TENTH EDITION

# Mechanical and Electrical Equipment for Buildings

TENTH EDITION

# Mechanical and Electrical Equipment for Buildings

**Benjamin Stein**
Consulting Architectural Engineer

**John S. Reynolds**
Professor of Architecture
University of Oregon

**Walter T. Grondzik**
Architectural Engineer

**Alison G. Kwok**
Professor of Architecture
University of Oregon

**WILEY**

John Wiley & Sons, Inc.

**Part opener pages are of the Lillis Business Complex at University of Oregon, designed by SRG Partnership, Portland, OR.**

This book is printed on acid-free paper. ∞

Published by John Wiley & Sons, Inc., Hoboken, New Jersey
Published simultaneously in Canada

For general information about our other products and services, please contact our Customer Care Department within the United States at (800) 762-2974, outside the United States at (317) 572-3993 or fax (317) 572-4002.

Wiley also publishes its books in a variety of electronic formats. Some content that appears in print may not be available in electronic books. For more information about Wiley products, visit our web site at www.wiley.com.

*Library of Congress Cataloging-in-Publication Data:*

Mechanical and electrical equipment for buildings / Benjamin Stein . . . [et al.].— 10th ed.
    p. cm.
    Rev. ed. of: Mechanical and electrical equipment for buildings / Benjamin Stein, John S. Reynolds. 2000.
    Includes index.
    ISBN 0-471-46591-7 (cloth)
    1. Buildings—Mechanical equipment.   2. Buildings—Electric equipment.   3. Buildings—Environmental engineering.   I. Stein, Benjamin.
    TH6010.S74   2006
    696—dc22        2005001253

Printed in the United States of America

10 9 8 7 6 5 4 3

# Contents

**v**

# PART III   ILLUMINATION   457

## PART IV   ACOUSTICS   727

## PART VI  FIRE PROTECTION  1065

# PART VII ELECTRICITY 1145

## PART VIII SIGNAL SYSTEMS 1335

## PART IX  TRANSPORTATION   1373

# PART X APPENDICES  1479

# Preface

SEVEN DECADES AND A FEW GENERATIONS HAVE passed since the first edition of *Mechanical and Electrical Equipment for Buildings* was published in 1935. At birth, this book was 429 pages long. Now, in the 10th edition, the page count is approximately 1700, a volumetric increase of 400%. Many new topics have been added, and a few have disappeared; computers are now routinely used in system design; equipment and distribution systems have undergone huge changes; mechanical cooling has become commonplace; fuel sources have shifted (coal provided half of the U.S. fuel in 1935). In recent editions, the book has increasingly added "why" to its long list of "how-tos."

Most of the systems presented in this book involve energy consumption. As North American society has moved from its initial reliance on renewable energy sources (wind, water, and horse power) to today's nonrenewable fossil fuels, it has also added vastly to its population and increased its per capita energy use. The resulting environmental degradation (primarily evident in air and water quality) has spurred efforts to reverse this decline. Governmental regulations are one part of such efforts, but this book emphasizes the investigation of alternative fuels and design approaches that go beyond the minimum acceptable to society.

It is becoming increasingly clear that gradual global warming is underway. It is less clear to what extent our hugely increased energy consumption is responsible, with its associated heat release and gaseous additions to the atmosphere. But it is very clear that the world's supply of fossil fuel is diminishing, with future consequences for all buildings that today rely so thoroughly on nonrenewable energy sources.

Buildings today contribute to negative global consequences of the future, and our approach to mechanical and electrical systems must consider how best to avoid negative environmental impacts. Thus, on-site resources—daylighting, passive solar heating, passive cooling, solar water heating, rainwater, wastewater treatment, photovoltaic electricity—share the spotlight with traditional off-site networks (natural gas, oil, the electrical grid, water and sewer lines). On-site processes can be area-intensive and labor-intensive and can involve increased first costs that require years to recover. Off-site processes are usually subsidized by society, often with substantial environmental costs. On-site energy use requires us to look beyond the building, to pay as much attention to a building's context as to the mechanical and electrical spaces, equipment, and systems within.

Throughout the many editions of this book, another trend has emerged. We have moved from systems that centralize all sources of heating, cooling, water, and electricity toward those that encourage more localized production and control. Increased sophistication of digital control systems has encouraged this trend. Further encouragement comes from multipurpose buildings whose schedules of occupancy are fragmented and from corporations with varying work schedules that result in partial occupancy on weekends. Another factor in this move to decentralization is worker satisfaction; there is increasing evidence that productivity increases with a sense of individual control of one's work environment. Residences are increasingly being used as office work environments. Expanding communications networks have made this possible. As residential designs thus become more complex (with office lighting, zones for heating/cooling, sophisticated communications, noise control), our nonresidential work environments become more attractive and individual.

Air and water pollution problems stemming from buildings (and their systems and occupants) are widely recognized and condemned. A rapidly increasing interest in green design on the part of clients and designers may help to mitigate such problems, although green design may be just an intermediate step toward survival through the mandatory consideration of sustainability. Another pervasive pollutant is noise. Noise impacts building siting, space planning, exterior and interior material selections—even the choice of cooling systems (as with natural ventilation). Air and water pollution can result in physical illness, but so can noise pollution, along with its burden of mental stress.

This book is written primarily for the North American building design community and has always emphasized examples from this region. Yet other areas of the world, some with similar traditions and fuel sources, have worthy examples of new strategies for building design utilizing on-site energy and energy conservation. Thus, some buildings from Europe and Asia appear in this 10th edition, along with many North American examples. Listings for buildings, researchers, and designers have been included in the index of this edition.

Building system design is now widely undertaken using computers, often through proprietary software that includes hundreds of built-in assumptions. This book is intended to encourage the designer to take a rational approach to system design: to verify intuitive design moves and assumptions and to use computers as tools to facilitate such verification, but to use patterns and approximations to point early design efforts in the right direction. Hand calculations have the added benefit of exposing all pertinent variables and assumptions to the designer. This in itself is a valuable rationale for conducting some portion of an analysis manually. Rough hand-calculated results should point toward similar results obtained with a computer; the greater

the disparity, the greater the need to check both approaches.

This book is written with the student, the architect- or engineer-in-training, and the practicing professional in mind. Basic theory, preliminary design guidelines, and detailed design procedures allow the book to serve both as an introductory text for the student and as a more advanced reference for both professional and student. We intend this work to be used as a textbook for multiple courses in architecture, architectural engineering, and building/construction management.

We are pleased to announce that a "MEEB 10" World Wide Web (WWW) site will provide supporting materials to enhance learning about and understanding the concepts, equipment, and systems dealt with in this book. The opportunity to provide color images via this medium is truly exciting. As with the 9th edition, an Instructor's Manual has been developed to provide additional support for this 10th edition. The manual, prepared by Walter Grondzik and Nick Rajkovich outlines the contents and terminology in each chapter; highlights concepts of special interest or difficulty; and provides sample discussion, quiz, and exam questions. The manual is available to instructors who have adopted this book in their courses.

*Mechanical and Electrical Equipment for Buildings* continues to serve as a reference for architectural registration examinees in the United States and Canada. We also hope to have provided a useful reference book for the offices of architects, engineers, and building managers.

BENJAMIN STEIN
JOHN S. REYNOLDS
WALTER T. GRONDZIK
ALISON G. KWOK

---

Visit **www.wiley.com/go/meeb**
for the expanding set of learning resources that accompany this book.

# Acknowledgments

Many people and organizations have contributed to the various editions of this book. We begin with those from whose work we have borrowed at length: J. Douglas Balcomb, Baruch Givoni, John Tillman Lyle, Murray Milne, William McGuinness, and Victor Olgyay; ASHRAE (the American Society of Heating, Refrigerating and Air-Conditioning Engineers), the American Solar Energy Society (ASES), the Illuminating Engineering Society of North America (IESNA), the National Drinking Water Clearinghouse, the National Small Flows Clearinghouse, the National Fire Protection Association (NFPA); and the many equipment manufacturers whose product information and photographs are used to illustrate the book.

Several professionals provided valuable assistance in assembling materials, and clarifying ideas and details. These include Craig Christiansen (NREL research reports), William Lowry (climate of cities), Daniel Panetta (AIWPS and recycling at California Polytechnic, San Luis Obispo), Dr. Jonathan Stein (computer applications), John A. Van Deusen (vertical transportation), and Martin Yoklic (cool-tower performance analysis).

In addition to drawings by Michael Cockram (whose work first appeared in the 8th edition), we are exceptionally pleased to include in this 10th edition illustrations by Dain Carlson, Amanda Jo Clegg, Eric Drew, Jonathan Meendering, and Erik Winter—students who embrace the principles and concepts of environmental technology in their design work and therefore clearly understand what they are drawing. We also acknowledge the many architects and engineers who provided illustrations of their buildings and processes throughout this book—citations to these firms and individuals are found throughout the book.

Testing in the classroom is a particularly valuable way to find needed improvements in any textbook. Students at the University of Oregon have, over many years, raised probing questions whose answers have resulted in changes. Valuable suggestions have come from the many graduate teaching fellows at the University of Oregon, particularly Christina Bollo, Alfredo Fernandez, Sara Goenner, Jeff Guggenheim, Susie Harriman, Angela Matt, Jonathan Meendering, Roger Ota, Therese Peffer, Troy Peters, David Posada, and Nick Rajkovich. Former Oregon students who helped with research include Troy Anderson, Daniel Irurah, Reza Javandel, Jeff Joslin, and Emily Wright.

We also extend special thanks to Nick Rajkovich who collaborated on the Instructor's Manual to accompany the 10th edition.

A large portion of the work involved in producing a manuscript is accomplished by supporting personnel. Among these, we wish particularly to thank Leesa Mayfield and Gina Livingston for coordinating illustrations with the text, maintaining a database to make all this manageable, and persevering throughout it all; Erik Winter for unfailing assistance with processing of illustrations; and Kathy Bevers for assistance with database troubleshooting. They also offered many helpful suggestions—as did the illustrators mentioned above.

Finally, we are indebted to the staff at John Wiley & Sons for their diligent and highly professional work, especially Amanda Miller, Vice President and Publisher; Paul Drougas, Acquisitions Editor; Lauren LaFrance, Assistant Developmental Editor; Donna Conte, Senior Production Editor; and Helen Greenberg, Copyeditor.

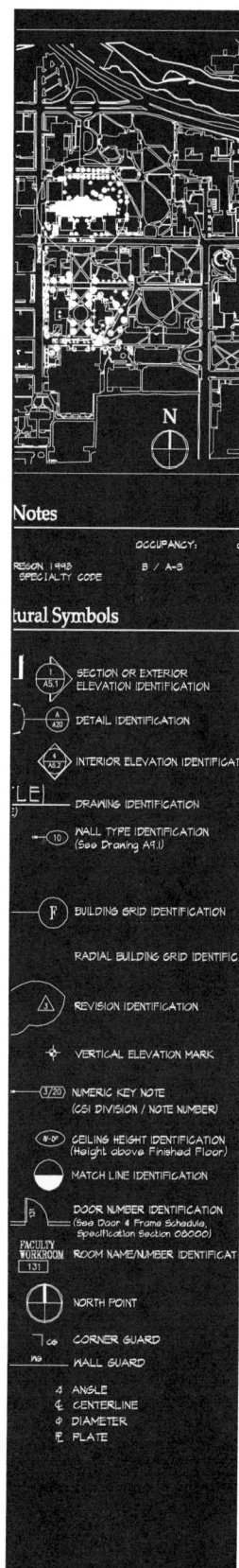

# DESIGN CONTEXT

Often the design of mechanical and electrical equipment for buildings is not considered until many important design decisions have already been made. In too many cases, such equipment is considered to have a corrective function, permitting a building envelope and siting to "work" in a climate that was essentially ignored.

Part I is intended to encourage designers to use the design process to full advantage and to include both climate and the key design objectives of comfort and indoor air quality in their earliest design decisions. Chapter 1 discusses the design process and the roles played by factors such as codes, costs, and verification in shaping a final building design. The critical importance of clear design intents and criteria is emphasized. Principles to guide environmentally responsible design are given. Chapter 2 discusses the relationship of energy, water, and material resources to buildings from design through demolition. The concept of environmental footprint is introduced as the ultimate arbiter of design decision making. Chapter 3 encourages viewing a building site as a collection of renewable resources, to be used as appropriate in the lighting, heating, and cooling of buildings. Chapter 4 discusses human comfort, the variety of conditions that seem comfortable, and implications of a more broadly defined comfort zone. It includes an introduction to design strategies for lighting, heating, and cooling. Chapter 5 introduces the issue of indoor air quality, which is currently a major concern of building occupants and the legal profession and an underpinning of green design efforts.

# Design Process

IN MARCH 1971 VISIONARY ARCHITECT MALCOLM WELLS published a watershed article in *Progressive Architecture.* It was rather intriguingly and challengingly titled "The Absolutely Constant Incontestably Stable Architectural Value Scale." In essence, Wells argued that buildings should be *benchmarked* (to use a current term) against the environmentally regenerative capabilities of wilderness (Fig. 1.1). This seemed a radical idea then—and remains so even now, over 30 years later. Such a set of values, however, may be just what is called for as the design professions inevitably move from *energy-efficient* to *green* to *sustainable* design in the coming decades. The main problem with Wells' "Incontestably Stable" benchmark is that most buildings fare poorly (if not dismally) against the environment-enhancing characteristics of wilderness. But perhaps this is more of a wakeup call than a problem.

As we enter the twenty-first century, *Progressive Architecture* is no longer in business, Malcolm Wells is in semiretirement, mechanical and electrical equipment has improved, simulation techniques have radically advanced, and information exchange has been revolutionized. In broad terms, however, the design process has changed little since the early 1970s. This should not be unexpected, as the design process is simply a structure within which to

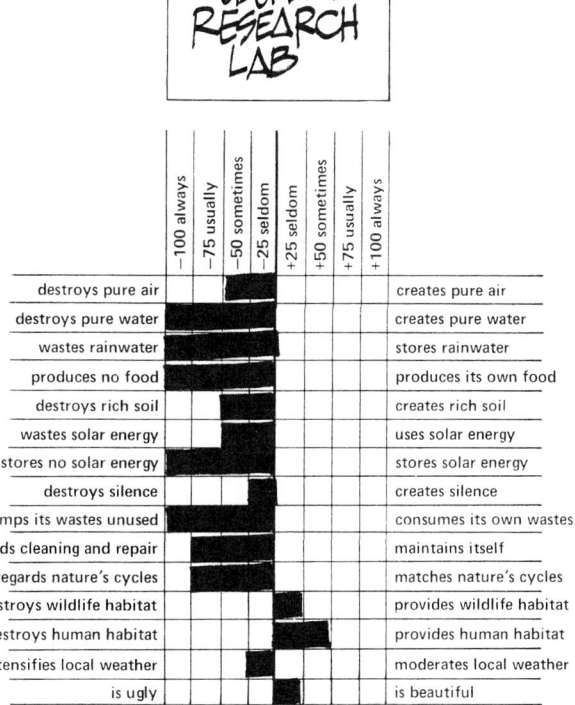

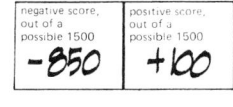

**Fig. 1.1** *Evaluation of a typical project using Malcolm Well's "absolutely constant incontestably stable architectural value scale." The value focus was wilderness; today it might well be sustainability. (© Malcolm Wells. Used with permission from Malcolm Wells. 1981. Gentle Architecture. McGraw-Hill, New York.)*

© Malcolm Wells 1969

develop a solution to a problem. What absolutely must change in the coming decades are the values and philosophy that underlie the design process. The beauty of Wells' Value Scale was its crystal-clear focus upon the values that accompanied his design solutions—and the explicit stating of those values. To meet the challenges of the coming decades, it is critical that designers consider and adopt values appropriate to the nature of the problems being confronted—both at the individual project scale and globally. Nothing less makes sense.

## 1.1 INTRODUCTION

The design process is an integral part of the larger and more complex building procurement process through which an owner defines facility needs, considers architectural possibilities, contracts for design and construction services, and uses the resulting facility. Numerous decisions (literally thousands) made during the design process will determine the need for specific mechanical and electrical systems and equipment and very often will determine eventual owner and occupant satisfaction. Discussing selected aspects of the design process seems a good way to start this book.

A building project typically begins with pre-design activities that establish the need for, feasibility of, and proposed scope for a facility. If a project is deemed feasible and can be funded, a multiphase design process follows. The design phases are typically described as conceptual design, schematic design, and design development. If a project remains feasible as it progresses, the design process is followed by the construction and occupancy phases of a project. In fast-track approaches (such as design-build), design efforts and construction activities may substantially overlap.

Predesign activities may be conducted by the design team (often under a separate contract), by the owner, or by a specialized consultant. The product of predesign activities should be a clearly defined scope of work for the design team to act upon. This product is variously called a *program*, a *project brief*, or the *owner's project requirements*. The design process converts this statement of the owner's requirements into drawings and specifications that permit a contractor to convert the owner's (and designer's) wishes into a physical reality.

The various design phases are the primary arena of concern to the design team. The design process may span weeks (for a simple building or system) or years (for a large, complex project). The design team may consist of a sole practitioner for a residential project or 100 or more people located in different offices, cities, or even countries for a large project. Decisions made during the design process, especially during the early stages, will affect the project owner and occupants for many years—influencing operating costs, maintenance needs, comfort, enjoyment, and productivity.

The scope of work accomplished during each of the various design phases varies from firm to firm and project to project. In many cases, explicit expectations for the phases are described in professional service contracts between the design team and the owner. A series of images illustrating the development of the Real Goods Solar Living Center (Figs. 1.2 and 1.3) is used to illustrate the various phases

**Fig. 1.2** *The Real Goods Solar Living Center, Hopland, California; exterior view. (Photo © Bruce Haglund; used with permission)*

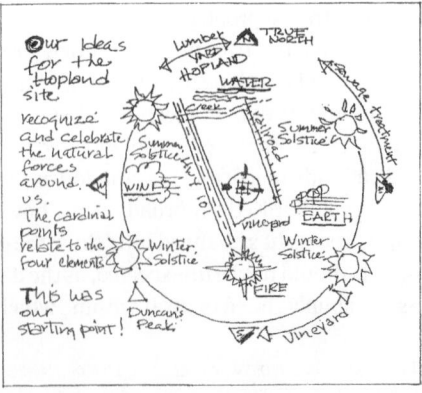

**Fig. 1.3** *Initial concept sketch for the Real Goods Solar Living Center, a site analysis. (Drawing by Sim Van der Ryn; reprinted from* A Place in the Sun *with permission of Real Goods Trading Corporation.)*

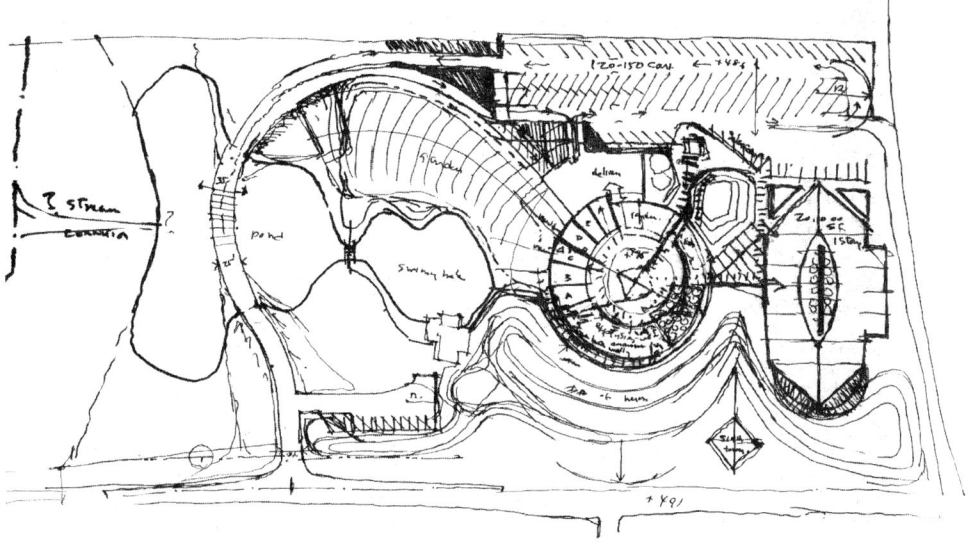

**Fig. 1.4** *Conceptual design proposal for the Real Goods Solar Living Center. The general direction of design efforts is suggested in fairly strong terms (the "first, best moves"), yet details are left to be developed in later design phases. There is a clear focus on rich site development even at this stage—a focus that was carried throughout the project. (Drawing by Sim Van der Ryn; reprinted from* A Place in the Sun *with permission of Real Goods Trading Corporation.)*

of a building project. (The story of this remarkable project, and its design process, is chronicled in Schaeffer et al., 1997.) Generally, the purpose of conceptual design (Fig. 1.4) is to outline a general solution to the owner's program that meets the budget and

captures the owner's imagination so that design can continue. All fundamental decisions about the proposed building should be made during conceptual design (not that things can't or won't change). During schematic design (Figs. 1.5 and 1.6), the

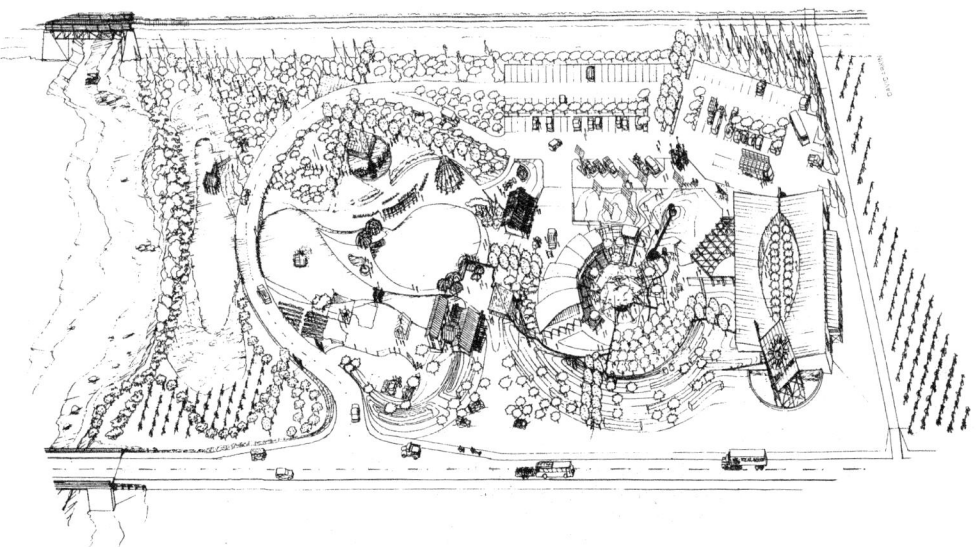

**Fig. 1.5** *Schematic design proposal for the Real Good Solar Living Center. As design thinking and analysis evolves, so does the specificity of a proposed design. Compare the level of detail provided at this phase with that shown in Fig. 1.4. Site development has progressed, and the building elements begin to take shape. The essence of the final solution is pretty well locked into place. (Drawing by David Arkin; reprinted from* A Place in the Sun *with permission of Real Goods Trading Corporation.)*

**Fig. 1.6** *Scale model analysis of shading devices for the Real Goods Solar Living Center. This is the sort of detailed analysis that would likely occur during design development. (Photo, model, and analysis by Adam Jackaway; reprinted from* A Place in the Sun *with permission of Real Goods Trading Corporation.)*

conceptual solution is further developed and refined. During design development (Fig. 1.7), all decisions regarding a design solution are finalized, and construction drawings and specifications detailing those innumerable decisions are prepared.

The construction phase (Fig. 1.8) is primarily in the hands of the contractor, although design decisions determine what will be built and may dramatically affect constructability. The building owner and occupants are the key players during the occupancy phase (Fig. 1.9). Their experiences with the building will clearly be influenced by design decisions and construction quality, as well as by maintenance and operation practices. A feedback loop that allows construction and occupancy experiences (lessons—both good and bad) to be used by the design team is essential to good design practice.

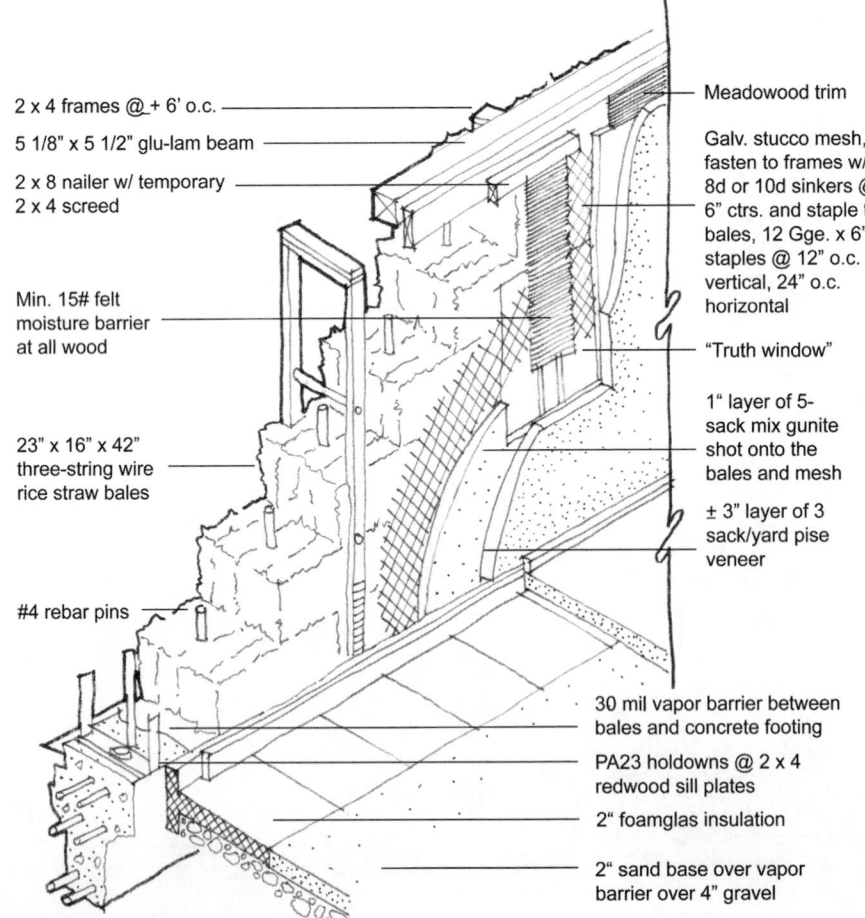

2 x 4 frames @ + 6' o.c.

5 1/8" x 5 1/2" glu-lam beam

2 x 8 nailer w/ temporary
2 x 4 screed

Min. 15# felt
moisture barrier
at all wood

23" x 16" x 42"
three-string wire
rice straw bales

#4 rebar pins

Meadowood trim

Galv. stucco mesh,
fasten to frames w/
8d or 10d sinkers @
6" ctrs. and staple to
bales, 12 Gge. x 6"
staples @ 12" o.c.
vertical, 24" o.c.
horizontal

"Truth window"

1" layer of 5-
sack mix gunite
shot onto the
bales and mesh

± 3" layer of 3
sack/yard pise
veneer

30 mil vapor barrier between
bales and concrete footing

PA23 holdowns @ 2 x 4
redwood sill plates

2" foamglas insulation

2" sand base over vapor
barrier over 4" gravel

**Fig. 1.7** *During design development the details that convert an idea into a building evolve. This drawing illustrates the development of working details for the straw bale wall system used in the Real Goods Solar Living Center. Material usage and dimensions are refined and necessary design analyses (thermal, structural, economic) completed. (Drawing by David Arkin; reprinted from* A Place in the Sun *with permission of Real Goods Trading Corporation. Redrawn by Erik Winter.)*

DESIGN CONTEXT

**Fig. 1.8** *Construction phase photo of Real Goods Solar Living Center straw bale walls. Design intent becomes reality during this phase. (Reprinted from* A Place in the Sun *with permission of Real Goods Trading Corporation.)*

## 1.2 DESIGN INTENT

Design efforts should generally focus upon achieving a solution that will meet the expectations of a well-thought-out and explicitly defined design intent. Design intent is simply a statement that outlines the expected high-level outcomes of the design process. Making such a fundamental statement is critical to the success of a design, as it points to the general direction(s) that the design process must take to achieve success. Design intent should not try to capture the totality of a building's character; this will come only with the completion of the design. It should, however, adequately express the defining characteristics of a proposed building solution. Example design intents (from among thousands of possibilities) might include the following:

**Fig. 1.9** *The Real Goods Solar Living Center during its occupancy and operations phase. Formal and informal evaluation of the success of the design solution may (and should) occur. Lessons learned from these evaluations can inform future projects. This photo was taken during a Vital Signs case study training session held at the Solar Living Center. (© Cris Benton, kite aerial photographer and Professor, University of California–Berkeley; used with permission.)*

- The building will provide outstanding comfort for its occupants.
- The building will use the latest in information technology.
- The building will be green, with a focus on indoor environmental quality.
- The building will use primarily passive systems.
- The building will provide a high degree of flexibility for its occupants.

Clear design intents are important because they set the tone for design efforts, allow all members of the design team to understand what is truly critical to success, provide a general direction for early design efforts, and put key or unusual design concerns on the table. Prof. Larry Peterson, former director of the Florida Sustainable Communities Center, has described the earliest decisions in the design process as an attempt to make the "first, best moves." Strong design intent will inform such moves. Weak intent will result in a weak building. Great moves too late will be futile. The specificity of the design intent will evolve throughout the design process. *Outstanding comfort* during conceptual design may become *outstanding thermal, visual, and acoustic comfort* during schematic design.

## 1.3 DESIGN CRITERIA

Design criteria are the benchmarks against which success or failure in meeting design intent is measured. In addition to providing a basis against which to evaluate success, design criteria will ensure that all involved parties seriously address the technical and philosophical issues underlying the design intent. Setting design criteria demands the clarification and definition of many intentionally broad terms used in design intent statements. For example, what is really meant by *green*, by *flexibility*, by *comfort?* If such terms cannot be benchmarked, then there is no way for the success of a design to be evaluated—essentially anything goes, and all solutions are potentially equally valid. Fixing design criteria for qualitative issues (such as *exciting, relaxing,* or *spacious*) can be especially challenging but equally important. Design criteria should be established as early in the design process as possible—certainly no later than the schematic design phase. As design criteria will define success or failure in a

specific area of the building design process, they should be realistic and not subject to whimsical change. In many cases, design criteria will be used both to evaluate the success of a design approach or strategy and to evaluate the performance of a system or component in a completed building. Design criteria might include the following:

- Thermal conditions will meet the requirements of ASHRAE Standard 55-2004.
- The power density of the lighting system will be no greater than 0.7 W/ft$^2$.
- The building will achieve a Silver *LEED®* rating.
- Fifty percent of building water consumption will be provided by rainwater capture.
- Background sound levels in classrooms will not exceed RC 35.

## 1.4 METHODS AND TOOLS

Methods and tools are the means through which design intent is accomplished. They include design methods and tools, such as a heat loss calculation procedure or a sun angle calculator. They also include the components, equipment, and systems that comprise a building. It is important that the right method or tool be used for a particular purpose. It is also critical that methods and tools (as means to an end) never be confused with either design intent (a desired end) or design criteria (benchmarks).

For any given design situation there are typically many valid and viable solutions available to the design team. It is important that none of these solutions be overlooked or ruled out due to design process short-circuits. Although this may seem unlikely, methods (such as fire sprinklers, electric lighting, and sound absorption) are surprisingly often included as part of a design intent statement. Should this occur, all other possible (and perhaps desirable) solutions are ruled out by direct exclusion. This does not serve a client or occupants well and is also a disservice to the design team.

This book is a veritable catalog of design guidelines, methods, equipment, and systems that serve as means and methods to desired design ends. Sorting through this extensive information will be easier with specific design intent and criteria in mind. Owner expectations and designer experiences will typically inform design intent. Sections of the book that address fundamental principles will provide

**TABLE 1.1 Relationships Between Design Intent, Design Criteria, and Design Tools/Methods**

| Issue | Design Intent | Possible Design Criterion | Potential Design Tools | Potential Implementation Method |
|---|---|---|---|---|
| Thermal comfort | Acceptable thermal comfort | Compliance with ASHRAE Standard 55 | Standard 55 graphs/tables or comfort software | Passive climate control and/or active climate control |
| Lighting level (illuminance) | Acceptable illuminance levels | Compliance with recommendations in the *IESNA Lighting Handbook* | Hand calculations or computer simulations | Daylighting and/or electric lighting |
| Energy efficiency | Minimal energy efficiency | Compliance with ASHRAE Standard 90.1 | Handbooks, simulation software, manufacturer's data, experience | Envelope strategies and/or equipment strategies |
| Energy efficiency | Outstanding energy efficiency | Exceed the minimum requirements of ASHRAE Standard 90.1 by 25% | Handbooks, simulation software, manufacturer's data, experience | Envelope strategies and/or equipment strategies |
| Green design | Obtain green building certification | Meet the requirements for a LEED gold rating | LEED materials, handbooks, experience | Any combination of approved strategies to obtain sufficient rating points |

assistance with establishment of appropriate design criteria. Table 1.1 provides examples of the relationship between design intent, design criteria, and tools/methods.

## 1.5 VALIDATION AND EVALUATION

To function as a knowledge-based profession, design (architecture and engineering) must reflect upon previous efforts and learn from existing buildings. Except in surprisingly rare situations, most building designs are generally unique—comprising a collection of elements not previously assembled in precisely the same way. Most buildings are essentially a design team hypothesis—"We believe that this solution will work for the given situation." Unfortunately, the vast majority of buildings exist as untested hypotheses. Little in the way of performance evaluation or structured feedback from the owner and occupants is typically sought. This is not to suggest that designers do not learn from their projects, but rather that little research-quality, publicly shared information is captured for use on other

projects. This is clearly not an ideal model for professional practice.

### (a) Conventional Validation/Evaluation Approaches

Design validation is very common, although perhaps more so when dealing with quantitative concerns than with qualitative issues. Many design validation approaches are employed, including hand calculations, computer simulations and modeling, physical models (of various scales and complexity), and opinion surveys. Numerous design validation methods are presented in this book. Simple design validation methods (such as broad approximations, lookup tables, or nomographs) requiring few decisions and little input data are typically used early in the design process. Later stages of design see the introduction of more complex methods (such as computer simulations or multistep hand calculations) requiring substantial and detailed input.

Building validation is much less common than design validation. Structured evaluations of occupied buildings are rarely carried out. Historically, the most commonly encountered means of validating building performance is the post-occupancy evaluation (POE). Published POEs have typically focused upon some specific (and often nontechnical) aspect of building performance, such as way-finding or productivity. Building commissioning and case studies are finding more application as building validation approaches. Third-party validations, such as the LEED rating system, are also emerging.

### (b) Commissioning

Building commissioning is an emerging approach to quality assurance. An independent commissioning authority (an individual or, more commonly, a team) verifies that equipment, systems, and design decisions can meet the owner's project requirements (design intent and criteria). Verification is accomplished through review of design documents and detailed testing of equipment and systems under conditions expected to be encountered with building use. Historically focused upon mechanical and electrical systems, commissioning is currently being applied to numerous building systems—including envelope, security, fire protection, and information systems. Active involvement of the design team is critical to the success of the commissioning process (ASHRAE, 2005).

### (c) Case Studies

Case studies represent another emerging approach to design/construction validation and evaluation. The underlying philosophy of a case study is to capture information from a particular situation and convey the information in a way that makes it useful to a broader range of situations. A building case study attempts to present the lessons learned from one case in a manner that can benefit other cases (future designs). In North America, the Vital Signs and Agents of Change projects have focused upon disseminating a building performance case study methodology for design professionals and students—with an intentional focus upon occupied buildings (à la POEs). The American Institute of Architects has developed a series of case studies dealing with design process/practice. In the United Kingdom, numerous case studies have been conducted under the auspices of the PROBE (post-occupancy review of building engineering) project.

## 1.6 INFLUENCES ON THE DESIGN PROCESS

The design process often appears to revolve primarily around the needs of a client and the capabilities of the design team—as exemplified by the establishment of design intent and criteria. There are several other notable influences, however, that affect the conduct and outcome of the building design process. Some of these influences are historic and affect virtually every building project; others represent emerging trends and affect only selected projects. Several of these design-influencing factors are discussed below.

### (a) Codes and Standards

The design of virtually every building in North America will be influenced by codes and standards. *Codes* are government-mandated and -enforced documents that stipulate minimum acceptable building practices. Designers usually interface with codes through an entity known as the *authority having jurisdiction.* There may be several such authorities for any given locale or project (fire protection requirements, for example, may be enforced separately from general building construction requirements or energy performance requirements). Codes essentially define the minimum that society deems acceptable. In no way is code compliance by itself likely to be adequate to meet the needs of a client. On the other hand, code compliance is undisputedly necessary.

Codes may be written in prescriptive language or in performance terms. A *prescriptive* approach mandates that something be done in a certain way. Examples of prescriptive code requirements include minimum R-values for roof insulation, minimum pipe sizes for a roof drainage system, and a minimum number of hurricane clips per length of roof. The majority of codes in the United States are fundamentally prescriptive in nature. A prescriptive code defines means and methods. By contrast, a performance code defines intent. A *performance* approach states an objective that must be met. Examples of performance approaches to code requirements include a maximum permissible design heat flow through a building envelope, a minimum design rainfall that can be safely drained from a building roof, and a defined wind speed that will not damage a roof construction. Some primarily prescriptive codes offer performance "options" for compliance. This is especially true of energy codes and smoke control requirements in fire protection codes.

Codes in the United States are in transition. Each jurisdiction (city, county, and/or state, depending upon legislation) is generally free to adopt whichever model code it deems most appropriate. Some jurisdictions (typically large cities) use home-grown codes instead of a model code. Historically, there were four model codes (the *Uniform Building Code,* the *Standard Building Code,* the *Basic Building Code,* and the *National Building Code*) that were used in various regions of the country. There is ongoing movement to development and use of a single model *International Building Code* to provide a more uniform and standardized set of code requirements. Canada recently adopted a major revision to its *National Building Code.* Knowledge of the current code requirements for a project is a critical element of the design process.

*Standards* are documents that present a set of minimum requirements for some aspect of building design that have been developed by a recognized authority (such as Underwriters Laboratories, the National Fire Protection Association, or the American Society of Heating, Refrigerating and Air-Conditioning Engineers). Standards do not carry the weight of government enforcement that codes do, but they are often incorporated into codes via reference. Standards play an important role in building design and are often used by legal authorities to define the level of care expected of design professionals. Typically, standards have been developed under a consensus process with substantial opportunity for external review and input. *Guidelines* and *handbooks* are less formal than standards, usually with less review and/or consensus. *General practice,* the least formalized basis for design, captures the norm for a given locale or discipline. Table 1.2 provides examples of codes, standards, and related design guidance documents.

## (b) Costs

Costs are an historic influence on the design process and are just as pervasive as codes. Typically, one of the earliest and strictest limits on design flexibility is the maximum construction budget imposed by the client. First cost (the cost for an owner to acquire the keys to a completed building) is the most commonly used cost factor. First cost is usually expressed as a maximum allowable construction cost or as a cost per unit area. Life-cycle cost (the cost for an owner to acquire and use a building for some defined period of time) is generally equally or more important than first cost, but is often ignored by owners and usually not well understood by designers.

Over the life of a building, operating and maintenance costs can far exceed the cost to construct or acquire a building. Thus, whenever feasible, design decisions should be based upon life-cycle cost implications and not simply first cost. The math of life-cycle costing is not difficult. The primary difficulties in implementing life-cycle cost analysis are estimating future expenses and the uncertainty naturally associated with projecting future conditions. These are not as difficult as they might seem, however, and a number of well-developed life-cycle cost methodologies have been developed. Appendix I provides basic information on life-cycle cost factors and procedures. The design team may find life-cycle costing a persuasive ally in the quest to convince an owner to make important, but apparently expensive, decisions.

## (c) Passive and Active Approaches

The distinction between passive and active systems may mean little to the average building owner but can be critical to the building designer and occupant. Development of passive systems must begin early in the design process, and requires early and continuous attention from the architectural designer. Passive system operation will often require the earnest cooperation and involvement of building occupants and users. Table 1.3 summarizes the identifying characteristics of passive and active systems approaches. These approaches are conceptually opposite in nature. Individual systems that embody both active and passive characteristics are often called hybrid systems. Hybrid systems are commonly employed as a means of tapping into the best aspects of both approaches.

The typical building will usually consist of both passive and active systems. Passive systems may be used for climate control, fire protection, lighting, acoustics, circulation, and/or sanitation. Active systems may also be used for the same purposes and for electrical distribution.

**TABLE 1.2 Codes, Standards, and Other Design Guidance Documents**

| Document Type | | Characteristics | Examples |
|---|---|---|---|
| Code | | Government-mandated and government-enforced (typically via the building and occupancy permit process); may be a legislatively adopted standard | Florida Building Code; California Title 24; Chicago Building Code; International Building Code (when adopted by a jurisdiction) |
| Standard | | Usually a consensus document developed by a professional organization under established procedures with opportunities for public review and input | ASHRAE Standard 90.1 (*Energy Standard for Buildings Except Low-Rise Residential Buildings*); ASTM E413-87 (*Classification for Rating Sound Insulation*); ASME A17.1—2000 (*Safety Code for Elevators and Escalators*) |
| Guideline | | Development is typically by a professional organization, but within a looser structure and with less public involvement | ASHRAE Guideline 0 (*The Commissioning Process*); *IESNA Advanced Lighting Guidelines;* NEMA LSD 12-2000 (*Best Practices for Metal Halide Lighting Systems*) |
| Handbook, design guide | | Development can vary widely—involving formal committees and peer review or multiple authors without external review | *IESNA Lighting Handbook; ASHRAE Handbook— Fundamentals;* NFPA *Fire Protection Handbook* |
| General practice | | The prevailing norm for design within a given community or discipline; least formal of all modes of guidance | System sizing approximations; generally accepted flashing details |

*Image Sources:* code—used with permission of the International Code Council; standard—used with permission of the American Society of Heating, Refrigerating and Air-Conditioning Engineers; guideline and handbook—used with permission of the Illuminating Engineering Society of North America; general practice—used with permission of John Wiley & Sons.

*Acronyms:* ASHRAE = American Society of Heating, Refrigerating and Air-Conditioning Engineers; ASME = American Society of Mechanical Engineers; ASTM = ASTM International (previously American Society for Testing and Materials); IESNA = Illuminating Engineering Society of North America; NEMA = National Electrical Manufacturers Association; NFPA = National Fire Protection Association.

## (d) Energy Efficiency

Some level of energy efficiency is a societally mandated element of the design process in most developed countries. Code requirements for energy-efficient building solutions were generally instituted as a result of the energy crises of the 1970s and have been updated on a periodic basis since then. As with all code requirements, mandated energy efficiency levels represent a minimum performance level that is considered acceptable—not an optimal performance level. Such acceptable minimum performance has evolved over time in response to changes in energy costs and availability and also in response to changes in the costs and availability of building technology.

In the United States, ANSI/ASHRAE/IESNA *Standard 90.1* (published by the American Society of Heating, Refrigerating and Air-Conditioning Engineers, cosponsored by the Illuminating Engineering Society of North America, and approved by the American National Standards Institute) is the most commonly encountered energy efficiency benchmark for commercial/institutional buildings. Some states (such as California and Florida) utilize state-specific energy codes. Residential energy efficiency requirements are addressed by several model codes and standards (including the *International Energy Code,* the *Model Energy Code,* and ANSI/ ASHRAE *Standard 90.2*). Appendix G provides a

**TABLE 1.3 Defining the Characteristics of Passive and Active Systems**

| Characteristic | Passive System | Active System |
|---|---|---|
| Energy source | Uses no purchased energy (electricity, natural gas, fuel oil, etc.)—example: daylighting system | Uses primarily purchased (and nonrenewable) energy —example: electric lighting system |
| System components | Components play multiple roles in system and in larger building—example: concrete floor slab that is structure, walking surface, and solar collector/storage | Components are commonly single-purpose elements— example: gas furnace |
| System integration | System is usually tightly integrated (often inseparably) with the overall building design— example: natural ventilation system using windows | System is usually not well integrated with the overall building design, often seeming an add-on— example: window air-conditioning unit |
| Passive and active systems represent opposing philosophical concepts. Design is seldom so straightforward as to permit the exclusive use of one philosophy. Thus, the hybrid system. Hybrid systems are a composite of active and passive approaches, typically leaning more toward the passive. For example, single purpose, electricity-consuming (active) ceiling fans might be added to a natural ventilation (passive) cooling system to extend the performance of the system and thus reduce energy usage that would otherwise occur if a fully active air-conditioning system were turned on instead of the fans. |||

sample of energy efficiency requirements from *Standard 90.1* and *Standard 90.2*.

Energy efficiency requirements for residential buildings tend to focus upon minimum envelope (walls, floors, roofs, doors, windows) and mechanical equipment (heating, cooling, domestic hot water) performance. Energy efficiency requirements for commercial/institutional buildings address virtually every building system (including lighting and electrical distribution). Most energy codes present a set of prescriptive minimum requirements for individual building elements, with an option for an alternative means of compliance to permit innovation and/or a systems-based design approach.

Technically speaking, *efficiency* is simply the ratio of system output to system input. The greater the output for any given input, the higher the efficiency. This concept plays a large role in energy efficiency standards through the specification of minimum efficiencies for many items of mechanical and electrical equipment for buildings. *Energy conservation* implies saving energy by using less. This is conceptually different from efficiency but is an integral part of everyday usage of the term. Energy efficiency codes and standards include elements of conservation embodied in equipment control requirements or insulation levels. Because of negative connotations some associate with "conservation" (doing without), the term *energy efficiency* is generally used to describe both conservation and efficiency efforts.

Passive design solutions usually employ renewable energy resources. Several active design solutions, however, also utilize renewable energy forms. Energy conservation and efficiency concerns are typically focused upon minimizing depletion of nonrenewable energy resources. The use of renewable energy sources (such as solar radiation and wind) changes the perspective of the design team and how compliance with energy efficiency codes/standards is evaluated. The majority of energy efficiency standards deal solely with on-site energy usage. Off-site energy consumption (for example, that required to transport fuel oil or natural gas, or the losses from electrical generation) is not addressed. This site-based focus can seriously skew thinking about energy efficiency design strategies.

## (e) Green Building Design Strategies

*Green design* considerations are increasingly becoming a part of the design process for many buildings. Green design goes well beyond energy-efficient design in order to address both the local and global impacts of building energy, water, and materials usage. Energy efficiency is a key, but not self-sufficient, element of green design. The concept, broadly called "green design," arose from concerns about the wide-ranging environmental impacts of design decisions. Although there is no generally accepted concise definition of *green*, the term is typically understood to incorporate concern for the health and well-being of building occupants/users and respect for the larger global environment. A green building should maximize beneficial impacts on its direct beneficiaries while minimizing negative impacts on the site, local, regional, national, and global environments.

Several green-design rating systems have found wide acceptance as benchmarks for design. These include the U.S. Green Building Council's LEED system (*Leadership in Energy and Environmental Design*) and an international evaluation methodology entitled GBTool. Somewhat similar rating systems are in use in the United Kingdom and Canada. The LEED system (Fig. 1.10) presents a palette of design options from which the design team can select strategies appropriate for a particular building and its context. Amassing points for selected strategies provides a means of attaining green building status—at one of several levels of achievement, via a formal certification procedure. Prerequisite design strategies (including baseline energy efficiency and acceptable indoor air quality) provide an underpinning for the optional strategies.

The emergence of green building rating systems has greatly rationalized design intent and design criteria in this particular area of architecture. Prior to the advent of LEED (or GBTool), anyone could claim greenness for his/her designs. Although green design is entered into voluntarily (no codes currently require it, although a number of municipalities require new public buildings to be green), there are now generally accepted standards against which performance can be measured. Appendix G provides an excerpt from the LEED green building rating system.

## (f) Design Strategies for Sustainability

Unlike green design, the meaning of "sustainability" in architecture has not yet been rationalized. The term *sustainable* is used freely—and often mistakenly—to describe a broad range of intents and performances. This is unfortunate, as it tends to make sustainability a meaningless term—and sustainability is far too important a concern to be meaningless. For the purposes of this book, sustainability will be defined as follows (paraphrasing the Brundtland Commission): *Sustainability involves meeting the needs of today's generation without detracting from the ability of future generations to meet their needs.*

Sustainability is essentially long-term survival. In architectural terms, sustainability involves the survival of an existing standard of living into future generations. From an energy, water, and materials standpoint, sustainability can be argued to require zero net use of nonrenewable resources. Any long-term removal of nonrenewable resources from the environment will surely impair the ability of future generations to meet their needs (with fewer resources available as a result of our actions). Because sustainability is so important a concept and objective, the term should not be used lightly. It is highly unlikely that any single building built in today's economic environment can be sustainable (yielding no net resource depletion). Sustainability

(a)                                           (b)

**Fig. 1.10** *(a) The Jean Vollum Natural Capital Center, Portland, Oregon. A warehouse from the industrial era was rehabilitated by Ecotrust to serve as a center for the conservation era. (b) LEED plaque on the front façade of the Vollum Center. The plaque announces the success of the design team (and owner) in achieving a key element of their design intent. (Photos © 2004 Alison Kwok; all rights reserved.)*

at the community scale is more probable; examples, however, are rare.

## (g) Regenerative Design Strategies

Energy efficiency is an attempt to use less energy to accomplish a given design objective (such as thermal comfort or adequate lighting). Green design is an attempt to maximize the positive effects of design while minimizing the negative ones—with respect to energy, water, and material resources. Sustainable design is an attempt to solve today's problems while reserving adequate resources to permit future generations to solve their problems. Energy efficiency is a constituent of green design. Green design is a constituent of sustainable design. *Regenerative design* steps out beyond sustainability.

The goal of energy efficiency is to reduce net negative energy impacts. The goal of green design is to reduce net negative environmental impacts. The goal of sustainability is to produce no net negative environmental impacts. The goal of regenerative design is to produce a net positive environmental impact—to leave the world better off with respect to energy, water, and materials. Obviously, if design for sustainability is difficult, then regenerative design is even more difficult. Nevertheless, there are some interesting examples of regenerative design projects, including the Eden Project in the United Kingdom and the Center for Regenerative Studies (Fig. 1.11) in the United States. Both projects involve substantial site remediation and innovative design solutions.

## 1.7 A PHILOSOPHY OF DESIGN

From a design process perspective, the operating philosophy of this book is that development of appropriate design intent and criteria is critical to the successful design of buildings and their mechanical and electrical systems. Passive systems should generally be used before active systems (this in no way denigrates active systems, which will be necessary features of almost any building); life-cycle costs should be considered instead of simply first cost; and green design is a desirable intent that will ensure energy efficiency and provide a pathway toward sustainability. Design validation, commissioning, and post-occupancy evaluation should be aggressively pursued.

John Lyle presented an interesting approach to design (that elaborates upon this general philosophy) in his book *Regenerative Design for Sustainable Development.* The following discussion presents an overview of his approach. The strategies provide design teams with varied opportunities to integrate site and building design with components and processes. Those strategies most applicable to the design of mechanical and electrical systems are presented here. This approach guided the design of the Center for Regenerative Studies at the California Polytechnic State University at Pomona, California (Fig. 1.11).

### (a) Let Nature Do the Work

This principle expresses a preference for natural/passive processes over mechanical/active processes.

*(a)*

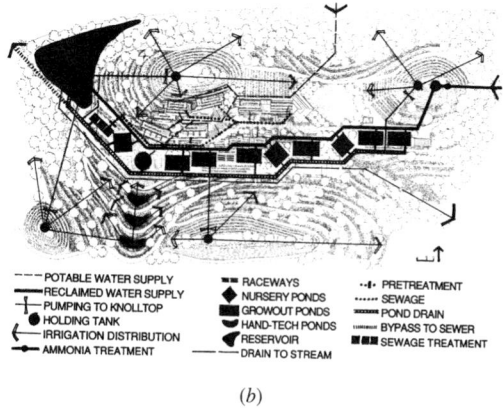

*(b)*

**Fig. 1.11** *(a) The Center for Regenerative Studies (CRS), California Polytechnic State University–Pomona. (b) Site plan for the CRS. It's not easy being regenerative—the highlighted elements relate only to the water reclamation aspects of the project. (Photo © 2004 Alison Kwok; drawing from Lyle, John Tillman. 1994. Regenerative Design for Sustainable Development. John Wiley & Sons, Inc., New York.)*

*Fig. 1.12* Letting nature do the work—via daylighting. Mt. Angel Abbey Library, Oregon, designed by Alvar Aalto. (Photo by Amanda Clegg.)

*Fig. 1.13* Consider nature as a model. Plants provide water treatment and generate biomass in an aquacultrural pond at the Center for Regenerative Studies, Cal Poly–Pomona. (Photo © 2004 Alison Kwok; all rights reserved.)

Designers can usually find ways to use natural processes on site (Fig. 1.12), where they occur, in place of dependence upon services from remote/nonrenewable sources. Smaller buildings on larger sites are particularly good candidates for this strategy.

### (b) Consider Nature as Both Model and Context

A look at this book reveals a strong reliance upon physical laws as a basis for design. Heat flow, water flow, electricity, light, and sound follow rules described by physics. This principle, however, suggests looking at nature (Fig. 1.13) for biological, in addition to the classical physical, models for design. The use of a Living Machine to process building wastes as opposed to a conventional sewage treatment plant is an example of where this strategy might lead.

### (c) Aggregate Rather Than Isolate

This strategy recommends that designs focus upon systems and not just upon the parts that make up a system—in essence, seeing the forest through the trees. The components of a system should be highly integrated to ensure workable linkages among the parts and the success of the whole. An example would be optimizing the solar heating performance of a direct gain system comprised of glazing, floor slab, insulation, and shading while perhaps reducing the performance of one or more constituent parts (Fig. 1.14).

### (d) Match Technology to the Need

This strategy seeks to avoid using high-grade resources for low-grade tasks. For example, it is obviously wasteful to flush toilets with purified water, but perhaps less obviously wasteful (but equally a mismatch) to use electricity (a very-high-grade energy form) to heat water for bathing. The corollary to this strategy is to think small, think simple, and think locally (Fig. 1.15).

### (e) Seek Common Solutions to Disparate Problems

This approach requires breaking out of the box of categories and classifications. An understanding of systems should lead to an increased awareness of systems capabilities—which will often prove to be multidisciplinary and multifunctional. Making a design feature (Fig. 1.16) serve multiple tasks (perhaps mechanical, electrical, and architectural in nature) is one way to counteract the potential problem of a higher first cost for green or sustainable design features. Solutions can be as simple and

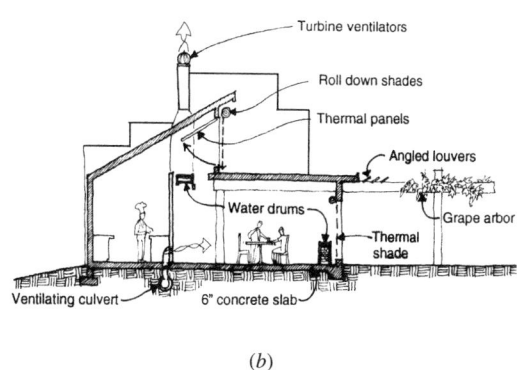

(*a*)  (*b*)

**Fig. 1.14** *Aggregating, not isolating. (a) The Cottage Restaurant, Cottage Grove, Oregon. (b) This section through the restaurant illustrates the substantial integration and coordination (aggregation) of elements typical of passive design solutions. (Photo by G.Z. Brown; drawing by Michael Cockram; © 1998 by John S. Reynolds, A.I.A., all rights reserved.)*

low-tech as using heat from garden composting to help warm a greenhouse.

## (f) Shape the Form to Guide the Flow

The most obvious examples of this strategy are solar-heated buildings that are shaped (Fig. 1.17) to gather winter sun or naturally ventilated buildings shaped to collect and channel prevailing winds. Daylighting is another obvious place to apply the "form follows flow" strategy, which can have a dramatic impact upon building design efforts and outcomes.

**Fig. 1.16** *Seek common solutions. The "atrium" of the Hood River County Library, Hood River, Oregon, provides a central hub for the library, daylighting, views (spectacular), and stack ventilation. (Photo © 2004 Alison Kwok; all rights reserved.)*

**Fig. 1.15** *Match technology to the need. Sometimes it's the simple things that count. (Photo © 2004 Alison Kwok; all rights reserved.)*

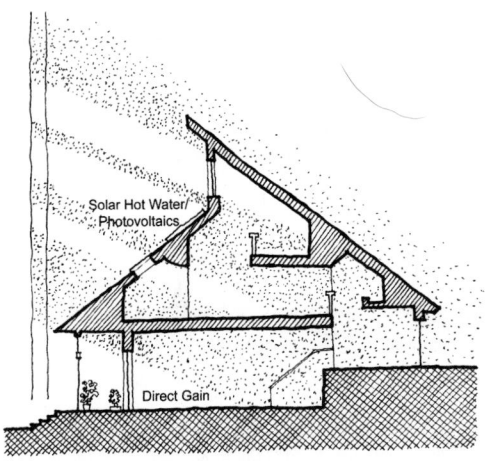

*Fig. 1.17 Shaping the form to the flow. Using a "band of sun" analysis as a solar form giver (see Chapter 3 for further details). (Redrawn by Jonathan Meendering.)*

## (g) Shape the Form to Manifest the Process

This is more than a variation on the adage "If you've got it, flaunt it." This strategy asks that a building inform its users and visitors about how it works both inside and out (Fig. 1.18). In passive solar-heated and passively cooled buildings, much of the thermal performance is evident in the form of the exterior envelope and the interior space, rather than hidden in a closet or mechanical penthouse. Prof. David Orr of Oberlin College addresses this issue succinctly by asking "What can a building teach?"

## (h) Use Information to Replace Power

This strategy addresses both the design process and building operations. Knowledge is suggested as a substitute for brute force (and associated energy waste). Designs informed by an understanding of resources, needs, and systems capabilities will tend to be more effective (successfully meeting intent) and efficient (meeting intent using less energy) than uninformed designs. Building operations informed by feedback and learning (Fig. 1.19) will tend to be more effective and efficient than static, unchangeable operating modes. Users of buildings can play a leading role in this approach by being allowed to make decisions about when to do what in order to maintain desired conditions. Reliance on a building's users is not so much a direct energy saver—most controls use very little power—as it is an education. A user who understands how a building receives and conserves heat in

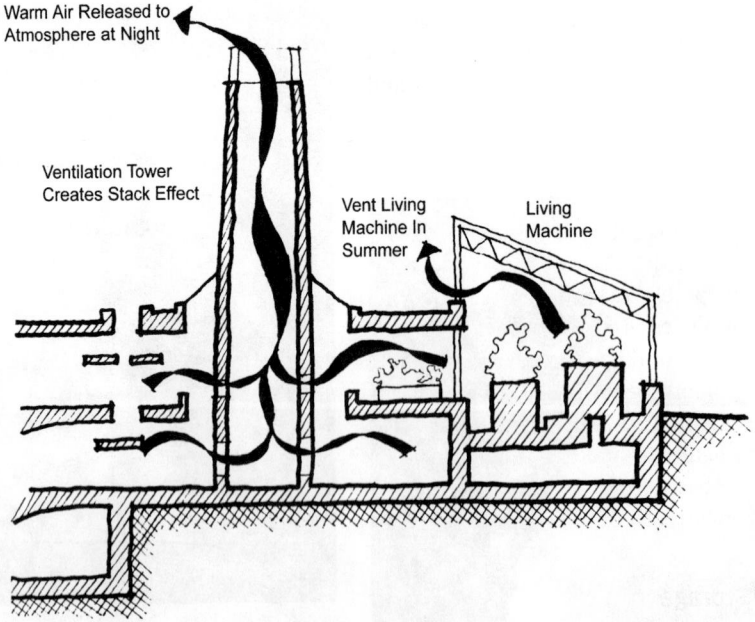

*Fig. 1.18 Shaping the form to the process. Stack effect ventilation is augmented by the building form in this proposal for the EPICenter project, Bozeman, Montana. (Courtesy of Place Architecture LLC, Bozeman, Montana, and Berkebile Nelson Immenschuh McDowell, Architects, Kansas City, Missouri. Redrawn by Jonathan Meendering.)*

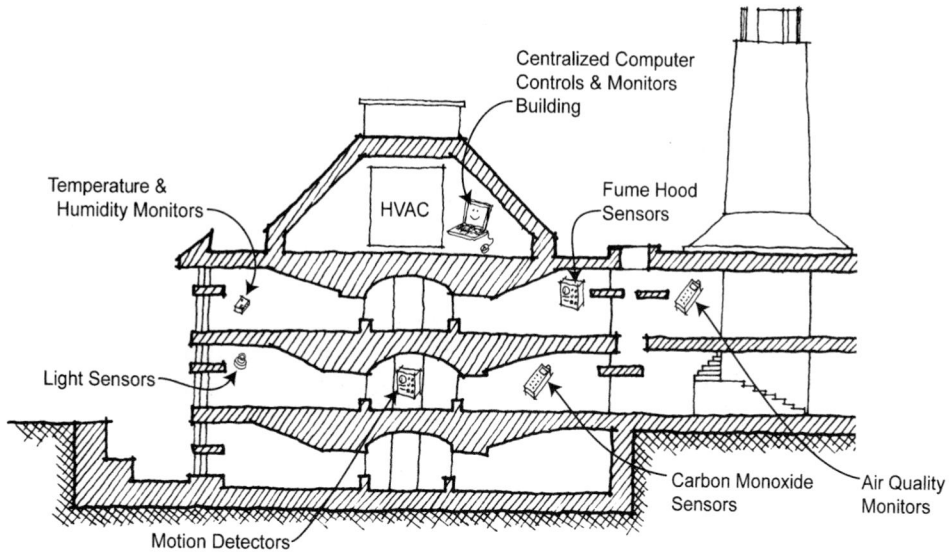

**Fig. 1.19** *Use information to replace power. Section showing intelligent control system components for the proposed EPICenter project, Bozeman, Montana. (Courtesy of Place Architecture LLC, Bozeman, Montana, and Berkebile Nelson Immenschuh McDowell, Architects, Kansas City, Missouri. Redrawn by Jonathan Meendering.)*

cold weather is likely to respond by lowering the indoor temperature and reducing heat leaks. Furthermore, some studies of worker comfort indicate that with more personal control (such as operable windows), workers express feelings of comfort across a wider range of temperatures than with centrally controlled air conditioning.

### (i) Provide Multiple Pathways

This strategy celebrates functional redundancy as a virtue—for example, providing multiple and separate fire stairs for emergency egress. There are many other examples, from backup heating and cooling systems, to multiple water reservoirs and piping pathways for fire sprinklers, to emergency electrical and lighting systems. This strategy also applies to climate–site–building interactions in which one site-based resource may temporarily weaken and can be replaced by another (Fig. 1.20).

### (j) Manage Storage

Storage is used to help balance needs and resources across time. Storage appears as an issue throughout this book. The greater the variations in the resource

supply cycle, the more critical storage management becomes. Rainwater can be stored in cisterns, balancing normal daily demands for water against variable monthly supplies. The high variability of wind-generated electricity output can be managed with hydrogen storage, providing a combustible fuel

**Fig. 1.20** *Providing multiple pathways. Three distinct sources of electricity are projected in this conceptual diagram for the proposed EPICenter project, Bozeman, Montana. (Courtesy of Place Architecture LLC, Bozeman, Montana, and Berkebile Nelson Immenschuh McDowell, Architects, Kansas City, Missouri. Redrawn by Jonathan Meendering.)*

that can be drawn on at a rate and time independent of wind speed.

On sunny winter days, a room's excess solar energy can be stored in its thermally massive surfaces (Fig. 1.21), to be released at night. On cool summer nights, *coolth* (the conceptual opposite of heat) can be stored in these same surfaces and used to condition the room by day. Most storage solutions will strongly impact building architecture.

**Fig. 1.21** *Manage storage. A concrete floor and barrels located high along the north wall provide thermal storage (for both heating and cooling) in the Cottage Restaurant, Cottage Grove, Oregon. (Photo by G.Z. Brown.)*

## 1.8 CASE STUDY—DESIGN PROCESS

### Gilman Ordway Campus of the Woods Hole Research Center

#### PROJECT BASICS

- Location: Falmouth, Massachusetts, USA
- Latitude: 41.3 N; longitude: 70.4 W; elevation: near sea level
- Heating degree days: 6296 base 65°F (3498 base 18.3°C); cooling degree days: 425 base 65°F (236 base 18.3°C); annual precipitation: 45.5 in. (1156 mm) (degree day data are for East Wareham; rainfall is for Woods Hole)
- Building type: remodeled and new construction; commercial offices and laboratory
- Building area: 19,200 ft² (1784 m²); four occupied stories
- Completed February 2003
- Client: Woods Hole Research Center
- Design team: William McDonough + Partners (and consultants)

**Background.** The Gilman Ordway Campus of the Woods Hole Research Center includes both new construction and extensive remodeling of a venerable old house to provide office and laboratory facilities. This recently opened building has generated a lot of interest. The clients are quite pleased with the facility and are using it as a vehicle to promote awareness of the environment and green design. The Research Center won an American

Institute of Architects/Committee on the Environment (AIA/COTE) Top Ten Green Project award and was recently the site of an Agents of Change POE training session. (The discussion that follows was extracted from information provided by William McDonough + Partners and the Woods Hole Research Center.)

**Context.** The work of the Woods Hole Research Center is focused upon the related issues of climate change and defending the world's great forests. When a new headquarters was considered, it was decided that the facility should reflect the Research Center's core values, support its research and education mission, and provide a healthy environment for building occupants and the outside world. Fund-raising was a major issue for this project and substantially impacted the design process and scheduling. Perhaps the most valuable lesson to be learned from this project is the inestimable value of perseverance and the benefit that a clearly enunciated set of objectives (design intent and criteria) can provide in seeing a donor-supported project through to completion.

**Design Intent.** The Woods Hole Research Center project sought to demonstrate that a modern building can "harmonize with a habitable earth" while providing a healthy, comfortable, and enjoyable workplace. Enhanced productivity and job satisfaction for employees was a key intent, as was far-beyond-code-minimum energy performance. In addition, the building was to serve as a teaching tool, providing an exemplar of a thoughtful approach to energy production and use, water quality and conservation, site design, and materials selection.

**Design Criteria and Validation.** The aggressive energy performance criteria set by the client and design team required the use of ENERGY 10 computer simulations and the ongoing services of an energy systems consultant. Interestingly, this same strong energy-related design intent allowed the retention of critical mechanical system elements during an extensive value engineering phase that cut approximately 15% from the construction budget. The owner retained an independent authority for building commissioning.

**Key Design Features**

- Extensive daylighting throughout the building
- Operable windows throughout the building
- An exceptionally tight and carefully detailed building envelope featuring triple-glazed windows and Icynene foam insulation (also an air barrier)

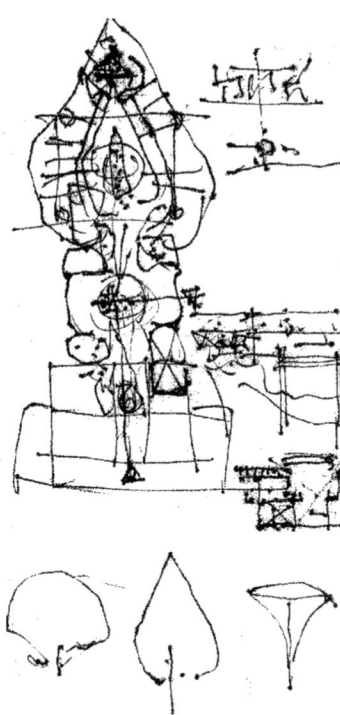

**Fig. 1.22** *Initial concept sketch for the Woods Hole Research Center (WHCR)—the "leaf." This is an exceptional example of a conceptual design phase product. (© William McDonough + Partners; used with permission.)*

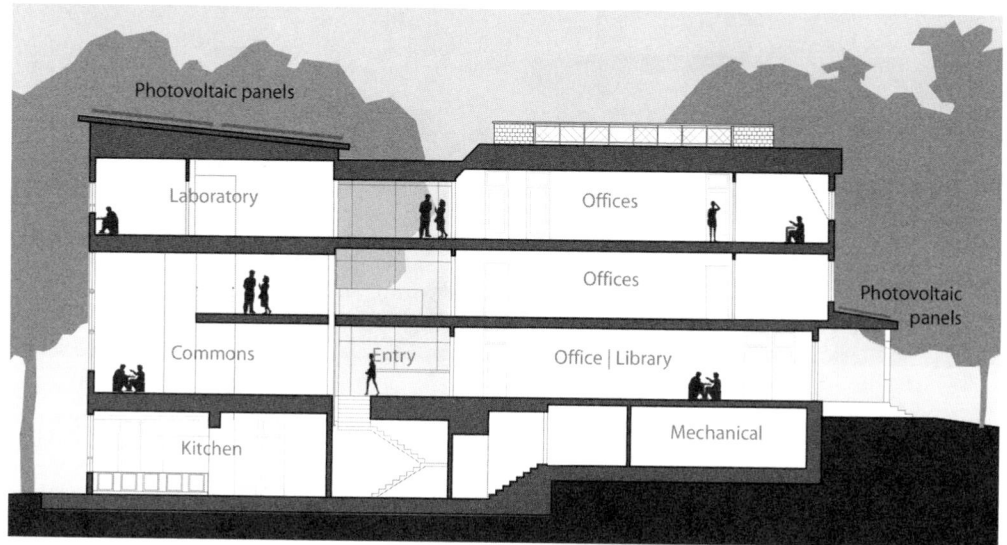

**Fig. 1.23** *Schematic design phase section through WHGC showing spatial organization and photovoltaic array locations. (© William McDonough + Partners; used with permission.)*

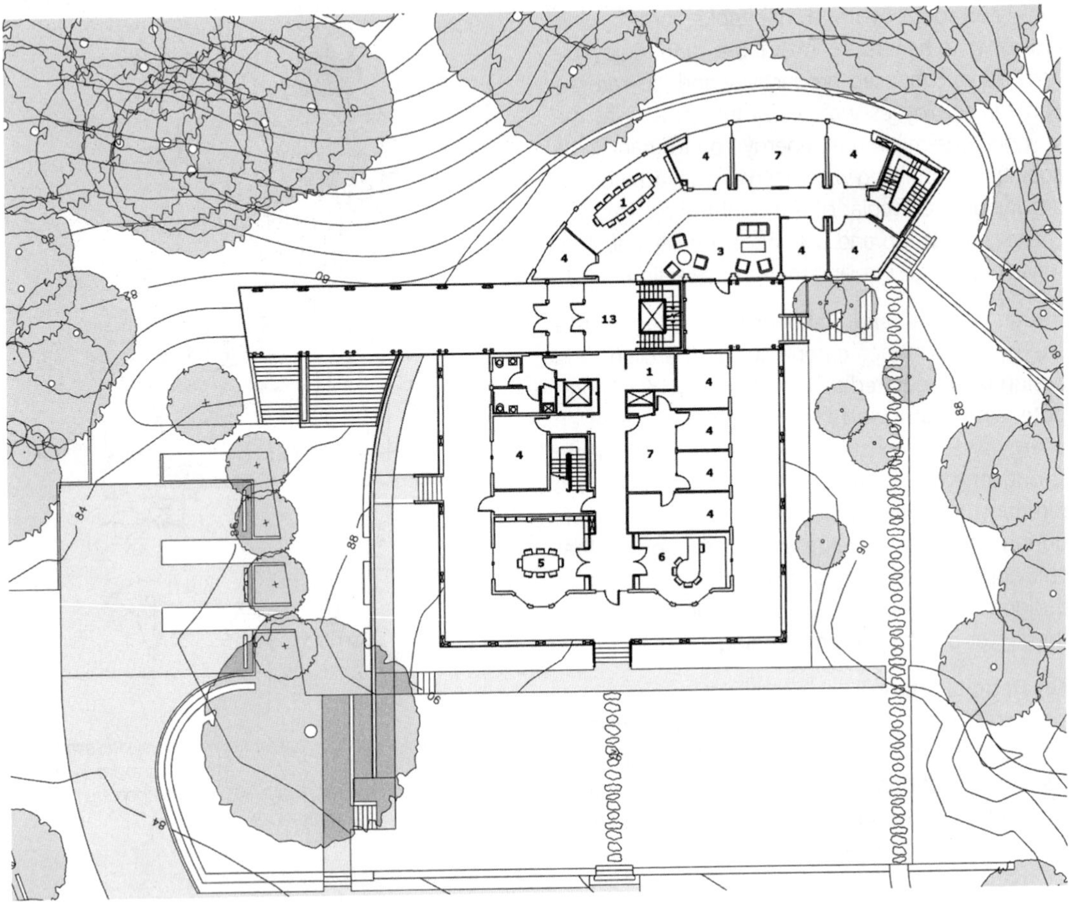

**Fig. 1.24** The site/floor plan of WHRC is representative of the evolution of a project as it moves into and through the design development phase. (© William McDonough + Partners; used with permission.)

(a)             (b)

**Fig. 1.25** Construction phase photos of WHRC; (a) showing the structure for the new addition and the existing house being remodeled; (b) showing the merger of new and remodeled parts of the building as the envelope enclosure is finalized. (© William McDonough + Partners; used with permission.)

**Fig. 1.26** *Exterior photo of the completed and occupied WHRC. (Photo © Alison Kwok; all rights reserved.)*

- A Ruck wastewater system, 95% on-site retention of stormwater, and collection of rainwater for site irrigation
- A ground source heat pump system for heating and cooling (coupled with a valence delivery system in office spaces)
- A rooftop, net-metered, photovoltaic array

**Post-Occupancy Validation Methods.** The client has installed an extensive energy monitoring and reporting system. Data collected by this system are available to the public via the World Wide Web (see below) and are also being used internally to optimize systems operations. Soils scientists from the Center are studying the effectiveness of the innovative septic system. In addition, the client has a very open and reflective attitude toward evaluation of the building and its systems. With a relatively small number of occupants, informal exchanges among Research Center users appear to be proving an effective means of POE.

**Fig. 1.27** *Bird's-eye view of the occupied WHRC building and site. Photovoltaic panels are a prominent feature on the roof. (© Cris Benton, kite aerial photographer and Professor, University of California–Berkeley; used with permission.)*

**Performance Data.** As this is a case study of design process as much as of a building, much of the following performance information relates to process outcomes.

- The building design received an AIA/COTE Top Ten Green Projects Award (2004).
- Measured data from the first year of occupancy show an energy consumption of about 20,000 Btu/ft$^2$ (227,200 kJ/m$^2$) per year; this is roughly 25% of the consumption of a typical office building and a 75% reduction from the energy density of the Research Center's previous facility.
- A grant from the Massachusetts Renewable Energy Trust allowed installation of a photovoltaic array consisting of 88 panels (each at 25 ft$^2$ [2.3 m$^2$]) that is expected to provide 37,000 kWh annually (about 40% of the building's power needs).

- All of the interior finish woodwork is a Forest Stewardship Council (FSC) certified sustainably harvested product; exterior wood finishes are also FSC certified, including cedar shingles and siding and Brazilian *ipé* wood for the extensive porch, deck, and entrance stairway.
- Paints and coatings meet low volatile organic compound (VOC) criteria; no carpet is used in the building.

### FOR FURTHER INFORMATION

Summary and real-time energy performance data for the Woods Hole Research Center building can be accessed at: http://www.whrc.org/building/education/performance.htm

A description of the building and design process can be found at http://www.aiatopten.org/hpb/overview.cfm?ProjectID=257

### References

Agents of Change. Department of Architecture, University of Oregon. http://aoc.uoregon.edu/

ANSI/ASHRAE/IESNA. 2004. *Standard 90.1: Energy Standard for Buildings Except Low-Rise Residential Buildings.* American Society of Heating, Refrigerating and Air-Conditioning Engineers, Inc. Atlanta, GA.

ANSI/ASHRAE. 2004. *Standard 55: Thermal Environmental Conditions for Human Occupancy.* American Society of Heating, Refrigerating and Air-Conditioning Engineers, Inc. Atlanta, GA.

ASHRAE. 2005. *Guideline 0: The Commissioning Process.* American Society of Heating, Refrigerating and Air-Conditioning Engineers, Inc. Atlanta, GA.

Fuller, S.K. and S.R. Peterson. 1995. *Life-Cycle Costing Manual for the Federal Energy Management Program.* National Institute of Standards and Technology. Gaithersburg, MD. http://fire.nist.gov/bfrlpubs/

ICC. 2003. *International Building Code.* International Code Council. Falls Church, VA.

ICC. 2003. *International Energy Conservation Code.* International Code Council. Falls Church, VA.

Lyle, J.T. 1996. *Regenerative Design for Sustainable Development.* John Wiley & Sons. New York.

McLennan, J.R. 2004. *The Philosophy of Sustainable Design.* ECOTone Publishing Company, LLC. Kansas City, MO.

NRC. 1995. *National Building Code of Canada.* National Research Council Canada, Institute for Research in Construction. Ottawa, ON.

Preiser, W.F.E. (ed.). 1989. *Building Evaluation.* Plenum Press. New York.

Schaeffer, J., et al. 1997. *A Place in the Sun: The Evolution of the Real Goods Solar Living Center.* Chelsea Green Publishing Company. White River Junction, VT.

USGBC. 2003. *LEED-NC 2.1: Leadership in Energy and Environmental Design (New Construction and Major Renovations).* U.S. Green Building Council. Washington, DC. Available online from http://www.usgbc.org/

Vital Signs. Center for Environmental Design Research, University of California–Berkeley. http://arch.ced.berkeley.edu/vitalsigns/

World Commission on Environment and Development. 1987. *Our Common Future* (the Brundtland Report). Oxford University Press. Oxford, UK.

CHAPTER 2

# Environmental Resources

THE DESIGNER OF TODAY'S BUILDINGS WILL ADD OR reshape spaces on a planet that has, for eons, evolved by using renewable energy arriving from the sun at a generally fixed and limited rate. Suddenly (in geologic time) our planet is experiencing population growth, nonrenewable resource depletion, and apparently global warming. Today's designers have available a vast, yet declining, reservoir of material resources, fossil fuel energy, and water. The building design professions thrive on a rapidly increasing population, each person consuming more resources than did her/his grandparents. Eventually, our planet must live within a fixed budget of renewable energy, water, and material resources. The question is: How can our present building designs best move toward this necessary accommodation with sustainability? How can building designers be environmentally proactive instead of simply reactive?

Population growth is both the source of much of our work as buildings professionals and the underlying source of our greatest problems. Our planet did not support a human population of 1 billion until about 1830, at which time the United States depended almost entirely upon the renewable energy sources of fuel wood and work animals; interior lighting was provided by burning oil or gas. In less than 200 years, 4 billion more people were added to our planet, and the United States shifted to almost total dependence upon nonrenewable fuels: coal, oil, and natural gas (Fig. 2.1). Another 1 billion in population is expected before the year 2010.

The building design process plays an active role in deciding where these people will live and work and how much of what kinds of resources they will use. The mechanical and electrical systems that support our new buildings can be part of a growing problem or an important start to a solution.

## 2.1 INTRODUCTION

Buildings depend upon energy and matter for their very existence and must pay heed to several fundamental rules of science. The First Law of Thermodynamics establishes the conservation of energy and matter (energy/matter can neither be created nor destroyed) and essentially states that you cannot get something for nothing. The Second Law of Thermodynamics expresses the tendency toward disorder that is part of the normal nature of things. Entropy is a measure of such disorder; as disorder increases, so does entropy. The Second Law is a declaration against perpetual motion, and essentially states that not only can't you get something for nothing, you can't even break even due to unavoidable losses (disorder) that contribute to increased entropy.

The construction and operation of buildings are fundamentally ordering processes. Materials are mined or harvested, refined or shaped, placed in manufactured products, transported to a building site, and assembled. All of these processes consume energy and materials as the various building systems are established. The operation and maintenance of a

**25**

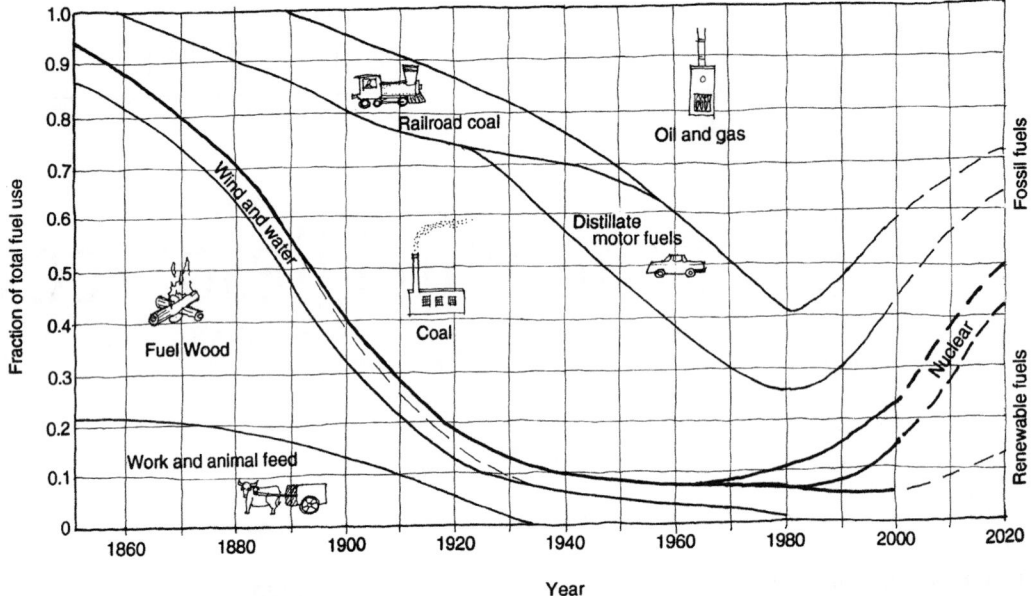

**Fig. 2.1** *U.S. fuel sources since 1850, showing a progression from dependence on renewable fuels (wood and work animals) to fossil fuels (coal, then oil and gas). Wind and water power were shifted from mills to electricity generation between 1890 and the present. Although not shown here, much fossil fuel is now converted to electricity before use. (Data 1850–1970 are from Fisher, 1974; data 1970–1980 are from Meyers, 1983; future projections are from Brower, 1990. Drawing by Michael Cockram; © 1998 by John S. Reynolds, A.I.A.; all rights reserved.)*

building consume further resources, as conditions (temperatures, light levels, flows of water) are established that would not otherwise occur under less ordered natural conditions. Maintaining building materials and systems over time against the forces of nature requires additional inputs of energy and materials. Buildings are inherently antientropic. This is not necessarily bad (learning and evolving are also antientropic), but it should be considered during the design process. As will be seen below, buildings have a substantial collective impact on our patterns of energy, water, and materials consumption.

The design process should consider various scales of concern relative to the impacts of buildings upon the environment. One such scale is geographic. Geographic scales of concern include the micro scale, the site scale, and the macro scale, with design focus historically being at the site scale. The terms *micro* (small) and *macro* (large) are not absolutely defined but are often referenced to a particular site. Thus, the area of influence of a microclimate is smaller than that of a macroclimate—but is usually also smaller than that of the site scale, often applying to one part of a site (perhaps with a steeper slope, lower elevation, or greater shading than other

parts). The site scale is normally self-explanatory, running from property line to property line. Energy efficiency issues are typically and historically addressed at the site scale and often ignore (unfortunately) energy consumption off site (such as electric power plant losses or natural gas transportation losses). Renewable/passive energy systems must consider microscale effects (such as orientation) in order to be successful. Nonrenewable/active systems often are oblivious to any scale of concern.

Time is another, and very interesting, scale of concern to building design. The scales typically addressed include *now* and *the future*—although *the past* is sometimes of concern with adaptive reuse and historic preservation projects. The concept of *the future* is usually left quite nebulous unless life-cycle costing is undertaken for a project, in which case the expected lifetimes of systems and equipment are explicitly estimated. It is clear that most buildings have a useful life of 25, 50, perhaps 100 years (or more). Stuart Brand provides an interesting look at buildings over time in *How Buildings Learn: What Happens After They're Built.* The problem with design for the future is that we don't know precisely what it holds. Nevertheless, design for sustainability requires the design team and the design process to consider the needs of future generations. This makes sustainability a very challenging concept and highly objective—but no less important than design for today.

## 2.2 ENERGY

Energy resources are broadly classified as renewable or nonrenewable. *Renewable* resources are those that are available indefinitely but are generally diffuse and arrive at a rate controlled by nature. For example, the influx of solar energy varies from day to day, but on average it should be available forever and at a generally predictable rate. Likewise, a woodlot produces a limited amount of wood per year but can do so for centuries if properly managed. An analogy for using renewable fuel sources is living on a fixed annual salary—with no hope for spectacular annual raises or unexpected bonuses but with long-term stability if wisely managed.

*Nonrenewable* energy resources are those that, once exhausted, cannot be replaced in a time frame that is meaningful to the human race. Coal, oil, and natural gas are examples of nonrenewable energy resources. Using nonrenewable fuel sources is analogous to living off a one-time lottery win that can be spent in 1 year or over 50 years, depending upon needs and planning, but that is gone for good when all spent.

The United States, like other industrialized countries, has spent the time since the mid-1800s in an energy transition (Fig. 2.1). This transition began with renewable energy sources, obtained locally, that did relatively low-grade work: animals pulled or pushed, and wood was burned to provide heated air, water, or steam. Buildings and people were directly affected by energy sources; work animals were fed, tended, and housed; fuel wood was cut nearby and stacked in large sheds (Fig. 2.2); fireplaces were social centers of buildings; and smoke from chimneys indicated activity inside. The side effects of energy use were also directly sensed—animal wastes, deforestation, and polluted air. Fuel use depended upon human labor. Architects of the era responded to the visual and spatial organization potentials of fireplaces and chimneys (Fig. 2.3).

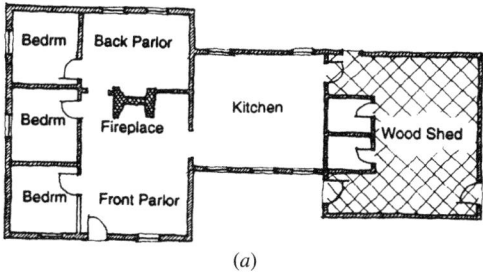

(a)

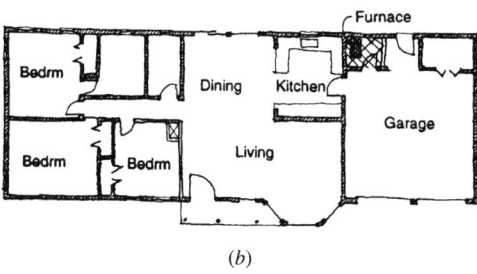

(b)

**Fig. 2.2** *Residential heating, past and present. (a) The house dependent on fireplaces or wood stoves also depends on someone to tend the fire. The warmer area near the fire in this early Oregon farmhouse was used for social purposes; the colder extremities served as sleeping areas and for storage of food and fuel. (Based upon a plan drawn by Philip Dole.) (b) The contemporary suburban home has either a small furnace area or electric heat built into each room. Heating equipment is no longer a major influence on building form.*

**Fig. 2.3** *The fireplace and the more efficient wood stove can inspire architectural form. This chimney symbolizes permanence as well as protection against the cold. The major social space of the house is marked both by the arched window and by the fireplace chimney. (Photo by William Johnston.)*

North America is now almost entirely dependent upon nonrenewable energy resources, an increasing proportion of which consists of imported oil and natural gas and electricity transported across substantial distances (Fig. 2.4). Buildings account for a good percentage of this energy demand. This trend is partly due to rapid growth in both population and per capita energy consumption. It is also due to the allure of highly concentrated energy available from fossil fuels, which encourages the use of high-quality energy sources such as electricity and natural gas for buildings and gasoline for transportation. People are now largely oblivious to their sources of energy: electricity is generated in far-off power plants; natural gas arrives through buried pipelines and fuel oil via supertankers. The experiential impact of energy use on building design and operation tends to be diluted. Energy consumption is regulated by automatic controls, and heating and cooling equipment is hidden from sight. Building occupants/users—who often do not personally control a thermostat, see climate control equipment, or pay a utility bill—have no direct contact with or concern about energy resources. Clients hire architects to provide for function and comfort. Architects then pay engineers to

design (and usually to successfully hide) mechanical and electrical equipment. Entropy, however, continues to increase.

For several reasons, buildings designed for today are likely to rely heavily upon electricity (Fig. 2.5), a situation that carries serious implications for resource depletion and environmental quality:

1. Consumption of electricity is expected to rise about twice as fast as overall energy demand, and we are more often using electricity in place of other energy forms. Part of the reason for this is that, for some primary energy sources (such as coal, heavy fuel oil, or nuclear), generation of electricity for subsequent (secondary) distribution to buildings is the only convenient usage option.
2. Other than daylighting (unfortunately still rare in today's buildings), electricity is the only source for building illumination. Heat produced by electric lighting may reduce a building's need for space heating, but it increases its need for space cooling—and mechanical cooling is almost universally provided by electric air-conditioning equipment.

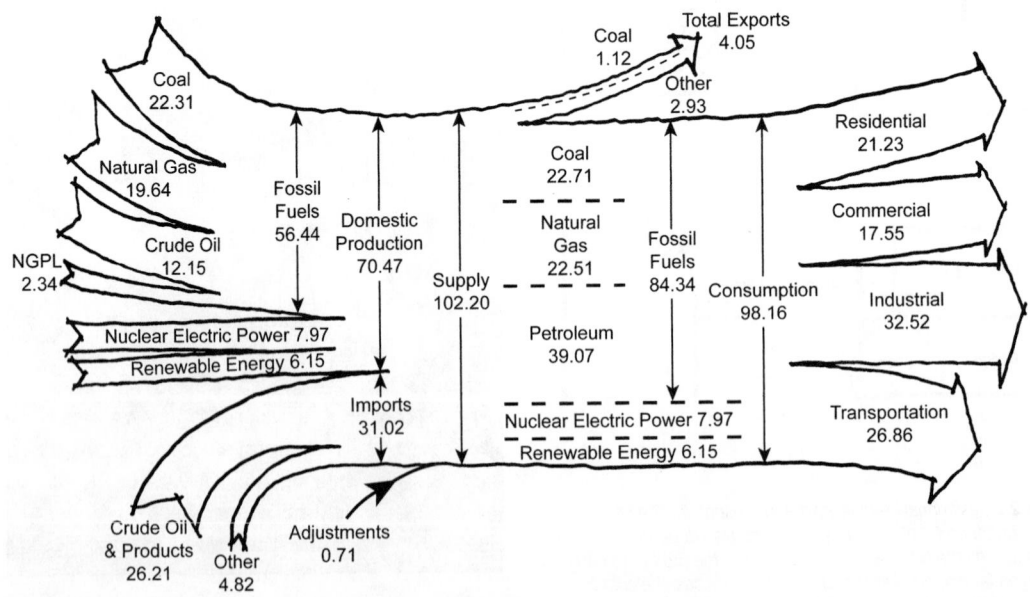

**Fig. 2.4** *U.S. energy flow, 2003: sources and end uses. Fuel types and sources are shown to the left and end use sectors to the right. Note the importance of residential and commercial consumption to total U.S. consumption—and the currently minuscule contribution of renewable energy sources to the whole. (Drawing by Jonathan Meendering using data from the Energy Information Administration, U.S. Department of Energy; Annual Energy Review, 2003.)*

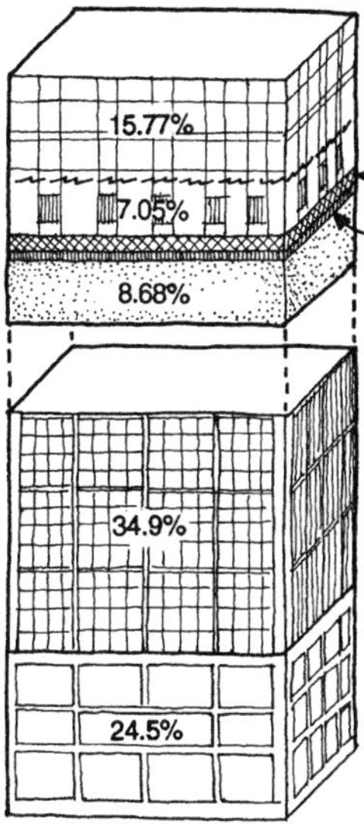

34.4 quads in U.S. buildings[1]

Electricity wasted
in production
15.77%

Electricity delivered
7.05%
2.24% Other[3]
0.65% Renewables[2]
Natural gas
8.68%

Industry
34.9%

Transportation
24.5%

**Fig. 2.5** *Sources of energy as consumed by end-use sector in the United States, 1996. Total: 94.0 quads (1 quad = 10^15 Btu). Not included are raw materials used in manufacturing. (1) Both residential and commercial buildings are included. (2) Renewables do not include passive solar energy use; hydroelectricity and PV are included in "electricity." (3) "Other" includes fuel oil, liquefied natural gas, coal, kerosene, and other petroleum products. (Data from U.S. Office of Building Technologies,* Core Databook, *1998. Drawing by Michael Cockram, © 1998 by John S. Reynolds, A.I.A., all rights reserved.)*

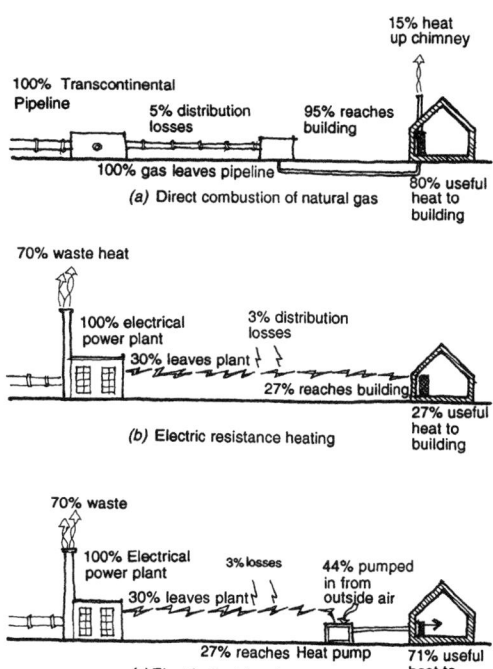

100% Transcontinental
Pipeline
5% distribution
losses
95% reaches
building
100% gas leaves pipeline
15% heat
up chimney
80% useful
heat to
building
*(a) Direct combustion of natural gas*

70% waste heat
100% electrical
power plant
3% distribution
losses
30% leaves plant
27% reaches building
27% useful
heat to
building
*(b) Electric resistance heating*

70% waste
100% Electrical
power plant
3% losses
44% pumped
in from
outside air
30% leaves plant
27% reaches Heat pump
71% useful
heat to
building
*(c) Electrically driven heat pump*

**Fig. 2.6** *Variations on higher-grade energy and lower-grade tasks. (a) Natural gas (a fossil fuel) is often burned in furnaces to provide low-grade space heating. With today's high-efficiency furnaces, over 80% of the gas' energy is delivered to the building as space heat. (b) However, when that natural gas is used instead to generate (higher-grade) electricity, and the electricity is then used for resistance space heating, the inefficiencies at the electric power plant cut deeply into the available energy: only about 27% is delivered to the building. (c) However, when the electricity generated by natural gas is used to drive a heat pump and the outdoor air is above freezing, about 71% of the gas' energy is delivered as space heat. (Drawing by Michael Cockram, © 1998 by John S. Reynolds, A.I.A., all rights reserved.)*

3. Electricity is a convenient and versatile energy form; it not only serves such high-quality and highly concentrated (or *high-grade*) tasks as lighting and providing drive power via electric motors, it can also serve such low-quality, low-temperature (or *low-grade*) tasks as cooking, water heating, and space heating (Table 2.1). All-electric buildings are commonplace, even though they are subject to paralysis in blackouts—as any building dependent upon a single energy source is vulnerable to disruptions. As shown in Table 2.1, of a total of 38 quads used for all building energy requirements, the primary energy used for electricity generation was 27 quads, equal to over two-thirds of the total.

4. Electricity generated by thermal processes (except for cogeneration) delivers to the end user less than one-third of the total energy that goes into its production; more than two-thirds is usually lost as waste heat at the generating plant (Fig. 2.6). (In Table 2.1, for every 1 quad of

electricity delivered, 3.22 quads were assumed used in generation.)

As we consume our planet's resources, including fossil fuels, many look to a return to renewable and sustainable energy. This vision will be partially implemented by solar energy converted directly to electricity (through photovoltaics [PV]) on or near the building requiring the electricity. Solar collectors (some for heating water, others for producing electricity) will shape the roofs and silhouettes of buildings. Building design professionals can choose to ignore or embrace such opportunities.

Hydrogen may be stored and distributed as a high-grade fuel, produced from water using electricity generated from renewable resources such as solar

**TABLE 2.1 Energy End Use in U.S. Buildings, by Fuel Type (Quads)[a]**

| End Use | Natural Gas | Fuel Oil[b] | LPG | Other Fuel[c] | Renewable Energy[d] | On-Site Electric | On-Site Total | On-Site % | Primary Electric[e] | Primary Total | Primary % |
|---|---|---|---|---|---|---|---|---|---|---|---|
| Space heating[f] | 4.96 | 1.02 | 0.30 | 0.19 | 0.40 | 0.69 | 7.55 | 38.6 | 2.21 | 9.08 | 23.7 |
| Space cooling | 0.01 | | | | | 1.43 | 1.45 | 7.4 | 4.62 | 4.63 | 12.1 |
| Ventilation[g] | | | | | | 0.31 | 0.31 | 1.6 | 1.01 | 1.01 | 2.6 |
| Water heating | 1.74 | 0.19 | 0.05 | | 0.05 | 0.55 | 2.58 | 13.2 | 1.77 | 3.79 | 9.9 |
| Lighting | | | | | | 2.12 | 2.12 | 10.9 | 6.84 | 6.84 | 17.8 |
| Refrigeration[h] | | | | | | 0.76 | 0.76 | 3.9 | 2.45 | 2.45 | 6.4 |
| Cooking | 0.47 | | 0.03 | | | 0.25 | 0.75 | 3.8 | 0.81 | 1.31 | 3.4 |
| Wet clean[i] | 0.07 | | | | | 0.29 | 0.36 | 1.8 | 0.94 | 1.01 | 2.6 |
| Computers | | | | | | 0.20 | 0.20 | 1.0 | 0.65 | 0.65 | 1.7 |
| Electronics | | | | | | 0.62 | 0.62 | 3.2 | 2.00 | 2.00 | 5.2 |
| Other[j] | 0.38 | 0.02 | 0.24 | 0.05 | 0.10 | 0.48 | 1.28 | 6.5 | 1.56 | 2.35 | 6.1 |
| Adjustments[k] | 0.64 | 0.22 | | | | 0.73 | 1.59 | 8.1 | 2.34 | 3.21 | 8.4 |
| Total | 8.27 | 1.46 | 0.62 | 0.24 | 0.54 | 8.45 | 19.58 | 100 | 27.20 | 38.33 | 100 |

*Source:* U.S. DOE (2004). Data are for the year 2002.

[a]Quad = $10^{15}$ Btu (1 Ej).

[b]Includes distillate fuel oil (1.38 quads) and residual fuel oil (0.08 quad).

[c]Kerosene (0.08 quad) and coal (0.11 quad) are assumed to be attributed to space heating; motor gasoline (0.05 quad) is assumed to be attributed to "other" end uses.

[d]Passive solar space heating is not included. It includes wood space heating (0.39 quad), geothermal space heating (<0.01 quad), solar water heating (0.05 quad), biomass (0.01 quad), and solar PV (<0.01 quad).

[e]Site-to-source electricity conversion = 3.22 due to generation and transmission losses.

[f]Includes electric furnace fans (0.25 quad).

[g]Commercial only (residential fan and pump energy use included proportionally in space heating and cooling).

[h]Includes refrigerators (1.37 quads) and freezers (0.43 quad) and commercial refrigeration.

[i]Includes clothes washers (0.10 quad), natural gas clothes dryers (0.07 quad), electric clothes dryers (0.76 quad), and dishwashers (0.08 quad).

[j]Includes commercial service station equipment, emergency electric generators, fuel oil cooking, natural gas-driven pumps, natural gas lighting, automated teller machines, telecommunications equipment, medical equipment, residential pool/hot tub heating, residential small electric devices, outdoor grilles, outdoor natural gas lighting, and the like.

[k]Energy Information Administration (EIA) adjustment to address discrepancies among data sources. Energy is attributable to the residential and commercial buildings sector but not directly to specific end uses.

energy and wind. The lifetimes of mechanical and electrical equipment specified today will probably overlap such a future. In this near future, lower-tech processes such as biomass conversion (combustion of wood and waste products) may develop faster than higher-tech processes such as PV. PV is growing very rapidly, though; its growth curve is as steep as that during the first 15 years of computer technology. One major oil company's energy scenario (in the late 1990s) anticipated 50% of world energy demand being met by alternatives to fossil fuels by the year 2050.

Today's fossil-fueled economy seems so entrenched as to defy a transition to renewable energy. A common question is often heard: Is solar energy adequate for our energy needs? Table 2.2 compares the Earth's receipt of solar energy at the surface in a single day with other energy phenomena. There appears to be adequate resource availability—given the will to move toward renewable and site-based resources. Energy efficiency efforts will play a critical role in making such a transition feasible.

To date, societal focus has been primarily upon using less energy—energy efficiency. The oil embargo of the 1970s spurred the development of energy efficiency standards, which have remained a fixture of building design ever since. In general, such standards seek to reduce building energy consumption—not to shift energy resources from nonrenewable to renewable. The green building design movement has provided momentum for a serious look at both reduced energy use and the use of energy from renewable resources (see Appendix G).

## 2.3 WATER

The building design profession's efforts toward a more resource-efficient product have, for the past 30-some years, focused primarily upon energy. This

**TABLE 2.2 Daily Arrival of Solar Energy on Earth Compared to Other Energy Quantities**

| | |
|---|---|
| Solar energy received each day | 1 |
| Melting of an average winter's snow during the spring | $1/10$ |
| A monsoon circulation between ocean and continent | $1/100$ |
| Use of energy by all mankind in a year | $1/100$ |
| A mid-latitude cyclone | $1/1,000$ |
| A tropical cyclone | $1/10,000$ |
| Kinetic energy of motion in earth's general circulation | $1/100,000$ |
| The first H bomb | $1/100,000$ |
| A squall line containing thunderstorms and perhaps tornados | $1/1,000,000$ |
| A thunderstorm | $1/100,000,000$ |
| The first A bomb | $1/100,000,000$ |
| The daily output of Boulder Dam | $1/100,000,000$ |
| A typical local rain shower | $1/10,000,000,000$ |
| A tornado | $1/100,000,000,000$ |
| Lighting New York City for one night | $1/100,000,000,000$ |

*Source:* Reprinted by permission from Lowry, W. 1988. *Atmospheric Ecology for Designers and Planners.* Peavine Publications, McMinnville, OR.

focus has been warranted by the limits upon nonrenewable energy sources imposed by the laws of thermodynamics. There is no option for the reuse or recycling of fossil fuel energy (as may be done with water and materials). Water concerns, however, are at crisis level in many parts of the United States, and water may well be the emerging limit to growth and development—especially locally and regionally—rather than energy.

Concerns about a viable supply of potable water have dominated politics and civil engineering in the arid western United States for a century. Surprisingly, Tampa, Florida, in the heavily rained-upon southeast, has a desalination plant to provide water for an otherwise underresourced region. Although water is a recyclable resource, it is not a renewable resource (no new daily supplies are being delivered to Earth). A quote generally attributed to *National Geographic* (October 1993) made this point succinctly: "All the water that will ever be is, right now" (UNH, 2004). In addition, where the water is, is not necessarily where it is wanted. Periodic water rationing is an unpleasant fact in many areas. Table 2.3 compares regional water resources with sustainable water usage capacity; it is clear that some areas of the United States now have serious water shortages. Accelerated depletion of underground water stocks (from aquifers, which are analogous to fossil fuel reserves) and maxed-out imports (both hydrologically and politically) suggest more trouble on the way. On a global scale, the disparities become even greater. Mostafa Tolba, former Executive Director of the UN Environment Program, speaking of the international picture, notes: "We used to think that energy and water would be the crit-

ical issues for the next century. Now we think water will be the critical issue" (UNH, 2004).

As with energy, per capita use of water involves more than consumption within a building. In the case of energy, transportation and industrial uses influence per capita consumption; with water, energy production and agricultural uses play a role. About half of all U.S. fresh and saline water withdrawals in 2000 were used in conjunction with thermoelectric power generation. Most of this was surface water used for once-through cooling at power plants. Withdrawals for this use have been relatively stable since 1985. (In a quirky turnabout, the California State Water Resources Control Board estimates that 6.5% of California's total electricity use is related to pumping and treating water.)

Irrigation remains the largest use of freshwater. Since 1950, irrigation has accounted for about 65% of total water withdrawals, excluding those for power generation. Historically, more surface water than groundwater has been used for irrigation. The percentage of total irrigation withdrawals from groundwater has continued to increase, from 23% in 1950 to 42% in 2000. Irrigated acreage more than doubled between 1950 and 1980, then remained constant before increasing nearly 7% between 1995 and 2000 (USGS, 2004).

Public water supply withdrawals in 1950 were 14 Bgal/day (53 GL/day); in 2000, more than 43 Bgal/day (163 GL/day). During 2000, about 85% of the U.S. population obtained drinking water from public suppliers compared to 62% during 1950. Surface water provided 63% of the total during 2000 compared to 74% during 1950. Potable

**TABLE 2.3  Comparison of Regional Water Use versus Resources for the Continental United States**

| Water Resources Region[a] | Consumptive Use | Renewable Water Supply[b] | Ratio (Use/Supply) |
|---|---|---|---|
| New England | 0.6 | 78.4 | 0.8% |
| Mid-Atlantic | 1.3 | 80.7 | 1.5% |
| South Atlantic/Gulf | 6.1 | 233.5 | 2.6% |
| Great Lakes | 1.9 | 74.3 | 2.6% |
| Ohio | 2.3 | 139.6 | 1.7% |
| Tennessee | 0.3 | 41.2 | 0.7% |
| Souris-Red-Rainy | 0.5 | 6.5 | 7.7% |
| Upper Mississippi | 2.3 | 77.2 | 3.0% |
| Lower Mississippi | 40.3[c] | 484.8 | 8.3% |
| Missouri | 17.5 | 52.9 | 33.1% |
| Arkansas-White-Red | 9.6 | 68.7 | 14.0% |
| Texas Gulf | 9.1 | 33.1 | 27.5% |
| Rio Grande | 3.5 | 5.4 | 64.8% |
| Upper Colorado | 4.2 | 13.9 | 30.2% |
| Lower Colorado | 10.6[c] | 10.3 | 103.0% |
| Great Basin | 3.5 | 10.0 | 35.0% |
| Pacific Northwest | 11.2 | 276.2 | 4.1% |
| California | 25.8 | 74.6 | 34.6% |

*Source:* Adapted from United States Geological Survey, 1984, with 1995 updates for water usage: http://water.usgs.gov/watuse/misc/consuse-renewable.html/

[a]These are water resource areas generally independent of state boundaries.

[b]Renewable water supply represents a long-term sustainable resource; precipitation plus imports less evaporation, exports, and water needed to maintain minimal stream flows.

[c]These values are for the entire river system.

water obtained from a surface source via a public supply system is becoming the norm; private well systems are less common than in the past. As with energy, population increases are increasing overall consumption, while efficiency efforts provide some counterbalancing effect.

Water-efficiency standards are not nearly as extensive or ubiquitous as energy-efficiency standards, although most building users are aware of low-flow toilet requirements, flow restrictors for showers, and self-closing bathroom faucets in public facilities. Surprisingly, given the few design restrictions that exist, per capita water use in the United States has remained flat for the past several years—and is currently 25% lower than in the late 1970s. This is partly because overall per capita use involves important sectors other than buildings (agriculture and power generation, for example) and partly due to increasing awareness of the value of water.

As with renewable energy sources, green building design efforts have also increased awareness of and design for water supply savings and alternatives. Part V: Water and Waste discusses many design alternatives that would likely be used in a green building. The role of water in the LEED rating system is outlined in Appendix G.

## 2.4  MATERIALS

A global or even countrywide perspective on materials resources is more difficult to obtain than is the case for energy and water. In the United States there appears to be a general upward trend in per capita consumption of materials. In the early 1900s, per capita consumption (in metric tons per person) was around 3.0, about 50% of which was construction materials; in 1950 about 6.0, with about 60% in construction materials; and from 1970 to 1990 about 10.0 (with ups and downs), with about 65% in construction materials.

Given the scarcity of quantitative data, a qualitative comparison with water must suffice. Many materials used in building construction and upkeep come from a generally fixed resource base. Like water, materials (at least many) may be recycled, but there is a fixed quantity of resources available on Earth. For many materials, what we have now is what we will have in the future, a marked exception being those organic materials (such as fiber products) that are renewable.

Finding adequate material resources for a building directly on the building site is rare: wood, straw, and earthen construction systems, for example, require large land areas to supply the materials for even a small building. The most commonly used construction systems involve materials brought to a site from some distance. A designer can generally select between imported renewable or nonrenewable materials and between imported virgin or recycled materials. The common thread is *imported*—not necessarily from overseas, but from a distance. Reducing the transportation distance and finding local materials when available are emerging design strategies.

Wood is the only renewable construction material currently in wide use in North America. It is easily worked, supports a wide variety of finishes, has moderate structural strength, requires regular maintenance for long life, burns easily, and has only moderate value either as thermal mass or as insulation. This common building material illustrates the impact a rapidly increasing population can have on a fixed, even if renewable, resource base. As huge old trees are harvested to the point of disappearance, growing demand for wood outstrips the supply available from younger, smaller trees. New production methods are devised (such as laminates, particle board, and engineered lumber) to allow large wood members to be constructed from smaller timber. The value of salvaged older wood members increases.

Nonrenewable materials are by far the most commonly used materials in mechanical and electrical systems; metals and plastics predominate. Their advantages include strength, durability, fire resistance, and conductivity or resistivity as required. Most such materials are obtained, however, at a significant energy cost to mine/manufacture, transport, and shape them for our use.

With rapidly increasing demand for both renewable and nonrenewable materials, how can a designer on a planet with fixed resources respond? What are the key materials issues for the design team to consider?

## (a) Embodied Energy

One issue is *embodied energy*, a complex and therefore elusive indicator of how much energy must be invested to mine/harvest/produce, fabricate, and transport a unit of building material. Table 2.4 summarizes information on embodied energy for common units of today's prevalent construction materials. Such apparently simple numbers are complicated by variations in availability of the raw

resource (more work needed to extract materials requires more energy), variations in distance from raw resource to manufacturing locations, and variations in the fuels used (and their efficiency of use) in the refining or fabricating processes.

Consider two alternatives for exterior wall cladding: wood siding and aluminum siding. Wood has embodied energy from a renewable resource—the sun. It takes the energy of humans and chain saws to cut the trees and fuel to haul them, perhaps 100 miles (160 km), to a mill. At the mill, more energy is invested as logs become lumber, and still more energy is used as lumber becomes finished siding—which is then transported to a construction site. Aluminum begins as bauxite, which requires energy to mine, then more energy to ship great distances to smelters, which use large quantities of electricity in the refining process. Cheap electricity (as in the Pacific Northwest, with its once-surplus hydropower) attracts bauxite mined thousands of miles away. Once aluminum ingots are formed at the smelter, they are shipped—again, sometimes thousands of miles—to factories that make products such as siding; the products are then transported to a construction site. For a given surface area of finished siding, the aluminum alternative represents about 100 times as much embodied energy as the wood. This, however, is not the end of the story—as a designer must consider the impacts of these two siding materials on building energy consumption and envelope maintenance and replacement needs and schedules.

## (b) Recycled or Virgin Material

It seems paradoxical that, while the world's population is increasing and its raw materials are decreasing, labor costs are growing so much faster than the costs of raw materials. One consequence is that labor-intensive practices become less economically attractive. Recycling is one such labor-intensive practice (see examples in Chapter 23), whether at the scale of a household, an office building, or an entire industry.

Building construction, renovation, and demolition involve many opportunities for recycling, but at present these activities represent a major source of waste. The U.S. Environmental Protection Agency's 1998 report *Characterization of Building-Related Construction and Demolition Debris in the United States*

DESIGN CONTEXT

### TABLE 2.4 Approximate Total Embodied Energy in Building Materials[a]

| Per Volume | | Per Weight | | Building Material | Per Area | | Per Unit | |
|---|---|---|---|---|---|---|---|---|
| Btu/Unit | MJ/Unit | Btu/lb | MJ/kg | | Btu/ft² | MJ/m² | Btu | MJ |
| | | | | MASONRY | | | | |
| | | 4,000 | 9.3 | Brick: clay fired[b] | | | 14,000/brick | 14.8/brick |
| | | | | ceramic glazed[c] | | | 33,400/brick | 35.2/brick |
| | | 123 | 0.3 | Adobe: semistabilized[b] | | | 3,700/block | 3.9/block |
| | | 730–960 | 1.7–2.24 | Concrete block[b] | | | 24,100–31,800/ | 25.5–33.6/block |
| | | | | Quarry tile[c] | 51,000 | 580 | block | |
| | | | | Ceramic tile[b] | 25,160 | 285 | | |
| | | | | Concrete | | | | |
| | | | | Ready-mix concrete[d] | | | | |
| 1,500,000/yd³ | 2,070/m³ | | | 3,000 psi | | | | |
| 1,700,000/yd³ | 2,346/m³ | | | 4,000 psi | | | | |
| 2,000,000/yd³ | 2,760/m³ | | | 5,000 psi | | | | |
| | | | | Ingredients | | | | |
| | | 2,400–4,000 | 5.6–9.3 | Portland cement[b] | | | | |
| 43,200/yd³ | 60/m³ | | | Sand for concrete, unprocessed[d] | | | | |
| 67,500/yd³ | 93/m³ | | | Sand, washed | | | | |
| 180,225/yd³ | 250/m³ | | | Crushed stone, dry, for concrete[d] | | | | |
| | | 2.2 | 0.005 | Slaked lime[e] | | | | |
| | | 12,680 | 29.5 | Reinforcing steel: 25 mm[e] | | | | |
| | | | | Reinforcing bars[c]: #2 | | | 2,600/ft | 0.84/m |
| | | | | #8 | | | 41,800/ft | 13.4/m |
| | | 3,900 | 9 | Welded wire mesh[c]: 2 × 4, 14/14 | | | | |
| | | 25,400 | 59 | 2 × 12, 8/8 | | | | |
| | | | | METAL FRAMING | | | | |
| | | 19,200 | 44.7 | Steel framing[b] | | | | |
| | | | | Steel shapes[c]: | | | | |
| | | | | W12 × 65, carbon | | | 1,217,800/ft | 390/m |
| | | | | W12 × 65, alloy | | | 1,749,200/ft | 560/m |
| | | | | WT6 × 27, carbon | | | 543,350/ft | 175/m |
| | | | | WT6 × 27, alloy | | | 780,400/ft | 250/m |
| | | 14,060 | 32.7 | Angles, 50 × 8 mm[e] | | | | |
| | | 16,830 | 39.1 | Joists, 203 × 152 × 52 mm (1 kg/m)[e] | | | | |
| | | | | Aluminum shapes[c]: | | | | |
| | | | | 8 I 8.81 | | | 811,800/ft | 260/m |
| | | | | 6 I 5.10 | | | 469,900/ft | 150/m |
| | | | | WOOD FRAMING | | | | |
| 91,620/ft³ | 3,400/m³ | | | Wood framing[b] | | | | |
| | | | | 2 × 4[b] | | | 3,750/ft | 1.2/m |
| | | | | Lumber[c] | | | 7,600–9,800/bd ft | (8–10.3/bd ft) |
| 160,800/ft³ | 6,000/m³ | 5,720 | 13.3 | Glue laminated timbers[b] | | | 13,400/bd ft | (14.2/bd ft) |
| | | | | METAL SHEETS | | | | |
| | | | | Steel[c]: 22 guage | 29,400 | 333 | | |
| | | | | 16 guage | 58,800 | 667 | | |
| | | | | Galvanized[c]: 22 guage | 49,800 | 565 | | |
| | | | | 16 guage | 98,500 | 1,118 | | |
| | | 40,330 | 93.8 | Corrugated galvanized iron (0.6 mm)[e] | | | | |
| | | | | Aluminum plate[c]: ¼ in. | 420,700 | 4,775 | | |
| | | | | 1 in. | 1,680,300 | 19,070 | | |
| | | 138,300 | 321.7 | Stainless steel[c]: cold rolled | | | | |
| | | 80,800 | 188 | hot rolled | | | | |
| | | 265,140 | 617 | Copper[e] | | | | |
| | | 105,620 | 245 | Lead[e] | | | | |

**TABLE 2.4 Approximate Total Embodied Energy in Building Materials[a] (Continued)**

| Per Volume | | Per Weight | | Building Material | Per Area | | Per Unit | |
|---|---|---|---|---|---|---|---|---|
| Btu/Unit | MJ/Unit | Btu/lb | MJ/kg | | Btu/ft² | MJ/m² | Btu | MJ |
| | | | | WOOD PRODUCTS | | | | |
| | | | | Shingles[c] | | | 7,300/bd ft | (7.7/bd ft) |
| | | | | Plywood: ⅜ in. softwood[c] | 5,000–5,800 | 56.8–65.8 | | |
| | | | | Flooring[c] | | | 10,300–14,300/bd ft | (10.9–15.1/bd ft) |
| | | | | Mouldings[c] | | | 17,900 bd ft | (18.9/bd ft) |
| | | | | Mineral surface insulating board[c] | 67,500 | 766 | | |
| | | | | ROOFING (SEE ALSO WOOD PRODUCTS) | | | | |
| | | 13,630 | 31.7 | Asphalt shingle[b]: | | | | |
| | | | | self-sealing | 29,730 | 337 | | |
| | | | | regular strip | 25,330 | 288 | | |
| | | | | Rolled roofing[c] | 7,800–11,000 | 89–125 | | |
| | | | | Saturated felt[b] | | | | |
| | | | | 15 lb | 1,840 | 21 | | |
| | | | | 30 lb | 3,680 | 42 | | |
| | | | | PLASTER AND LATH | | | | |
| | | 12,000 | 27.9 | Lath board[b] | 2,600 | 29.5 | | |
| | | | | Steel lath[b] | | | | |
| | | | | Gypsum board[b] | 2,600 | 29.5 | | |
| | | | | ⅜ in.[c] | 5,800 | 60.2 | | |
| | | | | GLASS | | | | |
| | | 6,750–7,500 | 15.7–17.4 | Glass[b,g] | | | | |
| | | | | Flat glass: | | | | |
| | | | | double-strength | 15,430 | 175 | | |
| | | | | tempered | 72,600 | 824 | | |
| | | | | Plate and float glass[c]: | | | | |
| | | | | ⅛–¼ in. | 48,000 | 545 | | |
| | | | | Laminated plate glass[c]: | | | | |
| | | | | ¼ in. | 212,500 | 2,412 | | |
| | | | | THERMAL INSULATION | | | | |
| | | 13,000 | 30.2 | Fiberglass[b] | | | | |
| | | 50,400 | 117 | Polystyrene[b] | | | | |
| | | 31,000 | 72.1 | Polyurethane[b] | | | | |
| | | | | Mineral wool, 4½ in.[c] | 8,300 | 94.2 | | |
| | | | | ACOUSTICAL CEILING SYSTEMS | | | | |
| | | 19,200 | 44.7 | Steel suspension systems[b] | | | | |
| | | 103,500 | 241 | Aluminum suspension systems[b] | | | | |
| | | | | FLOORING (SEE ALSO WOOD PRODUCTS) | | | | |
| | | 7,350 | 17.1 | Linoleum[b] | | | | |
| | | 22,560–27,730 | 52.2–64.5 | Vinyl[b] | | | | |
| | | 5,900 | 13.7 | Vinyl composition tile[b] | | | | |
| | | | | Modified resin vinyl tile[e] | 68,370 | 776 | | |
| | | | | FINISHES | | | | |
| 503,670/gal | 1,400/L | 33,303 | 76.8 | Water-based paint[b] | | | | |
| 437,000–508,500/gal | 1,220–1,420/L | | | Stains and warnishes[b] | | | | |
| | | | | Paints[c] | | | | |

[a]These numbers are rounded in most cases. Due to the quirky nature of many material dimensions, conversions to-from I-P and SI sizes are not attempted.

[b]From American Institute of Architects (1996).

[c]From Hannon et al. (1977).

[d]From Construction Technology Laboratories, Martha Van Geem, P.E. (correspondence, 1997). Metric conversions by the author.

[e]From Irurah (1997).

[f]Plaster information not yet available.

[g]Glass products currently available with 30% less embodied energy; under development with 60% less embodied energy.

estimated that 136 million tons per year (123 Mg) of such material is produced in the United States and that 65% to 85% of that total ends up in landfills.

Consider building demolition. If more of a demolished building can be recycled, more of the energy embodied in its material can be recovered and fewer virgin materials will be required for some other project. At present, the recovery of usable materials from demolition is limited because the cost of labor is high and the cost of energy and new products is relatively low. It is currently easier, quicker, and cheaper to reduce a building to rubble and haul it to a landfill than to recycle. As landfill capacity becomes scarce, design regulations concerning recycled material use can be expected.

An Atlantic City, New Jersey, project was able to recycle 90% of its demolition waste. Of a total of 1583 tons (1400 Mg) of demolition waste, only 152 tons (140 Mg) were nonrecyclable. Concrete and masonry became crushed aggregate for road building. Glass became "glasphalt" embedded in road surfaces as reflectors. Wood waste became mulch. At Fort Ord in California, four buildings totaling about 11,000 ft² (1022 m²) were dismantled rather than demolished, saving roofing boards, framing lumber, and tongue-and-groove wood flooring. Unpainted drywall was reclaimed for composting.

Construction recycling opportunities include crushed wallboard as a replacement for lime in agriculture, carpet ground up for attic insulation, plate glass crushed for use in glass fiber insulation, and pulverized wood as a composting aid at sewage sludge treatment facilities. Used acoustic ceiling tiles can become part of the slurry from which new acoustic tiles are made. Building materials are now increasingly being made from recycled materials:

reinforcing bars from ferrous scrap metal; cellulose insulation from newsprint; parking lot bumper strips, fence posts, and park benches from recycled plastics; and nonstructural concrete from incinerator ash. Even plastic yogurt containers, complete with scraps of aluminum foil, are made into a terrazzo-like floor tile.

Architect Pliny Fisk, codirector of the Austin, Texas–based Center for Maximum Potential Building Systems, has developed a wide array of such applications. His "Advanced Green Builder" home near Austin displays several applications of "ashcrete," made with coal fly ash and bottom ash, producing a 97% recycled-content concrete. This is used as ferro-cement for columns and beams and is foamed for hollow wall infill. Numerous other innovative recycled-material applications are showcased as well.

The most effective form of recycling involves reuse of a building or building shell. Audubon House in New York City is an excellent and well-publicized example of this level of reuse. The next most effective form of recycling involves the reuse of a building component as is. A residential demolition-by-hand salvage project in Portland, Oregon, recovered doors, windows, bathroom fixtures, framing lumber, plywood, siding, flooring, and bricks. The energy and economic summary for this project is shown in Table 2.5.

As with water and renewable energy sources, much of the current interest in lowering the embodied energy content of construction materials and increasing the recycling and reuse of products is attributable to green design efforts. The role of LEED in promoting a change in thinking about materials is outlined in Appendix G. The impact of building materials on occupant health and well-being is also an

**TABLE 2.5 Residential Salvage for Reuse**

| | Embodied Energy[a] Btu/ft² Floor Area (kJ/m² Floor Area) | Value U.S. $/ft² Floor Area (U.S. $/m² Floor Area) |
|---|---|---|
| Total for reusable salvage[b] | 46,890 (532,497) | 4.90 (52.74) |
| Demolition energy consumed[c] | −3,380 (−38,384) | |
| Value of energy embodied in salvage[d] | | +0.50 (5.38) |
| Value of avoided dumping fees | | +2.70 (29.06) |
| Total energy savings and value | 43,510 (494,112) | 8.10 (87.19) |

*Source:* Joslin et al. (1993).

[a]Based on Stein et al. (1981).

[b]Framing lumber alone constituted 38% of this embodied energy.

[c]Gasoline for transportation and hauling, plus human labor at 254.6 Btu/h (268.6 kJ/h).

[d]Assumed at $.04/kWh, very low rate typical of the Pacific Northwest.

emerging area of concern and interest. There is a direct link between the selection and maintenance of building materials and indoor air quality (see Chapter 5). Although beyond the scope of this book, LEED also looks at this aspect of materials use, and several reference texts provide fundamental information on design for healthy buildings. Many building products with substantially reduced heath impacts have been developed and marketed during the past 10 years.

## 2.5 DESIGN CHALLENGES

The buildings we design today are very likely, over their lifetimes, to experience major changes in the way they are used and in their sources of energy supply. The societal value of water and embodied materials will also probably change. The resource perspective of the future is unlikely to be that which we hold today. This poses some overall challenges to the designer.

### (a) Design for Building Recycling

Designing for the recycling of buildings is a two-part balancing act. First, the designer should provide enough flexibility to prolong the useful life of a building by enabling it to adapt easily to changed usage. Flexibility, however, can be expensive to implement physically and can result in a bland "sameness" throughout a building. The latter characteristic is easier for the designer to change than the former. Second, the design can allow for demounting of parts so that the structure can remain safely intact while reusable materials and components are removed. This can result in heavier buildings in which floor systems are not structurally integrated with beams. This approach also discourages integration of mechanical and structural systems, as discussed in Chapter 10. Furthermore, a demountable building may be especially subject to energy leaks, such as from cracks widening around self-contained components of the façade.

Some initial guidelines for recyclable buildings are as follows:

1. Design the structure to be separable from everything else and to be easily disassembled. Extensive remodeling is then possible without major structural modifications, and at the end of a building's life, elements of its structure can be reused elsewhere.
2. Design for "breathing room" where possible: between a building and its neighbors or between major spaces within a building. Some expansion is thus possible without rebuilding. This could include designing the columns and footings to support an extra floor or two for vertical expansion.
3. Maximize the utilization of on-site (natural) forces such as sun and wind. The less sophisticated the mechanical and electrical equipment, the less obvious will be the obsolescence of such equipment with the passing of time.
4. Use materials and components distinctly: avoid combinations that make recycling of these elements difficult. A steel or plastic pipe embedded in a concrete slab is neither easily repaired nor easily recycled; some "sandwiches" (manufactured building panels) do not allow metals, plastics, and other products they contain to be separated for reuse at the end of the panel's life.

Although maximum savings of embodied energy can be realized when a building component is reused as is, even the crushing and reprocessing of some (separated) building materials will save energy compared to their original manufacture from virgin material (see Chapter 23).

### (b) Design for Energy Transition

Two more challenges to designers arise:

1. To design buildings not only to save energy, but also so that they can eventually be weaned away from dependence on nonrenewable fuels. A transition away from electricity from the utility grid, to site-generated photovoltaic or fuel-cell electricity, seems easy enough given appropriate building orientation, collection surfaces, and equipment spaces.
2. To use energy wisely; to expect only a fair share of locally available renewable fuels, recognizing that such resources are limited even though they are continuously available. For example, in a high-density setting, it may be tempting to erect a large solar collector to intercept sunlight that would otherwise be utilized by a neighboring building. This temptation grows stronger as a building is designed to rely more heavily upon

the sun. The concept of a *solar envelope* to protect each building's fair share is discussed in Section 3.6.

## (c) Design for the Information Age

Controls for mechanical and electrical systems have become much more sophisticated, thanks to developments in information systems and electronics. With the advent of *smart houses, intelligent buildings,* and *smart appliances,* it is now possible to regulate an array of building systems collectively and across great distances to optimize performance and minimize resource consumption. For a building, with some zones requiring heating and other zones cooling, some zones with available daylight and others without, an automatic central control system can, without human intervention, integrate the flow of fresh air, sunlight through movable shading devices, and intensity of electric lighting to achieve maximum use of on-site renewable energy.

The Albany County Airport in New York State uses automated controls to regulate solar gain through a large central skylight (Fig. 2.7). A computer monitors indoor and outdoor temperatures, keeps track of solar altitude and azimuth, and then

(a)

*Fig. 2.7 The Albany County (New York) Airport features a central skylight (a) that provides 40% of the light and 20% of the heat for the building. (b) The insulated louvers are controlled by computer to admit or block the sun and to store heat within the building on winter nights. (Courtesy of Einhorn Yaffee Prescott, Architects, Albany, NY. Redrawn by Amanda Clegg.)*

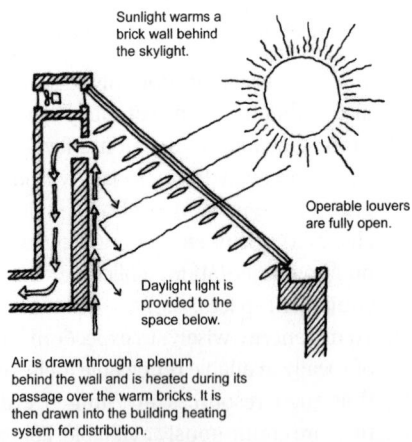

Sunlight warms a brick wall behind the skylight.

Operable louvers are fully open.

Daylight light is provided to the space below.

Air is drawn through a plenum behind the wall and is heated during its passage over the warm bricks. It is then drawn into the building heating system for distribution.

**Sunny Winter Day**
The sun is used to provide heat and light

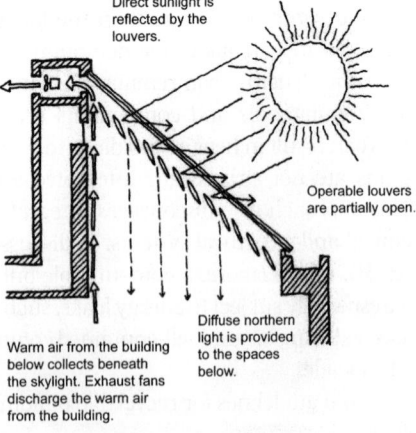

Direct sunlight is reflected by the louvers.

Operable louvers are partially open.

Diffuse northern light is provided to the spaces below.

Warm air from the building below collects beneath the skylight. Exhaust fans discharge the warm air from the building.

**Sunny Summer Day**
The sun is used to provide light, but is not allowed to penetrate the building and generate excessive heat

(b)

regulates the insulated shading louver position according to the building's heating or cooling needs. Solar gain is stored in a masonry wall that supports the skylight. A plenum behind the wall then heats air to be supplied to the vestibule areas of the airport, where winter heat losses are greatest. Photoelectric controls turn off electric lighting when daylight is adequate. The skylight provides 40% of the lighting and 20% of the heating needs of this 57,000 ft$^2$ (5295 m$^2$) building.

Information systems promise enormous energy conservation achievements while using very small amounts of energy themselves. They also require one of the least space-consuming distribution systems, or *distribution trees*, of all building service systems, especially compared to air ducts and plumbing pipes. In return for such agreeable characteristics, building designers must recognize that developments in information technology are so rapid that the nature of these systems is likely to undergo frequent and dramatic change. Information system distribution spaces may be quite small, but they must be highly accessible. Where information can be transmitted without wires or cables, the impact on building service space demands is even smaller. Adaptable information systems can make more feasible the renovation, rather than demolition, of older buildings for new tenants. These potentials notwithstanding, the need of occupants to play some role in the use and control of their environments should not be ignored.

## (d) Design for Transportation

There is clearly a link between design decisions at regional and neighborhood scales (urban planning and subdivision design), transportation, and resulting energy use for commuting, shopping, and recreation. This sphere of concern, however, is beyond the scope of this book. The link between transportation systems and building mechanical/electrical equipment may seem obscure, but consider the impact on buildings of the automobile and its internal combustion engine. Fresh air intakes at street level face significant pollution from engine exhausts. Parking lots below buildings compete for space with heavy mechanical and electrical equipment such as boilers, chillers, ice storage tanks, and switchgear and greatly complicate vertical transportation design. Sloped parking floors make future space use for other purposes quite difficult. Ventilating parking levels to remove fumes from automobiles requires large fans and energy to run them.

In a likely future of electrically powered vehicles, photovoltaic arrays over parking areas can provide electricity for a building and recharge the batteries of parked cars throughout the day (Fig. 2.8). In an alternative future of hydrogen fuel cell–powered vehicles, engine discharge consists simply of water. No fumes are emitted from such cars, saving fan space and energy. To the extent that cars become smaller, or are replaced by public transit or bicycles, significant space may be reclaimed

**Fig. 2.8** *The winning entry in a competition (sponsored by the New York Power Authority and the New York State Association of Architects) features a canopy of PV (solar) cells over a 19-acre (7.7-hectare) parking lot. In a future of electric-powered automobiles, such arrays can charge the batteries of parked cars, trucks, and buses while shading them. (Courtesy of Kiss Cathcart Anders Architects, New York, NY.)*

from parking for other uses. When entire buildings are now built as single-purpose parking garages, skillful design for reuse might allow such buildings to be renovated to serve new functions rather than demolished.

## 2.6 HOW ARE WE DOING?

From the discussion above and that in Chapter 1, it might seem that, environmentally, the building professions are doing pretty well. There are minimum standards for energy efficiency and plumbing fixture water consumption that affect virtually every North American building. Such standards are also common internationally. There is growing interest in green buildings, generally fueled by private sector and government owners seeking to set an example, which is moving design beyond the just-acceptable minimum requirements of codes and standards. Concern for energy consumption, renewable energy use, water resources and quality, and materials resources and consumption is an integral element of the green building movement. Per capita energy

and water use in the United States appear to be stable and/or deceasing.

From the perspective of yesterday, today's building designs (even the worst) are arguably more resource-efficient and respectful of the environment (this does not necessarily mean they are better designs). The question is: From the perspective of tomorrow, is today's good design good enough? The answer in one context is, unfortunately, no. That context is the *environmental footprint*. Environmental footprints are a concept promoted by Rees and Wackernagel (1995) that plot the gross resource demands of a geographic area as a *footprint* on the planet. Figure 2.9 provides an illustration of the environmental footprint concept applied specifically to water resources. The area in question may be a city, state or province, or country. If the footprint is larger than the geographic boundaries of the area in question, then the area is stepping on someone else's environmental toes. Such a city, state/province, or country needs more land to support itself than is available—thereby surviving through imports from other places. All is well as long as there are other places with surpluses; all is not well

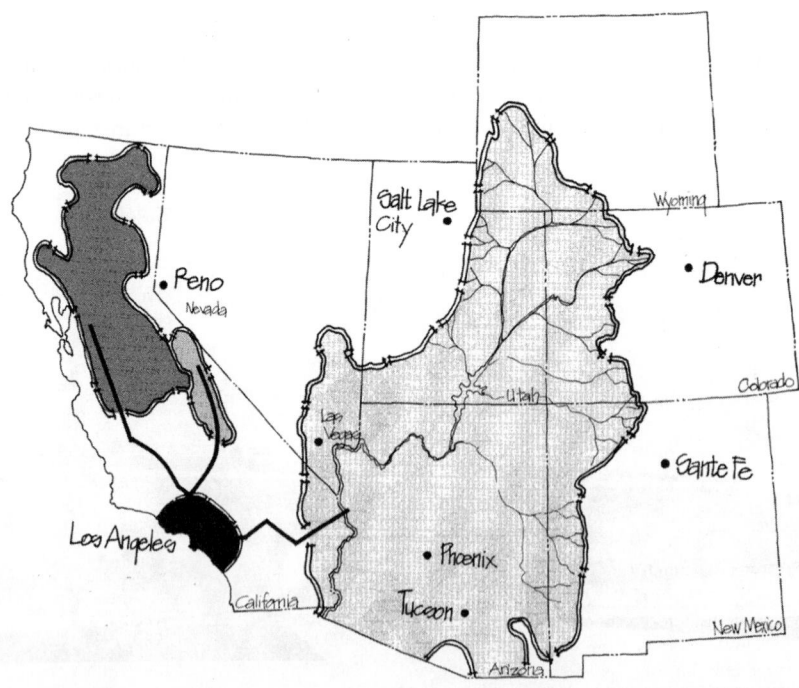

**Fig. 2.9** *The effective watershed of the greater Los Angeles area. The area (even if partial) needed to provide water to this metropolitan area (its water footprint) is vastly greater than the politically defined city limits. (From* Design for Human Ecosystems *by John Tillman Lyle. Copyright © 1999 by Harriet Lyle. Reproduced by permission of Island Press, Washington, DC.)*

when surpluses diminish or disappear. Table 2.6 shows estimated ecological footprints for several countries. It is clear that some countries are substantially overstepping their boundaries, while others are able to accommodate that overstep because of their less consumptive lifestyle. Continuing worldwide population growth makes the footstep balance tenuous. Table 2.7 provides similar environmental benchmarking for the same countries with respect to energy and water use and carbon dioxide ($CO_2$) emissions. $CO_2$ is becoming the key environmental metric in the United Kingdom and parts of Europe (Roaf, 2004).

The ecological footprint for the "world" shown

in Table 2.6 should lead to serious reflection regarding the meaning of sustainability. One unfortunate offspring of the growing interest in green design is a seemingly endless stream of one-upmanship that glibly promotes "sustainable" this and "sustainable" that—including buildings (virtually impossible in today's economic climate), communities (possible, but rare today), and states (perhaps necessary in the future). The term *sustainable* has lost almost any meaning through incessant misuse. This is unfortunate if one believes the story of the ecological footprint—sustainability is essentially keeping the Earth's footprint on the planet. A good idea, as it is the only planet we have right now.

**TABLE 2.6 Ecological Footprints[a] for Selected Countries**

| Country | 1997 Population | Footprint (ha/cap)[b] | Available Capacity (ha/cap) | Surplus (if +) or Deficit (if −) |
|---|---|---|---|---|
| Australia | 18,550,000 | 9.0 | 14.0 | 5.0 |
| Austria | 8,053,000 | 4.1 | 3.1 | −1.0 |
| Bangladesh | 125,898,000 | 0.5 | 0.3 | −0.2 |
| Brazil | 167,046,000 | 3.1 | 6.7 | 3.6 |
| Canada | 30,101,000 | 7.7 | 9.6 | 1.9 |
| China | 1,247,315,000 | 1.2 | 0.8 | −0.4 |
| Egypt | 65,445,000 | 1.2 | 0.2 | −1.0 |
| Germany | 81,845,000 | 5.3 | 1.9 | −3.4 |
| India | 970,230,000 | 0.8 | 0.5 | −0.3 |
| United States | 268,189,000 | 10.3 | 6.7 | −3.6 |
| WORLD | 5,892,480,000 | 2.8 | 2.1 | −0.7 |

*Source:* http://www.ecouncil.ac.cr/rio/focus/report/english/footprint/ranking.htm/
[a]Updated 1997.
[b]The ecological footprint, available ecological capacity, and surplus or deficit capacity are in hectares/capita (multiply hectares by 1.66 to obtain acres).

**TABLE 2.7 Per Capita Energy[a] and Water[b] Use and $CO_2$ Emissions[c] for Selected Countries**

| Country | 1997 Population | Energy Use[d] | Water Use[e] | $CO_2$ Emissions[f] |
|---|---|---|---|---|
| Australia | 18,550,000 | 5,975 | 1,250 | 16.8 |
| Austria | 8,053,000 | 3,790 | 261 | 7.9 |
| Bangladesh | 125,898,000 | 145 | 576 | 0.2 |
| Brazil | 167,046,000 | 1,064 | 345 | 1.8 |
| Canada | 30,101,000 | 8,000 | 1,494 | 16.2 |
| China | 1,247,315,000 | 887 | 494 | 2.7 |
| Egypt | 65,445,000 | 695 | 1,013 | 1.7 |
| Germany | 81,845,000 | 4,264 | 572 | 10.2 |
| India | 970,230,000 | 514 | 635 | 1.0 |
| United States | 268,189,000 | 7,921 | 1,682 | 19.8 |
| WORLD | 5,892,480,000 | 1,631 | 633 | 6.1 |

[a]*Source:* World Resources Institute, Earth Trends: The Environmental Information Portal; http://earthtrends.wri.org/searchable_db/
[b]*Source:* World Resources Institute, Earth Trends: The Environmental Information Portal; http://earthtrends.wri.org/searchable_db/
[c]*Source:* Nationmaster.com; http://www.nationmaster.com/; from World Resources Institute. 2003. Carbon Emissions from Energy Use and Cement Manufacturing, 1850 to 2000. Available on-line through the Climate Analysis Indicators Tool (CAIT) at http://cait.wri.org. Washington, DC: World Resources Institute.
[d]Units are thousand metric tons of oil equivalent per person per year. Data are for 2001. World per capita consumption has been stable over the past 10 years; that of the United States has increased slightly (7538 in 1990; 7921 in 2001).
[e]Units are cubic meters of water withdrawals per person per year. Data are for 2000.
[f]Units are thousand metric tons of carbon dioxide per 1000 people per year. Data appear to be for 2000.

## 2.7 CASE STUDY—DESIGN PROCESS AND ENVIRONMENTAL RESOURCES

### Philip Merrill Environmental Center, Chesapeake Bay Foundation

**PROJECT BASICS**

- Location: Annapolis, Maryland, USA
- Latitude: 38.9 N; longitude: 76.5 W; elevation: near sea level
- Heating degree days: 4381 base 65°F (2434 base 18.3°C); cooling degree days: 1271 base 65°F (706 base 18.3°C); annual precipitation: 42 in. (1063 mm)
- Building type: new construction; commercial offices and interpretive center
- 32,000 ft$^2$ (3000 m$^2$); two occupied stories
- Completed December 2000
- Client: The Chesapeake Bay Foundation
- Design team: SmithGroup (and consultants)

**Background.** The Philip Merrill Environmental Center was one of a half-dozen buildings certified as LEED Platinum at the time this case study was prepared. Platinum is the highest possible LEED rating. Elements of the design process for the Environmental Center are presented below in order to emphasize the critical importance of an appropriate design process to the development of high-performance buildings. Design team and client values were important to the success of this project—and led to the development of explicit and aggressive green design intent and criteria. Concern for energy efficiency and water conservation led to much of the distinctive form of the building—especially the signature water storage tanks on the entry façade. (The information that follows was provided by SmithGroup.)

**Context.** The Chesapeake Bay Foundation (CBF) is an environmental advocacy, restoration, and education organization headquartered in Annapolis, Maryland. Before the creation of the Philip Merrill Environmental Center, CBF's facilities included three properties in Annapolis and a small building outside of town. The functioning and unity of the organization suffered from the disparate locations and consequent separation of departments, justifying the creation of a new headquarters that could unify and house CBF in an optimum environment.

**Design Intent.** The new headquarters would not only house the Foundation, but would also be a reflection on CBF's mission. It would serve as a paragon for the Bay's watershed region of sustainable development—"walking the talk," "practicing what CBF preaches." The design was to emulate the regional vernacular and utilitarian functions of working on the Bay. The building was to respond to habitats, vegetation, soils, buffer zones, views, solar orientation, topography, prevailing wind direction, and functional requirements. The organization of the elements on the site would tell the story of CBF's mission to educate and involve the public in taking responsibility for the health of the Bay.

The leading principles behind the design were as follows:

- Set a precedent for sustainable development on the Chesapeake Bay.
- Provide for the functional needs of CBF.
- Create an effective work environment.
- Embody a sense of unity and connectiveness.
- Push the envelope of green building.
- Reflect the utilitarian nature of CBF.
- Mesh indoor and outdoor spaces.
- Create interactive spaces.
- Integrate CBF's departments while preserving distinction.
- Enhance public service.
- Facilitate an educational experience.

**Design Criteria and Validation.** The project was intended to achieve a LEED Platinum rating. At the time design commenced, LEED was a largely unknown rating system in its pilot phase of development. The LEED system was used both as a benchmark and as an assessment tool—a way of validating the design's sustainability. Energy modeling using Energy 10 software was performed during the preliminary design phases. The overall energy modeling during subsequent phases used Trace software.

**Post-Occupancy Validation Methods.** A full year of monitoring and POE was performed by the National Renewable Energy Laboratory (NREL). NREL provided the monitoring equipment to measure the resource consumption (water, electricity, propane) of the building and to measure the energy generated by the building. The Department of Energy performed a productivity analysis of the workers inside the building in an effort to evaluate how green buildings can not only save energy, but can create a healthier and more productive work environment.

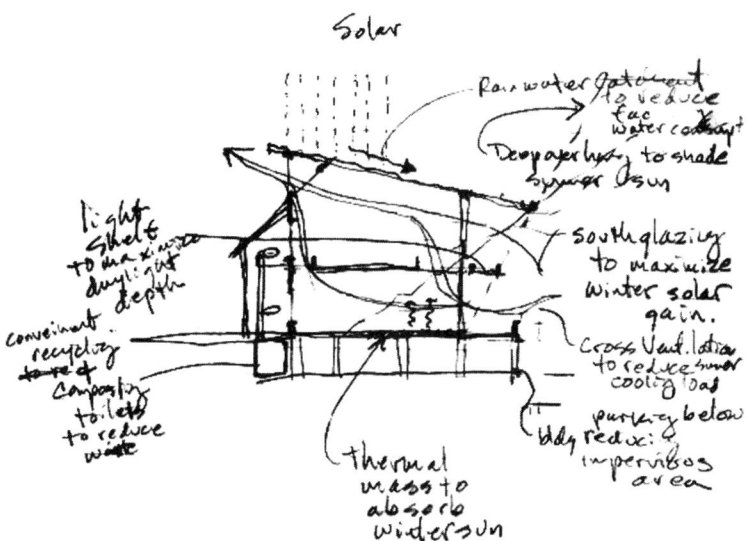

**Fig. 2.10** *Conceptual design sketch showing the earliest concept of the Philip Merrill Environmental Center and illustrating how the form of the building was directly related to the environmental goals for the project. This sketch is used by the design team as an ongoing example of how early goal setting allows designers to shape a building to respond to goals, thus creating an integrated design. (© SmithGroup; used with permission.)*

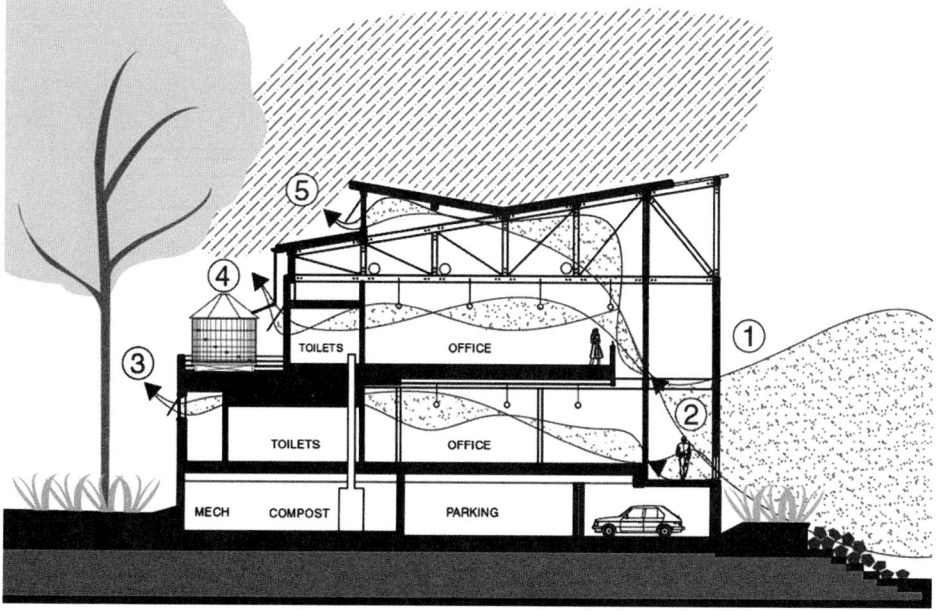

**Fig. 2.11** *Schematic design diagram illustrating how the conceptual design idea was refined, and the role that natural ventilation, passive solar heating, rainwater collection, and daylighting and views played in shaping the building. Energy and water are clearly focal elements. (© SmithGroup; used with permission.)*

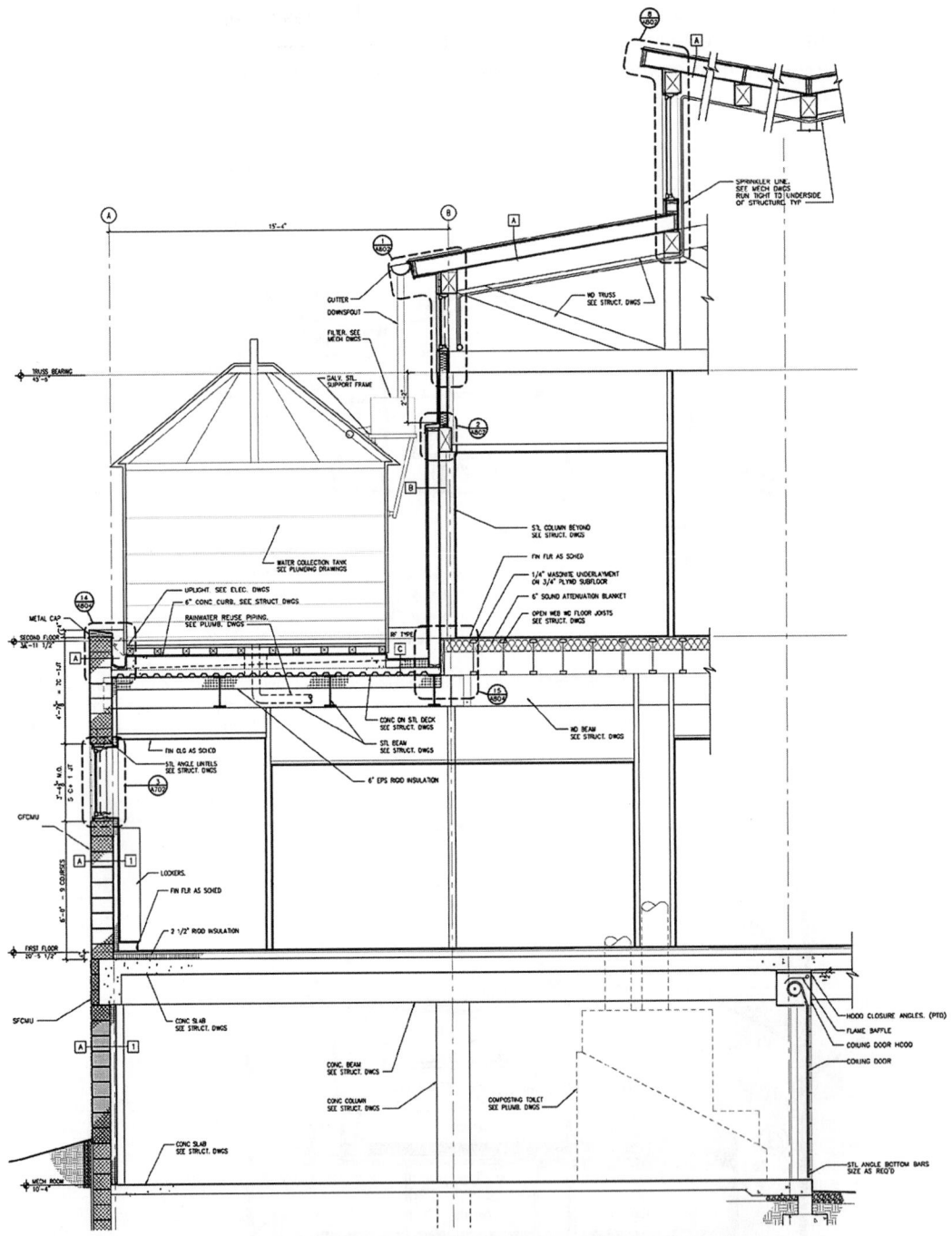

**Fig. 2.12** *Section through the Philip Merrill Environmental Center developed during the construction documents phase. The water storage tanks, which are a signature element of the final design, have evolved from concept to buildable artifact. (© SmithGroup; used with permission.)*

**Fig. 2.13** *North (inland) façade of the Philip Merrill Environmental Center showing the visual impact of rainwater collection intent and solution. Water conservation has informed this façade. (Photo © 2004 Walter Grondzik; all rights reserved.)*

**Fig. 2.14** *South (bay side) façade of the Philip Merrill Environmental Center showing PV panels, daylighting/solar apertures, and shading elements. Energy collection has informed this façade. (Photo © 2004 Walter Grondzik; all rights reserved.)*

**Performance Data.** Information available to date suggests substantial design team success in "pushing the envelope of green building"—particularly in the areas of material, water, and energy conservation:

- The building achieved a LEED Platinum rating.
- All wood was obtained from renewable resources; more than 50% of building materials were obtained from within a 300-mile (480-km) radius of the site.
- There is a projected 90% reduction in water use compared to that of a comparable (conventional) office building.
- There is a projected annual energy use of 350,000 kWh (90% electricity) with an anticipated contribution of 43,000 kWh equivalent from solar thermal systems and PV; plug loads account for roughly a third of the energy use, lighting another third, and climate control the remaining third.
- The project received a Grand Award, Building Team Project of the Year, from *Building Design & Construction* magazine in 2001.
- The building was named one of the AIA/COTE

Top Ten Green Projects in 2001 (American Institute of Architects/Committee on the Environment).

## FOR FURTHER INFORMATION

High Performance Buildings Database, U.S. Department of Energy, Office of Energy Efficiency and Renewable Energy: http://www.eere.energy.gov/buildings/highperformance/case_studies/index.cfm

National Renewable Energy Laboratory (NREL), High Performance Buildings Research: http://www.nrel.gov/buildings/highperformance/chesapeake.html

## References

American Institute of Architects. 1996. *Environmental Resource Guide.* John Wiley & Sons, New York.

Brand, S. 1994. *How Buildings Learn: What Happens After They're Built.* Viking, New York.

Brower, D. 1990. *Cool Energy.* Union of Concerned Scientists, Cambridge, MA.

The Earth Council. 2004. Ecological Footprints of Nations: How Much Nature Do They Use?—How Much Nature Do They Have? http://www.ecouncil.ac.cr/rio/focus/report/english/footprint/

Fisher, J. 1974. *The Energy Crisis in Perspective.* John Wiley & Sons, New York.

Hannon, B., R. Stein, B. Segal, and D. Serber. 1977. *Energy Use for Building Construction.* Center for Advanced Computation, University of Illinois, Champaign-Urbana, IL.

Irurah, D. 1997. "An Embodied Energy Algorithm for Energy Conservation in Building Construction as Applied to South Africa." Ph.D. thesis, University of Pretoria.

Joslin, J., et al. 1993. *The Waste Papers: Analysis and Discussion of the Potential for Salvage and Re-Use of Construction Materials from Residential Demolition.* Metro Solid Waste Department Publications, Portland, OR.

Lyle, J. T. 1999. *Design for Human Ecosystems: Landscape, Land Use, and Natural Resources.* Island Press, Washington, DC.

Meyers, R. 1983. *Handbook of Energy Technology and Economics.* John Wiley & Sons, New York.

Rees, W. and M. Wackernagel. 1995. *Our Ecological Footprint: Reducing Human Impact on the Earth.* New Society Publishers, Gabriola Island, BC.

Roaf, S. 2004. *Closing the Loop: Benchmarks for Sustainable Buildings.* RIBA Enterprises, Ltd., London.

Stein, R., et al. 1981. *Handbook of Energy Use for Building Construction.* U.S. Department of Energy, Washington, DC.

Sustainable Development Indicators. 2004. http://www.sdi.gov/indicators/lc_mater.htm (derived from the Materials Information Team, U.S. Geological Survey, U.S. Department of the Interior).

UNH. 2004. Watershed Protection, University of New Hampshire. http://www.unh.edu/marine-education/pages/watershed2004/ws-quotations.html/

USDOE. 2004. *2004 Buildings Energy Databook.* U.S. Department of Energy, Office of Energy Efficiency and Renewable Energy, Washington, DC.

USEPA. 1998. *Characterization of Building-Related Construction and Demolition Debris in the United States.* EPA530-R-98-010, U.S. Environmental Protection Agency, Washington, DC.

USGS. 2004. *Estimated Use of Water in the United States in 2000* (USGS Circular 1268), United States Geological Survey, Washington, DC.

CHAPTER 3

# Sites and Resources

SITE ANALYSIS TYPICALLY PRECEDES SITE PLANNING. The purpose of a site analysis is to understand the character of a given site. Such an analysis usually includes the collection of information on utility availability, noise sources, zoning, views, solar access, traffic and pedestrian patterns, climate, and the like. In some cases, long-term statistical data are available as an information resource (such as for climate, solar position, utility services); for other variables (such as noise, views, pedestrian circulation in an urban area) there is usually no existing database, and all information must be collected directly by the designer. To be useful and successful, a site analysis must do more than simply catalog information; it must place value on the collected information in the context of a proposed project and its design intent. For a given building, is solar radiation a desirable resource or a problem to be solved? Is wind a usable design element or an environmental force to be avoided? Understanding what resources are available for inclusion in a design solution, and what natural forces are potential problems to be mitigated by design, is the essence of site analysis—and a necessary precursor to green design.

A designer's early site planning decisions will, at a later date, influence available options for a building's climate control and lighting systems and affect a building's overall consumption of energy. When the site is seen as a collection of resources (sun, wind, water, plants) and also as part of the environment we all share, buildings can greatly reduce dependence upon nonrenewable fuels. They can also do this without limiting the availability of local energy resources for neighboring buildings. In addition to saving energy, the use of on-site resources can create outdoor spaces that become especially pleasant to be in. Such spaces can direct winter sun to a glass wall while blocking the wind or funnel summer breezes through shade to an open window. Site planning is greatly influenced by economic considerations, zoning regulations, and adjacent developments, all of which can interfere with the design of a site to utilize the sun, the sky, and the wind. Integration of all these concerns at the site-planning stage is the first step in adapting a building to its climate. This chapter looks briefly at some aspects of site–climate interactions.

## 3.1 CLIMATES

Climate is a long-term statistically derived picture of weather. Weather is what happened today or yesterday, while climate is what happened over the past 10, 15, or 20 years. Our most familiar names for climates describe their most severe season, as shown in Fig. 3.1. This is a convenient means of description, but it can be misleading for designers. "Cold" climates can have very hot, sometimes humid summer days; hot-arid climates can have bitterly cold winter conditions. Before designing buildings that will modify exterior conditions to provide indoor comfort, we should know when and how much

**DESIGN CONTEXT**

modifying is appropriate. This is not necessarily easy, as there are 8760 hours in a year and several climatic variables of interest for each hour (temperature, relative humidity, solar radiation, wind speed, etc.). A graphical means of portraying all these variables is helpful. A very useful set of graphic analysis tools for climates was developed by Olgyay (1963) under the name *bioclimatic* design. Bioclimatic design links comfort and climate. Chapter 4 provides details on thermal comfort, but a discussion of Olgyay's approach will provide an introduction to a few key issues.

Figure 3.2 shows a psychrometric chart with a superimposed comfort zone (bounding a range of conditions people will statistically find comfortable) and overlaid values of selected environmental variables. The "shading" and "overheated" lines correspond roughly to the comfort zone boundaries. The lines labeled "100, 200, and 300" (Btu/h ft²; equivalent to 315, 630, and 960 W/m²) indicate conditions where this much radiation is needed to remain comfortable outdoors without wind. These conditions correspond roughly to 61, 53, and 45°F (16, 12, and 7°C), respectively, although the higher the relative humidity, the lower the air temperature to which the lines correspond. The shading line corre-

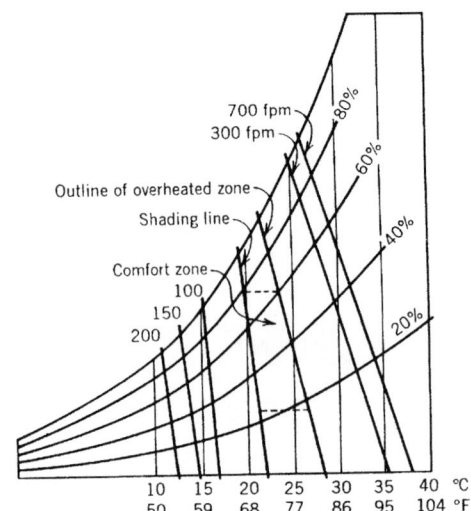

**Fig. 3.2** *Relationship of isolines to the thermal comfort zone. Isolines are lines with a constant value of a given property, such as wind speed. Several isolines are plotted on the timetables of climatic needs (see Fig. 3.3).*

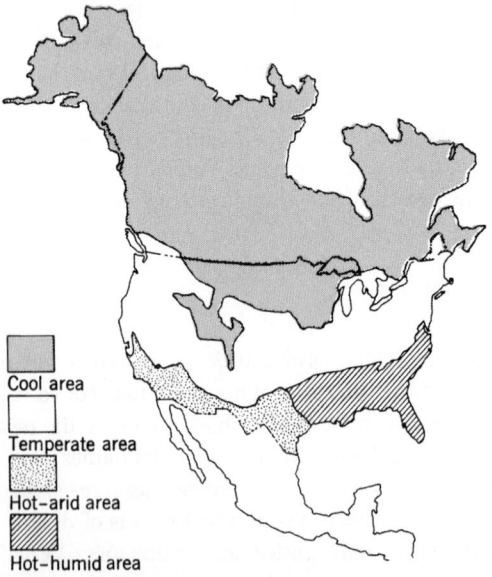

**Fig. 3.1** *Regional climate zones of the North American continent. (From Victor Olgyay,* Design with Climate: Bioclimatic Approach to Architectural Regionalism, *Copyright © 1963 by Princeton University Press. Reprinted by permission.)*

sponds to about 70°F (21°C) at the lowest relative humidity and to about 68°F (20°C) at the highest relative humidity.

Figure 3.3 shows Olgyay timetables for cities representative of two of the four climate zones shown in Fig. 3.1. These graphical views of thermal environmental needs are plotted across an entire year (the horizontal axis) for all hours of the day (the vertical axis). Those portions of the year within the bounds of the bold shading line require shade for thermal comfort. The darker-shaded (overheated) zones are bounded by an isoline that corresponds to a temperature of about 78–82°F (26–28°C), depending upon relative humidity. At higher temperatures, there is a need for air motion as well as shade to remain comfortable outdoors. Areas farther within the overheated zone correspond to thermal comfort needs for 300 and 700 fpm (1.5 and 3.6 m/s) air velocities. An occupancy schedule for a proposed building can be overlaid on such a timetable, providing insights into climate/building interactions (see Fig. 8.6 for an example).

The timetable for New York City (Fig. 3.3a) shows that about one-fifth of the year requires shade for outdoor comfort, with most of the hours from mid-July to mid-August falling in the overheated period. Little relief can be expected at night

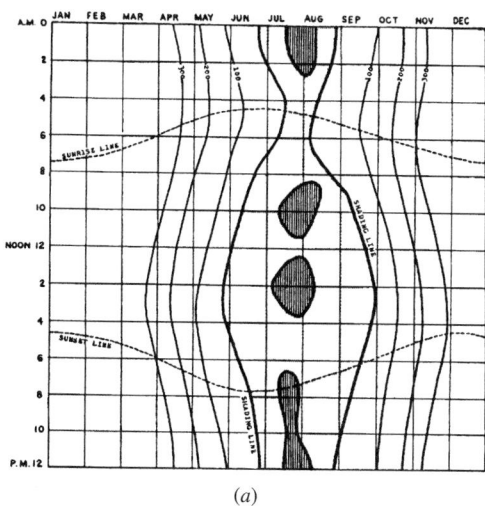 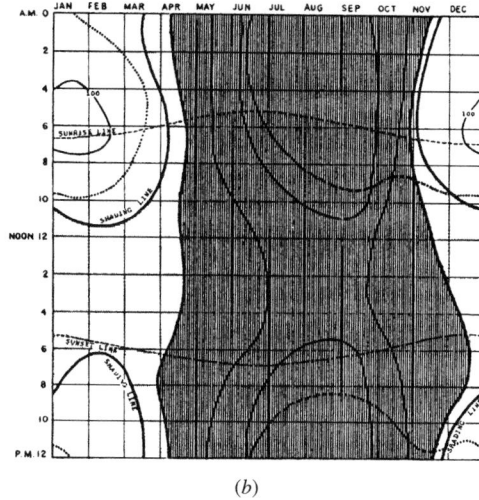

(a)                                                                                     (b)

**Fig.** *3.3* Timetables of climatic needs for (a) New York City and (b) Miami, cities representative of two of Olgyay's North American regional climate zones. Isolines within the shaded regions represent air speeds; the lines outside of the shaded areas represent solar radiation intensity. (From Victor Olgyay, Design with Climate: Bioclimatic Approach to Architectural Regionalism, Copyright © 1963 by Princeton University Press. Reprinted by permission.)

in summer. Perhaps one-third of the year is "too cold," with temperatures too low for shirtsleeve comfort outdoors and requiring more than 300 Btu/h ft² (960 W/m²) of radiation, an unlikely solar bounty during New York winters.

The timetable for Miami, Florida (Fig. 3.3*b*), shows that about four-fifths of the year demand shade for outdoor comfort, and about three-fifths are in the overheated period. A large portion of the year needs over 700 fpm (3.6 m/s) wind, with these hours occurring both during the day and at night. There is also a need for wind to counteract high vapor pressure (humidity), independently of high temperature. Clearly, wind is important to comfort in Miami in any unconditioned building. There is no such zone as "too cold" there.

Olgyay's climatic timetables provide a graphical view of outdoor conditions relative to thermal comfort for four key climate regions (Fig. 3.1). Of these four, the New York timetable suggests an emphasis on space heating but a need for cooling as well. In Miami, the focus is upon shading and cooling. Several strategies utilize the ambient climate as a heat source or sink for heating and for cooling. In Section 4.3, cooling strategies are outlined; in Section 4.4, solar heating strategies. Combinations of daylighting, heating, and cooling are considered in Section 4.5. Solar water heating is discussed in Chapter 21.

## 3.2 CLIMATES WITHIN CLIMATES

The climate at a particular site can be quite different from the climate data that are published as being representative of an entire region. This is particularly evident when visiting a site where a neighboring hill blocks wind or winter sun, or an adjacent lake cools summer breezes or adds a damp chill to the winter air. Such local variations constitute microclimates, with some characteristic distinctly different from those of the larger macroclimate. The characteristics of a *micro*climate are influenced by the interaction of the characteristics of both the site and the *macro*climate:

| *Site Characteristics* | *Climate Characteristics* |
|---|---|
| Soil type | Sun |
| Ground surface | Air temperature |
| Topography | Humidity |
| Vegetation | Precipitation |
| Water bodies/flows | Air motion |
| Views | Air quality |
| Human effects (heat, noise, etc.) | |

Most urban sites are under the influence of an urban subclimate that differs from the conditions of the surrounding countryside unaffected by urbanization. Probably the best-known urban climate feature is the *heat island.* Urban heat island effects are summarized in Table 3.1 and Fig. 3.4. Designers should note that city climatological stations are often located at nonurban sites, such as an outlying airport, masking the effects of a heat island.

The most obvious reason for a city's relative year-around warmth is its concentration of heat sources: the air conditioners, furnaces, and electric lighting in buildings and internal combustion engines in cars. This urban and industrial heat production is shown in Table 3.2 and Fig. 3.5. It appears that cities and industrial regions of the world release less internal heat per capita as people live and work closer together—although the heat density (temperature) is greater. Commercial and industrial cities release more heat for a given population density, and tropical cities release less, while still conforming to the pattern that greater compactness permits less energy use per capita.

## TABLE 3.1 Average Changes in Climate Effects Caused by Urbanization[a]

| Effect | Comparison with Rural Environment |
|---|---|
| Contaminants | |
| Condensation nuclei and particulates | 10 times more |
| Gaseous admixtures | 5 to 25 times more |
| Cloudiness | |
| Cover | 5 to 10% more |
| Fog, winter | 100% more |
| Fog, summer | 30% more |
| Precipitation[b] | |
| Totals | 5 to 10% more |
| Days with less than 2 in. (5 mm) | 10% more |
| Snowfall | 5% less |
| Relative humidity | |
| Winter | 2% less |
| Summer | 8% less |
| Radiation | |
| Global | 15 to 20% less |
| Ultraviolet, winter | 30% less |
| Ultraviolet, summer | 5% less |
| Sunshine duration | 5 to 15% less |
| Temperature | |
| Annual mean | 0.9 to 1.8°F (0.5 to 1.0°C) higher |
| Winter minima (average) | 1.8 to 2.6°F (1 to 2°C) higher |
| Heating degree days | 10% less |
| Wind speed | |
| Annual mean | 20 to 30% less |
| Extreme gusts | 10 to 20% less |
| Calms | 5 to 20% more |

*Source:* Landsberg (1970).

[a]These effects vary from city to city and from day to day.

[b]Research since 1970 has shown that it is not at all certain that urbanization causes increases in precipitation amount within a city.

Rain that falls on a city can be an effective evaporative cooling mechanism, especially as water evaporates from wet surfaces. Yet streets and buildings are usually designed to shed water quickly and thoroughly, sending rainwater into storm sewers instead of letting it evaporate slowly in the wind and sun. As a result, evaporative cooling from these surfaces is minimized.

A city also changes the overall cooling action of the wind by channeling it into narrow streets. The geometry of high vertical walls and narrow streets also increases summer heat collection in cities as the high sun is reflected downward to be absorbed and then reradiated by the often rocklike streets and building surfaces. In winter, however, this geometry puts most urban surfaces at a solar disadvantage because the low sun strikes only the upper portion of south-facing walls. A reduction in radiant heat loss at night, caused by this lack of access to the sky, is a key element in the formation of the urban heat island, as summarized in Fig. 3.6. Sky access is discussed in more detail in Section 3.6.

A more subtle urban climate influence is contaminated air. Small particles in a city's air can keep some sunlight from reaching the city, yet can also help to keep the city's heat from radiating outward. These particles also form additional nuclei for fog droplets. Table 3.1 suggests that fog may occur in a city in winter up to twice as often as in the surrounding countryside. Trees and greenery, which can act as crude filters of airborne dust, are not as available in a city.

The city thus modifies its climate from that of its surroundings. In winter, the city's factories, vehicles, surface materials, and geometry combine to increase temperature and reduce the amount of energy needed for heating buildings. The typical means of providing needed winter heating (fossil fuel–burning heating equipment and power plants) contribute to airborne particles and urban fog. Solar-assisted heating would diminish this pollution—and with less air pollution, more sun would reach a solar collector. However, the greater the population density, the more difficult is access to winter sun, especially at latitudes farther from the equator, where winter heating is most needed. High buildings readily block low sun altitudes.

In summer, the city's internal heat makes things worse. Air conditioners add their own process heat to the building heat that they pour into

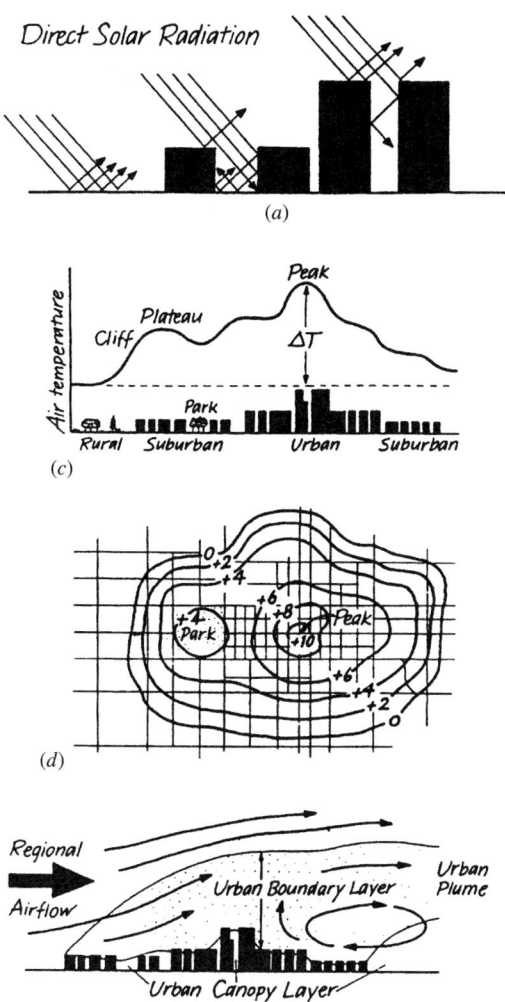

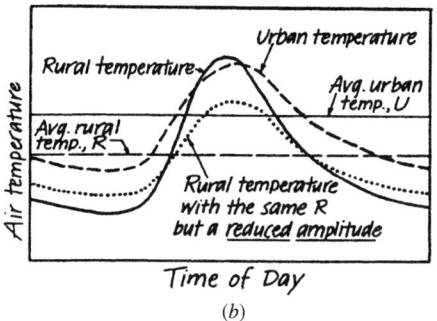

**Fig. 3.4** *Urban heat island: a densely occupied area with a temperature distinctly higher than that of the surrounding rural area. (a) Direct solar radiation is likely to be reflected within the city, thereby increasing solar heat gain in urban areas. (b) Temperature records at a rural site (solid line) and in the center of a city (dashed line) during a typical night and day. The city's heat-conducting materials and thin cloud of polluted air acting alone would not change the average air temperature but would reduce the day–night difference (the dotted line). In addition, the heat from increased solar gain and city-specific heat sources (cars, buildings) warms the air at all hours, producing the observed urban record (dashed line). (c) Idealized profile of the air temperature difference between urban and rural areas at times of peak differences—calm, clear nights. (d) Based upon (c), typical isotherms (lines of equal temperature) provide a "contour map" of the urban heat island. (e) The urban heat island can affect the countryside "downstream." (Reprinted by permission from Lowry, 1988.)*

the air; their appetite for electricity taxes power plants, in turn adding vastly more waste heat. Again, solar energy can help, this time by generating photovoltaic (PV) electricity to run air conditioners. High summer sun angles make solar access much less problematic. The stronger the sun, the more electricity is produced for air conditioners and other loads. PV arrays are currently high in first cost, but are noiseless, emit no exhaust gases, and add no internally generated heat to the summer air.

Site- and urban-planning responses to these urban climate characteristics can sometimes lead in contradictory directions. For example, the provision of greenways within cities would bring softer ground surfaces cooled in summer by shading, breezes, and evaporation. However, the winter impact of greenways might be an increase in fog via

evaporation of retained storm water; or perhaps fog would be discouraged locally by increased wind speeds in channels formed by the greenways. The summer impact of a greenway is also mixed; changes in comfort would depend upon the particular combination of increased humidity and cooler temperature a greenway produces. On balance, a greenway within a city seems to be a positive and aesthetic step in ameliorating the urban climate.

While considering how the sun, wind, and other climate elements can be utilized on a site for the benefit of a building, it is important to remember the need to protect the access of others to these same shared resources. In "The Tragedy of the Commons" (1968), Garrett Hardin writes about the commons as publicly owned meadows shared by many herders. Each herder realizes that

**TABLE 3.2 Heat Generated Within Cities**

| City or Region[a] | Population (millions) | Area (km$^2$) | Population Density ($10^3$/km$^2$) | Energy Use per Capita (kW/capita) | Energy Use per Unit Area[b] (W/m$^2$) |
|---|---|---|---|---|---|
| CITIES AND INDUSTRIALIZED REGIONS | | | | | |
| 3 Fairbanks[c] | 0.045 | 80 | 0.6 | 10.91 | 6.55 |
| 11 Vancouver | 0.6 | 111 | 5.4 | 3.55 | 19.2 |
| 12 Brussels | 1.3 | 163 | 8.0 | 3.5 | 28.0 |
| 14 West Berlin | 2.3 | 235 | 9.8 | 2.14 | 21.0 |
| 9 St. Louis | 0.75 | 250 | 3.0 | 5.3 | 15.9 |
| 10 Munich | 0.9 | 300 | 3.0 | 3.0 | 9.0 |
| 5 New Jersey suburbs of NYC | 4.7 | 6,100 | 0.8 | 9.1 | 7.3 |
| 1 Los Angeles County | 7.0 | 10,000 | 0.07 | 10.5 | 0.74 |
| 4 Nordheim–Westfalen[d] | 16.9 | 33,800 | 0.5 | 8.0 | 4.0 |
| 2 BosNyWash[e] | 33.0 | 87,000 | 0.38 | 11.2 | 4.3 |
| COMMERCIAL/INDUSTRIAL CITIES | | | | | |
| 15 Sheffield | 0.5 | 48 | 10.4 | 1.83 | 19.0 |
| 17 Manhattan | 1.7 | 59 | 28.8 | 5.52 | 159.0 |
| 16 Montreal | 1.1 | 78 | 14.1 | 7.02 | 99.0 |
| 13 Budapest | 1.3 | 113 | 11.5 | 3.74 | 43.0 |
| 8 Cincinnati (summer) | 0.6 | 222 | 2.7 | 9.3 | 25.1 |
| 7 Hamburg | 1.83 | 763 | 2.4 | 5.3 | 12.7 |
| 6 Chicago | 3.5 | 1,842 | 1.9 | 27.2 | 51.7 |
| TROPICAL | | | | | |
| 18 Hong Kong | 3.9 | 105 | 37.1 | 0.88 | 32.6 |
| 19 Singapore | 2.1 | 140 | 15.0 | 0.81 | 12.2 |

*Source:* Lowry and Lowry (1995) by permission.

[a]Index numbers refer to Fig. 3.5.

[b]$0.317$ W/m$^2$ = 1 Btu/ft$^2$; 1 km$^2$ = 0.386 miles$^2$.

[c]Data for Fairbanks, Alaska, include a sparsely populated incorporated area surrounding the central business district.

[d]Nordheim–Westfalen is the heavily urbanized and industrialized region in the lower Rhine valley near Düsseldorf–Dortmund.

[e]*BosNyWash* is the term often used by urbanists to refer to the megalopolis that stretches from Boston through New York to Washington, DC.

his personal wealth will increase as animals are added to his herd, so all herders increase their livestock. The meadow capacity, however, does not increase; overgrazing occurs, and as a result the commons become unable to support any animals. The following discussions of sun, air, and water resources available to a site are influenced both by the "private" needs of a building and the "public" patterns of resource availability, which should remain accessible to all.

## 3.3 BUILDINGS AND SITES

Buildings are temporary occupants of their sites. The arrival of a building usually produces rapid and dramatic changes to the biological systems on a site, to the microclimate, and often to the surface geology. Buildings are guests, sites are hosts; a funda-

mental design question is, how can the two most productively coexist?

A site offers a building earth for support, as a potential heat source and heat sink, and for the growth of plants where building density permits. Sounds on a site depend upon the context, urban or rural. Water is somewhere below a site, flows across it, falls on it as rain, and perhaps collects on its surface. Wind moves erratically across a site. Solar energy arrives in diurnal and seasonal cycles. Long-term patterns of sun, wind, and water are steady and generally predictable, although great variations can occur over shorter (daily, monthly, annual) time spans. At some level, an ecosystem of life is already established.

During construction a building arrives, bringing with it people and vehicles, a flow of materials into and out of the site, sounds of activity, and imported utility services such as electricity, water,

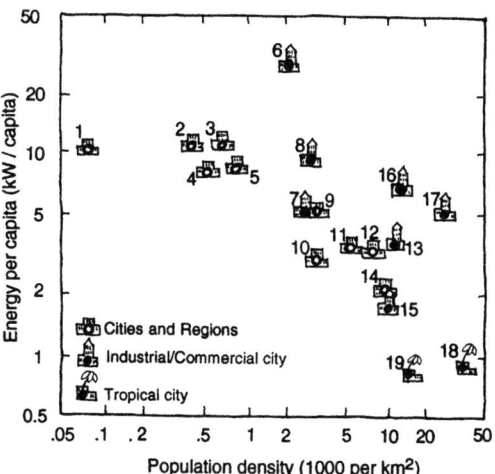

**Fig. 3.5** *Population density and energy use per capita for 19 cities and regions. Numbers refer to locations in Table 3.2. The heat island effect is influenced by both density and energy use. (Data with permission from Lowry and Lowry, 1995.)*

and natural gas. The building offers the site electric light by night, a continuous flow of heat (to or from), a radical change in water flows (both at and outside of the building envelope), and liquid and solid wastes (or nutrients, depending upon one's viewpoint). In our society, the waste outflows are most often whisked off for treatment "elsewhere."

When a site is larger than its building, both may be designed to improve building performance and user experiences. Large sites offer smaller buildings the opportunity to accept, filter, or block sun, wind, sound, and rainwater to make effective use of on-site resources. Vegetation, ground forms, and orientation, as well as roofs, walls, and floors, can play a role in this interaction.

When a building fills its site, opportunities for on-site resource use are more limited. Because buildings that completely fill sites are often located in densely built-up areas, less wind is available around the building walls and solar radiation may be blocked. The built-upon earth is less able to act as a heat sink or absorb water. The building roof often provides the major opportunity to receive sun and rain and discharge heat to the wind. It may offer the only opportunity to grow plants on the site. In dense urban areas, the future roofscape will be dominated by climate–building interactions: solar collection, rain collection, heat rejection, water treatment, and gardens. Compared to today's urban roofscape

dominated by horizontal black or gray surfaces behind parapet walls, this represents a dramatic design opportunity for aesthetic, social, and technical change.

## 3.4 ANALYZING THE SITE

After recognizing the resources that exist on and around a site, a designer then decides how best to integrate these resources into the building design while making the building and site a successful addition to the larger patterns of the surroundings. Schematic site plans are typically used as a kind of inventory; overlaid bubble diagrams can test possible design arrangements that relate rooms or functions to their surroundings in the plan view. Sun and wind conditions (in both summer and winter), noise sources, and water runoff patterns are often included in a schematic plan. It is particularly important to identify microclimates on the site, the places that have special characteristics differing from the regional climate. Microclimates can present opportunities for an expanded comfort zone: more sun in cool periods, more wind in warm periods. Microclimates can sometimes provide building sites where less energy is consumed because the winter is warmer or the summer

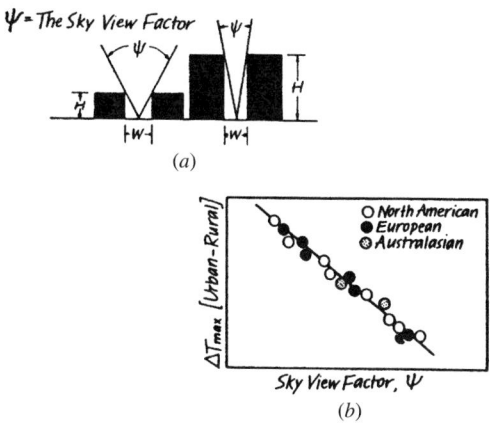

**Fig. 3.6** *The urban heat island effect is particularly strong on calm, clear nights. (a) With a greatly reduced "sky view factor" (Ψ) to the cold night sky, the walls and floors of urban canyons (the right part of the sketch) cannot lose heat as readily as can the open countryside or less dense suburban areas (the left part of the sketch). (b) The more narrow the Ψ, the more pronounced is the effect (ΔT) of the urban heat island in cities throughout the world. (Reprinted by permission from Lowry, 1988.)*

cooler. Microclimates can also present problem areas to be avoided for buildings or outdoor activity, or where special design measures need to be taken to correct their difficulties.

Microclimates on a site are not limited to those visible on the surface (in plan view). Conditions of privacy and accessibility, view, heat, light, air motion, sound, and water all change with vertical distance from the surface (Fig. 3.7a). To minimize energy consumption for constructing and using buildings and to integrate buildings with their surroundings, the conditions best suited to various functions should approach the characteristics of the layer of the site in which they are located. Consequently, both vertical and horizontal site analyses are needed. A lecture hall, requiring both an acoustically isolated and a closely controlled environment, is an obvious candidate for the subsurface layer. Electrical equipment, which benefits from a cool environment, is also suitable for the subsurface layer.

One building whose form responds to these layers is Boston's City Hall (Fig. 3.7b). Activities with the most frequent public interaction are located near the surface, such as the skylit, high-ceilinged lobby, whereas special ceremonial functions are elevated to distinctive forms in the near-surface layer above. The city offices with less frequent public contact occupy several floors in the sky layer, where daylight is plentiful. Storage and mechanical functions, as well as parking, are in the subsurface layer. An aesthetic equivalent of this horizontal layering is architect Louis Sullivan's concept of a façade as a "base, body and capital."

## 3.5 SITE DESIGN STRATEGIES

Sites can be organized to aid in heating, lighting, cooling, and controlling noise in buildings. A few of the most common site strategies are compared with their seasonal roles by Watson and Labs (1983) in Fig. 3.8. Later in this chapter, several of these strategies are discussed in more detail. The graphics in Fig. 3.8 suggest a focus upon housing but also apply to any function that can utilize an extended comfort zone.

One example of how the information in Figs. 3.7 and 3.8 can be applied (and manipulated) in the

planning of a building and its site is a house designed by Frank Lloyd Wright in the 1940s (Fig. 3.9). The direct gain of solar heat through its south-facing windows in winter makes this an early example of *passive solar heating*. (By contrast, *active solar heating* includes collectors and a single-purpose storage volume for solar heat, such as a tank of water or a bed of heated rocks.) This house, known as the Solar Hemicycle, was built in 1948 near Madison, Wisconsin, which lies between the cool and temperate climate zones. Winter heating is the dominant thermal influence in this area. The house stands on a rise in the prairie, particularly vulnerable to winter winds. During construction, earth was scooped from in front of the south face of the house and bermed against the entire curved north wall almost to roof level. Only a narrow strip of second-floor windows separates the berm from the roof. This berm and curve combination gives winter wind protection from the northeast to northwest and provides further insulation for the north wall. The north wall is made of stone, which absorbs and stores winter solar radiation (heat) that comes in through the floor-to-ceiling, southeast-to-southwest–facing glass. The concrete slab-on-grade first floor also stores solar heat in winter. The impression that this house and site are sun collectors is heightened by an entrance tunnel in the northeast end of the house that leads from the parking area through the berm and onto the sunny, protected south terrace overlooking a sunken "sun-trap" circular lawn. Passage from the cold north side through an even colder tunnel to the sunny south side prepares a visitor for the solar-heated interior.

This house plan is shorter in the north–south and longer in the east–west direction (about a 1:3 ratio). This elongation is typical of passive solar designs, which are able to store and use winter solar heat gain, and thus profit from having long south-facing glass walls to act as passive collectors. Had these large windows been well insulated at night and the roof and walls insulated to current standards, this house would be a very up-to-date case study of passive solar heating.

In the Solar Hemicycle house, protection from summer overheating is provided by an overhang along the south glass walls, as well as by the shaded and cool thermal mass of the north stone/berm

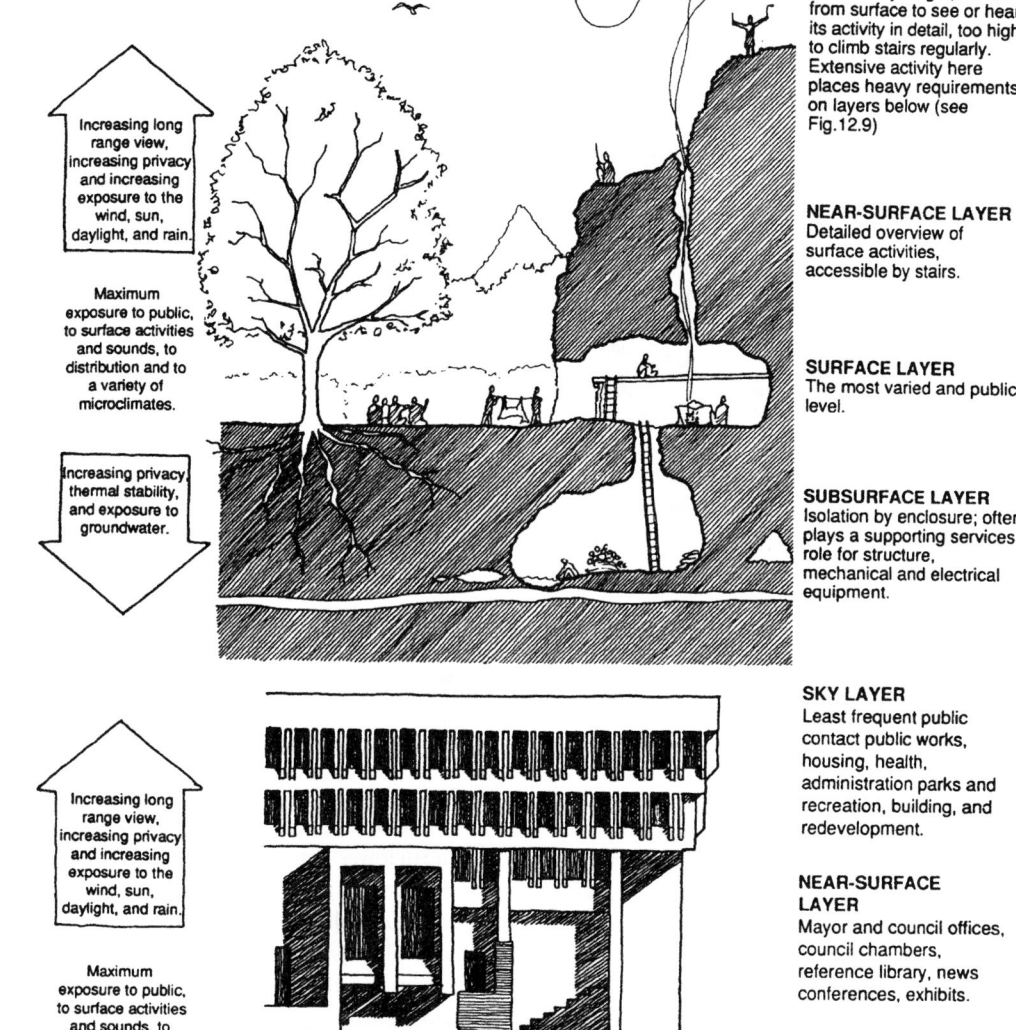

**SKY LAYER**
Isolation by height; too far from surface to see or hear its activity in detail, too high to climb stairs regularly. Extensive activity here places heavy requirements on layers below (see Fig.12.9).

**NEAR-SURFACE LAYER**
Detailed overview of surface activities, accessible by stairs.

**SURFACE LAYER**
The most varied and public level.

**SUBSURFACE LAYER**
Isolation by enclosure; often plays a supporting services role for structure, mechanical and electrical equipment.

Increasing long range view, increasing privacy and increasing exposure to the wind, sun, daylight, and rain.

Maximum exposure to public, to surface activities and sounds, to distribution and to a variety of microclimates.

Increasing privacy, thermal stability, and exposure to groundwater.

(a)

**SKY LAYER**
Least frequent public contact public works, housing, health, administration parks and recreation, building, and redevelopment.

**NEAR-SURFACE LAYER**
Mayor and council offices, council chambers, reference library, news conferences, exhibits.

**SURFACE LAYER**
Most frequent public contact; entry and lobby, complaints, elections, licensing, assessing, health registration

**SUBSURFACE LAYER**
Parking, mechanical equipment, data processing, inactive files, and storage.

Increasing long range view, increasing privacy and increasing exposure to the wind, sun, daylight, and rain.

Maximum exposure to public, to surface activities and sounds, to distribution and to a variety of microclimates.

Increasing privacy, thermal stability, and exposure to groundwater.

(b)

**Fig. 3.7** (a) Characteristics of horizontal layers of a site. (b) Horizontal layers and form: Boston City Hall, 1969. (Kallman, McKinnell and Knowles, Architects.)

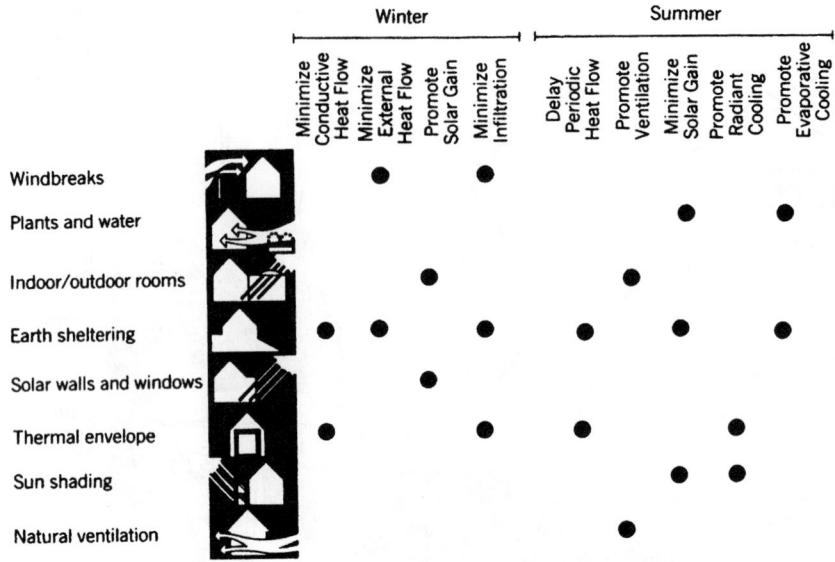

**Fig. 3.8** *Generic bioclimatic site design concepts and building strategies. (Reprinted from* Passive Cooling *by permission of the publisher, American Solar Energy Society.)*

wall combination. The high windows in both the north and south walls allow warm air to rise and escape.

## 3.6 DIRECT SUN AND DAYLIGHT

The Earth actually receives only a very small percentage of the sun's daily energy output. This small portion, however, is critically important to life on Earth. Almost all of today's energy sources have the sun as an ancestor. Fossil fuels are solar energy that has been concentrated by time and geologic conversion. Biomass, hydropower, and wind power represent shorter-term concentrations of solar energy. Even without concentration, the renewable solar resource is often more than adequate for building heating and lighting.

The amount of solar energy available at any given site varies both seasonally and daily. Typically, the closer the sun is to a position directly overhead, the more solar energy that reaches a horizontal site. Fortunately, sunlight's embodied heat energy is not its only resource; direct sun is our most intense light source. Indirect sun, such as on an overcast day or from the clear north sky, is a wonderfully diffuse and readily utilized light source.

### (a) Access to Light and Sun

The value of daylight (and fresh air) to buildings has long been recognized in zoning laws, which require that minimum distances (setbacks) be maintained between a building and the property line in lower-density areas. Height restrictions often accompany these setbacks, defining a maximum buildable volume that a building can fill (Fig. 3.10). As buildings become taller and density increases, daylight reflecting down between buildings is diminished; in response, the maximum buildable volume becomes narrower as it rises.

When direct sun in winter is desirable at the ground floor of all sites, the buildable volume is sharply reduced in height due to the low angle of the winter sun in northern latitudes. Protection of this most-restricted buildable envelope, called the *solar envelope,* is at present rarely mandated, but solar access ordinances have been enacted for residential zones in some cities. The most restrictive feature of the solar envelope is the low slope of its northern face, usually corresponding to the altitude of the sun above the horizontal for about 2 hours before and after noon on December 21. This feature allows 4 hours of access to direct sun for a neighbor on the site just to the north on even the shortest day.

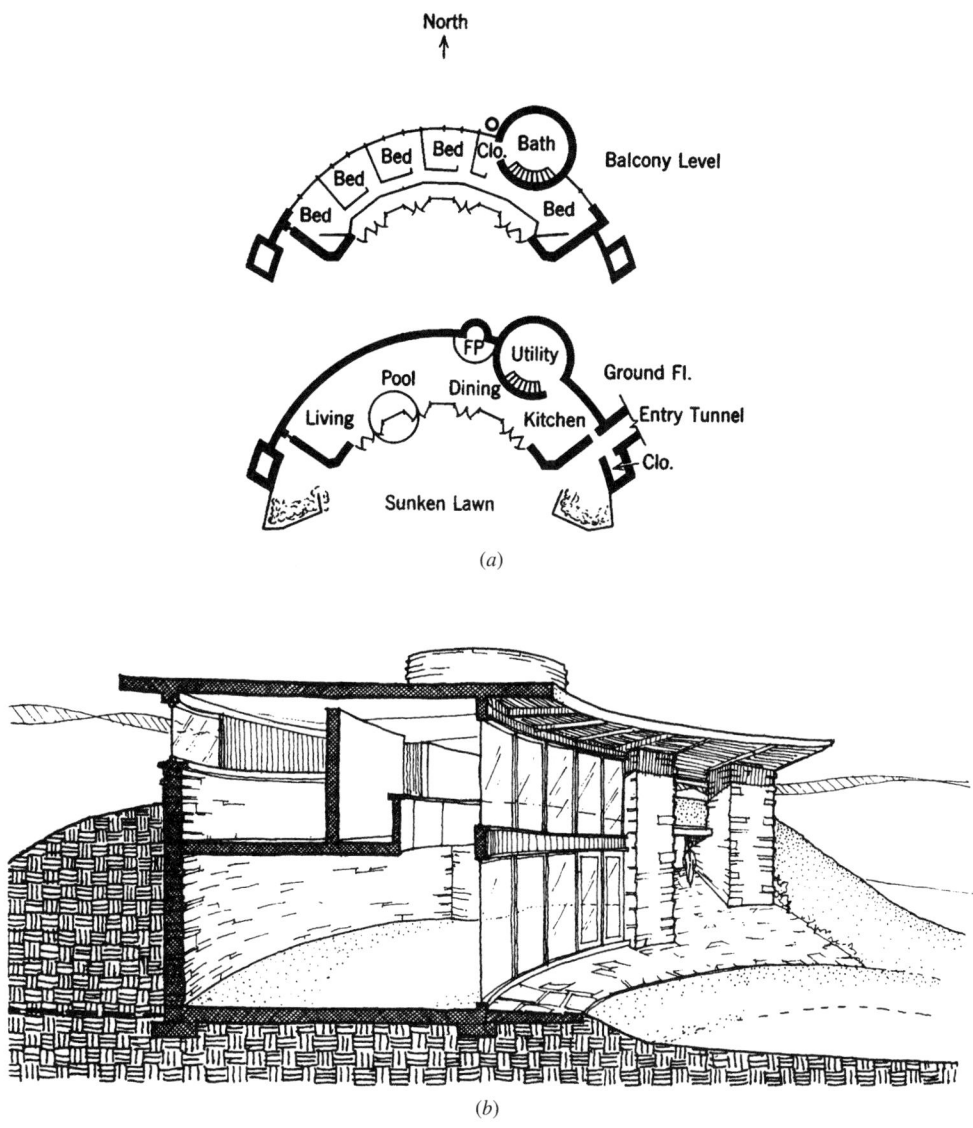

**Fig. 3.9** *An early passive solar-heated home, Frank Lloyd Wright's Solar Hemicycle (Jacobs House II) near Madison, Wisconsin. The house was designed in the early 1940s and built in 1948. (a) Floor plans. (b) Section-perspective, looking east toward the entry tunnel in the berm wall.*

A brief history of daylight and solar access regulations and an overview of current design guidelines are given in DeKay (1992). Figure 3.11 shows the application of six variations of regulations on an urban block at 40° N latitude, elongated east–west. The east–west streets are wider than the north–south streets. Further development of the solar envelope is shown in Fig. 3.12, based upon the work of Knowles (1981). Clearly, winter solar access is closely related to maximum density of development.

For daylight access, building surfaces can be almost as important as building geometry. The importance of light reflected from vertical and horizontal surfaces is apparent in the daylighting calculations presented in Chapter 14. Lighter-colored surfaces produce more internal daylighting, especially in crowded urban conditions, where a view of the sky from windows is not common. An increase in daylight results for all surfaces on an urban street if the building and site surfaces have light colors.

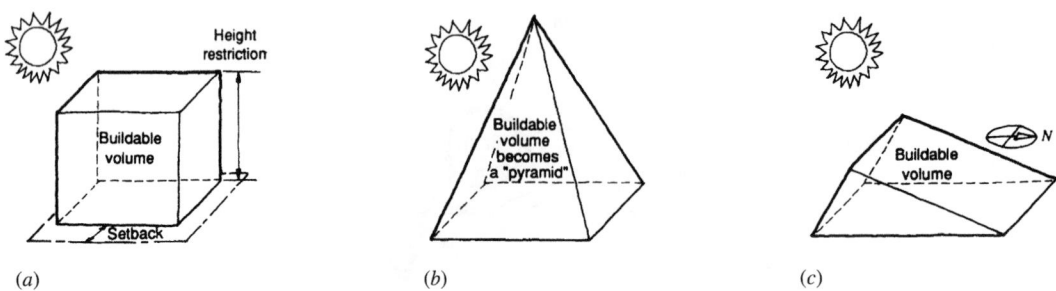

**Fig. 3.10** *Protecting access to light and solar radiation. Three regulatory approaches that compromise between private optima (e.g., maximum rentable floor space) and public optima (e.g., daylight at street level). (a) Simple daylight access, residential and low-rise commercial areas. (b) Daylight access in high-density areas. (c) Access to direct sun for winter heating.*

This can conflict with the use of planting for shading and evaporative cooling, although lighter-colored groundcover plants are available. Although daylight reflected in a diffused way from building and ground surfaces is a potential benefit, harsh specular reflections of direct sun from mirror surfaces are often unwelcome (see Section 3.6*f*).

## (b) Charting the Sun

For site planning, one of the earliest tasks is to determine when direct sun reaches a building or space—such as a playground, deck, or courtyard—on a site. Details of how a building can utilize the sun are found in later chapters on solar space heating (Chapter 8), solar water heating (Chapter 21), and

daylighting (Chapter 14). Information on solar geometry is presented in Chapter 6. Numerical data on sun position and intensity are found in Appendices C and E. Effective design for solar utilization, however, begins with an understanding of the basics.

The chart in Fig. 3.13 shows the sun's position for 40° N latitude and also shows, for the 6 hours of greatest insolation, what percentage of clear-day insolation is gained each hour by an unshaded south-facing vertical window. The emphasis on a southern orientation calls attention to its useful characteristic of receiving *more* sun in winter and *less* sun in summer than any other orientation. To convert these percentages to actual heat gains, see Tables C.1 through C.10 in Appendix C (for *clear*

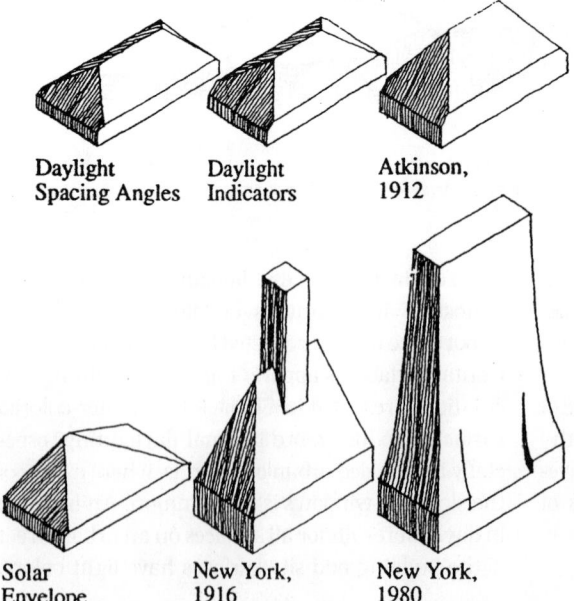

Daylight Spacing Angles    Daylight Indicators    Atkinson, 1912

Solar Envelope    New York, 1916    New York, 1980

**Fig. 3.11** *Various approaches to defining maximum allowable building envelopes for daylight access. These envelopes are applied to a 200-ft × 400-ft (61-m × 122-m) block at 40° N latitude. The east–west streets (along the longer side) are 65 ft (20 m) wide; the north–south streets are 45 ft (14 m) wide. In this case, "daylight spacing angles" and "daylight indicators" produce nearly identical envelopes. (From DeKay, 1992, with permission of the American Solar Energy Society.)*

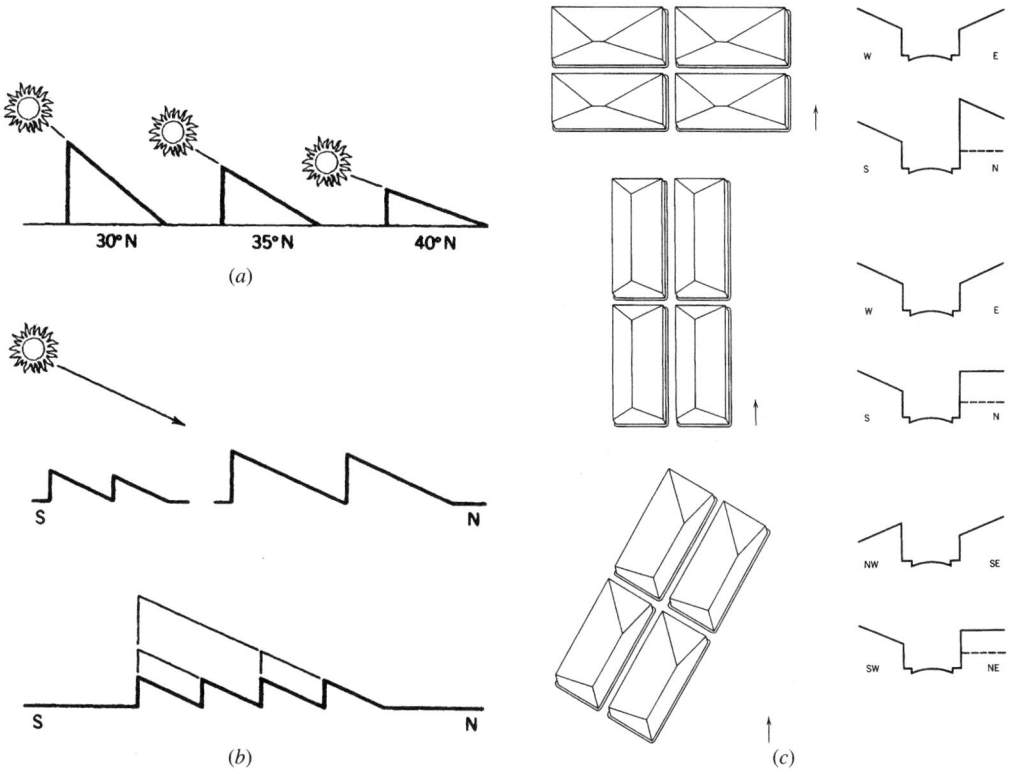

**Fig. 3.12** *These solar envelopes are refinements of the solar access "pyramid" of Fig. 3.10. (a) The slope of the solar envelope changes with latitude. (b) The larger the site, the greater the buildable volume of the solar envelope. (c) Solar envelopes for various orientations of individual sites. (d) Solar envelopes for east–west elongated blocks. (e) Solar envelopes for north–south elongated blocks. (Reprinted by permission of R. Knowles,* Sun, Rhythm, Form, *© 1981, MIT Press.)*

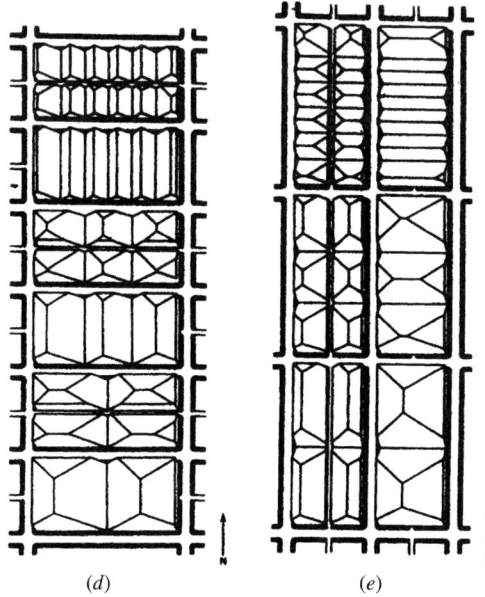

*(d)*  *(e)*

**Fig. 3.12** *(Continued)*

days at a given latitude) or Table C.15 for *average* day data by climate station (January and July only).

## (c) The "Band of Sun"

A building section drawing can be a powerful site analysis tool. A north–south section is drawn through a proposed building and its site. At three times of year (each solstice and the equinox), the noon solar altitude angle is determined, as in Fig. 3.14. Then the *band of sun* that strikes the building and site is drawn, incoming at the noon altitude. The portion of this available direct radiation that is used by the building and site can then be analyzed. The more of the band that is unused in each season, the higher the potential for redesign in order to better utilize the solar resource. This analysis helps to show how wintertime solar utilization depends upon lower sun altitudes and summertime uses depend upon higher altitude angles.

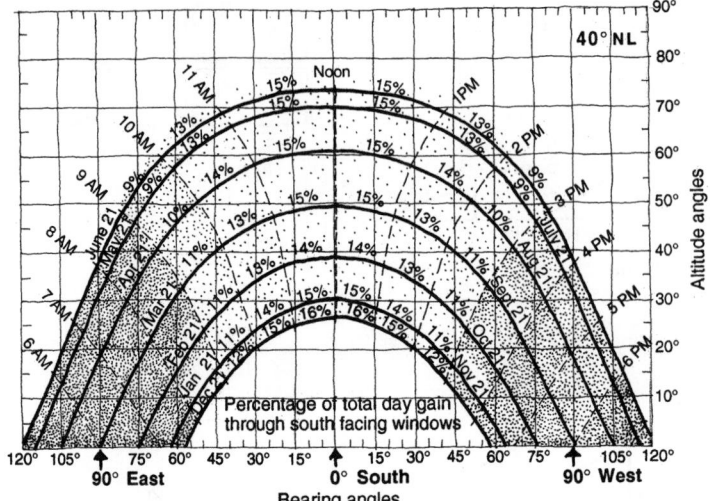

*Fig. 3.13* Sun chart for 40° N latitude showing the approximate percentage of clear day insolation for south-facing windows for each of the 6 maximum hours of sun each month. *(From Edward Mazria and David Winitsky,* Solar Guide and Calculator, *Center for Environmental Research, University of Oregon, 1976.)*

### (d) Skylines and Winter Sun

The skyline as actually seen from a given location on a site can be charted to determine access to direct sun at any time of the year (Fig. 3.15). Such an obstruction analysis should precede the placing of any solar collector, whether it is a south-facing window or a manufactured collector. Seasonal obstructions within the six "best" collection hours (9 A.M. to 3 P.M.) are particularly serious and should be minimized when siting a collector.

At the same time, consideration of neighbors' access to direct sun is appropriate. This can be checked by charting another skyline somewhere along the northern boundary of a site. The outline of the building being designed should be included on this skyline, and can then be modified, if necessary, to preserve solar access for the neighbor.

The location charted in Fig. 3.15 is quite a good one for a summer solar usage, such as a swimming pool heater, or for a PV array that serves summer loads such as air conditioners. It is also a possible location for a solar collector for domestic water heating, which will work well during the 6 months from March through September; obstructions are few within this period. However, it is quite poor for winter solar collection, allowing only about one-

fifth of the potential solar gain in the November to January period.

### (e) Sun and Shadows: Model Techniques

The graphic site analysis techniques described above have a limitation: each graph applies only to one particular location on a site. To study multiple locations, multiple graphs must be constructed. By contrast, a three-dimensional model used in conjunction with a sun shadow plot (Fig. 3.16) can yield three-dimensional sun penetration patterns as they change over time, for any position, indoors or out, in/on the model. Models are initially time-consuming to build but can save time when considering alternative locations on a site, alternative window and space combinations, alternative shading devices, and so on. Perhaps most important to a designer, models suggest three-dimensional solutions, because one is testing a volume rather than merely a plan or section. However, if obstructions to sun exist far from a site (such as nearby mountains), it might be better to rely on graphs rather than trying to include such large obstructions on a (quickly becoming huge) model.

A *sunpeg* chart (see Section 6.4), correctly attached to a model of any scale, allows a designer to quickly determine exact sun penetration and

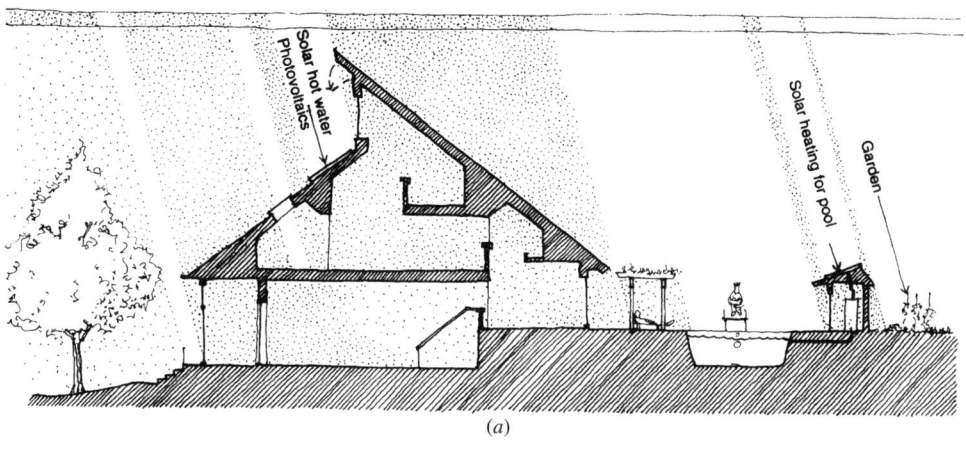

(a)

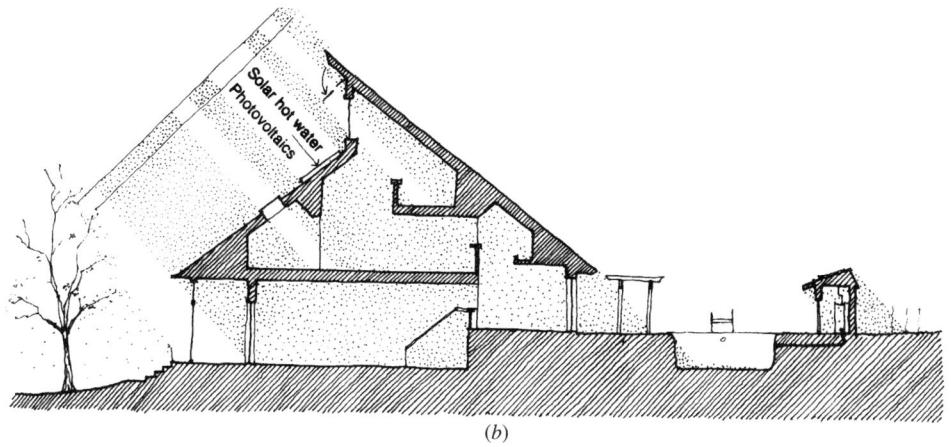

(b)

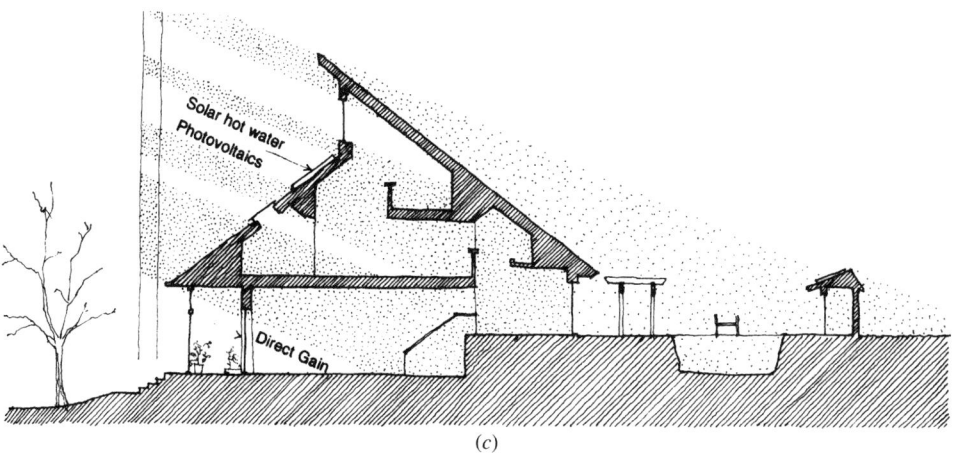

(c)

**Fig. 3.14** The band of sun available to a proposed building is charted at solar noon on a north–south section. (a) The summer solstice, where most-optimum collecting surfaces are at near-horizontal tilt angles. (b) The equinox. (c) The winter solstice, where most-optimum collecting surfaces are at near-vertical south-facing tilt angles.

DESIGN CONTEXT

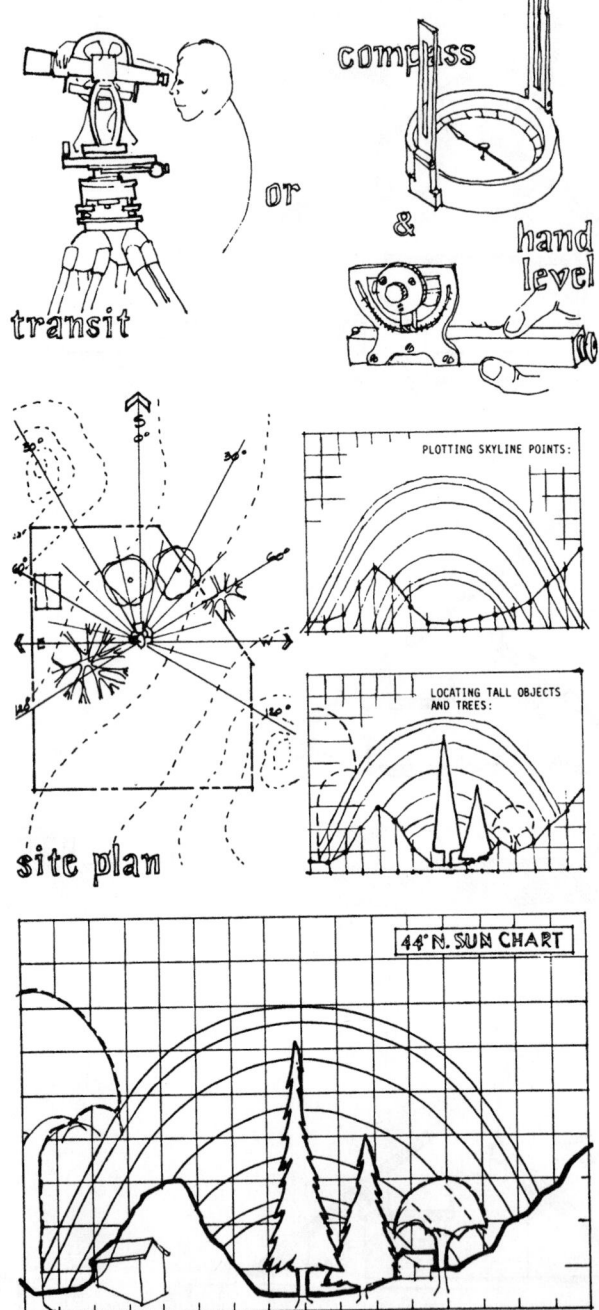

**Fig. 3.15** *Charting the skyline from a specific site position. (From Edward Mazria and David Winitsky,* Solar Guide and Calculator, *Center for Environmental Research, University of Oregon, 1976.)*

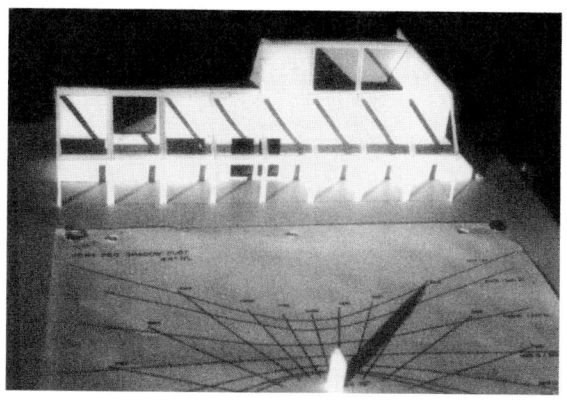

**Fig. 3.16** *A model of a small building with a glazed open-frame circulation space on the south side is observed at the sun's position at 3 P.M. on December 21 through the use of a sunpeg chart.*

shadow patterns at many times for any date. These plots are one of the most important tools for a concerned designer, both early and late in the design process.

In Fig. 3.16, a model is observed at a sun position corresponding to 3 P.M. on December 21. At this end of the 6-hour period of best collection, it can be seen that sun still fills the south-side glazed open-frame circulation space and enters spaces beyond through south openings. It can also be seen that passing from these spaces into the circulation space might involve considerable glare because the sun is so low in the sky. In such a case, vertical fins inside the south glazing would help intercept and

diffuse direct sun while still permitting solar heat gain.

### (f) Controlling Solar Reflections

The use of highly reflective (or "mirror") glass to reduce envelope heat gain in office buildings has increased the frequency of annoying solar reflections from buildings (Fig. 3.17). Large areas of nonmirror glazing, such as found on passive solar-heated buildings, can also cause reflection problems. The farther the sun's rays are from being perpendicular (or normal) to any surface, the more radiation is reflected, rather than absorbed or

**Fig. 3.17** *Mirror-glass windows in a newer office building (left) in Milwaukee, Wisconsin, cast strong reflections on the north-facing wall of an older building next door. Although this reflected radiation/heat might be welcome in winter, the glare can be intense. In summer, the older building is particularly disadvantaged by additional thermal loads.*

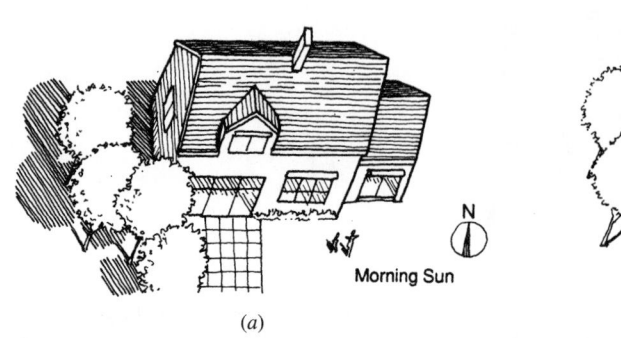

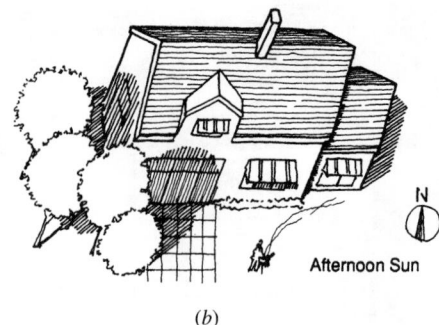

Morning Sun

*(a)*

Afternoon Sun

*(b)*

**Fig. 3.18** *Selective protection from reflections. (a) The trees standing west of this south window wall do not interfere with solar access during the best hours for solar collection (around noon), nor do they prevent early morning sun from entering the windows. Any reflections of the early morning sun are intercepted by the trees before they can annoy nearby buildings. (b) The late afternoon sun is blocked by the trees before either solar gain or reflections can occur.*

transmitted, by that surface. Thus, the intensity of reflection is greatest (and the transmission of solar gain is least) when the sun's rays are nearly parallel to a surface.

Because the most intense reflections occur

**Fig. 3.19** *The "eggcrate" shading devices shown on the southwest corner of this University of Oregon building prevent reflections by blocking acute sun angles from either side of the window. Different shading devices are appropriate for different façade orientations.*

when the sun's rays are nearly parallel to a wall, such reflections are fairly easily blocked or intercepted. For example, foliage (Fig. 3.18) can intercept reflected sunlight from a south-facing window without blocking solar collection during the best hours. Another common approach is to use external projections around windows, as in Fig. 3.19. The use of a model and a sunpeg chart is a good way to explore reflections. By using mirrored surfaces to represent reflective glass on a model, resulting reflections will be evident along their three-dimensional path.

## 3.7 SOUND AND AIRFLOW

Sound and airflow are considered together here because they are so difficult to separate. Many buildings that could be opened to ventilation or cooling by breezes rely instead on forced ventilation because of noise that would accompany breezes through an open window. Pollution is another potential deterrent to natural ventilation. Almost any object or device that reduces noise will also reduce the velocity of a breeze, as is true of most filtering devices used to remove dust particles.

### (a) Noise

Any sound that is unwanted becomes *noise* (see Chapter 17). The urban building in Fig. 3.20 is unusual both in providing an opportunity for wind-driven ventilation—air moves freely below it as well

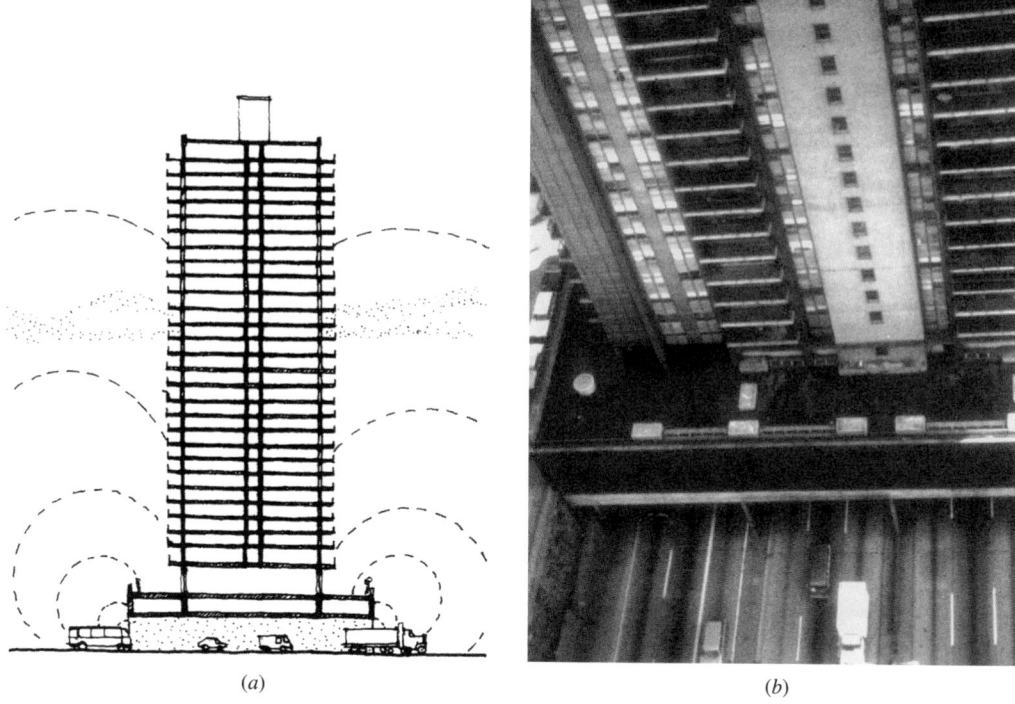

**Fig. 3.20** *Apartment buildings in series straddle the approach ramps to New York City's George Washington Bridge. (a) Section along the freeway. (b) Looking down to the freeway. These buildings were the scene of a study linking noise levels with reading disabilities for occupants of the apartments. (From Cohen et al., 1973.)*

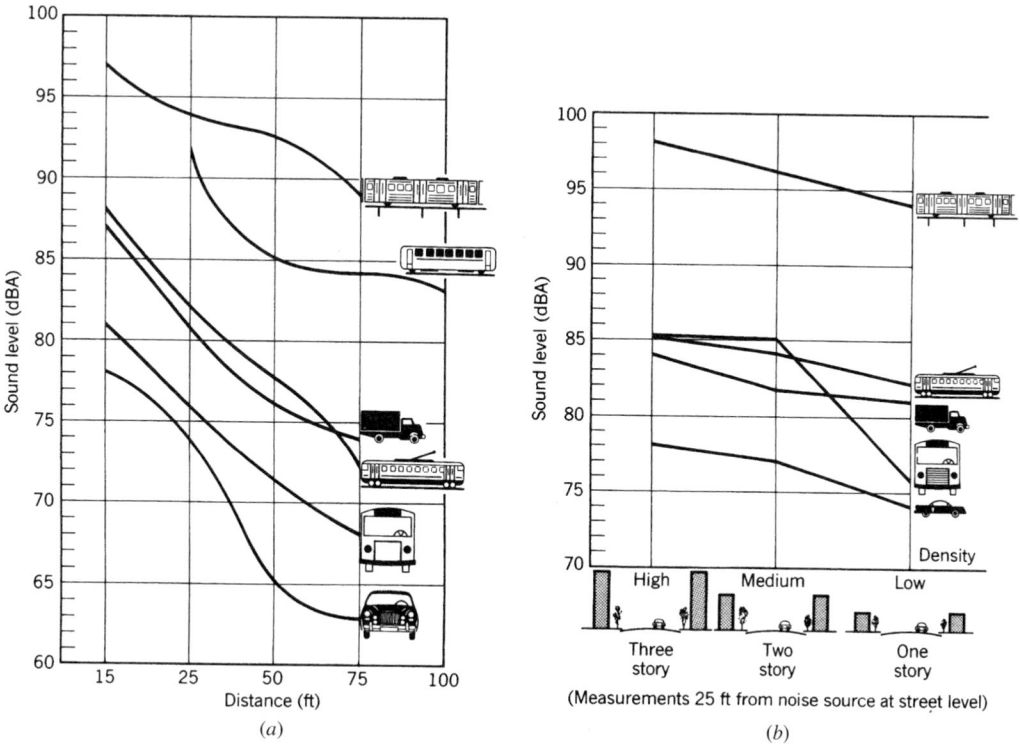

**Fig. 3.21** *Predicting noise levels outdoors. (a) Distance as a factor influencing sound pressure level. (b) Building height as a factor in noise propagation. (From Clifford R. Bragdon,* Noise Pollution: The Unquiet Crisis, *University of Pennsylvania Press, 1971. Reprinted by permission.)*

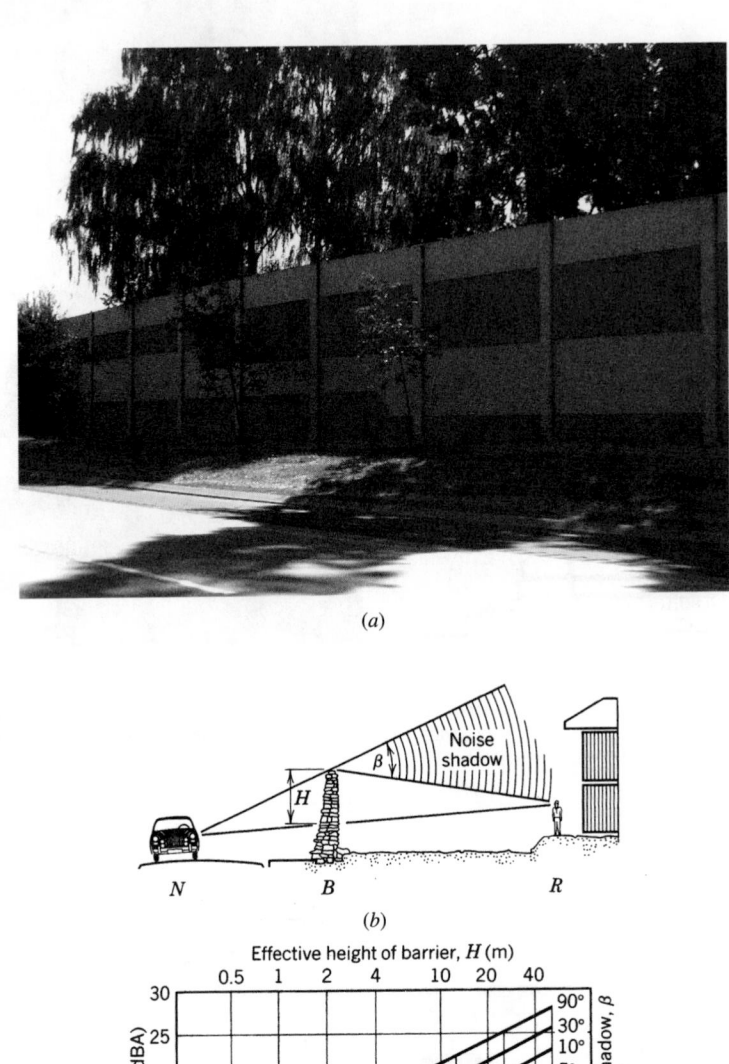

(a)

(b)

(c)

**Fig. 3.22** *Outdoor noise barriers. (a) A noise barrier abutting a freeway near Portland, Oregon. (Photo by Amanda Clegg.) (b) To determine the approximate noise reduction (in decibels) due to an outdoor barrier, construct a section locating the noise source (N), the solid barrier (B), and the receiver's location (R). On this section, determine the effective height (H) of the barrier and the diffraction angle (β) with the resulting "noise shadow." Enter graph (c) with H and β; where the lines intersect determines the noise reduction in dBA (left axis). A reduction of 10 dBA is perceived as half as loud as the original source. Note the perceptible noise reduction from simply breaking the line of sight (β = 1°). (From Doelle, 1972. Reprinted by permission.) For a more detailed procedure, see Fig. 19.33.*

as around it—and in the extraordinary intensity of traffic noise that is borne on the air.

Two characteristics of cities contribute to increased noise at street level: hard surfaces that reflect rather than absorb sound and parallel walls that intensify sound by interreflection rather than dissipating it. Increasing horizontal and vertical distances from urban noise sources affect outdoor noise levels in various ways, as shown in the graphs of Fig. 3.21. Unfortunately, obtaining much distance between sound source and receiver in dense urban areas is quite difficult.

Although building surfaces are generally made of hard materials for durability in weathering, softer and multiplaned materials (such as plants) are desirable from a public noise–reduction viewpoint. Their impact on measured sound levels may be slight, but visually softer surfaces reinforce a perception of acoustically softer environments (much as the sound of running water reinforces our perception of cool environments). Fountains are especially useful sources of *masking sound;* they can be kept flowing as long as a noise persists, and they enhance the cooling function of natural ventilation via evaporative cooling, especially in drier climates. Not all spaces benefit from masking sound; where a single sound source is expected to predominate over all others, masking sound will interfere and become unwanted "masking noise."

Where site conditions allow, barriers to street noise can be installed that cast *sound shadows* on a site (Fig. 3.22). Such barriers may do little to reduce noise levels at upper windows, but surface activities can be given much lower noise levels, especially near the barrier. Many cities now require such barriers between new housing developments and highways or railroads.

Another urban noise source is the mechanical equipment of buildings themselves. Many noise complaints against buildings involve air-conditioning equipment (the compressive refrigeration cycle and its year-round utilization as a heat pump are described in Chapter 9). When densely packed buildings are forced to rely upon mechanical cooling, such closeness makes the noise of the systems even more annoying. Noise is generated both by the compressor and by the great quantities of outdoor air that must be rapidly pushed through outdoor condenser coils. The condenser's need for outdoor air is so critical that attempts to surround it with noise shields can hinder its efficient opera-

tion and shorten its life. Fortunately, more recent energy-efficient equipment is quieter, although not silent.

In residential neighborhoods, the greater distance between buildings might be expected to lessen these difficulties. Yet the much lower ambient (or background) sound level of residential areas is one of their more appealing characteristics, and an intruding compressor on a hot summer night can irritate neighbors who formerly enjoyed cool—and quiet—night breezes.

## (b) Air Pollution

Problems of global importance are resulting from air pollution; these are summarized in Table 3.3. Building construction and operation are among the contributors to air pollution. The *greenhouse effect* (Fig. 3.23) threatens to produce global warming, a trend most scientists now firmly believe is underway. The greenhouse effect occurs because gases that block the outgoing flow of long-wave radiation (heat) from the Earth's surface are accumulating in the atmosphere. Energy production and use (especially of fossil fuels) are contributing heavily to these greenhouse gases, which include carbon dioxide, methane, nitrous oxide, ozone, and chlorofluorocarbons (CFCs). Another serious threat is stratospheric ozone depletion, with potentially devastating consequences to ecosystems due to increased ultraviolet radiation received at the Earth's surface.

The designers of buildings can influence these trends in several ways. First, they can help to greatly reduce the air pollution caused by electric power plants and by burning fuel in buildings, by designing for greater energy conservation and by utilizing clean and renewable energy sources within buildings. Chapter 8 is concerned largely with this topic. Second, they can specify materials and equipment that, through their manufacture or operation, lessen air pollution. This suggests avoiding fuel combustion (coal, oil, trash, wood, and natural gas, roughly in descending order of air pollution threat). This philosophy also encourages selection of refrigeration equipment that uses environmentally friendly refrigerants, as well as insulation and upholstery products made with non-CFC blowing agents. Many of these practices are mandated, but the designer can go further and seek out the most environmentally benign products.

DESIGN CONTEXT

### TABLE 3.3  Air Pollution: Sources and Effects

| Gas or Pollutant | Sources | Effects |
|---|---|---|
| Carbon monoxide (CO) | Gasoline-powered vehicles; industry using oil and gas; building heating using oil and gas; biomass burning | Enters human bloodstream rapidly, causing nervous system dysfunction and death at high concentrations; interferes with self-cleansing of atmosphere |
| Carbon dioxide ($CO_2$) | Fossil-fuel combustion, deforestation | Contributes to greenhouse effect |
| Methane ($CH_4$) | Rice fields, cattle, landfills, fossil-fuel production | Contributes to greenhouse effect |
| Sulfur oxides (sulfur dioxide [$SO_2$] and sulfur trioxide) | Industry using coal and oil; heating using coal and oil; power plants using coal, oil and gas; ore smelting | Acid rain, damaging plants and attacking building skin materials; irritates human respiratory tract and complicates cardiovascular disease; decreases visibility in atmosphere |
| Nitrogen oxides ($NO_x$) e.g. nitric oxide (NO) and nitrogen dioxide ($NO_2$) | Gasoline-powered vehicles; building heating using oil and gas; industry and power plants; biomass burning | Acid rain, damaging plants and attacking building skin materials; irritates human eyes, nose, and upper respiratory tract; triggers development of smog; decreases visibility in atmosphere |
| Nitrous oxide ($N_2O$) | Nitrogenous fertilizers, deforestation, biomass burning | Contributes to greenhouse effect |
| Hydrocarbons (compounds of hydrogen and carbon) | Petroleum-powered vehicles, petroleum refineries, general burning | Promotes smog; toxic to humans at high concentrations |
| Chlorofluorocarbons | Aerosol sprays, refrigerants, foams | Contributes to greenhouse effect and to stratospheric ozone depletion |
| Particulates (liquid or solid particles smaller than 500 micrometers) | Vehicle exhausts, industry, building heating, general burning, spore- and pollen-bearing vegetation | Promotes precipitation formation; some are toxic to humans; some pollens and spores cause allergic reactions in humans |

*Sources:* Adapted from Marsh (1991) and Graedel and Crutzen (1989).

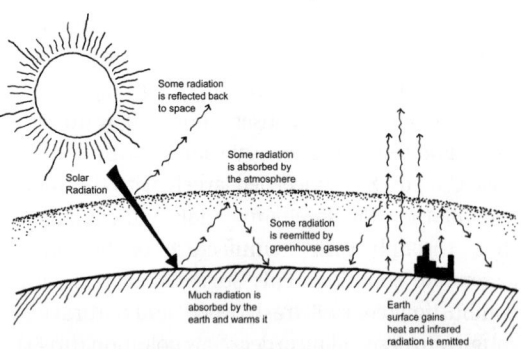

**Fig. 3.23** *The greenhouse effect traps heat in the Earth's upper atmosphere. Clouds and particles in the atmosphere reflect about one-fourth of incoming solar radiation while blocking about two-thirds of the heat that the Earth would otherwise lose to outer space. Currently, the atmosphere keeps the Earth about 33°C (60°F) warmer than it would be without this heat-trapping process. Increases in greenhouse gases theoretically will reflect more incoming solar radiation but block even more outgoing radiation, resulting in global warming. (Drawing by Amanda Clegg.)*

Buildings are substantial contributors to air pollution: the fuel combustion within, the power plants that supply electricity, the incinerators and landfills that receive waste from buildings. Buildings and power plants are major contributors to the greenhouse effect and the primary causes of acid rain (sulfur oxides) and smog (nitrogen oxides). The transportation that takes people to and from buildings is another major air pollution source. We design buildings to utilize "fresh" air, whether by natural or forced ventilation; we must then also design buildings to preserve our fresh air resource (see Fig. 1.1). The less energy buildings require, the cleaner the outdoor air will be.

On site, local sources of air pollution must be minimized and, as far as possible, isolated. Combustion gases pose a threat, especially where vehicles approach buildings (Fig. 3.24). Idling truck motors at loading docks or automobile motors at drive-up service windows can threaten building occupants

**Fig. 3.24** *Reactive protection of an outdoor air intake. A loading dock near an intake was a source of indoor air pollution from truck motor fumes, prompting the installation of a warning sign.*

through openings such as doors and windows; mechanical system fresh air intakes are particularly vulnerable, because outdoor air is intentionally drawn into them. Again, consider the neighboring buildings: will the location of some activity on your site threaten a neighbor's fresh air?

### (c) Wind Control

For most buildings, wind (like sun) changes from resource to detriment with the change of seasons. In many locations, wind also changes its prevailing direction with the seasons. Control of wind often means utilizing wind-sheltered areas in winter while encouraging increased wind speeds in summer. From spring through fall, outdoor spaces might benefit from this barrier-to-connector changeover on a daily basis: less wind during cool mornings, more during hot afternoons. Fortunately, winds are generally weakest in the early mornings and strongest in the afternoons because of the effect of the sun's heating on the surrounding environment.

The generalized patterns of wind flow around thin windbreaks and thicker buildings (Fig. 3.25) help us to understand where shelter and increased airflows occur. These patterns, however, are much more complicated than they first appear and are highly influenced by objects upstream, to the sides, and downstream of the wind-directing object being analyzed. Wind-tunnel tests using scale models are far more reliable than these generalized patterns; unfortunately, such tests are expensive and still fraught with opportunities

for misprediction. Nevertheless, a site can and should be analyzed graphically for seasonal wind utilization.

Wind ultimately returns to its original flow pattern after encountering an obstacle such as a windbreak or a building. Before it reaches the obstacle, it slows, builds (positive) pressure, and turns upward or sideways. As it passes the obstacle, it increases its speed, and reduced (negative) pressure results at the sides of and behind the obstacle. These pressure differences, flow patterns, and the size and shape of the wind-protected areas behind an obstacle are all usable for control of air motion, both inside a building and outside.

Windbreaks are commonly used to protect outdoor areas; these can be fences or plants. Figure 3.26 shows the relative reduction in wind velocity at the level of a windbreak as wind approaches and then passes. The distances (horizontal axis) are in units of the height of the windbreak. Note that the densest windbreak produces the greatest reduction in wind speed behind it—but that the wind recovers its full velocity closer to such a barrier compared to a less dense windbreak. Thus, the more dense the windbreak, the greater the reduction in wind speed but the smaller the area so affected.

Gaps in windbreaks can produce increased wind speeds through the gaps. Figure 3.27 shows a small area of wind speeds above a windbreak even greater than the undiverted wind speed. Although a gap is a threat to winter wind protection, it is an opportunity for summer wind speed enhancement. Given constant prevailing summer wind directions, a windbreak gap could provide a small area with above-average air motion.

Wind flow around buildings is a complex matter; nevertheless, some general patterns for shelter areas are shown in Fig. 3.25, whereas Figs. 3.28 and 3.29 show wind behavior to be expected in typical building combinations. Some patterns from the situations shown in Fig. 3.29 include:

- With the bar effect (*a*), the downward-spinning wind behind a building can reach 1.4 times the speed of the average wind.
- With the Venturi effect (*b*) and few obstructions upwind or downwind from the narrow neck of a building, wind speeds through the neck can reach 1.3 times the average, up to heights of 100 ft (30 m), and 1.6 times the average at about 165 ft (50 m) in height.

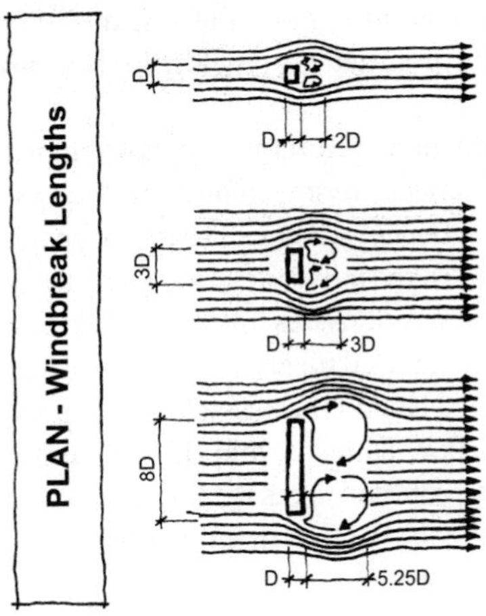

PLAN - Windbreak Lengths
D
D 2D
3D
D 3D
8D
D 5.25D

(a)

| Wind Speed Reduction (%) | | | | |
|---|---|---|---|---|
| Density of Belt | Average Over First: 50 Yd | 100 Yd | 150 Yd | 300 Yd |
| Very Open | 18 | 24 | 25 | 18 |
| Open | 54 | 46 | 37 | 20 |
| Medium | 60 | 56 | 48 | 28 |
| Dense | 66 | 55 | 44 | 25 |
| Very Dense | 66 | 48 | 37 | 20 |

(b)

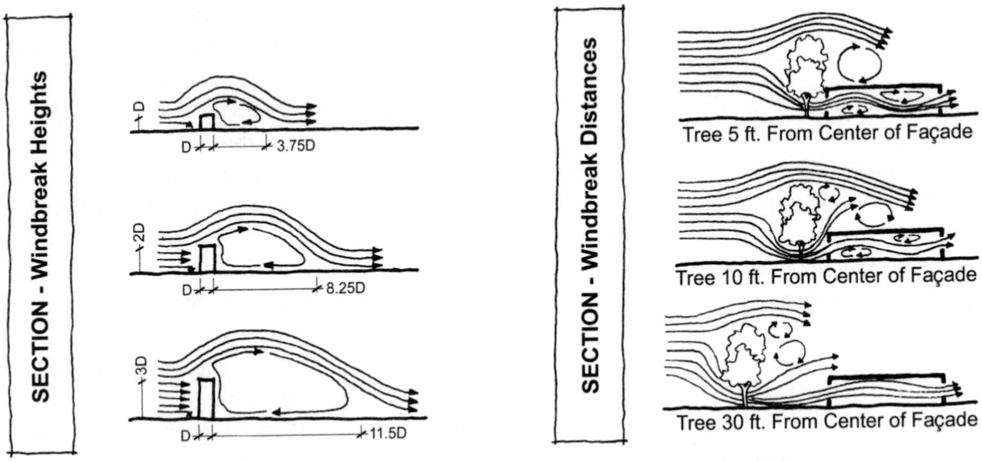

SECTION - Windbreak Heights
D
D 3.75D
2D
D 8.25D
3D
D 11.5D
SECTION - Windbreak Distances
Tree 5 ft. From Center of Façade
Tree 10 ft. From Center of Façade
Tree 30 ft. From Center of Façade

(c) (d)

***Fig. 3.25*** *Approximate patterns of wind around objects. (a) Effects of different barrier lengths (widths). (b) Reduction in wind speed due to windbreak density. (c) Effects of different barrier heights. (d) Wind flow through trees and buildings. (Reproduced with the permission of the American Institute of Architects, © 1981, AIA. Redrawn by Jonathan Meendering.)*

100%
80%
60%
40%
20%
0
10
5
0
5
10
15
20
25
30
Very permeable windbreak
Permeable windbreak
Impermeable windbreak

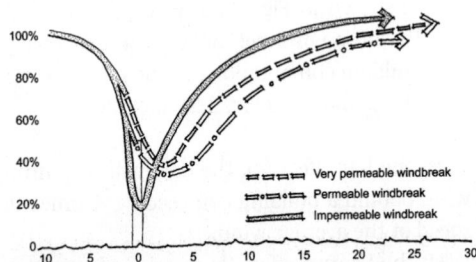

***Fig. 3.26*** *Wind speed reduction behind windbreaks of varying permeability. Solid (impermeable) barriers produce the lowest wind speeds, but these are effective for the shortest distance beyond the windbreak. Units of distance = heights of windbreak. (Redrawn by Erik Winter; from Brown and Gillespie, 1995.)*

- The gap effect (*c*) begins to occur with perpendicular winds and buildings of more than 5 stories (50 ft [15 m]) in height; by 7 stories, wind speeds 1.2 times the average can occur through the gaps; by 60 stories, gap wind speeds can be 1.5 times the average.
- For higher buildings (*d*), increased wind speeds occur at the corners (localized within a radius from the corner equal to the width of the building "d"); where height is 50 ft (15 m), wind speed can reach 1.2 times the average; for heights above 115 ft (35 m), wind speed can be 1.5 times the average. Where two towers approach each other, increased wind around corners and between the

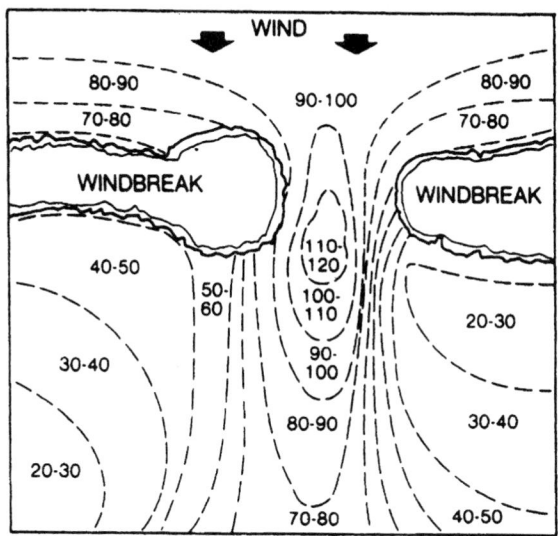

**Fig. 3.27** *Wind speeds accelerate through a gap in a windbreak. Numbers indicate the percentage of the incoming (unaffected) wind speed. (From Caborn, J.M. (1957).* Shelterbelts and Microclimate. *Edinburgh: H.M. Stationery Office; cited in McPherson, 1984.)*

towers can go as high as 2.2 times the average for towers 330 ft (100 m) high.

- Increased wind speed and turbulence within the wake of buildings (*e*) can be especially serious for towers at heights from 16 to 30 stories, where wind speeds can reach 1.4 to 2.2 times the average.

Beranek (1980) discussed methods for charting the shelter areas in such building groups. Localized

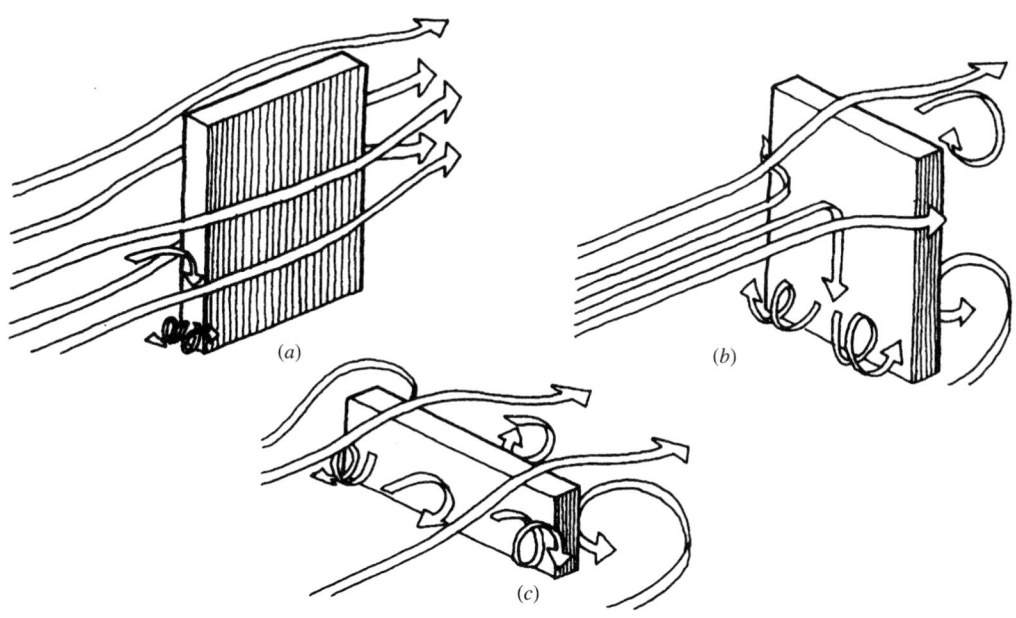

**Fig. 3.28** *Wind patterns around single buildings. (a) Tall, slender buildings: height greater than 2.5 times the width. (b) Tall, rather wide buildings; height between 2.5 and 0.6 times the width. (c) Long buildings; height less than 0.6 times the width. (From Beranek, W.J., "General Rules of the Determination of Wind Environment," in* Wind Engineering, *J.E. Cermak [ed.], Vol. 1, © 1980, Pergamon Press Ltd., reprinted by permission.)*

DESIGN CONTEXT

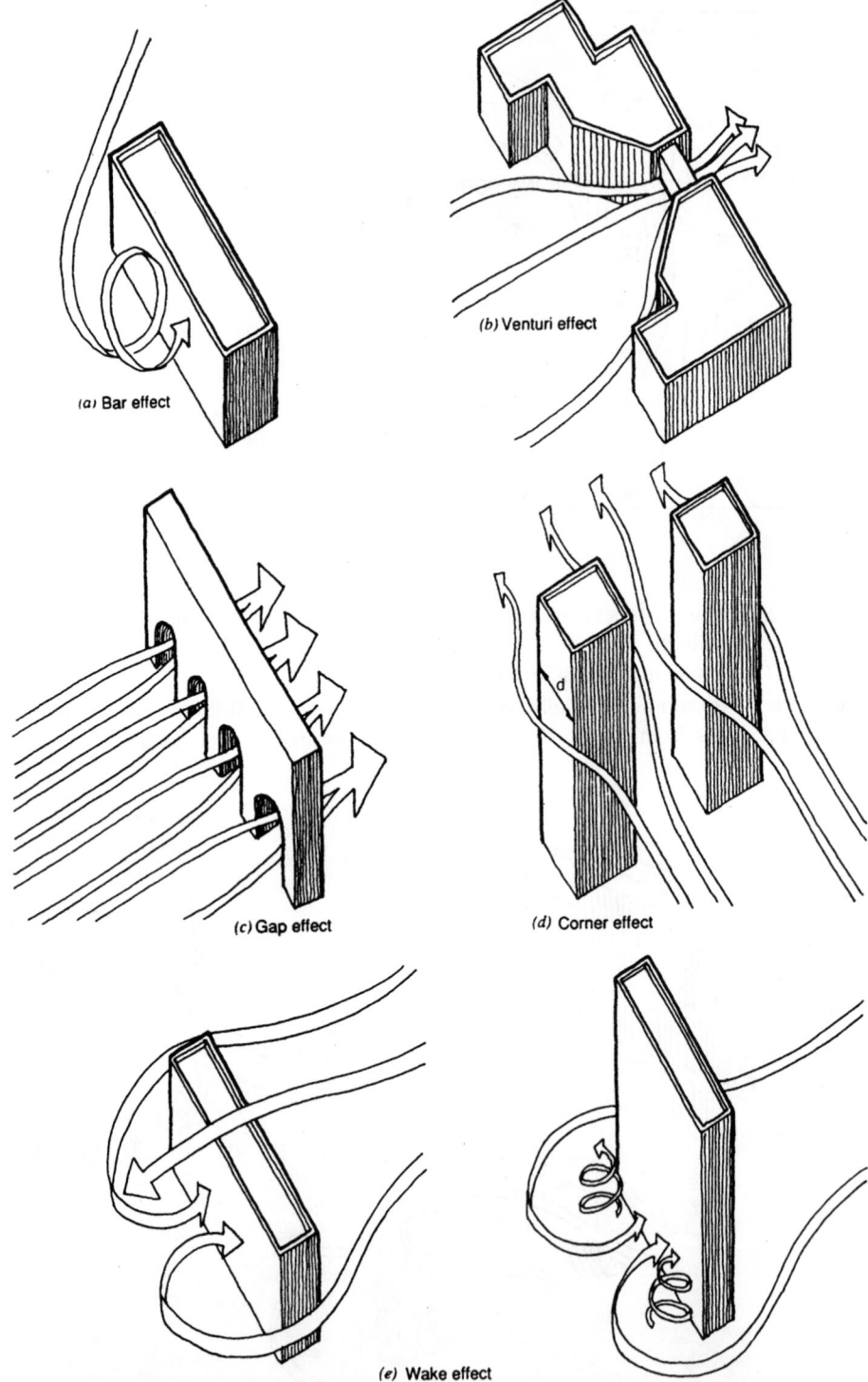

*(a)* Bar effect

*(b)* Venturi effect

*(c)* Gap effect

*(d)* Corner effect

*(e)* Wake effect

**Fig. 3.29** *Wind patterns among building clusters (see text for quantification). (From Gandemer, J. "Wind Environments Around Buildings: Aerodynamic Concepts," Wind Effects on Buildings and Structures, K.J. Eaton (ed.), © 1977, Cambridge University Press. Reprinted by permission.)*

effects of wind turbulence between buildings can have a particularly powerful impact on entryways. People leaving the controlled air motion of a building lobby are often unprepared for the speed and turbulence of channeled wind just beyond the doorway.

Wind flow for the cooling of buildings is discussed in detail in Chapter 8.

## (d) Ventilation and Cooling

Outdoor air is introduced to buildings for two distinct reasons. *Ventilation* involves the provision of fresh air to interiors to replenish the oxygen used by people and to help carry away their by-products of carbon dioxide ($CO_2$) and body odors. Ventilation is desirable year round; recommended minimum rates of fresh airflow are found in Tables E.25 and E.26 of Appendix E. *Passive cooling* (with outdoor air) replaces heated indoor air with cooler outdoor air. Cooling by breezes is a seasonal opportunity, limited to times when the outdoor air temperature is lower than the indoor air temperature and the outdoor humidity is at or below that desired indoors. When outdoor air temperatures are about the same as interior tem-

peratures, breezes might still be useful to increase interior air motion, thus extending the comfort zone, as quantified by Olgyay (1963) and discussed in Chapter 4.

Although airflows for indoor air quality control and for cooling are both commonly referred to as "ventilation," they are not identical design situations. Building codes/standards typically rigidly define ventilation. Airflow for passive cooling can require far greater quantities of air than airflow for control of air quality, and the influence of such requirements on building siting and window size and placement is considerable. Figure 3.30 illustrates the differing approaches to these two uses for airflow and suggests that window position and human occupants be considered together in design.

Two examples of contemporary buildings that use the wind appear in Figs. 3.31 and 3.32. The first served as a summer exhibition space in Montreal, Canada, and relied upon prevailing

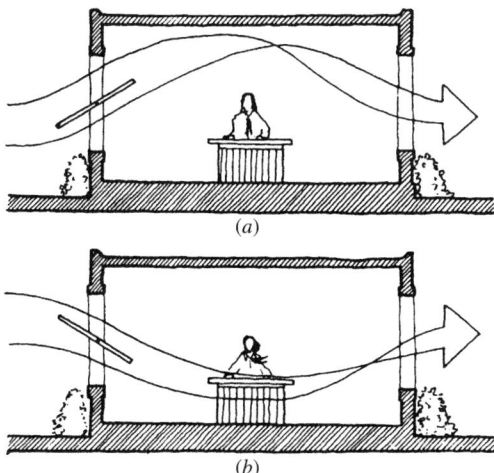

(a)

(b)

**Fig. 3.30** *Ventilation with and without occupant cooling. The size and position of a window will influence the flow of air within a spacer. (a) Ventilation: the window directs breezes upward, removing hot air at the ceiling. Airflow has minimum contact with occupants. (b) Space ventilation and people cooling: the window directs breezes toward the floor and across occupants and provides a direct people-cooling effect from air motion and fresh air for the space.*

**Fig. 3.31** *Cooling with the wind. The air inside this exhibition hall is heated by lights and crowds of people. It rises via the stack effect and is sucked out of the large clerestory opening by the prevailing winds. Note that winds do not blow into this high opening but create a negative pressure outside the opening, encouraging an outward flow of air from the exhibition space. (African Place at Expo '67 Montreal, Canada; John Andrews, architect.)*

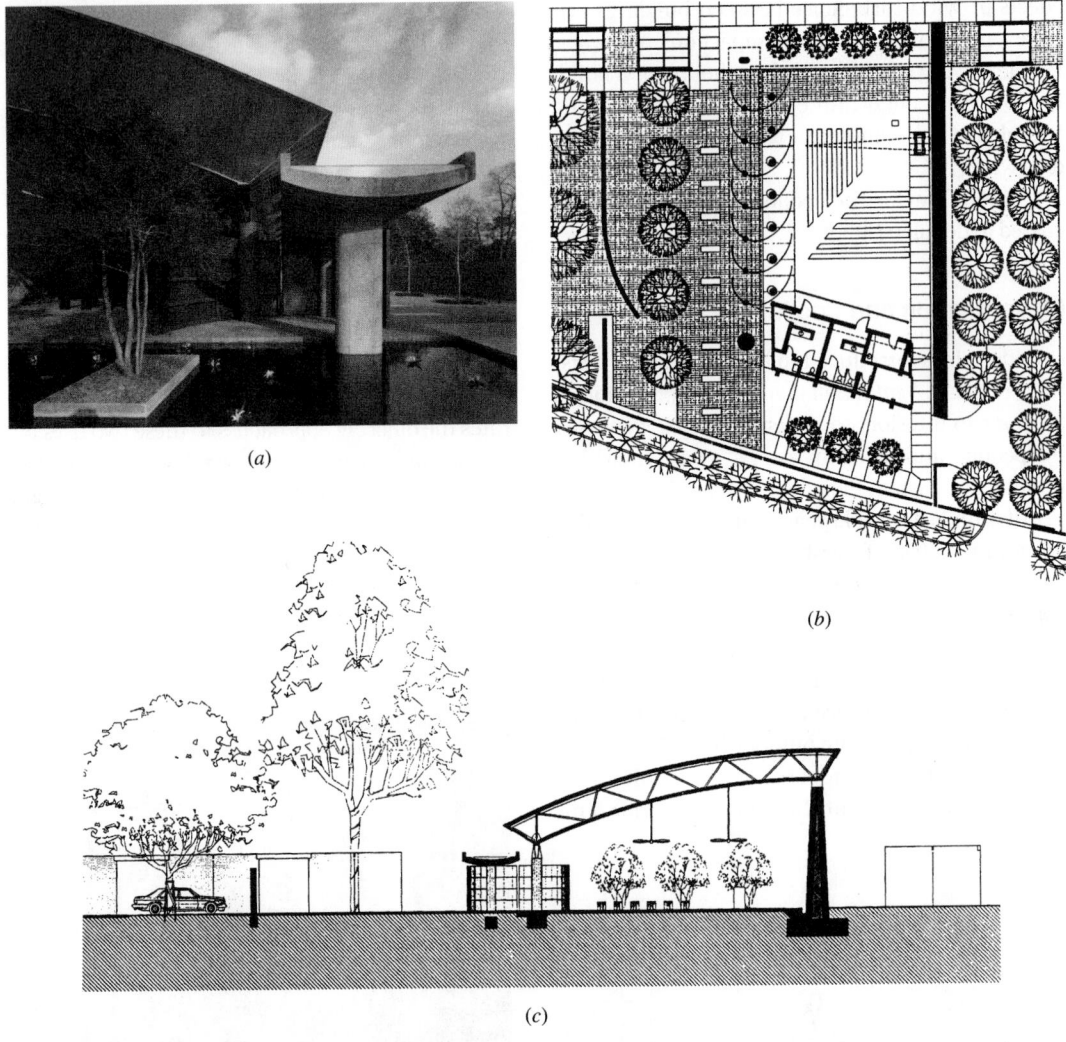

(a)

(b)

(c)

**Fig. 3.32** *The Beth Israel Chapel and Memorial Garden, Houston, Texas. (a) View from the west. The oversized gutter delivers rainwater to a pond that reflects daylight. Trees on islands in the pond provide evening shade. (b) Plan. Curved walls visually separate the open south courtyard from the roofed chapel but allow breezes to pass. A narrow triangular roof opening allows a shaft of direct sunlight to fall along the interior north wall, marking the passing of time. (c) Section, south to north. The curved roof sheds rainwater to an oversized gutter above the curved walls. Suspended ceiling fans can augment air motion. (Photo by Timothy Hursley. Courtesy of Solomon Inc., Architecture and Urban Design, San Francisco.)*

winds (and on rising hot air) to remove air heated by display lighting and crowds of people. The second is a chapel in a cemetery in the hot, humid climate of Houston, Texas. The architects intended this shaded, open-air space as a respite from the sealed, air-conditioned buildings and cars prevalent in Houston. In this building, a funeral is a time and place for reunion with physical phenomena; breezes and shade trees provide cooling, rainwater collected in a pool reflects skylight, and a narrow shaft of direct sun marks its

daily path along the rough stone interior north wall.

A building that relies on prevailing wind for cooling must be sited with attention to wind direction. One building should not be erected to obstruct another building's access to breeze. As seen previously, obstacles upstream from intake openings or downstream near an outlet can substantially reduce the velocity—and thereby the cooling effect—of the wind. Wind can cool people in hot weather (primarily by increasing

**TABLE 3.4  Beaufort Scale (Lower Speeds Only)**

| Beaufort Number | Speed 6 m (19.7 ft) Above Ground | | | Description of Effects Outdoors | |
|---|---|---|---|---|---|
| | m/s | fpm | mph | On Land | Over Water |
| 0 | 0.3 | <88 | <1 | Smoke rises; no perceptible movement | Smooth sea |
| 1 | 0.6–1.7 | 88–264 | 1–3 | Smoke drift shows wind direction; tree leaves barely move | Scale-like ripples |
| 2 | 1.8–3.3 | 352–616 | 4–7 | Wind felt on face; leaves rustle | Small wavelets |
| 3 | 3.4–5.2 | 704–968 | 8–11 | Leaves, twigs in constant motion; hair is disturbed; wind extends light flag | Large wavelets; occasional white foam crests |
| 4 | 5.3–7.4 | 1056–1408 | 12–16 | Small branches move; dust rises; hair disarranged | Small waves become longer |

*Source:* Reprinted from *Passive Cooling* (1981) by permission of the publisher, American Solar Energy Society, Inc.

evaporation from the skin), yet it can become an irritant at higher speeds (Tables 3.4 and 4.4). Manual controls for openings are a necessary part of a natural ventilation design. Finally, the proper size and placement of openings, with an unobstructed airflow path through the building, must be provided.

### (e) Wind, Daylight, and Sun

When daylight and wind-driven ventilation are desired, their design constraints combine to limit building width. Multistory hotels are an example; where such hotels fill entire blocks, they are often arranged around a central atrium, and the surrounding building is only two hotel rooms and one corridor deep. Multistory office buildings used to have a similar form, but increasing urban density and reliance on electric lighting and mechanical cooling have changed their form considerably (Fig. 3.33). There is now a trend back to arranging office workspaces closer to daylight openings and, in Europe, to including operable windows for fresh air as well.

When wind and sun are combined, they tend to be influential in opposite seasons. Fresh air, however, is desirable year round, so in winter a supply of tempered fresh air is necessary. Figure 3.34 captures some issues surrounding a design decision regarding whether to face a clerestory to the north or south for a given set of seasonal prevailing winds.

## 3.8  RAIN AND GROUNDWATER

Most buildings interact with four forms of water: rainwater and groundwater involve lightly controlled exterior interactions, whereas potable water and wastewater involve tightly controlled interior services. A detailed treatment of water within buildings is found in Part V, Water and Waste, and a closer look at rainwater utilization is found in Chapter 20.

Rain, like solar energy, is a generally diffuse, intermittent, and often seasonal resource. It is most often collected on site and used as a source of water where other water resources are scarce or of poor quality. Rain, like sun, has an influence on building design: heavy rains and pitched roofs have long been found in the same locales. Overhangs may extend further beyond walls exposed to storm winds; gutter and downspout details can become a design feature, as shown in the chapel in Fig. 3.32 and in several examples in Chapter 20. A building that reflects the combined influences of daylight, wind, and rain is shown in Fig. 3.35.

Rainwater's impact on site design can be thought of as similar to that of wind. Once on the surface, it will flow downhill in a wide path (on a wide plane), and buildings that obstruct this path must make provisions to divert it. Slight, shallow ditches called swales are frequently used in such diversion. The orientation of a building to the slope can also affect surface water diversion, as shown in Fig. 3.36.

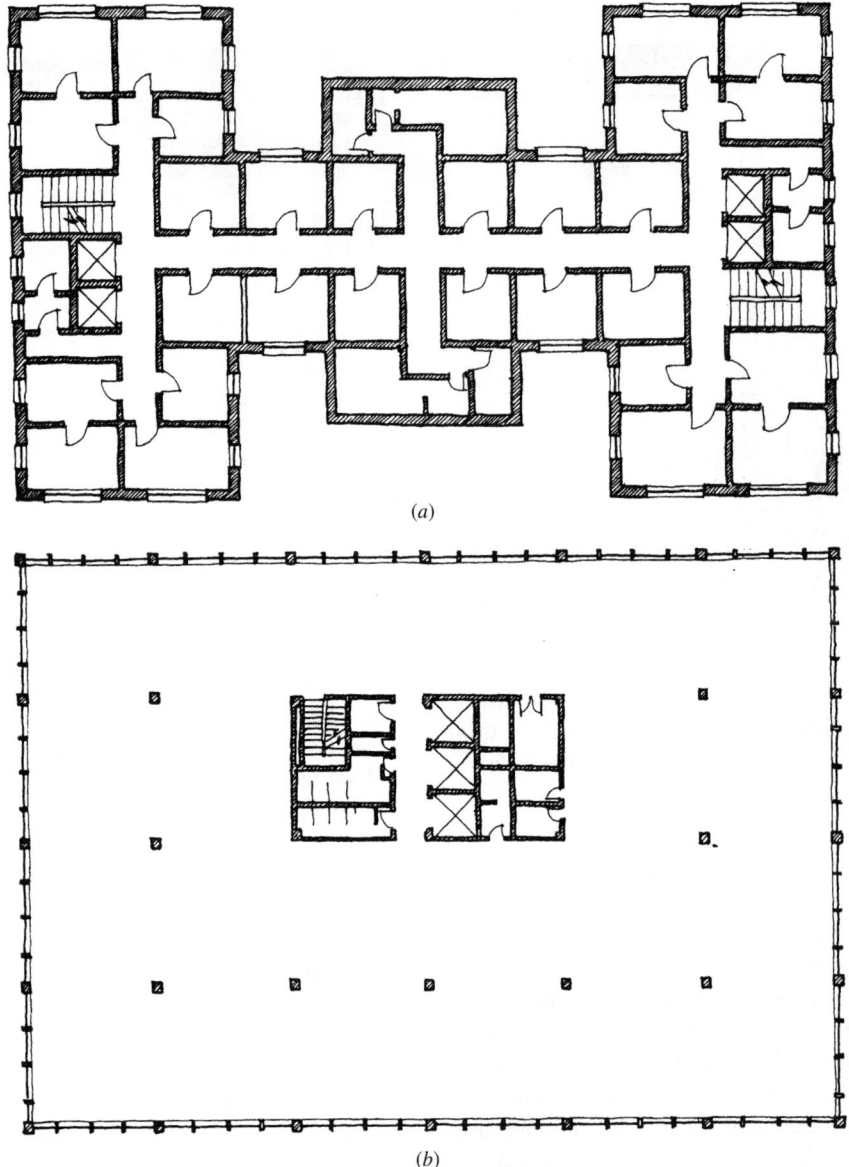

**Fig. 3.33** *In contrast to office building plan (a), which uses daylight and natural ventilation in each office, office building plan (b) receives cooled, filtered air, is less subject to exterior noise, typically provides both constant light and temperature throughout, and provides for more rentable floor space on its site. It also allows less daylight to reach the street level, requires much more electricity (though probably less heating fuel), and thus contributes more heat (and possibly more noise from mechanical equipment) to its surroundings year round.*

Surface water can be used to advantage in thermal, acoustic, and daylighting roles. Hot, dry breezes that pass over water surfaces (and especially through misty sprays above ponds) gain substantial moisture while undergoing a drop in dry-bulb temperature, as discussed under evaporative cooling in Section 4.3. Such conditioned air can provide improved comfort in hot, dry conditions. If a water feature also provides the sound of running water, it can serve to mask noises such as traffic or conversations in adjacent rooms.

Surface water plays a more complex role in daylighting due to the reflection characteristics of water. The surface of water is highly reflective to light striking at low angles of incidence; for example, the reflections of the setting sun from the small

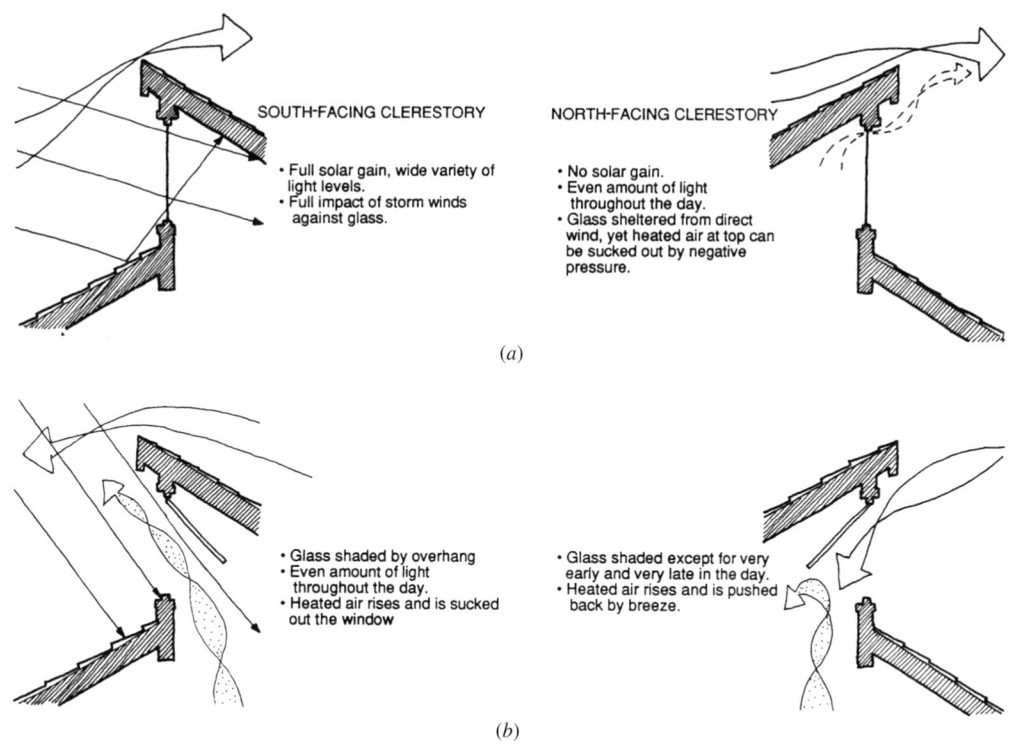

SOUTH-FACING CLERESTORY

- Full solar gain, wide variety of light levels.
- Full impact of storm winds against glass.

NORTH-FACING CLERESTORY

- No solar gain.
- Even amount of light throughout the day.
- Glass sheltered from direct wind, yet heated air at top can be sucked out by negative pressure.

(a)

- Glass shaded by overhang
- Even amount of light throughout the day.
- Heated air rises and is sucked out the window

- Glass shaded except for very early and very late in the day.
- Heated air rises and is pushed back by breeze.

(b)

**Fig. 3.34** *Some relative advantages of north versus south orientation for a clerestory window/shed roof combination. (a) Winter, with low sun and southerly storm winds. (b) Summer, with high sun and northerly breezes. (These wind directions are prevalent in the Pacific Northwest.)*

pond will illuminate the ceiling and east wall of the chapel in Fig. 3.32. On the coast, reflected sunlight from an infinitely huge surface can throw a blinding sheet of light across buildings on the shore. Conversely, sunlight near noon in summer strikes a water body at high angles of incidence—nearly perpendicular at southern U.S. latitudes. Water is highly absorptive and transmissive to radiation at these angles, thereby reflecting relatively little light.

Therefore, water bodies east, south, and west of buildings can provide increased reflected light on sunny winter days and somewhat decreased reflected light in summer relative to alternative grass surfaces. On heavily overcast days, however, when the sky is uniformly gray (and the least amount, but most glare-free quality, of daylight is available), water surfaces will not be particularly helpful.

The reflection of sunlight off water tends to be

**Fig. 3.35** *This roof over a covered outdoor tennis facility at the University of Oregon, Eugene, was designed to ward off rain-bearing south winds in winter. Direct glare from the sun is also unwelcome; instead, north skylight is admitted along with reflected light from roof surfaces. Gutters at the lower edge of each roof plane carry away rainwater; the courts stay dry for all but a few days each year. (Unthank Seder Poticha Architects. Photo by Amanda Clegg.)*

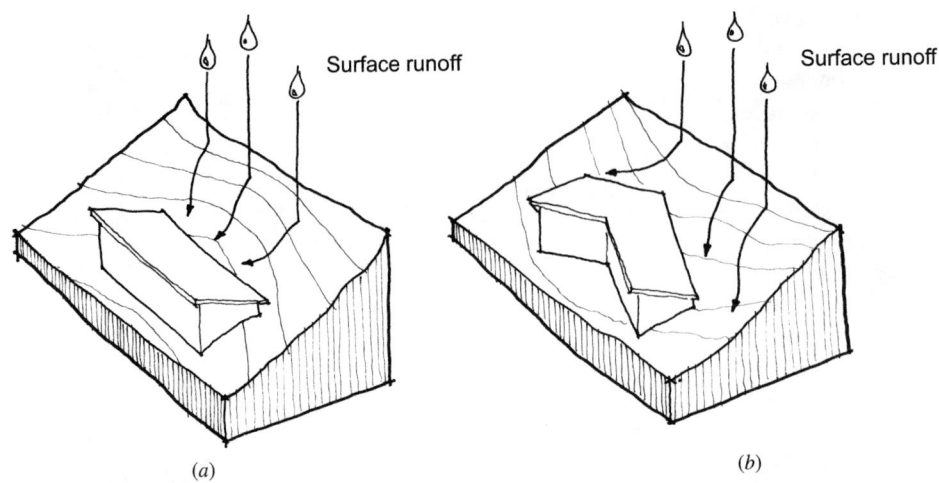

Surface runoff

Surface runoff

*(a)*

*(b)*

**Fig. 3.36** *Rain as surface flow. (a) Where buildings intercept surface water, provisions for diversion are necessary. A building sited as in (b) needs less elaborate provisions, as the form itself is a diverter. (Drawing by Dain Carlson.)*

in sparkling patches of always-changing patterns. This can provide a fascinating design feature or it can be annoying (either as glare or as a distraction from a visual task). Reflected off a matte-finish ceiling, such dancing light might be welcome; directly into eyes or onto a work surface, it easily becomes a problem. Sparkle for one viewer may be glare for another.

Groundwater is generally avoided by designers where possible, as it is a threat to foundations and below-ground spaces. This avoidance carries over into site planning; marshy places are usually unwelcome near buildings and represent an ecologically sensitive area as well. In urban areas, dry soil conditions are intensified as ponds are drained, streams are piped away, and hard surfaces impede water percolation into the soil.

Groundwater has interesting thermal potential as a heat sink, providing a place to discharge building heat in summer and providing heat from groundwater for winter building heating. The quantity of groundwater available to act as a heat sink varies with geographic location and subsurface conditions, and little information is readily available to indicate its potential on a given site. Geothermal systems using groundwater are becoming very common. As more buildings discharge and tap heat from this resource, eventually groundwater temperatures will change, but by how many degrees and for how long remains unpredictable. Groundwater pollution is another risk, through increased handling of groundwater or from refrigerants in the

December 21

June 21

September 21/ March 21

**Fig. 3.37** *Fixed overhang sun control. This south-facing exterior corridor in Oregon is open to the air year round. Low winter sun fills the corridor, and some of its heat is stored by the tile floor. In cold weather, however, the offices connected by the corridor are disadvantaged by unbuffered exposure to cold air. In summer, little of the corridor is exposed to sun, improving comfort. Sun control is identical in the (cool) spring and (warm) fall with such fixed shading.*

mechanical cooling equipment. As a result of such concerns, many localities now strictly regulate the use of groundwater as a heat sink/source.

The ready ability of water to conduct and store heat, which encourages its use as a heat sink, also makes it a thermal nemesis for heat storage tanks and bins in solar-heated buildings. There is little point in storing heat from solar energy in an underground tank for later use if surrounding groundwater is allowed to strip the tank of its heat. Groundwater, however, can help storage tanks for cooling systems. Such tanks are often used to provide cooling water to buildings during peak-load hours on hot days and can greatly lessen the demand for electricity to run conventional air conditioners.

*Snow* has special site design implications: it delays runoff, provides a blanket of thermal insulation, absorbs sound, and reflects more solar radiation than almost any other naturally occurring surface. Wind patterns can deposit snow much more thickly in some places, for better or worse. Snow hampers the movement of external control devices such as awnings or thermal shutters, can collect to excess on external light shelves, and can create disabling glare if it reflects low winter sun into windows at eye level.

## 3.9 PLANTS

Plants play several roles on a building site: they affect the absorptivity and emissivity of the Earth's surface; they are part of both the food and water cycles; by day they turn carbon dioxide into oxygen; they can provide organic matter suitable for building materials; and they help people mark time both by growth and by change with the seasons. Our associations with plants are mostly pleasant ones, and they contribute to the enjoyment of the places where they grow.

Plants are also of immediate practical value to building design because they enhance privacy, slow the winter wind, reduce glare from strong daylight, and/or prevent summer sun from entering and overheating buildings. In the last role, plants are particularly noteworthy, because they enhance a feeling of coolness when breezes rustle or sway their leaves. Most importantly, they respond more to cycles of outdoor temperature than to the cycles of sun position. Unlike fixed-position sunscreens on

*(a)*

*(b)*

**Fig. 3.38** *A deciduous tree as a naturally "smart" shading device. In hot August, the tree is in full, dense leaf, providing shade. In cold January, bare branches admit winter sun. These photographs were taken on the equinox at 4:45 P.M. solar time: (a) On September 21, the local average temperature is 75°F (24°C) and the tree functions as a shading device. (b) On March 21, the local average temperature is 53°F (12°C) and the tree has fortuitously not yet developed its leaves. (From Reynolds, 1976.)*

DESIGN CONTEXT

(a)　　　　　　　　(b)

(c)　　　　　　　　(d)

**Fig. 3.39** *Deciduous vines, temperature, and sun position. The sun's path through the sky is identical in late May (a) and late July (b). Paired—but lower—sun paths occur in late November (c) and late January (d). This deciduous vine responds more to the temperature of its Oregon climate than to the sun's position, which makes it particularly useful as a sun control device. (For termite control and wall longevity, it is best to keep vines on a trellis rather than on the wall surface.) (From Reynolds, 1976.)*

**TABLE 3.5 Deciduous Trees for Summer Shading and Winter Solar Collection**

| Botanical Name | Common Name | Transmissivity[a] Range % Summer | Winter | Foliation[b] | Defoliation[c] | Mature Height m | ft |
|---|---|---|---|---|---|---|---|
| Acer platanoides | Norway maple | 5–14 | 60–75 | E | M | 15–25 | 50–82 |
| Acer rubrum | Red maple | 8–22 | 63–82 | M | E | 20–35 | 65–115 |
| Acer saccharinum | Silver maple | 10–28 | 60–87 | M | M | 20–35 | 65–115 |
| Acer saccharum | Sugar maple | 16–27 | 60–80 | M | E | 20–35 | 65–115 |
| Aesculus hippocastanum | Horse chestnut | 8–27 | 73 | M | L | 22–30 | 72–98 |
| Amelanchier canadensis | Serviceberry | 20–25 | 57 | L | M | | |
| Betula pendula | European birch | 14–24 | 48–88 | M | M–L | 15–30 | 50–98 |
| Carcis canadensis[d] | Red bud | 62 | 74 | L | M | 12 | 40 |
| Carya ovata | Shagbark hickory | 15–28 | 66 | | | 24–30 | 78–98 |
| Catalpa speciosa | Western catalpa | 24–30 | 52–83 | L | | 18–30 | 60–98 |
| Cornus florida[d] | Dogwood | 43 | 53 | L | E | 11 | 36 |
| Fagus sylvatica | European birch | 7–15 | 83 | L | L | 18–30 | 60–98 |
| Fraxinus pennsylvanica | Green ash | 10–29 | 70–71 | M–L | M | 18–25 | 60–82 |
| Gleditsia tricanthos inermis | Honey locust | 25–50 | 50–85 | M | E | 20–30 | 65–98 |
| Juglans nigra | Black walnut | 9 | 55–72 | L | E–M | 23–45 | 75–148 |
| Liquidamber styraciflua[d] | Sweet gum | 33 | 47 | M | L | 24 | 78 |
| Liriodendron tulipifera | Tulip tree | 10 | 69–78 | M–L | M | 27–45 | 88–148 |
| Platanus acerifolia | London plane tree | 11–17 | 46–64 | M–L | | 30–35 | 98–115 |
| Populus detoides | Cottonwood | 10–20 | 68 | E | M | 23–30 | 75–98 |
| Populus tremuloides | Trembling aspen | 20–33 | | E | | 12–15 | 40–50 |
| Quercus alba | White oak | 13–38 | | | | | |
| Quercus palustris[d] | Pin oak | 45 | 47 | L | L | 23 | 75 |
| Quercus rubra | Red oak | 12–23 | 70–81 | | M | 23–30 | 75–98 |
| Robinia pseudoacacia[d] | Black locust | 38 | 40 | L | E | 21 | 69 |
| Tilia cordata | Littleleaf linden | 7–22 | 46–70 | L | E | 18–21 | 60–69 |
| Ulmus americana | American elm | 13 | 63–89 | M | M | 18–24 | 60–78 |

*Source:* Brown and Gillespie (1995), except as noted.

[a]Transmissivity to solar radiation; varies with instruments used by various researchers.

[b]Foliation: E = early (before April 30); M = middle (May 1–15); L = late (after May 15).

[c]Defoliation: E = early (before November 1); M = middle (November 1–30); L = late (after November 30).

[d]From Montgomery (1987). Transmissivity % = (100% − blockage %).

buildings, plants thus provide their deepest shade in the hottest weather.

To illustrate this mechanical-horticultural contrast, a fixed sunscreen (an overhang for south-facing windows) is shown in Fig. 3.37. Such sunscreens block the sun for some defined portion of the year, centered on June 21; that is, maximum shade is provided at the summer solstice. A typical approach to such sunscreens in the U.S. temperate zone is to shade at least half of a south window in a residence (or all of a window in an office with internal heat loads) from March 21 to September 21 (vernal equinox to autumnal equinox). Yet, March is (on the average) a colder month than September; March 21 is the last day of winter, whereas September 21 is the last day of summer. Indeed, average monthly temperatures for June and September can be quite similar. Full solar radiation is more welcome in early spring than in early fall, yet sun

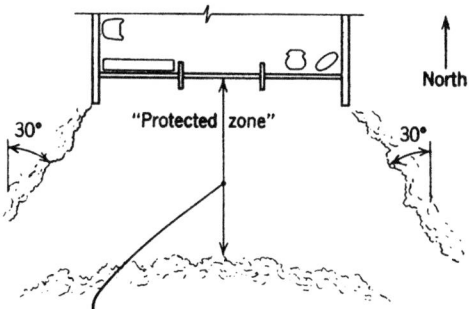

This distance is equal to at least twice the ultimate height of the trees or shrubs beyond.

*Fig. 3.40* Protecting access to winter sun, given a lawn or terrace of limited size to the south of solar collecting surfaces. Coniferous and even deciduous plants within the "protected zone" should be avoided unless they are very low growing or are a reliably early defoliating species (see Table 3.5). Summer sun protection for such south-facing windows is best provided by flexible architectural controls such as awnings or hanging screens.

position is identical at these times. In contrast to fixed sunscreens, deciduous plants do most of their shading from the middle of June to early October, giving windows access to solar radiation throughout much of the spring (Figs. 3.38 and 3.39).

Deciduous trees have a potential disadvantage for solar heating in that certain species (such as sweet gum and some oaks) hold on to their leaves well into the heating season, a tendency increased by fertilizing or irrigating near a tree. Other trees have a dense branch structure, blocking a surprisingly high percentage of solar radiation even when bare (Table 3.5). Agricultural extension services in most areas can provide information on the hardiness and average dates of defoliation for various tree species. For solar collection, avoiding trees or large shrubs within the area shown in Fig. 3.40 is recommended. In this *protected zone,* a deciduous vine on a trellis above a south-facing window is a better planting strategy.

Many large buildings have a relatively short heating season because they have constant internal heat gains. For such internal-load (heat-dominated) buildings, a tree or vine that foliates early in spring, defoliates late in fall, and has low transmissivity in summer is advantageous. The choice of trees, shrubs, and vines to improve a site and/or building microclimate can best be made after a month-by-month analysis of building heating and cooling needs.

## References

ASES. 1981. *Passive Cooling.* American Solar Energy Society, Boulder, CO.

Beranek, W.J. 1980. "General Rules for the Determination of Wind Environment," in *Wind Engineering,* J.E. Cermak (ed.). Pergamon Press, Elmsford, NY.

Bragdon, C. 1971. *Noise Pollution: The Unquiet Crisis.* University of Pennsylvania Press, Philadelphia.

Brown, R.D. and T.J. Gillespie. 1995. *Microclimatic Landscape Design: Creating Thermal Comfort and Energy Efficiency.* John Wiley & Sons, New York.

Cohen, S., D.C. Glass, and J.E. Singer. 1973. "Apartment Noise, Auditory Discrimination, and Reading Ability in Children," in *Journal of Experimental Social Psychology,* Vol. 9.

DeKay, M. 1992. "A Comparative Review of Daylight Planning Tools and a Rule-of-Thumb for Street Width to Building Height Ratio," in *Proceedings of the 17th National Passive Solar Conference,* Cocoa Beach, FL. American Solar Energy Society, Boulder, CO.

Doelle, L. 1972. *Environmental Acoustics.* McGraw-Hill, New York.

Gandemer, J. 1977. "Wind Environment Around Buildings: Aerodynamic Concepts," in *Wind Effects on Buildings and Structures,* K.J. Eaton (ed.). Cambridge University Press, New York.

Graedel, T.E. and P.J. Crutzen. 1989. "The Changing Atmosphere," in *Scientific American,* September.

Harden, G. 1968. "The Tragedy of the Commons," in *Science,* Vol. 162, Issue 3859, 13 December.

Knowles, R. 1981. *Sun, Rhythm, Form.* MIT Press, Cambridge, MA.

Landsberg, H.E. 1970. "Climates and Urban Planning," in *Urban Climates.* World Meteorological Organization Technical Note 108, Geneva.

Lowry, W.P. 1988. *Atmospheric Ecology for Designers and Planners.* Van Nostrand Reinhold, New York.

Lowry, W.P. and P.P. Lowry. 1995. *Fundamentals of Biometeorology, Volume 2: The Biological Environment.* Peavine Publications, McMinnville, OR.

Marsh, W.M. 1991. *Landscape Planning, Environmental Applications,* 2nd ed. John Wiley & Sons, New York.

Mazria, E. and D. Winitsky. 1976. *Solar Guide and Calculator.* Center for Environmental Research, University of Oregon, Eugene.

McPherson, E.G. (ed.). 1984. *Energy-Conserving Site Design.* American Society of Landscape Architects, Washington, DC.

Montgomery, D. 1987. "Landscaping as a Passive Solar Strategy," in *Passive Solar Journal,* Vol. 4, No. 1. American Solar Energy Society, Boulder, CO.

Olgyay, V. 1963. *Design with Climate: Bioclimatic Approach to Architectural Regionalism.* Princeton University Press, Princeton, NJ.

Reynolds, J.S. 1976. *Solar Energy for Pacific Northwest Buildings.* Center for Environmental Research, University of Oregon, Eugene.

Watson, D. and K. Labs. 1983. *Climatic Design.* McGraw-Hill, New York.

CHAPTER 4

# Comfort and Design Strategies

ONE OF THE EARLIEST REASONS FOR BUILDING was to create shelter from the climate and to enhance thermal comfort. This chapter is about the interrelationship among bodies, buildings, and climate. It begins by discussing bodily heat flow, then thermal comfort, and then design strategies that are appropriate to various climates.

## 4.1 THE BODY

Because we are alive, we are always generating body heat. Because the body's core must stay within a narrow temperature range, we nearly always need to lose this internally generated heat to our environment. The rate at which we produce heat changes frequently, as does the environment's ability to accept heat. To regulate our body's heat loss, we have available three common layers between our body cores and the environment: a first skin, which is our own; a second skin, our clothing; and a third skin, a building.

### (a) Metabolism

The rate at which we generate heat (our metabolic rate) depends mostly upon our level of muscular activity, partly upon what we eat and drink (and when), and partly on where we are in our normal daily cycle. Our heat production is measured in metabolic (MET) units (Table 4.1). One MET is defined as 50 kcal/h m² (equal to 18.4 Btu/h ft² or

58.2 W/m²). One MET is the energy produced per unit of surface area by a seated person at rest. Under these conditions, the total heat produced by a normal adult is about 360 Btu/h (106 W). The more active we are, the more heat we produce, and our own (first) skin is the most important regulator of heat flow.

Of the many interactions between our skin and the rest of our body, consider these three: touch, blood, and water. The sensation of touch includes pressure and pain as well as heat and cold. The sensations of heat and cold are produced by contact with surfaces or moving air as well as by radiation. They are frequently our signals for shifts in body heat regulation, which is controlled by the thermostat in our brains, called the *hypothalamus*.

In response to signals from our skin surface and to changes in our core temperature, the hypothalamus calls for changes in our blood distribution system. If we are too cold, we need to decrease our rate of heat loss, so the flow of blood from our core toward the surface of our skin decreases. Blood carries heat (in heating, ventilating, and air-conditioning [HVAC] terms, we are an "all-water" system): the less exposure to cool air at the skin surface, the less heat is lost. This decreased blood flow toward the surface is called *vasoconstriction,* and is triggered in part by cold signals from our skin. In this condition, less water is forced to the skin surface by our sweat glands, which reduces evaporation and therefore heat loss.

**83**

Note the implications of this zoning within the body. We strive to maintain, at all costs, a nearly constant core temperature for our vital organs. This most protected zone takes thermal precedence over the less vital zone of our extremities, such as the arms and legs; next in priority are our fingers and toes. The farther from our central body mass (fingers and toes) and the greater the surface area (ears), the faster the temperature will drop in cold conditions. The most variable thermal zone of all is our skin surface.

When cold conditions worsen, we get "goose bumps," symptoms of our skin's unsuccessful attempt to create insulation by fluffing up our body hair. Because we cannot add insulation this way, we soon increase our metabolic rate, or burn more fuel, by shivering, muscular tension, or increased muscular activity. At the point where shivering incapacitates us, we may reach 6 MET. Before this point, we seek help from our second and then our third skins of clothing and buildings.

The opposite occurs when we are too hot: first, blood flow toward the skin surface increases (*vasodilation*), triggered primarily by warm signals from our core. The sweat glands greatly increase their secretion of water and salt to the skin surfaces. This increases heat loss by evaporation (although salt accumulations impede evaporation by lowering the vapor pressure of water).

### (b) Heat Flow

Once blood and water transport our surplus heat to the skin surface, we have four ways to pass it to the environment: *convection* (air molecules contact our body, absorbing heat), *conduction* (we touch cooler surfaces, and heat is transferred), *radiation* (when our skin surface is hotter than other surfaces "seen" but not touched, heat is radiated to these cooler surfaces, and vice versa when other surfaces are hotter than our skin surface), and *evaporation* (a liquid can evaporate only by removing large quantities of heat from the surface it is leaving). The amount of heat we lose by each of these four methods depends upon the interaction between our metabolism, our clothing, and our environment. Figure 4.1 illustrates the typical situation of a person at rest as environmental conditions change.

**TABLE 4.1 Metabolic Rates for Typical Tasks**

| Activity | Metabolic Rate[a] | | |
|---|---|---|---|
| | MET Units[b] | Btu/h ft² | W/m² |
| Resting | | | |
|   Sleeping | 0.7 | 13 | 40 |
|   Reclining | 0.8 | 15 | 45 |
|   Seated, quiet | 1.0 | 18 | 60 |
|   Standing, relaxed | 1.2 | 22 | 70 |
| Walking (on the level) | | | |
|   2 mph (0.9 m/s) | 2.0 | 37 | 115 |
|   3 mph (1.2 m/s) | 2.6 | 48 | 150 |
|   4 mph (1.8 m/s) | 3.8 | 70 | 220 |
| Office activities | | | |
|   Reading, seated | 1.0 | 18 | 60 |
|   Writing | 1.0 | 18 | 60 |
|   Typing | 1.1 | 20 | 65 |
|   Filing, seated | 1.2 | 22 | 70 |
|   Filing, standing | 1.4 | 26 | 80 |
|   Walking about | 1.7 | 31 | 100 |
|   Lifting, packing | 2.1 | 39 | 120 |
| Driving/flying | | | |
|   Car | 1.0–2.0 | 18–37 | 60–115 |
|   Aircraft, routine | 1.2 | 22 | 70 |
|   Aircraft, instrument landing | 1.8 | 33 | 105 |
|   Aircraft, combat | 2.4 | 44 | 140 |
|   Heavy vehicle | 3.2 | 59 | 185 |
| Miscellaneous occupational activities | | | |
|   Cooking | 1.6–2.0 | 29–37 | 95–115 |
|   House cleaning | 2.0–3.4 | 37–63 | 115–200 |
|   Seated, heavy limb movement | 2.2 | 41 | 130 |
|   Handling 110-lb (50-kg) bags | 4.0 | 74 | 235 |
|   Pick and shovel work | 4.0–4.8 | 74–88 | 235–280 |
| Machine work | | | |
|   Sawing (table saw) | 1.8 | 33 | 105 |
|   Light (electrical industry) | 2.0–2.4 | 37–44 | 115–140 |
|   Heavy | 4.0 | 74 | 235 |
| Miscellaneous leisure activities | | | |
|   Dancing, social | 2.4–4.4 | 44–81 | 140–255 |
|   Calisthenics/exercise | 3.0–4.0 | 55–74 | 175–235 |
|   Tennis, singles | 3.6–4.0 | 66–74 | 210–270 |
|   Basketball | 5.0–7.6 | 90–140 | 290–440 |
|   Wrestling, competitive | 7.0–8.7 | 130–160 | 410–505 |

*Source:* Reprinted with permission of the American Society of Heating, Refrigerating and Air-Conditioning Engineers, Inc. From 1997 *ASHRAE Handbook—Fundamentals.*

[a]For average adult with a body surface area of 19.6 ft² (1.8 m²). For whole-body average heat production, see also Table F.8.

[b]One MET = 18.4 Btu/h ft² = 58.2 W/m².

As air and surface temperatures approach our own body temperature, we lose the options of convection, conduction, and radiation. Evaporation becomes essential, so access to dry, moving air is greatly appreciated. As air and surface temperatures fall, evaporation decreases while convection,

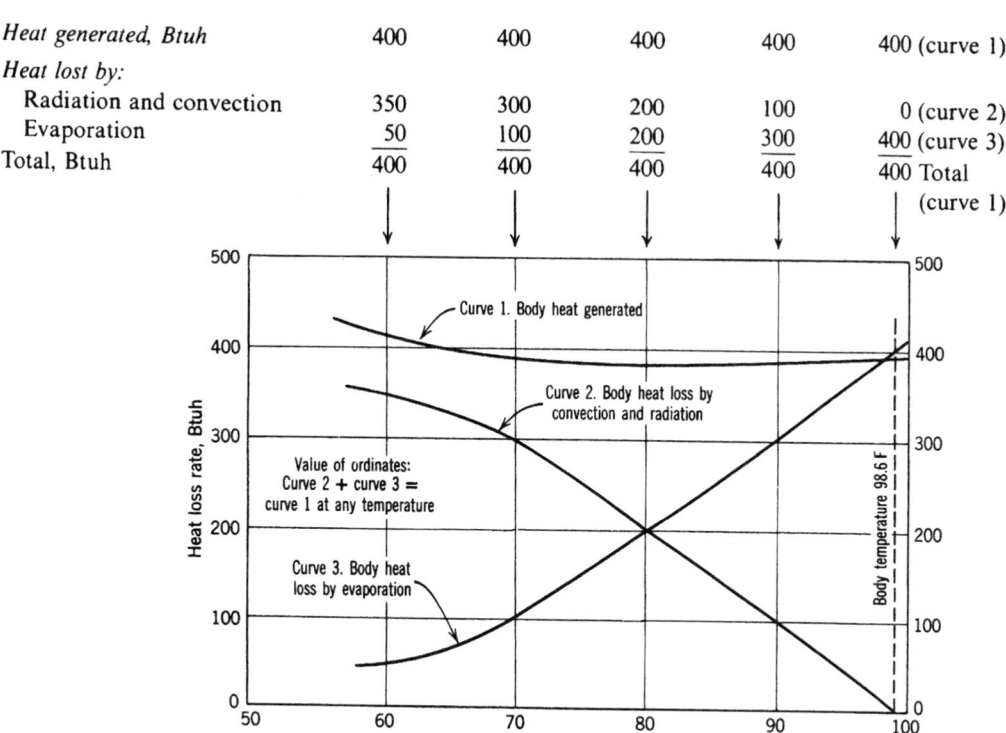

| Heat generated, Btuh | 400 | 400 | 400 | 400 | 400 (curve 1) |
|---|---|---|---|---|---|
| Heat lost by: | | | | | |
| Radiation and convection | 350 | 300 | 200 | 100 | 0 (curve 2) |
| Evaporation | 50 | 100 | 200 | 300 | 400 (curve 3) |
| Total, Btuh | 400 | 400 | 400 | 400 | 400 Total |
| | | | | | (curve 1) |

**Fig. 4.1** *Heat generated and lost (approximate) by a person at rest (RH fixed at 45%).*

conduction, and particularly radiation increase. Under the normally comfortable temperature in Fig. 4.1 of about 70°F (18°C), the proportions of body heat loss per hour are as follows:

| Radiation, convection, and conduction | 72% |
|---|---|
| Evaporation | |
| From skin surface: | 15% |
| From lungs (exhaled air): | 7% |
| Warming of air inhaled to lungs: | 3% |
| Heat expelled in feces and urine: | 3% |

### (c) Clothing

Usually clothing acts as an insulating layer and is particularly effective at retarding radiation, convection, and conduction. As air and surface temperatures in our environment fall well below our body's temperature, we adjust this second skin to provide increased insulation. However, in a hot, humid environment, our first skin needs exposure to moving air to encourage heat loss, yet needs protection from the sun's radiant heat. We need a simple sec-

ond skin that acts as a sunshade. In a hot, arid environment, our second skin may keep us from losing too much valuable water while also performing the vital role of shading.

The insulating value of clothing is measured in CLO units, 1 CLO being equivalent to the typical American man's business suit in 1941, when the concept of CLO was developed (1 CLO is equal to 0.88 ft² h F/Btu [0.155 m² K/W]). The total CLO of what you are now wearing can be estimated by assuming 0.15 CLO/lb (0.35 CLO/kg) of clothing weight. The total CLO of your attire may also be estimated by simply adding the CLO of each item from Table 4.2; this total will be a bit higher than the actual CLO value of the ensemble. Sometimes the position of clothing is as important as its CLO value; consider the role of socks as they separate our feet from contact with a cold floor.

Our second skin is just as likely as our third (building) skin to be dominated by considerations of style more than of thermal regulation; we cannot always count on our clothing—or buildings—to increase our thermal comfort.

**TABLE 4.2 Typical Insulation Values for Clothing Ensembles**

| Ensemble Description[a] | CLO[b] |
|---|---|
| Walking shorts, short-sleeve shirt | 0.36 |
| Trousers, short-sleeve shirt | 0.57 |
| Trousers, long-sleeve shirt | 0.61 |
| Same as above, plus suit jacket | 0.96 |
| Same as above, plus vest and T-shirt | 1.14 |
| Trousers, long-sleeve shirt, long-sleeve sweater, T-shirt | 1.01 |
| Same as above, plus suit jacket and long underwear bottoms | 1.30 |
| Sweat pants, sweat shirt | 0.74 |
| Long-sleeve pajama top, long pajama trousers, short ¾-sleeve robe, slippers (no socks) | 0.96 |
| Knee-length skirt, short-sleeve shirt, pantyhose, sandals | 0.54 |
| Knee-length skirt, long-sleeve shirt, full slip, pantyhose | 0.67 |
| Knee-length skirt, long-sleeve shirt, half slip, pantyhose, long-sleeve sweater | 1.10 |
| Same as above; replace sweater with suit jacket | 1.04 |
| Ankle-length skirt, long-sleeve shirt, suit jacket, pantyhose | 1.10 |
| Long-sleeve coveralls, T-shirt | 0.72 |
| Overalls, long-sleeve shirt, T-shirt | 0.89 |
| Insulated coveralls, long-sleeve thermal underwear, long underwear bottoms | 1.37 |

*Source:* Based on ASHRAE (1997).

[a]All ensembles include shoes and briefs or panties. All ensembles except those with pantyhose include socks unless otherwise noted.

[b]One CLO = 0.88 ft² h °F/Btu = 0.155 m²K/W.

## 4.2 THERMAL COMFORT

A positive definition of comfort is "a feeling of well-being." The more common experience of comfort is simply a lack of discomfort—thermally, of being unconscious of how you are losing heat to your environment. ASHRAE (2004) defines thermal comfort as "that condition of mind which expresses satisfaction with the thermal environment. . . ." There are three categories of factors that affect comfort: personal, measurable environmental, and psychological. Most *personal* factors are under your control: your metabolism and your clothing, as well as various adaptations such as migration to a more comfortable place, or drinking or eating warm or cold foods. *Measurable environmental* factors are the familiar tools of the designer: air temperature, surface temperature, air motion, and humidity. *Psychological* factors are also familiar designers' tools, but they are difficult to quantify for comfort: color, texture, sound, light, movement, and aroma. These factors are often overlooked as we strive to meet the numerical (therefore calculable) physical criteria

for thermal comfort. However, our primary design intent is to make people comfortable, and all aspects of buildings are our means to that end.

Consider the courtyard in a hot, dry climate (Fig. 4.2). Its fountain suggests coolness in the color and texture of its water; running water provides splashing sounds and sparkles of light and may generate some air motion. Vines provide shade and their leaves sway in the slightest breeze, evidence of at least some air motion. Blossoms of flowering plants yield a cool fragrance that blends with the aroma of moistened surfaces in hot, dry surroundings. The measured coolness of such a courtyard may be but a slight improvement over the environment beyond, but it seems cool.

Then consider a fireplace in a cold climate (Fig. 4.3). The fire's color is intensely warm, and it dances and casts a flickering light; it crackles, and it yields a smoky aroma (not appreciated by all). Few textures seem hotter than that of glowing coals. Yet a fireplace might add but slightly to the overall warmth of a room, because it draws in cold air to replace the air that passes up the chimney. We seek a place near the fire for the real radiant warmth, but also for the psychological comfort the fire offers.

In our society, designers are encouraged to take numerical data quite literally. The measurable environmental factors have been tested extensively in laboratories—but exclude other factors. A holistic view of designing for comfort considers numbers as a nonabsolute guide; common sense and a designer's own thermal experience play important roles as well. Lisa Heschong's *Thermal Delight in*

**Fig. 4.2** *Indicators of coolness in a courtyard include running water and shade from vines that move with the breeze. The senses of sight, sound, touch, smell, and taste all may be involved in a perception of coolness.*

**Fig. 4.3** *The open fire seems the very spirit of warmth, despite the heat lost by large quantities of exhaust air up the chimney. The senses of sight, sound, touch, and smell all may be involved in a perception of warmth, wrapped in this red brick environment. (Courtesy of the architect and photographer, Edward Allen.)*

*Architecture* (1979) is an excellent expression of these points of view.

Ultimately, our buildings will be expected to demonstrate success with regard to the measurable environmental factors of comfort, so it is necessary to understand how air and surface temperatures, air motion, and humidity are related to heat transfer.

| Heat Transferred By: | Primarily Dependent Upon: |
|---|---|
| Conduction | Surface temperature |
| Convection | Air temperature, air motion, humidity |
| Radiation | Surface temperature, orientation to the body |
| Evaporation | Humidity, air motion, air temperature |

From this comparison, it is evident that humidity is relatively unimportant in cold conditions,

where heat loss by convection, radiation, and conduction is dominant. However, humidity is of primary importance in hot conditions, dominated by evaporative heat loss. This is further evident in comfort studies, which show that skin temperature is an important factor in cold conditions, whereas skin wettedness (percentage covered by water) is most important in hot conditions.

## (a) Comfort Standards

The ASHRAE *comfort zone* (Fig. 4.5) represents combinations of air temperature and relative humidity that most often produce comfort for a seated North American adult in shirtsleeves (total 0.6 CLO) in the shade and without noticeable air motion. The concept of a comfort zone is diagrammed in Fig. 4.4, where surface temperatures are assumed to be not markedly different from air temperatures. Moving to the left of this zone, at lower air temperatures,

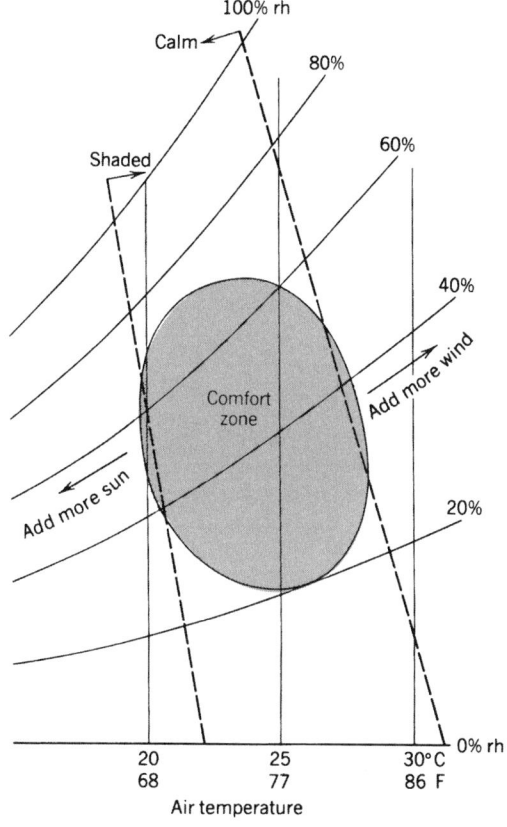

**Fig. 4.4** *Comfort zone defined by relative humidity and air temperature.*

DESIGN CONTEXT

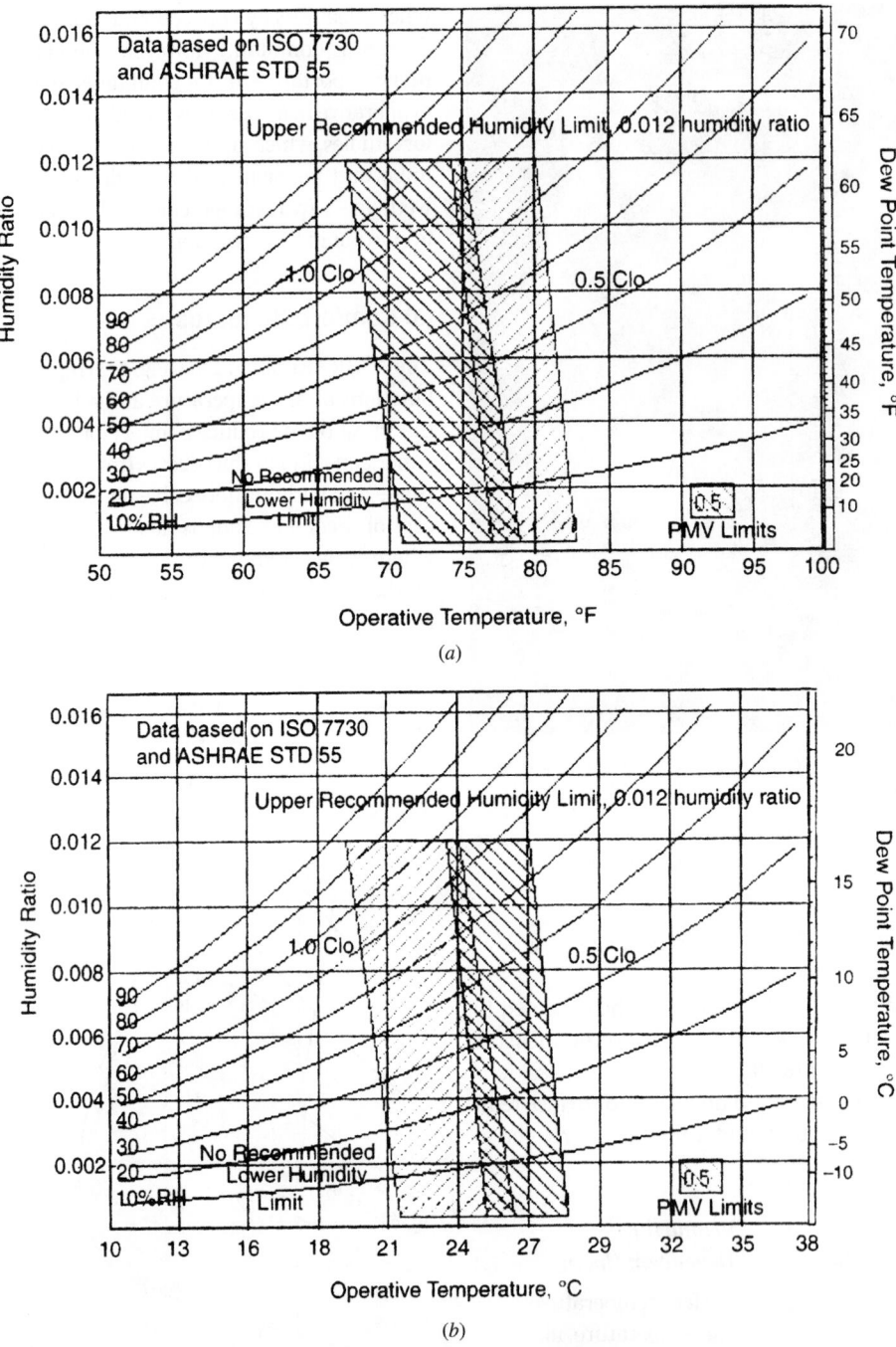

**Fig. 4.5** *(a, b) ANSI/ASHRAE Standard 55-2004 comfort zone (Reprinted with permission of the American Society of Heating, Refrigerating and Air-Conditioning Engineers, Inc., from ANSI/ASHRAE Standard 55-2004, Thermal Environmental Conditions for Human Occupancy.) (c) Relationships among preferred operative temperature, clothing, and activity. The shaded bands represent the degree to which departures from the optimum are tolerable; that is, nearly all people will find the departures only "slightly cool" or "slightly warm." (P.O. Fanger, Laboratory of Heating and Air Conditioning, Technical University of Denmark, Copenhagen.)*

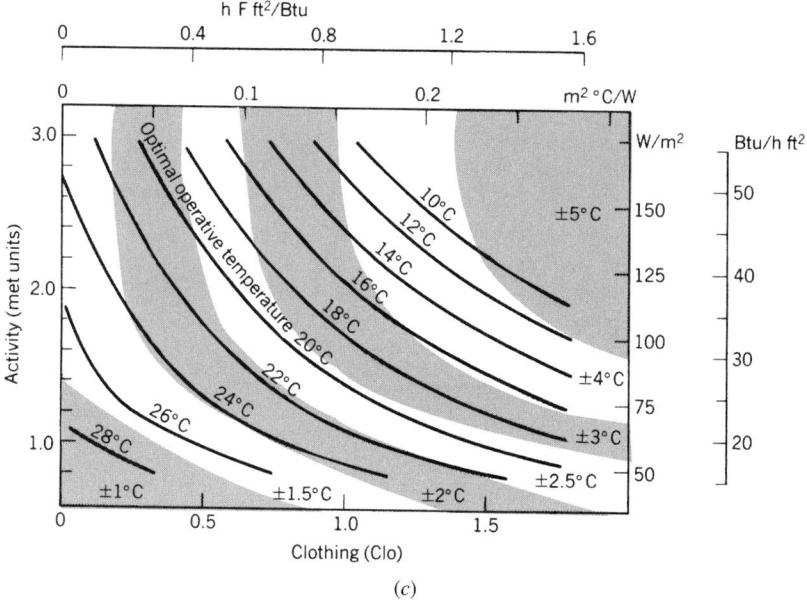

*(c)*

**Fig. 4.5** *(Continued)*

comfort is still attainable if added radiant heat is provided—such as by increasing surface temperatures or providing exposure to the sun. Higher activity and more clothing would also help achieve comfort. Similarly, by moving to the right of the comfort zone (at higher air temperatures), comfort is still attainable by increasing air motion (such as exposure to wind), lowering the activity level, and removing clothing. In both directions limits are soon reached, but the important point is that the basic comfort zone can be expanded by changing environmental variables and/or human behavior.

The human thermal response, perceived as thermal comfort or discomfort, is shaped by four environmental parameters: dry-bulb air temperature, relative humidity, radiant temperature, and air speed. These variables can be measured directly or derived from other measurements (using a psychrometric chart). Two nonenvironmental parameters, clothing and metabolism, are key personal variables.

Figure 4.5 shows the thermal comfort zone as defined by the American Society of Heating, Refrigerating and Air-Conditioning Engineers (ASHRAE), where acceptable ranges of operative temperatures ($T_{op}$) and relative humidity are given for sedentary (1.0–1.3 MET) persons wearing typical clothing (between 0.5 and 1.0 CLO of thermal insulation).

Figure 4.5 specifies the comfort zone parameters where there should be 80% occupant acceptability for the range of operative temperature and relative humidity, and where air speeds are not greater than 40 ft/min (0.20 m/s). Two zones are shown—one for 0.5 CLO of clothing insulation (assumed worn when the outdoor environment is warm) and one for 1.0 CLO of insulation (for when the outdoor environment is cool). The thermal comfort variables used here are defined as follows:

*Dry-Bulb (DB) Temperature.* DB temperature is the ambient air temperature as measured by a standard thermometer, thermocouple, or resistance temperature device. DB temperature can be used in combination with globe temperature and air velocity to calculate mean radiant temperature. Temperature affects comfort in a number of ways and, in combination with the other parameters described in this section, is a key factor in our energy balance, thermal sensation, comfort, discomfort, and perception of air quality.

*Operative Temperature.* This is the average of the dry-bulb temperature and the mean radiant temperature (MRT).

*Wet-Bulb (WB) Temperature.* WB temperature is measured by a thermometer with a wetted bulb rotated rapidly in the air to cause evaporation of its moisture—as with the sling psychrometer shown in Fig. 4.6. In dry air the moisture readily evaporates and draws heat out of the thermometer to produce a lower temperature reading, called the *wet-bulb depression* (the difference between DB and WB temperatures). A large depression is indicative of low relative humidity (RH). Slow evaporation, as when the air is already moisture-laden, results in a small wet-bulb depression and indicates high RH. Note that at 100% RH, DB and WB temperatures are equal.

*Relative humidity* may be measured directly or derived from DB and WB temperatures and is the ratio of the actual density of water vapor in air to the maximum density of water vapor that such air could contain, at the same temperature, if it were 100% saturated.

*Mean Radiant Temperature (MRT).* The radiant temperatures of surrounding surfaces influence human comfort. With respect to the human body at a particular location, *mean radiant temperature* is defined as the uniform temperature of an imaginary surrounding enclosure in which radiant transfer from the human body would equal the radiant heat transfer in the actual nonuniform enclosure (ASHRAE 2001). MRT is a calculated variable and cannot be directly measured. One calculation approach involves using a globe thermometer, which can be easily constructed using a hollow sphere (e.g., a ping pong ball) with a thermocouple or thermometer bulb at its center. Equation 4.1, for determining MR temperature from globe temperature, is as follows (ASHRAE, 2001):

I-P units: $\bar{t}_r =$

$$\left[(t_g + 460)^4 + \frac{4.74 \times 10^7 V_a^{0.6}}{\varepsilon D^{0.4}} (t_g - t_a)\right]^{0.25} - 460$$

$$(4.1)$$

SI units: $\bar{t}_r =$

$$\left[(t_g + 273)^4 + \frac{1.10 \times 10^8 V_a^{0.6}}{\varepsilon D^{0.4}} (t_g - t_a)\right]^{0.25} - 273$$

where

$\bar{t}_r$ = mean radiant temperature, °F (°C)
$t_g$ = globe temperature, °F (°C)
$V_a$ = air velocity, fpm (m/s)
$t_a$ = air temperature, °F (°C)
$D$ = globe diameter, ft (m)
$\varepsilon$ = globe emissivity (dimensionless)

Another calculation approach involves the geometry of the room relative to a specific point and uses Equation 4.2 (Egan, 1975):

I-P or SI units $\bar{t}_r = \dfrac{\sum t\alpha}{360}$

$$= \frac{t_1\alpha_1 + t_2\alpha_2 + t_3\alpha_3 + \ldots t_n\alpha_n}{360} \quad (4.2)$$

where

$\bar{t}_r$ = mean radiant temperature, °F (°C)

$\alpha$ = surface exposure angle (relative to the occupant) in degrees

DB air temperature by itself is usually not an adequate comfort indicator, especially in passively solar-heated or passively cooled spaces where radiant temperature or air motion may be more influential than air temperature.

The comfort zone in Fig. 4.5 purports to apply to both genders, any age, and any national origin; research by ASHRAE supports such a view. High humidity is avoided (to avoid mold and mildew as well as discomfort), and air motion is assumed to be unnoticeable. Occupants of naturally ventilated buildings, however, seem to find comfortable a combination of higher temperature and RH than do occupants of sealed,

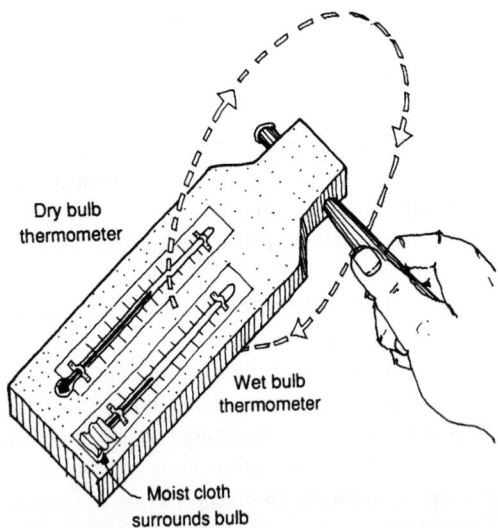

Dry bulb thermometer

Wet bulb thermometer

Moist cloth surrounds bulb

***Fig. 4.6*** *Sling psychrometer and its usage. Air motion encourages evaporation from the moist cloth, lowering the wet-bulb temperature below the surrounding air temperature, whereas the dry-bulb temperature stays constant at the surrounding air temperature. (At 100% RH, WB and DB temperatures will be equal.)*

air-conditioned buildings. Figure 4.7*a* shows a much higher incidence of these warmer, more humid environments in Hawaiian naturally ventilated classrooms than in similar air-conditioned classrooms. Yet, the majority of students and teachers in these Hawaiian schools voted these conditions as acceptable. These observations were made in both hot and cool seasons (see Kwok, 1998).

Does long-term *acclimatization* influence the sensation of thermal comfort? Researchers disagree. ASHRAE Standard 55 recognizes but two seasonal comfort zones, influenced by activity and clothing. Givoni (1998) advocates a higher temperature, higher humidity limit for "hot-developing" countries (Fig. 4.7*b*), based upon a combination of traditional and economic factors. In Fig. 4.7*b*, an interior air speed of 2 m/s (about 5 mph) is assumed.

An *adaptive model* of comfort recognizes acclimatization's influence as well as personal

actions to influence one's own comfort. ASHRAE Standard 55 recognizes that people can be comfortable in a wide range of temperatures and provides an optional method for determining acceptable thermal conditions for naturally conditioned spaces. Humphreys and Nicol (1998) elaborate on this approach, with occupant behaviors summarized in Table 4.3. They present a formula to determine the *indoor comfort* temperature, $T_n$, relative to an exponentially weighted running average of outdoor temperature and applicable to *free-running* buildings (without mechanically—narrowly—controlled indoor temperatures):

$$T_n = 0.534\, T_{rmo} + 12.9\ (°C) \qquad (4.3)$$

where $T_{rmo}$ is in °C and represents the mean outdoor air temperature during a period of hot weather. Note that relative humidity is not included in this relationship. Using this equation, with Phoenix, Arizona's, average July–August temperature of

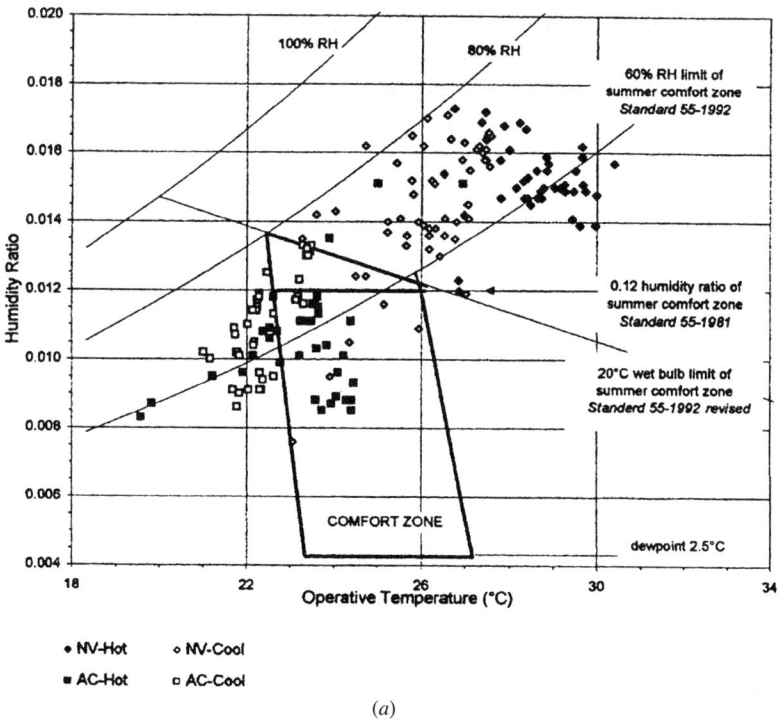

(a)

**Fig. 4.7** *Natural ventilation and comfort. (a) Measured conditions compared to the ASHRAE summer comfort zone (see Fig. 4.5b) for naturally ventilated and air-conditioned classrooms in Hawaii. The majority of the occupants voted these conditions acceptable. A higher temperature and humidity comfort zone for naturally ventilated buildings is supported by studies such as this (Kwok, 1998). (Reprinted with permission of the American Society of Heating, Refrigerating and Air-Conditioning Engineers, Inc. from ASHRAE Transactions, 1998, Vol. 104, Number 1.) (b) Suggested boundaries of outdoor air temperature and humidity within which indoor comfort can be provided by natural ventilation. Assumed air speed is 2 m/s (4.5 mph). The higher limits for "hot-developing countries" assume acclimatization by those cultures. (Based upon Givoni, 1998.)*

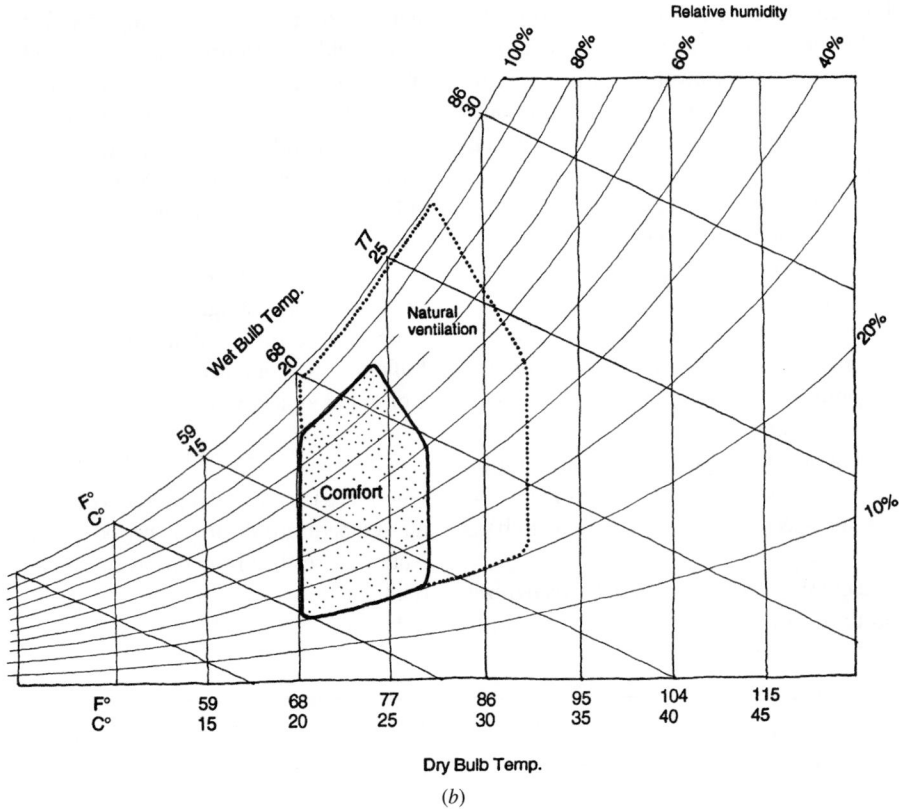

**Fig. 4.7** (continued)

32.3°C (90.2°F) (to approximate an exponentially weighted running average):

$$T_n = 0.534 (32.3) + 12.9 = 17.25 + 12.9$$

$$= 30.1°C (86.2°F)$$

Figure 4.5 specifies comfort zone boundaries for environments that meet specific operative temperature criteria, where the air speeds are not greater than 40 fpm (0.20 m/s), based upon specified clothing insulation values. Air speeds greater than 40 fpm (0.20 m/s) may be used to increase the upper operative temperature limit of the comfort zone in certain circumstances. For example, the borders of the summer comfort zone may be raised 1°F for each 30 fpm (1°C for each 0.275 m/s) increase in air motion up to a limit of 82.45°F at 160 fpm (28°C at 0.8 m/s). At this air speed, loose paper, hair, and other objects might start to be blown about (however, see Table 4.4). Note: Optimum summer operative temperature is

76°F (24.4°C). For each 0.1 CLO decrease in clothing, an increase of 1°F (0.6°C) is allowed in the borders of the summer comfort zone.

The upper and lower borders for relative humidity are based upon less precise data, but in general, the lower limit seeks to avoid problems such as coughs, nosebleeds, static electricity, and dust mites from excessively dry air; the upper limit tries to keep skin wettedness within acceptable levels, as well as discouraging the growth of mold and mildew.

### (b) Passive Building Comfort Standards

A somewhat different approach to comfort standards has been proposed for buildings that take a passive approach to heating and cooling in areas where acceptable humidity can be maintained. In these buildings, direct sun might add significantly to body warming in winter; strong air currents might be expected as a part of summer body cooling. These tend to be the kind of free-running

**TABLE 4.3 Adaptive Behavior for Thermal Comfort**

| Response Category | Actions in Response to Cold | Actions in Response to Heat |
|---|---|---|
| Regulating the rate of internal heat generation | Increasing muscle tension and shivering | Reducing one's level of activity<br>Drinking cold liquids (induces sweating)<br>Drinking hot liquids (induces sweating)<br>Eating less<br>Adopting the siesta routine (matching activity to the thermal environment) |
| Regulating the rate of body heat loss | Vasoconstriction (reduces blood flow to the surface tissues)<br>Curling or cuddling up (reduces exposed surface area)<br>Adding some clothing | Vasodilation (increases blood flow to the surface tissues)<br>Adopting an open posture (increases exposed surface area)<br>Taking off some clothing<br>Sweating |
| Regulating the thermal environment | Turning up the thermostat<br>Lighting a fire<br>Complaining to management (so that someone else will raise the temperature)<br>Insulating a loft or wall cavities<br>Improving windows or doors, weather stripping | Turning on the air conditioner<br>Switching on a fan<br>Opening a window<br><br>Shading a window from the sun |
| Selecting a different thermal environment | Finding a warmer spot (such as going to bed)<br>Visiting a friend (with a warmer place)<br>Visiting a heated public building<br><br>Building a new home<br>Emigrating: a long-term solution | Finding a cooler spot<br><br>Visiting a friend (with a cooler place)<br>Visiting a cooled public building (or going swimming)<br>Building a new home<br>Emigrating: a long-term solution |
| Modifying the body's physiological comfort conditions | Acclimatizing, letting the body and mind become more resistant to cold stress | Acclimatizing, letting the body and mind adjust so that heat is less stressful |

Source: Based on Humphreys and Nicol (1998).

buildings mentioned earlier as candidates for the adaptive model of comfort prediction.

Arens et al. (1980) demonstrated a wider range of comfort conditions and considerably more tolerance for summer air motion than ASHRAE (1995). A summary of these results is shown graphically in Fig. 4.8. Such graphs are called *bioclimatic* charts because of their interrelation of climate and human comfort factors. Where the users of buildings are expected to adjust to the wider temperature swings associated with passive buildings, these guidelines may be used. Some differences in the basic assumptions of Arens et al. from ASHRAE (1995) should be noted:

Activity: 1.3 MET (ASHRAE, 1.2 MET)
Winter: 0.8 CLO (ASHRAE, 1.0 CLO)
Summer: 0.4 CLO (ASHRAE, 0.5 CLO)

Note particularly that rather than operative temperature, ordinary air temperature (DB temperature) is used in Fig. 4.8. Added radiant heat for conditions below the shading line can be utilized to extend the comfort zone to lower temperatures; added air motion above the nearly still air line can extend the comfort zone to higher temperatures. *Radiant heat* to be added to a lower-extended comfort zone is shown in Fig. 4.8 using two quantities: the effective radiant field (ERF) is a measure of the net radiant heat flux to the body from all surfaces at temperatures other than air temperature. A more convenient quantity for designers may be the total solar radiation (or *insolation*) on a horizontal surface, termed $I_{TH}$. Insolation values are more readily available, and the added-radiation lines of Fig. 4.8 represent $I_{TH}$ converted to its approximate radiant impact on a human body's surface area when the sun is at an altitude of 45°. Arens et al. (1980) give further details on the effects of solar radiation on comfort at other sun altitudes, activity levels, and clothing combinations.

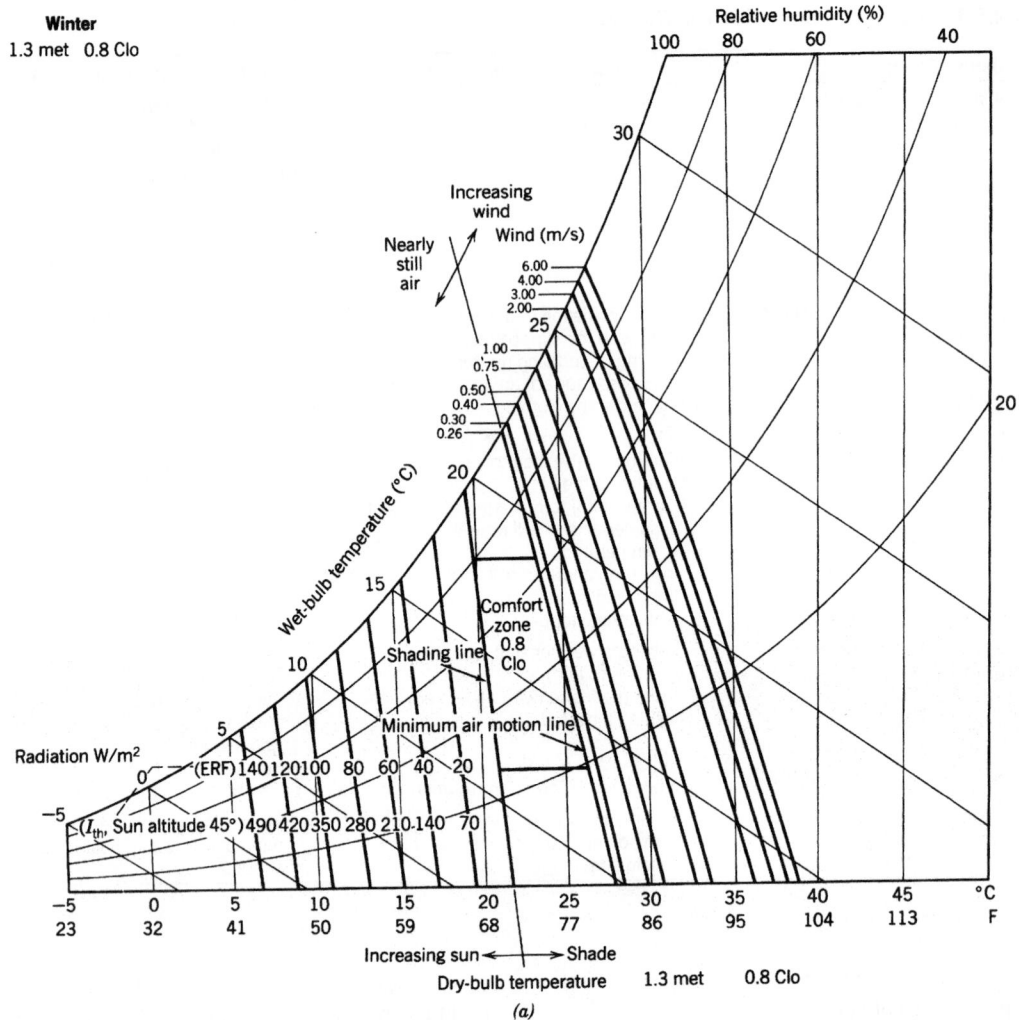

**Fig. 4.8** *Comfort zones that encourage passively heated and cooled buildings. The winter comfort zone (a) and the summer comfort zone (b) are combined for the year-round zone shown in (c). (Based upon Arens et al., 1980.)*

Added air motion is shown in Fig. 4.8 for quite a wide range of velocities. Table 4.4 indicates that people easily tolerate air motion outdoors up to about 600 fpm (3 m/s), but indoor studies have indicated that about 400 fpm (2 m/s) is the maximum tolerable air speed from overhead fans.

Caution is still advised when relative humidity is above 70% or below 20% for reasons cited earlier.

### (c) Localized Comfort

How can comfort be assured at each work station in an office or wherever people spend a lot of time? The location of heating, cooling, and ventilating components is an important detail. The human body is most affected by the thermal environment in its thermally sensitive places. Although we sweat and are sensitive to heat or cold over nearly all of our skin surface, we are most thermally sensitive in these places:

Heat receptors: fingertips, nose, elbows
Cold receptors: upper lip, nose, chin, chest, fingers

So, on a hot, humid day, cool air moving across the face is a particularly strong promise of comfort, whereas on a cold day, a burst of heat (such as

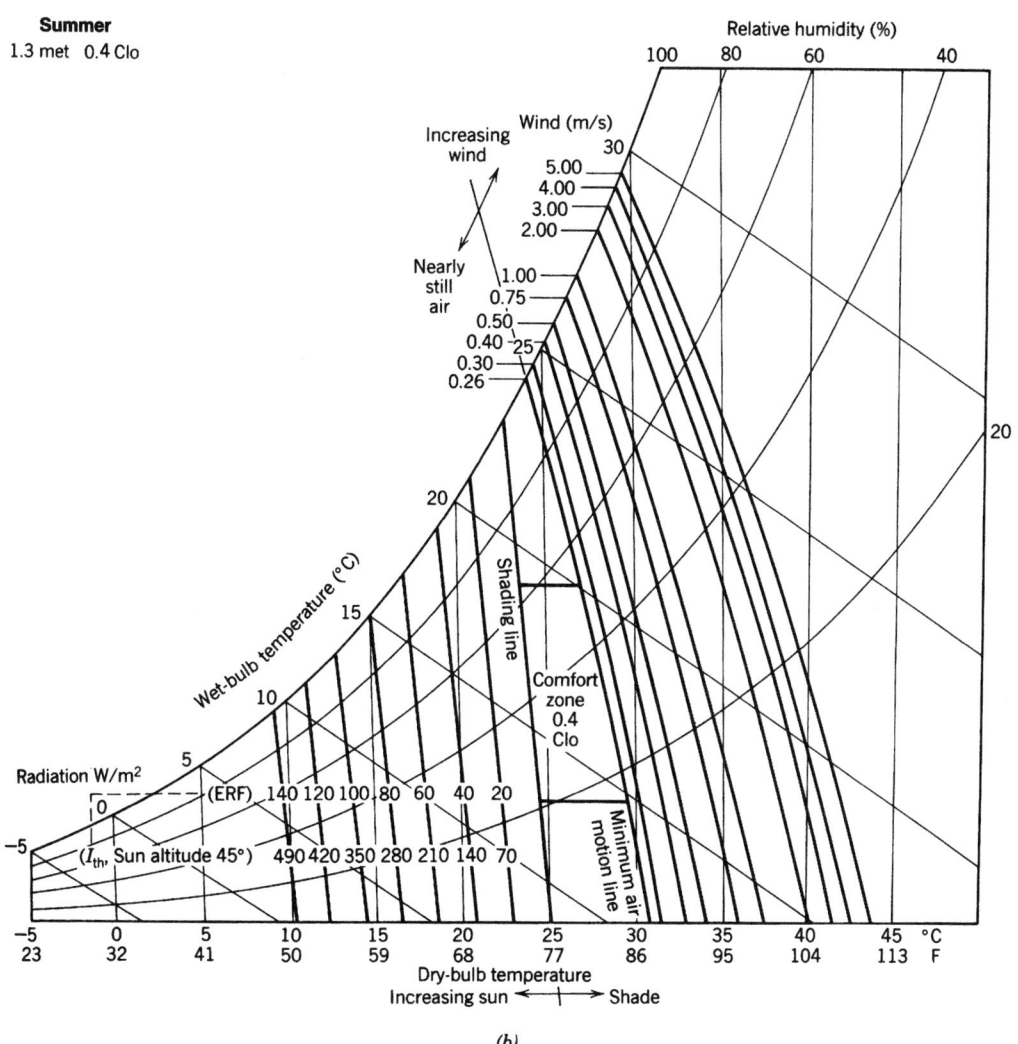

*Fig. 4.8* (Continued)

radiant heat from a window, a heater, or a cup of coffee) to the face and fingers is quickly effective.

Regarding our sense of touch: our fingertips are most sensitive to the *rate at which heat is being conducted* to colder objects or from hotter objects. The temperature of our skin at the fingertips under ordinary conditions is in the high 80s°F (high 20s°C), so our sense of touch works against many passively heated surfaces in winter, which feel cool even though they are warmer than room air temperature. These surfaces are made of materials that conduct heat rapidly so that they can soak up solar radiation without overheating

the room. This high *conductivity* makes them eager to accept human warmth as well, and persuades us as we touch them that they are cooler than, in fact, they are. Anyone who has walked barefoot from a rug to an unheated tile floor will sense that the tile is colder even when both have exactly the same surface temperature. This same characteristic works for passively cooled surfaces in summer.

Excessive air motion is called a *draft* and results in an undesirable local cooling of the body. It is a particularly serious threat in winter, yet summer drafts from very cool conditioned air are also

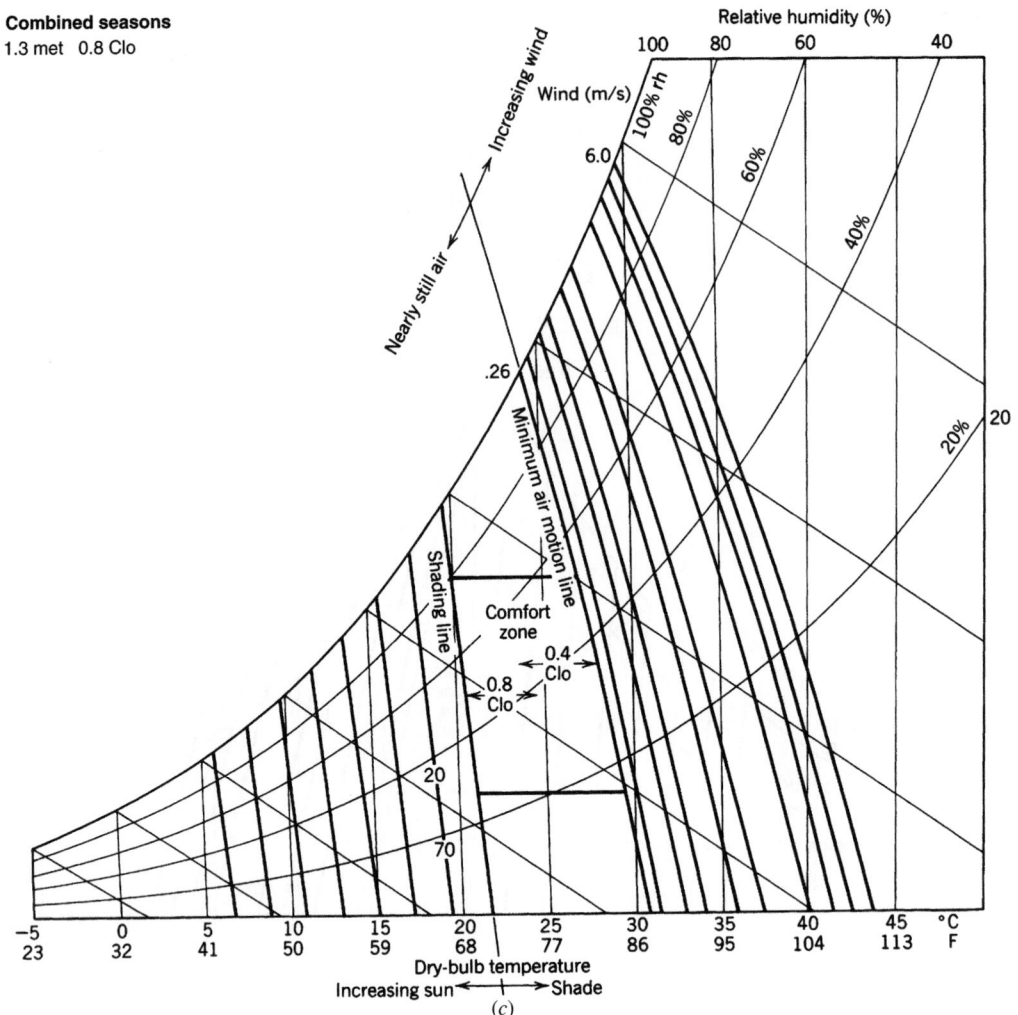

**Fig. 4.8** (Continued)

## TABLE 4.4 Indoor Air Velocity and Comfort

| Air Velocity | Possible Lower-Temperature Comfort Sensation (Between 80°F and 90°F; Larger Numbers Correspond to High-Humidity Areas) | Probable Impact |
|---|---|---|
| Up to 50 fpm (0.25 m/s) | No change in comfort sensation | Unnoticed |
| 50–100 fpm (0.25–0.51 m/s) | 2–3F° lower (1.1–1.7C°) | Pleasant |
| 100–200 fpm (0.51–1.02 m/s) | 4–5F° lower (2.2–2.8C°) | Generally pleasant but causing a constant awareness of air movement |
| 200–300 fpm (1.02–1.52 m/s) | 5–7F° lower (2.8–3.9C°) | From slightly to annoyingly drafty |
| Above 300 fpm (1.52 m/s) | More than 5–7F° lower (2.8–3.9C°) | Requires corrective measures if work is to be efficient and health secured |

*Source:* Adapted from Victor Olgyay, *Design with Climate: Bioclimatic Approach to Architectural Regionalism,* Copyright © 1963, Princeton University Press. Reprinted by permission.

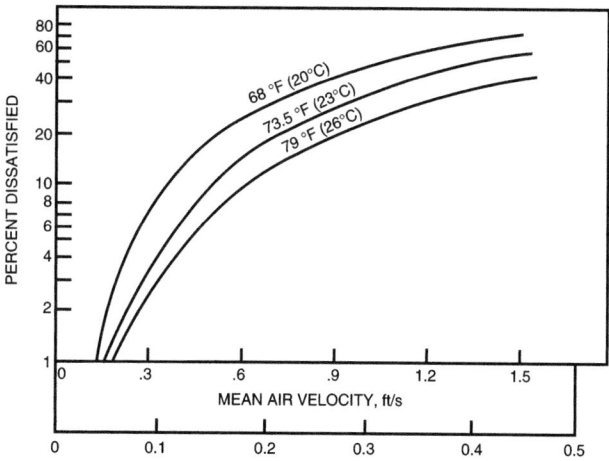

**Fig. 4.9** *Percentage of people dissatisfied as a function of mean air velocity. The warmer the air, the greater the tolerance. (Reprinted with permission of the American Society of Heating, Refrigerating and Air-Conditioning Engineers, Inc., from the 1997 ASHRAE Handbook—Fundamentals.)*

bothersome. Figure 4.9 shows the percentage of people dissatisfied due to a perceptible draft on their head, neck, shoulders, and back. Again, this applies to sedentary people wearing normal indoor clothing. Note the influence of air temperature: the lower the temperature, the higher the dissatisfaction. Many air-conditioning systems deliver summer supply air at temperatures as low as 55°F (13°C), well below the lowest temperature shown here. The standard summer 50 fpm (0.25 m/s) air speed at such supply air temperatures seems to threaten more than 40% dissatisfaction from drafts. However, ceiling fans with

the same air speed, for comfort at higher temperatures such as 80°F (27°C), should cause less than 15% dissatisfaction.

Another comfort factor is *radiant asymmetry*, the difference between the temperatures of two opposite surfaces as experienced by a body seated between them when one of these surfaces is markedly different in temperature from all other surfaces within a space. Figure 4.10 shows that the greatest dissatisfaction arises from "warm ceilings" (typical of radiantly heated ceilings in winter)—almost half of the experimental subjects experienced discomfort with ceilings 27°F (15°C) warmer

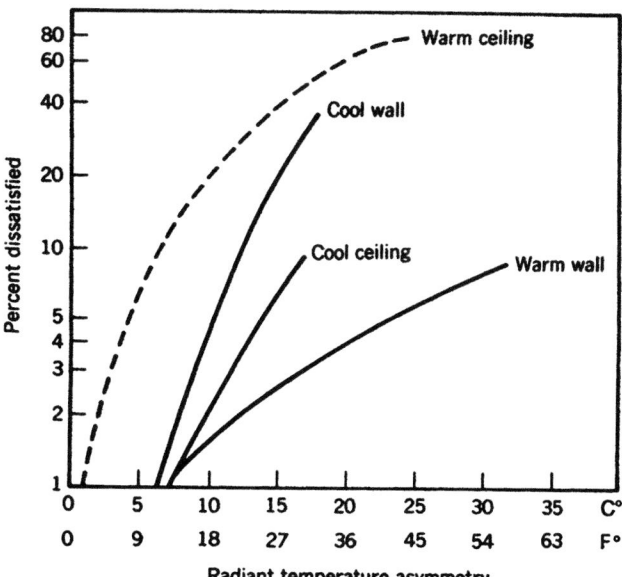

**Fig. 4.10** *Percentage of people expressing discomfort due to asymmetric radiation. A warmer ceiling produces the highest discomfort; a warmer wall, the least discomfort. (Reprinted with permission of the American Society of Heating, Refrigerating and Air-Conditioning Engineers, Inc., from the 1997 ASHRAE Handbook—Fundamentals.)*

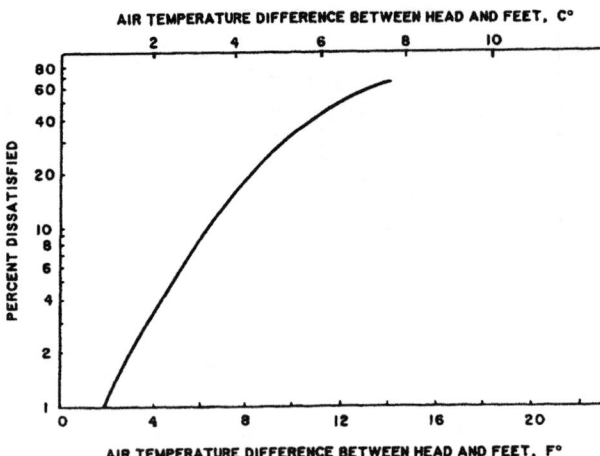

**Fig. 4.11** *Percentage of people dissatisfied as a function of the vertical air temperature difference between the head (higher temperature) and ankles (lower temperature). A cold floor proved uncomfortable. (Reprinted with permission of the American Society of Heating, Refrigerating and Air-Conditioning Engineers, Inc., from the 1997* ASHRAE Handbook— Fundamentals.*)*

than a space's average temperature. Less serious were "cool walls" (typical of a cold window in winter). Less than 10% dissatisfaction resulted from "cool ceilings" (such as radiant cooling panels in summer) and even less from "warm walls" (typical of a sunny window in summer or a passively heated mass wall in winter).

Because warm air is less dense than cold air, it rises; therefore, in most building spaces, air temperature is somewhat higher at the ceiling than at the floor, regardless of the season. Figure 4.11 shows experimental results of dissatisfaction with a vertical air temperature difference between the head at 43 in. (1.1 m) and the feet at 4 in. (0.1 m). (When the head level was cooler than the floor, much greater temperature differences were tolerated.)

### 4.3 DESIGN STRATEGIES FOR COOLING

The chart in Fig. 4.12 is a guide to various approaches to passive cooling in temperate climates. It is based on the *building bioclimatic chart* developed by Milne and Givoni (1979). It helps to assess the suitability of four approaches to passive cooling under summer conditions. Average monthly climate data (as in Table 4.5) are now available from software and various websites (see Appendix M). Such climate data can be plotted on this chart. If they do not exceed the "boundaries" of a strategy zone, then a match is indicated between the climate and that strategy. Although the edges of

each strategy zone are drawn as lines, these boundaries are more broad and vague than the lines suggest. When a climate surpasses the boundaries of all four strategy zones, conventional air conditioning is almost certainly desirable. The similarity between Figs. 4.8 and 4.12 is evident: the *comfort zones* include somewhat warmer conditions for naturally ventilated buildings.

To plot average monthly conditions for a climate on this strategy chart, assemble information such as that shown in Table 4.5. This is available for every climatological station maintained by the National Oceanic and Atmospheric Administration (NOAA) of the U.S. Department of Commerce. These climate summaries, often called LCDs (Local Climate Data), are available from the National Climatic Data Center (Federal Building, Asheville, NC 28801). Many libraries also carry such summaries in publications such as *Climates of the States.*

In addition to a typical hot summer day, it is useful to check for the *design condition*, that statistically hottest moment for which engineers would design a building's mechanical cooling equipment. The design condition shows the climate near its worst while avoiding the freakish conditions of extreme temperatures listed in the NOAA data. Design data are available in Appendix B as Summer Design Dry-Bulb and Mean Coincident Wet-Bulb (2.5%), meaning that such a combination will, in a typical summer, be exceeded during only 2.5% of the total hours in the 4-month period of June–September.

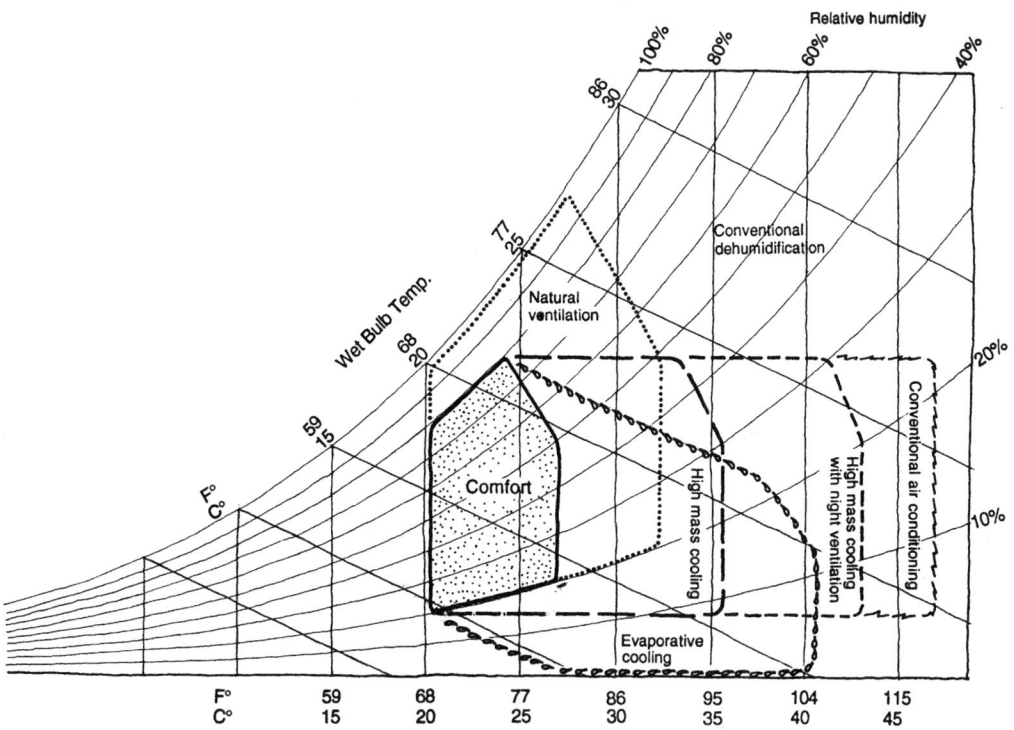

**Fig. 4.12** *Passive cooling design strategies by climate. Buildings usually contain sources of heat. The more heat that is generated within a building, the more an artificially warmer climate is created. Thus, after plotting the outdoor climate data on this chart, consider how shifting these plots would affect a design strategy. The more solar gains allowed inside, electric lights, business machines, and so on, the further the plots shift to the right. The more people, cooking, bathing, and other heat-plus-moisture sources, the further the plots move both upward and to the right. (Based upon Milne and Givoni, 1979.)*

**EXAMPLE 4.1** Plot the typical day of the hottest month for Dodge City, Kansas.

**SOLUTION**

By inspection of Table 4.5, the hottest month is July, but August is close and sun angles are lower (more gain through windows). To chart the typical August day, find the approximate RH for the coldest and hottest hours. Because RH is listed only at four times, first select the highest RH, occurring here at 6 A.M.: this will coincide approximately with the coldest hour. Thus, one end of the linear climate plot for August in Dodge City will be at the combination of 80% RH and 65.7°F. The other end of the plot will be at the lowest RH and the hottest hour, in this

**TABLE 4.5 Normal Data from Annual Summary of Local Climatological Data for Dodge City, Kansas[a]**

| Month | Daily Temperatures (°F) Normal | | Extreme[b] | | Relative Humidity (%) at Hour | | | | Wind | |
| | Max | Min | Max | Min | 00 | 06 | 12 | 18 | Mean Speed (mph) | Prevailing Direction |
|---|---|---|---|---|---|---|---|---|---|---|
| June | 86.0 | 61.4 | 108 | 41 | 63 | 75 | 44 | 38 | 14.4 | South |
| July | 91.4 | 66.9 | 109 | 47 | 66 | 78 | 46 | 42 | 12.9 | South |
| August | 90.4 | 65.7 | 107 | 47 | 71 | 80 | 50 | 46 | 12.7 | South |

[a]Records as of 1979.

[b]Extremes do not occur on the same day or series of days.

case at 6 P.M., a combination of 46% RH and 90.4°F. The line between these points is shown in Fig. 4.13.

We see from Fig. 4.13 that one passive cooling strategy—high mass with night ventilation—appears clearly adequate to meet the needs of the typical August day in Dodge City. (Using this strategy, the *low-temperature* end of the line is also important and should be below the comfort zone for best results.) Another strategy, high mass, appears just barely adequate. Two other strategies, natural ventilation and evaporative cooling, appear inadequate because the highest temperature/lowest RH combination falls outside their boundaries—slightly so for natural ventilation but greatly so for evaporation.

Now check on the *design condition* from Appendix B. The summer DB and mean coincident WB temperature is 97/69°F; this combination is shown as a circle around the point in Fig. 4.13. The 97°F DB temperature is plotted on the corresponding vertical line; the 69°F WB temperature is found by first locating the 69°F DB vertical line and following that line up to 100% RH, where 69°F DB and WB are

coincident. Then follow downward to the right along the constant WB line to its intersection with 97°F DB.

The design condition point falls just outside the zone of high-mass cooling and just within the zone of evaporative cooling. Because the design condition falls well within the zone of high mass with night ventilation, this strategy is thus confirmed as clearly the best of the passive cooling strategies for Dodge City. ∎

## (a) Natural Ventilation Cooling

This is the most obvious strategy suggested by the comfort charts presented earlier, in which higher air temperatures were offset by increased air motion. It may be the only passive strategy available in humid, hot climates in which temperatures are only slightly lower by night than by day. Buildings should be very open to breezes while simultaneously closed to direct sun. They may be thermally lightweight as well, because night air is not cool

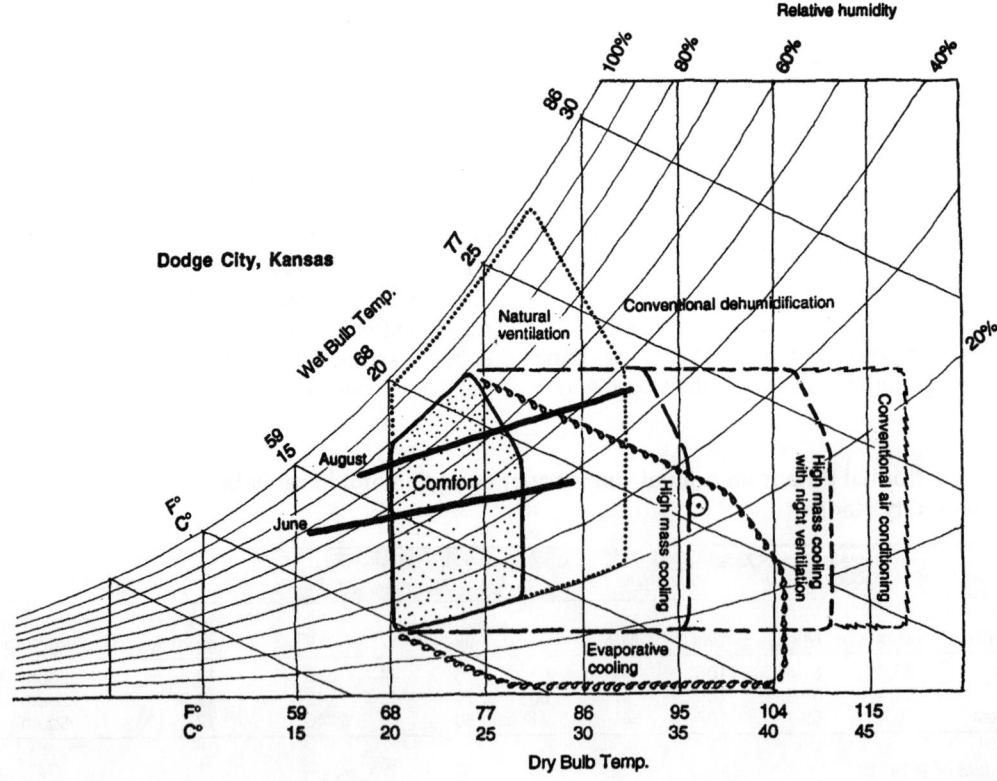

**Fig. 4.13** *Hot-month daily ranges and the summer design condition for Dodge City, Kansas, superimposed on the design strategy chart of Fig. 4.12.*

enough to remove much stored daytime heat. Very high humidity may be avoided only by sealing and air-conditioning buildings.

Natural ventilation has two variations: cross-ventilation and stack ventilation. *Cross-ventilation* is driven by wind and is accomplished with windows. It relies upon rather narrow plans with large ventilation openings on either side. Thus, it is naturally compatible with daylighting. *Stack ventilation* depends upon very low openings to admit outside air and very high openings to exhaust air; it is driven by the principle that hot air rises. Stack ventilation is generally weaker than cross-ventilation—except when there is no wind at all. Design guidelines for sizing apertures for ventilation are found in Chapter 8.

## (b) High-Mass Cooling

This strategy is for warm, dry summers, when the extremes of hot days are tempered by the still-cool thermal mass of a building. Cool nights then slowly drain away the heat that such mass accumulates during the day. The thermal mass can be in floors, walls, or roofs but will need a *sink* to which it can reject its heat by night. The roof has the advantage of radiating to the cold night sky, but it should be protected from exposure to sun by day. The masonry courtyard-type buildings of the Mediterranean are indigenous examples of this passive cooling strategy; their courtyard floors and roofs can be protected with movable shading devices (*toldos*) by day, then opened to night-sky radiation.

*Roof ponds* are a form of high-mass cooling for one- and two-story buildings. Because they require only the roof to be massive, they allow for considerable design freedom in walls and fenestration. Where cooling is the only objective, this approach uses water that is stored between the metal ceiling and the roof insulation; by night, the water is pumped (and/or sprayed) over the exposed roof surface and allowed to trickle back through the insulation to the storage pond. At lower latitudes (with high winter sun altitudes), roof ponds can be used for passive solar heating as well. This strategy uses sliding panels of insulation over bags of water; the panels slide open on winter days to collect sun and open on summer nights to radiate heat to the sky. A design guideline sizing procedure is found in Chapter 8.

Another variation on high-mass cooling depends upon earth contact. The earth acts as a heat sink, keeping walls and floors (even roofs when earth covered) cool. However, if the earth is allowed to continue to act as a heat sink in winter, heating needs could be greatly increased. Thus, a strategy for summer contact and winter isolation might be appropriate.

## (c) High-Mass Cooling with Night Ventilation

This hot–dry summer design strategy must use outside air at subcomfortable nighttime temperatures to flush away heat stored during the daytime. The fewer the subcomfortable night hours, the greater the area of thermally massive surface that must be provided to store the day's heat. Also, because there are more hours of daylight and fewer of nighttime, the ventilation must occur quickly and thoroughly, probably using fans. (Nightly wind velocities are typically lower than daytime velocities, because summer wind is often driven by regional solar overheating of the ground.) The building switches from a thermally closed condition by day (to exclude sun and hot outdoor air) to an open condition at night (to allow ventilation to cool the mass). Note: nighttime temperatures must be cooler than the comfort zone if this strategy is to be effective. Figure 4.14 shows the need to inspect both ends of the typical day's plot for this strategy.

This cooling strategy is highly compatible with passive solar heating strategies that rely on large areas of thermal mass, such as *direct gain* (Section 4.4a). It is suitable for large, high buildings, particularly those that have concrete (thermally massive) structural systems. A design guideline sizing procedure is found in Chapter 8.

## (d) Evaporative Cooling

This design strategy relies on the principle that when moisture is added to air, relative humidity increases while dry-bulb temperature decreases. (On the bioclimatic chart, this pattern exactly follows the constant wet-bulb line, upward and to the left.) In conditions that are more uncomfortably dry than uncomfortably hot, higher humidity is gladly exchanged for lower air temperature. However, large quantities of both water and outdoor air are

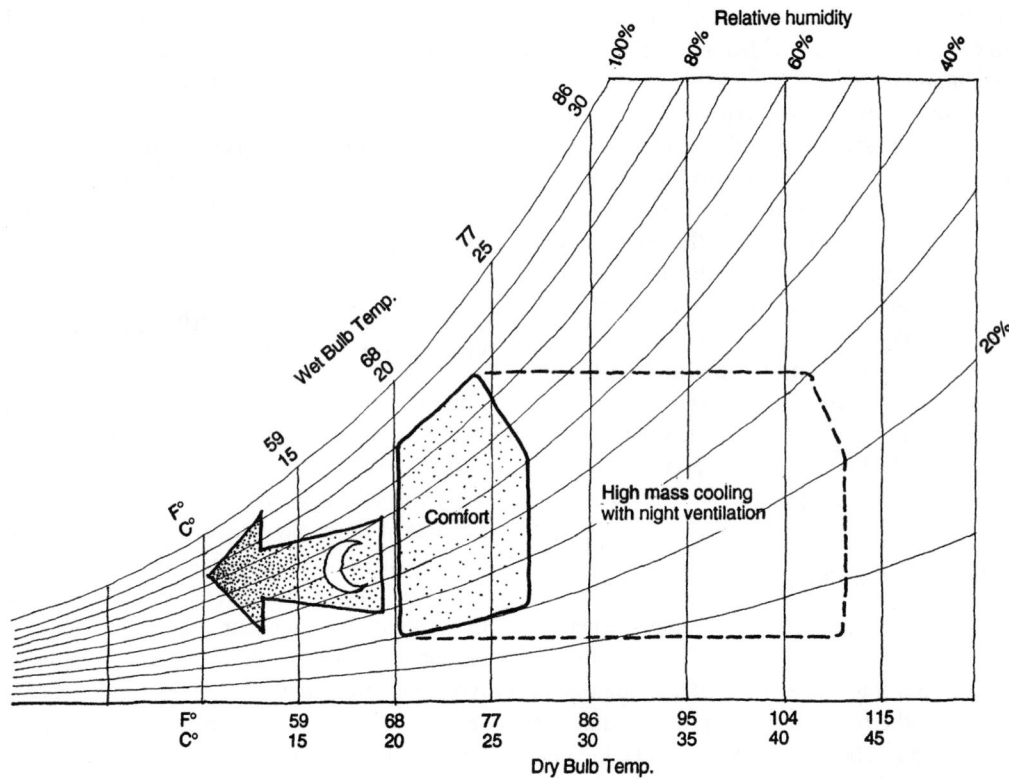

**Fig. 4.14** *Strategy chart segment for night ventilation of thermal mass. This strategy is most successful when cool nighttime temperatures are available, preferably below comfort zone temperatures for several hours.*

needed; fan-driven evaporative coolers are the most common way to provide this kind of cooling.

Evaporative cooling imposes few constraints on designers because the equipment resembles conventional HVAC systems. Rather high indoor air velocities and their associated sounds are typical of these systems, and the aroma of the wetted material of the cooler is often noticeable. The coolest air will be in the vicinity of the air inlet to the space, the warmest air at the outlet from the space. Design guideline sizing procedures are found in Chapter 8 for fan-driven coolers and for passive *cooltowers*. Indirect evaporative cooling, involving more mechanical equipment, is first discussed in Chapter 5.

## 4.4 DESIGN STRATEGIES FOR HEATING

Figure 4.15 indicates how greatly solar space heating may contribute to a building in the winter months. To use these charts, refer to Table C.15. For

your location, find the average January ambient (DB) air temperature, called *TA*, and the average January daily solar radiation on a vertical south-facing surface, called *VS*. On each chart, find these two data points to determine the approximate percentage contribution that solar energy can make to your building's winter seasonal fuel needs. In Fig. 4.15*a*, these points of TA and VS are plotted for seven North American cities. Sunny and cold Denver, Colorado, contrasts sharply with cloudy and cool Portland, Oregon. Yet, the subsequent charts show that even in Portland, solar energy can contribute to space heating.

The remainder of Fig. 4.15 compares two building envelope types with three common solar heating strategies. The left column (Fig. 4.15 b, d, f) assumes a building with insulation at code-required minimums and some south-facing glass (unshaded in winter). For this modestly solar building type, the seasonal fuel bill will be reduced, but rarely more than 40%, by using solar energy. The right column (Fig. 4.15 c, e, g) assumes a building that exceeds

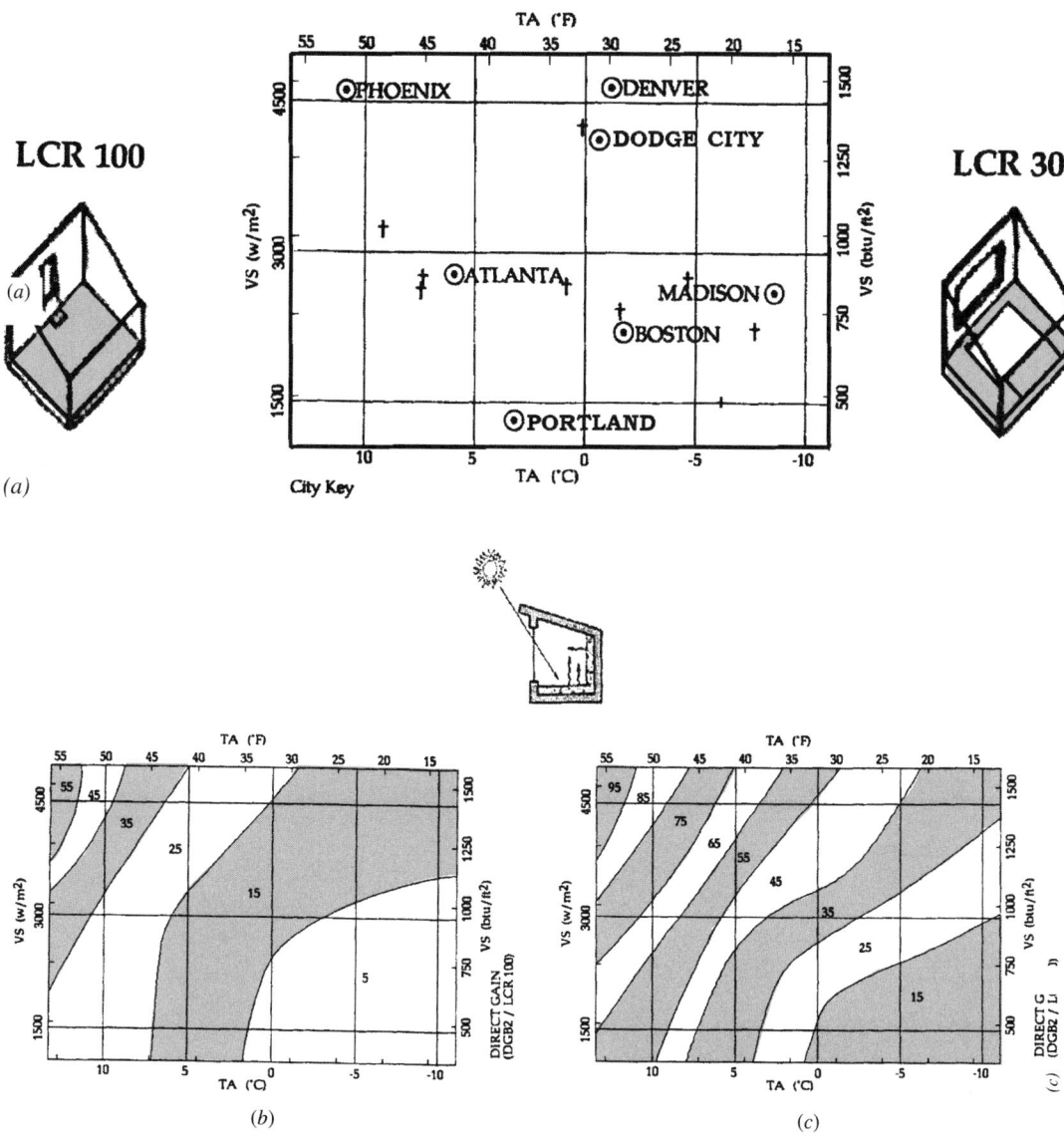

**Fig. 4.15** *Early assessment of passive solar heating potential. (a) From Table C.15, the January TA (average air temperature) and January VS (daily solar radiation on a vertical surface) are plotted for Atlanta, Georgia; Boston, Massachusetts; Denver, Colorado; Dodge City, Kansas; Madison, Wisconsin; Phoenix, Arizona; and Portland, Oregon. The Solar Savings Fraction (SSF), the percentage reduction in heating fuel achieved by a passive solar design, is approximated for modestly solar "LCR 100" (smaller south windows, moderately well insulated) and seriously solar "LCR 30" (larger south windows, very well insulated buildings.) († Indicates where SSF is approximately 6% higher than that shown on the graph for these locations: in California: Fresno, Mt. Shasta, Sacramento, and San Francisco; in Idaho: Boise and Pocatello; in Montana: Helena and Missoula; and in Reno, Nevada.) (b) Direct-gain systems in modestly solar and (c) seriously solar buildings (LCR 30). (d) Indirect-gain systems in modestly solar and (e) seriously solar buildings. (f) Isolated-gain systems in modestly solar and (g) seriously solar buildings.*

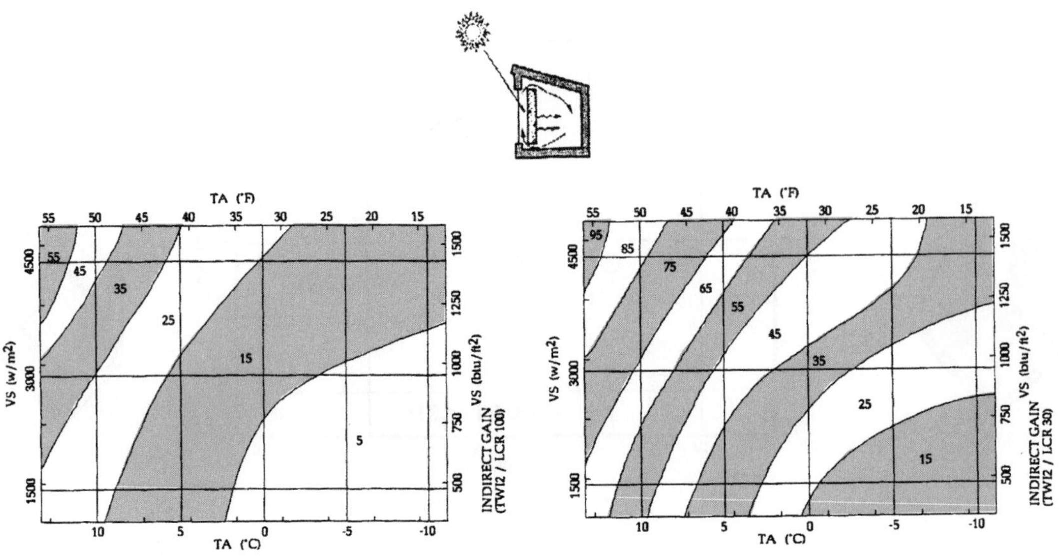

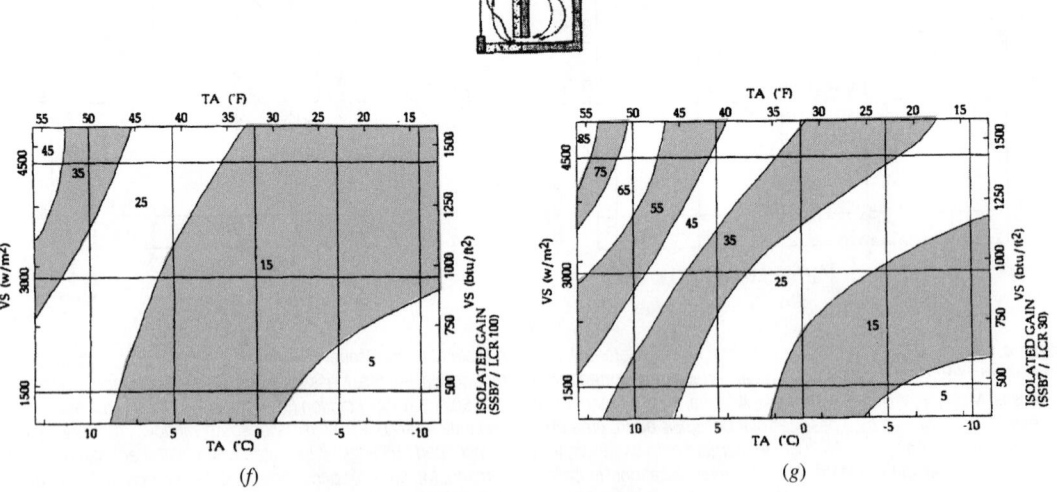

**Fig. 4.15** (Continued)

minimum insulation requirements and is designed with much greater areas of south-facing glass unshaded in winter. For this seriously solar building type, the seasonal fuel bill will be reduced by about twice as much as its modest counterpart.

The three solar heating strategies are *direct gain* (let sunlight into the space, where it warms exposed thermally massive surfaces), *indirect gain* (sun strikes the thermal mass first and is then passed only as heat to the space behind), and *isolated gain* (a sunspace or greenhouse heated greatly by the sun that then passes some of its heat to the space behind). Direct-gain systems carry the symbol DG; indirect-gain systems, TW (for Trombe wall); and isolated-gain systems, SS (for sunspace). For more details, see Appendix H. All the systems in this figure assume double-glazed windows (and for the direct and indirect ones, low-ε, argon-filled windows), but none use night insulation over windows. The direct gain is system type DGB2, the indirect gain is a vented TWD2, and the isolated gain is SSE1.

### (a) Direct Gain

Almost any building with some south-facing glass could claim to be a direct-gain building. The sun is admitted to the space to be heated, striking furnishings and room surfaces. In a well-designed direct-gain space, there are ample thermally massive surfaces (such as concrete slab; concrete block; or brick, quarry, or ceramic tile) that directly receive much of this incoming sun. Such massive surfaces should have at least three times the area of south-facing glass in order to keep the space from overheating in sunny hours. The mass should be thick enough, typically at least 4 in. (100 mm), to absorb and later reradiate a winter day's dosage of direct sun. In Chapter 8 and Appendix H, variations on direct-gain approaches involve glazing types, relative mass areas and thickness, and whether night insulation is used over windows.

Direct gain is popular because of its simplicity and its ample daylight and view to the south. Relative to the other passive solar approaches, it has problems of glare, overheating on sunny days, large radiant heat losses to glass areas by night (and thus a large diurnal difference in interior temperature), and fading of furnishings in direct sunlight. This winter heating approach is well matched with the summer cooling strategy of night ventilation of mass; both systems depend on large areas of internal exposed thermal mass, so such an investment pays off in both winter and summer.

### (b) Indirect Gain

This is the passive solar approach encountered least often, perhaps because a sheet of glass covering an opaque wall seems such a denial of "window." The mass wall behind the glass is usually 8 to 12 in. (200 to 300 mm) thick, and should be of a dense and highly conductive material such as standard-weight concrete or dense brick. Water in containers is also an option. An air space between the glass and the mass allows for the option to "vent" the mass wall, where cool air from the heated space behind the wall enters at the bottom, rises with solar heating, and then exits into the space behind near the ceiling. Sometimes this air space is made wide enough to admit a person for cleaning the inside surface of the glass, but operable casement or awning-type glazings are other cleanable options. The back surface of the mass (facing the heated space) should be kept clear of hangings or large furniture to facilitate radiant heat transfer to the space. In Chapter 8 and Appendix H, variations on indirect-gain approaches involve mass characteristics (water or masonry), thickness, and surface treatments; whether masonry walls are vented; glazing types; and whether night insulation is used behind the glazing.

Indirect gain is less popular than direct gain because it admits much less daylight and lacks a view to the south. Yet, relative to the other passive solar approaches, it has advantages of less glare, significantly less overheating on sunny days, and no fading of furnishings in direct sunlight. Another advantage is its large radiant heat contributions in the evening after sunny days. Many applications of indirect gain incorporate a smaller area of direct gain (window) within the larger indirect-gain wall surface.

### (c) Isolated Gain

This is a popular passive solar approach because it provides a uniquely sunny habitable space, alternately called a *sunspace, greenhouse, sun room,* or *winter garden.* This space experiences great variations in

# PASSIVE SOLAR
## DIRECT GAIN
### TOPLIGHTING    SIDELIGHTING

**PASSIVE COOLING STRATEGIES**

VENTILATIVE

RADIATIVE/EVAPORATIVE

EARTH COUPLING

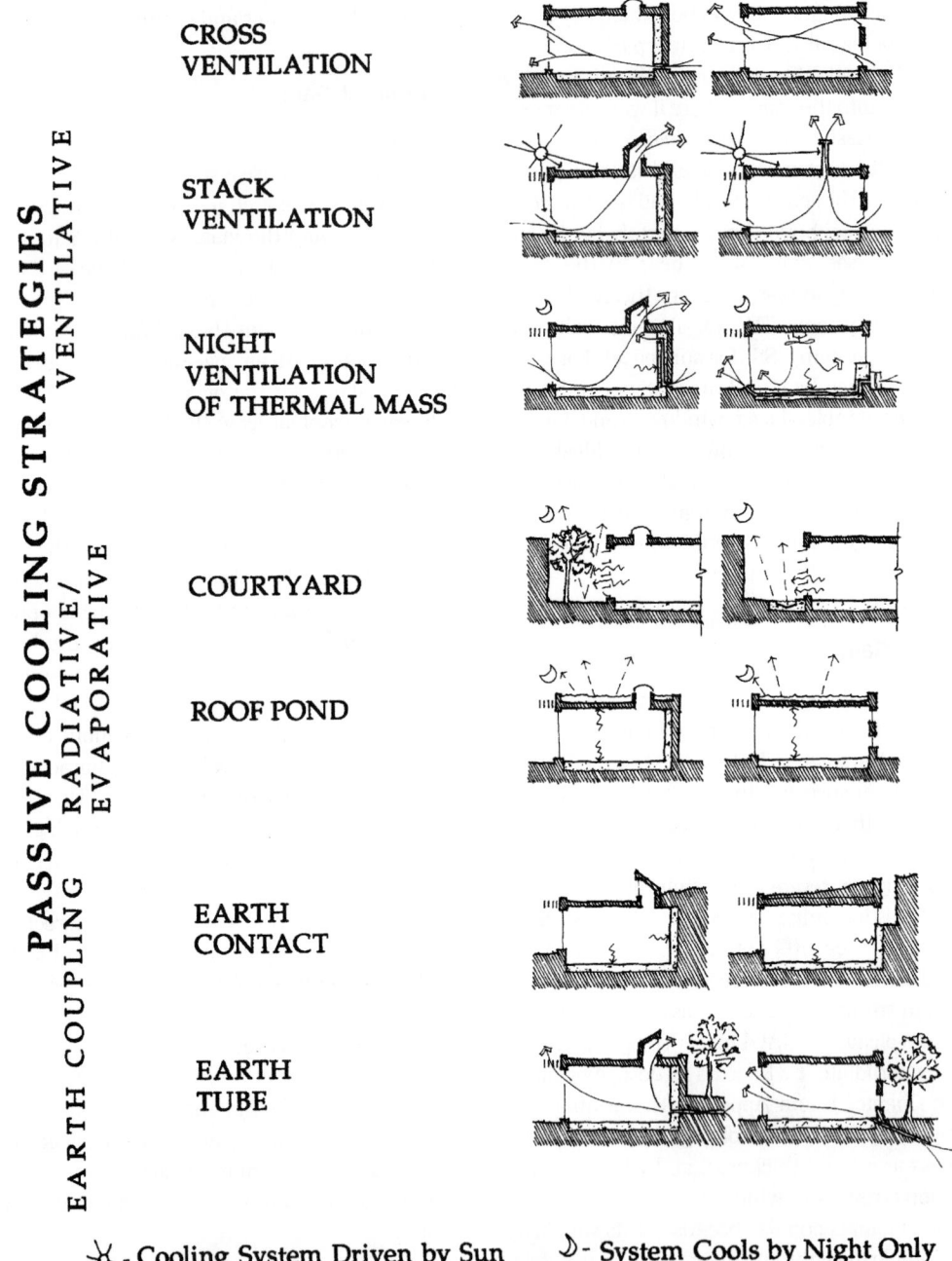

CROSS VENTILATION

STACK VENTILATION

NIGHT VENTILATION OF THERMAL MASS

COURTYARD

ROOF POND

EARTH CONTACT

EARTH TUBE

☼ - Cooling System Driven by Sun      ☾ - System Cools by Night Only

**Fig. 4.16** *Combining daylight, passive cooling, and passive solar heating opportunities. These schematic sections of simple one-cell buildings remind designers of the potential for site-based, renewable resources and how they might interact. (Sharon Shoshani and Alan Rutherford, University of Oregon class project.)*

# HEATING STRATEGIES

## ISOLATED GAIN
### TOPLIGHTING    SIDELIGHTING

## INDIRECT GAIN
### TOPLIGHTING    SIDELIGHTING

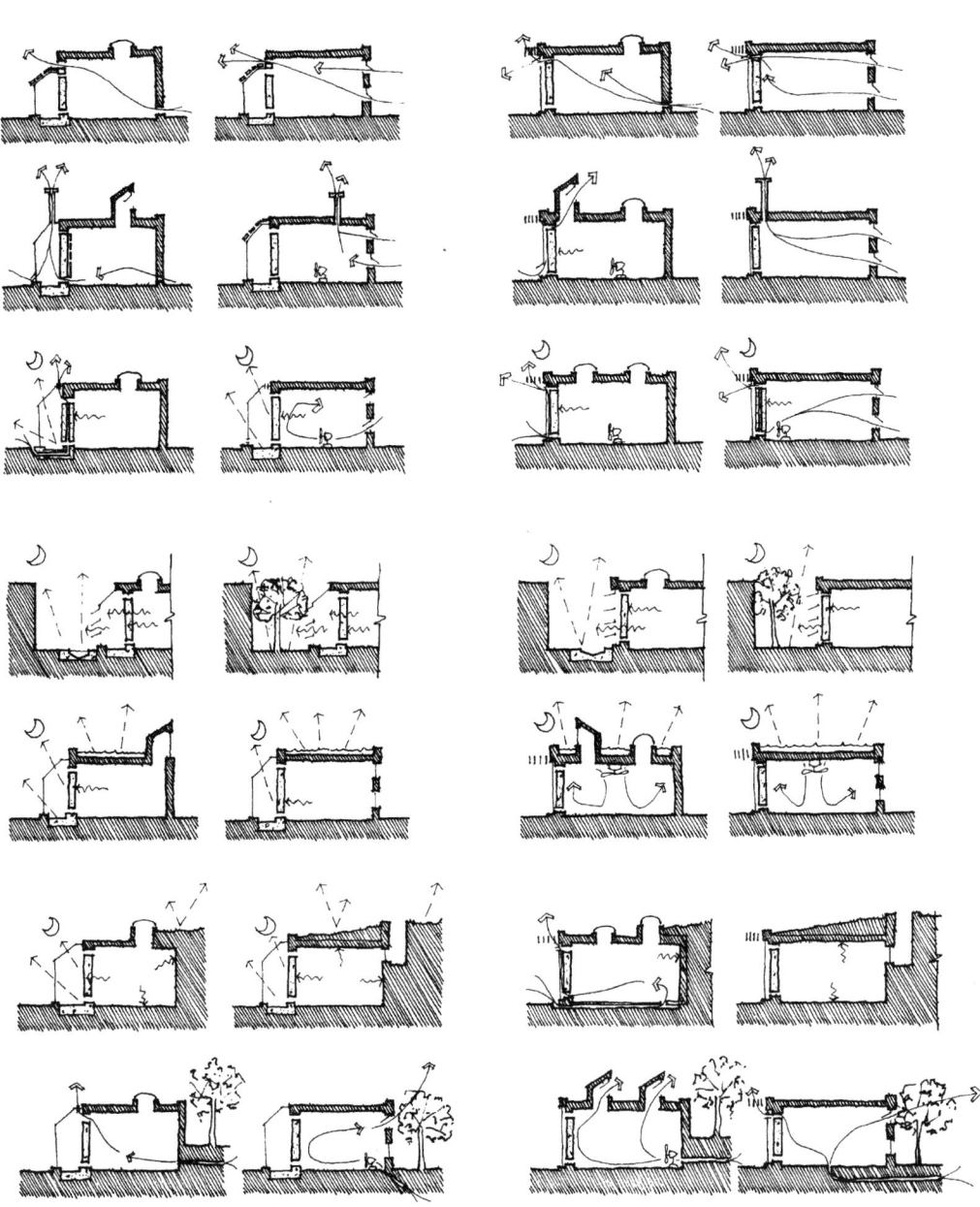

*Fig. 4.16* (Continued)

temperature—hot in the afternoon, cold before dawn—in order that the space behind it can be kept reasonably comfortable with solar heating delivered as needed. The sunspace usually has both south-facing vertical glazing and inclined glazing, increasing insolation in both winter and summer. In Chapter 8 and Appendix H, variations on isolated-gain approaches involve configurations of the glazings; whether the sunspace is surrounded by the building or is added on; whether the common wall between sunspace and building is masonry or insulated frame; whether the sunspace's end walls (facing east and west) are glazed or insulated; glazing types; and whether night insulation is used behind the glazing.

Chapter 8 presents much more detail on building insulation standards, south-facing window areas, thermal mass, and options for movable window insulation. This introduction to passive solar heating design is intended to help the designer find combinations of daylighting, passive cooling, and passive solar heating design approaches that are feasible in a given climate.

## 4.5 COMBINING STRATEGIES

Renewable sources of on-site energy for lighting, heating, and cooling allow a building to function with less need for imported energy sources. As building designs are shaped by the use of on-site sources, they become more regional in appearance. This works against the one-design-fits-all approach common to chain retailing (such as video stores or fast-food outlets), but works for the subtle integration of a building into its regional climate and culture while reducing the economic and environmental costs of imported energy.

### (a) Daylighting

This may be the most obvious and commonly used of all on-site energy sources. Two approaches are *sidelighting*, using windows in walls, and *toplighting*, using skylights in roofs. A hybrid approach is the clerestory window in very high walls; depending upon the clerestory window's relation to the floor plan, it could be considered either sidelighting or toplighting.

Sidelighting not only admits light but allows for a view of the exterior (and of the interior, especially at night). Sidelighting varies substantially by orientation and season, and direct sun can pour large quantities of heat through unshaded windows. With windows that face east and west, daylighting problems can occur due to the low sun altitudes in morning and evening that result in glare and unwanted summer heat gain.

Toplighting admits more light per unit area than does sidelighting, as skylights are directly exposed to the bright sky dome as well as direct sunlight. Toplighting less often results in glare problems because the light source is high in the space, unlike a sunny window that is in one's field of view. However, direct sun passing through a clear skylight may strike interior surfaces, with consequent glare and fading. Moreover, unshaded skylights (more than unshaded windows) are exposed to summer sun by day and to the cold winter sky. Although more efficient as admitters of light, they can represent a significant thermal penalty. Therefore, seasonal controls on skylights are even more important than on windows. Design guidelines for the sizing of windows and skylights are found in Chapter 14.

### (b) Daylighting, Cooling, and Heating

Although the relationships between these three aspects of environmental control are quite complex, in general the decision to extensively daylight a building tends to decrease its need for electricity and cooling and increase its need for heating. This is due to the influence of the large glass areas associated with daylighting. In summer, properly shaded daylighting openings contribute less heat gain than the electric lights that the daylight is replacing. This reduces the need for both electricity and cooling. In winter, some solar gain through daylight openings assists with heating (especially with south-facing windows), but the daylight still replaces electric lights that would otherwise be additional sources of heat. Thus, compared to a conventional building of the same size and function, the daylit building is likely to use more heating energy (unless passive solar heating is emphasized), less electricity for lighting, and less cooling energy.

The matrix of design strategies in Fig. 4.16 illustrates some opportunities for a simple one-space building across a range of on-site energy sources. *Ventilative* cooling strategies depend upon the air as a heat sink; *evaporative/radiative* cooling strategies depend upon both sky and air as heat sinks; *earth coupling* cooling strategies depend upon the earth as a heat sink. For a more extensive treatment of combined strategies by region for daylighting, cooling, and heating, see Lechner (2001) for 17 U.S. climate zones with bioclimatic charts of monthly conditions, vernacular design approaches, and recommended climate design priorities.

## 4.6 VISUAL AND ACOUSTICAL COMFORT

The state of being comfortable includes several environmental qualities—thermal, visual, and acoustic conditions that contribute to our health and well-being. As discussed in this chapter, thermal comfort is represented by a fairly unified theory in which the interrelationships of environmental and personal variables are defined and quantified. Visual and acoustical comfort variables are not so well developed or researched. The visual and acoustical comfort pictures are complex: if one variable changes, the impact on other variables is unclear and there are no guarantees of comfort. For example, if daylight and electric light are integrated, is glare necessarily eliminated by adequate illuminance and color rendering? Visual comfort can also mean providing a connection to the outdoors and visual stimulation through the use of view windows. Acoustic comfort in a classroom may mean that teachers and students can hear one another and external noises are prevented from reaching the classroom; or classrooms may simply have good reverberation time so that students can hear the teacher.

This chapter focused upon thermal comfort because of the highly developed state of thermal comfort information and standards. A focus on visual and acoustical comfort is no less important to building design, but the information available to the designer is not as well integrated or standardized. The designer can begin to understand the complexities of

these issues by examining issues individually and then integrating concerns via the design process.

### References

Arens, E., R. Gonzalez, L. Berglund, P. McNall, and L. Zeren. 1980. "A New Bioclimatic Chart for Passive Solar Design," in Vol. 5.2, *Proceedings of the Fifth National Passive Solar Conference,* American Section of the International Solar Energy Society, Boulder, CO, pp. 1202–1206.

ASHRAE. 1995. ANSI/ASHRAE Standard 55a-1995, addendum to ANSI/ASHRAE 55-1992, *Thermal Environmental Conditions for Human Occupancy.* American Society of Heating, Refrigerating and Air-Conditioning Engineers, Atlanta, GA.

ASHRAE. 1997. *Handbook—Fundamentals.* American Society of Heating, Refrigerating and Air-Conditioning Engineers, Atlanta, GA.

ASHRAE. 2001. *Handbook—Fundamentals.* American Society of Heating, Refrigerating and Air-Conditioning Engineers, Atlanta, GA.

ASHRAE. 2004. ANSI/ASHRAE Standard 55-2004, *Thermal Environmental Conditions for Human Occupancy.* American Society of Heating, Refrigerating and Air-Conditioning Engineers, Atlanta, GA.

Egan, M.D. 1975. *Concepts in Thermal Comfort.* Prentice-Hall, Englewood Cliffs, NJ.

Givoni, B. 1998. *Climate Considerations in Building and Urban Design.* Van Nostrand Reinhold, New York.

Heschong, L. 1979. *Thermal Delight in Architecture.* The MIT Press, Cambridge, MA.

Humphreys, M. and J.F. Nicol, 1998. "Understanding the Adaptive Approach to Thermal Comfort," in *ASHRAE Transactions,* Vol. 104, Number 1.

Kwok, A. 1998. "Thermal Comfort in Tropical Classrooms," in *ASHRAE Transactions,* Vol. 104, Number 1.

Lechner, N. 2001. *Heating, Cooling, Lighting: Design Methods for Architects* (2nd ed.). John Wiley & Sons, New York.

Milne, M. and B. Givoni, 1979. "Architectural Design Based on Climate," in *Energy Conservation Through Building Design,* D. Watson (ed.). McGraw-Hill, New York.

Olgyay, V. 1963. *Design with Climate: Bioclimatic Approach to Architectural Regionalism.* Princeton University Press, Princeton, NJ.

# Indoor Air Quality

Since the advent of modern **HVAC** systems, ventilation has generally been addressed via heating and cooling system design efforts. In centuries past, however, there were special systems to provide outside air to buildings, even at the residential scale. Banham (1969) describes both large and small buildings where outside air was deliberately introduced. In the home that Dr. John Hayward built for his family in 1867, the Octagon in Liverpool, England (Fig. 5.1), outdoor air was brought into the basement, slowed down to precipitate some particulates, then heated to help it rise throughout the four-story building. Ceiling vents just above gas lights drew "vitiated" air from each room, which was then vented to a "foul air chamber" in the attic. From there, a large shaft functioned using a combined siphon and stack effect. Powered at its low point by ever-present heat from the kitchen cooking range, it drew the foul air down, then up a very high chimney to discharge.

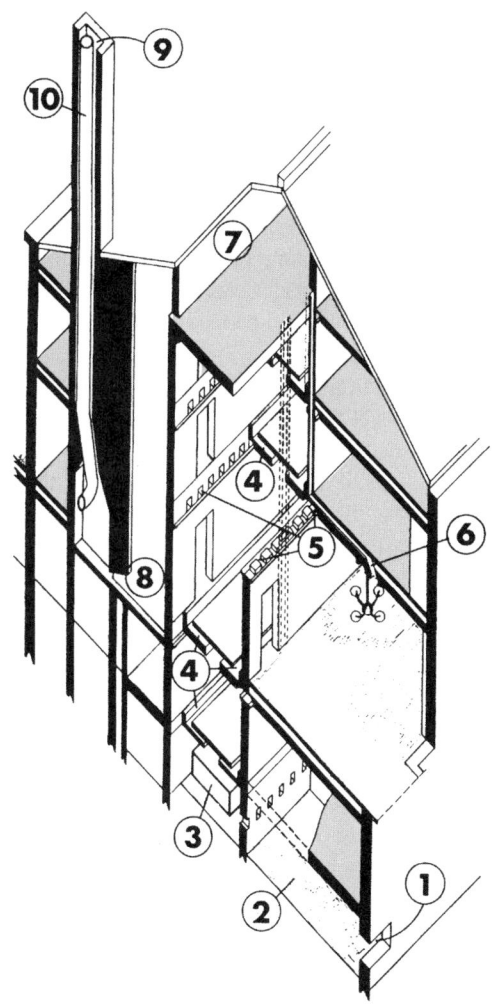

**Fig. 5.1** Fresh air intake, stale air exhaust at the Octagon, Grove Street, Liverpool, 1867. Dr. John Hayward built this exhaust system. Dirty air outside (coal-fired industries and buildings were common) and inside (gas lamps) probably produced an IAQ considerably worse than today's norms. 1. Fresh air intake; 2. Settling chamber in the basement; 3. Heating coils; 4. Air passages in lobby floors; 5. Air passages in the cornice; 6. Extract above a gas lamp; 7. Foul air chamber; 8. Foul air down a duct; 9. Foul air chimney; 10. Flue from a kitchen range. (Reprinted by permission from Banham, Reyner. 1969. The Architecture of the Well-Tempered Environment, *The Architectural Press, London.*)

## 5.1 INDOOR AIR QUALITY AND BUILDING DESIGN

Several trends have combined to bring indoor air quality (IAQ) concerns back into prominence. First, an increasingly large percentage of people's time is now spent indoors, and in more tightly controlled environments, as a service-based economy overtakes a manufacturing-based one. Second, the oil embargo of 1973 raised the world's consciousness regarding finite energy sources, producing a sudden and powerful rush toward energy-conserving designs. This, in turn, encouraged designers to limit the introduction of outdoor air that required cooling in summer and heating in winter. Third, a proliferation of chemicals in our environment has produced a vast array of potential air pollutants—from synthetic products permanently installed within buildings, from equipment used indoors, and from cleaning fluids used in maintenance. With more time spent in less fresh air and surrounded by more pollution sources, increasing numbers of buildings have seen experiences with *sick building syndrome* (SBS). SBS (by one definition) is a situation wherein more than 20% of the occupants complain of symptoms associated with SBS—such as headaches, upper respiratory irritation, and irritations of the eyes, among others. If these symptoms disappear after occupants leave the workplace (weekends are especially good periods of contrast), SBS is strongly indicated.

The building designer has an elusive task when air quality is at issue, because so little can be accurately predicted. Heat flow rates, occupancy schedules, and typical weather patterns can be combined to estimate with some confidence how much energy will be consumed by a building; construction types can then be altered in the design stage to yield predictably different results. Designers have few tables that provide rates of outgassing for various materials at given temperatures and no well-established data on design conditions for the quality of local or regional outdoor (fresh) air, even though there is readily available and reliable information for air temperature and humidity.

Controlling the quality of indoor air may be as important to building occupants as controlling for thermal or acoustical comfort. Designers know that saving energy for heating and cooling lowers the cost of maintaining a building; employers know that lost productivity from either on-the-job illness or sick leave can result in much greater costs. One estimate for a large office building compared the cost of increasing ventilation and improving air filtration to the value of projected health and productivity benefits. Initial improvements yielded estimated benefit-to-cost ratios of 50 to 1 (increased ventilation) and 20 to 1 (improved filtration). Acceptable thermal conditions and acceptable IAQ are *not* synonymous. Buildings that are thermally comfortable can still cause SBS when pollutants are sufficiently numerous.

ASHRAE (2004) has defined acceptable indoor air quality as: "air in which there are no known contaminants at harmful concentrations as determined by cognizant authorities and with which a substantial majority (80% or more) of the people exposed do not express dissatisfaction." There are two components to this definition— a comfort response and a health benchmark. Providing acceptable IAQ depends upon four major considerations, three of which depend largely on the designer:

1. Limiting pollution at the source (choosing materials and equipment carefully)
2. Isolating unavoidable sources of pollution
3. Providing for an adequate supply and filtering of fresh air (and recirculated air)
4. Maintaining a building and its equipment in a clean condition

Dealing with indoor pollution at its source, by choosing materials and equipment with care whenever possible, rather than increasing outdoor air flow rates (and related energy consumption) is the most rational IAQ strategy. The designer can further provide for improved IAQ by carefully locating a building on its site; zoning to isolate pollutant sources; and providing clean, adequate, and well-distributed outdoor air, air-cleaning devices, and building commissioning. Providing a *flush mode* following completion of construction might be considered. It is up to a building's managers to maintain IAQ by means of a regular equipment maintenance program, regular interior cleaning, and a careful selection of cleaning agents. Furthermore, a flush of the building after every unoccupied weekend or holiday period is helpful in removing accumulated pollutants from finishes and furnishings.

## 5.2 POLLUTANT SOURCES AND IMPACTS

Indoor air pollution can be described both in terms of the types of contaminants (gaseous, organic, or particulate) and the types of effects (odors, irritants, toxic substances) involved. People not only inhale contaminants, but also absorb and ingest some—the nose is not the only pollutant receptor/sensor. Table 5.1 summarizes some common indoor air pollutants, their effects, and simple strategies to ameliorate them. For some contaminants, the only method of avoidance is to design for their exclusion; equipment will not remove them, although increased ventilation can reduce their impact. Examples are asbestos, radon, and pesticides.

### (a) Odors

One of the most immediate indicators of IAQ problems is odor. People are sensitive to odors over an extraordinary range, whereas equipment to detect and classify odors is woefully lacking. Odors are perceived most strongly on initial encounter; then "fatigue" occurs and perception fades. Thus, visitors are more likely to detect odors than are the long-term inhabitants of a space. Odors may be simply unpleasant, with psychological consequences, or may be indicators of a more serious IAQ problem with physiological consequences. When an unfamiliar odor is detected, our reactions are positive, neutral, or negative, depending upon whether we perceive a threat (or enjoyment) through the odor.

Sometimes odors are directly traceable to a source, but in office environments odors are usually more complex. A typical office environment odor blend may include body odors, grooming products (perfumes, colognes), copy machines, food products, cleaning fluids, and outgassing from materials. More rarely, in a decreasing number of locales, tobacco smoke may also be present. This complexity produces an interesting reaction; people tend to become less sensitive to each of the component odors, with a resulting overall masking. However, an architecturally imposed masking approach—the deliberate introduction of a "perfume" to cover offending odors—is rarely successful. (Sweet-and-sour may work nicely for the palate, but olfactorily it can be uniquely nauseating.) Conversely, as the indoor environment is freed of multiple odors, people become more sensitive to the one or two odors that remain.

Often a simple measurement of carbon dioxide ($CO_2$) concentration is used as a first indicator (a surrogate) of potential IAQ problems related to occupancy, because the $CO_2$ concentration indoors is generally proportional to the human concentration.

Filtering odors from indoor air is usually accomplished with electronic or activated charcoal filters, described in Section 5.6.

### (b) Irritants

Unlike odors, which are immediately perceived and fade with prolonged exposure, irritants are often imperceptible at first but cause increasing distress over time. Symptoms of irritants include itching or burning eyes, sneezing, coughing, dry nose and throat, sore throat, and tightness of the chest. Most irritants are present in the form of particles and gas dispersoids (Table 5.2).

Sources of irritants typically include the building itself and the equipment and occupants within. New and newly renovated buildings are particularly prone to problems from outgassing of paints, adhesives, sealants, office furniture, carpeting, and vinyl wall coverings. *Volatile organic compounds* (VOCs) are chemicals containing carbon molecules that are volatile; that is, they off-gas or evaporate from material surfaces at room temperatures. The VOC list is long: methane, ethane, methylene chloride, trichloroethane, chlorofluorocarbons (CFCs), hydrochlorofluorocarbons (HCFCs), hydrofluorocarbons (HFCs), formaldehyde, and hydrocarbons such as styrene, benzene, and alcohols. All are now found frequently in new buildings.

Long-term occupancy brings other irritants. Ozone, valuable in the upper atmosphere but a smog component below, is produced by copy machines, high-voltage electrical equipment, and—ironically—electrostatic air cleaners. Mineral fibrous particles can be produced by the breakdown of duct liner/insulation and fireproofing. Hydrocarbon compounds come from copy machines and copy papers. Tobacco smoke is a mixture of gases and fine particles especially irritating to many individuals. Low humidity can exacerbate problems with irritants, producing symptoms similar to those from chemicals. Carpet shampooing yields organic solvents and ammonia; nighttime cleaning coinciding with reduced or nonexistent ventilation is especially problematic. In contrast, night

## TABLE 5.1 Common Air Pollutants

| Pollutant | Sources | Effects | Control Strategies |
|---|---|---|---|
| Excess moisture[a] | Cooking (heating open liquids), washing, exhaling | Increases growth of fungi, bacteria, and dust mites | Exhaust ventilation at source; dehumidification |
| Carbon dioxide ($CO_2$) | Human respiration | Minor discomfort at high concentrations; "stuffiness" | $CO_2$ is a good indicator of the ventilation rate in tightly enclosed spaces or where occupancy is high |
| Carbon monoxide (CO) | Incomplete combustion: furnaces, stoves, fireplaces; motor vehicle exhaust | Headaches, dizziness, sleepiness, muscle weakness, potentially lethal | Sealed combustion burners, adequate combustion air, safe exhaust flues |
| Nitrogen oxides | High-temperature combustion | Irritation, possible immune suppression | Safe exhaust flues, sealed combustion burners |
| Sulfur oxides | Combustion fuels containing sulfur (oil, coal) | Potential irritant, burning eyes, reduces lung function | Alternative fuels, safe exhaust flues, sealed combustion burners |
| Polynuclear aromatic hydrocarbons | Smoking, combustion of wood or coal, barbecuing, burnt food | Irritants and carcinogens | Prohibit smoking, lower-temperature in cooking, use clean fuels, burn wood in enclosed firebox with adequate oxygen supply |
| Ozone | Laser printers, photocopiers, small motors, electronic air cleaners | Inflammation of bronchia, wheezing and shortness of breath, dizziness, asthma attacks | Remove sources or exhaust at source, maintain electronic air cleaners |
| Volatile organic compounds (VOCs) Formaldehyde | Particle board, interior laminated panels, glues, fabric treatments, paints | Burning eyes and nose, skin rash, shortness of breath, headaches, nausea, dizziness, fatigue | Use alternative materials, seal particle board if used, ventilate |
| Others | Paints, solvents, carpets, soft plastics, adhesives, caulkings, softwoods, paper products, cleaning and maintenance products | Intoxication, burning eyes and nose, shortness of breath, headaches, nausea, dizziness, loss of judgment, panic | Use alternative materials, age materials before installing, ventilate |
| Lead | Pre-1970s paint, pre-1985 pipes and solder, dust and soil near roads (residue from leaded gas) | Neurotoxic, especially if ingested by young children; learning disabilities, nausea, trembling, numbness of extremities | Identify and remove or seal old paint, replace pipes and solder, avoid foods grown by roadside |
| Pesticide residues[b] | Treated basements and foundations, treated ceiling and wall cavities, treated cabinets and closets, treated soil outside foundation | Neurotoxic or long-term risk of liver, kidney, and other diseases, including cancers | Identification and removal by expert if history known, sealing in pesticide if possible |
| Asbestos fiber | Pre-1975 steam pipe and duct insulation, furnace and furnace parts, pre-1980 reinforced vinyl floor tile, and fiber cement shingles and siding | Long-term cancer risk from inhaling fibers | Leave material undisturbed, get expert identification and removal if required, seal with special sealant and cover with sheet metal if not crumbling |
| Mineral and glass fiber | Thermal insulation, pipe insulation, fire-resistant acoustic tile and fabrics | Potential irritant, burning eyes, itching skin, long-term risk of lung damage and cancer | Handle only with respirator and gloves, seal and enclose, do not disturb in place |
| Fungus particles, dust mites | Grow in basements, damp carpet, bedding, fabrics, walls and ceilings, closets | Very allergenic, burning eyes and nose, sneezing, skin rash, congestion, and shortness of breath | Keep surfaces dry and clean, cover bedding and upholstery with barrier cloth, ventilate, use borax treatments to retard fungus |
| Hazardous bacteria (e.g., *Legionella*) | Standing warm water, untreated hot tubs, air-conditioning drain pans, humidifier reservoirs | Severe respiratory illness, potentially lethal | Prevent standing water, clean and treat tubs and reservoirs |
| Radon gas | Natural radioactivity in soils | Increased lifetime lung cancer risk | Seal foundation and floor drains, ventilate subsoil |
| Methane and other soil gases | Decomposing garbage in landfills, leaking sewage lines, toxic waste | Possibly explosive or toxic, nuisance odors | Know site history before building, remove soil if necessary, seal foundation and floor drains, ventilate subsoil |

*Source:* Adapted from Rousseau and Wasley (1997).

[a]Too little moisture also adversely affects health.

[b]Pre-1980s treatments are more likely to leave residues and be toxic.

**TABLE 5.2 Characteristics of Particles and Particle Dispersoids**

| Particle Diameter, microns (μ) | | | | | | | |
|---|---|---|---|---|---|---|---|
| 0.0001 | 0.001 (1mμ) | 0.01 | 0.1 | 1 | 10 | 100 | 1,000 (1mm) | 10,000 (1cm) |

**Electromagnetic Waves:** X-Rays — Ultraviolet — Visible / Solar Radiation / Near Infrared — Far Infrared — Microwaves (Radar, etc.)

**Technical Definitions**
- Gas Dispersoids — Solid: Fume — Dust
- Liquid: Mist — Spray
- Soil: Atterberg or International Std. Classification System adopted by Internat. Soc. Soil Sci. Since 1934 — Clay — Silt — Fine Sand — Coarse Sand — Gravel

**Common Atmospheric Dispersoids:** Smog — Clouds and Fog — Mist — Drizzle — Rain

**Typical Particles and Gas Dispersoids:**
Rosin Smoke, Oil Smokes, Tobacco Smoke, Metallurgical Dusts and Fumes, Ammonium Chloride Fume, Sulfuric Concentrator Mist, Contact Sulfuric Mist, Paint Pigments, Zinc Oxide Fume, Colloidal Silica, Carbon Black, Aitken Nuclei, Atmospheric Dust, Sea Salt Nuclei, Combustion Nuclei, Viruses, Insecticide Dusts, Ground Talc, Spray Dried Milk, Alkali Fume, Milled Flour, Fertilizer, Ground Limestone, Fly Ash, Coal Dust, Cement Dust, Pulverized Coal, Flotation Ores, Plant Spores, Pollens, Beach Sand, Lung Damaging Dust, Nebulizer Drops, Pneumatic Nozzle Drops, Hydraulic Nozzle Drops, Red Blood Cell Diameter (Adults): 7.5μ ± 0.3μ, Bacteria, Human Hair

Gas Molecules (molecular diameters calculated from viscosity data at 0°C): O₂, CO₂, Cl₂, SO₂, C₂H₆, C₄H₁₀, F₂, HCl, CH₄, H₂O, N₂, H₂, CO

**Types of Gas Cleaning Equipment:**
Ultrasonics (very limited industrial application), High Efficiency Air Filters, Thermal Precipitation (used only for sampling), Electrical Precipitators, Cloth Collectors, Packed Beds, Common Air Filters, Liquid Scrubbers, Centrifugal Separators, Impingement Separators, Mechanical Separators, Settling Chambers

PREPARED BY C E LAPPLE

Source: Reprinted with permission of SRI International (formerly the Stanford Research Institute), Menlo Park, CA.

maintenance with *increased* ventilation rates—as with cooling by night ventilation of thermal mass—can reduce this threat.

As with odors, the impacts of irritants can be reduced with an increased outdoor air supply. Filters for the removal of irritants usually consist of particulate filters; less common are gaseous-removal filters, air washers, and electronic air cleaners. These are discussed in Section 5.6.

### (c) Toxic Particulate Substances

At the top of this list is asbestos, widely used in buildings until its toxicity was realized in the 1970s. Asbestos in tightly bound form is encountered as asbestos-cement and in vinyl-asbestos floor tiles, and in loosely bound form as sprayed-on asbestos insulation. The latter is particularly dangerous, readily releasing toxic asbestos fibers over the life of the material. With asbestos, neither increased ventilation nor filtering is acceptable; it must be either removed under stringent isolation controls or sealed and left in place.

Some of the respirable particles (see Table 5.2) that result from incomplete combustion are toxic. Incomplete combustion can occur from smoking, in woodstoves, fireplaces, gas ranges, and unvented gas or kerosene space heaters. Lacking control of combustion at its source, the remedies are to isolate the source insofar as possible, exhaust air from the immediate vicinity, increase the outdoor air supply to the area, and utilize particle filtering.

### (d) Biological Contaminants

Because living things inhabit both buildings and outdoor air, there will be biological contaminants such as bacteria, fungi, viruses, algae, insect parts, and dust within buildings. Moisture encourages both the retention and growth of these contaminants; standing water (which may occur in HVAC system components) and moist interior surfaces are likely trouble sites. Allergic reactions and infectious and noninfectious diseases can result. Outbreaks of Legionnaire's disease have occurred when improperly maintained HVAC systems incubated and then distributed disease-causing microorganisms. Now residential humidifiers, dehumidifiers, and air-conditioner drain pans are suspect.

Remedies for biological contaminants begin with good design and end with vigilant maintenance. Although exposure to ultraviolet radiation is sometimes used as a control strategy, filters are rarely an effective solution for these contaminants.

### (e) Radon and Soil Gases

Radon is a radioactive gas that decays rapidly, releasing radiation at each stage. It is colorless and odorless and thus is undetectable by humans. If we inhale radon, radiation release in the lungs can cause lung cancer. Other soil gases include methane (usually odiferous) and some pesticides that can volatize and enter buildings with soil gases. Effects on humans are not likely to be beneficial.

In many buildings with high levels of radon, the problem has been traced to exposure to soil. Radon penetrates through floor and wall cracks and openings around plumbing pipes; thus, below-ground spaces are particularly at risk. Penetrations of below-grade walls and floors should be both minimized and well sealed; under-slab ventilation (Section 5.5c) may be appropriate, especially in areas of high radon risk.

## 5.3 PREDICTING INDOOR AIR QUALITY

Assuming that pollutant sources have been minimized, designers essentially need to know how much outdoor air and what extent of filtering will produce acceptable indoor air quality. These questions are difficult to answer.

### (a) Ventilation Rate

The most common remedy for SBS (after controlling pollution sources) is to increase the rate of outdoor air ventilation. Recommended rates of ventilation are found in Tables E.25 (nonresidential) and E.26 (residential). Although very small amounts of outdoor air will provide sufficient oxygen, and although human body odor control is usually achievable at a rate of from 6 to 9 cfm (cubic feet per minute) (3 to 4.5 L/s) of outdoor air per occupant, outdoor air has more to do than provide oxygen and control odors. Defining minimum outdoor air supply rates has proven to be a controversial task. The current ASHRAE ventilation standard (Standard 62.1-2004—for other than low-rise residential buildings) establishes minimum rates (Table E.25) on the basis of an occupancy component and a building component in recognition of these distinct contaminant sources. Some feel that the outdoor air requirements are too high, others that they are too low.

Two units have been proposed to integrate the various indoor air pollutants in the same way that they are perceived by human beings. The *olf* is a unit of pollution (1 olf = the bioeffluents produced by the average person); the *decipol* is a unit of perceived air quality. These are related in this proposed comfort formula:

$$Q = 10 \frac{G}{C_i - C_o}$$

where

$Q$ = ventilation rate, L/s

$G$ = total pollution sources, olf

$C_i$ = perceived indoor air quality, decipol

$C_o$ = perceived outdoor air quality, decipol

At present, $C_i$ is recommended to be set at 1.4 decipol, which represents an expectation of 80% of occupants satisfied with IAQ. $C_o$ and $G$ may be roughly estimated from Table 5.3. These proposals are discussed in more detail in Fanger (1989).

The concept of *replacance* affects the design of ventilation systems. Table 5.4 shows that at the rate of 1 air change per hour (ACH) of outdoor air, an indoor space would have only 63% "new" air after 1 hour; about 8 hours at this rate is required for all the "old" air to be exhausted. There is, then, a difference between the fresh air input rate (ACH) and the

replacance—the fraction of air molecules at one specified time that was *not* in the indoor space at an earlier reference time. This relationship, along with details of air pollutants (and of heat exchanger design, for energy conservation), is thoroughly discussed in Shurcliff (1981).

The new campus for the Environmental Protection Agency (EPA) in Research Triangle Park, North

**TABLE 5.3 Estimating Indoor Air Quality**

| PART A. PERCEIVED OUTDOOR AIR QUALITY, $C_o$ | |
|---|---|
| During smog episodes | >1 decipol |
| In cities with moderate air pollution | 0.05–0.3 decipol |
| On mountains or at sea | 0.01 decipol |

| PART B. ESTIMATED OLF LOADS IN OFFICES PER M² FLOOR AREA | |
|---|---|
| **Pollution Source** | **olf/m²** |
| Occupants (10 m² per person) | |
|   Bioeffluents | 0.1 |
|   Additional load from 20% smokers | 0.1 |
|   40% | 0.2 |
|   60% | 0.3 |
| Materials and ventilation system | |
|   Average in existing buildings[a] | 0.4 |
|   Low-olf buildings[b] | 0.1 |
| Total load in office buildings | |
|   Average in existing buildings (40% smokers) | 0.7 |
|   Low-olf buildings (nonsmoking) | 0.2 |

*Source:* Fanger (1989). Reprinted by permission from the *ASHRAE Journal,* © 1989 by the American Society of Heating, Refrigerating and Air-Conditioning Engineers, Inc., Atlanta, GA.

[a]Based upon field studies of 15 randomly selected buildings in Copenhagen.

[b]Designed to contain low-outgassing materials and a frequently maintained ventilating system.

**TABLE 5.4 Air Replacance Compared to Input Air Changes per Hour (ACH)**

*Note:* Mixing is considered continuous and vigorous—as would be obtained in a forced-ventilation system.

| Time from Start of Run (h) | Rate of Air Input (ACH): | Replacance (%) | | | | | | | | |
|---|---|---|---|---|---|---|---|---|---|---|
| | | 0.06 | 0.12 | 0.25 | 0.5 | 1 | 2 | 4 | 8 | 16 |
| 1/16 | | 0.4 | 0.8 | 1.6 | 3.1 | 6.1 | 11.7 | 22.1 | 39.3 | 63.2 |
| 1/8 | | 0.8 | 1.6 | 3.1 | 6.1 | 11.7 | 22.1 | 39.3 | 63.2 | 86.5 |
| 1/4 | | 1.6 | 3.1 | 6.1 | 11.7 | 22.1 | 39.3 | 63.2 | 86.5 | 98.2 |
| 1/2 | | 3.1 | 6.1 | 11.7 | 22.1 | 39.3 | 63.2 | 86.5 | 98.2 | 99.9 |
| 1 | | 6.1 | 11.7 | 22.1 | 39.3 | 63.2 | 86.5 | 98.2 | 99.9 | 100 |
| 2 | | 11.7 | 22.1 | 39.3 | 63.2 | 86.5 | 98.2 | 99.9 | 100 | 100 |
| 4 | | 22.1 | 39.3 | 63.2 | 86.5 | 90.2 | 99.9 | 100 | 100 | 100 |
| 8 | | 39.3 | 63.2 | 86.5 | 98.2 | 99.9 | 100 | 100 | 100 | 100 |
| 16 | | 63.2 | 86.5 | 98.2 | 99.9 | 100 | 100 | 100 | 100 | 100 |

*Source:* Shurcliff (1981). Reprinted by permission.

Carolina, was designed with special emphasis on IAQ (Fig. 5.2). The designers considered several alternatives for fresh air provision, deciding that a simple variable air volume (VAV) system, set at a minimum of 3 ACH (of combined fresh and recycled air) would be acceptable. If the system were designed with a typical minimum, 1 ACH would have resulted during periods when neither heating nor cooling was required (typical spring and fall conditions). However, a 6 ACH alternative would have had dramatically increased energy consumption.

### (b) Testing

When a client is especially interested in IAQ, full-scale time tests can be used. At the new EPA campus (Fig. 5.2), the contractor was given a target for allowable contaminant concentrations (Table 5.5, Part B). Any material assembly deemed likely to contribute more than one-third of these allowable concentrations, and used in large quantities, was to be tested before acceptance. One desirable outcome of such testing and associated materials specifications is the possible avoidance (or shortening) of an anticipated flush-out of a completed building before occupancy.

### 5.4 ZONING FOR IAQ

After pollution control is implemented at the source (cleaner equipment, prohibiting smoking, careful material choices, etc.), remaining unavoidable pollutant sources should be identified. Then more sensitive building areas should be isolated from the key contaminators. This is sometimes difficult, as in "open offices" where walls are unwelcome but copying machines are essential. In such cases, the method is to erect as much of a barrier as possible around an offender, then *task ventilate* to remove the contaminated air immediately. Sometimes air pollution sources also produce unwanted sound, in which case the argument for a more complete barrier may become more compelling.

Many health care and laboratory buildings have "clean" and "dirty" zones, and even separate circulation pathways. Differential air pressures are often maintained to discourage air flow from dirty to clean zones—with higher pressure in clean areas, lower pressure in dirty areas. Lower-pressure areas can be created simply by exhaust air from such spaces, as well as by limiting the volume of supply air. Higher-pressure areas can be created by installing make-up air equipment, as well as increasing the volume of supply air from the HVAC system.

On a site-planning scale, try to locate air intakes upwind from pollution sources. Because winds frequently change direction, this may be more a matter of adequate separation distance than direction. The most obvious example is a major air intake for a central HVAC system, which should be as far as possible from parking areas, delivery docks, and streets—and from the exhaust outlets from that same HVAC

**Fig. 5.2** *The new campus of the U.S. Environmental Protection Agency at Research Triangle Park, North Carolina, features concentrated parking separate from the buildings. High exhaust stacks disperse air from laboratories; intakes are kept well away from exhaust. (Courtesy of Hellmuth, Obata + Kassabaum, Washington, DC.)*

**TABLE 5.5 Air Quality Standards**

| PART A. **NATIONAL AMBIENT-AIR QUALITY STANDARDS FOR OUTDOOR AIR** | | | | | | |
|---|---|---|---|---|---|---|
| | Long-Term Concentration Averaging | | | Short-Term Concentration Averaging | | |
| Contaminant | µg/m³ | ppm | | µg/m³ | ppm | |
| Sulfur dioxide | 80 | 0.03 | 1 year | 365 | 0.14 | 24 hours |
| Total particulate | 75[a] | — | 1 year | 260 | — | 24 hours |
| Carbon monoxide | | | | 40,000 | 35 | 1 hour |
| Carbon monoxide | | | | 10,000 | 9 | 8 hours |
| Oxidants (Ozone) | | | | 235[b] | 0.12[b] | 1 hour |
| Nitrogen dioxide | 100 | 0.055 | 1 year | | | |
| Lead | 1.5 | — | 3 months[c] | | | |

*Source:* U.S. Environmental Protection Agency.

| PART B. **MAXIMUM INDOOR AIR CONCENTRATION STANDARDS AT EPA CAMPUS** | |
|---|---|
| **Indoor Contaminants** | **Allowable Air Concentration Levels[d]** |
| Carbon monoxide (CO) | <9 ppm |
| Carbon dioxide (CO₂) | <800 ppm |
| Airborne mold and mildew | Simultaneous indoor and outdoor readings |
| Formaldehyde | <20 µg/m³ above outside air |
| Total VOC | <200 µg/m³ above outside air |
| 4 Phenyl cyclohexene (4-PC)[e] | <3 µg/m³ |
| Total particulates | <20 µg/m³ |
| Regulated pollutants | <National Ambient-Air Quality Standards (see Part A) |
| Other pollutants | <5% of TLV-TWA[f] |

*Source:* Hellmuth, Obata + Kassabaum (2001).

[a]Arithmetic mean.

[b]The standard is attained when the expected number of days per calendar year with maximum hourly average concentrations above 0.12 ppm (235 µg/m³) is equal to or less than 1.

[c]A 3-month period is a calendar quarter.

[d]These levels must be achieved prior to acceptance of building. They do not account for contributions from office furniture, occupants, and occupant activity.

[e]4-PC is an odorous contaminant constituent in carpets with styrene-butadiene-latex rubber (SBR).

[f]TLV-TWA is threshold limit value–time-weighted average.

system or outlets from other building systems. Even exhaust outlets should be located carefully, because there is a possibility that at times outdoor air can be drawn into these exhaust openings. A mechanical equipment room is the typical location for both intake and exhaust; energy conservation devices such as heat exchangers benefit from close proximity of intake and exhaust. Most animals use the same "ducts" to breathe in and exhale, obviously inviting air reentrainment. For a building, however, separation of these openings is prudent design.

The new campus for the EPA (Fig. 5.2) is an example of predesign planning for IAQ. This is a 1,000,000 ft² (92,900 m²) building complex serving a population of more than 2,000 on 133 acres (54 hectares) of farm land that has reverted to second-growth hardwoods. Table 5.6 gives a sum-

mary of design decisions and their impact on IAQ. Some of the more visible design consequences are the separation of parking and building, the concentration of parking in a structure (less impact on the existing landscape, more control of vehicle fumes), and the height of the exhaust stacks from the laboratories. Many other design decisions are hidden within the building's materials and HVAC system, as detailed in Table 5.6.

The topic of zoning includes decisions about local versus central equipment. Should individual exhaust fans be installed (creating selective lower-pressure areas) or a central exhaust fan (that can discharge up a very tall stack)? Should air cleaners be installed locally, where they can be selected according to the degree of pollution, or centrally, where they can be more easily and regularly maintained? What

## TABLE 5.6 Design Decisions and Impacts: EPA Campus, Research Triangle Park

| Item | Decision | Impact on IAQ |
|---|---|---|
| Siting of building | Locate exhaust downwind from air intakes, separate by >100 ft (30 m)<br>Maximize separation between parking areas and air intakes | Minimizes reentrainment of laboratory exhaust air at air intakes<br>Reduces the potential for vehicular exhaust entering building |
| Location of parking garages | Locate parking structure away from the building | Reduces the potential for vehicular exhaust entering building |
| Laboratory exhaust stacks | Increase stack height to 30 ft (9 m) based on wind tunnel testing | Minimizes reentrainment of laboratory exhausts into air intakes |
| Radon | Site-specific testing confirmed low levels of radon | Confirmed that radon levels are safe |
| Delivery/loading zone | Maintain negative pressure in loading area, positive pressure in building | Eliminates entrainment of delivery vehicle exhaust |
| Landscaping | Low maintenance and nonsporulating plants selected<br><br>Plants used as a barrier to vehicle exhaust | Intake of spores, fertilizer, or chemicals entering building is reduced<br>Minimizes entrainment of delivery vehicle exhaust |
| Laboratory fume hoods | Install flow gauges and alarms | Provides warning of air contaminants present in laboratory areas due to loss of flow |
| Acoustic insulation of ducts | Ductwork increased in size to reduce need for acoustical insulation; in select areas, mylar-coated duct silencers are used as ductwork transitions out of equipment rooms | Minimizes release of fibers into the airstream and possible contamination of the HVAC system (duct liners are difficult to monitor or to clean and can be sites of microbial growth) |
| Moisture accumulation | Install drain pans pitched toward drain pipe | Reduces moisture, which could result in introduction of bacterial contamination into HVAC system |
| Humidity control | No moisture carryover into system | Minimizes moisture in HVAC system and resultant bacterial contaminants due to moisture |
| Corrosion inhibitors | Inhibitors do not contain volatile amines | Eliminates exposure to certain air contaminants |
| System maintenance | Provide access panels at ductwork appurtenances and ample clearance around equipment | Maximizes ease of maintaining HVAC system |
| Outside ventilation rate | 100% outdoor air in laboratories; 20 cfm (10 L/s) per person in offices<br>Flexibility to increase ventilation rate 20% for unexpected sources | Maximizes occupant comfort and removal of air contaminants<br>Minimizes possible occupant exposure to contaminants |
| Airflow efficiency | Minimum airflow rate set at 3 ACH for VAV system in office areas<br>Flexible connections to room diffusers in open office areas | Increases air movement and ventilation effectiveness<br>Facilitates modifications to enhance airflow as necessary |
| Air cleaning | ASHRAE 30% efficiency pre-filters with 85% final filters<br>Flexibility to add scrubbers to laboratory exhaust<br>Use bird-proofing mesh screen | Minimizes dust and other aerosols entering indoor air via the HVAC system<br>Minimizes release of contaminants to ambient air<br>Eliminates bird droppings and possible microorganism infection in the HVAC system |
| Thermal control | Building Automated Control system<br>Fixed windows | Provides optimum control of temperature and pressure<br>Prevents unconditioned air from entering building; maintains positive pressure in laboratories |
| Exhaust system | 100% exhaust for photocopying rooms, laboratories, food preparation areas<br>Photocopiers located within 10 ft (3 m) of exhaust vent | Eliminates potential to recirculate contaminants and odors through the building via the HVAC system<br>Control potential source of air contaminants; recirculated air is filtered prior to its return |
| Smoking | Designate building as nonsmoking | Eliminates exposure to secondhand smoke and recirculation of tobacco smoke via the HVAC system |
| Building materials, finishes, furnishings | Materials selected to minimize release of contaminants from products | Minimize occupant exposure to contaminants as a result of off-gassing from building materials, finishes, furnishings |

*Source:* Hellmuth, Obata + Kassabaum, Architects, Washington, DC. This material was produced with U.S. government sponsorship through Order #70-2124-NTLX.

about heat exchangers for tempering incoming air: many smaller ones or one large one? The larger and more complex a building is, the more likely the development of a mix of local, specialized zones and a large, more general zone that is centrally served. Figure 5.3 explores the issue of the location of an office copier: at the edge, where task ventilation is easy but plentiful daylight and a view may be "wasted" on this function, or away from the edge, where a central exhaust system is more likely to be utilized.

## 5.5 PASSIVE AND LOW-ENERGY APPROACHES TO VENTILATION

This section deals with ventilation, an approach that assumes that "the solution to pollution is dilution." Another major approach, air cleaning, almost always involves forcing air through various filtering devices. Equipment for both ventilation and air cleaning is discussed in a following section.

### (a) Windows

Operable windows are one of the oldest and most common "switches" of all. Passive ventilation through windows and skylights is influenced by the position of the open window; if wind strikes the glass surface in its open position, it will be deflected. The direction of the wind approaching the window is generally unpredictable. Also, whereas for simple ventilation (without cooling) it is usually desirable to keep wind *away* from people, for cooling at temperatures above the standard comfort zone, wind *across* the body is helpful (see Fig. 3.30). For these reasons, a window that can be opened in a variety of positions can be useful; some examples are shown in Fig. 5.4.

Perhaps the best aspect of operable windows is that they give the building occupants some control over the source of outdoor air. Perhaps the worst aspect is that they rarely offer any means of filtering this incoming air. They also can confound attempts by a central HVAC system to regulate air flow and the resulting pressure. Sometimes they admit air (windward side); at other times, they exhaust it (leeward side). Note that the EPA Campus (Fig. 5.2) elected fixed, not operable, windows.

Some windows offer more free area of opening than others of the same size. Figure 5.5 compares

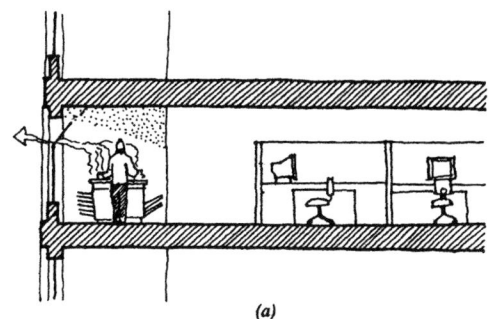

*(a)*

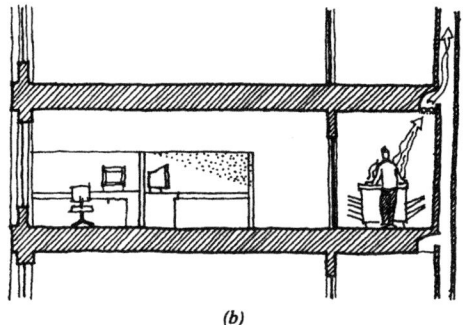

*(b)*

**Fig. 5.3** *Zoning for IAQ: the office copier. This notorious source of VOCs can be located at the perimeter (a), where access to fresh air and direct exhaust is simple; however, perimeter space is prime real estate for daylight and views. If the copier is located in less desirable interior space (b), exhaust air can create negative air pressure, drawing air from adjacent offices and containing the VOCs; longer runs of exhaust ducts are required.*

some common window types. The pattern of incoming flow is also highly influenced by the way in which these windows open. The outward flow is somewhat affected as well. Insect screens will reduce the flow of air. Details of windows and screens are discussed in Chandra et al. (1986). Estimating wind-driven air flow through window openings is discussed in Chapter 8.

Windows work best in the presence of wind. In calm conditions, they may still admit—or exhaust—air due to the stack effect. The taller the building, the more pronounced this effect. Operable windows in very tall buildings have been shunned by designers until recently. The Commerzbank in Frankfurt, Germany, is a 56-story tower. Each office's exterior window (Fig. 5.6) is operable in temperate weather, and the occupant decides the degree of openness; a lock-out is controlled by a building management system (BMS). The full-height office window's outer skin consists of fixed

(a)          (b)

**Fig. 5.4** *A window that can open in more than one position can enhance ventilation performance. (a) The window tilts from the top, directing the incoming air toward the ceiling; fresh air does not directly encounter workers within the space. (b) The window swings inward, allowing incoming air to move across people, enhancing warm-weather cooling. (Courtesy of Three Rivers Aluminum Company, Inc.)*

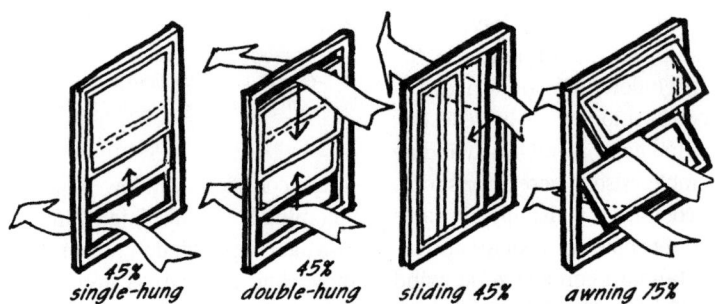

*45%*
*single-hung*      *45%*
*double-hung*      *sliding 45%*      *awning 75%*

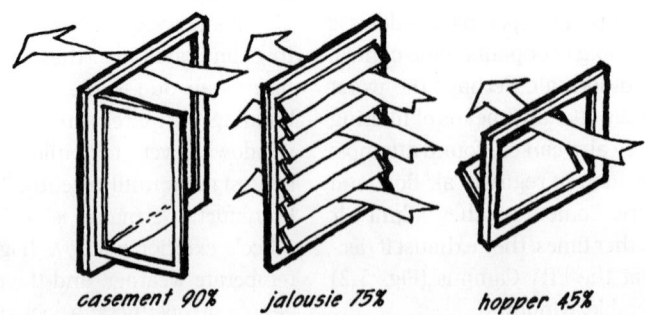

*casement 90%*      *jalousie 75%*      *hopper 45%*

**Fig. 5.5** *The percentage of actual openable area varies with the window type. (Reprinted by permission from Moore, Fuller. 1994.* **Environmental Control Systems:** *Heating, Cooling, Lighting. McGraw-Hill Company, New York.)*

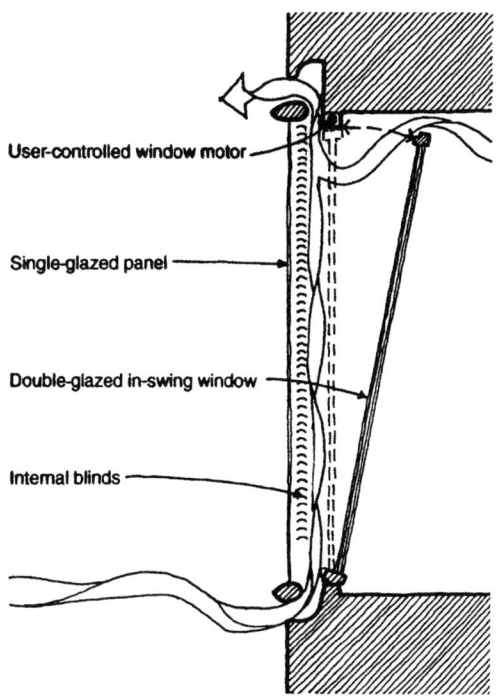

User-controlled window motor

Single-glazed panel

Double-glazed in-swing window

Internal blinds

*Fig. 5.6* *The Commerzbank is a recent addition to Frankfurt, Germany's skyline. Operable windows in this 57-story tower are tilted in from the top by occupants, while a fixed outer pane with continuous venting slots at the top and bottom keeps air currents under control. Shown here in exhaust mode, the window can also supply fresh air to the office. Under adverse outdoor conditions, the building management system (BMS) locks the inner window in the closed position.*

single-glazed safety glass, with 5-in. (125-mm) ventilation slots all across the top and bottom. These serve the 8-in. (200-mm) wide cavity between the outer and inner window skins. The inner window is double glazed, hinged at floor level, and has a motor-operated tilt-in mechanism at the top. Motorized blinds for solar shading are located within the cavity between the window skins. A central atrium provides a stack effect so that an open window is usually a source of incoming rather than exhaust air, although the top and bottom slots allow for a slight stack effect at each window.

## (b) Stack Effect

Several applications of the principle that hot air rises are applicable to IAQ. Estimating airflow due to the stack effect is discussed in Chapter 8, along with more detailed calculations. Devices can be used to enhance

the stack effect by creating suction when wind blows across the top of a stack. Probably the most common (available in chain-store catalogs) are wind gravity or turbine ventilators (Fig. 5.7); typical performance characteristics are listed in Table 5.7. This is probably not the most effective topping device, however. Figure

*Fig. 5.7* *Two wind-gravity (turbine) ventilators accent the skyline of the addition to Barton Hall at the University of New Hampshire. These draw hot exhaust air from the auditorium; cool outdoor air is admitted through low windows and shutters in the auditorium's north wall. (Courtesy of Banwell White Arnold Hemberger & Partners, Inc., Hanover, NH.)*

DESIGN CONTEXT

## TABLE 5.7 Turbine Ventilator Performance

*Note:* The combination of wind suction and stack effect produces the following exhaust capacities (cfm) for various throat diameters and stack heights of turbine ventilators. Recommended spacing between ventilators is 20 ft (6 m). (cfm × 0.472 = L/s)

| Outdoor Wind Velocity: mph (m/s) ΔT, indoors-outdoors | | 2 (0.9) | | | 4 (1.8) | | | 6 (2.7) | | | 8 (3.6) | | | 10 (4.5) | | |
|---|---|---|---|---|---|---|---|---|---|---|---|---|---|---|---|---|
| **F°** | | 10 | 20 | 30 | 10 | 20 | 30 | 10 | 20 | 30 | 10 | 20 | 30 | 10 | 20 | 30 |
| **C°** | | 5.6 | 11.1 | 16.7 | 5.6 | 11.1 | 16.7 | 5.6 | 11.1 | 16.7 | 5.6 | 11.1 | 16.7 | 5.6 | 11.1 | 16.7 |
| Turbine Throat Diameter | Height Above Intake ft (m) | Exhaust Capacity (cfm) | | | | | | | | | | | | | | |
| 6"  (150 mm) | 10 (3) | 114 | 125 | 130 | 210 | 221 | 226 | 314 | 325 | 330 | 426 | 437 | 442 | 534 | 545 | 550 |
| | 20 (6) | 122 | 135 | 144 | 218 | 231 | 240 | 322 | 335 | 344 | 434 | 447 | 456 | 542 | 555 | 564 |
| | 30 (9) | 129 | 144 | 156 | 225 | 240 | 252 | 329 | 344 | 356 | 441 | 456 | 468 | 549 | 564 | 576 |
| | 40 (12) | 135 | 152 | 166 | 231 | 248 | 262 | 335 | 352 | 366 | 447 | 464 | 478 | 555 | 572 | 586 |
| 10"  (250 mm) | 10 (3) | 209 | 222 | 274 | 370 | 383 | 435 | 545 | 558 | 610 | 728 | 741 | 793 | 915 | 928 | 980 |
| | 20 (6) | 234 | 269 | 301 | 395 | 430 | 462 | 570 | 605 | 637 | 753 | 788 | 820 | 940 | 975 | 1007 |
| | 30 (9) | 254 | 301 | 328 | 415 | 462 | 489 | 590 | 637 | 664 | 773 | 820 | 847 | 960 | 1007 | 1034 |
| | 40 (12) | 269 | 318 | 355 | 430 | 479 | 516 | 605 | 654 | 691 | 788 | 837 | 874 | 975 | 1024 | 1061 |
| 14"  (350 mm) | 10 (3) | 333 | 383 | 422 | 558 | 608 | 647 | 804 | 854 | 893 | 1062 | 1112 | 1151 | 1324 | 1374 | 1413 |
| | 20 (6) | 376 | 444 | 496 | 601 | 669 | 721 | 847 | 915 | 967 | 1105 | 1173 | 1225 | 1367 | 1435 | 1487 |
| | 30 (9) | 413 | 496 | 560 | 638 | 721 | 785 | 884 | 967 | 1031 | 1142 | 1225 | 1289 | 1404 | 1487 | 1551 |
| | 40 (12) | 444 | 539 | 614 | 669 | 764 | 839 | 915 | 1010 | 1085 | 1173 | 1268 | 1343 | 1435 | 1530 | 1605 |
| 18"  (450 mm) | 10 (3) | 476 | 564 | 623 | 755 | 843 | 902 | 1071 | 1159 | 1218 | 1399 | 1487 | 1546 | 1737 | 1825 | 1884 |
| | 20 (6) | 549 | 662 | 747 | 828 | 941 | 1026 | 1144 | 1257 | 1342 | 1472 | 1585 | 1670 | 1810 | 1923 | 2008 |
| | 30 (9) | 611 | 747 | 853 | 890 | 1026 | 1132 | 1206 | 1342 | 1448 | 1534 | 1670 | 1776 | 1872 | 2008 | 2114 |
| | 40 (12) | 662 | 819 | 941 | 941 | 1098 | 1220 | 1257 | 1414 | 1536 | 1585 | 1742 | 1864 | 1923 | 2080 | 2202 |
| 24"  (610 mm) | 10 (3) | 716 | 874 | 978 | 1101 | 1259 | 1363 | 1522 | 1680 | 1784 | 1963 | 2121 | 2225 | 2412 | 2570 | 2674 |
| | 20 (6) | 844 | 1046 | 1196 | 1229 | 1431 | 1581 | 1650 | 1852 | 2002 | 2091 | 2293 | 2443 | 2540 | 2742 | 2892 |
| | 30 (9) | 954 | 1196 | 1384 | 1339 | 1581 | 1769 | 1760 | 2002 | 2190 | 2201 | 2443 | 2631 | 2650 | 2892 | 3080 |
| | 40 (12) | 1046 | 1324 | 1542 | 1431 | 1709 | 1927 | 1852 | 2130 | 2348 | 2293 | 2571 | 2789 | 2742 | 3020 | 3238 |
| 30"  (760 mm) | 10 (3) | 1139 | 1385 | 1545 | 1719 | 1965 | 2125 | 2379 | 2625 | 2785 | 3068 | 3314 | 3474 | 3769 | 4015 | 4175 |
| | 20 (6) | 1342 | 1655 | 1890 | 1922 | 2235 | 2470 | 2582 | 2895 | 3130 | 3271 | 3584 | 3819 | 3972 | 4285 | 4520 |
| | 30 (9) | 1514 | 1890 | 2185 | 2094 | 2470 | 2765 | 2754 | 3130 | 3425 | 3443 | 3819 | 4114 | 4144 | 4520 | 4815 |
| | 40 (12) | 1655 | 2090 | 2430 | 2235 | 2670 | 3010 | 2895 | 3330 | 3670 | 3584 | 4019 | 4359 | 4285 | 4720 | 5060 |
| 36"  (910 mm) | 10 (3) | 1613 | 1967 | 2201 | 2475 | 2829 | 3063 | 3418 | 3772 | 4006 | 4414 | 4768 | 5002 | 5428 | 5782 | 6016 |
| | 20 (6) | 1901 | 2354 | 2692 | 2763 | 3216 | 3554 | 3706 | 4159 | 4497 | 4702 | 5155 | 5493 | 5716 | 6169 | 6507 |
| | 30 (9) | 2148 | 2692 | 3115 | 3010 | 3554 | 3977 | 3953 | 4497 | 4920 | 4949 | 5493 | 5916 | 5963 | 6507 | 6930 |
| | 40 (12) | 2354 | 2981 | 3470 | 3216 | 3843 | 4332 | 4159 | 4786 | 5275 | 5155 | 5782 | 6271 | 6169 | 6796 | 7285 |
| 42"  (1070 mm) | 10 (3) | 2183 | 2663 | 2998 | 3350 | 3835 | 4170 | 4645 | 5125 | 5460 | 6000 | 6480 | 6815 | 7365 | 7845 | 8180 |
| | 20 (6) | 2588 | 3203 | 3668 | 3760 | 4375 | 4840 | 5050 | 5665 | 6130 | 6405 | 7020 | 7485 | 7770 | 8385 | 8850 |
| | 30 (9) | 2928 | 3668 | 4243 | 4100 | 4840 | 5415 | 5390 | 6130 | 6705 | 6745 | 7485 | 8060 | 8110 | 8850 | 9425 |
| | 40 (12) | 3203 | 4058 | 4723 | 4375 | 5230 | 5895 | 5665 | 6520 | 7185 | 7020 | 7875 | 8540 | 8385 | 9240 | 9905 |
| 48"  (1220 mm) | 10 (3) | 2868 | 3500 | 3925 | 4412 | 5044 | 5469 | 6078 | 6710 | 7135 | 7843 | 8475 | 8900 | 9638 | 10270 | 10695 |
| | 20 (6) | 3378 | 4185 | 4785 | 4922 | 5729 | 6329 | 6588 | 7395 | 7995 | 8353 | 9160 | 9760 | 10148 | 10955 | 11555 |
| | 30 (9) | 3817 | 4785 | 5535 | 5361 | 6329 | 7079 | 7027 | 7995 | 8745 | 8792 | 9760 | 10510 | 10587 | 11555 | 12305 |
| | 40 (12) | 4185 | 5300 | 6175 | 5729 | 6844 | 7719 | 7395 | 8510 | 9385 | 9160 | 10275 | 11150 | 10955 | 12070 | 12945 |

*Source:* Reprinted courtesy of Western Ventilating Equipment, Inc., Los Angeles.

5.8 compares volumetric airflow results for a turbine and several other ventilators with those for a simple open stack. The tests were done in a wind tunnel at the Virginia Polytechnic Institute.

Because the stack effect works more forcefully with increased height, intakes should be as low as possible. When these openings are near the ground, precooling of summer intake air is possible. The Cottage Restaurant in Oregon (see Fig. 1.14) took some advantage of this principle. The Olivier Theatre at the Bedales School in rural Hampshire, England (Fig. 5.9), uses a gently sloping site to similar advantage. The tightly packed audience of a theater generates considerable heat, as do the lights. For

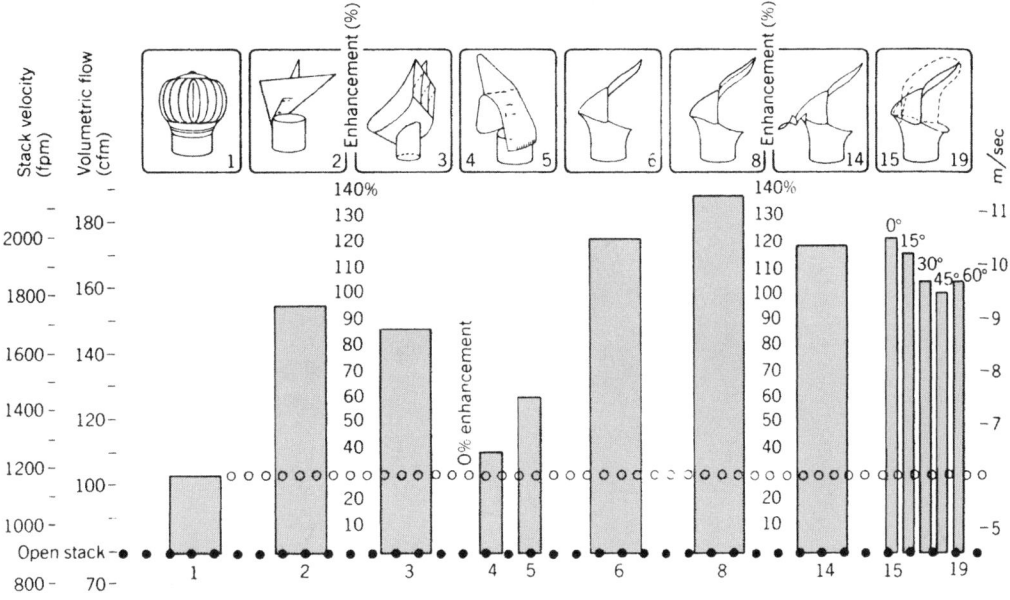

**Fig. 5.8** *Performance of some passive ventilators compared to a simple open stack. The popular turbine ventilator is identified as #1 but is outperformed by many other devices. (From the work of R.P. Schubert and Philip Hahn; reprinted from* Progress in Passive Solar Energy Systems, *© 1983, by permission of the publisher, American Solar Energy Society, Boulder, CO.)*

this theater, the maximum acceptable temperature around the audience is about 77°F (25°C); the heat produced by people and lights provides a temperature increase of about 12.6°F (7°C). Therefore, whenever the outdoor temperature is about 64.4°F (18°C), cooling of the incoming air is needed.

In this theater, air is introduced to an "undercroft" (crawl space) with a concrete floor, on which are built many concrete block walls, forming an indirect path for the incoming air. This undercroft is cooled by night ventilation, and thus is made ready for the next event's heat gains. The inlet openings are 5% of the theater floor area; the surface area of the undercroft is 32 ft² (3 m²) per person. The audience of 270 people is ventilated and cooled by the air rising from this undercroft, through openings that total 3.5% of the floor area. Gaining heat, the air rises toward the central cupola, aided when necessary by a "punkah" fan (see Fig. 9.4), and exits through four louvered sides of the cupola, with a total outlet area of 6% of the floor area. The overall height of this stack is 51 ft (15.5 m); a maximum of 15 ACH is expected at around 5°F (3°C) difference, warmer inside than outside. This ventilation–cooling system is silent and utilizes no refrigerant. In the heating season at partial occupancy, the stack out-

lets are closed, and the fan can be run in reverse to send collected warm air downward to the floor.

## (c) Underslab Ventilation

Although the theater uses the ground's coolth to advantage, there are some places where the ground contains radon (or other soil gases). A county map of the United States (Fig. 5.10) shows the relative risk from radon, a long-term harmful gas. Buildings on former industrial sites or landfills could be threatened by other dangerous soil gases. Even ordinary sites can be threatened by methane gas from a leaking sewer line. A precaution against soil gas is to design for a *passive sub-slab depressurization system*. This involves at least one 4-in. (100-mm) pipe open at both ends. The lower end is set into a layer of clean, crushed rock at least 4 in. (100 mm) thick that lies immediately below the floor slab. The object is to allow air within this rock layer to enter the open end of the pipe. The slab is poured and carefully sealed around the pipe, and the pipe is extended (through interior walls) through the roof, where it can vent radon and other soil gases to a safer place. Heat from the building drives the stack effect within this pipe.

(a)

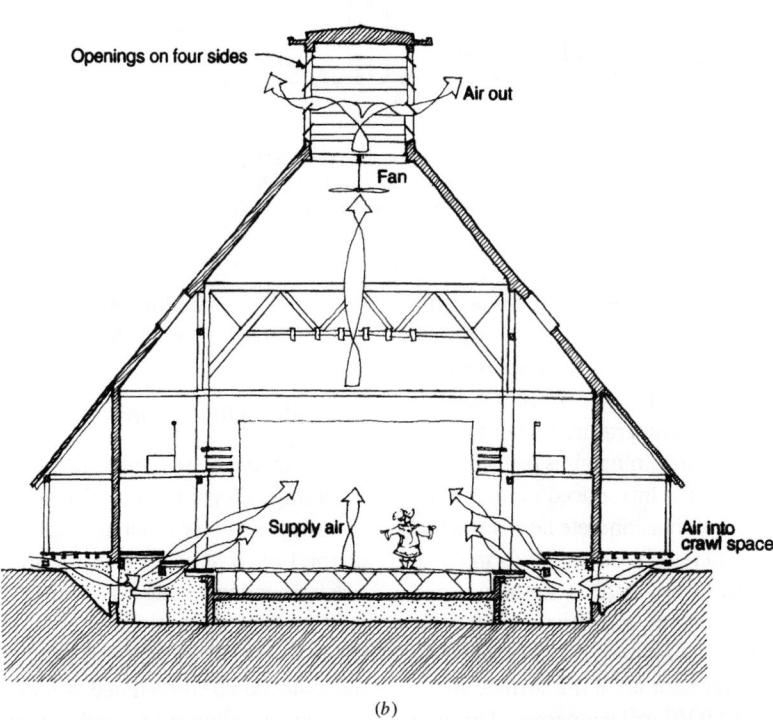

(b)

**Fig. 5.9** *Stack ventilation caps the Olivier Theatre at the Bedales School in rural Hampshire, England. (a) View from the east. (Photo by VIEW/Dennis Gilbert.) (b) Section shows the openings that admit outdoor air to the concrete "undercroft," where it is cooled, then admitted below seats to the auditorium; gaining heat, it rises (assisted when necessary by a punkah fan) and out the four-sided dampered openings in the cupola. (Designed by Fielden Clegg Architects, Bath, Avon; engineering by Max Fordham and Partners, London.)*

### (d) Preheating Ventilation Air

Fresh air brought directly into a space during the winter will improve IAQ, but at the expense of thermal comfort. Several passive or low-energy strategies to mitigate this problem are available. The office building in Fig. 5.11 is surrounded by a 4-ft (1.2-m)-wide cavity between the inner and outer glass surfaces. Air within this cavity is heated, both by the sun and by indoor heat sources, and rises out of a damper-controlled opening. Although this building does not utilize such heated air for ventilation, it demonstrates the possibility of such an approach. See the Comstock Center (Fig. 10.63) as another example.

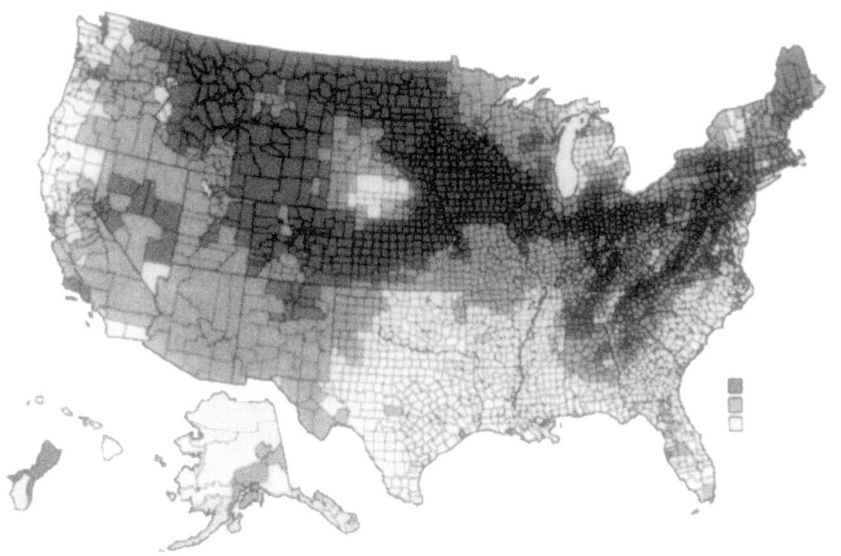

**Fig. 5.10** *Relative risks of high radon levels in the soil by county in the United States. The darker the zone, the higher the risk. (U.S. Environmental Protection Agency, courtesy of* Environmental Building News.*)*

The use of a south-facing wall as a winter pre-heating device with an (unglazed) *transpired collector* (available as Solarwall) is illustrated in Fig. 5.12. Aluminum sheeting, specially finished for solar absorption and penetrated by thousands of tiny holes, is the exterior surface. Behind this is a cavity kept under negative pressure by a fan. Outdoor air is drawn through the holes, heated by the dark outer surface, and drawn up the cavity to the fan and then on to the space. Insulation and the interior surface complete the south wall. Thus, heat loss from the building through the south wall is recaptured by the inflowing outside air. In summer, a fan draws air directly from the outside, bypassing the solar cavity.

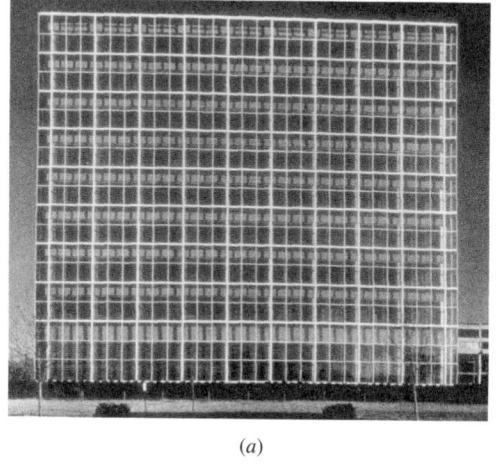

(a)

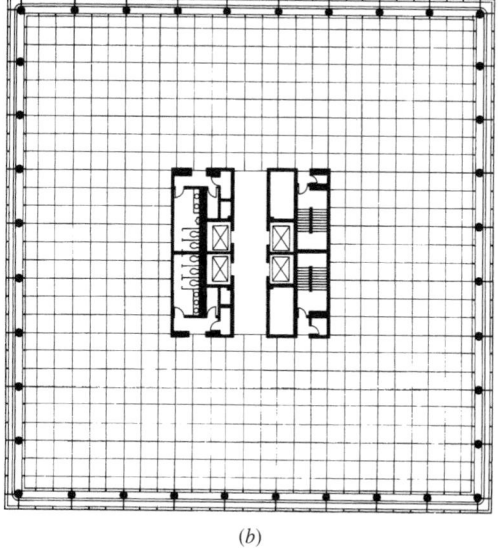

(b)

**Fig. 5.11** *The Occidental Chemical Corporate Office Building, Niagara Falls, New York. (Architects: Cannon Design, Inc.) (a) Fully exposed to sun and wind despite its downtown site, the building appears as a conventional curtainwall box. (b) An ordinary-looking plan with a central core and suspended ceilings. (c) On all four sides, a 4-ft (1.2-m) cavity allows for maintenance of movable daylight louvers and window washing, as well as for ample natural ventilation by the stack effect. Although cavity airflow is released at the roof, it could be utilized during winter as tempered fresh air to the interior.*

The holes at the top of the wall serve as outlets for the stack effect produced by the solar gains through the outer surface. There are numerous installations around the world—one of the largest, at 108,000 ft$^2$ (10,034 m$^2$), is for an aircraft manufacturer in Quebec, Canada. For the design procedure, see NREL (1988).

Another approach to both residential winter ventilation and heat exchange is the *breathable wall* combined with an exhaust air heat pump. This system depends upon a house being under negative pressure, which is assured by forced exhaust air. A heat pump then takes heat from the exhaust air and delivers that heat either for space heating or domestic water heating. The fresh air to replace that being expelled is drawn in through the outside walls by a unique combination of fiberglass lap siding board, fiberglass insulation batts, breathable sheathing, and *no* vapor barrier. This allows a slow, steady stream of cold air to enter, be warmed by the insulation, then enter the house. More information is available from the Canadian National Research Council.

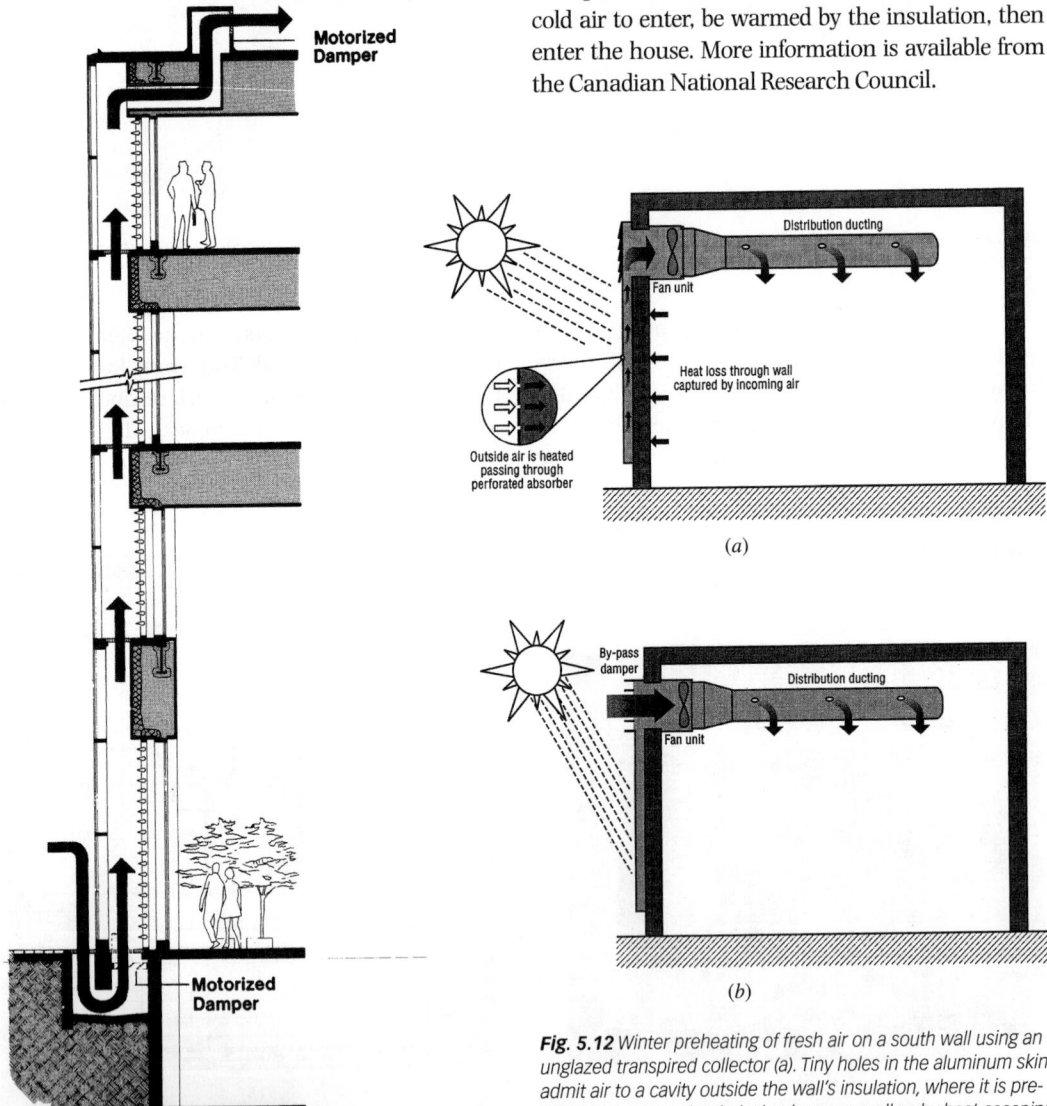

*(a)*

*(b)*

**Fig. 5.12** *Winter preheating of fresh air on a south wall using an unglazed transpired collector (a). Tiny holes in the aluminum skin admit air to a cavity outside the wall's insulation, where it is preheated by sun on the dark aluminum, as well as by heat escaping through the insulation. A fan draws this air to a supply plenum, after which it is distributed to the space. In summer (b), the fan draws directly from the outside, and the collector is self-ventilated by the stack effect. (Courtesy of the National Renewable Energy Laboratory.)*

Motorized Damper

Motorized Damper

Wall Section Detail

*(c)*

**Fig. 5.11** *(Continued)*

## 5.6 EQUIPMENT FOR CONTROL OF IAQ

This section concerns equipment that moves, heats or cools, humidifies or dehumidifies, and cleans air. A large range of capacities is involved, from room-sized to central whole-building air handling. A general note about heating and cooling system choice: acceptable IAQ will be easier to achieve if the heating and cooling systems utilize forced air motion, because some filtering is built into the air-handling equipment. However, separate air-cleaning systems are becoming increasingly common, so radiant heating systems with separate forced-air cleaning can yield high IAQ (given adequate outdoor air). For cooling, an economizer cycle (see Fig. 10.36) provides up to 100% outdoor air at times, and cooling by night ventilation of thermal mass provides many complete air changes during the nightly building maintenance activities that are so fume-producing. Evaporative cooling (see Chapter 8) provides a continuous flow of outdoor air.

### (a) Exhaust Fans

Exhaust fans remove air that is odorous and/or excessively humid before it can spread beyond bathrooms, kitchens, or process areas, creating a negatively pressured area that further limits the spread of undesirable air. In buildings with heating systems without air motion (radiant heating), exhaust fans are often the only built-in devices for moving air. They are often very noisy, which is sometimes useful for covering noises associated with bathrooms, but also noisy enough to discourage their use over long periods of time. In its most simple application, the exhaust fan is a stand-alone device, with no thought about where the replacement air will be drawn from, and rarely much concern about where this unwanted air will be discharged to. (Discharge into attics, basements, or crawl spaces is prohibited by code.)

ANSI/ASHRAE Standard 62.2-2004, *Ventilation and Acceptable Indoor Air Quality in Low-Rise Residential Buildings*, requires intermittent (user-controlled) exhaust fans of at least 50-cfm (25-L/s) capacity for bathrooms and 100 cfm (50 L/s) for kitchens. The intakes for these fans should be as close to the source of polluted air as possible so as not to drag such air across other locations before it leaves the space. In kitchens, this is directly above

the range (grease, odors, and water vapor); in bathrooms, in the ceiling above the toilet (and shower) in order to remove the warmest (therefore the most moist) air. Some options are shown in Fig. 5.13.

A more comprehensive approach in a residence includes the addition of a *principal exhaust fan*, which should be centrally located (drawing from the greatest space), quiet, and suitable for continuous use. Its exhaust capacity should be at least 50% of the entire system's capacity. In turn, this entire capacity is typically no less than 0.3 ACH. When this principal exhaust fan is operating, outdoor air must be brought in, tempered, and circulated throughout the residence. The tempering can be done by mixing with indoor air or by heating. Figure 5.14 shows two approaches to whole-house exhaust, one for forced-air systems and one without forced air motion.

This additional continuous exhaust capacity may cause problems of inadequate air for, or spillage of fumes from, combustion equipment. Such equipment (furnaces, stove-top barbecues, etc.) with a net exhaust greater than 150 cfm (75 L/s) must be provided with a make-up air fan that turns on/off with the equipment.

### (b) Heating/Cooling of Makeup Air

Where climates are mild and/or energy is inexpensive, special equipment (Fig. 5.15) other than heat exchangers can be used to heat and/or cool a particularly large quantity of makeup air. Especially common in factories or laboratory buildings with high exhaust air requirements, these simple devices often supplement the building's main heating/cooling system, which deals primarily with heat gains/losses through the building skin. In warm, dry climates, evaporative coolers are often used for makeup air because they are already designed to utilize 100% outdoor air. Even in hot and more humid climates, *indirect evaporative cooling* can help lower the temperature of makeup air. In Fig. 5.16, outdoor air is at 104°F (40°C) and 10% RH (point A). Two streams of outdoor air are involved. An evaporative cooler cools one air stream along a constant wet-bulb temperature line to point B, where it is now 70% RH but considerably cooler at 73°F (23°C): in the comfort zone, but warm and humid. At this point, it enters an air-to-air heat exchanger (see Section 5.5c). The other outdoor air stream

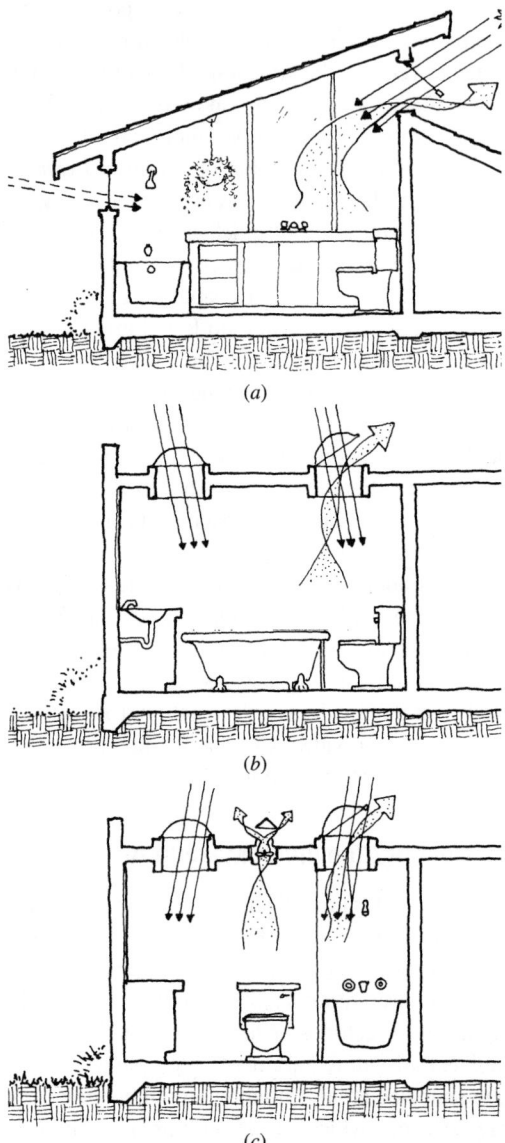

(a)

(b)

(c)

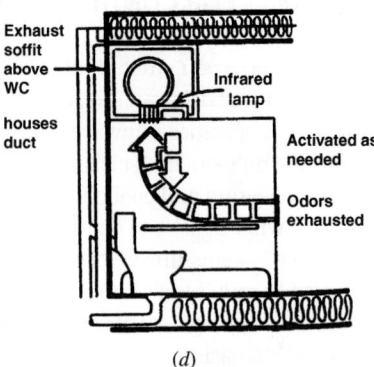

(d)

**Fig. 5.13** *Options for exhaust air from toilet rooms: (a) Daylight enters from the sides; incoming air from the lower window picks up heat and moisture near the ceiling and exits through a high clerestory. (b) Two skylights admit daylight; one exhausts hot, moist air at the ceiling. Makeup air will be needed from the adjacent rooms. (c) Adding an exhaust fan will ensure negative pressure in the bathroom, isolating moisture and odors from the adjacent rooms. (d) In public toilet rooms, exhaust directly above toilets is desirable. In cold weather, small radiant heat lamps with timer switches could add task heat and light.*

## (c) Heat Exchangers

As the tightness of construction increases and fewer air changes per hour (ACH) occur from infiltration (unintended air leaks), forced ventilation becomes more attractive as a means of reducing indoor air pollution. When a heat exchanger is used, it is possible to maintain an adequate supply of fresh air without severe energy consumption consequences. Figure 5.17 illustrates the basic principle of a simple air-to-air heat exchanger that is becoming increasingly common for tightly built small buildings. Note that the outgoing and incoming air streams must be adjacent.

Some commercially available heat exchangers are capable of extracting 70% or more of the heat from exhaust air. The lower the volume of airflow, the higher the efficiency. Table 5.8, Part A, shows representative sizes, airflows, and efficiencies for these devices. For the best diffusion of incoming fresh air through a building, the heat exchanger should be incorporated at the central forced-air fan (Fig. 5.17b). When a central forced-air system is not available, heat exchangers can be placed at various points in a building; typically, each heat exchanger is then equipped with its own fan. These may serve as makeup air units that are more energy efficient than the devices discussed in Section 5.6b.

enters the other side of this heat exchanger, again at point *A* conditions. As the two streams exchange heat, they move toward the same temperature: the evaporatively cooled stream moves from point *B* to point *C*, about 86°F (30°C) and 42% RH and is then exhausted. On the other side, outdoor air moves from point *A* to point *D*, about 91°F (33°C) and 16% RH. This second air stream is *indirectly evaporatively cooled.* Although it is still well above the comfort zone, it can then be either cooled by typical refrigerant systems or evaporatively cooled until it reaches the comfort zone.

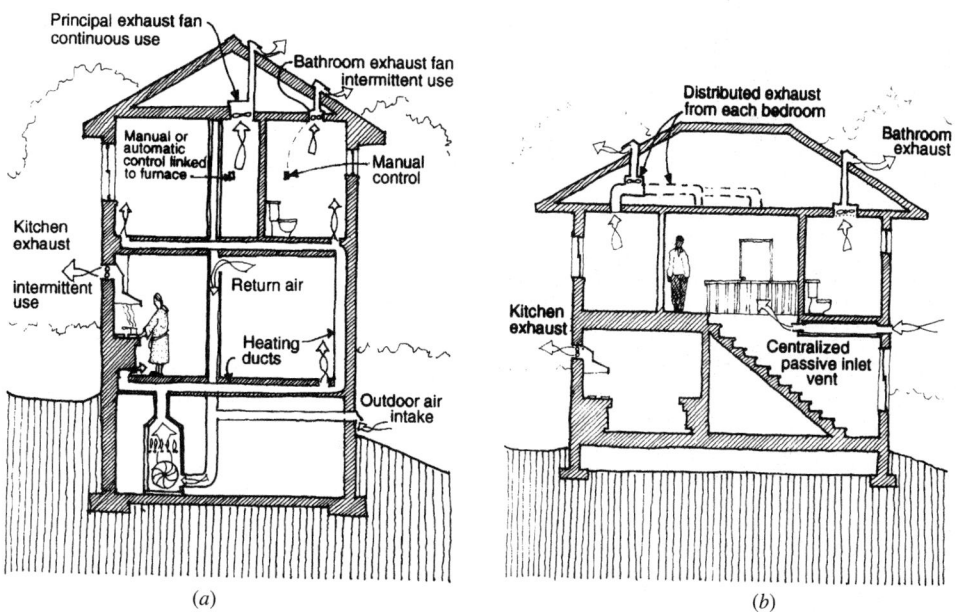

**Fig. 5.14** *Assured ventilation for residences. (a) Ventilation system with forced air heat. (b) Partially distributed exhaust system with a vent. (Based upon Canadian Research Council studies.)*

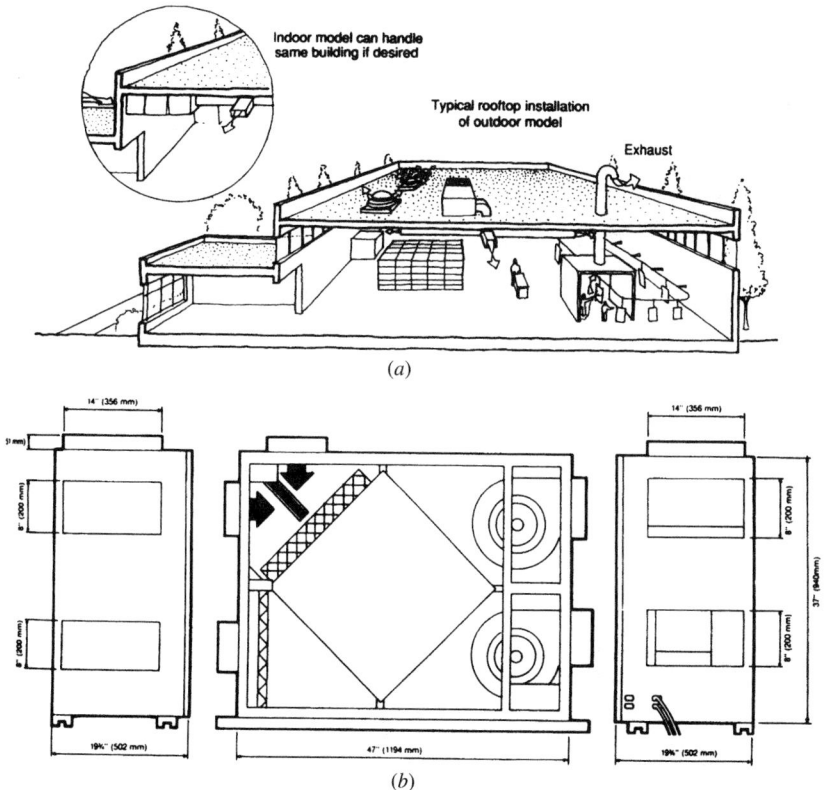

**Fig. 5.15** *Makeup air heaters/coolers. (a) Distributed on a factory roof, these units prevent negative indoor air pressures by providing replacement (makeup) air for the air that is exhausted from processes indoors. Often, a separate heating/cooling system will be used for gains or losses through the building skin. (b) Heat recovery ventilator, utilizing a cross-flow core, serves from 200 to 900 cfm (100 to 450 L/s), with supply air entering at a temperature 0.6 to 0.8 that of exhaust air temperature. (© Conservation Energy Systems, Minneapolis, MN.)*

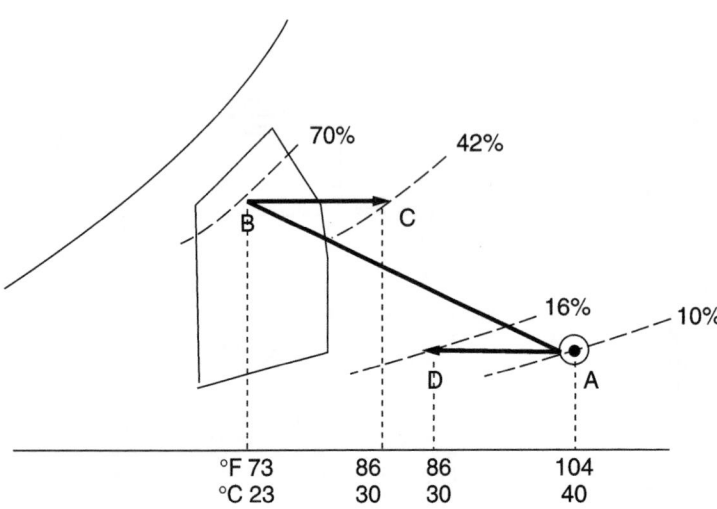

**Fig. 5.16** *Indirect evaporative cooling can precool fresh air. The psychrometrics of the process.*

Some cautions about air-to-air heat exchangers:

1. Avoid using them on exhaust airstreams that are contaminated with grease, lint, or excessive moisture (through cooking and clothes drying in particular) because clogging, frosting, and fire hazard problems can develop.

2. In colder winter conditions, a built-in defroster, which will consume energy, will be needed.

3. Carefully locate the outdoor fresh air intake. Keep this intake as far as possible from the exhaust air outlet to avoid drawing contaminated exhaust air back into the building. Keep the intake away from pollution sources such as vehicle exhaust, furnace flues, dryer and exhaust fan vents, and plumbing vents.

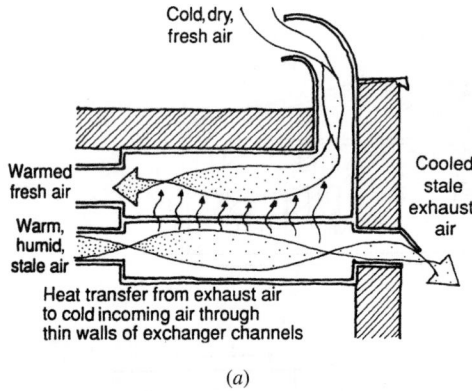

*(a)*

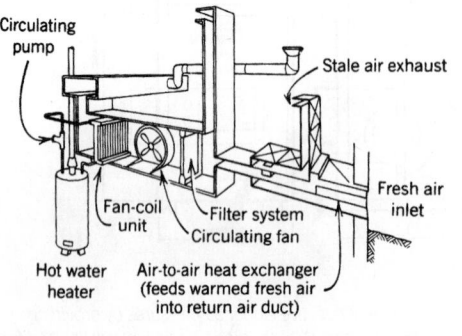

*(b)*

**Fig. 5.17** *Air-to-air heat exchangers are particularly helpful in cold weather. (a) The basic principle of operation. (b) A superinsulated home is heated by coils fed from the domestic water heater, plus a heat exchanger to preheat incoming cold air. (Reprinted by permission of Alberta Agriculture, Home and Community Design Branch, Low Energy Home Designs, © 1983 by Brick House Publishing Co.) (c) Fan-powered heat exchanger utilizing a heat/moisture exchange wheel (see also Fig. 5.20). This model is adjustable from 70 to 200 cfm (35 to 100 L/s), can be ceiling, wall, or floor mounted, and operates at 75% to 80% thermal efficiency. (Courtesy of Air Xchange, Inc., Rockland, MA.)*

**TABLE 5.8 Representative Heat Exchanger Data for Smaller Buildings**

| PART A. **AIR-TO-AIR SENSIBLE HEAT EXCHANGERS** | | | | |
|---|---|---|---|---|
| Approximate Size[a] L × W × H | | Airflow Range | | Temperature Recovery at 32°F (0°C) |
| in. | mm | cfm | L/s | |
| 21 × 13 × 25 | 535 × 330 × 630 | 50–120 | 25–60 | 70% |
| 34 × 17 × 21 | 865 × 435 × 530 | 75–225 | 35–110 | 80% |
| 53 × 17 × 21 | 1335 × 435 × 530 | 75–225 | 35–110 | 93% |
| 47 × 20 × 37 | 1195 × 505 × 940 | 350–700 | 165–330 | 65% |

*Source:* Conservation Energy Systems, Inc., Saskatoon and Minneapolis.

| PART B. **AIR-TO-AIR SENSIBLE + LATENT HEAT EXCHANGERS** | | | | |
|---|---|---|---|---|
| Approximate Size[b] L × W × H | | Airflow Range | | Wheel Efficiency |
| in. | mm | cfm | L/s | |
| 46 × 32 × 24 | 1170 × 815 × 610 | 500–1000 | 235–470 | 84–82% |
| 54 × 48 × 34 | 1375 × 1220 × 865 | 600–1600 | 285–760 | 87–80% |
| | | 1600–2300 | 760–1090 | 75–70% |
| 67 × 67 × 43 | 1705 × 1705 × 1095 | 1750–3250 | 830–1535 | 85–80% |
| | | 3000–4500 | 1420–2125 | 76–71% |
| 124 × 84 × 60[c] | 3150 × 2135 × 1525 | 3500–6500 | 1660–3070 | 85–80% |
| | | 6000–9000 | 2835–4250 | 76–71% |

*Source:* Greenheck Fan Corporation, Schofield, WI.

[a]Access to one face (L × W) is needed for servicing.

[b]Several arrangements (interior and rooftop) are available; dimensions do not include service clearance access, supply hoods, or exhaust hoods.

[c]This unit contains two wheels.

A student housing complex in Greensboro, North Carolina, utilizes heat exchangers on the exhaust air from bathrooms (Fig. 5.18). These exhaust devices are called *energy recovery ventilators* (ERVs). Each floor of the three-story complex has four apartments, each with 1078-ft² (100-m²) floor area and 8-ft (2-m) ceilings. Each apartment has its own air-to-air ERV that accepts air from the two small bathrooms in the apartment. Control is by individual switches in the bathrooms. Prefiltered outdoor air is drawn into the ERV, exchanging heat with the outgoing exhaust air. This fresh air is then fed directly into the air handler for each apartment's heat pump, adjacent to the ERV. Thus, the bathrooms are under negative pressure relative to the rest of the apartment. Outside, the fresh air intake is located high on the wall, the exhaust outlet lower. Intake and outlet locations on the walls are separated by a minimum of 8 ft (2 m).

A *heat pipe* (Fig. 5.19) also transfers sensible heat between adjacent airstreams. Within the heat pipe, a charge of refrigerant spends its life alternately evaporating, condensing, and migrating by capillary action through a porous wick. Because the only thing that moves is the refrigerant and it is self-contained, no maintenance and long life are likely. Efficiency ranges from 50% to 70%; modular sizes are available to 54 in. × 138 in. × 8 rows deep (1.4 × 3.5 m).

Heat pipes can assist in the dehumidification and cooling of incoming air. The typical cooling process using cooling coils is shown in Chapter 8. This potentially energy-intensive process often overcools the air to "wring out" (condense) water, then reheats the air. The heat pipe precools the air (subtracting heat) before the cooling coil, then warms it (adding heat) after the cooling coil. No energy input is required. A configuration of the heat pipe for this task is shown in Fig. 5.19b.

*Energy transfer wheels* (Fig. 5.20) go further than the two preceding devices in that they transfer latent as well as sensible heat. In winter, they recover both sensible and latent heat from exhaust air; in summer, they both cool and dehumidify the

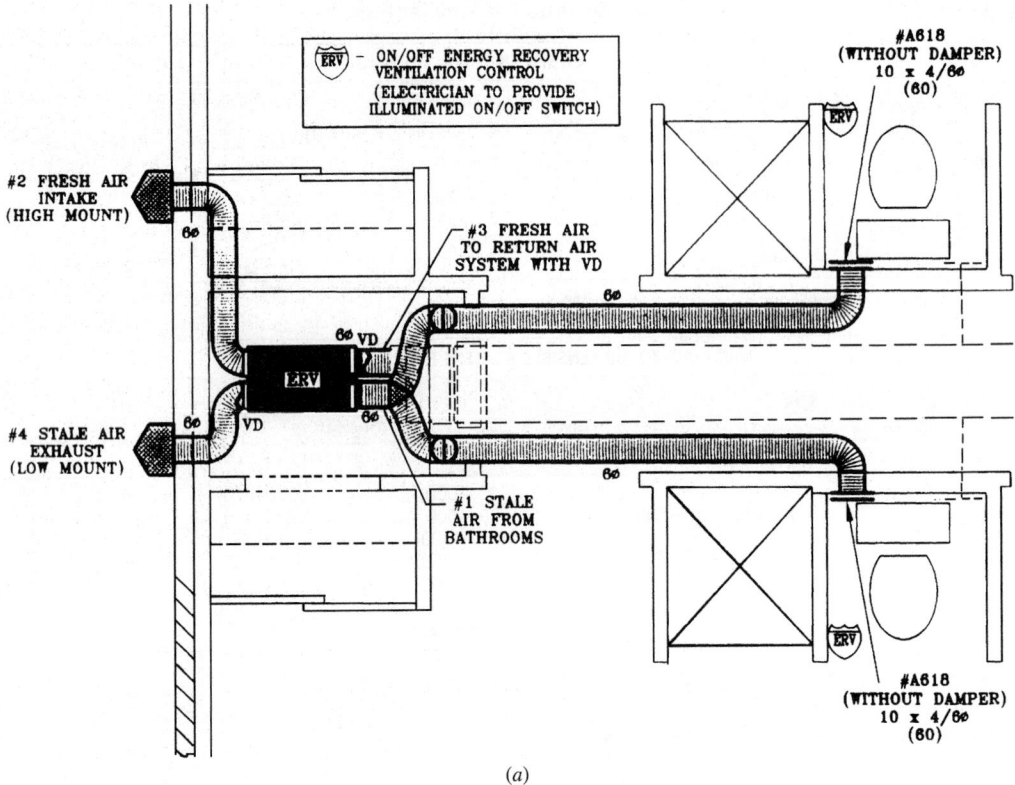

(a)

**Fig. 5.18** *Energy recovery ventilator (ERV) serves two small bathrooms (a) in a student housing complex in Greensboro, North Carolina. The ERV is activated by switches in either bathroom. Air is drawn out beside the water closet and exchanges heat (at about 85% effi-ciency) with incoming fresh air. This tempered fresh air is then mixed with some return air and fed directly to the indoor unit of a heat pump (b) located above the ERV. The supply air is then fed to other rooms, ensuring a negative pressure in the bathroom while the ERV operates. Each of 16 apartments has its own ERV and heat pump. (Design by Harry John Boody, Jamestown, NC.)*

incoming fresh air. Seals and laminar flow of air through the wheels prevent mixing of exhaust air and incoming air. A further precaution in the process purges each sector of the wheel briefly, using fresh air to blow away any unpleasant resid-ual effects of the exhaust air on the wheel surfaces. Carryover of exhaust air qualities, except those of heat and moisture, is between 4% and 8% without purging and less than 1% with purging. Efficiency ranges from 70% to 80%, and available sizes range up to 144 in. (3.6 m) in diameter. Table 5.8, Part B, shows some representative sizes, airflows, and effi-ciencies. A smaller example of this device is shown in Fig. 5.17c.

### (d) Desiccant Cooling

Another rotating-wheel process involves *desiccant cooling*. Desiccant cooling systems (Fig. 5.21) are

attractive because they use no refrigerants (that may contain CFCs), and they lower humidity with-out having to overcooling the air. The desiccants (such as silica gel, activated alumina, or synthetic polymers) in an *active* system must be heated to drive out the moisture they remove from the incom-ing air; at present, natural gas is typically used, but solar energy is a promising substitute due to its plentiful summer availability. Waste heat from other mechanical systems may also be used. In a *passive* desiccant system, heat from a building's exhaust air is enough to release and vent the moisture removed from incoming air.

Research on materials suitable for desiccant cooling and solar-driven regeneration should pro-duce improvements in the types available as well as their performance. For an extended discussion of both solid- and liquid-based desiccant systems, see Lorsch (1993).

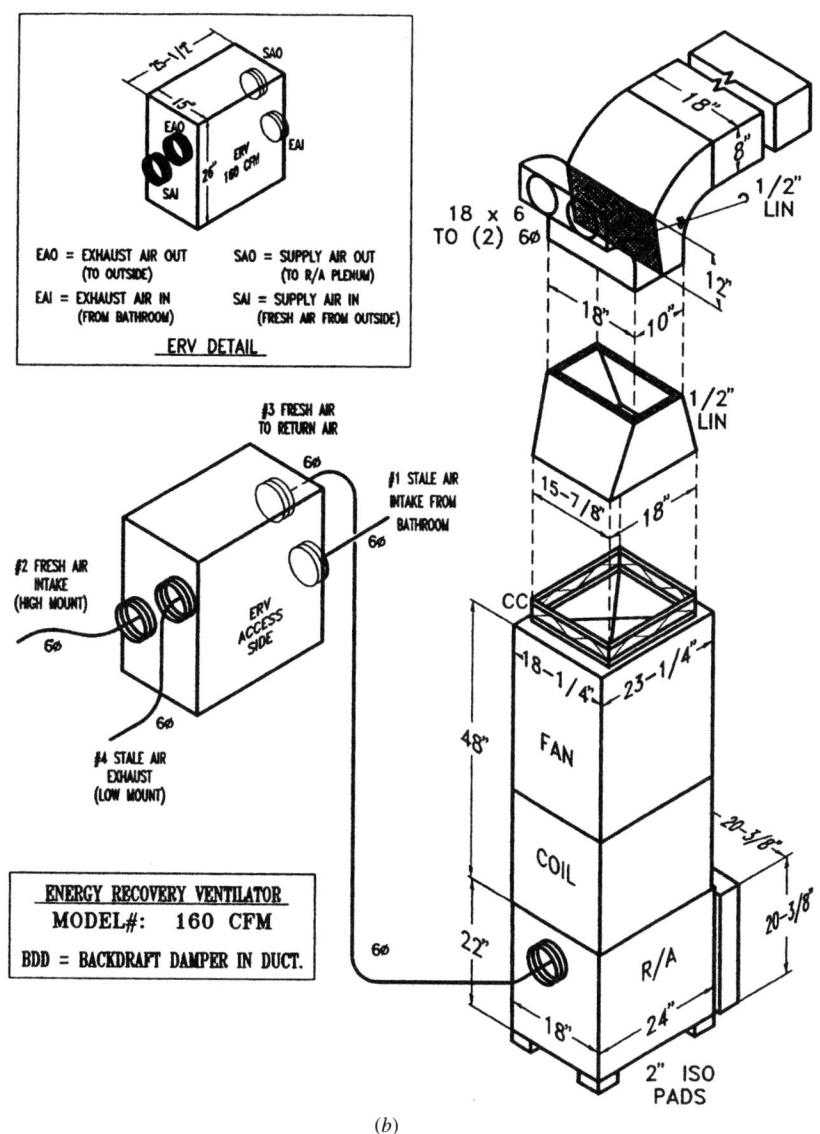

EAO = EXHAUST AIR OUT (TO OUTSIDE)
SAO = SUPPLY AIR OUT (TO R/A PLENUM)
EAI = EXHAUST AIR IN (FROM BATHROOM)
SAI = SUPPLY AIR IN (FRESH AIR FROM OUTSIDE)

ERV DETAIL

#3 FRESH AIR TO RETURN AIR
#1 STALE AIR INTAKE FROM BATHROOM
#2 FRESH AIR INTAKE (HIGH MOUNT)
#4 STALE AIR EXHAUST (LOW MOUNT)

ENERGY RECOVERY VENTILATOR
MODEL#:    160 CFM
BDD = BACKDRAFT DAMPER IN DUCT.

(b)

**Fig. 5.18** (Continued)

## (e) Task Dehumidification and Humidification

Humidity affects comfort; for a sedentary person, a 30% RH change will produce about the same comfort sensation as a 2°F (1°C) change in temperature. Higher humidity adds to hot air discomfort, making evaporation more difficult, increasing skin wettedness, and increasing the friction between skin and clothing or furniture surfaces. As RH exceeds 60%, problems with IAQ increase due to mold and mildew growth. Lower humidity produces cooler and drier sensations, but too-low humidity can irritate the skin.

For spaces that need only dehumidification rather than mechanical cooling, *refrigerant dehumidifiers* are commercially available (Fig. 5.22). Their advantage over desiccant dehumidifiers is that the air temperature remains essentially unchanged dur-

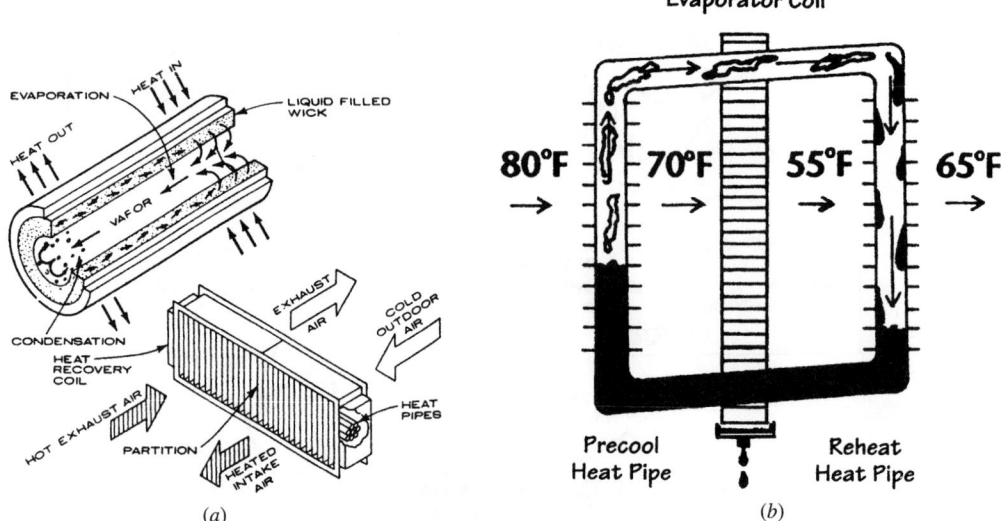

**Fig. 5.19** *The heat pipe is a self-contained device with no moving parts. (a) It silently transfers sensible heat between adjacent fresh-air intake and stale-air exhaust airstreams. (From AIA: Ramsey and Sleeper, Architectural Graphic Standards, 9th ed., © 1994 by John Wiley & Sons, Inc. Reprinted with permission of John Wiley & Sons, Inc.) (b) The heat pipe can increase moisture removal by first precooling incoming hot air before it reaches the cooling coils. Then the heat is returned to the cooled air, bringing it to a temperature–humidity combination acceptable for supply air to a space. (Courtesy of Heat Pipe Technologies, Inc., and* Environmental Building News.*)*

ing dehumidification (desiccant dehumidifiers raise the temperature of the dried air). These devices consume energy to run the refrigeration cycle, however, and this energy is added, as heat, to the space. Accumulated water must periodically be removed from these units and, if untended, could become a source of disease. Most refrigerant dehumidifiers encounter operating difficulties at air temperatures below 65°F (18°C), at which point frost forms on their cooling coils. This could cause problems in a tightly enclosed residence in winter.

Task humidifiers are widely available and often used to relieve symptoms of respiratory illnesses. Again, the problem of bacterial and mold growth in water reservoirs arises; some products add an ultraviolet (UV) lamp to counteract this threat.

### (f) Filters

There is a wide variety of particle air pollutants, as described in Table 5.2. The larger particulates are the easiest to remove, but much smaller respirable particles pose a greater threat to health. Not all pollutants can be removed by filters. Figure 5.23 compares filter groups and testing methods with associated filter efficiencies. Groups I, II, and III are widely used and are generally illustrated in Fig. 5.24. The highest efficiency filter, the high-efficiency particulate arrestance (HEPA), is most often found in special air cleaners for unusually polluted or IAQ-demanding environments. Air filter characteristics are summarized in Table 5.9. See ASHRAE Standards 52.1-1992 and 52.2-1999 for further details on air filter ratings. Specialists recommend using dust collectors rather than filters when the dust loading equals or exceeds 10 mg/m³. Again, it is wisest to remove pollutants at the source.

***Particulate Filters.*** Particulate filters are very common and come in several guises. *Panel filters* are furnished with HVAC equipment and function mainly to protect the fans from large particles of lint or dust. Because they are relatively crude, they are not really considered to be air-cleaning equipment. *Media filters* are much finer, using highly efficient pleated filter paper within a frame. They function both by straining and impaction. The larger particles are strained out by the closely spaced filter fibers, while some of the smaller particles that

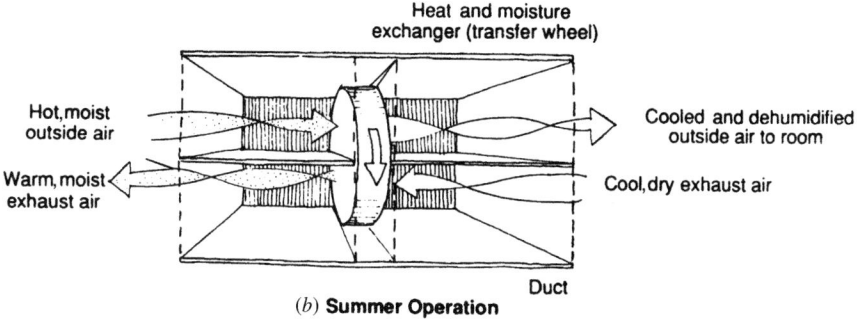

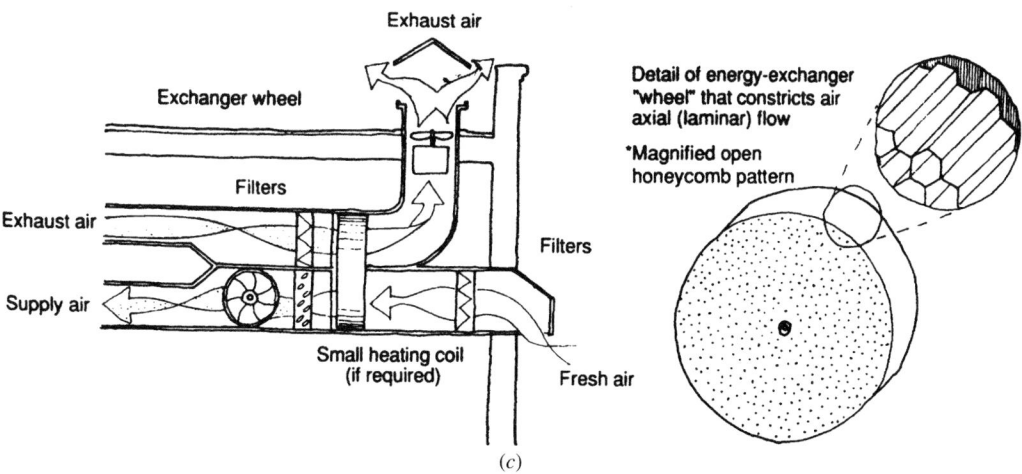

**Fig. 5.20** *Energy transfer wheels. The wheel surface is impregnated with lithium chloride (or another similar material), which absorbs moisture and transfers it to the other airstream. The wheel delivers moist air in winter (a) and dry air in summer (b). Cross section (c) through the wheel and the two airstreams it serves. Exhaust air may be filtered to help keep the wheel clean. (d) Multiple-unit installation. Room exhaust air passes through the upper chambers, and incoming fresh air passes through the lower chambers. Wheels rotate at 8 to 10 rpm. (e) Rooftop unit supplies up to 1000 cfm (500 L/s). (Courtesy of Air Xchange, Inc., Rockland, MA.)*

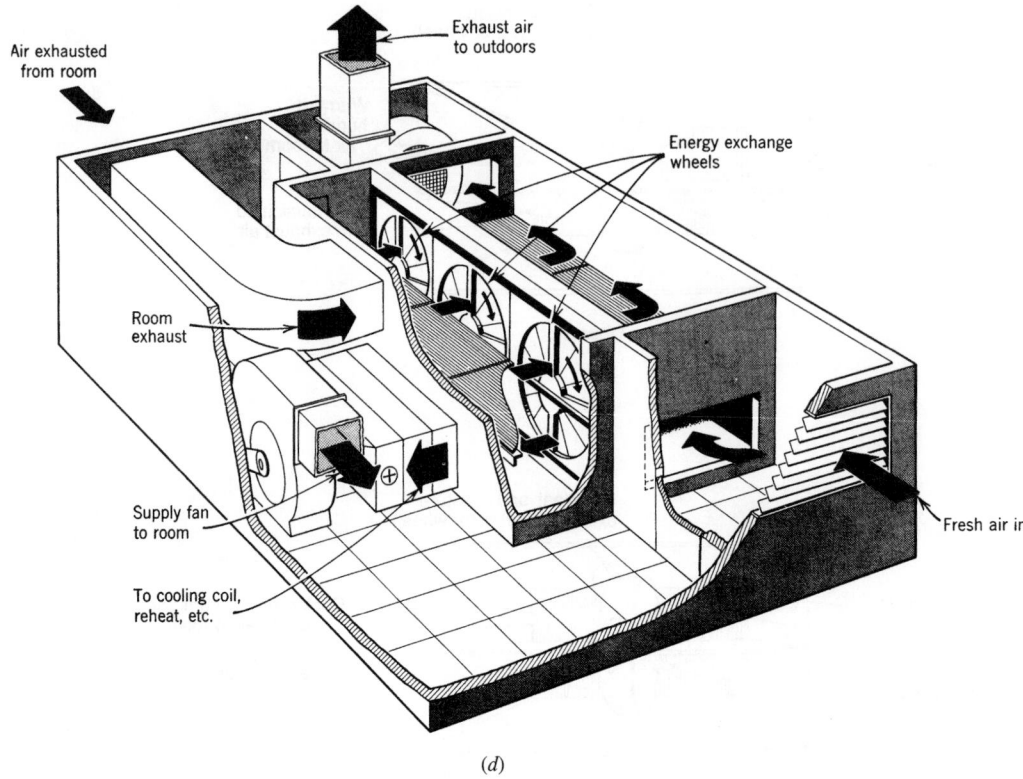

Exhaust air
to outdoors

Air exhausted
from room

Energy exchange
wheels

Room
exhaust

Supply fan
to room

To cooling coil,
reheat, etc.

Fresh air ir

*(d)*

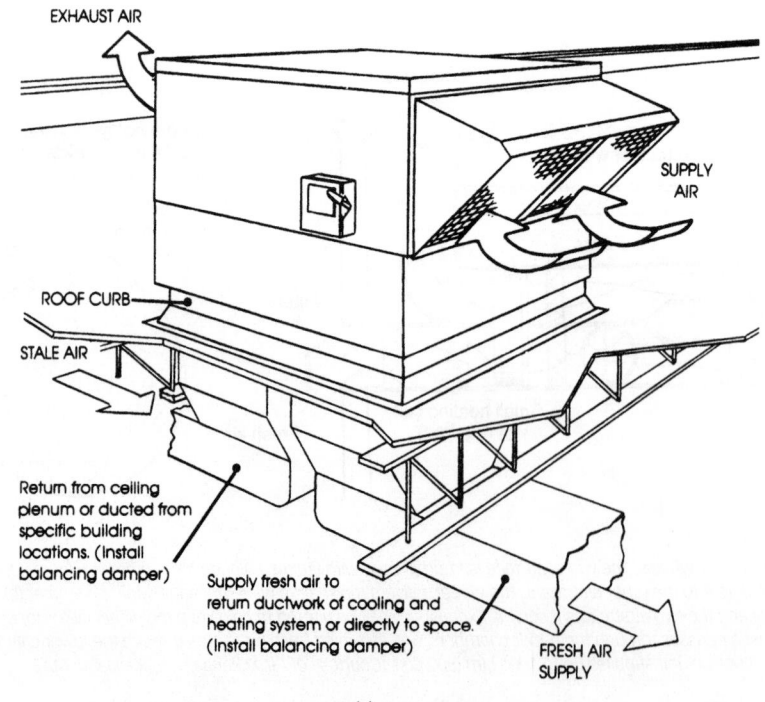

EXHAUST AIR

SUPPLY
AIR

ROOF CURB

STALE AIR

Return from ceiling
plenum or ducted from
specific building
locations. (Install
balancing damper)

Supply fresh air to
return ductwork of cooling and
heating system or directly to space.
(Install balancing damper)

FRESH AIR
SUPPLY

*(e)*

***Fig. 5.20*** *(Continued)*

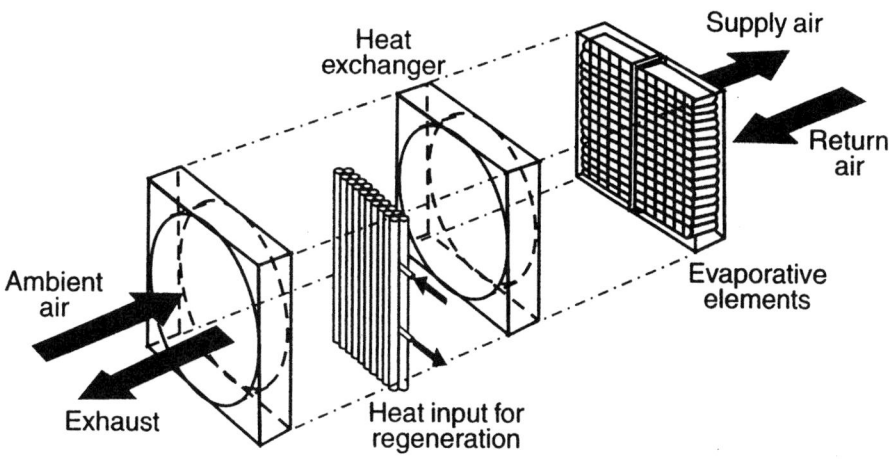

**Fig. 5.21** *Desiccant cooling utilizes a rotating wheel impregnated with either silica gel or Type 1M material. The wheel removes moisture from incoming hot outdoor air and delivers some of it to outgoing exhaust air. The remaining moisture must be driven from the wheel by a heat source; natural gas or, ideally, solar energy. (National Renewable Energy Laboratory.)*

would otherwise pass through are pushed into the fibers due to air turbulence. Particulate filters need regular maintenance, especially media filters, which can become blocked and cause damage to HVAC equipment and increased energy consumption if not replaced frequently enough. Media filters of high quality are expected to perform at an efficiency of 90% (minimum) and are typically at least 6 in. (150 mm) deep; this is for a 6-month minimum life cycle.

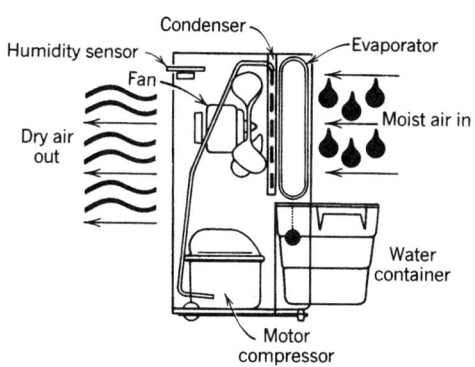

**Typical Performance Data**

| | | | | |
|---|---|---|---|---|
| Watts | 290 | 450 | 465 | 630 |
| Pints per day[a] | 14 | 20 | 25 | 30 |
| Unit dimensions (in.) | | | | |
| Height | 20½ | 20½ | 21½ | 21½ |
| Width, 11¾ | | | | |
| Depth, 16¾ | | | | |
| "Average" room size served | 20′ × 32′ | 35′ × 40′ | 40′ × 48′ | 40′ × 60′ |

[a]For air at 80 F, 60% rh.

**Fig. 5.22** *Refrigerant dehumidifiers typically are installed as free-standing units in smaller buildings.*

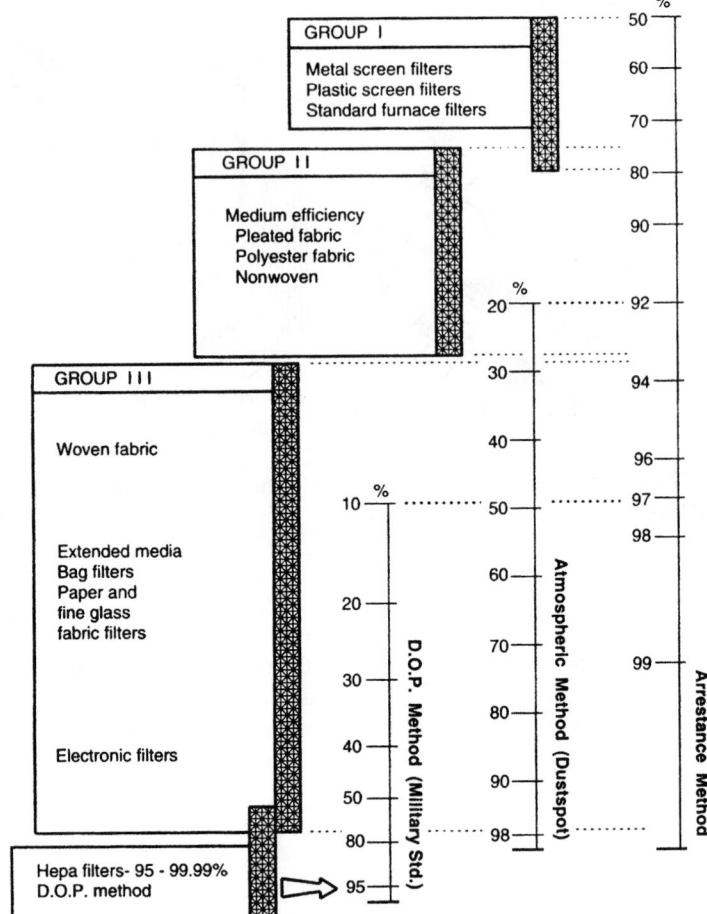

**Fig. 5.23** *Air filter efficiencies. Methods of determining efficiency vary; the Arrestance Method can be used across all groups, while the Atmospheric (Dust-Spot) Method should be used for medium- and high-performance filters. The Detection Operational Program (DOP) Method is a military standard.* (From Rousseau, D. and J. Wasley. 1997. Healthy by Design, *Hartley & Marks, Point Roberts, WA.)*

**Adsorption Filters.** Adsorption filters are for gaseous contaminant removal and vary according to the pollutant in question. *Activated-charcoal filters* are the most common of these types, absorbing materials with high molecular weights but allowing those of lower weights to pass. Other adsorption filters use porous pellets impregnated with active chemicals such as potassium permanganate; the chemicals react with contaminants, reducing their harmful effects. Adsorption filters must be regularly regenerated or replaced.

High-quality adsorption filters contain gas adsorbers and/or oxidizers with sufficient capacity to remain active over a full service cycle of 6 months at 24 hours per day, 7 days per week. Air velocity should be such as to allow the air to remain in the filter for about 0.06 second.

**Air Washers.** Air washers are sometimes used to control humidity and bacterial growth. The moisture involved can pose a threat if these devices are not well maintained.

**Electronic Air Cleaners.** Electronic air cleaners can pose a different threat due to ozone production but have the advantage of demanding less maintenance. Static electricity is produced in the *self-charging mechanical filter* by air rushing through it; larger particles thus cling to the filter. The more humid and/or higher the air velocity, the lower the filtering efficiency. In a *charged media filter,* an electrostatic field is created by applying a high dc voltage to the dielectric material of the filter. Many particles are not polarized, however, due to an insufficiently strong field. A *two-stage electronic air*

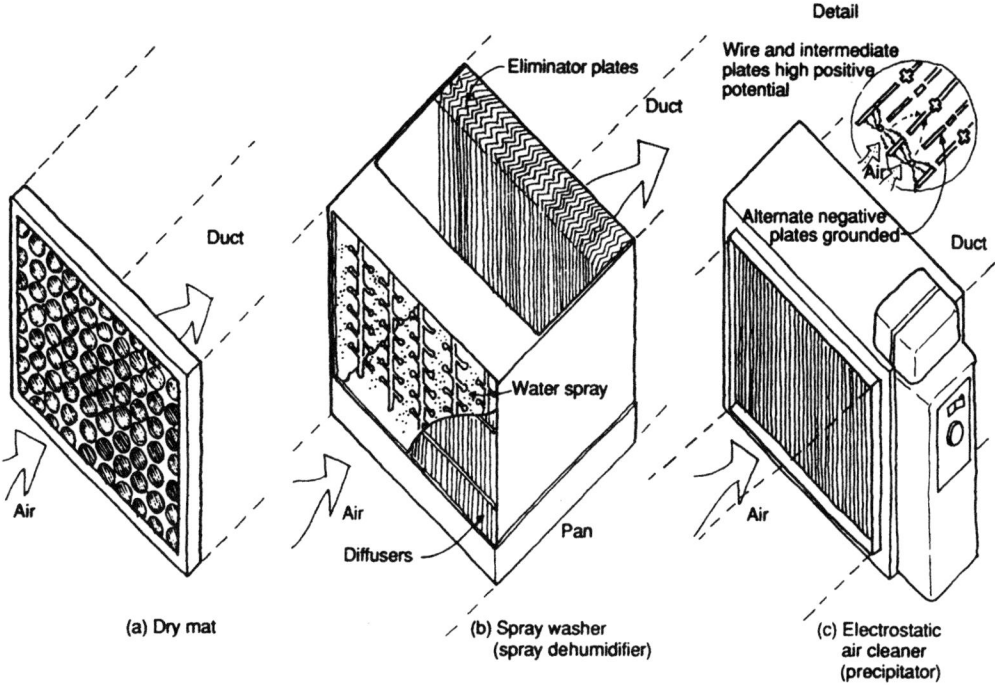

**Fig. 5.24** *Several air filter types; not shown are pleated and roll types.*

*cleaner* first passes dirty air between ionizing wires of a high-voltage power supply. Electrons are stripped from the particulate contaminants, leaving them positively charged. Then these ionized particles pass between collector plates that are closely spaced and oppositely charged. The particles are simultaneously repelled by the positive plates and attracted to the negative plates, where they are collected.

## (g) Locating Air-Cleaning Equipment

Before the advent of IAQ concerns, buildings were often designed with rather crude panel filters located only at the HVAC equipment; they were primarily intended to intercept materials that might adversely affect combustion or heat exchange. In addition to these equipment-protecting devices, a building requiring high IAQ will now have a combi-

**TABLE 5.9  Air Filter Characteristics**

| Media and Type | Percent Efficiency Range | | Dust-Holding Capacity | Airflow Resistance (In. Water)[a] |
|---|---|---|---|---|
| | Atmospheric Dust | Small Particles | | |
| Dry panel, throwaway | 15–30 | NA | Excellent | 0.1–0.5 |
| Viscous panel, throwaway | 20–35 | NA | Good | 0.1–0.5 |
| Dry panel, cleanable | 15–20 | NA | Superior | 0.08–0.5 |
| Viscous panel, cleanable | 15–25 | NA | Superior | 0.08–0.5 |
| Mat panel, renewable | 10–90 | 0–60 | Good to superior | 0.15–1.0 |
| Roll mat, renewable | 10–90 | 0–55 | Good to superior | 0.15–0.65 |
| Roll oil bath | 15–25 | NA | Superior | 0.3–0.5 |
| Close pleat mat panel | NA | 85–95 | Varies | 0.4–1.0 |
| High-efficiency particulate | NA | 95–99.9 | Varies | 1.0–3.0 |
| Membrane | NA | to 100 | NA | NA |
| Electrostatic with mat | 80–98 | NA | Varies | 0.15–1.25 |

*Source:* Reprinted by permission from AIA: Ramsey and Sleeper, *Architectural Graphic Standards,* 9th ed., © 1994 by John Wiley & Sons.
[a]Higher airflow resistance values will require increased fan energy; in. of water × 248.8 = Pa.

nation of high-efficiency particle filters and adsorption filters.

The panel filters provided with HVAC equipment are usually located *upstream* from the unit fan. High-efficiency particle and adsorption filtering systems should be located *downstream* from the HVAC cooling coils and drain pans to ensure that any microbiological contaminants from those wet surfaces are removed rather than being distributed throughout the building.

In buildings where a filtering system must be integrated with a central HVAC system (typical of small buildings), the HVAC system should be capable of continuously circulating air at the rate of 6 to 10 times per hour and of operating against the considerable static pressure that results from high-efficiency filters.

### (h) Ultraviolet Radiation (UV)

Since early in the twentieth century, UV radiation in the C band (200–280 nm wavelength) has been used to kill harmful microorganisms, but under tightly controlled conditions. Now there are UV lamp units that work within HVAC systems, promising to control fungi, prevent the development and spread of bacteria, and reduce the spread of viruses. As an additional benefit, cooling coils and drain pans stay cleaner. The germicidal output of these lamps is somewhat higher at room temperatures (and above) than at lower temperatures. These devices take up very little space (they are placed within ductwork) and generate no ozone or other chemicals. They are even more effective when installed in a UV-reflective duct interior: aluminum seems to be the best UV reflector commonly available. UV lamp life is 5000 to 7500 hours.

UV radiation also looks promising as a treatment for VOCs. The National Renewable Energy Laboratory is helping to develop a process that bombards polluted air with UV radiation in the presence of special catalysts. Pollutants including cigarette smoke, formaldehyde, and toluene are quickly broken down into molecules of water and carbon dioxide.

### (i) Individual Space Air Cleansing

Energy conservation considerations have reduced the air circulation rate in many central air-handling systems; moving less air reduces the energy used by fans. One result can be low distribution efficiency,

which causes poorly mixed air within the occupied spaces. With local (individual space) air-filtering equipment, both a high circulation rate and proper air mixing are achievable. Each such unit has its own fan that can operate either with or without the central HVAC fan. An example of an independent electrostatic air filter is shown in Fig. 5.25.

Air circulation through these filters should occur at rates between 6 and 10 times per hour. The air is then ducted to diffusers, hence circulating in a sweeping pattern across a space to return air intakes on the opposite side.

A variation on electrostatic air cleaning is shown in Fig. 5.26. In this equipment, a mixture of outdoor and indoor air is passed through a complex electrical field produced by both high-voltage dc and high-frequency ac. This supply air emerges with greatly reduced submicron particles that have coagulated into larger particles more easily carried by air currents. As this air moves through the occupied space, it picks up submicron particles and is returned to the equipment (some is exhausted), where it is filtered to remove the larger particles.

Portable air cleaners abound. One more elaborate model combines UV germicidal radiation with carbon/oxidizing media and HEPA. It is a counter-height rolling device about 15 in. (380 mm) square that emits cleansed air from the top. It claims to purify air in a space of about 12,000 ft$^3$ (340 m$^3$).

### (j) Controls for IAQ

A large number of air quality monitoring devices are now available, some of which can control the operation of IAQ-related equipment. One of the oldest and simplest devices measures the concentration of $CO_2$ in parts per million (ppm). Wherever there are indoor concentrations of people, elevated levels of $CO_2$ can be expected. Thus, $CO_2$ becomes a kind of "canary in the coal mine," an early indicator of pollutant buildups due to occupancy. In the Bedales school theater (Fig. 5.9), the operation of the exhaust air dampers in the cupola is controlled by a building management system. Under ordinary conditions, damper opening is regulated by information from the $CO_2$ monitor. This can be overridden by indoor and outdoor temperatures (summer night ventilation is encouraged), air velocity monitors below the seats (too much draft could produce discomfort), the presence of wind and rain (open-

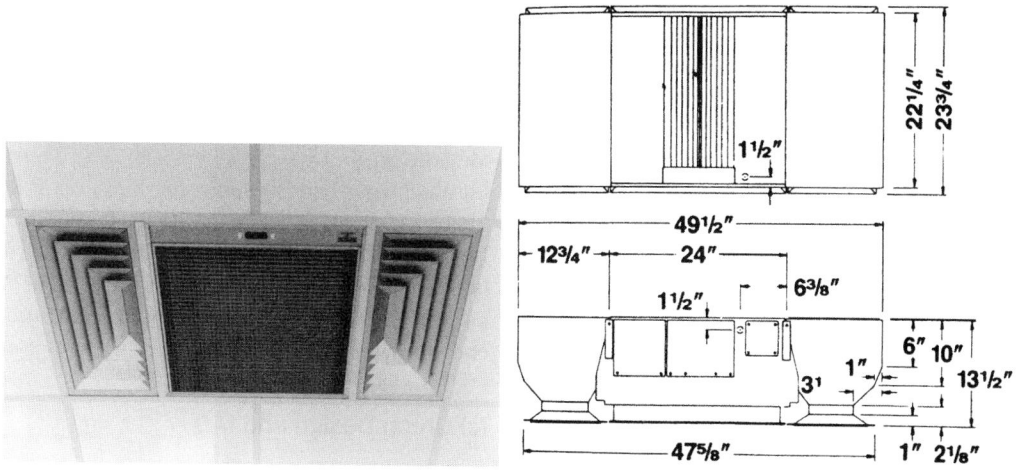

**Fig. 5.25** *A stand-alone electrostatic air filter operates independently of the central HVAC system; no ductwork is required. Airflow rates vary; the slower the flow, the more efficient the filtering. Flush mount (recessed in suspended ceiling), ceiling surface mount, wall mount, and portable models are available. (Courtesy of Tectronic Products Co. Inc., East Syracuse, NY.)*

ings limited), or a fire alarm (damper fully open to aid in smoke extraction).

Where contaminants are expected from sources in addition to humans (typically the case), monitors may be installed to detect carbon monoxide, or combinations of VOCs, or fuels such as propane, butane, or natural gas—or even for depletion of oxygen.

These monitors can be installed as stand-alone alarms or with additional relays that activate equipment. They are about the size of a programmable thermostat; the mounting height depends upon which gas is to be monitored.

Such devices can regulate ventilating heat exchangers, such as the ERV (shown in Fig. 5.18).

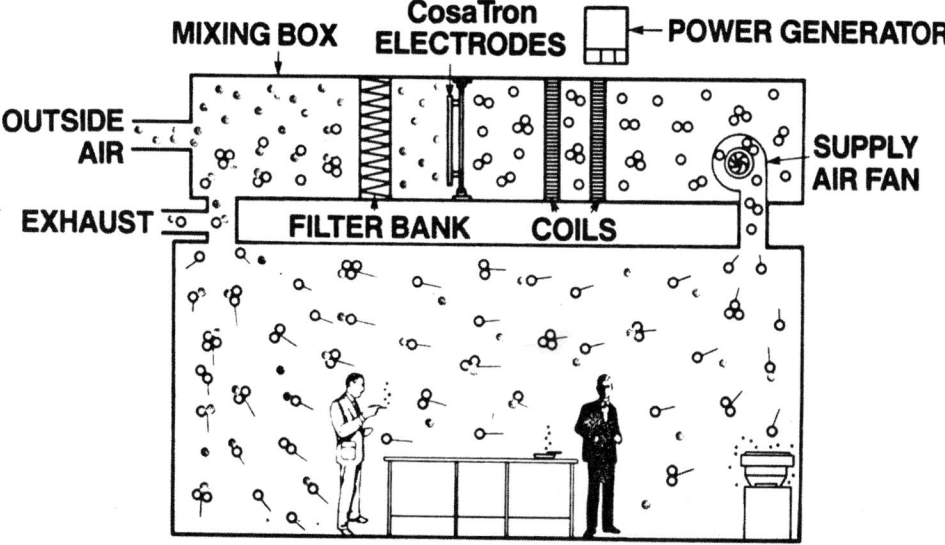

**Fig. 5.26** *This air-cleaning system uses a combination of high-voltage dc and high-frequency ac. The supply air emerges with greatly reduced submicron particles, which have coagulated into larger particles more easily carried by air currents. As this air moves through the occupied space, it picks up submicron particles and is either exhausted or returned to the equipment, where it is filtered to remove the larger particles. (Courtesy of CRS Industries Inc., Tampa, FL.)*

This could be especially useful during unoccupied periods, holidays, and weekends. Many HVAC systems are shut off during such extended periods; VOCs from finishes and furnishings continue to accumulate, however. Periodic flushing from ERVs, controlled by VOC monitors, will help maintain acceptable air quality and could eliminate (or greatly reduce) the need for a Monday morning preflush.

## 5.7 IAQ, MATERIALS, AND HEALTH

The effect of IAQ on occupant health and productivity seems to be an emerging building design issue. This concern stems partly from increasing interest in green design (with its focus upon environmental quality and the resulting impact on materials choices) and partly from continuing stories about the adverse health impacts of poor IAQ (especially cases involving mold and mildew).

### (a) Multiple Chemical Sensitivity

This is an unusual (and sometimes controversial) condition, also known as *environmental illness.* The controversy arises because the causes of this condition are poorly understood and likely involve numerous factors. When causes are mysterious, establishing targeted remedies becomes difficult. People with multiple chemical sensitivity are likely to see the provision of outstanding IAQ as a primary intent of the design process. People with this condition avoid environments with any known environmental risk factors.

Ecology House (in San Rafael, California) is an apartment complex for people with multiple chemical sensitivity. Its construction avoided plywood, using Douglas fir sheathing instead; the floors are tile instead of wall-to-wall carpet; cabinets are metal, not plywood or oriented strand board; the heating system is radiant hot water, not forced air. Barbecues and fireplaces are absent; painted surfaces are minimized; and any window coverings are alternatives to curtains. There is even an "airing room" where, for example, newspapers can be hung before they are read in order to evaporate ink odors.

### (b) Materials and IAQ

There is a rapidly growing availability of environmentally responsible building materials. Many such products address the issue of outgassing via low-VOC formulations. Others provide alternative materials to replace less benign products in common use. Even so, quantitative data are hard to find. Manufacturers of building materials are required to provide Material Safety Data Sheets (MSDS). These reports list all chemical constituents that make up at least 1% of a material (and are not deemed proprietary). Unfortunately, this information does not predict pollutant emission rates. A designer is left with the suspicion that the higher the percentage content of a chemical, the more likely its outgassing.

### (c) Green Design and IAQ

The U.S. Green Building Council's LEED rating system has helped to improve the visibility of acceptable IAQ as a design objective. LEED for new construction and major renovations, for example, establishes compliance with ASHRAE Standard 62 and control of tobacco smoke as prerequisites for LEED certification. Beyond these minimums, designers may choose to address $CO_2$ monitoring, low-emitting materials, or pollutant source control strategies (among others) to achieve rating points. See Appendix G.3 for further information

### References

AIA. 1994. Ramsey, C.G. and H.R. Sleeper. *Architectural Graphic Standards,* 9th ed. American Institute of Architects/John Wiley & Sons, New York.

ASHRAE. 1992. ANSI/ASHRAE Standard 52.1-1992: *Gravimetric and Dust-Spot Procedures for Testing Air-Cleaning Devices Used in General Ventilation for Removing Particulate Matter.* American Society of Heating, Refrigerating and Air-Conditioning Engineers, Inc., Atlanta, GA.

ASHRAE. 1999. ANSI/ASHRAE Standard 52.2-1999: *Method of Testing General Ventilation Air-Cleaning Devices for Removal Efficiency by Particle Size.* American Society of Heating, Refrigerating and Air-Conditioning Engineers, Inc., Atlanta, GA.

ASHRAE. 2004. ANSI/ASHRAE Standard 62.1-2004: *Ventilation for Acceptable Indoor Air Quality.* American Society of Heating, Refrigerating and Air-Conditioning Engineers, Inc., Atlanta, GA.

ASHRAE. 2004. ANSI/ASHRAE Standard 62.2-2004: *Ventilation and Acceptable Indoor Air Quality in Low-Rise Residential Buildings.* American Society of Heating, Refrigerating and Air-Conditioning Engineers, Inc., Atlanta, GA.

Banham, R. 1969. *The Architecture of the Well-Tempered*

*Environment.* University of Chicago Press, Chicago, and the Architectural Press, London.

Chandra, S., P.W. Fairey, and M. Houston. 1986. *Cooling with Ventilation,* Solar Energy Research Institute, Golden, CO.

Fanger, P.O. 1989. "The New Comfort Equation for Indoor Air Quality," in *ASHRAE Journal,* October.

Hellmuth, Obata & Kassabaum, P.C. 2001. *The Greening Curve: Lessons Learned in the Design of the New EPA Campus.* U.S. Environmental Protection Agency, EPA 220/K-02-001.

Lorsch, H. (ed.). 1993. *Air-Conditioning Systems Design Manual.* American Society of Heating, Refrigerating and Air-Conditioning Engineers, Inc., Atlanta, GA.

NREL. April 1988. *Federal Technology Alert: Transpired Collectors (Solar Preheaters for Outdoor Ventilation Air).* National Renewable Energy Laboratory, U.S. Department of Energy, Boulder, CO.

Rousseau, D. and J. Wasley. 1997. *Healthy by Design.* Hartley and Marks, Point Roberts, WA.

Shurcliff, W.A. 1981. *Air-to-Air Heat Exchangers for Houses.* Brick House, Andover, MA.

# P A R T   I I

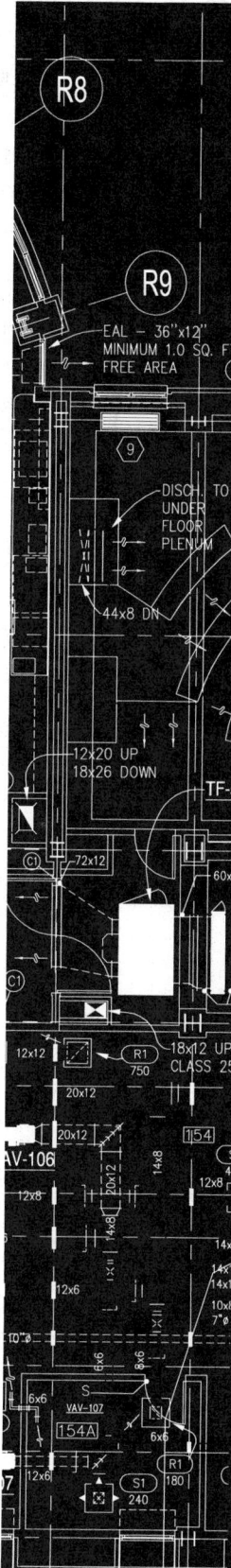

# THERMAL
# CONTROL

The next five chapters deal with solar geometry, heat flow fundamentals, and climate control systems for a range of building types and sizes. The underlying design intent is the comfort of people, for whom we heat, ventilate, and cool buildings, and the provision of acceptable indoor air quality. Thermal comfort criteria are discussed in Chapter 4 and indoor air quality in Chapter 5.

Chapter 6 presents information on solar geometry, a concern with direct impact upon the design of climate control and daylighting systems. Chapter 7 discusses the basic theory of heat flow; whether calculating heat gain in summer or heat loss in winter, the analysis process is based upon the same principles of heat flow. Chapter 8 presents extensive criteria and procedures for design and analysis of building climate control and associated daylighting systems. Initially, numerous strategies are presented as design guidelines, followed later by more detailed design and verification procedures. Chapter 9 addresses heating and cooling equipment for smaller buildings, and Chapter 10 deals with larger buildings and groups of buildings.

# Solar Geometry and Shading Devices

AN UNDERSTANDING OF THE SUN'S POSITION DURING the day is critical to site planning, daylighting design, passive solar design, and controlling unwanted heat gains. In early civilizations, most cultures revered the sun as a deity and understood its patterns and importance in the cycle of life. Today, such understanding is just as important when considering the relationships between the environment and technology in the design of buildings. We can no longer afford the resources to sustain the construction and operation of buildings with poorly designed glazed areas habitable only through the use of active climate control systems. We can rely on the patterns of the sun as a constant in the design process and create environments that work with these patterns, climate, and well-considered design intents.

## 6.1 THE SUN AND ITS POSITION

The sun is a giant star and the largest object in the solar system. The sun's energy is produced by nuclear fusion that occurs at temperatures in the range of 18–25 million °F (10–14 million °C). The energy from thermonuclear fusions at the sun's surface is released as electromagnetic radiation at approximately 10 million °F (5.6 million °C). The solar spectrum (Fig. 6.1) consists of approximately 5% ultraviolet (shortwave) radiation (less than 350 nm), 46% light—what we see—(350 to 750 nm), and 49% infrared (longwave) radiation

(greater than 750 nm). Though 93 million miles from Earth, the amount of the sun's radiation (heat energy) reaching the Earth's outer atmosphere is relatively stable and is called the *solar constant*—433 Btu/ft$^2$ hr (1.37 kW/m$^2$). Some annual variation in the solar constant occurs because of the elliptical orbit of the Earth around the sun, but it is so small that it has little effect on building design. The length of radiation's path through the atmosphere is the most important factor in determining how much radiation reaches the surface of the Earth. This received radiation is termed *insolation*.

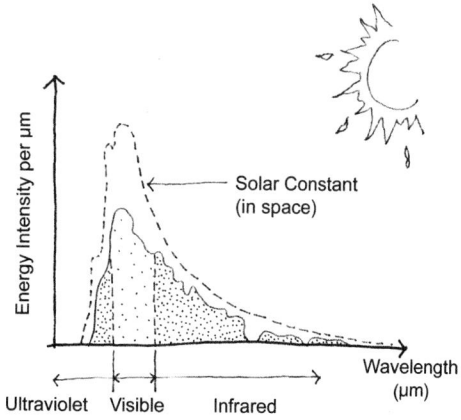

**Fig. 6.1** At the Earth's surface the solar radiation spectrum consists of ultraviolet radiation, visible radiation (light), and infrared radiation. (Drawing by Erik Winter; © 2004 Alison Kwok, all rights reserved.)

## (a) Earth's Rotation and Tilt

The Earth's axis of rotation is tilted at 23.5° (precisely 23.47°), also known as its *declination*. The Earth travels in the plane of the ecliptic (Fig. 6.2) in its orbit around the sun. The tilt of the Earth is toward the sun (June +23.5°) or away from the sun (December −23.5°). This tilt is responsible for the seasons. In the Northern Hemisphere in June, the sun's direct radiation is perpendicular to the Earth's surface; in December, there are fewer hours of sun and the radiation passes through a longer, nonperpendicular, length of atmosphere to reach the Earth. The rotation of the Earth (once every 24 hours) and its tilt determine the length of atmosphere that solar radiation passes through and consequently how much energy the Earth's surface receives.

## (b) Altitude and Azimuth

The position of the sun at any instant with respect to an observer on the ground is defined by its *altitude angle* and its *azimuth angle*. The *altitude* of the sun is the angle between the horizon and the sun's position above the horizon. The altitude varies during the day, beginning and ending at 0° at sunrise and sunset and reaching a daily maximum at *solar* noon. The altitude angle at noon varies from day to day, reaching a yearly maximum on June 21 (the summer solstice), a minimum on December 21 (the winter solstice), and a point halfway between the two on the vernal and autumnal equinoxes (March 21 and September 21). The *azimuth* (also called the solar bearing angle) is the angle along the horizon between the projected position of the sun and true (solar) south. Note that the azimuth is referenced from south herein, whereas in other sources it is sometimes referenced from north; there is no clear consensus on this issue. Refer to Fig. 6.3 and Tables C.11 to C.14 for solar position and insolation information.

Referring to the graphic relationships in Fig. 6.2, the height of the sun in the sky (altitude angle) depends upon the observer's position on the Earth (the latitude) and the seasonal changes (the tilt of the Earth). This can be expressed in Equation 6.1 as follows:

altitude angle at solar noon =
$$90° - \text{latitude} \pm \text{declination} \quad (6.1)$$

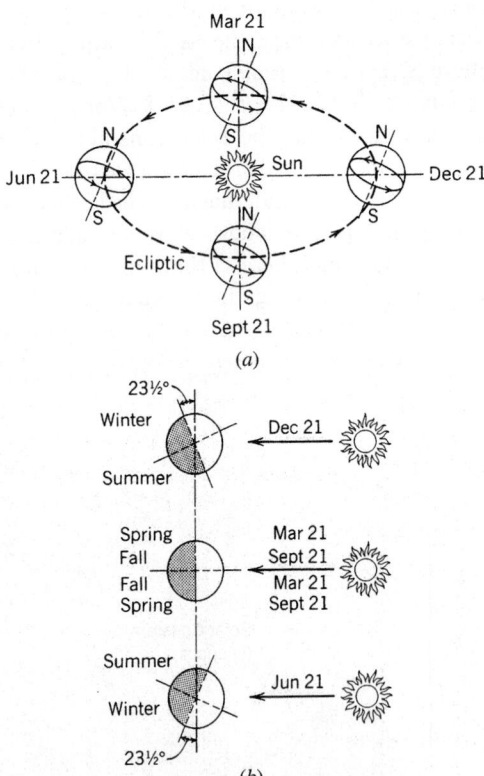

*(a)*

*(b)*

**Fig. 6.2** *(a) The ecliptic is the annual path of the Earth around the sun. (b) The tilt of the Earth's axis in the plane of the ecliptic results in the seasonal variations. In the Northern Hemisphere the Earth tilts away from the sun in December (−23.5°), resulting in winter's low sun altitude and cold weather. In June the effect is reversed.*

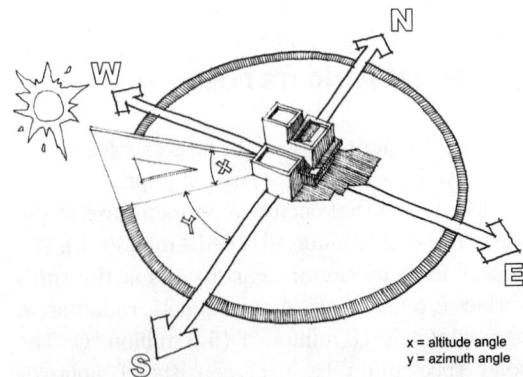

x = altitude angle
y = azimuth angle

**Fig. 6.3** *Altitude and azimuth angles. (Drawing by Erik Winter; © 2004 Alison Kwok; all rights reserved.)*

For example, the solar altitude for Minneapolis (44.9°N) for various seasons at solar noon is:

Maximum (June 21): = 90° − ~45° + 23.5° = 68.5°

Equinox (Mar/Sept 21) = 90° − ~45° − 0° = 45°

Minimum (Dec 21) = 90° − ~45° − 23.5° = 21.5°

These relations are shown graphically in Fig. 6.4.

The azimuth angle during its path from sunrise to sunset changes with the season. On the equinoxes (March 21 and September 21) the sun rises due east and sets due west. In the Northern Hemisphere, the sun rises north of east and sets north of west between March 22 and September 20. This is shown graphically in Fig. 6.5 for one specific latitude (32°N). Sunrise on June 21 is 28.5° north of east and sunset is correspondingly 28.5° north of west. On December 21 sunrise and sunset are 28.5° south of east and west.

A designer can use three basic solar pattern concepts when designing with the sun:

1. The altitude of the sun is highest in summer, lowest in winter, and in between in spring and

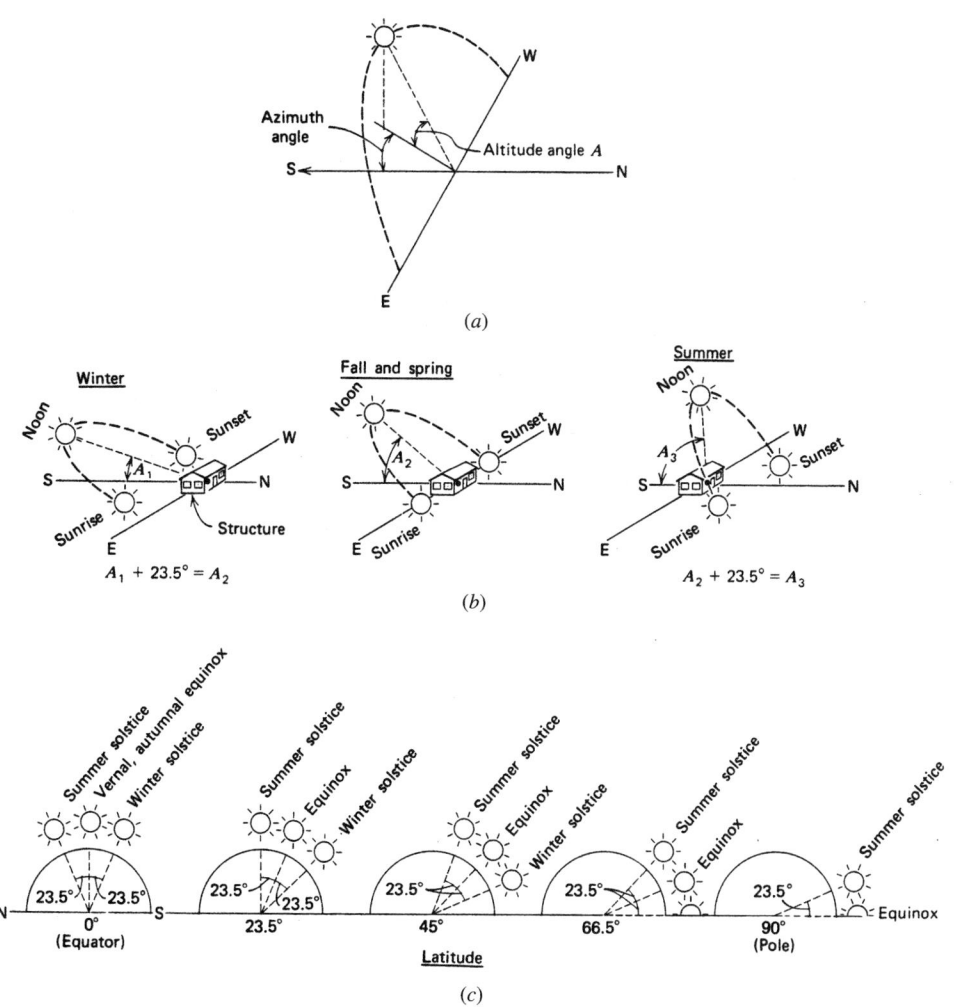

**Fig. 6.4** (a) The position of the sun is expressed in terms of vertical angle above the horizon (altitude) and horizontal angle (azimuth, measured from the south). (b) Approximate position of the sun in each of the seasons at a mid-northern latitude (approximately 45°). Note that the altitude angle is maximum in summer, minimum in winter, and in between in spring and fall. (c) Maximum sun altitude at various latitudes for both solstices and equinoxes.

THERMAL CONTROL

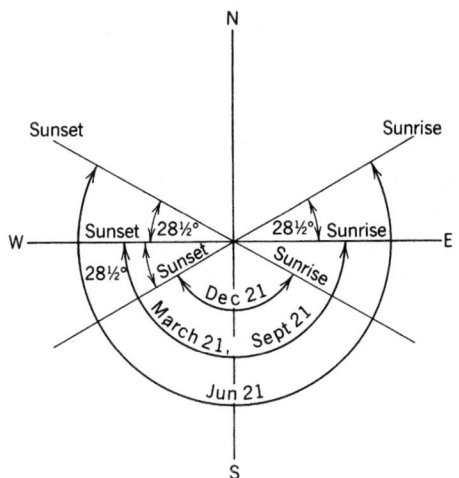

**Fig. 6.5** *A diagram of the sun's azimuth path from sunrise to sunset projected on a horizontal plane for solstices and equinoxes at latitude 32° N (Jerusalem, Savannah, Tucson). The sun's azimuth angle, like its altitude, varies with the time of day and the date.*

fall for all latitudes. The maximum difference in altitude angle between summer and winter solstices is about 47°.

2. The daily maximum altitude of the sun increases as a location approaches the equator (low latitudes, either north or south). The seasonal altitude *variation*, however, is the same for all latitudes (except at those extreme north and south latitudes where the sun is above or below the horizon for extended time periods). This factor not only has a pronounced effect on the design and efficacy of sun-shading devices, but also affects exterior daylight illuminance levels.

3. The azimuth angle of the sun is dictated by the time of day and by the season. The sun by definition traverses the sky between sunrise and sunset and in the Northern Hemisphere rises north of east during the summer, due east on the equinoxes, and south of east during the winter. The principal significance of the azimuth angle is its role as an important consideration when orienting a building on a site, establishing building exposures, and analyzing shading angles.

## 6.2 SOLAR VERSUS CLOCK TIME

The basis of all timekeeping is the length of the solar day, that is, one full rotation of the Earth on

its axis. The Earth's motion on the ecliptic is not uniform—it moves more rapidly when farther from the sun. As a result, the actual length of a solar day varies. Since it is impractical to use timepieces that need daily adjustment, we use an average, or *mean solar day*, as the basis for timekeeping. There are three reasons for the difference between solar time and clock time:

1. *Location within a time zone.* The 360° circumference of the Earth is divided into 24 one-hour time zones, each with a width of approximately 15° of longitude, which corresponds to approximately 1000 miles (1610 km). An observer located at any point other than directly on a standard time zone reference longitude (normally in the center of a time zone) must make a time correction for his/her distance from the reference longitude. (In some countries, a reference longitude is more than 30 minutes from time zone extremes. In such places, the time correction for longitude plus the equation of time can total well over an hour. This situation does not exist in the continental United States.)

2. *Equation of time.* The speed of the Earth in its orbit around the sun is nonuniform; it moves more rapidly when farther from the sun. As a result, the actual length of the solar day varies as defined by the *equation of time* (Fig. 6.6a). The extent of this variation can be determined from a curve called the *analemma* (Fig. 6.6b) or from tabulations in various sources. The analemma is the shape that results if the sun's position in the sky is recorded at the same time of day throughout the year.

3. *Daylight Saving Time (DST).* During DST, clocks are moved forward 1 hour to effectively give (from the morning) 1 more hour of daylight in the evening during the summer months. (Note: the official term is Daylight Saving Time, not Daylight Savings Time.) The purpose of DST is to make better use of daylight. In most of the United States (except Arizona, Hawaii, and Indiana), DST begins at 2:00 A.M. on the first Sunday of April and turns back to standard time on the last Sunday in October. In the European Union, Summer Time begins at 1:00 A.M. Universal Time (Greenwich Mean Time) on the last Sunday in March and ends the last Sunday in October.

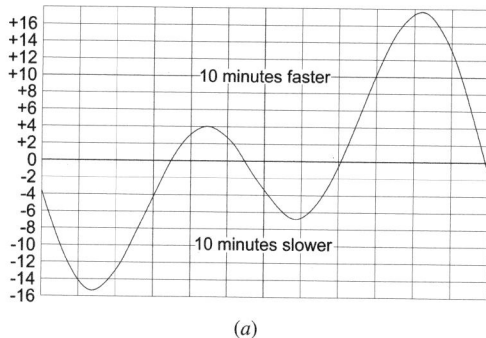

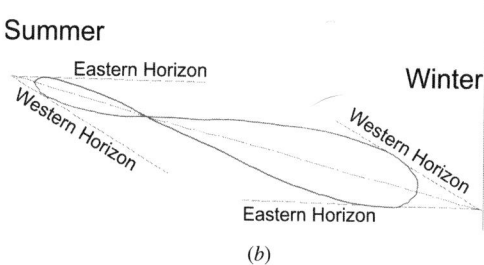

**Fig. 6.6** (a) Equation of time. (b) The analemma is a miniature almanac where the vertical coordinates of the path represent the sun's declination on a particular day of the year, while the horizontal coordinates tell how much the sun is ahead of or behind clock time on that day. (Drawings by Erik Winter; © 2004 Alison Kwok, all rights reserved.)

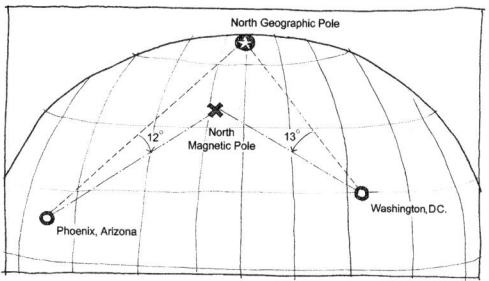

**Fig. 6.7** The concept of magnetic deviation (from true north), illustrated for Phoenix, Arizona (12° easterly) and Washington, DC (11° westerly). (Drawing by Erik Winter; © 2004 Walter Grondzik, all rights reserved.)

Unless otherwise clearly noted, most solar position tables and charts (including sunpath diagrams) are based upon solar time, not clock time. This must be considered when solar-driven loads are being added to clock-driven loads (for example, the heat gain from the western sun is being added to the heat gain from office occupants to determine the total load). Such an addition must use coincident values, and solar and clock times are rarely coincident.

## 6.3 TRUE SOUTH AND MAGNETIC DEVIATION

Working from a "true south" or "solar south" orientation (opposite of "true north," location of the Earth's axis of rotation) is the starting point in the planning and design of a building, particularly for passive solar strategies. When a compass is used on a building site to find north, the compass yields a reading toward magnetic north, determined by the Earth's internal magnetic field. This is usually (depending upon the locale) not the same as true north. Figure 6.7 illustrates this phenomenon. To correct between magnetic north (compass) and true north (solar), a magnetic deviation is applied to the compass reading. Depending upon the location on the Earth's surface, the deviation may be as large as 50°—but in most locations this is not the case. It should be noted that the magnetic field is not uniform, stationary, or perfectly aligned with the Earth's poles; thus, magnetic deviations can change and have changed over time.

## 6.4 SUNPATH PROJECTIONS

Design tools such as sunpath projection charts can be used to assess site conditions, orient a building, or provide information critical for passive solar design. Sunpath charts represent the sun's position throughout the year. There are two fundamental types of sunpath projections: vertical and horizontal. Horizontal projections are more commonly encountered. The basic idea behind the horizontal projection is shown in Fig. 6.8. The sky vault is represented as a hemisphere that projects as a circle, the center of which is the zenith. The observer looks down onto the hemispherical sky vault and projects the sunpaths from the vault onto the two-dimensional projected circle. The basic difference between the various horizontal projections is in the spacing between circles of equal altitude. Vertical projection charts are also available.

Four of the more common sunpath projections are (1) gnomonic, (2) equidistant, (3) rectilinear, and (4) stereographic.

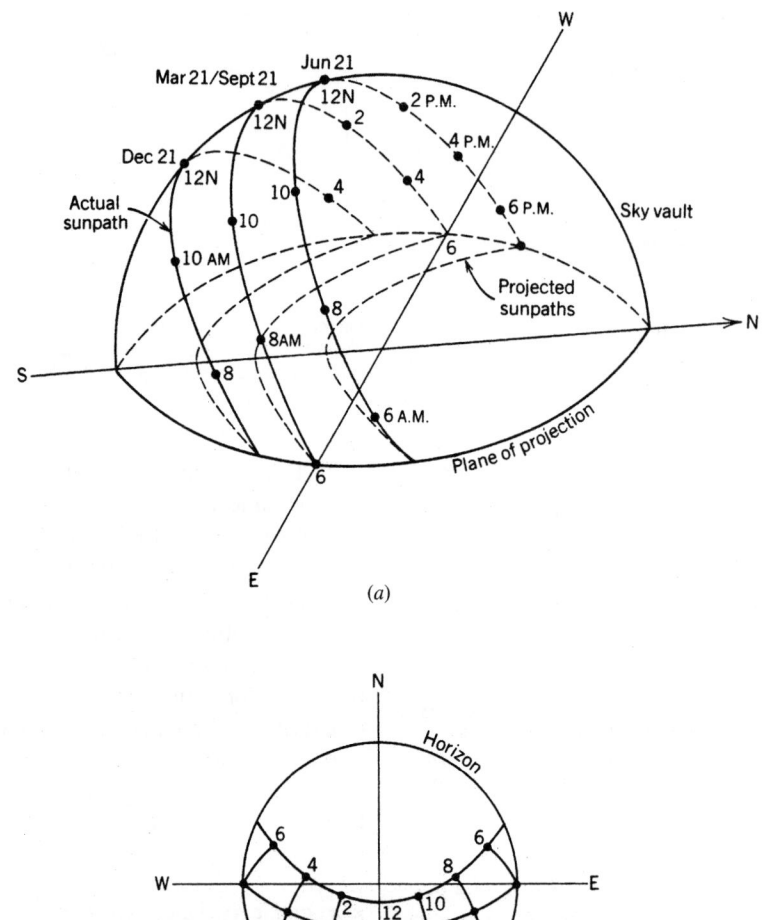

**Fig. 6.8** (a) The sky vault is represented as a hemisphere, which projects as a circle (b) on a projection plane. The exact shape of the sunpath chart depends upon the projection method. The hour points on the sunpath chart are connected to form hour lines that are useful for interpolation.

1. The *gnomonic projection* (used with sundials and sunpeg charts) is derived from the sundial. The observer is at the center of the projection; therefore low sun angles (sunrise, sunset) extend to infinity. For this reason, this method is rarely used for solar charts but is easily applied to building shadow studies. Sunpeg charts for specific latitudes (see Fig. 6.9 and Appendix D.2) will show the exact position of sun penetration and shadow on a model of any scale, on any date, at any time of day between shortly after sunrise and shortly before sunset. Use of the sunpeg chart is outlined below (reprinted by permission from Brown, Reynolds, and Ubbelohde, *Inside Out: Design Procedures for Passive Environmental Technologies,* John Wiley & Sons, © 1982) and shown in Fig. 6.10.

- *Find the sunpeg chart corresponding to the nearest latitude for a site.*
- *Make a copy of this chart. If the copier changes the chart size, the peg height line will also change; there should be no problem.*
- *Construct a "peg" (gnomon) whose finished height above the chart surface corresponds exactly to the "peg height" shown on the copy of the chart. This peg must stand and remain perfectly vertical relative to the model and not be bumped out of vertical alignment.*

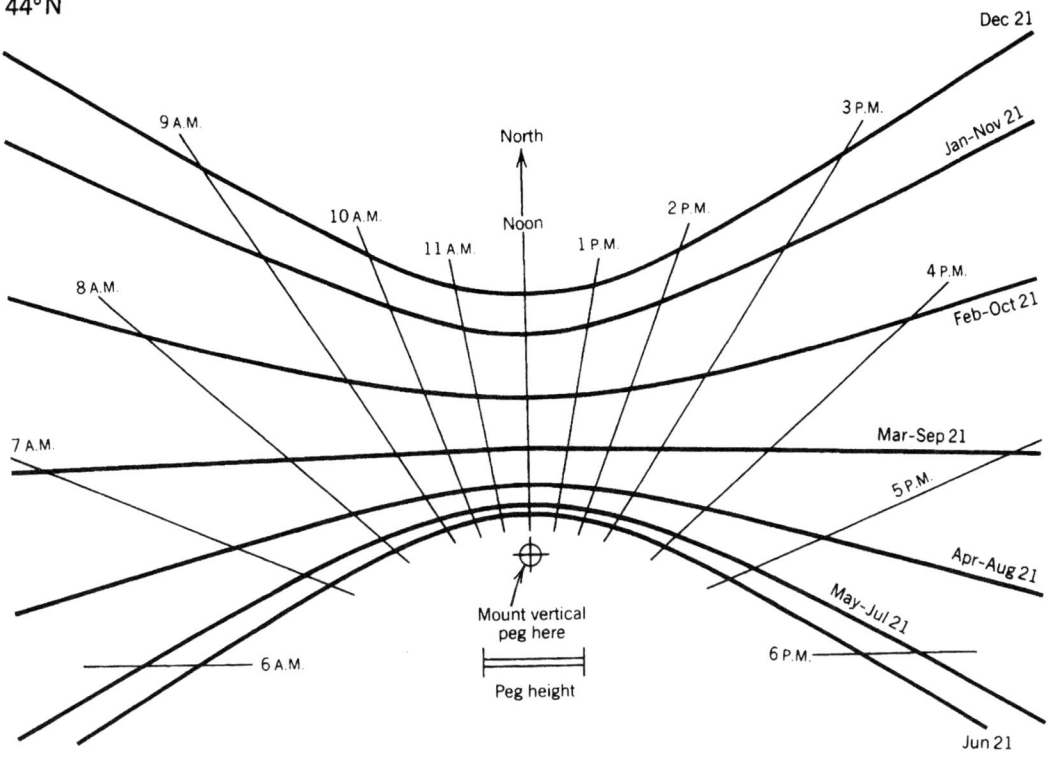

44°N

Dec 21

9 A.M.

North

3 P.M.

Jan-Nov 21

10 A.M.

Noon

2 P.M.

11 A.M.

1 P.M.

8 A.M.

4 P.M.

Feb-Oct 21

7 A.M.

Mar-Sep 21

5 P.M.

Apr-Aug 21

Mount vertical
peg here

May-Jul 21

6 A.M.

6 P.M.

Peg height

Jun 21

**Fig. 6.9** *Sunpeg chart for 44° N latitude.*

- *Mount the copy of the chart on the model to be tested. The chart must be perfectly horizontal over its entire surface, and the north arrow on the chart must correspond to true (solar) north on the model.*
- *Choose a test time and date. Take the model out into direct sunlight. (With a small model, a camera tripod might serve as a mount.) Tilt the model until the shadow of the peg points*

**Fig. 6.10** *A sunpeg chart attached to the base of a scale model to study solar access and shadow patterns in and about a building. (Photo by Jonathan Meendering; © 2004 Alison Kwok; all rights reserved.)*

*toward the intersection of the chosen time line and the chosen date curve. When the end of the peg's shadow touches this intersection, the model will show the same sun-shadow patterns as would occur on the time and date chosen.*

2. The *equidistant projection (horizontal, polar)* is used almost exclusively in the United States because of the ready availability and wide acceptance of the sunpath charts provided in the Pilkington (formerly Libbey Owens Ford) Sun Angle Calculator. The observer moves around the skydome, like the sun's view to the earth. Figure 6.11 shows a sample equidistant projection sunpath diagram. A full set of equidistant projections is provided in Appendix D.3. This projection method is equally applicable to all latitudes, and a latitude-specific sunpath is needed for a location under study. The Pilkington Sun Angle Calculator sunpath diagrams are available in steps of 4° of latitude between 24° and 52° latitudes.

3. The *rectilinear projection (vertical, cylindrical)* is a two-dimensional graph of the sun's

THERMAL CONTROL

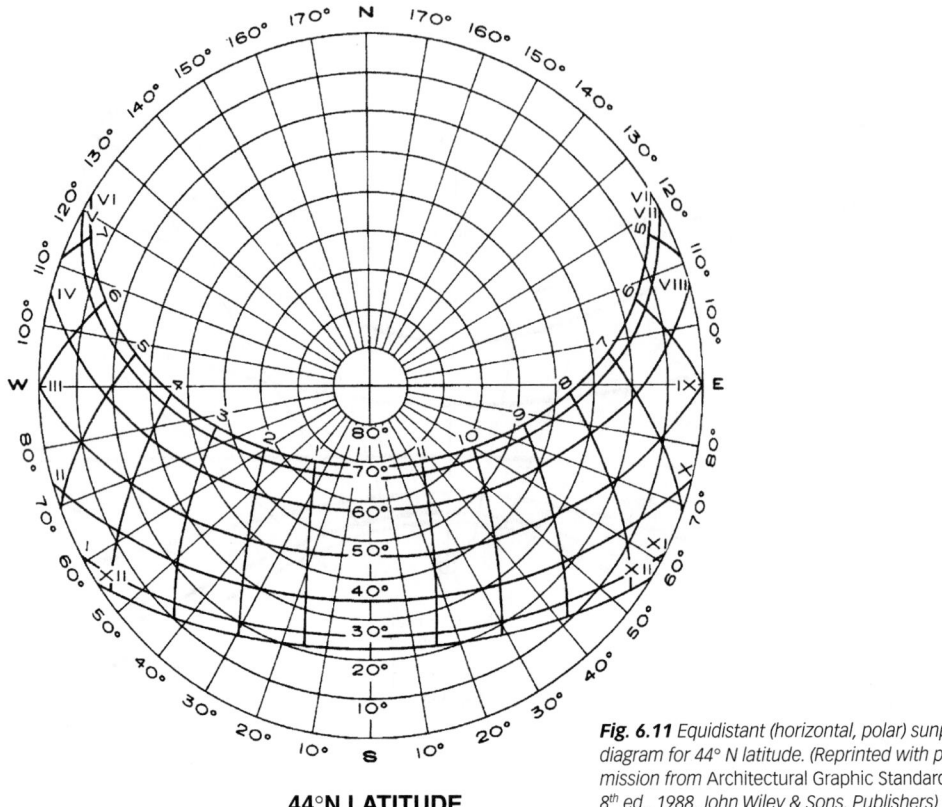

**44°N LATITUDE**

*Fig. 6.11 Equidistant (horizontal, polar) sunpath diagram for 44° N latitude. (Reprinted with permission from* Architectural Graphic Standards, *8th ed., 1988, John Wiley & Sons, Publishers)*

position in Cartesian coordinates. Azimuth is plotted along the horizontal axis and altitude on the vertical axis. The development of this type of chart is shown in Fig. 6.12. These charts are actually graphs superimposed on the eye of an observer. At the vertical center of each chart, the observer is looking due south. The horizon (a horizontal plane at the observer's eye level) is the line at the bottom of the chart. A set of rectilinear charts for several latitudes (from Mazria, 1979) is found in Appendix D.4.

These charts may also be used to easily determine the number of hours of greatest insolation, that is, what percentage of clear-day insolation is gained each hour by an unshaded *south-facing* vertical window (Fig. 6.13). This emphasis on south orientation calls attention to its useful characteristic of receiving *more* sun in winter and *less* sun in summer than any other orientation. To convert these percentages to actual heat gains, see Tables C.1 to C.10 (for *clear* days at a given latitude) or Table C.15 for *average* day data by climate station (January and July only).

4. The *stereographic projection* (circular, equal spacing) also represents the sun's changing position in the sky throughout the day and year. The format is like a fish-eye photograph of the sky pointed to the zenith directly above. The various paths of the sun throughout the year can be projected onto the sky, then "flattened." The principal advantage of this projection is that it is very simple to draw sunpaths at any scale and for any latitude without having to rely on published sunpaths. Thus, for any serious sunshading design, a designer can prepare a large-scale sunpath chart at the exact latitude of a project and do accurate graphical analysis. For a detailed description of the method of preparing these charts, see Szokolay (1980).

The design of sunshading devices necessitates projecting the three-dimensional solar path onto a two-dimensional surface. Stereographic and equidistant projections are the most popular for sunshading analysis.

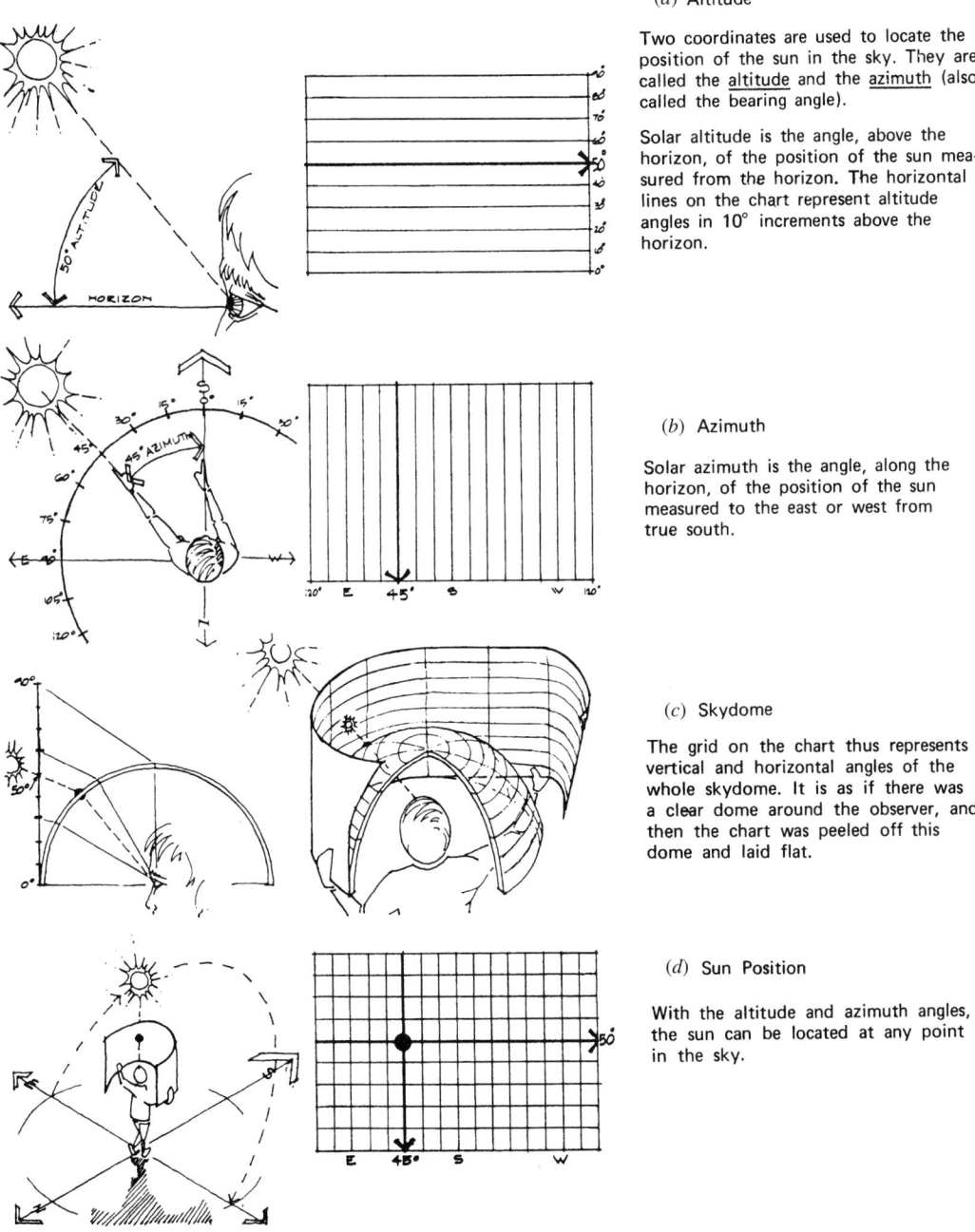

(*a*) Altitude

Two coordinates are used to locate the position of the sun in the sky. They are called the <u>altitude</u> and the <u>azimuth</u> (also called the bearing angle).

Solar altitude is the angle, above the horizon, of the position of the sun measured from the horizon. The horizontal lines on the chart represent altitude angles in 10° increments above the horizon.

(*b*) Azimuth

Solar azimuth is the angle, along the horizon, of the position of the sun measured to the east or west from true south.

(*c*) Skydome

The grid on the chart thus represents vertical and horizontal angles of the whole skydome. It is as if there was a clear dome around the observer, and then the chart was peeled off this dome and laid flat.

(*d*) Sun Position

With the altitude and azimuth angles, the sun can be located at any point in the sky.

**Fig. 6.12** *Illustrations (a) through (g) describe the development of the rectilinear sunpath chart. (From Edward Mazria and David Winitsky,* Solar Guide and Calculator, *Center for Environmental Research, University of Oregon, 1976.) (h) U.S. map showing cities along similar latitudes.*

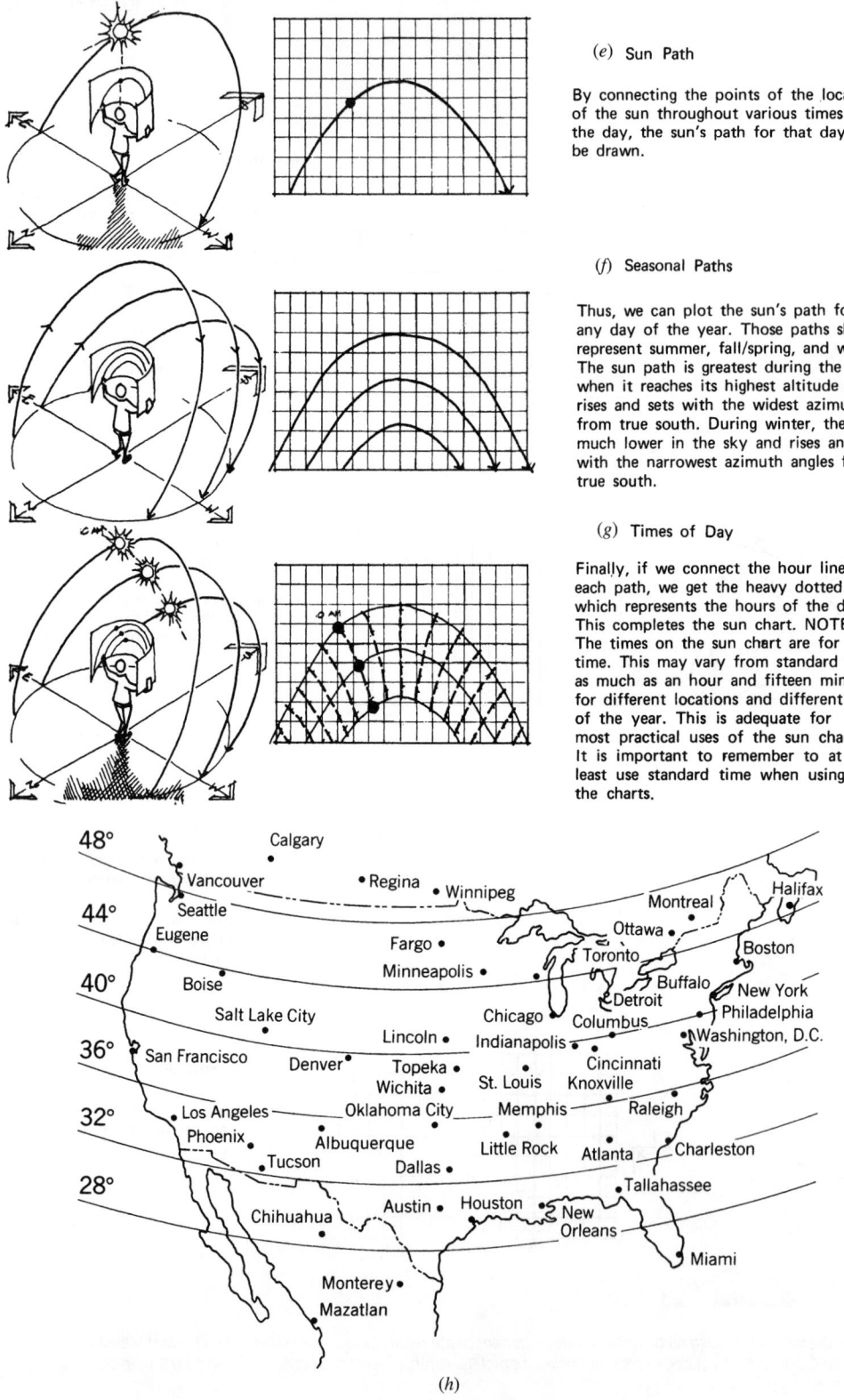

(e) Sun Path

By connecting the points of the location of the sun throughout various times of the day, the sun's path for that day can be drawn.

(f) Seasonal Paths

Thus, we can plot the sun's path for any day of the year. Those paths shown represent summer, fall/spring, and winter. The sun path is greatest during the summer when it reaches its highest altitude and rises and sets with the widest azimuth angle from true south. During winter, the sun is much lower in the sky and rises and sets with the narrowest azimuth angles from true south.

(g) Times of Day

Finally, if we connect the hour lines on each path, we get the heavy dotted line which represents the hours of the day. This completes the sun chart. NOTE: The times on the sun chart are for solar time. This may vary from standard time as much as an hour and fifteen minutes for different locations and different times of the year. This is adequate for most practical uses of the sun chart. It is important to remember to at least use standard time when using the charts.

(h)

**Fig. 6.12** (Continued)

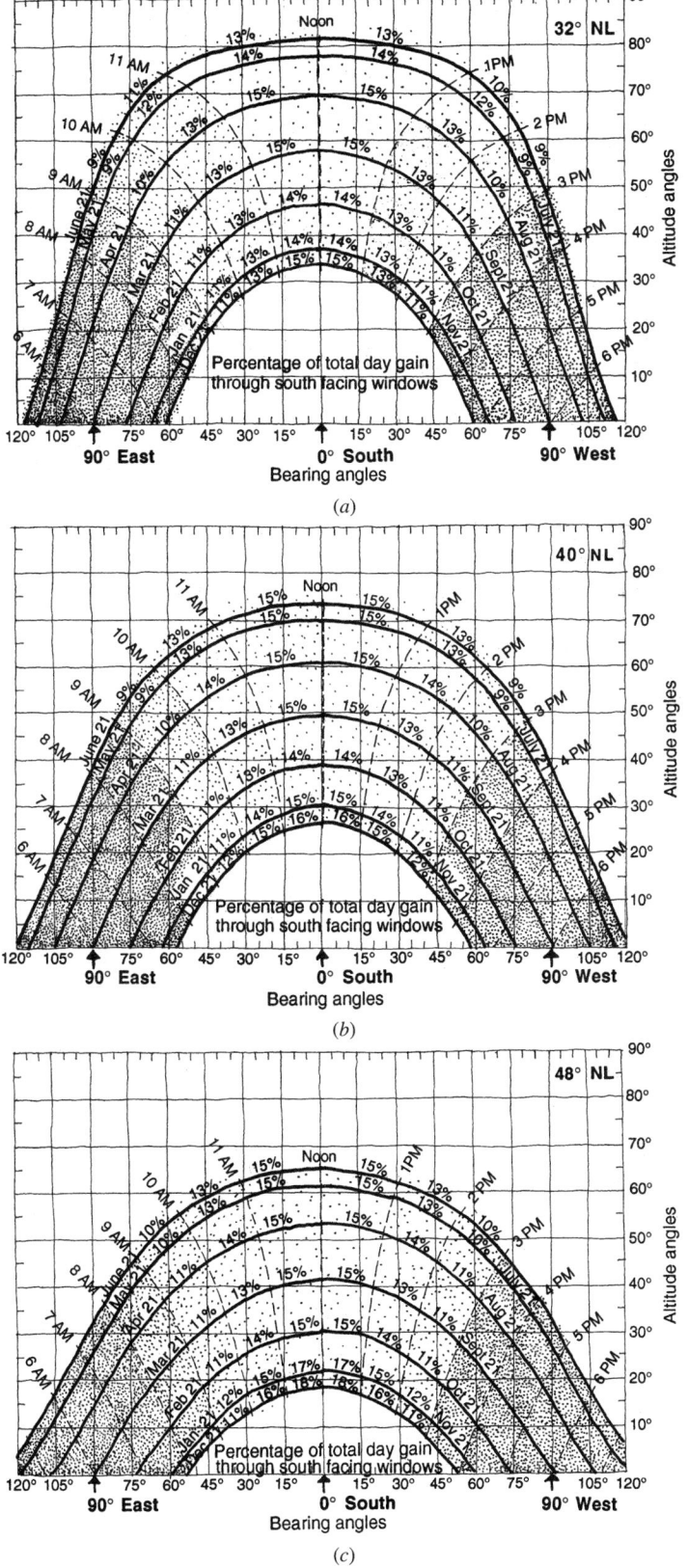

**Fig. 6.13** *Sunpath charts for three latitudes (32°, 40°, 48°) show the approximate percentage of clear-day insolation for south-facing windows for each of the maximum hours of sun each month.*

## 6.5 SHADING

Shading windows from solar heat gain is a key design strategy for passive cooling and to reduce cooling loads on active HVAC systems. Shading the opaque building envelope is also important, but since thermal resistance is usually greater through such elements than through glazing, the discussion in this section refers primarily to design strategies for shading windows oriented south, east, and west. Northern windows often also need shading devices, contrary to the myth that the north façade never receives direct beam radiation.

### (a) Shading for Orientation

Because of the high altitude of the sun, the most effective shading device for south-facing windows during the summer is a horizontal overhang. On east- and west-facing windows, a horizontal overhang is somewhat effective when the sun is at high positions in the sky but is not effective at low-altitude angles. A variety of shading devices are illustrated by the diagrams and examples in Figs. 6.14–6.18.

Direct solar gain through east- and west-facing windows can be an extraordinary heat gain liability and produce thermal and visual discomfort. During the early design stages, it is best to orient spaces to face north or south to avoid the east/west sun's low angle. If this is not possible, vertical fins are an effective strategy for east and west orientations. Eggcrate shading devices (a combination of overhangs and fins) provide optimal shading particularly in hot climates (Fig. 6.19). North-facing windows receive direct solar radiation in the summer in the early morning and near sunset, when the altitude of the sun is very low. For shading on the north side at these times, vertical fins are most effective.

### (b) Operable Shading Devices

Operable exterior shading devices are useful because they respond to daily and seasonal variations in solar and weather patterns in ways that fixed shading devices simply cannot do. The operation of a movable shading device can be as simple as twice-a-year adjustment—for example, manually extending roller shades, awnings, rotating fins, or louvers at the beginning of summer and

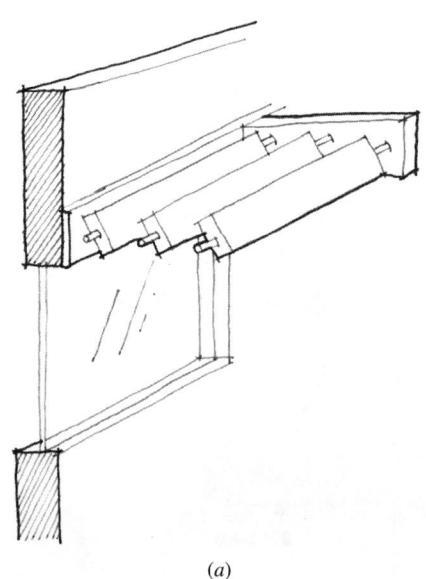

(a)

(b)

**Fig. 6.14** (a) Overhang with horizontal louvers. (b) Horizontal louvers allow free air movement—office building in Honolulu, Hawaii. (Drawing by Erik Winter, photo by Alison Kwok; © 2004 Alison Kwok, all rights reserved.)

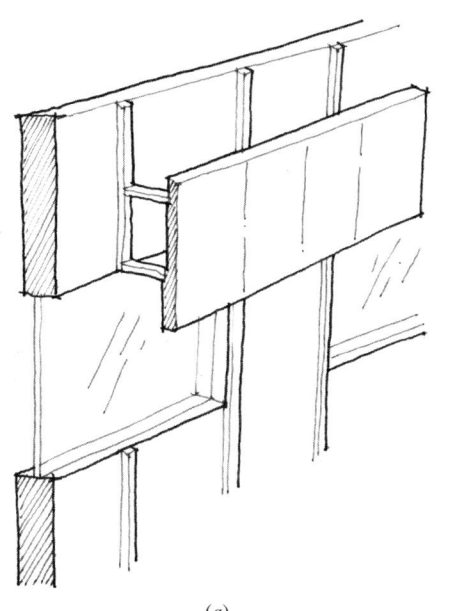

*Fig. 6.15* (a) Overhang with vertical panel. (b) Vertical panels at the IBM Tower in Kuala Lumpur, Malaysia. (Drawing by Erik Winter, photo by Alison Kwok; © 2004 Alison Kwok, all rights reserved.)

retracting the shade after the hot season has ended (in fall). These devices are very effective at blocking low sun angles from the east or west. More complex movable devices are typically on automated daily and seasonal programs. Although many facility managers are of the opinion that movable exterior shading devices require high maintenance and are prone to malfunctioning, the designer can apply appropriate technology to provide a low-maintenance solution.

*Fig. 6.16* (a) Vertical fins. (b) Fixed vertical fins on a government building in Honolulu, Hawaii. (Drawing by Erik Winter, photo by Alison Kwok; © 2004 Alison Kwok; all rights reserved.)

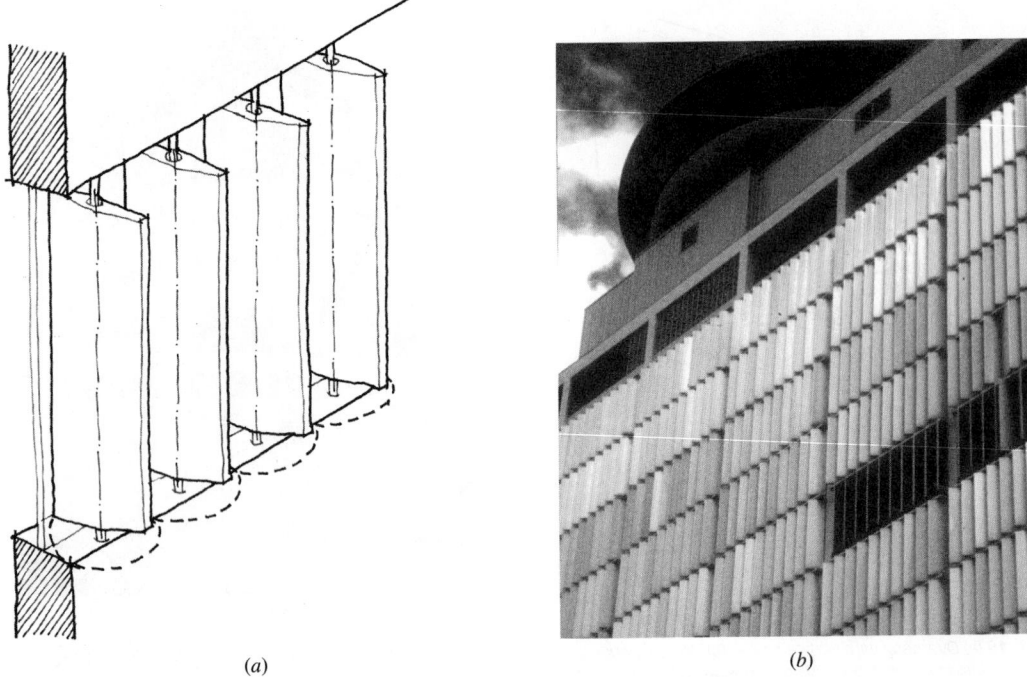

(a)  (b)

**Fig. 6.17** (a) Movable fins. (b) Movable fins are positioned according to the sun's position at the Ala Moana office building in Honolulu, Hawaii. (Drawing by Erik Winter, photo by Alison Kwok; © 2004 Alison Kwok; all rights reserved.)

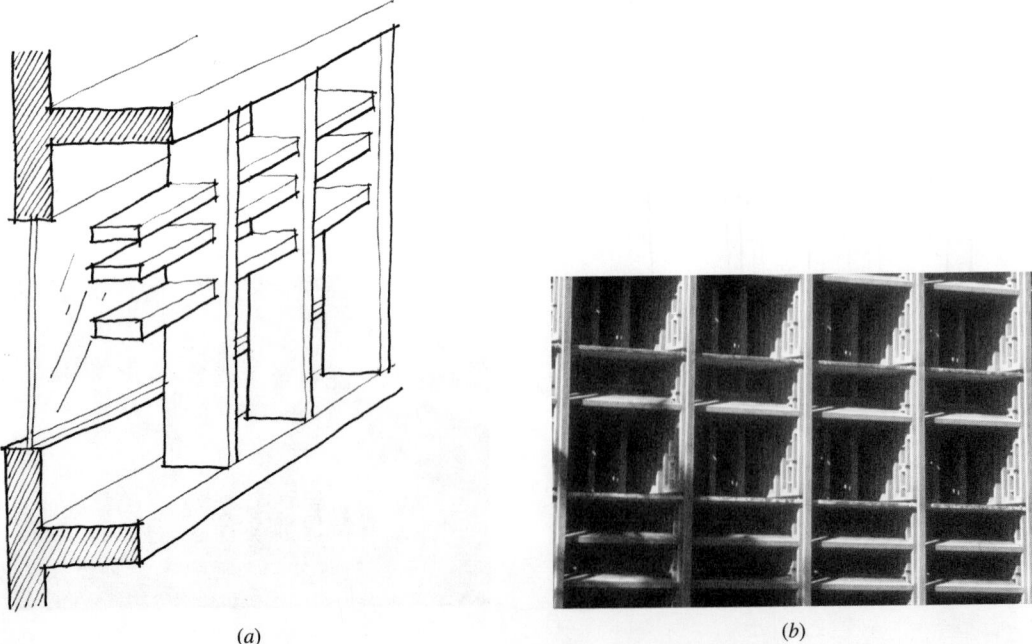

(a)  (b)

**Fig. 6.18** (a) Eggcrate shading device. (b) Modified eggrate to allow air movement and lighter structure at the Board of Water Supply in Honolulu, Hawaii. (Drawing by Erik Winter, photo by Alison Kwok; © 2004 Alison Kwok; all rights reserved.)

**162**

(a)                                                                 (b)

**Fig. 6.19** *Integration of eggcrate shading devices into the envelope at (a) the Hawaii Medical Services Association building in Honolulu, Hawaii and (b) the University of Arizona library in Tucson. (Photos by Alison Kwok; © 2004 Alison Kwok; all rights reserved.)*

**Fig. 6.20** *Vines grow on an exterior structure at the Finnish Embassy in Washington, DC. (Photo by Alison Kwok; © 2004 Alison Kwok; all rights reserved.)*

Designers can also situate deciduous trees and vines at key locations around a building, which will act as natural shading devices with a natural cycle for shade in the summer and loss of leaves in the winter (Fig. 6.20). Plants have the advantages of costing little, reducing glare, providing an attractive connection to nature, and reducing exterior surface (wall and ground) temperatures.

## 6.6 SHADOW ANGLES AND SHADING MASKS

Altitude and azimuth angles are very useful in understanding solar position and sunpath diagrams, but are much less useful in defining the shadow angles cast by a projection on a wall

exposed to the sun. More angles come into play with shading device design.

### (a) Shadow Angles

When designing shading devices, the geometry of the device itself and its relationship to the face of a building produce a number of angles relative to the desired shadow being cast. Since there is no universal nomenclature for these relationships, the angles involved in the design of shading devices will be very carefully defined. Refer to Fig. 6.21, which shows the shadow cast by a horizontal overhang on a wall exposed to sunlight. Note that the shadow is defined by two angles: the *vertical shadow angle* (VSA), which indicates the position of the leading edge of the shadow as defined from the leading edge of the overhang, and the *horizontal shadow angle* (HSA), which defines the leading edge of a shadow cast by a vertical element (indicated by a dashed line) as defined with respect to that element's leading edge. The terms *vertical shadow angle* and *horizontal shadow angle* are in use throughout the world, although in the United States the vertical shadow angle is also known widely as the *profile angle* because it is so designated on the Pilkington Sun Angle Calculator.

Figure 6.22 shows the same information in slightly different form; line *DA* is the shadow cast by line *DE*, which is the intersection between a horizontal projection and a vertical projection. This line is particularly important in determining the required extent of a shading element, as will be shown below.

THERMAL CONTROL

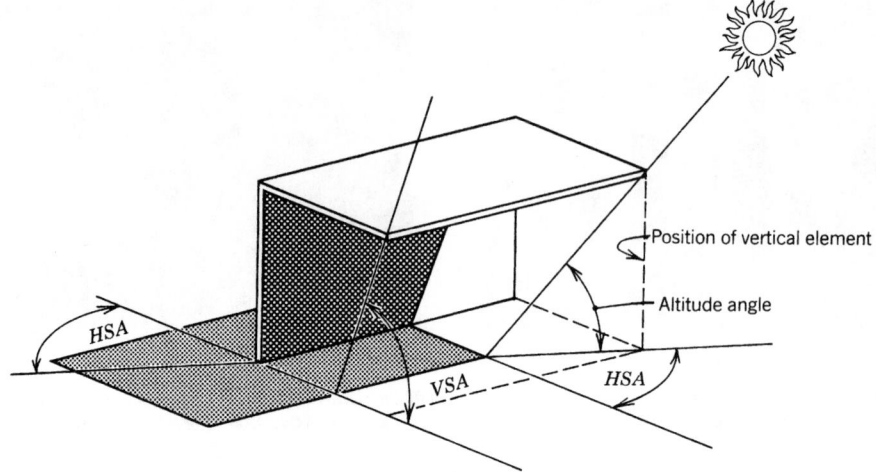

**Fig. 6.21** *The shadow cast by a horizontal overhang is best defined by the vertical shading angle (VSA) and the horizontal shading angle (HSA). The VSA is also known as the* profile angle *in the United States.*

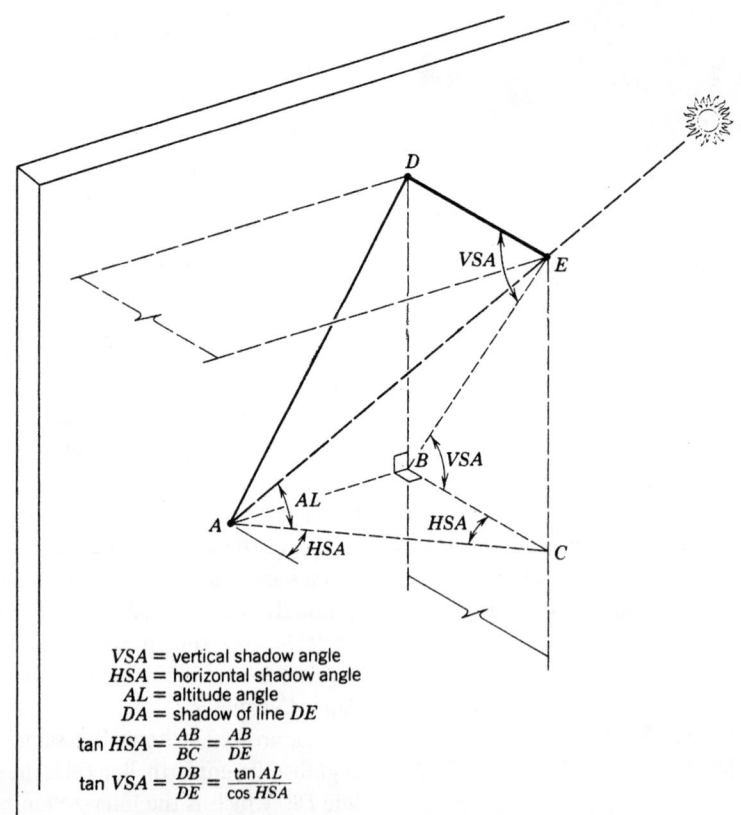

$$VSA = \text{vertical shadow angle}$$
$$HSA = \text{horizontal shadow angle}$$
$$AL = \text{altitude angle}$$
$$DA = \text{shadow of line } DE$$
$$\tan HSA = \frac{AB}{BC} = \frac{AB}{DE}$$
$$\tan VSA = \frac{DB}{DE} = \frac{\tan AL}{\cos HSA}$$

**Fig. 6.22** *The shadow DA cast by the intersection line DE between a horizontal and a vertical shading element defines the edge of each of these shadows. DE can also be thought of as a pin, normal to and extending from the wall, casting a shadow, DA, on the wall. The location and size of line DA are best defined in terms of angles VSA and HSA.*

## (b) Shading Masks

A shading mask is a sunpath chart (horizontal projection) that shows the shadow cast by a particular shading device, as shown in Fig. 6.23. For a horizontal shading device, the leading edge of the shadow cast by all horizontal elements with the same VSA (profile angle) projects as a segmental line, and when plotted on the sunpath chart, it is shown as a segmental mask. To draw this segment, a protractor is required. The Pilkington Sun Angle Calculator includes a profile angle protractor (an overlay to the sunpath chart) to draw the required segment. With no shading, there is no shadow, the VSA is 90°, and no segment exists. With a very deep overhang, the VSA approaches 0° and the *unshaded* area shrinks.

For vertical shading elements, such as fins (Fig. 6.23b), the leading edge of vertical shading elements forms a shadow, shown as radii projecting from the center, which forms an angle HSA, from a line normal to the wall. These radial lines can be drawn with the assistance of an ordinary protractor or by use of the protractor in Fig. 6.23c. The full segments and full radial lines of the shading masks are of *infinitely long elements*, which, of course, do not actually exist. Shading masks that appear in the literature (Olgyay and Olgyay, 1951; AIA, 1981) are frequently drawn as if for infinite elements, and the masks must be truncated for real-world shading design.

## (c) Use of Shading Masks

A shading mask is drawn on some transparent medium (e.g., paper or plastic sheet) and laid on top of a sunpath diagram for the proper latitude *drawn to the same scale* (see Fig. 6.24, which illustrates the placement of a shading mask). Its center point is placed on the center point of the sunpath diagram, and it is rotated until its facing direction (the direction of a normal to the wall) is aligned with the appropriate azimuth line on the sunpath diagram. Assuming that the shading mask has been correctly drawn to provide shading for the window (typically 50% or 100% coverage), the shaded hours for the window and device in question can then be read directly from the underlying sunpath diagram as summarized in Table 6.1. (This is why the shading mask must be drawn on a transparent medium.)

It is extremely important to note that a horizontal overhang shading element the same width as a window can provide full shade for the window only when the sun is exactly opposite the window (when the solar-window azimuth is 0°). This situation occurs for only an instant (Fig. 6.25). At all other times, some part of the window will be exposed to the sun. To provide full shading for more than an instant with a horizontal element and no vertical (side) elements, *an overhang must extend beyond the sides of the window.* The amount of such an extension can be determined both graphically and analytically. Since graphic solutions are amply treated in the literature (Harkness and Mehta, 1978; Lim et al., 1979; Cowan, 1980), the focus here is the analytic solution. In reviewing this discussion, keep in mind the requirement that shading masks for 100% coverage must be prepared so that the extremities of the opening are shaded.

## (d) Designing Finite Horizontal Shading Devices

Although any percentage of window shading coverage is possible, most shading devices are designed to give either 50% or 100% coverage. Deciding upon shading coverage is the first step in design. The second step is to establish the required *depth* of the shading device (the distance it projects from the wall). Figure 6.26 shows how the required depth and the corresponding segmental mask are determined. At times, because of window areas left unshaded by the original overhang, it is necessary to determine the required side extensions beyond the window's edges.

Due to the symmetry of solar motion about the ecliptic, the position of the sun is symmetrical on both sides of its maximum/minimum positions, that is, the solstices. Thus, the sun's position on May 21 is the same as on July 21, since both dates are 1 month from the summer solstice. As a result, any *fixed* shading device will give the same shade in the spring (before 21 June) as in the summer (after 21 June). Since in many locations spring is cool, desired late summer shading will produce spring shading that *may not* be desirable. Solutions to this apparent dilemma are either to use a movable or a variable-size fixed shading device or to compromise on the amount of shading, that is, to design for

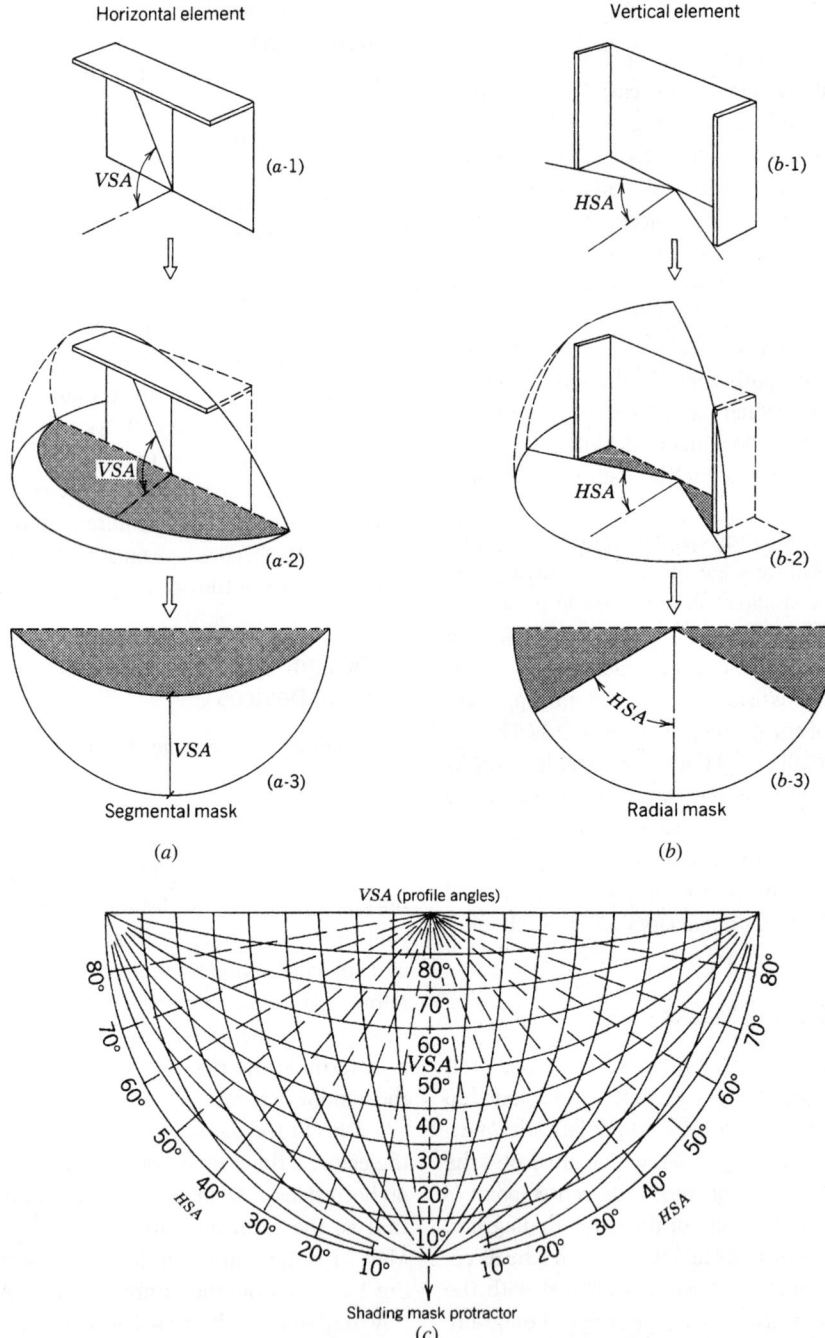

**Fig. 6.23** *A shading mask is the horizontal projection of the shadow cast by the elements with the same VSA or HSA. Thus, any infinitely long horizontal element with the VSA shown in (a-1) will have a shading mask, as shown in (a-3). Similarly, any infinitely high vertical elements with the HSA shown in (b-1) will have the shading mask shown in (b-3). A protractor (c) is required to draw the segmental horizontal element mask properly.*

THERMAL CONTROL

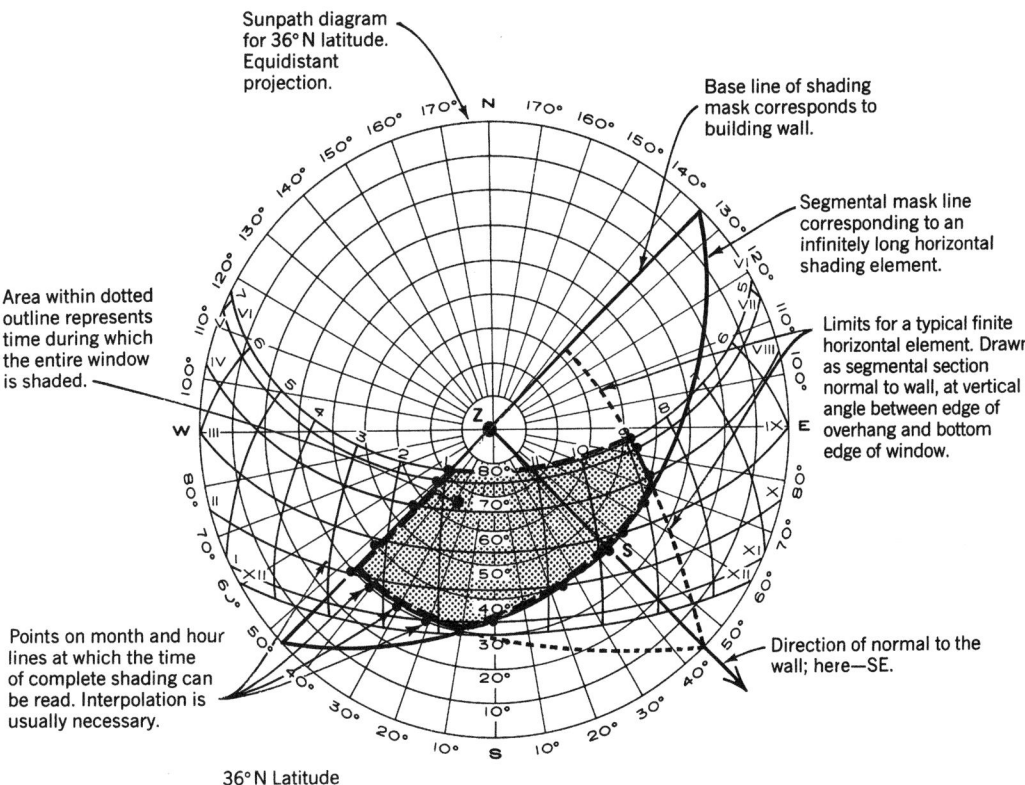

Sunpath diagram for 36° N latitude. Equidistant projection.

Base line of shading mask corresponds to building wall.

Segmental mask line corresponding to an infinitely long horizontal shading element.

Area within dotted outline represents time during which the entire window is shaded.

Limits for a typical finite horizontal element. Drawn as segmental section normal to wall, at vertical angle between edge of overhang and bottom edge of window.

Points on month and hour lines at which the time of complete shading can be read. Interpolation is usually necessary.

Direction of normal to the wall; here—SE.

36° N Latitude

**Fig. 6.24** *A shading mask for a typical horizontal overhang on a window is laid over a sunpath diagram to permit determination of the shaded hours. The mask is drawn to provide 100% shading for the entire window. The hours during which the entire window is shaded are determined by the intersection of the mask perimeter with the date/hour lines of the sunpath diagram and are tabulated in Table 6.1. Only during the hours falling within the dotted area is a window fully shaded by an overhang element that is as wide as the window. The shading element shown here is not symmetrical about the center, indicating that an extension to the left was used to provide a larger shade period after noon. This is typical for east-facing windows and is reversed for western exposures.*

**TABLE 6.1 Time of Day When Window Is Fully Shaded**

| | With Required Overhang Extensions | | With Element Same Width as Window |
|---|---|---|---|
| Date | From | To | Time |
| 21 June | 0900 | 1240 | 1115 |
| 21 July/May | 0850 | 1250 | 1100 |
| 21 Aug./Apr. | 0845 | 1330 | 1035 |
| 21 Sept./Mar. | 0940 | 1400 | 0955 |
| 21 Oct./Feb. | 1040 | 1410 | — |
| 21 Nov./Jan. | 12N | 1300 | — |
| 21 Dec. | 1230 | | — |

50% shading for late summer (and early spring) and 100% shading for early summer (and late spring).

### (e) Design Approaches

In addition to the approach described above, design might start with conceptual sketches of an appro-

priate shading device. As the sketches are developed into scaled drawings, a sectional drawing will show the geometry of the building. The VSA for 100% shading would be an angle from the window sill to the outer edge of the overhang. Once the VSA is established, a shading mask can be created and overlaid onto the appropriate sunpath chart to determine whether there is adequate shading during specific times of the day and year. A more detailed description of this design process is found in Olgyay and Olgyay (1951).

Shading devices can greatly enrich the aesthetics of a building as well as improve performance. Examples of shading device models for various orientations (Fig. 6.27) show a range of shading design solutions for a building in a temperate climate.

THERMAL CONTROL

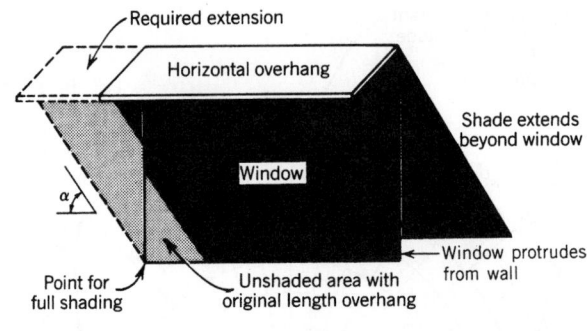

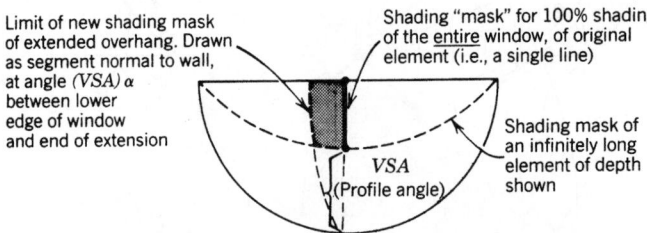

**Fig. 6.25** *A horizontal overhang with sufficient depth to provide 100% shading can provide full shading only when the sun is exactly opposite a window. At any other time, part of the window will be exposed to the sun. Full shading from low early morning or late afternoon sun is not possible with horizontal elements.*

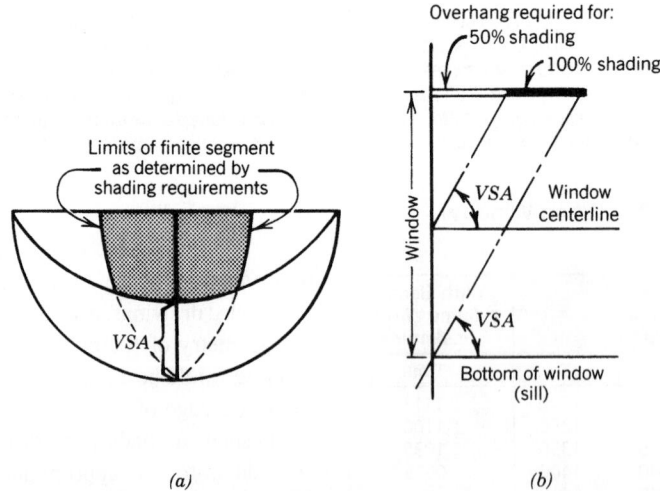

(a)

(b)

**Fig. 6.26** *One way to find the required depth of a horizontal overhang is to establish the required segmental shading mask (a) and read the VSA (profile angle) off the protractor that corresponds to the segment. Draw a wall section (b) to scale with the VSA. The VSA can be drawn for 50% shading coverage, 100% coverage, or any other value. The depth of the overhang can be measured directly from the section. Alternatively, if the overhang depth is known, the segment can be drawn.*

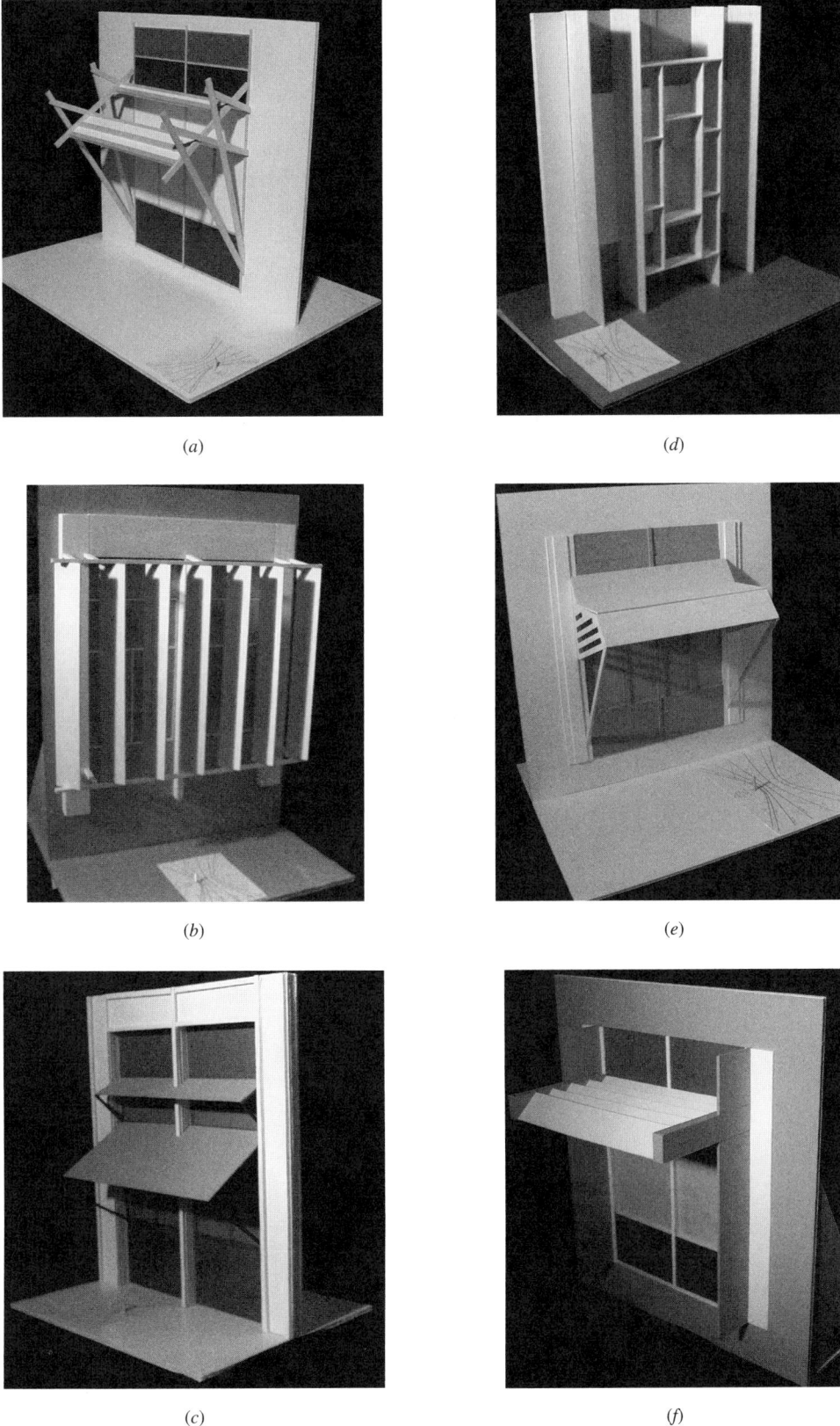

(a)

(d)

(b)

(e)

(c)

(f)

***Fig. 6.27*** *Examples of shading device models (a–f) illustrate a variety of design responses to orientation, shading requirements, and view for a university building located at 44° N latitude. (Photos by Fumiko Docker; © 2004 Alison Kwok; all rights reserved.)*

**169**

### References

AIA. 1981. Ramsey, C.G. and H.R. Sleeper. *Architectural Graphic Standards,* 8th ed. American Institute of Architects/John Wiley & Sons, New York.

Cowan, H.J. (ed.). 1980. *Solar Energy Applications in the Design of Buildings.* Applied Science Publishers, London.

Harkness, E. and M. Mehta. 1978. *Solar Radiation Control in Buildings.* Applied Science Publishers, London.

Lim, B.P., K.R. Rao, K. Tharmaratnam, and A.M. Mattar. 1979. *Environmental Factors in the Design of Building Fenestration.* Applied Science Publishers, London.

Mazria, E. 1979. *The Passive Solar Energy Book: A Complete Guide to Passive Solar Home, Greenhouse and Building Design.* Rodale Press, Emmaus, PA.

Olgyay, A. and V. Olgyay. 1951. *Solar Control and Shading Devices.* Princeton University Press, Princeton, NJ.

Pilkington Sun Angle Calculator. 1975. The Pilkington Sun Angle Calculator (formerly the Libbey Owens Ford Sun Angle Calculator) is available through the Society of Building Science Educators (SBSE).
http://www.sbse.org/resources/sac/index.htm/

Szokolay, S.V. 1980. *Environmental Science Handbook.* Halsted/John Wiley & Sons, New York.

THERMAL CONTROL

# Heat Flow

UNDERSTANDING HEAT FLOW IS FUNDAMENTAL TO ALL aspects of climate control. Chapter 4 addresses heat flows to and from the body that affect thermal comfort. Chapters 8, 9, and 10 deal in part with heat flows to, from, and within various elements of active and passive climate control systems. This chapter deals with heat flows through building envelopes, both through the materials of the building skin and by way of outdoor air that replaces conditioned indoor air. Basic concepts and calculations of heat flow are presented in this chapter, whereas applications of these concepts (passive solar heating, passive cooling, active HVAC system/equipment sizing, seasonal energy usage) are found in subsequent chapters. Numerous data tables that accompany this chapter are presented in Appendix E.

## 7.1 THE BUILDING ENVELOPE

From a building science perspective, the exterior enclosure (or envelope) of a building consists of numerous materials and components that are assembled on site to meet the intents of the owner and the design team. A building envelope typically includes some prefabricated components (such as windows and doors) that are available off the shelf and have well-defined and tested thermal performance characteristics. The typical envelope also includes materials in a variety of forms (sheets, blocks, bulk products, membranes, etc.)

that have been site-assembled to meet design requirements. These components and materials may be assembled into commonly used configurations with generally understood performance or into configurations unique to a given project and of uncertain performance. The job of envelope thermal analysis is to ensure that a proposed envelope will meet the design intent and criteria (including building codes).

From a functional perspective, the envelope of a building is not merely a two-dimensional exterior surface; it is a three-dimensional transition space—a theater where the interactions between outdoor forces and indoor conditions occur under the command of materials and geometries. Sun and daylight are admitted or rejected; breezes and sounds are channeled or deflected; and rain is repelled or collected. This transition space is where people indoors can experience something of what the outdoors is like at the moment, as well as where people outside can get a glimpse of the functions within. Figure 7.1 shows an example of an envelope that is a transitional space, not merely a surface. The more suited the outdoors is to comfort, the more easily indoor activity can move into this transition space. At building entries, a person will be especially aware of the difference between indoors and outdoors during the passage between the two conditions.

The envelope has a fourth dimension: it changes with time. Seasonal changes have a marked effect on the façade shown in Fig. 7.1 and a more subtle effect on the east-facing balconies

THERMAL CONTROL

**Fig. 7.1** *The envelope is more than a surface. This south-facing office façade in Oregon forms a microclimate zone that buffers the transition between indoor and outdoor conditions. Ground-cover plants at eye level for seated occupants and deciduous vines overhead give a seasonally changing view to the outdoors through a façade that also admits winter sun, year-round daylight, and summer breeze. (Photo by Amanda Clegg.)*

of the apartments in Fig. 7.2. The year-round usable volume of these apartments is increased by making the balcony into a sun porch. This allows sun to enter while blocking the wind—a response to Oregon's long, mildly cold winter. The more that users are involved in decisions about how much of the outside to bring inside, the more changeable the building envelope will be. Not all buildings encourage such flexibility and individual expression; an unchanging envelope can be a symbol of stability and may be considered appropriate for some governmental and religious monuments.

## 7.2 BUILDING ENVELOPE DESIGN INTENTIONS

Called by their familiar names, the basic components of a building envelope include windows, doors, floors, walls, and roofs. On closer inspection, windows can include skylights, clerestories, screens, shutters, drapes, blinds, diffusing glass, and reflecting glass—an array of components that determines how the envelope does its job of making the transition between inside and outside. Norberg-Schulz (1965) suggests that a component can more fundamentally be thought of in terms of its design intent relative to the exchange of energies: as a *filter, connector, barrier,* or *switch.*

In general, we define a *connector* as a means to establish a direct connection, a *filter* as a means to make the connection indirect (controlled), a switch as a regulating connector, and a *barrier* as a separating element. . . . An opaque wall thus serves as a filter to heat and cold, and as a barrier to light. Doors and windows have the character of switches, because they can stop or connect at will. (p. 113)

In addition to these historic envelope intents, a new option is emerging: the *transformer.* A transformer is intended to convert an environmental force (such as solar radiation) directly into a different and desirable energy form (such as electricity). A photovoltaic roof shingle is an example of a transformer.

The fundamental range of choices surrounding envelope intent and components can be illustrated by two opposite design concepts: the *open frame* and the *closed shell.* In harsh climates (or where unwanted external influences such as noise or visual clutter abound), the designer frequently conceives of the building envelope as a closed shell and proceeds to selectively punch holes in it to make limited and special contacts with the outdoors. Such an approach might be called *barrier-dominated.* When external conditions are very close to the desired internal ones, the envelope often begins as an open structural frame, with pieces of building skin selectively added to modify only a few outdoor forces. Such an approach might be called *connector-dominated.* (Note that a connector, filter, or barrier for one natural force may change its role for another force: glazing may be a connector to daylight but a barrier to wind.) The open-frame or the closed-shell approach to envelope design, when combined with material availability and the influence of local culture, can produce a distinct regional architecture, as shown in Fig. 7.3.

With a wide range of energy sources, building materials, and mechanical equipment available, it is possible to build connector-dominated buildings anywhere, regardless of the climate. The consequences of the resulting energy consumption can be severe. In contrast, if defending against outdoor conditions becomes an overriding consideration, barrier-dominated envelopes may be appropriate in any climate. The designer's use of connectors, filters, and barriers is basic to the design of building exteriors. With the addition of switches that allow

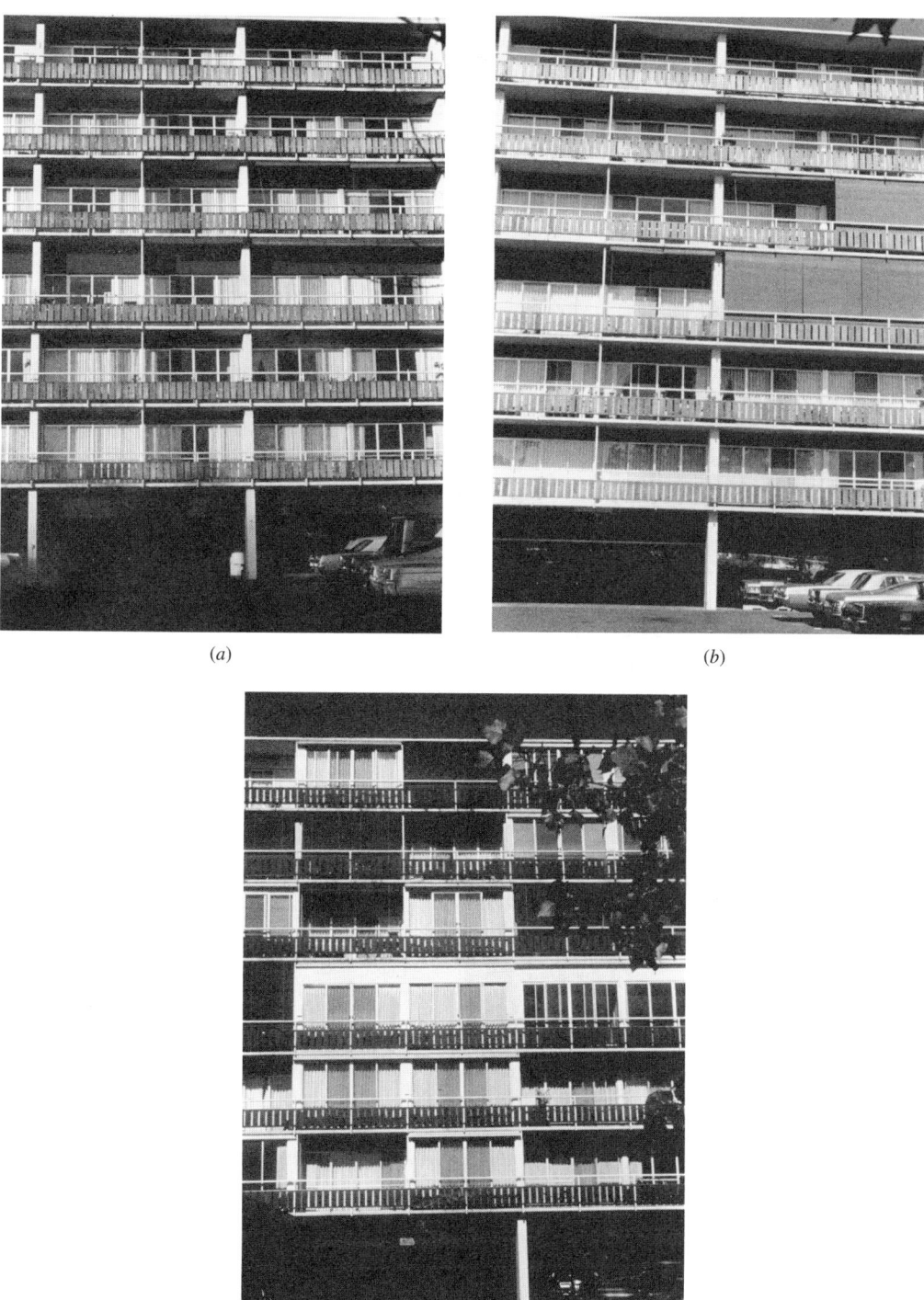

(a)

(b)

(c)

***Fig. 7.2*** *Envelopes change with time. This east-facing Oregon building façade changes by season and over the years. (a) December morning sun floods unshaded balconies in the building's early years. (b) Months later, June morning sun is blocked by shading devices on some balconies. (c) Years later, many balconies have been enclosed for greater winter comfort. (Cascade Manor, Eugene, OR; John Graham & Associates, Architects.)*

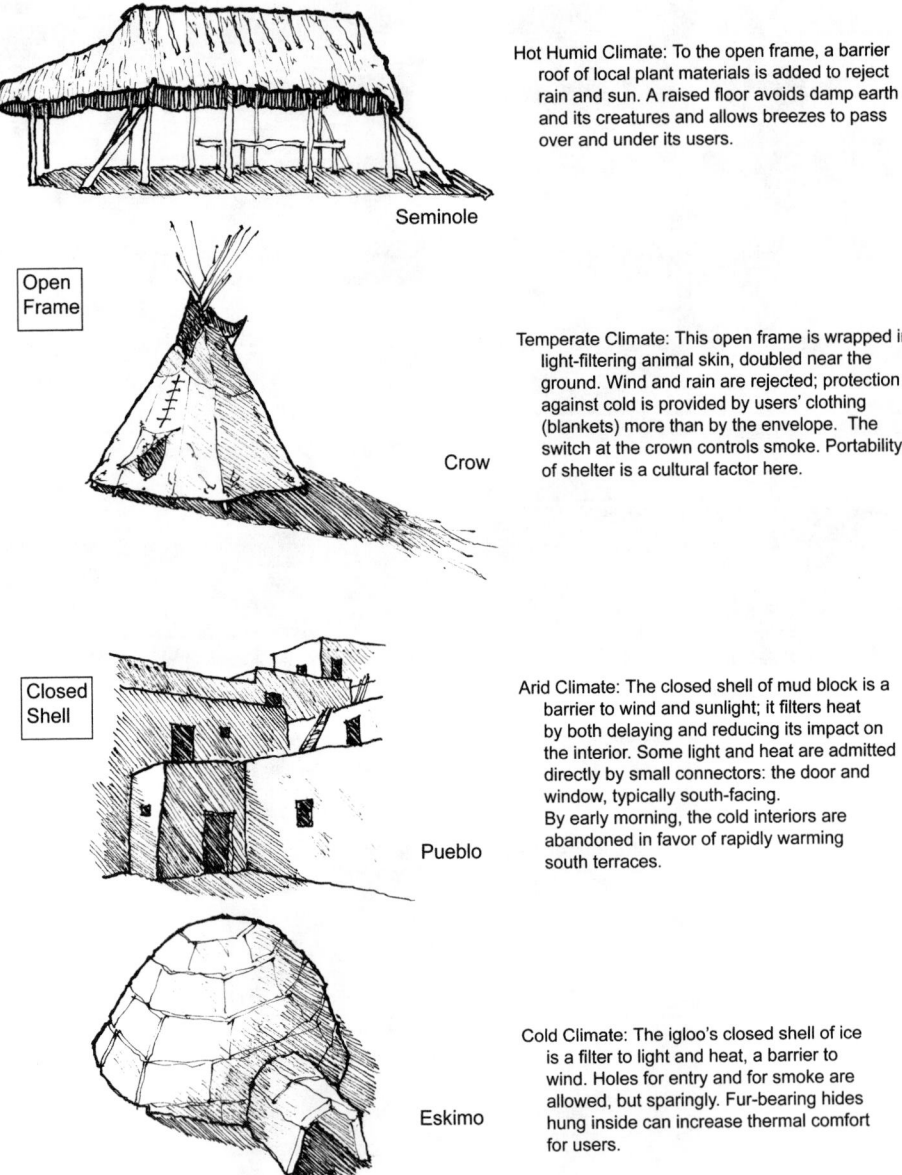

Hot Humid Climate: To the open frame, a barrier roof of local plant materials is added to reject rain and sun. A raised floor avoids damp earth and its creatures and allows breezes to pass over and under its users.

Seminole

Open Frame

Temperate Climate: This open frame is wrapped in light-filtering animal skin, doubled near the ground. Wind and rain are rejected; protection against cold is provided by users' clothing (blankets) more than by the envelope. The switch at the crown controls smoke. Portability of shelter is a cultural factor here.

Crow

Closed Shell

Arid Climate: The closed shell of mud block is a barrier to wind and sunlight; it filters heat by both delaying and reducing its impact on the interior. Some light and heat are admitted directly by small connectors: the door and window, typically south-facing.
By early morning, the cold interiors are abandoned in favor of rapidly warming south terraces.

Pueblo

Cold Climate: The igloo's closed shell of ice is a filter to light and heat, a barrier to wind. Holes for entry and for smoke are allowed, but sparingly. Fur-bearing hides hung inside can increase thermal comfort for users.

Eskimo

**Fig. 7.3** *Open frame (Seminole and Crow) and closed shell (Pueblo and Eskimo) envelope approaches are influenced by climate, materials, and culture. The influence of climate is dramatic, but material availability and cultural expectations also influence the envelope design solutions in these examples. (Drawing by Dain Carlson.)*

the envelope to respond to changing conditions and transformers that respond to building needs and site resources, a liveliness can result that makes a building an attractive addition to its environs.

Switches are a designer's way of having an envelope respond in a variable manner and/or giving building occupants some control of their own environment. Seasons and functions change; people are unpredictable. If the designer has carefully considered the range of choices that a switch will provide, successful user control of that switch is possible. Building skins plentifully supplied with switches become demonstrations that architecture is a performing art, not simply static sculpture. Figure 7.4 shows a remarkably integrated switch-dominated envelope in vernacular architecture.

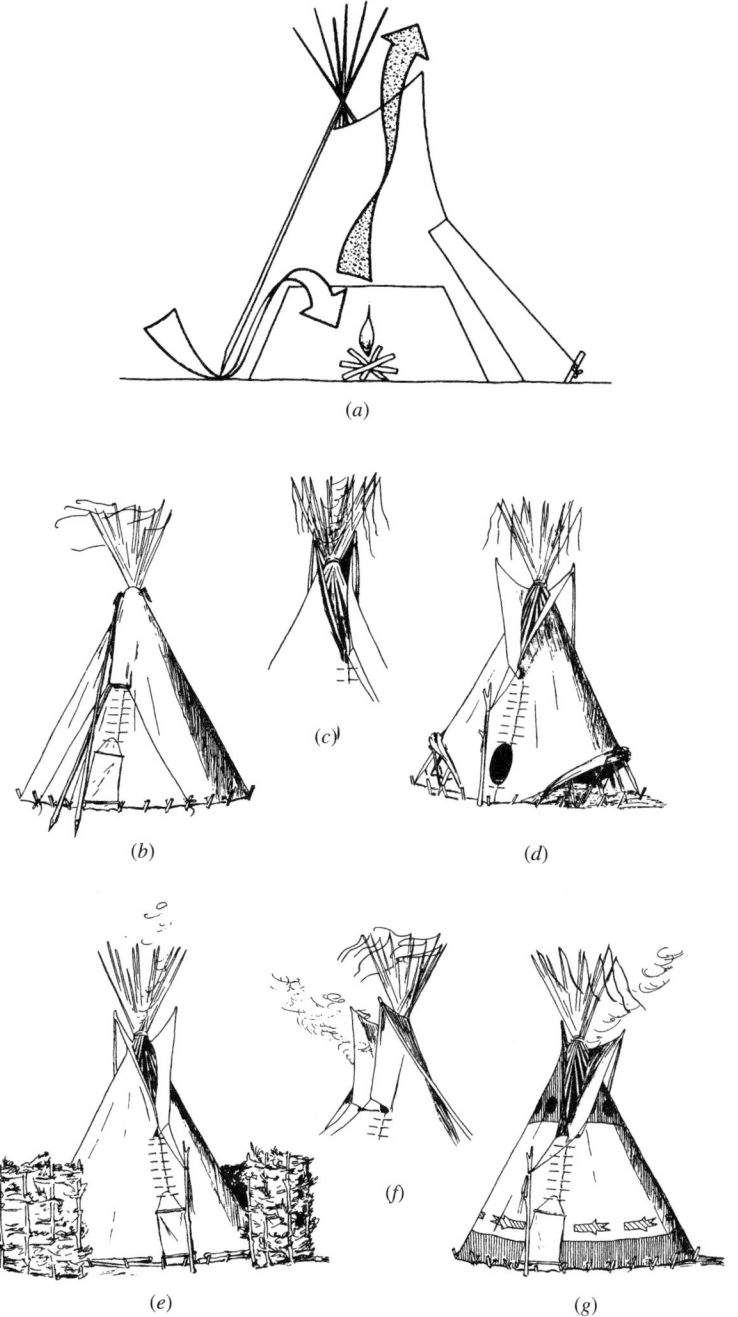

**Fig. 7.4** *Thermal switches on the Great Plains. (a) With an interior fire, smoke flaps at the top are adjusted to vent the smoke, and an interior liner forces the cold combustion air entering at ground level to first rise along the sides of the tipi, gaining some warmth, before it moves across the occupants on its way to the fire. Six adjustments to the lightweight, translucent, and portable tipi are shown: the tipis in these diagrams are facing east, with their backs to the prevailing westerly winds. (b) In severe rainstorms, the smoke flaps at the top can be closed. (c) For ordinary conditions of west wind, the smoke flaps block the wind, thus creating suction to draw out the smoke. (d) In hot weather, breezes are admitted under the cover at the ground. (e) In extremely cold weather, a temporary windbreak can be added. (f) For the unusual wind or (g) southwest wind, the smoke flaps are manipulated to block wind, again encouraging smoke draw-out, as in (c). (Parts b–g from* The Indian Tipi: Its History, Construction, and Use *by Reginald and Gladys Laubin. Copyright 1957 by the University of Oklahoma Press.)*

Daylighting/solar control devices are perhaps the most common and visible switches. Awnings block direct sun at some times and admit it at others. Opaque drapes can expose all of a window to incoming daylight on dark days or block out daylight entirely when desired. Translucent curtains can change bright sun into a diffuse light for the interior or be drawn back to allow strong direct sun and shadow to be cast inside a room. Passive solar heating systems rely on switches to control the incoming sun on warmer days, as seen in Fig. 7.2b. Operable windows are a commonly used thermal switch. Ventilating switches may be separate from windows.

Visually, switches are a particularly promising source of three- and four-dimensional interest on a building's exterior. For the office buildings in Fig. 7.5, on which façade would the daily and seasonal changes be most visually evident? If it is easier to imagine human beings working behind the windows of one building than of the other, might that suggest a more satisfactory work environment? Might more personal control of a window promote greater comfort?

Switches encourage interaction between users and their environments. This is usually satisfying to the users, who are able to select the desired exposure to the climate at a given moment. Yet without automation, supervision, or training in their use, switches can also be detrimental to system performance. Examples include a thermal shade left to cover a passive solar-collecting window on a cold, sunny morning; a vent left open during the hottest hours in a high-mass, night-ventilated building; or an awning rolled up to expose a window to summer sun. People in buildings often use switches as they feel appropriate, as demonstrated in Fig. 7.6. Conventionally air-conditioned buildings typically do away with user-operated ventilating switches (operable windows) so that the system will instead function with a closely and automatically controlled flow of filtered outdoor air. Thermally efficient as this practice may be, it can also be a source of widespread dissatisfaction with air-conditioned spaces. Sealed windows greatly curtail people's contact with sounds, smells, and breezes from the outdoors. This is frequently beneficial in urban areas, yet on beautiful days it can be very frustrating. A lack of switches can contribute to a feeling of helplessness about one's personal environment.

Passive heating and cooling systems are especially reliant upon switches, hence on the knowledge and cooperation of their users. These users often must base their actions in manipulating thermal switches at some point in time on what effect will be needed at a later time. This practice, called *thermal sailing,* is similar to the actions of outdoor workers in the far north, who learn to unbutton their coats in the cold early hours of the workday *before* they begin to sweat. Sweat would soak their insulating clothing, with harmful results later in the day as temperatures fall rapidly near dusk.

Misjudgments in passively solar-heated residences can result in extraordinarily high temperatures on a sunny day or uncomfortably cold nights without stored solar heat. For a building closely connected to its climate, the design of switches is also the design of an educational process for the users. The challenge is to involve, but not bind, the users in the management of their environment. Automated controls are a partial answer to this challenge; switches that are easy, fun, and obvious to use are another.

## 7.3 SENSIBLE HEAT FLOW THROUGH OPAQUE WALLS AND ROOFS

The flow of heat through a building envelope varies both by season (heat generally flows *from* a building in winter and *to* a building in summer) and by the path of the heat (through the materials of a building's skin or by way of outdoor air entering the interior through intentional and unintentional pathways). These complexities must be considered by a designer who intends to deliver comfort and energy efficiency. The following discussion of heat flow will focus first upon the building skin (opaque elements, then transparent elements), followed by heat flow via air exchange.

In the 1970s, designers began placing increasing emphasis on thermal performance of the building skin to conserve energy in the wake of energy scarcity. Tighter building envelopes resulted in decreased air leakage, leading to ongoing concerns about indoor air quality and "sick building syndrome." It is likely that today's building code requirements strike a fair balance between envelope energy efficiency and air quality requirements. Buildings have changed as a result of this balance.

(a)

(b)

(c)

**Fig. 7.5** *Sun control for Boston offices. (a) Shading switches both inside the glass (blinds) and outside the glass (awnings) are evident, and the windows themselves are operable. Awning use varies with window location and occupant needs and desires. (New England Merchants' Bank: Shepley Rutan & Coolidge, Architects; demolished 1966.) (b) Three-dimensional filters (overhangs) dominate the south façade (right) of this building at Boston University; the adjacent west windows (where overhangs are less effective) have only internal switches (blinds) deployed in a variety of positions. (Law and Education Building; Sert, Jackson and Gourley, Architects). (c) A two-dimensional filter of reflective glass equally sheathes all faces of this office tower, sending reflected sunlight to the neighborhood below. The "switches" in this case are thermostats; variability is removed from the envelope, and users are not involved in defining the outside–inside relationship. (John Hancock Headquarters; I.M. Pei and Partners, Architects; photo by Stephen Tang.)*

THERMAL CONTROL

(b)

(b)

**Fig. 7.6** *Which is spring, which is summer? The awnings for this south-facing office in Eugene, Oregon, are teamed with overhangs and side walls (or fins) to provide sun control for windows. (a) The altitude of the spring sun is low enough to threaten glare in the lower windows and cause awning deployment. (b) The summer sun, at a higher altitude, is readily blocked by the overhangs, and the awnings can be rolled up with less risk of glare. (Moreland-Unruh-Smith, Architects.)*

Today's typical new North American building loses somewhat more winter heat via incoming fresh air than it loses through its skin. In summer it gains somewhat more heat through its skin than it does via incoming fresh air (although this is climate dependent). This pattern is due in part to solar radiation being so important to envelope heat gains in summer and to the design outdoor–indoor temperature difference being much greater in winter than in summer (in most North American locations).

Sensible heat is a form of energy that flows whenever there is a temperature difference and that manifests itself as an internal energy of atomic vibration within all materials. Temperature is an indication of the extent of such vibration, essentially the "density" of heat within a material. Other forms of energy (such as solar radiation or sound) can be converted to heat and vice versa (within limits). Latent heat is sensible heat used to change the state of (evaporate or condense) water. *Power* refers to the instantaneous flow of energy (at a given time). In buildings, *energy* refers to power usage over time. Table 7.1 lists commonly encountered terms related to energy, power (the rate of energy use), and heat flow. Admittedly, some of the units used to quantify heat, power, and energy are quirky. Nevertheless, time spent understanding these units may be rewarded.

## (a) Static versus Dynamic; Sensible versus Latent

Although the general principles remain the same, analysis of heat flow under dynamic (rapidly changing) conditions is more complex than under static or steady-state (fairly stable over time) conditions. The effects of heat storage within materials become a greater concern under dynamic conditions than under static conditions. A static analysis requires consideration of fewer variables than a dynamic analysis and is therefore simpler. The key determinant of steady-state heat flow is thermal *resistance.* An analysis of heat flow under dynamic conditions involves more variables—including thermal *capacitance.* The following discussion begins with steady-state assumptions and then looks at the dynamic situation.

Heat flows are of two forms—sensible heat and latent heat. Sensible heat flow results in a change in temperature. Latent heat flow results in a change in moisture content (often humidity of the air). Total heat flow is the sum of sensible and latent flows. Materials react differently to sensible and latent heat flows; as such, the discussion below begins with sensible heat flow and then deals with latent heat flows.

## (b) Heat Flow Processes

Whenever an object is at a temperature different from its surroundings, heat flows from the hotter to the colder. Likewise, moisture flows from areas of greater concentration to areas of lesser concentration. Buildings, like bodies, experience sensible heat loss to, and gain from, the environment in three principal ways. In *convection,* heat is exchanged between a fluid (typically air) and a solid, with motion of the fluid due to heating or cooling playing a critical role in the extent of heat transfer. In *conduction,* heat is transferred directly from molecule to molecule, within or between materials, with proximity of molecules (material density) playing a critical role in the extent of heat transfer. In *radiation,* heat flows via electromagnetic waves from hotter surfaces to detached, colder ones—across empty space and potentially great distances. *Evaporation* can also be involved in envelope heat loss, carrying heat away from wet surfaces, but this is much less influential for most buildings than for our bodies. Moisture flow through envelope assemblies and via air leakage are the principal means of latent heat gain (or loss).

The combination of sensible heat flow by convection, conduction, and radiation through some typical combinations of materials is shown in Fig. 7.7. Heat flow through the various components of a building skin involves both heat flow through solids and heat flow through layers of air. Multiple air spaces and reflective surfaces are useful and inexpensive ways to slow the flow of heat via radiation from hot spaces where temperature differences are large. Some of the most effective insulating materials, therefore, combine dead-air spaces and layers of reflective films.

## (c) Thermal Properties of Components

Each material used in an envelope assembly has fundamental physical properties that determine how that material will interact with sensible heat. These key properties are described below.

*Conductivity.* Each material has a characteristic rate at which heat will flow through it. For homogeneous solids this is called *conductivity* (designated as $k$), and in the inch–pound (I-P) system it is the number of British thermal units per hour (Btu/h) that flow through 1 square foot (ft$^2$) of material that is 1 in. thick when the temperature difference across that material is 1°F (under conditions of steady heat flow). Thus the I-P units of conductivity are Btu in./h ft$^2$ °F.

The System International (SI) equivalent is the

**TABLE 7.1 Energy, Power, and Heat Terminology**[a]

| Concept | Terminology | Symbol | Discussion |
|---|---|---|---|
| ENERGY | | | The term *energy* implies a cumulative perspective, such as the potential in a barrel of oil or solar radiation collected during a heating season. |
| | British thermal unit | Btu | The British thermal unit is the fundamental I-P unit of energy. A Btu is the amount of energy (heat) required to raise the temperature of 1 lb of water by 1°F. A burning wooden match releases approximately 1 Btu. |
| | Joule | J | The Joule is the fundamental SI unit of energy. A Joule is a Newton-meter (a force of 1 Newton acting over a 1-m distance). A Joule is a fairly small unit of energy, so the kilo-Joule (1000 J—kJ) is commonly used in building design. |
| | Watt-hour | Wh[b] | The Watt-hour is a commonly used SI unit of energy. |
| | | | 1 W = 1 J/s<br>1 Btu = 1.055 kJ<br>1 kJ = 0.9478 Btu |
| POWER | | | The term *power* is used to describe the rate of energy usage, production, or flow. Power is always associated with a time frame (often an hour in building design). |
| | British thermal unit per hour | Btu/h or Btuh[b] | A rate of energy flow. Btu/h is used to express heat gains and losses and the heating and cooling capacity of equipment. |
| | Watt | W | The Watt is the SI unit of power and is used the same as Btu/h. |
| | | | 1 Btu/h = 0.0293 W<br>1 W = 3.412 Btu/h |
| | Horsepower | HP | An I-P expression of power, typically used to describe the capacity (size) of certain HVAC equipment (such as motors). |
| | | | 1 HP = 2545 Btu/h<br>1 HP = 746 W |
| HEAT FLOW | | | Heat flow is a form of power and in building design is typically expressed in Btu/h or W. There are a number of heat flow related properties of materials with specific definitions and uses as described below. These properties are based upon a unit area of material and a unit temperature difference. |
| | Conductivity | k | The rate of heat flow through a unit thickness of a homogeneous material. |
| | | | I-P units: Btu in./h ft² °F<br>SI units: W/m °C |
| | | | Btu in./h ft² °F × 0.1442 = W/m °C |
| | Conductance | C | The rate of heat flow through a specific nonhomogeneous object (such as a concrete masonry unit) or a defined thickness of a homogeneous material. |
| | | | I-P units: Btu/h ft² °F<br>SI units: W/m² °C |
| | | | Btu/h ft² °F × 5.678 = W/m² °C |
| | Film or surface conductance | h | The conductance of an air film; same concept and units as conductance. |
| | Overall coefficient of heat flow | U (or U-factor) | The rate of heat flow through an assembly (window, wall, etc.) bounded by air on both sides. Includes the effects of all materials, air films, and air spaces. |
| | | | I-P units: Btu/h ft² °F<br>SI units: W/m² °C |
| | | | Btu h ft² °F × 5.678 = W/m² °C |
| | Resistance | R | A measure of resistance to heat flow; the reciprocal of conductivity (or conductance). Essentially the force required to cause a unit flow of heat. |
| | | | I-P units: h ft² °F/Btu<br>SI units: m² °C/W |
| | | | h ft² °F/Btu × 0.176 = m² °C/W<br>(I-P) R per in. × 6.93 = (SI) R per m |
| | Permeance | M | Although permeance deals directly with water vapor flow, it is related to heat flow through the need to add or remove heat to humidify or dehumidify a building. |
| | | | I-P units: grains/ft² h in. Hg<br>SI units: ng/m² s Pa |
| | | | grains/ft² h in. Hg × 56.7 = ng/m² s Pa |

[a] A more complete listing of I-P and SI units and conversions is provided in Appendix A.

[b] Units oddity: the alternative symbol "Btuh" is shorthand for Btu/h and does not imply a product of energy and time; on the other hand, Wh is a product (1 Watt of power over a 1-hr period); furthermore, Btu (without a time unit) is a measure of energy, as is Wh (with a time unit).

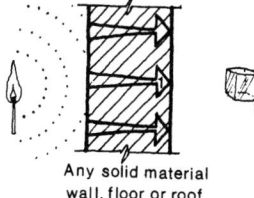

Any solid material
wall, floor or roof

A single solid material illustrates the transfer of heat from the warmer to the cooler particles by conduction (1).

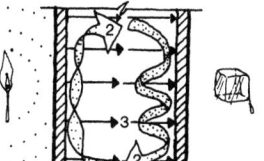

An air space in a wall

As air is warmed by the warmer side of the air space it rises. As it falls down along the cooler side it transfers heat to this surface (2). Radiant energy (3) is transferred from the warmer to the cooler surface. The rate depends upon the relative temperature of the surfaces and upon their emissive and absorptive qualities. Direction is always from the warmer to the cooler surface.

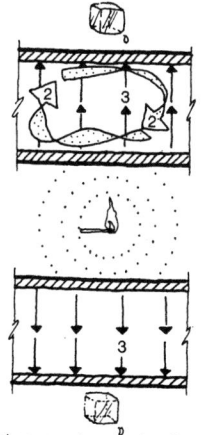

An air space in a roof or floor

The convective action (2) in the air space of a roof is similar to that in a wall, although the height through which the air rises and falls is usually less. The radiant transfer is up in this case because its direction is always to the cooler surface.

When the higher temperature is at the top of a horizontal air space, the warm air is trapped at the top and, being less dense than the cooler air at the bottom, will not flow down to transfer its heat to the cooler surface. This results in little flow by convection. The radiant transfer in this case is down because that is the direction from the warmer surface to the cooler one.

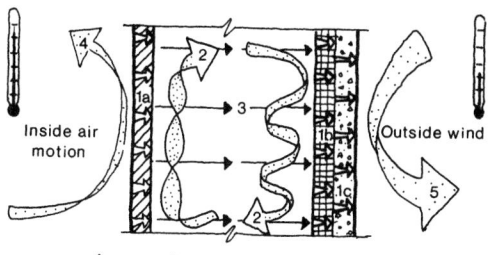

Inside air
motion

Outside wind

A composite assembly of building
materials in place at the site

This example of a wall in place illustrates the several methods by which heat is lost through a composite assembly of materials. Conduction at varying rates in different materials is accounted for in 1a, 1b, and 1c. Convection currents (2) and radiation (3) carry the heat across the air space.

**Fig. 7.7** *Heat flow through materials, across air spaces, and through construction assemblies. Means of heat flow include conduction, convection, and radiation.*

number of Watts that flow through 1 square meter (m²) of material 1 m thick when the temperature difference across that material is 1K (equal to 1C°) under conditions of steady heat flow. Thus, the SI units of conductivity are W/m K or W/m °C (these two terms are used interchangeably in the tables related to this chapter).

Conductivity is established by laboratory tests and is published as a basic property of homogeneous solids. Conductivity is an important factor in passive heating or cooling designs that depend heavily upon the rate at which heat is conducted into a material from its surface.

***Conductance.*** Many solids such as common brick, wood siding, batt or board insulation, gypsum board, and so on are widely available in standard thicknesses. For such common materials, it is useful to know the rate of heat flow for that standard thickness instead of the rate per inch. *Conductance*, designated as *C*, is the number of Btus per hour that flow through 1 ft² of a given thickness of material when the temperature difference is 1°F. The units are Btu/h ft² °F. SI units are W/m² K. Conductance is also used to describe the rate of heat flow through defined sizes of modular units of nonhomogeneous materials (such as a concrete masonry unit—composed of pockets of air surrounded by concrete).

**Resistance.** Conductivity and conductance are compared in Table 7.1 and Fig. 7.8, which also include another useful property, *resistance*. Designated as R, resistance indicates how effective any material is as an insulator. The reciprocal of conductivity (or conductance), R is measured in *hours* needed for 1 Btu to flow through 1 ft² of a given

thickness of a material when the temperature difference is 1°F. In the I-P system, the units are h ft² °F/Btu. SI units are m² K/W. Resistances and other important thermal properties are listed for many conventional building materials in Table E.1 (Appendix E). Table E.2 provides similar information for alternative (less commonly used and/or emerging) construction approaches.

Resistance is especially useful when comparing insulating materials, because the greater the R-value, the more effective the insulator. For this purpose, R is sometimes listed as "per inch" of thickness, in which case the units are h ft² °F/Btu-in. SI units are m K/W (indicating "per meter" of thickness across 1 m², even though the canceling units are odd).

**Emittance.** Radiation heat transfer is highly influenced by surface characteristics; shiny materials are much less able to radiate than common rough building materials. This characteristic is

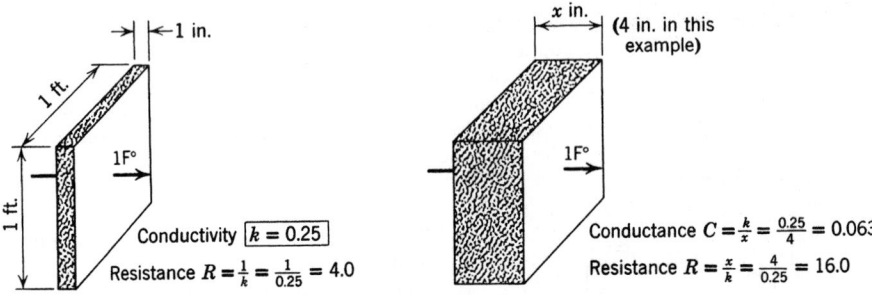

Glass Fiber Insulation Board

Conductivity $k = 0.25$

Resistance $R = \frac{1}{k} = \frac{1}{0.25} = 4.0$

Conductance $C = \frac{k}{x} = \frac{0.25}{4} = 0.063$

Resistance $R = \frac{x}{k} = \frac{4}{0.25} = 16.0$

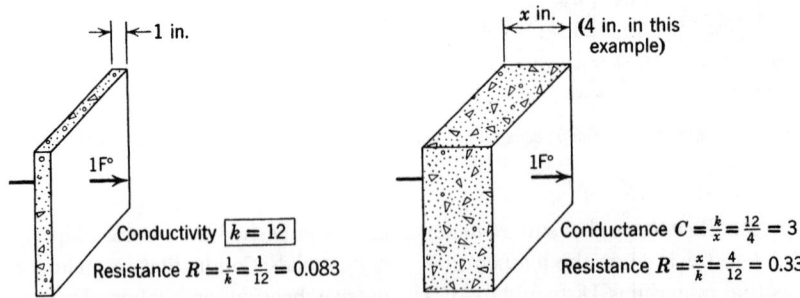

Sand and Gravel Concrete

Conductivity $k = 12$

Resistance $R = \frac{1}{k} = \frac{1}{12} = 0.083$

Conductance $C = \frac{k}{x} = \frac{12}{4} = 3$

Resistance $R = \frac{x}{k} = \frac{4}{12} = 0.33$

**Fig. 7.8** *Relationship between conductivity, conductance, and resistance for two typical materials. Glass fiber is a material of low conductivity (high resistance); concrete is a material of high conductivity (low resistance). Note: standard unit of area is 1 ft² (1 m²); standard air temperature differential is 1°F (1°C).*

called *emittance,* the *ratio* of the radiation emitted by a given material to that emitted by a blackbody at the same temperature. The impact of emittance (shiny vs. matte surfaces) is seen in Tables E.3, E.4, and E.5, which present properties of air layers and air spaces within construction assemblies; the lower the emittance, the lower the radiative heat exchange. For most materials, emittance is related to absorptance: a highly absorptive (low-reflectance) material will usually have a high emittance as well. *Selective surfaces* (sometimes used in solar collectors) are highly absorptive yet have very low emittance.

## (d) Thermal Classifications of Materials

Architectural materials generally interact with heat either as *insulators* that retard the flow of heat (useful for thermal barriers) or as *conductors* that encourage heat flow (useful for thermal storage materials). It is common to find both insulators and conductors in the same construction. For example, a wall can have an inner layer that is highly conductive and thermally massive (for thermal storage), with an outer layer that is also highly conductive and thermally massive (for durability and weathering) and a highly insulative, low-mass material in between.

*Insulations.* Materials used for insulation fall into three broad categories: (1) *inorganic* fibrous or cellular products (such as glass, rock wool, slag wool, perlite, or vermiculite), (2) *organic* fibrous or cellular products (such as cotton, synthetic fibers, cork, foamed rubber, or polystyrene), and (3) *metallic* or metalized organic *reflective* membranes (which must face an air space to be effective).

Insulating materials are available in a wide variety of forms. Form-fitting materials include *loose fill* (as above a ceiling on the floor of an attic); *insulating cement,* a loose material mixed with a binder and troweled onto a surface; and *formed-in-place* materials such as expanded pellets or liquid-fiber mixtures that are poured, frothed, sprayed, or blown in place. Less form-fitting but more common are batts and blankets of *flexible, semirigid* insulation, with varying degrees of compressibility and adaptability to substrates. *Rigid* insulation, with little on-site adaptability, is applied in blocks, boards,

or sheets and can be preformed to fit nonplanar surfaces such as pipes.

Exterior insulation and finish systems (EIFS) have become very popular (and controversial), both for retrofit and for new construction. Utilizing expanded polystyrene rigid boards applied to exterior gypsum, plywood, or cementitious substrates, then covered with fabric-reinforced acrylic, this construction method achieves slightly more than R-4 per inch of thickness (SI about R-28 per meter).

*Reflective* materials are available in sheets and rolls of either single or multiple layers, sometimes as preformed shapes with integral air spaces. When used without attachment to blanket or batt insulation, a reflective layer is called a *radiant barrier* and is especially applicable to roofs in warmer climates. Radiant barriers are also useful in east- and west-facing walls in such climates; details of applications in the southern United States are found in Melody (1987).

A combination of dead (still) air spaces and reflective surfaces produces some of the most effective insulating products, especially when made of lightweight materials of low conductivity. Glass fiber, cellular glass, expanded styrenes (foamed plastics), and mineral fibers all enclose vast numbers of dead-air spaces per unit volume. When they are bonded to reflective films and properly installed (the shiny film facing a dead air space), high resistance to heat flow is achieved.

For a summary of contemporary insulation materials, including environmental impacts and life-cycle considerations, see Wilkinson (1999). For a range of information on insulations see Bynum (2001).

*Conductors.* Materials used as conductors are typically dense, durable, and diffuse heat readily. Table E.1 lists the density, conductivity or conductance, and specific heat for many common materials. For a given material, the higher the numerical values of these three characteristics, the more successful that material's performance as a conductor. Some specifics on thermal mass are presented below and in Chapter 8.

*Air Films and Air Spaces.* Air films and spaces are interesting thermal components. Although they are actually void of material, they have potentially useful thermal properties and contribute substantially to the insulating capabilities of some construction assemblies. All above-ground envelope assemblies include at least two air films (interior and exterior), and many common assemblies include a substantial air space.

*Air Films.* At the exposed surfaces of solids, heat transfer takes place both by convection and by radiation. (Evaporation also can occur, occasionally with thermally significant results—see Fig. 8.44, the United Kingdom Pavilion, Seville, for an example of an entire wall used as an evaporative heat transfer element.) Convection is highly dependent upon air motion, so wind speed must be considered when estimating exterior surface convection. Also, because warm air rises and cold air falls, vertical surfaces that encourage surface airflow will exchange heat faster than similar surfaces placed horizontally, unless the direction of the heat flow is *upward* through the horizontal layer, as illustrated in Fig. 7.7.

When air motion along a surface is minimal, an *insulating layer* of air "attaches" itself to the surface. The resistance of this layer of still air along a vertical surface is equivalent to that of a thickness of ½-in. (12.7-mm) plywood. When this air layer is disturbed, however, its resistance drops quickly; with a 15-mph (6.7-m/s) wind, resistance drops to about one-quarter of the still-air value (see Table E.3). Similar drops in air film resistance occur when forced-air registers are located immediately above or below windows.

The insulation value provided by layers of air is often listed as conductance, the reciprocal of resistance. Surface conductances are designated $h_i$ for interior air layers and $h_o$ for exterior or outside air layers (sometimes the symbols $f_i$ and $f_o$ are used instead). Like other conductances, they are expressed in Btu/h ft² °F (in SI units, W/m² K). The variations in surface resistances and conductances are shown in Table E.3.

*Air Spaces.* An air space is a planar volume of air contained on two sides by some elements (drywall, brick, insulation, etc.) of an envelope assembly. Like the air films discussed above, air spaces can contribute to the overall thermal resistance of

a construction assembly. As indicated in Tables E.4 (I-P) and E.5 (SI), the resistance provided by an air space is a function of its width, position (vertical, horizontal, tilted), and surrounding emittances. To be effective in resisting heat flow, an air space must be relatively "dead"—without substantial air circulation.

### (e) Composite Thermal Performance

The variety of terms used so far to express thermal properties is potentially bewildering. These properties are, however, but components of a larger picture. Fortunately, there is *one* overall property that expresses the steady-state rate at which heat flows through architectural envelope assemblies. This property is the U-factor. U is the overall coefficient of thermal transmittance, expressed in terms of Btu/h ft² °F (in SI units, W/m² K). U-factors are commonly used to specify envelope thermal design criteria, as presented in Appendix G and Chapter 8. Many codes and standards specify maximum U-factors (or, for insulation alone, minimum R- or maximum C-values) for various components of the envelope. An example of such requirements for a small office is found in Example 7.7. Because U-factors are so important and so often encountered, data for typical wall, floor, roof, door, and window constructions are presented in many design resources (see, for example, Tables E.6 through E.16).

U-factors are calculated for a particular element (roof, wall, etc.) by finding the resistance of each constituent part, including air films and air spaces, then adding these resistances to obtain a total resistance. The U-factor is the reciprocal of this sum ($\Sigma$) of resistances:

$$U = \frac{1}{\sum R}$$

This procedure is shown in Example 7.1. "Precalculated" U-factors for many common constructions are found in Tables E.6 through E.16, typically based upon an outside wind speed of 15 mph (6.7 m/s), except for summer conditions as noted. Unfortunately, many common constructions are not so simple, as discussed in the sections that follow.

**EXAMPLE 7.1** What is the winter U-factor for the wall assembly shown in Fig. 7.9?

**SOLUTION**

| Component | R (I-P) | R (SI) | Data Source |
|---|---|---|---|
| Inside air film | 0.68 | 0.12 | Table E.3 |
| Gypsum board (0.375 in. [10 mm]) | 0.32 | 0.056 | Table E.1 |
| Plastic film vapor retarder | nil | nil | Table E.1 |
| Glass fiber batt insulation(nominal 6 in. [150 m]) | 19.00 | 3.35 | Table E.1 |
| Plywood (0.5 in. [12 mm]) | 0.62 | 0.11 | Table E.1 |
| Wood siding (1 in. [25 mm]) | 0.79 | 0.14 | Table E.1 |
| Outside air film | 0.17 | 0.03 | Table E.3 |
| Total resistance (R) | 21.58 | 3.81 | |

$$U \text{ (I-P)} = 1/\sum R = 1/21.58 = 0.046$$

$$U \text{ (SI)} = 1/\sum R = 1/3.81 = 0.262 \quad \blacksquare$$

There are several important points to remember about the U-factor. The U-factor is a *sensible* heat property—addressing heat flow resulting from a temperature difference but not addressing *latent* (moisture-related) heat flow. The U-factor is an *overall* coefficient of heat transfer, and includes the effects of all elements in an assembly and all sensible modes of heat transfer (conduction, convection, and radiation). The term *U-factor* should be used only where heat flow is from air to air through an envelope assembly. Air-to-ground heat flow (as through a slab-on-grade floor) is a different matter and is considered in Section 7.3f.

**Walls.** Compared to other elements of the building envelope, wall U-factors are quite straightforward. There are few complications such as ground contact, crawl spaces (as with floors), or intermediary attic spaces (as with roofs). There is, however, the issue of thermal bridging: where framing interrupts insulation, there are actually two different wall constructions and an averaged (an insulated portion with an uninsulated portion) U-factor must be found. Example 7.2 illustrates this procedure.

**EXAMPLE 7.2** Considering thermal bridging, what is the winter U-factor for the wall assembly shown in Fig. 7.10?

**SOLUTION**

For those portions of the wall where insulation is encountered:

| Component | R (I-P) | R (SI) |
|---|---|---|
| (6) Inside air film (still) | 0.68 | 0.12 |
| (5) Gypsum board (0.5 in. [12 mm]) | 0.45 | 0.079 |
| Plastic film vapor retarder | nil | nil |
| (4) Glass fiber batt insulation (6 in. [150 mm]) | 19.0 | 3.32 |
| (3) Vegetable fiber board (0.5 in. [12 mm]) | 1.32 | 0.23 |
| (2) Lapped wood siding (0.5 in. [12 mm]) | 0.81 | 0.14 |
| (1) Outside air film (15 mph [24 km/h]) | 0.17 | 0.03 |
| Total resistance through insulation | 22.43 | 3.92 |

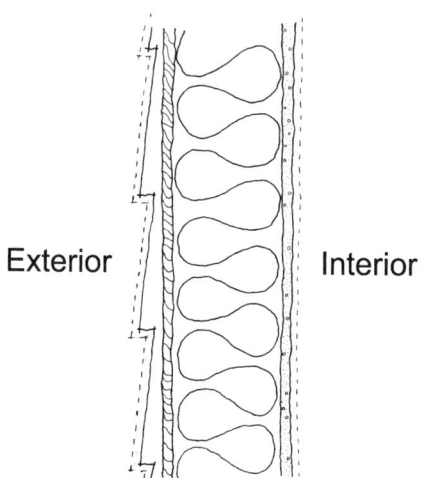

Exterior　　　　　Interior

**Fig. 7.9** *Wall section used in Example 7.1 to illustrate the process for determining a U-factor (overall thermal transmittance). (Drawing by Jonathan Meendering.)*

THERMAL CONTROL

For those portions of the wall where framing is encountered:

| Component | R (I-P) | R (SI) |
|---|---|---|
| (6) Inside air film (still) | 0.68 | 0.12 |
| (5) Gypsum board (0.5 in. [12 mm]) | 0.45 | 0.079 |
| Plastic film vapor retarder | nil | nil |
| (7) Wood studs (nominal 6 in. [150 mm]) | 6.82 | 1.20 |
| (3) Vegetable fiber board (0.5 in. [12 mm]) | 1.32 | 0.23 |
| (2) Lapped wood siding (0.5 in. [12 mm]) | 0.81 | 0.14 |
| (1) Outside air film (15 mph [24 km/h]) | 0.17 | 0.03 |
| Total resistance through framing | 10.25 | 1.80 |

Assuming that 12% of the surface area of the wall consists of framing (studs, sills, and plates), the area-weighted resistance and U-factor of the wall are as follows:

$R$ wtg = (0.88) (22.43) + (0.12) (10.25) = 20.97

$U$ wtg = 1/20.97 = 0.048 (compared to 1/22.43 = 0.045 if framing is ignored)

SI: $R$ wtg = (0.88) (3.92) + (0.12) (1.80) = 3.67

SI: $U$ wtg = 1/3.67 = 0.272 (compared to 1/3.92 = 0.255 if framing is ignored) ∎

The effects of framing with wood studs are nominal but noticeable. With metal studs, the detrimental effects of thermal bridging are more serious; Table E.8 provides information on correction factors and resulting R-values. The latest editions of ANSI/ASHRAE/IESNA Standard 90.1, *Energy Standard for Buildings Except Low-Rise Residential Buildings*, and ANSI/ASHRAE Standard 90.2, *Energy-Efficient Design of Low-Rise Residential Buildings*, present examples of whole-wall U-factors that account for thermal bridging. See Appendix G for excerpts from these standards.

To illustrate the complexity of thermal bridging, consider the following range of U-factors, all for nominal 4-in. (100-mm) thick framed walls with R-13 [SI: R-2.3] insulation located within the stud cavity:

| | |
|---|---|
| Metal building wall, nominal 4 in. (SI: 100 mm | U = 0.14 U = 0.79) |
| Steel frame wall, 3.5 in. at 16-in. o.c. (SI: 90 mm at 400-mm o.c. | U = 0.134 U = 0.76) |
| Steel frame wall, 3.5 in. at 24-in. o.c. (SI: 90 mm at 600-mm o.c. | U = 0.115 U = 0.65) |
| Wood frame wall, 3.5 in. at 16-in. o.c. (SI: 90 mm at 400-mm o.c. | U = 0.094 U = 0.53) |
| Wood frame wall, 3.5 in. at 24-in. o.c. (SI: 90 mm at 600-mm o.c. | U = 0.091 U = 0.52) |

For the same type of wall construction (wood frame) increasing stud spacing from 16-in. o.c. to 24-in. o.c. (400 mm to 600 mm) decreases (improves) the

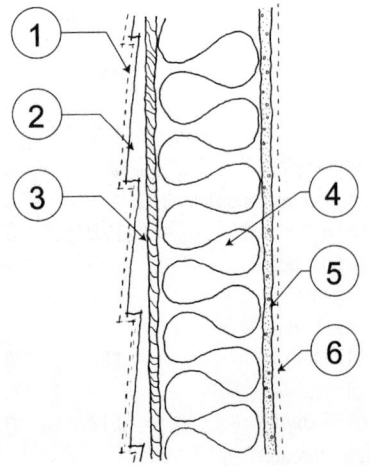
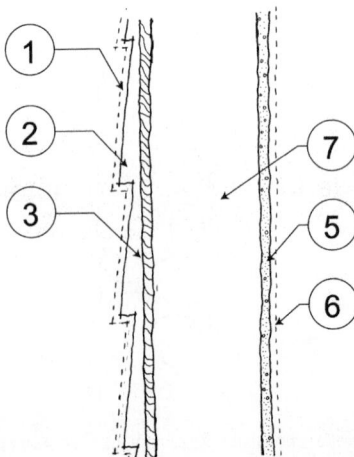

**Fig. 7.10** *Wall sections used in Example 7.2 to illustrate the process for determining a weighted-average U-factor to account for thermal bridging in a frame wall. (Drawings by Jonathan Meendering.)*

U-factor from 0.094 to 0.091 (SI: 0.53 to 0.52)— a nominal difference. Changing from wood to steel studs (at 16-in. o.c.; 400 mm), however, increases (degrades) the U-factor from 0.094 to 0.134 (SI: 0.53 to 0.76)—equivalent to an R-3 (SI: R-0.5) decrease in insulation. If a detailed calculation of thermal bridging effects is required, consult the 2005 *ASHRAE Handbook—Fundamentals*, Chapter 23.

Several new developments are promising to improve the thermal performance of wall systems. As discussed above, *structural insulated panels* (SIPs) are now available in a wide variety of surfaces, insulation types, and thicknesses. Because a single factory-built panel replaces site-built framing, savings in labor can be substantial. Thermal performance is considerably improved because no framing members penetrate the insulated core. The typical SIP consists of two structural surfaces (often oriented-strand board, or OSB) that enclose a core of either expanded polystyrene (EPS) (R-4 per inch; SI: R-27.7 per meter) or polyisocyanurate foam (R-6.5 per inch; SI: R-45 per meter). The panels are joined with plywood splines that connect the structural surfaces of adjacent panels in a manner that reduces the thickness of, but does not interrupt, the insulated core. Another joining method, shiplap joints, maintains full-thickness core insulation. Air tightness is generally greater than with stud framing. U-factors of framed walls and SIPs are compared in Table E.9.

*Insulated masonry systems* can offer lower U-factors while preserving a masonry exterior (for weathering and appearance) and a masonry interior (for thermal mass). Past approaches (filling hollow cores of concrete block or clay tile with insulation) left considerable thermal bridging through the solid masonry, with rather high resulting U-factors. If one of the masonry surfaces is not to be exposed, then a continuous layer of insulation can be applied to that face, with much lower resulting whole-wall U-factors. Even in cavity walls, the layer of insulation between wythes of block is usually penetrated by ties—with resulting thermal bridging. The most elementary improvement in this regard uses ties with lower thermal conductivity, such as fiber composite materials in place of steel. Another approach is to use precast blocks that can be stacked integrally with preformed rigid insulation; to the extent that the insulation layer can both be thick and unbroken by ties, thermal bridging is greatly reduced. Below is a sample of manufacturers' whole-wall U-factors for medium-density (120 lb/ft$^3$ [1920 kg/m$^3$]) concrete block insulation alternatives:

| | |
|---|---|
| 8-in. two-core concrete block, uninsulated | U = 0.39 |
| (SI: 200-mm two-core concrete block, uninsulated | U = 2.21) |
| 8-in. two-core concrete block with insulated cores | U = 0.19 |
| (SI: 200-mm two-core concrete block with insulated cores | U = 1.08) |
| 8-in. proprietary two-wythe concrete block and core | U = 0.11 |
| (SI: 200-mm proprietary two-wythe concrete block and core | U = 0.62) |

Where poured-in-place concrete is not to be used as a surface material, insulating concrete forms (ICFs) can be used to improve thermal performance. This system employs preformed rigid insulation as a formwork for poured concrete; the form/insulation remains in place after the wall is poured. Rigid insulation thus protects both faces of the structural wall (although it isolates the thermal mass); exterior and interior finishes are then applied to the insulation.

*Roofs.* In the simple case of a roof/ceiling separating a conditioned space from the outdoors, the U-factor is calculated as for walls. Where insulation is broken by framing, the effects of thermal bridging must be considered. Table E.8 shows correction factors to be applied to insulation/metal truss assemblies. Similar adjustments, although of lesser magnitude, are required with wood framing. ANSI/ASHRAE/IESNA Standard 90.1 presents several detailed tables of whole-roof U-factors that account for thermal bridging. If a detailed calculation of thermal bridging is required, consult the 2005 *ASHRAE Handbook—Fundamentals*, Chapter 23. Roof insulation, however, is often placed entirely above (or entirely below) the supporting structure, greatly reducing the effects of thermal bridging. There is a growing availability of above-the-roof-structure insulation materials, some penetrated by fastenings, others essentially unbroken. These options should be explored as a means of reducing roof heat loss and gain—with appropriate consideration of the environmental impacts of material selections.

**THERMAL CONTROL**

Where an insulated ceiling separates a space from an uninsulated attic, the simplest approach is to assume that the attic is at outdoor temperature. This assumption likely overestimates the winter heat flow rate because the temperature in a vented attic may be higher than the outdoor air temperature—but it simplifies analysis. If an accurate estimate of attic temperature is required, a more complex procedure is used, as presented in the 2005 *ASHRAE Handbook—Fundamentals,* Chapter 28. An attic is a space having an average distance of 1 ft (0.3 m) or more between the ceiling and the underside of the roof. A vented attic is often used to carry away moisture that may have migrated through the insulated ceiling. See Lstiburek and Carmody (1994) for detailed information on attic design practices.

*Floors.* When a floor is exposed to outdoor air (as with a cantilever or crawl space), the U-factor is calculated as for walls and roofs. When insulation is placed within framing cavities, the effects of thermal bridging must be included. It is becoming increasingly common to provide a continuous layer of insulation below the framing cavity, thus greatly reducing the detrimental effects of thermal bridging. ANSI/ASHRAE/IESNA Standard 90.1 presents several tables of whole-floor U-factors that account for thermal bridging.

Many codes require vented crawl spaces. In such situations, the simplest procedure is to insulate the floor above the crawl space and assume that the crawl space is at outdoor temperature. This assumption overestimates the heat flow rate, because a vented crawl space will rarely be as cold (warm) as outdoor air under winter (summer) design conditions. If a more accurate analysis of crawl space design temperatures is required, or if crawl space walls are insulated, a more complex procedure is used, as presented in the 2005 *ASHRAE Handbook—Fundamentals,* Chapter 29. It is rarely energy-conserving, however, especially in colder climates, to insulate the walls of a crawl space instead of insulating the floor above. See Lstiburek and Carmody (1994) for detailed information on crawl space design practices.

*Doors.* Table E.10 lists U-factors for solid (no glazing) doors in common use in North America. These precalculated values are convenient but should be used with care, as air film resistances are a relatively large factor in overall door thermal performance. Many doors are near forced-air supply registers or return grilles, reducing the resistance of the indoor air film. Conversely, the exterior air film may be more effective than anticipated because doors are typically protected somewhat by overhangs, porches, and so on. The wind conditions assumed in Table E.10 are typical of winter (not summer) conditions. Where significant door glazing is involved, U-factors for similar windows should be substituted or the specific manufacturer's product data used.

### (f) Special Envelope Heat Flow Conditions

The overall coefficient of heat transfer (U-factor) is a convenient property for analysis of above-ground envelope assemblies. It is only applicable, however, to assemblies exposed to air on both the interior and exterior surfaces. In other situations, such as slab-on-grade floors and below-ground walls and floors, U-factor is not a well-defined property (how much soil is part of the floor?) and other thermal analysis methods must be used. Typically, data derived from empirical studies are used in lieu of fundamental heat flow equations. Because the concept of U-factor is so simple to understand and apply, much of the empirical data are presented in terms of equivalent U-factors.

*Slab-on-Grade Floors.* Heat flow dynamics change when the lower surface of a floor is in direct contact with the ground. The ground is often at a temperature different from that of outdoor air, and earth is more conductive than air. Testing has shown that heat flow in this complex situation is strongly related to slab perimeter length. Table E.11 shows the heat flow rate through a concrete slab-on-grade, given as $F_2$ units (per foot or meter of perimeter length) rather than U-factors (per unit area). Four edge insulation conditions are shown, with corresponding values of $F_2$ for three climate zones. Interpolation can be used to estimate $F_2$ for other climates (correlated to heating degree days). This procedure relates all heat loss from the floor to the length of perimeter.

Slab-on-grade heat loss illustrates the principle of diminishing returns: an incrementally higher R-value does not produce an equally lower heat flow. An insulation truism is that the first few inches (millimeters) of insulation make a much greater impact on the rate of heat flow than do the last few inches (millimeters). Table E.11 lists a maximum slab-edge R-value of 5.4 h ft$^2$ °F/Btu (SI: R-0.95). If more slab-edge insulation seems desirable, consult Table E.12. Note how much more effective is vertical insulation at the exposed perimeter compared to insulation beneath the slab. The highest $F_2$ value listed in Table E.12 is for insulation at R-10 (SI: R-1.8) not only around the perimeter, but also under the entire area of the floor slab. This lowest rate of heat flow is a reduction of just 25% from the performance provided by the same insulation employed only around the perimeter to a depth of 4 ft (1.2 m).

**EXAMPLE 7.3** A warehouse will be built with a slab-on-grade floor. The warehouse is 80 ft (24.4 m) square in plan. What is the relationship between total slab insulation and rate of heat flow through the slab?

**SOLUTION**

From Table E.12, $F_2$ is related to total insulation as follows: building slab perimeter—four sides at 80 ft (24.4 m) each = 320 ft (97.5 m); building slab area: 80 ft (24.4 m) × 80 ft (24.4 m) = 6400 ft$^2$ (595 m$^2$).

I-P Units:

| Insulation Approach | Total Volume of Slab Insulation Used (ft$^3$) | $F_2$ (Btu/h ft °F) | $F_2 \times$ Perimeter = Rate of Heat Flow (Btu/h °F) | Reduction in Heat Flow (Btu/h °F) | Reduction per Unit (ft$^3$) of Insulation |
|---|---|---|---|---|---|
| Uninsulated slab | None | 0.73 | 233 | None | — |
| R-5 vertical, 2 ft, 1 in. | (2 ft) (320 ft) (1/12 ft) = 53.3 | 0.58 | 186 | 47 | 0.88 |
| R-10 vertical, 4 ft, 2 in. | (4 ft) (320 ft) (2/12 ft) = 213.3 | 0.48 | 154 | 32 | 0.20 |
| R-10 fully insulated, 2 in. | (2 ft) (320 ft) (2/12 ft) + (6400 ft$^2$) (2/12 ft) = 1173.4 | 0.36 | 115 | 39 | 0.04 |

SI Units:

| Insulation Approach | Total Volume of Slab Insulation Used (m$^3$) | $F_2$ (W/m K) | $F_2 \times$ Perimeter = Rate of Heat Flow (W/K) | Reduction in Heat Flow (W/K) | Reduction per Unit (m$^3$) of Insulation |
|---|---|---|---|---|---|
| Uninsulated slab | None | 1.26 | 123 | None | — |
| R-0.9 vertical, 0.6 m, 25 mm | (0.6 m) (97.8 m) (0.025 m) = 1.5 | 1.0 | 97.5 | 25.5 | 17.0 |
| R-1.8 vertical, 1.2 m, 50 mm | (1.2 m) (97.8 m) (0.051 m) = 6.0 | 0.83 | 80.9 | 16.6 | 2.8 |
| R-1.8 fully insulated, 50 mm | (0.6 m) (97.8 m) (0.051 m) + (595 m$^2$) (0.051 m) = 33.3 | 0.62 | 60.5 | 20.4 | 0.6 |

Thus, a minimal initial investment of 640 ft$^2$ (60 m$^2$) of 1-in. (25-mm) thick insulation cuts the uninsulated slab heat flow by one-fifth. However, a vastly greater investment of 7040 ft$^2$ (654 m$^2$) of 2-in. (50 mm) thick insulation (22 times as much!) cuts the less insulated heat flow only by one-fourth. ∎

*Basements.* Heat flow through basement walls and floors is complicated by an increasing length of the heat flow path with increasing depth, as shown in Figs. 7.11a and 7.11b. A further complication is that the temperature of the earth is not equal to ambient air temperature and becomes more and more out of phase with air temperature at increasing depths. To obtain a "design temperature" for below-ground heat loss, first estimate the mean winter temperature of the site location (for example: using Table B.15 in Appendix B, take the average of the ambient temperature (TA) for January and for the year. Then, from this mean winter temperature, subtract the value of the constant amplitude, Fig. 7.11c. This gives a design temperature. Table E.13 is then used to determine the heat flow rate.

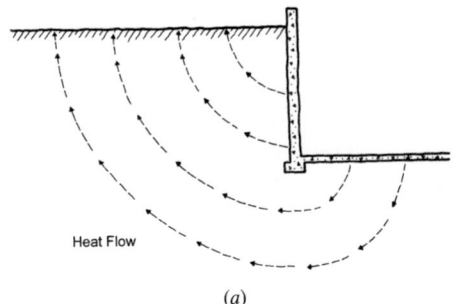

Heat Flow

(a)

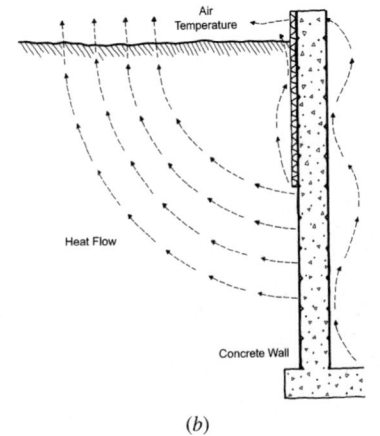

Air Temperature

Heat Flow

Concrete Wall

(b)

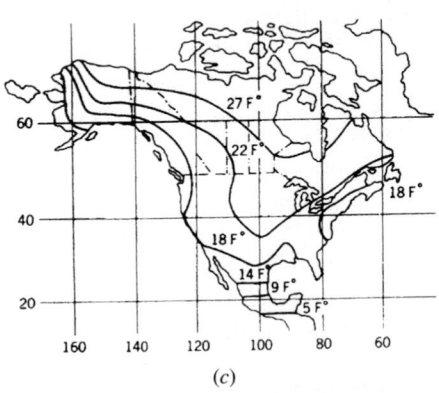

60

27 F°

22 F°

18 F°

40

18 F°

14 F°

9 F°

20

5 F°

160    140    120    100    80    60

(c)

**Fig. 7.11** *Heat flow below grade. (a) Heat flow through basement walls and floors follows radial paths (that are generally perpendicular to soil isotherms). (b) Heat flow paths of differing lengths for a partially insulated basement wall. (c) Lines of constant amplitude of ground temperature. Just below the surface, the ground temperature fluctuates around a mean annual temperature by this amplitude (5°F difference = 2.8°C difference; 9°F = 5°C; 14°F = 7.8°C; 18°F = 10°C; 22°F = 12.2°C; 27°F = 15°C).* (Reprinted with permission of the American Society of Heating, Refrigerating and Air-Conditioning Engineers, Inc. from the 2001 ASHRAE Handbook—Fundamentals; parts a and b redrawn by Amanda Clegg.)

**EXAMPLE 7.4** Calculate the design heat loss for a basement in Minneapolis, Minnesota, which is 28 ft (8.5 m) wide by 30 ft (9.1 m) long, sunk 6 ft (1.8 m) below grade. An insulation of R-8.34 (SI: R-1.5) is applied to this wall to a depth of 2 ft (0.6 m) below grade. An internal temperature of 70°F (21°C) is to be maintained.

**SOLUTION**

The soil-path design temperature is estimated from Table C.15; for Minneapolis, the TA in January is 12°F (−11°C); TA for the year is 44°F (6.7°C); the average of these is 28°F (−2°C). From Fig. 7.11c, the amplitude is about 24°F degrees (13.3°C) at Minneapolis. The design temperature at the ground surface is therefore 28 − 24 = 4°F (−2 − −13.3 = −15.3°C).

The heat loss through the basement wall is estimated using Table E.13, Part A, by determining the loss per degree of temperature difference per 1-ft (0.305-m)-high strip of wall and then summing for the full height of the wall:

1st ft below grade (insulated)      0.093 Btu/h ft °F

    SI: 1st 0.3 m: (0.53) (0.305) = 0.162 W/m K

2nd ft below grade (insulated)      0.079

    SI: 2nd 0.3 m: (0.45) (0.305) = 0.137

3rd ft below grade (uninsulated)      0.155*

    SI: 3rd 0.3 m: (0.88) (0.305) = 0.268*

4th ft below grade (uninsulated)      0.119

    SI: 4th 0.3 m: (0.67) (0.305) = 0.204

5th ft below grade (uninsulated)      0.096

    SI: 5th 0.3 m: (0.54) (0.305) = 0.165

6th ft below grade (uninsulated)      0.079

    SI: 6th 0.3 m: (0.45) (0.305) = 0.137

Total *per 1-ft length* of wal      0.621 Btu/h ft °F

Total *per 1-m length* of wall      1.073 W/m K

Basement perimeter = 2 (28 + 30 ft) = 116 ft

    SI: 2 (8.5 + 9.1 m) = 35.2 m

Total wall heat loss = 0.621 Btu/h ft °F × 116 ft = 72 Btu/h °F

    SI: 1.073 W/m K × 35.2 m = 37.8 W/K

---

*Note that although this is an increase in loss relative to the segment of wall directly above, it represents a design decision exhibiting understanding of the law of diminishing returns.

The heat loss through the basement floor is calculated using Table E.13, Part B.

Average heat loss per ft² floor
= 0.025 Btu/h ft² °F (for 28 ft width, 6 ft depth);
 SI: per m² floor = 0.14 W/m² K
 (for 8.5 m width, 1.8 m depth)

Floor area = 28 ft × 30 ft = 840 ft²
 SI: area = 8.5 m × 9.1 m = 77.4 m²

Total floor heat loss = 0.025 Btu/h ft² °F × 840 ft²
= 21 Btu/h °F
 SI: 0.14 W/m² K × 77.4 m² = 10.8 W/K

Total heat loss for the basement below grade: walls
72 + floor 21 = 93 Btu/h °F
 SI: walls 37.8 + floor 10.8 = 48.6 W/K

Design temperature difference: (70°F inside)
− (4°F at earth's surface) = 66°F
 SI: (21.1°C inside)
 − (−15.6°C at earth's surface) = 36.7°C

Design heat loss below grade = 93 Btu/h °F × 66 °F
= 6,138 Btu/h
 SI: design below grade heat loss
= 48.6 W/K × 36.7 °C or K = 1784 W  ■

## (g) Predicting Surface Temperatures and Condensation

Surface temperatures, as well as the temperature at any point within a wall, roof, or floor assembly, can be predicted if the inside and outside air temperatures and the thermal properties of the construction are known. The variance in temperature through a cross section of a construction assembly is called a *thermal gradient*. For any construction, the thermal gradient can be predicted by proportioning the collective thermal resistance at any point in the assembly to the overall difference in temperature across the assembly. This procedure is illustrated in Example 7.5:

**EXAMPLE 7.5** Determine the thermal gradient, under winter design conditions, through the wall assembly shown in Fig. 7.12. The interior air temperature is 68°F (20°C) and the exterior air temperature is 32°F (0°C).

## SOLUTION

I-P Units: resistance values are from Example 7.1; "reference point" is a shorthand means of describing a series of locations throughout the wall assem-

bly, each on the exterior side of a component; total temperature gradient spans (68 − 32) = 36°F:

| Thermal Component | Component Resistance | Cumulative Resistance | Reference Point | Temperature Difference to Reference Point (°F) | Temperature at Reference Point |
|---|---|---|---|---|---|
| Room air | — | — | A | — | 68.0 |
| Inside air film | 0.68 | 0.68 | B | (0.68/21.58) (36) = 1.1 | 66.9 |
| Gypsum board | 0.32 | 1.00 | C | (1.00/21.58) (36) = 1.7 | 66.3 |
| Vapor retarder | nil | 1.00 | D | (1.00/21.58) (36) = 1.7 | 66.3 |
| Batt insulation | 19.00 | 20.00 | E | (20.00/21.58) (36) = 33.4 | 34.4 |
| Plywood sheathing | 0.62 | 20.62 | F | (20.62/21.58) (36) = 34.4 | 33.6 |
| Wood siding | 0.79 | 21.41 | G | (21.41/21.58) (36) = 35.7 | 32.3 |
| Outside air film | 0.17 | 21.58 | H | (21.58/21.58) (36) = 36 | 32.0 |
| Outside air | — | — | I | — | 32.0 |

SI Units: resistance values are from Example 7.1; "reference point" is a shorthand means of describing a series of locations throughout the wall assembly, each on the exterior side of a component; total temperature gradient spans (20 − 0) = 20°C:

| Thermal Component | Component Resistance | Cumulative Resistance | Reference Point | Temperature Difference to Reference Point (°C) | Temperature at Reference Point (°C) |
|---|---|---|---|---|---|
| Room air | — | — | A | — | 20.0 |
| Inside air film | 0.12 | 0.12 | B | (0.12/3.81) (20) = 0.6 | 19.4 |
| Gypsum board | 0.056 | 0.18 | C | (0.18/3.81) (20) = 0.9 | 19.1 |
| Vapor retarder | nil | 0.18 | D | (0.18/3.81) (20) = 0.9 | 19.1 |
| Batt insulation | 3.35 | 3.53 | E | (3.53/3.81) (20) = 18.5 | 1.5 |
| Plywood sheathing | 0.11 | 3.64 | F | (3.64/3.81) (20) = 19.1 | 0.9 |
| Wood siding | 0.14 | 3.78 | G | (3.78/3.81) (20) = 19.8 | 0.2 |
| Outside air film | 0.03 | 3.81 | H | (3.81/3.81) (20) = 20 | 0.0 |
| Outside air | — | — | I | — | 0.0 |

■

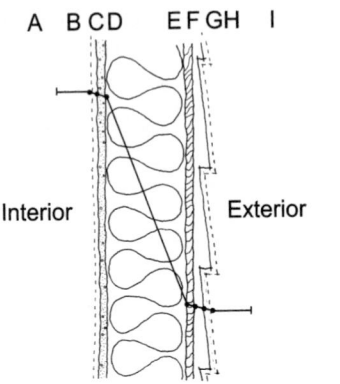

A  B CD    E F GH    I

Interior                    Exterior

**Fig. 7.12** *Wall section used in Example 7.5 to illustrate the proce-dure for calculating a thermal gradient through a construction assembly. (Drawing by Jonathan Meendering.)*

Understanding envelope surface temperatures is fundamental to predicting thermal comfort. How cold in winter, or hot in summer, will an interior sur-face be and how will this affect mean radiant temper-ature? Understanding temperature patterns within materials is critical to establishing where water vapor might condense within or on a wall, roof, or floor assembly; this is discussed further in Section 7.9.

### (h) Dynamic Thermal Effects

The above discussion of thermal properties of mate-rials and assemblies focused upon static (steady-state) conditions. Such conditions are usually presumed for analysis of winter heat loss (assumed to occur under the darkness of night). A more com-plex situation exists when thermal conditions expe-rience rapid changes—such as those brought about by solar radiation impinging upon various portions of the building envelope. Dynamic conditions are usually presumed for analysis of summer heat gain (assumed to occur during daylight hours). The dynamic situation requires consideration of more thermal properties.

Under static conditions, heat flow is primarily a function of temperature difference (the driving force) and thermal resistance (the resisting force). Under dynamic conditions, these two factors are still important, but heat storage in the envelope assembly itself becomes a compounding issue. Heat storage is a function of the *density* of a material and its *specific heat;* the product of these two properties is known as *thermal capacity.* Thermal capacity can (but will not always) reduce heat flow via storage.

Heat entering a wall construction during the day-time, for example, can be stored within the wall for several hours until it flows back out to the cool night air—assuming appropriate weather condi-tions and adequate thermal capacity. As a mini-mum, the effect of heat entering a massive wall at, say, 10 A.M. will not be seen on the interior surface until a few hours later.

***Density.*** Density is the weight of a material per unit volume. In the I-P system, density is given as lb/ft$^3$; in the SI system, it is given as kg/m$^3$. For a fixed volume of material, greater density will permit the storage of more heat. Table E.1 provides density data for common building materials.

***Specific Heat.*** Specific heat is a measure of the amount of heat required to raise the temperature of a given mass of material by 1°. In the I-P system, this is expressed as Btu/lb °F; in the SI system, it is expressed as kJ/kg K. It takes less energy input to raise the temperature of a low-specific-heat mater-ial than that of a high-specific-heat material. Table E.1 provides values of specific heat for common building materials.

***Thermal Capacity.*** Thermal capacity is an indi-cator of the ability of a fixed volume of material to store heat. The greater the thermal capacity of a material, the more heat it can store in a given vol-ume per degree of temperature increase. Thermal capacity for a material is obtained by taking the product of density and specific heat; for example:

Thermal capacity for concrete:

(140 lb/ft$^3$) (0.24 Btu/lb °F) = 33.6 Btu/ft$^3$ °F
SI: (2240 kg/m$^3$) (1.0 kJ/kg K) = 2240 kJ/m$^3$ K

Thermal capacity for water:

(62 lb/ft$^3$) (1.00 Btu/lb °F) = 62.0 Btu/ft$^3$ °F
SI: (992 kg/m$^3$) (4.18 kJ/kg K) = 4147 kJ/m$^3$ K

Thermal capacity for air:

(0.075 lb/ft$^3$) (0.24 Btu/lb °F) = 0.02 Btu/ft$^3$ °F
SI: (1.2 kg/m$^3$) (1.0 kJ/kg K) = 1.2 kJ/m$^3$ K

***Time Lag.*** Time lag is a measure of the delay in the flow of a pulse of heat through a material that results from thermal capacity. Units are hours. As an example, if the sun comes out from behind clouds and strikes the exterior surface of a mass wall at 10 A.M., the exterior surface temperature will rise quickly. It may be several hours, however, before this temperature "spike" is seen at the inside surface of the wall. The reason is that some heat is being stored in the wall material and that heat will not continue to flow through the wall. Time lags of several hours are possible with very heavy wall constructions.

The design effects of capacitive insulation are difficult to analyze without computer assistance. As a result, most hand calculation methods incorporate capacitive effects through the use of surrogate variables that attempt to simplify calculations for the end user. This approach is presented in Chapter 8.

## 7.4 LATENT HEAT FLOW THROUGH THE OPAQUE ENVELOPE

The information presented above focused upon sensible heat flow through various elements of the building envelope. Water also moves through building envelope assemblies—in both liquid and vapor states. The focus here is upon water vapor movement (assuming that proper architectural design of the envelope components will control the movement of liquid water). Water vapor will often need to be handled by a climate control system through the use of energy (termed *latent heat*). In the summer, moisture will typically flow into an air-conditioned building, increasing humidity and requiring dehumidification, which is often accomplished through removal of the latent heat of condensation of the added moisture. In the winter, it is not unusual to add water vapor to the air in a building to keep RH from dropping too low. This is often accomplished by evaporating water by adding latent heat of vaporization. In some building types and climates, dealing with latent heat may be as big a problem as dealing with sensible heat.

### (a) Moisture Control Fundamentals

A difference in *vapor pressure* is the driving force behind moisture flow through components of an intact building envelope assembly, while gaps in the envelope can provide a route for airflow that carries water vapor. Vapor pressure difference is to latent heat flow as temperature difference is to sensible

heat flow. The *permeance* of the materials of construction is the latent equivalent of sensible conductance. (As with conductance, *permeance* refers to the bulk properties of a material and *permeability* refers to unit thickness properties.) The less permeable a material is, the greater the resistance to water vapor flow. Materials with low permeance are termed *vapor retarders,* and are incorporated in envelope constructions as a means of reducing the flow of water vapor and subsequently the risk of condensation of the vapor within the envelope assembly. I-P units of permeance are grains/h ft$^2$ in. Hg; SI units are ng/s m$^2$ Pa. (Grains refers to grains of water, with 7000 grains per pound; in. Hg is inches of mercury, a pressure measurement; ng is a nano-gram or $10^{-9}$ of a gram.) The I-P units for vapor pressure are in. Hg; the SI units are Pa [Pascals].

From an architectural design perspective, reducing water vapor flow is accomplished using very thin materials (membranes) that must be carefully located to ensure that they work as intended. Although placement within an envelope assembly is critical, vapor retarders take up virtually no space—in drastic contrast to the thickness requirements of sensible heat retarders (insulations). The specific location of a vapor retarder within a wall, roof, or floor cross section will vary with climate and construction types. The fundamental principle, however, is that the retarder stop the flow of water vapor before the vapor can come in contact with its dew point temperature within the assembly.

### (b) Cold Climate Moisture Control

Most common building materials, including gypsum board, concrete, brick, wood, and glass fiber insulation, are easily permeated by moisture. Most surface/finish materials are also permeable. In cold climates, the winter outside air contains relatively little moisture, even though the RH may be high. By contrast, inside air contains much more moisture per unit of volume, despite its probably lower RH. The resulting differential vapor pressure drives the flow of water vapor from high to low vapor pressure (typically from warm to cold).

The primary problem with water vapor flow in winter occurs when the temperature somewhere within a wall, floor, or roof drops low enough for water vapor passing through the assembly to condense. The temperature at which this occurs is the dew point temperature of the air acting as the source for the water vapor. Such a temperature often occurs within the insulation layer. Insulation can then become wet and thereby less effective, because water conducts heat far better than the air pockets it has filled. If wet insulation compacts, the air pockets are then permanently lost. Worse yet, moisture damage can occur, such as dry rot in wood structural members.

The usual remedy for envelope condensation problems in cold climates is a *vapor retarder* installed within the building envelope assembly. Because very low permeability is desired, these retarders are commonly a plastic film installed with as few gaps and holes as possible. Because the moister air on the warm side of the envelope is the source of the problem, the vapor retarder needs to be installed *as close to the warm side as possible*—typically, just behind the interior surface (gypsum board, wood flooring, etc.) before the dew point can be encountered. However, with higher insulation R-values, it is becoming common to install the vapor retarder within the insulation at a point about one-third of the distance from the interior to the exterior. This allows the inner one-third of the wall to be used as a chase for wiring or plumbing without penetrating the retarder, yet with enough insulation beyond this point to prevent condensation by maintaining the temperature above the dew point.

Another approach is to use vinyl wallpaper or vapor retarder paints on interior surfaces of the envelope. These give less protection, however, than the around-the-corners wrap that can be achieved with properly installed plastic films. This performance disadvantage is shared by aluminum foil–faced insulation; it is very effective thermally but less effective as a vapor retarder. Also, the aluminum foil must face an air space if it is to be thermally effective as a radiant barrier.

A substantial benefit of plastic films is that they also reduce airflow through construction—acting as an air barrier. Outdoor air is always infiltrating a building, gradually replacing the indoor air. This unintended flow of cold air becomes a problem when temperatures outside are very different from those inside, especially when strong winds force outdoor air indoors fast enough to produce noticeably cold drafts. Therefore, the combined moisture- and infiltration-tight characteristics of plastic film vapor retarders are usually beneficial.

In cold climates, condensation on cold interior surfaces can also occur, especially at windows (with little resistance to reduce the slope of the thermal gradient). Occupant annoyance or material damage can result. Although the air indoors may not be particularly humid (perhaps 30% to 50% RH), it often contains enough moisture to support condensation on cold surfaces. Condensation can be readily predicted (see Fig. 7.12) and addressed during design.

### (c) Hot, Humid Climate Moisture Control

In hot, humid conditions, cool inside surfaces are often encountered—for example, a radiant cooling panel containing chilled water, a water pipe, or a supply air diffuser. If hot, humid air contacts such a surface, condensation can occur, with the moisture vapor in the air condensing to form visible droplets of water on the cool surface. The result can be mildly annoying if droplets of condensation fall on occupants or serious if water stains occur and, eventually, mold grows on damp surfaces.

In hot, humid climates, the object is to keep the moisture in the warmer outside air from penetrating to the cooler (and usually less humid) interior. One method is to use a *drainage plane*, installed just inside the exterior surface material, to block the flow of liquid water through the envelope. The drainage plane can consist of simple tar paper (building felt). If any moisture succeeds in penetrating this drainage plane, it should be allowed to wick to the interior; therefore, no vapor retarder such as plastic, vinyl wallpaper, or vapor retarder paints should be used on interior surfaces. Numerous severe problems with water vapor damage to building interiors—and related mold and mildew growth with associated IAQ problems—have recently been publicized, typically related to condensation of moisture on the exterior-facing (hidden) side of vinyl wall coverings used as an interior finish. See Lstibuek and Carmody (1994) for details on hot and cold climate moisture control.

## 7.5 HEAT FLOW THROUGH TRANSPARENT/TRANSLUCENT ELEMENTS

Heat flow through windows and skylights requires special attention for several reasons. Despite dramatic improvements, these transparent/translucent envelope components still usually have the lowest R (highest U) of all components of an envelope. Also, they are major contributors to infiltration of outdoor air, which adds to winter heating or summer cooling loads. Perhaps thermally most important, they admit solar heat, for better (winter) or worse (summer). By design, this radiation (heat) passes through a window or skylight with minimal resistance, which complicates thermal analysis. Transparent/translucent devices also admit daylight to buildings and often provide desired ventilation. The huge variety of such components and the several important roles they play require special attention from designers.

A standardized approach to rating window performance characteristics has been developed. This approach is coordinated by the National Fenestration Rating Council (NFRC). Each window or skylight manufactured in the United States bears an NFRC label certifying that the window has been independently rated. The label carries a brief description of the product, such as "Model #, Casement, Low-e = 0.2, 0.5" gap, Argon Filled" and lists the following information: U-factor, solar heat gain coefficient (SHGC), and visible light transmittance (VT). Figure 7.13 shows a sample NFRC label. Similar labels are available for site-assembled window units. Characteristics of some residential-sized windows are found in Table E.15.

The following sections provide information on fundamental thermal properties of transparent/translucent assemblies and approaches that can be taken to enhance the performance of such assemblies. A transparent material permits a generally undistorted view (as with clear glass); a translucent material permits at best a distorted view (as with glass blocks or milky plastic). Both types of material, however, permit some unimpeded transmission of radiation—as opposed to an opaque material/assembly that permits no direct solar radiation transfer.

**Fig. 7.13** *Certifying window thermal performance; a sample NFRC window label. (© National Fenestration Rating Council; used with permission.)*

### (a) U-factor

As with opaque envelope components, heat flow due to temperature differences through windows and skylights is a function of the U-factor. The determination of U-factors for windows and skylights is complicated by significant differences in heat flow rates between the center-of-glass, edge-of-glass, and frame portions of a unit. The NFRC "U-factor" melds these into a single representative value for an entire window or skylight unit. The size of the air gap between glazings, the coatings on the glazings, the gas fill between glazings, and the frame construction all influence the U-factor. The lower the U-factor, the lower the heat flow for a given temperature difference. Tables E.14, E.15, and E.16 list U-factors for various windows and skylights. Appendix G presents sample requirements for window U-factors and SHGC (see below) for nonresidential buildings excerpted from ANSI/ASHRAE/IESNA Standard 90.1-2004 and similar information for residential buildings from ANSI/ASHRAE Standard 90.2-2004.

### (b) Solar Heat Gain Coefficient (SHGC)

This thermal property is also generally based upon the performance of the entire glazing unit, not just that of the glass portion. SHGC represents the percentage of solar radiation (across the spectrum) incident upon a given window or skylight assembly that ends up in a building as heat. It is a measure of the ability of a window to resist heat gain from solar radiation. SHGC can theoretically range from 0 to 1, with 1 representing no resistance and 0 representing total resistance. SHGC values for real products typically range from about 0.9 to 0.2. SHGC is dimensionless. A high SHGC (meaning poor resistance to radiant gain) is desirable for solar heating applications, whereas a lower SHGC (good resistance) is better for windows where cooling is the dominant thermal issue. The SHGC depends upon the type of glass and the number of panes, as well as tinting, reflective coatings, and shading by the window or skylight frame.

NFRC tests and lists SHGC values only for glazing units; laboratory testing does not include auxiliary elements such as draperies, overhangs, trees, and the like. Prior to the development of the SHGC approach, the term *shading coefficient* (SC) was universally used to quantify the same concept. Shading coefficients, however, were based only upon the glass portion of a glazing unit—specifically excluding the frame. Although theoretical SC values also range from 0 to 1, the basis of SC is different from that of SHGC. SC is the ratio of radiant heat gain through a given type of glass relative to ⅛-in. (3-mm) thick single clear glass. SC is still a useful concept for comparing glass types and especially for expressing the effects of external or internal shading devices.

A wide variety of window and skylight U-factor, SHGC, and VT (see below) data can be found in Chapter 31 of the 2005 *ASHRAE Handbook—Fundamentals.* Tables E.17 through E.24 list SHGC, SC, and/or VT values for various types of transparent/translucent products and shading approaches.

A small sample of U-factors for skylights is shown in Table E.16. Note in this table the great thermal penalty paid when using smaller units: the "manufactured skylights" are 2 ft × 4 ft (600 mm × 1200 mm), whereas the "site-assembled glazing" unit size is 4 ft (1200 mm) square. For otherwise identical construction, the whole-component U-factors are almost twice as high for the smaller units. With these smaller units, a frame with better thermal performance is a smart investment.

## (c) Visible Transmittance (VT)

This thermal/optical property represents the percentage of incident light (only the visible spectrum) at a normal angle of incidence that passes through a particular glazing. VT is dimensionless. The higher the visible transmittance, the greater the daylight transmission. VT is influenced by the color of the glass (clear glass has the highest VT) as well as by coatings and the number of glazings. VT may be expressed relative to the glass portion only of a glazing unit or relative to the glass and frame. The appropriate expression will depend upon the nature of an analysis; in any case, noncomparable values should not be compared. All NFRC-certified VT values should be directly comparable.

It might seem intuitive that any glazing or coating that reduces SHGC (via lower radiation transmission) will also reduce VT (implying lower radiation transmission). This is not always the case, however, as SHGC deals with the full solar radiation spectrum (including light), whereas VT deals only with the visible (light) spectrum. Spectrally selective glazings and selective coatings are available that greatly reduce SHGC with little reduction in VT. The relationship between SHGC and VT is expressed as the *light-to-solar gain ratio* (LSG) obtained by dividing the VT by the SHGC. The greater the LSG, the more suitable a glazing is for daylighting in hot climates (or wherever cooling is the dominant thermal condition). Values of LSG are not listed on NFRC labels. LSG values for common glazing units (including the effects of the frame) are included in Table E.15.

## (d) Air Leakage

This is the rate of outdoor air infiltration between a new window and its frame measured under defined conditions—usually under pressure equivalent to that of a 25-mph (40-km/h) wind, with the window locked. *High-performance* windows may be tested/rated at even higher pressures. As weather stripping deteriorates with age, higher rates of infiltration may be expected with older and well-used windows. Design air leakage values are not listed on NFRC labels but may be obtained from manufacturers' catalog data.

### (e) Low-Emittance (low-ε) Coatings

These coatings are typically applied to one glass surface facing into the air gap between multiple glazings. A low-ε coating blocks a great deal of the radiant transfer between the glazing panes, reducing the overall flow of heat through the window and thus improving the U-factor. Indeed, one such coating is almost as effective as adding another layer of glazing. In Table E.15, compare the U-factor for window 5 with the U-factors for windows 7 and 12. An important added benefit of these films is their reduction of UV transmission, thus reducing fading of objects and surface finishes in rooms.

Two approaches to providing low-ε films are *hard-coat* (durable, less expensive, but less thermally effective) and *soft-coat* (better thermal performance but more expensive and subject to degradation by oxidation in the manufacturing stage).

Three common types of low-ε coatings are:

1. High-transmission low-ε: for passive solar heating applications, where a low U-factor is combined with a high SHGC; window 7 in Table E.15 is an example. The coating is on the inner glazing, where it traps outgoing infrared radiation that otherwise would be lost. Summer overheating can be avoided with external shading devices.
2. Selective-transmission low-ε: where winter heating and summer cooling are both important, requiring low U-factor and low SHGC, but with a relatively high VT for daylighting. Window 9 in Table E.15 is an example, with an LSG ratio of 1.65. The coating is on the outer glazing, where it blocks incoming infrared radiation, which as heat is then convected away by outdoor air.
3. Low-transmission low-ε: where the sun is the enemy, low U-factor, low SHGC, and even low VT seem warranted. Window 10 in Table E.15 is an example, again with the coating on the outer glazing, where it rejects more of the solar gain. With a tinted exterior glazing, even lower SHGC and VT could result.

### (f) Selective Transmission Films

Heat flow due to radiation can be greatly reduced by the introduction of selective transmitter (or *low-emittance*) film somewhere within the glazing cavity. As shown in Fig. 7.14, these films admit most of the incoming solar radiation in both the visible and near-infrared (short) wavelengths. Warm objects within a room emit far-infrared (long-wave) radiation. This long-wave radiation is reflected back into the room by the selective film. These selective films typically are available as separate sheets that can be inserted between sheets of glazing as a window is fabricated. As a separate sheet, a selective film could also be applied to existing windows—for instance, between storm windows and the ordinary windows they protect.

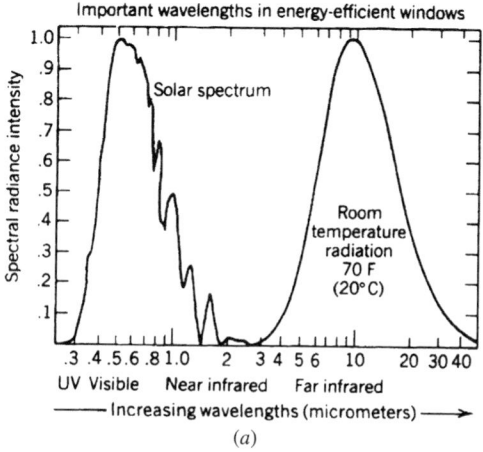

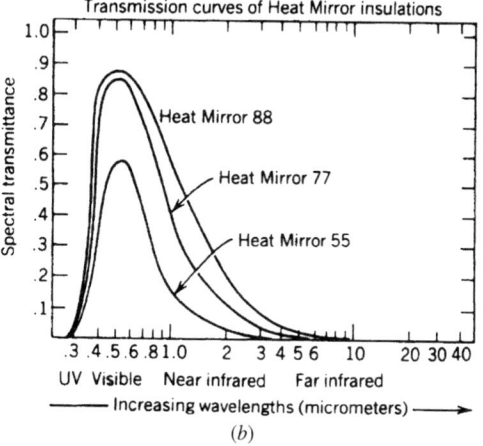

**Fig. 7.14** *Performance of selective transmitters. (a) Spectral characteristics of solar (short-wavelength) radiation and room-temperature (long-wavelength) radiation. (b) Transmission and reflection performance for several Heat Mirror selective transmitter films. Incoming solar radiation (both visible and near-infrared) is mostly transmitted, whereas heat radiated from room-temperature objects is reflected and thereby kept within heated spaces. (Courtesy of Southwall Corporation.)*

With continuing advances in glazing performance, an increasing range of selective transmission options is available. For example, products from Southwall Technologies range from Heat Mirror 77 (low reflectance, recommended for vertical glass) to Heat Mirror 22 (low transmission, recommended for sloping glass) and provide various colors. A clear Heat Mirror 88 has the lowest reflectance of all. Some skylight and greenhouse manufacturers offer this low-emittance product as a standard option.

### (g) Inert Gas in the Air Gap

Filling an enclosed air space with argon or krypton has thermal benefits. These less conductive gasses greatly reduce heat transfer by convective currents within the air gap between multiple glazings, producing lower U-factors. As a result, the inner surface of the glass is maintained at a temperature closer to that of the indoors, with greater comfort (because radiant heat to or from the window surface is reduced) and less chance of condensation on the inside surface. To preserve this gas fill over the life of the window, a very reliable edge seal is required.

For argon, the optimum air gap width is about ½ in. (12 mm). When a thinner window is needed, more expensive krypton can be used with an air gap of only ¼ in. (6 mm). Because the combination of inert gas and low-ε coatings is so effective at lowering the U-factor, most manufacturers offer them together (as with windows 7 through 11 in Table E.15) rather than separately.

### (h) Superwindows

When all of the currently available high-performance glazing options are combined in one product, it is called a *superwindow*. The combination of multiple glazings and/or suspended films, coatings, inert gas fill, and sealed/thermally broken frame construction yields a lower heat flow rate and a higher price. Window 11 in Table E.15 is a superwindow. An early example of a superwindow, a double-glazed assembly with two selective films, yielding three gas-filled cavities, is shown in Fig. 7.15. With a combination of superior thermal resistance and useful solar gain potential, superwindows can conceivably provide *better* heating season thermal performance than an insulated

opaque wall. The potential design consequences are enormous. Figure 7.16 compares the heating and cooling season performance, in a residential application, of several windows in three locations in the United States with quite different climates.

### (i) Shading

Perhaps the single most important energy-related component for passively cooled buildings is the sunshade. Because many solar-heated buildings will experience overheating in hot weather, sunshading is critical for passively heated buildings as well. If correctly implemented, sunshading rejects most solar heat gains yet aids in distributing daylight deep into buildings, where it reduces internal heat gains due to electric lights.

If a building is arranged to intercept the intense rays of the sun *before* they pass through its transparent envelope elements, instead of afterward, the cooling load can often be cut in half. In approximate terms, effective external shading rejects about 80% of the fierce attack of solar energy, whereas internal shading absorbs and reradiates 80% of it. Outside louvers have a chance to cool off in an occasional breeze, but inside draperies

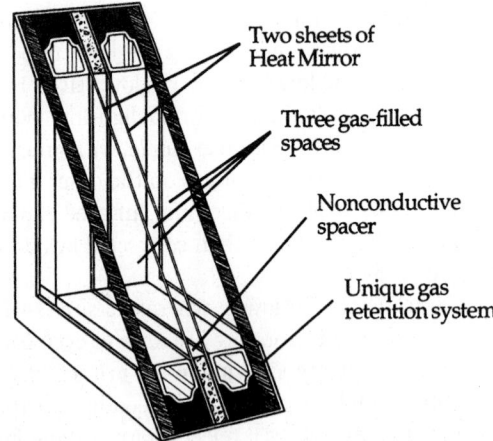

Two sheets of
Heat Mirror

Three gas-filled
spaces

Nonconductive
spacer

Unique gas
retention system

**Fig. 7.15** *The Superglass window system (an early superwindow) utilized two Heat Mirror films between two outer panes of glass. Three air spaces are thus created and filled with a nonreactive gas mixture to retard convection. Nonmetallic spacers reduce heat flow at window edges. The resulting window provided a center-of-glass insulating value of R-8.1 (SI: R-1.43). (ASHRAE lists for a similar window, nonoperable, 4 ft (1220 mm) square, an overall window U = 0.19.) Very low transmission of UV radiation is another characteristic. (Courtesy of Hurd Millwork Company, Medford, WI.)*

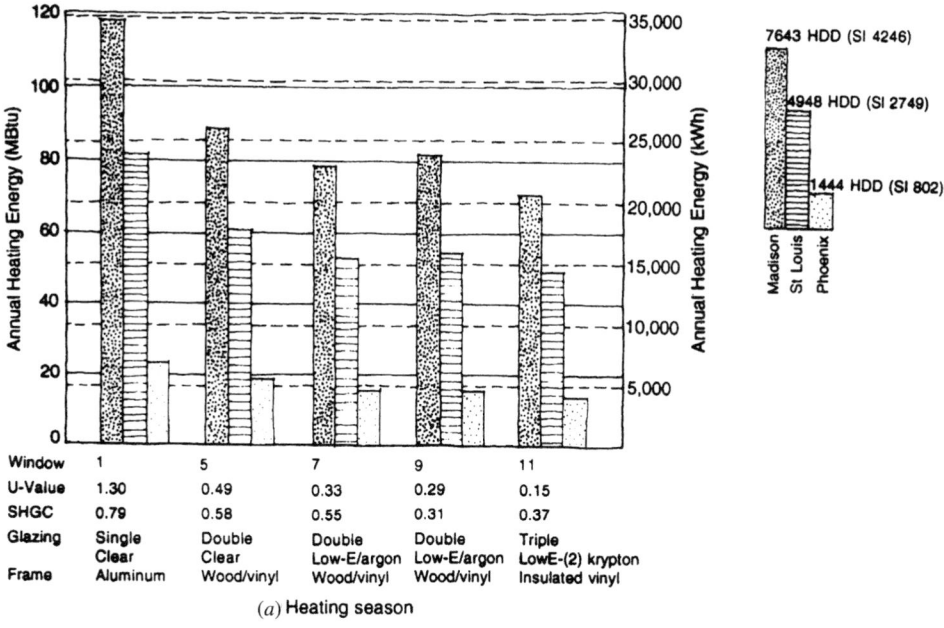

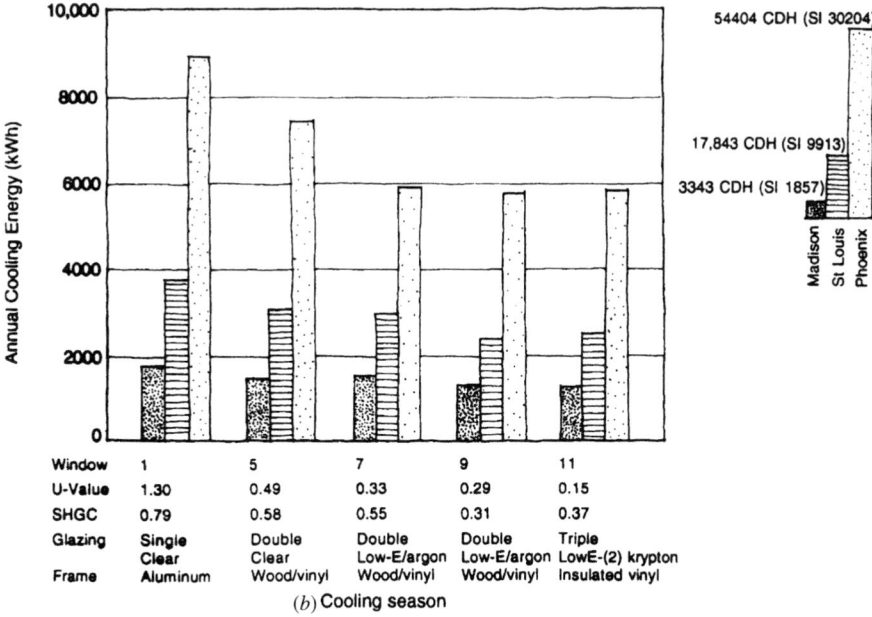

***Fig. 7.16*** *(a, b) Seasonal energy performance comparison of various windows—assuming a typical 1540-ft² (143-m²) residence with a glazing area of 15% of the floor area, equal glazing area on all four orientations, and no external or internal shading. Window numbers correspond to those in Table E.15. MBtu = millions of Btu; HDD = heating degree days, base 65°F (18.3°C); CDH = cooling degree hours, base 74°F (23.3°C). U-factors in figure are I-P values; multiply these by 5.678 for SI U-factors. (From* Residential Windows: A Guide to New Technologies and Energy Performance, *by John Carmody, Stephen Selkowitz, and Lisa Heschong. Copyright © 1996 by John Carmody, Stephen Selkowitz, and Lisa Heschong. Adapted by permission of W.W. Norton & Company, New York.)*

are part of a heat trap, and they constitute a system of hot-weather radiant heating that causes discomfort for those who work near perimeter surfaces.

To properly reject direct sun yet allow for a view and daylight, many sunshades project out from the windows they protect. These exterior projections become highly visible elements of façades, and they tempt some designers to impose formal aesthetic criteria that can be damaging to the solar control functions. A frequent example is the application of the same sunshade geometry to all façades of a building. When the sunshades are fixed in position, this tends to help one façade but not the others (see Fig. 3.19). Where they are movable, as with awnings, this same-sunshade effect is not so serious.

Fixed sunshading devices are very common, partly because they lack moving parts and controls that can be expensive and troublesome. They pose a dilemma in the spring and fall, however, because in order to block the sun on any elevation in September, they will also block it in March. For many buildings in temperate climates, March is a heating-need month, whereas September is a cooling-need month. A procedure for evaluating fixed shading device performance is presented in Chapter 6; some approximate shading effects, expressed as shading coefficients, are found in Tables E.20 and E.21.

Adjustable sunshading, once extremely common (it seemed as if every 1930s shop and office window had an awning) but later considered old-fashioned, has made a comeback with the increased interest in passive cooling. Some basic approaches to adjustable shading devices are shown in Table E.22. The first type shown, made of durable materials, has proved effective against both break-ins and hurricanes.

How are movable shading devices adjusted to the desired position? Most manufacturers offer three types of controls: manual, motorized, and automatic. Manual systems are cheap and relatively trouble-free, but they require thoughtful, timely action by the users of a building. So do motorized controls, but for adjusting large/heavy devices in remote places (clerestories, for example), motorized assistance is a practical necessity. Automatic systems have the advantages of freeing the users from adjustment tasks and of taking into account the thermal needs of the building as a whole when setting the sunshade's position. With computerized controls becoming commonplace for large build-

ings, the added costs of controls for automatic sunshading can easily be incorporated into the overall cost of controls.

Interior shading devices are less effective than exterior devices but are far more commonly used. There are several reasons for this seeming contradiction. They are not subject to weathering or dirt accumulation and generally are easier for users to adjust. The designer who prefers a clean, apparently unchanging façade appearance will rely on interior devices. Tables E.23 and E.24 list shading coefficients for several combinations of interior shading devices.

The overall performance of multiple shading devices/approaches can be estimated by taking the product of the shading coefficients for individual devices that are used in series. For example, the overall shading coefficient of a window with external shading with an SC of 0.75, integral shading from the glazing unit of 0.65, and internal shading providing an SC of 0.7 is $(0.75)(0.65)(0.7) = 0.34$. In general, shading coefficients lower than 0.2 should be used with caution. There is no direct conversion between SHGC values and SC. Nevertheless, it may be necessary to convert an SHGC value to SC in order to determine the overall shading performance of a glazing unit with external or internal shading devices. In this case, an approximate conversion of $SC = SHGC/0.87$ may be used.

## 7.6 TRENDS IN ENVELOPE THERMAL PERFORMANCE

As a result of the oil embargo of the early 1970s, there have been huge increases in installed component R (therefore, huge decreases in assembly U-factor) over the past 30-plus years. Sometimes this has occurred because of the use of increased thicknesses of insulation (as with many current wall and roof designs), sometimes because of substantially changed materials (as with windows). What might the next quarter century bring, and how might the designer of today's building envelopes anticipate— or participate in—these advances?

*Structural systems* may be changing due to the advent of SIP. They promise greatly improved insulation and air-tightness compared to the site-assembled wood or steel framing systems they would replace. SIP can incorporate the latest developments in both insulated cores and thermal storage surfaces, and do so with less thickness and lighter weight than stick-built alternatives. Since the early 1970s, walls have progressed from using R-7 (SI: R-1.23) batts between typical 2-in. × 4-in. (50-mm × 100-mm) studs to R-26 (SI: R-4.58) batts between 2-in. × 6-in. (50-mm × 150-mm) studs plus insulating sheathing, thereby providing a threefold reduction in heat flow. With insulated cores for SIP now approaching R-25 per inch (SI: R-173 per meter), the next quarter century could produce yet another threefold reduction in heat flow without an increase in wall thickness. Another potential development is an interior panel surface of phase-change wallboard, to store and release heat at room temperatures. This could increase thermal capacity considerably with very little increase in weight, with advantages for winter solar heating and summer night ventilation. Floors and roofs can similarly benefit from the use of SIP rather than framed construction.

*Insulation* developments include aerogel, gas-filled panels, powder-evacuated panels, and compact vacuum. Some of these products are suitable for retrofit applications in existing construction. Aerogel is both transparent and porous, one of the lightest solid materials. Silica aerogel can be foamed into cavities without ozone-depleting CFCs. In a 90% vacuum, silica aerogel has a resistance of R-20 per inch (SI: R-139 per meter). By adding carbon to absorb infrared radiation, R-32 per inch (SI: R-222 per meter) is possible.

Gas-filled panels are hermetically sealed plastic bags enclosing honeycomb baffles of thin polymer films and low-conductivity gas (argon, krypton, or xenon). Relative resistance values of these insulating cores are: argon, R-5.2 per inch; krypton, R-13.4 per inch; and xenon, R-19.3 per inch (SI: R-36, 93, and 134 per meter, respectively). Powder-evacuated panels contain a vacuum and compacted silica-based powder sealed within a multilayer gas barrier. Expensive to produce and threatened by punctures, this may be a development more suited for appliances than for buildings. The resistance is about R-20 to R-25 per inch (SI: R-139 to R-173 per meter).

*Windows* have undergone the most dramatic thermal improvement. Resistances for window units (glazing plus frame) have evolved from about R-1 (SI: R-0.18) to R-6.5 (SI: R-1.14) or more. Pro-totypes as high as R-20 (SI: R-3.52) exist. This improvement has been accomplished through a series of developments: a second layer of glass; a wider air space between glazing layers; tinted, reflective, photochromic, and low-emissivity coatings; low-conductivity gas between glazings; intermediate films between glazings; operable blinds between glazings; lower-conductivity glazing spacers and frames; and more air-tight weather stripping. Future developments are likely to include *smart windows* with on-demand variable light transmission. For example, in electrochromic glazing, an applied electric field switches the window from a clear state to one with a deep coloration; intermediate states of coloration are possible. The most likely application is in cooling-dominated climates where control of glare, solar gain, and visual privacy are paramount. Carmody et al. (1996) provide an excellent overview of residential window technologies, while Carmody et al. (2003) provide a similar overview of commercial window technologies.

Before the oil embargo of the early 1970s, windows represented the component with the greatest heat flow rate in a typical building envelope. Today, the highest rate of heat flow is more likely to be from outside air infiltration (or deliberate ventilation). Windows do so many things in addition to providing weather protection—admitting daylight, allowing views outdoors and in, admitting breezes, and admitting warming winter sun—that their relative thermal weakness compared to their desirability continues to challenge manufacturers. Over the next quarter century, windows may well continue to display the most dramatically improved resistance to heat flow of all the components of the envelope.

Roofs are of particular interest relative to heat gains and losses because they are subjected to such extremes. On a clear night, radiant losses to the sky can lower a roof's surface temperature below that of the outdoor air—beneficial in summer, but a problem in winter. In summer, elevated roof surface temperatures in sunlight are dramatic and problematic. Efforts are underway to disseminate information regarding the benefits of *cool roofs* and to incorporate credit for such roofs into national energy codes (CRRC, 2004). Both reflective roofs and green (or eco-roof) designs are being employed.

As discussed below, limiting heat flow via unwanted air movement is an important aspect of envelope design. The tightness of buildings has improved remarkably over the past 20 years, partly due to improved components (such as windows) and partly due to increased care in construction (involving caulking and air barriers/retarders). As a result of this improvement, building air tightness as a means of reducing energy usage is now in conflict with IAQ needs and concerns. It is likely that future improvements in this area will revolve around the use of heat recovery ventilators that permit reasonable ventilation rates without excess energy penalties.

Commissioning of the building envelope is an emerging trend. Commissioning goes beyond construction quality control to attempt to ensure (though design review, documentation, and testing) that envelope design, construction, and maintenance meet the owner's project requirements. The National Institute of Building Sciences is developing a building envelope commissioning guideline.

## 7.7  HEAT FLOW VIA AIR MOVEMENT

Outdoor air can enter a building by means of infiltration and/or ventilation. *Infiltration* is an unintended influx of outdoor air due to air leakage through the building skin. *Ventilation* is a deliberate, designed introduction of outdoor air. In either case, air at a different temperature and humidity interacts with the air inside a building. This interaction brings sensible and latent heat loads. Of the two means of air entry, infiltration is the more difficult to predict, since it is by definition unintended. Ventilation air quantities are intended and thus easily quantified. Although it is possible for infiltration and ventilation to occur simultaneously, usually it is reasonable to assume one of the following scenarios: only infiltration (which is typical of many smaller buildings with no ventilation) or only ventilation (typically required by code for larger buildings, and which usually pressurizes a building and blocks infiltration). Both infiltration and ventilation airflow rates are expressed in cubic feet per minute (cfm); in SI the units are liters per second (L/s).

### (a) Infiltration

The main problem with infiltration is estimating how much air will leak into a building once it is built. There are two primary means of calculating the infiltration airflow rate during design: the air-change method and the crack method.

The *air-change method* is very simple and quick but tends to overestimate. Air-change is a correlational method wherein observed performance of a set of existing buildings is matched (correlated) to key characteristics of a proposed building. The characteristics used are construction type and climate. Virtually no detailed information about a building or its spatial arrangement is required. Table E.27 lists estimated numbers of air changes per hour (ACH) as a function of these two characteristics. ACH is indicative of the "turnover" of air within a building or space. The greater the ACH, the greater the rate of outdoor airflow. A tight building might permit 0.25 ACH or so, a leaky building 1.0 ACH. ACH can be converted to airflow rate as follows:

$$V = \frac{(\text{ACH})(\text{volume, ft}^3)}{60 \text{ min/h}}$$

where $V$ is airflow rate in cfm.

In SI units:

$$V = \frac{(\text{ACH})(\text{volume, m}^3)}{3600 \text{ sec/h}}$$

where $V$ is in m³/s. Although both m³/s and L/s are used as SI units of airflow rate, L/s is more commonly encountered in building design work. L/s = m³/s × 1000.

The *crack method* requires more effort and assumes that data on window and door construction and wind velocities are known. It also assumes that all infiltration under design conditions is due to cracks around doors and windows. To estimate the airflow rate, calculate the length of cracks on the windward exposure(s) only and multiply this by the respective assembly air leakage rate(s) as follows:

$$V = (\text{total lin. ft of window/door edge})$$
$$\times (\text{air leakage rate(s), cfm per lin. ft})$$

where $V$ is in cfm.

In SI units:

$$V = (\text{total lin. m of window/door edge})$$
$$\times (\text{air leakage rate(s), L/s per lin. m})$$

where $V$ is in L/s.

See Table E.15 for generic air leakage data or get NFRC or manufacturer's ratings for a specific product. If air leakage rates are unknown, Table E.28 shows how much leakage per unit length of crack should be assumed, for the windward exposure(s) only, to arrive at the total infiltration airflow rate. Multiply the total length of crack by the infiltration rate from part B of Table E.28.

It is possible to measure the actual infiltration performance of a constructed building using a fairly simple device called a *blower door.* Some energy codes require that building tightness be verified by a blower door test. As a result, local utility companies may be able to provide reasonably accurate values for air change rates for buildings constructed using prevailing practices. Such local current data may provide better estimates of infiltration than the crack method.

### (b) Ventilation

It is important to provide some minimum amount of fresh air to indoor environments as a means of IAQ control. Odors and a sense of staleness can be uncomfortable, and dangerous buildups of pollutants such as formaldehyde and radon gas can occur within buildings. These pollutants are most effectively removed via fresh airflow. Tables E.25 and E.26 list recommended design outdoor airflow rates to produce acceptable IAQ. A more thorough discussion of IAQ is found in Chapter 5.

Many building codes require a minimum outdoor airflow rate based upon either building population or floor area. These requirements are generally based upon ASHRAE Standard 61.1 (non-residential buildings) or ASHRAE Standard 62.2 (residential buildings). This minimum is often (but not always) provided by mechanically induced ventilation. Since ventilation is intended, estimating such airflow rates is simple. With some simplification, when a building's required outdoor air (Table E.25) is based upon population:

$$V = \text{(cfm [or L/s] of outdoor air per person)} \\ \times \text{(number of people)}$$

When the required outdoor airflow rate is based upon floor area:

$$V = \text{(cfm [or L/s] of outdoor air per unit floor area)} \\ \times \text{(total floor area)}$$

## 7.8 CALCULATING ENVELOPE HEAT FLOWS

When the fundamentals of heat flow through a building envelope are understood, calculations to determine the magnitude of such flows can be undertaken. Three types of heat flow are usually of primary interest: (1) a design heat loss based upon "worst-hour" conditions, which is used to size heating systems; (2) a design heat gain (or cooling load) also based upon worst-hour conditions, which is used to size cooling systems; and (3) an annualized heat flow, based upon year-long climate conditions, which is used to predict annual energy usage and costs and/or demonstrate compliance with energy standards. Hand calculations can provide reasonably accurate estimates of design heat loss and gain—even though the use of computer programs for these analyses is very common. Although a rough estimate of annual energy usage can be made using hand calculations, it is much more likely that a computer program will be used for this purpose. Many such programs are readily available, although they generally have a moderately steep learning curve and require detailed input information. Regardless of whether a manual or computerized calculation approach is taken, understanding the concepts and issues involved is critical to good design.

### (a) Design Heat Loss

Calculation of design heat loss is an estimation of the worst likely hourly heat flow from a building to the surrounding environment. This value is used to size heating systems; the greater the design heat loss, the larger the required heating system capacity. The *design* heat loss is not the highest heat loss that can (or will) ever occur; rather, it is a statistically reasonable maximum heat loss based upon a chosen outside air temperature. By convention, design heat loss is assumed to occur at night (no sun), during the winter (coldest temperatures), with no occupants, lights, or equipment to offset heat lost through the envelope. These assumptions lead to an easy analysis based upon steady-state conditions.

To calculate the design (statistically-worst-hour) heat loss through a building's envelope, the following information is necessary:

1. The *rate* at which heat flows through each of the elements that make up the building envelope—the assembly U-factors
2. The *area* of each of these assemblies
3. The design *temperature difference* between inside and outside

With only the first two factors, a comparison of the thermal integrity of one building envelope versus another, in any location, can be made. These two factors are an outcome of the design process. The third factor, temperature difference, is a function primarily of climate (but also of comfort criteria). Design temperatures for various geographic locations (climates) are listed in Appendix B.

The design sensible heat loss through *all* types of above-ground elements of the building envelope can be calculated as follows:

$$q = (U)(A)(\Delta t) \quad \text{for each element}$$

I-P units: Btu/h = (Btu/h ft² °F) (ft²) (°F)

SI units: W = (W/m² K) (m²) (K)

where

$q$ = hourly heat loss through a specific envelope component (by conduction, convection, and radiation) under design conditions

$U$ = U-factor for a given envelope component

$A$ = surface area of the envelope component

$\Delta t$ = the design temperature difference between inside and outside air (see Appendix B)

(If a *representative* heat loss is desired, use average temperatures such as from the latter part of Appendix C or local climatological data.)

In addition to the above-ground components, heat loss through at- and below-grade elements and through airflow must be estimated. For slab-on-grade floors (see Example 7.3):

$$q = (F_2)(P)(\Delta t)$$

I-P units: Btu/h = (Btu/h ft °F) (ft) (°F)

SI units: W = (W/m K) (m) (K)

where

$q$ = hourly sensible heat loss through the slab-on-grade floor under design conditions

$F_2$ = heat loss coefficient for the floor slab/insulation configuration

$P$ = perimeter of the floor slab

$\Delta t$ = design temperature difference between inside and outside air

For below-grade basement walls and floors (see Example 7.4):

$$q = (F)(A)(\Delta t) \quad \text{summed across various depths below grade of the wall/floor}$$

I-P units: Btu/h = (Btu/h ft² °F) (ft²) (°F)

SI units: W = (W/m² K) (m²) (K)

where

$q$ = hourly sensible heat loss through the basement wall/floor under design conditions

$F$ = heat loss coefficient for the wall-floor/insulation configuration at a given depth below grade

$A$ = area of the floor or wall at a given nominal depth below grade

$\Delta t$ = design temperature difference between inside air and surface of the ground

For airflow due to infiltration or ventilation, design sensible heat loss is calculated as follows:

$$q = (V)(1.1)(\Delta t)$$

where

$q$ = sensible heat loss due to infiltration or ventilation (Btu/h)

$V$ = outdoor airflow rate in cfm

$\Delta t$ = temperature difference between outdoor and indoor air (°F)

1.1 = a constant derived from the density of air at 0.075 lb/ft³ under average conditions, multiplied by the specific heat of air (heat required to raise 1 lb of air 1°F, which is 0.24 Btu/lb °F) and by 60 min/h. The units of this frequently encountered, but quirky, constant are Btu min/ft³ h °F.

In SI units, this becomes

$$q = (V) (1.2) (\Delta t)$$

where $q$ is in Watts, $V$ is in liters per second, 1.2 is a constant (the density of air [1.20 kg/m$^3$] multiplied by the specific heat of air [1.0 kJ/kg K]), and $\Delta t$ is in degrees K.

The total design sensible heat loss for a building is the sum of the component losses (above-ground elements, on-ground and below-ground elements, and airflow). All the above-ground envelope elements could be lumped together (using an area-weighted U-factor) under the same equation (as orientation, tilt, and transparency are of no consequence in the absence of sun). It is very useful, however, to identify the loss of each component (window, floor, roof, etc.) so that design improvements can focus upon those envelope components with the biggest contribution to overall heat loss. Design for energy efficiency in the building envelope essentially consists of trying to reduce the value of any variable that occurs on the right-hand side of the above equations. This is the design palette; no other design change will affect design heat loss:

- Reduce component U-factors (within the constraints of the budget and constructability).
- Reduce areas of components (within the constraints of the program and design intent).
- Reduce the design temperature difference (usually by lowering indoor air temperature, perhaps by microclimate improvement).
- Reduce airflow into the building.

In practice, design latent heat loss is often ignored. There are two reasons for this: (1) humidity (directly affected by latent loss) is often allowed to float during the winter, and (2) there is no single system or piece of equipment that will handle both sensible and latent heat loss. Even so, it is possible to estimate design latent heat loss for the various envelope elements. The calculation of latent heat loss through above-ground components involves an equation similar to that used for sensible heat loss:

$$q = (M) (A) (\Delta p) (2500)$$

where

$q$ = latent heat exchange due to vapor pressure difference ($W$)

$M$ = permeance of the envelope element (ng/s m$^2$ Pa)

$A$ = surface area of the envelope component (m$^2$)

$\Delta p$ = difference in vapor pressure between outdoor and indoor air (Pa)

2500 = an approximate value of energy content (enthalpy in kJ/kg) under typical interior conditions

As it is common to install a reasonably effective (very-low-permeance) vapor retarder in cold climate constructions to minimize condensation problems, latent envelope heat loss is often assumed to be negligible. Because the on-grade and below-ground portions of a building are commonly waterproofed with very-low-permeability materials, these components are typically assumed to be impermeable and to support no latent heat loss. Substantial latent loss can occur, however, as a result of airflow.

The design latent heat loss due to the flow of outdoor air into a space is calculated as follows:

$$q = (V) (4840) (\Delta W)$$

where

$q$ = latent heat exchange due to ventilation (Btu/h)

$V$ = outdoor airflow rate in cfm

$\Delta W$ = difference in humidity ratio between outdoor and indoor air (lb of moisture/lb of dry air)

4840 = a constant derived from the density of air (0.075 lb/ft$^3$) multiplied by the heat content of water vapor (1076 Btu/lb) and by 60 min/h (the density and heat content values noted are for conditions typical of interior environments)

In SI units, this becomes

$$q = (V) (3010) (\Delta W)$$

where $q$ is in Watts, $V$ is in liters per second, $W$ is in kJ/kg, and 3010 is the density of air (1.20 kg/m$^3$) multiplied by the heat content of water vapor (2500 kJ/kg for typical indoor conditions).

## (b) Design Heat Gain

The calculation of design heat gain (more correctly called *cooling load*) is far more complex than the calculation of design heat loss. The simplifying assumptions made for winter conditions (nighttime with no sun and ignoring occupancy-related heat sources) are untenable for a summer situation. There is sun during the "worst" summer hour, and the highest temperatures are usually in the afternoon, when a nonresidential building would be in full work mode with occupants, lights, and equipment generating heat. In a nutshell, design cooling load calculations must deal with many more variables—including the compounding effects of solar radiation, which create the need for dynamic heat flow analysis. Things get very complicated.

There are hand calculation methods for estimating design cooling load and computer programs to do the same. The details of these methods will not be presented here (see Chapter 8 for more information). It is important, however, to establish the conditions under which such calculations are conducted. In addition to the design variables that affect design heat loss (envelope assembly U-factor, surface area, temperature difference, and airflow rate), the following variables will affect design cooling load:

- The orientation of an assembly (north, south, etc.)
- The tilt of an assembly (vertical, horizontal, inclined)
- The surface reflectance of an assembly
- The thermal capacity of an assembly
- The solar heat gain coefficient of a transparent/translucent assembly
- Shading for any envelope component
- Heat gain from occupants
- Heat gain from lights
- Heat gain from equipment

Design decisions influence all of these variables and therefore the resulting design cooling load. The above list is again the palette from which energy efficiency decisions may be selected. It is critical that this palette be fully appreciated early in the design process and not be delegated to a mechanical engineering consultant who may enter the design process too late to provide effective advice.

How does design *heat gain* differ from design *cooling load?* In the past, calculations for summer loads simply totaled all the heat flows at a given hour that entered or originated inside of a building envelope. This total instantaneous heat flow is

known as *heat gain.* It turns out, however, that some of this instantaneous gain does not immediately affect indoor air temperature—specifically, radiation heat flows that are absorbed and stored by the internal mass of a building. Those heat gains that affect air temperature at the hour of interest are collectively called the *cooling load.* It is only this distinct subset of heat gains that must be handled (at the time) by a cooling system. Conceptually, cooling load is equal to instantaneous heat gain minus any part of that heat gain that is stored within the building plus any previously stored gains that are now affecting air temperature. The distinction can be substantial, providing first cost savings in equipment capacity and life-cycle savings through improved operating efficiency.

Solar radiation plays an important role in building envelope heat gain, and not just via transparent/translucent components. The impact of solar radiation on opaque construction assemblies can be substantial. The concept of sol-air temperature helps to illustrate this impact. *Sol-air temperature* is the apparent outdoor air temperature that would produce the same heat flow experienced under the combined effects of temperature difference (based upon actual outdoor air temperature) and radiation. Essentially, sol-air temperature lumps the heat flow caused by radiation absorbed and retained by a surface with the heat flow caused by the air-to-air temperature difference. This is a convenient concept, as it produces a delta-*t* value that can be plugged into the conventional heat gain equation. The simplified sol-air temperature formula (in I-P units) is:

$$t_e = t_o + \frac{\alpha I}{h_o} - 7F°$$

where

$t_e$ = sol-air temperature

$t_o$ = outdoor (ambient) dry bulb temperature

$\alpha$ = absorptance of surface for solar radiation (for light-colored surfaces, usually assumed = 0.45; for dark-colored surfaces, usually assumed = 0.90; detailed values are listed in Table 9.1)

$I$ = total solar radiation incident on the surface, Btu/h ft$^2$ (see Appendix C for solar heat gain factors, which are approximately equivalent to $I$ for horizontal surfaces)

$h_o$ = coefficient of heat transfer by long-wave radiation and convection at the surface (usually assumed to be = 3.0 Btu/h ft$^2$ °F)

The simplified sol-air temperature formula in SI units is:

$$t_e = t_o + \frac{\alpha I}{h_o} - 3.9\text{C}°$$

where the differences from I-P units are as follows:

$I$ = total solar radiation incident on the surface, W/m$^2$

$h_o$ = usually assumed to be = 17.0 W/m$^2$ K

---

**EXAMPLE 7.6** What is the sol-air temperature for a horizontal white roof, compared to a dark roof, on a clear July 21 at noon, at 40° N latitude? Assume outdoor air temperature of 90°F (32°C).

**SOLUTION**

From Table C.3 of Appendix C, on July 21 at noon at 40° N latitude, the solar heat gain factor on a horizontal surface = 262 Btu/h ft$^2$ (827 W/m$^2$).

Sol-air temperature: $t_e = t_o + \dfrac{\alpha I}{h_o} - 7\text{F}°$

For a white roof:

$$t_e = 90°\text{F} + \frac{0.45 \times 262 \text{ Btu/h ft}^2}{3.0 \text{ Btu/h ft}^2 \text{ °F}} - 7\text{F}°$$

$$= 90 + 39.3 - 7 = 122.3°\text{F}$$

For a dark roof:

$$t_e = 90°\text{F} + \frac{0.90 \times 262 \text{ Btu/h ft}^2}{3.0 \text{ Btu/h ft}^2 \text{ °F}} - 7\text{F}°$$

$$= 90 + 78.6 - 7 = 161.6°\text{F}$$

Solar radiation has a marked impact upon surface temperature (and resulting heat flow). Elevated temperatures in full sun will drive considerably more heat through both roofs compared to shaded conditions. The white roof has a $\Delta t$ about 1.4 times greater in full sun than in shade. The dark roof has a $\Delta t$ about 1.3 greater than the white roof under the stated conditions. ∎

## 7.9 ENVELOPE THERMAL DESIGN STANDARDS

The building design team will find that there are code requirements that establish minimum thermal envelope performance requirements for virtually all jurisdictions in North America and for virtually all building types. In the United States, code adoption is a local matter and the specific energy code to be followed depends upon the jurisdiction. Even so, the energy code most likely to be encountered for nonresidential buildings is ANSI/ASHRAE/IESNA Standard 90.1. In California, Title 24 is the statewide counterpart of Standard 90.1. Residential buildings are likely to fall under the requirements of the *International Energy Conservation Code* (or its predecessor *Model Energy Code*—both of which are really standards until adopted as law). ASHRAE also publishes a residential energy standard, Standard 90.2. Although energy codes provide efficiency requirements for more than just the envelope (usually also dealing with HVAC equipment efficiency, hot water heating, and lighting), this discussion focuses upon envelope requirements.

It is typical for energy codes to provide two paths to compliance: (1) a prescriptive path, which provides precise statements of minimal acceptable component characteristics (for example, a minimum R for wall insulation or a maximum U for a window) and (2) a performance path, which sets a minimum level of overall energy performance that may be met using a wide range of solutions. Code compliance via the prescriptive path is usually straightforward and requires little analysis or creativity. Following the performance path may require detailed analysis (usually by way of a computer program) and opens the door to innovation and creativity.

Energy codes need to be placed in context. They establish a set of minimum requirements. Just barely meeting an energy code simply means that a building envelope is not illegal. In today's design environment, doing better—perhaps much better—than the minimum required by code is becoming a common design intent. The U.S. Green Building Council (USGBC) LEED rating system, for example, requires compliance with ANSI/ASHRAE/IESNA 90.1 as a prerequisite for green building status. To actually gain any points toward a green building rating, a design must exceed these minimum requirements (by 15% to 60%, depending upon the points sought). ASHRAE recently released its first *Advanced Energy Design Guide* (distinct from the 90.1 energy standard) to assist in the design of office buildings that will perform 30% better than a 90.1 compliant building.

Appendix G provides excerpts from the USGBC LEED program and ASHRAE energy-efficiency standards as follows:

- A sample of the prescriptive envelope requirements from ANSI/ASHRAE/IESNA Standard 90.1 (nonresidential buildings)
- A sample of the prescriptive envelope requirements from ANSI/ASHRAE Standard 90.2 (residential buildings)
- The criteria set used for LEED certification of new buildings

**EXAMPLE 7.7** A two-story medical office building is being designed for St. Louis, Missouri. What are the maximum permissible U-factors (and other envelope characteristics) for such a building in this location?

**SOLUTION**

Refer to Appendix B for climate data and to Appendix G (Table G.1) for sample provisions of ANSI/ASHRAE/IESNA Standard 90.1-2001.

- Climate: St. Louis, Missouri: HDD65: 4758 (SI: HDD18 = 2643); CDD50: 4283 (SI: CDD10 = 2379). (HDD and CDD values would be obtained from Standard 90.1 or another data resource.)

  These benchmark climate indicators (heating and cooling degree days; HDD and CDD, respectively—see Chapter 8 for more information) are used to establish which set of the Standard 90.1 prescriptive requirements apply to a building in St. Louis. Within the applicable set (Table G.1), note that one of the labeled columns of design values applies to a nonresidential (for this example, an office) building, while the other columns apply to other situations.

- Roof: maximum U-factor: 0.034 (SI: 0.193) (assuming a roof/attic as typical of this type of building).

  As noted in Table G.1, insulation of R-30 (SI: R-5.3) minimum is required. This compares to a total R of 1/0.034 = 29.4 (SI: 1/0.193 = 5.2) for all components of the roof, suggesting that the negative effects of thermal bridging and the positive effects other materials such as sheathing/shingles will generally cancel each other.

- Frame walls: if wood framed, the maximum U-factor is 0.089 (SI: 0.51); if steel framed, the maximum U-factor is 0.124 (SI: 0.70).

  Note from the above that the U-factor requirements vary with the type of construction. This flexiblility in requirements (depending upon the

construction approach) reflects the fact that Standard 90.1 is a consensus document attempting to satisfy many constituencies and is not based on scientific absolutes or idealized efficiencies. A further look at the wall requirements (Table G.1) shows that the minimum insulation resistance for both types of constructions is R-13 ("equivalent" to U = 0.077) (SI: R-2.3; "U" = 0.44). The oddly higher overall U-factor requirements for both construction types reflect the lower resistance of the framing and are especially reflective of the thermal short-circuiting that occurs with steel studs.

- Floor (assuming wood-framed, over a crawl space): maximum U-factor: 0.051 (SI: 0.29).

  Note from Table G.1 that in this situation (as opposed to the wall case above), the U requirements for wood framing and steel joists are virtually identical (0.051 [SI: 0.29] for wood, versus 0.052 [SI: 0.30] for steel joists). The minimum insulation resistance for both constructions is R-19 (SI: R-3.3), equivalent to U = 0.053 (SI: 0.30); thus, an insulation detail that is not breached by framing must be assumed (compared with the wall requirements above).

- Opaque doors: maximum U-factor: 0.7 (SI: 4.0) (assuming swinging doors).

  Doors are a substantial weak link in the thermal envelope resistance but are usually a small percentage of the overall wall area.

- Windows: If operable windows constitute 5% of the wall area, their maximum U-factor is 0.67 (SI: 3.8) and the maximum SHGC is 0.39 (except for north glazing at SHGC 0.49). If the windows constitute 25% or 35% of the wall area, then these maximums still apply; at 41% of wall area, however, the maximums reduce to U: 0.47 (SI: 2.7) and SHGC: 0.25 (north: SHGC 0.36).

  The window prerequisites are the most situation-specific of the envelope requirements, varying with the type of window (fixed versus operable), the extent of window glazing versus wall area, and the orientation (for SHGC). ∎

**References**

Publications of the American Society of Heating, Refrigerating and Air-Conditioning Engineers, Inc., Atlanta, GA:

Advanced Energy Design Guide for Small Office Buildings. 2004.

ASHRAE Handbook—Fundamentals. 2005.

ASHRAE Standard 62.1-2004: Ventilation for Acceptable Indoor Air Quality.

ASHRAE Standard 62.2-2004: Ventilation and

*Acceptable Indoor Air Quality in Low-Rise Residential Buildings.*

ANSI/ASHRAE/IESNA Standard 90.1-2004: *Energy Standard for Buildings Except Low-Rise Residential Buildings.*

ANSI/ASHRAE Standard 90.2-2004: *Energy-Efficient Design of Low-Rise Residential Buildings.*

Bynum, R.T., Jr. 2001. *Insulation Handbook.* McGraw-Hill, New York.

Carmody, J., S. Selkowitz, and L. Heschong, 1996. *Residential Windows: A Guide to New Technologies and Energy Performance.* W.W. Norton and Company, New York.

Carmody, J., S. Selkowitz, E. Lee, D. Arasteh, and T. Willmert. 2003. *Window Systems for High-Performance Buildings.* W.W. Norton and Company, New York.

CRRC. 2004. Cool Roof Rating Council: http://www.coolroofs.org/

ICC. 2003. *International Energy Conservation Code.* International Code Council, Falls Church, VA.

Lstiburek, J. and J. Carmody. 1994. *Moisture Control Handbook: Principles and Practices for Residential and Small Commercial Buildings.* John Wiley & Sons, New York.

Melody, I. 1987. *Radiant Barriers: A Question and Answer Primer.* Florida Solar Energy Center (FSEC-EN-15), Cocoa, FL. http://www.fsec.ucf.edu/pubs/energynotes/en15.htm/

National Fenestration Rating Council (NFRC). *Certified Products Directory.* NFRC Incorporated, Silver Spring, MD. http://www.nfrc.org/

Norberg-Schulz, C. 1965. *Intentions in Architecture.* MIT Press, Cambridge, MA.

Wilkinson, G. 1999. "Beyond R-Value—Insulating for the Environment," in *Environmental Design and Construction,* January–February 1999.

THERMAL CONTROL

CHAPTER 8

# Designing for Heating and Cooling

CHAPTERS 1 THROUGH 5 PROVIDE SOME PREPARATION for design: a perspective on energy, water, and material resources; an understanding of human comfort and indoor air quality; an analysis of the climate resources available on site; a list of the general design strategies appropriate to various climates; a discussion of the basics of heat transfer calculations; and some attention to requirements for the components of the building envelope. Hence, we now know much about the variables outside of a building, the desired conditions inside, and the individual components of a building's

skin. In this chapter, all these variables are integrated into the process of designing for heating and cooling with the important related factor of daylighting. The design of mechanical support systems is introduced in Chapter 9.

The information in this chapter is organized from the general to the particular. It also serves two distinct kinds of buildings—buildings using on-site energy resources and more conventional buildings using imported energy. The sections that pertain to one or the other building type—and to both—are outlined in Table 8.1.

**TABLE 8.1 Proceeding Through Chapter 8**

| Buildings Using On-Site Energy | Conventional Buildings Using Imported Energy |
|---|---|
| Prior Decisions: Chapter 3. Sites and Resources | |
| Chapter 4. Comfort and Design Strategies | |
| Chapter 7. Heat Flow | |
| ← Section 8.2 Zoning → | |
| ← Section 8.3 Daylighting Considerations → | |
| Section 8.4 Passive Solar Heating Guidelines | |
| Section 8.5 Summer Heat Gain Guidelines | |
| Section 8.6 Passive Cooling Guidelines | |
| Section 8.7 Reintegrating Daylighting, Passive Solar Heating, and Cooling | |
| ← Section 8.8 Calculating Worst-Hourly Heat Loss → | |
| | Section 8.9 Calculations for Heating-Season Fuel Consumption (Conventional Buildings) |
| Section 8.10 Passive Solar Heating Performance | |
| ← Section 8.11 Approximate Method for Calculating Heat Gain (Cooling Load) → | |
| ← Section 8.12 Psychrometry → | |
| | Section 8.13 Detailed Hourly Heat Gain (Cooling Load) Calculations |
| Section 8.14 Passive Cooling Calculation Procedures | |

## 8.1 ORGANIZING THE PROBLEM

How should the building envelope respond to the sometimes conflicting needs for heating, cooling, and daylighting? Heating and cooling design strategies were related to climate in Chapter 4, where it became evident that internal heat gains can shift the appropriate design strategies for a building toward cooling, perhaps eliminating heating needs entirely. Typically, much of this internal heat is provided by electric lighting. Daylighting can replace electric lighting for most of the typical working day in most building types—*if* the building is designed to allow daylight to reach most of the interior.

### (a) Fenestration

Codes and standards typically prescribe a relationship between floor area and fenestration area (residential buildings) or total wall area and fenestration area (nonresidential buildings). These prescriptions assume that a building will be designed conventionally, that is, to rely on imported energy for lighting, heating, and cooling. Thus, prescribed areas of fenestration tend to be rather small. If a designer wishes to rely on daylighting to a greater extent, then some proof of benefit will be needed.

Daylighting is accompanied by large glass areas, which increase a building's heating needs in winter; yet in such buildings, less heat from electric lighting is available to fill those needs. Another complication is that adequate daylight requires much larger glass areas under dim winter skies than under bright summer skies. If the glass area is sized for winter daylighting, then excessive daylight—and, along with it, excessive heat—might be admitted in summer. With proper controls, daylighting can reduce summer cooling loads, relative to electric lights, but it will usually increase winter heating loads. Where passive solar heating or surplus heat from another source is readily available, this trade-off is attractive.

One of the earliest and most difficult questions for the designer is how much fenestration is optimum for a building. Some of the major energy end uses in buildings are: 30% for space heating; 11% for space cooling and ventilation; and 14% for electric lighting. By building type and location, however, this proportion of energy use can change substantially. Daylighting and the related square footage of windows are considered first because it so largely determines whether space heating or space cooling will be the dominant need within the building.

### (b) Building Form

At its simplest, form can be reduced to questions of tall or short, thick or thin. Figures 8.1 and 8.2 compare these form variations to the relative importance of heating, cooling, and daylighting. It is apparent that thicker, taller buildings have more floor space away from climate influences; being electrically lit rather than daylit, they generate heat and need cooling all year. These buildings (Fig. 8.1) are called *internal load dominated* (ILD). In contrast, thinner buildings—in which nearly all spaces have an exterior wall—need heating in cold weather and cooling in hot weather; electric lights by day are largely unnecessary. These buildings (Fig. 8.2) are called *skin load dominated* (SLD).

The ultimate choice of building form is determined by a combination of design issues; the energy use issue can help in the selection process. Once the building form has been chosen, the functions can be distributed according to typical architectural criteria, including the thermal zoning considerations.

In the selection of a building form, some particularly important questions accompany each energy use. Because the question of daylighting versus electric lighting frequently is so influential in determining whether heating or cooling will be the dominant concern, we begin with daylighting.

DAYLIGHTING ISSUES

1. What will be the relative emphasis on sidelighting (characterized by uneven distribution and glare in the visual field but little glare on horizontal surfaces) and toplighting (the reverse characteristics)?
2. What role will direct sun play in daylighting? In winter, can solar heat without glare be admitted?
3. How can seasonal adjustments be made in the size of daylight openings?
4. To what extent will daily changes in daylighting control be necessary?
5. How can adequate daylight be admitted in an even way, such that unwelcome dark-appearing places are avoided?

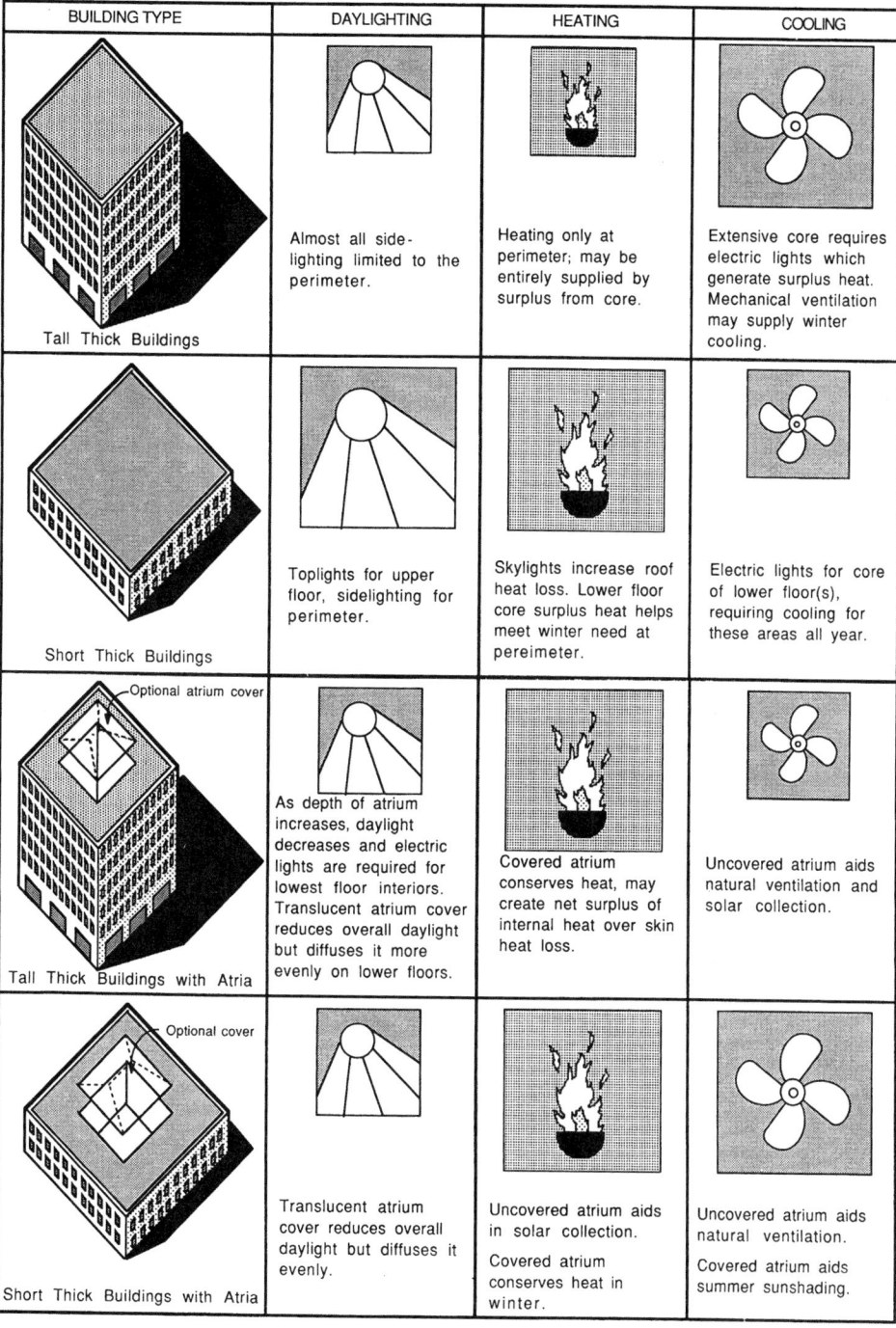

| BUILDING TYPE | DAYLIGHTING | HEATING | COOLING |
|---|---|---|---|
| Tall Thick Buildings | Almost all side-lighting limited to the perimeter. | Heating only at perimeter; may be entirely supplied by surplus from core. | Extensive core requires electric lights which generate surplus heat. Mechanical ventilation may supply winter cooling. |
| Short Thick Buildings | Toplights for upper floor, sidelighting for perimeter. | Skylights increase roof heat loss. Lower floor core surplus heat helps meet winter need at pereimeter. | Electric lights for core of lower floor(s), requiring cooling for these areas all year. |
| Tall Thick Buildings with Atria | As depth of atrium increases, daylight decreases and electric lights are required for lowest floor interiors. Translucent atrium cover reduces overall daylight but diffuses it more evenly on lower floors. | Covered atrium conserves heat, may create net surplus of internal heat over skin heat loss. | Uncovered atrium aids natural ventilation and solar collection. |
| Short Thick Buildings with Atria | Translucent atrium cover reduces overall daylight but diffuses it evenly. | Uncovered atrium aids in solar collection. Covered atrium conserves heat in winter. | Uncovered atrium aids natural ventilation. Covered atrium aids summer sunshading. |

**Fig. 8.1** *Internal-load-dominated buildings. The diagram shows how the relative importance of cooling, daylighting, and heating varies with building form and with climate. These buildings typically have energy-use patterns dominated by heat gains from lighting, equipment, and people and have limited opportunities for the use of on-site or natural energy sources.*

THERMAL CONTROL

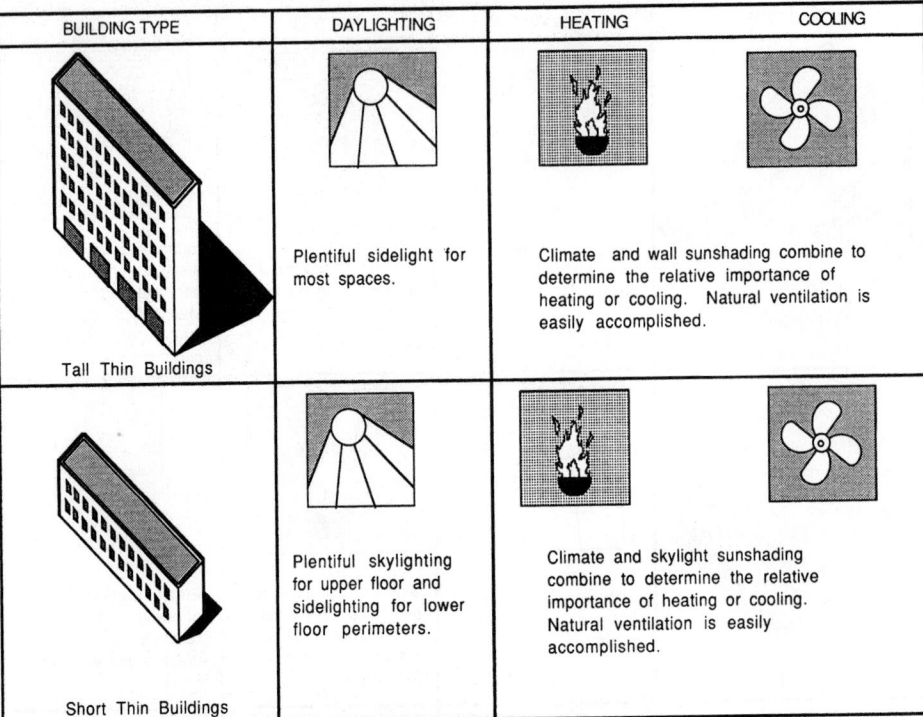

| BUILDING TYPE | DAYLIGHTING | HEATING | COOLING |
|---|---|---|---|
| Tall Thin Buildings | Plentiful sidelight for most spaces. | Climate and wall sunshading combine to determine the relative importance of heating or cooling. Natural ventilation is easily accomplished. | |
| Short Thin Buildings | Plentiful skylighting for upper floor and sidelighting for lower floor perimeters. | Climate and skylight sunshading combine to determine the relative importance of heating or cooling. Natural ventilation is easily accomplished. | |

**Fig. 8.2** *Skin-load-dominated buildings. The diagram shows how the relative importance of cooling, daylighting, and heating varies with climate and building. These building typically have energy use patterns dominated by heat gains and losses through the building envelope with opportunities for the use of on-site or natural energy sources.*

### HEATING ISSUES

1. Can the sun be used to heat spaces? If so, how will south-wall design be affected?
2. How can openings in walls facing other directions be kept to a minimum without daylight being shut out? Where such openings are desirable, how low a fenestration U-factor can be justified?
3. What role will direct sun through south glass or skylights play in daylighting?
4. How can daylight be admitted but the chilling effects of large, cold glass surfaces be minimized?
5. How can incoming fresh air be warmed before it chills the people sitting near the fresh air opening?
6. Is there surplus heat elsewhere in the building that can be used to help warm perimeter spaces?

### COOLING ISSUES

1. Will the strategy be to open the building to breeze or close the building for coolth retention, or to use a combination of these alternatives (open by night, closed by day)?

2. How can direct sun be kept out of the building? Can east and west windows be minimized and adequate daylight still be provided?
3. How can adequate daylight be admitted for winter conditions without overlighting (and thus overheating) for summer conditions?
4. When can cooling be provided by outdoor air rather than by a refrigeration cycle?
5. Can the operation of refrigeration machinery be concentrated during the coldest (nighttime) hours, when electric power is cheapest?
6. How can incoming fresh air be cooled before it warms the people sitting near the fresh air opening?
7. Can the structure of the building be used to absorb heat by day, then be flushed with night air in climates with cool nights?

### (c) Building Envelope

The next design step involves relating the climate to the design of the building's skin. Each skin element provides an opportunity for thermal and luminous

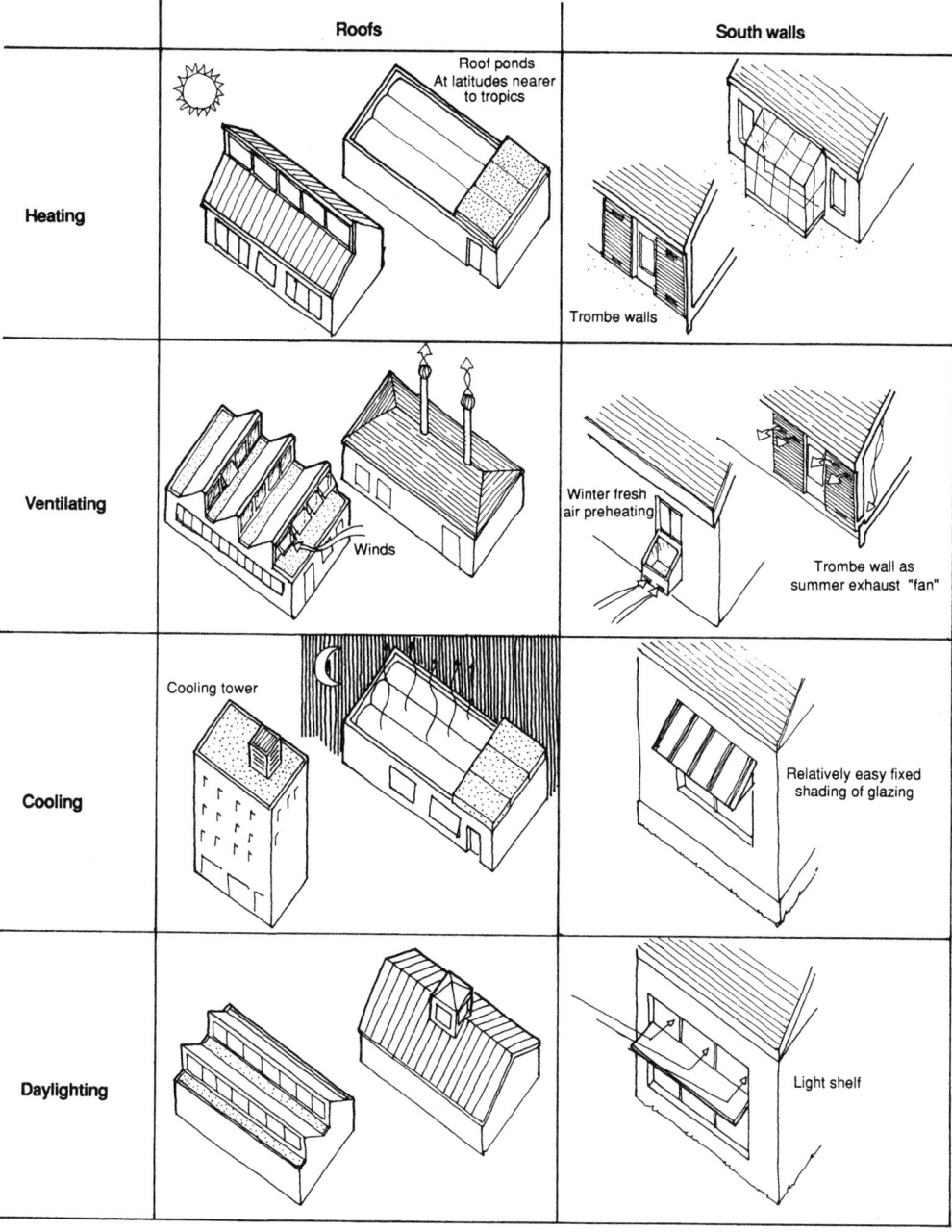

| | Roofs | South walls |
|---|---|---|
| **Heating** | | |
| **Ventilating** | | |
| **Cooling** | | |
| **Daylighting** | | |

(Roofs / Heating: Roof ponds — At latitudes nearer to tropics; Trombe walls)
(Ventilating: Winds; Winter fresh air preheating; Trombe wall as summer exhaust "fan")
(Cooling: Cooling tower; Relatively easy fixed shading of glazing)
(Daylighting: Light shelf)

**Fig. 8.3** *The components of a building's envelope can be used both to conserve energy and to admit on-site or natural energy sources.*

exchange between inside and outside; heating, cooling, ventilating, and daylighting devices can be mixed as needed. Figure 8.3 shows some of the most common of these devices for varying orientations. Sections in this chapter give numerical criteria for sizing these skin elements, with an emphasis on the use of on-site, renewable energy resources.

These criteria may conflict with the size relationships that may be prescribed by codes and standards for conventional buildings.

An example of how lighting, heating, and cooling considerations can combine with other criteria (structural and acoustic, in this example) is shown in Fig. 8.4. These design diagrams explore

THERMAL CONTROL

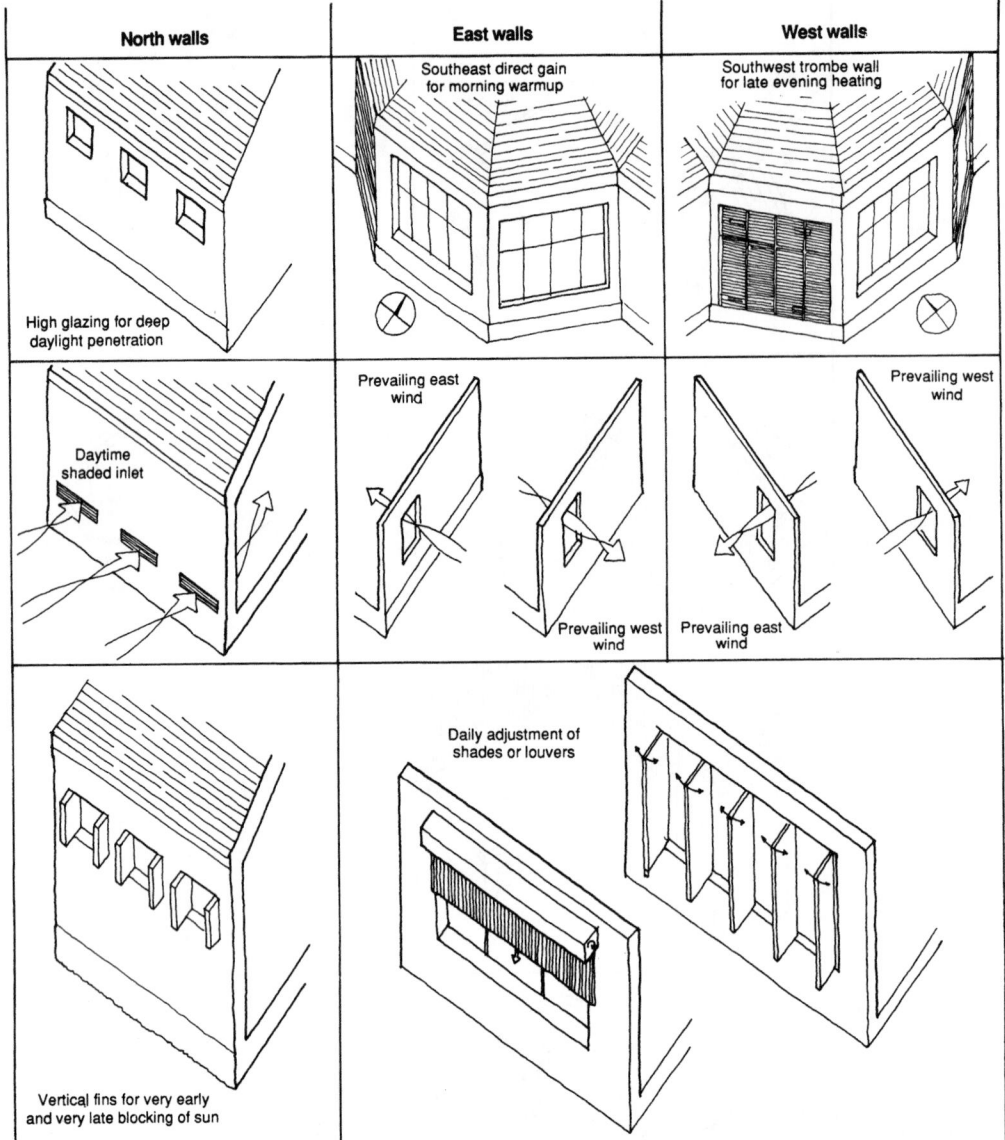

**Fig. 8.3** *(Continued)*

opportunities in an office building in a cool-winter, dry-summer climate; the actual resulting building is shown in Fig. 8.7.

## 8.2 ZONING

Before calculating winter heat loss and summer heat gain, zoning should be considered. Dividing the calculations into zones facilitates later decisions about sizes of equipment as well as fenestration and areas of thermal mass. The thermal and luminous zoning of a building recognizes that different envelopes and support systems may be required around and within the building. The more carefully zoning is considered in these early design stages, the better will be the lighting and thermal performance and the lower will be the annual energy consumption. (Also, the less likely it will be that all sides of a building will have an identical appearance.)

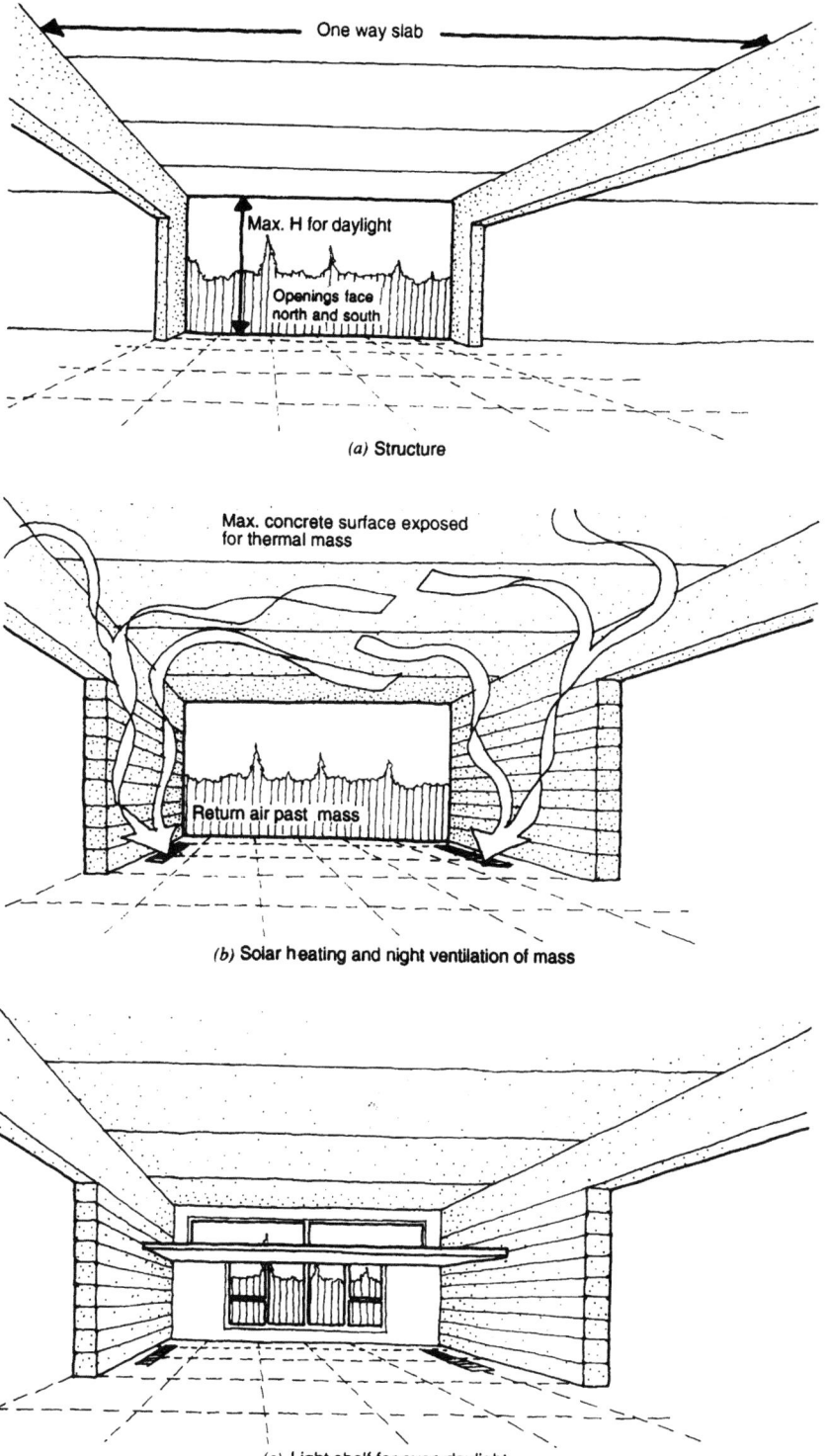

(a) Structure

(b) Solar heating and night ventilation of mass

(c) Light shelf for even daylight

**Fig. 8.4** *Design diagrams explore opportunities in an office building in western Oregon (see Fig. 8.7 for the resulting building). (a) Choosing a one-way structure allows windows to be placed as high as possible for deeper daylight penetration. (b) Maximizing exposed interior surfaces that are thermally massive facilitates both winter solar heat storage and summer night cooling. Carpet was installed at the insistence of the client. (c) Shaping the window with more area above a light shelf helps reduce the visual contrast near the window. (d) Indirect electric lighting is facilitated when higher ceilings are available; less glare on computer screens results. (e) With so much exposed mass surface, sound absorption is a must. This solution continues to expose the concrete ceiling and does not cast shadows from incoming daylight. (Drawings by Michael Cockram; © 1998 by John S. Reynolds, A.I.A.; all rights reserved.)*

THERMAL CONTROL

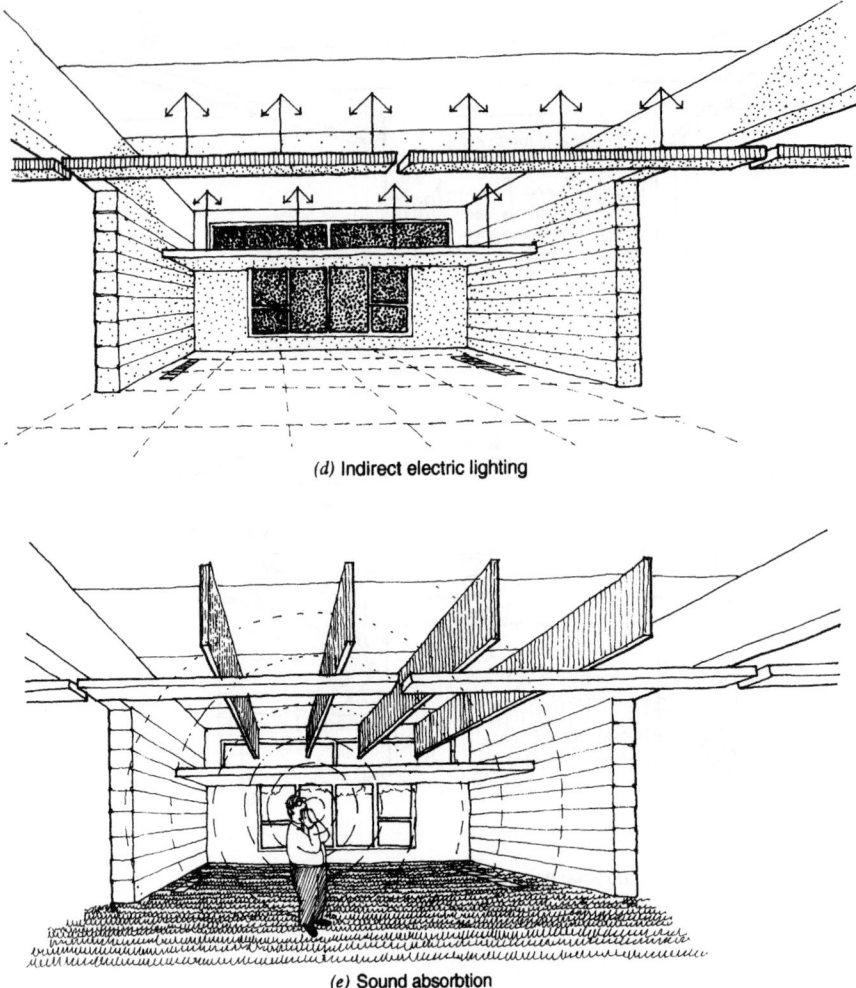

*(d)* Indirect electric lighting

*(e)* Sound absorbtion

**Fig. 8.4** *(Continued)*

Zoning is most often influenced by the following factors:

1. *Function.* Particularly important because of the variations in internal heat gains between functions, function may also influence the zonal organization of a building, as in Fig. 8.5. Comfort conditions may vary considerably between functions; air temperatures can be lower for a strenuous activity than for a sedentary activity, or heat tolerance may be greater for some activities (restaurant kitchens) than for others. Some functions thrive in daylight; others shun it. Some functions adversely affect the IAQ of other functions.

2. *Schedule.* Closely related to function, scheduling can influence both the envelope and the support system. An activity scheduled only between

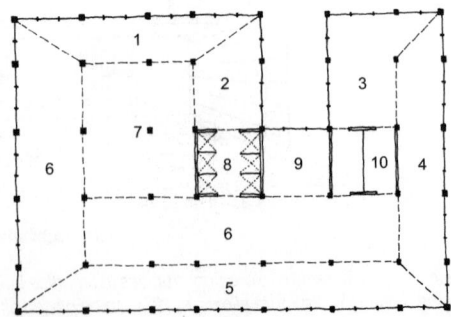

**Fig. 8.5** *Zoning for a mixed-use building with perimeter and internal zones as well as varying types of use. Scheduling and/or internal load differences within any one of these zones could require division into additional zones. (Drawing by Amanda Clegg.)*

9 A.M. and 4 P.M. can often be entirely daylit at a time when the outside temperatures are the warmest of the daily cycle. By contrast, an activity that takes place only from 9 P.M. to 4 A.M. will be entirely dependent on electric lighting, whose heat can be used to overcome the chill of the outside temperatures during these hours in winter. (In the summer, such heat can be flushed away with the cool outside night air in many U.S. locations.) Support systems are often divided by scheduling considerations: if one activity has operating hours different from those of the remainder of the building, a separate mechanical system is often provided. This saves energy, because large equipment will not be underused to provide heating or cooling for only one zone.

3. *Orientation.* The degree of exposure to daylight, direct sun, and wind is obviously important to zoning. Consider the block-square office building floors (Fig. 8.5) on a cold, sunny, and windy day. Perimeter spaces with direct sun through the windows may gain more heat than is lost and thus need cooling. This might be done by the opening of windows, but too much cold air (especially on the windy side of a building) may make the workers near the windows uncomfortable. Perimeter spaces without direct sun may have a net heat loss due to heat loss through glass, infiltration, and a lack of electric lights (because daylight is adequate). These spaces will need heat from a mechanical support system. Interior (no-daylight) spaces are overheated by electric lights because they cannot lose heat. These spaces will need cooling from the support system.

Zoning considerations are among the most important influences on building form and external design, along with the familiar aesthetic, social, legal, economic, and technical influences that combine in a tug of war familiar to the building designer.

## 8.3 DAYLIGHTING CONSIDERATIONS

Sizing of windows and analysis of daylighting performance will generally use the same guidelines for recommended daylight factor. Such guidelines are described in detail in Chapter 14, where they are discussed as a means of predicting daylighting performance. In previous chapters, win-

dows and skylights have been seen as a liability because of winter heat loss and summer heat gain. It is time to consider their benefits numerically. This section will use daylighting guidelines in a preliminary example of window sizing for an office building.

When a building is designed to rely on daylighting, a prime design concern is the *daylight factor* (DF), which is expressed as a percentage of the outdoor illuminance *under overcast skies* that is available indoors.

$$DF = \frac{\text{indoor illuminance from daylight}}{\text{outdoor illuminance}} \times 100\%$$

Some simple target overcast sky daylight factors and the simplest design guidelines that provide these target daylight factors are presented in Tables 14.2 and 14.5. These design guidelines consider two factors: *How high* is the window in the wall, and *how large* is the window or skylight area compared to the floor area for each daylit space?

The target daylight factors listed in Table 14.2 provide sufficient light during most of the daylight hours on overcast winter days. The relationship between office working hours, daylight, and outdoor temperature is explored in the *climatic timetables* of Fig. 8.6 for the building whose design diagrams appeared earlier (see Fig. 8.4). Obviously, much more light will be available on summer days—probably more light than is needed, bringing heat along with it. When sizing windows and skylights, remember that controlling direct sun is necessary and that less opening area is needed in summer than in winter.

**EXAMPLE 8.1** Many of the design guidelines presented in this chapter will be illustrated by applying them to a typical bay of the building in Fig. 8.7. This 24,000-ft$^2$ (2230-m$^2$) office building for an electric utility is located in Eugene, in western Oregon's wet-winter, dry-summer climate typical of the U.S. Pacific Northwest. Energy conservation and daylighting considerations played a major role in this building's design and were integrated with solar heating, night ventilation of mass, structure, and acoustics (see Fig. 8.4).

Example 8.1 is an ongoing illustration. Applications of various design guidelines will continue to appear in the context of this example throughout this

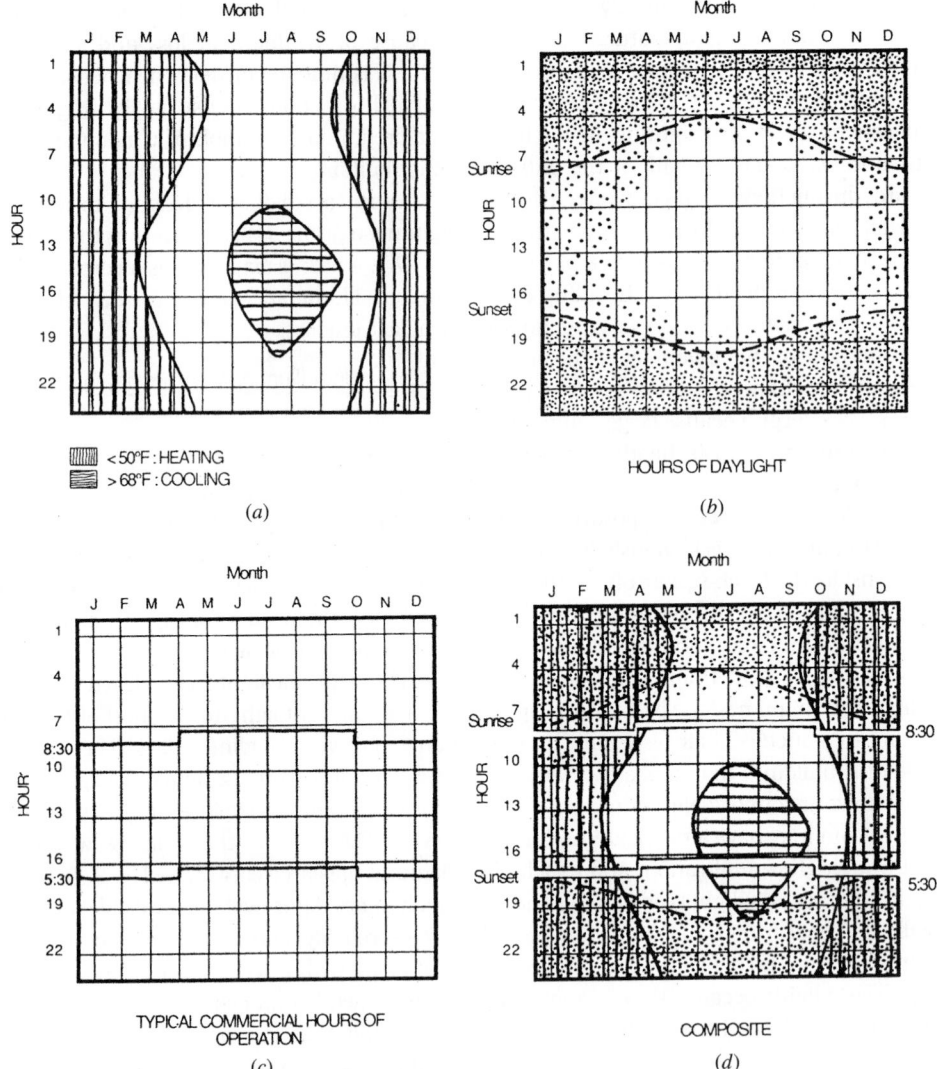

**Fig. 8.6** *A series of climate timetables were prepared while designing the office building for the Emerald People's Utility District near Eugene, Oregon, (Fig. 8.7). (a) Average outdoor temperatures show little daily variation in winter but a large daily range in summer. (b) Daylight peaks around noon near the summer solstice. (c) Work schedule shows the impact of summer's Daylight Savings Time. (d) A composite timetable indicates that daylight is available during almost all working hours and that half of the hottest hours occur after the building closes on summer afternoons—thanks to Daylight Saving Time. Cool nighttime temperatures make night ventilation of thermal mass an attractive cooling option. (Courtesy of Equinox Design, Inc., and WEGroup, PC, Architects, Eugene, OR.)*

chapter. Table 8.2 presents the type of information about a building that will be required in order to apply these design guidelines.

Part A. Daylight design

Part B. Overall rate of Btu/DD ft² heat loss

Part C. Approximate solar savings fraction (SSF)

Part D. Approximate heat gain

Part E. Cross-ventilation guidelines

Part F. Night ventilation of thermal mass guidelines

Part G. January balance point temperature

Part H. Annual SSF based upon the load to collector ratio (LCR)

Part I. Clear January day indoor temperature swing

Part J. Detailed night-cooling of mass calculation

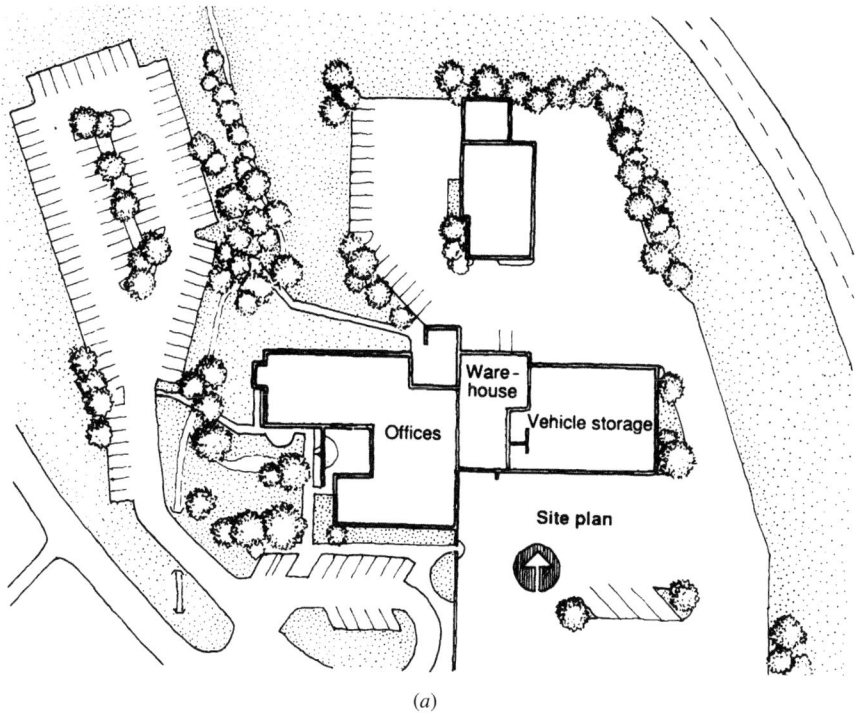

(a)

**Fig. 8.7** *The office building for the Emerald People's Utility District near Eugene, Oregon, is elongated east–west for maximum window areas facing north–south. (a) The site plan, (b) the ground floor plan, (c) the second floor plan, and (d) a section through a two-story wing. The cutaway axonometric (e) of a typical open-office two-story bay shows air flush through hollow cores of exposed precast concrete slabs, windows and shading, and suspended sound-absorbing baffles. The deciduous vines on trellises at the south façade change the building's appearance by season; (f) at the summer solstice, the shading stripes will soften as vines leaf out with age; at the fall equinox, the leaves will still shade, whereas at the spring equinox (g) the branches will be bare to allow sun to reach the windows; (h) at the winter solstice, most of the window is exposed to the sun. (Drawing c is courtesy of Virginia Cartwright. Courtesy of Equinox Design, Inc., and WEGroup, PC, Architects, Eugene, OR.)*

**EXAMPLE 8.1, PART A** Daylighting design for this office building began with design diagrams (Fig. 8.4) and an assessment of daylighting potential (Fig. 8.6). After it was established that daylighting was available during almost all normal working hours, the building plan was organized so that almost all windows faced either south or north (Fig. 8.7) to avoid the problems of low-altitude sun (year-round glare and summer heat gain) that accompany east- and west-facing windows. Then the building section was designed so that the height of the windows ($H$) was related to the depth of the floor plan served by those windows ($2.5H$).

After a generous target $DF_{av} = 4.0\%$ was chosen from Table 14.2, the next step was to size the windows and clerestories (as skylights), using the DF guidelines from Table 14.5. Applicable formulas for both sidelighting and for vertical monitor skylights:

$$DF_{av} = 0.2 \, \frac{\text{window (or skylight) area}}{\text{floor area}}$$

Applied to the entire typical bay,

$$DF_{av} =$$

$$0.2 \, \frac{97 \text{ ft}^2 \text{ north} + 97 \text{ ft}^2 \text{ south} + 132 \text{ ft}^2 \text{ clerstory}}{1440 \text{ ft}^2}$$

$$= 0.45, \text{ or } 4.5\%$$

slightly above the target $DF_{av}$ of 4.0%.

Note that in this example, $DF_{min}$ from the sidelighting occurs near the center of the building, at about the point (on the second floor) where the most light is available from the skylight. Therefore, relatively even daylighting distribution is expected on the upper floor. This is helped by the use of light shelves and the T-shaped windows shown in Fig. 8.7e.

The seasonal window size question (more

THERMAL CONTROL

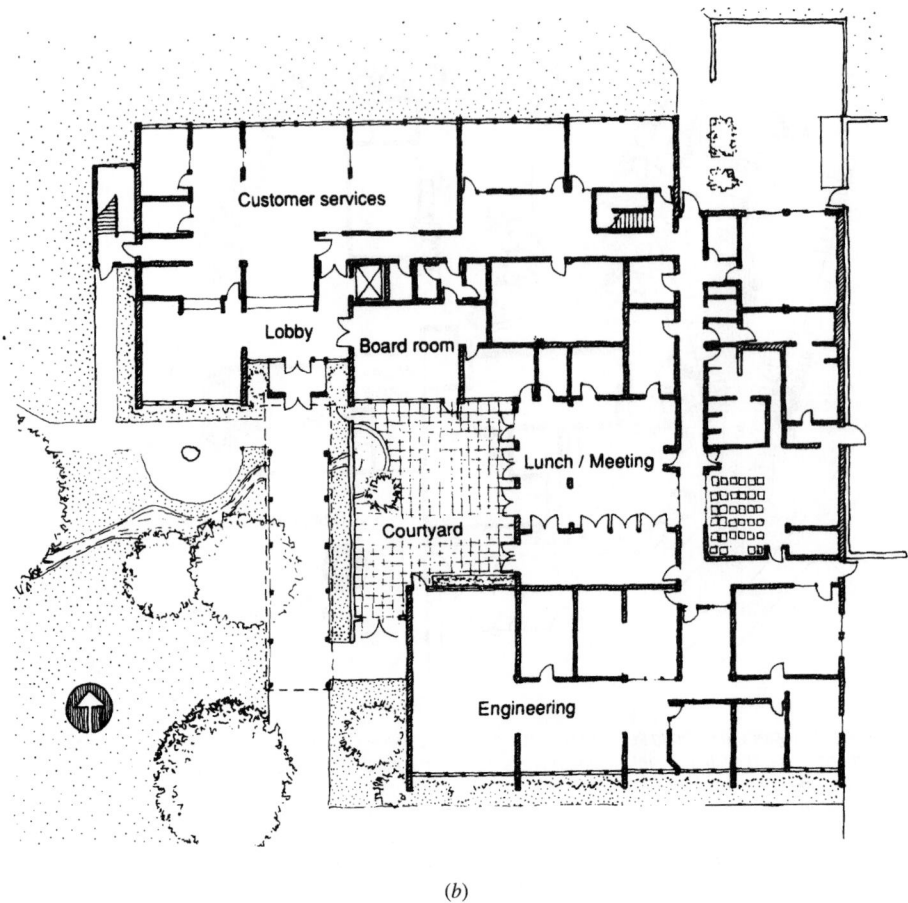

(b)

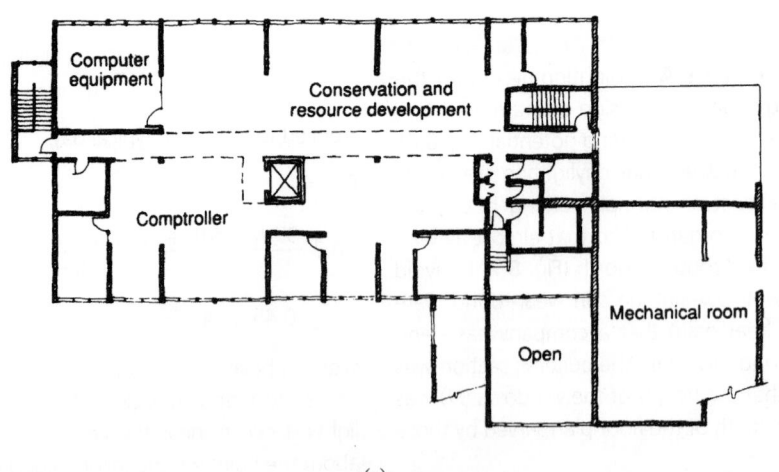

(c)

**Fig. 8.7** (Continued)

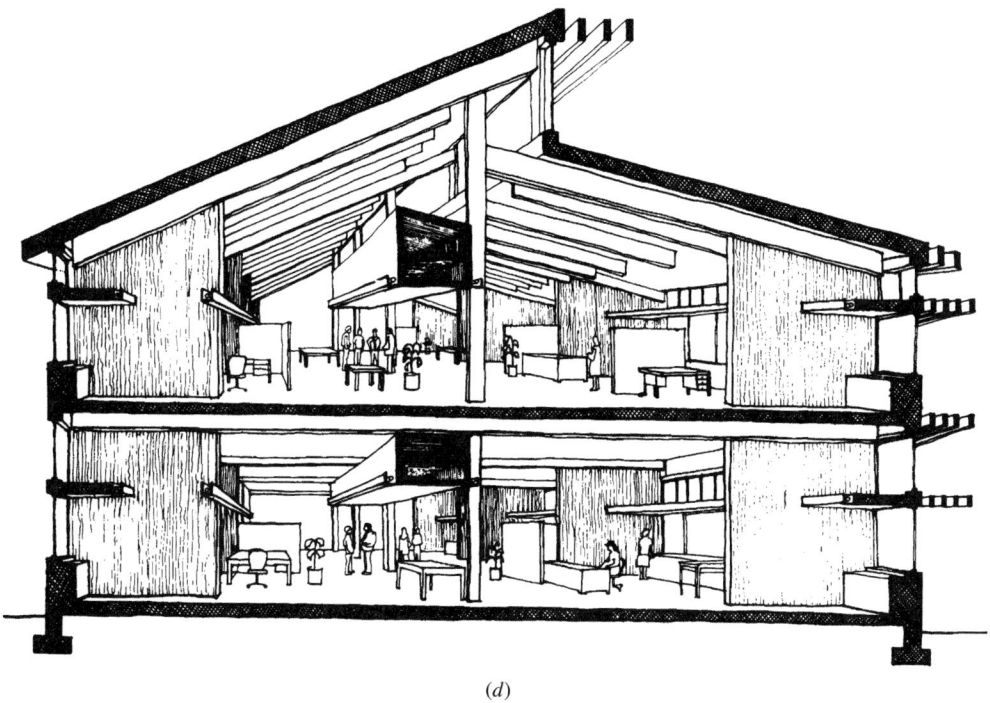

*(d)*

**Fig. 8.7** *(Continued)*

needed in winter, less in summer) was answered in this example by the use of deciduous vines outside the south windows. In winter and cool spring, the vines are bare of leaves. Warm weather brings leafy shade lasting well into the warm fall (Fig. 8.7).

## 8.4  PASSIVE SOLAR HEATING GUIDELINES

When the building's function, climate, and site are favorable, winter solar gains can contribute greatly to a properly designed building. One of the first criteria to be considered is energy conservation; there is little benefit in pouring solar heat into a leaky building. For solar designers, a motto should be "Insulate before you insolate."

### (a) Whole-Building Heat Loss Criteria

The recommended maximum rates of heat loss (in Btu per degree days [DD] per square foot) shown in Table 8.3 are the basis for development of the following passive solar heating guidelines. They are also useful as a quick check on overall envelope per-

formance in residential or small commercial buildings. (Some states have adopted similar criteria as part of their building codes; be sure to check the applicable code before relying on these numbers for code compliance.) The heat loss rates are shown for two conditions:

1. *Conventional (Nonpassively Solar-Heated) Small Buildings.* The overall rate of Btu/DD ft$^2$ is based on *total* heat loss, including all portions of the envelope *and* infiltration. To determine the whole-building heat loss rate, list for each envelope component (roof, walls, floor, windows, etc.) the U-factor (Chapter 7 and Appendix E) and the total exposed area $A$; then simply multiply $U \times A$. (For slab floors on grade, see Chapter 7 for determining perimeter heat losses.) For the special cases of walls below grade, such as berm walls, an approximation is needed. During the coldest weather, the temperature outside such walls will nearly always be higher than the outdoor air temperature, so any procedure based on Btu/DD will *overpredict* the heat loss through these walls. As a result, designers often calculate the UA of below-grade walls by using their actual U-factor but using only *half* of their actual area; this lesser UA

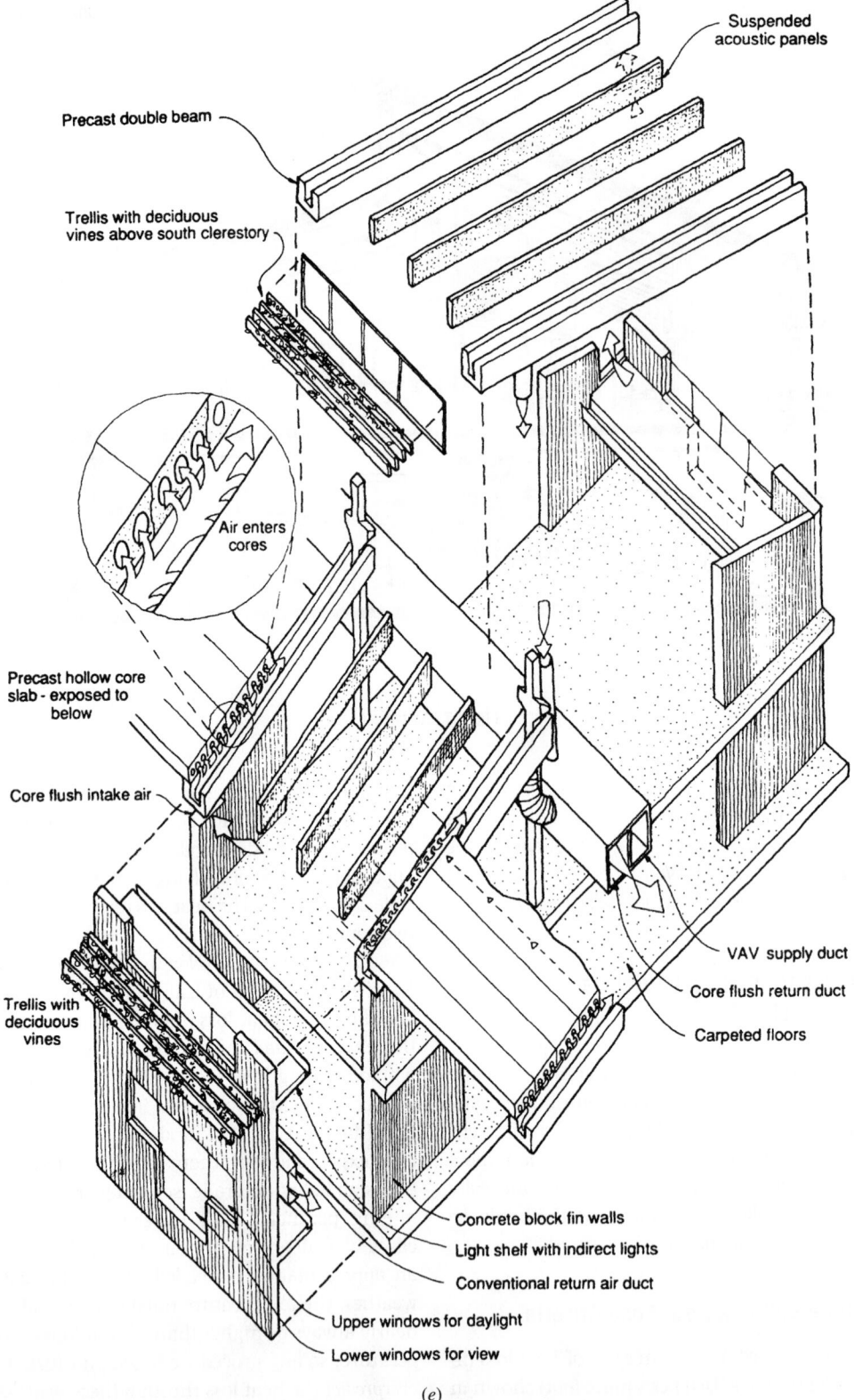

Suspended acoustic panels

Precast double beam

Trellis with deciduous vines above south clerestory

Air enters cores

Precast hollow core slab - exposed to below

Core flush intake air

Trellis with deciduous vines

VAV supply duct

Core flush return duct

Carpeted floors

Concrete block fin walls

Light shelf with indirect lights

Conventional return air duct

Upper windows for daylight

Lower windows for view

(e)

**Fig. 8.7** (Continued)

(f)

(g)

(h)

**Fig. 8.7** (Continued)

## TABLE 8.2 Design Data for a Typical Second-Floor Bay of the Office Building (Fig. 8.7e)

| Component of Envelope | U-Factor (Btu/h ft² °F) | Area (ft²) | U × A (Btu/h °F) |
|---|---|---|---|
| *Floor Area (basis for guidelines)* | | 1440 | |
| Exposed to Outside | | | |
| Roof[a] | 0.028 | 1512 | 42 |
| Exterior opaque walls[b] (north and south) | 0.084 | 296 | 25 |
| North window[c] | 0.74 | 97 | <u>72</u> |
| *Total for passive solar guidelines[d] UA$_{ns}$* | | | 139 |
| South window[c] | 0.74 | 97 | |
| South clerestory[c] | 0.74 | <u>132</u> | |
| *Total for passive solar guidelines A$_p$* | | 229 | |
| Interior Thermal Mass | | | |
| Roof underside | | 1512 | |
| Bearing walls, beams and columns | | <u>720</u> | |
| *Total for passive heating and cooling* | | 2232 | |
| Ventilation | | | |
| Volume of enclosed space, 22,900 ft³ | | | |
| Occupancy is eight persons | | | |
| Ventilation is provided at the rate of 20 cfm per person | | | |
| Openable window area: 16 ft² on north, 16 ft² on south | | | |

[a]R-40 over precast slab. Assume corrected R-35; therefore, U = 0.28.

[b]A 4-in. concrete block outside, 1-in. airspace, R-19 batt with metal studs at 24 in. and gypsum inside; approximate corrected R-11.9; therefore, U = 0.084.

[c]Clear double glazed, ⅜-in. airspace. nonthermal break aluminum, overall U = 0.74.

[d]Passive solar heating calculations do not directly include losses from south-facing glass. Also, there is no floor heat loss or gain, because this is a second-floor space.

roughly compensates for the lesser Δt through these walls.

For infiltration, determine the number of air changes per hour (ACH) under winter design conditions and multiply this infiltration (or fresh air) rate by a constant that accounts for density and specific heat:

$$\text{or} \left.\begin{array}{l} \text{ACH (volume, ft}^3\text{)} \times 0.018 = \\ \text{ACH (volume, m}^3\text{)} \times 0.33 = \end{array}\right\} \; UA \text{ for infiltration}$$

Add the envelope UA values to those for infiltration, multiply by 24 h/day to account for DD, and divide by the building's total heated floor area:

$$\frac{(UA_{\text{envelope}} + UA_{\text{infiltration}}) \times 24 \text{ h}}{\text{total heated floor area (ft}^2)} = \text{Btu/DD ft}^2$$

2. *Passively Solar-Heated Buildings.* Here the overall rate of Btu/DD ft² *excludes* the solar collecting portion(s) of the envelope; otherwise, it is also

## TABLE 8.3 Overall Heat Loss Criteria for Solar Guidelines

| *Maximum Overall Heat Loss* | | | | | |
|---|---|---|---|---|---|
| | | Btu/DDF ft² | | W/DDK m² | |
| **Annual Heating Degree Days (Base 65°F)** | **(Base 18°C)** | **Conventional Buildings** | **Passively Solar-Heated Buildings, Excluding Solar Wall[a]** | **Conventional Buildings** | **Passively Solar-Heated Buildings, Excluding Solar Wall[a]** |
| Less than 1000 | Less than 556 | 9 | 7.6 | 51 | 43 |
| 1000–3000 | 556–1667 | 8 | 6.6 | 45 | 37 |
| 3000–5000 | 1667–2778 | 7 | 5.6 | 40 | 32 |
| 5000–7000 | 2778–3889 | 6 | 4.6 | 34 | 26 |
| Over 7000 | Over 3889 | 5 | 3.6 | 28 | 20 |

*Source:* Balcomb et al. (1980). SI conversions approximated by the author.

[a]The guidelines in Table F.1 assume a solar building that meets this criterion.

based on total heat loss from all other portions of the envelope, and it includes infiltration. The equation used to determine the overall rate is as follows:

$$\frac{(UA_{\text{envelope, except south glass}} - UA_{\text{infiltration}}) \times 24\text{ h}}{\text{total heated floor area (ft}^2)}$$

$$= \text{Btu/DD ft}^2$$

One of the biggest unknowns in this procedure is the assumed rate of infiltration. A carefully designed and constructed small building can easily achieve a rate of 0.75 ACH; with increased attention to infiltration (vapor) retarder installation, caulking of all cracks, and so on, rates below 0.33 ACH have been demonstrated.

**EXAMPLE 8.1, PART B** Find the overall rate of heat loss, in Btu/DD ft$^2$, for the Oregon office building presented in Fig. 8.7.

**SOLUTION**
This office building is passively solar heated, so the overall rate of heat loss, in Btu/DD ft$^2$, is determined by using the total $U \times A$ of the non-south-glass envelope. From Table 8.2, this is 42 + 25 + 72 = 139 Btu/h °F.

To this must be added the effects of ventilation. From the same table, outdoor air is shown to be supplied at the rate of

20 cfm/person × 8 persons = 160 cfm

which can also be expressed as 160 cfm × 60 min/h = 9600 cfh. Comparing this hourly rate to the volume of the typical bay, we have

$$\frac{9600\text{ ft}^3/\text{h}}{22,000\text{ ft}^3} = 0.44\text{ ACH}$$

The $U \times A$ for ventilation is therefore

0.44 ACH × 22,000 ft$^3$ × 0.018 = 174 Btu/h °F

and the total rate of heat loss is

$$\frac{(139\text{ Btu/h °F} + 174\text{ Btu/h °F}) \times 24\text{ h/day}}{1440\text{ ft}^2}$$

$$= 5.2\text{ Btu/DD ft}^2$$

This building is located in Eugene, Oregon, which corresponds most closely to the location of Salem, Oregon, as listed in Table C.15, where we find DD65 in Salem = 4852 per year. From Table 8.3, the maximum recommended rate of heat loss for a passively solar heated building in a 3000- to 5000-DD

climate is 5.6 Btu/DD ft$^2$. This building's heat loss rate is under this maximum, thus enhancing energy conservation. ∎

Passive solar heating and energy conservation have a complex relationship. Relative to conventional buildings, passively solar-heated buildings usually conserve purchased energy; yet, buildings that aim at very high percentages of solar heating can use more *total* heating energy than is used by buildings with smaller window areas. Designers interested primarily in saving purchased energy may aim at lower solar percentages and more insulation; designers interested in buildings that closely relate to climate and climatic changes may aim at higher solar percentages (and more daylighting), along with higher thermal masses and, probably, greater ranges of indoor temperature.

### (b) Solar Savings Fraction (SSF)

The term *solar savings fraction* is used to evaluate a building's solar heating performance. The SSF is the extent to which a solar design *reduces a building's auxiliary heat requirement* relative to a "reference" building—one that has, instead of a solar wall, an energy-neutral wall that experiences neither solar gain nor heat loss; otherwise, the solar building and the reference building are identical. The SSF compares the auxiliary energy needed by the solar building to the auxiliary energy needed by the reference building, as illustrated in Fig. 8.8. Remember that the SSF is *not* the percentage of the solar building's heat supplied by the sun; typically, the sun provides a much *higher* fraction of a building's total space heat than does the SSF. Rather, the SSF is more a measure of the solar building's *conservation* advantage.

A starting point for passive solar preliminary design is Table F.1. For your location, both a range of SSF values and a range of ratios of areas of south glass/floor area can be determined. The table shows the SSF ranges for both "standard performance" (simple double-glazed windows) and "superior performance" (for either night-insulated or superwindow) solar openings. This same information is shown graphically for six geographically diverse locations in Fig. 8.9. (Add your own location's information from Table F.1 to a copy of this graph.)

The difference between standard and superior performance is explored in Fig. 8.10 for the

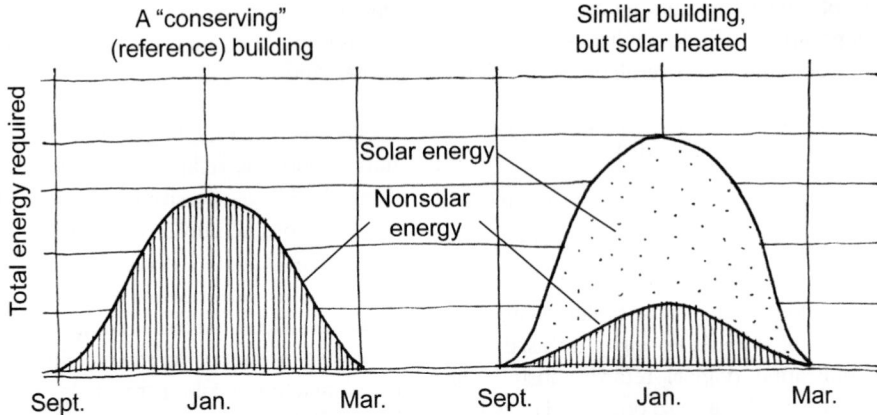

**Fig. 8.8** *The solar savings fraction compares the auxiliary heat needed by solar-heated buildings to that needed by a nonsolar but energy-conserving building that is otherwise similar, called the "reference" building. For example, if the solar building needs 25 units of auxiliary heat per year and the reference building needs 70 units, the difference is 70 − 25 = 45 units, or 64% of the reference 70 units. Therefore, SSF = 64%. (Note, however, that the solar building is 75% solar heated.) (From Brown, Reynolds and Ubbelodhe,* Inside Out: Design Procedures for Passive Environmental Technologies, *© 1982, John Wiley & Sons, Inc. Reprinted with permission of John Wiley & Sons, Inc. Redrawn by Erik Winter.)*

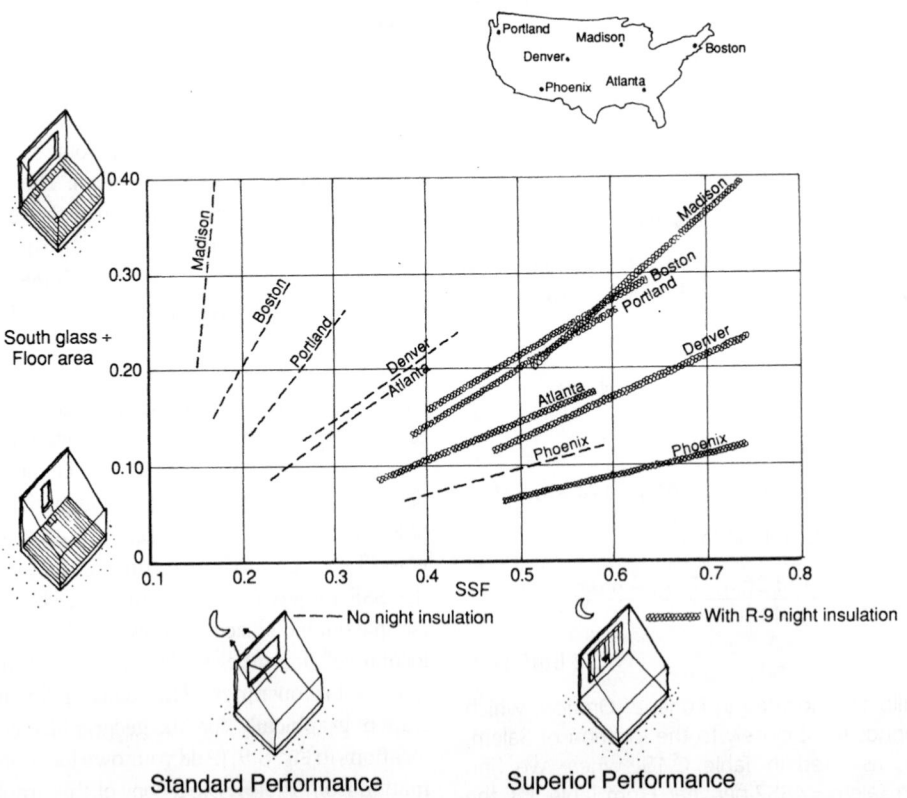

**Fig. 8.9** *Design guidelines for passive solar heating: ratio of south glass to floor area. The information in Table F.1 is presented graphically for six U.S. cities. "No night insulation" = standard performance; "With R-9 night insulation" = superior performance.*

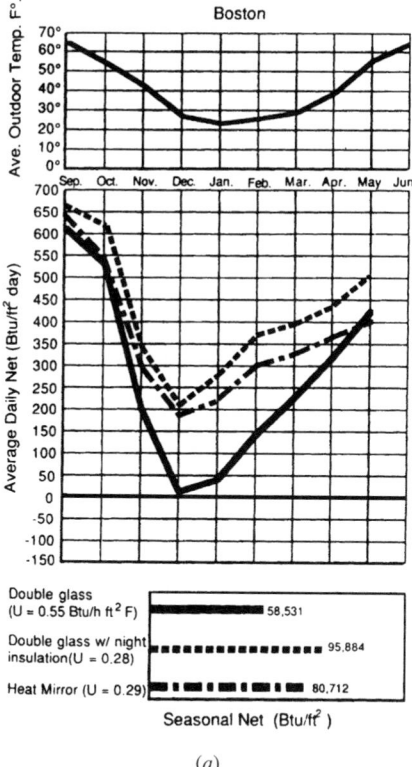

(a)

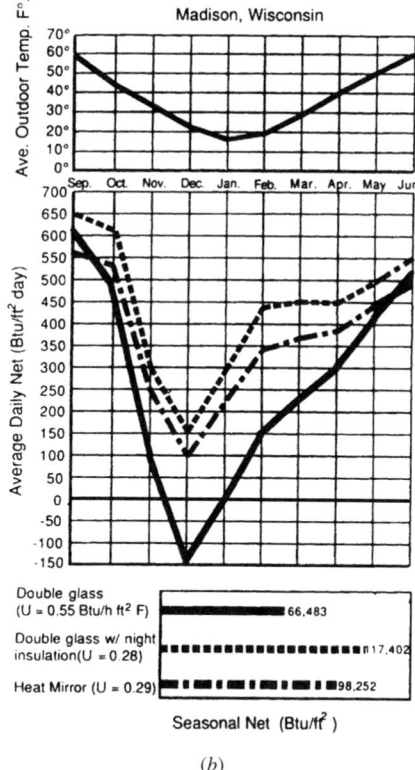

(b)

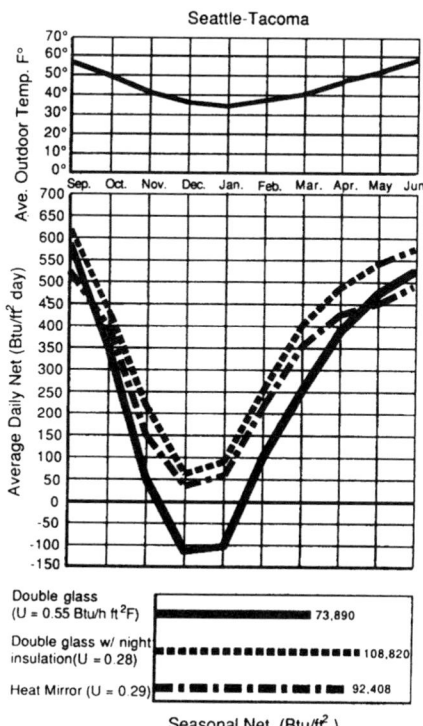

(c)

**Fig. 8.10** *Average daily net heating, per square foot of south window, for three glazing assemblies in three climates. (a) In Boston, ordinary double glazing is a slight energy plus even in December. (b) In Madison, ordinary double glazing is a net loser for almost 2 months. (c) In Seattle, ordinary double glazing is a net loser for more than 2 months. (From Johnson, 1981.)*

September–June period in Boston, Madison, and Seattle. The solid line charts the lower performance of simple double glazing (listed as U = 0.55 Btu/h ft² °F). The other two higher-performance alternatives (movable insulation and superwindows) show why the newer superwindows have made movable insulation nearly obsolete: without the need for any actions by the user, superwindows have a higher net heat gain than movable insulation in all three locations.

## (c) Thermal Mass

Another early design question involves the amount of thermal mass area necessary to store the solar heat admitted each day. This reveals the simplistic, early-design nature of Table F.1, because it does not distinguish between the various approaches to passive solar heating (*direct gain, indirect gain,* and *isolated gain*, as discussed in Chapter 4). Table F.2 details the simple relationship between SSF and the area (also the weight) of water or masonry that should be provided in direct-gain designs.

The *distribution* of the thermal mass is also important, however. Indirect gain systems (Trombe wall and water wall systems) usually place the thermal mass in full sun for the entire day, often just inside the glazing. In direct-gain systems, this thermal mass should be within (or should enclose) the direct-gain–heated space, and the exposed surface area of the mass should be *at least three times the glazing area*. Masonry surfaces are less thermally effective (on a daily basis) beyond a depth of about 4 to 6 in. (100 to 150 mm). Note that the thermal storage area is relatively unimportant at low SSF values, but as the SSF increases, so does the relative proportion of thermal mass area to solar glazing.

---

**EXAMPLE 8.1, PART C** Find the approximate SSF of a typical bay of the Oregon office building (Fig. 8.7).

### SOLUTION

Standard performance windows (simple double glazed) are used in this building, and the approximate SSF can be found from Table 8.2 (building data) and Table F.1 (SSF guidelines) as follows:

$$\frac{\text{south glass area } 97 \text{ ft}^2 + 132 \text{ ft}^2}{\text{floor area } 1440 \text{ ft}^2} = 0.16$$

For Salem, Oregon, with standard performance windows, a range of 0.12 to 0.24 ratio of glass/floor yields SSF 21 to 32%. So, with an actual ratio of 0.16, the approximate SSF is determined as follows:

Salem ratio range $0.24 - 0.12 = \Delta 0.12$
Salem SSF range $32\% - 21\% = \Delta 11\%$
(Actual ratio 0.16)
$-(\text{Salem minimum ratio } 0.12) = 0.04$,

$$\text{therefore} = \frac{0.04\ (\Delta 11\%)}{\Delta 0.12}$$

$$= 3.7\%, \text{ to be added to the minimum SSF;}$$

$$\text{Actual SSF} = \text{minimum } 21\% + 3.7\% = 24.7\%$$

This glass area/floor area ratio (0.16) is consistent with the previously calculated daylighting design considerations, although there is somewhat more south wall area that could be utilized for additional solar heat collection. (If more solar heat but not more daylight is desired, the additional collection area could be provided as a Trombe wall.)

Table F.2 shows that direct-gain thermal mass appropriate to SSF = 25% would be about 75 lb of masonry per square foot of south glass, or about 2 ft$^2$ of mass surface per square foot of south glass. From Table 8.2, the total exposed area of thermal mass in this typical bay is 2232 ft$^2$:

2232 ft$^2$ mass/229 ft$^2$ south glass
    $= $ almost 10 ft$^2$ mass/ft$^2$ glass

This is much more mass area than the minimum recommended. This excess of mass area will help prevent overheating on sunny winter days and contribute to thermal stability year round. For a more detailed analysis of this building's SSF, see Example 8.1, Part H.

At this point, one must wonder why standard rather than superior performance windows were used. For Salem, the approximate SSF would have increased to about 44%, with more than adequate internal areas of thermal mass. The simple answer is that the client did not want movable insulation, and superwindows were not available at an attractive cost when the building was being designed. ■

*Phase-change materials* are an alternative to simple thermally massive masonry surfaces. Flat bags of eutectic salts are shown in Fig. 8.11. Thin, horizontal tiles packed with these phase-change materials can store great quantities of heat with the phase change from solid to liquid. This change can be formulated to occur in the low 70s °F (20s °C) to prevent overheating of the space. As flat tiles enclosing bags of salt, they can form the finished surface in any horizontal application—floors, ceilings, or counter- and tabletops, for example. Tubes and trays of salts can be arranged as desired. Because the function of these materials is to keep room temperatures steady, their performance in preventing overheating on a sunny winter day will also be appreciated on hot summer days, provided they are taken below the phase change or melting temperature at night. For U.S. locations with large daily temperature ranges in summer (Mean Daily Range in Table B.1), thermal storage surfaces for passive solar heating in winter are potentially useful for night ventilation cooling in summer.

As a preliminary guide, *the phase-change tile surface area = one to three times the area of solar opening*. Additional information on these materials can be found in Johnson (1981).

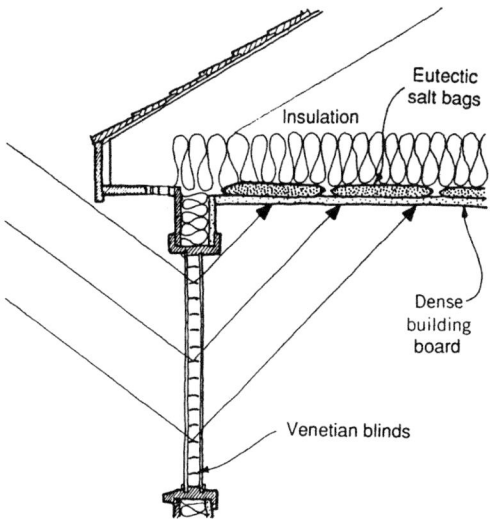

**Fig. 8.11** *Using bags of phase-change material (eutectic salt) for thermal storage above a flat ceiling. Mirrored-surface venetian blinds reflect direct sun to the ceiling, where it strikes a heavy (at least 90 lb/ft² [440 kg/m²]) board product with high conductivity. Metal ceilings or plaster are also suitable; ordinary gypsum board is not sufficiently conductive. Bags of salt weighing 5 lb/ft² (25 kg/m²) are in contact with the ceiling material and must be installed in a horizontal position. They are usually formulated to melt between 70 and 75°F (21 and 24°C). (Adapted by permission from Johnson, 1981.)*

The Society for the Preservation of New Hampshire Forests building utilizes phase-change storage above a metal ceiling (Fig. 8.12). It also uses water-filled tubes as an interior partition, masonry interior walls, and a concrete floor slab to help store the winter daily solar gain.

The New Canaan Nature Center (Fig. 8.13) is another innovative New England example using many passive/low-energy components. In addition to eutectic salts used in the south-facing railing on the upper level, the 4000-ft² (372 m²) building utilizes extensive clear, double-glazed south glass, movable insulating shades, operable vents at the skylight for stack-effect ventilation, ceiling fans, a woodstove, solar collectors placed inside the skylight monitor (their water is used for warming planting beds), warm air heat recovery ducts, a well-insulated envelope, and even a rainwater collection system. The manually operated switches for ventilation, insulation, and shading represent an unusual degree of user–building interaction. There is also unusual attention to microclimates and the transition between inside and outside in this award-winning building.

*Rock beds* are sometimes used to store the excess heat that indirect-gain (sunspace) systems can generate. They are typically placed directly beneath a concrete floor slab. The disadvantage of this approach is that any below-grade location raises suspicions of condensation and/or groundwater that may facilitate mold growth, yet cleaning of the rock bed is difficult.

The general guidelines cited by Mazria (1979) are as follows:

*Rock bed* **volume,** *ft³, per ft² of solar opening*

Cold climates: ¾ to 1½
Temperate climates: 1½ to 3

*Rock bed* **surface area** *in contact with floor above*

Cold climates: 75% to 100% of floor area above
Temperate climates: 50% to 75% of floor area above

### (d) Orientation

How important is it that the passive solar opening face due south? The general recommendation is that this orientation be *within 30° of south*. In *The Passive Solar Design Handbook* (Balcomb et al., 1980), the average penalties for off-south orientation are listed as follows.

5% decrease in SSF at 18° east or 30° west of true south

10% decrease in SSF at 28° east or 40° west of true south

20% decrease in SSF at 42° east or 54° west of true south

See Section 8.10 for more detailed coverage of passive heating performance that distinguishes between direct, indirect, and isolated gain systems and includes the expected annual auxiliary energy needed by a passively solar-heated building.

**EXAMPLE 8.2** A small office building in Omaha, Nebraska, is to be passively solar heated. How much superwindow south glass area should be provided?

**SOLUTION**

Table F.1 shows that an Omaha building with a south glass area equal to 20% of its floor area can expect an SSF of 51 if its solar openings are superior performance.

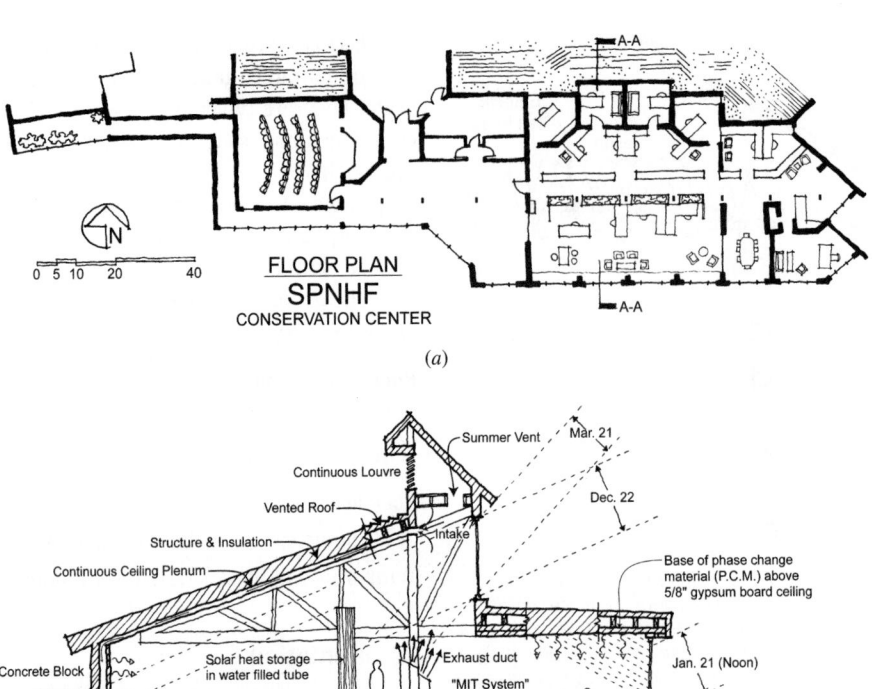

FLOOR PLAN
SPNHF
CONSERVATION CENTER

0  5 10  20       40

*(a)*

SECTION A-A

Summer Vent
Continuous Louvre
Vented Roof
Structure & Insulation
Continuous Ceiling Plenum
Mar. 21
Dec. 22
Intake
Base of phase change
material (P.C.M.) above
5/8" gypsum board ceiling
12" Concrete Block
Solar heat storage
in water filled tube
Exhaust duct
"MIT System"
minimizes glare and
temperature swings
at work space
Jan. 21 (Noon)
Berm
P.C.M. Tiles
Stratified direct gain heat
moved through masonry
wall & floor for low temp.
radiant heat
4" Conc. Floor
6" Deep Hypocaust
Duct/Mech. Tunnel
Fan Units
Light directing louvres
(adjustable) between double
glazed windows

*(b)*

*(c)*

**Fig. 8.12** *The Conservation Center for the Society for the Protection of New Hampshire Forests, Concord, Banwell White Arnold Hemberger & Partners, Architects. (a) The plan is elongated on the east–west axis to maximize southern exposure and facilitate daylighting. Direct gain serves the reception area, workroom, and offices, and a sunspace double envelope combination warms the lecture room. A wood-fired boiler provides backup heat. (b) Section with south glazing and thermal storage materials: translucent water tubes, masonry walls, and phase-change materials in ceiling and windowsill positions. The circulation of hot air that collects at the clerestory is also shown. In summer, the hot air is vented and an awning shades the clerestory. Daylighting is diffused through translucent tubes to the spaces on the north side. (c) At the workroom, mirror-finish venetian blinds reflect direct sun to the phase-change materials in bags above the dark metal deck ceiling. (Drawings a and b by Jonathan Meendering after original works by Banwell White Arnold Hemberger & Partners, Architects; photo c by C. Stuart White, Jr.)*

THERMAL CONTROL

(a)

(b)

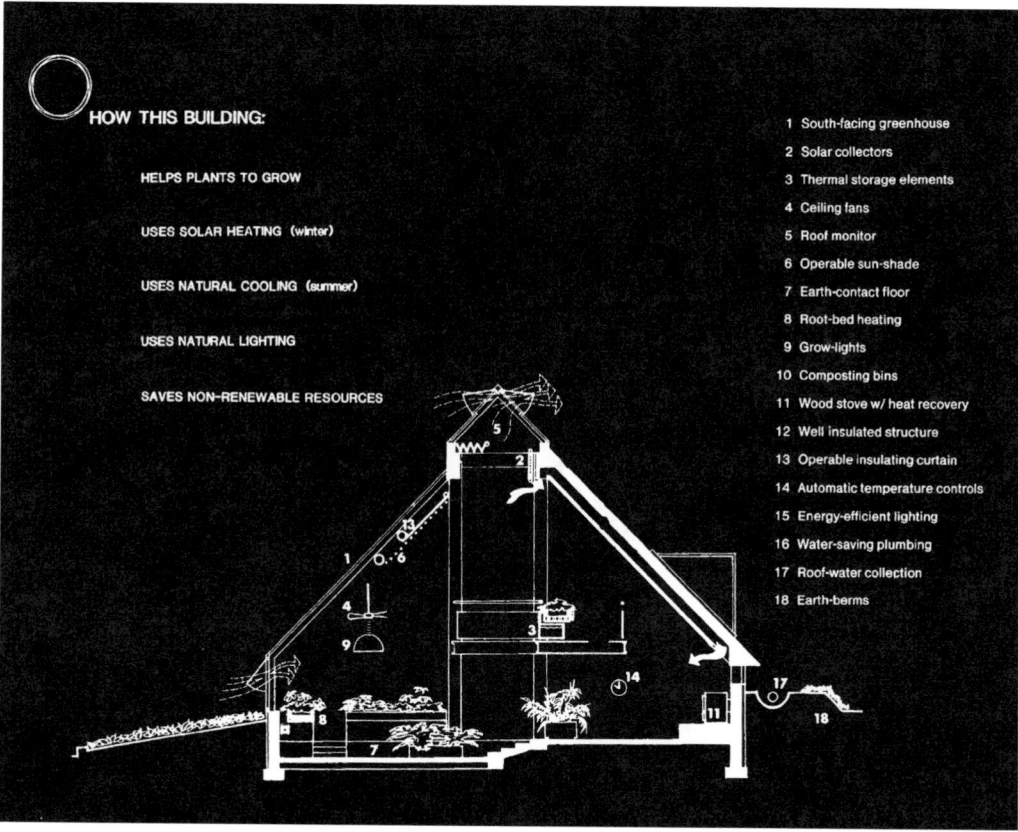

(c)

**Fig. 8.13** *The New Canaan (Connecticut) Nature Center (a,b) illustrates many passive and low-energy components. (c) Section illustrates the design features that utilize renewable energy and enhance energy conservation. The* thermal storage elements *are filled with eutectic salts that absorb excess solar heat via phase change (from solid to liquid) by day, then again change phase (from liquid to solid) at night to release their stored heat to the space. (Buchanan/Watson, Architects. Photo by Robert Perron. Courtesy of Donald Watson, FAIA, Troy, NY.)*

Select 20% of the floor area for south glass: SSF = 51. Thermal mass can be estimated from Table F.2; for SSF = 50, either 30 lb of water or 150 lb of masonry (per square foot of south glass) should be provided. Using brick as a thermally massive surface, 3.7 ft² of brick should be provided for each square foot of south glass. ∎

### (e) Roof Ponds

This is yet another approach to passive solar heating, one not covered by the preceding guidelines. In general, roof ponds are used in warmer, less humid areas of the southern United States, where snow will not impede the movement of roof insulating panels and the winter sun is higher in the sky than at northern latitudes. They are frequently sized for their *summer cooling* performance. In the U.S. southern latitudes, a pond sized for cooling will usually be adequate to absorb the needed winter sun. As a check for the pond's heating capacity, Mazria (1979) recommends the following guidelines:

Roof pond area = 85% to 100% of floor area for winter average outdoor temperatures of 25° to 35°F (−4° to +2°C)

Roof pond area = 60% to 90% of floor area for winter average outdoor temperatures of 35° to 45°F (2 to 7°C)

These guidelines are for roof ponds that have two layers of enclosing material between the water and the sky (i.e., are "double glazed") and movable night insulation.

### (f) Active Solar Heating

In contrast to passive systems, which incorporate sun collection and storage as part of a building's walls, floors, or ceilings, active solar heating uses mechanical equipment to collect and store solar energy. The most common early design questions for such systems involve the area of the solar collectors, their tilt and azimuth, and the size of the thermal storage. (Domestic hot water [DHW] solar heating information is given in Chapter 21.)

For active solar space-heating systems, the rules of thumb are more complex. Building heating needs vary by function and climate. The percentage of space heating that can economically be provided with active solar systems is another big variable.

Nevertheless, a rough guide is desirable as a design starting point:

Collector/floor area ratio = the smaller of the two window/floor area ratios listed in Table F.1

This should provide a portion of the annual heating load somewhere in the range of high-performance SSF from Table F.1. Larger arrays of collectors can be designed, of course, but they rarely will be economically attractive.

For collector tilt and azimuth optima:

Optimum tilt = latitude plus 10° to 15°; optimum azimuth is from due south to 15° W of south

where the tilt angle is measured up from horizontal. The orientation somewhat west of south is attractive in climates with frequent morning fog. Also, because air temperatures are higher in the afternoon, collectors lose less heat then and therefore operate more efficiently.

The design guidelines for storage size are as follows:
2 gal of water storage per ft² (81 L per m²) of collector

or

0.5 to 0.75 ft³ of rock bed per ft² (0.15 to 0.23 m³ per m²) of collector

The large arrays of collectors necessary for space heating must be served by correspondingly large pipes or ducts. Whereas pipe size rarely influences design, the air ducts for air-type collectors can consume large amounts of space. Therefore, the following typical flow rates are:

water flow rate of 0.25 to 0.5 gpm per ft² (0.17 to 0.34 L/s per m²) of collector

or

airflow rate of 2 cfm per ft² (10 L/s per m²) of collector

Approximate pipe sizes can be determined from Chapter 21, and approximate duct sizing is shown in Chapter 10.

### 8.5 SUMMER HEAT GAIN GUIDELINES

Design guidelines for cooling are complicated by the fact that cooling loads frequently are more closely related to individual building characteristics than to climate; sunshading and internal heat gains are particularly influential on cooling loads. Because passive cooling guidelines (Section 8.6) are expressed in heat to be removed per unit of floor area, it is first necessary to estimate the extent of the heat gain problem. Later (in Section 8.13), more detailed passive cooling procedures are shown. A comparison of the guidelines with these more detailed procedures is given in Table 8.4.

Detailed calculations for heat gain are presented in Sections 8.11 and 8.13. For now, Table F.3 gives a quick *approximation*. Many buildings (restaurants, factories, stores selling heating appliances, etc.) have special heat sources within. For these unusually heat-loaded situations, Table F.3 will be inadequate. As a starting point for preliminary passive-cooling sizing for typical buildings, however, it should be helpful.

Note that "open" buildings, such as those that

**TABLE 8.4 Comparing Passive Cooling Design Guidelines with Detailed Calculation Procedures**

| Design Guideline | Detailed Analysis Procedure |
|---|---|
| PART A. CROSS-VENTILATION | |
| *Section 8.6a*<br>• Assume 3F° [1.67C°] $\Delta t$<br>• Assume window orientation to wind | *Section 8.14a*<br>• Use actual $\Delta t$<br>• Use actual window orientation to wind |
| PART B. STACK VENTILATION | |
| *Section 8.6b*<br>• Assume 3F° (1.67C°) $\Delta t$ | *Section 8.14b*<br>• Use actual $\Delta t$ |
| PART C. NIGHT VENTILATION OF THERMAL MASS | |
| *Section 8.6c*<br>• Assume ratio of mass area/floor area<br>• Assume cooling during hour of max. $\Delta t$<br>• Assume max. $\Delta t$ for natural ventilation estimation | *Section 8.14c*<br>• Use actual exposed mass area<br>• Use actual mass heat capacity<br>• Use actual hourly chart of air and mass temperatures<br>• Find total cooling and required air flow rate |
| PART D. EVAPORATIVE COOLING (ACTIVE) | |
| *Section 8.6d*<br>• Assume 2.67 cfm/ft² floor area<br>• Assume 83°F exhaust air | *Section 8.14d*<br>• Use actual outdoor temperature<br>• Find actual indoor supply temperature<br>• Find allowable $\Delta t$ as air passes through indoors<br>• Then cfm = (Btu/h)/(1.1)($\Delta t$) |
| PART E. COOLTOWERS (PASSIVE, EVAPORATIVE) | |
| *Section 8.6e*<br>• Find approximate exit air temperature<br>• Find approximate flow rate<br>• Then Btu/h = (cfm) (1.1)($\Delta t$)<br>• Then Btu/h = (cfm)(1.1)($\Delta t$) | *Section 8.14e*<br>• Use actual outdoor temperature and wet-bulb depression<br>• Find actual exit air (supply) temperature<br>• Find actual exit airflow rate |
| PART F. ROOF PONDS | |
| *Section 8.6f*<br>• Assume pond max. temperature of 80°F<br>• Estimate pond minimum temperature<br>• Assume 30% gain through roof insulation<br><br>• Assume pond depth from 3 in. to 6 in.<br>• Find area of pond depth and area | *Section 8.14f*<br>• Actual outdoor temperature, resulting pond temperature<br>• Actual heat gain through roof insulation<br>• Actual hours of internal heat gains<br>• Choice of fan-forced or still-air layer at ceiling<br>• Evaporation or not on upper pond surface<br>• Find area of pond depth and area<br>• Find size of backup cooling |
| PART G. EARTH TUBES | |
| *Section 8.6g*<br>• Assume 65°F soil temperature<br><br>• Assume soil conductivity<br>• Does not specify depth<br>• Assume 85°F outdoor air<br>• Assume 500 fpm velocity<br>• Choose diameter and length to match cooling load | *Section 8.14g*<br>• Actual underground temperature; assume resulting tube temperature is within 4F°<br>• Actual soil conductivity<br>• Actual depth<br>• Actual outdoor air temperature<br>• Assume 500 fpm velocity<br>• Calculate actual cooling (assume 1.3 Btu/h F ft length $\Delta t$) |

are naturally ventilated, do not have heat gains from infiltration, because they assumedly maintain internal temperatures that are slightly *above* exterior temperatures. However, these buildings do experience heat gains through windows, walls, and roofs due to solar impacts on these surfaces. For "closed" buildings, heat gain from infiltration or ventilation must be added, because these structures maintain internal temperatures lower than outside temperatures.

---

**EXAMPLE 8.1, PART D** Find the approximate heat gain for the Oregon office building in Fig. 8.7.

**SOLUTION**
From Table F.3:

A. People and equipment (using the high end for Europe office buildings, based on assumptions about computer use)     5.8 Btu/h ft$^2$

B. Electric lighting (with a 4% DF from Example 8.1, Part A; this procedure assumes that most electric lights are off when daylight is available)
    0.5 Btu/h ft$^2$

C. Envelope (Eugene's design temperature, Appendix B, is 89°F)

South glass, shaded by vines:

$$\frac{(97 \text{ ft}^2 + 132 \text{ ft}^2) \times 16}{1440 \text{ ft}^2} \qquad 2.54 \text{ Btu/h ft}^2$$

North glass, unshaded (Table F.6 Part B; regular double glass, venetian blinds):

$$\frac{97 \text{ ft}^2 \times 14 \text{ Btu/h ft}^2}{1440 \text{ ft}^2} \qquad 0.94 \text{ Btu/h ft}^2$$

Walls:

$$\frac{296 \text{ ft}^2 \times 0.084 \text{ Btu/h ft}^2°\text{F} \times 15}{1440 \text{ ft}^2} \qquad 0.26 \text{ Btu/h ft}^2$$

Roof:

$$\frac{1512 \text{ ft}^2 \times 0.028 \text{ Btu/h ft}^2°\text{F} \times 35}{1440 \text{ ft}^2} \qquad 1.03 \text{ Btu/h ft}^2$$

D. Total heat gains, internal and through building skin:

    11.07 Btu/h ft$^2$

E. Heat gains from ventilation:

$$\frac{20 \text{ cfm/person} \times 8 \text{ persons} \times 16}{1440 \text{ ft}^2} \qquad 1.8 \text{ Btu/h ft}^2$$

Total approximate heat gains:    12.87 Btu/h ft$^2$ ∎

## 8.6 PASSIVE COOLING GUIDELINES

The design guidelines for passive cooling are much newer and less tested than those for passive heating. It is important that the designer *first* check the match between the climate and the cooling strategy, as was done in Chapter 4, before applying these guidelines to a building design.

### (a) Cross-Ventilation

One of the oldest methods known, this strategy provides plentiful fresh air but maintains a building at temperatures slightly *above* those outdoors. The cross-ventilation inlet (window) areas, expressed as percentage of total floor area, are related to wind speed and resulting heat removal in Fig. 8.14a. Remember that *an equal (or greater) area of outlet openings* must also be provided. Furthermore, any internal obstructions (such as partitions) must have a total area of openings at least equal to this required inlet area. The assumptions about wind direction and indoor–outdoor temperature differences that were used to produce these guidelines are explained in the figure. For more detailed wind speed and direction information, see Section 8.14.

The Δt of 3F° used in Fig. 8.14 is deliberately kept small to encourage open strategies in milder summer climates. Thus, an interior temperature of 83°F, which is comfortable with sufficient air motion, lower RH, and comfortable surface temperatures, would be obtainable with an outside temperature of 80°F. However, a greater Δt is often appropriate—for example, for spring or fall cooling of office buildings, or for summer cooling of factories or kitchens where internal temperatures may remain in the low 90s. In such cases, find the percentage of inlet area required; then multiply by the ratio

$$\frac{3\text{F}°}{\text{actual } \Delta t}$$

to obtain required cross-ventilation areas for any specific temperature difference (see Example 8.1, Part E).

See Table 8.4 for a comparison of these design guidelines with the more detailed calculations of Section 8.14 for both cross- and stack ventilation.

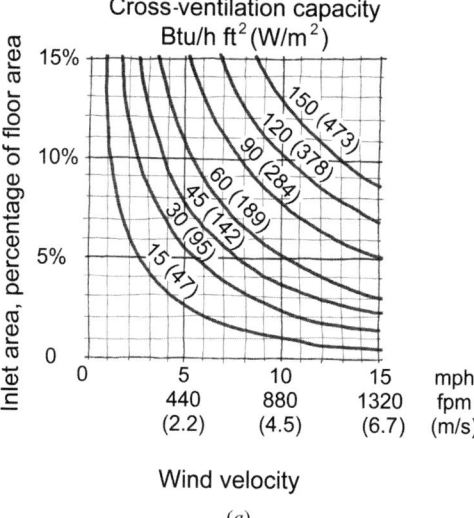

Cross-ventilation capacity
Btu/h ft² (W/m²)

*(a)*

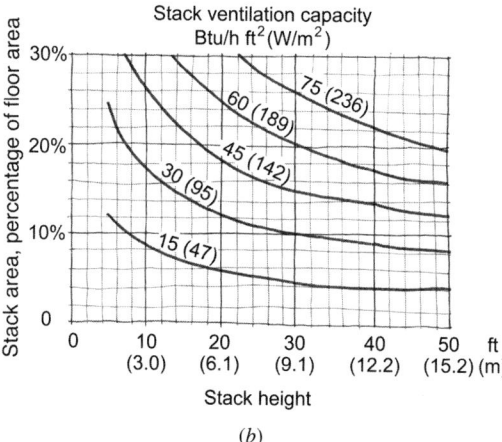

Stack ventilation capacity
Btu/h ft² (W/m²)

*(b)*

**Fig. 8.14** *(a) Cross-ventilation design guidelines for heat removed per unit floor area and relationship of inlet openings and wind speed. The total inlet opening area is expressed as a percentage of the total floor area served by cross-ventilation. Note: Outlet areas must also be at least equal to inlet areas. This graph is based on an internal temperature 3F° (1.7C°) above the exterior temperature and assumes that the wind is not quite perpendicular to the window openings, for a wind effectiveness factor of 0.4 (see Section 8.14). (b) Stack-ventilation design guidelines for heat removed per unit floor area and the relationship of stack height and inlet openings. The area of stack inlet/throat/outlet is expressed as a percentage of the total floor area served by stack ventilation. Note: Outlet areas and the stack throat area must also be at least equal to inlet areas. This graph is based on an internal temperature of 83°F and an exterior temperature 80°F, for a 3F° differential (28.3°C internal, 26.7°C external, for a 1.6C° differential). (Redrawn by Jonathan Meendering.)*

**EXAMPLE 8.1, PART E** Find the required inlet areas for cross-ventilation of a typical bay of the office building in Fig. 8.7.

**SOLUTION**

This office building is located in Eugene, Oregon, whose monthly average wind velocities may be obtained from NOAA Local Climatological Data. For July, the hottest month, the average wind velocity is about 8 mph from a nearly due-north average direction. From Table 8.2 the inlet area (= outlet area also in this case) is 16 ft². The inlet area, as a percentage of floor area, is therefore

$$16 \text{ ft}^2/1440 \text{ ft}^2 = 1\%$$

Entering Fig. 8.14a with 8-mph wind velocity and 1% inlet area, we find that somewhat less than 15 Btu/h ft² heat gain can be removed by cross-ventilation. In Example 8.1, Part D, we approximated a heat gain (excluding ventilation) of about 11 Btu/h ft². Therefore, under average July conditions, the window area is probably adequate to remove heat gains and maintain an indoor temperature 3F° *above* the outdoor temperature.

Because Eugene's average July daily high is 83°F, this would produce an interior daily high of 86°F. Such a temperature would require somewhat more than minimum air motion in order to feel comfortable. At Eugene's summer *design* temperature of 89°F, the interior temperature would be 92°F. For these reasons this building was designed to be night ventilated, relying on thermal mass rather than on cross-ventilation for hot-day comfort.

In milder weather, these windows can achieve substantial cooling because the temperature difference, Δt, is more than the 3F° assumed in Fig. 8.14. For example, when the temperature indoors is 75°F, outdoors 65°F, the fully opened windows with the 8-mph wind would result in a cooling rate of

$$\frac{\text{about 15 Btu/h ft}^2 \times 10\text{F}°}{3\text{F}°} = \text{about 50 Btu/h ft}^2$$

which is considerably more than the actual gain of 11 Btu/h ft². Therefore, under such conditions, the windows are only partially opened. ∎

## (b) Stack Ventilation

This is another historically useful strategy, which, like cross-ventilation, provides plentiful fresh air but maintains a building at temperatures slightly *above* outdoor temperatures. The stack inlet areas, expressed as percentage of total floor area, are related to stack height and the resulting heat removal in Fig. 8.14*b*. Remember that *an equal (or greater) area of stack outlet openings, as well as at least an equal cross-sectional area through the vertical stack*, is also required. Again, consider internal obstructions: partitions must have a total of openings at least equal to this required inlet area. The assumptions about indoor–outdoor temperature differences that were used to produce this guideline are explained in the figure.

Adjustment of the Δt for this guideline is similar to the procedure for cross-ventilation. It requires multiplication of the required percentage of stack area by the ratio

$$\frac{3F°}{\text{actual } \Delta t} \sqrt{\frac{3F°}{\text{actual } \Delta t}}$$

to obtain the required stack ventilation areas for any other specific temperature difference.

Until the advent of air conditioning, these methods of cross- and stack ventilation were used in virtually all commercial buildings. In some climates with mild summers, newer buildings are turning to these older methods (refer to Fig. 8.46). Figure 8.15 shows the Thoreau Center for Sustainability in San Francisco's Presidio National Park. An older military hospital was renovated for offices for nonprofit environmental organizations. To maximize daylighting and passive cooling, the original operable windows are utilized. Interior partitions are kept low enough to facilitate cross-ventilation, with integral indirect electric lighting to the higher white ceiling. The section indicates that both cross-ventilation and stack effect are anticipated. However, the stack openings to the attic and, subsequently, through the roof are less than 1% of the total floor area. The most likely benefit of the stack effect here is to keep the attic at cooler summer temperatures (with air drawn in at the eave vents, not flowing outward as shown).

(*a*)

**Fig. 8.15** *The Thoreau Center for Sustainability occupies renovated hospital buildings (a) in San Francisco's Presidio National Park. (b) Relatively narrow floor plans facilitate daylighting and natural ventilation. (c) Interior at the center corridor. (d) Section illustrates daylighting and cross-ventilation. Very small stack openings relative to the floor area served limit the stack effect to a minor role. (Drawings courtesy of Tanner Leddy Maytum Stacy Architects, San Francisco.)*

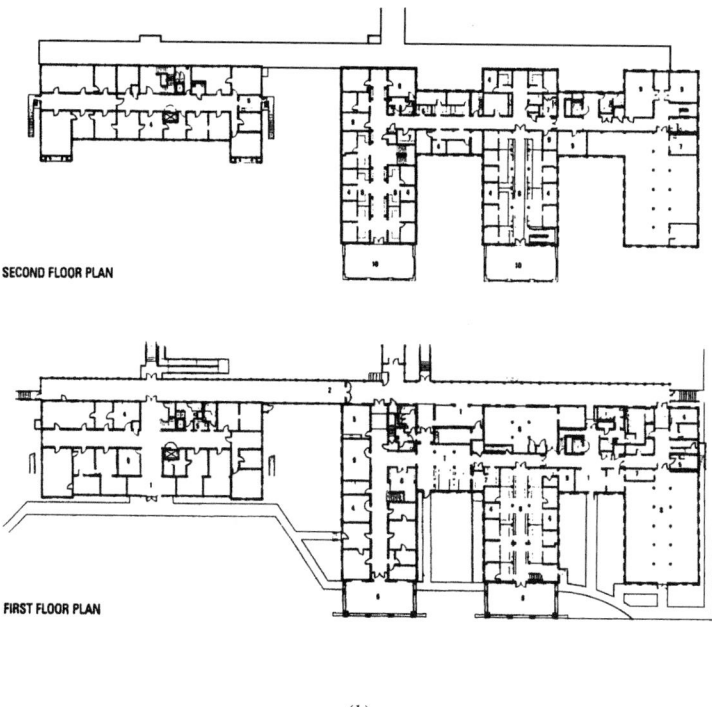

SECOND FLOOR PLAN

FIRST FLOOR PLAN

(b)

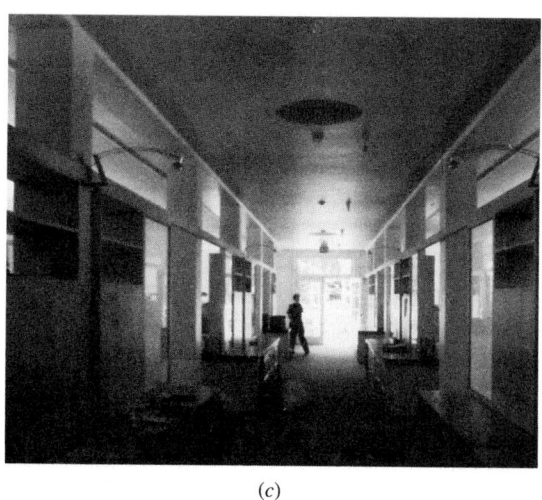

(c)

**Fig. 8.15** (Continued)

## (c) Night Ventilation of Thermal Mass

This strategy maintains a building at temperatures lower than those outside by day and flushes the building with plentiful fresh air by night. The sizing procedure is shown in Fig. 8.16, where climate data are related to two representative types of thermally massive building. For each type, the graphs show the *daily* Btu per square foot of floor area that can be stored. The "average" mass building is represented by a building with an exposed concrete floor 4 in. thick (or an exposed ceiling of equivalent construction). There is one unit area of exposed mass for each unit floor area. The "high" mass building is

THERMAL CONTROL

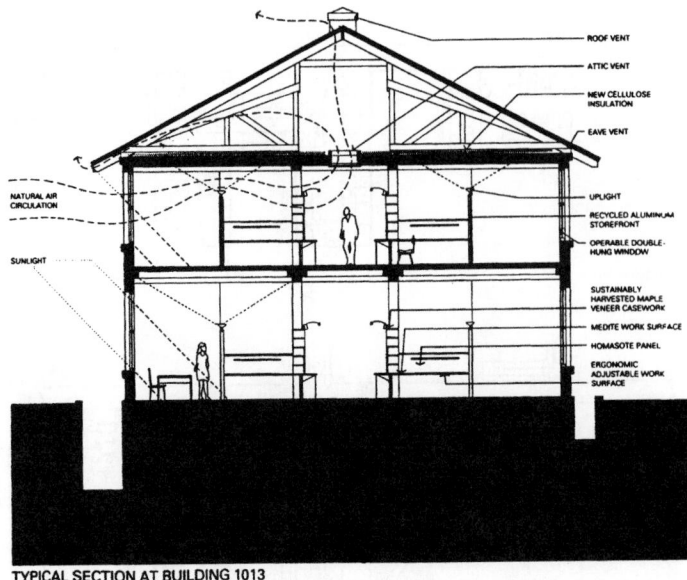

TYPICAL SECTION AT BUILDING 1013

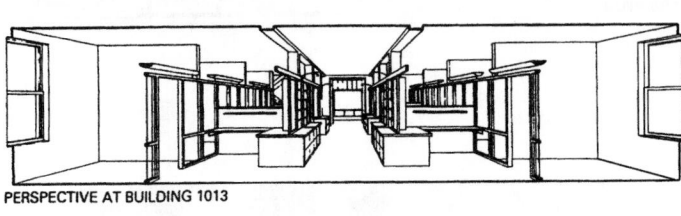

PERSPECTIVE AT BUILDING 1013

(d)

**Fig. 8.15** (Continued)

similar to the typical passively solar-heated, direct-gain building or to a multistory building with an exposed concrete structure, in which *both sides* of the floor slab are available for thermal storage, or an equivalent exposed mass area in walls, and so on. For this building, there are two units area of exposed mass for each unit floor area.

The climate data needed for Fig. 8.16 (maximum summer design DB temperature and mean daily range) are given in Appendix B. These data also allow calculation of the *minimum* summer design DB temperature; that is, maximum design DB temperature minus mean daily range. This minimum temperature is of interest here because the thermal mass of the building will be lowered toward (but not quite to) it during night ventilation. For high daily range climates, the lowest temperature obtained by the thermal mass will be *about one-fourth of the mean daily range above the minimum air temperature* (for lower daily range climates—30F° or less—*one-fifth of the mean daily range*).

**EXAMPLE 8.3** Compare the potential for night ventilation of mass in Sacramento, California, and Oklahoma City, Oklahoma.

**SOLUTION**

Sacramento, California:

Highest DB = 98°F (Appendix B)
Mean daily range = 36F° (Appendix B)
Lowest DB = 98 − 36 = 62°F

Approximate lowest mass temperature: 1/4 of 36 = 9F°; 62 + 9 = 71°F

Oklahoma City, Oklahoma:

Highest DB = 97°F
Mean daily range = 23F°
Lowest DB = 97 − 23 = 74°F

Approximate lowest mass temperature: 1/5 of 23 = 4.6F°; 74 + 4.6 = 78.6°F

From these data, it appears that Sacramento, with a comfortably low 71°F mass temperature, is a likely

Thus, the typical office building remains thermally closed during the 8 or 9 hours of summer occupancy. All the heat generated during the closed mode—8 to 9 daytime hours—must be stored in the structure:

Heat to be stored/ft² floor area =
hours occupied in closed mode
× heat gain, Btu/h ft² floor (from Table F.3)

Having found the amount of heat that can be stored each day, the designer must then solve the problem of removing the heat by night. Either natural or forced ventilation can be used, but in many locations wind speed is very low on summer nights. The ventilation rate is determined by the "best hour" of cooling during the night, that is, the hour during which the temperature difference, Δt, between inside mass and outside air is greatest, and hence the most heat is removed. This "best hour" information can be found in Fig. 8.16c and Fig. 8.16d. Then the cross- or stack ventilation rules of thumb can be used to size the openings.

See Table 8.4 for a comparison of these guidelines and more detailed calculation procedures for night ventilation of thermal mass.

---

**EXAMPLE 8.1, PART F** Find the potential for night ventilation of mass in the office building in Fig. 8.7.

**SOLUTION**

For Eugene, Oregon, from Appendix B, the summer design temperature is 89°F and the MDR is 31F°. From Table 8.2, the typical bay has a ratio of exposed thermal mass/floor area of

$$\frac{1512 \text{ ft}^2 + 720 \text{ ft}^2}{1440 \text{ ft}^2} = 1.55$$

which places this example halfway between average mass (ratio 1.0) and high mass (ratio 2.0). Entering Fig. 8.16a, with Eugene's climate data, we find that an average-mass building can store about 100 Btu/ft² day, whereas a high-mass building (Fig. 8.16b) can store about 170 Btu/ft² day. This building can therefore store about 135 Btu/ft² day.

Assuming a 9-hour typical working day (including the noon hour), at the average approximate heat gain rate of 13 Btu/h ft² (Example 8.1, Part D), the daily heat gain in this building is 9 h × 13 Btu/h ft² = 117 Btu/ft² day. This average heat gain is less than the storage capacity of 135 Btu/ft² day, so the

building should perform satisfactorily with this method of cooling.

Although this building was designed for nighttime forced ventilation because of summer night calm, we will check to see whether the operable windows are adequate for natural night cross-ventilation. In Fig. 8.16c, we find that during the hour of maximum night cooling, somewhat less than 14% of the total stored heat gains must be removed:

14% × 117 Btu/ft² day = 16 Btu/h ft²

From Fig. 8.14d, the resulting maximum hourly Δt is about 13F°.

In Example 8.1, Part E, we determine that less than 15 Btu/h ft² would be removed by cross-ventilation under average July conditions at a 3F° Δt. Because during this hour of maximum cooling the indoor–outdoor Δt is 13F°, cross-ventilation could remove

$$\left(\frac{13\text{F}°}{3\text{F}°}\right) \times 15 \text{ Btu/h ft}^2 = 65 \text{ Btu/h ft}^2$$

Cross-ventilation by night appears feasible for this building, because excess stored heat could be removed during this hour of maximum cooling. Unfortunately, Eugene's summer *nighttime* average wind velocity is nearly zero. Forced ventilation was therefore chosen, also in order to flush all interior mass surfaces thoroughly with cool air.

The rate of forced ventilation during the hour of maximum cooling can be estimated from

$$V = \frac{q_v \times 60 \text{ min/h}}{(1.1 \text{ Btu min/ft}^3 \text{ F}° \text{ h})(\Delta t)}$$

Expressed per square foot floor area in this building,

16 Btu/h ft² = V cfh [0.018 Btu min/ft³ °F h] 13F°

$$V = \frac{16}{0.018 \times 13}$$

= 68.4 cfh outdoor air/ft² floor area

This rate of forced outdoor air ventilation produces this air change: average ceiling height is 15.3 ft, so

$$\frac{68.4 \text{ cfh/ft}^2 \text{ floor}}{15.3 \text{ ft}^3/\text{ft}^2 \text{ floor}} = 4.47 \text{ ACH}$$

a rate considerably greater than that required for minimum outdoor air per person by day. This promises an attractive indoor air quality by the next morning. ∎

The administration building for the Fetzer Winery sits among grapevines in northern California's

Mendocino Valley (Fig. 8.17). The summer design condition is approximately 96/68, with MDR 30F° (35.6/20, MDR 16.7C°). As an average-mass structure, this building could store and remove about 60 Btu/ft$^2$ day (190 W/m$^2$ day); as a high-mass structure, it could store and remove about 100 Btu/ft$^2$ day (315 W/m$^2$ day).

### (d) Evaporative Cooling

The most common evaporative cooler is not strictly passive cooling, as it depends on a fan to force large quantities of outdoor air through a wet filter, thereby lowering the air's temperature and raising its RH before delivering the air to the space to be cooled. In hot and arid climates, the energy used by the fan in evaporative systems is less than the energy needed to achieve conventional cooling based on the compressive refrigeration cycle. Although this process requires quantities of water, it does not use refrigerants that pose a threat to the Earth's ozone layer.

Before using these guidelines, be sure to check the cooling strategy chart in Chapter 4 to determine whether evaporative cooling is appropriate for your climate. It is unlikely that evaporative cooling will be helpful in the humid southeastern United States.

The rates of evaporative cooling presented in Fig. 8.17 are based on a rather high airflow of 2.67 cfm per ft$^2$ of floor area. (In conventional cooling, an airflow rate closer to 1 cfm per ft$^2$ is more common.) With this amount of air motion, a highest indoor air temperature of 83°F is assumed; so the evaporatively cooled air, after absorbing the heat from the space, exits at 83°F. (Note: this may still be well below the outdoor temperature: might it do some useful cooling?) To use Fig. 8.17, first find in Table B.1 the summer design DB and mean coincident WB temperatures for your location. Enter the graph at these two data points, and at their intersection find the approximate amount of heat in Btu/h ft$^2$ that evaporative cooling can remove. As WB temperatures surpass 68°F (represented by dotted lines in the graph), indoor conditions become increasingly humid, producing almost certain discomfort indoors with a WB temperature of 75°F.

A more thorough method for evaluating evaporative cooling potential is presented in Section 8.14, where various airflow rates and temperatures of supply air and exit air can be examined. See Table 8.4 for a comparison of this guideline and the more thorough method.

---

**EXAMPLE 8.4** Evaluate the potential for evaporative cooling for a 3000-ft$^2$ retail store in Denver, Colorado. Due to large electric lighting and equipment display loads, the approximate heat gain is 30 Btu/h ft$^2$.

### SOLUTION
From Table B.1, Denver has a summer design DB temperature of 91°F, with a coincident WB temperature of 59°F. Checking back with Fig. 4.14, these design conditions fall well within the climate served by evaporative cooling. Next, enter Fig. 8.18 at the points of 91° DB and 59° WB; the intersection of these data lines shows that under such conditions, heat gains can be removed by evaporative cooling at a rate of about 37 Btu/h ft$^2$ of floor area.

37 Btu/h ft$^2$ capacity
> 30 Btu/h ft$^2$ approximate heat gains

Therefore, evaporative coolers can meet the need. Because Fig. 8.18 assumes an airflow rate of 2.67 cfm per ft$^2$, the cooler's approximate size will be

$$3000 \text{ ft}^2 \times 2.67 \text{ cfm/ft}^2 = 8000 \text{ cfm}$$

This cooler size could be reduced by the ratio of cooling capacity to heat removal need:

$$8000 \text{ cfm} \times \frac{30 \text{ Btu/h ft}^2 \text{ need}}{37 \text{ Btu/h ft}^2 \text{ capacity}}$$

$$= \text{about } 6500 \text{ cfm}$$

However, the larger capacity will allow a lower indoor air exit temperature than the 83°F assumed in Fig. 8.18, so the larger capacity is a safer approximation. ∎

### (e) Cooltowers

A more passive approach to evaporative cooling appears as a tower on the residence in Fig. 8.19. This University of Arizona experimental building near the Tucson Airport has a *cooltower* with wetted pads on all four faces at the top. Hot, dry air is cooled as it passes through the pads, dropping to the base of the tower and then into the house. Analysis by Givoni (1994) indicates that such a tower's delivery of wetter, cooler air is almost independent of wind speed and also is not dependent on the second tower at the opposite end of this building, a *solar chimney*

*(a)*

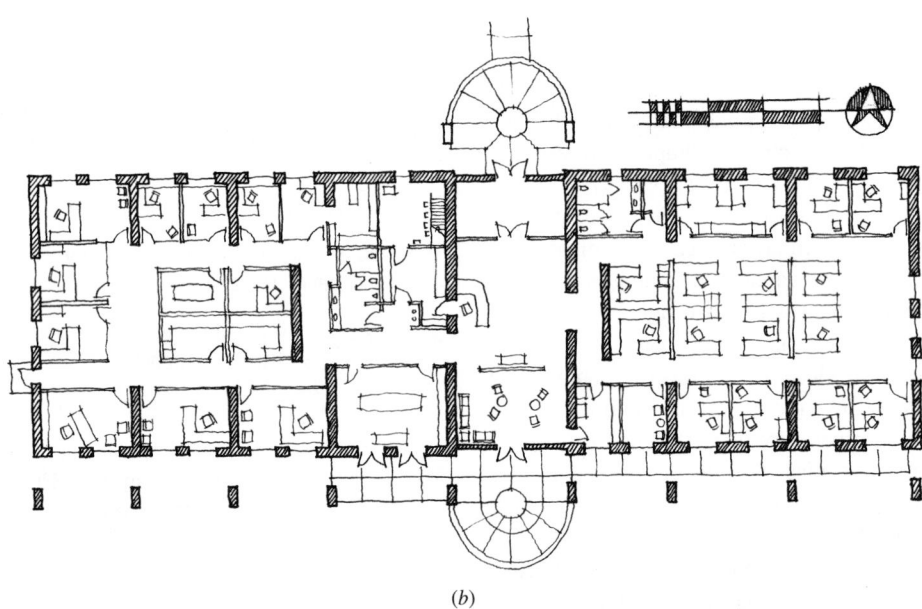

*(b)*

**Fig. 8.17** *The Fetzer Winery administration building (a) in its rural Mendocino County, California, setting. The south-facing clerestory and a deciduous vine and trellis are the more visible parts of the daylighting, passive solar heating, and passive cooling strategies. (b) Plan shows the relation of a workstation to north and south daylight. (c) Section with ducts for the night air flush as well as for conventional heating and cooling. (d) North-side dormers are for night air intake and exhaust. (Parts b and c redrawn by Dain Carlson from original drawings provided courtesy of Valley Architects, St. Helena, CA.)*

through which air from the house is discharged. Temperature and flow rate of air delivered to a building by a passive cooling tower can be approximated from Fig. 8.20, but such data are based on very little experimental work. A newer building using this principle, as well as Trombe wall passive solar heating, is the Zion National Park Visitor Center (Fig. 8.21).

**EXAMPLE 8.5** A 3000 ft² (279 m²) retail store in Las Vegas, Nevada, has a summer heat gain of 14 Btu/h ft²

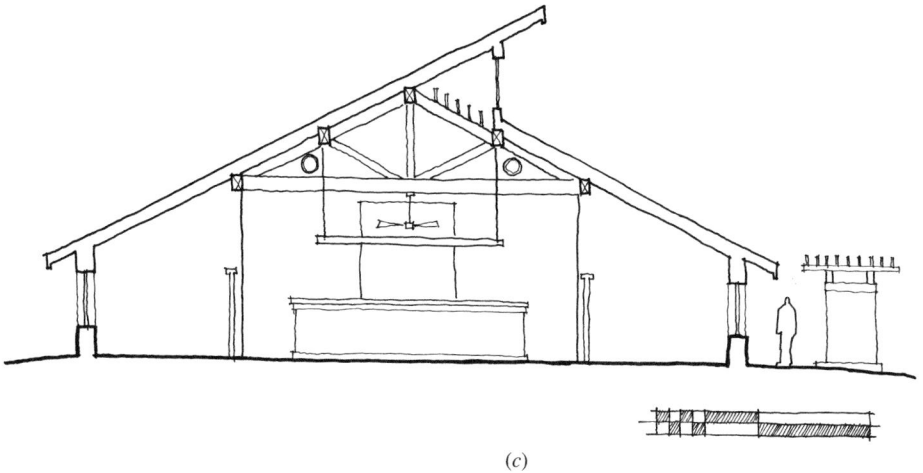

*(c)*

*(d)*

**Fig. 8.17** *(Continued)*

(44.2 W/m²). How large a passive cooling tower would be needed under summer design conditions?

**SOLUTION**

From Appendix B, Las Vegas has a summer DB = 106°F with a mean coincident WB = 65°F. This represents a *wet-bulb depression* (DB − WB) of 41F°. From Fig. 8.20*a*, at 106°F DB and 41F° (DB − WB), the air exiting a passive cooling tower will be about 69°F.

Next, determine the amount of supply air at 69°F needed to maintain this building at an indoor temperature of 82°F (probably a reasonable indoor temperature when it is 106°F outdoors, given the air motion provided by this passive cooling tower). From Chapter 7:

cfm = (Btu/h)/(1.1)(Δ*t*)
= (3000 ft² × 14 Btu/h ft²)/(1.1)(82 − 69)
= about 3000 cfm

At the Las Vegas design condition of 106 DB, 41F° DB − WB, from Fig. 8.20*c*, it appears that a passive cooling tower of 40 ft (12 m) in height, with a total wetted pad area of 32 ft² (3 m²), will provide about 3300 cfm (1558 L/s), somewhat more than the minimum just calculated. ∎

**(f) Roof Ponds**

This promising strategy has rarely been implemented, possibly due to water phobia among architects and clients alike. Yet it demonstrates the most stable interior temperatures of any of these techniques, needing electricity only to open and close the sliding insulation panels on the roof. Roof ponds sized for cooling will likely be nearly equal in area to the floors of the buildings they cool. Average pond depth is between 3 and 6 in. (75 and 150 mm). A more detailed and precise procedure for determining

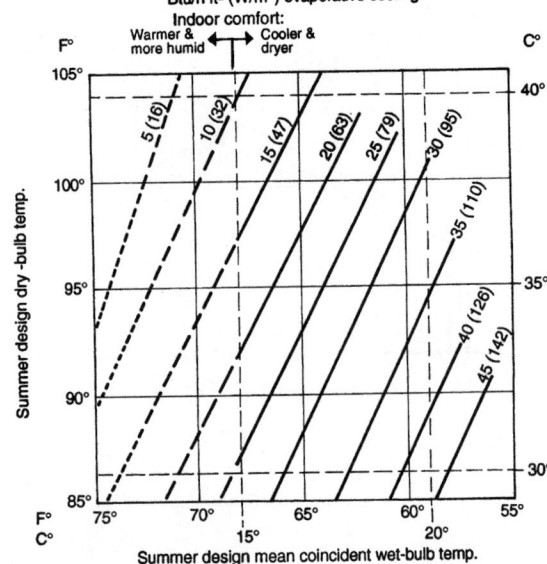

**Fig. 8.18** *Fan-forced evaporative cooling design guideline for Btu/h removed per ft² of floor area (W per m²). A forced airflow rate of 2.67 cfm per ft² of floor area (13.56 L/s per m² of floor area) is assumed, as is an exiting (exhaust) air temperature of 83°F (28.3°C). The dotted lines toward the left side of the graph represent increasingly humid indoor air, which may cause discomfort.*

pond area is presented in Section 8.14, but here is the quick approximation:

- Pond maximum* temperature: 80°F (*this temperature is based upon the assumption that any higher pond temperature would fail to produce cooling for the building interior)
- Pond minimum temperature = minimum nighttime DB (= max. daytime temp – mean daily range, from Appendix B)
- Pond $\Delta t$ = pond maximum of 80°F – pond minimum
- Pond's allowable daily heat stored (from build-

ing), Btu/day ft² = (0.7) (pond $\Delta T$) (pond depth, feet*) (62.5 lb/ft³ water)(1.0 Btu/lb °F) (*depth not to exceed 0.75 ft)

(This equation assumes 70% of the pond's daily heat gain from the building below and 30% through the insulated panels above the pond)

- Required size of pond (ft² pond per ft² floor area)

$$= \frac{\text{building total heat gain per day (Btu/day ft² floor area)}}{\text{pond allowable heat stored per day (Btu/day ft² floor area)}}$$

**Fig. 8.19** *An experimental residence in Tucson, Arizona, is passively and evaporatively cooled by a cooltower (at the right in the photograph). Hot, dry air passes through the wetted pads at the tower's top and undergoes a drop in DB temperature that causes it to fall, entering the residence at the bottom of the tower. The tower at the left is a solar chimney to help speed the flow of warmed exhaust air.*

THERMAL CONTROL

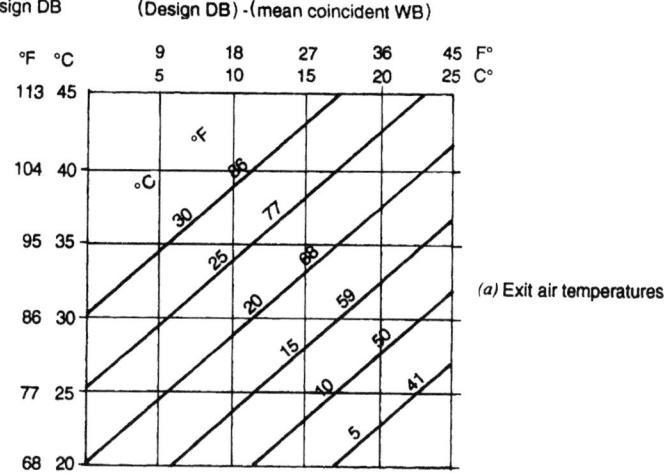

(a) Exit air temperatures

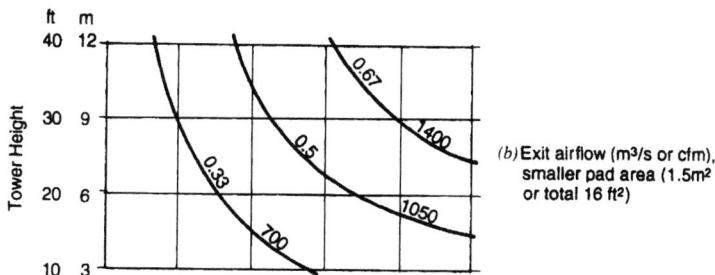

(b) Exit airflow (m³/s or cfm),, smaller pad area (1.5m² or total 16 ft²)

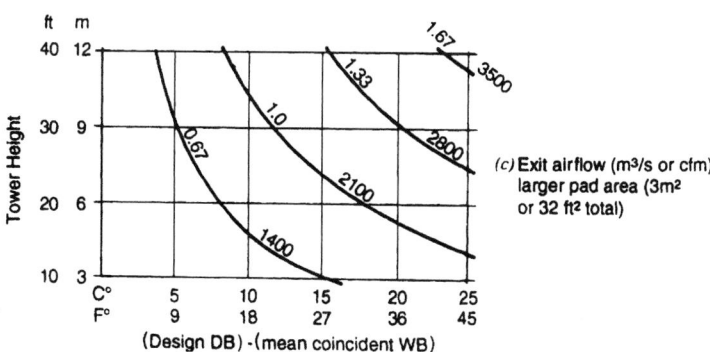

(c) Exit airflow (m³/s or cfm) larger pad area (3m² or 32 ft² total)

**Fig. 8.20** *Design guidelines for cooltowers. (a) The temperature of the air as it leaves the cooltower (enters the building) depends on the outdoor DB temperature, as well as the difference between the DB and WB temperatures. The flow rate of the exit air depends on the tower height, as well as the difference between the DB and WB temperatures. (b) Flow rate for a smaller area of wetted pads, 1.5 m² (16 ft²). (c) Flow rate for a larger area of wetted pads, 3 m² (32 ft² total). (Based upon the work of Givoni, 1994.*

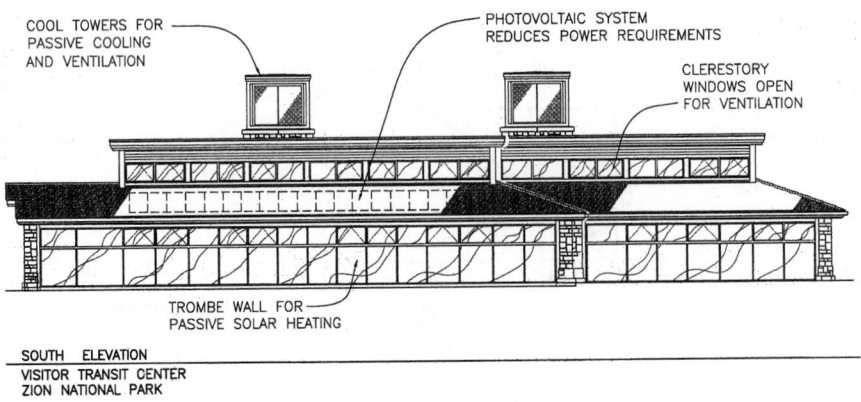

COOL TOWERS FOR
PASSIVE COOLING
AND VENTILATION

PHOTOVOLTAIC SYSTEM
REDUCES POWER REQUIREMENTS

CLERESTORY
WINDOWS OPEN
FOR VENTILATION

TROMBE WALL FOR
PASSIVE SOLAR HEATING

SOUTH   ELEVATION
VISITOR TRANSIT CENTER
ZION NATIONAL PARK

*(a)*

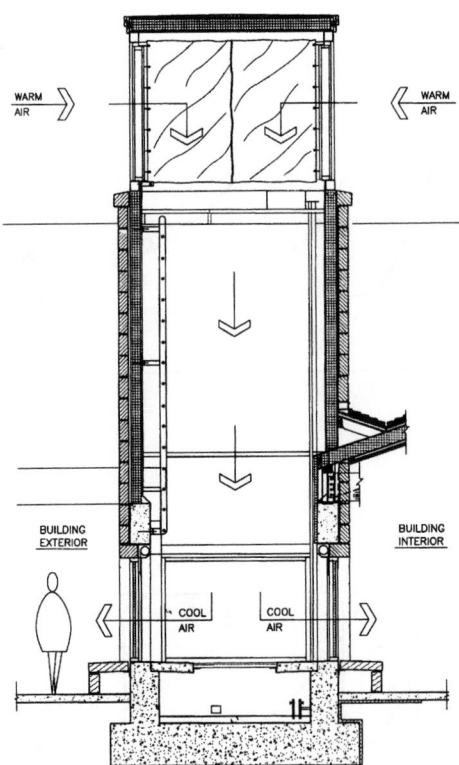

WARM
AIR

WARM
AIR

BUILDING
EXTERIOR

BUILDING
INTERIOR

COOL
AIR

COOL
AIR

SECTION THROUGH COOL TOWER
VISITOR TRANSIT CENTER
ZION NATIONAL PARK

*(b)*

**Fig. 8.21** *The new Visitor Center at Zion National Park (Utah) features both cooling by cooltowers and solar heating (a) by Trombe walls and a clerestory. Because many of the building's functions can be performed outside, each cooltower (b) has both an indoor and an outdoor outlet, either of which can be closed off when not needed. Air enters through wet media pads at the top of the tower (c). An X-baffle then diverts the air downward. The building also features PV and water harvesting. (Courtesy of the National Park Service, Denver Service Center.)*

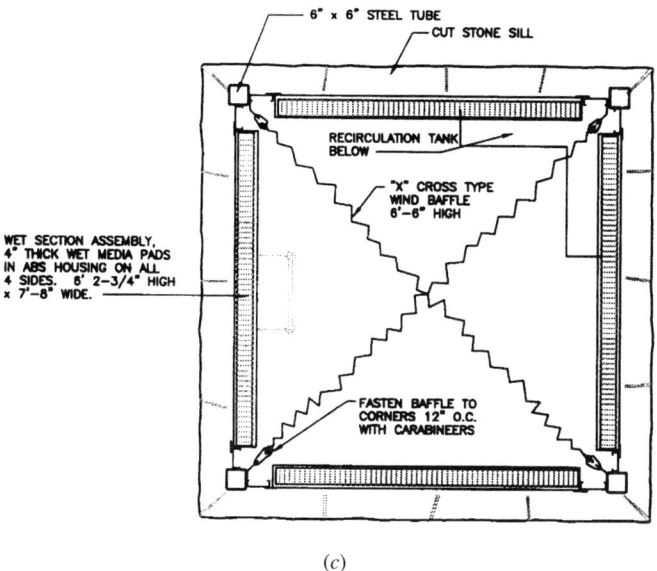

(c)

**Fig. 8.21** (Continued)

See Table 8.4 for a comparison of the design guideline and the more detailed calculations for roof ponds.

**EXAMPLE 8.6, PART A** An office building in Albuquerque, New Mexico, is to be cooled by a roof pond 4 in. deep. An hourly heat gain of 15 Btu/h ft² is assumed (excluding heat gain through the roof), somewhat more than the Oregon office building of Example 8.1.

**SOLUTION**

For Albuquerque:

Maximum temperature = 94°F (Appendix B)
Mean daily range = 27F° (Appendix B)
Minimum (night) temperature = 94 − 27 = 67°F

Therefore:

Pond $\Delta t$ = 80 maximum − 67 minimum = 13F°

Pond storage capacity for building heat
= 0.7(13°)(0.33 ft)(62.5 lb/ft³)(1 Btu/lb °F)
= 188 Btu/day ft²

The required pond size, then, is

$$\frac{15 \text{ Btu/h ft}^2 \times 9 \text{ h/day}}{188 \text{ Btu/day ft}^2} = 0.72 \text{ ft}^2/\text{ft}^2 \text{ floor area}$$

Therefore, a 4-in. pond covering 72% of the one-story building's floor area will be approximately large enough to cool the building. (If the entire ceil-ing is desirable as a cooled surface, then a slightly shallower pond of greater area could be used. However, the sliding insulation panels must be stacked somewhere beyond the edge of the roof pond.)

We revisit this building as Example 8.6, Part B, in Section 8.14. ∎

The residence in Fig. 8.22 is one of the first roof-pond structures. Located in Atascadero, California, at about 35°N latitude, it has a remarkable history of thermal stability, providing both passive solar heating in winter and passive cooling in summer (nearby Paso Robles has 2976 HDD65 and 5349 CDH74). Roof-pond inventor Harold Hay donated the building to California State Polytechnic at San Luis Obispo in the late 1990s. To emphasize the pond as the solar heating device, this house has almost no south-facing glass (most windows face east toward a small lake). The sliding insulation panels are stacked when open (by day in winter, by night in summer) over the carport on the north end of the building.

### (g) Earth Tubes

These provide a way to cool outdoor air before it enters a building. A fan is used to force sufficient quantities of air through these long tubes. Because earth tubes need to be well underground as well as rather long in order to cool outdoor air, it is rarely

THERMAL CONTROL

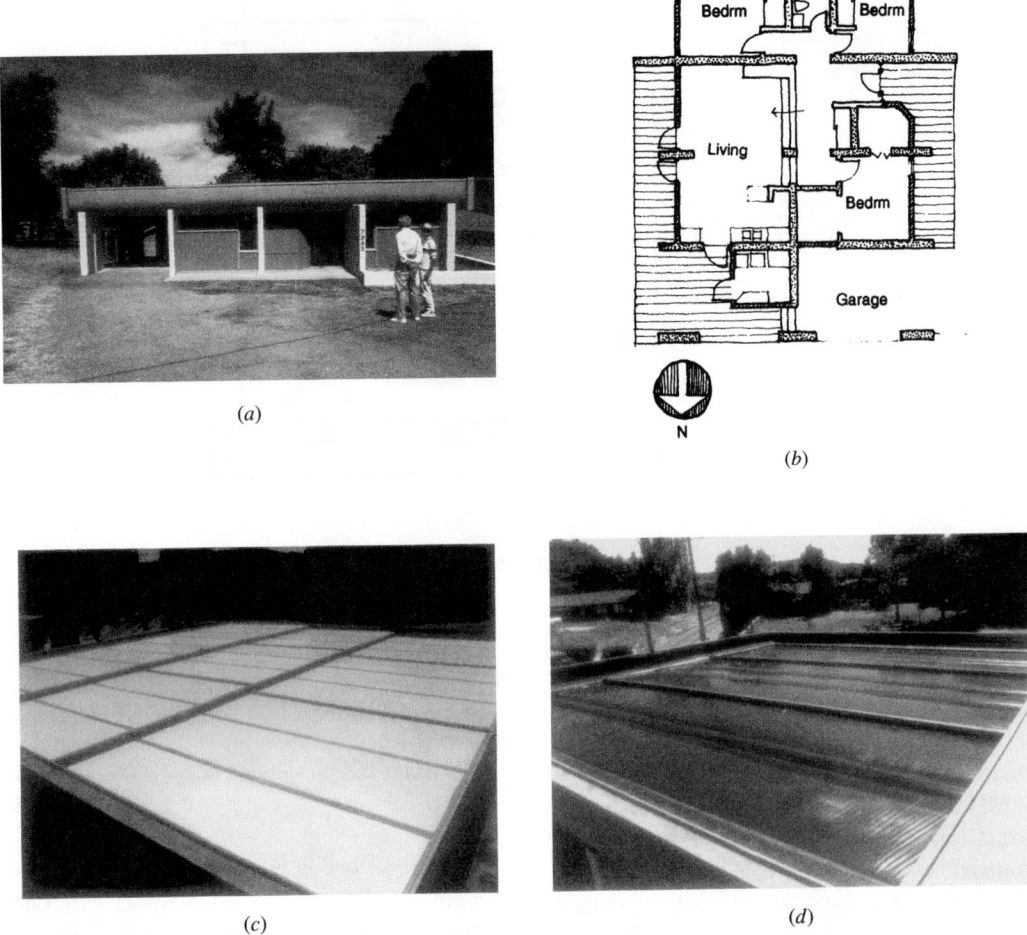

(a)

(b)

(c)

(d)

**Fig. 8.22** *One of the first roof pond buildings, this Atascadero, California, residence is now a research building for California State Polytechnic University at San Luis Obispo. (a) The exterior from the west. (b) Plan. South exposure is minimal; the roof is the surface of interest for heating and cooling. (c) The roof insulating panels in the closed position (as on a summer day). (d) The roof insulating panels in the open position (as on a summer night). Water is contained within plastic bags laid directly on the metal deck ceiling. (e) Sections showing summer day roof pond insulation and night exposure. (f) Sections showing winter day roof pond exposure and night insulation. (g) Year-long record of the indoor temperature range compared to the outdoor range. (Drawings based in part on Sandia National Laboratory, 1977.* Passive Solar Buildings: A Compilation of Data and Results, *Sand 77-1204.)*

economical to install enough earth tubes to completely meet a building's need for cooling. If long trenches are needed for another purpose (underground water lines, for example), an earth tube is more feasible. Where earth tubes are considered, the component of building heat gain that is represented by cooling fresh air (Table F.3, Part E) can sometimes be achieved using this strategy.

The estimate of cooling for the earth tubes described in Fig. 8.23 is based on Abrams (1986), who assumed a soil conductivity equal to that of heavy, damp earth. One of the most influential vari-

ables in earth tube performance is soil conductivity, so Fig. 8.24 (also from Abrams, 1986) compares the cooling performance of 6-in. (152-mm) diameter earth tubes when soil conductivity changes. Note that this design guideline presents total Btu/h cooling of *the outdoor air in the tube* rather than Btu/h ft² (for a typical ft² of building floor area). Therefore, two adjustments must be made: first, find the Btu/h cooling load of the entire building; second, adjust for the difference between hotter air outdoors and cooler air desired within the building.

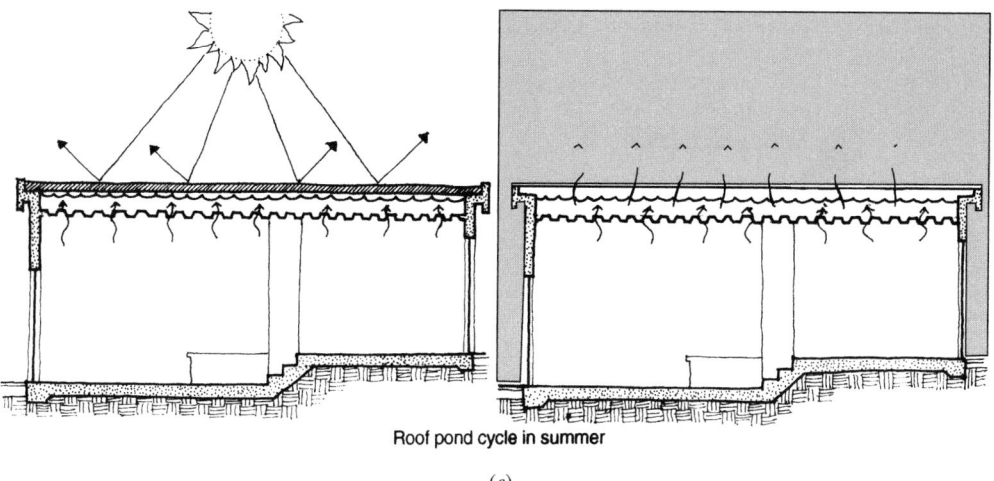

Roof pond cycle in summer

(e)

Roof pond cycle in winter

(f)

(g)

**Fig. 8.22** (Continued)

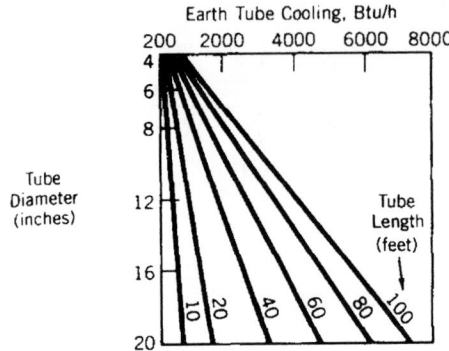

**Fig. 8.23** *Cooling of intake air provided by an earth tube in Btu/h. This tube is set in heavy, damp soil (k = 0.75 Btu/h °F ft) whose temperature around the tube is 65°F. The outdoor air entering the tube is at 85°F and is moved at a flow velocity of 500 fpm. (Adapted by permission from Donald W. Abrams,* Low Energy Cooling, *© 1986, Van Nostrand Reinhold Company.)*

See Section 8.14*g* for a more detailed calculation procedure for earth tubes in which tube depth, soil temperature, and soil conductivity are variable. See Table 8.4 for a comparison of the design guideline with the more detailed calculation procedure.

---

**EXAMPLE 8.7, PART A** A partially underground nature center of 4000 ft² in northern Pennsylvania, design DB = 91°F, has an hourly heat gain of 9.3 Btu/h ft², of which 2.2 Btu/h ft² are due to outdoor air required ventilation. If earth tubes are used only for cooling the fresh air, how many tubes at what length will be needed?

**SOLUTION**

First, the total building cooling needed from the earth tubes is (4000 ft²)(2.2 Btu/h ft²) = 8800 Btu/h. Next, adjust this figure to account for hot outdoor air versus cooler indoor air. If we accept 82°F indoors (the walls are earth sheltered and therefore cooler), then the total earth tube cooling needed is (8800 Btu/h)(91°F/82°F), or about 9770 Btu/h.

From Fig. 8.23, assuming heavy, damp soil, a 60-ft-long tube 20 in. in diameter will deliver about 4500 Btu/h; two such tubes will deliver about 9000 Btu/h.

We return to this example in Section 8.14*g*, Example 8.7, Part B. ∎

Table 8.4 compares the design guidelines for passive cooling with the more detailed calculation procedures that are given in Chapter 7 (basement walls and floor) and later in Section 8.14.

## 8.7 REINTEGRATING DAYLIGHTING, PASSIVE SOLAR HEATING, AND COOLING

At this point, preliminary investigations of envelope component U-factors and fenestration/floor area ratios have proceeded through various codes, standards, and design guidelines. It is quite likely that conflicting advice has resulted. Depending on the location, building type, and client's objectives, a very different mix of glass, insulation, and thermal

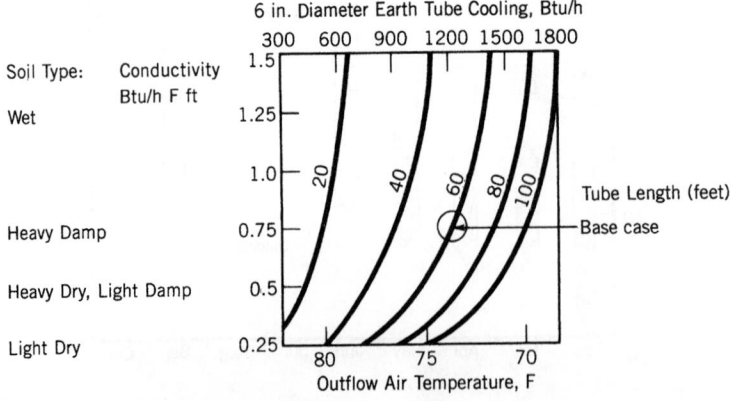

**Fig. 8.24** *Six-inch-diameter earth tube cooling capacities as soil conductivity varies. The* base case *is a 6-in.-diameter earth tube 60 ft long, shown in Fig. 8.23 to deliver 1200 Btu/h of cooling at soil conductivity of 0.75, with other conditions as specified in Fig. 8.23. (Adapted by permission from Donald W. Abrams,* Low Energy Cooling, *© 1986, Van Nostrand Reinhold Company.)*

mass could result from prescriptive conservation approaches compared to passive solar heating or any of several approaches to passive cooling.

When prescriptive standards (such as those presented in Table G.1 or G.2) appear to preclude either passive solar heating or passive cooling strategies in a building, the designer should remember that there are procedures for either window area trade-offs (in which more insulation in walls or roofs might allow more glass areas) or methods to compare whole-building annual energy consumption that can demonstrate the superiority of using well-designed passive strategies over a simple conventional prescribed approach.

---

**EXAMPLE 8.8** Consider a one-story residence in Billings, Montana. How does the fenestration area prescribed for code compliance compare with desirable fenestration areas for daylight, solar heating, and ventilation cooling?

**SOLUTION**

From ASHRAE Standard 90.2, windows that are ≥15% of the floor area must have a maximum U-factor of 0.36 and a maximum SC of 0.7. From Table E.15, windows #7 through #12 provide these values for U and SC (considered equivalent to SHGC for this purpose).

We now need to assume the floor area and dimensions in order to proceed. A plan 32 ft wide by 64 ft long (long walls to face north/south) yields a floor area of 2048 ft$^2$. Most windows can be placed in the long walls for both daylight distribution and cross-ventilation. However, the south wall will need the most window area to facilitate solar heating. Tentatively, assign window areas as follows:

| | |
|---|---|
| South window area | = 8% of floor area |
| North window area | = 5% of floor area |
| East and west window area | = 2% (1% each) of floor area |
| | 15% |

**Daylight:**

For sidelighting, DF$_{av}$ = 0.2 (window area/floor area), where all of the floor area is within 2.5 $H$ of the window wall. If we assume that $H$ = 7.5 ft (ceiling at wall 8 ft high), then the daylight area is 2.5 × 7.5 = 18.75 deep. Light from opposite exterior walls produces the maximum daylight width of 2 × 18.75 = 37 ft. With our assumed maximum 32-ft width, all floor area is within the daylight zone. For our total of

15% of floor area in fenestration, DF$_{av}$ = 0.2 [0.15] = .03, or 3%. This seems to be sufficient according to the target DFs in Table 14.2.

**Passive Solar Heating:**

Windows #7 through #12 (Table E.15) are under consideration. From Table F.1, footnote c, windows #7 and #12 qualify as superior performance glazings. For Billings, Montana, a south glass/floor area "low" ratio of 0.16 would produce SSF = 53%. From Table F.2, such an SSF would require at least 3.8 ft$^2$ of interior masonry surface for each square foot of south glass. However, only 8% south window area/floor area is allocated. The designer can:

- Reallocate window area so that essentially all windows face south (with daylight and cross-ventilation penalties), or
- Increase the total window area to more than 15%, with correspondingly better fenestration U-factors required (from Standard 90.2, for example, up to 21% of floor area would require U = 0.29), or
- Settle for a lower SSF with only 8% south window area/floor area.

**Cross-Ventilation:**

First, we need a quick estimate of heat gains. From Table F.3:

A. Internal gains from people and equipment: assume 4 people at 230 Btu/h plus 1200 Btu/h from appliances:

$$(4 \times 230) + 1200 = 2120 \text{ Btu/h}$$

$$2120 \text{ Btu/h}/2048 \text{ ft}^2 = 1.04 \text{ Btu/h ft}^2$$

B. Lights: assume with 3% DF a gain of about 2.0 Btu/h ft$^2$.

C. Gain through envelope: Billings' design temperature (Appendix B) is 91°F.

Externally shaded window gains = (0.15) × 16 = 2.4 Btu/h ft$^2$; gain through walls: total vertical area = [(2 × 32 ft) + (2 × 64 ft)] = 1536 ft$^2$. Of this, window area = (0.15 × 2048) = 307 ft$^2$; opaque wall area = 1536 − 307 = 1229.

To meet Standard 90.2, frame walls must have a U-factor no greater than 0.073, whereas masonry walls may have a maximum U-factor of 0.08. Therefore,

gain through walls = (1229 ft$^2$ wall area/2048 ft$^2$ floor area) × 0.08 × 15 = 0.72 Btu/h ft$^2$

Gain through roof: assume roof area equal to floor

area, and to comply with Standard 90.2, a maximum U-factor of 0.033. Therefore,

$$\text{gain through roof} = (1.0) \times 0.033 \times 35$$
$$= 1.16 \text{ Btu/h ft}^2$$

D. For an open building (using cross-ventilation)

$$\text{total heat gain} = 1.04 + 2.0 + 2.4 + 0.72 + 1.16$$
$$= 7.32 \text{ Btu/h ft}^2$$

We can assume a total operable area of perhaps 90% of the glazed area using fully openable case-ment windows. With 8% floor area in south win-dows and 5% in north windows, the maximum available ventilation area = $0.9 \times 5\% = 4.5\%$ of the floor area. From Fig. 8.14a, it appears that this rate of heat gain can be removed through this inlet area by a wind velocity of only about 2 mph. (However, when it is 91°F outdoors, the use of another cooling strategy is highly likely.)

**Summary:** Probably increase the south glass area (and the overall ratio window/floor area) and use windows with lower U-factors, with summer exter-nal shading. Floor and many interior wall surfaces will be faced with thermally massive materials. ∎

## 8.8 CALCULATING WORST-HOURLY HEAT LOSS

Chapter 7 showed how heat is exchanged by trans-fer through a building envelope ($q$) and through ventilation of a building ($q_v$). By combining these rates of heat exchange, we can obtain a building's total hourly heat loss in winter. (To calculate the total hourly heat gain in summer, other gains must be added; this more complicated procedure is discussed in Sections 8.11 and 8.13.) The total hourly heat loss of a building can be calculated under several different assumptions reflecting dif-ferent purposes.

### (a) Maximum Hourly Loss: Sizing Conventional Heating Equipment

The most typical use of $q + q_v$ is to determine the maximum amount of heat per hour that heating equipment must provide. Two important assump-tions are usually made:

1. No internal heat gains (lights, people, etc.) and no solar gains are present in the building.

2. The *design* lowest outdoor temperature (see Appendix B) is occurring.

These are conservative assumptions that lead to the installation of heating equipment that is rarely used to capacity. Such equipment does provide a safety margin for those rare times when even lower than design temperatures occur or when windows are inadvertently left open or other temporary and unexpected heat leaks occur in very cold weather.

Thus, to obtain design hourly heat loss, calculate

$$q_{total} = q + q_v$$

where

$$q = (\Sigma UA)\, \Delta t$$

with $\Sigma UA$ the sum, for all exposed components of the building's envelope, of $U \times A$; $\Delta t$ = (interior − exterior design temperature); and

$$q_v = 1.1\, V \Delta t \,(1.2V \Delta t; \text{ SI units})$$

where $V$ is the volume in cfm (L/s) of outdoor air introduced.

### (b) Maximum Hourly Heat Loss: Sizing Auxiliary Heating for Passive Solar Buildings

The one important difference between the maxi-mum heat loss calculations for conventional and passive solar buildings is the following assumption:

No internal heat gains (lights, people, etc.) are pre-sent, but there is sufficient stored solar energy to at least cancel out the heat losses through the south solar collection area.

Otherwise, the procedure is the same as that for conventional buildings:

$$q_{total} = q + q_v$$

where

$q = \Sigma\, UA_{ns}\, \Delta t$

$q_v = 1.1\, V \Delta t$ ($\Delta t$ again based on the design condition, Appendix B) ($1.2\, V \Delta t$; SI units)

$\Sigma\, UA_{ns}$ *excludes* the solar collector area

This can lead to occasional chilly interiors, as when several days of heavily overcast skies coincide with design-condition outdoor temperatures. In some locations, therefore, designers use the more conservative, conventional procedure for sizing the auxiliary heaters for passive solar buildings.

## (c) Maximum Hourly Loss: Checking Design Criteria

The calculations that produce $q$ and $q_v$ are also useful in reviewing a building's design. By showing where most of the heat loss is occurring, they can quickly pinpoint opportunities for energy conservation and increased comfort. If much of the building's heat loss is occurring through the large windows in one wall, for example, consider the following options:

1. Reduce the window size. (Architectural and daylighting considerations may override.)
2. Go to a lower window U-factor to reduce heat loss and increase the winter surface temperature of this large glass area. (Cost and detailing considerations may override.)
3. Add thermal shades or shutters to dramatically reduce heat loss and increase winter surface temperature. (Architectural, view, cost, and detailing considerations may override.)

If much of the building's heat loss occurs through ventilation air, consider these options:

1. Reduce infiltration by tighter construction (or reduce mechanical ventilation toward the code-required minimum).
2. Add a heat exchanger between outgoing and incoming air.

These calculations can also be used to check your building against published criteria for thermal performance (see Appendix G for example criteria). Redesign of building envelopes to meet such criteria is fairly common in the early stages of building design.

## (d) Hourly Rates of Fuel Consumption

When outdoor temperatures in winter drop below the building balance point (see Section 8.9), heating systems usually begin to operate. The hourly rate of fuel consumption depends on the hourly heat loss from the interior space. If the boiler (or furnace) is selected to run continuously at the outdoor, critical winter design temperature, then it will cycle

(run intermittently) at higher outdoor temperatures. The equipment, however, *is* selected on the basis of the maximum winter demand rate and therefore relates to the calculated heat loss at the design temperature. Sometimes your energy supplier will ask for the maximum hourly rate.

**EXAMPLE 8.9** Calculate the rates of burning for several fuels (or the rate of using electricity) to make up the hourly heat loss, under design conditions, of a mercantile store. Its maximum hourly heat loss is 159,840 Btu/h. For fuel values, refer to the data in Table 8.5.

**SOLUTION**

If, for instance, oil were used, the situation would be gal/h × Btu/gal heat value × efficiency = Btu/h capacity (= heat loss). Transposing, we have

$$\text{gal/h} = \frac{\text{Btu/h heat loss}}{\text{Btu/gal} \times \text{efficiency}}$$

(Other consumption statements are similar.) Applying values to this and to relationships for the other fuels, we obtain the following rates:

Coal: 159,840/14,600 × 0.70 = 15.6 lb/h

Oil: 159,840/141,000 × 0.75 = 1.51 gal/h

Gas (older, less efficient equipment): 159,840/1052 × 0.75 = 203 cfh

Electricity: 159,840/3.41 × 1.00 = 46,600 W (46.6 kW)

**TABLE 8.5 Approximate Heat Values of Fuels**

| Fuel | Heat Value | | Typical Seasonal Efficiency, % |
|---|---|---|---|
| | I-P Units | SI Units | |
| Anthracite coal | 14,600 Btu/lb | 33,980 kJ/kg | 65–75 |
| No. 2 oil | 141,000 Btu/gal | 39,300 kJ/L | 70–80[c] |
| Natural gas[a] | 1,050 Btu/ft³ | 39,100 kJ/m³ | 70–80[c] |
| Propane | 2,500 Btu/ft³ | 93,150 kJ/m³ | 70–80[c] |
| Electricity | 3,413 Btu/kW | 1 kW | 95–100 |
| Wood[b] | 7,000 Btu/lb | 16,290 kJ/kg | 30–50 |

*Note:* This table includes the thermal value of electricity used on site (but not losses in fuel energy at the electrical generating plant). Approximate seasonal efficiencies of typical burner–boiler equipment are also shown.

[a] Natural gas is frequently sold in therms; 1 therm = 100,000 Btu/h (SI units, 29.3 kW).

[b] At 20% moisture content.

[c] Higher-efficiency furnaces are now available.

These results are based on the assumption that the boiler and its piping are enclosed within the useful volume of the store. If they were in cold basements, or if the ducts or pipes ran through unheated space, more fuel would be used and system efficiency would decline. The rates established set the values by which the fuel-burning apparatus is selected. For instance, if oil is used, a nozzle that injects oil at the rate of about 1 gph should be tried. The above rates are for design (extreme) conditions and are not typical of the lower average rate of operation throughout the winter. ∎

## 8.9 CALCULATIONS FOR HEATING-SEASON FUEL CONSUMPTION (CONVENTIONAL BUILDINGS)

The following method of estimating the fuel used for space heating in a typical season best applies to residences and small commercial buildings that are skin-load dominated and not passively solar heated beyond SSF = 10%. To the extent that the combination of internal and solar gains can be predicted accurately, this method yields a reasonable estimate of annual fuel consumption for any building. (For passively solar-heated buildings, use instead the method given in Section 8.10.)

Internal and solar gains make almost any building warmer than the outdoors during the heating season. The furnace (or other space-heating device) is not needed until the outdoor temperature drops to the point at which these internal and solar gains are insufficient to heat the building by themselves, that is, when the heat lost through the building's skin and infiltration matches the heat gained through solar plus internal loads. This particular outdoor temperature is called the *balance point*; it represents the beginning of the need for space-heating equipment.

To estimate the annual energy needed for a building's space heating, it is necessary to know the following:

- The building's heat-loss rate (envelope and infiltration)
- The building's internal plus solar gain rate
- The building's balance point temperature
- The time period during which the outside temperature falls below the building's balance point temperature (DD)

## (a) Balance Point Temperature

When a building needs neither heating nor cooling, the internal gains equal the external losses:

$$Q_i = \text{balance point } q_{total}$$

where

$Q_i$ = internal gains plus solar gains (Btu/h or W)

balance point $q_{total} = UA_{total} (t_i - t_b)$

$t_b$ = balance point temperature

$t_i$ = average interior temperature over 24 hours, winter

$UA_{total}$ = total heat loss rate—envelope plus infiltration (Btu/h °F or W/°C)

Rewriting this equation to solve for the balance point temperature:

$$t_b = t_i - \frac{Q_i}{UA_{total}}$$

To determine the total heat loss rate $UA_{total}$, combine the envelope (or skin) losses and the infiltration (or ventilation) losses.

The quantity $Q_i$ cannot be determined so straightforwardly. The internal gains can be estimated as shown in Table F.3 (remember, these are summertime gains) or calculated more precisely from known building population, lighting, and equipment data. For residences, the following *daily total* internal gains are considered typical:

*People:* Two adults and two children (average times of occupancy): 23,000 to 24,500 Btu/day (6.7 to 7.2 kWh/day)

*Lights and equipment:* See Table 8.6 for individual heat sources, but if actual appliances are unknown, then:

53,000 Btu/day (15.5 kWh/day) for standard efficiency equipment

100,000 Btu/day (29.3 kWh/day) for old and inefficient equipment

For more details on internal gains in residences, see the 2005 *ASHRAE Handbook—Fundamentals*, Chapter 28.

The solar gains are elusive; each month has a different average gain. For simplicity in the heating season, use the average January daily insolation on a vertical surface (found in Appendix C) with this approach.

During the heating season:

$$Q_i = \frac{\text{internal gains (Btu/day)}}{24\ h}$$

$$+ \frac{\left[\begin{array}{l}\text{January insolation} \quad \text{area (ft}^2\text{),} \\ \text{(Btu/ft}^2 \text{ day average)}, \times \quad \text{south} \\ \text{vertical surface} \qquad \text{glass}\end{array}\right]}{24\ h}$$

The balance point temperature, $t_b$, can be used to do several things besides predict fuel consumption. By noting the $t_b$ on a graph of monthly outdoor temperatures, the designer can quickly see the relative importance of heating versus cooling for a specific building in a given climate. Also, it can be used to gain a better understanding of how zones in a building interrelate. Once the $t_b$ is calculated for each zone, it can be determined when the entire building needs heating (outdoor temperature below any zone's $t_b$) or cooling (outdoor temperature above any zone's $t_b$), or when one zone's surplus heat can be another zone's space heating source (outdoor temperature higher than one zone's $t_b$ but

## TABLE 8.6 Typical Residential Daily Internal Heat Gains from Appliances and Lighting

| PART A. ELECTRIC APPLIANCES | | |
|---|---|---|
| **Heat Source** | **Btu/day** | **kWh/day** |
| Frost-free refrigerator | 16,000 | 4.7 |
| Freezer | 10,900–15,700 | 3.2–4.6 |
| Dryer[a] | 8,900–11,600 | 2.6–3.4 |
| Range[a] | 6,500–11,300 | 1.9–3.3 |
| Television[b] | 3,400–4,400 | 1.0–1.3 |
| Dishwasher | 2,400 | 0.7 |
| Lights and miscellaneous[c] | 9,500–18,800 | 2.8–5.5 |
| Water heater[d] | 13,700–27,400 | 4.0–8.0 |
| PART B. GAS APPLIANCES | | |
| *Heat Source* | *Btu/day* | |
| Water heater | 13,700–27,400 | |
| Dryer[a] | 12,000–19,000 | |
| Range[a] | 12,000–27,000 | |

*Source: A New Prosperity: The SERI Solar Conservation Study.* Brick House Publishing, Andover, MA, 1981. Adapted by permission from the 1989 *ASHRAE Handbook—Fundamentals,* published by the American Society of Heating, Refrigerating and Air-Conditioning Engineers, Inc., Atlanta, GA.

[a]These are for the appliance's consumption per day; heat gain to a house is less, depending on the amount of heated exhaust air.

[b]Total daily use of TV per household.

[c]These figures are for an average 1350-ft$^2$ house; the rate per square foot of floor area may be extrapolated.

[d]Standby heat loss from water heater to house.

lower than that of another). If thermal exchange between zones occurs for a major portion of the typical year, the choice of either a heating or a cooling strategy could be influenced.

**EXAMPLE 8.1, PART G** Calculate the January balance point temperature for a typical bay of the office building in Fig. 8.7. Assume that the average indoor January temperature over a 24-hour period is an energy-conserving 68°F.

### SOLUTION

The (summer) internal gains from Example 8.1, Part D, may need some modifications. In darker winter, assume that the gain from electric lights is closer to the 2 < DF < 4 value of 2.0 Btu/h ft$^2$. The internal gain is then 5.8 + 2.0 = 7.8 Btu/h ft$^2$ from people, equipment, and electric lights. The average internal gains are as follows:

$$\frac{7.9\ \text{Btu/h ft}^2 \times 9\ h}{24\ h} = 2.96\ \text{Btu/h ft}^2$$

The January solar gain, *VS*, from Table C.15, for Salem, Oregon, is 471 Btu/ft$^2$ day per ft$^2$ of south glass:

$$\frac{471\ \text{Btu/h ft}^2 \times (97\ \text{ft}^2 + 132\ \text{ft}^2)\ \text{south glass}}{24\ h \times 1440\ \text{ft}^2}$$

$$= 3.1\ \text{Btu/h ft}^2$$

Therefore,

$$Q_i = 2.96 + 3.1 = 6.06\ \text{Btu/h ft}^2$$

*UA* for the typical bay (from Table 8.2 and Example 8.1, Part B) includes, for this calculation, all elements of the building envelope:

(42 + 25 + 72 + 72 + 98 + 174 infiltration)
$$= 483\ \text{Btu/h °F}$$

expressed per square foot of floor area:

$$483/1440 = 0.34\ \text{Btu/h °F ft}^2$$

Therefore,

$$t_b = 68°F - \frac{6.06\ \text{Btu/h ft}^2}{0.34\ \text{Btu/h °F ft}^2} = 68 - 17.8$$

$$= 50.2°F$$

Consulting NOAA data for Salem, Oregon, 50.2°F, although based on January insolation, is closest to the daily average temperature for the months of April and October. ■

**THERMAL CONTROL**

### (b) Degree Days (DD)

These data are published for each climate station and are calculated to various *base* temperatures (found in Appendix C). Heating degree days (HDDs) to a specified base temperature, such as HDD65, refer to average temperatures *below* the base temperature of 65°F. Cooling degree days (CDDs) to a specified base temperature, such as CDD65, refer to average temperatures *above* the base temperature of 65°F. (Cooling degree hours [CDH], such as CDH74, similarly refer to the average number of hours above a base temperature of 74°F.) For smaller buildings, such as residences, HDDs are in much wider usage than are CDDs (or CDHs), and many data sources simply list DD65; this notation is interchangeable with HDD65. Until recently, degree days were always based on 65°F, because older, indifferently insulated buildings, with low internal gains, had a typical balance point of about 65°F. The combination of much higher levels of insulation and much more electric equipment has shoved the average building's balance point temperature downward; hence DD50, DD55, and DD60 are included with the traditional DD65 in Appendix C.

To derive HDD for a particular climate and X base temperature (HDDX), each day's mean temperature (halfway between high and low) is subtracted from the base temperature; the result is the number of HDDX for that day. If the mean temperature equals or exceeds the base temperature, no HDDX are recorded. Then the HDDX are totaled for an average year.

For example, assume that a day in Troy, New York, had a high of 60°F and a low of 34°F. The mean temperature was $(60 - 34/2) + 34 = 47°F$.

$$65 - 47 = 18 \text{ DD65}$$
$$60 - 47 = 13 \text{ DD60}$$
$$55 - 47 = 8 \text{ DD55}$$
$$50 - 47 = 3 \text{ DD50}$$

Clearly, a building with a 65°F balance point will need more heat on a given day than will a building with a 50°F balance point temperature.

To convert DD I-P to DD SI, simply multiply DD I-P by ⁵⁄₉:

$$\text{DD SI} = 0.56 \text{ DD I-P}$$

To obtain the *DD balance point* needed to estimate a particular building's heating needs, interpolate between the various base DDs as required. (If the

balance point is below 50°F, get lower DD base figures from your local weather station. Do not extrapolate!)

### (c) Yearly Space Heating Energy

To estimate the energy needed over an average year for a building, E, calculate

$$E = \frac{(UA)(DD \text{ balance point})(24 \text{ h})}{(AFUE)(V)}$$

where

E is in units of fuel consumed per year (therms of gas or kWh of electricity)

UA is the total heat loss rate, envelope + infiltration (Btu/h °F, or W/°C)

DD balance point is obtained as just described

AFUE is the annual fuel utilization efficiency, displayed on all furnaces manufactured within the United States (Table 8.7)

V is the heating value of the fuel from Table 8.5

**TABLE 8.7 Some Typical Values of Annual Fuel Utilization Efficiency (AFUE)**

| Type of Furnace | AFUE (%) | |
| --- | --- | --- |
| | Indoor | ICS[a] |
| NATURAL GAS | | |
| Natural draft with standing pilot[b] | 64.5 | 63.9 |
| Natural draft with intermittent ignition[b] | 69.0 | 68.5 |
| Natural draft with intermittent ignition and auto vent damper[b] | 78.0 | 68.5 |
| Fan-assisted combustion with standing pilot or intermittent ignition | 80.0 | 78.0 |
| Fan-assisted combustion with improved heat transfer, with standing pilot or intermittent ignition | 82.0 | 80.0 |
| Direct vent with standing pilot, preheat[b] | 66.0 | 64.5 |
| Direct vent, fan-assisted combustion, and intermittent ignition | 80.0 | 78.0 |
| Fan-assisted combustion (induced draft) | 80.0 | 78.0 |
| Condensing | 93.0+ | 91.0+ |
| OIL | | |
| Standard[b] | 71.0 | 69.0 |
| Standard with improved heat transfer[b] | 76.0 | 74.0 |
| Standard, with improved heat transfer and auto vent damper[b] | 83.0 | 74.0 |
| Condensing | 91.0 | 89.0 |

*Source:* 1996 *ASHRAE Handbook—Systems and Equipment,* copyright © by the American Society of Heating, Refrigerating and Air-Conditioning Engineers, Inc., Atlanta, GA.

[a]Isolated combustion system: combustion air is drawn from outdoors, not indoors.

[b]Since 1992, all new furnaces are required to have an AFUE (ICS) level of at least 78.0%.

**EXAMPLE 8.10** A residence in Springfield, Illinois, has a total heat loss rate *UA* of 544 Btu/h °F and a balance point temperature of 55°F. It will have a natural gas condensing furnace, located indoors, for which *AFUE* = 0.93. For Springfield, DD55 = 3434. The approximate average annual energy used for space heating is

$$E_{therms} = \frac{(544 \text{ Btu/h °F})(3434 \text{ DD})(24 \text{ h/day})}{0.93 \times 100,000 \text{ Btu/therm}}$$

$$= 482 \text{ therms} \qquad \blacksquare$$

## 8.10 PASSIVE SOLAR HEATING PERFORMANCE

Section 8.4 presents design guidelines for determining solar opening size, thermal mass, and the solar savings fraction (SSF). As a building design takes shape, more detailed information becomes useful: Which passive system matches the architectural program? Which performs better thermally? For a given solar opening, what exactly is the resulting SSF? How much auxiliary fuel consumption per year must accompany that SSF? If a building overheats on sunny winter days, how hot will it get?

To this point, passive solar heating has been treated as a single approach; the design guidelines for SSF distinguished only between standard and superior system performance. Important architectural differences, however, characterize the various passive solar heating approaches, as summarized in Table 8.8. On the basis of the wider architectural implications presented in Table 8.8 and the detailed sizing information found in Appendix H, the designer can select a passive solar heating approach with some confidence in its applicability and its yearly need for auxiliary space heating.

### (a) Glazing Performance

In the reference systems of Table H.1, the glazing is assumed to face due south. The choices of glazing conditions include single, double, and triple glazing, and night insulation "no" or "yes." Single, double, and triple glazing with no night insulation are common approaches. Movable night insulation, once common (Fig. 8.25), has largely been supplanted by superwindows. Yet, what equivalent of the current superwindows might provide a

performance approximately equivalent to double glazing with R-9 night insulation?

The 24-hour averaged U-factor of the double-glazed fixed window with R-9 night insulation is found as follows: a double-glazed window, with a wood/vinyl frame, corresponds to window #5 in Table E.15, with an overall U = 0.49. "Night insulation" assumes adding R-9 insulation, in place from 5:30 P.M. to 7:30 A.M. This extra-insulated window should at night then have an overall R of at least

$$(1/0.49) + 9 = \text{R-11};$$

then nighttime U = 1/11 = 0.09.

$$U_{24av} = \left[\frac{(14 \text{ h})(0.09)] + [(10 \text{ h})(.49)]}{24 \text{ h}}\right] = 0.257$$

The equivalent superwindow would then have these characteristics, based on window #5's solar gain:

$$U = 0.26, SHGC = 0.58, VT = 0.57$$

Such a combination may be difficult to obtain. Here are some choices for superwindow substitutes for the "double/yes n.i." option within the Appendix H reference types:

- Where listed, choose the triple-glazing option instead of the double/yes n.i.
- Use window #7, Table E.15, with U = 0.33, SHGC = 0.55, VT = 0.52, with nearly identical solar gain characteristics, and interpolate between double/yes n.i. and double/no n.i. on the basis of U-factors:

  Double/no n.i., U = 0.49

  Double/yes n.i., U = 0.26

  Window #7, U = 0.33

This places window #7 at about one-third of the difference, closer to double/yes n.i.

- Use the clear triple-glazed window (#12, Table E.15) in place of double/yes n.i., and given its U = 0.34, SHGC = 0.52, VT = 0.53, interpolate for U-factor as with window #7.

Another option is *transparent insulation materials* (TIM), discussed earlier in Chapter 7, such as aerogel (one example). These materials transmit diffuse light and solar radiation but are opaque to thermal radiation. The four generic types of TIM are shown in Fig. 8.25. The absorber-parallel structures (Fig. 8.25a) are similar to the sheets of spectrally selective materials in superwindows; the more layers, however, the less the transmission of solar gain.

**THERMAL CONTROL**

### TABLE 8.8 Passive Solar Heating Systems Compared

| | Influence on Plan | Heating Characteristics |
|---|---|---|
| **Direct gain (DG)** 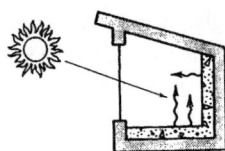 | Sun can enter through south windows or skylights; open plan can allow sun and stored heat to serve entire top floor of building. Large areas of thermal mass surface should be darker-colored and free of rugs, wall hangings, etc. Light-colored surfaces near glass reduce glare. Outdoor view and access to south are encouraged. | Quick to warm up in the morning; fast response to sun. Tendency to overheat at midday; large temperature swings. Much radiant loss to bare window by night; movable insulation encouraged (or triple glazing or selective film). Warmth spread throughout space along with thermal mass. |
| **Thermal storage wall** **Trombe (masonry)** **Wall (TW)** **Water wall (WW)**  | Needs to be on south wall of building. Inner wall of TW or WW should be kept clear of hangings, furniture, etc., but rest of space is unrestricted. Not much solar impact beyond about 25 ft from TW or WW. Outdoor view and access to south are discouraged. | Unvented TW, WW are slow to warm up by day, slow to cool by night; small temperature swings. Most radiant heat arrives in evening; comfort is most likely near the TW or WW surface. (Behavior of vented TW is midway between those of unvented TW and DG systems.) |
| **Sunspace (SS)**  | Same influences as TW and WW above; or SS can be insulated from building, becoming a less efficient heat source for a rock bed. Floor above rock bed should be kept free of rugs. SS becomes a special place with different characteristics from rest of building. Access to SS thereby encouraged. Access to south encouraged; view to south filtered. | SS thermally like DG, but with extreme temperature swings and accentuated radiant loss by night. Movable insulation often omitted and night use curtailed. Building beyond is thermally like TW or WW, depending on its connection with SS. Warm floor above rock bed in months with SS surplus heat. |
| **Roof pond** 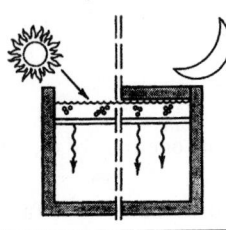 | Flat or nearly flat roof is desirable. Skylight is discouraged. Plan is completely unrestricted; sidelighting and views, access to outdoors encouraged | Low-temperature swings; steady temperatures in both summer and winter. Winter air stratification possible, since warmest surface is also highest surface in the space. |

**TABLE 8.8** *(Continued)*

| | Daylighting | Cooling |
|---|---|---|
| DG | Very high DF possible, with high glare potential at lower windows. Possible conflict between light-colored surfaces for glare reduction and dark-colored surfaces for solar absorption. Summer shading greatly reduces DF. Encourages both side- and toplighting. | Cross-ventilation: encouraged by large windows to south. Stack ventilation: helped when clerestories are used. Night ventilation/thermal mass: excellent potential, much mass surface and capacity. Other closed-building cooling: large south windows are big threat. Shading (and lower DF) necessary in overheating months. |
| TW, WW | No daylighting through TW unless interrupted by windows (high glare potential). Diffuse light possible through translucent WW. Discourages sidelighting. | Cross-ventilation: discouraged by solid TW, WW. Stack ventilation: TW or WW can itself produce a stack effect, but with risk of evening overheating. Night ventilation/thermal mass: limited interior mass surface exposure but a lot of mass capacity. Other closed-building cooling: with summer shading of TW or WW, good match with controlled cooling; thermal mass delays and reduces peak heat gains. |
| SS | Very high DF within SS; little or no daylight through common wall, except as encouraged by view and access to surface. Summer shading of SS reduces DF. | Cross-ventilation: only to extent that common wall is penetrated for view and access. Stack ventilation: SS can become a moderately effective stack. Night ventilation/thermal mass: both surfaces of common mass wall are available, but two spaces must be cooled. Other closed-building cooling: with summer shading of SS, good match with controlled cooling (except in SS itself). |
| Roof pond | Discourages toplighting, encourages sidelighting, with very light color on underside of roof. | Cross-ventilation: excellent potential. Stack ventilation: discouraged. Night ventilation/thermal mass: easily achieved with roof undersurface. Other closed-building cooling: roof pond night sky cooling is often sufficient by itself, making other cooling unnecessary. |

*Source:* Illustrations from AIA: Ramsey/Sleeper, *Architectural Graphic Standards* 7th ed., © 1981 John Wiley & Sons, Inc., New York. Reprinted by permission.

Absorber-vertical structures (Fig. 8.25*b*) include honeycomb or capillary materials that reflect and transmit the incoming solar radiation with fewer transmission losses. Combining these structures (Fig. 8.25*c*) can result in either transparent rectangular cross sections or transparent foam with bubble sizes of a few millimeters. Quasi-homogeneous layers (Fig. 8.25*g*) include aerogel. These scatter the incoming sunlight differently than do the types of structures shown in Fig. 8.25*c*.

For passive space heating, TIM use may be limited by the lack of *visual* transparency: these materials diffuse sunlight. They are thus appealing primarily for Trombe wall and water wall applications (Fig. 8.25*e*), where the view through the glazing is unimportant. Some information on U-factors and diffuse transmittance is presented in Table 8.9. Again, if we seek a substitute for movable insulation over double glazing, we would like U = 0.26, SHGC = 0.58, VT = 0.57; from Table 8.9, any of the honey-

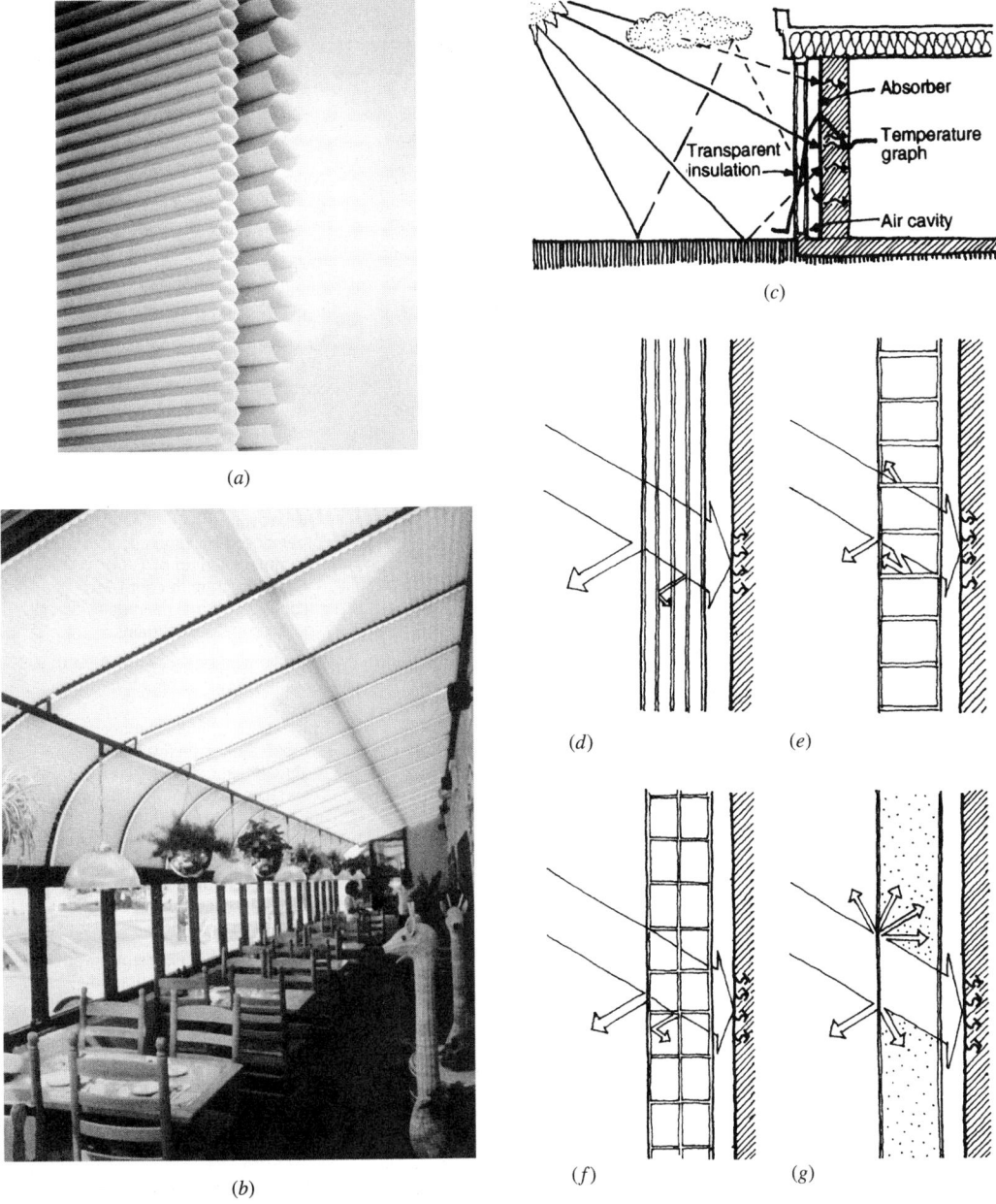

**Fig. 8.25** *Variations on insulation for solar apertures. Transparent insulation materials (TIMs) in most applications are more translucent than transparent. (a) The Duette window shade, when installed over double glazing, has winter R-values as follows: at ⅜ in. (9.5 mm) thick: translucent, R 3.23; ¾ in. (19 mm) thick: translucent R 3.57; opaque, R 4.2. (b) The shades can move either vertically or horizontally, cover both planar and curved surfaces, and be either motor or manually operated. (c) These translucent materials are particularly effective as insulation for Trombe walls, where the heated mass can be insulated from the cold winter air temperatures yet still receive strong sunlight. (Courtesy of the Fraunhofer Institute, Freiburg, Germany, and Solar Today, American Solar Energy Society.) (d) Absorber-parallel systems utilize multiple films, as in "superwindows." The more layers, the greater the optical losses. (e) Absorber-vertical structures include honeycomb or capillary materials. Some scattering and absorption of light occurs, but optical losses are lower than with type (d). (f) Cavity structures combine absorber-parallel and absorber-vertical structures. (g) Quasi-homogeneous layers are similar to cavity structures and include aerogel. (Courtesy of Hunter Douglas, Inc., Broomfield, CO.)*

**TABLE 8.9 Transparent Insulation Materials (TIM)**

| PART A. HONEYCOMB AND CAPILLARY STRUCTURES | | | |
|---|---|---|---|
| **TIM** | **τ Diffuse[b]** | **U-Factors[a,b] Btu/h ft² °F** | **W/m² K** |
| Honeycomb polycarbonate | | | |
| Thickness 1.97 in. (50 mm) | 0.85 | 0.35 | 2.0 |
| Thickness 3.94 in. (100 mm) | 0.78 | 0.19 | 1.07 |
| Capillaries, polycarbonate | | | |
| Thickness 3.94 in. (100 mm) | 0.73 | 0.17 | 0.98 |
| Capillaries, PMMA (acrylic glass) | | | |
| Thickness 3.94 in. (100 mm) | 0.80 | 0.16 | 0.91 |
| PART B. AEROGEL BETWEEN DOUBLE GLAZING | | | |
| **Aerogel Granule Diameter** | **τ Diffuse[b]** | **U-Factors[a] Btu/h ft² °F** | **W/m² K** |
| <0.079 in. (<2 mm) | 0.22 | 0.17 | 0.98 |
| 0.118–0.157 in. (3–4 mm) | 0.40 | 0.18 | 1.03 |
| 0.157–0.236 in. (4–6 mm) | 0.42 | 0.20 | 1.13 |
| 0.236–0.315 in. (6–8 mm) | 0.43 | 0.20 | 1.15 |

*Source:* Wittwer et al. (1991).

[a]Mean temperature 50°F (10°C), Δ*t* 18F° (10 K).

[b]Part *A* values are based on the TIM only, without the glass cover necessary in the actual application.

comb or capillary materials in Part A will be close if we remember to reduce the τ diffuse and slightly increase the U-factor to account for a single cover sheet of glass.

## (b) Direct-Gain (DG) Systems

These are the most commonly encountered systems because nearly all south-facing spaces have windows, and at least some have areas of internal thermal mass. In Table F.4, three of the four early solar examples utilize some DG collection. DG buildings include the Cottage Restaurant (Fig. 1.14) and the Oregon office in Fig. 8.7. The most common problems are overheating on sunny winter days and inadequate area of thermal mass.

For DG spaces, the thermal mass should be widely distributed around the room so that direct sun can strike the mass surface and/or be reflected to the mass surface as soon as possible on entering the window. Table F.2 gives design guidelines for mass; the DG reference systems (Appendix H) list

either a 3:1 or a 6:1 mass-to-glass area ratio. However, even more mass area will give a more thermally stable performance, as will become clear in Section 8.10*i*; a common recommendation for DG spaces above 50% SSF is for a thermal mass area five to seven times the area of glass (for no more than the optimum 4-in. masonry thickness). There are many common ways to provide such mass surfaces, including brick veneer and clay tile over a bed of grout. These surfaces can be applied to a frame construction. Floors (such as slab on grade) are easy ways to achieve large mass areas, but carpeting and rugs are very popular with occupants.

## (c) Sunspaces (SS)

These are the next most common systems, especially the prefabricated "add-on" type shown in the Kelbaugh house (Table F.4). Perhaps the most common problem is the expectation that an SS will be a greenhouse where exotic plants will thrive. But the SS is a "thermal servant" to the "thermal master" space behind it; comfortable temperatures in the master space are achieved at the cost of *very wide temperature swings* in the servant space. Consequently, while there are many times in a day when the SS is at a comfortable temperature, there are other times when it is not. In the First Village example (Table F.4 and Fig. 8.26), the SS contains the stairs. Although not always thermally comfortable, the passage between floors is at least rapid and visually pleasant. The "master" spaces beyond, on both floors, can choose the degree of interaction with the SS by opening doors and windows into it.

When the SS's common wall is masonry, the conductive flow of heat from servant to master is built in. Yet many of the SS systems in Appendix H show an insulated, lightweight frame common wall. It is important to remember that, with such insulated common walls, the SS is assumed to contain *a row of water containers* extending across the full east–west width of the SS. These theoretical containers are twice as high as they are wide and sit adjacent to (but not on) the floor and the common wall. They have a volume of 1 ft³ for every 1 ft² of common wall. This takes quite a bit of SS floor area.

Both masonry and insulated common wall SS systems assume available thermocirculation vents in the common wall; the top vents are 8 ft (2.4 m) above the bottom vents, and the top vents constitute

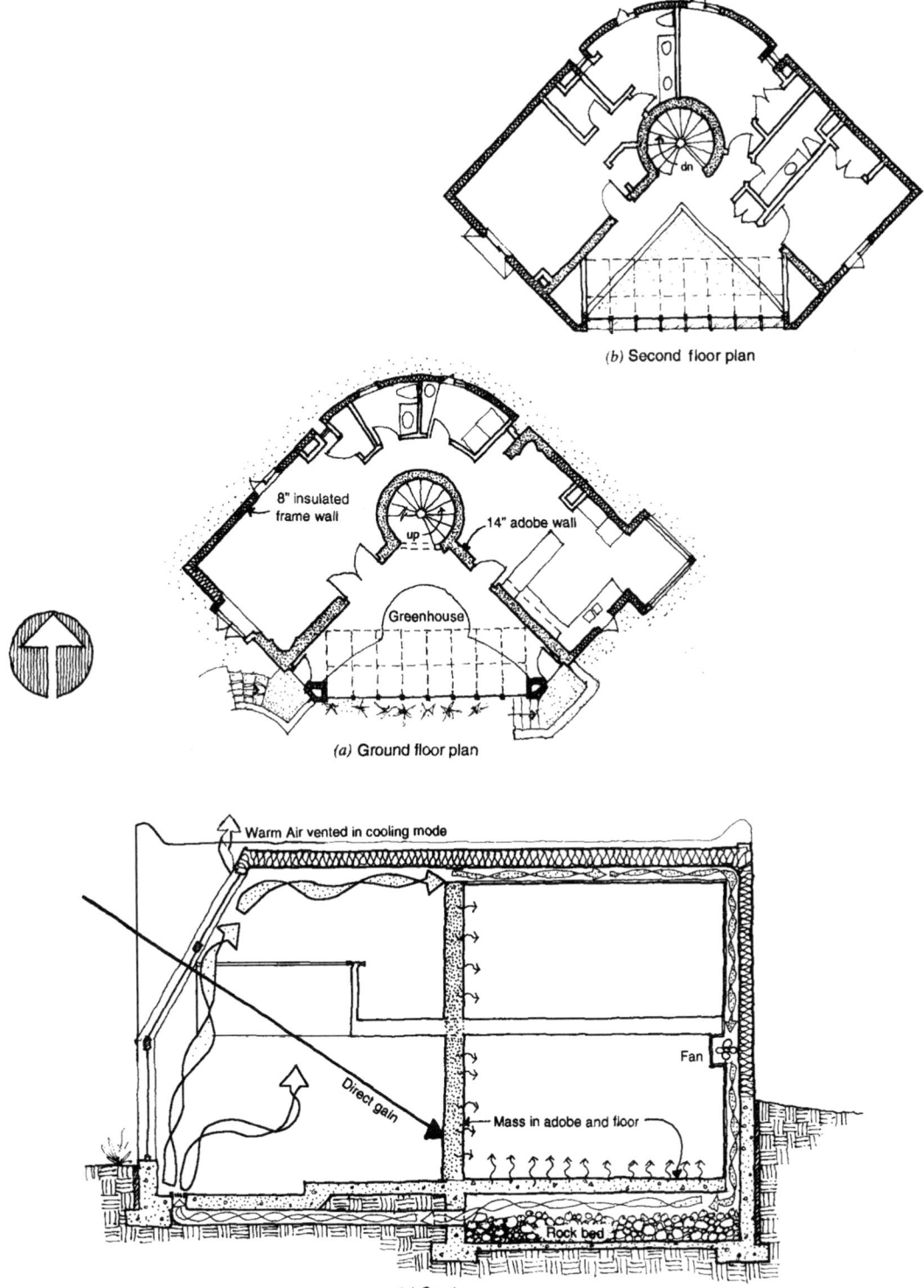

(b) Second floor plan

8" insulated frame wall

14" adobe wall

up

Greenhouse

(a) Ground floor plan

Warm Air vented in cooling mode

Direct gain

Mass in adobe and floor

Fan

Rock bed

(c) Section

**Fig. 8.26** *An early and well-known example of sunspace passive solar heating, the Unit 1, First Village residence borders the northeast and northwest sides of the enclosed sunspace. (a) Main floor plan. (b) Upper floor plan. (c) Section showing how the hottest air from the sunspace (midday to late afternoon) can be forced into a rockbed below the main floor for nighttime radiant heat. (Drawings based in part on Sandia National Laboratory, 1977.* Passive Solar Buildings: A Compilation of Data and Results. *Sand 77-1204.)*

3% of the area of the common wall, as do the bottom vents. All SS systems are assumed to have a thermally massive, perimeter-insulated floor slab on grade.

The SS system descriptions are quite explicit as to dimensions (Fig. H.1), but variations from these dimensions are not a problem *if the proportions of length-width-height are maintained.* It is the shape, rather than the dimensions, of each SS type that influences its performance.

## (d) Trombe Walls (TW)

These systems are considerably less common. When used, they often have rather large DG openings within the TW for daylight and view. This is the case with the Kelbaugh house (Table F.4), an unusual example that mixes three passive systems. The main advantage of TW systems is thermal stability; the diurnal temperature swings are less than with most other passive systems. They deliver a large portion of their heat by radiation to the space. The main disadvantages seem to be the loss of view and daylight, and keeping the air space clean between the Trombe mass wall and the glass. Objects that interfere with radiant heat transfer from the interior surface must also be minimized.

Two major choices are *vented* and *unvented.* The vented TW systems provide naturally moving air (via the stack effect between the mass and the glass) in addition to heat conducted through the mass wall. They deliver warmer air sooner than an identical-but-unvented mass wall, thanks to the flow of warmed air. They also introduce dust and dirt to the space between the mass and the glass.

The unvented TW systems deliver heat quite a bit later, with the result of quite low daily temperature variations and warmth arriving in the evening. They are somewhat less efficient than vented TW because the very high temperature at midday between the mass and glass results in more heat flow out through the glass.

The warehouse in Fig. 8.27 utilizes a very high vented TW in the New England winter. The detail shows a simple device to prevent reverse airflow at night. This Vermont warehouse utilizes ceiling fans to prevent stratification of warm air at the ceiling. Daylighting is provided by roof monitors; the TW is vented to the outside in summer.

## (e) Water Walls (WW)

These are the least common systems of all. Perhaps, as with roof ponds, this is another case of water phobia. They can be specially made, or made of corrugated galvanized steel culverts, steel drums, or fiberglass-reinforced plastic tubes (for which manufacturers provide suggested installation details). Within WW containers, some air space should be provided, as water expands when it heats. Either a rust inhibitor or a sacrificial anode should be added to the water within steel containers. Remember that WW systems are opaque, not transparent; the more transparent the tubes, the closer to DG will be their performance. This is discussed later in Section 8.10g.

Specially made WW can fit neatly below windows or anywhere else within exterior wall framing. The San Luis Solar Group Complex includes a house and an architectural studio (Fig. 8.28) near Santa Margarita, California. Both buildings utilize daylight (including clerestories and skylights), DG, PV, and domestic water panel collectors; there is also a small hydropower system adjacent to the complex. Of special interest here are the WW steel panels, 9 in. (230 mm) thick and painted on the outer surface with a black selective paint behind double glazing. The selective surface paint is an excellent absorber of short-wave solar radiation but a poor emitter of long-wave radiation from the heated black surface. The inside steel surface is painted and exposed to the space to be heated. These WW are used in the conference room of the studio, the dining area, and the upstairs bedroom. They are unobtrusive and correspond approximately to WWC2 in Appendix H.

## (f) Load Collector Ratio (LCR) Annual Performance

The following procedure is based on Balcomb et al., *Passive Solar Heating Analysis* (1984), published by ASHRAE and reprinted by permission. The reference offers a much wider variety of passive systems and a wider network of location listings than can be presented in this book. Along with more "sensitivity curves" to allow prediction for nonstandard passive systems, the reference also provides a much more time-consuming and detailed method for calculating the *monthly* SSF and auxiliary energy needs; the method presented here gives annual results only.

THERMAL CONTROL

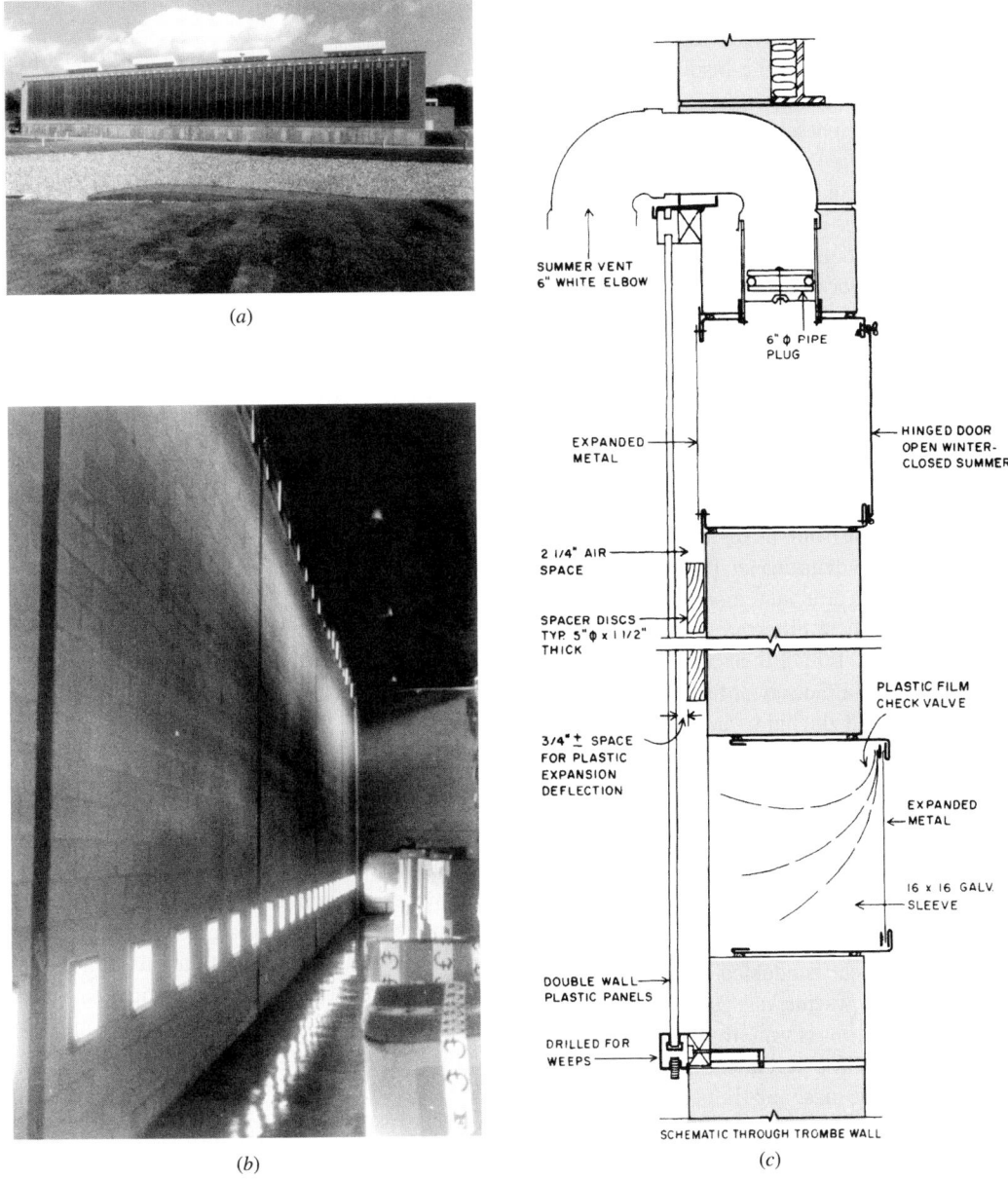

(a)

(b)

(c)

**Fig. 8.27** *The warehouse for the Famolare shoe company in Brattleboro, Vermont (a) uses a huge vented Trombe wall, 20 ft by 184 ft (6.1 m by 56.1 m). The Trombe wall faces 10° east of south. The pond in the foreground is an absorption basin for rainwater from the roof and pavement. (Photo by Robert Perron.) (b) Interior shows lower and upper vents. (c) Simple switches at the bottom prevent reverse thermosiphoning on cold nights; seasonal switches at the top provide building ventilation, via the Trombe wall, in summer. Spacer disks prevent plastic double glazing from deflecting too far inward when its inner surface overheats. (Reprinted from* Architectural Record, *November 1979, © 1979 by McGraw-Hill, Inc. All rights reserved. Reproduced with the permission of the publisher. Courtesy of Banwell White Arnold Hemberger & Partners, Architects, Hanover, NH.)*

(a)

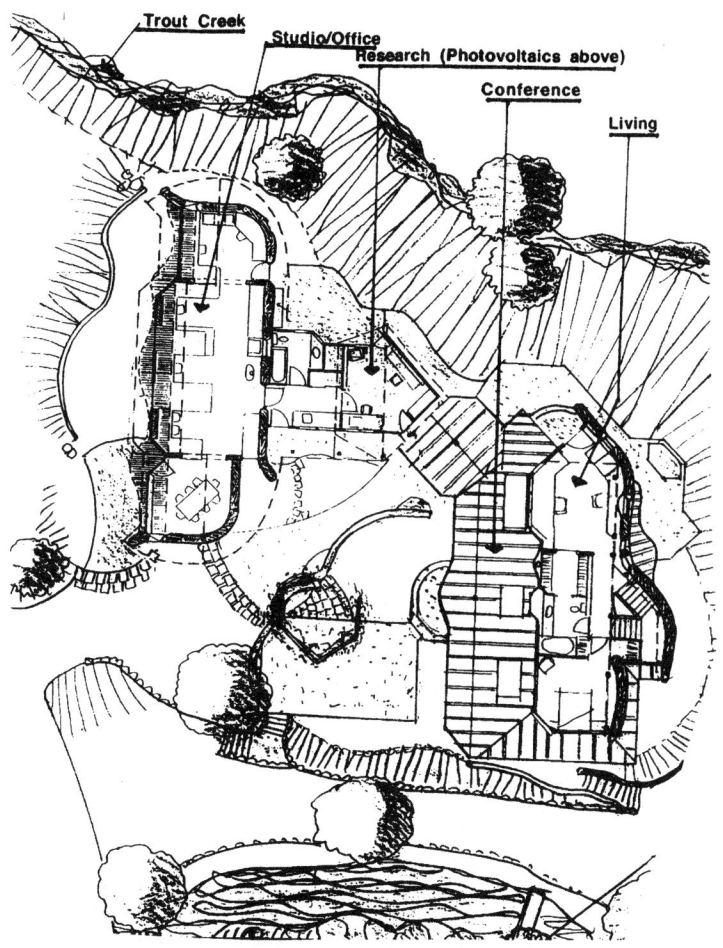

Trout Creek
Studio/Office
Research (Photovoltaics above)
Conference
Living

(b)

**Fig. 8.28** *The San Luis Solar Group Complex near Santa Margarita, California, includes a house and an architectural studio. PV and micro-hydroelectricity generation joins daylighting, passive solar heating, and passive cooling in this innovative installation. (a) South façade of the studio. (b) Site plan. (c) Section north–south. (d) Exterior of a specially built water wall below a window serving the dining room. (Drawings courtesy of San Luis Solar Group, Santa Margarita, CA.)*

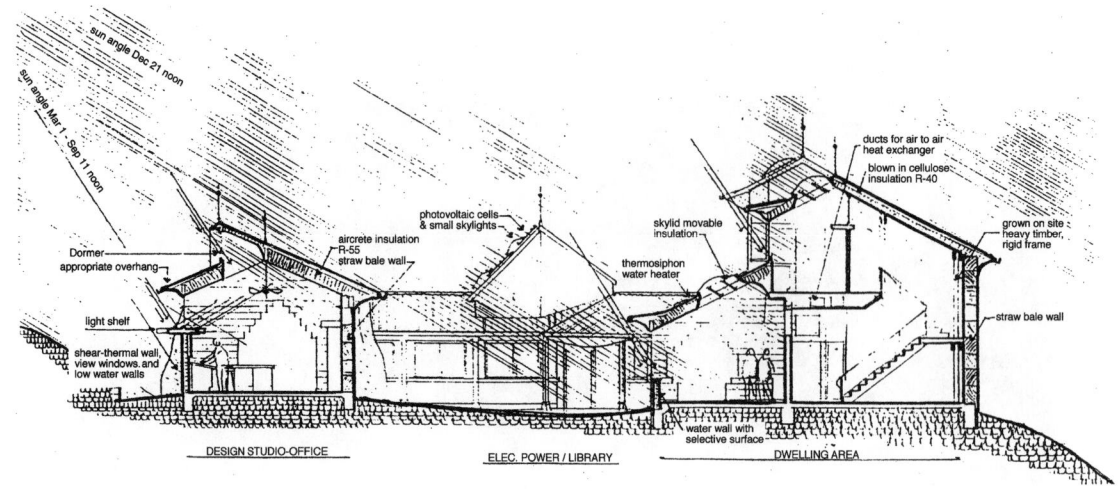

*(c)*

*(d)*

**Fig. 8.28** *(Continued)*

Thus, Balcomb et al. (1984) is an important and perhaps indispensable reference for the serious passive solar designer.

The method presented here, called the *load collector ratio* (LCR), yields the annual SSF and auxiliary energy needs for a building. This method has the following steps:

**STEP 1.** Choose the location and the reference passive system that most closely coincide with your building and its site. The locations listed in Appendices C and H are shown in Fig. C.1; the reference passive systems for which performance data are available (Appendix H) are summarized in Table H.1. If your system differs significantly from the closest reference system, see Section 8.10g.

**STEP 2.** Tentatively select a size for the solar openings, balancing the design guidelines for SSF (Table F.1) with those for daylighting (Table 14.5) and, if applicable, for ventilation (Fig. 8.14).

**STEP 3.** Calculate the "non-south" envelope heat loss rate, $UA_{ns}$, for the building design—one that *excludes* the solar openings but *includes* all other envelope losses, as well as the infiltration loss. Then multiply $UA_{ns}$ by 24 h to obtain Btu/DD; this is called the *building load coefficient* (BLC).

$$BLC = 24 \times UA_{ns}$$

**STEP 4.** Check your building's overall loss rate against the criteria from Table 8.3:

$$Btu/DD\ ft^2 = \frac{BLC}{floor\ area\ (ft^2)}$$

Does your building envelope conserve energy sufficiently, or do you need more insulation (or less nonsouth glass or less infiltration)?

**STEP 5.** Determine the *vertical projection* of the solar opening area $A_P$. (For a vertical solar opening, $AP$ is identical to the actual opening area; for a 45°-inclined solar opening, $A_P = 0.707$ actual area.)

**STEP 6.** Find the LCR, expressed in Btu/DD ft²:

$$LCR = \frac{BLC}{A_P}$$

**STEP 7.** For the reference system that most closely approaches your design, consult Table H.3 at the appropriate location. By interpolation, find the annual SSF that corresponds to your passive sys-

tem's listed LCR. Note also the annual DD65 listed in this table.

**STEP 8.** Finally, determine the approximate annual auxiliary heating Q required:

$$Q = (1 - SSF) \times BLC \times DD$$

Although this quick annual-results method is based on the DD65 listed in Table H.3, it is possible to adjust $Q$ to *approximately* account for higher internal gains or better-insulated envelopes. In this adjustment, use DD based on the balance point instead of DD65 in the previous equation.

**STEP 9.** Now compare the design guideline–predicted relationship between collector size and SSF to the actual one you have just calculated. If the SSF is *smaller* than you had hoped, can you decrease BLC (improve conservation) or increase collector size, or switch to another passive system with a more favorable SSF for the same LCR? If the SSF is *larger*, will you be happy with the increased fuel savings, or will you reduce the collector size or consider another, less efficient passive system that has some architectural advantage over the one for which you calculated SSF? Table F.4 shows approximate LCRs for four historic solar buildings.

---

**EXAMPLE 8.1, PART H** What is the annual SSF for the Oregon office building in Fig. 8.7?

**SOLUTION**

The reference passive system most closely corresponding to the DG in a typical bay is found in Table H.1, Part D: DGC1 has a high thermal storage capacity, based on an average thermal mass thickness of 4 in., and a ratio of mass to south glass area of 6:1. It has standard performance double-glazed windows with no night insulation. (From Table 8.2, the actual building's mass to south glass area ratio is 2232 ft²/229 ft² = almost 10:1.) The BLC of the typical bay is

24 h/day × (42 + 25 + 72 + 174 infiltration) Btu/h °F
= 7512 Btu/DD

The LCR of the typical bay is therefore

$$\frac{7512\ BLC}{229\ ft^2\ collector\ area} = 32.8$$

From Table H.3, for Salem, Oregon, with LCR = 33, system DGC1, SSF = about 27%, compared to the design guideline (Example 8.1, Part C) of SSF = 25%.

Note that if superior performance windows had been used throughout the typical bay, the BLC would be reduced. The north window *UA* would become 97 ft² × 0.30 Btu/h ft² °F = 29. Thus BLC =

24 h/day × (42 + 25 + 29 + 174 infiltration) Btu/h °F
= 6480 Btu/DD

The passive system now closest to the actual building is DGC2, with triple-glazed, thus lower U-factor openings. The new LCR is

$$\frac{6480 \text{ BLC}}{229 \text{ ft}^2 \text{ collector}} = 28.3$$

Therefore, with DGC23 in Salem, Oregon, with LCR of 28.3, find SSF as follows:

At LCR 25, SSF = 41; at LCR 30, SSF = 37 (from Table H.3)

Salem SSF range 41 − 37 = Δ4

Salem LCR range 30 − 25 = Δ5

(Actual LCR) − (Salem minimum LCR) = 28.3 − 25
= 3.3; therefore

= 3.3 (Δ4/Δ5) = 2.6, to be added to the minimum SSF

Actual SSF = 37 + 2.6% = 39.6%

This is a substantial improvement that increases the energy saved by about 50%. ∎

**EXAMPLE 8.11** We now take a more detailed look at the solar collecting area required for the passively solar heated building in Omaha, Nebraska, discussed in Example 8.2.

**SOLUTION**

From the design guidelines used in Example 8.2, we expected that 20% of the floor area in south glass with superior performance would yield a 51% SSF. More detailed characteristics of this building are shown in Fig. 8.29. From a program requirement of 2900 ft² of floor area, the 20% south glass area equals 0.20 × 2900 = 580 ft². For daylighting by sidelight only, $DF_{av}$ = 0.2 (window area/floor area); if all of the south glass area is available for daylighting (as with DG systems), $DF_{av}$ = 0.2 × 20%, or 4%. This is adequate for office work. However, because you want to avoid dark areas near the rear walls of these spaces and investigate Trombe walls, and so on, some north light is desirable. Choose about 3% of the floor area in north glass (say, 90 ft² glass); added $DF_{av}$ = 0.2 × 3% = 0.6% additional.

(Wall and roof insulation and overall percentage wall area in fenestration should be checked against local codes and the current ASHRAE Standard 90.1. The fenestration area:

south glass 580 ft² + north glass 90 ft² = 670 ft²

Total wall area:

north 1100 ft² + south 1100 ft² + east 600 ft²
+ west 600 ft² = 3400 ft²

Fenestration/wall area ratio = 670 ft²/3400 ft²
= 20%

Coincidentally, this is also the percentage floor area in south glazing.)

We now assume that the following maximum U-factors meet codes and standards and that ventilation at the rate of 20 cfm per person is provided. The building will house 2900 ft²/180 ft² per person = 16 people. Outdoor airflow (in place of infiltration) = 20 cfm per person × 16 people = 320 cfm.

**STEP 1.** Given the emphasis on daylight and the original assumption of night ventilation with the windows, choose system DGC2. This has triple-glazed windows and will require six times as much thermally massive surface area (minimum 4-in. thick) as a south glass area:

6 × 580 ft² south glass
= 3480 ft² exposed thermal mass, minimum

If all interior surfaces of exterior walls are exposed concrete block, we obtain

3400 ft² wall total − 670 ft² window
= 2730 ft² interior mass area

which is not sufficient. So, add at least 750 ft² exposed concrete slab floor area, such as a 10-ft-wide strip just inside the south windows, for the entire length of the building.

**STEP 2.** The solar opening was selected at 20% floor area, or 580 ft².

**STEP 3.** Calculate $UA_{ns}$.

North, east, and west opaque (mass) walls:
2210 ft² × 0.08 = 177

South opaque (mass) wall: 520 ft² × 0.08 = 42

Roof (insulation above deck): 1974 ft² × 0.048 = 95

Slab perimeter: 200 lin ft × 0.54* Btu/h °F lin ft
= 108

North glass: (triple glazed, U = 0.33): 90 ft² × 0.33
= 30

Ventilation: 320 cfm × 1.1 Btu min/h ft³ °F = 346

Total 798 Btu/h °F

---

*From Table E.11, assuming insulated construction type (a), with R = 5.4 insulation, and interpolating for 6601 DD65.

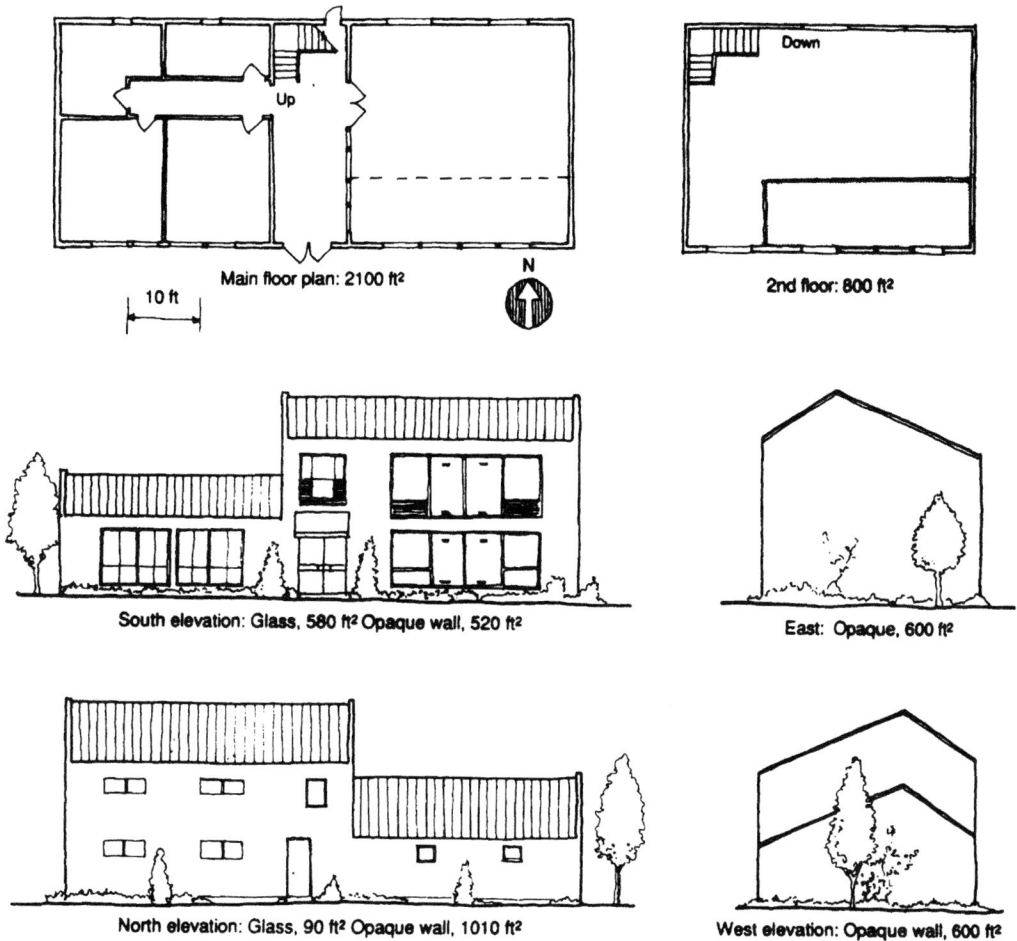

**Fig. 8.29** *Plans and elevations of the Omaha, Nebraska, office building discussed in Example 8.11.*

Then

$$BLC = 798 \text{ Btu/h } °F \times 24 \text{ h/DD}$$
$$= 19{,}152 \text{ Btu/DD}$$

**STEP 4.** Calculate overall heat loss rate:

$$\frac{19{,}152 \text{ Btu/DD}}{2900 \text{ ft}^2 \text{ floor}} = 6.6 \text{ Btu/DD ft}^2$$

This is considerably in excess of the Table 8.3 criterion of 4.6 Btu/DD ft² for this 6601 DD65 climate. Unless further conservation steps are taken, the annual SSF is likely to be much lower than was predicted by the design guideline. Promising targets: more insulation on the wall, slab edge, and roof; a heat exchanger for that 320 cfm of outdoor air.

**STEP 5.** Determine the vertical projection of the solar opening. The south windows are vertical; therefore, $A_p$ is identical to the actual area, 580 ft².

**STEP 6.** Find the LCR:

$$\frac{19{,}152 \text{ Btu/DD}}{580 \text{ ft}^2} = 33.0$$

**STEP 7.** Enter Table H.3 for Omaha, Nebraska, LCR = 33, and DGC2. Interpolate between LCR 30 and 40. The annual SSF is found to be about 29.2%.

**STEP 8.** The approximate annual auxiliary heating required is

$$Q = (1 - 0.292) \times 19{,}152 \times 6601 \text{ DD}$$
$$= \text{about } 89{,}500{,}000 \text{ Btu}$$

**STEP 9.** Compare the actual SSF with the predicted value: SSF = 29.2% is only about three-fifths of the 51% SSF predicted by the guideline due to inadequate conservation measures, as explained in Step 4. Obviously, a lower LCR yields a higher SSF; a larger collector area would lower the LCR. However,

Step 4 indicated that lower BLC, not higher $A_p$, is the better way to achieve a lower LCR in this case.

Table H.3 can be used to compare the relative efficiencies of many passive heating systems. Using the column of LCR = 30, for instance, the best-performing systems are:

*Omaha, Nebraska, LCR = 30*

| | |
|---|---|
| WWB4 | 47% SSF |
| WWC2 | 45% |
| SSE2 | 44% |
| DGC3 | 42% (movable insulation) |
| TWD4 | 42% |
| TWE2 | 42% |

If any of these systems are more architecturally compatible with the building program and/or design intent, they could be used instead of DGC2. In this case, DGC2 gives substantial daylighting with little impact on the exterior appearance; the large area of thermal mass necessary was relatively easily obtained. DGC2 appears to be a reasonable choice for this building in this climate if the client prefers not to use movable insulation over windows. ∎

What if two or more passive systems are used in the same building? In that case, first do calculations for the entire building, assuming that *one* system has *all* the solar area, and find the SSF. Next, do the calculations assuming that the second system has all the solar area, and find its SSF. Then average the SSF values according to the relative solar areas of the two systems.

---

**EXAMPLE 8.12** A veterinary clinic in Buffalo, New York, is using two systems: WW for examining rooms and DG for the waiting/reception area. The building's characteristics are

Balance point = 50°F

$UA_{ns}$ = 356 Btu/h °F

$A_p$ WW = 240 ft² (reference system WWB4)

$A_p$ DG = 150 ft² (reference system DGA3)

Total floor area = 1900 ft²

Predicted SSF from design guideline = about 36%, superior performance

Checking the overall heat loss criteria (Table 8.3),

$$\frac{356 \times 24}{1900 \text{ ft}^2} = 4.5 \text{ Btu/DD ft}^2$$

This is less than 4.6 (for 5000 to 7000 DD), so it is acceptable.

**SOLUTION**

First, calculate as though all $A_p$ (390 ft²) were system WWB4:

$$BLC = 24 \times 356 = 8544 \text{ Btu/DD}$$

$$LCR = \frac{8544}{390} = 22$$

For Buffalo, LCR 20 yields SSF = 0.37, and LCR 25 yields SSF = 0.31, so SSF = 0.35 by interpolation.

Next, calculate as though all $A_p$ were system DGA3. For LCR 22, Buffalo SSF = 0.28 by interpolation.

Now calculate the average SSF, given that the WW comprises 240/390 = 62% of the total solar opening and DG comprises 150/390 = 38%:

$$\frac{0.35(62\%) + 0.28(38\%)}{100\%} = 0.217 + 0.106$$

$$= 0.323; \text{ SSF} = 0.32$$

*Result:* The SSF is about 90% of what was predicted (0.32 as opposed to 0.36); because the glass area is already large, program requirements probably prevent further increases. The building already seems to conserve energy well, because it meets the Table 8.3 heat loss criteria.

The approximate annual auxiliary heating energy required is

$$Q = (1 - 0.323) \times 8544 \text{ Btu/DD}$$
$$\times 3322 \text{ DD50 (balance point)}$$
$$= 19 \text{ million Btu}$$

(This is equivalent to about 5600 kWh, or about 220 therms of natural gas burned at 85% efficiency.) ∎

## (g) Variations on Reference Systems

A particularly wide set of choices faces the designer of DG and SS systems, in which mass distribution and glass orientation can assume thousands of different combinations. The *sensitivity curves* furnished in Balcomb et al. (1984) give some guidance on how a predicted SSF might vary as an actual passive system departs from a reference system.

Sensitivity curves can serve as early general design guidelines. Looking at the curves for your location, which design changes yield dramatic results and which make little difference? The curves may also be used to adjust the SSF found for a reference design.

One passive solar example that departs radically from any reference case is the Class of 1959 Chapel at the Harvard Business School in Cambridge, Massachusetts (Fig. 8.30). The glazed "sunspace" resembles system type SSA in that it is attached to (rather than set into) the building behind it. However, this is manifestly *not* the simple rectangular SS plan, and the solar aperture is divided into southeast- and southwest-facing halves.

*(a)*

The aperture $A_p$ is doubly complicated: the vertical dimension is the vertical projection of the sloping wall, whereas the horizontal dimension is the due-south projection of the plan. From Table H.1, because of the glazed end walls and masonry common wall, it comes closest to either SSA3 (standard performance) or SSA4 (superior performance). This depends on the glazing chosen—in this case, clear double glazing with low-ε, air space approximately ½-in. (13-mm) thick. In Cambridge (Boston), for a small LCR (a huge solar aperture relative to the building behind) at 3320 DD65, the Balcomb reference shows:

$$\text{If LCR} \quad = \quad 25 \quad 20 \quad 15$$

$$\text{If SSA3, then SSF} \quad = \quad 37 \quad 41 \quad 47$$

$$\text{If SSA4, then SSF} \quad = \quad 55 \quad 61 \quad 69$$

Given the complex geometry of this SS, the safer assumption is standard performance.

Water walls are rather uncommon in buildings, despite their relatively high performance throughout Table H.3. These systems are based on opaque containers, but some designers are attracted by the idea of daylight filtering through translucent water containers. Algae growth in transparent plastic tubes, encouraged by the daylight that passes through them, is usually controlled by the addition of an algaecide to the water. Dyes may be added to change the color of daylight seen through tubes or water-filled glass block, as in Fig. 8.31. The "WW" of this Long Island library is actually a hybrid between

*(b)*

**Fig. 8.30** *The Class of 1959 Chapel at the Harvard Business School is entered (a) through a sunspace with a sunken garden. (b) South-facing glazing of the sunspace. (c) Plan; south is about 30° east of the point of the sunspace. (d) Section. This is not a typical sunspace type from Appendix B. (Drawings courtesy of Moshe Safdi and Associates, Inc., Somerville, MA.)*

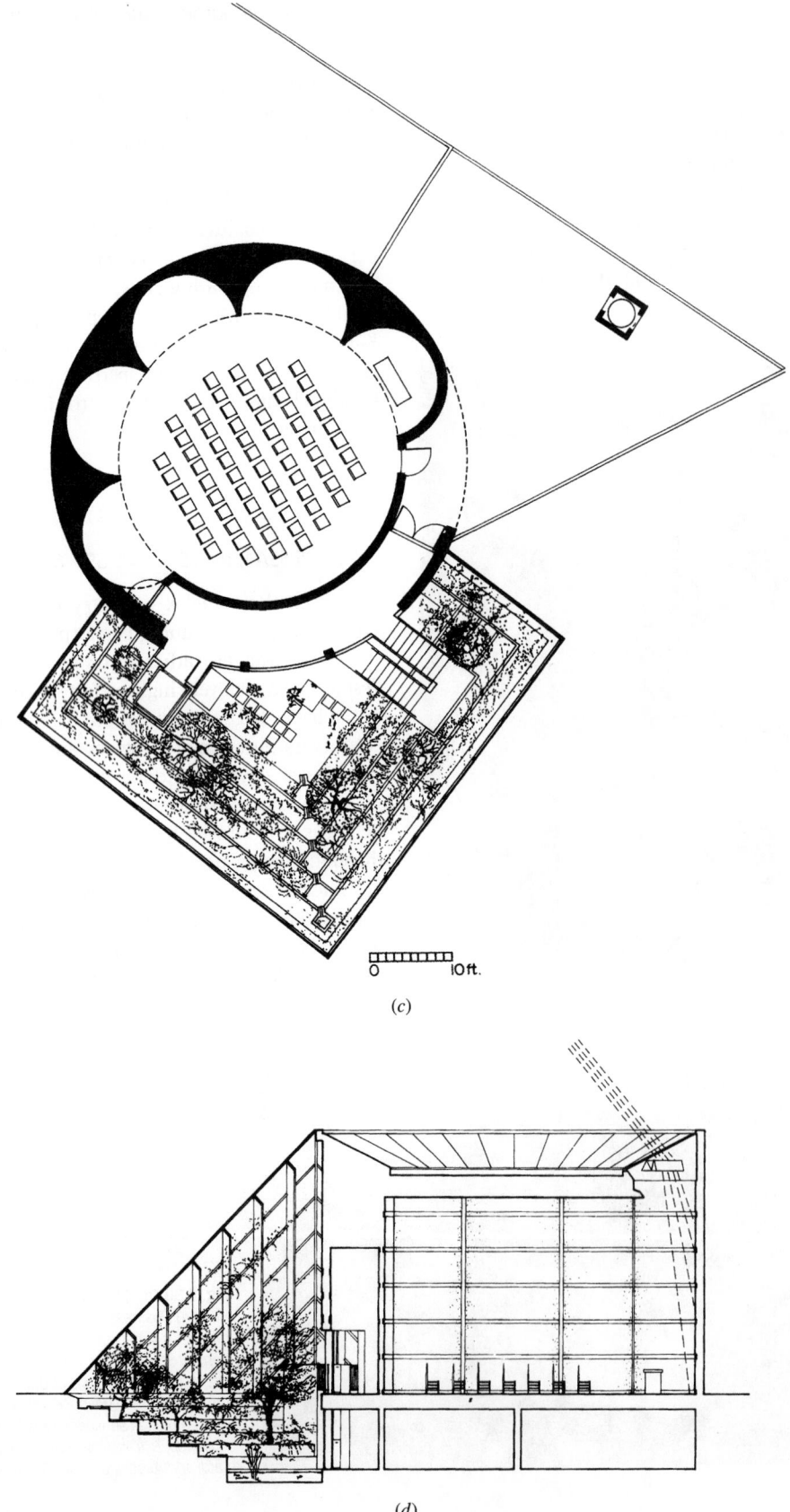

0       10ft.

(c)

(d)

      **Fig. 8.30** (Continued)

*(a)*

*(b)*

**Fig. 8.31** *The Center Moriches Free Public Library (Long Island, New York) utilized water-filled glass block as a heat storage and daylight-diffusing south wall (a). Because the water storage is translucent, it functions somewhere between a DG and a WW system. Dyes add color to selected blocks, forming the pattern shown (b). Lightshelves and stack ventilators add to its energy-conserving performance. (Courtesy of Banwell White Arnold Hemberger & Partners, Architects, Hanover, NH.)*

**TABLE 8.10 Time Lag Through Homogeneous[a] Walls**

| Material | Thickness (in.) | U-Factor[b] (Btu/h ft²) | Time Lag (h) |
|---|---|---|---|
| Stone | 8 | 0.67 | 5.5 |
| | 12 | 0.55 | 8.0 |
| | 16 | 0.47 | 10.5 |
| | 24 | 0.36 | 15.5 |
| Solid concrete | 2 | 0.98 | 1.1 |
| | 4 | 0.84 | 2.5 |
| | 6 | 0.74 | 3.8 |
| | 8 | 0.66 | 5.1 |
| | 12 | 0.54 | 7.8 |
| | 16 | 0.46 | 10.2 |
| Common brick | 4 | 0.60 | 2.3 |
| | 8 | 0.41 | 5.5 |
| | 12 | 0.31 | 8.5 |
| | 16 | 0.25 | 12.0 |
| Face brick | 4 | 0.77 | 2.4 |
| Wood | ½ | 0.68 | 0.17 |
| | 1 | 0.48 | 0.45 |
| | 2 | 0.30 | 1.3 |
| Insulating board | ½ | 0.42 | 0.08 |
| | 1 | 0.26 | 0.23 |
| | 2 | 0.14 | 0.77 |
| | 4 | 0.08 | 2.7 |
| | 6 | 0.05 | 5.0 |

*Source:* Victor Olgyay, *Design with Climate: Bioclimatic Approach to Architectural Regionalism,* Copyright © 1963 by Princeton University Press. Reprinted by permission.

[a]For composite constructions, add an estimated additional time lag to the sum of the individual materials' time lags as follows:

Two-layer, light construction: ½ hour more
Three or more layers: 1 hour more
Very heavy construction: 1 hour more

[b]The U-factor is based on outdoor surface conductance of 4.0 and an indoor surface conductance of 1.65 Btu/h ft² °F.

DG (light passes through the glass to strike massive surfaces beyond) and WW (all light is converted to heat within the opaque water containers, then passed to the space beyond). To assess the performance of such a system, interpolate between the entire aperture considered as the relevant DG system and same entire aperture considered as the relevant WW system. The higher the visible transmittance, the closer to DG performance will be the result.

## (h) Thermal Lag Through Mass Walls

The time necessary for solar heat to pass through various thermally massive materials is shown in Table 8.10. This time lag can be put to use when the time of maximum solar heat is different from the time of maximum internal heat need. A typical

example is for a residence's living room in winter, in which the late evening sedentary entertainment hours occur with cold temperatures outside, yet maximum warmth is desired inside. A Trombe wall, for instance, made of solid grouted concrete block and 12 in.-thick, will delay the arrival of maximum solar gain to the interior by almost 8 hours. If maximum solar heat gain occurs at about 1 P.M. (maximum sun at noon but highest temperatures at about 2 P.M.), then such a TW would deliver the maximum heat to the inner surface at about 9 P.M.

## (i) Internal Temperatures

Two quantities are of particular interest to passive solar designers: How much higher, compared to the outdoor temperature, will the average indoor temperature be from solar heating alone? Also, how widely will this internal temperature vary (swing) on a clear winter day?

THERMAL CONTROL

The approximate temperature difference between inside and outside on a *clear* January day, called Δt *solar,* can be estimated from Fig. 8.32; it varies with latitude and with the LCR. SS are not shown in the figure. Although the temperature within an SS cannot be easily approximated, the Δt solar for the room beyond the SS can be approximated by using Fig. 8.32a, if these spaces have an insulated common wall, or Fig. 8.30b if they have a masonry common wall.

To determine the average winter indoor temperature:

1. Find the average January ambient (outdoor) temperature, *TA,* from Appendix B.
2. Find Δt solar for the building and its site (Fig. 8.32).
3. Find the Δt due to internal heat sources:

$$\Delta t \text{ internal} = \frac{\text{total internal gains (Btu/day)}}{BLC + (UA_s \times 24)}$$

where $UA_s$ is for the solar area only. *Note:* Δt internal averages 5 to 7F° for residences.

4. Add the quantities from the first three steps to find the average January clear-day indoor temperature.

When internal gains are high, there is less need for Δt solar; the building is mostly "heating itself" (becoming an ILD rather than an SLD building). If the average indoor temperature is too high, smaller solar openings should be considered unless the climate is predominantly cloudy (clear days rare) in November through January.

The other important comfort question concerns the size of the *temperature swing* due to passive solar heating. Controlled by the sun and by the actions of users rather than by a thermostat, passive solar buildings typically experience larger daily variations (swings) in indoor temperature than do conventional buildings, especially on clear days. To estimate your building's clear-day January temperature swing, or Δt *swing,* see Table 8.11. The average indoor temperature determined previously will fall in the middle of this Δt swing.

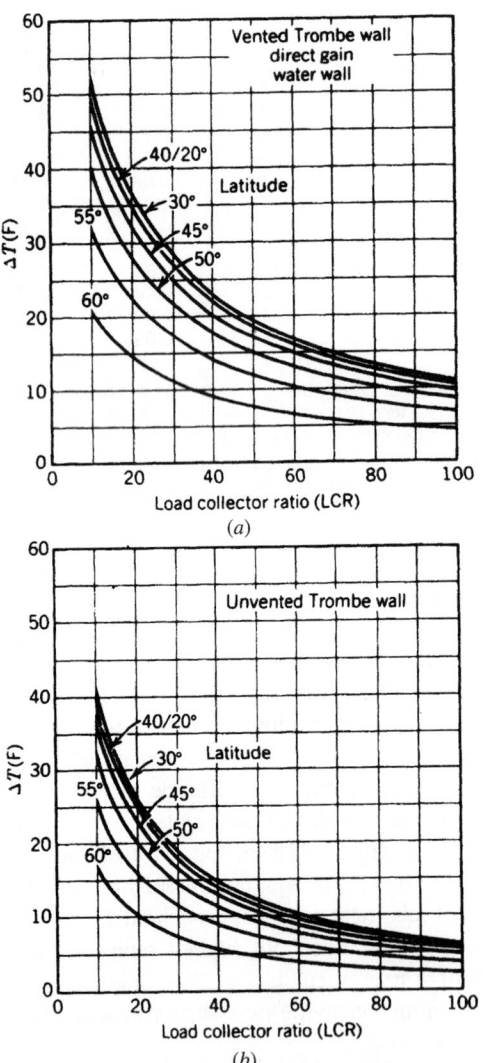

**Fig. 8.32** *Graphs of Δt solar, the temperature difference to be expected between the average inside temperature and the average outside temperature on a clear January day. The curve marked "40°/20°" applies to both 40° latitude and 20° latitude. (a) Δt solar for direct gain, water wall, vented Trombe wall, or sunspace (insulated common wall) systems. (b) Δt solar for unvented Trombe wall or sunspace (masonry common wall) systems. (From Balcomb et al., 1980.)*

**TABLE 8.11 Indoor Temperature Swing, Δt Swing**

| Passive Solar System | Δt Swing[a] |
|---|---|
| DG: $\dfrac{\text{mass area}}{\text{glass area}} = 1.5$ | 1.11 × Δt solar |
| = 3 | 0.74 × Δt solar |
| = 9 | 0.37 × Δt solar |
| WW | 0.39 × Δt solar |
| TW, vented for 3% of wall area | 0.65 × Δt solar |
| TW, unvented | 0.13 × Δt solar |

*Source:* Balcomb et al. (1980).

[a]These swings are based on a thermal storage mass capacity of 45 Btu/ft² °F; Δt solar can be found in Fig. 8.32.

THERMAL CONTROL

**EXAMPLE 8.1, PART I** Find the clear January day temperature swing in the typical bay of the office building in Fig. 8.7.

**SOLUTION**

1. (Appendix C) Salem's average January outdoor temperature = 39°F.
2. (Fig. 8.32a) At 44° N latitude, using a DG system, with LCR = 33, $\Delta t$ solar = 24F°.
3. $\Delta t$ internal = (from Example 8.1, Part G, winter, and Part H)

$$\frac{7.9 \text{ Btu/h ft}^2 \times 1440 \text{ ft}^2 \times 9 \text{ h/day}}{7512 \text{ Btu/DD} + (229 \text{ ft}^2 \times 0.74 \text{ Btu/h ft}^2 \text{ °F} \times 24 \text{ h/day})}$$

$$= \frac{102,384}{11,579} = 8.8\text{F°}$$

4. Average January clear-day temperature indoors is 39°F + 24F° + 8.8F° = 72.8°F (within the comfort zone).
5. (Table 8.11) Indoor temperature swing, for DG with a 9:1 ratio of mass to south glass area = 0.37 × $\Delta t$ solar:

$$0.37 \times 24\text{F°} = 8.9\text{F°} \ \Delta t \text{ swing}$$

Therefore, the clear January day interior will vary by 8.9°/2°F on either side of the average temperature:

$$72.8 + 8.9/2 = 77.3°F \text{ high}$$
$$72.8 - 8.9/2 = 68.4°F \text{ low}$$

This appears to be a quite comfortable range. Because the Pacific Northwest is largely overcast in winter, this daily range typically will be less wide, but the $\Delta t$ solar will also be smaller. Thus, the average January day temperature indoors will be lower. ■

## 8.11 APPROXIMATE METHOD FOR CALCULATING HEAT GAIN (COOLING LOAD)

Unlike the calculations for winter worst-hour heat loss, which simply assume nighttime conditions and few if any internal gains, summer worst-hour heat gain calculations are very complex. The difference in air temperature between inside and outside, which was so influential in winter, is much less important in summer. Solar gains and internal gains from lights, people, and equipment *must* be included. Summer calculations are complicated further by the

fact that the hourly change in summer load can be very great, both from changing sun position and from changing internal loads. Also, in summer the thermal mass of the building becomes influential, delaying the impact of the radiant component of heat gains from all sources. The design guideline approach represented in Table F.3 cannot respond to hourly changes or to thermally massive construction. Furthermore, Table F.3 deals only with sensible heat gains and ignores latent heat gains.

*Sensible heat* is the kind of heat that *increases the DB temperature of air*; the glowing coil of an electric range adds sensible heat to the air of the kitchen.

*Latent heat* is the kind of heat that is present within *increased moisture* in air; boiling water in a pan on top of an electric range is evaporating, and as it evaporates it increases the latent heat in the kitchen air. This latent heat will increase the WB temperature, but not the DB temperature, of the air. In conventionally cooled buildings, the mechanical cooling equipment must be sized so that latent as well as sensible heat will be removed.

A simplified heat gain procedure has been developed for residential buildings; with some risk and some judgment, it can also be used for a quick *approximation* of the conditions in commercial buildings. This simplified method was devised to be used with buildings that, like residences

1. Are occupied and air-conditioned (internal temperatures closely controlled) for 24 hours per day. (Separate calculations can be done for weekdays and weekends.)
2. Derive much of their gains through the building envelope and ventilation rather than internally.
3. Are tolerant of undersized cooling equipment, with the result that unusually hot weather means noticeably higher indoor temperatures (and that interior temperatures will vary during a typical summer day).

Because many commercial buildings do *not* have these characteristics, this method should not be applied if very accurate estimates are desired for such buildings. However, the method is rapid, if risky.

### (a) Gains Through Roof and Walls

The sensible heat gains through opaque parts of a building's envelope are calculated with the equation:

$$q = U \times A \times \text{DETD}$$

where

*U*-factors are for summer

*A* = area of the roof or wall

DETD (design equivalent temperature differences); values are listed for broad categories of construction in Table F.5

The DETD values are based on an average indoor temperature of 75°F (23.8°C) and the outdoor conditions listed. Note that for lightweight wall construction (and doors) the DETD varies by orientation. A means of correcting DETD for other temperatures is as follows.

Where the design temperature difference (outdoor design temperature minus assumed indoor temperature of 75°F (23.8°C) is not an even increment of 5F° (2.8C°), the equivalent temperature difference should be corrected 1F° (0.5C°) for each 1F° (0.5C°) difference from the tabulated values.

For rapid approximation, however, select the DETD directly from the table for the conditions nearest to your building/climate combination. The design temperature for your climate is listed in Appendix B.

Note that Table F.5 lists considerably higher DETD values for roofs than for any other component. This is due to the *sol-air temperature*, the effective temperature of outdoor air just above a solar-heated surface. When the sun strikes the surface, it adds its heat to that from the outdoor air. The darker the surface, the greater the resulting effective (sol-air) temperature. This elevated outdoor temperature thus increases the Δ*t* through the roof, resulting in greater heat gains. Sol-air temperature is discussed in the 2005 *ASHRAE Handbook—Fundamentals*, Chapter 30, with some tabulated values for July 21 at 40° N latitude. Uninsulated or poorly insulated roofs of darker colors are particularly impacted by sol-air temperature. The sol-air formula was discussed in Chapter 7.

### (b) Gains Through Glass

The quick way to approximate these gains is

$$q = A \times \text{DCLF}$$

DCLF (design cooling load factor) values are listed in Table F.6 and *include the U-factors* as well as the equivalent temperature differences. (These DCLF values do *not* correspond to the worst-hour gains; they were obtained by averaging the hours from 5:30 A.M. to 6:30 P.M. at both 30° N and 40° N latitudes.) Again, the DCLF values were based on an inside temperature of 75°F (23.8°C) and outside temperatures as listed in Table F.6. (For *rapid* approximation, ignore the corrections procedure shown in note *a*.) Glass protected by exterior shading devices that exclude all direct sun may be considered equal to the values listed in Table F.6 for "north glass protected by awnings."

### (c) Gains from Outdoor Air

In residences, outdoor air often enters by infiltration. In many other buildings, codes require the deliberate introduction of outdoor air by mechanical ventilation. Whichever way outdoor air enters your building in summer, its sensible heat gain can be calculated by either

$$q_{\text{infiltration}} = (A_{\text{exposed}}) \text{ (infiltration factor)}$$

where $A_{\text{exposed}}$ is the total area of exposed wall surface, including windows and doors, and the infiltration factor is found from Table F.7, or

$$q_{\text{mechanical ventilation}} = (Q) \text{ (ventilation factor)}$$

where *Q* is the volume of outdoor air (cfm or L/s; see Chapter 7) and the ventilation factor is found from Table F.7.

### (d) Gains from People

Only the sensible gains are tabulated, because a simple overall factor for latent gains is included later. The rate of heat gain from people in various activities is shown in Table F.8; values in the "Sensible Heat" column are generally used. (For residences, sensible heat gain per occupant is often assumed at 230 Btu/h [67.3 W]):

$$q_{\text{people}} = (\text{number of occupants}) \times (\text{sensible gain per occupant})$$

### (e) Gains from Lights

The power supplied to electric lamps (those that normally are on while cooling equipment is functioning) can be added directly to the sensible heat gain. Be sure to include ballast heat gains along with fluorescent lamps, usually done by taking *from 1.12 to 1.2 times the total bulb wattage* of such lights (use the lower figure with energy-efficient ballasts).

## (f) Gains from Equipment

In residences, a standard assumption is that 1200 to 1600 Btu/h (350 to 470 W) of sensible heat gain is produced by appliances. (Other residential heat loads are assumed to be vented.) For other buildings, see the gains of specific pieces of equipment given in Tables F.9 and F.10.

## (g) Latent Heat Gains

These vary a great deal with the type of occupancy, but this simple method assumes that the latent gains are closely associated with outdoor air infiltration. Figure 8.33 is a method to estimate *additional latent heat* as a percentage of *total sensible gains*. The design DB and mean coincident WB temperatures from Appendix B, as well as the relative tightness of building construction, determine the latent percentage of total sensible gain. A minimum of 10% and a probable maximum of 30% additional latent are recommended limits.

**EXAMPLE 8.13** (This example is adapted from the 1997 *ASHRAE Handbook—Fundamentals*. Reprinted with permission of the American Society of Heating, Refrigerating and Air-Conditioning Engineers, Inc., Atlanta, GA.)

A one-story office building (Fig. 8.34) is located "in the eastern United States near 40° N latitude." The adjoining buildings on the north and west are not conditioned, and their inside air temperatures are, for simplicity, assumed equal to the outdoor air temperature at any time of day. This is an unusual building both for its uninsulated walls and for very high lighting loads.

*Roof construction:* 4.5-in. (115-mm) flat roof deck of 2-in. (50-mm) gypsum slab on metal roof deck, 2-in. (50-mm) rigid above-deck roof insulation, surfaced with two layers of mopped felt vapor seal built-up roofing having dark-colored gravel surface, no false ceiling. Summer $U = 0.09$ Btu/h ft$^2$ °F (0.51 W/m$^2$ °C).

*South wall construction:* 4-in. (100-mm) face brick, 8-in. (200-mm) common brick, 0.625-in. (16-mm) plaster, 0.25-in. (6-mm) plywood panel glued on plaster. Summer $U = 0.24$ Btu/h ft$^2$ °F (1.36 W/m$^2$ °C).

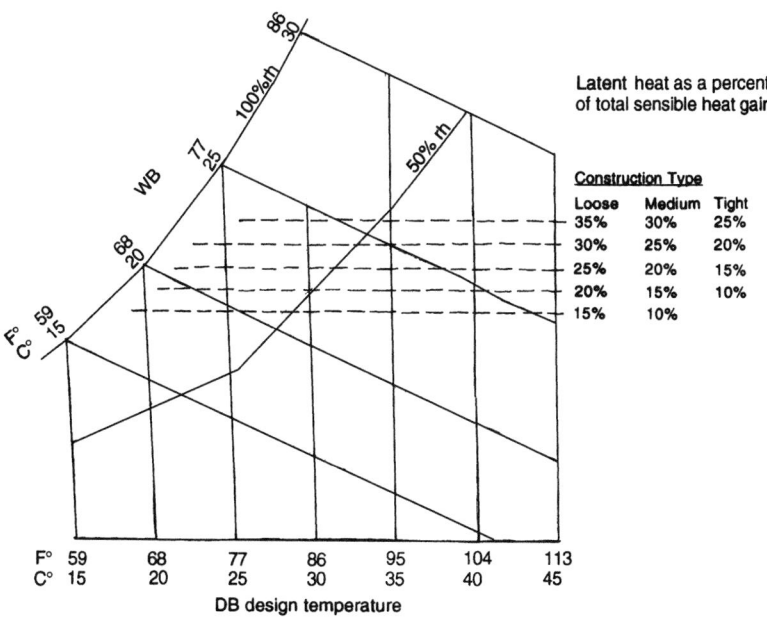

**Fig. 8.33** *Estimating the latent heat gain as a percentage of total sensible heat gain, assuming that these additional latent gains are closely associated with outdoor air infiltration. The design DB and mean coincident WB temperatures are from Appendix B. Relative tightness of building construction is also included.*

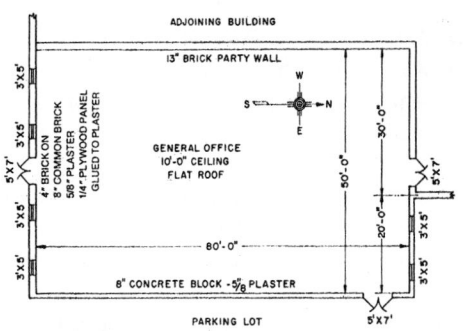

**Fig. 8.34** *Plan of the office building for which heat gain calculations are shown in Example 8.13 (and in Table 8.12). (Reprinted with permission of the American Society of Heating, Refrigerating and Air-Conditioning Engineers, Inc. from the 1997 ASHRAE Handbook—Fundamentals.)*

*West wall and adjoining north party wall construction:* 13-in. (325-mm) solid brick, no plaster. Interior partition U = 0.25 Btu/h ft² °F (1.40 W/m² °C).

*North exposed wall and east wall construction:* 8-in. (200-mm) concrete block and 0.625-in. (16-mm) plaster. Summer U = 0.48 Btu/h ft² °F (2.73 W/m² °C).

*Floor construction:* 4-in. (100-mm) concrete on ground.

*Fenestration:* 3-ft × 5-ft (1-m × 1.5-m) nonoperable windows of regular plate glass with light-colored venetian blinds. Summer U = 0.81 Btu/h ft² °F (4.6 W/m² °C).

*Front doors:* Two, 2.5 × 7 ft (1.5 m × 2 m).

*Side doors:* Two, 2.5 × 7 ft (1.5 m × 2 m).

*Rear doors:* Two, 2.5 × 7 ft (1.5 m × 2 m), interior.

*Door construction:* Light colored 1.75-in. (45-mm) steel door with solid urethane core and thermal break. Summer U = 0.18 Btu/h ft² °F (1.08 W/m² °C).

(U-factors for all doors and outside walls were calculated assuming a wind speed of 7.5 mph (12 km/h). For party and inside walls, still air was assumed.)

*Outdoor design conditions:* Dry-bulb temperature, 94°F (35°C); daily range, 20F° (11C°). Wet-bulb temperature, 77°F (25°C).

*Indoor design conditions:* Dry-bulb temperature, 75°F (24°C); wet-bulb temperature, 62.5°F (18°C).

*Occupancy:* 85 office workers from 8 A.M. to 5 P.M.

*Lights:* 17,500 W fluorescent, from 8 A.M. to 5 P.M.; and 4000 W tungsten, continuous.

*Equipment:* This example does *not* include any office equipment.

*Ventilation:* The ventilation rate is 15 cfm (7 L/s) per person, for a total of 1275 cfm (595 L/s).

The conditioning equipment is located in the adjoining structure to the north.

Determine the sensible, latent, and total space cooling load at design conditions.

**SOLUTION**

Before beginning the calculations, *estimate* the heat gain using the design guideline procedure of Table F.3.

People (omit equipment in this example):

$$2.3 \times 2^a = 4.6 \text{ Btu/h ft}^2$$

Lighting[b]:

$$21,500 \text{ W}/4000 \text{ ft}^2 = 5.4 \text{ W/ft}^2 \times 3.41 \text{ Btu/h W} = 18.4$$

Envelope gains:

| | |
|---|---|
| Windows: | 90 ft²/4000 ft² × 18[c] = 0.4 |
| Walls: | |
| South: | 405 ft²/4000 ft² × 0.24 × 19[c] = 0.5 |
| East, north | |
| exposed: | 935 ft²/4000 ft² × 0.48 × 19[c] = 2.1 |
| Party: | 1065 ft²/4000 ft² × 0.25 × 19[c] = 1.3 |
| Roof: | 4000 ft²/4000 ft² × 0.09 × 39[c] = 3.5 |
| Ventilation: | 1275 cfm/4000 ft² × 21[c] = <u>6.7</u> |

Total 37.5 Btu/h ft²

---

[a]4000 ft²/85 people = 47 ft² per person, more than double the heat gain at a lower density of 100 ft² per person that is assumed in this approximate procedure, so increase people gain by a factor of 2
[b]This is considerably in excess of the lighting loads on which this approximate procedure is based. Without a detailed knowledge of the actual installed lighting loads the design guideline would have suggested, based upon

$$DF = 0.2 \times \frac{90 \text{ ft}^2 \text{ windows}}{4000 \text{ ft}^2 \text{ floor}} = 0.5\% \text{ DF}$$

which is <1, therefore suggesting a load of only 5.1 Btu/h ft²
[c]Interpolate for 95°F, between 90°F and 100°F outdoor temperature

*Cooling load due to heat gain through building envelope:* The heat gains through roof, exposed walls, and doors shown in Table 8.12 were calculated by

$$q = U \times A \times DETD$$

for which DETD values are taken from Table F.5. In line with the approximate (and therefore rapid)

**TABLE 8.12 Cooling Load Through Building Envelope (Example 8.13)**

| Net Area (m²) | U-factor (W/m² K) | Δt (C°) | DETD (C°) | DCLF | Cooling Load (W) | Section | Reference | Net Area (ft²) | U-factor (Btu/h ft² °F) | Δt (F°) | DETD (F°) | DCLF | Cooling Load (Btu/h) |
|---|---|---|---|---|---|---|---|---|---|---|---|---|---|
| 371.6 | 0.51 | | 24.4 | | 4620 | Roof | Table F.5 | 4000 | 0.09 | | 44.0 | | 15,840 |
| 37.6ᵃ | 1.36 | | 9.1 | | 470 | South wall | Table F.5 | 405ᵃ | 0.24 | | 16.3 | | 1,580 |
| 71.1ᵃ | 2.7 | | 9.1 | | 1750 | East wall | Table F.5 | 765ᵃ | 0.48 | | 16.3 | | 5,990 |
| 15.8ᵃ | 2.7 | | 9.1 | | 390 | North wall exposed | Table F.5 | 170ᵃ | 0.48 | | 16.3 | | 1,330 |
| 98.9ᵃ | 1.41 | 11.2ᵇ | | | 1560 | Party walls | | 1065ᵃ | 0.25 | 20ᵇ | | | 5,330 |
| 3.25 | 1.08 | | 13.1 | | 50 | Doors: S | Table F.5 | 35 | 0.19 | | 23.6 | | 160 |
| 3.25 | 1.08 | 11.2ᵇ | | | 40 | N | Table F.5 | 35 | 0.19 | 20ᵇ | | | 130 |
| 3.25 | 1.08 | | 13.1 | | 50 | E | Table F.5 | 35 | 0.19 | | 23.6 | | 160 |
| 5.6 | (4.59)ᶜ | | | 97.9 | 550 | Windows: S | Table F.6 | 60 | (0.81)ᶜ | | | 31 | 1,860 |
| 2.8 | (4.59)ᶜ | | | 72.6 | 200 | N | Table F.6 | 30 | (0.81)ᶜ | | | 23 | 690 |
| | | | | Total: | 9680 | | | | | | | Total: | 33,070 |

ᵃCalculated from gross wall area less windows and door areas.

ᵇDesign temperature difference, inside to outside.

ᶜDCLF for glass *includes* the U-factor.

nature of this calculation method, *no corrections* were made to DETD values to adjust for either outside or inside design temperatures differing from those listed, and actual gains were rounded to the nearest 10. The climate's daily temperature range is medium; the roof is dark (typical of commercial buildings in air-polluted areas). An outside design temperature of 95°F (35°C) was used. To keep things simple, the temperature difference through party walls was made equal to that on which the table was based; in this case, $95 - 75 = 20F° (35 - 23.8 = 11.2C°)$.

*Cooling load due to heat gain through glass:* These heat gains were calculated by

$$q = A \times DCLF$$

for which the DCLF values (which include the U-factor) were taken from Table F.6. Again, *no corrections* were made, and gains were rounded to the nearest 10.

*Cooling load due to heat gain from lights, people, and equipment:* The rate of heat gain from the lighting can be approximated simply by taking the energy input (including extra power required by ballasts for fluorescents). For fluorescent lamps, 17,500 W × 1.2 ballast factor

$$= 21,000 \text{ W } (71,650 \text{ Btu/h}).$$

For incandescent lamps,

$$4000 \text{ W } (13,650 \text{ Btu/h})$$

The sensible gains from people can be determined from Table F.8. For office work, assume 75 W (250 Btu/h):

85 people × 75 = 6380 W (21,250 Btu/h)

Heat gains from office equipment are ignored in this example. (See Tables F.3 and F.9 for typical ranges of such gains.) The sensible gains from lights, people, and equipment thus total 106,550 Btu/h (31,380 W).

*Cooling load due to ventilation or infiltration:* Because this is a commercial building incorporating deliberate introduction of outdoor air, infiltration will be ignored. This example assumes 15 cfm (7 L/s) per person. Total mechanical ventilation is 1275 cfm (595 L/s). Sensible heat gains (from Table F.7):

Mechanical ventilation:

22.0 × 1275 cfm = 28,050 Btu/h
(13.6 × 595 L/s = 8090 W)

*Latent heat gains:* The climate description suggests a typical northeastern U.S. climate at 40° N latitude. Newark, New Jersey (91 DB, 73 WB), and Philadelphia, Pennsylvania (90 DB, 74 WB), are examples. From Fig. 8.33, estimate latent gain as about 20% of the total sensible load.

| | W | Btu/h |
|---|---|---|
| Sensible gains, envelope | 9,680 | 33,070 |
| Sensible gains, lights and people | 31,380 | 106,550 |
| Sensible gains, ventilation | 8,090 | 28,050 |
| Sensible gains, total | 49,150 | 167,670 |
| Additional latent gains (20%, this example) | 9,830 | 33,530 |
| Total latent and sensible heat gains | 58,980 | 201,200 |

■

**THERMAL CONTROL**

## 8.12 PSYCHROMETRY

Moisture, air, and heat interact with some consequences that are threats to, and other consequences that are opportunities for, building performance. In winter, condensation within insulation due to falling air temperatures can be disastrous. In summer, adding moisture to hot, dry air can lower its DB temperature while raising its humidity to more comfortable levels.

These moisture, air, and heat interactions are complex. As air temperature rises, its capacity to hold moisture rises also, and the warmer air becomes less dense. These combined interactions are described by *psychrometry*, the study of moist air. Fortunately, these interactions can be combined within a single chart—see Fig. F.1 (I-P units) and Fig. F.2 (SI units).

In Chapter 4, we encountered the terms *dry-bulb* (DB), *wet-bulb* (WB), and *relative humidity* (RH). These elements are combined in the schematic chart of Fig. 8.35, where the term *saturation line*, at 100% RH, is introduced. This is also called the *dew point* (DP) because dew forms (water vapor condenses) when saturated air touches any surface at or below the air's dew point temperature. This saturation is sometimes undesirable, as within walls or roofs, or on ceiling, air duct, or glass surfaces. However, it is often desirable, as on air-conditioner coils, where the resulting reduction of the moisture content in the air is deliberate.

The psychrometric chart may be used to graph a wide variety of processes, which are summarized in Fig. 8.36. To understand these processes, we must add to the basic chart of Fig. 8.35. The first addition is the *humidity ratio*, which indicates the amount of moisture by weight within a given weight of dry air. Air treatment processes that travel along these horizontal lines of constant humidity ratio (Fig. 8.37) are the familiar processes of simple heating (air passing through the heating coil of a furnace or through a solar collector) and simple (sensible) cooling (air passing through the cooling coil of an air conditioner before saturation). The humidity ratio is used in calculating latent heat gains from outdoor air.

The next addition shows how the *density* of air varies as its temperature and moisture content vary. These lines are those of *specific volume*, the reciprocal of density, a useful quantity in air-conditioning

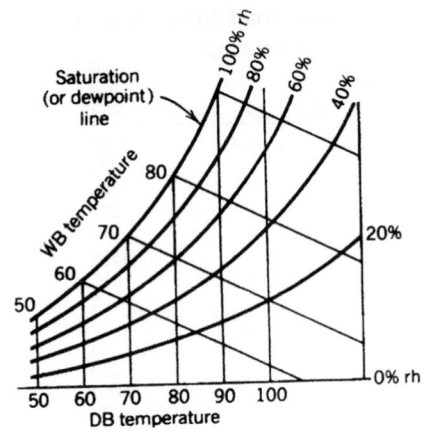

**Fig. 8.35** *Some basic components of the psychrometric chart: DB and WB temperatures and RH.*

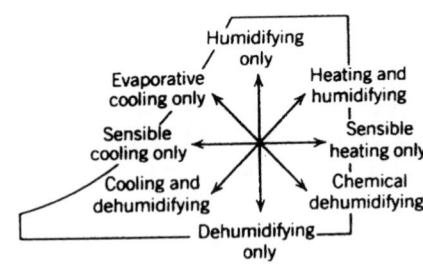

**Fig. 8.36** *Climatic-conditioning processes expressed on the psychrometric chart. (Adapted from "Architectural Design Based on Climate," by M. Milne and B. Givoni, in Watson (ed.), Energy Conservation in Building Design. Reprinted with the permission of the publisher, McGraw-Hill, Inc.)*

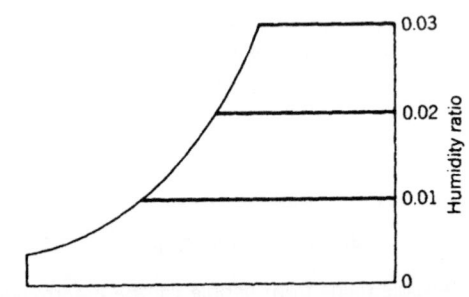

**Fig. 8.37** *Humidity ratio on the psychrometric chart: I-P units are lb moisture/lb of dry air; SI units are kg moisture/kg dry air.*

calculations and helpful in understanding the stack effect (Section 8.14*b*). The specific volume is given in ft³/lb (m³/kg) of dry air. It is evident from these lines in Fig. 8.38 that a pound of hot air is larger (has more volume) than a pound of cold air. This larger volume per unit of weight increases buoyancy; thus, hot air rises, whereas cold air sinks.

The next addition (Fig. 8.39) contains *enthalpy*, the sum of the sensible and latent heat content of an air–moisture mixture relative to the sum of the sensible and latent heat in air at 0°F (0°C in SI units) at standard atmospheric pressure. Enthalpy units are Btu/lb (kJ/kg) of dry air. Enthalpy lines are almost parallel to those of WB temperature. Perhaps the most familiar process to travel along the lines of constant enthalpy is evaporative cooling, where increased moisture and lower air DB temperature are obtained *without changing the enthalpy* (total heat content) of the air. There is indeed a drop in sensible heat as the temperature drops, but this is matched by an increase in latent heat as the moisture content increases. The opposite process is chemical (desiccant) dehumidifying, where decreased moisture

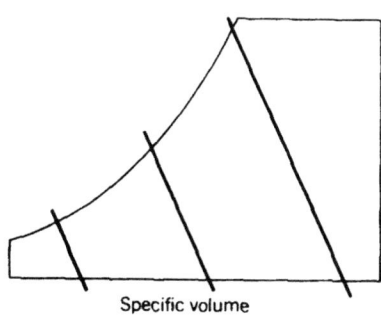

**Fig. 8.38** *Specific volume on the psychrometric chart: I-P units are ft³/lb dry air; SI units are m³/kg dry air.*

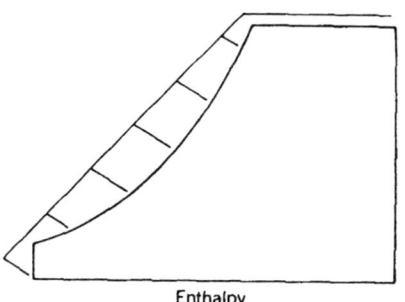

**Fig. 8.39** *Enthalpy on the psychrometric chart: I-P units are Btu/lb; SI units are kJ/kg.*

content is obtained at the price of increased air DB temperature; again, no change in enthalpy (total heat) occurs.

### (a) Cooling Process

The work to be done by mechanical air-conditioning equipment is measured by the total change in enthalpy that must occur within the air that is treated by such equipment. The psychrometric chart is used to size an air conditioner accurately.

First, consider the problem of determining the total change in enthalpy. Assume outdoor conditions of 90°F DB and 76°F WB, with desired indoor conditions at 75°F DB and 50% RH. From Fig. F.1, these indoor conditions are 75 DB/62.7 WB. In the simple (but relatively rare) case of cooling 100% outdoor air, the total heat to be removed is determined as shown in Fig. 8.40. (Once-through systems cooling 100% outdoor air require substantial energy. They may be found in hospital surgical rooms and laboratories with many fume hoods.) How much total heat will be removed in taking air from 90 DB/76 WB to 75 DB/62.7 WB?

For every pound of "dry" air (based on the weight of the air alone) that is cooled and dehumidified, 39.6 − 28.3 = 11.3 Btu must be extracted. Similarly, for every pound of air so treated, 0.0162 − 0.0093 = 0.0069 lb of condensed moisture must be disposed of.

The actual cooling process is more complex, as shown in Fig. 8.41. The conditioned air must be introduced to the space at both a lower DB temperature and lower RH than the desired indoor conditions so that such supply air can "soak up" heat and moisture, then leave through the return air grilles no worse than the desired indoor conditions (75 DB/50% RH). So another set of conditions, lower in DB and RH, will be established, depending on the rates at which air is introduced and the amounts of heat and moisture to be absorbed by the air passing through the space.

Within the cooling equipment, the outdoor air follows a complex path; the lines labeled 1, 2, and 3 in Fig. 8.41 trace the cooling and dehumidifying steps. Outdoor air is cooled without a loss of moisture (step 1) until it reaches saturation (its dew point). The cooling then continues; the air continues to lose sensible heat in step 2, and now loses moisture as well, extracting more heat. When step 3

THERMAL CONTROL

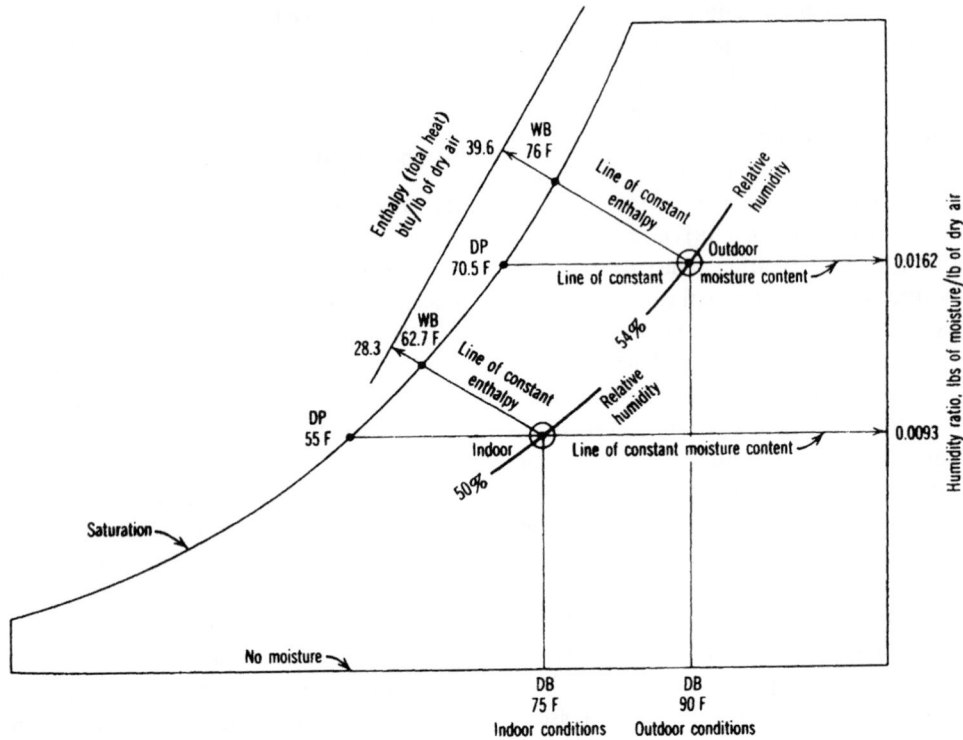

**Fig. 8.40** *Use of the psychrometric chart to determine the change in enthalpy between given outdoor and indoor conditions.*

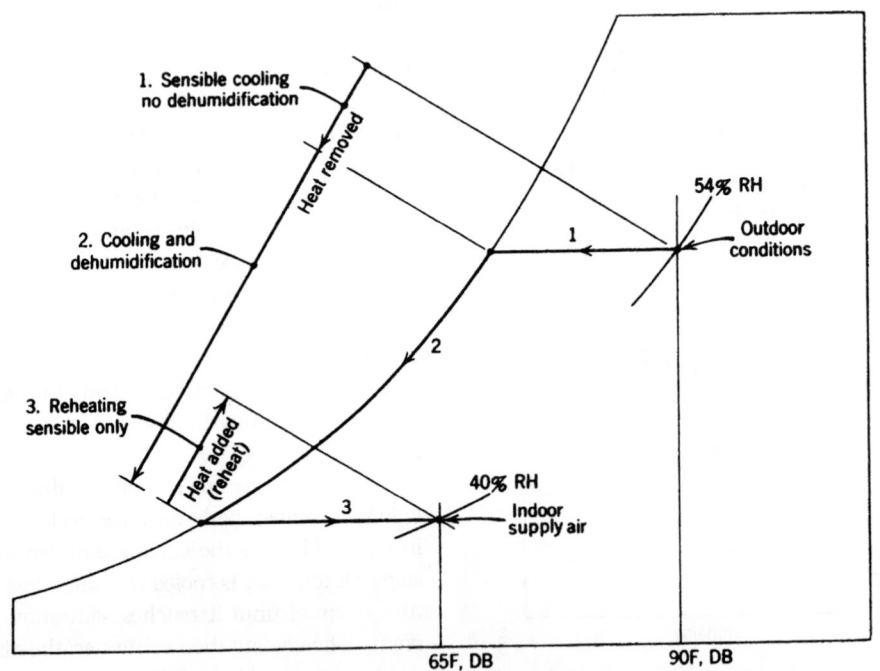

**Fig. 8.41** *Process of cooling and dehumidifying outdoor air, summer conditions.*

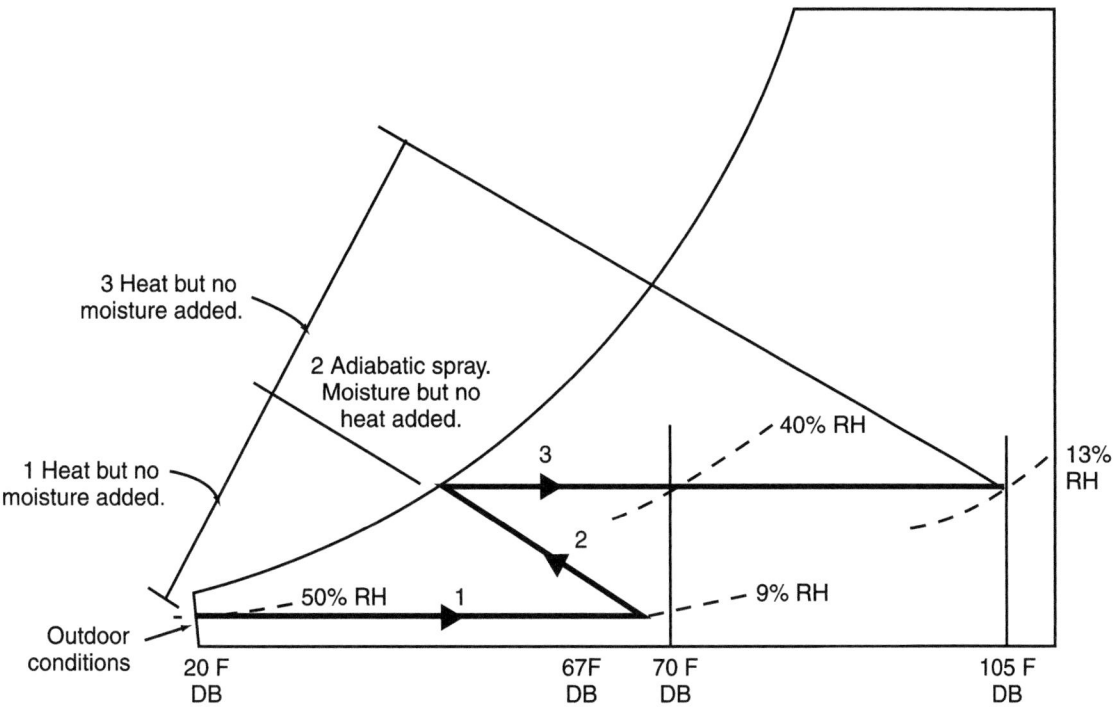

**Fig. 8.42** *Process of heating and humidifying outdoor air, winter conditions.*

begins (known as the *reheating process*), the remaining smaller amount of moisture and the air are both heated to 65°F, 40% RH—slightly below the conditions to be maintained in the space. The changes in the heat and moisture content of the air at the various stages are measured along the enthalpy scale.

Obviously, deliberately adding heat to just-cooled air sounds wasteful. Many heat-recovery processes can be incorporated in cooling equipment so that the reheat is provided by some of that just removed. Heat-recovery equipment is discussed in Chapter 5.

### (b) Heating Process

The process of heating outdoor air in winter is charted in Fig. 8.42. Air at low temperatures in winter often has a humidity ratio so low that this moisture content would be unacceptably low when the outdoor air is warmed (step 1). Therefore, moisture must be added. Often this is in the form of a warm water spray. However, in Fig. 8.42 an adiabatic spray is used—that is, no change in enthalpy—shown in step 2. Then, in step 3, the sat-

urated air is warmed to a higher temperature (105°F in this case) than the desired indoor conditions of 70 DB and 40% RH. This supply air then loses its heat to the space. Note that in this process, we start at desired indoor conditions and work backward (along step 3) to saturation, then downward along the constant enthalpy line (along step 2) until we reach the line of constant humidity ratio of the warming outdoor air—in this case, 67°F DB. As with cooling, the changes in heat can be read on the enthalpy scale.

## 8.13 DETAILED HOURLY HEAT GAIN (COOLING LOAD) CALCULATIONS

Ordinarily, the approximate method of calculating the peak cooling load (Section 8.11) is sufficient for a designer's preliminary estimates of cooling equipment size for a small building. However, much more detailed procedures are used by engineers to actually size the equipment and to assess the peak-load impact of various design options such as shading devices.

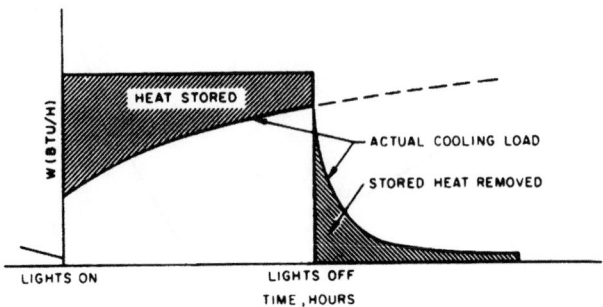

**Fig. 8.43** *Thermal storage effect in the cooling load caused by electric lights. (Reprinted with permission of the American Society of Heating, Refrigerating and Air-Conditioning Engineers, Inc. from the 1997* ASHRAE Handbook—Fundamentals.*)*

The 1997 *ASHRAE Handbook—Fundamentals* (Chapter 28) presents three detailed methods for calculating heat gains and equipment cooling loads. The transfer function method (TFM) and the total equivalent temperature differential, time-averaging method (TETD/TA) each yield hourly values over a 24-hour day. They account for thermal storage by building mass, which can significantly shift the impact of instantaneous heat gains on the actual cooling load for the HVAC equipment, as shown in Fig. 8.43. These detailed methods are tedious and/or complicated enough to require the use of a computer. Another more direct, one-step method, cooling load temperature difference/cooling load factor (CLTD/CLF), more quickly yields a 1-hour value that also accounts for the effects of thermal mass; however, considerable judgment is necessary to choose the correct hour for analysis. These methods are used by engineers, and rarely by architects, and are not presented here. However, a summary of the results of the calculations by each method for the office building of Example 8.13 is shown in Table 8.13.

The quick and uncomplicated heat gain methods shown in Table F.3 and Section 8.11 served the simple building in Example 8.13 rather well. Had the design guideline (Table F.3) been used to size an air conditioner, it would have matched the TFM worst-hour sensible gains, but was only about 84% of the other methods' worst-hour sensible gains. Latent gain estimates would have been necessary as well. The approximate method underpredicted latent gains, but the total (latent and sensible) gains of this quick, by-hand method ranged from 83% to 95% of the computer-based methods' results.

As a quick approximation for estimating the overall equipment size or passive cooling approach, these by-hand methods are adequate for smaller and simpler buildings. When utilizing these quicker

**TABLE 8.13 Comparison of Methods of Heat Gain Calculation (Example 8.13)**

| Method | Maximum Hourly Gain (W) | | | 24-h Average Gain (W/m²) | | | Maximum Hourly Gain (Btu/h) | | | 24-h Average Gain (Btu/h ft²) | | |
|---|---|---|---|---|---|---|---|---|---|---|---|---|
| | Sensible | Latent | Total | Sensible | Latent | Total | Sensible | Latent | Total | Sensible | Latent | Total |
| Design guideline (Table F.3) | (118/m²) | | | | | | (37.5/ft²) | | | | | |
| Approximate[a] (Section 8.11) | 49,150 (131/m²) | 9,830 | 58,980 | | | | 167,670 (41.9/ft²) | 33,530 | 201,200 | | | |
| Transfer Function[b] (worst hour 4 P.M.) | 41,415 (110/m²) | 15,040 | 56,455 | 68.3 | 32.0 | 100.3 | 149,623 (37.4/ft²) | 61,168 | 210,791 | 21.3 | 12.5 | 33.8 |
| CLTD/CLF[b] (for 4 P.M. only) | 53,499 (143/m²) | 15,040 | 68,489 | | | | 179,140 (44.8/ft²) | 63,208 | 242,348 | | | |
| TETD/TA[b] (worst hour 4 P.M.) | 50,096 (134/m²) | 15,040 | 65,136 | 68.3 | 32.0 | 100.3 | 169,082 (42.3/ft²) | 61,168 | 230,250 | 21.3 | 12.5 | 33.8 |

[a]See Table 8.12 and Example 8.13 for calculations.
[b]From 1997 *ASHRAE Handbook—Fundamentals*, Chapter 28.

methods, keep in mind that some judgments may require subdividing the calculations or even changing some of the multipliers to reflect unusual conditions. A designer's checklist of considerations for heat gain should include:

1. Characteristics of the building envelope: materials, sizes, external surface colors, and shapes
2. Building location and orientation, as well as the extent of external shading of the building by trees or adjacent structures
3. Outdoor design conditions
4. Indoor design conditions: DB, WB, and ventilation rate
5. The schedule of lighting, occupancy, equipment, and any other processes that contribute to internal heat gain
6. Thermal zoning requirements

## 8.14 PASSIVE COOLING CALCULATION PROCEDURES

When outdoor air is 85°F (30°C) or below, it is possible to cool buildings by simple ventilation, maintaining conditions indoors that are within the comfort zone. Design guidelines for sizing windows and stacks for either cross-ventilation or stack ventilation were presented in Section 8.6.

In the many climates in which the outdoor temperature is above 85°F (30°C) for a large number of working-day hours, buildings that are closed to the hot exterior during those hours are practical. Between 80° and 85°F (27° to 30°C), outdoor air can keep people within the comfort zone if it moves across the body fast enough. In areas with reliable winds, open buildings are feasible up to 85°F (30°C). Between 55° and 69°F (13° and 21°C), outdoor air is cool enough to use in place of, or to greatly supplement, mechanically cooled air if the humidity is sufficiently low. Passive cooling is even easier under these conditions; see Table F.11 for cooling data for a number of cities. Appendix B shows CDD74 for a wide variety of locations.

From the advent of air conditioning to recent years, the standard response of designers was to turn to mechanical cooling, utilizing the refrigeration cycle and forced air. More recently, some passive cooling alternatives have proved to be effective in climates that have clear or cool summer nights. The following procedures go beyond the quicker

design guidelines that appeared in Section 8.6 and allow the designer to better adjust a preliminary design to the program and the climate.

To introduce these more detailed methods, consider the United Kingdom Pavilion (now dismantled) at the 1992 Seville World Expo (Fig. 8.44). This was an unusually large passive building (213 ft × 105 ft × 82 ft high [65 m × 32 m × 25 m high]), with an unusually wide collection of passive cooling techniques, inspired by the traditional cooling methods of hot, dry southern Spain: shading, water, and mass. The Expo was an April to October event, with the greatest challenge in July–August, when average daily temperatures range from about 64° to 102°F (18° to 39°C).

*Shading* began with the flat roof, where standing wave-form racks carried both translucent fabric shades and PV cells facing south. Outside the south façade, projecting sail-like "fly-sheets" kept off direct sun. The east and west façades used different strategies.

*Water* became a theme of the exhibit, beginning with the wavelike silhouette on the roof and featured as a continuous sheet of water streaming down the east façade, dripping into a pool. The water was pumped, with some power (about 50%) provided by the PV cells on the roof racks. Waiting visitors standing in line outside were treated to the sight and sound of this controlled waterfall, then crossed a bridge over the pond, flanked by spray fountains, to enter the Pavilion. The pond continued below the east wall, exposed to the interior. Water also played a much less evident role on the west facade.

*Mass* was a particular challenge, as the building was designed to be a lightweight structure to be dismantled and largely reused elsewhere. The west wall consisted of white metal containers filled with water to reflect the greater part of the direct sun and store the remainder during the day, radiating by night to cooler outdoor air.

*Thermal zoning* was a vital part of this concept, recognizing that people were entering from very hot, unshaded conditions and that most visitors would then move through rather quickly. Passive cooling was given the task of keeping visitors cooler than the outside; mechanical refrigeration was responsible for maintaining even lower temperatures only where people gathered over extended periods of time, as in the "pods" that

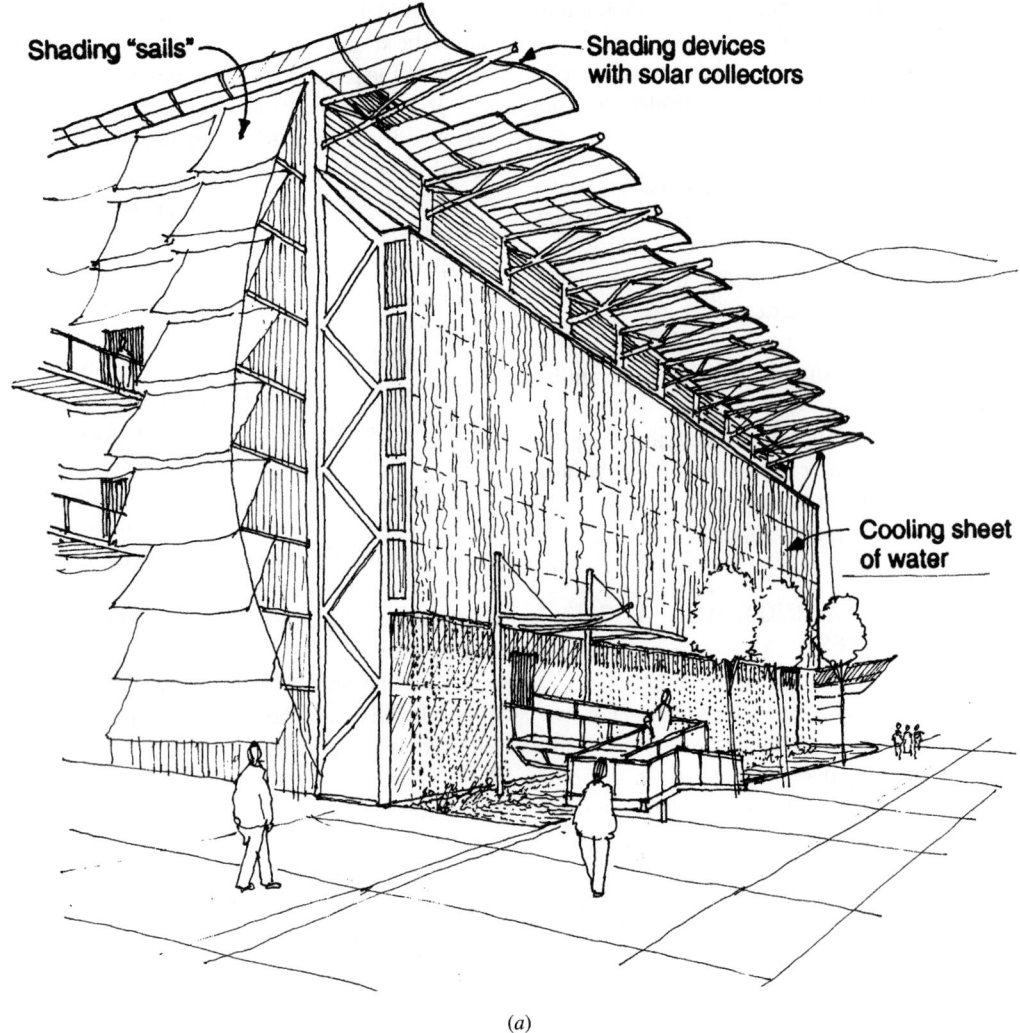

Shading "sails"

Shading devices
with solar collectors

Cooling sheet
of water

(a)

**Fig. 8.44** *The United Kingdom Pavilion at the Seville (Spain) World Expo, 1992. (a) Waveform shading devices on the roof also carry PV cells, whose electricity helps pump water from the pool that flows continuously over the east glass wall. (b) Section (east–west) with a combination of cooling strategies. (c) Section (east–west) with estimated maximum temperatures. (Sections are courtesy of Ove Arup Partnership, London.)*

contained the restaurant and the cafeteria. Gardner and Hadden (1992) describe how the typical July visitor left a sunny outdoor environment at about 100°F (38°C) for the cooler interior at about 90°F (32°C) supplemented by radiant coolth from the huge east glass wall at a maximum of 75°F (24°C) thanks to evaporation from the running water. Exhaust air from the pods lowered temperatures in their vicinity to about 82°F (28°C). The pods themselves, maintained at about 73°F (23°C), were entered only after the visitor underwent a lengthy transition through the Pavilion from the hot outdoors.

To some extent, both cross- and stack ventilation were also involved, despite the outdoor heat; the hottest air indoors rose to the ceiling and was replaced by outdoor air that was cooled slightly as it passed through the watery gap between the east glass wall and the pond. Very high vents at the top of both east and west walls admitted wind to sweep across the ceiling and remove the hottest air.

### (a) Cross-Ventilation

Once again, remember that cross-ventilation cooling works only when the *outside is cooler than the inside*. Otherwise, why bring in air that is warmer to

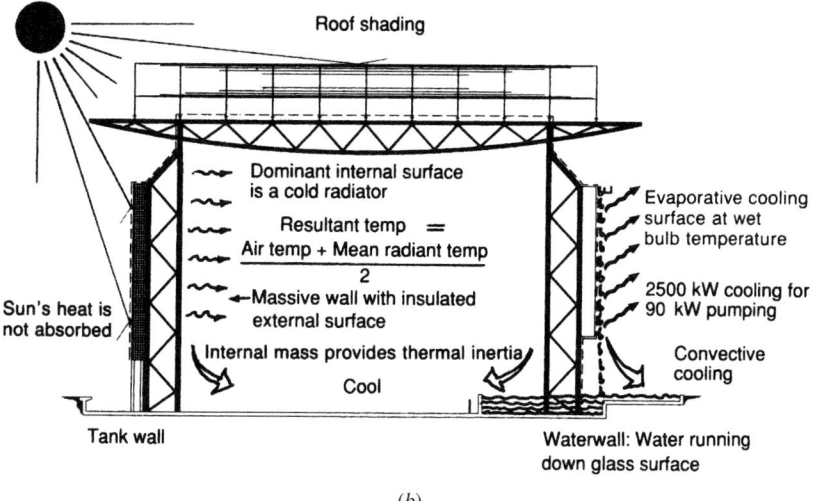

(b)

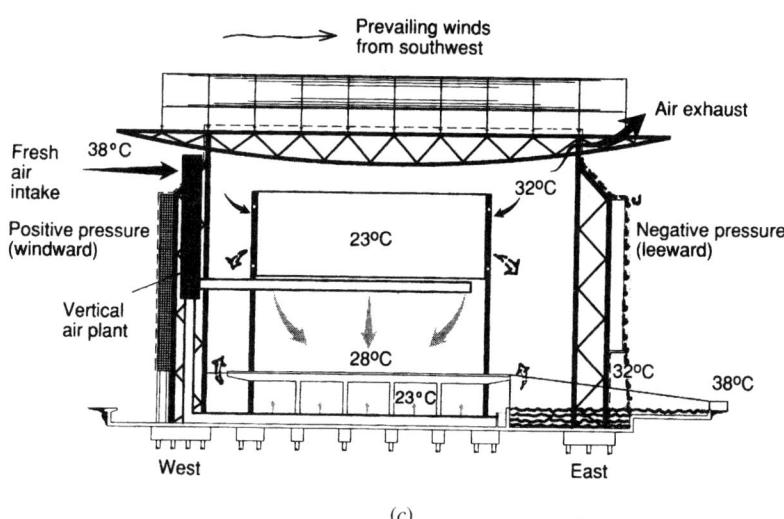

(c)

**Fig. 8.44** (Continued)

try to cool a space? For ventilation using windows, the quantity of outdoor air admitted is termed $V$. The flow of air from inlet windows through the building to outlet windows should have a minimum of obstacles. In I-P units,

$$V = C_v A v$$

where

$V$ = volume flow rate of air, cfm

$A$ = area of operable windows on inlet side or sides, ft$^2$

$C_v$ = effectiveness factor (dimensionless) that adjusts

for different wind orientations: 0.5 to 0.6 for winds perpendicular to the window openings; 0.25 to 0.35 for wind diagonal to the window openings

$v$ = velocity of wind, fpm (= mph × 88)

In SI units,

$$V = 1000 C_v A v$$

where

$V$ is in L/s (= m$^3$/s multiplied by 1000)

$A$ is in m$^2$

$v$ is in m/s

The sensible heat removed by this flow of outdoor air through indoor spaces was presented in Chapter 7:

$$q_v = (V) (1.1) (\Delta t)$$

where

$q_v$ = sensible heat exchange due to ventilation (Btu/h)

$V$ = volume flow rate, in cfm of outdoor air introduced

$\Delta t$ = temperature difference between outdoors and indoors (F°)

1.1 = a constant (the density of air multiplied by the specific heat of air); units are Btu min/ft³ h °F

In SI units, this becomes

$$q_v = (V) (1.2) (\Delta t)$$

where $q_v$ is in watts, $V$ is in L/sec, 1.2 is a constant (the density of air [1.20 kg/m³] multiplied by the specific heat of air [1.0 kJ/kg K]), and $\Delta t$ is in K.

A further modification to these cooling formulas is usually necessary because wind data are usually taken at airports at a height of about 30 ft (10 m) above open, unobstructed ground. Rarely will buildings that rely on cross-ventilation be so favorably situated. Figure 8.45 shows correction factors to be applied to such airport wind velocities for variations in terrain and height above the ground. The dimensionless correction factor should then be applied to the quantity $V$, usually resulting in a reduced volume flow rate.

A more detailed procedure for calculating cross-ventilation in residences given in Chandra et al. (1986) accounts for such factors as neighboring buildings and terrain, height of opening above grade, insect screens, and framing members within openings. To analyze natural ventilation in even more detail, wind tunnel tests are more likely to be useful than are detailed calculations.

## (b) Stack Ventilation

This is another reminder that stack ventilation cooling works only when the *outside is cooler than the inside*. The stack effect needs several conditions: warmer air indoors that can enter the bottom of the stack, cooler air outdoors, and low inlets to admit that cooler outdoor air to the building. This cooler outdoor air picks up heat from the building and enters the bottom of the stack. Within the stack, this now-warm air rises, because it is less dense—therefore lighter—than the cooler outdoor air that surrounds the top of the stack.

The following equation applies when there is no significant resistance to airflow within the building. To calculate ventilation cooling by the stack effect:

$$V = 60 \, KA \, \sqrt{gh(t_i - t_o)/t_i}$$

where

$V$ = airflow rate, cfm

$K$ = discharge coefficient for opening; assumed to equal 0.65 for multiple inlet openings

$A$ = in square feet, the smaller of either total free area of inlet or outlet openings or horizontal cross-sectional area (throat area) of the stack

$g$ = gravitational constant, 32.2 ft/s²

$h$ = height of the stack from inlet to outlet, ft

$t_i$ = temperature indoors ($>t_o$), °R

$t_o$ = temperature outdoors ($< t_i$), °R

Note: °R = °F + 459.67.

In SI units,

$$V = KA \, \sqrt{gh(t_i - t_o)/t_i}$$

where

$V$ = airflow rate, m³/s (× 1000 = L/s)

$K$ = discharge coefficient for opening, assumed to equal 0.65 for multiple inlet openings

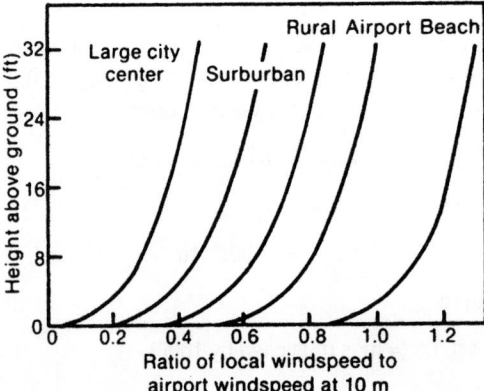

**Fig. 8.45** *Wind speed variation with height for various terrains. (From* Cooling with Natural Ventilation, *Florida Solar Energy Center, published by the Solar Energy Research Institute.)*

$A$ = in m², the smaller of total free area of inlet or outlet openings or horizontal cross-sectional area (throat area) of the stack

$g$ = gravitational constant, 9.81 m/s²

$h$ = height of stack from inlet to outlet, m

$t_i$ = temperature indoors ($>t_o$), K

$t_o$ = temperature outdoors ($< t_i$), K

Note: K = °C + 273.15.

The height of a stack and its cross-sectional area present an unusual opportunity for building form. The Inland Revenue Centre at Nottingham, England (Fig. 8.46), took advantage of this opportunity to provide cylindrical stair towers at corners (or at ends) of its buildings as stack ventilators. These are three- and four-story buildings, with a width of 44.6 ft (13.6 m) to encourage both daylight penetration and natural ventilation. The tower stacks serve all floors except the top. The fabric-covered top of each tower is 23 ft (7 m) above the highest floor served, and can be raised up to 3.3 ft (1 m) or lowered (closed) to vary the flow of exhaust air. The stacks are enclosed in glass block so that solar gain can assist the heating—and thus the flow—of air by day within the stack. Berry et al. (1995) report that the airflow increased from 4.8 to 6.2 ACH with the tower operating in solar mode.

Each of these lower floors' openings to each stack is 43 ft² (4 m²), with the doors held open magnetically unless fire alarms close them. This rather small opening to the stack (relative to the floor area served) is the limiting factor in the outward flow of heated air; to supplement the supply of outdoor air, small three-speed fans are installed in the floor below each window to draw in fresh air, then distribute it through a wide floor grille. The triple-glazed windows are also operable, both as sliding "doors" and as tilt-ins from the top.

This fan-assisted stack effect is used on summer nights to help precool the building for the following day; the lower floors feature very large areas of exposed structural mass in their precast concrete ceilings.

The top floor is a lighter-weight structure, with the stack effect culminating in a ridge vent. Monitoring reveals that this floor is usually a few degrees warmer than the massive night-ventilated lower floors.

## (c) Night Ventilation of Thermal Mass

Before doing these detailed calculations, be sure that (1) you have checked your summer climate against the passive cooling design strategies (Fig. 4.12) and that night ventilation of thermal mass is appropriate, and (2) your building and climate have been checked for approximate performance based on the design guidelines (Fig. 8.16) for this cooling strategy. Also have in mind a positive night ventilation strategy; Fig. 8.47 shows an example of a forced-air system integral with a concrete joist-and-girder structural system. The following procedure is adapted from one developed by Karen Crowther for a workshop at the Fifth National Passive Solar Conference in Amherst, Massachusetts, as presented in Miller (1980).

**STEP 1.** In column II of Table 8.14, list the hourly outdoor temperatures for the design condition (these may be approximated from the summer DB temperature and mean daily range given in Appendix B). This will give the worst-day performance. (To get average-day performance, list average hourly temperatures, which are available from local weather service records.) You need not list temperatures above 80°F (27°C) because outdoor air will not be used for cooling above that temperature.

**STEP 2.** Calculate the 24-hour heat gain for the building in Btu. Find the sum of all the hourly heat gains through the envelope, the minimum ventilation while "closed," and the internal gains while operating. List on line H of Table 8.14.

**STEP 3.** Find the total area of the thermal mass surface that is *exposed* (no rugs, etc.) both to the space to be cooled *and* to moving night air during the ventilation ("open") cycle; list on line A of Table 8.14. *Note:* The larger the mass area exposed, the better the performance. This is why direct-gain solar-heated buildings, with plentiful exposed mass, often make such good candidates for night ventilation cooling. Two additional comments on mass surface: place it where people "see" it so that it can readily receive their radiant heat, and keep direct sun off the mass (and out of the building) during the cooling season.

**STEP 4.** Find the mass heat capacity for the entire space to be cooled: mass volume × density × specific heat. Table E.1 lists both density and specific heat for most common building materials; Table 8.15 shows a quick way to get mass heat capacities for the most

*(a)*

*(b)*

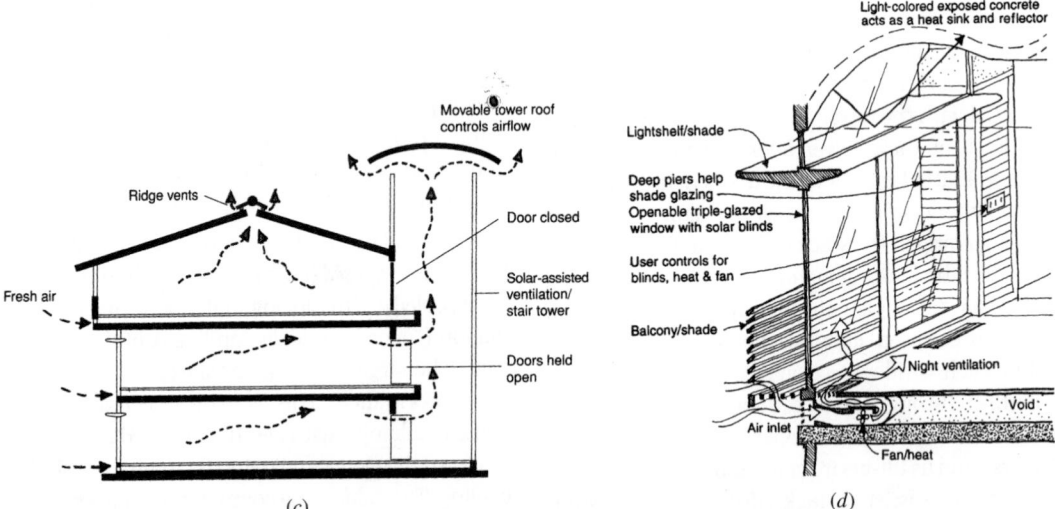

*(c)*

*(d)*

**Fig. 8.46** *The Inland Revenue Centre at Nottingham, England. (a) Cylindrical stair towers at the corners and ends of three- and four-story office blocks act as stacks; solar gain through glass helps speed the flow of exhaust air. (b) Tops of the towers are fabric and can be raised to increase the stack aperture and therefore increase the flow. (Photos courtesy of Vaughan Reynolds.) (c) Section of a three-story wing with a stack-ventilation stair tower. (Drawing courtesy of Ove Arup Partnership, London.) (d) Section perspective with fan-assisted intakes below each window and an exposed precast concrete ceiling that helps store heat for night ventilation cooling.*

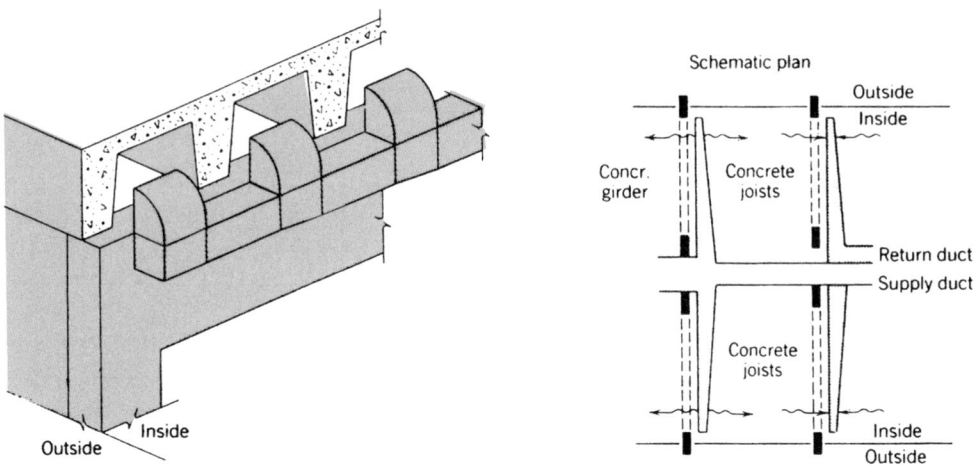

**Fig. 8.47** *Example of ductwork for night ventilation using the thermal mass of a concrete structural system.*

common thermal mass materials. Enter this total mass heat capacity on line B of Table 8.14.

**STEP 5.** For "supplementary" cooling due to surfaces *other* than those of the principal thermal mass, list the space's *floor area.* This step should be taken only for spaces with a significant amount of roof, wall, or floor area *in addition* to the thermal mass areas counted in step 3. For example:

- If your space has exposed concrete ceilings, walls, and floors—all counted already as thermal mass—skip this step.

- If all thermal mass is in the form of freestanding water containers, enter the entire floor area.

- If the entire ceiling is thermal mass but the walls or the floors are not, enter half of the floor area.

## TABLE 8.14 Night Ventilation of Thermal Mass

*Calculation Procedure*
A. Mass surface area (from step 3) _____ ft²
B. Mass heat capacity (from step 4) _____ Btu/°F
C. Floor area (supplementary cooling, step 5), and _____ ft²
   total building volume _____ ft³

| (I) Hour | (II) Outside Air Temperature (°F) | (III) Cooling (Btu) | (IV) Mass Temperature (°F) |
|---|---|---|---|
| 8 P.M. | _____ | _____ | _____ |
| 9 P.M. | _____ | _____ | _____ |
| 10 P.M. | _____ | _____ | _____ |
| 11 P.M. | _____ | _____ | _____ |
| Midnight | _____ | _____ | _____ |
| 1 A.M. | _____ | _____ | _____ |
| 2 A.M. | _____ | _____ | _____ |
| 3 A.M. | _____ | _____ | _____ |
| 4 A.M. | _____ | _____ | _____ |
| 5 A.M. | _____ | _____ | _____ |
| 6 A.M. | _____ | _____ | _____ |
| 7 A.M. | _____ | _____ | _____ |
| 8 A.M. | _____ | _____ | _____ |
| 9 A.M. | _____ | _____ | _____ |

D. Total mass cooling _____ Btu _____ °F
E. Final mass temperature _____ °F
F. Supplementary cooling (see steps 5 and 11) _____ Btu
G. Total cooling, D + F _____ Btu
H. 24-hr heat gain, from step 2 _____ Btu
I. Flow rate required for night ventilation _____ cfh   or _____ ACH

**TABLE 8.15 Common Mass Heat Capacities**

| I-P Units $ft^3 \times (Btu/ft^3 \,^\circ F) = Btu/^\circ F$ | | SI Units $m^3 \times (kJ/m^3 K) = kJ/K$ |
| --- | --- | --- |
| Volume × (62.4) | Water | Volume × (4181) |
| Volume × (22.5) | Ordinary concrete | Volume × (1507) |
| Volume × (18.7) | Masonry, grout-filled | Volume × (1253) |
| Volume × (15.6) | Brick | Volume × (1045) |

*Source:* Adapted from Crowther, "Night Ventilation Cooling of Mass," in Miller (1980).

- If the entire ceiling is thermal mass but the floor is not, and there are few or no walls (e.g., open office plan), enter one-third to one-fourth of the floor area.

**STEP 6.** Complete column III hour by hour after determining the mass temperature (column IV for the preceding hour):

$$
\text{cooling Btu/h} = \begin{bmatrix} \text{Previous} \\ \text{hour} \\ \text{mass} \\ \text{temp., }^\circ F \\ \text{(col. IV)} \end{bmatrix} - \begin{bmatrix} \text{outside} \\ \text{temp., }^\circ F \\ \text{(col. II)} \end{bmatrix}
$$

$$
\times \begin{bmatrix} \text{mass} \\ \text{surface} \\ \text{area, ft}^2 \\ \text{(line A)} \end{bmatrix} \times \begin{bmatrix} \text{surface} \\ \text{conductance,} \\ \text{Btu/h ft}_2 \,^\circ F \end{bmatrix}
$$

For the first hour, assume that the mass temperature is 80°F. The surface conductance is usually assumed as 1.0 Btu/h ft² °F. (See Table E.3 for other surface conductances under various conditions, many of which are considerably more than 1.0.)

**STEP 7.** Complete column IV, hour by hour, after calculating the cooling Btu/h (column III):

$$
\text{mass temp.} = \text{previous hour mass temp. (col. IV)}
$$

$$
- \frac{\text{cooling Btu/h (col. III)}}{\text{mass heat capacity, Btu/}^\circ F \text{ (line B)}}
$$

**STEP 8.** Continue this hourly process using columns III and IV *until* the falling temperature of the mass equals the rising temperature of the outdoor air. At that point, continuing with plentiful ventilation will only rob the mass of its coolth; the building therefore switches to (thermally) closed mode, with minimal ventilation.

**STEP 9.** Add all the hourly cooling Btu/h values (column III) to obtain the total mass cooling in Btu (line D).

**STEP 10.** Note the final mass temperature from column IV. This is probably at least 5F° *above* the *lowest* air temperature of the night (column II). If this lowest mass temperature is significantly higher, consider redesigning for more exposed mass surface area.

**STEP 11.** If supplementary cooling is appropriate (see step 5), calculate it as follows:

$$
\begin{matrix} \text{supplementary} \\ \text{cooling, Btu} \end{matrix} = \begin{bmatrix} 80^\circ F & - & \text{final mass} \\ & & \text{temperature} \\ & & \text{(line E)} \end{bmatrix}
$$

$$
\times 2.25 \times \begin{bmatrix} \text{floor area} \\ \text{(line C)} \end{bmatrix}
$$

The factor 2.25 assumes a modest role for the other, less thermally massive surfaces. Enter this supplementary cooling on line F.

**STEP 12.** Obtain total cooling by adding lines D and F; enter the total on line G.

**STEP 13.** In step 2, you entered the 24-hour heat gain for the building on line H.

Compare this cooling Btu needed to the total cooling provided (line G). If you have not provided enough cooling, and the final mass temperature is more than 7F° above the lowest nighttime outdoor temperature, consider redistributing the building mass over a wider surface (e.g., 3000 ft² of 4-in. slab rather than 2000 ft² of 6-in. slab) and trying again. If you do not have enough cooling, and the final mass temperature is 5 to 7F° above the lowest nighttime outdoor temperature, consider providing *both* more mass and more surface area (e.g., 3000 ft² of 4-in. slab rather than 2000 ft² of 4-in. slab) and trying again.

**STEP 14.** Determine and enter on line 1 the approximate flow required for night-ventilating air. Use the following formula:

$$cfh = \frac{Btu/h}{0.018\,\Delta t}$$

where

cfh (ft³/h) = the minimum required flow rate of night air

Btu/h = the cooling Btu for the hour of maximum cooling during the night (column III)

$\Delta t$ = the temperature difference between the final mass temperature (column IV) and the outdoor air (column II) for that same hour of maximum cooling

It is often useful to express this night ventilation flow rate in terms of ACH:

$$ACH = \frac{cfh\ required}{building\ volume\ (ft^3)}$$

**EXAMPLE 8.1, PART J** How much heat will be removed, and what is the final thermal mass temperature, in a typical bay of the Oregon office building in Fig. 8.7? In this procedure, assume that only the exposed surfaces of the thermal mass participate in night flush cooling. (In the actual building, the interior surfaces of the continuous cores in the precast floor slabs are also flushed with night air, yielding added cooling.)

**SOLUTION**

**STEP 1.** The design temperature for Eugene, Oregon, is 89°F, with a mean daily range of 31F°. This is distributed by hour (using the sine curve equation) as shown in Table 8.16, with the minimum daily temperature assumed at 4 A.M. and the maximum at 4 P.M.

**STEP 2.** The 24-hour heat gain is estimated from Example 8.1, Part D. During the 9-hour workday, the heat gains total 13 Btu/h ft². During the unoccupied hours, there are negligible gains from electric lights, people and equipment, windows, or ventilation. However, the *opaque* envelope gains remain because this envelope stores daily heat and releases it gradually. These unoccupied gains (from Table 8.2) total 0.26 Btu/h ft² from the walls, plus 1.03 Btu/h ft² from the roof, totaling 1.29 Btu/h ft².

Occupied:

$$9\ h \times 13\ Btu/h\ ft^2 = 117\ Btu/ft^2$$

Unoccupied:

$$15\ h \times 1.29\ Btu/h\ ft^2 = \underline{19.35}\ Btu/ft^2$$
$$136.35\ Btu/day\ ft^2$$

For the entire typical bay,

$$136.35\ Btu/day\ ft^2 \times 1440\ ft^2 = 196,344\ Btu/day$$

Enter this total on line H.

**STEP 3.** The area of exposed mass, from Table 8.2, totals 2232 ft². The concrete floor is covered by carpet and is therefore assumed to be disconnected from thermal storage for this space. The wallboard interior surfaces of the exterior walls are also ignored.

**STEP 4.** The mass heat capacity is estimated assuming that all elements are solid concrete at 150 lb/ft³ and 4 in. deep. The bearing walls are 8 in. thick, exposed on both sides, with concrete fill in the cores of all concrete blocks. The beams are 8 in. thick, but their inner faces are unexposed (they form the air duct for the night flush system). The precast ceiling slabs are 10 in. thick, but much of this thickness consists of air-filled cores; therefore, only the lower 4 in. are assumed to participate as "exposed" mass. From Table 8.15:

Mass heat capacity = 2232 ft² × 0.33 ft × 22.5 Btu/ft³ °F

= 16,573 Btu/°F

**STEP 5.** Little supplementary cooling is assumed, although the carpeted floor provides some (albeit insulated) mass, as does the wallboard surface of the exterior walls. Assuming that *only one-fourth* of the floor area contributes to cooling,

$$0.25 \times 1440\ ft^2 = 360\ ft^2$$

**STEPS 6–10.** See Table 8.16.

**STEP 11.** Determine supplementary cooling: (80°F − 67.2°F) × 2.25 × 360 ft² = 10,368 Btu.

**STEP 12.** See Table 8.16.

**STEP 13.** The building has more than adequate cooling capacity to meet this 89°F *design* condition; on *typical* summer days, with a high of only 83°F, there will be even more excess mass capacity. (A more conservative and very detailed analysis, assuming more electric lights left on, extremely high temperatures, and a less favorable coefficient of heat

**TABLE 8.16 Oregon Office Building (Fig. 8.7) Performance: Cooling by Night Ventilation of Thermal Mass for a Typical Second-Floor Bay**

| A. Mass surface area | 2,232 ft² |
| B. Mass heat capacity | 16,573 Btu/°F |
| C. Floor area (supplementary cooling) | 360 ft² |
| Total building (bay) volume | 22,900 ft³ |

| (I) Hour | (II) Outside Air Temperature (°F) | (III) Cooling$^a$ (Btu/h) | | (IV) Mass Temperature (°F) |
|---|---|---|---|---|
| P.M. | | | | |
| 8 | 80 | No heat removed | | 80 |
| 9 | 77 | (80 − 77) 2232 (line A) | = 6,696 | $80 - \dfrac{6696}{16{,}573 \text{ (line B)}} = 79.6$ |
| 10 | 74 | (79.6 − 74) 2232 | = 12,499 | $79.6 - \dfrac{12{,}499}{16{,}573} = 78.8$ |
| 11 | 71 | (78.8 − 71) 2232 | = 17,410 | $78.8 - \dfrac{17{,}410}{16{,}573} = 77.7$ |
| 12 | 68 | (77.7 − 68) 2232 | = 21,650 | $77.7 - \dfrac{21{,}650}{16{,}573} = 76.4$ |
| A.M. | | | | |
| 1 | 65 | (76.4 − 65) 2232 | = 25,445 | $76.4 - \dfrac{25{,}445}{16{,}573} = 74.9$ |
| 2 | 62 | (74.9 − 62) 2232 | = 28,793 | $74.9 - \dfrac{28{,}793}{16{,}573} = 73.2$ |
| 3 | 60 | (73.2 − 60) 2232 | = 29,462 | $73.2 - \dfrac{29{,}462}{16{,}573} = 71.4$ |
| 4 | 58 | (71.4 − 58) 2232 | = 29,909 | $71.4 - \dfrac{29{,}909}{16{,}573} = 69.6$ |
| 5 | 60 | (69.6 − 60) 2232 | = 21,427$^b$ | $69.6 - \dfrac{21{,}427}{16{,}573} = 68.3$ |
| 6 | 62 | (68.3 − 62) 2232 | = 14,062 | $68.3 - \dfrac{14{,}062}{16{,}573} = 67.5$ |
| 7 | 65 | (67.5 − 65) 2232 | = 5,580 | $67.5 - \dfrac{5580}{16{,}573} = 67.2$ |
| 8 | 68 | Stop flush: mass temperature is now below outdoor temperature | | |

| D. Total mass cooling | 212,933 Btu |
| E. Final mass temperature | 67.2°F |
| F. Supplementary cooling | 10,368 Btu |
| G. Total cooling (212,933 + 10,368) | 223,301 Btu |
| H. Compare to 24-h heat gain | 196,344 Btu |
| I. Flow rate required for night ventilation | About 5.5 ACH |

$^a$A surface conductance of 1.0 Btu/h ft² °F is assumed in this calculation.

$^b$At this point, enough heat has been removed to meet the typical bay's design-day heat gain.

THERMAL CONTROL

transfer between mass and air, led to the decision to also flush the precast cores at night.)

Because of this excess capacity, it appears advisable to begin the night flush a few hours *later,* when the $\Delta t$ between indoors and outdoors is greater and therefore more cooling is achieved per hour of fan operation. (Repeat the procedure above, beginning the flush at midnight with a $\Delta t$ of $80 - 68 = 12F°$, to see how similar results can be obtained with 3 fewer hours of fan operation.) The disadvantage is that the interior will remain hotter for a longer period in the evening, which could cause discomfort for a late worker. Alternatively, the flush could be stopped after 5 A.M., by which time the stored heat has been removed.

**STEP 14.** The airflow required will be checked at 4 A.M., the hour of maximum cooling:

$$\frac{29,909 \text{ Btu/h}}{0.018 \text{ Btu/ft}^3 \text{ °F h} \times (71.4 - 58)F°} = 124,000 \text{ cfh}$$

$$ACH = \frac{124,000 \text{ cfh}}{22,900 \text{ ft}^3} = 5.41$$

which is somewhat more than that predicted by the design guideline in Exercise 8.1, Part F, of 4.47 ACH. ∎

What might be the advantages of this approach compared to conventional mechanical cooling? For this Oregon building, detailed records of electricity end use are available (see Ashley and Reynolds, 1994). Figure 8.48 compares several August monthly records for the (conventional) compressors and the night-flush fans. In the first two Augusts, the building operated as a night-ventilated structure. In the third August (1991), due to a building-wide computer control changeover, the building operated only as a conventionally cooled structure. The sun's impact on the building should be relatively equal for these 3 months, but outdoor temperature varied as shown by the CDD18°C (= CDD65°F × 5/9). Not only was as much electricity necessary to cool the conventionally operated building, despite lower CDD than the previous August, but also all the fan energy use and almost all the compressor energy occurred during the daytime hours of high consumption. In many locations, utilities charge more for electricity used during the daytime. With night ventilation, the fans deliver huge quantities of fresh air off-peak by night, unlike the conventional daytime operation that recycles some air to save energy.

Another night-ventilation example is the California Highway Patrol offices at Gilroy, California, shown in Fig. 8.49. This one-story building sits near a freeway interchange south of San Jose (81 DB/65 WB, MDR = 26°F). In this building, the nighttime air is drawn in as far from the freeway as possible, taken underground a short distance (with just a bit of earth tube cooling), then distributed throughout the central corridor. At either end of a continuous skylight over the central corridor, large fans exhaust air by night. Corridor walls are concrete block, and the floor is a concrete slab on grade. The cool night air may carry a seasonal scent; Gilroy calls itself the garlic capital of the world.

## (d) Fan-Assisted Evaporative Cooling

Before beginning the following calculations, be sure that:

1. You have checked your summer climate against the passive cooling design strategies

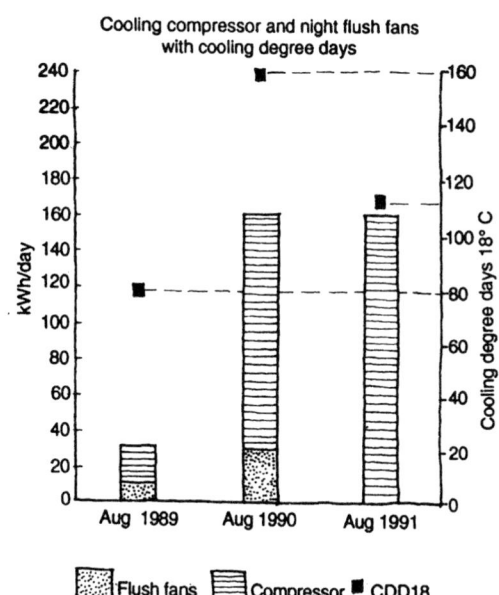

**Fig. 8.48** *Comparing night ventilation of a mass with conventional compressive cooling. The Emerald People's Utility District offices (Fig. 8.7 and Example 8.1, Part J) are monitored for end use of electricity. In August 1991 (117 CDD65 [65 CDD18]), with compressive cooling only, 160 kW/day of electricity was used, nearly all of it during periods of heavy use by day. Compare this to the hotter August 1990 (150 CDD65 [83 CDD18]), when 126 kW/day was used for compressive cooling by day, plus 34 kW/day for night-ventilation fan-forced cooling. (From Ashley and Reynolds, 1994.)*

THERMAL CONTROL

(a)

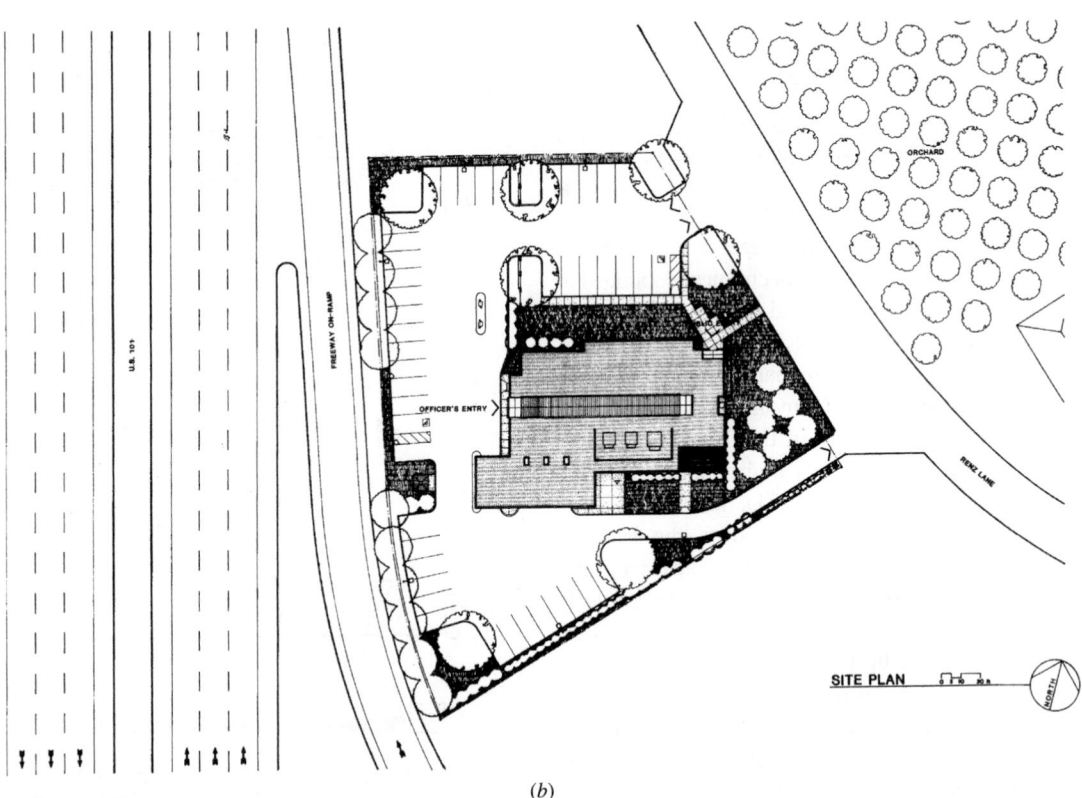

(b)

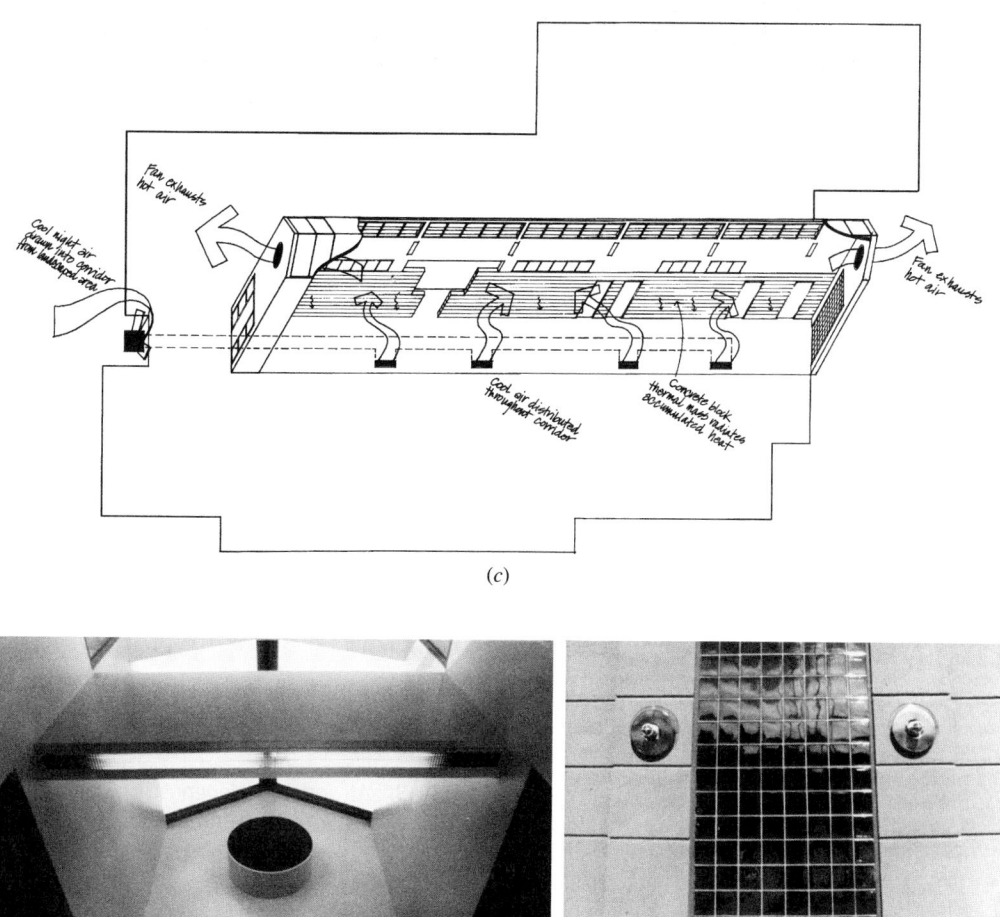

(c)

(d)   (e)

**Fig. 8.49** *The California Highway Patrol building near Gilroy, California (a) is flushed with night air for cooling. (b) Site plan. (c) Axonometric shows intake air through a very short earth tube; night flush exhausts are at either end of the central skylit corridor with thermally massive walls and floor. (d) Looking up at one night flush exhaust grille. (e) Exterior with night flush exhaust grille. (Drawings courtesy of The Colyer/Freeman Group, San Francisco.)*

(Fig. 4.12) and the evaporative cooling strategy is appropriate.

2. Your building and climate have been checked for approximate performance based on the design guidelines for evaporative cooling (Section 8.6*d*).

First, find the total sensible heat gain in Btu/h that is to be removed from your building by evaporative cooling. Generally, this is calculated at your climate's summer design DB and mean coincident WB temperatures (Table B.1). The psychrometric chart is then used to plot the progress of evapora-

tively cooled air, as shown in Fig. 8.50. (The complete psychrometric chart, in I-P units, is given in Fig. F.1.)

**STEP 1.** *Determine outdoor air conditions.* Enter the psychrometric chart at the summer design DB and mean coincident WB temperatures for your climate (point *A* in Fig. 8.50). As the air is blown through the evaporative cooler, it proceeds along the constant WB line toward saturation or 100% RH (from point *A* toward point *B* in Fig. 8.50).

**STEP 2.** *Determine supply air temperature.* The most efficient evaporative cooler will be able to cool and

THERMAL CONTROL

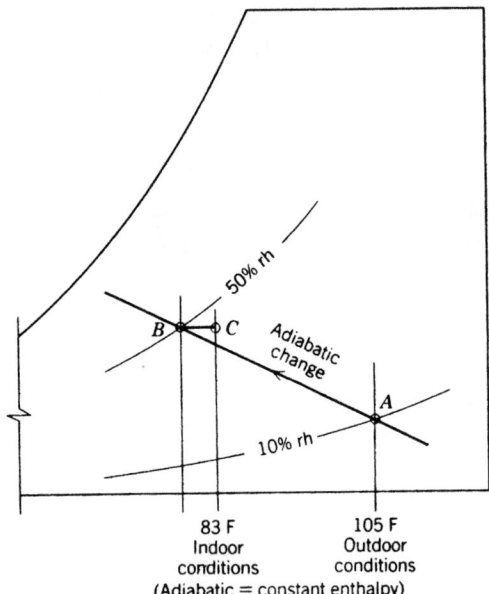

**Fig. 8.50** *Once-through cycle of outdoor air through a fan-forced evaporative cooler. Values are transcribed from the psychrometric chart (Fig. F.1). Hot, dry air can be humidified adiabatically (no change in total heat content) from point A to point B, reaching indoor conditions that fall within, or close to, the optimum summer comfort zone shown in Figs. 4.7 and 4.8. This is accomplished without the use of the high-energy-consuming compressive refrigeration cycle. As the humidified air picks up heat from the space, it moves from point B to point C.*

humidify the outdoor air to a point *about three-fourths* of the total distance on the chart from entering air conditions to saturation (the three-fourths point is reached at point *B* in Fig. 8.50); a more reasonable operating assumption may be two-thirds of that total distance.

**STEP 3.** *Determine Δt indoors.* At conditions of point *B*, the air leaves the evaporative cooler and enters the building. It immediately begins to pick up sensible heat (moving from point *B* toward point *C* in Fig. 8.50). When the air has picked up enough heat to reach an uncomfortably warm temperature, it is exhausted to the outdoors (point *C* in Fig. 8.50). Note, however, that this exhaust air is still well below outdoor temperatures, and therefore is capable of some cooling just outside the building.

**STEP 4.** *Calculate heat removed and airflow rate indoors.* The heat removed from the building by this exhaust air depends on two factors: (1) the DB temperature difference (Δt) between the supply air and the exhaust air (point *B* to point *C* in Fig. 8.50) and

(2) the airflow rate usually expressed in cubic feet per minute:

$$\text{Btu/h removed} = (\text{cfm airflow})$$
$$\times (1.1 \text{ Btu min/ft}^3 \text{ °F h})$$
$$\times (\Delta t, \text{F°})$$

The designer can vary the cfm airflow rate by choosing larger or smaller evaporative coolers; any point less than three-fourths of the total distance from outdoor air to saturation may also be chosen.

**EXAMPLE 8.14** A 4000-ft² retail store near Tucson, Arizona, has been calculated to have a total sensible heat gain of 100,000 Btu/h at summer design conditions (105 DB, 66 WB for this location).

**SOLUTION**

**STEP 1.** Enter Fig. 8.50 at 105 DB, 66 WB (point *A*).

**STEP 2.** Measure along the constant 66-WB line to saturation, and determine that three-fourths of this distance corresponds to 76 DB, 66 WB (point *B*).

**STEP 3.** Δt indoors depends on the assumed exhaust air DB temperature; assume 83°F in this example (point *C*, corresponding to 83 DB, 68 WB). Then, Δt = 83 − 76 = 7F°.

**STEP 4.** Calculate heat removed and airflow rate indoors.

$$100,000 \text{ Btu/h removed} = (\text{cfm airflow})$$
$$\times (1.1 \text{ Btu min/ft}^3 \text{ °F h})$$
$$\times (7\text{F°})$$

Rearranging the equation yields

$$\text{cfm} = \frac{100,000 \text{ Btu/h}}{1.1 \text{ Btu min/ft}^3 \text{ °F h} \times 7\text{F°}} = 13,277 \text{ cfm}$$

Note that this is a rate of 13,277/4000 = 3.3 cfm/ft² of floor area.

This airflow rate can be provided by one larger or several smaller cooling units, depending on possible zones within the building. For instance, two evaporative coolers at 7000 cfm each would meet the requirements for this retail store; one might serve a smaller-area zone in front with show windows and a lot of solar gain, while another might serve a larger-area interior zone with less solar gain. ∎

Some typical evaporative coolers are shown in Section 8.6.

## (e) Cooltowers

One of the first North American cooltowers was shown in Fig. 8.19, along with estimated cooling design guidelines in Fig. 8.20. To more closely estimate the performance of such a passive evaporative device with calculations, some very complex equations are involved (for example, see Thompson et al., 1994). For a shorter but less exact approach, consider this formula proposed by Givoni (1994) in either °F or °C:

$$t_{exit} = DB - 0.87 (DB - WB)$$

where

$t_{exit}$ = air entering the space (exiting the cooltower)

DB = dry-bulb temperature, outside

WB = coincident wet-bulb temperature, outside

This simple equation ignores the potential influences of both wind (driving more air through the evaporative cooling pads) and the optional solar chimney exhaust tower (seen in Fig. 8.19).

Cooling achieved by the tower depends both on the temperature of this exit air and on its flow rate. Givoni (1994) proposes, in SI units:

$$V = 0.033 \, A_{evap} \sqrt{H (DB - WB)}$$

where

$V$ = flow, in m³/s (× 1000 = L/s)

$A_{evap}$ = total area of wetted pads, m², in cooltower

$H$ = overall height of tower, m

DB = dry-bulb temperature, outside

WB = coincident wet-bulb temperature, outside

0.033 = accounts for pressure drop through the tower and the building

Note that this equation ignores the area of both the cooltower and the outlet. Knowing both the temperature and the flow rate, cooling achieved is

$$W = 1.2V[t_{int} - t_{exit}]$$

where

W = Watts removed from space by cooltower

$V$ = flow from cooltower, in m³/s (× 1000 = L/s)

$t_{int}$ = DB temperature to be maintained indoors

$t_{exit}$ = air entering the space (exiting the cooltower)

In I-P units, this becomes

$$V = 2.7 \, A_{evap} \sqrt{H (DB - WB)}$$

where

$V$ = flow, in cfm

$A_{evap}$ = total area of wetted pads, ft², in cooltower

$H$ = overall height of tower, ft

DB = dry-bulb temperature, outside

WB = coincident wet-bulb temperature, outside

2.7 = accounts for pressure drop throughout the tower and the building

Note that this equation ignores the area of both the cooltower and the outlet. Knowing both the temperature and the flow rate, cooling achieved is

$$Btu/h = 1.1 \, V [t_{int} - t_{exit}]$$

where

Btu/h = heat removed from space by cooltower

$V$ = flow from cooltower, in cfm

$t_{int}$ = DB temperature to be maintained indoors

$t_{exit}$ = air entering the space (exiting the cooltower)

In a remote desert west of Tucson, Arizona, the Agua Blanca ranch uses several cooltowers to serve a building around a T-shaped courtyard (Fig. 8.51). This off-grid building complex uses PV for lights and water pumping, propane for a refrigerator, and one wall heater and stand-by generator, and collects rainwater in a cistern below the courtyard. Wood stoves and fireplaces supplement direct solar gain in winter. The courtyard connects to the surrounding ranch by three passages, one covered with a roof and small open skylights, one covered with a deciduous vine on a trellis, and one open to the sky. Large rolling barn doors are located at the outer end of each passage; these provide both security and the ability to trap cool air from the cooltowers (or via the indoor spaces), thus increasing hot-weather comfort outdoors.

Three cooltowers, 22 ft (6 m) high, each with a total evaporative pad area of 64 ft² (5.9 m²), serve three units around the courtyard. At the outlet of each cooltower, switches allow the flow to be shared, sent to either of two locations, or closed off. One of the cooltowers can deliver air directly to the covered passage off the courtyard.

**THERMAL CONTROL**

(a)

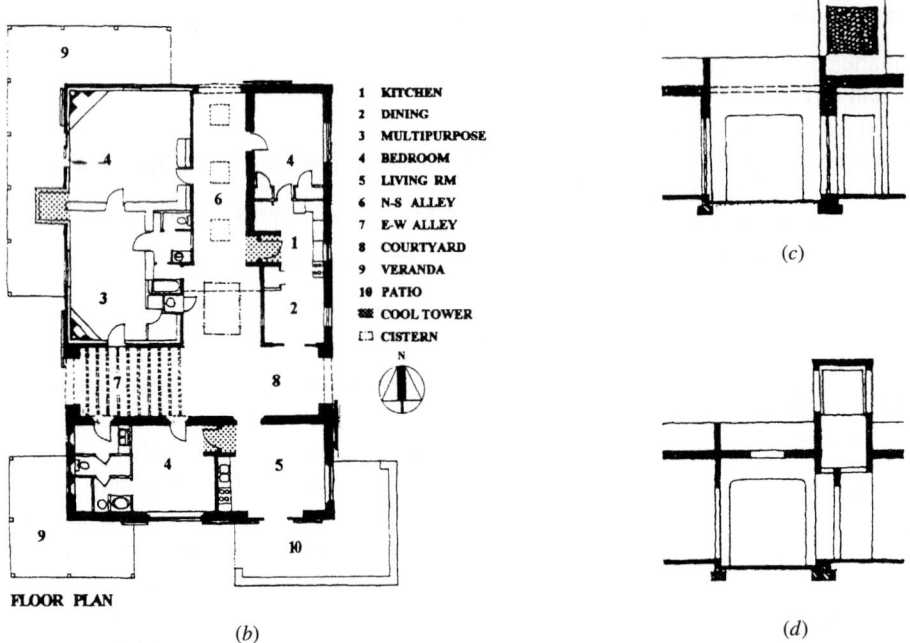

|   |   |
|---|---|
| 1 | KITCHEN |
| 2 | DINING |
| 3 | MULTIPURPOSE |
| 4 | BEDROOM |
| 5 | LIVING RM |
| 6 | N-S ALLEY |
| 7 | E-W ALLEY |
| 8 | COURTYARD |
| 9 | VERANDA |
| 10 | PATIO |
| ▨▨ | COOL TOWER |
| ⊏⊐ | CISTERN |

FLOOR PLAN

(b)

(c)

(d)

**Fig. 8.51** *Cooltowers dominate the silhouette of the Agua Blanca ranch in a remote desert near Tucson, Arizona. (a) West side, with one of the large rolling barn doors that allow the courtyard/outdoor spaces to be connected with cooler breezes or to be protected from hot winds. Air from the cooltowers, after use within the units, can be discharged to the courtyard, while (under the hottest conditions) it is still cooler than outdoor air. Thus, it helps to form a "cool pool" of air in the protected courtyard. (b) Plan of three units, each with a cooltower, around a T-shaped courtyard/covered passage outdoor space. (c) East–west section through the courtyard. (d) North–south section through one of the three cooltowers. (Courtesy of Martin Yoklic, Environmental Research Laboratory, University of Arizona.)*

Design simulations (Yoklic and Layseca, 1998) for a typical June in Tucson predict a flow from each cooltower of about 5000 cfm (2.36 m³/s), maintaining the indoor temperature at 74°F (23.3°C). In these quite small buildings, this translates into an ACH of about 60, considering the volume to include a height up to 6 ft (1.8 m), above which stratifica-

tion of warm air negates circulation through the upper potion of the space.

These computer-generated design simulations are summarized in Table 8.17. These results differ somewhat from the results of Givoni's (1994) equations, suggesting that perhaps in this case:

**TABLE 8.17 Cooltower Performance Hourly Simulations[a]**

| PART A. TOWER EFFECTIVE HEIGHT[b] = 10 FT (3 m) | | | | | | | | | | | | | | |
|---|---|---|---|---|---|---|---|---|---|---|---|---|---|---|
| Ambient (°F) | | | Tower Outflow | | | | | Ambient (°C) | | | Tower Outflow | | | |
| DB | WB | RH (%) | DB | RH (%) | Velocity (fpm) | Flow Rate (cfm) | Water Consumed (gal/h) | DB | WB | RH (%) | DB | RH (%) | Velocity (m/s) | Flow Rate[c] (m³/s) | Water Consumed (L/h) |
| 82 | 56 | 18.2 | 61.6 | 72.0 | 170 | 4254 | 9.8 | 27.8 | 13.3 | 18.2 | 16.4 | 72.0 | 0.86 | 2.00 | 37.1 |
| 87 | 59 | 18.0 | 65.1 | 71.0 | 175 | 4384 | 10.8 | 30.6 | 15 | 18.0 | 18.4 | 71.0 | 0.89 | 2.07 | 40.9 |
| 93 | 62 | 16.7 | 68.9 | 69.1 | 183 | 4572 | 12.3 | 33.9 | 16.7 | 16.7 | 20.5 | 69.1 | 0.93 | 2.16 | 46.6 |
| 98 | 64 | 15.0 | 71.8 | 67.0 | 190 | 4752 | 13.9 | 36.7 | 17.8 | 15.0 | 22.1 | 67.0 | 0.97 | 2.24 | 52.6 |
| 102 | 64 | 11.4 | 72.9 | 63.2 | 199 | 4987 | 16.1 | 38.9 | 17.8 | 11.4 | 22.7 | 63.2 | 1.01 | 2.35 | 60.9 |

| PART B. TOWER EFFECTIVE HEIGHT[b] = 15 FT (4.6 m) | | | | | | | | | | | | | | |
|---|---|---|---|---|---|---|---|---|---|---|---|---|---|---|
| Ambient (°F) | | | Tower Outflow | | | | | Ambient (°C) | | | Tower Outflow | | | |
| DB | WB | RH (%) | DB | RH (%) | Velocity (fpm) | Flow Rate (cfm) | Water Consumed (gal/h) | DB | WB | RH (%) | DB | RH (%) | Velocity (m/s) | Flow Rate[c] (m³/s) | Water Consumed (L/h) |
| 82 | 56 | 18.2 | 62.2 | 69.6 | 205 | 5133 | 11.5 | 27.8 | 13.3 | 18.2 | 16.8 | 69.6 | 1.04 | 2.42 | 43.5 |
| 87 | 59 | 18.0 | 65.7 | 68.7 | 212 | 5289 | 12.7 | 30.6 | 15 | 18.0 | 18.7 | 68.7 | 1.08 | 2.50 | 48.1 |
| 93 | 62 | 16.7 | 69.6 | 66.7 | 221 | 5516 | 14.4 | 33.9 | 16.7 | 16.7 | 20.9 | 66.7 | 1.12 | 2.60 | 54.5 |
| 98 | 64 | 15.0 | 72.5 | 64.5 | 229 | 5734 | 16.3 | 36.7 | 17.8 | 15.0 | 22.5 | 64.5 | 1.16 | 2.71 | 61.7 |
| 102 | 64 | 11.4 | 73.7 | 60.5 | 241 | 6017 | 18.9 | 38.9 | 17.8 | 11.4 | 23.2 | 60.5 | 1.22 | 2.84 | 71.5 |

| PART C. TOWER EFFECTIVE HEIGHT[b] = 20 FT (6.1 m) | | | | | | | | | | | | | | |
|---|---|---|---|---|---|---|---|---|---|---|---|---|---|---|
| Ambient (°F) | | | Tower Outflow | | | | | Ambient (°C) | | | Tower Outflow | | | |
| DB | WB | RH (%) | DB | RH (%) | Velocity (fpm) | Flow Rate (cfm) | Water Consumed (gal/h) | DB | WB | RH (%) | DB | RH (%) | Velocity (m/s) | Flow Rate[c] (m³/s) | Water Consumed (L/h) |
| 82 | 56 | 18.2 | 62.6 | 68.0 | 235 | 5864 | 12.9 | 27.8 | 13.3 | 18.2 | 17.0 | 68.0 | 1.19 | 2.77 | 48.8 |
| 87 | 59 | 18.0 | 66.2 | 67.0 | 242 | 6043 | 14.2 | 30.6 | 15 | 18.0 | 19.0 | 67.0 | 1.23 | 2.85 | 53.7 |
| 93 | 62 | 16.7 | 70.1 | 65.0 | 252 | 6302 | 16.1 | 33.9 | 16.7 | 16.7 | 21.2 | 65.0 | 1.28 | 2.97 | 60.9 |
| 98 | 64 | 15.0 | 73.0 | 62.7 | 262 | 6550 | 18.2 | 36.7 | 17.8 | 15.0 | 22.8 | 62.7 | 1.33 | 3.09 | 68.9 |
| 102 | 64 | 11.4 | 74.3 | 58.7 | 275 | 6874 | 21.1 | 38.9 | 17.8 | 11.4 | 23.5 | 58.7 | 1.40 | 3.24 | 79.9 |

*Source:* Excerpted from Yoklic and Layseca (1980). Additional simulations courtesy of Martin Yoklic. Metric conversions by the author.
[a]Tower has a total evaporative pad area of 64 ft² (5.95 m²); pad thickness is 4 in. (100 mm); tower outlet area is 25 ft² (2.32 m²); building outlet area is 50 ft² (4.65 m²). Ambient data are selected from Tucson, Arizona, typical June.
[b]Effective height is measured from the bottom of the evaporative pads to the top of the outlet to the building.
[c]Flow rate (m³/s) × 1000 = L/s.

$$t_{exit} = DB - 0.70\,[DB - WB]$$

and that the flow rate (in SI units) may be closer to

$$V = 0.04\,A_{evap}\,\sqrt{2z\,(DB - WB)}$$

where $z$ = effective height of the tower, that is, from the bottom of the evaporative pads to the top of the outlet to the building.

## (f) Roof Pond Cooling

Before doing the following detailed calculations, be sure that (1) you have checked your summer climate against the passive cooling design strategies (Fig 4.12), and that either the evaporative or the high thermal mass strategy is appropriate, and (2)

your building and climate have been checked for approximate performance based on the design guidelines. Note, however, that the design guidelines are based *only* on summer night DB temperature. With evaporative cooling (such as a light spray onto the surface of the roof pond containers), better performance can be expected, as shown by the following calculation procedure, based on Fleischhacker et al. (1982). This procedure can be used to check the size and depth required for your building's roof pond. It assumes an *optimum pond depth of 4 in. (100 mm)* but allows for other depths as well.

**STEP 1.** First, assemble the following data on your climate:

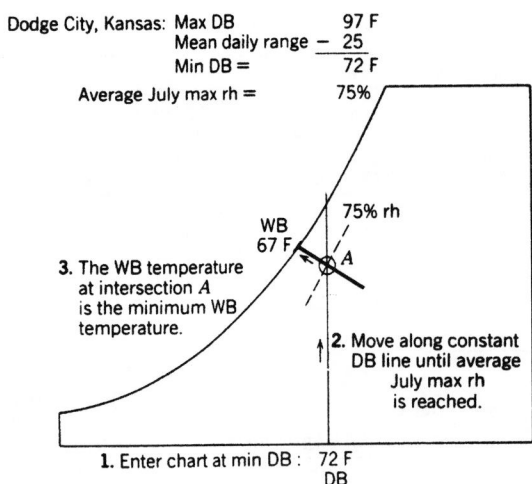

Dodge City, Kansas: Max DB          97 F
         Mean daily range −   25
         Min DB =            72 F
    Average July max rh =      75%

WB 67 F     75% rh

**3.** The WB temperature at intersection *A* is the minimum WB temperature.

*A*

**2.** Move along constant DB line until average July max rh is reached.

**1.** Enter chart at min DB :   72 F   DB

**Fig. 8.52** *Finding the minimum (nightly) WB temperature, summer design conditions, for Dodge City, Kansas.*

Maximum DB temperature (Appendix B)

Mean daily range (Appendix B)

Minimum DB temperature (= max DB − mean daily range)

Design WB (2½%) (Appendix B)

Average maximum RH for July (from local climatological data)

July average temperature (TA July in Appendix C)

From these data, determine two further characteristics for your climate: (1) minimum WB temperature and (2) average July operating hours for residential air conditioning, or *N*.

*Minimum WB temperature* can be determined from the psychrometric chart (shown in Figs. F.1 and F.2). Enter the chart with minimum DB and move vertically along the constant DB line until you reach the average maximum RH for July. At that intersection, refer to the diagonal WB temperature lines, from which minimum WB temperature can be determined (Fig. 8.52).

*Average July operating hours, N,* can be estimated from Fig. 8.53. Enter at July TA for your climate. Your climate's *N* will fall somewhere between the maximum *N* and minimum *N* lines; as a guide to estimating *N* for your climate, compare your climate's mean daily range and design WB to those of the cities shown in Fig. 8.53. The higher the mean daily range and the lower the design WB, the lower the *N*.

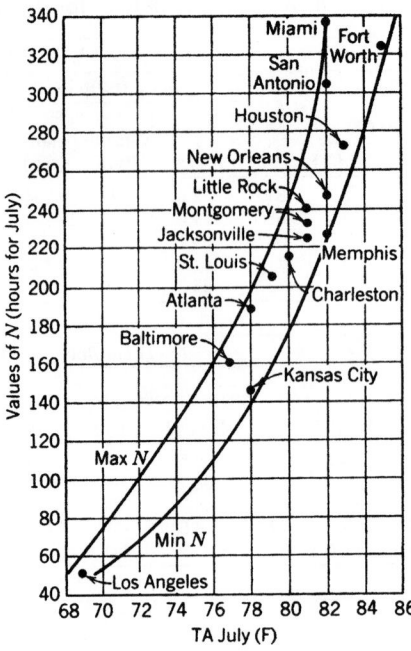

**Fig. 8.53** *Range of N, average July operating hours, for residential air conditioners. (Based on Fleischhacker et al., 1982.)*

**STEP 2.** Calculate the peak hourly heat gain *but exclude internal gains.* (The method given in Section 8.11 is appropriate for this roof–pond-sizing procedure.)

**STEP 3.** Approximate the daily total heat (excluding internal heat) to be stored in the roof pond, or $Q_E$, in Btu:

$$Q_E = \frac{(\text{Btu/h, peak hourly gain}) \times (N, \text{ hours for July})}{31, \text{ days in July}}$$

**STEP 4.** Determine the rate of the internal gain, in Btu/h, while the building is occupied.

**STEP 5.** Approximate the daily total internal heat to be stored in the roof pond, or, $Q_I$, in Btu.

$Q_I$ = hourly internal gains, in Btu/h
          × daily hours of building occupation

**STEP 6.** Calculate the daily heat gain directly to the roof pond through its insulated covers, or $Q_P$, in Btu. This formula assumes R-16 (R:SI − 2.8) insulating panels with white upper surfaces and foil faces on the under surface.

$$Q_P = 0.4(A_c)(4 \times \text{DB max, °F} - \text{DB min, °F} - 200)$$

where $A_c$ is the horizontal surface area of pond in square feet.

**STEP 7.** Consider whether fans will be used to stir the room air below the roof pond (to help the heat exchange between the pond and the room, as well as to provide comfort) and determine the value to be used for $h$, the overall heat transfer coefficient. A corrugated steel deck ceiling is assumed.

| Air Velocity | fpm (m/s) | 0 (0) | 44 (1.22) | 73 (0.37) | 115 (0.58) |
|---|---|---|---|---|---|
| Value of $h$ | | 1.25 | 1.47 | 1.53 | 1.70 |

**STEP 8.** Determine the highest comfortable internal air temperature, $t_{op}$, from the comfort criteria provided in Fig. 4.8.

**STEP 9.** Calculate the maximum allowable internal air temperature, $t_{imax}$, which will be higher than $t_{op}$ by a factor $F$ that is related to your climate's characteristics (see Table 8.18, Part A for values of $F$).

$$t_{imax} = t_{op} + F$$

**STEP 10.** Determine the allowable temperature swing of the roof pond.

(a) Calculate the maximum pond $t$:

max. pond $t =$

$$t_{imax} - \frac{\left( \substack{\text{peak hourly heat gain} \\ \text{including internal heat gains, Btu/h}} \right)}{h(A_c)}$$

where $t_{imax}$ is from step 9, $h$ is from step 7, and $A_c$ is the horizontal surface area of the pond in square feet.

(b) Calculate the minimum pond $t$, which depends on whether evaporative cooling will be used to help lower the pond's temperature and on several characteristics of the pond and the building.

1. For a *dry* pond surface:

min. pond $t_{dry} = DB_{min} + 1.5F° \pm$ corrections F°
(if any, from Table 8.18 Part B)

2. For a *wet* pond surface:

min. pond $t_{wet} = DB_{min} - \dfrac{DB_{min} - WB_{min}}{2}$

(c) Find the pond temperature swing $\Delta t_p$:

$$\Delta T_{pdry} = \text{max. pond } t - \text{min. pond } t_{dry}$$

$$\Delta T_{pwet} = \text{max. pond } t - \text{min. pond } t_{wet}$$

(Obviously, if neither minimum pond $t$ dry nor minimum pond $t$ wet is lower than maximum pond $t$, a roof pond cannot be used for cooling.)

**STEP 11.** Determine the required pond depth $D$ (in inches).

$$D = \frac{(0.19)(Q_E + Q_I + Q_p)}{(\Delta t_p)(A_c)}$$

where $Q_E$, $Q_I$, and $Q_P$ are the daily pond heat gains from steps 3, 5, and 6; $\Delta t_p$ is from step 10; and $A_c$ is the horizontal surface area of the pond in square feet.

*Note:* If $D$ is less than 4 in., consider reducing the pond size and recalculating. If $D$ is much more than 4 in., consider a larger pond area, more air motion indoors, or the use of a wet pond surface to more closely approach the optimum 4-in. depth, or see optional step 12.

**STEP 12.** (optional) Auxiliary mechanical air conditioning may offer a more economical alternative than increased pond size. The size, in tons (12,000 Btu/h) of air conditioning required, is determined as follows:

(a) desired $D$ is 4 in. optimum.

(b) $\Delta t_p = \left( \dfrac{0.19}{D} \right) \dfrac{(Q_E + Q_I + Q_P)}{A_c}$

(c) so max. pond $t = $ min. pond $t + \Delta t_p$

(d) and tons of AC at peak total hourly heat gain $=$

$$\frac{(\text{Btu/h} - [h(A_c)])(t_{i\,max} - \text{max pond } t)}{6000}$$

where the peak total hourly heat gain *includes* internal gains, $h$ is from step 7, $A_c$ is the horizontal surface area of the pond in square feet, $t_{imax}$ is from step 9, and max. pond $t$ is from step 10.

**EXAMPLE 8.6, PART B** Size the roof pond for the Albuquerque office building for which we predicted that a 4-in.-deep pond equal to three-fourths of the building's floor area would be sufficient. Assume that the one-story office is about 4000 ft² in area. Try a 4-in.-deep pond of 3000 ft².

## SOLUTION

**STEP 1.** Albuquerque, design data:

$DB_{max}$ is 94°F (Appendix B).

MDR is 27F° (Appendix B).

DB min. is $94 - 27 = 67°F$.

July RH *maximum* (nighttime) is approximately 50% (local data).

July TA is 79°F (Appendix C).

$WB_{min}$ is from Fig. F.1 at the intersection of the 67 DB line, and 50% RH is 56°F.

July operating hours $N$ for this dry, high area should be close to the minimum for TA = 79°; from Fig. 8.53, this is about 160 hours.

**STEP 2.** Determine the peak hourly gain. Earlier in this exercise, a total gain of 15 Btu/h-ft² was assumed as an average. Assume that 9 Btu/h ft² of this total represents the load from electric light, people, and equipment, and that 6 Btu/h ft² is due to envelope and ventilation gains. (No heat gains to the interior are assumed through the roof pond.) Then peak hourly gains (excluding internal)

$$= 6 \times 4000 \text{ ft}^2 = 24,000 \text{ Btu/h}$$

**STEP 3.**

$$Q_E = \frac{24,000 \text{ Btu/h} \times 160 \text{ h}}{31 \text{ days}} = 123,870 \text{ Btu}$$

**STEP 4.**

$$\text{internal gains} = 9 \text{ Btu/h ft}^2 \times 4000 \text{ ft}^2$$

$$= 36,000 \text{ Btu/ft}^2$$

**STEP 5.**

$$Q_I = (36,000 \text{ Btu/h})(9 \text{ hours of operation})$$

$$= 324,000 \text{ Btu}$$

**STEP 6.**

$$Q_p = 0.4(3000 \text{ ft}^2) (4 \times 94°F - 67°F - 200)$$

$$= 1200 \times (376 - 67 - 200)$$

$$= 1200 \times 109 = 130,800 \text{ Btu}$$

**STEP 7.** Fans will be used, at 115 fpm, for added comfort as well as for heat transfer. Therefore, $h = 1.7$.

**STEP 8.** From Fig. 4.4, with 115 fpm (0.58 m/s) air speed, it appears that 83°F is within the comfort zone. So, $t_{op} = 83°F$.

**STEP 9.** $t_{imax} = t_{op} + F$. In Table 8.18, Part A, Albuquerque's conditions appear to be somewhere between those of San Antonio and those of Phoenix, so assume $F$ to be 1.75.

$$t_{imax} = 83 + 1.75 = 84.75°F$$

**STEP 10.**

(a) max. pond $t = t_{imax} - \dfrac{\text{total hourly gain}}{h(A_c)}$

$$= 84.75 - \frac{24,000 + 36,000}{1.7 \times 3000}$$

$$= 84.75 - 11.75 = 73°F$$

(b) Assume a dry pond surface. Because this is a 4-in.-deep pond that is fully exposed to sky, with no night internal load, there is no correction factor from Table 8.18, Part B.

$$\text{min. pond } t_{dry} = \text{DB min.} + 1.5F°$$

$$= 67 + 1.5 = 68.5°F$$

For a wet pond,

$$\text{min. pond } t_{wet} = \text{DB}_{min} - \frac{\text{DB}_{min} - \text{WB}_{min}}{2}$$

$$= 67 - \frac{67 - 56}{2}$$

$$= 61.5°F$$

(c) The pond temperature swing is therefore

$$\Delta t_p \text{ dry} = 73 - 68.5 = 4.5F°$$

$$(\Delta t_p \text{ wet} = 73 - 61.5 = 11.5F°)$$

**STEP 11.**

$$D_{rqd} = \frac{0.19(123,870 + 324,000 + 130,800)}{4.5 \times 3000}$$

$$= \frac{109,947}{13,500} = 8.1 \text{ in. for a dry pond}$$

(For a wet pond, $D_{rqd} = 3.2$ in.) ■

It appears that a 4-in.-deep pond with a wet surface of somewhat less than 3000 ft² would be sufficient for this building in this climate.

Once the area and the depth of roof ponds have been determined, questions arise about the architectural integration of such large horizontal surfaces and their relative emphasis on heating or cooling. Figure 8.54 shows variations in the treatment of containers and the insulating panels to match the most important thermal role of roof ponds; Fig. 8.55 shows three typical approaches to the placement of roof ponds on buildings. A valuable source of information about the container bags, sliding insulation, roof decks, and other components needed for roof ponds is the *California Passive Solar Handbook* by Phillip Niles and Ken Haggard, written for the California Energy Commission in 1980. Commercially available products include the Skytherm® system, developed and pioneered by Harold R. Hay.

A variation of the roof pond (applying primarily to the cooling mode) is to move the water rather than move the insulation. Water-impervious insulation (such as Styrofoam) is secured in place above the metal ceiling at a distance sufficient to be a reservoir for the cooling water. At night, this water is pumped onto the top of the insulation, where it cools by radiation and evaporation, then trickles slowly back into the reservoir. This variation was tested at the School of Engineering Technology, University of Nebraska at Omaha campus. In winter, the reservoir of water above the metal ceiling acted as thermal mass for a DG passive system. However, cold winter rain must be excluded from that reservoir.

### (g) Earth Tubes

The use of the earth as a heat sink, in our culture at any rate, is still in an early development stage. The most direct application would be an underground building with uninsulated concrete walls set against the soil. The problem, of course, is that winter heat loss will likely exceed summer loss; if solar heat can be admitted (e.g., through skylights) to counterbalance the increased winter loss, uninsulated walls sized for the desired summer loss become more attractive. (However, except in dry climates, condensation on walls may pose a seasonal or even year-round problem.) A more thorough discussion of this option can be found in Watson and Labs (1983).

The long-term potential of the earth as a heat sink is lessened by the fact that soils are relatively slow heat conductors. For example, we ignore winter heat loss through concrete slabs to the earth below, calculating instead the heat losses through the slab's exposed perimeter.

Earth tubes (as discussed under design guidelines in Section 8.6g) are devices for cooling the incoming ventilating air through earth contact

before it enters the building. A small amount of air (perhaps equal to the minimum fresh air requirements—see Tables E.25 and E.26) is usually brought in through several tubes. These tubes are usually 8 to 20 in. in diameter, are buried at a depth of 5 to 10 ft, and are up to 200 ft long. Table F.12 summarizes several tested summertime applications.

The lowest temperature to which air can be cooled in such tubes will *approach* ground temperature; the more slowly the air moves through the tube, the more time there will be for cooling. The air temperature might come within 4F° (about 2C°) of the soil temperature under good conditions.

To approximate soil temperature around the tube:

1. Determine $t_{GW}$ the groundwater temperature (in °F) (assumed equal to deep-underground earth temperature) from the map in Fig. 8.56.
2. Determine $t_{amp}$, the average amplitude of surface temperature (in F°), using the map in Fig. 7.11.
3. Determine CF, the amplitude correction factor, depending upon whether the soil outside the earth tube is dry, average, or wet during the cooling period. The amplitudes, or seasonal variations in earth temperature, *diminish to near zero* at these depths (also called CF):

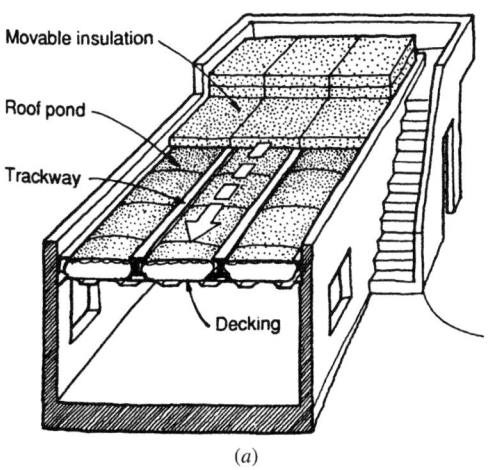

(a)

**Fig. 8.54** (a) Principal components of roof ponds with movable insulation. (b–g) Variations on roof ponds for optimization of heating or cooling performance. (Reprinted from Niles and Haggard, 1980, by permission of the California Energy Commission.)

THERMAL CONTROL

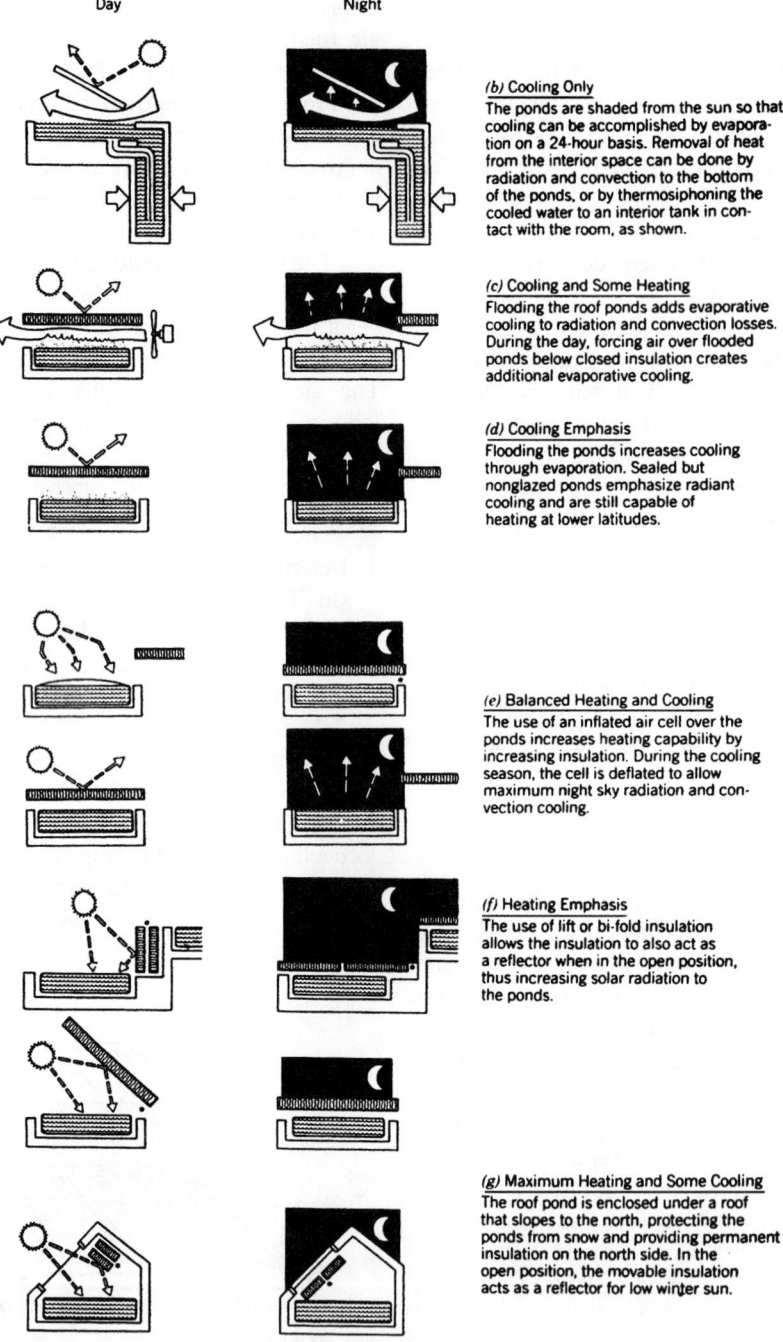

Day     Night

*(b)* Cooling Only
The ponds are shaded from the sun so that cooling can be accomplished by evaporation on a 24-hour basis. Removal of heat from the interior space can be done by radiation and convection to the bottom of the ponds, or by thermosiphoning the cooled water to an interior tank in contact with the room, as shown.

*(c)* Cooling and Some Heating
Flooding the roof ponds adds evaporative cooling to radiation and convection losses. During the day, forcing air over flooded ponds below closed insulation creates additional evaporative cooling.

*(d)* Cooling Emphasis
Flooding the ponds increases cooling through evaporation. Sealed but nonglazed ponds emphasize radiant cooling and are still capable of heating at lower latitudes.

*(e)* Balanced Heating and Cooling
The use of an inflated air cell over the ponds increases heating capability by increasing insulation. During the cooling season, the cell is deflated to allow maximum night sky radiation and convection cooling.

*(f)* Heating Emphasis
The use of lift or bi-fold insulation allows the insulation to also act as a reflector when in the open position, thus increasing solar radiation to the ponds.

*(g)* Maximum Heating and Some Cooling
The roof pond is enclosed under a roof that slopes to the north, protecting the ponds from snow and providing permanent insulation on the north side. In the open position, the movable insulation acts as a reflector for low winter sun.

**Fig. 8.54** *(Continued)*

**TABLE 8.18 Roof Pond Design Data**

| PART A. VALUES OF FACTOR "F" | | | | |
|---|---|---|---|---|
| | Design DB/ Mean Coincident WB | MDR | "F," Based on Interior Air Motion | |
| Location | (°F) | (F°) | None | Fans: 115 fpm (0.58 m/s) |
| Miami, Florida | 90/77 | 15 | 2.0 | 1.0 |
| San Antonio, Texas | 97/73 | 19 | 3.0 | 1.5 |
| Phoenix, Arizona | 107/71 | 27 | 4.0 | 2.0 |

| PART B. CORRECTIONS TO MINIMUM (DB) POND TEMPERATURE, F° | | | | | | | | | | | |
|---|---|---|---|---|---|---|---|---|---|---|---|
| Pond Depth (in.) | | | | Night Internal Load (Btu/h ft²) | | | | Pond Portion Exposed to Sky | | | |
| 2 | 4 | 6 | 10 | 0 | 2 | 4 | 8 | ⅓ | ½ | ⅔ | Fully |
| −1.5 | +0 | +0.7 | +1.0 | +0 | +0 | +0.8 | +2.5 | +3.3 | +2.1 | +1.2 | +0 |

*Source:* Fleischhacker et al. (1982).

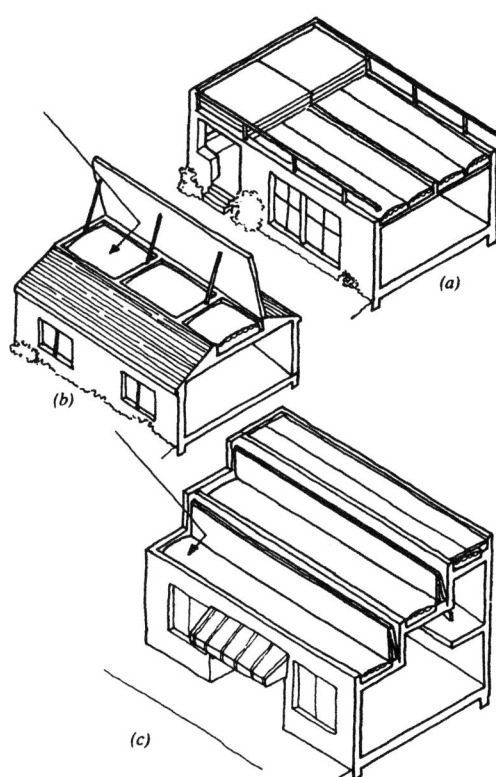

**Fig. 8.55** *Three examples of roof pond integration with building form. (a) Insulating panels slide open to stack over exterior (or untempered) space. See Fig. 8.22 for this system in use. (b) Hinged insulating panels with reflective undersurfaces can act as reflectors to increase winter insolation. However, some interference with summer night sky radiation can result, because the pond surface cannot "see" the entire sky dome. (c) Bifold insulating panels also act as reflectors, interfering less with diffuse solar radiation and allowing somewhat more exposure to the night sky than (b). (Adapted from Niles and Haggard, 1980, by permission of the California Energy Commission.)*

| | CF |
|---|---|
| Dry soil | 14 ft (4.25 m) |
| Average soil | 18 ft (5.5 m) |
| Wet soil | 22 ft (6.7 m) |

4. Determine D, the depth (ft) at which the earth tube will be installed.

5. Determine $t_{SG}$, the late summer ground temperature (°F) around the earth tube:

$$t_{SG} = (t_{GW}) + (t_{AMP}) [1 - (D/CF)]$$

(Note that when the tube is very deep and $D \geq CF$, $t_{SG} = t_{GW}$.)

6. Determine the temperature difference that makes cooling possible:

$$\Delta t = \text{desired indoor temperature} - t_{SG}$$

7. Determine the long-term cooling rate:

$$\text{Btu/h ft of tube length} =$$

$$(\Delta t)(C) \text{ (Area, tube surface, ft}^2\text{/ft of length)}$$

where $C$ is a factor that accounts for the long-term effects of heat flow from the tube to the soil:

$C = 0.11$ for dry soil
     (soil conductivity $k = 0.25$ Btu/h ft °F)
$C = 0.28$ for average soil ($k = 0.75$)
$C = 0.44$ for wet soil ($k = 1.5$)

  In SI units,

$C = 0.62$ for dry soil
$C = 1.59$ for average soil
$C = 2.50$ for wet soil

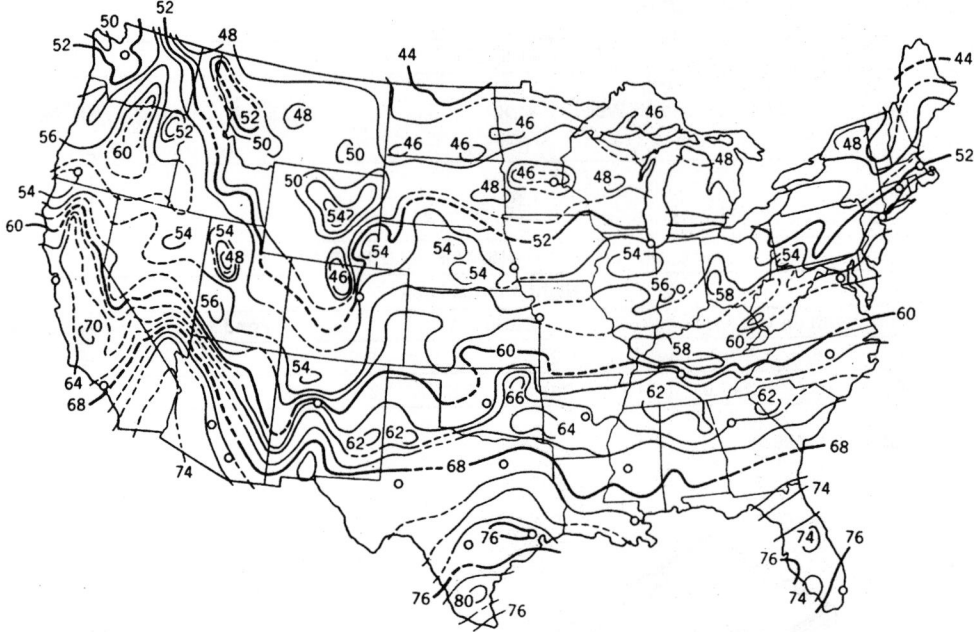

**Fig. 8.56** *Distribution of well-water temperature (°F) in the United States, adapted from the National Water Well Association. These temperatures may be assumed to be the yearly average ground temperatures. For seasonal temperature change amplitude, see Fig. 7.11. (Reprinted from* Passive Cooling, *1981, published by the American Solar Energy Society, Boulder, CO.)*

This represents an attempt to simplify a complex problem. The airflow rate is ignored, even though, with very slow flow, there is more time for heat exchange in the tube. The longer the tube is flooded with hot air, the higher will be the earth temperature. For more detailed information about and other approaches to earth tubes, see Givoni (1994), C.E. Frances, "Earth Cooling Tubes, Case Studies . . ." in Bowen et al. (1981), Abrams (1986), and Brown et al. (1992).

**EXAMPLE 8.7, PART B** A partially underground nature center of 4000 ft² (372 m²) in northern Pennsylvania has an hourly heat gain of 9.3 Btu/h ft², of which 2.2 Btu/h ft² are due to outdoor air–required ventilation. A long trench at a depth of 6 ft (1.8 m) in heavy, damp soil must be dug for a utility connection. If one earth tube, used only for cooling the fresh air, is installed at the bottom of the trench, what length and diameter will be needed?

**SOLUTION**

The total cooling needed from the earth tube is

$$(4000 \text{ ft}^2)(2.2 \text{ Btu/h ft}^2) = 8800 \text{ Btu/h}$$

(In Part A, using design guidelines, assuming heavy damp soil, a 60-ft-long tube 20 in. in diameter promised to deliver about 4500 Btu/h; two such tubes, about 9000 Btu/h.)

1. Determine $t_{GW}$: from Fig. 8.56, northern Pennsylvania has an average underground temperature of 52°F (11.1°C).

2. Determine $t_{AMP}$: the seasonal amplitude (from Fig. 7.11) is about 18F°.

3. Determine CF: in wet soil, this amplitude extends to 22 ft.

4. Determine D: the given trench depth is 6 ft.

5. Determine $t_{SG}$, the late summer ground temperature (°F) around the earth tube:

$$
\begin{aligned}
t_{SG} &= (t_{GW}) + (t_{AMP})\,[1 - (D/CF)] \\
&= (52) + (18)\,[1 - (6/22)] \\
&= (52) + (13) = \text{about } 65°F \text{ } at \text{ } most
\end{aligned}
$$

6. Determine the temperature difference that makes cooling possible:

$$\Delta t = \text{desired indoor temperature} - t_{SG}$$

In this case, we accept a maximum 82°F air temperature (the walls will be cooler, thanks to their underground location), so 82 − 65 = 17F°.

7. Determine the long-term cooling rate: Btu/h ft of tube length

$$= (\Delta t)(C)$$
(area, tube surface, ft²/ft of length)

Here, $C$ = about 0.44 for wet soil. Try first the 20-in.-diameter tube from the earlier design guidelines:

Btu/h ft of tube length

$$= (17F°)(0.44) \left[ \frac{20 \text{ in.} \times \pi \times 1 \text{ ft length}}{12 \text{ in./ft}} \right]$$
$$= 39 \text{ Btu/h ft}$$

To deliver the full 8800 Btu/h, the tube would need to be about 8800/39 = 225 ft in length. The design guideline promised considerably more, in this case, than could be delivered.

In early summer, the earth temperature around the tube would be at its yearly average temperature of 52°F. With this larger $\Delta t$, 82 − 52 = 30F°, the performance improves:

Btu/h ft of tube length

$$= (30F°)(0.44) \left[ \frac{20 \text{ in.} \times \pi \times 1 \text{ ft length}}{12 \text{ in./ft}} \right]$$
$$= 69 \text{ Btu/h ft}$$

and the tube length under those conditions would be about 8000/69 = 127 ft. ∎

## (h) Passive Cooling Summary

As a summary example of a passively cooled and solar-heated building, consider the Visitor Center for the Antelope Valley, California, Poppy Reserve, about 85 miles northeast of Los Angeles (Fig. 8.57). Set in a high desert where winter temperatures sometimes fall below 20°F, the building is 100% passively solar heated by a combination of direct-gain and Trombe wall strategies, interior thermal shades, and thermally massive ceiling, walls, and floor. An earth-sheltered, thermally massive building was chosen for its suitability to the environmentally sensitive site, as well as for its year-round thermal advantage in a desert climate. On summer days when air temperatures frequently exceed 100°F, a combination of an earth tube and evaporative cooling provides a modest supply of fresh air without overheating. Continuous power ventilating of the building at night draws the cool outdoor air through the tube, then over the thermally massive surfaces to provide added cooling capacity for the following day. The 2100-ft² building is also served by an 8-kW wind generator and has its own well.

**EXAMPLE 8.15** Use the psychrometric chart to investigate the cooling process at the Antelope Valley Visitor Center. The earth tube is 24 in (610 mm) in diameter and 150 ft (45.7 m) long, at an average depth of 6 ft (1.8 m).

### SOLUTION

First, determine the design conditions. Nearby Barstow, California, has a similar climate; the design condition (°F) is 104/68, with MDR 37 (and thus a nightly low of 104 − 37 = 67°F even at design conditions). Enter the psychrometric chart diagram (Fig. 8.58) at slightly more severe conditions, such as 105 DB, 70 WB (point A). Note that RH = about 17% in this desert location.

Next, examine the earth tube conditions:

1. Determine $t_{GW}$: from Fig. 8.56, the deep-earth temperature is about 64°F.

2. Determine $t_{AMP}$: the seasonal amplitude (from Fig. 7.11) is about 19F°.

3. Determine CF: in dry (desert) soil, this amplitude extends to 14 ft.

4. Determine D: the given trench depth is 6 ft.

5. Determine $t_{SG}$: the late summer ground temperature (°F) around the earth tube:

$$\begin{aligned} t_{SG} &= (t_{GW}) + (t_{AMP}) [1 - (D/CF)] \\ &= (64) + (19) [1 - (6/14)] \\ &= (64) + (10.9) = \text{about 75°F } at \text{ } most \end{aligned}$$

(Note that during the April poppy season, the earth temperature is likely to be at the yearly average of 64°F.)

The air in the earth tube will be about 4F° above this late-summer temperature, or 74.8 + 4 = approximately 79°F. (This may not sound very cool, but compare it to 104°F outside.) To follow what happens to the air in the tube, move horizontally on the chart in Fig. 8.58, from point A toward 79 DB (point B): at 79 DB, about 62 WB, the RH in the air has risen to about 34%.

At this point, the air enters the building through a small fountain consisting of water droplets. To chart this evaporative cooling, move from point B along the constant 62 WB line. If this fountain is a really good heat exchanger, we might achieve 80% of the distance from point B to saturation; call that point C (about 66 DB, 62 WB, RH now = 80%). This combination (of pessimistic air/earth temperatures and optimistic evaporative cooling expectations) is actually below the summer comfort zone!

The air now begins to pick up heat from the

*(a)*

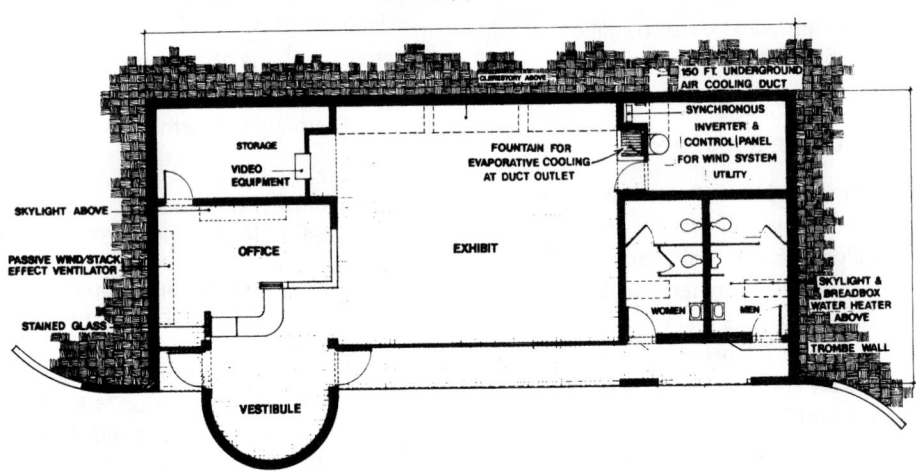

FLOOR PLAN  ↑ N

*(b)*

**Fig. 8.57** *Visitor Center for the Antelope Valley California Poppy Reserve near Palmdale, California. (a) The Visitor Center is set into the field of poppies; an 8-kW wind generator and the earth tube intake can be seen on the hillside above. (b) Plan shows earth sheltering for an elongated east–west axis, served both by direct gain and Trombe wall solar heating and by earth tube/evaporative cooling. (c) Solar energy enters mostly through the south glass; the glass block skylights and north clerestory are mostly for even distribution of daylighting. At rest rooms, a vented Trombe wall is used. Thermal shades on the interior protect the south glass and north clerestory on winter nights; a roll-down exterior sunscreen greatly reduces direct solar gain through south glass in summer. Concrete block walls and concrete roof and floor provide thermal mass, as useful for summer cooling as for winter solar heating. (d) In summer, the desert air first passes through an earth tube (assisted by a fan), then is evaporatively cooled within the building, in sight of visitors. This cooled air picks up the building's heat and is then exhausted from the stack ventilator. (Courtesy of The Colyer/Freeman Group, San Francisco.)*

**312**

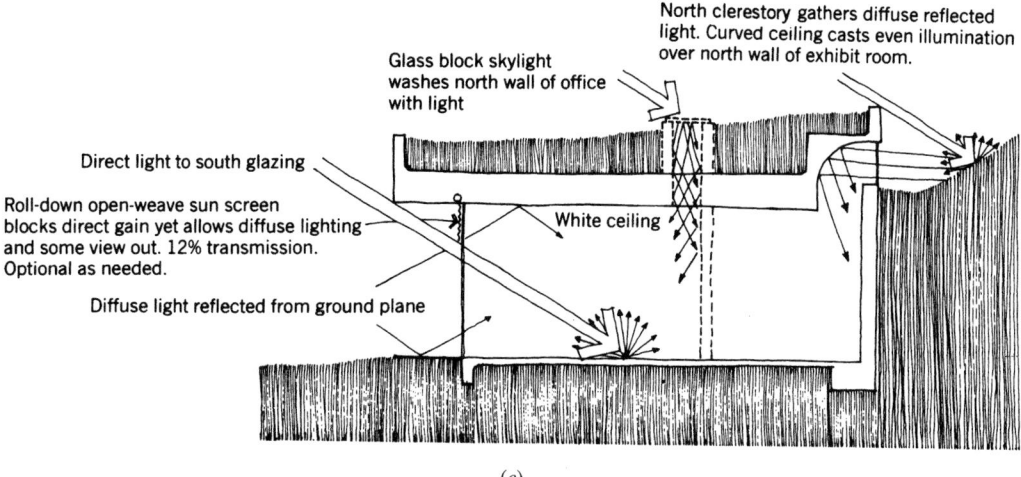

North clerestory gathers diffuse reflected light. Curved ceiling casts even illumination over north wall of exhibit room.

Glass block skylight washes north wall of office with light

Direct light to south glazing

Roll-down open-weave sun screen blocks direct gain yet allows diffuse lighting and some view out. 12% transmission. Optional as needed.

White ceiling

Diffuse light reflected from ground plane

*(c)*

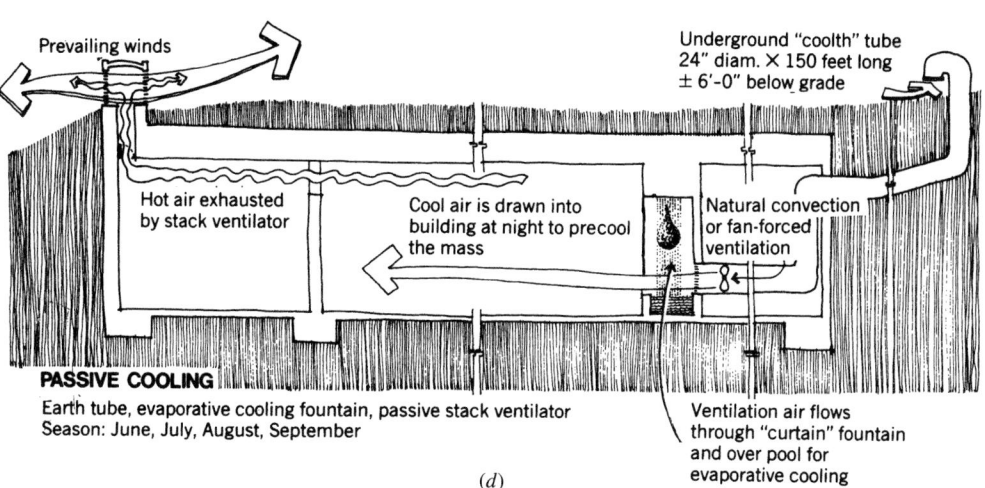

Prevailing winds

Underground "coolth" tube 24" diam. × 150 feet long ± 6'-0" below grade

Hot air exhausted by stack ventilator

Cool air is drawn into building at night to precool the mass

Natural convection or fan-forced ventilation

**PASSIVE COOLING**
Earth tube, evaporative cooling fountain, passive stack ventilator
Season: June, July, August, September

Ventilation air flows through "curtain" fountain and over pool for evaporative cooling

*(d)*

**Fig. 8.57** *(Continued)*

building's interior. With few sources of latent heat, we follow sensible heat gain horizontally to the right; if we are to maintain an interior temperature of, say, 82°F (with air motion and cool underground surfaces), the air can be heated to point *D:* 82 DB, 67 WB, RH = about 47%. In this condition, it is discharged through the stack ventilator at the west end of the building. (Note that it is much cooler than the outdoor air at point *A,* even at the point of discharge.)

But does this theoretical process actually occur? We must complete the calculation process:

**6.** Determination the temperature difference that makes cooling possible: For the psychometric equivalent process,

$$\Delta t = \text{outdoor temperature} - t_{SG}$$
$$= 105°F - 75°F = 30F°$$

**7.** Determine the long-term cooling rate:

$$\text{Btu/h ft of tube length} =$$
$$(\Delta t)(C)\ (\text{Area, tube surface, ft}^2/\text{ft of length});$$

Here, $C = 0.11$ for dry soil.
Btu/h ft of tube length

$$= (30F°)(0.11)\left[\frac{24\ \text{in.} \times \pi \times 1\ \text{ft length}}{12\ \text{in./ft}}\right]$$

$$= 20.7\ \text{Btu/h ft}$$

Note that this rate is similar to that of the tubes in Table F.12, most of which are smaller in diameter

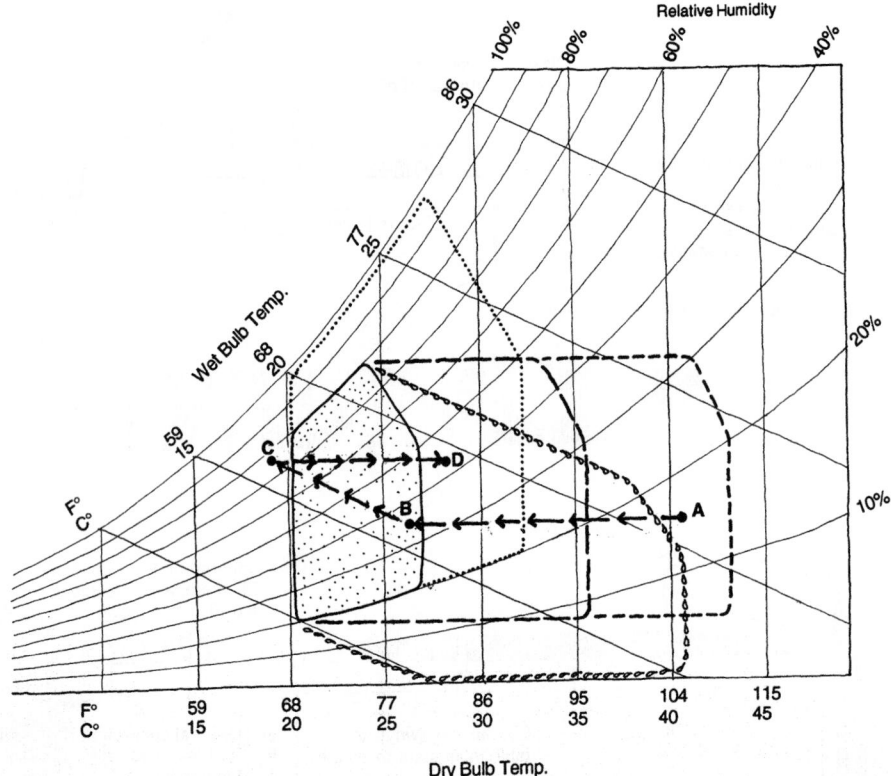

**Fig. 8.58** *Following the path of passively cooled air at the Antelope Valley Visitor Center. Under summer design conditions, air enters the earth tube at point A. It is sensibly cooled within the tube,\perhaps to the extent shown at point B. It then enters the building through a small fountain, perhaps reaching an evaporatively cooled state as far as point C. As the cooled air then picks up heat from the building, the air heads toward point D, where it is exhausted through the stack ventilator.*

but are located in moist soil. This tube will therefore cool outdoor air at the rate of (20.7 Btu/h ft) × (150 ft) = 3105 Btu/h under summer design conditions.

If the air in the tube is to drop from 105°F to 79°F assumed at point *B* in Fig. 8.58, at what rate must it move?

$$Btu/h = \text{flow rate, cfm} \times 1.1 \times \Delta t$$
$$3105 = \text{cfm} \times 1.1 \times [105 - 79]$$

The flow rate, 3105/28.1 = 110.6 cfm, is substantially less than the flow rates typical of Table F.12. Checking the velocity in this 2-ft.-diameter tube,

$$\text{ft/min} = \text{cfm/ft}^2 \text{ of cross section}$$
$$\text{velocity, ft/min} = 110.6/\pi(1\text{-ft radius})^2$$
$$= \text{about 35 ft/min}$$

again substantially less than the velocities typical of Table F.12.

It is likely that much more outdoor air will be pulled through the tube, thus entering the building at a higher temperature. Then evaporative cooling through the fountain will be necessary to deliver the air within the comfort zone.

For example, assume that for the cooling rate of 20.7 Btu/ft of length (based upon 105°F outside air, 75°F $t_{SG}$) and resulting total cooling of 3105 Btu/h, the fan provides a modest flow rate of 500 cfm. Then

$$3105 \text{ Btu/h} = 500 \text{ cfm} \times 1.1 \times \Delta t$$
$$\Delta t = 5.75 \text{ F}°$$

The outdoor air then leaves the tube (new point *B*) at about 99 DB, 68 WB and, with successful exposure to evaporative cooling in the fountain, might reach (new point *C*) 75 DB, 68 WB, quite humid but within the comfort zone and likely welcome in this desert environment. It then could acquire some heat from the building before it is exhausted. ∎

## References

Abrams, D.W. 1986. *Low Energy Cooling.* Van Nostrand Reinhold, New York.

Ashley, R. and J. Reynolds. 1994. "Overall and Zonal Energy End Use in an Energy Conscious Office Building," in *Solar Energy,* Vol. 52, No. 1, pp. 75–83.

ASHRAE. 2005. *ASHRAE Handbook—Fundamentals.* American Society of Heating, Refrigerating and Air-Conditioning Engineers, Inc., Atlanta, GA.

ASHRAE. 1997. *ASHRAE Handbook—Fundamentals.* American Society of Heating, Refrigerating and Air-Conditioning Engineers, Inc., Atlanta, GA.

Balcomb, J.D., D. Barley, R. McFarland, J. Perry, Jr., and W. Wray. 1980. *Passive Solar Design Handbook,* Vol. 2, *Passive Solar Design Analysis.* U.S. Department of Energy, Washington, DC.

Balcomb, J.D., R. Jones (ed.), R. McFarland, and W. Wray. 1984. *Passive Solar Heating Analysis.* American Society of Heating, Refrigerating and Air-Conditioning Engineers, Inc., Atlanta, GA.

Berry, B., J. Bown, J. Thornton, J. Turzynski, and M. Walton. 1995. "The New Inland Revenue Centre, Nottingham," in *The Arup Journal,* April. Ove Arup Partnership, London.

Bowen, A., E. Clark, and K. Labs. 1981. *Passive Cooling.* American Solar Energy Society, Boulder, CO.

Brown, G.Z. and M. DeKay. 2000. *Sun, Wind & Light,* 2nd ed. John Wiley & Sons, New York.

Brown, G.Z., B. Haglund, J. Loveland, J. Reynolds, and S. Ubbelohde. 1992. *Inside Out: Design Procedures for Passive Environmental Technologies,* 2nd ed. John Wiley & Sons, New York.

Chandra, S., P.W. Fairey, and M. Houston. 1986. *Cooling with Ventilation.* Solar Energy Research Institute, Golden, CO.

Fleischhacker, P., D. Bentley, and G. Clark. 1982. "A Simple Verified Methodology for Thermal Design of Roof Pond Cooled Buildings," in *Proceedings of the 7th National Passive Solar Conference.* American Solar Energy Society, Boulder, CO.

Gardner, I. and D. Hadden. 1992. "UK Pavilion, Expo '92, Seville," in *The Arup Journal,* Autumn. Ove Arup Partnership, London.

Givoni, B. 1994. *Passive and Low Energy Cooling of Buildings.* Van Nostrand Reinhold, New York.

Johnson, T. 1981. *Solar Architecture: The Direct Gain Approach.* McGraw-Hill, New York.

Mazria, E. 1979. *The Passive Solar Energy Book.* Rodale Press, Emmaus, PA.

Miller, H. (ed.). 1980. *Proceedings of the Passive Cooling Workshop, 5th National Passive Solar Conference.* American Solar Energy Society, Boulder, CO.

Niles, P. and K. Haggard. 1980. *California Passive Solar Handbook.* California Energy Commission, Sacramento.

Olgyay, V. 1963. *Design with Climate: Bioclimatic Approach to Architectural Regionalism.* Princeton University Press, Princeton, NJ.

SERI. 1981. *A New Prosperity: The SERI Solar Conservation Study.* Brick House Publishing, Andover, MA.

Thompson, T., N. Chalfoun, and M. Yoklic. 1994. "Estimating the Performance of Natural Draft Evaporative Coolers," in *Energy Conversion Management,* Vol. 35, No. 11. Elsevier Science Ltd., London.

Watson, D. and K. Labs. 1983. *Climatic Design: Energy Efficient Building Principles and Practices.* McGraw-Hill, New York.

Wittwer, V., W. Platzer, and M. Rommell. 1991. "Transparent Insulation," *Solar Today,* January–February.

Yoklic, M. and M. Layseca. 1998. "Earth Construction, Courtyards, Verandahs, and Cooltowers: An Arizona Ranch House with Middle East Vernacular," in *Proceedings of the 23rd National Passive Solar Conference.* American Solar Energy Society, Boulder, CO.

**THERMAL CONTROL**

# HVAC for Smaller Buildings

THE PRECEDING CHAPTERS DEALT WITH MOST OF THE early steps in the design process, such as the choice of thermal design strategy (Chapter 4), the choice of siting (Chapter 3), and the choice of the components and initial sizing of envelopes (Chapters 7 and 8). Chapter 5 dealt with strategies and equipment for indoor air quality. Chapters 9 and 10 carry the design process into the specifics of mechanical systems and equipment for heating and cooling.

There are various ways to organize this large collection of components and systems. The next two chapters are organized as follows:

Chapter 9:

1. A review of the relative thermal role of building envelopes versus their internal heating/cooling equipment
2. A look at the heating/cooling *system design process* begun by the architect and completed by the engineer
3. Equipment location and sizing for smaller buildings (one or a few thermal zones)
4. Controls for smaller buildings
5. Refrigeration cycles
6. Cooling-only systems
7. Heating-only systems
8. Heating and cooling systems
9. Psychrometry and refrigeration

Chapter 10:

10. The basics of the process by which me-

chanical systems are integrated into building design
11. The four generic HVAC system types for large buildings
12. Larger, centralized HVAC equipment
13. Air distribution considerations within spaces
14–17. Details of the four generic HVAC systems
18, 19. Large-scale system opportunities

Items 1, 2, and 10 are a more theoretical introduction to systems and equipment; hence, rather than reading the chapter in order, you could read those items first, then go on to the component-and-equipment sections.

## 9.1 REVIEW OF THE NEED FOR MECHANICAL EQUIPMENT

One of the most basic functions of buildings is to provide shelter from weather. In a carefully designed building, the roofs, walls, windows, and interior surfaces alone can maintain comfortable interior temperatures for most of the year in most North American climates. With appropriate scheduling, the most uncomfortable hours within buildings can often be avoided; for example, the siesta avoids the hottest afternoon hours within stores and office buildings. Several aspects of comfort and climate, however, pose difficult challenges for ordinary building forms and materials.

**317**

A building surface influences comfort primarily through its *surface temperature;* secondarily, it slowly changes *air temperature* (as when cool air moves across a warm surface). Important as these two determinants of comfort are, they sometimes are not sufficient by themselves. Especially in cooling situations, *air motion* and *RH* are significant comfort determinants; and for many indoor activities, *air quality* becomes an important issue in both heating and cooling.

Building form can work with climate to produce air motion for cooling, although the faster air speeds that extend the human comfort zone above 83°F (28°C) may be difficult to provide without mechanical assistance. Relative humidity is still controlled by mechanical (or chemical) means. Building form and materials may be able to keep spaces surprisingly cool, but without dehumidification, surfaces in many North American summer climates can become clammy and covered with mold.

It is difficult to filter air without a fan to force the air through the filtering medium. Electrostatic filtering, of course, is done within mechanical equipment.

Thus, whereas the desired air and surface temperatures can often be achieved by passive means (a combination of building form, surface material, and occasional user response), the comfort determinants of air motion, RH, and air quality often require mechanical devices. As the control of air properties—motion, moisture, particulate content—becomes more critical to comfort, the designer becomes more likely to respond with a sealed building, excluding outdoor air except through carefully controlled mechanical equipment intakes. In the recent past, this exclusion of outdoor air has often been accompanied by the exclusion of daylight, of view, of solar heat on cold days—in sum, by a general rejection of all aspects of the exterior environment. As designers come to terms with the role of mechanical equipment, they should also clarify the role of these devices in relation to the climate: are they occasional modifiers, permanent interpreters, or permanent excluders of the outdoors?

## 9.2 HEATING, VENTILATING, AND AIR CONDITIONING (HVAC): TYPICAL DESIGN PROCESSES

HVAC systems usually involve a minimum of three design stages. In the *preliminary design* phase, the most general combinations of comfort needs and climate characteristics are considered:

Activity comfort needs are listed.

An activity schedule is developed.

Site energy resources are analyzed.

Climate design strategies are listed.

Building form alternatives are considered.

Combinations of passive and active systems are considered.

One or several alternatives are sized by general design guidelines.

This level of analysis covers the process discussed in Chapter 1 and Sections 8.1 to 8.6. For smaller buildings, this analysis is often done by the architect alone. For innovative or unusual systems in smaller buildings, and especially for larger, multiple-zone buildings, consultants such as engineers and landscape architects often are included. The team approach is particularly valuable in assessing the strengths of various design alternatives. The architect and the consultants have very different perspectives, and when mutual goals can be clearly agreed on early in the design process, these perspectives are not only mutually supporting but can produce striking innovations whose benefits extend far beyond services to the clients of a particular building. By setting an example, the team can make available better environments for less energy for hundreds of subsequent buildings. With inspired teamwork, the distribution of HVAC services can enhance building form, as many examples in this chapter show.

By the time the *design development* phase is reached, one of the design alternatives has probably been chosen as the most promising combination of aesthetic, social, and technical solutions for the program. The consulting engineer (or architect, on a smaller job) is furnished with the latest set of drawings and the program. Typically, the architectural or mechanical engineer then:

I. Establishes design conditions.
   A. By activity, lists the range of acceptable air and surface temperatures, air motions, relative humidities, lighting levels, and background noise levels.
   B. Establishes the schedule of operations.
II. Determines the HVAC zones, considering:
   A. Activities.
   B. Schedule.

C. Orientation.

D. Internal heat gains.

III. Estimates the thermal loads on each zone:

A. For worst winter conditions.

B. For worst summer conditions.

C. For the average condition or conditions that represent the great majority of the building's operating hours.

D. Frequently, an estimate of annual energy consumption is made.

IV. Selects the HVAC systems. Often, several systems will be used within one large building because orientation, activity, or scheduling differences may dictate different mechanical solutions. Especially common is one system for the all-interior zones of large buildings and another system for the perimeter zones.

V. Identifies the HVAC components and their locations.

A. Mechanical rooms.

B. Distributions trees—vertical chases, horizontal runs.

C. Typical in-space components, such as under-window fan-coil units, air grilles, and so on.

VI. Sizes the components.

VII. Lays out the system. At this stage, conflicts with other systems (structure, plumbing, fire safety, circulation, etc.) are most likely to become evident. Because insufficient vertical clearance is one of the most common building coordination problems with HVAC systems (especially all-air systems—see Sections 10.2 and 10.5), the layouts must include sections as well as plans. Opportunities for integration with other systems also become more apparent at this stage: air ducts can also help distribute daylighting, act as sunshading devices, or fulfill other functions.

After the architect and the other consultants hold conferences in which HVAC system layout drawings are compared to those for other primary systems (structure, plumbing, electrical, etc.), *design finalizing* occurs. At this final stage, the HVAC system designer verifies the match between the loads on each component and the component's capacity to meet the load. Final layout drawings then are completed.

## 9.3 EQUIPMENT LOCATION AND SERVICE DISTRIBUTION

This chapter considers smaller buildings with one or a few thermal zones. For larger buildings, turn now to Chapter 10.

Smaller buildings are typically skin-load dominated: that is, for them, the climate dictates whether heating or cooling is the major concern. In some climates, only heating systems are needed; the building can "keep itself cool" during hot weather without mechanical assistance. In other climates, only cooling is needed. In still others, both heating and cooling are required.

### (a) Central or Local?

A skin-load-dominated building may have such differing but simultaneous needs that a room-by-room solution ("local") for heating, ventilating, and cooling is desirable. Consider the building with both north- and south-facing spaces on a cold, sunny winter day; one side gets ample solar heat, while the other side needs additional heat. The advantage of local systems is their ability to respond quickly to individual rooms' needs.

The central system also has advantages: the equipment is contained within its own space rather than taking up space within each room, and maintenance can be carried out without disrupting activities within those rooms.

### (b) Central Heating or Cooling Equipment

In the early design stages, determining an approximate size for the largest equipment is sometimes useful. Once the heating or cooling capacities are known, manufacturers' catalogs can be consulted for the dimensions of the heating and cooling equipment.

The critical decision in sizing the heating equipment is the *design temperature:* what is the lowest reasonable outdoor temperature for which a heating device can be sized if the desired interior temperature is to be maintained? These winter design temperatures are listed in Appendix B. When this is known, the next step is

design $\Delta t$ = inside temperature

– outside design temperature

In Section 8.4, the criteria for Btu/DD ft$^2$ were listed

in Table 8.5. To convert to the required capacity of a building's heating equipment, calculate

$$\frac{\text{Btu/DD ft}^2}{24 \text{ h}} \times \Delta t \times \text{ft}^2 \text{ floor area}$$
$$= \text{Btu/h heating capacity}$$

(*Note:* For passively solar-heated buildings, the backup heating unit is sometimes sized for the heat loss of the *entire* envelope and sometimes for the envelope *minus the solar wall*, just as the criteria were defined in Table 8.3. Similarly, buildings with *reliable* internal gains (lights, equipment, people, etc.) are sometimes designed with smaller heating units because internal gains supply a constant portion of the space heating needs. For a more complete discussion, see Section 8.8.)

The sizing of cooling (mechanical refrigeration) units is not so straightforward, as was evident when detailed hourly heat gain procedures were presented (Sections 8.11 and 8.13). However, a *very approximate* early estimate can be obtained from the estimated hourly gains in Table F.3. (*Warning:* This estimate is likely to be *lower* than that obtained using the peak heat gain hour, for which cooling equipment is often sized.)

sensible Btu/h cooling capacity
    = [approx. heat gain (Btu/h ft²)][floor area (ft²)]

Another common unit used for sizing mechanical refrigeration is tons of cooling capacity, 1 ton being equivalent to the useful cooling effect of a ton of ice, or 12,000 Btu/h (3516 W). Therefore, the required capacity in tons is

$$\frac{\text{heat gain, Btu/h}}{12{,}000} = \text{tons of cooling}$$

With this information about needed Btu/h (or Watts), equipment is then selected; physical size and service access areas are specified in catalogs. For a more general floor area estimate, see the sizing nomographs in Tables 10.3 and 10.4. Even more generally, equipment size may be estimated using these guidelines:

For ordinary equipment: 500 ft²/ton (46.5 m²/ton)

High-efficiency chillers: 1000 ft²/ton (93 m²/ton)

### (c) Distribution Trees

Central heating/cooling systems produce heating and cooling in one place, then distribute them to other building spaces according to their respective needs. The distribution tree is the means for delivering heating and cooling: the "roots" are the machines that provide heat and cold, the "trunk" is the main duct or pipe from the mechanical equipment to the zone to be served, and the "branches" are the many smaller ducts or pipes that lead to individual spaces.

For now, the questions to be answered about distribution trees for buildings are: How many? What kind? Where? A building can have one giant distribution tree, several medium-sized trees, or an orchard of much smaller trees. At one extreme, a large mechanical room is the scene of all heating and cooling production; leading from this room is a very large trunk duct with perhaps hundreds of branches. At the other extreme, each zone has its own mechanical equipment (such as a rooftop heat pump), with short trunks and relatively few branches on each tree.

What kind of distribution tree? Most simply, air (ducts) or water (pipes). Air distribution trees are bulky and therefore likely to have major visual impacts unless they are concealed above ceilings, below floors, or within vertical chases. Water distribution trees consume much less space (a given volume of water carries vastly more heat than does the same volume of air at the same temperature) and can be easily integrated within structural members such as columns. Both air and water trees can be sources of noise.

Where does the distribution tree fit in? On the exterior, it can lend a three-dimensional organizational structure to a façade. Exterior trees can take up a smaller amount of rentable floor space but require expensive cladding and are subject to considerable heat losses and gains, which could increase energy usage. Interior trees are often combined with other continuous vertical spaces, such as elevator shafts and stairways. If the choice is an exterior distribution tree, its potential contribution to façade performance should be considered. For example, the distribution tree might act as a sunshade or as a light reflector.

To carry the tree analogy to its logical conclusion, consider the "leaves," the points of interchange between the piped or ducted heating or cooling and the spaces served. One example is a large, bulky device such as a fan-coil unit on the exterior wall below windows. In contrast, a perforated ceiling system with thousands of small holes acting as a widely spread grille is essentially invisible.

## 9.4 CONTROLS FOR SMALLER BUILDING SYSTEMS

In the past, when buildings had mechanical equipment for heating only, thermostats were simple on–off devices; when they dropped below a setpoint, the heat was turned on. With the advent of mechanical cooling and then concerns about energy conservation, thermostats added a *deadband* separating heating need temperature from cooling need temperature and automatic setback provisions that allow buildings to change internal temperature between occupied and unoccupied hours. ANSI/ASHRAE Standard 90.2, *Energy-Efficient Design of Low-Rise Residential Buildings,* calls for thermostats capable of being set from 55°F to 85°F (13°C to 29°C) as well as an adjustable deadband, the range of which includes a setting of 10F° (5.6C°).

Building management systems (BMS) are now capable of regulating far more than temperature. They are capable of remote control, allowing systems to be activated in advance of the occupants' arrival; this is particularly useful for weekend and vacation homes, where heating/cooling and domestic hot water systems, if left on, could waste much energy during unoccupied periods. Door and window locks, security cameras, lighting, and appliances are potential partners in a comprehensive control system.

Future developments include *neural networks,* where automation systems are capable of learning while being used. They thus predict usage patterns, adjusting in advance without needing specific commands from occupants. When the building use pattern is highly predictable, as with many retail and commercial occupancies, these self-programming systems should learn very quickly how to anticipate needs while conserving energy.

In residences, usage patterns are likely more varied and less predictable. Experiments at the University of Colorado have revealed that even here, patterns may be more predictable than was first thought; for details, see Mozer (1998). The Adaptive Control of Home Environments (ACHE) system has two objectives: to anticipate inhabitants' needs and to save energy. Lighting, air temperature, venti-

lating, and water heating are controlled so that just enough energy, just when needed, is provided. Lights will be set to the minimum required, hot water maintained at the minimum temperature to meet demands, only occupied rooms will be kept at an optimum comfortable temperature, and so on.

The control framework compares energy costs to "misery" costs. Minimum settings and their time patterns are developed as the building is occupied over time. Whenever a preset minimum is overruled by the occupant, the system learns and readjusts accordingly; however, it occasionally tests lower minimum settings to be sure that energy conservation is not being unduly sacrificed to past desires for more.

## 9.5 REFRIGERATION CYCLES

Unlike the slow, diffuse heat transfer processes that characterize passive heating and cooling approaches, mechanical equipment can rapidly concentrate heating or cooling on demand. The refrigeration cycle is a particularly useful mechanical process in heating as well as cooling applications. The two types of heat transfer process commonly used in mechanical equipment for buildings are the *compressive* and the *absorption* refrigeration cycles. (Equipment that utilizes these cycles is discussed in the remainder of this chapter and in Chapter 10.)

### (a) Compressive Refrigeration

As shown in Fig. 9.1, the compressive refrigeration cycle is a scheme for transferring heat from one circulated water system (chilled water) to another (condenser water). This is done by the liquefaction and evaporation of a refrigerant, during which processes it gives off and takes on heat, respectively. The heat it gives off must be disposed of (except in the heat pump), but the heat it acquires is drawn out of the circulated water known as the *chilled water,* which is the medium for subsequent cooling processes.

Refrigerants are gases at normal temperatures and pressures, and must be compressed and liquefied to be of service later as heat absorbers. To be liquefied (see Fig. 9.1), the refrigerant must first be compressed to a high-pressure vapor; then,

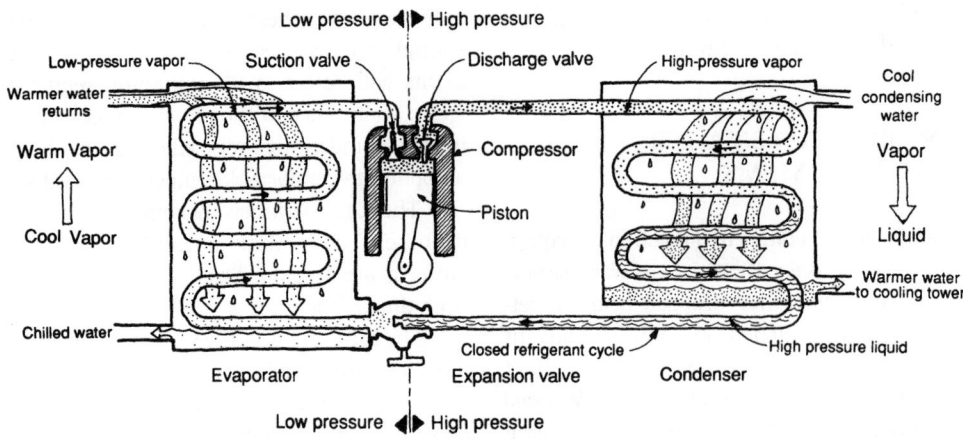

**Fig. 9.1** *Schematic arrangement of the compressive refrigeration cycle, providing chilled water for a building cooling system.*

by means of cool water, latent heat is extracted from the refrigerant, which condenses it to a liquid. This high-pressure liquid is a potential heat absorber because, when it is released through an expansion valve, it springs back mechanically to its gaseous form. In this change of state, it must take on latent heat by drawing heat out of the circulated water of the chilled water system.

It may be said that the refrigeration cycle pumps the heat out of the chilled water system into the condenser water system. Indeed, by special (reverse cycle) arrangements of the water systems, a *heat pump* is the result. (The compressive refrigeration cycle can be used to transfer heat between almost any media; whereas Fig. 9.1 illustrates a water–water cycle, Fig. 9.38 shows the heat pump in both air–air and water–air applications.)

The piston-type compressor in Fig. 9.1 can instead be one of several other types: *rotary, scroll,* or *screw* compressors, each with characteristics suitable to particular applications. See Chapter 10 for further discussion.

### (b) Alternative Refrigerants

Unfortunately, the most commonly used refrigerants of the early 1990s are a threat to our atmosphere, because they can escape from the equipment as chlorofluorocarbon (CFC) gases. The threats are stratospheric ozone depletion and global warming. A related global warming threat is the $CO_2$ released from the production of electricity that powers

chillers and other refrigeration machines. There is a potential conflict if lower-threat refrigerants are less efficient and therefore produce higher energy consumption. This conflict assumes continued power generation by fossil fuels; if photovoltaics are used to power these cooling devices, this aspect of the threat to global warming disappears.

Production of CFC refrigerants was banned in the United States by the mid-1990s. (A black market quickly developed for widely used CFC refrigerants such as Freon.) Fears that replacement refrigerants would be less efficient have largely disappeared; combinations of better chillers and new refrigerants produce energy savings.

Considerable efforts are underway both to reduce the likelihood of escaping refrigerant and to develop more efficient non-CFC refrigerants. The first alternative was hydrochlorofluorocarbon (HCFC) refrigerants—still a threat to our atmosphere but better than CFC. HCFC still contains chlorine, a major influence on ozone depletion. Thus, HCFCs themselves are due to be phased out in the first decades after 2000. A longer-term replacement is hydrofluorocarbons (HFCs); even this improvement still threatens our atmosphere. Yet another possibility is natural hydrocarbons (HC). In general, comparing these two alternatives, HFCs have higher global warming potential and long atmospheric lifetimes, but low toxicity and are nonflammable. HCs have negligible global warming potential and short atmospheric lifetimes, but are flammable and explosive.

One alternative is ammonia, used in the early years of compressive refrigeration but discontinued because of its acute toxicity and (relative to CFCs) its flammability. It is now returning in some absorption cooling cycles. Ammonia does have the advantage of having a strong and unmistakable odor; it warns of its leakage, unlike the other refrigerants. Use of ammonia will require regular maintenance, good ventilation, and good access/escape routes. It will encourage systems that keep the refrigerant loop within controlled spaces such as mechanical rooms.

The search for the ideal refrigerant continues and may never end. This is another area in which rapid developments may be expected in the near future. For a comprehensive review of refrigerants, see Calm (1994).

### (c) Absorption Refrigeration Cycle

This process is illustrated in Fig. 9.2. No CFCs or HCFCs are used here; the process uses distilled water as the refrigerant and lithium bromide (*salt solution*) as the absorber. In order to remove heat from chilled water, this cycle uses still more heat in regenerating the salt solution. Typically, it is less efficient than the simple compressive cycle and needs about twice the capacity for rejecting heat. However, the heat for salt solution regeneration may be provided by solar energy or by relatively high-temperature waste heat from another source, such as steam or hot water, such as from a fuel cell. Because the high-grade energy (electricity) needed to run a compressor is replaced by the lower-grade heat needed to run the generator, the absorption cycle can enjoy an energy advantage over the compressive cycle, even though it is less efficient.

There are several variations on the absorption cycle. When a fuel such as natural gas is used as the heat for the generator, it is called *direct fired*. When another heat source (such as waste heat) is used, it is called *indirect fired*. The relatively simple (!) cycle shown in Fig. 9.2 is a *single-effect* absorption cycle using one heat exchanger between the strong and weak salt solutions. A *double-effect* absorption cycle (powered by steam in Fig. 9.3) adds a second generator and condenser that operate at a higher temperature. It approximately doubles the efficiency of the single-effect cycle. A *triple-effect* cycle, in turn, pro-

vides a 50% efficiency improvement over the double-effect cycle.

## 9.6 COOLING-ONLY SYSTEMS

### (a) Fans

Before the advent of mechanical air conditioning, cooling was commonly achieved with simple air motion provided by fans. The summer comfort chart shown in Fig. 4.8 encourages increased air motion as a way to extend comfort into air temperatures in the mid-80s °F. As a general rule, people will perceive a 1F° decrease in air temperature for every 15 fpm increase in that air's speed past the body (about a 1C° decrease for every 1 m/s increase). Ceiling fans are often installed and run at slow speed to destratify warm air at the ceiling in winter; they can be run at higher speed in summer to provide added comfort through increased air motion. The air motion produced by ceiling fans will vary with the fan height above the floor, the number of fans in a space, and the fan's power, speed, and blade size. Figure 9.4 shows expected air speeds with one 48-in. (1220-mm) ceiling fan in a typical residential living room.

In the hot summer climate of Davis, California, an experimental house (sponsored by Pacific Gas & Electric, designed by the Rocky Mountain Institute) has eliminated a conventional cooling system through a series of alternative approaches. Fans play a major role: in addition to ceiling fans, a whole-house fan removes the hottest air from the central hallway, exhausting to the ventilated attic. More thermal mass (tile floors, double drywall), an attic radiant barrier, and low-ε, gas-filled windows are also used.

### (b) Unit Air Conditioners

The device shown in Fig. 9.5 is perhaps the most commonly seen piece of mechanical equipment in the United States. Perched in windows in full view of passersby, these window-box air conditioners noisily remind us that many of our buildings still are *not* centrally mechanically cooled. Mechanical cooling was considered a luxury until long after World War II.

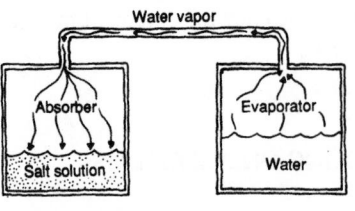

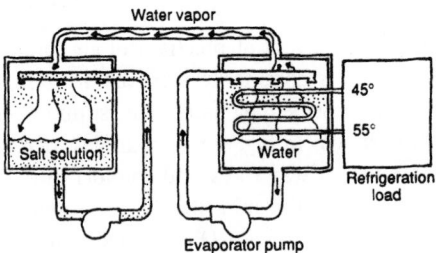

## 1 Evaporator and absorber

Consider two connected, closed tanks with a salt solution (lithium bromide) in one and water in the other. Just as common table salt absorbs water on a damp day, the salt solution in the absorber soaks up some of the water in the evaporator. The water remaining is thereby cooled by evaporation.

## 2 Evaporator coil and pump added

This refrigeration effect is utilized by putting a coil in the evaporator tank. Water from this tank is pumped to a spray header which wets the coil. The spray's evaporation chills water in the coil as it circulates to the refrigeration load. Solution pumped to spray in absorber raises efficiency.

## 3 Solution pumps and generator added

In an actual operating cycle, the salt solution is continuously absorbing water vapor. To keep the salt solution at the proper concentration, part of it is pumped directly to a generator where excess water vapor is boiled off. The reconcentrated salt solution is returned to the absorber tank, where it mixes with the solution sprayed to absorber in step 2.

## 4 Condenser and heat exchanger added

Water vapor boiled off from the weak solution is condensed and returned to the evaporator. A heat exchanger uses the hot, concentrated salt solution leaving the generator to preheat the cooler, weak solution coming from the absorber. Finally, condensing water circulating through the absorber and condenser coils removes the waste heat.

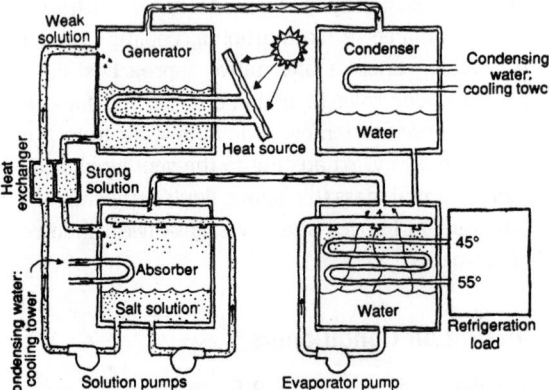

**Fig. 9.2** *The steps taken to build a* single-effect *absorption refrigeration cycle. The refrigeration load from this cycle (step 4) is a build-ing chilled water supply system. The heat source for the generator can be* indirect-fired *(steam, hot water, waste heat, or solar energy) or* direct-fired *(natural gas). (Adapted courtesy of the Carrier Corporation.)*

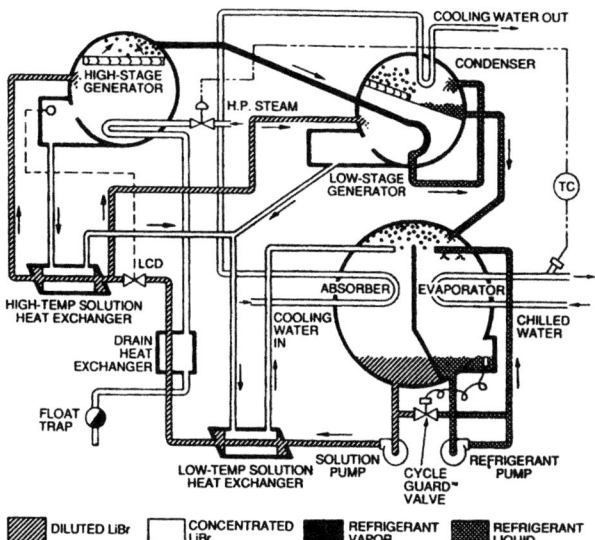

**Fig. 9.3** *Indirect-fired double-effect absorption cycle. Lithium bromide and water are used in this cycle. (Courtesy of the Carrier Corporation.)*

Built-in, through-wall air conditioners offer a low-first-cost way to provide separate zones for individual apartments, motel rooms, and so on (see Section 9.8). In noisy cities, the drone of these units masks street noise for the interior, thus potentially helping to promote relaxation. Unfortunately, such scattered units rarely afford the chance to conserve energy through the exchange of waste heat or the higher efficiencies that can accompany larger equipment. However, if turned on only when cooling is needed (i.e., when people are present), they can provide substantial savings over the larger always-on systems.

### (c) Evaporative Cooling: Misting

As explained in Section 8.12, the net effect of evaporative cooling is *no total change* in the heat content (enthalpy) of the treated air; its DB temperature is *lowered*, but there is an *increase* in RH. *People* feel cooler, although no change in total heat has occurred. One of the most direct approaches is a

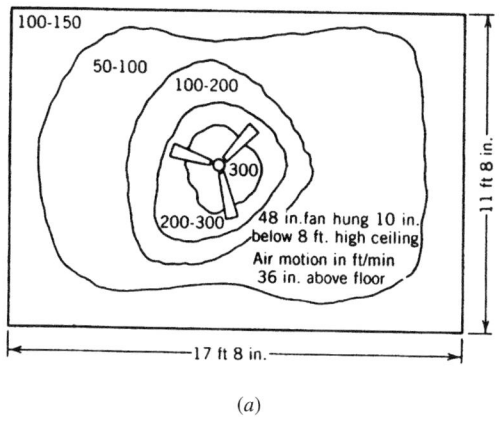

*(a)*

*(b)*

**Fig. 9.4** *Ceiling fans are useful in heating and cooling. In winter, at slow speeds, they destratify warm air at the ceiling. In summer (a) they extend the comfort zone by providing increased air motion. Room size is typical of residential living rooms. (Reprinted with permission of the American Society of Heating, Refrigerating and Air-Conditioning Engineers, Inc., from 1997 ASHRAE Handbook—Fundamentals.) (b) A ceiling fan is a visual feature in this North Carolina residence. (Christopher C. Morgan, AIA, Architect, Charlotte, NC.)*

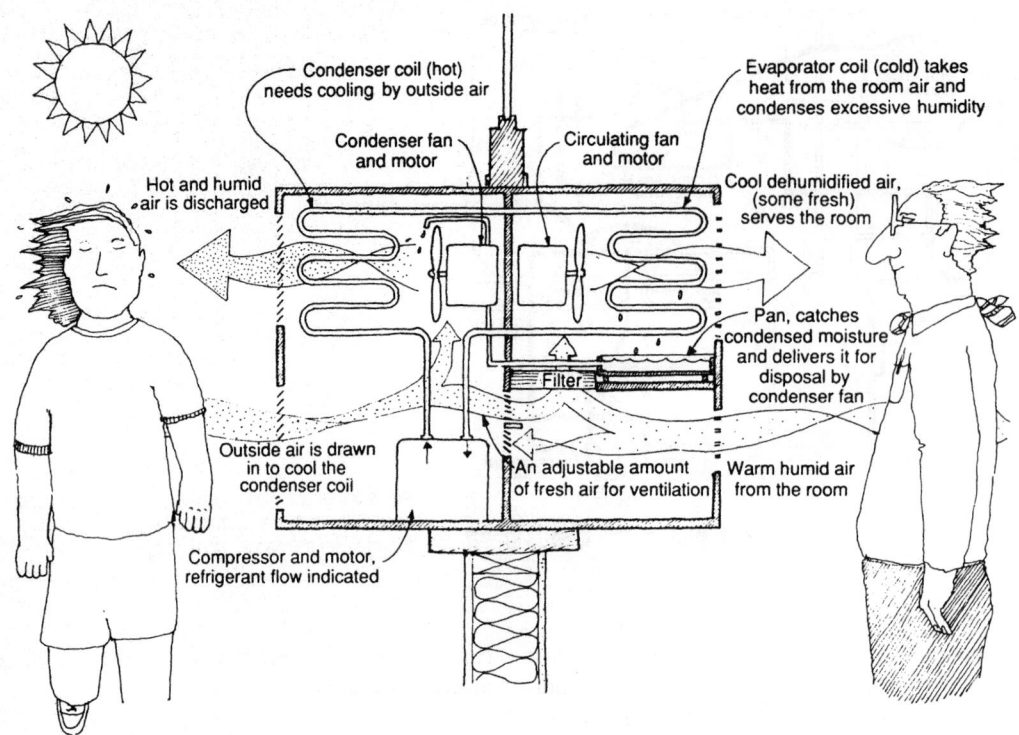

**Fig. 9.5** Diagram of a typical through-the-wall air-conditioning (cooling) unit. Direct heat exchange occurs between air and the process of evaporation (on one side) and condensation (on the other) of the refrigerant. The unit is self-contained, requiring access to outside air and electricity to power the compressor and the two fans. Typical capacities are 1 to 2 tons (12,000 to 24,000 Btu/h [3.5 to 7 kW]).

misting or fogging system whereby a fine spray of water droplets is blown into the air. A common application is for small outdoor areas—the team benches of football stadiums or refreshment pavilions. However, very large spaces in hot, dry climates can also benefit; Fig. 9.6 shows mist descending from a skylight diffuser in the Atocha railroad station in Madrid, Spain. A large space using fogging (in a Michigan conservatory) is shown later in Fig. 10.52.

### (d) Evaporative Cooling: Roof Spray

In the past, when roofs were poorly insulated, roof sprays were rather common ways to reduce heat gains. New variations promise new energy savings.

Roof color is the first consideration; white or near-white roofs are the first step toward energy savings through control of sol-air temperature. However, emissivity is also involved; the higher the emissivity, the faster a roof surface reradiates its heat to the sky. Table 9.1 compares the solar absorp-

tance, albedo (overall solar reflectance), and emittance of some common roofing materials. The combination of high albedo and high emittance resists solar heat gain most effectively. (It also promises to reduce the heat island effect in urban areas; see Chapter 3.) A solar reflectance index (SRI) is under development to allow quick comparisons between roofing products. Currently, the SRI scale = 0 at the combination of albedo 5% and emittance 90% (about equal to that of black asphalt shingles). The SRI scale = 100 at albedo 80%, emittance 90% (about equal to T-EPDM).

The *night roof spray thermal storage system* (NRSTS) cools water on a roof by night, using both night sky radiation and evaporation. The water is then stored for use the next day in building cooling. The water can be stored either on the roof (below floating insulation, above a structural ceiling) or in a tank below the roof. When stored on the roof, it becomes a variant of the roof pond cooling system. When stored in a tank, the water can be circulated through a cooling coil to precool air before it enters

**Fig. 9.6** *Mist is sprayed from a diffusing system located high in the Atocha railroad station in Madrid, Spain. It provides psychological reinforcement of evaporative cooling in a hot, dry summer.*

the chiller-fed cooling coils. It is thus an assistance to mechanical cooling. Three variations are shown in Fig. 9.7.

At a Nogales, Arizona, border patrol station, a 6500-ft² (604-m²) flat white roof was retrofitted with an NRSTS, utilizing a 10,000-gal (37,850-L) above-ground tank. Over the summer of 1997, it averaged 250 Btu/ft² day of cooling. Water use at various sites has averaged 4 to 5 gal/h (15 to 19 L/h) per 1000 ft² (per 93 m²) of roof area. Details are presented in Bourne and Hoeschele (1998).

### (e) Evaporative Coolers

These are also affectionately termed *swamp coolers* and *desert coolers* and are familiar devices in hot, arid climates (Fig. 9.8). (They are used in other climates for special high-heat applications such as restaurant kitchens.) They require a small amount of electricity to run a fan and some water to increase the RH of the air they supply to the building. This process of cooling is explained in Section 8.14.

The typical evaporative cooler (shown in Fig. 9.8) needs full access to outdoor air and is

thus often set on the roof; through-the-wall units are also available. Great quantities of dry, hot outdoor air are blown through pads kept moist by recirculated and makeup water. The "cooled" air is then delivered to the indoor space. The effect of the gently

### TABLE 9.1 Solar Performance of Roofing Materials

| Material | Absorptance (%) | Albedo (%) | Emittance (%) |
|---|---|---|---|
| White asphalt shingles | 79 | 21 | 91 |
| Black asphalt shingles | 95 | 5 | 91 |
| White granular-surface bitumen | 74 | 26 | 92 |
| Red clay tile | 67 | 33 | 90 |
| Red concrete tile | 82 | 18 | 91 |
| Unpainted concrete tile | 75 | 25 | 90 |
| White concrete tile | 27 | 73 | 90 |
| Galvanized steel (unpainted) | 39 | 61 | 4 |
| Aluminum | 39 | 61 | 25 |
| Siliconized white polyester over metal | 41 | 59 | 85 |
| Kynar white over metal | 33 | 67 | 85 |
| Gray EPDM | 77 | 23 | 87 |
| White EPDM | 31 | 69 | 87 |
| T-EPDM | 19 | 81 | 92 |
| Hypalon | 24 | 76 | 91 |

*Source:* Lawrence Berkeley National Laboratory.

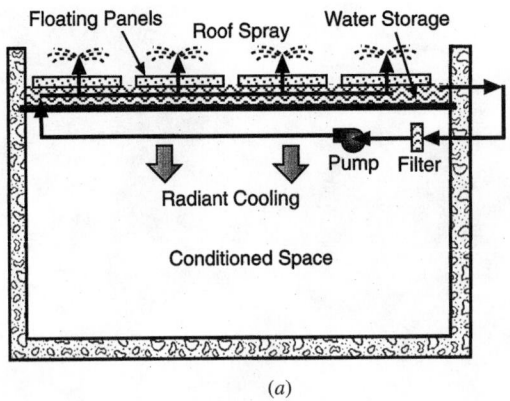

(a)

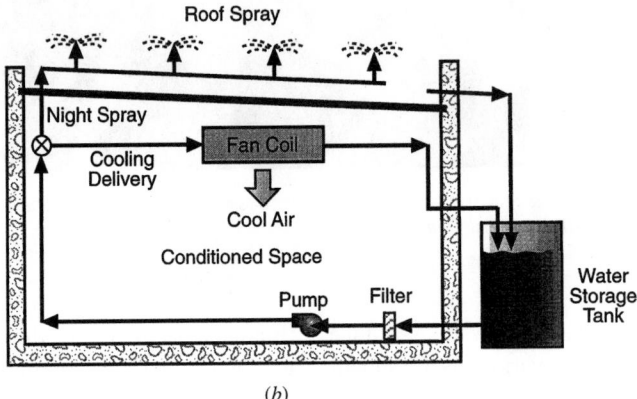

(b)

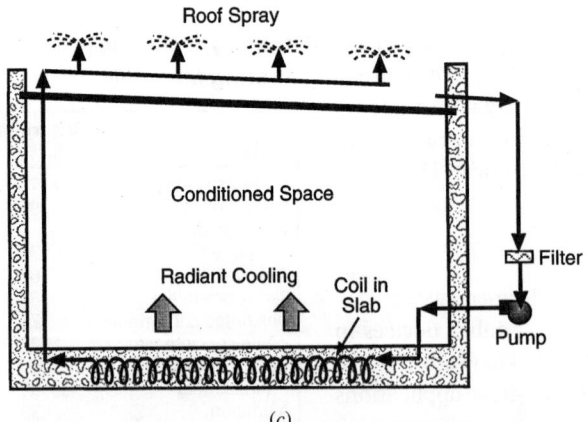

(c)

**Fig. 9.7** *Night roof spray thermal storage systems (NRSTS). (a) This version approximates the performance of the roof pond. (b) Remote water storage allows use of the cooled water at any time. (c) The floor slab is used as thermal storage in this version. (From* Technical Installation Review, *December 1997: WhiteCap Roof Spray Cooling System. Federal Energy Management Program, U.S. Department of Energy.)*

THERMAL CONTROL

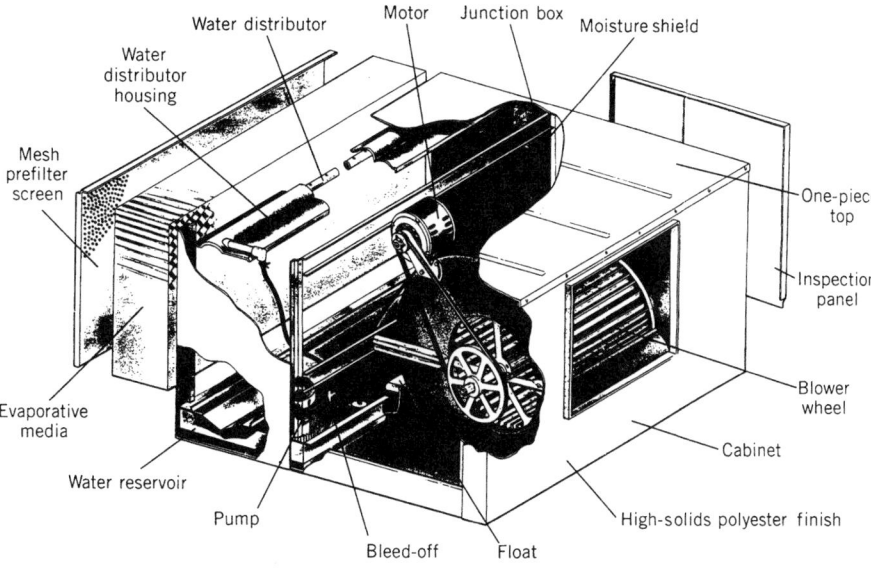

| Model | Dimensions (in.) | | | hp | Air Delivery | |
| | H | W | D | | cfm at 0.1 in. | cfm at 0.5 in. |
|---|---|---|---|---|---|---|
| AS50, AD50 | 28 | 42 | 45 | 1/3 | 2650 | 1600 |
| | | | | 1/2 | 3110 | 2130 |
| | | | | 3/4 | 3630 | 2806 |
| AS70, AD70 | 35 | 42 | 48 | 1/2 | 3750 | 2610 |
| | | | | 3/4 | 4350 | 3580 |
| | | | | 1 | 4820 | 4100 |

**Fig. 9.8** The UltraCool evaporative cooler has a low profile and uses internal components of plastic or stainless steel to reduce corrosion. Where ducts are short, use friction of cfm at 0.1 in. (Courtesy of Champion Cooler Corporation, El Paso, TX.)

moving cool air is to cool the body and to produce further cooling by evaporation of body moisture.

Air introduced into the indoor space must then be exhausted for the system to operate properly. By selecting the room through which the air is exhausted, one can route cool air as desired in any chosen path from unit to relief opening. The closer to the relief opening, the warmer the indoor air.

### (f) Indirect Evaporative Cooling

The preceding discussion concerns the simplest application of evaporative cooling, known as the *direct* process (once-treated outdoor air directly introduced to the space). Unfortunately, some areas have such hot summer daytime conditions that these simple evaporative approaches cannot pro-

duce comfort indoors. *Direct and indirect* processes have been combined to achieve more "real" cooling and better indoor comfort conditions; a psychrometric process diagram is shown in Fig. 5.16.

One of several such approaches is shown in Figs. 9.9 and 9.10. Warm, rather dry night outside air is evaporatively cooled and fed into a rock bed. The air's temperature is low enough to cool the rock bed, and its RH is moderate. (At the same time, the house is directly evaporatively cooled by a second cooling unit.) Figure 9.10 traces the process by day. Extremely hot, dry outdoor air (*A*) is drawn into the rock bed, where it is cooled by contact (*D*). It can then be passed through an evaporative cooler to achieve a better combination of RH and DB temperature (*E*). After picking up both sensible and latent heat, the air is exhausted (at approximately

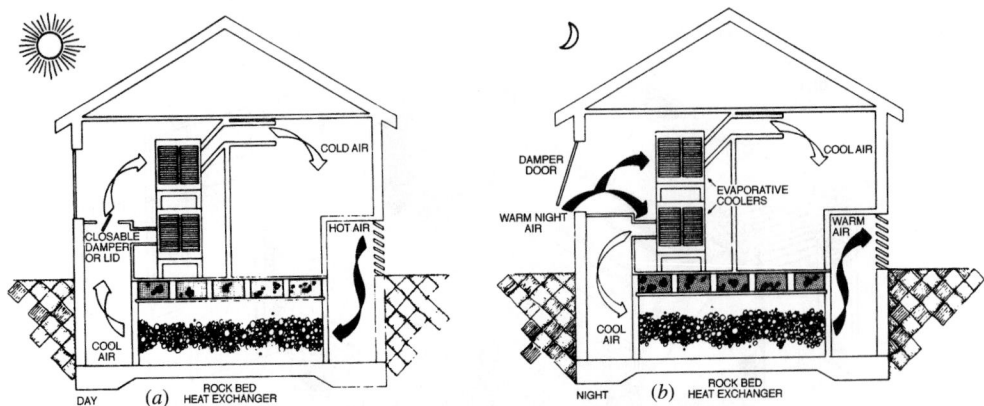

**Fig. 9.9** *(a,b) Direct–indirect evaporative cooling utilizing a rock bed for storage and heat exchange. (Reprinted by permission of the Environmental Research Laboratory, University of Arizona.)*

temperature *F*). Note that at condition *F*, it is still much cooler than outdoor air.

By comparison, simple direct evaporative cooling by day would have produced indoor supply air too hot and humid for comfort (*B*). Again, upon exhaust (*C*), it is still cooler than outdoor air.

Indirect evaporative cooling is combined with a direct refrigerant system in an innovative tent structure over a San Francisco department store (Fig. 9.11). Two layers of fiberglass, Teflon-coated fabric, separated by an average of 12 in. (300 mm), are supported by a network of cables hung from eight masts. The tent roof covers about 70,000 ft² (6500 m²) of sales floor; its 7% translucency to sunlight provides 450 to 550 footcandles of daylight. This greatly reduces the need for electric lighting, although some electric lamps, clipped onto exposed fire sprinkler pipes, are still used as accent lighting. About 3.5 W/ft² of solar gain, mostly in the form of this diffused daylight, penetrates the tent cover. When this solar gain is combined with heat gains from people and electric accent lights, a cooling load is always generated in San Francisco's mild climate, which rarely falls below 45°F (7°C). The roof's relatively high U-factor is thus advantageous in helping to lose heat. Because San Francisco overheats even more rarely, such low resistance to heat flow is not seriously disadvantageous in summer.

To remove this heat, four sets of direct refrigerant equipment are provided at the tent perimeter on grade. These feed into a perimeter plenum, from which the entire store is supplied with cooled air. The exhaust air is collected at the center and returned to help with the task of cooling. Indirect evaporative cooling (also called *sensible evaporative refrigeration*) units are used to cool air (see Fig. 9.11*b*). In this application, two units work in tandem. Air to be cooled (supply air) enters the heat exchanger of the first unit, which is cooled by evaporative cooling of outside air. During peak temperature periods, the supply air is only somewhat cooled by this process; sufficiently low temperatures for use on the interior are obtained by passing it through the second unit's heat exchanger. This second unit is cooled by evaporative cooling of the *exhaust* air from the store, which is cooler than outside air under summer conditions. Thus, the exhaust air does some work beyond the direct cooling of the tent's interior. This two-stage, indirect evaporative cooling process allows the supply air to be cooled without the RH being raised, as would be the case if direct evaporative cooling were used.

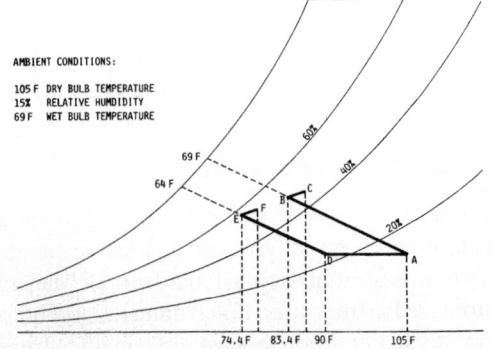

**Fig. 9.10** *Comparing direct (A to B to C) and indirect cooling (A to D to E to F) on the psychrometric chart.*

*(a)*

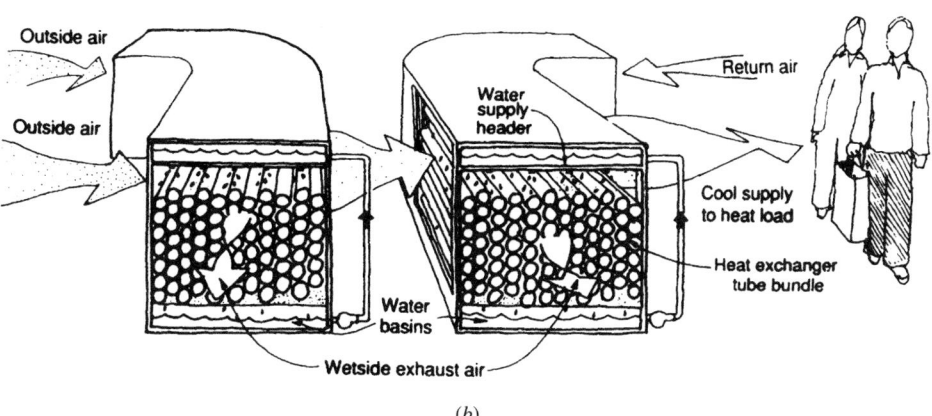

*(b)*

**Fig. 9.11** *Bullock's Department Store, San Mateo, California. (a) The eight-masted white fabric roof, highly reflective to ward off solar heat, transmits about 7% of daylight to provide ample diffuse daylight for sales areas. (Photo by Steve Proehl. Environmental Planning and Research, Inc., architects; Giampolo and Associates, Inc., mechanical and electrical engineers.) (b) Cooling is provided by a two-stage, indirect evaporative cooling system, which uses much less energy than does conventional compressive refrigeration.*

## 9.7 HEATING-ONLY SYSTEMS

There are substantial areas of North America in which the summers are so mild, but winters so cold, that heating systems are installed but not cooling systems. This is most common for residences and small commercial buildings.

Where should heating devices be placed within a space? The relative comfort in heated rooms from entire ceilings or floors as a heat source is discussed in Chapter 4. Figure 4.9 shows dissatisfaction with increased air velocity (the warmer the air, the higher the velocity tolerated), Fig. 4.10 shows dissatisfaction with asymmetric radiation (warmer ceilings were less tolerated), and Fig. 4.11 shows dissatisfaction with vertical air temperature differences (cold floors were less tolerated). Figure 9.12 shows why designers usually locate heat sources below windows, despite the fact that warmer temperatures just inside an exterior wall will drive more heat through the wall in cold weather (heat loss = $U \times A \times \Delta t$). As windows and walls become better insulated, their interior surface temperatures rise and the need for heat at the edge grows less. Indeed, with superinsulated components, the need for space heat disappears because internal gains from the sun, lights, appliances, and occupants can heat the space.

### (a) Wood Heating Devices

After the sun, the most ancient method of heating is the radiant effect of fire. With each step from campfire to fireplace to wood stove, more of the fuel's heat was captured for the room rather than wasted to the outdoors (Fig. 9.13). Although many people enjoy the sight, sound, and smell of the open fireplace, the tightly enclosed wood stove with a catalytic combustor is a substantially more efficient and less polluting approach to heating. Indoor as well as outdoor air pollution is a serious issue with fireplaces and stoves. Combustion generates carbon monoxide, breathable particulates, and, at times, nitrogen dioxide. Wood smoke can cause nose and throat irritation; it can remain in the lungs, and it can trigger asthmatic attacks. Keeping a clean chimney, burning small, hot fires rather than large, smoky ones, using seasoned wood, and ensuring adequate ventilation to the wood-burning device are strategies to minimize pollution and risk.

*Open fireplaces* may be lovely to look at, but the amount of air exhausted up the chimney can quickly cause more heat losses than heat gained from the fire. The colder the outside air, the greater the net heat loss. Masonry mass around fireplaces can store and release some heat; for energy conservation, the mass should be surrounded by the building rather than located on an exterior wall. ANSI/ASHRAE Standard 90.2 requires fireplaces to have a tight-fitting damper, firebox doors, and a source of outside combustion air within the firebox.

*Wood stoves* are available in a wide variety of styles and are made of several materials. The sizes of such stoves are often difficult to determine. Manufacturers rarely specify the Btu/h output, which depends on the density, moisture content, and burn time of the wood fuel. Wood that has been split, loosely stacked, and covered from rain for at least 6 months should achieve a moisture content of about 20% by weight. The following sizing procedure assumes no more than this 20% moisture content. For more details, see Volume 35 of *Alternative Sources of Energy Magazine* (1978).

The formula for the hourly heat output to a room from a wood stove is

$$\text{Btu/h} = \frac{(V)(E)(D)(7000)}{T}$$

where

$V$ = useful (loadable) volume of the stove ($ft^3$)

$E$ = percent efficiency, expressed as a decimal (<1.0); see Fig. 9.13

$D$ = density of the wood fuel (Table 9.2)

$T$ = burn time (hours) for a complete load of firewood; usually assumed at an 8-hour minimum

7000 = Btu/lb of firewood, 20% moisture content

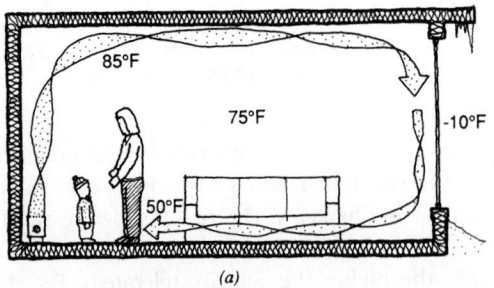
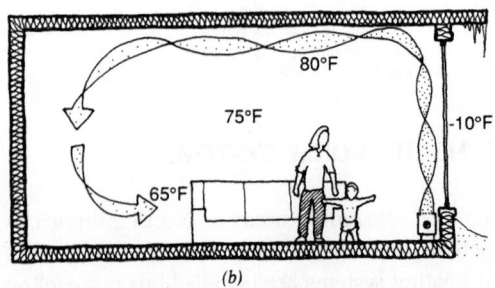

**Fig. 9.12** *Locating a heat source near an interior wall (a) encourages a cold draft along the floor in winter. Below a window (b) it evens the temperature throughout the room but also loses more heat through the window.*

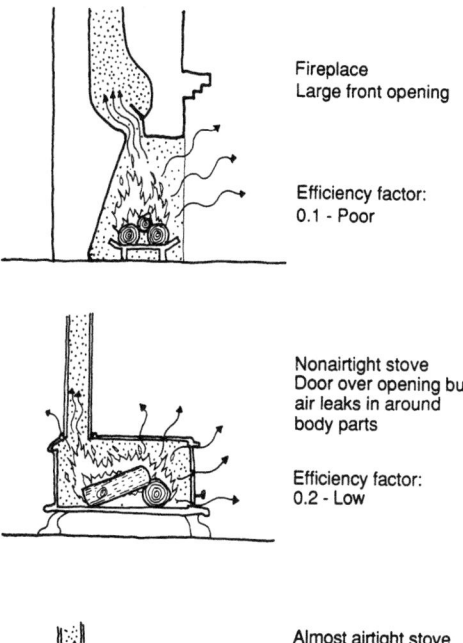

Fireplace
Large front opening

Efficiency factor:
0.1 - Poor

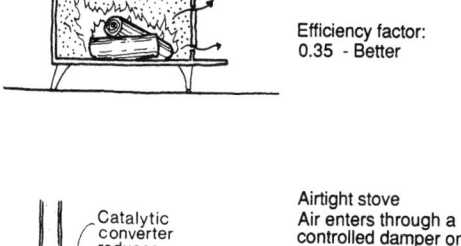

Nonairtight stove
Door over opening but
air leaks in around
body parts

Efficiency factor:
0.2 - Low

Almost airtight stove
Air leaks in around
door only

Efficiency factor:
0.35 - Better

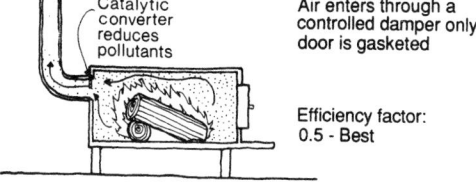

Catalytic
converter
reduces
pollutants

Airtight stove
Air enters through a
controlled damper only;
door is gasketed

Efficiency factor:
0.5 - Best

**Fig. 9.13** *Wood-burning devices have substantially increased in efficiency since the time of the open fireplace. (Copyright © 1978 by* Alternative Sources of Energy, *Issue 35.)*

**TABLE 9.2 Approximate Average Wood Density**

| Type | Density (lb/ft$^3$)[a] |
|---|---|
| Shagbark hickory | 40.5 |
| White oak | 37.4 |
| Red oak | 36.2 |
| Beech | 36.2 |
| Sugar maple | 34.9 |
| Yellow birch | 34.3 |
| White ash | 33.7 |
| Black walnut | 31.2 |
| American elm | 28.7 |
| Spruce | 25.6 |
| Hemlock | 23.7 |
| Aspen | 23.1 |
| Red cedar | 18.7 |
| White pine | 17.5 |

*Source: Alternative Sources of Energy Magazine,* #35, © 1978. Reprinted by permission.

[a]These values are approximate; they vary a great deal.

made from densified quality sawdust, a manufacturing by-product. The form and content of this fuel produce a highly efficient burn with less pollution emitted. The fuel is cleaner and takes less storage space than cordwood; an electric auger automatically feeds fuel into the burnplace to maintain a fire. From 10,000 to 50,000 Btu/h (2930 to 14,650 W) are produced, depending on the model and operating settings.

Wood stoves are frequently used as the sole mechanical heat source for an entire building, such as a residence or a small commercial building that is passively solar heated. Because radiant heat is the dominant form of heat output, the areas that "see" the stove get most of the benefit. However, *circulating* stoves convert a larger portion of their heat to convected heat, which produces a layer of hot air at ceiling level. By providing a path between rooms at the ceiling, this hot air will slowly spread throughout a building; it also easily finds its way upstairs, because warm air rises. The more thermally massive the ceiling construction, the longer it will store and reradiate the heat from this warm air mass.

The flue leading from a wood stove carries very hot gases that are a potential source of heat (and pollution). The flue can be exposed to a space, making its radiant heat available, or simple heat exchangers can be constructed (such as the preheating of domestic hot water). More elaborate heat recovery devices—for boiler flue heat recovery—are discussed in Section 10.3.

Catalytic combustors, a recent development, reduce the air pollution from wood burning. These

("Bone-dry" wood can be assumed to have 8600 Btu/lb.) Note that a *drop* in burn time *increases* the Btu/h output; it is evident that when the air supply is increased to the stove, the fire burns hotter, consuming the wood more quickly. To meet the design heat loss (worst condition) for a room, burn times of 8 to 10 hours should be assumed. Stoves rarely need relighting with a 10-hour burn time.

*Pellet stoves* were introduced in 1984 and have several desirable characteristics. The pellets are

devices are honeycomb-shaped, chemically treated disks as much as 6 in. in diameter and 3 in. thick (150 mm in diameter, 75 mm thick). They are either inserted in the flue or built into the stove itself. When wood smoke passes through the combustor, it reacts with the chemical and ignites at a much lower temperature; this causes gases to burn that otherwise would have gone up the flue. The result is more heat produced, less creosote buildup in the flue, and fewer pollutants in the atmosphere. Like the catalytic converters in autos, these devices impose limits on the fuel: plastic, colored newsprint, metallic substances, and sulfur are ruinous to combustors, which means that the stove must be used as a wood burner, not a trash incinerator.

Wood stoves have a larger impact on building design than do most other heating devices. Either noncombustible materials must be placed below and around them or minimum clearance to ordinary combustible building materials must be provided. Furniture arrangements and circulation paths must be designed with the very hot stove surfaces in mind. Hot spots occur near the stove; cold spots occur whenever visual access to the stove is blocked. Thermally massive materials near the stove are advantageous in leveling the large temperature swings that can accompany the on–off cycle of the stove; this affinity for thermal mass has made the wood stove a popular choice for backup heat in passively solar-heated buildings. Finally, the amount of space required for wood storage should not be overlooked; recall the impact of the wood storage space on the house shown in Fig. 2.2. A covered, well-ventilated, easily accessible, and quite large space is optimum.

*Masonry heaters* overcome many of the metal wood stoves' disadvantages. Their footprint is rather small compared to their height; typically, they are used to heat the entire building (such as a residence). An inner vertical firebox supports a hot, clean burn, resulting in efficient combustion; combustion gases then flow downward in outer chambers, transferring heat to exterior masonry surfaces. In Finnish masonry heaters, this is termed *contraflow* (Fig. 9.14). Cool air at the floor of the room flows upward as it is heated by contact with this masonry; the temperature difference between masonry and air remains fairly constant, with the highest temperatures at the top. Heat is gentle and even; dangerously hot surfaces are avoided. Fires

Warmed room air rises

Heating smoke sinks as it cools    To the chimney

**Fig. 9.14** *The Finnish contraflow masonry heater.*

may be built at 6 P.M., and combustion is completed by a family's bedtime; the heat continues to radiate all night, but no fire is burning while people sleep. Research at Finland's Tampere University of Technology has resulted in optimum masonry heater designs, described in Barden and Hyytiainen (1993).

## (b) Electric Resistance Heaters

These common devices carry the disadvantage of using high-grade energy to do a low-grade task, as shown in Fig. 2.6. Their advantages, however, are impressive: low first cost and individual thermostatic control that can easily be used to make each room a separate heating zone. Thus, the energy wasted at the electricity-generating plant (usually 60% to 70%) can be partially "recovered" at the building, where unused rooms can remain unheated. A few of the many types of electric resistance heaters are shown in Fig. 9.15. As in the case of metal wood stoves, surfaces can sometimes reach

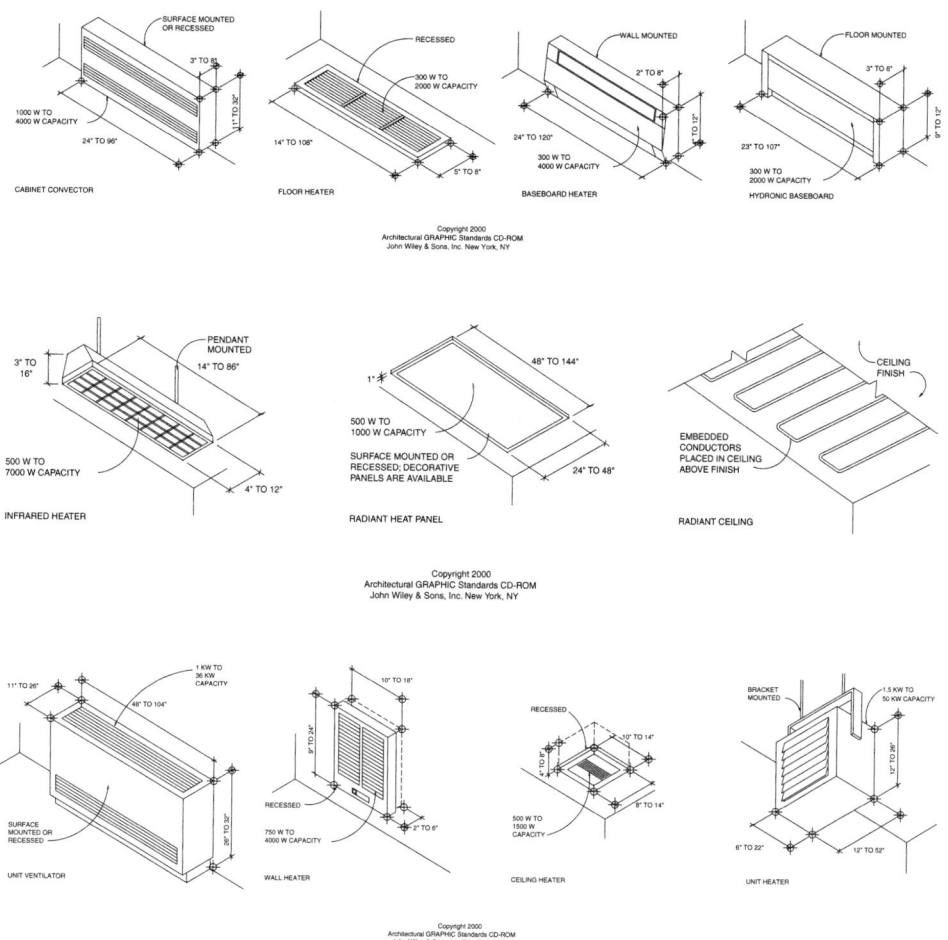

Copyright 2000
Architectural GRAPHIC Standards CD-ROM
John Wiley & Sons, Inc. New York, NY

Copyright 2000
Architectural GRAPHIC Standards CD-ROM
John Wiley & Sons, Inc. New York, NY

Copyright 2000
Architectural GRAPHIC Standards CD-ROM
John Wiley & Sons, Inc. New York, NY

**Fig. 9.15** *Varieties of electric resistance heating units. (Reprinted by permission from AIA: Ramsey/Sleeper,* Architectural Graphic Standards, *10th ed., © 2000 by John Wiley & Sons, Inc.)*

high temperatures, requiring care in the location of heaters relative to furniture placement, draperies, and traffic flow. Electric heaters are sized by their capacity in kilowatts (1 kW = 3413 Btu/h). The maximum watt density allowed is 250 W per linear foot of heater (820 W per linear meter).

### (c) Gas-Fired Heaters

These are often found in semioutdoor locations such as loading docks and repair shops. They are fired with either natural gas or propane. When vented, they can be used in more traditional environments such as the retail store in a remodeled warehouse (Fig. 9.16). Their advantage is that they heat surfaces first rather than air, so that comfort is obtained

without the need for high air temperatures. When high rates of air exchange are expected, high-intensity radiant heaters are often used.

Radiant heaters should be sized by the surface temperature change they produce. Many manufacturers specify this surface Δt for specified mounting heights and angles relative to the surface to be heated.

Gas-fired baseboard heaters are also available, using either natural gas or propane. They heat by both convection and radiation, as do their electric counterparts. At a steady-state efficiency of 80%, they use a lower-grade resource (than electricity) to do this lower-grade task. They are direct vented (using a built-in fan) to the outside, and therefore must be installed on or near an exterior wall; vents

THERMAL CONTROL

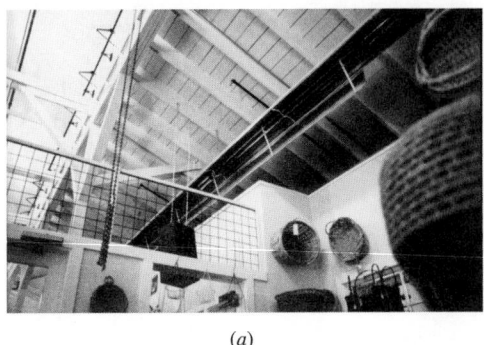

(a)

(b)

**Fig. 9.16** *A high-intensity infrared heater adds to the historic atmosphere of this retail store in an old warehouse (a). Exposed mechanical equipment includes the chains that operate the clerestory windows. The vented gas heater has an adjustable reflector that enables its radiant heat to be directed. (Clark-Ditton Architects, Eugene, OR.) (b) Vented gas high-intensity infrared heaters are available in both straight-line and U-shaped units. (Courtesy of Solaronics, Inc., Rochester, MI.)*

are 1½ in. in diameter, with a maximum length of 19 in. length (38 mm in diameter, 483 mm in length). With a cross section of 9 in. high by 5 in. deep (230 mm × 130 mm), a 48-in. (1220-mm) length will deliver 5800 Btu/h (1700 W); a 72-in. (1830-mm) length will deliver 9400 Btu/h (2755 W).

### (d) Ceiling Electric Resistance Heat

Ceilings can be constructed to include electric resistance in wiring (Fig. 9.17). Because the ceiling is

not touched, it can be safely heated to a rather high temperature. The primary disadvantage of ceiling heat is that hot air stratifies just below the ceiling, so that air motion is discouraged; remember also that Figs. 4.10 and 4.11 predict more discomfort with the warmer ceiling. Finally, the wires are hidden within the ceiling surface, and unwary occupants can puncture wires while installing hooks or additional light fixtures.

### (e) Hot Water Boilers

The remaining choices for whole-building heating systems are discussed in the following order: (1) hot water and (2) forced air. Such systems include a fuel, a heat source, a "mover" (such as a pump or fan), a distribution system, a heat exchanger or terminal within the space, and a control system.

Hot water boilers are rated according to heating capacity by several different categories. *Heating capacity* is the rate of useful heat output with the boiler operating under steady-state conditions, often expressed in MBh (1000 Btu/h). This "useful heat" assumes that the boiler is within the heated envelope of the building; thus, the heat that escapes from the boiler walls is available to help heat the building. AFUE, the annual fuel utilization efficiency, is defined as 100% minus the losses up the stack during both the on and off cycles, and the losses due to infiltration of outdoor air to replace the air used for combustion and for draft control. Finally, the net $I = B = R$ *rating* (a designation of the Institute of Boiler and Radiator Manufacturers) is published by the Hydronics Institute Division of The Gas Appliance Manufacturers Association (GAMA). The net $I = B = R$ rating load is lower than the heating capacity rating, because it consists only of the heating to be delivered to the spaces and excludes the heat loss of the boiler itself.

Select a boiler whose rating matches the calculated critical heat loss of the house or building; too small a boiler results in lower indoor temperatures at design conditions; too large a boiler costs more and is a waste of space. When using AFUE to select an efficient boiler, take care to see that the assumptions about "inches of water draft" and percentage $CO_2$ are similar for the boilers being compared. Minimum AFUEs are specified both in ANSI/ASHRAE Standard 90.2, *Energy-Efficient Design of Low-Rise Residential Buildings,* and in ANSI/ASHRAE/IESNA

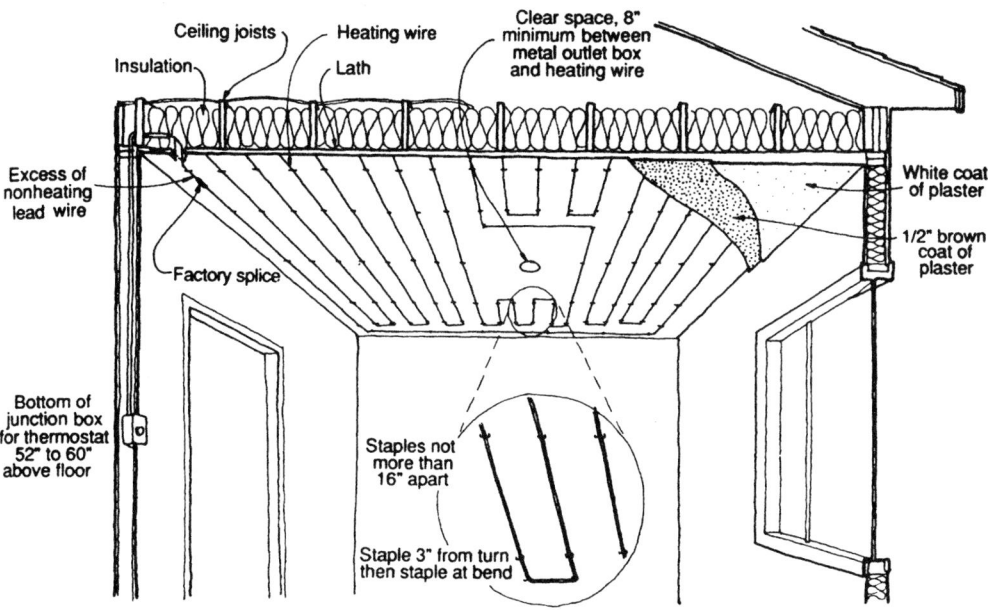

**Fig. 9.17** *Radiant heating electrical cable is installed prior to completion of a plaster ceiling.*

90.1, *Energy Standard for Buildings Except Low-Rise Residential Buildings.*

Boilers and their accessories comprise a wide inventory. A few selected types are discussed:

1. *Oil-fired steel boiler.* A refractory chamber receives the hot flame of the oil fire. Combustion continues within the chamber and the fire tubes. Smoke leaves through the breeching at the rear. Water, *outside* the chamber, receives the heat generated in the combustion chamber. If a domestic hot water coil is connected for use, a larger-capacity boiler is selected. An aquastat (water thermostat) turns on the burner whenever the boiler water cools off, thereby maintaining a reservoir of hot water ready for heating the building.

2. *Gas-fired cast iron hot water boiler* (Fig. 9.18). Cast iron sections contain water that is heated by hot gases rising through these sections. Output is related to the number of sections. Additional heat is gained from a heat extractor in the flue. With induced draft combustion, condensing unit in the flue, and intermittent electronic ignition instead of a pilot light, up to 90% AFUE is attainable. The American Gas Association (AGA) sets standards for gas-fired equipment.

3. *Oil-fired, cast iron hot water boiler.* Primary and secondary air for combustion may be regulated at the burner unit. Flame enters the refractory chamber and continues around the outside of the water-filled cast iron sections.

## (f) Hot Water Baseboard and Radiator Systems

Hot water heating circuits that serve baseboards or radiators come in four principal arrangements. Figure 9.19a shows the series loop system, usually run at the building's perimeter. The water flows to and *through* each baseboard or fintube in turn. Obviously, the water at the end of the circuit is a little cooler, but because in all hot water systems the water temperature drop seldom exceeds $20F°$ ($11C°$) in residences, the *average* temperature can usually be used to select the baseboard or other elements. Valves at each heating element are not possible, because any valve would shut off the entire loop. Adjustment is by a damper at each baseboard, which reduces the natural convection of air over the fins. This is a *one-zone* system—all elements on, or off, together. There is no general rule about the maximum allowable length of a water circuit, but for long runs, the pipe size can be increased or

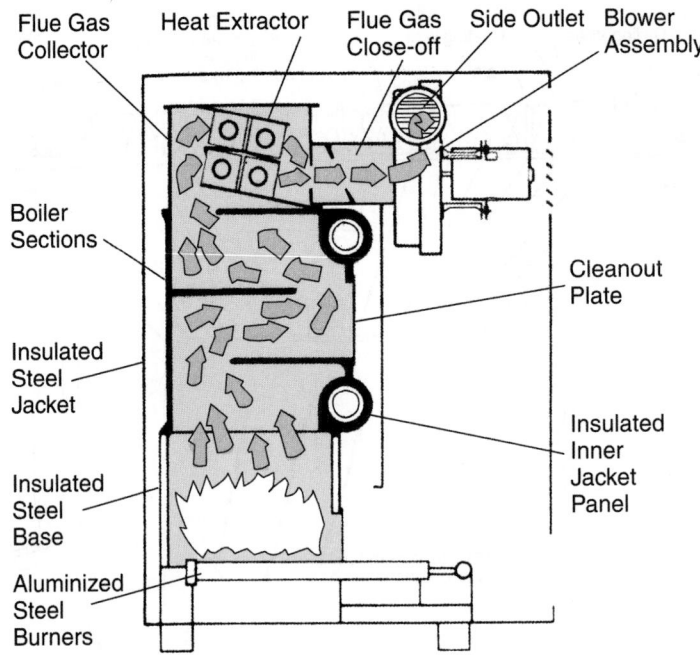

*Fig. 9.18* Gas-fired cast iron sectional boiler for hot water heating. Very high operating efficiency is possible; no chimney is required, as lower-temperature exhaust gases can be vented through the wall to the exterior.

*several* loops used in parallel to create more than one thermal zone.

The one-pipe system shown in Figs. 9.19*b* and 9.20 is a very popular choice. Special fittings act to divert part of the flow into each baseboard. A valve may be used at each one to allow for reduced heat or for a complete shutoff to conserve energy—an advantage that the loop system does not provide. The one-pipe system uses a little more piping and thus is not as economical to install as the loop system, in which piping is minimal. Again, the supply water temperature will be lower at the end of the run than at the beginning.

The two-pipe reverse–return shown in Fig. 9.19*c* provides the same supply water temperature to each baseboard or radiator, because it is not cooled either by passing through a previous baseboard or accepting the cooler return water. Equal friction, resulting in equal flow, is achieved through all baseboards (numbers 1 to 5) by *reversing* the return instead of running it directly back to the boiler. This equality is effected by equal lengths of water flow through any baseboard together with its lengths of supply-and-return main. More pipe is required for this system than for the systems shown in Fig. 9.19*a* or 9.19*b*.

Figure 9.19*d* shows an arrangement that is not usually favored because the path of water through baseboard number 1 is much shorter than that through the others, especially number 5. Baseboard number 5 could easily be undesirably cool, because it is short-circuited by the others.

*Pipe expansion* requires expansion joints in long runs of pipe and clearance around all pipes passing through walls and floors. Each time a hydronic system changes from room temperature to a heated condition, the piping will undergo the following expansion, assuming a 70°F (21°C) initial temperature:

| Pipe Expansion per 100 ft | | |
|---|---|---|
| Water Temperature (°F) | Iron Pipe (in.) | Copper Tubing (in.) |
| 160 | 0.7 | 1.0 |
| 180 | 0.9 | 1.3 |
| 200 | 1.0 | 1.5 |
| 220 | 1.2 | 1.7 |

## THERMAL CONTROL

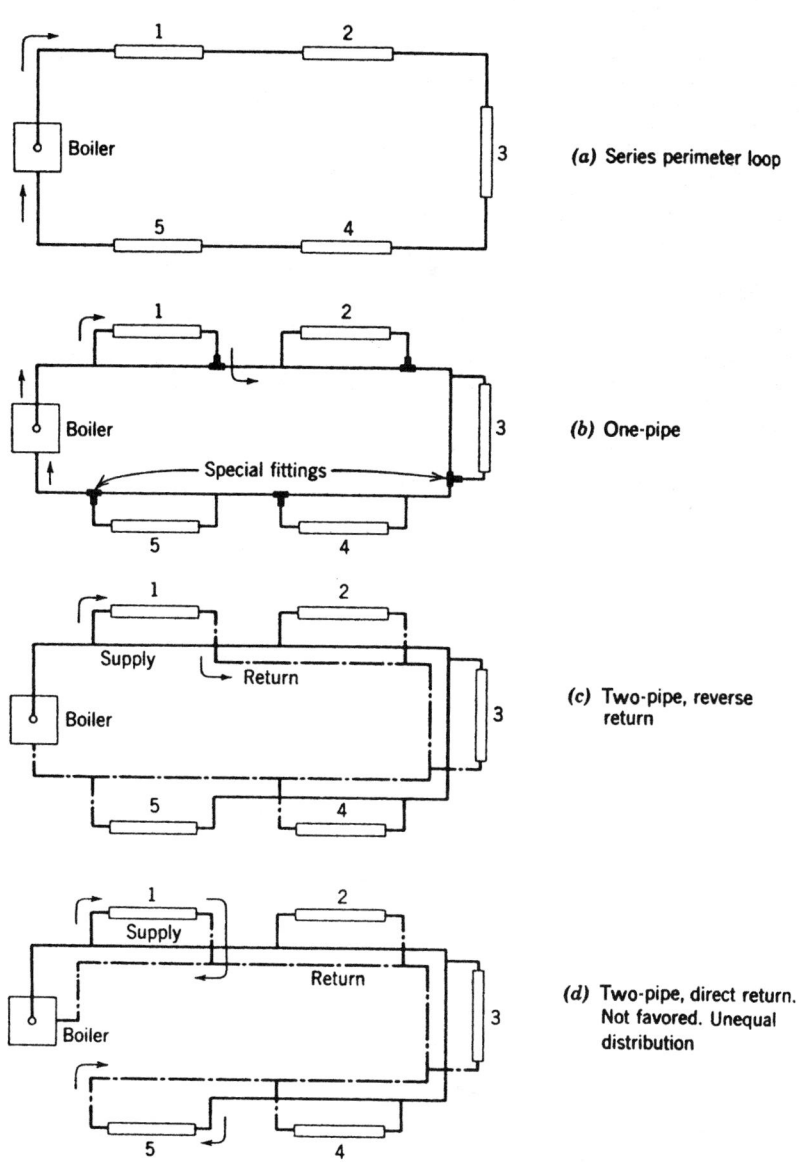

*(a)* Series perimeter loop

*(b)* One-pipe

*(c)* Two-pipe, reverse return

*(d)* Two-pipe, direct return. Not favored. Unequal distribution

**Fig. 9.19** *Diagrams of hydronic distribution options, seen in the plan. Baseboard convectors are shown here. Controls are not shown.*

An *air cushion tank,* compression tank, or expansion tank is a closed tank containing air, usually located above the boiler. When the water in the system is heated, it expands, compressing the air trapped in the tank. This tank allows for the usual range of temperatures within the system, including temperatures above the usual boiling point of water, with-

out the frequent opening of the pressure relief valve. One type of air cushion tank, called a *diaphragm tank,* separates the air and water with an inert, flexible material; this prevents reabsorption of the air by the water.

For a conventional (unpressurized) air cushion tank, allow 1 gallon of tank capacity for every

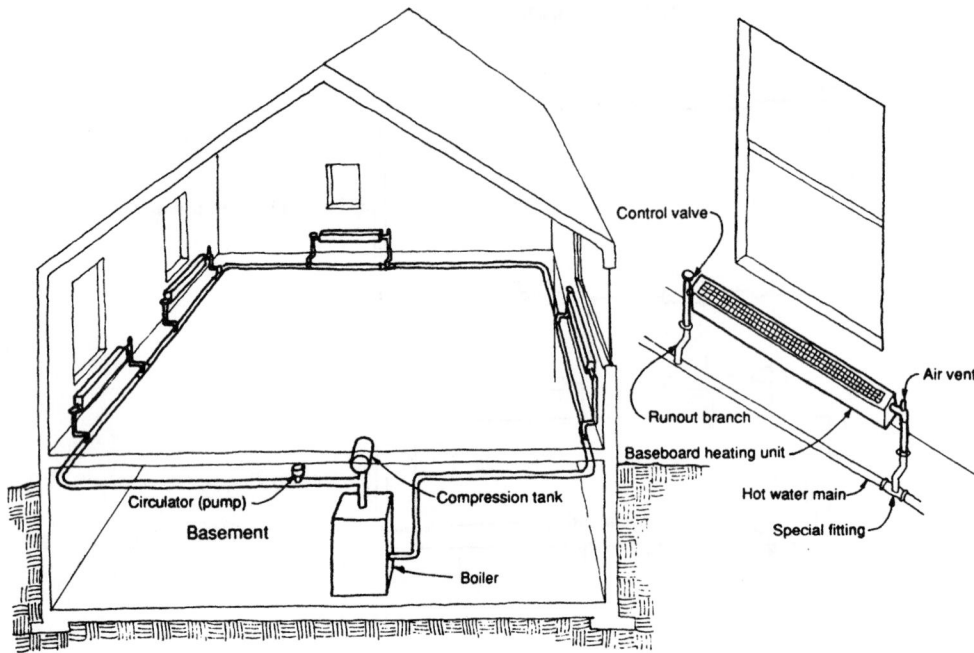

**Fig. 9.20** *One-pipe hydronic system.*

5000 Btu/h (1 L for every 385 W) of the total system heat loss. For a pressurized tank of at least 8 lb/in² (55 kPa), allow 1 gallon of tank capacity for each 7000 Btu/h (1 L for each 540 W) or see the manufacturer's recommendations.

*Air vents and water drains* are part of the distribution system. Except for the necessary air cushion in the upper part of the compression tank above the boiler, air must not be allowed to accumulate at high points in the piping or at the convector branches. Air vents at all high points relieve these possible air pockets that would otherwise make the system air-bound and inoperative.

If a system is drained and left idle in a cold house, water trapped in low points can freeze and burst the tubing or fittings. Operable drain valves must be provided at such locations and, of course, at the bottom of the boiler, as shown in Fig. 9.21.

*Hydronic and electrical controls* allow automatic operation, described in Fig. 9.21. There are two options for system control:

1. As in Fig. 9.21, the thermostat controls the circulating pump and the boiler. In colder weather, the system operates almost continuously, and the average temperature in the system gradually rises.

2. The thermostat controls only the boiler, and the circulating pump operates continuously. This uses more energy for the pump but minimizes system temperature variation and thus the possibility of expansion noises.

Makeup water is added as required, the air level in the tank is regulated by the air control fittings, and the circulator and burner operate as controlled by the aquastat and thermostat. If air vents in the *piping* are not automatic, they will require periodic manual "bleeding" of unwanted air.

*Circulating pumps* are used to overcome the friction of flow in the piping and fittings and to deliver water at a rate sufficient to offset the hourly heat loss of the house or building.

*Pipe insulation* is required whenever the pipes are outside the heated envelope of the building; ANSI/ASHRAE Standard 90.2 specifies insulation appropriate to the temperature of the heated water.

THERMAL CONTROL

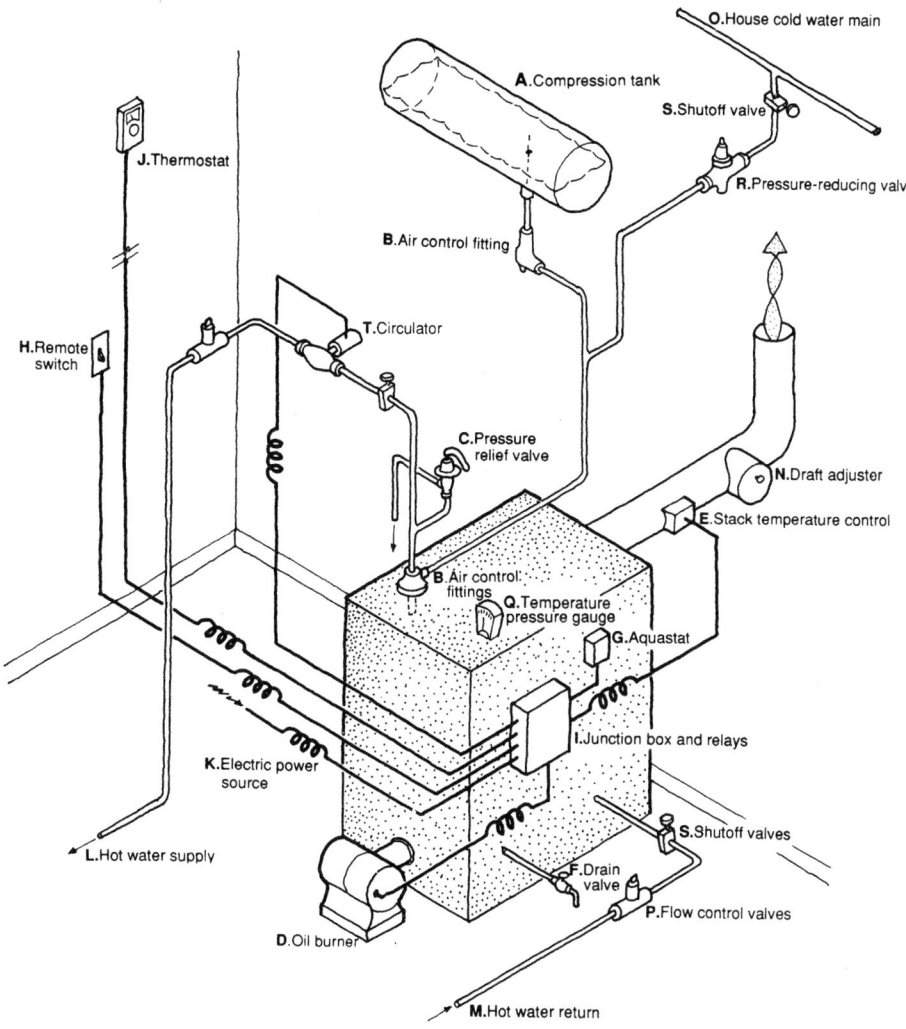

O.House cold water main

A.Compression tank

S.Shutoff valve

R.Pressure-reducing valve

J.Thermostat

B.Air control fitting

H.Remote
switch

T.Circulator

C.Pressure
relief valve

N.Draft adjuster

E.Stack temperature control

B.Air control
fittings

Q.Temperature
pressure gauge

G.Aquastat

I.Junction box and relays

K.Electric power
source

S.Shutoff valves

F.Drain
valve

L.Hot water supply

P.Flow control valves

D.Oil burner

M.Hot water return

(A) *Compression tank.* Accommodates the expansion of the water in the system.

(B) *Air control fittings.* Vent out unwanted air in the boiler and maintain the level in the compression tank.

(C) *Pressure relief valve.* Usually set for 30 psi. Initial cold pressure about 12 psi. Relieves excessive system pressure.

(D) *Oil burner.* Responds to aquastat or thermostat.

(E) *Stack temperature control.* Senses stack temperature and stops oil injection if ignition has not occurred.

(F) *Drain valve.* At low point in the water system.

(G) *Aquastat.* Maintains temperature of boiler water by starting the oil burner when temperature of water drops below the aquastat's setting. Sometimes set at about 180 F.

(H) *Remote switch.* At a safe distance from the boiler so that the plant can be turned off in case of trouble during which the boiler cannot be approached.

(I) *Junction box and relays.* General control center.

(J) *Thermostat.* When the room temperature drops below its setting, it turns on both the oil fire and the circulating pump.

(K) *Electrical power source.* Operates from a separate individual circuit at the power panel.

(L) *Hot water supply.* Copper tubing to convectors or baseboards.

(M) *Hot water return.* Copper tubing from convectors or baseboards.

(N) *Draft adjuster.* Regulates the draft (combustion air) over the fire.

(O) *House cold water main.* From which water is fed automatically into boiler.

(P) *Flow control valves.* Prevent casual flow of water by gravity when the circulator is not running.

(Q) *Temperature pressure gauge.* Indicates water temperature and pressure. Sometimes supplemented by immersion thermometers in supply and return mains.

(R) *Pressure-reducing valve.* Admits water into the system when the pressure there drops below about 12 psi. Has a built-in check valve to prevent backflow of boiler water into the water main.

(S) *Shutoff valves.* Normally open. Can be closed to isolate the system and permit servicing of components.

(T) *Circulator.* Centrifugal circulating pump that moves the water through the tubing and heating elements.

**Fig. 9.21** *An oil-fired boiler and its hydronic and electrical controls.*

## (g) Radiant Panels

Radiant floors have several comfort advantages over radiant ceilings (see again the introduction to Section 9.7). The components are largely the same as in the baseboard/radiator systems, except that now whole coils of pipes replace the individual radiators and baseboards. A balancing valve should be installed on each coil. Where uninsulated spaces underlie floors (or are above ceilings), special attention should be paid to adequate insulation, because the radiant panel will generate especially high temperatures, and thus a very high Δt through the floor (or ceiling).

Thanks to higher insulation, today's buildings often require panels smaller in area than the floor (or ceiling) area available. In a conventional radiant panel system, the panel is placed nearest the exterior walls, where the heat loss is greatest. In a solar-heated building, this is not so clear. If the panel heats the floor surface just inside south-facing glass, how much warming will be left to the sun? A preheated slab will absorb much less solar radiation. However, in cloudy cold weather, the area closest to the south glass could become uncomfortably cool.

In the past, copper tubing was widely used for the heating coils. This typically involved a number of connections within a coil; each represented a potential point of failure over the life of the panel. Today, coils are typically one-piece and are made of synthetic materials such as cross-linked polyethylene tubing. When the floor is concrete or other cast-in-place material, the coil is either directly embedded in the slab (tied down to resist floating during the pour) or stapled to an underfloor, over which the slab is poured. Radiant floors with coils underneath wood floors are increasingly popular.

Rugs or carpets over radiant floors are a mixed blessing; they are soft on the feet but interfere with the exchange of heat. Special undercarpet pads can facilitate heat transfer; higher water temperatures can be used in the coils, because skin contact with the floor is prevented by the carpet.

## (h) Hydronic Heating Sizing

The calculations for the sizing of a water distribution system are based upon the required flow and the friction in the piping. (For a domestic water supply, another factor is the vertical distance the water

must be raised. In these closed-loop heating systems, however, the weight of the cooler water falling back to the boiler essentially counterbalances the weight of the hot water being raised. Furthermore, gravity helps the hot—lighter—water rise and the cooler water fall.)

The key to pipe sizing is the overall required flow rate. Ordinarily, the temperature drop that occurs as the hot supply water gives up heat to the space (through the convector) is about 20F° [11C°] in residences; in commercial applications, 30, 40, or 50F° (17, 22, or 28C°) temperature drops are also common, as recommended by the manufacturers of unit heaters and convectors. Because the entire building's design heat loss is overcome by this system, in I-P units, total flow rate, gpm

$$= \frac{\text{design heat loss, Btu/h}}{20F° \times 60 \text{ min/h} \times 8 \text{ lb/gal} \times 1 \text{ Btu/lb F°}}$$

$$= \frac{\text{design heat loss}}{9600}$$

In SI units, total flow rate, L/s

$$= \frac{\text{design heat loss, W}}{11C° \times 1 \text{ kg/L} \times 4180 \text{ W sec/kg C°}}$$

$$= \frac{\text{design heat loss}}{45,980}$$

Then, using Section 21.11, we can account for friction through piping and fittings and can size the main supply and return pipes. The same procedure can be applied to branches, proportioned to the heat they must deliver. The total friction to be overcome in the most distant run is then converted from total psi to *feet of head*, with each foot of head = 0.433 psi (the pressure exerted by a foot-high column of water). In SI, 1 ft of water = 2.99 kPa.

With both the friction of the system expressed in feet of water, or head, and the flow rate established, a pump can be selected. Figure 9.22 shows typical performance curves for four pumps. The designer enters the curve with a desired flow rate, then selects the pump with a head capacity greater than or equal to the head required. Pump performance curves are provided by the manufacturer.

The critical choice, however, is not pipe size. It is relatively easy to distribute such small-diameter pipes within wall and floor/ceiling construction. Rather, the critical choice is the *hot-water supply*

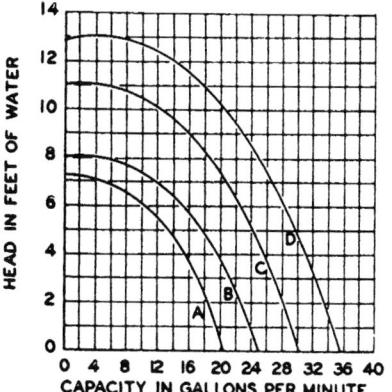

**Fig. 9.22** *Typical pump capacity (or performance) curves for four pumps used in hydronic heating systems. One foot of water = 2.99 kPa; 1 gpm = 0.0631 L/s. (Courtesy of the Hydronics Institute Division of GAMA, Berkeley Heights, NJ.)*

*temperature:* the higher this temperature, the smaller the convector units that discharge the heat to each space. However, higher temperatures endanger occupants, who may suffer skin burns if they touch exposed parts of the convectors or the distribution tree. Higher temperatures also can lead to steam within the boiler/distribution tree, although the system is under pressure and therefore the boiling point is greater than 212°F (100°C).

These systems are not designed to accommodate steam, and serious injury can sometimes result. A safer choice of average water distribution temperature is 180°F, even though this temperature results in larger convectors. A slightly lower annual average energy consumption should accompany the lower distribution temperature.

*Baseboard convector* selection is then made from the manufacturer's data tables, such as the one shown in Table 9.3. The two common baseboard types are *RC*, usually cast iron with a water-backed front surface and an extended rear heating surface, and *finned tube*, a metal tube with an extended surface in the form of fins, usually placed behind a metal enclosure.

**EXAMPLE 9.1** What length of baseboard convector (Table 9.3) is necessary along a 20-ft [6-m]-long living room wall? From heat loss calculations, the living room requires 9000 Btu/h (2635 W). The average water temperature is 180°F, and water flow is about 1 gpm (0.6 L/s).

**SOLUTION**
With 180°F supply water and 1 gpm flow rate, Table 9.3 shows that 580 Btu/h will be delivered for each

**TABLE 9.3 Hydronic Baseboard Convectors**

| Water Supply | | | | | |
|---|---|---|---|---|---|
| Temperature | | Flow Rate[a] | | Heat Delivered per Unit Length[b] | |
| °C | °F | L/s | gpm | Watts/m | Btu/h ft |
| 60 | 140 | 0.06 | 1 | 307 | 320 |
| | | 0.25 | 4 | 326 | 340 |
| 66 | 150 | 0.06 | 1 | 365 | 380 |
| | | 0.25 | 4 | 384 | 400 |
| 71 | 160 | 0.06 | 1 | 432 | 450 |
| | | 0.25 | 4 | 461 | 480 |
| 77 | 170 | 0.06 | 1 | 490 | 510 |
| | | 0.25 | 4 | 518 | 540 |
| 82 | 180 | 0.06 | 1 | 557 | 580 |
| | | 0.25 | 4 | 586 | 610 |
| 88 | 190 | 0.06 | 1 | 614 | 640 |
| | | 0.25 | 4 | 653 | 680 |
| 93 | 200 | 0.06 | 1 | 682 | 710 |
| | | 0.25 | 4 | 720 | 750 |
| 99 | 210 | 0.06 | 1 | 739 | 770 |
| | | 0.25 | 4 | 778 | 810 |

*Source:* Slant/Fin Ltd. Mississauga, Ontario.
[a]Use flow rate of 0.6 L/s (1 gpm) unless flow rate is known to be ≥0.25 L/s (4 gpm).
Pressure drop is as follows: flow 0.6 L/s (1 gpm) = 3.76 mm/m (47 mil.in./ft).
                flow 0.25 L/s (4 gpm) = 42 mm/m (525 mil.in./ft).
[b]Baseboard finished length will be 76 mm (3 in.) longer than the length required to meet the heating need.

lineal foot of baseboard convector; 9000 Btu/h ÷ 580 Btu/h ft = 15.5 ft. Choose a baseboard combination of 16-ft overall length, consisting of two 8-ft (finished length) pieces. The total active finned length will be about 15.5 ft. ∎

*Panel* (*radiant ceiling or floor*) heating design usually depends on a water temperature of 120°F (49°C) for heated floors and 140°F (60°C) for heated ceilings. An uncarpeted concrete floor slab using ¾-in.-diameter pipe or tube on 12-in. centers (20-mm diameter on 300-mm centers), with an average water temperature of 120°F, will deliver *50 Btu/h ft² (158 W/m²) of floor panel.*

A ceiling panel with nominal ⅜-in. tube on 6-in. centers (10-mm diameter on 150-mm centers), with an average water temperature of 140°F, will deliver *60 Btu/h ft² (189 W/m²) of ceiling panel.*

To determine the area of heated panel, divide the room's design heat loss by the rate of panel heat delivery. Although radiant ceilings deliver more heat per unit area, they also discourage air motion because the warmest air rises to lie against the warmest surface. In contrast, at a radiant floor the coolest air drops to contact the warmest surface, is then warmed, and rises to be cooled at the ceiling, drops to the floor, and repeats this cycle continually.

Each panel contains one or more coils. In floor panels, each coil delivers 10,000 Btu/h (2930 W) and should be no longer than about 200 linear feet (60 m). In ceiling panels (with smaller-diameter tubes), each coil delivers 3000 Btu/h (880 W) and should be no longer than about 100 linear feet (30 m).

Sizing of the main pipes and pump is based on the longest circuit, measured along the length of the supply pipe from the boiler to the coil and back along the return line to the boiler. Two guidelines apply:

1. Do not include the length of the coil itself in this circuit length.
2. No section of a floor panel main should be less than ¾ in. (20 mm) in diameter.

---

**EXAMPLE 9.2** Determine the panel area and coils required for a radiant floor in a living room 15 ft × 25 ft (= 375 ft²) with a design heat loss of 12,000 Btu/h (4.6 m × 7.6 m = 34.8 m², heat loss 3514 W).

**SOLUTION**
The panel area required is

$$\frac{12{,}000 \text{ Btu/h}}{50 \text{ Btu/h ft}^2} = 240 \text{ ft}^2 \text{ of floor panel}$$

The room has 375 ft² of floor available, so the heated panel is usually placed along the exterior walls, where the room heat loss is greatest (unless solar heat through south windows is involved). The number of coils in the panel is

$$\frac{12{,}000 \text{ Btu/h}}{10{,}000 \text{ Btu/h per coil}} = 1.2 \text{ coils}$$

Use one coil in this panel. ∎

## (i) Hydronic Zoning

Figure 9.23 shows that zoning is relatively easy to accomplish with hydronic systems. The installation shown in the figure is made up of three separately heated areas—the first, second, and third floors. Each can be heated to different temperatures as called for by thermostats in each separate apartment. For example, if only the thermostat serving the second floor (zone B) calls for heat, it turns on pump B. Flow-control valves B open, admitting hot water from the boiler header to main B. Flow-control valves A and C remain closed, preventing flow in mains A and C. Any or all of the zones may operate at one time. The boiler keeps a supply of hot water continually ready to supply any zone on demand. This is achieved by an aquastat immersed in the boiler water. When the boiler water drops below the prescribed temperature, it turns on the firing device, such as an oil burner or a gas burner, which brings the water up to the temperature setpoint. If an overhead main supplies downfeed, as in the first floor of this installation, special downfeed supply and return fittings are necessary. For the second- and third-floor zones, one special return tee is sufficient. If the designer also elects to use a special upfeed *supply* tee of the venturi type, higher outputs of the convectors will result.

Two of the more famous residences in the Midwest utilize hydronic systems. Frank Lloyd Wright's Robie House (Chicago) has wall radiators integrated below the north windows in the living room. Underfloor radiators with grilles in the floor were provided for below the full-height south windows but

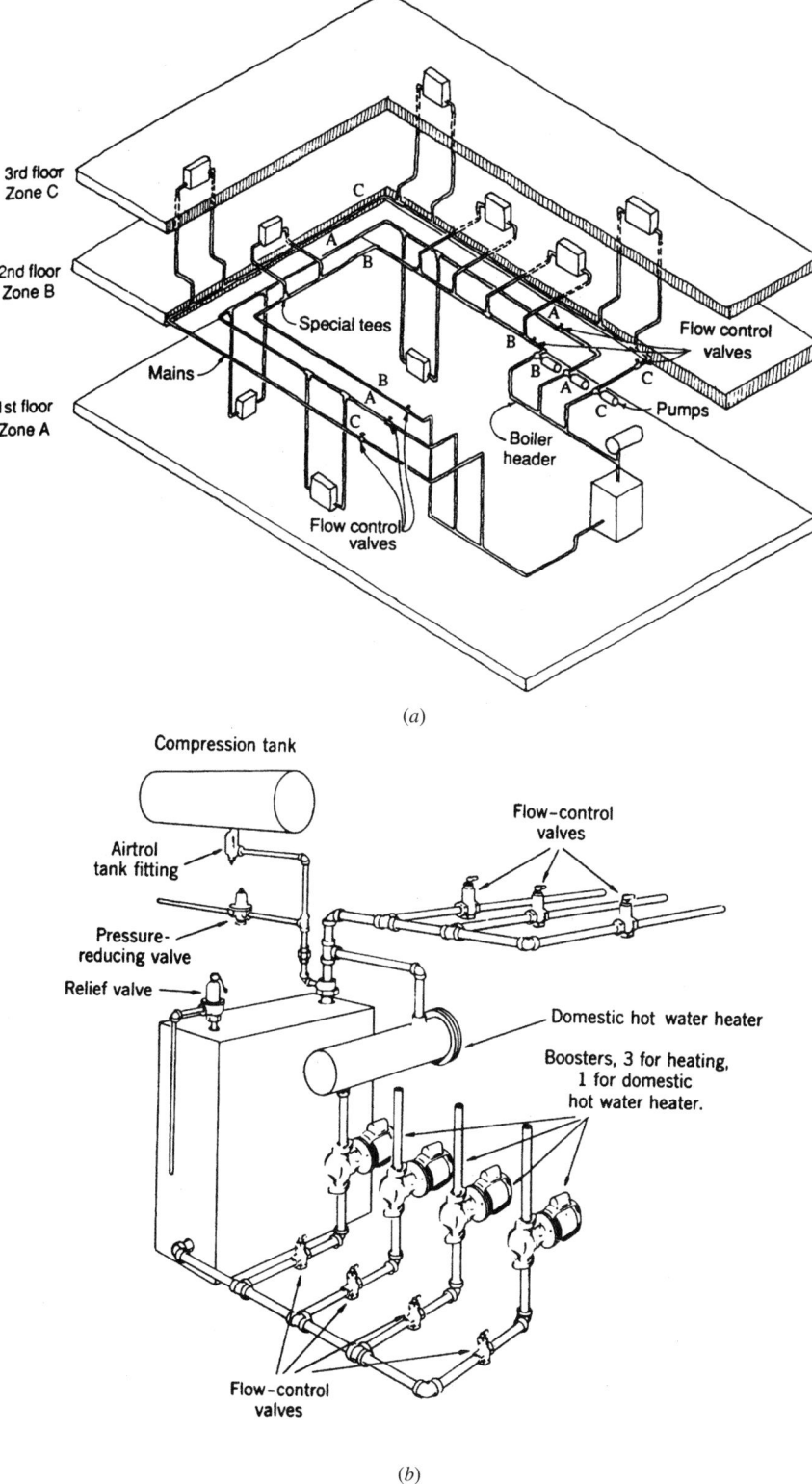

3rd floor
Zone C

2nd floor
Zone B

1st floor
Zone A

Special tees

Mains

Flow control
valves

Flow control
valves

Pumps

Boiler
header

Flow control
valves

*(a)*

Compression tank

Airtrol
tank fitting

Flow–control
valves

Pressure-
reducing valve

Relief valve

Domestic hot water heater

Boosters, 3 for heating,
1 for domestic
hot water heater.

Flow–control
valves

*(b)*

**Fig. 9.23** *Three-zone, multicircuit, one-pipe system. (a) Each convector has connections to the one pipe. (b) Boiler, piping, and water controls suitable for this three-zone, one-pipe system. Each one-pipe circuit should be provided with two flow-control valves and a circulator (also called a* booster *or pump) on the supply (or return) pipe. (Courtesy of ITT Bell and Gossett.)*

THERMAL CONTROL

were apparently never installed. The boiler sits in a basement room. Mies van der Rohe's Farnsworth House (Fox River, Illinois) preserves its four walls of ceiling-to-floor glass by concealing radiant heating pipes in the floor slab. The boiler sits within the central utility "closet."

Today's radiators are designed to reflect the sleek and simple lines of contemporary architecture. The radiator is getting new exposure with some colorful and pleasing products (Fig. 9.24). The Mayer Art Center at Phillips Exeter Academy in New Hampshire (Fig. 9.25) features new exposed radiators in some older buildings. These radiators are based on simple components (typically, 2¾ in. [70 mm] wide) that can be combined in many heights and widths, inviting the designer to feature them rather than to hide them in metal cabinets.

## (j) Heating Equipment Efficiency, Combustion, and Fuel Storage

As fuels burn to produce heat, they require oxygen to support the combustion. Because oxygen constitutes only about one-fifth of the volume of air, reasonably large rates of airflow are required. The air should be drawn in from outdoors at a position close to the fuel burner or (preferably) led to this location by a duct. ANSI/ASHRAE Standard 90.2 calls for 0.5 cfm/1000 Btu/h (1 L/s per 1242 W). This supply duct should be arranged to remain open at all times. Although permitted by Standard 90.2, this combustion air should *not* be drawn from the general building space. It is a waste of energy, and contemporary "tight" construction inhibits such airflow. A dangerous condition is created whenever stack flow is restricted.

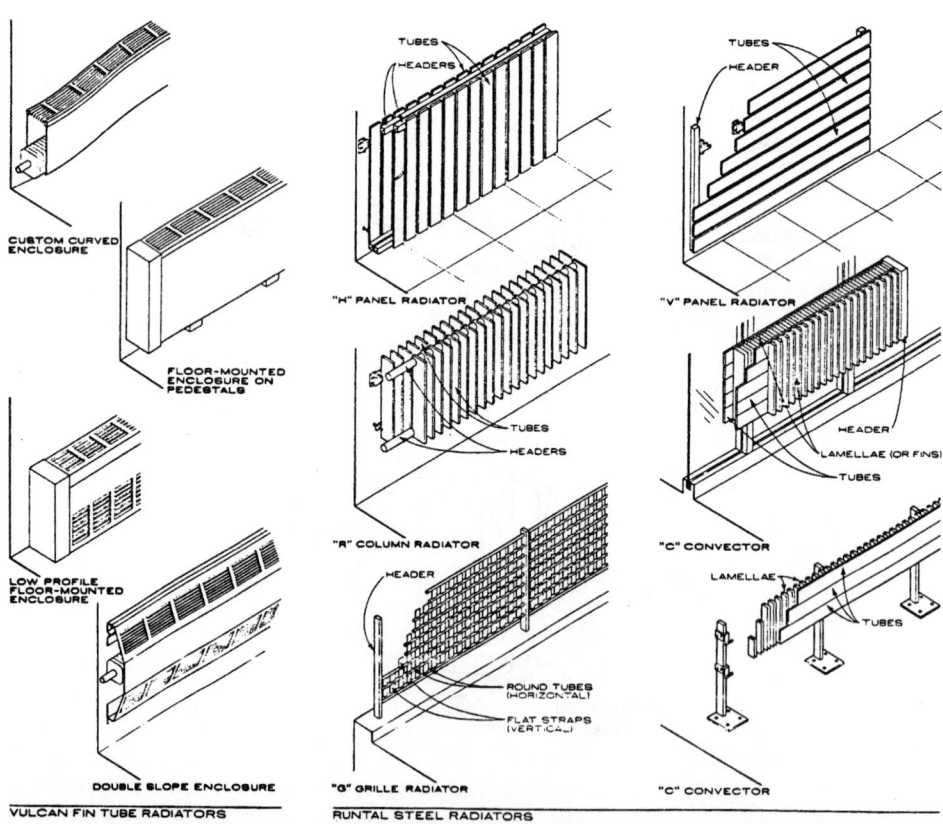

**Fig. 9.24** *Hot water radiators are available in bright colors and are based on simple components. (Reprinted by permission from AIA: Ramsey/Sleeper,* Architectural Graphic Standards, *10th ed., © 2000 by John Wiley & Sons, Inc.)*

THERMAL CONTROL

(a) (b)

**Fig. 9.25** *Hot water radiators fit below a window (a) at the Mayer Art Center, Phillips Exeter Academy, Exeter, New Hampshire. Amsler Hagenah MacLean, Architects. (Photo by Alex Beatty.) (b) Radiator formed by flat tubes. (Courtesy of Runtal/North American Energy Systems.)*

High-efficiency boilers and furnaces manage to remove so much heat from the exhaust gases that smaller flues at much lower temperatures result. These relatively small pipes can be vented through a wall to the exterior. Eliminating a chimney has lessened the impact on the building design of such boilers and furnaces.

For older or less efficient fuel-burning equipment, it is important that when chimneys carry high-temperature flue gases, they be safely isolated from combustible construction to prevent the possibility of fire. The size of the flue will depend upon the boiler or furnace selected. Flue height (Fig. 9.26a) had traditionally been 35 to 40 ft (11 to 12 m). The function of providing a draft, for which chimney height was an important consideration, is now provided by fans. For example, oil is injected under pressure, accompanied by air, and forced in by a fan. Often a draft adjuster in the breeching (smoke pipe) that carries the flue gases to the chimney is arranged to open slightly to *reduce* the normal stack draft. If increased draft is ever required, an induced draft fan that puts suction on the flue side of the fire is usually chosen instead of greater stack height. Draft hoods above gas burners prevent downdraft from blowing out the flames.

Prefabricated chimneys (Fig. 9.26b) are replacing with increasing frequency the bulkier and heavier field-built masonry. They offer a number of advantages and can be easily supported on a normal structure.

The storage space to be allowed for fuel oil depends on the proximity of the supplier and the space available at the building. For oil, when more than 275 gal (1040 L) was stored, it was common practice to use an outside tank buried in the ground. This practice eventually led to leaking tanks and contaminated soil and groundwater. Thus, several factors converged to discourage the use of oil: a cleaner and more efficient alternative (natural gas), few basements in new construction where oil tanks might be stored, and unsightly outdoor above-ground oil tanks.

Both ANSI/ASHRAE Standard 90.2 and ANSI/

THERMAL CONTROL

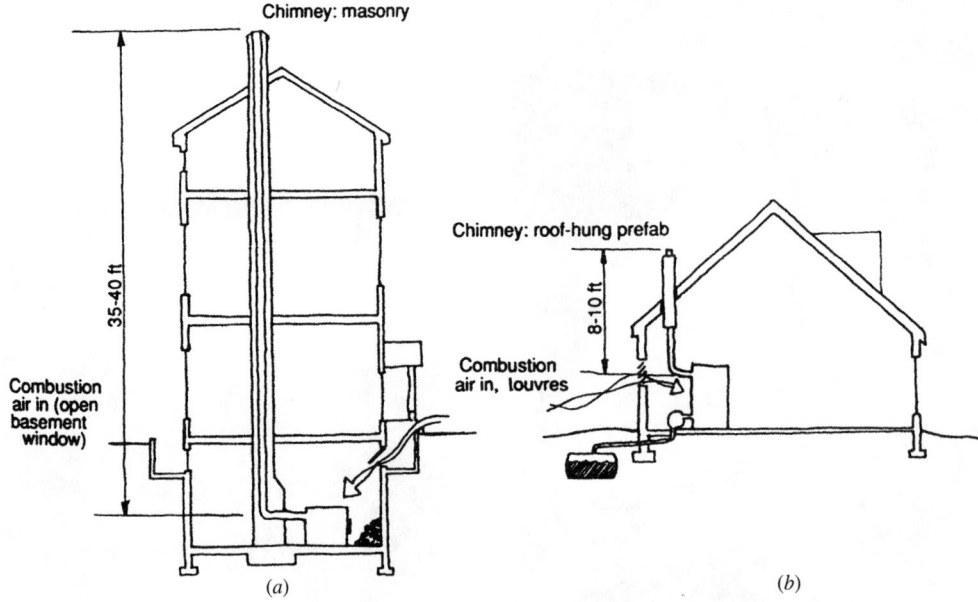

**Fig. 9.26** The need for 40-ft (12-m) chimneys (a) has been eliminated by controlled draft in burners (b) and by the use of high-efficiency heating equipment that can be directly vented to the exterior. Note the buried fuel oil tank, a potential environmental threat.

ASHRAE/IESNA 90.1 specify minimum efficiency ratings for heating and cooling equipment. Depending upon the size and the type, one of these terms will apply:

*Annual fuel utilization efficiency* (AFUE) is the ratio of annual fuel output energy to annual input energy, which includes any nonseason pilot input loss.

*Coefficient of performance* (COP) is defined slightly differently, depending upon the task. For *cooling*, it is the ratio of the rate of heat removal to the rate of energy input in consistent units, for a complete cooling system (or factory-assembled equipment), as tested under a nationally recognized standard or designated operating conditions. For *heating* (heat pump), it is the ratio of the rate of heat delivered to the rate of energy input in consistent units, for a complete heat pump system as tested under designated operating conditions. Supplemental heat is not included in this definition.

*Energy efficiency ratio* (EER) is the ratio of net equipment cooling capacity in Btu/h to the total rate of electric input in watts under designated operating conditions. (When consistent units are used, this ratio is the same as COP.)

*Integrated part load value* (IPLV) is a single-number figure of merit based on part-load EER or COP expressing part-load efficiency for air-conditioning and heat pump equipment on the basis of weighted operation at various load capacities for the equipment.

*Seasonal energy efficiency ratio* (SEER) is the total cooling output of an air conditioner during its normal annual usage period for cooling, in Btu/h, divided by the total electric energy input during the same period, in watt-hours.

### (k) Warm Air Heating Systems

These systems began to supersede the open fireplace in about 1900. Originally, an iron furnace that stood in the middle of the basement was hand-fired by coal. Surrounding it was a sheet metal enclosure. An opening in its side near the bottom admitted cool combustion air that gravitated to the basement. A short duct from the top of the enclosure delivered the warm air by gravity to a large grille in the middle of the floor of the parlor. Other rooms, including those in upper stories, shared a little of this warmth when doors were left open.

Very gradual changes had culminated by the middle of the twentieth century in systems essentially like the ones described in Fig. 9.27. The improvements included:

Automatic firing of oil or gas

Operational and safety controls

Ducted air to and from each room

Blowers to replace gravity

Filters

Adjustable registers

By the 1960s, the basement was beginning to disappear as subslab perimeter systems became popular for basementless houses (Fig. 9.28). The heat source was located centrally and fully *within* the insulated volume of the house; heat escaping from the unit itself merely helped heat the building. In general, air was delivered from below, upward across windows, to be taken back at a central high-return grille.

When electricity was used instead of oil and

gas for space heating, provisions for combustion, chimneys, and fuel storage were unnecessary. Horizontal electric furnaces began to appear in shallow attics or above furred ceilings. Air was delivered down from ceilings across windows and taken back through door grilles and open plenum space. Electric resistance furnaces use more electricity than heat pumps to do the same heating task (see again Fig. 2.6), so heat pumps have largely supplanted such furnaces.

As energy-saving design gained strength, insulated windows and well-insulated roofs, walls, and floors lessened the need for space heating. From a central furnace or heat pump, short ducts could deliver warm air to the inner side of each room, because warming at the insulated windows was less essential. Air returned to the unit through open grilles in doors and at the furnace or heat pump enclosure.

*Comfort* is one of warm air heating's advan-

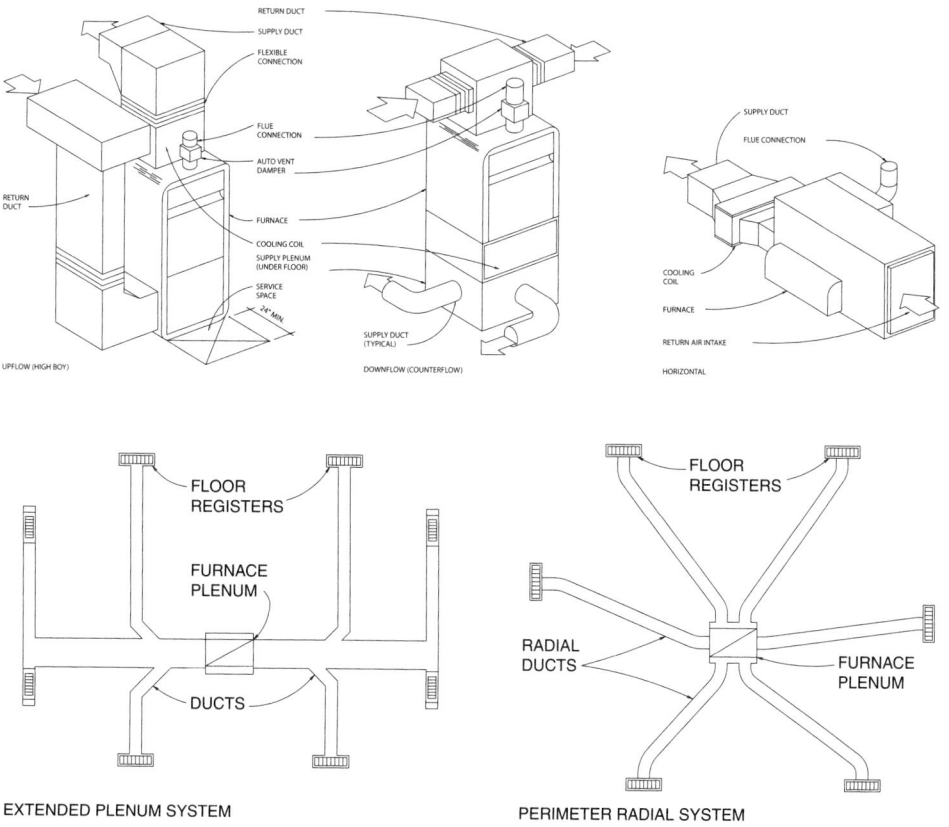

**Fig. 9.27** *Typical furnace types and duct distribution arrangements. (From AIA: Ramsey/Sleeper,* Architectural Graphic Standards, *10th ed., © 2000 by John Wiley & Sons, Inc. Reprinted with permission of John Wiley & Sons, Inc.)*

THERMAL CONTROL

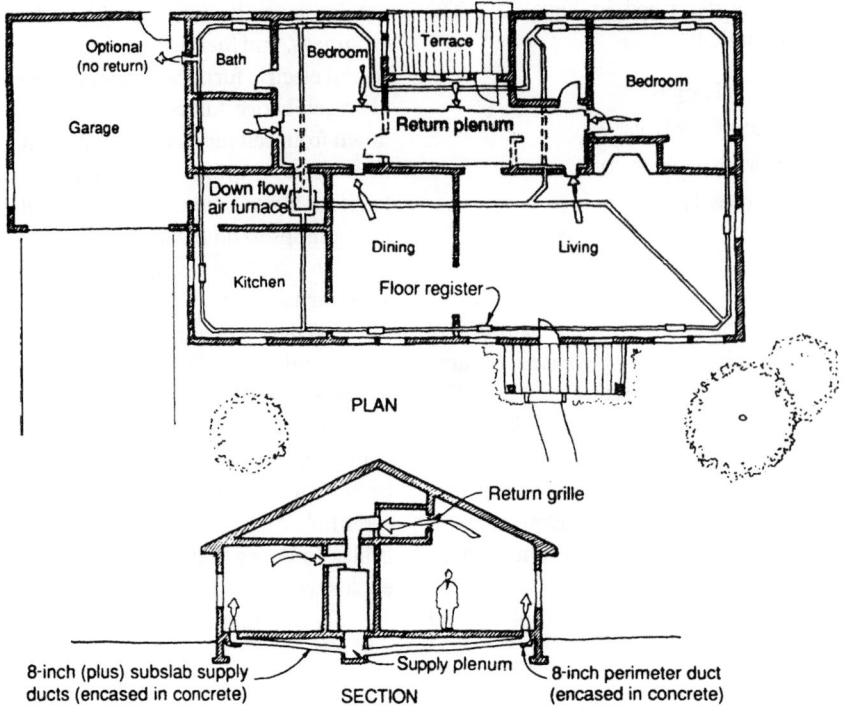

**Fig. 9.28** *Forced-warm-air, perimeter loop system, adaptable for a cooling coil at the furnace. No returns are taken from the kitchen, baths, or garage.*

tages. The motion of air in the space helps to assure uniform conditions and reasonably equal temperatures in all parts of a building. The building can quickly be warmed with a forced-air system. It is possible to clean both the recirculated air and the outdoor air by means of filters and other special air-cleaning equipment. Air may be circulated in non-heating seasons. Fresh air may be introduced to reduce odors and to make up the air exhausted by fans in kitchens, laundries, and bathrooms. Central cooling can be incorporated or introduced if ducts are designed originally to do so; cooling often calls for greater rates of air circulation. Humidification can be achieved by a humidifier in the air stream, and if cooling is included in the design, dehumidification can be accomplished in summer. For both heating and cooling, a common arrangement is to place the supply registers in the floor, below areas of glass. This is important for winter operation. With adequate attention to supply register placement, return grilles can be located so as to minimize return air ductwork. High return grilles pick up the warmer air for reheating at the equipment. In many

systems, air circulates at all times and is warmed or cooled as required.

*Planning* for warm air systems begins with the attempt to locate the furnace reasonably close to the center of the building. After the system is designed, a furnace must be selected. It should be capable of burning fuel at a rate suitable to make up the building's hourly heat loss. The rate of air delivery depends on the air temperature *rise* that is planned. Finally, the motor and blower must be powerful enough to overcome the friction of air against metal in both the supply and return duct systems, as well as the friction of air flowing through the furnace, filters, registers, and grilles (see later Fig. 9.31). Minor adjustments can be made at the furnace to adapt to the demands of the system and the building.

Some of the system components are discussed below.

*Furnaces* have become much more efficient in recent years, thanks to forced-draft chimneys and heat exchangers, as shown in Fig. 9.29. Seasonal efficiencies of up to 95% are possible, in contrast to

THERMAL CONTROL

(b)

(a)

**Fig. 9.29** Furnaces with greatly increased operating efficiencies are now available. (a) An Amana gas furnace. (b) The small, high-efficiency heat exchanger utilized by this furnace, which recovers heat from exhaust gases. (Courtesy of Amana Refrigeration, Inc.)

about 62% for older furnaces. The AFUE ratings for furnaces are based on an *isolated combustion system* that requires that all combustion and dilution air be drawn from outside.

Figure 9.30 shows the relationship between a furnace, the duct distribution tree, and some elements of the spaces they serve.

*Ducts* are constructed of sheet metal or glass fiber and are either round or rectangular. Ductwork will conduct noise unless these suggestions are followed:

Do not place the blower too close to a return grille.

Select quiet motors and cushioned mountings.

Do not permit connection or contact of conduits or water piping with the blower housing.

Use a flexible connection between bonnet and ductwork.

Ducts also can be lined with sound-absorbing material to further discourage noise transfer, but beware of materials that encourage mold and mildew growth.

*Duct sizes* may be selected on the basis of permissible air velocity in the duct (Table 9.4). (See Section 10.4 for more detailed duct sizing procedures.)

**EXAMPLE 9.3** The main duct in the low-velocity, warm air system of a residence delivers 1600 cfm (755 L/s). It is located above a drywall ceiling and has a rectangular cross section. Select a size for this duct.

**SOLUTION**

Table 9.4 indicates that for a residence, RC(N) should be between 25 and 35. Choose the quieter RC(N) = 25; then 1700 fpm would be a maximum

THERMAL CONTROL

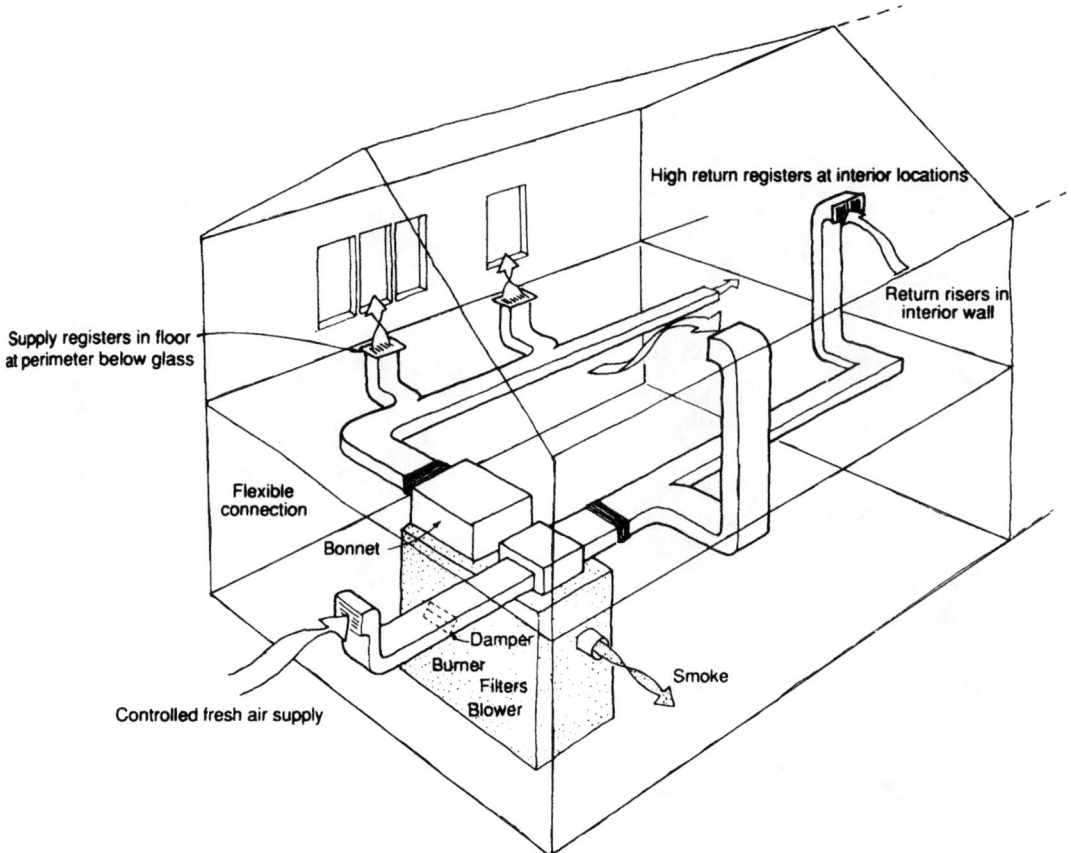

High return registers at interior locations

Return risers in interior wall

Supply registers in floor at perimeter below glass

Flexible connection

Bonnet

Damper

Burner

Filters

Blower

Smoke

Controlled fresh air supply

**Fig. 9.30** *Conventional warm-air furnace and ducts. Low supply registers under windows and high return grilles at interior walls help assure distribution within the spaces.*

acceptable velocity. The area of the duct in square inches would be

$$\frac{1600 \text{ cfm} \times 144 \text{ (in.}^2/\text{ft}^2)}{1700 \text{ fpm}} = 135 \text{ in.}^2$$

A 10 × 14 in. (140-in.²) duct is acceptable; a larger duct would produce less noise and require less fan power to overcome friction. ∎

Figure 9.31 illustrates a device that simplifies duct sizing. At a glance, it shows many duct cross-sectional configurations that will satisfy the combined requirements of friction, airflow, and air velocity.

*Dampers* are necessary to balance the system and adjust it to the desires of the occupants (Fig. 9.32). Splitter dampers are used where branch ducts leave the larger trunk ducts. The flow of each riser can be controlled by an adjustable damper in

the basement at the foot of the riser. Labels should indicate the rooms served. Some codes require dampers of fire-resistant material actuated by fusible links to prevent the possible spread of fire through a duct system (see Fig. 24.6). Figure 9.32*d* shows how turning vanes can be used to assist airflow at sharp turns in ductwork. Such assistance reduces friction within the ductwork, thus reducing the total static head (Fig. 9.33) against which the supply fan must work.

*Supply registers* (Fig. 9.34) should be equipped with dampers, and their vanes should be arranged to disperse the air and to reduce its velocity as soon as possible after it enters the room. This is commonly done by providing vanes that divert the air, half to the right and half to the left. When a supply register is in the corner of a room, it is best if the vanes deflect all the air away from the corner.

**TABLE 9.4 Air Velocities for Ducts and Grilles**

| PART A. MAIN DUCTS[a] | | | | | |
|---|---|---|---|---|---|
| | | **Maximum Airflow Velocity** | | | |
| | | **fpm** | | **m/s** | |
| **Duct Location** | **Design RC(N)** | **Rectangular** | **Circular** | **Rectangular** | **Circular** |
| In shaft or above drywall ceiling | 45 | 3500 | 5000 | 17.8 | 25.4 |
| | 35 | 2500 | 3500 | 12.7 | 17.8 |
| | 25 | 1700 | 2500 | 8.6 | 12.7 |
| Above suspended acoustic ceiling | 45 | 2500 | 4500 | 12.7 | 22.9 |
| | 35 | 1750 | 3000 | 8.9 | 15.2 |
| | 25 | 1200 | 2000 | 6.1 | 10.2 |
| Duct within occupied space | 45 | 2000 | 3900 | 10.2 | 19.8 |
| | 35 | 1450 | 2600 | 7.4 | 13.2 |
| | 25 | 950 | 1700 | 4.8 | 8.6 |

| PART B. FACE VELOCITIES AT SUPPLY AND RETURN OPENINGS[b] | | | |
|---|---|---|---|
| | | **Maximum Airflow Velocity** | |
| **Type of Opening** | **Design RC(N)** | **fpm** | **m/s** |
| Supply air outlet | 45 | 630 | 3.2 |
| | 40 | 550 | 2.8 |
| | 35 | 490 | 2.5 |
| | 30 | 430 | 2.2 |
| | 25 | 350 | 1.8 |
| Return air opening | 45 | 750 | 3.8 |
| | 40 | 670 | 3.4 |
| | 35 | 590 | 3.0 |
| | 30 | 490 | 2.5 |
| | 25 | 430 | 2.2 |

| PART C. RANGES OF RC (N)[c] | |
|---|---|
| **Function** | **HVAC System Noise in Unoccupied Spaces, RC (N)** |
| Residential, private | 25–35 |
| Hotels, individual rooms, meeting rooms | 25–35 |
|   Lobbies, corridors, service areas | 35–45 |
| Office buildings, private offices, conference rooms | 25–35 |
|   Teleconference rooms | 25 (max.) |
|   Open-plan offices | 30–40 |
|   Lobbies, circulation | 40–45 |
| Hospitals and clinics, private rooms, operating rooms | 25–35 |
|   Wards, corridors, public spaces | 30–40 |
| Performing arts, drama theaters, music teaching rooms | 25 (max.) |
|   Music practice rooms | 35 (max.) |
| Churches, mosques, synagogues | 25–35 |
| Schools, lecture halls | 35 (max.) |
|   Classrooms over 750 ft$^2$ (70 m$^2$) | 35 (max.) |
|   Classrooms up to 750 ft$_2$ (70 m$^2$) | 40 (max.) |
| Libraries | 30–40 |
| Laboratories (with fume hoods), group teaching | 35–45 |
|   Research, telephone use, speech communication | 40–50 |
|   Testing/research, minimal speech communication | 45–55 |
| Courtrooms, unamplified speech | 25–35 |
|   Amplified speech | 30–40 |
| Sports indoors, school gymnasiums and pools | 40–50 |
|   Large capacity, with amplified speech | 45–55 |

*Source: 1995 ASHRAE Handbook—HVAC Applications,* copyright © by the American Society of Heating, Refrigerating and Air-Conditioning Engineers, Inc., Atlanta, GA.

[a]Branch duct velocities should be about 80% of those listed, whereas velocities in final runouts to outlets should be 50% or less. Elbows and other fittings can increase noise substantially, so airflows should be reduced accordingly.

[b]These are "free" opening velocities. Diffusers and grilles can increase sound levels, in which case these velocities should be reduced accordingly.

[c]See also Table 19.8 for NC recommended levels.

THERMAL CONTROL

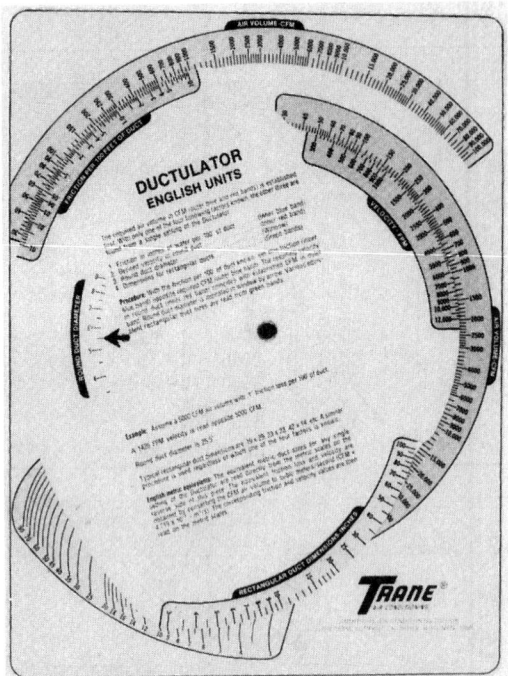

**Fig. 9.31** *The Ductulator is a duct-sizing device available from the Trane Company. The designer selects any two factors (e.g., friction and airflow volume), and the device yields all the other factors (e.g., air velocity, diameter of the round duct required, or combinations of rectangular-duct cross sections required).*

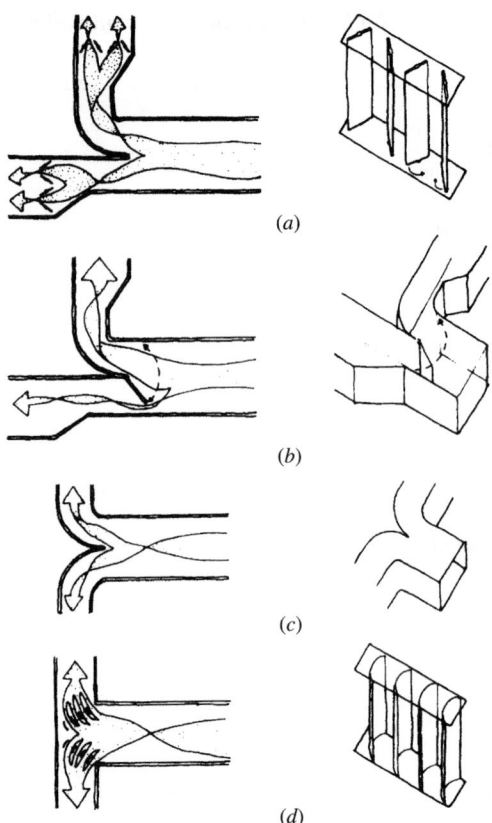

*(a)*

*(b)*

*(c)*

*(d)*

**Fig. 9.32** *Air controls in ducts. (a) Air adjustment by opposed-blade dampers. (b) Air adjustment by a splitter damper. (c) Conventional turns in ducts. (d) Right-angle turns with turning vanes—a more compact method.*

Return grilles are of the slotted type in walls and of the grid type in floors. All registers and grills should be made tight at the duct connection. See Tables 9.4 and 9.5 for selection of registers based on output and recommended face velocity.

*Controls:* The burner is started and stopped by a thermostat, which is usually placed in or near the living room at a thermally stable location that is protected from cold drafts, direct sunlight, and the warming effects of nearby warm air registers. A cut-in temperature of between 80° and 95°F (27° and 35°C) is selected for the fan switch in the furnace bonnet. After the burner starts, the fan switch turns on the blower when the furnace air reaches the selected cut-in temperature. Burner and blower then continue to run while heat is needed. When the burner turns off, the blower continues to run until the temperature in the furnace drops to a level a little below the cut-in temperature of the fan switch. If, during operation, the temperature unexpectedly exceeds 200°F

(93°C), a high-limit switch turns off the burner in the interest of comfort and safety. As in all automatically fired heating units, a stack temperature control in the breeching cuts off the fuel if ignition fails.

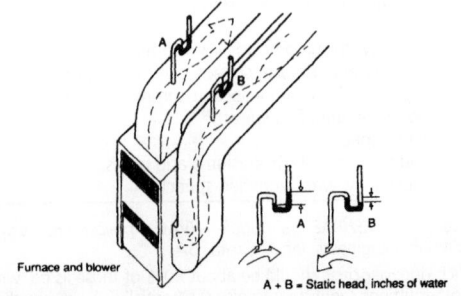

Furnace and blower

A + B = Static head, inches of water

**Fig. 9.33** *The static head is the pressure, measured in inches of water, available to overcome friction in the entire system.*

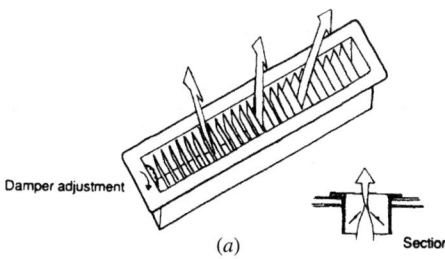

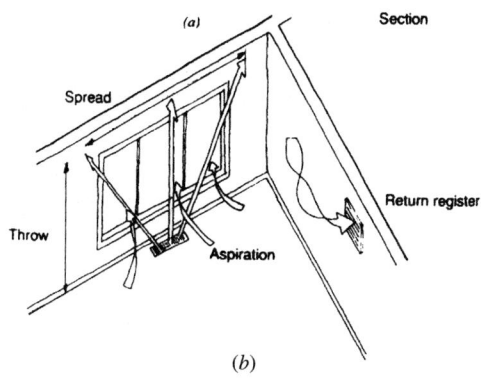

**Fig. 9.34** *The typical floor register (diffuser) in its 2¼-in. × 12-in. (60-mm × 300-mm) size (a). It has diverting vanes for spread and an adjustable damper. See Table 9.5 for characteristics. (b) Spread and throw; cooler room air is induced by aspiration to join the stream of warm air, resulting in a bland, pleasant air stream that crosses the room.*

## 9.8 HEATING/COOLING SYSTEMS

Mechanical cooling is a much more recent development than heating; in its early days, it was widely adopted as a retrofit to existing heating systems. The first two approaches to mechanical cooling reflect this early attitude.

### (a) Cooling Coils Added to Warm Air Furnaces

This common system utilizes a rather simple arrangement of the refrigeration cycle. Figure 9.1 illustrated the circuit of a refrigerant in compression, condensing, and evaporation, in which the condenser heat is carried away by water and the evaporation process draws heat out of water in another circuit to produce *chilled* water. Thus, the heat is *moved* to a heat rejection location outdoors. Figure 9.35 is a schematic diagram of an air-to-air (in contrast to a water-to-water) refrigeration device. Air instead of water can be used to cool the condenser, and indoor air can be cooled directly by being passed over the evaporator coil in which the refrigerant is expanding from a liquid to a gas. Thus, heat is moved from the indoor air to the outdoor air by the step-up action or heat-pumping nature of the refrigeration cycle. When indoor air is cooled directly in this manner, by the expanding refrigerant, the process is usually known as *direct expansion*. The cooling coils therefore are often

**TABLE 9.5 Typical Residential Forced-Air Register, 2¼ × 12 in. (60 × 300 mm)**

| PART A. I-P UNITS | | | | | | | | |
|---|---|---|---|---|---|---|---|---|
| Heating (Btu/h) | 3,045 | 4,565 | 6,090 | 7610 | 9,515 | 11,415 | 13,320 | 15,220 |
| Cooling (Btu/h) | 855 | 1,280 | 1,710 | 2,135 | 2,670 | 3,200 | 3,735 | 4,270 |
| Cfm | 40 | 60 | 80 | 100 | 125 | 150 | 175 | 200 |
| Vertical throw (ft) | 3 | 4 | 5 | 6 | 8 | 10 | 12 | 14 |
| Vertical spread (ft) | 6 | 8 | 10 | 11 | 14 | 17 | 22 | 25 |
| Face velocity (fpm) | 280 | 420 | 565 | 705 | 880 | 1,050 | 1,230 | 1,400 |

*Source:* Lima Register Company.

| PART B. SI UNITS | | | | | | | | |
|---|---|---|---|---|---|---|---|---|
| Heating (W) | 890 | 1,340 | 1,780 | 2,230 | 2,790 | 3,340 | 3,900 | 4,460 |
| Cooling (W) | 250 | 375 | 500 | 625 | 780 | 940 | 1,090 | 1,250 |
| L/s | 19 | 28 | 38 | 48 | 59 | 71 | 83 | 95 |
| Vertical throw (m) | 0.9 | 1.2 | 1.5 | 1.8 | 2.4 | 3.0 | 3.7 | 4.3 |
| Vertical spread (m) | 1.8 | 2.4 | 3.0 | 3.4 | 4.3 | 5.2 | 6.7 | 7.6 |
| Face velocity (m/s) | 1.4 | 2.1 | 2.9 | 3.6 | 4.5 | 5.3 | 6.2 | 7.1 |

SI conversions by the author.

referred to as *DX coils.* Figure 9.36 shows another popular arrangement in which the airflow through the furnace and coils is horizontal.

Meanwhile, the compressor–condenser unit is placed outdoors on a concrete slab or on the roof. The unit creates a noisy, hot microclimate in summer—an influence on both site and building planning.

### (b) Hydronic and Coils

This system combines a perimeter hot water heating pipe with an overhead air-handling system. A boiler with a tankless coil supplies domestic hot water. The boiler's heat output supplies both the perimeter loop and a coil in the air-handling unit of the duct system. The total heating load is met by the combination of radiant heat generated by the perimeter loop and heated air from the overhead air-handling system.

As installed in the Levittown standardized houses, the perimeter loop consists only of ½- or ¾-in. (13- or 19-mm) tubing embedded 4 in. (100 mm) below the top of the floor slab to overcome the cold slab effect. It has the capacity to maintain a 35F° (19C°) differential between the inside and outside temperatures at the perimeter.

The air-handling unit and overhead duct system, incorporating supply outlets in each room and central return, is used throughout the year. Its cooling coil is connected to an adjacent outdoor condensing unit (Fig. 9.37).

Because the heating load is shared by the slab loop and by warm air, the winter indoor temperature remains more constant. The air can be distributed at no more than about 120°F (49°C), or 20F° (11C°) less than with a conventionally ducted system.

### (c) Air–Air Heat Pumps

These use the refrigeration cycle to both heat *and* cool, thus eliminating the distinction between furnace (or boiler) and DX cooling coils. As shown in Fig. 9.38, heat is "pumped" from indoors to outdoors in summer (Fig. 9.38*a*) and from outdoors to indoors in winter (Fig. 9.38*b*). Heat pumps can transfer heat air–air, air–water, and water–water. The most common application for smaller buildings is the air–air heat pump, shown in Figs. 9.38 and

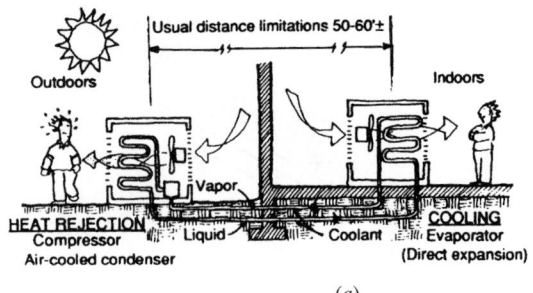

(*a*)

(*b*)

**Fig. 9.35** *The compressive refrigeration cycle (a) used in an exterior air-cooled heat rejection unit and an interior cooling coil suitable for placement in a central air stream. (b) Cooling/heating air handling unit. Cutaway view shows an upflow air pattern. At the lower left is the return air intake and a filter; adjacent are the fan and motor. Above these components are the gas-burning elements. At the top is the direct-expansion cooling coil. (Courtesy of American Furnace Division, Singer Company.)*

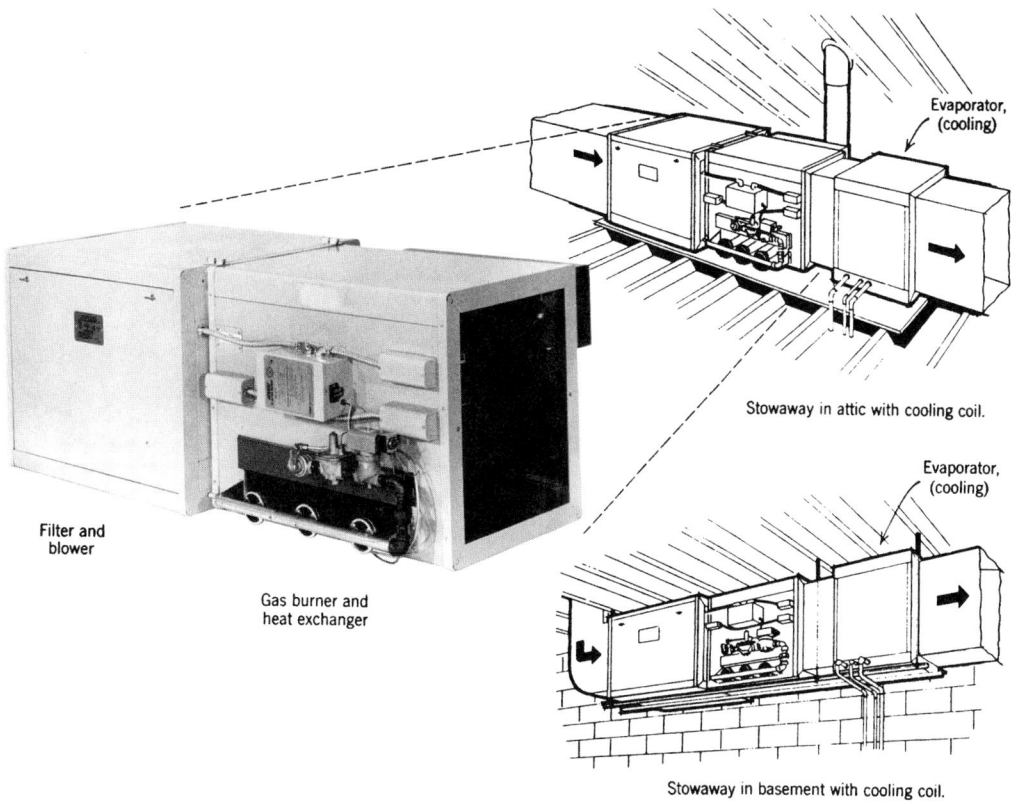

Filter and
blower

Gas burner and
heat exchanger

Evaporator,
(cooling)

Stowaway in attic with cooling coil.

Evaporator,
(cooling)

Stowaway in basement with cooling coil.

**Fig. 9.36** *Compact horizontal flow combinations for forced-air heating and cooling. The small pipe between the two refrigerant pipes of the cooling unit is a water drain that carries away the condensed moisture from the recirculated and outdoor air. The heating unit requires gas and flue or exhaust gas connections. Refrigerant pipes connect to the outdoor compressor–condenser unit.*

9.39. In a *single-package* (also called *unitary*) system (Fig. 9.39*a*), only one piece of equipment is involved. A single-package air–air heat pump moves heat between an outdoor airstream and an indoor airstream; although kept separate, both streams pass through a *single* outdoor unit. A system with both outside and inside components is called a *split* system (Fig. 9.39*b*). A split-system air–air heat pump moves heat via a refrigerant loop between the outdoor unit (which also contains the compressor), through which outdoor air passes, and the indoor unit (which usually contains backup heating coils) for the treatment and circulation of indoor air.

Single-package heat pumps are commonly located on roofs, where they have unlimited access to outdoor air and where their noise is less likely to annoy—provided they are sufficiently isolated from the building's structure. This approach is shown in the daylighted, passively solar-heated Mount Airy, North Carolina, library (Fig. 9.40). This 14,000-ft$^2$ (1300-m$^2$) building also has a solar preheating system for its hot water. The five air–air heat pumps utilize economizer cycles (up to 100% outside air when temperatures are favorable). The average annual building energy consumption has been monitored at about 17,000 Btu/ft$^2$ (53,635 W/m$^2$)—approximately one-third of nearby similar-function buildings.

Individual air–air heat pump units are especially common in building types with all-perimeter spaces with varying orientations and numerous thermal zones. Motels are a prime example. In Fig. 9.41, separate air–air heat pumps serve each motel room; at best, their constant noise helps mask the intermittent sounds from the adjacent parking lot/circulation

THERMAL CONTROL

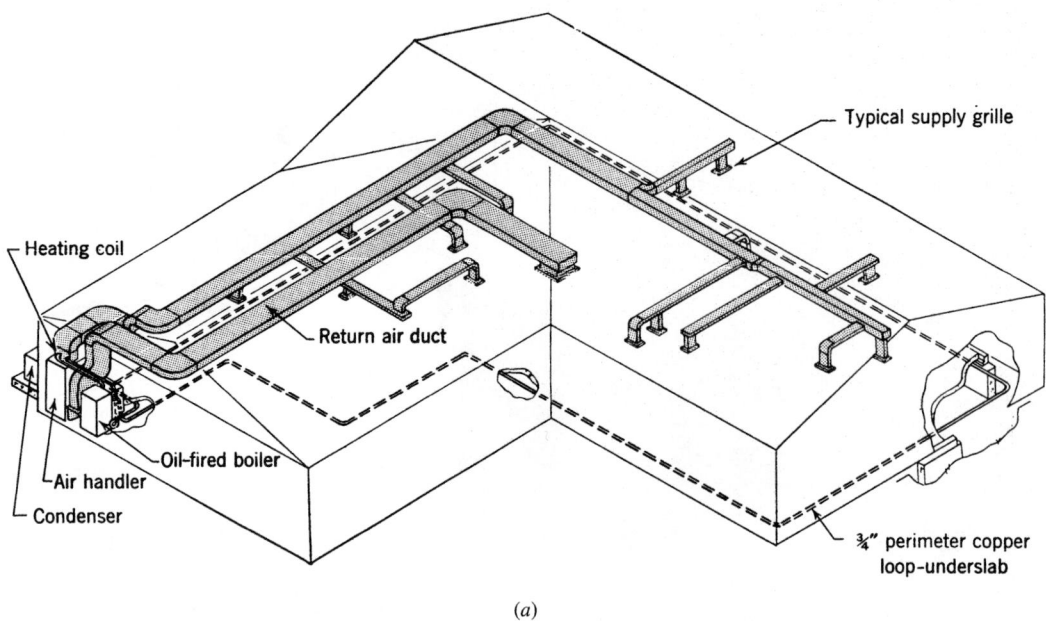

(a)

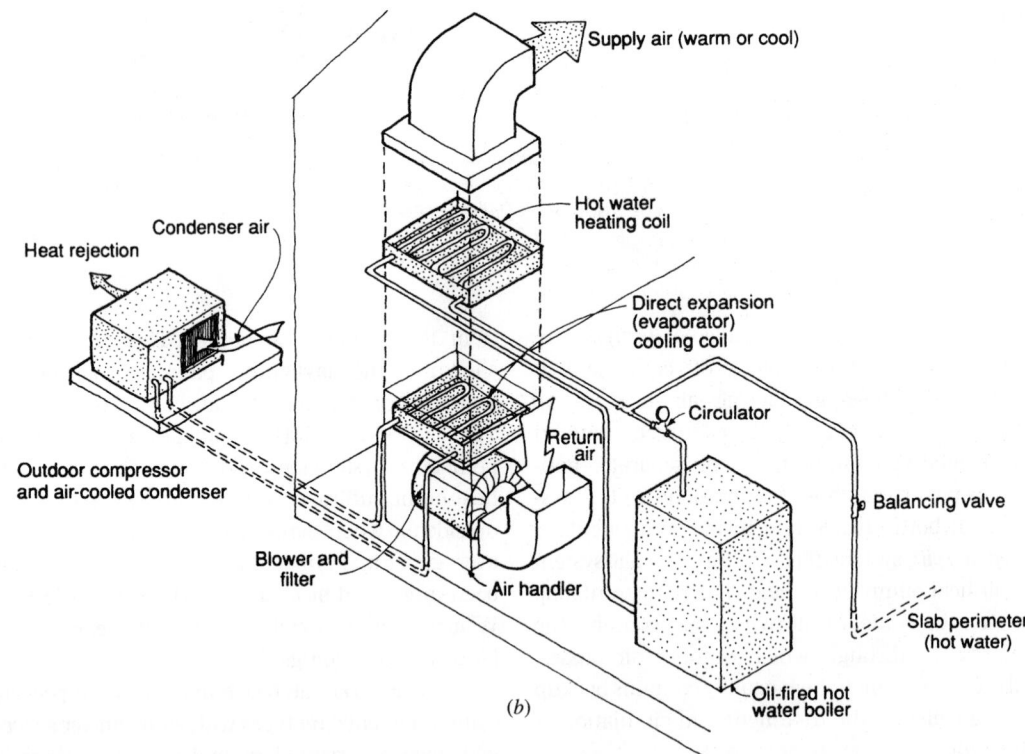

(b)

**Fig. 9.37** *Combination hydronic and forced-air system (a) using a perimeter hot water loop in the floor slab and an overhead air supply. (Courtesy of Levitt and Sons. Design by mechanical engineer John Liebl.) (b) Close-up of the heating and cooling components. The boiler heats both the slab perimeter and the coils in the air stream.*

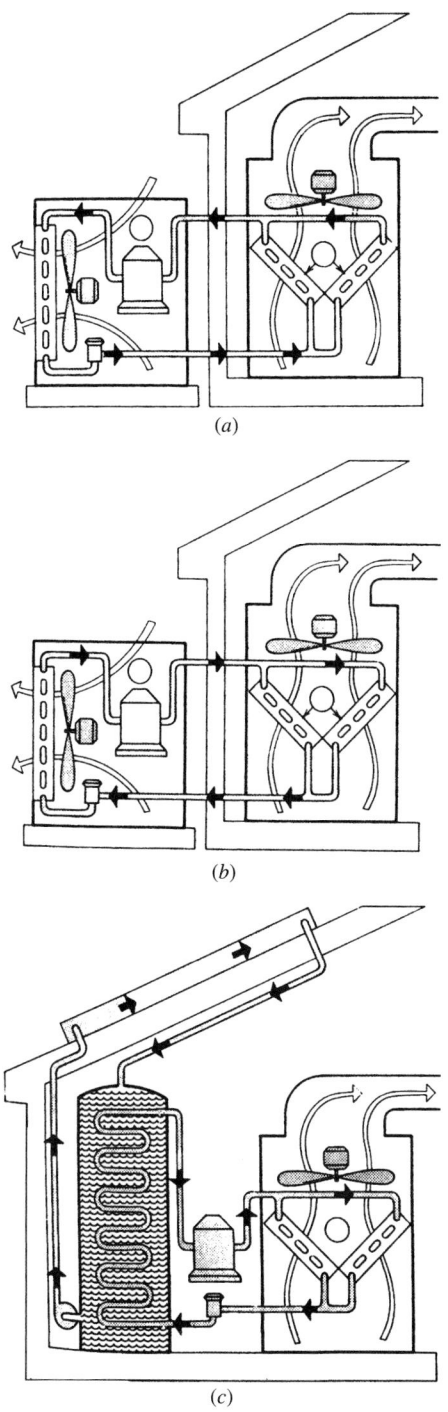

**Fig. 9.38** *Heat pump applications of the compressive refrigeration cycle. The simple air–air heat pump provides cooled air (a) or heated air (b). Teamed with a solar collector and water storage tank (c), the heat pump can yield usefully warm temperatures in the air stream while increasing the collector efficiency by lowering the storage tank temperature. (Reprinted with permission from* Popular Science, *© 1978 by Times Mirror Magazines, Inc.)*

space. Opportunities for heat exchange between these heat pumps are scarce; if a central water loop (see Section 9.8e) were substituted for outdoor air as the heat source/sink, energy costs would go down, although the first cost would rise.

Split systems are popular because the noise of the compressor and the outdoor air fan are removed from the interior, and the size of the indoor unit can be quite small. This indoor element is often mounted either high on the wall or on the ceiling. Such an indoor unit is available with automatically changing louvers; when it is the cooling mode, it delivers cool air along the ceiling, from where it sinks to the level of occupancy; cold air blowing directly on people is avoided. In the heating mode, the louvers shift to direct hot air steeply downward. The greater the distance between the indoor and outdoor units, the greater the strain on the refrigerant loop.

Heat pumps have a high initial cost, and they have shown a relatively high frequency of compressor failure. Noise from the compressor and the outdoor air fan may affect site planning, especially for residences.

One of the primary attractions of the heat pump is that in its heating mode it can give more energy than it receives (electrically). Although energy (usually electricity) is required to run the cycle, the pump draws "free" heat from a source such as outdoor air. The total heat delivered to the building is more than the heat (electricity) required to run the cycle. The measure of this heat advantage is called the *coefficient of performance* (COP), defined as

$$COP = \frac{\text{heat delivered to space}}{\text{necessary work input}}$$

(See the earlier discussion in Section 9.7j.)

In typical space-heating applications, a seasonal COP of 2 or more is common in mild-winter areas. Because the COP changes with outdoor conditions and indoor load, a *seasonal energy efficiency ratio* (SEER) rating system has been established. SEER measures the number of Btu/h removed for each watt of energy input, averaged over the cooling season. The higher the SEER, the more efficient the heat pump's seasonal performance. SEER ranges are roughly as low as 5 and as high as 15. The heating cycle of the heat pump has a similar rating system, called the *heating seasonal performance factor* (HSPF).

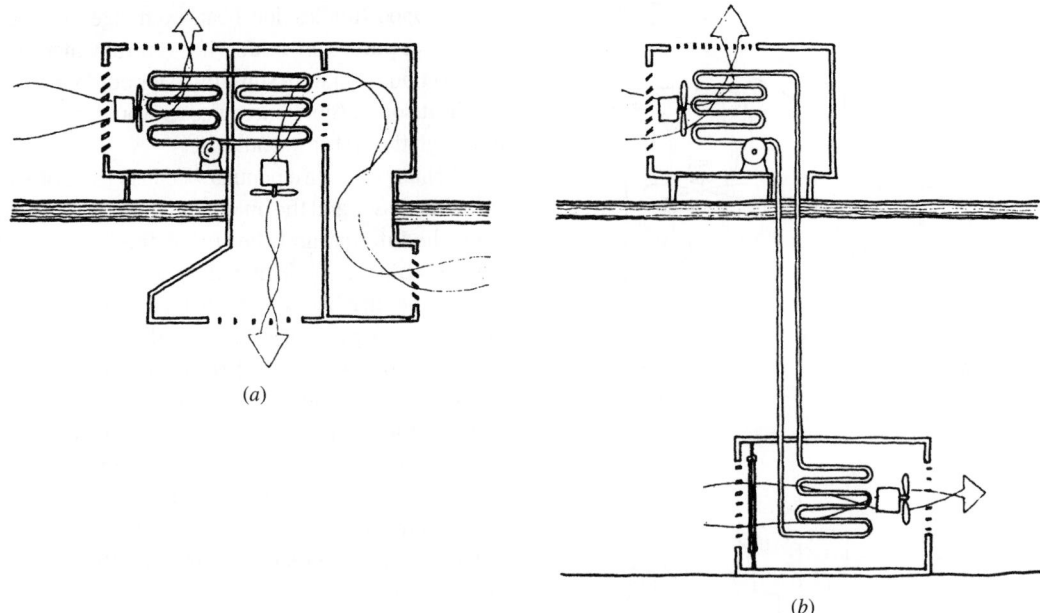

(a)

(b)

**Fig. 9.39** *The package unit (a) and the split system (b) are popular applications of the air–air heat pump. The package unit and the outdoor unit of the split system are also typically placed outside walls, as well as on roofs, as shown here.*

As might be expected from a device that draws heat from winter outdoor air, there are limitations to its heating performance. As outdoor temperatures approach 32°F (0°C), the COP drops and the outdoor coil tends to ice over. Built-in electric resistance coils must then be used; this, of course, ends the efficiency advantage that made the heat pump attractive. See Fig. 9.42 for a demonstration of falling performance with falling temperatures. Because of this characteristic, air–air heat pumps are less frequently used in cold-winter climates. They also generally make questionable backup choices for passively solar-heated buildings in colder climates, because backup sources are typically needed only in the coldest weather.

Heat pumps that pump heat from *water* sources, such as wells or solar-heated storage tanks, or from the *ground*, are much more dependable cold weather performers. Water–air heat pumps (shown in Fig. 9.38c) and solar collectors make an effective team. As these rather high-COP heat pumps remove heat from the solar storage tank (to deliver it to the indoor air), the resulting lower temperature of the solar-heated water *increases* the solar collector's performance. (See Fig. 21.36 for solar collector efficiencies.) Assume that on a cold, partly sunny day,

the collector being fed water from the tank at 90°F is able to raise its temperature by 4F° to 94°F (from 32° to 35°C). This improvement is slight because of the rather high heat loss that a 94°F collector experiences when surrounded by cold air. If, however, the collector were to be fed 59°F (15°C) water, its heat loss would be greatly reduced. The heat that the collector *does not* lose to the cold air can be invested in the 59°F water, which will thus leave at a considerably higher temperature than 59°F + 4 = 63°F. Thus, more solar energy is collected, and more is available for transfer to the building via the heat pump. The heat pump, in turn, can heat the 59°F tank water to much higher temperatures to serve the coils in the air handling unit.

Although most of today's heat pumps utilize electricity to drive the compressive refrigeration cycle, there are also absorption cycle heat pumps that utilize natural gas. The *GAX heat pump* (Fig. 9.43) not only uses this lower-grade resource for heating and cooling, it avoids CFCs and HCFCs and functions at outdoor temperatures much lower than those of the electric air–air heat pump. Eventually, solar energy can be used to drive the absorption cycle. Solar-driven refrigeration is a particularly elegant blend of energy source

*(a)*

*(b)*

**Fig. 9.40** *Mount Airy Library, Mount Airy, North Carolina. Mazria/Schiff & Associates, Architects. (a) View from the southwest. (Photo by Gordon H. Schenck, Jr.) (b) Plan. (c,d) North–south sections relate sunlight and natural heat flow. Individual package heat pumps are set on the flat-roof sections.*

**361**

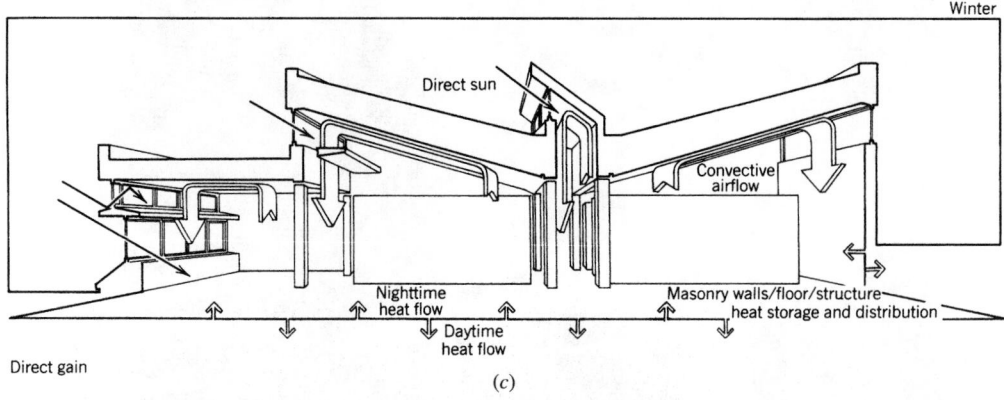

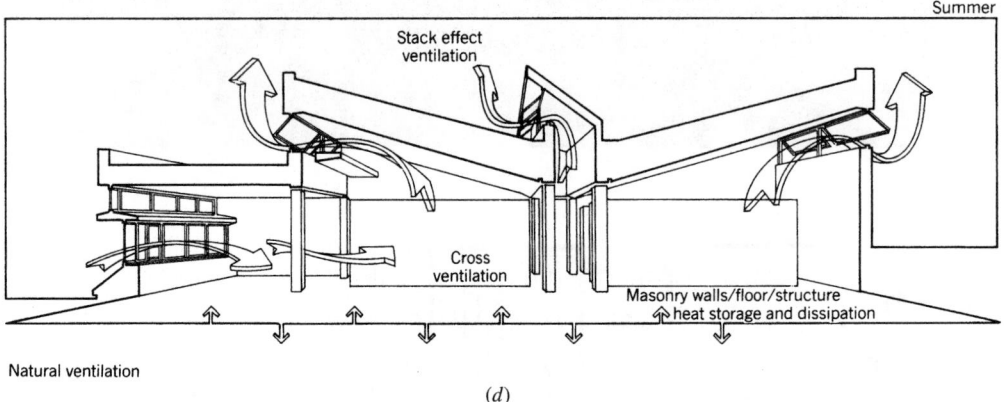

**Fig. 9.40** (Continued)

and task—the hotter the sun, the higher the cooling capacity.

### (d) Ground Source Heat Pumps

Ground–air heat pumps, also called *geothermal heat pumps* or *Geo Exchange systems,* are found in several configurations throughout North America. They often provide domestic hot water in addition to heating and cooling. An environmentally safe refrigerant is circulated through a loop installed underground (or in a pond or lake), taking heat from the soil in winter and discharging heat to the soil in summer. The loop is often high-density polyethylene (HDPE). Below the surface, soil temperatures are more stable year round than outdoor air temperatures, thus raising the COP relative to that of air–air heat pumps. Annual well-water (deep soil) temperatures were shown in Fig. 8.56. This

system is almost completely out of sight, with no maintenance or weathering of exterior equipment. Noise is confined to the compressor in a small indoor mechanical room.

Some common configurations are shown in Fig. 9.44. In the closed systems (*a–c*) the flow rate is typically 2 to 3 gpm/ton of refrigeration (0.3 to 0.5 mL/J), with lower flows in the open loop systems.

The horizontal ground source closed loop heat pump (Fig. 9.44*a*) requires trenches 3 to 6 ft (1 to 2 m) deep; typically, 400 to 650 ft (120 to 200 m) of pipe are installed per ton (12,000 Btu/h or 3.5 kW) of heating and cooling capacity. To squeeze more pipe length into a trench, a "slinky" coiled pipe is sometimes used. The trenches can be placed below parking lots or lawns and gardens.

The vertical ground source closed loop heat pump (Fig. 9.44*b*) is particularly applicable where the site area is limited. Vertical holes are bored from

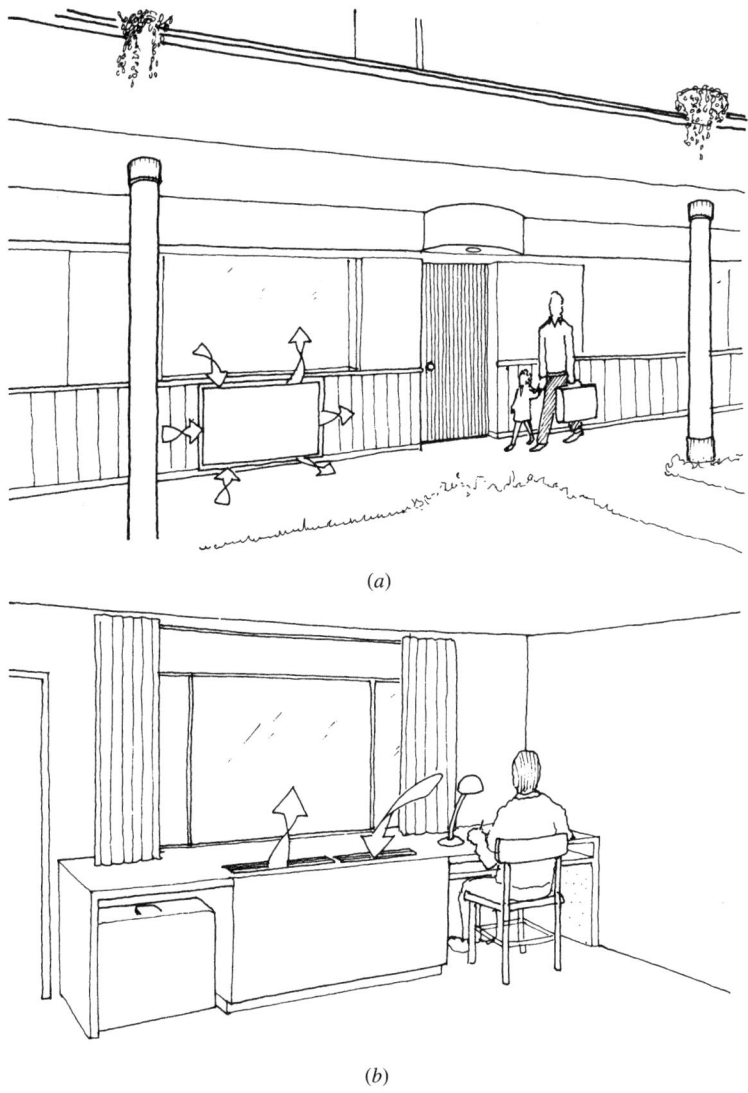

(a)

(b)

**Fig. 9.41** *Air–air heat pumps serve a motel. (a) A panel set just forward of the exterior wall allows the heat pump to inhale and exhale outdoor air around the panel edges. In summer it discharges warm air; in winter, cool air. (b) Interior cabinet contains the heat pump. Room air is taken in as shown, then discharged (after cooling or heating) upward across the glass surface—the likely point of maximum heat loss or gain.*

150 to 450 ft (46 to 137 m) deep. Each hole contains a single full-depth loop and is backfilled (or grouted) after the loop is installed. Because the temperature is much lower at greater depths, less pipe length is required than for horizontal loops. The distance between boreholes varies from a minimum of 15 ft (4.6 m) with high water table and low building cooling loads to as much as 25 ft (7.6 m) for buildings with high cooling loads. A minimum distance of 20 ft (6 m) is usually recommended.

The pond or lake closed loop heat pump (Fig. 9.44c) is sometimes used when a building is close to an adequately large body of water. The loop is submerged, and the surrounding water conducts heat far more rapidly than does soil. The resulting shorter length required, and the low cost of placing the coils in water, can make this attractive. However, the water level in the pond should never drop below a minimum of 8 ft (2.5 m) and must have sufficient surface area for heat exchange.

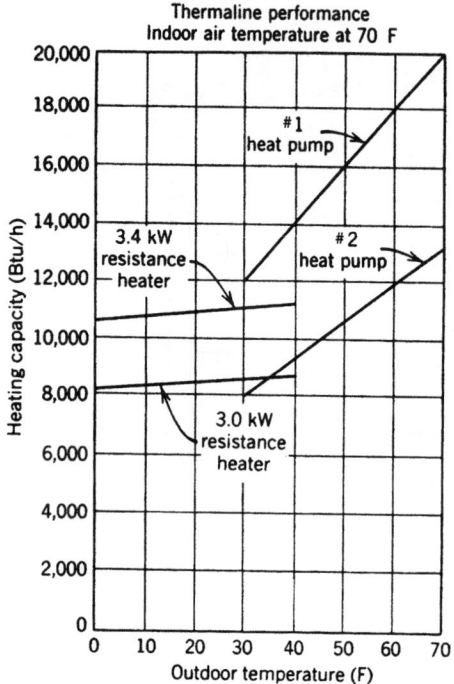

Thermaline performance
Indoor air temperature at 70 F

*Fig. 9.42 Typical air–air heat pump operating characteristics; heat delivery falls with falling outdoor temperature.*

The groundwater–source open loop heat pump ("pump and dump") is suitable only where groundwater is plentiful, and may be prohibited by local codes and environmental regulators. One variant of this system (Fig. 9.44*d*) takes water from one well, through a heat exchanger within the building, then discharges it to a second well. Another variant (Fig. 9.44*e*) takes water from the bottom and discharges back into the top of the same (standing) well, typically 6 in. (150 mm) in diameter and as deep as 1500 ft (460 m).

The Wildlife Center of Virginia at Waynesboro is a wildlife teaching–research hospital. Its 5700-ft² (530-m²) floor area is served by four geothermal heat pumps (two at 4 tons, two at 5 tons) connected to 11,350 ft (3460 m) of underground horizontal pipe, laid in a "slinky" configuration and thus fitting within about 2500 ft (760 m) of trench. The trench was dug around existing trees and placed under future roadways in its forest setting. Energy simulations estimate that annual heating, cooling, and hot water will use about 35,000 kWh. Had air–air heat pumps and an electric water heater been used instead, the estimate is 66,000 kWh, a 47% yearly savings advantage for the geothermal system.

Ground source heat pumps are often used in retrofits, especially in schools where site areas are plentiful, or historic structures where small-size interior mechanical equipment is highly desir-

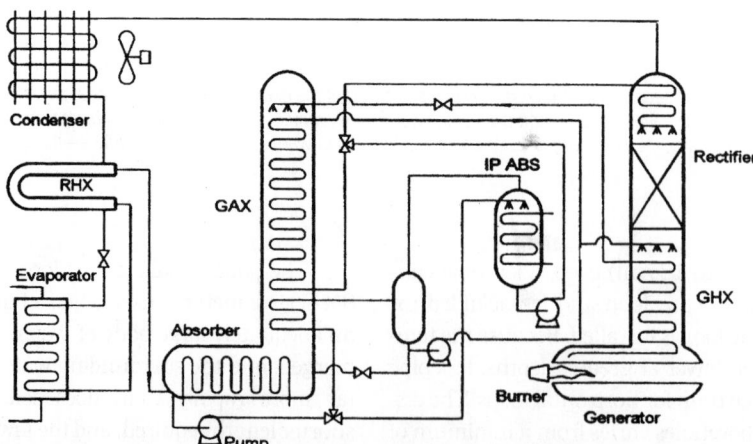

*Fig. 9.43 An air-cooled advanced GAX (generator-absorber heat exchange) heat pump uses aqua ammonia as the absorbent and is operational down to –10°F (–23°C). The target cooling COP is 0.95; heating COP is 1.55. Designed for light commercial applications, it ranges from 5 to 25 tons capacity. (RHX = refrigerant heat exchanger; IP ABS = intermediate pressure absorber.) (Courtesy of Energy Concepts Company, Annapolis, MD.)*

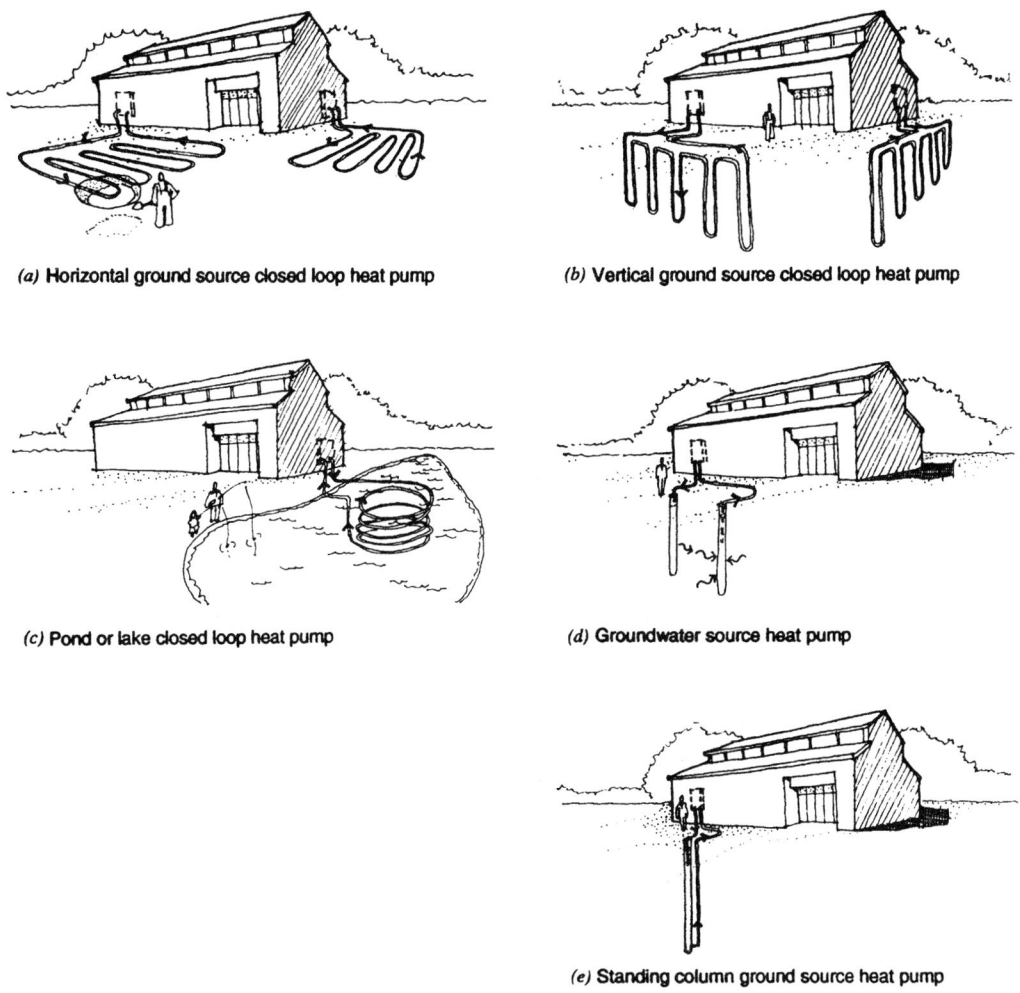

*(a)* Horizontal ground source closed loop heat pump

*(b)* Vertical ground source closed loop heat pump

*(c)* Pond or lake closed loop heat pump

*(d)* Groundwater source heat pump

*(e)* Standing column ground source heat pump

**Fig. 9.44** *Configurations of ground source heat pumps. (a) Horizontal ground source closed loop heat pump laid in trenches. (b) Vertical ground source closed loop heat pump placed in boreholes. (c) Pond or lake closed loop heat pump. (d) Groundwater source heat pump taking water from one well and discharging it to another. (e) Standing column groundwater source heat pump taking water from, then discharging to, the same well.*

able. The Daniel Boone High School near Johnson City, Tennessee, was built in 1971 with a two-pipe chilled water system and electric resistance heat. A 1998 ASHRAE Technology Award was won when this 160,000-ft$^2$ (14,864-m$^2$) school was retrofitted with a ground source vertical closed loop heat exchanger that is fed by 320 boreholes, each 150 ft (46 m) deep. Each borehole loop is 300 ft (91 m) of ¾-in. (19-mm) polyethylene pipe. The boreholes are arranged in sections of 20 holes each. The holes are 15 ft (5 m) on center. The 20-hole sections are separated by 20 ft (6 m). The loops all connect to an 8-in. (203-mm) supply and same size return line to the new heat exchanger within the existing mechanical room. Within the building, a new water loop heat pump system is installed, one heat pump in each zone.

### (e) Water Source Heat Pumps

The school described previously uses an interior closed water loop that connects all the heat pumps of all thermal zones. This system facilitates heating in one zone while another zone is being cooled, because simultaneous heat "deposits" and heat "withdrawals" actually help the system to function most efficiently. Figure 9.45 diagrams the typical water loop system, where a supplementary heat source (such as a boiler) in cold weather, and a supplementary heat rejector (such as a cooling tower) in hot weather, are installed to maintain usable water temperatures within the loop. A large office building in Pittsburgh that uses such a system is shown in Fig. 10.63.

Motels are good candidates for water loop heat pumps, because some rooms face sunny conditions, others are shaded; some are occupied, some unoccupied; and substantial domestic hot water needs all combine to make a heat-sharing water loop attractive.

## 9.9 PSYCHROMETRICS AND REFRIGERATION

The procedure for sizing cooling equipment is more complicated than for heating because latent as well as sensible heat must be considered. The example following refers to the psychrometric chart; a review of Figs. 8.37 to 8.41 may be helpful.

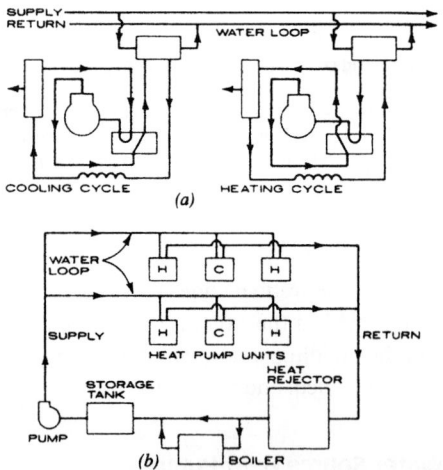

**Fig. 9.45** *Water source heat pumps. (a) Each water–air heat pump either deposits heat into the loop (while cooling) or withdraws heat (while heating). This system is particularly well suited to buildings where simultaneous heating and cooling needs occur. (b) Supplementary heat sources (boilers) and heat rejectors (cooling towers) are usually provided. (Reprinted by permission from AIA: Ramsey/Sleeper, Architectural Graphic Standards, 9th ed., © 1994 by John Wiley & Sons, Inc.)*

**EXAMPLE 9.4** Find the total heat to be removed, and thus the refrigeration capacity required, for a dance hall. The design conditions are:

| | |
|---|---|
| Room conditions (summer) | 75°F DB (24°C), 50% RH |
| Number of occupants | 80 people |
| Activity | Dancing |
| Ventilation provided | 35 cfm (18 L/s) per person |
| Conditions, outdoor air | 90°F DB, 75°F WB (32.2°C, 23.9°C) |

| Heat Gains in the Room | Sensible Heat, SH (Btu/h) | Latent Heat LH (Btu/h) |
|---|---|---|
| 80 people dancing (see Table F.8) | | |
| 80 × 305 Btu/h | 24,400 | |
| 80 × 545 Btu/h | | 43,600 |
| Total transmission and solar gain, lights, equipment, etc. | 67,600 | None |
| | Room sensible heat (RSH) | Room latent heat (RLH) |
| | = 92,000 | = 43,600 |
| Total heat gains in room: 135,600 Btu/h (RSH + RLH) | | |

## SOLUTION

First, determine the portion of the heat gain that is due to sensible heat gain, called the *sensible heat factor* (SHF):

$$SHF = \frac{RSH}{RSH + RLH} = \frac{92,000}{135,600} = 0.68$$

On the psychrometric chart (simplified in Fig. 9.46a), draw a line between the fixed "bull's-eye" (80°F DB, 50% RH) and the value of 0.68 on the SHF scale at the upper right edge of the chart. This is called the *SHF line*.

Point *A* is the room condition (of the "used" air) within the dance hall as it is returned for reprocessing: 75°F DB, 50% RH (62.5°F WB). Next, decide how much cooler the supply air should be than the return air. To avoid uncomfortable drafts, this supply temperature is usually 20F° (or less) below the space's air temperature. In this case, with such well-stirred air, choose 20F°. Then the quantity of air required to cool the room will be

$$cfm = \frac{RSH}{1.1\Delta t} = \frac{92,000 \text{ Btu/h}}{1.1 \, (20F°)} = 4260 \text{ cfm}$$

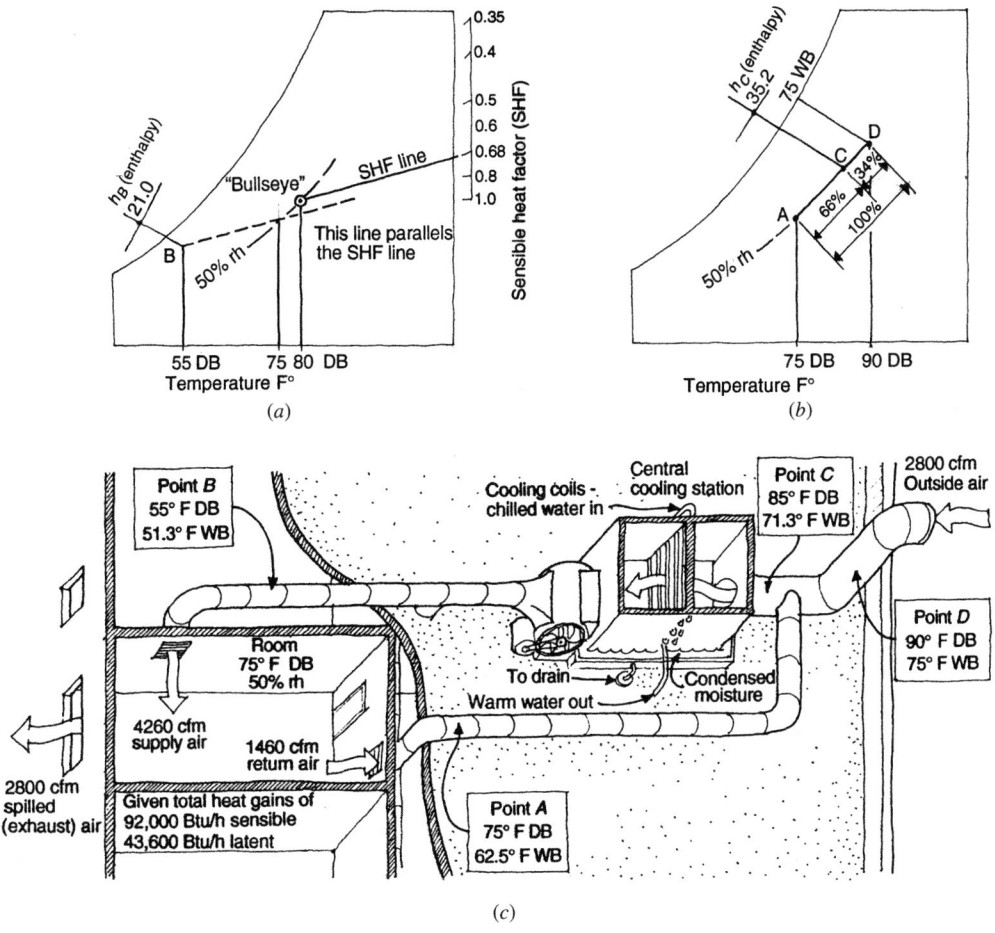

**Fig. 9.46** *Sizing cooling equipment using the psychrometric chart (see detailed charts in Appendix F). (a) Finding the conditions for the supply air. (b) Finding the conditions for the return air–outdoor air mixture. (c) Points A, B, C, and D, shown within the building and its cooling equipment.*

(The factor 1.1 is the constant explained in the ventilation formulas in Section 7.9.)

The portion of this supply air that is *outdoor* air is as follows:

80 people × 35 cfm/person = 2800 cfm

So the percentage of outdoor air is

$$\frac{2800}{4260} = 66\%$$

Now, several important points can be located on the chart. Point *B* is the condition of the air entering the rooms; it has been decided that it will be 20F° cooler than 75°F, which places point *B* somewhere on the 55° DB line. To determine exactly where, draw a dashed line through point *A*, *parallel* to the SHF line,

and extend it until it crosses the vertical 55° DB line. This is point *B*, and occurs at 55° DB, 51.3° WB; enthalpy ($h_B$) = 21.0 Btu/lb. Point *D* is the condition of the outdoor air, given (under design conditions) at 90°F DB, 75°F WB.

Point *C* (Fig. 9.46*b*) represents the mixture of 66% outdoor air and 34% return air that is brought to the cooling equipment for treatment and distribution back to the dance hall. First, connect points *A* (the return air) and *D* (outdoor air); then plot *C* at 66% of the distance from *A* to *D*. This occurs at 85°F DB, 71.3°F WB; enthalpy ($h_C$) = 35.2 Btu/lb.

The cooling equipment must remove the grand total heat (GTH) according to the formula

$$GTH = 4.5 \times cfm \times (h_C - h_B)$$

(where 4.5 is a constant = 60 min/h × 0.075 lb/ft³ average air density). So, in this example

$$GTH = 4.5 \times 4260 \text{ cfm} \times (35.2 - 21.0)$$
$$= 272{,}214 \text{ Btu/h}$$

The size of the required refrigeration unit is specified in tons, where 1 ton = 12,000 Btu/h. The refrigeration required:

$$= \frac{272{,}214 \text{ Btu/h}}{12{,}000 \text{ Btu/h ton}} = 22.7 \text{ tons}$$

*Note:* If *minimum* outdoor air requirements of, say, 25 cfm per person were provided,

$$80 \text{ people} \times 25 \text{ cfm/person} = 2000 \text{ cfm}$$

The percentage of outdoor air becomes 2000/4260 = 47%.

Point C then moves to about 82°F DB, 68.8°F WB, at which point $h_C$ = about 33 Btu/lb.

The refrigeration required then becomes

$$4.5 \times 4260 \times (33 - 21) = \frac{230{,}040 \text{ Btu/h}}{12{,}000 \text{ Btu/h ton}}$$
$$= \text{about 19 tons}$$

This represents a first-cost saving in equipment size and, of course, energy savings over the life of the dance hall. (Dancers would smell more sweat, however. It is that trade-off again—energy conservation versus IAQ.) ∎

### References

AIA, Ramsey, C.G., and H.R. Sleeper. 1994. *Architectural Graphic Standards*, 9th ed., Wiley, New York.

ASHRAE. 2001. *ASHRAE Handbook—Fundamentals.* American Society of Heating, Refrigerating and Air-Conditioning Engineers, Inc. Atlanta, GA.

ASHRAE. 2004. ANSI/ASHRAE/IESNA Standard 90.1-2004: *Energy Standard for Buildings Except Low-Rise Residential Buildings.* American Society of Heating, Refrigerating and Air-Conditioning Engineers, Inc. Atlanta, GA.

ASHRAE. 2004. ANSI/ASHRAE Standard 90.2-2004: *Energy-Efficient Design of Low-Rise Residential Buildings.* American Society of Heating, Refrigerating and Air-Conditioning Engineers, Inc. Atlanta, GA.

Barden, A.A., and H. Hyytiainen. 1993. *Finnish Fireplaces: Heart of the Home,* 2nd ed. Finnish Building Centre, Ltd., Helsinki.

Bourne, R. and M. Hoeschele. 1998. "Performance Results for a Night Roof Spray Storage Cooling System," in *Proceedings of the 23rd National Passive Solar Conference,* American Solar Energy Society, Boulder, CO.

Calm, J.E. 1994. "Refrigerant Safety," in *ASHRAE Journal,* July.

Hydronics Institute Division of GAMA. 1996. *Installation Guide for Residential Hydronic Heating Systems, No. 200.* Berkeley Heights, NJ.

Lorsch, H. (ed.). 1993. *Air-Conditioning Systems Design Manual.* ASHRAE, Atlanta, GA.

Mozer, M. 1998. "The Neural Network House: An Environment That Adapts to Its Inhabitants," in *Proceedings of the American Association for Artificial Intelligence, Spring Symposium on Intelligent Environments,* March 1998, Stanford University, Palo Alto, CA.

Natural Gas Cooling Equipment Guide (4th ed.). 1996. American Gas Cooling Center, Arlington, VA.

# Large Building HVAC Systems

THIS CHAPTER CONTINUES THE DISCUSSION THAT BEGAN in Chapter 9, with a focus now on more complex systems for larger buildings with many thermal zones. Those using this book for reference rather than as a text may have gone directly to this chapter without reading the preceding material. If so, pay attention to the comments directly below.

## Before Selecting an HVAC System for a Large Building:

The choice of an HVAC system should follow preliminary design decisions that are discussed in the preceding chapter and sections:

*Chapter 5, Indoor Air Quality,* discusses the relationship between comfort, zoning, and equipment that helps maintain acceptable air quality in buildings.

*9.1 Review of the Need for Mechanical Equipment* reviews the relative thermal role of building envelopes compared to their internal heating, cooling, and ventilating systems.

*9.2 Heating, Ventilating, and Air Conditioning (HVAC): Typical Design Processes* outlines the decisions made by the architect and the engineer.

*9.5 Refrigeration Cycles* reviews both the compressive and the absorptive refrigeration processes.

At the onset of the twenty-first century, large building HVAC is showing several trends. One is the increasing willingness to let mechanical equipment share its tasks with natural ventilation and daylighting. Building automation has made this easier to manage. Another trend is toward an underfloor plenum air supply (and the related *displacement ventilation*) rather than using ducts to diffusers and return grilles, both on the ceiling. Concern about air quality indoors and the environment outdoors is producing a variety of approaches to increased ventilation that avoid refrigerants containing CFCs and HCFCs. Fuel cells and photovoltaics are promising increased energy autonomy to larger buildings.

## 10.1 HVAC AND BUILDING ORGANIZATION

By this time, many decisions about a building's design have been made: design strategies appropriate to the climate and the building's activities have been identified, and the basic siting and overall form of the building have been determined from daylighting and thermal considerations, among others. This section begins by considering the internal yet broad issues of zoning and system choice and ends with a discussion of the more detailed consequences of system choice. A general guide to estimation of a building's thermal zone requirements was presented in Section 8.2, which discussed the importance of differences in function, schedule, and orientation.

## (a) Zoning

The minimum number of thermal zones for a conventionally designed multipurpose building is shown in Fig. 10.1. A need for more than these 16 zones could result from differences in scheduling within a zone, such as between offices and stores. As is true of the other occupied floors, apartments have a minimum of five zones (based on orientation); however, the emphasis on individual controls—and the variation in usage patterns—often produces as many zones as there are apartments. When the details of zoning are added to the other preliminary design decisions, the details of HVAC systems can be considered.

## (b) System Anatomy

Table 10.1 describes the basic organization of any HVAC system. Three kinds of common tasks (heating, cooling, and ventilating) are done by production components; usually, they require distribution and delivery components. Intake supplies and exhaust by-products accompany each task. Although the eventual choice of HVAC system should follow an analysis of the zone's needs, some early concepts underlie system choices.

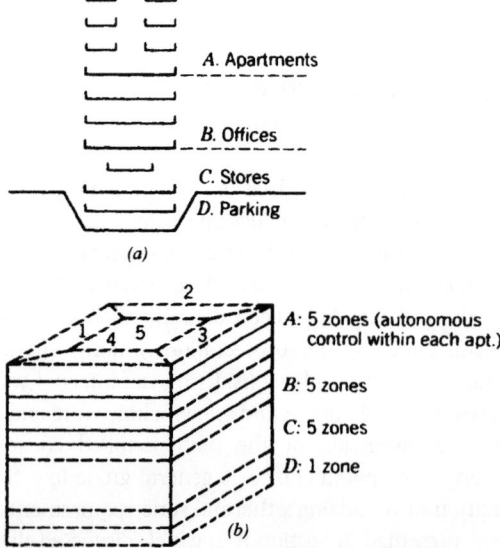

A. Apartments

B. Offices

C. Stores

D. Parking

*(a)*

A: 5 zones (autonomous control within each apt.)

B: 5 zones

C: 5 zones

D: 1 zone

*(b)*

**Fig. 10.1** *The minimum number of thermal zones for a rather large, conventionally designed, multipurpose building.*

## (c) Central versus Local Systems

This question was explored in Sections 5.5*g* for indoor air quality and 9.3*a* for smaller buildings; it is even more applicable to larger buildings. Central systems require one or several large mechanical spaces (often in basements and/or on roofs), sizable distribution trees, and complex control systems. The noise, heat, and other characteristics of such mechanical rooms can be controlled fairly easily, because the machinery is concentrated at a few locations. Similarly, maintenance is easy to perform without interrupting normal activities, although breakdowns in central equipment can paralyze the entire building. Air quality can be controlled by locating the air intakes high above the pollution at street level and by regular maintenance of the centralized air filtering equipment. Longer equipment life can be expected with regular maintenance. Energy conservation can be served by the recovery of one machine's heat by-product for a nearby machine's heat input. Although there are many ways to provide for the differing thermal needs of the many zones served by central systems, one important drawback of central systems is the size and length of the distribution trees necessary to carry centralized services to many local receivers. Another drawback is a difference in zone scheduling: when the entire system must be activated to serve one zone (such as computer operations in an office building on a weekend), energy is wasted.

*Local systems* therefore become increasingly attractive as scheduling differences multiply. Also, pronounced differences in other factors—function (with resulting comfort expectations) or placement within the building, for example—can lead to the choice of local systems. Large and centralized equipment spaces are not required with local systems; rather, production equipment is distributed throughout the building (or over the roofs of low-rise structures). Dispersal of equipment minimizes the size of distribution trees and greatly simplifies control systems. Moreover, system breakdowns affect only small portions of the building. However, noise and other by-products of multiple machines pose numerous potential threats to occupied spaces, and maintenance is demanding, because access to so many separate locations is often disruptive or constricted. Then, too, air qual-

**TABLE 10.1  Basic HVAC Systems: Tasks and Components**

| | Production/Motion<br>Movers, converters,<br>processors | Distribution<br>Supply and return<br>trees, delivery and<br>control components | Results |
|---|---|---|---|
| Heat: | Boilers<br>Furnaces<br>Pumps<br><br>Fans<br>Filters<br>Heat pumps | Pipes<br>Ducts<br>Electricity conduits<br><br>Diffusers<br>Grilles<br>Radiators<br>Thermostats<br>Valves, dampers | Warm air or surfaces<br>Air motion often controlled<br>Humidity control some-<br>times needed |
| Cool: | Evaporative coolers<br>Heat pumps<br>Chillers and cooling<br>towers<br>Coils<br>Pumps<br>Fans<br>Filters | Pipes<br>Ducts<br>Diffusers<br>Grilles<br>Radiators<br>Thermostats<br>Valves<br>Dampers | Cool air or surfaces<br>Air motion usually con-<br>trolled<br>Humidity control usually<br>provided |
| Vent: | Fans<br>Filters | Ducts<br>Diffusers<br>Grilles<br>Switches<br>Dampers | Fresh air<br>Air motion usually con-<br>trolled<br>Air quality control often<br>needed |

*Source:* Class notes developed by G.Z. Brown, University of Oregon.

ity depends on the regular cleaning of many filters scattered over the building, often within occupied spaces. The potential for energy conservation seems promising, because heating or cooling is produced only as locally needed, but there is little chance to use one zone's waste heat as another's needed source.

*Central heat/cool, local air distribution* has become a popular way to take advantage of the favorable characteristics of both central and local approaches. This is shown in Fig. 10.2*b*, with a central boiler/chiller space remotely located and fan rooms on each floor. This minimizes the bulky distribution tree for air; although the distribution tree for heated and chilled water is extensive, it is also of much smaller diameter and therefore is relatively easily accommodated. The central equipment room makes energy recovery systems from boilers and chillers more feasible.

## (d) Uniformity versus Diversity

How similar should be the interior environments of buildings? This question encompasses not only thermal experiences, but visual and acoustical ones as well.

THERMAL CONTROL

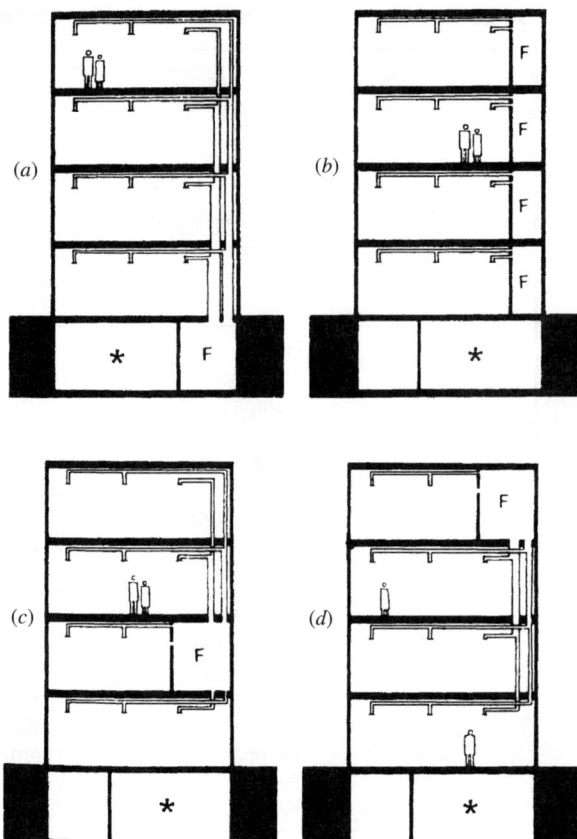

**Fig. 10.2** Fan rooms (F) can either be combined with or separated from boiler/chiller "equipment" rooms (*). (a) Common location for a central combined equipment room. (b) Increasingly common arrangement of a small fan room on each floor, with an equipment room in the basement. (c) An intermediate floor may be able to provide space for a central fan room, while the heavier and noisier equipment remains in the basement. (d) With a top floor central fan room, the equipment may be located either on the roof or in a mechanical penthouse, or may remain in the basement. (Adapted by permission from E. Allen and J. Iano, The Architect's Studio Companion, 2nd edition, © 1994, John Wiley & Sons, NY.)

* BOILER AND CHILLER EQUIPMENT ROOM

The advantages of uniformity are most evident in a rapidity of design and construction that, through mass production and speed, often brings lower first costs. Uniformity of ceiling heights, light fixture placement, grille locations, and so on promotes flexibility in office arrangements that can extend a building's usable life span. However, there are at least four types of offices, which may need to be interchangeable within such "flexible" space. The typical *enclosed office* has the privacy of four walls and a door. The *bullpen office* has repeated, identical workstations, with low dividers at about the height of the desk surface. The *uniform open plan office* resembles the bullpen, but with higher divider partitions for added privacy. The *free-form open plan office* has some individually designed workstations with divider partitions of varying heights (sometimes reflecting the varying status of workers). In the bullpen and uniform open plan office, the resulting uniformity is not always attractive to users, and

diversity is often encouraged at a more personal level—with office furnishings, for example. A more thorough approach to diversity can provide stimulus to the user who spends many hours away from the variability of the exterior climate.

If offices must be uniform in ceiling lighting, air handling, and size, the corridors that connect them and the lounges or other supporting service spaces can deliberately be made different. Diversity requires a complete and detailed design of places; it gives the builder a more complex and interesting task; and it can provide orientation and interest to the users. The attractiveness of diversity is evident in most collections of retail shops, in which light and sound—and sometimes heat and aroma—are used to distinguish one shop from the next.

Diversity in the thermal conditions to be maintained, such as warmer offices and cooler circulation spaces in the winter, can be used to enhance the comfort of the office users. Designers have long recognized that a space can be made to seem brighter and higher if it is preceded by a dark, low transition space. Thermal comfort impressions can be manipulated similarly. Less than comfortable conditions in circulation spaces or other less critical zones not only make the critical spaces seem more comfortable by contrast, but also save significant amounts of energy over the life of a building. Furthermore, such conditions can make passive strategies more attractive.

A large-scale demonstration of diversity in thermal zones is shown in Fig. 10.3 Passive solar heating can make a significant contribution, even through a shallow-sloped, single-glazed cover in cloudy Glasgow, Scotland, largely because the mall area and leisure areas are allowed a much wider thermal range than would be permitted in stores and offices. The overcast skies are quite suitable for daylight, and the addition of summer sunshading makes natural ventilation (through the stack effect, assisted by fans) possible during the cool summers. U.S. Pacific Northwest climate conditions are similar.

### (e) Comparing Systems and Zones

In the process of selecting systems from among the wide variety available, it is helpful to consider the match between the zones' characteristics and those of various systems. Among the considerations are zone placement (close to or away from the building skin), the zone's thermal loads, the comfort determinants based on the zone's activities, the space avail-

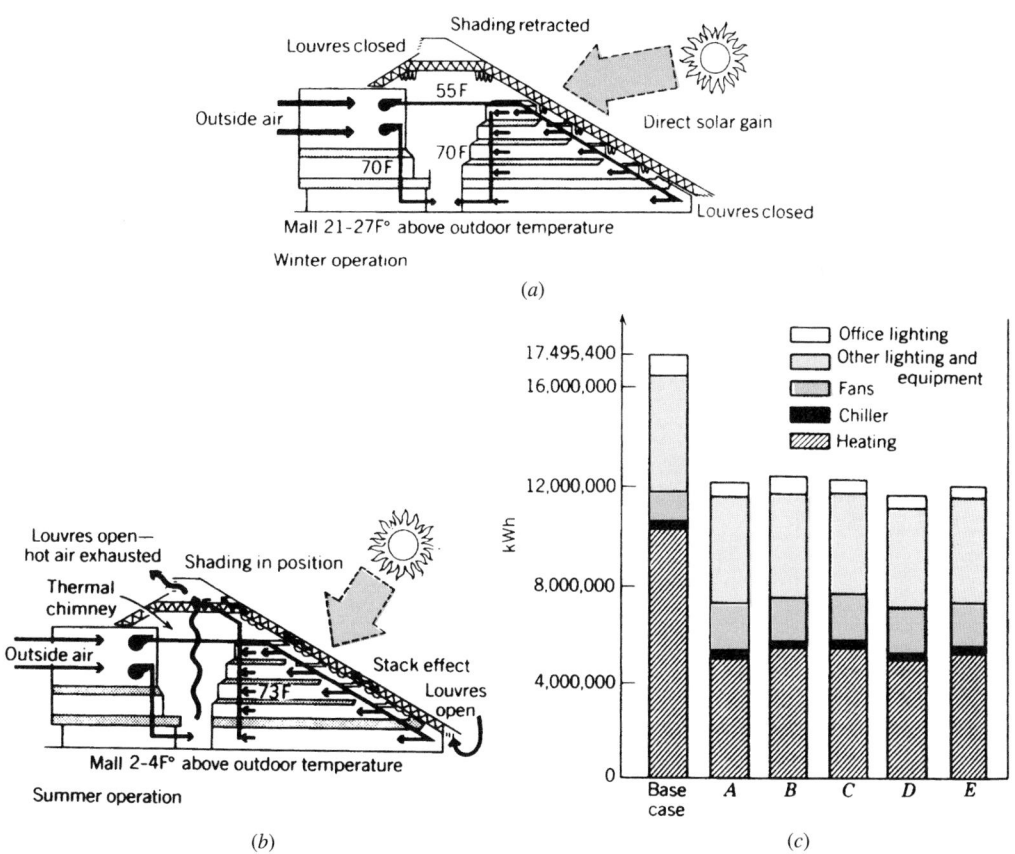

**Fig. 10.3** *St. Enoch's Square, Glasgow, Scotland: a proposal to use passive solar heating, daylighting, and natural ventilation. Reiach & Hall and GMW Partnership, architects (joint venture); Cosentini Associates, energy consultants; Princeton Energy Group, daylighting consultants. (a) Schematic section showing winter operation; the mall temperature varies around 63°F (17°C) during operating hours, while offices are kept near 70°F (21°C). (b) Schematic section showing summer operation; the mall temperature varies from about 68 to 74°F (20 to 23°C) during operating hours. (c) Estimates of annual energy consumption for a conventional-design base case and several alternative configurations. Note the significantly lower heating energy requirements, resulting in part from the lower winter temperatures allowed for the less critical zones such as the mall and the leisure areas in configurations A to E.*

able for system components within the zone, and the life-cycle costs of various system alternatives.

*Zone placement* will sometimes preclude local systems, which depend on easy access to outdoor air both for fresh air and for a heat source or sink. Local systems for interior (away-from-skin) zones are awkward. Relationships between zone placement and building forms are shown in Fig. 10.4.

The thermal loads on each zone determine the extent to which heating or cooling is the dominant problem—which, in turn, can influence the choice of system. A zone with little cooling load and low moisture production may be well served by a simple system of fresh air plus heating, with no humidity control. Zones that require cooling will usually also require more complete control of air motion and

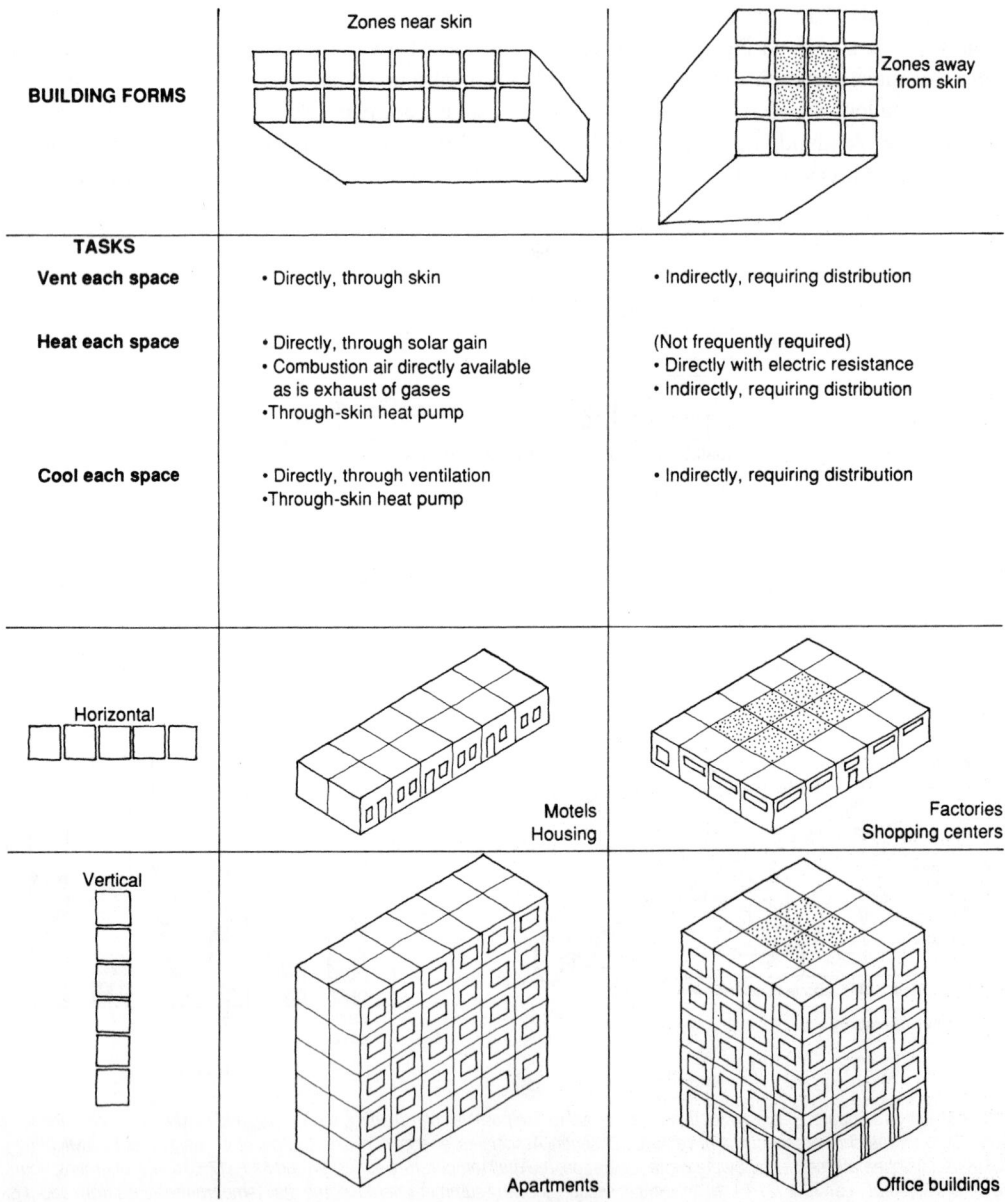

**Fig. 10.4** *Zone placement and building form are related to heating, cooling, and ventilating tasks; some applications take on typical building forms. (From class notes developed by G.Z. Brown, University of Oregon.)*

relative humidity. Although it is risky to generalize about which comfort determinants are most important (given the differences between activities and between individuals), it can generally be assumed that comfort and thermal tasks are related.

|  | For Heating of Spaces | For Cooling of Spaces |
|---|---|---|
| More important | Surface, air temperatures | Air motion |
| ↓ | Air motion | Relative humidity |
| Less important | Relative humidity | Air, surface temperatures |

Thus, the choice of systems can be based partly on whether the system provides good control of the more important comfort determinants.

## (f) Distribution Trees

Section 8.1 raised preliminary questions about distribution trees: how many, what kind (air or water),

and where to place them within buildings. HVAC system choice is influenced by the amount of space the system requires. In some cases, it is easy to provide small equipment rooms at regular intervals throughout a building, such that little or nothing in the way of a distribution tree will be required. In other cases, a network of distribution trees and central, large equipment spaces are easier to accommodate. These central systems typically fall into one of three classifications:

All-air (the largest distribution trees)

Air and water

All-water (the smallest distribution trees, with local control of fresh air)

The details of these systems can be found in Sections 10.5 to 10.7, along with typical applications and space requirements. Figure 10.5 shows the matrix of central–local, air–water influences on distribution trees.

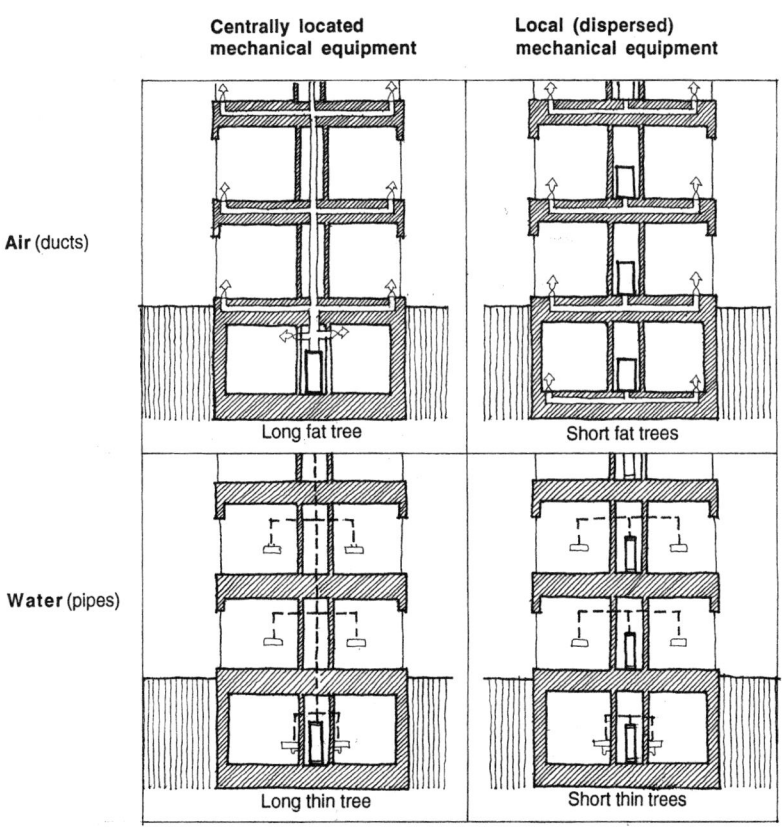

**Fig. 10.5** *Matrix of distribution trees. (© 1998 John S. Reynolds; drawing by Michael Cockram.)*

## TABLE 10.2 Procedure for Matching Zones and Systems

| CAPSULE DESCRIPTION |
|---|
| A multipurpose building (similar to that shown in Fig. 10.1) is situated in a cold winter–mild summer climate. |
| Apartments are on upper floors surrounding an open-air central court; they have adequate daylight and cross-ventilation. Floor heights are quite low. |
| Offices are rented to various tenants. Exterior offices have high ceilings to facilitate daylighting, and therefore have low internal gains but restricted clearance for horizontal ducts. Interior offices have lower ceilings; vertical chase space is limited because it reduces rentable area. |
| Shops are located around the perimeter on the high-ceilinged ground floor; some smaller shops are located on the mezzanine and ground floors in the interior zone. Space for vertical chases is severely limited on these highest rental floors. |
| The parking area is below grade, surrounded by air and light wells. Floor heights are very limited to reduce ramp length. |

| Activities (Program) | Apartments | Offices | | | Shops | | Restaurant | Parking |
|---|---|---|---|---|---|---|---|---|
| | | | Computer Center | | | | | |
| Schedule | 24 hours | 24 hours | 9 hours | | 9 hours | | 12 hours | |
| Placement | Exterior | Exterior for access | Exterior | Interior | Interior | Exterior | Exterior for access | Entire floor |
| Internal gains | Vary | High | Low | Medium | High | Medium | High plus moisture | Low with exhaust gas |
| Dominant HVAC task(s) | Heat-ventilate | Cool | Heat/cool | Cool | Cool | Cool | Cool | Ventilate |
| HVAC space available | | | | | | | | |
| Vert. (in plan) | Medium | Medium | Medium | Tight | Tight | Tight | Tight | Medium |
| Horiz. (in section) | Tight | Medium | Tight | Medium | Tight | Ample | Ample | Tight |
| System choices | | | | | | | | |
| Local | | A | | | | | B | C |
| Central | | | | | | | | |
| All-air | | | (D) | D | D | (D) | | |
| Air and water | | | E | | | E | | |
| All-water | F | | | | | | | |

| SUMMARY |
|---|
| A. The computer center's unique schedule and rate of internal gain usually requires a separate system equipped with humidity and air quality controls to protect the equipment. Some heat recovery for use in E and F seems advisable. |
| B. The restaurant's special problems of heat, moisture, and aroma, as well as its schedule, require a separate system. |
| C. The parking area needs only plenty of fresh air; it requires no tie with the other zones at all. |
| D. The always-cooling loads of interior zones are best served by all-air systems offering control of humidity and air quality. However, vertical chase size is tight, and high-velocity distribution may be required. (The exterior zones could also be served by all-air systems. But the need for heating, plus the likelihood of fresh air infiltration and the tight clearance for horizontal ductwork, suggest that the system for exterior zones should be separated from the system for interior zones.) |
| E. Quick changes from heating to cooling are best handled by water; some central air quality control is offered by air and water systems. |
| F. A central all-water system offers energy conservation advantages, recovering waste heat from system D (and potentially from A). Fresh air is easily and cheaply handled on a local basis, which also provides cooling. |

| MECHANICAL SPACE |
|---|
| Probably best located on the top office floor or on a floor of its own between offices and apartments. Distribution tree sizes will thereby be minimized on the high-rent ground floor. |

A simplified procedure for matching zones and systems is shown in Table 10.2, in which preliminary system choices are made for a building such as the multipurpose structure shown in Fig. 10.1. In this process, the original 16 zones are translated into three local systems and three central systems: one all-air, one air and water, and one all-water.

The Fox Plaza Building in San Francisco, which illustrates many of the matches between systems and zones, is shown in Fig. 10.6. This project includes four major building types in one structure:

1. Underground garage for the storage of cars
2. Commercial center at ground level, including a bank, a women's specialty store, and other commercial establishments
3. Ten floors of offices
4. Sixteen floors of apartments

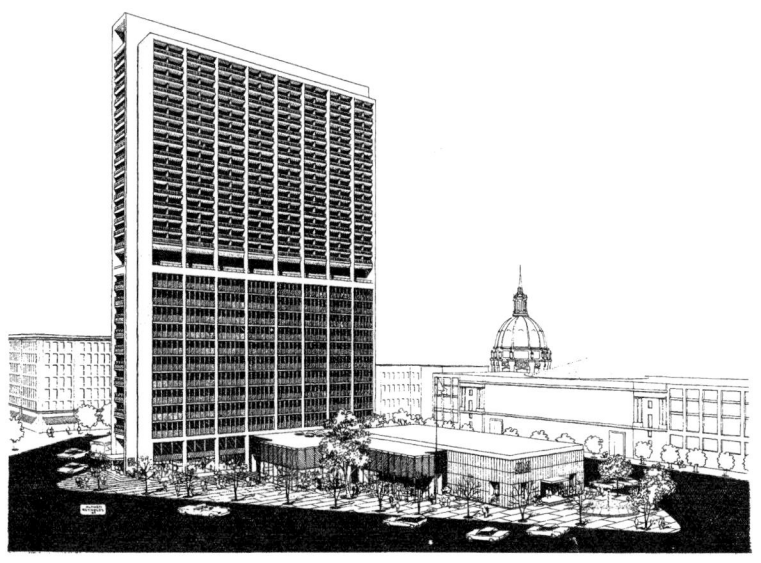

(a)

(b)

**Fig. 10.6** The Fox Plaza Building, San Francisco. Victor Gruen Associates, Inc., architects and engineers. (a) Elongated façade facing northeast shows the 16 floors of apartments above, the 10 floors of offices below, and the 13th-floor mechanical space. In that space are chillers, and pumps for cooling tower water and chilled water, as well as boilers and converters (steam to hot water) for the fan-coil units in the residential stories above and hot water coils in the office stories below. Air handling for the offices is also located here, downfed by high-velocity ducts. Residences are heated; offices are heated and cooled. (The roof has the cooling tower and the domestic hot water generator–storage units for the residential stories.) (Courtesy of Progressive Architecture.) (b) Construction photo with air-handling units visible on the 13th floor, and downfeed ducts that supply high-velocity hot and cold air to the office floors. (Photo by Morley Baer.)

The mechanical level is located between the office portion of the building and the apartments above. The distribution trees—heating, air conditioning, electrical, and so on—are thus directed both upward and downward, resulting in two shorter trees rather than one longer tree. The spatial requirements of offices and those of apartments are quite different; thus, the floor-to-floor heights, window treatment, and heating, cooling, electrical, elevators, and other services are different. The placement of the mechanical level between the offices and the apartments also provides for a definite visual separation between the two functions.

Quite unusual is the placement of the steam boilers on the 13th floor instead of in the conventional basement location. Only a small amount of auxiliary equipment is located on the roof and in a small portion of the garage. Residential areas have hot-water heating (residential cooling being rarely needed in San Francisco), offices have dual-duct, high-velocity heating/cooling, and commercial (ground-floor) tenants are supplied with hot and chilled water for individual climate control requirements.

### (g) Central Equipment Location

The Fox Plaza Building has an intermediate location for the heating and cooling production equipment—one that separates floors of apartments from floors of offices. Other typical locations for central equipment are in the basement (where machine noise is most easily isolated, utilities are easily accessed, and machine weight is little problem) and on the roof, where access to air as a sink for reject heat is easiest of all and headroom is unlimited. Very tall buildings may require several intermediate mechanical floors. Examples of these approaches are found throughout the rest of this chapter.

The equipment's considerable heat, moisture, air motion, noise, and vibration potentially annoy occupants on nearby floors (or even neighboring buildings). As shown in the Fox Plaza example, the equipment can be expressive of building services and can play a useful demarcation role between vertical layers of high-rise buildings. Moving the equipment off the roof also frees this prestigious view location for high-rent occupancy and allows a roof form much more expressive of great height than a flat roof with a cooling tower.

### (h) Concealment and Exposure

The pipes, ducts, and conduits that take the necessary resources to and from the interior are often carried within a network of spaces unseen by anyone except builders and repair people. The advantages of concealment include less noise from moving water and air, fewer surfaces requiring cleaning, less care necessary in construction (leaks, not looks, are important), and more control over the appearance of the interior ceiling and wall surfaces. Although maintenance access to such hidden supply lines is more difficult, a variety of readily removable covers is available, particularly in suspended ceilings.

However, the exposure of these supply networks provides an honest and direct source of visual (and occasionally acoustical) interest. Exposure in corridors and service areas and concealment in offices constitute an approach used in many office buildings. Flexibility is usually encouraged by exposure; changes can be easily made when there is no need for neatly cut holes in concealing surfaces. However, flexibility from full-height movable partitions requires constant ceiling heights—a feature of the suspended-ceiling approach.

One of the more spectacular examples of exposed mechanical (and structural) systems is shown in Fig. 10.7—the result of a design competition for a museum of modern art, reference library, center for industrial design, center for music and acoustic research, and supporting services in downtown Paris.

When users are invited to play an active role in adjusting conditions inside, exposure of the switches they manipulate is helpful. Visible mechanisms not only remind users of their opportunities but also encourage user interaction. In this way, adjustments are sometimes discovered that the designer had not anticipated.

### (i) Mechanical–Structural Integration or Separation

The similarity of these two technical support systems—structures and environmental controls—has intrigued designers ever since mechanical systems began to require substantial volume for distribution, as in air-duct systems. As the complexity and size of the mechanical distribution systems was *increasing* with technological development (typically, more air is required to cool a space than to

tower often moved to the roof, often taking the air-handling machinery with it. This further encouraged the merging of systems, for one system was growing wider as the other diminished (Fig. 10.8). Thus, a fixed-column cross section, consisting mostly of the structural column at the base and the air duct at the top, became possible.

Yet the functions of these systems differ widely: compared to the dynamic on–off air, water, and electrical distribution systems, the structural system is static—gravity never ceases. The moving parts in mechanical systems need maintenance far more frequently than the connections of structural components. Changes in occupancy can mean enormous changes in mechanical systems, requiring entirely different equipment; structural changes of such magnitude usually occur only at demolition. Mechanical systems can invite user adjustment; structural systems rarely do.

Thus, although it is possible to wrap the mechanical systems in a structural envelope, it is of questionable long-term value, given the differing life spans and characteristics of these systems. The probability of future change suggests that the *mechanical* system be the exposed one, despite the appeal to many designers of the structural system's cleaner lines.

## (j) Distribution Tree Placement Options

These options are summarized in Fig. 10.9. Vertical placement options are important because they affect floor space, influencing the flexibility of spatial layout and the availability of usable (or rentable) floor space. Horizontal placement options affect ceiling height—a particular issue in daylighting design and sometimes a critical factor when overall height limits are imposed yet maximum usable floor space is desired. (In Washington, DC, for instance, no building can rise higher than the Capitol.) Both vertical and horizontal distribution at the edges can have a dramatic impact on building appearance.

The history of distribution trees and high-rise buildings is one of trends and countertrends. Initially, multistory buildings relied upon daylight and cross-ventilation, so a thin, relatively high-ceiling plan with much perimeter was favored (refer to Fig. 3.33). The heat gain and loss was all at the perimeter, so perimeter distribution trees (carrying only steam or heated water, and of quite small diameter) were generally used. As electric

**Fig. 10.7** *Centre Georges Pompidou, Paris. A view of the mechanical support systems. Piano + Rogers, architects. (Photo by John Tingley.)*

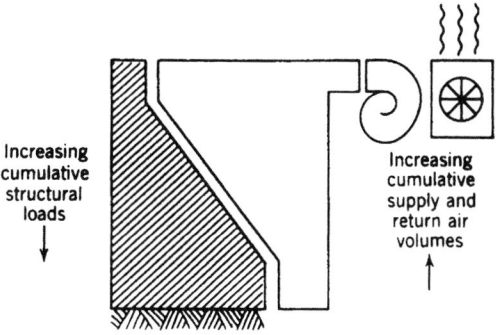

**Fig. 10.8** *Distribution trees: with rooftop centralized air handling, the supply and return air duct sizes decrease as they approach the ground. Conversely, the structural load increases toward the ground.*

heat it because of a lower Δt), the increased strength of materials was *reducing* the size of the structural system. The uncluttered floor areas between the more widely spaced columns became desirable for flexibility in spatial layout. With the mechanical systems at or within these columns, floor areas remained clear, thus giving mechanical–structural integration further impetus. With the new expectations for cooling, the refrigeration cycle's cooling

THERMAL CONTROL

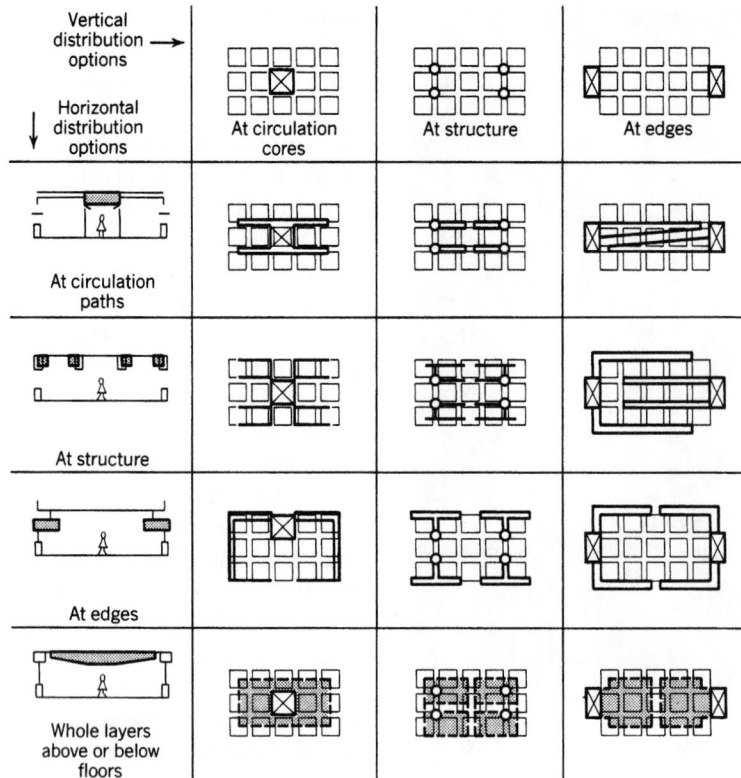

**Fig. 10.9** *Distribution tree placement options, both vertical (with impact on the plan) and horizontal (with impact on the section). (From class notes developed by G.Z. Brown, University of Oregon.)*

lighting and thus the need for air conditioning increased, so did the thickness of floor plans; large central internal areas needed a lot of forced, cooled air. Central boilers, chillers, and fan rooms were the norm. Thus, bulky air distribution trees appeared. At about the same time, the glass curtain wall and its slick, two-dimensional look of modernity became fashionable. The air distribution trees were so visually intrusive on the façades that they were pushed to the core, where cooling needs were relatively steady. However, the thin glass perimeter experienced extreme needs for both heating and cooling; getting from vertical trees at the core to the perimeter required larger cavities above suspended ceilings. This pushed the ceiling in the offices down to keep floor-to-floor distances economical. Vast office areas resulted that were visually dull, low-ceilinged, and without daylight.

Now, countertrends include decentralized air handling, with small fan rooms on each floor. Verti-

cal air distribution trees are shrinking, horizontal ones becoming more common. At the same time, daylighting is pushing office ceilings higher; so is a preference for indirect lighting and its compatibility with computer screen visual comfort. Night cooling utilizing thermal mass is encouraging the exposure of concrete structure and favoring raised-floor air supply/ventilation systems. A renewed interest in sun control is encouraging three-dimensional façades, replacing two-dimensional reflective glass façades (which merely redirect the sun toward someone else). With increased three-dimensionality at the façade, perimeter distribution trees are once again conceivable.

It is logical to place at the perimeter the parts of the system that deal with the effects of sun, shade, and temperature change in the several perimeter zones, leaving at the core a separate network to handle the more stable interior areas. The disadvantages of perimeter distribution include (usually) higher construction costs and an environment that is more thermally hostile due to the extremes of outdoor temperature.

*Vertical distribution within internal circulation cores* is very common, as it leaves a maximum of plan flexibility for the rest of each floor and does not disturb the prized floor areas nearest windows. However, one centralized vertical distribution trunk will require large horizontal branches near the core, so with this choice early thought must be given to the horizontal placement options.

An unusual example of vertical air distribution at the core is shown in Fig. 10.10. The Fox Plaza, Los Angeles, office building's unique features include both fan rooms on each floor *and* a large central vertical air shaft. This air shaft begins at the bottom as a fresh air intake to each floor and tapers to become, at the top, an exhaust (heated) air outlet from each floor. Thus, the stack effect is utilized to help supply fresh and exhaust stale air from a large building, with help from small supply fans at each floor.

*Vertical distribution integrated with structure* creates some intriguing possibilities where the structure–HVAC integration concept is suitable. Multiple HVAC trees are implied (because there are multiple columns with which they are integrated), so the horizontal branches tend to be small. However, these branches often join the vertical trunk at the same place where critical column-to-girder structural connections need to be made; interference is common and can be costly to correct. Vertical distribution at the *edges* is potentially dramatic in form but costly to enclose (if outside) or wasteful of prime floor space (if inside).

*Horizontal distribution above corridors* is very common, since reduced headroom here is more acceptable than in the main activity areas. Furthermore, corridors tend to be away from windows, so their lower ceilings do not interfere with daylight penetration. Because corridors connect nearly all spaces, horizontal service distribution to such spaces is also provided. Furthermore, exposure of these services above corridors can heighten the contrast between such serving spaces and the uncluttered, higher-ceilinged offices that are served. Horizontal distribution at the *structure* is sometimes chosen, particularly where U-shaped beams or box beams provide ready channels for HVAC distribution. However, the penetration of horizontal structure members by these continuous service runs must be coordinated. Horizontal distribution at the *edges*

can be integrated usefully with sunshading devices and light shelves; it can also act as a spandrel element that contrasts with the window strips. Horizontal distribution within *whole layers* below floors (or above ceilings) is often utilized, now increasingly common with displacement ventilation systems.

An example of supply at the edge for both vertical and horizontal distribution is found in the International Building in San Francisco (Fig. 10.11). Here the vertical shafts are prominently exposed at the corners; these shafts carry supply and return ducts serving the four perimeter air-conditioning zones. Air handling equipment and a 750-ton refrigeration plant are located on the floors just below the terrace level (those least desirable for renting). Each corner duct branches to serve two zones, which are separately controlled. Pressure reduction and blending are done by equipment in the hung ceiling, and from these points air flows to strip-grille diffusers directly above the glass on the four sides of the building. Local controls offer comfort to personnel in each area.

Interior zones on each floor are supplied by a riser duct in the building core, which branches at each floor to a loop just outside the line of elevators. The loop serves ceiling diffusers.

Between the perimeter loop and the interior loop, a return loop collects air for return to the central station (second and third floors). These return loops on the 11th to 21st floors are picked up by external return risers on alternate exterior corners. From the 10th floor down, the loops are picked up (as shown in Fig. 10.11) by an interior return riser that extends down through the core in front of the blank faces of the high-rise elevators. To provide a clear space between the elevator banks on the main floor (fourth or terrace), the two core ducts' risers are offset at the ceiling of that story.

In summary, perimeter air for all stories is supplied through corner ducts. Central air for all stories is supplied through a core duct. All return air above the 10th floor is carried down through the return ducts at the *other* two corners. Return air from the 10th floor and below is carried down by a return duct in the core.

Further examples of the relationships among HVAC systems, their distribution trees, and buildings are given in examples that accompany the more detailed descriptions of large-building HVAC systems in Sections 10.4 to 10.7.

THERMAL CONTROL

(a)

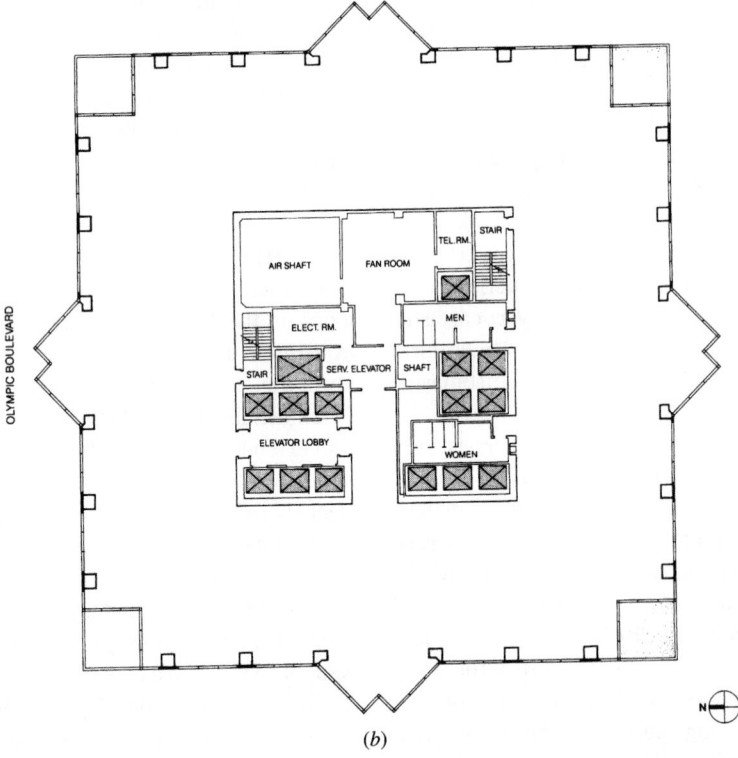

(b)

**Fig. 10.10** *The Fox Plaza, Los Angeles, office building is a 34-story, 800,000-ft² (74,320-m²) granite and glass tower (a) with an unusual vertical distribution tree. (b) Typical lower floor plan (floors 6 to 16) shows both a fan room and a large vertical air shaft. At this lower level, most of the shaft area is supplying outdoor air (from an intake in the bluff face below the building); the remainder is exhausting stale air toward the roof. Note the lack of columns between the core and perimeter, contributing to office layout flexibility. (c) Typical upper floor plan (floors 31 to 33) shows fewer elevators; by this level, most of the shaft area is exhausting stale air toward the roof. (d) Section shows the tapered interior of the constant-cross-section central air shaft, which relies upon the stack effect to bring in (usually cooler) outdoor air at the base and expels hotter exhaust air at the top. (Courtesy of Johnson Fain Pereira Associates, Architects, Los Angeles: and Kim, Casey and Harase, Inc., Engineers, Los Angeles. Photo by Wolfgang Simon.)*

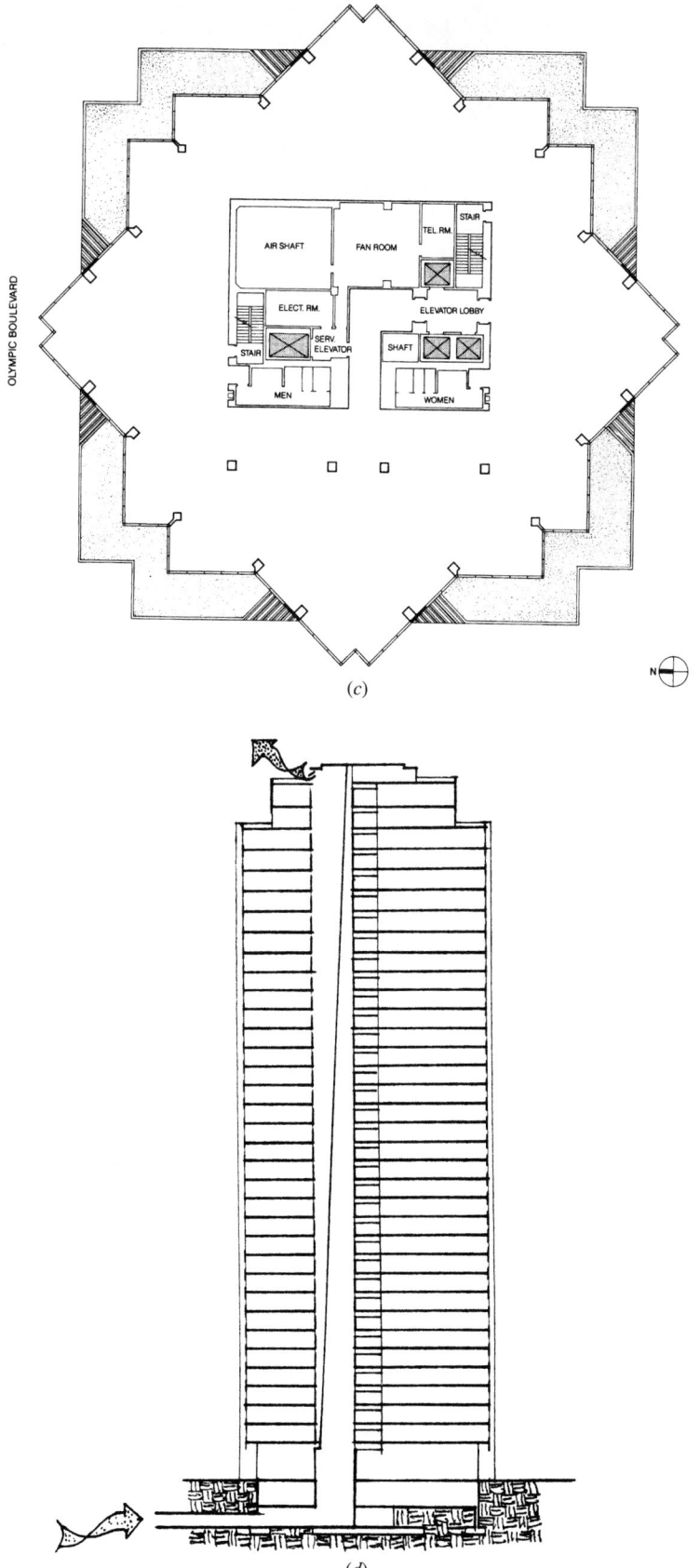

OLYMPIC BOULEVARD

AIR SHAFT

FAN ROOM

TEL. RM.

STAIR

ELECT. RM.

STAIR

SERV. ELEVATOR

MEN

ELEVATOR LOBBY

SHAFT

WOMEN

N

(c)

(d)

**Fig. 10.10** (Continued)

**383**

(a)

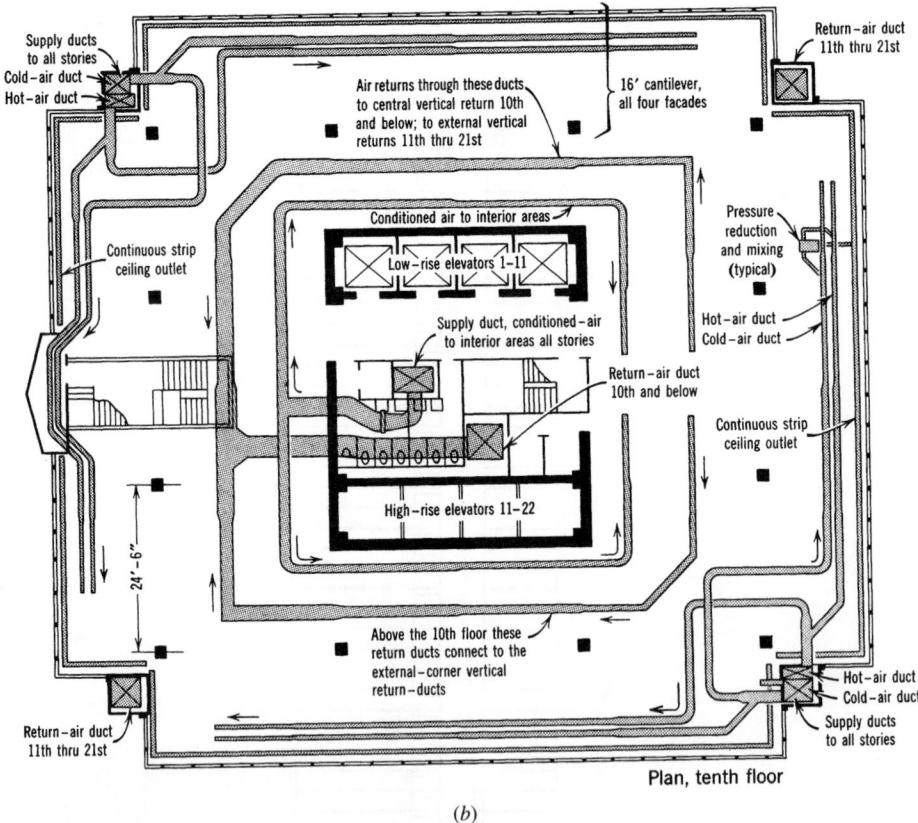

(b)

**Fig. 10.11** *The International Building, San Francisco. Anshen and Allen, architects; Eagelson, Engineers (Charles Krieger, E.E.) mechanical designers. (Courtesy of Progressive Architecture.) (a) Photo of one of the four corner main duct enclosures. (b) Tenth-floor plan. The major supply ducts (both hot and cold) to all 21 floors are located in two opposite corners. Each of these supply distribution trees serves two adjacent sides of perimeter offices. The conditioned air is supplied from a third-floor mechanical space. In the opposite two corners, return air from the upper 11 floors is collected and taken down to the mechanical space. The remainder of the return air is taken down through the core.*

## 10.2 HVAC SYSTEM TYPES

Large buildings have so many thermal zones, and there are so many ways to move heat from one place to another, that hundreds of HVAC system variations have been devised. A few of the most typical are introduced in this section; the following section treats in detail the major components of HVAC production and delivery. Finally, some common variations on each of the four main system classifications are presented.

One way to classify HVAC systems is by the media used to transfer heat. Although thousands of liquids and gases can be used as carriers of heat, the three most common in building applications are air, water, and refrigerant. Traditionally, there are four main system classifications:

Direct refrigerant systems

All-air systems

Air and water systems

All-water systems

In the last three cases, the heating/cooling production equipment typically is located centrally in a large building, often rather far from the thermal zones it serves. The air-handling components may be either centrally served or served floor by floor. Distribution tree size and placement thus become important issues when those systems are selected. In direct refrigerant systems, the heating/cooling machine usually is located adjacent to the zone(s) it serves; thus, the machine's environment—the microclimate it creates and its needs—relative to the zone's environment becomes an important consideration.

### (a) Direct Refrigerant Systems

These systems nearly eliminate the distribution trees of air or water, relying instead on a heating/cooling device adjacent to or within the space to be served. Thus, they are prevalent in skin-dominated buildings with extensive perimeter zones; these tend to be smaller buildings, so these systems are discussed in Chapter 9, Section 9.8.

### (b) All-Air Systems

The more common variations on all-air systems are shown in Fig. 10.12. Because air is the only heat transfer medium used between the mechanical room (central station) and the zones it serves, and because air holds much less heat per unit volume than water, the distribution trees for this class are quite thick. Sometimes, to reduce duct sizes, higher velocities are used for supply air. This generates more noise and higher friction, resulting in more energy used by fans; higher velocity should be used only sparingly, where space limitations are extreme.

For comfort, however, these systems are, overall, the best. The quantities of air moved through the central station(s) are heated or cooled, humidity-controlled, filtered, and freshened with outdoor air—all under controlled conditions. Within the zones, supply registers and return grilles allow a well-planned stream of conditioned air to thoroughly permeate all work areas. More details on air distribution are found in Section 10.4 and on this HVAC type in Section 10.5.

**_Single-Zone Systems_** (Fig. 10.12a). This is the common small-building forced air system controlled by a single thermostat.

**_Single-Duct, Variable-Air-Volume (VAV) Systems_** (Fig. 10.12c). This is the most popular large building system of recent years. Its single duct requires less building volume for distribution than do multiduct systems, and the variation of air _volume_ flow rate (rather than of air temperature) saves energy relative to the single duct with reheat (Fig. 10.12_e_). Depending on outdoor conditions and prevailing indoor needs, the central station supplies at normal velocity either a heated or a cooled stream of air. Automatic volume controls (linked to each zone's thermostat) adjust the volume admitted to that zone within an air terminal diffuser (often located above a suspended ceiling). When the central station is supplying cold air, a zone that needs more cooling will get more air; an unoccupied room with no internal gains, or a space with heat loss through an exterior wall, will get less air. Clearly, such a system is well suited to serve the interior, always-hot zones of internal-load-dominated buildings. Less clear is its suitability for the perimeter zones of buildings in cold, cloudy conditions.

**_Fan-Powered VAV Systems_** (Fig. 10.12f). This variation allows individual units to heat when the main supply system is cooling; it might therefore serve perimeter zones. In this case, the cool air is reduced to a minimum for IAQ, and the unit's fan draws additional air from a ceiling (or floor) plenum, heating it as required.

*Multizone Systems* (Fig. 10.12b). Because each zone has an individual centrally conditioned airstream, the total distribution tree volume grows to astonishing size with only a few zones. The central station produces both warm and cool airstreams, which are mixed at the central location to suit each zone. These systems are more likely to be found on medium-sized buildings or on larger buildings in which smaller central stations are located on each floor. The single-return airstream collects air from all zones (as is the case for the other systems in this class). Energy savings result when a "bypass" deck is added, allowing each zone to choose some unthermally treated return air as part of the supply air.

*Single-Duct with Reheat* (Fig. 10.12e). This system (along with VAV) has the smallest distribu-

tion tree of this class, because at each zone the only object added to the duct is a small reheat coil (heat provided by steam, hot water, or electric resistance). (Technically, this could also be called an *air and water system*.) The central station provides a single stream of cold air that must be cold enough to meet the maximum cooling demand of any one zone. All other zones *reheat* this air as needed. In cold weather, outdoor air at temperatures as low as 38°F (3°C) can be used; the colder this single central airstream, the less air need be circulated (and the smaller the ducts). For buildings with large interior zones in most U.S. climates, however, the central airstream must be *cooled* most of the time; then more energy must be spent to reheat the airstream at most zones. These systems thus are notorious for energy wastage, although careful engineering can make them attractive for some climates.

*Double-Duct, Constant-Volume Systems* (Fig. 10.12d). Two complete distribution trees are required; at the height of summer the cooling

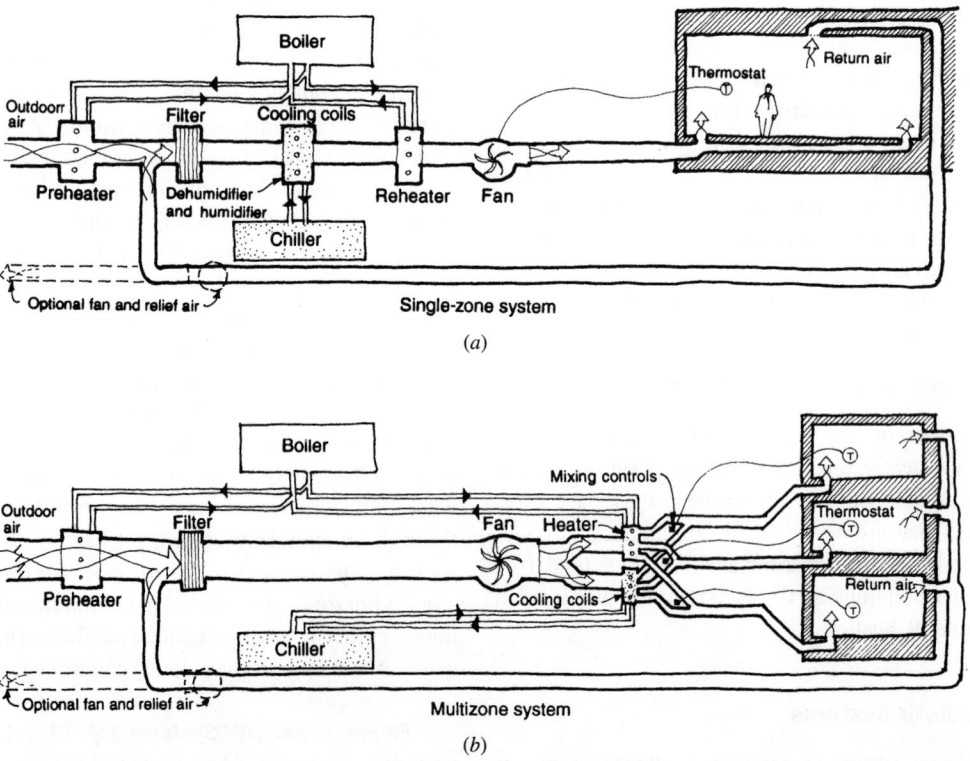

*Fig. 10.12* (a–f) All-air HVAC systems. An underfloor air supply is shown here, but a ceiling supply is more common.

THERMAL CONTROL

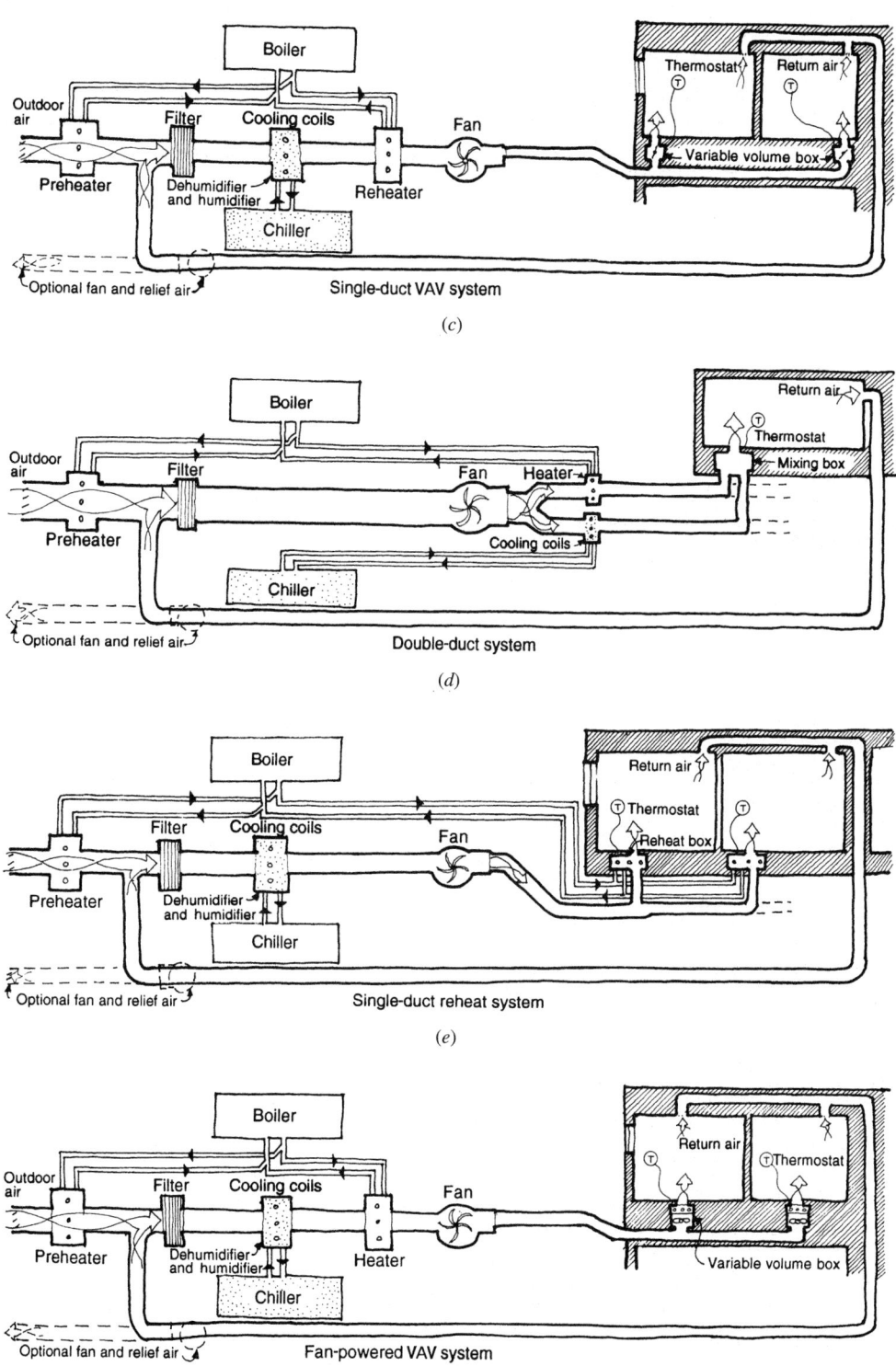

*(c)*

*(d)*

*(e)*

*(f)*

**Fig. 10.12** *(Continued)*

airstream does all the work, whereas in the coldest winter conditions the heating airstream carries the load. Most of the time, air from these two streams is mixed to order at each zone's air terminals. Because both temperature and volume can be controlled, this system offers better comfort under reduced load conditions (for example, an only partially occupied room) than does the single-duct VAV system. However, it is much more expensive to install, consumes much building volume for the two distribution trees, and usually consumes more energy than the single-duct VAV system that has largely replaced it.

### (c) Air and Water Systems

Several variations on air and water systems are shown in Fig. 10.13. Most of the heating and cooling of each zone is accomplished via the water

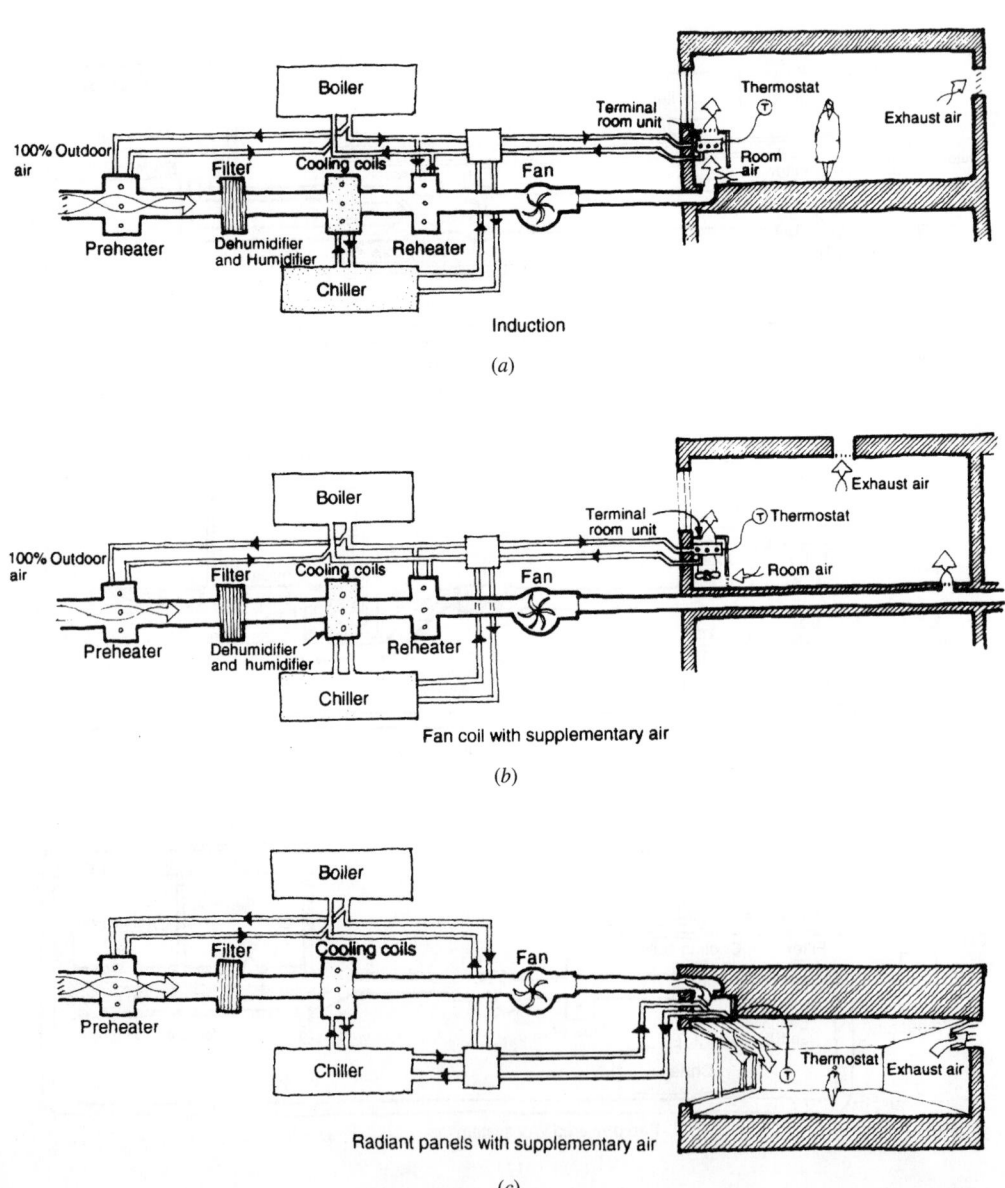

**Fig. 10.13** (a–c) Air and water HVAC systems. Where the underfloor air supply is shown here, a ceiling supply is also common. In (b) the supplementary air is often delivered directly to the fan-coil unit.

distribution tree, which is much thinner than the tree needed by air. For air quality—filtering, humidity, freshness—a small, centrally conditioned airstream, equal to the total fresh air required, is provided. Thus, several distribution trees are involved, yet the total space they require is almost always less than that required by all-air systems.

Exhaust air may be gathered in a return air duct system, making heat recovery possible. Or (a cheaper alternative) air can be exhausted locally to avoid the construction of yet another distribution tree. If the water distribution provides either heating *or* cooling only, it is called a *two-pipe system* (shown throughout Fig. 10.13). If it provides simultaneous heating *and* cooling, it is a *four-pipe system* or yet another heating and cooling variation utilizing heat pumps (see Fig. 9.45). (Three-pipe systems were a first-cost saving alternative that allowed simultaneous heating and cooling [from two supply pipes] but wasted energy by mixing the very warm hot return and the mildly warm cold return in only one return pipe. They are no longer permitted in most locales.)

This class of system frequently serves the perimeter zones of large buildings, whereas all-air systems (commonly, single-duct VAV) are used for the interior zones. More details on this HVAC class can be found in Section 10.6.

*Induction Systems* (Fig. 10.13a). This familiar system's air terminal may be found below windows throughout the United States. A high-velocity (and high-pressure), constant-volume fresh air supply is brought to each terminal, where it is forced through an opening in such a way that air already within the room (*bypass*, or secondary, air) is *induced* to join the incoming jet of air. A fairly thorough circulation of room air is thus accomplished with only a little centrally treated air. Air then passes over finned tubes for heating or cooling. Thermostats control the unit's output by controlling either the flow of the water or the flow of secondary air.

*Fan-Coil with Supplementary Air* (Fig. 10.13b). Another familiar piece of below-window equipment is the fan-coil, which moves the room air as it provides either heating or cooling. Centrally conditioned, tempered fresh air is brought to the space in a constant-volume stream; the fan moves both fresh and room air across a coil that either heats or cools the air, as required.

*Radiant Panels with Supplementary Air* (Fig. 10.13c). Either ceiling or wall panels contain the heated or cooled water to provide a large surface for radiant heat exchange. Centrally conditioned, tempered fresh air is brought to the space in a constant-volume stream. The "piece of equipment" within the space is replaced by a large surface, which must be kept clear of obstructions to radiant heat exchange.

*Water Loop Heat Pumps.* This variation on the two-pipe distribution system was shown in Fig. 9.45. Heat pumps (water–air) either draw heat from the loop (heating mode) or discharge heat to it (cooling mode). For a large building in cold weather, the excess heat of the interior zones is thus used to warm the perimeter zones. The loop's temperature ranges between 65° and 90°F (18° and 32°C); in hot weather, a central cooling tower disposes of the loop's excess heat, whereas in cold weather a central boiler adds the loop's needed heat. The loop is sized to carry 2 to 3 gpm (0.13 to 0.19 L/s) per ton, where the total tonnage equals the sum of all the individual units (often greater than the actual load).

Because individual heat pumps are used, this system is closely related to direct refrigerant systems; it is often considered an all-water system.

## (d) All-Water Systems

The more simple-appearing all-water systems are shown in Fig. 10.14. These systems only heat and cool; the distribution trees are indeed slim. Air quality is dealt with elsewhere—either locally, by means of infiltration or windows; or by a separate fresh air supply system; or simply by fresh air from an adjacent system, such as a ventilated interior zone. This ambiguity about fresh air leads to similar ambiguities about whether a system is air and water or all-water. A fan-coil terminal is often employed so that air motion occurs along with heating or cooling. (Sometimes the fan-coil unit is located against the exterior wall so that fresh air may be brought in and mixed with the room air through the fan.) Both baseboard and valence (above-window) units are also commonly available.

Because air is handled so locally, there is very little mixing of air from one zone to another, making this attractive where potential air contamination (or smoke from a fire) is a special concern. It is

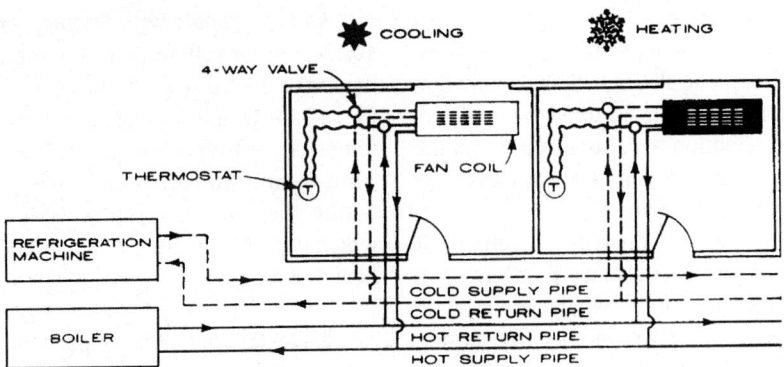

**Fig. 10.14** *All-water HVAC systems. Four-pipe distribution trees need smaller volumes than do those for air systems; however, less thorough conditioning of air is provided.*

also an easy system to retrofit. However, maintenance is high; filters in each fan coil must be cleaned, and drain pans are potentially problematic.

More details on this HVAC system class can be found in Section 10.7.

*Two-pipe* water distribution systems were shown throughout Fig. 10.13. They provide either heating or cooling. One pipe is for supply, the other for return. In a typical large-building application, they are used for heating in winter, cooling in summer. This raises the question of what changeover period will be required, a problem made much easier with the following alternative system.

*Four-pipe* systems are shown in Fig. 10.14. They allow quick changeover heating and cooling, utilizing two supply and two return pipes. A four-pipe system also allows for simultaneous cooling and heating in different zones within a single distribution system.

### (e) Equipment Space Allocations

An important early design decision is whether to integrate or separate the *heating/cooling* equipment and the *air handling* equipment (see Fig. 10.2). If they are integrated, one or a few central mechanical room(s) can serve many floors and each mechanical room will need area and height sufficient for both heating/cooling and air handling equipment. If separated, one (or several, in tall buildings) large space for heating/cooling equipment is typically located in the basement or the penthouse, with a smaller fan room on each floor.

Each mechanical room should have both a central location relative to the area it serves and

direct access to the outside—contradictory requirements in many cases. Central locations within the area served minimize the distribution tree size; access to the outdoors facilitates the use of outdoor air as a heat source (winter) or sink (summer) and allows equipment to be installed or removed in later remodeling. Mechanical rooms serving both heating/cooling and air handling equipment need relatively high ceilings; 12-ft (3.7-m) clear is a typical minimum, 20-ft (6-m) clear a typical maximum.

Tables 10.3 and 10.4 present the approximate space requirements for conventional mechanical systems. (For a more detailed look at equipment room space requirements, see Figs. 10.18, 10.29, and 10.42.)

The sizing graphs shown in Tables 10.3 and 10.4 are generous if buildings are designed with energy conservation in mind. For buildings with large heat gains or losses, these graphs may slightly undersize the areas needed. These graphs are intended to give a very fast approximation of areas; more detailed procedures for sizing are presented in Section 10.3 (for boilers, chillers, fan rooms) and Section 10.4 (for air ducts).

**EXAMPLE 10.1** Approximate the central equipment room needed to serve the Oregon office building (Fig. 8.7) that was analyzed for passive performance throughout Chapter 8. The total floor area of this office building is 24,000 ft² (2230 m²).

**SOLUTION**
Enter Table 10.3, spaces for heating and cooling, for an office building at 24,000 ft². The space for the

**TABLE 10.3 Approximate Space Sizes for Major Heating and Cooling Equipment**

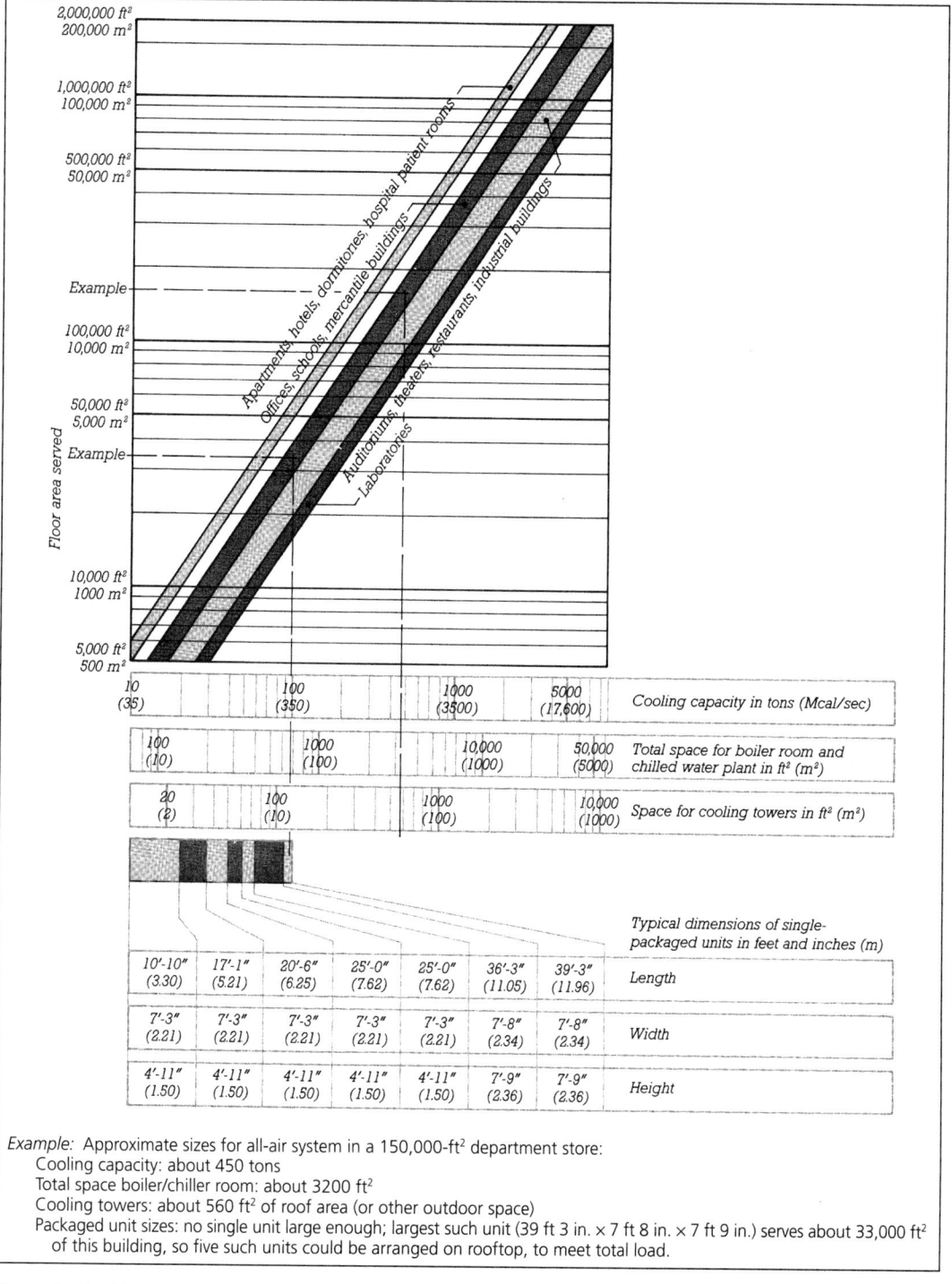

*Example:* Approximate sizes for all-air system in a 150,000-ft² department store:
Cooling capacity: about 450 tons
Total space boiler/chiller room: about 3200 ft²
Cooling towers: about 560 ft² of roof area (or other outdoor space)
Packaged unit sizes: no single unit large enough; largest such unit (39 ft 3 in. × 7 ft 8 in. × 7 ft 9 in.) serves about 33,000 ft²
of this building, so five such units could be arranged on rooftop, to meet total load.

*Source:* Reprinted by permission from Edward Allen, *The Architect's Studio Companion,* © 1989 by John Wiley & Sons, New York.

THERMAL CONTROL

**THERMAL CONTROL**

### TABLE 10.4  Approximate Space Sizes for Air-Handling Equipment

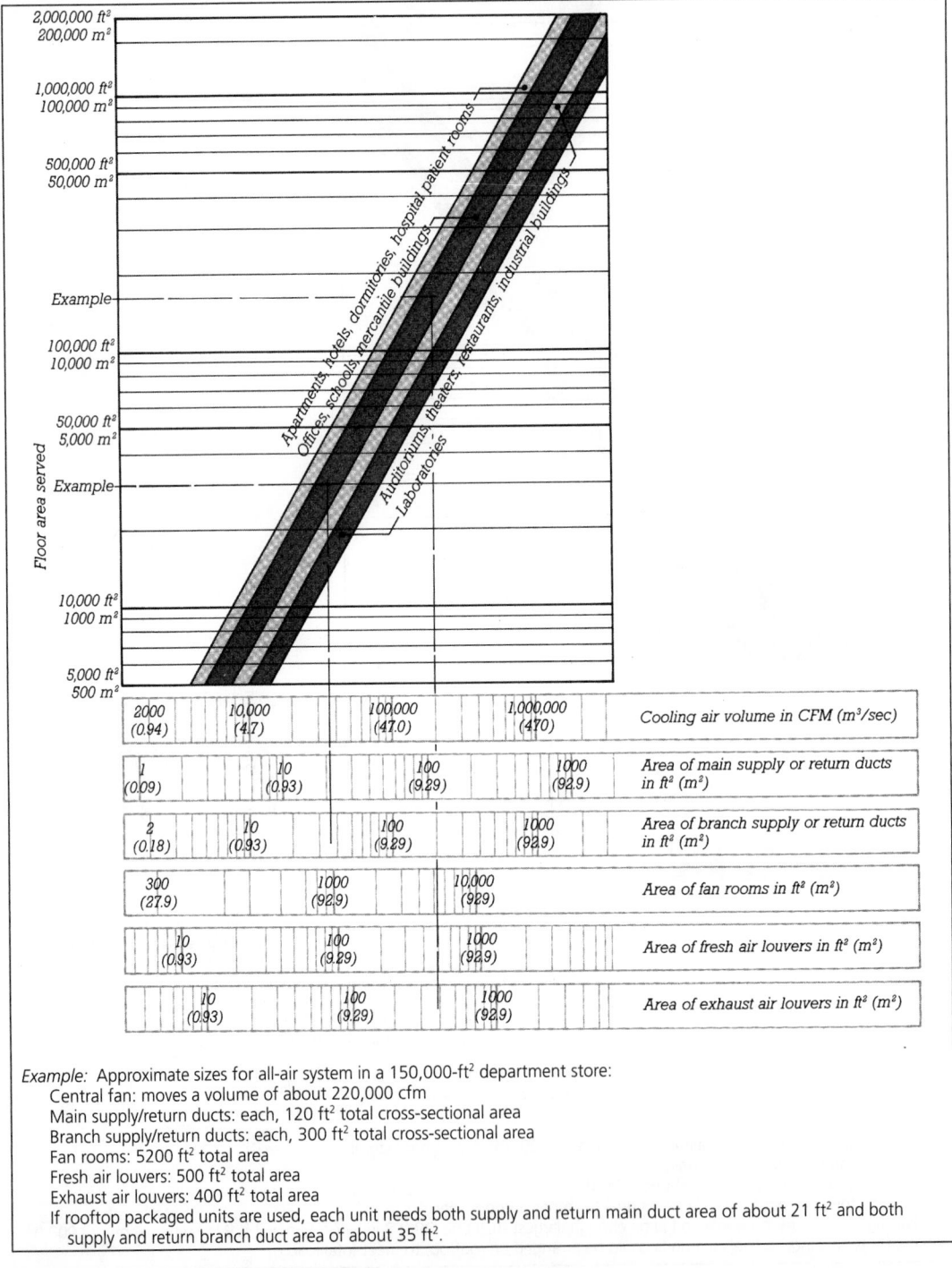

*Example:* Approximate sizes for all-air system in a 150,000-ft² department store:
Central fan: moves a volume of about 220,000 cfm
Main supply/return ducts: each, 120 ft² total cross-sectional area
Branch supply/return ducts: each, 300 ft² total cross-sectional area
Fan rooms: 5200 ft² total area
Fresh air louvers: 500 ft² total area
Exhaust air louvers: 400 ft² total area
If rooftop packaged units are used, each unit needs both supply and return main duct area of about 21 ft² and both
    supply and return branch duct area of about 35 ft².

*Source:* Reprinted by permission from Edward Allen, *The Architect's Studio Companion,* © 1989 by John Wiley & Sons, New York.

boiler and chiller is estimated at about 500 ft². Enter Table 10.4, spaces for air handling, for an office building at 24,000 ft². The space for the fan room is estimated at about 800 ft². Thus, a central mechanical room (boiler, chiller, and main fans) is estimated at 500 + 800 = 1300 ft² (121 m²). (The actual size of the mechanical room in this building is 1200 ft², a very close prediction.)

Now, approximate the largest duct sizes required. On the building's upper floor are five bays, each 1440 ft², for a total of 7200 ft² (669 m²). The building's mechanical room is beyond the east end of this upper floor, so all the HVAC system supply air must enter through one end of the central duct. Enter Table 10.4, spaces for air handling, for an office building at 7200 ft². The following ranges of sizes are indicated:

Volume of air: 7000–11,000 cfm

Area, main supply or return ducts: 4 to 6.5 ft²

(The actual central supply duct maximum size in this building's upper floor is 2.6 ft². The table thus overestimated the duct size by about 150% in this case of a well-insulated, shaded, and daylit building.) ∎

Engineers often refer to cooling equipment sizes in tons of refrigeration. The relationship between tons and floor area served is explored in Table 10.5.

For the thermally well-designed detached residence of today, a design guideline is 1000 ft² of floor area/ton.

## 10.3 CENTRAL EQUIPMENT

The many HVAC systems that are included in the categories of all-air, air and water, and all-water have in common a dependence on *central equipment* for the generation of heating and cooling, and/or air quality control. Figure 10.15 shows the basic relationships between some of the major pieces of central equipment and the spaces they serve. This section offers a general guide to some central equipment options and sizes. The consulting engineer chooses such equipment based upon a much more detailed analysis.

### (a) Boilers

These devices heat the recirculating hot water system used for building heating. The type of boiler selected depends on the size of the heating load, the heating fuels available, the desired efficiency of operation, and whether single or modular boilers are to be installed. Boiler sizes are commonly stated

**TABLE 10.5 Tons of Refrigeration: Design Guidelines**

| Type of Occupancy | Floor Area Served | | Assumptions |
|---|---|---|---|
| | ft²/ton | m²/ton | |
| General occupancy: | 350–550 | 32.5–51 | 400 cfm/ton (189 L/s per ton) |
|   Perimeter spaces | | | 1.5 cfm/ft² (0.7 L/s per m²) |
|   Interior spaces | | | 0.6 cfm/ft² (0.3 L/s per m²) |
| Offices | 500 | 46.5 | 200 ft²/person (18.6 m²/person) |
| | | | 3.5 W/ft² (38 W/m²) for lights, equipment |
| High-rise apartments | | | |
|   North-facing | 1000 | 93 | |
|   Other orientations | 500 | 46.5 | |
| Hospitals | 333 | 31 | 1000 ft²/bed (93 m²/bed) |
| Shopping centers | 400 | 37 | |
|   Department stores | | | 2 W/ft² (21.5 W/m²) |
|   Specialty stores | | | 5 W/ft² (53.8 W/m²) |
| Hotels | 350 | 32.5 | |
| Restaurants | 150 | 14 | |
| Central plants | | | |
|   Urban districts | 380 | 35.5 | |
|   College campuses | 320 | 29.5 | |
|   Commercial centers | 475 | 44 | |
|   Residential centers | 500 | 46.5 | |

*Source:* Lorsch (1993). Reprinted with permission of American Society of Heating, Refrigerating and Air-Conditioning Engineers, Atlanta, GA.
*Note:* 1 ton = 12,000 Btu/h (= 3.514 kW).

THERMAL CONTROL

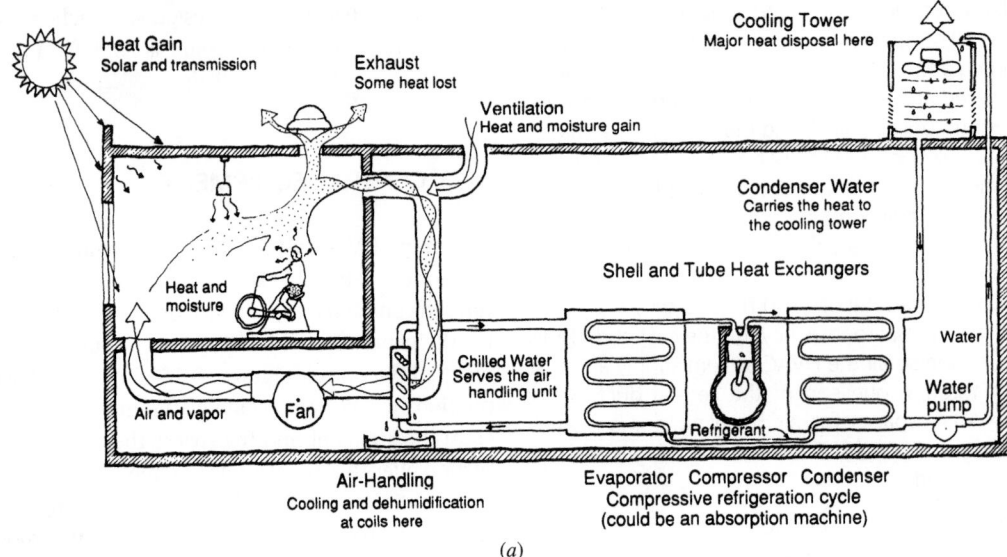

*(a)*

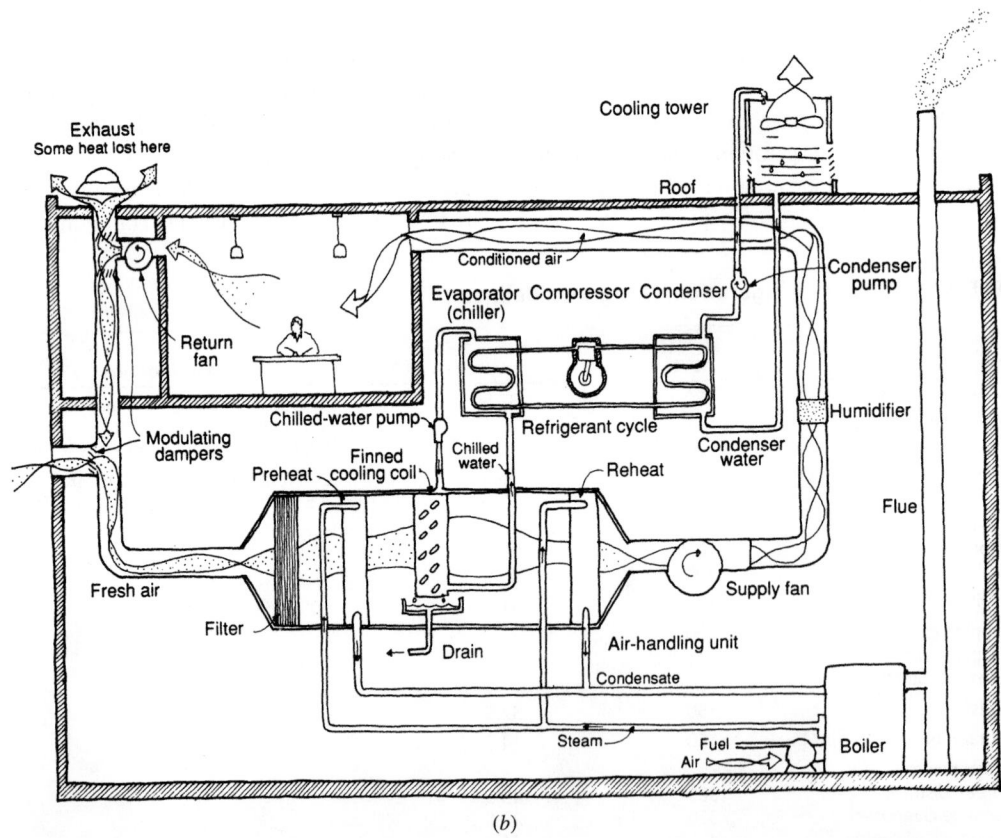

*(b)*

**Fig. 10.15** *Some basic components of HVAC central equipment. (a) A simplified diagram of a cooling cycle, in which chilled water is circulated to air handling coils and heat is disposed of through a cooling tower. (b) Schematic diagram of major components of central equipment for both heating and cooling.*

either in Btu/h of net output or in (gross) boiler horsepower, where, in I-P units,

$$\text{boiler horsepower} = \frac{\text{heating load (Btu/h)}}{\% \text{ boiler efficiency}}$$

$$\times 33,470 \text{ Btu/h per horsepower}$$

In SI units,

$$\text{boiler horsepower} = \frac{\text{heating load (kW)}}{\% \text{ boiler efficiency}}$$

$$\times 9.81 \text{ kW per horsepower}$$

Efficiency depends partly on the number of passes that the hot gases make through the water—the more passes, the higher the efficiency. It also depends on burner efficiency and on regular maintenance. Finally, efficiency is best when the equipment is operating near its capacity. Figure 10.16 compares typical boiler types, including two- and three-pass boilers.

*Fire Tube Boilers.* The hot gases of the fire are taken through tubes that are surrounded by the water to be heated. Firebox boilers place the boiler shell on top of the combustion chamber. Scotchmarine boilers feature multiple passes of the combustion gas through tubes. Fire tube boilers can be either dryback or wetback. Dryback designs have chambers outside the vessel to take combustion gases from the furnace to the tank. Wetback designs have water-cooled chambers that conduct the combustion gases.

*Water Tube Boilers.* The water to be heated is taken through tubes that are surrounded by the boiler's fire. They hold less water than the fire tube models, and so respond faster and can generate steam (where desired) at higher pressures.

*Cast-Iron Boilers.* Often used in residential and light-commercial applications, these are lower-pressure and lower-efficiency boilers. They do have the advantage of being modular.

In addition to the boilers themselves, there are choices of burner types (depending on the fuel(s) used), burner controls, and boiler feedwater systems. Consult the latest *ASHRAE Handbook—HVAC Systems and Equipment* for details.

Fossil fuel–burning boilers need flues for exhaust gases, fresh air for combustion, and required air pollution control equipment. The exhaust gas is usually first taken *horizontally* from the boiler; this horizontal enclosure, or flue, is called the *breeching.* The *vertical* flue section is called the *stack.* Guidelines for sizes and arrangements of breeching and stacks are shown in Fig. 10.17. Local codes determine the quantity of air required for combustion; local air pollution authorities set pollution control requirements. As a general rule, combustion air can be supplied in a duct to the boiler at an average velocity of 1000 fpm (5.1 m/s). The duct should be large enough to carry at least 2 cfm (1 L/s) per boiler horsepower. Furthermore, ventilation air to the boiler room should be provided; preferably, the inlet and outlet should be on opposite sides of the room. Minimum sizes: enough for 2 cfm (1 L/s) per boiler horsepower at a velocity of about 500 fpm (2.5 m/s).

Space requirements for boilers are summarized in Fig. 10.18, which shows multiple boilers. Note that clear space within the room must be provided so that the tubes of the boiler can be pulled when they must be replaced. Access for eventually replacing entire boilers must be considered.

Several types of single boilers are discussed here. The final boiler type discussed, the modular boiler, is preferred for energy conservation.

1. *High-output, package-type steel boiler.* For large buildings that use steam as a primary heating medium, one or several such boilers may be used. Direct use of steam can be seen in Fig. 10.15*b*, supplying preheat and reheat coils and also a humidifying unit. The relative lightness of this boiler type, compared to the older styles with ponderous masonry bases (boiler settings), makes it suitable for use on upper floors of tall buildings. Figure 10.6 shows two such boilers on the 13th floor of the Fox Plaza Building.

2. *Converter, steam to hot water* (Fig. 10.19). When, in a building that uses primary steam boilers, secondary circuits that use hot water for heating are required, a converter is used. It is considered a heat exchanger. In Fig. 10.6, there is downfeed steam supply for the two boilers on the 13th floor to two such converters, one for hot water heating in the apartments and one below the garage ceiling for hot water heating in the commercial area. A converter may also be used to transfer heat from steam to *domestic*

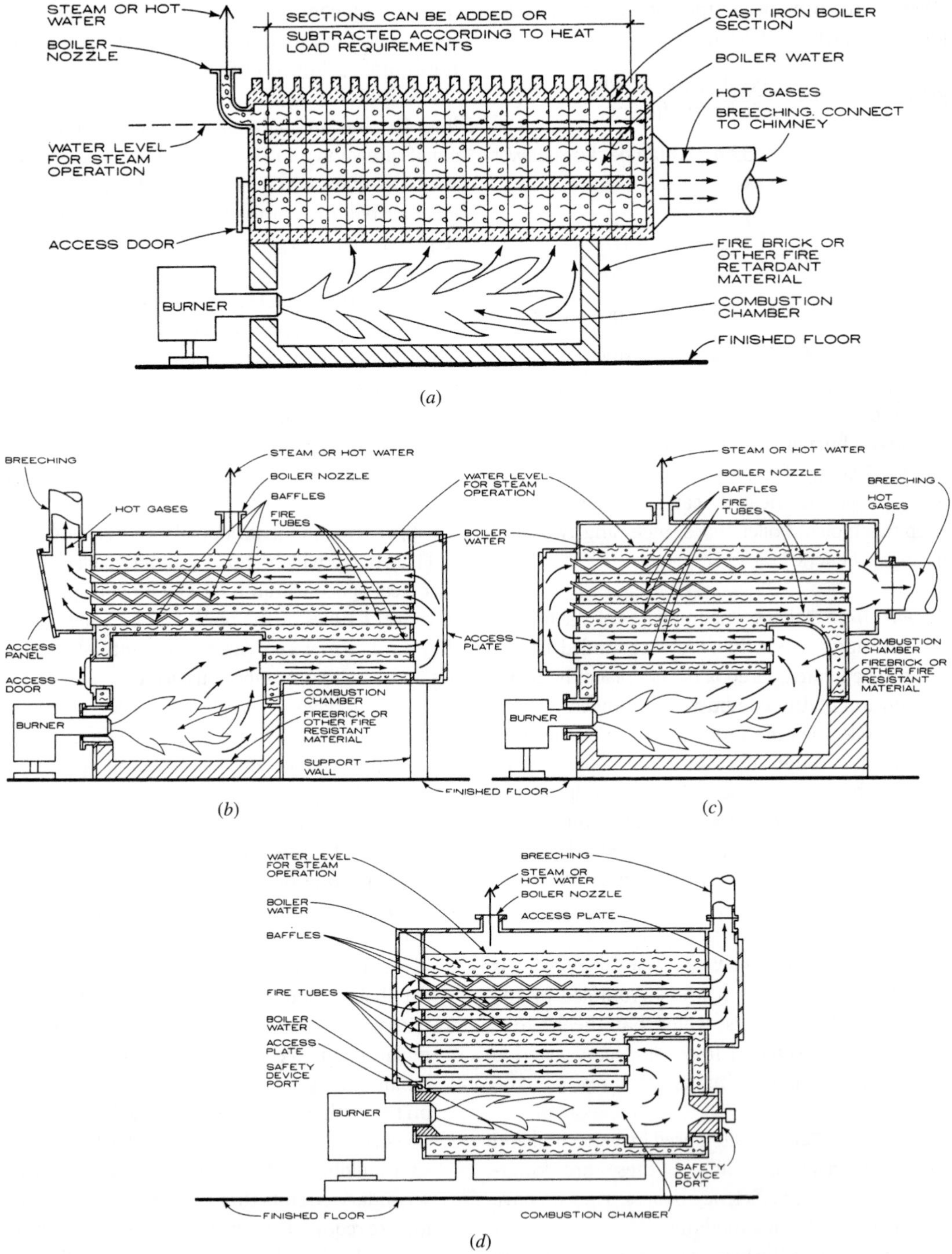

THERMAL CONTROL

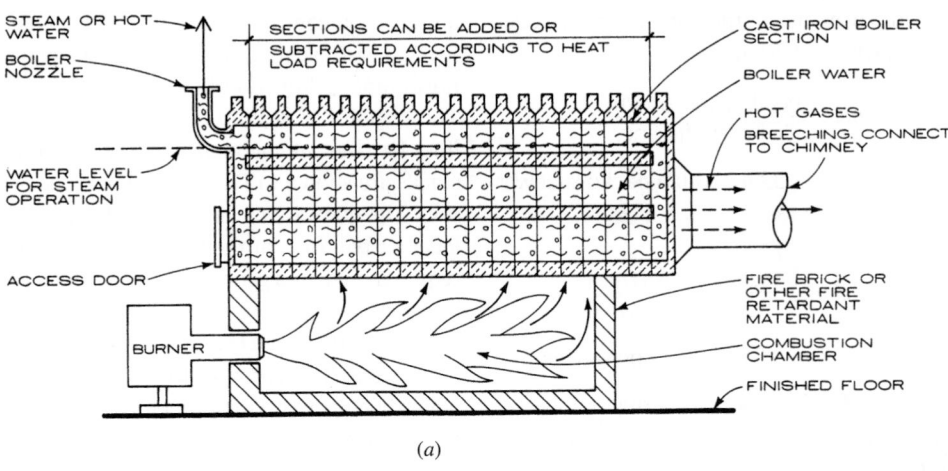

**Fig. 10.16** *Comparisons of boiler types. (a) Cast iron sectional type. (b) Two-pass fire tube. (c) Three-pass fire tube. (d) Three-pass wet-back Scotch marine. (From AIA: Ramsey/Sleeper,* Architectural Graphic Standards, *9th ed., © 1994 by John Wiley & Sons, Inc. Reprinted with permission of John Wiley & Sons, Inc.)*

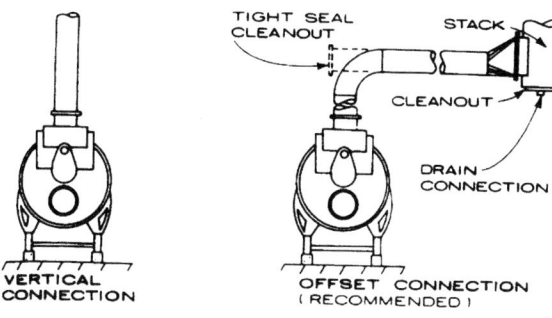

### STACK DIAMETER—SINGLE BOILER VENT OR STACK

| BOILER HORSE-POWER | STACK DIAMETER (IN.) | A (IN.) | B (IN.) | C (IN.) |
|---|---|---|---|---|
| 15-20 | 6 | 15 | 15 | 12 |
| 25-40 | 8 | 20 | 20 | 16 |
| 50-60 | 10 | 25 | 25 | 20 |
| 70-100 | 12 | 30 | 30 | 24 |
| 125-200 | 16 | 40 | 40 | 32 |
| 250-350 | 20 | 50 | 50 | 40 |
| 400-800 | 24 | 60 | 60 | 48 |

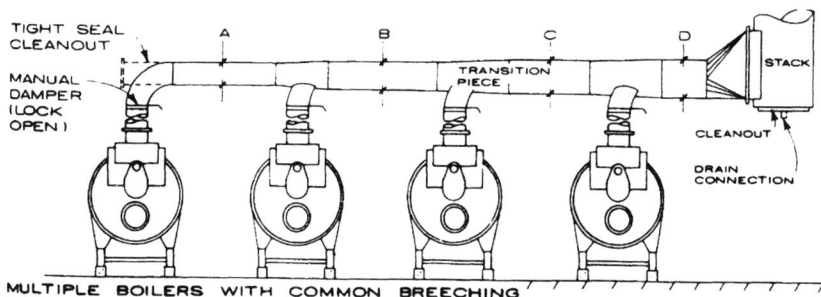

MULTIPLE BOILERS WITH COMMON BREECHING

### STACK DIAMETER—MULTIPLE BOILERS: COMMON BREECHING AND STACK

| BOILER HORSE-POWER | MINIMUM STACK DIAMETER (IN.) | | | | | |
|---|---|---|---|---|---|---|
| | NUMBER OF BOILERS | | | | | |
| | 2 | | 3 | | 4 | |
| | 100 FT | 200 FT | 100 FT | 200 FT | 100 FT | 200 FT |
| 25-40 | 11 | 12 | 13 | 14 | 14 | 16 |
| 50-60 | 13 | 14 | 15 | 16 | 17 | 18 |
| 70-100 | 16 | 17 | 19 | 20 | 21 | 23 |
| 125-200 | 21 | 22 | 24 | 26 | 28 | 30 |
| 250-350 | 26 | 28 | 32 | 34 | 34 | 40 |
| 400-600 | 32 | 34 | 38 | 40 | 42 | 46 |

### BREECHING DIAMETER— SINGLE AND MULTIPLE BOILERS

| BOILER HORSE-POWER | MINIMUM BREECHING DIAMETER (IN. OD) | | | |
|---|---|---|---|---|
| | A (IN.) 1 BOIL-ER | B (IN.) 2 BOIL-ERS | C (IN.) 3 BOIL-ERS | D (IN.) 4 BOIL-ERS |
| 15-20 | 6 | 8 | 9 | 9 |
| 25-40 | 8 | 10 | 11 | 12 |
| 50-60 | 10 | 12 | 14 | 15 |
| 70-100 | 12 | 15 | 17 | 18 |
| 125-200 | 16 | 20 | 22 | 24 |
| 250-350 | 20 | 25 | 28 | 30 |
| 400-600 | 24 | 30 | 33 | 36 |
| 700-800 | 24 | 34 | 38 | 42 |

Note: Stack diameter should be larger than breeching diameter.

**Fig. 10.17** *Breeching and stack size guidelines for fossil-fuel-fired boilers. (From AIA: Ramsey/Sleeper,* Architectural Graphic Standards, *10th ed., © 2000 by John Wiley & Sons, Inc. Reprinted with permission of John Wiley & Sons, Inc.)*

(service) water. Converters are frequently used where central steam supply systems are available, as in large-city downtown areas. The easier, quieter distribution of heat by hot water has largely replaced steam heating distribution trees within buildings.

3. *Electric boilers.* Where electricity costs are competitive with those of fossil fuels, electric boilers are sometimes used. Both hot water and steam electric boilers are available. The advantage of electric boilers is the elimination of combustion air, the flue, and air pollution at the building. The disadvantages are the use of a high-grade energy source for a relatively low-grade task and the pollution impact at the electric generating plant. In order to protect against high electric demand charges, a large number of control steps are desirable.

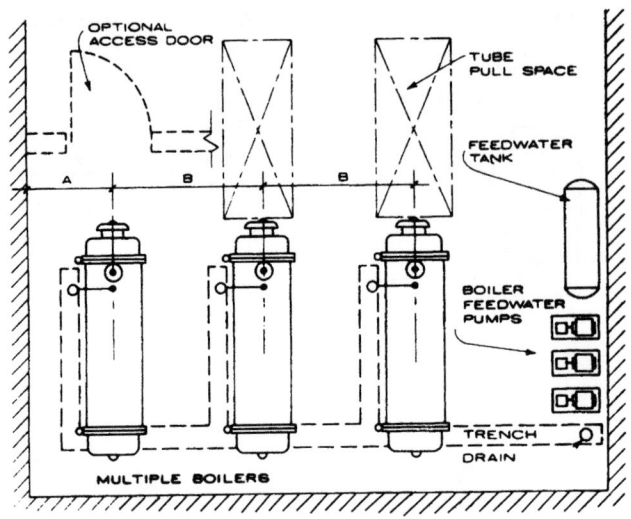

**BOILER ROOM SPACE REQUIREMENTS**

| BOILER HP | 15-40 | 50-100 | 125-200 | 250-350 | 400-800 |
|---|---|---|---|---|---|
| Dimension A | 5'-9'' | 6'-6'' | 6'-10'' | 7'-9'' | 8'-6'' |
| Dimension B | 7'-5'' | 8'-9'' | 9'-7'' | 11'-9'' | 14'-3'' |

**Fig. 10.18** Boiler room space requirements. Dimension A includes an aisle of 3 ft-6 in. (1 m) between the boiler and the wall. Dimension B between the boilers includes an aisle of at least 3 ft-6 in. (1 m), up to 5 ft (1.5 m) for the largest boilers. (From AIA: Ramsey/Sleeper, Architectural Graphic Standards, 10th ed., © 2000 by John Wiley & Sons, Inc. Reprinted with permission of John Wiley & Sons, Inc.)

4. *Compact boilers* (Fig. 10.20). Newer, smaller-dimension boilers with high thermal efficiencies are available. In addition to their space-saving footprint, they feature a variety of venting options that make them easily adaptable to smaller equipment rooms.

5. *Modular boilers* (Fig. 10.21). The primary advantage of modular boilers is efficiency. Boilers achieve maximum efficiency when they are operated continuously at their full rated fuel input. The single boilers discussed previously operate this way only under outside design conditions, which by definition occur, at most, during 5% of a normal winter. In a modular boiler design, each section is run independently. Therefore, only one section need be fired for the mildest heating needs; as the weather gets colder, more sections are gradually added. Because each section operates continuously at full-rated fuel input, efficiency is greatly increased (Fig. 10.22). Each module, being rather small, requires little time to reach a useful temperature and (unlike the larger single boilers) does not waste a lot of heat as it cools down. Thus, modular boilers

usually produce a 15% to 20% fuel savings for the heating season relative to single boilers. Their other advantages include ease of maintenance (one module can be cleaned while others carry the heating load) and small size (allowing easy installation and replacement in existing buildings).

Modular boilers also eliminate the initial cost of *oversizing* heating equipment. In cold climates, conventional boiler systems often use two or three large boilers to ensure that heat is available even if one large boiler fails. When two such boilers are used, it is common practice to size each boiler at two-thirds of the total heating load; an oversize of one-third results. When three such boilers are used, it is common practice to size each boiler at 40% of the total heating load; an oversize of 20% results. However, when a minimum of five modular boilers are used, oversizing can be eliminated because the failure of a single module will not have a crippling impact on the overall heat output.

*Gas-fired pulse boilers* are an even smaller and more energy-efficient choice for modular boilers.

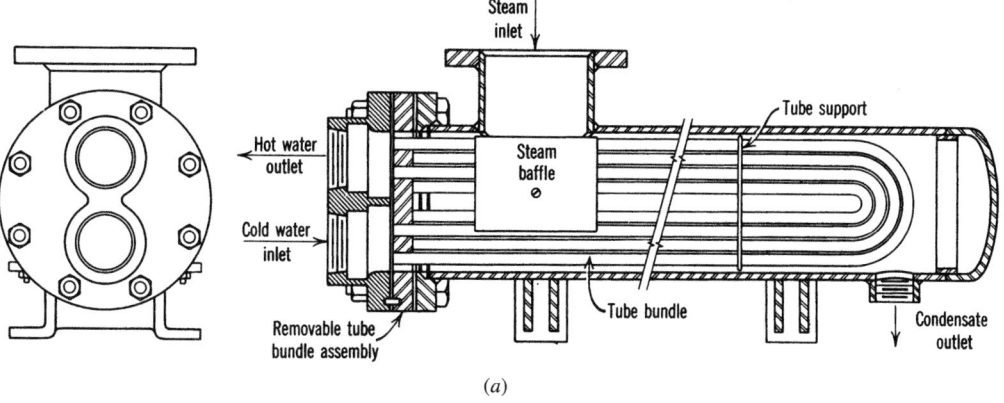

(a)

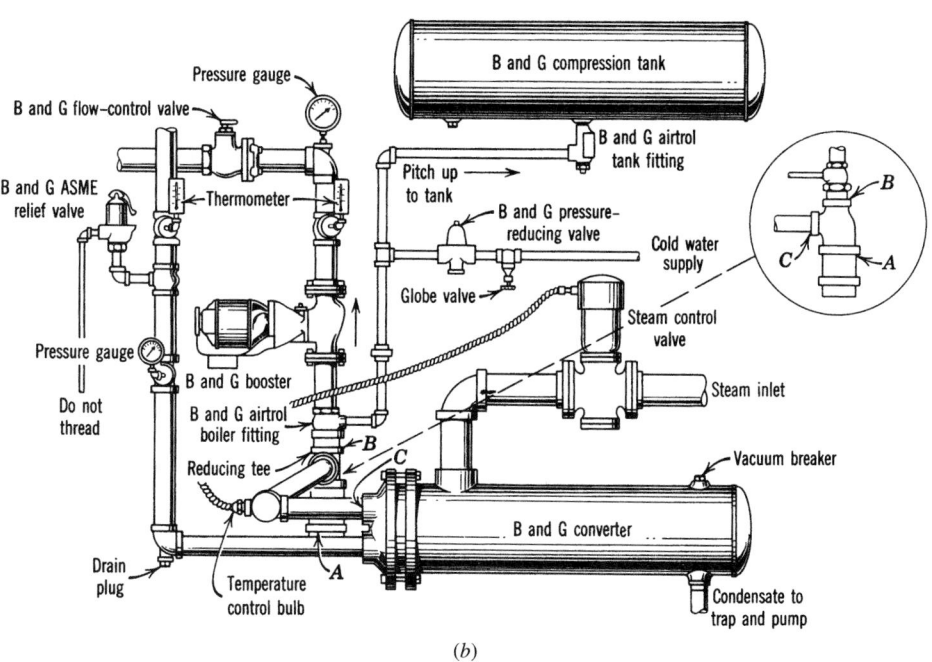

(b)

**Fig. 10.19** *Conversion unit that transfers heat from steam to hot water. (a) Section illustrating the principle of heat transfer from steam to water. (b) A converter connected to the steam supply and equipped with all devices necessary for a complete hot water heating system. (Courtesy of ITT Bell and Gossett.)*

Pulse boilers utilize a series of 60 to 70 small explosions per second, making the hot flue gases pulse as they pass through the firetube. This makes for very efficient heat transfer. Pulse boilers are available up to about 300,000 Btu/h (88 kW).

Pulse boilers operate with lower water temperatures so that water vapor in the flue gas can condense and drain. This change of state liberates additional heat, allowing these pulse boilers to achieve efficiencies up to 90%. They exhaust moist air, not hot smoke, so flues can be small-diameter plastic pipe rather than large-diameter, heat-resistant materials.

## (b) Chillers

These devices remove the heat gathered by the recirculating chilled water system as it cools the building. The selection of chillers depends largely on the fuel source and the total cooling load.

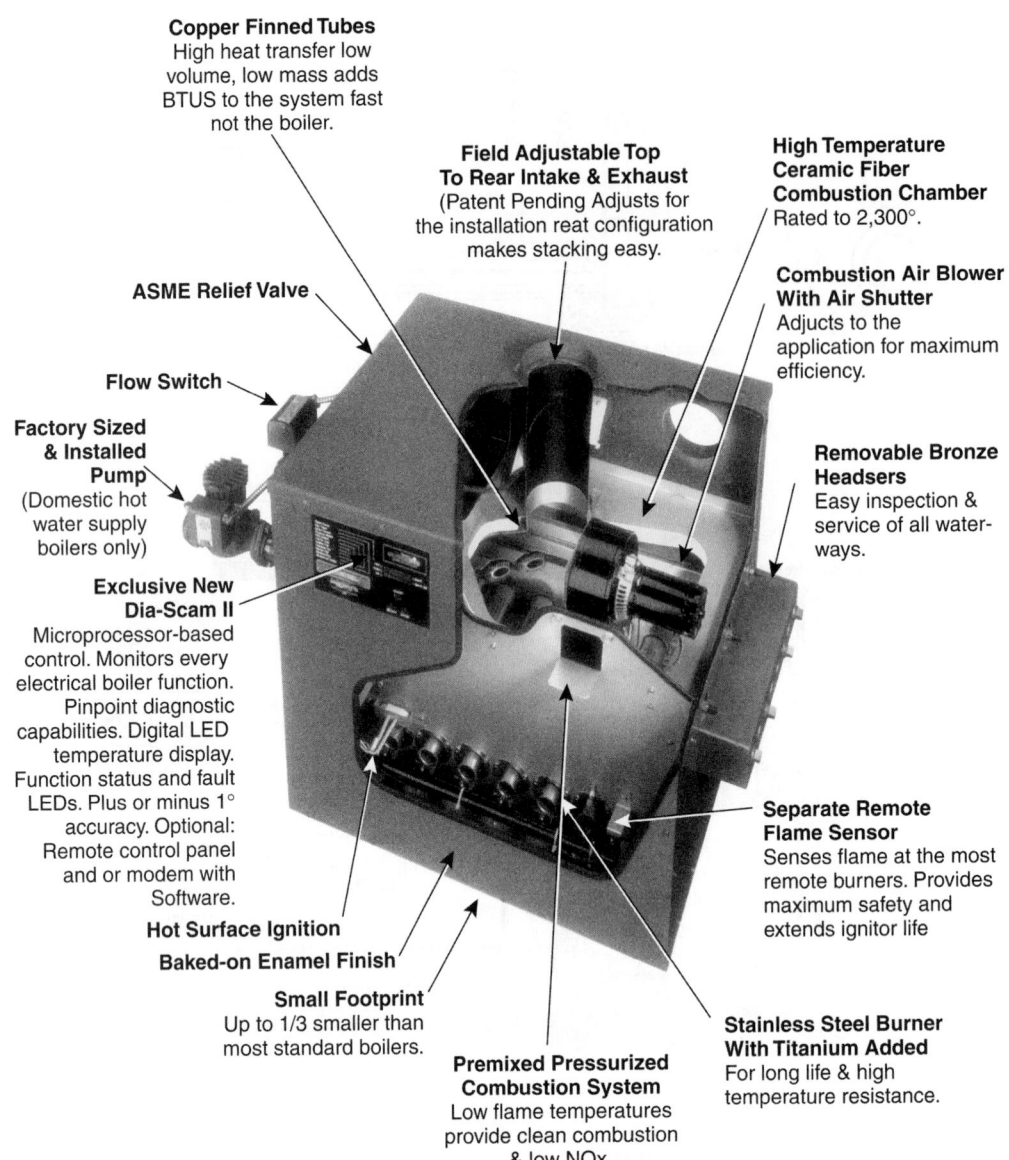

**Copper Finned Tubes**
High heat transfer low volume, low mass adds BTUS to the system fast not the boiler.

**Field Adjustable Top To Rear Intake & Exhaust**
(Patent Pending Adjusts for the installation reat configuration makes stacking easy.

**High Temperature Ceramic Fiber Combustion Chamber**
Rated to 2,300°.

**ASME Relief Valve**

**Combustion Air Blower With Air Shutter**
Adjucts to the application for maximum efficiency.

**Flow Switch**

**Factory Sized & Installed Pump**
(Domestic hot water supply boilers only)

**Removable Bronze Headsers**
Easy inspection & service of all water-ways.

**Exclusive New Dia-Scam II**
Microprocessor-based control. Monitors every electrical boiler function. Pinpoint diagnostic capabilities. Digital LED temperature display. Function status and fault LEDs. Plus or minus 1° accuracy. Optional: Remote control panel and or modem with Software.

**Hot Surface Ignition**

**Baked-on Enamel Finish**

**Separate Remote Flame Sensor**
Senses flame at the most remote burners. Provides maximum safety and extends ignitor life

**Small Footprint**
Up to 1/3 smaller than most standard boilers.

**Premixed Pressurized Combustion System**
Low flame temperatures provide clean combustion & low NOx.

**Stainless Steel Burner With Titanium Added**
For long life & high temperature resistance.

*Fig. 10.20* Burkay Genesis hot water boiler, fueled by either natural gas or propane, is available in ratings from 200,000 to 750,000 Btu/h (58,620 to 219,825 W). All units are 30 in. high × 24 in. deep (762 mm × 610 mm); the smallest boiler is 23 in. (584 mm) wide, and the largest is 57 in. (1454 mm) wide. The copper heat exchanger has an 83.7% thermal efficiency rating, and a variety of venting options are available. (Courtesy of A.O. Smith Water Products Company, Irving, TX.)

Chillers include both absorption and compressive refrigeration processes in a wide range of sizes.

New developments in chillers continue to result from a combination of concerns about the role of CFCs and HCFCs in global climate change and from changes in utility regulations that are producing unstable energy prices in many areas. Chillers capable of changing quickly between electricity and natural gas are becoming available as a result.

The single-effect, indirect-fired *absorption chiller* (Fig. 10.23) is attractive where central steam or high-temperature water (from solar collectors, as waste heat from an industrial process, a fuel cell, etc.) is available. This device uses the absorptive refrigeration cycle (explained in Fig. 9.2). Direct-fired absorption chillers use natural gas to power the cycle. In general, absorption equipment is less efficient than compressive refrigeration cycle equipment, although

| | | | |
|---|---|---|---|
| Shipping weight: | 644 lb | Floor loading | 179 lb/ft.$^2$ |
| Water content: | 7.25 gal | Pressure drop ( $\Delta T = 20F°$): | 1.7' WC |
| Fire side surface: | 49 ft.$^2$ | Water side surface: | 42 ft.$^2$ |
| Horsepower: | 9.1 hp | Pressure rating | 100 psi ASME |

(a)

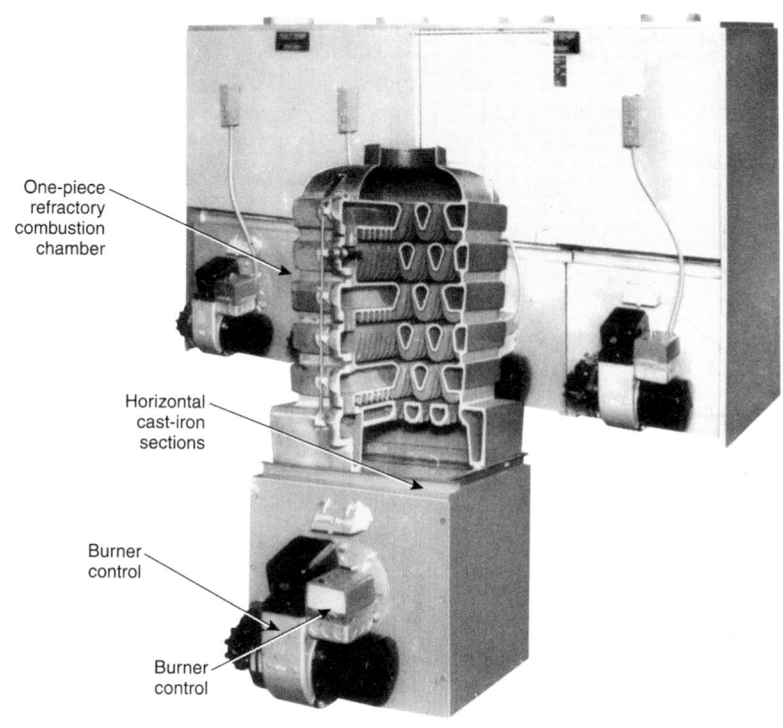

One-piece refractory combustion chamber

Horizontal cast-iron sections

Burner control

Burner control

Cutaway of Heating Module
385,000 Btu/h input
Factory assembled with burners and all oil controls
20" × 32" × 48 1/2" high

(b)

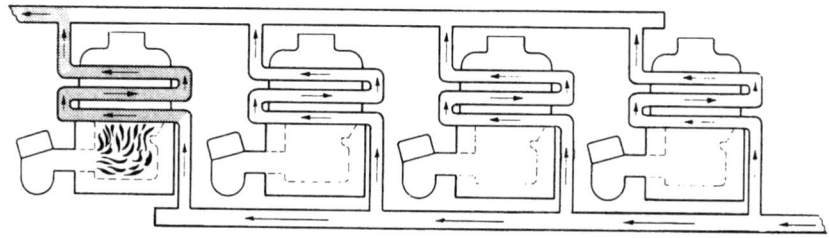

**SCHEMATIC FLOW DIAGRAM**
For 4-step 1,540,000 Btu/h Input Multi-Temp (MO-1540)

(c)

**Fig. 10.21** Modular boilers. (a) A bank of four modules—total input 1.5 million Btu/h (439 kW). (b) Details of one module. (c) Schematic of flow conditions in mild weather, with only one module in operation.

**401**

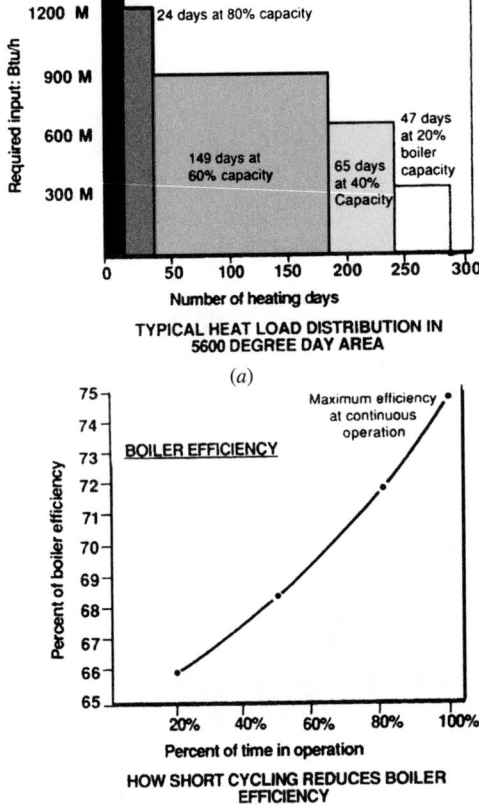

*Fig. 10.22* One large boiler versus many smaller ones. (a) Boilers rarely operate at full capacity; instead, they respond to part loads the majority of the time. (b) Under part load conditions, a boiler will often short-cycle, which on a single large boiler could drop the annual efficiency into the 66% to 75% range.

a cheap or even free heat source to power the cycle can rapidly overcome efficiency disadvantages. Absorption machines have fewer moving parts (and therefore require less maintenance) and are generally quieter than compressive cycle equipment. They are environmentally attractive, despite their much higher waste heat output (about 31,000 Btu/ton, compared to at most 15,000 Btu/ton for compressive cycle equipment), because they do not use CFCs or HCFCs and because they require far less electricity to operate. Newer developments include the double-effect absorption chiller (see Fig. 9.3) and the triple-effect chiller, each accompanied by an increase in efficiency.

The compressive refrigeration cycle (explained in Fig. 9.1) is used in the other types of chillers. Larger units are *centrifugal chillers* (Fig. 10.24), whose compressors either can be driven by an electric motor or can utilize a turbine driven by steam or gas. (When a steam-driven turbine is used, the exhaust steam is often used to run an auxiliary absorption cycle machine. These two devices make an efficient combination, and the steam plant that supplies them in summer can supply heating in winter.) Centrifugal chillers usually require about 1 hp/ton (0.57 kW, or 10 ft³ gas, or about 15 lb of steam per ton). These large chillers usually require a cooling tower. *Dual-condenser chillers* (Fig. 10.25) can choose whether to reject their heat to a cooling tower (via the heat rejection condenser) or to building heating (via the heat recovery condenser).

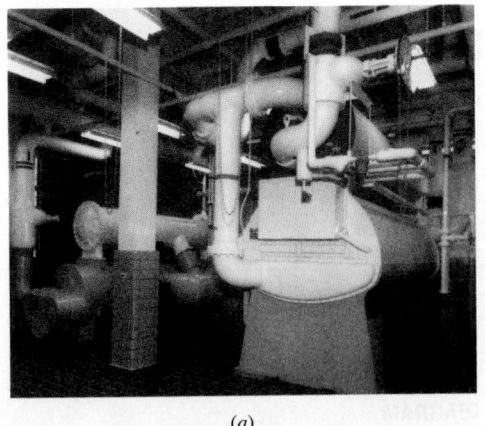

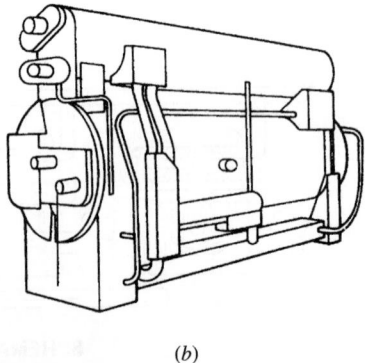

*Fig. 10.23* (a) An absorption chiller driven by heat to produce chilled water. (The Carrier Corporation; courtesy of Ingersoll-Rand.) (b) Two-stage absorption chiller utilizing steam, producing 200 to 800 tons of cooling. (From AIA: Ramsey/Sleeper, Architectural Graphic Standards, 10th ed., © 2000 by John Wiley & Sons, Inc. Reprinted with permission of John Wiley & Sons, Inc.)

Condenser
water in

Condenser
water out

Chilled water
out

Chilled water
in

Shell and tube
type condenser

Electric motor.
Could be steam
turbine or diesel

Centrifugal
compressor

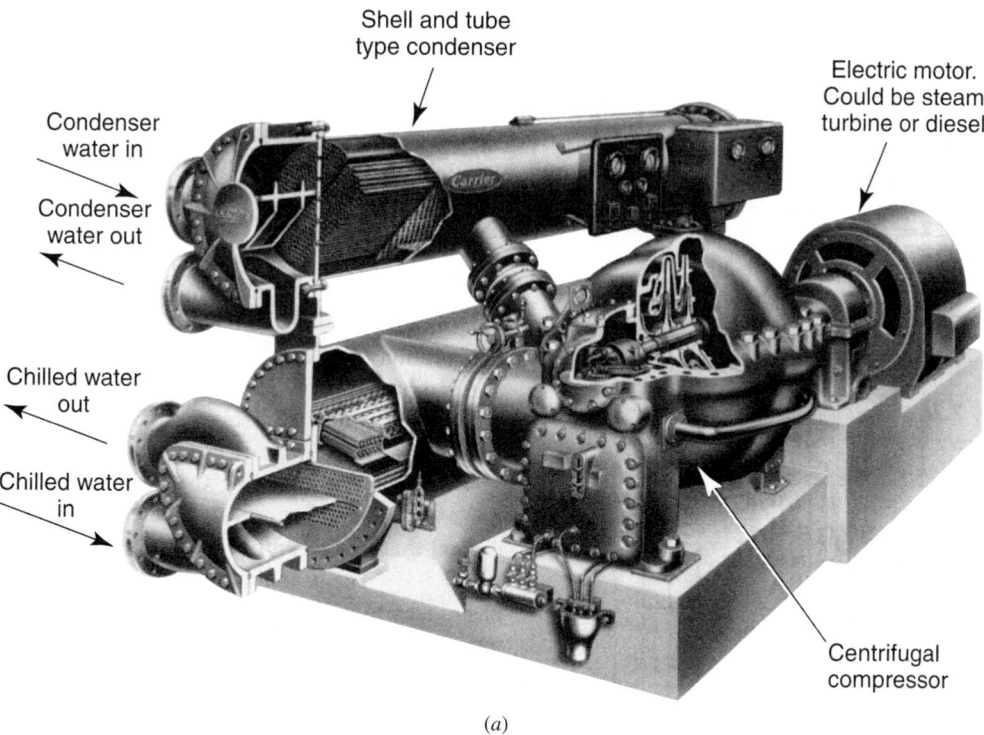

(a)

MOTOR
ROTOR

MOTOR

TO OVER TEMPERATURE
SAFETY SWITCH

VANE
MOTOR

MOTOR
STATOR

TO
PURGE
ORIFICE

BACK
PRESSURE
VALVE

FILTER

TRANSMISSION

COMPRESSOR

REFRIGERANT VAPOR
REFRIGERANT LIQUID
REFRIGERANT
LIQUID/VAPOR

CONDENSER

STRAINER

COOLER

SUBCOOLER COIL

CONDENSER
WATER

CHILLED
WATER

L 14'-4"   W 4'-9"   H 7'-7"
WT. 15,750 LB

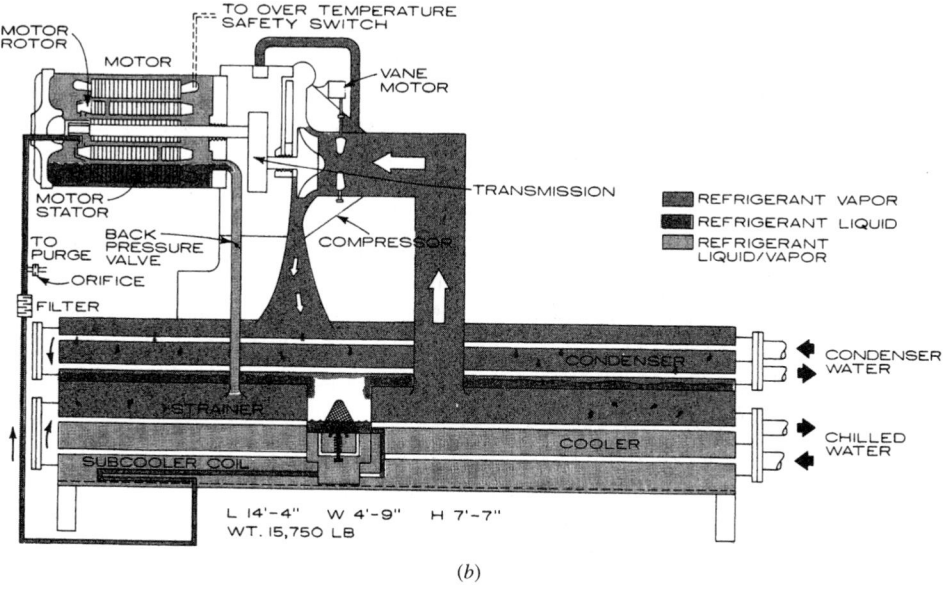

(b)

**Fig. 10.24** (a) A centrifugal chiller—a machine of large capacity using the compressive refrigeration cycle. (Courtesy of the Carrier Corporation.) (b) Centrifugal chiller with a flooded cooler and condenser within a single outer shell. This low-pressure unit typically produces 100 to 400 tons of cooling. (From AIA: Ramsey/Sleeper, Architectural Graphic Standards, 7th ed., © 1981 by John Wiley & Sons, Inc. Reprinted with permission of John Wiley & Sons, Inc.

THERMAL CONTROL

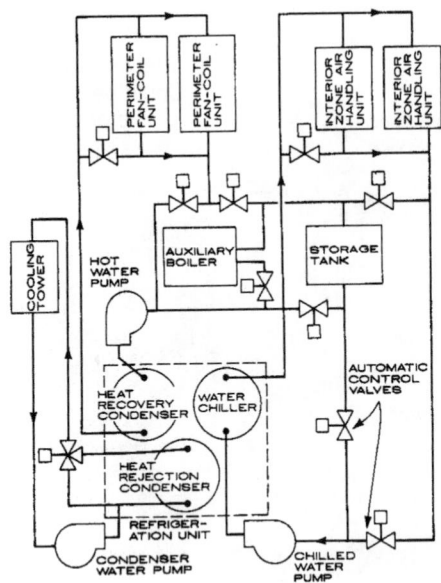

**Fig. 10.25** *Dual-condenser chiller. Heat drawn from the chilled water system is either rejected to the cooling tower or recovered for use in building heating. (From AIA: Ramsey/Sleeper,* Architectural Graphic Standards, *10th ed., © 2000 by John Wiley & Sons, Inc. Reprinted with permission of John Wiley & Sons, Inc.)*

Somewhat smaller chillers use either twin screws or a scroll in place of a piston in the compressor. The screw compressor (Fig. 10.26) has a pair of helical screws; as they rotate, they mesh and thus compress the volume of the gas refrigerant. They are small and quiet, with little vibration. The scroll compressor (Fig. 10.27) uses two inter-fitting spiral-shaped scrolls. Again, the refrigerant gas is compressed as one scroll rotates against the other fixed one. Gas is brought in at one end while the compressed gas is released at the other. Quiet and low-maintenance, they are also more efficient than reciprocating compressors.

Even smaller compressive-cycle machines are called *reciprocating chillers* (Fig. 10.28). Usually electrically driven, they are often combined with an air-cooled heat rejection process rather than a cooling tower. This makes them a closer relative of the smaller direct refrigerant machines discussed in Section 9.8.

Chilled water is usually supplied at between 40° and 48°F (4° and 9°C). When the chilled water is supplied cold and returns much warmer, the large rise in temperature reduces the initial size (cost) of equipment and increases its efficiency (thereby reducing the operating cost as well). Water treatment may be needed for chilled water to control corrosion or scaling.

Typical cooling capacities and space requirements of chillers are shown in Fig. 10.29. Each refrigeration machine in this illustration requires two pumps—one for the chilled water (to cool the building) and one for condenser water (to deal with reject heat). Typically, space is provided for future chiller additions, which may be required by building expansion and/or by higher internal gains from as-yet-uninstalled equipment, such as computer terminals within offices. Improved-efficiency chillers may replace older ones when energy costs and environmental regulations become compelling. Adequate clearance access to the equipment room is a major design issue.

### (c) Condensing Water Equipment

With chillers, there must be a way to reject the heat that is removed from the recirculating chilled water system. Reject heat is handled by the condensing water system, which serves the condensing process within refrigeration cycles. For larger buildings, the

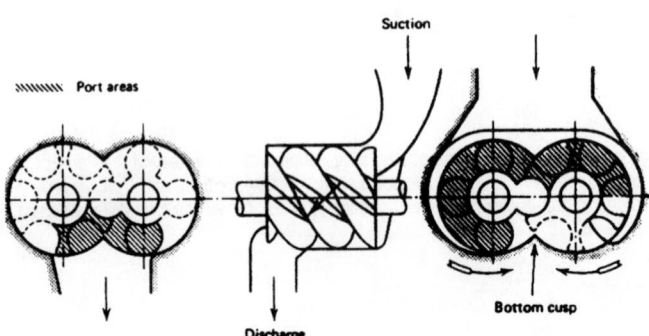

**Fig. 10.26** *A screw, or helical, compressor is a quieter, smaller machine with little vibration. (from Bobenhausen,* Simplified Design of HVAC Systems, *© 1994 by John Wiley & Sons, Inc. Reprinted with permission of John Wiley & Sons, Inc.)*

**Fig. 10.27** *Scroll compressor rotates one scroll form against another, with a quiet and efficient compression of the refrigerant. (From Boben-hausen,* Simplified Design of HVAC Systems, *© 1994 by John Wiley & Sons, Inc. Reprinted with permission of John Wiley & Sons, Inc.)*

condensing water requirement is most likely to be met by a *cooling tower.*

The cooling tower's place within the overall equipment layout was shown in Fig. 10.15*b;* a more detailed guide to sizes and types is given in Figs. 10.30 and 10.31. The object is to maximize the surface area contact between outdoor air and the heat condensing water. In crossflow towers, fans move air horizontally through water droplets and wet layers of fill (or packing), whereas in counterflow towers (prevalent in larger buildings), fans move the air up as the water moves down.

Cooling towers create a special—and usually unpleasant—microclimate. They demand huge quantities of outdoor air (approximately 300 cfm [142 L/s]), which they make considerably more humid. In cold weather, they can produce fog. They are typically very noisy—a natural consequence of forced-air motion. The condensing water flows are about 2.8 gpm (0.18 L/s) per ton of compressive refrigeration and about 3.5 gpm (0.22 L/s) per ton of absorption refrigeration.

The water that escapes as vapor from the tower is between 1.6 and 2 gph (1.7 and 2.1 mL/s). This water must be replaced, which is done automatically. The steady evaporation and exposure to the outdoors under hot and humid conditions spells trouble for the condensing water: controls for scaling, corrosion, and bacterial and algae growth are especially important. Ozone treatment systems have the advantage of reliable biological control and leave no chemical residue. Since the discovery of the link between Legionnaire's disease and cool-

ing towers, biological control has assumed greater importance.

The vapor that escapes the cooling tower should be kept from the vicinity of fresh air intakes, and from neighboring buildings or parked cars, where feasible. The floor space requirements can be approximated from Table 10.3, or use the average of 1/500 of the building gross floor area (for towers up to 8 ft [2.4 m] high) or 1/400 of the building gross floor area (for higher towers).

Although it is tempting to try to block the noise of cooling towers with solid barriers, it is critical that noise control not interfere with air circulation. The manufacturer's recommended clearances to solid objects near cooling towers must be consulted

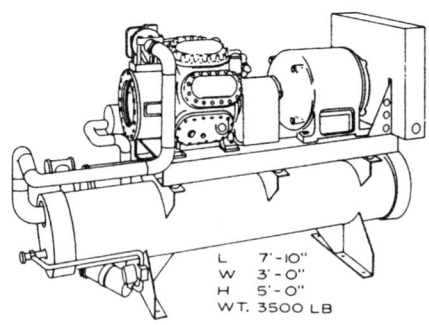

**Fig. 10.28** *A reciprocating chiller—a small-capacity machine that uses the compressive refrigeration cycle. Typically, this type of chiller produces less than 200 tons of cooling. (From AIA: Ramsey/Sleeper,* Architectural Graphic Standards, *10th ed., © 2000 by John Wiley & Sons, Inc. Reprinted with permission of John Wiley & Sons, Inc.)*

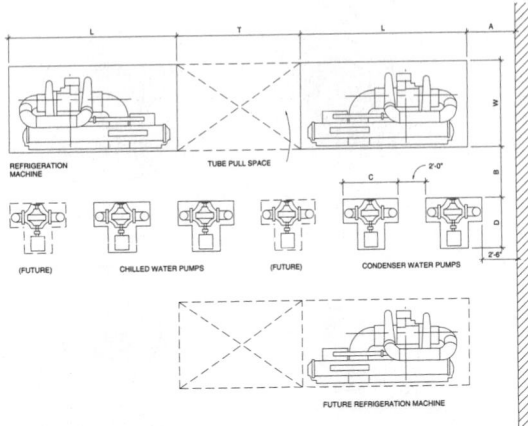

## General

The capacity of each refrigeration machine is equal to 50% of the peak cooling load. Each water pump provides the flow requirement of one refrigeration machine. Therefore, one pair of condenser and chilled water pumps is needed for each machine.

The cooling tower may be located on the roof of the refrigeration equipment room or on the ground adjacent to the equipment room. When located on ground, the condenser water outlet(s) on the cooling tower must be not less than 5 ft above the equipment room floor elevation for proper functioning of condenser water pumps.

See ASHRAE Standard 15–1992, "Safety Code for Mechanical Refrigeration," for required ventilation of chiller plant and monitoring of toxic refrigerants.

## Expansion of Equipment

For operational flexibility of a refrigeration plant, the size of the future refrigeration machine is generally planned to be the same as of the present machines. It may be economically advantageous to oversize some portions of the chilled and condenser waterpipes to handle the future flow rates.

Provision must also be made for expansion of the cooling tower capacity when the future refrigeration machine is installed.

## REFRIGERATION ROOM LAYOUT

### REFRIGERATION EQUIPMENT ROOM SPACE REQUIREMENTS

| EQUIPMENT (TONS) | DIMENSIONS | | | | | | | | MINIMUM ROOM HEIGHT |
|---|---|---|---|---|---|---|---|---|---|
| | L | W | HEIGHT | T | A | B | C | D | |
| **RECIPROCATING MACHINES** | | | | | | | | | |
| Up to 50 | 10'-0'' | 3'-0'' | 6'-0'' | 8'-6'' | 3'-6'' | 3'-6'' | 4'-0'' | 3'-0'' | 11'-0'' |
| 50 to 100 | 12'-0'' | 3'-0'' | 6'-0'' | 9'-0'' | 3'-6'' | 3'-6'' | 4'-0'' | 3'-6'' | 11'-0'' |
| **CENTRIFUGAL MACHINES** | | | | | | | | | |
| 120 to 225 | 17'-0'' | 6'-0'' | 7'-0'' | 16'-6'' | 3'-6'' | 3'-6'' | 4'-6'' | 4'-0'' | 11'-6'' |
| 225 to 350 | 17'-0'' | 6'-6'' | 7'-6'' | 17'-6'' | 3'-6'' | 3'-6'' | 5'-0'' | 5'-0'' | 11'-6'' |
| 350 to 550 | 17'-0'' | 8'-0'' | 8'-0'' | 16'-6'' | 3'-6'' | 3'-6'' | 6'-0'' | 5'-6'' | 12'-0'' |
| 550 to 750 | 17'-6'' | 9'-0'' | 10'-6'' | 17'-0'' | 3'-6'' | 3'-8'' | 6'-0'' | 5'-6'' | 14'-0'' |
| 750 to 1500 | 21'-0'' | 15'-0'' | 11'-0'' | 20'-0'' | 3'-6'' | 3'-6'' | 7'-6'' | 6'-0'' | 15'-0'' |
| **STEAM ABSORPTION MACHINES** | | | | | | | | | |
| Up to 200 | 18'-6'' | 9'-6'' | 12'-0'' | 18'-0'' | 3'-6'' | 3'-6'' | 4'-6'' | 4'-0'' | 15'-0'' |
| 200 to 450 | 21'-6'' | 9'-6'' | 12'-0'' | 21'-0'' | 3'-6'' | 3'-6'' | 5'-0'' | 5'-0'' | 15'-0'' |
| 450 to 550 | 23'-6'' | 9'-6'' | 12'-0'' | 23'-0'' | 3'-6'' | 3'-6'' | 6'-0'' | 5'-6'' | 15'-0'' |
| 550 to 750 | 26'-0'' | 10'-6'' | 13'-0'' | 25'-6'' | 3'-6'' | 3'-6'' | 6'-0'' | 5'-6'' | 16'-0'' |
| 750 to 1000 | 30'-0'' | 11'-0'' | 14'-0'' | 29'-6'' | 3'-6'' | 3'-6'' | 7'-0'' | 6'-0'' | 17'-6'' |

Note: Direct-fired absorption machines are roughly the same size as steam absorption machines.

**Fig. 10.29** Chiller room space requirements. Each refrigeration machine is served by two pumps (chilled water and condenser water). (From AIA: Ramsey/Sleeper, Architectural Graphic Standards, 10th ed., © 2000 by John Wiley & Sons, Inc. Reprinted with permission of John Wiley & Sons, Inc.)

before a tower is enclosed in any way. The roof is thus a favorite location for cooling towers, where wind can disperse the vapor and the noise and odor are remote from the street. However, the cooling tower can sometimes be featured; near downtown Denver, the cooling tower for the large performing arts complex sits in a forlorn stretch of grass bordered by arterial streets and away from pedestrians (Fig. 10.32). Its plume adds visual interest as it twists ghostlike above the equipment.

When fouling of the condensing water system cannot be tolerated, an alternative approach, called the *closed-circuit evaporative cooler*, is taken. Its schematic operation is described in Fig. 10.33. Usually used to cool the refrigerant directly, it can also be used for the condenser water, as well as on water loop heat pump systems (see Fig. 9.45). Either refrig-

erant or condenser water is protected within an always-closed loop, while a separate body of water is recirculated through the cooler, with steady evaporation and attendant problems. It requires much less makeup water than the cooling towers.

## (d) Energy Conservation Equipment

One big advantage of central equipment rooms is the opportunity they present for energy conservation. Regular maintenance is simplified when all the equipment lives in a generous space kept at optimum conditions; with regular maintenance comes increased efficiency of operation. Another conservation opportunity is that of heat transfer between various machines, or between distribution trees, where one's waste meets another's need.

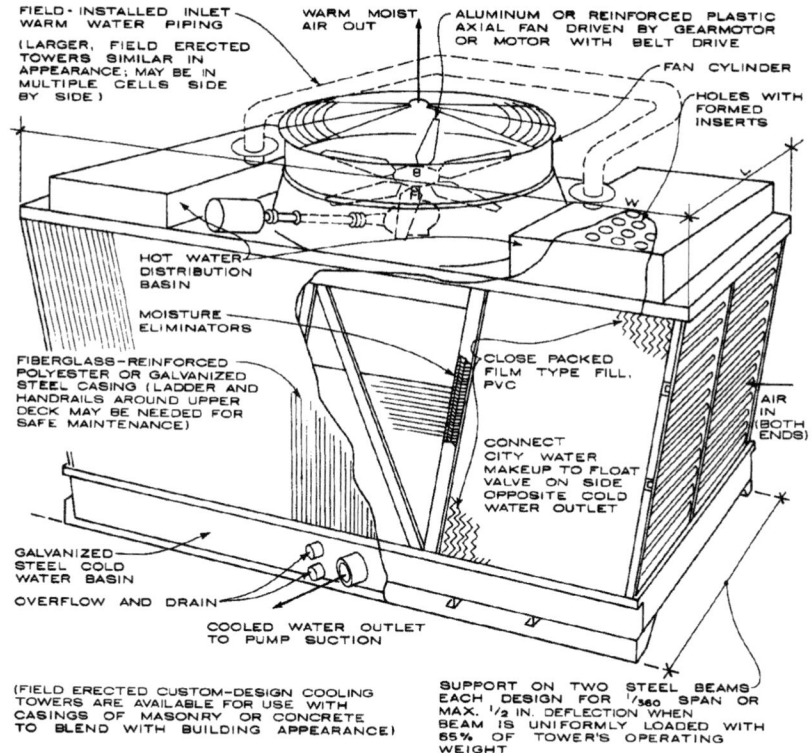

FIELD-INSTALLED INLET
WARM WATER PIPING

(LARGER, FIELD ERECTED
TOWERS SIMILAR IN
APPEARANCE; MAY BE IN
MULTIPLE CELLS SIDE
BY SIDE)

WARM MOIST
AIR OUT

ALUMINUM OR REINFORCED PLASTIC
AXIAL FAN DRIVEN BY GEARMOTOR
OR MOTOR WITH BELT DRIVE

FAN CYLINDER

HOLES WITH
FORMED
INSERTS

HOT WATER
DISTRIBUTION
BASIN

MOISTURE
ELIMINATORS

FIBERGLASS-REINFORCED
POLYESTER OR GALVANIZED
STEEL CASING (LADDER AND
HANDRAILS AROUND UPPER
DECK MAY BE NEEDED FOR
SAFE MAINTENANCE)

CLOSE PACKED
FILM TYPE FILL,
PVC

CONNECT
CITY WATER
MAKEUP TO FLOAT
VALVE ON SIDE
OPPOSITE COLD
WATER OUTLET

AIR
IN
(BOTH
ENDS)

GALVANIZED
STEEL COLD
WATER BASIN

OVERFLOW AND DRAIN

COOLED WATER OUTLET
TO PUMP SUCTION

(FIELD ERECTED CUSTOM-DESIGN COOLING
TOWERS ARE AVAILABLE FOR USE WITH
CASINGS OF MASONRY OR CONCRETE
TO BLEND WITH BUILDING APPEARANCE)

SUPPORT ON TWO STEEL BEAMS
EACH DESIGN FOR $^1/_{360}$ SPAN OR
MAX. $^1/_2$ IN. DEFLECTION WHEN
BEAM IS UNIFORMLY LOADED WITH
65% OF TOWER'S OPERATING
WEIGHT

(a)

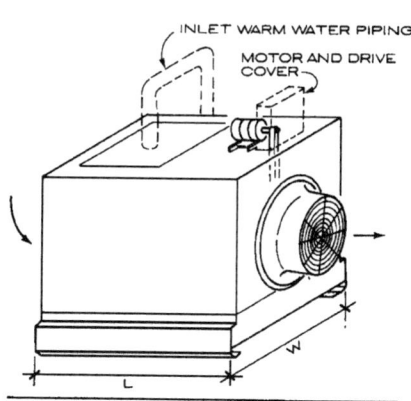

INLET WARM WATER PIPING

MOTOR AND DRIVE
COVER

NOTE
AVAILABLE
IN SINGLE
MODULES AS
SKETCHED, OR
END-TO-END OR
BACK-TO-BACK
DOUBLE INLET

WARM
WATER
INLET
TO SPRAY
NOZZLES

COOLED
WATER
OUTLET

AIR IN

| TONS 3 GPM/TON 95-85-78 | OVERALL DIMENSIONS (IN.) | | | OPERATING WEIGHT (LB.) | MOTOR (HP) |
|---|---|---|---|---|---|
| | L | W | HT. | | |
| 5 | 69 | 33 | 60 | 940 | ¼ |
| 25 | 75 | 46 | 80 | 1600 | 1 |
| 50 | 84 | 64 | 92 | 2500 | 3 |
| 100 | 93 | 100 | 92 | 4200 | 5 |
| 150 | 100 | 144 | 112 | 8000 | 7½ |

| TONS 3 GPM/TON 95-85-78 | OVERALL DIMENSIONS (IN.) | | | OPERATING WEIGHT (LB.) | MOTOR (HP) |
|---|---|---|---|---|---|
| | L | W | HT. | | |
| 20 | 36 | 36 | 78 | 950 | 2 |
| 50 | 72 | 36 | 96 | 1700 | 7½ |
| 150 | 144 | 56 | 122 | 4800 | 20 |
| 400 | 140 | 118 | 192 | 14,000 | 50 |

(b)

(c)

**Fig. 10.30** *Cooling towers that serve the condensing water system for large buildings. (a) Cutaway view of a large-capacity (200 to 700 tons) crossflow induced-draft package cooling tower. (b) Size ranges for crossflow induced-draft package cooling towers. (c) Size ranges for counterflow induced-draft package cooling towers. Note: "3 gpm/ton 95-85-78" refers to the cooling capacity in tons, with condensing water flow at 3 gpm per ton, the condensing water entering the tower at 95°F and leaving at 85°F, with outside air at no more than 78°F WB temperature. (From AIA: Ramsey/Sleeper,* Architectural Graphic Standards, *10th ed., © 2000 by John Wiley & Sons, Inc. Reprinted with permission of John Wiley & Sons, Inc.)*

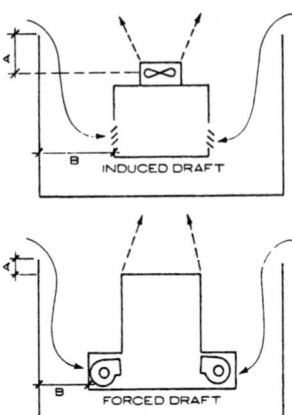

**Fig. 10.31** *The more wall clearance, the better the operation. A = maximum height of enclosure above the tower outlet; minimize this dimension. B = as large as possible, especially if walls have no air openings. (Consult the manufacturer for minimum dimensions.) (From AIA: Ramsey/Sleeper, Architectural Graphic Standards, 10th ed., © 2000 by John Wiley & Sons, Inc. Reprinted with permission of John Wiley & Sons, Inc.)*

*Boiler flue economizers* achieve heat transfer by passing the hot gases in a boiler's stack through a heat exchanger, thus preheating the incoming boiler water (Fig. 10.34).

*Runaround coils* (Fig. 10.35) can be used for heat transfer between intake and exhaust air ducts when these two airstreams are rather far apart. This circulating heat-transfer fluid usually contains antifreeze; it provides simple sensible heat transfer, with no restrictions on exhaust and intake location. No contamination of intake air by exhaust air is caused by this arrangement. The efficiency of such coils runs between 50% and 70%, and they are available in modular sizes up to 20,000 cfm (9440 L/s).

Heat exchange between incoming and exhaust airstreams was discussed in Sections 5.5c and 5.5d, where heat pipes (Fig. 5.19) and thermal transfer wheels (Fig. 5.20) can play major roles in energy conservation. The thermal transfer wheels can add to the central equipment space requirements. Other desiccant systems were discussed in Section 5.5d.

*Economizer cycles* (Fig. 10.36) use cool outdoor air, as available, to ease the burden on a refrigeration cycle as it cools the recirculated indoor air. The economizer cycle can thus be thought of as a central mechanical substitute for the open window; when it is cool enough (below the supply air temperature), 100% outside air can be provided and no chilled water is needed. When the outdoor air temperature is higher than the supply air temperature but lower than the return air temperature, 100% outdoor air is still brought in but chilled water is used to lower its temperature. Above the return air temperature, outdoor air is reduced to that volume required for IAQ.

Relative to open windows, this cycle has several advantages: energy-optimizing automatic thermal control, filtering of the fresh air, tempering of the cool outdoor air to avoid unpleasant drafts, and an orderly diffusion of fresh air throughout the building. Its disadvantages are the loss of personal control that windows offer and thus loss of awareness of exterior–interior interaction. In hot, humid climates, the moisture brought by 100% outdoor air may be unwelcome.

Economizer cycles are available as options on most direct refrigerant machines (such as single-package rooftop units) and are typically installed for large-building central air supply systems. Buildings with high internal gains (internal load dominated) are particularly good targets for economizer cycles because they need cooling even when the outside temperature is chilly. Economizer cycles lend themselves readily to a cooling strategy of night ventilation of thermally massive structures because they have a built-in option for 100% outdoor air.

### (e) GeoExchange Systems

Using the earth as a heat source and sink for small buildings was explored in Section 9.8d. Four typical applications were shown in Fig. 9.44. Larger buildings can also utilize such systems. A late 1990s building just east of Central Park, in New York City, utilizes two wells 1500 ft (457 m) deep; all but the top 50 ft (15 m) are lined by bedrock. Heat is taken from (or discharged to) water, which is pumped from one well and discharged to the other—a "groundwater source" system. The average year-round temperature in these deep wells (in an intensely urban area) is estimated at 56°F (13°C), about the temperature at which chilled air is delivered to a space in summer. Going to such depths is perhaps the only geothermal option in densely built-up areas.

The Cambridge (Massachusetts) Cohousing Project (Fig. 10.37) is also located in a densely settled urban neighborhood. The 41 living units feature passive solar heating, underground parking

***Fig. 10.32*** *A plume of mist hovers ghost-like above a cooling tower in full public view near the Denver performing arts complex.*

(this reserves some 20,000 ft² [1860 m²] of the surface for open green space), and centralized heating/cooling utilizing ground-source heat pumps with locally controlled thermal zones. In this urban setting, the relative quiet of the indoor heat pump's compressors and the lack of hot air (summer) or cold air (winter) noisy discharges are welcome amenities.

In England, the Hyndburn Borough Council decided that their new headquarters building should set an example as a "zero energy" building: that is, over a typical year, it should generate as much energy as it imports. The 38,750-ft² (3600-m²) building (Fig. 10.38) is elongated east–west and boasts an ambitious section that combines daylighting, PVs, a well-insulated skin, and even rainwater collection to use for flushing toilets. Windpower adds to electricity generation. Summer cooling is by

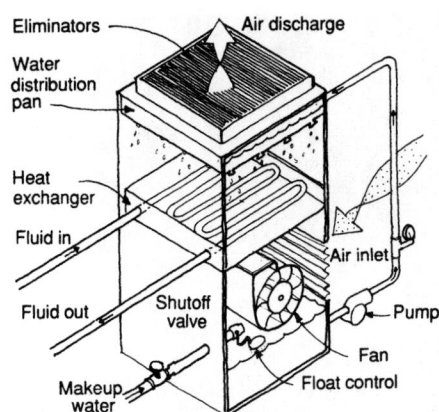

**Fig. 10.33** Closed-circuit evaporative coolers, which cool the condensing water system while protecting it from contact with outside air. A self-contained water system is circulated through the evaporative cooler; steady evaporation losses are replaced by makeup water. (Based upon AIA: Ramsey/Sleeper, Architectural Graphic Standards, 8th ed., © 1988 by John Wiley & Sons.)

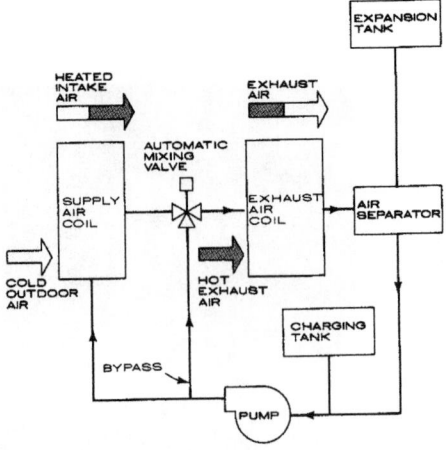

NOTE

Indirect sensible heat transfer between remote air streams with no cross-contamination. Exhaust airstream coil construction to suit application. Antifreeze fluid for low air temperatures. Bypass valve temperature control. Computerized equipment selection. Efficiency 50–70%. Modular coils to 20,000 cfm.

**Fig. 10.35** Runaround coils for heat transfer between fresh intake air and stale exhaust air, used where the airstreams are in separate locations. (From AIA: Ramsey/Sleeper, Architectural Graphic Standards, 10th ed., © 2000 by John Wiley & Sons, Inc. Reprinted with permission of John Wiley & Sons, Inc.)

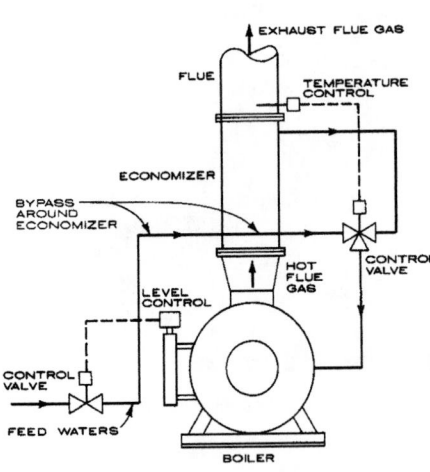

NOTE

Direct flue gas to feed water heat transfer for high pressure steam boilers. Boiler flue gas at 500°F leaving economizer at 325°F, heats feed water from 200 to 248°F. Mixing valve maintains minimum stack temperature leaving economizer to prevent moisture condensation in stack.

**Fig. 10.34** Heat recovery for boilers. Flue gas entering at 500°F (260°C) leaves the "economizer" at 325°F (163°C), a temperature still high enough to prevent condensation in the stack. The heat recovered here is added to incoming boiler water, raising its temperature from 200 to 248°F (93 to 120°C). (From AIA: Ramsey/Sleeper, Architectural Graphic Standards, 10th ed., © 2000 by John Wiley & Sons, Inc. Reprinted with permission of John Wiley & Sons, Inc.)

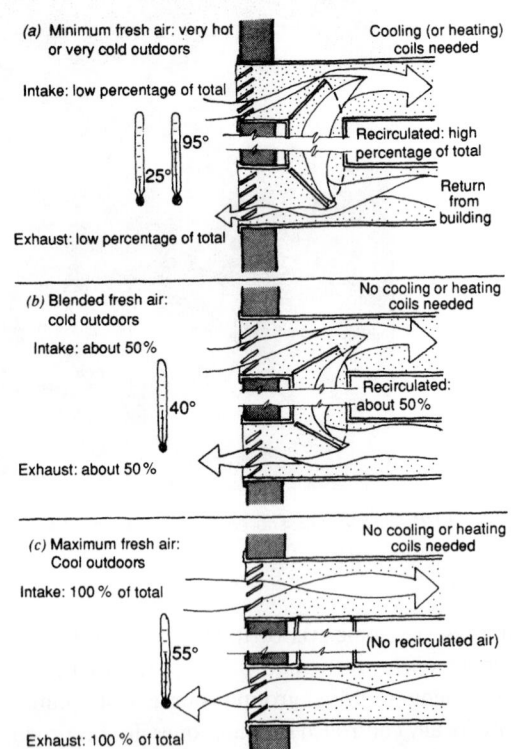

**Fig. 10.36** The economizer cycle controls the relationships between fresh, exhaust, and recirculated air. (a) When outside air is hot (or very cold), the economizer cycle is inactive, and minimum fresh air is introduced. (b) As very cold outside air gets warmer, it can be blended with recirculated air, and neither heating nor cooling coils are needed. (c) When outside air is cool, it can completely replace circulated air, making mechanical cooling inactive.

THERMAL CONTROL

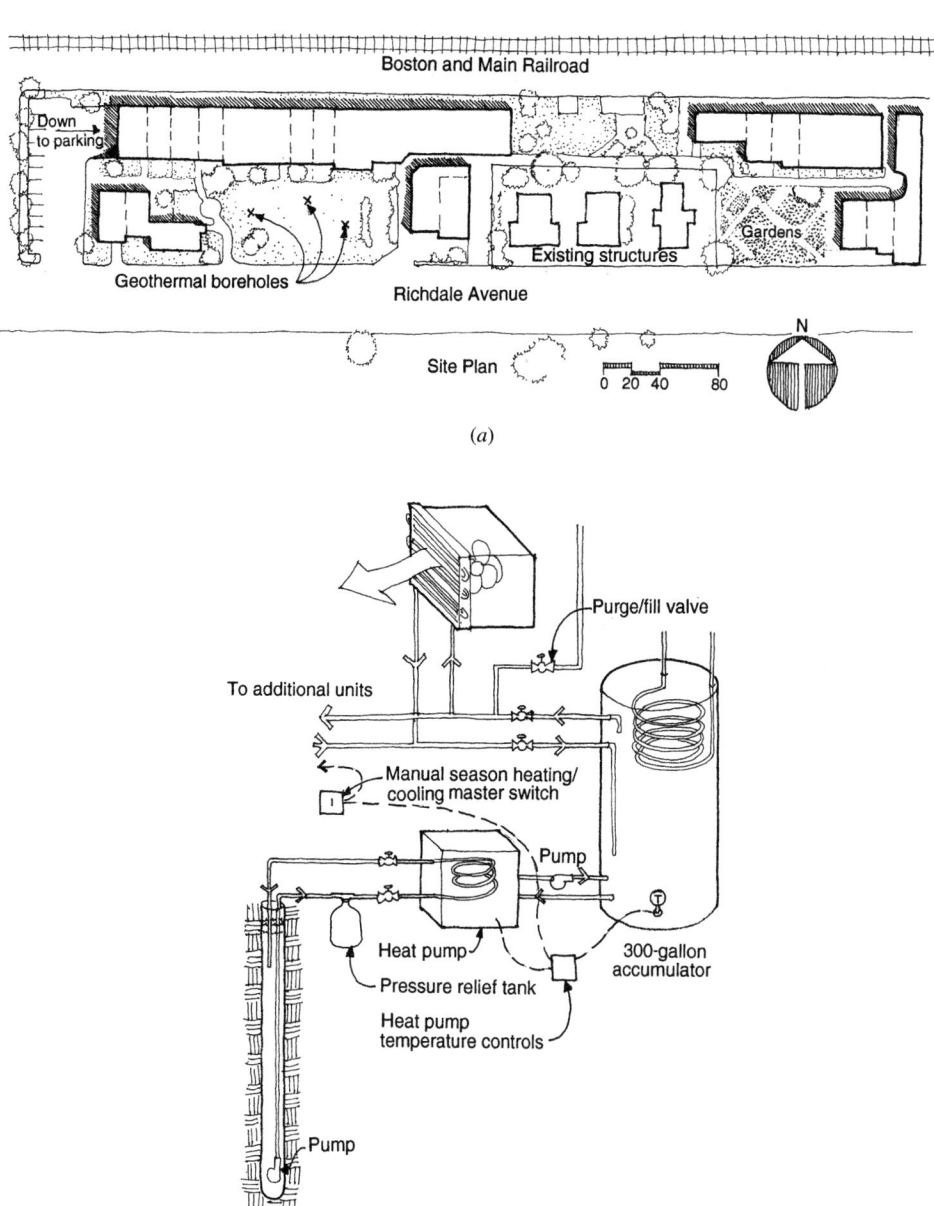

*(a)*

*(b)*

**Fig. 10.37** *Cambridge (Massachusetts) Cohousing development takes advantage of three boreholes on its urban site (a) to provide central heating and cooling, serving individual residential fan-coil units. It also preheats water (b) for the DHW (domestic hot water) system. The residences are sited to provide winter solar access, and a positive fresh air intake is located on the side sheltered from adjacent railroad tracks. Underground parking preserves open space for gardens and recreation. (Courtesy of Building Science Engineering, Harvard, MA.)*

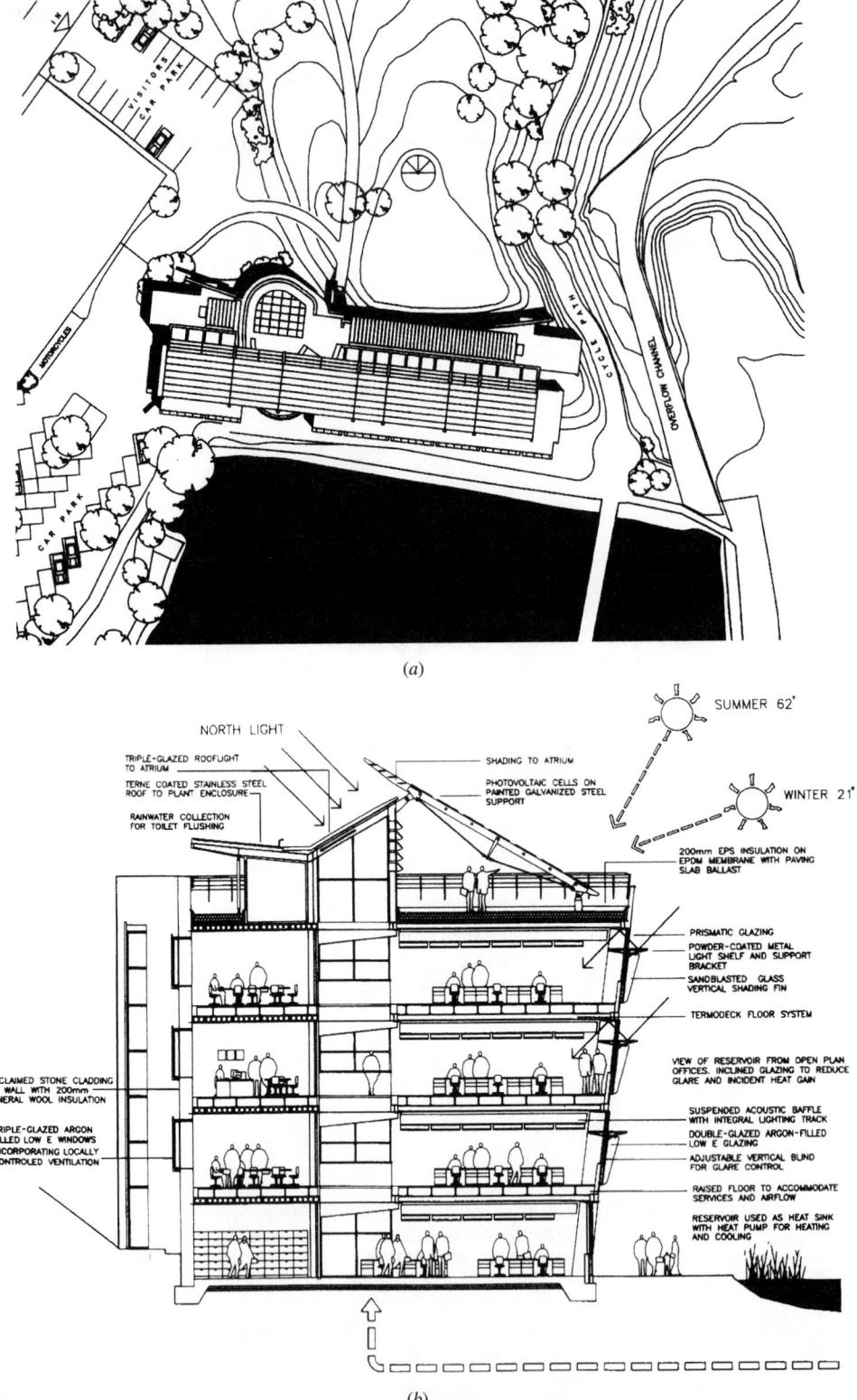

*(a)*

*(b)*

**Fig. 10.38** *The Hyndburn Borough Council (England) headquarters faces south toward a reservoir (a) that provides an evaporatively cooled microclimate and also acts as a heat source/sink for a water–water heat pump. (b) As suggested in the diagrammatic section, daylighting, passive solar heating, PVs, and a well-insulated shell are featured. Termodeck is a precast hollow-core slab that stores coolth on summer nights and precools ventilation air. The supply air then rises from the plenum created by raised floor. (Courtesy of Jestico + Whiles + Associates, architects, London.)*

ventilation (assisted by night ventilation of mass) through a raised "Termodeck" system. This is a pre-fabricated hollow-core slab through which cool night air circulates, storing coolth to assist the next day's hottest hours. (This is a concept explored by the Oregon office building in Fig. 8.7.) The raised floor then provides a distribution plenum, with a design rate of 4 ACH.

For more extreme future summer cooling, but now mostly for winter heating (beyond that provided by passive solar heating), an adjacent reservoir acts as a thermal sink. A "lake closed source heat pump" will serve the mechanical ventilation system. The lake is south of the building, providing some local evaporative cooling in summer and inviting a very climate-oriented south façade, behind which are open-office areas. The north façade faces the city and is more traditionally institutional in character, as are the individual offices behind it.

## (f) Energy Storage

We commonly experience daily changes from warmer to colder conditions, both in winter and in a hot, dry summer. Central storage equipment for large buildings can take advantage of this cycle to increase operating efficiency, save energy, and significantly reduce electricity demand charges. Some electric utilities offer incentives to install thermal storage in order to reduce the peak strain on their generating facilities.

*Water storage tanks* are one common approach to storage, such as those shown in Fig. 10.39. On typical winter days, the total internal heat generated by a large building can be somewhat greater than its total need for heating at the perimeter zones. Instead of being thrown away as exhaust air, this surplus heat is captured and stored in large water tanks, from which it can be withdrawn and used on cold winter nights and weekends. In the summer, chillers can work at night, when efficiency is high because cool outdoor air helps the refrigeration cycle reject its heat. By storing the coolth produced, less work need be done by chillers during the next day's peak, when electric rates are highest and operating efficiency is lowest.

Design guidelines for water storage are 0.5 to 1 gal/ft² (20 to 40 L/m²) of conditioned space. Such large, heavy tanks are frequently located in basements and underground parking facilities.

Water tanks require additional floor space for equipment, although they may allow somewhat smaller chillers to be installed. They may also contribute to lower fire insurance premiums because considerable water is stored and ready to use.

*Ice storage tanks* are another storage approach, as shown in Fig. 10.40. Because they take advantage of the latent heat of fusion (143.5 Btu/lb of water at 32°F [334 kJ/kg at 0°C)]), such units can store far more energy in a given-size tank than can water. Design guidelines for ice storage are 0.13 to 0.25 gal/ft² (5 to 10 L/m²) of conditioned space. A comparison of the relative sizes of storage tanks needed for an 18-story office building is shown in Table 10.6. Also, with ice storage there is less undesired mixing of hot and cold water within the tank.

Ice can be made in several ways. A common method is to form ice as a layer around a pipe that carries refrigerant in a closed circuit through the tank. Control of the thickness of this layer is important, because if too little ice is made at night, too little cooling will be available the following day. In another method, ice is formed and thawed by circulating a brine through coils in cylindrical water tanks. Other methods form ice on plates, then harvest it in an insulated bin.

The Iowa Public Service Building (Fig. 10.41) uses ice storage in six tanks that occupy about half of the floor space in the ground floor mechanical room. This installation serves a five-story, 167,000-ft² (15,515-m²) utility office building that utilizes solar collectors supplemented by small backup boilers. The six ice-making machines serve a 75,000-gal (283,900-L) ice storage pit with 90 million Btu (26,355 kWh) capacity. Winter heat exchange opportunities include reject heat from the ice-making machines, heat from the central toilet exhaust air, and heat from return air taken through luminaires. Solar collectors on the roof preheat the ventilation air, which is fed into the ceiling plenum at each floor. Fan-coil units then draw from this fresh air supply.

Water and ice storage tanks are sometimes located at the top of a building, especially when other related mechanical equipment is there also. Despite the weight of such tanks, they can become a structural advantage. The *tuned mass damper* method of reducing lateral vibration (or sway) in high-rise buildings utilizes a heavy, moving mass at the top; when the building begins to sway, the mass is moved

THERMAL CONTROL

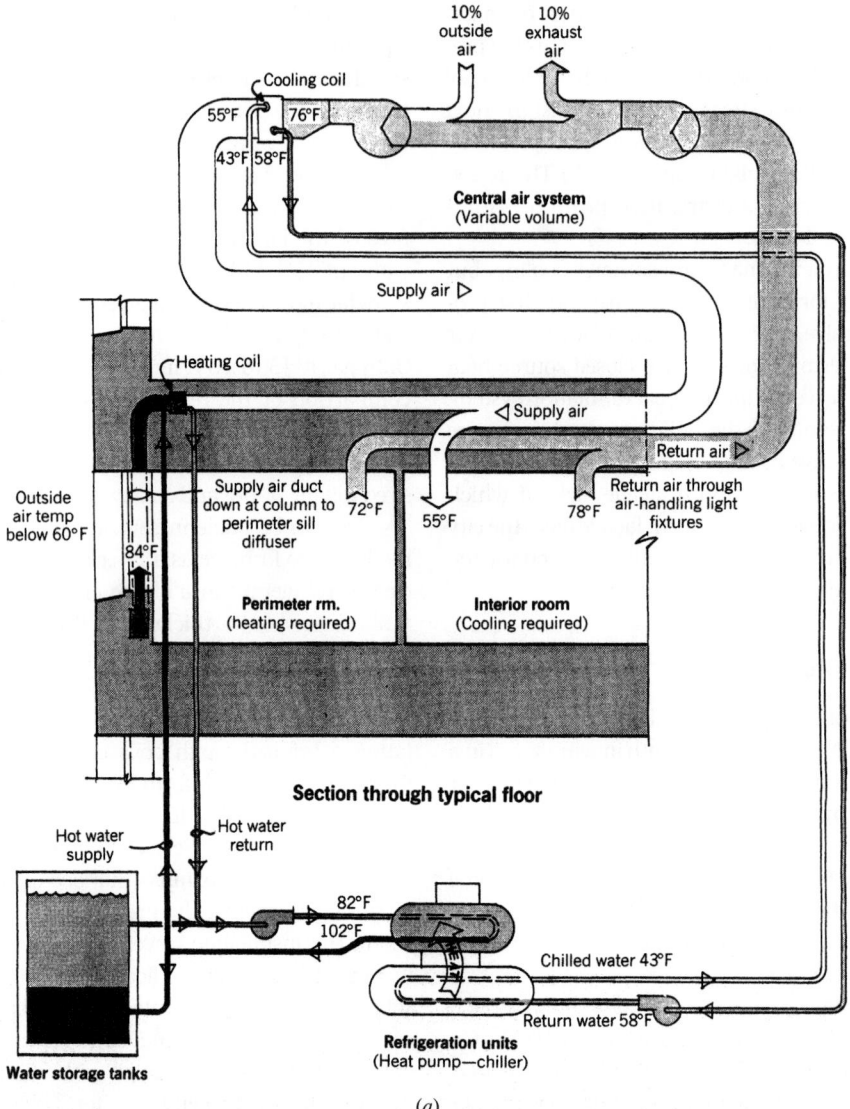

*(a)*

**Fig. 10.39** *Water tank heat storage. The 870,000-ft² (80,825-m²) transportation office building (Park Plaza, Boston), population 2000, uses a three-compartment insulated concrete tank storing 750,000 gal (2,838,990 L) of water. (a) At an outside air temperature of 40° to 50°F (4.4 to 10°C), the surplus internal heat is stored in the tanks rather than rejected as exhaust air. (b) By the time the outside air is about 50°F (10°C), the tanks are fully charged; up to an outside temperature of 60°F (15.5°C), the economizer cycle provides energy-conserving cooling. (c) At about 60°F (15.5°C) chillers must operate, but working all night when the outside air is cooler enables them to work less by day. Their nightly production is stored as cold water, available to help with the following day's peak. Smaller machines and more efficient operation are the result. (Courtesy of Shooshanian Engineering Associates, Inc.)*

in the opposite direction. In the Crystal Tower in Osaka, Japan, this mass is provided by ice storage tanks. This 515-ft (157-m), 37-story building has an ice thermal storage total of 25,400 ft³ (720 m³) divided into nine tanks. Six of the ice storage tanks provide the movable structural mass, suspended from roof girders, weighing 540 tons (489,880 kg),

including steel framing. The chiller and condenser are also in the equipment penthouse.

## (g) Air Handling Equipment

A given HVAC system type may have many variations. In some variations, all the air is passed

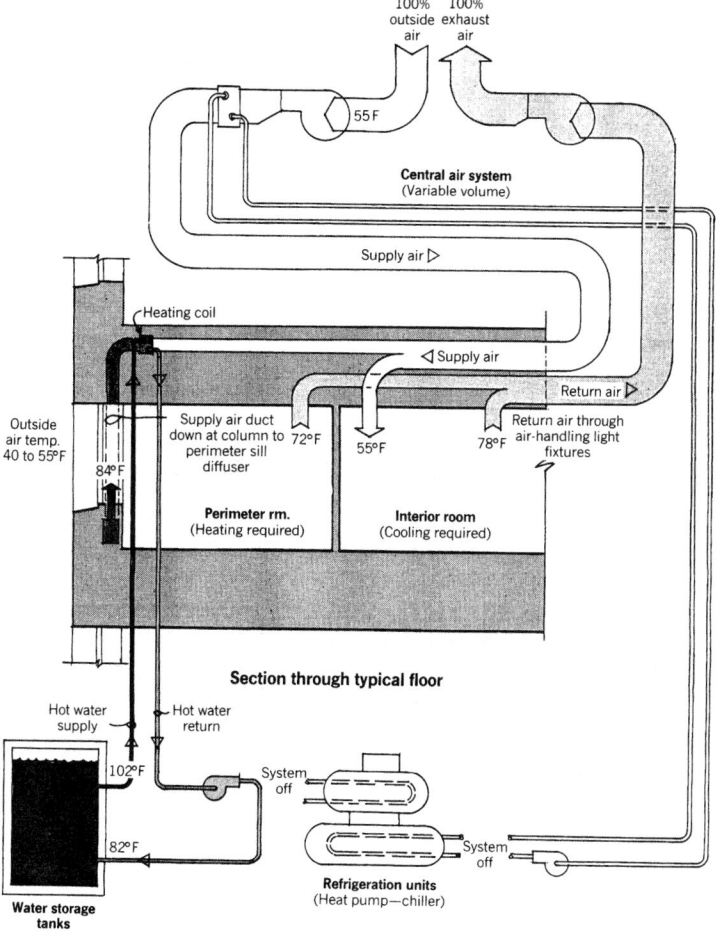

*(b)*

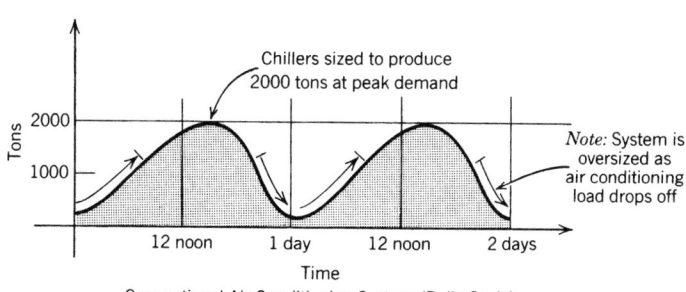

Conventional Air Conditioning System (Daily Cycle)

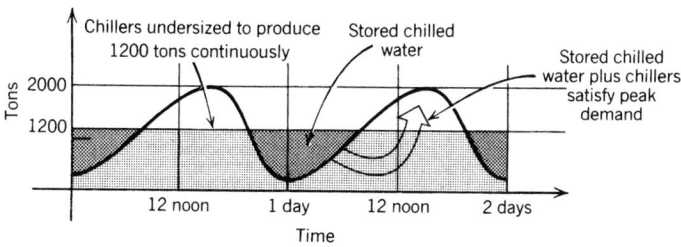

Transportation Building System (Daily Cycle)

*Note:* Charging of storage tanks with cold water during off hours provides added cooling media for peak air conditioning needs.

*(c)*

**Fig. 10.39** *(Continued)*

THERMAL CONTROL

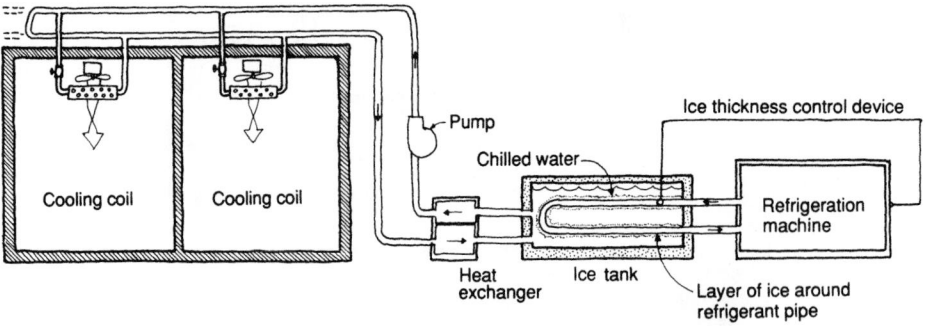

**Fig. 10.40** *Ice tank heat storage. Chilled water in an ice tank forms a layer of ice around the refrigerant pipe (supplied by the refrigeration machine). As chilled water at about 35°F (1.7°C) is removed from the tank, it enters a heat exchanger, so that its temperature is closer to 45°F (7.2°C) for distribution to cooling coils throughout the building.*

through one central equipment room. In others, air handling may be done in many separate and smaller rooms, whereas central heating and cooling require only one equipment room. Some detailed design guidelines for air handling equipment for either case are shown in Fig. 10.42. Air handling equipment room dimensions are shown in Table 10.7. Total air quantities (cfm) may be estimated from Table 10.4 or obtained more precisely from cooling load calculations.

### TABLE 10.6 Cooling Storage Comparison

| Eighteen-story office building: 375,000 gross ft² (34,840 m²), 800 tons peak load, 6250 ton-h/day at design, 75 million Btu (21,900 kWh) storage needed. | | |
|---|---|---|
| | **Water Storage** | **Ice Storage** |
| Btu/lb[a] (kJ/kg) | 15 (34.9) | 164 (381.5) |
| Pounds storage | 5,000,000 | 457,000 |
| (kilograms) | (2,267,960) | (207,290) |
| Gallons | 599,500 | 54,800 |
| (liters) | (2,269,300) | (207,435) |
| Storage efficiency | 0.90 | 1.0 |
| Percent ice | | 66% |
| Net gallons | 666,000 | 83,030 |
| (liters) | (252,100) | (314,295) |
| Cubic feet | 89,046 | 11,101 |
| (cubic meters) | (2,522) | (314) |
| Floor area of tank, approx. 8 ft deep | 80 by 150 ft | 30 by 50 ft |
| (2.5 m deep) | (24 by 46 m) | (9 by 15 m) |
| Storage ratio, water to ice | 8 : 1 | |

*Source:* Reprinted by permission from *Specifying Engineer,* January 1983. Metric conversions by the author.

[a]Btu/lb based on Δt = 15F° for water storage and on Δt = 20.5F° in the melted ice water (added to 143.5 Btu/lb at fusion).

Some common fan types include:

*Panel:* the most simple type on this list; for high-efficiency air delivery, the motor is generally mounted in the center of the propellers. It is likely the noisiest type. It is not designed for much static pressure from ductwork, filters, and so on.

*Fixed pitch vane axial:* capable of working against somewhat more static pressure than the panel fan; more common in industrial applications.

*Centrifugal:* with an airfoil bladed wheel, it has high efficiency over a wide operating range and is quieter than the previous two. Major changes in pressure result in only minor changes in the volume of air delivered.

*Vane axial adjustable pitch:* takes less space than centrifugal fans and can work against more static pressure. The pitch is adjustable occasionally, not continuously, for system balancing or seasonal changes in air volume.

*Vane axial controllable pitch:* the automatically controlled pitch responds to changes in temperature, humidity, airflow, and so on, depending on the sensors used. It is commonly used in VAV systems.

Filters and other methods of IAQ control were discussed in Chapter 5. There are numerous combinations of air handlers, filters, and coils for air heating/cooling, selected by the consulting engineer. For the architect, the primary issues include adequate space, adequate access (for present maintenance and future equipment replacement), and noise isolation.

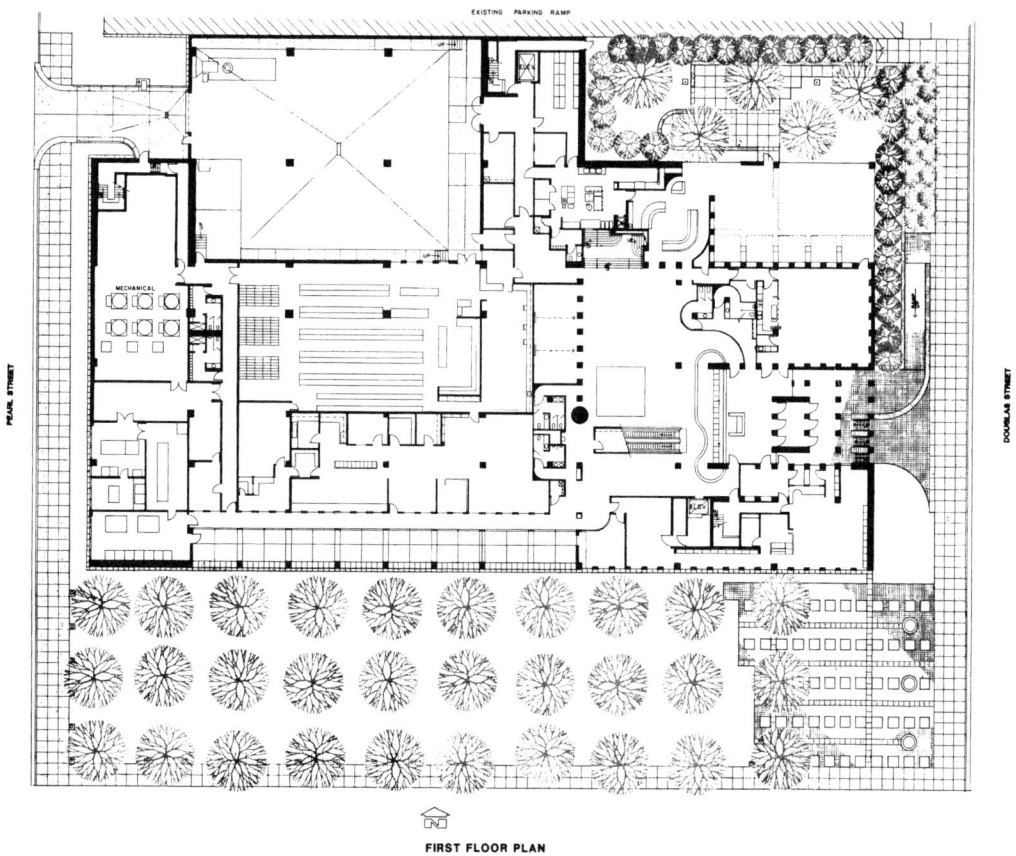

**Fig. 10.41** *The ground floor of the Iowa Public Service offices (Sioux City) contains six ice machines within the mechanical space adjacent to the loading dock. (Courtesy of Rossetti Associates and Foss, Engelstad, Heil Associates, joint venture architects.)*

## (h) Controls

Although the most obvious HVAC control function is to maintain desired comfort conditions, controls also increase fuel economy by promoting optimum operation and act as safety devices, limiting or overriding mechanical equipment. They also eliminate human error: controls fall asleep only during a power failure.

Although precise control of temperature and humidity everywhere in a building may be a tempting thought, controls can usually maintain only a *range* of conditions, not a setpoint. This range (within which neither heating nor cooling is called for) is called the *deadband*. Temperature variations occur vertically within a space—the higher the space, the greater the variation. Variations between horizontal positions within a zone are highly likely, especially where one or more walls are exterior and

where different rooms share a zone controlled by a single thermostat. Variations in time also occur: a building will always be warmer at the moment the heating system turns off than at the moment the system turns on.

Individual controls can be classified as follows: *controllers*, which measure, analyze, and initiate action; *actuators*, which are the controller's servants and in turn become the masters of pieces of equipment; *limit and safety controls*, which may function only infrequently, preventing damage to equipment or buildings; and *accessories*, a miscellaneous collection.

Systems of controls can be classified by power source: electric (both analog and direct digital control); pneumatic (in which compressed air is the motivating force); and self-contained, including "passive" controls such as those motivated by thermal expansion of liquids or metals. Another way to

THERMAL CONTROL

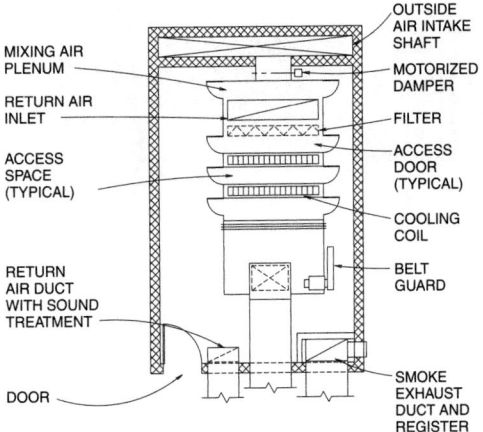

EQUIPMENT ROOM PLAN   *(a)*

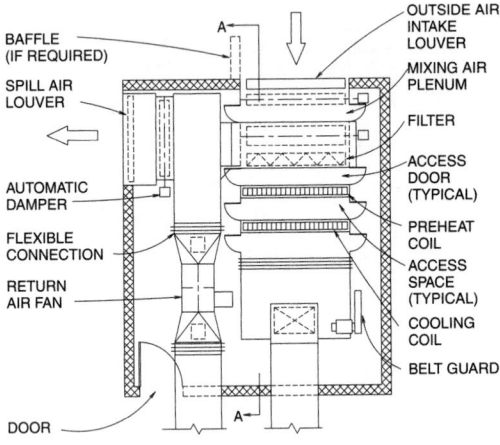

EQUIPMENT ROOM PLAN   *(b)*

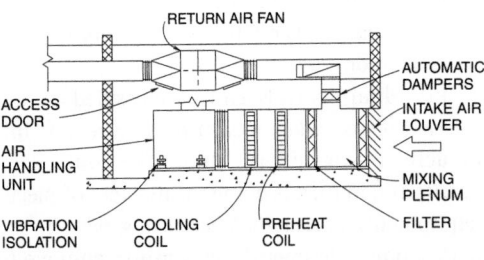

EQUIPMENT ROOM SECTION A-A   *(c)*

**Fig. 10.42** *Some variations on a huge variety of HVAC equipment rooms. (a) Plan with floor-mounted air-conditioning unit. Outdoor air is drawn from an adjacent shaft; return air ducts are above the equipment, and a remote exhaust fan expels what is not recirculated. (b) Plan on an exterior wall. Outdoor air is taken through the wall, and a return air fan is overhead. (c) Section through plan b. (From AIA: Ramsey/Sleeper,* Architectural Graphic Standards, *10th ed., © 2000 by John Wiley & Sons, Inc. Reprinted with permission of John Wiley & Sons, Inc.)*

classify control systems is by the motion of the controller equipment: *two-position* systems are of the simple on–off type; *multiposition* systems have several varieties of the on position, commonly used for separate operation of more than one machine; *floating* controls can assume any position in the range between minimum and maximum; *central logic control* systems can be programmed to integrate the many aspects of building control into one decision-making unit, and are now the prevalent approach to building management.

Control diagrams for two common HVAC applications are shown in Fig. 10.43. Single-duct VAV systems (Fig. 10.43*a*) are further described in Section 10.5; with the constantly varying flow rate, the fan must be regulated so as to maintain the minimum pressure (and therefore, flow) needed at the most demanding outlet. This outlet may be either the one most remote from the fan or the one needing the greatest flow (because it has the highest gain) at the moment. Economizer cycles (Fig. 10.43*b*) compare outside to inside positions, and vary the proportion of fresh (outdoor) air to return air, to provide "free" cooling (see again Fig. 10.36).

Most of today's large buildings are regulated by central logic control systems, usually called *building management systems* (BMS). The goal is a productive, cost-effective environment achieved by optimizing the interrelationships between the building's structure, systems, services, and management. The HVAC system is centrally regulated and interconnected with lighting, electric power (such as load shedding), elevators, service hot water, access control and security, telecommunications, and information management. Such systems not only maintain comfort with energy conservation, they also sound the alarm about malfunctions, can learn from past practice, and keep records of performance.

This integration and automation of the many building control systems is made possible through *direct digital control* (DDC), which can be applied to a wide variety of elements. They provide "dynamic control" in the case of HVAC anticipating time-based changes in heat flow patterns or in occupancy schedules. These actions depend on direct digital microcontrollers located on each piece of regulated equipment throughout the building. Three building types show applications of this comprehensive automated control opportunity.

**TABLE 10.7  Air Handling Equipment Rooms**

| PART A. I-P UNITS | | | | | | | |
|---|---|---|---|---|---|---|---|
| | Approximate Overall Dimension of Supply Air Equipment | | | | Recommended Room Dimensions | | |
| cfm Range | Width | × Height | × Length | Width | × Height | × Length | |
| 1000–1,800 | 4'–9" | 2'–9" | 14'–9" | 12'–6" | 9'–0" | 18'–9" | |
| 1,801–3,000 | 5'–0" | 3'–6" | 16'–0" | 13'–9" | 9'–0" | 20'–0" | |
| 3,001–4,000 | 6'–9" | 4'–6" | 16'–0" | 17'–6" | 9'–0" | 20'–0" | |
| 4,001–6,000 | 7'–6" | 4'–6" | 16'–9" | 18'–0" | 9'–0" | 20'–9" | |
| 6,001–7,000 | 7'–6" | 4'–9" | 18'–3" | 18'–6" | 9'–6" | 22'–3" | |
| 7,001–9,000 | 8'–0" | 5'–0" | 18'–9" | 19'–0" | 10'–0" | 22'–9" | |
| 9,001–12,000 | 10'–0" | 5'–6" | 21'–0" | 23'–0" | 11'–0" | 25'–0" | |
| 12,001–16,000 | 10'–3" | 6'–0" | 22'–0" | 13'–6" | 12'–6" | 26'–0" | |
| 16,001–19,000 | 10'–6" | 6'–6" | 23'–9" | 24'–0" | 13'–0" | 27'–9" | |
| 19,001–22,000 | 11'–9" | 7'–3" | 25'–0" | 26'–9" | 15'–0" | 29'–0" | |
| 22,001–27,000 | 11'–9" | 8'–6" | 26"–0" | 27'–0" | 16'–0" | 30'–0" | |
| 27,001–32,000 | 13'–0" | 9'–9" | 27'–9" | 29'–0" | 18'–0" | 31'–9" | |

| PART B. SI UNITS | | | | | | | |
|---|---|---|---|---|---|---|---|
| | Approximate Overall Dimension of Supply Air Equipment (m) | | | | Recommended Room Dimensions (m) | | |
| L/s Range | Width | × Height | × Length | Width | × Height | × Length | |
| 470–849 | 1.4 | 0.8 | 4.5 | 3.8 | 2.7 | 5.7 | |
| 850–1,415 | 1.5 | 1.1 | 4.9 | 4.2 | 2.7 | 6.1 | |
| 1,416–1,887 | 2.1 | 1.3 | 4.9 | 5.3 | 2.7 | 6.1 | |
| 1,888–2,831 | 2.3 | 1.3 | 5.1 | 5.5 | 2.7 | 6.3 | |
| 2,832–3,303 | 2.4 | 1.4 | 5.6 | 5.6 | 2.9 | 6.8 | |
| 3,304–4,247 | 2.4 | 1.5 | 5.7 | 5.8 | 3.0 | 6.9 | |
| 4,248–5,662 | 3.0 | 1.7 | 6.4 | 7.0 | 3.4 | 7.6 | |
| 5,663–7,550 | 3.1 | 1.8 | 6.7 | 7.2 | 3.8 | 7.9 | |
| 7,551–8,966 | 3.2 | 2.0 | 7.2 | 7.3 | 4.0 | 8.5 | |
| 8,967–10,381 | 3.6 | 2.2 | 7.6 | 8.1 | 4.6 | 8.8 | |
| 10,381–12,741 | 3.6 | 2.6 | 7.9 | 8.2 | 4.9 | 9.1 | |
| 12,742–15,100 | 4.0 | 3.0 | 8.5 | 8.8 | 5.5 | 9.7 | |

*Source:* AIA: Ramsey/Sleeper, *Architectural Graphic Standards,* 9th ed., © 1994 by John Wiley and Sons. SI conversions by the author.

*Laboratories* have proven to be especially difficult HVAC control problems due to their fume hoods. Conditioned air is provided from the central HVAC system, often by a VAV supply. Whenever a fume hood is exhausting air (frequently in huge quantities), the VAV supply and return systems are affected. Complicating this relationship is the nature of the laboratory work; where the experiment could be damaged by outside contaminants, the lab should be positively pressurized to minimize infiltration (and the VAV must therefore supply and return slightly *more* air than the fume hood exhausts). However, where the experiments involve diseased, toxic, or other hazardous substances, the lab must be negatively pressurized (and the VAV must supply and return slightly *less* air than the fume hood exhausts). DDCs tied to a central system can balance energy conservation, lab worker safety, and safety for the nonlab environment.

*Hotel rooms* can present serious energy loss problems from heating/cooling either an unoccupied room or a room with open windows. With DDCs inter-tied with the registration desk, an "unoccupied" mode of operating can be remotely controlled, with bare-minimum heating or cooling. When the room is occupied, the supply of either hot water heating or cooled air can be throttled back whenever the window is open. Also, a "purge" mode can enable a new arrival (or the front desk, in anticipation) to select a greatly increased flow of outdoor air for a limited time period to dilute cigarette smoke or other odors. Chapter 30 presents more detailed information on *intelligent buildings.*

*Offices* might be provided with DDC not only for

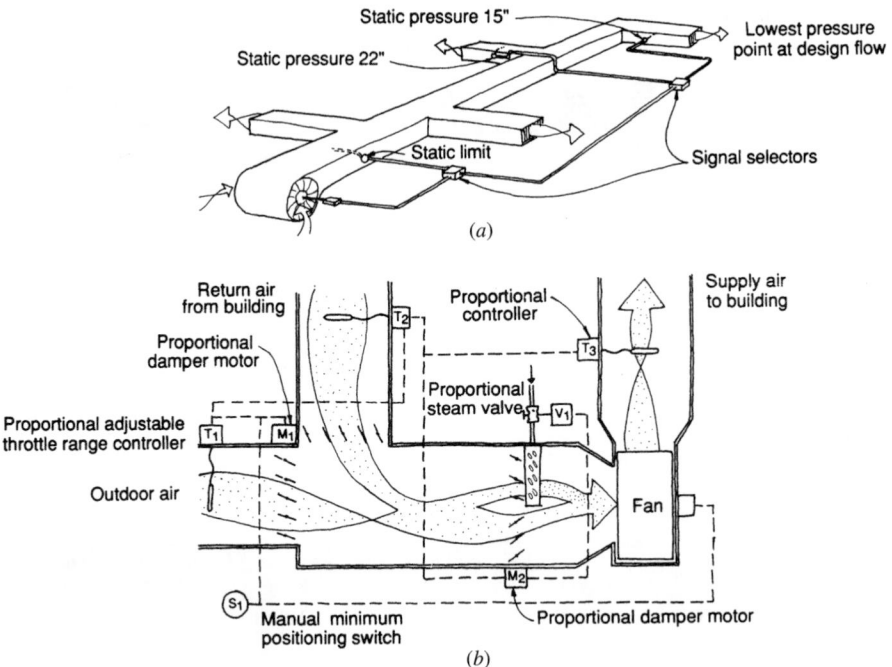

**Fig. 10.43** Controls for some common HVAC applications. (a) Single air duct with VAV. (b) Economizer cycle.

the VAV supply units, but also inter-tied DDCs for an interconnected ventilating window (preventing simultaneous open windows and treated forced-air delivery), daylight reflectors (mini-light shelves), venetian blinds, a radiant heater valve, an electric light switch, and an insulating shade. A control panel or "dashboard" gives the worker an opportunity to interact with the central control in operating these devices.

Office buildings that use more passive strategies also benefit from BMS. The British Research Establishment building for fire research, "Building 16" (Fig. 10.44), depends upon cross- and stack ventilation for its cooling and movable louvers on the south windows for sunshading and daylight penetration. A common network links this building's control systems; each worker has a TV-like controller that regulates lights and can override the programmed settings of the nearest high-level windows (for ventilation) and south-window louvers (for sun control and daylighting). This innovative office building uses a ground heat pump (borehole type) to supplement solar and internal winter heat gains or to provide supplementary cooling to the night ventilation system. Note the provisions for

cross-ventilation for individually enclosed offices on the ground floor, utilizing a cavity above the ceiling to carry the ventilation air on to exhaust on the other side of the building.

Rapid changes in centralized control systems are continuing. Initially, DDC systems were developed by individual companies, with little or no opportunity for communication between systems. Such proprietary systems left designers frustrated by an inability to specify a wide variety of products, all controlled by a single BMS. In the late 1990s, two competing systems had emerged that promised integration of products and systems from different manufacturers. BACnet was developed by ASHRAE committee members as a nonproprietary communication standard. LonMark was developed by the LonMark Interoperability Association, a user-funded organization of building owners, specifiers, system integrators, and product suppliers. *Open control system architecture* is an opportunity for a BMS in which components from several vendors interoperate over a BACnet-adapted Ethernet LAN (local area network). It makes possible a system combining BACnet, LonMark, and proprietary subsystems (Fig. 10.45).

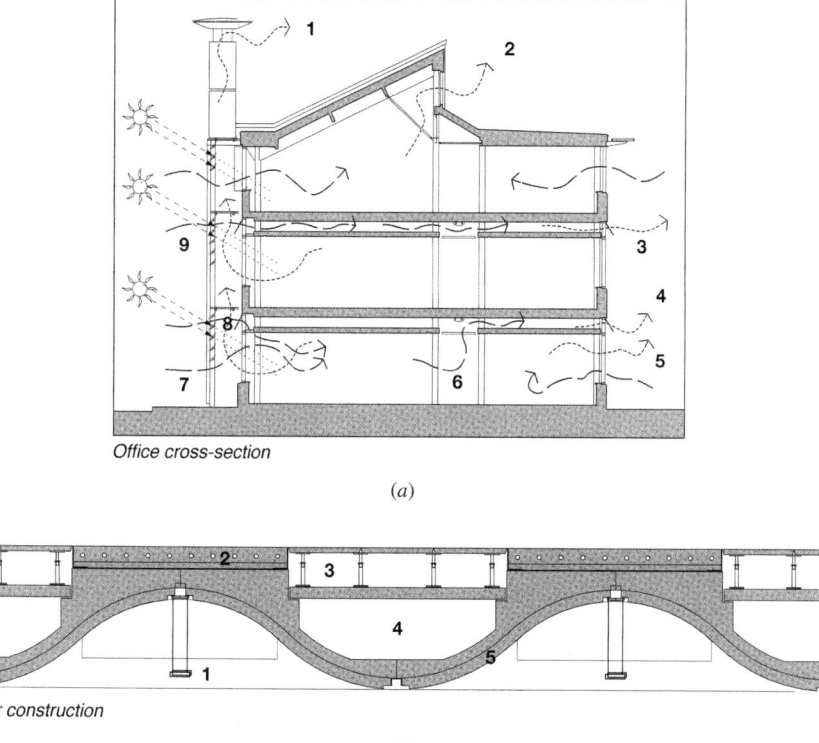

*Office cross-section*

(a)

*Floor construction*

(b)

**Fig. 10.44** *Natural ventilation and daylight strategies dominate the north–south section (a) of the British Research Establishment Building 16 at Garston. (1) Stack ventilation (hot, calm); (2) clerestory BMS-controlled ventilation; (3) night ventilation through slab, BMS-controlled; (4) cross-ventilation bypass over enclosed offices; (5) enclosed office single-sided ventilation; (6) corridor cross-over zone; (7) manually operated lower-level windows; (8) high-level BMS-controlled windows; (9) motorized external shading louvers, also BMS-controlled. (b) East–west section detail of precast floor structure. (1) Luminaire with integral photosensors; (2) heated/cooled screed using a geothermal source; (3) raised access floor for wiring; (4) cross-ventilation duct (night ventilation and/or cross-over ventilation for enclosed offices); (5) waveform precast concrete ceiling with poured-in-place topping slab. (Courtesy of Feilden Clegg Architects, Bath, England.)*

## 10.4 AIR DISTRIBUTION WITHIN SPACES

When large office buildings had far greater interior (core) areas than perimeter areas and were filled with less efficient lighting at high luminance levels, they were considered to always need cooling. Because cool air supplied at the ceiling would naturally fall toward the level of occupants and because suspended ceilings were ubiquitous, supply air from the ceiling was almost universal. The suspended ceiling was also tempting for the return air provisions, whether as a plenum or using another ducted system. With both supply and return air at the ceiling, the danger of short-circuiting arose: supply air heading quickly for the return opening, with resulting shortages of both cooling and IAQ. In this section, we first look at approximate duct sizing and

then consider three air distribution systems for multistory office buildings.

### (a) Air Ducts

Duct sizes (in cross section) are frequently of interest early in the design process. Duct depths can help determine floor heights; duct cross sections influence the sizes and shapes of the vertical cores that serve multistory buildings. An approximation of duct size can be obtained as follows:

1. Determine the quantity of air to be distributed through the largest duct, using Table 10.5, or the ACH, from a calculation of night ventilation of thermal mass. This will usually be expressed in cubic feet per hour (cfh).

THERMAL CONTROL

## Open Building Management and Control System

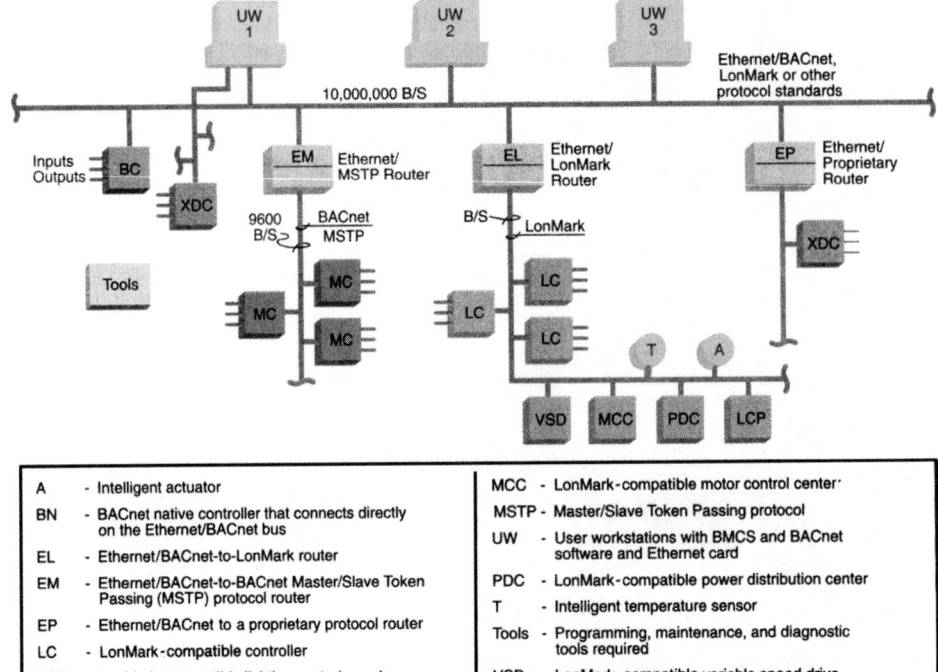

| | | | |
|---|---|---|---|
| **A** | - Intelligent actuator | **MCC** | - LonMark-compatible motor control center |
| **BN** | - BACnet native controller that connects directly on the Ethernet/BACnet bus | **MSTP** | - Master/Slave Token Passing protocol |
| **EL** | - Ethernet/BACnet-to-LonMark router | **UW** | - User workstations with BMCS and BACnet software and Ethernet card |
| **EM** | - Ethernet/BACnet-to-BACnet Master/Slave Token Passing (MSTP) protocol router | **PDC** | - LonMark-compatible power distribution center |
| **EP** | - Ethernet/BACnet to a proprietary protocol router | **T** | - Intelligent temperature sensor |
| **LC** | - LonMark-compatible controller | **Tools** | - Programming, maintenance, and diagnostic tools required |
| **LCP** | - LonMark-compatible lighting control panel | **VSD** | - LonMark-compatible variable speed drive |
| **MC** | - BACnet/MSTP compatable controller | **XDC** | - Proprietary digital controller |

**Fig. 10.45** *Building Management System (BMS) that is open to various protocol standards. (Courtesy of Honeywell, Inc.)*

2. If necessary, convert cfh to cfm:

$$\frac{\text{cfh} \times 1\text{ h}}{60\text{ min}} = \text{cfm}$$

3. Find the maximum velocity of this air within the duct from Table 9.4, expressed in feet per minute (fpm).

4. The approximate *minimum* required cross-sectional area of duct $A$ is then

$$A_{\text{in.}^2} = \frac{\text{volume of air (cfm)}}{\text{velocity (fpm)}}$$

$$\times\ 144\text{ in.}^2/\text{ft}^2 \times \text{friction allowance}$$

where the friction allowances are
- round ducts = 1.0 (may be neglected)
- nearly square ducts (ratio of width to depth, 1:1)
  small (<1000 cfm) = 1.10
  large (>1000 cfm) = 1.05
- thin rectangular ducts (ratio of width to depth, 1:5) = 1.25

Then check this against the *recommended* duct cross-sectional area from Table 10.4. Remember that the *minimum* duct area will carry a penalty of increased noise and friction. Techniques for noise suppression in ductwork are discussed in Chapter 19.

### (b) Ceiling Air Supply

This approach is so widespread that many lighting fixtures are made to either serve as diffusers for supply air or as intakes for return air. As return air fixtures, they are especially effective because they remove much of the heat from electric lighting before it can contribute to overheating the office space below. This makes less work for the cooling equipment.

When individual offices are enclosed, a supply diffuser and return grille ensure air circulation. However, as open office space proliferated, concerns grew about whether the workspace cubicles were being adequately served with conditioned air. Air diffusers and grilles (Fig. 10.46) on the ceilings are uncluttered by furniture and independent of re-arrangements of workstation cubicles. Canadian research in the early 1990s indicated that the most important variable was the quantity of the airflow, not the location of the diffusers relative to workstations or whether the cubicle partitions have a small gap where they meet the floor.

Are higher ceilings better for air distribution? There seems to be a trade-off between more vertical clear space (between ceiling and cubicle partitions) that would allow a wider range of air distribution and the increased distance from the actual occupant who should be receiving the conditioned air. The increased distance would allow colder air to be distributed, resulting in smaller ducts, and could also allow higher velocity of supply air, further reducing duct size. Higher ceilings allow deeper daylight penetration and represent a larger "pool" of air that is slower to become polluted by occupants or office equipment. Higher ceilings also encourage underfloor distribution, as seen in the next subsection.

When the noise from forced-air diffusers is critical, as in telecasting and recording studios, pin-hole-perforated diffusers (Fig. 10.47) can provide large quantities of low-velocity air. This is an especially challenging combination of high heat gains (from lights) and low background noise requirements. Again, ceiling distribution leaves the floor unencumbered.

## (c) Underfloor Supply with Displacement Ventilation

As open office spaces have grown larger in area and their ceilings higher (for daylight penetration and indirect lighting), and as concerns about IAQ have deepened, the floor has gained in popularity as the air source location. Displacement ventilation provides fresh air, cooled to just below the design room temperature, at a low velocity. Furthermore, fresh air supply rates are typically near 45 cfm (22 L/s), compared to about 30 cfm (15 L/s) often used in ceiling supply systems. This high-volume, low-Δt, low-velocity combination requires greatly increased duct sizes, provided by using the plenum below a

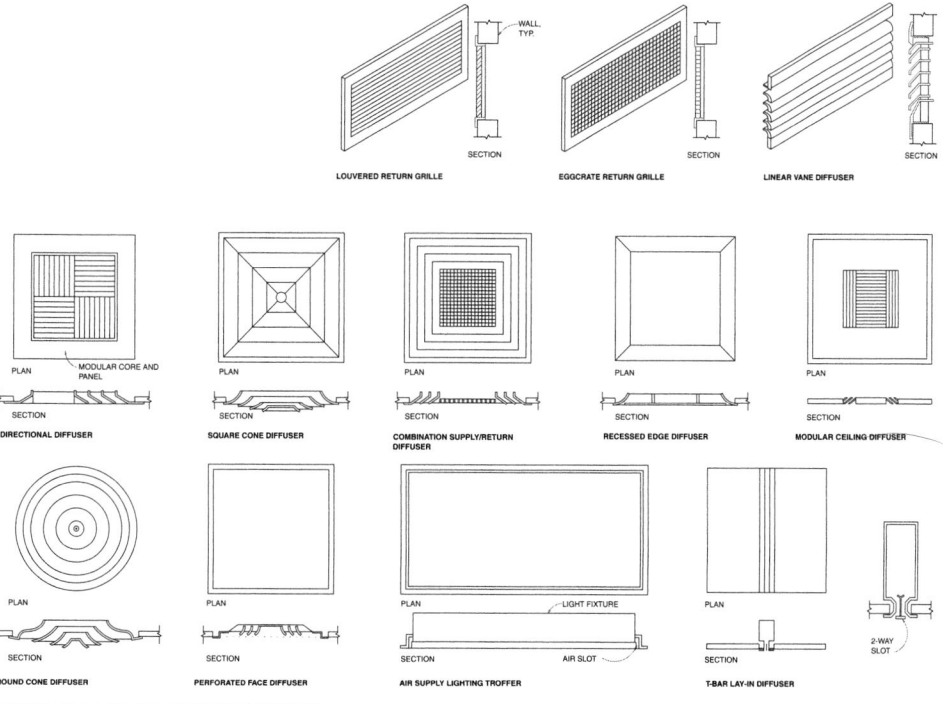

RETURN GRILLES AND SUPPLY DIFFUSERS

***Fig. 10.46*** *Common air distribution outlets. (From AIA: Ramsey/Sleeper,* Architectural Graphic Standards, *10th ed., © 2000 by John Wiley & Sons, Inc. Reprinted with permission of John Wiley & Sons, Inc.)*

**Fig. 10.47** *Pinhole perforated diffusers (above a lighting grid) provide large quantities of low-velocity air to the television studio at the Community Media Center of Santa Rosa, California. The background noise from more typical HVAC systems would be amplified by sound recording equipment and hence would be unacceptable. (© TLCD Architecture, Santa Rosa, CA.)*

raised floor system. The fresh air rises quickly to the occupant level, drawn upward by the stack effect of heat gain from lights, people, and office equipment. It continues to the ceiling, where it stratifies and where various approaches to return air collection are located. Freshest and coolest at the floor, hottest and most stale at the ceiling, this system promises better IAQ and thermal comfort than would be available from ceiling supply–return systems. Furthermore, the outlets at the floor are adjustable and easily reached by the office workers.

The raised floor is typically supported on a 2 × 2-ft (600 × 600-mm) module. The module that contains the diffusing outlet often also contains an outlet for power and cable. This diffuser is placed off-center within the modular unit, further expanding the variability of the floor opening relative to furniture placement because each plan rotation of the square module places the outlet slightly differently. Some codes restrict the height of the raised floor plenum; many codes require specially coated

wiring within the plenum because moving air will wash over the wires for the life of the building.

The overall floor-to-floor height needs to be great enough to accommodate both the raised floor and the higher office ceiling (at least 9 ft [2.7 m] to accommodate warm air stratification). However, elimination of the suspended ceiling and the supply air ductwork also eliminates the suspended ceiling cavity. This tends to expose the lighting, fire sprinkler, and public address systems below the structural ceiling, although the latter can be distributed in the raised floor cavity above and puncture the structural ceiling as required.

The structural slab below the raised floor can be of concrete, and this thermal mass can be night ventilated to store coolth to assist the ventilation air for the next day. The Inland Revenue Building (see Fig. 8.46) incorporates a night-ventilated raised-floor section within its precast waveform floor sections.

Library Square in Vancouver, British Columbia (Fig. 10.48), utilizes a raised floor to serve its interior zones of book stacks. The library has seven floors, totals 390,000 ft² (36,320 m²), and exposes its concrete structure as a finished ceiling. The floor-to-floor height is 16.4 ft (5 m). The underfloor supply is fed by VAV boxes and floor fan units (FFU) that blend low-temperature supply air (about one-third) with local filtered return air (about two-thirds) to provide 63°F (17°C) air to the pressurized plenum. Flush to the floor, high-induction swirl floor diffusers (8 in. [200 mm] in diameter) produce an upward air motion, taking the fresh-return air blend directly into the occupancy zone (the first 6.5 ft [2 m] of height on each floor). The stack effect of heat from people, computers, and lights takes hot air on to the stratification zone above. Thus, people-generated pollutants tend to rise to this hot, stale air zone, then into return air slots in the precast concrete ceiling. Return air is taken away in insulated ducts within the supply air plenum of the floor above. The plenum height is an unusually generous 2 ft (600 mm) high; it also contains the sprinkler distribution and all wiring systems. The warm, dry summer climate supports a night ventilation system, flushing

the plenum and storing coolth in the exposed concrete structure. This has reduced the use of an ice-storage system.

### (d) Workstation Delivery Systems

The air delivery system shown in Fig. 10.49 allows a variety of individual controls at each workstation. As developed by Johnson Controls, Inc., these systems are called Personal Environments® systems. A mixture of outdoor and recirculated indoor air, called *primary air,* is brought from the main duct (or floor plenum) to each Personal Environments system's mixing box in a duct carrying at most 120 cfm (56 L/s) but typically less. Each worker can adjust the supply air temperature, the mixture of primary and locally recirculated air, air velocity (and therefore volume) and direction, and radiant supplementary heat (below desk level). Task lights can be dimmed, and the background (masking) sound level is adjustable. The worker has almost as much environmental control at each office workstation as a driver has in an automobile's front seat.

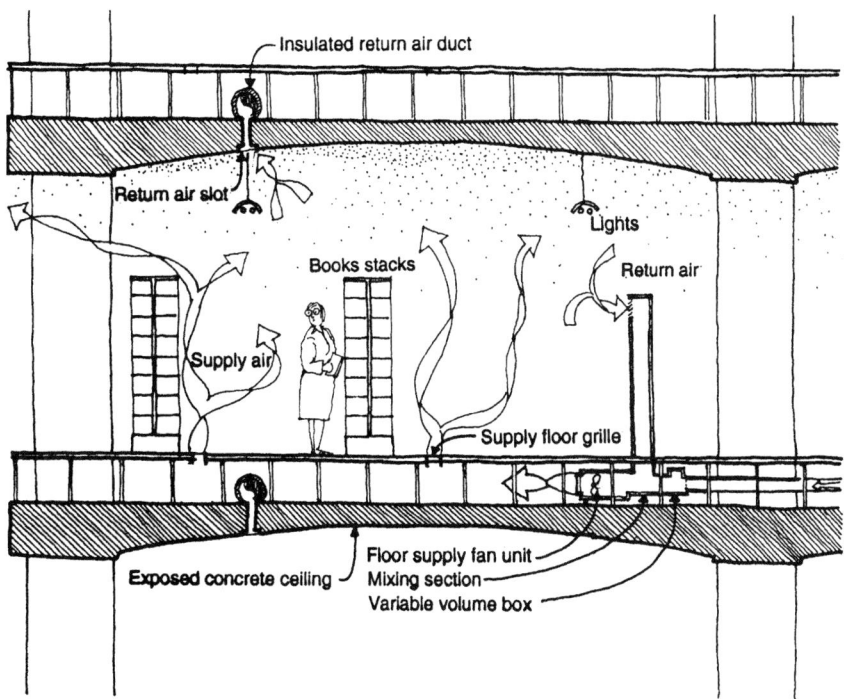

**Fig. 10.48** *Library Square, Vancouver, British Columbia, supplies conditioned air to the central book stacks via an underfloor plenum. This displacement ventilation results in return air taken through openings in the exposed structural ceiling. (Section adapted from information supplied by Blair McCarry, Keen Engineering, North Vancouver, BC.)*

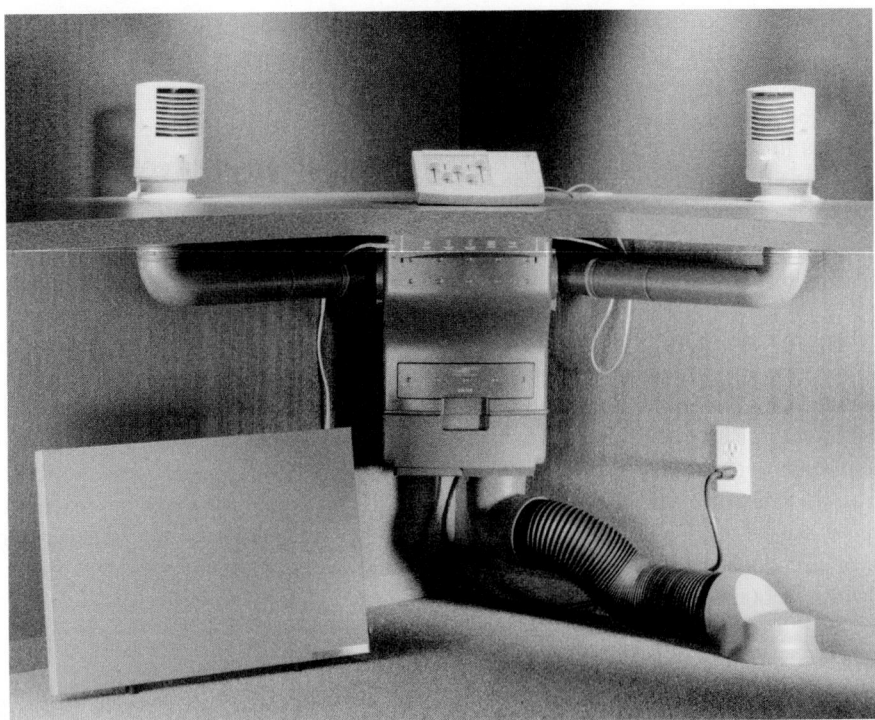

**Fig. 10.49** *The Personal Environments® system provides each workstation with a fan, air filters, an air mixing box, and a background sound (white noise) generator. The control panel allows adjustment of task lighting, background sound, fan speed, primary/recirculated air mixture, and radiant heating. Two diffusers distribute both air and background sound. Diffusers are adjustable about both horizontal and vertical axes. The below-desk radiant heating panel warms the lower body. (Courtesy of Johnson Controls, Inc., Milwaukee, WI.)*

When the workstation is unoccupied, an occupancy sensor shuts the system down, maintaining a minimum airflow of 12 cfm (5.7 L/s).

### (e) Alternative Supply/Return Systems

So far, the supply/return systems have been either both from the ceiling or supply from the floor and return at the ceiling. Supply from the ceiling, return through the floor is another possibility. The fresh air gets to the occupancy level later, and individual control is much more difficult, but debris from the floor falls into a return, not a supply, plenum. There seems less danger of debris falling into a floor supply diffuser while the system operates; it is during fan-off periods, which may include custodial maintenance, that this possibility increases.

An alternative approach with return at the perimeter is shown in Fig. 10.50. It is variously called an *air-extract window,* an *air curtain window,* or a *climate window.* Developed in Scandinavia in the 1950s, this is a triple-glazed window that passes room air between a typical outer double-glazed window and an inner single pane. The inner pane thus is kept at very nearly the same temperature as the room air, which greatly increases comfort near windows on very cold (or very hot) days. Venetian blinds are often inserted in this cavity, where they can intercept direct sun and redirect its light toward the ceiling. The solar heat intercepted by the blind is carried off by the room air to a plenum above the ceiling, where such air can be either exhausted or recirculated and its heat content either reclaimed or rejected. The U-factor of these windows is dependent upon the rate of airflow between the glazing (Fig. 10.50c); typical flow rates are 4 to 6 cfm per foot (6 to 9 L/s per meter) of window width.

The Ocosta Junior/Senior High School at Westport, Washington (Fig. 10.51), uses these windows on the south façade. (For an example of such windows in a larger U.S. building, see Fig. 10.63, which shows the Comstock Center in Pittsburgh.) The

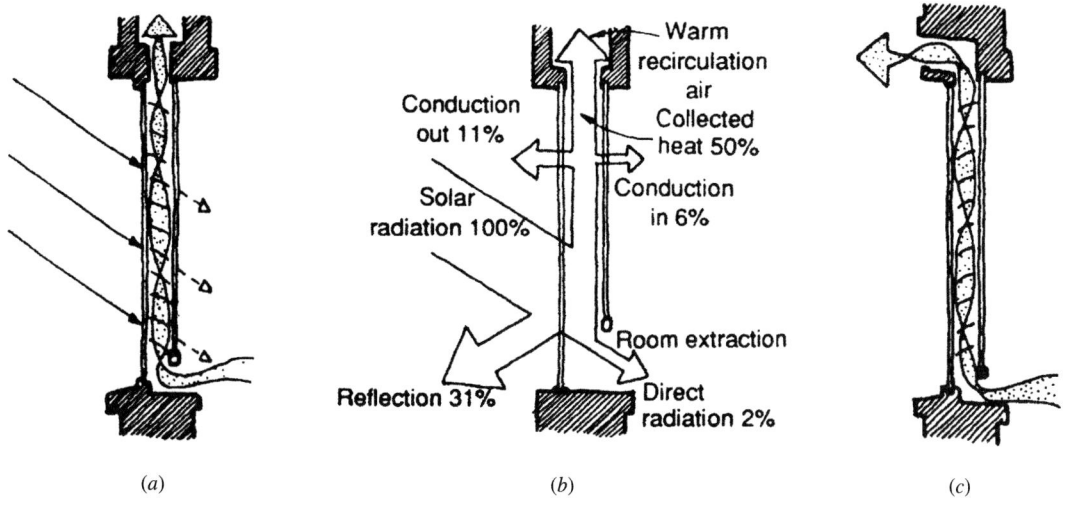

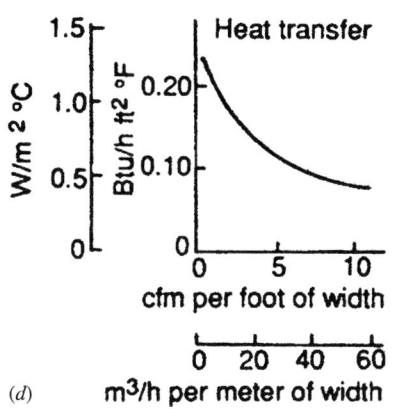

**Fig. 10.50** *Air extract window (also called an air curtain or climate window). (a) Cross section showing the window as a solar collector. (b) Distribution of solar heat in the collector mode. (c) The window as a "solar chimney" exhausting hot air. (d) The U-factor of these windows varies with the rate of airflow within the glazing: the typical flow rate is 4 to 6 cfm per foot (6 to 9 L/s per meter) of window width. (From articles by D. Aitken and O. Seppanen in* Proceedings of the Sixth National Passive Solar Conference. *© 1981 by the American Solar Energy Society, Boulder, CO.)*

classrooms have photocells that switch off some lights nearest the windows when daylight is adequate, and water loop heat pumps help to transfer passive solar heat from the south side to the sunless north side rooms. Excess heat in the Ocosta school is stored in an underground water tank. The compact plan, highly insulated envelope, and mass capacity within the school reduce the need for a boiler to supplement the winter temperature of the water loop. During consistently warm weather, the BMS exhausts the air from the window to the outdoors.

Supply openings in vertical risers are relatively rare but at times function effectively. The Frederick Meijer Gardens near Grand Rapids, Michigan, is a huge indoor tropical rain forest. This glass structure is more than 65 ft (20 m) high; it has more glass area than floor area—great for a high daylight factor, but imagine the heating loads! High humidity

must be maintained, yet water condensation on leaves is undesirable. A strong horizontal airflow could be helpful to the plants, acting as wind would to develop a stronger root structure.

The solutions here begin with a perimeter tunnel serving as an air supply duct, delivering air evenly around the glass walls, heated by finned tubes at planting level that are largely hidden by foliage. The tunnel insulates the planting beds from the extremes of the exterior, maintaining constant ground temperatures year round. A separate "wind" system continuously feeds air to vertical ducts with several diffusers (Fig. 10.52b), making leaves and branches sway. Humidity is provided by a high-pressure water system that creates fog (very small water particles). This is located high in the space (but also near waterfalls and a stream for special effects). This is also an evaporative cooling system to counteract summer

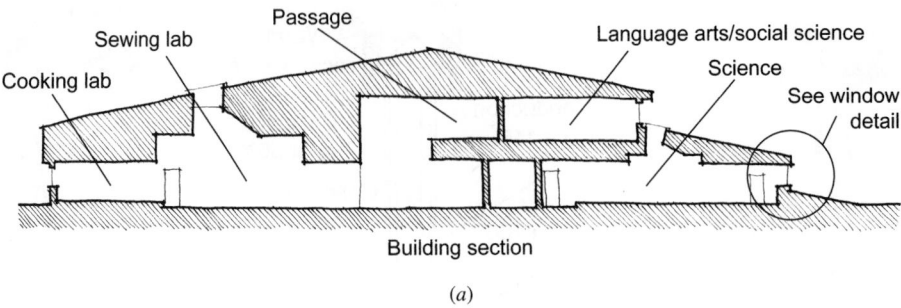

Building section

(a)

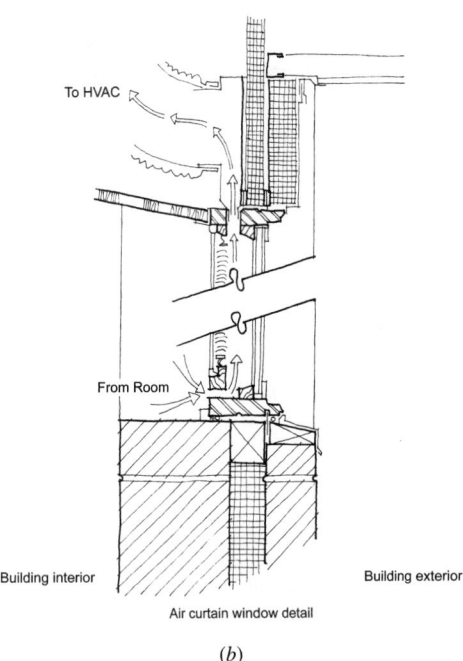

To HVAC

From Room

Building interior

Building exterior

Air curtain window detail

(b)

**Fig. 10.51** *The Ocosta Junior/Senior High School at Westport, Washington, uses air extract (air curtain) windows to help distribute solar energy from the south side (to the right in section a) to the north classrooms. (b) Detail through the south-facing windows. Photocells control the electric lights during daylight hours; venetian blinds direct sunlight toward the ceiling and deeper into the classroom. (Courtesy of Burr Lawrence Rising + Bates Architects, Tacoma, WA. Redrawn by Erik Winter.)*

overheating. Peak airflow is needed only for cooling; return air is taken through a large opening hidden behind a waterfall. About one-third of the peak supply air is returned; the remainder is exhausted at the top of the structure through operable vents high in the glass roof. A BMS controls the roof vents, the fogging system, the finned tube heating system, rolling shading screens, and the heating/cooling equipment in a basement, all in response to changing conditions outside this enormous glass house.

## 10.5 ALL-AIR HVAC SYSTEMS

This large, complex family of systems was introduced in Section 10.2. All-air HVAC systems require thick distribution trees—therefore more building volume—but can promise comfortable results because they effectively regulate IAQ, temperature, and humidity. This approach concentrates the mechanical equipment and its maintenance within a central mechanical room, freeing the rest of the building from repair interruptions and worries about water leaks from heating/cooling pipes. However, this family does less well with perimeter zones in cold climates.

With all-air systems, air quality can be manipulated through pressure control: *negative pressures* in odorous or excessively humid locations (kitchens, toilet rooms, pet shops in shopping centers, etc.), *positive pressures* in shopping malls, corridors of apartment houses, stair towers, and so on. The difference

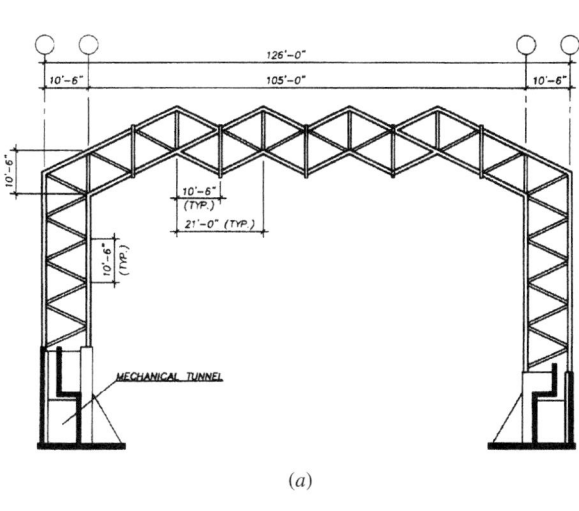

(a)

(b)

(c)

**Fig. 10.52** With more glass area than floor area, the Frederick Meijer Conservatory near Grand Rapids maintains a tropical environment in Michigan's winter. (a) A perimeter tunnel delivers air evenly around the walls, heated by finned tubes just above the planting level. (b) A separate "wind" system provides a continuous breeze from diffusers in the vertical ducts to strengthen plants' root systems. (c) A fogging system maintains tropical humidity while adding evaporative cooling under summer conditions. (Photos by Roger Van Vleck; courtesy of Fishbeck, Thompson, Carr & Huber, Ada, MI.)

in pressure sets up an overall direction of airflow that helps prevent the spread of odorous or otherwise contaminated air and can help manage smoke in a fire (see Chapter 24). Positive pressures in connecting spaces help each adjacent space keep its air to itself.

### (a) Single-Zone Systems (Fig. 10.12a)

Smaller buildings are often served by this least complicated all-air example (see examples in Section 9.8). Where an entire building is essentially one thermal zone, uncomplicated enough to be served by only one thermostat, this system offers a very low first cost. A closely related system is the multizone (see later in this chapter), which is a combination of single zones.

### (b) Single-Duct VAV Systems (Fig. 10.12c)

As described in Section 10.2, this system is an energy-saving option ideally suited to the internal zones of large buildings, where cooling is always needed. However, it can also be adapted to serve the entire building, with important savings over the constant-volume (CV) systems described later (Section 10.5e). The air handling units (fans, etc.) for each thermal zone are sized to meet the peak demand on that zone. In most CV systems, the fans always run at this peak condition speed, even though peaks are often of rather short duration. In VAV systems, fans run at peak speed only during peak hours. This obviously saves considerable energy needed to run fans. When the building uses only one or a few central fans, a VAV system will require smaller fans because the fans need meet only one peak condition at a time and do not have to be sized for all zones' peak flows. The variation in demands on the fan can be met either by selecting variable-pitch blades or (less expensively) by varying the speed of the fan.

Where a VAV system is used, provision must be made for at least code-minimum fresh air levels, sometimes resulting in overcooling or wasteful reheating. Although VAV systems are typically less noisy than CV systems (because less air is moving), the air motion noise *varies* with the volume, and variable noise sources are inherently more noticeable than steady ones.

Within (or near) the spaces it serves, the VAV system typically needs a mixing box or terminal; this is often placed above a suspended ceiling or below a raised floor. Of the several terminal types available, the standard and most simple one is shown in Fig. 10.53. This terminal serves not only to vary the quantity of air, but also to both attenuate the noise and reduce the velocity of air from the main trunk of the distribution tree. High velocity is commonly used in main ducts because it reduces the size (thickness) of these critical large portions of the tree.

The Utah Department of Natural Resources in Salt Lake City (Fig. 10.54) utilizes VAV throughout its three-story, 105,000-ft$^2$ (9755-m$^2$) office building. Air is supplied from linear ceiling diffusers located between the rows of indirect luminaires parallel to the windows. Return air is taken into the suspended ceiling plenum. Highly insulated walls, windows, and roof lessen the need for perimeter heating, although a supplementary warm water heating system is built in below the windows to offset perimeter

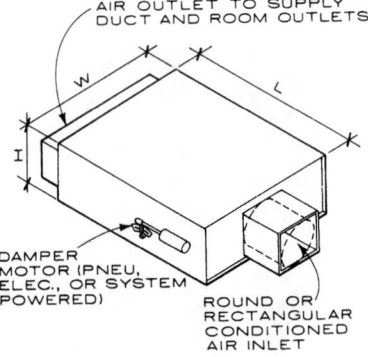

### RANGE OF DIMENSIONS

| CFM | HEIGHT | LENGTH | WIDTH |
|---|---|---|---|
| 400 | 8"- 9" | 24"-39" | 14"-30" |
| 800 | 10"-11" | 24"-53" | 18"-42" |
| 1600 | 14" | 30"-48" | 22"-44" |
| 2400 | 16" | 42"-60" | 26"-54" |
| 3200 | 18" | 42"-67" | 33"-54" |

**Fig. 10.53** *The most simple type of VAV terminal, which pinches back the volume of incoming air as thermal loads decrease. (From AIA: Ramsey/Sleeper,* Architectural Graphic Standards, *9th ed., © 1994 by John Wiley & Sons, Inc. Reprinted with permission of John Wiley & Sons, Inc.)*

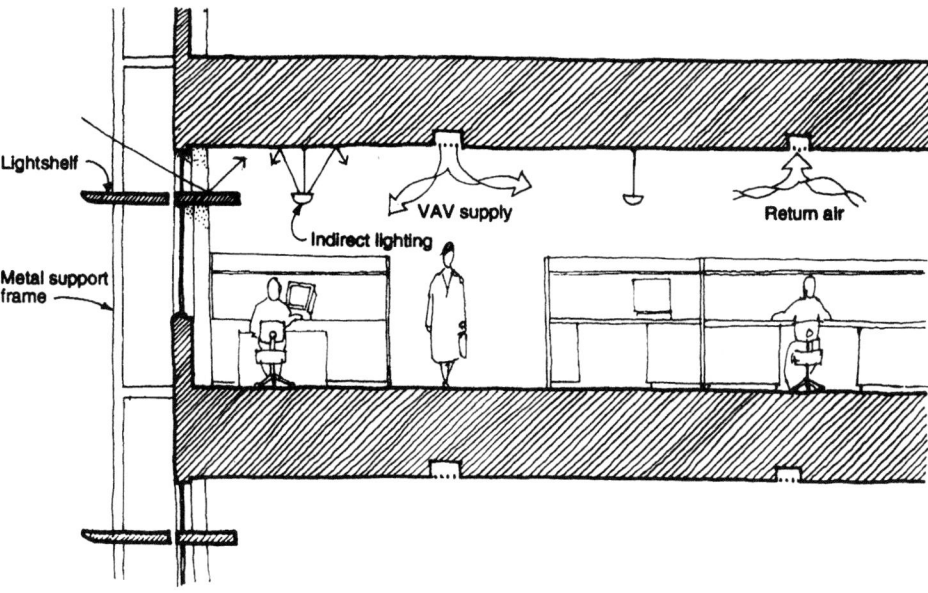

**Fig. 10.54** *The Utah Department of Natural Resources building features lightshelves, direct and indirect evaporative cooling, and an economizer cycle. Supply air is distributed by a VAV system with ceiling diffusers located between luminaires. (Based upon the design by Gillies Stransky Brems Smith Architects, Salt Lake City, UT.)*

heat losses. The lower 1 ft (300 mm) is a slot where room air is drawn in by convection, and a perforated metal grille at the top forms the window sill.

The building is elongated east–west for best daylight control and winter solar gain; lightshelves even the daylight penetration. With almost no east- or west-facing glass, the building relies on 4200 ft² (390 m²) of south glass and 3800 ft² (353 m²) of north glass to serve its width of 125 ft (38 m). Supplementary indirect lighting is controlled by photocells and dimming ballasts.

Utah's hot, dry summers permit a four-stage approach to cooling. First is the economizer cycle (see Fig. 10.36); at higher temperatures, direct evaporative cooling is used; then indirect evaporative cooling (see Figs. 9.10 and 9.11) utilizing an oversized cooling tower; and finally (estimated at about 10 days per year), a conventional chiller.

There are several variations on the basic VAV system, some of which respond to the problem of minimum airflows for rooms with little thermal load or to the desire to serve both interior and perimeter areas with the same HVAC system.

### (c) Fan-Powered VAV Systems (Fig. 10.12f)

To maintain minimum fresh air, VAV terminals are usually set so that they cannot be entirely closed off, ensuring some outdoor air at all times. If this fresh air minimum, entering at low velocity, does not provide the desired air motion and mixing within the room, VAV terminals can be equipped with fans, which are activated as needed with decreasing incoming air volume. These self-contained fans mix room air with incoming air to provide an airstream of the right temperature and velocity to maintain comfort.

This approach to VAV is sometimes taken when simultaneous heating and cooling are needed, such as at perimeter zones, which can generate sizable heating needs while the rest of the building needs cooling. Another approach is to utilize an induction-type VAV terminal, with which air heated by electric lights is induced to join the incoming cool airstream. Greater heating needs often require the use of reheat terminals supplied by a circulating hot water system or by electric resistance heating. This reheat applica-

(a)

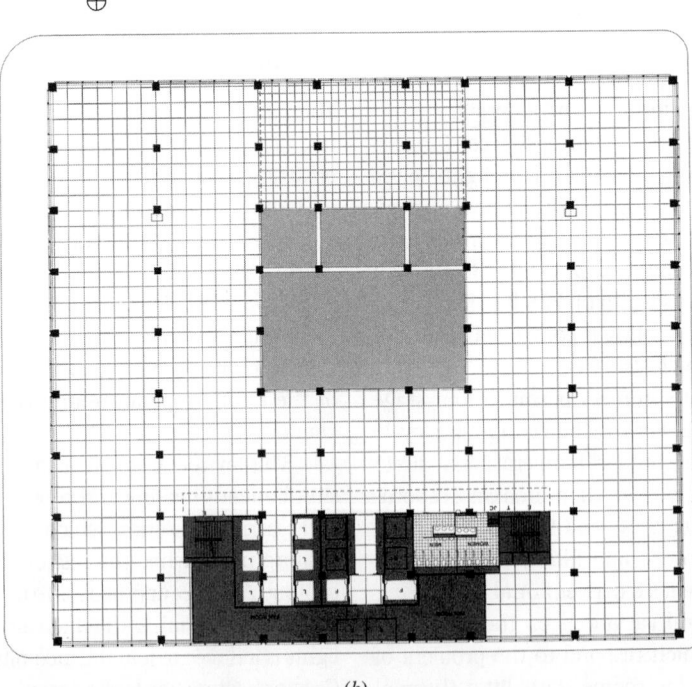

(b)

**Fig. 10.55** *Floor-by-floor fan rooms supply VAV for this Chicago office building. (a) Exterior view of 33 West Monroe. (Photo by Merrick, Hedrick-Blessing.) (b) Plan of a typical lower floor; two fan rooms occupy the space along the exterior wall. (c) Section perspective showing stacked atriums. (Courtesy of Skidmore, Owings & Merrill, Architects-Engineers, Chicago.)*

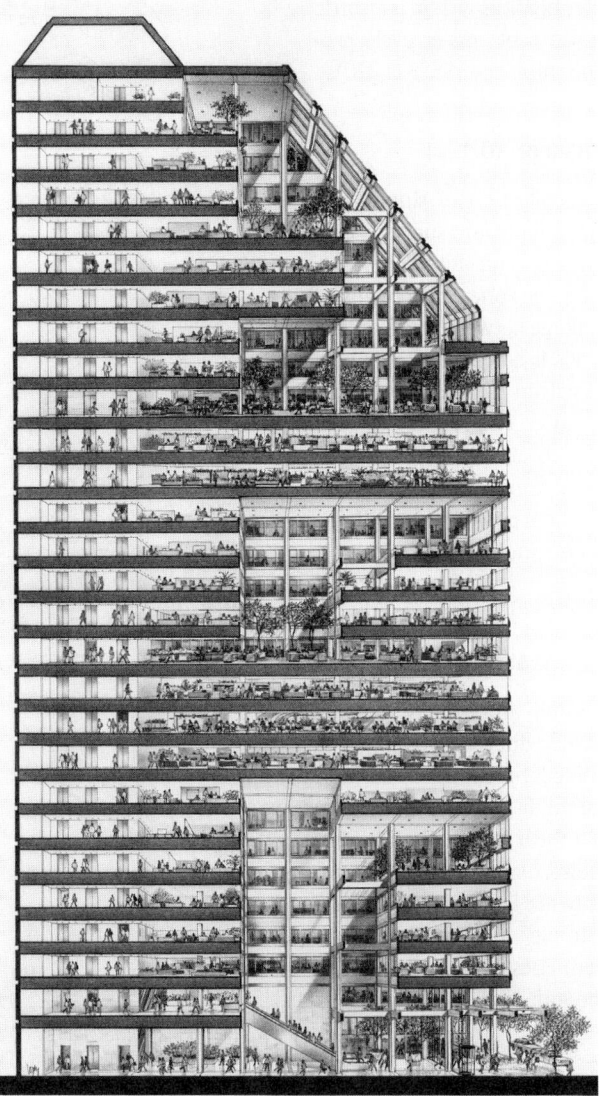

*(c)*

**Fig. 10.55** *(Continued)*

tion is more energy efficient than the standard CV reheat system described in Section 10.5*e*, because a much smaller volume of air is first cooled, then reheated. The water or electric coils can be incorporated either in the VAV terminal or in the ductwork between the terminal and the space it serves.

An example of a floor-by-floor VAV system (Fig. 10.55) is a 1-million-ft$^2$ (92,900-m$^2$) 28-floor Chicago office building designed by Skidmore Owings & Merrill. This mid-rise approach to office towers utilizes three "stacked" atriums to relieve the monotony of the wide interior floors. Another

result is lower structural and energy costs per square foot relative to conventional high-rise structures. The cubelike shape of the building exposes less skin area (38% of which is in insulating glass) to Chicago's cold winters; electric lighting at about 1.8 W/ft$^2$ holds down internal gains. To accommodate the differing schedules and comfort needs of a variety of tenants, each floor is provided with two VAV supply fans that can be operated at night and on weekends, independently of the rest of the building. Each floor's mechanical core has one exterior wall (on an alley) to facilitate fresh air intake/stale air exhaust. The perimeter heating system is electric resistance fin radiation; an economizer cycle provides cooling with outdoor air below 55°F (13°C).

### (d) Multizone Systems (Fig. 10.12b)

A multizone system is a collection of single-zone systems served by a single supply fan; such systems rarely exceed eight zones per air handling unit. Simultaneous heating of some zones and cooling of others is possible, but leakage between zones at the decks of hot and cold coils is common. Return air from all zones is mixed within one return duct; a bypass at the heating and cooling decks requires more space but allows such air to be mixed with supply without undergoing unnecessary heating or cooling. One multizone system per floor of medium- to high-rise buildings is an increasingly common application.

### (e) Single Duct with Reheat (Fig. 10.12e)

Formerly a widespread system and usually supplied at constant volume, single-duct reheat systems now are severely restricted by many codes and ASHRAE standards. For high-pressure and high-velocity main ducts, CV reheat boxes (Fig. 10.56a) are used to control the noise, temperature, pressure, and velocity of the supply air. For more simple all-low-velocity systems, simple duct insert heaters (Fig. 10.56b) can be used. These are sized to fit the duct, which needs to be enlarged only slightly to accommodate them.

### (f) Double-Duct Systems (Fig. 10.12d)

This is still considered the "Cadillac" of HVAC systems, not only because of its superior comfort control and flexibility for simultaneous heating/cooling zones, but also because of its high initial cost, large size, and high energy usage. The double-duct system is rarely installed now, except in hospitals. Building volumes needed for a double-duct system's three (two supply, one return) full-sized air distribution trees are harder to justify, given that VAV systems can provide acceptable comfort for most common spaces. An example of a double-duct office building was seen in Fig. 10.11, San Francisco's International Building.

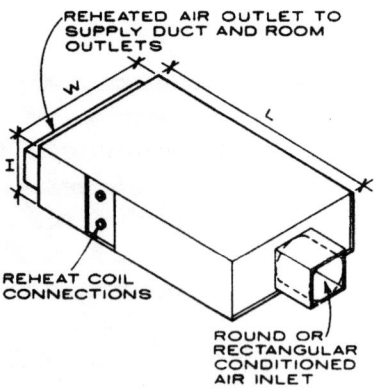

**RANGE OF DIMENSIONS**

| CFM | HEIGHT | LENGTH | WIDTH |
|-----|--------|--------|-------|
| 200 | 9"-11" | 30"-50" | 16"-22" |
| 400 | 9"-11" | 30"-51" | 18"-30" |
| 800 | 9"-11" | 30"-51" | 22"-42" |
| 1600 | 14"-16" | 48"-51" | 40"-44" |
| 2400 | 16"-18" | 60"-55" | 40"-54" |
| 3200 | 16"-18" | 60"-55" | 16"-66" |
| 5000 | 20"-18" | 60"-55" | 20"-80" |

(a)

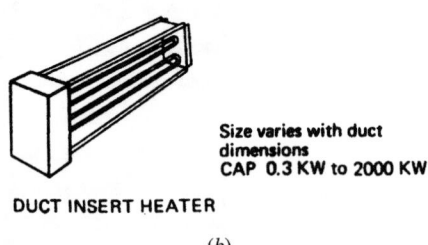

Size varies with duct dimensions
CAP 0.3 KW to 2000 KW

**DUCT INSERT HEATER**

(b)

**Fig. 10.56** Single duct with reheat coils (a) requires a simple terminal where velocity and pressure are reduced. (b) Electric coil within a duct. (From AIA: Ramsey/Sleeper, Architectural Graphic Standards, 9th ed., © 1994 by John Wiley & Sons, Inc. Reprinted with permission of John Wiley & Sons, Inc.)

The mixing boxes (terminals) of double-duct systems are similar to those of other all-air systems (Fig. 10.57) but are generally larger for the same airflow capacity. This adds still more to their impact on building volume. They are expensive and may require maintenance. Although most double-duct systems are CV, they can be VAV when the reduction in airflow is no more than 50% below the maximum. For details on other double-duct variations, see Lorsch (1993).

(a)

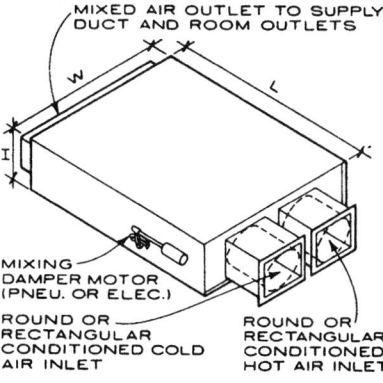

MIXED AIR OUTLET TO SUPPLY
DUCT AND ROOM OUTLETS

MIXING
DAMPER MOTOR
(PNEU. OR ELEC.)

ROUND OR
RECTANGULAR
CONDITIONED COLD
AIR INLET

ROUND OR
RECTANGULAR
CONDITIONED
HOT AIR INLET

High, medium, or low velocity systems. Inlet pressure
$^1/_4$ to $1^1/_2$ in. W.C. Capacity range from 150 to 2000
cfm per box (low velocity) to 5000 cfm (high velocity).
Box serves as converter from high to low velocity air
system, noise attenuator, and control device by mixing
hot and cold air streams.

### RANGE OF DIMENSIONS

| CFM | HEIGHT | LENGTH | WIDTH |
|---|---|---|---|
| 400 | 6''–10'' | 40''–51'' | 30''–19'' |
| 800 | 8''–11'' | 50''–51'' | 42''–24'' |
| 1600 | 12''–14'' | 48''–51'' | 44''–40'' |
| 2400 | 14''–18'' | 60''–55'' | 54''–44'' |
| 3200 | 14''–18'' | 60''–55'' | 54''–44'' |
| 5000 | 16''–18'' | 60''–55'' | 54''–66'' |

(b)

**Fig. 10.57** *High-velocity, double-duct terminal providing mixing and attenuation (pressure and sound reduction). (a) Pneumatically controlled from a thermostat, this unit blends and delivers air at the selected temperature. (Courtesy of Anemostat.) (b) Typical mixing box dimensions. (From AIA: Ramsey/Sleeper,* Architectural Graphic Standards, *9th ed., © 1994 by John Wiley & Sons, Inc. Reprinted with permission from John Wiley & Sons, Inc.)*

## 10.6  AIR AND WATER SYSTEMS

These systems, introduced in Fig. 10.13 and Section 10.2, have the design complexity—and first cost—of supply-and-return distribution trees for both water and air. This disadvantage is offset by the space-saving advantages of thinner water trees, the possibility of only one air tree (supply), and the superior comfort characteristics offered by air. When only fresh air is centrally treated and distributed, only an equal quantity of air need be exhausted, which rarely needs an extensive distribution tree. Because this exhaust air is not recirculated, these systems are attractive for hospitals and other buildings in which the mixing of air between zones is undesirable. When return air is used rather than exhausted, it forms a small percentage of the supply air.

These systems are most often used in perimeter zones, where extensive extra heating (winter) and cooling (summer) is readily provided by water. If only one water supply tree is used (two-pipe system), there is a problem of deciding when to change over between heating and cooling. In-space maintenance is required (filters in particular), and humidity is less tightly controlled than in all-air systems. In most air and water systems, the air is circulated in a cooled and dehumidified condition and heated by the water for most of the year.

Air and water systems can often be found in the perimeter zones of office buildings, hospitals, schools, apartments, and laboratories.

THERMAL CONTROL

## (a) Induction (Fig. 10.13*a*)

Centrally conditioned fresh air is supplied (at either high or medium pressures and velocities) to each induction terminal. Each terminal then mixes 20% to 40% incoming fresh air with the 80% to 60% room air that is induced to flow along with the fresh air, all passing over finned tubes for heating or cooling and circulating this mixture of air to the space.

In two-pipe systems, either hot or cold water—not both—is available to temper this air mixture ordered by the thermostat linked to each terminal. Because cool air is distributed for most of the year, the two-pipe system is largely in heating mode (in colder climates).

In four-pipe systems, the availability of both hot and chilled water makes it possible to switch instantly from heating to cooling for excellent thermal control.

The induction terminals typically are located either below perimeter windows (Fig. 10.58) or above a suspended ceiling. Condensation from the cooling coil in summer and the need to clean the filters of the induced room air make under-window

locations much easier to maintain, even if they intrude on the floor area available.

The CBS Tower in New York City (Fig. 10.59) uses a high-velocity induction system for its perimeter zones. The triangular black granite-faced exterior columns give the façade a three-dimensionality quite unlike the slick glass boxes of its contemporaries (it was designed by Eero Saarinen and built in 1962). These thick columns enclose the perimeter's distribution trees; every other column contains the high-velocity air supply; in between, every fourth column contains the supply water and every alternate fourth column the return water trees. These constant column sizes belie the thicker air, thinner water distribution trees. The interior zone of this tower is served by a VAV system; return air from both perimeter and interior zones is collected at the core.

## (b) Fan-Coil with Supplementary Air (Fig. 10.13*b*)

This system is closely related to the preceding induction system; however, this system uses a fan at each

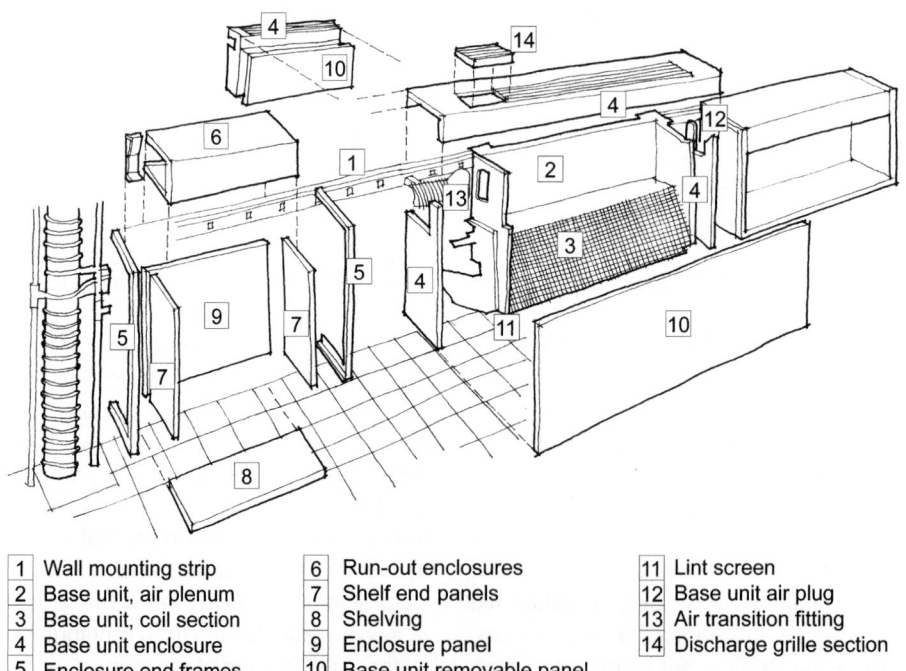

| | | | | | | | |
|---|---|---|---|---|---|---|---|
| 1 | Wall mounting strip | 6 | Run-out enclosures | 11 | Lint screen |
| 2 | Base unit, air plenum | 7 | Shelf end panels | 12 | Base unit air plug |
| 3 | Base unit, coil section | 8 | Shelving | 13 | Air transition fitting |
| 4 | Base unit enclosure | 9 | Enclosure panel | 14 | Discharge grille section |
| 5 | Enclosure end frames | 10 | Base unit removable panel | | |

**Fig. 10.58** *Installation of a high-velocity induction unit. Conditioned 100% outdoor air is brought in through a high-velocity duct to provide ventilation and to induce the circulation of room air. It is attenuated and silenced in the chamber (2), and then, through jets in the front of the plenum, it induces the flow of room air, which is heated or cooled at the fin coil (3). The lint screen (11) requires periodic maintenance for proper airflow. (Courtesy of Carrier Corporation. Redrawn by Erik Winter.)*

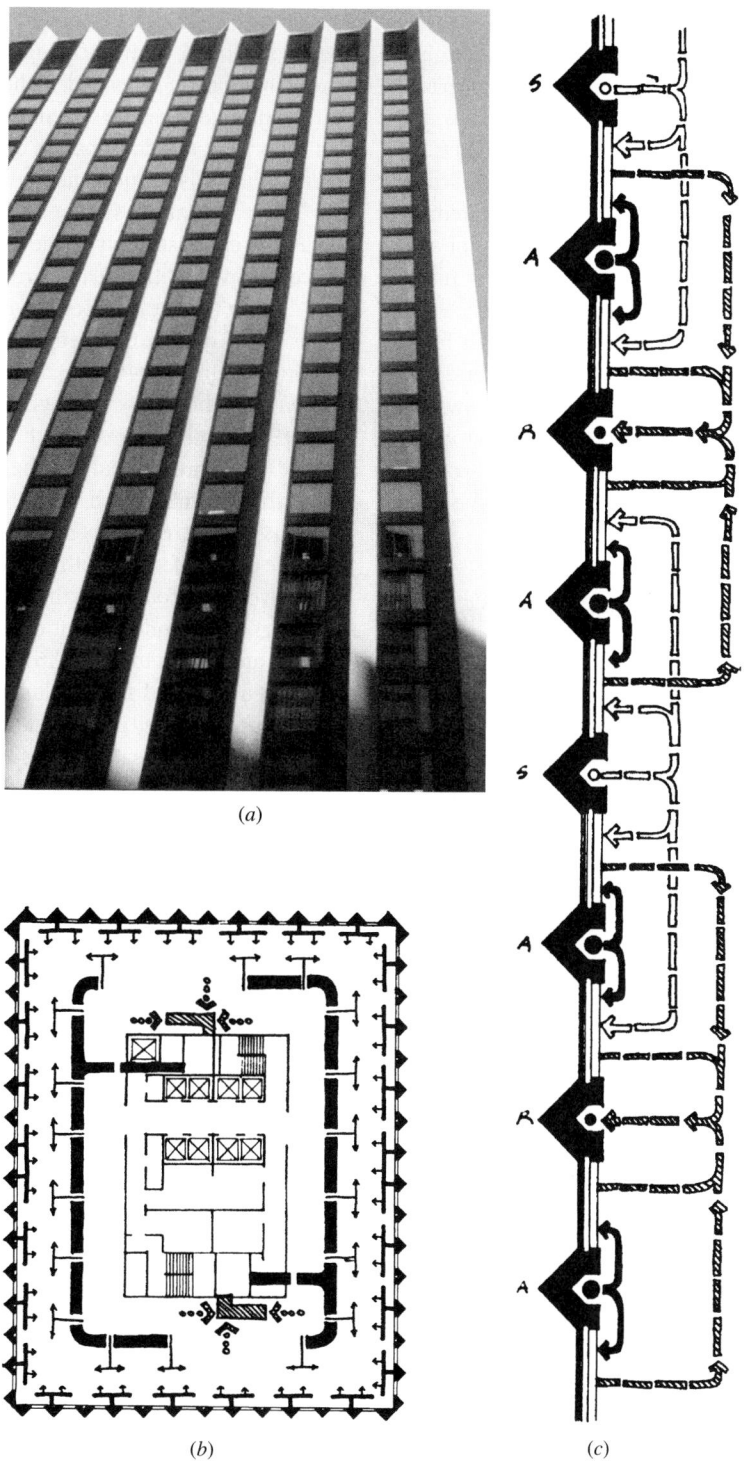

(a)

(b)                                                    (c)

**Fig. 10.59** *The identical triangular exterior columns of the CBS Tower in New York City (a) enclose a variety of distribution trees. (b) Plan of a typical floor; perimeter zones are served by a high-induction air and water system, interior zones by VAV. All return air is collected at the core. (c) At the perimeter, column type A contains an air supply tree feeding the high-velocity induction unit on either side of the column. Column type S contains the supply water tree, hot in winter and cold in summer. Column type R contains the return water tree. (From Guise, David, Design and Technology in Architecture, © 1985, John Wiley & Sons, Inc. Reprinted with permission of John Wiley & Sons, Inc.)*

unit, rather than high-velocity primary air, to move the mixture of supply and room air through the unit. A typical arrangement is a supply air plenum over a corridor, feeding horizontal fan-coils above the suspended ceilings of spaces on either side. These fan-coils draw some supply air and mix it with room return air to maintain desired temperatures. If condensation and resulting pans of standing water are to be avoided, the supply air must be sufficiently dehumidified.

Fan-coils are very widely used, both in conjunction with conditioned supply air (air and water) or as stand-alone units that take in their own fresh air at the perimeter (all-water), or even as room-air-only units. A building that uses both fan-coil units (at entries) and radiant panels (in offices) is shown in Fig. 10.60.

### (c) Radiant Panels with Supplementary Air (Fig. 10.13c)

Large areas of radiant surface can be used to offset large losses of bodily radiant heat, as in the case of large areas of cold glass on a winter day or when users are both scantily clothed and sedentary. In summer, such panels can help offset radiant gain from electric lights or large glass areas. The ceiling is often favored for the panel location, because it is uncluttered by the furniture, tackboards, and other items that cover floors and walls. At the 56-story Commerzbank in Frankfurt, Germany, chilled ceilings serve all office floors in conjunction with the ventilation system.

Green on the Grand (Fig. 10.60) is a 23,465-ft$^2$ (2180-m$^2$) two-story office complex in Kitchener, Ontario. Its design combines a very highly insulated envelope, daylighting, water conservation, a gas absorption (non-CFC, non-HCFC) chiller/heater with a heat rejection pond, a separate IAQ system, and heat recovery of exhaust air. It also features a dedicated air system for ventilation and a hydronic system for heating and cooling. The hydronic system uses fan-coils for the entryways and radiant panels on ceilings of office spaces. (The conditioned air in this example is provided by the ventilating system.) Fan-coils are better able to handle the extremes of entryway heat losses and gains. The radiant panels in the offices are used for both heating and cooling, but are sized to meet cooling loads and cover 30% of the ceiling area. Tenants were

given their choice of two designs: steel panels suspended below a drywall ceiling (painted to match the ceiling color) or extruded aluminum panels fit into a suspended ceiling system. To prevent condensation on these radiant panels during cooling, the ventilation air is dehumidified.

The ventilation system supplies low on the walls and exhausts high on the walls, a form of displacement ventilation. There are two rates of fresh air: 20 cfm (10 L/s) per person, and (when economizer cycles operate or when more fresh air or more cooling is required), 40 cfm (20 L/s) per person. The exhaust air then flows through two heat exchangers: the first reheats fresh air that has been greatly cooled for dehumidification; the second preconditions the incoming fresh air at the point of entry. The latter heat exchanger is a rotary-wheel ERV capable of transferring both heat and moisture.

Several approaches to the radiant cooling (and heating) panel, independent of a dehumidified air supply, are shown in Fig. 10.61. These approaches assume a minimal floor/ceiling thickness, which could contribute either to reducing the floor-to-floor height or to greater ceiling height (for daylight penetration or displacement ventilation). If a raised floor is used above these concrete slabs, even higher percentage heat gains come from the ceiling.

Radiant cooled floors are sometimes used, as in the Bleshman Regional Day School (Fig. 10.62). The designers of this school for the handicapped recognized that the majority of its users would spend much of their time quite close to the floor and that the colder air near the floor could be uncomfortable, especially in the New Jersey winter. The entire floor is warmed by the supply (ventilation) air in winter, which enters just below windows to counteract the downdraft off cold glass. In summer, the cool supply air first cools the floor and then cools the air in front of the warm glass. The concrete cellular "air floor" provides a thermal mass that helps maintain steady temperatures. Heated or cooled air is provided by rooftop air–air heat pumps, one of which is provided for each cluster of three to six classrooms. This HVAC example is therefore related to the direct refrigerant family, discussed in Section 7.8.

### (d) Water Loop Heat Pump

This approach was diagrammed in Fig. 9.45 and used in the school in Fig. 10.51. In the 175,000-ft$^2$

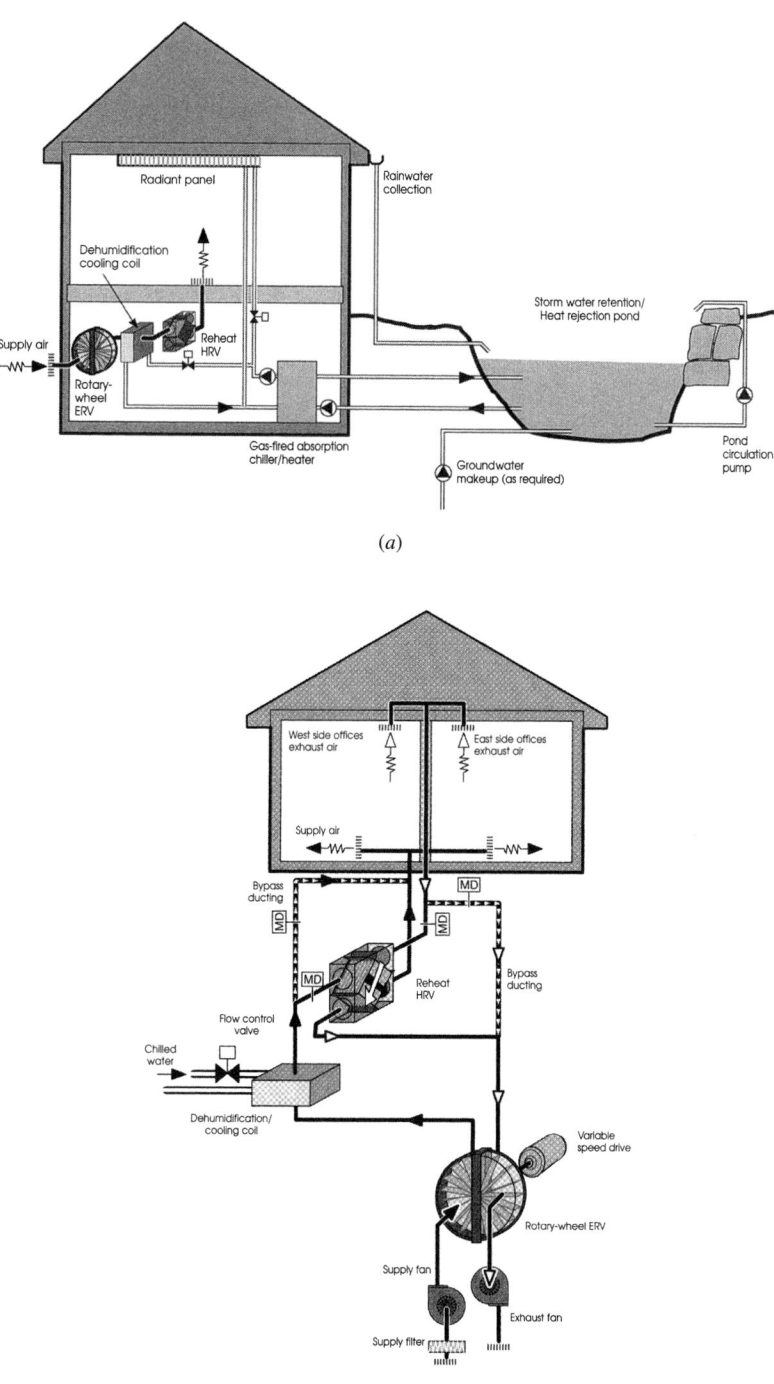

(a)

(b)

**Fig. 10.60** An innovative Canadian office building, Green on the Grand at Kitchener, Ontario. The cooling system combines a radiant panel system (a) with a dehumidified summer ventilation system (b). The building also features daylighting, a gas absorption chiller/heater, rainwater pond heat rejection, and two-stage heat recovery. (Courtesy of Enermodal Engineering, Ltd., Kitchener, ON.)

THERMAL CONTROL

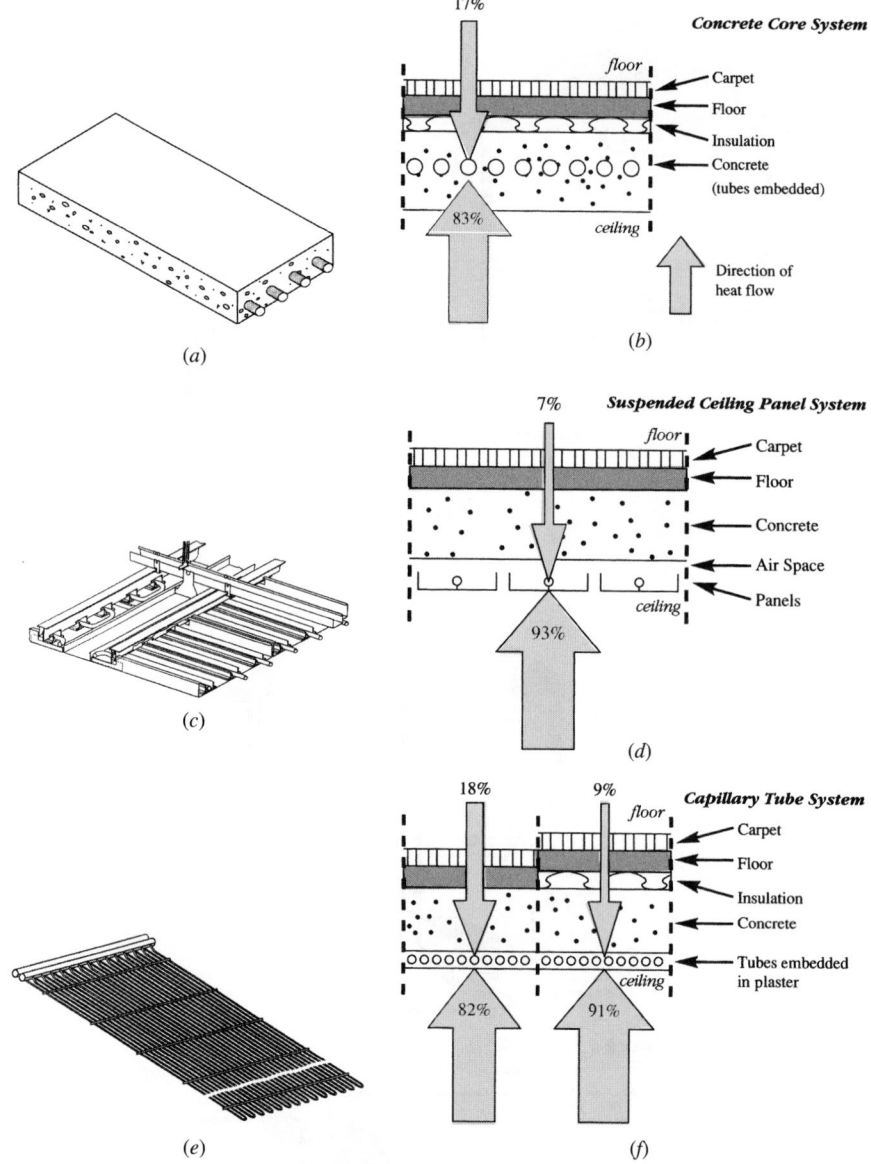

**Fig. 10.61** *Approaches to radiant cooling systems. (a) Concrete core system: water is circulated through plastic tubes embedded in the concrete floor/ceiling slab. (b) With carpet, pad, and insulation on the floor, most of the heat radiates to the ceiling. (c) Panel system, usually using aluminum facing, connected to metal tubes. (d) Panel system has the highest percentage of heat to the ceiling. (e) Cooling grid made of capillary tubes embedded in ceiling plaster (or in gypsum board or mounted on ceiling panels). This is the most even surface temperature distribution. (f) Again, a high percentage of the heat is radiated to the cooled ceiling, depending on the use of insulation below the carpet and pad. (From* Center for Building Science News, Fall 1994, Lawrence Berkeley Laboratory, CA.*)*

(16,260-m²) Comstock Center in Pittsburgh (Fig. 10.63), there are six to eight small heat pumps on each of 10 floors, located above the suspended ceiling. Their connecting loop doubles as the building's wet-pipe sprinkler system supply; this is possible because the heat pumps keep the loop between 65° and 85°F (18° and 29°C). Because neither hot nor chilled water supply and water return distribution trees are needed in addition to the sprinklers, there is a substantial first-cost saving, and relatively little building volume is consumed. A 23,000-gal (87,060-L) water storage tank allows daytime heat,

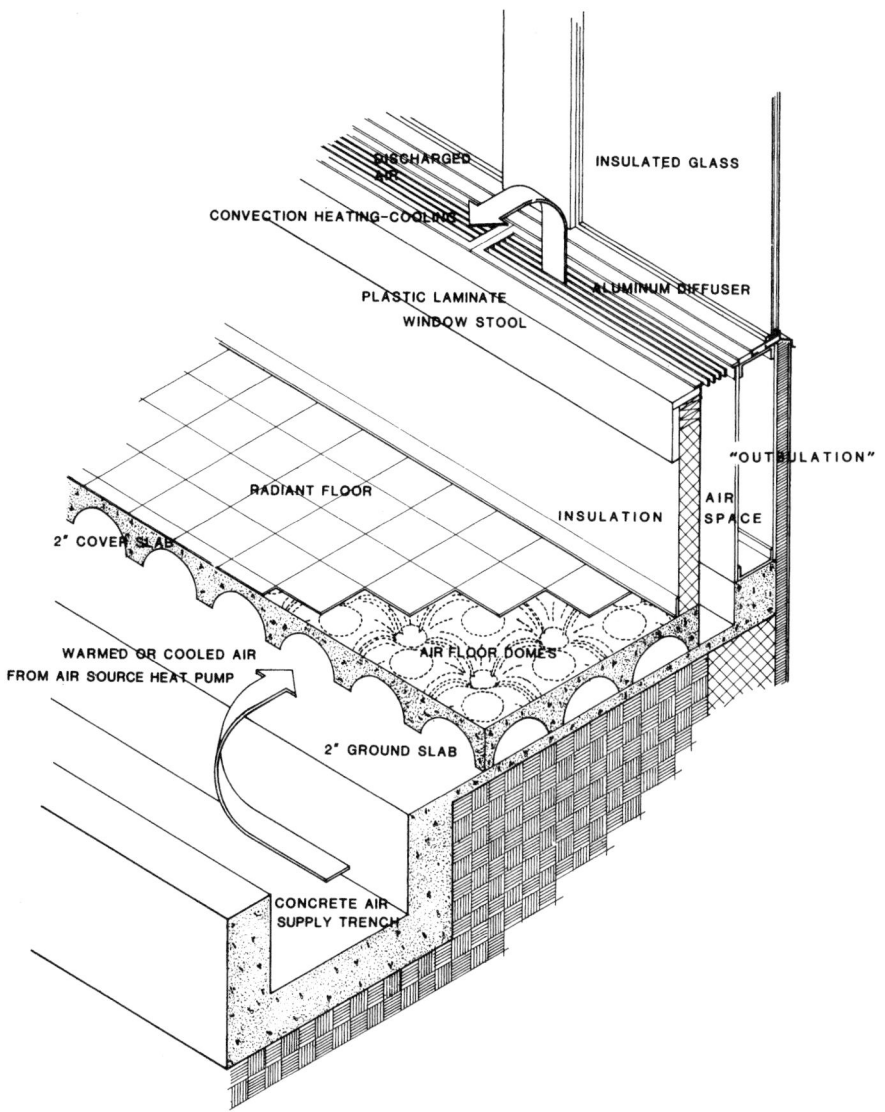

**Fig. 10.62** *Underfloor air distribution achieved with a cellular concrete air floor serves the Bleshman Regional Day School in Paramus, New Jersey. (Courtesy of Rothe-Johnson, architects.)*

rejected to the loop by the heat pumps, to be stored and recalled for nighttime heating. When necessary, penthouse equipment is used: two small boilers to maintain the 65°F minimum temperature in the water loop or an evaporative condenser to hold temperatures at 85°F.

Another feature of the Comstock Center is its use of air extract windows (see Fig. 10.50) to control infiltration and to moderate the perimeter zone's temperatures. After the return air passes up inside this window and arrives in the ceiling plenum, it is mixed with fresh air, then tempered and recirculated by the heat pumps.

The Comstock Center also utilizes a large daylighting atrium, whose temperature control is provided largely by exhaust air from the offices; the stack effect is utilized to provide natural ventilation on its west face in summer conditions.

THERMAL CONTROL

*(a)*

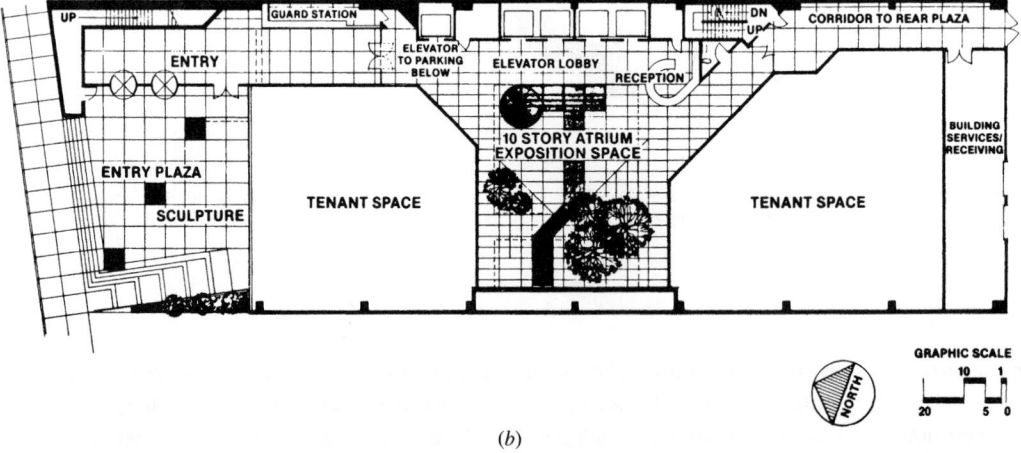

*(b)*

**Fig. 10.63** *The Comstock Center. Pittsburgh, Burt Hill Kosar Rittelmann Associates, architects. (a) The north and west faces. (b) Ground floor plan. (c) Water loop heat pumps are linked via the sprinkler system on each floor. This is also called a tri-water system. (d) Extract-air windows (see again Fig. 10.50) control infiltration and moderate the perimeter zone temperatures. Fresh air is ducted to the plenum, where it mixes with return air before being treated and recirculated through the heat pumps. (e) The central daylighting atrium is tempered by exhaust air from the offices; exhaust air then leaves via the stack effect. (f) The stack effect also controls summer overheating.*

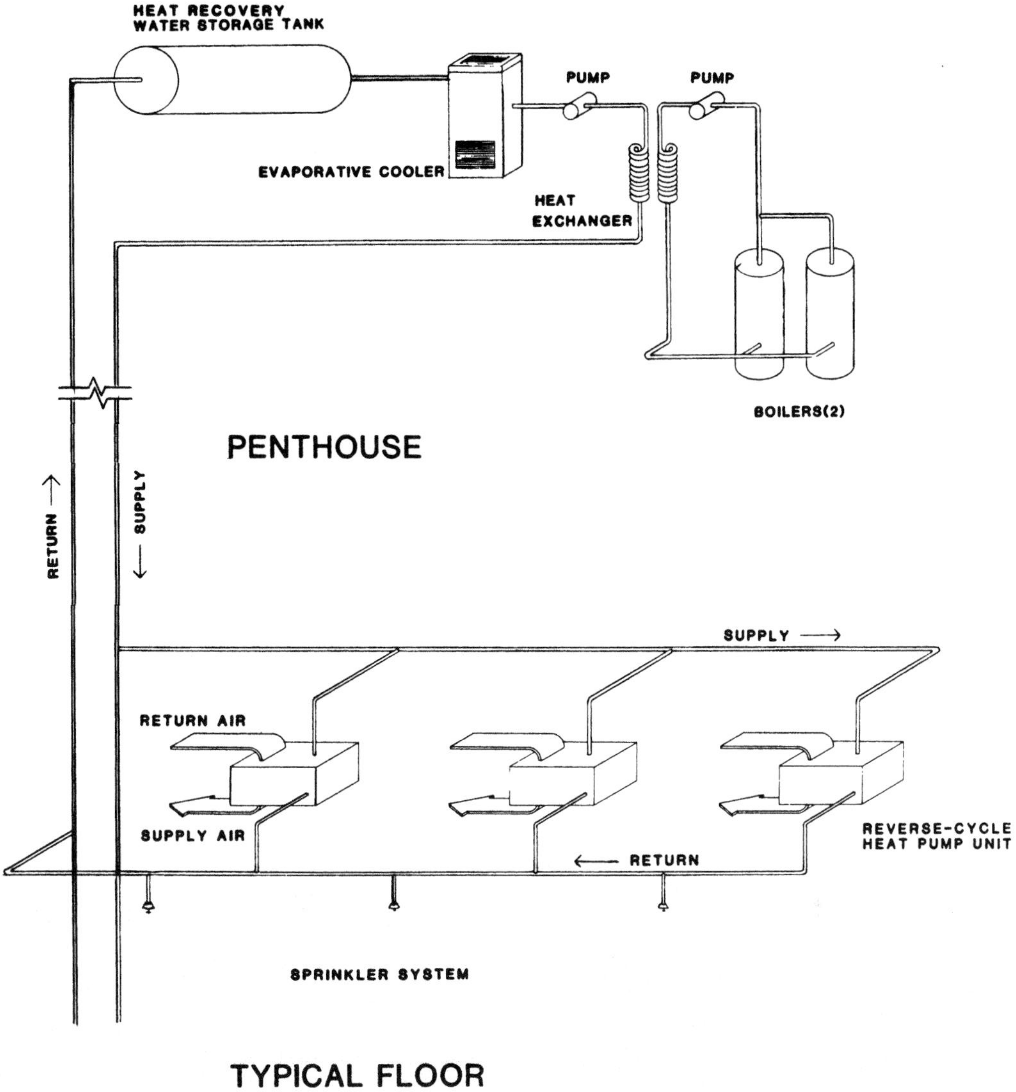

HEAT RECOVERY
WATER STORAGE TANK

PUMP

PUMP

EVAPORATIVE COOLER

HEAT
EXCHANGER

BOILERS(2)

RETURN →

← SUPPLY

PENTHOUSE

SUPPLY →

RETURN AIR

SUPPLY →

SUPPLY AIR

← RETURN

REVERSE-CYCLE
HEAT PUMP UNIT

SPRINKLER SYSTEM

## TYPICAL FLOOR

(c)

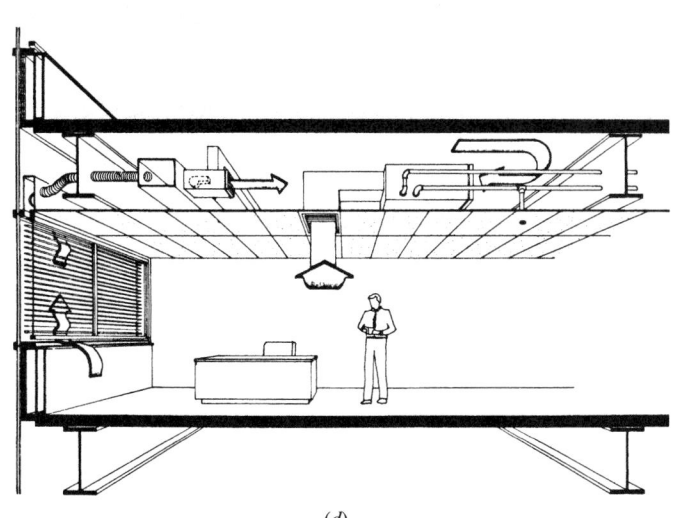

(d)

**Fig. 10.63** (Continued)

443

THERMAL CONTROL

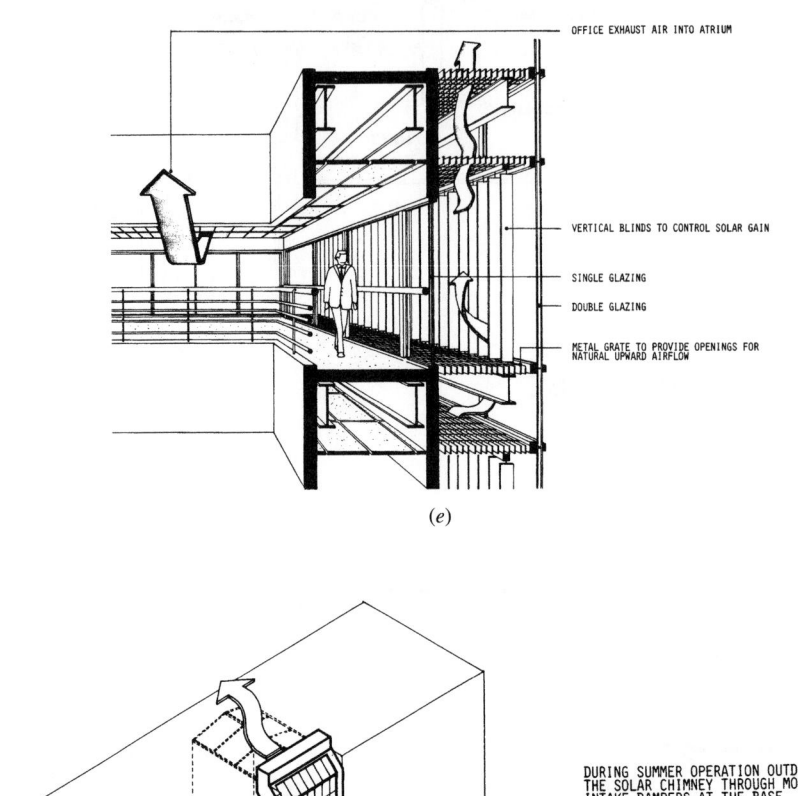

OFFICE EXHAUST AIR INTO ATRIUM

VERTICAL BLINDS TO CONTROL SOLAR GAIN

SINGLE GLAZING

DOUBLE GLAZING

METAL GRATE TO PROVIDE OPENINGS FOR
NATURAL UPWARD AIRFLOW

(e)

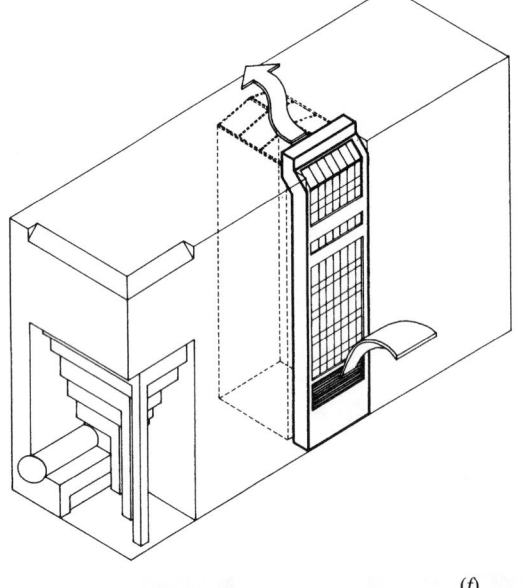

DURING SUMMER OPERATION OUTDOOR AIR ENTERS
THE SOLAR CHIMNEY THROUGH MOTORIZED AIR-
INTAKE DAMPERS AT THE BASE. SOLAR HEATED
AIR INDUCES A NATURAL UPWARD AIRFLOW
RISING APPROXIMATELY 100 FEET, WHICH IS
THEN EXHAUSED OUT THROUGH THE TOP OF THE
SHAFT. DURING WINTER OPERATION THE AIR
SHAFT IS SEALED TO CREATE A DEAD-AIR SPACE
BETWEEN THE INNER SINGLE AND OUTER DOUBLE
GLAZINGS.

(f)

**Fig. 10.63** (Continued)

## 10.7 ALL-WATER SYSTEMS

These systems (see Fig. 10.14 and Section 10.2d) typically deal only with temperature control; air quality is left to separate systems. An especially familiar all-water application is the simple *fan-coil unit* (Fig. 10.64). (Because units with the same name are used in air and water systems, this further blurs the distinction between these two HVAC families.) These units, which may be found above ceilings, below windows, or in corners, simply control the temperature (and, to a limited extent, the relative humidity) of the air already in the room, which is blown through the coils. Because water is often condensed from the room air when cooling is in progress, a drain line is required; water standing in drain pans is unhealthy. Exterior air intake grilles can easily be added when fan-coil units are located

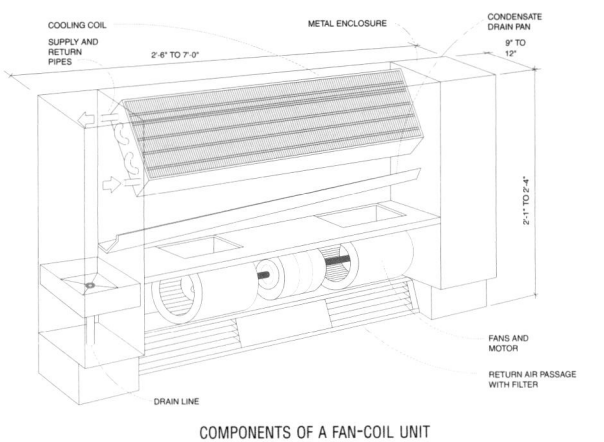

COOLING COIL

SUPPLY AND RETURN PIPES

METAL ENCLOSURE

CONDENSATE DRAIN PAN

2'-6" TO 7'-0"

9" TO 12"

2'-1" TO 2'-4"

FANS AND MOTOR

RETURN AIR PASSAGE WITH FILTER

DRAIN LINE

COMPONENTS OF A FAN-COIL UNIT

(a)

CONTROL PANEL INSIDE

IN-SPACE UNIT

INSULATED RISERS

INSULATED METAL ENCLOSURE

FLEXIBLE COPPER CONNECTIONS (TO ALLOW FOR RISER EXPANSION)

THERMOSTAT

COOLING COIL

CONDENSATE DRAIN PAN

MOTOR AND FAN

RETURN AIR SECTION WITH FILTER UNIT

RECESSED UNIT

CEILING FINISH

RECESSED UNIT

3'-5" TO 5'-8"

1'-2 1/2"

1'-0"

CEILING FINISH

NOTE

High-rise fan-coil units are available for stacked, vertical installations. Use of a stacked configuration can minimize the amount of piping required to connect the units to the distribution network and simplify collection of condensate. In this configuration the units are normally furred into the walls of the rooms being served (often in a corner).

HIGH RISE FAN-COIL

(b)

LOW PROFILE UNIT

VERTICAL FAN-COIL CONFIGURATIONS

(c)

IN-SPACE UNIT

HORIZONTAL FAN-COIL

(d)

**Fig. 10.64** Simple fan-coil units without fresh air. (a) Standard below-window unit. (b) High-rise unit for corners or cabinet locations. (c) Vertical units for below large windows. (d) Horizontal units for above-ceiling locations. (From AIA: Ramsey/Sleeper, Architectural Graphic Standards, 10th ed., © 2000 by John Wiley & Sons, Inc. Reprinted with permission of John Wiley & Sons, Inc.)

below windows to allow the local provision and tempering of ventilation air; this is especially common in motels, hotels, and apartments.

Fan-coil units are inherently noisy due to the fan, so careful attention to sound ratings is required when background sound levels must be low. In offices, a relatively high background sound level can help to maintain acoustic privacy at the workstation; the sound of the fan-coil provides reassurance that some kind of air treatment system is at work.

All-water perimeter systems with local fresh air can also take the simple form of *operable windows with hot water finned-tube radiation.* An expressive variation on this approach appears in the reading rooms and staff workrooms for the Seeley G. Mudd Library at Yale University (Fig. 10.65). Limestone spandrels are curved in to allow fresh air to enter these smaller-perimeter rooms just below the windows, where hot water finned-tube radiators are available when needed. The incoming fresh air replaces exhaust air, which flows out of the upper operable windows. The remainder of the building has conventional forced-air heating and cooling.

## 10.8 DISTRICT HEATING AND COOLING

Often, large projects made up of many large buildings are well served by *one* central station heating/cooling plant. The familiar economies of scale apply here; very large, efficient, and well-maintained boilers and chillers encourage energy recovery through heat exchange, reduce air pollution, and remove the noise and other nuisances associated with heating and cooling from the other buildings. This approach is called *district heating and cooling.*

District heating for residences and small commercial buildings is common in northern Europe; district steam systems serve the central areas of many U.S. cities. Most often, electrical generating plants are the heat (or steam) source; *cogeneration* (discussed in Section 10.9) allows the waste heat from the generating plant to be put to use either as space heating or steam (indirect-fired) absorption cooling. Ironically, the trend toward better-insulated buildings has reduced the market for distributed heat, making the installation less cost-effective. This discourages suppliers both from investing in a district system and from encouraging technical innovations in energy conservation by their customers. Smaller, densely built-up central heating districts tend to be more successful than large, sprawling ones; lengthy lines cost more to install and maintain and lose more energy.

### (a) High-Temperature Water and Chilled Water

Long-distance steam distribution has been used for more than a century; the development of *high-temperature water* (HTW) and chilled water distribution among buildings is a more recent development. Offering many advantages (although steam is still frequently chosen for city distribution), circulated high-temperature, high-pressure hot water and chilled water in closed systems are widely used in U.S. Air Force bases and airports, and for groups of buildings such as hospital complexes and college campuses. Increased efforts are now being made to install new district heating/cooling networks served by existing fossil-fueled electricity generating plants. Such plants waste more than half of their fossil fuel input (see Fig. 2.6), and district heating/cooling could intercept much of that waste; see also the next section on cogeneration.

Water will not flash into steam if kept at sufficiently high pressure. It may then be circulated by pumps through supply and return mains and through branches to heat exchangers, which operate conventional low-pressure hot water systems, generate steam, and perform numerous other thermal tasks. Pressures are on the order of 400 psig (pounds per square inch, gauge) and temperatures are about 300°F. During its circuit, the water will sometimes lose up to 150F° and 60 psig in pressure. The section shown in Fig. 10.66 illustrates a common arrangement.

High-temperature water has a number of advantages over steam for certain installations. It is a two-pipe system, and the temperature drop in the *supply* main is often as little as 10F°. With reasonably high water velocities, mains can be reduced to almost half the size of those required for steam distribution, with no need for steam traps and pressure-reducing valves. The pipes need not pitch to low points, as in the case of steam (to accommodate condensation), but can follow the contours of the ground. Although installation costs are greater, operational costs are less than those for steam. Feed

(a)

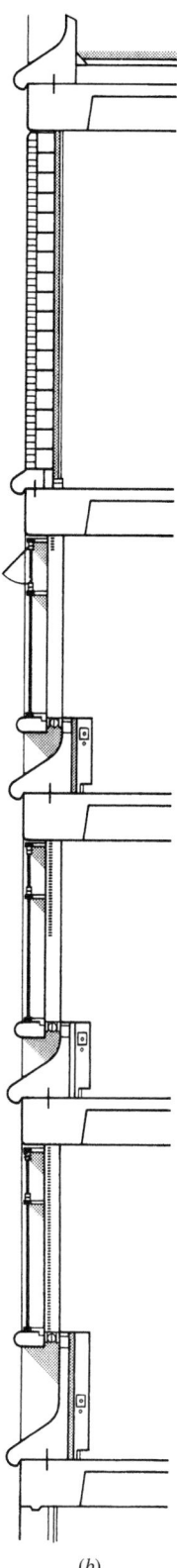

**Fig. 10.65** *Local fresh air and finned-tube radiation at the perimeter. The Seeley G. Mudd Library at Yale University; Roth and Moore, architects. (a) Exterior, showing curved limestone spandrels, along which flows incoming fresh air. (Photo © Steve Rosenthal.) (b) Section showing the fresh air intake, finned-tube radiation, and upper operable sash for exhaust air. This system is used for the smaller reading room and staff workrooms at the perimeter.*

water treatment is negligible and corrosion is minimal. Underground problems of expansion and insulation are the same as in other subterranean systems. Large sweep-type loops accommodate expansion between fixed points, and underground piping is embedded in special thermally efficient insulative fill.

District chilled water systems also offer advantages. The remote central chillers are likely candidates to use waste heat in a non-CFC absorption cooling cycle. Natural sources are possible; the

(b)

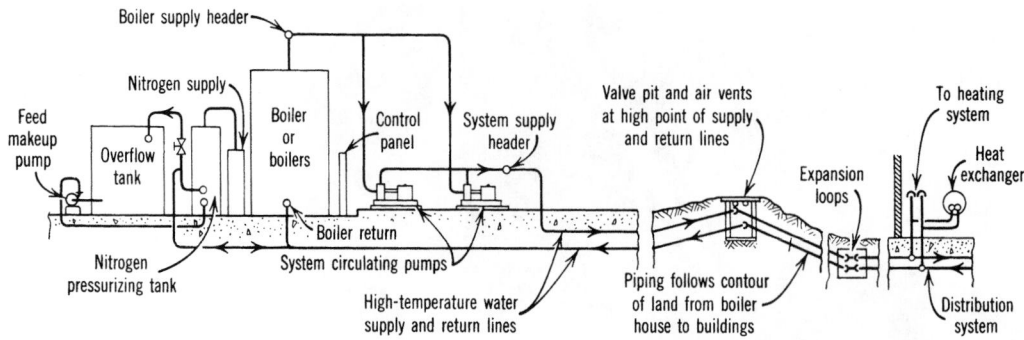

**Fig. 10.66** *Typical arrangement of a high-temperature water system. (Reprinted from* High-Temperature Water Systems, *Industrial Press. By courtesy of author Owen S. Lieberg, consulting engineer.)*

Toronto (Canada) District Heating Corporation uses Lake Ontario water, drawn from a 1.6-mile (2.6-km) intake at a depth of 200 ft (61 m) with a year-round temperature of 40°F (4.5°C). Passed through a heat exchanger (with the district chilled water) and then treated, the lake water then joins the city water supply.

With district heating/cooling, all facilities except air handling and ducts are located together in a remote central plant. This frees the buildings from the space requirements and visible impacts of stacks, boilers, fuel storage, water chillers, and cooling towers; the associated heat, humidity, fouled air, and noise are as remote as the central plant. When such a system serves individual customers, the heated or chilled water is metered. When it is owned by the group of buildings served (such as college campuses), it is usually not metered; this can become a problem when efforts to identify building energy waste and subsequent savings are being investigated.

## 10.9 COGENERATION

In the preceding chapters, electricity has repeatedly been called a *high-grade* source in reference to the high temperatures needed to produce electricity by conventional (fossil-fuel) means and to the large amount of waste heat (often twice the fraction of electricity) produced in the process. *Cogeneration* (also called *total energy*) is an attempt to recover some of the otherwise wasted lower-grade heat that accompanies the generation of electricity by steam turbines. Industrial cogeneration facilities are espe-cially cost-effective in the pulp and paper, petroleum, and chemical industries. Building-scale cogenera-tion is a less obvious energy bargain.

### (a) Electrical Power Generation at the Site

Where conditions are favorable, electricity for power and light can be generated economically by a system that also supplies the building with heating in winter and cooling in summer. Such a system uti-lizes a fuel such as gas or oil and is often supplemen-tal to the local electric utility company. Although installation costs are greater than those for the more conventional systems that use separate ser-vices of electricity and fuel, the savings in annual operating costs can sometimes pay for the excess installation cost in a reasonable time. Operational savings continue thereafter. This approach was developed largely in the 1960s and is used in hun-dreds of commercial and industrial buildings and in many schools. Cogeneration is particularly attrac-tive for district heating/cooling plants.

### (b) Early On-Site Power Generation

Before 1900 and for several years thereafter, nearly all large buildings and groups of buildings were supplied with direct current generated on or near the premises. The motive power was usually in the form of steam-driven reciprocating engines with belt connections to direct-current generators. Direct current cannot be transformed to different voltages and must be generated and distributed at the voltage used in the building. At these relatively low voltages, power loss in the distribution system is great, and distance adds greatly to the loss. This tended to keep electricity usage within a building.

With the development and use of alternating-current machines, utility companies were able to establish central power stations from which electricity could be transmitted great distances to the user at high voltage. There it was transformed down to domestic voltages for use. Because, at high voltage, power losses are very low, this system became universal. During the 1920s and 1930s, owners removed their private power generators and accepted utility service, with its savings in operating expense.

### (c) How Cogeneration Developed

Older buildings used steam produced from coal-burning boilers from which little or no energy salvage was possible. Today's fuels, used directly in reciprocating engines or turbines (including those that generate electricity), have residual heat value that can be recovered for purposes of heating or cooling. For cogeneration to be successful, there should be a reasonably steady demand in the building for the power generated and also for the heat recovered. Lighting and the demand for power by computers, electrical business machines, and other devices can create nearly constant demand for power throughout the year. Similarly, the exhaust heat recovery from the engines or turbines that power the generators is in demand for either heating or indirect-fired absorption cooling at most times of the year.

### (d) Turbines and Reciprocating Engines

Figure 10.67 shows two principal systems for total energy. In both systems—one using a turbine and the other a reciprocating engine to operate the generator that supplies electric power—heat is reclaimed to produce steam or hot water. The steam or hot water is then used for heating or, by use of an absorption chiller, for cooling. When a turbine is used, the fuels are natural gas or fuel oil. The heat is recovered by passing the hot turbine exhaust through a waste heat boiler to produce steam. Fuel for a reciprocating engine is natural gas or diesel fuel. Both the jacket cooling water and the hot engine exhaust are passed through heat exchangers that utilize the heat to produce steam or hot

water for heating or cooling. In both systems, an auxiliary boiler, fired directly by gas or oil, stands ready to help maintain a balance in the system.

Cogeneration offers a degree of electrical independence and a way to use otherwise wasted heat. The latter is essentially free energy because it would otherwise have to be purchased and paid for separately in the form of electricity or other fuels.

### (e) Cogeneration for Housing

The Harbortown apartment and townhouse complex in Detroit, Michigan, utilizes a year-round natural gas–fired primary energy and cogeneration/waste recovery system (Fig. 10.68). Each apartment has one or two stackable, upright fan-coil units using either chilled or heated water. This water is provided by two direct-fired chiller–heaters: each incorporates a natural gas boiler and a two-stage absorption chiller in the same unit, taking less space in the mechanical room. The natural gas–driven cogeneration electric plant serves rather constant loads such as corridor and outdoor lighting. The generator's waste heat is passed to a storage tank, where it preheats domestic hot water for the tenants. This meets about 87% of the hot water energy usage.

### (f) Fuel Cells

We first encountered fuel cells in connection with the explorations of outer space. Spaceships use stored oxygen and hydrogen to feed the fuel cell, yielding—seemingly miraculously—electricity, heat, and pure water. In effect, this reverses the process of electrolysis: using electricity to split water into its two components, hydrogen and oxygen. This theory dates back to Sir William Grove in 1839, long before the needs of space travel.

A fuel cell generates direct current power by converting the *chemical energy* of hydrogen and oxygen into electricity and heat. Because no combustion is involved, nitrogen oxides and carbon monoxide are nearly eliminated. For a building's fuel cell power plant, several components are added: first, a fuel processor (or fuel reformer) to prepare a hydrogen-rich stream of fuel, then the fuel cell stack, then a power conditioner (or inverter) to convert the fuel cell's direct current to alternating current.

THERMAL CONTROL

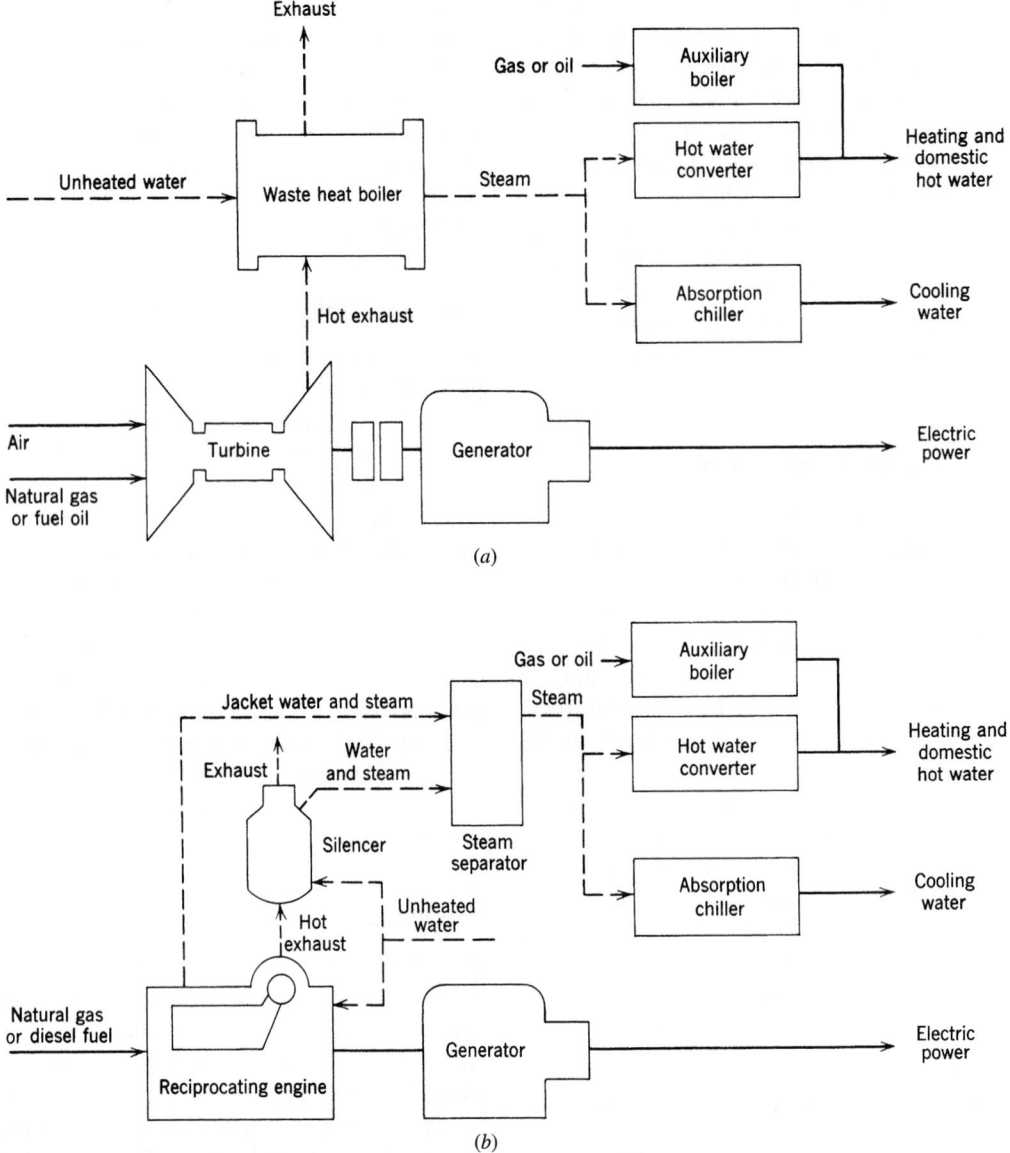

**Fig. 10.67** Cogeneration (total energy) systems. (a) Using a turbine. (b) Using a reciprocating engine. (Reprinted by permission from Total Energy, Educational Facilities Laboratories.)

Fuel cells are not, by themselves, that much more efficient than combustion processes; they are about 40% efficient for power production. When combined with a way to use that 60% waste heat, they climb toward a total of 90% efficiency, with very few harmful emissions.

Hydrogen is the ultimate fuel; already hydrogen-rich, it needs no processor and produces the least environmental impacts from the fuel cell.

Although hydrogen supply networks are not yet commonplace, scenarios exist for a "hydrogen economy" in which surplus electricity from windpower is used to split water into hydrogen and oxygen atoms and the hydrogen stored for later distribution. Windpower's extreme variability often presents power grid distribution challenges; storage solves this problem. Solar energy or biomass are other renewable sources for electrical generation.

(a)

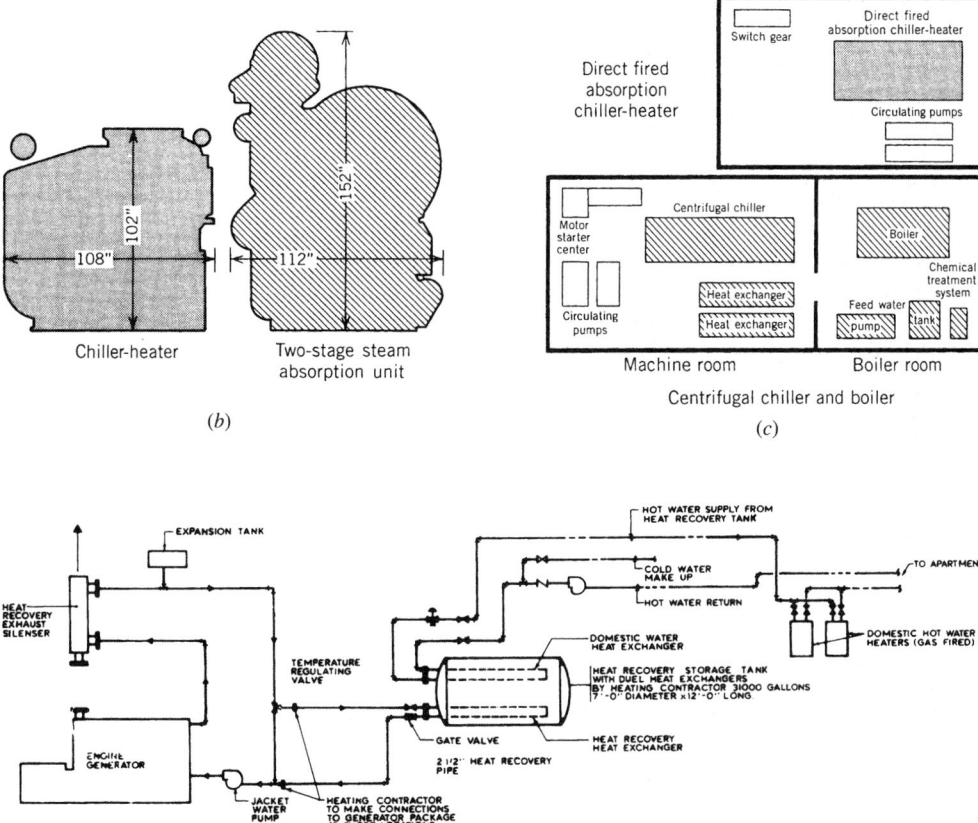

(b)

(c)

FLOW DIAGRAM OF HEAT RECOVERY SYSTEM

(d)

**Fig. 10.68** Harbortown is a 120,000-ft² (11,150-m²) apartment and townhouse complex (a) in Detroit that uses natural gas as a primary fuel year round. (Photo © 1989, William Kildow.) Chiller-heaters (Hitachi) combine a boiler and a two-stage absorption chiller in one compact unit. (b) Comparison of headspace and (c) floor space required by chiller-heaters versus conventional units of similar capacities. (d) Cogeneration plant sends its waste heat to preheat domestic hot water. (Courtesy of Skidmore, Owings and Merrill, Architects-Engineers, Chicago.)

However, other hydrocarbon fuels also can feed fuel cells: natural gas (most common for building fuel cell power plants), propane, methanol, and, with additional pretreatment, methane from landfills and anaerobic digester gas from sewage treatment plants. Even with these less hydrogen-rich fuels, the emissions from fuel cells are well below those of combustion processes; the fuel processor emits carbon dioxide and a trace of carbon monoxide.

The proton-exchange membrane (PEM) fuel cells (Fig. 10.69) are promising for transportation when zero-emission vehicles are the objective and hydrogen is available as the fuel. This process has building applications as well. It works at a temperature below the boiling point of water. Relative to other fuel cell types, it has a smaller size, lighter weight, and lower noise levels. The PEM fuel cell stack involves several subsystems, typically including water purifiers and pumps, air compressors, and heat rejection components (coolant pumps and a radiator). This complicates performance at subfreezing temperatures and adds maintenance requirements. But the smaller size promises smaller building applications in the future.

The phosphoric acid fuel cell power plant (Fig. 10.70) is at work in (or outside) a number of buildings and municipalities around the world. The ONSI Corporation's 200-kW fuel cell power plant measures 10 ft × 18 ft, is 10 ft high (3 m × 5.5 m, 3 m high), weighs 40,000 lb (18,144 kg), and needs a clearance of 8 ft (2.5 m) around the module for maintenance. It also needs a cooling module (or a building's cooling tower) for discharge of heat in excess of that recovered for building use. The acid electrolyte in the fuel cell stack works at about 390°F (200°C). At its rated power, it also produces 700,000 Btu/h (205 kW) to heat a water stream to 140°F (60°C). The rated sound level is 62 dBA at 30 ft (9 m) from the module.

The Durst Organization's high-rise office building at 4 Times Square in New York City (Fig. 10.71) is a demonstration of several future-oriented strategies: PV cells integral to the façade, direct-fired (natural gas) absorption chillers, increased fresh air for IAQ, a dedicated exhaust air shaft (for smoking and other polluting activities), and waste chutes to facilitate recycling, among others. It also utilizes two 200-kW fuel cell power packages fed by natural gas. The electricity is about 80% destined for external lighting by night; huge electrical signs will be a prominent part of the building's Times Square façades. By day, about 80% of the power goes to the building's base load. The hot water will be used in winter for perimeter heating. In summer, the water is wasted. Although there was hope that one of the absorption chillers could be fed by this water, a variety of considerations precluded this: the fuel cells are located on the fourth floor (accessible to maintenance and close to the huge signs), whereas the chillers are located in the penthouse.

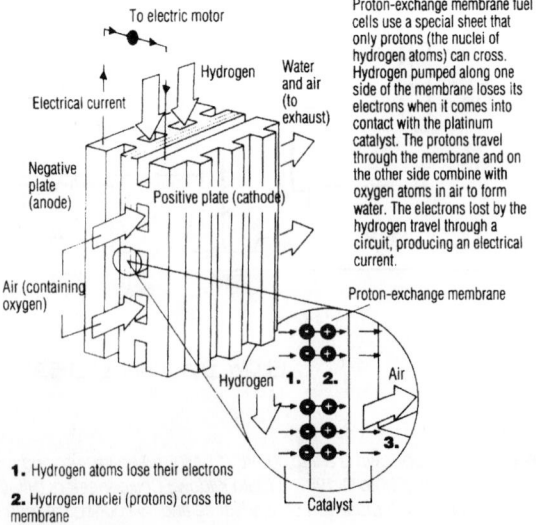

**Fig. 10.69** *Schematic diagram of a proton-exchange membrane (PEM) fuel cell with a description of its operation. (Courtesy of* Nucleus, *Fall 1994, © Union of Concerned Scientists, Cambridge, MA.)*

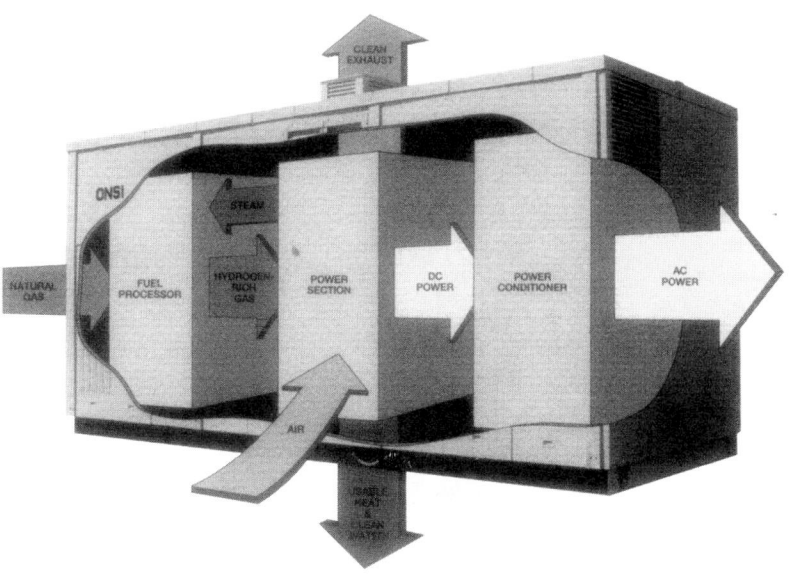

**Fig. 10.70** *A fuel cell power plant using natural gas (or other hydrocarbon fuel). The fuel processor (reformer) converts the natural gas to a hydrogen-rich stream that enters the fuel cell power section (or stack). The output is electricity (dc), water, carbon dioxide and considerable heat. Some heat is used by the fuel processor and some is rejected, but most is usable in a building or industry. A power conditioner (inverter) converts dc to ac power. (Courtesy of ONSI Corporation.)*

(a)

**Fig. 10.71** *The 48-story office building at 4 Times Square, New York City (a), under construction in 1998. (b) Many environmentally friendly features are included, such as façade-integrated PV and two fuel cell power plants like those shown in Fig. 10.70. Fed by natural gas, the fuel cells are expected to work without maintenance interruptions for up to 5 years and provide considerable hot water as well as electricity. (The fuel cells are actually installed at the fourth-floor level rather than at the penthouse.) (Courtesy of Fox and Fowle, Architects, P.C., New York. Consentini Associates, Mechanical Engineers; the Durst Organization, Developer.)*

**453**

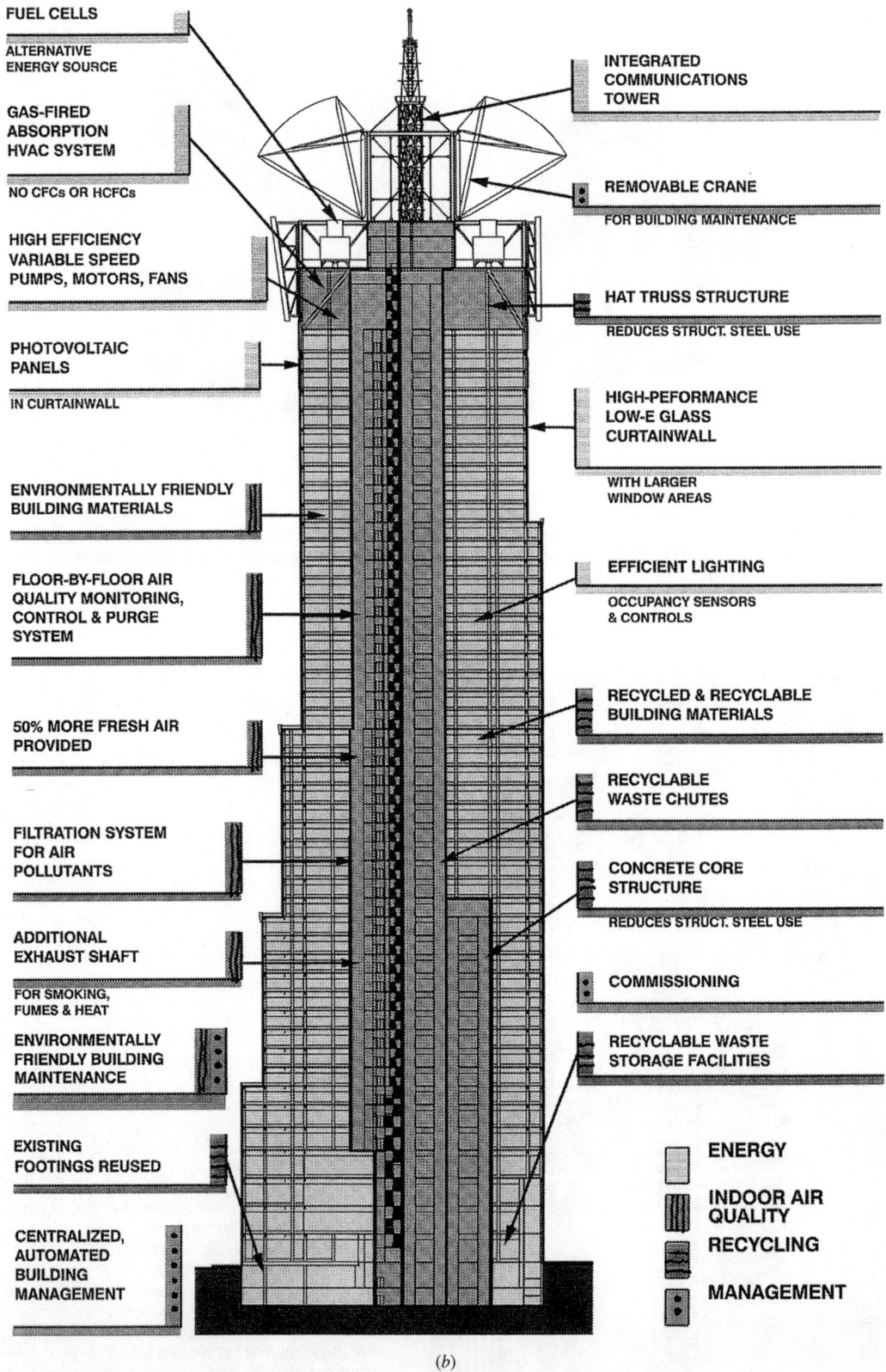

THERMAL CONTROL

FUEL CELLS
ALTERNATIVE
ENERGY SOURCE

GAS-FIRED
ABSORPTION
HVAC SYSTEM
NO CFCs OR HCFCs

HIGH EFFICIENCY
VARIABLE SPEED
PUMPS, MOTORS, FANS

PHOTOVOLTAIC
PANELS
IN CURTAINWALL

ENVIRONMENTALLY FRIENDLY
BUILDING MATERIALS

FLOOR-BY-FLOOR AIR
QUALITY MONITORING,
CONTROL & PURGE
SYSTEM

50% MORE FRESH AIR
PROVIDED

FILTRATION SYSTEM
FOR AIR
POLLUTANTS

ADDITIONAL
EXHAUST SHAFT
FOR SMOKING,
FUMES & HEAT

ENVIRONMENTALLY
FRIENDLY BUILDING
MAINTENANCE

EXISTING
FOOTINGS REUSED

CENTRALIZED,
AUTOMATED
BUILDING
MANAGEMENT

INTEGRATED
COMMUNICATIONS
TOWER

REMOVABLE CRANE
FOR BUILDING MAINTENANCE

HAT TRUSS STRUCTURE
REDUCES STRUCT. STEEL USE

HIGH-PEFORMANCE
LOW-E GLASS
CURTAINWALL
WITH LARGER
WINDOW AREAS

EFFICIENT LIGHTING
OCCUPANCY SENSORS
& CONTROLS

RECYCLED & RECYCLABLE
BUILDING MATERIALS

RECYCLABLE
WASTE CHUTES

CONCRETE CORE
STRUCTURE
REDUCES STRUCT. STEEL USE

COMMISSIONING

RECYCLABLE WASTE
STORAGE FACILITIES

ENERGY

INDOOR AIR
QUALITY

RECYCLING

MANAGEMENT

(b)

**Fig. 10.71** (Continued)

454

The 200-kW fuel cell in Fig. 10.70 is usually fed by natural gas, but other examples include hydrogen fuel (Hamburg, Germany: electricity to the grid, hot water to heat an apartment building), anaerobic digester gas (Yonkers, New York: wastewater treatment plant; electricity and hot water used within the facility), and landfill gas (Groton, Connecticut: closed landfill; electricity to the grid, plans for a hydroponic tomato–growing greenhouse to utilize the hot water). Plentiful hot water is a by-product of the fuel cell; it seems that finding a year-round use for this water is often a problem of excess supply.

Other fuel cell types include alkaline (operates at 480°F [250°C]), molten carbonate (works faster at 1110–1300°F [600–700°C]) and solid oxide (1200 to 1830°F [650 to 1000°C]) with a higher efficiency, and waste heat sufficient to run gas turbines. These higher-temperature processes are more likely at central station power plants.

Clearly, cogeneration and fuel cells have a promising future.

### References

AIA. 1994. Ramsey, C.G. and H.R. Sleeper. *Architectural Graphic Standards*, 9th ed. American Institute of Architects/John Wiley & Sons, New York.

Allen, E. 1989. *The Architect's Studio Companion.* John Wiley & Sons, New York.

Allen, E. and J. Iano. 1994. *The Architect's Studio Companion,* 2nd ed. John Wiley & Sons, New York.

ASHRAE. 2004. *ASHRAE Handbook—HVAC Systems and Equipment.* American Society of Heating, Refrigerating and Air-Conditioning Engineers, Inc. Atlanta, GA.

Lorsch, H. (ed.). 1993. *Air-Conditioning Systems Design Manual.* American Society of Heating, Refrigerating and Air-Conditioning Engineers, Inc. Atlanta, GA.

THERMAL CONTROL

# PART III

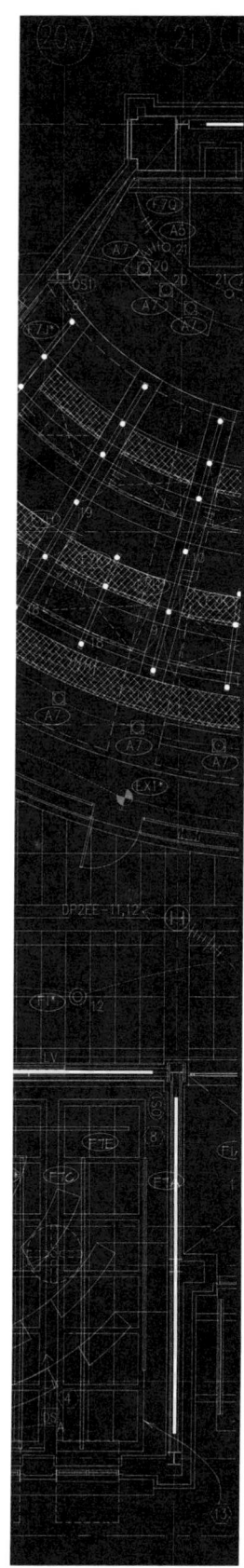

# ILLUMINATION

It has been estimated that 90% of the information we obtain from our senses is received via sight. Vision, in turn, is made possible by light, the proper provision of which in buildings is discussed in this part of the book. Because architecture is a uniquely visually oriented profession, its practitioners must thoroughly understand the art and science of illumination in order to be able to integrate enclosure and light into a working whole that will appear, and function, according to the design intent. A building designer must have a sufficient background, both technically and aesthetically, to be able to understand and apply the principles of light and lighting.

Chapter 11, Lighting Fundamentals, introduces the subject with terminology, definitions, basic characteristics, and measurements. A clear distinction is made between the study of light as a physical phenomenon and its study for visual photometric effect. The former considers light as a form of energy, whereas the latter views it as a visual phenomenon (i.e., the physiological response to a form of energy). Pursuing the subject of human response to light, the chapter continues with a discussion of factors in visual acuity. The chapter concludes with an overview of quantitative and qualitative indictors of lighting system performance and acceptability.

Chapter 12, Light Sources, covers the two principal light sources: daylight and electric light. The characteristics of daylight as a source are presented. The chapter continues with extensive information on electric light sources, covering in detail incandescent, fluorescent, mercury-vapor, metal–halide, and sodium-vapor lamps, plus a few recently developed electric light sources of potential interest to designers. The material includes a detailed description of lamp construction and operating characteristics, including accessories such as ballasts and lamp holders. Emphasis is placed on luminous efficacy and overall life-cycle costs.

Chapter 13, Lighting Design Process, leads the reader step-by-step through a recommended design process. Cost issues, power budgets, energy considerations, and approaches to providing appropriate illumination are discussed.

Chapter 14, Daylighting Design, provides a discussion of daylighting as a passive design solution. Means of providing daylighting to a building are presented, along with commonly used methods for daylighting design and analysis.

Chapter 15, Electric Lighting Design, discusses lighting fixture characteristics, illumination calculation techniques, and lighting control strategies and techniques. The chapter concludes with a discussion of lighting design evaluation. Illustrative examples are used to demonstrate and compare calculation techniques.

Chapter 16, Electric Lighting Applications, applies the knowledge and techniques from the preceding chapters to specific building occupancies (residential, educational, commercial, and industrial), each of which has its special needs. Special topics are discussed in detail, including emergency lighting, exterior lighting, and lighting for computer-intensive work areas.

# CHAPTER 11

# Lighting Fundamentals

Architecture is the masterly, correct and magnificent play of masses brought together in light. Our eyes are made to see forms in light; light and shade reveal these forms; cubes, cones, spheres, cylinders, or pyramids are the great primary forms which light reveals to advantage.

— Le Corbusier

## 11.1 INTRODUCTORY REMARKS

For many years, an unwarranted division existed in the field of lighting design, dividing it into two disciplines: architectural lighting and utilitarian design. The former found expression in design that took little cognizance of visual task needs and displayed an inordinate penchant for incandescent wall washers, architectural lighting elements, and form-giving shadows. The latter saw all spaces in terms of illuminance levels and cavity ratios, and performed its design function with footcandles (lux) and dollars as the ruling considerations. That both of these trends have generally been eliminated is due largely to the efforts of thoughtful architects, engineers, and lighting designers, assisted in part by the energy consciousness that followed the 1973 oil embargo. The last event spurred research into satisfying vision needs within a framework of minimal energy use. That research, and its resulting energy codes and continuing development of higher efficiency sources, are today motivated by environmental considerations.

Another positive factor in the rationalization of lighting design has been the work of the Illuminating Engineering Society of North America (IESNA).

Its activities in research, standardization, and publication have done much to place lighting design on a stable scientific basis while taking full cognizance of its essential artistic aspects. It is precisely this combination of science and art that makes lighting design an architectural-type discipline. For each project, a responsible lighting designer will consider *quantitatively:*

1. Daylight—its introduction and integration with electric light
2. The interrelationship between the energy aspects of electric and daylighting, heating, and cooling
3. The effect of lighting on interior space arrangement and vice versa
4. The characteristics, means of generation, and utilization techniques of electric lighting
5. Visual needs of specific occupants and of specific tasks
6. The effects of brightness patterns on visual acuity

and *qualitatively:*

7. The location, interrelationship, and psychological effects of light and shadow—that is, brightness patterns

**459**

8. The use of color, both of light and of surfaces, and the effect of the illuminant source on object color and sometimes the reverse
9. The artistic effects possible with patterns of light and shadow, including the changes inherent in daylighting, and so on
10. Physiological and psychological effects of the lighting design, particularly in spaces occupied for extended periods

The list is almost endless, because so much of the information we receive from our senses comes via our eyes, and what we see is a direct consequence of scene lighting.

As a result of the need to consider these and other interrelated factors, many of which are mutually incompatible, the lighting designer is faced with many difficult decisions. The purpose of Chapters 11 to 16 is then twofold: to provide the background that will help lighting designers make these decisions correctly and to make them proficient in the use of lighting as a design tool.

## PHYSICS OF LIGHT

### 11.2 LIGHT AS RADIANT ENERGY

The IESNA defines light as visually evaluated radiant energy or, more simply, as a form of energy that permits us to see. If light is considered as a wave, similar to a radio wave or an alternating current wave, it has a frequency and a wavelength. Figure 11.1 shows the position of light in the wave spectrum and its relation to other wave phenomena of various frequencies.

From Fig. 11.1 we see that even the longest-wavelength light (red) has a much higher frequency than radio waves and radar, and that light constitutes only a very small part of the wave energy spectrum. Color is determined by wavelength. Starting with the longest wavelengths (red), we proceed through the spectrum of orange, yellow, green, blue, indigo, and violet to arrive at the shortest visible wavelengths (highest frequency). Bordering the visible spectrum are infrared at the low-frequency (long wavelength) end and ultraviolet at the high-frequency end. Both are invisible to humans but not to some animals.

When a light source produces energy over the entire visible spectrum in *approximately* equal quantities, the combination appears white, whereas a source producing energy over only a small section of the spectrum produces its characteristic colored light. Examples are the blue–green clear mercury lamp and the yellow sodium lamp.

For our purposes, all light will be considered white unless specifically noted otherwise. This position is scientifically tenable because light sources with large differences in chromatic content all appear white after a short accommodation period,

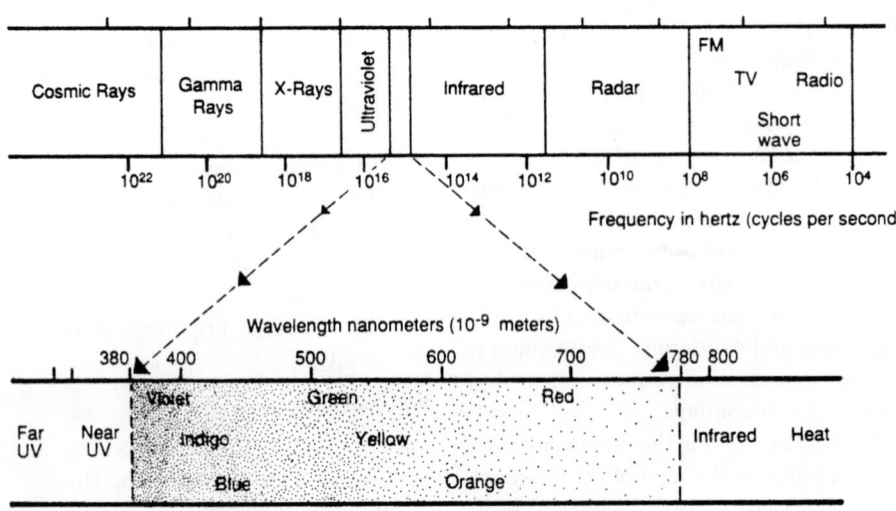

**Fig. 11.1** *Electromagnetic spectrum.*

and all standard commercial sources permit colors to be easily and correctly identified. It is only when sources differing widely in chromaticity are viewed side by side that the variation in whiteness can be noticed by the effect on colored objects and on neutral surfaces.

## 11.3 TRANSMITTANCE AND REFLECTANCE

Lighting design is possible because light is predictable; that is, it obeys certain laws and exhibits certain fixed characteristics. Although some of these are so well known as to appear self-evident, a review is in order.

The *luminous transmittance* of a material such as a luminaire lens or diffuser is a measure of its capability to transmit incident light. By definition, this quantity, known variously as *transmittance, transmission factor,* and *coefficient of transmission,* is the ratio of the total transmitted light to the total incident light. In the case of incident light containing several spectral components passing through a material that displays *selective* absorption, this factor becomes an average of the individual transmittances for the various components and must be used cautiously. A piece of frosted glass and a piece of red glass may both have a 70% transmission factor, but obviously they affect the incident light differently. In general, then, transmission factors should be used only when referring to materials displaying nonselective absorption—that is, those that transmit the various component colors equally. Clear glass, for instance, displays a transmittance between 80% and 90%, frosted glass between 70% and 85%, and solid opal glass between 15% and 40%. The remainder is absorbed and reflected. See Table 14.7 for typical transmission factors.

Similarly, the ratio of reflected to incident light is variously called *reflectance, reflectance factor,* and *reflectance coefficient.* Thus, if half the amount of light incident on a surface is bounced back, the reflectance is 50% (or 0.50). The remainder is absorbed, transmitted, or both. The amount of absorption and reflection depends on the type of material and the angle of light incidence because light impinging on a surface at grazing angles tends to be reflected rather than absorbed or transmitted (Fig. 11.2). An example of almost perfect reflection from an opaque surface would be that from a well-

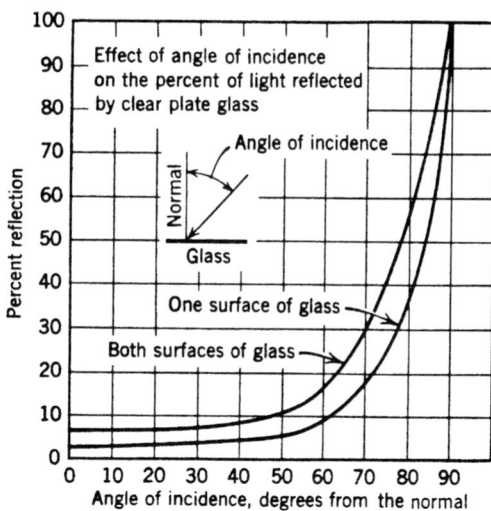

**Fig. 11.2** Relation between angle of incidence and percentage of reflectance. This effect is important when considering the penetration of sunlight into interior spaces and, conversely, the exterior glare produced by reflection of the sun from building windows.

silvered mirror, whereas almost complete absorption takes place on an object covered with lamp black or matte-finish black paint. The effect of the material's surface finish on reflection is shown in Fig. 11.3. See Table 14.9 for typical reflectance values. Reflectance measurement is discussed in Section 11.11.

The reflection that occurs on a smooth surface such as polished glass or stone is called *specular reflection,* as in Fig. 11.3a. If the surface is rough, multiple reflections take place on the many small surface projections and the light is diffused, as in Fig. 11.3b. Reflectance is a measure of total light reflected; it may be specular or diffuse, or a combination of both, as shown in Fig. 11.3c.

Diffuse transmission takes place through any translucent material such as frosted glass, white glass, milky Plexiglas, tissue paper, and so on. This diffusing principle is widely employed in lighting fixtures (luminaries) to spread the light generated by the source within the fixture. Diffuse and non-diffuse transmission are illustrated in Fig. 11.4a and Fig. 11.4b.

## 11.4 TERMINOLOGY AND DEFINITIONS

Before beginning any discussion of lighting studies, techniques, and effects, it is important to have a basic understanding of the physical concepts and

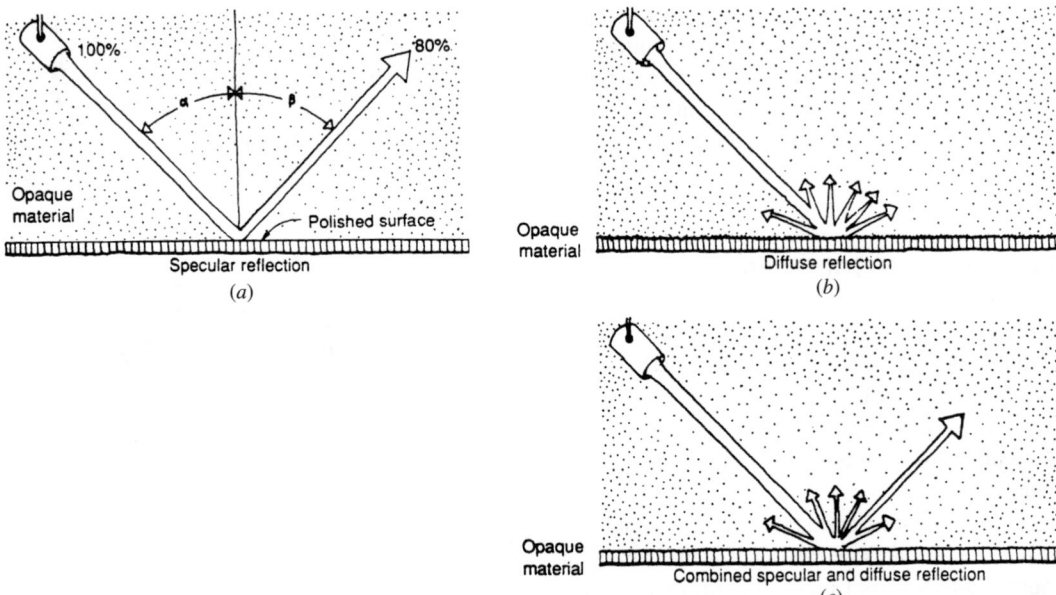

**Fig. 11.3** *Reflection characteristics. (a) In specular reflection, angle of incidence equals angle of reflection ($\alpha = \beta$). Because 80% of light is reflected, reflectance is 80%; 20% of light is absorbed. (b) In diffuse reflection, incident light is spread in all directions by multiple reflections on the unpolished surface. Such surfaces appear equally bright from all viewing angles. (c) Most materials exhibit a combination of specular and diffuse reflection. Such a surface mirrors the source while producing a bright background.*

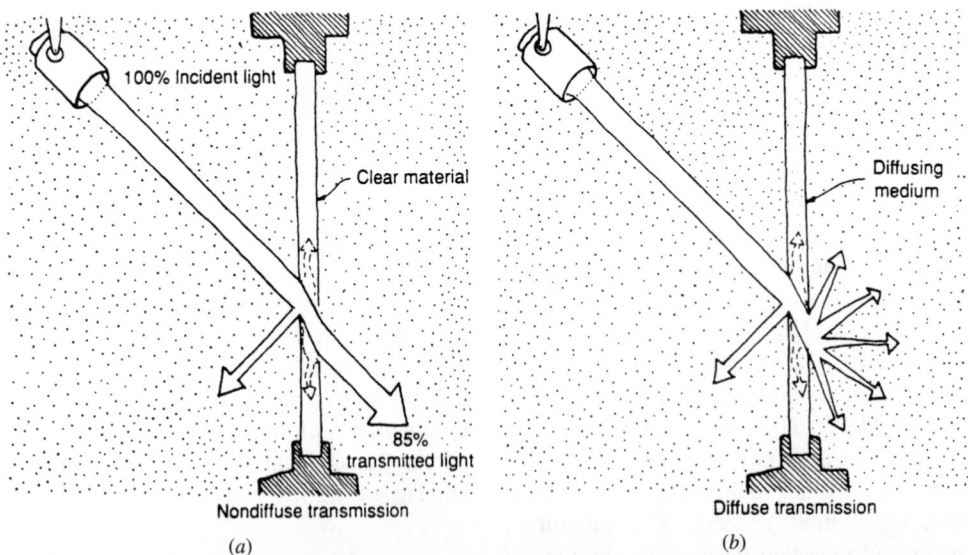

**Fig. 11.4** *Transmission characteristics. (a) In nondiffuse transmission, the light is refracted (bent) but emerges in the same beam as it enters. Clear materials such as glass, water, and certain plastics exhibit this type of transmission. In the instance illustrated, the transmittance is 85% (the remaining 15% is reflected and absorbed). The source of light is clearly visible through the transmitting medium. (b) With diffuse transmission, the source of light is not visible and, in the case of multiple sources, the diffusing surface exhibits generally uniform brightness if the spacing between the light sources does not exceed approximately 1½ times their distance from the material.*

terminology involved and their interrelations. The SI system of units is used as the basic system by the IESNA and in this book, whereas the lighting industry uses both the SI and I-P systems. As in other chapters, we frequently use dual units, with the second unit enclosed in parentheses.

## 11.5 LUMINOUS INTENSITY

The SI unit of *luminous intensity* is the candela (candlepower), abbreviated cd (cp), and normally represented by the letter *I*. It is analogous to pressure in a hydraulic system and voltage in an electric system, and represents the force that generates the light that we see. An ordinary wax candle has a luminous intensity horizontally of approximately 1 candela, hence the name. The candela and candlepower have the same magnitude. Luminous intensity is a characteristic of the source only; it is independent of the visual sense.

## 11.6 LUMINOUS FLUX

The unit of luminous flux, in both SI and I-P units, is the lumen (lm). If we take a 1-cd (candlepower)

source that radiates light equally in all directions and surround it with a transparent sphere of 1 m (ft) radius (Fig. 11.5), then *by definition* the amount of luminous energy (flux) emanating from 1 $m^2$ ($ft^2$) of surface on the sphere is 1 lm. Because there is $4\pi$ $m^2$ ($ft^2$) surface area in such a sphere, it follows that a source of 1 candela (candlepower) intensity produces $4\pi$, or 12.57, lm. The lumen, as luminous flux, or quantity of light, is analogous to flow in hydraulic systems and current in electric systems and is normally represented by the Greek letter $\phi$.

In physical terms, the lumen is a unit of power, like the watt. However, unlike the watt, which is a radiometric unit directly convertible to other power units such as Btu/h, the lumen is a measure of *photometric* power. This means light power as perceived by the human eye and therefore as a function of human physiology. Put another way, lumens (or luminous flux) is the time rate of flow of *perceived* luminous energy. Because the visual response of the eye is frequency dependent, the apprehended light power is therefore also frequency dependent, varying with the spectral content of the impinging light and the spectral sensitivity of the eye. Figure 11.6*a* shows the spectral content of the visible energy produced by a 500-W incandescent

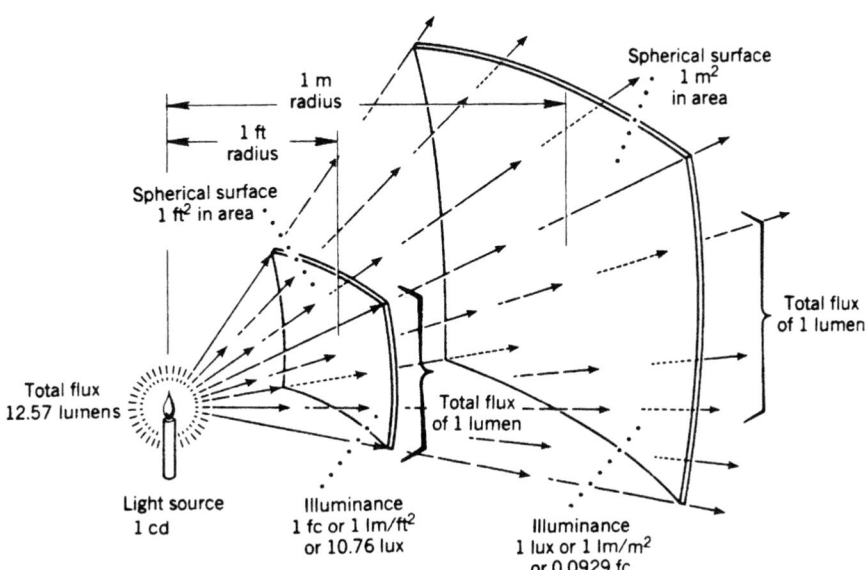

**Fig. 11.5** *A source of 1-cd intensity produces $4\pi$ (12.57) lumens of light flux. Thus, each square foot (square meter) of spherical surface surrounding such a source receives 1 lumen of light flux. This quantity of light flux produces an illuminance of 1 fc (lux) on the spherical surface.*

ILLUMINATION

lamp. Measured radiometrically, it amounts to 45 W. However, when passed through a selective filter (Fig. 11.6*b*), which is effectively what happens when the light enters the eye, the resultant "understood" light power appears as in Fig. 11.6*c*, and therefore can no longer be measured in watts. Instead, we use a unit of eye-perceived, or photometric, power called the *lumen*. If the spectral content curve in Fig. 11.6*a* were differently shaped, *even if the total radiometrically measured power were the same*, the resultant perceived power in Fig. 11.6*c* would be different.

Refer to Fig. 11.6*b*. A correlation can be made between photometric and radiometric power at the point of maximum response of the eye, which occurs at 555 nanometers (nm) wavelength—1 nm is $10^{-9}$ m. One watt of monochromatic light at that wavelength produces 683 lm. However, because common light sources such as incandescent, fluorescent, mercury, and so on, are not monochromatic but produce light in many parts of the spectrum (see Figs. 11.47 and 11.48 and Table 11.12), no single conversion factor between watts and lumens exists. Each source has its own luminous efficiency (lumens/watt), determined by its spectrum. For the 500-W lamp used as an illustration in Fig. 11.6, its luminous efficiency (efficacy) is 10,000 lm/500 W, or 20 lumens per watt (lm/W or lpw).

## 11.7 ILLUMINANCE

One lumen of luminous flux, uniformly incident on 1 m (ft²) of area, produces an *illuminance* of 1 lux (lx) (*footcandle* [fc]). Illuminance is normally represented by the letter *E*. Restated, illuminance is the density of luminous power, expressed in terms of lumens per unit area. If we consider a light bulb as analogous to a sprinkler head, then the rate of water flow would be the lumens and the amount of water per unit time per m² (ft²) of floor area would be the lux (footcandles). Thus, the SI unit, lux, is smaller than the corresponding I-P unit, footcandles, by the ratio of square meters to square feet. That is,

$$10.764 \text{ lux} = 1 \text{ fc}$$

or multiply footcandles by 10.764 to obtain lux. These relationships are shown in Fig. 11.5. Restating mathematically yields

$$\text{lux} = \frac{\text{lumens}}{\text{square meter area}} \tag{11.1}$$

$$\text{lx} = \frac{\text{lm}}{\text{m}^2}$$

and

$$\text{footcandles} = \frac{\text{lumens}}{\text{square foot area}} \tag{11.2}$$

$$\text{fc} = \frac{\text{lm}}{\text{ft}^2}$$

As an *approximation* (with 8% error)

$$10 \text{ lx} \cong 1 \text{ fc} \tag{11.3}$$

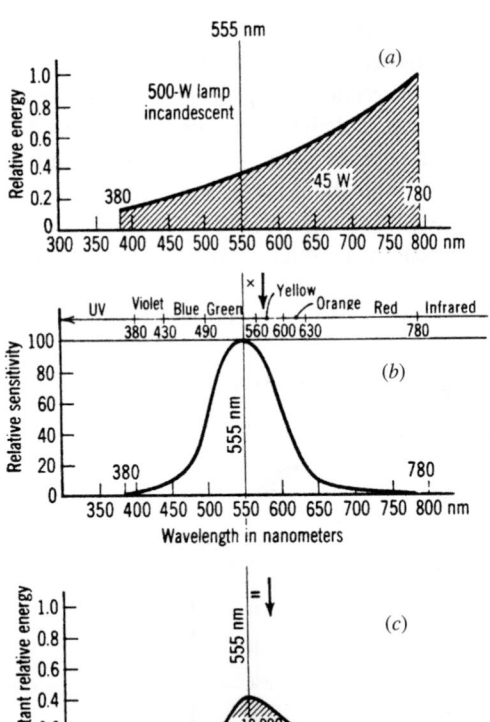

**Fig. 11.6** *Graphical demonstration of the method by which the unit of light flux is defined. (a) The spectrum of the light produced by a 500-W incandescent lamp. It amounts to approximately 45 W measured radiometrically. When filtered by the human eye, whose spectral sensitivity curve is given in (b), this light power is perceived as shown in (c). The new light power curve is expressed in lumens and indicates the quantity of light as perceived by the eye.*

**EXAMPLE 11.1** A 34-W, 425-mA (milliampere), 48-in. (122-cm) fluorescent tube produces 3200 lm.

What is the illuminance on the floor of a 3-m² room assuming 60% overall efficiency and uniform illumination?

**SOLUTION**

Useful lumens = 0.6 × 3200 = 1920

$$lx = \frac{1920}{3 \times 3} = 213.3 \ lx$$

$$fc = 213.3/10.76 = 19.8 \ fc$$

Calculating footcandles directly, we obtain (with 3 m = 9.84 ft)

$$fc = \frac{1920}{9.84 \times 9.84} = 19.8 \ fc$$

By approximation:

$$fc \cong lx/10 = 21.3$$

Note that this calculation gives *average* illuminance in the space. Illuminance at a *point* can be computed from intensity, as explained in Section 11.12. ∎

## 11.8 LUMINANCE, EXITANCE, AND BRIGHTNESS

An object is perceived because light coming from it enters the eye. The impression received is one of object *brightness*. This brightness sensation, however, is subjective and depends not only upon the object *luminance* (*L*), but also upon the state of adaptation of the eye (see Sections 11.18 and 11.19). For this reason, the physiological sensation is generally referred to in the literature as *subjective* or *apparent brightness*, or simply *brightness*, whereas the measurable, reproducible state of object luminosity is its *luminance* (formerly *photometric brightness*). Luminance is normally defined in terms of intensity; it is the luminous intensity per unit of *apparent* (projected) area of a primary (emitting) or secondary (reflecting) light source. Thus, its units are candela per area. Specifically, the SI unit of luminance is candela per square meter (cd/m²), sometimes referred to as the *nit*. Another unit formerly in common use in the I-P system is the footlambert. Conversion factors for SI and I-P units (plus other, obsolete units for the convenience of readers using older sources) are given in Table 11.1. Other luminance terms such as *stilb, apostilb, blondel, millilam-*

**TABLE 11.1 Lighting Units—Conversion Factors**

| Unit | Multiply | By | To Obtain |
|---|---|---|---|
| Illuminance (*E*) | Lux | 0.0929 | Footcandle |
| | Footcandle | 10.764 | Lux |
| Luminance (*L*) | cd/m² | 0.2919 | Footlambert |
| | cd/cm² | 10,000 | cd/m² |
| | cd/in.² | 1,550 | cd/m² |
| | cd/ft² | 10.76 | cd/m² |
| | millilambert | 3.183 | cd/m² |
| | Footlambert | 3.4263 | cd/m² |
| Intensity (*I*) | Candela | 1.0 | Candlepower |

*bert*, and *candela per square in.* are best avoided. In this book, the term *luminance* is used except where it is specifically intended to refer to the physiological sensation involved, in which case the terms *brightness, subjective brightness,* or *apparent brightness* will be used. Luminance has no readily conceivable mechanical or electrical analogy.

A word of caution at this point is in order. Although definitions and terminology are established for the specific purpose of accurate information exchange, the lighting literature is replete with articles, comments, and rebuttals that exist only because of the looseness of definitions and terminology. Some authors insist on applying *brightness* only to self-luminous surfaces, using *lightness* as the equivalent term for objects deriving their luminance from reflection. Thus, the sun has brightness; the moon, lightness. In this book, we use the term *brightness* for the subjective reaction to either source type. Other authors (constantly) point out that the luminance–brightness relation breaks down when light other than white light is used. Although this is demonstrable, it is of real interest only in theatrical lighting, where colored light is very frequently used. For our purposes we assume white light, and as pointed out previously, the color accommodation characteristic of our eyes recognizes as white (colorless) light of a large chromatic range. Within that range, and for a very large range of intensities, object color is readily recognizable and the fixed luminance-brightness ratio is maintained. Contrasting word usages such as *dim* and *dark, light* and *bright, clear* and *muddy, shallow* and *deep,* and so on as applied to lighting are best left to experienced lighting designers because the terms are almost entirely subjective and therefore unhelpful to novice designers.

Another concept that the lighting designer will encounter is know as *luminous exitance*, or simply as *exitance*, which, as the name implies, describes the

ILLUMINATION

total luminous flux density leaving (exiting) a surface, irrespective of directivity or viewer position. For instance, if a surface 1 m² emits 1 lumen, its luminous exitance is 1 lumen per square meter (1 lm/m²) or 0.093 lm/ft². A surface that is a perfect diffuser, whether by emitting light diffusely or reflecting light diffusely, is known as a *Lambertian surface*. It is fairly simple to demonstrate mathematically that the luminance of such a surface equals 1/π times its exitance. The importance of this relationship is its usefulness as an approximation. Although very few surfaces are truly Lambertian, many are approximately so, and this relationship can be used as an engineering-accuracy approximation in many such cases.

The concept of exitance is important in detailed photometric calculations such as those involved in determining coefficients of utilization, surface luminance coefficients, and in detailed point illuminance calculations. All of these are beyond the scope of this book because they are not usually performed by the lighting designer. Use of the derived coefficients is demonstrated in the referenced sections.

Detailed point calculations are today almost universally performed by computer, and the necessary mathematics is built into the computer program. Readers interested in further background on luminous exitance are referred to Murdoch (1985).

Because object luminance is that which is visually perceived and is a prime factor in visibility (and glare), it is important that the reader be able to perform basic luminance calculations. Although the eye does not differentiate between primary sources that generate and emit light and secondary sources that derive their luminance from reflection or transmission, the differentiation is important in calculation procedures. See Fig. 11.7 for a graphic representation of the basic relationships.

---

**EXAMPLE 11.2** Luminance of a light-emitting surface.
1. Calculate the luminance of an A-19 standard inside-frosted, 100-W incandescent light bulb with a maintained output of 1700 lm. Assume (for simplicity's sake) that the bulb is spherical.
2. Assume that an opal glass globe of 8 in. (20 cm) diameter and a transmittance of 35% surrounds the above bulb. Calculate the luminance of the globe. Use SI units throughout.

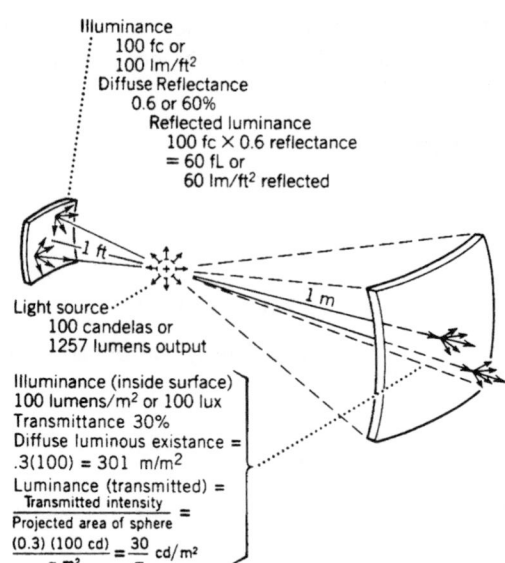

**Fig. 11.7** Luminance may be either reflected or transmitted. In the former case, it is calculated as the product of the incident lumens and the reflectance; in the latter case, as the transmitted intensity divided by the projected area.

## SOLUTION

1. Assume that the filament is a point source (an essentially valid assumption) and that the inside frosting of the glass does not reduce the output (it does by about 1%). The inside frosting serves to convert the point source filament to a uniformly emitting globe. The definition of a point source tells us that 1 candela produces 4π lumens distributed spherically.

Therefore:

$$\frac{1\ cd}{4\pi\ lm} = \frac{I\ cd}{1700\ lm}$$

The intensity, $I$, of the filament, and therefore also of the bulb (because the frosting is assumed not to reduce output), is

$$I = 1700\ lumens \times \frac{1\ cd}{4\pi\ lm} = 135\ cd$$

An A-19 bulb has a diameter of 19/8 in. (6 cm). The definition of luminance is

$$L = \frac{I}{A}\ \frac{(intensity)}{(projected\ area)}$$

The projected area becomes approximately a circle with a diameter of 6 cm (assuming a spherical bulb), whose area equals $\pi r^2$ or $\pi (0.03\ m)^2$. So

$$L = \frac{135\ cd}{\pi (0.03\ m)^2} = 47{,}750\ \frac{cd}{m^2}$$

This luminance is a potentially severe glare source if it is in the field of vision (depending on its distance from the eye; see Section 11.28).

2. We have already calculated the intensity of the source, but it is reduced to 35% by the globe. The projected area of the globe is larger than that of the bulb; that is,

$$A \text{ proj.} = \pi\, r^2 = \pi\, \frac{(0.2\text{ m})^2}{(2)}$$

So the expression for luminance becomes

$$L = \frac{135\text{ cd }(0.35)}{\pi\,(0.10\text{ m})^2} = 1500\,\frac{\text{cd}}{\text{m}^2}$$

This is no longer a potential source of direct glare. ∎

Note two important principles demonstrated by this example:

1. Intensity is not area or distance dependent; it varies only with transmission factors.
2. Luminance varies with both intensity and area.

For example, if a white lamp whose inside coating reduces lamp output by about 10% is used in lieu of an inside frost, both intensity and luminance would be reduced by 10%. However, if a clear glass lamp is used, intensity would remain the same but luminance would increase inversely with the ratio of filament area to bulb area. Therefore, a clear 100-W lamp, whose filament has an area of perhaps 0.3 cm², would have a luminance of

$$\frac{135\text{ cd}}{0.3(0.0004)\text{cm}^2} = 4{,}500{,}000\text{ cd/m}^2$$

which is so severe a glare source as to be disabling when in the near field of vision.

---

**EXAMPLE 11.3** Calculate the luminance of a 34-W, T12, 4-ft white fluorescent lamp. Assume a viewing angle normal to the long axis of the lamp and that the lamp is a diffuse (Lambertian) emitter. Use SI units.

**SOLUTION**

This problem can be solved in two ways: by using the relationship between exitance and luminance or by calculating intensity.

*By exitance/luminance:* Use the 2770 lm at 40% life as an average condition of lamp output. The luminous length is 4 ft (120 cm) and the diameter is

12/8 in. (3.8 cm). The luminous surface area of the tube is then

$$A = L \times d \times \pi = (120\text{ cm}) \times (3.8\text{ cm}) \times \pi$$
$$= (1.2\text{ m})\,(0.038\text{ m})$$
$$= 0.0456\,\pi\text{ m}^2$$

$$\text{exitance} = \frac{\text{luminous flux}}{\text{area}} = \frac{2770\text{ lm}}{0.0456\,\pi\text{ m}^2}$$

$$\text{luminance} = \frac{1}{\pi}\,(\text{exitance})$$

$$= \frac{2770}{0.0456\,\pi^2}\text{ cd/m}^2 = 6155\text{ cd/m}^2$$

*By intensity:* The equivalent spherical intensity of the lamp can be calculated by using the relationship that 1 cd produces $4\pi$ lumens. Therefore,

$$\text{equivalent intensity } I = \frac{2770/\text{m}}{4\pi}\text{ cd}$$

The radius of a sphere of equivalent surface area to the fluorescent tube would be

$$4\,\pi\, r^2 = 0.0456\,\pi\text{ m}^2$$

$$r^2 = \frac{0.0456}{4}\text{ m}^2 = 0.0114\text{ m}$$

$$r = 0.1067\text{ m}$$

The equivalent projected area of such a sphere is $\pi\, r^2$, or

$$A = 0.0114\,\pi$$

and its luminance is

$$L = \frac{\text{intensity}}{\text{area}}\,\frac{\text{cd}}{\text{m}^2} = \frac{2770/4\,\pi}{0.0114\,\pi}\,\frac{\text{cd}}{\text{m}^2}$$

$$= \frac{2770}{0.0456\,\pi^2}\,\frac{\text{cd}}{\text{m}^2}$$

$$= 6155\text{ cd/m}^2$$

This luminance can constitute a glare problem, depending on its position in the field of view. If we had used a 40-W lamp with 3200 lm output, its luminance would be higher in direct proportion to the ratio of output; that is,

$$L_{3200} = L_{2770} \times \frac{3200}{2770} = 6155\,\frac{\text{cd}}{\text{m}^2}\,\frac{(3200)}{(2770)}$$

$$= 7110\text{ cd/m}^2$$

This is the origin of the frequently used average figure of 7000 cd/m² as the luminance of a standard 48-in. fluorescent tube. It is a borderline glare source.

Because of their energy-saving characteristic, the 32-W, 48-in. T8 lamp with 3000 lm output has come into wide use. That it should not be used unshielded can be readily demonstrated by a similar luminance calculation:

Tube area = $L \times \pi \, d$

$$= (120 \text{ cm})(\pi)8/8 \text{ in.} \left(2.54 \ \frac{\text{cm}}{\text{in.}}\right)$$

$$= 958 \text{ cm}^2 = .0958 \text{ m}^2$$

$$\text{exitance} = \frac{\text{lumens}}{\text{area}} = \frac{3000 \text{ lm}}{.0958 \text{ m}^2} = 31{,}330 \text{ lm/m}^2$$

$$\text{luminance} = \frac{1}{\pi} \ (\text{exitance}) = \frac{31{,}330}{\pi} = 9972$$

or approximately 10,000 cd/m². Because this level of luminance is potentially a mild glare problem, these lamps should not be used in bare-bulb fixtures. ∎

---

**EXAMPLE 11.4** To demonstrate the usefulness of the luminance/exitance approximation, calculate the luminance, in SI units, of the page that you are now reading. Assume a uniform illuminance of 500 lx, a diffuse reflectance of 0.77, and a viewing angle normal to the page.

**SOLUTION**
From the definition of illuminance, 1 lx is produced by 1 lm falling on 1 square meter. Therefore, the exitance, or density of reflected lumens from the page, is

$$\text{exitance} = 500 \ \frac{\text{lm}}{\text{m}^2} \times 0.77 = 385 \ \frac{\text{lm}}{\text{m}^2}$$

and

$$L = \frac{1}{\pi} \times 385 \ \frac{\text{lm}}{\text{m}^2} = 122.5 \text{ cd/m}^2$$

(Typical luminances are given in Table 11.2; see Section 11.19.) ∎

## 11.9 ILLUMINANCE MEASUREMENT

Field measurements of illuminance levels are most commonly made with a portable illuminance meter, three of which are illustrated in Figs. 11.8 and 11.9. These devices contain a photoelectric material connected to a microammeter via electronic control circuitry and are calibrated in lux, footcandles, or both.

As explained in Section 11.6 and as shown in

**Fig. 11.8** Electronic, digital, color-corrected and cosine-corrected light (illuminance) meter from Minolta. (Photo by Jonathan Meendering.)

Fig. 11.6, the human eye is not equally sensitive to the various wavelengths (colors). Maximum sensitivity at high illuminance levels is in the yellow–green area (wavelength of 555 nm), whereas sensitivity at the red and blue ends of the spectrum is quite low. This effect is so pronounced that 10 units of blue energy are required to produce the same visual effect as 1 unit of yellow–green. Therefore, if a meter is to be useful, its inherent response, which is quite different from that of the human eye, must be corrected to correspond to the eye. For this reason, meters are "color corrected."

The cells (meters) must also be corrected for light incident at oblique angles that does not reach the cell due to reflection from the surface glass and shielding of the light-sensitive cell by the meter housing. This correction is known as *cosine correction*. A good meter must therefore be color and cosine corrected (and will plainly so indicate).

Modern photometers may have considerable

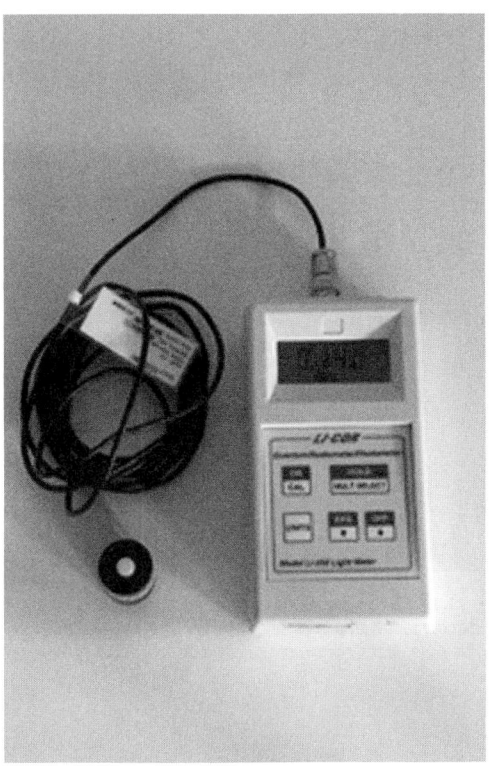

**Fig. 11.9** *Digital meter (from Li-Cor) has a variety of sensors that measure illuminance, solar irradiance, or photosynthetic radiation. Due to its small size, the illuminance sensor can be easily used for architectural model measurements. (Photo by Jonathan Meendering.)*

electronic circuitry, which provides such functions as automatic ranging, integration for flickering or time-varying sources, and connection facilities for data storage and transmission. For determining average room illuminance when using a conventional nonintegrating meter, a number of readings should be taken and an average computed. Where no definite heights is specified, readings are taken at 30 in. (75 cm) above the floor, a level known as the *working plane* because it is approximately normal desk height. The meter must always be held with the cell parallel to the plane of the test. Thus, to measure wall illuminance, the meter must be held with the cell parallel to the wall. If electric lighting readings are desired and the test is being conducted during daylight hours, readings should be taken with and without the electrical illumination and the results subtracted. Detailed instructions for conducting field surveys are contained in the IESNA publication *How to Make a Lighting Survey*. Briefly, a

survey of an existing indoor lighting installation should establish:

1. Type, rating, and age of sources.
2. Type, design, and model of luminaires.
3. Maintenance schedule.

It should also measure:

1. Mounting height of luminaires.
2. Spacing and pattern of luminaires.
3. Reflectances of walls, floor, ceiling, and major items of furniture and equipment.
4. Illuminance levels throughout the area plus levels at all working plane elevations. In addition, vertical plane illuminance at walls and other major vertical planes should be measured. The significance of vertical surface luminance is discussed in detail in subsequent chapters.

## 11.10 LUMINANCE MEASUREMENT

In terms of appreciation of the visual scene, including particularly considerations of glare, the measurement of luminance is more important and meaningful than that of illuminance (lux [fc]). This is so because it is luminance or, more accurately, subjective brightness and brightness contrasts caused by photometric luminance that we see, not illuminance. Light, as such, is invisible. That lux measurements are still more widely taken than luminance measurements and utilized as a gauge of the adequacy of a lighting installation is due to two factors:

1. Lux meters are cheaper and simpler to use than luminance meters.
2. Design recommendations for lighting levels are given in terms of illuminance. Lux measurements can therefore be used as a rapid, simple method of determining whether a particular lighting installation meets these design requirements (assuming that material reflectances and reflectance ratios are correctly chosen).

Luminance meters are available in a number of configurations, one of which is shown in Fig. 11.10. An approximation of the luminance of a reflecting or luminous source can be obtained using an illuminance (footcandle) meter of the type shown in

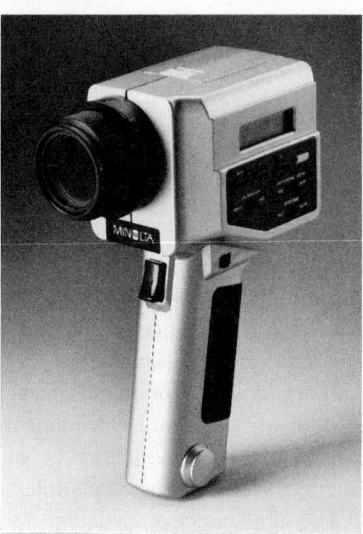

**Fig. 11.10** *Direct-reading, narrow-angle, spot-type luminance meter has an acceptance angle of 1°, a range of 0.001 to 299,000 cd/m² (0.001 to 87000 fL), a variable response speed to permit measurement of flickering sources, and a comparison mode that permits direct luminance comparison of two sources. Results are displayed digitally. (Courtesy of Minolta Corp.)*

**Fig. 11.11** *When the cell of a direct-reading illuminance meter is held in contact with a luminous source, the surface luminance can be read directly or simply calculated.*

Fig. 11.8. For diffuse *reflecting* surfaces, the cell of the meter is placed against the surface and then slowly retracted 2 to 4 in. (5 to 10 cm) until a constant reading is obtained. The luminance, in footlamberts, is then 1.25 times the reading in footcandles, the 1.25 factor compensating for wide-angle losses.

For a diffuse luminous source, the cell of an illuminance meter is placed directly against the surface (Fig. 11.11); the source luminance in footlamberts is equal to the reading on the meter in footcandles because footlamberts = lumens per area = footcandles. When using a meter calibrated in lux, the readings must be divided by $\pi$ to obtain the diffuse source luminance in cd/m².

## 11.11 REFLECTANCE MEASUREMENTS

It is often desirable to know the reflectance of a given surface because luminance can then be readily computed (see Fig. 11.7). Two methods of measuring diffuse (nonspecular) reflectance are shown in Fig. 11.12: the known-sample method and the light-ratio method. If a sample of known reflectance factor (RF) is available, this method should be used

because it yields more accurate results than the ratio method. The sample should be no smaller than 8 in. × 8 in. (200 mm × 200 mm).

It is a good idea for an inexperienced lighting designer to determine the reflectances, illuminance, and luminance levels of spaces and surfaces familiar to him or her, such as an office desk, adjoining wall, and the like—even to the extent of marking these figures on the respective surfaces in order to develop an appreciation of and a memory for these parameters. This enables the designer to visualize the result of a lighting design and should be of considerable assistance. See Table 14.9 for typical reflectance values.

## 11.12 INVERSE SQUARE LAW

We have already seen that, by definition, a point source of 1 cd intensity produces an illumination of 1 lux on the inside surface of a surrounding sphere of 1 m radius ($r$). Because the surface area of this sphere is $4\pi$ m², a 1-cd source produces $4\pi$ lm of luminous flux. Now, assume a sphere of 2-m radius surrounding this same source (Fig. 11.13). Because the same amount of flux is spread over a larger area,

## Reflected/Incident Light Method

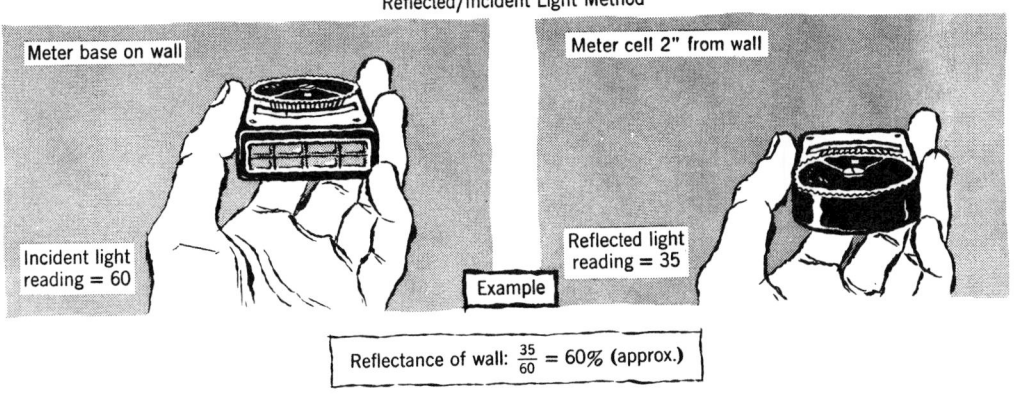

Meter base on wall

Incident light reading = 60

Meter cell 2" from wall

Reflected light reading = 35

Example

Reflectance of wall: $\frac{35}{60} = 60\%$ (approx.)

### Known Sample Comparison Method

Known reflectance sample (90% R.F.)

Meter 2" from sample

Unknown reflectance wall

Meter 2" from wall

Example

Meter reading with 90% reflectance sample = 55

Meter reading with test card removed = 35

Reflectance of unknown surface: $\frac{35}{55} \times 90 = 60\%$ (approx.)

**Fig. 11.12** *Two simple methods of measuring the diffuse reflectance of a surface.*

the illumination on the larger sphere is inversely proportional to the ratio of the sphere areas; that is,

$$\frac{lux_2}{lux_1} = \frac{area_1}{area_2} \quad (11.4)$$

or

$$lux_2 = lux_1 \times \frac{area_1}{area_2}$$

therefore,

$$lux_2 = lux_1 \times \frac{4\pi r_1^2}{4\pi r_2^2} \quad (11.5)$$

$$= lux_1 \times \frac{r_1^2}{r_2^2}$$

In other words, the illumination is inversely proportional to the square of the distance from the source. In general terms,

$$lux = \frac{cd\ intensity}{distance^2} \quad (11.6)$$

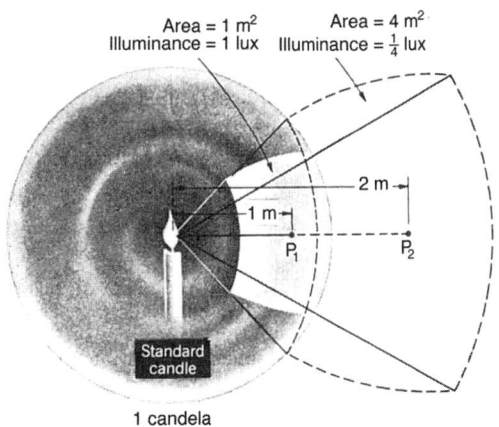

**Fig. 11.13** *Relationship between candelas, lumens, and lux defined with reference to a standard light source of 1 mean spherical cp (1 cd) located at the center of a sphere with a 1-m radius.*

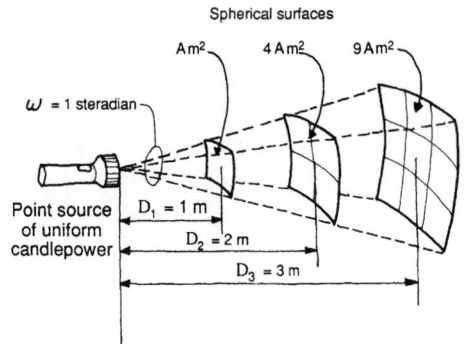

**Fig. 11.14** *Demonstration of inverse square law properties using a solid angle of unit size. Note that the surfaces are necessarily spherical because points on a planar surface are not equidistant from the source.*

where distance is expressed in meters (feet). (This holds true for surfaces normal to a source. For other situations, see Chapter 15.)

This relationship can also readily be derived by using any solid angle and the area it intercepts, as in Fig. 11.14. A glance at this figure shows clearly that the area intercepted is proportional to the square of the distance from the source; therefore, the illumination is inversely proportional, as stated previously.

## 11.13 LUMINOUS INTENSITY: CANDELA MEASUREMENTS

Luminous intensity (candela [candlepower]) cannot be measured directly but must be computed from its illumination effects. The simplest way of doing this is to use the inverse square relationship developed in the preceding section. Measure the illuminance produced on a plane at right angles to the source at a known distance and apply Eq. 11.6. For accurate measurement, the distance should be at least 5 and preferably 10 times the maximum dimension of the source because, for anything other than a point source, the equation is an approximation. The candela (candlepower) thus calculated is the luminous intensity in the direction being viewed. Because luminous intensity is not uniform in all directions for anything except an ideal point source, and because a single intensity figure for a source is desirable for calculation purposes, the average of a number of intensity figures taken from several directions is used. This average figure is called the *mean spherical candlepower* (mscp) and represents an equivalent point

source that produces $4\pi$ lm for every candela. Thus, a 10-cd lamp exhibits an average intensity of $\pm 10$ cd in all directions and produces $40\pi$ lm.

## 11.14 INTENSITY DISTRIBUTION CURVES

If the luminous intensity figures calculated in the preceding section are plotted on polar coordinate axes, the resultant figure is called a *candlepower distribution curve* (CDC) for the particular source involved. The procedure for making this curve is straightforward. A photo cell is rotated around the source in a single plane, illuminance measured, and intensity (cd) calculated. Alternatively, the photo cell can be fixed and the source rotated. If the source's distribution is symmetrical, as shown in Fig. 11.15, then only a single set of values is required, and the resultant plot is valid in all vertical planes through the source. Thus, for incandescent lamps, downlights, open circular reflectors, and the like, only a single CDC is required. For a nonsymmetric source such as a fluorescent luminaire, CDC curves in several planes are required to define the fixture's distribution characteristic. Normally, manufacturers will provide longitudinal and crosswise curves, plus a diagonal (45° plane) curve on request. This is illustrated in Fig. 11.16, where the three planes and typical resultant curves are shown.

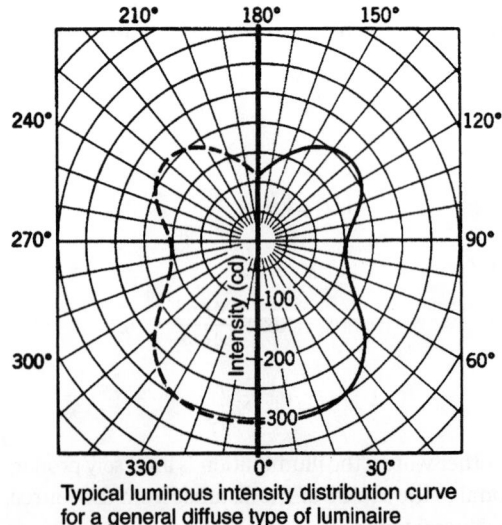

**Typical luminous intensity distribution curve for a general diffuse type of luminaire**

**Fig. 11.15** *Typical luminous intensity (cd) distribution curve for a general diffuse-type luminaire. Because the unit is symmetrical about its vertical axis, only one curve need be shown. Furthermore, only the right side of this curve need by shown due to symmetry.*

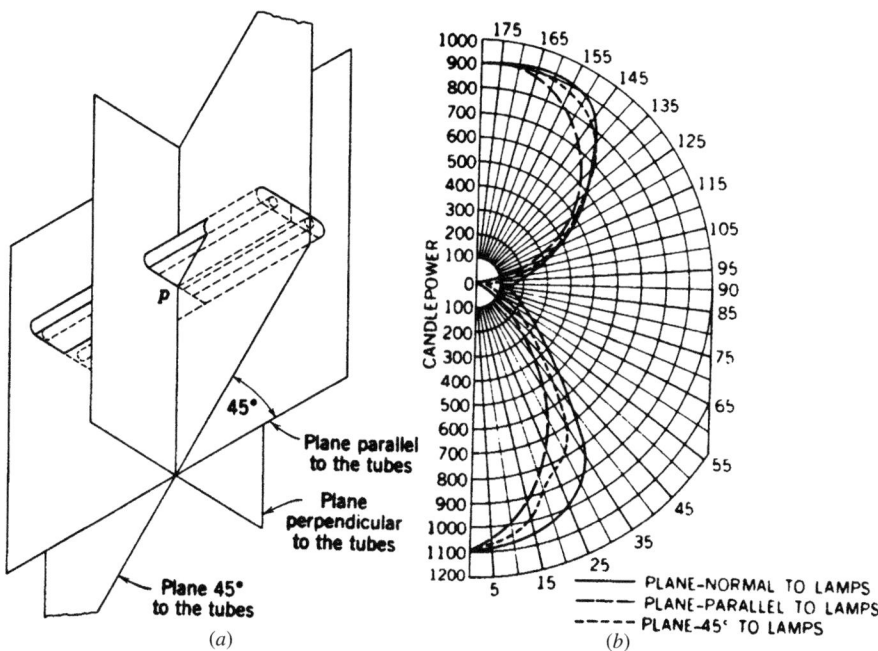

**Fig. 11.16** *(a) Due to the asymmetry of a fluorescent luminaire, intensity distribution curves in (at least) three planes are required. (b) Photometric distribution for this fixture is symmetrical in each individual plane; therefore, only one side of a curve is required. By convention, the right side is used.*

Most CDC plots are made on polar coordinates because such a plot clearly shows directions and magnitudes. Nevertheless, polar plots tend to crowd near the nadir and accurate magnitude readings at the cutoff angle are difficult to make. For this reason, it is occasionally desirable to obtain a plot on rectangular coordinates. One such plot is shown in

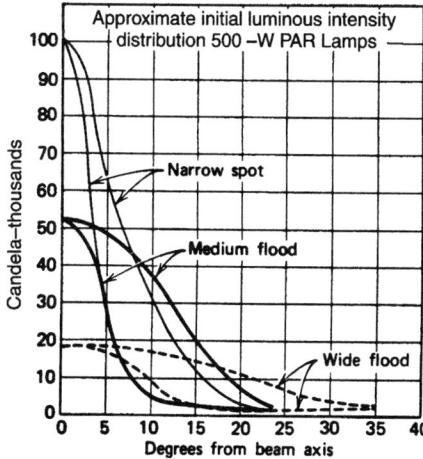

**Fig. 11.17** *Luminous intensity distribution curves plotted in rectangular coordinates. Note that candela values near the cutoff angles are easily read, which is not the case in polar plots.*

Fig. 11.17. The usefulness of intensity distribution curves will become clear in our subsequent discussions on lighting fixture diffusers, point-by-point calculations, and direct and reflected glare. It should be noted that the area of the CDC curve is *not* a measure of the lumen output.

## LIGHT AND SIGHT

### 11.15 THE EYE

Because any discussion of light and lighting techniques is irrelevant to our purposes unless ultimately related to vision, we turn to a cursory examination of the human eye before proceeding further with discussions of lighting.

Light impinging on the eye enters through the pupil, the size of which is controlled by the iris, thereby controlling the amount of light entering the eye. The lens focuses the image on the retina, from which the optic nerve conveys the visual message by electric impulse to the brain. Figure 11.18

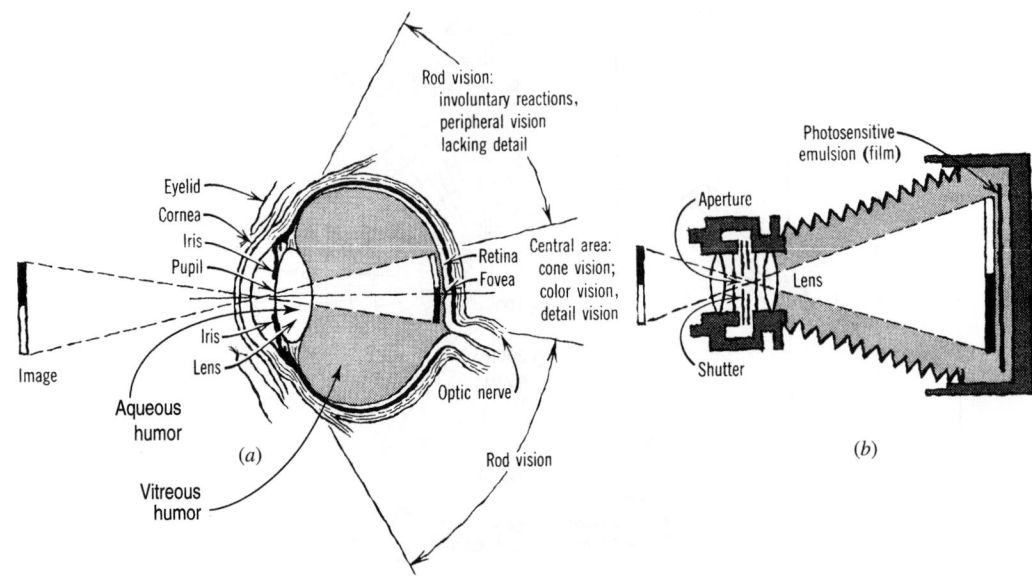

**Fig. 11.18** *The human eye (a) and the camera (b) operate on similar optic principles. The cornea acts as an outer refracting lens that introduces light into the iris. The iris and pupil control the f-stop, or opening of the eye, and correspond roughly to a range of f2.1 to f11. The lens, which acts as a perfectly smooth automatic zoom lens, can focus from about 2 in. to infinity.*

shows the structure of the eye and the parallel structure of a camera.

Light is focused on the retina, which contains in all some 150 million light-sensitive cells of two types: rod and cone cells The central portion of the eye, near the fovea, is an area of pinhead size containing about 100,000 *cone* cells, which accounts for the extreme precision of foveal (center-focus) vision. The cones are responsible for the ability to discriminate detail and also give us our sensation of color and detect *luminances in the range 3 to 1,000,000 cd/m²*. Proceeding outward from the fovea, a second type of cell is encountered called a *rod* cell. Rods can detect luminances from 1/1000 cd/m² to approximately 120 cd/m² and are extremely light sensitive, giving a response to light 1/10,000 as bright as that required by cone cells. However, rod cells lack color sensitivity, thus accounting for the fact that in dim light (rod vision), we have no color perception and all colors appear as varying shades of gray. Rod cells also lack detail discrimination, making "night vision" quite coarse. Finally, rod cells are slower acting than cone cells and therefore have a low degree of flicker fusion; stated conversely, they are highly motion sensitive. Because these cells occur at the outer portions of the retina, their motion sensitivity results in our

being best able to detect movement when looking out of the "corner of the eye." Looking at a fluorescent tube directly and then obliquely demonstrates this effect.

Figure 11.19 is a sketch illustrating the angles involved in the field of vision. Of particular interest is the extreme narrowness of the cone of central (foveal) vision, in which acute perception of detail takes place. This area is so small that the eye must refocus on each dot in a colon (:) if you wish to examine each individually. Surrounding this central area is a cone of binocular vision of 30° half-angle, called the *near field* or *surround*, in which area most of the coarser sight information is gathered. Beyond this cone we have far field and peripheral, primarily horizontal, monocular vision. It is the far field and peripheral areas that largely give us our subjective, ambience-type reactions.

## 11.16 FACTORS IN VISUAL ACUITY

The three components of any seeing task are the object or task itself, the lighting conditions, and the observer. Listed below are the variables affecting each of these three components. Based upon the results of many investigations, they

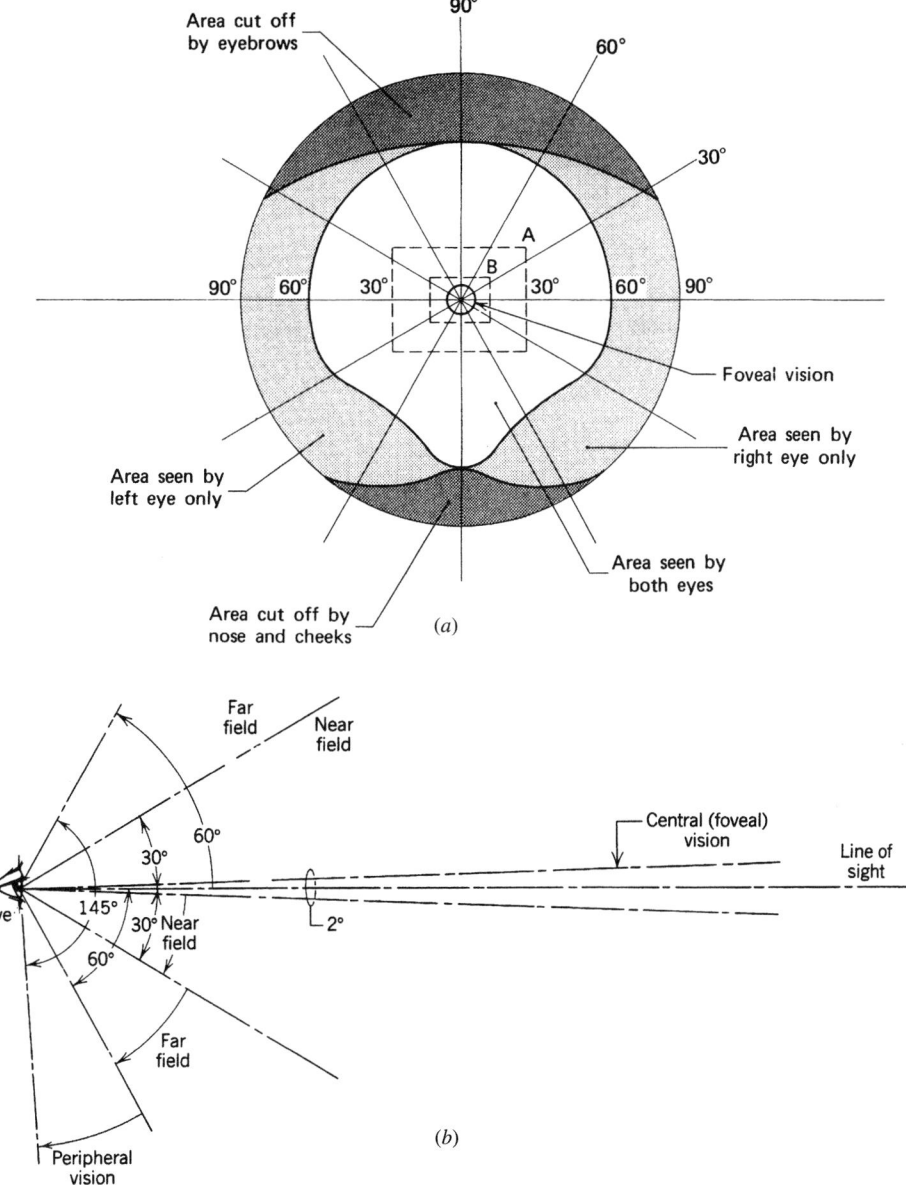

**Fig. 11.19** *The fields of vision of a normal pair of human eyes (a) and the subtended angles (b). The rectangles A and B superimposed on the field of vision in (a) represent a large magazine and a small book, respectively.*

can be categorized as of primary or secondary importance.

I. The Task

*Primary Factors*

a. Size

b. Luminance (brightness)

c. Contrast, including color contrast

d. Exposure time—needed or given

*Secondary Factors*

e. Type of object—required mental activity; familiarity with the object (in reading, familiarity is so important as to become the primary factor)

    *f.* Degree of accuracy required
    *g.* Task—moving or stationary
    *h.* Peripheral patterns
  II. The Lighting Condition

    *Primary Factors*
    *a.* Illumination level
    *b.* Disability glare
    *c.* Discomfort glare

    *Secondary Factors*
    *d.* Luminance ratios
    *e.* Brightness patterns
    *f.* Chromaticity
  III. The Observer

    *Primary Factors*
    *a.* Condition of the eyes (both health and age)
    *b.* Adaptation level
    *c.* Fatigue level

    *Secondary Factors*
    *d.* Subjective impressions; psychological reactions

Although in the following discussions these factors are considered individually, many are interrelated. Thus luminance I*b* and adaptation III*b* result from the presence of illumination II*a*; subjective impressions III*d* are dependent on brightness patterns II*e* and chromaticity II*f*; fatigue III*c* results from a combination of many of the factors, and so on.

    In the literature, it is common to find reference to the *quantity* and *quality* of the lighting environment. In terms of these factors, the quantity of light has reference to item II*a* and the quality to items II*b* through II*f*.

    The basic visual tasks are the perception of low contrast, fine detail, and brightness gradient. Assuming a good lighting environment—that is, low glare, acceptable luminance ratios, and white light plus a normal pair of unfatigued eyes—visual acuity is primarily dependent on items I*a* to I*d*, the interrelated effects of which have been determined by a large number of field tests. Remember that the seeing task under discussion involves foveal vision (i.e., focusing and concentrating on small-area detail). This is a vastly different task than normal reading, where the eye rapidly scans familiar images without focusing on details and the brain immediately understands even when much of the information is missing, as in poor reading copy. The task discussed next is quite different and could be com-

pared to studying mathematical equations or reading an unfamiliar language or even proofreading spelling. All of these tasks require detailed examination of each symbol individually.

## 11.17 SIZE OF THE VISUAL OBJECT

Visual acuity is generally proportional to the physical size of the object being viewed given fixed brightness, contrast, and exposure time. Because the actual parameter is not physical size but subtended visual angle, visual ability can be increased by bringing the object nearer the eye (Fig. 11.20). It is assumed that we are dealing with a pair of *young eyes*, because at ages above 40, the accommodation ability of the eye becomes limited and bringing the object closer blurs the focus.

## 11.18 SUBJECTIVE BRIGHTNESS

The sensation of vision, as explained previously, is caused by light entering the eye. This light may be thought of as a group of convergent rays, each ray coming from a different point in space and therefore carrying different visual information. The composite of these rays comprises the entire visual picture that the eye sees and the brain comprehends. The individual rays differ from each other in intensity and chromaticity, depending on the part of the viewed object from which they were reflected. The

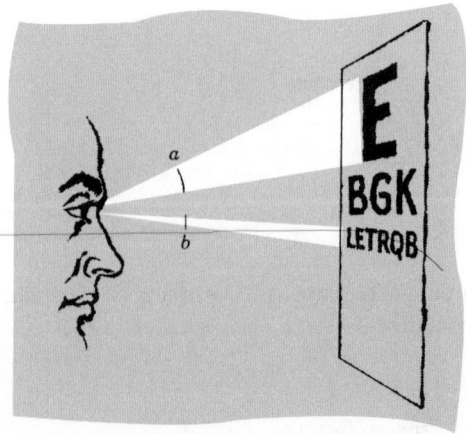

**Fig. 11.20** *Relationship between object size and visibility is demonstrated by comparison of subtended angles* a *and* b.

intensity of these cones of light determines and describes the perceived brightness of the object being viewed.

The human eye detects luminance over an astonishing range of more than 100 million to 1, the lower levels being accomplished after an adjustment period called *adaptation time.* This period varies from 2 minutes for cone vision to up to 40 minutes for rod vision for dark adaptation but is much faster for both types for light adaptation (going from dark to light). The effects of adaptation on apparent (photometric) brightness are discussed in the following section. Tables 11.2 and 11.3 list some measured luminances of everyday visual tasks.

An interesting characteristic of light-level adaptation is a shift in the sensitivity curve of the eye (Fig. 11.6b). Whereas for the light-adapted eye (photopic vision) maximum sensitivity occurs at 555 nm in the yellow–green region, the dark-adapted eye (scotopic vision) peaks at 520 nm in the blue–green region. This means that as the light dims, the warm colors—yellow, orange, red—become grayed and the blues and violets stand out. This phenomenon can be important in the lighting design of restaurants, where light levels generally vary inversely with restaurant quality. Very few foods are blue or violet.

Returning then to the primary consideration

## TABLE 11.3 Preferred and Permissible Luminances

| Item | Luminance in cd/m² |
|---|---|
| Recommended road luminance | 1–2 |
| Minimum discernible, chromatic | 2–3 |
| Clearly discernible human features | 15–20 |
| Preferred wall luminance | 25–150 |
| Preferred ceiling luminance | 50–250 |
| Preferred task luminance | 100–500 |
| Permissible luminaire luminance (depending on position in field of vision) | 1000–7000 |

of visual acuity as affected by luminance, we can state that, in general, visual performance increases with object luminance. However, a great deal depends on the background against which an object is viewed and the consequent contrast in brightness between the object being viewed and its surroundings.

## 11.19 CONTRAST AND ADAPTATION

The discussion that follows assumes full-spectrum white light and ignores the effects of chromaticity, which is considered separately. Many researchers in the area of visibility have concluded that contrast is the single most important factor in visual acuity. This is self-evident when we realize that in fact, the eye sees only contrast. This can readily be demonstrated by viewing a large, evenly lighted, monochromatic, diffuse-finish surface (preferably white) that encompasses the entire visual field. The eye is unable to focus on such a surface because it sees no contrast, but only a single luminance. Therefore, the eye itself attempts to provide the missing contrast by seeing an internally reflected view of the retina in yellow.

To properly evaluate the effect of contrast (luminance ratio) on visibility, we must first determine the nature of the visual task or, more simply, exactly what it is that we are trying to see. As stated before, the basic visual tasks are detail discrimination and detection of low contrast. Examples of the former are drafting, industrial product inspection, or something as simple as discriminating between the numbers 3, 6, and 8, which are similarly shaped. Detection of low contrast includes reading faint copy, sewing a black fabric with black thread, and the like.

## TABLE 11.2 Typical Luminance Values[a]

| Object | Luminance | |
|---|---|---|
| | cd/m²[b] | Footlamberts |
| Black glove on a cloudy night | 0.0003 | 0.0001 |
| Wall brightness in a well-lighted office | 100 | 30 |
| This sheet of paper in an office | 120 | 35 |
| Green electroluminescent lamp | 150 | 45 |
| Asphalt paving— overcast day | 1,300 | 380 |
| North sky | 3,500 | 1,000 |
| Moon, candle flame | 4,000–5,000 | 1,300 |
| Fluorescent tube | 6,000–8,000 | 2,200 |
| Kerosene flame | 8,500 | 2,500 |
| Hazy sky or fog | 15,000 | 4,400 |
| Snow in sunlight | 25,000 | 7,300 |
| 100-W inside frost incandescent lamp | 50,000 | 14,600 |
| Sun | 2.3 E9 | 0.67 E9 |

[a]Values are rounded off.

[b]To obtain footlamberts, divide by 3.42.

ILLUMINATION

Contrast is a dimensionless ratio, defined as

$$C = \frac{L_T - L_B}{L_B} \text{ or } \frac{L_B - L_T}{L_B} \text{ or } \left| \frac{L_B - L_T}{L_B} \right| \quad (11.7)$$

where $L_T$ and $L_B$ are the luminance of the task and background, respectively, in any units. Thus, $C$ varies from 0 for no contrast to 1.0 for maximum contrast. In most situations, the illumination on the task and background is the same. Therefore, because luminance is the product of illuminance (lux) and reflectance, contrast can also be expressed as

$$C = \left| \frac{R_B - R_T}{R_B} \right| \quad (11.8)$$

where $R_T$ and $R_B$ are the reflectances of the task and the background, respectively. From this equation, we can conclude that *contrast is generally independent of illumination* (ignoring specularity). Thus, for the black-on-white lettering that you are now reading,

$$C = \frac{0.77 - 0.045}{0.77} = 0.94$$

which accounts for the excellent legibility. (We are assuming no specularity.) Reflectance figures are taken from Table 11.9.

It is obvious that high contrast is the critical factor in visual appreciation of outline, silhouette, and size, which are the factors involved in the task of reading. Thus, black-on-white print can be read with ease even in moonlight, which is at best 0.1 lux illuminance, because the contrast is so high (94%). An important conclusion can then be drawn about a *reading task*: with high contrast (clear, legible print), visibility is essentially *independent* of illumination above a certain minimum. Indeed, high illuminance values can be detrimental because they generally go hand-in-hand with high luminance sources, and these in turn can cause veiling reflections (see Section 11.29).

Now refer to Fig. 11.21. Note that as the contrast between the letters in the word "performance" and the background diminishes, the individual letters become harder to read. The end letters of the word require an illuminance of up to 1000 lux, and that suffices only because we expect the letter *e* at the end. Were it an unknown sign, illuminance of a magnitude of 10,000 lux or more would be needed. These latter letters are an example of the second type of visual task mentioned previously—

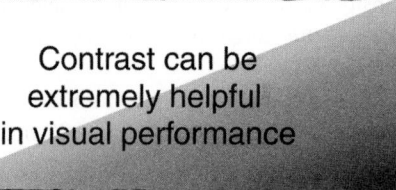

**Fig. 11.21** High contrast is helpful when the seeing task involves detection of silhouette detail.

low-contrast poor copy requiring surface detail study.

The *e* at the end of "performance" in Fig. 11.21 is printed with the same density as the *C* in "Contrast." It exists, and with enough lighting, the negative effect of lack of contrast can be overcome. This is not so with copy from a used-up printer cartridge or a washed-out photocopy. There, the data simply do not exist, and increased lighting only makes this fact more evident.

High background luminance makes an object look darker, and therefore assists in outline detail discrimination, which is precisely the visual task involved in reading. For this reason, black-on-white is desirable for reading. (This is a special case of lateral adaptation, which is discussed in Section 11.36.) Conversely, high background luminance makes surface examination more difficult. A simple experiment demonstrates this effect. Stand near a window and hold your hand in front of you with the floor as background. The skin surface detail is perfectly clear—in rough proportion to its luminance. Now hold your hand up against the window with the daytime sky as background. The hand outline is clear, but the skin surface appears dark—the brighter the sky, the darker the skin surface. The reason for this is that the eye automatically adapts to the average brightness of the entire scene.

It is well known that when using an automatic exposure control camera to photograph a dark object on a light background (such as a person in a snow scene) it is necessary to manually increase the camera aperture in order to obtain additional light to photograph the detail of the darker object. (In doing this, we overexpose the rest of the scene.) Because we cannot easily control the aperture of our eyes, we must compensate for the detrimental effect of high background luminance in another way—for example, by increasing the surface luminance of the visual task. Indeed, this method is frequently employed (see Section 11.31*b*). Limited visual compensation can be made by squinting; this reduces the field of vision and the overall scene brightness. For maximum visual acuity, *the luminance of a surface-type task should be the same as, or slightly higher than, that of the background,* but ratios of 3:1 are acceptable in most circumstances.

Another way of understanding this is to consider the adaptation characteristic of the human eye (Fig. 11.22). As stated, the eye adapts to the brightness level of the *overall* scene and sees each object in the scene in the framework of that adaptation level. Thus, at an adaptation level of 1 fL (3.4 cd/m$^2$), a measured luminance ratio of 1:10 (horizontal scale on Fig. 11.22) appears to be only approximately 1:4 (vertical scale); that is, the apparent ratio is *smaller* than the actual one. Put another way, the low level of eye adaptation causes the eye to diminish the difference between high brightnesses. This effect becomes smaller as the adaptation level rises, until at an adaptation level of 1000 fL (3400 cd/m$^2$) (daylight conditions), the apparent and actual ratios correspond; that is, smaller ratios are recognizable. Because visual acuity is, by definition, the ability to distinguish between different levels of luminance, we have in effect demonstrated that *visual acuity increases with increased adaptation level.*

The second important conclusion that can be drawn is that at high adaptation levels, apparent brightness is lower than actual brightness and vice versa. Thus, a shadowed object near a window looks *darker* than it actually is; contrary to first expectation, it must be better lighted than a similar object further inside the room for equal visibility. That this effect (high-level adaptation) is primarily important in daylight situations is also apparent from the curves. At a 100 fL (340 cd/m$^2$) adaptation level, which is approximately that of a brightly lighted interior space, apparent and actual luminance levels coincide. The reverse effect, resulting from low adaptation levels, can be very important in design situations where low lighting levels are found, such as theaters, lecture halls, restaurants, and storage spaces. Sources of light that would be entirely acceptable at a higher adaptation level can easily become an annoying glare at a low level. Good examples are a theater usher's flashlight or the blinding glare of an oncoming car's headlights. Reduction of contrast due to veiling reflections and the measurement of contrast and contrast reduction are discussed in Section 11.29.

The foregoing discussion of contrast deliberately avoided any discussion of *object* colors and the

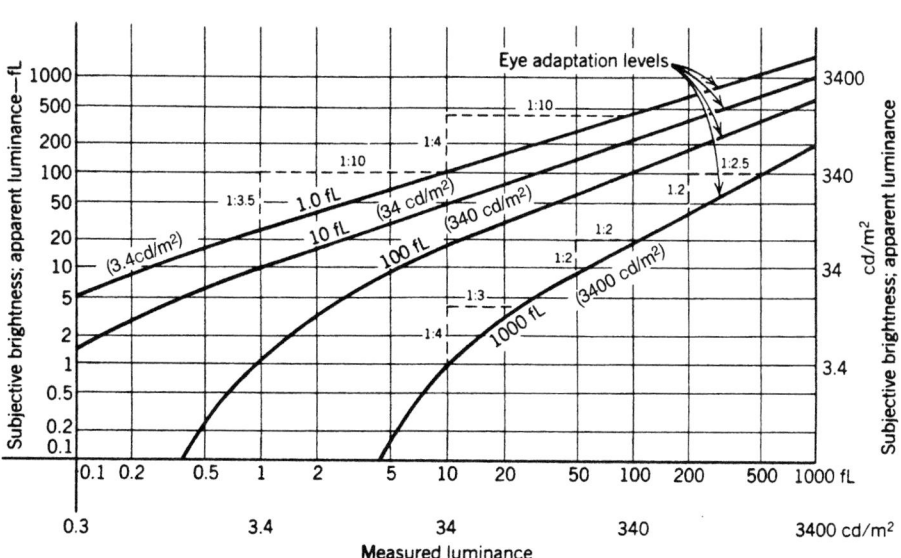

**Fig. 11.22** *The effect of an eye's adaptation level on perceived (subjective) brightness is clearly shown. (Adapted from H. Cotton,* Principles of Illumination, *John Wiley & Sons, New York, 1960.)*

effect of color contrast on visual acuity for several reasons:

- Office-type tasks (paperwork) are most often black-on-white tasks.
- Most work tasks involving colored objects deal with unsaturated colors, where the pronounced effects of color contrast are minimal.
- The effect of object color on visual acuity is very complex because it involves the color characteristic not only of the object, but also of the background and the surround plus the chromaticity of the illuminant.

That being said, we must, however, at least mention a number of important object color phenomena that bear on visual acuity.

- The subjective brightness of a colored (heterochromatic) object is greater than that of an achromatic object for the same photometric luminance. This effect varies with hue and saturation. It is more pronounced with saturated colors than with those of low chroma and more in the blue–purple–red area than the yellow–green area.
- Colored objects on a dark-to-black background appear light and desaturated. Conversely, colors on a light-to-white background appear darker and more saturated.
- Adjacent complementary colors produce a pale-to-white border between them. The effect also appears when the task and background colors are complementary, and is most pronounced with saturated (high-chroma) colors.

These remarks deal with *object* color only. As stated at the beginning of this section, we are assuming white, full-spectrum light, so that neither object color nor visual acuity is affected. Effects on visual acuity resulting from unbalanced illuminants, such as those from high-pressure sodium lamps and, to a lesser extent, from halide and fluorescent sources, are the subject of intensive and ongoing research. Some of the results are adduced in the discussion of illuminant chromaticity at the end of this chapter. These effects are most pronounced and therefore most important in relation to elderly persons and persons with visual defects. We must again, therefore, emphasize that unless otherwise specifically stated, our discussions of vision, light, and

lighting assumes full-spectrum white light and young, healthy eyes.

## 11.20 EXPOSURE TIME

Registering a meaningful visual image is not an instantaneous process, but one that requires finite amounts of time. Just as a photograph can be taken in dim light by using a longer exposure, so can the human eye better distinguish and discriminate fine detail in poor light given time (and neglecting eyestrain). Of course, the time needed depends on the type of task, but the principle of shorter time at higher illumination, within limits, remains the same. This is particularly true when the object being viewed is not static but is in motion.

The phenomenon, however, is not linear. For one specific task tested, increasing the luminance by a factor of 6 halved the seeing time, whereas a further sixfold increase in luminance reduced the time only another 20%. Thus, as in the case of improved contrast with increasing background brightness, we have a case of diminishing returns.

With the parameter of time, as with other parameters of visual acuity, the same qualification applies. When dealing with material that does not require detail discrimination, improved performance does not necessarily result from improved illumination. It has been amply demonstrated that speed of reading and comprehension are substantially independent of illumination levels above a minimum but are very much dependent on the contrast quality of the material.

## 11.21 SECONDARY TASK-RELATED FACTORS

Refer again to the list of factors in visual acuity in Section 11.16. We have discussed at some length the primary task-oriented factors Ia–d, and at this point wish to consider secondary factors e–h of that list. These items refer essentially to the level of concentration required. Thus, spray painting a large metal object or packing fruit are very different from inspecting the painted object for defects or the fruit for bruises. The former tasks are largely mechanical and repetitious, whereas the latter tasks require continuous judgmental decisions based on visual information. Because both of the latter tasks are frequently moving (assembly-line type of work) and both involve penalties for inaccuracy (rejection at a later inspection state or negative feedback from the purchaser), the lighting required for these tasks is several orders of magnitude better than that for related but largely mechanical tasks. Indeed, extrapolation from laboratory condition tests yields only a range of lighting recommendations. Thereafter, considerable field testing and adjustment is required.

An observer performing a lab test has a different level of concentration and performance than that of a person at an 8-hour-a-day task. The latter compensates for an unsatisfactory seeing condition by:

1. Moving the work to a better viewing angle.
2. Moving the head and eyes to a more comfortable position.
3. Reducing the distance between the eyes and the task to the extent that the eyes can accommodate.
4. Complaining about a poor contrast task so that something is done about it (such as fixing the photocopying machine).
5. Taking more time to perform the seeing task involved. (This item, if it affects production, frequently spurs management to make appropriate alterations in the work environment.)

The last item in the task list—peripheral patterns—deals with the visual surround rather than the immediate area of the work. Other than glare sources, which are discussed separately, there are many items that, although outside the central field of vision, can disturb the viewer's concentration and therefore the task performance of a worker. These include movement (vehicles, machines, persons), to which peripheral vision is particularly sensitive; large variations in the brightness pattern of the background caused by such activities as periodic opening of an outside door or welding; and even nonvarying patterns, which are disturbing because of their very nature, such as checkerboard light–dark patterns, or devices on which it is difficult to focus the eyes, such as crossed patterns of wires and bars. None of these items is strictly lighting oriented, but they are noted here to demonstrate that an adequate lighting design necessarily includes adjustments for particular field conditions.

## 11.22 OBSERVER-RELATED VISIBILITY FACTORS

It is a well-documented fact that the visual performance of *healthy* eyes decreases with age. This reduction is demonstrated principally in two areas: an increase in minimum focusing distance caused by increasing lens rigidity and a decrease in sensitivity caused by clouding of the cornea, lens, and vitreous humor. Both of these effects can be compensated for, the former with external lensing (eyeglasses) and the latter by increased task size, luminance, contrast, and exposure time, as explained previously. The point is, of course, that the age of the worker is an important parameter in the vision equation.

Furthermore, maximum performance may not be synonymous with maximum comfort or minimum fatigue. Indeed, the reverse may sometimes be true. Most experts agree that what is normally referred to as *eyestrain* is a condition of the eye muscles resulting from extensive and intensive eye use. Thus, excellent performance under excellent lighting conditions can still produce fatigue because of the demanding nature of the task. In addition, as discussed later, discomfort glare or even excessive lighting can cause fatigue without affecting performance.

The lighting designer must be concerned not only with providing adequate uniform lighting levels, but also with all of the factors involved. Many, indeed, are beyond his or her control. Some, such as task contrast, previously thought to be outside the lighting designer's province, should be examined by the designer in the framework of an overall lighting plus task plus observer problem and recommendations made. Acceptance of these recommendations is a management decision.

Recent work in field testing of visibility situations in real-work conditions considers all of these parameters in arriving at a visibility judgment and, in addition, goes far toward identifying any difficulty as lying with the task, the lighting condition, or the observer. The entire field of visibility and visual performance is undergoing very active continuing research, and definitive answers to the elusive question of how to design for optimal viewing have not yet been found, nor indeed is there any assurance that they will be. What has emerged over the years have been various

ILLUMINATION

parameters and criteria for judging aspects of visibility, such as ESI, which is a contrast-related visibility criterion; RVP, which rates a vision situation in terms of speed and accuracy of performance; VCP, which judges the visual comfort of an overall scene; and a somewhat ephemeral and entirely subjective impression called *visual clarity*. Other research has concentrated on the relationship between eye pupil size and visual acuity and, based on results from this research, the relationship between pupil size and various aspects of light and lighting. What has been confirmed to date is the complexity of human work-oriented vision plus our inability thus far to adequately quantify the interrelationships between the many factors involved—quantification being the basis of reliable, duplicatable design.

## 11.23 THE AGING EYE

The past few decades have seen a remarkable increase in life expectancy in modern Western countries with a resultant sharp increase in the aged population. In the United States at this writing, about 15% of the population is above 65 years of age, a proportion that is expected to reach 20% by the year 2020. As a result, lighting design must take cognizance not only of special requirements in buildings specifically intended for use by the aged but also, increasingly, of general use public buildings. To this end, a brief review of these special requirements is presented here.

Refer to Fig. 11.18. Light enters the eye through the cornea, passes through the aqueous humor, and enters the lens through the pupil. After being focused by the lens, it continues through the vitreous humor and finally projects the viewed image, reversed, on the retina. As the eye ages, a whole spectrum of physiological changes may occur; some are usual and are therefore classified as normal; others, such as cataracts, are less common but are still considered by ophthalmologists to be an expected development. The unusual developments are classified as *pathologies*, only indirectly or partially related to aging and therefore outside the purview of normal lighting design. The normal developments and their influence on lighting design are briefly described in the following subsections.

### (a) Cornea

This perfectly clear outer lens tends to become cloudy, with corresponding reduction of visual clarity and acuity. This results in a requirement for more light to overcome the reduction in light intensity on the retina. The overall effect is very similar to that of a neutral density filter on a camera lens, the difference being that such a filter is most often used to reduce excessive ambient light without excessively reducing the shutter opening (increasing the *f*-stop). The need for additional light for aging eyes is recognized in most modern systems of illumination specification. See Table 11.6 for the factors used in the IESNA system.

### (b) Lens

The lens, which begins life as a very lightly yellow–tinted flexible crystalline body, gradually thickens and yellows. As a result of the thickening, flexibility is reduced, resulting in the well-known inability to close-focus. The yellowing both reduces the overall light intensity in the eye and selectively filters the blue portion of the spectrum. Research seems to indicate improved visual acuity through pupil size control when the incident light is rich in the blue area of the spectrum. Because lens yellowing *reduces* the blue frequencies, the overall effect is again to require additional light.

A second and more important degenerative phenomenon of the lens is its gradual clouding. When the opacity is confined to the perimeter, its effect is negligible because vision is unaffected. When small, opaque areas appear within the visual axis through the lens, vision is affected in two ways:

1. The viewed image is dimmed and blurred due to opacities in the field of view.
2. Light entering the lens is scattered by interreflections from the opaque particles, resulting in a subjective impression of glare. This effect is particularly severe outdoors, where light enters the lens from all angles, and both indoors and outdoors as a result of concentrated highluminance (glare) sources.

The net result of these reactions is a requirement for more light but an even more pressing requirement that sources of glare and peripheral light be eliminated. (People with this condition frequently wear eyeglasses and sunglasses with large, opaque earpieces to block peripheral light and thus reduce glare.) Also, because short-wavelength (blue) light interreflects and scatters more readily than does long-wavelength (yellow–red) light, these people are more comfortable with incandescent sources and low-color temperature fluorescent lamps (2700–3100° K) than with sources rich in the blue–green spectrum. (Ophthalmologists frequently prescribe yellow-tinted eyeglass for people with this lens condition to filter out blue light.)

Finally, a less common but still prevalent condition of the aging lens is the development of fluorescent particles, called *fluorigens*, in the vision path. In the presence of UV radiation, such as exists in daylight, fluorescent, and high-intensity discharge (HID) sources, these particles fluoresce, causing scatter, blur, and glare. The solution to this problem is a combination of yellow-tinted eyeglass lenses and a reduction of light sources containing appreciable quantities of UV.

## (c) Pupil

The pupil controls the amount of light entering the eye and is therefore intimately involved in the constantly changing accommodation level of the eye. The pupil muscles react more slowly as they age, thus lengthening accom.modation time. Dark-to-light accommodation is very rapid in the young eye, and is barely noticed except for extreme changes such as exiting a cinema into sunlight. With an aging eye, the slower pupil results in severe glare sensations with even much smaller brightness changes.

The net result of all of the normal conditions of the aging eye is a heightened sensitivity to glare, an intolerance to the blue–UV end of the spectrum, and an overall requirement for higher illuminance levels. For the lighting designer, these needs translate into requirements for very careful selection and placement of luminaires, increased use of indirect lighting, and particular attention to the spectrum of the light sources used. Because some of these requirements are not only mutually incompatible but also contrary to energy-efficient design practice, it may be particularly difficult to satisfy all the requirements in spaces occupied by persons with a wide range of ages. In work areas of this type, it may be wise to provide for the possibility of readily

changing lighting conditions in a limited area to accommodate older occupants rather than attempt an overall design, keeping in mind constantly the glare and color factors discussed previously.

## QUANTITY OF LIGHT

### 11.24 ILLUMINANCE LEVELS

Returning to the list of factors in Section 11.16, and having discussed the task-oriented and observer-oriented items (except for item IIId, psychological reactions, which is covered in Section 11.36), we turn now to item II, the *lighting condition*. This is frequently, if somewhat inaccurately, divided into two groups—quantity and quality of lighting—with item IIa representing quantity and items IIb to IIf representing quality. That such a division is not accurate becomes clear in our discussion of glare in Sections 11.28 to 11.31.

An understanding of the factors involved in visual acuity, as discussed previously, does not answer the most basic lighting design question, which is "How much light must I provide for the specific visual task at hand?" That this question is extremely difficult to answer is evidenced by the fact that even today, recommendations for similar tasks vary by ratios as high as 10:1 among countries with highly developed technologies. Because this is obviously an unsatisfactory situation in an era of global markets, international construction, the European Union, and international cooperative lighting research, the trend since the late 1980s has been to attempt a degree of standardization.

The North American (IESNA) recommendations were originally developed analytically by extrapolation from extensive laboratory tests. The function of these tests was to determine the conditions under which small differences in contrast could be detected for specific degrees of accuracy, with variable parameters of task luminance, size, and exposure time. The idea behind the tests was that visual acuity could be defined as the ability to distinguish differences in contrast.

The British (and, to a large extent, the European) approach was to study specific tasks in actual and simulated field conditions. To these results, modifying factors of size, contrast, accuracy requirement, speed, and duration were added. The disadvantage of this approach was the necessity to study a large number of visual tasks individually. In the absence of a very specific visual task description, a detailed listing of visual categories was developed, as is discussed in the next section. Recently, IESNA recommendations have been modified by a graded system—more in line with British and CIE (Commission Internationale de l'Eclairage) recommendations—that establishes a median or average requirement for a task, within a range, and then modifies the median up or down to specific conditions of speed, accuracy, error importance, task duration, background reflectances, and viewer visual capability.

In addition, because of the moral and legal pressure (ANSI/ASHRAE/IESNA Standard 90.1 is mandated by many codes) for energy conservation, the IESNA has taken additional salutary steps toward rationalization of its very influential illumination standards. They include recognition of fatigue and task familiarity (e.g., reading) as factors in determining illuminance levels, establishment of lighting power budgets, and energy standards (see Section 13.5) that encourage use of daylight as a normal component of a space's illumination and use of task/ambient lighting design as the preferred technique where high levels of task lighting are required.

## 11.25 ILLUMINANCE CATEGORY

Before discussing illuminance (lux) recommendations, it is necessary to understand the basis of their derivation, applicability, and shortcomings. As noted above, most of the IESNA task illuminance recommendations are derived by extrapolation from threshold contrast visibility tests that yield a required task luminance. Assuming uniform, diffuse task reflectance and uniform illuminance, it is then a simple step to calculate required illuminance since luminance is simply the product of illuminance and reflectance. In SI units,

$$L = \frac{E \times RF}{\pi}$$

and in I-P units,

$$fL = fc \times RF$$

where

$L$ = luminance in cd/m$^2$

$E$ = illuminance in lux

$RF$ = reflection factor

$fL$ = luminance in foot-lamberts

$fc$ = illuminance in footcandles (see Example 11.4 and Fig. 11.7)

This simple relationship is used to derive much of the current IESNA tables, with modifying factors as explained previously (see Table 11.6).

A number of reservations about this method have been voiced by respected authorities. One is based on research that indicates that suprathreshold visibility requirements are more readily related to eye brightness adaptation levels than to threshold contrast luminance levels (see Fig. 11.22 and the associated text discussion). Another objection is that deriving suprathreshold luminances from threshold values depends on applied criteria and can therefore vary considerably. Still another objection is based on the readily demonstrable fact that the sensation of vision is not mathematically related to photometric luminance. Thus, a black surface of 10% reflectance illuminated with 900 lux and a white surface of 90% reflectance illuminated with 100 lux both have exactly the same luminance, yet the eye always sees the white surface as lighter than the black one by a large margin. All of these reservations add up to a recommendation to the lighting designer to use the IESNA illuminance recommendations as one aspect of the overall design and not dogmatically.

Returning to recommendations, visual task studies indicate that assuming good contrast, the required luminances, categorized by type of task, are roughly as follows:

| Category of Visual Task | Required Luminance (cd/m²) |
|---|---|
| Causal | 10–20 |
| Ordinary | 20–100 |
| Moderate | 100–200 |
| Difficult | 200–400 |
| Severe | Above 400 |

Dependance of required illumination on task reflectance (RF) can be seen by a glance at the following tabulation, which shows quantitatively the illuminance requirements in the previous categories for tasks of radically different reflectance.

| Category of Visual Task | Required E (lux)[a] | |
|---|---|---|
| | RF = 50% | RF = 10% |
| Casual | 62–125 | 300–625 |
| Ordinary | 125–625 | 625–3,125 |
| Moderate | 625–1,250 | 3,125–6,250 |
| Difficult | 1,250–2,500 | 6,250–12,500 |
| Severe | >2,500 | >12,500 |

[a]Lux figures rounded.

This illustrates that a single illumination scheme is often inadequate for an area containing widely differing visual tasks. Note that a 10% RF makes all tasks difficult and that casual seeing comprises only outline recognition.

## 11.26 ILLUMINANCE RECOMMENDATIONS

Because American architects and engineers are involved in many construction projects outside of the United States, this section presents the essentials of British illuminance recommendations, which are similar to those of the CIE and many European countries, in addition to the American IESNA recommendations. In any specific design, the standards of that country should be consulted. Most developed nations, including Australia, Brazil, China, France, Germany, Japan, and others, have their own published lighting standards.

### (a) British Lighting Standards

These standards are published by the Chartered Institution of Building Services Engineers (CIBSE: http://www.cibse.org/). The particular publication in which illuminance recommendations appear is the *Code of Interior Lighting*, 1994, amended 1997 and 2004. The method of use is to determine the recommended average illuminance level, called the *standard maintained illuminance,* from either the very detailed, extensive listing of specific tasks in the previously mentioned publication (a small sample of which is reproduced in Table 11.4) or, if only the representative type of task is known, from the task category chart in Table 11.5. Having established this recommendation, the designer then modifies it (if necessary) by using the flow chart shown in

**TABLE 11.4 Typical Illuminance Recommendations, CIBSE (UK)**

| Offices and Shops | | |
|---|---|---|
| | Standard Maintained Illuminance (lux) | Notes |
| OFFICES | | |
| General offices | 500 | Local lighting may be appropriate |
| Computer workstations | 300–500 | See ref. 25 |
| Conference rooms, executive offices | 300–500 | Dimming or switching to permit use of visual aids may be necessary |
| Computer and data preparation rooms | 500 | See ref. 25 |
| Filing rooms | 300 | Vertical surfaces may be especially important |
| DRAWING OFFICES | | |
| General | 500 | |
| Drawing boards | 750 | Local lighting may be appropriate |
| Computer-aided design and drafting | 300–500 | Special lighting is required; see ref. 25 |
| Print rooms | 300 | |

*Source:* Reproduced with permission from the CIBSE *Code for Interior Lighting* (1994).
Reference 25: *CIBSE Lighting Guide LG3: Areas for Visual Display Terminals*

*Notes:*

The information in the table will be influenced by reference to the "core recommendations" in Sections 4.3 to 4.5 of the Code.

Check that installations designed to meet the needs of visual display screen tasks also have the task-to-wall and task-to-ceiling ratios recommended in Section 4.4 of the Code.

Where air conditioning or mechanical ventilation is required, air handling luminaires may be appropriate.

**TABLE 11.5 Examples of Activities/Interiors Appropriate for Each Maintained Illuminance**

| Standard Maintained Illuminance (lux) | Characteristics of Activity/Interior | Representative Activities/Interiors |
|---|---|---|
| 50 | Interiors used rarely with visual tasks confined to movement and casual seeing without perception of detail | Cable tunnels, indoor storage tanks, walkways |
| 100 | Interiors used occasionally with visual tasks confined to movement and casual seeing calling for only limited perception of detail | Corridors, changing rooms, bulk stores, auditoria |
| 150 | Interiors used occasionally or with visual tasks not requiring perception of detail but involving some risk to people, plant, or product | Loading bays, medical stores, plant rooms |
| 200 | Interiors occupied for long periods or for visual tasks requiring some perception of detail | Foyers and entrances, monitoring automatic processes, casting concrete, turbine halls, dining rooms |
| 300[a] | Interiors occupied for long periods, or when visual tasks are moderately easy (i.e., large details [>10-min arc]) and/or high contrast | Libraries, sports and assembly halls, teaching spaces, lecture theaters, packing |
| 500[a] | Visual tasks moderately difficult (i.e., details to be seen are of moderate size [5–10 min arc] and may be of low contrast); also, color judgment may be required | General offices, engine assembly, painting and spraying, kitchens, laboratories, retail shops |
| 750[a] | Visual tasks difficult (i.e., details to be seen are small [3–5 min arc] and of low contrast); also, good color judgments or the creation of a well-lit, inviting interior may be required | Drawing offices, ceramic decoration, meat inspection, chain stores |
| 1000[a] | Visual tasks very difficult (i.e., details to be seen are very small [2–3 min arc] and can be of very low contrast); also, accurate color judgments or the creation of a well-lit, inviting interior may be required | General inspection, electronic assembly, gauge and tool rooms, retouching paintwork, cabinet making, supermarkets |
| 1500[a] | Visual tasks extremely difficult (i.e., details to be seen are extremely small [1–2 min arc] and of low contrast); optical aids and local lighting may be of advantage | Fine work and inspection, hand tailoring, precision assembly |
| 2000[a] | Visual tasks exceptionally difficult (i.e., details to be seen exceptionally small [<1 min arc] with very low contrast); optical aids and local lighting will be of advantage | Assembly of minute mechanisms, finished fabric inspection |

*Source:* Reproduced with permission from the CIBSE *Code for Interior Lighting* (1994).

[a]One minute of arc (min arc) is one-sixtieth of a degree. This is the angle of which the tangent is given by the dimension of the task detailed to be seen divided by the viewing distance.

Table 11.6, which either increases or decreases the recommended illuminance to suit the task size, contrast, duration, and error risk. Explanatory usage notes accompany Table 11.6.

### (b) North American Illuminance Recommendations (IESNA)

This (1981) system is patterned, to an extent, on CIE recommendations. In lieu of the single illumi-

**TABLE 11.6  Design Maintained Illuminance Flow Chart[a]**

| Standard Maintained Illuminance (lux)[b] | Task Size and Contrast[c] | | Task Duration[d] | | Error Risk[e] | Design Maintained Illuminance (lux)[f] |
| | Unusually Difficult to See? | Unusually Easy to See? | Unusually Long Time? | Unusually Short Time? | Serious for People, Plant, or Product? | |
|---|---|---|---|---|---|---|
| 200 | Yes-200 | 200 | Yes-200 | 200 | Yes-200 | 200 |
| | 250 | 250 | Yes-250 | 250 | Yes-250 | 250 |
| 300 | Yes-300 | Yes-300 | Yes-300 | Yes-300 | Yes-300 | 300 |
| | 400 | 400 | Yes-400 | Yes-400 | Yes-400 | 400 |
| 500 | Yes-500 | Yes-500 | Yes-500 | Yes-500 | Yes-500 | 500 |
| | 600 | 600 | Yes-600 | Yes-600 | Yes-600 | 600 |
| 750 | Yes-750 | Yes-750 | Yes-750 | Yes-750 | Yes-750 | 750 |
| | 900 | 900 | Yes-900 | Yes-900 | Yes-900 | 900 |
| | 1000 | 1000 | Yes-1000 | Yes-1000 | Yes-1000 | 1000 |
| | | | 1300 | | 1300 | 1300 |
| | | | 1500 | | 1500 | 1500 |

*Source:* Reproduced with permission from the CIBSE *Code for Interior Lighting* (1994).

[a]To use the chart, follow the horizontal path from the "standard" maintained illuminance in the schedule until the answer to a question is "yes." If the "yes" is strong, follow the solid arrow; if moderate, follow the dashed arrow.

[b]The flow chart should be used for all standard maintained illuminance recommendations from 200 to 750 lux for general activities and interiors in the lighting schedule. For recommendations of 150 lux or less, the modifying factors are not relevant (see Table 11.5). Where a standard maintained illuminance of more than 750 lux is recommended, this always applies to a stated task for specific industries or activities where the modifying factors have usually been taken into account.

[c]The standard maintained illuminance given in the schedule assumes that the task is representative of its type. If the task is much more visually difficult than usual (e.g., smaller size, lower contrast), then an increase in the maintained illuminance is appropriate. Reduced contrast may arise from the use of safety lenses or safety screens because they reduce the transmission of light. Increase illuminance to take account of the age or eyesight of the operator. Conversely, if the task detail is such that the task is easier to see than usual (e.g., larger size, higher contrast), a reduction in maintained illuminance can be made.

[d]The standard maintained illuminance given in the schedule assumes that the task is to be undertaken over a conventional working period. If the work is to be undertaken continually for a much longer period than usual, the maintained illuminance should be increased in order to diminish the risk of visual fatigue. Conversely, if the work is to be carried out over a much shorter period than usual, the maintained illuminance may be reduced.

[e]The schedule assumes that the consequences of any errors are typical of the activity. However, if errors have unusually serious consequences for people, plant, or product, an increase in illuminance may be appropriate.

[f]If the design maintained illuminance is more than two full steps (e.g., 300–750 lux) on the illuminance scale above the standard maintained illuminance, consideration should be given to whether changes in the task details or organization of the work are more appropriate than substantial increases in the maintained illuminance.

nance recommendation that characterized the early IES tables, this system provides a luminance range determined by task difficulty (contrast, size), within which a specific illuminance is selected based on three weighting factors—the age of the observer, the importance of speed and/or accuracy, and the reflectance of the background on which the task is seen. The procedure is as follows:

1.  An illuminance category (A through I) is initially selected, based either on a general description of the activity involved (Table 11.7) or, if known, on a specific activity in a specific setting.

(The extensive tables for this latter selection are found in the IESNA *Lighting Handbook*, 8th ed., 1993, and are not reproduced here.) The category selection gives a three-number range of illuminances that corresponds to minimum significant subjective illuminance changes.

2.  In the next step, three weighing factors are introduced. Age under 40 reduces the light requirement, unusual demand for speed or accuracy increases it, and particularly low or high background reflectance (below 30% or above 70%) increases or reduces it, respectively. This step is not a mechanical one in that at the

**TABLE 11.7 Illuminance Categories and Illuminance Values for Generic Types of Activities in Interiors**

| Type of Activity | Illuminance Category | Ranges of Illuminances | |
|---|---|---|---|
| | | Lux | Footcandles |
| *General lighting throughout spaces* | | | |
| Public spaces with dark surroundings | A | 20–30–50 | 2–3–5 |
| Simple orientation for short temporary visits | B | 50–75–100 | 5–7.5–10 |
| Working spaces where visual tasks are only occasionally performed | C | 100–150–200 | 10–15–20 |
| *Illuminance on task* | | | |
| Performance of visual tasks of high contrast or large size | D | 200–300–500 | 20–30–50 |
| Performance of visual tasks of medium contrast or small size | E | 500–750–1,000 | 50–75–100 |
| Performance of visual tasks of low contrast or very small size | F | 1,000–1,500–2,000 | 100–150–200 |
| *Illuminance on task, obtained by a combination of general and local (supplementary) lighting* | | | |
| Performance of visual tasks of low contrast and very small size over a prolonged period | G | 2,000–3,000–5,000 | 200–300–500 |
| Performance of very prolonged and exacting visual tasks | H | 5,000–7,500–10,000 | 500–750–1,000 |
| Performance of very special visual tasks of extremely low contrast and small size | I | 10,000–15,000–20,000 | 1,000–1,500–2,000 |

*Source:* Courtesy of the Illuminating Engineering Society of North America.

higher luminances (categories D through I), the designer is expected to become thoroughly familiar with the importance, duration, and visual difficulty of the specific task involved, because the design aims at *task* lighting and not simply general room illumination (see Table 11.5). Having determined the weighting factors, the designer can then select the recom-mended (target) illuminance from Tables 11.8 and 11.9.

The resultant recommended illuminance is "raw" or conventional illuminance, that is, average maintained lux (fc), *on the task* for categories D through I, and *in the room* for categories A through C. Therefore, the calculation procedure for the target

**TABLE 11.8 Effect of Weighting Factors on General Lighting Illuminance Values, for Illuminance Categories A, B, and C[a]**

| Weighting Factors | | Illuminance (lux) | | |
|---|---|---|---|---|
| | | Categories | | |
| Average Age of Occupants | Average Room Surface Reflectance (%)[b] | A | B | C |
| Under 40 | Over 70 | 20 | 50 | 100 |
| | 30–70 | 30 | 75 | 150 |
| | Under 30 | 30 | 75 | 150 |
| 40–50 | Over 70 | 30 | 75 | 150 |
| | 30–70 | 30 | 75 | 150 |
| | Under 30 | 30 | 75 | 150 |
| Over 55 | Over 70 | 30 | 75 | 150 |
| | 30–70 | 30 | 75 | 150 |
| | Under 30 | 50 | 100 | 200 |

[a]Based on IESNA recommendations.

[b]Average weighted surface reflectances, including wall, floor, and ceiling reflectances, if they encompass a large portion of the task area or visual surround. For instance, in an elevator lobby, where the ceiling height is 25 ft, neither the task nor the visual surround encompasses the ceiling, so only the floor and wall reflectances would be considered.

**TABLE 11.9 Effect of Weighting Factors on Task Lighting Illuminance Value
for Illuminance Categories D through I[a]**

| | | | ILLUMINANCE ON TASK | | | | | |
| --- | --- | --- | --- | --- | --- | --- | --- | --- |
| *Weighting Factors* | | | *Illuminance Categories* | | | | | |
| Average of Workers' Ages | Demand for Speed and/or Accuracy[1] | Task Background[2] Reflectance (%) | D | E | F | G[b] | H[b] | I[b] |
| Under 40 | NI[c] | Over 70 | 200 | 500 | 1,000 | 2,000 | 5,000 | 10,000 |
| | | 30–70 | 200 | 500 | 1,000 | 2,000 | 5,000 | 10,000 |
| | | Under 30 | 300 | 750 | 1,500 | 3,000 | 7,500 | 15,000 |
| | I[c] | Over 70 | 200 | 500 | 1,000 | 2,000 | 5,000 | 10,000 |
| | | 30–70 | 300 | 750 | 1,500 | 3,000 | 7,500 | 15,000 |
| | | Under 30 | 300 | 750 | 1,500 | 3,000 | 7,500 | 15,000 |
| | C[c] | Over 70 | 300 | 750 | 1,500 | 3,000 | 7,500 | 15,000 |
| | | 30–70 | 300 | 750 | 1,500 | 3,000 | 7,500 | 15,000 |
| | | Under 30 | 300 | 750 | 1,500 | 3,000 | 7,500 | 15,000 |
| 40–55 | NI | Over 70 | 200 | 500 | 1,000 | 2,000 | 5,000 | 10,000 |
| | | 30–70 | 300 | 750 | 1,500 | 3,000 | 7,500 | 15,000 |
| | | Under 30 | 300 | 750 | 1,500 | 3,000 | 7,500 | 15,000 |
| | I | Over 70 | 300 | 750 | 1,500 | 3,000 | 7,500 | 15,000 |
| | | 30–70 | 300 | 750 | 1,500 | 3,000 | 7,500 | 15,000 |
| | | Under 30 | 300 | 750 | 1,500 | 3,000 | 7,500 | 15,000 |
| | C | Over 70 | 300 | 750 | 1,500 | 3,000 | 7,500 | 15,000 |
| | | 30–70 | 300 | 750 | 1,500 | 3,000 | 7,500 | 15,000 |
| | | Under 30 | 500 | 1,000 | 2,000 | 5,000 | 10,000 | 20,000 |
| Over 55 | NI | Over 70 | 300 | 750 | 1,500 | 3,000 | 7,500 | 15,000 |
| | | 30–70 | 300 | 750 | 1,500 | 3,000 | 7,500 | 15,000 |
| | | Under 30 | 300 | 750 | 1,500 | 3,000 | 7,500 | 15,000 |
| | I | Over 70 | 300 | 750 | 1,500 | 3,000 | 7,500 | 15,000 |
| | | 30–70 | 300 | 750 | 1,500 | 3,000 | 10,000 | 15,000 |
| | | Under 30 | 500 | 1,000 | 2,000 | 5,000 | 10,000 | 20,000 |
| | C | Over 70 | 300 | 750 | 1,500 | 3,000 | 7,500 | 15,000 |
| | | 30–70 | 500 | 1,000 | 2,000 | 5,000 | 10,000 | 20,000 |
| | | Under 30 | 500 | 1,000 | 2,000 | 5,000 | 10,000 | 20,000 |

Notes

1. In determining whether speed and/or accuracy is not important, important, or critical, the following questions need to be answered: What are the time limitations? How important is it to perform the task rapidly? Will errors produce an unsafe condition or product? Will errors reduce productivity and be costly? For example, in reading for leisure there are no time limitations, and it is not important to read rapidly. Errors will not be costly and will not be related to safety; thus, speed and/or accuracy is not important. If, however, prescription notes are to be read by a pharmacist, accuracy is critical because errors could produce an unsafe condition and time is important for customer relations.

2. The task background is that portion of the task upon which the meaningful visual display is exhibited. For example, on this page the meaningful visual display includes each letter, which combines with other letters to form words and phrases. The display medium, or task background, is the paper, which has a reflectance of approximately 85%.

[a]Based on data from the *IESNA Handbook*, 8th ed. (1993).

[b]Obtained by a combination of general and supplementary lighting.

[c]NI = not important; I = important; C = critical.

illuminance is a lumen method for categories A through C and a point-by-point method for the *task* illuminance of categories D through I.

For a space with several tasks of varying visual difficulty, the designer is expected to so design the lighting and controls that task requirements are met without overlighting. A uniform layout keyed to the most severe task is tedious and energy wasteful, and is strongly to be discouraged.

The IESNA-recommended illuminance values are *not* applicable to installations where a visual task is not the deciding factor. Such installations include merchandising spaces, displays of all sorts, theatrical and artistic lighting, lighting for mood, lighting for safety, light used as part of an industrial process, and so on. Recommended illuminances for exterior spaces are found in the same group of tables in the IESNA handbook referred to previously.

ILLUMINATION

### (c) Evolution of IESNA Illuminance Recommendations

The IESNA illuminance selection system has been amended from the above described system as follows:

1.  The traditional nine illuminance categories (A through I) have been collapsed to seven categories (A through G). There is a single target illuminance (see below) for each category.

| Orientation and Simple Visual Tasks (lux) | | Common Visual Tasks (lux) | | Special Visual Tasks (lux) | |
|---|---|---|---|---|---|
| A | 30 | D | 300 | G | 3,000–10,000 |
| B | 50 | E | 500 | | |
| C | 100 | F | 1000 | | |

2.  A matrix called the IESNA Lighting Design Guide lists a wide range of lighting applications and tasks and provides a specific illuminance category for each listing. In addition to the lighting quantity recommendations inherent in the illuminance categories, the Design Guide provides recommendations regarding lighting quality issues.

3.  Additional factors related to the lighting condition can be factored into the choice of illuminance. These include scene geometry, shadowing and modeling, luminance ratios, presence of daylight, glare considerations, and others. All those concerned with lighting design should use the current *IESNA Lighting Handbook* and the illuminance recommendations presented therein.

## QUALITY OF LIGHTING

### 11.27  CONSIDERATIONS OF LIGHTING QUALITY

*Quality of lighting* is a term used to describe all of the factors in a lighting installation not directly connected with quantity of illumination. Certainly it is obvious that if two identical rooms are lighted to the same *average* illuminance, one with a single bare bulb and the other with a luminous ceiling, there is a vast difference in the two lighting systems. This difference is in the *quality* of the lighting, a term

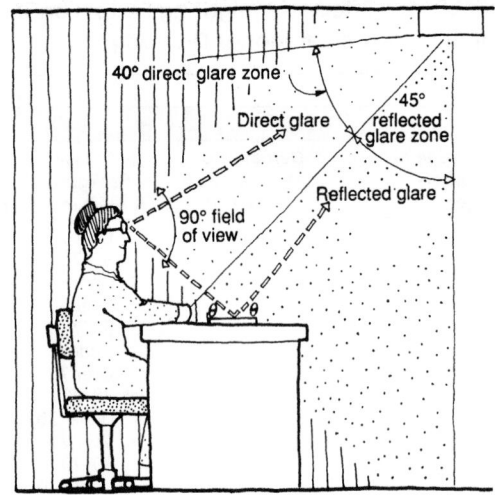

**Fig. 11.23** *Glare zones. Direct glare presupposes a head-up position, whereas reflected glare assumes eyes down at a reading angle.*

that describes the overall scene—that is, the luminances, diffusion, uniformity, and chromaticity of the lighting.

Excessive luminances and/or excessive luminance ratios in the field of vision are commonly referred to as *glare*. The quality of the lighting system must also include the visual comfort of the system, that is, the absence of glare. When the glare is caused by light sources in the field of vision, it is known as *direct* or *discomfort glare*. When the glare is caused by reflection of a light source in a viewed surface, it is known as *reflected glare* or *veiling reflection* (see Fig. 11.23) Factors affecting the severity of glare are adaptation level of the eyes, the apprehended size of the glare source, luminance ratios, room size and surface finishes, and size and position of lighting fixtures and windows. Light sources in the far-field and peripheral-vision areas beyond the central 90° cone are less troublesome as glare sources.

### 11.28  DIRECT (DISCOMFORT) GLARE

The factors involved in producing discomfort glare are the luminance, size, and position of each light source in the vision field plus the adaptation level of the eye. The discomfort of direct glare stems from two facts: first, the eye adapts (rapidly) to the average brightness of the overall visual scene; and second, the eye is attracted to the highest luminance in that scene. (The latter fact is used effectively in merchandising displays.) Thus, if an area of high brightness, such as a window or a lighting fixture, exists in the visual scene and we are looking at an area of lower brightness, such as a work task, three visually disturbing things occur:

1. The eye adapts to a higher luminance level, thus effectively reducing the subjective brightness of the task—or, put more simply, making it harder to see what we are looking at (see Fig. 11.22). This is readily demonstrated by alternately blocking and unblocking a direct glare source with one's hand while trying to perform a moderately difficult visual task and noting the immediate improvement in visibility when the glare source is obscured.
2. The eye is drawn simultaneously in two directions: involuntarily to the source of high luminance and volitionally to the object we are looking at. The resultant tension causes considerable visual discomfort.
3. The adaptation level is continuously varying as the eye is drawn to the glare source and away again.

Glare is proportional to a source's luminance and its apprehended solid angle. Therefore, a small, bright source is usually not a problem, whereas a large, low-brightness source (such as a luminous ceiling) may be. Indeed, a small, bright source adds sparkle to the field of vision, and many observers find it a pleasant addition in a monotonous lighting environment. Although discomfort glare from a scene is cumulative, source luminance is more important than the number of sources. For instance, if the luminance of a number of sources is halved, the reduction in glare is greater than is achieved by reducing the number of such sources by half. Indeed, the latter procedure has little effect on discomfort glare.

The remaining two factors are less self-evident. Glare decreases rapidly as the brightness source is moved away from the direct line of vision; thus, the glare produced depends on the source's position in the field of view. The amount of *discomfort glare* produced by a source is inversely proportional to the background luminance (eye adaptation level). Thus, a ceiling fixture with a luminance of $4000$ cd/m$^2$ at

$65°$ might easily constitute a source of discomfort glare in a space with an eye adaptation level of $150$ cd/m$^2$. The same fixture would not be objectionable in a daylight condition, where the eye adaptation level might be $1500$ cd/m$^2$. A more striking example is that of an automobile's headlights, which at night are so severe a source of glare as to constitute *disabling glare*, whereas in daylight, with its concomitant high eye adaptation level, these lights are very noticeable but not usually disturbing.

Keeping in mind the dependence of direct glare on eye adaptation level, a useful rule of thumb is that the luminance of large sources should not exceed $2500$ cd/m$^2$ and that of small sources should not exceed $7500$ cd/m$^2$. The former is roughly the luminance of blue sky; the latter approximates that of a fluorescent lamp. The terms "large" and "small" depend not only on the actual physical dimensions of the source but also on the distance from the observer. That is, the actual criterion is apprehended size, or subtended visual angle, as shown in Fig. 11.20.

The glare effect of a number of individual glare source contributions in an interior space can be quantified by a criterion called *visual comfort probability* (VCP), which is defined as the percentage of normal-vision observers who will be comfortable in that specific visual environment. The IESNA has established a set of standard conditions for which VCP of sources can be calculated. These include a 1000-lux illuminance, representative room dimensions, fixture height and observer position, and a head-up field of view limited to $53°$ above and directly forward from the observer (Fig. 11.24).

With these conditions, direct glare will not be a problem if all three of the following conditions are satisfied:

1. The VCP is 70 or more.
2. The ratio of maximum-to-average luminaire luminance does not exceed 5:1 (preferably 3:1) at $45°$, $55°$, $65°$, $75°$, and $85°$ from the nadir, crosswise and lengthwise.
3. Maximum luminaire luminances crosswise and lengthwise do not exceed the following:

| Angle Above Nadir (degrees) | Maximum Luminance (cd/m$^2$) |
|---|---|
| 45 | 7710 |
| 55 | 5500 |
| 65 | 3860 |
| 75 | 2570 |
| 85 | 1695 |

ILLUMINATION

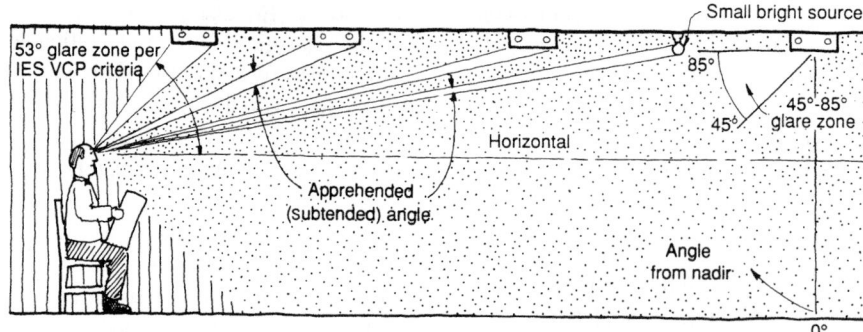

**Fig. 11.24** *Determination of direct glare. The glare contribution of each source depends on its size (subtended or apprehended solid angle), luminance, and location in the field of view. Note that the apprehended solid angle of a small source is such that even with high luminance, it is not objectionable. Such sources are normally called sparkle. Glare is much more objectionable with a dark background than with a light one; therefore, light-colored paints on ceilings and upper walls are recommended.*

A typical set of manufacturer's luminance and VCP data is shown in Fig. 11.25 for a ceiling-mounted fluorescent fixture with 4-40WT12 lamps. Note that all VCP values are considerably above the 70 minimum criterion. If full VCP data of this type are not available, they can be calculated with almost any lighting calculation program, given the luminaire luminance data. Despite the usefulness of the VCP criterion, it is inherently limited by its own standard conditions, which are not easily applied to other situations:

1. In small spaces, VCP has little significance.
2. Tabulated VCP figures are given for the *worst* viewing position in the room. Because VCP varies dramatically with observer position, the VCP values given are always lower than the space's *average* VCP.

In view of these (and other) reservations, most of which tend to make the actual direct glare situation better than the VCP calculation would indicate, it is recommended that layouts giving a VCP somewhat below 70 not be discarded out of hand. Instead, they should be examined carefully and, if possible, several observer positions calculated using one of the many readily available PC-based computer programs. With these as a guide, the designer can usually rearrange and substitute equipment to obtain the desired condition.

See Section 15.4 for a comparison of the direct glare characteristics of lighting fixture diffusers.

## 11.29 VEILING REFLECTIONS AND REFLECTED GLARE

Although there is no generally accepted convention with respect to nomenclature, many people refer to *reflected glare* when dealing with specular (polished or mirror) surfaces and to *veiling reflections* when considering source reflections in dull or semi-matte finish surfaces, which always exhibit some degree of specularity. We use the terms interchangeably.

### (a) Nature of the Problem

The problem of veiling reflections is much more complex than that of direct glare because it involves both the source *and* the task and is inherent in the act of seeing (Fig. 11.26). Vision is produced by light being reflected from the object seen. Thus, if the object being viewed were replaced by a mirror, we would see the source(s) of light clearly (Fig. 11.26a). In commercial spaces there are usually one or more lighting fixtures near the observer that furnish most of the light by which to see. These *principal* sources are the main contributors to reflected glare. Other, more remote fixtures in the room are lesser sources of veiling reflections (Fig. 11.26b).

To the extent that the sources can be seen in the vision task, glare exists. It is imperative to an understanding of this problem to appreciate the importance of the reflection characteristic of the object being viewed. If the object were perfectly absorbent—that is, if it had a reflection coefficient

**4–40-W Lamps**
**Prismatic Lens Diffuser**

*Average Luminance Data cd/m²*

| Vertical Angles | Across Axes | Along Axes |
|---|---|---|
| 60° | 2000 | 1750 |
| 65° | 1060 | 1075 |
| 70° | 500 | 560 |
| 75° | 410 | 380 |
| 80° | 480 | 343 |
| 85° | 560 | 420 |

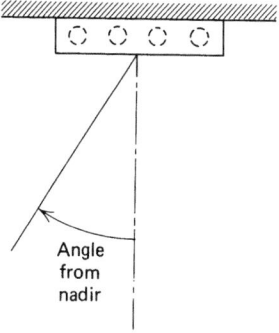

Angle from nadir

*IES Visual Comfort Probability Data*

| Room Size | | Luminaires Lengthwise | | | Luminaires Crosswise | | |
|---|---|---|---|---|---|---|---|
| W (ft) | L (ft) | Ceiling Height (in ft) | | | | | |
| | | 8.5 | 10.0 | 13.0 | 8.5 | 10.0 | 13.0 |
| 20 × | 20 | 80 | 77 | 76 | 79 | 75 | 72 |
| | 30 | 80 | 78 | 76 | 79 | 76 | 73 |
| | 40 | 82 | 79 | 77 | 79 | 77 | 74 |
| | 60 | 80 | 80 | 78 | 79 | 77 | 75 |
| 30 × | 20 | 84 | 80 | 77 | 82 | 78 | 74 |
| | 30 | 83 | 80 | 77 | 81 | 79 | 74 |
| | 40 | 82 | 80 | 78 | 81 | 79 | 75 |
| | 60 | 82 | 80 | 78 | 80 | 79 | 75 |
| | 80 | 82 | 80 | 78 | 80 | 78 | 76 |

Reflectances:
Wall 50%
Ceiling cavity 80%
Floor cavity 20%
Work plane illumination: 1000 lux

**Fig. 11.25** *A typical set of manufacturer's published VCP and luminance data.*

of 0%—it would appear completely black, as no light would be reflected into the eye (Fig. 11.26*c*). Conversely, if the object were perfectly specular, like a clean mirror, and no light source were within the geometry of reflection, it too would appear black (Fig. 11.26*d*). Thus, if we took a clean mirror out on a moonless night and shined a light on if from over our shoulder, it would be practically invisible because no light would be reflected back into our eyes.

The reader might try this experiment: In an inside space with a single overhead luminaire, try to examine the surface of a very clean, dust-free mirror. You will find that the best angle to hold it is *almost* at the angle at which the light source is seen. This is because the mirror is *almost* completely specular, and it is the slight diffuse reflection near the viewing angle that permits us to see the surface. This means that reflected glare is due to task surface specularity, whereas object definition, that is, the ability to see the task itself, is due to task surface diffuseness. A corollary of this conclusion is that veiling reflections, which are caused by mirroring of a source in the task, are proportional to source luminance and substantially independent of illuminance level. The brighter the source, the more troublesome its reflection.

Glare sources within the geometry of reflected vision are shown in Fig. 11.27, and the effects are shown in Fig. 11.28. Figure 11.27 clearly shows that although large sources are difficult to avoid, small sources can usually be easily avoided by a small change in the source-task-eye geometry, such as by moving the head or tilting the task. Table 11.10 lists a few sample reflectance figures to demonstrate that most materials exhibit both specular and diffuse reflectance. In studying Fig. 11.27,

**ILLUMINATION**

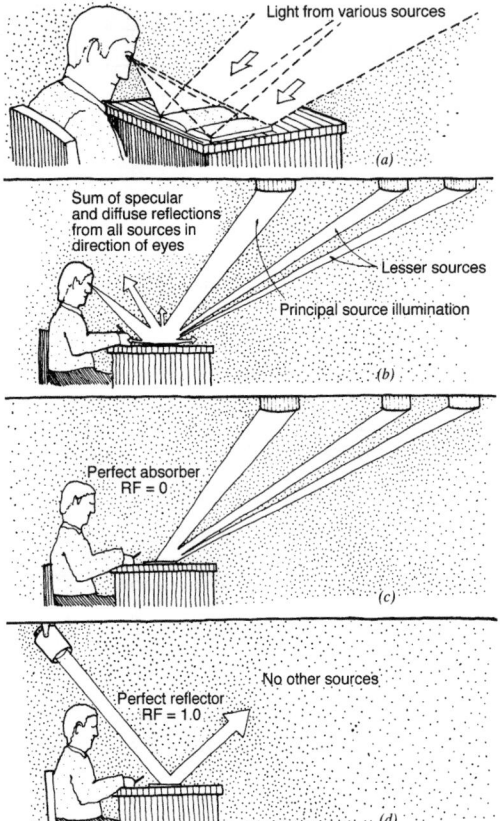

**Fig. 11.26** (a) The nature of the seeing process requires that light from the source(s) be reflected by the task into the eye. (b) The light entering the eye is the sum of all of the reflected light, specular and diffuse, from all sources in the direction of the eye. If the task is specular, all of the sources will be seen reflected in the task. (c) A perfectly absorptive object is jet black because it reflects nothing. (d) A perfectly reflective object positioned as shown is also black because geometrically it cannot reflect light into the eyes.

it is important to note that a majority of visual work is done in the zone of 20° to 40° from the vertical, below the eye, with a maximum at the 25° reading angle (Fig. 11.29).

### (b) Contrast Reduction

The principal effect of the reflection of a light source in a visual object is to reduce contrast between the object and its background, thus reducing visibility. It is as if a bright veil were spread over the object being viewed, which accounts for the term *veiling reflection*. As the angle of the incident light approaches the viewing angle, the specularly

reflected component of this light becomes more and more pronounced, and task contrast drops. This is clearly visible in Figs. 11.28 and 11.29. The worst situation occurs when the incident angle equals the viewing angle. When the specular reflectance of the task and background is high, as with the glass screen of a visual display terminal, for instance, an image of the source is superimposed on the object, making viewing impossible (Fig. 11.29). However, even with the highly specular finish of "slick" magazine paper, vision is still possible because of the very high contrast between black ink and white paper, although with much reduced clarity and considerable annoyance.

When considering specular *and* diffuse reflectance, the equation for contrast given in Section 11.19 must be rewritten as

$$C = \frac{(L_{BD} + L_{BS}) - (L_{TD} + L_{TS})}{L_{BD} + L_{BS}} \quad (11.9)$$

where $L_B$ and $L_T$ are background and task luminances caused by diffuse ($D$) and specular ($S$) reflectance; that is, $L_{BS}$ is the background luminance due to its specular reflectance, and so forth. If we were to rework the calculation of contrast of Section 11.19, including specularity, the result would be much different.

**EXAMPLE 11.5** Assume an interior space lighted to an average illuminance of 75 fc (750 lux) using bare bulb fluorescent fixtures (luminance = 7000 cd/m² [2000 fL]). The task is drafting with India ink on vellum. Reflectances are:

|  | Specular | Diffuse |
|---|---|---|
| Ink | 0.021 | 0.038 |
| Paper | 0.018 | 0.71 |

Calculate the task contrast without and with reflection of the 2000 fL source on the work.

**SOLUTION**

1. Without specularity, using Eq. 11.6 for diffuse reflection only:

$$C = \frac{R_B - R_T}{R_B} = \frac{0.71 - 0.038}{0.71} = 0.947$$

2. With specularity, using Eq. 11.9,

$$C = \frac{(L_{BD} + L_{BS}) - (L_{TD} + L_{TS})}{L_{BD} + L_{BS}}$$

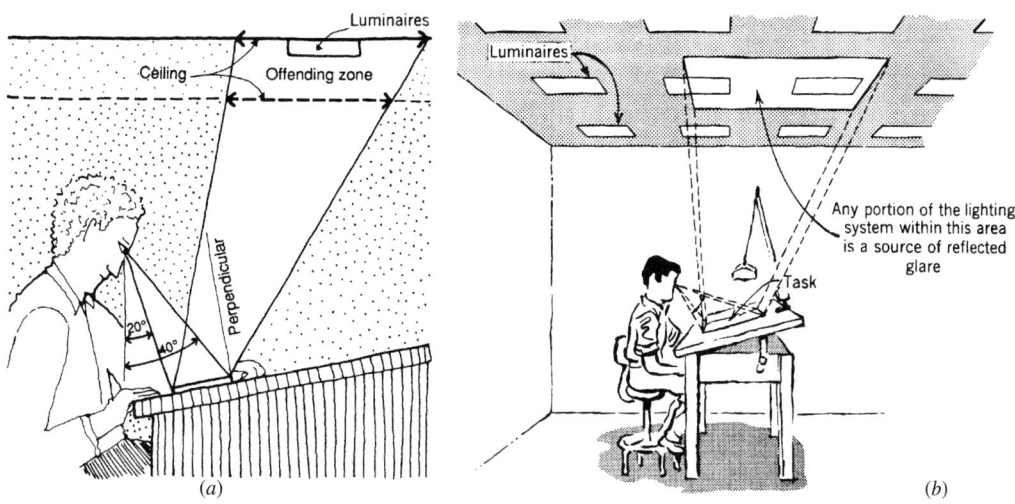

**Fig. 11.27** Geometry of reflected glare. (a) Because normal desktop, head-down viewing angles vary from 20° to 40° from the vertical, the offending zone is the area on the ceiling corresponding to specular reflection between these two angles. Note that the higher the ceiling, the larger this area becomes. (b) In an office situation, the draftsman would see ceiling fixtures in the offending zone reflected in his instruments, parallel straightedge, and work. Note the important fact that the offending zone moves back and becomes smaller as the table tilts up.

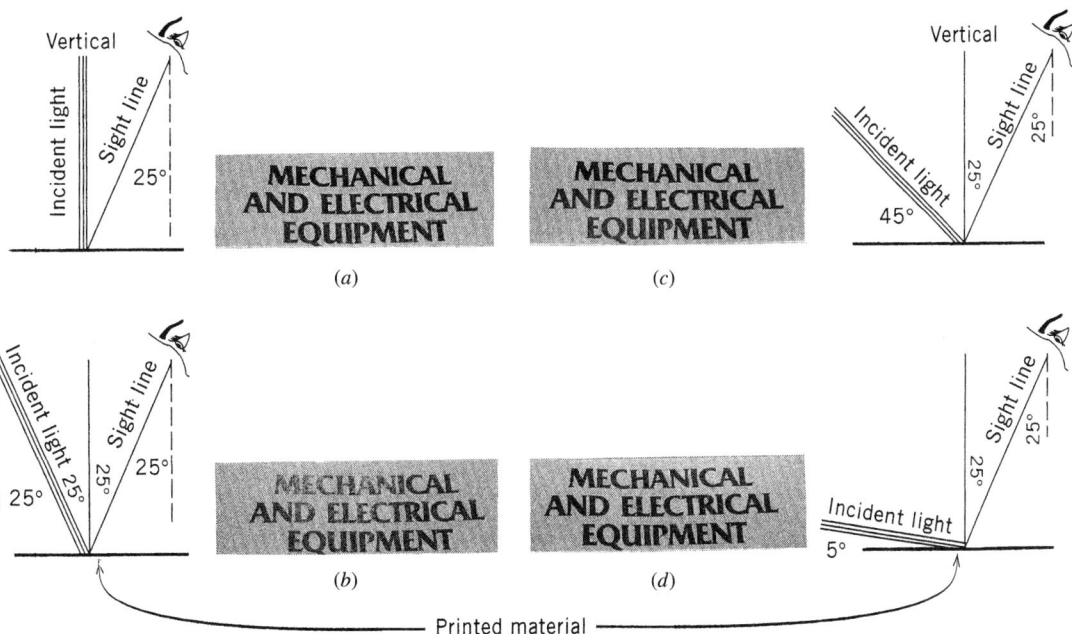

**Fig. 11.28** Note the effect of light in the area of the ceiling that causes reflected glare; contrast is reduced, and the print washes out. The usual viewing angle to a horizontal surface is between 20° and 40° from the vertical; we show 25° because it is the most common viewing angle. With vertical incident light on a diffuse surface (a), such as the pages of a textbook, the print is dark and clear. When the angle of light incidence is equal to the viewing angle (b), we have a mirror reflection situation. Even with diffuse paper, the print is light at best and almost invisible at worst. As the angle of incidence becomes larger (c), reflected glare decreases. When the incident light is at a very low angle (d), there is little reflected glare but all the print appears lighter. (Photos by B. Stein.)

ILLUMINATION

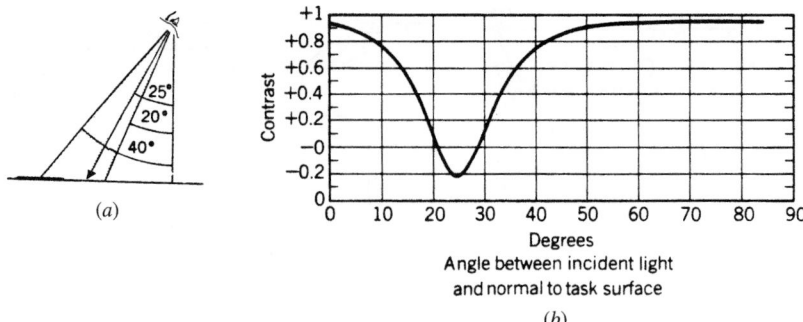

**Fig. 11.29** *(a) Normal viewing angle of a task on a horizontal surface ranges from 20° to 40° from the vertical. Most common viewing angle is 25°. (b) Graph showing contrast reduction of a task with a specular background (such as a sheet of glossy paper) as a function of the angle between incident light and the normal to the task surface. Viewing angle is assumed to be 25° from normal. Note that between 22° and 27° the contrast is negative. This indicates that background luminance exceeds that of the task, making the task essentially invisible. What is visible, clearly, is a reflected (mirrored) image of the source.*

where

$$L_{BD} = 75 \text{ fc} \times 0.71 = 53.25$$
$$L_{BS} = 2000 \text{ fL} \times 0.018 = 36.0$$
$$L_{TD} = 75 \text{ fc} \times 0.038 = 2.85$$
$$L_{BD} = 2000 \text{ fL} \times 0.021 = 42.0$$

and

$$C = \frac{(53.25 + 36) - (2.85 + 42)}{53.25 + 36} = 0.497$$

which is just over half of the previous contrast! If the contrast is normalized to the maximum contrast (as is usually done), the contrast reduction $R$ can be expressed as

$$R = 1 - \frac{C}{C_{MAX}} \qquad (11.10)$$

Thus, in this case, contrast reduction would be

$$R = 1 - \frac{0.497}{0.947} = 0.47$$

That is, the contrast reduction would be 47%. ∎

### TABLE 11.10 Typical Reflectances

| Material | Reflectance | |
| --- | --- | --- |
| | Specular | Diffuse |
| Matte black paper | 0.0005 | 0.04 |
| Matte white paper | 0.0030 | 0.77 |
| Newspaper | 0.0065 | 0.68 |
| Very glossy white photo paper | 0.048 | 0.83 |
| Metallic paper—copper | 0.11 | 0.28 |
| Dull black ink | 0.006 | 0.045 |
| Super gloss black ink | 0.039 | 0.016 |

*Source:* Courtesy of the IESNA.

A similar calculation for clear, black typewritten material on good white bond paper yields a contrast reduction from 94% to 77%, or $R = 17\%$. This 77% figure simply serves to emphasize the fact that with high task to background contrast, effective seeing is possible almost regardless of the lighting condition. However, this most emphatically does not relieve the lighting designer of the responsibility to provide a comfortable and efficient lighting environment in which pronounced veiling reflections do not exist. In general, any contrast reduction of more than 15% is undesirable.

Because both specular and diffuse reflectances frequently vary with the angle of view and exact figures are rarely available, accurate calculation is difficult. If a lighting system exists or a mock-up can be made, measurements of contrast reduction can be made accurately with a contrast/luminance meter of the type shown in Fig. 11.30. A standard contrast device that is designed to correspond to a normal office task (black typeface and white paper background) is positioned on the work surface and exposed to the ambient illumination. The task contrast is then measured at the same angle at which it would normally be viewed. Contrast reduction is automatically calculated and displayed.

A contrast reduction map of the work surface can thus easily be made (Fig. 11.30c). Thereafter, changes can be made in the position of the work, viewer, or illumination sources to minimize contrast reduction. Note the pronounced effect of simply shifting the source out of the ceiling glare source zone (offending zone) (Figs. 11.30d and 11.27).

ILLUMINATION

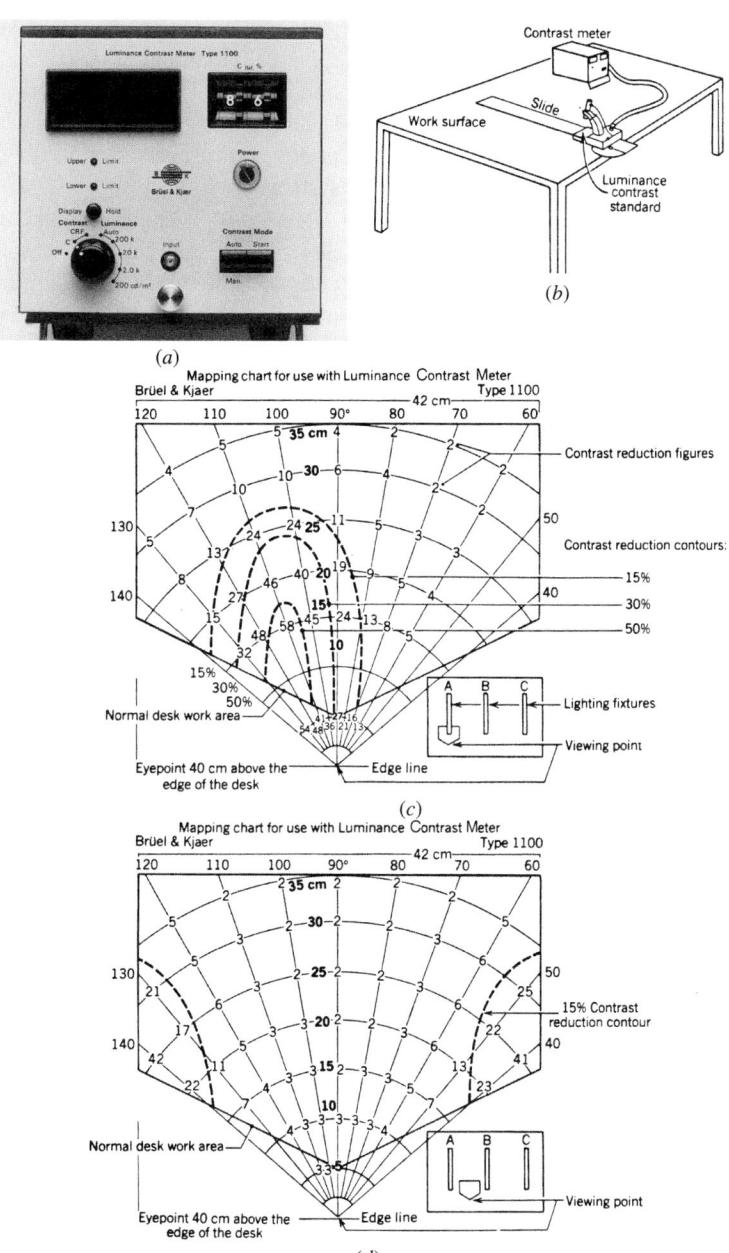

**Fig. 11.30** (a) This unit is a high-precision electronic digital instrument that measures luminance, luminance ratio, contrast, and contrast reduction. Use of the unit is shown in (b), (c) and (d). Luminance measurement range is 0 to 200,000 cd/m² (56,400 fL) at high accuracy. (b) The meter measures contrast reduction at a specific viewing angle. (c) Note that with a lighting fixture directly in the offending zone—that is, above and in front of the viewer—a severe loss of contrast occurs over much of the work surface. (d) By shifting the relative position of the viewer and the source so that no source exists in the offending zone, contrast reduction is held to 3% to 4% over most of the work surface. Contrast reduction in the normal work area should not exceed 15%. (Courtesy of Brüel & Kjaer.)

Contrast reduction in the primary work area should not exceed 15% for good task visibility with specular work items such as glossy papers.

## 11.30 EQUIVALENT SPHERICAL ILLUMINATION AND RELATIVE VISUAL PERFORMANCE

### (a) Equivalent Spherical Illumination

Another way of approaching the problem of contrast reduction is to define a reference lighting system that is effectively free of veiling reflections and then relate actual lighting systems to it with a figure of merit. Conversely, one can measure the effectiveness of a given lighting system in terms of the equivalent glare-free system. Both of these ideas, which are essentially the same, are the basis of the concept of *equivalent spherical illumination* (ESI).

In order to achieve a lighting system almost free of reflected glare, it is necessary to construct an enclosed volume whose surfaces are uniformly diffusely reflective and whose primary source is obscured to the maximum extent possible. As illustrated (Fig. 11.31), the integrating sphere is such a device. Light is introduced from the outside, split by a deflector, and evenly distributed throughout the sphere by the multiple reflections from the white-painted walls. The result is an evenly illuminated volume. When a task is introduced, the illumination falling on it is *entirely* uniform; that is, there are no high-luminance sources reflected in it. It is therefore termed *spherically* illuminated. (Note the parallel to sky illumination.) The extent to which any other illumination system can duplicate this

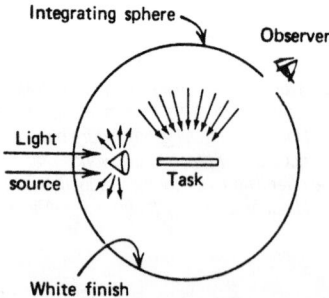

**Fig. 11.31** *Spherical illumination is produced by illuminating an object by diffuse reflection from the inside walls of an integrating sphere. The light source and observer are normally external.*

glare-free environment is that system's *equivalent spherical illumination* (ESI), representing the portion of its total illumination that is spherical—that is, diffuse and glare-free. ESI is determined by comparing contrast rendition in the spherical and test systems.

A study of school lighting (Fig. 11.32) gave the illustrated results for four viewing positions in a classroom lighted with ceiling-mounted continuous rows of 2-ft × 4-ft, 4-lamp, 40-W fluorescent fixtures with lens-type wraparound diffusers on 10-ft centers. Carefully note that:

1. ESI depends entirely on the viewing position and viewing angle, other factors in the space being equal.
2. In an ostensibly very well lighted (215 fc) position (*M1*), the glare-free illuminance is only 28 fc! This does not mean that visual work in this position is impossible. It does mean, emphatically, that in position *M1*, a pronounced veiling reflection exists on all specular objects. (Because of the size, orientation, and location of the glare source, this reflected glare is difficult to avoid.) It further means that a large amount of energy is being utilized (effectively wasted) to produce essentially negative results.

The results could have been anticipated, at least qualitatively by examination of the observer positions vis-à-vis the layout. Positions *M1* and *M3* have bright sources in the offending zone—*M1* more so than *M3*—as is borne out by the results. *M2* is an excellent position in that it receives light contributions from the two sides, its illuminance value being lower than the others due to wide row spacing. *M4* is ideally placed; no glare sources are in the offending zone and a row of fixtures is positioned *behind* it, which makes it geometrically impossible to act as a glare source. The ESI analysis gives quantitative expression to our qualitative judgment and, as such, is a valuable design tool. Note particularly that the ESI results shown in Fig. 11.32 clearly correspond to the results of a similar test made with the contrast meter, as shown in the charts of Fig. 11.30.

As with VCP criteria for direct glare, so with ESI; there are ameliorating factors that generally make a given lighting system better than these criteria figures would indicate. Some of these factors have already been mentioned but bear repetition.

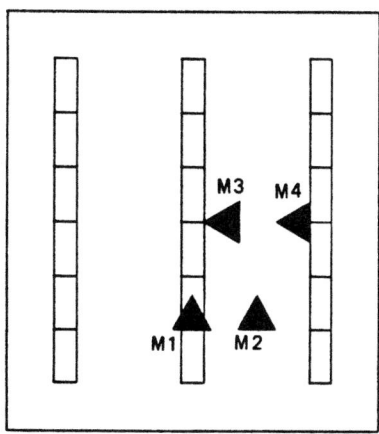

|  |  | M1 | M2 | M3 | M4 |
|---|---|---|---|---|---|
| TI | 2L | 108 | 92 | 125 | 118 |
|  | 4L | 215 | 185 | 250 | 235 |
| CRF | 2L | .75 | 1.00 | .82 | 1.01 |
|  | 4L | .76 | 1.00 | .83 | 1.03 |
| ESI | 2L | 17.8 | 91.9 | 31.5 | 127.8 |
|  | 4L | 28.4 | 185.3 | 58.1 | 308.3 |

*Observer Position*

TI–Task illumination
2L–2 lamps (inside pair)
4L–4 lamps
CRF–Contrast Rendition Factors

ILLUMINATION

**Fig. 11.32** *A test classroom illuminated by three widely spaced rows of four-lamp fixtures with lens-type wraparound diffusers. Observer positions are shown by arrows. The row of fixtures in front of position M4 is too far forward to be in the offending zone. (From Sampson, 1970.)*

1. ESI is critically dependent on observer position and viewing angle. Although position is generally fixed by chair location, observers can and do change their viewing angle and head aspect to correct for glare situations.
2. The nature of the task (i.e., its specularity) is assumed to be fixed and unique. In some situations the task nature varies, and thus also the contrast. When tasks are constant, severe veiling reflections frequently lead to measures being taken that improve the task, the lighting, or both.
3. The lighting distribution characteristic of the fixture involved is a critical factor in glare production. The characteristic of a wraparound lens diffuser is such (see Section 11.31c[4]), Fig. 20.23, and Section 15.2) that considerable light falls in the glare zone. Other diffuser characteristics yield different results.

The concept and use of ESI have come under considerable criticism in the professional lighting literature because ESI addresses contrast, which is not identical with visibility, it requires recalculation for even the small geometry changes that can change contrast dramatically, it tracks raw footcandles, and it is based on extrapolation from threshold conditions, the efficacy of which has been called into doubt. These published reservations and criticisms of ESI have disturbed the lighting profession

sufficiently to result in wide abandonment of its use and its disappearance from most modern computer lighting programs.

The ESI procedure, however, does exactly what it is designed to do: to point out locations of poor lighting geometry immediately and quantitatively, and to flag luminaires with unsuitable distribution characteristics for the proposed use. That is, as stated in the 1981 *IES Lighting Handbook*, it is "used as a tool in determining the effectiveness of controlling veiling reflections and as part of the evaluation of lighting systems."

### (b) Relative Visual Performance

In more recent years, a metric called *relative visual performance* (RVP) has appeared in the literature and has gained considerable acceptance. It tests (also via computer calculation) the effectiveness (i.e., the relative [to perfection] visual performance of a given visual environment) of task accomplishment in regard to speed and accuracy. Like ESI, it is based on luminance and contrast, but unlike ESI, it judges the relative performance of a task rather than simply contrast reduction. It seems to ignore the discomfort of veiling reflections if the task can be performed efficiently. The author has computed RVP for a number of common lighting layouts with various types of lighting

ILLUMINATION

fixtures. The results give uniformly high values for RVP varying between 0.95 and 0.99, which to our mind makes effective judgment of glare situations very difficult. The reader is encouraged to use the available lighting programs to calculate ESI and RVP for proposed lighting layouts and, where possible, to compare the results with completed construction.

## 11.31 CONTROL OF REFLECTED GLARE

Because the causes of veiling reflections are well understood, it would seem that a solution to the problem should long since have been developed. Unfortunately, this is not the case. Although there is no known lighting method or material that completely eliminates veiling reflections, there are a number of techniques that minimize contrast loss due to veiling reflections while maintaining adequate illumination. These are:

- Physical arrangement of sources, task, and observer so that reflected glare is minimal (see Section 11.31a)
- Adjusting brightness (eye adaptation level) so that objectionable brightness is minimized (see Section 11.31b)
- Design of the light source so that it causes minimal reflected glare (see Section 11.31c).
- Changing the task quality (see Section 11.31d).

### (a) Physical Arrangement of System Elements

Arrange the lighting geometry to avoid sources of high luminance at reflection angles when dealing with specular tasks. This is often difficult to accomplish in modern offices, which frequently have both horizontal and vertical work surfaces, the latter being the specular screen surface of a visual display terminal (VDT). The latter problem is so widespread that today, it effectively governs the design of office lighting. As should be clear from Figs. 11.30 and 11.32 and the related discussion, in a space using multiple sources, particularly in continuous rows, placing the work between rows with the line of sight parallel to the long axis of the units is an effective technique (see Fig. 11.32, position *M2*). Position *M4* is dangerous in that the center row can be a source of reflections. In this case it is not, due to the 10-ft row spacing. Note that the offending zone for horizontal tasks depends on the tilt of the desk. Thus, for a horizontal desk, the offending zone is forward of the desk, as in Fig. 11.33*a*; with an elevated table, the ceiling glare source zone may well be behind the source, as in Fig. 11.33*b*.

All of the geometric solutions mentioned presuppose a detailed, fixed furniture layout, a situation that obtains in many but certainly not all cases. In the absence of such data, two alternatives are possible: a uniform layout with furniture adjusted to it or vice versa. In practice, a

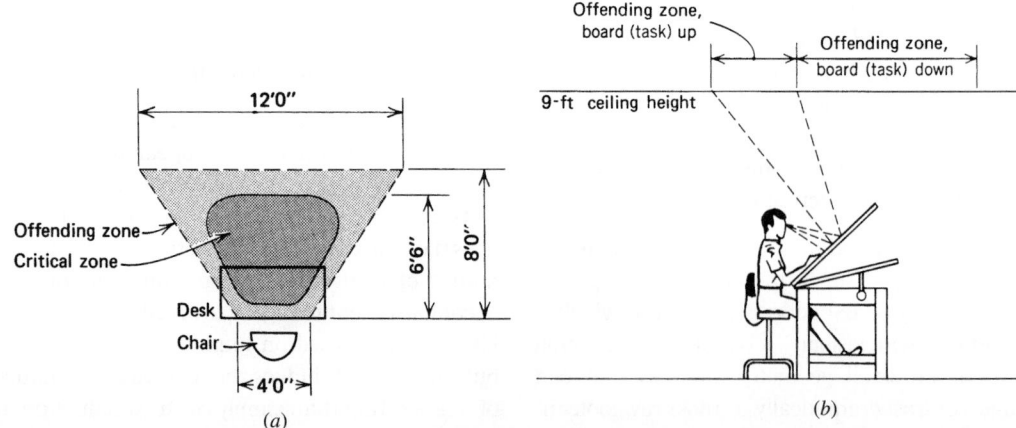

**Fig. 11.33** (a) If luminaires are kept out of the trapezoidal offending zone, contrast will be excellent. If the bulk of one or more luminaires projects into this zone, and in particular into the critical zone, contrast will drop sharply. The dimensions shown are for a flat desk 3 ft × 5 ft and a 9-ft ceiling height. (b) The dependence of the glare zone on table tilt is illustrated. The offending zone becomes smaller as the table is raised, so that with a table near the vertical position, glare is all but eliminated. (a from Ross and Baruzzini, Inc., 1975.)

combination of both is the most practical approach. Because low watts per square foot budgets have made ducted lighting fixture heat removal systems (air troffers) much less prevalent, fixtures are easily shifted. This mobility is further enhanced by the extreme flexibility of lighting fixtures fed from ceiling plug-in raceways. Figure 11.34 shows such a rearrangement, which results in saving five fixtures, a load reduction of 800 W, and an *improvement* in visibility.

## (b) Control of Area Brightness and Eye Adaptation Level

As discussed in Section 11.19, loss of contrast can be compensated for (and glare reduced) by increased overall nonglare illumination. In so doing, we are simply making the task brighter to override the detrimental veiling reflection. The problem with this technique, however, is that a large increase in illuminance is required to overcome the glare. This increase can, in many instances, be most practically accomplished not by increasing overall room illumination, with the associated extremely high energy consumption, but by adding a supplementary task-lighting source so arranged as to be free of reflected glare. By making this supplementary source's position adjustable (as in Fig. 11.27b), we accomplish three things:

1. Veiling reflection is overcome.
2. The high level of illumination needed for exacting tasks is provided with minimum energy expenditure.
3. The observer is granted complete control, with resultant optimum lamp placement plus psychological satisfaction that generally prevents worker complaint. (The optimum position is generally to the left and slightly forward of the task).

We can demonstrate the effectiveness of a supplemental desk lamp by returning to Example 11.5.

---

**EXAMPLE 11.6** Recalculate the contrast reduction of the ink-on-vellum visual task of Example 11.5, assuming that a desk lamp raises the illuminance to 200 fc (2000 lux) and is positioned so as to be glare-free. (*Note:* An adjustable lamp with 2 at 15-W fluorescent tubes produces approximately that luminance on the desk.)

**SOLUTION**
Contrast from Eq. 11.9:

$$C = \frac{(L_{BD} + L_{BS}) - (L_{TD} + L_{TS})}{L_{BD} + L_{BS}}$$

where

$$L_{BD} = 200 \text{ fc} \times 0.71 = 142$$
$$L_{BS} = 2000 \text{ fL} \times 0.018 = 36.0$$
$$L_{TD} = 200 \text{ fc} \times 0.038 = 7.6$$
$$L_{BD} = 2000 \text{ fL} \times 0.021 = 42.0$$

and

$$C = \frac{(142 + 36) - (7.6 + 42)}{142 + 36} = 0.72$$

With this contrast (0.72), the contrast reduction from the original no-glare situation has been improved from the original 47% reduction (0.947 to 0.497) to 24% reduction (0.947 to 0.72). However, because even a contrast reduction of 24% is undesirable, a change in task-source geometry or a change in source luminance would be required, assuming that the task itself must remain unchanged. ∎

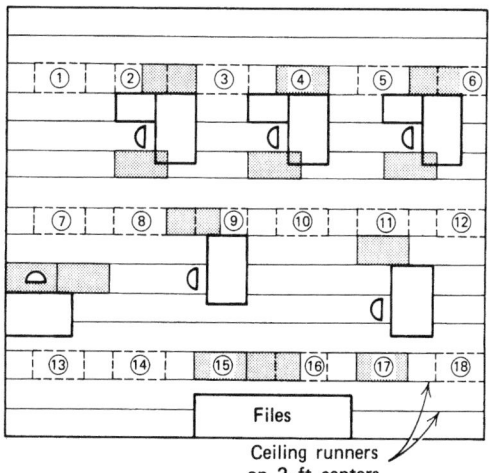

Files

Ceiling runners on 2-ft centers

*Fig. 11.34 The original uniform fixture layout utilized three rows of six 2 × 4, four-lamp fixtures, giving a total load of 2880 W, a load density of 2.6 W/ft², and a uniform illumination level of approximately 90 (raw) fc (900 lux). The original layout is shown dotted and numbered. The rearranged layout uses 13 fixtures (shown shaded) for a total 2080 W, a load density of 1.9 W/ft², and more than 100 ESI fc (1000 lux) on each work surface. In addition, five fixtures are saved. Note: This level of illumination is justified only for difficult visual tasks.*

## (c) Control of Source Characteristics

The reflected luminance that causes loss of contrast is proportional to the luminaire's luminance at that viewing angle, and therefore may be reduced by reducing luminaire luminance at that angle. This can be accomplished in four ways:

Dimming or switching lamps (see number 1 following)

Using luminaires with lower overall luminance (see number 2 following)

Using the luminaire as a primary source to illuminate a large, low-brightness secondary source (see number 3 following)

Reduce the luminaire luminance *only at the offending angles* (see number 4 following)

1. Reducing the total output of a fixture also reduces its output in the critical portion of the ceiling glare zone and can actually *increase* the ESI illuminance (i.e., improve task contrast).
2. In lieu of using a few small high-output sources, utilize larger-area, low-output sources (Fig. 11.35). This has the effect of reducing the source luminance in the ceiling glare zone while increasing the illumination contribution from outside the glare zone, resulting in better contrast for the same or lower illuminance level (lux). The disadvantage of this technique is an increased lighting fixture cost.
3. To overcome the economic disadvantage of multiple low-output, low-luminance sources, the ceiling can be used as a secondary source illuminated from high-output indirect or semi-indirect fixtures. These sources, which can be fluorescent or HID (e.g., metal–halide), have the advantage of high efficiency. The space's ceiling height must be sufficient to permit suspending the unit while avoiding "hot spots" on the ceiling. The minimum suspension length depends on the luminaire characteristics and is normally provided by the manufacturer. To assure high efficiency, the ceiling should be painted with a high-reflectivity matte white paint and kept clean. Results obtained from a semi-indirect installation using 1500-mA, very-high-output lamps are shown in Fig. 11.36.
4. Because most horizontal task vision takes place between 20° and 40° from the vertical (see Figs. 11.27 and 11.30), any fixture that emits little or no light below 40° from the horizontal *cannot* produce veiling reflection regardless of its position in the field of view (Fig. 11.37). As a result, diffuser manufacturers produce prismatic diffusers the output of which is diminished below 30° and above 60° in order to minimize both reflected and direct glare. Due to the characteristic shape of the distribution curve, they are known industrywide as *batwing* diffusers. For observers positioned so that their sight lines are parallel to the longitudinal axis of the ceiling fixtures, lenses with linear (side-to-side) batwing characteristics perform well. If the observing position varies in aspect with respect to the fixture, a radial batwing curve (in all directions) is required. Note carefully that these diffusers have only limited usefulness in reducing reflected glare in specular vertical surfaces (VDT screens). All types of diffusers and their characteristics are discussed in Section 15.4.

**Fig. 11.35** A concentration of light in the glare zone (a) produces the largest amount of reflected glare. As the number of light sources is increased (b) in the glare zone and luminance is decreased, reflected glare is decreased. The least glare is from an all-luminous ceiling, which also has the lowest luminance (c).

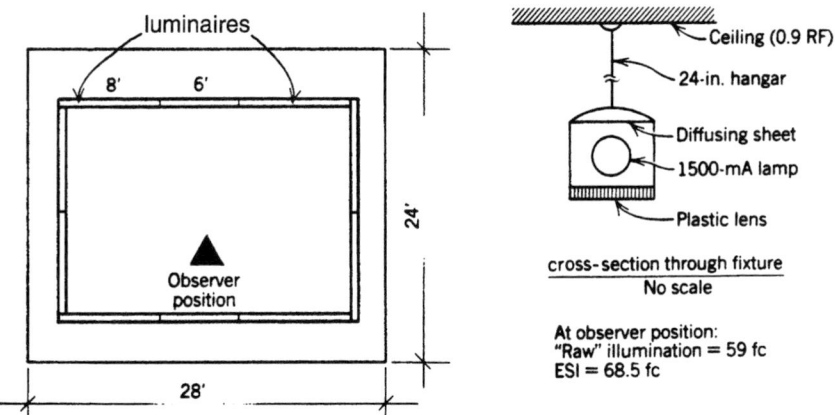

*Fig. 11.36* With a high-reflectance diffuse finish ceiling, this semidirect perimeter lighting installation yields a higher ESI illuminance than raw illuminance at the viewing location illustrated, indicating excellent contrast rendering. The plastic lens at the bottom of the fixture serves to provide perceived light source luminance and to avoid an impression of gloominess, despite the satisfactory overall luminance level. (From Sampson, 1970.)

## (d) Changing the Task Quality

At this point, it should be clear that reducing the task specularity is at least as effective a means of reducing veiling reflections as changing the lighting system characteristics, if not more so. It is therefore recommended that task contrast and specularity be actively considered and recommendations made in a framework of energy and cost effectiveness. Thus, to produce adequate visibility it is often cheaper and always more energy economical to upgrade the task (in the visibility sense) than to change the lighting system.

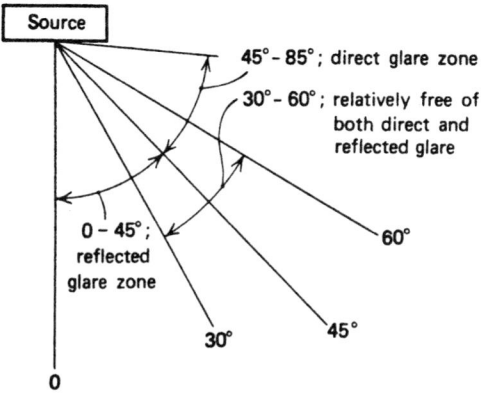

*Fig. 11.37* Glare zones are 0° to 45° and 45° to 85° for reflected and direct glare, respectively. Therefore, a diffuser that emphasizes the 30° to 60° zone will be least objectionable on both counts.

## 11.32 LUMINANCE RATIOS

As explained previously, other factors being equal, visual performance increases with contrast—that is, with the difference in luminance between the object being viewed and its immediate surroundings. Conversely, however, the difference between the average luminance of the visual field (task) and the remainder of the field of vision should be low to avoid the discomfort of large, rapid changes in eye adaptation level. Restated, *contrast is desirable in the object of view but undesirable in the wider surrounding field of view.*

Providing reflectances of 50%, 30%, and 80% for walls, floor, and ceiling, respectively, and 35% for furniture, establishes a fairly high eye adaptation level so that direct glare (which results from excessive luminances in the field of view) is minimized. Recommendations for maximum luminance ratios to achieve a *comfortable* environment are presented in Table 11.11. Effective visual performance is entirely possible in environments with much higher ratios. It is simply not as visually comfortable and may be fatiguing.

To achieve the recommended luminance ratios, it is obviously necessary to carefully control the reflectances of the major surfaces in a room. The reflectance figures given are averages for work-type commercial and educational spaces. The marked difference between a background with proper

**TABLE 11.11 Recommended Maximum Luminance Ratios[a]**

*Note:* To achieve a comfortable brightness balance, it is desirable to limit luminance ratios between areas *of appreciable size* as seen from normal viewing positions as follows:

| | |
|---|---|
| 1 to one-third | Between task and adjacent surroundings |
| 1 to one-tenth | Between task and more remote darker surfaces |
| 1 to 10 | Between task and more remote lighter surfaces |
| 20 to 1 | Between luminaires (or fenestration) and surfaces adjacent to them |
| 40 to 1 | Anywhere within the normal field of view |

[a]These ratios are recommended as maximums; reductions are generally beneficial.

reflectance and one with excessive brightness ratios caused by the low surrounding reflectances is shown in Fig. 11.38.

## 11.33 PATTERNS OF LUMINANCE: SUBJECTIVE REACTIONS TO LIGHTING

In the list of characteristics in Section 11.16, we included among the secondary factors in illumination *patterns of luminance*—that is, the patterns of

**Fig. 11.38** *The reflected glare from luminaires disappears when a piece of light, diffuse linoleum is placed over the dark, polished desktop. Light-colored desktops with 35% to 50% reflectance result in task-to-background ratios within the 3:1 recommended range. Before the linoleum was placed, a reflection similar to the one seen on the right also existed on the left of the desk due to another luminaire. (Courtesy of IESNA.)*

light and shadow in a space resulting from the illumination. Thus, a single source may produce sharp shadows, whereas a luminous ceiling or a completely indirect illumination system produces almost completely diffuse light. *Diffusion* is the degree to which light is shadowless and is therefore a function of the number of directions from which light impinges on a particular point and their relative intensities.

Perfect diffusion, rarely obtainable (or desirable), would have equal intensities of light impinging from all directions, therefore yielding no shadows. The only naturally occurring example of perfectly diffuse lighting is a daytime fog, which we know to be extremely disturbing to the eye, demonstrating that some directivity is desirable. Diffusion can be judged by the depth and sharpness of shadows. A room with well-diffused illumination resulting from multiple sources and high room surface reflectances yields soft multiple shadows that do not obscure the visual task. Because purely diffuse lighting is monotonous and not entirely conducive to extended periods of work effectiveness, some directional lighting is often introduced as an adjunct to diffuse general lighting to lend interest by producing shadows and brightness variations. Where texture must be examined or surface imperfections detected by grazing angle reflections, highly directional lighting is required. Indeed, as seen in Fig. 11.39, directional light is what creates shape and is precisely the characteristic best used to influence architectural space and form.

Sections 13.10 through 13.15, which deal with systems of lighting, illustrate a few of the light/dark patterns produced by different lighting arrangements. The combinations of uplighting and downlighting are legion; each produces its own shadows and modeling, and each has a quality of its own. It is very much in the interest of the lighting designer to be familiar with these effects so that he or she can mentally visualize them as the design progresses. Indeed, it would be well for a designer to prepare a reference sketchbook of such shadow diagrams. It is these patterns of light and darkness that give the ambience and the subjective reactions of sociability/isolation, clarity/fuzziness, spaciousness/crampedness, simplicity/clutter, formality/informality, boredom/excitement, definition/shapelessness, and so on. Indeed, many lighting designers *begin* a lighting design by sketching pictorially the area to

**Fig. 11.39** *Totally diffuse lighting (a) destroys texture, whereas a combination of diffuse and directional lighting (b) produces the required modeling shadows. (Courtesy of Holophane.)*

be lighted, showing the patterns of brightness and shadow desired to achieve their objective. This technique is obviously most useful in nonoffice areas such as lobbies, waiting areas, all type of recreational spaces, restaurants, merchandising spaces, and so on. In office areas the technique can also be applied to provide the points of visual interest referred to in our discussion.

Color has a great deal to do with subjective reactions and is discussed separately. The subject of psychological reactions to the lighting environment is extensive and complex, and can be only touched on here to the extent of mentioning a few of the salient lighting techniques and their usual subjective responses.

In addition to modeling and texture accent, small, high-brightness sources, usually called *sparkle*, create points of interest and visual excitement. Lighting installations generally yield a sense of vividness or activity proportional to the level of illumination. This is not the case with very diffuse lighted areas, which, even at high illuminance levels, are tedious. This is particularly noticeable in large, luminous ceiling installations that are especially oppressive when the ceiling is low. Small, exposed incandescent lamps, a brightly lighted, rough-textured wall, and pendant fixtures with pierced reflectors are some of the techniques used to create visual interest.

Visual attention can be drawn by high brightness. This well-known fact is used constantly in displaying merchandise. Note the following usual reactions:

- A 3:1 luminance ratio between object and surround will be noticed, but usually will not affect behavior or draw attention.
- A 10:1 luminance ratio will attract attention and, if interesting, hold it.
- A 50:1 luminance ratio or higher will highlight the object thus illuminated, practically to the exclusion of all else in the field of view.

Because areas of high luminance draw the eye's attention, all of the individual brightness sources in the field of view produce an overall impression. If there is some form or order or pattern to them (as a pattern of lighting fixtures), then the overall impression is not disturbing—it can be thought of as visually harmonious. If, however, they are in disarray, they produce a discordancy in the eye precisely as noise produces discordancy in the ear. This visual "noise" is frequently referred to as *visual clutter* and can be very disturbing. The designer is well advised to keep this important fact in mind when arranging light sources that are the primary sources of luminance in an enclosed space. Aspects of fixture patterning are shown in Section 13.16.

Other subjective reactions to lighting on which there is wide consensus are:

1. Bright walls (about 25% of the horizontal light level) increase the impression of spaciousness. Conversely, dark walls diminish a space. As a corollary, high fixture luminance attracts the eye away from the walls and diminishes spaciousness.
2. Worker-adjustable task lights increase the feeling of control and therefore comfort.
3. Downlights (and color highlights) increase feelings of relaxation and comfort.
4. Hidden-source indirect lighting and very-low-brightness lighting fixtures cause discomfort because of the inability to locate the source of light.

## FUNDAMENTALS OF COLOR

The subject of color is vast. Here we consider only a few aspects of color that it is imperative that the lighting designer understand. Furthermore, because it is difficult to discuss color without actually using it, the coverage is brief.

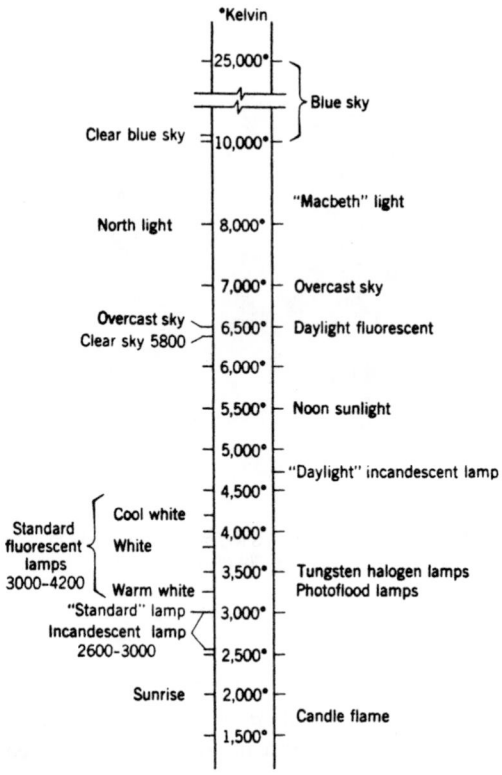

*Fig.* **11.40** *Approximate color temperatures of common illuminants.*

### 11.34 COLOR TEMPERATURE

A light source is often designated with a *color temperature*, such as 3400 K for halogen lamps, 4200 K for certain fluorescent tubes, and so on. This nomenclature derives from the fact that when a light-absorbing body (called a *black body*) is heated, it first glows deep red, then cherry red, then orange, until it finally becomes blue–white hot. The color of the light radiated is thus related to its temperature. Therefore, by developing a black-body color temperature scale, we can compare the color of a light source to this scale and assign to it a *color temperature*—that is, the temperature to which a black body must be heated to radiate a light similar in color to the color of the source in question. Temperature is measured in Kelvin, which is a scale that has its zero point at −460°F. Figure 11.40 shows the assigned color temperature of some common light sources.

Strictly speaking, a color temperature can be assigned only to a light source that produces light by heating, such as the incandescent lamp. Other sources, such as fluorescent lamps, produce light by processes that are detailed in Chapter 12. Such sources are assigned a *correlated color temperature* (CCT), which is the temperature of a black body whose chromaticity most nearly matches that of the light source. For such sources there is *no relation whatever* between their operating temperature and the color of the light produced.

Any nonspectral color illuminant is composed of two or more component color illuminants. When such a composite light—for example, white—falls on a surface other than black or white, selective absorption occurs. The component colors are absorbed in different proportions so that the light reflected or transmitted is composed of a new combination of the same colors that had impinged on the surface. Thus, a white light reflected from a red wall acquires a red tint because the component colors of the white light other than red were absorbed in greater proportion than the red. When reflected, the red light takes prominence, thus giving the reflected light a red tint. This is illustrated in Fig. 11.41*a*.

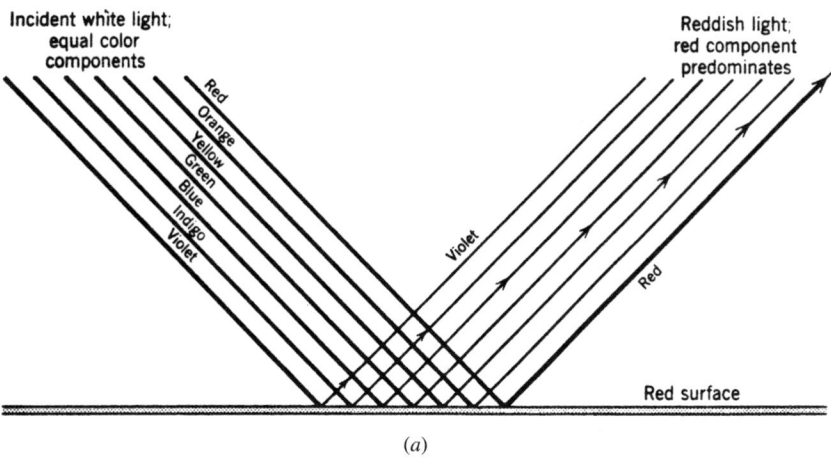

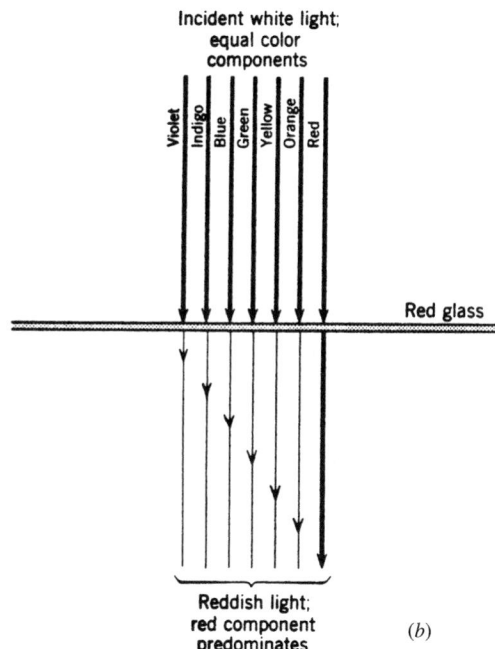

**Fig. 11.41** *Selective absorption of (a) reflected light and (b) transmitted light.*

Similarly, a white light passed through a piece of red glass emerges as a reddish light because the other components are absorbed in much greater proportions than the red. This well-known phenomenon is illustrated in Fig. 11.41*b*.

It is this phenomenon that allows us to see color at all; the individual object pigmentation absorbs other colors of light and reflects or transmits to the eye only its own hue in greater concentration than in the incident light.

## 11.35 OBJECT COLOR

The color of the illuminant (light) and, correspondingly, the coloration of the objects within a space constitute an important facet of the lighting quality. The two factors, however, must not be considered separately because by definition the color of an object is its ability to modify the color of light incident on it by selective absorption. The color reflected or transmitted is apprehended by the eye as the color of the object. An object is technically said to be "color-less" (not transparent) when it does not exhibit selective absorption, reflecting and absorbing the various components of the incident light nonselec-tively. Thus, white, black, and all shades of gray are colorless, neutral, achromatic, or, more precisely, lack hue.

*Hue* is defined as that attribute by which we recognize and therefore describe colors as red, yel-low, green, blue, and so on. Just as it is possible to form a series from white to black with the interme-diate grays, it is possible to do the same with a hue.

The difference between the resultant colors of the same hue so arranged is called *brilliance* or *value.* White is the most brilliant of the neutral colors and black the least; pink is a more brilliant red hue than ruby; and golden yellow is a more brilliant (lighter) yellow hue than raw umber.

Colors of the same hue and brilliance may still differ from each other in *saturation,* which is an indi-cation of the vividness of hue or the difference of the color from gray. Thus, pure gray (black plus white) has no hue; as we add color, we change the saturation without changing the brilliance. The three characteristics then that define a particular coloration are *hue, brilliance,* and *saturation.* Using these terms, we may define "bay" as a color red–yellow in hue of low brilliance and low saturation,

whereas carmine is a color red in hue, of low bril-liance and very high saturation.

Various systems of color classification have been devised, including the ISCC-NBS (Inter-Society Color Council—National Bureau of Standards) color sys-tem, the Munsell Color System, the Ostwald Color System, CIE, and the Chromaticity Diagram. In the well-known and widely used Munsell Color System (Fig. 11.42), brilliance is referred to as *value* and sat-uration as *chroma;* thus, a color is defined by hue, value, and chroma. The brilliance (value) of a pig-ment or coloration is related to its reflectance to white light. The higher the brilliance or value, the higher the reflectance factor, as might be expected when one considers that white and black are the poles of brilliance. Chroma or saturation may be thought of as either the difference from gray or the purity of the color. Spectral colors have 100% purity and therefore maximum chroma.

When white is added to a pigment, it produces a *tint;* adding black produces a *shade.* When pigments are mixed to produce a particular color, the color is created by a subtractive process. That is, each pig-ment absorbs certain proportions of full-spectrum white light; when mixed, the absorptions com-bine to subtract (absorb) various colors of the white spectra and leave only those colors that finally constitute the hue, value, and chroma of

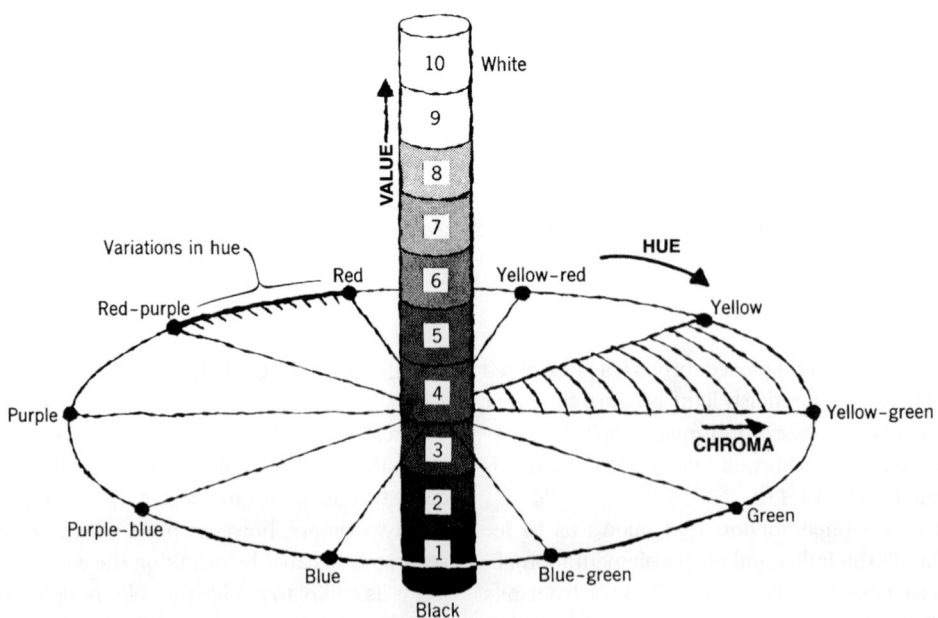

**Fig. 11.42** *The Munsell Color System defines a color by three characteristics: hue (color), chroma (saturation), and value (grayness).*

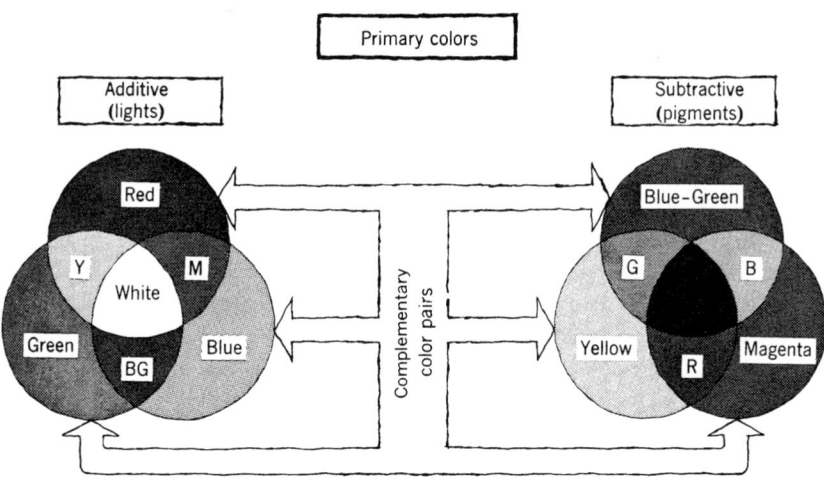

**Fig. 11.43** *Primary and complementary colors. Complementary color pairs are shown by arrows. Pigments from color by an absorptive (subtractive) process; colored lights from colors by an additive process.*

the pigment. This subtractive effect is also utilized when producing colors by filtering white light. Each filter selectively absorbs component colors, transmitting only the component desired. Thus, a blue filter transmits only blue, and so on (see Fig. 11.41*b*). Conversely, when lights of the three primary colors—red, green, and blue—are combined, they form white by an additive process (Fig. 11.43).

The additive and subtractive primary colors are complementary; they combine to give a white or neutral gray, respectively. Thus, red and blue–green, blue and yellow, and green and magenta are complementary. Therefore, if a red object is illuminated with blue–green light, the object color appears gray because the red pigment absorbs the blue–green and reflects nothing; hence the gray. This accounts for the once common "lost red car" in parking lots illuminated with clear mercury lamps with their characteristic blue–green color. This phenomenon is today rare, as clear mercury lamps have fallen into disuse in favor of metal–halide and sodium lamps.

## 11.36  REACTIONS TO COLOR

Light of a particular hue (other than white) is rarely used for general illumination except to create a special atmosphere. When a space is lighted with colored light, the eye adapts by a phenomenon known as *color constancy* so that it can, to a considerable degree (depending on the chromaticity of the light), recognize colors of objects despite the spectral quality of the illuminant. Thus, even when wearing heavily tinted sunglasses, we can still distinguish the color of objects quite easily. Indeed, after only a very short while, we no longer notice the green, yellow, blue, amber, or other color cast caused by the tinted lenses. However, the eyes become more sensitive to the missing colors that would make up white light. This phenomenon can be used to make meat look redder on a butcher's counter by using blue-rich, red-poor, cool white lighting in the remainder of the store.

A similar phenomenon occurs when the eye is exposed to a monochromatic scene, where the chromaticity is due to coloration of the objects rather than the illumination. The eye in such a situation becomes sensitized to the complementary color; thus, if after looking at a green surface one shifts the gaze to a white surface, one sees the complementary red color. Returning to our meat market, the use of green paint on the walls also enhances the redness of the meat. This effect in reverse also partly accounts for the extensive use of green for paints, linens, gowns, and so on in operating rooms. The eyes of the surgeons and nurses when diverted from the redness of the surgical area are more comfortable seeing green on a green background than on a white one.

By a process known as *lateral adaptation,* the apparent color of an object changes when the

ILLUMINATION

background color is changed. Thus, a green object looks somewhat blue–green on a yellow background because the eye is supplying the complementary color to yellow—that is, blue. Similarly, the same green object looks slightly yellow–green when on a blue background, the eye supplying the yellow.

Apparent brightness of a color is a function of its hue, in that light colors appear lighter than dark colors even when measured luminance is the same. Thus, spaces may be defined by color within an area of equal illumination. Also, all colors tend to appear less saturated; that is, they appear "washed out" when illumination is high. Thus, pigments of high saturation (chroma) must be used in well-lit spaces if they are to be effective, although extensive use of saturated colors is generally best avoided.

Other well-known psychological effects of colors are the coolness of blues and greens and the warmth of reds and yellows. Thus, cool colors might well be used in a fur salon and warm colors in a display of summer wear. Red and yellow are "advancing" colors because objects lit with them tend to advance toward the observer, giving the appearance of becoming larger. The opposite effect is noted with blue and green, accounting for their being known as "receding" colors.

A practical, energy-saving application of these color phenomena would be to use warm colors to compensate somewhat for lowered thermostats in the winter and cool colors for the opposite effect in

## Fluorescent Lamps

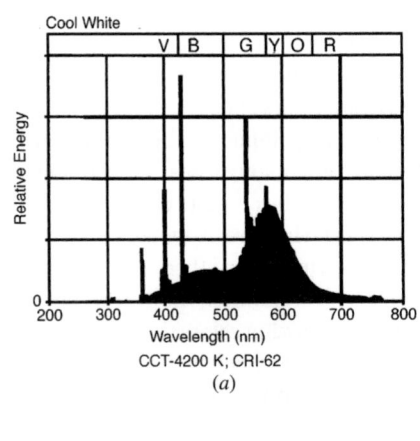

CCT-4200 K; CRI-62

(a)

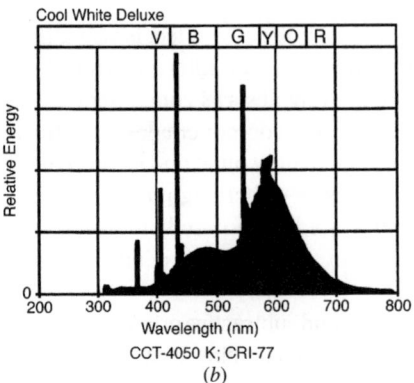

CCT-4050 K; CRI-77

(b)

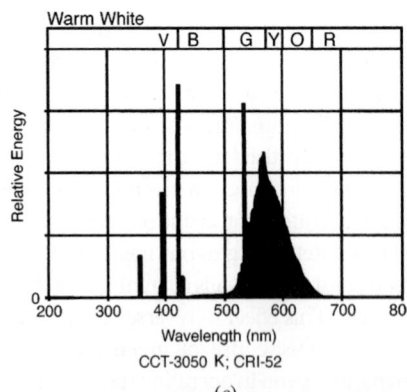

CCT-3050 K; CRI-52

(c)

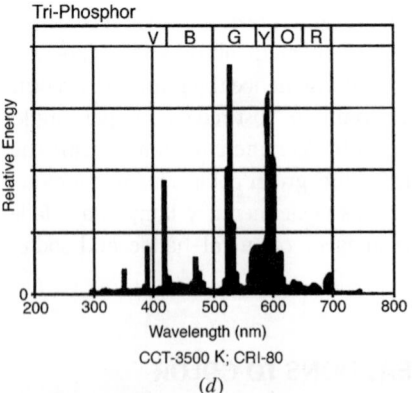

CCT-3500 K; CRI-80

(d)

**Fig. 11.44** *Spectral energy distribution of specific fluorescent lamp types with their correlated color temperatures (CCT) and color rendering indices (CRI). In actual fact, the curves are not continuous but consist of individual color lines. They are shown connected for simplicity. Because only a radiating black body has a true color temperature, a source with mixed color illuminants is assigned a CCT, which is the temperature of a blackbody radiator whose chromaticity most nearly matches that of the light source.*

summer. How to accomplish this without the expense of repainting twice a year is left to the ingenuity of the architect and interior designer. In an atmosphere designed to be calm and restful, greens should generally predominate either in illuminant color, object color, or both, except in eating areas, which should be lighted with reds and yellows because cool colors are generally unappetizing. Yellows and browns emphasize motion sickness, whereas blues and greens tend toward the reverse. Warm and saturated colors produce activity; conversely, cool, unsaturated colors are conducive to meditation. Cool colors also seem to shorten time passage and are well applied in areas of dull, repetitive work.

A further discussion of color control, illuminant colors, color measurement, and color matching is found in the next sections, dealing with spectral energy distribution of sources and the color rendering index.

## 11.37 CHROMATICITY

The CIE color system is the internationally accepted standard for designating illuminant color. In this system, the relative proportions of each of the three primary colors (red, green, and blue) required to produce a given illuminant color are calculated. These values are called the *tristimulus values* for that color and are designated by capital letters: X (red), Y (green), and Z (blue). See Fig. 11.47 for an example of a chromaticity diagram.

## 11.38 SPECTRAL DISTRIBUTION OF LIGHT SOURCES

In addition to providing sufficient light of adequate quality, the lighting designer must be concerned with the spectral content of the selected illuminant,

ILLUMINATION

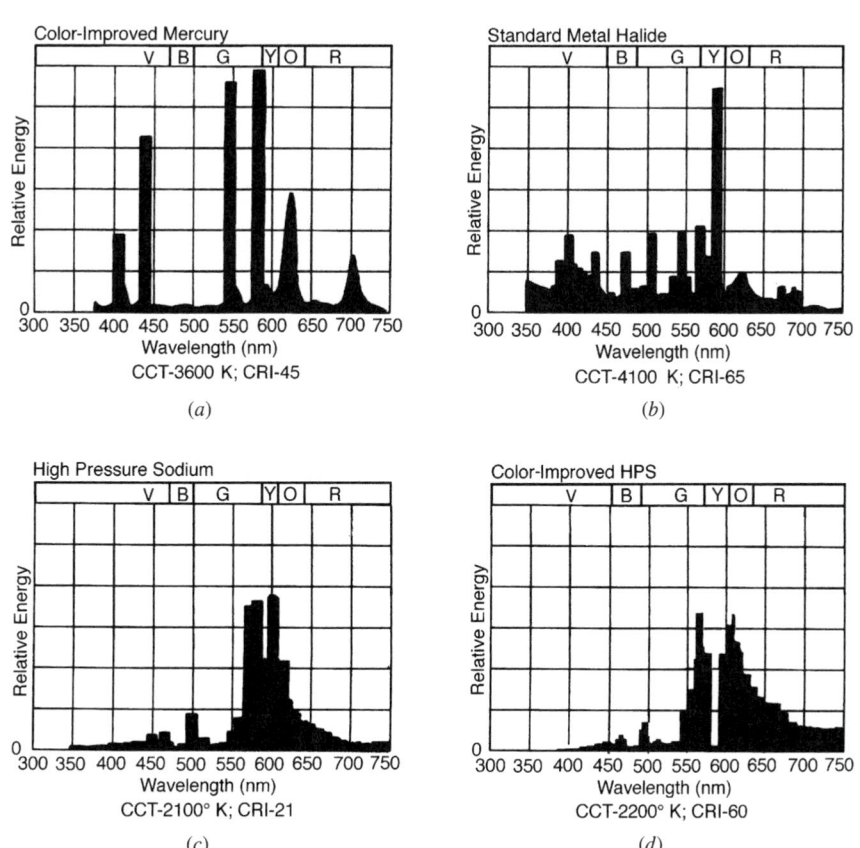

High Intensity Discharge (HID) Lamps

Color-Improved Mercury
CCT-3600 K; CRI-45
(a)

Standard Metal Halide
CCT-4100 K; CRI-65
(b)

High Pressure Sodium
CCT-2100° K; CRI-21
(c)

Color-Improved HPS
CCT-2200° K; CRI-60
(d)

**Fig. 11.45** *Spectral energy distribution of typical high-intensity discharge (HID) lamps with their CCT and CRI.*

because perceived object color depends heavily on the illuminant. As discussed earlier, perceived object color is the result of selective absorption and reflection of components of the illuminating light by the pigments of the object being viewed. It is therefore necessary that the *illuminant* contain the color of the *object* in order for us to see the object's color. It is not so obvious that the relative energy of an illuminant at a particular wavelength determines the saturation and brilliance with which we see a color. To understand this, refer to Fig. 11.44. In graphic form, the relative spectral energy distributions of a few common light sources have been plotted (as a function of wavelength—that is, color). If we compare the graphs in Figs. 11.44*a* and 11.44*c*, which

show the spectral content of two of the most common light sources, cool white and warm white fluorescent lamps, respectively, we note that the principal difference lies in the amount of blue in their spectrum. As a result, a blue object will be bright under cool white light and dull (grayed) under warm white light. The situation is more pronounced with the standard high-pressure sodium (Fig. 11.45*c*) compared to the clear metal halide lamp (Fig. 11.45*b*). A blue object will be gray under the sodium lamp and unrecognizable as blue, whereas under the metal halide lamp its blue color will show clearly, and so on.

This concern for perceived object color, which relates not only to furnishings but also to paints and

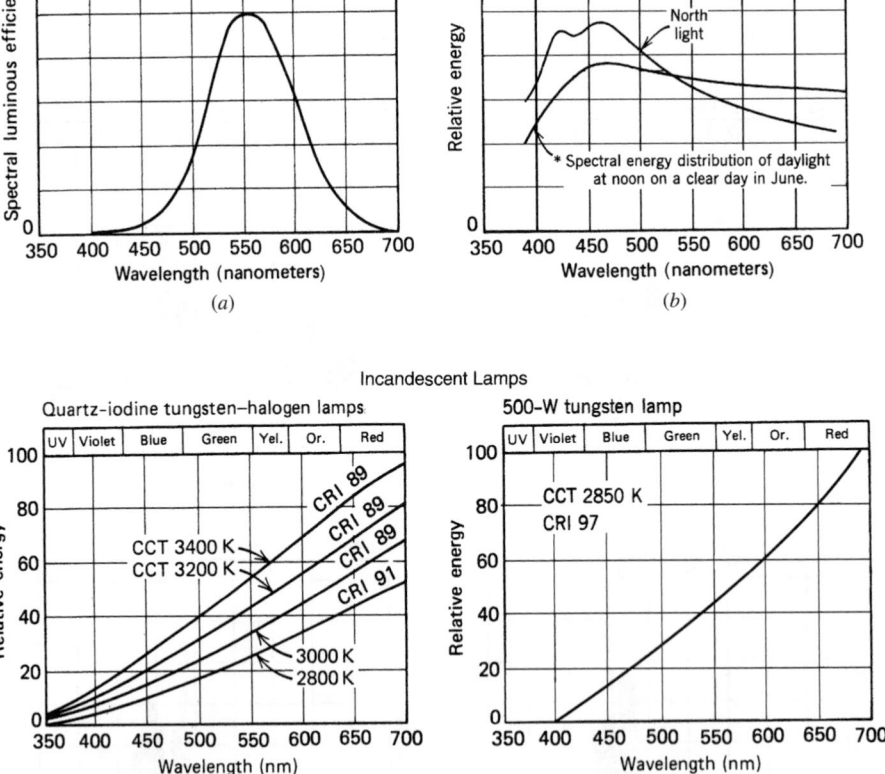

**Fig. 11.46** *(a) Standard photopic (cones) eye sensitivity curve. Note that maximum sensitivity occurs in the daylight range of 500–750 nm. (b) Spectral energy distribution of two specific types of daylight. North light, with a color temperature of 8000–10,000 K, peaks in the blue range, whereas noon daylight contains all spectral colors in roughly equal proportions. (c) Tungsten–halogen lamps are incandescent light sources and therefore contain all spectral colors. As wattages increase, the color changes from orange–red to white and the CRI drops slightly. (d) A simple filament-type incandescent lamp is very close to being a blackbody radiator (i.e., its actual temperature and CCT are almost the same). This is indicated by its high CRI (97).*

prefinished construction materials such as carpets or floor tiles, is quite properly the province of the architect and lighting designer, who in turn must possess the necessary knowledge and information to make the appropriate choices of both illuminant *and* object color. Spectral composition graphs of the types shown in Figs. 11.44 through 11.46 are available from manufacturers for all light sources, and they should be examined when considering the characteristics of a particular light source.

One of the best ways to compare illuminants is first to expose a dull white surface to the illuminants, side by side but separated by an opaque divider, to get an impression of the illuminant color and then expose a series of colored chips—again, side by side—to see which colors are bright-

ened and which are grayed. The intensity of illumination also influences the appearance of colors, and it must be considered in choosing object colors. As intensity is increased, reflection increases, particularly with pale tints (high value) that contain much white pigment and thus tend to wash out color. Therefore, with high-intensity lighting, saturation of colors should be high for true, brilliant color rendition.

Refer now to Fig. 11.46. Note from diagrams *b*, *c*, and *d* that the spectrum of a light source that produces light as a result of heating is continuous. Sunlight is equal in spectrum to a blackbody radiator at 5500 K; north light equal to one at about 8000 to 10,000 K; a 500-W incandescent lamp approximately equal to one at 2850 K; and so on.

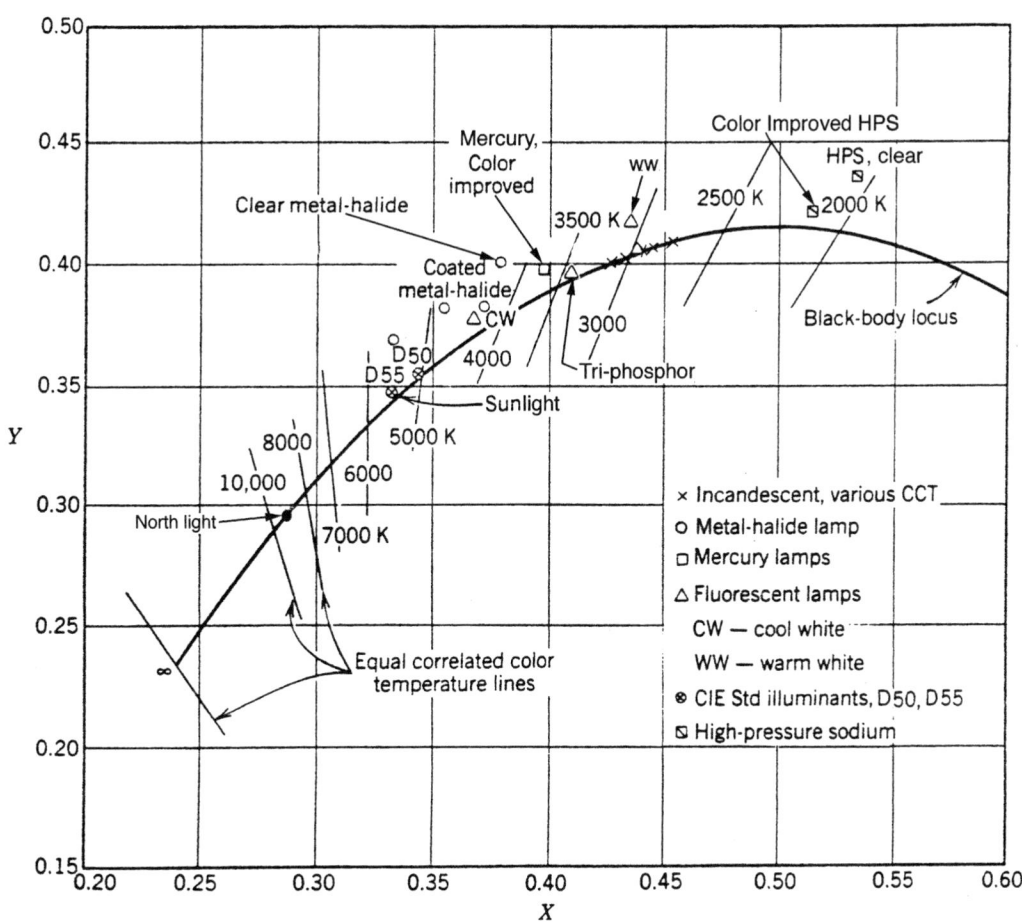

**Fig. 11.47** Portion of a chromaticity diagram showing the relation of common illuminant chromaticities to that of the blackbody locus. Illuminants whose coordinates fall on the same line crossing the blackbody locus have the same CCT but may have entirely different component colors.

If the spectrum of a blackbody radiator is plotted on a chromaticity diagram, its locus is a continuous curved line, as seen in Fig. 11.47. The chromaticity of all true blackbody radiators falls exactly on this line, with the location depending on temperature. Daylight, for most purposes, falls on this locus, although because of selective atmospheric absorption and other phenomena, it is actually slightly off. Incandescent lamp chromaticity is very close to this locus because it is also a heat–light radiator.

A source that produces light by means of individual phosphors can also have chromaticity on this locus if the phosphors are selected to produce a continuous spectrum similar to that of a blackbody radiator. Thus, we see in Fig. 11.47 that triphosphor fluorescent lamps (Fig. 11.44d) and metal halide lamps (Fig. 11.45d) have spectral components over the entire spectrum, yielding chromaticities fairly close to the blackbody locus. For other sources, the CCT is established by their chromaticity locus in relation to the diagonal lines crossing the blackbody locus, as seen in Fig. 11.47. Each of these lines is isothermal—that is, all chromaticities on it have the *same CCT*. Thus, the reader can see that two sources with widely differing spectral content and therefore object color rendering can have the same CCT.

**TABLE 11.12 Effect of Illuminant on Object Colors**

| Lamp | CRI (approximate) | CCT (K) | Whiteness | Colors Enhanced | Colors Grayed | Notes |
|---|---|---|---|---|---|---|
| FLUORESCENT | | | | | | |
| Warm white | 52 | 3050 | Yellowish | Orange, yellow | Red, blue, green | |
| Cool white | 62 | 4200 | White | Yellow Orange Blue | Red | |
| Cool white deluxe | 77 | 4050 | White | Green Orange Yellow | Red | |
| Triphosphor | 75 | 2800 | Yellowish | Red Orange | Deep red, blue | |
| | 80 | 3000 | Pale yellowish | Red Orange Green | Deep red | |
| MERCURY | | | | | | |
| Clear | 20 | 7000 | Blue–Green | Blue Green | Red Orange | Poor overall color rendering |
| Deluxe | 45 | 3700 | Pale purplish | Deep blue, red | Blue–green | Shift over life to greenish |
| METAL–HALIDE | | | | | | |
| Clear | 65 | 4000 | White | Blue Green Yellow | Red | May shift to pinkish over life |
| Phosphor-coated | 80 | 4200 | White | Blue Green Yellow | None | Shifts to pinkish over life |
| HIGH-PRESSURE SODIUM | | | | | | |
| Standard | 21 | 2100 | Yellowish | Yellow Green | Red, blue | |
| Color-corrected | 60 | 2200 | Yellowish-white | Red Green Yellow | Blue | CRI decreases slightly over life |
| Incandescent | 99+ | 2900 | Yellowish | Red, orange, yellow | Blue, green | |

## 11.39 COLOR RENDERING INDEX

*Color rendering* is defined as the degree to which perceived colors of objects illuminated by a test source conform to the colors of the same objects as illuminated by a reference source. The *color rendering index* (CRI) of a source is a two-part concept, comprising a color temperature that establishes the reference standard and a number that indicates how closely the illuminant approaches the standard. *The standard is always daylight at that color temperature.* Therefore, the CRI of a lamp is really a measure of how closely it approximates daylight of the same color temperature. Two sources cannot be compared unless their color temperatures are equal or quite close. A CRI of 100 indicates an illuminant whose spectral content is equal to daylight of that temperature. CRIs for typical common lamps are given in Figs. 11.44 through 11.46.

Table 11.12 lists the color characteristics of a few of the major sources. An illuminant's own color appearance on a neutral surface depends on its own spectral content, but if the observer is placed in a space illuminated with this source, after a short exposure time the eye becomes adapted to the source color and detects only a degree of whiteness rather than an actual tint.

Where it is necessary to detect small color differences between two objects, a light poor in object color or complementary to the object color should be used at a relatively high illumination level, followed by a light high in object color at the same illumination level. If this is not possible, two widely different but broad-spectrum illuminants should be used, preferably at the same illumination level. Another technique is the use of a special, fixed-color source. For a full discussion, see the current *IESNA Lighting Handbook.*

It should be remembered in all considerations of color, comparison, matching, and rendering that object color depends on the spectral energy distribution of the light source (illuminant), and therefore any change in the spectral content changes the object's appearance. Two sources of the same color temperature and, therefore, apparent whiteness can have quite different spectral content and therefore render object colors differently. A case in point would be a 3000 K warm white fluorescent tube and an incandescent lamp (500-W photoflood) of approximately the same color temperature. Color

**Fig. 11.48** *A hand-held lightweight (≈8 oz) chromaticity meter for determining the chromaticity and color temperature of a light source. The unit reads X,Y coordinates, color temperature in degrees Kelvin, and illuminance in lux. (Courtesy of Minolta Corp.)*

temperature is an expression of dominant color, not spectral distribution.

A convenient hand-held chromaticity meter is illustrated in Fig. 11.48. This unit measures the $X, Y$ coordinates of an illuminant, which can then be plotted on a standard CIE color diagram to determine absolute chromaticity. This is very useful in comparing illuminants to predict the color response and avoid color metamerisms. The meter also reads color temperature in kelvin and illuminance in lux.

### References

CIBSE. 1994. *CIBSE Code for Interior Lighting.* Chartered Institution of Building Services Engineers, London.

CIBSE. 1996. *The Visual Environment for Display Screen Use* (CIBSE Lighting Guide LG03). Chartered Institution of Building Services Engineers, London.

CIBSE. 2004. *CIBSE Code for Lighting.* Chartered Institution of Building Services Engineers, London.

Cotton, H. 1960. *Principles of Illumination.* John Wiley & Sons, New York.

ILLUMINATION

IESNA. 1963. "How to Make a Lighting Survey." *Illuminating Engineering* 57(2):87–100.

IESNA. 1993. *IESNA Lighting Handbook,* 8th ed. Illuminating Engineering Society of North America, New York.

IESNA. 2000. *IESNA Lighting Handbook,* 9th ed. Illuminating Engineering Society of North America, New York.

Murdoch, J.B. 1985. *Illumination Engineering.* Macmillan, New York.

Ross and Baruzzini, Inc. 1975. "Energy Conservation Applied to Office Lighting" (Conservation Paper No. 18). Federal Energy Administration, Washington, DC.

Sampson, F.K. 1970. *Contrast Rendition in School Lighting.* Educational Facilities Laboratory, New York.

# CHAPTER 12

# Light Sources

DESIGNING THE LUMINOUS ENVIRONMENT TODAY involves a complex balance of considerations, such as illuminance, luminance, view, visual comfort, shading, and glazing types—and associated concerns such as thermal comfort and energy efficiency. Historically, human activities and tasks were relegated to daylight hours and often within proximity of tall windows. Candles and oil lamps were expensive and a fire hazard, in addition to providing poor illumination for certain tasks. Electrical lighting began around 1870 with the development of commercially usable arc lamps and was given greater impetus nine years later by Edison's first practical incandescent lamp. Development of the fluorescent lamp (and other electric discharge

lamps) has revolutionized the workplace. Buildings such as offices, shopping centers, and factories can operate during evening hours. The many new technological developments in the lighting industry offer the designer a variety of energy-efficient and environmentally responsible sources and controls to fully integrate daylight and electric light into the design process (Fig. 12.1).

## 12.1 BASIC CHARACTERISTICS OF LIGHT SOURCES

Daylight and electric light sources are discussed in this chapter. Daylight sources may be categorized as

(a)

(b)

**Fig. 12.1** (a) A daylit space; sunlight streams through a window at the Santa Anna Monastery in Santa Anna, Italy. (Photo by Alison Kwok, © 2004; all rights reserved.) (b) An electrically lit space—a wall sconce at the Westin Peachtree Plaza hotel in Atlanta, GA. (Photo by Walter Grondzik, © 2004; all rights reserved.)

ILLUMINATION

direct (direct sunlight or diffuse skylight) or indirect (light reflected or modified from its primary source). Electric light sources today fall generally into two generic classifications: incandescent lamps (including tungsten-halogen types); and gaseous discharge lamps (including fluorescent, mercury vapor, metal halide, high-pressure and low-pressure sodium lamps, and the induction lamp).

*Efficacy* is a basic characteristic common to both daylight and electric light sources—measured in lumens per watt (lm/W). Efficacy is the ratio of lumens provided to watts of heat produced by a light source. Table 12.1 lists efficacies of common light sources. Due to its high efficacy, daylight introduces less heat per lumen than electric sources, making use of daylight an attractive strategy for reducing cooling loads in buildings caused by lighting (assuming effective, balanced distribution and utilization of illumination).

The efficiency of a standard incandescent lamp in converting electrical energy to light is approximately 7%; the other 93% is released as heat. Fluorescent lamps are approximately 22% efficient, and although they are a great improvement over incandescents, the generally low efficacy of lighting in buildings accounts for a large proportion of building energy use.

## 12.2 SELECTING AN APPROPRIATE LIGHT SOURCE

Choosing light sources for buildings—whether daylight or electric light (or, more likely, a combination of both)—involves simultaneous lighting and thermal considerations. Because electric lighting in American nonresidential buildings consumes 25% to 60% of the electric energy utilized, any attempt to reduce this quantity must necessarily include integration of the cheapest (insofar as energy is concerned), most abundant, and, in many ways, most desirable form of lighting available—daylight. In selecting appropriate light sources for buildings, understanding the characteristics of the light sources will allow a designer to use them appropriately for energy efficiency and to provide visual and thermal comfort. For resource efficiency, a designer should first optimize daylight sources through building geometries and material finishes, and then design the electric lighting system to supplement and enhance illumination and effect. This chapter describes the characteristics of daylight and electric light sources to assist a designer in understanding the limits and capabilities of each source.

**TABLE 12.1 Efficacy of Various Light Sources**

| Source | Efficacy (lm/W) |
|---|---|
| Candle | 0.1 |
| Oil lamp | 0.3 |
| Original Edison lamp | 1.4 |
| 1910 Edison lamp | 4.5 |
| Incandescent lamp (15–500 W) | 8–22 |
| Tungsten-halogen lamp (50–1500 W) | 18–22 |
| Fluorescent lamp (15–215 W)[a] | 35–80 |
| Compact fluorescent lamp[b] | 55–75 |
| Mercury vapor lamp (40–1000 W)[a] | 32–63 |
| Metal halide lamp (70–1500 W)[a] | 80–125 |
| High-pressure sodium lamp (35–100 W)[a] | 55–115 |
| Induction lamp[c] | 48–70 |
| Sulfur lamp[c] | 90–100 |
| Direct sun (low altitude = 7.5°) | 90 |
| Direct sun (high altitude > 25°) | 117 |
| Direct sun (mean altitude) | 100 |
| Sky (clear) | 150 |
| Sky (average) | 125 |
| Global (average) | 115 |
| Maximum source efficacy predicted by the year 2010 | 150 |
| Maximum theoretical limit of source efficacy | 250 (approximate) |

[a]Includes ballast losses (with electronic ballasts, lumens per watt become much higher). Losses vary between ballasts and manufacturers.

[b]With electronic ballasts.

[c]With a power supply.

# DAYLIGHT SOURCES

## 12.3 CHARACTERISTICS OF DAYLIGHT

The most prominent characteristic of daylight is its variability. The source of all daylight is the sun. Exterior illumination, at a particular place and time, depends upon (1) solar position, which can be determined if the latitude, date, and time of day are given; (2) weather conditions (e.g., cloud cover, smog); and (3) effects of local terrain (natural and built obstructions and reflections). The position of the sun in the sky is expressed in terms of its altitude above the horizon and its azimuth angle. For all latitudes in the Northern Hemisphere, the sun's altitude is highest in summer, lowest in winter, and in between in spring and fall. Azimuth angle is defined as the sun's horizontal position angle, measured from the *south*. Solar position is absolutely predictable for any given time and location.

Cloud cover, unlike solar position, is only statistically predictable, on the basis of extensive U.S. Weather Service observations at numerous weather stations throughout the United States. At locations other than those for which recorded data are available, an educated guess is necessary. Outside the United States, a designer must rely on locally available data, which are often difficult to obtain.

The third factor—local terrain and construction conditions that either reduce illumination by shadowing or increase it by reflection—can be considered only on a case-by-case and site-specific basis.

For manual calculation procedures, it is sufficient to establish four basic sky conditions. These are:

1. Solid overcast sky
2. Clear sky without sun (in the field of view)
3. Clear sky with sun
4. Partly cloudy sky

## 12.4 OVERCAST SKY

This condition, which occurs for much of the year in northerly climates such as England, Scandinavia, and the Pacific Northwest, is called the *CIE sky*, because it was adopted by the Commission Internationale de l'Eclairage (CIE) as the standard design sky for daylighting calculations (CIE, 1970). This sky, as defined by the CIE, has a nonuniform brightness distribution, increasing from horizon to zenith in approximately a 1:3 ratio. Sky luminance at any altitude angle above the horizon is defined in Eq. 12.1 as

$$L_A = L_Z \frac{1 + 2 \sin A}{3} \qquad (12.1)$$

where

$L_A$ = luminance at $A°$ above the horizon (in any direction)

$L_Z$ = luminance at the zenith

Thus at the horizon, where $A = 0°$,

$$L_A = \frac{L_Z}{3}$$

The illuminance (density of light in lux) on unobstructed exterior horizontal and vertical surfaces produced by this luminance distribution has an approximate ratio of 2.5:1 (Fig. 12.2).

There is agreement among all sources that with an overcast sky, exterior horizontal illuminance varies directly with the sun's altitude, *irrespective of*

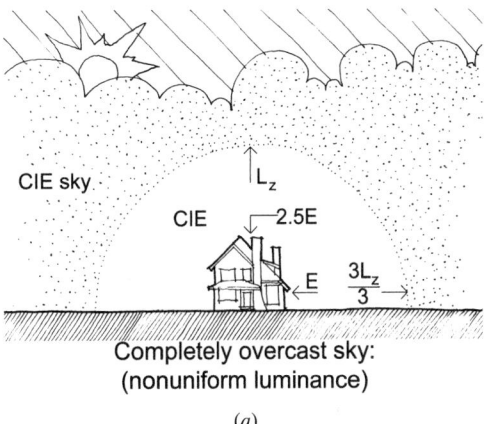

Completely overcast sky:
(nonuniform luminance)

*(a)*

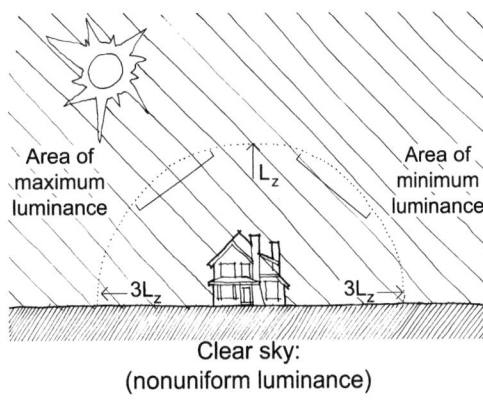

Clear sky:
(nonuniform luminance)

*(b)*

**Fig. 12.2** *(a) The completely overcast sky has a zenith luminance $L_Z$, which is three times the horizon luminance. With such a sky, illuminance on unobstructed exterior horizontal surfaces ($E_H$) is about 2½ times that on similar vertical surfaces ($E_V$). (b) The clear sky has the area of brightest luminance around the sun. The area opposite the sun is darkest and can be considered as essentially uniform at approximately 3500 cd/m² (1000 fL). (Redrawn by Erik Winter.)*

*azimuth.* Various formulations for this relationship have been put forward. One formulation that gives good agreement with observations is by Krochman (1963), shown in Eq. 12.2:

$$E_H = 300 + 21,000 \sin A \qquad (12.2)$$

where $E_H$ is exterior horizontal illuminance (lux) and $A$ is the solar altitude, in degrees. Solar altitude (and azimuth) for various times of day can be obtained from Table D.1. Figure 12.3 is a plot of year-round averages for both vertical and horizontal illuminance from an overcast sky as a function of solar altitude, based on U.S. Weather Service observations.

It is interesting to compare the exterior horizontal illuminance obtained from the two sources given: Krochman's formula (Eq. 12.2) and the observation-based data of Fig. 12.3 for a few typical conditions. Solar altitude is obtained from Table D.1.

Latitude:      38°
Solar Time:   10 A.M.
Dates:        Dec. 21, March/Sept. 21, June 21

|            | Eq. 12.2            | Fig. 12.3            |
|------------|---------------------|----------------------|
| Dec 21     | 8,500 lux (790 fc)  | 8,608 lux (800 fc)   |
| Mar/Sept 21| 14,623 lux (1,359 fc)| 15,923 lux (1,480 fc)|
| June 21    | 18,669 lux (1,735 fc)| 23,134 lux (2,150 fc)|

The degree of agreement is generally satisfactory, and either source will yield suitable results.

One of the most convenient ways of expressing the quantity of daylight illuminance during the schematic design of buildings is the concept of *daylight factor* (primarily intended for overcast skies). Daylight factor is the ratio of indoor illuminance to available outdoor illuminance. Daylight factor is discussed in Chapter 14 as a means of setting criteria for, and determining the effectiveness of, a daylighting design.

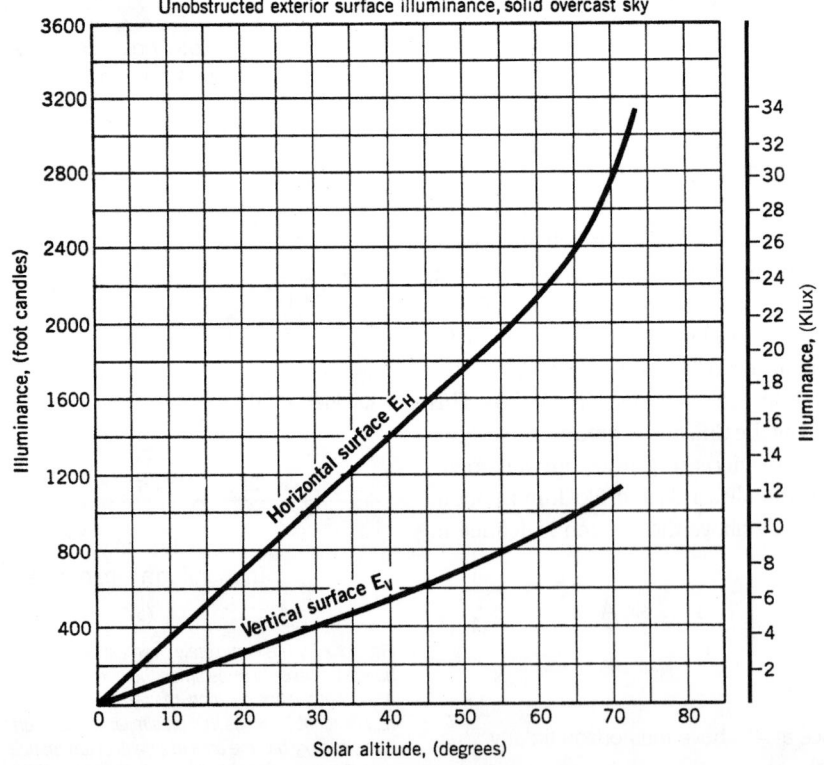

**Fig. 12.3** *Curves giving unobstructed exterior surface illuminance directly from an overcast sky. (To obtain illuminance in lux, multiply footcandles by 10.76 [or 10 as an approximation].) (Data based on U.S. Weather Service observations; courtesy of Libbey-Owens-Ford.)*

## 12.5  CLEAR SKY

### (a) Horizontal Illuminance

Exterior horizontal illuminance on a cloudless day consists of two source components: diffuse illumination from the entire sky plus the much larger component of direct sunlight. As with overcast sky, various empirical formulas for both components have been proposed, and here too, all sources agree that the total illumination, diffuse plus direct, varies directly with solar altitude.

Figure 12.4 gives values for both components of exterior horizontal illuminance based upon observations. The *sky only* values are used to determine shaded skylight illuminance or daylong ground illuminance outside a shaded window—that is, a north-facing window, or an east/west window when the sun is on the opposite side of the building. In determining ground illuminance, the values given in Fig. 12.4 must be reduced somewhat, as they represent unobstructed horizontal illuminance, whereas the area outside a building window is partially obstructed from sky light by the building itself. If a building is so large that the ground outside the shaded window effectively receives diffuse radiation only from the half of the sky away from the sun, an average figure for $E_H$ of 1000 fc (~10,000 lux) can be used. This is because the luminance of the half of the sky away from the sun varies from a minimum of approximately 300 fL (1031 cd/m$^2$) for the deep-blue patch directly opposite the sun to about 2000 fL (6874 cd/m$^2$) at the sides, giving an average half-sky luminance of about 1000 fL (~3400 cd/m$^2$). This, in turn, gives a horizontal illuminance $E_H$, diffuse, of about 1000 fc (~10,000 lux) (see Fig. 12.2).

Figure 12.4 also gives horizontal illuminance from the sun only, as a function of solar altitude. This value, when combined with the proper portion of diffuse illuminance, as discussed previously, is useful in determining ground illuminance outside a sunny building exposure or illuminance on an unshaded skylight. The light incident on an

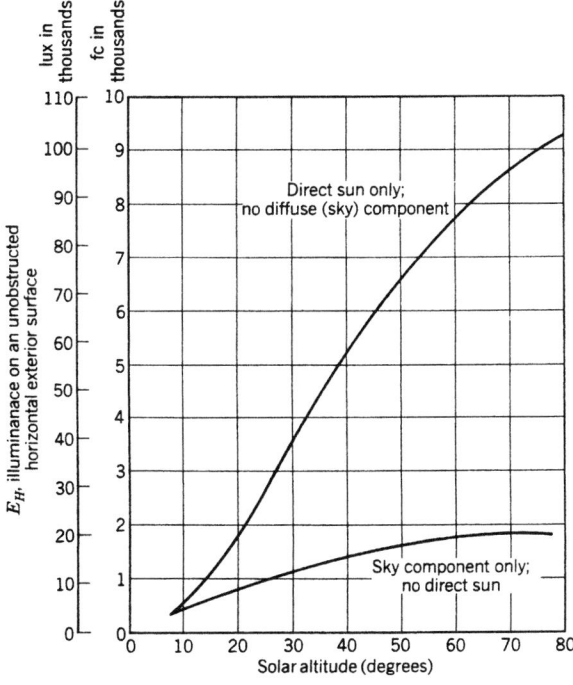

**Fig. 12.4** *Components of the exterior horizontal illuminance on an unobstructed surface, from a clear sky, as a function of solar altitude. Total illuminance $E_H$ is the sum of the two components. (From data in Rennhackkamp, 1967.)*

external reflector or light shelf at a window can also be determined from these figures.

### (b) Vertical Surface Illuminance

Inasmuch as most daylighting is accomplished via vertical fenestration, vertical surface illumination is the major component of interior daylight. It is also important for determining the daylight contribution of vertical elements in skylights. There is no simple relationship between horizontal and vertical illuminance from a clear sky, as there is for an overcast sky, because the illumination on a vertical surface depends upon solar azimuth as well as altitude. More specifically, it depends upon the *bearing angle* Fig. 12.5, which is defined as the horizontal angle between a vertical plane containing the sun and a plane perpendicular to the vertical surface in question. A bearing angle of 0° indicates that the sun plane is perpendicular to the vertical surface. Like $E_H$, $E_V$ (vertical illuminance) is divided into two components: sky only and direct sun only, which are plotted in Fig. 12.6 as a function of solar altitude and bearing angle. The *sky only* component is effectively for the half-sky

because a vertical surface can be exposed to a maximum of only half of the full sky. Solar radiation data may be translated into illuminance by using average "efficiency" figures for solar energy in units of lumens per watt of received radiation (Fig. 12.6).

## 12.6 PARTLY CLOUDY SKY

The luminance of a partly cloudy sky is difficult to express mathematically because of its infinite variability of conditions. However, statistical data on cloud cover are available from observations at many weather stations, and these data should be used in computer-calculated, hour-by-hour energy analysis programs. For the purpose of lighting design, it is important to note that the illumination from a partly cloudy sky is *higher* than that from a clear sky by 10% to 15% because of additional reflected sunlight from cloud edges. Several attempts have been made to account for this type of sky in terms of the effect on the daylight factor within a room, but none have received general acceptance.

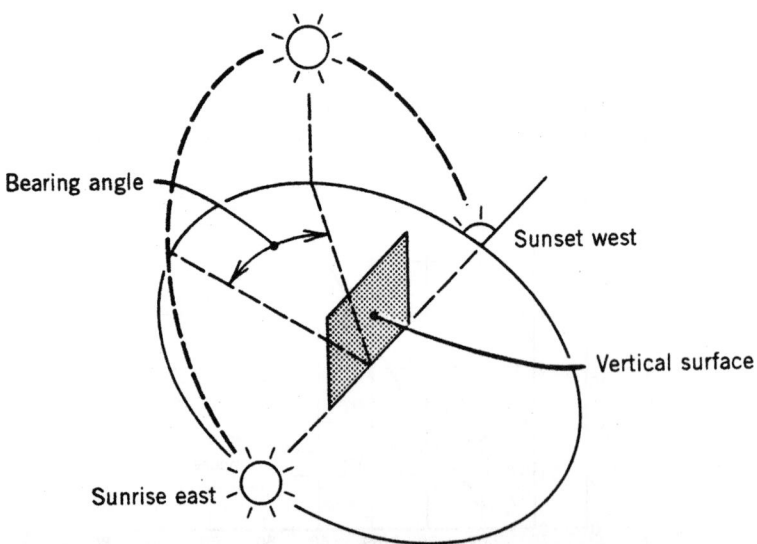

**Fig. 12.5** *The bearing angle of a vertical surface—the angle between a hypothetical vertical plane perpendicular to the surface (say, a window) and a hypothetical vertical plane containing the sun. (Other sources refer to this angle as the* window-to-sun azimuth angle *or* surface azimuth.*)*

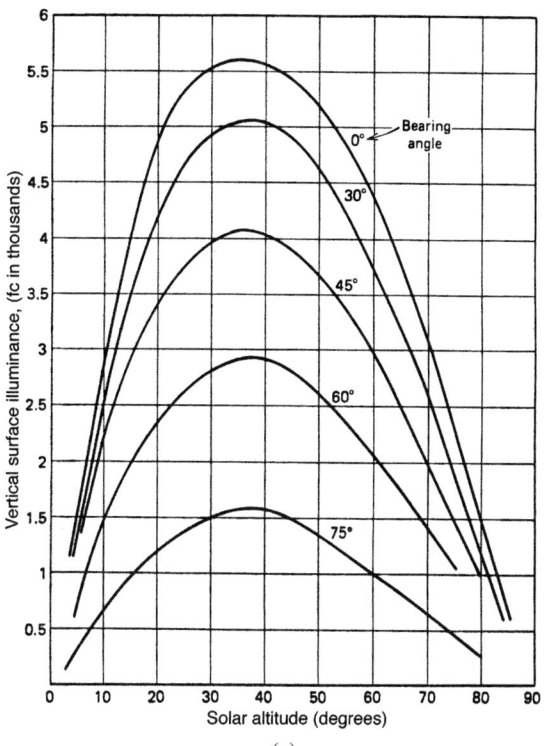

*(a)*

## ELECTRIC LIGHT SOURCES

## Incandescent Lamps

### 12.7 THE INCANDESCENT FILAMENT LAMP

### (a) Construction

The standard incandescent lamp consists simply of a tungsten filament inside a gas-filled, sealed glass envelope (Fig. 12.7). Current passing through the high-resistance filament heats it to incandescence, producing light. Gradual evaporation of the filament causes the familiar blackening of

**Fig. 12.6** *(a) Vertical surface illuminance, year-long average, sun only, no sky contribution. (b) Vertical surface illuminance, clear summer sky, no sky contribution. (c) Vertical surface illuminance, clear sky during various seasons, no sky contribution. (Courtesy of Libby-Owens-Ford.)*

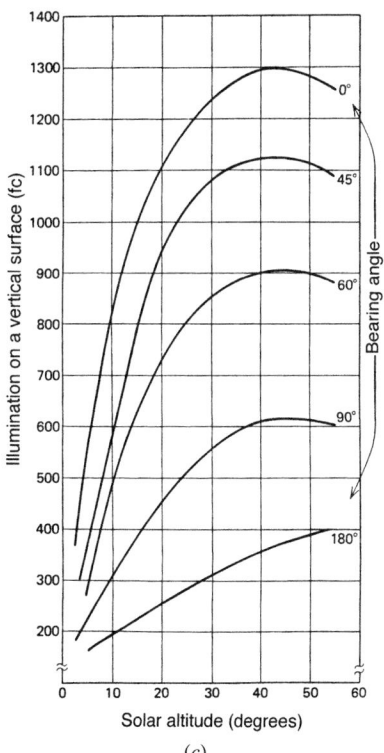

*(b)*

*(c)*

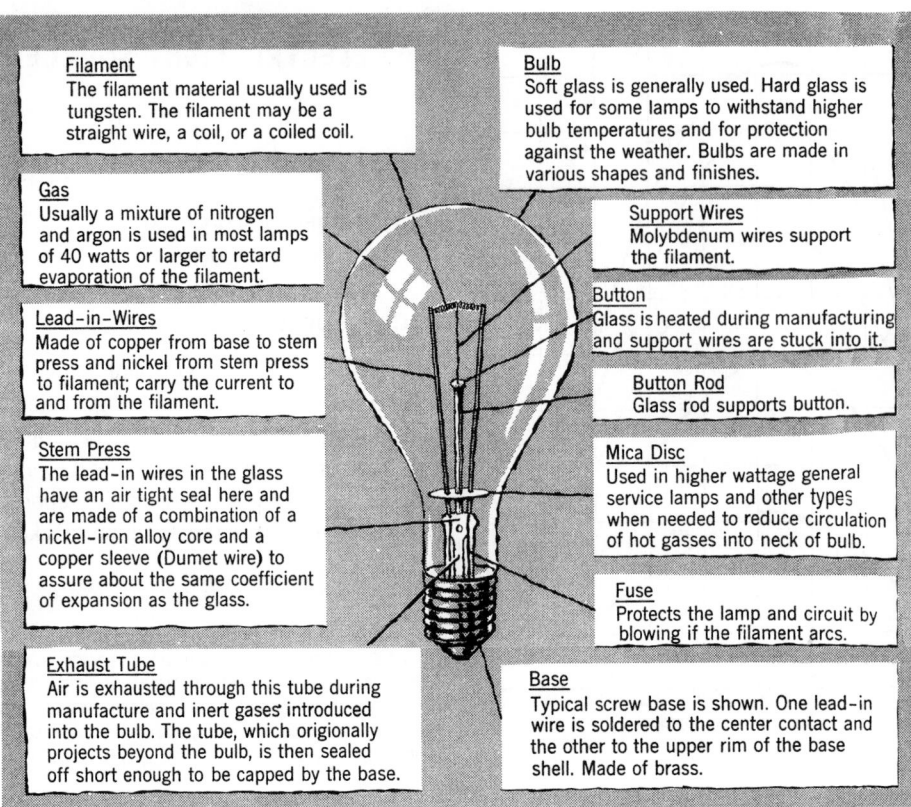

**Filament**
The filament material usually used is tungsten. The filament may be a straight wire, a coil, or a coiled coil.

**Gas**
Usually a mixture of nitrogen and argon is used in most lamps of 40 watts or larger to retard evaporation of the filament.

**Lead-in-Wires**
Made of copper from base to stem press and nickel from stem press to filament; carry the current to and from the filament.

**Stem Press**
The lead-in wires in the glass have an air tight seal here and are made of a combination of a nickel-iron alloy core and a copper sleeve (Dumet wire) to assure about the same coefficient of expansion as the glass.

**Exhaust Tube**
Air is exhausted through this tube during manufacture and inert gases introduced into the bulb. The tube, which originally projects beyond the bulb, is then sealed off short enough to be capped by the base.

**Bulb**
Soft glass is generally used. Hard glass is used for some lamps to withstand higher bulb temperatures and for protection against the weather. Bulbs are made in various shapes and finishes.

**Support Wires**
Molybdenum wires support the filament.

**Button**
Glass is heated during manufacturing and support wires are stuck into it.

**Button Rod**
Glass rod supports button.

**Mica Disc**
Used in higher wattage general service lamps and other types when needed to reduce circulation of hot gasses into neck of bulb.

**Fuse**
Protects the lamp and circuit by blowing if the filament arcs.

**Base**
Typical screw base is shown. One lead-in wire is soldered to the center contact and the other to the upper rim of the base shell. Made of brass.

*Fig. 12.7* Construction of a standard incandescent lamp.

the bulb and eventual filament rupture and lamp failure. Incandescent lamps are available in many bulb and base types, with special designs for particular applications (Fig. 12.8 and Fig. J.1). In order to diffuse the light output, most bulbs are coated inside with white silica providing almost complete light diffusion at a cost of approximately 2% to 3% of the light output. Colored light is also available from either coated bulbs or bulbs of colored glass.

The incandescent lamp base is the means by which a connection is made to the socket and thereby to the source of electric current. Most lamps are made with screw bases of various sizes, the most common being the medium screw base. General service lamps of 300 W and larger use the mogul screw base. When lamps are placed in precise reflectors or in lens systems where exact positioning of the filament is important, one of the special bases illustrated in Fig. 12.8 is used.

**(b) Operating Characteristics**

Critically dependent upon the supplied voltage, the life, output, and efficiency of a lamp can be markedly altered by even a small change in operating voltage, as illustrated Fig. J.2. For example, operating a 120 V lamp at 125 V or 115 V (Table 12.2) affects lumen output and, in particular, lamp life. In installations where lamp replacement is difficult and/or expensive, and use of an incandescent lamp is indicated, lamps may be operated slightly undervoltage to prolong life, thereby decreasing the frequency of replacement. Because luminous efficacy (the number of lumens emitted for each watt of electricity used) is decreased by this procedure, and because recognizing that energy cost is normally a major consideration over the life of any lighting installation, a detailed life-cycle cost analysis (Appendix I) should be made by the design professional involved. Conversely, where lamps are replaced

Bulb shapes

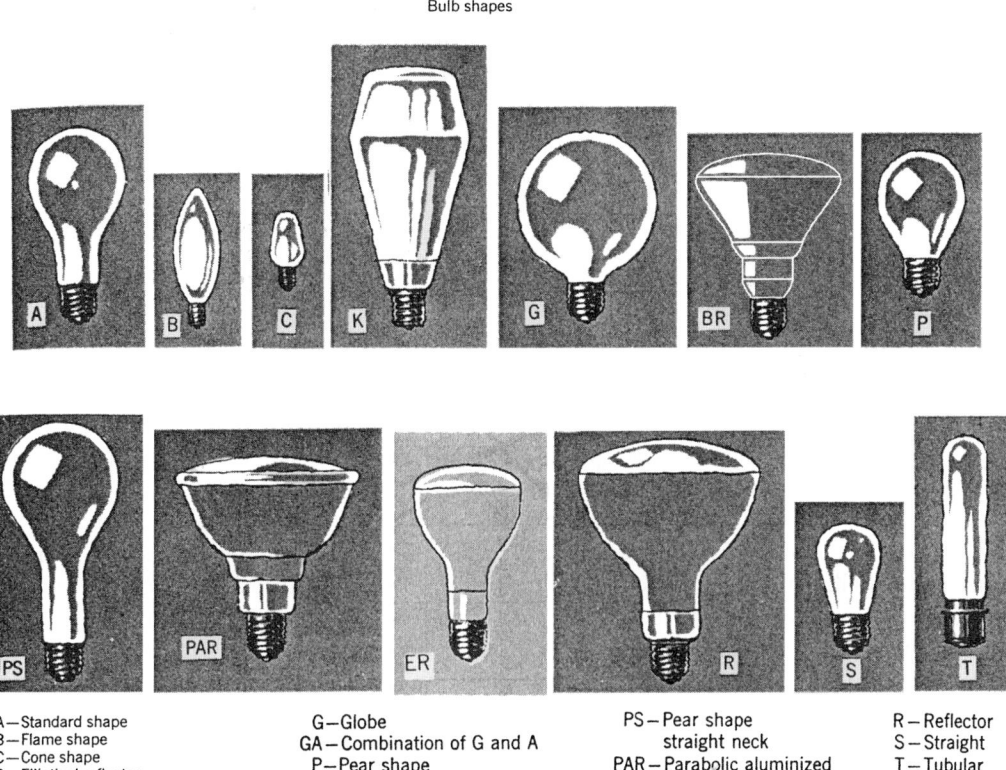

A—Standard shape
B—Flame shape
C—Cone shape
ER—Elliptical reflector

G—Globe
GA—Combination of G and A
P—Pear shape
K—Arbitrary designation

PS—Pear shape
  straight neck
PAR—Parabolic aluminized
  reflector

R—Reflector
S—Straight
T—Tubular

Base types

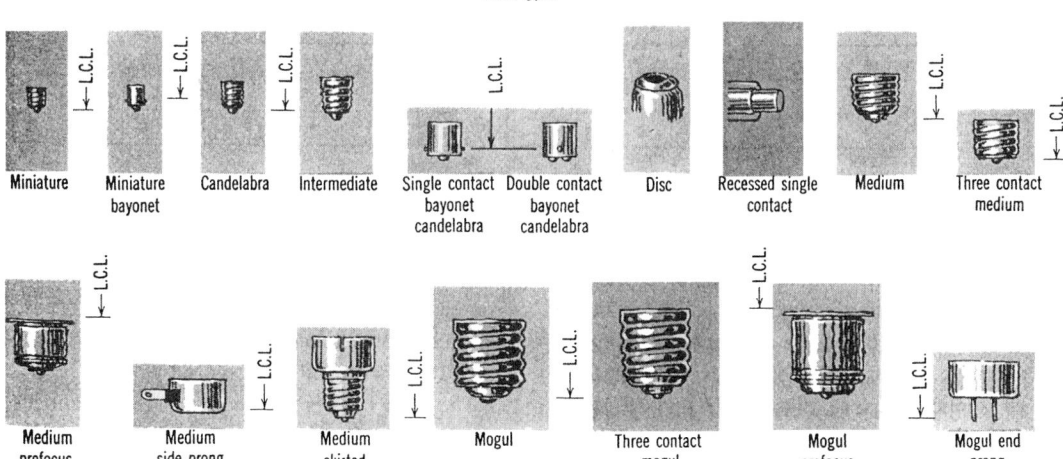

Fig. 12.8 Common incandescent lamp bulb and base types with nomenclature. The complete bulb nomenclature indicates type and size. The letter is an abbreviation of the shape, and the number is the diameter in eighths of an inch. A PS-52 is a pear-shaped bulb 52/8 (6½) in. in diameter. A PAR-38 is a parabolic reflector lamp 38/8 (4¾ in.) in diameter.

ILLUMINATION

**TABLE 12.2 Comparison of Operating Characteristics**

| Operation of Lamps | 120-V lamp at 125 V (104.2%) | 120-V lamp at 115 V (95.8%) |
|---|---|---|
| Amount of light (lumens) | 16% more | 15% less |
| Power consumption (watts) | 7% more | 7% less |
| Efficacy (lumens per watt) | 8% higher | 8% lower |
| Life (hours) | 42% less | 72% more |

before burnout using a group replacement system and initial installation cost per lux and/or energy costs are high, lamps may be operated overvoltage, thereby increasing their output and efficacy but shortening their life. This procedure is normal in sports-lighting installations because of the high installation cost of tower-mounted floodlights, making it mandatory to extract the maximum light from each unit. In stadium installations with yearly lamp operation schedules averaging less than 200 hours, 10% overvoltage operation doubles the light output but still allows a once-a-year, off-season relamping and is therefore a highly economical procedure. Generally, however, it is advisable to operate incandescent lamps at the rated voltage, accepting balanced efficacy, output, and life.

## (c) Other Characteristics

1. *Lumen maintenance.* Light output decreases slowly with lamp life as an incandescent bulb blackens. Lamp position (vertical or horizontal) during operation and the resulting bulb temperatures affect this characteristic.
2. *Color.* Incandescent light has a large yellow–red component and is therefore highly flattering to the skin. The spectral content of the light produced by a heated source depends upon its temperature: high-wattage lamps are bluer, low-wattage lamps are yellower. Dimmed lamps give yellow–red light.
3. *Surroundings.* Generally, incandescent lamps are impervious to surrounding heat, cold, or humidity. Starting is completely unaffected by ambient conditions. Bulbs, however, must be appropriately selected if exposure to water is expected.

4. *Luminous efficacy.* Incandescent lamps produce light as a by-product of heat; as a result, they are inherently inefficient. Luminous efficacy increases with wattage. Thus, a 60-W general-service lamp produces 890 initial lumens, or 14.8 lm/W, whereas an A-21 100-W lamp produces only slightly less output than two 60-W lamps, but the higher wattage results in an 18% energy savings.

## (d) Summary

The principal advantages of incandescent lamps are low cost; instant start and restart; simple, inexpensive dimming; simple, compact installation requiring no accessories; cheap fixtures; focusability as a point source; high power factor; lamp life independent of the number of starts; and skin-flattering, full-spectrum color. From a human factor perspective, the full-spectrum quality of light, with higher amounts of light in the red wavelengths, is best for rendering skin tones and facial expressions.

The principal disadvantages are low efficacy (see Table 12.1), short lamp life, and critical voltage sensitivity. Low efficacy means more fixtures and larger heat gain than more efficient alternatives. Short lamp life results in high lamp replacement labor costs. Voltage sensitivity may require careful and expensive circuit design. Also, light concentration at the filament (point source) requires careful fixture design in order to avoid glare and, if undesirable, sharp shadows. Because of its poor energy characteristics, incandescent lamp use should be limited to the following applications:

- Where use is infrequent
- Where there is frequent short-duration use
- Where low-cost dimming is required
- Where the point source characteristic of the lamp is important, as in focusing fixtures
- Where minimum initial cost is essential
- Where its characteristically good color rendering is desired

A brief list of conventional incandescent lamps and their general physical and operating characteristics is provided in Table J.1. Specific lamp data for use during design development should be taken from current manufacturers' literature.

## 12.8 SPECIAL INCANDESCENT LAMPS

Beyond the tungsten-halogen lamp, which is discussed separately, numerous special types of incandescent lamps are available. Some of the more important types are covered briefly in the following pages.

*Rough service* and *vibration* lamps are built to withstand rough handling and continuous vibration, respectively. Both conditions are extremely hard on general-service lamp filaments. Neither of these types is intended for general use, and both have lower luminous efficacy than a general-service lamp (see Table J.1).

*Extended-service* lamps are designed for 2500-hour life. They are useful in locations where maintenance is irregular and/or relamping is difficult. Such lamps are really designed for slightly higher voltage than that which is applied, and therefore efficacy is reduced (see Table J.1 and Fig. J.2). So-called long-life lamps, which are guaranteed to burn for 2, 3, or 5 years, are actually just lamps designed for higher voltages than those listed. As they normally sell at a high cost and are very inefficient, their use is seldom advisable. Before using such lamps, a life-cycle cost comparison, including the cost of lamps, energy, and relamping, should be made (see Appendix I).

### (a) Reflector Lamps

These are made in "R," "BR," "ER," and "PAR" shapes (see Fig. 12.8). They contain a reflective coating on the inside of the glass envelope that gives the entire lamp accurate light-beam control. Many reflector lamp types are available in narrow or wide beam design, commonly called *spot* and *flood*, respectively. R lamps are generally made in soft glass envelopes for indoor use, whereas PAR lamps are hard glass, suitable for exterior application. When using R and PAR lamps, the fixture acts principally as a lamp holder since beam control is built into the lamp. These lamps have an improved reflector design that increases their efficiency.

The Energy Policy Act (EPACT) of 1992 made a number of incandescent reflector lamps obsolete because of the act's minimum efficacy requirements. These requirements state that R and PAR lamps rated 115 V to 130 V with a medium screw base, a bulb diameter greater than 2.75 in., and nominal wattages between 40 and 205 shall have minimum efficacies, as shown in Table 12.3. All major manufacturers produce incandescent reflector lamps that meet EPACT requirements. Among these are elliptical reflector (ER) and bulge reflector (BR) lamps that use a more efficient reflector design. These lamps are normally catalog listed as "energy-saving" lamps.

### (b) Energy-Saving Lamps

Major national energy legislation including the National Appliance Energy Conservation Act of 1987, the Energy Policy Act of 1992, and the comprehensive energy legislation passed by the U.S. House of Representatives and U.S. Senate in 2003 have created requirements to conserve lighting energy. The American Council for an Energy-Efficient Economy (ACEEE), the U.S. Department of Energy (DOE), and the Environmental Protection Agency (EPA) have developed and support lighting energy efficiency programs, new research, and initiatives. For example, the Energy Star® program (http://www.energystar.gov/), established by the EPA in 1992 to improve and provide energy-efficient products (appliances, lighting, and heating and cooling equipment) and practices, is aimed at reducing emissions from power plants, avoiding the need for new power plants, and reducing energy bills. Energy Star encourages every U.S. household to change the five fixtures used most often at home (or the lamps in them) to Energy Star–qualified lighting to save more than $60 every year in energy costs. If every household did this, it would keep more than 1 trillion pounds of greenhouse gases out of our air—a $6 billion energy savings equivalent to the annual output of more than 21 power plants.

Every major manufacturer is producing a line of energy-saving lamps. Energy-efficient lamps are frequently known by trademarked names. They are rated at a wattage lower than that of the standard

**TABLE 12.3 Minimum Required Efficacy of R and PAR Lamps**

| Nominal Lamp Wattage (W) | Minimum Efficacy (lm/W) |
|---|---|
| 40–50 | 10.5 |
| 51–66 | 11.0 |
| 67–85 | 12.5 |
| 86–115 | 14.0 |
| 116–155 | 14.5 |
| 156–205 | 15.0 |

lamps they are intended to replace and are generally more efficient. Any additional first cost should be analyzed by a life-cycle cost analysis and the payback period calculated. The designer will find that a proper control system is often more economically attractive than low-wattage lamps and that energy-saving incandescent lamps are primarily useful in retrofit work.

## 12.9 TUNGSTEN–HALOGEN (QUARTZ–IODINE) LAMPS

This lamp type, illustrated in Fig. 12.9, is similar to the standard incandescent lamp in that it produces light by heating a filament. It differs in that a small amount of halogen gas (iodine or bromine) is added to the inert gas mixture that fills a small capsule constructed of quartz glass that surrounds the filament within the bulb of the lamp. This addition results in retardation of filament evaporation, which is the usual cause of incandescent lamp failure, and thereby extends lamp life (Fig. 12.10).

Although the tungsten–halogen lamp has only slightly higher efficacy than an equivalent standard incandescent lamp, it has the advantages of longer life, lower lumen depreciation (98% output at 90% life), and a smaller envelope for a given wattage (see Fig. 12.9). The last characteristic is due to the high temperature required by the halogen cycle, which in turn requires a compact, high-temperature filament. As a result, the lamp is effectively a point source, making it ideal for use in precision reflectors. Indeed, it is in this area, as discussed later, that most of the recent developments in tungsten–halogen lamp technology have occurred.

Due to the lamp's high filament temperature, the bulb envelope is generally made of quartz or a special high-temperature glass, which can withstand high temperatures better than glass; this, in turn, gave rise to the alternative name—*quartz–iodine*—that is sometimes applied to this lamp type. Another result of high filament temperature is that the gas pressure inside the quartz envelope is elevated and the lamps have been known to rupture violently, spraying hot quartz fragments over a wide area. As a consequence, all manufacturers now provide a cautionary notice with their lamps. The wording varies but essentially states that due to the possibility of rupture, lamps should be handled carefully, guarded

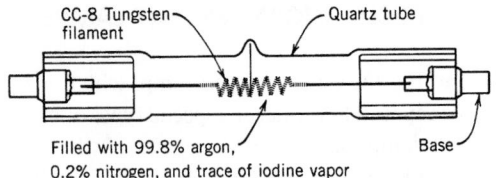

Double-ended tungsten halogen lamp 120 V, 50-1500 W

(*a*)

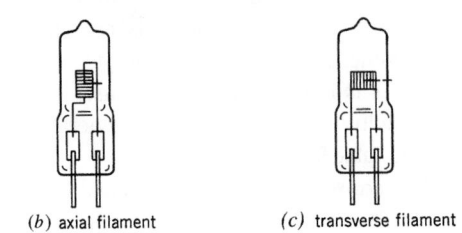

(*b*) axial filament  (*c*) transverse filament

Single ended pin terminal lamps; 12 V, 120 V  20-65 W

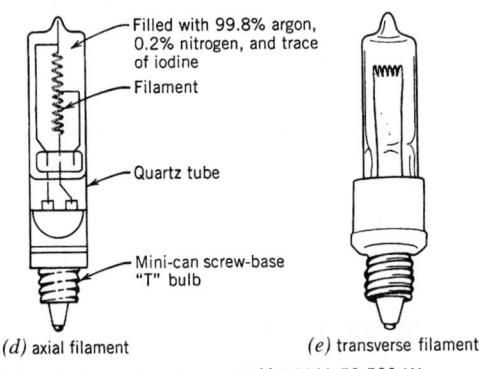

(*d*) axial filament  (*e*) transverse filament

Single ended screw base lamps; 12 V, 120 V  50-500 W

**Fig. 12.9** *Common tungsten–halogen lamps are available in a variety of designs.*

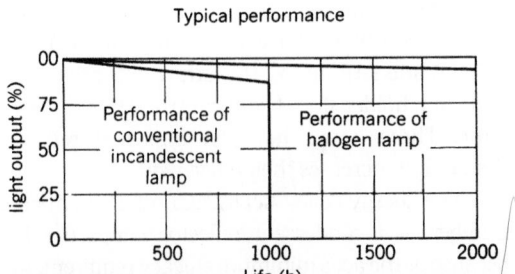

**Fig. 12.10** *The self-regenerative halogen cycle slows the evaporation of the tungsten filament, and consequently lowers light depreciation and lengthens the lamp life compared to a standard incandescent lamp.*

against abrasion and overvoltage operation, and, most importantly, adequately shielded or screened. The shield can be a reflector cover, a fixture lens, a screen, or other devices that will contain hot flying fragments in the event that a lamp shatters or high temperatures cause a fire hazard.

A number of standards organizations, and professional societies have adopted and published cautionary notices, including the *IESNA Lighting Handbook*, ANSI Standard C78.1451-2002, the Canadian Standards Association, and the International Electrotechnical Commission (IEC). Exceptions to these precautions exist when a lamp is protected by encapsulation inside a sealed envelope. Such encapsulated construction is now common in R, PAR, MR, and modified A-lamp shapes. These encapsulated lamps are intended for direct replacement of standard incandescent lamps of the same bulb shape.

Other halogen lamp characteristics are similar in all respects to those of the standard incandescent lamp. Color temperature ranges between 2000 K and 3400 K; spectral energy distribution is typical of blackbody radiation, and dimming characteristics are similar to those of standard incandescent lamps.

## 12.10 TUNGSTEN–HALOGEN LAMP TYPES

The basic lamp is a small gas-filled quartz tube, as shown in Fig. 12.9. Because the lamp must be used with some sort of reflector, it is manufactured with different terminations to suit the fixture reflector or secondary lamp envelope in which it is placed. All the lamp types shown in Fig. 12.9 can be used where lamp-only replacement is intended, such as in floodlights or reflector fixtures, by using appropriate

ILLUMINATION

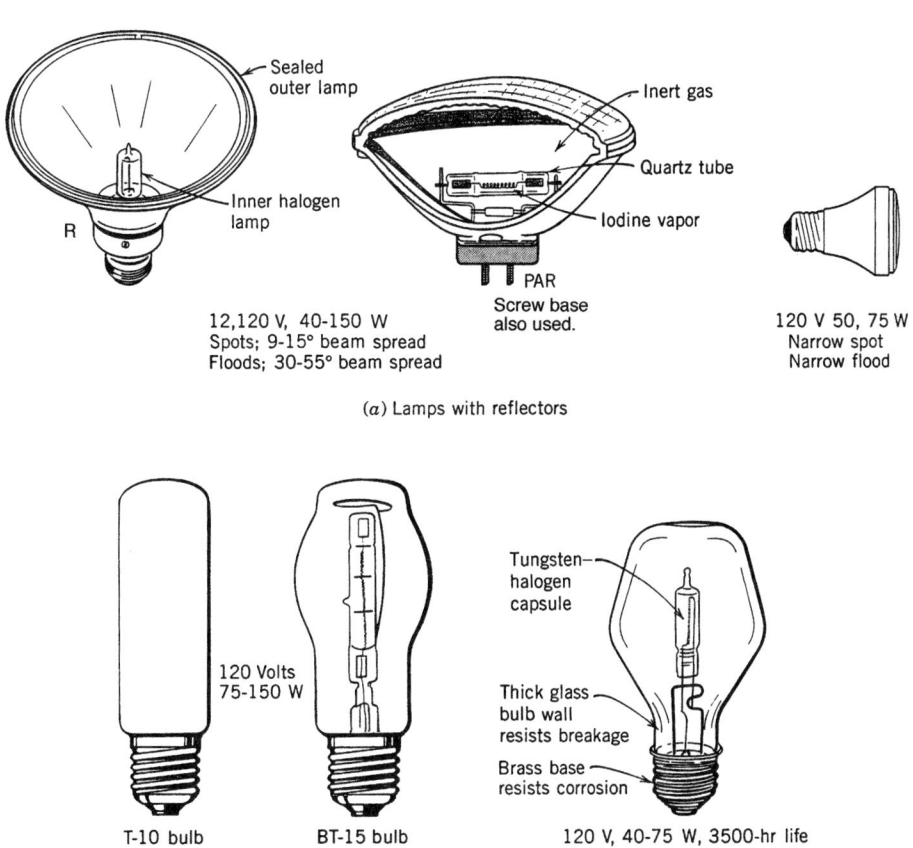

(a) Lamps with reflectors

(b) General service encapsulated lamps

**Fig. 12.11** *Tungsten–halogen lamps can be mounted in a variety of ways inside an enclosing glass envelope. (a) Lamp is used either horizontally or vertically to the reflector or (b) used without a reflector in a protective glass envelope and an Edison medium screw base.*

bases—slide contacts for double-ended lamps, screw bases for screw-base lamps, and special ceramic pin-hole bases for pin-type lamps.

## (a) Encapsulated Lamps

These lamps (Figs. 12.11 and 12.12) are sealed units intended for direct replacement of either a corresponding distribution-type incandescent lamp in the case of reflector units or general service incandescent lamps for reflectorless units. As a sealed unit, the halogen lamp is not replaceable, and the entire unit is discarded on burnout. Reflector units are available in a wide variety of beam patterns detailed in manufacturers' catalogs. Typical data are given in Fig. 12.13.

Reflector lamps are also available with a variety of filters: so-called cool lamps that direct much of the heat out through the back of the lamp; high-efficiency units that reflect and concentrate the heat back on the lamp filament; lamps with ultraviolet (UV) filters for use in displays of UV-sensitive objects; and others. For complete design information, access to current manufacturers' catalogs is a necessity.

**Fig. 12.12** PAR halogen lamp with standard medium screw bases in sizes PAR 16, 20, 30, and 38 with a short or elongated lamp neck. Lamp wattages range from 45 to 90 W. (Courtesy of Philips Lighting Co.)

**Fig. 12.13** Typical PAR halogen lamp data. The beam of the PAR lamp is conical in shape. Each type of PAR lamp has a distinct illumination pattern (a–f) that varies in size and light intensity, depending on the angle at which the lamp is aimed and its distance from the area illuminated. (g) The round lighting pattern changes to oval or elliptical when the lamp is aimed at an angle, making illuminance calculations much more complex. (Data in a–f courtesy of Philips Lighting Co.)

## Light Distribution – Candlepower Distribution

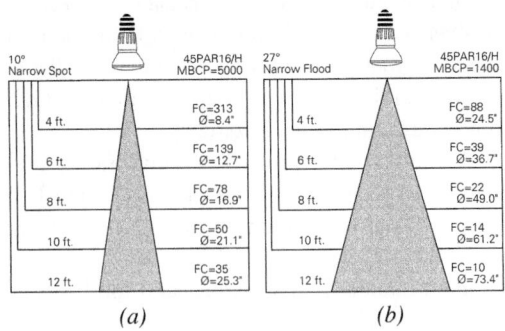

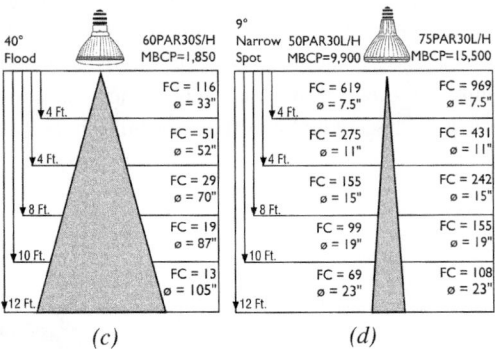

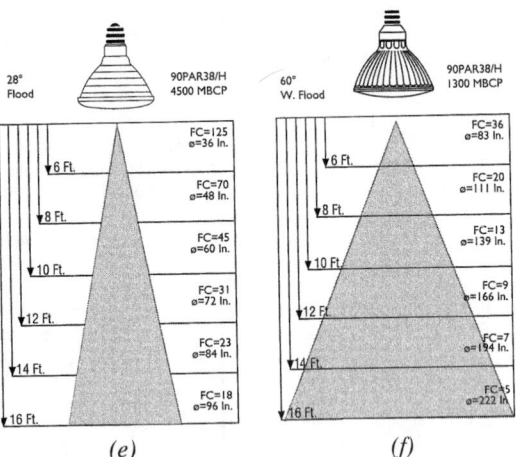

MBCP = maximum beam candlepower
ø = diameter of beam spread in inches
FC = footcandle measured at 0°

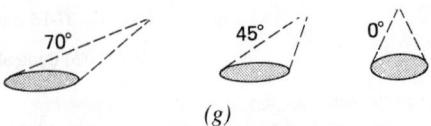

## (b) MR-16 Precision Reflector Units

Miniature single-ended 12-V lamps of 2-in. diameter (and smaller), with multifaceted dichroic (heat-ejecting) reflectors and a bipin base, have found very wide acceptance in all types of display and accent lighting applications. These reflector units, illustrated in Fig. 12.14, essentially comprise an entire lighting fixture, like R and PAR lamps, requiring only a base for electrification. The lamps are known by the generic name *MR-16*, after an early 2-in.-diameter model, although each major manufacturer utilizes its own trade name. A multi-mirror-faceted dichroic reflector produces a "cool" precision light beam by ejecting approximately two-thirds of the lamp heat (long-wave radiation) through the back of the reflector. Luminaires must provide adequate means to dissipate this heat to avoid early lamp failure or creation of a fire hazard. The lamps are rated from 20 to 75 W. Beam characteristics of a few MR-16 lamps are given in Fig. 12.14.

(a)

**FOOT CANDLE CONES**

**TRU-AIM TITAN™**

**10° NSP**

| Distance from Source (in ft.) | Diameter (in ft.) | Center FC | | | |
|---|---|---|---|---|---|
| | | 20W | 35W | 50W | 65W |
| 0' | | | | | |
| 3' | 0.5 | 556 | 922 | 1278 | 1556 |
| 6' | 1.0 | 139 | 231 | 319 | 389 |
| 9' | 1.6 | 62 | 102 | 142 | 173 |
| 12' | 2.1 | 35 | 58 | 80 | 97 |
| 15' | 2.6 | 22 | 37 | 51 | 62 |

**25° NFL**

| Distance from Source (in ft.) | Diameter (in ft.) | Center FC | |
|---|---|---|---|
| | | 50W | 65W |
| 0' | | | |
| 3' | 1.3 | 356 | 444 |
| 6' | 2.7 | 89 | 111 |
| 9' | 4.0 | 40 | 49 |
| 12' | 5.3 | 22 | 28 |
| 15' | 6.7 | 14 | 18 |

**40° FL**

| Distance from Source (in ft.) | Diameter (in ft.) | Center FC | | | |
|---|---|---|---|---|---|
| | | 20W | 35W | 50W | 65W |
| 0' | | | | | |
| 3' | 2.2 | 78 | 139 | 222 | 233 |
| 6' | 4.4 | 19 | 35 | 56 | 58 |
| 9' | 6.6 | 9 | 15 | 25 | 26 |
| 12' | 8.7 | 5 | 9 | 14 | 15 |
| 15' | 10.9 | 3 | 6 | 9 | 9 |

**60° VWFL**

| Distance from Source (in ft.) | Diameter (in ft.) | Center FC | | | |
|---|---|---|---|---|---|
| | | 20W | 35W | 50W | 65W |
| 0' | | | | | |
| 3' | 3.5 | 39 | 72 | 111 | 117 |
| 6' | 6.9 | 10 | 18 | 28 | 29 |
| 9' | 10.4 | 4 | 8 | 12 | 13 |
| 12' | 13.9 | 2 | 5 | 7 | 7 |
| 15' | 17.3 | 2 | 3 | 4 | 5 |

(b)

**Fig. 12.14** (a) Photo of a particular design of an MR-16-type lamp. (b) Typical illumination cones for the lamp shown in (a). Abbreviations: NSP, narrow spot; NFL, narrow flood; Fl, flood; VWFL, very wide flood. (Courtesy of Osram-Sylvania Products, Inc.)

## Gaseous Discharge Lamps

Lamps in this category include fluorescent and high-intensity discharge (HID) lamps (mercury vapor, metal halide, high-pressure sodium), which function by producing an ionized gas in a glass tube or container rather than heating a filament. Discharge lamps are known for their long life and high efficacy. This section describes the function of ballasts and the various types of gaseous discharge lamps.

### 12.11 BALLASTS

All gaseous discharge lamps require a ballast to trigger the lamp with a high ignition voltage and to control the amount of electric current for proper operation. Ballasts discussed in this section primarily apply to fluorescent systems. Refer to manufacturers' information for details regarding HID ballasts. Matching of ballast to lamp is critical to successful lamp operation.

The function of a ballast is threefold:

- To supply controlled voltage to heat the lamp filaments in preheat and rapid-start circuits
- To supply sufficient voltage to start the lamp by striking an arc through the tube
- To limit the lamp current once the lamp is started

Organizations involved with ballast standards and testing include:

ANSI—American National Standards Institute: originates standards on a national level

CBM—Certified Ballast Manufacturers: a group of fluorescent ballast manufacturers who produce ballasts that conform to certain ANSI specifications

ETL—Electrical Testing Laboratories, Inc.: a private, independent organization and recognized authority in measurement and testing of lamps and lighting equipment

UL—Underwriters Laboratories, Inc.: an independent, nonprofit organization that certifies electrical products to ensure public safety from fire

In the United States, ballasts should be UL labeled and CBM/ETL certified (for a limited number of fluorescent ballasts). The UL label assures intrinsic safety. CBM establishes high-quality design criteria, and ETL tests ballasts to determine that design standards have been met.

### (a) Ballast Characteristics

Because of the considerable energy that is lost in inefficient ballasts, manufacturers and standards organizations have established criteria by which ballast energy efficiency can be judged. These characteristics allow comparisons of lighting system operation and performance parameters.

*Ballast Factor.* Ballast factor is the measured ability of a ballast to produce light from a connected lamp. It is the ratio of the light output of a lamp when operated on a tested ballast to the light output of the same lamp operated by a standard laboratory reference ballast (using ANSI test procedures). Ballasts with extremely high or low ballast factors can reduce lamp life because of inconsistencies in lamp current. ANSI Standard C82.11 prescribes a minimum ballast factor for CBM certification for a certain number of ballast types. A ballast may have different ballast factors for different lamps—for example, one ballast factor for operating standard lamps and another for operating energy-saving lamps. A lamp with a low ballast factor uses less energy but light output is also less. A lamp with a high ballast factor uses more energy and provides more light output. Energy savings with high ballast factors may be achieved by using lower-wattage lamps and fewer fixtures. Ballast factor is not a measure of energy efficiency. Although a lower ballast factor reduces lamp lumen output, it also consumes proportionately less input power. As such, careful selection of a lamp-ballast system with a specific ballast factor will allow a designer to better minimize energy use by "tuning" the lighting levels in a space. For example, in new construction, high ballast factors are generally best, since fewer luminaires will be required to meet the light level requirements. In retrofit applications or in areas with less critical visual tasks, such as aisles and hallways, lower ballast factors may be more appropriate (Eley Associates, 1993).

*Ballast Efficacy Factor.* Ballast efficacy factor is the ratio of ballast factor (as a percentage) to power (in watts). The ballast efficacy factor is an expression of lumens per watt for a given lamp–ballast combination. Comparisons using the ballast effi-cacy factor are valid only when comparing ballasts for equivalent systems in terms of lamp type and number of lamps.

Because ballast factor is an indication of the amount of light produced by a ballast–lamp combi-nation and input watts is an indication of power consumed, the ballast efficacy factor is an expres-sion of lumens per watt for a given lighting system. This measurement is generally used to compare the efficiency of various lighting systems. For example, a ballast with a ballast factor of 0.88 using 60 watts of input power has a ballast efficacy factor of 1.466 ($0.88 \times 100 \div 60 = 1.466$). Another ballast utilizing the same input power with a ballast factor of 0.82 has a ballast efficacy factor of 1.366. The first bal-last therefore offers greater efficacy because it has a higher ballast efficacy factor (1.466 vs. 1.366).

*Power Factor.* Ballast power factor is a mea-sure of how effectively a ballast converts the voltage and current supplied by a power source into watts of usable power delivered to the lamp. In general, the power factor is determined from the ballast design and is considered high (if above 0.90), low (below 0.79), or "corrected" (0.80 to 0.90). Power factors pertain only to the effective use of the power sup-plied to a ballast and not to how well the ballast provides light through a lamp. High power factor ballasts are more expensive, but the additional cost is readily repaid by lower line losses, smaller circuit conductors in long runs, and a larger number of fix-tures per circuit. Energy conservation and eco-nomic considerations dictate the use of power factor–corrected and high power factor ballasts.

### (b) Ballast Types

There are three basic types of ballasts: magnetic, hybrid, and electronic. Ballast technology has greatly changed in the past 10 years because of energy pol-icy changes developed by the U.S. DOE, state energy offices, the ACEEE, the Alliance to Save Energy, the Natural Resources Defense Council, and light-ing manufacturers.

*Magnetic.* Magnetic ballasts (core-and-coil) contain a magnetic core of several laminated steel plates wrapped with copper windings and operate at line frequency (60 Hz). These ballasts (Fig. 12.15)

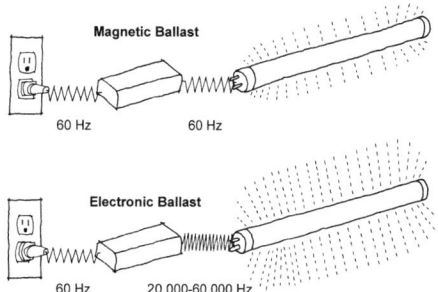

**Fig. 12.15** *Ballasts for fluorescent lamps have traditionally been of the electromagnetic type, operating at a frequency of 60 Hz. Electronic ballasts operate at frequencies of 20,000–60,000 Hz and cause lighting systems to convert electric power to light more efficiently than systems run by electromagnetic ballasts. (Drawing by Jonathan Meendering; © 2004 Alison Kwok; all rights reserved.)*

have become obsolete, although they are found in existing buildings.

*Hybrid.* Also called *cathode-disconnect ballasts,* hybrid ballasts use a magnetic core-and-coil trans-former and an electronic switch for the electrode-heating unit. Like magnetic ballasts, they operate at 60 Hz. The ballast disconnects the electrode-heating unit after starting the lamp.

*Electronic.* Solid-state electronic ballasts oper-ate lamps at 20–60 kHz and have half the power loss of magnetic ballasts. Lamp efficacy increases by approximately 10–15% compared to operation at 60 Hz. Electronic ballasts are more expensive but more energy-efficient, generate less heat, are virtu-ally silent, lighter, and start and operate without flicker. They are also available as dimming ballasts, which allow light output to be controlled between 1% and 100%.

*Special.* Lamps operating at other than 430 mA, including low-current and high-current units, require matching ballasts to supply the required current, waveform, and circuitry. Use of one manufacturer's low-current lamp with another's low-current ballast is not suggested without prior testing or a specific manufacturer's recommenda-tion. The principal varieties of nonstandard ballasts are as follows:

1. *Low-current ballasts* are intended to match spe-cific low-current lamps, including T8 triphos-phor units, slimline lamps, and others.

2. *High-current ballasts* are intended to be used with high-output lamps. The purpose of this combination is either to increase output in an existing installation or to reduce the number of fixtures in a new installation.

3. *Energy-saving ballasts* are designed to reduce the total wattage of the lamp–ballast combination. Part of this power reduction is produced by more efficient design of the ballast itself. Another part is due to a lower current rating for the lower wattage of the lamp itself. A third portion is frequently a switching arrangement in the ballast that disconnects the lamp filaments after an arc is struck (after the lamp ignites). This technique can save 4 to 8 W per lamp–ballast pair.

4. *Multilevel ballasts* are useful when it is desired to change lighting levels evenly. The usual unit is two-level—that is, full output and 50%—but three-level units are available for full, two-thirds, and one-third output.

In addition to these ballast types, there are special units for low or high ambient temperature, weatherproof units, and low leakage-to-ground units for hospital applications.

### (c) Ballast Performance

*Heat.* Ballast heat is usually transferred to the luminaire body by direct metal-to-metal contact (which must be unimpeded) and is then dissipated by radiation and convection from the fixture. The location and method of fixture installation affect the heat transfer from the fixture and, consequently, the ballast temperature. Operating temperature directly affects ballast life. At normal operating temperature, a ballast life of 12 to 15 years can be expected. Generally, ballast life is halved for every 50°F above the 194°F (27.8°C above 90°C) operating temperature and, conversely, is doubled for every 50°F (27.8°C) reduction in operating temperature below 194°F (90°C). Electronic ballasts will usually start a lamp at 50°F (24°C) minimum. A special ballast is required for starting at temperatures to 0°F (−18°C). The cooler operation of electronic ballasts reduces air-conditioning costs. Not only do the ballasts operate cooler, but lamps operated by electronic ballasts produce the same light output with lower losses. Therefore, overall energy costs for an electronic ballast installation are reduced because the fixtures use less energy and produce less heat for the same light output.

*Noise.* All electromagnetic and many electronic ballasts make a humming sound that originates from the inherent magnetic action causing vibrations in the steel laminations of the core and coil assembly. Because electronic ballasts have a small (or no) core-and-coil assembly, they have the lowest noise output. Most electronic ballasts make almost no sound. Ballast noise, if any, may become amplified because (1) of the method of mounting the ballast in the fixture; (2) of loose parts in the fixture; or (3) ceilings, walls, floors, and hard furniture reflect the noise. Ballasts are sound-rated by a letter, A through F, which indicates not actual sound developed, but performance in a space. A rating of A designates the quietest ballast. Selection should be made on the basis of the ballast sound rating and the requirements of the installation.

*Flicker.* Flicker is caused by extinguishment and reignition of the arc within a fluorescent tube and is visible only where long-persistence phosphors are thin or entirely absent (i.e., at the extreme ends of a lamp). Electromagnetic ballasts are designed to condition 60-Hz input voltage to the electrical requirements of a lamp. A magnetic ballast alters the voltage but not the frequency. Thus, the lamp voltage crosses zero 120 times each second, resulting in 120 light output oscillations per second. This results in about 30% flicker for standard phosphor lamps operated at 60 Hz. This flicker is generally not noticeable—but there is evidence that flicker of this magnitude can cause adverse effects, such as eyestrain and headache. Most electronic ballasts use high-frequency operation, which reduces lamp flicker to an essentially imperceptible level. Manufacturers typically specify the flicker percentage of a particular ballast. For a given ballast, the flicker percentage is a function of lamp type and phosphor composition.

*Dimming Control.* Dimming of electronically ballasted lamps is accomplished within the ballast itself. The dimming process uses energy that should be accounted for in lighting system energy-use calculations. Electronic ballasts alter the output power to the lamps by a low-voltage signal into the output circuit. High-power switching devices to condition the input power are not required. This arrangement allows control of one or more ballasts independent of the electrical distribution system.

*Radio Noise.* Occasionally, a defective ballast will cause radio noise (commonly referred to as *radio frequency interference* [RFI]). In general, however, RFI is *not* produced by a ballast, but by the arc discharge in a fluorescent tube. To minimize RFI, ballasts are available with integral RF noise suppressors. In extreme cases, additional suppression can be obtained by installation of RF noise attenuators in a lighting fixture.

## Fluorescent Lamps

The second major category of electric light sources is gaseous discharge lamps, of which the fluorescent lamp is the best-known and most widely used type. Since their introduction in 1937, fluorescent lamps have almost completely supplanted incandescent lamps in all fields except specialty lighting and residential use. The typical linear fluorescent lamp comprises a cylindrical glass tube sealed at both ends and containing a mixture of an inert gas, generally argon, and *low-pressure* mercury vapor. Built into each end of the tube is a cathode that supplies the electrons to start and maintain an electric arc, or gaseous discharge. Short-wave UV radiation, which is produced by the mercury arc, is absorbed by phosphors coating the inside of the tube, causing a reaction that emits visible radiation (light). The particular mixture of phosphors used governs the quantity and spectral quality of the light output. Light from fluorescent sources radiates from a larger lamp surface area than is the case with incandescent sources. The light is diffuse, which is suitable for illuminating or washing large areas such as ceiling planes.

## 12.12 FLUORESCENT LAMP CONSTRUCTION

Rapid-start and instant-start fluorescent lamps are commonly used today. Preheat lamps are a legacy type. Lamp families include linear and compact. Linear lamps are tubular in shape, with the most

popular versions being T8 and T5 (26 mm and 16 mm) in standard and high output (HO) and the legacy T12 (38 mm) lamp. Compact fluorescents include dedicated socket versions of single-tube, double-tube, and triple-tube lamps. The descriptions that follow cover *standard* lamps and circuits. Special lamps, accessories, and circuits, including low-wattage lamps, triphosphor lamps, and special-shape lamps, are discussed separately.

### (a) Preheat Lamps

Older fluorescent fixtures use a preheat technology that heats the gas in order to start the lamp and use a mechanism called a *starter.* Preheat fixtures either have an automatic starter or require a manual starting action. The original T12 (38 mm) fluorescent lamp was a preheat design. Construction of a typical hot cathode lamp (used with both preheat and rapid-start types) is shown in Fig. 12.16. All preheat lamps have bi-pin bases (see Fig. 12.16).

This lamp circuit utilizes a separate starter, a small cylindrical device that plugs into a preheat fixture. When the lamp circuit is closed, the starter energizes the cathodes; after a 2- to 5-second delay, it initiates a high-voltage arc across the lamp, causing it to start. Most starters are automatic, although in desk lamps preheating is accomplished by depressing the start button for a few seconds and then releasing it. This closes the circuit and allows the heating current to flow; releasing the button causes the arc to strike. Preheat lamps are no longer the industry standard but are included here as a point of comparison.

### (b) Rapid-Start Lamps

Today, the most popular fluorescent lamp design is the rapid-start lamp, shown in Fig. 12.17. This design functions similarly to the traditional preheat lamp, but without a starter switch. Instead, the lamp's ballast constantly channels current through both electrodes. This current flow is configured so that there is a charge difference between the two electrodes, establishing a voltage across the tube. Most fluorescent fixtures with two or more lamps are known as *rapid start.* Instead, the ballast keeps a low flow of current running through the filaments at all times or during the start-up period, eliminating the delay inherent in a preheat circuit. When the lamp circuit is energized, the arc is struck immediately. No external starter is

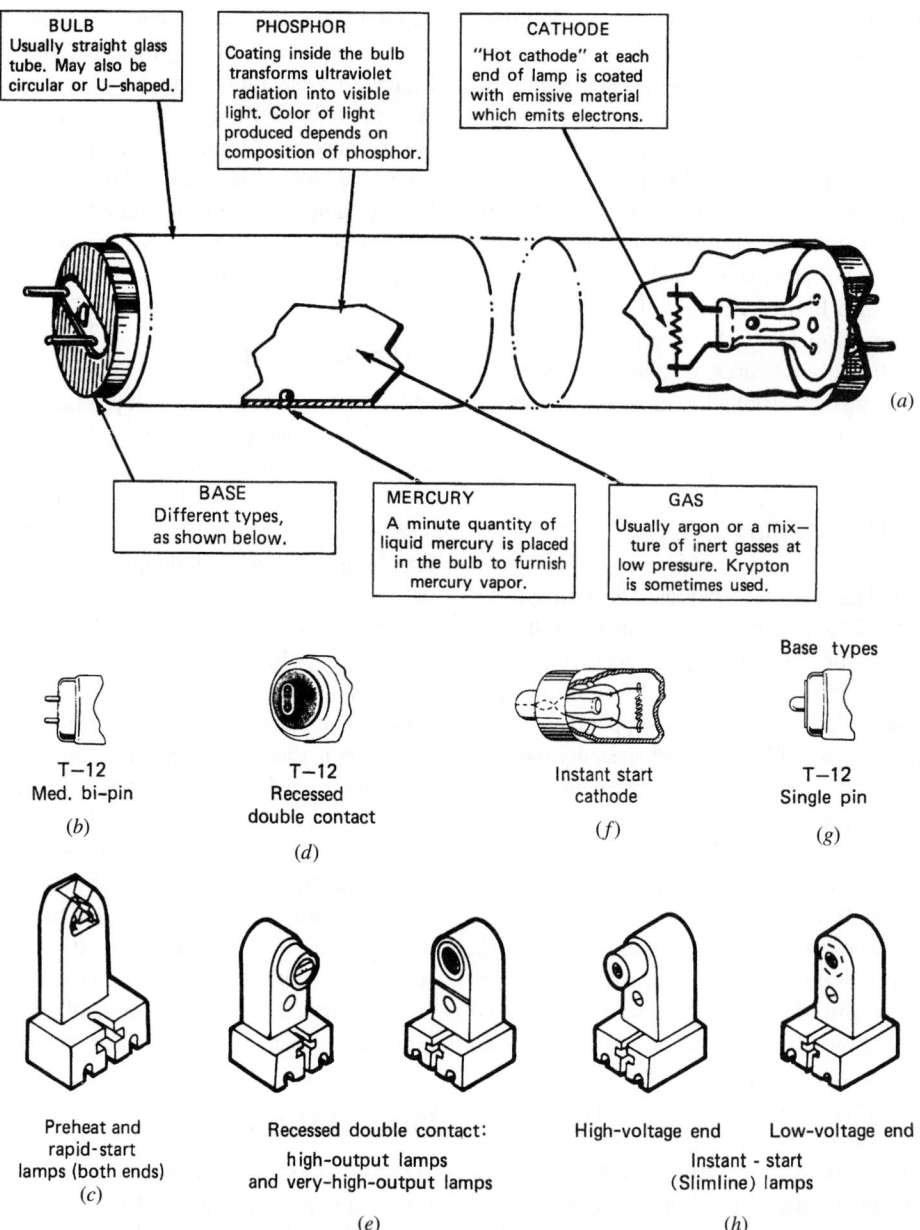

**BULB**
Usually straight glass tube. May also be circular or U—shaped.

**PHOSPHOR**
Coating inside the bulb transforms ultraviolet radiation into visible light. Color of light produced depends on composition of phosphor.

**CATHODE**
"Hot cathode" at each end of lamp is coated with emissive material which emits electrons.

(a)

**BASE**
Different types, as shown below.

**MERCURY**
A minute quantity of liquid mercury is placed in the bulb to furnish mercury vapor.

**GAS**
Usually argon or a mix—ture of inert gasses at low pressure. Krypton is sometimes used.

Base types

T—12
Med. bi–pin
(b)

T—12
Recessed
double contact
(d)

Instant start
cathode
(f)

T—12
Single pin
(g)

Preheat and
rapid-start
lamps (both ends)
(c)

Recessed double contact:
high-output lamps
and very-high-output lamps
(e)

High-voltage end     Low-voltage end
Instant - start
(Slimline) lamps
(h)

**Fig. 12.16** Details of typical fluorescent lamps and associated lampholders. (a) Construction of preheat–rapid-start bi-pin base lamp. (Courtesy of GTE/Sylvania, Inc.) This type of lamp has type (b) base and is held in type (c) lampholder. High-output HO and VHO rapid-start lamps use a recessed dc base (d) and lampholders (e). Instant-start lamps are similar in construction to (a) except with cathode construction (f), have a single-pin base (g), and use single-pin lampholders (h), which are different at each end.

required. Because of the similarity of operation, rapid-start lamps will operate satisfactorily in a pre-heat circuit. The reverse is not true, because the pre-heat lamp requires more current to heat the cathode than the rapid-start ballast provides.

Rapid-start T12 lamps operate at 425 to 430 mA. If the current is increased, the output of the lamp also increases. Two generic types of higher output rapid-start lamps are available. One operates at 800 mA and is called simply *high output* (HO). The second, which operates at 1500 mA (1.5 A), is called (by different manufacturers) *very high output*

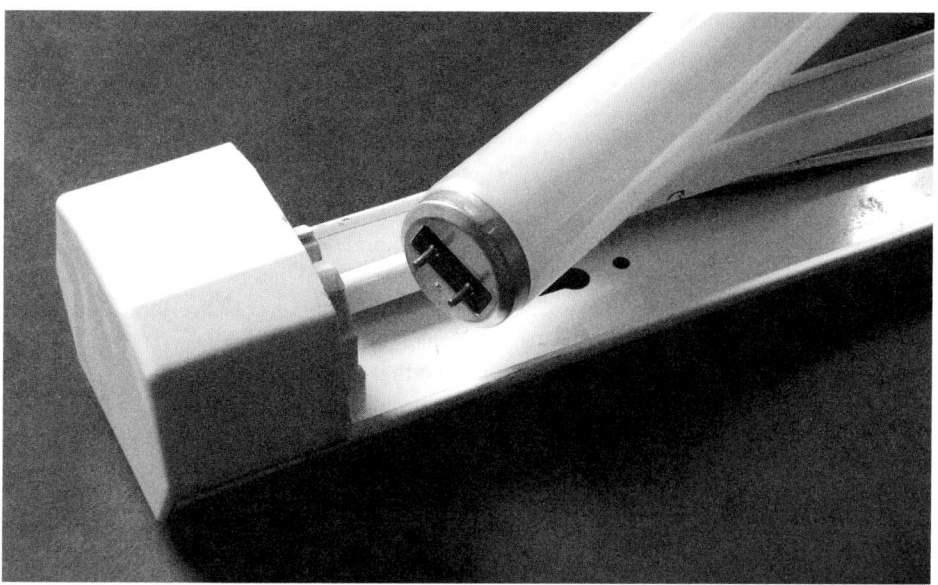

**Fig. 12.17** *Rapid-start fluorescent lamps have two pins that slide against two contact points in an electrical circuit. The ballast constantly channels current through both electrodes, creating a charge difference between the two electrodes and establishing a voltage across the tube. (Photo by Jonathan Meendering; © 2004 Alison Kwok; all rights reserved.)*

(VHO), *superhigh output,* or simply the *1500-mA, rapid-start lamp.* There is also a 1500-mA lamp that uses what looks like a dented or grooved glass tube. This lamp has a somewhat higher output than a standard VHO tube. All HO lamps use double-contact bases and special ballasts (see Fig. 12.16). HO lamps are used in applications such as outdoor sign lighting, street lighting, and merchandise displays where high output is required from a limited size source. Because of the serious heat problems involved, VHO lamps are frequently operated without enclosing fixtures. Conversely, HO and VHO lamps are frequently used in cold environments that prevent proper operation of standard output 430-mA lamps. Most HO and VHO lamps have slightly lower luminous efficacy than a standard 430-mA, rapid-start lamp and have a considerably shorter life. It should be noted that only rapid-start lamps are to be used with motion sensors or in conjunction with sequential repetitive dimming of lamps. Use of instant-start lamps in this application will overload the cathodes and lamp life will be reduced.

#### (c) Instant-Start Fluorescent Lamps

Instant-start fluorescent lamps use a high-voltage transformer to apply a very high initial voltage to the cathodes. An excess of electrons on the cathode surface forces some electrons into the fill gas, which ionizes the gas. This creates an instant voltage difference between the cathodes, establishing an electric arc. These lamps have only a single pin at each end that also acts as a switch to break the ballast circuit when the lamp is removed, thus lessening the shock hazard (see Figs. 12.13 and 12.16). The lamps are generally operated in two-lamp circuits at various currents; normal currents are 200 and 430 mA. The high-voltage starting characteristic of instant-start circuits lowers the lamp life to about half that of a corresponding rapid-start lamp. Instant-starts have the advantage of being able to start at much lower ambient temperatures (below 50°F [10°C]) than rapid-start circuits. This starting characteristic makes the instant-start lamp and circuit particularly applicable to outdoor use.

### 12.13 FLUORESCENT LAMP LABELS

Standard fluorescent lamp labels are printed on the end of a lamp and are identified by several letters and numbers, as shown in Table 12.4. The typical labeling is in the form FSWWCCC-TDD

### TABLE 12.4 Fluorescent Lamp Label Designations

| Label | Explanation |
|---|---|
| F | Fluorescent lamp. "G" means germicidal shortwave UV lamp. |
| S | Style—no letter indicates a normal straight tube; "C" means Circline. |
| W | Nominal power in watts: 4, 5, 8, 12, 15, 20, 30, 40, etc. |
| CCC | Color. W = white, CW = cool white, WW = warm white, BL/BLB = black light, etc. |
| T | Tubular bulb |
| DD | Diameter of tube in eighths of an inch. T8 is 1 in., T5 is ⅝ in., etc. |

(each manufacturer has variations on this format). Depending upon the type of fluorescent lamp, designations for color rendering index and color temperatures are also included on the label.

## 12.14 FLUORESCENT LAMP TYPES

*T8 Fluorescent Lamps.* Over the past 20 years, T8 lamps have afforded designers (and their clients) cost-effective and energy-efficient lighting systems that are visually comfortable and have a high degree of flexibility in their application. T8 lamps are 1 in. (26 mm) in diameter; they are available in wattages of 17, 25, 32, and 40 W at 2-ft, 3-ft, 4-ft, and 5-ft (600-mm, 900-mm, 1200-mm, and 1500-mm) lengths, respectively. There are two color rendering categories of T8 lamps: 700 (75 CRI)

and 800 (85 CRI) series, which relate to the color rendering properties of the triphosphor coatings used. Lamp manufacturers have developed a standard designation to indicate the color temperature of a lamp. For example, "30" indicates a 3000 K lamp. T8 lamps are available in 3000 K, 3500 K, and 4100 K color temperatures, designated "30", "35", and "41," respectively. Table 12.5 shows a comparison of lamp technologies as fluorescent lamp manufacturers have increased efficacy and color rendering while decreasing diameter and wattage.

*T5 Fluorescent Lamps.* A new line of T5 lamp technology was developed in Europe and introduced in North America in 1996. Although it was still expensive, the introduction of a T5 HO line in 1998 offered about twice the lumen output in the same length as its T8 counterpart with an efficacy that is attractive in meeting project energy goals. The T5 is the first "metric" lamp introduced in the United States, yet it is commonly called T5 because of industry nomenclature. Standard and HO T5 lamps are available in 22-in., 34-in., 46-in., and 58-in. (560-mm, 864-mm, 1163-mm, and 1473-mm) lengths. The standard T5 and the T5 HO lamps are the same diameter and width. The 46-in. T5 (nominal 4 ft) is rated at 2900 lumens, similar to the lumen per watt output of a T8 lamp (2950 lumens). By contrast, the 46-in. T5 HO lamp is rated as high as 5000 lumens, offering twice the maintained light output of a T8 lamp. Because of its smaller ⅝-in. (15-mm) diameter construction,

### TABLE 12.5 Comparative Characteristics of Tubular Fluorescent Lamps[a]

| | T12 | T8 | T5 | T5HO |
|---|---|---|---|---|
| Initial rated light output | 3350 lumens | 2950 lumens | 2900 lumens | 5000 lumens |
| Nominal lamp watts | 40W | 32 W | 28W | 54W |
| Initial lamp efficacy[1] | 84 lm/W | 92 lm/W | 104 lm/W | 93 lm/W |
| Initial system efficacy[2] | 88 lm/W | 90 lm/W | 89 lm/W | 85 lm/W |
| Lumen maintenance[1] | 78% | 93% | 97% | 95% |
| Maintained system efficacy | 69 lm/W | 84 lm/W | 86 lm/W | 81 lm/W |
| Rated life[3] | 20,000 hr | 20,000 hr | 16,000 hr | 16,000 hr |
| CRI | 80 | 85 | 85 | 85 |
| Optimum operating temperature | 77°F [25°C] | 77°F [25°C] | 95°F [35°C] | 95°F [35°C] |

[a]Figures are representative; for exact figures, consult current catalogs.
[1]Based on 4-ft nominal length, CRI 85 lamps.
[2]Based on 4-ft nominal length, CRI 85, two-lamp rapid-start, electronic ballast.
[3]Varies with manufacturer and phosphor coating technology.

significantly less glass, mercury, and high-quality phosphors are needed for its construction. T5 lamps also allow a designer to use fewer lamps (and fixtures), thus providing certain savings on installation and long-term maintenance. The narrow lamp diameter has provided an opportunity for the design of new fixtures and for use in low-profile, indirect luminaires. The color rendering quality of light from T5 lamps (CRI 85) is excellent, although the potential for glare problems exists, which can be addressed by sophisticated shielding techniques. Utilizing T5 (and particularly T5 HO) lamps in direct lighting installations requires special attention to glare control.

## 12.15 CHARACTERISTICS OF FLUORESCENT LAMP OPERATION

Five characteristics define the operation of fluorescent lamps:

- *Efficacy*—light output per unit of power input
- *Lumen maintenance*—the decreasing output of light as a lamp ages
- *Lamp life*—average (statistically defined) lamp life expectancy
- *Temperature and humidity*—how a lamp responds to extreme environmental operating conditions.
- *Dimming*—output reduction of a fluorescent lamp

### (a) Efficacy

$$\text{luminous efficacy} = \frac{lumens \ (light \ output)}{\text{watts (power consumed including ballast losses)}}$$

$$= \text{lumens per watt (lm/W)}$$

The design efficacy (lumens per watt) of a fluorescent lamp depends upon the operating current and the phosphors utilized. Fluorescent lamp efficacy is further dependent upon the lamp length, ambient temperature, frequency of the electricity supply, and ballast operation. *Wattage*, in itself, is a meaningless quantity unless it is associated with a lumen output figure. Thus, energy-saving low-wattage or high-output lamps with their special ballasts are seldom the indicated choice in new design because their efficacy does not usually jus-

tify the cost premium. These special lamps are useful in retrofit work, in which case field measurements of illuminance and lamp temperature are required before selecting a replacement lamp–ballast combination.

### (b) Lumen Maintenance

The lumen output of a fluorescent tube decreases rapidly during the first 100 hours of burning and thereafter much more slowly. Phosphors deteriorate, typically blackening at the ends of a lamp, thereby blocking some light. Most product catalogs list "initial lumens," which is the lamp output under *laboratory conditions* after 100 hours of burning, and "mean" or "design" lumens, which is lamp output at 40% of life. Lighting levels gradually drop as a system ages until, somewhere in the middle of the effective life of the lamps (Fig. 12.18), the intended "maintained" illuminance of a system is (temporarily) achieved.

### (c) Lamp Life

The life of a standard fluorescent lamp is defined as the period of time an average lamp is expected to

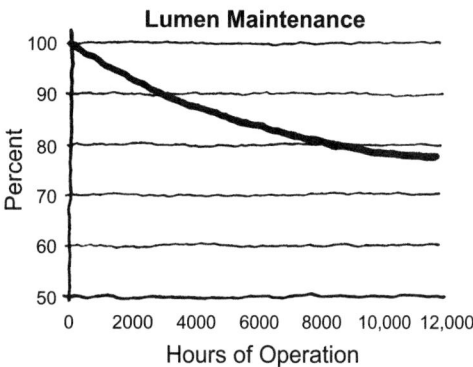

**Fig. 12.18** *Lumen maintenance curve for fluorescent lamps. Lumen maintenance is the ability of a lamp to retain its lumen output over time. Greater lumen maintenance means that a lamp will remain brighter longer. The opposite of lumen maintenance is lumen depreciation, which represents the reduction of lumen output over time. In the United States, mean lumens is a measure taken at 40% of the rated lamp life. In the United Kingdom, lighting levels are based upon maintained illumination, and it is necessary to determine the minimum lighting level in the installation when replacement of lamps is due, taking into account all possible reasons for deterioration. (Drawing by Jonathan Meendering.)*

last, depending upon the burning hours per start. It is expressed as "rated average life" in hours of operation. The values listed in lamp catalogs for life are based upon a burning cycle of 3 hours per start (and 20 minutes of "off" status) and represent the average life of a group of lamps; that is, half of the lamps in any group will have burned out at that time. Typical lamp mortality curves are shown in Fig. 12.19, and the effect of burning hours per start is shown in Fig. 12.20. Average rated lamp life is not the same as the time at which lamps are typically replaced, which is usually well before 50% failures occur in a batch. Several factors affect fluorescent lamp life. Longer burning hours per start will extend lamp life. Lamp life is shortened by improper lamp current, improper voltage to the ballast, or improper cathode heating.

From an energy utilization viewpoint, if a lighted space is not utilized for 10 minutes or more, fluorescent lamps should be shut off. This takes into account both direct energy consumption and the resource energy required to replace a lamp as a result of shortening its life. From a cost viewpoint, the breakeven point depends upon these factors: (1) lamp life reduction as a function of burning hours per cycle, (2) cost of energy, (3) cost of lamp and lamp replacement, (4) amount of time the lamp remains off when shut off, (5) cost of switching equipment (if any), and (6) life of the building.

With this number of variables it is not possible to give general solutions, and an individual analysis is required. However, several analyses for ordinary office conditions, using lamp life data as given in Fig. 12.20 (a 20-year fixture life, $0.085/kWh energy cost escalating 3% annually, $1.25 lamp cost, 15-minute relamping time, and $8 per lamp to provide the necessary switch [one switch per two 2-lamp fixtures]), have shown that lamps should be switched off any time they are not in use for 5 to 8 minutes or more. (The spread is caused primarily by variation in local labor rates.) It is thus clearly an economic fallacy to leave lamps burning to achieve longer lamp life.

## (d) Effect of Temperature and Humidity

Fluorescent lamps are affected by extremes in ambient temperature and by high humidity. Outside of the optimal operating temperature range, 41–77°F (5–25°C), there is a rapid drop in light output and difficulty in starting. High humidity causes electrical leakage along the lamp surface, lowering the starting voltage provided by the ballast. Lamps are precoated with silicone to break up moisture films and prevent such leakage.

The temperature of the coolest point on the lamp bulb wall determines a lamp's mercury-vapor pressure, which in turn determines the lamp lumen output, wattage, and color. Maximum output for standard lamps occurs at a *bulb* temperature of 104°F (40°C). The bulb wall temperature itself is affected by room ambient temperature, airflow over

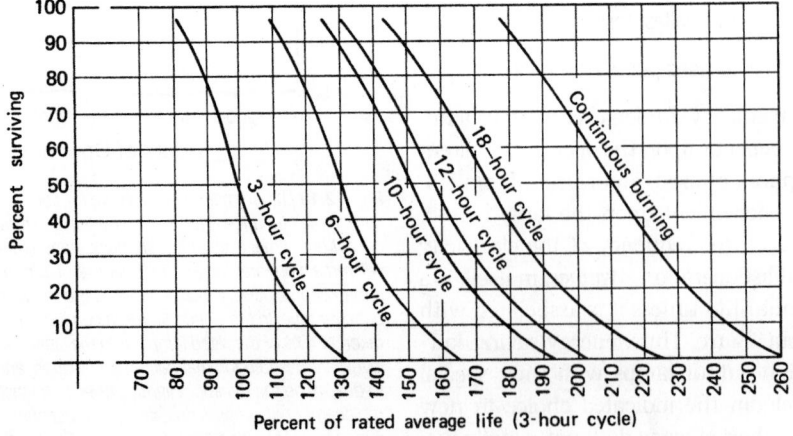

**Fig. 12.19** *Typical mortality curves of standard fluorescent lamps.*

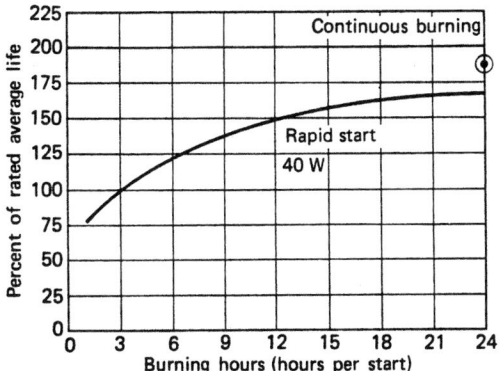

**Fig. 12.20** *Effect of burning hours on fluorescent lamp life. Note that at 3 burning hours per start, the average lamp life is 100% of the nominal catalog figure.*

the lamp (as with air-return fixtures), and the temperature of adjacent surfaces such as a ballast enclosure. Thus, catalog data on lamp output and wattage, based upon laboratory tests of bare tubes at 77°F (25°C) ambient temperature in still air, may be very far from actual field performance.

## (e) Dimming

Dimming of a fluorescent lamp system reduces energy consumption, can correct overlighting, can balance illumination through integration with daylighting, and allows flexibility when full lighting output is required. Unlike incandescents, which can be dimmed with just a wallbox device, fluorescent lamps require dimming ballasts. The dimming range differs greatly among ballasts. With most electronic dimming ballasts, output can vary between full and a minimum of about 10% of full output. However, electronic, full-range dimming ballasts are also available for some lamp types that operate lamps down to 1% of full lumen output.

A ballast can be configured so that it (1) receives a signal from a control device and subsequently (2) changes the current flowing through a lamp, thereby achieving a gradual, controlled reduction in lamp output. The characteristics of the ballast circuitry affect the duration and extent of the change in current and subsequent lamp output.

Electronic dimming ballasts for fluorescent lamps are designed to respond to either an analog or digital signal to achieve the dimming effect. Because a dimming ballast must be able to communicate with connected controllers, the method becomes the basis for a protocol (common operating parameters adopted by all manufacturers of dimming ballasts and controllers that use that method). This assures interchangeability between a ballast made by a particular manufacturer and various controllers made by controls manufacturers. Typical applications for dimming include both new construction and retrofit installations: auditoriums and training areas, conference rooms and boardrooms, department and specialty stores, education and healthcare institutions, hotels, houses of worship, private and executive offices, and restaurants.

The primary dimming methods are:

- *Analog:* An analog electronic dimming ballast includes components that perform these functions: electromagnetic interference filtering, rectification, power factor correction, and ballast output to power a lamp. There are several analog methods, including 0-10VDC, two-wire phase-control, three-wire phase-control, and wireless infrared, with 0-10VDC being most often used.
- *Digital:* The digital electronic dimming ballast includes components that perform these functions: electromagnetic interference filtering, rectification, power factor correction, ballast output to power a lamp, and control (as a micro-controller). The micro-controller functions as a storage, receiver, and sender of digital information. The micro-controller can store the ballast address, receive control signals, and send status information.
- *Wireless infrared:* This method uses an infrared transmitter to control the signal and does not require additional wires. The dimmer is either contained in the ballast or provided as an additional component in the light fixture. Wireless infrared control is a good retrofit solution and allows for occupant fixture control. Wireless infrared control is ideally suited for spaces where individual control is desired without additional wiring, such as conference rooms, boardrooms, and open and private offices.

## 12.16 FEDERAL STANDARDS FOR FLUORESCENT LAMPS

Enacted in October 1992, the National Energy Policy Act (EPACT) was designed to reduce the U.S. energy bill by approximately $250 billion over a 15-year period. EPACT mandates minimum standards for lamps in terms of efficacy (lumens per watt) and color rendering index (CRI; the ability of a light source to render colors accurately). EPACT standards eliminated the manufacture and distribution of several major fluorescent lamp types and incandescent lamps that provided the least amount of light for the highest use of energy. The major lamps eliminated are 40 W F40T12 (CW and WW); 75 W F96T12 (CW and WW), and 110 W F96T12/HO (CW and WW). Lamps with very good CRI and special service fluorescent lamps are excluded from the act.

## 12.17 SPECIAL FLUORESCENT LAMPS

### (a) Low-Energy Lamps

The need for energy conservation, the discontinuance of certain lamp types due to EPACT regulations, and a desire to reduce lighting levels in existing overlighted spaces has resulted in the development of a complete line of low-energy lamps. Wattage ratings for these lamps are lower than those of standard lamps because they are intended primarily as lower-energy direct replacements for existing lamps. All such lamps are clearly marked by the manufacturer. They require special matching ballasts for maximum effectiveness and have an efficacy equal to, or somewhat higher than, that of standard lamps and ballasts. They have the disadvantage of higher cost, the need for special ballasts where maximum energy reduction is desired, generally shorter life, inability to be used in most dimming circuits, and problems with inventory and proper lamp replacement. Their use is indicated only where other light-output and wattage-reduction schemes, such as circuit dimming or reduced-wattage ballasts, are inapplicable.

### (b) U-Shaped Lamps

U-shaped lamps were developed to answer a need for a high-efficacy fluorescent source that could be utilized in a square fixture. The U lamp is basically a standard fluorescent tube bent into a U shape and available with 3⅝- or 6-in. leg spacing; the former can be accommodated three to a 2-ft² × 2-ft² fixture. (The narrower T8 envelope of triphosphor lamps

permits a tighter bend; these U lamps have a 1⅝-in. leg-to-leg spacing.) U lamps operate on standard ballasts and have slightly lower output than a corresponding straight tube. In all other respects, a U lamp has the same characteristics as a straight lamp of similar type.

### (c) Ecologically Friendly Lamps

ALTO® lamps were developed in 1995 by the Phillips Lighting Company to support the reduction of mercury at the source and provide users with environmentally responsible methods for disposal. The ALTO family of lamps includes a broad selection of TCLP-compliant lamps: linear and compact fluorescents, high-pressure sodium lamps, metal halide, U-bent fluorescent, and the lead free MasterLine™ ALTO lamps, all of which can be recycled (always the preferred method) or disposed of conventionally. TCLP is the *Toxicity Characteristic Leaching Procedure*—a test developed by the EPA in 1990 to measure hazardous substances that might dissolve into the ecosystem and that is used by the federal government and by most states to determine whether old fluorescent lamps should be characterized as hazardous waste. All fluorescent lamps contain mercury; however, ALTO lamps have the lowest mercury doses available (on average, 70% less mercury than the 2001 industry average) on the market. This product development encouraged other manufacturers to reduce the mercury content in their products—as in the Osram Sylvania (Ecological®) and General Electric (Ecolux®) lines.

### (d) UV Lamps

UV lamps emit radiation in the UV spectrum, which includes all electromagnetic radiation with wavelengths in the range of 10–400 nm.

- The UVA range includes wavelengths from 315 to 400 nm. Wavelengths from about 345 to 400 nm are used for "blacklight" effects (causing many fluorescent objects to glow) and are usually slightly visible if isolated from the more visible wavelengths. Shorter UVA wavelengths from 315 to 345 nm are used for sunning.
- UVB refers to wavelengths from 280 to 315 nm. These wavelengths are more hazardous than UVA wavelengths and are largely responsible for sunburn.

- UVC refers to shorter UV wavelengths, usually from 200 to 280 nm. Wavelengths in this range, especially from the low 200s to about 275 nm, are especially damaging to exposed biological cells. Such short-wave UV radiation is often used for germ killing purposes.

Although UV lamps are not commonly used in architectural lighting, they are included in this section as specialty lamps because they address a wide range of applications in industrial, technological, laboratory, and medical settings. UV lamps include fluorescent black lights, fluorescent tanning and medical UV lamps, "RS" reflector ("floodlamp") sunlamps, and germicidal and EPROM (erasable programmable read-only memory) erasing lamps.

## 12.18  COMPACT FLUORESCENT LAMPS

*Compact fluorescent lamps* (CFLs) offer a comparable (in brightness and color rendition), energy-efficient alternative to incandescent lamps. Unlike standard fluorescent lamps, they can directly replace standard incandescent bulbs.

CFLs are simply folded fluorescent tubes with both ends terminating in a common base. Some compact fluorescent lamps have the tubes and ballast permanently connected with a screw-in medium base. Others have separate tubes and ballasts, allowing the tubes to be replaced without changing the ballast. As a result, an exhausted lamp is simply replaced in the existing ballast, resulting in considerable economy. A CFL produces a diffuse light, unlike single-point incandescent lamps. This is an important factor to consider when replacing incandescents with CFLs in high ceiling applications.

CFLs are manufactured in a variety of styles or shapes: two, four, or six tubes; circular or spiral tubes (Fig. 12.21). They are efficient at lower wattages and can produce light output equivalent to that of higher-wattage incandescents (e.g., a typical 60-W incandescent lamp with a 900-lumen output could be replaced by a 15- to 19-W CFL). The total surface area of the tube(s) determines how much light is produced. The efficacy of

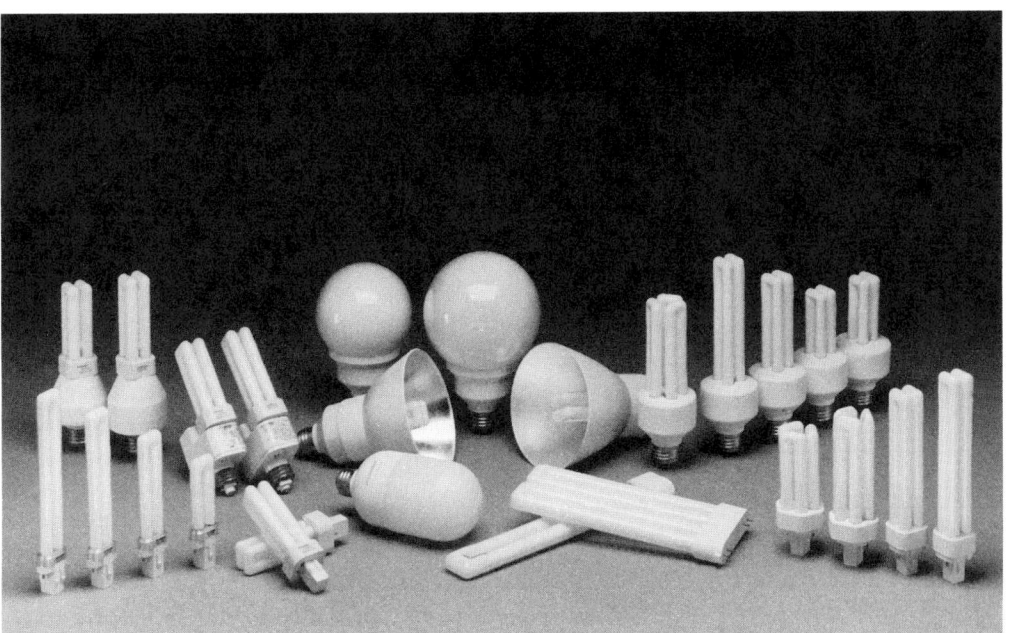

**Fig. 12.21** *Family portrait of compact fluorescent lamp designs. Pin base lamps designed for use with a separable ballast are shown in the foreground, the same lamps mounted in their screw base ballasts are in the left background, and the one-piece combined lamp–ballast design stands in the right background. Globe and reflector-type replacements for incandescent lamps are in the center of the photo. (Courtesy of Osram-Sylvania.)*

**TABLE 12.6  Equivalent Wattage of Common Incandescent Lamps and Compact Fluorescents**

| Incandescent Watts | Compact Fluorescent Watts |
|---|---|
| 50 | 9 |
| 60 | 15 |
| 75 | 20 |
| 100 | 25 |
| 120 | 28 |
| 150 | 39 |

*Source:* U.S. Department of Energy, Energy Efficiency and Renewable Energy—http://www.eere.energy.gov/consumerinfo/factsheets/ef2.html/

lamp–ballast combinations ranges from 55 to 75 lm/W, assuming an electronic ballast. Lamps with magnetic ballasts are available but are not favored because of excessive heat, weight, flicker, and reduced efficiency. Lamp colors are similar to straight lamps (i.e., 3000 K, 3500 K, 4200 K, and 5000 K). All CFLs have a CRI of 80 or higher. Their life is 10,000 to 12,000 hours based on 3 hours per start. Table 12.6 compares the wattage of commonly available incandescent lamps to that of a CFL that provides similar light output.

A major advantage of using CFLs is saving money, as shown in Table 12.7. This table assumes that a lamp is on for 6 hours per day and that the electric rate is 8 cents per kilowatt-hour.

**TABLE 12.7  Cost Comparison for Operation of an Incandescent Lamp and a Compact Fluorescent Lamp**

| | Incandescent 100 W 1750 lumens | Compact Fluorescent 27 W 1750 lumens |
|---|---|---|
| Lamp cost ($) | $0.50 | $20.00 |
| Rated life (hours) | 750 | 10,000 |
| Efficacy (lumens per watt) | 17 | 64 |
| Energy cost (@8¢/kwh for 10,000 hrs) | $80 | $22 |
| Total cost (lamps + energy) | $85 | $42 |

*Sources:* U.S. Department of Energy, Energy Efficiency and Renewable Energy—http://www.eere.energy.gov/consumerinfo/factsheets/ef2.html and Southface Energy Institute—http://www.southface.org/home/sfpubs/techshts/13e_lite.pdf/

# HIGH-INTENSITY DISCHARGE LAMPS

High-intensity discharge (HID) lamps (Fig. 12.22) produce light by discharging electricity through a high-pressure vapor. Lamps in this category include mercury vapor (CRI range 15–55), metal halide (CRI range 65–80), and high-pressure sodium (CRI range 22–75). These lamps are characterized by high efficacy, rapid warm-up time, rapid restrike time, and historically poor color rendering capabilities. Mercury vapor lamps were the first commercially available HID lamps and originally produced a bluish-green light. Today they are available with a color-corrected whiter light, but because of inefficiency and potential hazards they are being replaced by the newer, more efficient metal halide and high-pressure sodium lamps. Standard high-pressure sodium lamps have the highest efficacy of all HID lamps, but they produce a yellowish light. High-pressure sodium lamps that produce a whiter light are now available, but their efficiency is somewhat lower.

HID lamps are typically used when high illuminance is required over large areas and when energy efficiency and/or long life are desired. Typical applications include gymnasiums, large public areas, warehouses, outdoor activity areas, roadways, parking lots, and pathways.

## 12.19 MERCURY VAPOR LAMPS

The mercury vapor lamp was the first HID lamp to be developed, and for many years it was the only HID lamp commercially available. It has been largely supplanted by the metal halide lamp because of the latter's better color and efficacy. Though many installations of mercury vapor exist, this technology is no longer specified for new buildings. For the same energy efficiency, metal halide lamps have better color rendering properties and are now often preferred for indoor applications. Mercury vapor lamps, most often used to light streets, gymnasiums, and sports arenas, must be maintained properly to be safe.

The mercury vapor lamp operates by passing an arc through *high-pressure* mercury vapor contained in a quartz arc tube (Fig. 12.23). This produces

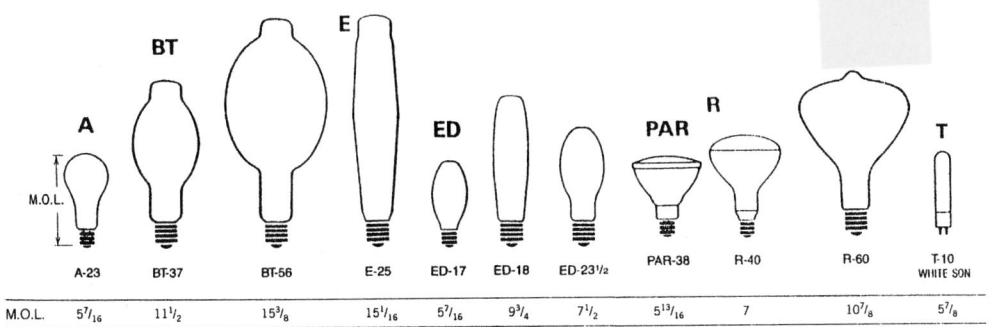

*Fig. 12.22* Bulb shapes for most HID lamps, with their maximum overall length (M.O.L.).

radiation in both the UV region (as in the low-pressure fluorescent lamp tube) and the visible region, principally in the blue–green band. This color is characteristic of a clear mercury lamp.

### (a) UV Radiation

A considerable portion of a mercury lamp's energy spectrum is in the UV range. This does not normally constitute a hazard to persons exposed to even a clear lamp because the outer glass bulb absorbs most of the UV radiation while transmitting light. However, if the outer bulb is broken, the quartz arc tube will continue to burn and the UV radiation emitted constitutes a safety hazard, particularly to the skin and eyes of persons exposed to it. As a result, manufacturers include a warning to this effect with all mercury lamps sold. Users are also generally informed that a safety-type mercury lamp is readily available that will self-extinguish if the outer glass envelope is broken. An alternative is to use mercury vapor lamps in an enclosing fixture designed to both prevent lamp breakage from external sources, such as vandalism, and protect users of the lamp in the (unlikely) event of a spontaneous lamp fracture. In the interest of safety, it is suggested that mercury lamps be shut off at least once a week for at least 30 minutes to allow them to cool completely.

In connection with protection from the injurious effects of UV radiation, it is well to note two facts:

1. The shorter the UV wavelength, the more potentially irritating it is to the skin and eyes. Germicidal UV radiation is in the short-wave range (200 to 300 nm).
2. White plaster and polished metal are good reflectors of UV radiation. As a result, UV radiation reflected from such surfaces is almost as dangerous as that from line-of-sight exposure to a UV source.

### (b) Lamp Life

Lamp life is extremely long, averaging 24,000 hours or more based upon 10 burning hours per start. Mercury vapor lamps are not suitable for applications that are subject to constant switching. Their life is affected by ambient temperature, line voltage, and ballast design.

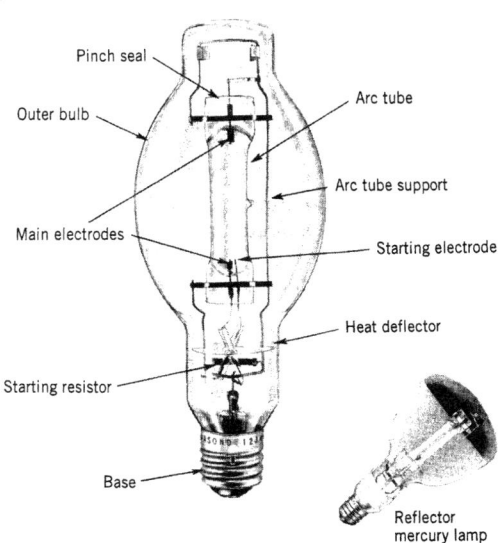

*Fig. 12.23* Typical construction of a clear mercury vapor lamp.

## (c) Lumen Maintenance

This depends upon the specific type of lamp and its burning position. Manufacturers publish data for *each* of their lamp types. In general, clear lamps have the best lumen maintenance, followed by color-improved and phosphor-coated units.

## (d) Color Correction and Efficacy

Color correction is normally required because the blue–green light from a clear lamp distorts almost all object colors. (Mercury vapor lamps are frequently used to illuminate outdoor gardens because the blue–green light enhances the green of trees and vegetation.) Color correction is achieved by adding phosphor to the inside of the outer bulb. The phosphors convert UV radiation to light exactly as in fluorescent lamps, and the stain on the glass acts as a filter to some of the blue–green radiation. The phosphors reradiate generally in the red band, which is entirely absent in the basic lamp color. Depending upon the arc tube design and the phosphors used, the color of the emitted light can be corrected to make it acceptable for general indoor use. Lamps are available in clear, white, color-corrected, and deluxe white in ascending order of color improvement. Efficacy, including ballast loss, ranges from 25 lm/W for a 50-W lamp to a maximum of 55 lm/W for a 1000-W color-corrected lamp. Note that, in general, efficacy is lower than for fluorescent lamps. CRI ranges from a low of 20 for a clear lamp to a high of only 50 for a deluxe white lamp. A short list of representative lamp data is given in Table 12.8.

## (e) Ballasts and Lamp Starting

Ballasts are required, as with all arc discharge lamps, to start a mercury vapor lamp and thereafter to control the arc. From 3 to 6 minutes are required for the lamp to reach full output because heat must be generated by electron flow to vaporize the mercury in the arc tube before the arc will strike. Once extinguished, the lamp must cool before restrike is possible. This restart delay amounts to 3 to 8 minutes, depending upon the ballast type, and is an important consideration in design, as a momentary power outage will extinguish all lamps, leaving an

**TABLE 12.8 Typical Data for Mercury Vapor Lamps**

| Watts | Type[a] | Bulb[b] | Initial Lumens | Efficacy[c] (lm/W) |
|---|---|---|---|---|
| 100 | DX | A-23 | 4,300 | 39 |
| 100 | DX | ED-23 ½ | 4,400 | 40 |
| 100 | DX | R-40 | 2,800 | 25 |
| 250 | DX | ED-28 | 13,000 | 47 |
| 250 | DX/SB | E-28 | 6,000 | 24 |
| 400 | DX | R-57 | 23,000 | 54 |
| 450 | W/SB | BT-37 | 9,700 | 28 |
| 700 | DX | BT-46 | 43,000 | 55 |
| 750 | W/SB | R-57 | 14,000 | 19 |

*Note:* All lamps using an external ballast have a life of 24,000+ hours, based upon 10 hours burning per start. Life of self-ballasted lamps: 250 W—12,000 hours; 400 W and 700 W—16,000 hours.

All deluxe lamps with external ballast have a CRI of 45 and a CCT of 3700 K. All self-ballasted lamps have a CRI of 50 and a CCT of 3300 K.

[a]Type abbreviations: DX = deluxe; W = white; SB = self-ballasted.

[b]For bulb shape and dimensions, see Fig. 12.22.

[c]Efficacy includes an estimated loss in a magnetic ballast. For self-ballasted lamps, efficacy is as shown.

interior area in the dark. Mercury vapor lamp luminaires are available that utilize small halogen lamps to supply light during such outages. Alternately, some incandescent lighting can be utilized to maintain minimum illumination.

The principal mercury vapor ballast types are reactor, regulating, and electronic. Magnetic mercury ballasts are large, heavy, and quite noisy. Where this may be a problem, remote ballast mounting should be considered or lighter, quieter, and more expensive electronic ballasts used. Because lamp-operating characteristics depend heavily upon the type of ballast and because the choice of an appropriate ballast involves highly technical electrical considerations, selection should be left to an electrical or lighting consultant.

## (f) Self-Ballasted Lamps

These have been available for some years and consist of a screw-base color-corrected lamp with an internal resistive/reactive ballast. They have a CRI of 50, an efficacy of 20 to 25 lm/W, and a correlated color temperature (CCT) of 3500 K to 4000 K. Their great advantage is their long life, which can be used to advantage in applications involving burning periods of 8 to 10 hours minimum, relative inaccessibility, limited space that precludes ballast installation, and indifferent color requirements.

## (g) Application

Mercury vapor lamps are applicable to indoor and outdoor use with proper attention to color and fixture luminance. The most common exterior application is for parking lots. Indoor application is generally limited to mounting heights of 10 ft AFF (above finish floor) or higher to avoid direct glare and permit adequate area coverage. Their use in industrial spaces and stores was once common, but today retrofitting with metal halide lamps is typical. Warehouses and non-color-sensitive industrial areas continue to use mercury vapor lamps.

## 12.20 METAL HALIDE LAMPS

This lamp began its life in the early 1960s as a modified mercury vapor lamp. Major advances in miniaturization, color rendering, color temperature, and consistency—by the addition of halides such as thallium, indium, and sodium to the arc tube—resulted in changes in the output, efficacy, color, and life of the lamp. Metal halide lamps have excellent color characteristics and therefore almost unlimited applicability. The number of types and sizes is so large, and changes so rapidly, that any abbreviated tabulation would be inadequate at best and misleading at worst. As with all lamps, but more so with lamps undergoing intensive development, a current manufacturer's catalog should be consulted for accurate lamp data. Pulse-start metal halide lamps utilize a glass arc tube to contain the arc. Pulse-start technology includes a new family of lamps (it has been used with high-pressure sodium lamps), and improves the start system, efficacy, and lumen maintenance and yields faster warmup and restrike.

Ceramic metal halide lamps were introduced to the market a number of years ago and have become an industry standard, offering a high CRI of 80–90, a color temperature of 3000 K or 4100 K, improved lumen maintenance, and stable color consistency. Typical metal halide lamp characteristics and types are discussed in the following subsections.

## (a) Lamp Configurations

Construction details for a basic metal halide lamp are shown in Fig. 12.24 and are similar to those of its "parent" mercury vapor lamp, illustrated in Fig. 12.23. In addition to this design (in a BT-shaped bulb envelope), metal halide lamps are manufactured in elliptical bulbs, PAR reflector lamps, and single- and double-ended tubular shapes (Figs. 12.22, 12.24 to 12.26).

## (b) Safety

Being essentially a modified-vapor mercury lamp, the metal halide lamp carries the same safety warning as mercury lamps. An additional warning, however, refers to the fact that metal halide lamp arc tubes have a tendency to explode; therefore, the lamp must be used in an approved enclosing luminaire. All major manufacturers also make lamps with internal shields that will contain the flying pieces of a ruptured arc-tube without damaging the outer bulb. Such lamps may be used in open lighting fixtures. One such lamp is illustrated in Fig. 12.26.

As with mercury vapor lamps, manufacturers also produce a line of safety metal halide lamps that self-extinguish within 15 minutes of an outer bulb fracture, thus limiting exposure to harmful UV radiation. Both of these safety designs are clearly noted

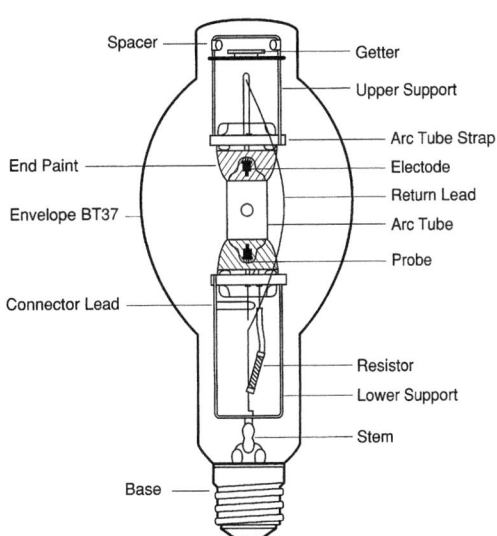

**Fig. 12.24** *Construction details of a 400-W standard metal halide lamp, which can be mounted either horizontally or vertically. (Courtesy of Osram-Sylvania.)*

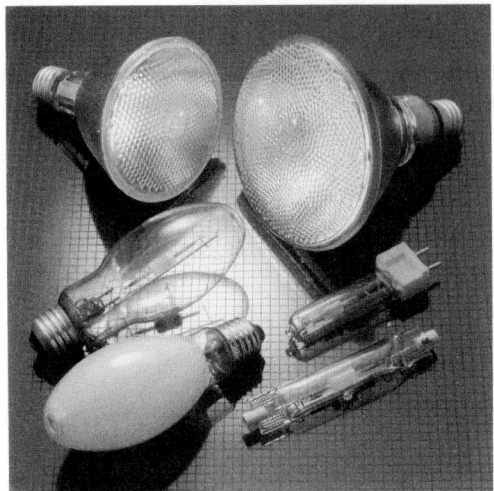

**Fig. 12.25** *Various configurations of metal halide lamps. Clockwise from the bottom left: phosphor-coated and clear elliptical bulbs; PAR 30 and 38 reflector lamps; single-ended and double-ended tubular lamps—all have ceramic arc tubes and a CRI greater than 80. (Courtesy of GE Lighting.)*

on the lamps by trade name and sometimes by description.

## (c) Designs, Shapes, and Ratings

The metal halide lamp designs available as of this writing are:

- Standard lamps, available in ED, BT, and PAR shapes, in wattages from 50 to 1500 W, efficacies of 75 to 105 lumens per watt with magnetic ballasts, and slightly higher with electronic ballasts

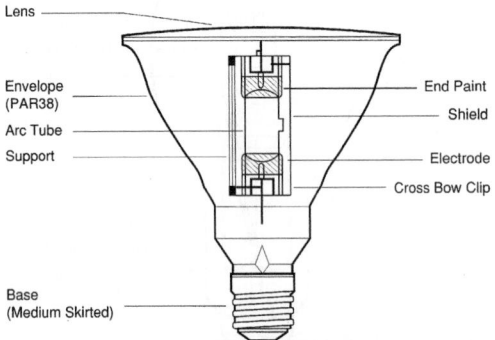

**Fig. 12.26** *Construction details of a PAR enclosure for a metal halide lamp. The lamp itself is tubular and constructed with a surrounding protective shield designed to contain arc tube fragments in the event of a violent rupture. (Courtesy of Osram-Sylvania.)*

(efficacy also increases with wattage). Lamps are clear or phosphor-coated, with a CCT ranging from 3000 K to 4200 K and a CRI ranging from 65 to 85. Their life, at 10 burning hours per start, varies from 10,000 to 20,000 hours depending upon lamp type, size, and burning position.

- Safety-shielded lamps with integral shields to contain a ruptured arc tube.
- Self-extinguishing lamps designed to shut down automatically upon a break in the outer glass envelope.
- High-output lamps designed for a specific burning position (that is specified on each lamp). Output is 5% to 8% higher than that of standard lamps, but the color is somewhat poorer, with a CRI range of 65 to 70.
- Single-ended and double-ended tubular lamps. These lamps are characterized by a very high CRI (80–93); a somewhat shorter life than standard lamps, particularly for the single-ended units (6000–10,000 hours); and slightly lower efficacy (70–85 lm/W). They are intended for applications requiring very high color rendering.

## (d) Operating Characteristics

Like mercury vapor lamps, metal halide lamps are not instant starting, requiring approximately 2 to 3 minutes on initial startup and 8 to 10 minutes for restrike. (Tubular lamps require only about half of these times.) As a result, when they are used for indoor installations, a secondary instant-start source must be available. A number of manufacturers produce special hot-restrike ballasts that provide immediate restrike of lamps on restoration of power after an outage. Lamp output on restrike is inversely proportional to the duration of a power outage.

It is important to note that the spectrum of light produced by a metal halide lamp *changes* as a lamp ages. The change is gradual, definite, but usually unnoticed and depends upon the particular design of lamp. Where color rendering is important, or where the lamp is used with other light sources, a designer should choose metal halide lamps that are specially made for color stability. These lamps are designed not to vary in CCT more than 200 K over the lamp life. Finally, dimming or reduced output operation of metal halide lamps is not normally recommended because of the very noticeable color shift that occurs when a lamp is dimmed.

## (e) Lamp Ballasts

Metal halide lamps operate satisfactorily on a simple reactor ballast, although a separate ignitor is usually required to start the lamp. These ballasts have a low power factor, of about 50%, which is undesirable from the perspective of energy conservation, wiring economy, electrical losses, and component heating. High-power factor magnetic ballasts (pf > 90%) are also available. Magnetic ballasts are large, heavy, and tend to be noisy. The last characteristic can be improved by the use of a potted (epoxy-filled) ballast. Electronic ballasts are also readily available, along with dimming and multilevel ballasts. However, as noted previously, dimming ballasts are not frequently used because of the large shift in lamp color that they cause.

## 12.21 SODIUM VAPOR LAMPS

The highest-efficacy general-purpose HID source available is the high-pressure sodium lamp (HPS). The basic construction of this type of lamp is illustrated in Fig. 12.27, which shows schematic drawings of the design. Typical performance data for various types of HPS lamps are given in Table 12.9. Construction is quite different from that of mercury vapor and metal halide lamps. The characteristic color of HPS lamps stems from the spectral absorption phenomenon of the sodium contained in the arc tube—with a resultant pronounced yellow tinted light.

### (a) Primary Characteristics of HPS Lamps

*Standard HPS Lamps.* The extremely low CRI (20–22) is not acceptable where any degree of color rendition is required, thus limiting the standard HPS lamp to exterior areas and road lighting.

*Color-Corrected HPS Lamps.* Color can be improved considerably by increasing the pressure inside the arc tube. This causes some of the sodium in the arc to be reabsorbed, and the radiated light widens its spectrum into the red range (at the expense of efficacy and lamp life).

*"White" HPS Lamps.* A still greater increase in lamp pressure improves lamp color and yields a

**TABLE 12.9 Typical Data for Clear[a] High-Pressure Sodium Lamps**

| Watts | Bulb[b] | Life (h) | Initial Lumens[c] | Lamp Efficacy (lm/W)[d] |
|---|---|---|---|---|
| "WHITE" LAMPS: CRI: 85, CCT: 2700 K | | | | |
| 35 | T10 | 10,000 | 1,250 | 36 |
| 50 | ED-17 | 10,000 | 2,000 | 40 |
| 100 | ED-17 | 10,000 | 4,200 | 42 |
| COLOR-CORRECTED LAMPS: CRI: 60, CCT: 2200–2300 K | | | | |
| 100 | ED-17 | 15,000 | 7,300 | 73 |
| 250 | ED-18 | 15,000 | 22,000 | 88 |
| 400 | ED-18 | 15,000 | 37,000 | 93 |
| STANDARD HPS LAMPS: CRI: 22, CCT: 1900–2100 K | | | | |
| 50 | ED-17 | 24,000+ | 4,000 | 80 |
| 100 | ED-17 | 24,000+ | 9,500 | 95 |
| 250 | ET-18 | 24,000+ | 28,000 | 112 |
| 400 | ET-18 | 24,000+ | 48,000 | 120 |
| 750 | BT-37 | 24,000+ | 110,000 | 147 |
| 1,000 | E-25 | 24,000+ | 133,000 | 133 |

*Note:* Data extracted from current manufacturers' catalogs.

[a]Data are identical for coated lamps except for bulb shapes and lumen output.

[b]Other bulb shapes are available in some sizes. See Fig. 12.22 for bulb data.

[c]Initial lumens for coated lamps are 6% to 9% lower.

[d]Based on initial lumens. Efficacy with ballast is approximately 10% lower.

"whiter" lamp color in a limited wattage range. These low-wattage, reduced-efficacy, shortened-life lamps are normally operated with small, lightweight electronic ballasts.

Because of its high output and narrow linear arc tube, any HPS lamp is a potential glare source, and wattages of 150 W and higher must be either completely shielded or mounted at sufficient height (if in an open reflector) to be above the near field of vision. These glare-prevention strategies also apply to metal halide lamps.

If a diffusing coating is added to a sodium vapor lamp, the entire glass envelope becomes the light-emitting source. This reduces lamp luminance, and therefore glare potential, drastically but also reduces output (and efficacy) by 6% to 8%. Light distribution from a coated lamp in an open reflector is vastly improved, as can be seen in Fig. 12.28.

### (b) Other Operating Characteristics

In contrast to both mercury vapor and metal halide lamps, HPS lamps do not emit any appreciable UV radiation, do not tend to rupture violently, and can

**ILLUMINATION**

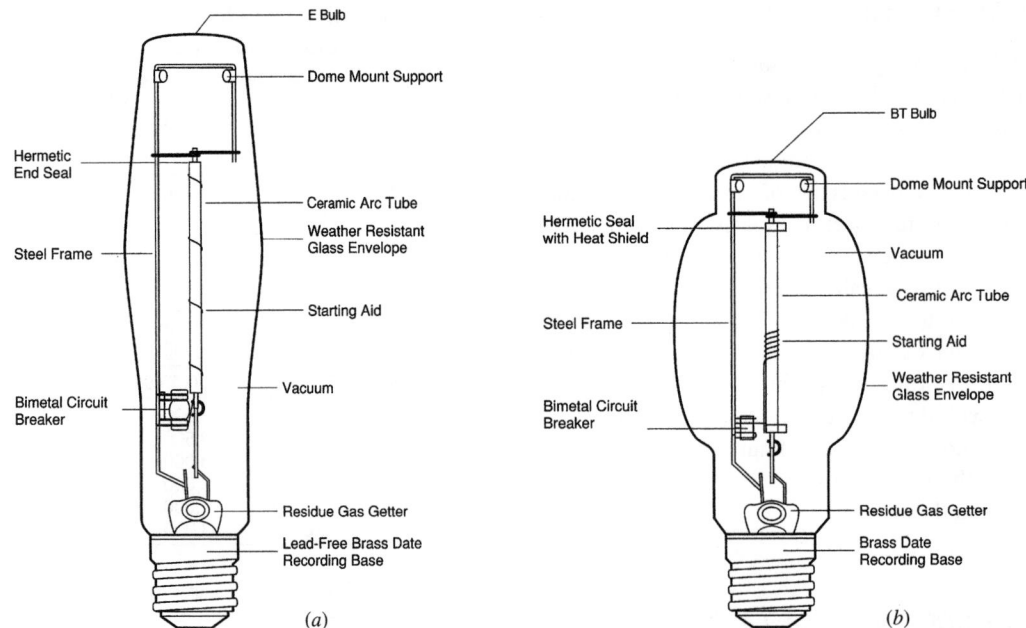

**Fig. 12.27** *Simplified drawings of the internal construction of two high-pressure sodium (SON) lamp designs. (a) Noncycling lamp (in an E-shaped glass bulb) designed to indicate by a special color when the lamp has reached the replacement stage. Unlike standard HPS lamps, this lamp will not cycle on and off at the end of its useful life. (b) Retrofit HPS (SON) lamp in a BT bulb, intended for direct replacement of an existing mercury lamp. This lamp operates efficiently on a mercury lamp ballast. (Drawings courtesy of Osram-Sylvania.)*

be installed any position without affecting operating characteristics.

Like all discharge lamps, HPS lamps require a ballast for ignition and arc control. Due to the extremely high voltage required for lamp ignition, both magnetic and electronic ballasts contain an electronic ignition circuit. Because of this, an HPS lamp *must* be used with a compatible ballast that carries the same ANSI designation as found on the outer glass bulb. If it is used with an incompatible ballast (and fixture), the lamp may rupture and constitute a serious safety hazard. As with metal halide lamps, ballasts are available for HPS lamps that provide instantaneous restrike after a power interruption. Light output on restrike is inversely proportional to the length of the outage; for example, after a 10-second outage, restrike light output will be 85% of full capacity, whereas after a 2-minute outage, lamp output will be only 10%. It then takes approximately another minute to regain full output.

### (c) Lamp Design Types

In addition to standard HPS lamps, including the color-corrected types, three additional special lamps

have been developed to solve specific problems. They are:

1. *Noncycling lamp.* As an HPS lamp ages, its arc voltage rises. Eventually, the ballast is unable to sustain the arc and the lamp extinguishes. After cooling, the lamp lights to full brightness and soon thereafter extinguishes again. This on–off cycling is characteristic of an HPS lamp at the end of its life. To eliminate this condition, a special noncycling lamp was developed that uses very little sodium amalgam in the arc tube and is more environmentally friendly because of this reduced content. To enhance this environmental aspect, these lamps are made with a lead-free base and lead-free solder. Photometric characteristics are similar to those of standard lamps. The E-shaped lamp in Figs. 12.27 and 12.28 is of this design.

2. *Standby lamps.* As noted previously, HPS lamps require a minute or more to restrike after being extinguished. A crowded public area plunged into complete darkness is a recipe for disaster; therefore, such areas must always be furnished with instant-on emergency lighting. Standby

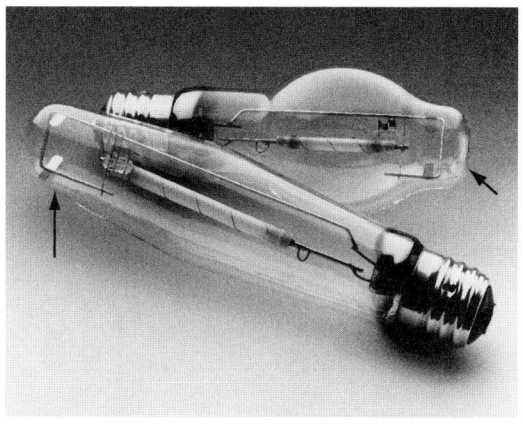

(a)

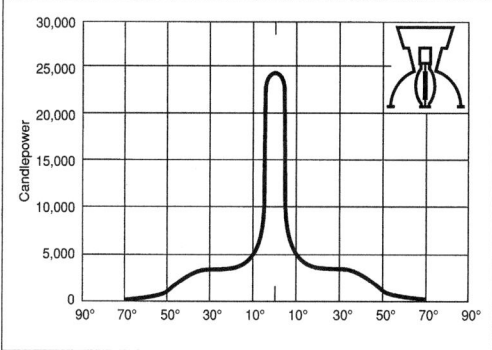

Clear HPS lamp operating at 12,000 lumens in open-bottomed fixture.

(b)

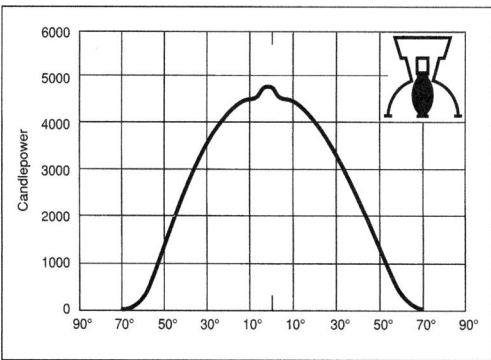

Coated HPS lamp operating at 12,000 lumens in open-bottomed fixture.

(c)

***Fig. 12.28*** *(a) HPS retrofit and noncycling lamps. (b) A narrow linear arc-tube of an HPS lamp is not suitable for use in an open reflector designed for a larger source such as a phosphor-coated mercury lamp. (c) A coated HPS lamp creates a large, diffuse source with a highly improved light distribution. (Drawings courtesy of Osram-Sylvania.)*

lamps have two arc tubes. When the lighted one is extinguished due to a momentary power loss, the second (cool) arc tube immediately begins to glow—assuming that voltage has returned. This arrangement *may* be acceptable to some code authorities as a substitute for instant-on replacement lighting following a momentary power outage. It is *not* acceptable as emergency lighting as required by NFPA 101: Life Safety Code.

3. *Retrofit lamps.* These are HPS lamps designed as a direct replacement for mercury vapor lamps of the same wattage. They are enclosed in BT-shaped envelopes of the same size as the lamp being replaced, and they operate properly on mercury vapor lamp ballasts. A retrofit lamp is illustrated in Fig. 12.28.

## 12.22 LOW-PRESSURE SODIUM LAMPS

The low-pressure sodium (SOX) lamp produces light characteristic of sodium's monochromatic saturated yellow color, making it inapplicable for general lighting. Because of its very high efficacy of more than 150 lm/W, *including* ballast loss (but with a CRI of 0), it can be applied wherever color rendition is not an important criterion but energy efficiency is. Thus, SOX lamps are used for street, road, parking lot, and pathway lighting. SOX lamps are used around astronomical observatories because the yellow light can be filtered out of a telescope. Another desirable aspect of SOX lamps is their 100% lumen maintenance. This, coupled with a discharge lamp's typically long life (18,000+ hours), makes SOX lamps fairly economical in terms of life-cycle cost.

## OTHER ELECTRIC LAMPS

## 12.23 INDUCTION LAMPS

The induction lamp is filled with low-pressure mercury vapor. When ionized by a high-frequency induction coil inside the lamp, the mercury vapor produces UV radiation, which then strikes a phosphor coating on the inside of the lamp, producing

ILLUMINATION

light. Similar to the light-producing process used by standard fluorescent lamps, the difference is that the gas is ionized by an induction coil (rather than an electron stream)—thus the name *induction lamp.* Two such designs are shown in Figs. 12.29 and 12.30.

Characteristics of the design shown in Fig. 12.29 are as follows:

| | |
|---|---|
| Wattage | 85 W total, including external devices |
| Initial lumens | 6000± |
| System efficacy | 70 lm/W |
| CCT | 3000 K or 4000 K |
| CRI | 80+ |
| Ignition time | Under ½ second |
| Time to 75% output | Up to 1 minute |
| Hot restrike time after outage | Less than ½ second |
| Life (50% survival) | 100,000 hours |
| Lumen maintenance | 70% at 60,000 hours |
| Burning position | Any |

The extraordinarily long life, excellent lumen maintenance, and instantaneous restrike time make the induction lamp suitable for illuminating public areas. As with all light sources, a comparison with other sources can only be made for a given proposed usage based on project-specific considerations.

A self-contained induction lamp of lower wattage is illustrated in Fig. 12.30. This lamp is built into

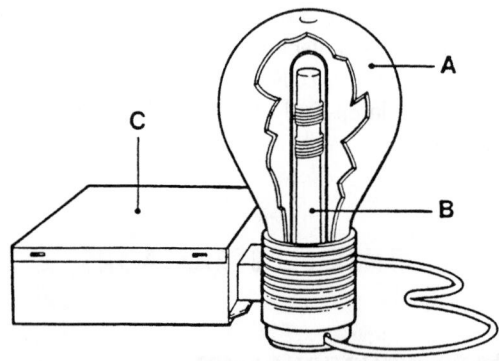

**Fig. 12.29** *Schematic diagram of an induction lamp (rated 85 W, including all losses) shows its operating principles. The lamp is 4.3 in. (109 mm) in diameter and 7.5 in. (191 mm) high overall. The high-frequency generator (C) produces a high-frequency current, which circulates in the coil on the power coupler (B). This ionizes the mercury vapor inside the lamp (A), producing UV radiation. The UV radiation strikes the fluorescent coating inside the lamp, producing light. (Illustration courtesy of Philips Lighting Company.)*

a modified R-shaped envelope with a standard Edison base, and is therefore readily applied as a direct replacement for a 100-W incandescent reflector lamp. The lamp's published performance figures are:

| | |
|---|---|
| Total wattage | 23 W |
| Initial lumens | 1100 |
| Efficacy | 48 lm/W |
| CCT | 3000 K |
| CRI | 82 |
| Life | 10,000 hours |
| Lumen maintenance at 70% life | 75% |

## 12.24 LIGHT-EMITTING DIODES

Light-emitting diodes (LEDs) (or light-emitting semiconductors) have a wide range of applications, such as in lamps (medical instruments, bar coders, fiber optic communication), mobile technologies (phones, digital cameras, laptops), consumer appliances, automotive (instrument panels, signal lights, courtesy lighting), and signals (traffic, rail, aviation). LED uses in architectural illumination include signage, retail displays, emergency lighting (exit and emergency signs), and accent lights for pathways.

LEDs are available in a full range of colors, are small (⅛ in. [3mm] across), have a very low light output, use very little power, and have a fast response. They emit light in proportion to the forward current through the diode. The color of the light emitted by a LED depends upon the band gap material of the semiconductor, such as gallium arsenide phosphide (GaAsP) or gallium phosphide (GaP). GaAsP emits from red to amber, depending upon the concentration of phosphorus. GaP emits green light and can emit red light.

## 12.25 SULFUR LAMPS

The principle of microwave energy excitation has been applied successfully to another recently developed lamp that consists of a golf ball–sized globe filled with an inert gas and a few milligrams of sulfur. In contrast to the induction lamp described previously, no mercury vapor is used in this lamp, and

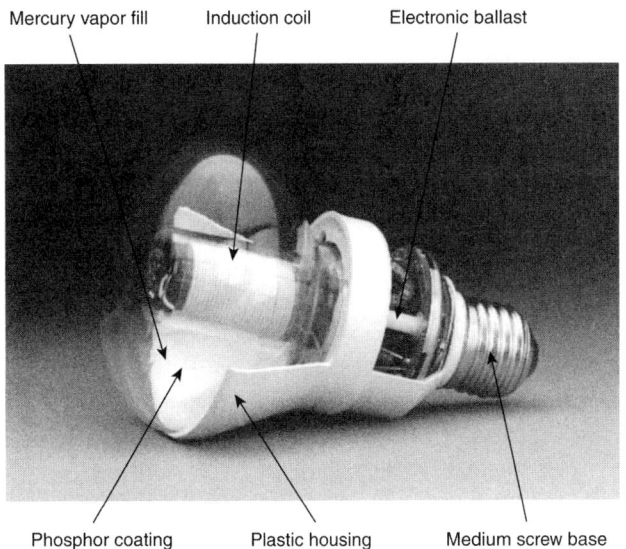

Mercury vapor fill    Induction coil    Electronic ballast

Phosphor coating    Plastic housing    Medium screw base

**Fig. 12.30** *Cutaway of the GE induction lamp showing the essential elements: induction coil, phosphor-coated bulb with mercury-vapor fill, and electronic ballast. This 23-W lamp in a modified R (reflector)-shaped bulb has a height just under 5 in. (127 mm) overall and a 3-in. (76-mm) maximum bulb diameter. (Photo courtesy of GE Lighting.)*

the radiation is full-spectrum light with very little UV or infrared radiation. The originally developed lamp was a 6-kW unit that emitted more than 400,000 lumens. The most recent (1998) version of the lamp has the following characteristics:

| | |
|---|---|
| Total power input | 1320 W |
| Initial lumens | 130,000 |
| Efficacy | 101 lm/W |
| Diameter of globe | 29 mm (1.1 in.) |
| CCT | 5600 K |
| CRI | 80 |
| Life of lamp (estimated) | 60,000 hours |
| Life of exciting magnetron | 15,000 hours |
| Lumen maintenance | Very high |
| Color constancy with aging | Excellent |

The current sulfur lamp technology provides approximately the same output as a 1000-W HPS lamp but with much better color characteristics. It also compares to about 1200 W of metal halide lamp(s) with similar efficacy, but again, because of its full-spectrum radiation, it has superior color characteristics. Initial trials of the lamp as a driver for a "light pipe" have been highly successful. Because the lamp is still in development at this time, its eventual commercial usage is difficult to predict.

## 12.26 FIBER OPTICS

Although optical fibers have been available since the 1920s, practical applications, in the medical field, were not developed until the late 1950s and early 1960s. Bundled fibers used as a diagnostic tool can deliver light to remote regions of the body and carry coherent (understandable) images back to a doctor. In recent years, fiber optics have made their most significant advances in the communications field: long distance telephone cables or thousands of paired wires have been replaced by a single-fiber cable. Architectural applications of fiber optics in recent years have included alternatives to directly replace recessed ceiling downlights, track and display case lighting in museums, pools/spas, supermarkets, and other commercial buildings.

Illuminators for fiber optic systems utilize a variety of lamp types. The primary advantages of using a fiber optic system are that no heat is produced where the light exits the fiber and no UV radiation is transmitted through the fiber.

### References

ANSI. 2004. ANSI C82.11: *Lamp Ballasts—High Frequency Fluorescent Lamp Ballasts—Supplements.* American National Standards Institute, New York.

ILLUMINATION

ANSI. 2002. ANSI C78.1451: *Electric Lamps—Use of Protective Shields with Tungsten-Halogen Lamps—Cautionary Notice.* American National Standards Institute, New York.

CIE. 1970. *Daylight, International Recommendations for the Calculation of Natural Daylight* (Publication No. 16 E-3.2). Commission Internationale de l'Eclairage, Paris.

Eley Associates. 1993. *Advanced Lighting Guidelines: 1993* (EPRI TR-101022s Rev. 1). Electric Power Research Institute, Palo Alto, CA.

IESNA. 2000. *IESNA Lighting Handbook,* 9th ed. Illuminating Engineering Society of North America, New York.

Krochman, J. 1963. "Uber die horizontal Beleuchtungsstarke der Tagesbeleuchtung." Lichtechnik, 15(11).

NFPA. 2003. NFPA 101: *Life Safety Code.* National Fire Protection Association, Quincy, MA.

Rennhackkamp, W.M.H. 1967. "Sky Luminance Distribution on Warm Arid Climates." *Proceedings 16th International Conference on Illumination,* Washington, DC.

CHAPTER **13**

C H A P T E R

# Lighting Design Process

## 13.1 GENERAL INFORMATION

LIGHTING DESIGN IS A COMBINATION OF APPLIED ART
and applied science. There can be many solutions to
the same lighting problem, all of which will satisfy
the minimum requirements, yet some will be dull
and pedestrian, whereas others will display ingenuity and resourcefulness. The competent lighting
designer approaches each problem afresh, bringing
to it knowledge of current technology and years of
background and experience, yet rarely being satisfied with a carbon copy of a previous design. It is
these years of background with their successful and
not so successful designs, coupled with a constant
striving for improvements, that are the characteristics differentiating the competent lighting consultant, designer, or engineer from the person who
attempts to force each new job into the unwilling
mold of a previous design.

Because of the large number of interrelated
factors in lighting, no single design is the correct
one; for this very reason, it is not entirely desirable
to solve a lighting problem with a step-by-step technique. However, since experience has shown this
technique to be a good approach for the uninitiated
who lack the experience necessary to view an entire
solution, we have adopted it.

## 13.2 GOALS OF A LIGHTING DESIGN

Simply stated, the goal of lighting is to create an
efficient and pleasing interior. These two requirements—that is, utilitarian and aesthetic—are not
antithetical, as is demonstrated by every good lighting design. Light can and should be used as a primary architectural material.

1. Lighting levels should be adequate for efficiently
   seeing the particular task involved. Variations
   within acceptable luminance ratios in a given
   field of view are desirable to avoid monotony
   and to create perspective effects.
2. Lighting equipment should be unobtrusive but
   not necessarily invisible. Fixtures (luminaires)
   can be chosen and arranged in various ways to
   complement the architecture or to create dominant or minor architectural features or patterns. Fixtures may also be decorative and thus
   enhance the interior design.
3. Lighting must have the proper quality, as discussed previously. Accent lighting, directional
   lighting, and other highlighting techniques increase the utilitarian as well as the architectural quality of a space.
4. The entire lighting design must be accomplished efficiently in terms of capital and energy
   resources, the former determined principally by
   life-cycle costs and the latter by operating-
   energy costs and resource-energy usage. Both
   the capital and energy limitations are, to a large
   extent, outside the control of the designer, who
   works within constraints in these areas. Obviously, these constraints are maxima.

With these goals before us, we can write a
lighting design procedure, keeping in mind that the

order of steps shown is not necessarily the same in each lighting problem and that since all factors are closely interrelated, it is often necessary to address several of the stages simultaneously before arriving at a decision.

It is appropriate to note at this point that the lighting design approach and procedure that we explain in this chapter is primarily an analytic one; that is, the design procedure establishes requirements primarily in numerical form and then manipulates the variables of sources, fixtures, placement of units, and so on, to arrive at a design conclusion. There exists an alternative approach, frequently referred to as *brightness design*, in which the designer labels surfaces on a perspective plan of a space with desired luminances as established by a mental picture (or by other means) and designs the lighting accordingly. This approach, which can be very effective, is highly intuitive (i.e., requires considerable prior experience on the part of the designer). For this reason, and because that approach requires a good deal of hands-on, field trial-and-error work, we feel that it is not appropriate to a textbook and therefore is not used here. The only exception to this statement occurs in our discussion of daylight models in Chapter 12. There the entire purpose of the model is to give the designer a visual/mental picture of the space's brightness patterns on the basis of which windows, window treatments, and possible exposures can be varied to achieve the desired light and shadow patterns (luminances and luminance ratios).

## 13.3 LIGHTING DESIGN PROCEDURE

### (a) Project Constraints

The flow chart in Fig. 13.1, which represents the design procedure and its interactions, should be referred to throughout the necessarily lengthy discussion that follows in order to maintain perspective. It is important that the reader be aware of job constraints and of the interactions between the lighting designer and the remainder of the design group. We deliberately emphasize this to demonstrate the interdisciplinary nature of lighting design in general and its connection with HVAC and daylight (fenestration) in particular. This approach, which is most often referred to as the *systems design approach*, is followed throughout the discussion.

Item 4 of the list in Section 13.2 referred to constraints. These can be related to the owner-designer-user team and/or the jurisdictional authorities. In some detail, these are:

1. *Owner-designer-user group.* The owner establishes the cost framework, both initial and operating. A part of both of these may be a rent structure, which in turn determines and is determined by the space usage. If the owner is also the occupant, the cost factors change somewhat but remain in force. The architect determines the amount and quality of daylighting and the architectural nature of the space to be lighted. Many of these data are detailed in the building program. Obviously, the architect and lighting designer (who may be the same person) should interact in this aspect of building design.

2. The jurisdictional authorities *may* include:

   DOE—U.S. Department of Energy

   GSA—General Services Administration

   NFPA—National Fire Protection Association (codes)

   ASHRAE—American Society of Heating, Refrigerating and Air-Conditioning Engineers

   IESNA—Illuminating Engineering Society of North America

   NIST—National Institute of Science and Technology (formerly National Bureau of Standards)

Most of these are jurisdictional by reference; that is, the actual authorities may specify that the lighting system meet the requirements of ASHRAE, IESNA, and so on. If federal funds are involved, DOE/GSA standards will probably be involved. The principal areas of involvement are energy budgets and lighting levels, both of which affect every aspect of lighting design including source type, fixture selection, lighting system, fixture placement, and even maintenance schedules. For this reason, the first step in the lighting design procedure is to establish the *project lighting cost framework and the project energy budget.*

### (b) Task Analysis

As shown in Fig. 13.1, this step essentially determines the needs of the task. Factors to be considered in addition to the nature of the task are its repetitiveness, variability, who is performing the task (i.e.,

Lighting Design Procedure

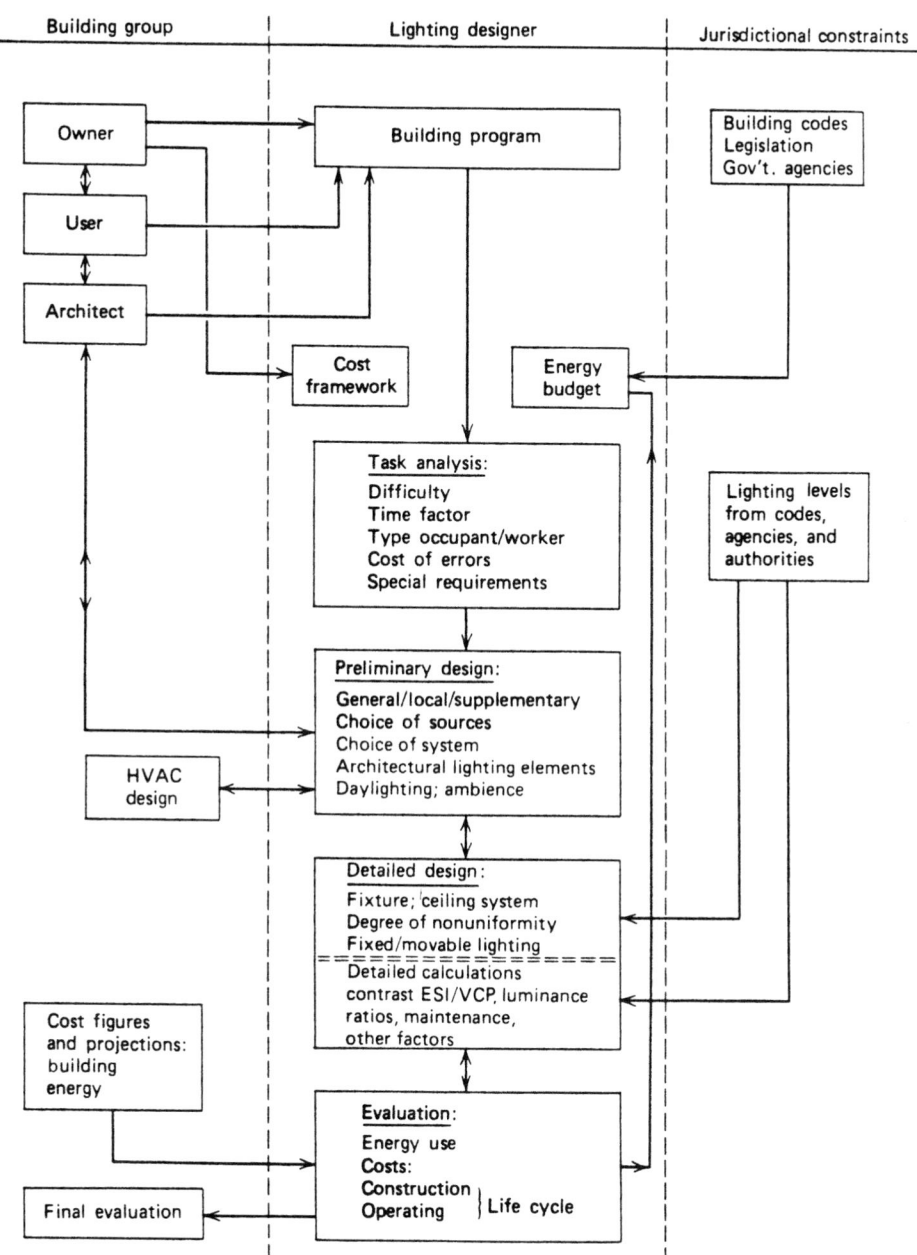

**Fig. 13.1** *Lighting design procedure chart.*

physical condition of the occupant), task duration, cost of errors, and special requirements. Several of these factors have been discussed in the preceding sections dealing with quality of light. These are cross-referenced in the appropriate sections in the following analysis.

**(c) Design Stage**

This is the active consideration stage during which detailed suggestions are raised, considered, modified, accepted, or rejected. This is also the most interactive stage, as is clearly seen in Fig. 13.1. At its completion, a detailed, workable design is in hand. The critical interactions here are with the architect in daylighting and with the HVAC group in power loads. The former may result in relocating a space within the building, the latter in making a change in a lighting system or HVAC system. In brief, this stage consists of the following steps:

1. Select the lighting system. Select the type of light source and the distribution characteristic of fixture(s) or the area source and consider the effects of daylighting, economics, and electric loads.
2. Calculate the lighting requirements. Use the applicable calculation method and establish the fixture pattern, considering the architectural effects.
3. Design the supplemental decorative and architectural (built-in) lighting.
4. Review the resultant design. Check the design for quality, quantity, aesthetic effect, and originality.

### (d) Evaluation Stage

With the design on paper, it can now be analyzed for conformance to the principal constraints of cost and energy. If the design stage has been carefully completed, with due attention to these factors, the results of the final evaluation should be gratifying. The results of this stage are fed to the architectural group for use in the final overall project evaluation. In the following sections, we consider in detail each of the steps in the design procedure.

## 13.4  COST FACTORS

This is a particularly difficult item for a novice lighting designer because it requires experience in the field and an acquaintance with commercially available equipment. Also, the inevitable trade-offs between first cost and operating cost cannot be made intelligently unless the cost structure is clearly understood. The following guidelines should be of considerable assistance both in avoiding unpleasant surprises when a job is estimated and in preparing cost analyses:

1. Decide at the outset what cost criteria will be applied—that is, the relative importance of first

cost, operating costs, annual owning costs, and life-cycle costs.
2. Trade-off decisions are required between first cost and operating costs. For example, incandescent lamps and fixtures are low in first cost and high in operating cost, and so on. Dimming and control equipment falls into this area of decisions.
3. Manufacturers' catalog items are *always* cheaper than specials and can be priced more readily.
4. Compare the annual owning costs of two systems or methods. Conversion of these data to life-cycle cost comparisons is straightforward (see Appendix I).
5. The impact of lighting energy on the operating cost of the entire building must be studied and the apportionment of costs determined. The only practical means of accomplishing this is by using a computer program. Programs can be readily adjusted to reflect the effect of the lighting system on building costs, and in particular on HVAC first cost and operating costs. It is incorrect to artificially separate the lighting system from the HVAC system with which it intimately interacts.

The lower the lighting system's energy level, the lower the building's overall operating cost. The argument that heat from a lighting system is fully utilized to heat the building and is therefore not wasted is a specious one that has been refuted on many counts:

HVAC system first cost is higher.

HVAC year-round cost is higher.

Lighting energy cost is higher.

Life-cycle costs are higher.

Energy resource use is higher.

## 13.5  POWER BUDGETS

The requirement to establish a project's lighting power budget in accordance with a specified procedure has now been incorporated into the building codes of many states.

The purpose of this budget determination procedure is not to dictate the design procedure. Indeed, all standards explicitly so state. Instead, the purpose is to develop an overall maximum power budget

*within* which the designer is free to do as he or she wishes. Obviously, extravagance in one area is necessarily at the expense of another area, as maximum power is inflexible, and the entire power budget is based on reasonable design techniques. Still, there is enough leeway in the budget and enough exceptions so that the designer is not overly restricted.

The nationally accepted standard that defines the establishment of a lighting power budget is ANSI/ASHRAE/IESNA Standard 90.1: *Energy Efficient Design of New Buildings Except Low-Rise Residential Buildings,* published by ASHRAE and regularly updated.

ASHRAE 90.1 sets forth design requirements for the efficient use of energy in new buildings intended for human occupancy. Specifically excepted in the current edition are single-family and multi-family residences of three or less stories above grade and buildings whose primary function is directed to a specific purpose, for which human occupancy may be required but is secondary. These include industrial buildings, buildings with very low energy use, and very small buildings (<100 ft² area). The standard encompasses energy use requirements for all of a building's environmental systems (i.e., HVAC, lighting, electrical, power, water, envelope, and energy management).

The standard recognizes the advances that have taken place in the field of lighting controls, as a result of which it is able to focus on *energy use* limitation (which is its specific purpose) without necessarily limiting *power use.* For details on how this is accomplished, the reader is referred to the current issue of Standard 90.1.

## 13.6 TASK ANALYSIS

Refer to Fig. 13.1. This is the stage at which the quantity and quality of lighting required for the tasks are decided. The factors affecting this choice, as shown in Fig. 13.1, are difficulty, time factor, occupant, cost of errors, and special requirements.

### (a) Difficulty

The components of visual difficulty were discussed at length in Sections 11.16 to 11.22, and the results in terms of lighting levels appear in Tables 11.4 to 11.9. Essentially, the designer examines the type of

task involved, and after determining the applicable authority, he or she selects the required illuminance.

In the absence of specific instructions or reasons to the contrary, the designer uses IESNA recommendations. If there are several tasks to be performed at the same point and the most difficult one occurs infrequently, it may be reasonable to provide supplementary portable lighting or even to suggest moving to another brighter location. If it is the major task, lighting should be based on it and provision made for intensity reduction for less demanding work.

Variation in task difficulty is particularly common in spaces in public buildings. Thus, a school gym can be used for athletics, band concerts (despite the acoustics), and town meetings—activities with totally disparate lighting requirements. In these and similar instances, it is common practice to treat the space as essentially three different spaces and design lighting for each, with a careful eye to maximum common equipment usage. Similar problems are encountered in basements, multipurpose rooms, and conference/meeting/lecture/exhibition rooms. Fortunately, most such spaces do not have severe seeing tasks.

The task variation referred to here is the variation that occurs in one very specific location and is not to be confused with task variation in an area, however restricted. Thus, a small private office of, say, 8 × 8 ft has a desk, file cabinet, and circulation space, involving three tasks of differing but constant difficulty in one small space. The corresponding lighting for these is also fixed and varies with the task severity. The values listed in Tables 11.4 to 11.9 represent the required illumination on the surface in question, whether horizontal, vertical, or in between. Inasmuch as the flux method of calculating illuminance normally yields the 30-in., horizontal-plane illuminance level, it is helpful in the early design stages to know the ratio of horizontal to vertical illuminance for various lighting systems. This ratio is approximately

| | |
|---|---|
| Narrow distribution (direct and semi-direct) | 3:1 |
| Wide distribution (direct and semi-direct) | 2.5:1 |
| General diffuse (indirect) | 1.5:1 |

Once the design is advanced, computer calculation will yield exact illuminance data on any desired surface, including important vertical surfaces.

As the illuminance values listed assume adherence to both recommended luminance ratios and reflectances (see Section 11.32), it is necessary to select, in conjunction with the interior designer, finishes and reflectances for surfaces within the area. If, for instance, in a private office a dark wall finish of 10% reflectance is chosen, it will be necessary for the lighting designer to compensate for this by additional wall lighting to maintain the recommended maximum 10:1 brightness ratio (see Table 11.11 and the discussion of point $g$ in Section 13.7). The atmosphere created by vertical surface luminances is discussed later. Table 13.1 lists the reflectances of some common interior paint finishes.

### (b) Time Factor

As discussed in Section 11.20, the length of time in which the task must be accomplished is important in exacting work. Beginning with moderately difficult tasks—that is, luminance of about 170 cd/m² (50 fL)—prolonged intensive application or rapidly changing tasks would require illumination to be raised one (or more) levels. Alternatively, the quality could be improved by increasing daylight or task contrasts.

### (c) Occupant

Inasmuch as the age and other specific characteristics of the worker are usually not known, a standard distribution is assumed and the recommendations as tabulated take account of this. However, if there is a high percentage of older workers, as is the case in certain industries, lighting should be raised one level. This compensates for the inability of aging eyes to accommodate and for the tendency to tire easily (see Section 11.23).

### (d) Cost of Errors

This involves an economic trade-off between savings resulting from improving visual accuracy against the cost of the improved lighting. Performance can be brought close to perfection, but the cost of so doing increases much more rapidly than the proportional increase in performance. Lighting is usually designed to provide approximately 90% work accuracy. Thus, this step is basically an economic calculation, the criteria for which must come from the owner or user. Tasks in which this problem is encountered include inspection, proofreading, textile matching, very fine machining, and jewelry manufacture.

### (e) Special Requirements

These include any nonstandard task lighting requirements. Some of these are specific illuminant color, directionality for shadowing, reflections as required for inspection, polarization, and controlled variations, as required in a space with varied tasks or a varying daylight factor. In addition to these, the physical dimensions of the task often create special requirements of their own. We tend to assume a small object in the horizontal plane because that is the normal office task. However, there are exceptions, such as a drafting board, a large machine and an inspection bench, or a cutting table. Consequently, these special requirements arise:

1. *Large tasks.* In large tasks, the angle of seeing varies from 20° to 70° from the vertical, resulting in radically changing glare angles and reflections from the task.
2. *Three-dimensional tasks.* These tasks shadow themselves, particularly when containing undercuts and reveals. An architect's model shop presents such tasks. When it is necessary to see into an opening, an intense narrow beam is required.
3. *Tools.* Tools cast shadows below and in front when lighted from above and behind. A fabric cutter must see ahead of and below the cutting machine.

**TABLE 13.1  Approximate Reflection Factors**

| Medium-Value Colors | Reflection Factor |
|---|---|
| White | 80–85 |
| Light gray | 45–70 |
| Dark gray | 20–25 |
| Ivory white | 70–80 |
| Ivory | 60–70 |
| Pearl gray | 70–75 |
| Buff | 40–70 |
| Tan | 30–50 |
| Brown | 20–40 |
| Green | 25–50 |
| Olive | 20–30 |
| Azure blue | 50–60 |
| Sky blue | 35–40 |
| Pink | 50–70 |
| Cardinal red | 20–25 |
| Red | 20–40 |

4. *Nonhorizontal tasks.* These must be calculated for the plane in which they stand. As noted in Section 13.6a, the ratio between horizontal and vertical illumination varies between 1.5:1 and 3:1, depending upon the system. Task lighting requirements are stated in the plane of the task. This can have a pronounced effect on the lighting system selected and its arrangement.

5. *Task observed from various positions.* There are instances in which a fixed task is observed from several angles, such as a drawing in a conference room or a wall display. Illumination must be adequate for all viewing angles.

## 13.7 ENERGY CONSIDERATIONS

Energy considerations must pervade every aspect of the design process. Some background material is in order here to place the lighting energy subject in proper perspective. Best current estimates indicate that lighting consumes approximately 25% of the electric power generated in the United States. In terms of *resources*, this amounts to approximately 4 million barrel (bbl) of oil per day. The same sources indicate usage by occupancy as approximately:

| | |
|---|---|
| Residential | 20% |
| Industrial | 20% |
| Stores | 20% |
| Schools and offices | 15% |
| Outdoor and other | 25% |

In commercial buildings, lighting consumes about 20% to 30% of the building's electric energy, more in residences and less in industrial facilities. By judicious design, a reduction of 40% to 50% in lighting energy is attainable. Translated into resources, this reduction can readily amount to more than 1 million bbl of oil per day. Few will disagree that such a goal is well worth the effort. Every 1.0 W/ft$^2$ reduction in lighting energy results in at least 1.25 W/ft$^2$ (13.5 W/m$^2$) savings in air-conditioned buildings. It has been demonstrated by actual designs that offices and schools can be *well* lighted with less than 1.5 W/ft$^2$ (16.1 W/m$^2$). The question to be answered then is: What design guidelines can be followed to effect this energy-conscious design?

The following are general recommendations regarding the design of energy-efficient lighting systems. For specific design requirements and details,

refer to Standard 90.1 (ASHRAE, 2004), the Advanced Energy Design Guide (ASHRAE, 2004), and LEED (USGBC, 2005).

I. Conceptual-level approaches to energy-conscious lighting design include:

A. *Design lighting for expected activity.* This is the task lighting approach. It is wasteful of energy to light any surface to a higher level than it requires. Nonuniform lighting is recommended where high illuminance levels are required for selected tasks in multitask spaces. One way to accomplish this for areas where an exact furniture layout is not available is to use readily movable fixtures. Providing overall high-level illumination with provision for switching to reduce lighting levels is not advisable because of the increased first cost and the psychological impetus to operate at maximum levels. Another solution is to use fixed luminaires for general low-level lighting and supplementary task lighting. Other factors and techniques to be borne in mind are:

1. Group tasks with similar lighting requirements.
2. Place the most severe seeing tasks at the best daylight locations.
3. Improve the quality of difficult visual tasks. This is more energy-economical than providing additional light.
4. The advantages of nonuniform lighting increase as the space between workstations increases.
5. When using the task-ambient design approach, keep in mind that a nonuniform ceiling layout may give a chaotic appearance to a space. Therefore, the preferred approach is uniform ambient lighting and local task lighting.

B. *Design with effective, high-quality, efficient, low-maintenance, thermally controlled luminaires. Effective* means providing useful light and minimum direct glare. In cases where much of the viewer's time is spent in a head-up position, as in schools, or where the viewer can compensate for veiling reflections, the decision should lean toward high VCP (see Section 11.28). Where work and viewing position are fixed and most of the

viewer's time is spent head down, the decision should lean toward low reflected glare (see Sections 11.29–11.31).

1. A *high-quality* luminaire is made with permanent finishes such as Alzac or multicoat baked enamel or any of the high-quality permanent aluminum finishes. This ensures that its performance after 8 to 10 years of service will be comparable to the original.

2. An *energy-efficient* luminaire is one with a high luminaire efficacy rating (LER). This metric is explained in detail in Section 15.12. It combines photometric efficiency (a high coefficient of utilization [CU]; see Section 15.10) with energy efficiency of the luminaire's components to obtain a high overall *efficacy* for the luminaire, expressed in lumens per watt.

3. A *low-maintenance* luminaire remains clean for extended periods and is designed so that all reflecting surfaces can be easily and rapidly cleaned without demounting. Enclosed fixtures should be gasketed. Nongasketed units collect and retain dust and cause rapid output depreciation. Relamping should be simple and rapid to encourage group relamping programs that are energy efficient and cost effective. A 20% increase in maintained light is possible if lamps are replaced at the end of their *useful* life—that is, when output is down to 70% of initial maintained lumens—and if fixtures are cleaned and maintained on a fixed schedule. No cost trade-off is generally involved because periodic maintenance and relamping are normally cheaper than one-at-a-time maintenance and burnout replacement, *and* yield 20% higher average lumens. Fixtures in relatively inaccessible locations such as high ceilings must be designed for low maintenance, and maintenance should be on a fixed schedule.

4. A *thermally controlled* luminaire is one that controls the heat generated by the light source. This item depends to a large extent on the type of HVAC system, the lighting heat load, and the types of fixtures employed. Detailed analysis of this point involves HVAC considerations and the overall impact of lighting energy on the building. In the late 1980s, the thermal problem was principally directed at removing fixture heat. Today, the use of electronic ballasts and low-wattage lamps has reduced the seriousness of this problem (Fig. 13.2).

C. *Use efficient light sources and accessories.* This point is self-explanatory. The ready availability of high-efficacy, high color rendering index (CRI), compact sources has made this item far less problematic than previously. The only cost trade-off involved is between relatively expensive, high-CRI sources and cheaper, low-CRI units. The cost differential is not large, and the choice is frequently made on other than an economic basis.

Spill light and borrowed light are often neglected sources. Glass in upper wall sections can provide sufficient corridor lighting from borrowed office lighting. Sources with high lumen *maintenance* should be given preference.

D. *Select the appropriate lighting system.* As detailed in Sections 13.10 to 13.15, there are five general approaches to light delivery, each with its particular distribution characteristics and, therefore, applicability. In addition, all luminaires must be carefully located to provide uniformity of ambient lighting and task lighting, where applicable. When using indirect luminaires, correct spacing and hanging-stem length will avoid ceiling "hot spots," which can cause direct and reflected glare. Properly designed indirect lighting allows fully effective use of a space in that users are not restricted to facing in any particular direction to avoid direct or reflected glare from fixtures. The mounting height of suspended fixtures must be coordinated with cavity sizes and finishes to decrease light loss and maximize coefficient of utilization.

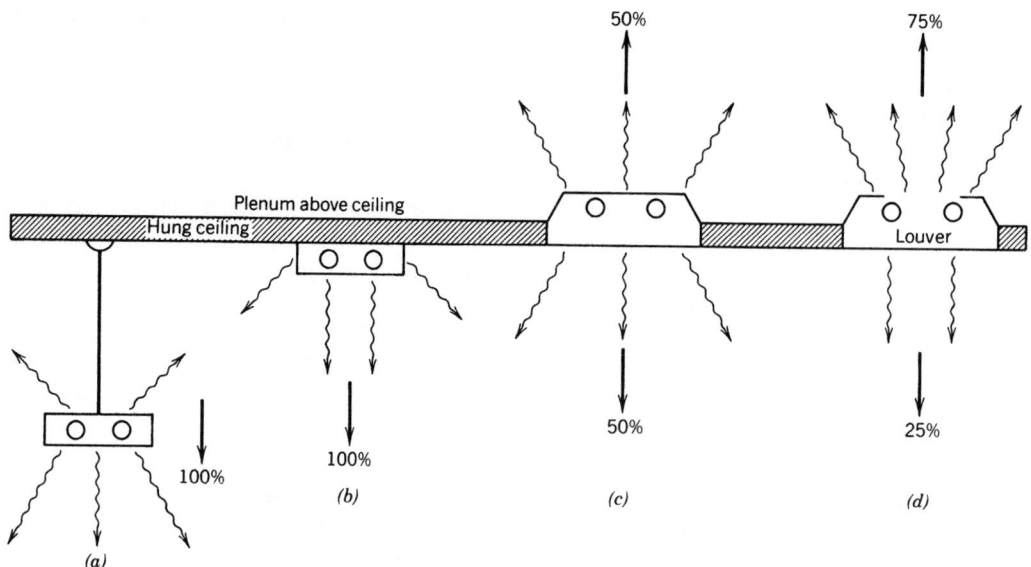

**Fig. 13.2** *Method of fixture installation controls the transfer of a luminaire's heat. (a) Suspended units contribute all their heat to the space while remaining fairly cool. (b) Surface-mounted fixtures also place all their heat in the space but, because of blocked transfer upward, run hot. (c) Completely enclosed recessed units transfer about 50% of their heat to the plenum. (d) Open-louvered, baffled units transfer about 75% of their heat. When they are ducted, heat transfer up can be as high as 85%.*

E. *Use daylight, properly.* Daylight must be considered as a normal light source subject to weather variations and time of building use. Obviously, a three-shift industrial plant cannot use daylight on all shifts, but it can for at least one shift, and design should address this fact. Part of proper daylight design is control of window luminance, which can cause severe and even disabling glare. A corollary of excessive luminance is excessive heat gain. Both are manageable with common control devices, both manual and automatic. Window control devices must also be designed to reflect light back into the space at night to avoid light loss via windows.

F. *Use energy-efficient lighting control strategies.* The subject of lighting control, including manual and automatic switching, dimming, sensing, and intensity control, is covered in Sections 15.13 to 15.15 for both new installations and retrofit work. Proper design of controls can reduce energy consumption over a noncontrolled installation by as much as 60% without reducing lighting effectiveness.

Appropriate controls are mandated in ANSI/ASHRAE/IESNA Standard 90.1.

G. *Use light finishes on ceilings, walls, floors, and furnishings.* This point is self-explanatory and is examined in a number of sections. A brief summary of recommended ranges is:

| | |
|---|---|
| Ceilings | 80–92% |
| Walls | 49–60% |
| Furniture, office machines, and equipment | 25–45% |
| Floors | 20–40% |

In addition to producing higher illumination levels in the room, high reflectances minimize uncomfortable luminance ratios, as between luminaire and upper wall or between task and background (see Table 11.11). For luminance ratios:

1. Between task and near surround—aim for 3:1.
2. Between task and immediate area—aim for 10:1.
3. Between luminaires and their background—aim for 20:1.
4. Anywhere in the normal field of view—aim for 40:1.

Note that these targets themselves are *maximums* and that values above maximum should be accepted only with excellent justification.

II. Detailed requirements, generally from ANSI/ASHRAE/IESNA Standard 90.1, include:

    A. *Mandatory requirements* (summarized)

        1. All interior and exterior lighting must conform to the stated energy limitations. Trade-offs between the two are not permitted.

        2. All interior and exterior lighting systems in buildings larger than 5000 ft$^2$, except emergency and exit lighting, must be equipped with some form of automatic shutoff control.

        3. Where applicable, occupancy sensors and automatic daylight compensation control should be used.

        4. Separate tasks or spaces must have separate controls.

        5. Areas with lighting power densities above 1.1 W/ft$^2$ should have facilities for at least two lighting levels.

    B. *Recommendations*

        1. Where task/ambient lighting is used, the ambient level should not be lower than a third of the task level to avoid uncomfortable luminance ratios.

        2. Accent lighting should not exceed five times the ambient level. Therefore, in merchandising areas where the contrast between ambient and task levels is critical, reduce ambient levels as much as is practical.

        3. When specifying superreflective aluminum in fluorescent (or other) luminaires, determine that the material's high reflectance will be maintained in the specific application and that the required luminaire maintenance procedures will be available.

        4. Utilize self-luminous exit lights or those utilizing LED displays where permitted by local codes.

The energy-efficiency benchmarks in ANSI/ASHRAE/IESNA Standard 90.1 are updated on a regular basis. The current version of this standard should always be used for design. The accompanying User's Manual (ASHRAE 2004) is recommended as a valuable supporting document to assist with interpreting and implementing Standard 90.1.

## 13.8 PRELIMINARY DESIGN

Again referring to Fig. 13.1, the preliminary design phase is the time during which ideas crystallize, but in terms of areas and patterns as well as light and shadow, and not yet in terms of hardware. At this stage the quality of the system is decided on; that is, the luminances, diffuseness, chromaticity, and proportion of vertical to horizontal lighting are determined. The last factor establishes in large measure a room's "mood" or lighting ambience. In preceding sections, these items were discussed in some detail. In the sections that follow on lighting systems (direct, indirect, etc.), the quality of each is considered and applications are suggested. In the overall view, however, the ultimate quality of the lighting system, its visual pleasantness, centers of visual attention, highlights and shadows, as well as texture and forms, are a deft and perhaps artistic combination of the previously mentioned considerations and establish, as the term implies, the quality of the lighting design. A few observations, not covered elsewhere, are mentioned in the following paragraphs.

Planes other than the working plane must always be considered. The ratio of vertical to horizontal illumination of the chosen lighting system determines wall luminance, which in turn greatly influences the overall impression of the space's brightness (see Fig. 13.12). The floor finish has a pronounced effect on the ceiling illumination for direct lighting systems because in direct systems, ceiling illumination derives only from room interreflectances, with floor reflectance being particularly important in large spaces (see Fig. 15.36).

The chromaticity of a room's lighting depends primarily on the source but secondarily on the luminaire and surface finishes. A "white" source can be tinted slightly by the use of a colored reflector in the luminaire. Of course, the effect on luminaire output of such a change must be considered. In the case of semi-indirect and indirect lighting, this same effect can be accomplished by the use of colored ceiling and upper wall surfaces, which serve as secondary reflectors and become the actual luminous source for the room. Recommendations in Table 11.12 cover the choice of source as affected by its chromaticity.

## 13.9 ILLUMINATION METHODS

The discussion that follows is primarily addressed to electric lighting systems. The descriptions apply directly to lighting fixtures and fixture arrangements. Similar considerations, however, can be seen to apply to daylighting design—and the implications of these considerations on daylighting design should not be overlooked during the design process.

There are three broad methods of illumination: general, local/supplementary, and combined general and local.

### (a) General Lighting

This is a system designed to give uniform and generally, although not necessarily, diffuse lighting throughout the area under consideration. The method of accomplishing this result varies from the use of luminous ceiling to properly spaced and chosen downlights, but the resultant lighting on the *horizontal working plane* must be the same, that is, reasonably uniform. It may be, but is not necessarily, task lighting.

### (b) Local/Supplementary Lighting

These are two terms that are used interchangeably. By definition, both *local lighting* and *supplementary lighting* provide a restricted area of relatively high intensity. A desk lamp, a high-intensity downlight on a merchandising display, and a track light illuminating wall displays in practice are all referred to as *local, supplementary,* or *local–supplementary* lights. Typical of this genre are the units illustrated in Fig. 13.3.

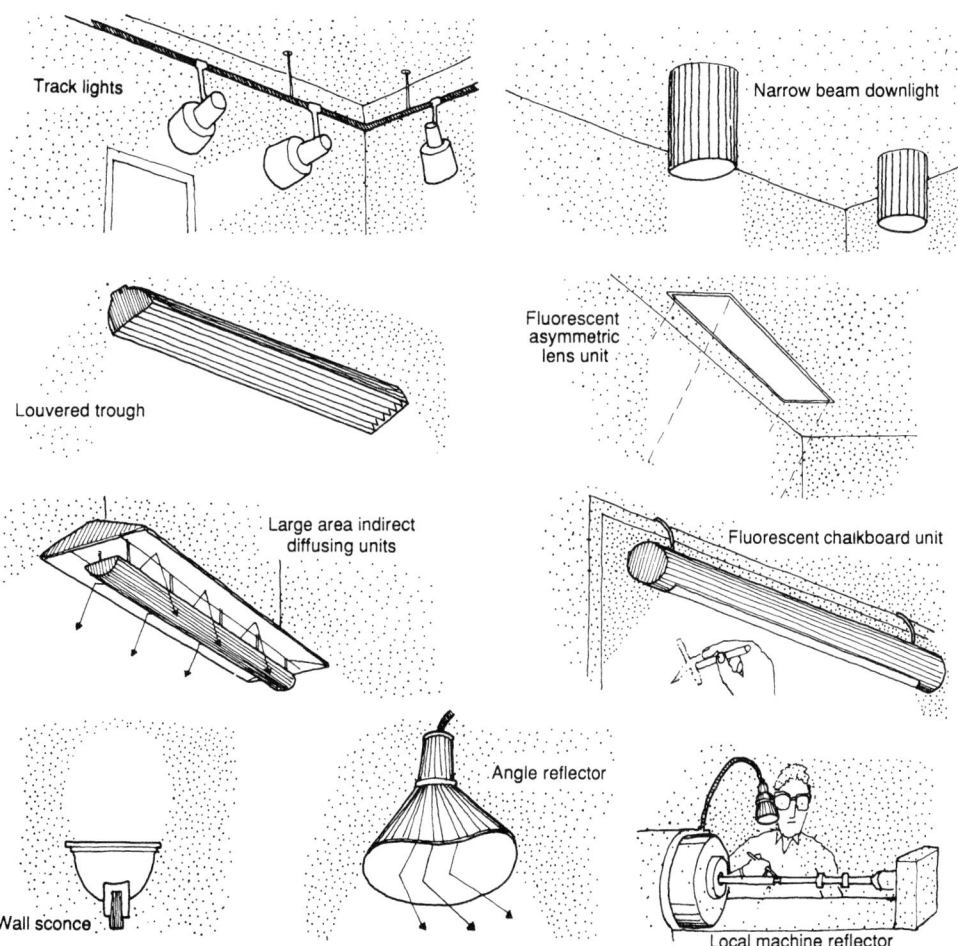

*Fig. 13.3* Typical supplementary lighting units for incandescent, CFL, and linear fluorescent sources.

### (c) Combined General and Local Lighting

This illumination method is used in spaces where the general visual task is low, but supplementary lighting is required in a limited area for a particular task.

These three *methods* of illumination can be accomplished in many ways by the use of luminaires and luminous sources of different types, because the illumination method is a function of *both* luminaire arrangement and the luminaire's inherent lighting distribution. The term used to describe the effect of the combination of a particular fixture type applied in a particular way is the *lighting system*. Thus, a reflector-type fixture, when aimed down, gives *direct* light. The same fixture beamed up at the ceiling gives *indirect* light. The following section describes the systems that constitute the vast majority of lighting installations.

### 13.10 TYPES OF LIGHTING SYSTEMS

No single lighting system can be said to be the only choice in a given instance; on the contrary, the designer normally has a choice of at least two systems that, if utilized properly, yield illumination of adequate quantity and good quality. However, other factors, such as harmonization with the architecture and economics, usually tip the balance in favor of one or the other.

The five generic types of lighting systems are indirect, semi-indirect, diffuse or direct-indirect, semi-direct, and direct.

### 13.11 INDIRECT LIGHTING

See Fig. 13.4*a*. Between 90% and 100% of the light output of the luminaires is directed to the ceiling and upper walls of the room. The system is called *indirect* because practically all the light reaches the horizontal working plane indirectly, that is, via reflection from the ceiling and upper walls. *Therefore, the ceiling and upper walls in effect become the light source*, and if these surfaces have a high-reflectance finish, the room illumination is highly diffuse (shadowless). Because the source must be suspended at least 12 in. (30 cm) and preferably 18 in. (45 cm) or more from the ceiling (depending on the unit's output) to avoid ceiling "hot spots," this system requires a minimum ceiling height of 9 ft, 6 in. (~3 m). If luminaires are correctly spaced, the resultant illuminance is uniform, and direct and reflected glare potentials are both low.

To avoid an excessive luminance ratio between the luminaire and its surrounding field, the luminaire can be made translucent on the bottom, the sides, or both. Approximately 750 lux (75 fc) is the maximum horizontal-plane illuminance attainable without exceeding an *overall* ceiling luminance of about 2500 cd/m² (730 fL) (see Section 11.28). With practically no veiling reflections, this illuminance level is sufficient for all but the most difficult tasks. The lack of shadow, low source brightness, and highly diffuse quality created by indirect lighting give a very quiet, cool ambience to the space, suitable for private offices, lounges, and plush waiting areas. Areas having specular visual tasks such as an office with visual display terminals (VDTs) use this system to advantage. In such spaces, indirect fixtures *without* luminous bottoms or sides should be specified.

When properly designed, particularly when the source of light is architectural coves (see Fig. 13.4*c*), the ceiling has a floating sky quality, which is pleasant and can be used to give an impression of height in a low-ceilinged room. (This system is not to be confused with a transilluminated ceiling, which is really a direct lighting system of entirely different quality and effect.) A further characteristic of the indirect lighting system is loss of texture on vertical surfaces, as is common to all fully diffuse lighting.

Indirect lighting is inherently inefficient, as much of the useful light reaches the working plane only after double reflection—within the luminaire and off the ceiling. Although to a considerable extent this inefficiency is offset by the glare-free lighting, applications to difficult seeing tasks frequently require an additional light source. Thus, an indirectly lighted architect's drafting room having tables equipped with supplementary lamps would take advantage of both systems—the local high-intensity light at a maximum of 200 fc (~2000 lux) for the restricted area being worked on and overall table lighting of 40 to 50 fc (~400 to 500 lux) of high-quality lighting free of veiling reflections. The latter would also solve any reflected glare problems arising from the many viewing angles required by large tasks such as drawings.

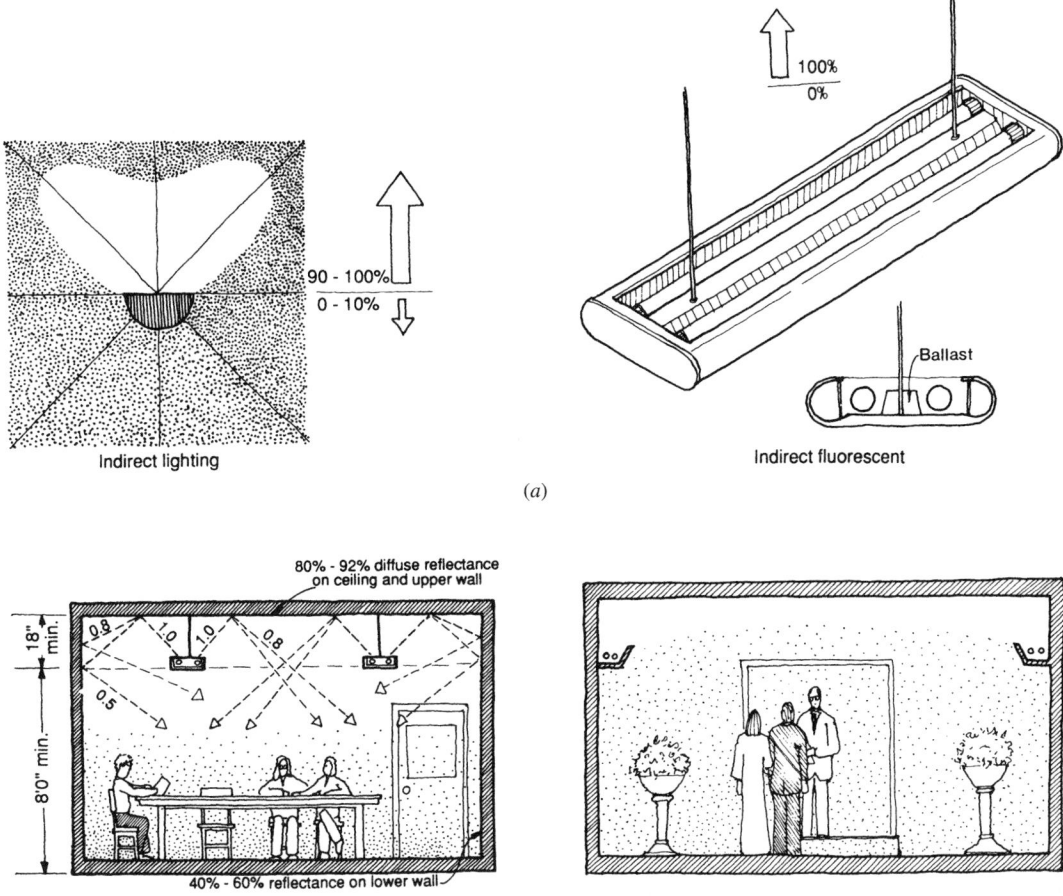

**Fig. 13.4** *Indirect lighting. (a) The luminaires deliver 90–100% of their output above their own horizontal plane. (b) The ceiling and upper wall surfaces of the space are directly illuminated, and by reflection become large secondary sources that illuminate the space below. When properly designed, this type of installation yields a substantially uniformly bright ceiling. (c) Use of architectural coves gives an acceptable luminance gradient on the ceiling and, if properly designed, nearly uniform, glareless illumination in the room. This system of illumination is particularly useful in spaces with VDTs.*

## 13.12  SEMI-INDIRECT LIGHTING

Between 60% and 90% of the light is directed upward to the ceiling and upper walls. This distribution is similar to that of indirect lighting, except that it is somewhat more efficient and allows higher levels of illumination without undesirable brightness contrast between the luminaire and its background, along with lower ceiling brightness. A typical fixture, illustrated in Fig. 13.5, employs a translucent diffusing element through which the downward light component passes. The ceiling remains the principal radiating source, and the diffuse character of room lighting remains. Direct and reflected glare are both very low, as they are with indirect lighting (see Fig. 11.36). In both indirect and semi-indirect systems, it is often desirable to add accent lighting or downlighting to break the monotony inherent in these systems and establish visual points of interest or to create required modeling shadows.

In both indirect and semi-indirect lighting systems, the light undergoes a number of ceiling and wall reflections before reaching the horizontal working plane. The use of colored paints, particularly on the ceiling, can serve to tint the room illumination slightly by selective absorption.

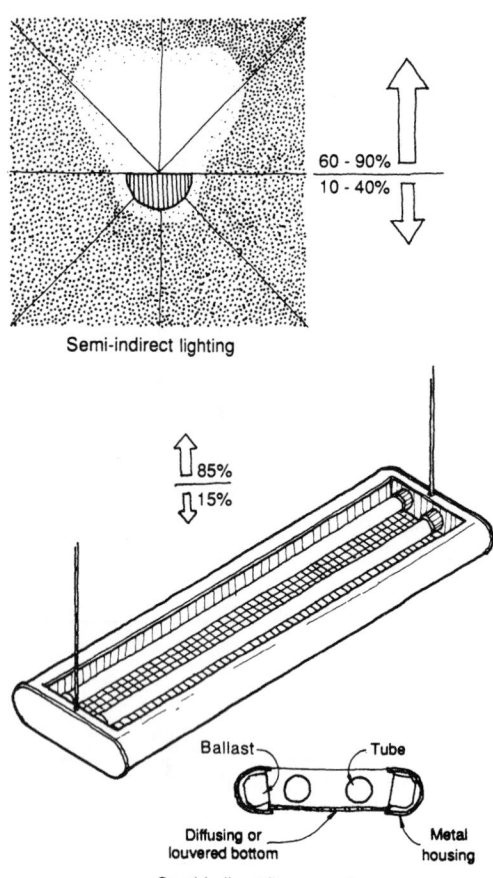

Semi-indirect lighting

Semi-indirect fluorescent

Ballast — Tube

Diffusing or louvered bottom — Metal housing

60 - 90%
10 - 40%

85%
15%

*Fig. 13.5 Semi-indirect lighting.*

## 13.13 DIRECT–INDIRECT AND GENERAL DIFFUSE LIGHTING

Direct–indirect lighting provides an approximately equal distribution of light upward and downward, resulting in a bright ceiling and upper wall (Fig. 13.6). For this reason, luminance ratios in the upper-vision zone are usually not a problem. As the ceiling is an important although secondary source of room illumination, diffuseness is good, with resultant satisfactory vertical-plane illumination.

Diffusing fixtures (Fig. 13.7) give light in all directions, whereas direct–indirect fixtures have little horizontal component. Stems for both types should be of sufficient length to avoid excessive ceiling brightness, generally not less than 12 in. (30 cm).

Because the impression of illumination depends to a large extent on *wall luminance* (as this is the surface we see most often), a space with general diffuse illumination *appears* lighter than one with direct–indirect illumination due to the darker walls in the latter (see Figs. 13.7b and 13.11).

By avoiding excessively bright luminaires and giving attention to the positioning of sources and viewing angles, direct and reflected glare can both be kept low. Furthermore, because the luminaire (like any other luminous source in the field of view whose luminance is higher than that of the average scene) draws the eyes' attention, particular care must be taken to limit its luminance and to avoid disturbing fixture patterns (see Fig. 13.14).

The efficiency of these two systems is good. Both are well applied in spaces requiring overall uniform lighting at moderate levels such as classrooms, standard office work spaces, and merchandising areas.

## 13.14 SEMI-DIRECT LIGHTING

With this type of lighting system, 60% to 90% of the luminaire output is directed downward and the remaining upward component serves to illuminate the ceiling (Fig. 13.8). If the ceiling has a high reflectance, this upward component is normally sufficient to minimize direct glare from the luminaires, depending on eye adaptation level. The degree of diffuseness depends largely on the reflectances of room furnishings and of the floor. Shadowing should not be a problem when upward components are at least 25% and ceiling reflectance not less than 70%. With smaller upward components, the system is essentially direct lighting (see Section 13.15). The system is inherently efficient. Reflected glare can be controlled by the methods discussed in Section 11.29. With adequate wall illumination, the quality of the lighting gives a pleasant working atmosphere. It is applicable to offices, classrooms, shops, and other working areas.

## 13.15 DIRECT LIGHTING

In this system essentially all the light is directed downward. As a result, ceiling illumination is

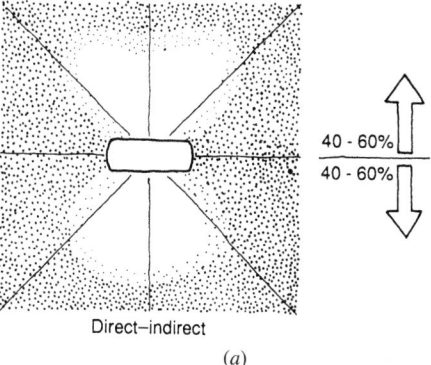

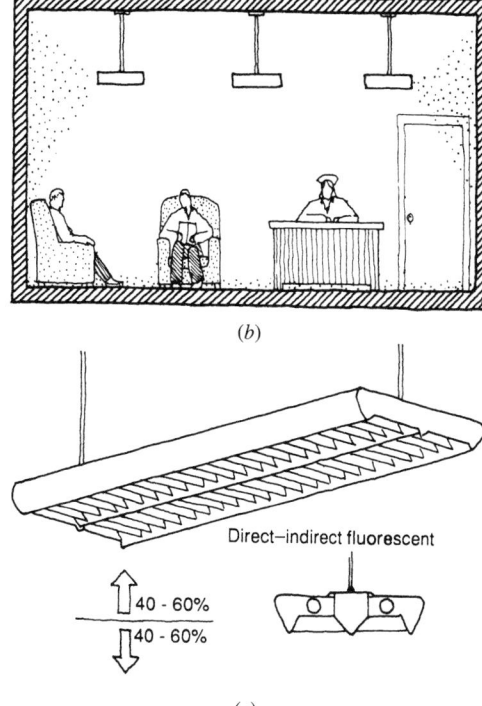

**Fig. 13.6** (a) Direct–indirect lighting. Upper and lower room surfaces are luminous (b), but the center of walls is not because of the lack of horizontal light from fixtures (c). Principal light on the working plane comes directly from the luminaire.

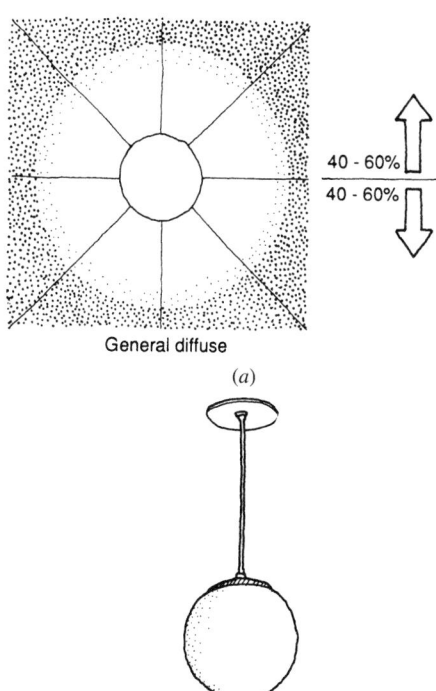

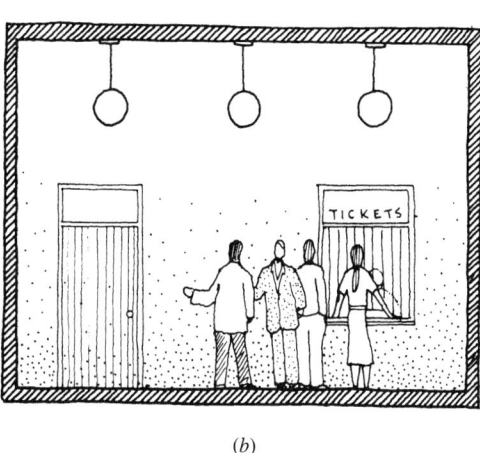

**Fig. 13.7** (a) General diffuse lighting. (b) Note that all room surfaces are illuminated and become secondary sources, although those closest to the fixtures (ceiling, upper wall) are the brightest secondary sources. The primary source of illumination is the direct radiation from the fixture. The floor contribution is low due to its normally low reflectance.

ILLUMINATION

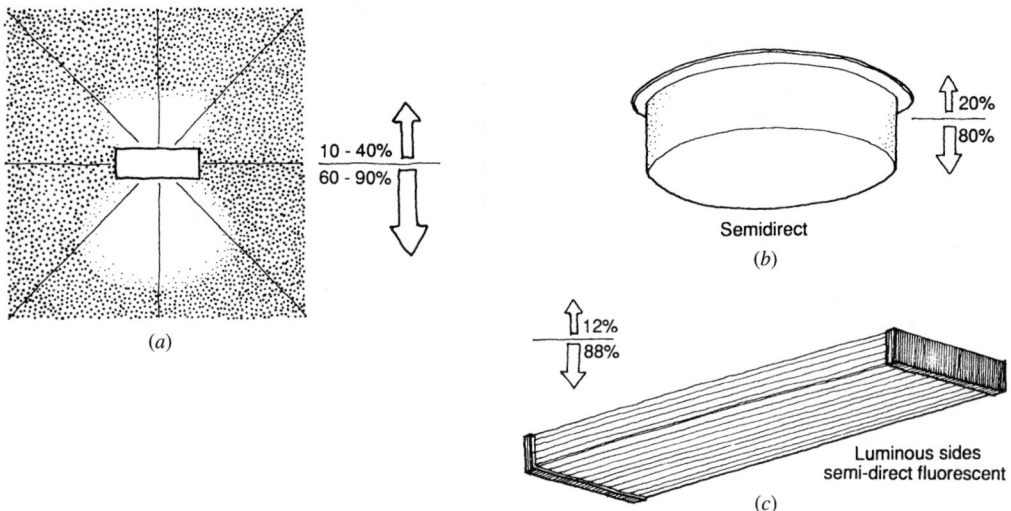

10 - 40% ⇧
60 - 90% ⇩

(a)

⇧20%
⇩80%

Semidirect
(b)

⇧12%
⇩88%

Luminous sides
semi-direct fluorescent

(c)

**Fig. 13.8** *Semi-direct lighting provides its own ceiling brightness (a), with surface-mounted fixtures (b) or pendant/surface units (c). Other characteristics are similar to those of direct lighting.*

entirely due to light reflected from floor and room furnishings. This system, then, more than any other, requires a light, high-reflectance, diffuse floor unless a dark ceiling is desired from an architectural or decorative viewpoint. Occasionally, the ceilings are deliberately painted a dark color and pendant direct fixtures used to lower the apparent ceiling of a poorly proportioned room or hide unsightly piping, ductwork, and so on.

The effect of direct lighting depends greatly on whether the luminaire light distribution patterns are spread or concentrating (Figs. 13.9 and 13.10). In the former cases, considerable diffusion of light results from reflections on the floor, furniture, and walls. The result is a working atmosphere with slightly darkened walls and ceiling. This type of lighting, which is most widely represented by the recessed fluorescent troffer in a suspended ceiling, is common for general office lighting. The luminaires themselves form a ceiling surface of light and dark areas, and the quality of the entire system is not unpleasant. Difficulties associated with direct glare and veiling reflections can be controlled by proper use of reflectances, use of low-brightness units, and judicious arrangement of viewing positions. When direct lighting units are used in a uniform pattern, this latter option disappears, and the need for particularly low-brightness units and high ceiling reflectivity, or

specialty diffusers such as those with a batwing distribution, is increased.

Direct lighting gives little vertical surface illumination, requiring the addition of perimeter lighting in business atmospheres (Fig. 13.11).

Concentrating downlights create sharp shadows and a theatrical atmosphere that are not appropriate to a working commercial space. They can be used in restaurants and other areas where the privacy type of atmosphere generated by limited-area horizontal illumination and minimal vertical-surface illumination is desired. When a lighting fixture is designed with a black cone or baffle or another device that is nonreflecting at the viewing angle, it appears dark even when lighted. It is our opinion that installations providing high-horizontal surface illumination, with no apparent source of brightness, such as those using black-cone downlights, are disturbing to our normal bright-sun-and-sky orientation, and should be used cautiously and only in limited areas. This same comment, but to a lesser extent, is applicable to very low brightness diffusers such as the parabolic wedge type (see Fig. 15.11). (There, however, the unit has the redeeming characteristic of low reflected glare, which is not the case with downlights.) Stated otherwise, the psychological impression of a space without a visible source of light is gloomy and cavelike. When very low brightness

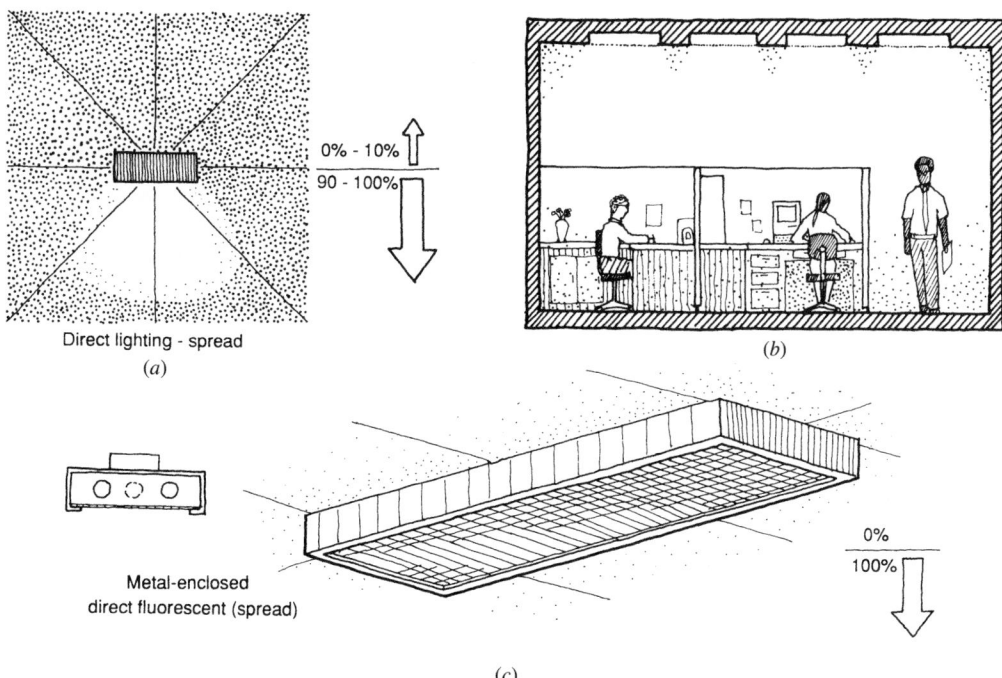

**Fig. 13.9** *Spread-type direct lighting (a) illuminates all room surfaces except the ceiling (b), which is only illuminated by reflection from the floor. Some diffuseness is evident. The most common type of unit in this category is the direct fluorescent unit, either surface-mounted (c) or troffer type, recessed in the ceiling.*

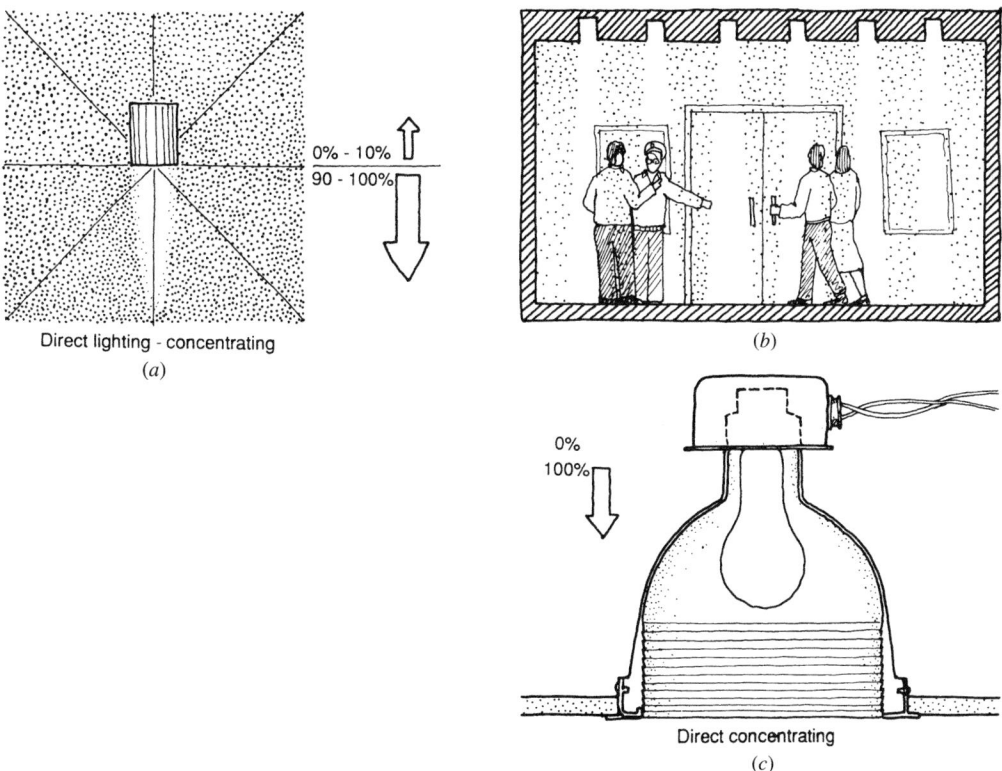

**Fig. 13.10** *With concentrated direct distribution (a), the floor is the only luminous surface (b) other than the ceiling fixture. Diffuseness is absent. Walls are dark. Incandescent downlights (c) and, to a lesser extent, CFL downlights are of this type unless equipped with spread-type lenses.*

ILLUMINATION

**571**

(a)  (b)

**Fig. 13.11** *Large-dimension lighting fixtures may be used in a low-ceilinged room if the apparent size of the unit is reduced. Here at a mounting height of 7 ft-6 in. (2.3 m), 4 × 4-ft (1.2 × 1.2-m) units are acceptable because the lattice on the face of each unit gives the impression of reduced fixture size. Note also that the apparent illumination in (b) is greater than in (a) (although both are exactly equal on the table surface) due to the wall wash in the background. The eye perceives vertical surface illumination more readily than horizontal illumination and retains the impression for the entire space.*

sources are used, as in VDT areas, this negative impression can be alleviated by the addition of luminous surfaces in nonreflecting areas or points of light sparkle.

In summary then, spread-type direct lighting is suitable for general lighting, whereas concentrated direct lighting, which reduces vertical illumination, is appropriate for highlights, local and supplementary lighting, and specialized privacy-atmosphere installations.

## 13.16 SIZE AND PATTERN OF LUMINAIRES

Because of its luminance, each luminaire or other luminous source is a point of visual attention. To the extent that luminaires are numerous, large, very bright, or arranged in striking patterns, attention is drawn to them and away from other surfaces. Furthermore, color elements or accent lighting can be added deliberately to draw attention. Rigid

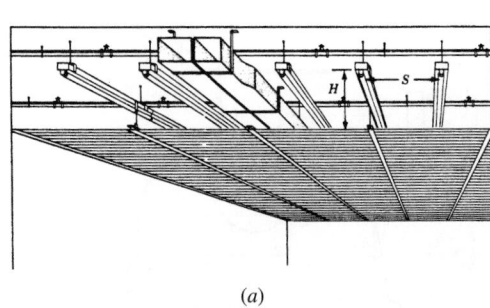 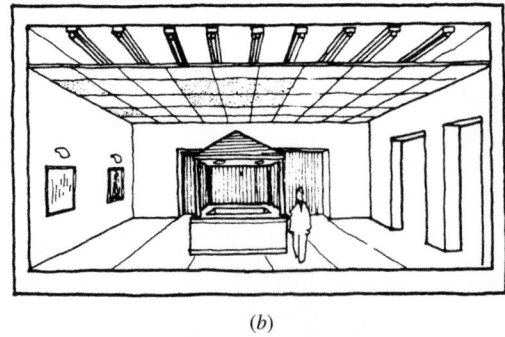

(a)  (b)

**Fig. 13.12** *(a) Luminous ceiling. When properly designed, piping and ductwork above the suspended diffusing material are not seen and do not affect the installation's light distribution. As a rule of thumb, the strip fluorescent fixture spacing S should not exceed 1.5 times the fixture height above the diffusing element for uniform illumination. (b) A luminous ceiling provides low brightness and highly diffuse, uniform illuminance, generally at levels exceeding 500 lux (45 fc). It is particularly useful for specular tasks where supplemental lighting is impractical. To relieve the monotony of large, unbroken expanses of diffusers, as in (a), designers frequently use clearly defined diffuser panels, as shown.*

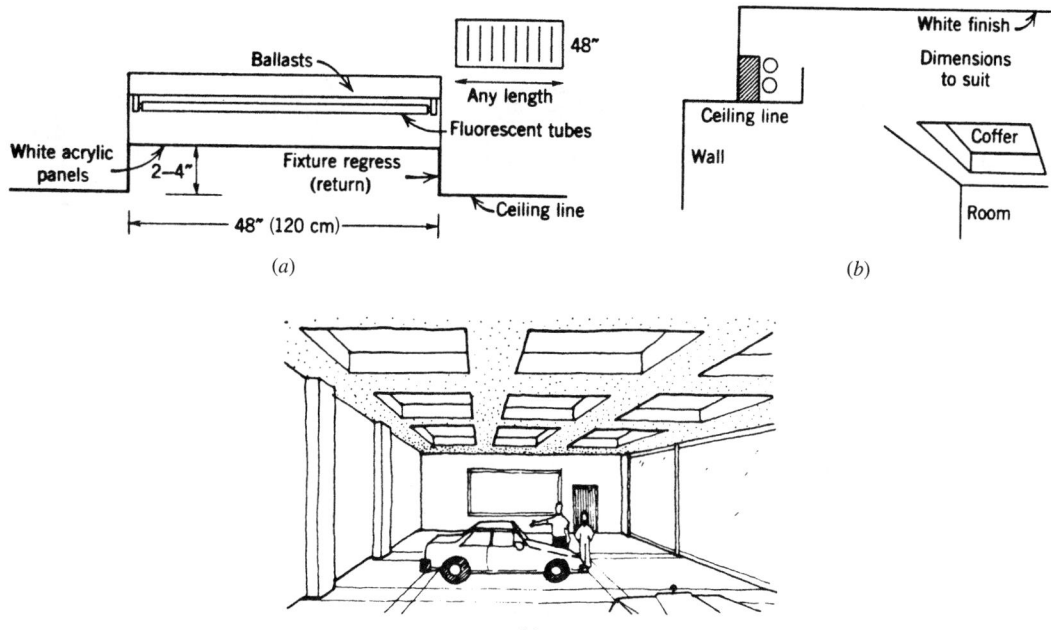

**Fig. 13.13** *Coffer-type light sources can be manufactured (a) or architectural (b). Manufactured coffers (a) are large direct lighting fixtures generally available in a 48-in. (120-cm) width with standard-length 4-ft (1.2-m) fluorescent lamps installed across the width, as shown. Length is variable, as required. Architectural coffers (b) can be constructed in any shape or size using hidden linear fluorescent lamps or CFL units to illuminate a white (plaster) flat or domed surface, which in turn illuminates the space beneath. Both types of coffer can give an illusion of great depth and of a floating illuminated surface. (c) Multiple coffers create a dominant architectural effect and, when designed in conjunction with skylights, can furnish soft, glare-free illumination throughout the day and night.*

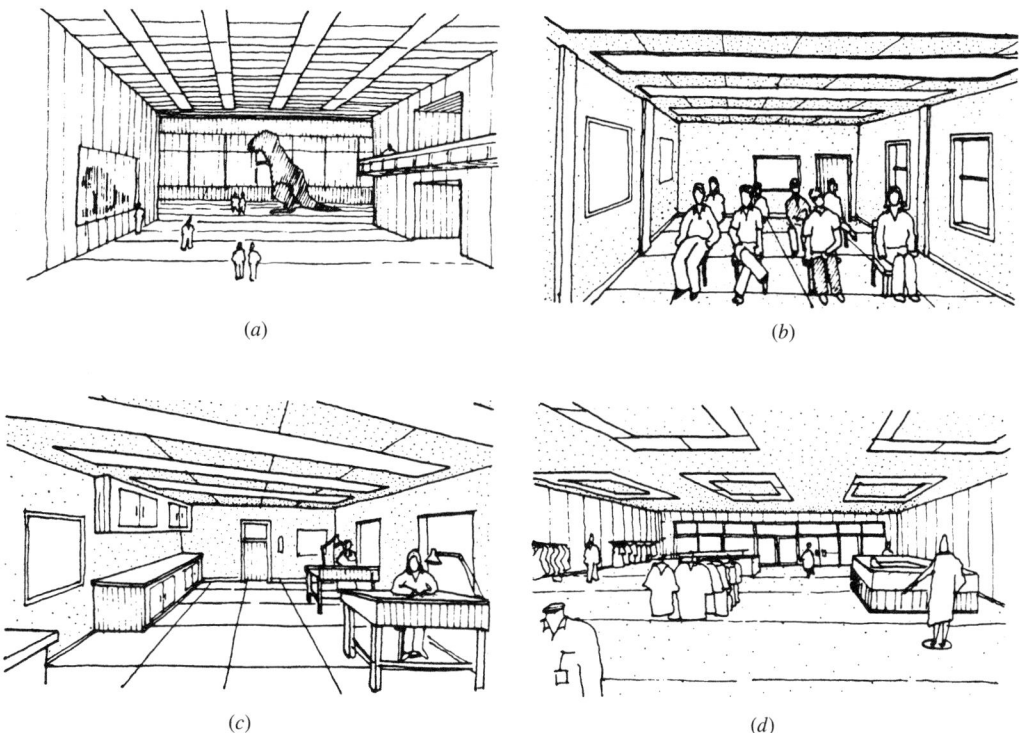

**Fig. 13.14** *(a) Longitudinal lines in the direction of the sight line increase apparent length, direct traffic flow, and decrease direct glare. (b) Lines perpendicular to the line of sight shorten and widen a space but also increase direct glare. (c) Diagonal lines minimize shadows and break rectangular patterns. (d) Rectangular pattern is architecturally dominant.*

ILLUMINATION

(a)

(b)                                                              (c)

***Fig. 13.15*** *(a) Lighting design in various spaces with high ceilings. The fixtures in (a) and (b) follow structural beams. (Photos by M.B. Warren.) (b) Floor reflection and daylight provide ceiling and wall illumination. (Photo by L. Reens.) (c) Lighting in this space was handled by recessing fixtures into the lattice ceiling pattern. Metal halide HID units and tungsten–halogen units were used. (Courtesy of GTE/Sylvania, Inc.)*

**Fig. 13.16** *Downlights are unobtrusive light sources. They can be spaced evenly throughout a room (a) or unevenly (b).*

rules cannot be established to cover these criteria, but examples can demonstrate the principles involved.

Luminaire size should correlate with room size and ceiling height. Fluorescent fixtures larger than $2 \times 4$ ft ($60 \times 120$ cm) should not be used in ceilings lower than 10 ft (3.1 m) unless their size is minimized by some sort of surface pattern (see Fig. 13.11). Transilluminated (luminous) ceilings (Fig. 13.12) are *totally* lighting fixtures and require a minimum of 12 ft (3.66 m) mounting height. When they are installed below this level, particularly in large rooms, the effect is oppressive, as if the sky were lowered on us. To offset this effect, the use of colored, shaped, or dark panels is of some help. In place of a luminous ceiling, large-area, coffer-type fixtures can be utilized that give the impression of depth (Fig. 13.13).

To achieve the uniformity of illumination desirable for general lighting, regular spacing is required. However, various effects may be obtained within the regularity to accomplish an architectural purpose, as shown in Fig. 13.14*a–d.* The pattern of lights must never be at cross-purposes with any dominant architectural pattern; rather, it should either reinforce an architectural form or be neutral. If a strong architectural element is absent, a dominant lighting pattern may be desirable. Conversely, a strong architectural element can either be reinforced (Fig. 13.15*a*) or utilized to carry a neutral lighting pattern (Fig. 13.15*b,c*).

Continuous row installations eliminate the dominant checkerboard effect of (closely spaced) individual luminaires and are cheaper. Coves and cornices give the ceiling a floating or lightness

effect. Geometric patterns can be used to add interest or break the monotony of large areas, such as those found in department stores. Generally, downlights are not dominant, and regularity of placement is not essential (Fig. 13.16). However, when downlights are surface mounted (a generally inadvisable procedure), the result can be very far from visually neutral (Fig. 13.17). Nonuniform layouts with large sources create a distinct pattern problem inasmuch as they are too large to be neutral, and the nonuniformity can create visual confusion (Fig. 13.18). The only cure for this problem is to minimize the source brightness by using low-brightness luminaires (see Fig. 16.16 and Section 15.11).

In spaces where circulation is the primary "seeing task," yet they contain isolated areas requiring

**Fig. 13.17** *The large, can-shaped, surface-mounted downlights dominate the area's appearance despite the high ceiling.*

ILLUMINATION

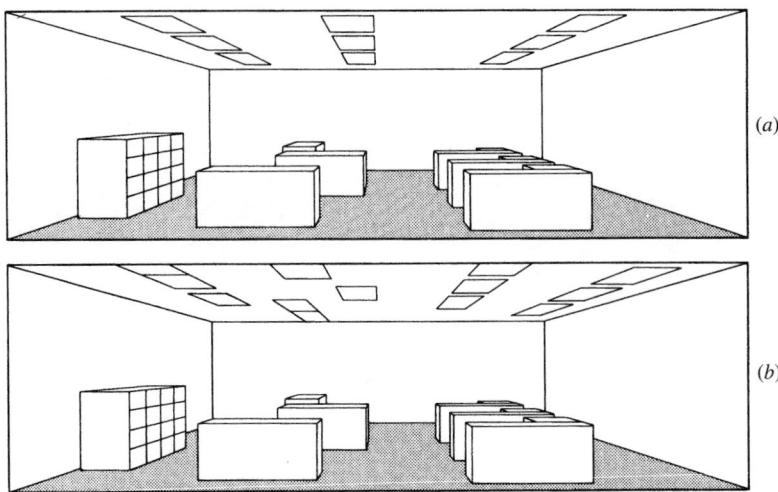

**Fig. 13.18** *(a) The layout of the lighting fixtures is economical and may provide uniform illuminance. (b) The nonuniform layout lacks integration with the functions below and may not provide the illuminance needed for the task below.*

greater illumination, such as waiting areas in transportation terminals, a perimeter lighting system with additional local lighting as required is a viable solution (Fig. 13.19). In addition to their illumination function, lighting patterns are often used as directional markers. This is particularly useful in transportation terminals, where signs only partially serve this purpose. Figure 13.20 illustrates this principle, as well as the use of a lighting pattern to emphasize an architectural aspect of the space.

A frequently neglected consideration is the appearance of a luminaire when deenergized. With proper daylight and energy-conserving design, many sources are unlighted during the normal-use hours of the space. (Obviously, low-brightness sources change least in appearance.) Cognizance of the visibility and appearance of luminaires in daylight, regardless of whether they are illuminated, from inside or outside a building can be used to advantage, as shown in Figs. 13.21 and 13.22.

**Fig. 13.19** *Cornices, valances, and coves are luminous ceiling borders. In large rooms, suspended coves achieve a uniform ceiling brightness gradient and, when designed with a downward component or combined with local lighting, as illustrated, create a pleasant, intimate atmosphere.*

**Fig. 13.20** *The lines of lights in the center converge optically to produce a directional flow of traffic toward the escalator. (Penn Station, New York; photo by B. Stein.)*

(a)           (b)

*Fig. 13.21* (a) Lighting can be utilized as a medium to connect the inside and outside of a building. The simple maneuver of continuing the lighting pattern beyond the window visually connects the inside and outside spaces. Care must be exercised to avoid fixture placement that reflects in the glass. (b) As fixtures are readily visible even when unlit during daylight hours, their outline can be accentuated and the resultant pattern utilized as an architectural motif. (Courtesy of Welton Becket & Assoc.)

*Fig. 13.22* Electric lighting fixtures can be seen through the glazing even during the daytime. (Photo by B. Stein.)

## 13.17 OTHER DESIGN CONSIDERATIONS

Because the phenomenon of vision as affected by lighting is the core of our discussion, it is appropriate to present here some miscellaneous yet important observations that are not readily subsumed under any of the major headings in our discussion.

1. As shown in Figs. 13.14*a* and 13.14*b*, the impression of room length and width can be emphasized by the direction of lines of lighting. An even wall wash of light can also be used to shorten and widen a hallway or corridor.

2. As can be seen in Fig. 13.11, a light wall in the line of sight increases the room's apparent size. Therefore, both light and low-chroma, high-value paint hues can be used to expand a space (lighted walls, light colors) or contract it (unlighted walls, dark color).

3. Vertical surface illuminance should be approximately 25% to 35% of horizontal illuminance for a space to appear dimensionally undistorted. Because high luminance attracts the eye, fixtures with sparkle draw the eye away from the walls and thereby shrink the space's horizontal dimensions.

4. As shown in Figs. 13.13*c*, 13.14*c,d*, 13.16, and 13.17, luminaire patterns can be dominant (i.e., can become a focus of attention) by virtue of their size or arrangement. The same is true of wall-lighting patterns to an even greater extent, as walls are always in the direct line of sight. Therefore, wall lighting that creates meaningless scallops, spots, irregular gradients, and points of sparkle become a dominant visual element. In most cases, this is neither intended nor desired. Unscalloped, even-gradient wall illumination is readily accomplished with linear sources, with elliptical reflector lamps or luminaire reflectors, utilizing proper luminaire spacing.

5. Concentrated pools of light in an overall low-ambient-light space effectively isolate the illuminated areas from each other (i.e., define a limited specific geographic space). This can be used to advantage in restaurants and work or school areas where it is desired to define individual "territories" in a single large space.

## References

ASHRAE. 2004. ANSI/ASHRAE/IESNA Standard 90.1: *Energy Standard for Buildings Except Low-Rise Residential Buildings.* American Society of Heating, Refrigerating and Air-Conditioning Engineers, Inc., Atlanta, GA.

ASHRAE. 2004. *Standard 90.1-2004 User's Manual.* American Society of Heating, Refrigerating and Air-Conditioning Engineers, Inc., Atlanta, GA.

ASHRAE. 2004. *Advanced Energy Design Guide For Small Office Buildings.* American Society of Heating, Refrigerating and Air-Conditioning Engineers, Inc., Atlanta, GA.

USGBC. 2005. *LEED (Leadership in Energy and Environmental Design).* U.S. Green Building Council. http://www.usgbc.org/

# CHAPTER 14

# Daylighting Design

**DESIGNING WITH DAYLIGHT IS BOTH AN ART AND A** science. The designer in concert with the appropriate combination of building geometries, materials, and light produced by site conditions can produce health and well-being, visual delight (Fig. 14.1), intended ambience and reduce dependency on

electrical energy use. Studies have demonstrated that daylighting improves indoor environmental quality for occupants. Daylighting design is often mistakenly understood to mean that an abundance of light should fill a space; successful design, however, involves a careful balance and control of heat

(a)

(b)

**Fig. 14.1** (a) East Wing, National Gallery, Washington, DC (I.M. Pei, 1978). (b) Hong Kong International Airport, Hong Kong, China (Norman Foster, 1998). (Photos by Alison Kwok; © 2004; all rights reserved.)

ILLUMINATION

gain and loss. A variety of strategies are available to control and enhance daylight through shading devices, light shelves, glazing, atria, courtyards, and material finishes (both interior and exterior). Daylighting is an easily achievable LEED strategy if a building can provide a minimum daylight factor of 2% in 75% of all occupied spaces.

## 14.1 THE DAYLIGHTING OPPORTUNITY

### (a) Importance of Daylighting Design

Historically, the architectural form of buildings, placement of windows, and location of rooms were guided by the availability of daylight as the primary source of illumination. Daylight was the only source of abundant light for buildings, provided through deep windows and thick walls and perhaps replaced (although inadequately) in the evening by the flicker of a candle flame or an oil lamp. Building form changed dramatically with the development of fluorescent lighting technologies, allowing interiors to be uniformly lit by electric lighting systems and to function at cooler temperatures (as opposed to the higher temperatures produced by heat-intensive incandescent lamps). Designing with daylight can improve energy efficiency by minimizing the use of electricity for lighting as well as reducing associated heating and cooling loads. Daylighting is a critical design factor to those concerned about global warming, carbon emissions, and sustainable design—in addition to visual comfort.

Research has found daylight to be an important factor influencing human behavior, health, and productivity. Windows admitting daylight provide occupants with a view and a temporal connection with the outdoors. Daylight renders the environment in a vivid range of experiences and delight. It is important for basic visual requirements to view tasks and to perceive space. How daylight is delivered is in the hands of the designer at the beginning stages of design. The option of ignoring daylight in our high-energy-cost and rapidly-diminishing-natural-resource world is no longer available. This chapter describes daylight strategies to increase occupant satisfaction, control glare, provide appropriate vertical and horizontal illumination, and address the potential for energy savings to enable the designer to create a proper visual environment.

### (b) Planning for Daylight Throughout Design

Designing buildings for daylighting is a complex systems integration process. Daylighting design begins with situating a building on its site and continues through each phase of design; making the best use of daylight continues throughout a building's occupancy. While overall design goals remain generally fixed throughout each design phase, there are key concerns associated with each of the phases. For example, in the conceptual design phase, building form, orientation, layout, and major apertures might be primary elements. Further into design development, there would be specification of materials and interior finishes, zoning for integration with electric lighting and other services, control systems, and commissioning procedures set in place. During occupancy, fine-tuning and maintenance of the system would occur, and postoccupancy evaluation to determine satisfaction, visual comfort, and lighting system performance.

### (c) Energy Savings with Daylighting

To obtain lighting energy savings in a building, six "essential" ingredients for daylighting design are recommended by the Illuminating Engineering Society of North America (IESNA) in RP-5-99 *Recommended Practice of Daylighting*:

1. Plan interior space for access to daylight.
2. Minimize sunlight in the vicinity of critical visual tasks.
3. Design spaces to minimize glare.
4. Zone electric lighting for daylight-responsive control.
5. Provide for daylight-responsive control of electric lighting.
6. Provide for commissioning and maintenance of any automatic controls.

Energy savings from reduced electric lighting will be compromised if any of these factors are overlooked. If any one of the first three is missing, daylight will make little contribution to the illumination of the space. Each of these factors is discussed in this chapter.

### (d) Goals of Daylighting

Improved aesthetics, provision of human biological needs (circadian rhythms and visual relief), and reduction of electric lighting energy usage are the most important advantages of daylighting a building. Key goals in daylighting design are to provide sufficient illuminance, minimize the perception of glare, and provide for overall visual comfort.

## 14.2 HUMAN FACTORS IN DAYLIGHTING DESIGN

The following human-related factors (as opposed to the physical aspects of light) are described briefly to illustrate the importance of considering daylighting and especially these factors in the design of spaces.

### (a) Windows and View

There is a common belief that if a window is placed in a wall, there will be sufficient view and daylight. The view function of a window, however, is very different from the daylighting function. The most preferred views from a window include the sky, the horizon, and the ground. In offices, people enjoy having windows in their work space because of the views. The functional advantage is that people can look into the distance to reduce eye fatigue after doing close desk tasks. Depending upon the type of facility, the designer should be aware of special circumstances, such as in care facilities to ensure that occupants who are bedridden can have a view from their vantage point, or providing lower sills in facilities for children (depending upon safety), or accommodating people in wheelchairs by providing low-sill windows in bedrooms and other areas.

### (b) Productivity and Satisfaction

Productivity is a complex issue that is difficult to isolate or attribute to a single parameter such as daylighting. The connection to the temporal qualities of daylight improves our psychological well-being and productivity. In studies of classrooms, windows, daylight, and performance, researchers found that students with more daylighting in their classrooms progressed faster on math and reading tests than students with less daylighting. Also, sources of glare negatively impact student learning, and the issues of control of windows, blinds, sun penetration, and acoustic conditions are important for teachers. In another study, retail stores were found to have a "daylight effect on increased monthly sales" (Heschong Mahone Group, 1999–2003).

### (c) Controlling Daylight in Interior Spaces

Daylight, whether diffuse light or direct sunlight, provides significant benefits associated with psychological well-being. On the other hand, there are potential problems—such as glare or substantial cooling loads—caused by uncontrolled quantities and qualities of light. Direct sunlight is not, however, always a liability. In nontask areas, a momentary sunny patch, a streak of sunlight against a wall, or a series of multiple shapes provides visual interest and dynamicism to a space. Sunlight in task areas can be controlled in a number of ways:

- Provide exterior fixed shades that exclude sunlight for all sun positions.
- Use systems that diffuse the incident sunlight sufficiently to eliminate glare potential.
- Provide occupant controlled adjustable shades.

### (d) Minimize Glare

Glare is a difficult problem to overcome when balancing daylight and view. Any window (including north exposures) can produce problematic glare if the window is within the field of view. High contrast ratios between a window and adjacent surfaces can occur unless the window is designed to reduce luminance ratios through the use of sunshading devices, light shelves, high-reflectance interior surfaces, light-colored window surrounds and mullions, and low-transmittance glazing (though such glazing will reduce light flux through the window). Furniture may be oriented to introduce daylight from another side of a space (as opposed to having an occupant face a window).

## 14.3 SITE STRATEGIES FOR DAYLIGHTING BUILDINGS

Optimal daylighting opportunities depend upon a building's position on a site relative to available daylight, horizon obstructions, orientation, and building form. The quality of daylight, its effects on illumination, and solar position for passive heating and control of cooling loads are of particular importance to the designer. Site obstructions such as neighboring buildings, trees, and landforms will determine the maximum available daylight on a site and the maximum project envelope that will preserve daylight access to adjacent properties (Fig. 14.2).

*Orientation.* In many locations, an ideal orientation for buildings is an elongated, narrow plan allowing the north and south façades of the building maximum exposure to more easily controllable daylight. From a daylighting standpoint, this is desirable because direct solar radiation received by the south façade is easier to control to prevent excess solar gain, is relatively uniform, and is necessary for passive solar heating strategies. The nearly constant diffuse skylight availability on the north façade is advantageous for uniform and soft daylighting. Figure 14.3 shows how orientation affects cooling, lighting, and heating energy for a building.

*Form.* Establishing an appropriate building form in the early stages of design is critical to daylighting performance. The width of the long, narrow plan previously described will determine how much of the floor area will have access to usable daylight. Generally, a 15-ft (4.5-m) perimeter zone can be completely daylit, a 15- to 30-ft (4.5- to 9.0-m) area can be partially daylit, and an electrically lit area beyond 30 ft (9 m) can be used to determine the width of a building.

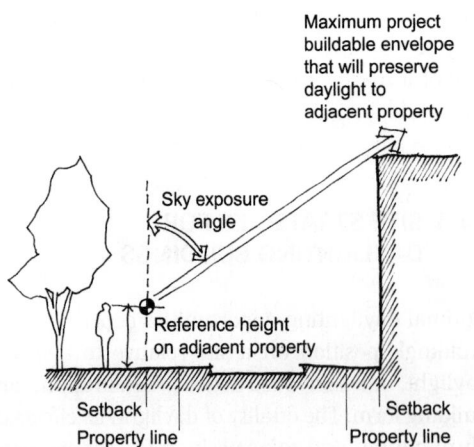

**Fig. 14.2** *Protecting a site from obstructions to daylight. (Drawing by Erik Winter.)*

## 14.4 APERTURE STRATEGIES: SIDELIGHTING

Sidelighting systems admit light from apertures in window walls, and light sweeps across the space from one or more sides. The distance to which usable daylight penetrates a space and falls onto a work plane is the variable that designers work with to provide for sufficient illuminance. Generally, sidelighting is best for desk tasks because there are no veiling reflections, provided there is proper orientation of the worker. A number of generalized sidelighting strategies provide greater illuminance further into a space and improve visual comfort. The variety of approaches and components available to the designer is extensive. A complete discussion is provided in *Daylight in Buildings, A Source Book on Daylighting Systems and Components* (International Energy Agency, 2000). Schematic examples of a few typical design strategies are described below and in Figs. 14.4–14.11.

- *Design for bilateral lighting.* Daylight within a space is generally most evenly distributed when a space is lit from two walls (bilateral lighting), as shown in Fig. 14.4. Bilateral lighting from "opposite" walls produces the most evenly distributed lighting condition. Unilateral lighting (windows on one wall) can increase the potential for glare.
- *Place windows high on a wall.* In general, for a given window area, daylight will penetrate further into a space and have more uniform distribution when windows are placed high on a window wall (Fig. 14.5). If possible, raise the ceiling height to accommodate a higher window position. Use the ceiling as a reflecting surface by placing window heads as close as possible to the ceiling.
- *Use adjacent walls as reflectors.* Interior walls become reflectors when windows are placed adjacent to them, thus reducing the contrasting edge around the window (Fig. 14.6). This arrangement can also bring visual delight if the reflecting wall is a light color, which will reveal patterns and colors from sunlight and reflect diffuse light further into the space.
- *Splay the walls of an aperture.* This strategy is similar to the reflector strategy described previously, where light washes across a longer or rounded surface area around the window. When the edges of window openings are splayed (Fig. 14.7) or rounded, these illuminated surfaces surrounding the window reduce contrast and are more

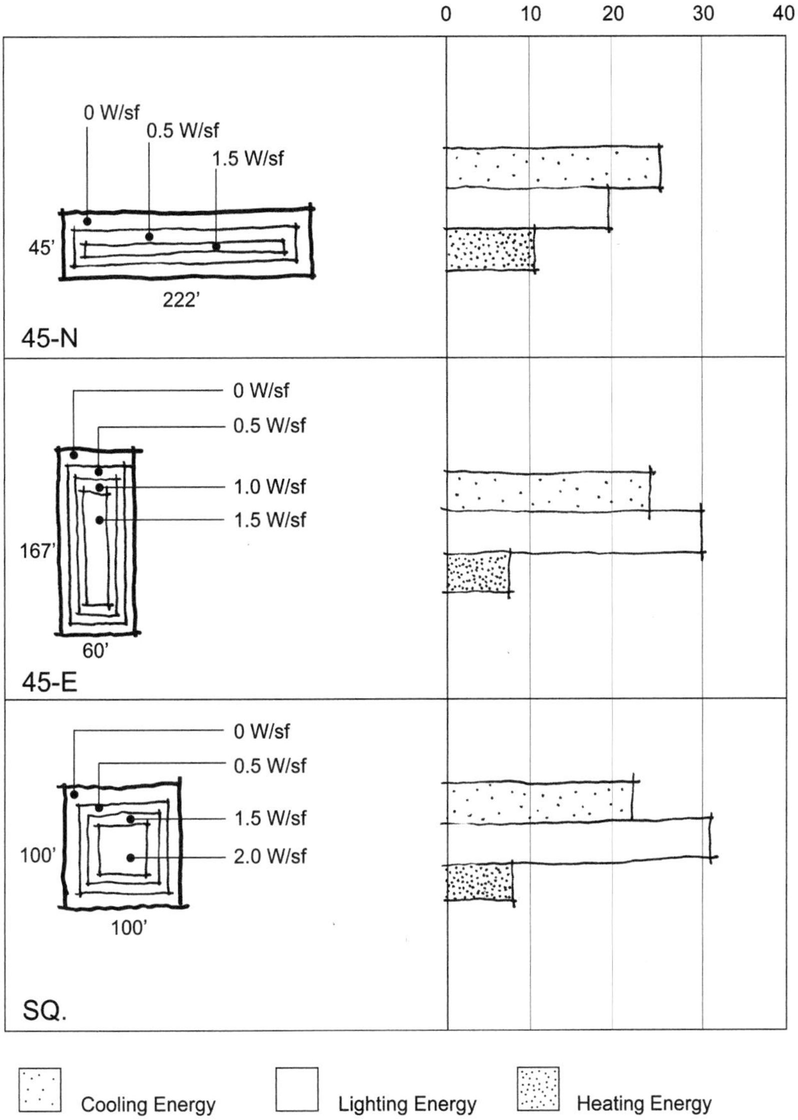

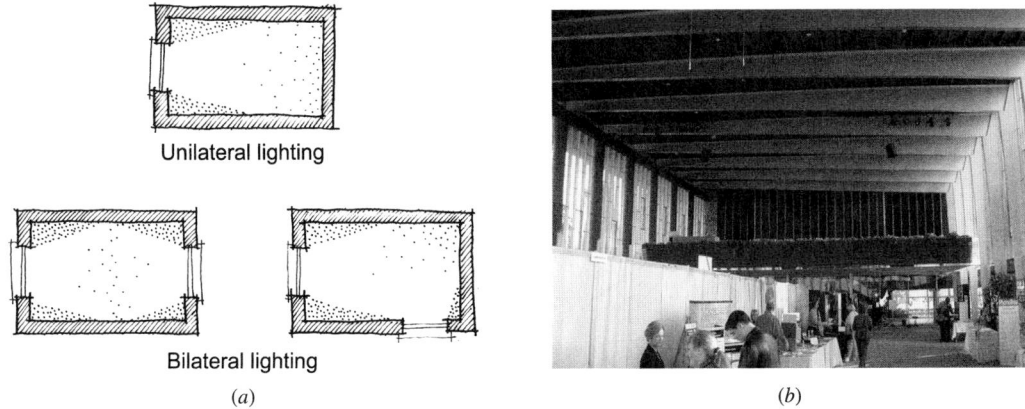

0    10    20    30    40

0 W/sf
0.5 W/sf
1.5 W/sf

45'
222'

**45-N**

0 W/sf
0.5 W/sf
1.0 W/sf
1.5 W/sf

167'
60'

**45-E**

0 W/sf
0.5 W/sf
1.5 W/sf
2.0 W/sf

100'
100'

**SQ.**

⬚ Cooling Energy    ☐ Lighting Energy    ⬚ Heating Energy

**Fig. 14.3** *Effect of building orientation on energy consumption. (Drawing by Erik Winter after Moore,* Environmental Control Systems, *1993.)*

Unilateral lighting

Bilateral lighting

*(a)*

*(b)*

**Fig. 14.4** *(a) Plan diagrams of unilateral and bilateral daylighting. (b) Windows on two sides (a bilateral approach) at the Crystal Cathedral campus, Anaheim, California. (Drawing by Erik Winter, photo by Alison Kwok; © 2004 Alison Kwok; all rights reserved.)*

**583**

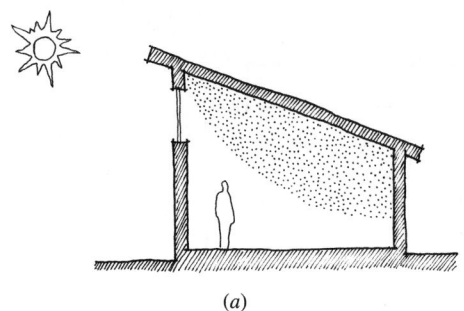

(a)

(b)

**Fig. 14.5** (a) With higher windows, daylight extends farther into a space. (b) High windows in a study space at Mt. Angel Library, Mt. Angel, Oregon. (Drawing by Erik Winter; photo by Fuller Moore; © 2004 The Society of Building Science Educators; used with permission.)

visually comfortable, and therefore reduce the potential for glare.

- *Provide daylight filters.* Daylight may be modified (either blocked or diffused) by a number of elements, which include trees, vines, and trellises (Fig. 14.8) on the exterior of a building; filters for the interior of a building may include blinds, drapes, or translucent glazing.

- *Provide summer shading.* Depending upon passive solar heating and cooling design strategies, in certain instances direct sunlight should be blocked before it enters a space at certain times of the year. Figures 14.9–14.11 show examples of exterior louvers (horizontal or vertical), overhangs, trellises, trees, and light shelves that can block direct sunlight, reflect diffused sunlight into a

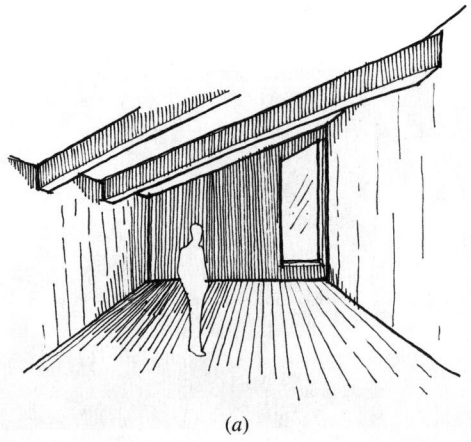

(a)

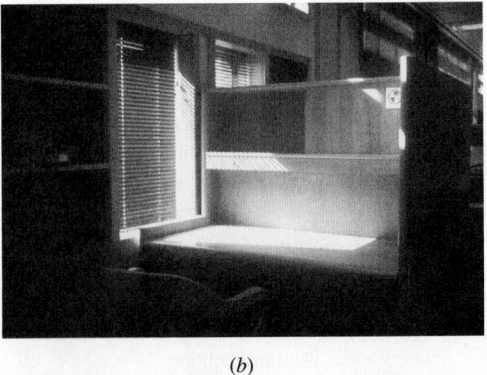

(b)

**Fig. 14.6** (a) Windows adjacent to a wall provide an additional reflecting surface. (b) Reading carrel adjacent to a window at the Graduate Theological Union Library, Berkeley, California. (Drawing by Erik Winter; photo by Alison Kwok; © 2004 Alison Kwok, all rights reserved.)

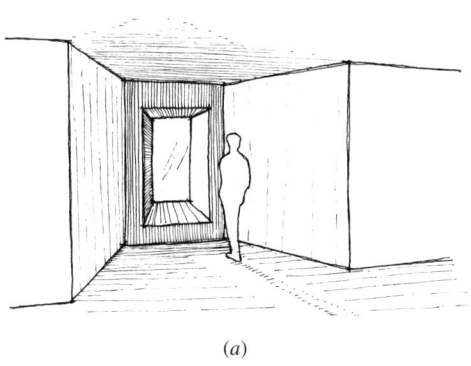

(a)

(b)

**Fig. 14.7** (a) Splayed window provides additional reflecting surfaces. (b) Deeply set window within a thick wall of a room in a medieval village in Il Borro, Italy. (Drawing by Erik Winter; photo by Alison Kwok; © 2004 Alison Kwok; all rights reserved.)

(a)

(b)

**Fig. 14.8** (a) Trellis at Westcave Environmental Center, Round Mountain, Texas. (b) Trees outside of a window can filter light and shade a window. (Photos by Walter Grondzik; © 2004, all rights reserved.)

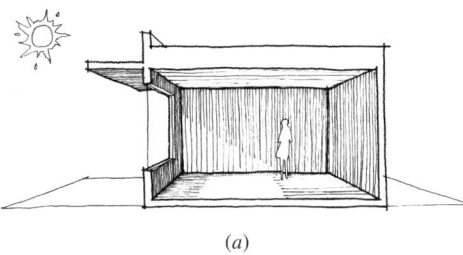

(a)

**Fig. 14.9** (a) Horizontal overhangs block light but also act as a reflector for light from the ground plane. (b) Horizontal shading devices at Ash Creek Intermediate School, Monmouth, Oregon. (Drawing by Erik Winter; photo by Alison Kwok; © 2004 Alison Kwok, all rights reserved.)

(b)

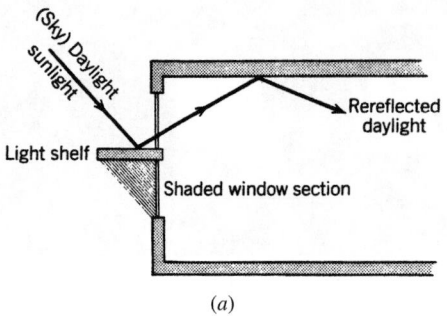

*(a)*

*(b)*

**Fig. 14.10** *(a) Light shelf reduces the daylight factor near a window and increases it at greater depths. Shelf material (opaque, translucent) and angle of installation (horizontal, sloped up) markedly affect performance. (b) Opaque white surface at the top of a light shelf in the Emerald People's Utility District office building, Eugene, Oregon. (Photo © 2004 by Alison Kwok; all rights reserved.)*

space, and provide solar control. Light-colored materials or finishes on these components will reduce contrast between the element and the view or sky beyond. Light shelves that also serve as a shading device can be designed as a horizontal (integrated or attached) component positioned inside and/or outside a window. Typically above eye level, they divide a window into a lower

**Fig. 14.11** *Fixed horizontal louvers on the south façade at the Phoenix Public Library, Phoenix, Arizona. (Photo © 2004 by Walter Grondzik; all rights reserved.)*

view portion and an upper area exclusively for daylight.

## 14.5 APERTURE STRATEGIES: TOPLIGHTING

Skylights, roof monitors, and clerestories are suitable aperture strategies for the top floor of a building, particularly for interior locations of large floors that are far from perimeter windows. To prevent veiling reflections or direct glare situations, toplighting components should be away from the *offending zone* (areas with a direct view from an occupant) or use a baffle or interior reflector to diffuse and control daylight. Most of the strategies for sidelighting also apply to toplighting, several of which are discussed here.

- *Splay the "walls" of an aperture.* By splaying the sides of a skylight, the skylight appears larger because light washes along a larger surface area and reflects diffuse light into the space (Fig. 14.12). This strategy reduces the potential for glare similar to the way splayed windows function.

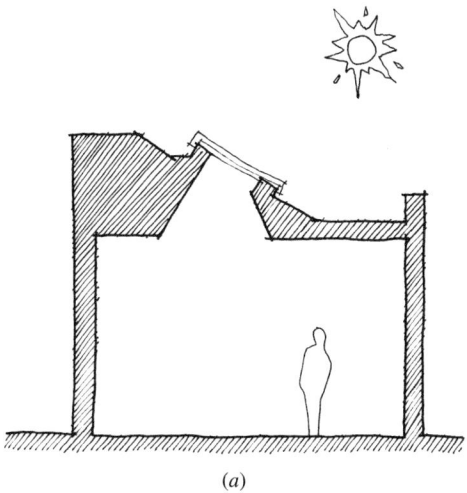

(a)

(b)

(a)

(b)

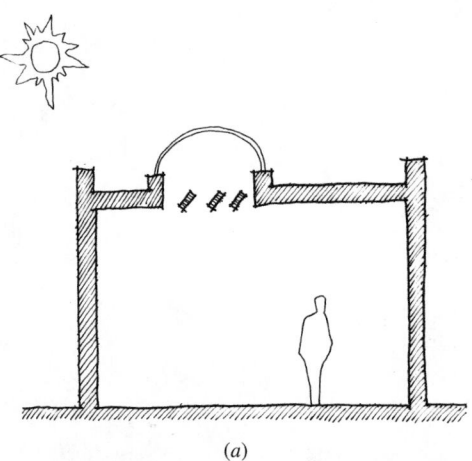

(a)           (b)

**Fig. 14.14** (a) Skylight with baffles that block direct solar radiation. (b) Baffled skylight daylighting design at Mt. Airy Public Library, Mt. Airy, North Carolina (Edward Mazria, 1984). (Drawing by Erik Winter; photo by Fuller Moore, © 2004 The Society of Building Science Educators; used with permission.)

**Fig. 14.15** Clerestory skylights with louvers at the U.S. Holocaust Memorial Museum in Washington, DC (Pei, Cobb, Freed & Partners, 1993). (Photo © 2004 by Alison Kwok; all rights reserved.)

- *Place toplights high in the space.* Higher ceilings with skylights allow more surface area for light to diffuse upon, virtually becoming a larger source (Fig. 14.13). This strategy works well in spaces where the skylight is well above the field of view.
- *Use interior devices to block, baffle, or diffuse light.*

Direct sunlight can be redirected by a reflector below a skylight, clerestory, or roof monitor that, depending upon the surface material, diffuses the light onto another surface within the space (Figs. 14.14 and 14.15).

## 14.6 SPECIALIZED DAYLIGHTING STRATEGIES

A number of innovative daylighting systems can be categorized as experimental, yet they have tremendous potential. Some of these strategies include laser-cut or prismatic panels, fiber optics, solar tubes, and heliostats. More advanced systems use concentrating reflectors and lenses to introduce concentrated luminous energy into some type of light-conducting device. These may be fiber optic bundles, prismatic light pipes, or some type of mirrored channel. The problem of heat rejection becomes more severe as the degree of solar energy concentration increases. Thus, light pipes are much less critical in this regard than are optical-fiber systems. The efficiency and economic feasibility of these systems are interdependent because of the materials used; the farther the light is transmitted, the lower the system's overall efficiency and the higher the cost to make the system technically feasible.

*Light pipes.* This term refers to several strategies: daylight pipes, electric light pipes, and fiber optic pipes (Fig. 14.16). The light pipe operates by

(a)

(b)

**Fig. 14.16** (a) Heliostat on the rooftop of a building tracks the sun and directs light into an 8-ft. (2.4-m)-wide atrium and down 14 floors. (b) Solar Light Pipe is a 120-ft (37-m)-long, 6-ft (1.8-m)-diameter, 12-sided steel-and-aluminum frame—enclosing laminated glass panels and surrounded by fabric—within the atrium at the Morgan Lewis building in Washington, DC. (Photos © 2004 Carpenter Norris Consulting; used with permission.)

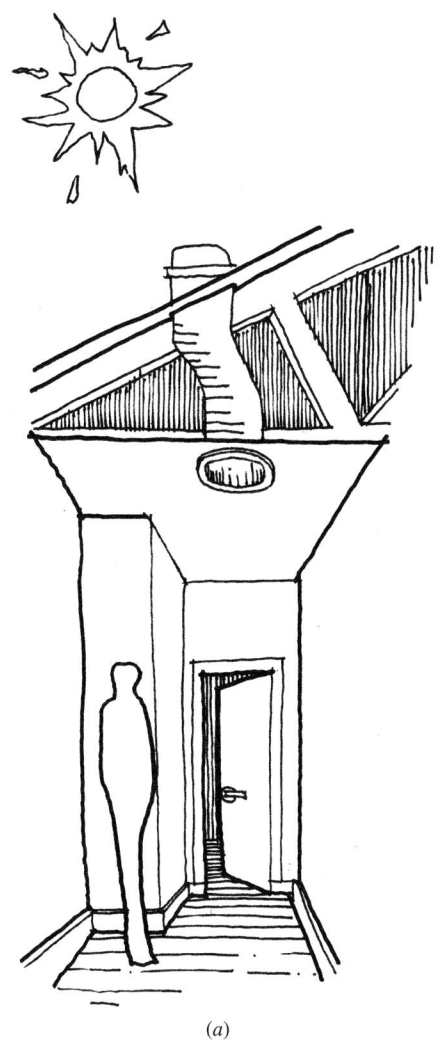

(a)

(b)

**Fig. 14.17** (a) Skylight construction through a roof structure. (b) Top of the Solatube® skylights on the roof at Ash Creek Intermediate School in Monmouth, Oregon. (Drawing by Erik Winter; photo by Alison Kwok; © 2004 Alison Kwok, all rights reserved.)

collecting light through a heliostat; channeling daylight (or electric light) through a reflective tube made of prismatic glass, plastic film, and mirrors; and diffusing light at the end of the pipe. It is an exciting new strategy because of the length it can transport—65 ft (20 m)—from a single point light source.

*Tubular skylights.* These light shafts have highly reflective surfaces and are capped by a clear skylight (Fig. 14.17). The amount of light transmitted and delivered varies with the diameter of the shaft. They are convenient and economical for supplemental illumination in hallways, closets, and areas without a need for a lot of control. The light quality is comparable to that of a ceiling-mounted fluorescent fixture (often indistinguishable).

## 14.7 DAYLIGHT FACTOR

Daylight factor (DF) is defined as the ratio of interior illuminance ($E_i$) to available outdoor illuminance

$$DF = \frac{E_i, \text{ indoor illuminance, at a given point}}{E_H, \text{ outdoor illuminance}}$$

$$\times \, 100\% \qquad (14.1)$$

where $E_H$ is the unobstructed horizontal illuminance. The daylight factor concept is applicable only where the sky luminance distribution is known or can reasonably be estimated. The Commission Internationale de l'Eclairage (CIE) defines an overcast sky and a clear sky whose luminance distributions are fixed for the purpose of calculations. Daylight factor cannot be used with skies with constantly changing luminance (partly cloudy and direct sun) because under such conditions the daylight factor at a given point also varies continuously, making the concept useless as a calculation tool *for absolute daylight values.*

Daylight factor as a means of expressing interior daylight illuminance is both absolute and relative. With a given sky luminance distribution, variations in daylight illuminance *inside* correspond exactly to variations *outside* (i.e., the daylight factor remains the same). This assumes a minimal effect from obstructions and ground reflections. Thus, the daylight factor allows de-

termination of interior daylight distribution for varying fenestration, spatial arrangement, and building orientation.

Daylight factor is constant for a given space and window configuration. Interior illuminance can easily be calculated by knowing the daylight factors for locations in a given space and the exterior illuminance derived from sky luminance data. Daylight design analysis can use a combination of minimum exterior illuminance and corresponding minimum daylight factor requirements to predict daylight sufficiency under almost all exterior conditions.

## 14.8 COMPONENTS OF DAYLIGHT

The general characteristics of daylight as a source of light (versus electric lamps) were discussed in Chapter 11. Understanding the components of daylight is important to the design of apertures and the selection of materials. Daylight in a building consists of three components (see Fig. 14.18):

1. Sky component (SC)
2. Externally reflected component (ERC)
3. Internally reflected components ($IRC_1 + IRC_2$)

DF is the sum of these three components, each calculated individually for each location being considered. DF is a ratio, but the value of a given DF is based upon contributions from these components: DF = SC + ERC + IRC.

The *sky component* (SC) is that portion of total daylight illuminance at a point received directly from the area of the sky *visible through* an aperture.

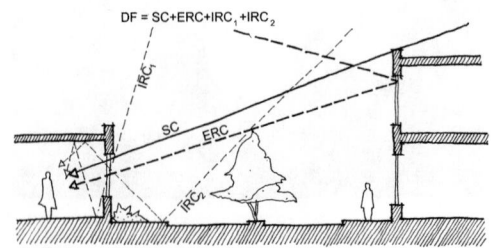

**Fig. 14.18** *Total daylight factor (DF) is composed of the SC, ERC, and IRC. The IRC, in turn, is subdivided into reflected sky light and reflected ground light components. Note that surfaces deep in the room are illuminated with rereflected light. (Redrawn by Erik Winter.)*

As the SC represents *received* light, it takes into account reductions due to window obstructions (mullions, etc.) and losses in transmission; that is,

$$SC = \text{incident skylight} - \text{window losses}$$

The *externally reflected component* (ERC) represents light reflected from exterior obstructions onto the point under consideration. This *does not include ground-reflected light.* ERC is of significance only in built-up areas (where there are structures opposite an aperture) and can be estimated as the portion of the SC for that area of obstructed sky, reduced by the percentage of the sky obstructed (RD) and the *reflectance factor* (RF) of the obstruction; that is,

$$ERC = SC \times RD \times RF$$

Thus, if 25% of the sky is obstructed by a building with a 20% RF, we have

$$ERC = SC \times 0.25 \times 0.20$$

For this particular example, then

$$ERC = 5\% \text{ of } SC$$

to be added to the remaining 75% of SC (25% of the sky was obstructed).

The *internally reflected component* (IRC) represents the light received at the point under consideration that has been reflected from interior surfaces. IRC is subdivided into reflected skylight ($IRC_1$) and reflected ground light ($IRC_2$). $IRC_2$ is generally small, and $IRC \cong IRC_1$. IRC is, therefore, primarily dependent upon interior surface reflectances and upon the amount of window glazing, and becomes a large portion of DF deep

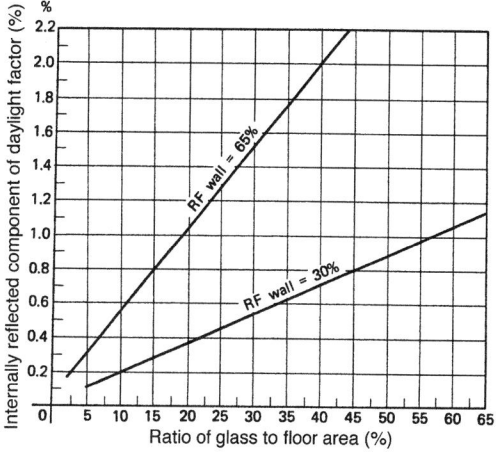

**Fig. 14.19** *Plot of the IRC of the daylight factor as a function of amount of glazing, expressed as per DF (i.e., as a percentage of exterior illuminance). As expected, the effect of a lighter wall finish becomes more pronounced as the fenestration area increases.*

within an interior space (see Table 14.1 for wall reflectance factors and Fig. 14.19 illustrating IRC as a function of the amount of glazing). IRC is normally calculated using published interreflectance tables, as direct calculation is extremely complex.

Typical curves for both horizontal and vertical daylight factors for a room with single (unilateral) sidelighting (windows on one side) are shown in Fig. 14.20. These curves are produced by a long-hand daylight-protractor-aided technique (Building Research Station, London). Any change in parameters, such as window dimensions or height above the working plane, ceiling height, surface reflectance, ground reflection, and obstructions, alters these

**TABLE 14.1 Effect of Wall Reflectance Factor on the Proportion of IRC in the DF**

| Distance from Window in ft (m) | 30% Wall Reflectance | | 60% Wall Reflectance | |
|---|---|---|---|---|
| | Total DF | $\frac{IRC}{DF}$ (%) | Total DF | $\frac{IRC}{DF}$ (%) |
| 0 | 30 | 1 | 31 | 3.5 |
| 5 (1.5) | 16 | 1.9 | 17 | 6.5 |
| 10 (3.0) | 5.5 | 5.5 | 6.3 | 16.9 |
| 15 (4.5) | 2.1 | 14.3 | 2.9 | 37.9 |
| 20 (6.0) | 1.3 | 23 | 2.1 | 52.4 |

*Room data:*

Room 24 × 28 ft (7.3 × 8.5 m) 70% ceiling reflectance

Window on 28-ft wall—one side only 20% floor reflectance

Window area = 20% of floor area

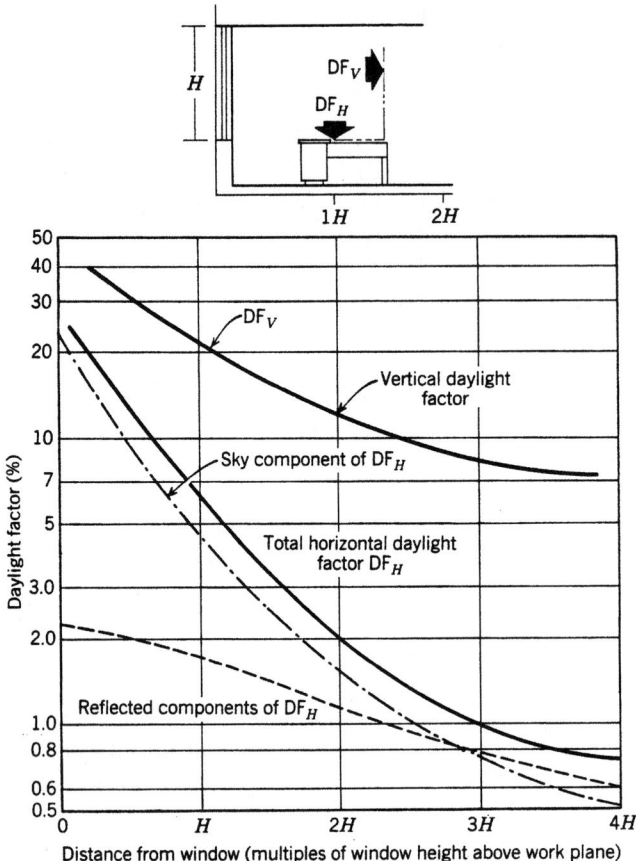

**Fig. 14.20** *Typical daylight factor curves for horizontal (DF$_H$) and vertical (DF$_V$) illuminance for a room with large windows on one side only. Note that the SC represents almost the entire DF near the window, but its proportion reduces at greater depths. There, inter-reflected light constitutes 50% of the available daylight.*

curves and requires recalculation and replotting. Exact calculation of even a few variants for a space is a tedious and time-consuming procedure.

The following manual methods describe alternative approaches available to save time and increase accuracy:

1. Use of simplifications, such as standard curves, tabular data, or the CIE method.
2. Use of a library of graphic light distribution plots with varying parameters.
3. Use of a less laborious manual calculation procedure. One such technique is known commonly as the *lumen method* or the *IES method*.
4. Use of computer simulation software.

Designers may use daylight factor criteria as a starting point for daylight design, translating the DF values (such as those given in Table 14.2) into actual illuminances in footcandles (lux) and comparing the results to recommended illuminance values. As an example, consider two cities in the United States that have overcast skies for appreciable portions of the year—Columbus, Ohio (40° N latitude) and Seattle, Washington (48° N latitude). Table 14.3 compares illuminance values calculated by the DF method with those recommended by IESNA and the Chartered Institution of Building Services Engineers (CIBSE). For most of the year (with the exception of winter), daylight provides all the light necessary for the tasks in Table 14.2. In this case, where the available exterior daylight is as low as 5000 to 7000 lux (465 to 650 fc), supplemental electric lighting would be required for all interior areas beyond H feet

**TABLE 14.2 Recommended Daylight Factors**

| Task | DF[a] |
|---|---|
| Ordinary seeing tasks, such as reading, filing, and easy office work | 1.5–2.5% |
| Moderately difficult tasks, such as prolonged reading, stenographic work, normal machine tool work | 2.5–4.0% |
| Difficult, prolonged tasks, such as drafting, proofreading poor copy, fine machine work, and fine inspection | 4.0–8.0% |

*Source:* Millet and Bedrick (1980).

[a]Use the smaller DF values for southern latitudes with plentiful winter daylight.

(that is, one window height from the window—see Fig. 14.20).

In addition to the recommendations in Table 14.2, the ratio between the minimum and average daylight factor in a space, which relates to contrast ratios, should be no less than 30%:

$$\frac{DF_{min}}{DF_{avg}} \geq 0.3$$

The minimum daylight factor in any portion of a space should not drop below 0.5%, which is sufficient for circulation.

## 14.9 GUIDELINES FOR PRELIMINARY DAYLIGHTING DESIGN

Guidelines provide the designer with a variety of broadly based rules useful during the conceptual and schematic stages of design. Based upon design experience and lighting research, these guidelines assume overcast sky conditions. During design development, they may be used as a starting point for performance analyses that include other parameters such as sky conditions, orientation, and wall color, using computer simulation software, physical models, and calculations.

### (a) The 2.5H Guideline

This longstanding guideline in the lighting design field (Fig. 14.21) assumes that there will be sufficient workplane illuminance from a window up to a

**TABLE 14.3 Horizontal Illuminances ($E_H$) from Overcast Sky, at Selected Times, in Columbus, Ohio, and Seattle, Washington, Corresponding to the Recommended DF**

| Location | 10 AM Solar Altitude[a] | Available Daylight, $E_H$ fc (lux)[b] | DF Recommendation (%)[c] | Illuminance (fc, lux) Calculation from DF | Illuminance (fc, lux)[d] Recommendation |
|---|---|---|---|---|---|
| Columbus 40° N latitude | June 21 60° | 2100 (22,500) | 1.5–2.5 | 31–52 (338–563) | 28–47 (300–500) |
| | | | 2.5–4 | 52–84 (563–900) | 47–70 (500–750) |
| | | | 4–8 | 84–167 (900–1800) | 70–93 (750–1000) |
| | Mar./Sept. 21 41° | 1400 (15,500) | 1.5–2.5 | 21–35 (225–375) | 28–47 (300–500) |
| | | | 2.5–4 | 35–56 (375–600) | 47–70 (500–750) |
| | | | 4–8 | 56–112 (600–1200) | 70–93 (750–1000) |
| | Dec. 21 21° | 700 (7500) | 1.5–2.5 | 11–18 (113–188) | 28–47 (300–500) |
| | | | 2.5–4 | 18–28 (188–300) | 47–70 (500–750) |
| | | | 4–8 | 28–56 (300–600) | 70–93 (750–1000) |
| Seattle 48° N latitude | June 21 56° | 1950 (21,000) | 1.5–2.5 | 29–49 (315–525) | 28–47 (300–500) |
| | | | 2.5–4 | 49–78 (525–840) | 47–70 (500–750) |
| | | | 4–8 | 78–156 (840–1680) | 70–93 (750–1000) |
| | Mar./Sept. 21 36° | 1220 (13,000) | 1.5–2.5 | 18–30 (195–325) | 28–47 (300–500) |
| | | | 2.5–4 | 30–48 (325–520) | 47–70 (500–750) |
| | | | 4–8 | 48–97 (520–1040) | 70–93 (750–1000) |
| | Dec. 21 14° | 500 (5400) | 1.5–2.5 | 8–13 (81–135) | 28–47 (300–500) |
| | | | 2.5–4 | 13–20 (135–216) | 47–70 (500–750) |
| | | | 4–8 | 20–49 (216–532) | 70–93 (750–1000) |

[a]From Appendix D.1.

[b]From Fig. 12.3.

[c]From Table 14.2.

[d]From CIBSE and IESNA recommendations (Tables 11.4, 11.5, 11.7, and 11.8).

ILLUMINATION

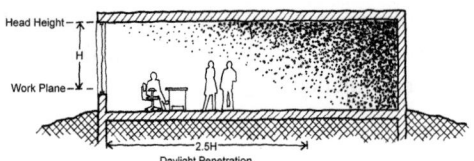

**Fig. 14.21** *Section shows the 2.5H guideline, which assumes that sufficient daylight for the desk plane will be delivered at a depth 2.5 times the height of the window above the desk plane. (Drawing by Jonathan Meendering; © 2004 Walter Grondzik; all rights reserved.)*

distance of 2.5 times the head height of the window above the workplane—assuming clear glazing, overcast skies, no major obstructions, and a total window width that is approximately half that of the exterior perimeter wall.

### (b) The 15/30 Guideline

This preliminary design guideline assumes that a 15-ft-wide (4.6-m) zone from a window wall (Fig. 14.22) can be daylit sufficiently for office tasks. The next 15-ft (4.6-m) zone can be partially daylit and supplemented with electric lighting. Zones farther than 30 ft (9.1 m) from the window would receive very little daylight. In schematic design, these areas might ideally be allocated to circulation.

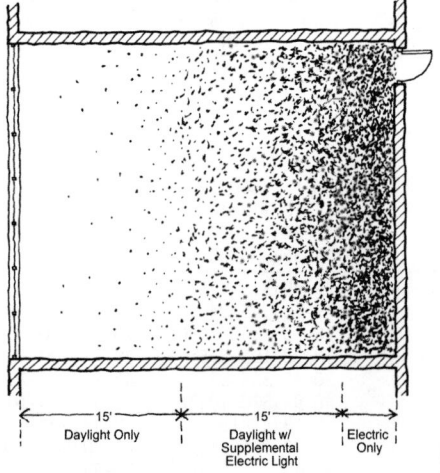

**Fig. 14.22** *Section shows the 15/30 guideline, which assumes that sufficient daylight will be delivered the desk plane at a 15-ft (4.8-m) distance from the window wall. The 15- to 30-ft (4.8- to 9.0-m) daylight zone will need supplementary electric lighting, and the zone beyond 30 ft (9 m) will receive virtually no daylight. (Drawing by Jonathan Meendering; © 2005 Walter Grondzik; all rights reserved.)*

### (c) The Sidelighting and Toplighting Daylight Factor Guideline

The size of windows, clerestories, or skylights may be estimated by using the simple formulas in Table 14.4, Parts *A* and *B*, which provide target daylight factor values. These design guidelines consider two factors: the height of the window in the wall and the window or skylight area compared to the floor area for each daylit space.

For an example of such a calculation, see the example previously discussed in Chapter 8: the two-story office building in Oregon (shown in Fig. 14.23c) of the Emerald People's Utility District. Briefly, after selecting a target value of $DF_{av} = 4.0\%$ from Table 14.2, the windows and clerestories (as skylights) were sized, using the DF guidelines for sidelighting and for the vertical monitor skylight from Table 14.4:

$$DF_{av} = 0.2 \ \frac{\text{window (or skylight) area}}{\text{floor area}}$$

Applied to the entire typical bay,

$$DF_{av} =$$
$$0.2 \ \frac{97 \text{ ft}^2 \text{ north} + 97 \text{ ft}^2 \text{ south} + 132 \text{ ft}^2 \text{ clerestory}}{1440 \text{ ft}^2}$$
$$= 0.45, \text{ or } 4.5\%$$

slightly above the target $DF_{av}$ of 4.0%.

Note that in this example, $DF_{min}$ from the sidelighting occurs near the center of the building, at about the point (on the second floor) where the most light is available from the skylight. Therefore, a relatively even daylighting distribution is expected on the upper floor. This is helped by the use of light shelves and the T-shaped windows shown in Fig. 14.23. The seasonal window size question (more needed in winter, less in summer) was answered in this example by the use of deciduous vines outside the south windows. In winter and cool spring, the vines are bare of leaves. Warm weather brings leafy shade lasting well into the warm fall.

Table 14.4, Part *C*, shows design guidelines for buildings with an atrium that provides daylighting to surrounding offices. Because the lowest daylight factor will occur in offices on the lowest floor (deepest within the atrium), the designer might find the required atrium *aspect ratio* first for the lowest floor and then size the atrium on that basis. This design approach would provide a higher daylight factor in all the offices higher in the atrium. The atrium

## TABLE 14.4 Daylight Factor Design Estimates for Overcast Sky Conditions

| PART A. **SIDELIGHTING**[A,B] |
|---|

$$DF_{av} = 0.2 \left( \frac{\text{window area}}{\text{floor area}} \right)$$

$$DF_{min} = 0.1 \left( \frac{\text{window area}}{\text{floor area}} \right)$$

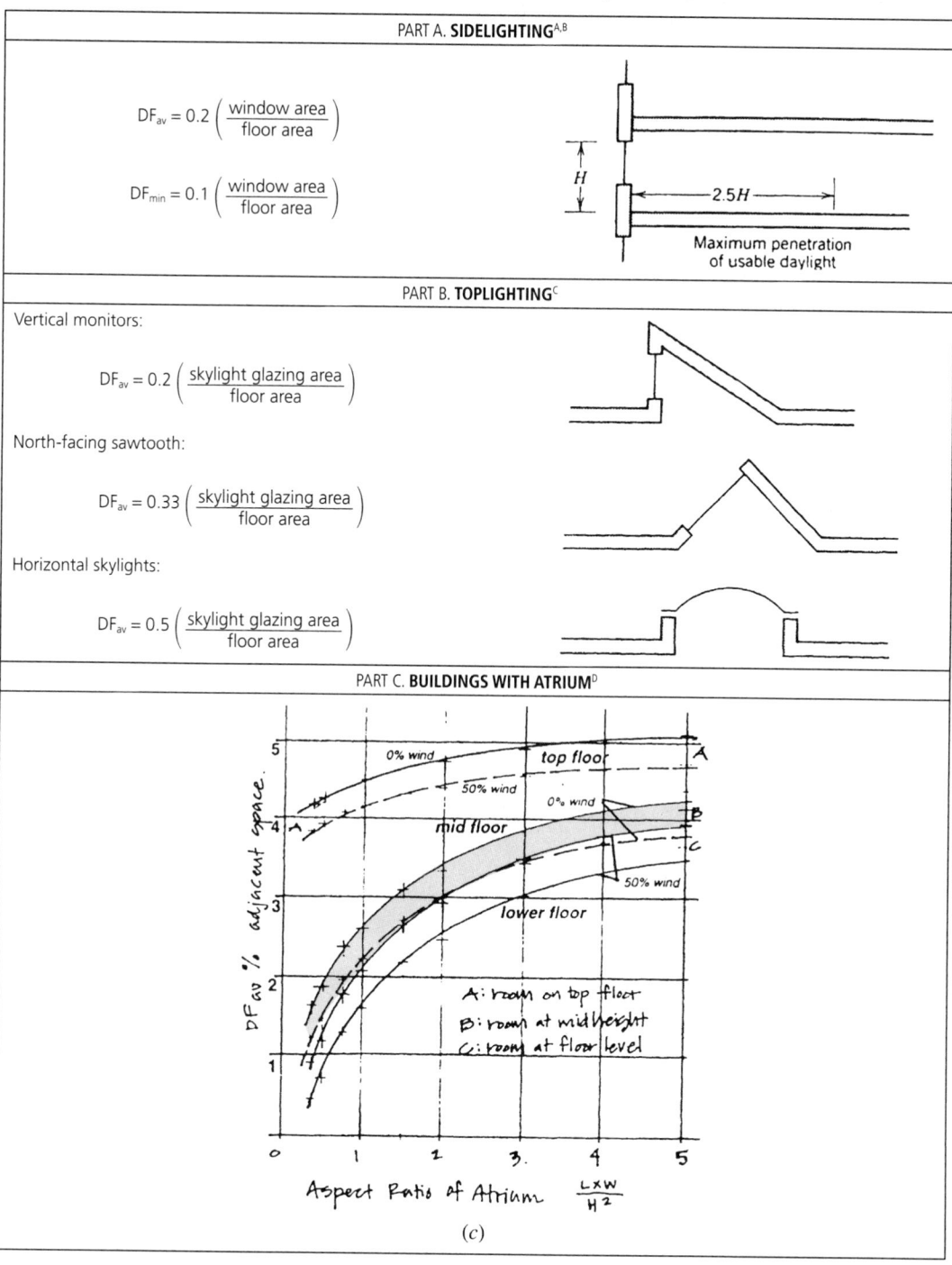

| PART B. **TOPLIGHTING**[C] |
|---|

Vertical monitors:

$$DF_{av} = 0.2 \left( \frac{\text{skylight glazing area}}{\text{floor area}} \right)$$

North-facing sawtooth:

$$DF_{av} = 0.33 \left( \frac{\text{skylight glazing area}}{\text{floor area}} \right)$$

Horizontal skylights:

$$DF_{av} = 0.5 \left( \frac{\text{skylight glazing area}}{\text{floor area}} \right)$$

| PART C. **BUILDINGS WITH ATRIUM**[D] |
|---|

(c)

*Source:* Parts A and B: Millet and Bedrick (1980). Part C: Brown and DeKay (2000).

[a]Assumes windows in one wall of a room with relatively light-colored surfaces.

[b]Window height/room depth relationships based on the works of R.G. Hopkinson (1966) and others at the British Research Station.

[c]Assumes an even distribution of such skylights in the roof so that an even distribution of light results in the room below: thus, only average DF, no minimum, is listed.

[d]Based on model tests of a square atrium with white walls open to the sky. "No windows" average atrium wall reflectance = 70%. "50% windows" average atrium wall reflectance = 40%. DF values are for an office of 9 × 9 × 3 m (30 × 30 × 10 ft). The window opening to the atrium is 1.5 m high × 9 m long (5 ft high × 30 ft long) and the sill height is 0.85 m (2.8 ft).

ILLUMINATION

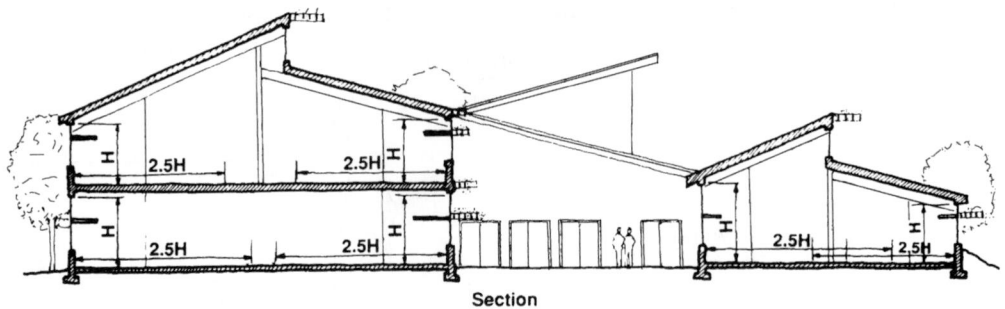

**Section**

Daylighting cross section (south to right)

(a)

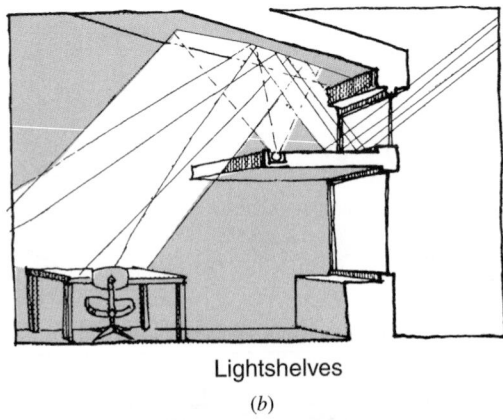

Lightshelves

(b)

North Façade

(c)

**Fig. 14.23** (a) Diagrammatic daylighting section of the office building for the Emerald People's Utility District near Eugene, Oregon. Light shelves (b) reduce the contrast between a lot of daylight near windows and too little daylight farther in. The resulting windows (on north façade) are T-shaped (c), with more glass area above the light shelves and less glass below.

aspect ratio = [(length × width)/height$^2$]. For a detailed discussion of the relationship of atrium size, rentable office floor area, and latitude see DeKay (1992).

## 14.10 DESIGN ANALYSIS METHODS

Because of the variability of daylight, the designer may provide a balance of illumination to save electric energy and reduce utility costs while addressing issues of glare, direct sunlight, and heat gain. The art and science of daylighting is largely about understanding how to control the admission of daylight into buildings.

In the following sections, several interior daylighting analysis methods are described. The manual methods range from hand calculations that address only minimum, maximum, and average conditions to physical scale models where surfaces and apertures are easily changed. The manual and graphic calculation methods are inexpensive but limited to simple spatial geometries. Computer simulation programs can produce detailed and realistic presentations in three-dimensional graphical form. Software is widely available, but its use is dependent upon cost and training, and the user must understand daylighting concepts and principles in order to interpret the results and to overcome the limitations of simulation. Physical models still offer the designer an economical, realistic, and accurate alternative. Additionally, the intuitive understanding provided by scale models may increase the client's understanding of lighting phenomena.

### (a) CIE Method

This method resulted from a search for a simple, rapid, straightforward, and reasonably accurate daylighting calculation method that would yield reliable results without the time-consuming constructions and calculations necessitated by other manual methods. After a study of considerable length and intensity, the CIE adopted and adapted a system developed in Australia by Dresler (*Daylight Design Diagrams*, 1963). The current CIE method was published in *Daylight*, 1970.

This system is based upon the daylight factor described previously as applied to the *standard overcast CIE sky*. Dresler developed a set of more than 100

curves covering rooms of varying proportions and fenestration. A typical curve is shown in Fig. 14.24. The curves relate minimum daylight factor (at a point 2 ft (0.6 m) from the wall opposite a window) to the maximum permissible room depth, for given reflectances and a standard window design, thus establishing the room's proportions. Depth, or width, is the dimension at right angles to the window wall.

The curves imply that the number of design variables is so large and daylight itself is so variable that a simple routine method can be based only on minimal conditions for a given (selected) daylight duration. Therefore, the diagrams give the *lowest* level of daylight that can reliably be expected for a given percentage (percentile) of normal working hours in sidelighted rooms and the *average* level in toplighted spaces.

*Advantages* of the system are:

1. Consideration of obstructions, exterior reflections, and interior reflections.
2. Applicability to a very wide range of side and top fenestration designs.
3. Establishment of required room proportions is architecturally more useful than solving for specific dimensions.

*Limitations* of the system are:

1. Inapplicable to clear sky and direct sun conditions.
2. Inapplicable to other than rectangular rooms.
3. Unusable with sunshading devices or high-reflectance ground.
4. Results give points of minimum, twice minimum, and four times minimum daylight only. Other points must be interpolated or extrapolated.
5. Window proportions and position in a wall are fixed.

Overall, the system accomplishes what it intended. The limitations listed are inherent in any quick, simplified daylight calculation technique.

The CIE system is usable in two modes:

1. Given complete architectural dimensional data, find interior illuminance.
2. Given incomplete architectural dimensional data and required interior illuminance, find maximum room depth and/or other room proportions that satisfy the illuminance requirement.

ILLUMINATION

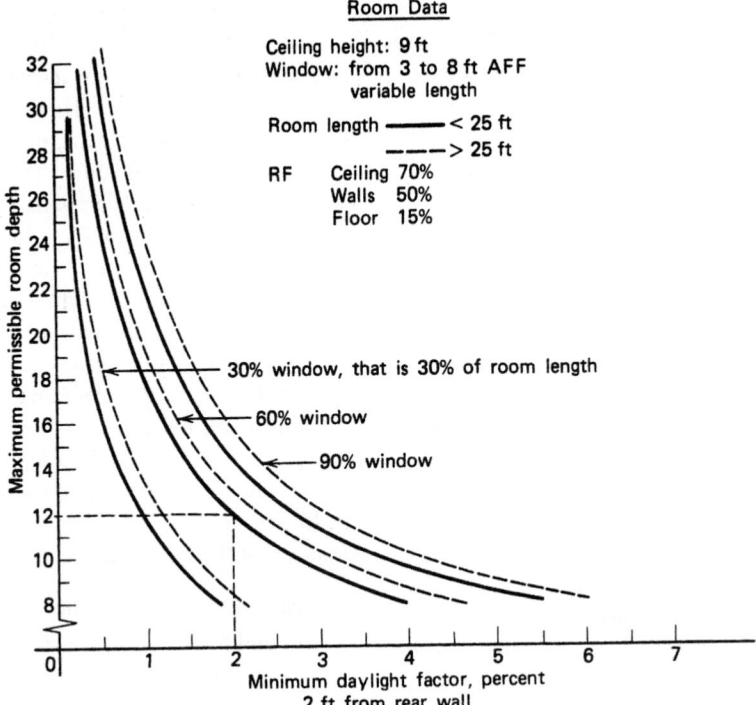

**Fig. 14.24** *Maximum room depth required to maintain a minimum daylight factor is proportional to window size. Thus, for a room less than 25 ft (7.6 m) long with a 5-ft-high window for 60% of the room's length, the depth cannot exceed 12 ft (3.7 m) if 2% DF is to be maintained at a point 2 ft (61 cm) from the rear wall. (From* Daylight Design Diagrams, *1963.)*

Mode 1 is simpler because it leads directly to an answer. For this reason, the designer should set the room length (window wall dimension) and percentage of fenestration of the window wall, leaving the room depth (perpendicular to window wall) as a variable. Alternatively, room length and depth may be set, with percentage of fenestration as the variable. Ceiling height is usually fixed. See Fig. 14.25 for sketches showing room parameters.

**EXAMPLE 14.1** An example of the CIE method in Mode 1 uses a classroom in a single-story Seattle elementary school, 25 ft (7.6 m) long, 18 ft (5.5 m) deep, and with a 9-ft, 6-in. (3-m) ceiling. It receives daylight unilaterally from windows totaling 18 ft (5.5 m) in length (see the room sketch in Fig. 14.26). Window glazing is wired glass having a transmittance of 80%. The school is situated in a dense residential area. Determine the portion of the year during which tasks requiring a minimum illuminance of 14 fc (150

lux) can be carried out by daylight throughout the room. Also, determine what illuminance levels can be maintained for 85% of daylight hours and at what distances from the window. (Note that 14 fc [150 lux] corresponds to a DF of 2% applied to an $E_H$ of 70 fc [7500 lux].)

*Calculation.* The latitude of Seattle is 47.6° N. The design condition for Seattle is solid overcast sky for 85% of the hours between 0900 and 1700. From Fig. 14.27, the minimum unobstructed horizontal illuminance $E_H$ during these hours is 67 fc (7200 lux).

**STEP 1.** *Determine the room depth in terms of window height.* Window height H is 5 ft (1.5 m) (see Fig. 14.26). In plan, the room depth is expressed as multiples of window height above sill level:

$$\frac{18\text{-ft depth}}{5\text{-ft window height}} = 3.6H$$

**STEP 2.** *Determine window coverage.* This variable is expressed as a percentage of the total room length based upon the width of the glazing used in the

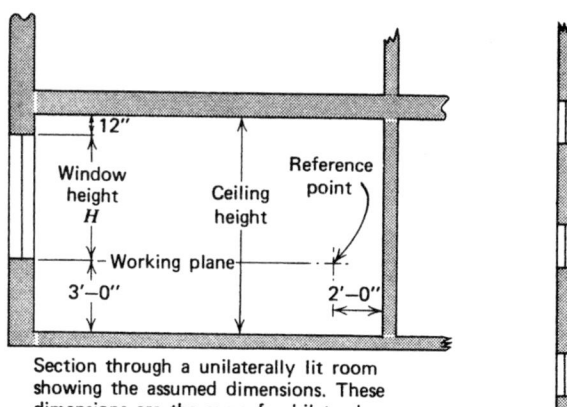

Section through a unilaterally lit room showing the assumed dimensions. These dimensions are the same for bilateral lighting except that the reference point is midway between the window walls.

(a)

Plane of window walls showing window width expressed as a percentage of total room length.

(b)

**Fig. 14.25** *Sketches indicating the parameters of the CIE calculation system. (a) A vertical section through a room with dimensional data relevant to this system. Note that the sill height has been selected to coincide with a working plane at 90 cm (3 ft). The height of the working plane usually varies between 76 and 91 cm (30 and 36 in.), the former being more common in North America, the latter in Europe. A lower sill contributes only ground-reflected light at the working plane. Where the window sill is significantly above the working plane (i.e., short windows high on a wall), this design system is inapplicable. (b) Calculation size (length) of windows with respect to overall room length. (From* Daylight, International Recommendations for the Calculation of Natural Daylight, *1970, Commission Internationale de l'Eclairage; reproduced with permission.)*

room. It is assumed that the window head is 12 in. (300 mm) below the ceiling. This distance was selected as the representation of best practice "without being duly optimistic."

$$\frac{3 \times 6 \text{ ft}}{25 \text{ ft}} = 72\%$$

**STEP 3.** *Determine the* design *daylight factor and the* service *daylight factor.* The design daylight factor is a point 2 ft (0.6 m) from the wall opposite the window—on the window centerline, which is the minimum daylight factor relative to the maximum permissible room depth. From Fig. 14.28, for a room length of 25 ft (7.6 m), ceiling height 9 ft, 6 in. (2.9 m), window coverage 72%:

design daylight factor at 3.6H = 1.3

The service daylight factor takes into account correction factors such as glazing transmission and dirt accumulation. The service daylight factor is a product of the design daylight factor and correction factors.

$$DF_{service} = DF_{design} \times \text{correction factors}$$

Correction factors from Table 14.5 and Fig. 14.29 are:

Glass transmission 0.95

Glass cleanliness 0.8

Therefore

$$DF_{service} = 1.3 \times 0.95 \times 0.8 \cong 1.00$$

**STEP 4.** *Determine the required exterior illuminance.* Use the service daylight factor in Eq. 14.2 to obtain required exterior illuminance.

$E_H$, required exterior illuminance

$$= \frac{\text{required interior illuminance}}{DF_{service}} \quad (14.2)$$

$$E_H = \frac{150 \text{ lux min}}{1.00} \times 100 = 15,000 \text{ lux}$$

**STEP 5.** *Obtain the percentage of hours between 0900 and 1700 during which the required illuminance is maintained.* From Fig. 14.27 with the given conditions (roughly 48 °N latitude, exterior illumi-

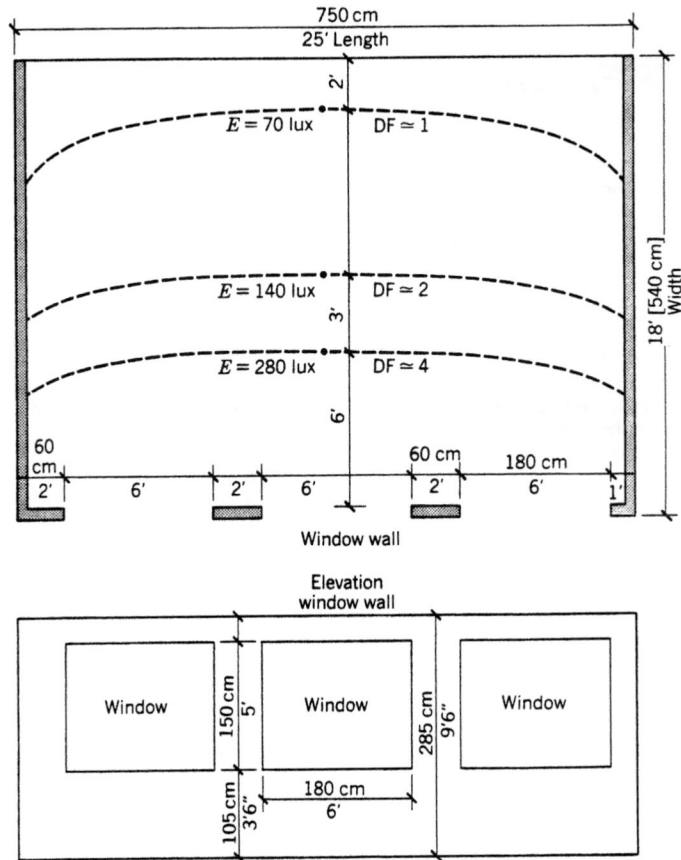

**Fig. 14.26** *Plan and window wall elevation of the Seattle classroom calculation example using the CIE method. The three daylight contours are estimated based upon the calculated center point. They represent the levels maintained for 85% of daylight hours. Levels twice as high are maintained for 60% of the daylight hours.*

nance 15,000 lux), an illuminance of 150 lux will be maintained for less than 60% of the hours between 0900 and 1700 (for time periods other than this, see Table 14.5c). The level that *is* maintained for 85% of the hours is

$$E_{min} = 7200 \text{ lux} \times 0.01 \text{ (DF of 1.0)} \approx 72 \text{ lux}$$

**STEP 6.** *Determine the locations in the room that will receive adequate illuminance.* Figure 14.30 shows the distance from the window at which the daylight factor is twice or four times the minimum. Using Fig. 14.30a, the room depth from Step 1 is 3.6H. This point intersects with the $\alpha = 0°$ angle of obstruction line at a 1.8H distance (1.8 × 5 ft = 9 ft [2.7 m]) from the window at which the daylight factor doubles.

$$7200 \text{ lux} \times 0.02 \text{ (a DF of 2.0)} \approx 144 \text{ lux}$$

Using Fig. 14.30b, the room depth from Step 1 is 3.6H. This point intersects with the $\alpha = 0°$ angle of obstruction line at a 1.2H distance (1.2 × 5 ft = 6 ft [1.8 m]) from the window at which the daylight factor quadruples.

$$7200 \text{ lux} \times 0.04 \text{ (a DF of 4.0)} \approx 288 \text{ lux}$$

These calculated illuminance levels are accurate only at the centerline of the window wall. By visual estimate and extrapolation, rough contours can be drawn and are shown in Fig. 14.28 and Fig. 14.31 (for comparison with the graphic method discussed in the next section). For bilateral sidelighting, top-lighting (skylights, sawtooth roofs, and monitor roofs), and calculations for other $\alpha$ angles of obstruction, additional information is found in the CIE document (*Daylight,* 1970). ∎

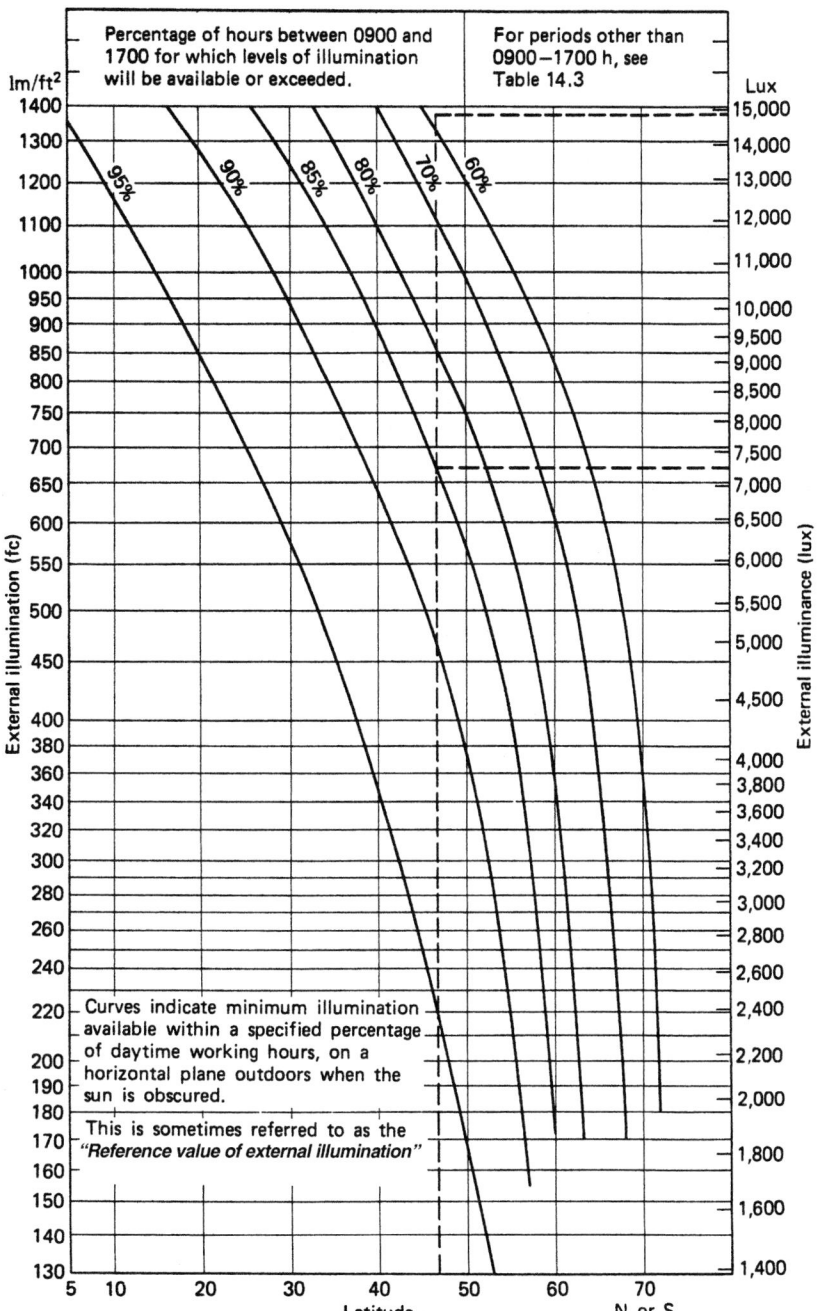

**Fig. 14.27** *Minimum maintained external illuminance as a function of latitude for a given percentage of the normal working day.* (From Daylight, International Recommendations for the Calculation of Natural Daylight, *1970, Commission Internationale de l'Eclairage; reproduced with permission.*)

ILLUMINATION

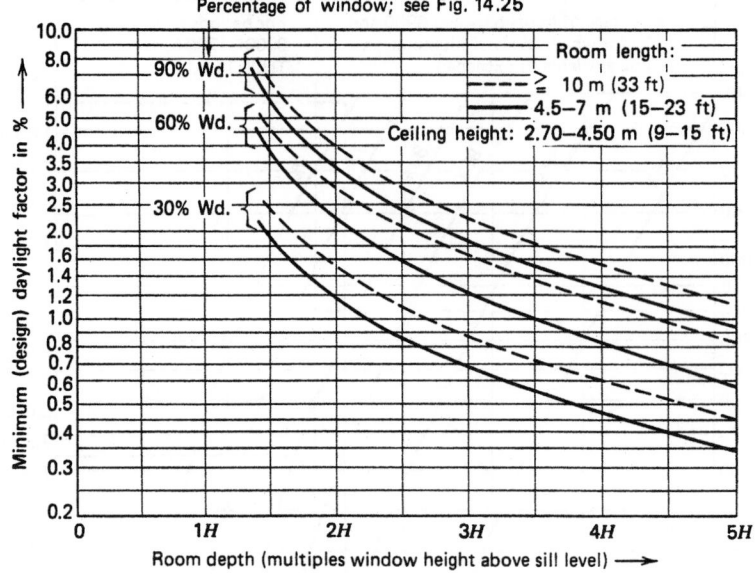

Relation between room depth and minimum (design) daylight factor
(For various room lengths and window widths)

Percentage of window; see Fig. 14.25

Room length:
------ ≥ 10 m (33 ft)
——— 4.5–7 m (15–23 ft)
Ceiling height: 2.70–4.50 m (9–15 ft)

90% Wd.
60% Wd.
30% Wd.

Minimum (design) daylight factor in %

Room depth (multiples window height above sill level) ⟶

**Fig. 14.28** *Basic design diagram that relates minimum daylight factor to room depth. Inasmuch as room depth is expressed in terms of window height, the curves effectively relate minimum daylight factor (2 ft [61 cm] from the back wall) to room proportion. (From* Daylight, International Recommendations for the Calculation of Natural Daylight, *1970, Commission Internationale de l'Eclairage; reproduced with permission.)*

## TABLE 14.5 Correction Factors to Be Used in CIE Daylight Calculations

| A. CORRECTION FACTOR TO ACCOUNT FOR GLASS TRANSMITTANCE | |
|---|---|
| **Diffuse Transmittance of Glass (%)** | **Correction Factor** |
| 80 | 0.95 |
| 70 | 0.80 |
| 60 | 0.70 |
| 50 | 0.60 |
| 40 | 0.45 |
| 30 | 0.35 |

| B. CORRECTION FACTORS TO ACCOUNT FOR DIRT ACCUMULATION ON GLASS | | | | |
|---|---|---|---|---|
| | | **Angle of Slope (Measured to the Horizontal)** | | |
| **Locality** | **Class of Industry** | **90–75°** | **60–45°** | **30–0°** |
| Country or outer-suburban area | Clean | 0.9 | 0.85 | 0.8 |
| | Dirty | 0.7 | 0.6 | 0.55 |
| Built-up residential area | Clean | 0.8 | 0.75 | 0.7 |
| | Dirty | 0.6 | 0.5 | 0.4 |
| Built-up industrial area | Clean | 0.7 | 0.6 | 0.55 |
| | Dirty | 0.5 | 0.35 | 0.25 |

| C. PERCENTAGES TO USE WHEN FIGURE 14.27 CURVES ARE APPLIED TO PERIODS OTHER THAN 09.00–17.00 | | | | | | |
|---|---|---|---|---|---|---|
| **Curve in Figure 14.27** | **95%** | **90%** | **85%** | **80%** | **70%** | **60%** |
| Alternative period | | | Percentage of alternative period | | | |
| 07.00–15.00 | 95 | 90 | 85 | 80 | 70 | 60 |
| 08.00–16.00 | 100 | 100 | 95 | 85 | 70 | 60 |
| 07.00–17.00 | 95 | 85 | 75 | 65 | 55 | 45 |
| 06.00–18.00 | 75 | 70 | 65 | 60 | 50 | 40 |

*Source:* Daylight, International Recommendations for the Calculation of Natural Daylight *(1970).*

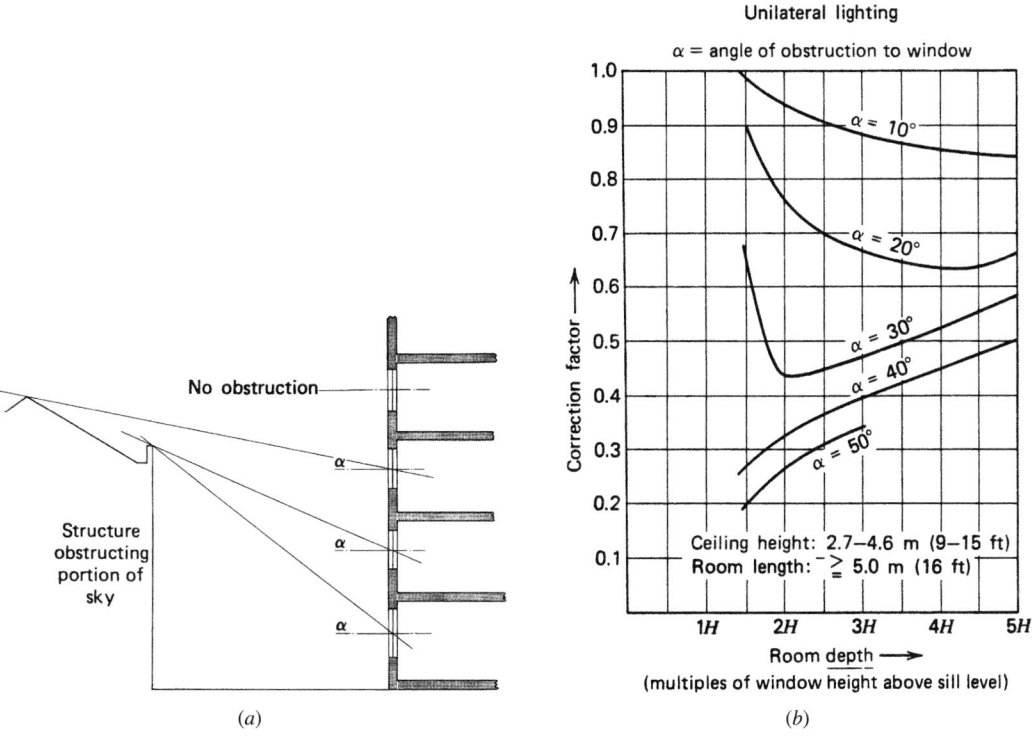

**Fig. 14.29** (a) Angle of obstruction α of an external object. (b) Correlation factors to account for the influence of external obstructions on the minimum daylight factor. (From Daylight, International Recommendations for the Calculation of Natural Daylight, 1970, Commission Internationale de l'Eclairage; reproduced with permission.)

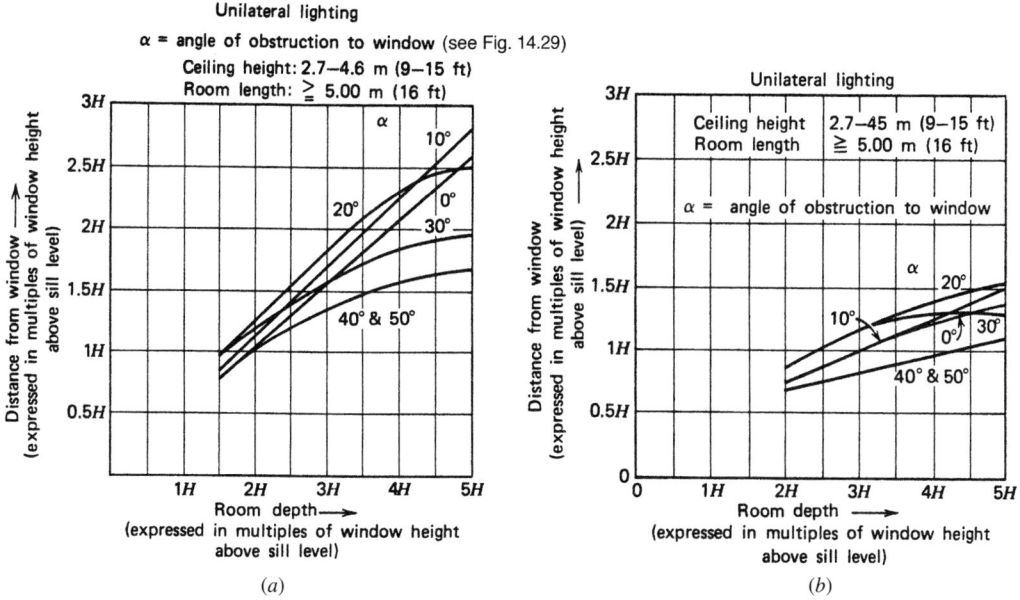

**Fig. 14.30** (a) Distance from the window at which the daylight factor is twice the minimum daylight factor. (b) Distance from the window at which the daylight factor is four times the minimum daylight factor. (From Daylight, International Recommendations for the Calculation of Natural Daylight, 1970, Commission Internationale de l'Eclairage; reproduced with permission.)

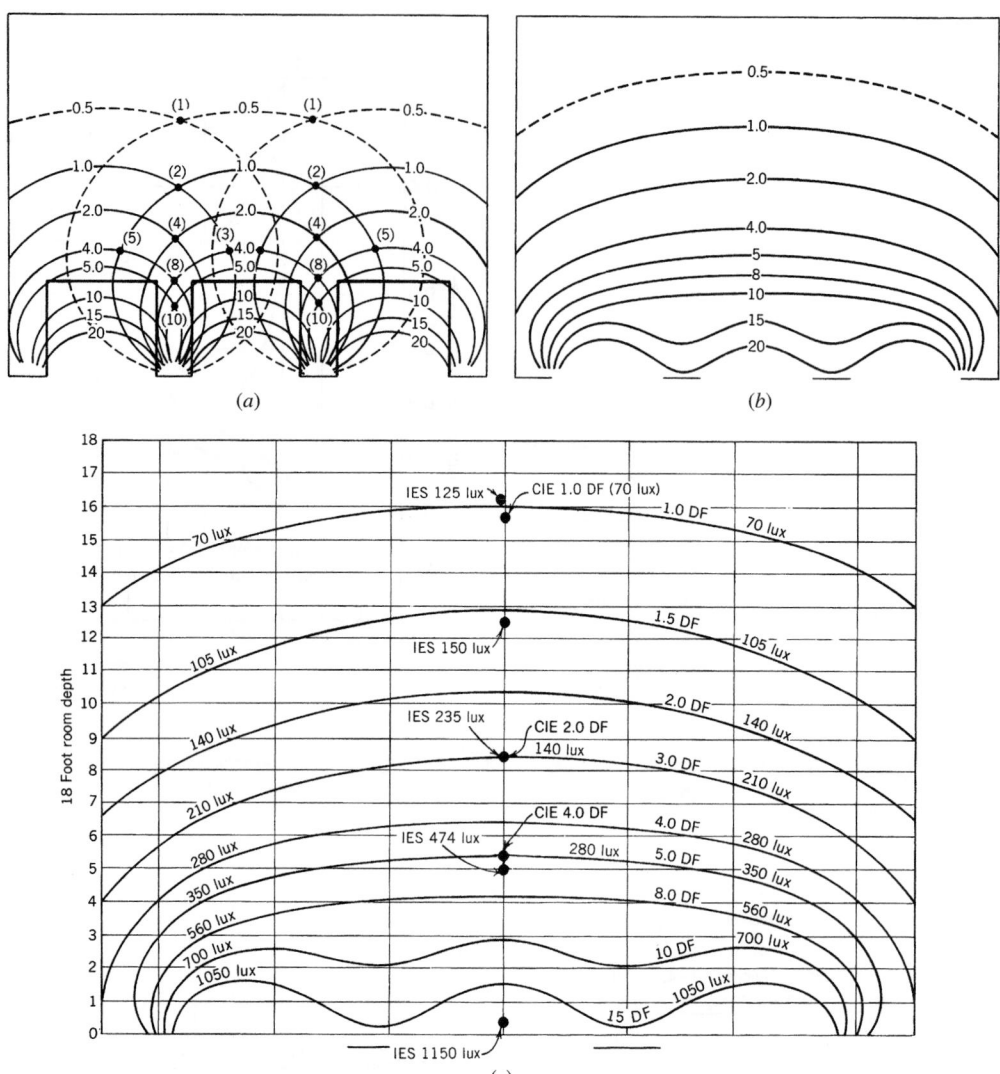

**Fig. 14.31** *(a) Daylight contours for each window of Fig. 14.26 are plotted on the floor plan of the room being studied. Numbers in parentheses are combined SCs. (b) The isolux contours of (a) are combined to form new isolux contours that represent the total SC of daylight within the room. (c) The final isolux contours are calculated, including correction factors. The numbers represent daylight factors. Note the variance between these contours and the points calculated by the CIE method. The five design points calculated by the IESNA method are also shown. A comparison of the results of the three methods described indicates that on the room centerline (a location where comparison of all methods is possible), agreement is within engineering accuracy (see text discussion).*

In summary, the CIE method is relatively simple but provides only limited data on predicted performance. In this example, its exterior illuminance data (7250 lux from Fig. 14.27) seem to agree well with a measured average value of 7200 lux for Seattle. From the rough contours shown in Fig. 14.26, an integrated daylight and electric lighting strategy should be designed for areas further from the window wall.

(*Note:* The term *design* illuminance as used with the CIE method differs from common lighting system design usage, where *design illuminance* is used to identify the criteria (benchmark) illuminance established for a space or position; *initial illuminance* identifies the illuminance actually provided at system startup, and *maintained illuminance* is the illuminance provided after some defined time period.)

## (b) Graphic Daylighting Design Method (GDDM)

This method, which applies to overcast sky conditions and shows results as daylight factor (isolux) contours within a room (rather than individual daylight factors at specific points), was developed by Millet and Bedrick (1980). Its primary advantage over the CIE method is that its results are a family of daylight factor contours that are more useful to a lighting designer than is numerical output. The disadvantages of this method are that it is not readily applicable to clear-sky conditions and it requires that a designer acquire a "library" of 200 or so patterns that cover most design situations. An outline of the method is presented here.

A computer simulation program (UWLIGHT) developed the daylight distribution patterns resulting from either sidelights or skylights. To generalize the system, windows are identified by height to width proportion ($H/W$), and the position of the isolux contours on the plan is determined by the ratio of the height of the sill above the work plane to the window height. The GDDM method can account for high windows, clerestories, and other designs intended to introduce daylight deep into a space—something that the CIE method cannot do because it is restricted to a sill height *at* the work plane.

**EXAMPLE 14.2** Figure 14.32 shows a typical isolux pattern for a window whose *H/W* ratio is 0.8 and whose sill is at the work plane. This particular window pattern was selected because it corresponds to the window in the previous example, Fig. 14.26, enabling graphical comparison.

*Calculation.* Employing the GDDM method with the dimensions of the Seattle classroom used in the last section:

**STEP 1.** *Determine the window proportion.*
Referring to Fig. 14.26, the window proportion is

$$\frac{H}{W} = \frac{5}{6} = 0.83$$

**STEP 2.** *Select the appropriate window pattern.* In this case, Fig. 14.32 (selected from a library of isolux patterns developed by Millet and Bedrick) is the closest pattern to match the example. *S/H = 0* indicates that the pattern begins at the window wall, as shown.

**STEP 3.** *Develop an isolux pattern for each window of the space.* On a plan of the room, trace the isolux pattern for each window (Fig. 14.31*a*). The patterns overlap because the windows are close together. Where contours meet, the daylight factors of the contours are added together, producing values for the new combined contours. The combined

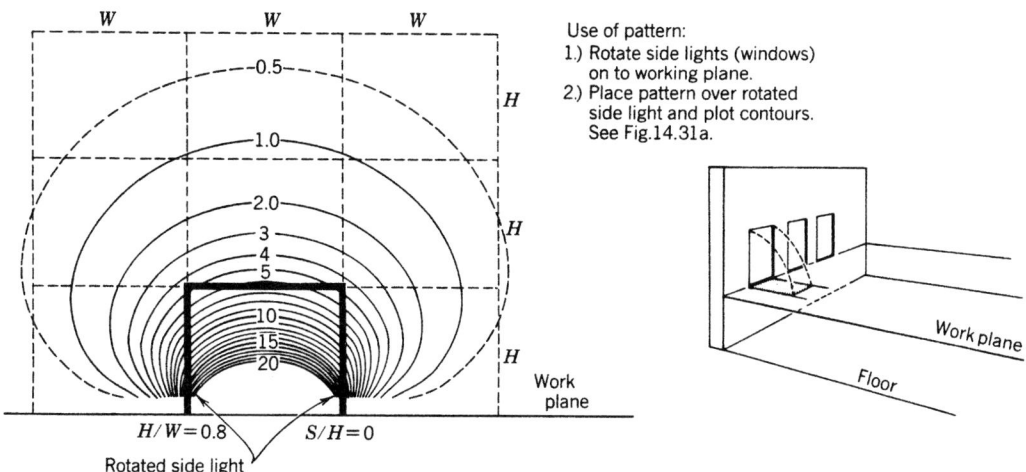

**Fig. 14.32** *Typical isolux contour map for a window with a height-to-width ratio of 0.8 and a sill at the working plane elevation. Numbers represent the SC of daylight factors for an overcast sky condition. The rectangles are the window outlines rotated (projected) onto the working plane. See insert. (From Millet and Bedrick, 1980, p. 191.)*

contours and their daylight factors are shown in Fig. 14.31*b*.

**STEP 4.** *Make corrections to the isolux pattern.* The value of the combined contours is corrected to account for internally reflected components of daylight plus light reduction due to glazing. The final contours are shown in Fig. 14.31*c*. Note that this diagram gives the designer a much more complete picture of the daylight contours than do the results of the CIE method. For the purpose of comparison, the three calculated daylight factors from the CIE method are shown in Fig. 14.31*c*. ∎

## (c) IESNA Lumen Method

IESNA developed the lumen method for calculating daylight availability (published as RP-23-89, *Recommended Practice for the Lumen Method of Daylight Calculations*). Although inexpensive, like many manual methods it is limited in application—in this case, to rectilinear spaces with flat ceilings. A trade-off between usability, learning curve, and cost, however, is often made when selecting a design method.

The calculation procedure for sidelighting is discussed in this section, as this is a more frequently encountered strategy than toplighting. The method, as fully described in RP-23-89, consists of four detailed steps. In the discussion that follows, the same notation and terms found in RP-23-89 are used *except* for *bearing angle*, which is referred to in the IESNA procedure as *solar window azimuth*. The term *bearing angle* is commonly used in international sources.

*Characteristics of the Method.* The IESNA method is probably the most flexible manual technique available. It has the following major characteristics for sidelighting:

1. It takes into account reflected light from the ground and adjacent structures, as well as the reduction in sky light due to such structures.
2. It cannot accommodate direct sunlight, but conversely, it readily accommodates the shading devices normally used to block direct insolation.
3. Provision is made for various types of glazing, as well as common window controls such as horizontal and vertical blinds.

4. The principles of the zonal cavity calculation approach for interior lighting (see Chapter 15) are applied. The window height determines the cavities (i.e., the floor cavity extends to the window sill height and the ceiling cavity from the top of the window to the ceiling). The room cavity is therefore the window height.
5. The work plane is always at the sill height of the window. Where this is decidedly not the case (a difference of up to 1 ft is usually negligible), such as when a clerestory or a floor-to-ceiling window is used, work plane illuminance can be calculated by superposition. For instance, with a clerestory, subtract a work-plane-to-clerestory sill height window from a work-plane-to-top-of-clerestory window to obtain the desired result. A degree of inaccuracy is unavoidable in the calculation when the work plane is *above* the sill.
6. Cavity reflectances are fixed (Fig. 14.33) at 70%-50%-30% for ceiling, room, and floor cavities, respectively.
7. The system calculates only five points in a room on the window centerline. As noted with reference to the three points calculated by the CIE method, this is not normally sufficient to give a picture of the interior daylight distribution.
8. The method is usable in only one mode; that is, given location and full dimensional data, daylighting can be calculated. It cannot readily be used to determine desirable room proportions, given the other data, as can the CIE method.

**EXAMPLE 14.3** To again use the Seattle classroom as an example, here are the conditions of the problem:

*Location:* Seattle, Washington.

*Latitude:* 47.6° N.

*Room:* 25 ft (7.6 m) long, 18 ft (5.5 m) deep, with a 9-ft, 6-in. (3-m) ceiling.

*Window:* height 5 ft (1.5 m) above a 30-in. (760-mm) working plane; total length, 18 ft (5.5 m).

*Transmittance:* 85%; net glass area 92%.

*Ground Reflectance:* not previously specified. Assume an extensive area of mixed grass and concrete walk-

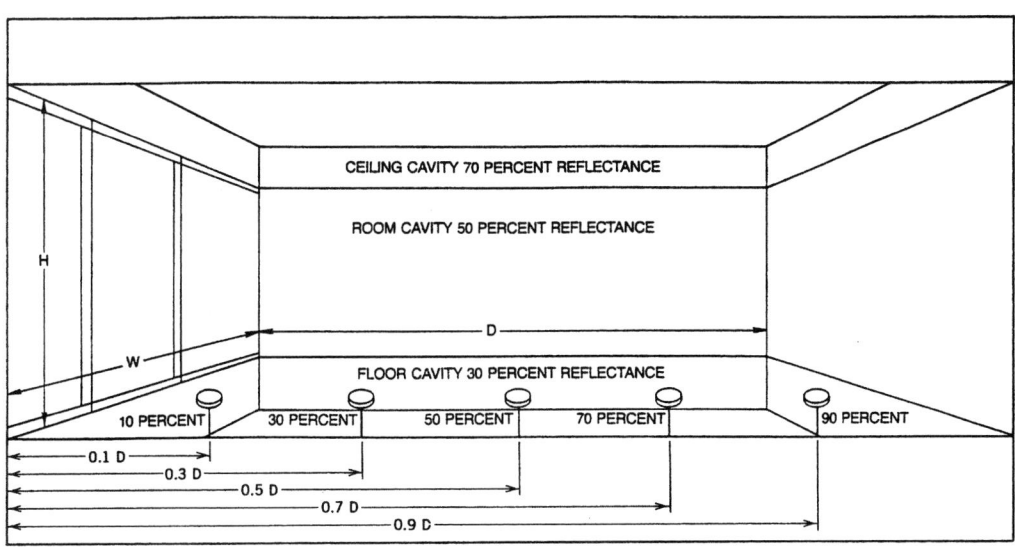

**Fig. 14.33** *Standard conditions in a room for daylighting calculations: sidelighting. (IESNA Recommended Practice for the Lumen Method of Daylight Calculations, RP-23-89.)*

ways, with an average overall reflectance of 20% (see Table 14.3). For a more accurate calculation of reflectance, see the method described in the IES *Recommended Practice for the Calculation of Daylight Availability* (1994).

**Calculation.** Illuminance $E_i$ will be found at the five points shown in Fig. 14.33 for a spring day and a winter day at 10 A.M. and 2 P.M. Assume that the sky is overcast so that a direct comparison with other methods can be made. Reflectances are assumed to be 70% for the ceiling, 50% for the wall, and 30% for the floor to correspond to the IESNA method standard conditions.

**STEP 1.** *Determine the vertical and horizontal illuminance on the exterior of the window.*

Using the solar altitude for the spring and winter day (Table D.1, solar data), the vertical and horizontal illuminances can be found from Fig. 12.3. Calculate $E_{xhk}$, the half-sky illuminance (a vertical window sees only half of the sky). The results are compiled in Table 14.6.

Vertical illuminance from the ground is

$$E_{xvg} = RF_g \times \frac{E_{xHk}}{2} \qquad (14.3)$$

Dec. 21: $E_{xvg} = RF_g \times \frac{E_{xHk}}{2} = 0.2\,(5380) = 1076$ lux

Mar. 21: $E_{xvg} = 0.2\,(13,340) = 2670$ lux

**STEP 2.** *Determine net transmittance of the window.* A number of factors affect the transmittance of light through glazing. The net transmittance is the product of the *glazing* transmittance (Table 14.7) and a *light loss factor* (Table 14.8), which represents the cleanliness of the window. The net transmittance should also account for the net glazing area (i.e., gross window area less mullions, glazing bars, and so on) and any other factor (such as insect screens) that would reduce actual transmittance.

Glass transmittance is 85% (Table 14.7), light loss factor is 0.9 (Table 14.8), and the net glass area

**TABLE 14.6 Vertical and Horizontal Illuminance Values for Spring and Winter, Seattle, Washington**

| Solar Altitude (Appendix D.1) | Vertical Window Illuminance, $E_{xvk}$ (Fig. 12.3) | Horizontal Illuminance from Full Sky, $E_{xHk}$ (Fig. 12.3) | Horizontal Illuminance from Half Sky, $E_{xhk}$ |
|---|---|---|---|
| Dec. 21: 14° | 200 fc (2,150 lux) | 500 fc (5,380 lux) | 250 fc (2,690 lux) |
| Mar. 21: 36° | 500 fc (5,380 lux) | 1,240 fc (13,340 lux) | 620 fc (6,670 lux) |

## TABLE 14.7 Transmittance Data of Glass and Plastic Materials

| Material | Approximate Transmittance (%) |
|---|---|
| Polished plate/float glass | 80–90 |
| Sheet glass | 85–91 |
| Heat-absorbing plate glass | 70–80 |
| Heat-absorbing sheet glass | 70–85 |
| Tinted polished plate | 40–50 |
| Figure glass | 70–90 |
| Corrugated glass | 80–85 |
| Glass block | 60–80 |
| Clear plastic sheet | 80–92 |
| Tinted plastic sheet | 9–42 |
| Colorless patterned plastic | 80–90 |
| White translucent plastic | 10–80 |
| Glass-fiber-reinforced plastic | 5–80 |
| Double glazed—two lights clear glass | 77 |
| Tinted plus clear | 37–45 |
| Reflective glass[a] | 5–60 |

*Source: IES Recommended Practice for the Lumen Method of Daylight Calculations RP-23-1989; reprinted with permission.*

[a]Includes single glass, double-glazed units, and laminated assemblies. Consult manufacturer's material for specific values.

## TABLE 14.8 Typical Light Loss Factors for Daylighting Design

| Location | Light Loss Factor Glazing Position | | |
|---|---|---|---|
| | Vertical | Sloped | Horizontal |
| Clean areas | 0.9 | 0.8 | 0.7 |
| Industrial areas | 0.8 | 0.7 | 0.6 |
| Very dirty areas | 0.7 | 0.6 | 0.5 |

*Source: IES Recommended Practice for the Lumen Method of Daylight Calculations RP-23-1989; reprinted with permission.*

is given as 92%. The net transmittance ($\tau$) of the window is a product of these factors:

$$\tau = (0.85)(0.9)(0.92) = 0.70$$

**STEP 3.** *Select coefficients of utilization* for the five calculation locations (10%, 30%, 50%, 70%, and 90%) (from Tables C.16–C.21) on the basis of room dimensions and the portion of the sky seen by the window. The SC seen by the window is determined by the ratio of vertical to horizontal illuminance at the window.

SC $E_{xvk}/E_{xhk}$:

Dec. 21: 2150/2690 = 0.8;

Mar. 21: 5380/6670 = 0.8 (Table C.16)

The $E_{xvk}/E_{xhk} = 0.8$ value corresponds most closely to the $E_{xvk}/E_{xhk} = 0.75$ value of Table C.16. Use the following values to find the coefficients of utilization, $CU_k$:

$$\frac{\text{room depth}}{\text{window height}} = 18/5 = 3.6;$$

use 4.0 (first column variable)

$$\frac{\text{window length}}{\text{window height}} = 18/5 = 3.6;$$

use 4.0 (first row variable)

The ground component vertical illuminance ($E_{xvg}$) is the product of the ground reflectance (Table 14.9) and half of the horizontal illuminance (Eq. 14.3). The sky and ground component values are com-

## TABLE 14.9 Reflectances of Building Materials and Outside Surfaces

| Material | Reflectance (%) |
|---|---|
| Aluminum | 85 |
| Asphalt (free from dirt) | 7 |
| Bluestone, sandstone | 18 |
| Brick | |
| Light buff | 48 |
| Dark buff | 40 |
| Dark red glazed | 30 |
| Red | 15 |
| Yellow ochre | 25 |
| White | 75 |
| Cement | 27 |
| Chromium | 65 |
| Concrete | 55 |
| Copper | 40 |
| Earth (moist cultivated) | 7 |
| Granolite pavement | 17 |
| Glass | |
| Clear | 7 |
| Reflective | 20–30 |
| Tinted | 7 |
| Grass (dark green) | 6 |
| Gravel | 13 |
| Granite | 40 |
| Marble (white) | 45 |
| Macadam | 18 |
| Marble | 45 |
| Paint (white) | |
| New | 75 |
| Old | 55 |
| Plaster | |
| Smooth | 80 |
| Rough | 40 |
| Stippled | 40 |
| Slate (dark clay) | 8 |
| Snow | |
| New | 74 |
| Old | 64 |
| Vegetation (mean) | 25 |

*Source:* Values compiled from Lam (1986), Stein and Reynolds (1992), and the Lighting Design Lab (© 2005, used with permission).

piled in Table 14.10 for each of the five reference locations.

**STEP 4.** *Calculate the illuminances for each of the five reference locations.* (Table 14.11 tabulates the illuminance values from this calculation.) The basic equation for each location is

$$E_i = \tau\,(E_{xvk} \times CU_k + E_{xvg} \times CU_g) \qquad (14.4)$$

where

$E_i$ = interior illuminance at a specific reference point

$\tau$ = net transmittance of the window

$E_{xvk}$ = exterior vertical illuminance at the window from half of the sky (an unobstructed vertical window sees only half of the sky)

$E_{xvg}$ = exterior vertical illuminance at the window from the ground

$CU_k$ = coefficient of utilization for sky light

$CU_g$ = coefficient of utilization for ground light

To compare the lumen method results with the CIE and GDDM methods, the values calculated for December 21 are plotted on the room plan of Fig. 14.31. The December values are minimum values and correspond most closely to the 85th per-

centile figures of the CIE and GDDM methods. The agreement is excellent for the half of the room nearest the window. For the deeper half, where there is significant ground contribution, the IESNA method yields higher illuminances because it considers the rereflected ground light contribution. Overall, the agreement among the three methods is excellent. The figures for March 21 correspond most closely to the 50th to 60th percentile of Fig. 14.27, that is, somewhat more than double the minimum figures—and here again the agreement is good.

To demonstrate the use of the IESNA method for *clear-sky* conditions, we can work through a brief example where five IESNA point illuminances are calculated for the same Seattle classroom, on June 21 at 10 A.M. for clear-sky conditions. Assume that the window faces southwest (azimuth angle = 45° west of south).

Solar azimuth at 10 A.M. on June 21 at 48° N latitude is 56° (Table D.1). The bearing angle is therefore 45° + 56° or 101° (i.e., no direct sunlight enters the window, which is a necessary condition of the IESNA clear-sky method). Solar altitude is 55°.

### Calculation

**STEP 1.** *Determine the horizontal illuminance* (refer to Fig. 12.4):

| Sun only: | 80,000 lux (7432 fc) |
|---|---|
| Sky only: | 18,000 lux (1672 fc) full sky |
| | 9,000 lux (836 fc) half sky $(E_{xhk})$ |

*Determine the vertical illuminance* (refer to Fig. 12.6b):

At a solar altitude of 55° and a bearing angle of 101°, the $E_{xvk}$ = 700 fc (7500 lux).

From Eq. 14.3, vertical illuminance on the exterior of the window resulting from ground light is

$$E_{xvg} = 0.2 \times \frac{80{,}000 + 9000}{2} = 8900 \text{ lux (827 fc)}$$

**STEP 2.** *Determine net transmittance of the window:*

$$\tau = 0.70, \text{ as previously}$$

**STEP 3.** *Select the coefficients of utilization* (refer to Tables C.16–C.21):

$$\frac{E_{xvk}}{E_{xhk}} = \frac{7500}{9000} = 0.83$$

**TABLE 14.10 Coefficients of Utilization for Sky and Ground Components for Five Interior Locations**

| Location | $CU_k$ (Sky) | $CU_g$ (Ground) |
|---|---|---|
| 10 | 0.673 | 0.183 |
| 30 | 0.235 | 0.159 |
| 50 | 0.104 | 0.103 |
| 70 | 0.065 | 0.071 |
| 90 | 0.053 | 0.060 |

**TABLE 14.11 Illuminance Values for Winter and Spring in Seattle, Washington, at Five Reference Locations**

| Location | $E_i$ Dec. 21, lux (fc) | $E_i$ Mar. 21, lux (fc) |
|---|---|---|
| 10 | 1151 (107) | 2877 (267) |
| 30 | 474 (44) | 1182 (110) |
| 50 | 234 (22) | 585 (54) |
| 70 | 151 (14) | 378 (35) |
| 90 | 125 (12) | 312 (29) |

ILLUMINATION

**TABLE 14.12 Illuminance Values for Clear Sky Conditions on June 21 at Five Reference Locations**

| Location | $E_i$ June 21, lux (fc) |
|---|---|
| 10 | 4674 (434) |
| 30 | 2225 (207) |
| 50 | 1188 (110) |
| 70 | 784 (73) |
| 90 | 652 (61) |

**STEP 4.** *Calculate the illuminance:*

$$E_i = \tau \, (E_{xvk} \times CU_k + E_{xvg} \times CU_g)$$

$$= 0.7 \, (7500 \times CU_k + 8900 \times CU_g)$$

Table 14.12 tabulates the illuminances at the five reference locations. ∎

## 14.11 DAYLIGHTING SIMULATION PROGRAMS

Until recently, daylighting simulation tools were too expensive and complex to use for day-to-day designs, and they were primarily used by lighting consultants or researchers. Computer rendering tools have been developed so that many simulation programs now provide realistic visual daylighting output with varying degrees of accuracy. Computational approaches can simulate the distribution of light from both daylight and electric sources, for any selected season, time of day, and building location (orientation and latitude). Many of these programs use *radiosity* techniques. Radiosity-based renderings are produced by dividing all the surfaces in a scene into a mesh of small polygons. Each polygon takes on a different value of light absorption/reflection, depending upon its relationship to a light source and its surface parameters. These values simulate the light distribution throughout the scene. The following are some commonly used programs; selected characteristics and features are compared in Table 14.13.

- *Desktop Radiance: http://radsite.lbl.gov/deskrad/* This program is a Windows version of Radiance that integrates a realistic rendering package (Fig. 14.34) with a computer-aided design (CAD) input environment. Libraries of materials, glazings, luminaires, and furnishings facilitate data entry. Lawrence Berkeley National Laboratory, Pacific Gas and Electric, and the California Institute for Energy and Environment developed this program.
- *ECOTECT: http://www.squ1.com/daylight/df-ecotect. html/* ECOTECT is a building analysis software program offering a range of modeling and analysis features such as visualization, shading, shadows, solar analysis, lighting, thermal performance, ventilation, and acoustics. It can export to

**TABLE 14.13 Comparison of Processing Features for Daylighting Simulation Programs**

| Key:<br>3 = excellent<br>2 = good<br>1 = satisfactory<br>— = not applicable | Simulating/Rendering | | | | | Output | | | | | | User Interface | Online Help |
|---|---|---|---|---|---|---|---|---|---|---|---|---|---|
| | RADIOSITY | RAYTRACING | ITERATIVE PROCEDURE | VIEW POINTS | MEMORY REQUIREMENTS | ISOLUMEN CONTOURS | NUMERIC OUTPUT | REALISTIC RENDERING | ANIMATION | GLARE ANALYSIS | DAYLIGHT FACTORS | EASE OF USE | EXTENSIVE HELP |
| Desktop Radiance | 3 | 3 | — | 3 | 1 | — | 3 | 3 | — | 3 | 3 | 3 | 3 |
| FormZ | 3 | 3 | — | 3 | 3 | — | — | 3 | 3 | — | — | 2 | 2 |
| Lightscape | 3 | 3 | 3 | 3 | 2 | 3 | 3 | 3 | 3 | — | — | 2 | 3 |
| Lumen Micro | 3 | — | — | 3 | | 3 | 3 | 1 | — | — | — | 3 | 1 |

*Source:* From Bryan, H. and S.M. Autif, 2002; used with permission of the American Solar Energy Society.

*Note:* Recently developed software—in addition to those programs listed above—include ECOTECT, Autodesk VIZ Render, Lighting Designer, and AGI32; these programs include capabilities such as daylight factor analysis and glare analysis but have not been rigorously evaluated on the parameters established for this comparison table.

(a)

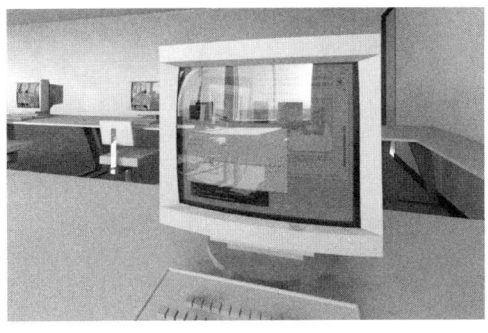

(b)

**Fig. 14.34** (a) Rendering of shading provided by the blinds at the New York Times building. (b) Rendering study of veiling reflections on a computer monitor. (© 2004; example a by Greg Ward & Judy Lai; example b by Chas Erlich, Lawrence Berkeley National Laboratory, rendered by Radiance; used with permission.)

Radiance for higher-level ray-tracing techniques. Daylighting capabilities can model shadows and reflections on the surfaces of other buildings at a single point in time; show an entire year's shadow patterns for a single surface; model surface solar radiation relative to the effects on thermal mass; and calculate daylight factor (Fig. 14.35).

- *FormZ RadioZity: http://www.formz.com/products/formz_radiozity.html/* This program is a version of FormZ, a general-purpose solid and surface modeling program. FormZ RadioZity provides radiosity-based rendering and can simulate the distribution of light in a space. Although familiar to many as an architectural rendering program,

it has the ability to accurately show shadows and radiosity rendering (Fig. 14.36).

- *Autodesk® VIZ 2005: http://usa.autodesk.com/* AutoDesk purchased the Lightscape software and incorporated features with improved interoperability from the company's other products (such as AutoCAD and 3d MAX). VIZ 2005 (Fig. 14.37) is a specialized visualization and rendering program that includes lighting effects from indirect illumination and shadows under varying conditions of daylight and electric light.

- *Lumen Micro 2000 and Lumen Designer: http://www.lighting-technologies.com/* Lumen Micro 2000 (Fig. 14.38) operates in a Windows

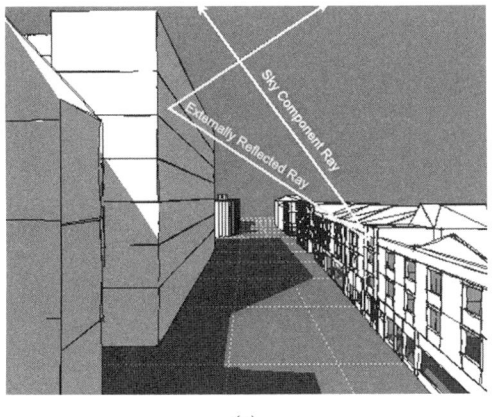

(a)

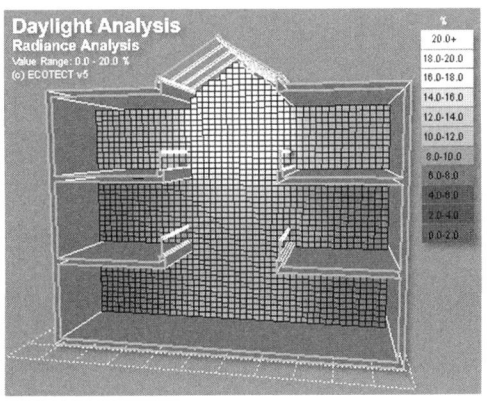

(b)

**Fig. 14.35** (a) Schematic output showing exploratory rays passing through a window and hitting an external obstruction. (b) Daylight factor distribution analysis applies a ray-tracing technique using the Building Research Establishment (BRE) Daylight Factor method. (© 2004 Dr. Andrew J. Marsh, of Square One research; created with ECOTECT v5 [ecotect.com]; used with permission.)

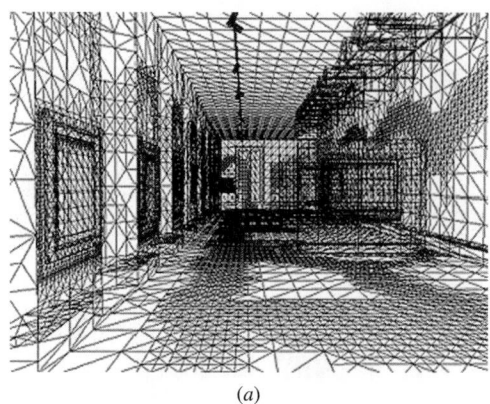

*(a)*      *(b)*

***Fig. 14.36*** *(a) Radiosity rendering of polygons for a gallery space. (b) Rendering of the same gallery space. (© 2004, modeled and rendered with form.Z RadioZity by Markus Bonn, auto.des.sys, Inc.; used with permission.)*

*(a)*      *(b)*

***Fig. 14.37*** *(a,b) Rendering of daylight in a residence Autodesk VIZ. (Certain images reproduced from Autodesk® VIZ Render 2004 software with permission of Autodesk, Inc.; © 2002, all rights reserved.)*

*(a)*      *(b)*

***Fig. 14.38*** *(a) Lumen Micro rendering of daylighting in a church. (b) Lumen Designer rendering of electric lighting in an office. (© 2005 Lighting Technologies, Inc., Lumen Micro Software; used with permission. © 2005 Lighting Technologies, Inc., Lumen Designer Software; used with permission.)*

environment and can provide detailed daylighting analysis. There is an extensive product library with luminaire data from over 70 manufacturers. Lumen Designer is the most recent generation of software (although Lumen Micro 2000 is still available). It can create any geometry, produce realistic renderings, interface with product databases from industry, and calculate daylighting.

## 14.12 PHYSICAL MODELING

Physical models are a useful and indispensable tool for the investigation of complex daylighting phenomena. Simple physical models can give both the designer and the client a visual understanding of a daylighted space. Physical models can duplicate the lighting phenomena that would occur in a full-scale space and, when placed under identical sky conditions, will yield accurate results relative to brightness, shadows, and daylight factor. By changing window design or orientation, adding light shelves, reflectors, or shading devices, and/or modifying surface materials, a designer can quickly produce a three-dimensional visual image that displays qualitative and quantitative performance results for a proposed daylighting design.

The advantages of physical models include:

- The opportunity for accurate daylight measurements and for qualitative evaluation
- Easy construction (for most designers)
- Crude models that can yield critical information
- Easy comparisons of various schemes (e.g., interchangeable wall or ceiling elements)
- Realistic visualization for clients

The principal disadvantage of using physical models is the need to expose them to the desired sky conditions. For example, waiting for suitable sky conditions in order to view a particular space under both overcast and clear sky conditions, or at different seasons of the year, or during different times of day is not always practical.

Constructing scale models is relatively simple, using corrugated cardboard, mat board, and colored paper—mounted on a base for ease of manipulation (Fig. 14.39). The model should be made modularly so that alternative design proposals can be interchanged. For example, to compare various skylight configurations, several replaceable roof

(a)

(b)

(c)

**Fig. 14.39** *These photographs illustrate the effectiveness of even a crude model in daylighting studies. (a) A faculty office at the University of California–Berkeley served as an exercise for a daylighting study. (b) The scale model for this office was constructed of cardboard. Significant reflecting surfaces such as desk surfaces and window sills were carefully modeled. (c) A quick modification to the model introduced light washing along the wall from a skylight above (upper left of photo) so that the rear of the office would receive more light. All photos were taken on site at midday on an overcast day. (Photos © 2004 by Alison Kwok; all rights reserved.)*

configurations can be constructed. Model size depends upon the size of photometers used to measure interior illuminance, the size of the space, and the need to accommodate a camera viewport. Considering ease of construction and visualization opportunities, bigger is usually better—although larger models are often preceded by smaller/cruder study models.

Cardboard is an ideal material for daylighting models because it is opaque, unlike foam core board, which is translucent and transmits some light. Unintentional light leaks must be prevented, typically by sealing the joints of a model with black electrician's (or duct) tape or by using strips of black cardboard to close gaps. "Portholes" in one or both of the long sides of a model (approximately 2 in. [50 mm] in diameter) will accommodate visual inspection and insertion of a camera lens to photograph the distribution of light. Model surface reflectances (both interior surfaces and exterior surfaces that contribute to daylight distribution) should be the same as those proposed for the actual building (see Table C.22 for reflectances and mat board colors). Special care should be taken to accurately replicate details around daylight openings—the size and depth of mullions, the depth and reflectivity of the sill, louvers, shading devices, and surfaces just outside daylight openings. Any major furnishings that might have a significant impact upon light distribution should also be included.

Daylight model testing may be conducted under a real or an artificial sky and may also involve heliodon studies.

1. Use of a real sky with daylighting models is logical but often difficult to coordinate (as described above).
2. *Artificial sky or mirror box.* Carefully designed and controlled artificial sky domes or a mirror box can duplicate overcast sky conditions with a high degree of accuracy and are ideal for testing physical models. A number of such units exist in major universities and lighting laboratories around the world. Sky domes are usually illuminated by interior perimeter lamps with the model located in the center. A mirror box is essentially a room with a luminous ceiling (using fluorescent lamps) and mirrored walls to create a sky with an "infinite" horizon. For construction details, see Moore (1985).

3. *Heliodon.* The heliodon, as shown in Fig. 14.40, is a sophisticated device that allows the study of shading and solar access for a specific latitude, time of day, and time of year using architectural scale models. It operates by rotating and tilting a building model with respect to the real sky or an "artificial sun" (a narrow-beam electric light source) until the desired solar altitude and azimuth are reached. Over the years, heliodons (*sun machines, sun tables, helioluxes, sun emulators*) have been built using a variety of configurations to simulate the sun's position relative to an architectural scale model. In all cases, the device establishes a geometric relationship for three variables: site location (latitude), solar declination (time of year), and the Earth's rotation (time of day). By adjusting any one of these variables, a heliodon can simulate sunlight penetration and shading for any combination of site location and time. Other types of heliodons keep the position of the model fixed and use a band of lamps that move along three axes to simulate the sun's position for different times and seasons.

**Fig. 14.40** *ShadowTracker heliodon at the Baker Lighting Lab at the University of Oregon. Automated adjustments for latitude, elevation, and tilt permit exposure of the model to desired clear sky conditions. (Photo by Sara Goenner; © 2004 Alison Kwok; all rights reserved.)*

## 14.13 CASE STUDY—DAYLIGHTING DESIGN

### Audubon House

### PROJECT BASICS

- Location: New York City, New York, USA
- Latitude: 40.5° N; longitude: 74° W; elevation 87 ft (26.5 m)
- Heating degree days: 4805 base 65°F (2672 base 18.3°C); cooling degree days: 1140 base 65°F (634 base 18.3°C); annual precipitation: 47 in. (1200 mm)
- Building type: adaptive reuse
- 98,000 ft² (9104 m²); eight occupied stories
- Completed December 1992
- Client: National Audubon Society
- Design team: Croxton Collaborative, Architects (and consultants)

**Background and Context.** Audubon House is a restored and remodeled century-old romanesque building in New York City. The National Audubon Society wanted a headquarters to reflect their environmental mission to restore ecosystems and ensure a healthy environment for people, wildlife, and natural resources. The design team and the client worked together to create a place that served as a model for energy efficiency and environmental responsibility and to enforce the model. The Croxton Collaborative brought in many design-lessons-learned about energy efficiency and environmental performance from the redesign and renovation of the Natural Resources Defense Council (NRDC, another New York–based environmental group) building in 1988, just prior to the beginning of the Audubon project. Placing environmental criteria as a priority for the design in terms of resources, energy efficiency, air quality, and occupant well-being set the stage for the development of many new green office buildings. A case study of Audubon House won first place in the 1998 Vital Signs Case Study Competition. The following information is extracted from *Audubon House* (National Audubon Society, 1994) and information sheets.

**Design Intent.** The design of Audubon House took the approach of developing a "living model" in four environmental dimensions: energy conservation and efficiency, creating a healthy indoor environment, resource conservation and recycling, and reducing negative environmental impacts—an

antidote to the plague of standard-issue, grossly inefficient office buildings that dot urban and suburban landscapes. The motto was "Green design is affordable." Compared to code-compliant buildings in New York City at that time, Audubon House was designed to use 62% less energy overall energy and deliver 30% more outdoor air to its occupants.

**Design Criteria and Validation.** Benchmarks were set for building systems, materials, and strategies based upon previous experience. The criteria included a target of 0.97 W of electricity per square foot (conventional buildings at that time used as much as 2.4 W/ft²); an in-house recycling system that would capture 80% of the office refuse (mostly paper); 26 cfm (12.3 L/s) of outdoor air per person, which would exceed the existing standards and guidelines (then 10–20 cfm); pendant ceiling fixtures reflecting 88% of the light from the fixture up to the ceiling; and occupancy sensors that would turn off lights in unoccupied zones.

**Post Occupancy Validation Methods.** Even with recent developments in environmental technology, Audubon House continues to serve as a model. Several New York City projects have applied the principles used in this project. Five years after the renovation was completed, an occupant survey found a high degree of satisfaction with daylight quality and availability in the workspaces. The farther a workstation was located from a window, the more often occupants used their task lights. Several nonfunctioning occupancy and daylight sensors reduced the potential of the daylight integrated lighting system; the lighting power density, however, was calculated to be 0.82 W/ft², which exceeded the design goal of 0.97 W/ft² and the ASHRAE Standard 90.1 recommendations. In 1996 a separate detailed study on indoor air quality was conducted as part of a larger study of air quality in green buildings.

### Key Design Strategies

- A gas-fired absorption chiller-heater uses no CFCs, emits no sulfur dioxide, and emits 60% less nitrogen oxides than conventional units. Additionally, no CFCs are used in the insulation.

(a)

(b)

**Fig. 14.41** (a) Audubon House, headquarters for the National Audubon Society, located in New York City, is a reused and renovated building (Croxton Collaborative, 1992). (b) Lobby space uses daylight from high windows and skylight. (Photos © 2004 by Walter Grondzik; all rights reserved.)

(a)

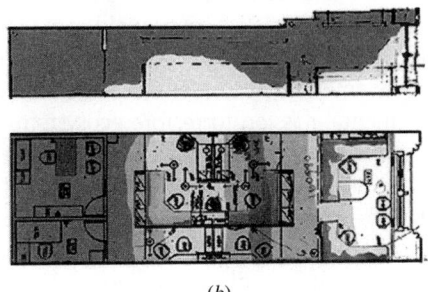

(b)

**Fig. 14.42** (a) Office space along the south-facing façade of Audubon House. (b) Daylighting section showing illuminance measurements. (Photo © 2004 by Alison Kwok; all rights reserved.)

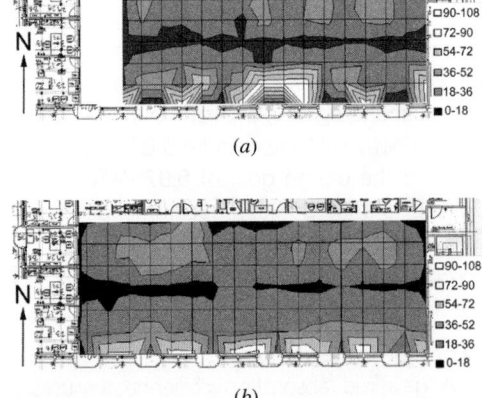

(a)

(b)

**Fig. 14.43** (a) Illuminance contours in Audubon House during an overcast day. (b) Illuminance contours during a clear day. (© 2004 Alison Kwok; all rights reserved.)

**Fig. 14.44** Recycling bins in the Audubon House basement receive materials sorted by categories of waste (paper, glass, and newspaper). (Photo © 2004 by Alison Kwok; all rights reserved.)

- To promote a healthy environment, natural and recycled materials were used for the furnishings and renovation materials, such as Air-Krete™ wall insulation, wool rugs without adhesive backing, wood and fabric furniture, and recycled content countertops and tiles.
- A daylight integrated lighting system producing lowered electricity use during daylight hours through the use of light-colored furnishings and interior surfaces, a layout of interior work areas to ensure that daylight is not blocked by walls and corridors, pendant uplight ceiling fixtures with fluorescent lamps, daylight dimming sensors, and occupancy sensors in all work spaces (producing what is affectionately known as the "Audubon flap").
- There were low VOC emissions from building materials and furnishings, high-efficiency filters to remove particles, and air intakes on the roof rather than at street level.
- Recycling the entire building by adaptive reuse reduced embodied energy costs, as did the use of recycled materials (Homasote™ subfloor from 50% recycled newsprint, tiles from light bulbs, bathroom countertops from plastic containers, etc.) and the provision of an in-house recycling system.

### FOR FURTHER INFORMATION

National Audubon Society. 1994. *The Audubon House: Building the Environmentally Responsible, Energy-Efficient Office.* John Wiley & Sons, New York.

Fact sheet about Audubon House: http://www.audubon.org/nas/ah/

Vital Signs Case Study: http://arch.ced.berkeley.edu/vitalsigns/bld/case_studies.html/

### References

Brown, G.Z. and M. DeKay. 2000. *Sun, Wind & Light: Architectural Design Strategies,* 2$^{nd}$ ed. John Wiley & Sons, New York.

Bryan, H. and S.M. Autif. 2002. "Lighting/Daylighting Analysis: A Comparison," in *Proceedings of the National Solar Energy Conference.* American Solar Energy Society, Boulder, CO.

CIE. 1970. *Daylight, International Recommendations for the Calculation of Natural Daylight* (Publication No. 16 E-3.2). Commission Internationale de l'Eclairage, Paris.

DeKay, M. 1992. "Volumetric Implications and a Rule-of-Thumb for Thickness of Atria Buildings," in *Proceedings of the 17$^{th}$ National Passive Solar Conference.* American Solar Energy Society, Boulder, CO.

Dresler, A. 1963. *Daylight Design Diagram.* Service Division, Commonwealth of Labor and National Service, Melbourne, Australia.

Heschong Mahone Group. 1999–2003. Executive summaries and reports available from http://www.h-m-g.com/projects/daylighting/projects-PIER.htm/

Hopkinson, R.G., R. Pethebridge, and J. Longmore. 1966. *Daylighting,* Heinemann, London.

IESNA. 1994. *Recommended Practice for the Calculation of Daylight Availability* (RP-21-84; reaffirmed 1994), Illuminating Engineering Society of North America, New York.

IESNA. 1999. *Daylighting* (RP-5-99), Illuminating Engineering Society of North America, New York.

IESNA. 1989. *Recommended Practice for the Lumen Method of Daylight Calculations* (RP-23-89), Illuminating Engineering Society of North America, New York.

International Energy Agency. 2000. *Daylight in Buildings: A Source Book on Daylighting Systems and Components.* International Energy Agency, Solar Heating and Cooling Programme, Energy Conservation in Buildings and Community Systems.

Lam, W. 1986. *Sunlighting as Formgiver for Architecture.* Van Nostrand Reinhold, New York.

Millet, M.S., and J.R. Bedrick. 1980. *Graphic Daylighting Design Method.* Lawrence Berkeley Laboratory/U.S. Department of Energy, Washington, DC.

Moore, F. 1985. *Concepts and Practice of Architectural Daylighting,* Van Nostrand Reinhold, New York.

National Audubon Society and the Croxton Collaborative. 1994. *Audubon House: Building the Environmentally Responsible, Energy-Efficient Office.* John Wiley & Sons, New York.

Stein, B. and J. Reynolds. 1992. *Mechanical and Electrical Equipment for Buildings,* 8th ed., John Wiley & Sons, New York.

# CHAPTER 15

# Electrical Lighting Design

## LUMINAIRES

### 15.1 DESIGN CONSIDERATIONS

#### (a) General

REFER TO FIG. 13.1 TO SEE THE context for this chapter. At this point in the lighting design process, where design development of the electrical lighting system occurs, lighting hardware is chosen based upon considerations brought forth from earlier design stages and appropriate validation calculations are performed. Some spaces require overall uniform illumination. These spaces are calculated by the lumen method, which yields average illuminance. Other spaces utilize local lighting alone, or local lighting in addition to general lighting, requiring point-by-point illuminance calculations or some other method for restricted-area calculation. Additional considerations at this design stage are the control strategy (see Sections 15.13–15.15); type of ceiling system (e.g., modular, movable fixture, and integrated service); and ancillary considerations of ballast noise, luminaire heat distribution, and maintenance. Also decided here is whether to utilize workstation-mounted or built-in lighting, both of which are principally applicable to open-plan spaces.

#### (b) Luminaire Characteristics

The purpose of a *luminaire* or *lighting fixture* (the terms are synonymous) is twofold: physically, to hold, protect, and electrify the light source; and photometrically to control the lamp output (i.e., to redirect the light produced) because most common light sources emit light in substantially all directions. The means by which this beam shaping is accomplished are well known. The characteristics of reflectors, baffles, lenses, and louvers that perform these functions are discussed in some detail in the following sections. However, the problem in luminaire selection is that the requirements are, in many respects, incompatible, and therefore a trade-off between various luminaire characteristics must be made. Thus, for instance, high efficiency frequently entails high fixture luminance with resultant glare. Desirable wall lighting means high-angle luminaire output with resultant direct glare (see Figs. 11.23 and 11.24). Low-angle light means minimum direct glare but possible veiling reflections. A high shielding angle (>35°) means good visual comfort but reduced efficiency, and so on. It is therefore obvious that

- No single luminaire design is ideal for even a majority of applications and that
- To make an intelligent selection among the hundreds of lighting fixtures commercially available, it is absolutely necessary that the designer understand both the specific requirements of the

application and the light control characteristics of the luminaire being considered.

## 15.2 LIGHTING FIXTURE DISTRIBUTION CHARACTERISTICS

At this point, review of Section 11.14 by the reader would be useful. The two distribution curves shown in Fig. 15.1a are actual test results of two 2-lamp, 1-ft-wide by 4-ft-long semi-direct fluorescent fixtures with prismatic enclosure. The flat bottom of the curve in Fig. 15.1a indicates even illumination over a wide area, therefore permitting a high spacing to mounting-height ratio (S/MH) (1:4) for uniform illumination. The rounded bottom of the curve in Fig. 15.1b indicates uneven illumination and closer required spacing for horizontal uniformity as defined in Section 15.5.

The straight sides of the curve in Fig. 15.1a show a fairly sharp cutoff, and the small amount of light above 45° means high efficiency, probably insufficient wall lighting, barely adequate diffuseness, and very little direct glare potential but a distinct possibility of veiling reflections. Conversely, the curve in Fig. 15.1b shows a large amount of horizontal illumination (above 45°) with resultant direct glare potential, diffuseness, and relative inefficiency, because horizontal light output is attenuated by multiple reflections before reaching the horizontal working plane. Here, however, low out-

put below 45° minimizes reflected glare potential. The uplight component of luminaire (a) is directed outward to cover the ceiling and will not cause hot spots; the corresponding light from fixture (b) is concentrated above the fixture and gives uneven illumination of the ceiling.

These conclusions were reached based on the following observations:

1. *Uniformity of illumination* requires that the intensity at angles above the nadir (0° from the vertical) be greater than the intensity at 0° so that points distant from the fixture centerline obtain the same illumination as those below the fixture (because illuminance varies inversely with the square of distance). This is exactly the case with the flat-bottom characteristics of Fig. 15.1a. Therefore, such fixtures can be spaced more widely than the units of Fig. 15.1b.

2. *High efficiency* is achieved by directing the luminaire output to the work plane (i.e., from 0° to 45° from the vertical). Light above 45° is directed to the walls and reaches the working plane only after multiple attenuating interreflections.

3. *Diffuseness* exists when light reaches the work plane from multiple directions. This requires that light be reflected from walls and ceilings to the work plane, which in turn requires luminaire light output above 45° from the vertical.

4. *Direct glare* is caused by light output at high angles (i.e., above 45° from the vertical). Direct

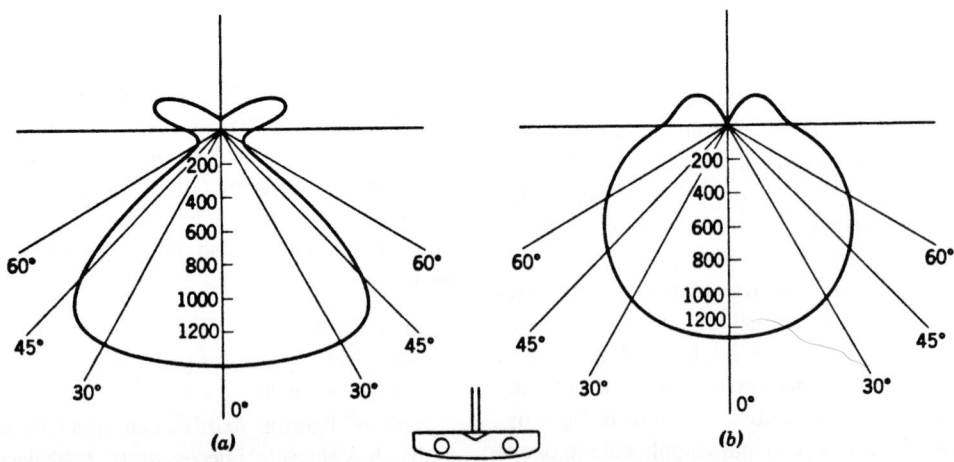

**Fig. 15.1** Semi-direct fluorescent fixture crosswise distribution (two lamps, 32 W each, prismatic enclosure). Note the sharp cutoff and wide, horizontally even distribution of (a) in contrast to the diffuse, broad, and horizontally uneven distribution of (b).

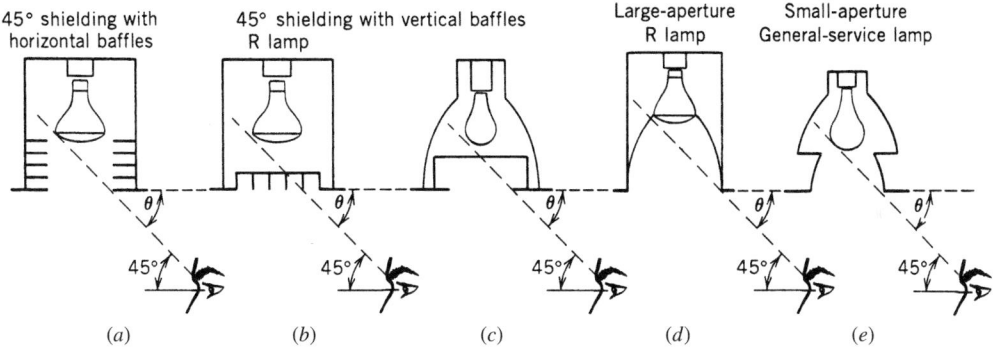

**Fig. 15.2** *Methods for shielding downlights using circular shields for vertically symmetrical sources such as incandescent and HID lamps. Halogen lamps mounted vertically in R lamp envelopes are symmetrical; horizontally mounted units in PAR or other reflectors are not. Baffled downlights (a–c) control unwanted high-angle light by cutoff as illustrated. Black baffles aid by absorbing light and appearing dark. Other colors give a ring of light at the baffled edge. Cones (d, e) control brightness by cutoff and by redirection of light due to their shape. They are either parabolic or elliptical. A light specular finish such as aluminum appears dull; a black specular finish appears unlighted. Black finishes require high-quality maintenance because dust shows as a bright reflection. CFL lamps in reflectors are not normally a serious direct glare concern and are considered to be vertically symmetrical. A shielding angle of 45° minimum is recommended for high-luminance lamps.*

glare from linear fluorescent fixtures can be minimized by placing the long axis parallel to the line of sight, because such fixtures normally have low *endwise* high-angle output.

5. *Reflected glare* is caused by reflection of low-angle output from the task. Therefore, fixtures with control means that limit output between 0° and 45° minimize the potential for veiling reflections (see Section 15.4*b*).

6. *Shielding* is a function of the shape of the fixture housing plus any additional lamp concealment means, such as louvers or baffles. The *shielding angle* is defined as the angle between a horizontal plane through the louvers or baffles and the inclined plane at which the lamp first becomes visible as one approaches the fixture (Figs. 15.2 and 15.3). *Cutoff angle* is usually defined as being synonymous with the shielding angle. However, because some sources define it as the complement of the shielding angle, it is best to avoid the term and use only *shielding angle.*

7. *Ceiling illumination* is produced by light above the horizontal. As with light below the horizontal, a spread characteristic (Fig. 15.1*a*) means good ceiling coverage, no hot spots, and, ultimately, good diffuseness. Concentrated uplight

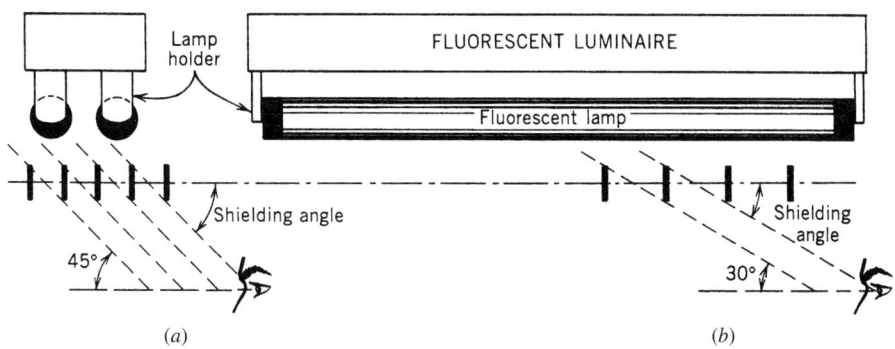

**Fig. 15.3** *Shielding of fluorescent lamps is less critical due to lower lamp luminance. For T12 lamps, 45° × 35° crosswise/lengthwise shielding as shown is excellent and 35° × 30° is satisfactory. For T8 lamps, 45° × 35° should be used. Because fixture luminance is higher in the transverse direction (a) than lengthwise (b), a better cutoff angle is required. The shielding elements may be louvers or baffles. Opaque shielding elements have a higher visual comfort rating than translucent plastic units.*

ILLUMINATION

means a potential hot spot if the fixture suspension hanger is too short, and in any event it yields uneven ceiling illumination.

Thus, we see that a rapid inspection of a fixture curve performed by an informed person can yield a large amount of information on the fixture's performance. The reader is encouraged at this juncture to review the comments on the two distribution curves of Fig. 15.1 and then analyze similarly other distribution curves in manufacturers' catalogs.

## 15.3 LUMINAIRE LIGHT CONTROL

### (a) Lamp Shielding

Except where it is desired to use a bare lamp as a source of sparkle, such as in chandeliers and other decorative fixtures, all lamps in interior fixtures should be shielded from normal sight lines (i.e., sight lines in a head-up, eyes-straight-ahead position; see Fig. 11.19). The reason is obvious; bare lamps are so bright (see Table 11.2) that they usually constitute a source of direct or even disabling glare, depending on the apprehended angle (close-

ness to the eye and size of the lamp) and eye adaptation level. The range of permissible luminaire luminances (listed in Table 11.3) of 1000 to 7000 cd/m² depends upon these two variables (apprehended angle and adaptation level).

As a general rule, exposed incandescent lamps, 6 W and larger, are sources of direct glare and should be avoided. Note that the upper direct-glare limit corresponds to the luminance of a bare 34-W T12 fluorescent tube, which accounts for the fact that bare-lamp fluorescent fixtures are well tolerated. However, when such a fixture is relamped (and reballasted) with a more efficient, better CRI, T8 lamp whose luminance exceeds 10,000 cd/m², it becomes a source of annoying direct glare that actively impairs visual ability.

Shielding of lamps is accomplished with the fixture housing/reflector or with baffles and louvers, as mentioned previously (see Figs. 15.2, 15.3, and 15.4). Fluorescent fixtures require shielding most when placed crosswise to the line of sight, thus exposing the entire length of the lamp to the field of view. Such fixtures require longitudinal baffles, deep housings, or louvers to provide the necessary shielding. Alternatively, when at all possible, place fixtures with their long axis in the direction of sight lines.

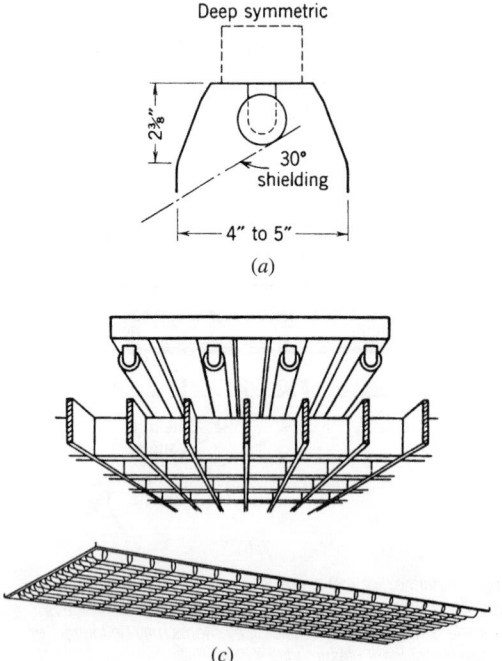

(a)

(c)

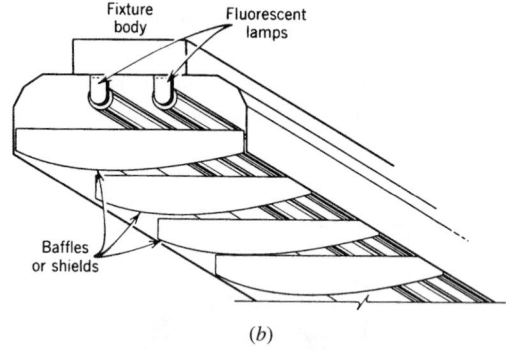

(b)

**Fig. 15.4** Shielding elements. (a) The most basic shielding element is the lamp reflector, which may double as the fixture body. (b) Shields perpendicular to the long axis of a linear fluorescent lamp are normally called baffles. They are less important than lengthwise shields (louvers) because endwise lamp luminance is lower than crosswise luminance. (c) Two-way shielding is most effective but seriously lessens luminaire efficiency. (From Stein, B., Building Technology, Mechanical and Electrical Systems, 2nd ed., 1997; reproduced by permission of John Wiley & Sons.)

## (b) Reflectors

It is important to understand the action of luminaire reflectors. The basic shapes and beam patterns are illustrated in Figs. 15.5 through 15.7. Note from Fig. 15.6 that the so-called pinhole downlight requires an elliptic reflector to focus the light through this hole at point $f2$ in order to maintain even minimal fixture efficiency. Elliptic reflectors are large, and frequently the space above the ceiling is too restricted for their use. Lamps with integral elliptic reflectors can be utilized with a standard baffled reflector to achieve roughly the same effect.

## (c) Reflector Materials

Until fairly recently, reflector materials were of two types: white gloss paint for portions of fixture body interiors that acted as reflectors, and formed anodized aluminum sheet for the shaped reflectors of the types shown in Figs. 15.5 through 15.7. The reflectances (reflection factors) of both of these materials are approximately the same, varying between 0.84 and 0.88 *when new and clean.* Neither, however, is truly specular; the paint finish is actually primarily diffuse, whereas the aluminum is principally specular. Where shaped reflectors are not used, as in the case of a fluorescent troffer, the lack of specularity is essentially immaterial because, at worst, the diffuseness will reduce luminaire output slightly by increasing the number of interreflections within the fixture body (Fig. 15.8a). The idealized specular reflections for shaped reflectors shown in Figs. 15.5 through 15.7 are just that; in reality, the reflectances are considerably more diffuse and become increasingly so with reflector aging and dirt accumulation.

Painted fixture body interiors lose their high reflectance by rapid aging due to elevated temperatures and accumulation of dust and dirt. This

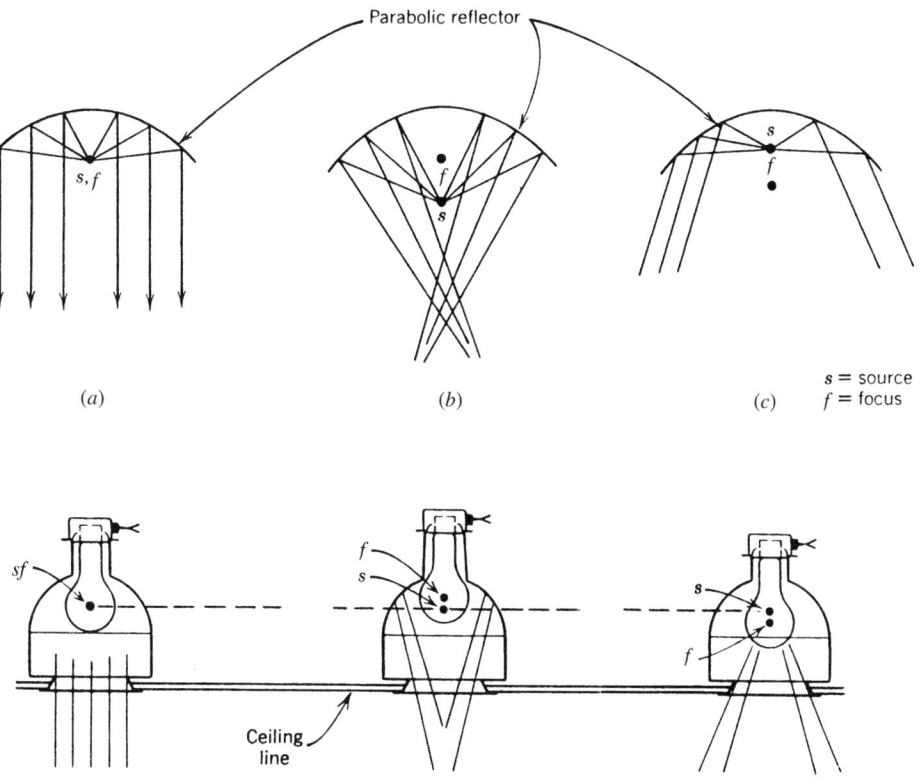

**Fig. 15.5** *Parabolic reflector action shown with the fixture below: (a) with the source at the focal point, rays are parallel; (b) with the source below the focal point, they converge; (c) with the source above the focal point, they diverge. This focusing action is illustrated by fixtures correspondingly designated. Note that type (c) requires a large ceiling opening to achieve even minimal efficiency.*

ILLUMINATION

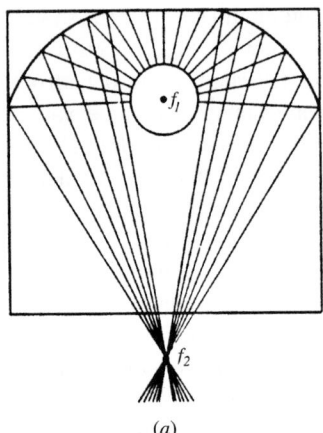

(a)

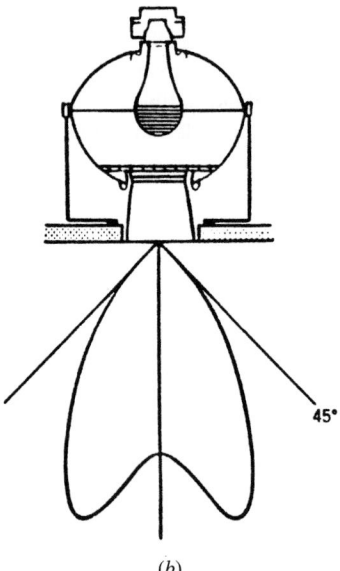

(b)

**Fig. 15.6** (a) Action of an elliptical reflector section. With the light source at focal point f₁ the light converges at the other focal point, f₂. This effect is useful in fixture design, as in (b). By projecting light up only (through the use of a silvered bowl lamp) the output light can be redirected through a constricted aperture at the other focal point, with little loss. This design is the basis of high-efficiency "pinhole" downlights.

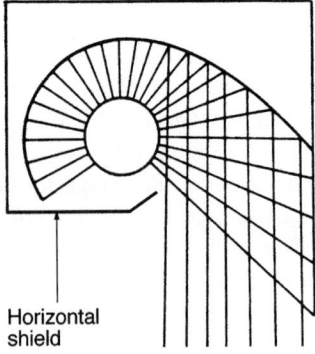

**Fig. 15.7** The extended section reflector allows the source to be concealed (shielded) while projecting its light directly down, but horizontally displaced, from the source.

causes a decrease in overall reflectance, which is compensated for by initial overdesign, as explained in Section 15.20. However, because a rectangular fixture body is not an accurately shaped reflector, the result of overall reflectance reduction is simply an overall reduction in output while maintaining the same photometric *distribution* characteristics.

Energy conservation programs and utility re-

bates (now generally terminated) led to the introduction in retrofit of very-high-reflectance auxiliary reflectors that were added to existing fluorescent troffers approximately as shown in Fig. 15.8*b*. Unfortunately, many of the claims of highly increased efficiency were based on retrofitting aged, dirty luminaires, and their results can be very misleading. A reasonable estimate of the possible improvement in luminaire efficiency can be made by considering these facts:

- Approximately 40% of a lamp's output in an open luminaire is directed downward and is therefore completely independent of any reflector action.
- The difference in reflectance between a new, clean, painted surface and an old, dirty surface is, *at most,* 50%. That means that the *maximum* light loss of an open fixture due to poor maintenance is 50% of 60% (the maximum reflected light component), or 30% of the overall light output. Reference is always to an open-bottom fixture.
- The maximum reflectance of the best (and most expensive) silver reflectors is about 95%, comprising 93% specular and 2% diffuse. This is only 10% higher than the original *minimum* 85% paint reflectance. Therefore, at most, retrofit of a very dirty, old fixture with a high-quality (expensive) silver reflector improves performance by the 30% lost to dirt plus 10% of the 60% reflected light for a total of 36% *maximum*. Relamping, of course, also improves output, but that is not connected with fixture body reflectance.

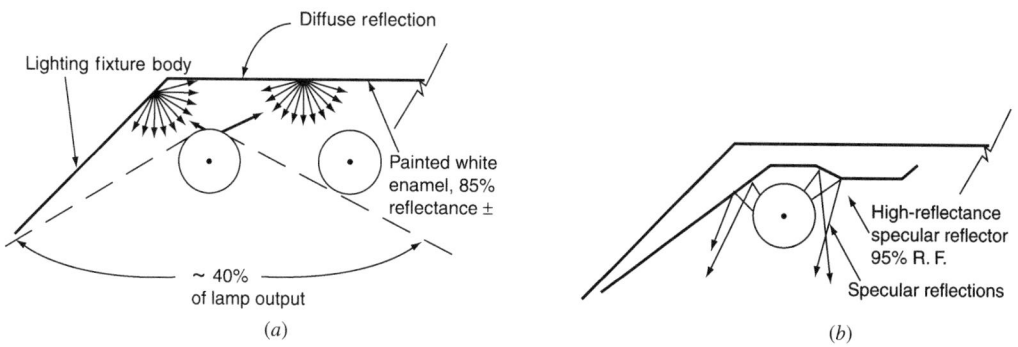

**Fig. 15.8** (a) Approximately 40% of lamp output in an open linear fluorescent fixture is unrestricted. The remainder leaves the fixture after one or more reflections. (b) A mirrored reflector narrows the distribution pattern of the luminaire by specular reflection and increases output somewhat.

- Simple cleaning of a very dirty fixture body restores 20% to 25% of the light loss. The remaining loss is due to aging of the paint. A cost analysis is required to determine whether the 10% to 15% differential in light output between simple cleaning and silvered reflector addition is economically feasible.

A very important consideration in retrofit and new construction is the photometric characteristic of high-reflectance linear fluorescent luminaires. In general, they are shaped like the curve of Fig. 15.1b. This means, as explained in Section 13.16, that light is concentrated downward, resulting in a requirement for closer luminaire spacing to obtain uniform illumination. In new construction, this can be considered in design, although it can be a serious economic penalty. In retrofit work, it can result in unacceptable lack of illuminance uniformity, requiring additional luminaires and expensive relocation of existing units.

Another factor to be considered is the degree of maintenance required to keep "super" reflectors in pristine condition in order to achieve the 15% ± maximum output differential. To determine this, designers should request an aging test and inspect a previously retrofited installation with ambient conditions and cleaning schedules similar to those of the area under consideration.

## 15.4 LUMINAIRE DIFFUSERS

Subsumed under this heading, in trade parlance, are all of the devices placed between the lamp(s) and the illuminated space, whose function it is to diffuse the light, control fixture brightness, redirect the light, and obscure (hide) and shield the lamps. As most of these devices perform multiple functions, they are discussed individually.

### (a) Translucent Diffusers

Because these do not redirect the light but merely diffuse it, the distribution characteristic is circular, as seen in Fig. 15.9a. Typical examples of this type are white opal glass, frosted glass, and white plastics such as plexiglass, polystyrene, vinyl, and polycarbonates. The distribution is basically the same as it would be for bare lamps. Lamp hiding power is good. Depending on the material, direct glare can be a problem (VCP is poor). Veiling reflections are high. The S/MH does not exceed 1.5. The fixture is generally inefficient. Wall illumination is good because of a large component of high-angle light (which reduces VCP). The net result of using this type of diffuser is to lower lamp luminance by distributing the lamp output over a larger diffusing area. Applications include corridors, stairwells, high-ceilinged spaces, and other areas without demanding visual tasks.

A special type of flat plastic panel that polarizes the transmitted light was introduced some years ago. These panels held great promise because they produce a marked decrease of veiling reflection at an angle of 60°, but much less at other angles. Because most viewing is in the range 20° to 40°, using these panels does not result in any appreciable reduction in reflected glare in normal office

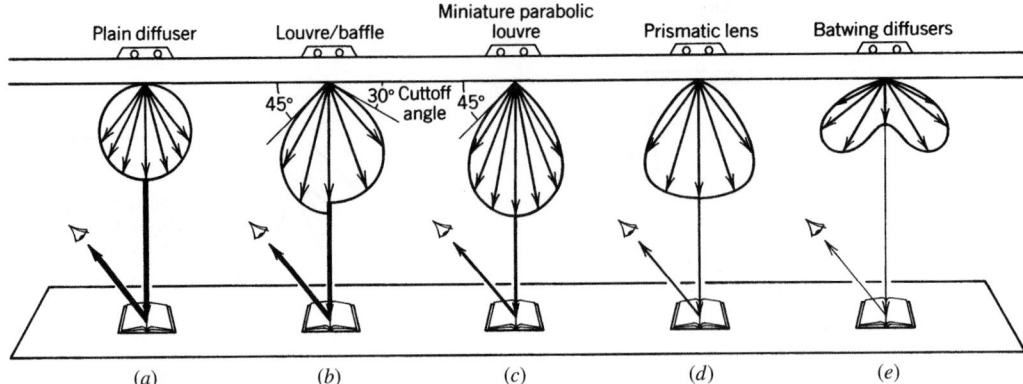

**Fig. 15.9** *Comparison of typical candlepower distribution curves for common linear or PL lamp fluorescent luminaire diffuser elements (for a full description of types a–d, see text). Note that for a given geometry of viewer and luminaire, the severity of veiling reflections depends entirely on the fixture's photometric characteristic. In the individual figures, the potential to produce reflected glare is indicated by the weight of the line representing fixture output and reflectance from the work task. The batwing distribution (e) concentrates its output in the 30° to 60° range, which minimizes both direct and reflected glare.*

work situations. From experience in a drafting room equipped with luminaires utilizing high-efficiency multilayer polarizers, it can be stated that visual discomfort from reflected glare, as personally experienced and as reported by a large staff, is not noticeably reduced.

### (b) Louvers and Baffles

These are generally rectangular section, metal or plastic, and serve primarily to shield the source (see Figs. 15.2, and 15.3) and to diffuse the output, particularly when plastic translucent louvers are used. Candlepower distribution curves are shown in Fig. 15.9*b*. The exact curve shape depends on the shielding angle, design of the louver, and its finish. Louvers finished in specular aluminum or dark colors exhibit low direct glare. The large downward light component can cause serious veiling reflections. Overall fixture efficiency is average.

The S/MH, a luminaire metric that indicates the maximum spacing permissible for a given luminaire mounting height that yields uniform illumination and is given as a dimensionless ratio, is fully explained in the next section. For this diffuser type, it does not exceed 1.5 and varies inversely with the shielding angle. This is because the basic circular distribution is changed to an egg shape by cutoff and redirection, reducing the high-angle light. Thus,

a 45° shielding angle has lower direct glare but requires closer spacing.

A special design in this category is the miniature egg-crate parabolic wedge louver shown in Figs. 15.10 and 15.11. These units redirect a large portion of the light directly downward, and because of this redirection and their specular finish, they appear completely dark—darker, indeed, than the unlighted portion of the ceiling when viewed obliquely (see Fig. 15.11). Fixtures using these louvers have low efficiency due to trapped light, with a maximum coefficient of utilization of about 0.5 (see Section 15.20). VCP is very high, but veiling reflections can be troublesome; S/MH varies between 1.0 and 1.5. The shielding angle is usually 45°. A typical candlepower distribution curve is shown in Fig. 15.9*c*. When these units are used, additional wall lighting is almost always required.

Because of the low efficiency of the parabolic louver design shown in Fig. 15.10*a* caused by the wide light-trapping tops of the louvers, an improved version was developed with the tops shaped to reflect incident light efficiently. This improved design, which is shown in Fig. 15.10*b*, increases luminaire efficiency by about 20%.

### (c) Prismatic Lens

Many designs are available with varying distribution characteristics. Figures 15.1*a* and 15.9*d* can be

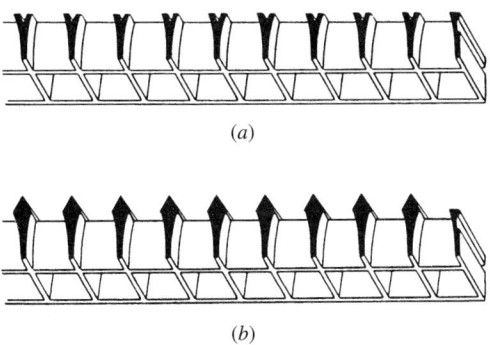

*(a)*

*(b)*

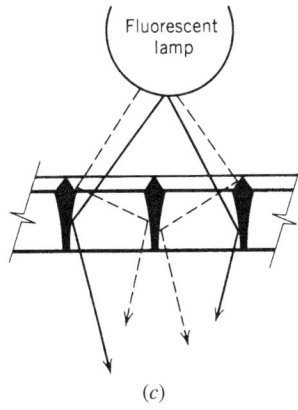

*(c)*

**Fig. 15.10** *(a) Section through a conventional, miniature parabolic wedge, eggcrate type of louver. These units give exceptionally low brightness when seen at a normal viewing angle. Most such units are made of aluminized plastic. Fixtures equipped with these units exhibit low overall efficiency due to the large amount of light trapped by the broad top of each parabolic wedge. (b) A modified wedge design uses a curved top on each wedge to redirect and utilize light striking the top. (c) Solid lines represent light rays redirected by the bottom curve, whereas dotted lines show light redirected by the top curve, which was lost in the design of (a). Typical louver cell dimensions are a ½-in. cube for design (a), with a consequent 45° shielding angle, and ⅝- to ¾-in. square by ½-in. high for design (b), giving a 35° to 45° shielding angle.*

taken as typical of this genre. They produce an efficient fixture (high coefficient of utilization), good diffusion, wide permissible spacing—an S/MH as high as 2.0—and low direct glare (high VCP). Veiling reflections can be troublesome, depending upon viewing angles and position (see Fig. 11.32).

### (d) Fresnel Lens

The action of this lens is similar to that of a reflector. Lamp hiding power is poor, but efficiency is high

**Fig. 15.11** *In the illustration the designer is checking the luminance of a fluorescent fixture equipped with a miniature parabolic wedge louver, which has very low luminance above 45°. This characteristic is readily apparent in comparison to the adjacent fixture, which utilizes a prismatic lens diffuser.*

and visual comfort is good. S/MH is seldom more than 1.5 (Fig. 15.12).

### (e) Batwing Diffusers

The theory behind this type of diffuser is covered in Section 11.31c. A typical characteristic is shown in Fig. 15.9e. There are two fluorescent luminaire

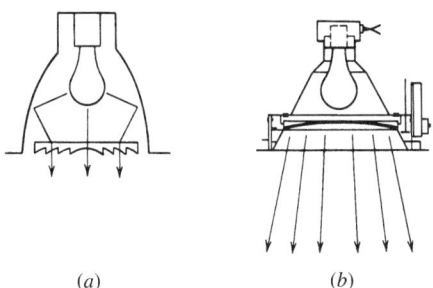

*(a)*      *(b)*

**Fig. 15.12** *Action of a Fresnel lens. With a Fresnel lens fixture, a smaller housing without a reflector can be used while still maintaining beam control. The lens performs the same function as a reflector, controlling the beam as a function of source placement. By utilizing a lens fixture, the curved reflector (a) can be largely eliminated, yielding a smaller fixture while maintaining accurate beam control. A common design (b) uses a regressed lens to provide shielding, although lens brightness is not normally objectionable.*

designs that produce the batwing shape distribution characteristic—a prismatic lens and parabolic reflectors and baffles.

1. *Prismatic batwing diffusers.* These are either linear or radial; that is, they produce the batwing distribution either in one direction or in all directions. Typical characteristics of both types are shown in Fig. 15.13. Note that the characteristic shape is more pronounced in the linear diffuser, which indicates better control of veiling reflections in that direction (usually crosswise). Fixtures equipped with these diffusers have good

efficiency, low direct and reflected glare, and good diffusion. As with all enclosed ungasketed fixtures, the lens acts as a dust trap, necessitating frequent cleaning to maintain high output.

2. *Deep parabolic reflectors.* These luminaires (Fig. 15.14) produce modified versions of the characteristic batwing distribution in the normal (crosswise) direction. Distribution in the parallel or lengthwise direction is circular (diffuse), indicating minimum beam control in that direction. These fixtures, like the batwing lens–type diffuser units, have high efficiency, high S/MH, low reflected glare, and low to very low surface brightness, making them usable in VDT areas. They are normally applied with the long axis in the direction of sight lines.

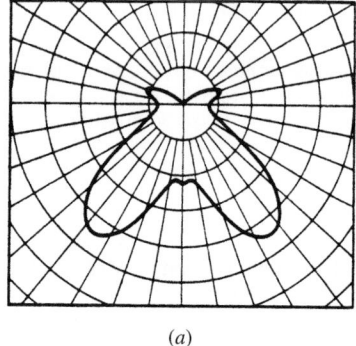

(a)

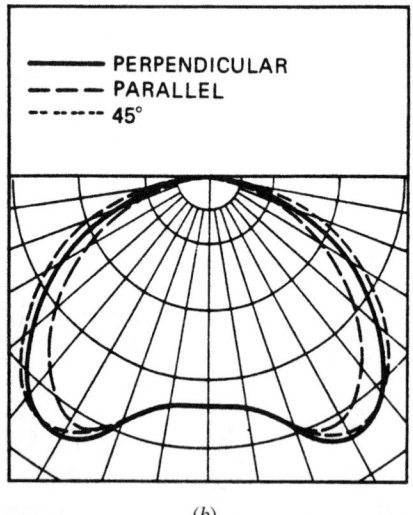

—— PERPENDICULAR
– – – PARALLEL
·· —·· — 45°

(b)

**Fig. 15.13** (a) Linear batwing distribution with extremely sharp cutoff in the upper and lower ranges. The curve is taken across the lamp axis for a single-lamp unit. (b) Distribution curves for a radial batwing distribution lens. Note that the perpendicular, parallel, and diagonal curves are almost identical. Zonal flux is maximum in the 30° to 60° range and drops off at both extremes, as desired.

## 15.5 UNIFORMITY OF ILLUMINATION

In any space intended to be lighted uniformly with multiple, discrete, ceiling-mounted direct-lighting system light sources, it is necessary to establish a fixture spacing that gives acceptable uniformity of illumination. A ratio of maximum to minimum illuminance on the working plane of 1.2 to 1.3 is readily acceptable because lesser ratios are not easily noticed. For general background or circulation lighting, a ratio of up to 1.5 is acceptable. The recommended S/MH (above the working plane) given by manufacturers (see the figures immediately above the distribution curves for each fixture in Table 15.1) are generally based upon a 1.0 illuminance ratio (Fig. 15.15). Therefore, the S/MH recommendation may be exceeded somewhat without seriously affecting uniformity.

When the luminaire's distribution characteristic is symmetrical in all directions, as is generally the case with small source lamps such as incandescent, CFL, and HID, only a single S/MH figure is required. However, with the asymmetrical distribution of most fluorescent fixtures, an S/MH ratio is required both crosswise and lengthwise. Due to the characteristic of the lamp itself, the transverse (crosswise, perpendicular) ratio is almost always considerably higher than the longitudinal (parallel, endwise, lengthwise) ratio. See, for instance, fixtures 26, 28, and 42 in Table 15.1. The S/MH (also

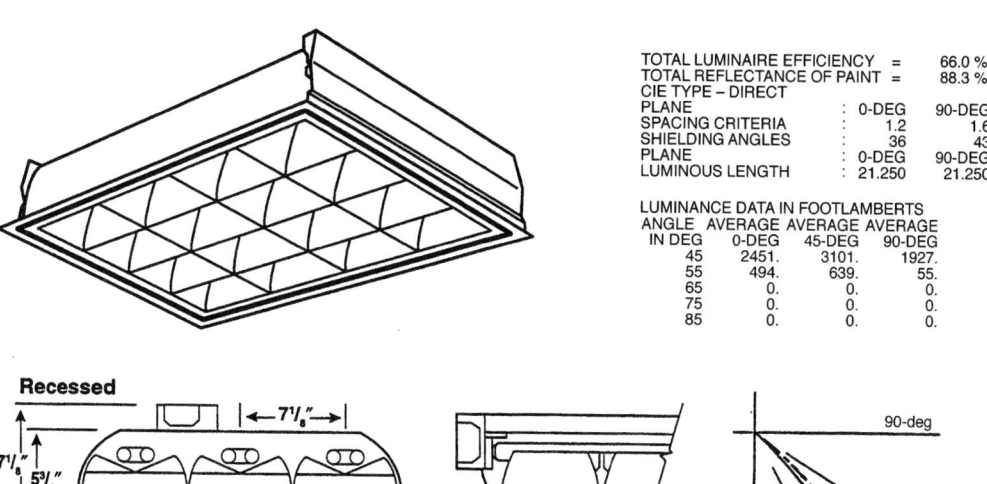

TOTAL LUMINAIRE EFFICIENCY = 66.0 %
TOTAL REFLECTANCE OF PAINT = 88.3 %
CIE TYPE – DIRECT

| PLANE | : | 0-DEG | 90-DEG |
|---|---|---|---|
| SPACING CRITERIA | : | 1.2 | 1.6 |
| SHIELDING ANGLES | : | 36 | 43 |
| PLANE | : | 0-DEG | 90-DEG |
| LUMINOUS LENGTH | : | 21.250 | 21.250 |

LUMINANCE DATA IN FOOTLAMBERTS

| ANGLE IN DEG | AVERAGE 0-DEG | AVERAGE 45-DEG | AVERAGE 90-DEG |
|---|---|---|---|
| 45 | 2451. | 3101. | 1927. |
| 55 | 494. | 639. | 55. |
| 65 | 0. | 0. | 0. |
| 75 | 0. | 0. | 0. |
| 85 | 0. | 0. | 0. |

**Recessed**

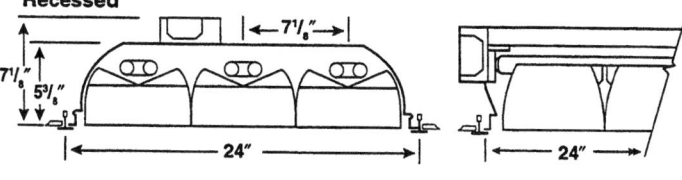

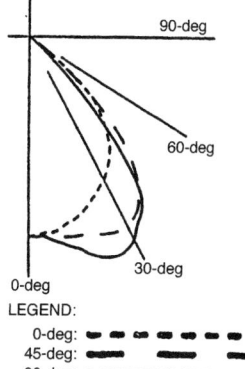

LEGEND:
0-deg: – – – –
45-deg: — — —
90-deg: —————

ILLUMINATION

**Visual Comfort Probability**

Reflectance - 80, 50, 20
Work Plane Illumination - 100 FC @ 2.5 ft.    **VCP**

| Room Room | | | Luminaires Lengthwise | | | Luminaires Crosswise | | |
|---|---|---|---|---|---|---|---|---|
| W | L | Ht. | 8.5 | 10.0 | 13.0 | Ht. 8.5 | 10.0 | 13.0 |
| 20 | 20 | | 68 | 81 | 77 | 90 | 90 | 87 |
| 20 | 40 | | 69 | 81 | 77 | 91 | 90 | 84 |
| 30 | 30 | | 69 | 81 | 77 | 91 | 90 | 84 |
| 30 | 60 | | 69 | 82 | 78 | 91 | 90 | 84 |
| 40 | 40 | | 69 | 82 | 78 | 91 | 90 | 84 |
| 40 | 60 | | 70 | 82 | 78 | 91 | 90 | 84 |
| 60 | 30 | | 69 | 82 | 78 | 91 | 90 | 84 |
| 60 | 40 | | 70 | 82 | 78 | 91 | 90 | 84 |
| 60 | 60 | | 70 | 82 | 78 | 91 | 91 | 84 |

**Coefficients of Utilization**

| | Coefficients of Utilization | | | | | | |
|---|---|---|---|---|---|---|---|
| Zonal Cavity Method | | | | | Floor Reflectance - .20 | | |
| RC RW | 80 | | | | 50 | | |
| | 70 | 50 | 30 | 10 | 50 | 30 | 10 |
| 1 | 75 | 73 | 71 | 69 | 69 | 67 | 66 |
| 2 | 71 | 67 | 65 | 62 | 64 | 62 | 60 |
| 3 | 67 | 62 | 59 | 56 | 59 | 57 | 54 |
| 4 | 63 | 57 | 54 | 50 | 55 | 52 | 50 |
| 5 | 59 | 53 | 49 | 46 | 51 | 48 | 45 |
| 6 | 55 | 49 | 44 | 41 | 47 | 43 | 41 |
| 7 | 51 | 44 | 40 | 37 | 43 | 39 | 36 |
| 8 | 48 | 41 | 36 | 33 | 39 | 35 | 33 |
| 9 | 44 | 37 | 32 | 29 | 36 | 32 | 29 |
| 10 | 41 | 34 | 29 | 26 | 33 | 29 | 26 |

**Energy Data**

**LER: FP-42**          **Energy Cost: $5.71\***
**Input Watts: 134**    **BF: .90**

The above energy calculations were conducted using a specific lamp/ballast combination. Actual results may vary depending upon the lamp and ballast used. Lamp and ballast specifications are subject to change without notice.

\*Comparative annual lighting energy cost per 1000 lumens based on 3000 hours and $0.08 per KWH.

***Fig. 15.14*** *Modern 2-ft square, deep parabolic reflector luminaire designed for three F40, 22.5-in. twin-tube CFL lamps, with a rated output of 3150 lm each. Total power input is 134 W and ballast factor (BF) is 0.9, giving a luminaire efficacy rating (LER) of 42. The typical modified batwing crosswise distribution and circular lengthwise distribution are clearly shown. Photometric data of interest to lighting designers are also shown. (Courtesy of Columbia Lighting.)*

# TABLE 15.1 Coefficients of Utilization for Typical Luminaires with Suggested Maximum Spacing Ratios

To obtain a coefficient of utilization (CU):
1. Determine the cavity ratios for the room, ceiling, and floor.
2. Determine the effective ceiling and floor cavity reflectances from Table 15.2. Use initial ceiling, floor, and wall reflectances.
3. Obtain the CU for a 20% effective floor cavity reflectance from the appropriate table below for the luminaire type to be used. Interpolate, when necessary, to obtain the CU for the exact room cavity ratio for the nearest effective ceiling cavity reflectances above and below the reflectance obtained in step 2; interpolate between these CUs to obtain the CU for the step 2 ceiling cavity reflectance.
4. If the effective floor cavity reflectance differs significantly from 20%, obtain the multiplier from Table 15.3 and apply this to the CU obtained in step 3.
5. To obtain the CU for a ceiling cavity reflectance ($\rho_{CC}$) of 30 or 10%, multiply the figure for $\rho_{CC}$ = 50% by 0.85 or 0.70, respectively. This is an approximation. For exact figures, see the IESNA *Lighting Handbook* (2000).
6. Use the figure in the last column ($\rho_{CC}$ = 0; $\rho_w$ = 0) for outdoor lighting, i.e., there are no walls or ceiling.
7. Legend:

$$\rho_{CC} = \text{percent effective ceiling cavity reflectance}$$
$$\rho_w = \text{percent wall reflectance}$$
$$RCR = \text{room cavity ratio}$$
$$\text{Maximum S/MH guide} = \text{ratio of maximum luminaire spacing to mounting above work plane}$$

*Note:* In some cases, luminaire data in this table are based on an actual typical luminaire; in other cases, the data represent a composite of generic luminaire types. Therefore, whenever possible, specific luminaire data should be used in preference to this table.
The polar intensity sketch (candlepower distribution curve) and the corresponding spacing-to-mounting height guide are representative of many luminaires of each type shown.

| Typical Luminaire | Typical Distribution and Percent Lamp Lumens / Maintenance Category | Maximum S/MH | RCR | $\rho_{CC}$: 80 | | | 70 | | | 50 | | | 0 |
| --- | --- | --- | --- | --- | --- | --- | --- | --- | --- | --- | --- | --- | --- |
| | | | $\rho_w$: | 50 | 30 | 10 | 50 | 30 | 10 | 50 | 30 | 10 | 0 |
| | | | | *Coefficients of Utilization for 20% Effective Floor Cavity Reflectance ($\rho_{FC}$ = 20)* | | | | | | | | | |
| 1 Pendant diffusing sphere with incandescent lamp | V / 5% / 45% | 1.5 | 0 | .87 | .87 | .87 | .81 | .81 | .81 | .69 | .69 | .69 | .44 |
| | | | 1 | .71 | .67 | .63 | .66 | .62 | .59 | .56 | .53 | .50 | .31 |
| | | | 2 | .61 | .54 | .49 | .56 | .50 | .46 | .47 | .43 | .39 | .23 |
| | | | 3 | .52 | .45 | .39 | .48 | .42 | .37 | .41 | .36 | .31 | .18 |
| | | | 4 | .46 | .38 | .33 | .42 | .36 | .30 | .36 | .30 | .26 | .15 |
| | | | 5 | .40 | .33 | .27 | .37 | .30 | .25 | .32 | .26 | .22 | .12 |
| | | | 6 | .36 | .28 | .23 | .33 | .26 | .21 | .28 | .23 | .19 | .10 |
| | | | 7 | .32 | .25 | .20 | .29 | .23 | .18 | .25 | .20 | .16 | .09 |
| | | | 8 | .29 | .22 | .17 | .27 | .20 | .16 | .23 | .17 | .14 | .07 |
| | | | 9 | .26 | .19 | .15 | .24 | .18 | .14 | .20 | .15 | .12 | .06 |
| 3 Porcelain-enameled ventilated standard dome with incandescent lamp | IV / 10% / 85% | 1.3 | 0 | .99 | .99 | .99 | .97 | .97 | .97 | .92 | .93 | .93 | .83 |
| | | | 1 | .88 | .85 | .82 | .86 | .83 | .81 | .83 | .80 | .78 | .72 |
| | | | 2 | .78 | .73 | .68 | .76 | .72 | .67 | .73 | .69 | .66 | .61 |
| | | | 3 | .69 | .62 | .57 | .67 | .61 | .57 | .65 | .60 | .56 | .52 |
| | | | 4 | .61 | .54 | .49 | .60 | .53 | .48 | .58 | .52 | .48 | .45 |
| | | | 5 | .54 | .47 | .41 | .53 | .46 | .41 | .51 | .45 | .41 | .38 |
| | | | 6 | .48 | .41 | .35 | .47 | .40 | .35 | .46 | .39 | .35 | .32 |
| | | | 7 | .43 | .35 | .30 | .42 | .35 | .30 | .41 | .34 | .30 | .28 |
| | | | 8 | .38 | .31 | .26 | .38 | .31 | .26 | .37 | .30 | .26 | .24 |
| | | | 9 | .35 | .28 | .23 | .34 | .27 | .23 | .33 | .27 | .23 | .21 |
| 7 EAR-38 lamp above 51-mm (2 in.)-diameter aperture (increase efficiency to 54½% for 76-mm (3 in.)-diameter aperture) | IV / 0% / 43½% | 0.7 | 0 | .52 | .52 | .52 | .51 | .51 | .51 | .48 | .48 | .48 | .44 |
| | | | 1 | .49 | .48 | .48 | .48 | .48 | .47 | .47 | .46 | .46 | .42 |
| | | | 2 | .47 | .46 | .45 | .46 | .45 | .44 | .45 | .44 | .43 | .41 |
| | | | 3 | .45 | .44 | .43 | .45 | .43 | .42 | .44 | .42 | .42 | .40 |
| | | | 4 | .43 | .42 | .41 | .43 | .41 | .40 | .42 | .41 | .40 | .38 |
| | | | 5 | .42 | .40 | .39 | .41 | .40 | .38 | .41 | .39 | .38 | .37 |
| | | | 6 | .40 | .39 | .37 | .40 | .38 | .37 | .39 | .38 | .37 | .36 |
| | | | 7 | .39 | .37 | .36 | .39 | .37 | .36 | .38 | .37 | .35 | .35 |
| | | | 8 | .37 | .36 | .34 | .37 | .35 | .34 | .37 | .35 | .34 | .33 |
| | | | 9 | .36 | .34 | .33 | .36 | .34 | .33 | .35 | .34 | .33 | .32 |
| 18 "High-bay" wide distribution ventilated reflector with clear HID lamp | III / 1½% / 77½% | 1.5 | 0 | .93 | .93 | .93 | .91 | .91 | .91 | .87 | .87 | .87 | .78 |
| | | | 1 | .85 | .82 | .80 | .83 | .81 | .79 | .79 | .78 | .76 | .70 |
| | | | 2 | .77 | .73 | .70 | .76 | .72 | .69 | .73 | .70 | .67 | .63 |
| | | | 3 | .70 | .65 | .61 | .68 | .64 | .60 | .66 | .62 | .59 | .56 |
| | | | 4 | .63 | .58 | .53 | .62 | .57 | .53 | .60 | .56 | .52 | .49 |
| | | | 5 | .57 | .51 | .47 | .56 | .51 | .47 | .55 | .50 | .46 | .44 |
| | | | 6 | .51 | .45 | .41 | .51 | .45 | .41 | .49 | .44 | .40 | .38 |
| | | | 7 | .46 | .40 | .35 | .45 | .39 | .35 | .44 | .39 | .35 | .33 |
| | | | 8 | .41 | .35 | .31 | .41 | .35 | .31 | .40 | .34 | .31 | .29 |
| | | | 9 | .37 | .31 | .27 | .37 | .31 | .27 | .36 | .30 | .27 | .25 |

# TABLE 15.1 (Continued)

| Typical Luminaire | Typical Distribution and Percent Lamp Lumens | Maintenance Category | Maximum S/MH | RCR | ρcc → 80 | | | 70 | | | 50 | | | 0 |
|---|---|---|---|---|---|---|---|---|---|---|---|---|---|---|
| | | | | | ρw → 50 | 30 | 10 | 50 | 30 | 10 | 50 | 30 | 10 | 0 |
| | | | | | Coefficients of Utilization for 20% Effective Floor Cavity Reflectance (ρFC = 20) | | | | | | | | | |
| 26 — Diffuse aluminum reflector with 35° crosswise shielding | 17% ↑, 66% ↓ | II | 1.5/1.3 | 0 | .95 | .95 | .95 | .91 | .91 | .91 | .83 | .83 | .83 | .66 |
| | | | | 1 | .85 | .82 | .80 | .82 | .79 | .77 | .75 | .73 | .72 | .59 |
| | | | | 2 | .76 | .72 | .68 | .74 | .70 | .66 | .68 | .65 | .62 | .52 |
| | | | | 3 | .69 | .63 | .59 | .66 | .61 | .57 | .62 | .58 | .54 | .46 |
| | | | | 4 | .62 | .56 | .51 | .60 | .54 | .50 | .56 | .51 | .47 | .41 |
| | | | | 5 | .55 | .49 | .44 | .53 | .48 | .43 | .50 | .45 | .41 | .36 |
| | | | | 6 | .50 | .43 | .39 | .48 | .42 | .38 | .45 | .40 | .36 | .31 |
| | | | | 7 | .45 | .38 | .34 | .43 | .37 | .33 | .41 | .36 | .32 | .27 |
| | | | | 8 | .40 | .34 | .29 | .39 | .33 | .29 | .37 | .31 | .28 | .24 |
| | | | | 9 | .36 | .30 | .25 | .35 | .29 | .25 | .33 | .28 | .24 | .20 |
| 28 — Diffuse aluminum reflector with 35° crosswise, 35° lengthwise shielding | 17% ↑, 56½% ↓ | II | 1.5/1.1 | 0 | .83 | .83 | .83 | .79 | .79 | .79 | .72 | .72 | .72 | .56 |
| | | | | 1 | .75 | .72 | .70 | .72 | .69 | .67 | .65 | .64 | .62 | .50 |
| | | | | 2 | .67 | .63 | .60 | .65 | .61 | .58 | .59 | .57 | .54 | .45 |
| | | | | 3 | .61 | .56 | .52 | .58 | .54 | .51 | .54 | .50 | .48 | .40 |
| | | | | 4 | .55 | .49 | .45 | .53 | .48 | .44 | .49 | .45 | .42 | .36 |
| | | | | 5 | .49 | .44 | .40 | .47 | .42 | .39 | .44 | .40 | .37 | .31 |
| | | | | 6 | .45 | .39 | .35 | .43 | .38 | .34 | .40 | .36 | .33 | .28 |
| | | | | 7 | .40 | .35 | .31 | .39 | .34 | .30 | .36 | .32 | .29 | .25 |
| | | | | 8 | .36 | .31 | .27 | .35 | .30 | .26 | .33 | .28 | .25 | .21 |
| | | | | 9 | .33 | .27 | .23 | .32 | .26 | .23 | .29 | .25 | .22 | .19 |
| 33 — Luminous bottom-suspended unit with very-high-output lamp | 66% ↑, 12% ↓ | VI | N.A. | 0 | .77 | .77 | .77 | .68 | .68 | .68 | .50 | .50 | .50 | .12 |
| | | | | 1 | .67 | .64 | .62 | .59 | .57 | .54 | .44 | .42 | .41 | .10 |
| | | | | 2 | .59 | .54 | .50 | .52 | .48 | .45 | .38 | .36 | .34 | .09 |
| | | | | 3 | .51 | .46 | .42 | .45 | .41 | .37 | .34 | .31 | .28 | .07 |
| | | | | 4 | .45 | .40 | .35 | .40 | .35 | .31 | .30 | .27 | .24 | .06 |
| | | | | 5 | .40 | .34 | .30 | .35 | .30 | .27 | .26 | .23 | .20 | .05 |
| | | | | 6 | .36 | .30 | .26 | .32 | .27 | .23 | .24 | .20 | .18 | .05 |
| | | | | 7 | .32 | .26 | .22 | .28 | .23 | .20 | .21 | .18 | .15 | .04 |
| | | | | 8 | .29 | .23 | .19 | .25 | .21 | .17 | .19 | .16 | .13 | .03 |
| | | | | 9 | .26 | .20 | .17 | .23 | .18 | .15 | .17 | .14 | .12 | .03 |
| 35 — Two-lamp prismatic wraparound; multiply by 0.95 for four lamps | 11½% ↑, 58½% ↓ | V | 1.5/1.2 | 0 | .81 | .81 | .81 | .78 | .78 | .78 | .72 | .72 | .72 | .59 |
| | | | | 1 | .71 | .69 | .66 | .69 | .66 | .64 | .64 | .62 | .60 | .50 |
| | | | | 2 | .64 | .59 | .56 | .61 | .58 | .54 | .57 | .54 | .51 | .44 |
| | | | | 3 | .57 | .52 | .48 | .55 | .50 | .47 | .51 | .48 | .45 | .38 |
| | | | | 4 | .51 | .46 | .41 | .49 | .44 | .41 | .46 | .42 | .39 | .34 |
| | | | | 5 | .46 | .40 | .36 | .44 | .39 | .35 | .41 | .37 | .34 | .29 |
| | | | | 6 | .41 | .35 | .31 | .40 | .35 | .31 | .38 | .33 | .30 | .26 |
| | | | | 7 | .37 | .31 | .27 | .36 | .31 | .27 | .34 | .29 | .26 | .23 |
| | | | | 8 | .33 | .28 | .24 | .32 | .27 | .23 | .30 | .26 | .22 | .19 |
| | | | | 9 | .30 | .24 | .20 | .29 | .24 | .20 | .27 | .23 | .19 | .17 |
| 38 — Four-lamp, 610-mm (2-ft)-wide troffer with 45° plastic louver | 0% ↑, 50% ↓ | IV | 1.0 | 0 | .60 | .60 | .60 | .58 | .58 | .58 | .56 | .56 | .56 | .50 |
| | | | | 1 | .54 | .52 | .50 | .52 | .51 | .49 | .50 | .49 | .48 | .44 |
| | | | | 2 | .48 | .45 | .43 | .47 | .44 | .42 | .45 | .43 | .41 | .39 |
| | | | | 3 | .43 | .40 | .37 | .42 | .39 | .37 | .41 | .38 | .36 | .34 |
| | | | | 4 | .39 | .35 | .32 | .38 | .35 | .32 | .37 | .34 | .32 | .30 |
| | | | | 5 | .35 | .31 | .28 | .35 | .31 | .28 | .34 | .30 | .28 | .26 |
| | | | | 6 | .32 | .28 | .25 | .32 | .28 | .25 | .31 | .27 | .25 | .23 |
| | | | | 7 | .29 | .25 | .22 | .29 | .25 | .22 | .28 | .25 | .22 | .21 |
| | | | | 8 | .26 | .22 | .20 | .26 | .22 | .20 | .25 | .22 | .20 | .18 |
| | | | | 9 | .24 | .20 | .17 | .24 | .20 | .17 | .23 | .20 | .17 | .16 |
| 42 — Fluorescent unit with flat prismatic lens, four-lamp, 610 mm (2 ft) wide | 0% ↑, 63% ↓ | V | 1.4/1.2 | 0 | .75 | .75 | .75 | .73 | .73 | .73 | .70 | .70 | .70 | .63 |
| | | | | 1 | .67 | .65 | .63 | .66 | .64 | .62 | .63 | .62 | .60 | .55 |
| | | | | 2 | .60 | .57 | .54 | .59 | .56 | .53 | .57 | .54 | .52 | .49 |
| | | | | 3 | .54 | .50 | .47 | .53 | .49 | .46 | .52 | .48 | .45 | .43 |
| | | | | 4 | .49 | .44 | .40 | .48 | .44 | .40 | .47 | .43 | .40 | .37 |
| | | | | 5 | .44 | .39 | .35 | .43 | .38 | .35 | .42 | .38 | .34 | .33 |
| | | | | 6 | .40 | .34 | .31 | .39 | .34 | .31 | .38 | .34 | .30 | .29 |
| | | | | 7 | .36 | .30 | .27 | .35 | .30 | .27 | .34 | .30 | .27 | .25 |
| | | | | 8 | .32 | .27 | .23 | .32 | .27 | .23 | .31 | .26 | .23 | .22 |
| | | | | 9 | .29 | .24 | .20 | .28 | .23 | .20 | .28 | .23 | .20 | .19 |

(continued)

ILLUMINATION

## TABLE 15.1 *(Continued)*

| Typical Luminaire | Typical Distribution and Percent Lamp Lumens | Maintenance Category | Maximum S/MH | RCR | ρcc → 80 | | | 70 | | | 50 | | | 0 |
|---|---|---|---|---|---|---|---|---|---|---|---|---|---|---|
| | | | | | ρw → 50 | 30 | 10 | 50 | 30 | 10 | 50 | 30 | 10 | 0 |
| | | | | | Coefficients of Utilization for 20% Effective Floor Cavity Reflectance (ρFC = 20) | | | | | | | | | |
| 44 — Radial batwing distribution—louvered fluorescent unit | 0%↑ 60↓ 45° | IV | N.A. | 0 | .71 | .71 | .71 | .70 | .70 | .70 | .66 | .66 | .66 | .60 |
| | | | | 1 | .65 | .63 | .61 | .63 | .62 | .60 | .61 | .59 | .58 | .54 |
| | | | | 2 | .59 | .55 | .53 | .58 | .55 | .52 | .55 | .53 | .51 | .48 |
| | | | | 3 | .53 | .49 | .46 | .52 | .48 | .45 | .50 | .47 | .45 | .42 |
| | | | | 4 | .47 | .43 | .40 | .47 | .43 | .40 | .45 | .42 | .39 | .37 |
| | | | | 5 | .42 | .38 | .34 | .42 | .37 | .34 | .41 | .37 | .34 | .32 |
| | | | | 6 | .38 | .33 | .30 | .38 | .33 | .30 | .37 | .33 | .30 | .28 |
| | | | | 7 | .34 | .29 | .26 | .33 | .29 | .26 | .33 | .28 | .25 | .24 |
| | | | | 8 | .30 | .25 | .22 | .30 | .25 | .22 | .29 | .25 | .22 | .20 |
| | | | | 9 | .27 | .22 | .18 | .26 | .22 | .18 | .26 | .21 | .18 | .17 |
| 45 — Radial batwing distribution—four-lamp, 610-mm (2 ft)-wide fluorescent unit with flat prismatic lens and overlay | 0%↑ 48%↓ 45° | V | N.A. | 0 | .57 | .57 | .57 | .56 | .56 | .56 | .53 | .53 | .53 | .48 |
| | | | | 1 | .50 | .48 | .47 | .49 | .47 | .46 | .47 | .46 | .44 | .41 |
| | | | | 2 | .44 | .41 | .38 | .43 | .40 | .38 | .41 | .39 | .37 | .34 |
| | | | | 3 | .39 | .35 | .32 | .38 | .34 | .31 | .37 | .33 | .31 | .29 |
| | | | | 4 | .34 | .30 | .27 | .33 | .29 | .26 | .32 | .29 | .26 | .24 |
| | | | | 5 | .30 | .25 | .22 | .29 | .25 | .22 | .28 | .24 | .22 | .20 |
| | | | | 6 | .26 | .22 | .19 | .26 | .22 | .18 | .25 | .21 | .18 | .17 |
| | | | | 7 | .23 | .19 | .16 | .23 | .19 | .16 | .22 | .18 | .16 | .14 |
| | | | | 8 | .21 | .16 | .13 | .20 | .16 | .13 | .19 | .16 | .13 | .12 |
| | | | | 9 | .18 | .14 | .11 | .18 | .14 | .11 | .17 | .14 | .11 | .10 |
| 46 — Bilateral batwing distribution—one lamp, surface-mounted fluorescent with prismatic wraparound lens | 12%↑ 63½%↓ 45° | V | N.A. | 0 | .87 | .87 | .87 | .84 | .84 | .84 | .77 | .77 | .77 | .64 |
| | | | | 1 | .76 | .73 | .70 | .73 | .70 | .67 | .67 | .65 | .63 | .53 |
| | | | | 2 | .66 | .61 | .57 | .64 | .59 | .56 | .59 | .56 | .52 | .44 |
| | | | | 3 | .59 | .53 | .48 | .56 | .51 | .47 | .53 | .48 | .44 | .38 |
| | | | | 4 | .52 | .45 | .40 | .50 | .44 | .40 | .47 | .42 | .38 | .32 |
| | | | | 5 | .46 | .39 | .34 | .44 | .38 | .33 | .41 | .36 | .32 | .27 |
| | | | | 6 | .41 | .34 | .29 | .39 | .33 | .29 | .37 | .31 | .27 | .23 |
| | | | | 7 | .36 | .30 | .25 | .35 | .29 | .24 | .33 | .27 | .23 | .20 |
| | | | | 8 | .32 | .26 | .21 | .31 | .25 | .21 | .29 | .24 | .20 | .17 |
| | | | | 9 | .29 | .22 | .18 | .28 | .22 | .18 | .26 | .21 | .17 | .14 |
| 47 — Radial batwing distribution—four-lamp, 610-mm (2-ft)-wide fluorescent unit with flat prismatic lens — see note 2 | 0%↑ 59½%↓ | V | 1.7 | 0 | .71 | .71 | .71 | .69 | .69 | .69 | .66 | .66 | .66 | .60 |
| | | | | 1 | .62 | .60 | .58 | .61 | .59 | .57 | .59 | .57 | .55 | .51 |
| | | | | 2 | .55 | .51 | .47 | .53 | .50 | .47 | .51 | .48 | .46 | .42 |
| | | | | 3 | .48 | .43 | .39 | .47 | .43 | .39 | .45 | .41 | .38 | .36 |
| | | | | 4 | .42 | .37 | .33 | .41 | .37 | .33 | .40 | .36 | .32 | .30 |
| | | | | 5 | .37 | .32 | .27 | .36 | .31 | .27 | .35 | .30 | .27 | .25 |
| | | | | 6 | .33 | .27 | .23 | .32 | .27 | .23 | .31 | .26 | .23 | .21 |
| | | | | 7 | .29 | .24 | .20 | .29 | .24 | .20 | .28 | .23 | .20 | .18 |
| | | | | 8 | .26 | .21 | .17 | .25 | .20 | .17 | .25 | .20 | .17 | .15 |
| | | | | 9 | .23 | .18 | .14 | .23 | .18 | .14 | .22 | .17 | .14 | .13 |
| 48 — Two-lamp fluorescent strip unit | 20½%↑ 68%↓ | I | 1.6/1.2 | 0 | 1.01 | 1.01 | 1.01 | .96 | .96 | .96 | .87 | .87 | .87 | .68 |
| | | | | 1 | .85 | .81 | .77 | .81 | .77 | .73 | .73 | .70 | .67 | .53 |
| | | | | 2 | .73 | .66 | .61 | .69 | .63 | .58 | .63 | .58 | .54 | .42 |
| | | | | 3 | .63 | .56 | .50 | .60 | .53 | .48 | .55 | .49 | .44 | .35 |
| | | | | 4 | .56 | .47 | .41 | .53 | .46 | .40 | .48 | .42 | .37 | .29 |
| | | | | 5 | .49 | .40 | .34 | .46 | .39 | .33 | .42 | .36 | .31 | .24 |
| | | | | 6 | .43 | .35 | .29 | .41 | .34 | .28 | .38 | .31 | .26 | .20 |
| | | | | 7 | .39 | .31 | .25 | .37 | .29 | .24 | .34 | .27 | .23 | .17 |
| | | | | 8 | .34 | .27 | .21 | .33 | .26 | .21 | .30 | .24 | .19 | .15 |
| | | | | 9 | .31 | .23 | .18 | .30 | .23 | .18 | .27 | .21 | .17 | .12 |
| 49 — Two-lamp fluorescent strip unit with 235° reflector fluorescent lamps | 12½%↑ 85%↓ | I | 1.4/1.2 | 0 | 1.13 | 1.13 | 1.13 | 1.09 | 1.09 | 1.09 | 1.01 | 1.01 | 1.01 | .85 |
| | | | | 1 | .96 | .92 | .88 | .93 | .89 | .85 | .87 | .83 | .80 | .68 |
| | | | | 2 | .83 | .76 | .70 | .80 | .74 | .68 | .75 | .69 | .65 | .55 |
| | | | | 3 | .73 | .65 | .58 | .70 | .63 | .57 | .66 | .59 | .54 | .46 |
| | | | | 4 | .64 | .55 | .49 | .62 | .54 | .48 | .58 | .51 | .46 | .39 |
| | | | | 5 | .56 | .47 | .41 | .55 | .46 | .40 | .51 | .44 | .38 | .33 |
| | | | | 6 | .50 | .41 | .35 | .49 | .40 | .34 | .46 | .38 | .33 | .28 |
| | | | | 7 | .45 | .36 | .30 | .44 | .35 | .30 | .41 | .34 | .28 | .24 |
| | | | | 8 | .40 | .32 | .26 | .39 | .31 | .25 | .37 | .30 | .25 | .21 |
| | | | | 9 | .36 | .28 | .22 | .35 | .27 | .22 | .33 | .26 | .21 | .18 |

ILLUMINATION

**TABLE 15.1** *(Continued)*

| Typical Luminaire | Maintenance Category | Maximum S/MH | RCR | $\rho_{cc}$: → 80 | | | 70 | | | 50 | | | 0 |
|---|---|---|---|---|---|---|---|---|---|---|---|---|---|
| | | | $\rho_w$: → | 50 | 30 | 10 | 50 | 30 | 10 | 50 | 30 | 10 | 0 |
| | | | | Coefficients of Utilization for 20% Effective Floor Cavity Reflectance ($\rho_{FC}$ = 20) | | | | | | | | | |
| 50 Single-row fluorescent lamp cove without reflector (multiplied by 0.93 for two rows and by 0.85 for three rows) | | | 1 | .42 | .40 | .39 | .36 | .35 | .33 | .25 | .24 | .23 | Coves are |
| | | | 2 | .37 | .34 | .32 | .32 | .29 | .27 | .22 | .20 | .19 | not recom- |
| | | | 3 | .32 | .29 | .26 | .28 | .25 | .23 | .19 | .17 | .16 | mended- |
| | | | 4 | .29 | .25 | .22 | .25 | .22 | .19 | .17 | .15 | .13 | for lighting |
| | | | 5 | .25 | .21 | .18 | .22 | .19 | .16 | .15 | .13 | .11 | areas |
| | | | 6 | .23 | .19 | .16 | .20 | .16 | .14 | .14 | .12 | .10 | having |
| | | | 7 | .20 | .17 | .14 | .17 | .14 | .12 | .12 | .10 | .09 | low reflec- |
| | | | 8 | .18 | .15 | .12 | .16 | .13 | .10 | .11 | .09 | .08 | tances |
| | | | 9 | .17 | .13 | .10 | .15 | .11 | .09 | .10 | .08 | .07 | |

For luminaire 53 (below), $\rho_{cc}$ = 10%; $\rho_w$ = 50, 30, 10 (two groups shown):

| Typical Luminaire | RCR | 50 | 30 | 10 | 50 | 30 | 10 |
|---|---|---|---|---|---|---|---|
| 53 $\rho_{cc}$ from below ~45% — Louvered ceiling. Ceiling efficiency ~50; 45° shielding opaque louvers of 80% reflectance. Cavity with minimum obstructions and painted with 80% reflectance paint—use $\rho_{cc}$ = 50. | 1 | .51 | .49 | .48 | .47 | .46 | .45 |
| | 2 | .46 | .44 | .42 | .43 | .42 | .40 |
| | 3 | .42 | .39 | .37 | .39 | .38 | .36 |
| | 4 | .38 | .35 | .33 | .36 | .34 | .32 |
| | 5 | .35 | .32 | .29 | .33 | .31 | .29 |
| | 6 | .32 | .29 | .26 | .30 | .28 | .26 |
| | 7 | .29 | .26 | .23 | .28 | .25 | .23 |
| | 8 | .27 | .23 | .21 | .26 | .23 | .21 |
| | 9 | .24 | .21 | .19 | .24 | .21 | .19 |
| | 10 | .22 | .19 | .17 | .22 | .19 | .17 |

*Source:* Data extracted from *IES Lighting Handbook Reference Volume,* (1981); with permission.

Notes:

1. Refer to the manufacturer's catalog data for more precise values when a specific luminaire type is proposed for use.

2. Multiply coefficients by 1.05 for three lamps and by 1.1 for two lamps.

called *spacing criteria* [SC]) for a specific luminaire is determined by measuring the distance between two test luminaires that yields the same illuminance on the working plane midway between them as directly under each one. This ignores (deliber-ately) any contributions from other fixtures in a multifixture installation and from interreflections, accounting only for the direct component of illuminance from the two test fixtures. Therefore, in an actual installation with several rows of fixtures, the illuminance at point $P_1$ in Fig. 15.16 is *higher* than the average by 20% to 30% because of the other fixtures and interreflections, and the illuminance at point $P_2$ is approximately equal to the room average. The illuminance levels along the walls, assuming a distance between the last row and the wall equal to one-half of the side-to-side spacing, range from 60% of average at point $P_3$ down to 50% at point $P_4$, assuming light-colored walls.

When walls are dark due to paint or aging, bookshelves, dark wood paneling, and the like, the illuminance levels drop to less than 50% of average, which is obviously insufficient as task lighting. To counteract this effect, particularly when placement of furniture is such that visual

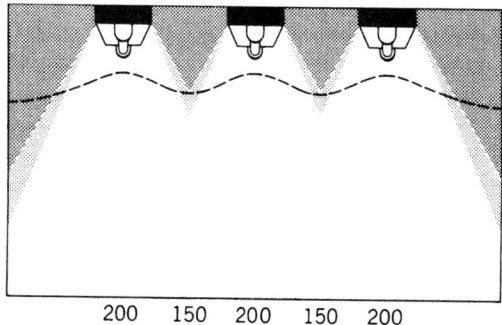

200 150 200 150 200

**Fig. 15.15** *The ratio of maximum to minimum illuminance should not exceed 1.3 in areas requiring uniform illumination.*

ILLUMINATION

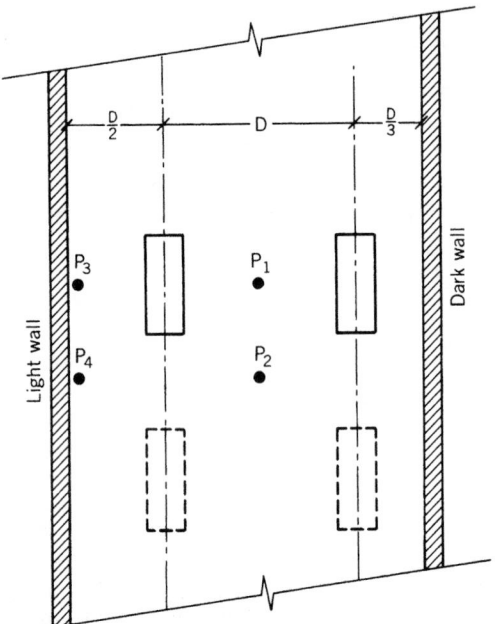

**Fig. 15.16** The diagram shows lighting fixtures installed according to the manufacturer's recommended spacing criteria (spacing to mounting height above the workplane ratio), with a row-to-row spacing, D and a row-to-wall spacing, D/2, as shown at the left wall. Assuming a high-reflectance finish on the wall (light color), illuminances $P_3$ and $P_4$ are at least one-half of the illuminance directly below a fixture. At a dark wall, as on the right, illuminance would fall below this value. As a consequence, the designer would move the right row of luminaires closer to the wall, as shown.

tasks *will* occur near walls, the designer has three choices:

1. Reduce the distance between the last row of fixtures and the wall to a third or less of the row-to-row spacing. This also provides required wall lighting.
2. Provide some type of continuous perimeter lighting or wall-wash units, both of which increase illuminance levels at the walls.
3. A combination of choices 1 and 2.

Because endwise illumination from linear fluorescent fixtures is considerably lower than crosswise illumination, end walls have lower illuminance than side walls and greater illuminance variation. It is, therefore, particularly important to provide some additional illumination, as discussed previously, and terminate fixture rows no more than 1 ft from an end wall. This is all the more important

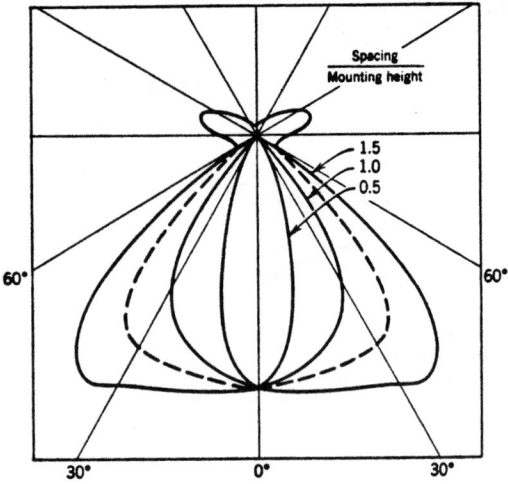

**Fig. 15.17** Typical distribution curves for approximating the ratio of fixture spacing to mounting height (S/MH) above the working plane for direct lighting luminaires. Although these curves were developed for direct lighting point sources such as incandescent (and HID), they can also be used with asymmetric distribution luminaires such as fluorescent ones. The permissible S/MH is somewhat higher than the curves indicate because this is a semi-direct distribution and the ceiling light component permits wider spacing between units. (After Odle and Smith, from IESNA Journal, January 1963.)

where visual tasks without supplementary task lighting occur at these walls.

We mentioned previously that the fixture in Fig. 15.1*a* had a high S/MH because of its flat-bottomed curve. This ratio, when not given by the manufacturer, may be approximated from Fig. 15.17. An accurate method of calculating maximum to minimum illumination ratios is available (see the *IESNA Lighting Handbook*, 1993).

The foregoing discussion of illumination uniformity was concerned with uniformity on a horizontal work plane. Occasionally, it is necessary to know the degree of uniformity vertically—that is, on horizontal planes at different elevations *directly below the fixtures*. Four different lighting situations are normally encountered. They are point sources, such as point source downlights; line sources, such as continuous-row fluorescent fixtures; infinite sources, such as luminous ceilings—whether transilluminated or indirect; and parabolic reflector beams, such as from PAR lamps. The vertical uniformity of each type is shown graphically in Fig. 15.18.

ILLUMINATION

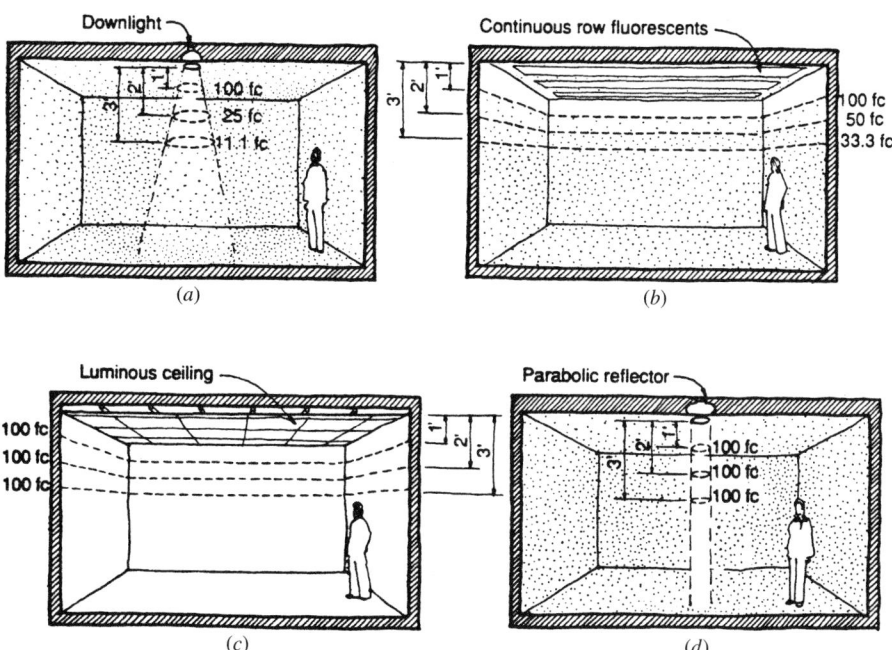

**Fig. 15.18** *Variation of illuminance vertically, directly below the fixtures, for different source types. (a, b) Illuminance directly below the fixture varies inversely with the square of the distance for a point source and inversely with the distance for a line source. (c, d) Illuminance remains constant at all distances from either an infinite (or nearly) source or a parabolic reflector.*

## 15.6 LUMINAIRE MOUNTING HEIGHT

The mounting height of luminaires is normally established before their spacing. In arriving at a mounting height for fixtures with an upward component, a balance must be struck between low mounting, which controls ceiling brightness and gives good utilization of light, and a reluctance to dominate an area, particularly a large room, by using such a low mounting height that the apparent ceiling height is affected (Fig. 15.19). General rules for mounting height are:

1. Indirect and semi-indirect luminaires should normally be suspended no less than 18 in. from the ceiling and preferably 24 to 36 in. Single-lamp luminaires with a very wide distribution (inverted batwing) may be suspended as little as 12 in. from the ceiling. Manufacturers' recommendations should be sought on this point.
2. Direct–indirect and semi-direct fluorescent fixtures should be suspended not less than 12 in. for two-lamp units and 18 in. for three- and four-lamp units.

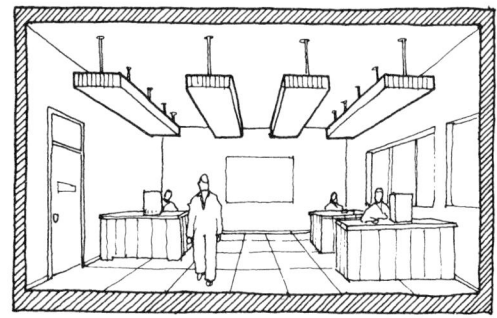

**Fig. 15.19** *Mounting height of fixtures may be lower in a small room than in a large room because of the illusion of lowness created in a large room.*

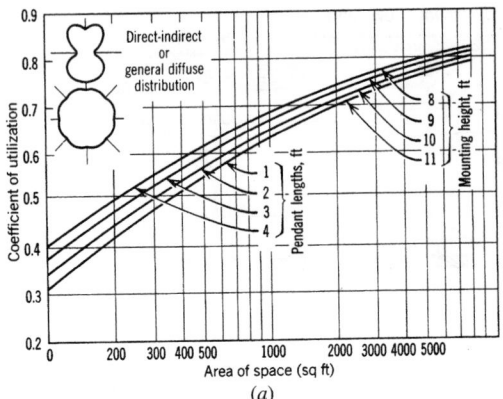

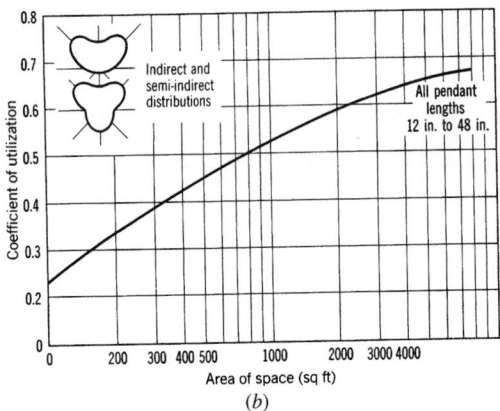

**Fig. 15.20** *Coefficient of utilization (CU; lighting system efficiency) as a function of pendant length for various distributions. With a substantial downward component, as in direct-indirect or general diffuse lighting (a), system efficiency rises slowly as the fixture descends (pendant length increases). Maximum differentials occur in small rooms and can reach 20%. Where the ceiling is the light source, as in indirect and semi-indirect systems (b), the pendant length does not change the room illumination. This curve can be used to estimate CU for indirect and semi-indirect luminaries in the absence of manufacturer's data.*

The effect of pendant length on the coefficient of utilization (efficiency; see Section 15.11) is given in Fig. 15.20.

## 15.7 LIGHTING FIXTURES

Before proceeding further with design, we will discuss the principal item of lighting hardware: the luminaire itself. This section and the sections that follow cover luminaire construction, installation,

and appraisal. The architect should simply stop to consider that lighting fixtures constitute 25% to 30% of the electrical budget or 4% to 5% of the overall building budget to appreciate their importance. Because the difference between a quality unit and an inferior one is often not readily visible to the casual observer, particular care must be taken in the specification of lighting fixtures and in the examination of shop drawings and *samples*. All fixtures, if applied properly, give a sufficient quantity of light, but only a good unit combines quantity with good quality, ease of installation, ease of maintenance, and indefinite life. In addition, installation must be proper to ensure mechanical rigidity and safety, electrical safety, freedom from excessive temperatures, and requisite accessibility of component parts and of the fixture outlet box. The following material is a combination of National Electrical Code (*NEC*) minimum requirements and factors beyond these minima that the authors have found important.

## 15.8 LIGHTING FIXTURE CONSTRUCTION

1. All fixtures should be wired and constructed to comply with local codes, *NEC* (Article 410), and the Underwriters Laboratories (UL) Standard for Luminaires, and should bear the UL label where label service is available. Reflector Luminaire Manufacturers (RLM) standards should be adhered to for all porcelain-enameled fixtures.
2. Fixtures should generally be constructed of 20-gauge (0.0359-in.)-thick steel minimum. Cast portions of fixtures should be no less than $\frac{1}{16}$ in. thick.
3. All metals should be coated. The final coat should be a baked-enamel white paint of at least 85% reflectance, except for anodized, aluminum, or silvered surfaces.
4. No point on the outside surface of any fixture should exceed 90°C after installation and on continuous operation. For an exception, see *NEC* Article 410 M.
5. Each fixture should be identified by a label carrying the manufacturer's name and address and the fixture catalog number.
6. Glass diffuser panels in fluorescent fixtures should be mounted in a metal frame. Plastic

diffusers should be suitably hinged. "Lay-in" plastic diffusers should not be used.

7. Plastic diffusers should be of the slow-burning or self-extinguishing type with a low smoke-density rating and low heat-distortion temperatures. The latter should be low enough so that the plastic diffuser distorts sufficiently to drop out of the fixture before reaching ignition temperature.

8. It is *imperative* that plastics used in air-handling fixtures be of the noncombustible, low-smoke-density type. These requirements also apply to other nonmetallic components of such fixtures.

9. All plastic diffusers should be clearly marked with their composition material, trade name, and manufacturer's name and identification number. Results of ASTM combustion tests should be submitted with fixture shop drawings. The characteristics of many plastic diffusers change radically with age and exposure to UV radiation.

A brief survey of the most common standard transparent and translucent lighting fixture diffusers follows.

*Glass.* Transparent to translucent, scratch resistant, easily cleaned, nonflammable, available in all grades of impact resistance, nonyellowing, usable with all sources, unaffected by and effective in blocking UV, readily formed into desired patterns. Heavy and expensive.

*Acrylic (Plexiglass).* Clear to translucent, easily scratched, slow burning, low impact resistance, resistant to yellowing, and available in a special low-yellowing composition (UV grade). Usable indoors and outdoors. Not usable at the elevated temperatures (>90°C) found in some HID applications. Does not readily embrittle, warp, or craze. Molds well.

*Polycarbonate.* Initially very clear and highly impact resistant, but with a tendency to opacity and strength loss with age. Good scratch resistance and excellent thermal resistance. Usable indoors and outdoors with all sources. Readily molded to prismatic forms. Self-extinguishing (burning rate). Expensive.

*Polystyrene.* Usable only indoors because of rapid yellowing when exposed to exterior UV radiation. Slow burning, but smoke generation properties problematic with some fire codes. Not usable in the long term because of discoloration, particularly with UV-producing sources. Good thermal resistance. Readily molded. Cheap. Not scratch resistant.

As can readily be seen, no ideal diffuser material exists. A designer must select the material that best suits the use and budget, considering both initial and replacement costs as well as long-term optical properties.

10. Ballasts should be mounted in fixtures with captive screws on the fixture body to allow ballast replacement without fixture removal.

11. All fixtures mounted outdoors, whether under canopies or directly exposed to the weather, should be constructed of appropriate weather-resistant materials and finishes, including gasketing to prevent entrance of water into wiring, and should be marked by the manufacturer "Suitable for Outdoor Use."

## 15.9 LIGHTING FIXTURE STRUCTURAL SUPPORT

Although some codes allow fluorescent fixtures weighing less than 40 lb to be mounted directly on the horizontal metal members of hung-ceiling systems, experience has shown that vibration, member deflection, routine maintenance operation on equipment in hung ceilings, and poor workmanship can cause such fixtures to fall, endangering life. It is therefore strongly recommended that all fixtures—surface, pendant, or recessed—whether mounted individually or in rows, be supported from the ceiling system support (purlins) or directly from the building structure, but in no case by the ceiling system itself. This is particularly important in the case of an exposed "Z" spline ceiling system.

## 15.10 LIGHTING FIXTURE APPRAISAL

The intense competition in the lighting products field necessitates close scrutiny of the characteristics of luminaires and all accessories. To compare the relative merits of similar lighting fixtures manufactured by different companies, complete test data plus a sample in a regular shipping carton from a normal manufacturing run are needed.

The following list should be used as a basic guide, with additional items added according to job requirements:

1. *Photometric and design data.* Manufacturers should furnish complete test data, including candlepower distribution curve(s), coefficients of utilization, wall and ceiling luminance coefficients, luminance data from 45° to 85°, a table of VCP, energy data including LER (see Section 15.12), and recommended S/MH (SC). These data should come from a reliable independent testing laboratory, not from the manufacturer's test facilities. In addition, many manufacturers either regularly publish or make available on request various design aids such as isolux (isocandle) curves and point-by-point computer printouts for different layouts. These are very useful, and their applications are covered in Section 15.11.

2. *Construction and installation.* The designer should check the sample for workmanship; rigidity; quality of materials and finish; and ease of installation, wiring, and leveling. Installation instruction sheets should be sufficiently detailed. Results of actual operating temperature tests in various installation modes should be included. Air-handling fixtures should be furnished with heat-removal data, pressure-drop curves, air-diffusion data, and noise criteria (NC) data for different airflow rates.

3. *Maintenance.* Luminaires should be simply and quickly relampable, resistant to dirt collection, and simple to clean. Replacement parts must be readily available.

## 15.11 LUMINAIRE–ROOM SYSTEM EFFICIENCY: COEFFICIENT OF UTILIZATION

Because of internal reflections inside a luminaire, some of the generated lumen output of the lamp is lost. The ratio of output lumens to lamp (input) lumens, expressed as a percentage, represents the luminous efficiency of the fixture. This characteristic has little meaning by itself, however, as the efficiency of a luminaire in doing a particular lighting job depends on the space in which it is used.

To illustrate, let us consider the case of a large room with a high, dark ceiling. If we were to use a high-efficiency (say, 80%) indirect lighting unit in such a room, most of the light directed upward would be lost (absorbed) and the actual illuminance on the working plane would be very low. If, however, the same room were illuminated with 50% efficiency direct-lighting units utilizing the same wattage, the illuminance on the working plane would be considerably higher than in the first case.

Similarly, if we consider a small room with dark walls and ceiling, lighted alternatively by diffuse lighting and direct lighting units of the same wattage and unit efficiency, the horizontal-plane illumination is higher for the direct units because of the large loss of the horizontal and upward components of the diffuse lighting on the walls and ceiling. Fixture efficiency *alone* is not sufficient; *the overall luminous efficiency of a particular unit in a particular space* is the required figure of merit. This number, inasmuch as it describes the utilization of the fixture output in a specific space, is known as the *coefficient of utilization* (CU). It is defined as the ratio between the lumens reaching the horizontal work plane and the generated lumens. As each luminaire has a different coefficient for every different space in which it is used, a system of standardization has evolved utilizing room cavities (explained later) of certain proportions and various surface reflectances. The fixture coefficients are then computed and tabulated as shown in Table 15.1. It should be emphasized that the figures given in this table are for generic fixture types only; in an actual job, luminaire data as found in manufacturers' catalogs should be used. To summarize, CU is a factor that combines fixture efficiency and distribution with room proportions, mounting height, and surface reflectances.

## 15.12 LUMINAIRE EFFICACY RATING

As a result of an EPACT mandate calling for an industrywide testing and information program designed to improve lighting fixture energy efficiency, a collaborative effort produced NEMA standard LE5, *Procedure for Determining Efficacy of Luminaires.* Unlike the CU discussed in the previous section, which defines the illumination efficiency of a particular luminaire in a particular space, standard LE5 determines the energy efficiency of the luminaire *alone*. Because this efficiency is expressed in lumens per watt (lumens output per watt input), it uses the same descriptive term used for light sources (i.e., *efficacy*). This metric takes into account all power used by a luminaire, including ballast, and includes the ballast factor, which is itself a ballast energy efficiency metric. The expression used to calculate the luminaire efficacy rating (LER) is

$$LER = \frac{photometric\ efficiency \times ballast\ factor}{luminaire\ input\ watts}$$

An LER metric applies to a specific type of fluorescent luminaire and is identified by an abbreviation as

FL = fluorescent lensed

FP = fluorescent parabolic

FW = fluorescent wraparound

FI = fluorescent industrial

FS = fluorescent strip light

Thus, the energy figures shown in Fig. 15.14 are specifically labeled "FP" to denote a fluorescent parabolic luminaire. This enables a designer to compare LER figures for different fixtures on a common basis. This commonality also extends to ballast type (i.e., magnetic or electronic). In addition, standard LE5 lists benchmark LER figures that are considered to represent an acceptable luminaire. An additional item of data useful in economic comparisons that is included in the NEMA standard is the yearly cost per 1000/lm, based on 3000 burning hours and $0.08 per kilowatt-hour (see Fig. 15.14.). The actual cost is easily calculated from this figure.

The LER approach has been expanded beyond fluorescent fixtures to include "Commercial, Non-residential Downlight Luminaires" (NEMA LE 5A) and "High-Intensity Discharge Industrial Luminaires" (NEMA LE 5B).

## LIGHTING CONTROL

### 15.13 REQUIREMENT FOR LIGHTING CONTROL

The term *lighting control* means all the techniques by which a lighting system can be operated, and

covers both manual and automatic controls. The control strategy must be decided on simultaneously with the lighting design because the control scheme must be appropriate to the light source. In turn, the system's accessories and arrangement depend on the control scheme. For instance, if dimming is decided on, using a fluorescent light source, then the range of dimming determines the type of ballast, the ballast switching points, and the degree of dimming flexibility.

The primary purposes of lighting control are flexibility and economy: flexibility to provide the modifications of luminances and patterns desired by the designer, and economy of both energy resources and monetary resources (see Appendix I for treatment of economic analyses). A properly designed lighting control system can reduce energy usage up to 60% over a simple on–off system installation. In addition, financial operating economies result from

- Reduced energy use
- Reduced air-conditioning costs as a result of lower lighting waste heat
- Longer lamp and ballast life due to lower operating temperatures and lower output
- Lower labor costs due to control automation

As noted previously (Chapter 13), lighting in most new nonresidential construction is designed within the energy constraints of ANSI/ASHRAE/IESNA Standard 90.1. This standard has a system of lighting power credits for lighting control systems designed with automatic energy-conserving controls. These credits permit the effective connected load to be reduced by factors that increase with energy conservation effectiveness. Thus, for example, circuits with a simple on–off mode initiated by a daylight sensor have a smaller power credit than daylight sensing with continuous dimming because the latter is more energy economical (but much more expensive initially). To avoid confusion, particularly in view of overlapping and sometimes inaccurate terminology, it is necessary to differentiate between control functions, control devices, and control systems.

For lighting, the only control *functions* are switching and dimming. The control *devices* are the means by which the switching and dimming functions are accomplished. They are numerous, ranging from a simple wall switch to time switches and dimmers of all sorts. Generally also included in this

ILLUMINATION

category are control *initiation* devices, such as occupancy sensors and photocells. The control *system* is the entire assembly of control and signal-initiating equipment together with their interconnections. Included here also are microprocessors and programmable controllers. The system can be a stand-alone arrangement or, alternatively, as is the case in large facilities, it can be part of an energy management system (EMS), a building automation system (BAS), or both. The difference in operation in these instances lies in the control algorithm, which is primarily energy oriented for an EMS and overall building function oriented for a BAS. In the discussion that follows, the control criteria are energy conservation, cost reduction, and operating flexibility.

## 15.14 LIGHTING CONTROL: SWITCHING

There are two basic control functions—switching and dimming. Switching is an on–off function. By selecting the number of lighting elements to be switched in each switching action, the designer can establish the number of control levels. The more levels, the finer the control. Thus, in a space requiring several levels of uniform illumination for different functions, the designer has many control alternatives. He or she can switch entire fixtures, but this adversely affects uniformity. Taking three-lamp fluorescent fixtures as an example, the designer can obtain better uniformity and four

levels of illumination by switching the ballasts (assuming one two-lamp and half of a two-lamp ballast per fixture):

| | |
|---|---|
| All ballasts on | 100% illumination |
| Two-lamp ballast on | 66% illumination |
| Half of two-lamp ballast on | 33% illumination |
| All ballasts off | 0% illumination |

(Magnetic ballasts have been preferred in this arrangement because a single electronic ballast is often used for up to four lamps as one means of obtaining its benefits while reducing its cost per lamp.)

This type of switching has the advantage of light reduction in relatively small steps at low cost. A typical arrangement is shown in Fig. 15.21. Use of split-wired two-lamp units is advantageous from the cost and energy viewpoints. Even more uniform light reduction and finer control are possible with two-level ballasts (at increased cost). There, each lamp remains lighted but at either full or half output. Thus, the designer could have 0%, 50%, and 100% levels or, by combining alternate ballast switching, with two-level ballasts could have 0%, 17%, 33%, 50%, 67%, 83%, and 100% levels. However, if that degree of control is desirable, dimming is probably preferable. The choice depends on the type of space and on the situation economics, as discussed later. An alternative method of achieving lower lighting levels in discrete steps, by switching, is to introduce impedance into the lighting circuit. This acts to reduce circuit current and light output.

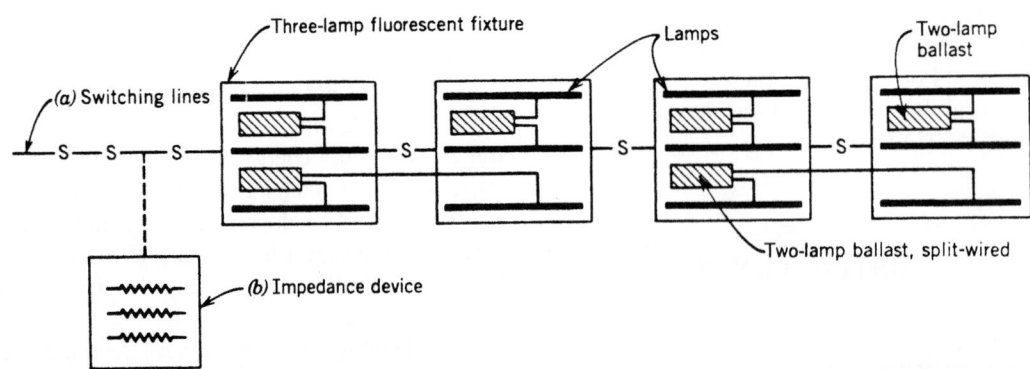

**Fig. 15.21** *Schematic diagram of switching arrangements to achieve multiple discreet lighting levels with three-lamp fluorescent lighting fixtures. Two-lamp ballasts are used in the interest of energy conservation and financial economy. In scheme (a) ballasts are switched, thus removing either one or two lamps from service. Finer control is achieved by using two-level ballasts or by introducing impedance (b) into the circuit, either in a block for an entire circuit or distributed in each fixture.*

Such devices are readily available for control of fluorescent lamps (see Fig. 15.26).

Recognition of the fact that an increased number of control (switch) points makes possible finer control and therefore energy conservation led to requirements in ANSI/ASHRAE/IESNA Standard 90.1 relating to the number of control points in a space and their types. Many field studies have demonstrated conclusively that reliance on manual switching is not an effective long-term energy conservation strategy, regardless of the good intentions of the space occupants. Indeed, studies have shown that space "ownership" affects even the low level of conservation possible with manual switching (i.e., lighting in private offices, conference rooms and small storage spaces may be switched off, whereas lighting in multioccupancy and common-use spaces such as libraries, large office spaces, break areas, and the like may not). As a result, Standard 90.1 first specifies the minimum number of control points required in a space and then awards automatic switching or dimming a higher number of equivalent control points.

The basic requirement is for one control point for every 450 ft$^2$ or fraction thereof of enclosed lighted space plus one control point for each task (or group of tasks) located in the space. Thus, the classroom in Example 15.1 would require two control points for its 517 ft$^2$ of area plus one control point for all similar tasks grouped in the room, for a total of three control points. This requirement could be met with three wall switches, or one wall switch and an occupancy sensor, or an automatic dimming system probably initiated by daylight sensors. The point is, of course, to encourage use of automatic controls, which, as has been pointed out repeatedly, is the only proven method of attaining significant energy conservation.

## 15.15 LIGHTING CONTROL: DIMMING

The techniques and equipment required for dimming each of the different light sources, as well as the effect on the color of the light produced and on the lamp, are discussed in Chapter 12. Figure 15.22 shows typical lumen output versus power input curves for common light sources. Note that for fluorescent lamps, *even with conventional magnetic ballasts,* dimming down to approximately 40% of

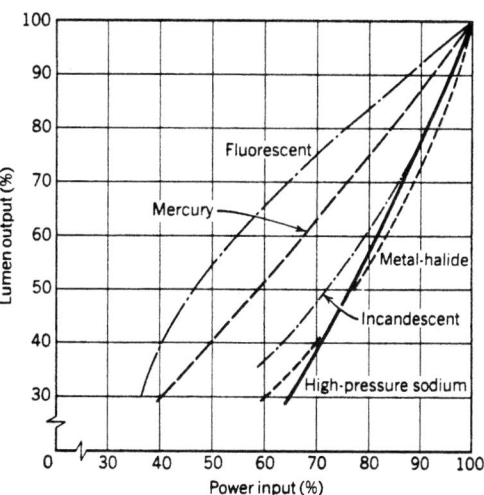

**Fig. 15.22** Typical dimming curves for generic light source types. Note that fluorescent lighting is efficient and approximately linear down to 40% output. All other sources have reduced efficacy when dimmed.

output is possible without reducing efficacy. This desirable characteristic can be exploited in control schemes where it is desired to change light output gradually without sacrificing efficiency. Due to efficacy reduction below 40% output, an economical and efficient control scheme combines dimming and switching of multilamp fluorescent fixtures to yield an almost stepless output range of 13% to 100% output. Continuous dimming over a 10% to 100% range is practical with special individual magnetic dimming ballasts or with electronic ballasts. As discussed in Chapter 12, electronic ballasts are much more energy efficient than conventional ones and must be considered for all *new* installations, dimmed or not. For retrofit work, silicon-controlled rectifier (SCR) or triac dimmers give excellent results with existing conventional core-and-coil ballasts down to 40% output.

## 15.16 LIGHTING CONTROL: CONTROL INITIATION

Control initiation is either manual, automatic, or a combined manual-automatic function. The last is usually in the form of an override function: manual override of an automated procedure to assure adequate control for special or unusual situations and automatic reset of the manual function to reestablish normal or steady-state operation.

### (a) Manual Control Initiation

Numerous studies dating back at least 50 years have indicated increased employee satisfaction when at least a degree of control of the working environment is in his or her hands. This satisfaction is frequently accompanied by increased work output, at least in the short term. Unfortunately, manual control of lighting levels has been demonstrated to be wasteful of energy due to the tendency to leave lights on at maximum level even when daylight is supplying more than enough light or when leaving a room for an extended period. A modicum of energy conservation in the latter instance is possible with the installation of *time-out* switches in spaces normally used for short periods, such as supply closets (see Fig. 26.48). However, for normal working spaces, even installation of manual dimmers in private offices is not effective because of the need to go to the dimmer control location on the wall to readjust. Manual dimming in multiple-occupancy spaces is effective only in creating personnel dissatisfaction and friction.

Modern electronics has made practical a remote-control dimming system that can control single or multiple luminaires, making it applicable to all occupancies. Figure 15.23 illustrates the use of the system in an open multiple-occupancy space. This enables individual workers to adjust the output of luminaire(s) closest to their workstation without disturbing other employees. This adjustment, which is simply accomplished (see Fig. 15.23a) by remote control, can alleviate direct and/or reflected glare or can be a temporary expedient suitable to the task at hand, as, for instance, work with a visual display terminal (VDT). Figure 15.23b–d shows the system components. Wiring of the system can be arranged so that a single receiver-dimmer can control up to 20 ballasts or, conversely, multiple dimmers are connected on a single circuit. The latter arrangement would be used in wiring a group of small single- or double-occupancy rooms. Such rooms are frequently wired with a wall-mounted dimmer, which can then be activated by the remote control, as shown in Fig. 15.24.

### (b) Automatic Control Initiation

Automatic controls are of two types: an open-circuit type and a closed-loop feedback type, also known as *static* and *dynamic control*, respectively. The former initiates a control function that is independent of the actual lighting situation, whereas the latter reacts to the condition of the lighting situation it controls via a feedback loop.

1. *Static control.* The most common type of open-circuit lighting control is the programmable time-base controller. These vary from small, relatively simple units that replace wall switches and fit into a common device box to the more sophisticated units shown in Figs. 26.13 through 26.15. These devices are available in myriad designs and capacities, but all perform the same basic function—(remote) control of loads and circuits on a preprogrammed time basis. The programming, in turn, is determined after analysis of operating schedules, task requirements, and field conditions. With "tight" programming, energy savings of up to 50% over an uncontrolled installation are possible.

Because these devices act only on a time base and are insensitive to actual field conditions, an override feature must be incorporated to permit accommodation of special conditions. Thus, if a timer is set to shut off a row of fixtures adjacent to windows between 10 A.M. and 3 P.M., local override must be provided to accommodate dark rainy days and the like. Similarly, if lighting is shut off during nonworking hours, provision must be made for persons working overtime. The override arrangement can be entirely local, in which case it may lead to energy waste because it depends on local cancellation; it can be local with time-out, which can be a nuisance to a person working for an extended period; or it can incorporate an override feedback link to the controller, usually operated by telephone lines. In general, programmable time controls are best applied to facilities with regular, repetitive schedules and few exceptional situations.

2. *Dynamic control.* The second type of automatic control initiation responds to sensor-indicated field conditions via an information feedback loop. It is frequently referred to as *dynamic control* because the initiation of a control function depends not on a fixed programmed parameter such as time, but on real-time field parameters (i.e., as measured at that instant). In the case of lighting control, these parameters may be ambient illuminance, time, system

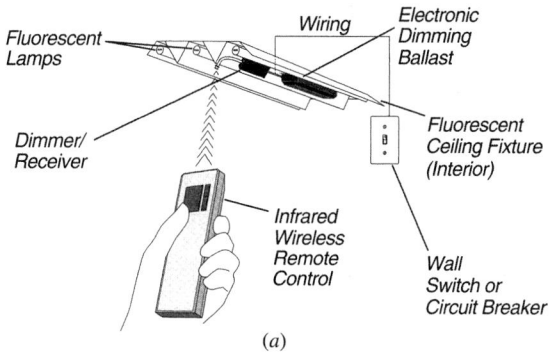

*(a)*

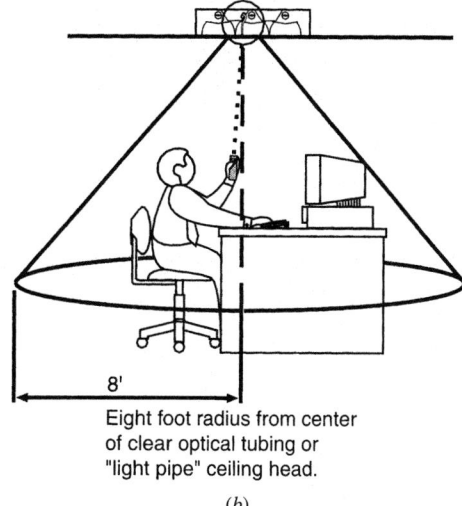

8'

Eight foot radius from center
of clear optical tubing or
"light pipe" ceiling head.

*(b)*

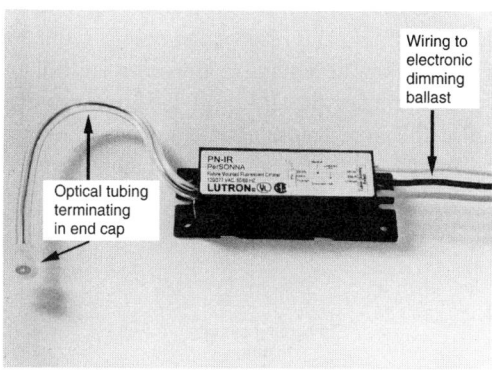

*(c)*

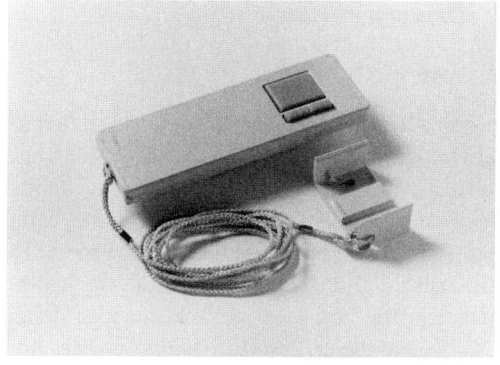

*(d)*

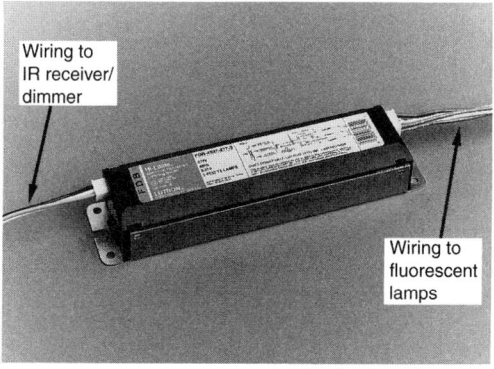

*(e)*

**Fig. 15.23** *(a, b) Ceiling-mounted fluorescent fixtures within an 8-ft radius of the optical sensor can be dimmed by remote control and restored almost instantaneously. (c) The hand-held IR remote control device (shown with a permanently mounted holster and a keeper-cord) arranged for continuous dimming and for dimming override for maximum or minimum light. (d) High-frequency electronic dimming ballast for one to four lamps can provide full-range dimming down to 1–5% of output or down to 10% of output, depending on the type. (e) The IR-activated dimmer/receiver, which fits into a 4-in. square standard outlet box, can be mounted inside the luminaire as in (a) or adjacent to it. The optical tube that carries the IR signal must be exposed on the ceiling. It may be mounted on the underside of an open luminaire or close by. (Courtesy of Lutron Electronics Co.)*

kW demand, kWh usage in a time period, or space occupancy, singly or in combination, depending on the programming algorithm (i.e., how the controller's microprocessor has been programmed). The control device in its entirety is called a *programmable controller* (see Section 26.16), which, in combination with the field sensors and the interconnecting wiring, constitute the control system. Some systems are wireless, using high-frequency signals impressed on the power wiring system to transmit control signals. This arrangement is known as a *power line carrier* (PLC) system and is discussed in Section 26.34. Other systems are completely wireless, using radio frequencies and a system of wireless transmitters and receivers.

In addition to its CPU (microprocessor), a programmable controller contains input/output (I/O) interfaces, memory, and means for programming (and reprogramming). An operational block diagram is shown in Fig. 26.16. The controller accepts not only the usual time-based signal function, but

ILLUMINATION

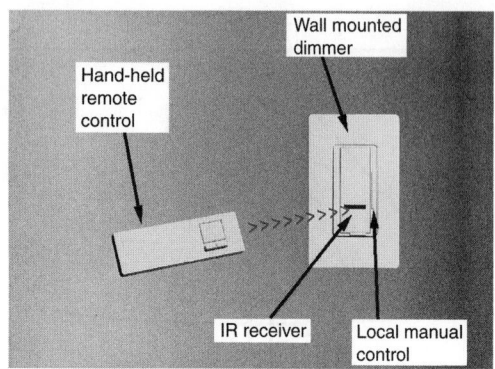

**Fig. 15.24** *Remote controller may be used to operate a wall-mounted dimmer. (Courtesy of Lutron Electronics Co.)*

also information (feedback) from field devices via its I/O device. It then "processes" the signal in its CPU, which consists of logic and storage memory, and sends out a resultant processed control signal.

Large lighting control systems use a computer in lieu of a programmable controller and usually have additional control facilities, such as telephone interfaces and local control/relay/switching centers, which process local sensor input and control local lighting blocks (see Figs. 15.25 and 26.56). Identical systems are used for building automation (Chapter 30), HVAC control, energy management, and the like, the difference being the types of sensing devices and the control algorithm.

## 15.17  LIGHTING CONTROL STRATEGY

It is apparent that a good lighting control system varies the lighting supplied to match the lighting required, *as the requirement varies;* thus, overlighting and underlighting are avoided. In addition, the control system must be capable of permitting initial adjustments and of accommodating external, non-lighting-connected constraints such as commands from a peak-demand controller. The common lighting system situations addressed by the control system follow.

### (a) System "Tuning"

In every lighting installation there is a difference between the design intent and the field result. This is due to assumptions and imprecision in calculation, differences between specified and installed equipment, equipment location changes, and so on. The responsible lighting designer "tunes" the lighting system in the field to attain the intended design. This usually means *reducing lighting levels* in non-task areas because spill light is frequently sufficient for circulation, rough material handling, and the like. This tuning can result in an energy reduction of 20% to 30%, depending on the control technique.

The smaller the group of light sources controlled, the more accurate the tuning and the larger the energy saving proportionally. Lighting system

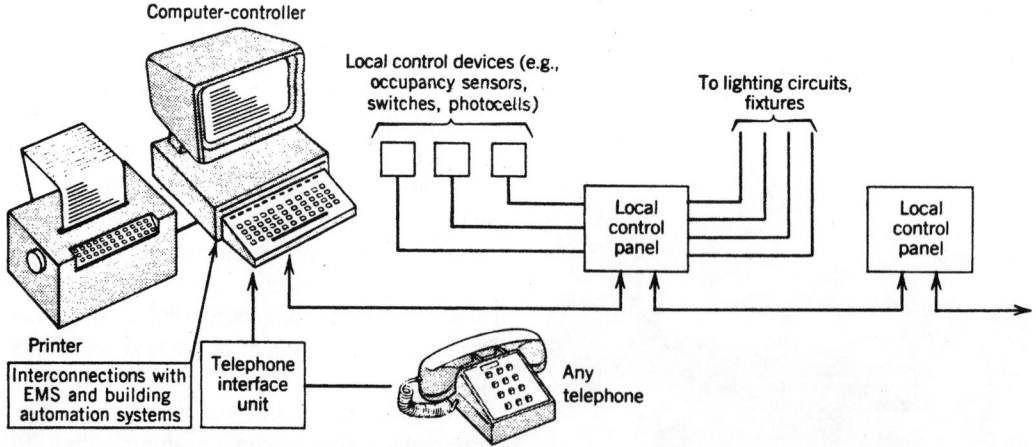

**Fig. 15.25** *Schematic arrangement of a large lighting control system. The controller can schedule and supervise thousands of control points via the local control panels. The local control panel accepts coded commands from the controller and operates individual devices and circuits. It also accepts local control signals from manual (switches) and automatic devices. Override is provided via a telephone command to the controller, and all functions, including overrides, appear as hard copy on the printer.*

retuning is also required when the function of an entire space is changed, or when a single area is altered by a furniture move or by a task change. An ancillary benefit of field tuning is glare reduction, which frequently improves task visibility dramatically. Tuning is a one-step function in the sense that, once accomplished, it does not require change unless the space function changes. Therefore, it should also be reversible to accommodate such changes. Tuning should not be confused with *lumen maintenance*, described in Section 15.17*d*, which is a control strategy designed to compensate for normal system output decline and its corollary—initial system overdesign.

As stated, the tuning action most often required is a reduction in illuminance. This can be accomplished by:

1. Making appropriate field modifications to adjustable fixtures (aiming, lamp position, etc.).
2. Replacing lamps with others of lower wattages and replacing fluorescent tubes with low-wattage tubes. These changes reduce light output reversibly.
3. Replacing fluorescent ballasts with low-current ballasts, thereby lowering output.
4. Adding current-limiting impedances to luminaires or lighting circuits, as mentioned in Section 15.14*a* (Fig. 15.26).
5. Ballast switching or use of multilevel ballasts (see Section 15.14*a*).
6. Dimming by adjustment of a potentiometer at individual fixtures so equipped.

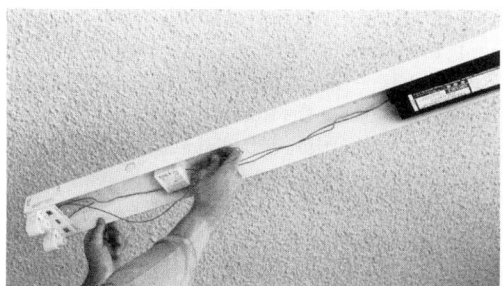

**Fig. 15.26** *This type of compact solid-state electronic circuit reduces ballast power input by approximately 30% and lamp output by about 28%, resulting in a net gain in efficacy. The power factor remains above 90%. Similar units are available for larger power decrease. They can be mounted on the lamp end or in the fixture channel, as shown. An ancillary benefit is cooler ballast operation, resulting in extended life. (Photo courtesy of Remtec Systems.)*

7. Replacing standard wall switches with time-out units (see Fig. 26.48), programmable units, or dimmer units.

### (b) Variable Time Schedule

No normal task area has a constant 24-hour, 365-day lighting requirement. In commercial and industrial spaces, work areas have regularly scheduled periods during which task lighting is not required. These include coffee and lunch breaks, cleaning periods, shift changes, and unoccupied periods. Programmed time controls can readily save 10% to 25% of the energy use compared to relying on occupants to manually operate controls. The action of such a controller in an actual installation is shown in Fig. 15.27. "Tight" scheduling took account of lunch hour and provided lighting only in restricted areas being cleaned rather than whole floors. The payback period for the investment in control equipment varies between 1½ and 5 years. Note that static control that is insensitive to actual field conditions has only limited ability to conserve energy.

### (c) Occupancy Sensing

Within a normal 9 A.M. to 5 P.M. working schedule, offices in commercial spaces are unoccupied for 30% to 60% of the time. The reasons are manifold—coffee breaks, conferences, work assignments, illness, vacations, and reassignment to a different work location are a few. Occupancy sensors can operate relays to turn off lights after a preset minimum period of about 10 minutes or can dim the light level to a minimum in areas such as corridors, which always require some light. (They can also turn off other energy consumers such as fan-coil units, air conditioners, and fans.) Reestablishment of the original lighting level can be instantaneous, delayed, or manual on the action of the occupant. Another useful function of an occupancy sensor is to provide an automatic override in schedule systems, thus both relieving the occupant of the necessity of using a manual override and limiting the energy use to actual occupancy time. It can also light the occupant's way into a space and shut off the system after he or she has left.

Occupancy sensors that react to a human presence are of three types—passive infrared (IR), ultrasonic, and a hybrid of both technologies. The

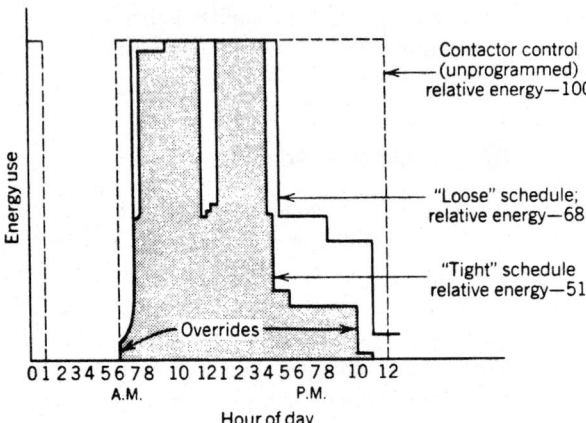

**Fig. 15.27** *Actual plot of energy usage on a typical floor of a high-rise building in New York City, with three degrees of lighting control. Note the importance of override by occupants when tight scheduling is used. Override was provided on a 1000-ft²-zone basis. (Data extracted from Peterson and Rubenstein, 1982.)*

IR sensor (Fig. 15.28) reacts to the motion of a heat source within its range. It operates by creating a pattern of beams, and alarms when a heat source (such as a person) moves from one beam to another. It will *not* alarm to a stationary heat source. Although the IR sensor is quite sensitive, it has several disadvantages:

- Small movements may not be detected, as they may not cross from one beam (zone) to another. As a result, a person sitting quietly may not be detected and the sensor may shut down/off the lights.
- Very slow movements may not be detected even when they cross zones.
- The IR detector must "see" the heat source. Therefore, a heat source blocked by furniture will not be detected.
- The beams have a discrete width and depth. Therefore, there may be "dead" spots under the beams if the units are not carefully selected to have adequate multilevel beams or properly located to give the desired coverage.

Ultrasonic sensors emit energy in the 25- to 40-kHz range, which is well above the range of human hearing. The waves immediately fill a space by reflecting and rereflecting off all hard surfaces, establishing a pattern that is detected by the sensor. Any movement within the space disturbs this pattern and is immediately detected by the sensor. Ultrasonic sensors have distinct advantages over IR sensors in that they do not require a direct line-of-sight exposure to the movement and they detect small movements. The latter characteristic, how-

ever, is also a disadvantage inasmuch as movement of curtains and even air movement can trigger a sensor. It is frequently necessary to adjust (reduce) a sensor's sensitivity to avoid false sensing. Unfortunately, this also decreases its coverage.

Hybrid (dual-technology) sensors combine the characteristics of both sensors by usually requiring both sensors to react to turn lights on, but once on, a reaction in either sensor keeps the lights on. In addition, sophisticated electronic circuitry "learns" a space's occupancy patterns and is programmed to react accordingly.

Placement of sensors is very important and should be tested before final installation. Studies have shown that reducing the minimum on period below 10 minutes is counterproductive and frequently causes space occupants to shut off the sensors. Depending on their type and mounting position, sensors cover a maximum area of 250 to 1000 ft² per unit. The payback period for this equipment runs between 6 months and 3 years, depending on the type of space and the degree of control already existing. Occupancy sensors are best applied in areas that are divided into individual rooms and work spaces. Sensors can be wall or ceiling mounted or mounted on a wall-outlet box in a combined sensor/wall-switch configuration. A few designs are shown in Fig. 15.29. (See Section 30.2, which discusses the security application of motion detectors and illustrates their operation.)

ANSI/ASHRAE/IESNA Standard 90.1 recognizes the great effectiveness of these devices by giving both connected power credit and a high control point equivalency.

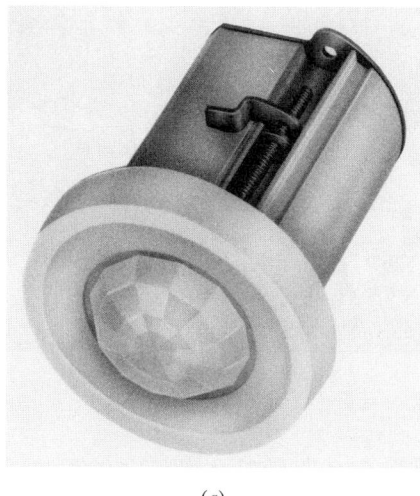

(a)

(b)

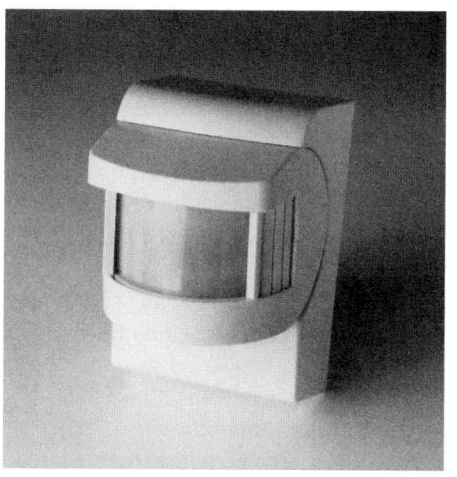

(c)

**Fig. 15.28** *Passive infrared (PIR) occupancy sensors. All sensors can be equipped with an adjustable override to prevent turning on lights when ambient light is sufficient. All units have adjustable delayed-off timing and a flashing LED that indicates sensor operation. (a) Flush-mounted ceiling unit. Note the 360° circular pattern of the lens, which indicates omnidirectional coverage. The unit is approximately 4 in. in diameter. (b) Flush-mounted combination wall switch and PIR occupancy sensor. Operation of the switch overrides the sensor function. (c) Surface-mounted wall PIR occupancy sensor. Note the semicircular shape of the lens, which indicates wide horizontal coverage. (Photos courtesy of Leviton Manufacturing Co.)*

ILLUMINATION

### (d) Lumen Maintenance

Referring to Section 15.20, we see that in order to maintain a minimum lighting level to the end of a maintenance period, we must deliberately overdesign initially. The extent of the overdesign is the reciprocal of the light loss factor (LLF). With an average LLF of 0.60, this initial overdesign amounts to (1/0.60), or 66%. Assuming a linear light falloff over a 2-year maintenance period, this overdesign results in an average of 33% annual energy waste. (In the next 2-year maintenance cycle, the actual overdesign is slightly less due to a small amount of unrecoverable loss.) Because the light depreciation is a continuous and very gradual process over the maintenance period, the most appropriate control strategy is one that reduces the initial overlighting by the required amount (as measured in the field) and gradually restores it as the system ages. This control strategy, known as *lumen maintenance*, is accomplished by a dimming system operating in conjunction with local light sensors (photocells). The photocells measure ambient light, and in response to their signals the controller(s) operates the dimming units to raise (or lower) the light output. Depending on the size of the installation, the dimmers can either be dispersed or centralized. The modulating action of such a system over the maintenance period is shown in

ILLUMINATION

(a)

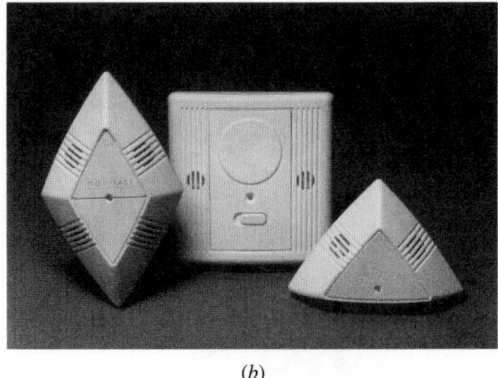

(b)

**Fig. 15.29** *Ultrasonic occupancy detectors. (a) Ceiling-mounted sensors, bidirectional on the left and unidirectional on the right. (Courtesy of Leviton Manufacturing Co.) (b) Sensors in various designs, intended for differing applications. Left to right: Ceiling-mounted two-way sensor intended for large rooms of up to 2800 ft². Dual-function sensor and wall switch designed for rooms of up to 300 ft², wall mounted on a single- or double-gang wall box. One-way ceiling-mounted sensor designed for use in rooms of up to approximately 1250 ft². (Courtesy of Novitas Inc.)*

Fig. 15.30. This strategy, in a purely manual mode (periodic maintenance, initial light reduction in accordance with the length of the maintenance period), gives only about a 10% energy reduction, whereas in an automated system almost all of the 33% annual energy waste can be eliminated.

In a new installation, the choice of whether to use electronic ballasts and full-range dimming, conventional ballasts and partial dimming, or a system of multilevel switching is an economic one and depends on many factors, one of the most important of which is the cost of energy. Often a combined system is advisable. In interior zones, initial lighting reduction does not exceed 30% to 40%, and full-range dimming is not required. In perimeter zones, daylight often provides all of the required light, and either full-range dimming, dimming plus switching, or a multilevel switching system is required (see Section 15.14a and the following discussion of daylight compensation). The payback period for a lumen maintenance installation of this type varies from 1 to 5 years. Shorter payback periods can be obtained by using multilevel switching rather than dimming because of its lower first cost.

An additional favorable effect of initial light reduction is the lengthening of effective lamp life and a reduction in the rate of its lumen depreciation. When a lamp is operated at the rated voltage, its lumen output drops during its life (according to the type of lamp). However, if lamps are operated at reduced output, as is the case if lamps are dimmed to compensate for initial overlighting, the lamp life cycle is greatly extended, lumen depreciation is reduced, and lamp energy consumption is linearly reduced.

Typical life extension (to economic replacement) figures are:

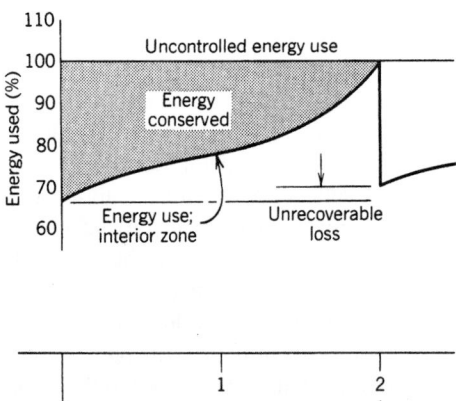

**Fig. 15.30** *Graph of energy use by a system that reduces the initial lighting level to compensate for initial overdesign and gradually increases the level as system output depreciates. In subsequent cycles the energy savings are reduced slightly because of unrecoverable output loss.*

| | |
|---|---|
| Fluorescent | 80% |
| Metal halide | 40% |
| High-pressure sodium | 20% |

These figures are for interior zones; for perimeter zones with ambient daylight compensation, they are higher.

## (e) Daylight Compensation

It should be apparent that a control system arranged for continuous ambient light compensation, as described in the preceding subsection, is automatically also arranged to compensate for ambient daylight. The difference is that ambient compensation for lumen maintenance due to light loss factors is a very gradual process of *increasing* output, whereas daylight compensation can be a minute-by-minute variation and generally in the direction of *decreased* electric lighting. Because of these possible rapid variations, switching systems are undesirable, as the constant on–off or level switching of lamps can be very annoying to occupants and deleterious to lamps.

Automatic dimming is the system of choice. Recognizing that in perimeter areas daylight often supplies all the required light, the system for fluorescent installations must be either full-range dimming with electronic ballasts or partial dimming with magnetic or electronic ballasts. Figure 15.31 shows the action of both types of dimming systems in a fluorescent installation with daylight compensation. The crucial design element in a daylight-compensating system is the establishment of zone areas. Depending on the latitude and climate, the southern and possibly the eastern and western exposures can have an interior (second) perimeter zone that receives sufficient daylight for a large enough portion of the year to economically warrant dimming. The northern exposure has only a narrow perimeter zone (Fig. 15.32). As a starting point, the size of the zones is established by determining the maximum room depth that receives at least half of its illuminance from daylight for several hours a day. A computer daylight and dimming cost study with perimeter zone depth as a variable is an effective approach. Several such programs are available.

Placement of the control photocells depends on the control system. Where daylight compensation is desired in conjunction with lumen maintenance (see the preceding subsection), area photocells are desirable, as they give a feedback control signal for the specific area involved. Alternatively, a daylight-factor map of a space can be made, preferably after

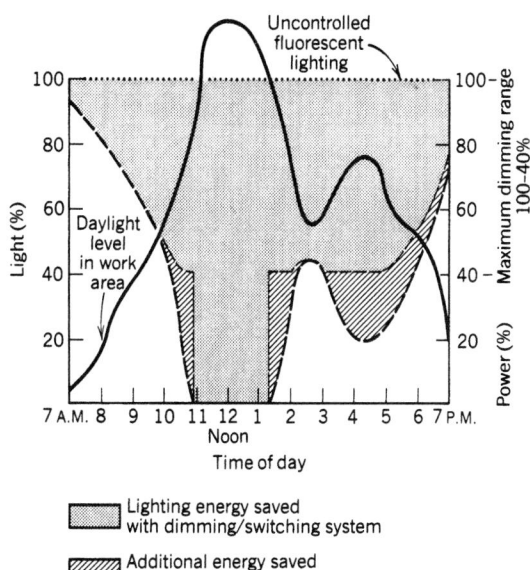

**Fig. 15.31** *Typical graph of energy savings with daylight-compensating lighting control. A full-range dimming system is more effective than one that dims down to only 40% because daylight often supplies most of the required lighting. An economic analysis is required to determine whether the additional cost of such a system is justified.*

the installation is complete and the furniture in place, and photocells located at one point, based on this map (Fig. 15.33). We know from our study of daylight in Chapter 14 that the daylight factor at any inside point is constant. Therefore, by measuring actual daylight level at a point either immediately inside or outside the window, shielded from direct sun and from inside electric lighting, we can relate it to any area in the room by the ratio of

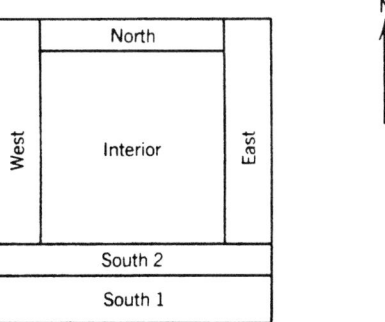

**Fig. 15.32** *Typical building plan showing approximate daylight perimeter zones. Exact delineation of zones depends upon latitude, climate, window design, and cost of electric energy.*

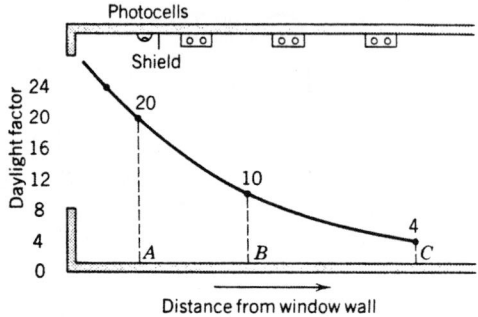

**Fig. 15.33** *Typical daylight factor curve plotted on a room section with one side window. The photocell control technique for daylight compensation described in the text assumes a constant DF distribution indoors—that is, fixed luminance sky and no direct sun.*

daylight factors and establish a switching level at which the lighting for that inside area is switched, partially or fully.

An example should make this clear. In Fig. 15.33, the photocells are mounted at point *A*. The ratio of daylight between points *A* and *B* is the ratio of their daylight factors—that is, 20/10, or 2.0. Therefore, if 500 lux of daylight is required at point *B* before switching lights in that area, twice that amount (2.0 × 500 lux) or 1000 lux is required at point *A*. Similarly, a 500-lux requirement at point *C* corresponds to (20/4)(500) or 2500 lux at point *A*. Therefore, switching can be initiated by two single-level photocells or a multilevel unit at point *A*, with settings at 1000 and 2500 lux. Other switching arrangements can be made on the same principle. Dimming initiation can also be arranged in this fashion.

Daylight compensation can reduce energy use in perimeter areas by up to 60%, depending on latitude, climate, depth of perimeter zone, hours of building use, initial power density, and so on. The amount saved for the entire building obviously depends on the building's configuration—that is, the ratio of perimeter to total area. Payback time is usually in the range of 3 months to 3 years.

In summary, a well-designed lighting control system can provide energy conservation of up to 60%, extremely long lamp life, reduced cooling costs, extended ballast life, and reduced maintenance costs. When all these factors are considered, the investment payback period (see Appendix I) is always short and therefore financially attractive. An aspect of a centralized lighting control system

not discussed previously, because it is not directly concerned with lighting, is its use in connection with peak demand reduction (see Chapter 25). When interconnected with a demand controller, this approach acts to reduce the electric lighting load in accordance with a predetermined preferential-load schedule and can achieve significant savings.

## DETAILED DESIGN PROCEDURES

### 15.18 CALCULATION OF AVERAGE ILLUMINANCE

Once a luminaire has been selected on the basis of the foregoing criteria, it remains only to calculate the number of such fixtures required in each space, for uniform *general* illuminance, and to arrange them properly. Although a number of calculation methods are available, the lumen (flux) method is simplest and most applicable to our need for area-lighting calculations. Illuminance calculation from point, line, or area sources is covered in Sections 15.27 to 15.30. Luminance (photometric brightness) calculations are covered in Section 15.32.

Before beginning a detailed description of the zonal cavity calculation method, a general comment on precision is in order. The precision of any calculation should not exceed either the accuracy required or the precision of available data. Thus, there is no point in working to three decimal places if a ±10% rounding of the result is common and acceptable or if the data available are accurate to only one decimal place. As the reader will see, the lumen (lighting flux) method of *average* illuminance calculation is replete with assumptions and estimates. Among these are:

1. It is assumed that the space is empty. This is not normally the case.
2. It is assumed that all surfaces are perfect diffusers. This is not the case.
3. All surface reflectances are estimates, ±10%.
4. Maintenance conditions are estimates, at best ±10%.
5. No allowance is made for deviation of the performance of an individual product from its specification.

Any attempt to account accurately for these approximations would enormously complicate the calculation and would serve no useful purpose for this type of *average* illuminance calculation. For this reason, the procedure presented here introduces some approximations (which we feel are well justified) in the interest of simplification. These approximations are noted wherever used.

## 15.19 CALCULATION OF HORIZONTAL ILLUMINANCE BY THE LUMEN (FLUX) METHOD

The lumen method of calculation is a procedure for determining the *average* maintained illuminance (footcandles, lux) on the working plane in a room. The method presupposes that luminaires will be spaced so that uniformity of illumination is provided in order that an *average* calculation will have validity. The method is based on the definition of 1 lux (footcandle) of illuminance as 1 lumen incident on 1 meter (square foot) of area; that is,

$$\text{lux or (fc)} = \frac{\text{lumens}}{\text{area m}^2 \text{ or (ft}^2)}$$

As explained previously, the ratio between the lumens reaching the working plane in a specific space and the lumens generated is the coefficient of utilization, CU. Or

lumens on the working plane = lamp lumens × CU

Therefore,

$$\text{illuminance } E = \frac{\text{lamp lumens} \times \text{CU}}{\text{area}}$$

The coefficient CU is selected from tables provided by the manufacturer of a selected luminaire by a technique known as the *zonal cavity method* (explained in Section 15.21). In the absence of specific CU data, an approximation can be made by using the generic fixture types in Table 15.1.

The illuminance figure so calculated is the *initial average* illuminance. This initial level is reduced by the effect of temperature and voltage variations, dirt accumulation on luminaires and room surfaces, lamp output depreciation, and maintenance conditions. All of these effects are cumulatively referred to as the *light loss factor* (LLF):

$$\text{maintained } E = \text{initial } E \times \text{LLF}$$

(This factor was previously termed the *maintenance factor*, MF.) The procedure required to arrive at this factor is explained in the following section.

Our final expression for maintained illuminance $E$ as calculated by the lumen method is, therefore,

$$E = \frac{\text{lamp lumens} \times \text{CU} \times \text{LLF}}{\text{area}} \quad (15.1)$$

where $E$ is lux if area is expressed in square meters or footcandles if the area is in square feet.

Lamp lumens is the total within the space and is equal to

number of fixtures × lamps per fixture
× initial lumens per lamp

The formula then becomes

$$E = \frac{\begin{array}{c}\text{number of luminaires} \times \text{lamps/luminaire} \\ \times \text{lumens/lamp} \times \text{CU} \times \text{LLF}\end{array}}{\text{area}} \quad (15.2)$$

or, conversely, solving for the number of luminaires required to achieve a target maintained illuminance $E$:

number of luminaires =

$$\frac{E \times \text{area}}{\begin{array}{c}\text{lamps per luminaire} \times \text{lumens} \\ \text{per lamp} \times \text{CU} \times \text{LLF}\end{array}} \quad (15.3)$$

For large areas, a much more useful figure is the area illuminated per luminaire:

area per luminaire =

$$\frac{\text{lamps per luminaire} \times \text{lumens per lamp} \times \text{CU} \times \text{LLF}}{E}$$

$$(15.4)$$

For instance, it is much more convenient to know that to maintain, say, 60 fc within a space with a given luminaire, 70 ft$^2$ per unit is required than it is to know that for an 18,000-ft$^2$ floor, 257 fixtures are necessary. The former figure allows us to establish a pattern, say $7 \times 10$; the latter figure is too large to be immediately useful. Therefore, for rooms requiring more than a small number of luminaires, the latter calculation should always be used.

## 15.20 CALCULATION OF LIGHT LOSS FACTOR

The light loss factor (LLF) is composed of elements that can be categorized as *recoverable* and *nonrecoverable*. The former can be improved by maintenance; the latter cannot. The total LLF is the product of all the individual factors. The overview that follows includes approximations. For more precise data, see the IESNA *Lighting Handbook* (2000).

Among the nonrecoverable loss factors are the following:

### (a) Luminaire Ambient Temperature

Light output changes when a *fluorescent* fixture operates at other than its design temperature. With normal indoor installation use 1.0—that is, no depreciation. For other conditions, refer to technical data on the luminaire involved.

### (b) Voltage

When a lamp operates at the rated voltage, use 1.0. Details of source sensitivity to voltage are given in Chapter 12.

### (c) Luminaire Surface Depreciation

This factor is proportional to age and depends upon the type of surface involved. The designer must estimate this factor based on knowledge of the luminaire materials.

### (d) Components

Losses due to components include ballast factor, ballast-lamp photometric factor, equipment operating factor, and lamp position (tilt) factor. Air troffers also introduce a thermal factor. In the absence of specific data for component factors, use a total of 0.92.

In the absence of reliable data for any of the foregoing nonrecoverable factors, use an overall factor representing the product $a \times b \times c \times d$ of 0.88.

The factors below are *recoverable*—that is, they can be returned to their initial state by maintenance.

### (e) Room Surface Dirt

This factor is self-explanatory. Obviously, lighting approaches that depend heavily on surface reflections, such as indirect systems, are more seriously affected than systems that deliver most of their useful light directly. Assuming a 24-month cleaning cycle and normal conditions of cleanliness, use the appropriate factor in the following list. Alter it for other conditions such as infre-

quent maintenance and unusual cleanliness or dirtiness.

Direct lighting: $0.92 \pm 5\%$
Semi-direct lighting: $0.87 \pm 8\%$
Direct–indirect lighting: $0.82 \pm 10\%$
Semi-indirect lighting: $0.77 \pm 12\%$
Indirect lighting: $0.72 \pm 17\%$

### (f) Lamp Lumen Depreciation

This factor depends upon the type of lamp and the replacement schedule. Use the following when exact data are unavailable:

|  | Group Replacement | Replacement on Burnout |
|---|---|---|
| Incandescent | 0.94 | 0.88 |
| Tungsten–halogen | 0.98 | 0.94 |
| Fluorescent | 0.90 | 0.85 |
| Mercury vapor | 0.82 | 0.74 |
| Metal halide | 0.87 | 0.80 |
| High-pressure sodium | 0.94 | 0.88 |

### (g) Burnouts

This factor accounts for lamps that produce no output but have not been replaced. It depends upon maintenance schedules and method of replacement. Use the following as a general rule:

Group replacement procedures: 1.0
Individual replacement on burnout: 0.95

### (h) Luminaire Dirt Depreciation

This factor depends upon luminaire design, atmosphere conditions in the space, and maintenance schedule. The luminaire maintenance category is obtained from the manufacturer's data or from Table 15.1. The type of atmosphere is determined by considering the space involved. Assuming a 12-month cleaning schedule and normal room cleanliness, use the base number in Fig. 15.34 and change it to match the conditions of dirt and maintenance. The categories correspond to those used by the IESNA.

Total LLF is the product of all the depreciation factors:

$$LLF = a \times b \times c \times d \times e \times f \times g \times h$$

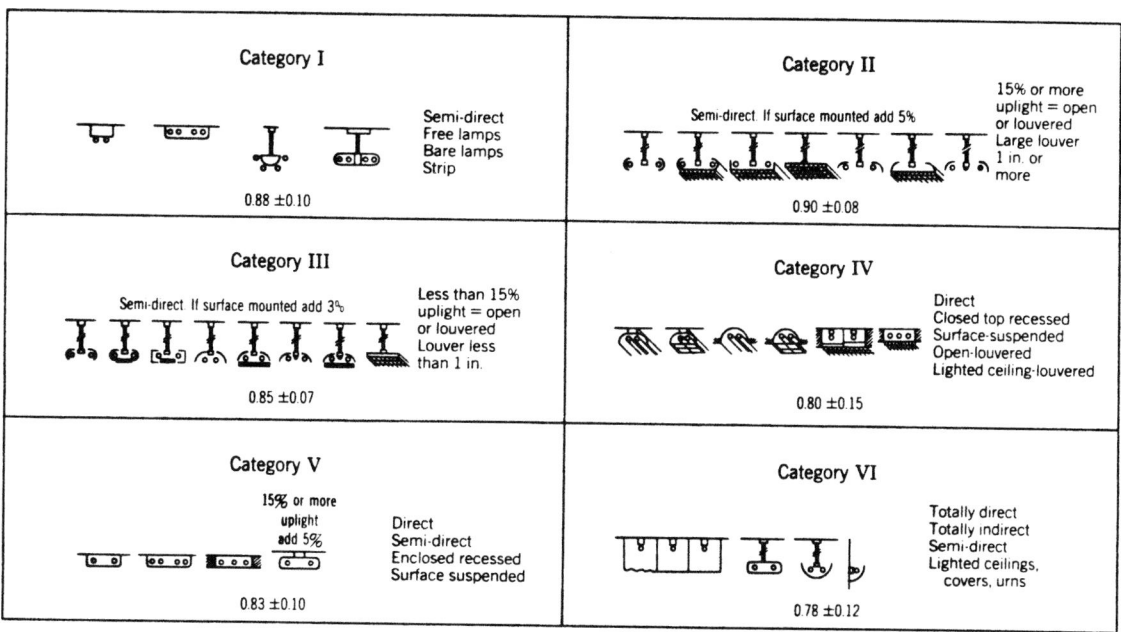

**Fig. 15.34** *The LDD factor is determined from the category of luminaire (which is an indication of its proneness to dirt accumulation) plus knowledge of room ambient conditions.*

For example, a fluorescent air troffer in a regularly maintained group lamp-replacement, air-conditioned office might typically have an LLF of 0.8. The same fixture in the same office, but with walls and fixture cleaned only when burned-out lamps are replaced, would typically have an LLF of 0.55. Thus, if in the first case the maintained illumination is $E$ fc, in the second case it is 0.55/0.80 or 0.69 $E$ fc, that is, a reduction of 31% as a result of poor maintenance. When a detailed determination of LLF is not possible, use the factors given in Section 15.22 (they are somewhat conservative).

## 15.21 DETERMINATION OF THE COEFFICIENT OF UTILIZATION BY THE ZONAL CAVITY METHOD

The coefficient of utilization (CU) connects a particular fixture to a particular space by relating the luminaire's light distribution characteristic to the room's size and its surface reflectances. To account for the luminaire's mounting height and its relationship to the working plane, the space is divided into three cavities: a ceiling cavity above the fixture, a floor cavity below the working plane, and a room cavity between the two (Fig. 15.35). Given the surface reflectances, the effective reflectances of the floor and ceiling cavities can be obtained. With these, the CU can be selected from tables (either Table 15.1 or from manufacturer's data) and the lumen formula (Eq. 15.3) applied to arrive at average illuminance. A step-by-step explanation of the method plus illustrative examples demonstrates the procedure. The reader should follow the steps with the flow chart in Fig. 15.36 and the calculation form in Fig. 15.37 in hand.

STEP 1. First, dimensional data are established. In offices, schools, and many other occupancies, the work plane is 30 in. above the finished floor (AFF). In drafting rooms it is 36 to 38 in.; in shops, 42 to 48 in.; in carpet stores and sail-cutting rooms at floor level. The three $h$ terms are the heights of the various cavities. Also identify the initial reflectance of the room surfaces and fill in the sketch in Fig. 15.37. Utilize the reflectance closest to those given in Table 15.2.

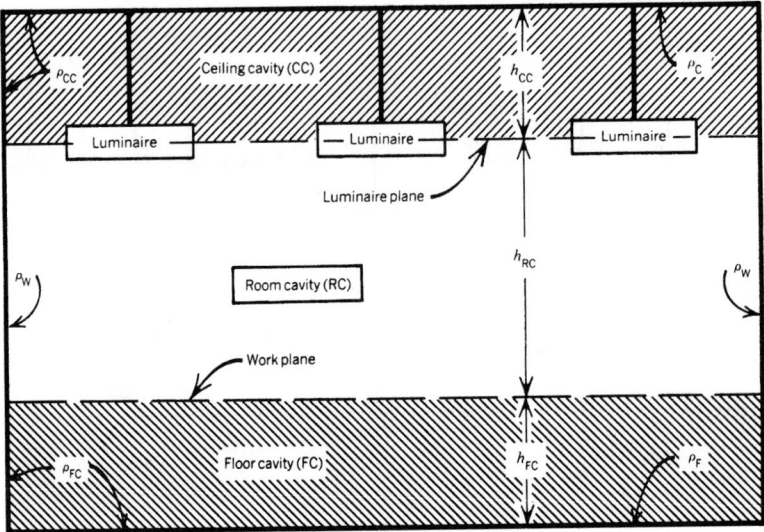

Legend:

$\rho_C$ = ceiling reflectance
$\rho_{CC}$ = ceiling cavity reflectance
$\rho_W$ = wall reflectance
$\rho_F$ = floor reflectance
$\rho_{FC}$ = floor cavity reflectance
$h$ = height in feet or meters
$h_{RC}$ = height of room cavity

**Fig. 15.35** *Room cavities as used in the zonal cavity method.*

**STEP 2.** See Fig. 15.37. This step involves determining the cavity ratios of the room by calculation. The basic expression for a cavity ratio (CR) is

$$CR = 2.5 \times \frac{\text{area of cavity wall}}{\text{area of work plane}} \quad (15.5)$$

In a rectangular space, the area of the cavity wall is $h \times (2l \times 2w)$ or $2h(l + w)$; therefore,

$$CR = \frac{2.5 \times 2h \times (l \times w)}{\text{area of work plane}}$$

or

$$CR = 5h \times \frac{l + w}{l \times w} \quad (15.6)$$

For other than rectangular rooms, the area can be calculated as required by geometry. For instance,

in a circular room, the cavity wall area $= h \times 2\pi r$ and the work plane area is $\pi r^2$. Thus,

$$CR = \frac{2.5 \times h \times 2\pi r}{\pi r^2} = \frac{5h}{r} \quad (15.7)$$

For each of the cavities in a rectangular room we have:

Room cavity ratio

$$RCR = 5h_{RC} \frac{l + w}{l \times w} \quad (15.8)$$

Ceiling cavity ratio

$$CCR = 5h_{CC} \frac{l + w}{l \times w} \quad (15.9)$$

Floor cavity ratio

$$FCR = 5h_{FC} \frac{l + w}{l \times w} \quad (15.10)$$

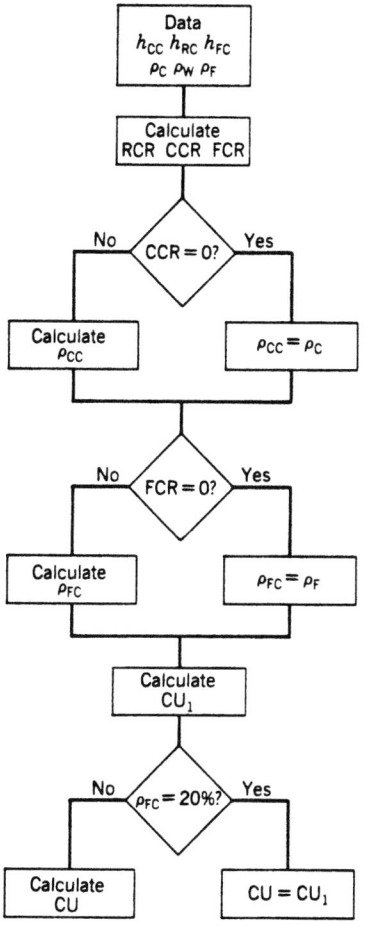

**Fig. 15.36** *Zonal cavity method flowchart.*

For reference, because CR values for a space are related, once one has been determined, the others can be obtained by ratios

$$CCR = RCR\,\frac{h_{CC}}{h_{RC}} \qquad (15.11)$$

$$FCR = RCR\,\frac{h_{FC}}{h_{RC}} \qquad (15.12)$$

and

$$CCR = FCR\,\frac{h_{CC}}{h_{FC}} \qquad (15.13)$$

**STEP 3.** See Table 15.2 and Figs. 15.36 and 15.37. This step involves obtaining the effective ceiling cavity reflectance ($\rho_{FC}$) from Table 15.2. Note that the wall reflectance remains as selected in step 1. If

the fixtures are surface mounted or recessed, then CCR = 0 and $\rho_{CC}$ = selected ceiling surface reflectance.

**STEP 4.** See Table 15.2 and Figs. 15.36 and 15.37. This step involves obtaining the effective floor cavity reflectance $\rho_{FC}$, as in step 3 for $\rho_{CC}$. If the floor is the working plane, FCR = 0 and $\rho_{FC}$ = selected floor surface reflectance.

**STEP 5.** Select the CU from the manufacturer's data. Note that interpolation may be necessary for CCR ($\rho_{CC}$) if it is between the figures in the CU table. See Example 15.1 in the next section. CU correction factors for $\rho_{FC}$ other than 20% (the standard value in CU tables) are given in Table 15.3.

**STEP 6.** Calculate the illuminance and the number of fixtures or area per luminaire as in Section 15.19.

Illustrative examples and shortcut methods are demonstrated below. CU coefficients are listed in Table 15.1 for generic luminaire types.

## 15.22 ZONAL CAVITY CALCULATIONS: ILLUSTRATIVE EXAMPLES

**EXAMPLE 15.1** It is suggested that the reader photocopy Fig. 15.37 and fill it in as the solution to this example is developed.

*Given.* Classroom: 6 m × 8 m × 3.70-m height, elementary school. Initial reflectances: ceiling 80%, entire wall 50%, floor 20%. (Note that the sketch in Fig. 15.37 can accommodate different reflectances for the upper, center, and lower wall sections.) Provide adequate illuminance using fluorescent fixtures. Assume yearly maintenance, lamp replacement at burnout, proper voltage and ballasts, and a medium clean atmosphere.

**SOLUTION**

*(a) Illuminance Criterion.* Refer to Table 11.7. Depending on the class grade, the illuminance category could be either D or E. We use E, which gives an illuminance range of 500–750–1000 lux. In Table 11.9 under category E we find: age—under 40, speed/accuracy are important, and background is more than 70% (white paper). This yields a recommendation of 500 lux. The daylight contribution

**GENERAL INFORMATION**

1. Project identification: _____
   *(Give name of area and/or building and room number)*

2. Average maintained illumination for design: _____ lux [footcandles]

Luminaire data:                           Lamp data:
3. Manufacturer: _____      5. Type and color: _____
4. Catalog number: _____      6. Number per luminaire: _____
                                          7. Total lumens per luminaire: _____

**SELECTION OF COEFFICIENT OF UTILIZATION**

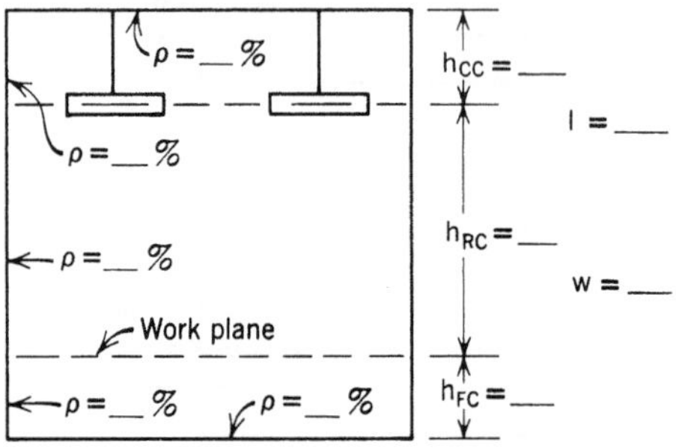

8. Step 1: Fill in sketch at right.

9. Step 2: Determine Cavity Ratios by formulas.

9a. Room cavity ratio, RCR  = _____
9b. Ceiling cavity ratio, CCR = _____
9c. Floor cavity ratio, FCR = _____

10. Step 3: Obtain effective ceiling cavity reflectance ($\rho_{CC}$) from Table 15.2.   $\rho_{CC}$ = _____

11. Step 4: Obtain effective floor cavity reflectance ($\rho_{FC}$) from Table 15.2.   $\rho_{FC}$ = _____

12. Step 5: Obtain coefficient of utilization (CU) from manufacturer's data.   CU = _____

**SELECTION OF LIGHT LOSS FACTORS**

Unrecoverable                                    Recoverable
13. Luminaire ambient temperature _____       17. Room surface dirt depreciation _____
14. Voltage to luminaire _____                18. Lamp lumen depreciation _____
15. Luminaire surface depreciation _____      19. Lamp burnouts factor _____
16. Other factors **(components)** _____       20. Luminaire dirt depreciation LDD _____

21. Total light loss factor, LLF (product of individual factors above): _____

**CALCULATIONS**

(Average maintained illumination level)

$$\text{Number of luminaires} = \frac{(\text{Illuminance}) \times (\text{Area})}{(\text{Lumens per luminaire}) \times (\text{CU}) \times (\text{LLF})}$$

22.                $$= \frac{\rule{4cm}{0.4pt}}{\rule{4cm}{0.4pt}} =$$

$$\text{Lux [footcandles]} = \frac{(\text{number of luminaires}) \times (\text{lumens per luminaire}) \times (\text{CU}) \times (\text{LLF})}{(\text{area})}$$

23.                $$= \frac{\rule{6cm}{0.4pt}}{\rule{6cm}{0.4pt}} =$$

24. Calculated by: _____ Date: _____

**Fig. 15.37** *Zonal cavity method calculation form. (Courtesy of IESNA.)*

ILLUMINATION

**TABLE 15.2  Percent Effective Ceiling or Floor Cavity Reflectance ($\rho_{cc}$, $\rho_{Fc}$) for Various Reflectance Combinations**

| Percent Ceiling $\rho_c$ or Floor Reflectance $\rho_F$: | | 90 | | | | 80 | | | | 70 | | | 50 | | | 30 | | | | 10 | | |
|---|---|---|---|---|---|---|---|---|---|---|---|---|---|---|---|---|---|---|---|---|---|---|
| **Percent Wall Reflectance $\rho_w$:** | | 90 | 70 | 50 | 30 | 80 | 70 | 50 | 30 | 70 | 50 | 30 | 70 | 50 | 30 | 65 | 50 | 30 | 10 | 50 | 30 | 10 |
| CEILING OR FLOOR CAVITY RATIOS—CCR OR FCR | 0 | 90 | 90 | 90 | 90 | 80 | 80 | 80 | 80 | 70 | 70 | 70 | 50 | 50 | 50 | 30 | 30 | 30 | 30 | 10 | 10 | 10 |
| | 0.2 | 89 | 88 | 86 | 85 | 79 | 78 | 77 | 76 | 68 | 67 | 66 | 49 | 48 | 47 | 30 | 29 | 29 | 28 | 10 | 10 | 9 |
| | 0.4 | 88 | 86 | 83 | 81 | 78 | 76 | 74 | 72 | 67 | 65 | 63 | 48 | 46 | 45 | 30 | 29 | 27 | 26 | 11 | 10 | 9 |
| | 0.6 | 88 | 84 | 80 | 76 | 77 | 75 | 71 | 68 | 65 | 62 | 59 | 47 | 45 | 43 | 29 | 28 | 26 | 25 | 11 | 10 | 9 |
| | 0.8 | 87 | 82 | 77 | 73 | 75 | 73 | 69 | 65 | 64 | 60 | 56 | 47 | 43 | 41 | 29 | 27 | 25 | 23 | 11 | 10 | 8 |
| | 1.0 | 86 | 80 | 74 | 69 | 74 | 71 | 66 | 61 | 63 | 58 | 53 | 46 | 42 | 39 | 29 | 27 | 24 | 22 | 11 | 9 | 8 |
| | 1.2 | 86 | 78 | 72 | 65 | 73 | 70 | 64 | 58 | 61 | 56 | 50 | 45 | 41 | 37 | 29 | 26 | 23 | 20 | 12 | 9 | 7 |
| | 1.4 | 85 | 77 | 69 | 62 | 72 | 68 | 62 | 55 | 60 | 54 | 48 | 45 | 40 | 35 | 28 | 26 | 22 | 19 | 12 | 9 | 7 |
| | 1.6 | 85 | 75 | 66 | 59 | 71 | 67 | 60 | 53 | 59 | 52 | 45 | 44 | 39 | 33 | 28 | 25 | 21 | 18 | 12 | 9 | 7 |
| | 1.8 | 84 | 73 | 64 | 56 | 70 | 65 | 58 | 50 | 57 | 50 | 43 | 43 | 37 | 32 | 28 | 25 | 21 | 17 | 12 | 9 | 6 |
| | 2.0 | 83 | 72 | 62 | 53 | 69 | 64 | 56 | 48 | 56 | 48 | 41 | 43 | 37 | 30 | 28 | 24 | 20 | 16 | 12 | 9 | 6 |
| | 2.2 | 83 | 70 | 60 | 51 | 68 | 63 | 54 | 45 | 55 | 46 | 39 | 42 | 36 | 29 | 28 | 24 | 19 | 15 | 13 | 9 | 6 |
| | 2.4 | 82 | 68 | 58 | 48 | 67 | 61 | 52 | 43 | 54 | 45 | 37 | 42 | 35 | 27 | 28 | 24 | 19 | 14 | 13 | 9 | 6 |
| | 2.6 | 82 | 67 | 56 | 46 | 66 | 60 | 50 | 41 | 53 | 43 | 35 | 41 | 34 | 26 | 27 | 23 | 18 | 13 | 13 | 9 | 5 |
| | 2.8 | 81 | 66 | 54 | 44 | 66 | 59 | 48 | 39 | 52 | 42 | 33 | 41 | 33 | 25 | 27 | 23 | 18 | 13 | 13 | 9 | 5 |
| | 3.0 | 81 | 64 | 52 | 42 | 65 | 58 | 47 | 38 | 51 | 40 | 32 | 40 | 32 | 24 | 27 | 22 | 17 | 12 | 13 | 8 | 5 |
| | 3.5 | 79 | 61 | 48 | 37 | 63 | 55 | 43 | 33 | 48 | 38 | 29 | 39 | 30 | 22 | 26 | 22 | 16 | 11 | 13 | 8 | 5 |
| | 4.0 | 78 | 58 | 44 | 33 | 61 | 52 | 40 | 30 | 46 | 35 | 26 | 38 | 29 | 20 | 26 | 21 | 15 | 9 | 13 | 8 | 4 |
| | 4.5 | 77 | 55 | 41 | 30 | 59 | 50 | 37 | 27 | 45 | 33 | 24 | 37 | 27 | 19 | 25 | 20 | 14 | 8 | 14 | 8 | 4 |
| | 5.0 | 76 | 53 | 38 | 27 | 57 | 48 | 35 | 25 | 43 | 32 | 22 | 36 | 26 | 17 | 25 | 19 | 13 | 7 | 14 | 8 | 4 |

*Source:* Extracted from the *IESNA Lighting Handbook* (1993); reprinted with permission. For more complete data, see the current *IESNA Lighting Handbook*.

**ILLUMINATION**

**TABLE 15.3[a]  Factors for Effective Floor Cavity Reflectances Other Than 20% (Any Wall Reflectance)[b]**

For 30% effective floor cavity reflectance, *multiply* by the appropriate factor below.
For 10% effective floor cavity reflectance, *divide* by the appropriate factor below.

| | Percent Effective Ceiling Cavity Reflectance, $\rho_{cc}$ | | | |
|---|---|---|---|---|
| **Room Cavity Ratio** | **80** | **70** | **50** | **10** |
| 1 | 1.08 | 1.06 | 1.04 | 1.01 |
| 2 | 1.06 | 1.05 | 1.03 | 1.01 |
| 3 | 1.04 | 1.04 | 1.03 | 1.01 |
| 4 | 1.03 | 1.03 | 1.02 | 1.01 |
| 5 | 1.03 | 1.02 | 1.02 | 1.01 |
| 6 | 1.02 | 1.02 | 1.02 | 1.01 |
| 7 | 1.02 | 1.02 | 1.01 | 1.01 |
| 8 | 1.02 | 1.02 | 1.01 | 1.01 |
| 9 | 1.01 | 1.01 | 1.01 | 1.01 |
| 10 | 1.01 | 1.01 | 1.01 | 1.01 |

[a]Extracted from the *IESNA Lighting Handbook* (1993); reprinted with permission.

[b]For more precise data, for varying $\rho_w$, see the current *IESNA Lighting Handbook*.

in much of the space is usually considerable because of the hours of use. It frequently exceeds this 500-lux level, but to demonstrate the calculation procedure, we assume an absence of daylight. Lines 1, 2, and 8 (room sketch) of Fig. 15.37 can be filled in.

*(b) Luminaire Selection.* The criteria we have previously developed for a classroom situation require an installation that yields

1. Low direct glare (high VCP) because schoolchildren spend a large proportion of their time in a heads-up position.

2. Low veiling reflections because much of the seeing task involves high-reflectance materials, occasionally specular.

3. High efficiency and low energy use to meet ANSI/ASHRAE/IESNA Standard 90.1 and most governmental requirements.

4. Minimum required maintenance in view of the poor cleaning and maintenance situation that exists in many schools.

Although the ceiling height is sufficient to permit use of indirect lighting (e.g., luminaire 33 in Table 15.1) with all of its distinct advantages, it is not chosen because

5. The luminaire maintenance category is inappropriate, and given the type of maintenance expected, light reduction would be serious.
6. Indirect lighting depends on a highly reflective ceiling, requiring yearly cleaning and repainting at intervals not exceeding 5 years. This is not generally the case in public schools.

As a result, we select a two-lamp version of luminaire No. 44 from Table 15.1. This unit is a parabolic aluminum reflector with louvers and exhibits a 45° cutoff and a crosswise batwing distribution. (Fig. 15.14 shows a unit of this design.) This fixture meets requirements 1 through 4.

Although its distribution curve (see Table 15.1) shows no upward component, most commercial units of this basic design do have slots in the top of the reflector and show 5% to 10% uplight. In practice, we would select such a unit. This avoids an excessively dark ceiling and undesirable luminance ratios between fixture and background. Furthermore, fixtures with top slots stay cleaner because of the upward air movement through the fixture caused by the warm lamps and ballast. The mounting height should be about 270 cm (~9 ft) to permit easy maintenance and good row spacing. For this luminaire the recommended maximum SC can be estimated at between 1.5 and 2.0. The work plane height is 75 cm (30 in.).

*(c) Calculations.*

**STEP 1.** The required data that should appear in the sketch in Fig. 15.37 are

$h_{CC} = 1.0$ m          $h_{RC} = 1.95$ m

$h_{FC} = 0.75$ m          $l = 8$ m

$\rho_C = 80\%$          $\rho_w = 50\%$

$\rho_F = 20\%$          $w = 6$ m

**STEP 2.** From Eqs. 15.8, 15.9, and 15.10

$$\frac{l + w}{l \times w} = 0.29$$

$$RCR = 5 h_{RC} \frac{l + w}{l \times w}$$

$$RCR = 5(1.95)(0.29) = 2.84$$
$$CCR = 5 (1)(0.29) = 1.46$$
$$FCR = 5 (0.75)(0.29) = 1.09$$

**STEP 3.** From Table 15.2 obtain effective reflectances: For $\rho_{CC}$ use $\rho_C = 0.8$, $\rho_w = 0.5$, and CCR = 1.46. Therefore,

$$\rho_{CC} = 0.61$$

**STEP 4.** For $\rho_{FC}$ use $\rho_F = 0.2$, $\rho_w = 0.5$, and FCR = 1.09. Therefore,

$$\rho_{FC} = 0.18 \text{ by interpolation}$$

**STEP 5.** Interpolation between RCR of 2.8 and 3.0 is necessary. The CU for the selected luminaire—No. 44 of Table 15.1—can now be obtained by double interpolation.

| $\rho w = 0.50$ | CU | | |
|---|---|---|---|
| $\rho_{CC} \rightarrow$ | 0.70 | 0.61 | 0.50 |
| RCR | | | |
| 2. | 0.58 | | 0.55 |
| 2.84 | | ? | |
| 3. | 0.52 | | 0.50 |

Design CU = 0.52. No correction from Table 15.3 is required, as $\rho_{FC}$ is close to 20%. At this stage, lines 9 through 12 of Fig. 15.37 can be filled in.

**STEP 6.** The LLF (see Section 15.20) results from establishing items 13 to 20 of Fig. 15.37. These are:

Items 13–16          0.88

Item 17          0.95

Item 18          0.85

Item 19          0.95

Item 20          0.80

Item 21: LLF = (0.88)(0.95)(0.85)(0.95)(0.80) = 0.54

**STEP 7.** *Lumen calculation.* A typical classroom is normally divided into a large student seating area and a teacher's area. The illuminance requirement is approximately the same for both, so the room can be treated as a single unit for visual task purposes. A good lighting design includes chalkboard lighting, the spill light from which is usually sufficient to illuminate the front of the room. Treating the entire space as one visual task area, and assuming two nominal 40-W, 4100 K fluorescent lamps per fixture, the number $N$ of luminaires required is, from Eq. 15.3,

$$N = \frac{(500 \text{ lux})(6 \times 8 \text{ m})}{2(3200 \text{ lm})(0.52)(0.54)} = 13.35$$

The remaining lines of Fig. 15.37 can now be completed.

Refer now to Fig. 15.38 for the layout. Two lengthwise rows of fixtures, spaced 300 cm apart,

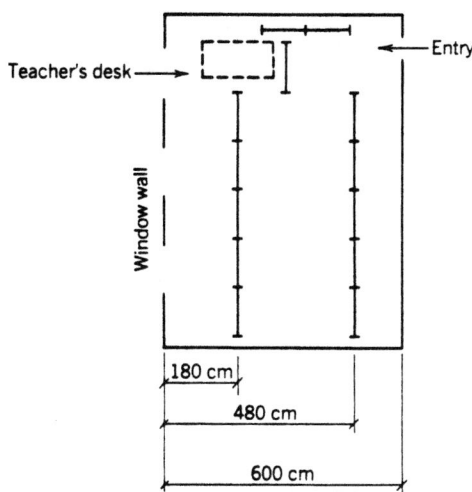

*Fig. 15.38* Layout of pendant parabolic aluminum reflector luminaires in a typical classroom. The units have a modified batwing distribution in the crosswise direction, which mandates their being hung with their long axis parallel to the line of sight, as shown. Furthermore, the fixtures have lower brightness in that direction, that is, VCP for students in the head-up position is excellent. Note that the distance between the outside fixture row and the window wall is more than one-half of the side-to-side spacing because of daylight contribution during school hours, whereas the inside fixture row to inside wall spacing is less than one-half of the side-to-side spacing in order to maintain sufficient illuminance near that wall. The single fixture to the right of the teacher's desk (to the teacher's left) was so placed to avoid the veiling reflections that would result from a fixture directly above the desk and to take advantage of the fixture's transverse batwing distribution characteristic.

give excellent coverage, as shown. The resultant spacing-to-mounting-height ratio

$$\frac{300 \text{ cm spacing}}{270 \text{ cm mounting height}} = 1.1$$

is well below the maximum of 1.5 to 2.0 estimated by inspection of the candlepower distribution curve. The sixth fixture in the outside (window) row provides illumination for the teacher's desk. Two *single-lamp* batwing distribution or asymmetric distribution luminaires provide chalkboard and front-of-room lighting. The window wall row is separately switched and is farther from the wall than the inside row due to the usual presence of daylight. Point-by-point illuminance can be computed by one of the available computer programs. The daylight contribution can also be included. Actual *average* illuminance in the room would be simply the ratio of the number of fixtures used to

the number required (calculated) times the target illuminance:

$$E = \frac{11 + 1/2 + 1/2}{13.35} \text{ (500 lux)} \approx 450 \text{ lux}$$

assuming similar characteristics for the one- and two-lamp fixtures. The designer must decide if this illuminance is appropriately close to the 500 lux target illuminance. ∎

**EXAMPLE 15.2** large modern business office, 60 × 100 × 8-ft hung ceiling height. Initial reflectances are 0.80, 0.50, and 0.30. Provide general lighting using fluorescent lamps with high-efficiency ballasts in recessed troffer luminaires. The space is fully air-conditioned. Lamps are replaced on a burnout basis, and the fixture is then cleaned.

**SOLUTION**

From Table 11.7 we obtain a recommended target illuminance for routine office tasks of 500 lux. As almost every desk in a modern business office is equipped with a visual display terminal (VDT), we must select a fixture that is designed to minimize the extremely disturbing reflection of luminaires in VDT screens. (See Section 16.19 for a full discussion of this ubiquitous problem.) Luminaires with parabolic diffusers, reflectors, and/or baffles, which exhibit very low high-angle luminance, meet this requirement, as explained in Section 15.4.

A second consideration is that in an office of this size, which may well be "landscaped" using half-height partitions, viewing directions vary greatly, negating any benefits inherent in a luminaire's directionality. As a result, we selected a specular parabolic baffled 2-ft unit, which is directionally and architecturally neutral and displays the required low levels of luminance at high angles. The unit's characteristics are given in Fig. 15.39 when equipped with two 32-W, T8, U-shaped lamps with 2800 initial lumens. Note that the luminaire exhibits a modified batwing intensity distribution in the transverse direction.

*Calculations.* The reader should fill in a copy of Fig. 15.37 as we proceed. The working plane is taken as the desktop height (i.e., 30 in. or 2.5 ft).

**STEP 1.** $h_{CC} = 0$; $h_{RC} = 5.5$; $h_{FC} = 2.5$.

**STEP 2.** CCR = 0; RCR = 0.73; FCR = 0.33.

**STEP 3.** $\rho_W = 50\%$; $\rho_{CC} = 80\%$ (for a recessed fixture we use the ceiling reflectance).

**STEP 4.** From Table 15.2, $\rho_{FC} = 29\%$.

## REK D/RIK-D  2/31-U

Recessed mounted luminaire with parabolic mirrored louver
(without mirrored reflector)

Shielding angle :  Parl.: 60  -  Norm.: 60

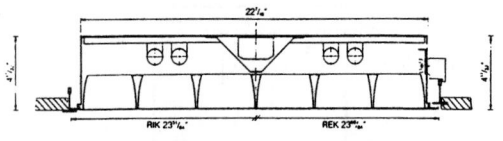

| CANDLEPOWER | DEG. | Parl. | 22.5 | 45 | 67.5 | Norm. | Flux |
|---|---|---|---|---|---|---|---|
| | 0 | 1609 | 1609 | 1609 | 1609 | 1609 | |
| | 5 | 1603 | 1603 | 1607 | 1608 | 1605 | 177 |
| | 10 | 1577 | 1579 | 1589 | 1598 | 1600 | |
| | 15 | 1534 | 1534 | 1551 | 1565 | 1564 | 506 |
| | 20 | 1473 | 1480 | 1490 | 1522 | 1529 | |
| | 25 | 1400 | 1412 | 1431 | 1481 | 1496 | 773 |
| | 30 | 1312 | 1327 | 1365 | 1464 | 1502 | |
| | 35 | 1212 | 1235 | 1311 | 1448 | 1501 | 964 |
| | 40 | 1102 | 1141 | 1263 | 1328 | 1352 | |
| | 45 | 970 | 1034 | 1138 | 1135 | 1149 | 892 |
| | 50 | 539 | 572 | 617 | 621 | 635 | |
| | 55 | 122 | 129 | 139 | 140 | 144 | 223 |
| | 60 | 14 | 14 | 15 | 15 | 16 | |
| | 65 | 12 | 6 | 11 | 7 | 10 | 10 |
| | 70 | 3 | 3 | 2 | 3 | 4 | |
| | 75 | 2 | 2 | 1 | 2 | 2 | 2 |
| | 80 | 2 | 1 | 1 | 1 | 2 | |
| | 85 | 1 | 1 | 1 | 2 | 2 | 2 |
| | 90 | 2 | 2 | 2 | 2 | 2 | |
| | 95 | 0 | 0 | 0 | 0 | 0 | |
| | 100-180 | 0 | 0 | 0 | 0 | 0 | 0 |

2500 cd (based on 2 x 2900 lm)

### COEFFICIENTS OF UTILIZATION

LUMINAIRE EFFICIENCY :  0.61

| | CEILING | 80 | | | 70 | | | 50 | | | 30 | | | 10 | | | 0 |
|---|---|---|---|---|---|---|---|---|---|---|---|---|---|---|---|---|---|
| REFL. FACT. | WALLS | 50 | 30 | 10 | 50 | 30 | 10 | 50 | 30 | 10 | 50 | 30 | 10 | 50 | 30 | 10 | 0 |
| | FLOOR | 20 | 20 | 20 | 20 | 20 | 20 | 20 | 20 | 20 | 20 | 20 | 20 | 20 | 20 | 20 | 20 |
| RCR 1 | | 64 | 61 | 59 | 62 | 60 | 58 | 60 | 58 | 56 | 57 | 56 | 54 | 55 | 54 | 53 | 52 |
| 2 | | 57 | 54 | 51 | 56 | 53 | 50 | 54 | 51 | 49 | 52 | 50 | 48 | 50 | 48 | 47 | 45 |
| 3 | | 53 | 49 | 46 | 53 | 49 | 46 | 51 | 48 | 45 | 49 | 47 | 44 | 48 | 45 | 44 | 42 |
| 4 | | 49 | 45 | 42 | 49 | 45 | 41 | 47 | 44 | 41 | 46 | 43 | 40 | 45 | 42 | 40 | 39 |
| ROOM CAVITY RATIO 5 | | 45 | 41 | 38 | 45 | 41 | 37 | 44 | 40 | 37 | 42 | 39 | 37 | 41 | 39 | 36 | 35 |
| 6 | | 42 | 37 | 34 | 41 | 37 | 34 | 40 | 36 | 33 | 39 | 36 | 33 | 38 | 35 | 33 | 32 |
| 7 | | 38 | 34 | 30 | 38 | 33 | 30 | 37 | 33 | 30 | 36 | 33 | 30 | 35 | 32 | 30 | 29 |
| 8 | | 35 | 31 | 28 | 35 | 31 | 28 | 34 | 30 | 27 | 33 | 30 | 27 | 33 | 30 | 27 | 26 |
| 9 | | 33 | 29 | 26 | 33 | 28 | 26 | 32 | 28 | 25 | 31 | 28 | 25 | 31 | 28 | 25 | 24 |
| 10 | | 30 | 26 | 23 | 30 | 26 | 23 | 30 | 26 | 23 | 29 | 26 | 23 | 29 | 25 | 23 | 22 |

#### ZONAL SUMMARY / AVG [ cd/m² ]

| ZONE | LUMENS | LAMP | FIXT | DEG. | NORM. | 45° | PARL. |
|---|---|---|---|---|---|---|---|
| 0 - 30 | 1456 | 25.1 | 41.0 | 45 | 5091 | 5041 | 4295 |
| 0 - 40 | 2420 | 41.7 | 68.2 | 50 | 3092 | 3009 | 2624 |
| 0 - 50 | 3312 | 57.1 | 93.3 | 55 | 786 | 759 | 665 |
| 0 - 60 | 3535 | 61.0 | 99.6 | 60 | 98 | 94 | 85 |
| 0 - 70 | 3545 | 61.1 | 99.9 | 65 | 75 | 81 | 86 |
| 0 - 90 | 3549 | 61.2 | 100.0 | 70 | 35 | 16 | 24 |
| 90 - 180 | 0 | 0.0 | 0.0 | 75 | 28 | 14 | 21 |
| 0 - 180 | 3550 | 61.2 | 100.0 | 80 | 31 | 21 | 31 |

**Fig. 15.39** *Complete photometric data on a 2-ft square fixture with a 36-cell mirrored parabolic louver. The unit uses two T8 U-shaped 32-W fluorescent lamps. Note that the transverse distribution characteristic has a modified batwing shape typical of this type of louver. The louver's specular finish yields very low brightness at high angles, as can be seen from both the distribution curve and the luminance (cd/m²) data. (Courtesy of Zumtobel-STAFF.)*

**STEP 5.** We find CU for $\rho_{FC}$ = 20% by extrapolation,

| RCR | CU | |
|---|---|---|
| 0.73 | ? | |
| 1.0 | 0.64 | CU = 0.66 |
| 2.0 | 0.57 | |

**STEP 6.** From Table 15.3, the multiplier for $\rho_{FC}$ = 30% (close to 29%) is 1.085.

**STEP 7.** Final CU = 1.085 (0.66) = 0.72.

**STEP 8.** LLF per Fig. 15.37:

| | |
|---|---|
| 1.0 | 0.92 |
| 1.0 | 0.85 |
| 0.9 | 0.95 |
| 0.92 | 0.88 |

LLF = 0.54

**STEP 9.**

$$\text{area/luminaire} = \frac{\text{lamps} \times \text{lumens/lamp} \times \text{CU} \times \text{LLF}}{\text{illuminance}}$$

$$= \frac{2(2800)(0.66)(0.54)}{500/10.76 \text{ lux per fc}} = 43 \text{ ft}^2$$

With mounting height 5.5 ft above the working plane, estimated spacing criteria of 1.8 transverse and 1.2 parallel yield maxima of 9.9 ft transverse and 6.6 ft parallel for fixture centerlines. We would test grids of 9 × 5 ft and 8 × 6 ft on a computer to see which yielded better results for uniformity of illumination and lower reflected glare in *all* directions because, as noted previously, viewing directions will probably vary. In the absence of a computer program, we would select the 9 × 6 grid to take advantage of the batwing characteristic. Note the important fact that a maintained level of 500 lux (50 fc of high-quality illuminance) for VDT areas is achieved with a power density of less than 1.6 W/ft² (17.54 W/m²). ∎

## 15.23 ZONAL CAVITY CALCULATION BY APPROXIMATION

Although the foregoing zonal cavity calculations are straightforward and essentially simple, they can become tedious if more than a few areas are involved.

Step 1: **GENERAL INFORMATION**

(a) Project identification: _____

*(Give name of area and/or building and room number)*

(b) Average maintained illumination for design: _____ lux [footcandles]

Luminaire data: _____ Lamp data: _____
(c) Manufacturer: _____ (e) Type and color: _____
(d) Catalog number: _____ (f) Number per luminaire: _____
(g) Total lumens per luminaire: _____

**SELECTION OF COEFFICIENT OF UTILIZATION**

Step 2: Determine the equivalent square-room RCR (room cavity ratio):

$$W_{sq} = W + \frac{L-W}{3}$$

$$RCR = \frac{10 \text{ hrc}}{W_{sq}}$$

Step 3: Determine ceiling cavity equivalent reflectance by equivalent square room size.

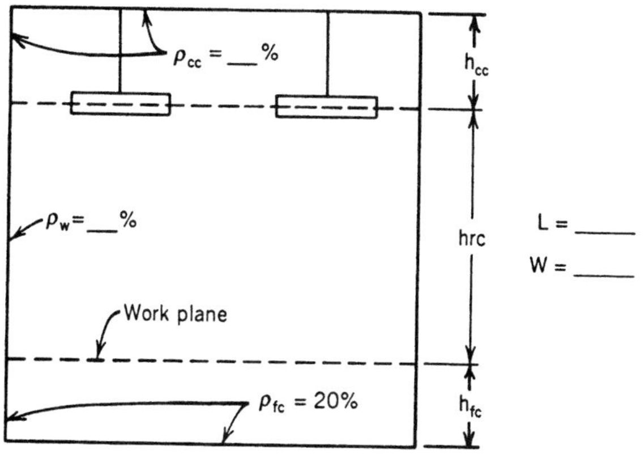

$\rho_{cc} = \_\_ \%$     $h_{cc}$

$\rho_w = \_\_ \%$     hrc     $L = \_\_\_\_$     $W = \_\_\_\_$

Work plane

$\rho_{fc} = 20\%$     $h_{fc}$

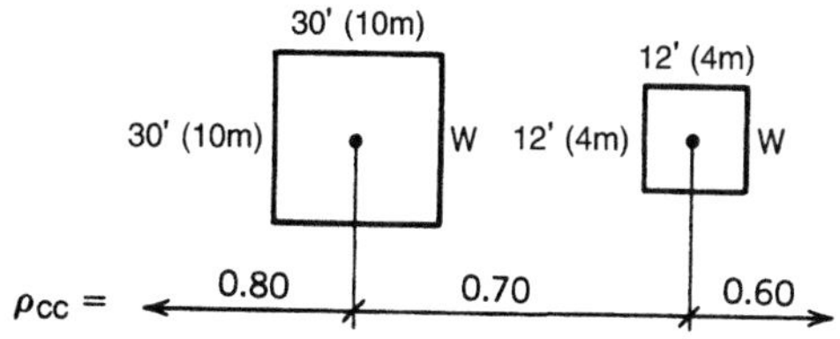

30' (10m)

30' (10m)  W     12' (4m)  W

12' (4m)

$\rho_{cc} = $     0.80     0.70     0.60

Step 4: Fill in sketch of room above.
Step 5: Assume floor cavity reflectance $\rho_{fc} = 20\%$.
Step 6: Obtain coefficient of utilization CU from manufacturers' data: CU = _____

Step 7: Light loss factor LLF     Good conditions: 0.65
Average conditions: 0.55
Poor conditions: 0.45

Step 8: **CALCULATIONS**

$$\text{Number of luminaires} = \frac{\text{Recommended illuminance} \times \text{area}}{(\text{Lumens per luminaire}) \times (\text{CU}) \times (\text{LLF})}$$

$$\text{OR} \quad \text{Lux [footcandles]} = \frac{(\text{Number of luminaires}) \times (\text{lumens per luminaire}) \times (\text{CU}) \times (\text{LLF})}{\text{area m}^2 \text{ (ft}^2)}$$

***Fig. 15.40*** *Sheet for illuminance calculation using an approximate zonal cavity method (based on a method developed by B.F. Jones).*

Two alternatives exist: to utilize one of the many computer programs available or simply to shorten the calculations with reasonable approximations. Computer assistance is discussed in Section 15.26 and Appendix M. An effective computational method using approximations is demonstrated in this sec-

tion, which is based upon a method developed by B. F. Jones.

Fill in the calculation sheet for illuminance calculation by approximation, as shown in Fig. 15.40. Assume that all rooms are square. To do this for a rectangle, take one-third of the difference in dimensions and add to the smaller dimension to obtain the equivalent width w. Then, for square rooms

$$RCR = \frac{10h_{RC}}{w}$$

Then, using the calculated side dimension of the square equivalent room, assume:

1. $\rho_{CC} = 0.80$ for a large room, that is, equal to or larger than $30 \times 30$ ft $(10 \times 10$ m).

   $\rho_{CC} = 0.70$ for a medium room, that is, between $30 \times 30$ ft $(10 \times 10$ m) and $12 \times 12$ ft $(4 \times 4$ m).

   $\rho_{CC} = 0.60$ for a small room, that is, equal to or smaller than $12 \times 12$ ft $(4 \times 4$ m).
2. Assume that $\rho_{FC} = 0.20$.
3. Assume that LLF = 0.65 for good conditions, 0.55 for average conditions, and 0.45 for poor conditions.

---

**EXAMPLE 15.3.** Classroom as in Example 15.1 by approximation.

1. "Square" the room.

$$w_{SQ} = 6 + \frac{8-6}{3} = 6 \ 2/3$$

2. Assume

$$\rho_{CC} = 70; \ \rho_{W} = 50; \ \rho_{FC} = 20.$$

3. Calculate RCR

$$RCR = \frac{10 \times 1.95}{6 \ 2/3} = \frac{19.5}{6.66} = 2.93$$

4. Obtain CU from Table 15.1, fixture No. 44. CU = 0.52 by visual inspection.
5. Calculate fixtures

$$N = \frac{500(6 \times 8 \ m)}{2(3200)(0.52)(0.55)}$$

$$= 13.11 \ fixtures$$

Thus, the result is substantially the same as the accurate calculation. Let us also check Example 15.2.

---

**EXAMPLE 15.4.** Office as in Example 15.2 by approximation.

1. $w_{SQ} = 60 + \dfrac{100-60}{3} = 73$ ft

2. $\rho_{CC} = 80; \ \rho_W = 50; \ \rho_{FC} = 20$

3. $RCR = \dfrac{10 \times 5.5}{73} = \dfrac{55}{73} = 0.75$

4. CU = 0.66 by inspection (mental interpolation)

5. LLF = 0.55

6. area/luminaire $= \dfrac{2 \times 2800 \times 0.66 \times 0.55}{50 \ fc}$

   $= 40.6$ sq ft

This result is within 5.5% of the accurate calculation. Thus, we see that these simple approximations give answers sufficiently accurate for most uses and are therefore recommended. ∎

In conclusion, then, with respect to zonal cavity calculations, we can make the following statements:

1. For preliminary and routine calculations of rectangular rooms, with assumed reflectances, use the assumptions listed previously. A modified calculation form is provided in Fig. 15.40 to assist with this method.
2. For rooms where a high degree of accuracy is desired and actual reflectances are known, use the long method with visual interpolation.
3. For rooms of unusual shape or rooms with special conditions, such as coffered ceilings, mixed-material walls, and partial height partitions, use computer assistance.
4. For spaces in which a number of different solutions are to be tried, use a computer.

## 15.24 EFFECT OF CAVITY REFLECTANCES ON ILLUMINANCE

The reflectances of the various room cavities have a marked effect on the CU because of light reflections within the room. To demonstrate this graphically, we have plotted in Figs. 15.41, 15.42, and 15.43 the effect of varying cavity reflectances on the three principal types of fixture distribution: semi-indirect, direct–indirect, and direct-spread. Note that as expected, ceiling cavity reflectance has the most pronounced effect with indirect fixtures and floor reflectance with direct units. Because lighting costs amount to 3% to 5% of the total construction cost for many types of buildings such as offices, a 20% differential in lighting fixtures can amount to as much as 1% of the total cost of a facility. This amount would not only pay for the increased cost of higher-reflectance finishes and materials but would

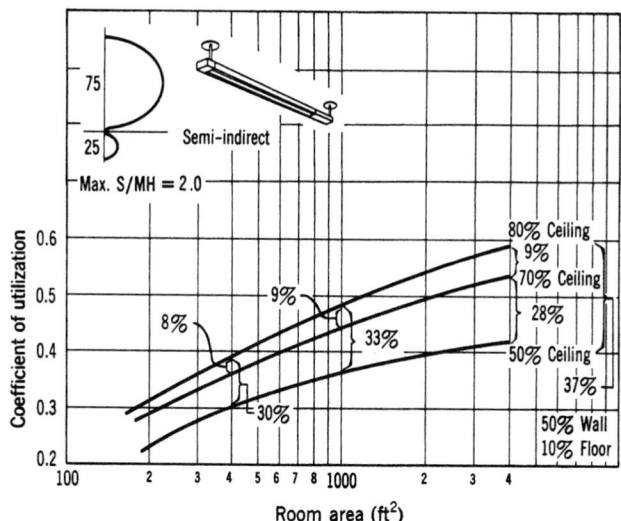

**Fig. 15.41** *Effect of surface reflectances on the CU of a luminaire with semi-indirect distribution. As expected, because the ceiling becomes the light source, its reflectance has the most pronounced effect. With this particular unit having a 25% downward component, the floor finish also has an appreciable effect, increasing the CU by an average of 10% for a 30% reflectance floor. The effect of wall reflectance naturally increases as rooms become smaller and the proportion of wall surface becomes larger. The change in CU between a 30% and a 50% reflectance wall varies from 15% for a 400-ft² room to 5% for a 4000-ft² room.*

also reduce both initial and operating costs. These data clearly indicate the necessity for the lighting designer to have considerable influence on the selection of room materials and finishes, a situation that, unfortunately, does not usually occur.

## 15.25 MODULAR LIGHTING DESIGN

An increasingly large number of buildings are being designed around a modular system, resulting in a need for flexible lighting to fit the module utilized. In such buildings, once the general lighting scheme and luminaire are established, it is convenient to draw a family of curves for the fixture chosen, thereby facilitating the utilization of the modular unit in various spaces. "Area" may readily be replaced with multiples of modular areas, as shown in Fig. 15.44.

## 15.26 CALCULATING ILLUMINANCE AT A POINT

The lumen (flux) method of horizontal illuminance calculation explained previously is appropriate for

spaces in which illuminance is essentially uniform throughout. However, even in such a space, illuminance varies at least ±10% and, near columns, walls, windows, bookcases, and the like, considerably more. Therefore, to answer the often asked question "How much light will I have on my desk?" the designer must turn to other methods. Three are available:

1. Calculation of illuminance at selected points by computer, as explained in Section 15.31
2. Utilization of one of the design aids explained in Section 15.27
3. Longhand calculation by one of the methods presented in Sections 15.28 through 15.30

These methods also yield results where the lumen method is simply inapplicable. Among such situations are layouts that are intentionally nonuniform; calculation of illuminance on planes other than horizontal (e.g., wallwashers); calculation of illuminance resulting from architectural lighting elements such as coves, valances, and the like; and illuminance calculations for nonstandard light sources for which CU data of the type given in Table 15.1 are not available.

ILLUMINATION

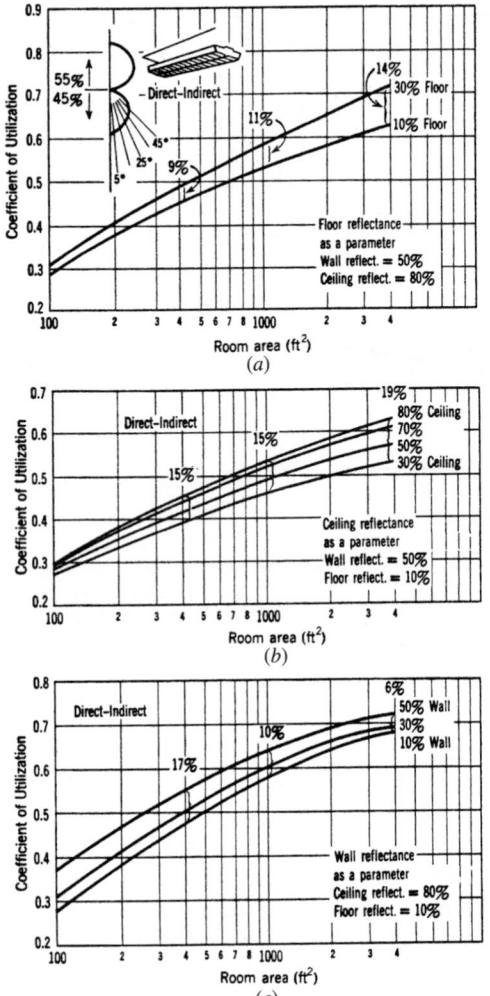

**Fig. 15.42** *Effect of surface reflectances on the CU of a luminaire with direct–indirect distribution. With this distribution, the effects of the ceiling and floor are most pronounced, with an appreciable wall effect only in small rooms.*

## 15.27 DESIGN AIDS

By *design aids* we mean any of the various curves, charts, plots, or tables either prepared by the designer or made available by luminaire manufacturers, the purpose of which is to simplify and speed lighting design when using a particular lighting fixture. The reliability of data so obtained depends entirely on the manufacturer involved, and their use should be governed accordingly. More recently, it has become customary for major manufacturers to provide computer output charts and tables based on the designers' proposed layout(s). When using these, it behooves the designer to carefully study the data input to the computer program, as the fixture supplier is certainly not a disinterested party and the program may be "weighted" accordingly. A brief description of common design aids follows.

### (a) Isolux Charts

These charts, also called *isofootcandle charts*, are based on the type traditionally supplied by manufacturers of outdoor lighting equipment, such as street lights and floodlights, but are equally applicable to interior lighting. Their use is illustrated in Fig. 15.45. The basic tool is an isolux diagram for a single luminaire. This is either calculated (see Sections 15.28–15.29), measured from a full-scale mockup (the most accurate if not the most practical method), or obtained from the manufacturer. Inasmuch as the relative positions of the

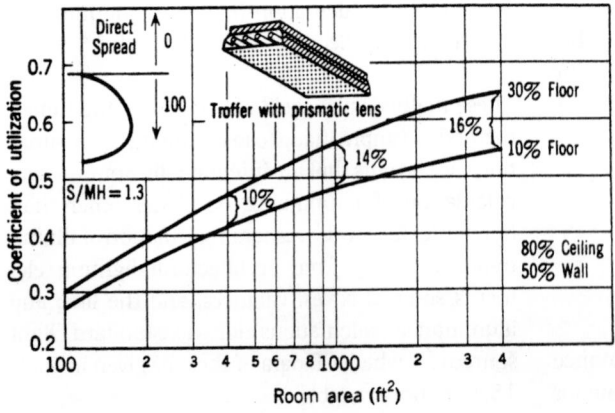

**Fig. 15.43** *Effect of surface reflectances on the CU of a luminaire with direct (spread) distribution. Floor finish is most important, with wall reflectance important only in small rooms. As these fixtures have no upward component, all ceiling illumination is derived from reflection. Thus, in a room with floor reflectance of less than 20%, ceiling finish has no effect on room illumination.*

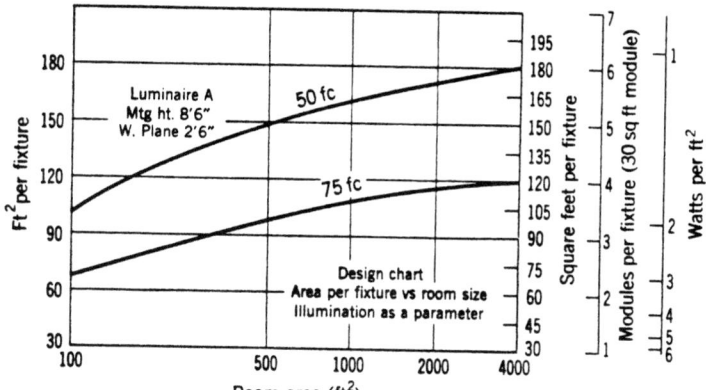

**Fig. 15.44** *Luminaire design chart. For frequently used fixtures, this type of chart gives an easy design figure for various size rooms. As seen from the ordinates, the figures can be translated into number of modules and watts per square foot.*

source and illuminated point are reversible—that is, if a source at *A* causes illuminance *E* at point *B*, then the source at *B* will cause the same illuminance *E* at point *A*—placing the center of the isolux chart at the point in question permits direct reading of the illuminance contribution of every other luminaire. It then remains simply to sum the individual contributions to obtain the (scalar) illuminance at the desired point. An example is shown in Fig. 15.45.

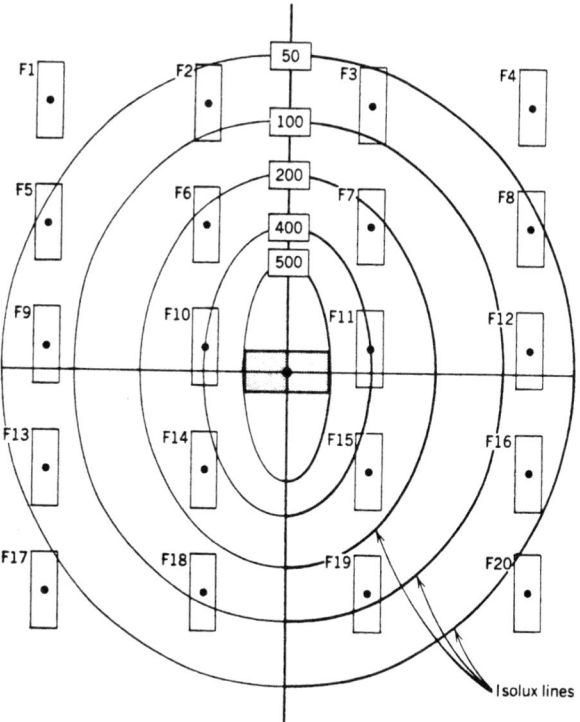

| Luminaire | Contribution (Lux) |
|---|---|
| F2, F3 | 75 × 2 |
| F5, F8 | 50 × 2 |
| F6, F7 | 225 × 2 |
| F9, F12 | 75 × 2 |
| F10, F11 | 400 × 2 |
| F13, F16 | 60 × 2 |
| F14, F15 | 300 × 2 |
| F18, F19 | 100 × 2 |
| | 2570 lux total |

**Fig. 15.45** *The ellipses represent isolux lines for a single luminaire at a given height above the work plane. They are centered on the point (the work area of a desk) for which the illuminance must be determined. The total illuminance at that point is the sum of the individual luminaire contributions. The center of the luminaire is the point of reference. Therefore, when two or more isolux lines pass through a fixture, its contribution is determined by the interpolated isolux line passing through its center. Note the symmetry around the vertical axis, necessitating a plot of only half of the ellipses.*

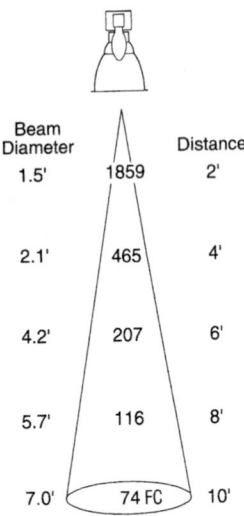

| Beam Diameter | | Distance |
|---|---|---|
| 1.5' | 1859 | 2' |
| 2.1' | 465 | 4' |
| 4.2' | 207 | 6' |
| 5.7' | 116 | 8' |
| 7.0' | 74 FC | 10' |

**Fig. 15.46** *For downlights with symmetrical circular distribution, a "cone of light," as shown, can be drawn. The illuminance at varying distances on the beam centerline directly below the luminaire is given in the center column. A circle at the circumference of which the illuminance is half of this maximum is drawn at each distance from the downlight (2 ft, 4 ft, etc.). The numbers in the left column show the diameter of this (beam) circle. (Courtesy of Zumtobel-STAFF.)*

### (b) Illuminance "Cone" Charts

See Fig. 15.46. When the light distribution of a direct downlight is symmetrical, as is generally the case, a cone can be drawn showing the illuminance directly under the fixture at various distances. The projected circles are defined by maximum illuminance at the center and half of this illuminance at the edge. This projected circle can be used in the same fashion as the isolux chart in the preceding section, except that only two values are given—that at the center and that on the circumference.

### (c) Illuminance Tables and Charts

These take various forms but all give specific illuminance data, in numerical form, for specific points. The values are obtained from a computer printout or an actual test. Figure 15.47 shows the illuminance pattern on a wall produced by the wallwash version of the downlight shown in Fig. 15.46. The difference between the two fixtures is the addition of an interior reflector in the wallwasher version.

## 15.28 CALCULATING ILLUMINANCE FROM A POINT SOURCE

It is well to note at the outset that all of the following methods calculate illuminance at a *point*. The answer to the constant query "How much light will I have on my desk from a luminaire at this location?" is arrived at by taking several points on the desk and calculating illuminance at each one. Shortcuts can be made for symmetry and so on. The reader should also understand that most of these calculations are laborious and are generally performed by a consulting engineer or lighting specialist rather than the architect. They are presented here as background material for the technically oriented designer.

**Fig. 15.47** *Addition of an interior reflector to the downlight of Fig. 15.46 converts it to a dual-purpose downlight and wallwash unit. The wall illuminances produced by multiple units spaced 3 ft apart and ceiling-mounted 3 ft from the wall are given in the chart. Similar charts are available for other luminaire spacings. (Courtesy of Zumtobel-STAFF.)*

## Wallwash Lighting Data Chart

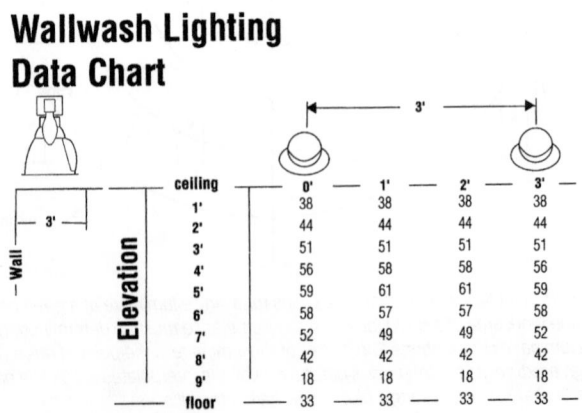

| Elevation | 0' | 1' | 2' | 3' |
|---|---|---|---|---|
| ceiling | | | | |
| 1' | 38 | 38 | 38 | 38 |
| 2' | 44 | 44 | 44 | 44 |
| 3' | 51 | 51 | 51 | 51 |
| 4' | 56 | 58 | 58 | 56 |
| 5' | 59 | 61 | 61 | 59 |
| 6' | 58 | 57 | 57 | 58 |
| 7' | 52 | 49 | 49 | 52 |
| 8' | 42 | 42 | 42 | 42 |
| 9' | 18 | 18 | 18 | 18 |
| floor | 33 | 33 | 33 | 33 |

The basis of point source calculations is the inverse square law developed in Section 11.12:

$$fc = \frac{cp}{D^2}$$

where $fc$, $cp$, and $d$ are footcandle illuminance, candlepower intensity, and distance, respectively. Refer to Fig. 15.48. The horizontal illuminance at a point $P$ as shown in Fig. 15.48 is

$$\text{horizontal } E = \frac{cp}{D^2} \cos\theta \qquad (15.14)$$

and the vertical illuminance at that same point is

$$\text{vertical } E = \frac{cp}{D^2} \sin\theta \qquad (15.15)$$

However, because

$$\cos\theta = \frac{H}{D} \quad \text{and} \quad \sin\theta = \frac{R}{D}$$

we have then at point $P$

$$\text{horizontal illuminance} = \frac{cp}{H^2} \cos^3\theta \quad (15.16)$$

$$\text{vertical illuminance} = \frac{cp}{R^2} \sin^3\theta \quad (15.17)$$

Inasmuch as the candlepower intensity in the direction of point $P$ is taken from a candlepower distribution curve, and $\theta$ is known, these expressions are readily usable. Very few commercial light sources are actually point sources. However, *when the maximum dimension of the source is less than five times the distance to point P*, the equations give satisfactory results. Note that these equations can be used to calculate and plot isolux diagrams for point sources of the type shown in Figs. 15.45 and 15.46.

---

**EXAMPLE 15.5** Referring to Fig. 15.48 and the candlepower distribution curve of Fig. 15.49, find the horizontal and vertical illuminance at point $P$, which is 10 ft below and 12 ft horizontally distant from the source.

**SOLUTION**

$$H = 10 \text{ ft} \qquad R = 12 \text{ ft}$$

$$\theta = \tan^{-1}\frac{12}{10} = 50°$$

$$\sin\theta = 0.766 \qquad \cos\theta = 0.643$$

$$cp \text{ at } 50° = 6600 \text{ (from Fig. 15.49)}$$

$$\text{horizontal illuminance} = \frac{6600}{10^2}(0.643)^3 = 17.5 \text{ fc}$$

$$\text{vertical illuminance} = \frac{6600}{12^2}(0.766)^3 = 20.8 \text{ fc} \quad \blacksquare$$

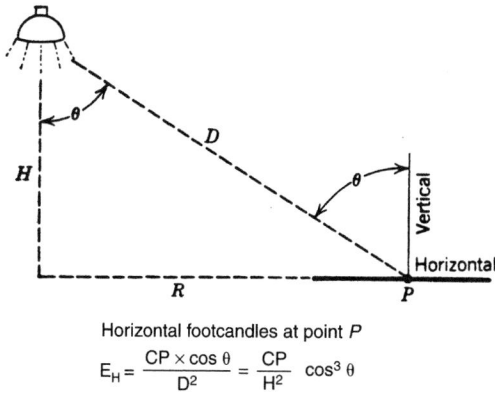

Horizontal footcandles at point $P$

$$E_H = \frac{CP \times \cos\theta}{D^2} = \frac{CP}{H^2}\cos^3\theta$$

Vertical footcandles at point $P$

$$E_V = \frac{CP \times \sin\theta}{D^2} = \frac{CP}{R^2}\sin^3\theta$$

**Fig. 15.48** *Relationship between intensity in candlepower (cp) and illuminance when the source can be considered a point source—that is, when the inverse square law applies. Source major dimension must not exceed 0.2D to be considered a point source. Measurement in feet yields fc; distances in meters yield lux.*

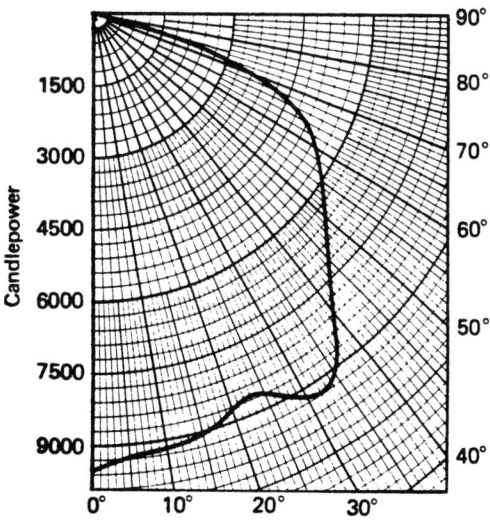

**Fig. 15.49** *Typical candlepower distribution plot for use in inverse square law calculation.*

ILLUMINATION

## 15.29 CALCULATING ILLUMINANCE FROM LINEAR AND AREA SOURCES

When the source is too large to be considered a point source (the definition is relative and depends on the distance to the illuminated surface), it is referred to as either a *linear source* or an *area source*. The *direct component* of the illuminance at a point, resulting from such sources, can be calculated by manual graphical or analytical means, both of which are based on an assumed distribution, generally lambertian (diffuse). Inasmuch as most lighting fixtures being applied today do *not* have lambertian characteristics (e.g., parabolic reflectors, prismatic diffusers), the results of such calculations are necessarily approximate. (Skylights and luminous ceilings *do* have a lambertian distribution, and for these sources these calculation methods do give reliable results.) In addition to the direct component of illuminance, a *reflected component* must be added that depends on the point's location in the room and the room characteristics. The calculations involve charts, diagrams, and tables. The interested reader will find a full description of these manual methods in the IESNA *Lighting Handbook* (2000).

Because these manual methods are laborious and frequently less than reliable, they are not presented here. The ready availability of desktop computers and computer programs that can handle detailed input for a specific light source, without broad approximations, has made these manual procedures obsolete. We recommend that when point-by-point illuminance calculation is desired, such a program be used. Alternatively, one of the design aids described previously, based on a specific light source, can be used.

## 15.30 COMPUTER-AIDED DESIGN

As pointed out in previous sections, the use of computers in lighting design is a practical necessity if really useful results are to be obtained. Once the desired luminance patterns and illuminance levels have been established, and luminaires selected and located by using either average illuminance (zonal cavity) calculations or one of the design aids previously demonstrated and/or the manufacturer's assistance, the responsible designer will confirm the preliminary design solution with accurate calcula-

tions. Furthermore, only a computer analysis will give useful point-by-point illuminance figures plus valuable data on VCP and reflected glare for selected work locations and viewing directions.

In addition, computer analysis gives the designer a degree of flexibility not otherwise possible in that:

1. The calculations are performed accurately and rapidly.
2. The designer is freed for other, less routine work.
3. The designer has the ability to change parameters repeatedly without making the analysis excessively burdensome, as would be the case with hand calculations.

It is this last characteristic that gives the designer greatest flexibility. The ability to run a series of calculations for a pendant fixture with varying pendant lengths, or to change paint colors and reflectances for various surfaces and note the effect, or to test different layout patterns, as mentioned in Example 15.2, gives the designer a very powerful and extremely useful design tool. In addition, computer analysis can consider related items, such as first costs, energy use, operating costs, and impact on HVAC systems—items whose complexity because of interrelations puts them well beyond the pencil-and-hand-calculator's ability. A few lighting analysis programs currently available are listed in Appendix M.

## 15.31 COMPUTER-AIDED DESIGN: ILLUSTRATIVE EXAMPLE

In order to demonstrate the huge amount of accurately calculated data available at a designer's fingertips when using any of the comprehensive lighting design programs, we have chosen to run Example 15.2 with a typical program: Lumen-Micro 7 (© Lighting Technologies, Boulder, CO). The first step, of course, is to input the data and the target illuminance. For convenience they are repeated here:

- Room: 60 ft × 100 ft × 8 ft
- Reflectances: 0.80 ceiling, 0.50 wall, 0.30 floor
- Required uniform illuminance: 50 fc (500 lux)
- Luminaire: recessed parabolic troffer

Most lighting analysis programs (like this one) have a large library of luminaires that is constantly updated, plus an efficient search program. This obviates the tiresome task of physically searching through several dozen catalogs and the even more tiresome task of keeping the catalogs current. We decided to use a 2-ft square luminare in order to reduce the possibility of both direct and reflected glare to persons viewing the fixture crosswise, and further chose to use U-shaped lamps. A very rapid search of the software's extensive library of lighting fixtures resulted in the selection of the luminaire shown in Fig. 15.50 (as printed from the program's library). On the basis of a preliminary calculation, an area of approximately 75 ft$^2$ per fixture seemed appropriate (because in the original longhand calculation an area of about 50 ft$^2$ was used for a two-lamp fixture). A row spacing of 7.5 ft and a column spacing of 10 ft was "drawn" on the computer, and a complete point-by-point calculation was run for horizontal and vertical illuminance, VCP and RVP. The numerical results are shown in Figs. 15.51–15.53. As the room is very large and symmetrical and is being calculated with no daylight contribution, a calculation for one-quarter of the space is sufficient. Figure 15.51 gives a summary of the results. If daylighting calculations had been done, their results would be shown as well. The *average* illuminance in the space is 48 fc, compared to the 50-fc target illuminance, and is therefore well within the desired range. Unit power density (UPD) is only 1.29 W/ft$^2$, which is within the requirements of the major energy codes. The Luminaire Summary shows the fixture chosen; had several units been used, all would be listed. Similarly, a list of luminaires in their positions in the space appears at the bottom of the figure. Inasmuch as only one type was used in this space, the long, repetitive list was truncated. Figure 15.52 is a printout of the point-by-point horizontal illuminance on the working plane (30 in. AFF), and it immediately demonstrates the value of a point-by-point calculation. Although the *average* illuminance is 48 fc and therefore satisfies the target requirement, the illuminance drops to as low as 38.5 fc and rises to 65 fc in the body of the space, away from the walls. (The calculation grid is 2.5 ft by 2.5 ft.) In practice, this situation might cause the designer to select a closer fixture spacing or perhaps change the luminaire, because 38.5 fc is 23% below the target illuminance.

On another printout of horizontal illuminance (Fig. 15.53) we have plotted luminaire locations, desk locations, and one drafting table. (The plan is not scalable because the printout row and column spacings are different.) Note that on desks against the wall, the *average* illuminance is near 50 fc, whereas on the drafting table it is about 45 fc because of its location. A staggered arrangement might improve uniformity, and therein lies one of the great advantages of computer calculation. With an additional half hour of work, an experienced designer who is accustomed to working with a given program could try three luminaires, each in three different arrays, and compare the results to arrive at an optimum lighting/economic solution.

Figure 15.54 shows partial printouts of three other metrics of the lighting design. Vertical illuminance (Fig. 15.54*a*) is useful for visual tasks on other than horizontal surfaces. In conventional offices where the majority of work is done on the horizontal plane, vertical illuminance is not critical. As pointed out repeatedly in previous sections, diffuse lighting is required to achieve good vertical surface illumination from overhead sources. In this space, with direct lighting only and (effectively) no walls to reflect light and increase diffusion, vertical illumination comes primarily as a component of direct lighting, with a small contribution from light rereflected from the floor (see Fig. 16.29).

Figure 15.54*b* indicates uniformly high VCP when looking north (directly ahead). This is to be expected from a parabolic reflector luminaire, which is noted for low luminance and therefore low direct glare. Relative visual performance (RVP) is the metric that replaced equivalent spherical illumination (ESI) as an indication of suprathreshold visual response, considering task size, contrast, and background luminance. The task we selected for these calculations is a difficult one: HB pencil on vellum, on a white table. Nevertheless, the calculations indicate perfect RVP. In the discussion of reflected glare, ESI and RVP in Sections 11.29 and 11.30, we expressed serious reservations about the usefulness of RVP as a visual performance metric when considering veiling reflections; the results shown in Fig. 15.54*c* seem to reinforce these reservations.

Graphical output is often easier to grasp than a raft of numbers. Using the same calculation grid as previously, we printed out horizontal illumi-

ILLUMINATION

Photometric Summary:

Luminous size: 2.0 x 2.0 x 0.0 ft
No. of Lamps: 3
Lamp Lumens: 2800
Input Wattage: 97
Efficiency:     60.0%
Max Candela: 2916

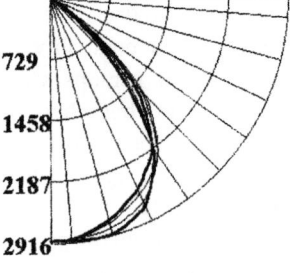

729
1458
2187
2916

| No. of Lamps: | 3 | Lamp Lumens: | 2800 |
|---|---|---|---|
| Multiplier: | 1 | Input Wattage: | 97 |
| Ballast Factor: | 1 | Ball/Lamp Factor: | 1 |
| Horz. Angles: | 5 | | |
| Vert. Angles: | 11 | | |

| | 0° | 22.5° | 45° | 67.5° | 90° |
|---|---|---|---|---|---|
| 0° | 2888 | 2888 | 2888 | 2888 | 2888 |
| 5° | 2881 | 2880 | 2882 | 2889 | 2893 |
| 15° | 2738 | 2771 | 2837 | 2898 | 2916 |
| 25° | 2515 | 2580 | 2685 | 2750 | 2745 |
| 35° | 2188 | 2300 | 2297 | 2186 | 2241 |
| 45° | 1494 | 1653 | 1578 | 1272 | 1089 |
| 55° | 247 | 351 | 296 | 48 | 17 |
| 65° | 3 | 8 | 1 | 10 | 1 |
| 75° | 0 | 10 | 0 | 0 | 1 |
| 85° | 0 | 0 | 0 | 0 | 1 |
| 90° | 0 | 0 | 0 | 0 | 0 |

COEFFICIENTS OF UTILIZATION - Zonal Cavity Method:

Floor Refl(%) = 20

| Ceiling Refl(%) | 80 | | | | 70 | | | | 50 | | | 30 | | | 10 | | | 0 |
|---|---|---|---|---|---|---|---|---|---|---|---|---|---|---|---|---|---|---|
| Wall Refl(%) | 70 | 50 | 30 | 10 | 70 | 50 | 30 | 10 | 50 | 30 | 10 | 50 | 30 | 10 | 50 | 30 | 10 | 0 |
| RCR | | | | | | | | | | | | | | | | | | |
| 0 | 71 | 71 | 71 | 71 | 70 | 70 | 70 | 70 | 67 | 67 | 67 | 64 | 64 | 64 | 61 | 61 | 61 | 60 |
| 1 | 68 | 66 | 64 | 62 | 66 | 64 | 63 | 61 | 62 | 61 | 59 | 60 | 59 | 58 | 58 | 57 | 56 | 55 |
| 2 | 63 | 60 | 57 | 55 | 62 | 59 | 56 | 54 | 57 | 55 | 53 | 55 | 53 | 52 | 53 | 52 | 51 | 50 |
| 3 | 59 | 55 | 51 | 48 | 58 | 54 | 51 | 48 | 52 | 50 | 47 | 51 | 48 | 47 | 49 | 47 | 46 | 45 |
| 4 | 56 | 50 | 46 | 43 | 54 | 49 | 46 | 43 | 48 | 45 | 42 | 47 | 44 | 42 | 46 | 43 | 41 | 40 |
| 5 | 52 | 46 | 42 | 39 | 51 | 45 | 41 | 39 | 44 | 41 | 38 | 43 | 40 | 38 | 42 | 40 | 37 | 36 |
| 6 | 49 | 42 | 38 | 35 | 48 | 42 | 38 | 35 | 41 | 37 | 34 | 40 | 37 | 34 | 39 | 36 | 34 | 33 |
| 7 | 45 | 39 | 35 | 32 | 45 | 38 | 34 | 31 | 38 | 34 | 31 | 37 | 34 | 31 | 36 | 33 | 31 | 30 |
| 8 | 43 | 36 | 32 | 29 | 42 | 36 | 31 | 29 | 35 | 31 | 29 | 34 | 31 | 28 | 33 | 30 | 28 | 27 |
| 9 | 40 | 33 | 29 | 26 | 39 | 33 | 29 | 26 | 32 | 29 | 26 | 32 | 28 | 26 | 31 | 28 | 26 | 25 |
| 10 | 38 | 31 | 27 | 24 | 37 | 31 | 27 | 24 | 30 | 26 | 24 | 30 | 26 | 24 | 29 | 26 | 24 | 23 |

$$RCR = \frac{5 * H * (\text{Room Length} + \text{Room Width})}{(\text{Room Length} * \text{Room Width})}$$

H = distance between workplane and luminaire plane

**Fig. 15.50** *Photometric report obtained from the computer library of Lumen-Micro 7© as a result of a search for a suitable 2-ft square, recessed parabolic luminaire for the example in Section 15.31. (Courtesy of Lighting Technologies, Boulder, CO.)*

Output Generated by Lumen Micro 7 for Windows. Copyright 1996 Lighting Technologies, Inc., Boulder, Colorado

**Room / Site Summary**

**Room 10 (Floor 1)**

**Calculation Summary**

| | Attained |
|---|---|
| Zonal Cavity Illum.*: | 48 fc |
| Unit Power Density: | 1.29 W/sq. ft. |

*Zonal cavity illuminance does not take into account objects within the space or daylighting.

**Room Summary**

| | |
|---|---|
| Overall Size: | 100 ft. x 60 ft. x 8 ft. |
| Reflectances: | Ceiling: .8 |
| | Walls: .5 |
| | Floor: .3 |

**Daylighting Information**

| | |
|---|---|
| Latitude: | 0 |
| Longitude: | 0 |
| Ground Refl.: | 0 |
| Bldg. Rotation: | 0 |

**Luminaire Summary**

| Type | Catalog Number |
|---|---|
| 2PMO 3U | Lithonia "Optimax" 2PMO 3U31 12LS |

**Luminaire Positions**

| Type | X | Y | Z | Rotate |
|---|---|---|---|---|
| 2PMO 3U | 5.00 | 3.75 | 8.00 | 0.00 |
| 2PMO 3U | 15.00 | 3.75 | 8.00 | 0.00 |

***Fig. 15.51*** *Summary of the results of a single layout plan. (Courtesy of Lighting Technologies, Boulder, CO.)*

nance contours overlaid with the point-by-point calculations in Fig. 15.55. (They can just as easily be printed separately.) The rectangular boxes in the figure are closed 40-fc contour lines.

## 15.32 AVERAGE LUMINANCE CALCULATIONS

The basic equations relating luminance to candela intensity and to illuminance are covered in Section 11.8, which also deals with the calculation of source luminance and reflected luminance when the illuminance is known. Thus, once horizontal illuminance has been calculated by any of the methods described previously and the reflectance of an object is known, its horizontal luminance can readily be calculated (see Section 11.8). However, as explained in detail in Sections 11.32 and 11.33 (which discuss lighting quality) and in Sections 13.10 through 13.16 (which deal with lighting systems and patterns) the luminance impression of

**Calculation Grid: Horizontal Grid #1 (1 of 1)**

**Room 10 (Floor 1)**

**Grid Summary**

Grid Properties
Grid Type:       Horizontal Illuminance
Grid Height:     2.50 ft
Units:           English

**Calculation Grid**

| | 1.3 | 3.8 | 6.3 | 8.8 | 11.3 | 13.8 | 16.3 | 18.8 | 21.3 | Dis 23 |
|---|---|---|---|---|---|---|---|---|---|---|
| 28.8 | 35.9 | 59.2 | 59.2 | 38.6 | 38.7 | 59.4 | 59.5 | 38.7 | 38.9 | 59 |
| 26.3 | 38.2 | 65.0 | 64.9 | 41.5 | 41.6 | 65.1 | 65.2 | 41.6 | 41.8 | 65 |
| 23.8 | 36.0 | 59.3 | 59.2 | 38.6 | 38.8 | 59.5 | 59.5 | 38.7 | 38.9 | 59 |
| 21.3 | 36.1 | 59.4 | 59.4 | 38.8 | 38.9 | 59.6 | 59.7 | 38.9 | 39.0 | 59 |
| 18.8 | 38.1 | 64.9 | 64.8 | 41.4 | 41.6 | 65.1 | 65.1 | 41.5 | 41.7 | 64 |
| 16.3 | 36.1 | 59.4 | 59.3 | 38.8 | 38.9 | 59.6 | 59.6 | 38.9 | 39.0 | 59 |
| 13.8 | 35.8 | 59.1 | 59.1 | 38.5 | 38.6 | 59.3 | 59.4 | 38.6 | 38.7 | 59 |
| 11.3 | 38.2 | 65.1 | 64.9 | 41.5 | 41.7 | 65.2 | 65.2 | 41.6 | 41.7 | 65 |
| 8.8 | 35.8 | 59.1 | 59.1 | 38.5 | 38.7 | 59.4 | 59.4 | 38.6 | 38.7 | 59 |
| 6.3 | 36.5 | 59.5 | 59.5 | 38.9 | 39.1 | 59.8 | 59.8 | 39.0 | 39.2 | 59 |
| 3.8 | 37.1 | 63.5 | 63.3 | 40.6 | 40.9 | 63.8 | 63.8 | 40.7 | 40.9 | 63 |
| 1.3 | 30.0 | 47.0 | 46.8 | 32.1 | 32.9 | 47.4 | 47.6 | 32.1 | 32.5 | 47 |

Distance from Grid Origin Point

**Fig. 15.52** Point-by-point horizontal illuminance results. (Courtesy of Lighting Technology, Boulder, CO.)

a visual environment is affected more by vertical than by horizontal surface luminance. (See also Table 11.2 and Fig. 13.11.) For this reason, it is important to be able to calculate *average* vertical surface (wall) luminance in the same simple, straightforward fashion used to calculate *average* horizontal illuminance. In addition, it is useful to know the average luminance of the ceiling cavity in a space in order to judge the contrast between all luminous objects, including luminaires, which have the ceiling cavity as background.

Straightforward calculation of both wall and ceiling cavity luminance ($L_W$ and $L_{cc}$) is possible through the use of luminance coefficients that are similar in concept and application to coefficients of utilization. These coefficients are listed in Table 15.4 for some of the generic fixture types listed in Table 15.1. Others are listed in the *IESNA Lighting Handbook* (2000). For actual design calculations, it is preferable to obtain coefficients from luminaire manufacturers. The average luminance calculations are parallel to those for illuminance.

Average initial wall luminance ($cd/m^2$):

$$L_W = \frac{\text{lamp lumens} \times \text{wall luminance coefficient}}{\pi \times \text{floor area in m}^2}$$

(15.18)

Output Generated by Lumen Micro 7 for Windows. Copyright 1996 Lighting Technologies, Inc., Boulder, Colorado

### Calculation Grid: Horizontal Grid #1 (1 of 1)

## Room 10 (Floor 1)

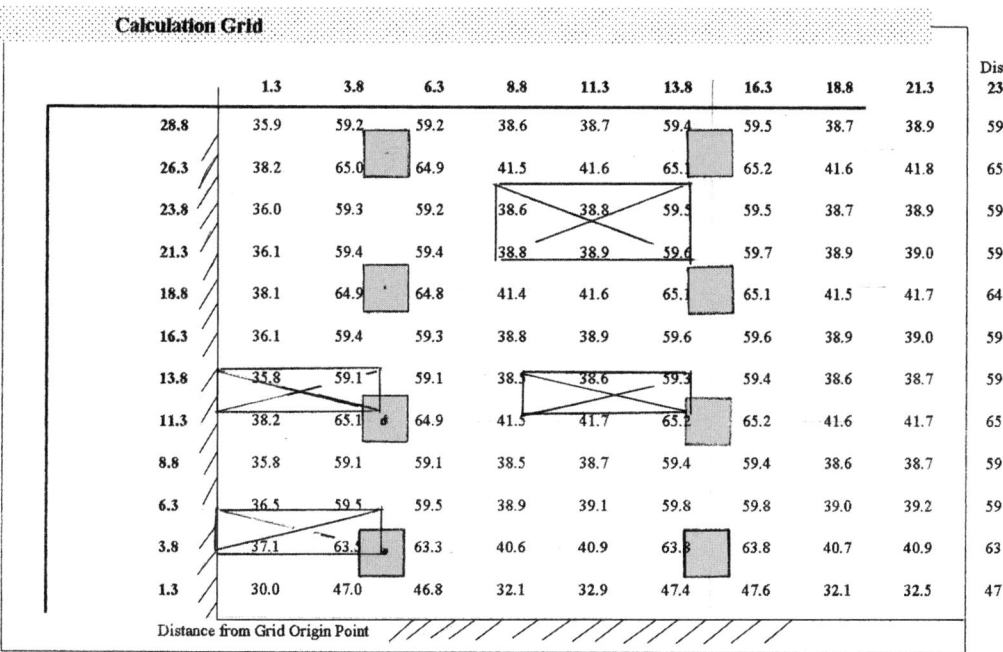

**Fig. 15.53** *Illuminance chart with luminaires and proposed furniture layout plotted on the chart. (Chart courtesy of Lighting Technologies, Boulder, CO.)*

ILLUMINATION

and average initial ceiling cavity luminance in cd/m$^2$:

$L_{CC} =$

$$\frac{\text{lamp lumens} \times \text{ceiling cavity luminance coefficient}}{\pi \times \text{floor area in m}^2}$$

$$(15.19)$$

If area is expressed in square feet and $\pi$ is omitted from these equations, $L$ will be expressed in footlamberts.

To obtain *maintained* values, an LLF similar to that explained in Section 15.20 is introduced. It is calculated similarly, except that item 17 (see Fig. 15.37), room surface dirt, is calculated using the following figures:

| Lighting System | Wall Luminance | Ceiling Luminance |
|---|---|---|
| Direct | $0.82 \pm 10\%$ | $0.75 \pm 10\%$ |
| Semi-direct | $0.87 \pm 7\%$ | $0.82 \pm 10\%$ |
| Direct–indirect | $0.92 \pm 5\%$ | $0.85 \pm 8\%$ |
| Semi-indirect | $0.87 \pm 7\%$ | $0.88 \pm 7\%$ |
| Indirect | $0.82 \pm 10\%$ | $0.90 \pm 5\%$ |

For ceiling-mounted or recessed luminaires, $L_{CC}$ is the average luminance of the ceiling between luminaires. For pendant luminaires, the calculated

## Grid Summary

| Grid Properties | |
|---|---|
| Grid Type: | Vertical Illuminance North |
| Grid Height: | 2.50 ft |
| Units: | English |

### Calculation Grid

(a)

| | 1.3 | 3.8 | 6.3 | 8.8 | 11.3 | 13.8 | 16.3 | 18.8 | 21.3 | Dis 23 |
|---|---|---|---|---|---|---|---|---|---|---|
| 28.8 | 15.8 | 22.1 | 22.4 | 16.9 | 17.3 | 23.0 | 23.1 | 17.2 | 17.3 | 22 |
| 26.3 | 11.2 | 13.4 | 13.7 | 13.0 | 13.3 | 14.3 | 14.4 | 13.3 | 13.4 | 14 |
| 23.8 | 15.5 | 23.8 | 24.1 | 16.7 | 17.1 | 24.8 | 24.9 | 17.0 | 17.1 | 24 |
| 21.3 | 15.8 | 22.5 | 22.9 | 17.5 | 17.8 | 23.6 | 23.7 | 17.7 | 17.8 | 23 |
| 18.8 | 12.6 | 13.9 | 14.2 | 13.5 | 13.9 | 14.9 | 15.0 | 13.8 | 13.9 | 14 |
| 16.3 | 15.8 | 24.2 | 24.5 | 17.1 | 17.4 | 25.1 | 25.3 | 17.3 | 17.4 | 25 |
| 13.8 | 11.4 | 19.4 | 20.0 | 14.6 | 14.8 | 20.6 | 20.7 | 14.9 | 15.0 | 20 |
| 11.3 | 13.1 | 14.4 | 14.7 | 14.0 | 14.4 | 15.4 | 15.6 | 14.3 | 14.5 | 15 |

### Calculation Grid                           VCP – NORTH

(b)

| | 1.3 | 3.8 | 6.3 | 8.8 | 11.3 | 13.8 | 16.3 | 18.8 | 21.3 | Dis 23 |
|---|---|---|---|---|---|---|---|---|---|---|
| 28.8 | 91.1 | 86.0 | 82.4 | 92.2 | 93.3 | 89.3 | 89.3 | 93.3 | 93.3 | 89 |
| 26.3 | 97.8 | 94.5 | 96.7 | 97.4 | 97.5 | 97.2 | 97.2 | 97.5 | 97.5 | 97 |
| 23.8 | 100.0 | 100.0 | 100.0 | 100.0 | 100.0 | 100.0 | 100.0 | 100.0 | 100.0 | 100 |
| 21.3 | 91.1 | 86.0 | 82.4 | 92.2 | 93.3 | 89.3 | 89.3 | 93.4 | 93.4 | 89 |
| 18.8 | 97.8 | 94.5 | 96.7 | 97.4 | 97.5 | 97.2 | 97.2 | 97.5 | 97.5 | 97 |
| 16.3 | 100.0 | 100.0 | 100.0 | 100.0 | 100.0 | 100.0 | 100.0 | 100.0 | 100.0 | 100 |
| 13.8 | 91.1 | 86.0 | 82.4 | 92.2 | 93.3 | 89.3 | 89.3 | 93.4 | 93.4 | 89 |
| 11.3 | 97.8 | 94.5 | 96.7 | 97.5 | 97.5 | 97.2 | 97.2 | 97.5 | 97.5 | 97 |

### Calculation Grid                           RVP – NORTH

(c)

| | 1.3 | 3.8 | 6.3 | 8.8 | 11.3 | 13.8 | 16.3 | 18.8 | 21.3 | Dis 23 |
|---|---|---|---|---|---|---|---|---|---|---|
| 28.8 | 1.00 | 1.00 | 1.00 | 1.00 | 1.00 | 1.00 | 1.00 | 1.00 | 1.00 | 1.0 |
| 26.3 | 1.00 | 1.00 | 1.00 | 1.00 | 1.00 | 1.00 | 1.00 | 1.00 | 1.00 | 1.0 |
| 23.8 | 1.00 | 1.00 | 1.00 | 1.00 | 1.00 | 1.00 | 1.00 | 1.00 | 1.00 | 1.0 |
| 21.3 | 1.00 | 1.00 | 1.00 | 1.00 | 1.00 | 1.00 | 1.00 | 1.00 | 1.00 | 1.0 |
| 18.8 | 1.00 | 1.00 | 1.00 | 1.00 | 1.00 | 1.00 | 1.00 | 1.00 | 1.00 | 1.0 |
| 16.3 | 1.00 | 1.00 | 1.00 | 1.00 | 1.00 | 1.00 | 1.00 | 1.00 | 1.00 | 1.0 |
| 13.8 | 1.00 | 1.00 | 1.00 | 1.00 | 1.00 | 1.00 | 1.00 | 1.00 | 1.00 | 1.0 |
| 11.3 | 1.00 | 1.00 | 1.00 | 1.00 | 1.00 | 1.00 | 1.00 | 1.00 | 1.00 | 1.0 |

*Fig. 15.54* Partial printout report for the computer analysis of Example 15.2. (a) Point-by-point calculation of vertical illuminance at the working plane level. Note that it averages less than half of the horizontal illuminance. (b) VCP chart of the room. (c) RVP chart indicating perfect seeing conditions in the entire area looking north. (Courtesy of Lighting Technologies, Boulder, CO.)

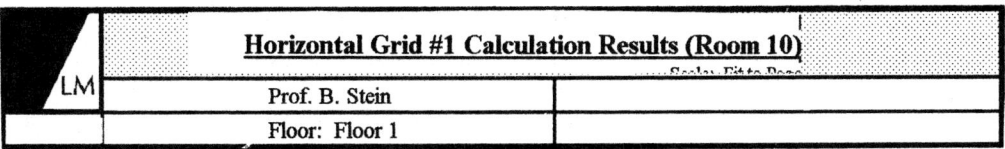

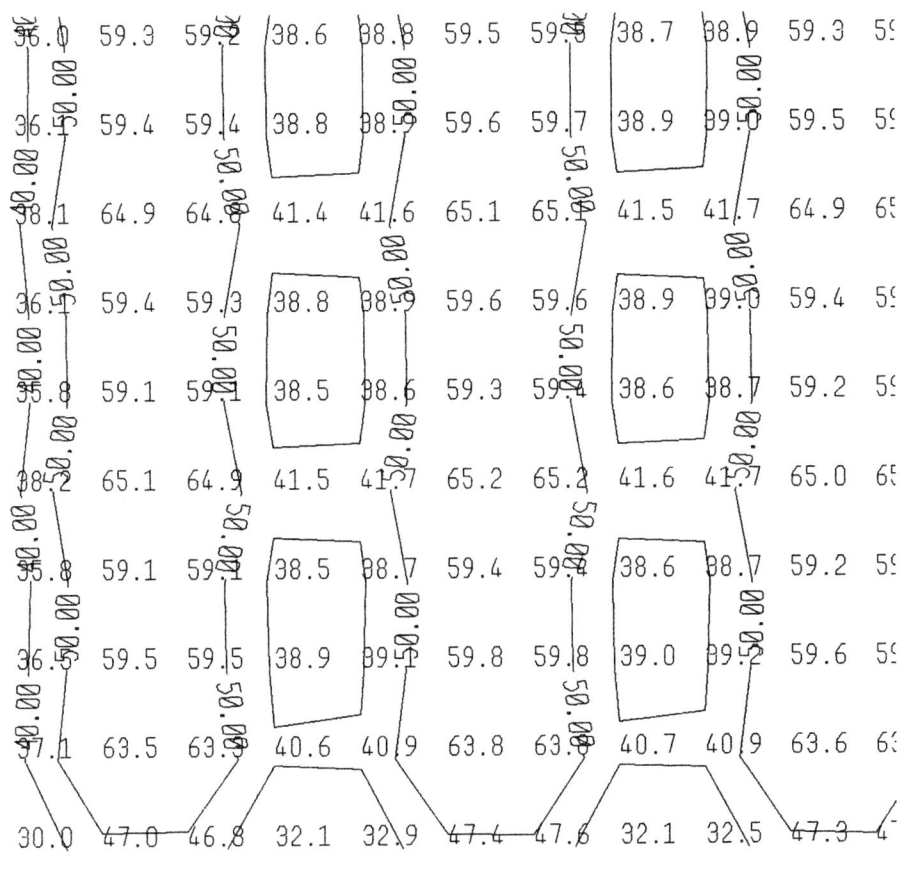

**Fig. 15.55** *A combined graphic (contours) and point-by-point printout of horizontal illuminance for lighting design. The almost rectangular figures represent closed contours at 40 fc. (Courtesy of Lighting Technologies, Boulder, CO.)*

$L_{CC}$ is that of an imaginary plane at the height of the luminaires. $L_{CC}$ is useful in determining the brightness ratios when compared to luminaire luminance at the seeing angle involved. The ceiling cavity, like the wall, is assumed to have a lambertian characteristic—that is, perfect diffuseness—making luminance independent of viewing angle.

It would be instructive to calculate the wall luminance of the office in Example 15.2. The photometric data in Fig. 15.39 do not include the wall luminance coefficient because luminance coefficients are not normally published by luminaire

manufacturers. However, based on other available data, a figure of 0.22 for wall luminance is a good estimate given RCR, $\rho_{CC}$, and $\rho_W$ of 0.66, 80%, and 30%, respectively. Initial wall luminance is then

$$L_W = \frac{3200 \times 0.22}{32\ ft^2} = 22\ fL = 75.5\ cd/m^2$$

This is within the preferred range of 25 to 150 cd/m² (see Table 11.3). In actuality, the average wall luminance would probably be higher because of the practice of placing the last row of luminaires quite close to the wall.

## TABLE 15.4 Wall Luminance Coefficients and Ceiling Cavity Luminance Coefficients for Typical Luminaires

To obtain a luminance coefficient, follow the procedure detailed at the beginning of Table 15.1 for finding a CU. More precise design data should be obtained from the manufacturers of the actual luminaires used.

| Typical Luminaire | RCR | $\rho_{CC}$: 80 | | | 50 | | | 80 | | 50 | |
|---|---|---|---|---|---|---|---|---|---|---|---|
| | | $\rho_w$: 50 | 30 | 10 | 50 | 30 | 10 | 50 | 30 | 50 | 30 |
| | | Wall Luminance Coefficients for $\rho_{FC}=20$ | | | | | | Ceiling Cavity Luminance Coefficient $\rho_{FC}=20$ | | | |
| 1 | 0 | | | | | | | .42 | .42 | .25 | .25 |
| | 1 | .32 | .18 | .06 | .27 | .15 | .05 | .42 | .40 | .25 | .23 |
| | 2 | .27 | .15 | .05 | .23 | .13 | .04 | .42 | .38 | .24 | .23 |
| | 3 | .24 | .13 | .04 | .20 | .11 | .03 | .41 | .37 | .24 | .22 |
| | 4 | .21 | .11 | .03 | .17 | .09 | .03 | .41 | .36 | .24 | .22 |
| | 5 | .19 | .10 | .03 | .16 | .08 | .02 | .40 | .35 | .24 | .21 |
| | 6 | .18 | .09 | .03 | .14 | .07 | .02 | .39 | .34 | .23 | .21 |
| | 7 | .16 | .08 | .02 | .13 | .07 | .02 | .39 | .34 | .23 | .21 |
| | 8 | .15 | .07 | .02 | .12 | .06 | .02 | .38 | .34 | .23 | .20 |
| Pendant diffusing sphere | 9 | .14 | .07 | .02 | .12 | .06 | .02 | .38 | .33 | .23 | .20 |
| with incandescent lamp | 10 | .13 | .06 | .02 | .11 | .05 | .01 | .37 | .33 | .22 | .20 |
| 3 | 0 | | | | | | | .15 | .15 | .09 | .09 |
| | 1 | .23 | .13 | .04 | .21 | .12 | .04 | .15 | .13 | .09 | .08 |
| | 2 | .22 | .12 | .04 | .21 | .11 | .04 | .14 | .11 | .08 | .07 |
| | 3 | .21 | .11 | .03 | .20 | .11 | .03 | .13 | .10 | .08 | .06 |
| | 4 | .20 | .10 | .03 | .19 | .10 | .03 | .13 | .08 | .08 | .05 |
| | 5 | .19 | .09 | .03 | .18 | .09 | .03 | .12 | .08 | .07 | .05 |
| | 6 | .18 | .09 | .03 | .17 | .09 | .02 | .12 | .07 | .07 | .04 |
| | 7 | .17 | .08 | .02 | .16 | .08 | .02 | .11 | .06 | .07 | .04 |
| Porcelain-enamaled | 8 | .16 | .08 | .02 | .15 | .07 | .02 | .11 | .06 | .06 | .04 |
| ventilated standard dome | 9 | .15 | .07 | .02 | .14 | .07 | .02 | .10 | .05 | .06 | .03 |
| with incandescent lamp | 10 | .14 | .07 | .02 | .13 | .07 | .02 | .10 | .05 | .06 | .03 |
| 7 | 0 | | | | | | | .08 | .08 | .04 | .04 |
| | 1 | .06 | .03 | .01 | .05 | .03 | .01 | .08 | .07 | .04 | .04 |
| | 2 | .05 | .03 | .01 | .05 | .03 | .01 | .07 | .06 | .04 | .04 |
| | 3 | .05 | .03 | .01 | .04 | .02 | .01 | .06 | .05 | .04 | .03 |
| | 4 | .05 | .02 | .01 | .04 | .02 | .01 | .05 | .04 | .03 | .03 |
| | 5 | .05 | .02 | .01 | .04 | .02 | .01 | .05 | .04 | .03 | .02 |
| | 6 | .04 | .02 | .01 | .04 | .02 | .01 | .05 | .03 | .03 | .02 |
| Reflector downlight with | 7 | .04 | .02 | .01 | .04 | .02 | .01 | .04 | .03 | .03 | .02 |
| battles and inside-frosted | 8 | .04 | .02 | .01 | .04 | .02 | .01 | .04 | .03 | .02 | .02 |
| lamp (see note on this unit | 9 | .04 | .02 | .01 | .04 | .02 | .01 | .04 | .02 | .02 | .02 |
| in Table 15.1) | 10 | .04 | .02 | .00 | .04 | .02 | .01 | .04 | .02 | .02 | .01 |
| 28 | 0 | | | | | | | .27 | .27 | .16 | .16 |
| | 1 | .17 | .10 | .03 | .14 | .08 | .03 | .26 | .24 | .15 | .14 |
| | 2 | .16 | .09 | .03 | .14 | .08 | .02 | .25 | .23 | .15 | .14 |
| | 3 | .15 | .08 | .02 | .13 | .07 | .02 | .24 | .21 | .14 | .13 |
| | 4 | .15 | .08 | .02 | .13 | .07 | .02 | .24 | .21 | .14 | .12 |
| | 5 | .14 | .07 | .02 | .12 | .06 | .02 | .23 | .20 | .14 | .12 |
| | 6 | .13 | .07 | .02 | .11 | .06 | .02 | .23 | .19 | .14 | .12 |
| | 7 | .13 | .06 | .02 | .11 | .06 | .02 | .23 | .19 | .13 | .11 |
| Diffuse aluminum reflector | 8 | .12 | .06 | .02 | .11 | .05 | .02 | .22 | .18 | .13 | .11 |
| with 35° crosswise and 35° | 9 | .12 | .06 | .02 | .10 | .05 | .01 | .22 | .18 | .13 | .11 |
| lengthwise shielding | 10 | .11 | .05 | .01 | .10 | .05 | .01 | .22 | .18 | .13 | .11 |
| 33 | 0 | | | | | | | .65 | .65 | .38 | .38 |
| | 1 | .20 | .12 | .04 | .13 | .08 | .02 | .65 | .63 | .38 | .37 |
| | 2 | .19 | .10 | .03 | .12 | .07 | .02 | .64 | .61 | .38 | .37 |
| | 3 | .17 | .09 | .03 | .11 | .06 | .02 | .64 | .60 | .37 | .36 |
| | 4 | .16 | .08 | .02 | .11 | .06 | .02 | .63 | .59 | .37 | .36 |
| | 5 | .15 | .08 | .02 | .10 | .05 | .02 | .63 | .59 | .37 | .36 |
| | 6 | .14 | .07 | .02 | .09 | .05 | .01 | .60 | .58 | .37 | .35 |
| | 7 | .13 | .07 | .02 | .09 | .04 | .01 | .62 | .58 | .37 | .35 |
| Luminous bottom- | 8 | .12 | .06 | .02 | .08 | .04 | .01 | .61 | .57 | .37 | .35 |
| suspended unit with extra- | 9 | .12 | .06 | .02 | .08 | .04 | .01 | .61 | .57 | .36 | .35 |
| high-output lamp | 10 | .11 | .05 | .01 | .07 | .04 | .01 | .61 | .57 | .36 | .35 |

# TABLE 15.4 *(Continued)*

| Typical Luminaire | RCR | ρcc: 80 ρw: 50 | 30 | 10 | 50 ρw: 50 | 30 | 10 | 80 ρw: 50 | 30 | 50 ρw: 50 | 30 |
|---|---|---|---|---|---|---|---|---|---|---|---|
| | | Wall Luminance Coefficients for $\rho_{FC}$ = 20 | | | | | | Ceiling Cavity Luminance Coefficient $\rho_{FC}$ = 20 | | | |
| 44 | 0 | | | | | | | .114 | .114 | .066 | .066 |
| | 1 | .137 | .078 | .025 | .125 | .072 | .023 | .105 | .094 | .061 | .055 |
| | 2 | .131 | .072 | .022 | .121 | .067 | .021 | .097 | .079 | .057 | .047 |
| | 3 | .127 | .068 | .020 | .118 | .064 | .019 | .092 | .068 | .054 | .041 |
| | 4 | .123 | .064 | .019 | .115 | .061 | .018 | .087 | .060 | .052 | .036 |
| | 5 | .119 | .060 | .018 | .112 | .058 | .017 | .084 | .053 | .050 | .032 |
| | 6 | .114 | .057 | .016 | .108 | .055 | .016 | .080 | .048 | .048 | .029 |
| | 7 | .110 | .054 | .015 | .104 | .053 | .015 | .078 | .044 | .046 | .027 |
| Radial batwing | 8 | .106 | .052 | .015 | .101 | .050 | .014 | .075 | .041 | .045 | .025 |
| distribution—louvered | 9 | .102 | .049 | .014 | .097 | .048 | .014 | .073 | .038 | .043 | .023 |
| fluorescent, unit | 10 | .097 | .047 | .013 | .093 | .046 | .013 | .070 | .036 | .042 | .022 |
| 46 | 0 | | | | | | | .236 | .236 | .138 | .138 |
| | 1 | .234 | .133 | .042 | .208 | .119 | .038 | .229 | .210 | .134 | .124 |
| | 2 | .213 | .117 | .036 | .190 | .106 | .033 | .222 | .193 | .130 | .115 |
| | 3 | .195 | .104 | .031 | .175 | .095 | .029 | .216 | .180 | .127 | .108 |
| | 4 | .181 | .094 | .028 | .162 | .086 | .026 | .211 | .170 | .124 | .102 |
| | 5 | .170 | .087 | .025 | .153 | .080 | .023 | .206 | .163 | .122 | .098 |
| Bilateral batwing | 6 | .159 | .080 | .023 | .143 | .073 | .021 | .201 | .157 | .119 | .095 |
| distribution—one lamp, | 7 | .149 | .074 | .021 | .135 | .068 | .020 | .197 | .152 | .117 | .092 |
| surface-mounted | 8 | .141 | .069 | .019 | .128 | .064 | .018 | .193 | .148 | .115 | .090 |
| fluorescent, with prismatic | 9 | .134 | .065 | .018 | .121 | .060 | .017 | .189 | .144 | .113 | .088 |
| wraparound lens | 10 | .126 | .061 | .017 | .115 | .056 | .016 | .185 | .141 | .111 | .086 |
| 35 | 0 | | | | | | | .22 | .22 | .13 | .13 |
| | 1 | .19 | .11 | .03 | .17 | .10 | .03 | .21 | .20 | .12 | .12 |
| | 2 | .18 | .10 | .03 | .15 | .09 | .03 | .21 | .18 | .12 | .11 |
| | 3 | .16 | .09 | .03 | .14 | .08 | .02 | .20 | .17 | .12 | .10 |
| | 4 | .15 | .08 | .02 | .14 | .07 | .02 | .19 | .16 | .11 | .10 |
| | 5 | .14 | .07 | .02 | .13 | .07 | .02 | .19 | .15 | .11 | .09 |
| | 6 | .14 | .07 | .02 | .12 | .06 | .02 | .18 | .15 | .11 | .09 |
| | 7 | .13 | .06 | .02 | .12 | .06 | .02 | .18 | .14 | .11. | .09 |
| Two-lamp prismatic | 8 | .12 | .06 | .02 | .11 | .05 | .02 | .18 | .14 | .11 | .08 |
| wraparound—multiply by | 9 | .12 | .06 | .02 | .11 | .05 | .01 | .17 | .13 | .10 | .08 |
| 0.95 for four lamps | 10 | .11 | .05 | .01 | .10 | .05 | .01 | .17 | .13 | .10 | .08 |
| 42 | 0 | | | | | | | .12 | .12 | .07 | .07 |
| | 1 | .16 | .09 | .03 | .15 | .09 | .03 | .11 | .10 | .06 | .06 |
| | 2 | .15 | .08 | .03 | .14 | .08 | .02 | .10 | .08 | .06 | .05 |
| | 3 | .15 | .08 | .02 | .14 | .07 | .02 | .10 | .07 | .06 | .04 |
| | 4 | .14 | .07 | .02 | .13 | .07 | .02 | .09 | .06 | .05 | .04 |
| | 5 | .13 | .07 | .02 | .12 | .06 | .02 | .09 | .06 | .05 | .03 |
| Fluorescent unit with flat | 6 | .12 | .06 | .02 | .12 | .06 | .02 | .09 | .05 | .05 | .03 |
| prismatic lens; four-lamp, | 7 | .12 | .06 | .02 | .11 | .06 | .02 | .08 | .05 | .05 | .03 |
| 2-ft-wide—multiply by 1.05 | 8 | .11 | .05 | .02 | .11 | .05 | .02 | .08 | .04 | .05 | .03 |
| for three lamps and 1.10 | 9 | .11 | .05 | .01 | .10 | .05 | .01 | .08 | .04 | .05 | .02 |
| for two lamps | 10 | .10 | .05 | .01 | .10 | .05 | .01 | .07 | .04 | .04 | .02 |
| 47 | 0 | | | | | | | .114 | .114 | .066 | .066 |
| | 1 | .175 | .100 | .032 | .163 | .094 | .030 | .107 | .093 | .063 | .055 |
| | 2 | .168 | .092 | .028 | .157 | .087 | .027 | .102 | .079 | .060 | .047 |
| | 3 | .157 | .083 | .025 | .147 | .080 | .024 | .097 | .068 | .057 | .041 |
| | 4 | .147 | .076 | .022 | .138 | .073 | .022 | .093 | .060 | .055 | .036 |
| | 5 | .139 | .071 | .021 | .131 | .068 | .020 | .090 | .054 | .053 | .033 |
| Radial batwing | 6 | .130 | .065 | .019 | .123 | .063 | .018 | .086 | .049 | .051 | .030 |
| distribution—four-lamp, | 7 | .122 | .060 | .017 | .116 | .059 | .017 | .082 | .045 | .049 | .027 |
| 610-mm (2-ft)-wide | 8 | .115 | .056 | .016 | .110 | .055 | .016 | .079 | .042 | .047 | .025 |
| fluorescent unit with flat | 9 | .109 | .053 | .015 | .104 | .052 | .015 | .076 | .039 | .045 | .024 |
| prismatic lens—see note 2 | 10 | .103 | .049 | .014 | .099 | .048 | .014 | .072 | .037 | .043 | .022 |

*Source:* Data extracted, with permission, from the *IESNA Lighting Handbook, Reference Volume* (1993). Coefficients for fixtures 1, 3, 7, 28, 33, 35, and 42 have been rounded to two decimal places.

Notes:

1. Refer to the manufacturer's catalog data for more precise values when a specific luminaire is proposed for use.

2. Multiply coefficients by 1.05 for three lamps and by 1.1 for two lamps.

## EVALUATION

### 15.33 DESIGN EVALUATION

The final step in lighting design is evaluation of the design relative to three key aspects—lighting, costs, and energy. The lighting aspects include quantity, quality, luminance ratios, mood, ambience, texture, color, variation, psychological impressions, orientation, and daylight use—in short, a review of all the lighting factors previously discussed in detail. A good deal of experience is required to visualize actual lighting results from design drawings. The novice designer would do well to have someone with such experience assist in doing the review. The other two aspects of evaluation, cost and energy, can be evaluated readily with the aid of the contractor's estimating figures for cost and a straightforward calculation for energy. The estimates are compared to the cost and energy budget figures developed at the preliminary design stage.

As we have repeatedly stressed, the important cost figures are life cycle cost, annual operating cost, and first cost for economic comparisons, operating budgets, and construction budgets, respectively. In Chapter 16 we present lighting recommendations for specific occupancies accompanied by actual cost studies and energy analyses. Detailed cost studies including the impact of lighting on air conditioning, the proportional cost of the wiring system, and the proper apportionment of costs involve the entire building and can be accurately performed only by computer. Studies of this type are generally made by consulting engineers rather than architects, and then only after initial, operating, and total costs have been set in proper perspective for a particular job by the architect and client. This is necessary because often, as in the case with speculative construction, the client's overriding consideration is first cost, thereby rendering a complete cost analysis unnecessary. Any attempt to completely separate costs for lighting, HVAC, structure, and so on is arbitrary because of the intimate interactions between these elements. Lighting designers are well advised to keep themselves and the construction team aware of this if they are to fulfill their responsibility.

### References

ASHRAE. 2004. ANSI/ASHRAE/IESNA Standard 90.1-2004: *Energy Standard for Buildings Except Low-Rise Residential Buildings.* American Society of Heating, Refrigerating and Air-Conditioning Engineers, Inc., Atlanta, GA.

IESNA. 1981. *IES Lighting Handbook,* 7th ed. Illuminating Engineering Society of North America, New York.

IESNA. 1993. *IESNA Lighting Handbook,* 8th ed. Illuminating Engineering Society of North America, New York.

IESNA. 2000. *IESNA Lighting Handbook,* 9th ed. Illuminating Engineering Society of North America, New York.

NEMA. 1998. NEMA LE 5B-1998: *Procedure for Determining Luminaire Efficacy Ratings for High-Intensity Discharge Industrial Luminaires.* National Electrical Manufacturers Association, Rosslyn, VA.

NEMA. 1999. NEMA LE 5A-1999: *Procedure for Determining Luminaire Efficacy Ratings for Commercial, Non-Residential Downlight Luminaires.* National Electrical Manufacturers Association, Rosslyn, VA.

NEMA. 2001. NEMA LE 5-2001: *Procedure for Determining Luminaire Efficacy Ratings for Fluorescent Luminaires.* National Electrical Manufacturers Association, Rosslyn, VA.

NFPA. 2005. NFPA 70-2005: *National Electrical Code.* National Fire Protection Association, Quincy, MA.

UL. 2004. UL 1598: *Standard for Luminaires.* Underwriters Laboratories, Inc, Northbrook, IL.

ILLUMINATION

# CHAPTER 16

# Electric Lighting Applications

## 16.1 INTRODUCTION

CHAPTERS 11 THROUGH 15 EXAMINED lighting fundamentals, sources, and design procedures. This final chapter on lighting considers the application of lighting principles to specific situations—with a primary focus upon electric lighting systems. The facilities covered in some detail include residential, educational, commercial, institutional, and industrial occupancies. Each is examined from the viewpoint of its special requirements, and design approaches are suggested. The latter include lighting materials and sources as well as comparative economics and energy considerations. The chapter concludes with a consideration of special types of indoor lighting, plus a short section on exterior lighting.

## RESIDENTIAL OCCUPANCIES

## 16.2 RESIDENTIAL LIGHTING: GENERAL INFORMATION

The buzzwords in modern residential construction are *automation* (convenience), *environmental considerations* (energy conservation, "green" design), and the *home as a working area* (home office, special communications considerations). These underlying considerations color decisions on sources, control, energy use, and budget, as discussed later.

Residential lighting offers to the lighting designer a great opportunity for originality and ingenuity because a residence combines more diverse functions and needs than almost any other building. Furthermore, it often requires that all work be done at minimal cost and that the result please persons with a range of tastes. The designer approaches the problems with a list of requirements, a perception of the space, and two basic tools: the lighting fixture and the architectural lighting element. The former was discussed at length in Chapter 15.

There are numerous books and countless periodical articles devoted to residential lighting design. We would recommend using IESNA publication RP-11 (1995), *Design Criteria for Interior Living Spaces*, as a reliable reference and guide.

## 16.3 RESIDENTIAL LIGHTING: ENERGY FACTORS

Although residences are generally excluded from the requirements of ANSI/ASHRAE/IESNA Standard 90.1 and lighting is not seriously addressed by ANSI/ASHRAE Standard 90.2, there may be local code requirements that deal with lighting energy efficiency. In any case, there are energy implications associated with residential lighting design decisions. General energy efficiency recommendations include:

1. Provide means for multiple light levels in all areas. A kitchen during food preparation does not have the same lighting requirements as a kitchen being entered for a "refrigerator raid." Low-level lighting provisions should be made in *all* rooms, including bathrooms. To accomplish this, use high-low switches, simple dimmers, multilevel ballasts, and multilevel switching. An ancillary benefit is that ambience can be changed thereby in multiuse rooms such as dining rooms, family and recreation rooms, and finished basements.

2. Provide local task lighting for areas where relatively difficult visual tasks are performed, such as the kitchen location where menus are planned and accounts are handled.

3. Provide dimming and switching for accent lighting.

4. Use programmable timers with photocell override for exterior lights.

5. In large residences, consider low-voltage or wireless control for ease of remote control and energy savings (see Section 16.7).

6. Use daylight in areas normally occupied during daylight hours, such as kitchens and living rooms. Consider skylights with built-in electric lighting for these areas.

Although automatic daylight compensation is generally not justified economically, area switching is and should be considered in all spaces, including work areas of children's rooms and home offices.

## 16.4 RESIDENTIAL LIGHTING SOURCES

Incandescent sources have traditionally found very wide use in residences despite their inefficiency because of their desirable characteristics, which include flattering skin color, low first cost, small size, focusability, and simple, economical dimmability. Other sources with at least some of these characteristics and much higher efficiency should be considered:

- Fluorescent sources with color temperatures between 3000 and 3500 K can be used in kitchens and other work areas. Where linear fluorescent lamps are not desirable, PL lamps and folded CFL units can readily be used. Keep in mind, however, that the lamp life of *all* fluorescent sources is shortened by switching.

- Architectural elements such as coves and cornices can readily use CFL lamps.

- Tungsten–halogen lamps should be restricted to highlighting and specialty requirements. They are incandescent lamps and, in addition to having low efficacy, must be carefully applied because of their concentrated heat.

- Consider use of low-wattage (9–12 W) PL fluorescent lamps in frequently used corridors and stairwells rather than incandescents. These small lamps can be left on for extended periods to provide the required low-level lighting without the constant switching and lamp replacement required for incandescents. This practice is even more practical in difficult access areas such as stairwells.

- For closets, pantries, and other small areas with frequently switched lighting, incandescent lamps remain the source of choice.

- For home offices and dual-purpose areas that serve for both work and recreation, design the lighting for each use individually, with maximum common use. Fluorescent sources are recommended for work purposes.

- Bathrooms can use 3000 to 3500 K fluorescent sources for general lighting, a separate incandescent source for short-time use, and low-brightness globe-shaped incandescents for mirror lighting.

- Although many manufacturers produce downlights and wall washers for CFL sources, none that we have seen deliver the "punch" of incandescents. Where such brightness is not required, CFL downlights are a good choice when burned for at least 3 hours per use.

- HID sources are appropriate for all exterior lighting.

## 16.5 RESIDENTIAL LIGHTING: DESIGN SUGGESTIONS

The following are a few general design suggestions:

- Use a general/task lighting approach, with the levels recommended in Tables 16.1 and 16.2.

**TABLE 16.1 Illuminance Recommendations for General Lighting**

| Activity or Area | Average Lux |
|---|---|
| Conversation and relaxation | 50–100[a] |
| Passage areas | 50–100[a] |
| Areas other than kitchen | 200–500 |
| Kitchen | 500–1000 |

[a]General lighting in these areas need not be uniform.

- Provide luminance ratios as in Fig. 16.1.
- Provide general lighting sufficient for movement and casual seeing in all spaces. Hallways require little lighting; stairs require more. Light stairs from directly above or ahead to create a shadow directly below the tread front. Lighting from the front eliminates shadows and can create a safety hazard.
- Do *not* avoid ceiling light sources, as is so frequently done. Wide-profile ceiling fixtures provide general lighting; switch-controlled table lamps do not.

- Lighting in areas specifically intended for use by older occupants, such as "grandma apartments" that are part of larger residences and residential buildings intended for older occupants, should take cognizance of the special requirements listed in Section 11.23.

## 16.6 RESIDENTIAL LIGHTING: LUMINAIRES AND ARCHITECTURAL LIGHTING ELEMENTS

Guidelines that can assist the designer in selecting luminaires from the huge variety available commercially are as follows:

1. Utilize diffuse distribution for general lighting; narrow-distribution downlights for area and furniture accents; and narrow-distribution, ceiling-recessed incandescent wallwashers for accenting textured surfaces such as brick.
2. Use built-in lighting to the extent possible,

**TABLE 16.2 Illuminance Recommendations for Specific Residential Visual Tasks[a]**

| Seeing Task | Typical North American Recommendation: Average Lux[b] | Other Authorities: Average Lux[b] |
|---|---|---|
| Dining | 100–200 | 100–150 |
| Grooming, makeup | 200–500 | 500 |
| HANDCRAFT | | |
| Ordinary seeing tasks | 200–500 | 200–500 |
| Difficult seeing tasks | 500–1000 | 500–750 |
| Critical seeing tasks | 1000–2000 | >1250 |
| KITCHEN DUTIES | | |
| Food preparation and cleaning involving difficult seeing tasks | 500–1000 | 750–1000 |
| Serving and other noncritical tasks | 200–500 | 200–300 |
| LAUNDRY TASKS | 200–500 | 100–300 |
| READING AND WRITING | | |
| Handwriting, reproductions, and poor copies | 500–1000 | 750 |
| Books, magazines, and newspapers | 200–500 | 300 |
| SEWING, HAND OR MACHINE | | |
| Dark fabrics | 1000–2000 | >1250 |
| Medium fabrics | 500–1000 | 700–1000 |
| Light fabrics | 200–500 | 300–500 |
| Table games | 200–300 | 300 |

[a]Selection of illuminance within the given range is based on the criteria given in Section 11.24.
[b]Divide by 10 to get footcandles. Due to the range of values, use of the exact 10.76 figure is unnecessary.

ILLUMINATION

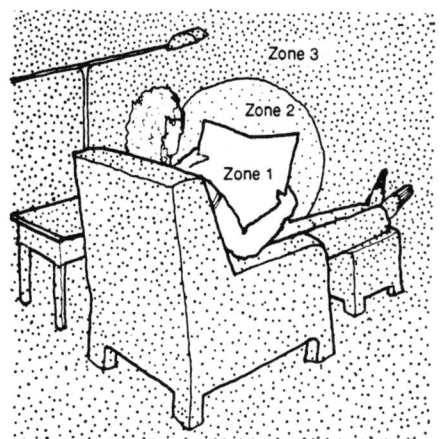

| Zone 2 | The immediate surroundings (area adjacent to the visual task) |
|---|---|
| Desirable ratio | 1/3 to equal to task* |
| Minimum acceptable ratio | 1/5 to equal to task* |

| Zone 3 | The general surroundings (not immediataly adjacent to task) |
|---|---|
| Desirable ratio | 1/5 to 5 times task* |
| Minimum acceptable ratio | 1/10 to 10 times task* |

*Typical task luminance range is 40 to 120 cd/m$^2$ |12-35 fL| and seldom exceeds 200 cd/m$^2$ |60 fL|.

**Fig. 16.1** *Seeing zones and recommended luminance ratios for residential visual tasks.*

frequently this can best be accomplished by original designs.

Architectural lighting elements include coves, cornices, valances, coffers, skylights, and other luminous constructions not normally comprising a lighting fixture. Although such units are normally less efficient than lighting fixtures, their use is often indicated by architectural considerations. Empirical design data are given in Figs. 16.3 and 16.4.

When using fluorescent lamps in architectural lighting elements, dark spots between lamps can be avoided by placing lamps at a slight angle rather than end to end, thus enabling ends to overlap. Similar overlapping is readily accomplished with PL lamps as well. Reflectors increase efficiency of an installation. When used in coves, reflectors should be aimed 15° to 25° above the horizontal and field-adjusted for the best ceiling coverage. When two-lamp strips are used, they should be arranged vertically, as shown in Fig. 16.4. Light output for a double-lamp installation rarely exceeds 1.75 times the single-lamp output.

## 16.7 RESIDENTIAL LIGHTING: CONTROL

In large residences, remote control of lighting becomes more of a necessity than a convenience from the viewpoints of control, safety, and status awareness. We will discuss two wiring and control systems that perform the necessary functions satisfactorily: low-voltage switching in Section 26.32 and power line carrier systems in Section 26.34. Completely wireless, radio wave–controlled lighting is particularly applicable to retrofit work because it

including architectural lighting elements. We believe that this demonstrates integrity of concept. For this reason, we recommend that the flexibility of track lighting be utilized for accent and task lighting but not for general lighting (Fig. 16.2).

3. Private residences are the exception to the rule of selecting off-the-shelf items in preference to specials. The lighting should complement the architecture and furnishings, and

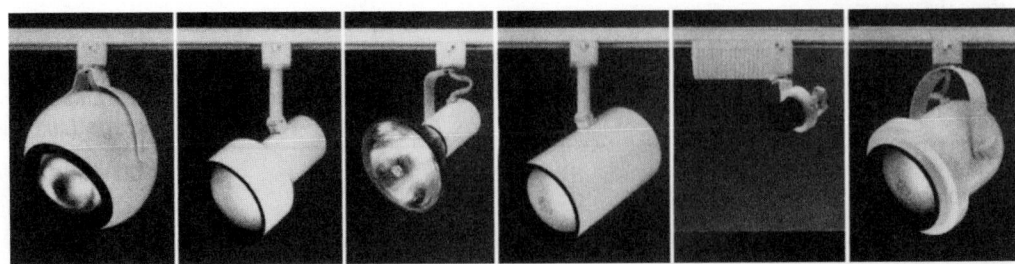

**Fig. 16.2** *Track lighting is available with a wide variety of luminaire bodies. Those illustrated include (left to right) spheres, step cylinders, swivel reflector-lamp holders, flat-back cylinders, external transformer low-voltage spots, and gimbal-ring cylinders. Not illustrated are wallwashers, adjustable and filtered spots, framing projectors, barndoor shutter units, and so on. In addition, a variety of single-circuit and multicircuit track designs are available. (Photos courtesy of Ruud Lighting.)*

*(a)* Lighted Cornices

Cornices direct all their light downward to give dramatic interest to wall coverings, draperies, murals, etc. May also be used over windows where space above window does not permit valance lighting. Good for low-ceilinged rooms.

*(b)* Lighted Valances

Valances are always used at windows, usually with draperies They provide up-light which reflects off ceiling for general room lighting and down-light for drapery accent. When closer to ceiling than 10 inches use closed top to eliminate annoying ceiling brightness.

*(c)* Lighted Coves

Coves direct all light to the ceiling. Should be used only with white or near-white ceilings. Cove lighting is soft and uniform but lacks punch or emphasis. Best used to supplement other lighting. Suitable for high-ceilinged rooms and for places where ceiling heights abruptly change.

*(d)* Lighted High Wall Brackets

High wall brackets provide both up and down light for general room lighting. Used on interior walls to balance window valance both architecturally and in lighting distribution. Mounting height determined by window or door height.

*(e)* Lighted Low Wall Brackets

Low brackets are used for special wall emphasis or for lighting specific tasks such as sink, range, reading in bed, etc. Mounting height is determined by eye height of users, from both seated and standing positions. Length should relate to nearby furniture groupings and room scale.

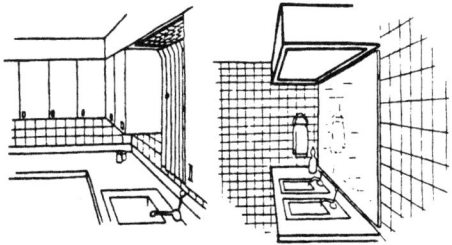

*(f)* Lighted Soffits

Soffits over work rease are designed to provide higher level of light directly below. Usually they are easily installed in furred-down area over sink in kitchen. Also are excellent for niches over sofas, pianos, built-in desks, etc.

Bath or dressing room soffits are designed to light user's face. They are almost always used with large mirrors and counter-top lavatories. Length usually tied to size of mirror. Add luxury touch with attractively decorated bottom diffuser.

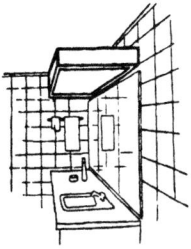

*(g)* Lighted Canopies

The canopy overhang is most applicable to bath or dressing room. It provides excellent general room illumination as well as light to the user's face.

*(h)* Luminous Wall Panels

Luminous wall panels create pleasant vistas; are comfortable background for seeing tasks; add luxury touch in dining areas, family rooms and as room dividers. Wide variety of decorative materials available for diffusing covers.

**Fig. 16.3** *Residential lighting elements. (Courtesy of IESNA.)*

### (a) Typical Valance

This typical dimensional drawing applies only to commonly encountered window valance situations. Obviously, other window treatments could necessitate modifications in these critical dimensions; i.e., vertical blinds, double-track situations, curved bay windows, etc.

The same "job-tailored" variations can occur in the design of any type of structural lighting device. Therefore, no other dimensional drawings have been included here.

Wood blocking locates lamp out from wall to minimize upper wall brightness and approximately 4" in front of drapery track. For good spread of light down draperies.

For good spread of light on ceiling, keep shielding in line with top of channel approximately 12" below ceiling and bevel top inside edge 45°

Lamp approximately 2" behind shielding for easy removal.

4" Min — 2"

Inside flat white to redirect light.

Shielding size determined by proportions of interior space may vary from 6" to 10" for good light spread and adequate shielding.

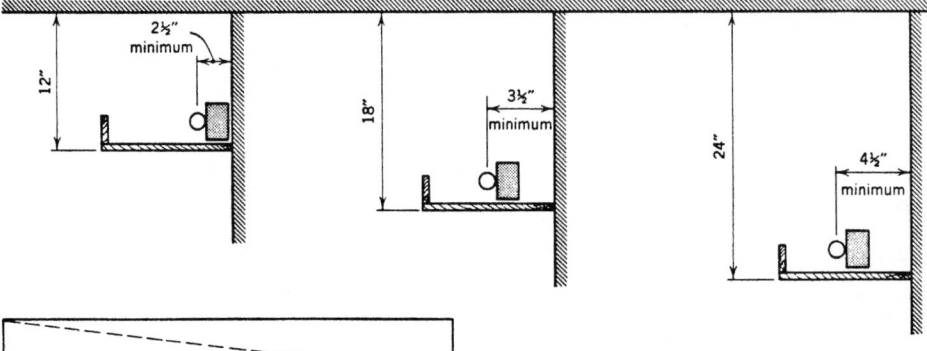

### (b) and (c) Cove Installations

Proper cove proportions: Height of front lip of cove should shield cove from the eye yet expose entire ceiling to the lamp. Orientation of fluorescent strip as shown is preferable. Cove interiors should be painted with high reflectance matte-finish paint. *Westinghouse Lighting Handbook. (Out of print)*

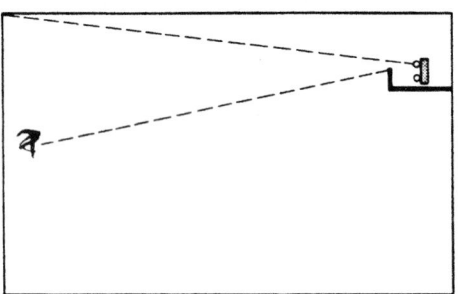

### (d) Typical Cornices

Wallwashing equipment mounted in valances and cornices provides improved brightness ratios and may be used for lighting desks against walls or vertical illumination of walls and objects mounted thereon. *Westinghouse Lighting Handbook. (Out of print)*

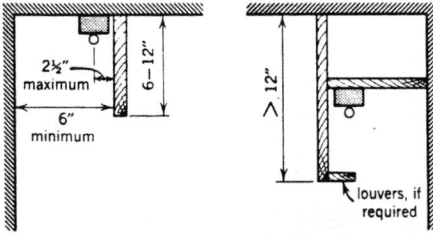

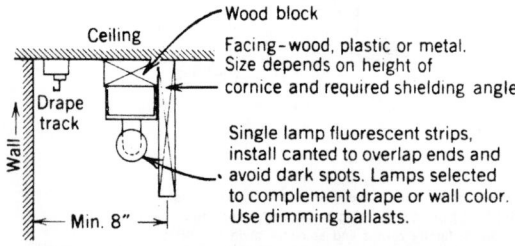

Wood block

Facing–wood, plastic or metal. Size depends on height of cornice and required shielding angle

Single lamp fluorescent strips, install canted to overlap ends and avoid dark spots. Lamps selected to complement drape or wall color. Use dimming ballasts.

Paint all surfaces matte white.

**Fig. 16.4** *Selected lighting elements: construction details.*

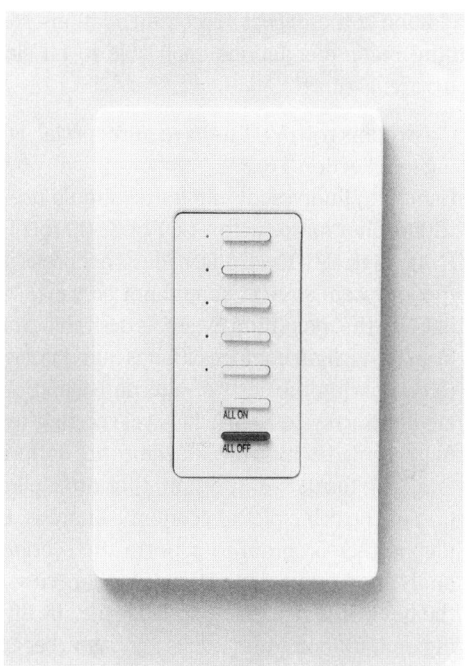

**Fig. 16.5** *Typical wall-mounted master station has five control buttons in addition to "All On" and "All Off" on every master. The control buttons can be arranged to control a single outlet or a group to establish lighting of a scene. Thus, the entry unit might have homecoming, extended absence, pathlighting, nightlighting, and garage lighting scenes, whereas the kitchen master may have lighting scenes for breakfast, dinner, formal and informal entertaining, and night. Each scene can be arranged to turn on, brighten or dim, and shut off selected lighting. (Courtesy of Lutron Electronics Co., Inc.)*

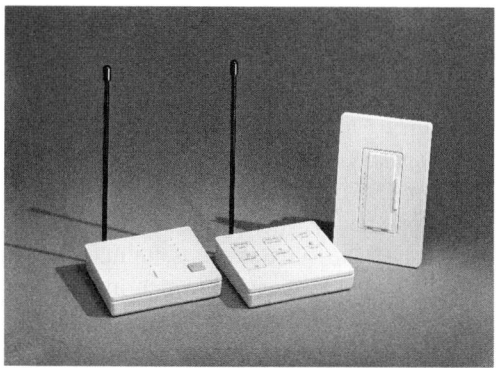

**Fig. 16.6** *Typical components of a wireless control system.* Left to right:
- *Bedside master station has five scene buttons and 10 individual outlet controls in addition to "All On" and "All Off" buttons.*
- *Repeater, which is used in large buildings to strengthen signals.*
- *Typical radio-controlled wall dimmer with local manual control. (Courtesy of Lutron Electronics Co., Inc.)*

requires only that control modules be substituted for existing switches. A conventional system might use a wall-mounted master at each entrance in the kitchen and in the home office (Fig. 16.5), a bedside tabletop master in the master bedroom (Fig. 16.6), and radio-operated wall switches and dimmers in all rooms.

## EDUCATIONAL FACILITIES

### 16.8 INSTITUTIONAL AND EDUCATIONAL BUILDINGS

The lighting requirements for some spaces in educational facilities coincide with the requirements for commercial (office) and institutional buildings. To that extent, the remarks herein are applicable there as well. Generally, educational buildings (excluding private colleges and universities) are maintained by operating funds obtained from taxes, and the budget is *always* tight. Therefore, all equipment in these public buildings must be extremely hardy, vandalproof, as maintenance-free as possible, and low in energy consumption. Maintenance in such buildings is generally poor and is performed on a repair rather than preventive basis. With these factors in mind, the following remarks apply to lighting equipment:

1. Use sources with the highest possible efficacy. Remember that daylight has the highest efficacy, followed by HPS, fluorescent, and metal halide sources.
2. Where specific color lamps are called for, such as 3500 K T8 and the like, this requirement should be permanently stenciled in large letters on the lighting fixture to ensure proper relamping.
3. Long-life sources should always be given preference because of their lower maintenance. Thus, corridor and stair lighting should use fluorescent or HID lamps. This is also important in locations where relamping is difficult, as in high-ceiling rooms such as gyms and assembly rooms.
4. In calculating illuminance, low LLF figures should be used to allow for aging of paints and dirt accumulation. Cleaning of lighting fixtures in schools is virtually unknown. An LLF figure of

0.5 is reasonable. This being so, provision should be made to reduce initial overlighting.

5. Many schools are not air-conditioned. With the masking of air noise absent, careful control must be exercised over noise and vibration from ballasts, diffusers, and so on. Electronic ballasts are preferred if the construction budget tolerates the additional cost. Ballast noise increases with current rating. Therefore, 800-mA, high-output and 1500-mA, very-high-output lamps must be used with caution, particularly in locations that amplify sounds, or where low noise criteria (NC) obtain.

6. Lighting equipment must be designed for absolutely minimum maintenance, and those fixtures within easy reach should be vandal-proof. This means using captive screws, rust-preventive plated parts, captive-hinged diffusers whose cleaning requires only one person, ballast replacement without demounting fixtures (plug-in ballasts are available), nonyellowing plastics, and high-quality finish and assembly.

For recommended illuminances for the various visual tasks in an educational facility, see Sections 11.24 to 11.26 and IESNA RP-3 2000, *Lighting for Educational Facilities.* In the following sections, we discuss the lighting requirements of specific school building areas.

## 16.9 GENERAL CLASSROOMS

The classroom is the basic space in a school. Unlike the classic schoolroom with fixed seats, a single viewing direction, and a fixed teacher location, many modern classrooms, at all grade levels, utilize multiple-student groups, teacher mobility, multiple tasks in the same overall space, and movable seating arrangements. Such spaces require:

1. Controls that permit subdivision of lighting, and level control within the subdivision (see Sections 15.13–15.15).
2. A lighting system with an appreciable indirect component and good diffusion to minimize the problem of veiling reflections due to viewing direction (see Sections 11.27–11.32).
3. Low-brightness luminaires with high VCP in all viewing directions inasmuch as a considerable portion of the students' time is spent in a head-up position (see Section 11.28).

In addition to these special recommendations, some lighting recommendations applicable to all classrooms are:

4. Classrooms with VDT units require special lighting (see Section 16.19).
5. Generally, fluorescent luminaires should use T8 triphosphor lamps with a CCT of 3500 to 4100 K and high CRI. Electronic ballasts are preferred. Incandescent sources should not be used. Daylight, to the maximum extent, is desirable. Some form of daylight compensation is mandatory.
6. In comparing the costs of alternative appropriate lighting systems, use life-cycle costing techniques only, because these facilities are nonprofit and long-term. See Appendix I for an explanation of the principles of economic analysis. Use a computer program to perform the economic analysis where possible. Do *not* compare costs on the basis of footcandles per dollar (i.e., by dividing maintained illuminance by cost), because this leads to a preference for higher illumination levels and, consequently, higher energy usage.
7. Illuminance category (see Section 11.26) and other lighting recommendations for the various types of classrooms and their activities are:
   a. *Reading.* C, D, E, depending upon the visibility characteristics of the reading material.
   b. *Mechanical Drawing and Drafting.* E, F. Drafting tables should be equipped with an adjustable fluorescent lamp. Incandescents are not recommended because of their heat, low efficacy, and both direct and reflected glare. Proper positioning of the drawing lamp provides glareless illumination even when working with pencil on vellum.
   c. *Typing.* See the special considerations of lighting for VDT units in Section 16.19.
   d. *Sight Saving.* These are general classrooms for visually impaired students. Illuminance category F is recommended, which indicates the need for supplemental lighting. These units should be individually adjustable by the user for both psychological and physiological reasons.

Figure 16.7 shows a perimeter classroom lighting layout using a two-lamp T8 louvered luminaire, and calculated initial horizontal and vertical illuminance values. A single-lamp version of the luminaire and its photometry are shown in Fig. 16.8. Note the crosswise batwing distribution that assists in limiting reflected glare.

**Fig. 16.7** *Computer-generated image of a 24-ft-W × 28-ft-L classroom using a perimeter layout of 16 4-ft luminaires in a 14-ft × 22-ft rectangle suspended 18 in. from the ceiling and 9 ft AFF. The luminaires use T8 triphosphor lamps and high-frequency electronic ballasts. Average horizontal luminance on the desks is 74 fc (800 lux), and average vertical luminance on the chalkboards is 39 fc (420 lux). Room power density is 1.8 W/ft² (Courtesy of Zumtobel-STAFF Lighting Co.)*

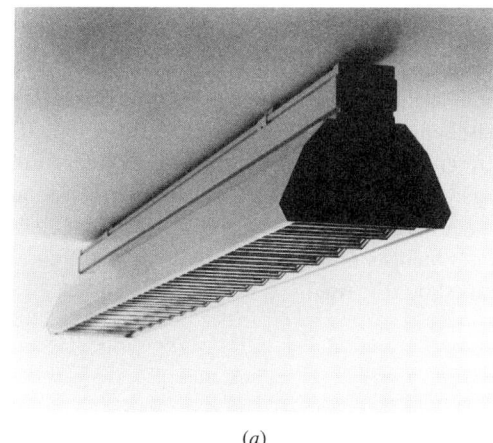

(a)

### ZX-ST 1/32W T8
BIVERGENT STEPPED VANE LOUVER

Total Luminaire Efficiency 65%

0% Uplight        100% Downlight

Spacing Criteria
Lateral Plane     0           90
                 1.2        1.5
TOTAL LAMP LUMENS = 2900
INPUT WATTS = 32

### Candela Distribution

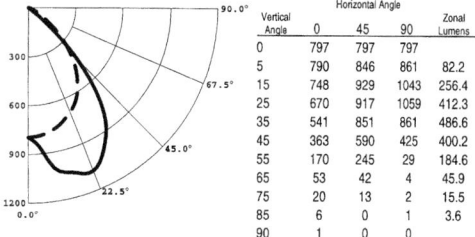

| Vertical Angle | Horizontal Angle | | | Zonal Lumens |
|---|---|---|---|---|
| | 0 | 45 | 90 | |
| 0 | 797 | 797 | 797 | |
| 5 | 790 | 846 | 861 | 82.2 |
| 15 | 748 | 929 | 1043 | 256.4 |
| 25 | 670 | 917 | 1059 | 412.3 |
| 35 | 541 | 851 | 861 | 486.6 |
| 45 | 363 | 590 | 425 | 400.2 |
| 55 | 170 | 245 | 29 | 184.6 |
| 65 | 53 | 42 | 4 | 45.9 |
| 75 | 20 | 13 | 2 | 15.5 |
| 85 | 6 | 0 | 1 | 3.6 |
| 90 | 1 | 0 | 0 | |

### Luminance Data in Candela / Sq. Meter

| Angle in Vertical | Average 0 | Average 45 | Average 90 |
|---|---|---|---|
| 45 | 2763 | 4491 | 3235 |
| 55 | 1595 | 2299 | 272 |
| 65 | 675 | 535 | 51 |
| 75 | 416 | 270 | 42 |
| 85 | 371 | 0 | 62 |

### Coefficients of Utilization

Effective Floor Cavity Reflectance = 20%

| pcc | 0.8 | | | | 0.7 | | | | 0.5 | | | 0.3 | | |
|---|---|---|---|---|---|---|---|---|---|---|---|---|---|---|
| pw | 0.7 | 0.5 | 0.3 | 0.1 | 0.7 | 0.5 | 0.3 | 0.1 | 0.5 | 0.3 | 0.1 | 0.5 | 0.3 | 0.1 |
| 0 | 78 | 78 | 78 | 78 | 76 | 76 | 76 | 76 | 73 | 73 | 73 | 69 | 69 | 69 |
| 1 | 73 | 71 | 69 | 67 | 71 | 69 | 67 | 66 | 67 | 65 | 64 | 64 | 63 | 62 |
| 2 | 68 | 64 | 61 | 58 | 67 | 63 | 60 | 57 | 61 | 58 | 56 | 59 | 56 | 55 |
| 3 | 63 | 58 | 54 | 50 | 62 | 57 | 53 | 50 | 55 | 52 | 49 | 53 | 51 | 48 |
| 4 | 59 | 52 | 48 | 44 | 57 | 52 | 47 | 44 | 50 | 46 | 43 | 49 | 45 | 43 |
| 5 | 55 | 48 | 43 | 39 | 53 | 47 | 42 | 39 | 46 | 42 | 39 | 45 | 41 | 38 |
| 6 | 51 | 44 | 39 | 35 | 50 | 43 | 38 | 35 | 42 | 38 | 35 | 41 | 37 | 34 |
| 7 | 48 | 40 | 35 | 31 | 46 | 39 | 35 | 31 | 38 | 34 | 31 | 38 | 34 | 31 |
| 8 | 44 | 37 | 32 | 28 | 43 | 36 | 31 | 28 | 35 | 31 | 28 | 35 | 31 | 28 |
| 9 | 42 | 34 | 29 | 26 | 41 | 33 | 29 | 26 | 33 | 28 | 26 | 32 | 28 | 25 |

(b)

**Fig. 16.8** *(a) This luminaire uses a very-high-reflectance silvered reflector and an unusual stepped mirrored baffle that provide low brightness for direct glare control, a batwing crosswise distribution for reflected glare control, and a sparkle on each baffle step for interest. (b) Photometrics of the luminaire shown in (a). (Courtesy of Zumtobel-STAFF Lighting Co.)*

**687**

## 16.10 SPECIAL-PURPOSE CLASSROOMS

### (a) Shops

Illuminance categories for all types of bench and machine work are

| Rough | D |
|---|---|
| Medium | E |
| Fine | F, G |
| Extra fine | G, H |

As for other tasks requiring levels above 500 lux, a task-ambient design involving an adjustable supplemental task fixture is recommended. As with drafting, adjustability of the task lighting is particularly important because of the frequently specular quality of the visual task—such as in a metal machining process. Another aspect of lighting design in shops that does not exist to the same extent in other school environments is that of safety lighting (Table 16.3).

### (b) Music Rooms

The illuminance category here is D for clear printed music and E or F for older copy or handwritten scores. These requirements are for vertical surface illuminance, as music scores are normally held in that position. Direct–indirect luminaires provide a high degree of diffusion that is helpful in this regard. For all spaces with vertical tasks, a computer analysis of the design is recommended to determine compliance with recommendations.

### (c) Art Rooms

The primary requirement here is for constant-color daylight. Thus, north windows and skylights are highly desirable. For electric lighting, because color is so important, high-CRI fluorescent lamps are required. General illumination should be augmented by user-adjustable supplementary lighting. If use of models is anticipated, adjustable accent lights are advisable. For display of artwork, adjustable wall illumination is required. Ceiling track-mounted units are an excellent choice (see Figs. 16.2 and 16.9).

## 16.11 ASSEMBLY ROOMS, AUDITORIUMS, AND MULTIPURPOSE SPACES

The varied activities in these rooms make flexible lighting imperative. For performances, low-level dimmed incandescent lighting is required. Here, incandescent lamps are the recommended source because of the lower cost of dimming and the short burning periods. For assembly rooms, this can be augmented by architectural elements along walls and draperies and in the ceiling. For study, additional ceiling fluorescents or HID units can be switched on. The combinations are legion; the different usages are the critical considerations (Figs. 16.10 and 16.11). Acoustic considerations are acute because of the low NC criteria. Thus, electronic ballasts for HID sources as well as fluorescents should be used.

**TABLE 16.3 Illuminance Levels for Safety[a]**

| *Hazard Requiring Visual Detection:* | *Slight* | | *Slight* | | *High* | | *High* | |
|---|---|---|---|---|---|---|---|---|
| **Normal Activity Level:** | **Low** | | **High** | | **Low** | | **High** | |
| Areas | Normal classrooms Lounges Small offices Dorm rooms Washrooms | | Shops Business classrooms Large offices Corridors Drafting rooms Lecture rooms Large classrooms Parking area Exterior walkways | | Stairs Libraries Reading rooms Cafeteria Swimming pools Locker rooms Interior sports Bleachers | | Boiler rooms Auditoriums Exits Exitways Laboratories Kitchens | |
| Lux (footcandles) | 5.4 (0.5) | | 11 (1.0) | | 22 (2.0) | | 5.4 (0.5) | |

*Source:* IESNA RP-3-1988; reproduced with permission.

[a]Minimum illuminance for safety or personnel, absolute minimum at any time and at any location where safety is related to seeing conditions.

**Fig. 16.9** *Art exhibition room illustrating good and bad lighting techniques. The upper wall fenestration is excellent for deep daylight penetration. Track lighting is ideal for display of art. The mixture of incandescent downlights for general lighting is excessive and an eyesore. Also, the positioning of the track lights can create both direct and reflected glare problems and annoying shadows unless the sources are selected properly and ceiling height is at least 10 ft.*

**Fig. 16.10** *Schools frequently utilize spaces for multiple functions. This space, normally used as a dining area, doubles as an assembly room. High-intensity, indirect tungsten–halogen units, tucked into concrete beam junctures, provide sufficient light for both uses.*

**Fig. 16.11** *Institutional cafeteria illuminated by cove lighting in a deep pyramidal coffer. Lighting is even, glare-free, soft in quality, and pleasant, yet of sufficient intensity to permit using the cafeteria as a working-meeting space. (Photo by B. Stein.)*

An additional consideration is step lighting. These units should be mounted to the side, or in risers, to illuminate the tread, and particularly its leading edge. Stage lighting is too highly specialized to be discussed here. Obviously, however, some form of stage lighting is required, and the building designer is advised to consult an expert.

## 16.12 GYMNASIUM LIGHTING

Gyms present a situation similar to auditoriums in that they have widely varying usages. All fixtures should be sturdy and guarded. Phosphor-coated mercury, HPS, and high-CRI metal halide are excellent choices for color, life, control, and efficacy. Multiple lighting levels should be available by switching or dimming. For dance and social events requiring low-level general lighting, other fixtures can be provided with long-life incandescent or tungsten–halogen lamps, which provide good color for low-intensity lighting as well as illumination during HID startup or restart after an outage. All fixtures should be designed for relamping from the floor. Locker rooms should use guarded-strip fluorescents.

## 16.13 LECTURE HALL LIGHTING

Lecture hall lighting is similar, with respect to sources and other considerations, to illumination for classrooms. Adjustable level fluorescent lighting is necessary for demonstrations, videos, and the like. Auxiliary lighting for demonstration tables and chalkboards completes the design. High-ceiling installations can utilize metal halide for general lighting. Controls for lighting should be located at the demonstration table (Fig. 16.12).

## 16.14 LABORATORY LIGHTING

Laboratories differ from classrooms in that tables are fixed, bench surfaces are frequently very dark, many of the items used exhibit specular reflection, vertical surface illumination is important, and visual tasks are not normally prolonged or severe. With low ceilings, use direct fixtures with an uplight component run crosswise to the tables. Luminaires with a batwing distribution minimize reflected glare from specular equipment. If the ceiling height is sufficient, indirect lighting is highly desirable for the same reason. Indirect lighting also provides a high

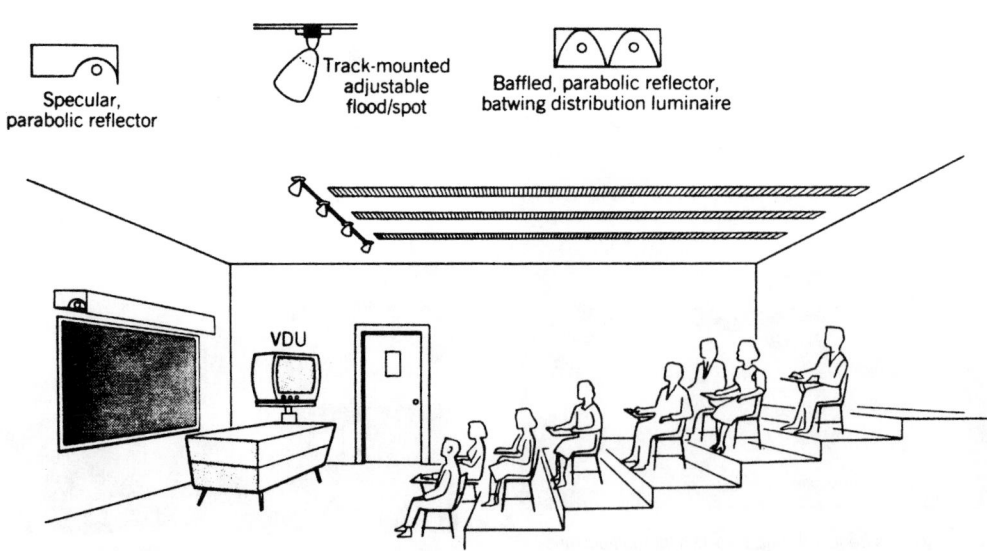

**Fig. 16.12** *Typical lecture room lighting utilizes 45° cutoff baffled parabolic reflector troffers for minimum direct and reflected glare, adjustable track lights for demonstration table illumination, and an asymmetric reflector for chalkboard lighting. The large visual display unit in the room will not have veiling reflections with the illustrated lighting arrangement.*

degree of diffuseness necessary for vertical surface illumination. See Fig. 16.13 for suggested layouts.

## 16.15 LIBRARY LIGHTING

Libraries comprise several different seeing tasks, each of which requires its own lighting solution.

### (a) General Reading Room

Here two solutions are possible, and both are in common use. In the first, general lighting is supplied over the entire area, which is sufficient for reading tasks. For this purpose, fluorescent or metal halide sources are normally applicable, the latter

with ceiling heights of at least 10 ft. The long life and high efficacy of these sources are suited to the long burning hours found in libraries. The second and more energy-efficient solution involves low-level general lighting supplemented by local reading lighting on the tables or in carrels. This solution is consonant with task-lighting orientation and is preferred. Reading lights should be fluorescent, user adjustable if possible, and arranged to avoid veiling reflections when not user adjustable.

Wherever HID sources are used, an instant restart source must be available to supply minimal lighting after an outage. Many commercial HID luminaires contain a small tungsten–halogen source for this purpose. Ballast noise can be a problem in low-NC-criteria spaces such as libraries. Electronic

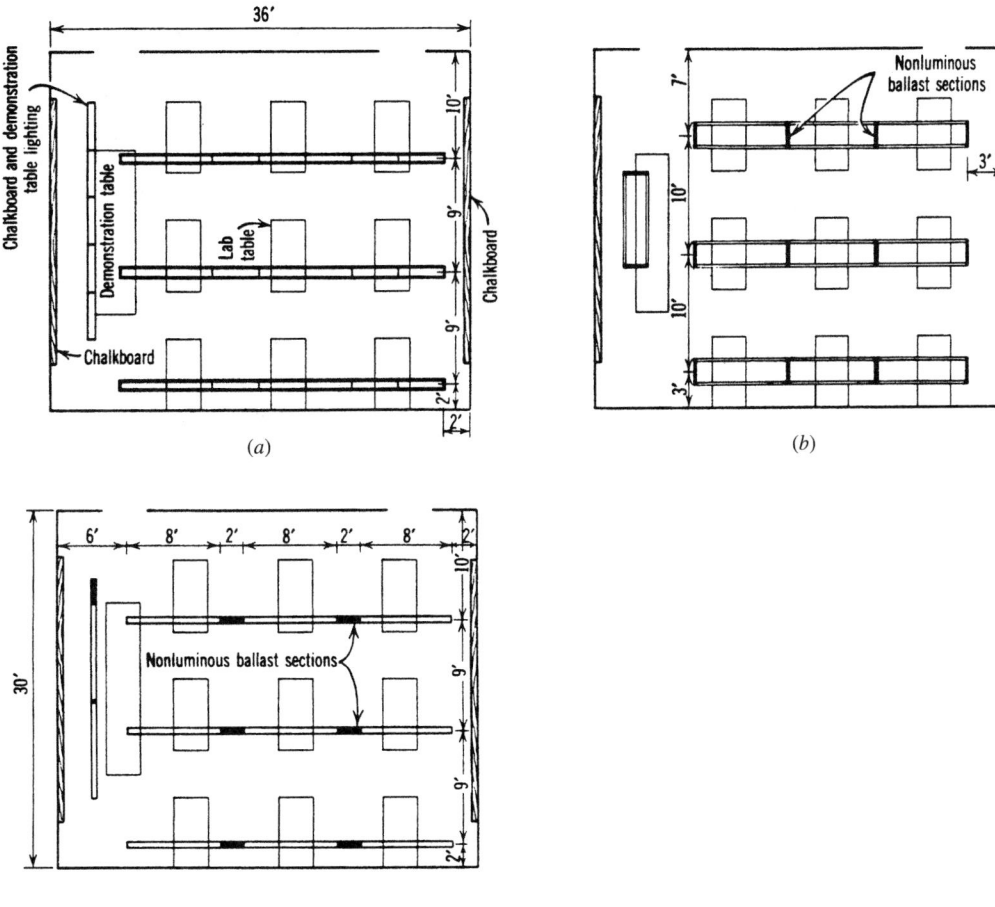

**Fig. 16.13** *Laboratory lighting schemes. Running fixtures across tables or in aisles is preferable to fixtures in the transverse direction in terms of reflected glare. (a) Pendant direct–indirect units. (b, c) Variations of the single semi-direct HO design.*

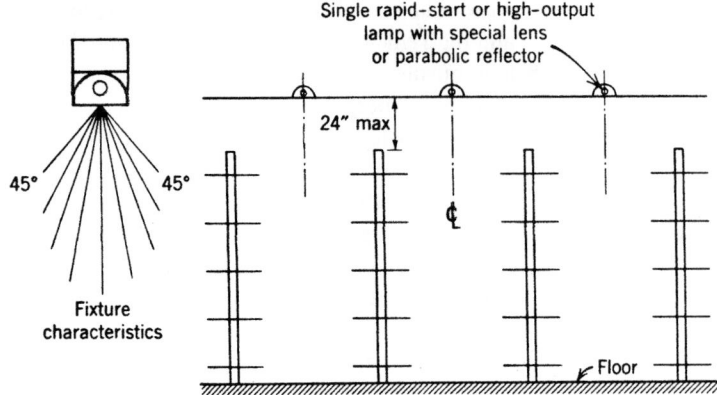

**Fig. 16.14** *Stack lighting is best accomplished by fixtures with lenses specifically designed for the purpose. Fixtures with baffles and plastic diffusers generally do not give adequate vertical surface illumination.*

ballasts for fluorescent and HID sources are available and should be employed.

### (b) Stack Areas

Here the required vertical surface illumination is best supplied by a special fluorescent unit designed for this purpose. These are mounted between stacks, and no higher than 24 in. above them, for best results (Fig. 16.14).

## 16.16  SPECIAL AREAS

Most schools contain areas devoted to functions not covered previously. Some lighting recommendations for a few of these areas are as follows:

1. *Spaces where color rendering and color matching are important,* such as sewing rooms and textile and art work spaces, must be lighted with particular attention to illuminant color. Consistent color

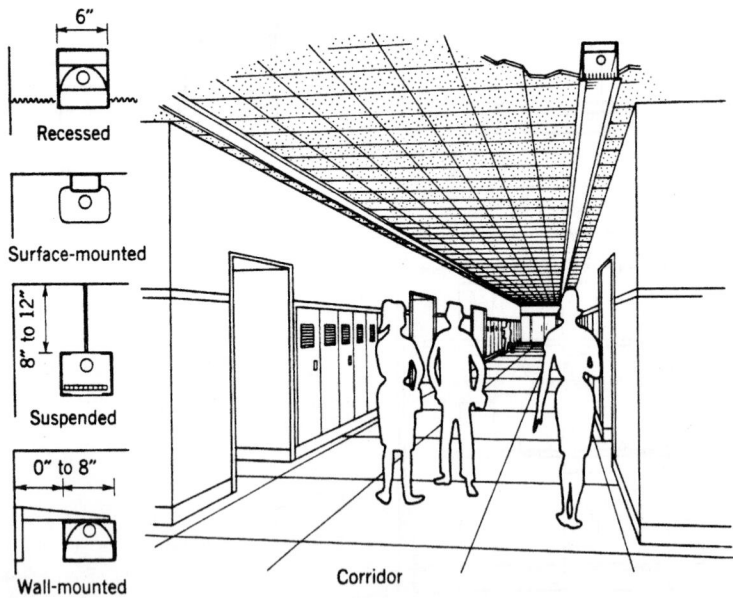

**Fig. 16.15** *Lighting of school corridors. High-reflectance walls, floor, and ceiling improve utilization of light and increase the feeling of cheerfulness. The lighting technique illustrated is appropriate for school corridors. The rows of luminaires at each sidewall illuminate bulletin boards, special displays, and the faces and interiors of lockers more effectively than do units centered in the ceiling.*

rendering without disturbing metamerisms requires a continuous-spectrum source such as incandescent (including tungsten–halogen) and, of course, daylight.

2. *Food service areas* must be well lighted to emphasize cleanliness and food attractiveness. Color rendering of food is particularly important to enhance its appetizing appearance. Use 3100 K fluorescent or metal halide lamps in preparation and serving areas, and eating areas, but in the latter at lower illuminance levels.

3. *Medical attention spaces* require high-level, good color rendering light for diagnosis and conventional office lighting for records and desk tasks.

4. *Offices, storage spaces, industrial spaces, and outdoor facilities* have the same requirements as similar spaces in other buildings and are covered in the discussion that follows.

5. *Corridors and stairways in all types of buildings* require special lighting. Corridors intended only for circulation need be lighted to only ±10 fc unless a specific seeing task requires higher levels—for example, bulletin boards and lockers (Fig. 16.15). Lighting can also be used to give direction by longitudinal arrangement. Wall-mounted or recessed wall lighting is particularly effective in corridors, providing walking illumination plus lighting for posters, bulletins, and so on. Fluorescent luminaires mounted across corridors, particularly when corridors are long, are effective in reducing the "tunnel" impression. Incandescent sources are not recommended because of their low efficacy, high maintenance, and frequency of relamping. Fluorescent and HID sources are also suggested for stairwells. Care must be exercised here, however, to avoid direct glare, which causes attention to shift from the stairs to the light source and may thereby create a hazard.

## 16.17 OTHER CONSIDERATIONS IN SCHOOL LIGHTING

### (a) Controls

See Sections 15.13 to 15.17. Because schools operate for the most part on fixed time schedules and during daylight hours, lighting controls of the preprogrammed time-base and daylight compen-

sation types are readily applicable. Other energy-conserving control strategies can be utilized as applicable.

### (b) Safety Lighting

See Table 16.3 for recommended illuminance levels. As pointed out, designers must be aware of the requirements of all jurisdictional codes.

### (c) Emergency Lighting

See Sections 26.39, 28.21, and 16.31. The local, state, and National Fire Protection Association (NFPA) codes establish minimum requirements. Exits must be clearly identified with lighted signs and a lighted path to these exits provided. As with safety lighting, it is not sufficient that average illuminance meet requirements. Illuminance in any given area must be free of large level differences, which can cause disabling glare, particularly in view of the relatively high adaptation levels of occupants' eyes immediately before a lighting outage.

## COMMERCIAL INTERIORS

## 16.18 OFFICE LIGHTING: GENERAL INFORMATION

The following information applies primarily to offices in commercial buildings and secondarily to similar spaces in other occupancies, such as educational and industrial buildings. In the latter cases, the general remarks applicable to facilities of those types take precedence. The special problems associated with VDTs (computer monitors) are discussed in Section 16.19. Task-ambient (nonuniform) lighting is covered in Sections 16.20 and 16.26. The reader is referred to IESNA RP-1 (2004), *American National Standard Practice for Office Lighting,* for a full discussion of office lighting.

### (a) Light Sources

In the interest of energy economy and good color, T8 3500 to 4000 K triphosphor linear lamps or equivalent CFL units are recommended, along with high-frequency electronic ballasts. HID can be used in indirect installations with sufficient ceiling height (minimum 9 ft, 6 in. clear). Color-corrected HPS and metal halide units of high CRI are both suitable.

Source color must be coordinated with the color scheme of room surfaces and furnishings. In areas with a large daylight contribution, the source correlated color temperature should be at least 4000 K. Incandescents may be used for storage areas, closets, and other short-burning-period uses. Incandescent and tungsten–halogen track lighting is used to advantage to illuminate displays of all sorts.

### (b) Illuminance Levels

These are discussed in Section 11.26 for the particular type of activity involved. See also Standard RP-1. Recommended reflectances for room surfaces are:

| | |
|---|---|
| Ceiling | 80% minimum |
| Walls | 50–70% |
| Partitions | 50–70% |
| Floor | 20–40% |
| Desktops, furniture | 25–45% |
| Window blinds | 40–60% |

In landscaped offices, the typical half-height partitions block 30% to 80% of the light from ceiling luminaires, depending on the furniture arrangement. It is therefore all the more important that partition finishes (including fabrics) be light-colored. It is also desirable to have upper-wall sections painted to match the ceiling—that is, with a lighter finish than the remainder of the wall. This serves the dual function of increasing ceiling cavity brightness, particularly with suspended fixtures, and increasing vertical illumination due to reflection from this surface.

### (c) Vertical Surface Illumination

This is required for many visual tasks in offices, such as those involving files, desk drawers, card files, and copy stands. Large area luminaires and a high degree of diffusness are desirable. This is especially true in large offices where wall reflections are absent. Light-finished furniture surfaces, luminaires that yield an illuminated ceiling, and high-reflectance floors are also helpful.

## 16.19 LIGHTING FOR AREAS WITH VISUAL DISPLAY TERMINALS

The computer era has resulted in an explosive proliferation of VDTs in business offices. These devices (also called monitors, video displays, or simply computers) have become standard desktop items. Because of its specular face, the VDT creates a special problem for office lighting that becomes the deciding consideration in selecting the lighting system. Simply stated, the primary problem is to avoid reflection on the screen of any luminous source in the area, including luminaires, windows, illuminated walls, and even light-colored clothing. Any such reflection makes reading data on the screen difficult and sometimes impossible.

The problem of reflections on the VDT screen is not only one of work difficulty. With a *sharply defined* reflection on the screen, such as of that of a lighting fixture, the problem becomes physiological (Fig. 16.16). The eye is naturally drawn to any bright areas in the field of vision. On a VDT screen with a clear reflected luminaire image there are two bright items in the vision field: the luminaire reflection and the VDT text or graphics. These two images are separated in depth by the thickness of the glass on the face of the VDT. The eyes are drawn to and attempt to focus on both images, but because of the difference in depth, the eye focus changes continuously, causing eyestrain and severe fatigue. This effect is

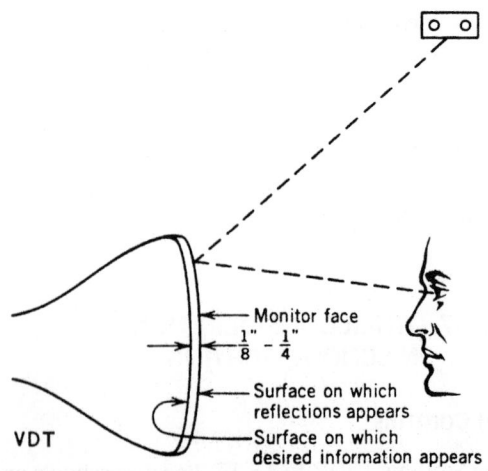

**Fig. 16.16** *A clear reflection on the face of a VDT is physiologically disturbing because it causes the eye's focus to move continuously between the plane of the monitor image (inside) and the plane of the reflection (monitor face). This constant refocusing causes eye fatigue.*

pronounced only with a clear reflected image; a fuzzy or blurred image appears only as a bright veiling reflection. As a result of this and associated problems, a number of governmental agencies have already issued guidelines for employees using VDTs that limit work periods and monitor physical effects. The health department of the state of New Jersey, for instance, has issued guidelines for all state employees that call for 15-minute breaks every 2 hours and periodic eye examinations. In addition, these guidelines call for all VDTs to have tiltable swivel mounts that are user adjustable and *separate* adjustable-angle keyboards. Fixed-position keyboards are not permitted. The reasoning here is obviously that each employee's situation is unique because of differences in work position, posture, and the body's physical dimensions. Therefore, the VDT user must be given the wherewithal to adjust and thereby maximize visual (and physical) comfort at his or her working position.

A secondary problem is avoidance of reflections on the usually specular keyboard and other specular objects in the vicinity. Matte-finish keyboards are available, as is semispecular glass. The problem with the latter is that it tends to dull and blur the information on the monitor, which, especially when working with complex graphics, is highly undesirable. This is even more true of retrofit glare filters placed over the VDT screen.

The VDT problem (Fig. 16.17) is one of geometry, complicated by the fact that the screen is often up to 20° from the vertical and the keyboard is up to 30° from the horizontal. A further difficulty lies in the contradictory nature of adjacent visual tasks. Viewing of the screen calls for a low (vertical) ambient light level (75 to 125 lux), whereas the reading and writing work done in conjunction with viewing data on the screen requires a much higher (horizontal) level (300 to 700 lux). A full solution to the problem therefore involves careful attention to the selection and location of the VDT equipment and control of room surface reflectances in addition to proper lighting design. The following recommendations refer to all these considerations, with the knowledge that some may not be in the purview of the architect/lighting designer.

### (a) Equipment

1. The screen should be recessed as deeply as possible into the VDT. This reduces the ambient light level at the screen and makes reading easier. It also reduces glare by restricting the room surface area that is reflected in the screen. (If the VDT is not constructed in that fashion, a simple hand-made shade does almost as well.)
2. Convex screens reflect a larger ceiling area than flat screens and therefore should be avoided.
3. All parts of a VDT or computer, including keys, should have a matte finish.
4. The VDT should be mounted on an adjustable tilting swivel that permits changing the geometry of reflection.
5. The keyboard should be separate from the monitor, with an adjustable tilt angle.

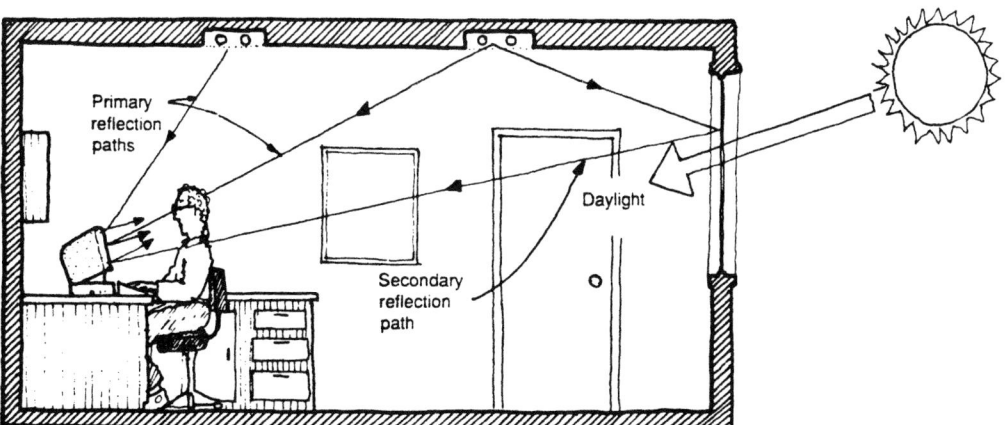

**Fig. 16.17** *Primary sources of reflection in the VDT screen and keyboard are luminous areas such as lighting fixtures and windows. Specular surfaces such as window glass can become secondary reflection sources via rereflection of light from luminaires (or other light sources), as shown.*

ILLUMINATION

### (b) Location

1. Avoid locations where the VDT *viewer* faces a window or another large, bright area.
2. Equip windows within the geometry of reflection with low-reflectance vertical blinds. Blackout curtains are not required. Remember that windows can become a secondary source of reflection—that is, luminaire to window to VDT screen (see Fig. 16.17).

### (c) Luminance Ratios

Throughout the course of the chapters on lighting, we repeatedly emphasize the need to limit the luminance ratios in the visual field, with the stringency diminishing as we recede from the center of the vision field. Thus, in Section 11.32, Table 11.11, we recommend a maximum luminance ratio of 3:1 between task and immediate surround and 1:10 between task and remote lighter surfaces. These ratios become essentially impossible to maintain when the "task" involved is a VDT because of its low luminance.

The luminance of a color monitor varies from 20 to 85 cd/m$^2$ and averages about 50 cd/m$^2$ (~20 fL). A 3:1 task-to-surround luminance ratio would be impossible to achieve without almost completely darkening the room. A sheet of white paper with 85% reflectance would have to be illuminated with approximately 60 lux to achieve this ratio, and that level of illuminance, in almost any conceivable office environment, is unacceptably low. As a result, with respect to VDT area lighting design, IESNA RP-1 recommends these maximum luminance ratios between a VDT screen and its surroundings:

VDT to lighter immediate surround 1:3
VDT to darker immediate surround 3:1
VDT to lighter remote surround *
VDT to darker remote surround 10:1

*The permissible ratio of screen luminance to higher remote luminance depends on the size and location of the luminance, as explained in Section 11.28. The recommendation is that no point on the ceiling of the room containing a VDT exceed 850 cd/m$^2$ (17 times the average screen luminance) on any 60-cm square and that lighting fixture luminance preferably be limited to:

850 cd/m$^2$ at 55° from the vertical
350 cd/m$^2$ at 65° from the vertical
175 cd/m$^2$ at 75° or more from the vertical

The limitation becomes more severe as the angle increases because high-angle luminance is reflected in the VDT screen from luminaires behind the VDT.

### (d) Light Delivery

1. *Indirect lighting.* With a ceiling height of 10 ft or more, an indirect lighting system with the 850-cd/m$^2$ luminance constraint can provide a uniform ambient illuminance level of about 30 fc (300 lux) horizontal and approximately half of that vertically. If these illuminance levels are not sufficient, portable desktop task fixtures can be utilized. It is advisable to design for somewhat higher levels and incorporate switching or dimming controls for field tuning of the installation. A ceiling height of less than 10 ft would yield unacceptably high ceiling luminance in a totally indirect installation.

2. *Semi-indirect and direct–indirect lighting.* Spaces with a ceiling height of 9 to 11 ft can utilize these systems to advantage, using semi-indirect lighting in the taller rooms and direct–indirect lighting in lower-ceiling areas. Uplight can vary between 40% and 80%, depending on the pendant stem length and the degree of spread of the upward component. Ceiling and fixture luminance should meet the criteria given previously. The ratio between fixture luminance and background ceiling luminance and between areas of ceiling of varying luminance should not exceed 5:1 at any viewing angle within the reflection geometry of the VDT. Inasmuch as manufacturers provide only fixture luminance data, the only way to determine ceiling luminances with different pendant lengths is with a full-scale mock-up. A few manufacturers provide *near-field* candela intensity data with which ceiling luminances can be calculated using a sophisticated computer program. However, because the majority of manufacturers use only far-field intensity measurement techniques, luminances of surfaces close to the luminaire, such as the ceiling, cannot be calculated accurately and only a full-scale mock-up gives reliable data. When designed properly, a direct–indirect system with approximately half of the luminaire output in each direction can provide a very effective VDT environment. Figures 16.18 and 16.19 show a well-designed semi-indirect unit with its characteristics and a typical installation. Figure 16.20 shows the application of a direct–indirect luminaire in an open-plan office with a relatively low ceiling. The batwing uplight characteristic prevents excessive ceiling brightness.

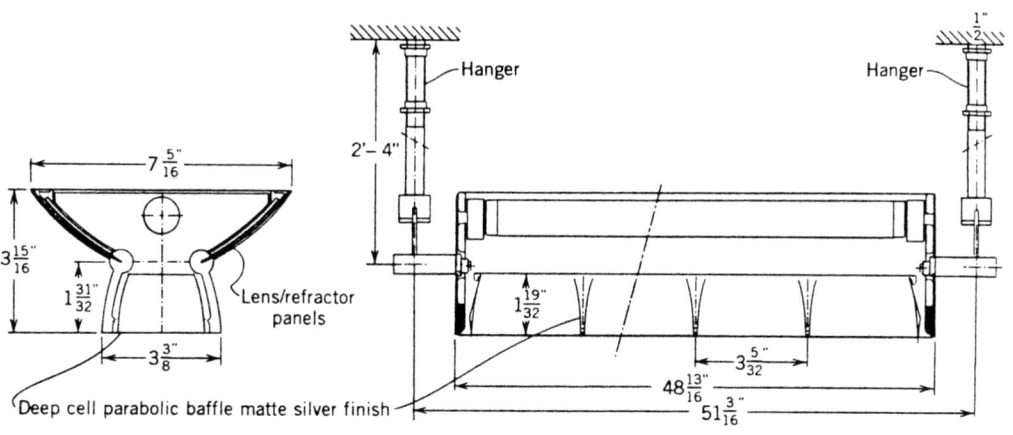

($a$)

## Photometrics

### ID-VM 1/32W T8
4' DIRECT/INDIRECT

Total Luminaire Efficiency  79%
76% Uplight          24% Downlight
Spacing Criteria
Lateral Plane          0          90
                       1.2        0.7
TOTAL LAMP LUMENS = 2900
INPUT WATTS = 31

### Candela Distribution

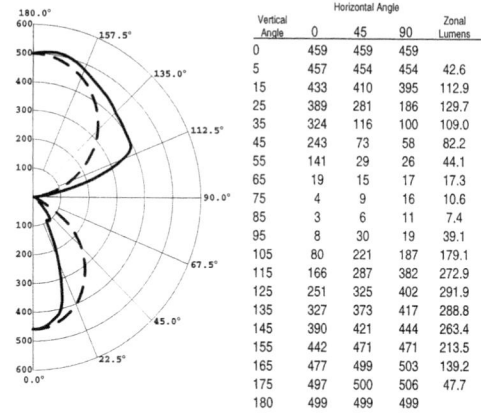

| Vertical Angle | Horizontal Angle 0 | Horizontal Angle 45 | Horizontal Angle 90 | Zonal Lumens |
|---|---|---|---|---|
| 0 | 459 | 459 | 459 | |
| 5 | 457 | 454 | 454 | 42.6 |
| 15 | 433 | 410 | 395 | 112.9 |
| 25 | 389 | 281 | 186 | 129.7 |
| 35 | 324 | 116 | 100 | 109.0 |
| 45 | 243 | 73 | 58 | 82.2 |
| 55 | 141 | 29 | 26 | 44.1 |
| 65 | 19 | 15 | 17 | 17.3 |
| 75 | 4 | 9 | 16 | 10.6 |
| 85 | 3 | 6 | 11 | 7.4 |
| 95 | 8 | 30 | 19 | 39.1 |
| 105 | 80 | 221 | 187 | 179.1 |
| 115 | 166 | 287 | 382 | 272.9 |
| 125 | 251 | 325 | 402 | 291.9 |
| 135 | 327 | 373 | 417 | 288.8 |
| 145 | 390 | 421 | 444 | 263.4 |
| 155 | 442 | 471 | 471 | 213.5 |
| 165 | 477 | 499 | 503 | 139.2 |
| 175 | 497 | 500 | 506 | 47.7 |
| 180 | 499 | 499 | 499 | |

### Luminance Data in Candela / Sq. Meter

| Angle in Vertical | Average 0 | Average 45 | Average 90 |
|---|---|---|---|
| 45 | 1516 | 455 | 362 |
| 55 | 1084 | 223 | 200 |
| 65 | 198 | 157 | 177 |
| 75 | 68 | 153 | 273 |
| 85 | 152 | 304 | 557 |

($b$)

($c$)

**Fig. 16.18** (a) Semi-indirect, low-brightness, deep-cell parabolic baffle fluorescent fixture. The parabolic baffle provides the required low luminance in the parallel direction (see text), as does the fixture body in the normal plane, so the fixtures should not cause glare. (See luminance figures in the photometric table.) (b) Photometry. Maximum ceiling luminance and ceiling luminance ratios depend on hanger length, ceiling reflectance, and fixture spacing. (c) Photograph of a semi-indirect luminaire showing construction and suspension hardware. (Courtesy of Zumtobel-STAFF Lighting Co.)

**Fig. 16.19** *Semi-indirect lighting installation in a New York stock brokerage office utilizing the fixture shown in Fig. 16.18. Every desk is equipped with a VDT. The translucent sides of the fixture are within the acceptable 5:1 luminance ratio to the ceiling recess luminance and therefore should not create a reflected glare problem. (Courtesy of Zumtobel-STAFF Lighting Co.)*

3. *Direct lighting.* In spaces with a ceiling height under 9 ft, and where otherwise required, a low-brightness direct-lighting luminaire can be used. The CIE standard for direct luminaires in areas with heavy VDT use calls for a maximum luminance of 200 cd/m$^2$ (58 fL) at 50° from the vertical and above, and the same luminance at 60° and above in areas with moderate VDT use. Such low luminance prevents both annoying reflections on monitor screens and direct glare, which can be problematic in VDT areas because the working posture is in a head-up rather than the traditional head-down

**Fig. 16.20** *A direct–indirect luminaire (see insert) with batwing uplight distribution can be used in a space with ceilings as low as 9 ft and even less, without excessive ceiling brightness, as shown in this open-plan landscaped office. (Courtesy of Zumtobel-STAFF Lighting Co.)*

position used for horizontal tasks. Very few North American manufacturers meet such stringent criteria, adhering instead to the preferred maxima given in IESNA RP-1: 850 cd/m² (~250 fL) at 55°, 350 cd/m² (~100 fL) at 65°, and 175 cd/m² (~50 fL) at 75°, all measured from the vertical.

Furthermore, these luminances, which can be met by many one- and two-lamp low-brightness fixtures, avoid the dingy, cave-like atmosphere created by very-low-brightness luminaires. The reader should note, however, that these luminances refer to the direction causing the reflection. Therefore, if the viewing direction is not fixed, it is necessary to obtain and check parallel, normal, and 45° figures from the manufacturer.

To attain the required low luminance, fluorescent troffers can be equipped with miniature parabolic wedge louvers (Fig. 15.10a) or the more efficient large-cell specular parabolic louvers (Fig. 16.21). One problem with the latter is that although a good shielding angle (minimum 35°, preferably 45°) avoids lamp image reflection on the VDT and yields high VCP (minimum of 80 recommended), the lamps are reflected in the specular louvers, causing bright spots, which in turn can reflect from a VDT screen. To avoid this, some designers place a diffusing material (translucent overlay) *above* the louvers, which makes the louver luminance uniform but reduces efficiency. Another alternative is to use semispecular louvers, which increase overall louver luminance but without the highlights of specular louvers. A third alternative is to use specially made high-grade (and expensive) louvers whose curvature is carefully controlled to eliminate the ripple reflections seen in Fig. 16.21. Finally, specular louvers must be very well maintained, as every speck of dust shows up as a bright spot on a dark background.

### (e) Finishes

Walls, floors, and furniture should be finished in low-chroma, low-value colors, with a maximum reflectance of 50%. Very dark and very light desktops are to be avoided because of excessive luminance ratios and the latter also because of reflections. Similarly, users should be advised to avoid light-colored clothing and specular clothing accessories. In general, the standard recommended maximum luminance ratios of 1:3 in the near field and 1:10 in the far field should be the design goal here as well.

**Fig. 16.21** *The illustrated installation shows a newspaper office with a VDT on every desk and a variety of viewing directions. The lighting fixtures utilized are deep-cell specular parabolic louver units, which typically have low surface brightness and crosswise batwing distribution. Note particularly that the VDT screens are deeply recessed into the monitor case to minimize glare and that each VDT is mounted with tilt and swivel capability. (Courtesy of Zumtobel-STAFF Lighting Co.)*

## 16.20 OFFICE LIGHTING GUIDELINES

### (a) Private Offices

In these spaces, a task-ambient approach is frequently appropriate because there is usually only one primary visual task location, with the remainder of the space devoted to circulation and storage. Often, sufficient illumination for the latter is supplied by spill light from the task area, particularly when the task fixture is a pendant unit with an uplight component, as seen in Fig. 16.22. In large rooms, provide downlighting in sitting areas and some type of wall illumination using wallwashers, sconces, or a recessed perimeter unit to brighten the often dark wood-paneled walls. Accent lights are required for pictures and other displays.

### (b) General Offices

The two basic approaches to general office lighting are a uniform layout that provides task-level lighting in the entire area or a task-ambient design. The former, discussed in this section, is most appropriate

ILLUMINATION

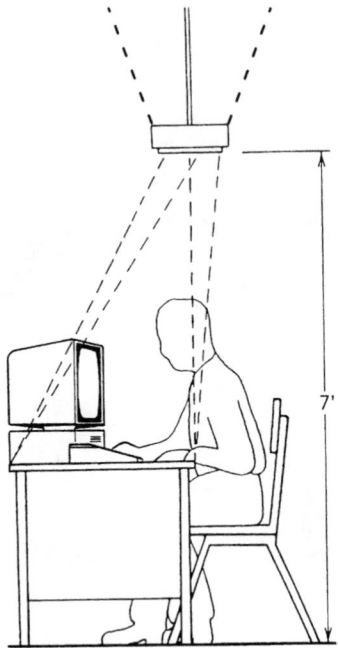

**Fig. 16.22** *Geometry of a pendant task-ambient fixture. This unit is intended not only to supply task illumination, but also a measure of ambient lighting via its 20% uplight component. In a small room, this component is sufficient for general circulation. When installed as shown, veiling reflections are minimized and glare on the VDT screen is essentially eliminated.*

to speculative-type construction, where the furniture layout is unknown but a complete job is required. (Frequently, however, the construction contractor provides only sufficient electric power for lighting and miscellaneous power; the lighting for the space is designed *after* the space has been rented as "tenant" work. In such cases, a task-ambient design is possible because tenant work is tailored to the user's needs.)

A task-illuminance overall layout is wasteful of energy and increases building energy budgets. Therefore, a uniform layout, when used, should be designed so that levels can be lowered easily in areas not requiring task lighting (see Sections 15.13–15.18). Furthermore, because no control of the geometry of direct and reflected glare is possible in such layouts, the luminaire selected must give minimum average glare in *all* viewing directions. The system that best meets these requirements is one that uses a large number of low-brightness luminaires in a dense layout. This is a very expensive solution, however, unless the ceiling is either modular or coffered.

Modular ceilings, with a single-lamp luminaire in each module, are frequently a feasible and economical approach. A structurally coffered ceiling with a luminaire in alternate coffers (Fig. 16.23) is another possible alternative, depending on coffer size. The higher cost of the luminaires may be offset by the absence of a hung ceiling. For a flat ceiling, a low-brightness troffer would be selected. An unusual luminaire that creates a direct–indirect low-brightness unit in a troffer enclosure is shown in Fig. 16.24. A batwing distribution is generally not sought because of high direct glare (low VCP) and strong veiling reflections at work locations whose line of vision is crosswise to a lighting fixture. Where the viewing direction is established but the exact furniture layout is unknown, a uniform layout with low-brightness parabolic reflectors with a batwing characteristic can be used to advantage. Luminaire selection criteria for VDT

**Fig. 16.23** *A coffer-ceilinged room (a) is illuminated with 1 × 4 two-lamp units in alternate coffers. The appearance both unlighted (a) and lighted (b) is symmetrical. (Photos by B. Stein.)*

ILLUMINATION

(a)

(b)

**Fig. 16.24** *Luminaire construction giving a direct–indirect appearance (a) using a direct troffer construction (b). The medium-brightness luminaire interior reflects light, giving a deep, airy appearance. The translucent lamp enclosure maintains an acceptable luminance against the background (interior) reflector. The photometric characteristic is essentially circular in both longitudinal and transverse planes. (Courtesy of Zumtobel-STAFF Lighting Co.)*

areas were given in the preceding section. Where VDTs are not the overriding consideration, the following criteria are helpful:

1. In terms of appearance, a 2 × 4 fixture is appropriate in a 2 × 2 tile ceiling; a 1 × 4 fixture is suitable in a 1 × 1 ceiling; and a square 2 × 2 fixture is suitable in all tile patterns.
2. Two-inch-deep parabolic louvers permit a shallow fixture, but lamp shielding is poor. A 3-in.-minimum louver depth is suggested for all large-cell parabolic louver luminaires.
3. Three-lamp parabolic louver units, which are usually too bright for VDT areas, can be used to advantage. Two-lamp units give better uniformity and VCP at a premium price.
4. Two-foot-square deep cell parabolics with 31-W T8 U lamps or biaxial compact lamps are an excellent choice in terms of VCP, uniformity of illumination, and appearance.
5. Premium-quality specular parabolic louvers that eliminate hot (bright) spots are available.

6. In relatively small offices with 1 × 1 ceiling tiles, 1-ft$^2$ troffers with deep miniature parabolic wedge louvers and 16-W U lamps or compact fluorescents are an excellent all-purpose lighting solution (Fig. 16.25).
7. Direct–indirect shallow luminaires with a wide, inverted batwing uplight distribution can be

**Fig. 16.25** *One-foot-square, 9-cell luminaire with deep parabolic louvers. This unit is applicable to small offices and/or areas with 1 ft × 1 ft ceiling-tile arrangements. (Courtesy of Zumtobel-STAFF Lighting Co.)*

pendant mounted as low as 7 ft, 6-in. AFF (Fig. 16.20).

### (c) Office Lighting Equipment

Office lighting equipment is generally not treated roughly. Fixtures with touch latches, light hinges, and adjustable devices may be selected without fear of breakage or vandalism.

### (d) Maintenance

In most offices, maintenance is provided on a trouble-call basis. Lamps are replaced on burnout, and the fixture is then cleaned. Because of the long life of fluorescents and HID sources, this generally means a 3- to 5-year cleaning cycle. An LLF of 0.65 is reasonable in air-conditioned spaces; a lower LLF is appropriate in open-window offices.

### (e) Fenestration

When fenestration is absent, a lighted valance around the room is recommended. This removes the wall-ceiling line and partially compensates for the lack of windows. It also brightens the walls and increases illumination on desks placed adjacent to the walls.

### (f) Control

The control strategy (see Sections 15.13–15.17) should *minimally* provide (after tuning):

1. Daylight compensation.
2. Possibility of operating individual small groups of lights while the remainder are off to permit off-hours work.
3. Path lighting through large spaces to permit traversing without turning on all lights.
4. Careful scheduling with supervised local override. With these general guidelines in mind, the following sections discuss specific topics in office lighting.

### 16.21 TASK-AMBIENT OFFICE LIGHTING DESIGN USING CEILING-MOUNTED UNITS

Efficient modern office lighting, like other work-area lighting, is often predicated on a task-ambient

design. This approach has the advantages of minimum contrast reduction at the task and minimum power use, the latter resulting in low operating cost. The method involves designing a uniform ceiling layout that provides low-level lighting for circulation and miscellaneous easy visual tasks, plus workstation–mounted task lighting. The latter can either be integral with the furniture (see the following section) or a separate, generally adjustable unit on or adjacent to the desk. The former is particularly useful for very severe seeing tasks when adjustability of angle and distance between light and work is vital. Integral furniture units are usually not adjustable, and are therefore limited to a maximum of about 750 lux on the task. Higher levels would generate excessive heat and, almost certainly, glare.

### 16.22 TASK-AMBIENT OFFICE LIGHTING USING FURNITURE-INTEGRATED LUMINAIRES

In lieu of ceiling-mounted luminaires to provide ambient lighting, indirect HID fixtures can be mounted on the top of furniture or can be freestanding. Task lighting would be integral with the office furniture. Advantages of this arrangement include the following:

1. The problem of furniture layout and layout changes is eliminated.
2. Initial construction cost is reduced.
3. Energy requirements are lowered because of short distances between light source and task.
4. Each occupant has on-off control of his or her task lighting, including, in some designs, positioning control.
5. Maintenance is greatly simplified because fixtures are readily accessible from the floor.
6. Floor-to-floor height can frequently be reduced.
7. Tax advantages normally accrue due to higher depreciation rates on furniture than on a building.

Disadvantages include the following:

1. Difficulty in dissipating heat and minimizing magnetic ballast noise due to proximity of sources to the user. For this reason, electronic ballasts should be used.
2. Veiling reflections are *always* present and can be severe.

3. Luminance ratios in the near and far surround may exceed recommended levels.
4. Difficulty in lighting a free-standing open desk, because most of the fixture types are under-counter or sidewall mounted.
5. Difficulty in evenly lighting large tables or L-shaped desk areas because of the concentrating nature of the lighting units.
6. Not readily applicable to automatic switching and dimming schemes.
7. For satisfactory operation with fixed task lighting, the desk must be usable by both left- and right-handed people. This requires a considerable degree of duplication, as shown by lamps on both sides of the drawing in Fig. 16.26.

Figure 16.26 graphically shows the problem of local desk lighting. Experience suggests that only systems that permit user positioning adjustment of the light source have a high degree of user acceptability.

## 16.23 INTEGRATED AND MODULAR CEILINGS

The cost, appearance, and design-flexibility advantages of an integrated ceiling design over field-assembled and coordinated systems have long been known. As a result, ceiling systems with integrated lighting, acoustic control, and air-handling capabilities are commercially available in modular sizes (including 60-in. square, 48-in. square, and $30 \times 60$ in.). Modules are made in flat and pyr-

amidal shapes, the latter having several distinct advantages:

1. More interesting and aesthetically pleasing.
2. More acoustic absorbency due to ceiling angles and a larger surface area.
3. The recessed center provides visual baffling, permitting the use of higher-brightness sources while maintaining high VCP.

Possible luminaire arrangements for both flat and pyramidal shapes are given in Fig. 16.27. In addition to the design flexibility available, an electrified track can be integrated into the system runners to supply both the lighting fixtures and power poles.

## 16.24 LIGHTING AND AIR CONDITIONING

The reduction of lighting power densities to below 2 $W/ft^2$ in all but special areas considerably reduces the impact of lighting-generated heat on a building's HVAC system. In non-air-conditioned buildings, the lighting heat contribution is partially applicable to building heating. Fixture efficiency is directly affected by its temperature. Fluorescent units operate at an optimum temperature of 77°F. Temperatures above and below this decrease output and fixture efficiency. Thus, heat removal from units is desirable even at low lighting-energy levels. The most effective method of fixture heat removal is duct connection to the unit itself. This method, however, is relatively expensive and immobilizes the fixture. Alternatively, the plenum can be exhausted

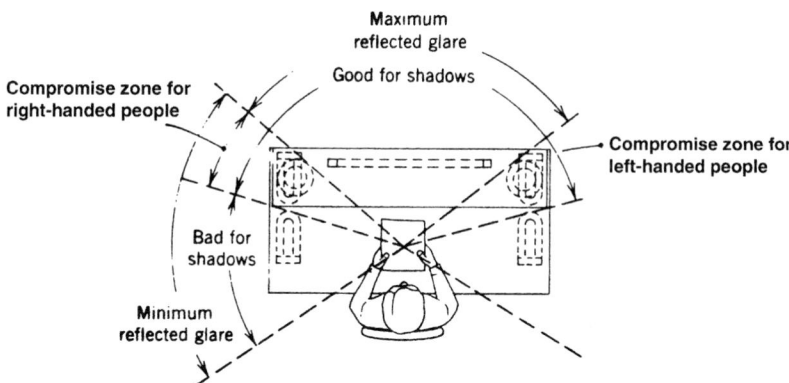

**Fig. 16.26** *Furniture-mounted task lighting can create severe reflected glare or hand shadows if the luminaire location is not selected carefully. Further constraints are location in elevation—for shielding (needs to be low) and for good distribution (needs to be high), appearance considerations, and compatibility with a worker's physical movements. Light source position adjustability removes most user objections to this type of task lighting.*

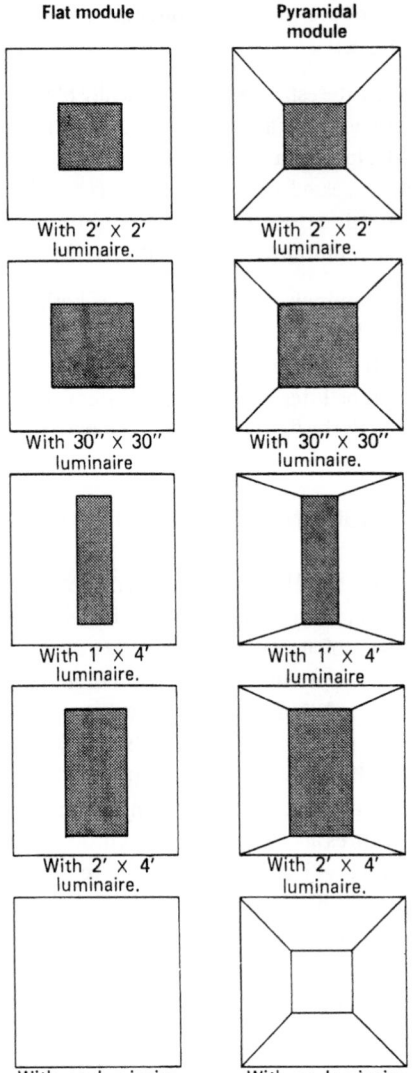

**Fig. 16.27** *Various configurations of 5-ft² modules.*

with air passing over the fixtures, picking up excess heat. These details are essentially part of HVAC system design and are covered in Part II.

## INDUSTRIAL LIGHTING

### 16.25 GENERAL INFORMATION

In industrial lighting, the primary consideration is economic. Given acceptable standards of comfort and safety for the working staff, additional costs for lighting must be self-justifying economically. In one case, a good lighting installation was improved at considerable cost. Production jumped 15%, of which 3% was sufficient to amortize the cost of the lighting alteration. In another case, an outlay for new inspection lighting reduced product failures and proved economically sound. In a third, improved lighting reduced accidents, improved employee morale, and consequently improved production. The cases studied are far too numerous to mention; general principles are adduced instead. For control strategies, see Sections 15.13–15.17.

### 16.26 LEVELS AND SOURCES

Illuminance requirements are detailed in IESNA RP-7-2001, *Lighting Industrial Facilities.* The standard describes in detail a method for prescribing illuminance for specific industrial occupancies and tasks. Where levels higher than 50 to 75 fc are required, general illumination must be supplemented by local illumination. Industrial facilities lend themselves readily to daylighting because many are one-story structures. Thus, roof monitors, skylights, and clerestories are readily applicable and extremely desirable. However, inasmuch as industrial facilities are frequently sited in industrial areas with attendant heavy atmospheric soot and dirt, a frequent cleaning and maintenance program is necessary if the LLF is to be kept at reasonable levels. This observation is also applicable to indoor lighting systems.

Light sources for industrial applications should exhibit high efficacy, good lumen maintenance, and long life. Of the sources available today, fluorescent and HID lamps meet these criteria. Induction and sulfur lamps may join this group after additional development and field testing. Where color is not critical, HPS is the recommended source. Adaptation to its warm yellow color is rapid, and if it is mixed with metal halide or mercury sources, no problem should be encountered. HID lamps are easier to maintain, store, clean, and relamp than fluorescent lamps and have equal or better efficacy, but have the disadvantages of delay and lower output on restrike. Because of their relatively small size, HID sources are used in focusing reflectors that produce intensity (cp) distribution characteristics designed for specific illumination objectives. Thus,

HID sources are generally used for high-bay (>25 ft) and medium-bay (15–25 ft) installations. For low-bay lighting, industrial reflector fluorescent luminaires and low-bay HID reflectors are both applicable.

One of the most common industrial lighting tasks is warehouse aisle lighting that must provide adequate vertical surface illumination on racks on both sides of an aisle. The required cp distribution to perform this task efficiently is a modified batwing curve. This is more easily accomplished in low-bay lighting with a continuous row fluorescent installation than with discrete HID units.

## 16.27 INDUSTRIAL LUMINANCE RATIOS

For reasons explained at length in preceding sections, luminance ratios must be controlled. Recommendations are given in Table 11.11. In many situations it is difficult to control the surrounding brightness. Ceilings, which frequently are covered with piping, ducts, and other equipment, should be light. Therefore, this mechanical equipment must be painted with matte, light unsaturated colors; maintenance and cleaning must be good; and fixtures should have an upward component of light to avoid more than a 20:1 ratio of task-to-ceiling luminance.

Use of bright saturated colors for general surface painting should be avoided because they draw attention and frequently have special significance. In addition to color-coded piping (banding is preferable), red frequently means fire equipment; green, first aid; orange, danger; and so on. White is also to be avoided, being excessively bright and susceptible to dirt. Recommended minimum reflectances are:

| | |
|---|---|
| Ceiling | 75–85% |
| Walls | 40–60% |
| Equipment | 25–45% |
| Floors | 20% |

## 16.28 INDUSTRIAL LIGHTING GLARE

The problem of direct glare can be acute in low-bay installations, and that of reflected glare in high-bay designs, when either uses a point source. One method of reducing direct glare is the use of low-brightness prismatic lens units with a black reflec-

tor behind the lens. Methods of minimizing veiling reflection from all sources were discussed previously.

## 16.29 INDUSTRIAL LIGHTING EQUIPMENT

The cost of maintenance increases with labor rates. For this reason, high-quality lighting equipment yields the lowest owning and life-cycle costs. For instance, the cost of replacing a ballast for HID lighting units frequently exceeds the cost of the ballast. It is thus obvious that it is more economical to utilize long-life, high-quality ballasts, particularly where luminaires are not readily accessible.

Other suggestions for lowering costs, both initial and operating, include using ventilated luminaires that tend to be self-cleaning by convection (Fig. 16.28) in addition to giving the needed upward light component, using bus-mounted fixtures for rapid installation and repair (see Fig. 27.12), using lowering mechanisms on high-bay units to avoid catwalk or platform relamping with a concomitant *extremely* high cost, using fixtures arranged for "stick" relamping from the floor in medium- and low-bay work, and generally incorporating modern equipment into the plant.

Proper maintenance is of paramount importance in industrial facilities because of the prevalence of dirt, vibration, and rough service. Maintenance includes cleaning, relamping, inspection, and preventive maintenance. Relamping on a burnout basis is extremely uneconomical because of disruption of production and lowered production due to lumen

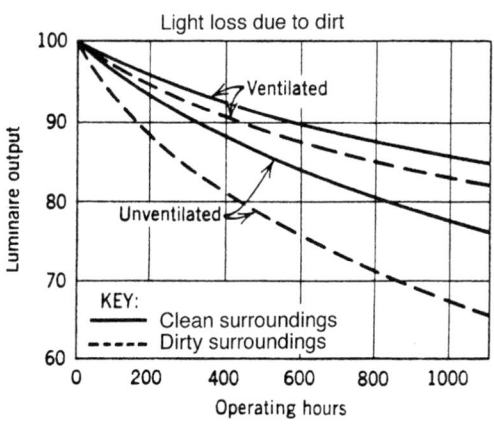

**Fig. 16.28** *Graph demonstrating the advantage of ventilated fixtures.*

depreciation before burnout. Relamping should be done on a planned group basis. Similarly, if the specific facility has a high dirt accumulation rate, cleaning must also be done on a planned group basis rather than only at relamping time.

Ballast noise, including the high levels of HID ballasts, is not usually a factor in industrial facilities because of high ambient noise. In relatively quiet installations and/or where fluorescent fixtures are mounted a short distance above a work bench, as in inspection and fine assembly, this is not true and electronic ballasts are needed.

## 16.30 VERTICAL-SURFACE ILLUMINATION

In industrial facilities more than any other occupancy, the illumination of vertical surfaces is crucial. This is a result of the nature of the work: machines, storage, gauges, and so on all require high vertical-surface illuminance (Fig. 16.29). The illuminance on a vertical surface is the result of the horizontal component of the lighting. This is

$$fc = \frac{cp}{D^2} \sin \theta = \frac{cp \times R}{D^3} = \frac{cp \times \cos^2\theta \times \sin \theta}{H^2}$$

To maximize the horizontal component, we set to zero the derivative of fc with respect to $\theta$. Thus

$$\frac{dfc}{d\theta} = \frac{cp}{H^2} (-2\cos\theta \times \sin^2\theta + \cos^3\theta) = 0$$

or

$$2 \sin^2 \theta = \cos^2 \theta$$

$$\tan^2 \theta = 1/2$$

$$\tan \theta = 0.707 = \frac{R}{H}$$

$$\theta \cong 35°$$

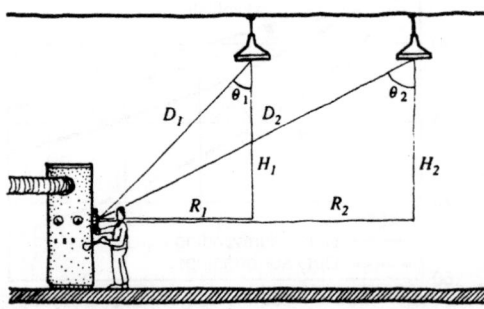

**Fig. 16.29** *Vertical surface illuminance is maximum where θ is approximately 35° (see derivation in text).*

Therefore, maximum vertical illuminance (illumination resulting from the horizontal-lighting component) is obtained when the angle between the fixture's vertical axis and the work is approximately 35°. Hence, we should select a fixture whose candlepower distribution curve demonstrates a high value at that angle. Of course, the derivation is for a single location and fixture. For good vertical and angular illumination over a large area, arrange fixtures with considerable overlap.

## 16.31 EMERGENCY LIGHTING

Emergency lighting is required when the normal lighting is extinguished, which can occur for any of three reasons:

1. General power failure
2. Failure of the building's electrical system
3. Interruption of current flow to a lighting unit, even as a result of inadvertent or accidental operation of a switch or circuit disconnect

As a result of the third reason, sensors must be installed at the most localized level—that is, at the lighting fixture (voltage sensor) or in the lighted space (photocell sensor).

### (a) Codes and Standards

Because emergency lighting is a safety-related item, it is covered by various codes, several of which may have jurisdiction. In addition, there are widely accepted technical society and industry standards whose recommendations normally exceed the minima required by codes.

1. *Life Safety Code (NFPA 101, 2003)*. This code defines the locations within specific types of structures requiring emergency lighting and specifies the level and duration of the lighting.
2. *National Electrical Code (NFPA 70, 2005)*. This code deals with system arrangements for emergency light (and power) circuits, including egress

and exit lighting. It discusses power sources and system design.

3. *Standard for Health Care Facilities (NFPA 99, 2005).* This code deals with special emergency light and power arrangements for these facilities.

4. *OSHA regulations.* These are primarily safety oriented and, in the area of emergency lighting, discuss primarily exit and egress lighting requirements.

5. *Industry standards.* These include the publications of the IESNA and the IEEE, in particular *Recommended Practice for Emergency and Standby Power Systems for Industrial and Commercial Applications* (IEEE Standard 446, 1995).

Because codes and standards are constantly being revised and updated, the designer for an actual project must determine which codes have jurisdiction, obtain current editions, and design to fulfill their requirements. The following material provides general information and focuses on good practice but is not intended to take the place of applicable construction and safety codes.

## (b) Minimum Illumination Levels and Duration of Emergency Lighting

Most codes and authorities require a *minimum average illuminance at floor level,* throughout the means of egress, of 1.0 fc (10 lux). This is considered sufficient, after eyes accommodate, to permit orderly egress. No point along the path of egress should have an illuminance of less than 0.1 fc (1 lux), and the maximum-to-minimum ratio of illuminances along the egress path should not exceed 40:1. The language of the codes is precise, and its consequences must be completely understood in order to properly design an emergency lighting system.

The *Life Safety Code* relates to life safety—that is, safe egress during an emergency that often involves fire. In such instances, smoke normally obstructs vision at eye level, and it is therefore extremely important that the required illumination be available at floor level. Ceiling-mounted emergency lighting may not illuminate the floor in smoky areas and may even worsen a situation by creating a bright, fog-like condition. It is therefore widely recommended that adequate egress lighting be provided at *baseboard level.* Some codes already mandate

baseboard egress lighting. The requirement that the *average minimum* illuminance be 1.0 fc (10 lux) along with the mandated minimum of 0.1 fc (1 lux) at any point and a maximum 40:1 ratio of maximum to minimum illuminance, should effectively eliminate faulty design that utilizes several widely separated bright sources.

Other authorities recommend that escape routes be lighted to not less than 1% of their normal illuminance, but in no case less than 0.5 fc (5 lux). Where the 1% of normal requirement recommendation falls below 1 fc (10 lux) (normal illuminance less than 100 fc), we would abide by the 1-fc minimum required by the *Life Safety Code.*

Duration of the 1-fc (10 lux) level of emergency lighting is normally specified to be a minimum of 90 minutes (for egress) and 0.6 fc thereafter. Facilities that cannot be evacuated quickly require higher levels for indefinite periods.

## (c) Central Battery Emergency Lighting Systems

Sections 26.39 and 28.21 cover emergency power systems, including generators and central battery systems. Generators are normally arranged to replace the utility during outages, using the building's normal electrical distribution system, with the difference that only essential loads remain connected. This obviously does not meet the requirement that emergency lighting be supplied on power failure, *even if the fault is local* (i.e., within the building).

Central battery systems are of three types:

1. Those supplying low-voltage dc (6 to 48 V) to dedicated emergency lighting units (Fig. 16.30*a*). This type of central battery installation consists of a battery, charger, and switching equipment. The emergency lighting units are generally not part of the normal lighting system, consisting instead of dedicated incandescent fixtures. The system has the same limitation as a generator, and therefore is not applicable to most emergency lighting situations except when equipped with multiple downstream sensors. As such sensing would be highly uneconomical, this system is limited to a single (large) space with common outage sensing. This arrangement has the further disadvantage that

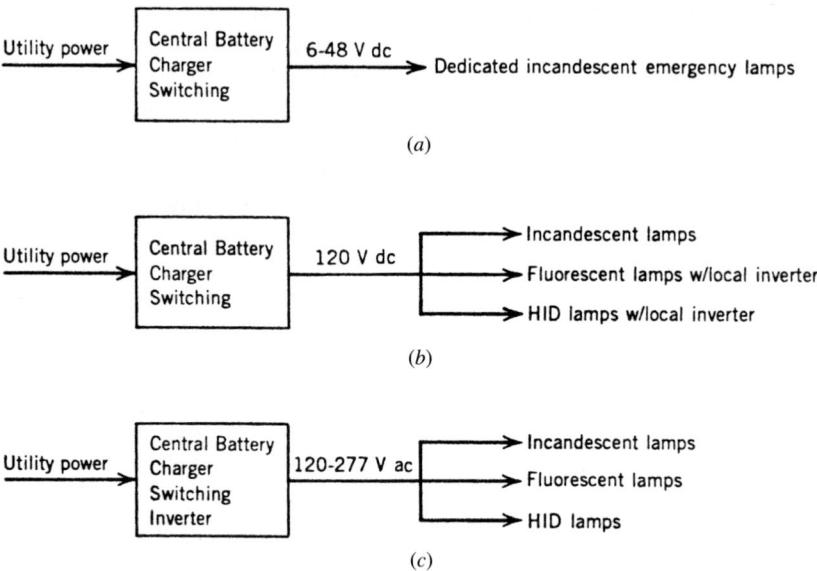

**Fig. 16.30** *Schematic diagrams of an emergency power system using central batteries. In (a) the emergency system is completely separate from the normal lighting system and utilizes dedicated lighting units. (b) This arrangement furnishes dc to the luminaire, where it is inverted to ac (except for incandescent lamps). The fixtures are utilized during both normal and emergency operation. As switching is instantaneous, HID lamps are not extinguished. (c) An arrangement that is similar to (b) except that a central inverter supplies ac continuously. Here, too, the lamps are used in both normal and emergency modes.*

the emergency lighting units may obtrude on the building's decor and architecture inasmuch as they are not part of the normal lighting system. Also, because the lighting units are isolated devices, it is difficult to obtain a satisfactory degree of emergency illumination uniformity, as explained previously (see also Section 16.31*e*).

2. Systems supplying 120-V dc (Fig. 16.30*b*). This arrangement supplies line voltage dc to emergency lighting fixtures, which are also used as part of the normal system. HID and fluorescent luminaires are furnished with integral inverters to change the incoming dc to ac. This arrangement has the advantage of a central, well-maintained battery supply but the disadvantage of requiring a full dc distribution system.

3. Systems supplying line voltage ac (Fig. 16.30*c*). Here the central unit contains a battery, charger, central dc/ac inverter, and switching equipment. Such a system can supply power to incandescent, fluorescent, and HID systems in spaces where a single voltage-loss sensing point is suf-

ficient. The emergency lighting units are part of the normal lighting system. This arrangement is most frequently used where a distributed system with local batteries (discussed later) would not be readily applicable due to the size of the load, as would be the case with high-wattage HID luminaires. A typical 10-kVA on-line UPS cabinet designed to supply emergency ac power to HID lamps measures approximately 40 in. W × 50 in. H × 20 in. D and weighs about 1000 lb.

## (d) Distributed (Local) Emergency Lighting Arrangements

As a result of the inability of a central system to respond to a localized outage, a frequently used emergency lighting arrangement is one where the source of power as well as all the voltage sensing and switching equipment are installed at the emergency lighting fixture, which is usually part of the normal lighting system. The emergency pack consists of a rechargeable battery and charger, voltage sensing and switching equipment, and, in the case

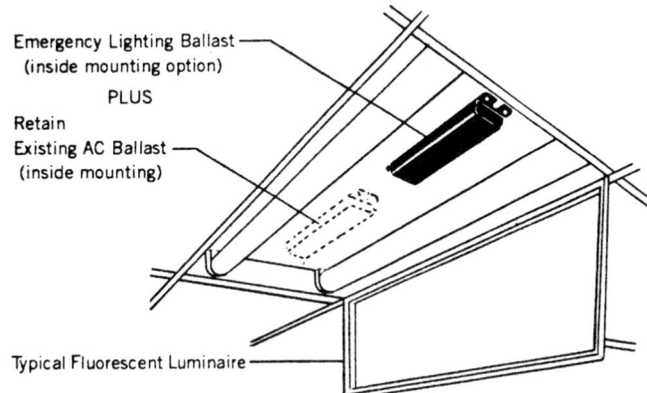

Emergency Lighting Ballast
(inside mounting option)

PLUS

Retain
Existing AC Ballast
(inside mounting)

Typical Fluorescent Luminaire

**Fig. 16.31** *The emergency lighting ballast package contains a 90-minute-capacity high-temperature battery, inverter, sensing and switching equipment, and electronic ballast circuitry to operate the lamp(s) at full or reduced output. The emergency ballast is energized on loss of normal power. The entire package measures 9½ × 2⅜, × 1½ in. (Courtesy of Big Beam Emergency System, Inc.)*

of fluorescent lamps, an electronic package (ballast) that operates the lamp at high frequency and generally at reduced output (Fig. 16.31). These packages, which are designed to supply the code-required 90 minutes of emergency lighting, are maintenance free for periods of up to 5 years. A second type of distributed emergency lighting unit is the familiar packaged unit with integral incandescent lamps. Several designs are shown in Figs. 16.32 and 16.33.

### (e) Emergency Lighting Design Considerations

The levels and duration discussed in Section 16.31*b* refer only to egress lighting. However, general emergency lighting is required to avoid distress and even panic that unfortunately may rapidly ensue. The required emergency illuminance in specific areas should be related to the area's normal illuminance and the degree of hazard in the area. Therefore, we would suggest:

| | |
|---|---|
| Exit area | 5 fc (50 lux) |
| Stair | 3.5–5 fc (35–50 lux) |
| Hazard areas, such as | |
| machinery room | 2–5 fc (20–50 lux) |
| Other spaces | 1.0 fc (10 lux) |

These levels should be essentially uniform. When the illuminance in an interior space drops instantaneously from 30 to 100 fc to 1 to 5 fc, the eyes require up to 5 minutes to fully accommodate.

During this long period, occupants are partially sightless—a condition that lends itself readily to panic. For this reason, bright, spotlight-type heads must be *very carefully arranged.* Otherwise they can create disabling glare and distorting shadows and impede eye accommodation.

Although it is customary to furnish the required emergency illumination from ceiling- or wall-mounted fixtures, consideration must be given to the code requirement, as noted in Section 16.31*b*, that specified emergency lighting levels be maintained at floor level. Because heavy smoke can readily obscure light from overhead fixtures, it is advisable to install some (preferably directional) emergency lighting fixtures near floor level.

### (f) Exit Lighting

Most codes require 5 fc (50 lux) on nonilluminated exit signs, internally illuminated signs with the same visibility as an externally illuminated sign, or self-luminous signs with a specified luminance (usually at least 0.21 cd/m$^2$). Some exit signs are equipped with a battery and controls that provide 1½ hours of illumination on loss of utility power. Others are arranged to illuminate the exit area beneath the sign (Fig. 16.34), and still others are equipped with a flasher and/or an audible beeper that assists people in finding the exit in a light-obscuring, smoke-filled room. Finally, some exit signs are nonelectrical self-illuminating, requiring continuous illumination, and are part of the emergency lighting system.

ILLUMINATION

(a)                                        (b)

**Fig. 16.32** *Commercial emergency lighting units. (a) This decorative unit contains a sealed, maintenance-free lead–acid battery, an automatic charger, and an automatic low-battery-voltage disconnect. The battery is designed to supply a minimum of 90 minutes of operation of fully adjustable 5-W sealed beam lamps. The unit measures 14 in. W × 5 in. H × 5 in. D. (b) Conventional design emergency unit contains a lead–acid or lead–calcium battery, which is housed with its required electronic control and safety equipment in a poly-carbonate housing. The lamps are rated 7.2 W each. The entire unit measures approximately 11 in. L × 12 in. H × 10 in. D; exact dimensions vary with battery type, voltage, and capacity. (Photos courtesy of Dual-Lite, a unit of General Signal.)*

(a)                                        (b)

**Fig. 16.33** *Industrial-type emergency lighting units. (a) This traditional-design unit is available in a variety of battery voltages and uses a pure lead or a lead–calcium maintenance-free battery. The cabinet conforms to a NEMA 1 enclosure. The unit can be arranged to sup-ply additional remote 7.2-W sealed-beam heads. The unit measures 14 in. W × 8 in. D × 19 in. H (over the lamps), and is provided with an external charge indicator light and a test switch. (b) This compact polyester enclosure unit is sealed and gasketed for rugged indus-trial application. It measures approximately 10 in. W × 11 in. D × 14 in. H and contains the usual electronics for control and charging of its lead battery. A charge indicator and a test switch (to simulate power failure) are mounted on the side. (Photos courtesy of Dual-Lite, a unit of General Signal.)*

*Fig. 16.34* Typical application of an exit light with a built-in battery, charger, and controls. Note that the bottom of the unit is designed to illuminate the area immediately in front of the exit. (Photo by B. Stein.)

## 16.32 FLOODLIGHTING

Floodlighting, both interior and exterior, is used extensively for the diverse locations listed in Table 16.4, in addition to the more common sports facility lighting, which is not listed. At the designer's disposal are a variety of sources with respect to output, color, life, efficacy, and wattage (see Chapter 12).

Although a detailed floodlighting design involves complex calculations beyond the scope of this book, it is often sufficient for the designer to utilize a watts per square foot table such as Table 16.4 to determine the approximate floodlighting requirements. The designer should also consult the unit power allowances for exterior lighting listed in ANSI/ASHRAE/IESNA Standard 90.1.

Thus, if one is concerned with lighting a self-service parking lot at a neighborhood shopping cen-

ter, and metal halide is selected, Table 16.4 tells us that approximately 0.055 $W/ft^2$ will suffice. This is well within the limits of Standard 90.1. If the lot is $200 \times 500$ ft ($100,000$ $ft^2$), then $0.055 \times 100,000$ $= 5500$ W is required. This figure is a good first estimate. Obviously, the actual power level depends on the specific equipment and design. This may vary from the initial estimate by $\pm 20\%$.

The arrangement and choice of equipment must be determined before the problem can be considered solved. Considerable assistance on this score can be obtained from either the lighting engineer involved or from representatives of equipment manufacturers.

Although most floodlight installations use a single type, the installation shown in Fig. 16.35*b* used a combination of metal halide and mercury to obtain the desired effect.

## 16.33 STREET LIGHTING

Although detailed street-lighting calculations and design considerations are beyond the scope of this book (see appropriate IESNA standards), a few remarks are in order. New installations now use HID sources almost exclusively. The low efficacy and short life of incandescent sources and the bulkiness of linear fluorescents make them obsolete. Furthermore, high street-lighting levels reduce vandalism and crime, improve night merchandising, and add to an area's attractiveness. A typical walkway luminaire is shown in Fig. 16.36.

## 16.34 LIGHT POLLUTION

An oft-neglected corollary to the principle of placing light where it is required is not to place light where it is not required. *Light pollution* is frequently defined as unwanted light in public places, whereas *light trespass* is the intrusion of unwanted light on private property. The latter includes light intrusion in windows and on private property, whereas the former covers excessive brightnesses everywhere, plus stray light that finds its way into the night sky to the distress not only of astronomers, but also of anyone who simply wishes to enjoy the beauty of a star-filled sky. At this writing, no standards exist

**TABLE 16.4 Lighting Application Guide**

| Application | Minimum Footcandles Maintained[a] | Watts per Square Foot Generally Required | | | |
|---|---|---|---|---|---|
| | | Tungsten–Halogen | Mercury | Metal Halide | High-Pressure Sodium |
| **Automobile Parking** | | | | | |
| Attendant parking | 2 | 0.38 | 0.17 | 0.11 | 0.075 |
| Industrial lots | 1 | 0.13–0.15 | 0.06–0.07 | 0.037–0.044 | 0.026–0.03 |
| Self-parking lots | 1 | 0.13–0.15 | 0.06–0.07 | 0.037–0.044 | 0.026–0.03 |
| **Shopping Centers** | | | | | |
| Neighborhood | 1 | 0.13–0.19 | 0.06–0.09 | 0.037–0.055 | 0.026–0.038 |
| Average commercial | 2 | 0.26–0.3 | 0.12–0.135 | 0.075–0.087 | 0.052–0.06 |
| Heavy traffic | 5 | 0.65 | 0.29 | 0.19 | 0.13 |
| **Automobile Sales Lots** | | | | | |
| Front row (front 20 ft) | 50 | 10. | 4.5 | 2.9 | 2.0 |
| Remainder | 10 | 1.5–1.8 | 0.68–0.81 | 0.44–0.52 | 0.3–0.36 |
| **Building** | | | | | |
| Construction | 10 | 1.5–1.8 | 0.68–0.81 | 0.44–0.52 | 0.3–0.36 |
| Excavation | 2 | 0.26–0.3 | 0.12–0.14 | 0.075–0.09 | 0.052–0.06 |
| **Buildings Up to 50 ft High** (Adj. Area Light / Dark) | | | | | |
| Light surfaces | 15 / 5 | 3.3  1.2 | 1.5  0.54 | 0.96  0.35 | 0.66  0.24 |
| Medium light surf. | 20 / 10 | 4.3  2.2 | 1.94  1.0 | 1.25  0.64 | 0.86  0.44 |
| Dark surfaces | 50 / 20 | 10.0  4.3 | 4.5  1.94 | 2.9  1.2 | 52.0  0.86 |
| **Billboards and Signs** (Adj. Area Light / Dark) | | | | | |
| Good contrast | 50 / 20 | 10.0  4.3 | 4.5  1.94 | 2.9  1.25 | 2.0  0.86 |
| Poor contrast | 100 / 50 | 20.0  10.0 | 9.0  4.5 | 5.8  2.9 | 4.0  2.0 |
| **Protective Lighting** | | | | | |
| Gates and vital area | 5 | 1.2 | 0.54 | 0.35 | 0.24 |
| Building surrounds | 1 | 0.15–0.19 | 0.07–0.09 | 0.044–0.055 | 0.03–0.04 |
| **Roadways** | | | | | |
| Along buildings | 1 | 0.24 | 0.11 | 0.07 | 0.05 |
| Open areas | 0.5 | 0.08–0.1 | 0.036–0.045 | 0.023–0.029 | 0.02 |
| Storage yards | 20 | 3.6–4.3 | 1.6–1.94 | 1.04–1.25 | 0.72–0.86 |
| Storage yards (inactive) | 1 | 0.15–0.19 | 0.07–0.09 | 0.044–0.055 | 0.03–0.04 |
| **Shopping Centers** | | | | | |
| Parking areas (attraction) | 5 | 0.65 | 0.29 | 0.19 | 0.13 |
| Buildings (attraction) | | (See Buildings) | | | |
| Used Car Lots | | (See Automobile Parking) | | | |

[a]All footcandle levels for ground area applications are *horizontal* values.

that define the light levels that constitute an intrusion or a nuisance glare. However, a few simple guidelines can assist the designer in avoiding the creation of a nuisance:

1. Light all exterior vertical surfaces from above, not below, wherever possible. This reduces skylight pollution.
2. Use luminaires with sharp cutoff beyond the illuminated area. Shields can be added to standard luminaires to accomplish this.
3. After an installation is complete, inspect it at night to determine whether any nuisance has been created.

Further information on this subject can be found in IESNA technical memoranda TM-10-00 and TN-11-00, dealing with light trespass.

## 16.35 REMOTE SOURCE LIGHTING

Every experienced lighting designer has at some time felt the need for a remote source luminaire to fill a specific lighting need. Among the most common situations are:

• Display lighting for light- and heat-sensitive objects such as old books, fabrics, drawings and

ILLUMINATION

*(a)*

*(b)*

**Fig. 16.35** *(a) Floodlighted section of a wall surrounding the Old City of Jerusalem, Israel, adjacent to the Jaffa Gate. Light sources are 400-W HPS units giving an average illuminance level of 50 lux. A sodium source was chosen to enhance the yellow-red color of the stone. (b) Church of All Nations, Mount of Olives, Jerusalem, Israel. Floodlight sources are 250- and 400-W mercury and metal halide units, giving an average illuminance of 70 lux. Sources were selected to complement the colors in the mosaic at the top of the façade. (Photos courtesy of City of Jerusalem and J. Stroumsa, Chief Engineer.)*

paintings, and, in general, objects containing organic materials, dyes, and coloring. Of particular importance in this category is the acute sensitivity of such objects to UV radiation, which exists in daylight and light from most electrical sources.

• Installations where relamping is a major logistic and financial problem. These include high-ceiling auditoriums and public assembly spaces

**Fig. 16.36** *A "lollypop" fixture, even if aesthetically pleasing to some, gives poor illumination downward (note the large collar). (Photo by B. Stein.)*

of all sorts, high-bay lighting areas and other difficult-access locations, clean rooms, spaces with security entry limitations, and rooms that cannot tolerate disturbances or interruptions, such as air-traffic control rooms, continuous process manufacturing control areas, and the like.

• Installations where lamp heat is highly undesirable and its removal is difficult, expensive, or both. Among these are store show windows, refrigerated showcases, and conditions where lamp heat is felt by the space occupant(s), such as halogen and other incandescent lighting at workstations and low-ceiling spaces.

• Installations where the presence of electrical wiring is undesirable, such as patient-controlled hospital bed lighting or light sources in devices used by children.

• Spaces classified by the NEC as electrically hazardous. These areas typically use large, heavy, expensive, and inefficient explosionproof lighting.

• Installations where the electric and magnetic fields produced by fluorescent and HID lighting fixtures are unacceptable.

• Applications where the light source must be very small and effectively invisible.

Two basic designs have been developed over the years to fill these lighting requirements: arrangements using optical fibers and those using light guides of various types. Applications of these two techniques do not generally overlap, and they are therefore discussed separately in the following sections.

## 16.36 FIBER-OPTIC LIGHTING

Fiber optics have been in use for years, principally in instruments designed to permit illumination and observation of essentially inaccessible locations. In medical instruments designed for intrusive applications, fiber-optic (FO) cable has the additional advantages of absence of heat and elimination of electrical wiring, both of which could be at the very least uncomfortable and frequently dangerous when inserted into the body.

The physical principle by which an optical fiber conducts light is that of total internal reflection, illustrated in Fig. 16.37. The fiber is constructed of an inner core of transparent material (silica, glass, optical plastics) and an outer coating (cladding) of another material of *lower* refractive index. Because of this difference in refractive indices, light rays within the acceptance cone angle ($2\theta$) are reflected at the interface of the two materials and thus proceed almost unimpeded down the core by a series of 100% reflections. Although there is (theoretically) no loss at the reflection, the core material itself exhibits an optical impedance that varies with the type of material. Ultraclear glass fibers, used in communications, have losses as low as 1 dB per kilometer, whereas the best plastic optical fiber used today has losses in the area of ⅛ dB per *meter*—that is, several hundred times as high. However, because of the short distances involved, these plastic fibers are satisfactory for lighting work (1-dB loss is the equivalent of 8.6% loss per meter, or 2.6% loss per foot).

Air, with a refractive index of 1.0, can also be used as a "cladding" material (unclad core), resulting in a fiber that emits light over its entire length. This light is shown in Fig. 16.37 as light "lost" in the cladding. In practice, a bare core is not used because it would be damaged at supports and dirt accumulation on the surface would result in high losses. Sidelight fibers (lateral mode) are constructed commercially of a core, cladding, and sheathing, all of which are transparent plastic materials with decreasing refractive index from core to sheath. Note also in the figure that the angle $\theta$, and therefore the size of the acceptance cone, increases with the difference in the refractive indices between cone and cladding. It is desirable to have an acceptance cone as large as possible

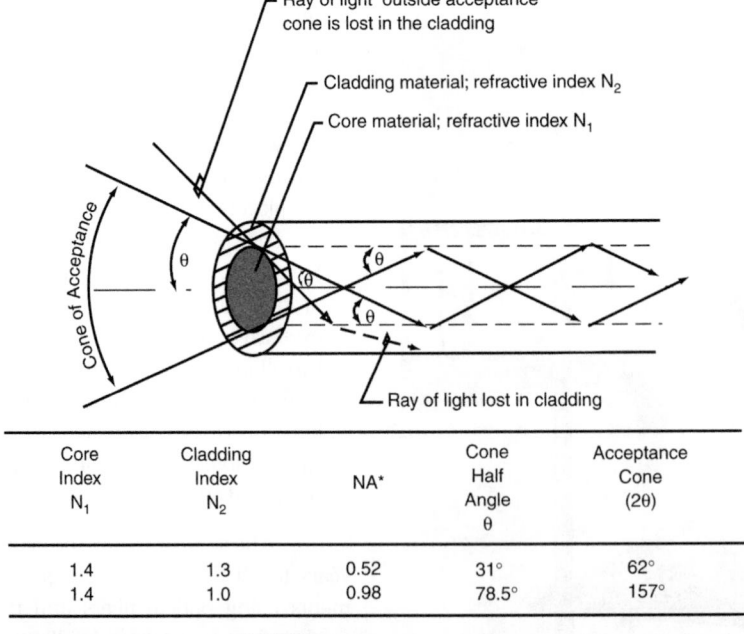

| Core Index $N_1$ | Cladding Index $N_2$ | NA* | Cone Half Angle $\theta$ | Acceptance Cone ($2\theta$) |
|---|---|---|---|---|
| 1.4 | 1.3 | 0.52 | 31° | 62° |
| 1.4 | 1.0 | 0.98 | 78.5° | 157° |

*NA (Numerical Aperture) $= \sqrt{N_1{}^2 - N_2{}^2} = \sin\theta$

**Fig. 16.37** *Schematic representation of the phenomenon of total internal reflection (TIR) in optical fibers (OF). Light rays entering the end of an OF at an angle to the fiber centerline no greater than $\theta$ are totally reflected at the interface between core and cladding. Angle $\theta$ is determined by the difference between the refractive indices of the core and cladding materials and increases as that difference increases. Light rays entering within the acceptance core (solid angle $2\theta$) travel along the fiber; all other light rays are lost.*

**TABLE 16.5 Typical Characteristics of Plastic Optical Fiber Cables**

| Type | Construction | Diameter Fiber/Core (mm) | OD (mm) | NA[a] | Cone Angle (degrees) | Attenuation Percent per Meter |
|---|---|---|---|---|---|---|
| End-light Fiber bundle | See Illustration (a) | 0.15–2.0 | 6–18 | 0.5 | 60 | 4.7 |
| Side-light Fiber bundle | | 0.15–2.0 | 6–14 | 0.5 | 60 | 4.7 |
| End-light Solid core | See Illustration (b) | 3–13 | 5–16 | 0.62 | 76 | 8.2 |
| Side-light Solid core | | 7–13 | 9–16 | 0.57 | 70 | 23.7 |

[a]NA (Numerical Aperture) = $\sqrt{N_1^2 - N_2^2}$ = sin $\Theta$; see Fig. 16.37.

*Note:* Sheathing is opaque for end-light cables and transparent for side-light cables.

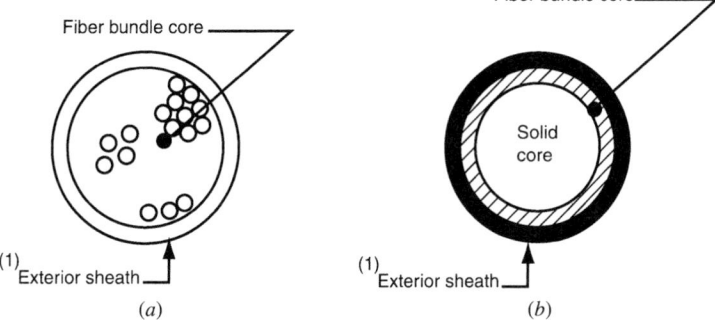

(high numerical aperture) so that maximum light from the source (illuminator) can be introduced (coupled) into the fiber. Glass fibers are generally not used in lighting applications because of their high cost, small light-carrying capacity, and large bending radius. Table 16.5 shows the construction and major characteristics of typical lighting-use optical fiber cables.

## 16.37 FIBER-OPTIC TERMINOLOGY

Because FO cable is becoming common in lighting work, a lighting designer should be familiar with FO terminology. A brief glossary of the most important terms follows.

**Acceptance angle (also acceptance cone).** The maximum solid interior angle of a cone that defines the spread of light rays that will enter a single fiber. As this angle is determined only by the fiber and the cladding materials, it is also the cone of light ray acceptance of a fiber bundle.

**Attenuation.** The degree of light reduction as it travels along the fiber. Because attenuation varies with frequency, the light color exiting an

optical fiber is different from that entering, and the change becomes more pronounced with increasing fiber length.

**Axial mode (also end-light).** The mode of conduction from one end of the fiber (cable) to the other, with no deliberate loss by emission over the cable length.

**Core.** The center of an optical fiber that carries the light.

**Fiber-optic port.** A factory- or field-applied terminating connector that enables coupling of an FO cable to a light source or other item in the system. Glass FO is factory terminated; plastic fibers can be either.

**Glass optical fiber.** The type always used in communication work and seldom in lighting application. It is expensive, requires a large bending radius, and exhibits very low attenuation.

**Illuminator.** The light source in an FO lighting installation. It usually consists of a metal box containing the light source (usually metal halide or halogen) and its accessories, plus required optical devices that collect and concentrate the light from the source and couple it optically to an optical port to which an optical cable is connected. The illuminator may also contain color filters, local and remote-control connections, and, almost always, a fan or blower for forced-air cooling of the source lamp.

**Lateral mode (also sidelight(ing)).** A light conduction and distribution system whereby light is emitted over the entire length of the fiber, simulating neon tubing. This is accomplished by using transparent cladding and sheathing, with refractive indices selected to increase light rays entering (and leaving) the cladding.

**Numerical aperture.** A numerical metric based on the refractive indices of the core and cladding materials, indicating the angle of the acceptance cone (i.e., that portion of the light that is conducted by the fiber core).

**Plastic optic fiber.** The fiber material normally used in lighting work, usually a clear acrylic-type compound as the core, with various plastic compounds as cladding or sheathing, depending on the design and mode of operation. The fibers are classified by size as small core (up to 2 mm in diameter) and large core (up to 20 mm in diameter). Small-core fibers are manufactured in continuous and essentially unlimited lengths. Large-core fibers and sheathed fiber-bundle cables do not usually exceed 150 ft in length.

**Refractive index.** The ability of a light-conducting (nonopaque) material to bend a light ray entering from another medium, expressed numerically.

**Tail.** A single optical fiber or a bundle of fibers extending from an illuminator to an output point. A single fiber separated from a bundle for a specific small, isolated illumination purpose (often decorative) is also referred to as a tail.

## 16.38 FIBER-OPTIC LIGHTING— ARRANGEMENTS AND APPLICATIONS

The development of large, efficient plastic optical fibers coupled with a concomitant price reduction has produced a lighting tool whose applications are limited only by the imagination of designers and manufacturers. In one form or another, optical fibers are applicable where:

- A single remote source can supply a large number of relatively small point-source lights.
- Burial of the light carrier (fiber bundle) in almost any substrate is desired or required.
- The heat, UV content, and electrical fields associated with most sources are absent.
- The presence of electrical wiring and its associated hazards are undesirable or unacceptable.

A few of the many current configurations are described in the following subsections.

### (a) Axial-Mode Linear Devices

In this arrangement, a bundle of fibers is placed in a longitudinal enclosure and individual fibers or small groups of fibers are separated from the bundle(s) and brought out of the enclosure as a light-emitting point (Fig. 16.38a). Most of these *light bars* are custom-made for a specific application, enabling the designer to specify all of the parameters including spacing of points, intensity of each point, dimensions of the bar, individual fiber size and bundle size, color characteristics, and so on. Light bars are used in retail display lighting, accent lighting,

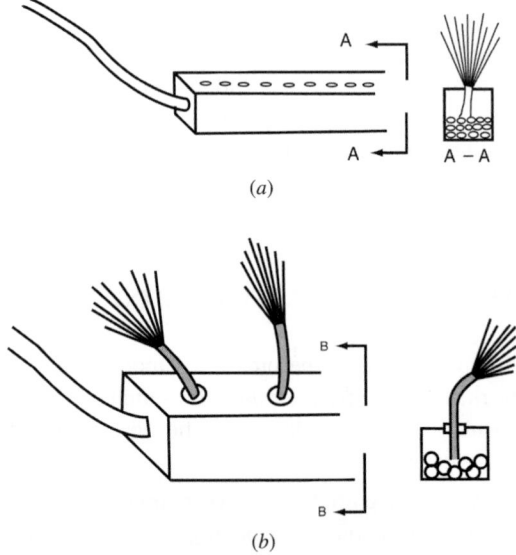

(a)

(b)

**Fig. 16.38** *Linear constructions using end-light fibers. (a) A light bar is simply a box containing multiple tails that are brought out of the box-type container at intervals, usually equal. By selecting fiber size, exit spacing, and illuminator size, a low-intensity linear lighting fixture is created. (b) By using large fibers and/or bundles brought out in adjustable groups, a linear bar with adjustable multihead spotlights can be readily constructed.*

directional devices, decorative applications, and other applications requiring a low-intensity, linear, "sparkling" and/or lighting device.

The essential design of a light bar can be adapted to the use of large-core fibers or multiple fiber bundles to produce a linear string of focusable, medium-intensity spotlights (Fig. 16.38*b*). In lieu of the closely spaced tiny light points of Fig. 16.38*a*, large fibers or bundles can be extracted. They can be used as light sources or can be coupled to optical plastic-fiber rods or other devices. These can then be "aimed" as desired to illuminate individual areas in a cabinet, store window, and the like. Here, too, the units are commonly custom-designed to perform a specific lighting task.

### (b) Axial-Mode Discrete Sources

By using large groups of end-light FO bundles combined to produce a point light source, semiconventional lighting fixtures can be constructed with a variety of common diffusing elements (Fig. 16.39). The advantages of these discrete sources compared to common electrically powered lighting fixtures are the same as those listed previously: absence of heat and electricity, and low maintenance and higher efficacy of high-output lamps in the illuminators. One disadvantage common to all plastic-fiber FO lighting is the necessity of keeping runs short due to the attenuation of the fibers. Lengths in excess of 20 ft are inefficient and therefore uncommon.

### (c) Lateral-Mode Fiber-Optic Lighting

As explained previously, lateral mode fibers emit light throughout their length. This makes them par-

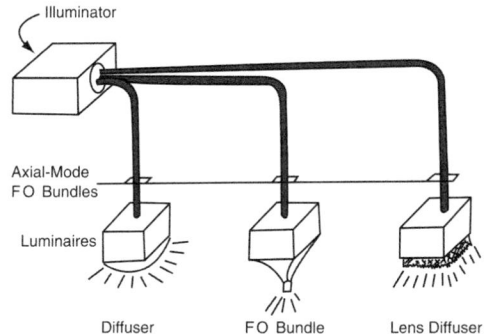

**Fig. 16.39** *Relatively large end-light FO bundles can be utilized as point sources in conventional types of lighting fixtures.*

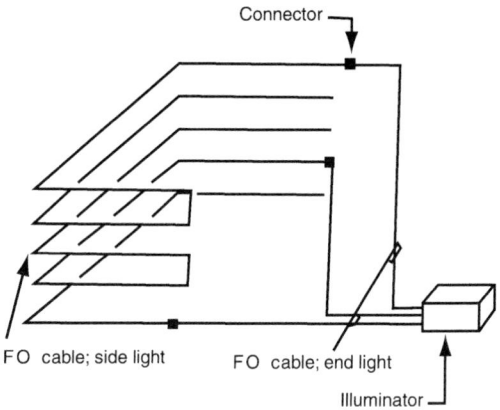

**Fig. 16.40** *Side-light-emitting FO cables are ideal for linear lighting tasks such as stair-edge illumination, under-shelf lighting, and outline lighting of all sorts. The illustrated 3-tail arrangement is representative of this type of FO lighting.*

ticularly suitable for linear lighting tasks such as illuminating stair nosings, path lighting, and all sorts of decorative trim lighting. Figure 16.40 shows a possible arrangement for under-shelf lighting, illumination under stair nosings, and the like. The practicability of burying parts of the cable in concrete is useful in giving the remaining lighted cable a disconnected, floating appearance. Color filters in the illuminator, when applied to multiple cable runs as in Fig. 16.40, can produce dramatic outlining effects.

### 16.39 HOLLOW LIGHT GUIDES

The idea of conducting light from one place to another apparently arose several millennia ago from an architectural desire to provide daylight deep in interior spaces. The interior court is one example of an architectural light guide that is only recognizable as such in multistory buildings. Figure 16.41 illustrates the use of an interior light shaft in a six-story building. Notice the rapid attenuation as daylight descends by multiple reflections, accounting for the increasing size of the windows at lower levels. Indeed, attenuation is the principal problem with light pipes and guides that depend upon multiple internal reflections to conduct light any appreciable distance. The best metallic mirrors available today specularly reflect only about 95% of the impinging light. It is therefore

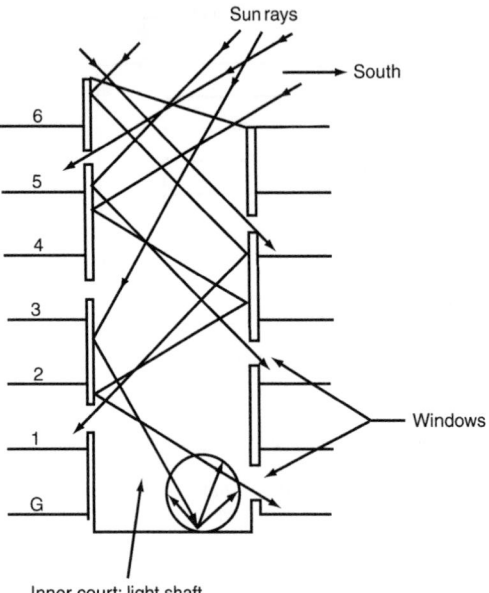

*Fig.* **16.41** *An interior court or light well (shaft) with a reflective wall surface acts as a light guide to introduce daylight at lower floors. Buildings in the Northern Hemisphere are oriented as shown. Due to attenuation of rereflected sunlight beams traveling down the shaft, lower floor windows are larger than those at upper floors to capture more (of the attenuated) light. (Reprinted from NASA Tech. Brief LAR-12333.)*

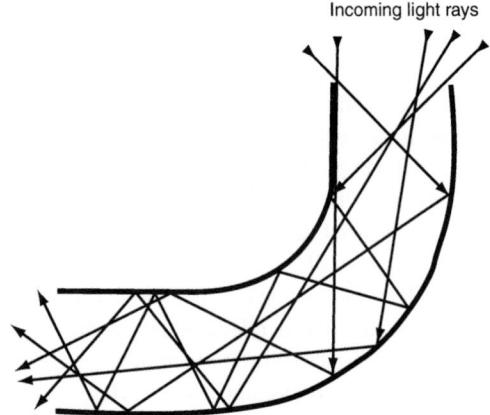

*Fig.* **16.42** *Curvature in a light guide increases the number of reflections that a light ray makes over the guide length, thereby increasing attenuation. (Reprinted from NASA Tech. Brief LAR-12333.)*

relatively easy to calculate the number of 5% loss reflections that reduce the output light from a light pipe with random nondirectional input to an uneconomical level (where the "free" light received is more expensive due to construction and maintenance costs than alternative electric light). Furthermore, open-ended light pipes of this sort (Fig. 16.42) are notoriously difficult to keep clean of dust and surface-reflectance degeneration, so that the 95% initial specular reflectance rapidly decreases to 85% semidiffuse reflection, thereby increasing ray rereflection and consequently overall light attenuation. (Sealing the upper end of the pipe with a transparent medium is not advisable due to the rapid accumulation of dust and condensation.) In order to capture an appreciable amount of sunlight (sky light is insufficient), an open-ended pipe at roof level must be very large. This, in turn, increases construction and maintenance costs and decreases (in the case of a commercial building) the usable/rental area, making the light-guide idea generally impractical except for very short guides. Such com-

mercially available units are essentially small skylights with an elongated, reflective collar.

One solution to the problem of rapid attenuation in a light guide caused by multiple interior wall reflections is to collimate the incoming light. This has the additional advantage of permitting reduction of the light guide's cross-sectional area. Such a system, however, requires a sun-tracking arrangement plus mirrors and lenses in a collimating optical train. A typical arrangement is shown schematically in Fig. 16.43. The increased efficiency of arrangements of this sort is offset by losses in the optical train and increased equipment cost (although the light pipe itself is smaller and therefore cheaper). Systems of this type are in use today, with generally satisfactory results, although system economics is marginal except in arid, sunny climates.

## 16.40 PRISMATIC LIGHT GUIDES

The first major improvement in hollow light guide design occurred in 1981, with the development and patenting of a prismatic rectangular hollow acrylic light guide (Whitehead, 1981, 1982). The patentable novelty of this device is that the facets of the prismatic exterior transparent acrylic walls of the pipe act as total internal reflection "mirrors," thus preventing light from escaping from the transparent box guide. As a result, light is guided

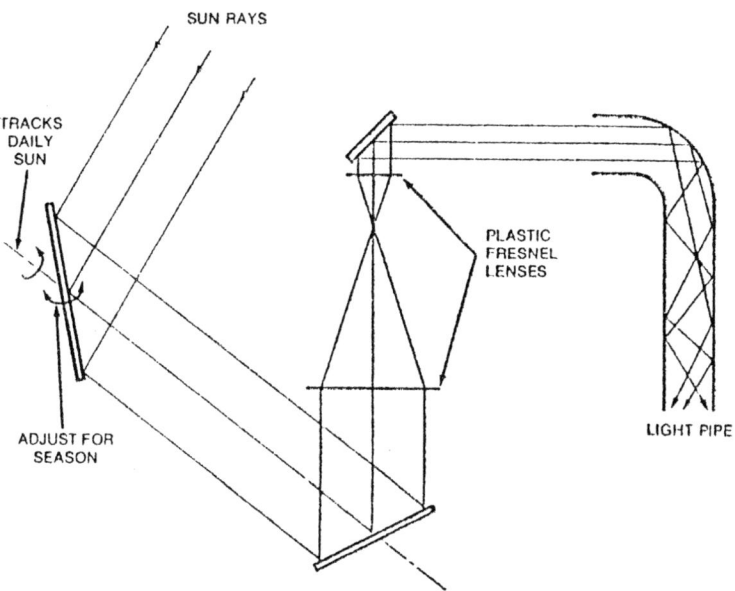

**Fig. 16.43** *Concentration and collimation of sunlight can be accomplished with a sun-tracking mirror and an optical train of mirrors and lenses. Full tracking of the sun requires altitude and azimuth drives on the collection mirror. To reduce costs without severely reducing collection efficiency, a full azimuth drive tracks the sun from east to west daily, while the mirror tilt around a horizontal axis (altitude tracking) is adjusted only seasonally to the mid-season position. (The sun's maximum altitude varies 23.5° from equinox to solstice.) (Reprinted from NASA Tech. Brief LA12333.)*

along the length of the device with very low losses, because reflection at the prismatic walls is theoretically perfect and losses in the air space within the guide are very small. In practice, it was found that at each reflection about 2% of the light was lost by absorption and, because of microscopic imperfec-

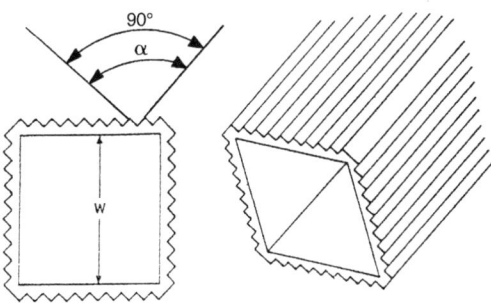

**Fig. 16.44** *Views of the hollow, clear acrylic, prismatic light guide developed by L.A. Whitehead. The cross section measures 13-cm square, and the prism angle α is 90°. (Reprinted from L.A. Whitehead et al.,* A New Efficient Light Guide for Interior Illumination, *1981.)*

tions in the prisms, about 6% escaped through the wall. This 6% "loss" converted the light guide into a long, rectangular lighting fixture emitting light *uniformly* over its entire length. By placing a mirror at the end of the guide, all the light reaching the end is reflected back into the guide, thereby increasing the light output over its length. Overall efficiency of this extended "lighting fixture" is high; this fact, combined with very low maintenance, produced a highly desirable lighting product. Its first application was in an electrically hazardous area previously illuminated by explosionproof lighting fixtures (Fig. 16.44).

## 16.41 PRISMATIC FILM LIGHT GUIDE

In 1988, the 3M Company developed a thin plastic prismatic film that utilizes the principle of total internal reflection at the prism face. Tests have shown that this material loses only about 1% of the light at each reflection. As a result, tubular hollow light guides using this material can extend

ILLUMINATION

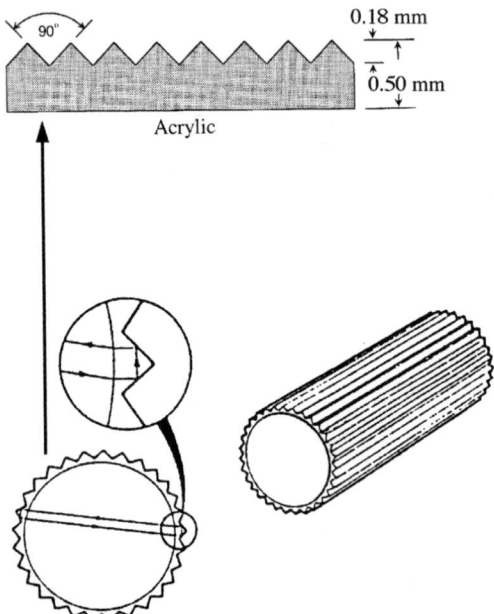

**Fig. 16.45** *Circular hollow light guide. If the prismatic plastic film is formed into a cylindrical shape as shown, it will act as a light guide with very low losses. (Reprinted from a paper presented at Globalcon '96 by K.G. Kniepp of the 3M Company.)*

for several hundred feet if the input light is collimated (Figs. 16.45 and 16.46). When used as an extended lighting fixture, this light guide exhibits good photometric characteristics.

Prismatic light pipes have all the advantages of FO lighting but fewer limitations, although in practice the two systems do not compete because they address entirely different needs. In addition to having all the advantages of remote-source FO lighting, including absence of heat, minimum UV radiation, absence of electrical wiring and the associated electromagnetic fields and electrical wiring hazards, and usability at any temperature, interior or exterior, prismatic light guides (and lighting fixtures) are constructed to handle very large quantities of light and do not produce color distortion (as do optical fibers). As a result, a luminous light pipe (light fixture) can be coupled to a very-high-output source such as an arc lamp, Xenon lamp, or the relatively new sulfur lamp, and the pipe can be used to illuminate large areas, either interior and exterior. The economies in wiring, luminaires, installation, and maintenance, in addition to the high efficacy of these large

sources, more than offset the cost of the prismatic material.

Because of the very high light conduction efficiency of optical lighting film (OLF), using it as a lighting fixture requires special techniques and materials to extract light from the pipe. One method is to apply another prismatic material to the inside of the light pipe with a different prism angle that causes incident light to change direction and exit the pipe. When this material is placed judiciously, selected areas of the pipe circumference emit light, in effect producing a longitudinal lighting fixture with the desired directionality. Typical lighting applications of prismatic film light-pipe fixtures are high-bay industrial and commercial installations, large exterior signs, tunnel illumination, highway signs, continuous rail-type guidance illumination, and exterior architectural building lighting. Figure 16.47 shows a circular prismatic light guide designed for exterior use, with several configurations and the associated photometric data. Figure 16.48 shows an application of the prismatic light-guide fixture.

## 16.42 REMOTE SOURCE STANDARDS AND NOMENCLATURE

Due to the relative newness of long, remote-source tubular lighting, industrywide standards do not exist and even nomenclature varies. Clear differentiation between devices intended primarily to carry light from one location to another and devices intended to illuminate along their length does not exist. Nor is terminology differentiated between light guides operating on the principle of internal reflection from specular surfaces and prismatic light guides. All these devices are variously and interchangeably referred to as

- Hollow light guide
- Light pipe
- Hollow light pipe
- Prismatic light guide
- Prismatic light pipe
- Hollow prismatic light guide
- Remote source hollow light guide

and various other names. See NEMA LSD 4-1999: *Glossary of Terms Pertaining to Remote Illumination Systems* for further information.

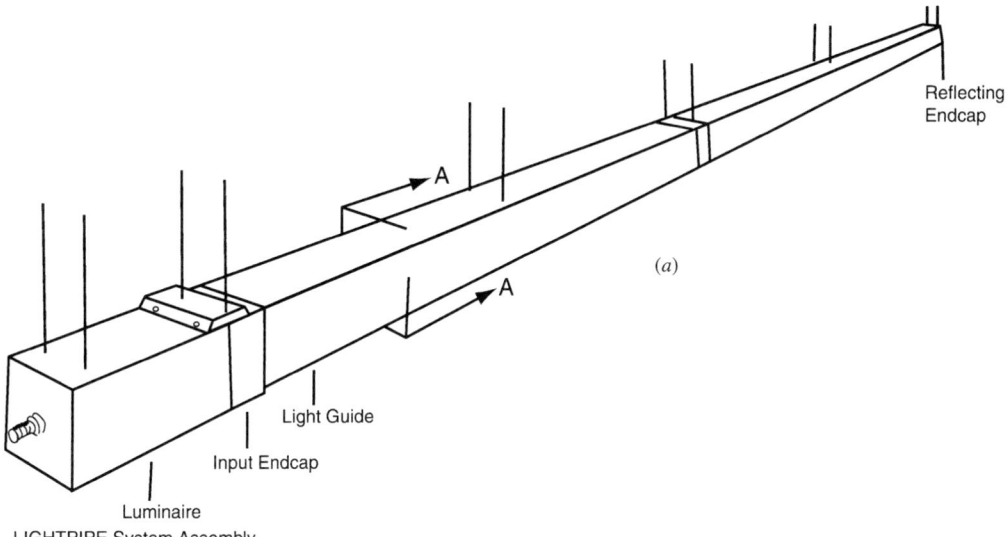

(a)

Reflecting
Endcap

Light Guide

Input Endcap

Luminaire
LIGHTPIPE System Assembly

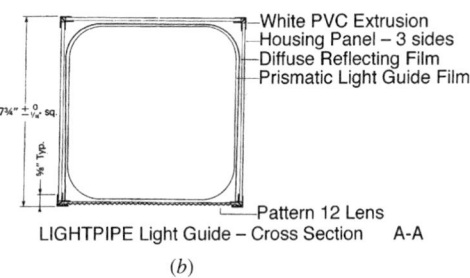

White PVC Extrusion
Housing Panel – 3 sides
Diffuse Reflecting Film
Prismatic Light Guide Film

7¾" ± ⁰⁄₁₆ sq.

¼" Typ.

Pattern 12 Lens
LIGHTPIPE Light Guide – Cross Section   A-A

(b)

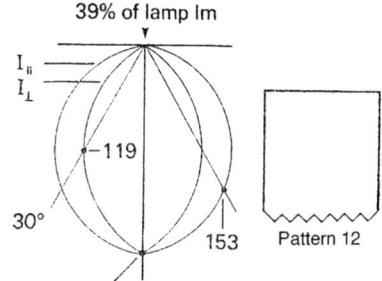

39% of lamp lm

$I_{\shortparallel}$

$I_{\perp}$

—119

30°

153

Pattern 12

200 cd / 1000 lamp lm

| ρcc | 80 | | 50 | | 30 | |
|---|---|---|---|---|---|---|
| ρw | 50 | 30 | 50 | 30 | 50 | 30 |
| RCR | | | | | | |
| 0 | .46 | .46 | .43 | .43 | .41 | .41 |
| 1 | .41 | .40 | .39 | .38 | .37 | .37 |
| 2 | .37 | .35 | .35 | .33 | .34 | .32 |
| 3 | .33 | .31 | .32 | .30 | .31 | .29 |
| 4 | .30 | .27 | .29 | .26 | .28 | .26 |
| 5 | .27 | .24 | .26 | .23 | .25 | .23 |
| 6 | .25 | .22 | .24 | .21 | .23 | .21 |
| 7 | .22 | .19 | .21 | .19 | .21 | .19 |
| 8 | .20 | .17 | .19 | .17 | .19 | .17 |
| 9 | .18 | .15 | .18 | .15 | .17 | .15 |
| 10 | .17 | .14 | .16 | .14 | .16 | .13 |

**CU Table — One Side Emitting**

**Fig. 16.46** *Details of a rectangular light-guide fixture. (a) The entire lighting unit comprises a "luminaire" at one end containing the light source (250- or 400-W metal–halide lamp) and its accessories connected to the hollow prismatic light guide. The end of the guide is sealed with a mirror that reflects the light back into the light guide. (b) The light guide itself is enclosed on three sides and equipped with a prismatic lens diffuser on the open (bottom) surface through which light is emitted. This specific design is usable to a length of 40 ft. Longer units are equipped with light sources at both ends. (c) Photometric characteristics of the luminaire-light guide assembly. (Courtesy of TIR Systems, LTD.)*

(c)

ILLUMINATION

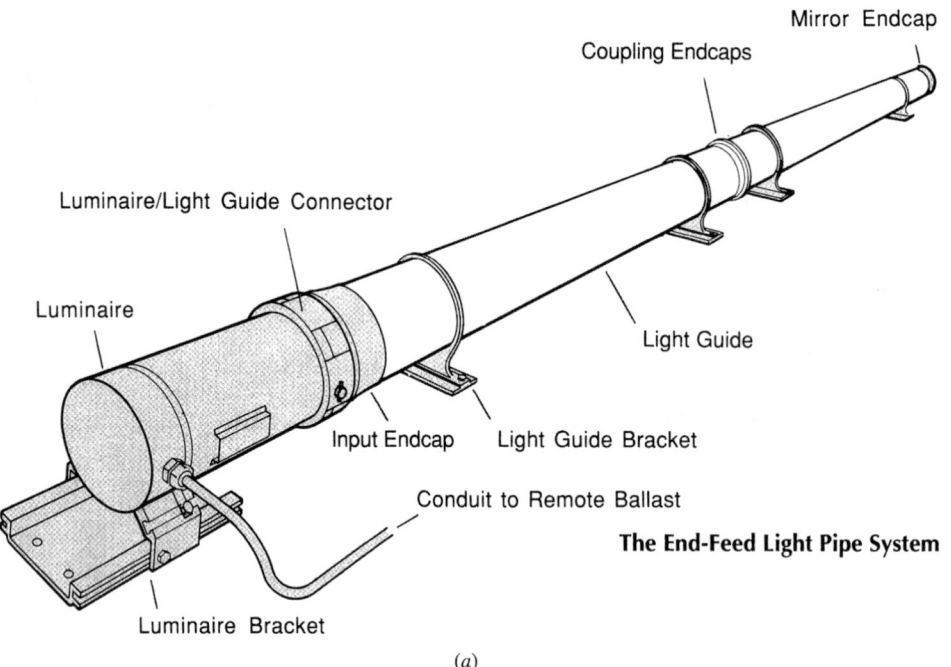

Mirror Endcap

Coupling Endcaps

Luminaire/Light Guide Connector

Luminaire

Light Guide

Input Endcap    Light Guide Bracket

Conduit to Remote Ballast

**The End-Feed Light Pipe System**

Luminaire Bracket

(a)

**Fig. 16.47** Details of a 7-in.-D hollow prismatic end-feed light-guide fixture. (a) The entire unit consists of an opaque "luminaire" section containing a 250-W metal halide lamp connected to a 7-in.-D acrylic tube up to 44 ft in length containing prismatic optical film and terminating in a reflective (mirrored) end cap. (b) The portion of the tube circumference that emits light (uniformly) is controlled by placement of a diffuse reflecting film between the inner prismatic film and the outer transparent acrylic envelope (see Fig. 16.46b). The reflective film is shown as the black portion of the tube circumference. (c) Photometric data for an end-fed system with a single 250-W metal halide source. (Courtesy of TIR Systems, Ltd.)

| Emitting Sector | Luminous Intensity Curve |
|---|---|
| 90° Emitting Sector | |
| 120° Emitting Sector | |
| 180° Emitting Sector | |
| 240° Emitting Sector | |
| * Available for end-feed system only | |

(b)

ILLUMINATION

# CU Tables — End-Feed System

| $\rho_{cc}$ | 70 | | 50 | |
|---|---|---|---|---|
| $\rho_w$ | 50 | 30 | 50 | 30 |
| RCR | | | | |
| 0 | .34 | .34 | .32 | .32 |
| 1 | .30 | .29 | .29 | .28 |
| 2 | .26 | .25 | .25 | .24 |
| 3 | .24 | .21 | .23 | .21 |
| 4 | .21 | .19 | .20 | .18 |
| 5 | .19 | .16 | .18 | .16 |
| 6 | .17 | .14 | .16 | .14 |
| 7 | .15 | .13 | .15 | .12 |

**CU Table - 90° Emitting**

| $\rho_{cc}$ | 70 | | 50 | |
|---|---|---|---|---|
| $\rho_w$ | 50 | 30 | 50 | 30 |
| RCR | | | | |
| 0 | .36 | .36 | .34 | .34 |
| 1 | .30 | .29 | .28 | .27 |
| 2 | .26 | .24 | .24 | .23 |
| 3 | .23 | .20 | .21 | .19 |
| 4 | .20 | .17 | .19 | .16 |
| 5 | .17 | .15 | .17 | .14 |
| 6 | .15 | .13 | .15 | .12 |
| 7 | .14 | .11 | .13 | .11 |

**CU Table - 120° Emitting**

| $\rho_{cc}$ | 70 | | 50 | |
|---|---|---|---|---|
| $\rho_w$ | 50 | 30 | 50 | 30 |
| RCR | | | | |
| 0 | .39 | .39 | .36 | .36 |
| 1 | .33 | .31 | .30 | .29 |
| 2 | .28 | .25 | .26 | .24 |
| 3 | .24 | .22 | .23 | .20 |
| 4 | .21 | .18 | .20 | .17 |
| 5 | .19 | .16 | .18 | .15 |
| 6 | .17 | .14 | .16 | .13 |
| 7 | .15 | .12 | .14 | .11 |

**CU Table - 180° Emitting**

| $\rho_{cc}$ | 70 | | | 50 | |
|---|---|---|---|---|---|
| $\rho_w$ | 50 | 30 | 10 | 50 | 30 |
| RCR | | | | | |
| 0 | .39 | .39 | .39 | .35 | .35 |
| 1 | .32 | .30 | .28 | .29 | .27 |
| 2 | .27 | .24 | .22 | .24 | .22 |
| 3 | .23 | .20 | .18 | .21 | .18 |
| 4 | .20 | .17 | .15 | .18 | .16 |
| 5 | .18 | .14 | .12 | .16 | .13 |
| 6 | .16 | .13 | .10 | .14 | .11 |
| 7 | .14 | .11 | .09 | .13 | .10 |

**CU Table - 240° Emitting**

| Length of Run (feet) | | 10 ft. | 20 ft. | 30 ft. | 40 ft. |
|---|---|---|---|---|---|
| **Lumens/foot** | | 500 lm/ft. | 270 lm/ft. | 165 lm/ft. | 110 lm/ft. |
| **Peak Luminous Intensity (cd)\ (Mean Exitance (lm/ft$^2$))** | 90° emitting | 1569\(1070) | 1708\(583) | 1569\(357) | 1435\(245) |
| | 120° emit. | 1143\(850) | 1243\(462) | 1143\(283) | 1043\(194) |
| | 180° emit. | 1091\(631) | 1187\(343) | 1091\(210) | 998\(144) |
| | 240° emit. | 696\(497) | 757\(270) | 696\(166) | N/A |

*Note:*   *All values listed are based on the maintained output of a single T250 luminaire.*
**Photometric Data - End-Feed System**

(*c*)

**Fig. 16.47** (Continued)

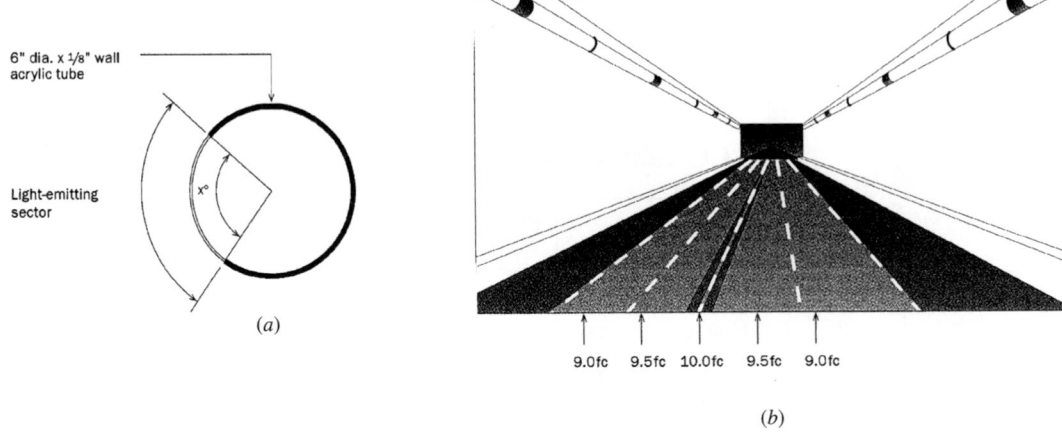

9.0fc  9.5fc  10.0fc  9.5fc  9.0fc

*(b)*

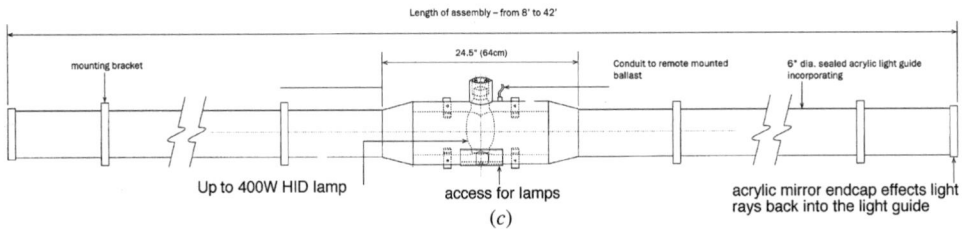

*(c)*

**Fig. 16.48** *Application of a prismatic tubular light guide to tunnel lighting. (a) Section demonstrating how light is emitted from the light guide. (b) Perspective drawing showing tubular lights on both sides of the tunnel and the resultant uniform tunnel illuminance levels. (c) Drawing of one section of the center-feed light guide. (Courtesy of TIR Systems, Ltd.)*

## References

ASHRAE. 2004. ANSI/ASHRAE/IESNA Standard 90.1-2004: *Energy Standard for Buildings Except Low-Rise Residential Buildings.* American Society of Heating, Refrigerating and Air-Conditioning Engineers, Inc., Atlanta, GA.

ASHRAE. 2004. ANSI/ASHRAE Standard 90.2-2004: *Energy-Efficient Design of Low-Rise Residential Buildings.* American Society of Heating, Refrigerating and Air-Conditioning Engineers, Inc., Atlanta, GA.

IEEE. 1995. Standard 446: *IEEE Recommended Practice for Emergency and Standby Power Systems for Industrial and Commercial Applications,* Institute of Electrical and Electronics Engineers, Inc., New York.

IESNA. 1995. RP-11-95. *Design Criteria for Interior Living Spaces.* Illuminating Engineering Society of North America, New York.

IESNA. 2000. TM-10-00: *Addressing Obtrusive Light (Urban Sky Glow and Light Trespass) in Conjunction with Roadway Lighting.* Illuminating Engineering Society of North America, New York.

IESNA. 2000. TM-11-00: *Technical Memorandum on Light Trespass: Research, Results, and Recommendations.* Illuminating Engineering Society of North America, New York.

IESNA. 2001. RP-7-01. *Lighting Industrial Facilities.* Illuminating Engineering Society of North America, New York.

IESNA. 2003. RP-3-2003. *Lighting for Educational Facilities.* Illuminating Engineering Society of North America, New York.

IESNA. 2004. RP-1-04. *American National Standard Practice for Office Lighting,* Illuminating Engineering Society of North America, New York.

NASA. NASA Tech. Brief. PB80-974440 (LAR-12333): Optics for Natural Lighting. National Aeronautics and Space Administration, Langley Research Center, Hampton, VA.

NEMA. 1999. LSD-4-1999: *Glossary of Terms Pertaining To Remote Illumination Systems.*

National Electrical Manufacturers Association, Rosslyn, VA.

NFPA: 2003. NFPA-101: *Life Safety Code.* National Fire Protection Association, Quincy, MA.

NFPA: 2005. NFPA-70: *National Electrical Code.*

National Fire Protection Association, Quincy, MA.

NFPA: 2005. NFPA-99: *Standard for Health Care Facilities.* National Fire Protection Association, Quincy, MA.

ILLUMINATION

# PART IV

# ACOUSTICS

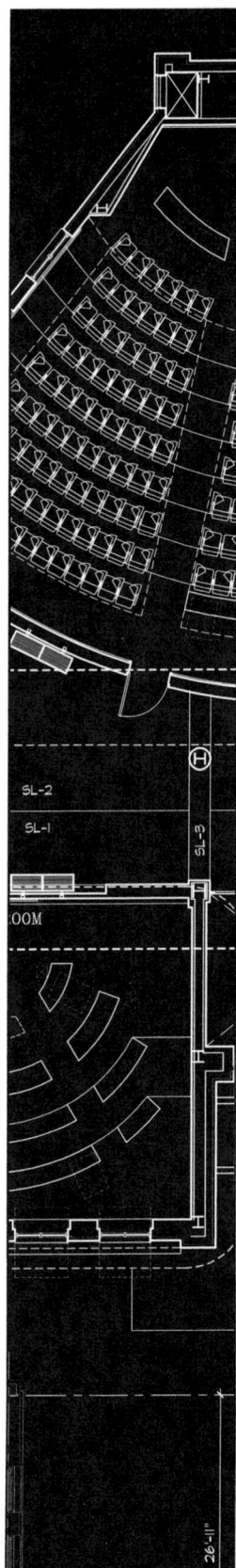

The distinction between the terms *sound* and *noise* is a simple and totally subjective one—sound is desirable, noise is not. This differentiation does not consider the specific content of an acoustical signal. For example, speech, which is a desirable sound in most instances, can become noise when coming from a neighbor's apartment at 2 A.M. or from an adjoining office. An air-handling unit, which is often considered a noise producer, can in some situations provide a desirable sound that can mask intruding speech (noise). The function of architectural acoustics is simply to follow this logic: to enhance desired sounds and to attenuate noises.

It is important to note that there is no unified theory of acoustical comfort, such as is available to guide design efforts dealing with thermal comfort. Design for good acoustics involves addressing a number of acoustical issues correctly—but individually. What is missing—because it does not yet exist— is a holistic view of what effect changing one acoustic parameter (say, reverberation time) will have on design criteria for another variable (such as sound pressure level). It becomes the designer's job to qualitatively balance the interactions between acoustic properties of spaces when establishing design intent and criteria.

Chapter 17 introduces the subject of architectural acoustics with a discussion of physical sound theory and physiological hearing phenomena. The latter include the negative effects of noise, which are primarily psychological impressions (involving annoyance) at low levels but become physical effects and can result in hearing damage at high levels. The chapter provides a description of the types of sounds typically encountered in architectural spaces and criteria for acceptable background noise levels in enclosed spaces. These are framed as noise criteria (NC) and room criteria (RC) curves.

Chapter 18 moves to the heart of architectural acoustics—the topic of room acoustics—with an explanation of absorption and reverberation, and develops acoustic design criteria for various indoor activities. The chapter concludes with a description of sound reinforcement systems.

Chapter 19 is devoted to noise control. It begins with a discussion of the use of absorptive materials for room noise reduction. It treats the problem of interspace noise conduction, dividing the problem into two parts (airborne noise and structure-borne noise) because the solutions are different. Relevant criteria, including sound transmission class (STC) and impact isolation class (IIC), are introduced and explained, and solutions to noise transfer problems of different types are suggested. The problem of speech privacy for both enclosed spaces and open (often office) areas is treated in detail. The chapter offers information on mechanical system noise control and acoustic recommendations and criteria, ending with a discussion of exterior acoustics. Acoustical reference information is also provided.

ACOUSTICS

# Fundamentals of Architectural Acoustics

THE ACOUSTIC ENVIRONMENT PLAYS AN IMPORTANT role in supporting (or disturbing) an overall sense of comfort in many of the spaces we occupy on a daily basis—including both residential and commercial/institutional spaces. For several reasons, many design solutions seem to shortchange the acoustical environment. This is partly due to the perceived complexity of architectural acoustics, partly due to lack of coverage of the topic area in many architecture programs, and partly due to people's amazing ability to overlook less than desirable acoustical situations. Good acoustics is not required by most building codes and is not a key element of green building rating systems. Nevertheless, providing acceptable acoustical conditions is a fundamental part of good design practice.

## 17.1 ARCHITECTURAL ACOUSTICS

Architectural acoustics may be defined as the design of spaces, structures, and mechanical/electrical systems to meet hearing needs. With proper design efforts, wanted sounds can be heard properly and unwanted sounds (noise) can be attenuated or masked to the point where they do not cause annoyance. Achieving good acoustics, however, has become increasingly difficult for a variety of reasons. To lower construction costs, the weight of various materials used in many of today's buildings has been reduced from those prevalent 25–50–100 years ago. Since light structures generally transmit sound more readily than heavy ones, lightweight buildings bring the potential for major acoustical problems. Population density in office spaces has steadily increased, thus raising the amount of noise generated. Worse yet—from the acoustics point of view—many offices today are designed as open areas with, at best, only thin partial-height dividers (cubicles) separating workers. Forty percent or more of a building budget may be allocated for mechanical systems—most of which generate noise. Outside noise sources, such as cars, trucks, trains, and airplanes, can also present problems and require isolation of interior spaces from exterior sounds.

Building owners and tenants are aware that quality acoustic environments are required for high productivity and comfort in buildings and hence competitive rental or purchase values. The architect is expected to provide such acoustic quality. A clear understanding of the principles explained in this and the following chapters will assist the architect in preparing straightforward designs alone and, in more complex instances, by cooperating knowledgeably with an acoustic consultant. Proper acoustic design responses early in the design process are critically important, as after-the-fact acoustic "repair" is often difficult (and, therefore, costly) and frequently impossible without substantial structural alterations (very costly).

All acoustical situations have three common elements—a sound source, a sound transmission path or paths, and a receiver of the sound. Through design, a source can be made louder or quieter and a path can be made to transmit more or less sound. Working with sources and paths throughout the various design phases (and into construction and often occupancy) is the bread and butter of architectural acoustics. The listener's reception of sound also may be influenced, although this is not normally an "architectural" solution. This chapter presents the fundamental bases of architectural acoustics to assist a designer in defining appropriate acoustic intents and criteria. Moreover, it describes basic methods for reaching such intents through the design process.

**TABLE 17.1  Speed of Sound Propagation in Various Media**

| Medium | Speed Meters per Second | Speed Feet per Second |
|---|---|---|
| Air | 344 | 1130 |
| Water | 1410 | 4625 |
| Wood | 3300 | 10,825 |
| Brick | 3600 | 11,800 |
| Concrete | 3700 | 12,100 |
| Steel | 4900 | 16,000 |
| Glass | 5000 | 16,400 |
| Aluminum | 5800 | 19,000 |

*Note:* These figures are approximate, since the listed materials vary in density. An average frequency is assumed.

## 17.2 SOUND

Sound can be defined in a number of different ways, depending upon the aspect of most interest or concern. Thus, sound can be described as a physical wave, or as a mechanical vibration, or simply as a series of pressure variations in an elastic medium. For airborne sounds, the medium is air. For structure-borne sounds, the medium may be concrete, steel, wood, glass, or combinations of these materials. A much more limited definition of sound, more appropriate to architectural acoustics, is that it is simply an audible pressure variation. This establishes that architectural acoustics is concerned with the building occupant. It also suggests that there may be inaudible pressure variations that cannot be heard. This is the case with *vibration*, which is a pressure variation that can be felt but not heard.

### (a) Speed of Sound

Sound travels at different speeds, depending upon the medium. In air, at sea level, sound velocity is 1130 fps (344 m/s). This is 770 miles per hour (1239 kilometers per hour)—slow indeed when compared to light, which has a speed of 186,000 miles per *second* (299,338 km per second). Since sound travels not only in air but also through parts of a structure, it is of interest to know the speed of sound in other media (Table 17.1). For architectural design purposes, speed variations due to changes in temperature and altitude (atmospheric pressure) may be ignored and, for most calculations, 1130 fps (344 m/s) may be used as the speed of sound in air (usually within 3% error). From a

practical standpoint, the speed of sound in air is slow enough that the travel time of a sound signal can be a key design issue.

### (b) Wavelength

The *wavelength* of a sound is defined as the distance between similar points on successive waves, which is the distance a sound travels in one cycle. The relationship between wavelength, frequency, and speed of sound is expressed as

$$\lambda = \frac{c}{f} \tag{17.1}$$

where

$\lambda$ = wavelength, ft (m)

$c$ = velocity of sound, fps (m/s)

$f$ = frequency of sound, Hz

Low-frequency sounds are characterized by long wavelengths and high-frequency sounds by short wavelengths. Sounds with wavelengths ranging from 0.5 in. to 50 ft (12 mm to 15 m) can be heard by humans. A simple nomograph (see Fig. 18.18) permits rapid determination of wavelength, given sound frequency, and vice versa.

### (c) Frequency

The number of times that a cycle of compression and rarefaction of air occurs in a given unit of time is described as the *frequency* of a sound. For example, if there are 1000 such cycles in 1 second, the frequency of the sound is 1000 cps—1000 Hertz

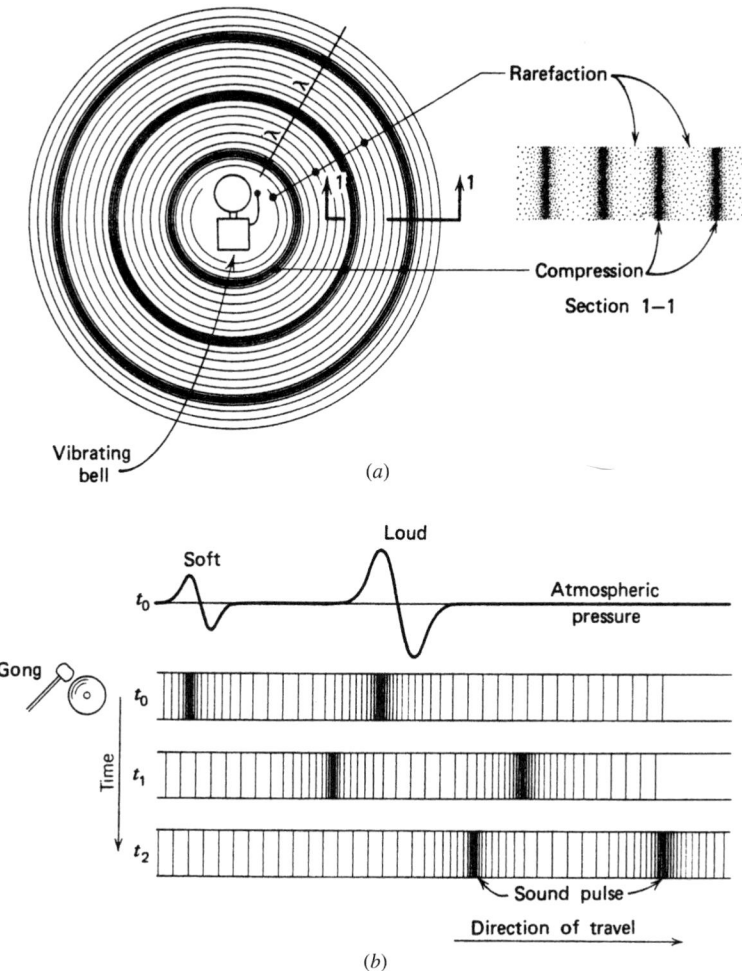

**Fig. 17.1** *Sound pressure waves. (a) The continuous vibration from the bell causes a series of compressions and rarefactions of the air to travel outward longitudinally from the source. Amplitude information is carried by pressure; that is, greater amplitude means greater compression and greater rarefaction. (Compression and rarefaction are shown diagrammatically as line density, although they are actually molecular phenomena, as shown in the upper drawing.) (b) Two single impulses of different magnitude (amplitude) traveling away from the source. Note how amplitude information is carried by the difference in pressure.*

(Hz) in standard nomenclature. Thus, in Fig. 17.1, higher frequencies would be shown by compressions and rarefactions that are closer together and lower frequencies by those that are farther apart. In architectural acoustics, frequency is sometimes referred to using a term borrowed from music—*pitch*. The higher a sound's frequency, the higher its pitch, and vice versa.

Sound frequency is integrally linked to speech and hearing. The approximate frequency range of a healthy young person's hearing is 20 to 20,000 Hz. The human speaking voice has a range of approximately 100 to 600 Hz in *fundamental frequencies*, but

*harmonics* (overtones) reach to approximately 7500 Hz. Most speech information is carried in the upper frequencies, whereas most of the *acoustic energy* exists in the lower frequencies. The critical frequency range for speech communication is 300 to 4000 Hz. Overtones outside these core frequencies, however, give the voice its characteristic sound and specific identity. Telephone and radio communication are accomplished using a considerably narrower frequency band by sacrificing some voice quality and intelligibility (see Fig. 17.2).

A sound composed of only one frequency is called a *pure tone*. Except for the sound generated by

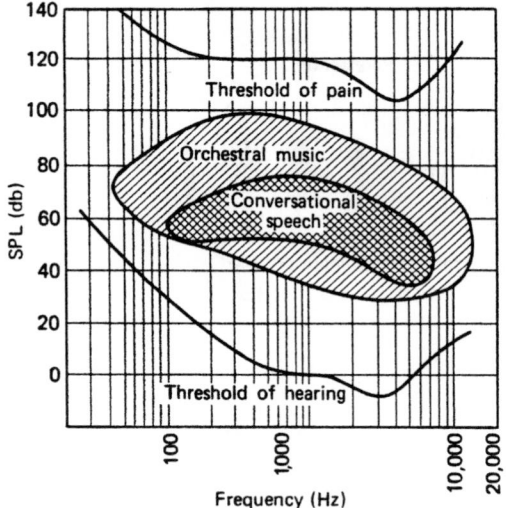

**Fig. 17.2** *The positions of speech and wide-range music in the aural field of the human ear are illustrated. Speech is in the nominally linear response area of the ear (see Fig. 17.7), as is most music. Beyond these frequencies, the ear effectively attenuates the incoming signal.*

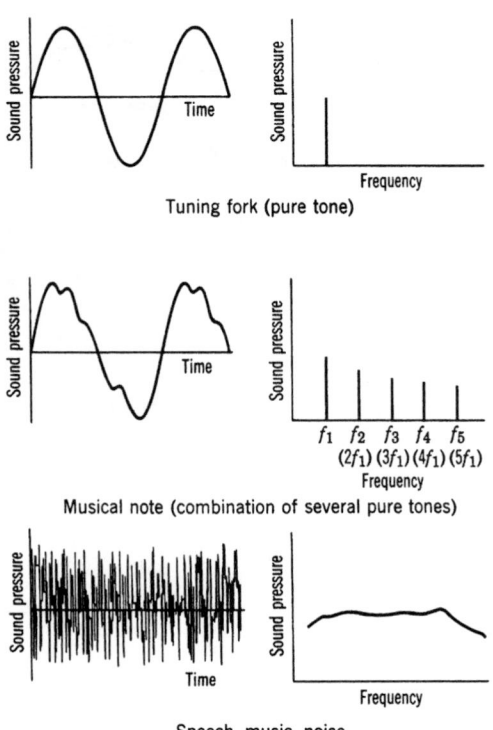

**Fig. 17.3** *Schematic representations of a pure tone, a musical note, and a more complex sound (such as speech, music, or noise) showing the variation of sound pressure with time and frequency.*

a tuning fork, few sounds are truly pure. Musical sounds (tones) are composed of a fundamental frequency and integral multiples of the fundamental frequency (harmonics). Most common sounds are complex combinations of frequencies. Figure 17.3 shows examples of pure tones, musical notes, and common sounds, while Fig. 17.4 shows the frequency ranges of some common devices and phenomena.

### (d) Octave Bands

The frequencies in the scale of Fig. 17.4 all stand in the ratio of 2:1 to each other—that is, 16:32:63: 125:250, and so on. Borrowing again from musical terminology, they are one *octave* apart. These particular frequencies are also accepted internationally as the center (reference) frequencies of octave bands used for the purpose of sound specification. For technical reasons, a geometric mean is used. Thus, 250 is the center frequency of an octave band ranging from $250/\sqrt{2}$ to $250\sqrt{2}$, with that particular octave being known as the *250-Hz octave*. If a finer division is required for analysis purposes (unusual in architectural acoustics), ½-octave and ⅓-octave bands are used. Because frequency is so important to architectural acoustics, and because there are 19,980 dis-

crete whole frequencies in the normal range of hearing, octave bands are used repeatedly during design as a way of capturing frequency-specific information about sounds without becoming buried in detail.

### (e) The Concept of Sound Magnitude

The magnitude of a sound signal is a relatively complex concept because there are a number of different metrics (and associated terminology) in common use and because of the great range of values involved in day-to-day acoustic situations. Sound magnitude is often equated with loudness, which is a subjective, receiver-oriented response not linearly related to the power of a sound (in watts). The physical magnitude of sound is variously described as *sound power, sound power level* (PWL), *sound pressure, sound pressure level* (SPL), *sound intensity,* and *sound intensity level* (IL). Each of these metrics has a place, and each differs from the others and from subjective loudness. To clearly understand these concepts—and it is imperative that they be understood—a comprehension

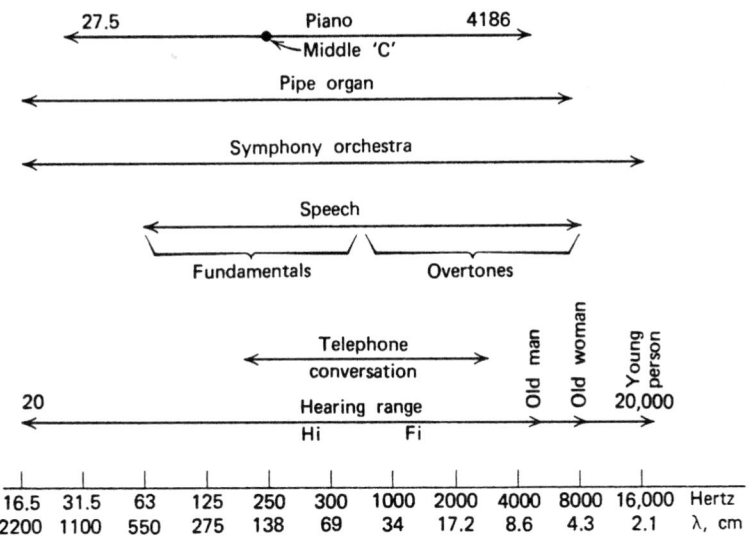

**Fig. 17.4** *Frequency ranges of common instruments. Wavelength (λ) is calculated on an assumed propagation velocity of 344 m/s (1130 fps).*

of how we hear and how sound is propagated in free space is necessary.

### (f) Sound Propagation

For simplicity, it is probably best to view sound as a series of pressure variations. In air, these pressure variations take the form of periodic compressions and rarefactions. The bell in Fig. 17.1 radiates a tone in all directions equally—that is, it creates a circular wavefront. As the material of the bell vibrates, it sets up vibrations of the same frequency in the air, which can best be visualized in a sectional view. Notice that the pressure changes containing the sound information travel in the same direction as the wavefront—longitudinally. Sound is therefore a *longitudinal* mechanical wave motion. This is unlike (for example) a radio signal, in which the wave travels longitudinally but the information— that is, the modulation—is transverse.

### 17.3 HEARING

As noted above, the approximate frequency response of a healthy young person's hearing is 20 to 20,000 Hz. The upper limit decreases with age as a result of a process called *presbycusis* (Fig. 17.5). The loss is more pronounced in men than women. Recogni-

tion of this phenomenon can be of importance in schools, since very high-pitched sounds that are inaudible to most adults can be a source of extreme annoyance to young students. For example, dentists report that high-speed drills and tooth-cleaning devices cause extreme auditory discomfort in many young patients. These devices produce sounds in the 15- to 20-kilohertz (kHz) range.

### (a) The Ear

Referring to Fig. 17.6, the outer ear is funnel-shaped and serves as a sound-gathering input device for the auditory system. Sound energy travels through the auditory canal (outer ear) and sets in motion the components of the middle ear, comprising the eardrum, hammer, anvil, and stirrup. The stirrup acts as a piston to transmit vibrations into the fluid of the inner ear. The motion of this fluid causes movement of hair cells in the cochlea, which, in turn, stimulates nerves at the bases of the hairs. The nerves, in turn, transmit electrical impulses along the eighth cranial nerve to the brain. These impulses we understand as sound.

It is often assumed that the ear ignores phase differences and combines frequencies. This may not always be the case, however, particularly when the frequencies are very far apart. For this reason, a single-number representation of a complex sound

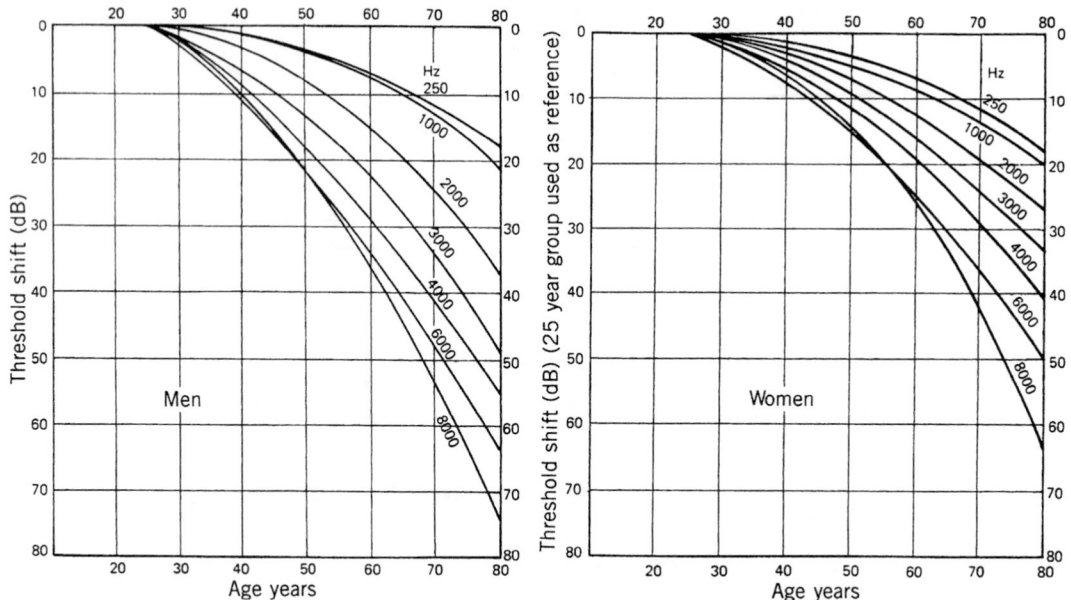

**Fig. 17.5** *These curves illustrate the average hearing threshold shift for men and women, with increasing age relative to a 25-year-old reference group. Note that at age 60 the male threshold shift at upper speech frequencies (4000 Hz) is 30 dB compared to 20 dB for women, a difference of 10 dB or one-half of the subjective loudness (see Table 17.3). For men older than 50 and women older than 60, frequencies above 10 kHz are effectively inaudible, and even at normal speech frequencies of 1000 to 2000 Hz, subjective loudness is reduced to one-half of that of a 25-year-old listener. (Reproduced with permission from F. A. White,* Our Acoustic Environment, *Wiley, New York, 1975.)*

(the type of information found in Table 17.5) can be misleading and must be used with caution.

At the threshold of hearing (approximately 0 dB), the displacement of air molecules impinging on the eardrum, and the eardrum excursion, are approximately one angstrom unit (1 Å = $10^{-8}$ cm), which is approximately the diameter of an *atom*. Were the ear an order of magnitude more sensitive, it would hear thermal noise. The human ear is thus operating close to the practical limit of sensitivity. At the other end of the magnitude spectrum, the threshold of pain corresponds to a sound pressure level of 130 dB and to an eardrum motion of approximately 0.25 mm (0.01 in.)—an astonishing range indeed.

Movement of the eardrum (and thus hearing) is caused directly by air pressure variations. Therefore, the magnitude quantity of most interest to architectural acoustics is sound pressure, which is a *force density,* and sound pressure level, which is the ratio of a given sound pressure to a base level, expressed in decibels.

## (b) Equal Loudness Contours

The human ear is not uniformly sensitive over its entire frequency range of 20 Hz to 20 kHz. The 120- to 130-dB upper limit (pain threshold) occurs at all frequencies. At the lower limit, however, the 0-dB threshold occurs only at 1000 Hz. The ear is in fact most sensitive at 3000 to 4000 Hz, at which frequencies the threshold is about −5 dB (relatively speaking). This type of nonlinear response exists throughout the ear's hearing range. To determine the nature of this nonlinearity, a large number of tests were conducted with pure tones of different frequencies, in which listeners were asked to equate the subjective loudness of signals. These test results produced a family of curves called *equal loudness level contours* (sometimes called *Fletcher–Munson equal loudness contours* after two of the principal researchers). These curves (Fig. 17.7) are internationally recognized and standardized, and are used as the reference for normal hearing response. They are also used to "weight" measuring devices,

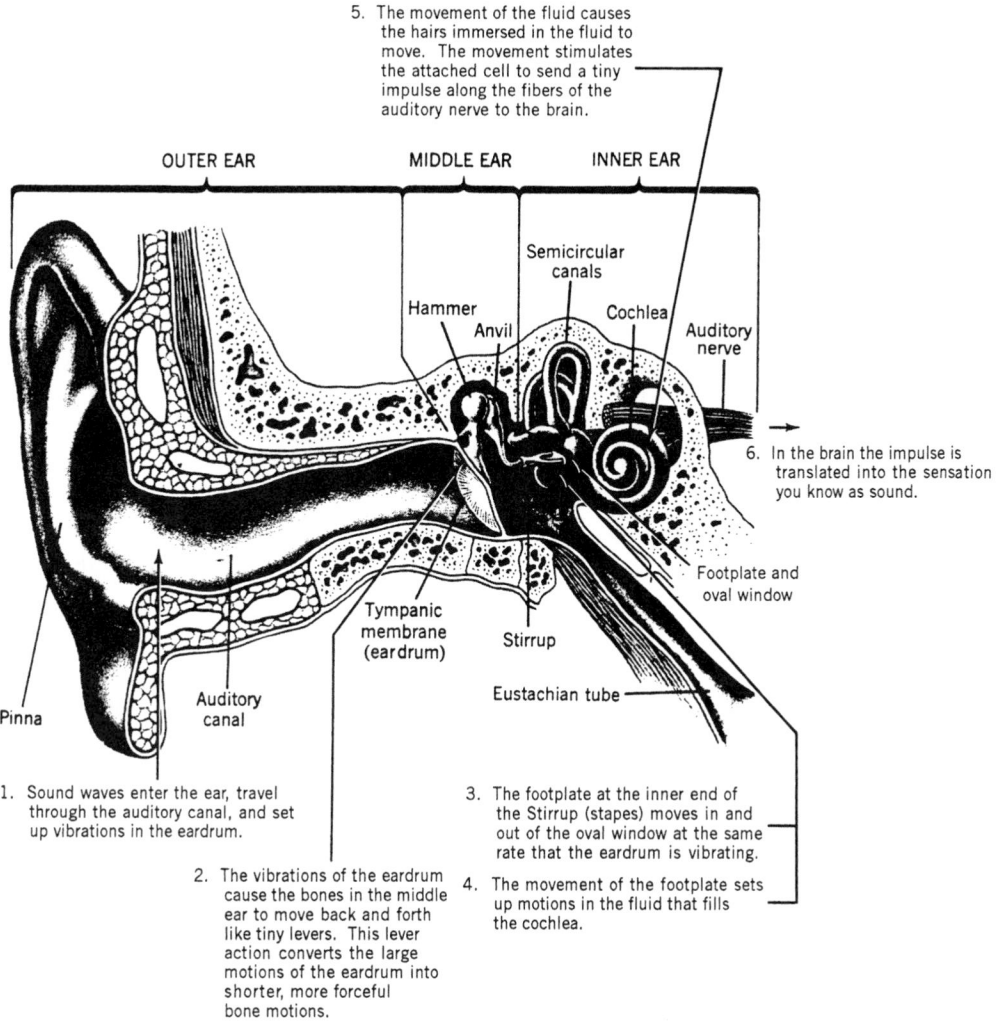

5. The movement of the fluid causes the hairs immersed in the fluid to move. The movement stimulates the attached cell to send a tiny impulse along the fibers of the auditory nerve to the brain.

OUTER EAR          MIDDLE EAR          INNER EAR

Semicircular canals

Hammer          Cochlea

Anvil          Auditory nerve

6. In the brain the impulse is translated into the sensation you know as sound.

Footplate and oval window

Tympanic membrane (eardrum)          Stirrup

Eustachian tube

Pinna          Auditory canal

1. Sound waves enter the ear, travel through the auditory canal, and set up vibrations in the eardrum.

2. The vibrations of the eardrum cause the bones in the middle ear to move back and forth like tiny levers. This lever action converts the large motions of the eardrum into shorter, more forceful bone motions.

3. The footplate at the inner end of the Stirrup (stapes) moves in and out of the oval window at the same rate that the eardrum is vibrating.

4. The movement of the footplate sets up motions in the fluid that fills the cochlea.

**Fig. 17.6** *Drawing of the human ear showing the principal parts and its function as a sound receptor. (Drawing reproduced with permission from F. A. White,* Our Acoustic Environment, *Wiley, New York, 1975. Notes are reproduced from* Quieting: A Practical Guide to Noise Control, *1976.)*

ACOUSTICS

as explained later in this chapter. Note that *by definition:*

1. All points on a single contour have the same *subjective* sensation of loudness.
2. The loudness level in *phons* (the number shown in the center of each curve in Fig. 17.7) of the *entire* contour is defined by the decibel level of that contour at 1000 Hz.

The phon scale was constructed as a way of defining perceived sound (subjective loudness impressions) in terms of the ear's nonlinear response. Assigning a single number—in a unit called the

*phon*—to an equal loudness contour makes it possible to compare subjective loudness impressions of two sounds, regardless of frequency or actual sound pressure level. Thus, in Fig. 17.8, the subjective loudness of a whisper in the ear (20 phons) is the same as that of distant pounding surf, despite the fact that the sound pressure level (SPL) of the former is 20 dB and that of the latter is 80 dB. The 60-dB differential is due to the ear's sharp drop in sensitivity at low frequencies. However, because the phon scale is nonlinear in terms of perceived loudness (a given phon differential does not correspond to the same perceived loudness change throughout

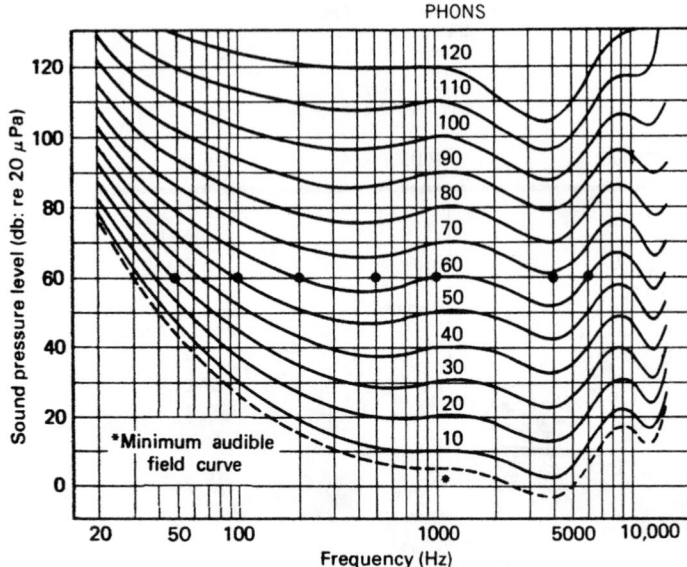

**Fig. 17.7** *Standard equal-loudness contours. These curves are accurate for a listener with normal binaural hearing situated in the near field of a source producing pure tones directly ahead of the listener. The subjective (perceived) loudness of each contour is quantified in a unit called the phon. The phon value of each curve is shown in the center of the figure. Note that a sound pressure level of 60 dB corresponds to 30 phons at 50 Hz, 50 phons at 100 Hz, 63 phons at 500 Hz, 60 phons at 1 kHz, 68 phons at 4 kHz, and 60 phons at 6 kHz. This indicates the relative flatness of the ear's response in the central frequency range and the sharp drop at low frequencies.*

the phon scale) and because phons cannot be combined arithmetically to give a resultant subjective loudness (60 phons plus 50 phons is *not* 110 phons), the scale has found few uses (one of which is to rate the noisiness of bathroom fans).

The Fletcher–Munson contours demonstrate some interesting phenomena:

1. Sensitivity drops off sharply at low frequencies, particularly at low dB levels. (For this reason, most stereo amplifiers provide automatic bass boost at low volume levels.)
2. Maximum sensitivity occurs between 3 and 4 kHz—precisely the frequencies that convey the most information in human speech (see Fig. 17.2).
3. In a normal listening range of 45 to 85 dB, and in the most often used frequency range of 150 Hz to 6 kHz, the contour is substantially flat; that is, the ear's response is effectively linear in this zone. It is only at extremes of sound level and frequency that nonlinearity occurs.

The ear "averages" sounds over some minimum time period, and sound impulses of shorter duration than this period sound quieter than they would as a steady-state sound. This minimum sound length (i.e., the time required for the sound to achieve full loudness) varies within the range of 50 and 200 milliseconds (ms) (½₀ to ⅕ s) and depends to an extent upon the frequency content of the sound. The figure of 70 ms (¼₄ s) is frequently found in the literature. This threshold time becomes important when considering the effect of echoes (see Sections 18.11–18.14).

## (c) Masking

When two separate sound sources are perceived simultaneously, the perception of each is made more difficult by the presence of the other. This effect is known as *masking,* which is defined technically as the number of decibels by which the threshold of audibility of one sound is raised by the presence of another sound. The masking effect is greatest when two sounds are close in frequency or frequency content, since the ear has greater difficulty separating like frequencies. Also, a low frequency will mask a high frequency more effectively than the reverse for the same decibel levels. With broad-frequency sounds, the masking effect is difficult to predict, since

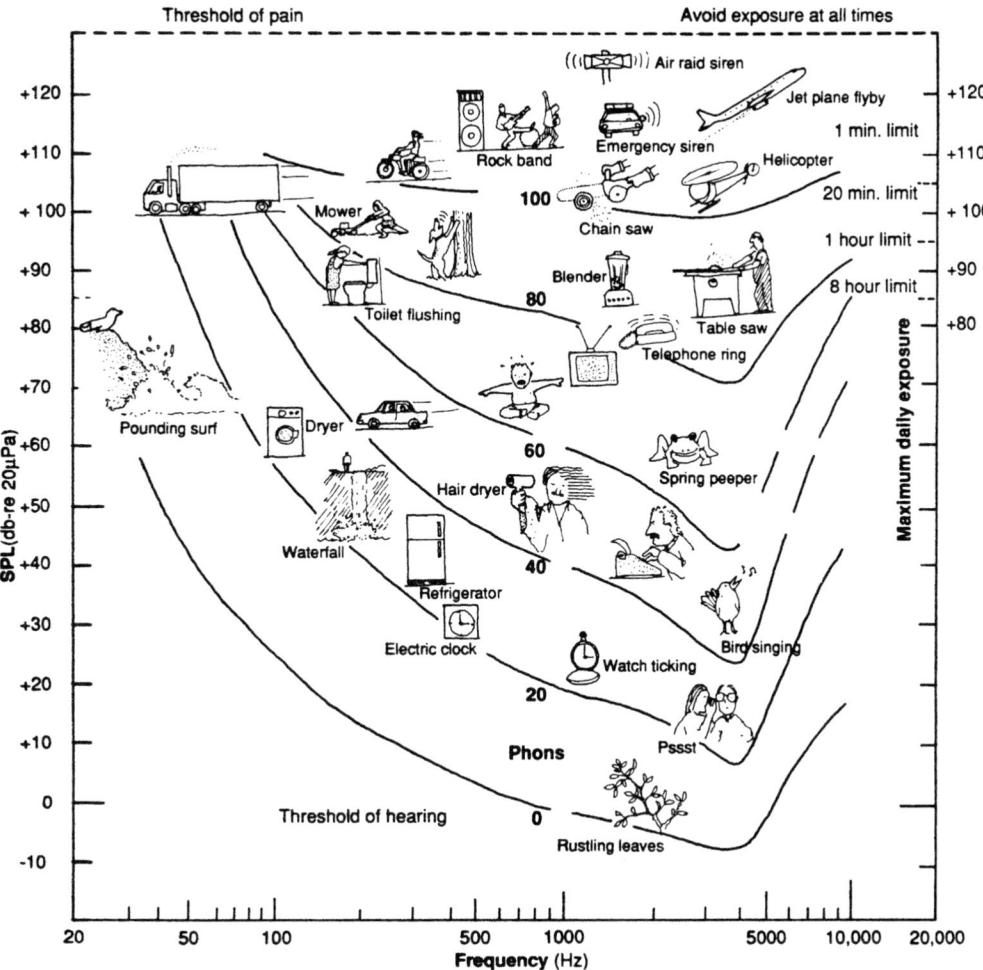

**Fig. 17.8** *Common sound sources plotted at their dominant frequencies and levels as typically heard by the observer. The equal-loudness contours (see Fig. 17.7) show why certain sounds seem louder than others, despite the pressure levels that would indicate the contrary (Reprinted with modification from* Quieting: A Practical Guide to Noise Control, 1976.)

it depends in part upon how "hard" a listener is listening. Masking is an extremely important and useful technique for noise control, wherein background sound levels are deliberately manipulated to mask other unwanted sounds (see Section 18.21). The background sounds used for this purpose are of a broadband continuous nature, such as "pink" noise (see the Glossary at the end of Chapter 19), that are non-information-bearing. They serve to depress the intelligibility of lower-magnitude, information-bearing sounds that would otherwise cause annoyance, and are particularly useful in open (landscaped) office plans where few noise control alternatives exist. Frequently, mechanical equipment

sound (such as that from air diffusers and fans) can be utilized to good advantage as masking sound.

### (d) Directivity

The exact mechanism by which the binaural aspect of hearing detects direction is not entirely understood. It is clear, however, that the brain can sense the direction of a sound source in many environments. This is a useful trait in both social and dangerous situations. In enclosed spaces, echoes from surface reflections (reverberation) will blur most directivity, and any "stereo" information will be almost completely dependent upon high frequencies in the near field (the sound field that is close to the source and therefore minimally affected by reverberation). Since high frequencies (with short wavelengths) travel in a relatively straight line, they will reach a listener before their reflections and in so doing will cue the hearer as to their origin. Low frequencies (with long wavelengths) reach a room's perimeter rapidly and even in the near field are already mixed with their reflections, thus disguising the location of their origin. This is the reason that low-frequency noise in a room is so difficult to locate; it seems to come from everywhere.

### (e) Discrimination

Frequency recognition is accomplished in the cochlea by the basilar membrane. This membrane resonates at one end at about 20 Hz and at the other at 20,000 Hz, giving the ear its frequency range. The ear can hear and recognize distinct frequencies, yet the hearing mechanism has the ability, apparently as directed by the brain, either to hear individual frequencies or to combine them into a single more complex sound. Thus, when we hear a string quartet we can, generally *at will*, hear either the entire quartet or each instrument individually. With concentration (vision helps in this regard), a trained ear can pick out a single instrument in an orchestra of 120 pieces, even if there is more than one such instrument in the same section. Conductors do this regularly. Similarly, the ear can perform the discrimination feat known as the cocktail party effect, that is, pick out one voice among background sounds that may be substantially louder than the wanted signal. In effect, the ear in this case attenuates the unwanted sound signals. Normally, however, the ear does precisely the opposite. It combines sounds that are clearly distinct from each other in frequency and phase. The three tones in a chord struck on a piano are different in frequency and, if played as a very rapid triplet, out of phase. Yet the ear combines them and hears a single sound, despite the fact that the maxima of the three tones do not occur simultaneously.

### 17.4 SOUND SOURCES

Building occupants encounter a wide range of sounds in the course of a day. Some of these sounds are produced by humans, others by mechanical equipment or natural phenomena. Some of the sounds are information-bearing, while others convey nothing intelligible. Speech and music are sounds of particular interest to building designers, as are those sounds deemed to be noise.

### (a) Speech

As can be seen from Fig. 17.2, the ear's sensitivity is highest in the speech frequency and the normal energy range. Individual speech sounds vary in duration between 30 and 300 ms, and the ear normally perceives these individually and clearly. Speech is comprised of *phonemes*, which are individual and distinctive sounds that, to an extent, vary from language to language. Certain phonemes exist in one language and not in another. Since some phonemes carry more information than others, it is these that good architectural acoustics must be particularly careful to preserve and support in order to maintain intelligibility. In English, consonants carry much more information than vowels, as can readily be demonstrated by writing a sentence (1), then rewriting it without consonants (2), and then without vowels (3):

1. Most speech energy is concentrated in the 100- to 600-Hertz range.
2. o ee ee i oeae i e 100 o 600 e ae.
3. Mst spch nrgy s cncntrtd n th 100 t 600 Hrtz rng.

The male voice centers its energy at around 500 Hz, the female voice at around 900 Hz. It is, however, in the high frequencies that consonants have most of their energy. Phonemes such as *s* and *sh*, for example, have most of their energy above 2 kHz, and both are particularly important in conveying intelligible content.

Normal speech averages between 40 and 50 dBA sound pressure level at 3 to 4 ft (0.9 to 1.2 m) from the source, with a dynamic range from about 30 dBA for soft speech to about 65 dBA for loud speech (at the same distance). Extremes of speech are 10 dBA for a soft whisper and 80 dBA for a shout, but in both of these instances intelligibility is sharply reduced because of lack of consonant power. Indeed, in shouting, emphasis is actually on vowels, so that it is generally accepted that a 70-dBA sound pressure level is about the upper limit of fully intelligible human speech. Singers who frequently exceed 90 dBA do so at great loss of intelligibility.

Another result of the high-frequency content of consonants, and therefore intelligibility, is its directiveness. The higher the frequency, the greater a sound's directivity and the less its diffraction (ability to be heard beyond a partial barrier). Therefore, intelligibility of speech is greatest directly in front of a speaker and least behind him/her. High-frequency tones are most easily absorbed and least diffracted.

## (b) Other Sounds

Instrumental music is much broader in dynamic range and more complex in frequency than speech. It has no direct parallel to intelligibility. A person's "reception" of music is a combination of physiological and psychological phenomena. As such, it is an experience beyond the scope of this book to cover in depth, but it is briefly examined in the subsequent discussion of room acoustics, auditoriums, and music halls. Noises are dealt with in Section 17.6.

## 17.5 EXPRESSING SOUND MAGNITUDE

Six different quantitative measures of sound magnitude are commonly encountered in architectural acoustics. Three of these—sound power, sound pressure, and sound intensity—are absolute measures; the other three—sound power level, sound pressure level, and sound intensity level—are ratio values that compare an absolute measure to a baseline reference value. Use of the qualitative metric of loudness is also common.

## (a) Sound Power

Sound power is an independent property of a sound source that quantifies the source's acoustical output. Sound power is constant for any given source operating under defined conditions (a certain level of speech effort or a rotating speed on a motor) and is not influenced by the nature of the surroundings into which a source is placed. Thus, the sound power (output) of a chiller or orchestra is not changed by the distance to a receiver or the characteristics of a mechanical room or auditorium.

Sound power is expressed in watts (of acoustical power) and varies widely from source to source. Some selected sound power values are: a jet engine, 100,000 W; a symphony orchestra, 10 W; a loud radio, 0.1 W; normal speech, 0.000010 W. Note the wide range of values just in this sample: from $10^5$ to $10^{-6}$ W. Sound power values for natural phenomena need to be obtained from an appropriate information source. Sound power data for manufactured equipment and devices can be obtained from the manufacturer. In fact, a manufacturer can only provide information on the sound power of a product—and cannot give precise data regarding sound pressure since there are so many environmental variables that will influence the possible pressures that may result from a given source.

## (b) Sound Pressure

Sound pressure is the deviation around an ambient air pressure that is caused by sound waves. Sound pressure is caused by the acoustic power output of a sound source but is influenced by the nature of the environment between the source and the receiver, including the distance between source and receiver. Sound power is a source property; sound pressure is an environment-source property. Sound pressure must be referenced to a particular location in a space, as pressure will often (usually) vary from location to location in a room.

Sound pressure is expressed in Pascals (Pa) (or microbars in the I-P system). It is common practice to use SI units for all architectural acoustics metrics—even in the United States. Sound pressures for some common situations include: near a jet plane, 200 Pa; the threshold of pain, 20 Pa; a loud nightclub, 2 Pa; next to a highway, 0.2 Pa; and normal speech, 0.02 Pa. As with sound power, note the substantial range of values in these sample situations.

## (c) Sound Intensity

The threshold of hearing—that is, the minimum sound intensity ($I$) that a normal ear can detect—is $10^{-16}$ W/cm$^2$. (The ear actually responds directly to pressure variations, but such pressures involve various energy densities.) The maximum sound intensity that the ear can accept without damage is approximately $10^{-3}$ W/cm$^2$. This gives an intensity range of

**TABLE 17.2 Comparison of Decimal, Exponential, and Logarithmic Statements of Various Acoustic Intensities**

| Intensity (W/cm²) | | Intensity Level, | |
|---|---|---|---|
| **Decimal Notation** | **Exponential Notation** | **Logarithmic Notation (dB)** | **Examples** |
| 0.001 | $10^{-3}$ | 130 | Painful |
| 0.0001 | $10^{-4}$ | 120 | |
| 0.00001 | $10^{-5}$ | 110 | 75-piece orchestra |
| 0.000001 | $10^{-6}$ | 100 | |
| 0.0000001 | $10^{-7}$ | 90 | Shouting at 5 ft (1.5 m) |
| 0.000000001 | $10^{-9}$ | 70 | Speech at 3 ft (0.9 m) |
| 0.00000000001 | $10^{-11}$ | 50 | Average office |
| 0.0000000000001 | $10^{-13}$ | 30 | Quiet, unoccupied office |
| 0.00000000000001 | $10^{-14}$ | 20 | Rural ambient |
| 0.000000000000001 | $10^{-15}$ | 10 | |
| 0.0000000000000001 | $10^{-16}$ | 0 | Threshold of hearing |

$10^{13}$, or 10 trillion to 1 (10,000,000,000,000:1). Table 17.2 gives an idea of the physical significance of these numbers. As an interesting comparison of energy densities, note that the maximum (painful) acoustic intensity in Table 17.2 is 0.001 W/cm², or 10 W/m²; this is only 1% of the average clear sky solar radiation of 1000 W/m².

A point sound source of constant power radiating in free space—that is, at a location far from the effects of any reflecting surface—is represented in the drawing of Fig. 17.9. The *sound intensity* at any (defined) distance from the source is expressed as

$$I = \frac{P}{A} \qquad (17.2)$$

where

$I$ = sound (power) intensity, W/cm²

$P$ = acoustic power, W

$A$ = area, cm²*

Since the sound radiates freely in all directions.

$$I = \frac{P}{4\pi r^2} \text{ W/cm}^2 \qquad (17.3)$$

---

*It is traditional in architectural acoustics to express area in square centimeters, although the SI system requires area to be stated in square meters. Conversion data for units are given in Table 19.15. See also Table 19.16 for a listing of symbols and abbreviations.

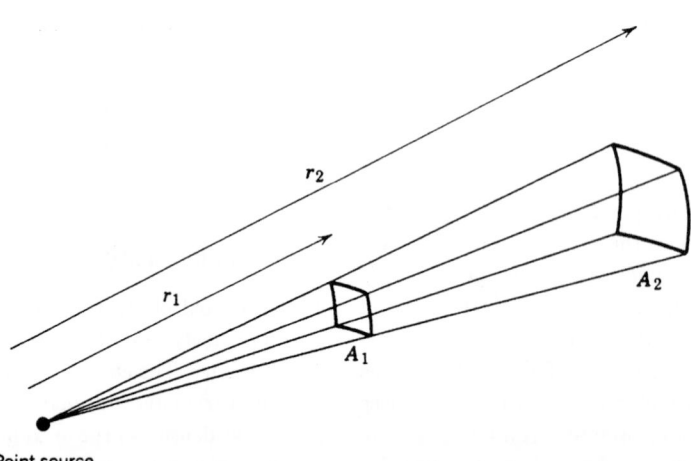

Point source

**Fig. 17.9** *The same total energy passes through areas A₁ and A₂. Since A is proportional to the square of r, the energy density or intensity is inversely proportion to r².*

where $r$ is the radius of an imaginary sphere enclosing the sound source. This is an implementation of the classic *inverse square law*, stating that *intensity is inversely proportional to the square of the distance from the source.* (In I-P units, this is

$$I = \frac{P}{930 \times 4\pi r^2} \text{ W/ft}^2 \qquad (17.4)$$

since $1 \text{ ft}^2 = 930 \text{ cm}^2$.) Using Eq. 17.3 to determine the intensities $I_1$ and $I_2$ at distances $r_1$ and $r_2$ from point source $P$, and dividing the equations, we find that the intensities at distance $r_1$ and $r_2$ from the source stand in the ratio

$$\frac{I_1}{I_2} = \frac{r_2^2}{r_1^2} \qquad (17.5)$$

Note the exact correspondence of these relations to the derivations for illuminance from a point source, found in Section 11.12 and Fig. 11.14. Figure 17.10 shows graphically how a sound pulse is attenuated in *strength* (but not in waveform) as it travels outward from a source by the action of distance.

The preceding derivation is based upon a *point source*—that is, a source that is small relative to the wavelength of the sound produced. This type of source produces spherical waves. Line sources, such as strings, produce cylindrical waves. Large vibrating surfaces, such as walls, produce plane waves. The importance of these distinctions will become clear in the discussion of sound barriers and diffraction in Chapter 19.

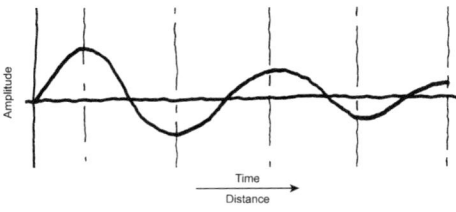

**Fig. 17.10** *Attenuation of a sound signal in air as it travels away from the source. The shape (information) remains constant when traveling in a nondispersive medium such as air. This is not the case with travel in solids, where different frequencies travel at different velocities, causing a wave-shape change with time and distance. Since velocity of propagation is constant, time and distance are linearly related and can therefore be plotted on the same axis. (Drawing by Jonathan Meendering. ©2004 by Walter Grondzik; all rights reserved.)*

## (d) The Decibel

Two problems arise immediately when dealing with quantities of the type encountered with sound power, pressure, and intensity. The numbers themselves are very small or very large. Furthermore, the human ear responds logarithmically, not arithmetically, to sound pressure (or intensity); that is, doubling the intensity of a sound does not double its loudness—such a change is barely perceptible. To address these problems, it would be much more convenient if there were a scale that:

1. Started at zero for the minimum sound (intensity or pressure) that can be heard.
2. Used whole numbers rather than powers of 10.
3. Had some fixed relationships between an arithmetic difference and a loudness change, say, 10 units equals a doubling (or halving) of loudness. Thus, on such a scale, the difference between 20 and 30, between 60 and 70, would always be a doubling of loudness.

Such a scale exists. It is the decibel scale.

The word *level* when appended to power, pressure, or intensity indicates a quantity expressed relative to a base quantity—in decibels. *Intensity level* is thus the ratio between a given intensity and a base intensity. If intensity level is expressed as

$$IL = 10 \log \frac{I}{I_0} \qquad (17.6)$$

where

$IL$ = intensity level, dB

$I$ = intensity, W/cm$^2$

$I_0$ = base intensity (i.e., $10^{-16}$ W/cm$^2$, the threshold of hearing)

log = logarithm to base 10

then a scale is established that satisfies the three conditions set forth above. The quantity $IL$ is dimensionless, since it indicates simply a ratio between two numbers. It is expressed in decibels, however, to clarify its status as a ratio quantity. This proves a convenient way to express the large range of sound magnitudes encountered. Table 17.2 shows the great convenience of using the logarithmic decibel scale compared to either decimal notation or exponential notation. Table 17.3 gives a short listing of subjective loudness changes expressed in decibels.

**TABLE 17.3 Intensity Level Changes and Corresponding Subjective Loudness Changes**

| Change in Intensity Level (dB) | Subjective Change in Loudness |
|---|---|
| 3 | Barely perceptible |
| 6[a] | Perceptible |
| 7 | Clearly perceptible |
| 10 | Twice (or half) as loud |
| 20 | Four times (or one-quarter) as loud |

[a]Six decibels corresponds to the change encountered when the distance to the source in a free field is doubled (halved).

Note that *10 dB indicates a doubling of loudness*, and 20 dB is loudness doubled twice—that is, a situation four times as loud. The *difference* ($\Delta$) between any two intensity levels can be expressed as

$$\Delta IL = IL_2 - IL_1 = 10 \log \frac{I_2}{I_0} - 10 \log \frac{I_1}{I_0}$$

Therefore,

$$\Delta IL = 10 \log \frac{I_2}{I_1} \text{ dB} \qquad (17.7)$$

A few examples using decibel notation and logarithmic calculations should help establish this useful system. By the way, the *bel* in decibel is in honor of Alexander Graham Bell (thus the capital B in dB).

**EXAMPLE 17.1** Two sound sources ($I_1$ and $I_2$) produce intensity levels of 60 and 50 dB, respectively, at a point. When these sources are operating simultaneously, what is the total sound intensity level? (Assume identical frequency content and random phasing—that is, the phase relationship between the two sources changes in a random manner.)

**SOLUTION**
Note that this example deals with intensity level, not intensity, since intensity itself has little significance for architectural acoustics. The technique involved in adding two sound intensity levels has three steps:

1. Convert both to actual intensity.

$$IL = 10 \log \frac{I}{I_0}$$

so

$$60 = 10 \log \frac{I_1}{10^{-16}}$$

or

$$6.0 = \log \frac{I_1}{10^{-16}}$$

Then, using the definition of a base 10 logarithm:

$$10^6 = \frac{I_1}{10^{-16}}$$

$$I_1 = (10^{-16})(10^6) = 10^{-10} \text{ W/cm}^2$$

By similar calculation,

$$I_2 = 10^{-11} \text{ W/cm}^2$$

2. Add the intensities arithmetically.

$$I_1 + I_2 = 10^{-10} + 10^{-11}$$
$$= (10 \times 10^{-11}) + 10^{-11}$$
$$I_{tot} = 11 \times 10^{-11} \text{ W/cm}^2$$

3. Reconvert to decibels. To find the intensity level (*IL*) corresponding to the combined or total intensity $I_1 + I_2$, simply apply Eq. 17.6:

$$IL_{tot} = 10 \log \frac{I_{tot}}{I_0}$$

$$= 10 \log \frac{11 \times 10^{-11}}{10^{-16}}$$

$$= 10 (\log 11 + \log 10^5)$$

$$= 10 (1.04 + 5)$$

$$= 60.4 \text{ dB}$$

which is only a fraction larger than the original 60 dB of the stronger of the two sounds. As demonstrated in this example, decibels cannot be added arithmetically. ∎

**EXAMPLE 17.2** Assume two sounds of 60 dB each. What is the combined sound intensity level in decibels?

**SOLUTION**
One method would be to calculate levels as in Example 17.1. A shorter method is to find the difference between the sum and either of the (equal) signals and add it to either individual signal. Using Eq. 17.7:

$$\Delta IL = IL_{comb} - IL_1 = 10 \log \frac{I_{comb}}{I_1}$$

$$= 10 \log \frac{2I_1}{I_1}$$

$$= 10 \log 2$$

$$= 10 (0.30)$$

$$= 3 \text{ dB}$$

This answer (which is independent of any particular sound level) yields the extremely important and useful fact that *doubling a signal's intensity raises the intensity level by 3 dB*. (In this case, the combined intensity level would be 60 dB + 3 dB, or 63 dB.)

Similarly, quadrupling a signal's intensity raises the resulting level by 6 dB. This is because

$$\Delta IL = 10 \log \frac{4I}{I}$$

$$= 10 \log 4$$
$$= 10 (0.60)$$
$$= 6 \text{ dB}$$

Therefore, quadrupling 60 dB gives 66 dB (or, alternatively, 60 dB + 60 dB = 63 dB and 63 dB + 63 dB = 66 dB). This technique is very useful when combining a large number of identical sound levels, as in the following example. ∎

**EXAMPLE 17.3** A factory will contain 20 identical machines, each of which generates a sound intensity level of 80 dB. What will be the combined sound intensity level? (Ignore issues of frequency content, phase, and sound fields.)

**SOLUTION**

$$\Delta IL = IL_{\text{tot}} - IL_{\text{single}}$$

$$= 10 \log \frac{I_{\text{tot}}}{I_{\text{single}}}$$

$$= 10 \log \frac{20 \, I_{\text{single}}}{I_{\text{single}}}$$

$$= 10 \log 20$$
$$= 10 (1.3) = 13 \text{ dB}$$

Therefore, the total sound intensity level will be
$$IL_{\text{tot}} = 80 \text{ dB} + 13 \text{ dB} = 93 \text{ dB}$$ ∎

A chart for combining the decibel levels of two sources is given in Fig. 17.11 that eliminates the somewhat lengthy procedure detailed in Example 17.1. Referring to Table 17.3, note that the human ear is not responsive to fractional decibel changes; indeed, even a 3-dB change is barely perceptible. This being so, it is recommended that the detailed calculations, and even the chart, be reserved for situations where a high degree of accuracy is required.

For everyday calculations, the following approximations may be used to combine the decibel levels of two sources:

- When the difference between two sources is 1 dB or less, add 3 dB to the higher decibel level to obtain the total.
- When the difference is 2 to 3 dB, add 2 dB.
- When the difference is 4 to 8 dB, add 1 dB.
- When the difference is 9 dB or more, ignore the lower-level source (add 0 to the higher).

A comparison in Table 17.4 of addition using these rules and a more accurate method shows that at usual levels, the error resulting from simplification is always less than 1%.

Returning to the inverse square law expressed in Eq. 17.5, it is now possible to determine the effect

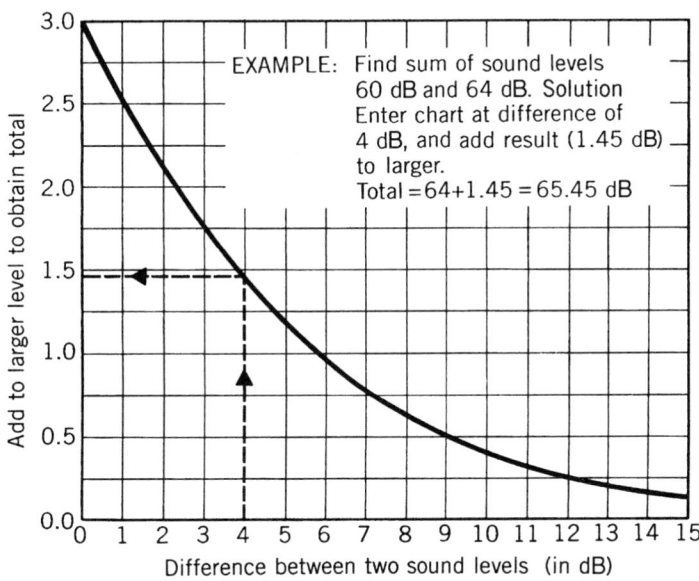

**Fig. 17.11** *Chart for adding two uncorrelated sound pressure levels when both are expressed in decibels. (Reprinted by permission from E.B. Magrab,* Environmental Noise Control, *Wiley, New York, 1975.)*

**TABLE 17.4 Addition of Uncorrelated Sound Pressure Levels**

| dB Levels | | Sum | |
|---|---|---|---|
| Lower | Higher | Approximate[a] | Accurate[b] |
| 60 | 60 | 63 | 63.0 |
| 60 | 62 | 64 | 64.4 |
| 60 | 64 | 65 | 65.5 |
| 60 | 66 | 67 | 67.0 |
| 60 | 68 | 69 | 68.7 |
| 60 | 70 | 70 | 70.5 |

[a]See text for approximation rules.
[b]In architectural acoustics, decimal values of decibels are not warranted and these values would be rounded off to a whole number.

on sound intensity level of moving away from a sound source.

---

**EXAMPLE 17.4** Given a sound source that produces an intensity level *IL* at a distance $d_1$ from a source (substitute any numbers desired or follow the problem with symbols), what is the intensity level at twice the distance? Four times?

*Note:* A sound intensity distribution that obeys the inverse square law on which these calculations are based results from a source in an open (unenclosed), obstruction-free space (e.g., outdoors). Also, intensity measurement $I_1$ must be taken at a sufficient distance $d_1$ from the source that a free field has developed. See the discussion in Section 18.7 for an explanation of sound fields.

**SOLUTION**
From Eq. 17.7 it is known that

$$\Delta IL = \frac{I_2}{I_1}$$

and from Eq. 17.5 that

$$\frac{I_1}{I_2} = \frac{d_2^2}{d_1^2}$$

Therefore, since $d_2 = 2d_1$

$$\frac{I_2}{I_1} = \frac{(d_1)^2}{(2d_1)^2} = \frac{1}{4}$$

Substituting in Eq. 17.7, we have

$$\Delta IL = 10 \log \frac{I_2}{I_1}$$
$$= 10 \log \frac{1}{4}$$
$$= 10 (-0.6)$$
$$= -6 \text{ dB}$$

which tells us that sound intensity level (not pressure) is reduced by 6 dB. Similarly, if the distance is quadrupled, it is reduced by 12 dB. ∎

To summarize, *the intensity level changes by 3 dB with every doubling or halving of power and changes by 6 dB with every doubling or halving of the distance from a point source.* Figures 17.12 and 17.13 illustrate the latter relationship.

### (e) Sound Power Level

Sound power levels may be derived from sound power values using the ratio process described above. The

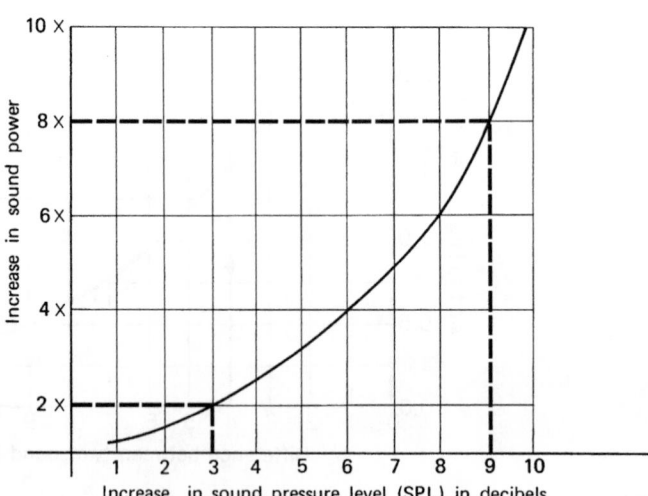

*Fig. 17.12* Decibel level increase as a function of power (intensity) increase.

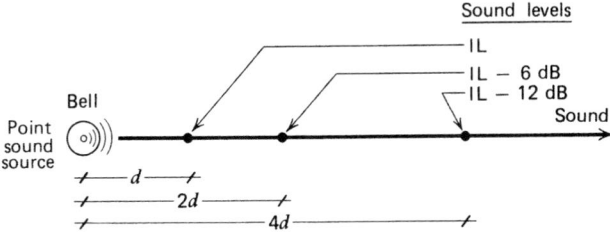

**Fig. 17.13** *Sound energy levels at varying distances from a source. Each doubling of distance reduces the intensity level by 6 dB. These relationships hold true only in a free field.*

reference sound power is $10^{-12}$ W. Sound power level is expressed in decibels.

### (f) Sound Pressure Level

Sound pressure levels are derived from sound pressures using the ratio approach described above. The usual reference sound pressure corresponds to the threshold of hearing and is taken to be 20 µPa or $2 \times 10^{-5}$ Pa ($2 \times 10^{-4}$ microbars [µbar]) (see Table 19.15). As with intensity, this sound pressure reference is established as 0 dB for the purpose of calculating sound pressure *level*. Since the ear responds logarithmically to intensity and since pressure varies as the square root of intensity, we can write the expression

$$SPL = 10 \log \frac{p^2}{p_0^2}$$

or

$$SPL = 20 \log \frac{p}{p_0} \qquad (17.8)$$

where

$SPL$ = sound pressure level, dB

$p$ = pressure, Pa or µbar

$p_0$ = reference base pressure, 20 µPa or $2 \times 10^{-4}$ µbar

Since the 0-dB base corresponds to the hearing threshold for both sound intensity level and sound pressure level, the decibel scales for sound pressure level and sound intensity level have been equalized *and the decibel values of the two can be used interchangeably.* The actual intensity and the actual pressure corresponding to a particular decibel level, however, are different—completely different—in magnitude and units. For instance, 70 dB may equal $10^{-9}$W/cm² intensity or 0.063 Pa pressure. The important fact, though, is that 70 dB corresponds approximately to a particular sound magnitude. It is necessary to say

"approximately" because assigning a single-number decibel level to a sound presents two difficulties:

1. Sound pressure level varies with time, except for a pure steady tone.
2. The different components of most common (complex) sounds vary in pressure level.

Two techniques are used to overcome this problem. If a sound has a dominant frequency, that frequency's level can be used (Fig. 17.8). This would be the case for a relatively constant sound such as that of a motor, fan, or pump. Other sounds that vary widely in constituent level and frequency can be plotted on an octave-band chart using maximum level for minimum percentage of time (Fig. 17.14). Where the position of the listener is not specified in the table, it is assumed to be at normal *close* distances: that is, 10 to 20 ft (3 to 6 m) from a train, 3 to 5 ft (0.9 to 1.5 m) from a radio, and the like.

As suggested previously, the combined effect of two sounds depends upon their frequency content. In the examples above, we assumed signals either of identical frequency and random phase or of a very-wide-frequency spectrum—so wide that phase phenomena are not significant. In architectural acoustics work, such an assumption is generally valid.

### (g) Measuring Sound

The need for a means of measuring sound levels in built projects to confirm that design criteria have been met should be obvious. One very useful instrument is the integrating sound-level meter illustrated in Fig. 17.15. To correlate meter readings with subjective loudness impressions, most such instruments that provide a single-number output are furnished with weighting networks, the characteristics of which are given in Fig. 17.16. The A network corresponds to an inverted 40-phon contour and

ACOUSTICS

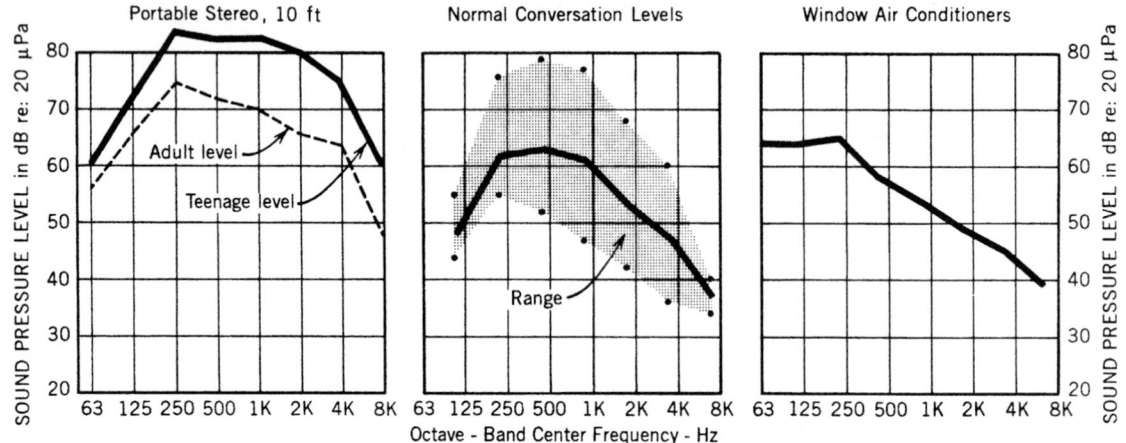

**Fig. 17.14** *Sound pressure level curves of common noise sources plotted according to octave band frequency content. Values shown are averages of multiple measurements. (Reprinted from* A Guide to Air-borne, Impact, and Structure-Borne Noise Control in Multifamily Dwellings, *1968.)*

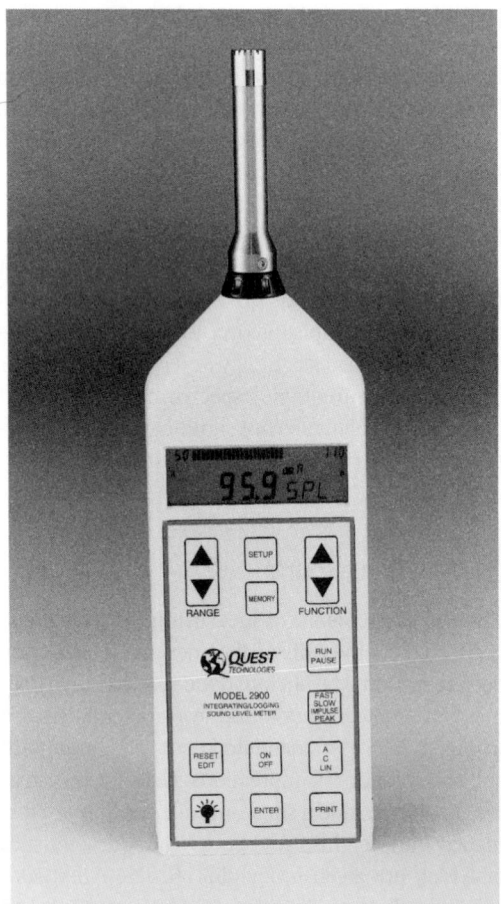

**Fig. 17.15** *Modern general-purpose (type 2) integrating, data logging sound level meter. This unit will measure and record isolated, nonrepetitive noise events in addition to continuous, fluctuating, and impulsive sounds, and will give both instantaneous and equivalent (L_eq) sound levels. The latter are used when an equivalent noise level of varying sound conditions over a selected measurement time is required. The meter will measure dBA, dBC, and linear with fast, slow, peak, or impulse response. When equipped with a filter set, the meter can be used for octave or ½-octave analysis. (Photo courtesy of Quest Technologies, Inc.)*

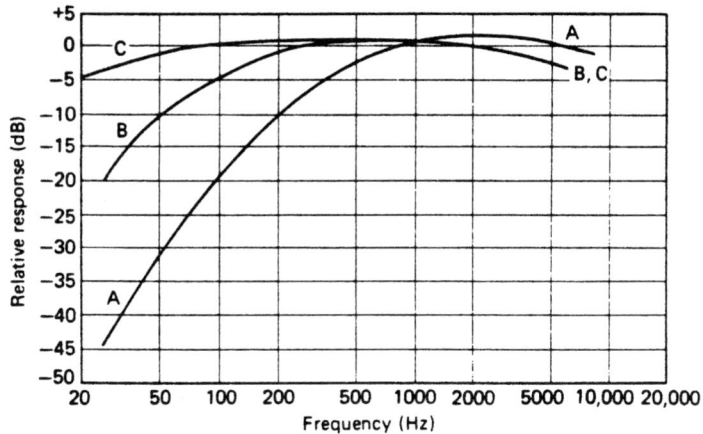

**Fig. 17.16** *Internationally standardized* A, B, *and* C *weighting curves.*

discriminates against low frequencies (see Fig. 17.7), as does the human ear. The B and C networks correspond to the 70-phon and 100-phon contours, respectively. In addition, a completely linear response is usually available. The reasoning behind the use of these weighting networks is evident when they are compared to the equal loudness curves of Fig. 17.7. The original intention was to use the A network at levels of up to 55 phons, the B network to 85 phons, and the C network at higher levels. In practice, however, it was found that only the A network corresponded fairly well to subjective loudness reports. As a result, the B network has fallen into disuse and the A network is used today for all measurements, regardless of loudness. The discrepancy at high loudness levels is apparent when the 40 phon–A weighting network curve is compared to the equal loudness curves above 80 phon. All measurements should be identified with the weighting network used, such as 50 dBA or 100 dBC.

More accurate measurements of complex sounds than are possible with a standard sound level meter are made with sophisticated instruments that measure intensity in octave bands and also often plot the results, as per the graphs in Fig. 17.14. Such measurements are necessary for accurate application of sound absorption and attenuation materials whose characteristics are nonlinear over the frequency spectrum.

Single-number dBA readings are known as *overall* levels and are useful as preliminary data and for broad-spectrum design. Table 17.5 shows a list of common sound levels as measured by the dBA scale. Such single-value numbers are useful to establish a mental-aural comparison base and for use in maximum noise exposure calculations, as discussed below.

## 17.6 NOISE

*Noise* is variously defined as unwanted sound, sound with no intelligible content, and/or broadband sound, depending upon the listener and the situation. Each definition is appropriate for various times and situations. It should generally be assumed that any sound can at some point be referred to as noise by someone.

Although noise effects and their control are the specific subject of Chapter 19, noise criteria and their development are discussed here as part of an overall view of hearing and sound sources. There are two basic approaches to the negative effects of noise; a psychological-practical one and a purely physiological one. The latter is concerned with the physical impact of noise on the body, including hearing loss and other deleterious conditions. The former is concerned with noise levels that cause annoyance and disturbance to daily activities, including work, relaxation, and rest.

### (a) Annoyance

Research has developed accurate data on perceptions of loudness. The concept of annoyance,

**TABLE 17.5 Common Sound Levels**

| Sound Pressure Level (dBA) | Typical Sound | Subjective Impression |
|---|---|---|
| 150 | | (Short exposure can cause hearing loss) |
| 140 | Jet plane takeoff | |
| 130 | Artillery fire, riveting, machine gun | (Threshold of pain) |
| 120 | Siren at 100 ft (30 m), jet plane (passenger ramp), thunder, sonic boom | Deafening |
| 110 | Woodworking shop, hard-rock band, accelerating motorcycle | Sound can be felt (threshold of discomfort) |
| 100 | Subway (steel wheels), loud street noise, power lawnmower, outboard motor | |
| 90 | Noisy factory, unmuffled truck, train whistle, machine shop, kitchen blender, pneumatic jackhammer | Very loud, conversation difficult; ear protection required for sustained occupancy |
| 80 | Printing press, subway (rubber wheels), noisy office, supermarket, average factory | (Intolerable for phone use) |
| 70 | Average street noise, quiet typewriter, freight train at 100 ft (30 m), average radio, department store | Loud, noisy; voice must be raised to be understood |
| 60 | Noisy home, hotel lobby, average office, restaurant, normal conversation | |
| 50 | General office, hospital, quiet radio, average home, bank, quiet street | Usual background; normal conversation easily understood |
| 40 | Private office, quiet home | |
| 30 | Quiet conversation, broadcast studio | Noticeably quiet |
| 20 | Empty auditorium, whisper | |
| 10 | Rustling leaves, soundproof room, human breathing | Very quiet |
| 0 | | Intolerably quiet Threshold of audibility |

however, being primarily subjective and psychological, is much more elusive. Tests have shown that *in general*, annoyance as a result of noise is:

1. Proportional to the loudness of the noise
2. Greater for high-frequency than low-frequency noise
3. Greater for intermittent than continuous noise
4. Greater for pure-tone than for broadband noise
5. Greater for moving or unlocatable (reverberant) noise than for fixed-location noise
6. Much greater for information-bearing noise (a neighbor's radio) than for nonsense noise (foreign language on a neighbor's radio)

### (b) Noise Criteria

To establish criteria for *acceptable background noise* (i.e., noise whose extent of annoyance is considered acceptable), certain of these effects must be neglected at this point for the sake of simplicity. (They can be and are considered in design and in establishing levels of masking noise [see Chapter 19].) Thus, ignore for the time being:

Factor 3, assuming, instead, continuous sounds.

Factor 4, as broadband noise is assumed.

Factor 5, as the noise source is assumed to be fixed in location.

Factor 6, assuming that we can only consider general noise level, not actual content.

Thus, the particular and special characteristics of noises such as a barking dog (3), a whistle (4), a single passing vehicle (5), and intelligible sounds (6) are not considered when establishing conventional noise criteria.

In order to quantify the concept of acceptable background noise, it was necessary to remove it from the purely psychological arena and relate it to a physical phenomenon. This was done by studying the effect of the two remaining annoyance factors (1 and 2) on speech communication. This study resulted in two design concepts: the Articulation Index (AI) and the Speech Interference Level (SIL). These are both determined by reading a carefully selected set of phonetically balanced nonsense syllables to a test audience in the presence of different levels and compositions of background noise. The ratio of correctly identified syllables to total syllables read is the Articulation Index. An AI in excess of 0.5 was considered indicative (Beranek, 1988a) of a condition in which acceptable intelligibility could be expected for *male* voices. A simplified version of the AI, called the Speech Interference Level SIL, was devised by Beranek. It consists simply of the arithmetic average in decibels of the background sound pressure levels in the four octave bands centered on 500, 1000, 2000, and 4000 Hz, for which acceptable intelligibility could be expected, for a given

voice effort, at a given distance between the speaker and listener.

### (c) Noise Criteria Curves

Beranek also developed the well-known and widely accepted and used noise criteria (NC) curves shown in Fig. 17.17. The NC curves take cognizance of the field-determined fact that most people prefer to speak at a level no greater than 22 phons above the background noise level. The NC curves are derived by combining the SILs in decibels with this fact; they

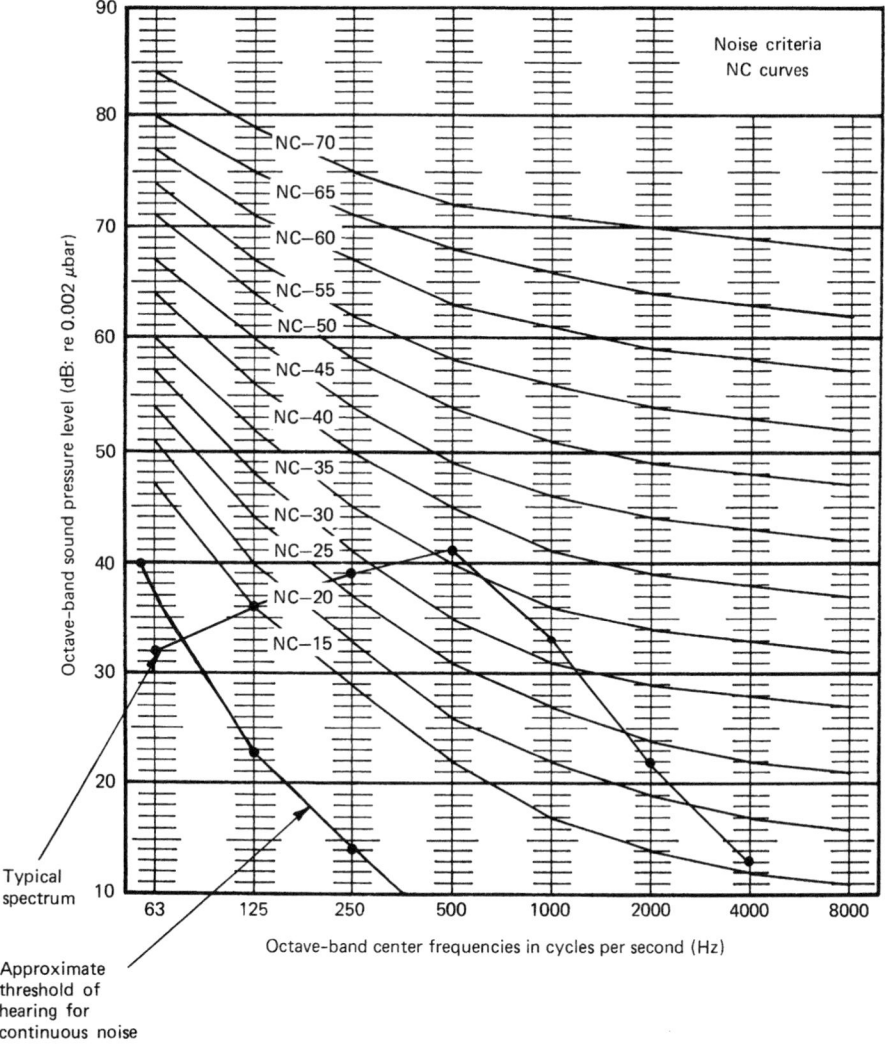

**Fig. 17.17** *Application of NC curves. The typical spectrum plotted would be rated NC-36, as it exceeds NC-35 at 500 Hz by 1 dB. (See Table 19.7 for specific NC recommendations for interior spaces.)*

represent a loudness level 22 phons higher than the SIL in dB. These contours then represent the maximum *continuous* background noise that will be considered acceptable in the environment specified and correspond fairly accurately to background noise levels in commercial environments. A similar set of curves called *noise rating* (NR) curves find considerable application outside the United States. They are less stringent than NC curves in low frequencies but more stringent in high frequencies.

To apply the NC curves, the spectrum of a specific noise being studied, using octave bands over the range of 63 to 8000 Hz, is measured and plotted on a sheet with the NC curves. The lowest NC curve that is not exceeded by any portion of the plot is the NC rating of the particular noise. Thus, specifying a maximum noise level of NC-30 for a space means that no portion of the sound pressure level curve of any continuous background noise in the space may cross the NC-30 contour. A piece of equipment rated NC-35 has an octave-band spectrum completely below NC-35. A fan rated NC-53 indicates that at some point in its frequency spectrum the fan exceeded NC-50 by 3 dB.

The NC rating of a noise falls between 5 and 10 dB below the measured dBA for the noise. The virtue of the NC curves is that they provide a single-number specification for sound across an entire frequency spectrum. Their disadvantage is that they were derived for, and are most accurate with, speech conversation conducted against a backdrop of continuous, non-intelligence-bearing noise. This is not the situation that is typically most troublesome in an office; indeed, continuous equipment noise may be *helpful* in masking unwanted speech. Nevertheless, NC curves remain the most commonly used criteria for establishing acceptable continuous *nonintentional and nonintelligible* background noise levels.

### (d) Room Criteria Curves

Due to the recognized shortcomings of the criteria inherent in the NC curves, specifically that they are undefined in the very low frequencies (16 and 31.5 octave bands) and are not sufficiently stringent at frequencies above 2 kHz, a similar approach to setting criteria was developed—called *room criteria* (RC) curves. These curves, shown in Fig. 17.18, were adopted by ASHRAE (the American Society of Heating, Refrigerating and Air-Conditioning Engineers)

as the suggested noise limitation benchmark in preference to NC curves. The curves were used to evaluate the acceptability of background mechanical system noise for typical space types. RC curves differ from NC curves in a number of aspects:

- They are straight lines.
- Their slope is constant at −5 dB (determined from extensive tests, mostly in the range of 40–50 dB).
- Regions labeled A and B as in Fig. 17.18 address the problem of very low frequencies and high sound pressure levels, an issue that is ignored in the NC criteria. This addition then deals with rumble and vibration that can cause extreme annoyance for many occupants.

Referring to Fig. 17.18, the procedure for determining the RC value of a specific item of equipment whose noise spectrum is known is as follows:

1. Calculate the arithmetic average of the sound pressure levels (SPL) in the 500-, 1000-, and 2000-Hz octave bands. This number is the RC value of that noise spectrum.
2. Draw a straight line at a −5 dB slope through this value of RC at 1000 Hz.
3. Plot the SPL values for the octave band center frequencies on the RC curve sheet and compare the plot to the RC line drawn in Step 2.
4. Classify the equipment sound quality from this comparison as follows:
   a. *Neutral.* If the octave band data plotted in Step 3 do not exceed the RC line drawn in Step 2 by more than 5 dB at or below 500 Hz and more than 3 dB at or above 1000 Hz, then the sound is considered neutral (bland, uncharacteristic) and the designator letter *N* is placed after the RC level. A piece of mechanical equipment with an *N* designation is then understood to have a neutral tone quality that most people would classify as unobtrusive and lacking specific character. ASHRAE design guidelines for HVAC system noise levels are listed as RC (*N*).
   b. *Rumble.* If the octave band plot *does* exceed the RC line by more than 5 dB at 500 Hz or below, the spectrum is classified as "rumbly" and the descriptor letter *R* is appended to its RC level number.
   c. *Hiss.* If the octave band plot *does* exceed the RC line by more than 3 dB at or above

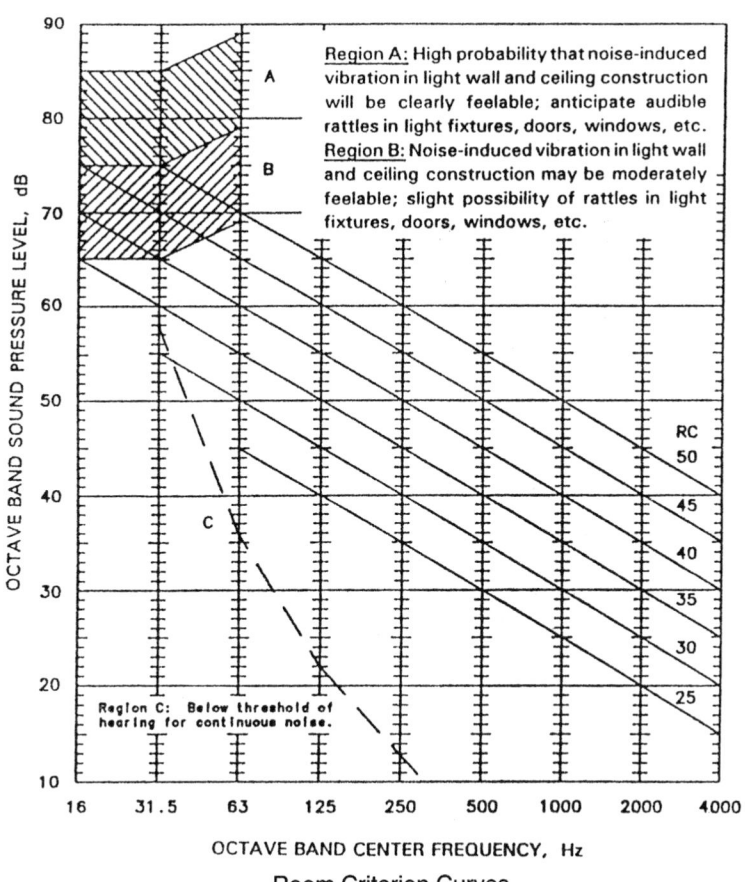

**Room Criterion Curves**

**Fig. 17.18** RC curves. These curves were adopted by ASHRAE in lieu of NC curve criteria. See text for an explanation of their use (Reprinted with permission from ASHRAE Handbook—Applications, 1995.)

ACOUSTICS

1000 Hz it is classified as "hissy" and the descriptor letter *H* is appended to its RC level.

d. *Vibration.* The shaded areas in Fig. 17.18 labeled *A* and *B* represent high sound pressure levels in the 31.5- and 63-Hz octave bands. In the *A* area, vibration will likely be felt in light construction and furniture, and rattling may occur in loosely constructed devices, cabinets, glassware, and the like. In the lower-energy *B* area, rattling will be less likely but vibration may still be felt.

The result of this classification is to give the designer not only an average SPL for an item of equipment, but also a sense of subjective sound quality that should be of considerable assistance in determining the noise abatement measures to be taken.

ASHRAE has updated the RC curve concept to what is now called the *RC Mark II* method. This updated method is a bit more complex than the original RC approach. For further information, see the ASHRAE *Handbook—HVAC Applications* (2003). The original NC curves have also been modified (by Beranek) and issued as balanced noise criteria (NCB) curves. They amended a failing of the NC approach by adding very-low-frequency coverage and the NCB curves were straightened, so that they resemble RC curves above 125 Hz, except that the slope angle is −3.33 dB compared to −5 dB for the RC curves. The slope at the higher frequencies has also eliminated the hiss that characterizes equipment noise conforming to the NC criteria. These adjustments have yielded a "balanced" neutral sound, which reflects the chart's name. Application of these curves to mechanical equipment noise is similar to that for RC curves, except that letter descriptors are not used. For a full discussion of the construction and application of BNC curves, see Beranek (1988b). Given these competing options for criteria, selection of NC, RC, RC Mark II, or NCB criteria for background noise is a function of the design intent. The approach that best meets the needs (and budget) of the owner should be chosen.

## (e) High Noise Levels and Hearing Protection

It has long been recognized that continuous exposure to high noise levels causes a degree of temporary deafness in most people and that long periods of such exposure, even on an intermittent 8-hour workday basis, can produce permanent hearing impairment. Most experts place the safe 8-hour

---

### § 1910.95 Occupational noise exposure.

(a) Protection against the effects of noise exposure shall be provided when the sound levels exceed those shown in Table G-16 when measured on the A scale of a standard sound level meter at slow response. When noise levels are determined by octave band analysis, the equivalent A-weighted sound level may be determined as follows:

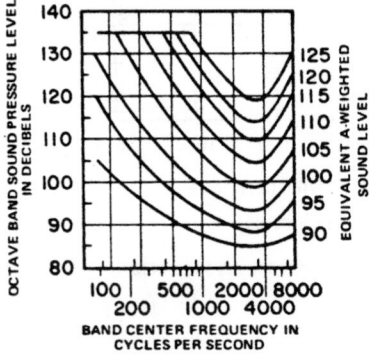

Equivalent sound level contours. Octave band sound pressure levels may be converted to the equivalent A-weighted sound level by plotting them on this graph and noting the A-weighted sound level corresponding to the point of highest penetration into the sound level contours. This equivalent A-weighted sound level, which may differ from the actual A-weighted sound level of the noise, is used to determine exposure limits from Table G-16.
[1910.95 amended at 39 FR 19468, June 3, 1974]

(b)(1) When employees are subjected to sound exceeding those listed in Table G-16, feasible administrative or engineering controls shall be utilized. If such controls fail to reduce sound levels within the

levels of Table G-16, personal protective equipment shall be provided and used to reduce sound levels within the levels of the table.

(2) If the variations in noise level involve maxima at intervals of 1 second or less, it is to be considered continuous.

(3) In all cases where the sound levels exceed the values shown herein, a continuing, effective hearing conservation program shall be administered.

TABLE G-16—PERMISSIBLE NOISE EXPOSURES[1]

| Duration per day, hours | Sound level dBA slow response |
|---|---|
| 8 | 90 |
| 6 | 92 |
| 4 | 95 |
| 3 | 97 |
| 2 | 100 |
| 1½ | 102 |
| 1 | 105 |
| ½ | 110 |
| ¼ or less | 115 |

[1] When the daily noise exposure is composed of two or more periods of noise exposure of different levels, their combined effect should be considered, rather than the individual effect of each. If the sum of the following fractions: $C_1/T_1 + C_2/T_2 + \cdots C_n/T_n$ exceeds unity, then, the mixed exposure should be considered to exceed the limit value. Cn indicates the total time of exposure at a specified noise level, and Tn indicates the total time of exposure permitted at that level.

[1910.95 Table G-16 amended at 39 FR 19468, June 3, 1974]

Exposure to impulsive or impact noise should not exceed 140 dB peak sound pressure level.

---

**Fig. 17.19** *The standard for exposure to noise in the workplace. (From OSHA, July 1988; current as of late 2004.)*

upper limit at 85 dBA. In addition, studies have indicated that continual exposure to noise levels as low as 75 to 85 dBA can produce or contribute to numerous physical and psychological ailments, including headache, digestive problems, tachycardia, high blood pressure, anxiety, and nervousness—an extensive catalog of human illnesses. Since continuous noise exposure is most severe in industry, regulatory legislation in the United States has been directed at this area.

In 1969 the Walsh–Healy Public Contracts Act was passed, and thereafter its provision for maximum permissible exposure to noise levels was incorporated into the Occupational Safety and Health Act. Both the act and the associated regulatory agency, the Occupational Safety and Health Administration, are known as OSHA. The relevant provisions of this act are reproduced in Fig. 17.19. To avoid overly complex regulations, limitations on exposure are given as single-number dBA values. Since workers rarely remain in a single acoustic environment for 8 hours, their total daily exposure, or *time-weighted average* (TWA) exposure, can be calculated from timed dBA measurements using formulas and tables given by OSHA and then compared to permissible levels. Alternatively, a dosimeter (Fig. 17.20) can be used, which automatically integrates the noise to which a person is exposed over a given time period and reads out the permissible TWA exposure directly.

Typical industrial noise levels are given in Table 17.6. When permissible levels are exceeded, management must take steps to reduce the exposure, either by reducing the noise or by providing hearing protectors. Typical characteristics of a few types of ear protectors are given in Fig. 17.21. Photographs of three of the most common ear protectors in industrial use are shown in Fig. 17.22, and their laboratory-measured mean attenuation spectra are given in Table 17.7.

OSHA does not deal extensively with impulse noises, except to state that they shall not exceed a 140-dB peak sound pressure level. Impulse noise is quite different from continuous noise, since apprehended noise levels depend on duration, and specification is difficult. Much work has been done in this area by the military for obvious reasons. The

**Fig. 17.20** *A noise dosimeter. The unit is carried by a person whose noise exposure is being tested, with the clip-mounted microphone at shoulder (ear) level. The meter will display, store, and calculate maximum, minimum, and peak levels of SPL, $L_{eq}$, TWA, exposure levels, and dosage. Noise level histories are stored and can be printed out as desired. The unit shown measures 5.5 × 2.8 × 1.4 in. (140 × 70 × 40 mm) and weighs 15.5 oz (440 g). (Photo courtesy of Quest Technologies.)*

**TABLE 17.6 Typical Industrial Noise Levels**[a]

| Equipment | dBA |
|---|---|
| Printing press plant (medium-sized automatic) | 86 |
| Heavy diesel-propelled vehicle (about 25 ft [7 m] away) | 92 |
| Heavy-duty grinder | 93 |
| Air compressor | 94 |
| Plastic chipper | 96 |
| Cutoff saw | 97 |
| Multiple spot welder | 98 |
| Turbine condenser | 98 |
| 15-cu-ft (425-L) air compressor | 100 |
| Drive gear | 103 |
| Banging of steel plate | 104 |
| Magnetic drill press | 106 |
| Air chisel | 106 |
| Positive displacement blower | 107 |
| Air hammer | 107 |
| Vacuum pump | 108 |
| Jolt squeeze hammer | 122 |

[a]These are approximate values for some typical generic equipment types and should not be used as design values.

ACOUSTICS

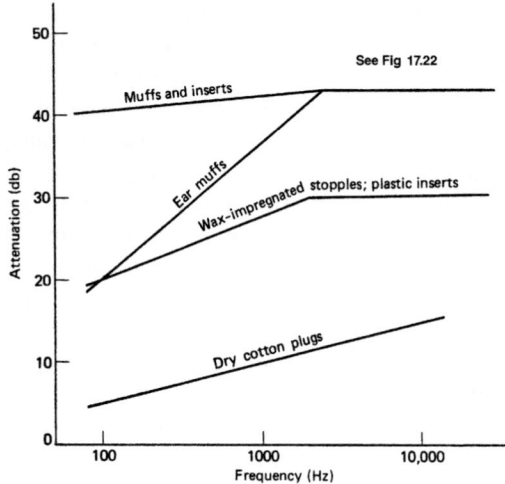

**Fig. 17.21** *Sound attenuation characteristics typical of various types of ear protectors. For design purposes, specific attenuation characteristics must be used and the attenuation figures modified to reflect field usage. Three common types of ear protectors are shown in Fig. 17.22. (From* Quieting: A Practical Guide to Noise Control, *1976.)*

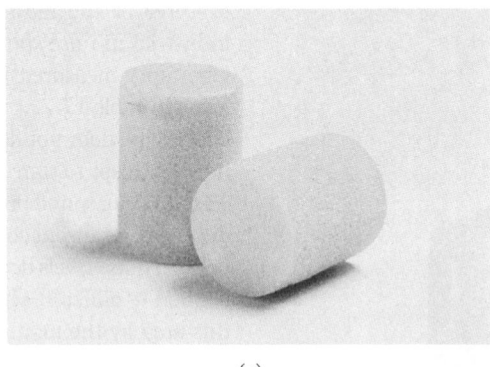

(*a*)

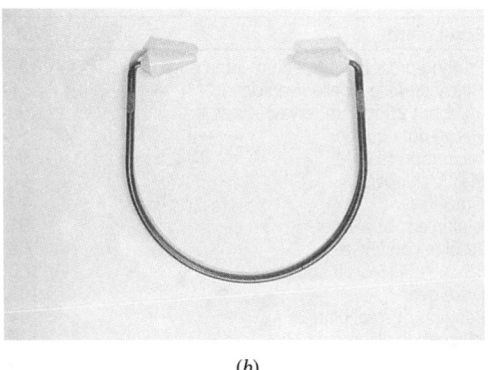

(*b*)

(*c*)

**Fig. 17.22** *Three common types of ear protection devices: (a) PVC foam ear plugs are rolled and squeezed to reduce their diameter before being placed in the ear canal, where they reexpand to their previous size. The devices have an EPA noise reduction rating (NRR) of 20 when tested to ANSI 53.19-1974. They are generally discarded after a single use. (b) Ear protector for intermittent use consists of (replaceable) foam pads mounted on a flexible neck band. The foam does not require rolling. The unit is rated NRR 20. (c) Spring-type head band and foam-filled ear cushions provide an NRR rating of 20 for this earmuff design. Mean attenuation characteristics for the three types of ear protectors are given in Table 17.7. (Photos courtesy of E-A-R Specialty Composites, a Div. of AEARO Company.)*

**TABLE 17.7 Mean Attenuation Data (dBA) for Ear Protectors**[a]

| 125 | 250 | 500 | 1000 | 2000 | 3150 | 4000 | 6300 | 8000 | NRR[c] |
|------|------|------|------|------|------|------|------|------|------|
| FOAM EARPLUGS (FIG. 17.22a) | | | | | | | | | |
| 37.4 | 40.9 | 44.8 | 43.8 | 36.3 | 41.9 | 42.6 | 46.1 | 47.3 | 29 |
| SEMIAURAL HEARING PROTECTOR (FIG. 17.22b) | | | | | | | | | |
| 28.7 | 28.3 | 28.0 | 28.6 | 32.2 | 42.1 | 44.3 | 47.2 | 44.6 | 20 |
| EARMUFF (FIG. 17.22c) | | | | | | | | | |
| 11.7 | 16.2 | 26.5 | 31.6 | 32.5 | 35.0 | 38.1 | 41.8 | 41.8 | 20 |

[a]Data extracted with permission from published AEARO catalog material.
[b]Test frequency (Hz).
[c]NRR, noise reduction rating.

interested reader is referred to the literature since this subject is substantially outside the realm of architectural acoustics.

## 17.7 VIBRATION

Sound is heard; vibration is felt. If sound is defined as a pressure variation that is audibly received, then vibration is a tactilely received pressure variation. Vibrations of concern to building design have frequencies that begin just below the range of human hearing at around 20 Hz. For someone with a hearing threshold above 20 Hz, vibration frequency might be higher. The means of reception of the energy is more important than specific frequency cutoffs. In general, vibration in a building is always an unwanted experience. Most mechanical and electrical equipment produces energy output that will be sensed as noise and output that will be sensed as vibration. Various means of vibration control are presented in Chapter 19. As vibration is not received through the sense of hearing, vibration mitigation approaches are quite different from noise mitigation approaches.

## References

ASHRAE. 1995. *ASHRAE Handbook—HVAC Applications.* American Society of Heating, Refrigerating and Air-Conditioning Engineers, Atlanta, GA.

ASHRAE. 2003. *ASHRAE Handbook—HVAC Applications.* American Society of Heating, Refrigerating and Air-Conditioning Engineers, Inc., Atlanta, GA.

Beranek, L.L. 1988a. *Acoustical Measurements*, revised ed. Acoustical Society of America, Melville, NY.

Beranek, L.L. 1988b. *Noise and Vibration Control,* revised ed. Institute of Noise Control Engineering, Washington, DC.

HUD. 1968. *A Guide to Airborne, Impact, and Structure-Borne Noise Control in Multifamily Dwellings.* U.S. Department of Housing and Urban Development, Washington, DC.

Magrab, E.B. 1975. *Environmental Noise Control.* John Wiley & Sons, New York.

NBS. 1976. *Quieting: A Practical Guide to Noise Control.* National Bureau of Standards (now NIST) (Handbook 119), Washington, DC.

OSHA. 2004. http://www.osha.gov/pls/oshaweb/

White, F.A. 1975. *Our Acoustic Environment,* John Wiley & Sons, New York.

**ACOUSTICS**

# Sound in Enclosed Spaces

CHAPTER 17 PRESENTED THE FUNDAMENTALS OF SOUND theory, some characteristics of human hearing and speech, and information on the negative effects of unwanted and excessive sounds. This chapter will discuss in some detail the interreaction between an enclosing space and a sound generated within it, which is essentially the definition of room acoustics.

## 18.1 SOUND IN ENCLOSURES

When a continuous sound is generated in an enclosure, fields are set up as described in Section 18.7. When the sound is not a continuous tone or noise but a series of discrete sounds, following one another and containing information as in speech or music, the room must be designed to maintain and enhance the information's intelligibility. That is what is meant by *design of room acoustics*.

Generated sound radiates out from a source until it strikes a room boundary or another large surface. Before reaching this surface, the sound intensity is attenuated by distance (the inverse square law—Section 17.5) and by absorption in the air. The latter is appreciable only in large rooms and at frequencies above 2000 Hz. When the sound reaches a wall, it is partially reflected and partially absorbed, and a small portion is transmitted into adjoining spaces. The energy transmitted has little effect on the space within which the sound originates, although, as discussed in Chapter 19, it may be very important in the surrounding spaces.

The ratio between the energy absorbed and the energy reflected by a surface will significantly affect what one hears within a space. Specifically, if little energy is absorbed and much is reflected, two effects will be noticeable. Intermittent sounds will be mixed together (which may make speech *less* intelligible or music more pleasant), and steady sounds will accumulate into a reverberant field, making the space noisy. Conversely, if much energy is absorbed and little reflected, the room will sound quiet for speech and "dead" for music. Quantification of these two primary characteristics of an enclosed space—absorption and reverberation (echo)—is the subject of the next six sections.

## ABSORPTION

## 18.2 SOUND ABSORPTION

When sound energy impinges on a material, part is reflected and the remainder is absorbed (in the sense that it is not reflected). Some of the "absorbed" energy is transmitted, although that part is so small that for our discussion here it will be ignored. Materials are neither perfect reflectors nor perfect absorbers. The term used to define a material's sound absorption characteristic is its *coefficient of absorption*, which is usually represented by the lowercase Greek letter alpha ($\alpha$). This sound absorption coefficient is defined as

$$\alpha = \frac{I_a}{I_i} \qquad (18.1)$$

where

$I_a$ = sound power density (intensity) absorbed by the material, W/cm$^2$

$I_i$ = intensity impinging on the material, W/cm$^2$

$\alpha$ = absorption coefficient, with no units since it represents a ratio

This absorption coefficient $\alpha$ can also be be thought of as a measure of absorption effectiveness; the larger the absorption coefficient, the more effective a sound absorber the material is. Thus, an absorption coefficient of 1.0 indicates 100% absorption and zero reflection of the impinging sound energy. We have already said that no material is a perfect absorber, but an absence of material, that is, an open space, transmits (absorbs) all the impinging energy and therefore can be considered a perfect absorbing "material."

Since open space has this characteristic, $\alpha$ has also been defined as the ratio between the absorption of a given material and that of an *open window* of the same area. By definition, then, for an opening (open window, open door, etc.)

$$\alpha = 1.0$$

The total absorption $A$ of a given quantity of material is proportional to its area and its absorption coefficient, that is,

$$A = S\alpha \qquad (18.2)$$

where

$A$ = total absorption, sabins

$S$ = surface area, square feet or square meters

$\alpha$ = coefficient of absorption

Since $\alpha$ is a ratio and thus unitless, and $S$ is a unit of area, $S\alpha$ should be in units of area as well. Instead, sound absorption units are called *sabins* in honor of W.C. Sabine, a pioneer in architectural acoustics. One sabin (m$^2$) is the sound absorption equivalent of an open window 1 m$^2$ in area. Similarly, 1 sabin (ft$^2$) is equivalent to 1 ft$^2$ of open window. As expected, 1 sabin (m$^2$) equals 10.76 sabins (ft$^2$). In

this text, when not otherwise specified, a sabin (ft$^2$) is intended.

All rooms are constructed of several materials, each having a different absorption coefficient. In addition, most rooms contain furnishings, which have their own individual coefficients of absorption. Thus, to determine the total absorption of a room, it is necessary to sum the component absorptions, that is,

$$\Sigma\, S\alpha = S_1\alpha_1 + S_2\alpha_2 + \cdots + S_n\alpha_n$$

or

$$\Sigma\, A = A_1 + A_2 + \cdots + A_n \qquad (18.3)$$

where

$\Sigma\, S\alpha$ = total absorption in the room, sabins

$S_1$, $S_2$, etc. = area of each material

$\alpha_1$, $\alpha_2$, etc. = absorption coefficient of each material

$A_1$, $A_2$, etc. = total absorption of each material

If $S$ is expressed in square feet, then $A$ is in sabins (ft$^2$); if $S$ is expressed in square meters, then $A$ is sabins (m$^2$).

Absorption coefficients for some common materials, sound-absorbing materials, and auditorium furnishings are tabulated in Table 18.1. It is important to note that for most common materials, absorption (and therefore $\alpha$) varies with frequency. Thus, accurate absorption calculations must be made individually for the frequencies being studied.

## 18.3 MECHANICS OF ABSORPTION

At this point, it is appropriate to examine absorption as an acoustic phenomenon so that we may understand the use of absorptive materials. In an untreated room of normal construction (Fig. 18.1a), when the sound waves strike the walls or ceiling, a small portion of the sound is transmitted, a small portion is absorbed, and most of it is reflected. The exact proportions depend upon the nature of the construction. When acoustical treatment is applied to the room surfaces as in Fig. 18.1b, some of the energy in the sound waves is dissipated before the sound reaches the wall. The transmitted portion is slightly reduced, but the reflection is greatly reduced.

## TABLE 18.1  Octave Band Average Sound Absorption Coefficients[a]

| General Building Materials and Furnishings[b] | | Absorption Coefficients ($\alpha$) | | | | | | |
|---|---|---|---|---|---|---|---|---|
| | | 125 Hz | 250 Hz | 500 Hz | 1000 Hz | 2000 Hz | 4000 Hz | NRC[c] |
| Brick, unglazed | | 0.03 | 0.03 | 0.03 | 0.04 | 0.05 | 0.07 | 0.005 |
| Brick, unglazed, painted | | 0.01 | 0.01 | 0.02 | 0.02 | 0.02 | 0.03 | 0.00 |
| Carpet, heavy, on concrete | | 0.02 | 0.06 | 0.14 | 0.37 | 0.60 | 0.65 | 0.29 |
| Carpet, heavy, on 40-oz hairfelt or foam rubber | | 0.08 | 0.24 | 0.57 | 0.69 | 0.71 | 0.73 | 0.55 |
| Concrete block, coarse | | 0.36 | 0.44 | 0.31 | 0.29 | 0.39 | 0.25 | 0.35 |
| Concrete block, painted | | 0.10 | 0.05 | 0.06 | 0.07 | 0.09 | 0.08 | 0.05 |
| Fabrics | | | | | | | | |
|   Light velour, 10 oz/yd$^2$, hung straight, in contact with wall | | 0.03 | 0.04 | 0.11 | 0.17 | 0.24 | 0.35 | 0.15 |
|   Medium velour, 14 oz/yd$^2$, draped to half area | | 0.07 | 0.31 | 0.49 | 0.75 | 0.70 | 0.60 | 0.55 |
|   Heavy velour, 18 oz/yd$^2$, draped to half area | | 0.14 | 0.35 | 0.55 | 0.72 | 0.70 | 0.65 | 0.60 |
| Floors | | | | | | | | |
|   Concrete or terrazzo | | 0.01 | 0.01 | 0.015 | 0.02 | 0.02 | 0.02 | 0.00 |
|   Linoleum, asphalt, rubber, or cork tile on concrete | | 0.02 | 0.03 | 0.03 | 0.03 | 0.03 | 0.02 | 0.05 |
|   Wood | | 0.15 | 0.11 | 0.10 | 0.07 | 0.06 | 0.07 | 0.10 |
| Glass | | | | | | | | |
|   Large panes of heavy plate glass | | 0.18 | 0.06 | 0.04 | 0.03 | 0.02 | 0.02 | 0.05 |
|   Ordinary window glass | | 0.35 | 0.25 | 0.18 | 0.12 | 0.07 | 0.04 | 0.15 |
| Gypsum board, ½ in. nailed to 2 × 4's 16 in. o.c. | | 0.10 | 0.08 | 0.05 | 0.03 | 0.03 | 0.03 | 0.05 |
| Marble or glazed tile | | 0.01 | 0.01 | 0.01 | 0.01 | 0.02 | 0.02 | 0.00 |
| Openings | | | | | | | | |
|   Stage, depending on furnishings | | | | | 0.25–0.75 | | | |
|   Deep balcony, upholstered seats | | | | | 0.50–1.00 | | | |
|   Grilles, ventilating | | | | | 0.15–0.50 | | | |
| Plaster, gypsum or lime, smooth finish on tile or brick | | 0.013 | 0.015 | 0.02 | 0.03 | 0.04 | 0.05 | 0.05 |
| Plaster, gypsum or lime, on lath | | 0.14 | 0.10 | 0.06 | 0.05 | 0.04 | 0.03 | 0.05 |
| Plywood paneling, ⅜ in. thick | | 0.28 | 0.22 | 0.17 | 0.09 | 0.10 | 0.11 | 0.15 |
| Rough wood, as tongue-and-groove cedar | | 0.24 | 0.19 | 0.14 | 0.08 | 0.13 | 0.10 | 0.14 |
| Slightly vibrating surface (e.g., hollow core door) | | 0.02 | 0.02 | 0.03 | 0.03 | 0.04 | 0.05 | 0.03 |
| Readily vibrating surface (e.g., thin wood paneling on 16-in. studs) | | 0.10 | 0.07 | 0.05 | 0.04 | 0.04 | 0.05 | 0.05 |
| Water surface, as in a swimming pool | | 0.008 | 0.008 | 0.013 | 0.015 | 0.020 | 0.025 | 0.00 |

| Absorption of Seats and Audience[d] | 125 Hz | 250 Hz | 500 Hz | 1000 Hz | 2000 Hz | 4000 Hz | NRC[c] |
|---|---|---|---|---|---|---|---|
| Audience, in upholstered seats, per ft$^2$ of floor area | 0.60 | 0.74 | 0.88 | 0.96 | 0.93 | 0.85 | — |
| Unoccupied cloth-upholstered seats, per ft$^2$ of floor area | 0.49 | 0.66 | 0.80 | 0.88 | 0.82 | 0.70 | — |
| Wooden pews, occupied, per ft$^2$ of floor area | 0.57 | 0.61 | 0.75 | 0.86 | 0.91 | 0.86 | — |
| Students in tablet-arm chairs, per ft$^2$ of floor area | 0.30 | 0.42 | 0.50 | 0.85 | 0.85 | 0.84 | |

| Acoustic Absorptive Materials | Mtg[e] | 125 Hz | 250 Hz | 500 Hz | 1000 Hz | 2000 Hz | 4000 Hz | NRC[c] |
|---|---|---|---|---|---|---|---|---|
| High-performance vinyl-faced fiberglass ceiling panels | | | | | | | | |
|   1 in. thick | E405 | 0.73 | 0.88 | 0.71 | 0.98 | 0.96 | 0.77 | 0.90 |
|   1.5 in. thick | E405 | 0.79 | 0.98 | 0.83 | 1.03 | 0.98 | 0.80 | 0.95 |
| Painted nubby glass cloth panels | | | | | | | | |
|   ¾ in. thick | E405 | 0.81 | 0.94 | 0.65 | 0.87 | 1.00 | 0.96 | 0.85 |
|   1 in. thick | E405 | 0.78 | 0.92 | 0.79 | 1.00 | 1.03 | 1.10 | 0.95 |
| Random fissured ¾-in.-thick panels | E405 | 0.52 | 0.58 | 0.60 | 0.80 | 0.92 | 0.80 | 0.70 |
| Perforated metal panel with infill 1 in. thick | E405 | 0.70 | 0.86 | 0.74 | 0.88 | 0.95 | 0.86 | 0.85 |
| Typical averages, mineral fiber tiles and panels | | | | | | | | |
|   ¾ in. fissured | E405 | 0.47 | 0.50 | 0.52 | 0.76 | 0.86 | 0.81 | 0.65 |
|   ¾ in. textured | E405 | 0.49 | 0.55 | 0.53 | 0.80 | 0.94 | 0.83 | 0.70 |
|   ⅝ in. fissured | E405 | 0.28 | 0.33 | 0.66 | 0.73 | 0.74 | 0.75 | 0.60 |
|   ⅝ in. textured | E405 | 0.29 | 0.35 | 0.66 | 0.63 | 0.44 | 0.34 | 0.50 |
|   ⅝ in. perforated | E405 | 0.27 | 0.29 | 0.55 | 0.78 | 0.69 | 0.53 | 0.60 |
|   3 in. thick × 16 in. square on 24-in. centers | A | 0.40 | 0.61 | 1.92 | 2.54 | 2.62 | 2.60 | |

[a]This table will be primarily useful in making preliminary calculations. Complete tables of coefficients of the various materials that normally constitute the interior finish of rooms may be found in books on architectural acoustics.

[b]Selected data courtesy of Owens-Corning Fiberglass.

[c]Noise reduction coefficient: the arithmetic average of the $\alpha$ values at 250, 500, 1000, and 2000 Hz.

[d]When the audience is randomly spaced, use an average of 5.0 sabins (ft$^2$) per person.

[e]See Fig. 18.4 for mounting methods.

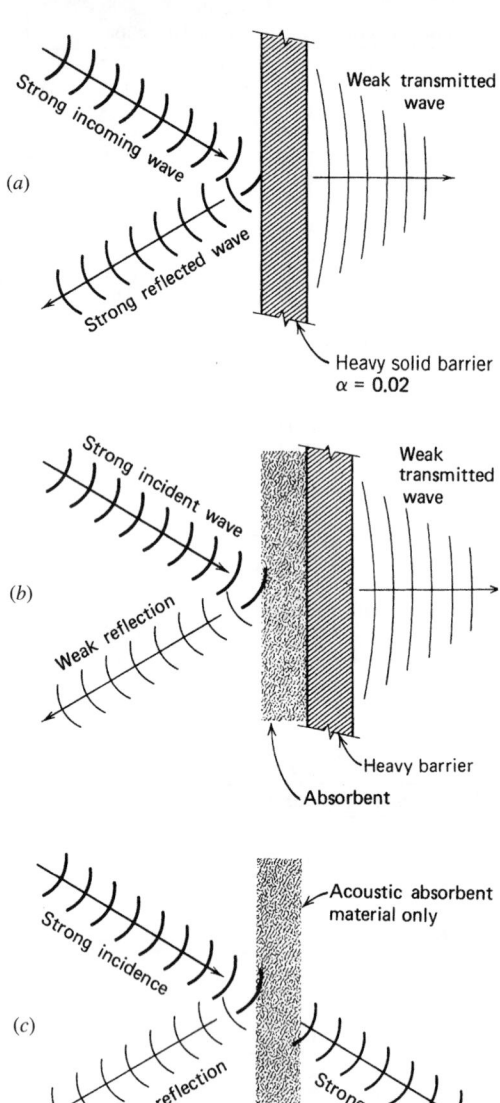

**Fig. 18.1** *(a) Action of an incoming sound wave striking a heavy barrier. Much of the energy is reflected, some is absorbed, and a little is transmitted. (b) When absorbent material is applied to a heavy wall, it "traps" sound, preventing reflection, while the wall mass acts to reduce transmission. (c) Action of acoustic absorbent material alone. Very little energy is reflected, some is absorbed, and most is transmitted.*

It is important to understand, however, that the principal effect of absorptive material is on the reflected sound. The transmitted sound energy is essentially determined by the mass of the solid airtight barrier between the two spaces. This is graphically represented in Fig. 18.1*c*. The effect of added

acoustic absorption in a space is shown in Fig. 18.2 and is calculated in Section 18.9.

## 18.4 ABSORPTIVE MATERIALS

We will now examine acoustic materials and the effect of varying type, quantity, thickness, and installation methods. There are three families of devices for sound absorption—fibrous materials, panel resonators, and volume resonators. All three types absorb sound by changing sound energy into heat energy. Only fibrous materials and panel resonators are used commonly in buildings. Volume resonators, also known as *Helmholtz resonators,* after their originator, are used principally as devices for absorbing a narrow band of frequencies. The discussion in this section refers to fibrous absorbers; the other two types are discussed in Section 19.2.

Fibrous materials absorb acoustic energy by the frictional drag of air moving in the tiny spaces between the fibers. The absorption provided by a specific material depends upon its thickness, density, porosity, and resistance to airflow. Since the action depends upon absorbing energy by "pumping" air through the material, the air paths *must extend from one side to the other.* A fibrous material with sealed pores is almost useless as an acoustic absorbent. (Therefore, painting will generally ruin a porous absorber.) A simple test is to blow smoke through the material. If the smoke passes through freely and the material is porous, fibrous, and thick, it should be a good sound absorbent. Absorbency increases with increasing material porosity up to approximately 70% porosity; above that figure absorbency remains fairly constant.

Table 18.1 gives absorption coefficients for fibrous absorbent materials and for some other building materials and furnishings. Several important conclusions can be drawn from examination of this table and Fig. 18.3:

1.  For absorbent materials, absorption is normally higher at high frequencies than at low ones.
2.  Absorption is not always proportional to thickness, but depends upon the type of material used and the method of installation (see Fig. 18.3). It is clear from this figure that beyond a nominal thickness, little is to be gained by additional thickness except at very low frequencies,

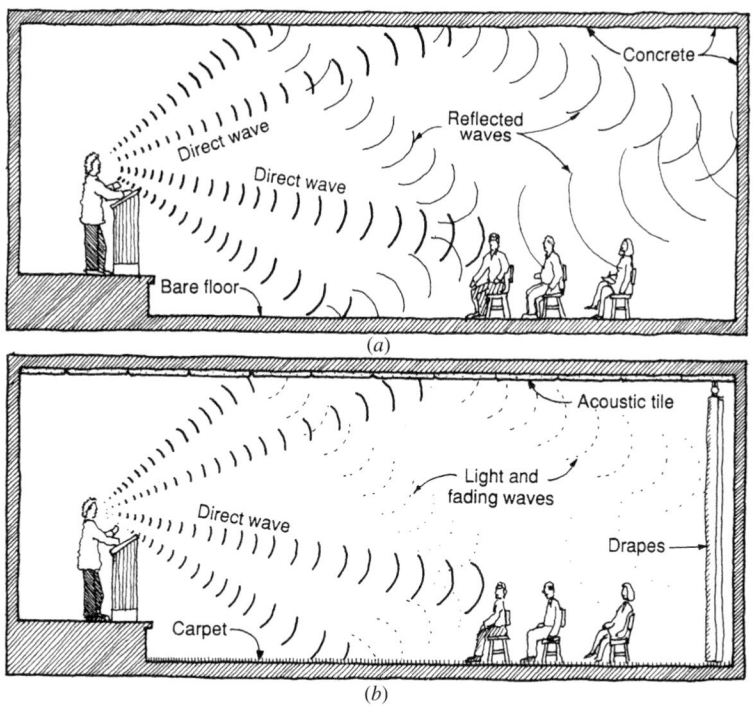

**Fig. 18.2** *In an untreated space (a) reverberant (reflected) sound constitutes a large portion of received sound in much of the room. These reflections are largely eliminated in (b) by wall and ceiling absorption. Note that direct wave sound is completely unaffected.*

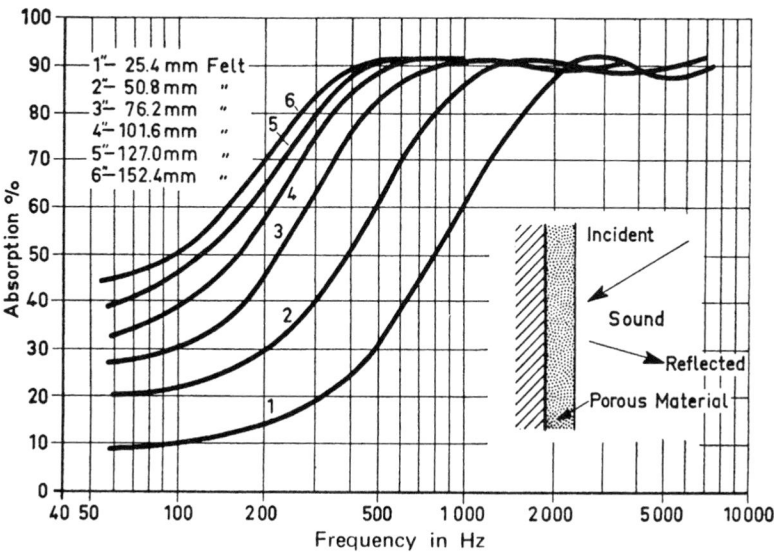

**Fig. 18.3** *Variation of the absorption coefficient with the thickness of felt absorbent. Note particularly that beyond 1 kHz, all thicknesses give the same* α, *whereas at low frequencies absorption is proportional to thickness. Note also that a very heavy layer is required to give appreciable absorption at low frequency. (Courtesy of Brüel and Kjaer.)*

or when installed discontinuously, as described in the next item.

3. It is possible to obtain an α greater than 1.0 by using very thick blocks. As they are installed at a distance from each other, and their edge absorption is very large (particularly at high frequencies), they exhibit an α greater than 1.0.

4. Installation methods have a pronounced effect, as discussed in the following section.

5. All other factors remaining constant, the thicker an absorbent material, the better its low-frequency absorption characteristic.

## 18.5 INSTALLATION OF ABSORPTIVE MATERIALS

Coefficient ratings for absorptive materials are always given with mountings corresponding to ASTM (2000) requirements. The most common standardized mounting methods are shown in Fig. 18.4. Installation of absorbent material directly on a wall or ceiling is the least effective means, since exposure to sound energy is minimal. When an air gap is left between the porous layer and the rigid surface, the combination acts almost as well in midfrequencies as an absorbent layer equivalent in thickness to the air plus porous material (Fig. 18.5). One problem with

this technique, however, is that at the λ/2 node of a standing wave there is a severe drop in absorption, as can be seen in Fig. 18.5c. At 1000 Hz, one-half wavelength is approximately 7 in. At that distance, α drops severely but is a maximum at λ/4 or 3.5 in.

For ceiling tile hung at 16 in. below the slab (Fig. 18.4, method E405), the drop in absorption occurs at

$$\lambda/2 = 16 \text{ in.}$$

$$\lambda = 32 \text{ in.} = 2.67 \text{ ft}$$

$$f = \frac{1128}{2.67} = 422 \text{ Hz}$$

which is midfrequency. This factor should be considered in applying absorptive material. Avoid a spacing corresponding to a drop in absorption at a sensitive frequency. To obtain good low-frequency absorption, it is essential that a deep air space is provided behind the absorbent material and that walls are treated in addition to the ceiling.

In increasing order of effectiveness, absorbent material can be applied:

1. Directly to the room surface
2. Hung below the ceiling and supported away from the walls
3. Hung from the ceiling as louvers or baffles
4. Made up into shapes such as cubes or tetrahedrons and suspended from the ceiling

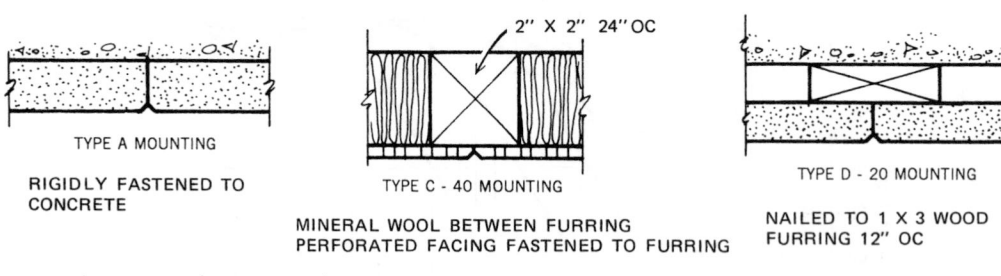

TYPE A MOUNTING

**RIGIDLY FASTENED TO CONCRETE**

TYPE C - 40 MOUNTING

**MINERAL WOOL BETWEEN FURRING PERFORATED FACING FASTENED TO FURRING**

TYPE D - 20 MOUNTING

**NAILED TO 1 X 3 WOOD FURRING 12" OC**

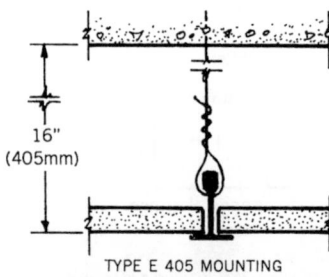

16" (405mm)

TYPE E 405 MOUNTING

**STANDARD HUNG–CEILING CONSTRUCTION**

*Fig. 18.4 Standard test-mounting methods for absorptive material, in accordance with which absorptive coefficients are given by manufacturers. The number following the letter E represents the distance in millimeters from the mounting surface.*

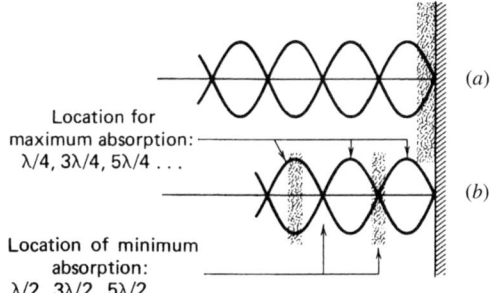

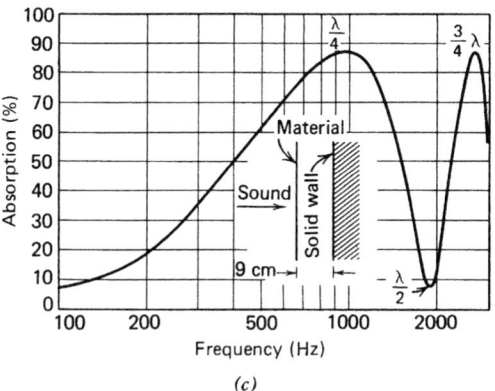

**Fig. 18.5** *Sound waves striking a surface of a large mass will create standing waves at certain frequencies (a), depending upon the room dimensions. Only insulation placed at the peaks (b) of these waves (λ/4, 3λ/4 . . . ) is effective, as seen in (c). (Part c reproduced with permission of Brüel and Kjaer.)*

The last two techniques are extremely effective because they expose a very large surface of porous material—much larger than could be obtained with wall or ceiling covering. Of course, these suspended objects become architectural elements and must be handled accordingly. In contrast, surface coverings are relatively architecturally neutral. In general, treatment should not be limited to one room surface such as the ceiling. All three principal surfaces in the direction of sound propagation—that is, the ceiling, floor, and back wall—should be treated *approximately equally* for best results. The common practice of treating the ceiling only is generally inadvisable, since high frequencies are highly directive and may not reach the ceiling until the third reflection.

In order to fully understand the effects of absorbent materials in an enclosed space, it is necessary to introduce two new concepts to the reader:

*reverberation* and *sound fields.* This will be done in the next two sections, after which we will return to our discussion of the use and effects of acoustic absorptive material in enclosed spaces.

# ROOM ACOUSTICS

## 18.6 REVERBERATION

Reverberation is the persistence of sound after the sound source has ceased. Such persistence is a result of repeated reflections in an enclosed space. Reverberation time ($T_R$) is defined as the time required for the sound level to decrease 60 dB after the sound source has stopped producing sound. For rooms of usual size and shape, the reverberation time at a specific frequency may be found by the formula

$$T_R = K \times \frac{V}{\Sigma A} \quad \text{seconds} \quad (18.4)$$

where

$K$ = a constant, equal to 0.05 when measurements are in feet and 0.16 when in meters

$V$ = room volume, ft³ or m³

$\Sigma A$ = total room absorption, sabins (ft² or m²) at the frequency in question

(For spaces of unusual shapes, see Beranek, 1988.) Reverberation is one of the most pronounced hearing reactions in an enclosed space. It is the ear's reaction to echoes, giving a subjective impression of "liveness" or "deadness" to a space.

A space with highly reflective surfaces, and therefore a low average absorption coefficient ($\overline{\alpha} < 0.2$), sounds live. Conversely, a highly absorptive nonreflective environment ($\overline{\alpha} > 0.4$) sounds dead. ($\overline{\alpha}$ is the average absorption coefficient of the entire space, as explained in detail in Section 18.8.) Since room absorption is related to total surface area, which in turn is related by room proportions to room volume, it is possible to relate all three factors in a single diagram. Figure 18.6 is drawn using room proportions of 2:1.5:1, for L:W:H, based on Fig. 18.22 for preferred room proportions. These proportions (2:1.5:1) represent the average of the extremes of Fig. 18.22. The maximum differential

**ACOUSTICS**

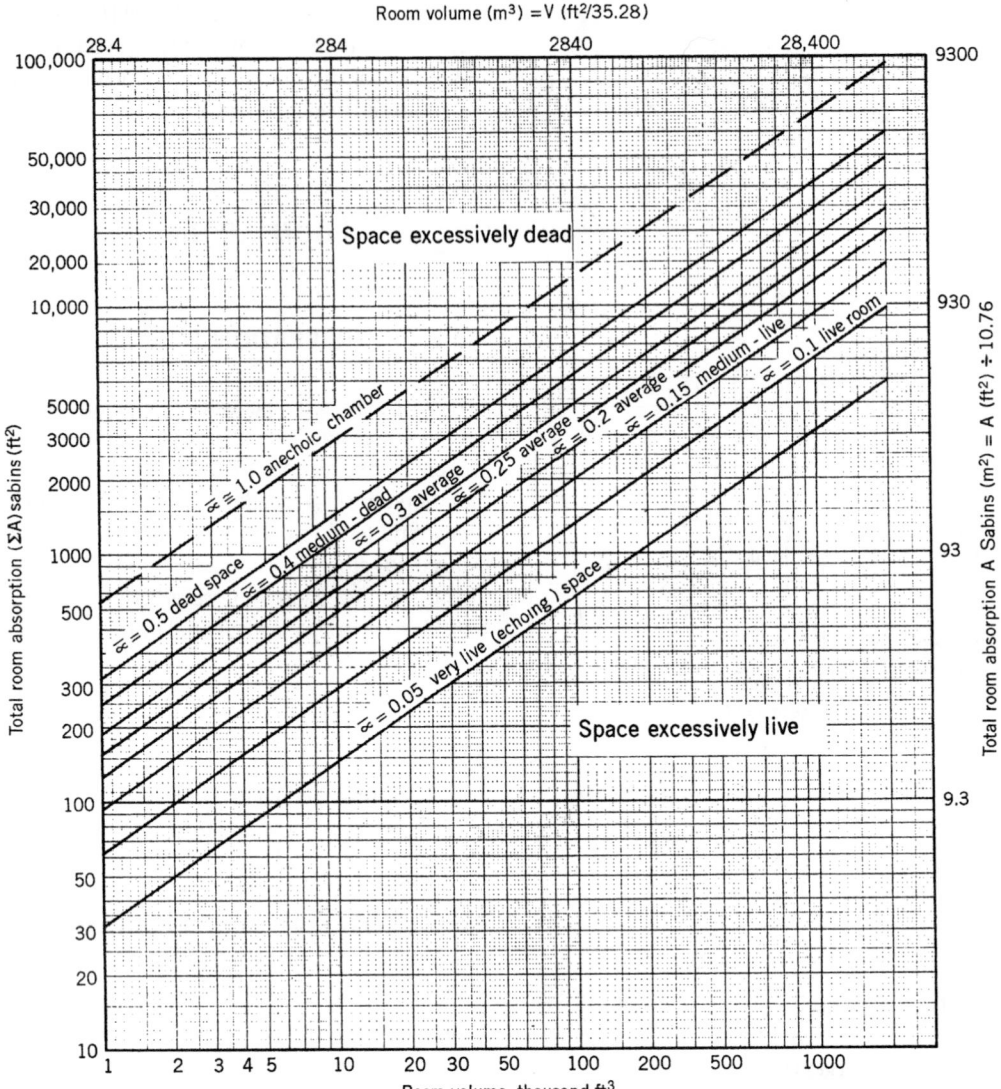

**Fig. 18.6** *Chart indicating a room's "liveness" as a function of its volume and total Sabin absorption. The L-W-H room proportions chosen—2H:1.5H:H—represent the average of the three extreme points of the recommended room proportion triangle in Fig. 18.22. For these proportions, $S = 6.25\,V^{2/3}$ and therefore $A = S\,\bar{\alpha} = 6.25\,\bar{\alpha}\,V^{2/3}$.*

introduced by using average proportions is 2½%—well within engineering accuracy.

The ultimate dead space is one where there is no reflection at all—as is the case outdoors in a flat open area. Indoors, we receive an auditory cue in the form of reflected sound (feedback) that helps us regulate our sound power output (voice level). Outdoors, the absence of this cue, to which we are so accustomed by our largely indoor existence, automatically causes us to raise our output. For this reason, most people tend to speak excessively loudly outdoors. Conversely, a highly reflective indoor condition gives a large feedback signal, which usually results in lowered vocal output.

In common room acoustics studies, reverberation times are calculated at 125, 500, 1000, and 2000 Hz. The midfrequency (500 to 1000 Hz) range is generally the reference used in specifying the reverberation time of a room when studying the speech characteristic of the space.

Reverberation can be considered as a mixture of previous and more recent sounds. The converse of *reverberation* or *reverberance* is *articulation*. An articulate environment keeps each sound event separate rather than running them together. Spaces for speech activities should be less reverberant—more articulate—than those designed for performance of music. See Sections 18.11 and 18.12 for reverberation criteria for speech and music rooms, respectively.

## 18.7 SOUND FIELDS IN AN ENCLOSED SPACE

The inverse square law described in Section 17.5 holds true for the acoustic *far field*, which is a sound field sufficiently far from a source that intensity is proportional to power and inversely to distance. This type of acoustic field is developed in open, obstruction-free space. Propagation in an enclosed space is quite different. There, when a sound reaches a wall or another large (with respect to wavelength) obstruction, part of the sound energy is reflected and part absorbed. As a result, the sound at any point in the room is a combination of direct sound from the source plus reflected sound from walls and other obstructions. If the reflections are so large that the sound level becomes uniform throughout the room, the acoustic field within the room is termed a *diffuse* one (no shadows), and intensity measurements with respect to a specific source are meaningless. Of course, if it is our intention to measure sound pressure level at a specific point, such as a seat in an auditorium, the type of acoustic field in the room is irrelevant.

Most indoor spaces do not have such a high level of reflection that a diffuse field is created. Instead, there is a *near field* near the source, a *free field* beyond the near field, and a *reverberant field* near the walls (Fig. 18.7). These can be recognized as follows:

1. The *near field* is generally within one wavelength of the lowest frequency of sound produced by the source. Within this distance, sound pressure level measurements vary widely and are not meaningful. (The maximum wavelength for the human male voice is about 11 ft.)
2. Close to large obstructions such as walls, the *reverberant field* is dominant and approaches a

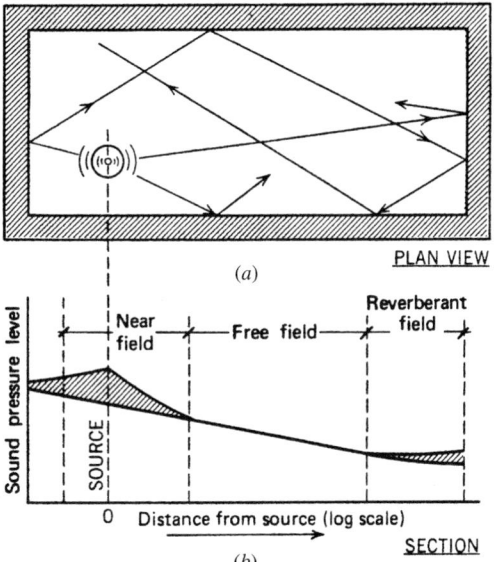

*Fig. 18.7* The type of sound field in an enclosed space depends largely on reflections (reverberation) and absorption. In a typical room there is a near field adjacent to the sound, a free field beyond that, and a reverberant field adjacent to the walls. In a large hall or auditorium, the reverberant field dominates and the sound pressure level remains approximately constant.

diffuse condition. In well-designed music auditoriums, the reverberant (diffuse) field predominates and the sound pressure level remains relatively constant beyond the free field area.
3. The *free* (far) *field* exists between the near and reverberant fields, and there intensity varies directly with pressure and inversely with distance squared. In this field, sound pressure level drops 6 dB with each doubling of distance from the source, and it is in this field that meaningful sound pressure level measurements can be made with respect to a specific small source.

## 18.8 SOUND POWER LEVEL AND SOUND PRESSURE LEVEL

Sound *power level* (PWL) is a measure of the amount of sound generated by a source, independent of its environment. Sound *pressure level* (SPL) can be thought of as the noise or sound in an enclosed space resulting from a source in that space, as affected by the characteristics of the space and the position of the listener. It is thus an end effect.

Since it is desirable to be able to predict the SPL in a space during the design stage, *before* construction, and also because of the difficulty of SPL measurement *after* construction in rooms with various types of sound fields (as explained above), it is useful to have equations and/or a graphic means of relating PWL to SPL. PWL data are supplied by manufacturers of equipment referenced to either octave or one-third octave bands. In free space, SPL and PWL are simply related by the inverse square law, whereas in enclosed spaces the room characteristics come into play. Roughly speaking, an analogy to lighting can be drawn: SPL corresponds to illuminance (that is, footcandles or lux) and PWL corresponds to the lumen output of the source causing the illuminance.

The basic relationship between SPL and PWL for a single small source in a room large enough to have both free and reverberant fields is:

I-P units:

$$SPL = PWL + 10 \log \left( \frac{Q}{4\pi r^2} + \frac{4}{R} \right) + 10.5 \qquad (18.5)$$

SI units:

$$SPL = PWL + 10 \log \left( \frac{Q}{4\pi r^2} + \frac{4}{R} \right) + 0.2 \qquad (18.6)$$

where

$Q$ is a directivity constant

$SPL$ = sound pressure level, dB

$PWL$ = sound power level, dB

$r$ = distance from source, ft (m)

$R$ = room factor, ft² (m²)

The factor $R$ can be calculated from

$$R = \frac{\Sigma\, S\bar{\alpha}}{1 - \bar{\alpha}}$$

where

$\Sigma S$ = total room surface area, ft² (m₂)

$\bar{\alpha}$ = average absorption coefficient of all materials in the room

that is,

$$\bar{\alpha} = \frac{\Sigma\, A \text{ (total room absorption)}}{\Sigma\, S \text{ (total room surface area)}} \qquad (18.7)$$

or

$$\bar{\alpha} = \frac{S_1\alpha_1 + S_2\alpha_2 + \cdots + S_n\alpha_n}{S_1 + S_2 + \cdots + S_n} \qquad (18.8)$$

The directivity constant $Q$ is either inherent in the sound source, and as such will be part of the given data, or can be obtained from Fig. 18.8 for a nondirectional source made directional by adjacent reflecting surfaces.

Thus, for a source suspended or supported at least ½ wavelength (at its lowest frequency) from a ceiling or floor, $Q = 1$; a source near a floor and distant from the walls uses $Q = 2$; and so on. In a space containing more than one sound source, the SPLs can be combined as explained in Section 17.5. In the two extreme cases (i.e., rooms with only far [direct] fields and rooms with only reverberant [diffuse] fields, corresponding to dead and live rooms, respectively) for a nondirectional source located on the floor and away from walls ($Q = 2$), Eqs. 18.5 and 18.6 reduce to:

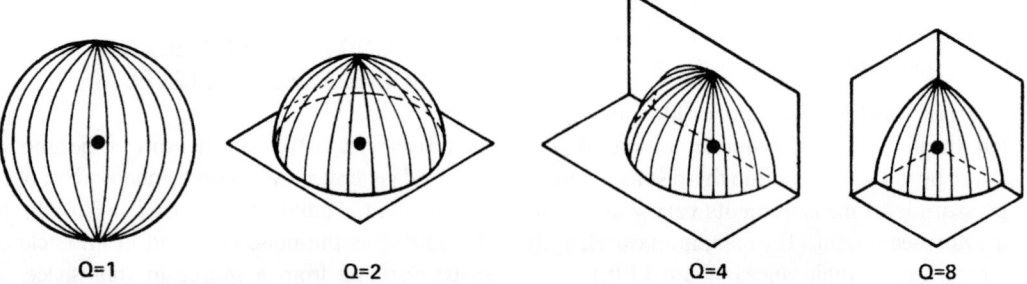

Q=1    Q=2    Q=4    Q=8

**Fig. 18.8** *Diagrams illustrating directivity factors for either inherently directive sources or nondirective sources placed adjacent to large reflecting surfaces. (Courtesy of Barry Blower Co.)*

ACOUSTICS

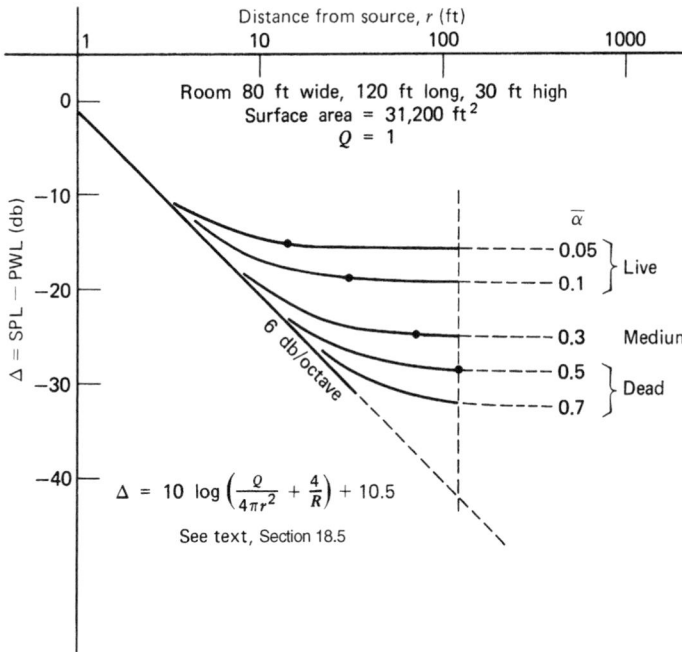

**Fig. 18.9** *Calculated curves show that the development of a reverberant field (constant SPL, or flat curve) is dependent upon a room's absorption characteristics. On each curve, the point at which a reverberant field begins is indicated by a dot on the curve. In a live room ($\overline{\alpha} = 0.1$), the reverberant field begins 20 ft from the source. In a dead room ($\overline{\alpha} = 0.5$), it begins 120 ft from the source, that is, at the back wall. Effectively, then, there is no reverberant field in such a room (or in any room with $\overline{\alpha} = 0.5$). The slope of the asymptote is 6 dB per octave, that is, inverse square attenuation.*

Dead room, direct field:

I-P units:

$$SPL = PWL - 20 \log r + (Q - 0.6)$$
$$= PWL - 20 \log r + 1.4 \qquad (18.9)$$

SI units:

$$SPL = PWL - 20 \log r + (Q - 10.9)$$
$$= PWL - 20 \log r - 8.9 \qquad (18.10)$$

Live room, diffuse-reverberant field:

I-P units:

$$SPL = PWL - 10 \log \Sigma A + 16.3 \qquad (18.11)$$

SI units:

$$SPL = PWL - 10 \log \Sigma A + 6.0 \qquad (18.12)$$

To understand the effect of room absorption on the development of the reverberant field in a large space, refer to Fig. 18.9. The field is plotted from Eq. 18.5, for a room 80 ft × 120 ft × 30 ft high, using average absorptions ($\overline{\alpha}$) ranging from 0.05 to 0.7, that is, from very live to very dead. Note particularly that the addition of sufficient absorbent material can, by drastically reducing sound reflections from room surfaces, entirely prevent the development of a reverberant (diffuse) field. This is particularly important in noisy industrial interiors where the building materials commonly employed are gener-

ally nonabsorbent, and the noise sources are both numerous and loud. An example of the use of absorptive material to effect a considerable degree of in-room noise reduction is demonstrated in the next section. We emphasize that the effect is almost exclusively in the room where the absorptive material is used since the acoustic energy transmitted to adjacent spaces is only minimally affected. Sound transmission from one enclosed space to an adjoining space through a common barrier depends almost entirely upon the type of barrier. This is discussed in detail in Chapter 19.

## 18.9 NOISE REDUCTION BY ABSORPTION

Equation 18.6 relates SPL and PWL as a function of distance from the source and room absorption. The following example shows the application of this equation and the result of adding absorption to the space.

**EXAMPLE 18.1** An open blower is installed on the floor, away from the walls, in a large enclosed space that is 6 m long, 12 m wide, and 4 m high. The floor is concrete ($\alpha = 0.01$); the walls and ceiling are painted block ($\alpha = 0.07$). The PWLs supplied by the manufacturer are 90 dB at 500 Hz and 87 dB at 2000 Hz. Calculate the SPL at distances of 5 m and 10 m from the blower outlet: (a) in the original room; (b) with double the absorption; and (c) with quadruple the absorption.

### SOLUTION

1. From Eq. 18.6, we have the expression

$$SPL = PWL + 10 \log_{10} \left( \frac{Q}{4\pi r^2} + \frac{4}{R} \right) + 0.2$$

In our example, $Q = 2$ (see Fig. 18.8).

2. The first step is to calculate the room factor $R$ for the three situations (a, b, and c) of absorption.

$$R = \frac{S\bar{\alpha}}{1 - \bar{\alpha}}$$

$$S = 2(48) + 2(24) + 2(72) = 288 \text{ m}^2$$

From Eq. 18.8:

$$\bar{\alpha} \text{ original} = \frac{3(48)(0.07) + 1(72)(0.07) + (72)(0.01)}{288}$$

$$= 0.055$$

$$\begin{array}{ll} \bar{\alpha}_a = 0.055 & R_a = 16.76 \\ \bar{\alpha}_b = 2(0.055) = 0.11 & R_b = 35.6 \\ \bar{\alpha}_c = 4(0.055) = 0.22 & R_c = 81.23 \end{array}$$

3. Calculating SPL values and tabulating, we obtain

| SPL for: | 500 Hz | | 2000 Hz | |
|---|---|---|---|---|
| | 5 m | 10 m | 5 m | 10 m |
| (a) Original room | 84.1 | 84.0 | 81.1 | 81.0 |
| (b) Double $\bar{\alpha}$ | 81.0 | 80.8 | 78.0 | 77.8 |
| (c) Quadruple $\bar{\alpha}$ | 77.7 | 77.3 | 74.7 | 74.3 |

The results indicate the very important fact that *doubling the absorption decreases the noise level by only 3 dB*. Therefore, it requires a quadrupling of absorption to make the decrease noticeable (see Table 17.3). This is obviously an expensive procedure with diminishing returns.

If the entire space is considered to be a reverberant field, then from Eq. 18.11 the sound intensity level (IL) can be expressed as follows:

$$IL = PWL - 10 \log \Sigma A + 16.3 \text{ dB} \quad (18.13)$$

where

$$\Sigma A = \text{total absorption in room, sabins (ft}^2)$$
$$IL = \text{intensity level, dB}$$
$$PWL = \text{sound power level, dB}$$

Although increasing absorption decreases the sound/noise level, the level cannot be reduced below the free field level for that distance from the source because the free field situation corresponds to outdoors, where $\bar{\alpha} = 1.0$.

The amount of noise reduction provided by additional absorption may be determined by noise reduction (NR):

$$NR = IL_1 - IL_2$$
$$= 10 \log \Sigma A_2 - 10 \log \Sigma A_1$$

Therefore,

$$NR = 10 \log \frac{\Sigma A_2}{\Sigma A_1} \quad (18.14)$$

where

$$NR = \text{noise reduction, dB}$$
$$\Sigma A_2 = \text{total absorption, final condition}$$
$$\Sigma A_1 = \text{total absorption, initial condition}$$

From Eq. 18.14 it is seen that doubling the absorption results in a noise reduction of 3 dB, since $10 \log_{10} 2 = 3$ (dB). ∎

It is of interest to work out a practical example using Eqs. 18.13 and 18.14 and compare the results with the more precise relation given in Eq. 18.5.

**EXAMPLE 18.2** Referring to Fig. 18.10, calculate the original sound level and the subsequent noise reduction by three steps of sound absorption treatment, assuming a completely reverberant field in the space, that is, SPL independent of location. Data:

Original condition: Painted concrete block chamber, $10 \times 10 \times 10$ ft

Fan sound power level:

At 500 Hz = 88 dB

At 2000 Hz = 78 dB

### SOLUTION

| Frequency (Hz) | Area (ft²) | $\alpha$ | Total Absorption ($\Sigma S\alpha$) |
|---|---|---|---|
| 500 | 600 | 0.06 | 36 Sabins (ft²) |
| 2000 | 600 | 0.09 | 54 Sabins (ft²) |

a. *Sound Intensity Level before Treatment*

At 500 Hz:

$IL$ = sound power level – 10 log $\Sigma A$ + 16.3 dB

= 88 dB – 10 log 36 + 16.3 dB

= 88 dB – 15.6 dB + 16.3 dB

= 88.7 dB

At 2000 Hz:

$IL$ = 78 dB – 10 log 54 + 16.3 dB

= 78 dB – 17.3 dB + 16.3 dB

= 77 dB

b. *Ceiling Treatment Only*

At 500 Hz:

$\alpha = 0.82$

additional absorption = 100 (0.82 – 0.06)

= 76 sabins

$NR = 10 \log \dfrac{76 + 36}{36}$

= 4.9 dB

At 2000 Hz:

$\alpha = 0.94$

$\Delta A = 100\ (0.94 - 0.09) = 85$

$NR = 10 \log \dfrac{85 + 54}{54}$

= 4.1 dB

c. *Ceiling and Half-Wall Treatment*

At 500 Hz:

added absorption = 300 (0.82 – 0.06)

= 228 sabins

$NR = 10 \log \dfrac{228 + 36}{36}$

= 8.7 dB

At 2000 Hz:

added absorption = 300 (0.94 – 0.09)

= 255 sabins

$NR = 10 \log \dfrac{255 + 54}{54}$

= 7.5 dB

d. *Ceiling and Full-Wall Treatment*

At 500 Hz:

$\Delta A = 500\ (0.82 - 0.06)$

= 380 sabins

$NR = 10 \log \dfrac{380 + 36}{36}$

= 10.6 dB

At 2000 Hz:

$\Delta A = 500\ (0.94 - 0.09)$

= 425 sabins

$NR = 10 \log \dfrac{425 + 54}{54}$

= 9.5 dB

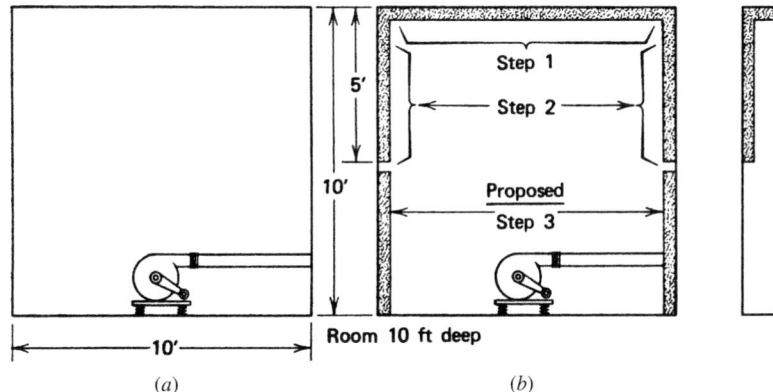
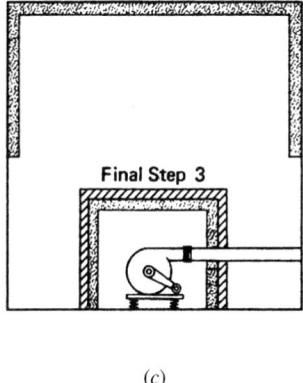

**Fig. 18.10** *Quieting a room (a) by addition of absorptive material is cost effective only through Step 2 (b). Further quieting would be accomplished locally (c), which might obviate the necessity for wall treatment (Step 2). Ceiling treatment should remain.*

## Summary

| | IL (SPL)[a] | |
|---|---|---|
| | 500 Hz | 2000 Hz |
| Bare room | 88.7 | 77 |
| Ceiling treatment | −4.9 dB | −4.1 dB |
| | (−4.9)[a] | (−4.2)[a] |
| Half-wall treatment | −8.7 dB | −7.5 dB |
| | (−8.7)[a] | (−7.7)[a] |
| Full-wall treatment | −10.6 dB | −9.5 dB |
| | (−10.7)[a] | (−9.6)[a] |

[a] The numbers in parentheses are the SPL differences as calculated from Eq. 18.5 using $Q = 2$ and $r = 5$ ft.

Two important conclusions can be drawn from these results:

1. The third step of adding absorptive material is not worthwhile since negligible additional room quieting is accomplished. This clearly demonstrates the law of diminishing returns as applied to room quieting with absorptive materials. Starting with a live room, the initial application is effective. Beyond that, additional quieting by absorption is not economical, and the same outlay would be better used in quieting the machine itself, probably with a machine enclosure, as indicated in Fig. 18.10.

2. If we compare the figures in parentheses that resulted from a full calculation (Eq. 18.5) to those arrived at using the much simpler equations applicable only to a fully diffuse field (Eq. 18.11), we see that the differences are negligible. This means that the shortcut method can always be used in preliminary calculations and frequently for final results as well. The diffuse (reverberant) field calculation also lends itself nicely to a graphical solution.

Using the diffuse field Eq. 18.11:

$$SPL = PWL - 10 \log \Sigma A + 16.3$$

and remembering that reverberation time is expressed in terms of total absorption and room volume, that is,

$$T_R = \frac{KV}{\Sigma A} \quad (18.15)$$

where $K$ is 0.05 when room volume is in cubic feet and 0.16 when it is in cubic meters. Using these

equations, we can then rewrite Eqs. 18.11 and 18.12 in terms of reverberation time and room volume:

For room volume in ft³:

$$SPL - PWL = -10 \log \frac{0.05V}{T_R} + 16.3 \quad (18.16)$$

For room volume in m³:

$$SPL - PWL = -10 \log \frac{0.16V}{T_R} + 6.0 \quad (18.17)$$

These two equations are plotted graphically in Figs. 18.11 and 18.12. The charts are accurate for live room, diffuse (reverberant) fields and provide a close approximation for other situations. An illustrative example will demonstrate use of the charts. ∎

**EXAMPLE 18.3** A generator in a 14,000-ft³ room has a PWL of 95 dB (re: $10^{-12}$ W) at 400 Hz.

a. Using the chart in Fig. 18.11, find the SPL at that frequency. Assume that the sound field in the room is reverberant. The room's reverberation time is 1.0 s at 400 Hz.

b. Check the result using the chart in Fig. 18.12

**SOLUTION**

a. Enter the chart in Fig. 18.11 at 14,000 ft³ and draw a vertical line until it intersects the 1.0-s sloping line. Extend a line from that intersection point to the X-axis and read −12 dB.
Then, since

$$SPL - PWL = -12 \text{ dB}$$

and PWL is given as 95 dB,

$$SPL = 95 - 12 = 83 \text{ dB}$$

b. The 14,000-ft³ room volume is equal to 400 m³, within engineering accuracy (it is actually 396.7 m³). Enter the chart in Fig. 18.12 at 400 m³, and extend a line upward to $T_R$ of 1 s and left to the X-scale. Read −12 dB. The two answers check, as they should. ∎

## 18.10 NOISE REDUCTION COEFFICIENT

The last column in Table 18.1 is labeled NRC—noise reduction coefficient. This figure is the arithmetic average of the absorption coefficients at 250, 500, 1000, and 2000 Hz. The name is ill

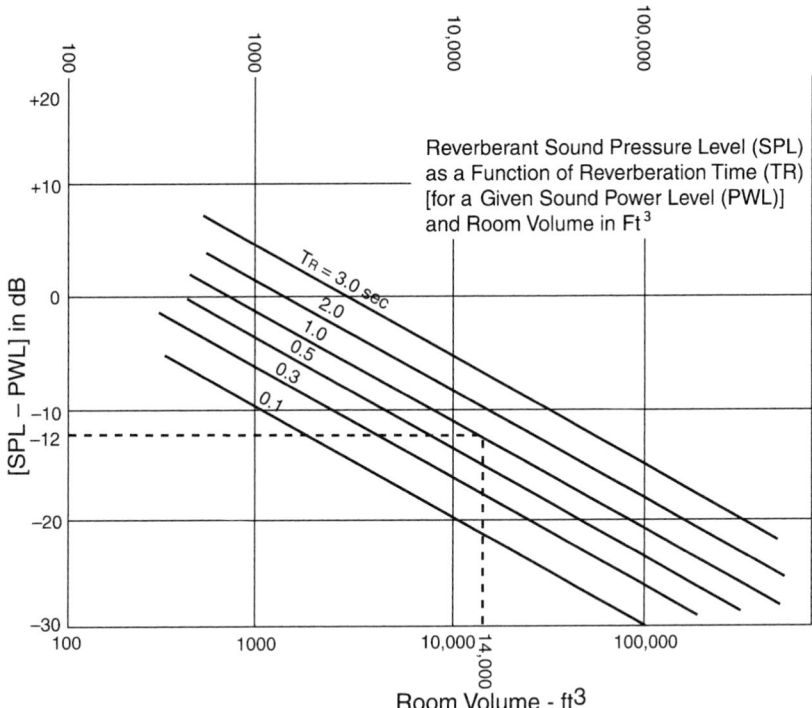

**Fig. 18.11** I-P chart for determining the reverberant sound pressure level (SPL) when the sound power level (PWL) of an item of equipment is known. Room volume in cubic feet and reverberation time in seconds are the variables.

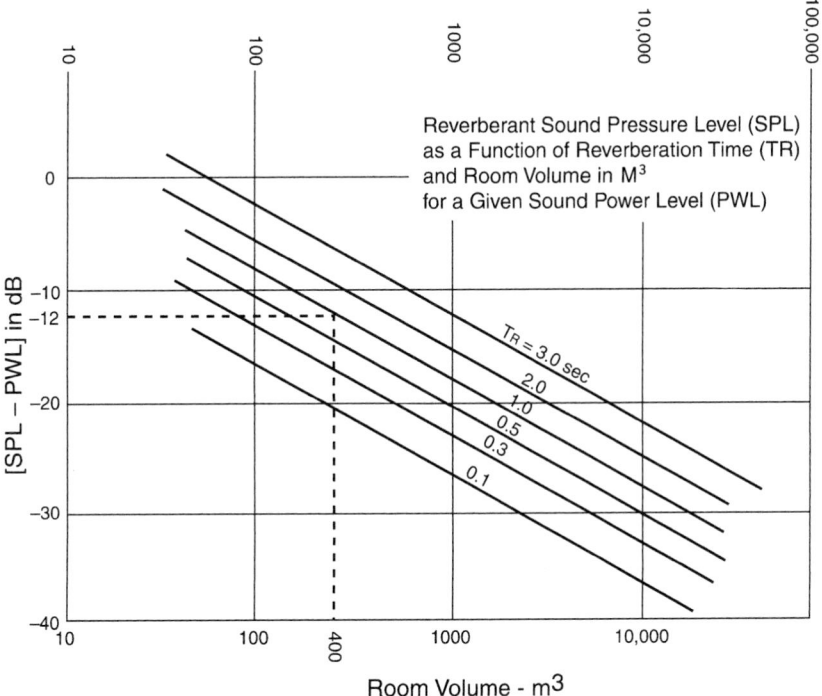

**Fig. 18.12** SI chart for determining the reverberant sound pressure level (SPL) when the sound power level (PWL) of an item of equipment is known. Room volume in cubic meters and reverberation time in seconds are the variables.

chosen inasmuch as it cannot be used as the words seem to imply. It is simply useful as a single-number criterion for the midband effectiveness of a porous absorber. If careful design is required or high- and low-end frequencies are of interest, NRC is nearly useless, and detailed calculations over the entire frequency range must be made. Similarly, two materials with the same NRC can perform quite differently, since the NRC is an average and few materials have a flat characteristic. Further, absorption is not directly equivalent to noise reduction.

## ROOM DESIGN

### 18.11 REVERBERATION CRITERIA FOR SPEECH ROOMS

The overriding criterion for speech is intelligibility. Since speech consists of short, disconnected sounds 30 to 300 ms in length (see Section 17.4), among

which are high-frequency, low-energy phonemes, the ideal room must ensure the ear's undistorted reception of these phonemes. This requires keeping reverberation to a minimum. We can obtain a good approximation of the subjective feeling of liveness of a room, for purposes of speech, from the relation

$$T_R \text{ (speech)} = 0.3 \log \frac{V}{10} \qquad (18.18)$$

where

$T_R$ (speech) = optimum reverberation time in seconds, for speech

$V$ = room volume, m³

For instance, a typical classroom might have a volume of 150 m³ (5300 ft³). Optimum reverberation time is

$$T_R = 0.3 \log 15 = 0.35 \text{ s}$$

Reverberation times longer than this would sound live, shorter ones dead and flat. Indeed, an increase of 20% in reverberation time would make the room excessively live and boomy and would negatively affect speech intelligibility. Figure 18.13 gives

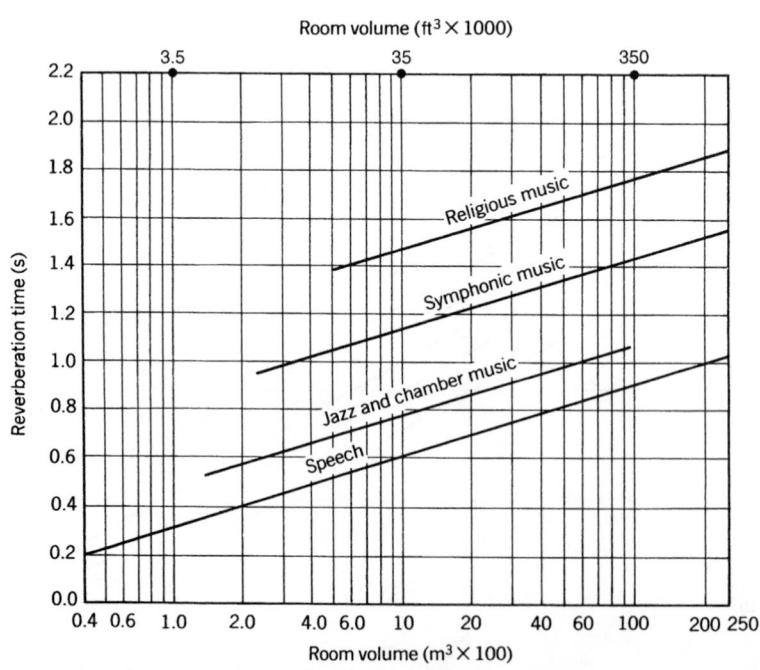

**Fig. 18.13** Optimum reverberation times for the frequency range 500 to 1000 Hz as a function of room volume. (The cubic foot volume figures are based upon the approximate conversion factor of 35 ft³/m³). (Reprinted by permission from E.B. Magrab, Environmental Noise Control, Wiley, New York, 1975, p. 206.)

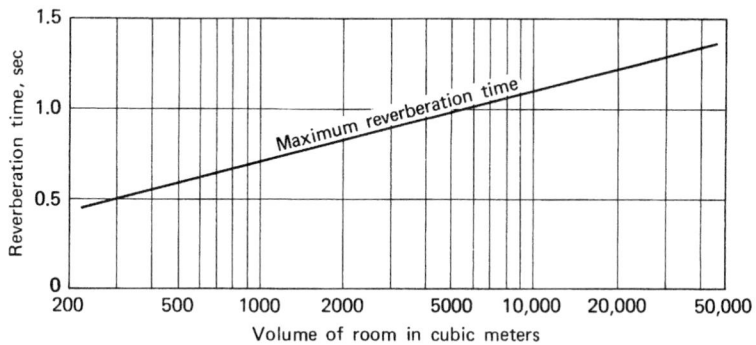

**Fig. 18.14** *Maximum recommended reverberation time for speech in auditoriums and lecture halls. (From P.V. Brüel,* Sound Insulation and Room Acoustics, *Chapman & Hall, London, 1951.)*

ACOUSTICS

*optimum* midfrequency reverberation times as a function of room size and use. Figure 18.14 gives maximum reverberation times for speech in large rooms. For good speech intelligibility, reverberation time should remain essentially flat down to 100 Hz.

The reflected sounds associated with reverberation can have either a salutary or a deleterious effect. The ear cannot distinguish between sounds that arrive within a maximum of 50 ms ($\frac{1}{20}$ s) of each other (some authorities use 40 ms, i.e., $\frac{1}{25}$ s). Sounds arriving within this time *reinforce* the direct path signal and appear to come from the source. Sounds arriving after this time are apprehended as a fuzzy echo or elongation of the sound, reducing intelligibility and directivity.

Since the range of 40 to 50 ms corresponds at 344 m/s to 13.7 to 17.2 m, a speech room should be so arranged that the difference between the first reflection path and the direct path is no greater than 17 m and preferably 14 m or less at midfrequency (Fig. 18.15). For more details concerning this and related factors on reflected paths, refer to Section 18.13.

Too *low* a reverberation time (very high absorption, minimum reflection) is also undesirable because:

1. It limits the size of the room to that which can be covered by direct sound only.
2. It is disturbing to the speaker, since absence of reflection prevents him or her from gauging the proper voice level and tends to cause excessive effort (shouting, as when outdoors).

Thus, proper design of a room for speech is a compromise between the need for some reflection and the desire to minimize reflection to preserve intelligibility.

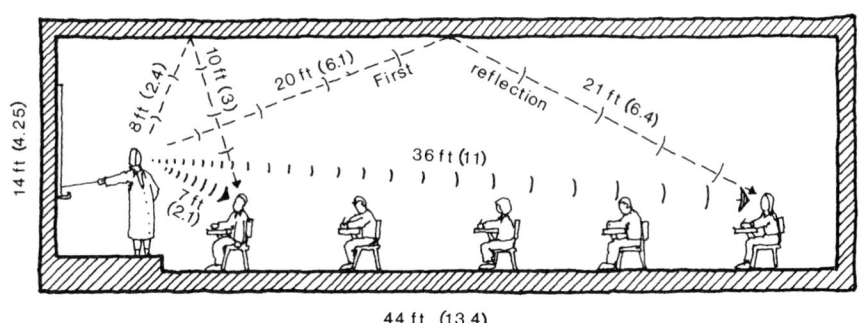

**Fig. 18.15** *Sound paths in a typical medium-sized lecture room. Note that for both extremes of listener position, the maximum path-length difference between direct and first reflection is 11 ft (3.4 m). Thus signal is reinforced, and intelligibility should be excellent if room absorption is provided to limit reverberation time to about ½ second maximum (see Fig. 18.14). Numbers in parentheses are dimensions in meters.*

## 18.12 CRITERIA FOR MUSIC PERFORMANCE

Adequate design for a music space requires recognition of the following:

1. Large-volume spaces require direct-path sound reinforcement by reflection.

2. Relatively long reverberation time is needed to enhance the music—the exact amount depending upon the type of music (Figs. 18.13 and 18.16). Designers should keep in mind that reverberation time recommendations vary as much as 100% among respected sources.

3. It is generally agreed that reverberation time should vary inversely with frequency (i.e., $T_R$ should be longer at lower frequencies [than the midfrequency recommendation] and shorter at higher frequencies). The longer $T_R$ at low frequencies adds fullness to music and "body" to speech. Thus, $T_R$ at

100 Hz should be, according to most researchers, 35% to 75% longer than $T_R$ at the center frequencies.

4. Short $T_R$ at upper frequencies adds directivity to the music. With large ensembles, directivity gives the sense of depth and instrument location necessary for proper appreciation. This is often referred to as *clarity* or *definition* in music. With a solo instrument, this problem is diminished.

5. Brilliance of tone is primarily a function of high-frequency content. Since these frequencies are most readily absorbed, a good direct path must exist between sound source and listener. Since our eyes and ears are close together, a good sound path exists when a good vision path exists. At the other end of the spectrum, lack of sufficient bass expresses itself as a loss of "fullness," which is often caused by resonant absorption.

The actual design of a music performance space is a very complex procedure involving exten-

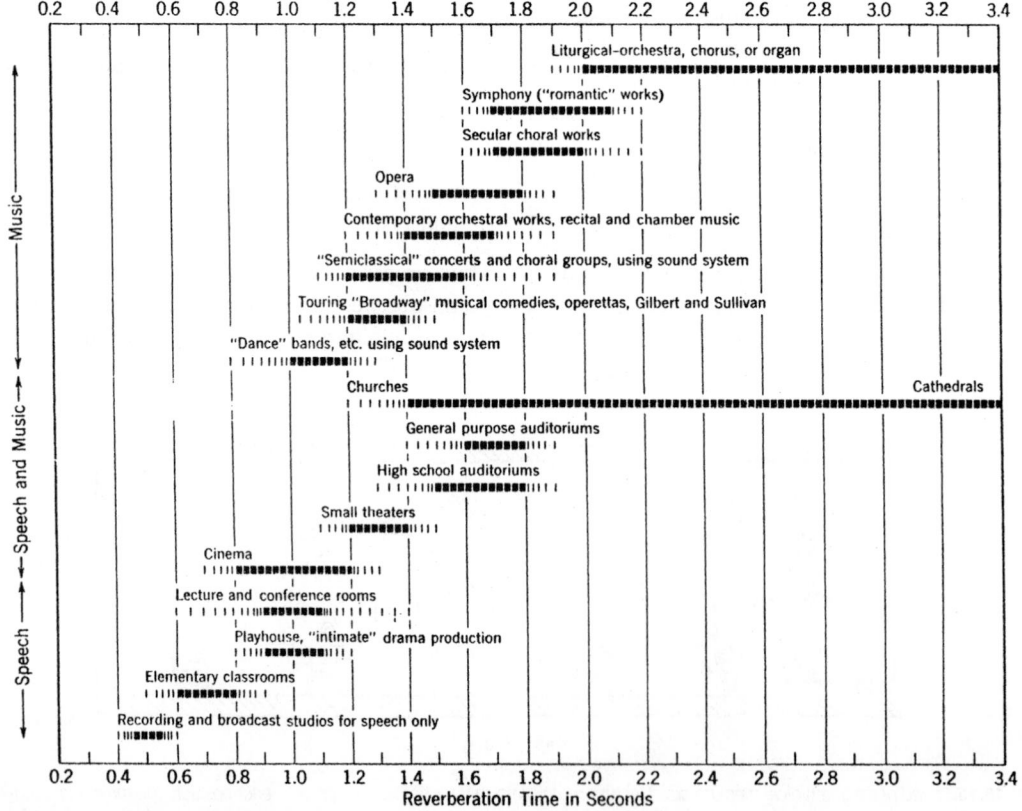

**Fig. 18.16** Optimum reverberation times at mid-frequencies (500–1000 Hz) for various types of facilities. The wide range shown reflects the effects of room proportions, shape, and volume on the optimum $T_R$, as well as differences of opinion among designers.

sive calculations of absorption, reverberation time, and ray diagramming, as well as juggling of materials, dimensions, and wall angles. Simulation techniques and acoustic models are also often employed.

Recent research and simulation studies of concert and recital halls have demonstrated that the sensation of fullness of music, or what is today referred to as *sound envelopment*, is enhanced by lateral reflections that reinforce the direct signal. It has also been found that the subjective judgment of reverberance is more strongly affected by *early decay time* (the time required for a 10-dB decrease in signal strength) than by the conventional 60-dB decay time. Finally, crispness or clarity of the music (particularly important in recital halls and for chamber music) depends upon reflections arriving within 40 to 70 ms. All of these factors are considered both in the original design and in the often lengthy "tuning" process of a space in-

tended for music performance. Most modern design solutions also use movable reflector panels and other active variables. After construction is completed, extensive tests are conducted and field adjustments are made.

## 18.13 SOUND PATHS

Ideally, every listener in a lecture hall, theater, or concert hall should hear the speaker or performer with the same degree of loudness and clarity. Since this is obviously impossible by direct-path sound, the essential design task is to devise methods for reinforcing desirable reflections and minimizing and controlling undesirable ones. Normally only the first reflection is considered in ray diagramming (discussed in the next section) since it is strongest. Second and subsequent reflections

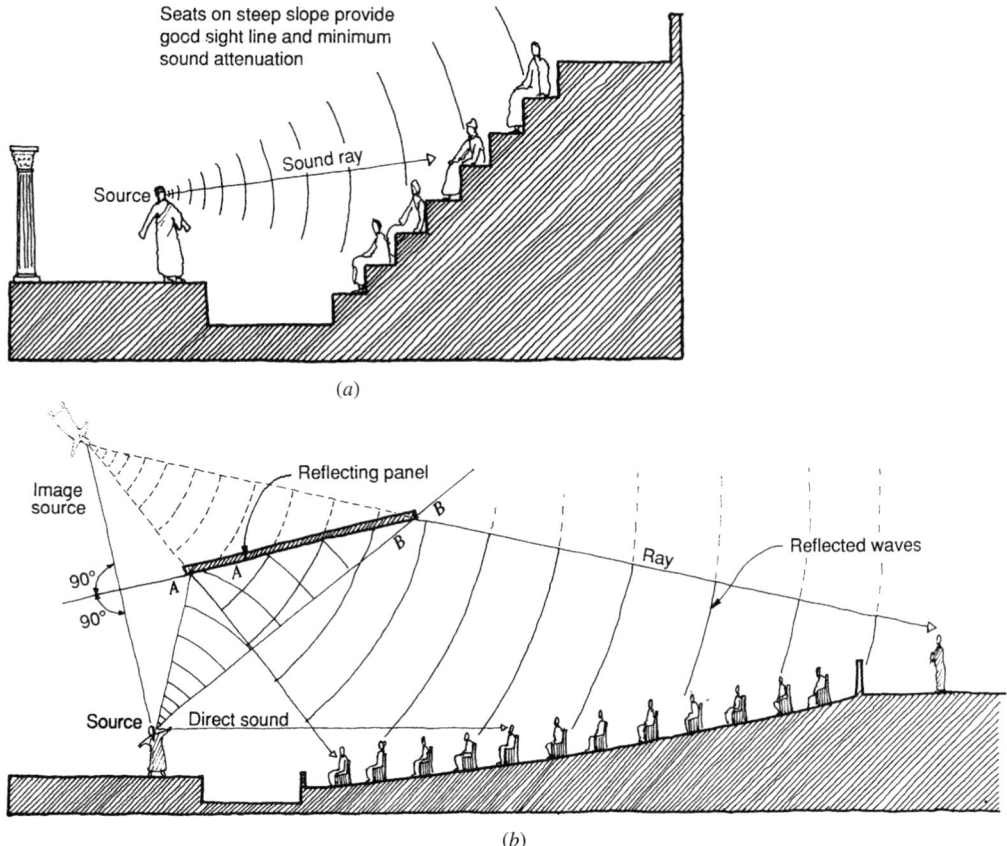

**Fig. 18.17** *Use of an angled reflector panel (b) creates an image source that stands in approximately the same relation to the audience as does the performer in the classic Greek theater (a).*

are usually attenuated to the point that they need not be considered except for the special situations of flutter, echoes, and standing waves, discussed below.

### (a) Specular Reflection

Specular reflection occurs when sound reflects off a hard, polished surface. This characteristic can be used to good advantage to create an effective image source. In ancient Greek and Roman theaters, seats were arranged on a steep conical surface around the performers. The virtue of this arrangement (Fig. 18.17a) is that the sound energy travels to each location with minimal attenuation. The same effect can be accomplished by placing the sound source above the seats. This is not practical physically, but it can be accomplished effectively by the use of a reflecting panel (Fig. 18.17b). The panel dimension must be at least one wavelength at the lowest frequency under consideration. Figure 18.18 is a chart for converting from frequency to wavelength in feet and meters.

### (b) Echoes

As explained in Section 18.11, a clear echo is caused when reflected sound *at sufficient intensity* reaches a listener more than 50 ms after he or she has heard the direct sound. (Some authorities place this figure as high as 80 ms.) Echoes, even if not distinctly discernible, are undesirable. They make speech less intelligible and make music sound "mushy." The relative undesirability depends upon the time delay and loudness relative to the direct sound, which, in turn, are dependent upon the size, position, shape, and absorption of the reflecting surface.

Typical echo-producing surfaces in an auditorium are the back wall and the ceiling above the proscenium. Figure 18.19 shows these problems and suggests remedies. Note that the energy that produced the echoes can be redirected to places where it becomes useful reinforcement. If echo control by absorption alone were used on the ceiling and back wall, that energy would be wasted. The rear wall, since its area cannot be reduced too far, may have to be made more sound absorptive to reduce the loudness of the reflected sound.

### (c) Flutter

A flutter, perceived as a buzzing or clicking sound, comprises repeated echoes traversing back and forth between two nonabsorbing parallel (flat or concave) surfaces. Flutters often occur between shallow domes and hard, flat floors. The remedy for a flutter is either to change the shape of the reflectors or their parallel relationship or to add absorption. The solution chosen will depend upon reverberation requirements, cost, and aesthetics.

### (d) Diffusion

This is the converse of focusing and occurs primarily when sound is reflected from convex surfaces. A degree of diffusion is also provided by flat horizontal and inclined reflectors (Fig. 18.20). In a diffuse sound field the sound level remains relatively constant throughout the space, an extremely desirable property for musical performances.

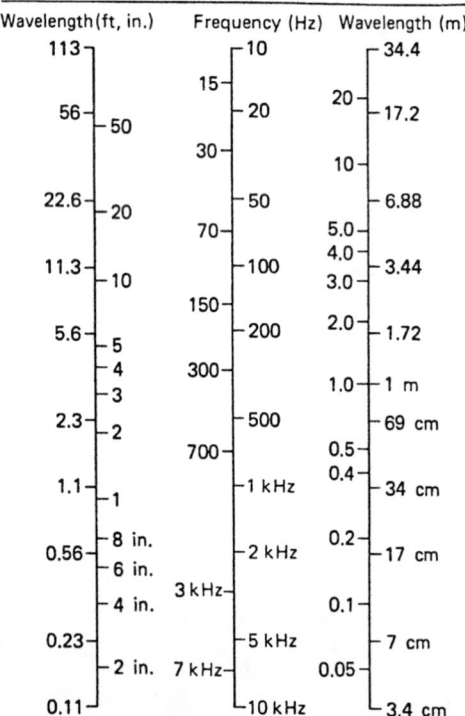

**Fig. 18.18** *Nomograph for determining wavelength in feet or meters, given frequency in Hertz, or vice versa. Speed of sound is taken as 1128 ft/s (344 m/s). To use it, hold a straightedge horizontally across the nomograph and read the figures directly.*

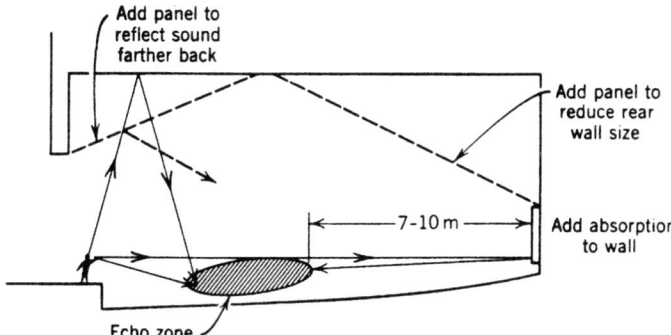

*Fig. 18.19* Auditorium section showing the causes and remedies for two typical echoes.

### (e) Focusing

Concave domes, vaults, or walls will focus reflected sound into certain areas of rooms. This has several disadvantages. For example, it will deprive some listeners of useful sound reflections and cause hot spots at other audience positions (Fig. 18.21a).

### (f) Creep

This describes the reflection of sound along a curved surface from a source near the surface. Although the sound can be heard at points along the surface, it is inaudible away from the surface. Creep is illustrated in Fig. 18.21b.

### (g) Standing Waves

Standing waves and flutter are very similar in principle and cause but are heard quite differently. When an impulse (such as a hand clap) is the energy source, a flutter will occur between two highly reflective parallel walls. It is perceived as a slowly decaying buzz. When a steady, pure tone is the source, a standing wave will occur, but only

when the parallel walls are spaced apart at some integral multiple of a half-wavelength.

When parallel walls are exactly one-half wavelength apart, the tone will sound very loud near the walls and very quiet halfway between them. This is because at the center, the reflected waves traveling in one direction are exactly one-half wavelength away from those traveling in the other direction, and are thus equal *but opposite* in pressure, which results in total cancellation. In other rooms, standing waves are noted as points of quiet and loudness in the room. Standing waves are important only in rooms that are small with respect to the wavelengths generated (smallest room dimension <30 ft for music or <15 ft for speech).

Another effect of standing waves, called *resonance*, is the accentuation of a particular frequency that will cause a standing wave. Thus, if one speaks (or plays a musical instrument) while standing near a wall of a room about 8 ft × 8 ft in size, one will notice an abnormal and sometimes unpleasant loudness in the sound at about 280 Hz.

Similarly, when a musician plays a scale, one note may seem far louder than the adjacent ones,

ACOUSTICS

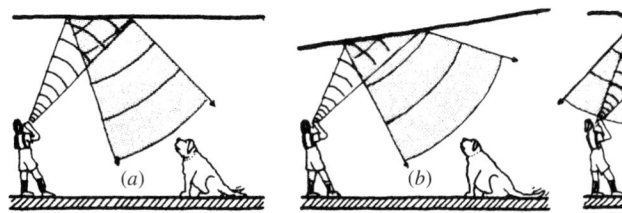

*Fig. 18.20* Sound diffusion can be created with reflectors of different shapes, including horizontal flat (a), inclined flat (b), and convex (c) reflectors. Diffusion improves from (a) to (c).

ACOUSTICS

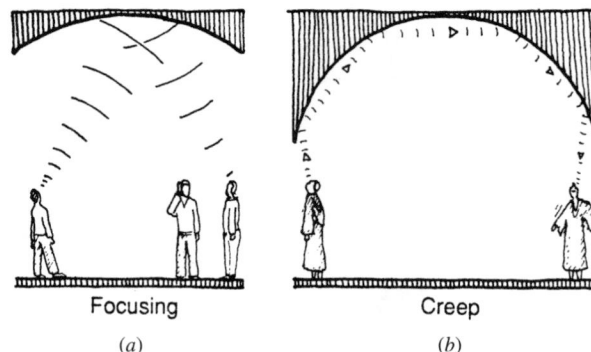

Focusing            Creep

(a)                    (b)

*Fig. 18.21 Two undesirable phenomena in room acoustics.*

and listeners in one section of the room will hear a quality of sound different from that heard by those in other sections. This effect *must* be avoided for music performance but is merely an annoyance in rooms designed for speech use. This is one of the reasons that one finds music rehearsal rooms, broadcast studios, and the like with nonparallel walls and undulating ceilings. These irregularities prevent most of the undesirable effects described above from occurring. However, since rooms with nonparallel walls and undulating ceilings are for most applications unacceptable, and since standing waves and other resonant phenomena are related to room geometry,

it is possible to calculate room proportions with conventional geometries that will minimize these effects. Figure 18.22 shows such room proportions, which are applicable to bass frequencies (i.e., below 100 Hz). These are the problematic frequencies, since at higher frequencies (and in large rooms) many standing waves occur, and the total effect is much less disturbing and frequently hardly noticeable.

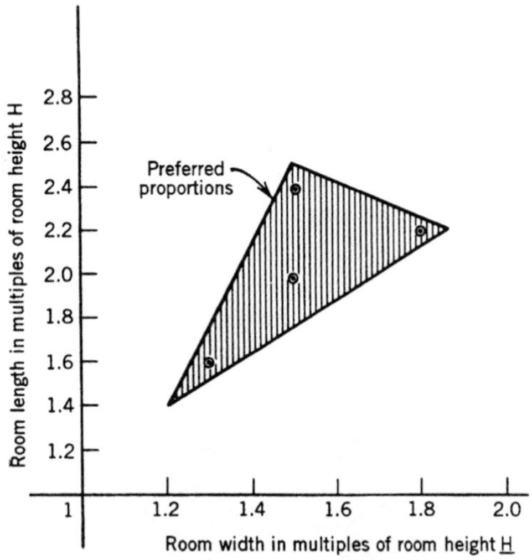

## 18.14 RAY DIAGRAMS

Ray diagramming is a design procedure for analyzing the reflected sound distribution throughout a hall using the first reflection only. Figure 18.23 shows a ray diagram. The rays are drawn normal (perpendicular) to the spherically propagating sound waves. Specular reflection is assumed, that is, at reflecting panels the angles of the incident and reflected rays are always equal. Thus, in addition to direct sound, each listener is receiving reflected sound energy. It is as though there were additional sound sources, the real one and numerous image sources. Figure 18.23 shows the application of a ray diagram to the design of a lecture hall. In Fig. 18.23a, the stage height and seating slope are arranged to provide good sight lines, and the ceiling height is established by reverberation requirements, aesthetics, cost, and so on. It can be seen that less than half of the ceiling is providing useful reflection. By dividing the ceiling into two panels (Fig. 18.23b), people in the rear of the room perceive the direct source plus two image sources, and the useful reflecting area is increased by 50%. In Fig. 18.23c, the shape has been further refined to include a lighting slot and a loudspeaker grille.

As mentioned previously, the sensation of musical envelopment that is so satisfying to a listener,

*Fig. 18.22 The triangular figure contains the room proportions recommended for avoidance of disturbing low-frequency acoustic phenomena such as flutter, standing waves, and resonances, particularly at frequencies below 100 Hz. The points shown at the three apexes were averaged to obtain the center point with a proportion of 2H:1.5H:H.*

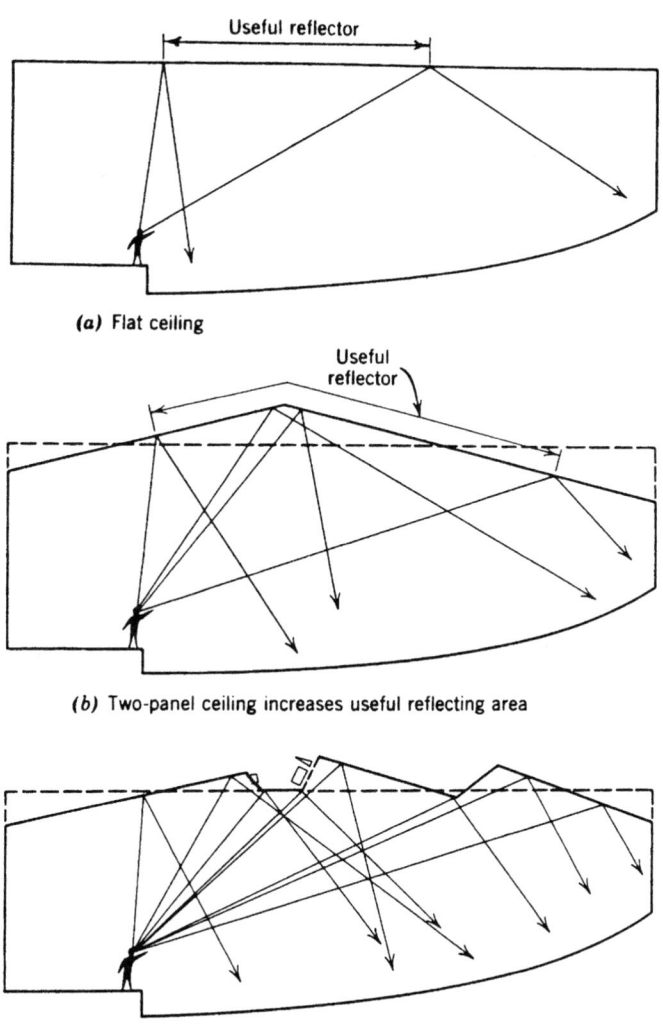

*(a)* Flat ceiling

*(b)* Two-panel ceiling increases useful reflecting area

*(c)* Multifaceted ceiling incorporates lights and loudspeakers

***Fig. 18.23*** *Section through a typical lecture room showing the use of ray diagrams.*

particularly when dealing with large groups playing symphonic music, is due in large measure to reflections received from the side (lateral reflections). This being so, ray diagramming, which was originally performed on room *sections* to determine ceiling shapes and the proper placement of ceiling-hung reflector panels, is now done in *plan* as well. This is particularly important in fan-shaped auditoriums, where a canted wall does not provide useful lateral reflections. The solution to this problem is to build a sawtooth wall or to use reflector panels along the wall that will provide the desired lateral reflection. Nonhorizontal ceiling reflector panels can also provide a measure of lateral reflection, particularly in balconies.

Although they are a useful design tool, ray diagrams have certain restrictions. Design solutions always will require compromise between ray diagram results for various speaking positions on a stage. Thus, a paraboloid may be a perfect shape for one source position but a very poor shape for other positions.

## 18.15 AUDITORIUM DESIGN

*Auditorium* is a general term used to describe a space where people sit and listen to speech or music. Acoustical design of an auditorium includes room acoustics, noise control, and sound system design.

Noise control is covered in Chapter 19, and sound systems are discussed in Sections 18.16 through 18.18. If the program for the auditorium includes activities that require different acoustical environments (as is frequently the case), then it must be decided early whether the acoustics will be a compromise between the program extremes or adjustable for various activities. Acoustical environments can be altered by changing the space volume, moving reflecting surfaces, and adding or subtracting sound-absorbing treatment. Figure 18.24 illustrates several examples of acoustical adjustability.

Factors that influence acoustical design include audience size, range of performance activities, and sophistication of the potential audience. A small school auditorium and a professional theater will have widely divergent demands from both audiences and performers. The audience size determines the basic floor area of an auditorium, assuming no balconies. Once this area has been fixed, the volume of the room is developed according to reverberation requirements of the space.

Figure 18.25 shows a typical auditorium in plan and section. The shape of the wall and ceiling surfaces is developed to provide proper distribution of sound and eliminate focusing or echoes. Essential characteristics of the design include:

1. Ceiling and side walls at the front of the auditorium distribute sound to the audience. These surfaces must be close enough to the perform-

ers to minimize time delays between direct sound and reflected sound.
2. Ceiling and side walls provide diffusion.

Acoustics must be considered in the selection of materials used in an auditorium. All auditoriums use both sound-reflecting and sound-absorbing materials. Since the largest area of sound-absorbing material in any auditorium is the audience, the difference in acoustical characteristics that occurs without an audience may be minimized by using fully upholstered seating.

Chairs with fully upholstered seats and backs, covered in an open-weave material, will have absorption characteristics approximating those of an audience. Using the auditorium in Fig. 18.25 as an example, the reverberation characteristics of an auditorium with various materials may be examined. In the first example, the room use is assumed to be for music performance. The only sound absorption is that provided by the audience and seating. In the second set of calculations, absorptive curtains were installed along the rear wall and a portion of the side wall. This configuration might be used for lectures in a room that is adjustable between speech and music configurations. A third configuration might use permanent sound-absorbing treatment installed on the ceiling and rear and side walls. Because of its low reverberation time, this configuration would be appropriate only for movies and lectures, not for music activities.

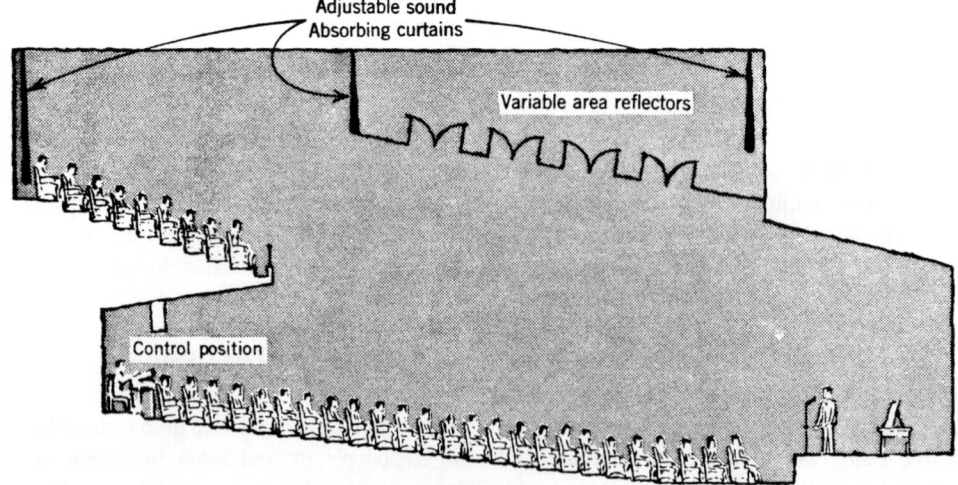

**Fig. 18.24** *Adjustable acoustic elements in an auditorium.*

ACOUSTICS

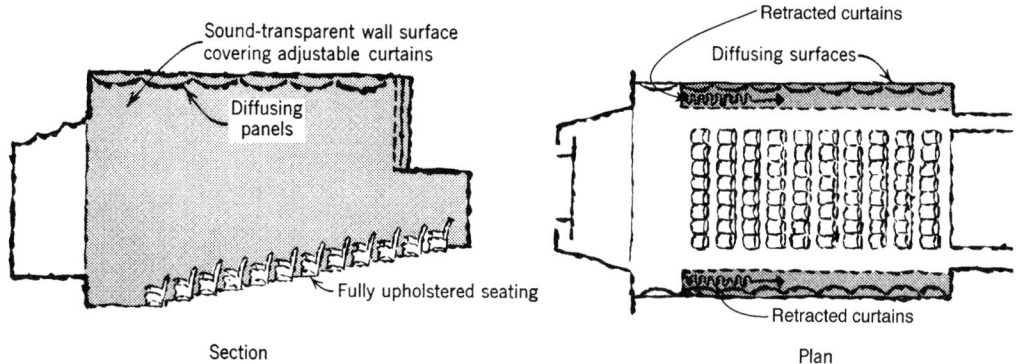

Section      Plan

### Simplified Calculations of Midfrequency (500 and 1000 Hz) Average Reverberation Times

$$\text{Reverberation time } (T_R) = \frac{0.05 \times \text{volume (ft}^3)}{\text{total absorption (sabins)}} \text{ seconds}$$

$$\text{volume} = 155{,}500 \text{ ft}^3$$

**More Reverberant Condition**
*(curtains retracted)*

**Less Reverberant Condition**
*(curtains exposed)*

| | Area | $\propto$ | Absorption | | Area | $\propto$ | Absorption |
|---|---|---|---|---|---|---|---|
| Seating and stage (with audience and performers) | 3323 | .92 | 3060 | Seating and stage (with audience and performers) | 3323 | .92 | 3060 |
| Wall area Concrete block | 8000 | .2 | 1600 | Wall area Concrete block (balance covered by curtains) | 3600 | .2 | 720 |
| | | | | Curtains | 4400 | .45 | 1970 |
| Lower rear wall Permanent sound-absorbing treatment | 450 | .88 | 396 | Lower rear wall Permanent sound-absorbing treatment | 450 | .88 | 396 |
| Total absorption | | | 5056 | Total absorption | | | 6146 |

More reverberant condition

Less reverberant condition

$$T_R = \frac{0.05 \times 155{,}500}{5056} = 1.5 \text{ s}$$

$$T_R = \frac{0.05 \times 155{,}500}{6146} = 1.2 \text{ s}$$

**Fig. 18.25** *Auditorium illustrating the use of surface treatments for control of reflections and reverberation.*

These simple examples indicate the effect of changes in the amount of absorption on the characteristics of a room. Adjustable treatments permit the characteristics of the room to be modified to any point between the extremes to meet the acoustic program requirements of a multipurpose hall.

Existing spaces may require remedial treatment to eliminate unwanted phenomena such as focusing and echoes, as shown in Fig. 18.26. In the first example, the surface of the dome was covered with sound-absorbing material to eliminate focusing; in the second, sound-absorbing treatment was

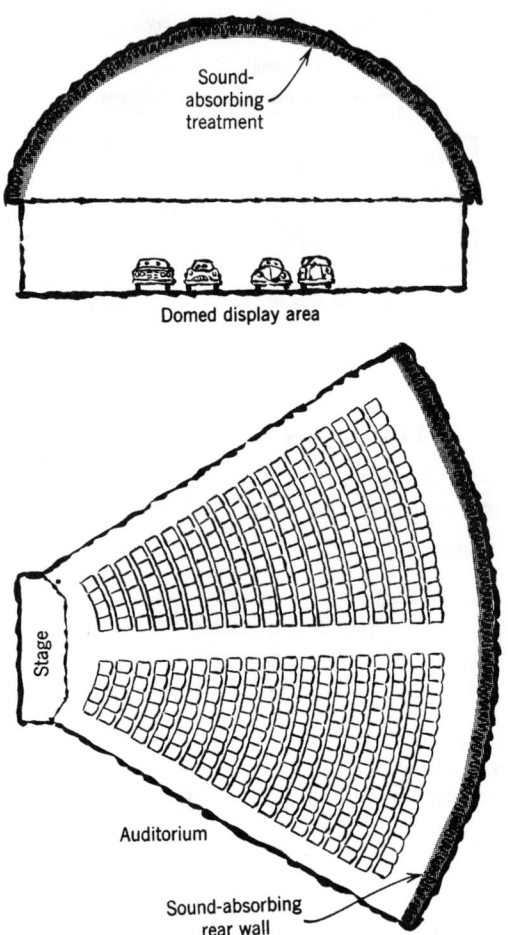

**Fig. 18.26** *Sound-absorbing treatment used to eliminate focusing from the dome and the curved auditorium wall.*

applied to a curved rear wall to eliminate an echo. Such treatment also will affect the reverberation characteristics.

## SOUND REINFORCEMENT SYSTEMS

### 18.16 OBJECTIVES AND CRITERIA

The purpose of a sound reinforcement system is just what the name indicates—to reinforce the sound, which would otherwise be inadequate. Thus, an ideal sound system will give the listener the same loudness, quality, directivity, and intelligibility as if the source of sound were immediately adjacent—a distance of 2 to 3 ft for speech and farther for music, depending upon the type and number of instruments. This situation must obtain for every position in the space within $\pm 3$ dB. The other factors should remain constant. Of these factors, loudness and intelligibility have previously been discussed. By *quality* we mean that the frequency response should be linear so that reproduced sound bears the same relation between its frequency components as the original sound. (Quality is then field-adjusted by voicing or equalization, as discussed below.)

*Directivity* is the characteristic whereby the sound appears to be coming from the originating source, that is, the loudspeakers should be directionally "invisible," and the listener must have the impression of actually hearing the source. It should be emphasized that sound systems cannot correct a poor acoustic design completely, although they can improve a bad situation.

Generally, sound systems will be required in spaces larger than 50,000 ft$^3$ ($\approx 1400$ m$^3$). In terms of occupancy, this volume translates into 550 persons in lecture rooms (15-ft average ceiling height and 6 ft$^2$ per person) and 325 persons in theaters (20-ft average ceiling height and 7.5 ft$^2$ per person). In such a room (50,000 ft$^3$) a normal speaking voice can maintain a volume level of only 55 to 60 dB, depending upon room design and voice strength. With background noise at NC 30 (see Table 19.8), a speaker will be heard; at higher noise levels, intelligibility will suffer.

### 18.17 COMPONENTS AND SPECIFICATIONS

All sound systems consist of three basic elements: input devices, amplifier(s), and loudspeaker systems.

#### (a) Input

Input usually means a microphone, a source of commercial broadcast material of various types, and means for reproducing recorded material in all common commercial formats. Connections to local computers and computer networks are available in sophisticated systems.

## (b) Amplifier and Controls

Amplifiers must be rated to deliver sufficient power to produce intensity levels of 80 dB for speech, 95 dB for light music, and 105 dB for symphonic music. This assumes a *maximum* background noise level of 60 dBA. Thus, 80-dB speech intensity will be 20 dB higher—or four times as loud as the noise level. If the noise level is *known* to be below 60 dB maximum, amplifier and loudspeaker power ratings can be reduced accordingly. The amplifier should carry technical specifications for signal-to-noise ratio, linearity, and distortion. Exact values depend upon the application and are left to the acoustics specialist or sound engineer to supply.

In addition to the usual volume, tone mixing, and input-output selector controls, the amplifier *must* contain special equalization controls for signal shaping. These are highly critical filter networks that, by selective amplification and attenuation of portions of the overall audio frequency spectrum, *voice* or *equalize* a system after installation. Equalization is the sine qua non of a good sound system; without it, the system will howl, sound rough, give insufficient and poorly distributed gain and sound level, and generally bad sound. Essentially, voicing tailors the system to the acoustic properties of the space. A system not equipped for equalization is not a professional system, and results will verify it. Furthermore, the specification must provide for the services of a competent sound engineer to perform the equalization after installation and construction is complete.

Another control frequently required in theater systems is a delay mechanism or circuit that can introduce a time delay into a signal being fed to a loudspeaker. Figure 18.27 shows a sound system that covers a majority of an auditorium from a central loudspeaker cluster. The under-balcony seating areas are hidden from the central cluster and receive reinforced sound from distributed loudspeakers in the under-balcony soffit. To provide directional realism, the signal to the under-balcony loudspeakers must be delayed to allow the weaker signal from the central speakers to arrive first. Delay is necessary because electrical signals travel at the speed of light, whereas sound is much slower (one-millionth of light speed, approximately). With this arrangement, sound will seem to come from the source, and the directivity so necessary to realism is maintained.

## (c) Loudspeakers

These are the heart of any sound system and obviously must be of the same high quality as the remainder of the system. Indeed, system economies will show up much more quickly in loudspeaker performance than in any other component. Selection of speakers is a complex technical task beyond the scope of our discussion. Nevertheless, a few general remarks are in order. The best systems with traditional components use central-speaker arrays consisting of high-quality, sectional (multicell), directional, high-frequency horns and large-cone

**Fig. 18.27** *Loudspeaker system using a delayed signal to the under-balcony area.*

woofers. These assemblies are frequently very large, and the architect should be aware of the dimensions that must be accommodated. Smaller units with folded horns can be used, at a sacrifice in low-frequency response. If only speech is to be reproduced, these units will perform adequately. Distributed systems use small (4- to 12-in.-diameter) low-level speakers, ceiling-mounted and firing directly down.

Recent developments in loudspeaker design have produced units much smaller than those previously required for high-power, high-quality, low-frequency sound reproduction. Here again, the considerations are highly technical and speakers should be supplied under a performance specification that guarantees user satisfaction subject to specified test procedures.

## 18.18  LOUDSPEAKER CONSIDERATIONS

Loudspeaker system design and placement must be coordinated with the architectural design. The two principal types of loudspeaker systems are central and distributed. The loudspeakers in a conventional central system are a carefully designed array of directional high-frequency units combined with less directional low-frequency units placed above and slightly in front of the primary speaking position. In most theaters, this location is just above the proscenium on the centerline of the room. Located in this position, the system provides directional realism and is simple in its design.

A distributed loudspeaker system consists of a series of low-level loudspeakers located overhead throughout the space. Each loudspeaker covers a

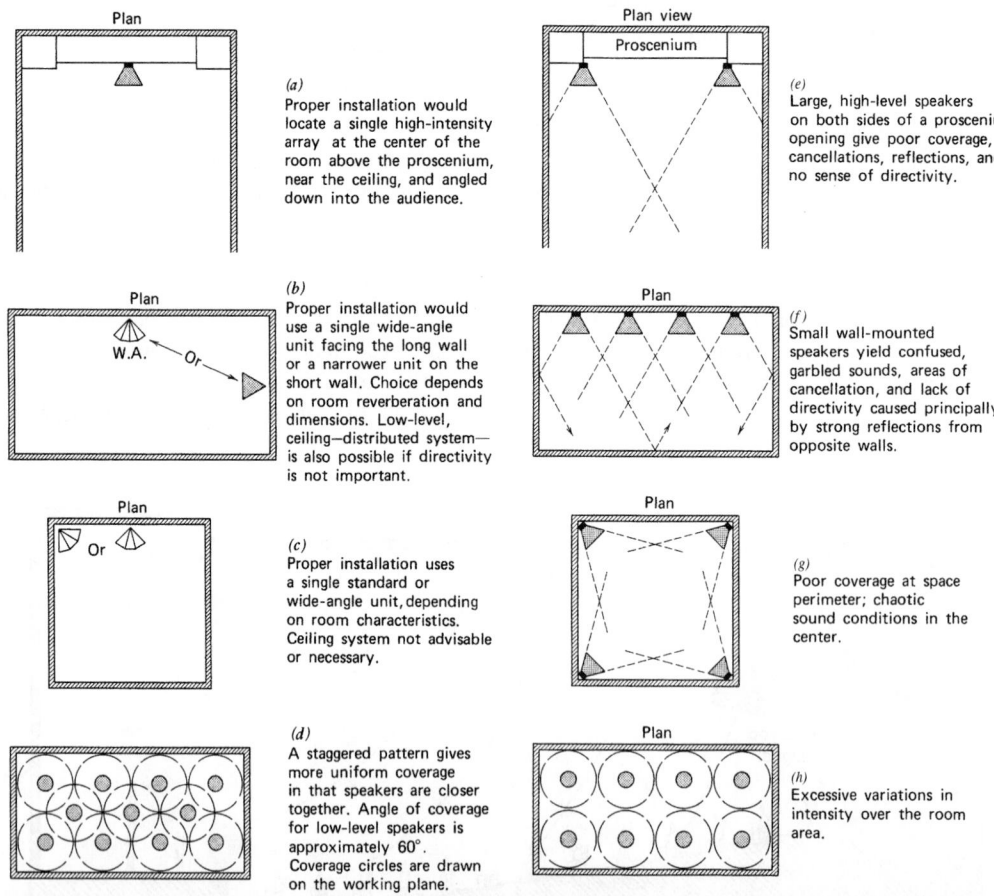

*(a)*
Proper installation would locate a single high-intensity array at the center of the room above the proscenium, near the ceiling, and angled down into the audience.

*(b)*
Proper installation would use a single wide-angle unit facing the long wall or a narrower unit on the short wall. Choice depends on room reverberation and dimensions. Low-level, ceiling–distributed system— is also possible if directivity is not important.

*(c)*
Proper installation uses a single standard or wide-angle unit, depending on room characteristics. Ceiling system not advisable or necessary.

*(d)*
A staggered pattern gives more uniform coverage in that speakers are closer together. Angle of coverage for low-level speakers is approximately 60°. Coverage circles are drawn on the working plane.

*(e)*
Large, high-level speakers on both sides of a proscenium opening give poor coverage, cancellations, reflections, and no sense of directivity.

*(f)*
Small wall-mounted speakers yield confused, garbled sounds, areas of cancellation, and lack of directivity caused principally by strong reflections from opposite walls.

*(g)*
Poor coverage at space perimeter; chaotic sound conditions in the center.

*(h)*
Excessive variations in intensity over the room area.

**Fig. 18.28** *Good speaker layouts (a–d) and poor layouts (e–h). Wide-angle speakers are available to fit most needs. Ceiling speakers give effective coverage on a 60° cone. The working plane is taken to be 4 ft (1.2 m) AFF for seated audience and 6 ft (1.8 m) AFF for standing listeners.*

small area, in a manner similar to downlights. This type of system is used in low-ceiling areas where a central loudspeaker cluster cannot provide proper coverage. It also can be used for public address functions if directional realism is not essential, in spaces such as exhibition areas, airline terminals, and offices. In public areas such as transportation terminals, where the absence of absorptive material makes such spaces highly reverberant, particular care must be taken in speaker positioning and volume levels. Failure to do so will result in the unfortunately very common condition of extremely loud yet unintelligible speech.

Distributed loudspeaker systems provide flexibility for use in spaces where source and listener locations vary according to the use of the space, since loudspeakers can easily be switched to provide proper coverage. In general, a listening position should receive sound from only one loudspeaker. Systems that cover seating areas with signals from several scattered loudspeakers will increase the loudness of the sound but tend to produce garbled speech. This rule is the principal reason that the arrangements shown in Fig. 18.28*e–h* will guarantee a bad job. The common practice of placing one loudspeaker on either side of a proscenium opening (Fig. 18.28*e*) or rows of speakers on one or both sides of a room (Fig. 18.28*f*), is particularly to be deplored.

Location and design of the sound system control position can create problems for the architect. The sound system operator must be within the coverage pattern of the loudspeakers. For proper operation, he or she should be able to hear the sound as it is heard by the audience. Some current auditorium designs locate sound system controls within the audience seating pattern (see Fig. 18.24). Other designs place a control room with a completely open wall or a large window at the rear of the auditorium. Monitor loudspeakers and earphones are inadequate substitutes for actual listening within the auditorium. In churches the control equipment can usually be located at the rear of the congregation area.

### References

ASTM. 2000. Standard E795-00: *Standard Practices for Mounting Test Specimens During Sound Absorption Tests.* ASTM International, West Conshohocken, PA.

Bruel, P.V. 1951. *Sound Insulation and Room Acoustics.* Chapman and Hall, London.

Magrab, E.B. 1975. *Environmental Noise Control.* John Wiley & Sons, New York.

CHAPTER **19**

# Building Noise Control

ACOUSTICS

## NOISE REDUCTION

NOISE CONTROL IN BUILDINGS IS COMPRISED OF three components:

1. Reduction of noise generation at the source by proper selection and installation of equipment
2. Reduction of noise transmission from point to point (along the transmission path) by proper selection of construction materials and appropriate construction techniques
3. Reduction of noise at the receiver through acoustical treatment of the relevant spaces to meet the noise criteria developed in Chapter 17

Speech privacy is achieved by manipulation of all of the above plus the use of masking noise where necessary.

Noise reduction is essentially the science of converting acoustical energy into another, less disturbing form of energy—heat. Since the amounts of energy involved are minute—130 dB corresponds to 1/1000 of a watt, or 0.003 Btu/h—the heat produced is completely negligible. This energy conversion is accomplished by absorption of sound energy by the room contents and wall coverings and also by the structure itself. The former controls noise levels *within* a space and the latter noise transmission between spaces. The reasons for this will become clear as our discussion proceeds.

## ABSORPTION

### 19.1 THE ROLE OF ABSORPTION

Absorptive noise control treatment of a room will affect the reverberant noise level *within* that room but will have a minimal effect on the noise level in adjoining spaces. Refer to Fig. 19.1 for a graphic presentation of this fundamental fact. The best that can be accomplished with acoustic room treatment is elimination of the reverberant field, that is, making the intensity at the room boundaries what it would have been in free space, as in Fig. 19.1*d*. (Even this is extremely difficult; the actual field at the wall would be above 72 dB, except in a completely anechoic chamber.) Adding further wall (or other) acoustic absorbent (as in Fig. 19.1*e*) does nothing in the room itself and has a minimal effect on the overall transmission loss, since the transmission loss in the acoustic material itself is very low, as can be seen in Fig. 18.1.

The subject of acoustic energy absorption and absorptive materials, and their effect on room acoustics including noise reduction, is treated extensively in Sections 18.1 through 18.10 for porous absorptive materials. Two other types of absorptive material are in use, although much less commonly: panel resonators and cavity resonators.

**787**

(a) TV set in free space produces 75-dB sound level, which drops 6 dB for each doubling of distance. Attenuation by inverse square law (see Eq.17-5).

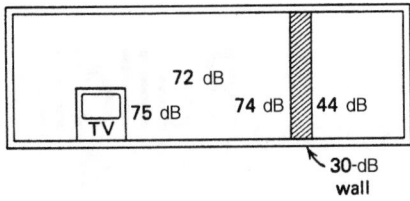

(b) TV still produces 75 dB. In the free field, sound drops to 72 dB but builds up to 74 dB at the wall due to reverberant field reinforcement (see Fig. 18.7). Wall attenuation is 30 dB. Sound on other side of the wall is 74 − 30 = 44 dB.

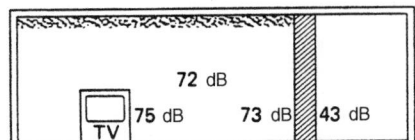

(c) Acoustic tile ceiling acts to reduce room reverberant field. Free field is extended. Level at wall is 73 dB. Level in second space is 73 − 30 = 43 dB.

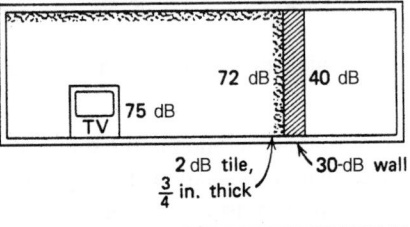

(d) Entire room is acoustically treated, effectively eliminating reverberant field. Room is "dead." Level on second side of wall is 72 dB less acoustic tile loss, less wall loss (that is, 72 − 2 − 30 = 40 dB).

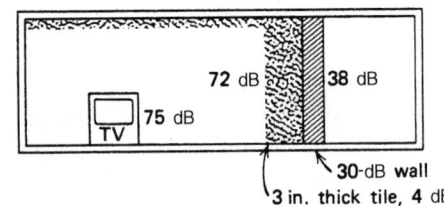

(e) Add another 2¼ in. of acoustic wall treatment. Room is "dead." Level at wall 72 dB. Level in second space = 72 − 4 − 30 = 38 dB.

**Fig. 19.1** *Graphic description of the effect of porous absorptive material on fields and SPL in adjoining spaces.*

## 19.2 PANEL AND CAVITY RESONATORS

*Panel resonators* are built with a membrane such as thin plywood or linoleum in front of a sealed air space generally containing absorbent material. The panel is set in motion by the alternating pressure of the impinging sound wave. The sound energy is converted into heat through internal viscous damping. Panel resonators are used where efficient low-frequency absorption is required and middle- and high-frequency absorption is unwanted or provided by another treatment (Fig. 19.2). Panel resonators are often used in recording studios.

A *volume* or *cavity resonator* (Helmholtz resonator) is an air cavity within a massive enclosure connected to the surroundings by a narrow neck opening. The impinging sound causes the air in the neck to vibrate, and the air mass behind causes the entire construction to resonate at a particular frequency. At that frequency absorption approaches unity, but drops fairly sharply above and below this frequency (see Fig. 19.2). By adjusting the neck opening and cavity dimensions, a unit can be tuned to resonate at different frequencies. This makes it extremely useful when a major single frequency is present, as with 120-Hz transformer

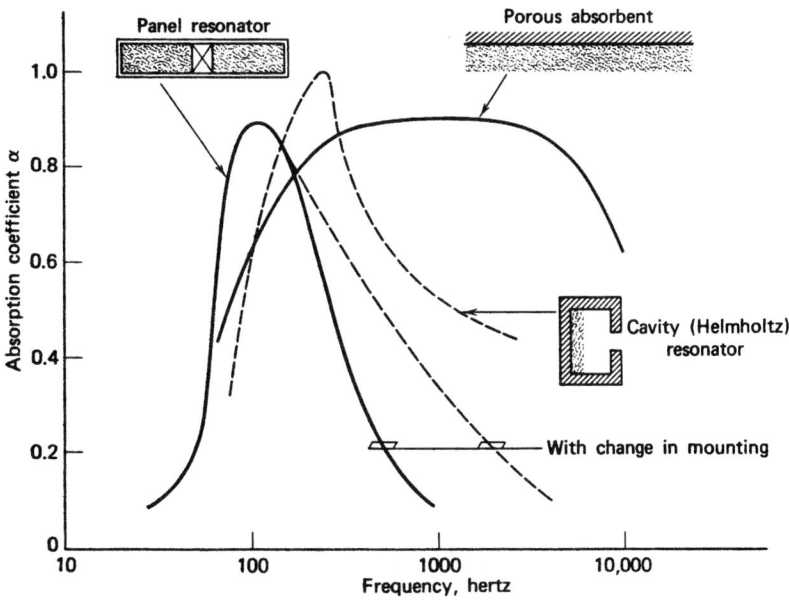

**Fig. 19.2** *Typical absorption curves for the three major types of sound absorbers. The absorption characteristic of each type can be changed by varying the design, as discussed in the text.*

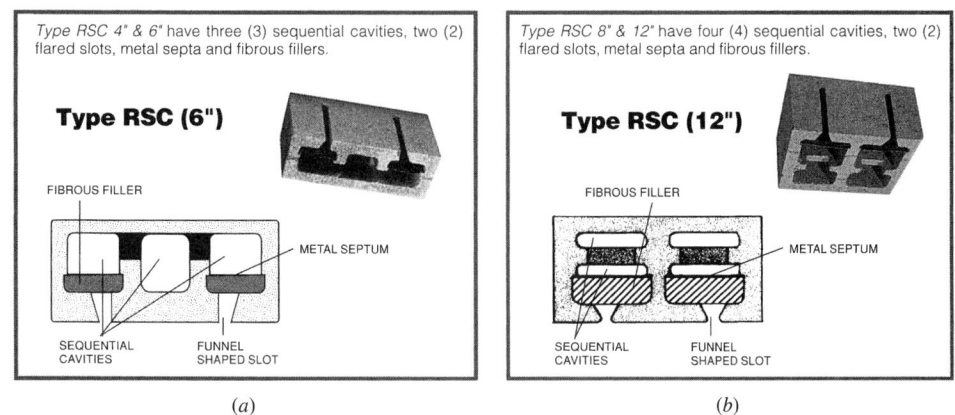

(a)          (b)

## Sound Absorption Coefficients — Type RSC

| Size | Type | Surface | Exposed Slots/ Cavities | FREQUENCY – Hertz | | | | | | | | | | | | | | | | | NRC |
|------|------|---------|------|-----|-----|-----|------|------|-----|-----|-----|-----|------|------|------|------|------|------|------|------|------|
| | | | | 125 | 160 | 200 | 250 | 315 | 400 | 500 | 630 | 800 | 1000 | 1250 | 1600 | 2000 | 2500 | 3150 | 4000 | 5000 | |
| 6" | RSC | PAINTED | 2/3 | .48 | .70 | .93 | 1.14 | 1.05 | .97 | .91 | .84 | .75 | .76 | .77 | .70 | .67 | .68 | .56 | .51 | .59 | **.85** |
| 8" | RSC | PAINTED | 2/4 | .48 | .85 | 1.17 | .99 | .90 | .88 | .98 | .79 | .62 | .58 | .60 | .61 | .70 | .69 | .70 | .64 | .51 | **.80** |
| 12" | RSC | PAINTED | 2/4 | .57 | * | * | .76 | * | * | 1.09 | * | * | .94 | * | * | .54 | * | * | .59 | * | **.85** |

(c)

**Fig. 19.3** *Details of a Helmholtz resonator design using concrete blocks, with tuned block cavities as the resonating chambers. (a) A 6-in. block with three sequential cavities tuned with a metal divider (septum). (b) A 12-in. block with four cavities and added absorptive filler. (c) Sound absorption coefficients. (Courtesy of The Proudfoot Co., Inc.)*

ACOUSTICS

*Fig. 19.4 Application of concrete block resonators of the type shown in Fig. 19.3 in a school gymnasium. (Courtesy of The Proudfoot Co., Inc.)*

hum. Concrete blocks can be used as resonators by tuning their cavities. Their absorption characteristic over the entire frequency band is improved by adding absorptive material in the cavities. Fibrous filler can be used in the block to increase high-frequency absorption. The blocks also serve as standard concrete construction blocks. See Fig. 19.3 for typical block details and Fig. 19.4 for a common application.

## 19.3 ACOUSTICALLY TRANSPARENT SURFACES

The soft, porous material of which acoustic absorbents are constructed may be covered with perforated metal or other materials to provide physical protection and act as stiffeners. These coverings are generally acoustically transparent except at higher frequencies. The frequency at which a noticeable reduction in absorption occurs for a perforated metal cover with circular holes can be estimated as

$$f = \frac{40p}{d} \qquad (19.1)$$

where

$f$ = frequency, Hz

$p$ = percentage of open area

$d$ = diameter of holes, in.

Thus, for ¼-in. holes and 60% open area, which is a typical commercial material,

$$f = \frac{40(60)}{0.25} = 9600 \text{ Hz}$$

which is very high and generally not of major concern. It is always preferable, given a fixed percentage of open area, to use covers with small holes rather than large ones since, as is clear from the formula, this raises the frequency at which absorption drops.

It is also desirable to stagger the holes, as this improves absorption. An open-weave fabric is almost completely transparent to sound and is often used as a decorative cover on absorbent wall coverings.

## 19.4 ABSORPTION RECOMMENDATIONS

To summarize, absorption techniques are useful and effective:

1. To change room reverberation characteristics.
2. In spaces with distributed noise sources such as offices, schools, restaurants, and machine shops.
3. In spaces with hard surfaces and little absorptive content.
4. Where listeners are in the reverberant field. (No amount of absorptive material can reduce intensity levels in the free field.)

Concentrated noise sources are better handled by individual equipment enclosures than by room treatment, since enclosures reduce the amount of sound emitted into a room, which room surface treatment cannot do. In addition to ceiling tiles and wall panels, acoustic absorbent material can be applied in the form of a sprayed-on finish (acoustic plaster), suspended unit absorbers, carpets, draperies, and the like.

## 19.5 CHARACTERISTICS OF ABSORPTIVE MATERIALS

1. *Acoustic tile* is available in size multiples of 12 in., from 12 in. × 12 in. up to 48 in. × 96 in., in a huge variety of patterns and finishes, including units with fire ratings. Installation methods include lay-in, nailing to furring strips, and gluing. Tile materials are generally mineral fiber or faced fiberglass, with NRC absorption ratings in the range of 0.45 to 0.75 for mineral fiber tiles and up to 0.95 for fiberglass. The latter, which are frequently used in open-office applications, have articulation class (AC) ratings of 170 to 210 (see Section 19.20*d*). Tiles for use in high-humidity areas should be certified by the manufacturer for that application (see Fig. 19.4).

2. *Perforated metal-faced units* are usually installed in a lay-in suspension ceiling, although

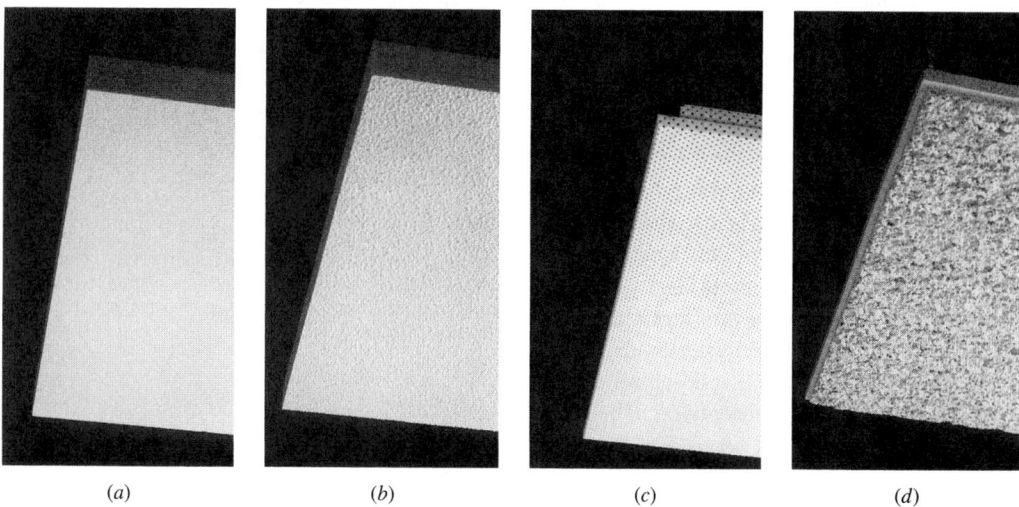

(a)      (b)      (c)      (d)

**Fig. 19.5** *Four of the many types of acoustic ceiling tiles available are illustrated. (a) and (b) are 2 ft × 4 ft × 1½ in. fiberglass ceiling tiles with an acoustically transparent plastic humidity shield facing and a foil back for good sound dispersion in the plenum of open-plan offices. (a) Tile has 1.0 NRC, 200 AC, 0.89 light reflectance and a thermal R value of 4.0. (b) Tile has similar acoustic and thermal characteristics, but due to its coarser surface, its light reflectance is 0.85. Both tiles are applicable to open-plan offices and are usable with indirect lighting systems. (c) Perforated metal cassette panels are used in lobbies, corridors, entries, outdoor canopies and soffits, and areas with high humidity. They are usable in HVAC systems for air returns and, when supplied with an acoustic infill (mineral fiber or fiberglass), have good acoustic properties. (d) General-purpose, coarse-textured mineral fiber ceiling tile is rated 0.70 NRC, 25 CAC, and 0.73 light reflectance. It can be used acoustically in offices, libraries, restaurants, and public spaces. (Photos courtesy of Armstrong World Industries.)*

ACOUSTICS

nailing to furring strips is possible. Units range in size from 12 in. × 24 in. to 24 in. × 96 in. (and larger on special order). The fill material is either wrapped mineral wool or fiberglass, with NRC ratings somewhat lower than those of acoustic tile of the same material. The metal finish is generally baked enamel in a range of colors. The units are applicable to all spaces and have the advantages of easy cleaning, high luminous reflectivity, and incombustibility. With the acoustic backing removed, a perforated unit can be used for air return (Fig. 19.5).

3. *Acoustic panels* (boards) are made of treated wood fibers, bonded with an inorganic cement binder, in sizes ranging from 12 in. × 24 in. to 24 in. × 120 in., with a thicknesses from 1 to 3 in. and a smooth or "shredded" finish. Panels are used in ceiling suspension systems, or nailed or glued when applied to walls and structural ceilings. Their principal advantage is their high structural strength, which makes them applicable to installations requiring acoustic treatment combined with strength and abuse resistance. A second advantage is their excellent flame-spread rating. Typical applications include full-span corridor ceilings, long-span direct-attached ceiling finish, wall panels in school gyms and corridors, and the like. NRC ratings range from 0.40 to 0.70. Acoustic panels are usually resistant to humidity, but usage in high-humidity spaces should be confirmed with the product manufacturer. This is particularly true for panels with "reveal" edges (Fig. 19.6).

4. *Acoustic plaster,* a material comprising a plaster-type base into which is introduced fibrous or light aggregate, is useful for application to curved and other nonlinear surfaces, in thicknesses of up to 1.5 in. Its advantages are ease of application and a high fire rating; its disadvantages are its inability to resist even mild abuse and its inapplicability to humid atmospheres. The noise absorption characteristics of acoustic plaster vary widely with composition, thickness, and application technique, and are generally below those of acoustic tile and panels.

5. *Sound blocks, baffles* and *hanging panels* are simply masses of absorptive acoustic material that achieve absorption coefficients in excess of 1.0 by exposing more than a single absorptive surface to the impinging sound. In all cases, because of their prominent appearance, necessitated by their shape,

*(a)*

*(b)*

**Fig. 19.6** *Acoustic panel material made of pressed organic fiber with an inorganic binder can be used as ceiling tile (a) or wall panels (b). The latter has cloth covering as an outer finish. (Photos courtesy of Tectum.)*

they obtrude into the space and frequently become a major architectural element. This is especially true of hanging baffles (Fig. 19.7).

6. *Wall panels* consist of a wood or metal backing on which is mounted a mineral fiber or fiberglass substrate and a fabric covering. NRC coefficients vary from 0.5 for direct-mounted 1-in. mineral fiber substrate to as high as 0.85 for strip-mounted 1½-in. fiberglass substrate panels. Wall panels are available in widths ranging from 18 to 48 in. and lengths

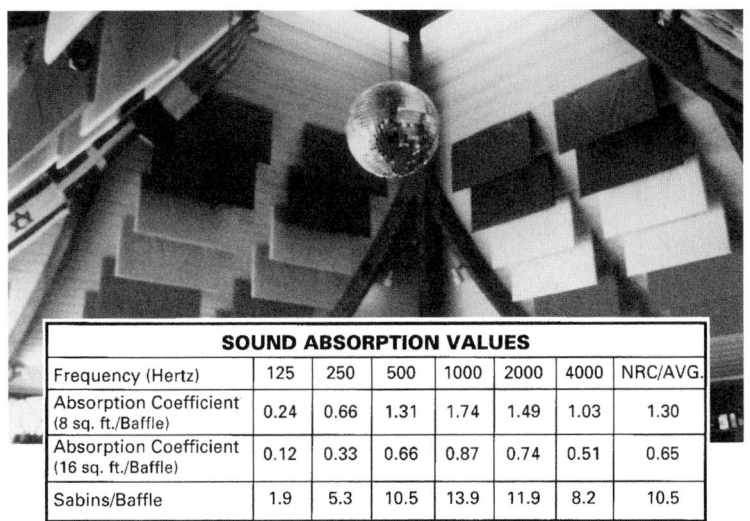

| SOUND ABSORPTION VALUES | | | | | | | |
|---|---|---|---|---|---|---|---|
| Frequency (Hertz) | 125 | 250 | 500 | 1000 | 2000 | 4000 | NRC/AVG. |
| Absorption Coefficient (8 sq. ft./Baffle) | 0.24 | 0.66 | 1.31 | 1.74 | 1.49 | 1.03 | 1.30 |
| Absorption Coefficient (16 sq. ft./Baffle) | 0.12 | 0.33 | 0.66 | 0.87 | 0.74 | 0.51 | 0.65 |
| Sabins/Baffle | 1.9 | 5.3 | 10.5 | 13.9 | 11.9 | 8.2 | 10.5 |

*Fig. 19.7* Application of hanging acoustic baffles to reduce reverberation in the high pyramid-shaped ceiling of a restaurant. The baffles are 1½-in.-thick fiberglass with a tough fire-rated cover. Sound absorption values are given in the accompanying table. (Courtesy of The Proudfoot Co., Inc.)

ACOUSTICS

*Fig. 19.8* Acoustic wall panel consists of a rigid backing covered with either mineral fiber or fiberglass and finished with any of a wide selection of fabrics. NRC averages 0.6 for mineral fiber substrate and 0.8 for fiberglass for contact mounting (A mounting) and 0.7/0.9 when the panels are mounted on 2-cm furring strips (D-20 mounting). The latter mounting is particularly effective in increasing absorption below 500 Hz. (Photo courtesy of Armstrong World Industries.)

of up to 120 in. Fabric coverings generally carry a fire-spread rating. These panels are frequently used in offices, conference rooms, auditoriums, theaters, teleconferencing centers, and educational facilities (see Fig. 19.8).

7. *Resonator sound absorbers* are available in a wide variety of sizes and shapes. Although some designs are available off the shelf, most are tailored to the specific acoustic needs of a project, using one of several standard designs. One fairly common type is shown in Fig. 19.9. In general, resonators are large and must therefore be integrated into the architectural design of a space. Exterior shapes can be altered, and the units can be installed less obtrusively to fulfill this architectural requirement.

8. *Carpeting and drapery* are used to cover large acoustically reflective surfaces in a space. Carpeting can be selected in almost any degree of density, looping, and depth, plus an additional depth of padding, to produce a high degree of absorption in middle and high frequencies (see Table 18.1). In general, absorption is proportional to pile height and density and increases when the carpet is installed on a thick, fibrous pad. Where drapery is not feasible and wall panels are impractical, carpeting can be installed on walls. In such instances, installation on furring strips with an enclosed air space behind will increase absorption over the entire acoustic spectrum and especially

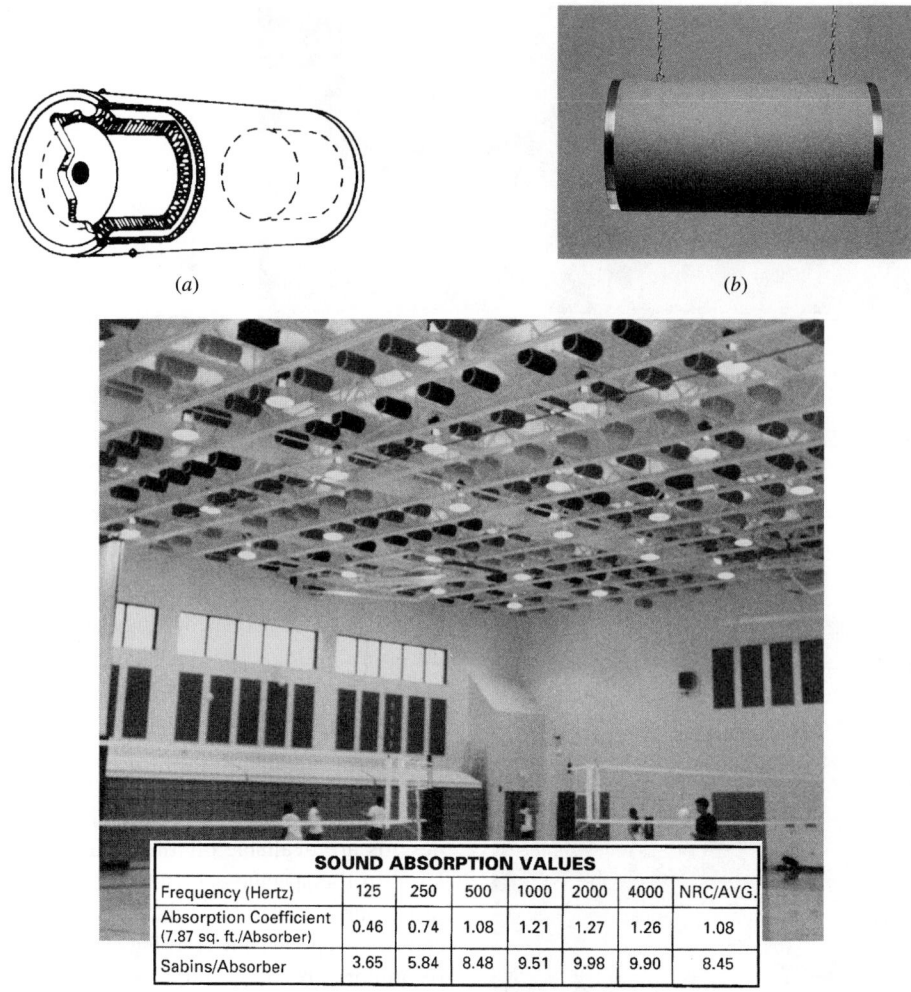

*(a)*

*(b)*

| SOUND ABSORPTION VALUES | | | | | | | |
|---|---|---|---|---|---|---|---|
| Frequency (Hertz) | 125 | 250 | 500 | 1000 | 2000 | 4000 | NRC/AVG. |
| Absorption Coefficient (7.87 sq. ft./Absorber) | 0.46 | 0.74 | 1.08 | 1.21 | 1.27 | 1.26 | 1.08 |
| Sabins/Absorber | 3.65 | 5.84 | 8.48 | 9.51 | 9.98 | 9.90 | 8.45 |

*(c)*

***Fig. 19.9*** *(a) Resonator/absorber unit consists of a molded fiberglass cylinder, 12-in. diameter and 24-in. length, containing a metal resonator at each end for low-frequency absorption and a fill of acoustic fiberglass for full-spectrum absorption. (b) The unit, which weighs about 6 lb, is suspended on thin wires or chains that are essentially invisible when the unit is mounted 10 ft AFF or higher. (c) A typical application in a school gymnasium. Because of their large size, the units constitute a major architectural element. By color selection and hanging patterns, they can be effectively used as such. The absorption characteristics of these units are listed in the table as a function of frequency. (Courtesy of The Proudfoot Co., Inc.)*

at low frequencies (where direct contact installation exhibits poor absorption).

Draperies are essentially acoustically transparent and provide appreciable absorption only in the middle and upper frequencies with heavy, dense, fuzzy fabrics, particularly when draped with a high degree of fold. As with carpeting, absorption increases over the entire spectrum when a heavy folded drapery forms an air space between itself and the wall. Approximate absorption figures are given in Table 18.1.

## SOUND ISOLATION

### 19.6 AIRBORNE AND STRUCTURE-BORNE SOUND

In contrast to the preceding material, which was concerned with in-room sound reduction by absorption, the following sections will discuss the characteristics of sound *transmission* between enclosed spaces. A distinction is often made between airborne and structure-borne sound, although in reality they differ only in the origin of the sounds. *Airborne sound* originates in a space with any sound-producing source, and although it changes to structure-borne sound when the sound wave strikes the room boundaries, it is still referred to as airborne because it originated in the air. *Structure-borne sound* is generally understood as energy delivered by a vibrating or impacting source directly contacting the structure. Hence, a child crying in an adjoining apartment is contributing airborne sound; the same child bouncing a ball on the floor is creating structure-borne sound, in this case by impact. Pumps that were installed without proper damping mounts create structure-borne sound by vibration.

In reality, all sound transmission is both airborne and structure-borne since, once having entered the structure, the sound travels along the structure and causes the structure to vibrate, in turn generating airborne sound. Figure 19.10 should assist in understanding this action. In Fig. 19.10*a*, the sound is airborne, originating in the air on one side of the partition. The incidence of sound energy causes the partition to vibrate, generating sound on the other side. Sound does not "pass through" unless an air path exists. If the partition is airtight, then the sound energy causes the structure to become a secondary source. The partition vibrates primarily in the direction of the sound, that is, in the vertical plane. It also vibrates in other modes, causing some sound energy to pass into the floor and ceiling, depending upon the details of attachment. This energy becomes structure-borne sound.

In Fig. 19.10*b*, the process is similar but reversed. Energy is introduced into the structure directly (and efficiently) by mechanical contact, that is, by vibration and impact. Sound travels along the structure, as shown, and, by causing the structure to vibrate, creates *airborne sound*. In a structure with rigid wall-to-floor connections, these sounds are clearly heard throughout the building. It is a common misconception that in a (relatively) massive concrete structure with masonry walls, such as a multistory residential building, light impacts such as footfalls will not be transmitted. On the contrary, the rigidity of the structure, and in particular the

rigid, airtight connections between partitions, floors, and ceilings, provide an excellent path for structure-borne sound. Only impact absorption by heavy carpeting will attenuate the sound of footsteps, and resilient floor-wall connections will attenuate structure-borne sound.

Airborne sound (originating in the air) is generally much less disturbing than structure-borne sound, since its initial energy is very small and it attenuates rapidly at boundaries. Structure-borne sound generally has a much higher initial energy level and attenuates slowly as it travels through a structure, thereby causing disturbance over large sections of a building. This disturbance is magnified by the "sounding board effect."

We are all familiar with the fact that a tuning fork must be held up to the ear to be heard directly, but if its handle is placed on a table the sound is amplified. This action is not really amplification but an increase in the efficiency of energy transfer. In general, the efficiency of a radiator is proportional to the ratio of its surface dimensions to the sound wavelength. A tuning fork vibrating at concert A (440 Hz) with a wavelength of 2½ ft cannot efficiently couple its energy into the air. It is simply too small. By placing the instrument on a table whose dimensions are approximately one wavelength, we permit it to transfer its energy efficiently, hence the amplification. The same effect can be extremely troublesome in structure-borne sound. A vibrating pump itself makes little sound. However, it transfers a large amount of energy into the structure, which will appear as audible sound at each partition, floor, and wall that is rigidly coupled to the structure. Soft (damping) connections prevent energy transfer, thereby greatly attenuating the transmission of sound energy into connecting efficient radiating surfaces—hence the desirability of such flexible connections.

Airborne sound changes direction easily (diffracts), with low frequencies being most flexible in this regard. Structure-borne sound travels much more rapidly than airborne sound (see Table 17.1) and with attenuation as low as 1 dB per kilometer. A sound traveling along a massive structure will radiate outward from the structure only minimally (although enough to be very annoying) because the large mass minimizes vibration in that direction. Thus, in Fig. 19.10*b*, noise from impact on the floor above will be louder (for equal impact energy) than

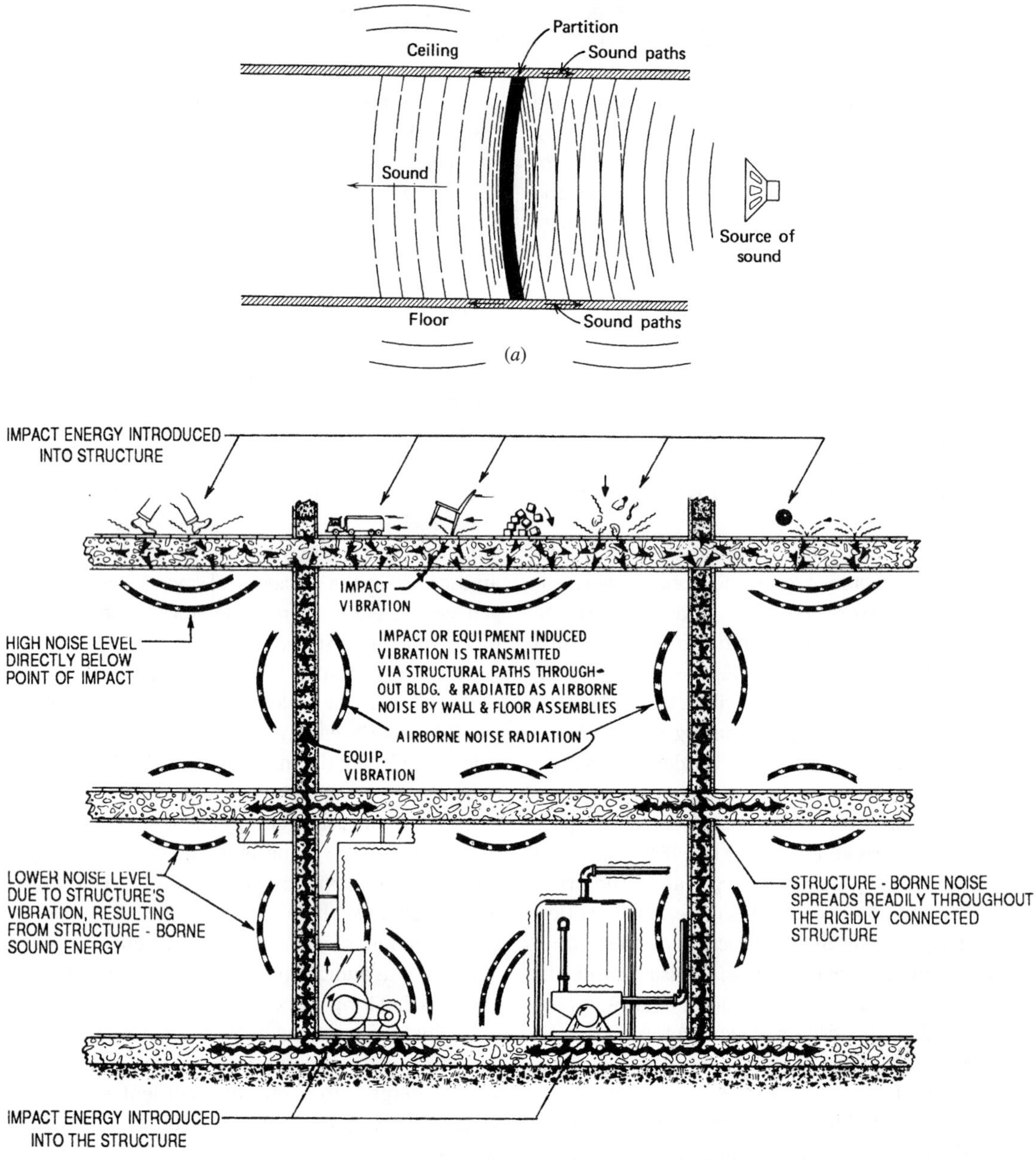

ACOUSTICS

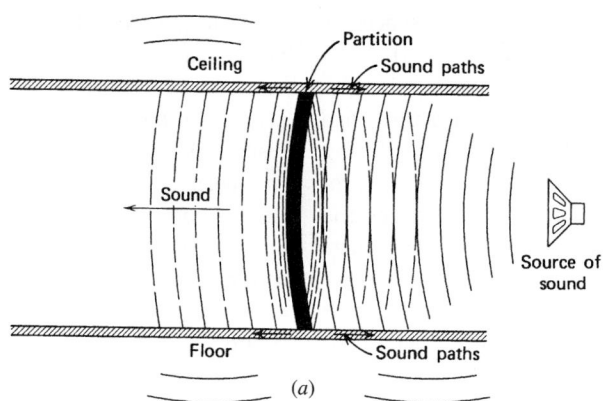

(a)

IMPACT ENERGY INTRODUCED INTO STRUCTURE

IMPACT VIBRATION

IMPACT OR EQUIPMENT INDUCED VIBRATION IS TRANSMITTED VIA STRUCTURAL PATHS THROUGH-OUT BLDG. & RADIATED AS AIRBORNE NOISE BY WALL & FLOOR ASSEMBLIES

AIRBORNE NOISE RADIATION

EQUIP. VIBRATION

HIGH NOISE LEVEL DIRECTLY BELOW POINT OF IMPACT

LOWER NOISE LEVEL DUE TO STRUCTURE'S VIBRATION, RESULTING FROM STRUCTURE - BORNE SOUND ENERGY

STRUCTURE - BORNE NOISE SPREADS READILY THROUGHOUT THE RIGIDLY CONNECTED STRUCTURE

IMPACT ENERGY INTRODUCED INTO THE STRUCTURE

(b)

**Fig. 19.10** (a) Airborne sound is so called because it originates in air; its energy level is low. Here the loudspeaker is the source of acoustic energy, which is converted to a vibration of the partition (shown greatly exaggerated), which in turn becomes a secondary source of airborne sound for the adjoining space. A small amount of energy is reradiated by the ceiling and floor of both spaces directly, and indirectly via the rigid partition-to-floor and partition-to-ceiling connections. (b) Structure-borne sound originates from mechanical contact between the structure and vibrating or impacting sources. As such, its energy level is usually much higher than that of airborne sound. This energy is transmitted with little attenuation throughout the structure via the rigid partition-to-floor and partition-to-ceiling connections. The entire structure is then set into vibration as shown, converting the (large) structure-borne sound energy to noise throughout the building.

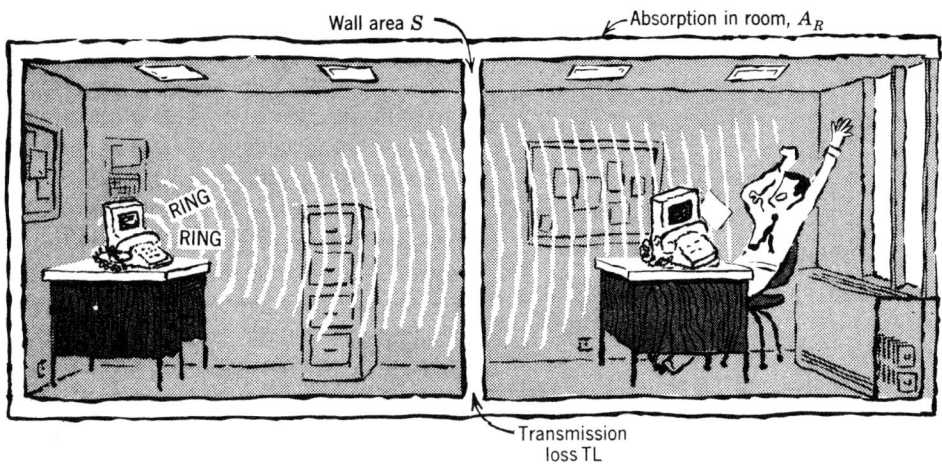

**Fig. 19.11** *A simple case of airborne sound transmission between adjacent rooms through a common barrier. With a sound source in one room, the transmitted sound level is dependent not only on the transmission loss of the barrier but also on the area of the barrier and the receiving-room absorption. The background noise level in the receiving room determines whether the transmitted sound will be noticed.*

ACOUSTICS

noise from machines below because the former generates sound directly downward, while the latter introduces energy into the entire network of parallel paths.

The sections immediately following deal with airborne sound and the means for controlling it (Fig. 19.11). Impact noise (a form of structure-borne sound) is covered in Sections 19.22 to 19.34.

## AIRBORNE SOUND

### 19.7 TRANSMISSION LOSS AND NOISE REDUCTION

The transmission loss ($TL$) of a barrier is the ratio, expressed in decibels, of the acoustic energy reradiated by the barrier to the acoustic energy incident on it. This number is a figure of merit for the sound-isolating quality of the wall itself and is obtained from controlled laboratory tests. (In Europe, transmission loss is referred to as *sound reduction index, R.*) However, the number that is of greater importance to the building designer is the actual noise reduction ($NR$) between two spaces separated by a barrier, that is, the action of the barrier in context. This noise reduction is defined as the difference

between the sound intensity levels in the two rooms, that is,

$$NR = IL_{room\ 1} - IL_{room\ 2} \qquad (19.2)$$

and is related to the TL of the barrier by the expression

$$NR = TL - 10 \log \frac{S}{A_R} \qquad (19.3)$$

where

$NR$ = noise reduction, dB

$TL$ = barrier transmission loss, dB

$S$ = area of the barrier, ft$^2$ (m$^2$)

$A_R$ = total absorption of the *receiving* room, sabins, ft$^2$ (m$^2$)

We see, therefore, that noise reduction and transmission loss are not equal but are related by the size of the dividing barrier $S$ and the absorption characteristic of the receiving room, $A_R$. A moment's thought will confirm the logic of this relation. When sound energy impinges on the barrier, the barrier in turn becomes the sound source, radiating into the receiving room. Therefore, the amount of sound energy transferred is proportional to the (log of) area $S$ of the common barrier between the two spaces.

The sound level in the receiving room is related to its own reverberance (absorption characteristic

$A_R$), as we have seen repeatedly. Thus, if the receiving room is a reverberant, live space, $A_R$ is low and NR is less than TL. Conversely, if the receiving room is dead, $A_R$ is large and NR can be greater than TL, depending upon the ratio of the barrier wall size to the room area (see Fig. 19.11). In lieu of precise calculations, the following generalizations can be used:

1. For a live receiving room,

$$NR = TL - 1 \text{ dB}$$

2. For a medium receiving room,

$$NR = TL + 4 \text{ dB}$$

3. For a dead receiving room,

$$NR = TL + 7 \text{ dB}$$

The extreme case of "deadness" of a receiving room is one with no walls, that is, sound transmission from inside to the exterior. In such cases, NR exceeds TL by 10 to 15 dB, depending upon the size of the exterior opening and the point outside where IL is measured. To acquire facility with sound-isolation techniques, the designer must become familiar with the relationship of transmission loss to the barrier's physical characteristics, its mass, rigidity, material of construction, and method of construction and attachment. These considerations are the subject of the following sections.

*Note:* The reader is cautioned to be careful when encountering the term *noise reduction* (NR), since a similar and completely unrelated term—*noise reduction coefficient* (NRC)—also exists. The latter is very poorly named.

## 19.8 BARRIER MASS

Sound transmission between spaces requires that a barrier be set into vibration by the incident sound energy. Although this was stated above, we repeat it here to emphasize the fundamental importance of this simple statement. (We are assuming a barrier that is impervious to air—i.e., a solid barrier. Otherwise, the moving air molecules bearing the sound will simply pass through with minimal transmission loss.) The impinging sound energy acts as a force on the wall. Since $F = MA$, the larger the mass, the less it will vibrate. When other factors (particularly angle of incidence) are taken into account, the

resultant acoustical relationship is known as the *mass law*. It states that for a nonporous, homogeneous structure of low stiffness, the sound transmission loss is proportional to the logarithm of the surface mass (the weight of the wall per unit of surface area) and to the frequency of vibration. Thus, doubling the mass (or frequency) will, theoretically, cause an increase of 6 dB in the transmission loss; stated otherwise, the slope of TL versus the frequency times mass (*fM*) curve is 6 dB. Figure 19.12 is a graphic representation of mass law operation. With sound incident at 9°, maximum energy is imparted to the barrier and the entire mass resists, resulting in maximum transmission loss. In practice, however, sound is incident from 0° to 80° (called *field* or *random incidence*), reducing the mass effect but keeping the slope at 6 dB per octave. Due to nonhomogeneity, porousness, and stiffness, actual field results indicate transmission losses closer to 4 dB per octave, as shown by the lower curve in Fig. 19.12.

## 19.9 STIFFNESS AND RESONANCE

The *stiffness* of a barrier is a function of its material composition and the rigidity of its mounting. The former depends upon its internal cohesiveness—that is, its modulus of elasticity—and the latter depends upon its boundary restraints—whether the barrier is tightly or loosely held. A homogeneous material of high Young's modulus (such as steel) has great cohesiveness between its molecules. As soon as one molecule is set in motion by incident sound energy, the motion is passed to the next molecule, and so on, making the material an excellent sound conductor. Homogeneous materials with a low modulus of elasticity have high internal damping (the motion of molecules is not transmitted well), and they are good sound insulators. Composite materials such as concrete and organic materials such as wood do not conform to these general rules.

Rigidity of mounting can be likened to a drumhead—the tighter it is stretched, the better it resounds. Rigidity (stiffness) in a panel barrier resists damping and assists vibration, making it a good transmitter and, conversely, a poor noise transmission insulator. As a result, a material such as lead, which has a high mass and a low modulus of elasticity and resists rigid mounting, is an excellent sound attenuator.

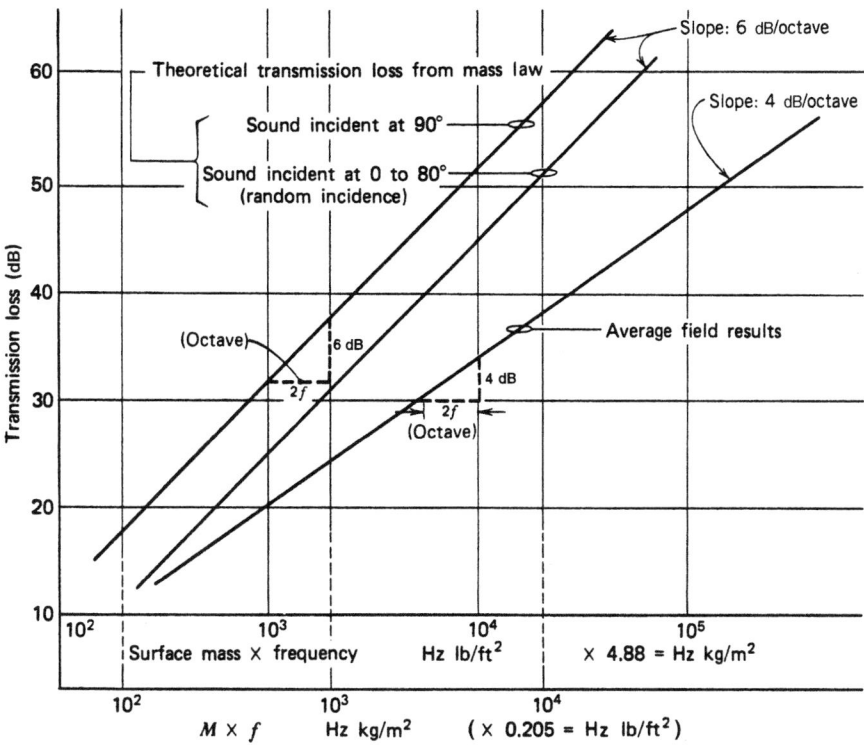

**Fig. 19.12** *Graphic representation of mass law action in attenuation of transmitted sound. Perpendicular (90°) sound incidence results in maximum transmission loss. Field or random incidence (0° to 80°) is approximately 6 dB lower but maintains the 6-dB-per-octave slope. Field results are lower due to flanking and stiffness effects and the slope of the curve averages only 4 dB per octave (frequency doubling).*

The effects of stiffness and mass both vary with frequency, unfortunately in opposite directions. Stiffness acts to reduce transmission loss as frequency increases, while, as we have seen, mass acts to increase it; the combined effect is shown in Fig. 19.13. Therefore, stiffness is most effective at low frequencies and mass at high frequencies. At very low frequencies the mass and stiffness effects negate each other, giving the resonance dips shown. Beyond approximately 200 Hz, most common wall construction enters the mass law range and continues with it until the critical frequency. Deviations from a smooth 4 to 6 dB per octave slope are due to the non-homogeneous nature of most wall constructions. At the *critical frequency* the phase of incident sound waves corresponds to or coincides with the phase of vibration (shear wave) of the barrier in such a way as to pass a large portion of the incident energy. See the insert in Fig. 19.13, which shows this effect as the coincidence dip. This effect is most pronounced in thin, homogeneous partitions and light, stiff ones.

Critical frequency, $f_c$, as a function of panel thickness for common materials, is plotted in Fig. 19.14. To avoid a coincidence dip in the audible range, partitions can be either very heavy and/or very stiff (which greatly decreases the critical frequency) or heavy and limp (resilient—which greatly increases the critical frequency). In practical terms, cost effectiveness is heavily in favor of the latter alternative. Thus, for instance, the transmission loss of a wood partition can be improved by grooving it to increase its flexibility, thereby increasing the critical frequency. The dramatic improvement in transmission loss, resonances, and coincidence dip achieved by the use of resilient mounting of a simple masonry partition is shown in Fig. 19.15. Both walls have the same weight—$21 + lb/ft^2$. The solid wall A has better attenuation below 200 Hz in the stiffness-controlled range. Above that frequency the resilient-mounted partition is 10 dB better, which means that the transmitted noise is only one-half as loud. The sound transmission class (STC) (which is a figure of merit

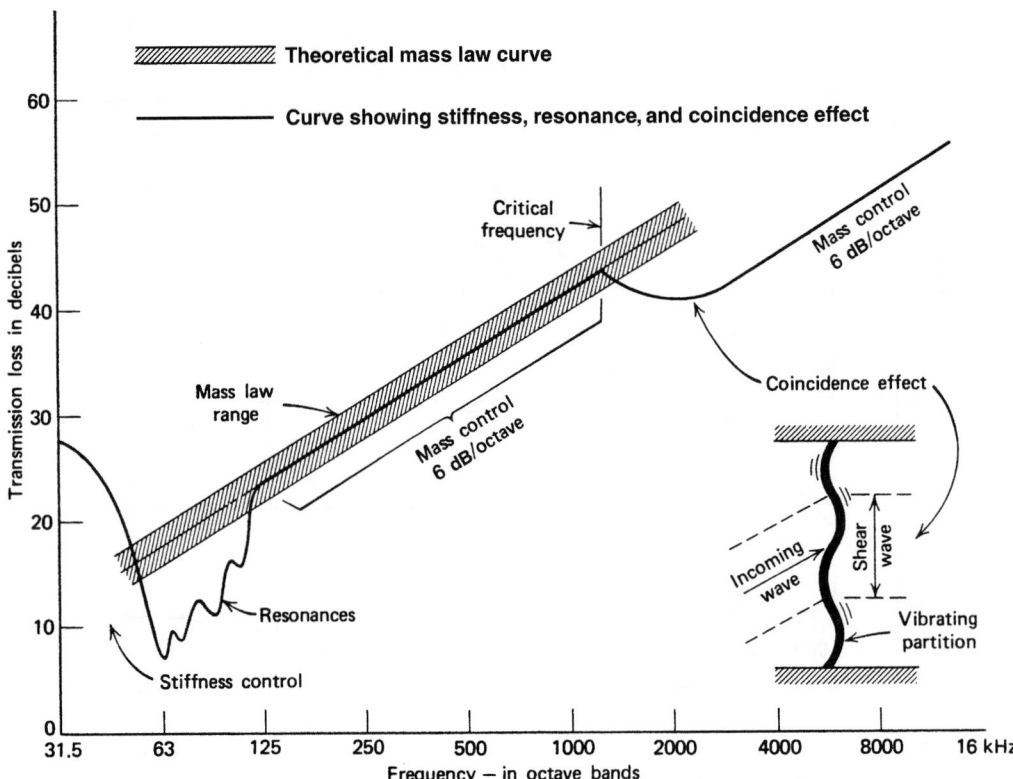

**Fig. 19.13** *Partition transmission loss as a function of frequency showing the effects of stiffness, resonance, and coincidence. At frequencies below resonance, control is almost purely a stiffness function; at frequencies above critical, control is almost purely a surface mass function.*

for partition sound transmission, as explained in Section 19.11) of wall A is 40; that of wall B is 51. Both show a coincidence dip at approximately 250 Hz (see Fig. 19.14), but that of B is shallow, whereas that of A is deep and wide. Furthermore, wall B is consistently better than mass law attenuation would predict, wall A consistently worse (due to stiffness).

## 19.10 COMPOUND BARRIERS (CAVITY WALLS)

Since the maximum theoretical increase in transmission loss with an increase of mass is 6 dB per doubling of mass, it is apparent that this method of transmission loss improvement rapidly reaches the limits of practicality. Indeed, as we have seen, the transmission loss of actual single homogeneous walls fall below the mass law curve. This is because mass increase brings with it stiffness increase, which acts to *reduce* transmission loss. If, however, a barrier

is constructed of two separate layers without rigid interconnection, its performance exceeds the calculated transmission loss based on mass alone. Note that even the nonrigid wire ties of wall B in Fig. 19.16 lower the STC by five points. At low frequencies, where stiffness controls the transmission loss (see Fig. 19.13), the cavity in Fig. 19.16 *C* acts as a rigid connection between the layers, adding stiffness and increasing transmission loss. At higher frequencies, in the mass law range, the air in the cavity acts as a damping coupling to reduce stiffness. The net result is an improvement in performance throughout the frequency range.

Transmission loss for the entire cavity wall increases with the width of the air space at the rate of approximately 5 dB per doubling. Performance can be improved still further by filling the void with porous, sound-absorbent material. This acts to further decrease the stiffness of the compound structure *and* to absorb sound energy that reflects back and forth between the two inside surfaces. The performance of

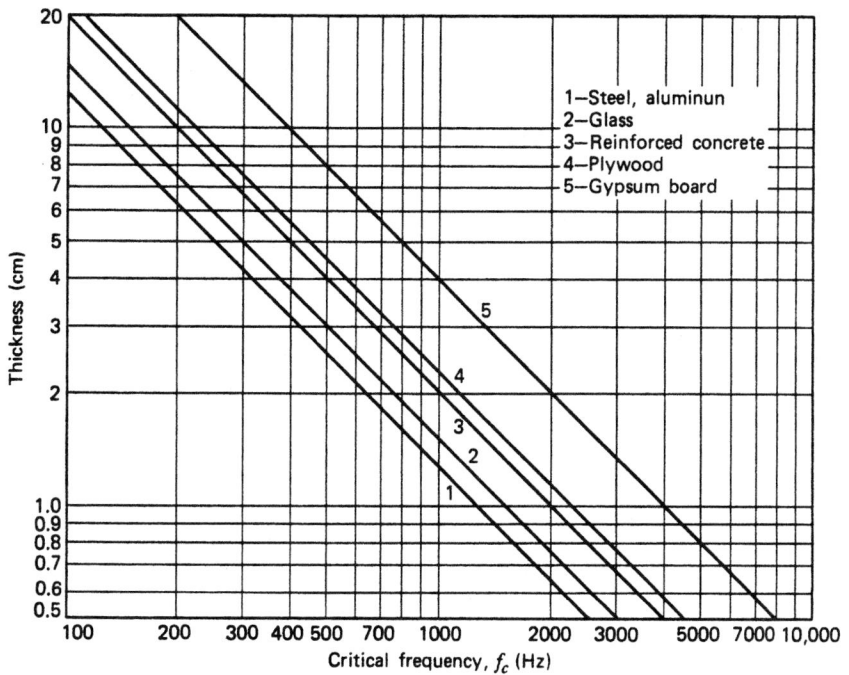

**Fig. 19.14** *Critical frequency as a function of thickness for several common materials. (Reprinted with permission from E.B. Magrab, Environmental Noise Control, Wiley, New York, 1975.)*

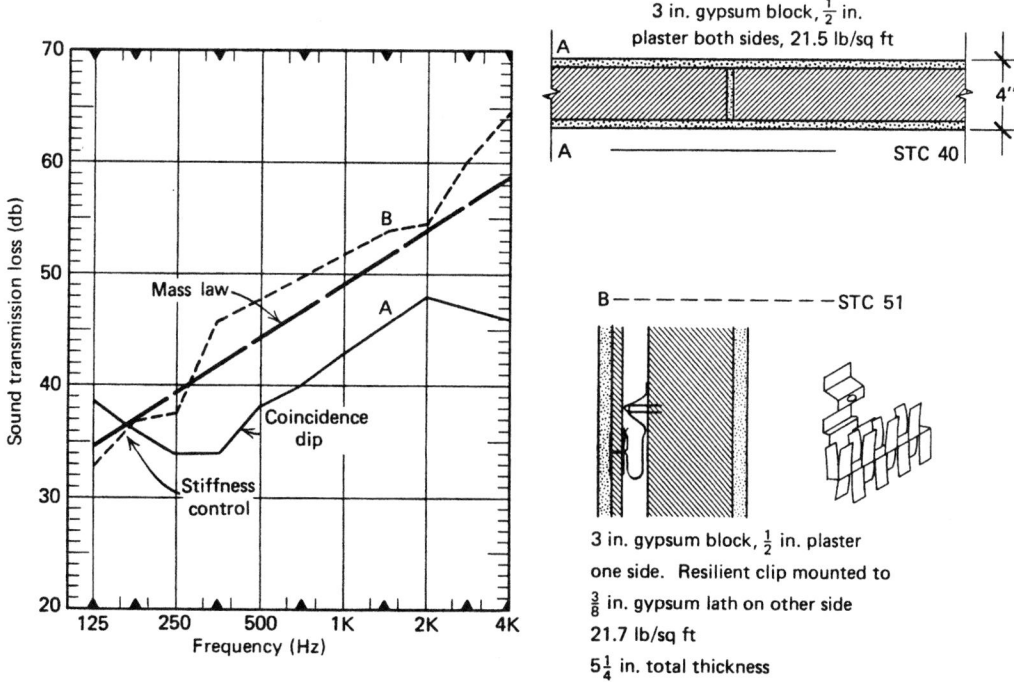

**Fig. 19.15** *Transmission loss characteristics of two equal-weight partitions with similar boundary constraints. The solid partition A performs worse than the mass law due to stiffness. The resilient-mounted wall B performs better than the mass law and much better than wall A, except at the lowest frequencies. (Data extracted from A Guide to Airborne, Impact, and Structure-Borne Noise Control in Multifamily Dwellings, 1968.)*

ACOUSTICS

ACOUSTICS

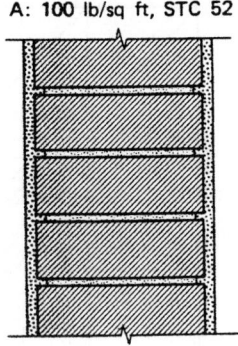

A: 100 lb/sq ft, STC 52

A: _Single 9-in. brick wall_

B: 100 lb/sq ft, STC 49

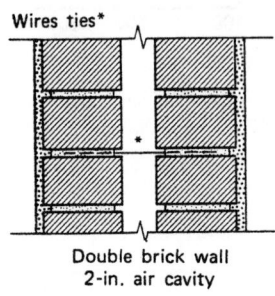

Wires ties*

Double brick wall
2-in. air cavity

B: _12-in. total thickness_
*Without wire ties, STC rises to 54

C: 120 lb/sq ft, STC 62

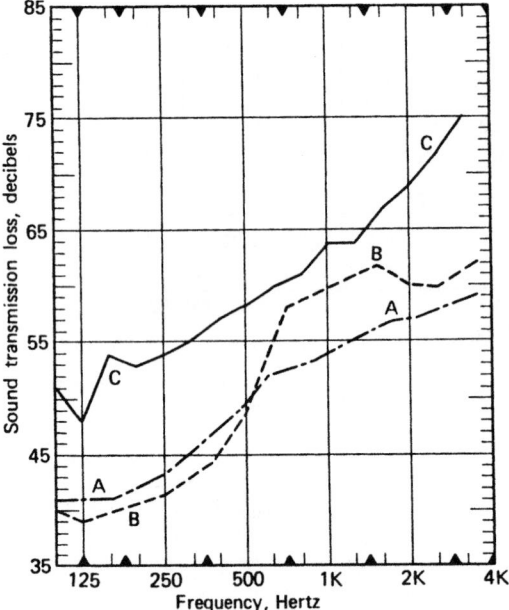

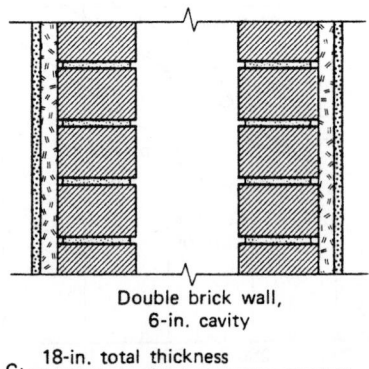

Double brick wall,
6-in. cavity

C: 18-in. total thickness

**Fig. 19.16** *Transmission loss curves showing the effect of air space on heavy wall construction. All three walls are approximately the same mass. The 2-in. air space of wall B is not significant until the higher frequencies, whereas the large 6-in. air space of wall C is effective throughout the frequency spectrum. (Data extracted from* A Guide to Airborne, Impact, and Structure-Borne Noise Control in Multifamily Dwellings, *1968.)*

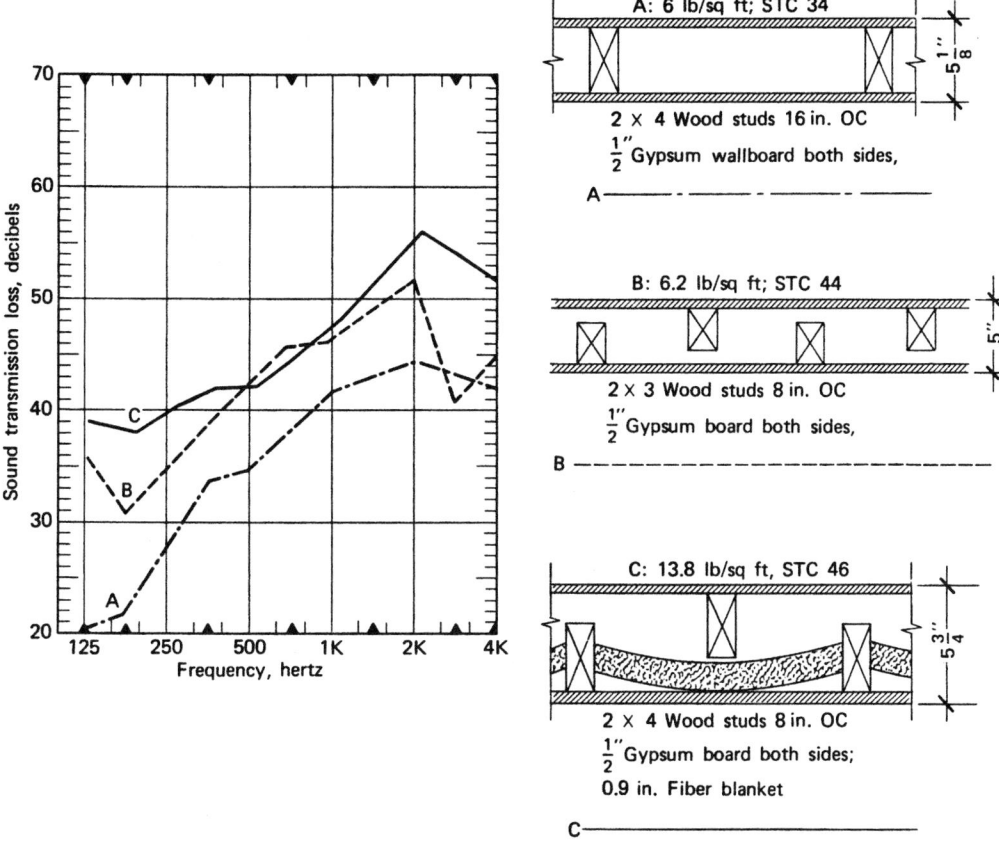

**Fig. 19.17** *Transmission loss curves illustrating the effect on lightweight walls of stiffness reduction and addition of absorptive material in the cavity. Curve A is a standard stud wall typically found in frame construction. Curve B shows the advantage of staggered studs over the entire frequency range. The dip at 3 kHz is a coincidence dip for a single leaf. Addition of absorptive material (Curve C) improves the attenuation characteristic at both ends of the spectrum and is particularly useful for its low end improvement. (Data extracted from* A Guide to Airborne, Impact, and Structure-Borne Noise Control in Multifamily Dwellings, *1968.)*

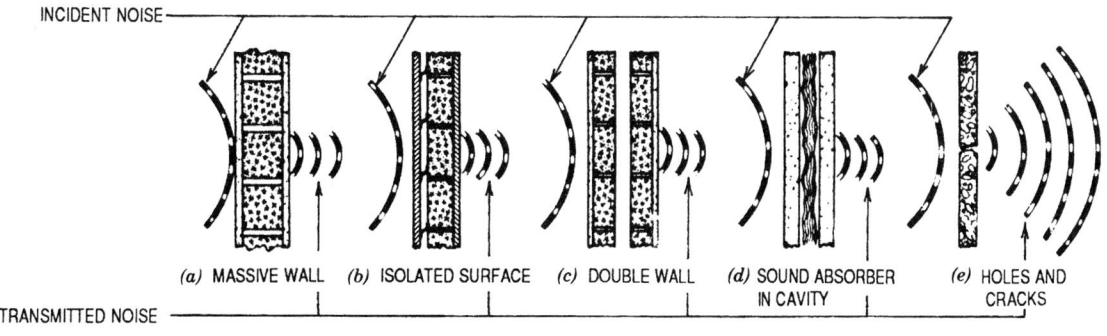

**Fig. 19.18** *The various techniques used to increase the transmission loss of a partition are shown in (a–d). In contrast, even a very small air path through the partition (e) can effectively destroy its effectiveness as a sound barrier. (Data extracted from* A Guide to Airborne, Impact, and Structure-Borne Noise Control in Multifamily Dwellings, *1968.)*

cavity walls is reduced by any rigid interconnections between leaves. Thus, a common stud wall with frequent rigid interconnections acts little better than a single homogeneous wall. However, a stud wall with staggered studs exhibits greatly improved performance over a single-material wall or a common stud wall. The above effects are illustrated in Figs. 19.16 and 19.17 (see also Appendix K). The effects of mass (Section 19.8), stiffness (Section 19.9), and com-

pound barriers with and without filler are shown qualitatively and graphically in Fig. 19.18.

## 19.11 SOUND TRANSMISSION CLASS

Various attempts at using a single-number average transmission loss to describe a barrier's characteristics have been made, with only limited success.

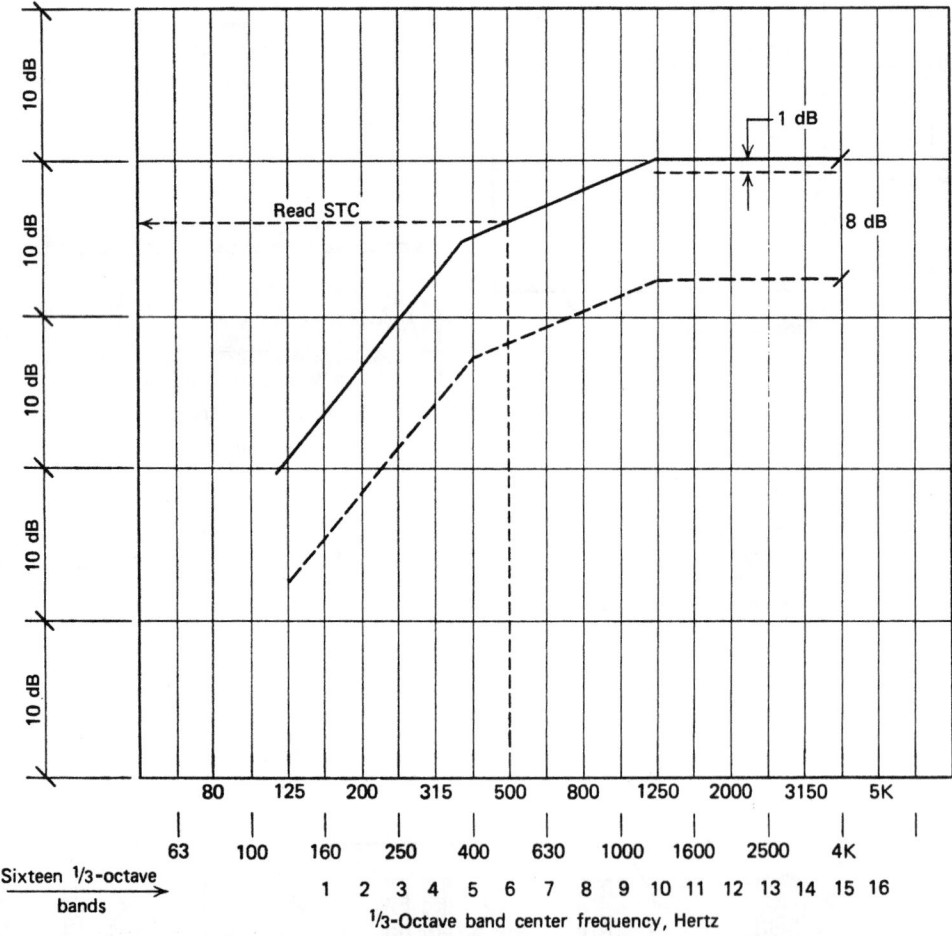

The STC is determined by comparison with a transparent overlay of this graph on which the STC contour is drawn. The STC contour is shifted vertically, relative to the test curve, until some of the measured *TL* values for the test specimen fall below those of the STC contour (the solid line) and the following conditions are fulfilled:

1. The sum of the deficiencies (i.e., the deviations below the contour) shall not be greater than 32 dB.
2. The maximum deficiency at a single test point shall not exceed 8 dB [the broken (dashed) line beneath the STC contour].

When the contour is adjusted to the highest value (in integral db) that meets the above requirements, the sound transmission class for the specimen is the *TL* value corresponding to the intersection of the contour and the 500-Hz ordinate.

*Fig. 19.19* Overlay from which sound transmission class (STC) is determined graphically.

Indeed, such averages can be misleading, since they ignore both deficiencies and proficiencies at particular frequencies. Their use, therefore, in all but rough work is to be discouraged.

To avoid the shortcomings of averages and yet to benefit from the indisputable convenience of single-number ratings, a system of standard contours was developed in the United States called *sound transmission class* (STC) contours. (A similar system, conforming to ISO 717-1, 1996, is used in Europe, involving a weighted sound reduction index $R_W$. In practice, the STC number for a particular barrier construction is derived by comparing actual test results measured in a series of sixteen ⅓-octave bands to the standard STC contours according to a fixed procedure. The technique is illustrated in Fig. 19.19. Figure 19.20 shows two transmission loss curves and STC ratings of each. Because STC fails to give credit for performance *above* the established requirements, octave-band transmission loss data, rather than STC ratings, should be used in all critical areas such as music rooms or mechanical rooms where certain particular frequencies may be dominant.

Figure 19.21 gives three standard STC contours that are of interest because they are used by the Federal Housing Administration (FHA) to specify grades of construction. The criteria for their application are found in Section 19.33. An appreciation of the degree of speech sound isolation provided by walls with different STC ratings is given in Section 19.17 (see Table 19.5). Since the subjective reaction on the quiet side depends upon the background sound level, the table gives this reaction for two NC curve levels. To assist the designer, extensive sound transmission testing has been performed on most types of standard wall and partition construction and the results published. Tables 19.1 and 19.2 and Appendix K give descriptions of constructions with typical details, transmission loss data, STC ratings, and other pertinent data.

## 19.12 COMPOSITE WALLS AND LEAKS

It is frequently necessary to determine the transmission loss of a composite wall, that is, a wall with a window, door, louver opening, and the like. It should be clearly appreciated that the two elements are "in parallel," to borrow an electrical concept, and the behavior is similar to that situation. That is, the overall performance will be strongly affected by the poorer of the two, with some tempering of the

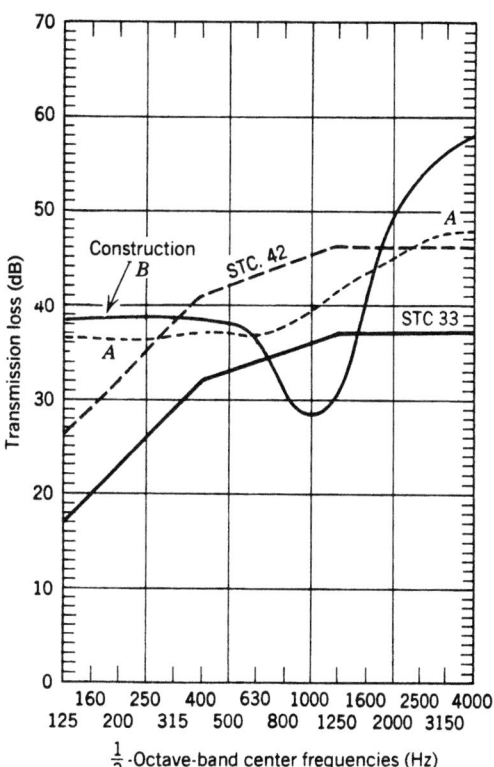

**Fig. 19.20** Curves A and B are two different construction types with the same average transmission loss. However, application of the STC curve criteria (given in Fig. 19.19) yields an STC of only 33 for construction B because of its deep center dip compared to STC of 42 for construction A. (From E.B. Magrab, Environmental Noise Control, Wiley, New York, 1975.)

degradation when the poorer barrier is much smaller in area than the other barrier element. Figure 19.22 enables us to analyze situations of this type.

Since an opening in a wall is effectively a second material of $TL = 0$, the curves in Fig. 19.22 can be replotted for this situation as in Fig. 19.23. Note that the curves very rapidly flatten out; thus, any wall with a 1% open area will have a *maximum* transmission loss of 20 dB, which is all but useless as a sound barrier. For this reason, it is imperative that all openings be completely sealed, particularly those around doors and windows. A hairline crack degrades a wall 6 dB, a keyhole degrades a door 3 dB, and so on. Special considerations for doors and windows are discussed below. Care must also be taken with such common acoustic leaks as back-to-back electric outlets, pipes passing through walls, and medicine cabinets—in fact any break in the integrity of a partition. All such openings must be

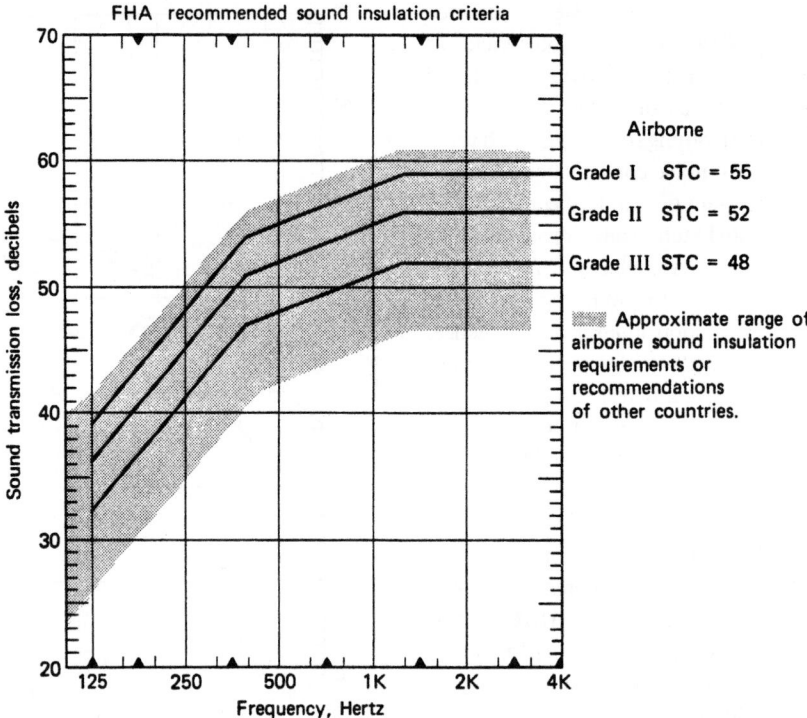

**Fig. 19.21** *Sound insulation criteria recommended by the FHA. (From* A Guide to Airborne, Impact, and Structure-Borne Noise Control in Multifamily Dwellings, *1968.)*

## TABLE 19.1 Improvements in STC Rating of Stud[a] Partitions[b]

| Description | STC[c] |
|---|---|
| Basic partition: single wood studs, 16 in. on centers, ½-in. gypsum board on both sides, air cavity | 35 |
| Add to basic partition | |
| Double gypsum board, one side | +2 |
| Double gypsum board, both sides | +4 |
| Single-thickness absorbent material in air cavity | +3 |
| Double-thickness insulation | +6 |
| Resilient channel supports for gypsum board | +5 |
| Staggered studs | +9 |
| Double studs | +13 |

[a]For application to metal stud partitions, use adders as in note *b*, but begin with STC = 40 for a 3⅝-in. basic partition.

[b]When using two improvements, add an additional +2; for three improvements, add +3.

*Example:* Improvements to 35 STC basic partition:

| | |
|---|---|
| Staggered wood studs | +9 |
| Double gypsum board, one side | +2 |
| Single-thickness insulation | +3 |
| Adder (3 improvements) | +3 |
| Total | +17 |
| Total STC | 35 + 17 = 52 |

[c]The STC figures are conservative. Other sources list the same constructions with 1 to 5 points higher STC.

## TABLE 19.2 STC Ratings of Masonry Walls

| Description | STC[a] |
|---|---|
| 4-in. lightweight[b] hollow block | 36 |
| 4 in. dense hollow block | 38 |
| 6-in. lightweight hollow block | 41 |
| 6-in. dense hollow block | 43 |
| 8-in. lightweight hollow block | 46 |
| 8-in. dense hollow block | 48 |
| 12-in. lightweight hollow block | 51 |
| 12-in. dense hollow block | 53 |
| 4-in. brick | 41 |
| 6-in. brick | 45 |
| 8-in. brick | 49 |
| 12-in. brick | 54 |
| 6-in. solid concrete | 47 |
| 8-in. solid concrete | 50 |
| 10-in. solid concrete | 53 |
| 12-in. solid concrete | 56 |

[a]See note *c*, Table 19.1.

[b]All ratings of lightweight block assume sealing with paint. Note that this reduces absorption.

Modifications

| | |
|---|---|
| Add sand to cores of hollow blocks | +3 |
| Add plaster to one side | +2 |
| Add plaster to both sides | +4 |
| Add furring strips, lath and plaster: | |
| One side | +6 |
| Two sides | +10 |
| Add plaster via resilient mounting: | |
| One side | +10 |
| Two sides | +15 |

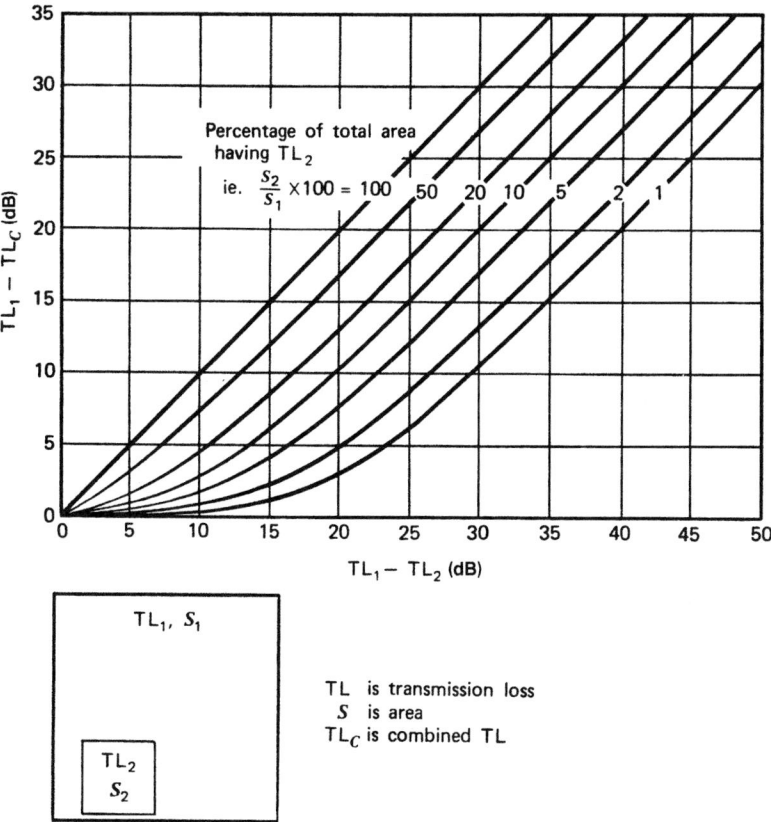

**Fig. 19.22** *Transmission loss of a two-element composite barrier as a function of the relative transmission loss of the components. (From E.B. Magrab,* Environmental Noise Control, *Wiley, New York, 1975.)*

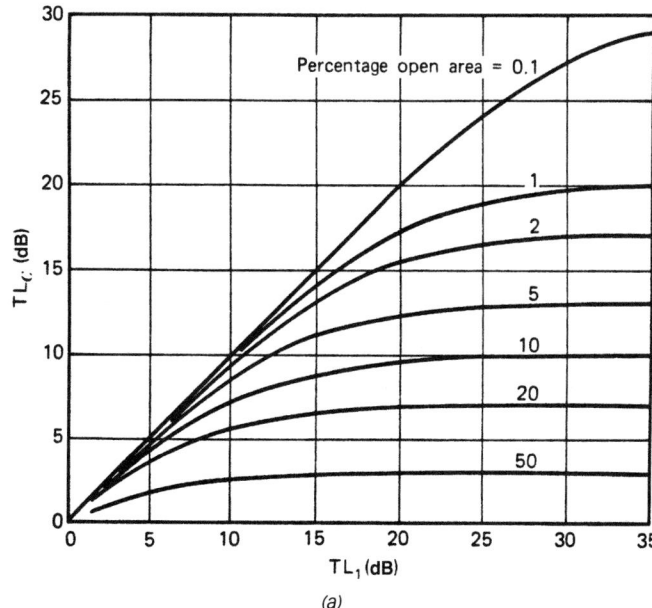

*(a)*

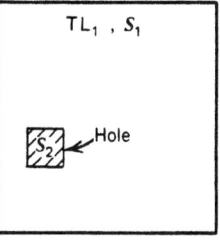

*(b)*

**Fig. 19.23** *Effect of a hole of a given size (as a percentage of the total area) on a partition with a given transmission loss. (From E.B. Magrab,* Environmental Noise Control, *Wiley, New York, 1975.)*

caulked to make an airtight joint if any appreciable degree of sound isolation is to be maintained.

Examples 19.1 and 19.2 show how barriers of high transmission loss are degraded by standard openings. To maintain the integrity of a barrier, special care must be taken in the design of windows and doors, as explained below.

**EXAMPLE 19.1** Given a 9-ft × 18-ft wall with a transmission loss of 52 dB at 1000 Hz, containing a 3-ft × 7-ft, 6-in. hollow core door of 22 dB transmission loss at that frequency, find the overall transmission loss of the composite wall.

**SOLUTION**

Refer to Fig. 19.22.

$$TL_1 - TL_2 = 30 \text{ dB}$$

$$\frac{S_2}{S_1} = \frac{3 \times 7.5}{9 \times 18} \times 100 = 13.9\%$$

From the curves in Fig. 19.22:

$$TL_1 - TL_c = 21.5$$

$$TL_c = 52 - 21.5 = 30.5 \text{ dB}$$

That is, a door with an area of only 14% of the entire wall reduces the transmission loss of the structure from 52 to 30.5 dB—that is, from excellent to very poor. ■

**EXAMPLE 19.2** An exterior brick/frame wall having a transmission loss of 54 dB at 1000 Hz, measuring 8 ft × 16 ft, is pierced by two wood frame windows, each of area 3 ft × 4 ft, with single ⅛-in. glass, with a transmission loss of 34 dB at 1000 Hz. Find the combined transmission loss.

**SOLUTION**

$$TL_1 - TL_2 = 54 - 34 = 20 \text{ dB}$$

$$\frac{S_2}{S_1} = \frac{2 \times 3 \times 4}{8 \times 16} \times 100 = 18.8\%$$

From Fig. 19.22:

$$TL_1 - TL_c = 12.5 \text{ dB}$$

$$TL_c = 54 - 12.5 = 41.5$$

Again, the result is a reduction from an excellent wall to a poor one. ■

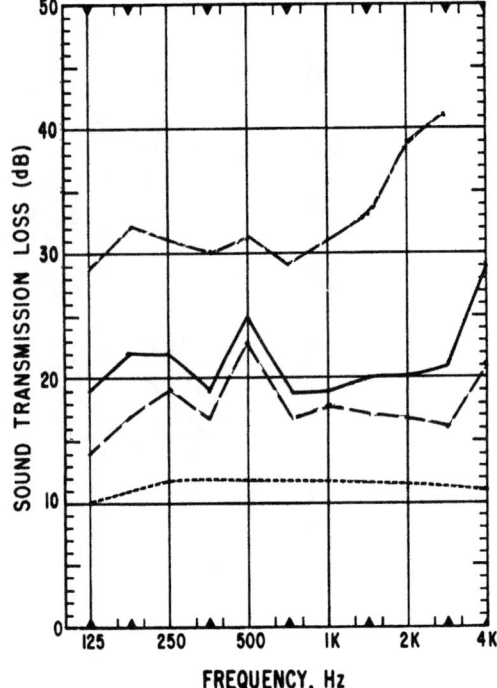

*(a)* Sound Transmission Loss of Doors

-··-· 1 3/4″ solid wood core door with gaskets and drop closure

——— 1 3/4″ hollow wood core door with gaskets and drop closure

-·—·— Same hollow door, no gaskets or closure, 1/4″ airgap at sill

········· Louvered door, 25-30% open area

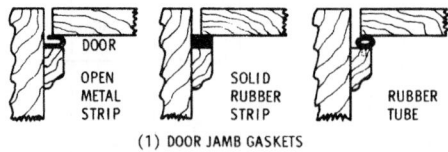

(1) DOOR JAMB GASKETS

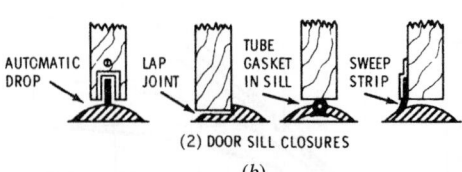

(2) DOOR SILL CLOSURES

*(b)*

**Fig. 19.24** *(a) Chart of typical transmission loss values for representative door constructions as a function of frequency. (b) Method for gasketing a door edge enclosure (1) and sealing the gap between the bottom of the door and the door saddle (2). (Chart and drawings extracted from* A Guide to Airborne, Impact, and Structure-Borne Noise Control in Multifamily Dwellings, *1968.)*

## 19.13 DOORS AND WINDOWS

As can be appreciated from the preceding section, doors and windows can in large measure determine the overall transmission loss of a wall. Since in almost every instance doors and windows have a lower acoustic transmission loss than the wall in which they are mounted, particular care must be taken not to degrade performance further with air leaks.

### (a) Doors

Figure 19.24 gives nominal transmission loss data for the most common types of doors as a function of frequency. Average transmission loss values for doors, that is, the arithmetic average of the octave band transmission losses in the range of 150 Hz to 3000 Hz, are not useful for two reasons:

- The very important low-frequency attenuation data are absent.
- Sharp peaks and valleys in the curves (see, for instance, the 6-dB peaks at 500 Hz in Fig. 19.24*a*) are unrecognized. As a result, a particularly troublesome frequency may not be sufficiently attenuated. In the absence of a complete frequency analysis, the STC rating of a door is a better indication than an average transmission loss figure. Typical STC values are given in Table 19.3.

Conclusions that can be drawn from inspection of Fig. 19.24*a* are:

1. Louvered doors (and doors undercut to permit air movement) are useless as sound barriers.
2. The most important step in soundproofing doors is complete sealing around the opening. A door in the closed position should exert pressure on gaskets, making the joints airtight (see Fig. 19.24*b*).

When a single door does not provide sufficient attenuation and specially constructed high-attenuation commercial acoustic doors are not practical, a simple and very effective technique is the construction of a sound lock consisting of two doors, preferably with sufficient space between them to permit full door swing (see Fig. 19.25). All surfaces in the sound lock should be covered completely by absorbent material and the floor carpeted. Such an arrangement will increase attenuation across the board by at least 10 dB and by as much as 20 dB at some frequencies, depending upon the shape of the sound lock and the type, amount, and mounting of absorptive material in the sound lock. The two doors of the sound lock must be gasketed, as explained above.

Another important consideration with respect to sound intrusion via doors is the location of a door with respect to sources of unwanted sound. This is particularly important in multiple-resident buildings of all types including private homes, apartment houses, dormitories, hotels, and rooming

### TABLE 19.3 Typical STC Values for Doors

| Door Construction | STC |
|---|---|
| Louvered door | 15 |
| Any door, 2-in. undercut | 17 |
| 1½-in. hollow core door, no gasketing | 22 |
| 1½-in. hollow core door, gaskets and drop closure | 25 |
| 1¾-in. solid wood door, no gasketing | 30 |
| 1¾-in. solid wood door, gaskets and drop closure | 35 |
| Two hollow core doors, gasketed all around, with sound lock | 45 |
| Two solid core doors, gasketed all around, with sound lock | 55 |
| Special commercial construction, with lead lining and full sealing | 45–65 |

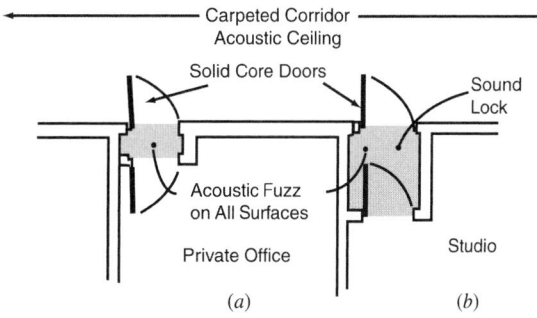

**Fig. 19.25** *A two-door sound lock should increase the transmission loss of a door assembly by a minimum of 10 dB for the small lock (a) and 15 dB for the larger one (b), depending upon the type, thickness, and mounting technique of the absorptive material in the lock. The solid-core doors must be sealed by one of the techniques shown in Fig. 19.24(b).*

houses. The same principle applies to commercial spaces where numerous private spaces such as offices open onto a common lobby or foyer. Figure 19.26 shows the effect of improvements in door placement.

## (b) Windows

Windows are critically important to block exterior noise, and all the more so, since exterior wall construction is generally of high STC, making the window the deciding factor in the composite exterior wall transmission loss. Sound leaks through cracks in closures of operable windows will normally establish a window's rating, regardless of the type of glazing. Fortunately, the attention now given to the sealing of windows for thermal purposes has had a salutary effect on their acoustic properties. As with doors, the importance of proper gasketing and sealing cannot be overemphasized. Double glazing is effective only when the two panes are separated by a wide air gap (Fig. 19.27). A narrow air gap acts as a stiff spring between the panes and transmits sound energy almost unattenuated. The result is approximately that of a single pane of double weight. Note that here too, as with absorptive material, the requirements of acoustic and thermal insulation are opposed. A small sealed air space between panes is desirable for thermal insulation, because a large space allows convection currents to transfer heat. For acoustical purposes, a small sealed space is not very useful, as explained above, whereas a large space traps acoustic energy and is an effective noise barrier, as is clearly seen in Fig. 19.28.

In addition to a window's sound transmission characteristic when closed, it is important to consider the transmission loss when open because of ventilation and passive cooling requirements. The sound attenuation between the center of a room with a clear-through open window and a point some distance outside is 5 to 15 dB. This drops to about 5 dB as the receiver-observer approaches the open window. By making the path from inside to outside indirect, the open window attenuation can be increased to as much as 25 dB, but with considerable reduction of airflow and hence ventilation capacity. Several possible arrangements with approximate midfrequency transmission loss figures are given in Fig. 19.19. This principle can be applied advantageously when exterior noise reduction is important but

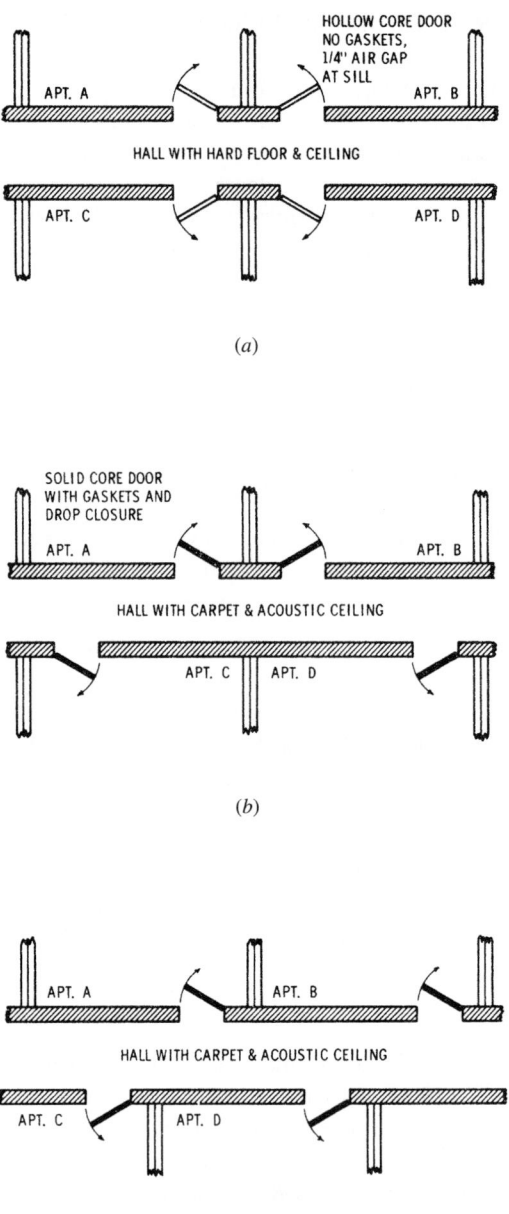

**Fig. 19.26** *Proper arrangement of doors to rooms on a common corridor can diminish noise transfer in the area. (a) Poor arrangement because any noise emanating from one of the rooms or from the corridor has a very short and unattenuated path into the remaining rooms. (b) Better arrangement than plan (a) because noise from any source must travel a minimum of a room width along an absorbent corridor to reach any other room. Noise to the remaining rooms is further attenuated. The weak point of this plan is a noise short circuit via adjacent doors for Apt. A and Apt. B. Plan (c) is best because there are no short circuits for sound travel. Although the A–C and B–D paths are slightly shorter than in plan (b), the difference would be unnoticeable. (Extracted from* A Guide to Airborne, Impact, and Structure-Borne Noise Control in Multifamily Dwellings, *1968.)*

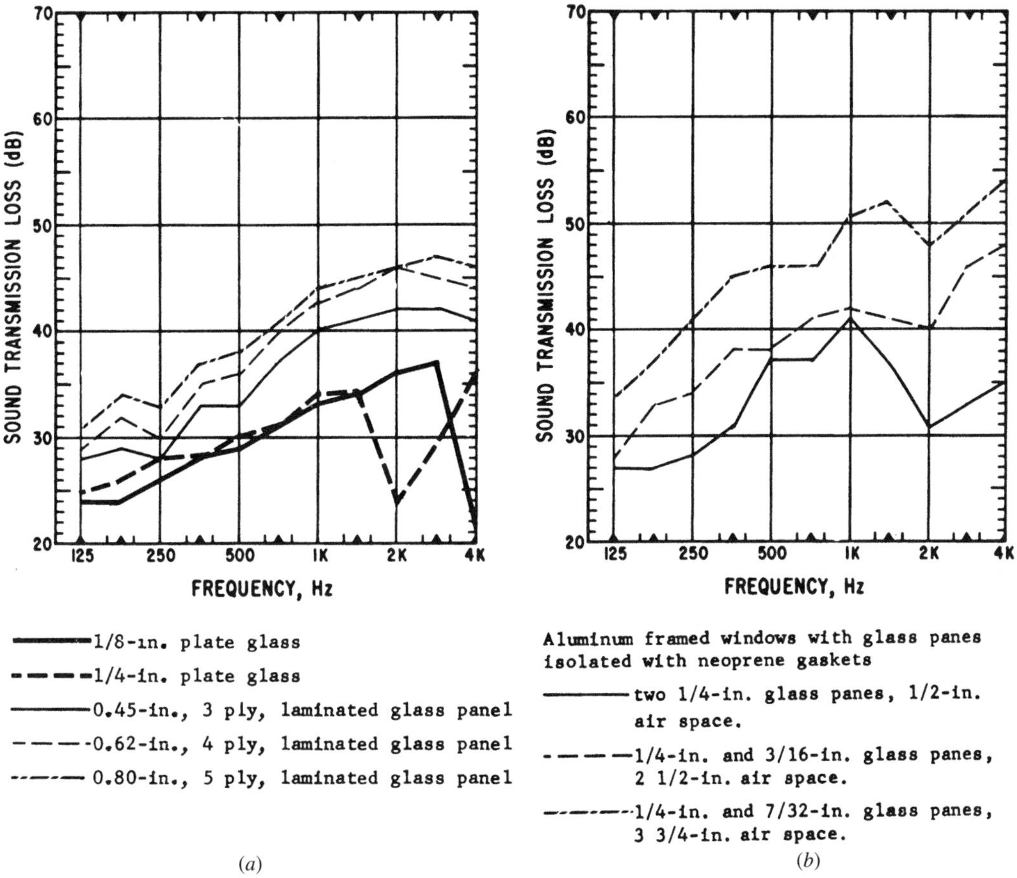

ACOUSTICS

**Fig. 19.27** (a) Sound transmission loss frequency spectrum shows the effect of thickening glass. The ¼-in. plate shows a very sharp (10-dB) coincidence dip at 2kHz, making it less effective than ⅛-in. plate between 1500 Hz and 2500 Hz. Further thickening with laminated glass eliminates the coincidence drop but reaches a practical limit at about ½-in. thickness. (b) Note that two ¼-in. panes with a ½-in. air space act as a stiff, thick pane and exhibit the sharp coincidence drop at 2 kHz. Larger sealed inter-pane air spaces markedly improve the acoustic isolation performance.

*Legend for (a):*
———— 1/8-in. plate glass
– – – – 1/4-in. plate glass
———— 0.45-in., 3 ply, laminated glass panel
– – – – 0.62-in., 4 ply, laminated glass panel
–··–··– 0.80-in., 5 ply, laminated glass panel

(a)

*Legend for (b):*
Aluminum framed windows with glass panes isolated with neoprene gaskets
———— two 1/4-in. glass panes, 1/2-in. air space.
– · – · – 1/4-in. and 3/16-in. glass panes, 2 1/2-in. air space.
–··–··– 1/4-in. and 7/32-in. glass panes, 3 3/4-in. air space.

(b)

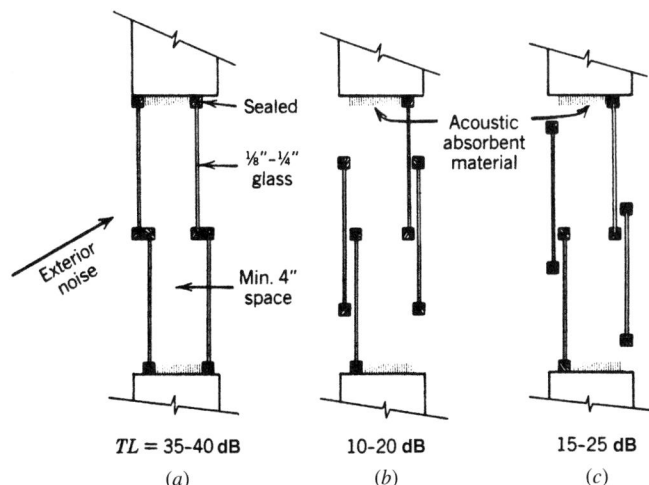

$TL$ = 35-40 dB
(a)

10-20 dB
(b)

15-25 dB
(c)

**Fig. 19.28** The degree of attenuation of external noise can be regulated with acoustic sealant and absorbent materials when using pairs of double-hung (or horizontally sliding) windows. Ventilation airflow varies inversely with transmission loss.

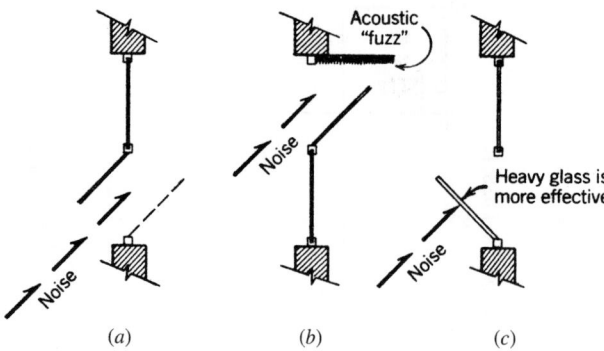

*Fig. 19.29* Alternative arrangements of the same basic "hopper" window design can yield results differing by as much as 10 dB. Design (a) is entirely open, and the noise path is unobstructed deep into the room. Design (b) is about 5 dB better than (a) at frequencies above 1 kHz because of higher absorption and less diffraction. Lower frequencies diffract readily around the window leaf and are less affected by absorptive material. Design (c) can be 10 dB better than (a), particularly at high frequencies, because it interposes a rigid barrier into the noise path. In this arrangement, the glass thickness is important.

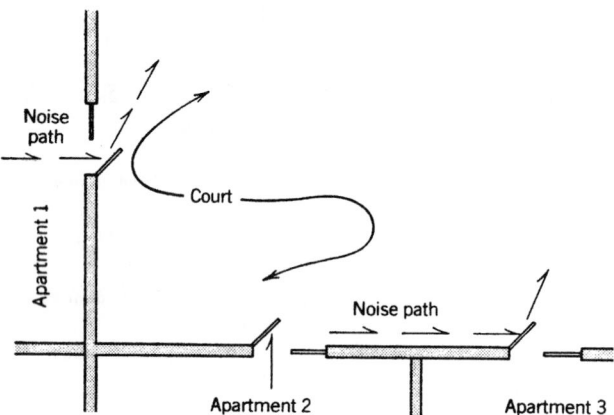

*Fig. 19.30* Noise transfer between contiguous corner spaces, as in Apartments 1 and 2, can be particularly severe if windows are improperly designed. Swinging windows, as shown, are preferable to double-hung or hopper windows because they reflect sound away from the adjacent space. Similarly, adjacent spaces on the same wall, such as Apartments 2 and 3, can benefit from this type of swinging arrangement, which is preferable to sliding or double-hung designs.

sealed windows are undesirable. Window-opening style and placement can also have an effect on the amount of exterior noise admitted, as shown in Figs. 19.29 and 19.30. Typical STC ratings of common window constructions are given in Table 19.4.

### TABLE 19.4 Typical STC Values for Windows

| Window Construction | STC |
|---|---|
| Operable wood sash, ⅛-in. glass, unsealed | 23 |
| Operable wood sash, ¼-in. glass, unsealed | 25 |
| Operable wood sash, ¼-in. glass, gasketed | 30 |
| Operable wood sash, laminated glass, unsealed | 28 |
| Operable wood sash, double-glazed, ⅛-in. panes, ⅜-in. air space, gasketed | 29 |
| Fixed sash, double ⅛-in. panes, 3-in. air space, gasketed | 44 |
| Fixed sash, double ⅛-in. panes, 4-in. air space, gasketed | 48 |

## 19.14 DIFFRACTION: BARRIERS

The physical process by which sound passes around obstructions and through very small openings is called *diffraction*. Simply stated, diffraction is a process whereby any point on a sound wave establishes a new wave when passing an obstacle. Thus, although much of a sound wave is blocked by a small opening, the portion that does get through establishes a new wave front (see Fig. 19.18e). The *amplitude* of the diffracted wave is determined by the relationship between the size of the opening and the wavelengths of the signal components. For a small hole, short wavelengths (high frequencies) are attenuated less than long wavelengths (low frequencies). See Fig. 19.31.

When sound encounters a finite-length barrier, it diffracts around and over it, approximately

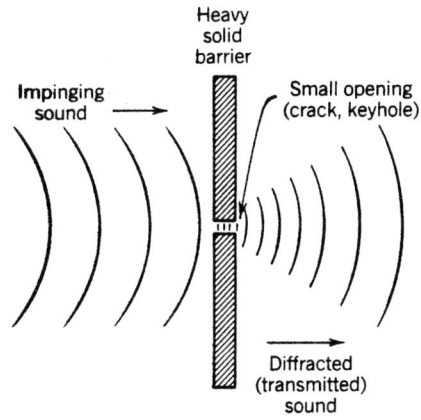

**Fig. 19.31** *Sound passes through small openings by diffraction. The intensity of the transmitted sound is proportional both to its frequency and to the size of the opening. It is always less than the intensity of the impinging sound.*

as shown in Fig. 19.32. The attenuation of the diffracted sound depends upon the frequency, type of source, and dimensions of the barrier. For a *point source*, with only a single *practical* path around the obstruction (barrier), the noise reduction in deci-

bels can be calculated from Maekawa's empirical equation:

$$NR = 20 \log \left[ \frac{\sqrt{2\pi N}}{\tanh \sqrt{2\pi N}} \right] + 5 \text{ dB} \quad (19.4)$$

where

$$N = (f/565)\,(A + B - d)$$

$NR$ = noise reduction, dB

$f$ = frequency, Hz

$A + B$ = shortest path length around the barrier, ft (over or around)

$d$ = straight-line distance, source-to-receiver, ft

Note that this equation:

1. Is applicable only to exterior barriers where sound passing over the barrier is partially diffracted and partially attenuated by distance. In an interior situation, sound passing over a partial height barrier (see Fig. 19.40) strikes the ceiling and is reflected down, increasing the received sound and effectively reducing barrier attenuation. Maximum exterior barrier attenuation is 24 dB as compared to about 15 dB for a partial-height interior partition.

**Fig. 19.32** *Comparison of the effect of a barrier on sources of different frequencies. The low-frequency sound (a) diffracts more readily over the barrier than the high-frequency sound (b) because of its longer wavelength. Thus, the lower the frequency, the smaller the acoustic shadow and the lower the barrier attenuation. The shadow is not as sharply defined as shown; it represents increasing attenuation closer to the barrier.*

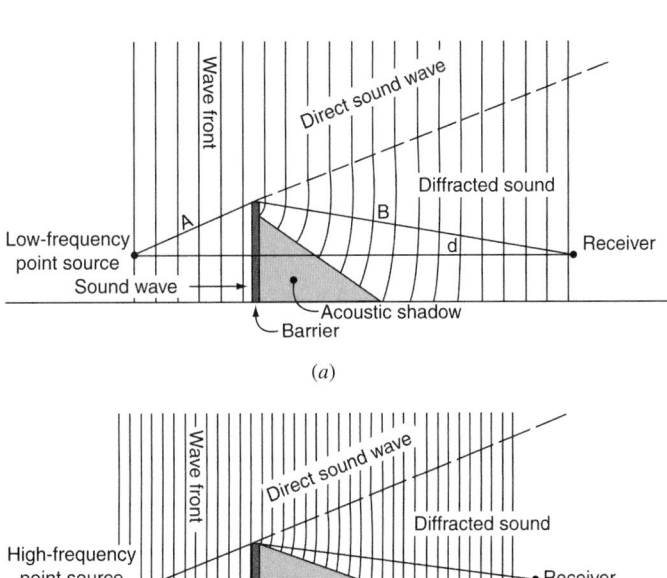

ACOUSTICS

2. Assumes that the barrier is very long (or very high), so that only one sound path exists. In practice, a barrier whose length (height) is at least four times the distance between the source and the wall is sufficient if *the barrier is close to the source.* If the barrier is close to the receiver, it must be longer (higher) still.

3. Assumes a point source. Line sources (such as traffic) show 20 to 25% less attenuation for the same barrier.

The equation will, however, give reliable, usable results when the dimensions of the source are small with respect to the barrier, as is the case for speech; individual motors, fans, engines, and other mechanical devices; and individual motor vehicles. The chart in Fig. 3.22 relates barrier dimensions and position to traffic noise reduction. Note that frequency is not a variable on the chart, since it has been plotted for an average attenuation at 220 Hz, which is the center frequency for random car and truck traffic.

It should be apparent that the best location for a barrier is either very close to the source or very close to the receiver. The worst position for attenuation is halfway between them. All effective barriers are assumed to be opaque and to have a minimum surface density of 5 lb/ft² (~25 kg/m²). The inherent transmission loss of the barrier need not be very high; a massively thick barrier has only marginally higher attenuation than one with the minimum surface weight specified above. Absorptive material placed on the source side of a barrier will reduce the noise reflected back toward the source but will not effectively increase the barrier's attenuation with respect to the receiver. Although the maximum theoretical noise reduction of an *exterior* barrier is about 24 dB, in practice it rarely exceeds 20 dB. Figure 19.33 is a nomograph based upon Eq. 19.4.

## 19.15 FLANKING

Just as sound will pass through the acoustically weakest part of a composite wall, it will also find parallel or flanking paths, that is, an acoustic short-circuit. Proper design of window locations to avoid flanking paths has already been shown in Fig. 19.30. The same situation obtains with respect to doors, as shown in Fig. 19.26, and any other openings between spaces. Thus, in Fig. 19.34, a high-STC wall between the two spaces is in large measure defeated by flanking paths F5, F6, and F7. In other spaces the most common flanking path is via the plenum, as in Fig. 19.34 (path F1) and in Figs. 19.35b and 19.35d. Ductwork (with registers or grilles in various rooms) acts as an excellent intercom system unless it is completely lined with sound-absorptive material (see Section 19.27). Even then, low-frequency sound is only minimally attenuated, and special measures must be employed if good transmission loss is required. This subject is discussed further below in Sections 19.25–19.27.

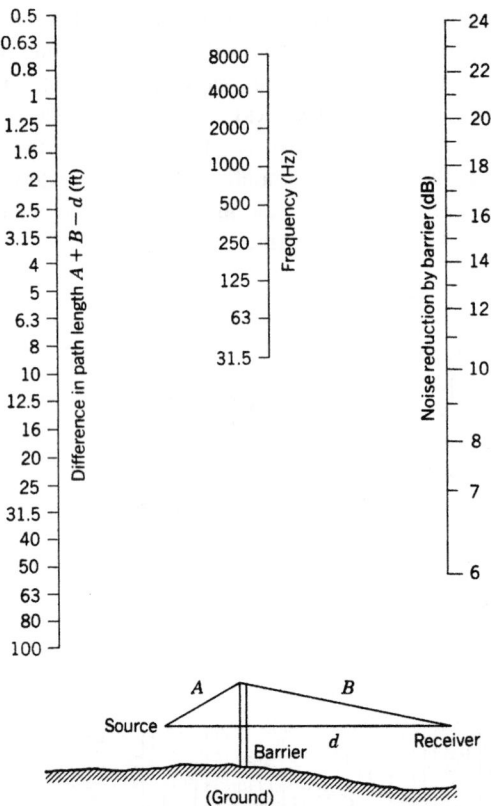

**Fig. 19.33** *This nomograph for estimating the noise reduction afforded by a barrier is based on Eq. 19.4 and assumes a point (small) source and only a single path around the barrier. The dimensions A, B, and d are taken from the insert sketch. Dimensions A plus B represent the shortest path around the barrier—which may be over or around it. (Reprinted with permission from B. Fader,* Industrial Noise Control, *Wiley, New York, 1981.)*

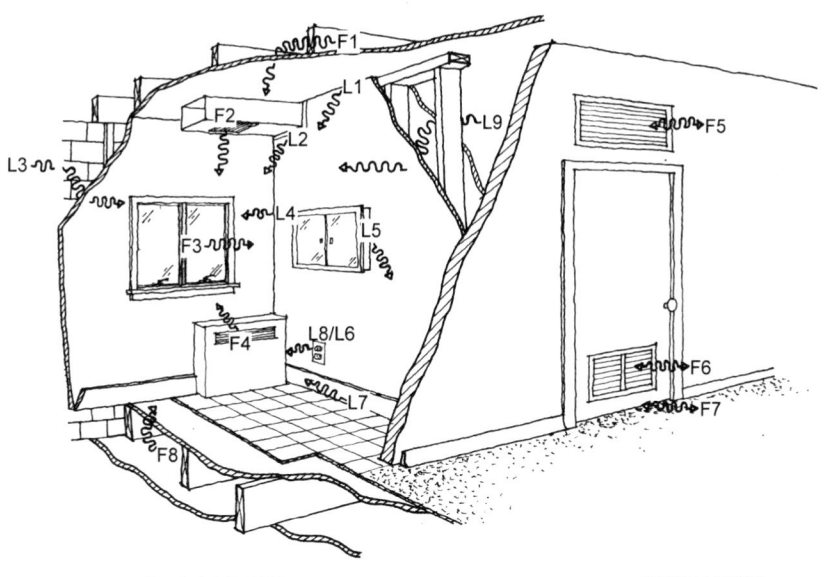

**FLANKING NOISE PATHS**

F1 OPEN PLENUMS OVER WALLS, FALSE CEILINGS
F2 UNBAFFLED DUCT RUNS
F3 OUTDOOR PATH, WINDOW TO WINDOW
F4 CONTINUOUS UNBAFFLED INDUCTOR UNITS
F5 HALL PATH, OPEN VENTS
F6 HALL PATH, LOUVERED DOORS
F7 HALL PATH, OPENINGS UNDER DOORS
F8 OPEN TROUGHS IN FLOOR-CEILING STRUCTURE

**NOISE LEAKS**

L1 POOR SEAL AT CEILING EDGES
L2 POOR SEAL AROUND DUCT PENETRATIONS
L3 POOR MORTAR JOINTS, POROUS MASONRY BLK
L4 POOR SEAL AT SIDEWALL, FILLER PANEL, ETC.
L5 BACK-TO-BACK CABINETS, POOR WORKMANSHIP
L6 HOLES, GAPS AT WALL PENETRATION
L7 POOR SEAL AT FLOOR EDGES
L8 BACK-TO-BACK ELECTRICAL OUTLETS

OTHER POINTS TO CONSIDER, RE: LEAKS ARE (A) BATTEN STRIP A/O POST CONNECTIONS OF PREFABRICATED WALLS, (B) UNDER-FLOOR PIPE OR SERVICE CHASES, (C) RECESSED, SPANNING LIGHT FIXTURES, (D) CEILING & FLOOR COVER PLATES OF MOVABLE WALLS, (E) UNSUPPORTED A/O UNBACKED WALL BOARD JOINTS, (F) EDGES & BACKING OF BUILT-IN CABINETS & APPLIANCES, (G) PREFABRICATED, HOLLOW METAL EXTERIOR CURTAIN WALLS.

**Fig. 19.34** *Flanking transmission of airborne noise. (Reprinted from* A Guide to Airborne, Impact, and Structure-Borne Noise Control in Multifamily Dwellings, *1968. Redrawn by Jonathan Meendering.)*

## SPEECH PRIVACY

### 19.16 PRINCIPLES OF SPEECH PRIVACY BETWEEN ENCLOSED SPACES

The subject of speech privacy has always been of paramount importance in office design. Numerous studies have demonstrated that productivity and noise are related inversely when the noise carries information. When noise does not carry information, it can be annoying and therefore counterpro-ductive or it can be useful as a masking sound, depending upon its frequency content, intensity level, and constancy. Referring to Section 19.7, which discusses the noise reduction of an *airtight* barrier between two spaces, we saw that the sound intensity levels in the source room (1) and the receiving room (2) are related by the expression

$$IL_2 = IL_1 - NR$$

where $NR$ is noise reduction and $IL_2$ and $IL_1$ are sound intensity levels in the receiving and source rooms, respectively. If the receiving room is completely quiet and has no sound source other than the transmitted sound (essentially $IL_2$), then that

ACOUSTICS

ACOUSTICS

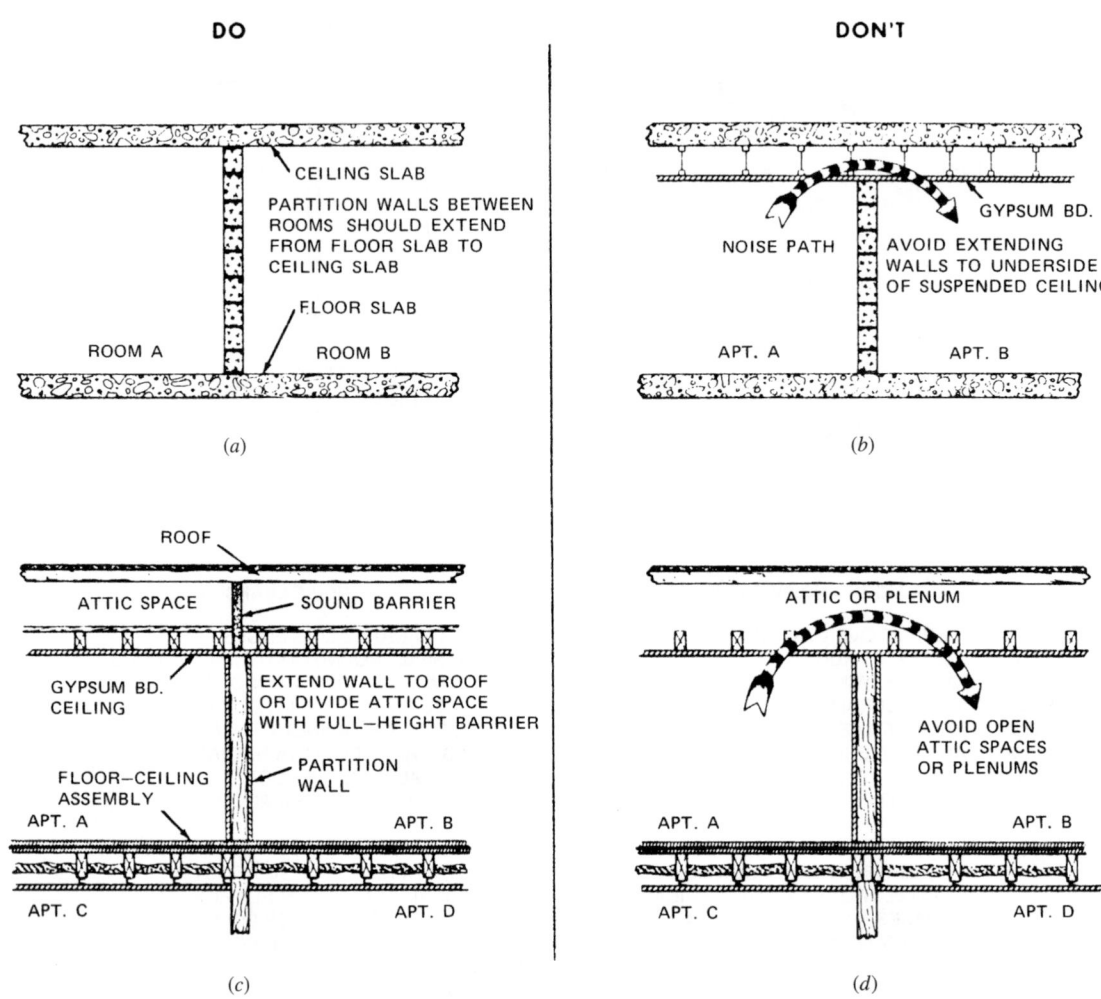

**Fig. 19.35** *Construction technique recommendations to avoid flanking paths. (Reprinted from* A Guide to Airborne, Impact, and Structure-Borne Noise Control in Multifamily Dwellings, *1968.)*

sound will always be a potential source of annoyance to the occupant of the receiving room as long as its intensity level is above the hearing threshold. If, however, there is a *constant* ambient sound level in the receiving room, then, depending upon its characteristics, it may mask the transmitted sound $IL_2$, even to the extent of making it completely inaudible. In most instances, however, it simply reduces or eliminates annoyance without completely masking the source. What we hear (and therefore what can potentially be a source of disturbance) depends upon our level of attention both to what we are doing and to the intrusive sound. (A remarkable exception is the ability of some, generally young, students to study in the presence of very loud, familiar—and therefore

information-bearing—music. Indeed, some claim that they can *only* study that way.)

Tests have shown that a majority of adults will not consider an intruding noise level $IL_2$ to be annoying if the intensity of a properly designed background sound is either greater than or no more than 2 dB less than $IL_2$. Thus, a transmitted $IL_2$ of 40 dBA will not be considered annoying by most people if the level of the background sound is at least 38 dBA. The upper level of usable background masking sound is usually taken to be about 50 dBA. Any higher intensity level will itself become a source of annoyance. Figure 19.36 gives a graphic representation of the relation between transmitted and background sound levels in a receiving room.

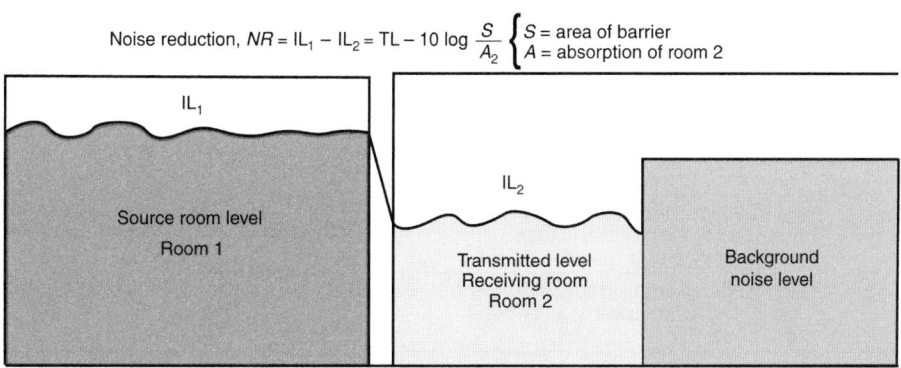

Noise reduction, $NR = IL_1 - IL_2 = TL - 10 \log \dfrac{S}{A_2}$ $\begin{cases} S = \text{area of barrier} \\ A = \text{absorption of room 2} \end{cases}$

$IL_1$

Source room level
Room 1

$IL_2$

Transmitted level
Receiving room
Room 2

Background
noise level

**Fig. 19.36** *The background noise level determines whether a transmitted sound will actually be heard. In this case, because the background noise level is considerably higher than the transmitted level from the adjacent room, the transmitted noise will not be noticeable (i.e., it will be masked).*

In summary, then, of the above discussion, we can simply state that the degree of *speech privacy* in a space is a function of two factors:

- The degree of sound isolation provided by the barriers between rooms
- The ambient sound level in the receiving room.

## 19.17  SOUND ISOLATION DESCRIPTORS

Since the degree of intrusiveness of extraneous noise in a space varies, depending as it does upon both transmitted and ambient sound, a descriptive scale of some sort is obviously necessary. If we restrict our discussion to speech sounds, since ultimately we are interested in office design, where speech is the primary sound source, then a descriptive scale, as shown in Table 19.5, can be established. With these absolute descriptions in mind,

and remembering that the hearing condition in a receiving room can be altered by changing either the barrier characteristics or the background sound level, *or both*, we can express the effectiveness of a construction as a speech sound barrier in terms of its STC for a given ambient sound level. Since the ambient sound (noise) level is frequency dependent, it can be approximated by an NC value. This is particularly useful when the ambient sound level is generated by machinery or by the sound of an air-conditioning system rather than by a shaped signal from an electronic masking sound system.

Table 19.6 shows the hearing conditions in a receiving room with an NC-25 background noise as a function of the barrier STC rating. If the background noise level were raised to NC-30, then each descriptor would roughly increase one level in quality (i.e., poor would become fair, fair would become good, and so forth). Stated otherwise, the *apparent* isolation provided by a barrier may be increased by

**TABLE 19.5  Relative Quality of Sound Isolation**

| Ranking | Descriptor | Hearing Condition[a] |
|---|---|---|
| 6 | Total privacy | Shouting barely audible. |
| 5 | Excellent | Normal voice levels not audible. Raised voices barely audible but not intelligible. |
| 4 | Very good | Normal voice levels barely audible. Raised voices audible but largely unintelligible. |
| 3 | Good | Normal voice levels audible but generally unintelligible. Raised voices partially intelligible. |
| 2 | Fair | Normal voice levels audible and intelligible some of the time. Raised voices generally intelligible. |
| 1 | Poor | Normal voice audible and intelligible most of the time. |
| 0 | None | Normal voice levels always intelligible. |

[a]Hearing condition in the presence of ambient noise, if any.

**TABLE 19.6 Relation Between Barrier STC and Hearing Condition on the Receiving Side, Background Noise Level at NC-25**

| Barrier STC | Hearing Condition | Descriptor and Ranking[a] | Application |
|---|---|---|---|
| 25 | Normal speech can be understood quite easily and distinctly through the wall. | Poor/1 | Space divider |
| 30 | Loud speech can be understood fairly well. Normal speech can be heard but not easily understood. | Fair/2 | Room divider where concentration is not essential |
| 35 | Loud speech can be heard but is not easily intelligible. Normal speech can be heard only faintly, if at all. | Very Good/4 | Suitable for offices next to quiet spaces |
| 42–45 | Loud speech can be faintly heard but not understood. Normal speech is inaudible. | Excellent/5 | For dividing noisy and quiet areas; party wall between apartments |
| 46–50 | Very loud sounds (such as loud singing, brass musical instruments, or a radio at full volume) can be heard only faintly or not at all. | Total Privacy/6 | Music room, practice room, sound studio, bedrooms adjacent to noisy areas |

[a]See Table 19.5.

raising the background (masking) sound level in the receiving room. Figure 19.37 shows two conditions of adjacent spaces. Although the source room level is uniform and partitions on both sides of the source room are identical, the background sound in the two receiving rooms is different. In *A*, the background is NC-35; in *B*, it is NC-25. The occupant of room *A* is not disturbed by the little heard from the source room. The occupant of room *B* hears clearly. Occupant *A* will probably praise the partition, whereas occupant *B* will complain. Although the levels of reradiated sound are identical in the two receiving rooms, the intruding signal is masked by the background sound in *A* but it is clearly audible in *B*. Thus, the apparent noise reduction is substantially higher in *A* than in *B*. This clearly demonstrates the effectiveness of masking sound in providing *apparent* sound isolation and speech privacy.

Sound isolation can also be improved by careful planning. Storage and circulation areas can serve as buffers for noise-sensitive areas. Physical separation of noisy areas from quiet ones often eliminates the need for complicated and expensive compound barriers.

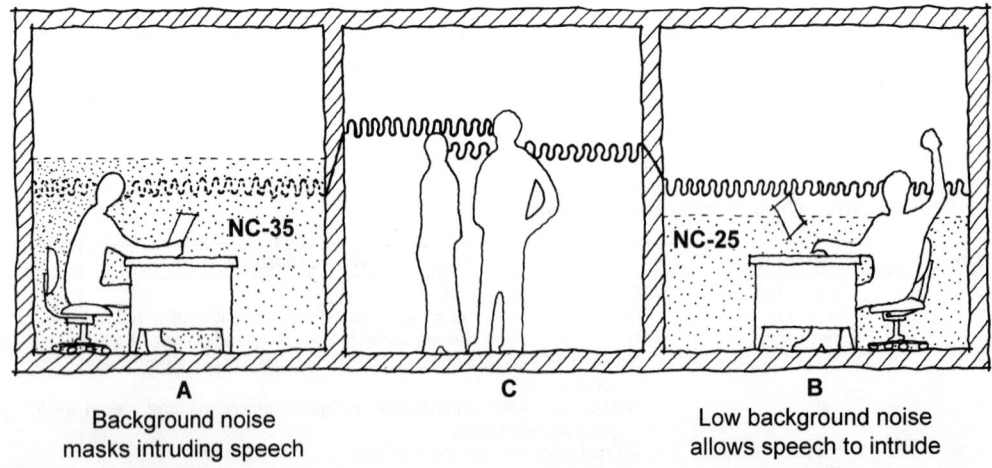

| A | C | B |
|---|---|---|
| Background noise masks intruding speech | | Low background noise allows speech to intrude |

**Fig. 19.37** *The occupant in room A with background noise NC-35 (≈ 45 dBA) is unaware of the noise (loud speech from room C) that is so disturbing to occupant B, whose NC-25 (≈ 36 dBA) is insufficient to mask the transmitted noise. (Drawing by Jonathan Meendering.)*

## 19.18 SPEECH PRIVACY DESIGN FOR ENCLOSED SPACES

The study of speech privacy received considerable emphasis with the advent of open-plan offices (office landscaping), although the same problem prevails with both open and enclosed office designs.

Essentially, the problem was to determine the factors affecting speech privacy and to quantify them with a degree of accuracy sufficient for design purposes. It rapidly became evident that although the physical principles are the same, the solutions to speech privacy design problems are radically different for closed spaces and for open-plan offices. In

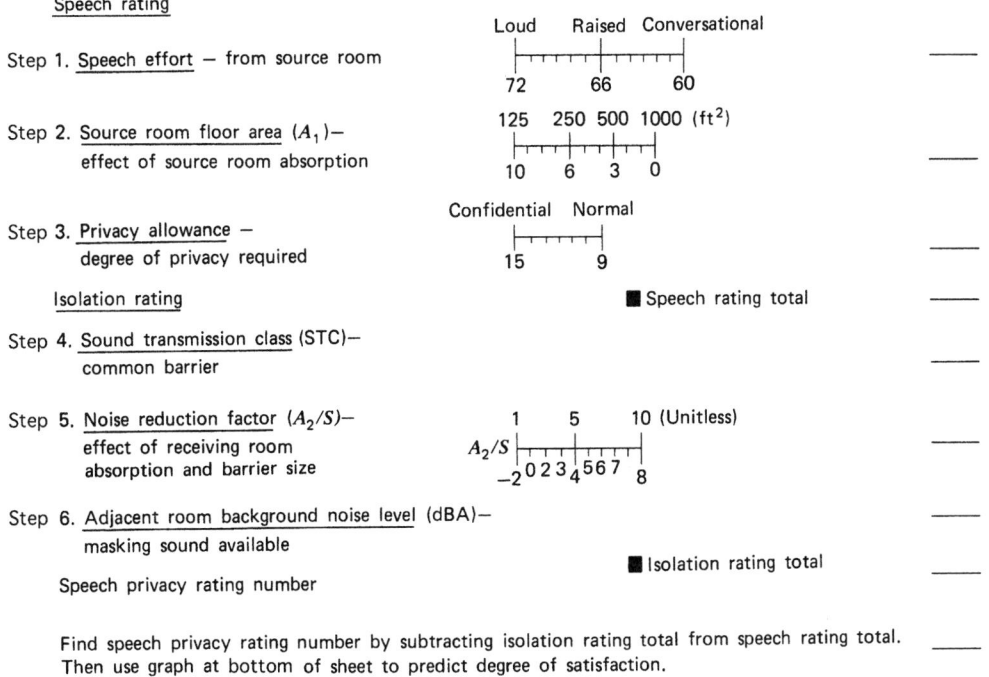

**a. Procedure for determining speech privacy rating**

Speech rating

Step 1. Speech effort — from source room

    Loud   Raised  Conversational
    72      66      60

Step 2. Source room floor area $(A_1)$ — effect of source room absorption

    125  250 500 1000 (ft$^2$)
    10   6   3   0

Step 3. Privacy allowance — degree of privacy required

    Confidential  Normal
    15      9

   Isolation rating

■ Speech rating total

Step 4. Sound transmission class (STC) — common barrier

Step 5. Noise reduction factor $(A_2/S)$ — effect of receiving room absorption and barrier size

    1    5    10 (Unitless)
$A_2/S$   $-2\ 0\ 2\ 3\ 4\ 5\ 6\ 7$  $8$

Step 6. Adjacent room background noise level (dBA) — masking sound available

   Speech privacy rating number

■ Isolation rating total

Find speech privacy rating number by subtracting isolation rating total from speech rating total. Then use graph at bottom of sheet to predict degree of satisfaction.

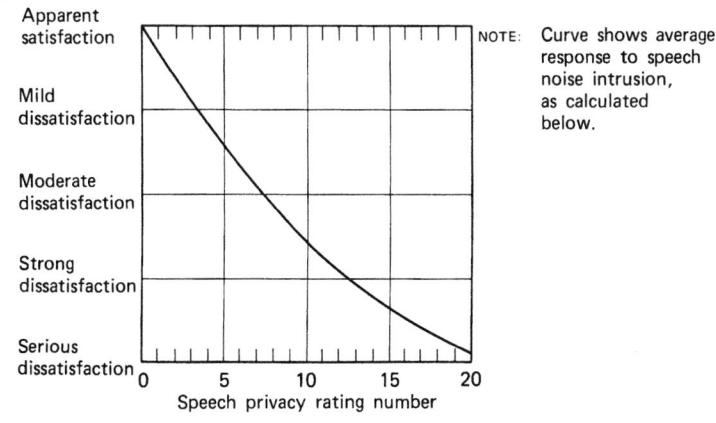

**b. Anticipated response to privacy situation**

Apparent satisfaction

Mild dissatisfaction

Moderate dissatisfaction

Strong dissatisfaction

Serious dissatisfaction

NOTE: Curve shows average response to speech noise intrusion, as calculated below.

0   5   10   15   20
Speech privacy rating number

***Fig. 19.38*** *Speech privacy analysis sheet. (Based on Cavanaugh et al., 1962.)*

ACOUSTICS

the former, the acoustic character of an airtight barrier between two spaces is the crucial element in the design, because all of the interspace sound transmission is through this barrier. In contrast, sound transmission between adjacent "cubicles" in an open office is primarily the result of *reflected* and *refracted* sound, with the direct component passing through the barrier being of secondary importance. As a result, the acoustic transmission characteristics of the barrier also become secondary in importance.

Studies indicate that six factors are involved in enclosed-space speech privacy, which can be subsumed under two headings (Fig. 19.38):

1. Speech rating of the source room (Room No. 1)
   a. Speech effort—a measure of the loudness of speech.
   b. Source room factor—gives the approximate effect of room absorption on the speech level in the source room. The scale in Fig. 19.38 is drawn for average absorption. For live rooms, raise the factor by 2 points; for dead rooms, lower it by 2 points. Factors *a* + *b* give the approximate source-room voice level.
   c. Privacy allowance—what is the measure of privacy required?

The privacy criteria definitions used in this step (Fig. 19.38, Step 3) are:

*Normal privacy*—such that the receiving-room occupant can understand a small portion of normal voice conversation in the source room by listening intently. This corresponds to "Good" and a ranking of 3 in Table 19.5. It was found that most occupants can work normally with this level of speech intrusion.

*Confidential privacy*—assumes that only a few words will occasionally be intelligible. This privacy level corresponds to "Excellent" and a ranking of 5 in Table 19.5. The six speech-rating points between Normal Privacy (9 points) and Confidential (15 points) in Fig. 19.38 correspond roughly to a 5-dB difference. If we remember that a sound intensity differential of 3 dB is barely perceptible, 6 dB is clearly noticeable, and 10 dB is a doubling of perceived sound (Table 17.3), we can appreciate that the difference between *normal* and *confidential* privacy is small and calls for accurate design,

**TABLE 19.7 Typical STC Ratings of Interior Partitions**

| Type of Partition | STC |
|---|---|
| Demountable partition | STC 20–30 |
| Drywall partition up to acoustical ceiling | STC 30 |
| Drywall partition extending 12 in. above acoustical ceiling tile system into ceiling plenum | STC 35 |
| Drywall partition with cavity insulation, full height to the underside of slab above | STC 40–45 |
| Two-layer drywall partition with insulation, erected full height to underside of slab above | STC 50 |

plus a measure of field adjustment of masking sound.

2. Isolation rating of the receiving room (Room No. 2)
   d. The STC rating of the barrier. Table 19.7 gives some typical STC ratings for office partitions.
   e. The noise reduction factor $A_2/S$ is an indication of receiving-room absorption, that is, the difference between $NR$ and $TL$, where $A_2$ is the area of the receiving room and $S$ is the area of the barrier between the rooms. Absorption is assumed to be average. For live rooms, lower this factor 2 points; for dead rooms, raise it 2 points.
   f. For the recommended background noise level in the receiving room, use Table 19.8.

An analysis sheet for enclosed spaces is provided in Fig. 19.38 (see also Cavanaugh et al., 1962, and Young, 1965.) The two examples of this analysis that follow should clarify its use. The reader should follow the analysis with Fig. 19.38 in hand. The numbered steps in the examples correspond to the numbers in the figure.

**EXAMPLE 19.3.** Evaluate the effectiveness of a partition.
Source room:

General clerical office, 40 × 60 × 9 ft, average $\bar{\alpha}$, 16-ft-long full-height partition, STC 40

Receiving room:

Conference room, 16 × 24 ft, medium-dead room

Background noise level, 40 dBA (NC-30) (from Table 19.8)

**TABLE 19.8 Suggested Noise Criteria Ranges for Steady Background Noise**

| Type of Space (and Acoustical Requirements) | NC Curve | Equivalent[a] dBA |
|---|---|---|
| Concert halls, opera houses, and recital halls (for listening to faint musical sounds). | 10–20 | 20–30 |
| Broadcast and recording studios (distant microphone pickup used). | 15–20 | 25–30 |
| Large auditoriums, large drama theatres, and houses of worship (for excellent listening conditions). | 20–25 | 30–35 |
| Broadcast, television, and recording studios (close microphone pickup only). | 20–25 | 30–35 |
| Small auditoriums, small theatres, small churches, music rehearsal rooms, large meeting and conference rooms (for good listening), or executive offices and conference rooms for 50 people (no amplification). | 25–30 | 35–40 |
| Bedrooms, sleeping quarters, hospitals, residences, apartments, hotels, motels, and so forth (for sleeping, resting, relaxing). | 25–35 | 35–45 |
| Private or semiprivate offices, small conference rooms, classrooms, libraries, and so forth (for good listening conditions). | 30–35 | 40–45 |
| Living rooms and similar spaces in dwellings (for conversing or listening to radio and TV). | 35–45 | 45–55 |
| Large offices, reception areas, retail shops and stores, cafeterias, restaurants, and so forth (for moderately good listening conditions). | 35–50 | 45–60 |
| Lobbies, laboratory work spaces, drafting and engineering rooms, general secretarial areas (for fair listening conditions). | 40–45 | 50–55 |
| Light maintenance shops, office and computer equipment rooms, kitchens, and laundries (for moderately fair listening conditions). | 45–60 | 55–70 |
| Shops, garages, power-plant control rooms, and so forth (for just acceptable speech and telephone communication). Levels above PNC-60 are not recommended for any office or communication situation. | — | — |
| For work spaces where speech or telephone communication is not required, but where there must be no risk of hearing damage. | — | — |

*Source:* Extracted with permission from E.B. Magrab, *Environmental Noise Control,* Wiley, New York, 1975.

[a]For information only. These data are not part of the NC information and do not appear in the source.

**ACOUSTICS**

Privacy analysis (see Fig. 19.38):

(a) 1. Speech effort: raised     66
     2. Area $A_1 > 1000$ ft$^2$     0
     3. Privacy—normal     9
          Speech rating $a = 75$

(b) 4. STC (given)     40
     5. $A_2/S$     3
        $(16 \times 24)/(16 \times 9) = 2.6$, corresponding to 2 on the scale; add 1 for higher than average $\bar{\alpha}$
     6. Background noise level     40
          Isolation rating $b = 83$
        Speech privacy rating $a - b = -8$

The partition performance is acceptable. In fact, the STC rating of the partition can be reduced to 32 without affecting speech privacy. ∎

**EXAMPLE 19.4.** Evaluate speech privacy.

Source room:

Drafting room 20 × 30 ft, medium-live

Common wall 12 × 8 ft high

STC: 26 (half glass, with door)

Receiving room:

Supervisor's office, 12 × 14 × 8 ft, average absorption

Background noise level, 35 dBA

Privacy analysis:

1. Speech effort—conversation     60
2. Source room factor:
    For area     +2
    For liveness     +1     Total     +3
3. Privacy—(almost) confidential     13
          Speech rating $a = 76$
4. STC (given)     26
5. $A_2/S$
    $(12 \times 14)/(12 \times 8) = 1.8$, corresponding to a reduction factor of 1.6; gives 2.0 on the scale since the method uses whole numbers only.) No adder required for average absorption.
          Therefore, $2 + 0 = 2$
6. Background noise level     35
        Isolation rating: $b = 63$
    Speech privacy rating $a - b = 13$
    which indicates strong dissatisfaction

The suggested corrections are to increase the wall STC to 36 by gasketing the door and to increase the background noise level in the receiving room to 40 dBA (NC-30). This would give a speech privacy rating of −2, as follows:

| | |
|---|---|
| STC of barrier | 36 |
| $A_2/S$ | 2 |
| Background noise | 40 |
| Isolation rating | 78 |

Speech privacy rating = $a − b = 76 − 78 = −2$

This result should be satisfactory according to the chart in Fig. 19.38. ∎

## 19.19 PRINCIPLES OF SPEECH PRIVACY IN OPEN-AREA OFFICES

The huge increase in the service sectors of the world economy has brought with it a corresponding increase in desk jobs, each of which is often equipped with a computer console. This trend has also necessitated increased space density for office workers, made possible by the general elimination of paper storage (files) and the corresponding elimination of the necessity for employees to continually move about. The increased density problem has been largely solved by open-office plans with ever-smaller "cubicles," usually with single occupancy, but recently also with dual occupancy. This situation has obviously aggravated the serious problem of annoyance due to the intrusion of speech sounds from neighboring workers, that is, a lack of *speech privacy*.

Since production is adversely affected by the inability of a worker to concentrate because of annoyance with speech intrusion—an annoyance that usually *increases* over time—the proper design of open office plans can have major economic benefits.

### (a) Sound Paths in Open Offices

In contrast to the single, or at most dual, sound paths that exist with full-height enclosures (Fig. 19.39), the sound paths in an open-plan arrangement (including first reflections) are direct, diffracted, ceiling reflected, and laterally reflected (Figs. 19.40 and 19.41). Careful study of these illustrations shows a number of important facts affecting speech privacy in open offices.

1. The angles of reflection of sound waves from the ceiling depend upon the location and height of the source and the ceiling height (Fig. 19.40a). Measurements have shown that these angles vary from a minimum of 30° for a standing speaker in the center of a cubicle to a maximum of 60° for a speaker close to, and facing, a partition for ceiling heights of up to 9 ft. Since much of the sound energy reaching an adjacent occupant does so after reflecting off the ceiling at these angles (30°–60°), a

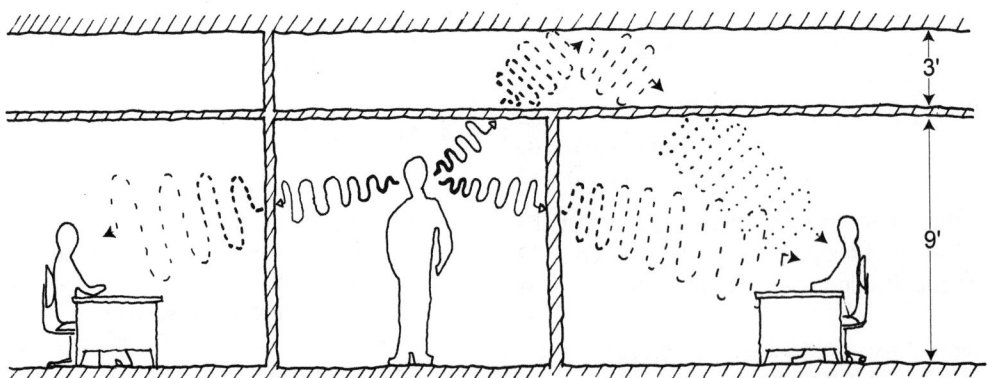

**Fig. 19.39** Sound paths between enclosed spaces are determined by the type of barrier separating them. In the case of a full-height barrier reaching to the underside of the ceiling slab, the only sound path is through the barrier and its STC determines the noise level in the receiving room. In the case of a ceiling-height partition with an overhead plenum, most of the sound energy will travel the upper, less attenuating path. Factors affecting the level of received sound are the ceiling's CAC rating (transmission characteristic) and the acoustic characteristics of the plenum, including all of its contents. In all cases, sound within a reasonably absorptive space attenuates with increasing distance. (Drawing by Jonathan Meendering.)

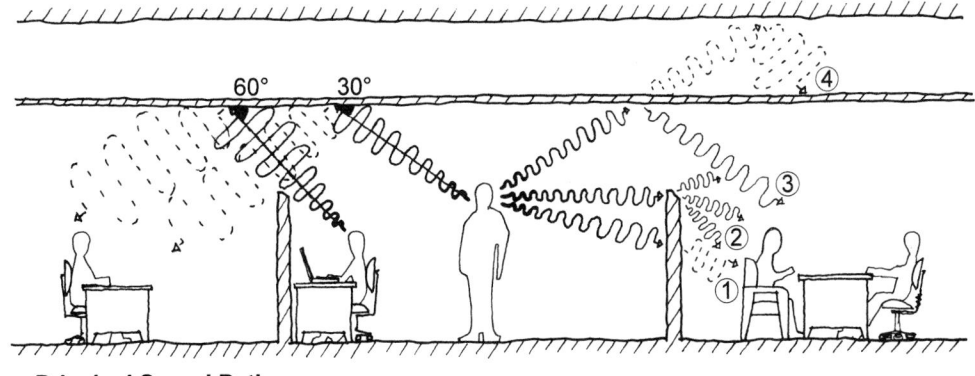

**Principal Sound Paths**

① Direct path through 5'6"-high partition
② Diffracted path over partition
③ Reflected path over partition
④ Sound absorbed and diffused

*(a)*

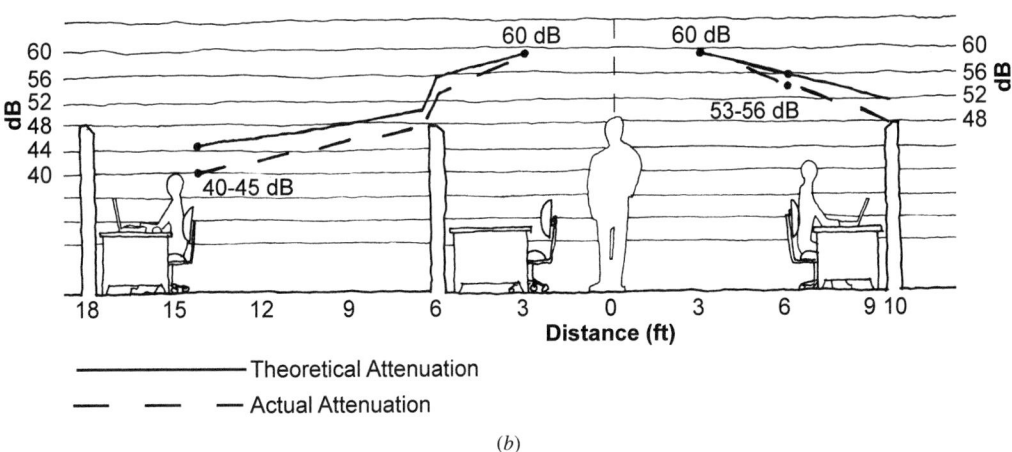

———————— Theoretical Attenuation

— — — Actual Attenuation

*(b)*

**Fig. 19.40** *(a) The principal sound paths between occupancies in an open office plan are (1) direct, (2) diffracted, (3) ceiling reflected, and (4) reflected from the slab above the plenum. Only paths (2) and (3) are problematic from the speech privacy viewpoint, requiring masking sound. Note that the angle of incidence of the sound wave at the ceiling varies from 30° for a standing speaker to 60° for a seated speaker, requiring a ceiling material of high absorption between these angles. Large VDT screens (17–19 in.), now in common use, create another path of strongly reflected sound. (b) Sound is attenuated by distance, dropping 4–6 dB for each doubling of the distance from the source (6 dB in the open; 4 dB in enclosed spaces due to interreflections). Average SPL of conversational speech is 60 dBA at 3 ft from a speaker. Attenuation of a diffracted signal at a partial-height partition is 4–8 dB. Note that the diffracted sound (without a contribution from ceiling-reflected sound) drops to 40–45 dB in the adjacent cubicle. In a two-person open office, the received signal from a standing speaker is at least 53–56 dBA (i.e., perfectly intelligible, even in the presence of maximum [50 dB] masking noise). (Drawings by Jonathan Meendering.)*

ceiling material with high absorption at these angles of incidence is required for speech privacy. This effectively negates the use of the noise reduction coefficient (NRC) as a useful factor to describe a ceiling material's absorption characteristic in an open-office design, since the NRC averages absorp-

tion at all angles. All major ceiling material manufacturers have tested, and will supply, accurate angular absorption data for their products. A single number descriptor that relates to this characteristic, called the *Articulation Class,* is discussed in detail in Section 19.20.

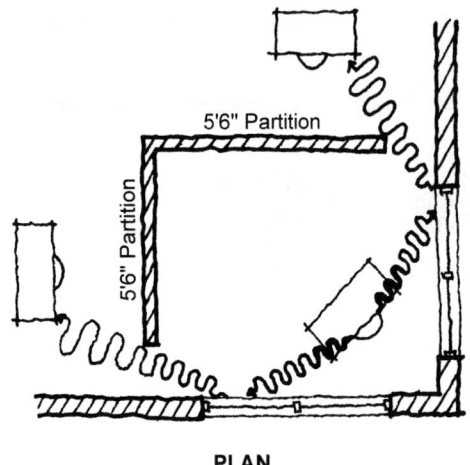

**PLAN**

**Fig. 19.41** *Confidentiality in offices on exterior building walls is difficult to maintain because of the lack of attenuation of sound reflected from windows. The problem is compounded by the custom of placing management personnel in such offices.*

2. Since absorption at the ceiling increases with the angle of incidence of a sound wave, it is always desirable for speech privacy to have maximum ceiling height. Most authorities recommend a *minimum* ceiling height of 9 ft.

3. Absorption efficiency of ceiling tiles varies inversely with the STC, which is to be expected, since sound transmission depends upon mass and absorption at air pockets. Light fiberglass ceiling tiles will typically have an absorption coefficient $\alpha$ of 0.95 at voice frequencies, with a *Ceiling Attenuation Class* (CAC) of 22–24 compared to more massive mineral fiber tiles with a maximum $\alpha$ of 0.8–0.85 but an STC of 34–36. Since we are interested in maximum absorption at the ceiling, the tile of choice is always that with the highest $\alpha$ at voice frequencies. The fact that more energy will pass through a tile of lower CAC (STC) does not affect the acoustic result, since most of the sound energy enters the plenum above the ceiling through openings in the far from airtight suspension system, and is then largely absorbed and dissipated by the spray fireproofing, sound-absorbing material, ductwork, and structural members typically found in a plenum.

4. Since sound will always find the path of least resistance to travel, very little sound energy will pass through partitions between cubicles; the paths over the partition by diffraction and reflection are much less resistant. That being so, the STC of these partitions need not be high. Where the source is close to the partition, as is the case with a seated speaker facing a partition and delivering sound energy directly at the partition at a height of approximately 44–48 in. from a distance of about 3 ft, the STC of the partition should be 25–26. For speakers at greater distances and heights, an STC of 20–22 is usually sufficient.

An exception to this general rule *may* occur when a large (17 in. to 19 in.) computer console is interposed between a seated speaker and an absorbent partition. Large VDTs are becoming increasingly common. Their smooth, highly reflective glass surface creates a strong sound path to the rear of the speaker and decreases the sound energy absorbed by the partition behind the VDT. The variables, however, are so numerous that conservative design will use the higher STC rating. (Contrast these values with those required of a full-height, fixed-partition construction typical of enclosed spaces, as listed in Table 19.7.)

The absorption coefficient $\alpha$ of a partial-height partition at voice frequencies should be a minimum of 0.8 and preferably 0.85–0.95. Some manufacturers have assigned an Articulation Class rating to their partition products, although that descriptor is usually reserved for ceiling tiles as explained above. The recommended ratings for partitions range from 180 to 220.

5. Partitions must be tall enough to block direct line-of-sight voice transmission, since such a path is unattenuated except by distance. The median mouth height of a standing American male is 63 in. This is the basis of the widely accepted recommendation that partitions between adjoining cubicles should not be lower than 65 in. and preferably 66–72 in. Since a 72-in.-high partition blocks vision for all but the tallest people, giving a subjective closed-in sensation to the occupant of a (small) cubicle, this height is normally used only between departments, with intradepartment partitions being 63–66 in. Increasing the height of a partition from 65 to 72 in. will increase path attenuation to an adjacent cubicle by 1–3 dB, depending upon ceiling height and speaker height and location.

6. Refer to Fig. 19.40*b*. This figure indicates signal attenuation due to distance in three different paths. Speech intensity at a conversational level is approximately 60 dBA at 3 ft from the

speaker. Using the fact that sound in a free field attenuates 6 dB for every doubling of distance, and making the assumption that the sound field in a cubicle approaches that of a free field because of the large amount of highly absorptive material in the area, we obtain a sound intensity level of 54 dBA at 6 ft from the speaker and 48 dBA at 12 ft. In practice, the received sound level is several decibels higher because, despite the high $\alpha$ of the space, there exist intraspace reflections that increase the sound level.

Referring now to Fig. 19.40*b*, we see that the minimum sound level at a receiver *within* the cubicle would be 55–56 dB, a level that no practical amount of background sound can mask. Thus, two occupants of a single large cubicle will always hear each other quite clearly. The attenuation of a partition in the diffracted paths (over and around a partition) depends upon the location and height of the speaker and varies from 4 to 8 dB. The attenuation of the signal in the transmitted path (through the partition) will be at least 10 dB. Based on these figures, it is recommended that the *minimum* horizontal distance between occupants of adjoining cubicles when seated at their workstations be 10 ft for minimum speech privacy. (Degrees of speech privacy are discussed in the next section.) Speech levels in teamwork areas readily reach 66 dB. This necessitates either locating such areas away from normal working spaces or the use of full-height fixed or demountable partitions to completely enclose such areas. The spaces that require careful siting or complete enclosure because of raised voice levels (64–66 dB) include videoconferencing rooms, telecommunications spaces, and areas where workers use speakerphones or voice-activated computers. The latter two devices are usually forbidden in densely populated work areas where a reasonable degree of speech privacy is required for the conduct of regular business tasks.

7. Refer to Fig. 19.41. As pointed out previously, sound will be received via paths of least resistance. These are often flanking paths that do not become evident except in plan view. In Figure 19.41*a*, the flanking paths are particularly important because the first reflection occurs at a window. Glass has negligible absorption and, because of its smoothness, exhibits specular reflection. As a result, the corner office occupant's voice will be heard clearly via the flanking paths shown, thus destroying the

confidentiality of conversation in that office. Since offices on the building perimeter are usually reserved for middle and upper management, and since the large windows in such offices act to minimize the speech privacy so important to managerial personnel, the space designer has several options to ameliorate this condition:

a. Use full-height fixed partitions, with fixed glass vision panels if required, and doors rather than openings.

b. Use heavy drapes over the "offending" glass windows, although this option defeats the visual and daylighting purpose of the windows.

c. Locate spaces requiring confidentiality in groups, sound-buffered from open-office spaces by unoccupied areas such as storage rooms.

It is also important to note that although the arrow signifying a sound path in Fig. 19.41 shows reflection from the windows, the sound energy will also strike the exterior walls, which are usually plastered and therefore highly reflective. Here, the placement of absorbent acoustic material on all walls, to ceiling height, is not impractical, as it is with windows, but it does entail considerable expense.

8. Refer to Fig. 19.42. The furniture arrangement establishes the source location of speech energy and consequently all of the sound paths that contribute to speech privacy—or, more accurately, the lack of speech privacy. In layout (*a*), sound power, unattenuated except by distance, reflects off the *back* of the opposing aisle partition and travels to the occupant of the neighboring cubicle. Since the back of an acoustic partition is usually metallic and nonabsorbent, this arrangement would entail the additional expense of an absorbent rear surface on the corridor panels to maintain a degree of speech privacy between cubicles.

Changing the desk location in the same-shaped cubicle to that shown in Fig. 19.42*b* improves speech privacy considerably by reducing the SPL of both the reflected and flanking paths, since the speakers face a highly absorptive surface. It may be unnecessary to use absorbent material on the rear of the corridor panels. A disadvantage of this arrangement lies in the short distance between speakers and neighbors if the same pattern of cubicles is continued longitudinally. A second possible disadvantage may be employee dissatisfaction with a working position facing a blank wall.

The arrangement in Fig. 19.42*c* uses the same

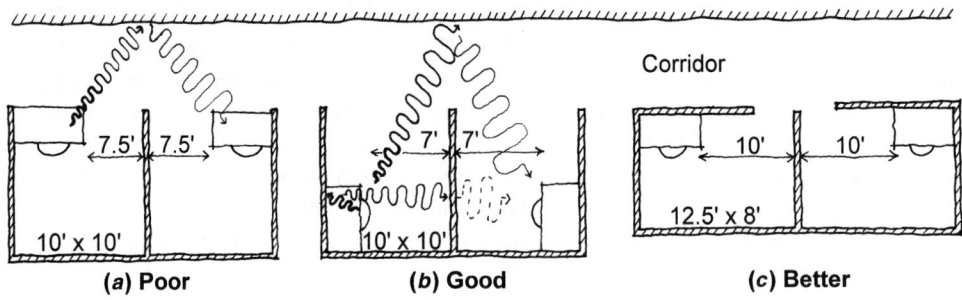

**Fig. 19.42** *Open-office shapes and furniture arrangements have a marked effect on speech privacy. (a) A strong, unattenuated signal reflects from the opposite corridor wall directly into the adjoining office. The side-to-side 15-ft total spacing with an intervening partition is barely satisfactory. (b) The sound reflected from the corridor wall is much weaker than in (a) because of increased distance and decreased voice intensity from the side. Lateral sound is also weak due to the attenuation at the first reflection off the absorbent partition. (c) Reflection from the corridor is eliminated by the use of a front closure partition with glass at the upper 18 in. for visual purposes. Side-to-side sound is weak due to wide spacing and the intervening partition. (Drawing by Jonathan Meendering.)*

area per cubicle. By changing its shape and adding a 6-ft-high acoustic partition whose top 18 in. is glass, the designer has reduced the sound energy levels in all paths while overcoming any employee resentment related to working in a blank, unexposed corner. Furthermore, the problematic flanking path in Fig. 19.42a has been eliminated, as has probably any requirement for exterior acoustic absorption on the corridor partitions.

## 19.20 OPEN-OFFICE SPEECH PRIVACY LEVELS AND DESCRIPTORS

### (a) Factors

The factors involved in determining the level of acoustic privacy that can be expected at a specific open-office location, as a result of neighboring speech sources, should be understood at this point. It may be helpful, however, to restate them. They are:

1. Loudness of the source(s).
2. Acoustic characteristics of the source(s). These include location, height above the floor, directivity, frequency spectrum, and information content. The last factor refers to the ability of a listener to make sense of what is being said or heard. Therefore, most types of non-information-bearing sound would not be considered intrusive sources of noise. A completely foreign language should also fall into this category, although in practice that is so only when the

words and syllables are muffled. Clearly heard words in a foreign language do constitute a source of speech annoyance, although to a far lesser extent than those in a comprehended language. Speech privacy calculation procedures ignore foreign languages and assume that all speech is a possible source of annoyance.

3. Signal attenuation along each path between the source and the receiver. This factor is different for every normal position of even a single source, as, for instance, for a seated or standing source person.
4. Level and frequency content of background sound (deliberate masking, HVAC noise, continuous machinery noise, and the like).
5. Degree of privacy required. This is discussed in detail in the next subsection. It is important to keep in mind, however, that the three classifications—confidential, normal, and transitional (minimal, marginal)—are assumed to remain constant as long as the physical factors involved do not change. This ignores the oft-demonstrated fact that intrusive noise causes psychological and physiological reactions in some people that tend toward aggravation and increased severity with passage of time. A speech privacy situation originally classified as confidential may deteriorate to normal as the hearer's increasing sensitivity to noise causes him or her to strain to hear, and thereby to actually hear noise that ordinarily is masked. This is, in effect, another

aspect of the cocktail party effect discussed in Chapter 17.

## (b) Levels (Degrees) of Speech Privacy

Speech privacy is often achieved by masking intruding speech with background sound. By consensus among acousticians, the definitions of the three levels of speech privacy in an open office are:

1. *Confidential privacy.* Normal voice levels are audible but generally unintelligible. Raised voices are partially intelligible. Noise level is minimal. To achieve this level of privacy, the background sound level must be no more than 2 dB less than the intruding sound, and no more than about 3 dB more than the intruding sound, to satisfy the minimal noise level requirement. In this acoustic situation, approximately 95% of people will not sense any sound-intrusive disturbance and will be able to concentrate on most types of work.

2. *Normal privacy.* At this level, normal voice levels from adjacent spaces are heard but are not intelligible without concentration (i.e., by straining to hear and catch every syllable). Raised voices are generally intelligible. The overall noise level is low. This level of speech privacy is achieved when the background sound level is within 6 dB of (less than) the intruding speech level. This corresponds roughly to an intruding speech level of 50 to 54 dB, and a background sound level of 44 to 45 dB, that together give a range of 51 to 55 dB, levels that can be considered to meet the low-noise requirement.

3. *Transitional (minimal, marginal) privacy.* At this speech privacy level, speech at normal voice levels in adjacent open offices is readily understood most of the time, and the overall noise level is average. This noise level occurs when the intruding speech level is 10 dB or more than the background sound level. Since background sound is limited to about 50 dB, this privacy level would mean an intruding speech intensity of 60 dB or more. Since 60 dB is approximately normal speech at 3 ft, this "privacy" level would occur with two occupants in a single office or a single occupant receiving intruding noise from at least three neighboring offices. This speech intrusion level would be considered intolerable by about 40% of people and would negatively affect the work efficiency of a higher percentage.

In summary, it is interesting to compare these three open-office speech privacy levels with the "absolute" grading given in Table 19.5:

| Open Office Class | Table 19.5 Rating |
|---|---|
| Confidential | Good |
| Normal | Fair |
| Marginal | Poor |

It should therefore be apparent that, by its very nature, an open office cannot achieve the top three grades of privacy listed in Table 19.5. If those levels of privacy are desired or required, fully enclosed spaces are necessary.

## (c) Articulation Index (AI)

In order to quantify speech privacy for an open-office design, a single number metric called the Articulation Index (AI) was developed in the 1970s by the acoustics consulting firm of Bolt, Beranek and Newman. This work was based in part upon studies of speech intelligibility by Bell Labs in the 1940s. Essentially, the AI relates speech intelligibility, speech intensity, and background sound level at the center of the five octave-band frequencies that encompass the spectrum of the human voice: 250, 500, 1000, 2000, and 4000 Hz. AI is determined by measuring the percentage of individual words that can be understood under specific speech and background sound levels. An AI of 0% indicates zero intelligibility and therefore ideal speech privacy. This, of course, does not mean silence; it means that with a specific combination of speech intensity level and background sound level, intelligibility is nil, and therefore speech privacy, defined as a lack of intelligible intruding speech, is ideal. At the other end of the scale, a high percentage of intelligible words yields a high AI and a correspondingly low speech privacy rating. In practice, the resultant AI figures are related to listener satisfaction and speech privacy descriptors, as shown in Table 19.9.

The calculation procedure for AI involves the use of weighting factors that are applied to intensity level differences between speech and background

ACOUSTICS

**TABLE 19.9 Articulation Index (AI) and Speech Privacy**

| AI | Persons Satisfied with Speech Privacy (%) | Open Office Speech Privacy Descriptor |
|---|---|---|
| 0–0.05 | 95–92% | Confidential |
| 0.06–0.2 | 90–80% | Normal |
| 0.21–0.3 | 79–65% | Minimal |
| >0.3 | <65% | Unacceptable |

sound levels at different frequencies in order to reflect the connection between intelligibility and frequency. As pointed out in Chapter 17, most of the information in English words is carried by consonants, whose frequencies are generally above 2 kHz. Thus, the problem frequently encountered in telephone conversations, of distinguishing between f and s, b and v, t and d, and so on, is due to excessive attenuation of the high frequencies that distinguish these letters from each other. The AI calculation emphasizes the importance of high frequencies to intelligibility by using the following weighting factors:

| Octave Band Center Frequency (Hz) | Relative Weighting Factor |
|---|---|
| 250 | 1.0 |
| 500 | 2.5 |
| 1000 | 3.5 |
| 2000 | 5.0 |
| 4000 | 4.0 |

The calculation of AI values for an actual open-office design is complex and laborious, since it considers all sound paths between each source and each receiver, including the acoustic characteristics of all reflective and absorptive surfaces in the path. The results are a specific AI factor for every receiver location. Because of the very large number of calculations involved, the analysis is done by computer. Changes in materials, plan arrangements, and dimensions can be made if the calculated AI does not satisfy the space's speech privacy requirement.

These changes can be predicted fairly accurately, since it has been demonstrated that a 3-dB change in the relative level of an intruding speech signal with respect to the background sound level will result in a 0.1-level change in AI. Thus, an increase of 3 dB in the background sound level or a decrease of 3 dB in the intruding signal (by increased absorption or path length) will have the effect of

increasing the AI between the involved workstations by 0.10, which is the difference between normal and poor speech privacy. Since 3 dB is a barely perceptible change in intensity (see Table 17.3), we can appreciate how sensitive speech privacy is to small changes in intensities.

### (d) Articulation Class

Numerous measurements in actual open-office installations have indicated that the absorption characteristics of the ceiling are the most important factors in speech privacy design. As noted in the previous section, the angles of incidence of speech sound on the ceiling range between 30° and 60°, with the majority of sound energy falling at the top of this range. A figure of merit for absorption, called the *Articulation Class* (AC), was established that indicates absorption effectiveness at angles of incidence between 45° and 55°. The usual range of AC is between 180 and 220 (no units), with higher numbers representing better absorption.

## 19.21 DESIGN RECOMMENDATIONS FOR SPEECH PRIVACY IN OPEN OFFICES

### (a) General Factors

The architectural arrangement of spaces in an open-office design has a marked influence on speech privacy. Areas should be grouped according to their speech privacy requirements. Spaces rated as "confidential" should be placed on the perimeter of the open area to limit their exposure to speech intrusion, with the caveats relating to exterior windows and walls, discussed above, observed. The design emphasis for these areas should be not only an AI between 0.0 and 0.05, but also on a low overall sound level, *including* background noise. Similarly, high-noise-producing areas should be grouped and placed on the perimeter at a maximum distance from confidential speech privacy areas. Use of demountable full-height partitions for such spaces should be considered.

Because of reflection from perimeter walls, open-area spaces should be as large as practical, with absorbent perimeter walls. Ceiling height should be no less than 9 ft clear, with a 3-ft plenum above. Extreme care must be taken with air-conditioning ductwork, which, if untreated, will act as an excellent speech and noise conduit via multiple ceiling outlets. Furthermore, these outlets, which were once relied upon to produce an even level of background sound, generally no longer do so. Most HVAC systems today are variable air volume (VAV) designs, whose noise levels vary $\pm 10$ dB, making them useless as a reliable source of masking sound.

## (b) Individual Office (Cubicle) Design

Offices should be designed for maximum closure and maximum partition length. Separation between occupants of adjacent offices should never drop below 10 ft, with a 12-ft minimum as a design target for normal privacy and 16 ft for confidential privacy. Minimum office area should be 80 ft$^2$, with a design target of 100 to 120 ft$^2$ for normal privacy and 200 ft$^2$ for confidential privacy. Desk arrangements should be checked for optimum speech paths (for privacy), recognizing that office furniture arrangements need not be uniform in all offices (see Section 19.19).

## (c) Ceilings

As the ceiling is the most important design element in speech privacy, care must be taken to avoid unintentional strongly reflective speech paths, as from metal pan air diffusers, flat lighting fixture diffusers, and the like. If the use of such or similar items is unavoidable, highly absorptive vertical baffle strips may be placed on their perimeter to block sound paths. In general, ceiling tiles should have an Articulation Class rating of 220 minimum, and minimum absorption coefficients ($\alpha$) at incidence angles of 30°–60° as follows:

| Frequency (Hz) | $\alpha$ |
|---|---|
| 250 | 0.65 |
| 500 | 0.65–0.75 |
| 1000 | 0.85 |
| 2000 | 0.90 |
| 4000 | 0.90 |

## (d) Partitions

As explained above, the minimum height of partitions should be 65 in., with 72-in.-high units separating offices from aisles and dividing departmental groups. The AC rating, if available, should range

from 200 to 220. STC ratings, as explained above, depend to an extent upon speaker locations and vary from 20 to 26. Joints between partitions should be carefully sealed, as even small openings can seriously compromise a partition's already limited efficiency. All partitions should reach the floor, although the lower portion is not always absorptive in low-speech-privacy areas.

## (e) Floors

Although carpeted floors do not seriously affect overall sound absorption, they do drastically reduce chair movement and footfall sounds. For this reason, all floors in open office areas should be carpeted. The difference in effectiveness of shallow-pile carpet compared to deep-pile carpet is minimal, and the same differential can be achieved by using a polyurethane cushion backing in lieu of the more common jute pad. The principal purpose of carpeting is to cushion the footfall impact so that its energy is not introduced into the structure. This subject is covered in detail in the discussion of structure-borne sound that follows.

## (f) Lighting Fixtures

Flat-bottom lighting fixtures must never be used. Fixtures should not be placed directly over partitions in order to avoid an interoffice speech reflection path. Experience has shown that the best lighting fixture (from the speech privacy point of view) is one with deep parabolic reflector cells and overall dimensions of 1 ft $\times$ 4 ft or 2 ft $\times$ 4 ft.

## (g) Masking Sound

It is imperative that the level of masking sound be uniform throughout an open office area, and at as low a level as will yield the desired speech privacy. Nonuniformity will immediately be noticed as people move about, and the masking sound itself will become a source of auditory annoyance. For a similar reason, loudspeakers should not be visible. Visible units become themselves a source of interest initially and then a source of annoyance. Speakers should be placed *in* the plenum, preferably facing up to increase dispersion and improve uniformity. Speakers mounted in the ceiling and facing down should be avoided. Most ceiling tiles in open-office spaces have a low CAC so that sound will easily penetrate into the office area below the ceiling.

A masking sound system should comprise a signal (noise) generator, a sophisticated equalizer for shaping the signal, an amplifier with appropriate controls, and a distribution system to feed the speakers. Speakers are normally 12 in., and are installed in a grid on 12- to 16-in. centers. The amplifier should be arranged so that volume levels can be remotely controlled. This permits time control so that the background sound volume can be reduced automatically after working hours. This is necessary so that the few people working late are not annoyed by relatively loud background sound in the absence of intruding speech.

The noise produced by a masking sound system is variously described as white noise, noise of air rushing through an opening (whoosh sound), noise of water in piping, and the like. The actual sound can be tailored to the user's preference by adjusting the filters in the system's equalizer. Generally, masking sound emphasizes low frequencies, because higher frequencies are immediately noticed as an annoying hiss. As noted above, the background sound level should not exceed 48 to 50 dBA. In some installations, the masking sound system doubles as a public address system, although this practice is not recommended because the sound/noise stops during an announcement, and when it returns it is noticed. Background sound must be designed so as to blend into the ambience of the background, and anything that disturbs the hidden quality of masking sound is to be avoided.

## (h) Design Procedure

Unfortunately, due to the large number of variables involved in open-office speech privacy design, a straightforward manual design method that will yield reliable results does not exist. However, a number of computer programs are available that will calculate the AI for any location as a result of a specified speech source intrusion. On the basis of these calculations, changes can be made to the design and the program rerun to achieve improvements where the calculated AI is excessive. The most effective way to perform a complete design is then, on the basis of the above calculations, to construct a full-scale mock-up that can then be field tested and "tuned." Although this is expensive and time-consuming, it is frequently far cheaper than making the requisite changes after construction of an unacceptable solution.

One of the distinct advantages of this two-step design procedure is that it enables a designer to equalize the acoustic absorption "strength" of various paths so that the attenuations of major paths are equal. There is no economic or engineering sense in a system that is much more effective for one path than for another, since sound will always choose the path of least acoustic resistance. To accomplish this balancing, the designer has many variables to juggle. They include barrier height and material, ceiling material, baffle sizes and positions (if used), distances between the source and receiver, position and directions of sources, and level of background sound.

## (i) Standards

The acoustic design and testing of open offices is covered by a group of American Society for Testing and Materials (ASTM) standards that should be in the hands of anyone engaged in open-office design. The standards are available from ASTM, 100 Barr Harbor Drive, West Conshohocken, PA 19428-2959.

ASTM E1573-02—*Standard Test Method for Evaluating Masking Sound in Open Offices Using A-Weighted and One-Third Octave Band Sound Pressure Levels.* This test method specifies the procedures that can be used to evaluate the spatial and temporal uniformity of masking sound in open offices using A-weighted sound levels. It also specifies the procedure for evaluating the masking sound spectrum and level using ⅓-octave band sound pressure levels.

ASTM E1110-01—*Standard Classification for Determination of Articulation Class.* This classification provides a single figure rating that can be used for comparing building systems and subsystems for speech privacy purposes. Excluded from this classification are applications involving female speakers and children, languages other than English, and sound spectra other than speech.

ASTM E1111-02—*Standard Test Method for Measuring the Interzone Attenuation of Ceiling Systems.* This test method is intended to provide measurements of the sound-reflective characteristics of ceiling systems when used in conjunction with partial-height space dividers.

ASTM E1130-02e1—*Standard Test Method for Objective Measurement of Speech Privacy in Open Offices Using Articulation Index.* This method describes a field test for measuring speech privacy objectively between locations in open offices. It relies upon acoustical measurement, published information on speech levels, and standard methods for assessing speech communication. This test method does not measure the performance of individual open-office components that affect speech privacy; it measures the privacy that results from a particular configuration of components. This method relies upon the AI, which predicts the intelligibility of speech for a group of talkers and listeners.

ASTM E1179-87(2003)—*Standard Specification for Sound Sources Used for Testing Open Office Components and Systems.* This specification states the requirements for sound sources used for measuring the speech privacy between open offices or for measuring the laboratory performance of acoustical components. The sound source is a loudspeaker located in an enclosure and driven with an appropriate test signal.

ASTM E1264-98—*Standard Classification for Acoustical Ceiling Products.* This classification covers ceiling products that provide acoustical performance and interior finish in buildings. It classifies acoustical ceilings by type, pattern, and certain ratings for acoustical performance, light reflectance, and fire safety.

ASTM E1374-02—*Standard Guide for Open Office Acoustics and Applicable ASTM Standards.* This guide discusses the acoustical principles and interactions that affect the acoustical environment and acoustical privacy in an open office. In this context, it describes the application and use of the series of ASTM standards that apply to open offices.

ASTM E1375-90(2002)—*Standard Test Method for Measuring the Interzone Attenuation of Furniture Panels Used as Acoustical Barriers.* This test method covers the measurement of the interzone attenuation of furniture panels used as acoustical barriers in open-plan spaces to provide speech privacy or sound isolation between working positions.

ASTM E1376-90(2002)—*Standard Test Method for Measuring the Interzone Attenuation of Sound Reflected by Wall Finishes and Furniture Panels.* This laboratory test method measures the degree to which reflected sound is attenuated by the most commonly found vertical surfaces in open-plan spaces. The vertical surfaces covered by this test method include wall finishes such as sound-absorbent panels and furniture panels or screens. It does not cover such items as window finishes or furniture other than panels.

## STRUCTURE-BORNE NOISE

### 19.22  STRUCTURE-BORNE IMPACT NOISE

The term noise will be used in lieu of sound in the following discussion of structure-borne, impact, and equipment noise control. Although use of the term noise assumes that a decision has been made that a particular sound is unwanted, this is a very reasonable assumption for the situations to be discussed. Use of the term noise gets to the point.

Structure-borne noise is at least as serious a problem as airborne noise for the following reasons:

1. There is no air cushion between the source and the structure; thus, high-intensity energy is introduced into the structure, through which it travels with minimum attenuation and at great speed.

2. Sound, once introduced into the structure, is attenuated well only by discontinuities in the structure. Since the structure must have structural integrity to carry the loads, discontinuities of the type that will stop noise are complex and expensive.

3. The entire structure constitutes a network of parallel paths for sound. Therefore, partial solutions are useless, since sound will find flanking paths. The entire structure must be soundproofed to yield good results.

4. Unlike the case of airborne noise, additional mass does not usually block structure-borne noise, particularly in long spans where a floor can act as a diaphragm, thereby improving the structure-to-air noise transfer efficiency (like a drum).

5. The increasing use of exposed structural ceilings eliminates the attenuation that can be introduced by a plenum above a hung ceiling. This is particularly bad, since most structure-borne noise

is carried by floor structures (rather than walls), which radiate sound up and down. The discussion that follows will be limited to impact noise. Refer to Section 19.26 for a brief treatment of vibration, which is felt rather than heard and is, in effect, a very low-frequency noise. Many of the practices and techniques that will minimize impact noise will also reduce vibration.

## 19.23 CONTROL OF IMPACT NOISE

Impact noise problems can be controlled in two ways—by preventing or minimizing the impact and by attenuating it once it has occurred. Prevention is discussed first; attenuation is covered in Section 19.25. Impact on floors is more serious than wall impact because the latter is partially attenuated at the wall/floor joint, whereas the former is introduced directly into the building framework. The discussion below addresses each of the solutions shown in Fig. 19.43.

### (a) Cushion the Impact

See Fig. 19.43a. This obvious solution will frequently eliminate all but severe problems. Resilient cushioning materials in common use are floor tile of rubber and cork, or carpeting on pads, in ascending order of impact insulation. See Section 19.24 and Appendix L for quantitative data on impact isolation.

### (b) Float the Floor

See Fig. 19.43b. Since the key to elimination of structure-borne sound is *isolation*, separating the

impacted floor from the structural floor by a resilient element is extremely effective. This element can be rubber or mineral wool pads, or blankets, or special spring metal sleepers. The effectiveness depends upon the mass of the floating floor, compliance of the resilient support, and degree of isolation of the floating floor. The last element is extremely important, since flanking paths via end contacts with walls can short-circuit the floating element's sound impedance and defeat the system. With floating floors it is important that:

1. The mass of the floating floor be large enough to spread the loads properly. Otherwise, the pad will compress and deform sufficiently to transmit the impact.
2. Total construction must be airtight. Airtight is soundtight.
3. Particular care be exercised where partitions rest on the floating floor (see Fig. 19.44a).
4. Short-circuits at walls or by penetrations be avoided; see Fig. 19.10b. Details of proper construction techniques are given in *A Guide to Airborne, Impact and Structure-Borne Noise Control in Multi-family Dwellings* (1968).
5. Construction throughout be consistent. Mixed construction types invite flanking noise paths (see Fig. 19.44b).

### (c) Suspend the Ceiling—and Use an Absorber in the Cavity

See Fig. 19.43c,d. As stated, the most disturbing noise is that radiated down from the ceiling. A flexibly suspended ceiling with an acoustic absorbent layer suspended in it can be very effective if not

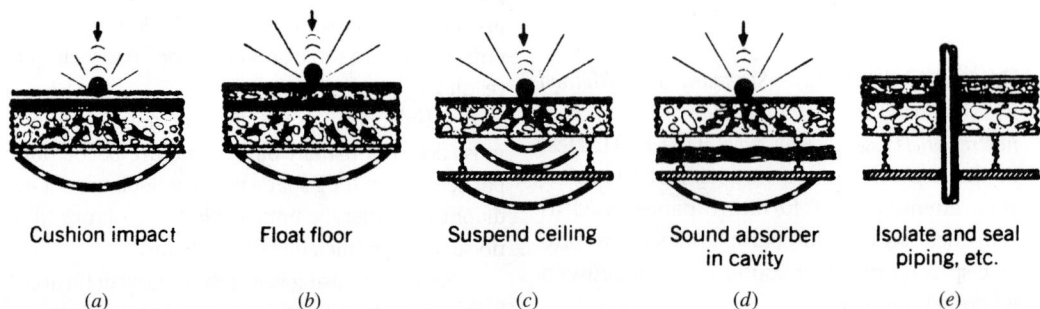

| Cushion impact | Float floor | Suspend ceiling | Sound absorber in cavity | Isolate and seal piping, etc. |
| (a) | (b) | (c) | (d) | (e) |

**Fig. 19.43** *Methods of controlling impact sound transmission through floors. (Reprinted from* Quieting: A Practical Guide to Noise Control, *1986.)*

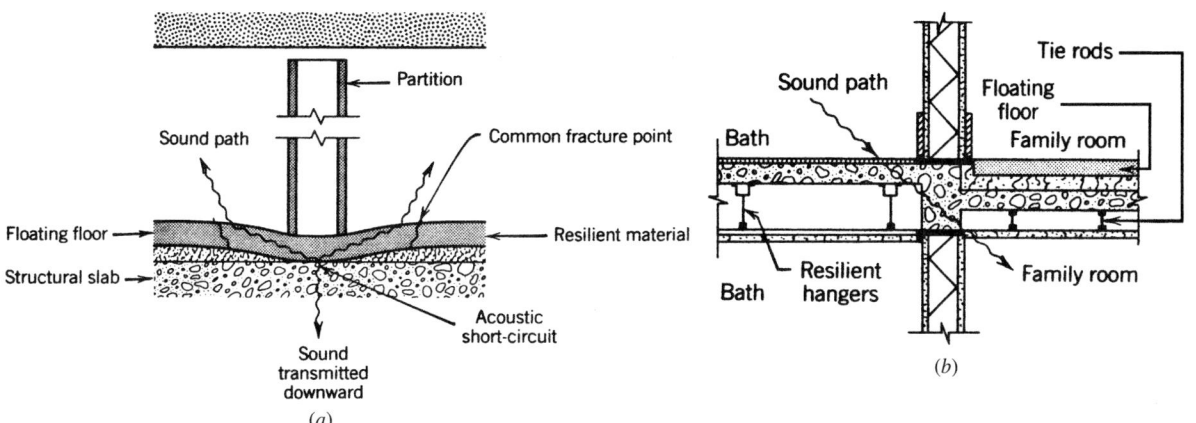

**Fig. 19.44** (a) Caution must be exercised when supporting partitions on floating floors to prevent structural failures or short-circuiting of the floating element, as illustrated. (b) Flanking paths in mixed-construction type floors. The FHA does not recommend mixing construction types unless provisions have been made to prevent flanking (e.g., expansion joints or breaks in all structural paths between each space). (Reprinted from A Guide to Airborne, Impact, and Structure-Borne Noise Control in Multifamily Dwellings, 1968.)

flanked by paths leading into the walls and from there reradiating into the space below. It is imperative that the entire floor slab above be decoupled from the walls below by resilient separators.

### (d) Isolate All Piping

See Fig. 19.43e. All rigid structures such as piping must be isolated so as not to form a flanking path, and penetrations must be caulked with resilient sealant so as not to constitute an air–sound leakage path.

### 19.24 IMPACT ISOLATION CLASS

The impact isolation class (IIC) is a single-number, impact isolation rating for floor construction, similar in intent and derivation to STC wall ratings. Tests are made with a standard tapping machine and noise levels measured in $\frac{1}{3}$-octave bands. These are plotted and compared to a standard contour, approximately as with the sound transmission class. Details of typical floor constructions along with IIC ratings are given in Appendix L. Resilient floor finishes on any of the floor constructions not specifically provided with them will add to the IIC ratings approximately as follows:

| | |
|---|---|
| $\frac{1}{16}$-in. vinyl tile | 0 |
| $\frac{1}{8}$-in. linoleum or rubber tile | $4 \pm 1$ |
| $\frac{1}{4}$-in. cork tile | $10 \pm 2$ |
| Low-pile carpet on fiber pad | $12 \pm 2$ |
| Low-pile carpet on foam rubber pad | $18 \pm 3$ |
| High-pile carpet on foam rubber pad | $24 \pm 3$ |

## MECHANICAL SYSTEM NOISE CONTROL

### 19.25 MECHANICAL NOISE SOURCES

Mechanical devices make noise. And generally, the more power they consume, the more noise they make. In many of today's buildings, 40% of the total construction budget is spent on mechanical systems located throughout a building.

In most buildings, the primary sources of mechanical noise are the components of the air-conditioning and air-handling systems such as fans, compressors, cooling towers, condensers, ductwork, dampers, mixing boxes, induction units, and diffusers. The curve of Fig. 19.45 depicts typical air-handling system noise and indicates the portions of the spectrum produced by each group of components. Pumps are another source of mechanical noise, which (along with the noise of flowing liquid) is transmitted along pipes to locations throughout the building.

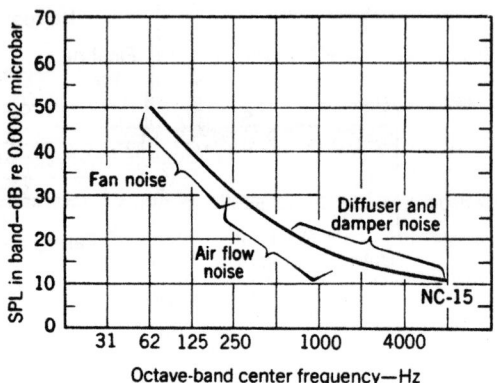

**Fig. 19.45** *Sound pressure level frequency spectrum of noise from HVAC system components.*

Elevators, escalators, and freight elevators also introduce mechanical noise into buildings. Escalators and freight elevators pose few problems, since they are localized in a specific area and have low operation speeds. Passenger elevator car operation, however, is rapid, and it affects large areas. In addition, the motors and controls are located on or above the prime upper floors of a building. Motor,

shaftway, and other equipment noise must be properly controlled to prevent annoyance to building tenants located near the shaftways or elevator penthouses and mechanical equipment rooms. Vibration isolation of these major components is a specialized problem beyond the scope of this book.

## 19.26 QUIETING OF MACHINES

Machines cause noise by vibration. This noise is imparted directly to the surrounding air and by vibrational contact to the surrounding structure. Therefore, there are three ways to reduce this noise:

1. Reduce the vibration itself.
2. Reduce the airborne noise by decoupling the vibration from efficient radiating sources.
3. Decouple the vibrating source from the structure.

Refer to Fig. 19.46. Items 1, 3, and 4 reduce vibration; items 4, 5, 6, and 7 reduce and decouple the vibration from the radiating cabinet; and items 2 and 8 decouple the vibrating source from the structure. Once a noise becomes airborne or

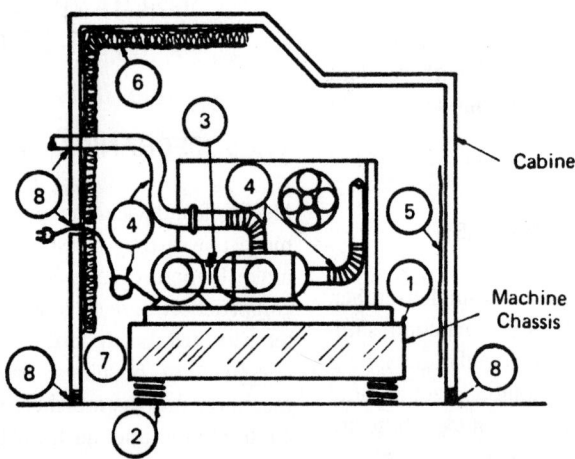

1. Install motors, pumps, fans, etc. on most massive part of the machine.
2. Install such components on resilient mounts or vibration isolators.
3. Use belt drive or roller drive systems in place of gear trains.
4. Use flexible hoses and wiring instead of rigid piping and stiff wiring.
5. Apply vibration damping materials to surfaces undergoing most vibration.
6. Install acoustical lining to reduce noise buildup inside machine.
7. Minimize mechanical contact between the cabinet and the machine chassis.
8. Seal openings at the base and other parts of the cabinet to prevent noise leakage.

**Fig. 19.46** *Techniques used to reduce the transmission of airborne and structure-borne noise from machines and appliances. (Reprinted from* A Guide to Airborne, Impact, and Structure-Borne Noise Control in Multifamily Dwellings, *1968.)*

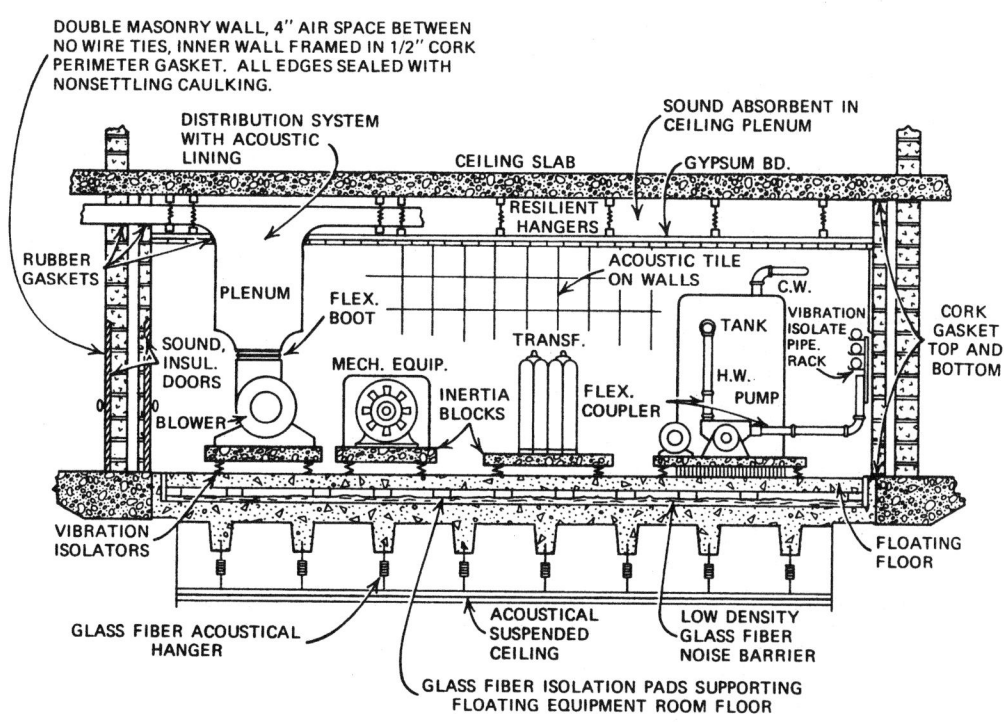

**Fig. 19.47** *Soundproofing a mechanical equipment room. Additional noise reduction to the space below can be achieved by inserting a layer of highly absorbent material in the space above the suspended acoustical ceiling. (Reprinted from* A Guide to Airborne, Impact, and Structure-Borne Noise Control in Multifamily Dwellings, *1968.)*

structure-borne, the isolation techniques presented above are employed.

*Vibration reduction* takes two forms, damping and isolation. One form of damping is accomplished by rigidly coupling the vibrating source to a large mass, frequently called an *inertia block*. Much of the energy is absorbed and dissipated as friction; the remainder results in lower-amplitude vibration (Fig. 19.47). Isolation is accomplished by supporting the vibrating mass on resilient supports. These take many forms and can be used in combinations. Thus, machines are supported on fibrous, rubber, or spring steel vibration isolators, and the entire mass can be supported on a floating floor, which in turn rests on resilient vibration isolators, as in Fig. 19.47. Large machines are supported on special commercial "sandwiches" of asbestos, lead, cork, and other strong, resilient materials. Piping is supported on cork pads and hung on resilient hangers.

Use of a diesel-driven electric generator for load peaking or cogeneration can cause very serious noise and vibration problems. The ideal solution to this situation is to completely isolate the unit in a separate outbuilding that is designed specifically to contain the very high noise level produced. If that is not practical and an inside location is necessary, a complete enclosure may be required to ameliorate the noise problem.

Vibration damping can be an even more serious problem, which can be solved satisfactorily with sufficient mass and proper vibration isolation. In the case of vibrating sheet metal, soft foam-type damping material glued directly to the metal is effective in damping. Flexible joints in all pipes and ducts connected to vibrating machines are mandatory. This includes flexible conduit connectors to all motors, transformers, and lighting fixtures using magnetic ballasts.

## 19.27 DUCT SYSTEM NOISE REDUCTION

Design of a quiet duct system entails more than specifying an absorptive duct lining. Air turbulence generates noise. Turbulence increases as the velocity of airflow increases and anywhere in the duct system where smooth laminar flow is disturbed, such as at sharp bends. The permissible "sharpness" of a bend depends in turn upon air velocity; the higher the air velocity, the more aerodynamic the duct system must be to prevent turbulence and, therefore, noise. Table 19.10 demonstrates this principle. The farther one proceeds upstream from a point of turbulence, the higher the air velocity may be from a noise perspective, since truly laminar flow is essentially noiseless. In principle, velocities should be as low as practical, since air turbulence noise increases exponentially with velocity.

Sound travels as easily against as with the airflow in ductwork. Therefore, both supply and return systems must be lined to control transmission of fan noise. Maximum fan noise reduction occurs at bends in the ductwork. For maximum fan noise reduction in short runs, a pair of 90° bends is sometimes deliberately inserted. However, because 90° bends also introduce turbulence that generates noise, introducing bends as fan noise attenuators can be counterproductive at air velocities above 600 fpm. Another disadvantage of bends is added system friction and the additional energy and cost required to move air. This point will be discussed further in the next section.

Other design approaches that create a quiet system include smooth transitions at changes of duct size and large-radius bends with turning vanes, the purpose of which is to reduce turbulence. Attenuation drops rapidly as duct size increases; therefore, ducts should not be deliberately oversized. Cross-talk between rooms and between ducts can be minimized by using lined ducts, separating adjacent ducts as much as possible, and gluing damping material on the outside and lining on the inside. Damping material is particularly effective in preventing the thin metal walls of ducts from resonating. Mufflers and silencers are effective in reducing the high-frequency components of fan noise but much less so with low frequencies. The pressure drop these devices introduce, which can be considerable, must be compensated for in the fan selection.

The ASHRAE *Handbook—HVAC Applications* should be consulted for recommendations on noise control in air-handling units, plenums, housings, and ducts. Figure 19.48 shows some of the ways in which cross-talk and flanking noises can be reduced. Figure 19.49 shows some techniques employed for quieting duct noise. Active noise cancellation (see Section 19.28) is particularly useful in duct systems since it does not reduce airflow, as do liners, baffles, and other mechanical silencing devices, and it is effective at low frequencies, whereas these devices are not.

The increased use of variable air volume (VAV) systems has introduced some noise problems that should not be neglected. VAV system noise can be minimized by following a few basic design rules. Maintain minimum system static pressure since fan noise increases exponentially with static pressure. Select the air volume modulating device at the fan with care, as it can be a noise source. Since outlet air volume control involves duct area restriction with attendant velocity increase and resultant noise, such a design must include some sort of downstream silencing equipment. Ceiling diffuser acoustic characteristics must be coordinated with design air velocity and with any requirement for masking sound. Finally, avoid the use of throttling dampers on ceiling diffusers since a partially closed damper can generate very high noise levels.

**TABLE 19.10 Maximum Air Speeds in Ducts to Yield NC-15 or NC-25 Background Levels[a]**

| Location | Supply | | Return | |
|---|---|---|---|---|
| | **NC-15** | **NC-25** | **NC-15** | **NC-25** |
| Slot speed at min. ½-in. opening | 250 fpm | 350 fpm | 300 fpm | 420 fpm |
| 10 ft of duct before opening | 300 | 420 | 350 | 490 |
| Next 20 ft | 400 | 560 | 450 | 630 |
| Next 20 ft | 500 | 700 | 570 | 800 |
| Next 20 ft | 640 | 900 | 700 | 980 |
| Next 20 ft | 800 | 1120 | 900 | 1260 |
| Next 20 ft | 1000 | 1400 | 1100 | 1540 |
| Next 20 ft | 1300 | 1820 | 1450 | 2030 |
| Next 20 ft | 1600 | 2240 | 1800 | 2520 |

[a]Ducts with 1- to 2-in. inside duct lining, all duct sizes.

**Good**             **Poor**

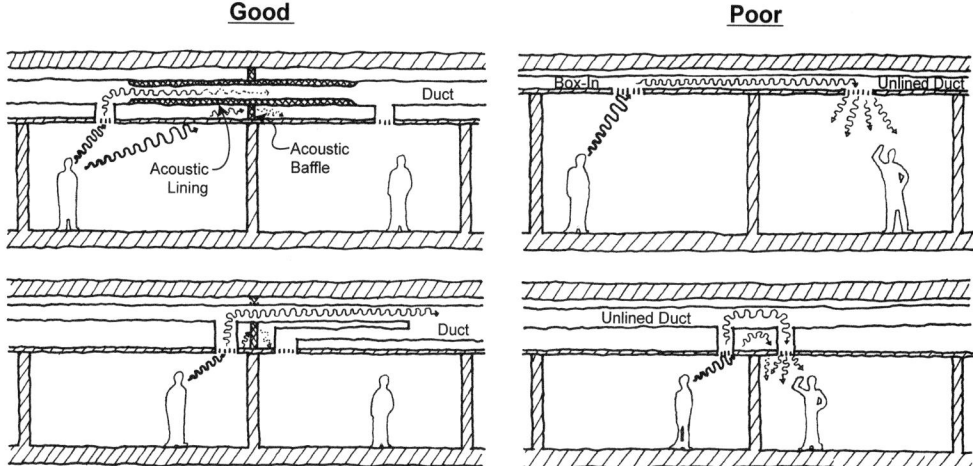

**Fig. 19.48** *Since ducts are efficient sound transmission paths, precautions must be taken to avoid cross-talk, ventilation air noise, and equipment noise. Avoid running ducts as a common supply or return between rooms unless they are properly baffled and lined with sound-absorbing material. The common practice, in wood frame structures, of using troughs between joists as a common return duct between rooms and between separate dwelling units results in serious noise transmission problems. Caulk or seal around ducts at all points of penetration through partitions. Use double-wall ducts, acoustical lining, flexible boots, and resilient hangers where required. Dwelling units should be serviced by separate supply and return ducts that branch off a main duct system. (Reprinted from A Guide to Airborne, Impact, and Structure-Borne Noise Control in Multifamily Dwellings, 1968. Redrawn by Jonathan Meendering.)*

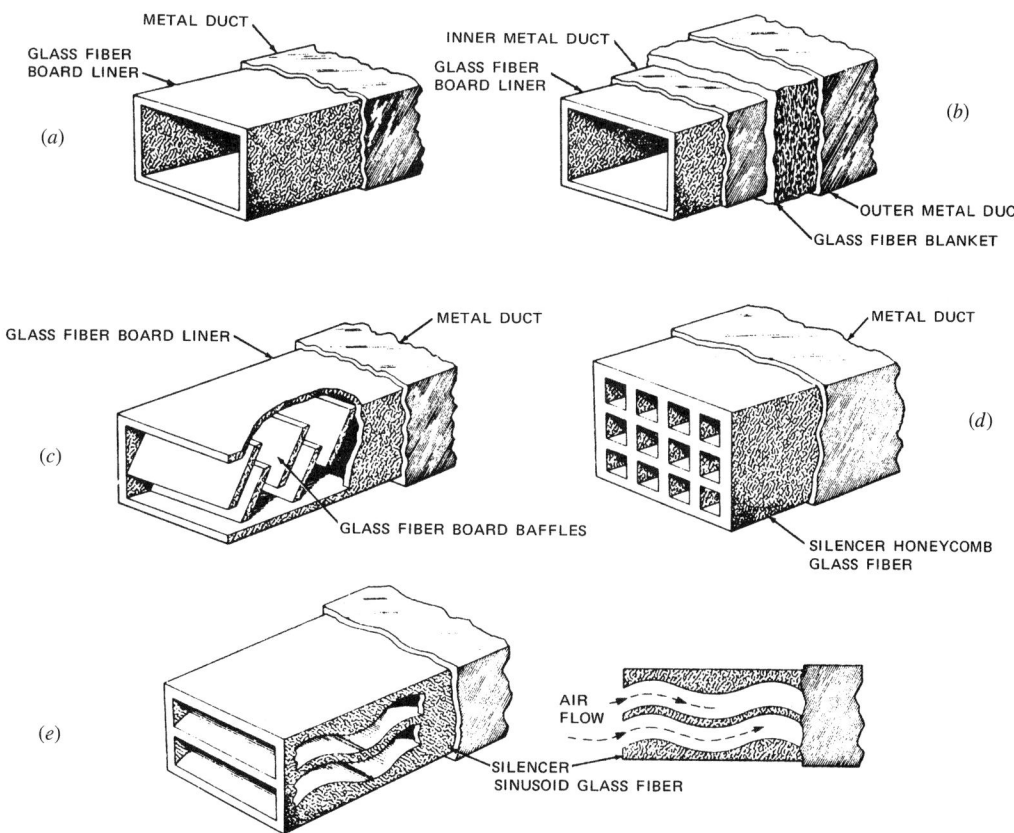

**Fig. 19.49** *Unlined duct has negligible sound attenuation. Inside lining (a) gives 2–3 dB attenuation per foot in the range 1–2 kHz, dropping rapidly above and below those frequencies and giving negligible low-frequency attenuation. Double lining (b) gives higher attenuation and reduces cross-talk between ducts. Duct silencers and baffles (c–e) give high broadband attenuation: a maximum of 10–12 dB/ft in the range 1–2 kHz and lower above and below. They are useful to reduce fan noise in short runs but cause considerable pressure drop. (Reprinted from A Guide to Airborne, Impact, and Structure-Borne Noise Control in Multifamily Dwellings, 1968.)*

**837**

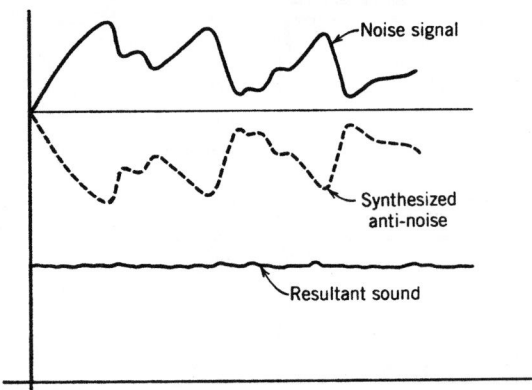

*Fig. 19.50* Silencing of noise by introduction of a synthesized noise signal exactly out of phase with the original signal. The resultant sound is effectively zero—that is, silence.

## 19.28 ACTIVE NOISE CANCELLATION

Since noise is a phenomenon consisting of acoustic wave energy transmitted at certain frequencies, it is theoretically possible to *eliminate* (not mask) noise by simultaneous transmission of identical wave energy exactly out of phase with the noise. The sum of the two energy waves is zero—hence silence (Fig. 19.50). The phase matching, however, must be exact; if the injected out-of-phase noise is not precisely a negative image of the original noise, then not only will the signals not cancel but they may even increase the noise level. The technique for accomplishing noise cancellation is straightforward in theory but much less so in practice. A microphone samples a noise source and feeds that signal into an analysis/synthesis device. This, as the name implies, analyzes the frequency and amplitude content of the noise and synthesizes the anti-noise, which is then fed to a loudspeaker in the original noise path. The resultant residual noise is detected by a downstream microphone and fed back into the controller as a feedback correction

signal. This entire procedure is shown in block diagram form in Fig. 19.51. The technical problems that must be overcome are formidable. Without delving deeply into the physics involved, we can describe them qualitatively.

1. It takes a finite amount of time to analyze the frequency content of a noise and to synthesize the anti-noise. If the noise is random it will have changed by the time the anti-noise signal is injected and will therefore not be canceled. As a result, *the only type of noise that can effectively be attenuated by anti-noise is one that is continuous and/or predictable.* Random noise, such as that of a barking dog, cannot be actively attenuated with the present technology. Candidates for noise cancellation are sounds like those produced by operating machinery, that is, continuous and repetitive sounds. This includes very-low-frequency noise.

2. The analysis and synthesis process is achieved by a device called an *adaptive digital filter.* For technical reasons, the higher the frequencies (pitch) involved, the more complex and expensive are the required digital electronics, microphones, and

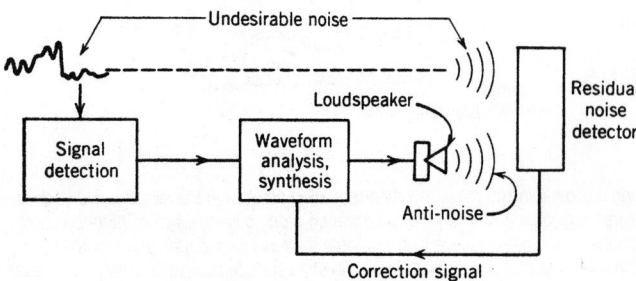

*Fig. 19.51* Block diagram of an active noise cancellation system. The noise signal is detected, its periodicity determined and its waveform are analyzed, and an out-of-phase noise is synthesized and injected into the acoustic environment. A residual noise detector provides a feedback signal that acts to improve noise cancellation.

loudspeakers. This further narrows the range of practical noise-cancellation candidates to those producing low frequencies, such as blowers, fans, rotors, internal combustion engines and their exhausts, transformers (hum), air movement in ducts, fluid movement in pipes, and the like.

3. The waveform of the noise, its modes, and its dispersion in space impact heavily on the type and amount of equipment required to effect attenuation economically (i.e., commercially). The simplest type of noise to treat is one that is confined by some sort of waveguide, at low frequency, and exhibits constant sound pressure and phase. Such a wave is known as a *plane wave*, and it can be "treated" with a single sampling microphone, a single loudspeaker, and a single processor.

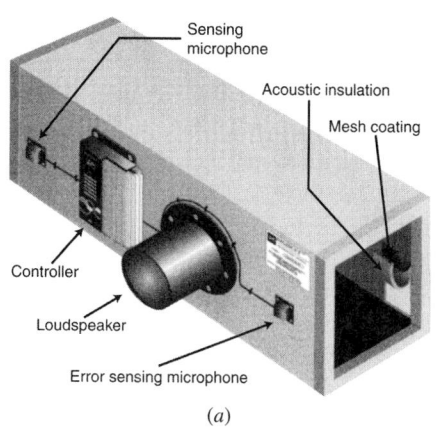

(a)

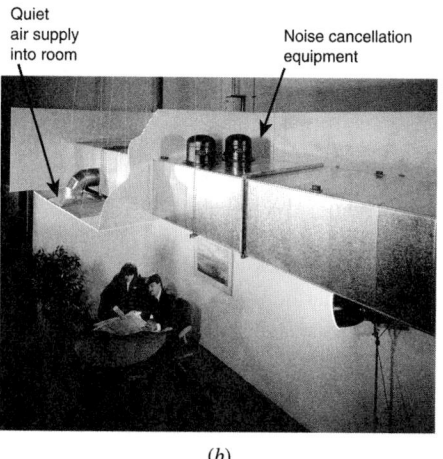

(b)

(c)

**Fig. 19.52** *Active duct noise cancellation equipment. (a) The system components, consisting of the sensing and error-signal microphones, controller, and a loudspeaker are built into a duct section of the required size, which is inserted into the duct system. (b) A mock-up of an office installation showing the relative size and positions of a dual-channel system. (c) Photograph of a multichannel duct-noise cancellation system in a new office building (designed with large, unlined HVAC ducts). (Courtesy of Digisonix.)*

Considering the above three criteria, one of the ideal candidates for economic noise cancellation treatment is duct noise, which is low frequency, continuous, and a plane wave. Figure 19.52 shows the construction of a typical commercial duct noise suppression unit, plus application photographs. Practical duct noise cancellation equipment in use today can reduce levels from NC-50 to NC-35 and is particularly effective at the "rumble" frequencies below 200 Hz.

Important auxiliary advantages of active duct noise cancellation are energy conservation and economic benefits. Passive duct noise reduction, particularly at low frequency, requires massive noise absorbers, requiring 7 to 10 ft of duct for installation. See Fig. 19.49c–e. In addition to the high first cost of these devices and their installation, they introduce a static pressure loss of ½ in. to 1½ in. w.g., depending upon airflow and speed. This, in turn, requires a higher horsepower (and noisier) fan. Economic analysis of such designs generally shows an advantage for active noise cancellation equipment. Such an analysis should be performed for every design where duct silencers are desired or required.

Another important application of active noise cancellation, already in wide use, is in the area of hearing conservation for people exposed to high noise levels at work. Here the out-of-phase noise is introduced into miniature loudspeakers (earphones) in an acoustically transparent headset. This allows the wearer to hear random sounds (such as speech) clearly, while repetitive cyclic noise from engines and the like is attenuated (Fig. 19.53).

Other areas where active noise cancellation is already in use include engine exhausts, heavy machine vibration, interiors of luxury automobiles, military and space vehicles, and selected shipboard spaces. Applications will undoubtedly increase with advances in digital signal processing technology, equipment miniaturization, and reduction in equipment costs.

## 19.29 PIPING SYSTEM NOISE REDUCTION

As with airflow, noise increases exponentially with liquid flow velocity. Piping is not a major noise source normally, since the radiating diameter is small, except for flow velocities much in excess of 8 fps where a pipe is in contact with the structure. This is, of course, most serious where a pipe passes through NC-15 to NC-25 areas (see Table 19.8). Domestic water system mains should be limited to 50 psi in other than tall buildings and pressure in branches to 35 psi. In high-rise structures, pressure-reducing valves will be required in high-pressure mains to meet these recommendations. Piping must be designed to prevent water hammer, and noise sources must be located away from quiet areas.

Pumps, like all rotating equipment, are sources of vibration and noise and should be treated as described in Section 19.26. Figure 19.54 shows a typical pump installation with appropriate noise reduction measures. For at least a distance of 100 pipe diameters beyond the pump, resilient pipe hangers should be used. With centrifugal pumps, as with fans and blowers, machine sound concentrates in narrow bands and, if extremely disturbing, can be attenuated with resonant filters. Reciprocating pumps are more difficult to control, as the pulsations are more vibration than noise. Flexible connections and U-joints in the piping will absorb much of this vibration.

## 19.30 ELECTRICAL EQUIPMENT NOISE

Electrical equipment is generally overlooked as a noise source, and this is unwise. Most electrical noise is a 120-Hz hum. This can be very disturbing because the frequency is so low and, as we have noted repeatedly, low-frequency noise is difficult to attenuate passively. Transformer noise levels

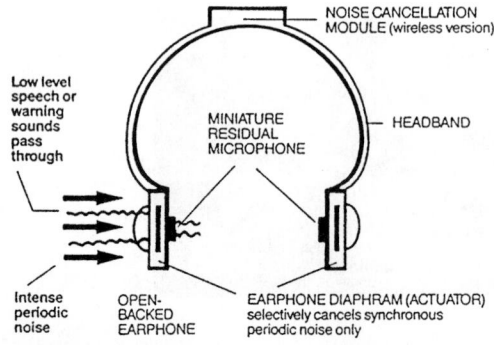

**Fig. 19.53** Schematic diagram of a headset with active noise cancellation. Because of the physical proximity of all the elements, only a single microphone is used. (Courtesy of Noise Cancellation Technologies.)

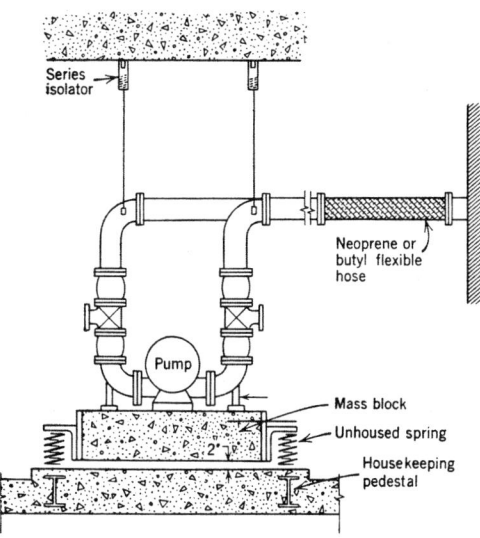

**Fig. 19.54** *Typical pump installation with appropriate vibration isolation and damping measures.*

are dictated by National Electrical Manufacturers Association (NEMA) and American National Standards Institute (ANSI) standards. For a premium price, lower-noise units are obtainable. Table 19.11 lists maximum noise levels for dry-type units. Most manufacturers warranty noise below these levels. Oil- and silicone-filled units are normally quieter than dry-type transformers, as are units designed for lower temperature rise. Transformer noise can be minimized by these steps:

1. Mount the unit on vibration isolators.
2. If the transformer is wall-hung, use resilient hangers. If it is floor-mounted, place it on a massive slab if possible.
3. Locate the unit so that reflections do not amplify the sound. Sound-absorbent material on the walls behind the units is not useful at 120 Hz. Only cavity resonators will absorb appreciable amounts of sound at that frequency.

**TABLE 19.11  Maximum Sound Levels: Dry-Type Transformers**

| kVA | Decibels (NEMA Standard) |
|---|---|
| 0–9 | 40 |
| 10–50 | 45 |
| 51–150 | 50 |
| 151–300 | 55 |
| 301–500 | 60 |

4. Use only flexible conduit connections.
5. Avoid locating transformers adjacent to, or immediately outside, quiet areas. A common error in this regard is placing a transformer pad immediately below the window of an NC 15–25 area.

The second major source of 120-Hz hum is conventional core-and-coil discharge lamp ballasts. These include magnetic ballasts for fluorescent and all HID sources. Fortunately, electronic ballasts, which are practically noiseless, are rapidly replacing core-and-coil ballasts in fluorescent fixtures, and to a lesser extent in HID units. Table 19.12 lists recommended applications of nonelectronic fluorescent ballasts. Conventional coil-type HID ballasts can be very noisy, and care must be exercised in their placement. With all ballasts, the method of mounting has a marked effect on the radiated noise. As pointed out earlier, when a small vibrating source is coupled rigidly to a larger body, noise is amplified because of increased source-to-air coupling. Since core-and-coil fluorescent ballasts for linear fluorescent lamps are necessarily closely coupled to large metal fixtures for heat dissipation purposes, the sound radiation is greatly amplified. A large number of such fluorescent fixtures mounted in a plenum can create a serious noise problem. Solution of the problem lies either in ballast replacement or in the use of absorptive material in plenums,

**TABLE 19.12 Acoustic Criteria for Selection of Conventional Core-and-Coil Fluorescent Lamp Ballasts**

| For an Installation in: | Use of Ballasts with This Rating Will Usually Be Satisfactory |
|---|---|
| TV or radio station, church, synagogue | A |
| Office, residence, library, reception or reading room, school study hall | B |
| Noisy office, doctor's or dentist's office, classroom | C |
| Industrial applications | D |

flexible conduit connection to fixtures, and resilient fixture hanging. In severe cases, ballasts can be remote-mounted. Coil-type HID ballasts are inherently noisier than fluorescent ones but, like ballasts for compact fluorescent lamps, are generally less troublesome, being coupled to small radiating bodies.

### 19.31 NOISE PROBLEMS DUE TO EQUIPMENT LOCATION

Roof-mounted HVAC units have proven to be very economical and very noisy. Vibration, short duct runs, and sound reflections are serious problems that can be solved with vibration isolators, sound mufflers, and careful location of equipment. Roof-mounted cooling towers are a particular problem when they are located adjacent to a taller building. This problem has led to a spate of lawsuits and noise control legislation in many cities. For this reason, particular attention should be paid to all exterior equipment during the design process.

In high-rise buildings, problems are caused by conflicts between the stringent noise requirements of the prime upper floor space and the near presence of elevator machine rooms, mechanical equipment rooms, and cooling towers. These problems are almost impossible to solve after construction and require the services of an acoustics expert during design.

### 19.32 SOUND ISOLATION ENCLOSURES, BARRIERS, AND DAMPING

In buildings with concentrated high-level noise sources, such as certain types of machinery, it is always more desirable to reduce the noise at its source than to attempt to treat the larger enclosing space. This is most effectively accomplished by enclosing the noise source with materials that provide a combination of reverberant noise reduction by absorption (as explained in Section 18.9 and Fig. 18.10) and blocking of airborne sound with high transmission loss (as detailed in Sections 19.7 to 19.12). These materials are available in the form of curtains, panels, and prefabricated partial and full enclosures tailored to the specific characteristics of the noise source (Fig. 19.55). Such

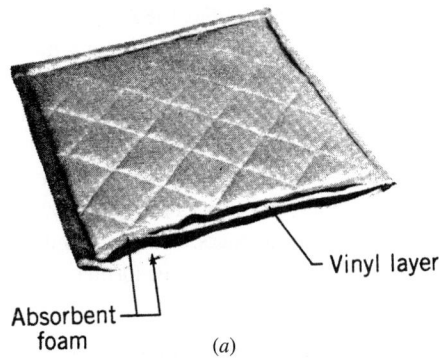

Vinyl layer

Absorbent foam

(a)

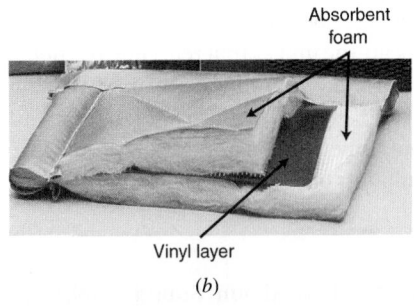

Absorbent foam

Vinyl layer

(b)

**Fig. 19.55** (a) Typical flexible nonmetallic composite acoustical barrier material that can be used as a free-hanging curtain or applied to a rigid surface. It consists of a layer of flexible vinyl sandwiched between two layers of fiberglass or other acoustic foam. The construction is seen clearly in the cutaway (b). The material is most effective when applied with an air space between it and any rigid support barrier. (Courtesy of E-A-R Specialty Composites, division of AEARO Co.)

ACOUSTICS

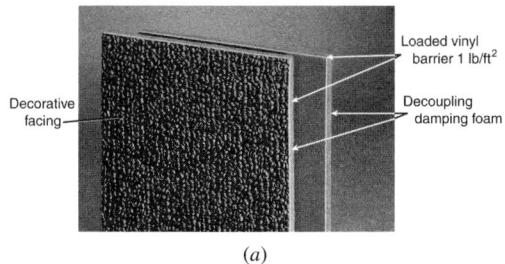

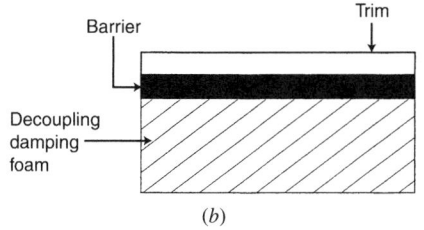

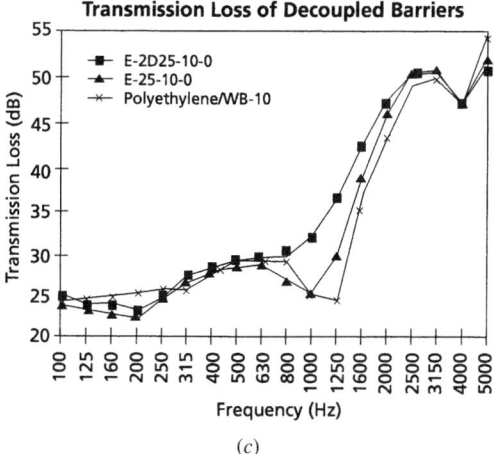

*Fig.* **19.56** *Composite damped, decoupled barrier material comprising a layer of damping foam plus an integral layer of loaded vinyl barrier material. The exterior may be covered (a) with a decorative cover layer. The foam layer acts to damp the vibrating surface to which it adheres (b) and to decouple the barrier layer. Preliminary attenuation test data (c) for a composite consisting of ¼ in. foam and 1 lb/ft². (Courtesy of E-A-R Specialty Composites, div. of AEARO Co.)*

vibrating (sheet) metal enclosures, as for instance on laundry machines, mixers, bins, chutes, polishing drums, and the like, a very effective noise control technique is to damp the vibration. This can be done by permanent attachment of a layer of foam to the vibrating metal, which, by minute flexing of the foam mass, converts the vibration energy to heat. The noise reduction can be further enhanced by adding a heavy, limp barrier material to the outside of the foam. Such combinations are called *composite damping barrier materials,* several of which are illustrated in Fig. 19.56. The foam layer also acts to decouple the barrier and thereby increase the noise attenuation.

## STC AND IIC RECOMMENDATIONS AND CRITERIA

Recommendations for background noise levels (NC criteria) are given in Table 19.8. Criteria for partition isolation (STC) and impact isolation (IIC) are given in the following sections and in Table 19.13.

### 19.33 MULTIPLE-OCCUPANCY RESIDENTIAL STC/IIC CRITERIA

The most important acoustical design criteria for residential work in the United States are issued by the Department of Housing and Urban Development in conjunction with the Federal Housing Administration (HUD/FHA). The reader is referred to the latest issue of *A Guide to Airborne, Impact and Structure-Borne Noise Control* and to subsequent HUD/FHA publications. Tables 19.14 and 19.15 give the essential data presented in the current edition of the *Guide.*

The recommendations are divided into grades I, II, and III. Grade II is the most important category and is applicable primarily in residential urban and suburban areas considered to have an average noise environment. The nighttime exterior noise levels may be about 40 to 45 dBA, and the permissible interior noise environment characteristics should not exceed NC 25–30. Grade I is suburban, with a quiet noise environment characterized by a nighttime exterior noise level of about 35 to 40 dBA. *Grade I STC/IIC criteria are 3 points higher than those of grade II.*

enclosures are not normally the responsibility of the building designer. It is, however, important to know that they will be used, as well as their characteristics, so that appropriate isolation can be designed into the building for the residual sound that is radiated from the enclosures.

Where a noise is at least partially the result of

**TABLE 19.13 Recommended STC for Partitions; Specific Occupancies**

| Type of Occupancy | Wall, Partition, or Panel Between | | | Sound Isolation Requirement: Background Level in Room Being Considered | |
|---|---|---|---|---|---|
| | Room Being Considered | and | Adjacent Area | Quiet | Normal |
| Normal school buildings without extraordinary or unusual activities or requirements | Classrooms | | Adjacent classrooms | STC 42 | STC 40 |
| | | | Corridor or public areas | STC 40 | STC 38 |
| | | | Kitchen and dining areas | STC 50 | STC 47 |
| | | | Shops | STC 50 | STC 47 |
| | | | Recreation areas | STC 45 | STC 42 |
| | | | Music rooms | STC 55 | STC 50 |
| | | | Mechanical equipment rooms | STC 50 | STC 45 |
| | | | Toilet areas | STC 45 | STC 42 |
| | Music practice rooms | | Adjacent practice rooms | STC 55 | STC 50 |
| | | | Corridor and public areas | STC 45 | STC 42 |
| Executive areas, doctors' suites; confidential privacy requirements | Office | | Adjacent offices | STC 50 | STC 45 |
| | | | General office areas | STC 48 | STC 45 |
| | | | Corridor or lobby | STC 45 | STC 42 |
| | | | Washrooms and toilet areas | STC 50 | STC 47 |
| Normal office; normal privacy requirements; any occupancy using rooms for group meetings | Office | | Adjacent offices | STC 40 | STC 38 |
| | | | Corridor, lobby, exterior | STC 40 | STC 38 |
| | | | Washrooms, kitchen, dining | STC 42 | STC 40 |
| | Conference rooms | | Other conference rooms | STC 45 | STC 42 |
| | | | Adjacent offices | STC 45 | STC 42 |
| | | | Corridor or lobby | STC 42 | STC 40 |
| | | | Exterior of building | STC 40 | STC 38 |
| | | | Kitchen and dining areas | STC 45 | STC 42 |
| Large offices, drafting areas, banking floors, etc. | Large general office areas | | Corridors, lobby, exterior | STC 38 | STC 35 |
| | | | Data-processing area | STC 40 | STC 38 |
| | | | Kitchen and dining areas | STC 40 | STC 38 |
| Motels and urban hotels, Hospitals and dormitories | Bedrooms | | Adjacent bedrooms[a] | STC 52 | STC 50 |
| | | | Bathroom[a] | STC 50 | STC 45 |
| | | | Living rooms[a] | STC 45 | STC 42 |
| | | | Dining areas | STC 45 | STC 42 |
| | | | Corridor, lobby, or public spaces | STC 45 | STC 42 |

*Source:* Courtesy of U.S. Gypsum.

[a]Separate occupancy.

The fundamental criteria for airborne sound insulation between dwelling units are, for grade II:

Wall partitions $\qquad$ STC > 52
Floor-ceiling assemblies $\qquad$ IIC > 52

These apply where similar function spaces are contiguous, such as bedroom to bedroom and living room to living room. Where this is not the case, the isolation must be increased to meet the higher sensitivity requirement.

Grade III recommendations are minimal and can be characterized as noisy, with an average nighttime exterior noise level of about 55 dBA or higher. *Grade III STC/IIC recommendations are 4 points lower than those of grade II.*

## 19.34 SPECIFIC OCCUPANCIES

### (a) Schools

School buildings house spaces of many kinds—classrooms, auditoriums, gymnasiums, cafeterias,

**TABLE 19.14 Criteria for Airborne Sound Insulation of Partitions Between Dwelling Units**

| Partition Function Between Dwellings | | | |
|---|---|---|---|
| Apt. A | | Apt. B | Grade II STC |
| Bedroom | to | Bedroom | 52 |
| Living room | to | Bedroom[a] | 54 |
| Kitchen[b] | to | Bedroom[a] | 55 |
| Bathroom | to | Bedroom[a] | 56 |
| Corridor | to | Bedroom[a,c] | 52 |
| Living room | to | Living room | 52 |
| Kitchen[b] | to | Living room[a] | 52 |
| Bathroom | to | Living room | 54 |
| Corridor | to | Living room[a,c,d] | 52 |
| Kitchen | to | Kitchen[e] | 50 |
| Bathroom | to | Kitchen | 52 |
| Corridor | to | Kitchen[a,c,d] | 52 |
| Bathroom | to | Bathroom | 50 |
| Corridor | to | Bathroom[a,c] | 48 |

*Source:* Reprinted from *A Guide to Airborne, Impact, and Structure-borne Noise Control in Multifamily Dwellings* (1968). For Grade I, add 3 points; for Grade III, subtract 4 points.

[a]Whenever a partition wall may serve to separate several functional spaces, the highest criterion must prevail.

[b]Or dining or family or recreation room.

[c]It is assumed that there is no entrance door leading from the corridor to the living unit.

[d]Criterion applies to the partition. Doors in corridor partitions must have the rating of the partition, not vice versa.

[e]Double wall construction is recommended to minimize kitchen impact noises.

**TABLE 19.15 Criteria for Airborne and Impact Sound Insulation of Floor-Ceiling Assemblies Between Dwelling Units**

| Assembly Function Between Dwellings | | | Grade II | |
|---|---|---|---|---|
| Apt. A | | Apt. B | STC | IIC |
| Bedroom | above | Bedroom | 52 | 52 |
| Living room | above | Bedroom[a] | 54 | 57 |
| Kitchen[b] | above | Bedroom[a,c] | 55 | 62 |
| Family room | above | Bedroom[a,d] | 56 | 62 |
| Corridor | above | Bedroom[a] | 52 | 62 |
| Bedroom | above | Living room[e] | 54 | 52 |
| Living room | above | Living room | 52 | 52 |
| Kitchen | above | Living room[a,c] | 52 | 57 |
| Family room | above | Living room[a,d] | 54 | 60 |
| Corridor | above | Living room[a] | 52 | 57 |
| Bedroom | above | Kitchen[c,e] | 55 | 50 |
| Living room | above | Kitchen[c,e] | 52 | 52 |
| Kitchen | above | Kitchen[c] | 50 | 52 |
| Bathroom | above | Kitchen[a,c] | 52 | 52 |
| Family room | above | Kitchen[a,c,d] | 52 | 58 |
| Corridor | above | Kitchen[a,c] | 48 | 52 |
| Bedroom | above | Family room[e] | 56 | 48 |
| Living room | above | Family room[e] | 54 | 50 |
| Kitchen | above | Family room[e] | 52 | 52 |
| Bathroom | above | Bathroom[c] | 50 | 50 |
| Corridor | above | Corridor | 48 | 48 |

*Source:* Reprinted from *A Guide to Airborne, Impact, and Structure-borne Noise Control in Multifamily Dwellings* (1968). For Grade I, add 3 points; for Grade III, subtract 4 points.

[a]This arrangement requires greater impact sound insulation than the converse, where a sensitive area is above a less sensitive area.

[b]Or dining or family or recreation room.

[c]It is assumed that plumbing fixtures, appliances, and piping are installed with proper vibration isolation.

[d]The airborne STC criteria in this table apply as well to vertical partitions between these two spaces.

[e]This arrangement requires equivalent airborne sound insulation and perhaps less impact sound insulation than the converse.

shop areas, swimming pools, and music suites—that pose acoustics problems.

1. *Auditoriums.* All auditoriums require a sound system for some of the activities accommodated (see Figs. 18.24 and 18.27). The most difficult aspect, architecturally, is integration of a loudspeaker system into the design. To provide proper sound reinforcement, loudspeakers must be located properly without large obstructions. To accomplish this, the loudspeaker system should be incorporated in the earliest design stages.

In general, a school auditorium is a multipurpose facility. It should be designed to meet speech requirements and also should be suitable for the school's music activities. Often a modified gymnasium (gymnatorium) or cafeteria (cafetorium) functions as an auditorium. Obviously, acoustic compromises occur in such facilities. Large areas of sound-absorbing treatment in either kind of space make them unsuitable as auditoriums and for most events that require speech amplification.

2. *Classrooms.* Typical classrooms are approximately 30 ft square with 10-ft ceilings. Adequate speech communication is easily achieved in a room of this size. Classroom acoustic design usually involves:

a. Locating sound-absorbing treatment to reduce classroom noise levels

b. Ensuring adequate privacy between adjacent spaces

c. Control of air-handling system noise

Acoustic tile ceilings provide adequate sound absorption for most classrooms. An NRC of 0.7 is recommended (see Section 18.10).

Partition systems must produce sufficient isolation to prevent disturbance from activities in other classrooms and corridors. Such partitions should

ACOUSTICS

run full height from floor to ceiling slab or roof construction. If return air transfer ducts are needed, their noise reduction characteristics must be as good as those of the walls or doors that they penetrate (for NC data, see Table 19.8). Unit ventilators commonly used for classrooms produce approximately the required level of background sound.

3. *Music Suites.* School music programs usually range from individual instruction to band and choral concerts. The teaching spaces required for such a program include practice rooms, ensemble rooms, and large rehearsal spaces. Both room acoustics design and sound isolation are important in music suites. Privacy between adjacent spaces is critical, since simultaneous use is necessary.

4. *Dining Areas.* The activity in cafeterias or lunchrooms usually generates a great deal of noise. The kitchen and serving areas should be separated from the eating spaces. Ceilings and wall areas in the cafeteria should be treated with sound-absorbing material. Unless the ceiling is completely treated with a highly efficient sound-absorbing material, the environment will be unsatisfactory due to its high noise level. The minimum NRC of this material should be 0.8.

5. *Gymnasiums.* Activities in gymnasiums create so much noise that even extensive treatment will not quiet these spaces. A quiet gymnasium probably would be unsatisfactory in any case, since spectators are conditioned to consider the noise as an enjoyable aspect of athletic events. However, to provide a proper environment for normal sports activities, the ceiling area should absorb sound. In addition, if a sound amplification system is to be used, sound-absorbing wall treatment may be required to eliminate echoes that would reduce the intelligibility of announcements. An NRC of 0.7 is suggested with sound-absorbent material to be ceiling-mounted.

If a gymnasium will also serve as an auditorium, loudspeaker system placement requires special consideration. For example, the loudspeakers should be located above the source location for speeches and plays.

6. *Swimming Pools.* The acoustic environment of swimming pools is often chaotic. Most sound-absorbing materials disintegrate in the high-humidity conditions prevalent in pool areas. Use

special sound-absorbing units that have moisture-resistant properties.

7. *Shops.* Metal, woodworking, and scenery shops in schools contain many noise sources—saws, planers, drill presses, and manual tools. Each generates high airborne and structure-borne noise levels. Consolidating noisy areas and maximizing the distance between them and quiet spaces are essential. Ceiling and wall absorptive treatment with an NRC of at least 0.75 is recommended.

### (b) Houses of Worship

The basic activities of these buildings usually combine speech and music. Thus, the worship environment must be acoustically hospitable to both. The architectural plan also must respond to religious requirements including the relative positioning of pulpits, lecterns, the altar (if any), and the choir.

Successful acoustics can be achieved by designing the overall environment for music and providing special assistance for speech. Large congregational spaces frequently include a sound-reflecting canopy over the pulpit to direct the speaker's voice to the congregation. In some large buildings, a loudspeaker located above the canopy further reinforces speech from the pulpit. The choir and organ communicate with the entire volume of the building and, therefore, benefit from a reverberant environment.

### (c) Offices

Although office buildings may contain public spaces, auditoriums, and restaurants, prime occupancy is in office areas. Most acoustics problems in office buildings relate to privacy—either between spaces within a single firm or between adjacent firms. Speech privacy has been discussed at length in Sections 19.16 through 19.21, including consideration of open-plan offices. Mechanical and electrical equipment noise problems are discussed in Sections 19.25 through 19.32.

### (d) Apartment Buildings

Large apartment buildings house hundreds and even thousands of residents. Privacy and freedom from annoyance are high on the list of tenant requirements. See the HUD/FHA criteria in Section 19.33.

The performance of partitions is compromised in many designs by careless planning of convenience outlets, medicine cabinets, and mechanical services. Direct-exhaust duct connections between apartments and back-to-back placement of medicine cabinets result in loss of privacy. Back-to-back convenience outlets must be avoided.

Installation of rugs or carpeting provides the best protection against footfall noise. Many leases now require that a tenant provide such impact-reducing floor covering over most of the floor area in an apartment. Good design also dictates that similar spaces in adjacent apartments be grouped—bedrooms next to bedrooms, for example. Absorptive material in bedrooms should be ceiling mounted. A minimum NRC of 0.6 is recommended.

Apartment house site selection seldom includes consideration of acoustics. Nevertheless, truck routes, superhighways, and airports can be annoying "neighbors." Cooling towers serving adjacent buildings must be considered during the planning stages.

## OUTDOOR ACOUSTIC CONSIDERATIONS

### 19.35 SOUND POWER AND PRESSURE LEVELS IN FREE SPACE (OUTDOORS)

The equations in Section 18.8 are not applicable to outdoor sound propagation, in which the large reflective component of the indoor condition is absent. Although the propagation of sound outdoors may not appear to be of immediate importance in architectural acoustics, outdoor noise sources such as traffic, cooling towers, and aircraft are frequently loud enough to disturb activities within or immediately adjacent to a building. Conversely, the noise made by building equipment such as cooling towers, heat pumps, and even window air conditioners may be loud enough to disturb

neighbors in a nearby building. For this reason, it is desirable to have some basic understanding of outdoor sound propagation.

For preliminary evaluation of an outdoor noise problem, assuming a small nondirectional source on the ground, the SPL can be determined from Eqs. 18.5 and 18.6. For large sources such as cooling towers and traffic, which do not exhibit inverse square properties, sound level estimates are best made on the basis of experience and empirical data beyond the scope of this book (see Magrab, 1975; Schaudinischky, 1976). For small outdoor sources, the equipment power level can be estimated by measuring the sound pressure level at 5 ft and adding 15 dB. Other factors (such as moisture in the air, the presence of trees, wind, and temperature gradients) will affect outdoor sound propagation to some extent, but they can be ignored except when great distances (i.e., over 1000 ft) are involved. Barriers, which are the most effective outdoor attenuators, were discussed in Section 19.14.

### 19.36 BUILDING SITING

Building siting, vis-à-vis exterior noise sources, is as important as interior structural design. Since this subject is somewhat beyond our scope of concern, the discussion is brief. Buildings should be sited, with respect to noise sources:

1. To use natural terrain noise barriers (Fig. 19.57a).
2. Regarding trees as noise barriers, rely only on thickly wooded areas (Fig. 19.57b).
3. To avoid naturally poor sites (Fig. 19.57c).
4. To avoid sound reflection from other buildings (Fig. 19.57d).

Point 4 above is also important in a multiwing building, in avoiding U-shapes or other configurations where a central court becomes an echo chamber.

Where avoidance of an exterior noise source is impossible, quiet zones can be buffered from noises by placing higher-noise areas on the noisy side of a building. Thus, in a school, classrooms and offices can be buffered by a cafeteria and gym; in a residence, bedrooms by living rooms and corridors; in an office building, private offices by noisier clerical offices; and so on.

ACOUSTICS

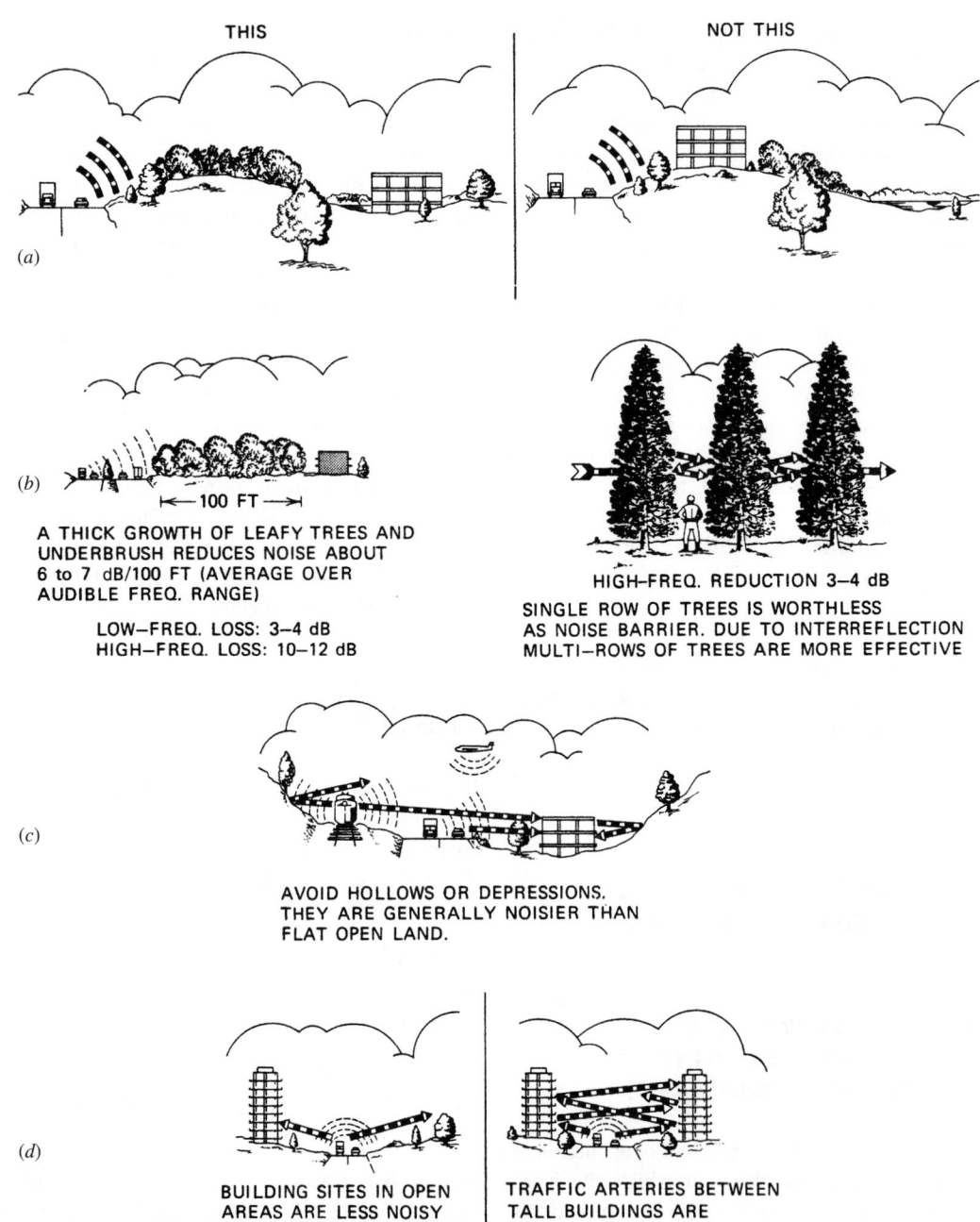

**Fig. 19.57** (a) Use of natural noise barriers. (b) Effectiveness of wooded areas as noise barriers, showing noise reduction of trees. (c) An example of a poor building site. (d) Building site issues near traffic arteries and other buildings. (Reprinted from A Guide to Airborne, Impact, and Structure-Borne Noise Control in Multifamily Dwellings, 1968.)

# REFERENCE MATERIAL

## 19.37 GLOSSARY

**A-Scale.** A filtering system with characteristics that roughly match the response characteristics of the human ear. Referred to as dBA.

**Absorption Coefficient α.** The ratio of the sound absorbed to the sound incident on a material or device.

**Anechoic Room.** A room that provides a free field acoustic testing environment like the outdoors. All the sound emanating from a source is essentially absorbed at the surfaces of the room.

**ANSI.** American National Standards Institute, a nonprofit national technical association that publishes standards covering definitions, test methods, recommended practices, and specifications of materials. Formerly the American Standards Association (ASA) and the United States of America Standards Institute (USASI).

**Articulation Class (AC).** A figure of merit for acoustic absorption indicating the absorption efficiency of a material for angles of incidence of the sound wave between 45° and 55°. The range is 150–250 (no units), with higher numbers indicating better absorption.

**Articulation Index (AI).** A numerical index in the range of 0.0–1.0 indicating the degree of speech intelligibility in an open-office design, measured with background noise, including masking sound, if any. Also a measure of speech privacy. An index of 0.0 indicates no speech intelligibility, hence ideal speech privacy; an index of 1.0 indicates perfect speech intelligibility and therefore no speech privacy.

**ASTM.** Formerly the American Society for Testing and Materials, now ASTM, a nonprofit national technical society that publishes definitions, standards, test methods, recommended installation practices, and specifications for materials. ASTM is a consensus group of the entire building materials industry. It sets standards for products and establishes methods for testing.

**Baffle or Barrier, Sound.** A shielding structure or partition used to increase the effective length of a sound transmission path between two locations.

**Ceiling Attenuation Class (CAC).** A single number or range of numbers for evaluating the effectiveness of an acoustical ceiling construction in isolating audible airborne sound transmission, tested at 16 one-third-octave frequencies. Higher numbers indicate more effectiveness in preventing noise between rooms. Tested in accordance with ASTM E1414. Previously described as the Ceiling Sound Transmission Class (CSTC) or Sound Transmission Class (STC).

**Critical Frequency.** The lowest frequency at which the wavelength of a bending wave, traveling in a structure, is the same as the wavelength in air at that frequency.

**Damping.** Dissipation of structure-borne noise. This is usually accomplished by using a material with a high internal energy-absorbing capacity (i.e., high internal damping).

**Decibel (dB).** A measurement approach adopted for convenience in representing vastly different sound quantities.

**Diffraction.** The tendency of sound waves to flow readily around obstacles that are small in comparison to the wavelength of the sound.

**Diffuse Sound Field.** A region where sound at any given point is made up of sound waves with all angles of incidence.

**Direct Sound Field.** A region in which all or most of the sound arrives directly from the source without reflection.

**FIIC.** Field Impact Insulation Class, which is determined by an actual field test. Also see *FSTC*, Field Sound Transmission Class, also determined by an actual field test.

**Flanking Sound Path.** The transmission of sound or noise from one room to another by indirect paths rather than directly through an intervening partition.

**Flutter.** A multiple echo set up between parallel reflecting surfaces.

**Free Sound Field (Free Field).** A region in a homogeneous medium free from boundaries.

In a free field, the sound pressure level decreases 6 dB for a doubling of the distance from a point source.

**FSTC.** Field Sound Transmission Class, which is determined by an actual field test performed per ASTM E336. Also see *FIIC*, Field Impact Insulation Class, also determined by an actual field test.

**Impact Isolation (Insulation) Class (IIC).** A single-figure rating that provides an estimate of the impact sound-isolating performance of a floor-ceiling assembly.

**Intensity.** The amount of sound energy per second that is carried across a unit area.

**Intensity Level (IL).** A measure of the acoustic power passing through a unit area expressed in the decibel scale and referenced to some standard base (usually $10^{-12}$ W/cm$^2$).

**Interference.** The destructive or reinforcing action of two or more waves arriving at the same position simultaneously.

**Loudness.** A subjective human definition of the intensity of a sound.

**Masking.** The presence of a background sound increases the level to which a sound signal must be raised in order to be heard or distinguished. If the level of the background sound is significantly higher than that of the sound signal, the signal cannot be heard. This effect is known as masking.

**Mass Law.** States that the transmission loss of walls (in part of the frequency range) is controlled entirely by the mass per unit area of the panel. It also states that the transmission loss increases 6 dB for each doubling of frequency or each doubling of the panel mass per unit area.

**Noise.** Any undesired sounds, usually of different frequencies, resulting in an objectionable or irritating sensation.

**Noise Reduction (NR).** (1) The reduction in sound pressure level caused by making an alteration to a sound source. (2) The difference in sound pressure level measured between two adjacent rooms caused by the transmission loss of an intervening barrier.

**Noise Reduction Coefficient (NRC).** The average sound absorption coefficient (to the nearest .05) measured at the four one-third octave bands

centered on frequencies of 250, 500, 1000, and 2000 Hz.

**Octave Band.** A range of frequency where the highest frequency of the band is double the lowest frequency. The band is usually specified by its center frequency.

**Phon.** Loudness level, at a particular frequency, equal to the 1000-Hz decibel level of that equal-loudness contour.

**Pink Noise.** Wide spectrum noise whose amplitude drops 3 dB per octave with increasing frequency (equal energy per octave). Useful for masking.

**Random Noise.** A noise whose magnitude and/or frequency cannot be predicted precisely at any given time. A rough approximation of random noise is the static heard on a radio between stations (see *Noise, Pink Noise, White Noise*).

**Reverberation.** A persistence or echoing of previously generated sound caused by reflection of acoustic waves from the surfaces of enclosed spaces.

**Reverberation Time.** The time required for a sound to decay to a value one-millionth of its original intensity or to reduce 60 dB after the sound source has stopped.

**Sabin.** The unit of acoustic absorption. One Sabin (ft$^2$) (m$^2$) is the absorption of 1 ft$^2$ (m$^2$) of perfect sound-absorbing material or open space with no reflecting surfaces.

**Sound Absorption Coefficient.** The fraction of the incident energy absorbed (not reflected) by a material when a sound wave strikes it is the sound absorption coefficient of that material. Usually represented by the Greek letter alpha ($\alpha$).

**Sound Barrier.** A material installed to prevent the passage of sound from one area to another. Sound-deadening board and lead sheet or special insulation make good sound barriers.

**Sound Level Meter.** An instrument for the direct measurement of sound pressure level. Sound level meters may also incorporate octave-band filters for measuring sound directly in octave bands.

**Sound Power Level (PWL).** A measure of the total airborne acoustic power generated by a noise source, expressed in the decibel scale and referenced to some standard base (usually $10^{-12}$ W).

**Sound Pressure Level (SPL).** A measure of the air pressure change caused by a sound wave. Expressed in the decibel scale and referenced to some standard base (usually 0.0002 μ bar).

**Sound Transmission Class (STC).** A single-number rating of a building element's efficacy in blocking the transmission of sound compared to a standard transmission attenuation/frequency curve. See also *Ceiling Attenuation Class.*

**Transmission Loss (TL).** The reduction of airborne sound power caused by placing a wall or barrier between the reverberant sound field of a source and its receiver. Transmission loss is a property of the wall or barrier.

**White Noise.** Noise of a wide frequency range in which the amplitude of the noise is essentially the same in all frequency bands (equal energy per frequency band).

## 19.38 REFERENCE STANDARDS

See Section 19.11*h* for a description of the contents of these standards.

ASTM C423-02a. *Standard Test Method for Sound Absorption and Sound Absorption Coefficients by the Reverberation Room Method*

ASTM E90-04. *Standard Test Method for Laboratory Measurement of Airborne Sound Transmission Loss of Building Partitions and Elements*

ASTM E336-97e1. *Standard Test Method for Measurement of Airborne Sound Insulation in Buildings*

ASTM E413-04. *Classification for Rating Sound Insulation*

ASTM E795-00. *Standard Practices for Mounting Test Specimens During Sound Absorption Tests*

ASTM E1110-01. *Standard Classification for Determination of Articulation Class*

ASTM E1111-02. *Standard Test Method for Measuring the Interzone Attenuation of Ceiling Systems*

ASTM E1130-02e1. *Standard Test Method for Objective Measurement of Speech Privacy in Open Offices Using Articulation Index*

ASTM E1179-87(2003). *Standard Specification for Sound Sources Used for Testing Open Office Components and Systems*

ASTM E1264-98. *Standard Classification for Acoustical Ceiling Products*

ASTM E1374-02. *Standard Guide for Open Office Acoustics and Applicable ASTM Standards*

ASTM E1375-90(2002). *Standard Test Method for Measuring the Interzone Attenuation of Furniture Panels Used as Acoustical Barriers*

ASTM E1376-90(2002). *Standard Test Method for Measuring the Interzone Attenuation of Sound Reflected by Wall Finishes and Furniture Panels*

ASTM E1414-00a. *Standard Test Method for Airborne Sound Attenuation Between Rooms Sharing a Common Ceiling Plenum*

ASTM E1573-02. *Standard Test Method for Evaluating Masking Sound in Open Offices Using A-Weighted and One-third Octave Band Sound Pressure Levels*

ISO 717-1, 1996. (International Organization for Standardization). *Acoustics—Rating of Sound Insulation in Buildings and of Building Elements— Part 1: Airborne Sound Insulation*

ISO 717-2, 1996. *Acoustics—Rating of Sound Insulation in Buildings and of Building Elements— Part 2: Impact Sound Insulation.*

## 19.39 UNITS AND CONVERSIONS

See Table 19.16.

**TABLE 19.16 Acoustic Units and Conversions**

| Variable | MKS units | CGS units |
|---|---|---|
| Force | kilogram-meter/s$^2$ = newton | gram-cm/s$^2$ = dyne |
| Intensity | watts/meter$^2$ | watts/cm$^2$ |
| Pressure | newton meter$^2$ = pascals | dynes/cm$^2$ = microbars |

| | In Conversion: | | |
|---|---|---|---|
| Quantity | Multiply | By | To Obtain |
| Force | newtons | $10^5$ | dynes |
| | dynes | $10^{-5}$ | newtons |
| Intensity | watts/cm$^2$ | $10^4$ | watts/m$^2$ |
| | watts/m$^2$ | $10^{-4}$ | watts/cm$^2$ |
| Pressure | pascals | 10 | microbars |
| | microbar | $10^{-1}$ | pascals |

*Note:* One atmosphere = 1 bar = $10^6$ μbar.

## 19.40 SYMBOLS

See Table 19.17.

**TABLE 19.17 Symbols and Abbreviations Used Commonly in Acoustics**

| | |
|---|---|
| $A$ | Total absorption, Sabins; area in unit being used |
| $A_R$ | Absorption in receiving room, Sabins |
| $A_{1,2...}$ | Total absorption of each material in a space, Sabins |
| AC | Articulation Class |
| $c$ | Velocity of sound, feet per second |
| CAC | Ceiling Attenuation Class |
| $d$ | Distance from source, meters or feet |
| dB | Decibel |
| $f$ | Frequency of sound, Hertz (Hz) |
| $I$ | Intensity, W/cm$^2$ |
| $I_a$ | Absorbed energy, W/cm$^2$ |
| $I_i$ | Incident energy, W/cm$^2$ |
| $I_0$ | Reference intensity, $10^{-16}$ W/cm$^2$ |
| IIC | Impact Insulation Class, no units |
| IL | Intensity level, decibels |
| NC | Noise criterion, no units |
| NRC | Noise reduction coefficient, no units |
| $NR$ | Noise reduction, decibels |
| $p$ | Pressure, pascals or microbars |
| $p_0$ | Reference base pressure, $2 \times 10^{-5}$ Pa |
| $P$ | Acoustic power, Watts |
| Pa | Pascal, unit of pressure (SI) |
| $PWL$ | Sound power level, decibels |
| $R$ | Room constant, square feet (meters) |
| $r$ | Distance from source, meters or feet |
| $S$ | Surface area, in unit being used |
| $SPL$ | Sound pressure level, decibels |
| STC | Sound Transmission Class, no units |
| $T_R$ | Reverberation time, seconds |
| $TL$ | Transmission loss, decibels |
| $V$ | Volume (geometric) |
| $W, P$ | Sound power, watts |
| $W_0$ | Reference base sound power, $10^{-12}$ W |
| $\alpha, \alpha$ | Absorption coefficient (no units) |
| $\bar{a}, \bar{\alpha}$ | Average absorption coefficient (no units) |
| $\lambda$ | Wavelength, feet or meters |
| $\Sigma$ | Sum of, or total (no units) |
| $\Sigma S\alpha = \Sigma A$ | Total absorption, Sabins |
| $\Delta$ | Change in a quantity or difference between two quantities |

*Note.* Where definitions are expressed in feet, centimeters or meters are also understood, with proper conversion factors, and vice versa.

### References

ASHRAE. 2003. *ASHRAE Handbook—HVAC Applications.* American Society of Heating, Refrigerating and Air-Conditioning Engineers, Atlanta, GA.

Cavanaugh, W.J., W.R. Farrell, P.W. Hirtle, and B.G. Watters. 1962. "Speech Privacy in Buildings," in *Journal of the Acoustical Society of America,* Vol. 34, No. 4.

Fader, B. 1981. *Industrial Noise Control.* John Wiley & Sons, New York.

HUD. 1968. *A Guide to Airborne, Impact, and Structure-Borne Noise Control in Multifamily Dwellings.* U.S. Department of Housing and Urban Development, Washington, DC.

Magrab, E.B. 1975. *Environmental Noise Control.* John Wiley & Sons, New York.

Schaudinischky, L.H. 1976. *Sound, Man and Building.* Applied Science Publishers, London.

Young, R.W. 1965. "Revision of the Speech Privacy Calculation," in *Journal of the Acoustical Society of America,* Vol. 38, No. 4.

ACOUSTICS

# WATER AND WASTE

W ater and its appropriate use are becoming an increasingly important part of design, especially the design of green buildings. Almost every North American building designed today is supplied with potable (drinkable) water. In the vast majority of these buildings, *most* of this clean water is used to carry away organic waste. The result is a wide range of design impacts covering everything from the detailed arrangement of bathroom fixtures and interior surfaces to the overall plans of very large, complex water and sewage treatment facilities.

This part of the book presents information on the following topics: basic planning information and rainwater (Chapter 20); water supply (Chapter 21); organic waste disposal, both waterborne and waterless (Chapter 22); and solid waste design issues (Chapter 23).

Fire protection, which often requires the most extensive water supply piping network within a building, is covered in Part VI.

# Water and Basic Design

"**THE NEXT GREAT WORLD CRISIS WILL BE** water supply." This prediction is becoming increasingly frequent as countries downstream threaten war over water rights with countries upstream, and water withdrawals in coastal areas allow saltwater intrusions into aquifers. State against state, county against county, water rights continue to cause problems. With a finite planetary water supply pitted against an increasing population and, worse yet, an increasing *per capita* consumption of water, we see again (as with fossil fuels) the problem of limited resources versus growing demand. At least in this case the amount of water is fixed, not diminishing. However, the problem of fair allocation remains (Fig. 20.1). Countries have fought wars over oil; must they also wage war over water?

Agricultural and industrial use of water may dwarf that of buildings, but designers still have a role in this dilemma. We can use water beautifully where it is appropriate and avoid its use when possible.

## 20.1 WATER IN ARCHITECTURE

Throughout history, in nearly all climates and cultures, the designer's major concern about water was how to keep it *out* of a building. Only since the end of the nineteenth century has a water supply *within* a building become commonplace in industrialized countries. In the rest of the world today, running water is still not available within most buildings. Water's potential contributions to lifestyle

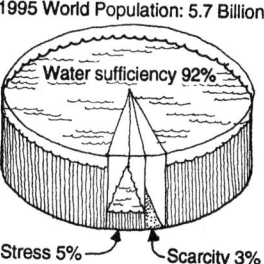

1995 World Population: 5.7 Billion

Water sufficiency 92%

Stress 5% — Scarcity 3%

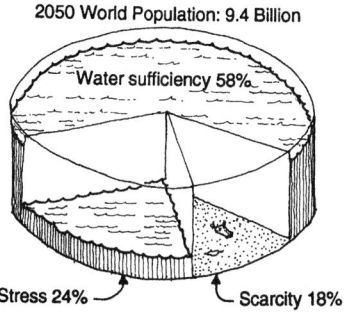

2050 World Population: 9.4 Billion

Water sufficiency 58%

Stress 24% — Scarcity 18%

**Fig. 20.1** *Earth's present and projected future annual per capita fresh water supply for all purposes. The relative size of the graphs reflects population growth. A "sufficient" supply is 1700 m³ (449,000 gal) per person; the "stress" supply is between 1000 and 1700 m³ (264,180 and 449,000 gal) per person; the "scarcity" supply is less than 1000 m³ (264,180 gal) per person. Currently, worldwide water use is 70% irrigation, 20% commercial-industrial, and 10% residential. (*Worldwatch*, State of the World, 1999; courtesy of Population Action International.)*

and architecture are as numerous and varied as appropriate design responses to the supply, use, and return of this versatile commodity.

## (a) Nourishment

Much of the human body is water, the most abundant chemical in our bodies as well as in our diet. The amount of pure (potable) water that we need for drinking and cooking is very small—only about 3 g/cd (gallons per capita per day) (11.4 L/cd) in the United States. The most common supply system throughout history has been the central municipal fountain or well (Fig. 20.2), whose technical importance to the community has often been emphasized by the aesthetics of both the fountain's overall sculptural composition and the elegance of detail in water spouts, basins, and other elements. The social importance of the fountain is evident from its loca-

**Fig. 20.2** *The fountain in the central piazza at Tarquinia, Italy, still serves as a gathering place and as a symbol of the town's provision of a basic need.*

tion in the central plaza, from where small amounts of water are carried daily by townspeople to homes or workplaces. Opportunities abound for conversations around the fountain.

As potable water became available on demand within buildings, water-related social opportunities diminished. However, employees still converse around the drinking fountain—for example, when it is located in a place that invites lingering or relaxing.

## (b) Cleansing and Hygiene

Water is a nearly ideal medium for the dissolution and transport of organic waste, and because of its high heat-storage capacity, it easily retains comfortable temperatures for bathing. Much larger quantities of water are used for cleaning than for nourishment: in the average U.S. home, about 14 g/cd (53 L/cd) is used for clothes washing and dishwashing, and another 21 g/cd (79.5 L/cd) is used for bathing and personal hygiene.

In the past, water for cleaning was carried to the home infrequently; the Saturday night (only) bath was typical well into the twentieth century in the United States. Bathing vessels were usually portable and sometimes were combined with other pieces of furniture (a couch that sat over a tub, a metal tub that folded up inside a tall wooden cabinet, etc.). Thus, a *bathplace* rather than a bathroom was the common design response.

Although the physical constraints of water carrying were important design influences on bathing, cultural attitudes were at least as strong an influence throughout history. Perhaps the most startling contrast is between the medieval concept of bathing as an almost sinful indulgence and the earlier Roman attitude toward public baths as the social centers of cities.

Today bathing facilities are commonly designed to be used on a personal scale, in privacy. There also are welcome opportunities to design more social bathing (recreational rather than cleansing) places, such as swimming pools, bath houses, and hot tubs. The characteristics of the water supply (spouts, jets, cascades) can be matched with those of the water body (mirror-smooth, gently flowing, rippled, rolling, foaming) to obtain the desired atmosphere.

## (c) Ceremonial Uses

Largely through its associations with cleaning, water acquired a ceremonial significance that remains particularly evident in religious services. Examples of the ceremonial use of water include vessels containing holy water at entrances to Catholic churches, pools in the forecourts of mosques, and full-immersion baptismal fonts at the altars of some Protestant churches. The opportunities for aesthetic expression are particularly rich in these ceremonial applications.

## (d) Transportation Uses

In stark contrast to its uses in nourishing, cleansing, and celebrating, water is used in our buildings principally to transport organic waste. There is perhaps no more flagrant example of a mismatch in architecture than the high-grade resource of pure water being used for the low-grade task of carrying away a cigarette butt. The typical U.S. home uses 32 g/cd (121 L/cd) just to flush toilets.

In the past, table scraps were commonly fed to animals or composted, and human waste was thrown out of windows (accompanied by warning cries) or deposited in holes below outhouses. Organic waste disposal was thus dependent on either portable vessels or special structures set apart from the typical building.

As water supplies were developed, water's advantages over the foul smell and inconvenience of these methods became irresistible. A typical sequence of events unfolded on Manhattan Island. In the 1700s, Manhattan was farm country that, like all other areas that later developed into large cities, had minimal water needs. Potable water was available in shallow wells and from some springs and streams. These sources were largely unaffected by the minor ground pollution from widely separated dry-pit privies (outhouses) that received human wastes. Paved city streets appeared in the 1800s, at which time the natural streams were enclosed in pipes called *storm sewers*. These pipes led the rainfall to the many waterways surrounding the island. Then in the later 1800s, flush toilets appeared. It seemed natural to connect the toilets to the already established storm sewers and to rename the pipeways *combined sewers*, which now carried both storm water and so-called sanitary drainage to the rivers (sanitary for the building but not for the rivers). Fast-flowing rivers are natural sewage treatment plants, and, surprisingly, for many decades they did a fair job of keeping pollution reasonably in check. With the prospect of future sewage treatment plants, separate *sanitary sewers* were built. Also, there were some remaining (and some newly built) storm sewers that did not carry the wastes from toilets.

In cities where this confused pattern of sewer systems still exists (including most larger and older cities), it is now extremely difficult and expensive to sort out and reroute sewers so that *only* sanitary drainage goes to treatment plants and *all* storm drainage goes to waterways or into the ground. It seems particularly ironic that in most U.S. locations the rainwater that falls on a home's roof is adequate in both quality and quantity to supply a family's cleansing needs (21 + 14 = 35 g/cd) (132 L/cd). In this chapter and in Chapter 21, these possibilities for rainwater are developed further.

As the human waste disposal place became a room within a building, the design issues grew more complex. Physically, there was a need for running water and for large-diameter pipes that sloped downward continuously from the toilet to a sewer or septic tank. As sewer gas became a recognized problem, an elaborate system of traps and vents became necessary. Again, cultural attitudes were also influential: How private an activity was this elimination from the body? To what extent could/should one plumbing fixture accommodate both body cleansing and waste elimination? If males insisted on standing rather than sitting while urinating, how to devise a toilet that would also properly accommodate defecation, which requires a low seat? Some designers' responses to these questions can be found in Chapter 21.

## (e) Cooling

Water has a remarkable cooling potential: it stores heat readily, removes large quantities of heat when it evaporates, and vaporizes readily at temperatures commonly found at the human skin surface. In hot-dry climates, designers can place water surfaces (or sprays) upwind from the place to be cooled or resort to the evaporative coolers discussed in Section 4.3.

Cooling towers (Section 10.3) are familiar components of large-building cooling systems.

Because all of us have experienced the physical cooling of the skin by water, we all carry psychological associations between water and cooling that can enhance our comfort on hot days. The sight of sunlight reflected on a water surface, with its characteristic "dancing" quality, connotes coolness, as does the sound of running or splashing water. Thus, even when water does not physically cool people, it can make an important psychological contribution to human comfort (Fig. 20.3).

### (f) Ornamental Uses

In almost any landscaping application, indoors or out, water becomes a center of interest. Our association of water with nourishing, cleansing, and cooling make water a very powerful design element—a fact recognized by landscape designers throughout history. In arid regions, water is often used sparingly, in small, tightly controlled channels and at lower flow rates. The gardens of Islamic architecture are especially effective demonstrations of such design restraint. Where water is more plentiful, it has been used lavishly, as at the Villa d'Este in Tivoli, outside Rome, where much of a river's flow is diverted through the gardens.

Especially useful design characteristics of water include its *reflectivity*, which sets it apart from most plant and ground materials in a garden; its *liquidity*, which attracts attention to its motion and creates unique sounds wherever it is moved; and its *life-sustaining* potential, which allows the addition of both water plants and animals to a garden.

### (g) Protective Uses

Every designer dreads water's ability to penetrate a roof and damage a building and its contents. However, we all depend on water as the best fire protection medium available in most buildings. The vast quantities of water potentially required for firefighting must be delivered quickly; the result is pipes of large diameter regulated by very large valves. Because this system's distribution tree must be immediately obvious to firefighters, some degree of exposure is prudent. Despite its size and guarantee of at least partial exposure in public places, a fire protection water supply system is rarely treated as a visually integral design element. This mismatch of potential and actuality is discussed further in Chapter 24.

Another protective use of water has been as a means to control circulation; moats around castles may seem quaint today, but designers still sometimes use water as a means of directing traffic over a bridge to an entry.

### 20.2 THE HYDROLOGIC CYCLE

There is a finite quantity of water in the Earth and its atmosphere. (This statement is challenged by a controversial theory of a steady rain of small water-bearing comets from outer space; see Gleick, 1998.) The process whereby this water constantly circulates, powered by about one-fourth of the Earth's solar energy, is called the *hydrologic cycle* (Fig. 20.4). More than 99% of this water is "inaccessible"—

**Fig. 20.3** Water as an aesthetic feature of a courtyard during the hot-arid season in Colima, Mexico. The sound and sight of running water add to a psychological impression of coolness.

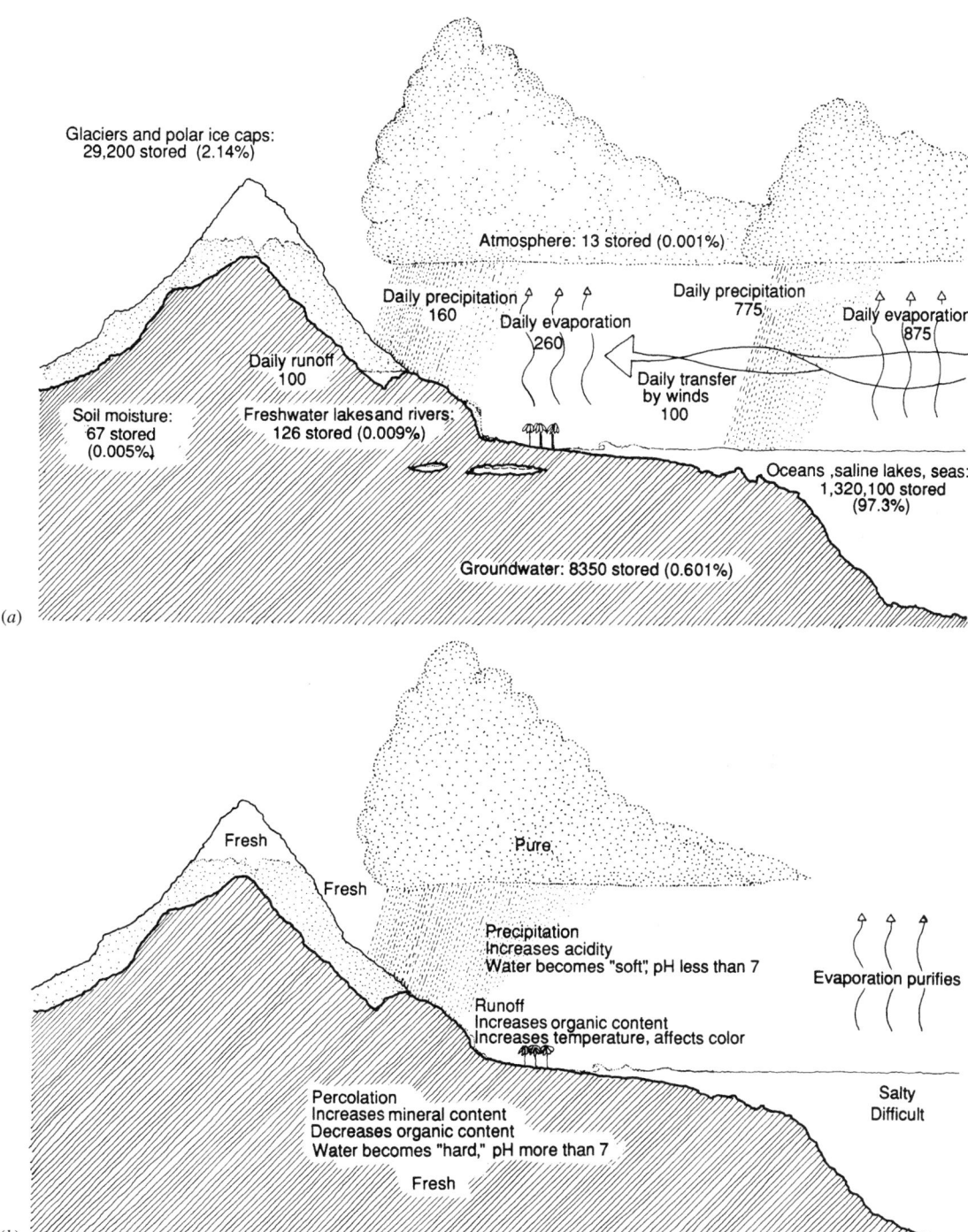

**Fig. 20.4** (a) Hydrologic cycle. The figures given are in cubic kilometers. "Evaporation" includes transpiration from plants as well as evaporation from surfaces. "Precipitation" can be rain, hail, sleet, or snow. The vast majority of stored water is in the ocean. (b) Quality of water at various stages within this cycle.

either because it is saltwater or because it is frozen in glaciers or polar ice caps. The most accessible sources of water are precipitation and runoff.

*Precipitation* has the advantage of relative purity, although acid rain is a growing threat in many parts of the world, including much of the United States and Canada. Like solar energy, precipitation is a very large but very thinly spread resource; its capture is therefore likely to take place on an individual basis. Until we experience a water crisis similar to the energy crisis that began in the 1970s, rainwater capture will remain a mostly untapped resource in the United States.

*Runoff* enjoys the advantage of a concentrated flow of water, which permits easy capture of large quantities. Its most serious disadvantage is the possibility of pollution—organic, chemical, and radioactive—depending on what is upstream from the point of capture. In some regions of North America, river water is reused 50 times on its way to the ocean. Further discussion of water sources and treatments occurs in Chapters 21 and 22.

That part of daily precipitation that neither evaporates nor joins the runoff becomes part of *soil moisture.* Much soil moisture is used by growing plants and is soon transpired (evaporated) by the plant to the atmosphere. As water works downward below the root zone of plants, it eventually reaches a zone of saturation, where all voids in the ground are filled with water. This zone of saturation is called *groundwater;* the upper surface of groundwater is called the *water table.* Wells are commonly sunk to a point well below the water table so that the latter's seasonal fluctuations will not interrupt the well's access to groundwater.

## 20.3  BASIC PLANNING

After considering the relationship between a building and the roles that water plays within it, the designer must do basic sizing—of the quantities of water needed, of the areas in which water will be used, and of the areas and equipment associated with water's return to the hydrologic cycle.

However, before discussing basic planning for the amounts of water used daily within buildings and for the treatment of that water, we should note that water is also an important component of building construction. The production of 1 ton (907 kg)

of bricks requires 580 gal (2200 L) of water; 1 ton of steel requires 43,600 gal (165,000 L); 1 ton of plastic requires 348,750 gal (1.32 million L). Water is one of the main ingredients of concrete; a typical 94-lb (42.6-kg) bag of cement requires about 6 gal (23 L) of water.

Another way in which buildings contribute to water consumption is through electricity consumption. Most power plants require very large quantities of water, which they quickly return to the hydrologic cycle warmer in temperature (and perhaps as vapor). A large nuclear power plant that utilizes a cooling tower can evaporate daily the approximate equivalent transpiration from about 9 mi$^2$ (23 km$^2$) of forest.

## (a) Water Supply

The task of estimating water needs is complicated by the conflict between *current practice* and *conservation.* Current practice tends toward the use of large amounts of water for very-low-grade tasks. Conservation reserves high-quality water for high-grade tasks and emphasizes recycling as well as diminished overall usage of water.

Water supply often is first estimated in terms of gallons per capita per day (liters per capita per day). Table 20.1 shows some common terms used in measuring the water supply. Typical quantities are matched with nourishing, cleansing, and other usages in the United States in Table 20.2. To gain an appreciation of how the rate of urban water usage has changed over time, consider the following figures (from Milne, 1976):

| | |
|---|---|
| Imperial Rome | 38 g/cd (144 L/cd) |
| London, 1912 | 40 g/cd (151 L/cd) |
| American cities just before World War II | 115 g/cd (435 L/cd) |
| Los Angeles, mid-1970s | 182 g/cd (689 L/cd) |

Compare these usage rates to the "Basic Water Requirement" for four domestic needs—drinking, sanitation, bathing, and cooking—of about 13 g/cd (50 L/cd) proposed by Gleick (1998).

Although the historical trend has clearly been toward higher per capita use of water, the recent emphasis on conservation has resulted in significant changes in this pattern. In Section 22.1, the influences of various conservation practices will be considered in more detail. Table 20.2 can be used to

### TABLE 20.1 Water Measurement Terms and Conversions

| QUANTITY | |
|---|---|
| 1 gal | = 8.33 lb |
| | = 231 in.$^3$ |
| | = 0.134 ft$^3$ |
| | = 3.785 L |
| 1 liter | = .0263 gal |
| 1 cubic foot | = 7.48 gal |
| | = 62.4 lb |
| | = 0.028 m$^3$ |
| 1 cubic meter | = 1000 L |
| | = 35.32 ft$^3$ |
| 1 ton | = 240 gal |
| 1 acre | = 0.4 hectare |
| | = 40 km$^2$ |
| 1 acre-foot (a-f) | = 325,851 gal |
| | = 43,560 ft$^3$ |
| | = 12 acre-inches (ac-in.) |
| 1 million gallons | = 3.07 a-f |
| **FLOW** | |
| 1 cfs | = 1.98 a-f/day |
| | = 0.028 m$^3$/s |
| 1 a-f/year | = annual water supply for five people at 180 gal/cd |
| 1000 gpm | = 2.23 cfs |
| | = 4.42 a-f/day |
| 1 million gpd | = 694.4 gpm |
| | = 1.55 cfs |
| | = 1120 a-f/yr |
| **LEAKS** | |
| Slow drip | = 170 gpd |
| | = 62,050 gal/yr |
| ⅛-in. (3.2-mm) diameter stream | = 3600 gpd |
| | = 4 a-f/yr |
| **COST** | |
| 10¢ per 1000 gal | = 7.48¢ per 100 ft$^3$ |
| | = $32.59 per a-f |
| 10¢ per 100 ft$^3$ | = $43.56 per a-f |
| | = $13.40 per 1000 gal |
| 10¢ per ton | = $136 per a-f |

*Sources:* Milne (1976) and Ferguson (1998).

estimate the daily indoor usage of water in various facilities if *current practices* are anticipated. For a very rough approximation of *conservation* effects:

- Reduce Table 20.2 values by 25%, assuming simple conservation measures such as flow controls.
- Reduce Table 20.2 values by 50%, assuming partial recycling.

The more water used for flushing toilets within the building types listed in Table 20.2, the greater the potential for savings through conservation.

In urban areas, public water mains usually provide the necessary quantities of water at the pressures and rates of flow required to operate typical plumbing fixtures. For isolated buildings or

those independent of public networks, the water supply can come from individual sources: wells, springs, cisterns, lakes, and so forth. For the minimum pressures and flow rates necessary from these sources (and/or the storage vessels associated with them), see Table 21.15.

### (b) Cisterns

Where rainwater is to be utilized, a rough approximation of catchment area and cistern storage volume is initially needed. This procedure is detailed later in this chapter. For now:

1. From Table 20.2, find the quantity of rainwater to be used daily:

$$g/cd \times population = gpd$$
$$(L/cd \times population = L/d)$$

2. Convert this quantity to the yearly need for water:

$$gpd \times 365 \text{ days} = gal/yr$$
$$(L/d \times 365 \text{ days} = L/yr)$$

3. Assume, conservatively, that a dry year will have two-thirds of the precipitation of an average year; this measurement is the *design precipitation.* (Average annual precipitation is available from NOAA annual summaries.)

Average annual precipitation × ⅔ = design precipitation

4. From Fig. 20.5, determine the catchment area required.

5. Roughly size the cistern (storage) capacity by finding the longest dry period (in days of negligible rainfall, from NOAA local climatological data):

cistern capacity = gpd × days of dry period

6. Convert capacity to volume by the formula

1 ft$^3$ stores 7.48 gal of water
(1 m$^3$ stores 1000 L water)

**EXAMPLE 20.1, PART A** A 20,000-ft$^2$ one-story factory near Salem, Oregon, will use roof-collected rainwater to flush its toilets. Water-conserving toilets using 3 gal/flush and serving 20 workers are to

### TABLE 20.2 Planning Guide for Water Supply[a]

| Building Usage | Per Capita (as Listed) Daily Usage | |
|---|---|---|
| | Gallons | Liters |
| Airports (per passenger) | 3–5 | 11–19 |
| Apartments, multiple-family (per resident) | 60 | 227 |
| Bath houses (per bather) | 10 | 38 |
| Camps | | |
| Construction, semipermanent (per worker) | 50 | 189 |
| Day with no meals served (per camper) | 15 | 57 |
| Luxury (per camper) | 100–150 | 378–568 |
| Resorts, day and night, with limited plumbing (per camper) | 50 | 189 |
| Tourist, with central bath and toilet facilities (per person) | 35 | 132 |
| Cottages with seasonal occupancy (per resident) | 50 | 189 |
| Courts, tourist, with individual bath units (per person) | 50 | 189 |
| Clubs | | |
| Country (per resident member) | 100 | 378 |
| Country (per nonresident member present) | 25 | 95 |
| Dwellings | | |
| Boardinghouses (per boarder) | 50 | 189 |
| Additional kitchen requirements for nonresident boarders | 10 | 38 |
| Luxury (per person) | 100–150 | 378–568 |
| Multiple-family apartments (per resident) | 40 | 151 |
| Rooming houses (per resident) | 60 | 227 |
| Single family (per resident) | 50–75 | 189–284 |
| Estates (per resident) | 100–150 | 378–568 |
| Factories (per person per shift) | 15–35 | 57–132 |
| Highway rest area (per person) | 5 | 19 |
| Hotels with private baths (two persons per room) | 60 | 227 |
| Hotels without private baths (per person) | 50 | 189 |
| Institutions other than hospitals (per person) | 75–125 | 284–473 |
| Hospitals (per bed) | 250–400 | 946–1514 |
| Laundries, self-service (per washing) | 50 | 189 |
| Livestock (per animal) | | |
| Cattle (drinking) | 12 | 45 |
| Dairy (drinking and servicing) | 35 | 132 |
| Goat (drinking) | 2 | 8 |
| Hog (drinking) | 4 | 15 |
| Horse (drinking) | 12 | 45 |
| Mule (drinking) | 12 | 45 |
| Sheep (drinking) | 2 | 8 |
| Steer (drinking) | 12 | 45 |
| Motels with bath, toilet, and kitchen facilities (per bed space) | 50 | 189 |
| With bed and toilet (per bed space) | 40 | 151 |
| Parks | | |
| Overnight, with flush toilets (per camper) | 25 | 95 |
| Trailer, with individual bath units, no sewer connection (per trailer) | 25 | 95 |
| Trailer, with individual baths, connected to sewer (per person) | 50 | 189 |
| Picnic | | |
| With bathhouses, showers, and flush toilets (per picnicker) | 20 | 76 |
| With toilet facilities only (per picnicker) | 10 | 38 |
| Poultry | | |
| Chickens (per 100) | 5–10 | 19–38 |
| Turkeys (per 100) | 10–18 | 38–68 |
| Restaurants with toilet facilities (per patron) | 7–10 | 26–38 |
| Without toilet facilities (per patron) | 2½–3 | 9–11 |
| With bar/cocktail lounge (additional quantity per patron) | 2 | 8 |
| Schools | | |
| Boarding (per pupil) | 75–100 | 284–378 |
| Day, with cafeteria, gymnasium, and showers (per pupil) | 25 | 95 |
| Day, with cafeteria but no gymnasiums or showers (per pupil) | 20 | 76 |
| Day, without cafeteria, gymnasiums, or showers (per pupil) | 15 | 57 |
| Service stations (per vehicle) | 10 | 38 |
| Stores (per toilet room) | 400 | 1514 |
| Swimming pools (per swimmer) | 10 | 38 |
| Theaters | | |
| Drive-in (per car space) | 5 | 19 |
| Movie (per auditorium seat) | 5 | 19 |
| Workers | | |
| Construction (per person per shift) | 50 | 189 |
| Day (school or office, per person per shift) | 15 | 57 |

*Source*: U.S. Environmental Protection Agency (1975).

[a]These values may be reduced as follows: with flow controls, up to 25% reduction; with water recycling, up to 50% reduction.

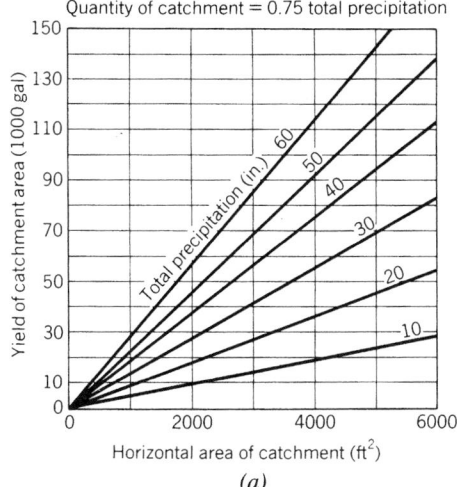

Quantity of catchment = 0.75 total precipitation

*(a)*

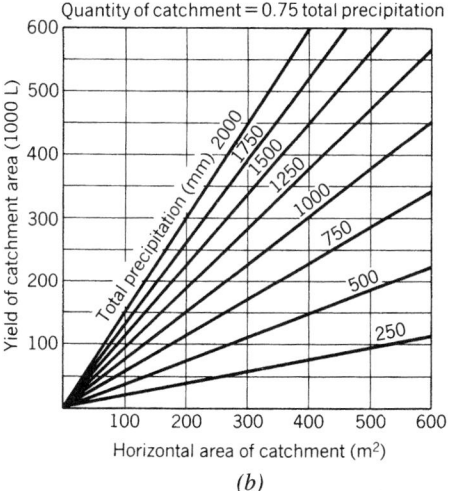

Quantity of catchment = 0.75 total precipitation

*(b)*

**Fig. 20.5** *Yields of rainfall catchment areas (roofs) in terms of total precipitation. In these graphs, 75% of the total precipitation is assumed to be catchable; the remainder is lost to evaporation or spillage. (a) I-P units from the U.S. EPA's* Manual of Individual Water Supply Systems, *1975. (b) SI units (1 m³ = 1000 L).*

be used. Table 20.2 shows that factory workers use between 15 and 35 g/cd. Because usage at this factory will be low—no showers, for example—assume 15 gal/cd:

$$15 \text{ g/cd} \times 20 \text{ workers} = 300 \text{ gpd for } \textit{all} \text{ usages}$$

Because low-flush toilets are to be used, reduce this figure by 25%:

$$0.75 \times 300 = 225 \text{ gpd}$$

Toilets will probably account for most of this 225 gpd: for example, at three flushes per day per worker,

$$3 \text{ flushes/day} \times 3 \text{ gal/flush} \times 20 \text{ workers} = 180 \text{ gpd}$$

Assume, then, that up to 200 gpd of rainwater will be utilized.

*Catchment Area:* Salem's average annual rainfall is 41 in. Design rainfall is $\frac{2}{3} \times 41 = 27.3$ in. in a dry year; the yearly need is 200 gpd × 365 days = 73,000 gal. (The combination of 73,000 gal and 27.3 in. is off the chart in Fig. 20.5a, so divide 73,000 gal by 2 to obtain 36,500 gal and then double the resulting catchment area.) At 27.3 in. of precipitation, 36,500 gal will be caught by 2800 ft². The catchment area for 73,000 gal is, therefore, 2 × 2800 = 5600 ft². (So about 28% of the 20,000-ft² factory's roof area will suffice for a catchment area.)

*Cistern Capacity:* Salem normally has very dry summers; average monthly rainfall is as follows:

| | |
|---|---|
| May | 2.1 in. |
| June | 1.4 in. |
| July | 0.4 in. |
| August | 0.6 in. |
| September | 1.5 in. |
| October | 4.0 in. |

The dry period, then, runs from mid-June to mid-September—about 90 days. Thus, capacity = 200 gpd × 90 days = 18,000 gal:

$$\text{volume} = \frac{18,000 \text{ gal}}{7.48 \text{ gal/ft}^3} = 2406 \text{ ft}^3 \qquad ■$$

(For example, 5 ft deep × 22 ft square = 2420 ft³.) A more detailed sizing procedure for this building's cistern is presented later in this chapter.

## (c) Required Facilities

Another important early design question is, how many plumbing fixtures should be provided? Table 20.3 lists the minimum plumbing facilities required for various types and sizes of building occupancies. (Note that local requirements may differ somewhat from this particular guide, which is taken from the 1997 *International Plumbing Code.* There are many plumbing codes in use in North America, and they sometimes disagree.) Because these are considered minimal requirements, more generous provisions are sometimes appropriate. Chapter 21 discusses the design of the spaces in which these services are used.

**WATER AND WASTE**

## TABLE 20.3 Minimum Number of Plumbing Facilities[a]

| Occupancy | Water Closets Urinals[b] Male | Female | Lavatories | Bathtubs/ Showers | Drinking Fountains[c] | Others |
|---|---|---|---|---|---|---|
| Assembly | | | | | | |
| Theaters | 1 per 125 | 1 per 65 | 1 per 200 | — | 1 per 1000 | 1 service sink |
| Nightclubs | 1 per 40 | 1 per 40 | 1 per 75 | — | 1 per 500 | 1 service sink |
| Restaurants | 1 per 75 | 1 per 75 | 1 per 200 | — | 1 per 500 | 1 service sink |
| Halls, museums, etc. | 1 per 125 | 1 per 65 | 1 per 200 | — | 1 per 1000 | 1 service sink |
| Coliseums, arenas | 1 per 75 | 1 per 40 | 1 per 150 | — | 1 per 1000 | 1 service sink |
| Churches[d] | 1 per 150 | 1 per 75 | 1 per 200 | — | 1 per 1000 | 1 service sink |
| Stadiums, pools, etc. | 1 per 100 | 1 per 50 | 1 per 150 | — | 1 per 1000 | 1 service sink |
| Business[e,f,g] | 1 per 25 | | 1 per 40 | — | 1 per 100 | 1 service sink |
| Educational | 1 per 50 | | 1 per 50 | — | 1 per 100 | 1 service sink |
| Factory and industrial | 1 per 100 | | 1 per 100 | [h] | 1 per 400 | 1 service sink |
| High hazard[e,f] | 1 per 100 | | 1 per 100 | [h] | 1 per 1000 | 1 service sink |
| Institutional | | | | | | |
| Residential care | 1 per 10 | | 1 per 10 | 1 per 8 | 1 per 100 | 1 service sink |
| Hospitals, ambulatory nursing home patients[i] | 1 per 50 | | 1 per room[j] | 1 per 15 | 1 per 100 | 1 service sink per floor |
| Day nurseries, sanitariums, nonambulatory nursing home patients, etc.[i] | 1 per 15 | | 1 per 15 | 1 per 15[k] | 1 per 100 | 1 service sink |
| Employees, other than residential care[i] | 1 per 25 | | 1 per 35 | — | 1 per 100 | — |
| Visitors, other than residential care | 1 per 75 | | 1 per 100 | — | 1 per 500 | — |
| Prisons[i] | 1 per cell | | 1 per cell | 1 per 15 | 1 per 100 | 1 service sink |
| Asylums, reformatories, etc.[i] | 1 per 15 | | 1 per 15 | 1 per 15 | 1 per 100 | 1 service sink |
| Mercantile[e,f,g] | 1 per 500 | | 1 per 750 | — | 1 per 1000 | 1 service sink |
| Residential | | | | | | |
| Hotels, motels | 1 per guestroom | | 1 per guestroom | 1 per guestroom | — | 1 service sink |
| Lodges | 1 per 10 | | 1 per 10 | 1 per 8 | 1 per 100 | 1 service sink |
| Multiple family | 1 per dwelling unit | | 1 per dwelling unit | 1 per dwelling unit | — | 1 kitchen sink per dwelling unit; 1 automatic clothes washer connection per 20 dwelling units units |
| Dormitories | 1 per 10 | | 1 per 10 | 1 per 8 | — | 1 service sink |
| One- and two-family dwellings | 1 per dwelling unit | | 1 per dwelling unit | 1 per dwelling unit | — | 1 kitchen sink per dwelling unit; 1 automatic clothes washer connection per dwelling unit[l] |
| Storage[e,f] | 1 per 100 | | 1 per 100 | [h] | 1 per 1000 | 1 service sink |

*Source:* Copyright © 1997, International Code Council, Inc., Falls Church, Virginia. *International Plumbing Code.* Reprinted with permission of the author. All rights reserved.

[a]Based on one fixture being the minimum required for the number of persons indicated or any fraction of the number of persons indicated. The number of occupants shall be determined by the building code. Unless otherwise shown, the required water closets, lavatories, and showers or bathtubs shall be distributed evenly between the sexes based on the percentage of each sex anticipated in the occupant load. The occupant load shall be composed of 50% of each sex, unless statistical data approved by the code official indicate a different distribution.

[b]In each bathroom or toilet room, urinals shall not be substituted for more than 50% of the required water closets.

[c]Drinking fountains shall not be installed in public restrooms. Where water is served in restaurants or where bottled water coolers are provided in other occupancies, drinking fountains shall not be required.

[d]Fixtures located in adjacent buildings under the ownership or control of the church shall be made available during the periods the church is occupied.

[e]Separate employee facilities shall not be required in occupancies in which 15 or fewer people are employed. Separate facilities for each sex shall not be required in structures or tenant spaces with a total occupant load, including both employees and customers, of 15 or fewer in which food or beverage is served for consumption within the structure or tenant space.

[f]Access to toilet facilities in occupancies other than assembly or mercantile shall be from within the employees' regular working area. The

required toilet facilities shall be located not more than one story above or below the employees' regular work area, and the path of travel to such facilities shall not exceed a distance of 500 ft (152 m).

$^g$In mercantile and assembly occupancies, employee toilet facilities shall be either separate facilities or public customer facilities. Separate employee facilities shall not be required in tenant spaces of 900 ft$^2$ (84 m$^2$) or less where the travel distance from the main entrance to a central toilet area does not exceed 500 ft (152 m) and where such central facilities are located not more than one story above or below the tenant space.

$^h$Emergency showers and eyewash stations shall be provided with a supply of cold water as required by the manufacturer. Waste connection shall not be required for emergency showers and eyewash stations.

$^i$Toilet facilities for employees shall be separate from facilities for inmates or patients.

$^j$A single-occupant room with one water closet and one lavatory serving not more than two adjacent patient rooms shall be permitted where such room is provided with direct access from each patient room and with provisions for privacy.

$^k$For day nurseries, a maximum of one bathtub shall be required.

$^l$For attached one- and two-family dwellings, one automatic clothes washer connection shall be required per 20 dwelling units.

## (d) Sewage

Where public sewers are available (as in most urban areas), the designer usually is not concerned with estimating the flow of sewage. However, where private or on-site sewage treatment is required or where public sewage treatment facilities are over-taxed, total sewage flow is an early design concern. Table 20.4 lists these sewage flows by type of occupancy, again in g/cd (L/cd). Note that sewage flow may differ from supply flow, especially where supply water is used for irrigation, for car washing, or in evaporative processes.

Once the daily flow of sewage is established, some early guidelines are needed for determining the suitability of and the area required by some treatment processes. One of the most common reasons for the rejection of a potential rural building site is a lack of suitability for sewage disposal. A geologic analysis of structural and sewage disposal potential is one of the first documents needed by the designer. Chapter 22 describes the sizing of some common treatment methods. At this earlier stage, the following design guidelines can be useful:

*Septic Tank Drainfields.* In I-P units (minimum 750-ft$^2$ area for any system):

- For shallow trenches in poorly draining soil: drainfield area = total sewage flow in gpd × 3.6 ft$^2$/gal
- For deep trenches in well-draining soil: drainage area = total sewage flow in gpd × 0.4 ft$^2$/gal

In SI units (minimum 70-m$^2$ area for any system):

- For shallow trenches in poor soil: L/day × 0.087 m$^2$/L
- For deep trenches in good soil: L/day × 0.01 m$^2$/L

(These guidelines allow for an expansion area equal to the original size of the drainage field in case of field failure.)

*Mounds.* These are built-up leaching fields on top of the existing grade (see Fig. 22.39). For a single-family dwelling, allow for a 4-ft-high (1.2-m-high) mound whose bottom area is a square 44 ft (13.4 m) on each side and whose sides slope at a 1:3 vertical-to-horizontal ratio.

*Package Sewage Plant Drainfields.* In these, sewage is treated to a much greater extent than in septic tanks, and effluent is filtered:

- In poorly draining soil: total sewage flow in gpd × 0.49 ft$^2$/gal (L/day × 0.012 m$^2$/L).
- In well-draining soil: total sewage flow in gpd × 0.23 ft$^2$/gal (L/day × 0.006 m$^2$/L).

*Sewage Lagoons.* These consist of at least two open treatment ponds (primary and secondary) and are sized on the basis of pounds of biochemical oxygen demand (BOD) rather than gallons of sewage flow. A typical assumption for estimating BOD is

- 0.2 lb BOD/person for ordinary domestic sewage
- 0.3 lb BOD/person where garbage grinders or other devices contribute added organic material to domestic sewage

The *total* acreage (ac) needed for the two ponds can be estimated as

- 20 lb BOD/ac for the (colder) northern United States
- 35 lb BOD/ac for the (drier, warmer) southern United States

(The primary pond is usually sized for 50 lb BOD/ac.)

WATER AND WASTE

**TABLE 20.4 Estimated Sewage Flow Rates**

| Type of Occupancy | Unit Gallons (Liters) per Day |
|---|---|
| Airports | 15 (56.8) per employee |
| | 5 (18.9) per passenger |
| Auto washers | Check with equipment manufacturer |
| Bowling alleys (snack bar only) | 75 (283.9) per lane |
|   Camps, campgrounds with central comfort station | 35 (132.5) per person |
|   Campgrounds with flush toilets, no showers | 25 (94.6) per person |
|   Day camps (no meals served) | 15 (56.8) per person |
|   Summer and seasonal | 50 (189.3) per person |
| Churches (sanctuary) | 5 (18.9) per seat |
|   With kitchen waste | 7 (26.5) per seat |
| Dance halls | 5 (18.9) per person |
| Factories, no showers | 25 (94.6) per employee |
|   With showers | 35 (132.5) per employee |
|   Cafeteria, add | 5 (18.9) per employee |
| Hospitals | 250 (946.3) per bed |
|   Kitchen waste only | 25 (94.6) per bed |
|   Laundry waste only | 40 (151.4) per bed |
| Hotels (no kitchen waste) | 60 (227.1) per bed (2 person) |
| Institutions (resident) | 75 (283.9) per person |
|   Nursing home | 125 (473.1) per person |
|   Rest home | 125 (473.1) per person |
| Laundries, self-service | 50 (189.3) per wash cycle |
|   Commercial | Per manufacturer's specifications |
| Motel | 50 (189.3) per bed space |
|   With kitchen | 60 (227.1) per bed space |
| Offices | 20 (75.7) per employee |
| Parks, mobile homes | 250 (946.3) per space |
|   Picnic parks (toilets only) | 20 (75.7) per parking space |
|   Recreational vehicles, without water hook-up | 75 (283.9) per space |
|     With water and sewer hook-up | 100 (378.5) per space |
| Restaurants/cafeterias | 20 (75.7) per employee |
|   Toilet | 7 (26.5) per customer |
|   Kitchen waste | 6 (22.7) per meal |
|   Add for garbage disposal | 1 (3.8) per meal |
|   Add for cocktail lounge | 2 (7.6) per customer |
| Kitchen waste-disposable service | 2 (7.6) per meal |
| Schools—staff and office | 20 (75.7) per person |
|   Elementary students | 15 (56.8) per person |
|   Intermediate and high | 20 (75.7) per student |
|     With gym and showers, add | 5 (75.7) per student |
|     With cafeteria, add | 3 (11.4) per student |
|   Boarding, total waste | 100 (378.5) per person |
| Service station, toilets | 1000 (3785) for 1st bay |
| | 500 (1892.5) for each additional bay |
| Stores | 20 (75.7) per employee |
|   Public restrooms, add (per unit of floor space) | 1 per 10 ft$^2$ (4.1 per m$^2$) |
| Swimming pools, public | 10 (37.9) per person |
| Theaters, auditoriums | 5 (18.9) per seat |
|   Drive-in | 10 (37.9) per space |

Source: *Uniform Plumbing Code,* copyright © 1997. Printed by permission of the International Association of Plumbing and Mechanical Officials.

## 20.4 RAINWATER

There is a striking similarity between rainwater and solar energy. Both are essential for agriculture, and both have been well understood and utilized by farmers since agrarian societies first emerged. Both can be very beneficial to architecture and were utilized as needed by anonymous builders for centuries. Both fell out of favor with designers as plentiful supplies of pure, centrally treated water and concentrated, centrally controlled fuels became commonplace. Both are thinly yet relatively evenly spread over the world's population, so they are at least seasonally available to help meet nearly every building's needs for water and heat. Yet both are difficult to utilize in industrialized societies, because they require *individual preoccupancy* expenditures.

Consider the typical public water and electric utility in a U.S. city. It can raise the funds to build large water treatment plants, electricity-generating plants, and the network of pipes and wires that bring these commodities to every building. The utility's costs, including interest on its construction debts, will be passed on to its consumers on a monthly basis, along with a margin of profit that is usually controlled by state governments. Thus, our society has a well-established method for encouraging central suppliers of water and power.

Now consider the individual building owner. To build a cistern and a solar-heated building, she must borrow money at an interest rate higher than that which the utility pays; and both options cost more initially than simple connections to the utility's pipes and lines. Even though she is willing to flush her toilets with rainwater rather than with chlorinated and filtered potable water from the util-

ity, and even though she is willing to heat with lower-grade solar energy than with higher-grade electricity, she must pay a substantial first-cost penalty—with interest—to do so.

The overall public good could be well served by a mixture of public networks of pure water and electricity and individual cisterns and solar applications. The environmental benefits would be substantial—less water withdrawn from rivers, lakes, and underground aquifers; less energy and chemicals used to treat and deliver such water; less storm water discharged to pollute rivers; less fuel used to generate electricity; and less environmental damage from power plants. Yet we continue to economically discourage the individual who uses the rain and the sun.

Architects Fernau and Hartman's design for an invited competition for a "low-entropy kindergarten" in Frankfort, Germany (Fig. 20.6), celebrates sun and rainwater as teaching elements. In their words, "The building reveals natural phenomena

**Fig. 20.6** *This low-entropy kindergarten competition entry for Frankfurt, Germany, features sun, adjustable shade, and a roof that collects rainwater and leads it to an interior stone-filled catchment. The rainwater downspout can be seen in the center of the drawing; the stone-filled catchment is below the stair landing. The overflow then goes outside, through a layer of stones and into a below-ground cistern. (Courtesy of Fernau & Hartman, Architects, Berkeley, CA.)*

such as the play of light and shadow over a wall during the course of a year, the warmth of a wintergarden in January, the cooling shade from deciduous vines on a trellis overhead, water pouring off the roof into an interior stone catchment during a downpour."

## 20.5 COLLECTION AND STORAGE

In terms of both quality and quantity, rainwater is an attractive alternative. Figure 20.4b shows that rainwater is close to the purest state in the hydrologic cycle. More recently, it is true, air pollution has begun to threaten the quality of rainwater in some areas as acid rain has become widespread in the northeastern section of the North American continent and in Europe. In some particularly air-polluted locations, lead poses a threat to rainwater quality. Also, on any catchment surface, dust and bird droppings are common pollutants that must be considered. Other factors that bear on rainwater quality are roofing materials and the form of the roof. The appropriate health authority should be consulted for a list of roofing materials that will have no toxic effects on rainwater. Steeper roofs are

scoured by winds, and thus collect less dust and give cleaner runoffs. Devices to discourage roosting birds are strongly recommended, as are periodic checks of cistern water for bacteria. Fungicides (for moss control) should be scrupulously avoided, as should roofing paints containing lead. For these reasons, urban rainwater commonly is not used for drinking and cooking. (Bottled water or on-site water distilling can supply potable water.) For the typical residence, however, that still leaves about 95% of indoor water usage that could be provided by rainwater. Also, in those cities that have very "hard" public water supplies, rainwater's "soft" characteristics make it particularly attractive.

The quantity of rainwater available in most U.S. locations could meet a high percentage of typical home or business needs. Milne (1976) pointed out that the 42 in. (1067 mm) of rain that falls annually on the streets and roofs of Manhattan Island could, if collected and stored, provide 148% of the residential needs of its 1.7 million inhabitants. For the typical U.S. suburban house (with a roof area of 1500 ft$^2$ [139 m$^2$]), the annual catchment can be estimated by combining the rainfall quantities (Fig. 20.7) with the resulting catchment yield (Fig. 20.5). Even at a "dry" rate of only

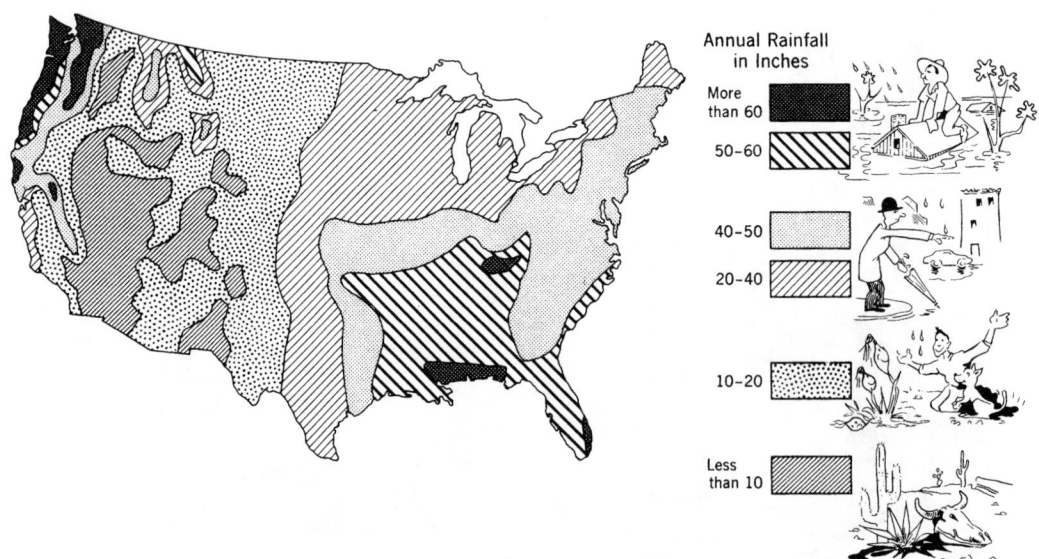

**Fig. 20.7** *Annual rainfall in the United States. Cistern volume conversion factors:*

*1 ft$^3$ = 7.48 gal = 0.02832 m$^3$*
*1 m$^3$ = 1000 L = 264.2 gal*

20 in. annually, a 1500-ft² roof would yield about 12,000 gal, or 33 gpd—nearly enough to meet the clothes-cleaning and dishwashing needs of the family. (In SI units, a 140-m² roof in an area with annual precipitation of 500 mm will yield about 50,000 L, or 137 L/d.)

Unfiltered rainwater seems particularly well suited to the irrigation of small lawns or gardens, both because it lacks additives unneeded by plants, such as chlorine and sodium fluoride, and because it can reduce the user's demand on the public water supply on the hottest summer days. Cisterns located above an irrigated area have the advantage of replacing a pump with simple gravity flow. A rain barrel at the bottom of a downspout is an example of this approach—although one of limited capacity.

In many of the world's drier areas, small cisterns within a home are common. Such cisterns, which can be fed both by rainwater and by the public supply, are frequently used for all domestic purposes, including drinking. The presence of such a large water volume also can be advantageous in the event of fire. Sometimes these storage cisterns are required because the public supply is diminished or cut off at peak usage hours due to insufficient water main capacity. Figure 20.8 shows a cistern in a typical outdoor location, along with various options by which pollution from dirty roofs can be minimized.

When cisterns are taken seriously as water storage devices, their function and size can become strong design-form determinants. The country home of architect John Andrews in the dry ranchland of New South Wales, Australia, offers a particularly striking example of cisterns and form (Fig. 20.9). These demonstrate the design image of "pregnant downspouts" storing water above the area where it will be used rather than belowground. The corner rainwater collectors also help deflect cooling summer breezes into the house along the diagonal walls at each corner. The central skylight then vents the breezes, along with the house's heat. From the corner cisterns, rainwater is pumped up to the central tower's storage tank. It then can be heated by solar collectors (to be installed on the sloping top of the tower), or used to supply the house's plumbing fixtures, or even used to sprinkle the metal roof surface, whose surface temperature could quickly be lowered by evaporation. Unevaporated roof water simply runs back into the cisterns.

As a final design-with-water gesture, the shower (off bedroom 1) has been made into a true celebration place for cleansing and refreshing, with a sweeping view of the ranch.

In this passively solar-heated home, the living areas are placed on the elongated north side—the warmer side in the cold season for this Southern Hemisphere house. Passive cooling is aided by pergolas on the west and south, which are covered with vines for hot season shading of windows.

***Sizing.*** The procedures described in Section 20.3 are for rough sizing of both the catchment area and the storage capacity for cisterns. When rainwater is to be a primary, as opposed to merely a supplementary source, a closer look must be taken at rainfall deposits and user withdrawals from a cistern.

This procedure depends on the monthly average rainfall (from NOAA Local Climatological Data), the monthly water usage, and the catchment area yield (from Fig. 20.5).

**EXAMPLE 20.1, PART B** Take a closer look at the cistern that was approximately sized in Part A. Daily usage for this cistern, to be used for toilet flushing in a factory near Salem, Oregon, was estimated at 200 gal and the catchment area at 5600 ft². The cistern capacity was estimated at 18,000 gal. Begin the process in the midst of the wettest months.

From Table 20.5, the following conclusions can be drawn:

1. For an 18,000-gal cistern, when the end-of-December cumulative capacity is added to January's surplus, the cistern will be at capacity from November through April.
2. With no allowances for abnormally dry months, we could reduce the cistern size by about 3000 gal (the size of the smallest cumulative capacity in September).
3. A larger cistern—one that could utilize everything from the catchment area—could be built. Its approximate size would be the year-end surplus of 38,510 gal plus the maximum spring monthly cumulative capacity of 25,410 gal in April.
4. Alternatively, the surplus could be devoted to additional usage of rainwater, beyond mere toilet flushing, from November through April. ■

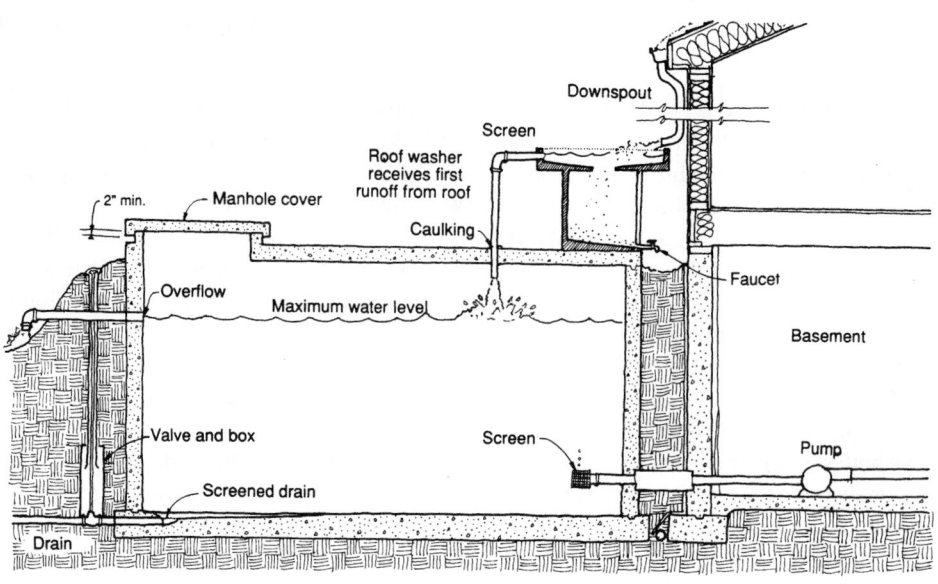

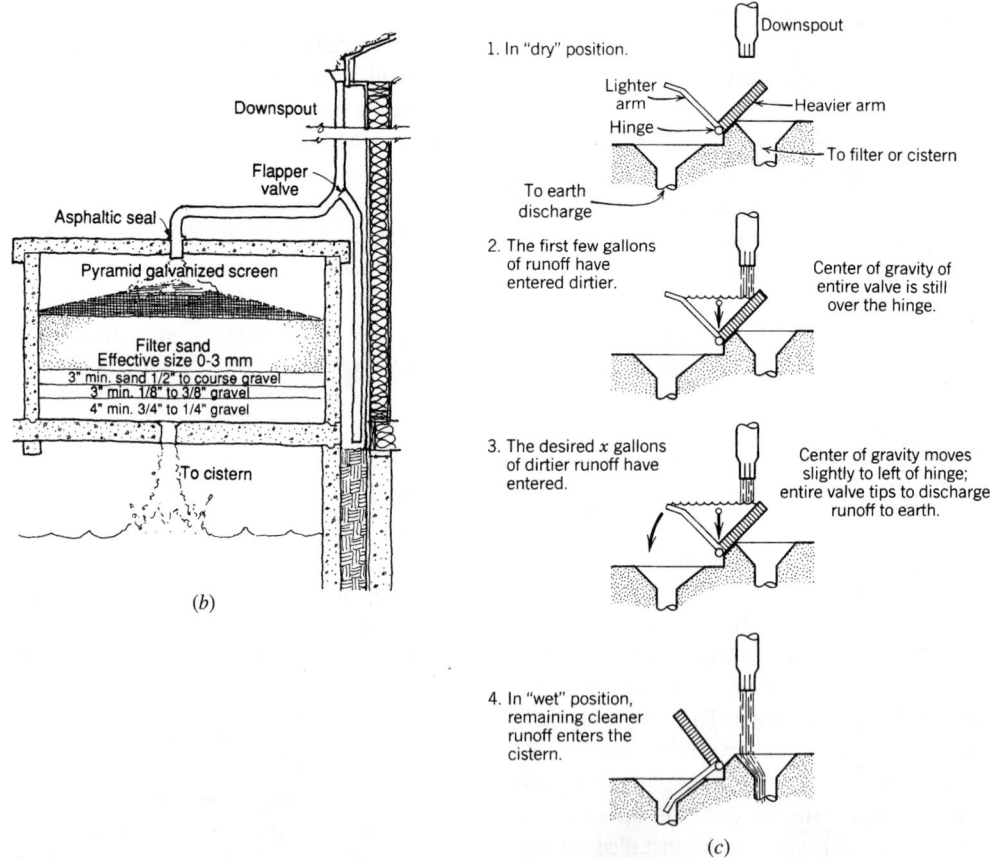

**Fig. 20.8** Section through a typical outdoor cistern. (a) A "roof washer" gets the dirtiest first runoff from the roof. It can later be emptied either by opening the faucet wide or by leaving the faucet slightly open so that it will slowly drain. (b) In place of the roof washer, a sand filter may be used. The "flapper valve" is used rarely, to divert the first rainfall after a prolonged dry spell. After this, the valve is left in a position to divert all rainwater to the sand filter. (Based on the U.S. EPA's Manual of Individual Water Supply Systems, 1975.) (c) Another roof-washing option is a "tipping valve," which dumps the first few gallons of each rainfall. After each rainfall, the valve must be manually (or spring) reset to the "dry" position if it is to again intercept dirty water. In an extended wet period, it would probably be left in the "wet" position.

(a)

(b)

**Fig. 20.9** *Country home of architect John Andrews, Eugowra, New South Wales, Australia. (a) View from the southwest showing corner cisterns, the central daylight/ventilation barrel vault, and the "energy tower" with a water storage tank. (b) View from the south showing the shading pergola to be covered by vines. (c) Floor plan showing the fireplace at the center and cisterns at all corners. (d) Section through the living area. (Courtesy of John Andrews International Pty, Ltd.)*

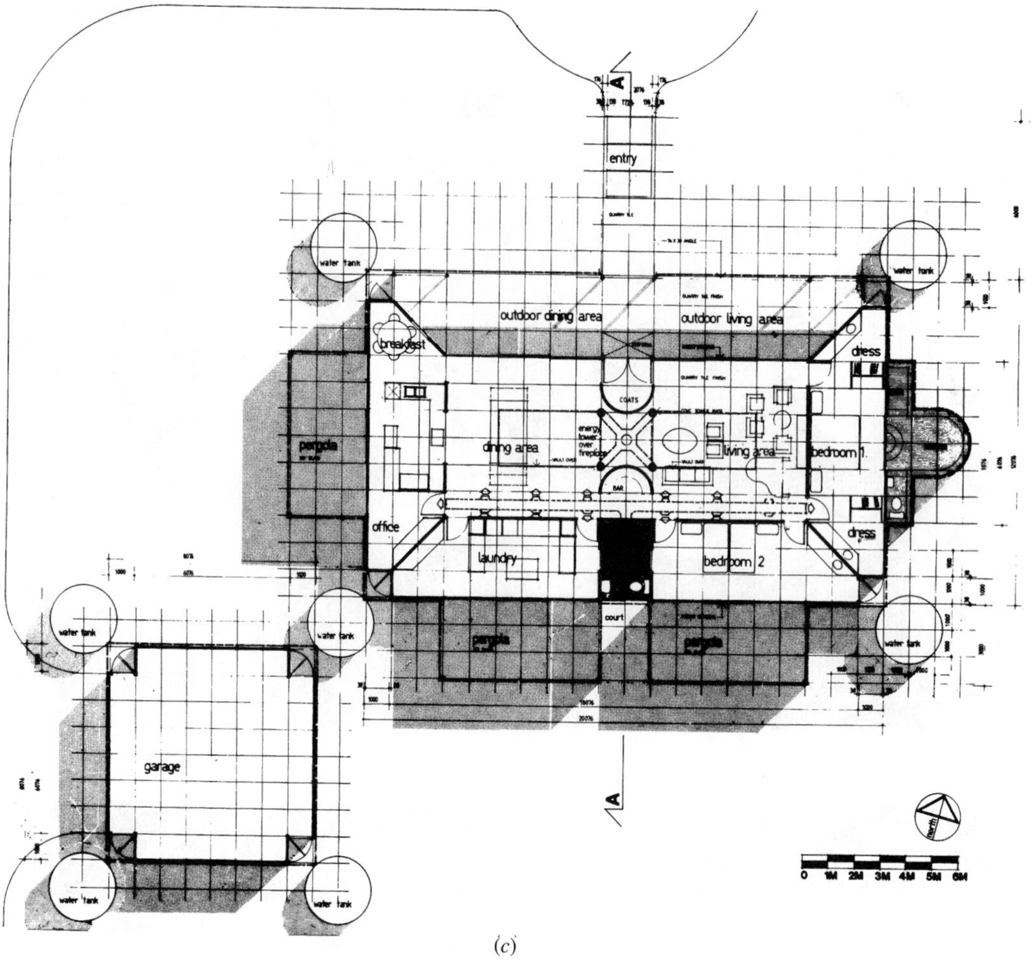

entry

water tank

water tank

outdoor dining area

outdoor living area

breakfast

dress

COATS

pergola

dining area

energy tower over fireplace

living area

bedroom 1

office

BAR

dress

laundry

bedroom 2

court

water tank

water tank

pergola

pergola

water tank

garage

water tank

water tank

north

0 1M 2M 3M 4M 5M 6M

(c)

GARAGE       PERGOLA    BEDROOM 2   LIVING AREA  ENERGY TOWER   OUTDOOR LIVING  PERGOLA

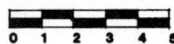

0 1 2 3 4 5

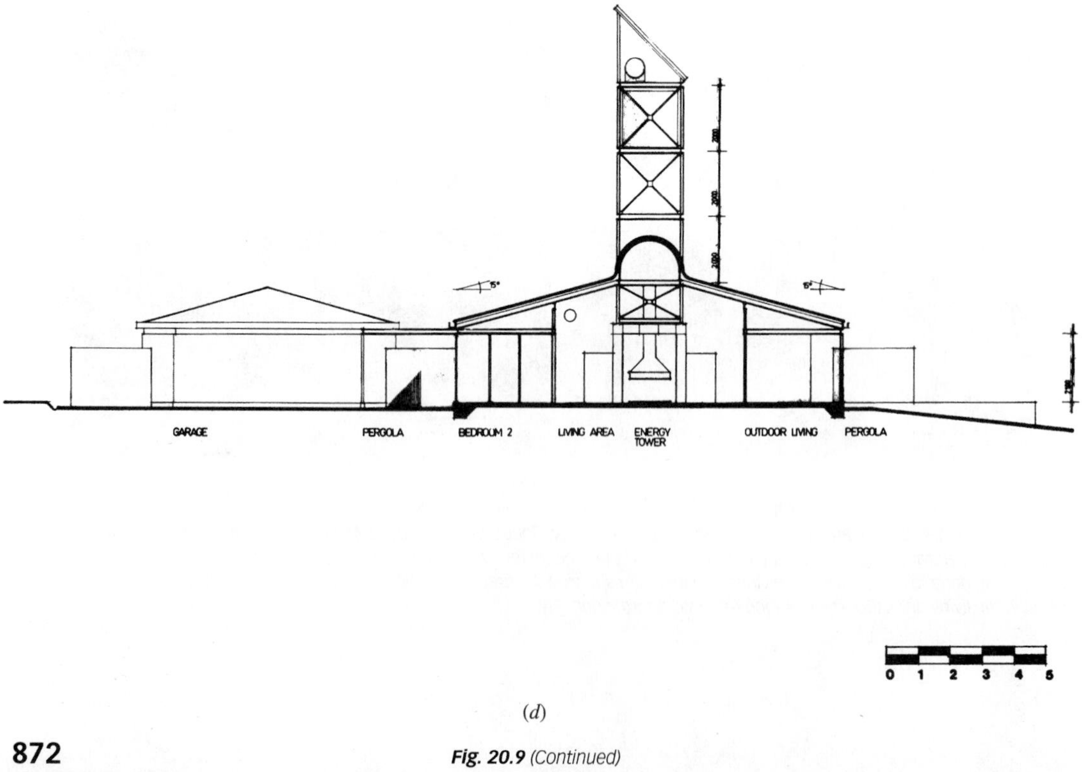

(d)

*Fig. 20.9* (Continued)

**TABLE 20.5  Rainfall Cistern Sizing Procedure**

| I<br>Month and Rainfall[a]<br>(in.) | | II<br>Catchment<br>Yield[b]<br>(gal) | III<br>Usage[c]<br>(gal) | IV<br>Net[d]<br>(gal) | V<br>Cumulative<br>Capacity[e]<br>(gal) | VI<br>Cumulative<br>Capacity<br>Adjusted for<br>Actual Size[f]<br>(gal) |
|---|---|---|---|---|---|---|
| January | 6.9 | 18,630 | 6,200 | 12,430 | 12,430 | 12,430 |
| February | 4.8 | 12,960 | 5,600 | 7,300 | 19,790[g] | 18,000 |
| March | 4.3 | 11,610 | 6,200 | 5,410 | 25,200[g] | 18,000 |
| April | 2.3 | 6,210 | 6,000 | 210 | 25,410[g] | 18,000 |
| May | 2.1 | 5,610 | 6,200 | −530 | 24,880[g] | 17,470 |
| June | 1.4 | 3,780 | 6,000 | −2,220 | 22,660[g] | 15,250 |
| July | 0.4 | 1,080 | 6,200 | −5,120 | 17,540 | 10,130 |
| August | 0.6 | 1,620 | 6,200 | −4,580 | 12,960 | 5,550 |
| September | 1.5 | 4,050 | 6,000 | −1,950 | 11,010 | 3,600 |
| October | 4.0 | 10,800 | 6,200 | 4,600 | 15,610 | 8,200 |
| November | 6.1 | 16,470 | 6,000 | 10,470 | 26,080[g] | 18,000 |
| December | 6.9 | 18,630 | 6,200 | 12,430 | 38,510[g] | 18,000 |

*Source:* Based upon procedure from Brown, Reynolds, Ubbelohde, *InsideOut: Design Procedures for Passive Environmental Technologies,* © 1982 by John Wiley & Sons.
[a]NOAA data for Salem, Oregon.
[b]From Fig. 20.5, according to which 10 in. of precipitation yields about 27,000 gallons for this 5600-ft² catchment area.
[c]The factory uses 200 gpd times days in each month.
[d]Col. II minus col. III.
[e]Values in col. IV, added month by month.
[f]Set at 18,000 gal for the factory in Example 20.1.
[g]The cumulative capacity exceeds the actual storage capacity of 18,000 gal.

## 20.6  RAINWATER AND SITE PLANNING

Prior to the spread to rural areas of buildings, streets, roads, and paved parking lots, water from rainfall and melting snow found its own way to natural destinations. Surface flow to creeks, streams, and rivers accounted for part of this drainage. Underground flow aided the general runoff. Outcropping of flowing groundwater created springs and artesian wells. Low, dished areas formed lakes that in turn overflowed to outlet streams. Flat areas sometimes developed into swamps or marshes.

At a time when there was a choice of locations for towns and villages, sites next to rivers were usually chosen. The waterways provided transportation, and water was supplied from the river or from adjacent wells. As streets and roads were built, slopes could be arranged whereby the rain falling on these areas and flowing onto them from roofs of buildings could run to the river. At interior parts of the country, high ground was favored for building sites and growing communities. Swampy or marshy ground would not be chosen, but did provide terminal locations for the storm water that ran off the high ground. In the course of this natural flow, much of the water was drawn by evaporation to the clouds. The rest, conforming to topographic river basins, continued to seek its way to the sea.

As building increased, desirable locations grew scarce. The possibility of selecting high, dry ground diminished. Great areas, formerly low and marshy, were filled in and buildings constructed, often on piles. From such locations storm water could not be disposed of by drainage to some adjacent lower area or even recharged to the earth through dry wells. Moreover, extensive grids of paved streets and sidewalks in these level developments caught and held the water, resulting in considerable "ponding." Storm sewers had to be built and the water transported great distances, often having to be lifted at intermediate pumping stations before reaching its destination, which might be a remote river.

This emphasis on the removal of storm water has led to an expensive and elaborate system based on quick disposal of rainfall. By decreasing the time between precipitation and runoff, quick disposal increases the peak flows within such systems, thereby increasing flooding of rivers during storms but reducing the rivers' flow between storms. It also contributes pollutants to waterways that otherwise would have been filtered by the soil, as detailed in Table 20.6.

The overloading of storm sewers not only causes minor flooding, but also can influence building design. The designers of the New Orleans Convention and Exhibition Center, a building with 610,000 ft² (56,670 m²) of roof area, spared the city's storm sewers from the impact of a 14-acre (5.7-hectare) runoff by carrying the rainwater over the roofs of adjoining wharfs and discharging it directly into the Mississippi River.

As urban storm sewers reach capacity and suburban groundwater levels drop, designers have begun to emphasize storm water infiltration (or recharge of groundwater) rather than quick runoff. Three design strategies for encouraging such recharge have emerged: roofs that will retain water and slowly release it, porous pavement, and onsite infiltration of runoff.

### (a) Roof Retention

If storm water is to be sent to storm sewers or to soak into the ground, a *slow flow* from roofs will help by diminishing peak flows in sewers and giving soaked

soil more time to absorb still more runoff. A nearly flat roof with specially designed drains (see Fig. 20.23) forms a temporary pond that permits slower discharge, yet eventually drains completely dry (to discourage mosquito breeding, etc.). (For cisterns, however, sloped roofs should be used, as they stay cleaner than flat roofs.) Another possibility is a sod roof that retains water for longer periods than ponded flat roofs (Fig. 20.10). Another option is to store the cistern water on the roof itself (Fig. 20.11).

However achieved, this temporary pond on top of a building will clearly add to structural requirements. Another problem could be posed by high winds blowing sheets of water onto people below. In summer, however, a flooded roof can provide a significant thermal advantage by greatly lowering daytime roof surface temperatures. Rainfall-retaining roofs are now required in some urban areas with overtaxed storm sewers. Sewer districts frequently charge for storm water runoff; by ponding storm water on site, such charges can be reduced or avoided.

### (b) Porous Pavement

To retain rainfall on site, many builders use either porous asphalt (Fig. 20.12), porous concrete,

**TABLE 20.6  Some Constituents of Stream Water**

| Constituent | Source in Nature | Role in Natural Ecosystem | Source of Urban Excess | Role of Excess |
|---|---|---|---|---|
| Sediment | Banks of meandering channels | Maintain stream profile and energy gradient; store nutrients | Construction sites; eroding stream banks | Abrade fish gills; carry excess nutrients and chemicals in adsorption; block sunlight; cover gravel bottom habitats |
| Organic compounds | Decomposing organic matter | Store nutrients | Herbicides; pesticides; fertilizers | Deprive water of oxygen by decomposition |
| Nutrients | Decomposing organic matter | Support ecosystems | Organic compounds; organic litter; fertilizers; food waste; sewage | Unbalance ecosystem; produce algae blooms; deprive water of oxygen by decomposition |
| Trace metals | Mineral weathering | Support ecosystem | Cars; construction materials; all kinds of foreign chemicals | Reduce resistance to disease; reduce reproductive capacity; alter behavior |
| Chloride | Mineral weathering | Support ecosystems | Pavement deicing salts | Sterilize soil and reduce biotic growth |
| Bacteria | Native animals | Participate in ecosystems | Pet animals; dumpsters; trash handling areas | Cause risk of disease |
| Oil | Decomposing organic matter | Store nutrients | Cars | Deoxygenate water |

*Source:* Ferguson, B. (1998). *Introduction to Stormwater.* © John Wiley & Sons, New York.

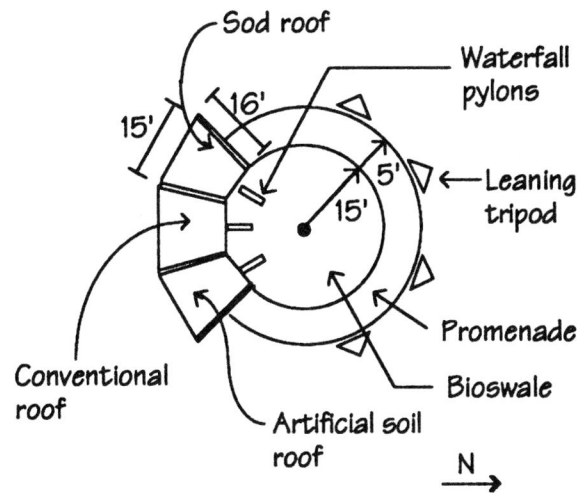

WATER AND WASTE

**Fig. 20.10** *The Eco-roof Pavilion is a demonstration structure at the confluence of the Columbia and Willamette rivers. Overlooking wetlands that further treat outflows from Portland's sewage treatment, this three-segment roof compares the rain retention characteristics of a sod roof, a conventional roof, and an artificial soil roof. Waterfall pylons will offer direct runoff comparisons. (Courtesy of Energy Studies in Buildings Laboratory, University of Oregon.)*

"incremental" paving, or open-celled pavers (alternation of paving materials with grass or ground-cover plants) for parking lots and roadways.

Porous concrete has been used for many years in building construction as a low-strength, high-porosity material that has some insulating properties (R-5 for 10-in. thickness). Patented porous concrete pavement now in use in Florida has a strength of more than 3800 psi and a permeability of 2.3 gallons of water per minute per square foot. (A 2500-psi mix has a permeability of 18.5 gallons per minute per square foot.) In cold weather areas, the freeze–thaw cycle could be destructive to porous concrete.

*Incremental paving* features many small, adjacent paving units; the resulting many joints allow

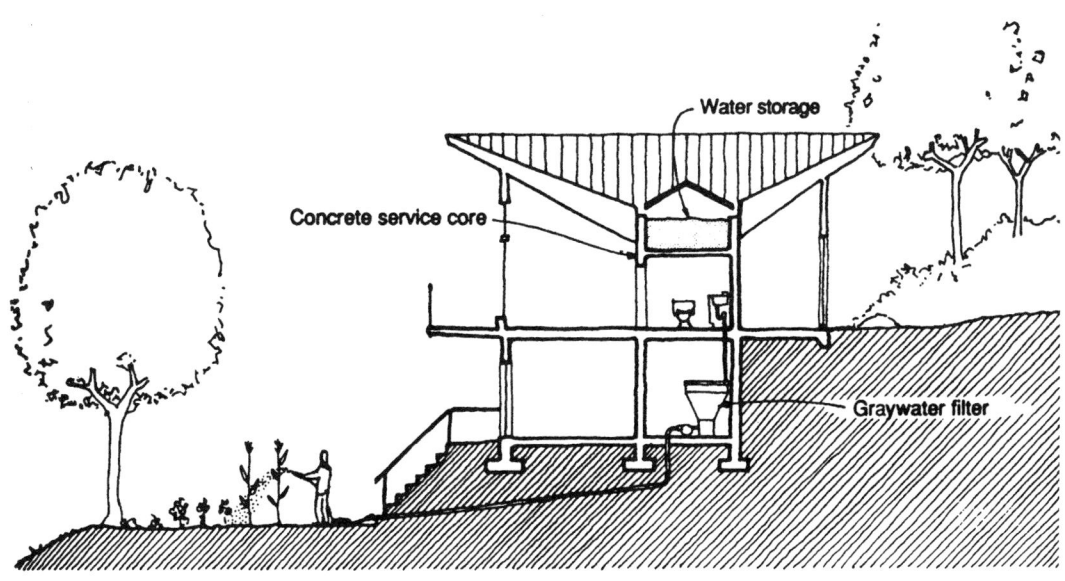

**Fig. 20.11** *Retaining rainwater so that gravity flow can encourage its usage.*

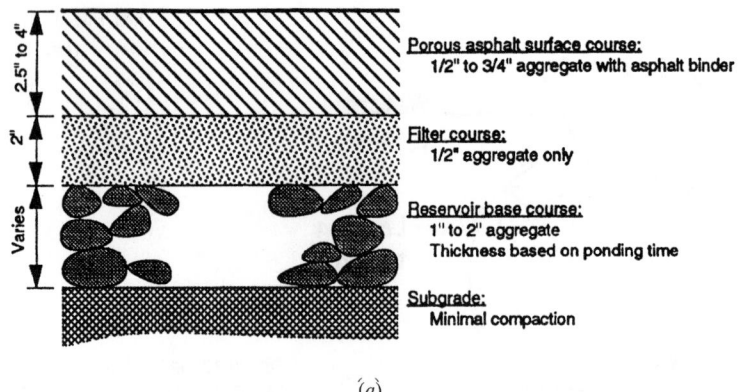

(a)

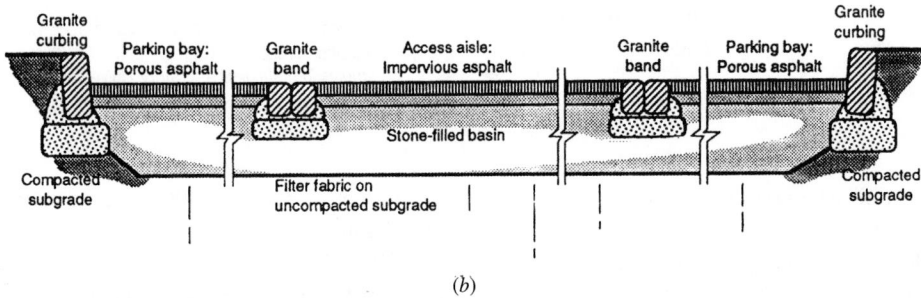

(b)

**Fig. 20.12** *Porous paving encourages groundwater recharge rather than storm runoff. (a) Standard porous asphalt pavement as used in Rockville, Maryland. (b) Subsurface basin allows required retention ponds to serve as parking lots; this example is at the Morris Arboretum in Philadelphia. (Reprinted by permission from Ferguson, B., 1998,* Introduction to Stormwater, © *John Wiley and Sons, New York.)*

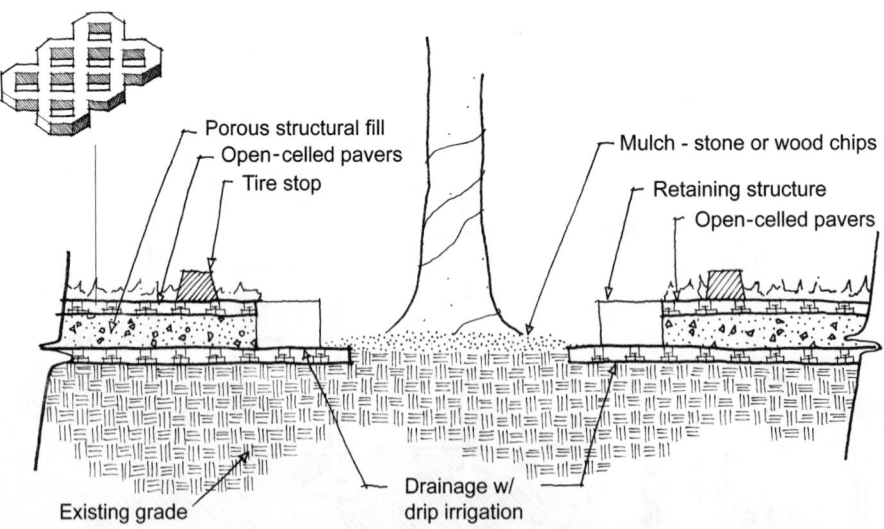

**Fig. 20.13** *Open-celled pavers on a porous fill allow grass or hardy ground cover to grow within the units. (Drawing by Erik Winter of GrassPave from Invisible Structures, Inc. Information used with permission.)*

water to pass through more readily than do solid, unbroken surfaces.

*Open-celled pavers* alternate small concrete or plastic paving units with grass or ground cover, as shown in Fig. 20.13. These are particularly applicable to short-term parking, as for a sports stadium, or for more remote areas of retail mall parking lots. They are also possible for roadway shoulders and emergency vehicle–only access areas.

If storm water is not retained, its runoff can be polluted to the extent that treatment is required. Using strips of planting for such treatment (phytoremediation) is discussed at the end of Chapter 22.

### (c) Site Design for Recharging

This tactic is especially advisable for suburban-density developments in drier climates with absorptive soil (sand, gravel, etc.). The first option is at each house; design of entire subdivisions can also be influenced.

The simplest design approach is the gutterless sloped roof illustrated in Fig. 20.14, which is applicable to one-story, basementless homes with wide, overhanging roofs. A gravel-filled trench skirting the perimeter directly below the edge of the eaves catches the water flowing off the roof.

Some designers do not like the appearance of conventional gutters and leaders; other designers celebrate them (Fig. 20.15). There are many ingenious ways to avoid or modify their use and yet provide proper drainage. In many cases, however, gutters and leaders will be required, either to collect rainwater for cisterns or to control conditions around the perimeter of a house. Several options for storm water recharge can be used with the gutter-leader combination.

A splash pan at the foot of each leader (Fig. 20.16a) offers the simplest method. It will lead the water a few feet from the house but will accommodate only a relatively low rate of flow. A gravel-filled pipe is somewhat more effective (Fig. 20.16b). When the soil is not very permeable (as, for instance, with clay), it is best to use a dry well with an extended area and many perforations through which the water can be discharged to the ground (Fig. 20.16c).

Footing drains are often used to collect and lead away groundwater that accumulates around foundations. This reduces the likelihood of basement wall leakage. These drains are most necessary

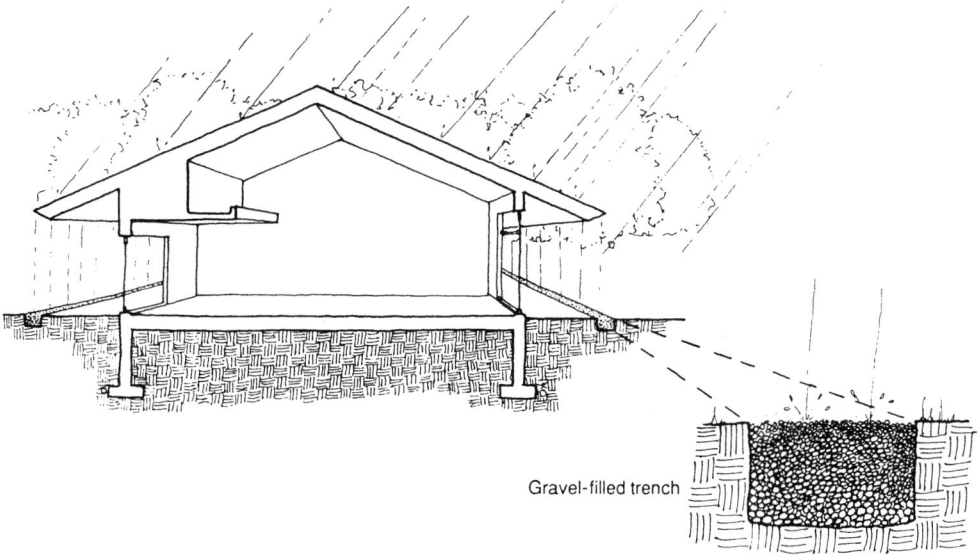

Gravel-filled trench

**Fig. 20.14** *Gutters and leaders are not always essential, provided that doorways, walls, foundations, and landscaped lawns are not subjected to rain concentrations.*

(a)

(b)

**Fig. 20.15** *Falling rain is featured at this Oregon residence, where the gutter is held below the metal flashing at the roof's edge. Runoff is seen from indoors as it falls into the suspended gutter. (Unthank Poticha Waterbury, Architects, Eugene, OR.)*

when higher ground near a building increases the flow of groundwater against underground walls. Figure 20.17 illustrates this and also shows how storm water from drains and roofs may be led to a surface absorption area of rock and gravel beyond a head wall where the general storm drain outcrops.

This method can be chosen where there is sufficient property area and slope. It has the advantages of easy maintenance and service. Also, one can observe whether it is functioning correctly.

In new suburban developments, for which there are no storm sewers, recharge basins (Fig. 20.18)

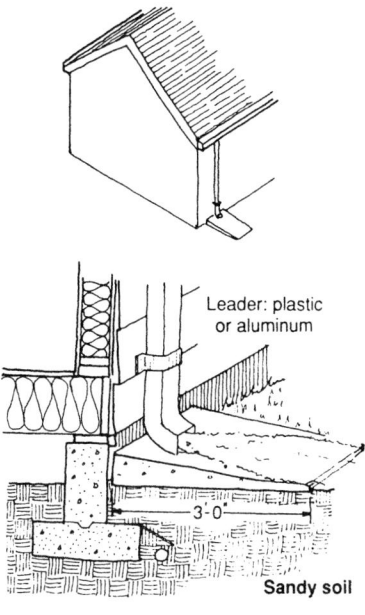

(a) Concrete splash pan

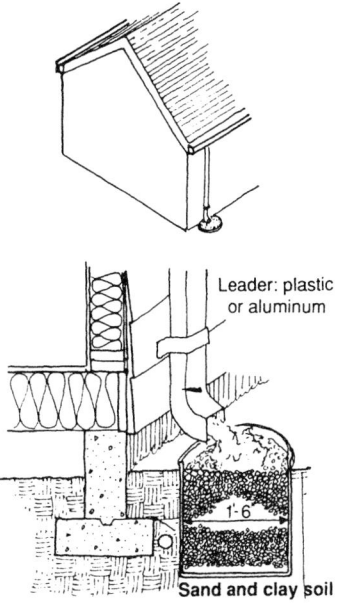

(b) Gravel-filled pipe, vitrified
clay or reinforced concrete

**Fig. 20.16** *Roof drainage for houses. Method (a) is suitable for low rates of flow introduced into very pervious soil. When denser soil is encountered, method (b) is used to get the water into the ground and thus avoid surface erosion. For heavy flow or to lead the water farther from the structure, method (c) may be used with one or several dry wells.*

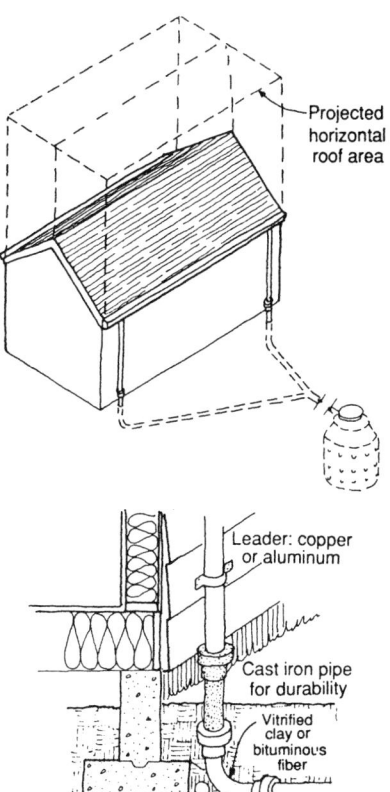

(c) Dry well

are sometimes required to deliver storm water to the ground. Water from numerous roofs, paved areas, and curb catch basins is collected and piped to an open, unpaved pit, where it sinks into the earth. This method is not recommended in areas of dense, impervious clay soil. A particularly effective example of this approach is offered by the community of passively solar-heated residences known as Village Homes, outside Davis, California. This area receives only about 20 in. (about 500 mm) of rain annually, so the recharge of groundwater was to this garden-oriented community a far more attractive (and less expensive) option than loss of the rainwater to a storm sewer. Storm water flows from leaders to dry, rockbed channels, along which are gardens and bicycle paths (Fig. 20.19). Occasionally, small dams across these channels create temporary holding ponds in case the runoff has not yet soaked through the channel bottom. In the event of extraordinarily heavy rainfall, an inlet to the public

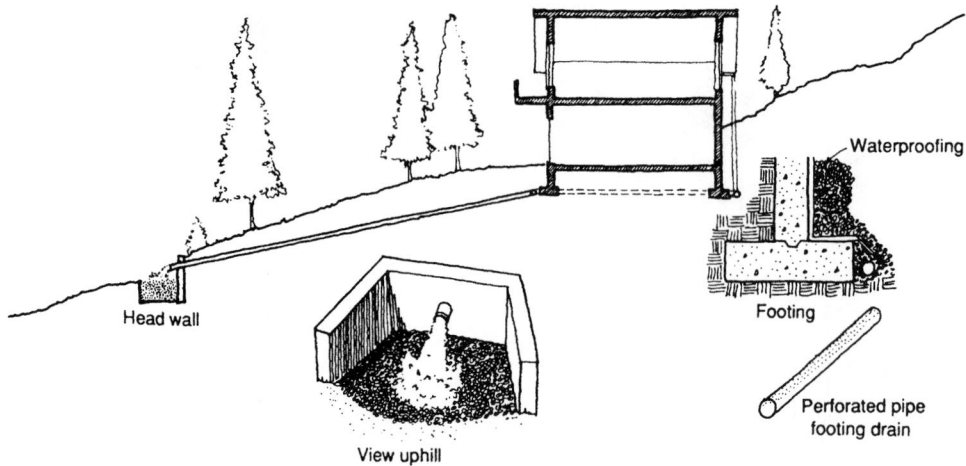

Waterproofing

Head wall

Footing

Perforated pipe
footing drain

View uphill

**Fig. 20.17** *Disposal of storm water on the site but remote from the house or building. When a wall is against a hill, it is usually subjected to the pressure of groundwater during storms. Open-joint clay, plastic, or fiber tile accepts this water and carries it away. Footing drains are tight-joint clay tile or bituminous fiber pipe. Flow through stone and gravel returns the water to the earth. A head wall is appropriate in lieu of a dry well if the site permits.*

storm sewer is available beyond the final holding pond; this inlet is needed approximately once every 5 years.

The National Wildflower Research Center (Fig. 20.20) is built over an endangered aquifer southwest of Austin, Texas. This complex of small buildings and larger gardens includes two cisterns, one of which forms the base of a central observation tower. Water gardens and irrigation are both served by the cisterns. The buildings' roofs are designed for rainwater collection, and a small channel (aqueduct) carries water over a wall to the outer cistern.

The planning of the landscape around buildings may closely follow such considerations as irrigation. The *hydrozone* concept of landscape planning is shown in Fig. 20.21. To minimize water consumption, exotic plant species are kept to a minimum and located near the house, where storm runoff and irrigating water are readily available. Native and adapted plantings, which can survive on normal rainfall, are utilized elsewhere. Landscape that requires no additional water is often called a *xeriscape;* see Fig. 21.68, and see Section 21.12 for a discussion of irrigation systems and an example in California's Napa Valley.

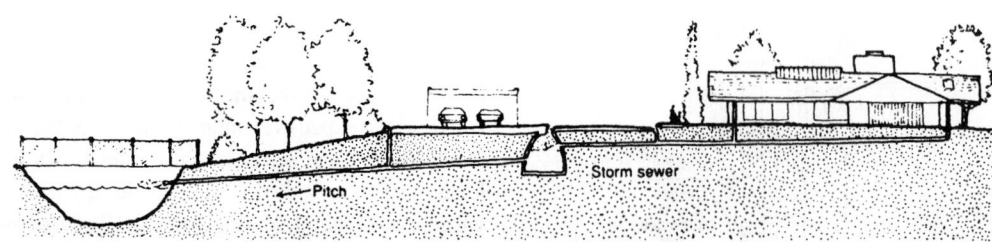

Pitch

Storm sewer

**Fig. 20.18** *Recharge basins in suburban communities. When topography, groundwater level, and porosity of the soil permit, developers are sometimes required to install systems that collect stormwater and carry it from catch basins at street curbs, the roofs of all houses, and paved areas to a recharge basin that receives the water and returns it to the ground. For the safety of children, a fence is sometimes required to prevent unauthorized access to the basin.*

(a)

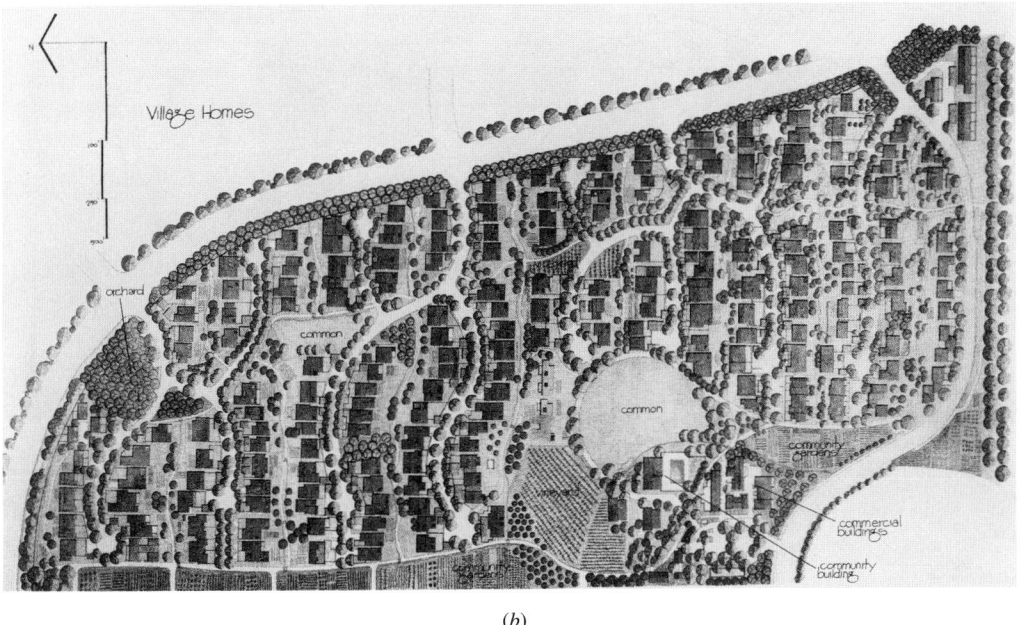

(b)

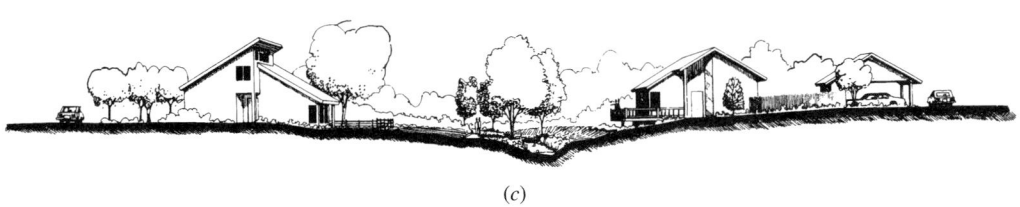

(c)

**Fig. 20.19** *Village Homes, a northern California subdivision of solar homes and storm water recharge areas. (a) Photo illustrates the open drainage ways and pedestrian paths. (Photo by Alan Butler.) (b) Plan shows the emphasis on bicycle paths, narrow streets, and widespread community-maintained garden space through which the recharge streambeds are led. (c) Site section explains the gradual drainage from leaders at houses to recharge stream. (Reprinted by permission from M. Corbett,* A Better Place to Live, *© 1981 by the Rodale Press.)*

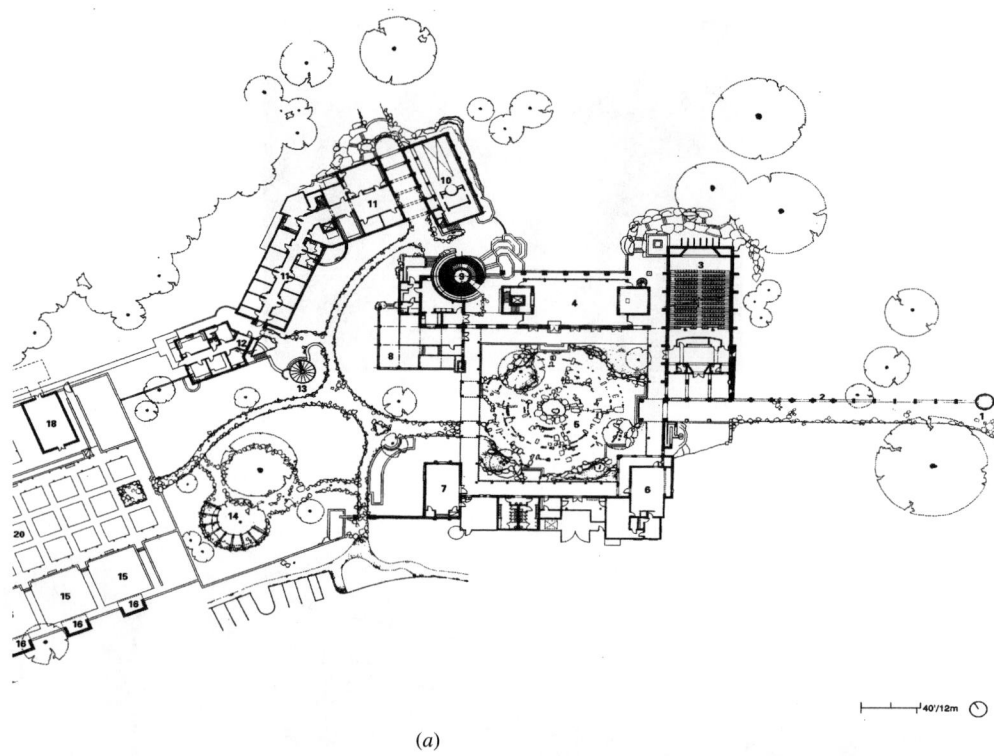

(a)

**Fig. 20.20** *The National Wildflower Research Center is designed to capture and utilize rainwater. (a) Site plan shows two cisterns, one at the far right (1), the other in the central observation tower (9). (b) Section through the observation tower/cistern. (Courtesy of Overland Partners, architects, San Antonio, TX.) 1. Cistern; 2. Aqueduct; 3. Auditorium; 4. Gallery; 5. Main plaza; 6. Gift shop; 7. Children's discovery room; 8. Kitchen; 9. Observation tower/cistern; 10. Library/ board room; 11. Administration; 12. Research laboratories; 13. Seed silo; 14. Pergola; 15. Comparative garden; 16. Pavilion; 17. Display greenhouse; 18. Research greenhouse; 19. Shade house; 20. Demonstration garden.*

1   WATER CISTERN
2   STAIRWELL
3   CISTERN ROTUNDA
4   OCULUS
5   OBSERVATION
      PLATFORM

TOWER SECTION

(b)

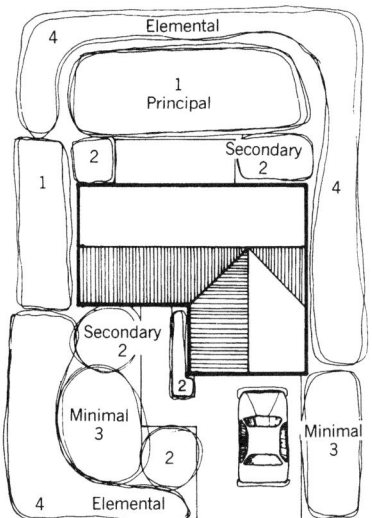 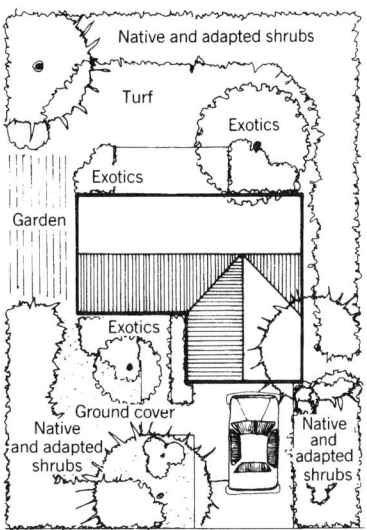

**Fig. 20.21** *The Hydrozone concept of landscape planning restricts exotic planting to special, easily watered areas. Native and adapted plantings are used elsewhere. Hydrozonics are here shown on a typical suburban lot. (Reprinted by permission from* Energy-Conserving Site Design, *E. G. McPherson, ed., © 1984 by the American Society of Landscape Architects.)*

## 20.7 COMPONENTS

The first storm water system design decision to be made involves the establishment of "watersheds" on a building's roof. To what edges, or at what points, will runoff be directed? To what depths will it accumulate before it leaves the roof? Because the answers to these questions depend on the intensity of storms as well as on the roof's geometry, it is necessary to find the maximum hourly rainfall for each location. This figure is available from local building code officials, or from Fig. 20.22.

*Gutters* and *leaders* (downspouts) can be sized through the use of Tables 20.7 and 20.8. The sizing of gutters and leaders depends both on the horizontal projected area of the roof, as shown in Fig. 20.16, and on the maximum hourly rainfall.

were designed, then at ½-in. slope per foot (4%), only a 4-in. diameter gutter would be required.)

Because two leaders will be used, each will drain 350 ft². Table 20.8 shows that a 2-in. leader can be used. For this gutter-leader combination, specify the detail of Fig. 20.23a.

Storm gutters and leaders can have an important impact on a building's appearance (Fig. 20.24). Alternatively, leaders can be set within buildings, and gutters can be built into a roof's surface to minimize the visual impact.

Where routing of storm water inside a building is preferable, drains and leaders can be sized from Table 20.8, and horizontal piping can be sized from Table 20.9. Care should be taken to insulate such lines; cold rainwater inside pipes can cause condensation to form on the outside pipe surface, sometimes resulting in staining and other water damage. ∎

**EXAMPLE 20.2** Select a gutter and two leaders for the front half of a house, as shown in Fig. 20.16. The rainfall rate is 4 in./h. The projected roof area is 700 ft², and the slope of the gutter is ⅟₁₆ in. in 1 ft of length (0.5%).

### SOLUTION

From Table 20.7, choose a semicircular gutter with a 6-in. diameter. (Note that if a steeper gutter slope

**EXAMPLE 20.3** Select the sizes for vertical conductors and horizontal storm drains for the building shown in Fig. 20.25. The roof, balcony, and courtyard areas are as shown, the rainfall rate is 4 in./h, and the pitch of horizontal drains is ¼-in. slope in 1 ft of run (2%).

### SOLUTION

The sizes selected and shown in Fig. 20.25 may be verified in Tables 20.8 and 20.20. ∎

*(a)*

**Fig. 20.22** *Maximum 100-year, 1-hour rainfall (in inches). (a) eastern United States; (b) central United States; (c) western United States. (Note: Alaska, range 0.4 to 1.4 in.; Hawaii, range 1.5 to 8 in.)* (International Plumbing Code. *Reprinted with permission. All rights reserved.* © 1997, International Code Council, Inc., Falls Church, VA.)

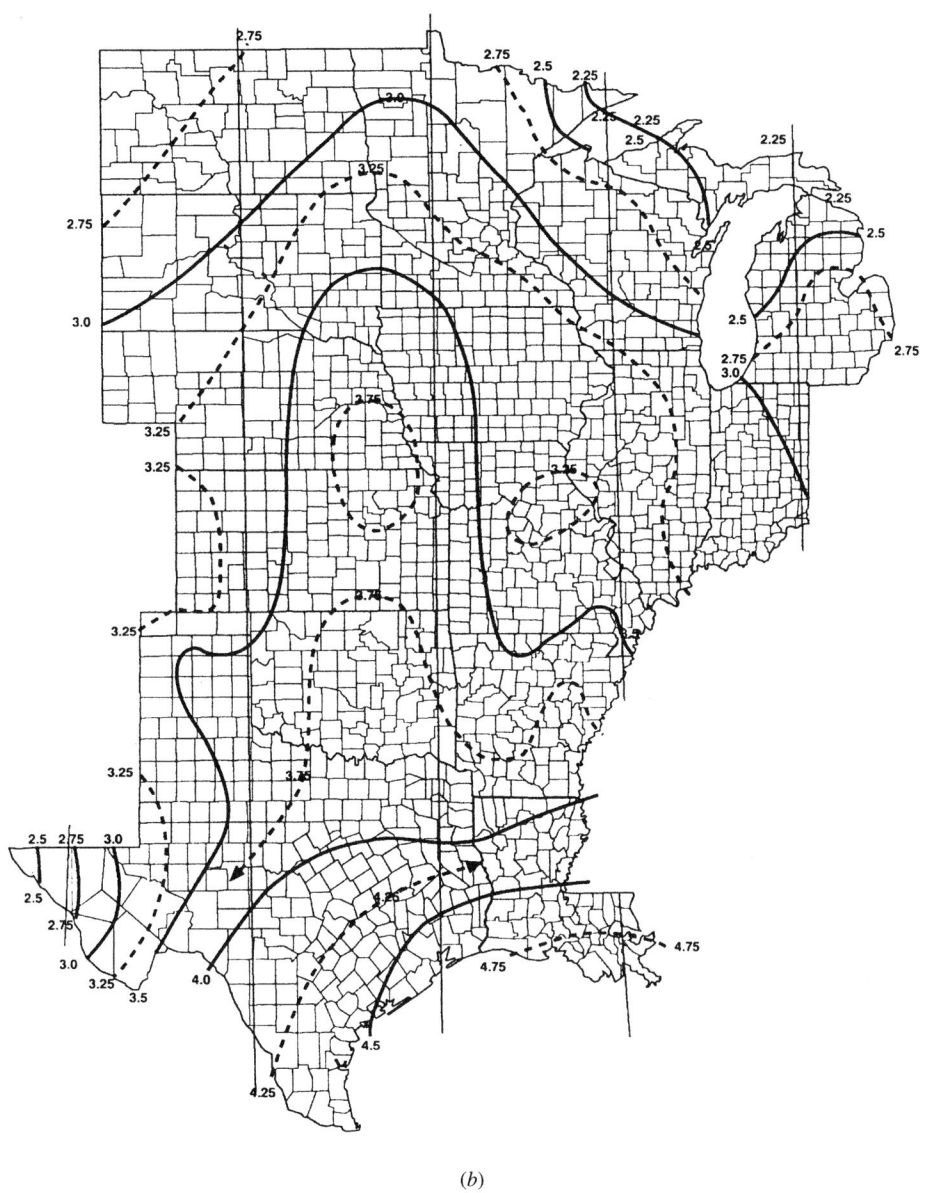

(b)

***Fig. 20.22*** (Continued)

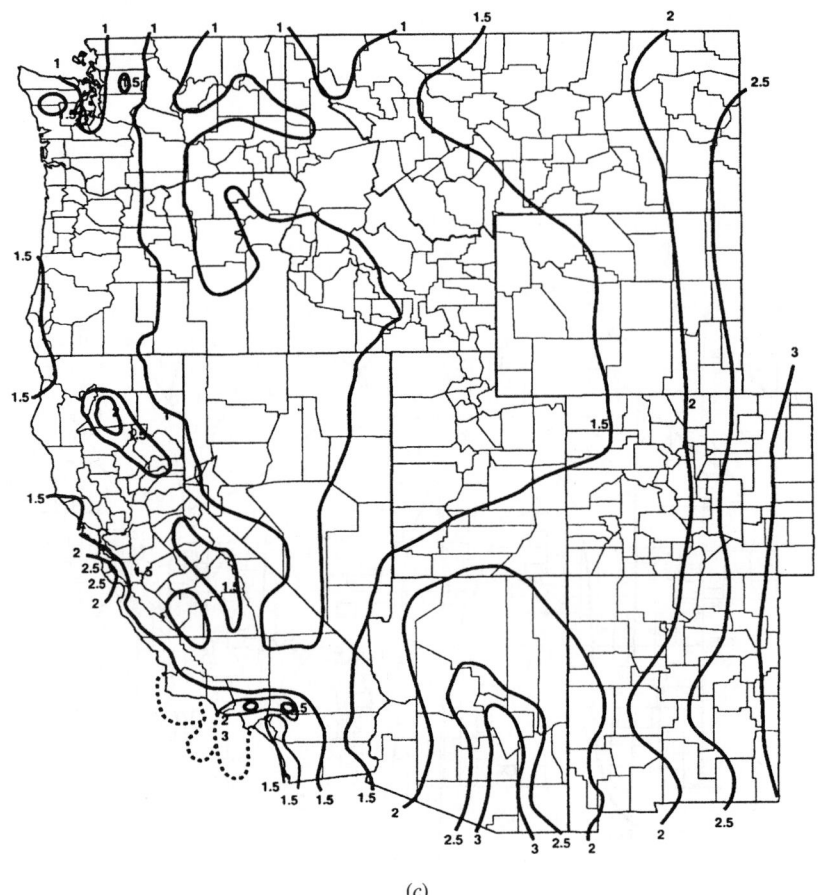

(c)

**Fig. 20.22** (Continued)

## TABLE 20.7  Size of Gutters

| PART A. I-P UNITS | | | | | |
|---|---|---|---|---|---|
| | **Maximum Rainfall (in. per h)** | | | | |
| Diameter of Gutter (in.) ¹⁄₁₆ in./ft Slope | **2** | **3** | **4** | **5** | **6** |
| 3 | 340 | 226 | 170 | 136 | 113 |
| 4 | 720 | 480 | 360 | 288 | 240 |
| 5 | 1250 | 834 | 625 | 500 | 416 |
| 6 | 1920 | 1280 | 960 | 768 | 640 |
| 7 | 2760 | 1840 | 1380 | 1100 | 918 |
| 8 | 3980 | 2655 | 1990 | 1590 | 1325 |
| 10 | 7200 | 4800 | 3600 | 2880 | 2400 |
| | **Maximum Rainfall (in. per h)** | | | | |
| Diameter of Gutter (in.) ⅛ in./ft Slope | **2** | **3** | **4** | **5** | **6** |
| 3 | 480 | 320 | 240 | 192 | 160 |
| 4 | 1020 | 681 | 510 | 408 | 340 |
| 5 | 1760 | 1172 | 880 | 704 | 587 |
| 6 | 2720 | 1815 | 1360 | 1085 | 905 |
| 7 | 3900 | 2600 | 1950 | 1560 | 1300 |
| 8 | 5600 | 3740 | 2800 | 2240 | 1870 |
| 10 | 10,200 | 6800 | 5100 | 4080 | 3400 |

## TABLE 20.7  (Continued)

| Diameter of Gutter (in.) ¼ in./ft Slope | Maximum Rainfall (in. per h) | | | | |
|---|---|---|---|---|---|
| | 2 | 3 | 4 | 5 | 6 |
| 3 | 680 | 454 | 340 | 272 | 226 |
| 4 | 1440 | 960 | 720 | 576 | 480 |
| 5 | 2500 | 1668 | 1250 | 1000 | 834 |
| 6 | 3840 | 2560 | 1920 | 1536 | 1280 |
| 7 | 5520 | 3680 | 2760 | 2205 | 1840 |
| 8 | 7960 | 5310 | 3980 | 3180 | 2655 |
| 10 | 14,400 | 9600 | 7200 | 5750 | 4800 |

| Diameter of Gutter (in.) ½ in./ft Slope | Maximum Rainfall (in. per h) | | | | |
|---|---|---|---|---|---|
| | 2 | 3 | 4 | 5 | 6 |
| 3 | 960 | 640 | 480 | 384 | 320 |
| 4 | 2040 | 1360 | 1020 | 816 | 680 |
| 5 | 3540 | 2360 | 1770 | 1415 | 1180 |
| 6 | 5540 | 3695 | 2770 | 2220 | 1850 |
| 7 | 7800 | 5200 | 3900 | 3120 | 2600 |
| 8 | 11,200 | 7460 | 5600 | 4480 | 3730 |
| 10 | 20,000 | 13,330 | 10,000 | 8000 | 6660 |

PART B. SI UNITS

| Diameter of Gutter (mm) 5.2 mm/m Slope | Maximum Rainfall (mm per h) | | | | |
|---|---|---|---|---|---|
| | 50.8 | 76.2 | 101.6 | 127.0 | 152.4 |
| 76.2 | 31.6 | 21.0 | 15.8 | 12.6 | 10.5 |
| 101.6 | 66.9 | 44.6 | 33.4 | 26.8 | 22.3 |
| 127.0 | 116.1 | 77.5 | 58.1 | 46.5 | 38.7 |
| 152.4 | 178.4 | 119.1 | 89.2 | 71.4 | 59.5 |
| 177.8 | 256.4 | 170.9 | 128.2 | 102.2 | 85.3 |
| 203.2 | 369.7 | 246.7 | 184.9 | 147.7 | 123.1 |
| 254.0 | 668.9 | 445.9 | 334.4 | 267.6 | 223.0 |

| Diameter of Gutter (mm) 10.4 mm/m Slope | Maximum Rainfall (mm per h) | | | | |
|---|---|---|---|---|---|
| | 50.8 | 76.2 | 101.6 | 127.0 | 152.4 |
| 76.2 | 44.6 | 29.7 | 22.3 | 17.8 | 14.9 |
| 101.6 | 94.8 | 63.3 | 47.4 | 37.9 | 31.6 |
| 127.0 | 163.5 | 108.9 | 81.8 | 65.4 | 54.5 |
| 152.4 | 252.7 | 168.6 | 126.3 | 100.8 | 84.1 |
| 177.8 | 362.3 | 241.5 | 181.2 | 144.9 | 120.8 |
| 203.2 | 520.2 | 347.5 | 260.1 | 208.1 | 173.7 |
| 254.0 | 947.6 | 631.7 | 473.8 | 379.0 | 315.9 |

| Diameter of Gutter (mm) 20.9 mm/m Slope | Maximum Rainfall (mm per h) | | | | |
|---|---|---|---|---|---|
| | 50.8 | 76.2 | 101.6 | 127.0 | 152.4 |
| 76.2 | 63.2 | 42.2 | 31.6 | 25.3 | 21.0 |
| 101.6 | 133.8 | 89.2 | 66.9 | 53.5 | 44.6 |
| 127.0 | 232.3 | 155.0 | 116.1 | 92.9 | 77.5 |
| 152.4 | 356.7 | 237.8 | 178.4 | 142.7 | 118.9 |
| 177.8 | 512.8 | 341.9 | 256.4 | 204.9 | 170.9 |
| 203.2 | 739.5 | 493.3 | 369.4 | 295.4 | 246.7 |
| 254.0 | 133.8 | 891.8 | 668.9 | 534.2 | 445.9 |

| Diameter of Gutter (mm) 41.7 mm/m Slope | Maximum Rainfall (mm per h) | | | | |
|---|---|---|---|---|---|
| | 50.8 | 76.2 | 101.6 | 127.0 | 152.4 |
| 76.2 | 89.2 | 59.5 | 44.6 | 35.7 | 29.7 |
| 101.6 | 189.5 | 126.3 | 94.8 | 75.8 | 63.2 |
| 127.0 | 328.9 | 219.2 | 164.4 | 131.5 | 109.6 |
| 152.4 | 514.7 | 343.2 | 257.3 | 206.2 | 171.9 |
| 177.8 | 724.6 | 483.1 | 362.3 | 289.9 | 241.4 |
| 203.2 | 1040.5 | 693.0 | 520.2 | 416.2 | 346.5 |
| 254.0 | 1858.0 | 1234.4 | 929.0 | 743.2 | 618.7 |

*Source:* Reprinted by permission from the 1997 *Uniform Plumbing Code.* © by the International Association of Plumbing and Mechanical Officials, Walnut, CA.

**WATER AND WASTE**

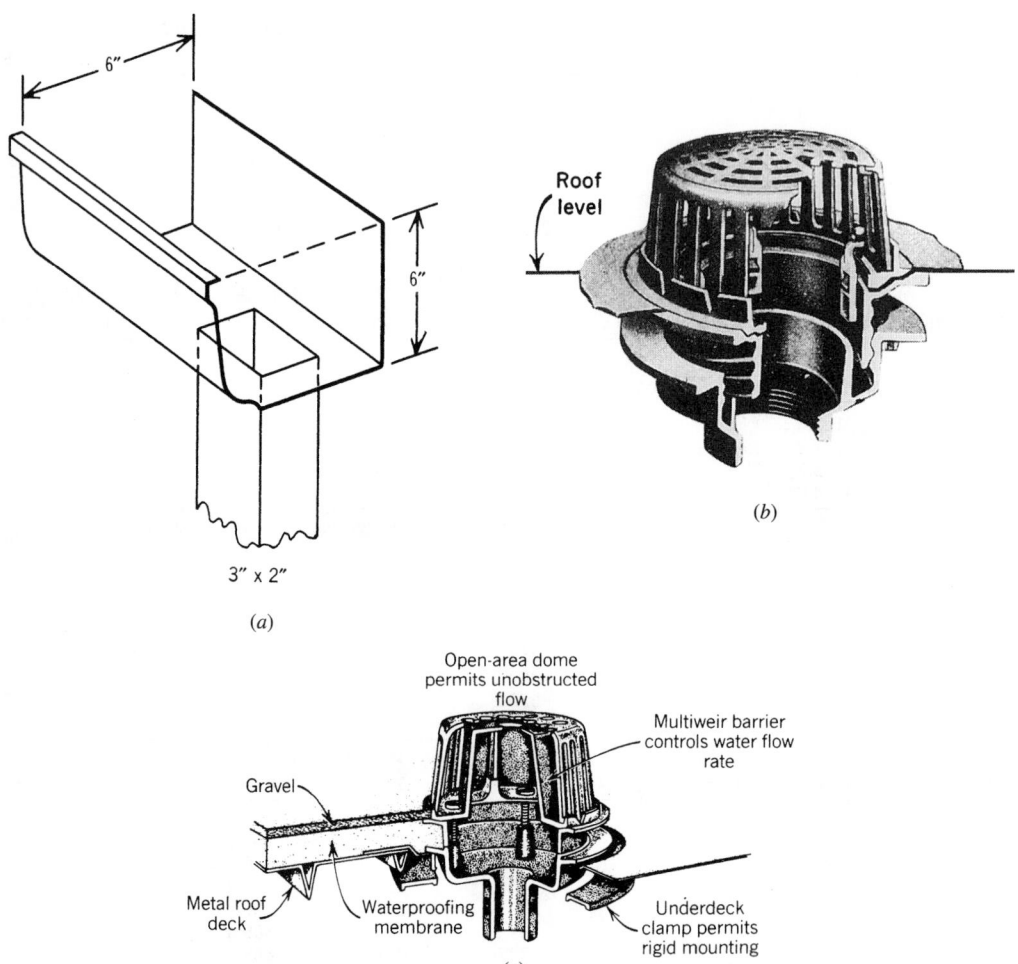

**Fig. 20.23** *Storm drainage components. (a) Conventional gutter and leader for houses; sizes vary by manufacturer. (b) Ordinary roof drain (Josam Manufacturing Co.). (c) Roof drain for controlled flow. (From* Specifying Engineer, *November 1982.)*

### TABLE 20.8 Sizing Roof Drains, Leaders, and Vertical Rainwater Piping

NOTES: The sizing data for vertical conductors, leaders, and drains is based on the pipes flowing ⁷⁄₂₄ full. For rainfall rates other than those listed, determine the allowable roof area by dividing the area given in the 1 inch/hour (25 mm/hour) column by the desired rainfall rate. Vertical piping may be round, square, or rectangular. Square pipe shall be sized to enclose its equivalent round pipe. Rectangular pipe shall have at least the same cross-sectional area as its equivalent round pipe, except that the ratio of its side dimensions shall not exceed 3 to 1.

| PART A. I-P UNITS | | | | | | | |
|---|---|---|---|---|---|---|---|
| Size of Drain, Leader or Pipe (in.) | Flow (gpm) | Maximum Allowable Horizontal Projected Roof Areas Square Feet at Various Rainfall Rates | | | | | |
| | | 1 in./h | 2 in./h | 3 in./h | 4 in./h | 5 in./h | 6 in./h |
| 2 | 23 | 2176 | 1088 | 725 | 544 | 435 | 363 |
| 3 | 67 | 6440 | 3220 | 2147 | 1610 | 1288 | 1073 |
| 4 | 144 | 13,840 | 6920 | 4613 | 3460 | 2768 | 2307 |
| 5 | 261 | 25,120 | 12,560 | 8373 | 6280 | 5024 | 4187 |
| 6 | 424 | 40,800 | 20,400 | 13,600 | 10,200 | 8160 | 6800 |
| 8 | 913 | 88,000 | 44,000 | 29,333 | 22,000 | 17,600 | 14,667 |

| PART B. SI UNITS | | | | | | | |
|---|---|---|---|---|---|---|---|
| Size of Drain, Leader or Pipe (mm) | Flow (L/s) | Maximum Allowable Horizontal Projected Roof Areas Square Meters at Various Rainfall Rates | | | | | |
| | | 25 mm/h | 50 mm/h | 75 mm/h | 100 mm/h | 125 mm/h | 150 mm/h |
| 50 | 1.5 | 202 | 101 | 67 | 51 | 40 | 34 |
| 75 | 4.2 | 600 | 300 | 200 | 150 | 120 | 100 |
| 100 | 9.1 | 1286 | 643 | 429 | 321 | 257 | 214 |
| 125 | 16.5 | 2334 | 1117 | 778 | 583 | 467 | 389 |
| 150 | 26.8 | 3790 | 1895 | 1263 | 948 | 758 | 632 |
| 200 | 57.6 | 8175 | 4088 | 2725 | 2044 | 1635 | 1363 |

*Source:* Reprinted by permission from the 1997 *Uniform Plumbing Code.* © by the International Association of Plumbing and Mechanical Officials, Walnut, CA.

***Fig. 20.24*** *A downspout terminates the roof with flair at the Pleasanton BART (Bay Area Rapid Transit) station near San Francisco. (Photo by Jane Lidz.)*

WATER AND WASTE

### TABLE 20.9 Sizing of Horizontal Rainwater Piping

NOTES: The sizing data for horizontal piping is based on the pipes flowing full. For rainfall rates other than those listed, determine the allowable roof area by dividing the area given in the 1 inch/hour (25 mm/hour) column by the desired rainfall rate.

| PART A. I-P UNITS | | | | | | | |
|---|---|---|---|---|---|---|---|
| Size of Pipe (in.) | Flow at ⅛ in./ft Slope (gpm) | Maximum Allowable Horizontal Projected Roof Areas Square Feet at Various Rainfall Rates | | | | | |
| | | 1 in./h | 2 in./h | 3 in./h | 4 in./h | 5 in./h | 6 in./h |
| 3 | 34 | 3288 | 1644 | 1096 | 822 | 657 | 548 |
| 4 | 78 | 7520 | 3760 | 2506 | 1880 | 1504 | 1253 |
| 5 | 139 | 13,360 | 6680 | 4453 | 3340 | 2672 | 2227 |
| 6 | 222 | 21,400 | 10,700 | 7133 | 5350 | 4280 | 3566 |
| 8 | 478 | 46,000 | 23,000 | 15,330 | 11,500 | 9200 | 7670 |
| 10 | 860 | 82,800 | 41,400 | 27,600 | 20,700 | 16,580 | 13,800 |
| 12 | 1384 | 133,200 | 66,600 | 44,400 | 33,300 | 26,650 | 22,200 |
| 15 | 2473 | 238,000 | 119,000 | 79,333 | 59,500 | 47,600 | 39,650 |
| Size of Pipe (in.) | Flow at ¼ in./ft Slope (gpm) | Maximum Allowable Horizontal Projected Roof Areas Square Feet at Various Rainfall Rates | | | | | |
| | | 1 in./h | 2 in./h | 3 in./h | 4 in./h | 5 in./h | 6 in./h |
| 3 | 48 | 4640 | 2320 | 1546 | 1160 | 928 | 773 |
| 4 | 110 | 10,600 | 5300 | 3533 | 2650 | 2120 | 1766 |
| 5 | 196 | 18,880 | 9440 | 6293 | 4720 | 3776 | 3146 |
| 6 | 314 | 30,200 | 15,100 | 10,066 | 7550 | 6040 | 5033 |
| 8 | 677 | 65,200 | 32,600 | 21,733 | 16,300 | 13,040 | 10,866 |
| 10 | 1214 | 116,800 | 58,400 | 38,950 | 29,200 | 23,350 | 19,450 |
| 12 | 1953 | 188,000 | 94,000 | 62,600 | 47,000 | 37,600 | 31,350 |
| 15 | 3491 | 336,000 | 168,000 | 112,000 | 84,000 | 67,250 | 56,000 |
| Size of Pipe (in.) | Flow at ½ in./ft Slope (gpm) | Maximum Allowable Horizontal Projected Roof Areas Square Feet at Various Rainfall Rates | | | | | |
| | | 1 in./h | 2 in./h | 3 in./h | 4 in./h | 5 in./h | 6 in./h |
| 3 | 68 | 6576 | 3288 | 2192 | 1644 | 1310 | 1096 |
| 4 | 156 | 15,040 | 7520 | 5010 | 3760 | 3010 | 2500 |
| 5 | 278 | 26,720 | 13,360 | 8900 | 6680 | 5320 | 4450 |
| 6 | 445 | 42,800 | 21,400 | 14,267 | 10,700 | 8580 | 7140 |
| 8 | 956 | 92,000 | 46,000 | 30,650 | 23,000 | 18,400 | 15,320 |
| 10 | 1721 | 165,600 | 82,800 | 55,200 | 41,400 | 33,150 | 27,600 |
| 12 | 2768 | 266,400 | 133,200 | 88,800 | 66,600 | 53,200 | 44,400 |
| 15 | 4946 | 476,000 | 238,000 | 158,700 | 119,000 | 95,200 | 79,300 |
| PART B. SI UNITS | | | | | | | |
| Size of Pipe (mm) | Flow at 10 mm/m Slope (L/s) | Maximum Allowable Horizontal Projected Roof Areas Square Meters at Various Rainfall Rates | | | | | |
| | | 25 mm/h | 50 mm/h | 75 mm/h | 100 mm/h | 125 mm/h | 150 mm/h |
| 75 | 2.1 | 305 | 153 | 102 | 76 | 61 | 51 |
| 100 | 4.9 | 700 | 350 | 233 | 175 | 140 | 116 |
| 125 | 8.8 | 1241 | 621 | 414 | 310 | 248 | 207 |
| 150 | 14.0 | 1988 | 994 | 663 | 497 | 398 | 331 |
| 200 | 30.2 | 4273 | 2137 | 1424 | 1068 | 855 | 713 |
| 250 | 54.3 | 7692 | 3846 | 2564 | 1923 | 1540 | 1282 |
| 300 | 87.3 | 12,375 | 6187 | 4125 | 3094 | 2476 | 2062 |
| 375 | 156.0 | 22,110 | 11,055 | 7370 | 5528 | 4422 | 3683 |

## TABLE 20.9 (Continued)

| Size of Pipe (mm) | Flow at 20 mm/m Slope (L/s) | Maximum Allowable Horizontal Projected Roof Areas Square Meters at Various Rainfall Rates | | | | | |
|---|---|---|---|---|---|---|---|
| | | 25 mm/h | 50 mm/h | 75 mm/h | 100 mm/h | 125 mm/h | 150 mm/h |
| 75 | 3.0 | 431 | 216 | 144 | 108 | 86 | 72 |
| 100 | 6.9 | 985 | 492 | 328 | 246 | 197 | 164 |
| 125 | 12.4 | 1754 | 877 | 585 | 438 | 351 | 292 |
| 150 | 19.8 | 2806 | 1403 | 935 | 701 | 561 | 468 |
| 200 | 42.7 | 6057 | 3029 | 2019 | 1514 | 1211 | 1009 |
| 250 | 76.6 | 10,851 | 5425 | 3618 | 2713 | 2169 | 1807 |
| 300 | 123.2 | 17,465 | 8733 | 5816 | 4366 | 3493 | 2912 |
| 375 | 220.2 | 31,214 | 15,607 | 10,405 | 7804 | 6248 | 5202 |

| Size of Pipe (mm) | Flow at 40 mm/m Slope (L/s) | Maximum Allowable Horizontal Projected Roof Areas Square Meters at Various Rainfall Rates | | | | | |
|---|---|---|---|---|---|---|---|
| | | 25 mm/h | 50 mm/h | 75 mm/h | 100 mm/h | 125 mm/h | 150 mm/h |
| 75 | 4.3 | 611 | 305 | 204 | 153 | 122 | 102 |
| 100 | 9.8 | 1400 | 700 | 465 | 350 | 280 | 232 |
| 125 | 17.5 | 2482 | 1241 | 827 | 621 | 494 | 413 |
| 150 | 28.1 | 3976 | 1988 | 1325 | 994 | 797 | 663 |
| 200 | 60.3 | 8547 | 4273 | 2847 | 2137 | 1709 | 1423 |
| 250 | 108.6 | 15,390 | 7695 | 5128 | 3846 | 3080 | 2564 |
| 300 | 174.6 | 24,749 | 12,374 | 8250 | 6187 | 4942 | 4125 |
| 375 | 312.0 | 44,220 | 22,110 | 14,753 | 11,055 | 8853 | 7367 |

*Source:* Reprinted by permission from the 1997 *Uniform Plumbing Code.* Copyright © by the International Association of Plumbing and Mechanical Officials, Walnut, CA.

WATER AND WASTE

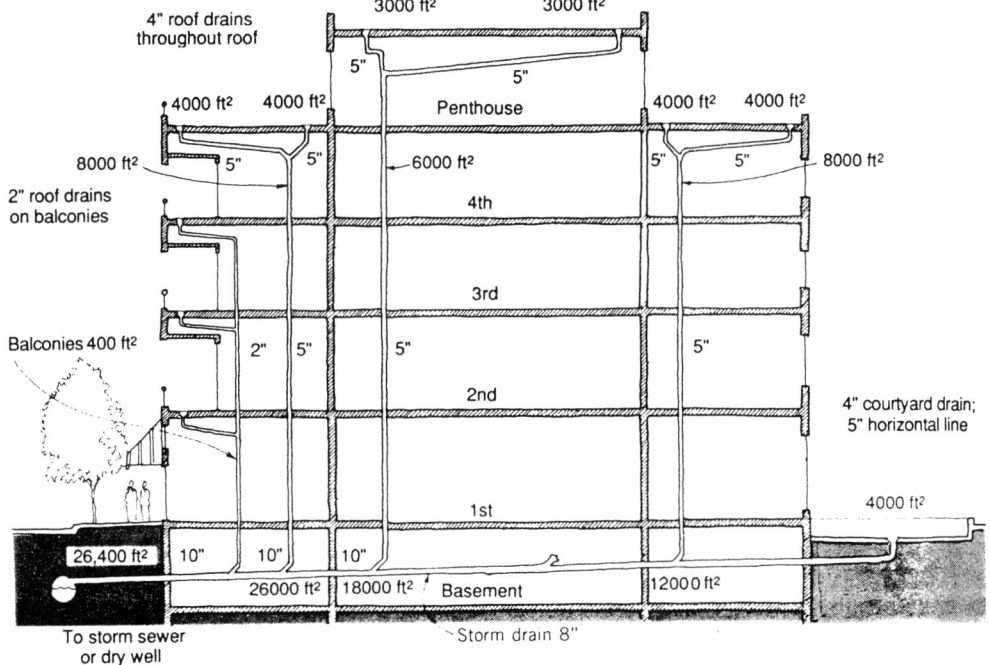

**Fig. 20.25** *Separate storm drainage. Areas drained and corresponding sizes of vertical leaders and horizontal drains are from Tables 20.8 and 20.9. Storm drain piping within a building needs insulative covering with a vapor retarder on the outside. This prevents condensation (sweating) on the pipes when, in winter, warm, moisture-laden air in the building could otherwise reach the pipe surface (which would be cold from carrying icy water), condense there, and lead to wet, dripping conditions on the pipes. Each roof has two drains in case one is temporarily blocked.*

WATER AND WASTE

## References

Brown, G.Z., J. Reynolds, and S. Ubbelohde. 1982. *InsideOut: Design Procedures for Passive Environmental Technologies.* John Wiley & Sons, New York.

Corbett, M. 1981. *A Better Place to Live.* Rodale Press, Emmaus, PA.

Ferguson, B. 1998. *Introduction to Stormwater.* John Wiley & Sons, New York.

Gleick, P. 1998. *The World's Water.* Island Press, Washington, DC.

International Association of Plumbing and Mechanical Officials. 1997. *Uniform Plumbing Code.* Walnut, CA.

International Code Council. 1997. *International Plumbing Code.* Falls Church, VA.

McPherson, E.G. (ed.). 1984. *Energy-Conserving Site Design.* American Society of Landscape Architects, Washington, DC.

Milne, M. 1976. *Residential Water Conservation,* U.S. Office of Water Research and Technology, Department of Commerce, NTIS, Springfield, VA.

U.S. Environmental Protection Agency. 1975. *Manual of Individual Water Supply Systems,* U.S. EPA, Washington, DC.

# CHAPTER 21

# Water Supply

WITH WATER, ONE OF THE DESIGNER'S FIRST concerns is to match the quality of the water to the task it performs. As this becomes a more serious design issue, designers will provide for the recycling of water within and around buildings, as well as specify plumbing fixtures that use less water. Table 21.1 shows typical relationships between water quality and usage. Water recycling and conservation are discussed in detail in Chapter 22. In this chapter we deal primarily with *potable* (drinkable) water: first with issues of water quality and then with the matter of assuring an adequate supply of water throughout a building.

## 21.1 WATER QUALITY

A summary of water quality at the various stages of the hydrologic cycle was shown in Fig. 20.4. As precipitation, water contains few impurities: almost no bacterial content is present, and only small amounts of minerals and gases can be expected. To collect this nearly pure water, surfaces are needed—and at this point, foreign substances can readily contaminate the water. These pollutants can affect water's physical (mostly organic), chemical (mostly inorganic), biological, or radiological characteristics. Both surface water and groundwater are subject to pollution.

The U.S. National Drinking Water Clearinghouse is a source of information on water supply and waste-water treatment systems. Its brief his-

tory of waterborne diseases worldwide begins with a major cholera epidemic that began in Calcutta, India, in 1817, killing thousands and spreading by 1832 to New York City, emptying the streets in panic over the disease. In 1854, a London physician demonstrated that local cases of cholera could be traced to one pump contaminated with sewage from a nearby house, but how the disease was transmitted remained a mystery. In summer 1859, London's Thames River, carrying combined storm and sanitary wastes, stank so badly that Parliament was suspended. In 1892, the bacterium causing cholera was identified during an epidemic in Hamburg, Germany, proving the relationship between contaminated water and the disease. In 1939, an outbreak of typhoid fever killed 60 people at an Illinois mental hospital; typhoid and gastroenteritis resulted from an accidental pumping of polluted river water into supply mains in Rochester, New York, in 1940. Another worldwide cholera epidemic began in Indonesia in 1961, eventually reaching Latin America by 1991. In 1993, an outbreak of cryptosporidiosis in the public water supply of Milwaukee, Wisconsin, shook the public's faith in water systems across the United States; that outbreak killed 104 and infected more than 400,000 people.

In response to this history, we disinfect huge quantities of water, expending energy and adding chemicals so that every drop of water entering North American buildings is potable. Again, the high-grade resource/low-grade use question arises;

**TABLE 21.1 Water Use and Quality in Buildings**

| Use | Desired Quality |
|---|---|
| A. CONSUMED | |
| 1. Drinking and cooking | Potable |
| 2. Bathing | Potable |
| 3. Laundering | Soft |
| 4. Irrigation and watering of livestock | Unpolluted |
| 5. Industrial processes | As required |
| 6. Vapor to increase the relative humidity of air | |
| B. CIRCULATED | |
| 1. Hot water for heating | *Note:* Make-up water should be soft or neutral and, for swimming, potable |
| 2. Chilled water for cooling | |
| 3. Condenser cooling water | |
| 4. Swimming pool water | |
| 5. Steam for heating, later condensed | |
| C. GENERALLY STATIC | |
| 1. Water stored for fire protection | No special requirement |
| 2. Water in fire stand pipes | |
| 3. Water in sprinkler piping | |
| D. CONTROLLED | |
| 1. Vapor condensed to reduce relative humidity of air | |

*Note:* For water uses in Section A, flow is often continuous. Section B comprises uses for which flow other than circulation is intermittent or at a relatively low rate, the water added to the systems being known as *make-up water.* Items C2 and C3 call for piping to provide adequate, though infrequent, flow in emergencies. Item D1 relates only to moisture condensed out of the air and involves no design for supply.

recycled water for lower-grade use is discussed in Chapter 22.

## (a) Physical Characteristics

Some of the most noticeable aspects of water quality fall within this category. Water from surface sources (roof runoff, streams, rivers, lakes, ponds) is particularly subject to physical pollutants.

*Turbidity* is easy to see and thus is a likely source of dissatisfaction for the would-be consumer. It is caused by the presence of suspended material such as clay, silt, other inorganic material, plankton, or finely divided organic material. Even those materials that do not adversely affect health are usually aesthetically objectionable.

*Color,* another visible alteration, is often caused by dissolved organic matter, as from decaying vegetation. Some inorganic materials also color water,

as do microorganisms. Like turbidity, such color changes usually do not threaten health, but they are often psychologically undesirable.

*Taste* and *odor* can be caused by organic compounds, inorganic salts, or dissolved gases. This condition can be treated only after a chemical analysis has identified which source is responsible.

*Temperature* is another characteristic of psychological importance—we expect drinking water to be cool. In general, water supplied between 50° and 60°F (10° and 16°C) is preferred.

*Foamability* is usually caused by concentrations of detergents. The foam itself does not pose a serious health threat, but it may indicate that other, more dangerous pollutants associated with domestic waste are also present. Because of increased foaming in water in the 1960s, today's detergents must use linear alkylate sulfonate (LAS), which biodegrades rapidly—except in the absence of oxygen. Because this lack of oxygen is characteristic of some septic tank drainage fields, foam in drinking water should be investigated promptly.

## (b) Chemical Characteristics

Groundwater is particularly subject to chemical alteration because as it moves downward from the surface it slowly dissolves some minerals contained in rocks and soils. A chemical analysis (such as that given in Table 21.2) is usually required for individual water supply sources. Such analysis will indicate (1) the possible presence of harmful or objectionable substances, (2) the potential for corrosion within the water supply system, and (3) the tendency for the water to stain fixtures and clothing. Concentrations are expressed in milligrams per liter (mg/L), which is essentially equivalent to parts per million (ppm).

Some general terms commonly used to describe chemical characteristics of water are as follows:

**Alkalinity.** Caused by bicarbonate, carbonate, or hydroxide components. Testing for these components of water's alkalinity is a key to determining which treatments to use.

**Hardness.** A relative term (see Fig. 21.1). Hard water inhibits the cleaning action of soaps and detergents, and it deposits scale on the inside of hot water pipes and cooking utensils, thus wasting heating fuel and making utensils

## TABLE 21.2 Example of Chemical Analysis of Water

| Quality | | Parts per Million (ppm)[a] |
|---|---|---|
| Total hardness | As $CaCO_3$ | 30 |
| Calcium hardness | As $CaCO_3$ | 20 |
| Alkalinity (methyl orange) | As $CaCO_3$ | 27 |
| Alkalinity (phenolphthalein) | As $CaCO_3$ | 0 |
| Free carbon dioxide | As $CO_2$ | 13.5 |
| Chlorides | As Cl | 6 |
| Sulfates | As $SO_4$ | 4 |
| Silica | As $SiO_2$ | 19 |
| Phosphates—normal | As $PO_4$ | 0 |
| Phosphates—total | As $PO_4$ | 0.5 |
| Iron—total | As Fe | 1.6 |
| Total dissolved solids | | 66 |
| Turbidity or sediment | | Present |

*Source:* A report by Olin Water Service for a private well in Virginia.

[a]Note that ppm and mg/L are essentially equivalent terms for concentration in water.

unusable. Hardness, which is caused by calcium and magnesium salts, can be classified as temporary (carbonate) or permanent (noncarbonate). Temporary hardness is largely removed when the water is heated—it forms the scale just described. Permanent hardness cannot be removed by simple heating (see Section 21.4*c*).

**pH.** A measure of the water's hydrogen ion concentration, as well as its relative acidity or alkalinity (Fig. 21.1). A pH of 7 is neutral. Measurements below 7 indicate increasing acidity (and corrosiveness); water in its natural state can have a pH as low as 5.5, with 0 being the ultimate acidity. Measurements higher than 7 indicate increasing alkalinity. A pH as high as 9 can be found in water in its natural state, with 14 representing the ultimate alkalinity. The pH value is the starting point for determining treatments for corrosion control, chemical dosages, and disinfection.

Unintentional chemical additions to water supplies most commonly include the following elements:

*Toxic substances* are occasionally present in water supplies. Local health authorities can provide information about acceptable concentrations of such substances as arsenic (As), barium (Ba), cadmium (Cd), chromium ($Cr^{6+}$), cyanides (CN), fluoride (F), lead (Pb), selenium (Se), and silver (Ag). Although limited amounts of *fluoride* are frequently added to water supplies to help prevent tooth decay, fluorides in excess of such optimum concentrations can produce mottling of teeth. *Lead* poses a dangerous threat, even in relatively small amounts, because it is a cumulative poison. Lead in water usually comes from lead piping (in older buildings or cities) or from corrosive water on lead-painted roofs. The maximum recommended concentration is 0.05 mg/L.

Unfortunately, the list of toxic chemicals is expanding, and as chemical waste dumps have been abandoned or mismanaged, groundwater has become contaminated. The U.S. Environmental Pro-

**WATER AND WASTE**

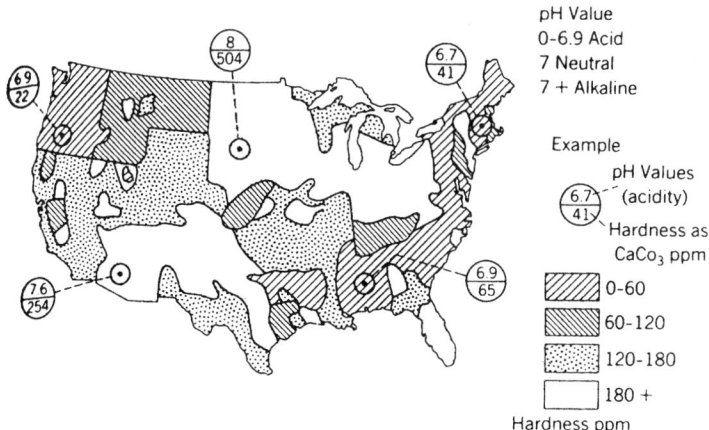

**Fig. 21.1** *Approximate groundwater chemical characteristics across the United States. Treatment may be needed when the pH is less than 7.0 (acidity results in corrosion) or when hardness as $CaCO_3$ exceeds 65 ppm (ppm and mg/L are essentially identical measures). (Courtesy of Progressive Architecture.)*

tection Agency (EPA) has estimated that 75% of both active and abandoned chemical waste dumps—some 51,000 in all—are leaking. In addition to the following list of inorganic chemicals, we are becoming aware of many new organic chemicals as well, some of which are suspected of causing 5% to 20% of U.S. cancers. The threat to our groundwater supplies is illustrated in Fig. 21.2. Once polluted, aquifers are extremely difficult to clean. This is one reason why ground source heat pumps are tightly regulated.

*Chlorides* can enter water as it passes through geologic deposits formed by marine sediment, or because of pollution from seawater, brine, or industrial or domestic wastes. A noticeable taste results from chloride in excess of 250 mg/L.

*Copper* can enter water from natural copper deposits or from copper piping that contains corrosive water. Concentrations of copper in excess of 1.0 mg/L can produce an undesirable taste.

*Iron* is frequently present in groundwater. Corrosive water in iron pipes will also add iron to water. At concentrations above 0.3 mg/L, iron can lend a brownish color to washed clothes and can affect the taste of the water.

*Manganese* can both pose a physiological threat (it is a natural laxative) and produce color and taste effects similar to those produced by iron. The recommended limit is 0.05 mg/L.

*Nitrates* in high concentrations pose a threat to infants, in whom they can cause "blue baby"

disease. In shallow wells, nitrate concentrations can indicate seepage from deposits of livestock manure.

*Pesticides*, a growing threat to water supplies, are particularly common in wells near homes that have been treated for termite control. Avoid using pesticides near wells.

*Sodium* is primarily dangerous for people with heart, kidney, or circulatory ailments. For a low-sodium diet, the sodium in drinking water should not exceed 20 mg/L. Salts spread on roadways for ice control can leach into the soil and enter groundwater. Note that some water softeners (discussed in Section 21.4c) can raise sodium concentrations in water.

*Sulfates*, which have laxative effects, can enter groundwater from natural deposits of Epsom salts (magnesium sulfate) or Glauber's salt (sodium sulfate). Concentrations should not exceed 250 mg/L.

*Zinc* sometimes enters groundwater in areas where it is found in abundance. Although not a health threat, it can cause an undesirable taste at concentrations above 5 mg/L.

### (c) Biological Characteristics

Potable water should be kept as free as possible of disease-producing organisms—bacteria, protozoa, and viruses. These organisms are not easily identified; a thorough biological water test is complex and time-consuming. For this reason, the standard

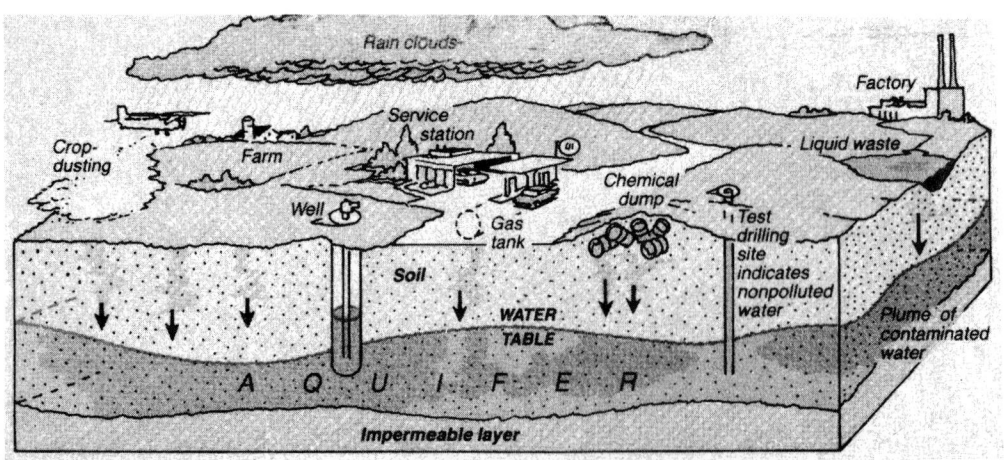

**Fig. 21.2** *How groundwater becomes contaminated. The "plume" formed by contaminants can often go undetected. (© 1982 by Newsweek, Inc., all rights reserved. Reprinted by permission.)*

test is for *one* kind of bacteria—the coliform group (*Escherichia coli*, better known as *E. coli*), which is always present in the fecal wastes of humans (as well as those of many animals and birds) and which outnumbers all other disease-producing organisms in water. The recommended maximum concentration of coliform bacteria is one organism per 100 mL (about ½ cup) water.

For biological activity to be kept to a minimum in drinking water, a water source should be chosen that does not normally support much plant or animal life, hence the popularity of groundwater rather than surface water as a source. In addition, the supply should be protected from subsequent biological contamination. Where cities depend on small lakes for water, humans are frequently excluded from the watersheds. Organic fertilizers and nutrient minerals should also be kept out of the water supply to further discourage biological activity. For the same reason, stored water should be kept dark and at low temperatures. Finally, organisms (or their by-products) are commonly destroyed at treatment facilities.

### (d) Radiological Characteristics

The mining of radioactive materials and the use of such materials in industry and power plants have produced radiological pollution in some water supplies. Because radiological effects are cumulative, concentrations of radioactive materials should be low indeed. "Safe" minimum concentrations have

continually been revised downward for other radiation exposures; consult the local public health service for current recommendations.

## 21.2 FILTRATION

In the preceding section, water pollution was broken down into physical, chemical, biological, and radiological categories. The various forms of treatment for such pollutants do not necessarily fall into the same categories, as one treatment may be effective for several different polluted conditions. A general look at common domestic water quality problems and treatments is provided in Table 21.3.

We begin with filtration, because so often it is the first treatment in a series; it is also one of the oldest and simplest methods. This very common treatment removes suspended particles, some bacteria, and color or taste by passing water through a permeable fabric or a porous bed of materials. The more common approaches are listed below, beginning with filtering to remove suspended particles and then moving on to more specialized applications for the removal of iron and/or manganese, tastes, and odors.

### (a) Sedimentation

Before entering a filter, this process removes some suspended matter from water simply by allowing

WATER AND WASTE

**TABLE 21.3 Common Water Quality Problems and Treatment in Small Systems**

| Item | Cause | Bad Effect | Correction |
|------|-------|-----------|-----------|
| Hardness | Calcium and magnesium salts from underground flow | Clogging of pipes by scale, burning out of boilers, and impaired laundering and food preparation | Ion-exchanger (Zeolite process) |
| Corrosion | Acidity, entrained oxygen and carbon dioxide (low pH) | Closing of iron pipe by rust, leaking connections, destruction of brass pipe | Raising the alkaline content (neutralizer) |
| Biological pollution | Contamination by organic matter or sewage | Disease | Chlorination by sodium hypochlorite or chlorine gas; or ozonation |
| Color | Iron and manganese | Discoloration of fixtures and laundry | Chlorination or ozonation and fine filtration |
| Taste and odor[a] | Organic matter | Unpleasantness | Filtration through activated carbon (purifier); aeration |
| Turbidity[a] | Silt or suspended matter picked up in surface or near-surface flow | Unpleasantness | Filtration |

[a]These problems are not common in private systems that use deep wells.

time, the inactivity of the water, and gravity to do the work of settling out heavier suspended particles. Simple basins, ponds, or tanks constructed for this purpose are large enough to retain the water for at least 24 hours and are equipped with baffles to slow the water flow. To clean out the sediment, water usually is diverted to an identical second basin while the first is being cleaned.

## (b) Coagulation

This process also removes suspended matter, along with some coloration. A chemical such as alum (hydrated aluminum sulfate) is added to water made turbulent by baffles or static mixers to distribute the chemicals evenly.

## (c) Flocculation

The water is then held in a quiet condition in which the suspended particles will combine with the alum to form floc. These heavy particles then settle out in a process similar to sedimentation. Some adjustment of the pH may be necessary.

See the National Drinking Water Clearinghouse, *Tech Brief* (September 1996) *Filtration* for details on the following methods of filtering.

## (d) Slow Sand Filters

These are common in small-scale water supply systems (Fig. 21.3; also see Fig. 20.8 for a rainwater application). Not suitable for water with high turbidity, they do not usually require coagulation/flocculation and may not even require sedimentation. Water should not be chlorinated before entering this filter, because it will interfere with the subsequent biological activity. These filters are able to remove up to 99.9% of *Giardia* cysts.

Slow sand filters are low-maintenance, easily constructed devices that should be cleaned as often as the turbidity of the water demands—from once a day to perhaps once a month. They are cleaned by removal and replacement of about the top 1 in. (25 mm) of sand, which has formed a layer of biological slime called (descriptively) the *schmutzdecke*, which traps small particles and degrades organic material in the water. This sand is then either washed for reuse or discarded.

The approximate rate of flow is slow, requiring a rather large surface area: 0.03 to 0.10 gpm per ft$^2$ (0.02 to 0.07 L/s m$^2$) of filter bed surface; in other units, 40 to 140 gallons per day per square foot (1630 to 5720 liters per day per square meter) of filter bed surface. Overall thickness is usually 30 to 48 in. of sand over 12 in. of gravel (900 to 1200 mm of

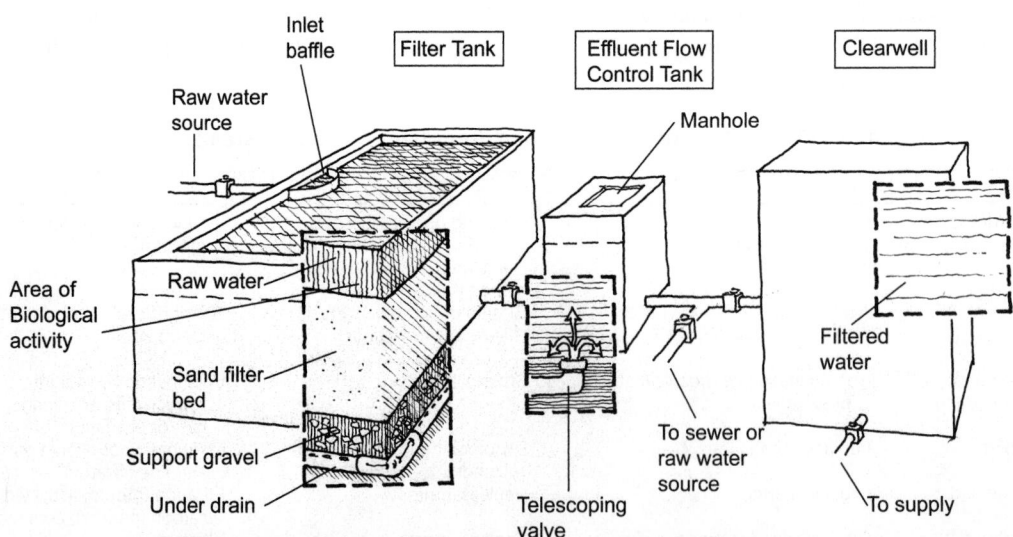

**Fig. 21.3** *Slow sand filter effective in removing cysts of* Giardia. *Water should not be chlorinated before it enters the filter. (From National Drinking Water Clearinghouse, Tech Brief [September 1996]* Filtration. *U.S. Environmental Protection Agency, Washington, DC. Redrawn by Dain Carlson.)*

sand over 300 mm of gravel) with an underdrain system.

In freezing temperatures, slow sand filters must be housed; if they develop an ice layer, this prevents cleaning.

### (e) Diatomaceous Earth Filters

Also known as *precoat* or *diatomite filters,* these can be of either the vacuum or the pressure type. They rely on a layer of diatomaceous earth, a minimum of ⅛ in. (3 mm) thick placed on a septum or filter element (for *Giardia* removal, the thickness should be increased to about ⅕ in. [5 mm]). They are most suitable for water with low bacterial counts and low turbidity. Simple to operate and effective in removing cysts, algae, and asbestos, they require periodic attention to remain effective, including backwashing every 1 to 4 days.

### (f) Direct Filtration

Intended for water supplies of high quality and seasonally consistent flow, these systems omit sedimentation but should include coagulation for most effective *Giardia* removal. These are often used with steel pressure tanks to maintain pressure in the water supply line.

*Packaged filtration* combines features such as chemical addition, flocculation, and sedimentation, along with filtration, in one compact unit. Most often used for small community water supplies, these systems treat surface water to remove turbidity, color, and coliform organisms.

### (g) Membrane Filtration

Also called *microfiltration* or *ultrafiltration,* this rapidly developing technique can remove bacteria, *Giardia,* and some viruses. It does not require coagulation as pretreatment. Using hollow fiber or spiral-wrapped membranes, it is able to exclude all particles greater than 0.2 micron from the water stream. It is best used on water supplies of low turbidity because of fouling of the fibers or membranes.

Water is forced at high pressure through these filters. The contaminants trapped on the inflow side must be frequently removed by reversing the flow and flushing the waste; calcium and other persistent contaminants must be periodically removed with chemical cleaning.

Nanofiltration, using much smaller pores, is discussed in Section 21.3.

### (h) Cartridge Filtration

Increasingly popular on lavatory faucets as well as on small supply systems, these systems are easy to operate and maintain. They require water of low turbidity and last longer when some prefiltering by more crude means is performed upstream. They can exclude particles of 0.2 micron (or even smaller). A disinfectant can prevent surface-fouling microbial growth on the cartridge filters; some periodic chemical cleaning will likely be required.

### (i) Other Filters

*Activated carbon filters* are particularly effective for removing tastes and odors. The water is passed through granular carbon, which attracts large quantities of dissolved gases, soluble organics, and fine solids.

*Porous stone, ceramic, or unglazed porcelain filters* (also called *Pasteur filters*) are usually made in small sizes so that they can be attached to water faucets. They are used widely in some countries, such as Mexico, but poor maintenance or hairline cracks often lead to bacterial infiltration, complicating the filtration process. A more positive approach to the disinfection of drinking water thus is desirable.

### 21.3 DISINFECTION

Disinfection is the most important health-related water treatment, because it destroys microorganisms that can cause disease in humans. Disinfection is required of water supply systems that rely on surface water or groundwater sources under the influence of surface water. Initially, primary disinfection achieves the desired level of microorganism kill (inactivation); then secondary disinfection maintains a disinfectant residual in the treated water that prevents microorganism regrowth.

Although chlorination has become the standard approach to removing harmful organisms from water, there are alternatives: nanofiltration, ultraviolet (UV) radiation (unsuitable for water with high turbidity because it cannot easily penetrate), bromine, iodine, ozone, and heat treatment, among

**WATER AND WASTE**

others. Chlorine continues to disinfect after the initial application. It is this continuing secondary disinfection that has made it universally relied on, despite dangers such as that posed by deadly chlorine gas. Although chlorine affects the taste and odor of water, it is also effective in removing less desirable tastes and odors. Unfortunately, chlorine can react with organic materials in water to form halogenated by-products. It is easier to either remove the organic materials before treatment, or to use another disinfectant strategy, than to try to remove these halogenated by-products after chlorine treatment.

See the National Drinking Water Clearinghouse, *Tech Brief* (June 1996), *Disinfection* for details on methods (*a*) through (*d*).

### (a) Chlorination

Factors that affect chlorine's ability to disinfect include:

1. *Chlorine concentration.* The higher the concentration, the faster and more complete the rate of disinfection.
2. *Contact time.* The longer the chlorine contacts the organisms in water, the more complete the disinfection. At a minimum, 0.4 mg/L of chlorine should contact water 30 minutes before use.
3. *Water temperature.* The higher the temperature during contact, the more complete the disinfection.
4. *pH.* The lower the pH, the more effective the disinfection.

There are three common forms in which chlorine is used to disinfect water supplies. *Chlorine gas* is stored in a cylinder (Fig. 21.4) as a liquid under high pressure and is released (as a gas) by a regulator to an injector attached to a water pipe or tank. The injector passes highly pressurized water through a venturi orifice, creating a vacuum that draws the chlorine into the water stream.

*Sodium hypochlorite solution* is easier to handle than deadly chlorine gas but is very corrosive and decomposes rather quickly. It should be stored in a cool, dark, dry area for no more than a month. *Hypochlorinators* automatically pump (or inject) a sodium hypochlorite solution into water. They are usually no larger than the pumps used in small water systems. Some hypochlorinators are specially

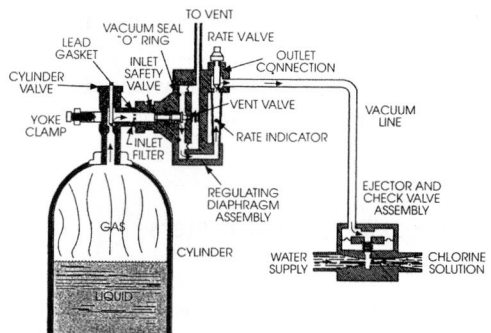

*Fig. 21.4* Cylinder-mounted chlorinator for automatic disinfection of supply water. (From National Drinking Water Clearinghouse, Tech Brief [June 1996] Disinfection. U.S. Environmental Protection Agency, Washington, DC.)

designed for low and fluctuating water pressures or for use where electricity is not available.

*Solid calcium hypochlorite* is a white solid containing 65% available chlorine that dissolves easily in water. It is corrosive, with a strong odor, but very stable and can be stored for up to a year. However, it readily absorbs moisture, forming chlorine gas; also, reactions between calcium hypochlorite and organic materials (wood, cloth, petroleum products) can generate enough heat to cause a fire or explosion. Again, hypochlorinators are used to deliver the disinfectant to water.

### (b) Chloramine

This is generated on site by adding ammonia to water containing chlorine or when water containing ammonia is chlorinated. This is a weaker disinfectant against viruses or protozoa than the chlorination processes, but it produces fewer disinfection by-products. It is most often used as a secondary rather than a primary disinfectant. Again, hypochlorinators are used to inject chlorine, after which ammonia is added.

### (c) Ozonation

This was first used in full-scale drinking water treatment in 1906. It is a powerful oxidizing and disinfecting agent, destroying most bacteria, viruses, and other pathogenic organisms. It requires a shorter contact time and dosage than chlorine and leaves no chlorine taste. Ozone is formed by passing dry air (or pure oxygen) through a system of high-voltage electrodes. It is an unstable gas and must be generated on site. When ozone reacts with an organic, it produces oxygen and an oxidized form of the organic. Ozone not used in this process quickly decays to oxygen.

In the United States, ozone is commonly used in cooling tower water treatment (Fig. 21.5), where its effectiveness against *Legionella pneumophila* is especially appreciated, as well as its control of algae and scale that can greatly reduce cooling efficiency. Ozone is also used in food processing, waste water cleanup, smoke removal, swimming pools and spas, bottled water, and pulp and paper bleaching.

Equipment includes an ozone generator, a contactor, and a destruction unit, plus instrumentation and controls. Operation and maintenance are relatively complex; electricity accounts for 26% to 43% of the operating costs for small systems. Because it acts only as a primary disinfectant, a secondary disinfectant (often chlorine) is usually required.

## (d) Ultraviolet Radiation

Special lamps are used within a reactor (Fig. 21.6), whose radiation disrupts the genetic material of the cells of organisms, making them unable to reproduce. Although effective against bacteria and viruses, UV radiation does not inactivate either *Giardia* or *Cryptosporidium* cysts. Otherwise, it is an effective primary disinfectant system, requiring a short contact time and without halogenated by-products. Yet again, a secondary disinfectant system is usually necessary. This system is not suitable for water that contains high levels of suspended solids, turbidity, color, or soluble organic matter.

## (e) Nanofiltration

These filter membranes start with pore sizes of 0.2 to 0.3 micron and are then dipped into a polymer that leaves a thin film, decreasing the pore size to 1 nanometer. This pore size removes bacteria, viruses, pesticides, and organic material. It also gives the membranes an affinity for calcium, contributing to water softening. However, it also means that the membranes need periodic acid cleaning to remove the calcium deposits. Adding phosphates to nanofiltered water reduces its capacity to dissolve lead.

With such extremely small pore sizes, this process requires very high water pressures, in turn requiring energy. Yet again, a secondary disinfectant system is usually necessary.

WATER AND WASTE

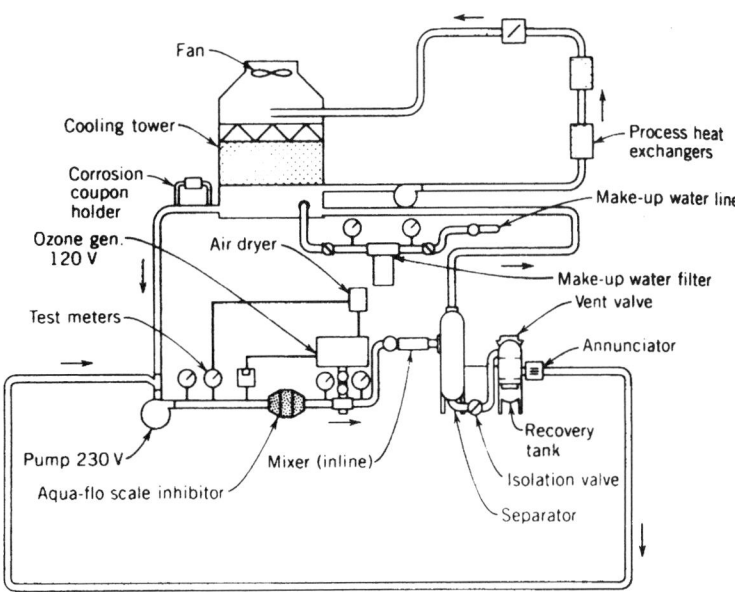

**Fig. 21.5** *Recycled cooling tower water is treated by an ozonator, a magnetic descaler, and a filtration system. This controls scale formation, algae and slime, corrosion, and sludge buildup. (Courtesy of Aqua-Flo, Inc., Baltimore.)*

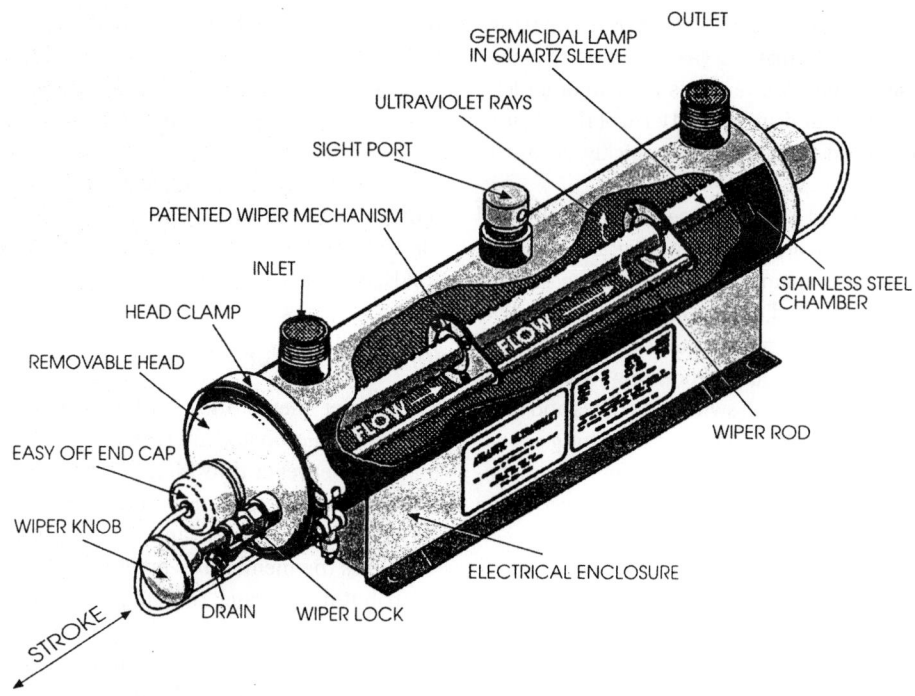

**Fig. 21.6** *Ultraviolet water purifier, effective against bacteria and viruses (but not against* Giardia *or* Cryptosporidium *cysts). (Courtesy of Atlantic Ultraviolet Corporation, Hauppauge, NY.)*

## 21.4 OTHER WATER TREATMENTS

### (a) Aeration (Oxidation)

This simple process can improve the taste and color of water and help remove iron and manganese. In aeration, as much of the water's surface as possible is exposed to air. The methods used are rich in aesthetic possibilities—the spraying of water into air, the fall of a turbulent stream of water over a spillway, and *flowforms,* sculptural waterfalls designed to carry water in a rhythmical, pulsating, figure-eight pattern. The Real Goods headquarters in Hopland, California, features these as a waterfall (Fig. 21.7) within a recycled water irrigation system.

To guard against contamination, these processes are often enclosed; if exposed, they must be kept clean. For aeration within tanks, water is passed through a series of perforated plates in streams or droplets.

Aeration improves the flat taste of distilled water and cistern water by adding oxygen. It also oxidizes iron or manganese, which then can more easily be removed by filtration. It also removes odors caused by hydrogen sulfide and algae.

Because aeration raises the level of dissolved oxygen in water, it should be avoided as a treatment when corrosion is a threat.

### (b) Corrosion Control

It is important to control corrosion both to keep water systems operating freely and to prevent corrosive water from increasing the concentration of hazardous materials (as from copper pipes). Corrosion also imparts a taste and/or odor to water that is objectionable. Corrosion is a slow degradation of a metal by a flow of electric current from the metal to its surroundings. Some factors involved in corrosion control are:

1. *Acidity.* The more acid (low pH, less than 6.0), the more corrosive the water.
2. *Conductivity.* As dissolved mineral salts increase the water's conductivity, they encourage the flow of the electrical current of corrosion.

WATER AND WASTE

(a)

(b)

**Fig. 21.7** *The Real Goods Solar Living Center at Hopland, California, illustrates photovoltaic (PV) power, passive solar heating, daylighting, and passive cooling, as well as creative water recycling. Water from on-site irrigation ponds is pumped by PV to a tank on a small hill (a) where, as flow varies with strong sunlight, it overflows and splashes down a series of flowforms to oxygenate the water. It passes through a plaza in front of the retail store, eventually (b) returning to the ponds. Landscape design by Land and Place.*

**WATER AND WASTE**

3. *Oxygen content.* Dissolved oxygen destroys the thin protective hydrogen film on immersed metals, thus promoting corrosion.
4. *Carbon dioxide content.* Carbon dioxide forms carbonic acid, which attacks metal surfaces.
5. *Water temperature.* Increased temperature increases corrosion.
6. *Lower flow rates.* Reduced turbulence means reduced erosion of the protective layers that form on the inner surfaces of pipes.

The products of corrosion often contribute to scale formation. Scale then lines surfaces, eventually clogging openings.

See the National Drinking Water Clearinghouse, *Tech Brief* (February 1997) *Corrosion Control,* for details on this subject.

*Acid neutralizers* can be installed on water supplies with low pH; their function is often combined with those of hypochlorinators. Typically, neutralizing solutions are mixtures of lime, soda ash, and water. However, pH adjustment should be made just before water delivery, *after* treatment processes such as coagulation and disinfection.

*Corrosion inhibitors* cause protective coatings to form on pipes. They are commonly fed into the water, as are other chemicals. Inorganic phosphates, sodium silicates, and mixtures of phosphates and silicates are the more commonly used additives.

Other corrosion control strategies include commercial pipe coatings/linings, installing dielectric or insulating unions (to avoid complications from dissimilar pipe metals), and avoiding metal piping and fixtures altogether.

### (c) Softening

Water hardness is caused primarily by calcium and magnesium deposits; when they are removed, water will be soft. Where water hardness is mild enough to affect only laundering, cisterns may be used to collect soft rainwater to use for washing clothes. Where water hardness produces scale in pipes and water heating appliances and cisterns are not feasible, water-softening equipment is used. Demineralization of water is accomplished with one of three methods: ion exchange, reverse osmosis, or electrodialysis.

See the National Drinking Water Clearinghouse, *Tech Brief* (May 1997) *Ion Exchange and Demineralization,* for details on the following methods of treatment.

*Ion exchange* is popular for small systems, and is effective not only with hardness ions but also with radionuclides. It can be used with fluctuating flow rates but requires pretreatment of most surface waters, and its waste is highly concentrated (requiring careful disposal). A large variety of resins are used for the exchange medium, each effective for specific contaminants.

On the exchange medium's charged surface, one (contaminant) ion is exchanged for another (regenerant) ion. Eventually, saturation occurs; the contaminants are flushed and the medium is regenerated—once per day is the common shortest cycle. Sodium chloride is often used to regenerate the exchange medium, resulting in a rather high sodium residual—an undesirable development for people on low-sodium diets. Another regenerant material is potassium chloride.

Equipment includes prefiltration, ion exchange, disinfection, storage, and distribution elements; in smaller systems, single-package units incorporate all of these processes.

*Reverse osmosis* (RO), like ion exchange, is popular for small systems and can be used with fluctuating flow rates, but it requires pretreatment of most surface waters and its waste is also highly concentrated. RO is effective not only with hardness ions but also with radium, natural organic substances, pesticides, and microbiological contaminants. RO units used in series can remove an even higher percentage of contaminants.

An inert, semipermeable membrane has a high-pressure supply water on one side; as the pressure slowly forces water through this filtering membrane, most of the contaminants are removed. Water must be used to flush the membrane so that mineral buildup is avoided; this produces a brine requiring careful disposal. Membranes are available in varying types and pore sizes; they are prone to fouling.

Commercial RO units are available in sizes ranging from a 1 gpm (3.9 L/m) water delivery rate (using two membranes and a 3-hp [2.2 kW] motor, requiring 4.5 gpm [17 L/m] feedwater for a 22% recovery rate) to a water delivery rate of 12.5 gpm (47.3 L/m) (using 12 membranes and a 15-hp [11.2 kW] motor, requiring 19.2 gpm [72.8 L/m] for a 65% recovery rate).

*Electrodialysis* effectively removes fluoride and nitrate, and can also remove barium, cadmium and selenium. It is relatively insensitive to levels of flow and of total dissolved solids but has an enormous appetite for water. From 10% to 80% of the total water supplied to an RO filtering unit is delivered to the user; the remainder is a waste stream. It also requires a higher level of pretreatment.

In this process, membranes adjacent to the inflowing stream are charged (either positively or negatively), attracting counter-ions to these membranes. The membranes allow either positively or negatively charged ions to pass through; thus, the ions leave the inflow stream and enter the waste streams (on the other side of each membrane). High water pressure and a source of dc power are needed in this process.

Membranes can become fouled when the pores are clogged by salt precipitation or blocked by suspended particulates. For either of these water conditions, pretreatment is essential. Reversing the charge on the membranes, *electrodialysis reversal,* helps to flush the attached ions from the membrane surface and can extend the time between membrane cleanings.

## (d) Nuisance Control

Some organisms may not be injurious to health but can multiply so rapidly that piping or filters become clogged or the water's appearance, odor, and taste are affected. This situation is most common with surface water sources, and it is within surface reservoirs that these treatments are most often applied. Algae growths, the most prevalent nuisance, can usually be controlled by applying copper sulfate (blue stone or blue vitriol) to the water body. Sudden and massive algae kills can have adverse impacts on other life forms within the water, because the decomposing algae rob the water of oxygen. As a further precaution, stored water should be shielded from sunlight whenever possible.

Cooling towers present an especially difficult

water treatment problem. The murky water with high turbidity and a high bacterial count leads to clogged passages (thus inefficient performance), deteriorated surfaces, and the growth of potentially lethal bacteria (*L. pneumophila*). As a result, enormous quantities of water are commonly passed once-through a cooling tower, and then dumped into storm sewers rather than being recycled in a closed system. (As late as 1989, San Francisco's City Hall was reported to be using 96,000 gallons [363,000 L] per summer day for once-through cooling—in a year of drought.) To treat cooling tower water successfully, a method is needed for microbial control, removing organics, and precipitating inorganics. Ozonation (Fig. 21.5) is a common answer to this widespread treatment problem.

### (e) Fluoridation

A heated controversy continues over the addition of fluoride to drinking water. The advantage of fluoridation is that children who drink fluoridated water have lower rates of tooth decay; and because everyone drinks water, all children benefit, not just those who can afford fluoride pills. Its disadvantage is that only children need the fluoride, not adults. Also, in amounts *above* those used in water treatment, fluoride is toxic and can cause mottled teeth. Opponents of fluoridation suggest that because sugar is a leading cause of tooth decay in children, sugar, rather than water, should be fluoridated. Small water systems can be equipped with fluoridation units. However, fluoride levels in the water supply must be carefully monitored.

### (f) Distillation

This is a simple, low-technology approach to purification that produces the equivalent of bottled water for drinking, cooking, and laboratory uses. In one process, it promises the removal of suspended solids, salts, bacteria, and (apparently) halogenated hydrocarbons. When water pollution is extreme, as in the case of sea (salt) water, distillation may be the best treatment. Water is heated to encourage evaporation. As the water turns to vapor, virtually all pollutants are left behind. When this vapor encounters a cooler surface, it condenses, and pure water (although flat in taste) can be collected from this surface.

**Fig. 21.8** *A solar still can be used to provide a small daily quantity of pure water; this installation serves a laboratory. (Courtesy of McCracken Solar Co., Alturas, CA.)*

Any heat source can be used in the distilling of water; solar stills (Fig. 21.8) are gaining in popularity because the energy used is free. Solar distillation of cistern water is an autonomous approach. In semiarid, rather sunny climates, a solar still should produce about ½ gal per square foot of collector surface area (4 L/m²) per day. This rate of production suggests that only the water used directly for drinking or other specialized purposes is usually feasible for distillation. Another factor to consider is that the cleaning of the still is generally accomplished by flushing it with twice as much water as was delivered. If not excessively brackish, this flush water could be used for irrigation or other non-potable-quality tasks.

## 21.5  WATER SOURCES

This section focuses on the equipment used to capture and store groundwater from wells. Other water sources are less often used for smaller systems, either because they require much more extensive treatment (surface water from lakes or rivers) or because they provide water intermittently (cisterns). A multistage treatment (flocculation, sedimentation, etc.) may be inappropriate for small water systems that receive only occasional maintenance. Cisterns were discussed in Chapter 20. Another increasingly important source of city water, the treated effluent of sewage treatment plants, is discussed in Chapter 22.

### (a) Wells

Farms and remote housing developments usually have private water systems. In rural and suburban areas where the progress of building is faster than the development of municipal water supplies, private sources may also be sought. Driven or drilled wells are preferable; water from these sources usually has at least the advantages of purity, coolness, and freedom from turbidity, odor, and unpleasant taste—any of which may be encountered in addition to either acidity or hardness.

*Bored wells,* which are dug with earth augers, are usually less than 100 ft (30 m) deep. They are used when the earth to be bored through is boulder-free and will not cave in. The diameter range is 2 to 30 in. (50 to 760 mm). The bored well is then cased with metal, vitrified tile, or concrete.

*Driven wells* are the simplest and usually the least expensive type. A steel drive-well point is fitted on the end of pipe sections and driven into the earth. The drive point is usually 1¼ to 2 in. (32 to 50 mm) in diameter. The materials and design of drive-well points vary according to the expected characteristics of the earth in which the well is driven. First, a pilot hole is bored (frequently with a simple hand auger), and the drive-well point and pipe sections are lowered into it. Then the well is driven to a point well below the water table.

*Jetted wells* require a source of water and a pressure pump. A washing well point is supplied with water under pressure; this loosens the earth and allows the point and pipe to penetrate.

*Drilled wells* require more elaborate equipment of several types, depending upon the geology of the site. The percussion (or cable-tool) method involves the raising and dropping of a heavy drill bit and stem. Having thus been pulverized, the earth being drilled is mixed with water to form a slurry, which is periodically removed. As drilling proceeds, a casing is also lowered (except when drilling through rock).

Rotary drilling methods (either hydraulic or pneumatic) utilize a cutting bit at the lower end of a drill pipe; a drilling fluid (or pressurized air) is constantly pumped to the cutting bit to aid in the removal of particles of earth, which are then brought to the surface. After the drill pipe is withdrawn, a casing is lowered into position.

Another method is the down-the-hole pneumatic (air) hammer method, which combines the percussion effect with a rotary drill bit.

Local well drillers, who usually use the method most suited to the prevailing geology of a region, can offer useful advice about well construction methods. When clients plan to build in a remote location, the architect and engineer should advise them about water problems. Quality-corrective measures can always be taken and pumping equipment purchased, but the amount of water that can be obtained from the ground, and the depth and cost of wells, are important considerations. There are some problem areas where wells several hundred feet in depth will yield as little as 5 gpm (0.3 L/s) or nothing. The cost of drilling a number of exploratory wells may be excessive. Unfortunately, when such difficulties occur, there is often no easy solution. Conferences with neighboring owners, state and federal geologists, and local well drillers can all be helpful.

A low-yield well can be combined with storage tanks so that a pump can run all night, slowly filling tanks to meet the next day's demands.

### (b) Pumps

The Hydraulic Institute, a nonprofit trade association of pump manufacturers, has several recommendations for reducing the amount of electricity consumed by pumps. Some of these apply to small water supply systems.

1. Design systems with lower capacity and total head requirements. (Using larger pipe sizes and minimizing the elevation of tanks are examples.)
2. Avoid excessive capacity. It is typically less expensive to add pumping capacity later on if needs increase. Operating a smaller pump closer to its capacity saves energy compared to a larger pump operating well under its capacity.
3. Select the most efficient pump type and size, even if its first cost is greater; life-cycle costs are likely to be lower.
4. Use two (or more) smaller pumps instead of one large one so that excess pump capacity can be turned off.
5. Maintain pumps and system components in virtually new condition to avoid efficiency loss.

Three common types of pumps used in well water supply are the positive displacement, the centrifugal, and the jet pump.

***Positive Displacement Pumps.*** There are two principal types of positive displacement pumps. In a *reciprocating pump,* a plunger moves back and forth

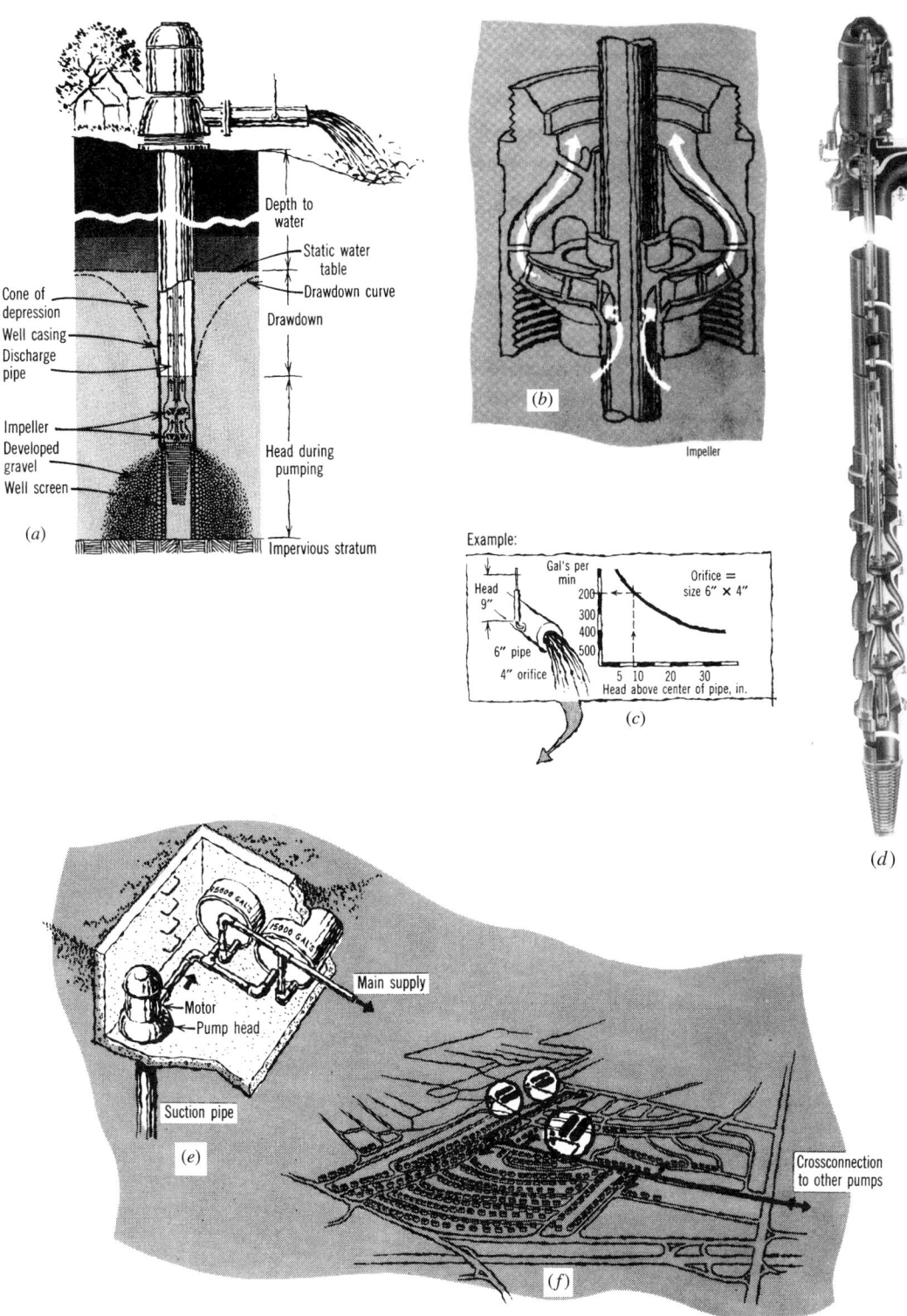

**Fig. 21.9** (a) A turbine well-pump. (b) Its operation. (c) Measurement of its capacity. (d) A Jacuzzi multistage lineshaft turbine well-pump. (Jacuzzi Bros., Inc.) (e,f) Its use in supplying a small community with groundwater. Capacities of turbine pumps range from 50 to 16,000 gpm (3 to 1010 L/s). (By permission of Progressive Architecture.)

within a cylinder equipped with check valves. The cylinder is best located near or below the ground-water level. Water enters the cylinder through an initial check valve (which allows flow in only one direction). As the plunger moves toward this check valve, the water is forced through the second check valve, located within the plunger itself. Then, as the piston returns to its original position, the water is forced upward toward the surface.

A *rotary pump* has a helical or spiral rotor—a turning vertical shaft within a rubber sleeve. As the rotor turns, it traps water between it and the sleeve, thus forcing the water to the upper end of the rotor.

**Centrifugal Pumps.** This type of pump contains an impeller mounted on a rotating shaft. The rotating impeller increases the water's velocity while forcing the water into a casing, thus converting the water's velocity into higher pressure. Each impeller and casing is called a *stage;* many stages can be combined in a multistage pump. The number of stages depends upon the pressure needed to operate the water supply system, as well as the height to which the water must be raised. The most common centrifugal pumps are those used in deep wells.

The *turbine pump* has a vertical turbine located below groundwater level and a driving motor located higher up, usually over the well casing at grade. A long shaft is thus required between the motor and the turbine. Substantial head clearances for this shaft's removal may be required. Figure 21.9 shows a turbine pump installation for a small community on Long Island, New York. The water is taken from the ground by multistage turbine pumps at depths of several hundred feet. It is delivered to submerged hydropneumatic tanks at a pressure of about 80 psi (about 550 kPa). As water is demanded in the houses, the air under pressure in the upper part of the tanks forces water through the mains.

*Submersible pumps* are designed so that the motor can be submerged along with the turbine (Fig. 21.10). The lengthy pump shaft is thus eliminated.

**Jet (or Ejector) Pumps.** In a jet pump, a venturi tube is added to the centrifugal pump. A portion of the water that is discharged from a centrifugal pump at the wellhead is forced down to a nozzle and the venturi tube (Fig. 21.11). The lower pressure within the venturi tube induces well water to flow

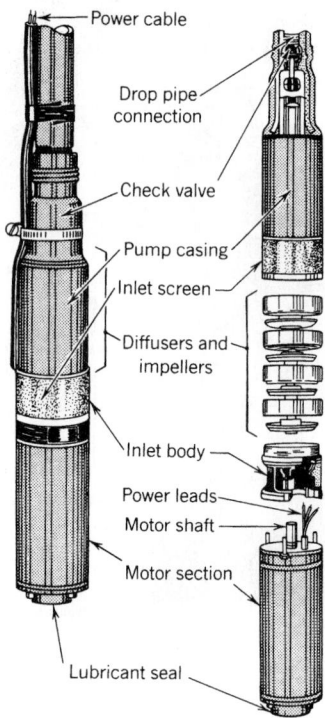

**Fig. 21.10** *Submersible pump (centrifugal type) in exploded view. (From the U.S. EPA's* Manual of Individual Water Supply Systems, *1975.)*

in, and the velocity of the water from the nozzle pushes it up toward the centrifugal pump, which can then lift it more easily by suction.

**Pump Selection.** Pump variations and characteristics are summarized in Table 21.4. The type of pump selected depends upon many factors, including the rate of yield of a well, the daily flow (and maximum instantaneous flow rate) needed by the users, the size of the storage or pressure tank used, and the total operating pressure against which the pump works (including the height to which water must be raised within the well). First cost, maintenance, and reliability are also factors, as is the energy used by the pump. In cold climates, a pump and water supply system must be protected from freezing.

Of these factors, the two critical selection determinants are the flow rate (volume per minute or per hour to be delivered) and the total pressure (or *head*). The flow rate depends upon the number of fixtures to be served (Fig. 21.12). The total pressure

*(a)*

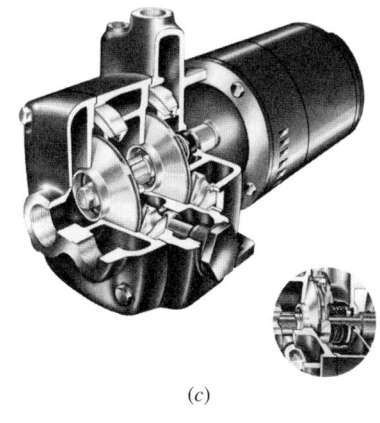

*(c)*

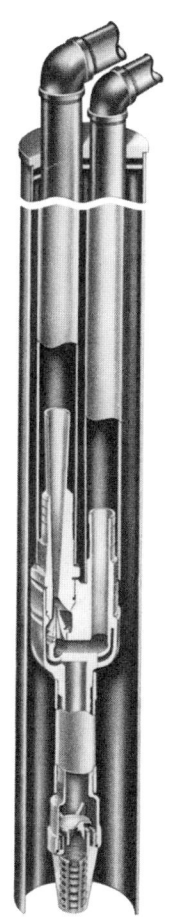

*(b)*

CHOOSE THE CORRECT JH FROM THESE CHARTS
DEEP WELL (Down to 120 feet)

The JH in the ½, ¾, 1 and 1½ horsepower rating, matched with the appropriate injector and pipe sizes, will pump this amount of water at the indicated lift or depth to the water in the well:

| If suction lift or depth to water is | JH ½ HORSEPOWER | JH ¾ HORSEPOWER | JH 1 HORSEPOWER | JH 1½ HORSEPOWER |
|---|---|---|---|---|
| | Produces these gallons per hour between 20-50 lbs. discharge pressure | | Produces these gallons per hour between 30-60 lbs. discharge pressure | |
| 30 feet | 795 G.P.H. | 990 G.P.H. | 1140 G.P.H. | 1620 G.P.H. |
| 40 | 680 | 875 | 1000 | 1470 |
| 50 | 575 | 735 | 875 | 1300 |
| 60 | 445 | 630 | 745 | 1200 |
| 70 | 360 | 495 | 620 | 920 |
| 80 | 310 | 385 | 530 | 820 |
| 90 | 255 | 315 | 435 | 700 |
| 100 | 220 | 275 | 340 | 590 |
| 110 | 195 | 240 | 295 | 540 |
| 120 | | 205 | 250 | 480 |

*(d)*

**WATER AND WASTE**

**Fig. 21.11** Details of a deep-well jet pump. (a) Photograph of a multistage jet pump housing and equipment. At the top is the on/off electrical switch activated by pressure settings. It controls the direct-connected electric motor to the left. Impellers are enclosed in the pump housing at the right. Circulating connections to and from the jet can be seen to the right, the pump discharge at the top. The pump can be set to operate up to 100 psi. (b) Well casing and circulating lines. Jet element can be seen at the bottom of the left-hand (larger) pipe. (c) Cutaway section of the pump. (d) Pumping capacity in gph under various conditions and discharge pressure ranges of 20 to 50 psi and 30 to 60 psi. (Jacuzzi Bros., Inc.) (e) Jet-type (also known as venturi or ejector) deep-well pump and storage tank for a house or small building (for well lifts greater than 25 ft [7.6 m]). Reduced pressure at (f); the jet nozzle, induces the flow of groundwater into the circulated flow.

WATER AND WASTE

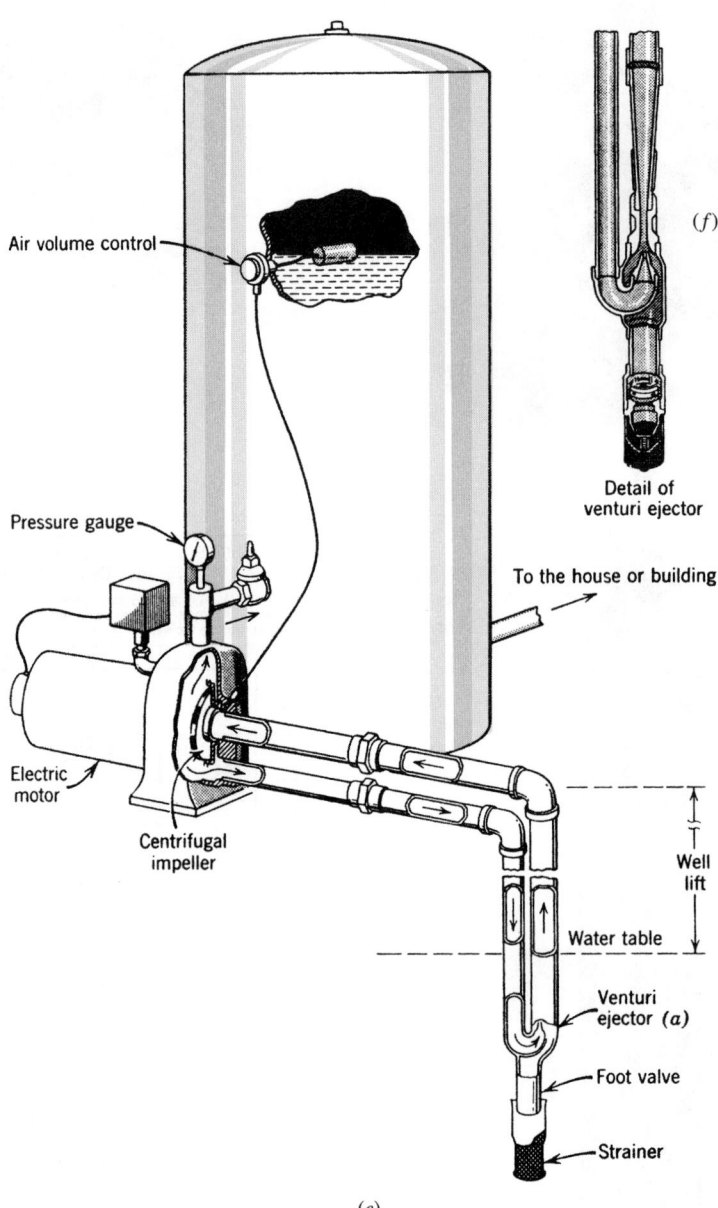

Air volume control

Pressure gauge

Electric motor

Centrifugal impeller

To the house or building

Detail of venturi ejector

(f)

Well lift

Water table

Venturi ejector (a)

Foot valve

Strainer

(e)

**Fig. 21.11** (Continued)

(Fig. 21.13) includes the suction lift, static head, and friction loss plus the pressure head. This relationship will be explained in detail in Section 21.11.

### (c) Pressure Tanks

Serving also for water storage, these tanks are frequently used both to maintain a constant pres-

sure on a pump-supplied water system and to allow for temporary peaks in water supply rates that exceed the capacity of the pump. (Elevated tanks offer one alternative to pressure tanks, cisterns another—although the latter usually are not located high enough to provide pressure to the supply system.)

Pressure tanks are often housed in outbuildings, along with the pump and any water treatment

**TABLE 21.4  Pumps for Water Supply**

| Type of Pump | Practical Suction Lift[a] | Usual Well-Pumping Depth | Usual Pressure Heads | Advantages | Disadvantages | Remarks |
|---|---|---|---|---|---|---|
| POSITIVE DISPLACEMENT | | | | | | |
| 1. Reciprocating<br>(a) Shallow well<br>(b) Deep well | 22–25 ft<br>22–25 ft | 22–25 ft<br>Up to 600 ft | 100–200 ft<br>Up to 600 ft above cylinder | 1. Positive action.<br>2. Discharge against variable heads.<br>3. Pumps water containing sand and silt.<br>4. Especially adapted to low capacity and high lifts. | 1. Pulsating discharge.<br>2. Subject to vibration and noise.<br>3. Maintenance cost may be high.<br>4. May cause destructive pressure if operated against closed valve. | 1. Best suited for capacities of 5–25 gpm against moderate to high heads.<br>2. Adaptable to hand operation.<br>3. Can be installed in very-small-diameter wells (2-in. casing).<br>4. Pump must be set directly over well (deep well only). |
| 2. Rotary<br>a. Shallow well (gear type) | 22 ft | 22 ft | 50–250 ft | 1. Positive action<br>2. Discharge constant under variable heads.<br>3. Efficient operation. | 1. Subject to rapid wear if water contains sand or silt.<br>2. Wear of gears reduces efficiency. | |
| b. Deep well (helical rotary type) | Usually submerged | 50–500 ft | 100–500 ft | 1. Same as shallow well rotary.<br>2. Only one moving pump device in well. | 1. Same as shallow well rotary except no gear wear. | 1. A cutless rubber stator increases life of pump. Flexible drive coupling has been weak point in pump. Best adapted for low capacity and high heads. |
| CENTRIFUGAL | | | | | | |
| 1. Shallow well<br>a. Straight centrifugal (single stage) | 20 ft max. | 10–20 ft | 100–150 ft | 1. Smooth, even flow.<br>2. Pumps water containing sand and silt.<br>3. Pressure on system is even and free from shock.<br>4. Low-starting torque.<br>5. Usually reliable and good service life | 1. Loses prime easily.<br>2. Efficiency depends on operating under design heads and speed. | 1. Very efficient pump for capacities above 60 gpm and heads up to about 150 ft. |
| b. Regenerative vane turbine type (single impeller) | 28 ft max. | 28 ft | 100–200 ft | 1. Same as straight centrifugal except not suitable for pumping water containing sand or silt.<br>2. They are self-priming. | 1. Same as straight centrifugal except maintains priming easily. | 1. Reduction in pressure with increased capacity not as severe as straight centrifugal. |

(cont'd)

WATER AND WASTE

**TABLE 21.4 (Continued)**

| Type of Pump | Practical Suction Lift[a] | Usual Well-Pumping Depth | Usual Pressure Heads | Advantages | Disadvantages | Remarks |
|---|---|---|---|---|---|---|
| CENTRIFUGAL | | | | | | |
| 2. Deep well<br>a. Vertical line shaft turbine (multistage) | Impellers submerged | 50–300 ft | 100–800 ft | 1. Same as shallow well turbine.<br>2. All electrical components are accessible above ground. | 1. Efficiency depends on operating under design head and speed.<br>2. Requires straight well large enough for turbine bowls and housing.<br>3. Lubrication and alignment of shaft critical.<br>4. Abrasion from sand. | |
| b. Submersible turbine (multistage) | Pump and motor submerged | 50–400 ft | 50–400 ft | 1. Same as shallow well turbine.<br>2. Easy to frost-proof installation.<br>3. Short pump shaft to motor.<br>4. Quiet operation.<br>5. Well straightness not critical. | 1. Repair to motor or pump requires pulling from well.<br>2. Scaling of electrical equipment from water vapor critical.<br>3. Abrasion from sand. | 1. 3500 RPM models, although popular because of smaller diameters or greater capacities, are more vulnerable to wear and failure from sand and other causes. |
| JET | | | | | | |
| 1. Shallow well | 15–20 ft below ejector | Up to 15–20 ft below ejector | 80–150 ft low heads. | 1. High capacity at as lift increases.<br>2. Simple in operation.<br>3. Does not have to be installed over the well.<br>4. No moving parts in the well. | 1. Capacity reduces<br>2. Air in suction or return line will stop pumping. | |
| 2. Deep well | 15–20 ft below ejector | 25–120 ft 200 ft max. | 80–150 ft | 1. Same as shallow well jet.<br>2. Well straightness not critical. | 1. Same as shallow well jet.<br>2. Lower efficiency, especially at greater lifts. | 1. The amount of water returned to ejector increases with increased lift—50% of total water pumped at 50-ft lift and 75% at 100-ft lift. |

*Source:* U.S. Environmental Protection Agency, *Manual of Individual Water Supply Systems,* 1975.

[a]Practical suction lift at sea level. Reduce lift 1 ft for each 1000 ft above sea level.

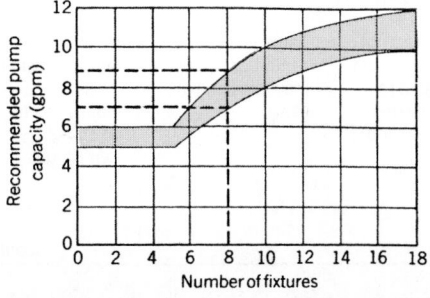

**Fig. 21.12** *The relationship between the recommended flow-rate capacity of a pump and the number of domestic plumbing fixtures it supplies. For details, see Section 21.9a. (From the U.S. EPA's* Manual of Individual Water Supply Systems, *1976.)*

equipment (Fig. 21.13). The temperature of the outbuilding must be kept above freezing, and its roof or walls should be removable to allow for replacement of parts over time. One type of pressure-storage tank is shown in Fig. 21.14.

The capacity of pressure tanks usually is small in comparison to the daily *total* water consumption; they provide short-term responses to peak flow demands. As a general rule, the pressure tank should be sized to deliver about 10 times the pump's capacity in gpm (L/m). For a typical residence, allow 10 to 15 gal (38–57 L) tank capacity per person served.

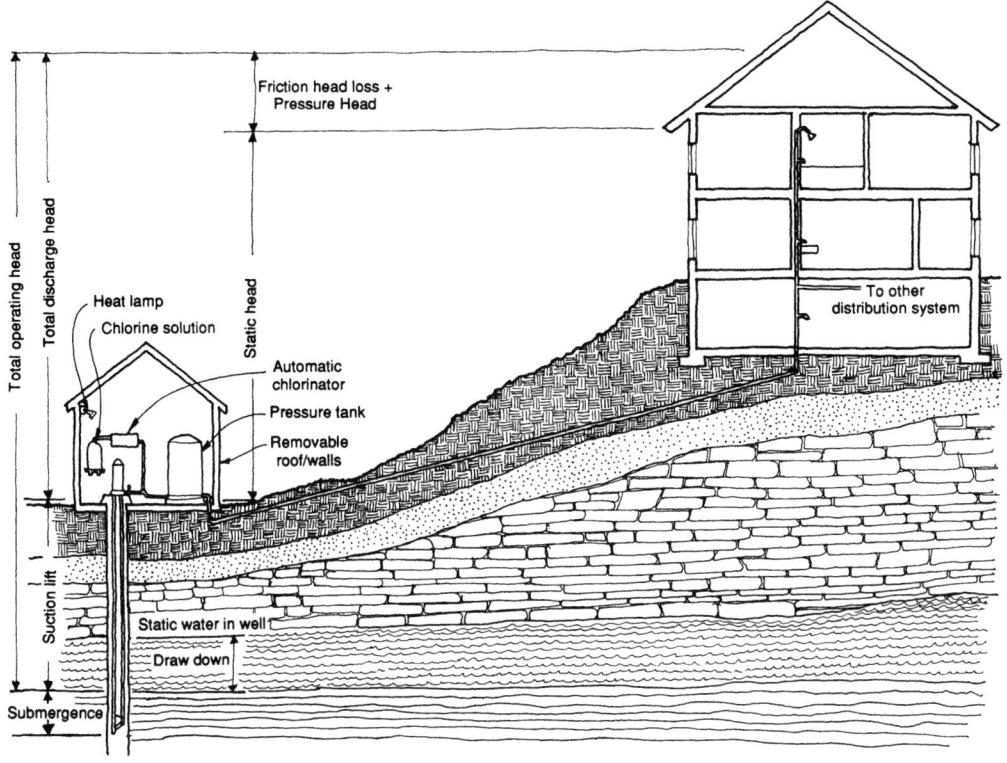

**Fig. 21.13** *The components of the total operating pressure (or head), a critical determinant of pump size. For details, see Section 21.9a. The pumphouse is usually a separate structure with water treatment and storage components. (From the U.S. EPA's* Manual of Individual Water Supply Systems, *1975.)*

For larger installations, the size of a pressure storage tank can be calculated by

$$Q = \frac{Qm}{1 - \dfrac{P_1}{P_2}}$$

where

$Q$ = tank volume (gal)

$Qm$ = 15 minutes of storage at peak usage rate (gal)

$P_1$ = minimum allowable operating pressure (psi) plus atmospheric pressure (14.7 psi)

$P_2$ = maximum allowable operating pressure (psi) plus atmospheric pressure (14.7 psi)

The ranges of allowable pressures are discussed in Section 21.9. An alternative tank-sizing procedure is shown in Table 21.5.

---

**EXAMPLE 21.1** An office building in a remote location has a water supply system served by a pump and

well. The peak demand is 50 gpm. The fixtures will operate at a minimum of 50 psi; a maximum of 70 psi should not be exceeded. Therefore,

$$Qm = 15 \text{ min} \times 50 \text{ gpm} = 750 \text{ gal}$$

and

$$Q = \frac{Qm}{1 - \dfrac{P_1}{P_2}} = \frac{750}{1 - \dfrac{50 + 14.7}{70 + 14.7}} = \frac{750}{1 - 0.76}$$

$$= 3125 \text{ gal} \qquad\blacksquare$$

The capacity of elevated tanks usually is equal to at least 2 days of average water usage. For fire-fighting or other special requirements, the capacity may have to be even greater.

A summary of the quality, treatment, and supply issues raised thus far is provided in the example of a rural estate shown in Fig. 21.15 and Table 21.5. On this large estate, water was required for an estimated demand of 30,000 gpd (113,560 Lpd). This was for domestic use only; irrigation was served by a separate installation pumping from a lake. Wells

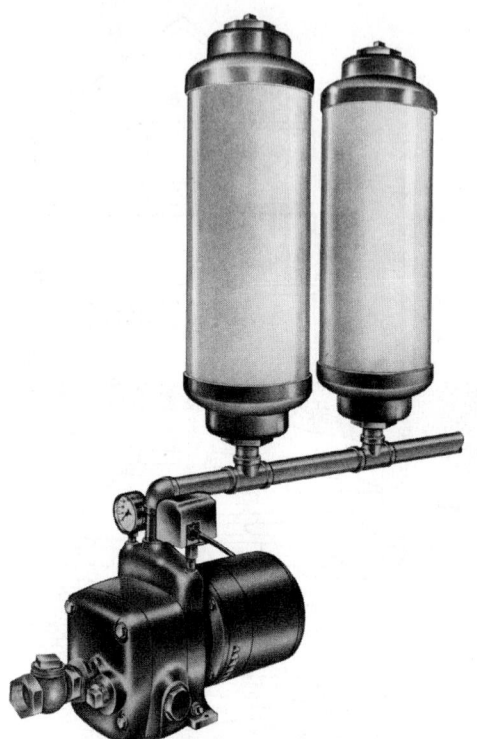

**Fig. 21.14** *Small pressure-storage tanks are installed primarily to keep a water supply system at constant pressure. These Hydrocel models are small (8½ in. in diameter, 27 in. long [216 mm in diameter, 686 mm long]), so they can be installed almost anywhere along the supply system. (Courtesy of Jacuzzi Bros., Inc.)*

were dug for the domestic supply. It was quickly evident that despite the great depths of the wells drilled, the available flow would be small. Four wells yielded a total rate of only 25 gpm.

Calculations for the amount of water to be stored to supplement this meager supply are shown in Table 21.5. The table shows that during the 14 daytime hours under conditions of peak demand, the well pumps would run continuously. Concurrently, an additional 9000 gal would be drawn from the tanks. At night, the well pumps would run for 6 hours to restore the 9000 gal drawn from the tanks during the day.

The operation of the system is illustrated in the diagram and notes of Fig. 21.15a. The pressure of 75 psi in the hydropneumatic tank is sufficient to raise the water to the greatest height in the distribution system, overcome friction in the piping, and leave a residual pressure available at each fixture of about 10 to 15 psi. Excessive pressure can always be moderated by a valve in the branch supply of the fixture.

Pressure in this tank is assured by the air compressor, activated by a pressure switch. A float switch starts the centrifugal pump to deliver water from the storage tanks as needed. The storage tanks, piped together and acting as a single reservoir, are fed by the four wells. The wells operate in unison, singly, or in groups, depending upon the level in the storage tanks. If the level drops rapidly, all well pumps can run. Although they are arranged to operate all day when needed, during periods of minimum demand only one or two may be called upon. Each has its own power supply, but controls emanate from panel N.

Swimming pools should be supplied with pure, potable water. Biological tests showed that this well water was safe. Thus, the fill line to the swimming pool is connected from this system rather than from the lake water supply used for irrigation. Because

**TABLE 21.5 Calculations to Establish Capacity of Water Storage Tanks at Kinloch (Fig. 21.15)**

| | |
|---|---|
| 1. Potable domestic water demand per day | 30,000 gpd |
| 2. Assumed hours of use | 7 A.M. to 9 P.M. (14 hours) |
| 3. Assumed hours of virtual nonuse | 9 P.M. to 7 A.M. (10 hours) |
| 4. Well-yield rate per hour | 25 gpm × 60 min = 1500 gph |
| 5. Total well yield in 14 hours | 1500 gph × 14 hr = 21,000 gal |
| 6. Total water needed in 14 hours | 30,000 gal |
| 7. Well yield in 14 hours | 21,000 gal (see 5 above) |
| 8. Net-tank capacity required, minimum | 30,000 − 21,000 = 9000 gal |
| 9. Net-tank capacity as designed | |

$20{,}000 \times 0.80$ (80% full) $= 16{,}000$ (2 tanks at 10,000 gal each)

$$2000 \times 0.70 \text{ (70\% full)} = \frac{+1400 \text{ (1 tank at 2000 gal)}}{17{,}400 \text{ gal (OK, greater than 9000)}}$$

| | |
|---|---|
| 10. Well operation at night to restore 9000 gal to tanks | 9000 ÷ 1500 = 6 hours of the 10 night hours available |

Washington State Ferries
Bainbridge Auto # 2

# 10/05/18 21:38

XXXXXXXXXXXXX7086
HEATHER F CRAIN
Approval 005744

Purchased

| ty | Description | PLU | Amount |
|----|-------------|-----|--------|
| 1 | Ad Veh U22' | 202881211A00AT | 15.35 |

Seattle — Bainbridg

| | |
|---|---|
| Total | 15.35 |
| VISA | 15.35 |
| Change | 0.00 |

10173118503139                    1193

## CUSTOMER COPY

10173 1185

Total includes the following:
* $0.25 per Fare Capital Surcharge

************************************************
Disputed fares must be submitted within 30 days.
Please retain receipt as proof of payment.
************************************************
Don't wait! Save a Spot
Reserve your travel to the San Juans
More info at TakeAFerry.com
************************************************

Washington State Ferries
Bainbridge Auto # 2

10/05/18 21:38

XXXXXXXXXXXXX7095
HEATHER T CRAIN
Approved 005744

Purchased

| Qty Description | PLU | Amount |
|---|---|---|
| 1 Ad Veh U22 | 2028213160001 | 15.35 |

Seattle - Bainbridge

| Total | | 15.35 |
| VISA | | 15.35 |
| Change | | 0.00 |

1017371850393                    1193

CUSTOMER COPY

1017371785

Total includes the following:
• $0.25 per Fare Capital Surcharge

++++++++++++++++++++++++++++++++++++
Discounted fares must be submitted within 30 days
Please retain receipt as proof of payment
++++++++++++++++++++++++++++++++++++

Don't wait! Save a Spot
Reserve your Travel to the San Juans
More info at TakeAFerry.com
++++++++++++++++++++++++++++++++++++

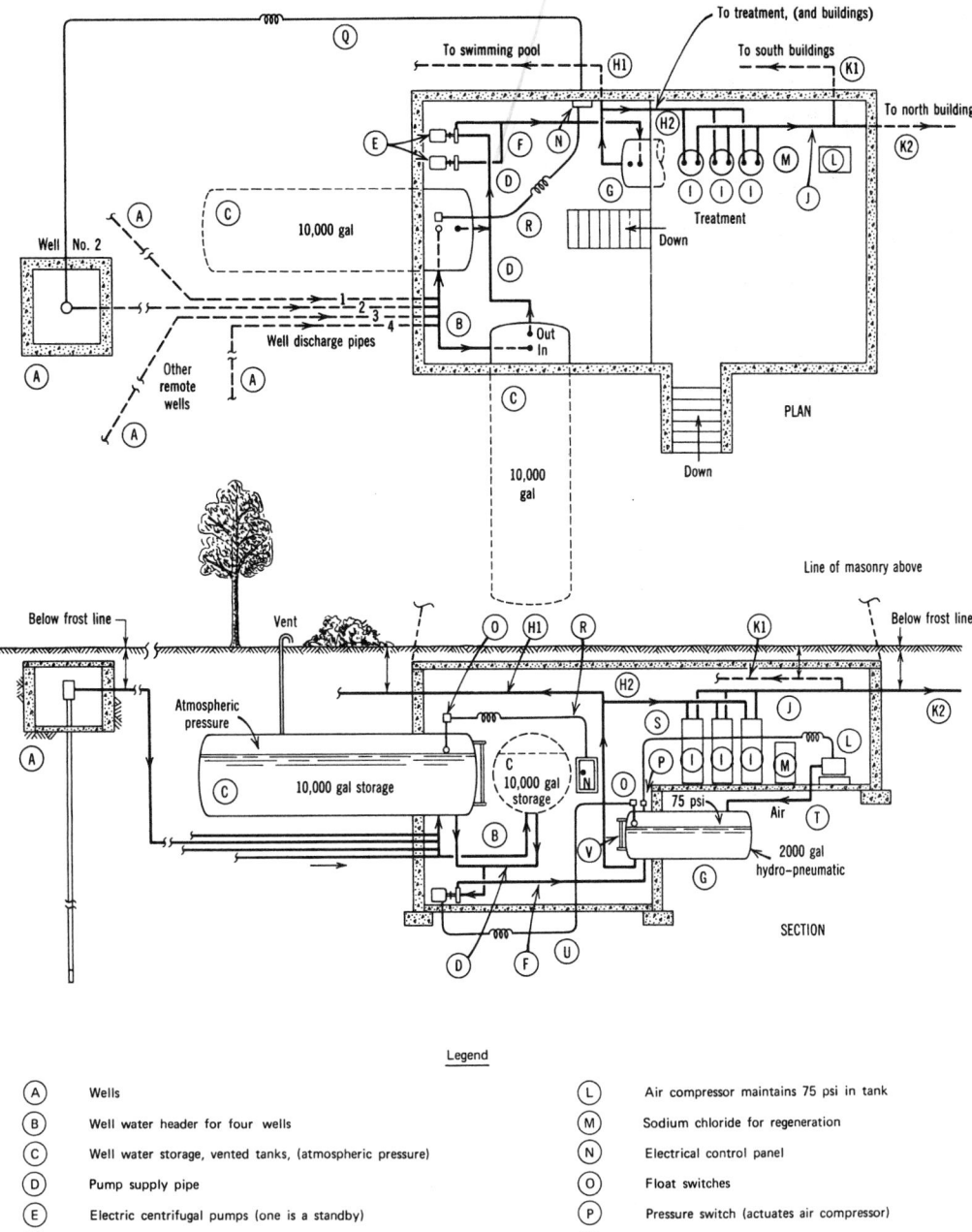

Legend

| | | | | |
|---|---|---|---|---|
| (A) | Wells | (L) | Air compressor maintains 75 psi in tank |
| (B) | Well water header for four wells | (M) | Sodium chloride for regeneration |
| (C) | Well water storage, vented tanks, (atmospheric pressure) | (N) | Electrical control panel |
| (D) | Pump supply pipe | (O) | Float switches |
| (E) | Electric centrifugal pumps (one is a standby) | (P) | Pressure switch (actuates air compressor) |
| (F) | Pump discharge to hydro-pneumatic tank | (Q) | Control wiring to well pumps |
| (G) | Hydro-pneumatic tank, 75 lbs per in.² | (R) | Control wiring, float switch to panel |
| (H1) (H2) | Discharge pipes, pressurized well water | (S) | Control wiring, pressure switch to compressor |
| (I) | Water softeners, (ion exchangers) | (T) | Compressed air piping |
| (J) | Treated well water under pressure | (U) | Control wiring, float switch to centrifugal pump |
| (K1) (K2) | Branch mains to buildings (treated well water) | (V) | Gauge glass indicates water level |

This schematic diagram shows system components and connections. For clarity, details (valves, drains, checkvalves etc.) are not shown. All control wiring, here simply indicated, operates through control panel (N). See text for operation.

(a)

**Fig. 21.15** (a) Schematic diagram of the water control and distribution center for the Kinloch Estate. (Bentel & Bentel, Architects, FAIA.) The plant is located on a hillside below the bathing and dressing pavilion. (b) One of two pavilions adjacent to the swimming pool that serves residents and guests at Kinloch. Below this pavilion is the control center. Exterior tubing to and from the control center is all below grade.

**915**

*(b)*

**Fig. 21.15** *(Continued)*

swimming pool water is separately recirculated and provided with its own purification treatment at a location adjacent to the pool, the supply line for make-up water to the pool would have infrequent and off-hour use. Water for domestic use in the buildings is passed through one of three ion exchangers to provide softening and to make the water more suitable for washing, bathing, and cooking. Periodically, the calcium precipitate can be flushed out and the tanks regenerated by sodium chloride.

## 21.6 HOT WATER SYSTEMS AND EQUIPMENT

There are many ways to provide the "domestic" or "service" hot water needs within a building—the hot water used *not* for space heating, but for bathing, clothes washing, dishwashing, and other related functions. Whereas much of the world either heats such water on a cookstove or does without it, North Americans enjoy an array of choices for domestic hot water (DHW) supply systems. In this section, rough supply estimates are considered first, then basic design choices, conventional heater tanks, solar water heaters, and finally, energy recovery heaters.

With today's better-insulated buildings, the demand for space heating can fall to a level about equal to DHW. As a result, a trend is developing toward using one rather larger water heater designed to meet the need both for space heating and for DHW. This saves both floor space and equipment cost compared to the more conventional boiler for space heating and a separate water heater for DHW. Some restrictions apply; the 1997 International Plumbing Code restricts the maximum water outlet

temperature to 160°F (71°C) and requires water in such a heating system to remain potable.

### (a) Water Temperature

Higher water supply temperatures have several advantages but carry the risk of scalding. Although water becomes uncomfortably hot to the touch above 110°F (43°C), much higher temperatures are often used for some commercial processes, as shown in Table 21.6. When determining the temperature at which DHW is to be supplied, consider the following factors.

*High Temperatures*

- Allow the installation of smaller storage tanks (less hot water is mixed with cold water to achieve a final usage temperature at the shower, sink, or lavatory) but require larger heating units.
- Can be achieved at the point of use (rather than in the tank) by heaters built into equipment such as dishwashers.
- Can cause scale to form on heating coils and within piping (above 140°F [60°C] in areas of hard water quality).
- May be required by code for some applications but limited by code for others.
- Limit the potential for growth of *Legionella pneumophila* bacteria (above 140°F [60°C]).

**TABLE 21.6 Representative Hot Water Temperatures**

| Use | Temperature °F | °C |
|---|---|---|
| Lavatory | | |
| Hand washing | 105 | 40 |
| Shaving | 115 | 45 |
| Showers and tubs | 110 | 43 |
| Therapeutic baths | 95 | 35 |
| Commercial and institutional laundry (based on fabric) | Up to 180 | Up to 82 |
| Residential dish washing and laundry | 140 | 60 |
| Surgical scrubbing | 110 | 43 |
| Commercial spray-type dishwashing[a] | | |
| Wash | 150 minimum | 65 minimum |
| Final rinse | 180 to 195 | 82 to 90 |

*Source:* Reprinted with permission of the American Society of Heating, Refrigerating and Air-Conditioning Engineers, Inc., from the 1995 *ASHRAE Handbook—HVAC Applications.*

[a]For other types of commercial dishwashers, see the source listed above.

*Lower Temperatures*

- Are less likely to cause burns but may not achieve desired sanitation.
- Mean less energy consumed, because storage and pipe heat losses are lower.
- Allow the installation of smaller heating units but require larger storage tanks.
- Make possible the use of lower-grade heat sources for DHW, such as solar energy or waste heat recovery. Table 2.1 shows that around 13% of annual U.S. energy usage is for DHW, which requires the lowest-grade heat source of all the common usages.

## (b) Heat Sources and Methods

These include familiar concentrated-energy, high-grade sources such as natural gas and electricity. Oil- and coal-fired boilers are frequently equipped with DHW coils as well. Buildings served by steam may use steam as a DHW heat source. Cogeneration provides heat for DHW, as in Fig. 10.67. Because of the relatively low-grade final temperatures needed, DHW can also be readily provided by wood-burning equipment, incinerators, solar energy equipment, heat pumps, and heat recovery devices (as in commercial ice-making machines whose discharged heat is contributed to DHW).

*Heating Methods.* There are two basic methods:

1. *Direct* heating brings water in contact with directly heated surfaces: electric-resistance elements or other electrically warmed surfaces within tanks, or surfaces directly exposed to fire or hot gases.
2. *Indirect* heating can be accomplished in several ways. Coils containing steam or fluids can be submerged within water tanks or (Fig. 21.16) set within boilers, whose primary function usually is to provide space or industrial process heating. Alternatively, coils containing DHW can be placed outside a boiler but within a casing containing steam, hot exhaust gases, or very hot water.

Direct and indirect methods can be utilized in a variety of equipment:

*Storage tank water heaters,* the type most commonly used for residential and small commercial purposes (see Section 21.6h)

*Circulating storage water heaters,* in which the water is first heated by a coil, then circulated through a storage tank (as in some solar heaters—see Section 21.6i)

*Tankless (instantaneous) heaters,* in which the water is very quickly raised to the desired temperature within a heating coil and immediately sent to the point of usage

## (c) Tankless Water Heaters

Also called *instantaneous water heaters,* these are available in larger sizes for central hot water systems (Fig. 21.16) or in smaller sizes for wall mounting adjacent to remote plumbing fixtures that occasionally need hot water, or for isolated bathrooms, laundry rooms, and so forth. Decentralized tankless units can be as small as "instant hot water taps" for kitchen or bar sinks—electric-resistance heaters capable of generating perhaps 3 gph at up to 200°F (2 L/h at up to 93°C). Bathroom groups can use either electric resistance or gas-fired units (Fig. 21.17); electric heaters require very high amperage and convert only about one-third of the primary energy into electricity; gas heaters must be vented and are most efficient without continuously burning pilots.

These tankless heaters can be very small; one unit of but 2.65 ft$^3$ produces up to 4 gpm (0.24 L/s) with a temperature rise of 45°F (20°C). With a greater temperature rise of 100°F (55.5°C), the flow rate drops to somewhat less than half. Because tankless heaters often consist of fairly long coils through which the water passes as it is heated, they may add considerable friction to the total supply system design requirements (see Section 21.11).

To gain a better understanding of hot water used per activity and the role of instantaneous heaters, see Table 21.7. Domestic water system design usually concentrates on the storage tank size (see Section 21.6h).

## (d) Energy Factors

The U.S. Department of Energy developed this standardized measure of annual overall efficiency. Standard storage tank water heaters that are gas-fired may reach an energy factor (EF) of 0.60 to 0.64. Because they have virtually none of the stand-by

WATER AND WASTE

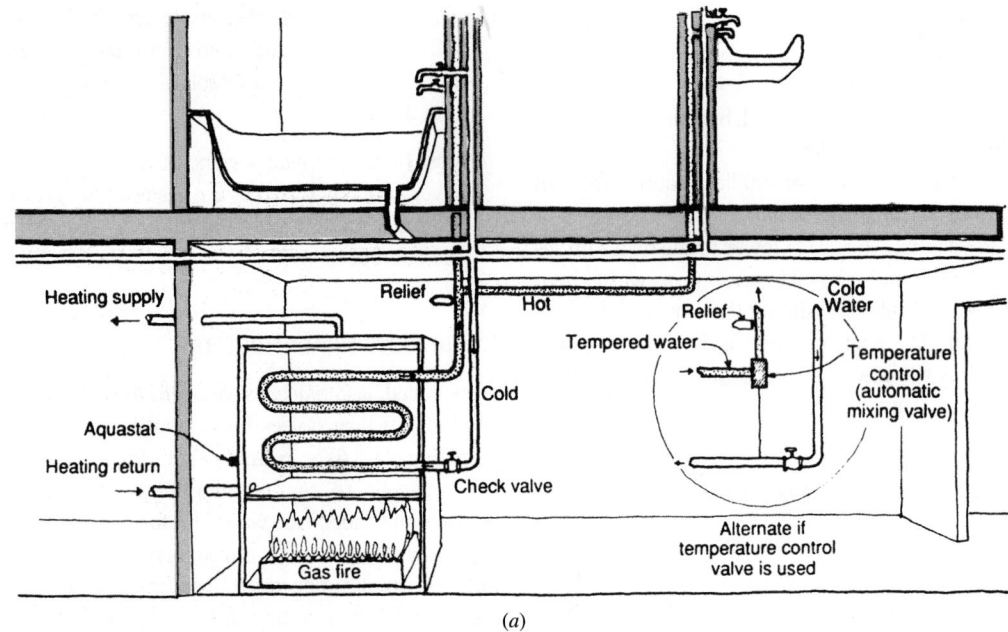

(a)

(b)

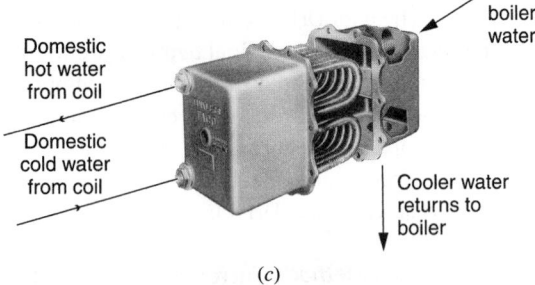

(c)

**Fig. 21.16** *(a) Example of an indirect, tankless heater for DHW utilizing a boiler that provides hot water for space heating. Because no storage is used, the point of DHW use should be very close to the boiler. (b) Internal tankless heating coil for DHW immersed in the jacket water of a gas-fired hot water heating boiler. The approximate capacity range is 3 to 15 gpm at a 100F° rise (0.2 to 0.9 L/s at a 55C° rise). (c) External-type tankless heater for DHW. Boiler water is piped to the unit and circulates by gravity, transferring heat to the coil. The approximate capacity range is 3 to 15 gpm at a 100F° rise (0.2 to 0.9 L/s at a 55C° rise). Note: Because it can be highly inefficient to provide summertime DHW with a boiler designed for winter space-heating loads, codes may alter or prohibit this arrangement.*

losses of storage tank heaters, gas-fired tankless water heaters offer EF of up to 0.69 with continuous pilots and as high as 0.93 with electronic ignition (although this minor usage of electricity is not included in the EF rating).

Some gas-fired instantaneous heaters are designed to operate with battery-powered pilot light

ignition. These are useful in remote installations with propane but without electricity.

### (e) Central versus Distributed Equipment

This choice can be particularly complex. In the United States, a central water-heater storage tank is

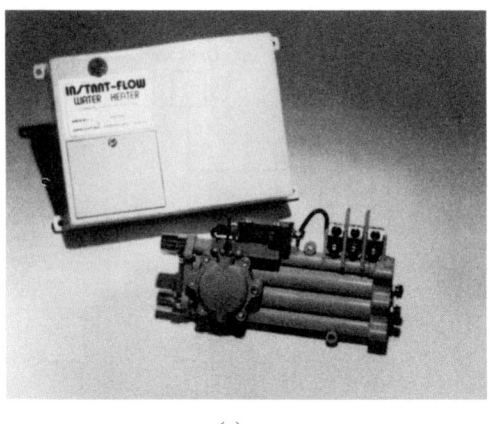

(a)

(b)

*Fig. 21.17* Instantaneous or tankless water heater. (a) In this instant-flow water heater, a series of coils heats water as it flows through. (b) Installation below a typical lavatory. A wide range of sizes is available; the 4.6-kW unit raises water by 31F° (17.2C°) at 1 gpm (0.06 L/s); the 9-kW unit raises water by 61F° (33.9C°) at 1 gpm (0.06 L/s). (Courtesy of Chronomite Laboratories, Inc.)

standard equipment in homes and small stores. However, the growing use of solar water heating, along with awareness of the heat losses from water-heater storage tanks, has created new interest in *combinations* of central and distributed DHW.

Consider a large residential DHW application such as that shown in Fig. 21.18. Here, two areas where hot water is used are separated by some 50 ft (15 m). If the simple, centralized water-heater storage tank is used (Fig. 21.18a), it would probably be placed nearest the maximum-use fixtures—dishwashers and clothes washers. (Also, floor space for the water-heater storage tank is usually more plentiful there.) Codes typically require that at least the first 8 ft (2.4 m) of the hot water pipe be insulated, as it leaves the water heater. However, each

time hot water is needed in the remote bathroom serving the bedrooms, the previously heated water in at least 50 ft (15 m) of supply pipe will almost certainly have cooled—despite insulation even over its entire length—and must be drained off before hot water finally arrives. (A ¾-in.-diameter pipe, for example, will hold 4.6 gal (17.4 L) in 50 ft (15 m) of pipe. If the water heater is set at an energy-conserving temperature of 120°F (49°C) and the incoming city water is 50°F (10°C), the 4.6 gal (17.4 L) of wasted water will also waste the 2682 Btu (2830 kJ) invested during its heating.)

Another way to conserve water is with a recirculating hot water system (Fig. 21.18b). The primary disadvantages of this system are the increased heat loss through the hot water pipe (now kept at

## TABLE 21.7 Domestic Hot Water Consumption—Residences

| | Hot Water Required in Gallons (Liters) per Use | |
|---|---|---|
| *Clothes Washing Machine* | *14-lb (6.4-kg) Machine* | *18-lb (8.2-kg) Machine* |
| Hot wash/hot rinse | 38 gal (144 L) | 48 gal [182 L] |
| Hot wash/warm rinse | 28 gal (106 L) | 36 gal [136 L] |
| Hot wash/cold rinse | 19 gal (72 L) | 24 gal [91 L] |
| Warm wash/cold rinse | 10 gal (38 L) | 12 gal [45 L] |
| *Dishwashing* | *Small* | *Large* |
| Dishwashing machine | 10 gal (38 L) | 15 gal (57 L) |
| Sink washing | 4–8 gal (15–30 L) | |
| *Personal Hygiene* | | |
| Tub bathing | 12–30 gal (45–134 L) | |
| Wet shaving/hair washing | 2–4 gal (8–15 L) | |
| Showering | 2–6 gpm (13–38 L/s) | |

*Source:* Reprinted by permission from Russell Plante, *Solar Domestic Hot Water,* copyright © 1983 by John Wiley & Sons.

WATER AND WASTE

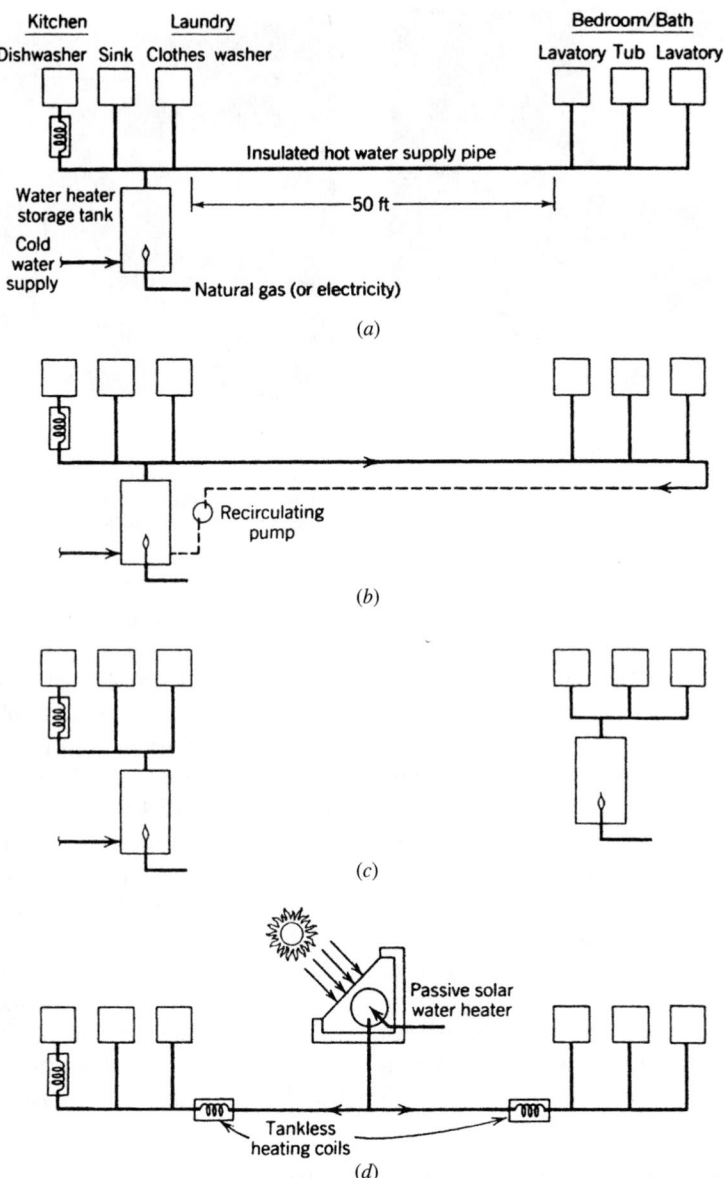

**Fig. 21.18** Options for DHW in a larger residence. (a) Typical centralized water-heater storage tank; water and energy are wasted in the 50–ft (15-m) run. (b) Recirculating pump added; this saves water but requires continuous energy to operate the pump and make up the pipe's heat losses. (c) Decentralized approach using two water-heater storage tanks; this roughly doubles the energy lost from storage but eliminates water and energy loss from the 50-ft (15-m) run. (d) Central solar water heater/decentralized tankless heating coil combination; it saves on water and energy but not on initial cost.

120°F [49°C] for 24 hours per day rather than for just a few minutes) and the energy required to run the circulating pump (again for 24 hours daily). Codes typically require a conveniently located cutoff switch for the circulating pump and pipe insulation with a minimum $k = 0.3$ Btu in./h ft² °F (0.04 W/m K) over the entire length of the circulating hot water system.

A decentralizing option (Fig. 21.18c) is to install two water-heater storage tanks, one for each group. The first cost will be greater and more floor space required. The daily waste heat from *two* heaters is clearly a disadvantage that will at least partially off-set the energy saved by eliminating the 50-ft (15-m) run of pipe. (Furthermore, in warm weather this waste heat will add to discomfort within the house.)

A decentralized-centralized mix (Fig. 21.18d) will save energy, water, and floor space, but it will almost certainly be costlier to install. A central solar water heater—either passive (shown) or active—brings the water up to warm (winter) or very hot (summer) temperatures. At each fixture group, a tankless heating coil instantaneously heats the water as needed. Almost no water is wasted, and the energy lost in the 50 ft (15 m) of pipe will be solar energy, not purchased energy from natural gas or electricity.

Another option is to omit the solar water heater and simply install two tankless heaters. The primary disadvantage of this setup is the need for larger-capacity heaters, whose instantaneous demand for energy could lead to electric cost penalties in areas where peak-load rates are high. Compared to central storage tanks, however, less heat is lost.

### (f) Distribution Trees

The choice of a distribution tree for a central water heating system must take into account the gradual cooling of water within the distribution system once it has left the central heater. If the heat losses associated with constantly circulating hot water (looped trees) are preferable to the heat and water wastage associated with simple, single hot water distribution trees, such loop systems can be achieved in either of two ways.

*Thermosiphon* hot water circulation depends upon the fact that water expands and becomes lighter when heated, as can be seen in Fig. 21.19a. If heat is applied to the lower loop of a glass tube, both ends of which have been inserted in an inverted bottle containing water, the water moves from A to B and rises through tube BC into the bottle. There it becomes cooled and drops through tube DA to A, is again heated, and rises in tube BC—thus completing the circulation. Because the movement depends upon the difference in weight between the two columns of water, the velocity and consequent effectiveness of the circulating system increase as both the temperature of the water and the height of the circuit increase. Hot water supply systems therefore usually consist of a heater with a storage tank, piping to carry the heated water to the farthest fixture,

and a continuation of this piping to return the unused cooled water back to the heater. A constant circulation is thereby maintained, and hot water may be drawn at once from a fixture without first draining off through the faucet the cooled water that would be standing in the supply pipe if there were no recirculation. Because heat increases the corrosive action of water on metals, copper tubing or rated polyvinyl chloride (PVC) pipe is often chosen for use in hot water and hot water circulating systems. Thermosiphon circulating systems are particularly effective in multistory buildings because of the beneficial effect of the increased circuit height on circulation.

*Forced circulation* of hot water is often utilized where a height advantage is unavailable. Low, long, rambling buildings, such as some large one-story residences, schools, and factories, lack the height needed to set up good hot water circulation by gravity. Also, flow is diminished by friction in long pipe runs. For such buildings, the forced-circulation scheme shown in Fig. 21.20 offers one option. Three independent aquastats—devices that create an electrical signal when water temperature drops—control this system. Aquastats A, B, and C, respectively, sense the temperatures of the water in the

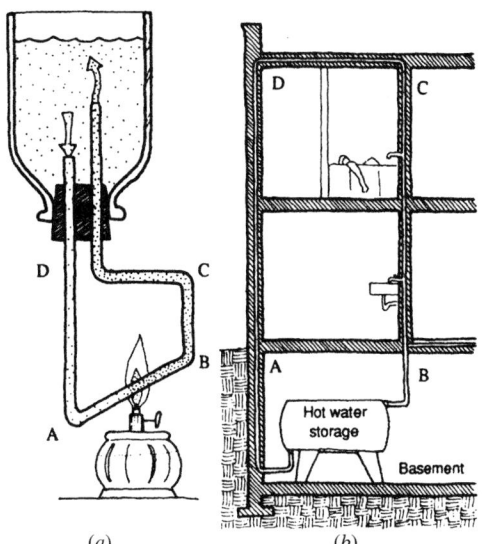

**Fig. 21.19** (a) Principle of hot water circulation by gravity (thermosiphon). (b) Its application to hot water service. During periods of no demand, there is sufficient incidental cooling between C and D so that dense (less warm) water at A forces the lighter (hot) water at B to rise for speedy availability at each faucet.

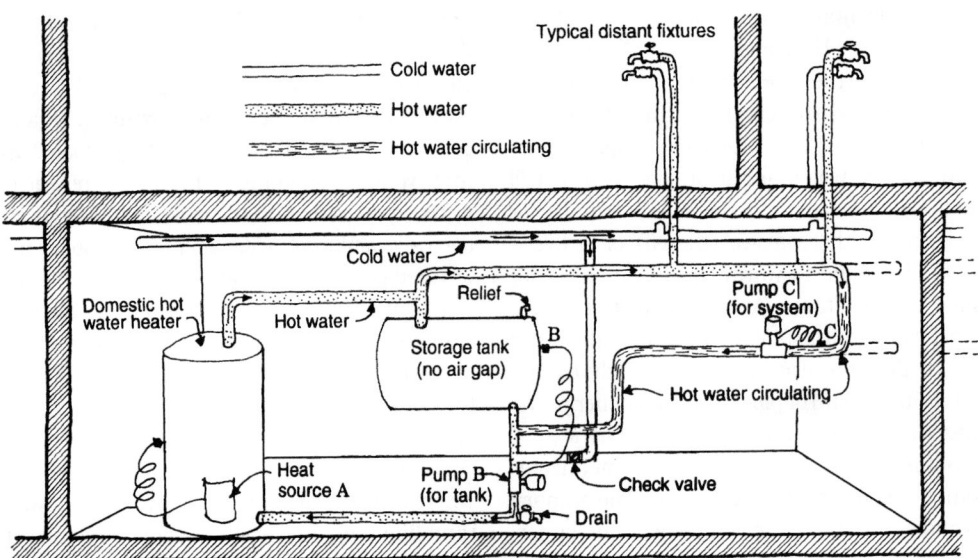

**Fig. 21.20** *Forced circulation of DHW for a long, low building. (Note that the use of space-heating boilers for DHW may cause system inefficiencies in warm weather.)*

heater, the tank, and at the end of the circulation-return main. As needed, they turn on the oil or gas burner, the tank-circulating pump, and the system-circulating pump. Fixtures remote from the tank are as close to hot water as the length of their hot water runout pipes. Water is usually available at full temperature in 5 to 10 seconds. Trial aquastat settings in °F could be *A* 180, *B* 160, *C* 120 (°C: *A* 82, *B* 71, *C* 49).

### (g) Variable Storage Temperature

An energy-saving computer control is available for hotels, motels, apartments, and larger commercial buildings. This device (Fig. 21.21) varies the supply temperatures of hot water so that the hottest water is supplied at the busiest hours. In hours of low usage, much lower supply temperatures mean that more hot water will be mixed with less cold water at showers, lavatories, and sinks. This increased hot water quantity poses no problem at off-peak hours, and the lower temperature means significant decreases in heat loss from the recirculating supply water system. A typical payback period is 1 year. The device stores a memory (adjusted weekly) of the typical daily patterns of usage and varies the supply temperature accordingly.

### (h) Conventional Water Heater Selection

One of the most common appliances used in the United States is the water-heater storage tank, an energy-conserving model of which is shown in Fig. 21.22.

For residences, the capacity of hot water heating/storage equipment can be taken from the values in Table 21.8. Note that for the familiar tank-type direct water heaters (gas, electric, or oil), there is a stated relationship among storage size, rate of hourly heat input, rate of draw (demand) over 1 hour, and recovery rate. Example 21.2 uses Table 21.8 for the sizing of a residential water heater.

---

**EXAMPLE 21.2** Select a natural gas water heater for a five-bedroom house with three baths. The *minimum* requirements of HUD-FHA are shown in Table 21.8: 50 gal, 47,000 Btu/h, 90-gal draw per hour, and 40 gph recovery.

**SOLUTION**
From Fig. 21.22 and Table 21.9, a model BTH 120 is selected; it exceeds all the minimums. ∎

Estimation of the hot water demand for commercial and institutional buildings is not so simple, and design guidelines are less reliable. For larger

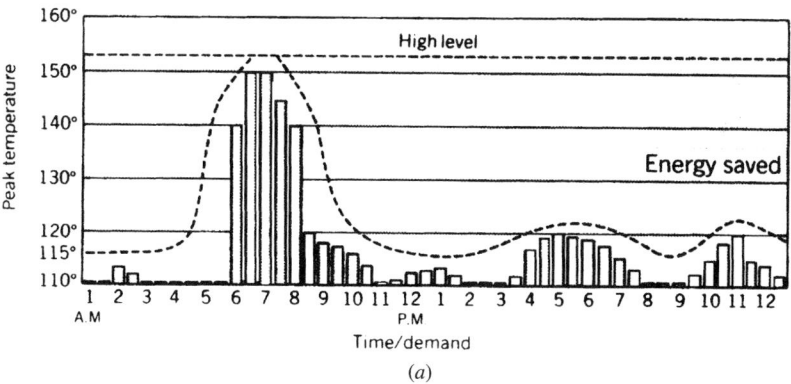

*(a)*

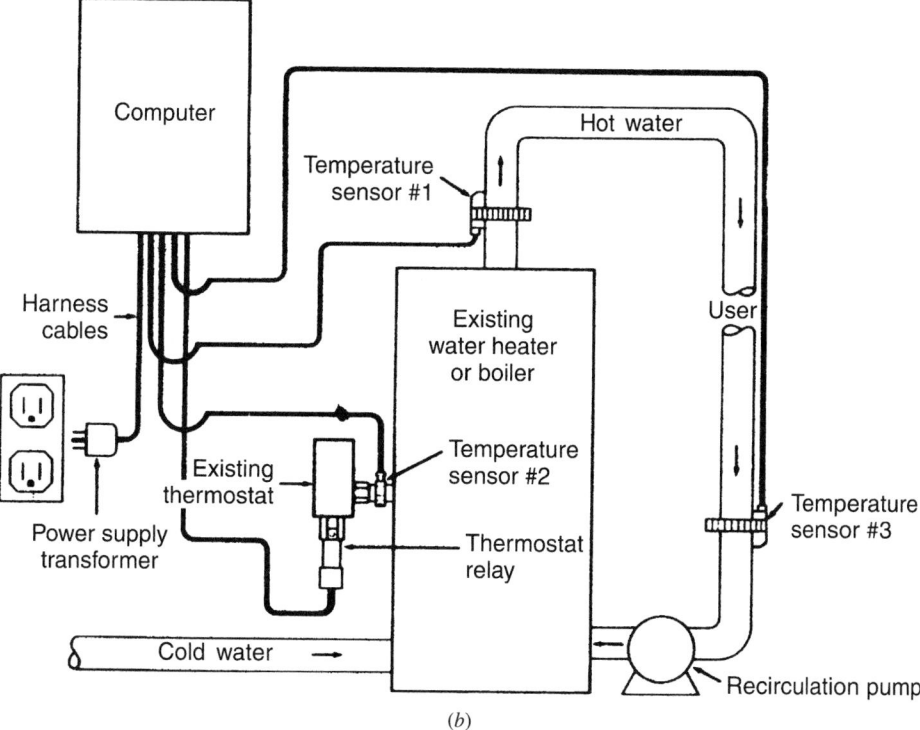

*(b)*

**Fig. 21.21** *Variable temperature DHW supply. (a) Energy savings are possible when the supply temperature of hot water is varied. Heat losses from supply pipes are greatly reduced for most of the typical day. A computer controls the supply temperature. (b) Schematic of computer-controlled, variable-temperature DHW system. (Courtesy of Fluidmaster, Inc., Anaheim, CA.)*

buildings, there is a trade-off between quick recovery (high heating capacity) and storage size; big tanks have small heaters, and vice versa. Another variable is storage temperature, as discussed earlier in Section 21.6a.

Small shops and very small office buildings can be treated like residences for hot water sizing. For larger buildings, the trade-off between heaters

and storage tanks should be explored. To begin this process, consult Table 21.10, which shows the maximum hourly and daily demands and contrasts them to an average day (which should be used to estimate monthly energy consumption).

The relationship between heater size and tank size is graphed in Fig. 21.23. The advantage of a *larger heater* is its smaller tank, which consumes less

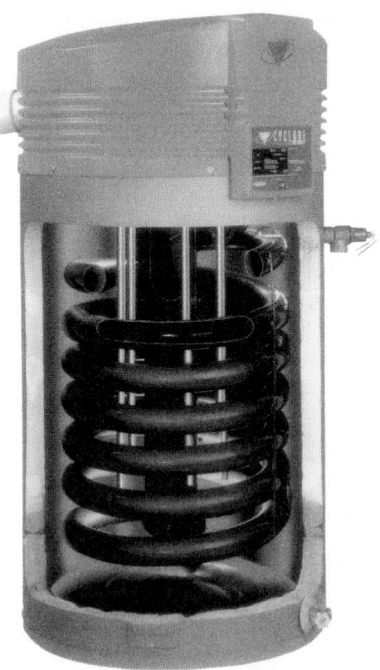

*Fig. 21.22* "Cyclone XHE" water-heater storage tank using natural gas (or propane). This model achieves 94% thermal efficiency with a burner located at the top, forcing blower-driven air into a countercurrent contact with rising fuel. A precise mix of air and fuel at the point of ignition results in high combustion efficiency, and the swirling flame continues downward in a submerged central combustion chamber. The heated gas is then forced at high velocity through the spiral heat exchanger coil, with a swirling action that maximizes heat exchange with the water. (Courtesy of the A.O. Smith Water Products Company, Irving, TX.)

*Cyclone XHE Data:*

[All models are 27¾ × 30 in. (705 × 762 mm) with 3-in. (76-mm) PVC vents]

| Model | Storage Capacity, gal (L) | Btuh Input/Output | Recovery (gal/h) at 94% Thermal Efficiency | | Height | |
|---|---|---|---|---|---|---|
| | | | 80F° Δ$t$ | 100F° Δ$t$ | in. | mm |
| BTH-120 | 60 (227) | 120,000/112,800 | 171 | 137 | 55½ | (1410) |
| BTH-150 | 100 (378) | 150,000/141,000 | 214 | 171 | 74½ | (1892) |

space and volume, weighs less, and probably has a lower first cost. The advantage of a *larger tank* is that its smaller heater tends to work steadily rather than in spurts; this lower, steadier demand for energy will lead to lower utility rates in the case of electric heating and thus to money savings over the life of the system. When the fuel supply is solar energy or waste heat, a larger tank is usually best suited to the characteristics of the fuel supply.

**EXAMPLE 21.3** (Adapted from the 1995 *ASHRAE Handbook—HVAC Applications.*) A women's dormitory housing 300 students, with a cafeteria serving 300 meals in 1 hour, is to be built. Find the required hot water storage size for two conditions: (1) assum-

ing a minimum recovery rate for both dorm and cafeteria and (2) assuming a dorm recovery rate of 2.5 gph (2.6 mL/s), which is half of the maximum hourly value given in Table 21.10, and a cafeteria recovery rate of 1.0 gph (1.1 mL/s), which is two-thirds of the maximum hourly value given in Table 21.21.

**SOLUTION**

*Minimum recovery:*
From Fig. 21.23a, the minimum recovery rate for women's dormitories is 1.1 gph. For 300 students,

$$300 \times 1.1 = 330 \text{ gph recovery}$$

At this rate, again from Fig. 21.23a, the minimum usable storage capacity is 12 gal/student. Assume that 70% of the total capacity is usable capacity.

This means that after 70% of the stored hot water is withdrawn, the remaining water has been cooled (by incoming water) to an unusably low temperature. Storage size must be increased by

$$\frac{100\%}{70\%} = 1.43$$

Thus

$$12 \times 300 \times 1.43 = 5150 \text{ gal storage}$$

From Fig. 21.23e, the minimum recovery rate for the cafeteria (serving full meals, type A) is 0.45 gph. For 300 meals,

$$300 \times 0.45 = 135 \text{ gph recovery}$$

At this rate, the minimum usable storage capacity is 7 gal/meal. Thus

$$300 \times 7 \times 1.43 = 3000 \text{ gal storage}$$

Combining these requirements for dorm and cafeteria,

$$\text{recovery} = 330 + 135 = 465 \text{ gph}$$
$$\text{storage} = 5150 + 3000 = 8150 \text{ gal}$$

*Faster recovery:*
At the specified dorm recovery rate of 2.5 gph,

$$300 \times 2.5 = 750 \text{ gph}$$

and the minimum usable storage required is 5 gal/student. Thus

$$300 \times 5 \times 1.43 = 2150 \text{ gal}$$

At the specified cafeteria recovery rate of 1.0 gph,

$$300 \times 1.0 = 300 \text{ gph}$$

and the minimum usable storage required is 2 gal/meal. Thus

$$300 \times 2.0 \times 1.43 = 860 \text{ gal}$$

Combining these requirements for dorm and cafeteria,

$$\text{recovery} = 750 + 300 = 1050 \text{ gph}$$
$$\text{storage} = 2150 + 860 = 3010 \text{ gal}$$

For this example, an increase in heater size of 225% for faster recovery allows the size of the tank to be reduced to only 37% of original size.

*Note:* This sizing procedure does not account for *system losses,* that is, heat lost from the storage tanks and from the hot water piping. Recovery capacities are usually increased because of these losses, which are simple to calculate when the U-factors of the tank and pipe insulation are known: Btu/h = $U \times A \times \Delta t$. ∎

**WATER AND WASTE**

## TABLE 21.8 HUD-FHA Minimum Water Heater Capacities, Residential

| | PART A. I-P UNITS | | | | | | | | | | |
|---|---|---|---|---|---|---|---|---|---|---|---|
| **Number of Baths** | **1–1.5** | | | **2–2.5** | | | | **3–3.5** | | | |
| **Number of Bedrooms** | 1 | 2 | 3 | 2 | 3 | 4 | 5 | 3 | 4 | 5 | 6 |
| *Gas[a]* | | | | | | | | | | | |
| Storage, gal | 20 | 30 | 30 | 30 | 40 | 40 | 50 | 40 | 50 | 50 | 50 |
| Input, 1000 Btu/h | 27 | 36 | 36 | 36 | 36 | 38 | 47 | 38 | 38 | 47 | 50 |
| 1-h draw, gal | 43 | 60 | 60 | 60 | 70 | 72 | 90 | 72 | 82 | 90 | 92 |
| Recovery, gph | 23 | 30 | 30 | 30 | 30 | 32 | 40 | 32 | 32 | 40 | 42 |
| *Electric[a]* | | | | | | | | | | | |
| Storage, gal | 20 | 30 | 40 | 40 | 50 | 50 | 66 | 50 | 66 | 66 | 80 |
| Input, kW | 2.5 | 3.5 | 4.5 | 4.5 | 5.5 | 5.5 | 5.5 | 5.5 | 5.5 | 5.5 | 5.5 |
| 1-h draw, gal | 30 | 44 | 58 | 58 | 72 | 72 | 88 | 72 | 88 | 88 | 102 |
| Recovery, gph | 10 | 14 | 18 | 18 | 22 | 22 | 22 | 22 | 22 | 22 | 22 |
| *Oil[a]* | | | | | | | | | | | |
| Storage, gal | 30 | 30 | 30 | 30 | 30 | 30 | 30 | 30 | 30 | 30 | 30 |
| Input, kW | 70 | 70 | 70 | 70 | 70 | 70 | 70 | 70 | 70 | 70 | 70 |
| 1-h draw, gal | 89 | 89 | 89 | 89 | 89 | 89 | 89 | 89 | 89 | 89 | 89 |
| Recovery, gph | 59 | 59 | 59 | 59 | 59 | 59 | 59 | 59 | 59 | 59 | 59 |
| *Tank-type indirect[b,c]* | | | | | | | | | | | |
| I-W-H-rated draw, gal in 3 h, 100F° rise | | 40 | 66 | | 66 | 66[e] | 66 | 66 | 66 | 66 | 66 |
| Manufacturer-rated draw, gal in 3 h, 100F° rise | | 49 | 49 | | 75 | 75[e] | 75 | 75 | 75 | 75 | 75 |
| Tank capacity, gal | | 66 | 66 | | 66 | 66[e] | 83 | 66 | 82 | 82 | 82 |
| *Tankless-type indirect[c,d]* | | | | | | | | | | | |
| I-W-H-rated draw, gpm, 100F° rise | | 2.75 | 2.75 | | 3.25 | 3.25[e] | 3.75 | 3.75 | 3.75 | 3.75 | 3.75 |
| Manufacturer-rated draw, gal in 5 min, 100F° rise | | 15 | 15 | | 25 | 25[e] | 35 | 25 | 35 | 35 | 35 |

**TABLE 21.8** *(Continued)*

| | PART B. SI UNITS | | | | | | | | | | |
|---|---|---|---|---|---|---|---|---|---|---|---|
| **Number of Baths** | **1–1.5** | | | **2–2.5** | | | | **3–3.5** | | | |
| **Number of Bedrooms** | **1** | **2** | **3** | **2** | **3** | **4** | **5** | **3** | **4** | **5** | **6** |
| *Gas[a]* | | | | | | | | | | | |
|   Storage, L | 76 | 114 | 114 | 114 | 150 | 150 | 190 | 150 | 190 | 190 | 190 |
|   Input, kW | 7.9 | 10.5 | 10.5 | 10.5 | 10.5 | 11.1 | 13.8 | 11.1 | 11.1 | 13.8 | 14.6 |
|   1-h draw, L | 163 | 227 | 227 | 227 | 265 | 273 | 341 | 273 | 311 | 341 | 350 |
|   Recovery, mL/s | 24 | 32 | 32 | 32 | 32 | 36 | 42 | 34 | 34 | 42 | 44 |
| *Electric[a]* | | | | | | | | | | | |
|   Storage, L | 76 | 114 | 150 | 150 | 190 | 190 | 250 | 190 | 250 | 250 | 300 |
|   Input, kW | 2.5 | 3.5 | 4.5 | 4.5 | 5.5 | 5.5 | 5.5 | 5.5 | 5.5 | 5.5 | 5.5 |
|   1-h draw, L | 114 | 167 | 220 | 220 | 273 | 273 | 334 | 273 | 334 | 334 | 387 |
|   Recovery, mL/s | 10 | 15 | 19 | 19 | 23 | 23 | 23 | 23 | 23 | 23 | 23 |
| *Oil[b]* | | | | | | | | | | | |
|   Storage, L | 114 | 114 | 114 | 114 | 114 | 114 | 114 | 114 | 114 | 114 | 114 |
|   Input, kW | 20.5 | 20.5 | 20.5 | 20.5 | 20.5 | 20.5 | 20.5 | 20.5 | 20.5 | 20.5 | 20.5 |
|   1-h draw, L | 337 | 337 | 337 | 337 | 337 | 337 | 337 | 337 | 337 | 337 | 337 |
|   Recovery, mL/s | 62 | 62 | 62 | 62 | 62 | 62 | 62 | 62 | 62 | 62 | 62 |
| *Tank-type indirect[f,g]* | | | | | | | | | | | |
|   I-W-H-rated draw, L in 3 h, 55 K rise | | 150 | 150 | | 250 | 250[e] | 250 | 250 | 250 | 250 | 250 |
|   Manufacturer-rated draw, L in 3 h, 55 K rise | | 186 | 186 | | 284 | 284[e] | 284 | 284 | 284 | 284 | 284 |
|   Tank capacity, L | | 250 | 250 | | 250 | 250[e] | 310 | 250 | 310 | 310 | 310 |
| *Tankless-type indirect[g,h]* | | | | | | | | | | | |
|   I-W-H-rated draw, mL/s, 55 K rise | | 170 | 170 | | 200 | 200[e] | 240 | 200 | 240 | 240 | 240 |
|   Manufacturer-rated draw, L in 5 min, 55 K rise | | 57 | 57 | | 95 | 95[e] | 133 | 95 | 133 | 133 | 133 |

*Source:* Reprinted with permission of the American Society of Heating, Refrigerating and Air-Conditioning Engineers, Inc. from 1995 *ASHRAE Handbook—HVAC Applications.*

[a]Storage capacity, input, and recovery requirements are typical and may vary with individual manufacturers. Any combination of these requirements to produce the stated 1-hour draw will be satisfactory.

[b]Boiler-connected water heater capacities (180°F boiler water, internal, or external connection).

[c]Heater capacities and inputs are the minimum allowable. Variations in tank size are permitted when recovery is based on 4 gph/kW at 100F° rise for electrical, AGA recovery ratings for gas, and IBR ratings for steam and hot water heaters.

[d]Boiler-connected heater capacities (200°F boiler water, internal, or external connection).

[e]Also for 1.5 baths and 4 bedrooms for indirect water heaters.

[f]Boiler-connected water heater capacities (82°C boiler water, internal, or external connection).

[g]Heater capacities and inputs are the minimum allowable. Variations in tank size are permitted when recovery is based on 4.2 mL/s kW at 55 K rise for electrical, AGA recovery ratings for gas, and IBR ratings for steam and hot water heaters.

[h]Boiler-connected heater capacities (93°C boiler water, internal, or external connection).

For these larger DHW applications, central steam is often used as a heat source. Figure 21.24 shows an indirect storage tank system. Such systems may cost more than tankless systems, but they are usually cheaper to operate, both because the peak prices for the instantaneous fuel demands of tankless heaters are avoided and because low-grade steam or waste high-temperature water sources can be utilized.

In any system that heats water under pressure,

**TABLE 21.9  Procedure for Sizing a Residential Water Heater Storage Tank**

| | *Table 21.8: HUD-FHA* | *Fig. 21.22 Data: Cyclone XHE* |
|---|---|---|
| **Characteristic** | **5 Bedrooms, 3 Baths** | **Model BTH-120** |
| Storage (gal) | 50 | 60 |
| Btuh input | 47,000 | 120,000 |
| 1-h draw (= tank capacity + 1-hr recovery) | 90 | 231 at 80F° |
| Recovery | 40 | 171 |

**TABLE 21.10 Domestic Hot Water, Commercial/Institutional**

| Type of Building | Maximum Hour | Maximum Day | Average Day |
|---|---|---|---|
| Men's dormitories | 3.8 gal (14.4 L)/student | 22.0 gal (83.4 L)/student | 13.1 gal (49.7 L)/student |
| Women's dormitories | 5.0 gal (19 L)/student | 26.5 gal (100 L)/student | 12.3 gal (46.6 L)/student |
| Motels: no. of units[a] | | | |
| 20 or less | 6.0 gal (23 L)/unit | 35.0 gal (132.6 L)/unit | 20.0 gal (75.8 L)/unit |
| 60 | 5.0 gal (20 L)/unit | 25.0 gal (94.8 L)/unit | 14.0 gal (53.1 L)/unit |
| 100 or more | 4.0 gal (15 L)/unit | 15.0 gal (56.8 L)/unit | 10.0 gal (37.9 L)/unit |
| Nursing homes | 4.5 gal (17 L)/bed | 30.0 (114 L)/bed | 18.4 gal (69.7 L)/bed |
| Office buildings | 0.4 gal (1.5 L)/person | 2.0 gal (7.6 L)/person | 1.0 gal (3.8 L)/person |
| Food service establishments: | | | |
| Type A—full meal restaurants and cafeterias | 1.5 gal (5.7 L)/max meals/h | 11.0 gal (41.7 L)/max meals/h | 2.4 gal (9.1 L)/avg meals/day[b] |
| Type B—drive-ins, grilles, luncheonettes, sandwich and snack shops | 0.7 gal (2.6 L)/max meals/h | 6.0 gal (22.7 L)/max meals/h | 0.7 gal (2.6 L)/avg meals/day[b] |
| Apartment houses: no. of apartments | | | |
| 20 or less | 12.0 gal (45.5 L)/apt. | 80.0 gal (303.2 L)/apt. | 42.0 gal (159.2 L)/apt. |
| 50 | 10.0 gal (37.9 L)/apt. | 73.0 gal (276.7 L)/apt. | 40.0 gal (151.6 L)/apt. |
| 75 | 8.5 gal (32.2 L)/apt. | 66.0 gal (250 L)/apt. | 38.0 gal (144 L)/apt. |
| 100 | 7.0 gal (26.5 L)/apt. | 60.0 gal (227.4 L)/apt. | 37.0 gal (140.2 L)/apt. |
| 200 or more | 5.0 gal (19 L) | 50.0 gal (195 L)/apt. | 35.0 gal (132.7 L)/apt. |
| Elementary schools | 0.6 gal (2.3 L)/student | 1.5 gal (5.7 L)/student | 0.6 gal (2.3 L)/student[b] |
| Junior and senior high schools | 1.0 gal (3.8 L)/student | 3.6 gal (13.6 L)/student | 1.8 gal (6.8 L)/student[b] |

*Source:* Reprinted with permission of the American Society of Heating, Refrigerating and Air-Conditioning Engineers, Inc. from 1995 *ASHRAE Handbook—HVAC Applications.*

[a]Interpolate for intermediate values.

[b]Per day of operation.

safety precautions are necessary. To minimize the dangers of excessive pressure (which can damage the system) and of superheated water (which can severely injure people), most codes require *pressure and temperature (P/T) relief valves* to be installed on top of all water heaters. Because these valves are designed to release hot water whenever a danger-ous pressure or temperature is reached, they should be attached to a length of pipe that will conduct the released water to a drain.

### (i) Solar Water Heating

Solar energy is most attractive to a designer when it will do most of its work during the time that it is most available. On a seasonal basis (Fig. 21.25), solar energy is especially attractive for outdoor swim-ming pool heating, as it will be used only in sunny, warm months. Throughout the United States, solar energy can easily meet most of the summertime demand for DHW as well. In the northern part of the country, a much smaller solar contribution can be expected in winter. An especially difficult problem

is the winter space-heat mismatch between greatest need and least solar energy supply. For both solar DHW and space-heating systems, winter brings the added complication of the need to protect against freezing.

Again, there are many ways in which solar energy can be used to heat water. Solar water-heating systems are usually classified by the means of fluid circulation (passive or active), the means by which heat from the collector piping is transferred to the DHW itself (direct, indirect, or closed-loop), and the means of protection against freezing.

*Passive systems* rely upon gravity for circula-tion, as explained in Section 21.6f. Hence, the stor-age tank usually must be placed *above* the collector (Fig. 21.26) and the number of bends in the system supply and return piping minimized to reduce fric-tion. Heavy storage tanks located high in a building can cause structural problems. The advantages of this passive approach include lower cost of compo-nents, high mechanical reliability (no pump, etc.), and low operational costs.

*Active systems* use pumps to force the fluid

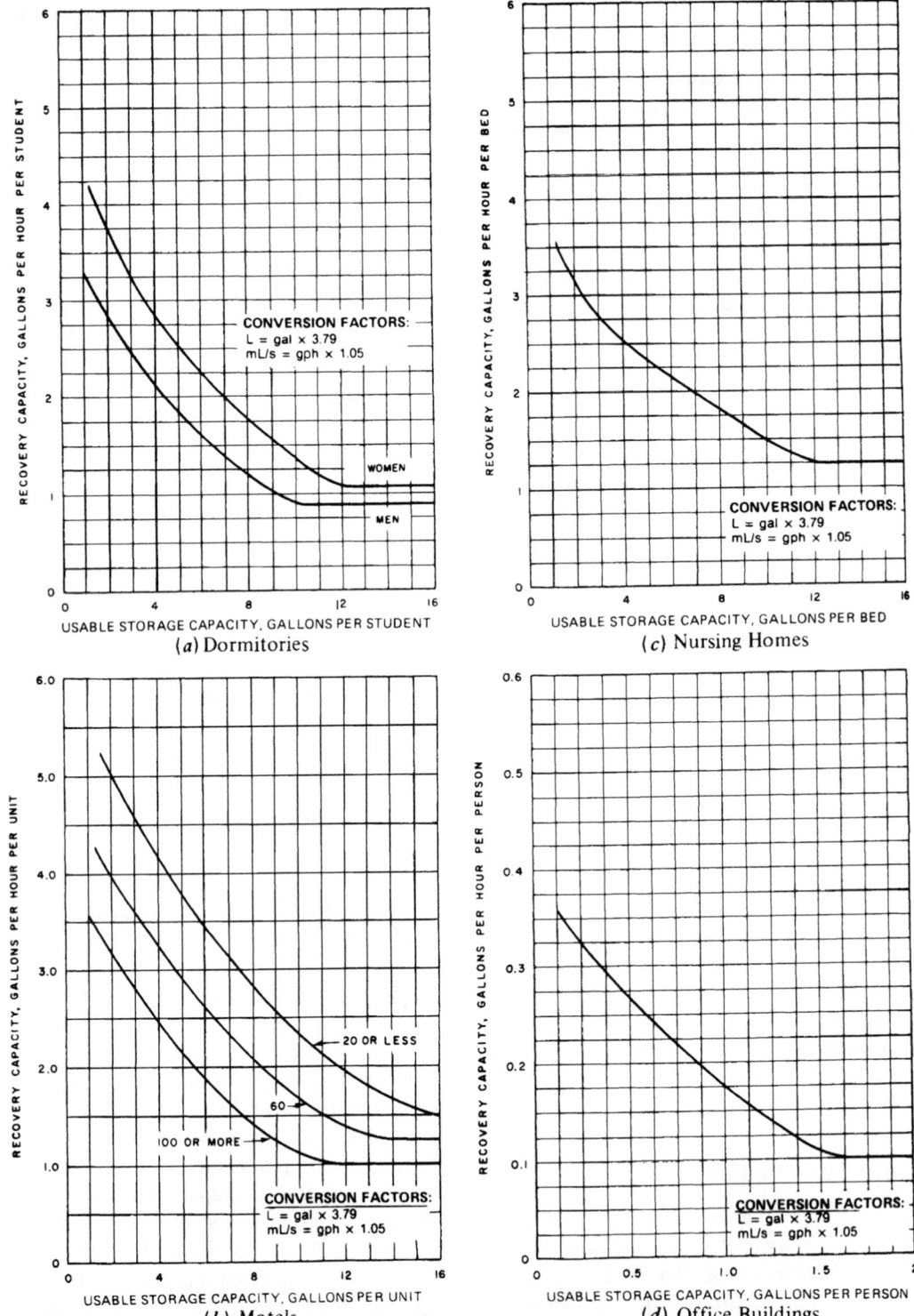

**Fig. 21.23** Domestic hot water sizing; the trade-off between recovery (heater) capacity and storage capacity. Usable storage capacity is usually considered to be between 60% and 80% of the total tank capacity. For (b) Motels, and (f) Apartments, curves are based upon the number of dwelling units. For (e) Food service, Types A, B, see Table 21.10. (Reprinted by permission of the American Society of Heating, Refrigerating and Air-Conditioning Engineers, Inc. from the 1995 ASHRAE Handbook—HVAC Applications.)

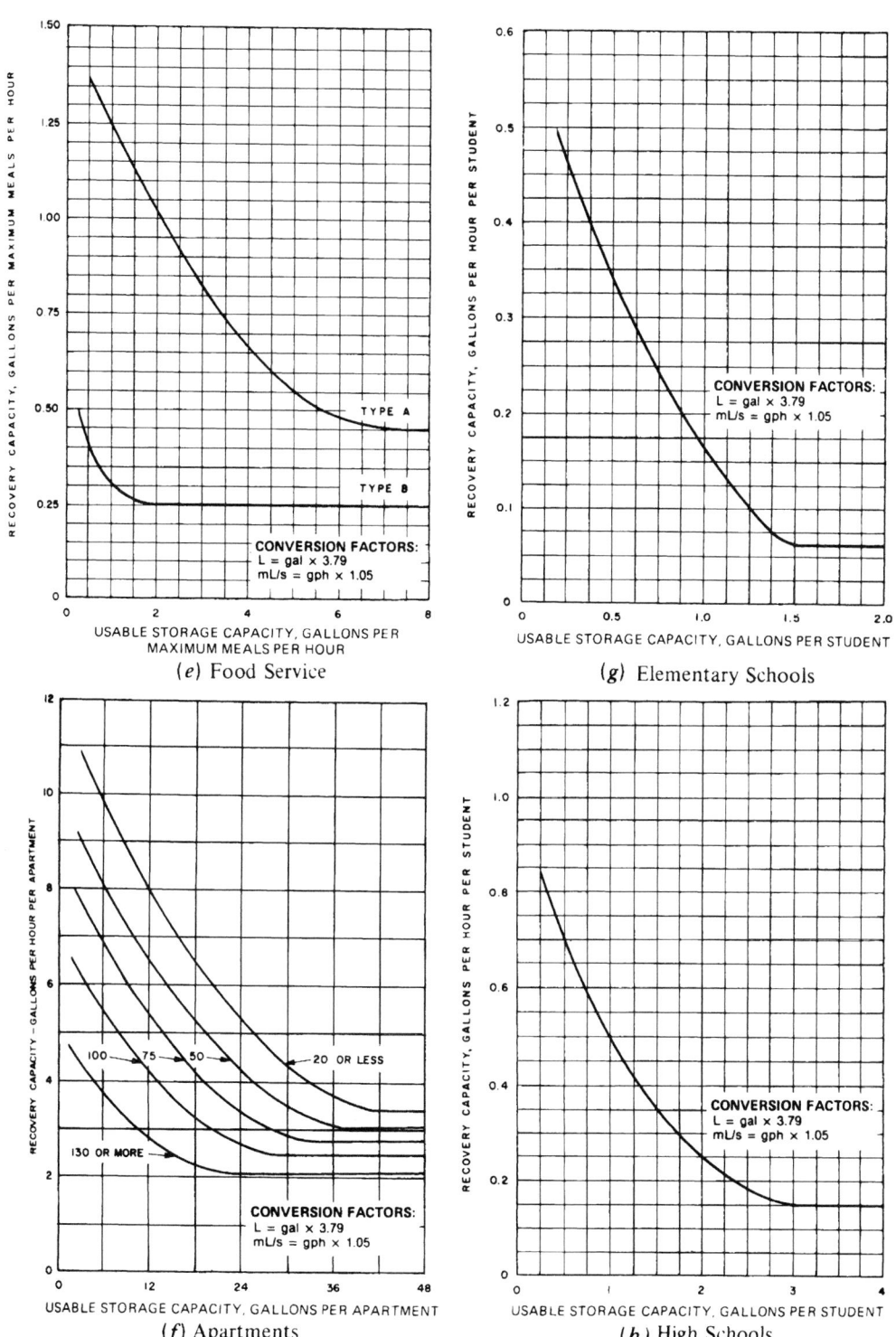

Fig. 21.23 (Continued)

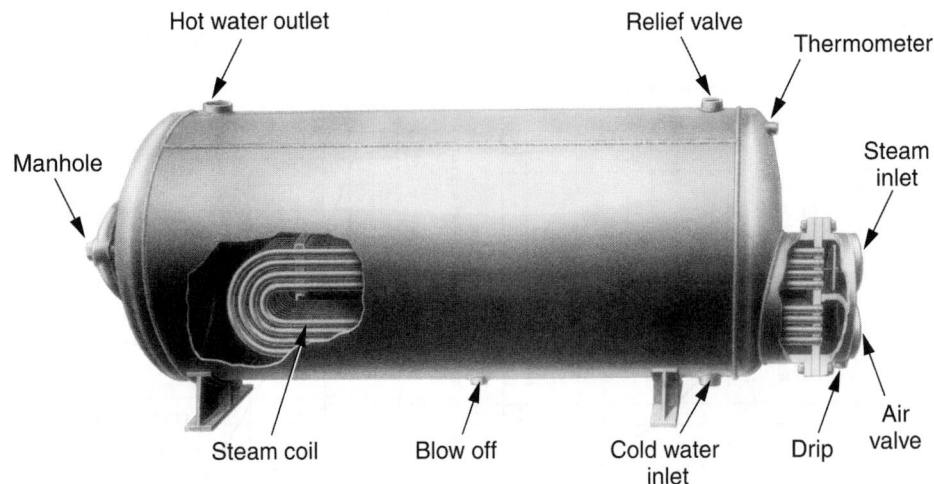

**Fig. 21.24** *Storage tank and heater for DHW for large-demand applications. A steam coil is submerged in the tank. Capacities from 100 to 10,000 gph (380 to 37,850 L/h), varying by length of coil, for a 140F° (from 40° to 180°F) (78C°) temperature rise.*

into the collector. Although this setup allows the collector and storage to be located anywhere that is convenient for the designer, it introduces the complications of mechanical breakdown, increased maintenance, and the cost of the energy needed to

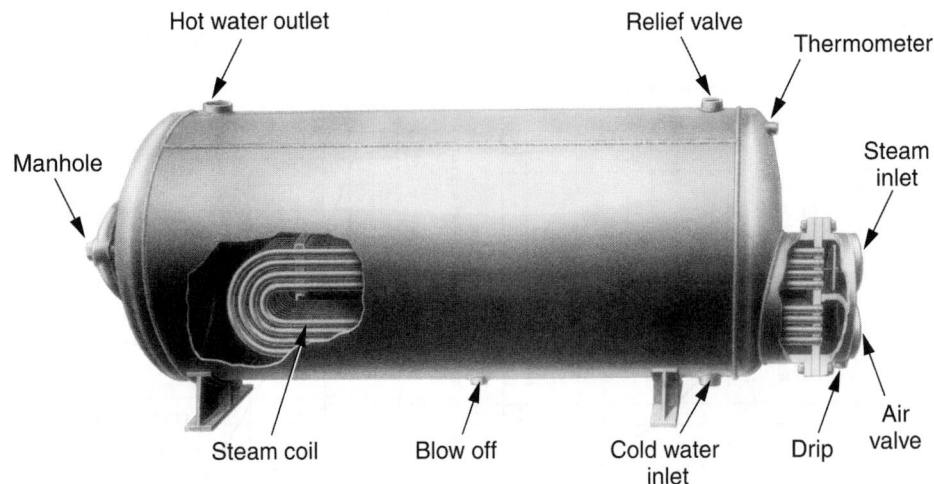

**Fig. 21.25** *Comparison of the pattern of solar energy supply (on a horizontal surface) to various patterns of heating needs.*

run the pump. Active DHW systems are widespread in North America; see Figs. 21.27 and 21.28 for typical systems.

*Direct systems* utilize only one fluid: the water to be heated for use in the building is circulated through the solar collector. Such a system has the advantages of simplicity and efficiency, as it does not require a separate fluid loop and the attendant piping complications and inefficiencies of heat exchange.

*Indirect systems* use a closed loop containing a fluid that circulates through the collector and storage tank. The fluid is not mixed with the DHW itself; rather, heat is passed from one fluid to the other through a heat exchanger. One advantage of this system is that it allows nonfreezing solutions to be used in the collector loop. It also allows collectors to be operated at low pressure rather than at the high pressures typical of urban public water systems. The choice of fluid is determined by its freezing and boiling points, its specific heat, and its level of toxicity.

As Table 21.11 shows, these design elements (passive and active, direct and indirect) are combined in a number of typical solar DHW systems. Brief descriptions of some of the more common systems follow.

1. *Batch systems.* In these, the simplest of all such systems, a black-painted storage tank is exposed to the sun within a glazed collector box (Fig. 21.29).

**TABLE 21.11  Solar Domestic Hot Water Systems**

| Type | Main Features | Advantages | Disadvantages |
|---|---|---|---|
| Batch system | Batch tank inside collector box.<br>Potable water within collector/tank. | No external power.<br>Few components.<br>Collector/tank at any location. | Seasonal; dependent on freezing locations.<br>Heat loss at night from storage. |
| Thermosiphon system[a] | Flat plate liquid collectors.<br>Normally open loop but no pump or external power (passive DHW system).<br>Storage tank higher than collector. | No external power.<br>Few components.<br>High performance. | Seasonal; dependent on freezing locations (if water is collector fluid).<br>Needs structural support for high storage tank. |
| Geyser pumping system | Flat-plate liquid collectors.<br>Methanol-water solution passively pumped by solar heat to a heat exchanger below. | No freezing damage.<br>No mechanical or electric parts.<br>No liquid service. | Temporarily stops operation at subzero temperatures, as solution freezes to a slush. |
| Closed-loop freeze-resistant system[a] | Flat plate liquid collectors.<br>Closed loop of piping from collectors to storage tank.<br>Uses external energy (circulator and differential controller).<br>Uses nonfreezing collector fluid.<br>Pressurized stonelined storage tank. | Can be used in coldest climates.<br>More and better-established competition.<br>Circulator; small consumption of external energy.<br>High performance. | Liquid; service/maintenance required.<br>More components required. |
| Drain-back system[a] | Flat plate liquid collectors.<br>Water is collector fluid (open loop).<br>Potable water circulates through the heat exchanger in the storage tank (not through the collectors).<br>Large heat exchanger.<br>Pitched headers. | Can be used in the coldest climates.<br>No antifreeze used.<br>Most simple of active flat plate systems (no valves). | Larger pump; larger consumption of external energy.<br>System must drain thoroughly.<br>Use of corrosion inhibitor recommended. |
| Drain-down system[a] | Flat plate liquid collectors.<br>Potable water circulated through collectors.<br>Line pressure feeds collectors (open loop).<br>Has automatic drainage valves.<br>Pitched headers. | No heat exchanger or extra storage tank needed.<br>High performance. | In some instances a larger pump; larger consumption of external energy.<br>System must drain thoroughly.<br>No corrosion inhibitor possible.<br>Freeze danger with valve failure. |
| Air-to-liquid system[a] | Flat plate air collectors.<br>Air-to-water heat exchanger.<br>Ductwork and blower.<br>Pipes and circulator.<br>Larger collector area than liquid system. | Won't freeze (dependent on exchanger location).<br>Air leaks won't cause damage.<br>Integrates well with space heating. | Hard to detect leaks.<br>More space required for ducts.<br>Blower and circulator required.<br>More carpentry involved.<br>Less efficient than other systems. |
| Phase-change system[a] | Flat plate liquid collectors.<br>Freon 114 or R12.[b]<br>Storage tank higher than collectors (passive type).<br>Closed-loop refrigerant-grade piping from collectors to storage tank (passive type) or to condenser (subambient type). | No external power (passive type).<br>Can be mounted at any location (subambient type). | Very hard to detect leaks.<br>Special equipment to install.<br>More components required (subambient type). |

[a]Reprinted by permission from Russell H. Plante, *Solar Domestic Hot Water: A Practical Guide*, © 1983 by John Wiley & Sons.
[b]Hydrochlorofluorocarbon or hydrocarbon refrigerants must be used in new systems.

WATER AND WASTE

**Fig. 21.26** *Solar water heaters are widely used. This house in Miami, Florida, incorporates a storage tank above the collector, which is enclosed in the chimney-like form. This permits passive (thermosiphon) circulation without a pump. (Photo by M. Steven Baker.)*

The price of simplicity is the inefficiency of exposing a tank of hot water to the cold nighttime conditions of the collector. Movable insulation, however inconvenient, can be used to reduce this problem. These heaters are also referred to as *breadbox* water heaters and *integral passive solar water heaters.* The batch system was used at the Antelope Valley California Poppy Reserve (Fig. 8.57), as shown in Fig. 21.30. Because of their simplicity, they are favored by do-it-yourself builders, especially in mild winter climates.

A more recent batch collector design utilizes eight copper tubes in a well-insulated, double-glazed (glass out, Teflon in) frame. The water flows in series, allowing the colder replacement water to enter at the bottom of the collector, with the hottest water at the top, ready for use. Insulation between the tubes helps maintain this temperature difference. Collectors are available in storage sizes of approximately 30, 35, 40, and 50 gal, weighing when filled 425 to 664 lb (114, 133, 152, and 190 L, weighing 193 to 301 kg).

2. *Thermosiphon systems.* The sun acts as both the pump and the heat source for these systems (Fig. 21.31). With no moving parts, maintenance needs are low. The collector, however, must be lower than the tank, and piping must be kept as simple as possible. Because the coldest water remains in the lower collector at night, the hot water in the upper tank is not threatened with undue heat loss. However, freezing conditions pose a severe threat to the collector. Accordingly, indirect (closed-loop) systems containing a nonfreezing fluid are frequently used; *phase-change systems* are a promising cold-winter option for this type of passive system.

3. *Closed-loop, freeze-resistant systems.* In addition to being used in thermosiphoning arrangements, these systems are commonly used in active systems (Fig. 21.32). A small pump circulates nonfreezing fluid to the collector when there is sufficient sun. This process is governed by a differential controller. Alternatively, a PV-(photovoltaic) driven circulation pump can be used (see Fig. 29.16); when there is sufficient solar energy to activate the pump, there should be enough to also heat water in the collector.

The price of this freeze protection is the inefficiency of the heat exchanger, used between the

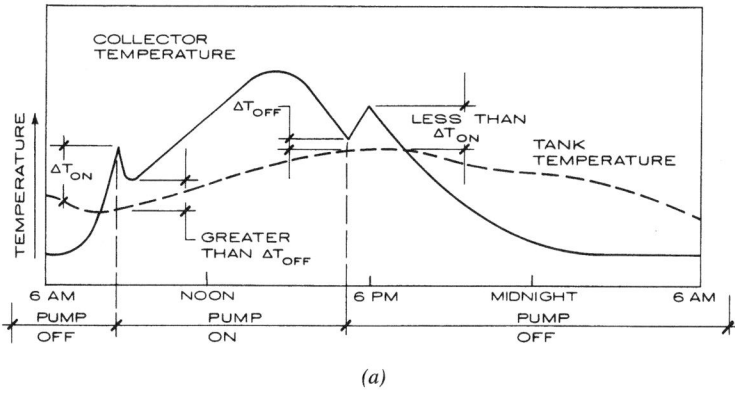

(a)

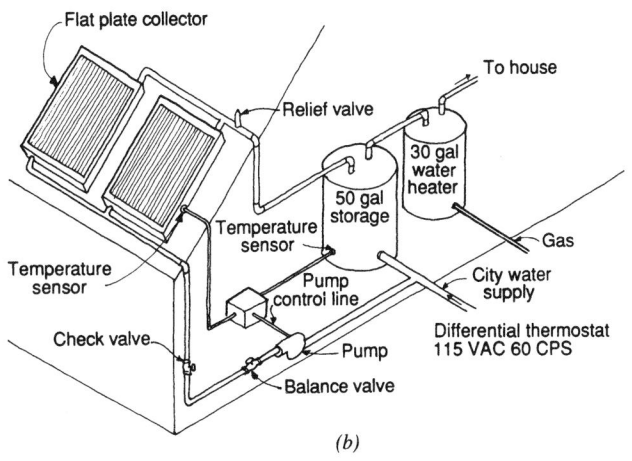

(b)

**Fig. 21.27** *Active solar water heating utilizes differential controllers (a) that compare the collector temperature to that of the storage tank. (b) Typical DHW system with separate solar and conventional heater tanks. (The alternative of only one storage tank produces less standby heat loss, but may reduce solar collection by feeding the collector with gas-heated water.)*

collector fluid and the potable hot water. Some codes require a *double* wall heat exchanger between any toxic nonfreezing fluid and the potable water, which further reduces efficiency.

4. *Drain-back systems.* Although these systems (Fig. 21.33a) use water as the fluid pumped from tank to collector, this water is not the potable hot water itself. Instead, the potable water passes through a heat exchanger in the solar storage tank. With the potable water kept out of the collector, the solar collector can operate at lower water pressure. This arrangement, however, requires a large heat exchanger, with attendant inefficiency. It also requires care in the design and installation of piping between the collector and tank so that the collector will drain thoroughly. When the controller senses that no solar energy can be gathered, it cuts

off the pump and water drains back into the tank. Therefore, the collector will be filled with air, not water, during all nighttime and cloudy-cold daytime hours. Corrosion inhibitors should be added to the collector/tank's water because the piping is frequently exposed to air.

5. *Drain-down systems.* These are the only active systems that do not utilize heat exchangers (Fig. 21.33b); DHW is circulated directly through the collector. Both higher pressure and higher efficiency result. In this system, the collector is usually filled with water that moves only when a differential controller activates the pump. Whenever the outside temperature drops near freezing, the controller activates solenoid valves and the water in the collector is drained down (dumped). In cold-winter areas, this process can result in several gallons of water

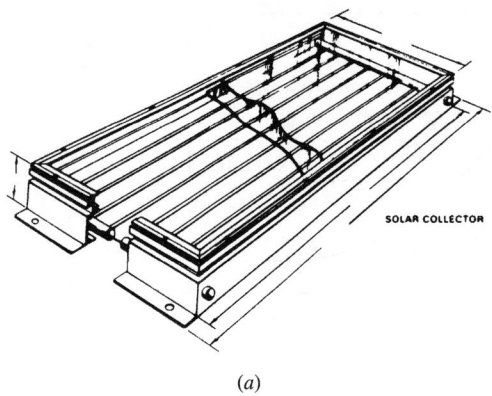

(a)

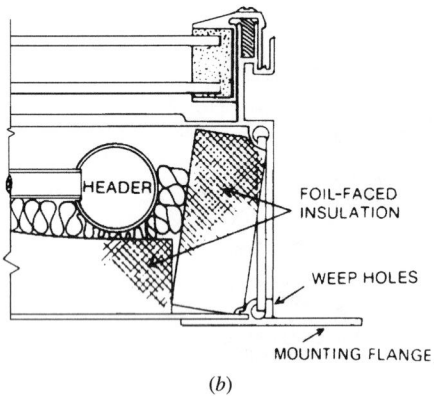

FOIL-FACED INSULATION

WEEP HOLES

MOUNTING FLANGE

(b)

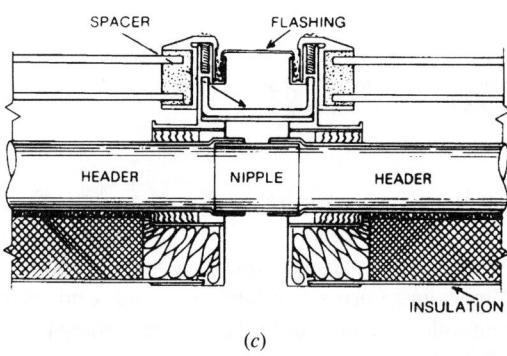

INSULATION

(c)

**Fig. 21.28** (a) Cutaway section of a typical active solar collector. (b) Cross section at the header. (c) Method of connecting manifolds of adjacent solar collectors.

wasted per winter day. Although the lack of a heat exchanger is attractive from the standpoint of cost savings and thermal efficiency, the set of electrically controlled solenoid valves is not foolproof. When malfunctions occur in pressurized systems, the loss

of water can be enormous, and there is great potential for water damage to the building.

6. *Air-to-liquid systems.* These systems use air collectors and rock storage beds; a heat exchanger transfers heat from the collector-heated air to the hot water. Much lower efficiencies result, and the ductwork is much more space-consuming than are pipes for water collectors. Leaks in air collectors or storage beds are very difficult to detect. However, air collectors are not damaged by freezing.

7. *Phase-change systems.* In any of the above-discussed systems in which potable water is kept out of the collector, the fluid within the collector not only can be freeze resisting, but also can take advantage of latent heat (the considerable amount of heat stored when fluid vaporizes and released when a fluid condenses). The primary disadvantages of a phase-change system (Fig. 21.34) are the high first cost, the difficulty in detecting leaks, and the resulting threat of chlorofluorocarbon or hydrochlorofluorocarbon refrigerant fluids to the environment.

*Passive* approaches to phase-change materials were outlined in the discussion of thermosiphoning systems. *Subambient* approaches utilize a heat pump and can glean heat even from cold-cloudy ("subambient") conditions. They thus represent a form of solar-assisted heat pump, closely related to those discussed in Section 21.6j.

The *sizing* of solar water heaters usually begins with simple design guidelines. For batch heaters, such a sizing guideline (Fig. 21.35) is:

0.45 to 0.65 $ft^2$ glazing per gallon of water stored
(0.011 to 0.016 $m^2$ per liter of water stored)

For the flat-plate solar collectors used in all the solar DHW systems listed above, the sizing guidelines are:

12 to 25 $ft^2$ collector/person (residential)

(1.1 to 2.3 $m^2$/person)

Optimum tilt (up from the horizontal) equal to latitude (or less)

1 to 1.5 gal of storage per square foot of collector area

(40.7 to 61 L per square meter of collector area)

This collector-sizing guideline is for *residential* hot water usage, including cooking and bathing. For warmer climates with ample insolation, use the lower figure; this will supply about half of the hot water on an annual average basis. (For *nonresidential*

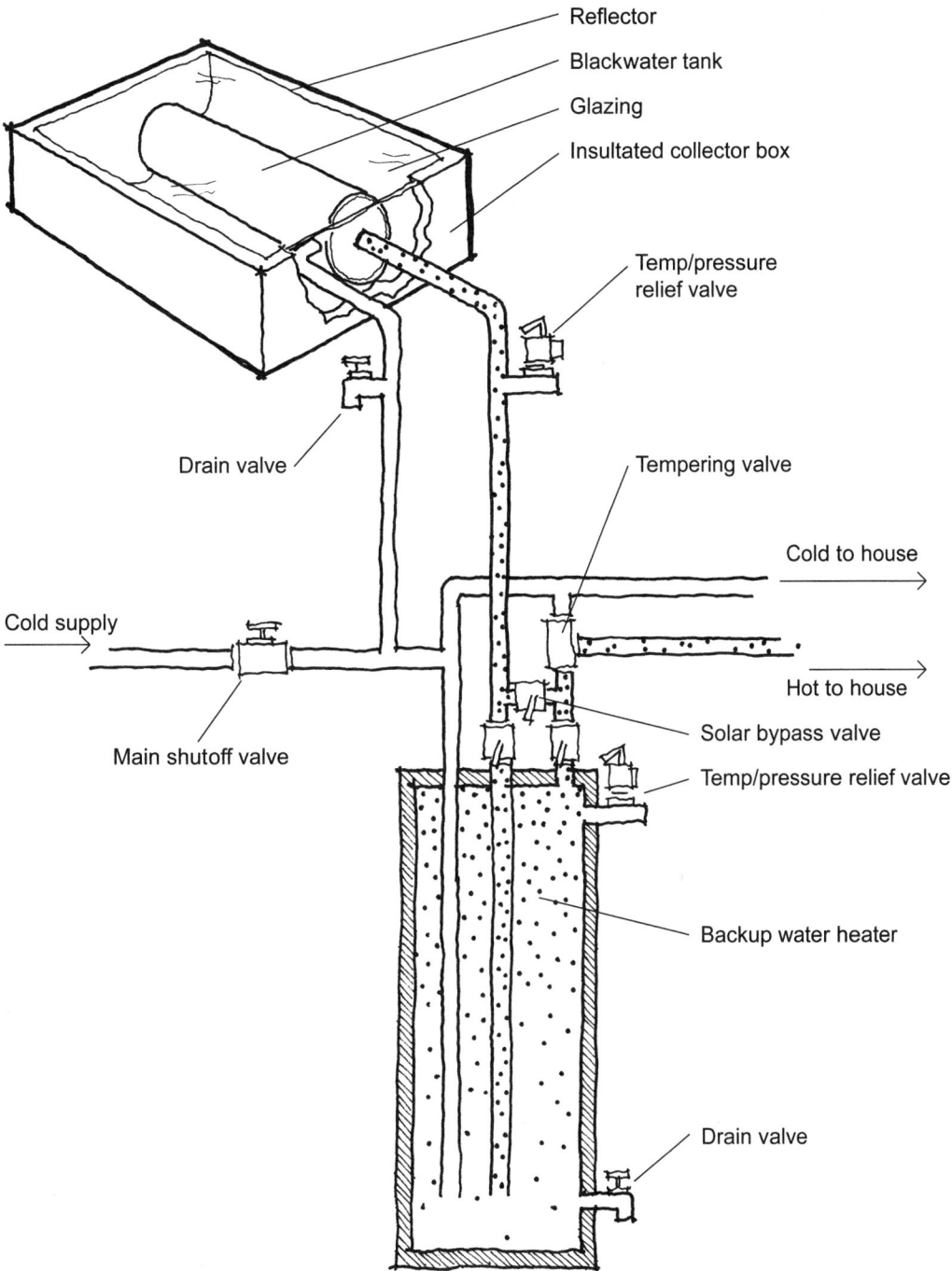

Reflector

Blackwater tank

Glazing

Insultated collector box

Temp/pressure relief valve

Drain valve

Tempering valve

Cold to house

Cold supply

Hot to house

Main shutoff valve

Solar bypass valve

Temp/pressure relief valve

Backup water heater

Drain valve

**Fig. 21.29** *Batch solar water-heating system. (Drawn by Dain Carlson; © 2004 Alison Kwok, all rights reserved.)*

Hot water from tank
Cold water to tank
Hot water to fixture
Mixing valve
Max. output 120°F
Water supply from well

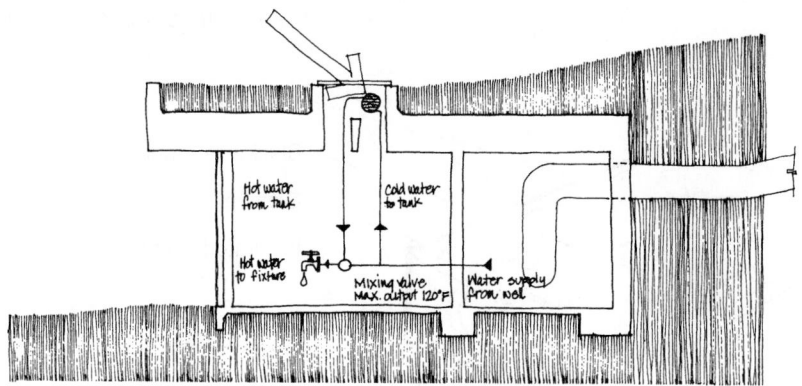

***Fig. 21.30*** *A batch solar water heater supplies the hot water for the Antelope Valley California Poppy Reserve (described in Fig. 8.57). (Courtesy of the Colyer/Freeman Group, Architects, San Francisco.)*

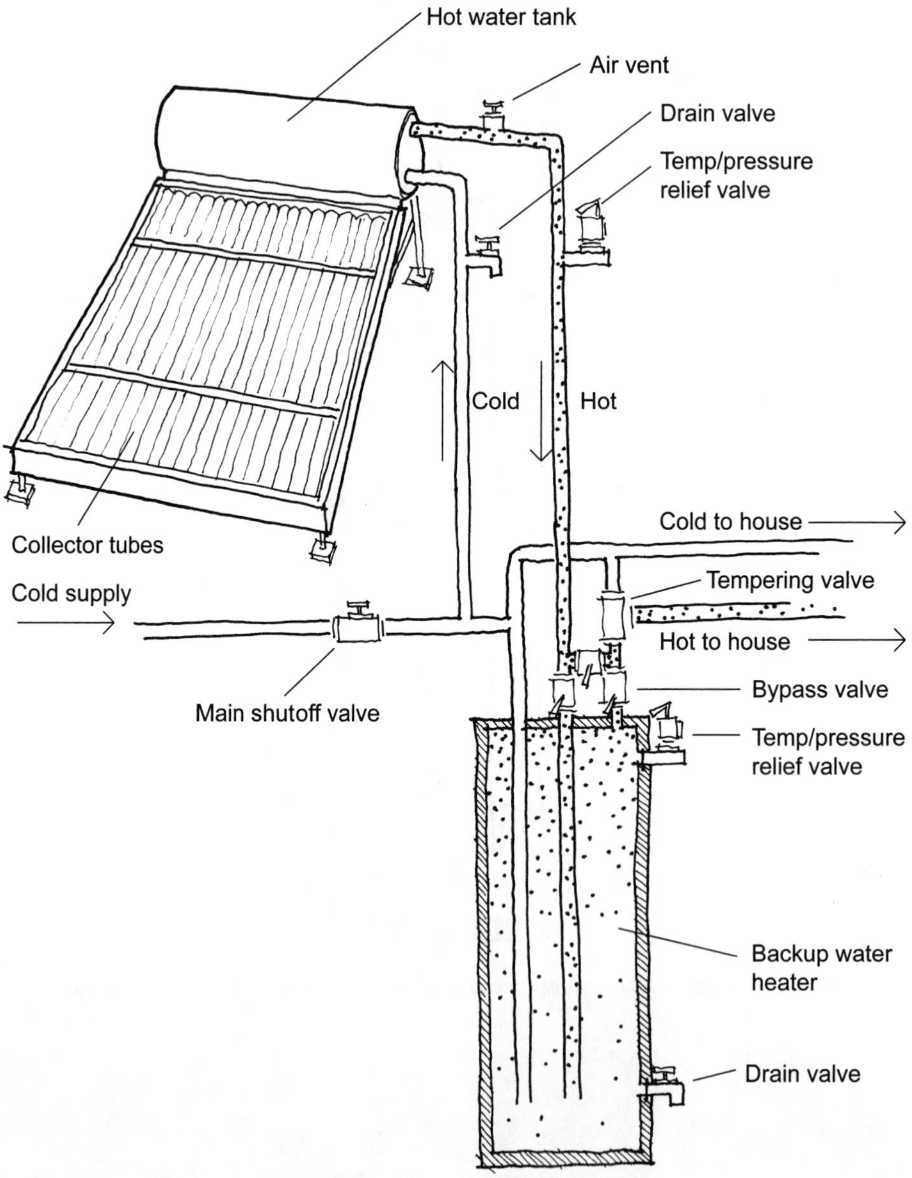

Hot water tank
Air vent
Drain valve
Temp/pressure relief valve
Cold
Hot
Collector tubes
Cold supply
Cold to house
Tempering valve
Hot to house
Main shutoff valve
Bypass valve
Temp/pressure relief valve
Backup water heater
Drain valve

***Fig. 21.31*** *Thermosiphon solar water-heating system. (Drawn by Dain Carlson; © 2004 Alison Kwok, all rights reserved.)*

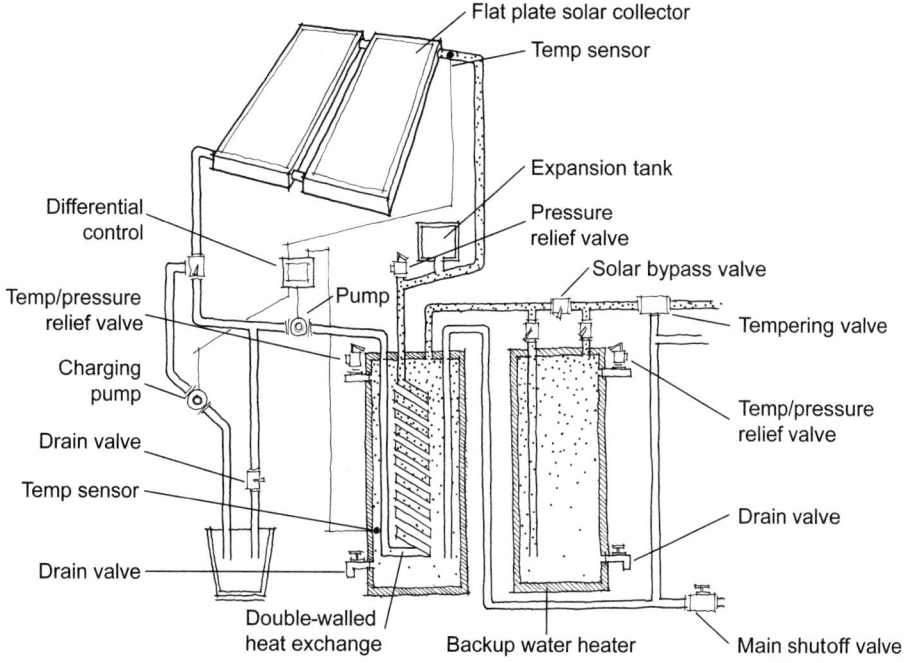

**Fig. 21.32** *Closed-loop, freeze-resistant solar water heating system. (Drawn by Dain Carlson; © 2004 Alison Kwok, all rights reserved.)*

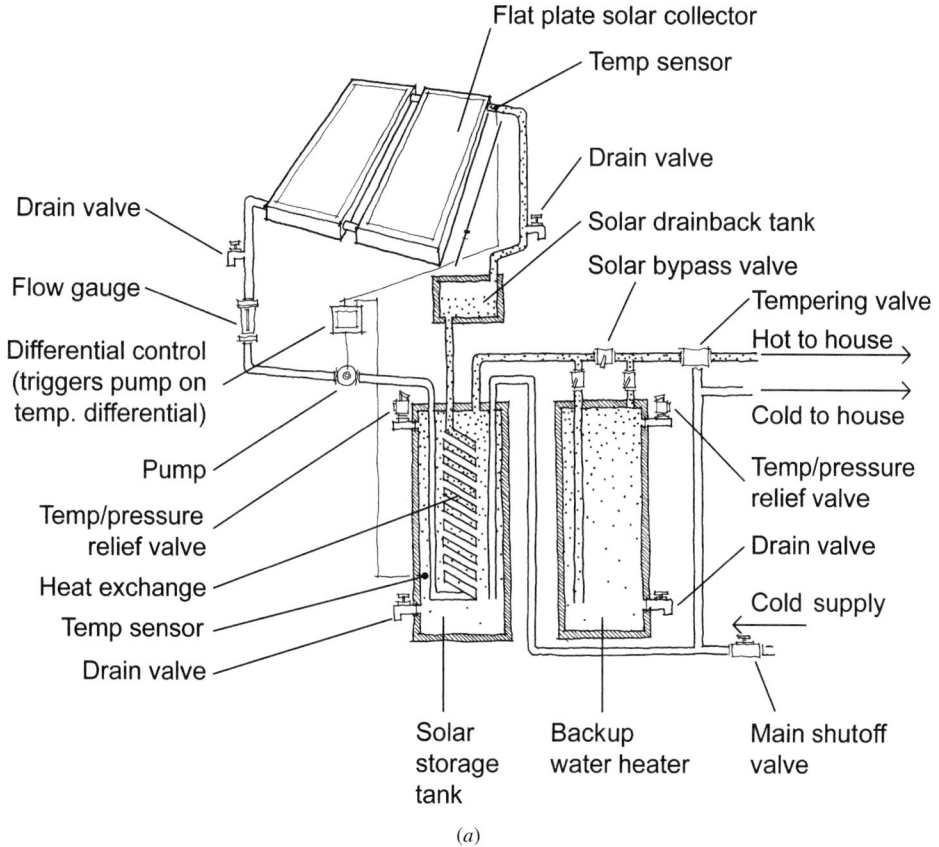

(a)

**Fig. 21.33** *Comparison of (a) a drain-back solar water-heating system with (b) a drain-down system. The drain-back system's performance can be increased by the addition of a check valve that allows the colder water at the bottom of the storage tank to circulate into the heat exchanger. The drain-down system is best used in mild-winter climates with infrequent freezing temperatures, since the water in the collector is dumped with each freezing threat. (Drawings by Dain Carlson; © 2004 Alison Kwok, all rights reserved.)*

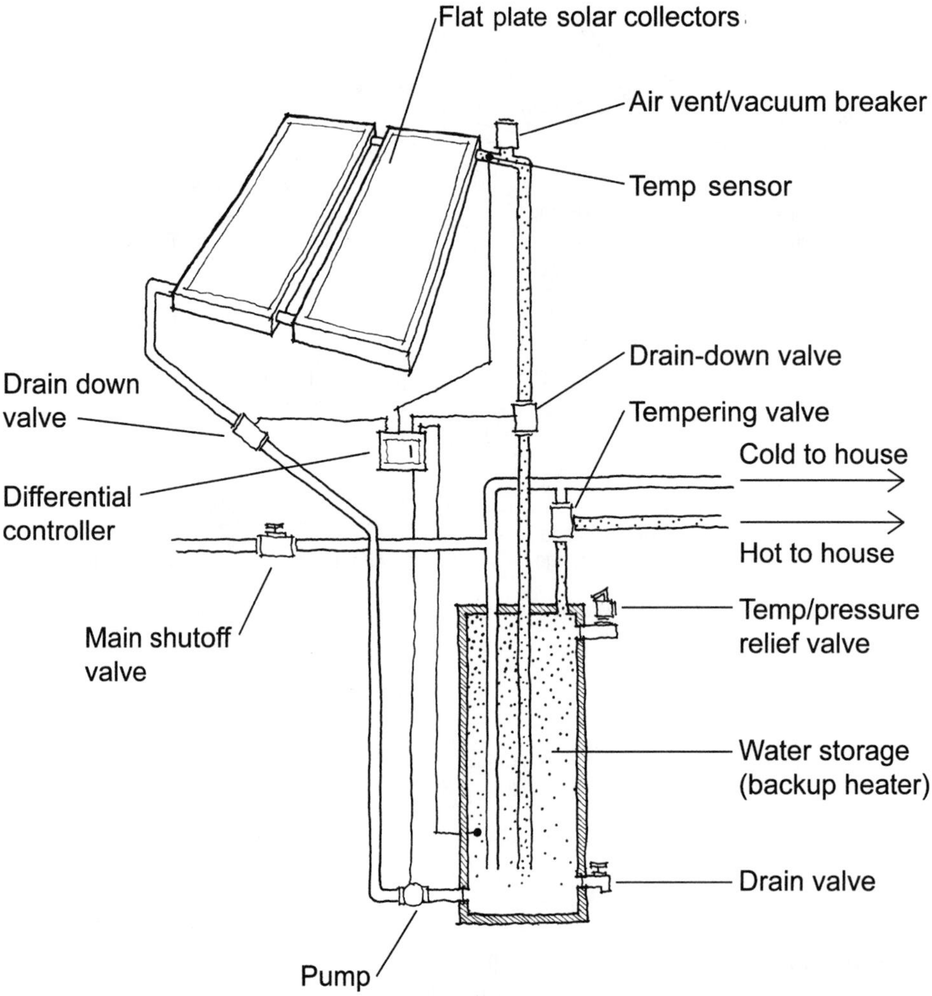

Flat plate solar collectors
Air vent/vacuum breaker
Temp sensor
Drain-down valve
Tempering valve
Cold to house
Hot to house
Drain down valve
Differential controller
Main shutoff valve
Temp/pressure relief valve
Water storage (backup heater)
Drain valve
Pump


(*b*)

***Fig. 21.33*** *(Continued)*



Cold water in
Coolant heat exchanger
PTR valve
Hot water in
Hot water out
Backup water heater


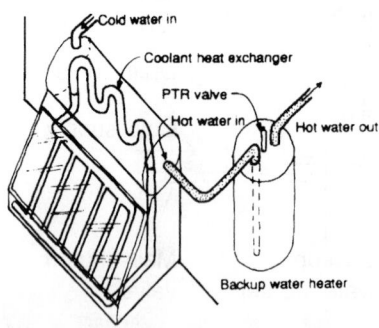

***Fig. 21.34*** *In a phase-change system, small heat exchangers can be built into the tops of thermosiphoning collectors; DHW is drained down when the threat of freezing arises. (Adapted by permission from* Solar Age, *November 1983.)*

DHW, adjust the sizing guideline by comparing the gallons of hot water per person per day used in residences to those used in the nonresidential function under design.)

A more detailed sizing procedure would consider the collector's expected heat contribution for some typical months. Such a procedure requires both climate data and assumptions about water temperatures and other values. As a result, detailed answers are best obtained from computer programs, such as those listed in Appendix M. For those interested in something more than sizing guidelines but less than computer programs, consider the following approach to collector sizing (adapted from

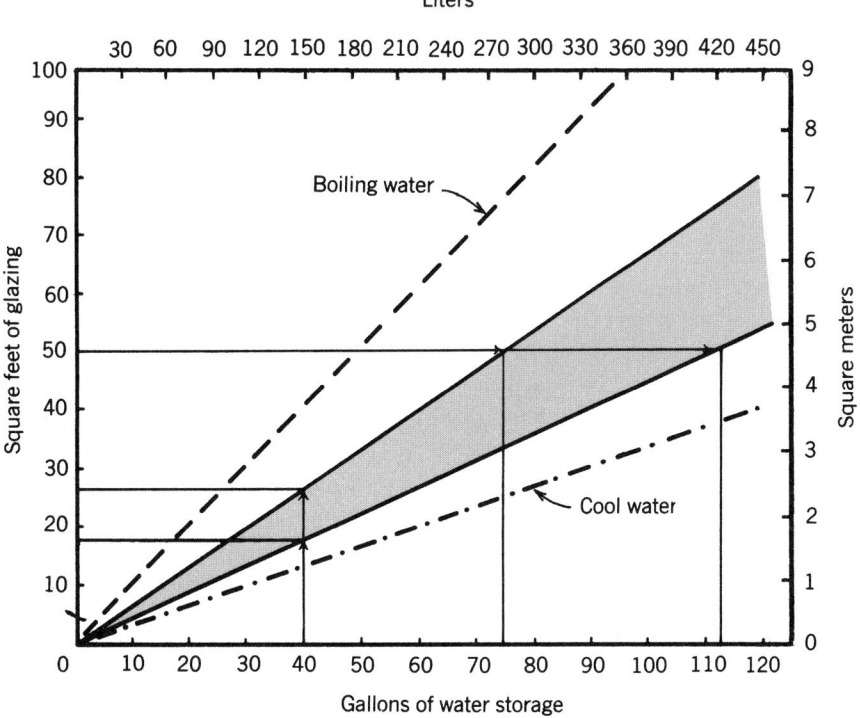

**Fig. 21.35** *Design guideline for the sizing of batch solar water heaters. In this example, a 40-gal (150-L) storage tank should be contained within a collector that has between 18 and 27 ft² (1.7 and 2.5 m²) of south-facing glazing, or a 50-ft² (4.7 m²) collector box should contain a tank of between 75 and 113 gal (285 and 430 L). (Adapted from Daniel K. Reif,* Passive Solar Water Heaters, *copyright © 1983 by Brick House Publishing Company, Inc. Reprinted by permission.)*

Brown, Reynolds, and Ubbelohde, *InsideOut: Design Procedures for Passive Environmental Technologies,* © 1982 by John Wiley & Sons).

**STEP 1.** *Select the collector tilt angle.* This can be done by consulting Tables C.11 to C.14, which present clear-day insolation values for various tilt angles. Appendix C, Table C.15, presents average insolation values on horizontal surfaces, which should also be checked. The tilt angle selected should be close to the optimum angle for the best month for total insolation.

**STEP 2.** *Check the collector efficiency.* This should be done at least twice: for the best insolation month and for the worst month. This step requires the following data:

Hourly insolation on tilted surface (Appendix C, Tables C.11 to C.14)

Outdoor average temperature (Appendix C, Table C.15)

The hourly insolation values on tilted surfaces in Tables C.11 to C.14 are for *clear* days. For most locations, this should be adjusted for *average* conditions, which are shown for vertical (south) and horizontal surfaces in Table C.15. Because for most of North America the optimum yearly DHW tilt angle will be closer to horizontal than to vertical, a simple correction to insolation must be made:

Hourly average day insolation on tilted surface

$$
= \begin{bmatrix} \text{hourly clear-day} \\ \text{insolation on} \\ \text{tilted surface} \end{bmatrix} \times
\begin{bmatrix} \dfrac{\text{average-day total}}{\text{horizontal insolation}} \\[4pt] \dfrac{\text{clear-day total}}{\text{horizontal insolation}} \end{bmatrix}
$$

This step also requires assumptions about the input temperature of the water supplied to the collector

WATER AND WASTE

($T_i$). In summer, this can be quite high—even higher than the thermostat setpoint temperature of the hot water tank. In winter, it is likely to be lower, by perhaps 10 to 20F° (5.5 to 11C°), than the thermostat setpoint temperature. Finally, this step requires a choice of collector type so that efficiency can be determined. Figure 21.36 shows the efficiency curves for a variety of solar collectors.

Note that the simplest of all, the unglazed flat black collectors, have the highest efficiency of all collectors at combinations of very low $T_i - T_a$ and very high $I_o$. These conditions are typical of summer days, on which the water needs to be heated only a few degrees. This corresponds to swimming pool applications and is one reason such collectors are so popular for that purpose; low cost is another reason.

For the opposite conditions—those of winter space heating—more elaborate collectors are appropriate. Evacuated-tube collectors have the advantage of nearly eliminating collector heat loss by convection. Fresnel lens and/or tracking-concentrating collectors increase the incoming solar energy per unit area, whereas heat loss increases only slightly (due to higher $\Delta t$). Both of these approaches are expensive,

and tracking collectors require added maintenance for the tracking mechanism. Selective surfaces represent a simple, relatively cheap improvement over flat black collectors (compare D to B); they absorb solar energy just as well but emit radiant energy at a vastly lower rate, thus cutting collector heat loss by radiation.

**STEP 3.** *Approximate system efficiency.* Once the hot water leaves a collector, it must be led back to storage. In indirect systems, it must go through a heat exchanger. The tank will lose some heat; heat losses also occur as the cooled water is led back to the collector. Another "loss" to be considered is noncollection time. Estimates of collector performance are based on total daily insolation, yet the first few and last few hours of daylight generally will *not* bring enough insolation to warm the collector. Therefore, some reduction should be made in daily insolation totals when performance is estimated. All these factors can be accounted for by assuming a lower system efficiency. A perhaps optimistic assumption is that system efficiency = (0.8) × (collector efficiency).

**STEP 4.** *Determine the total heat needed for DHW.* Estimates of the quantity of hot water needed were made in Section 21.6h. Once the total gallons (or liters) per day are known and the desired storage temperature determined, the heat needed for the daily supply of DHW is easy to calculate. In I-P units:

$$Q = 8.33 \,(gpd)\,(t_s - t_g)$$

where

  $Q$ = daily heat need, Btu
  8.33 = weight of water, pounds per gallon ×
        Btu/lb °F
  $t_s$ = storage temperature, °F (see Table 21.7)
  $t_g$ = groundwater temperature, °F (see Fig. 8.56,
        well-water temperatures)

In SI units:

$$Q = 1.16 \,(L/d)\,(t_s - t_g)$$

where

  $Q$ = daily heat need, Wh
  1.16 = the SI equivalent of the 8.33 factor
  $t_s$ = storage temperature, °C
  $t_g$ = groundwater temperature, °C

**STEP 5.** *Determine the desired percentage solar contribution to DHW.* There are several approaches to this

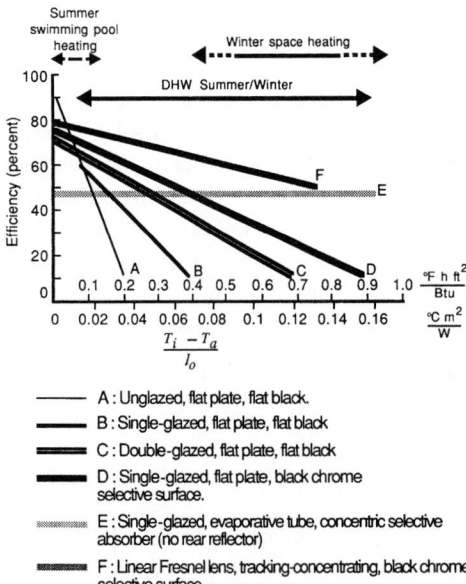

**Fig. 21.36** *Collector efficiencies depend upon insolation ($I_o$), water temperature entering the collector ($T_i$), and ambient (air) temperature ($T_a$). (Based on the U.S. Dept. of Housing and Urban Development's* Intermediate Minimum Property Standards Supplement, *1977.)*

problem. One common strategy is to have solar provide 100% of the hot water needs in the best insolation month. A closely related strategy is to choose the *yearly* percentage of solar heat desired and then provide that percentage in the average month (such as March). Whatever strategy is chosen, a month and its percentage of solar heat must be identified.

**STEP 6.** *Size the collector.* This is done by combining the preceding steps as follows:

collector area

$$= \frac{\text{daily heat need} \times \text{percent solar desired}}{\text{daily insolation} \times \text{system efficiency}}$$

This may be done for several months as a check on optimum collector size.

**EXAMPLE 21.4** Determine the approximate size of solar collectors needed to serve the women's dormitory described in Example 21.3. Assume the location to be Springfield, Illinois.

**STEP 1.** *Tilt angle.* Refer to Appendix C (Table C.15). Springfield's latitude is 39.8° N. In Table C.15, average insolation is given for July. Use this month as the best month.

From Table C.12, the optimum tilt angle for July at 40° N latitude is horizontal. However, at 30° tilt (latitude minus 10°), much better performance will be obtained in average months, such as March and September. Choose a 30° tilt angle.

**STEP 2.** *Collector efficiency.* For the best month (July), the best hourly clear-day insolation is 307 Btu/h ft$^2$, at 30° tilt, in the hour centered at noon. To adjust this for *average* hourly value, calculate

$$307 \times \frac{2058 \text{ Btu/day average horizontal (Table B.15)}}{2534 \text{ Btu/day clear horizontal (Table B.12)}}$$

$$= 249 \text{ Btu/h ft}^2$$

Outdoor average temperature (TA) is approximately obtainable from Table C.15; TA for July is 76°F. This is quite conservative, since it is the average daily temperature; the temperature at noon should be higher. From Appendix B, the mean daily range for Springfield is 21F°. Thus, 76 + 21/2 = about 86°F. The thermostat setpoint temperature of the water storage tank is assumed to be 115°F. This allows adequately hot water for most dormitory uses, although a higher temperature will be required for the cafeteria's dishwashing.

Assume, because it is summer, that $T_i$ is at the setpoint of 115°F. (In winter, it would be lower.) Therefore, the quantity $(T_i - T_a)/I_o$ can be calculated at

$$\frac{115°F - 86°F}{249 \text{ Btu/h ft}^2} = 0.116$$

Enter Fig. 21.36 with this number, and assume a single-glazed, flat plate selective surface collector (type D). The efficiency will be about 70% under these conditions.

**STEP 3.** *System efficiency.* Use the design guideline

0.8 (70% collector efficiency) = 56%

**STEP 4.** *Total heat needed.* The average usage in gpd can be estimated from Table 21.21. For women's dormitories, the rate is 12.3 gal/student × 300 students = 3690 gal. For the cafeteria, the average rate (type A) is 2.4 gal/meal = 2.4 × 300 = 720. Total gallons needed: 3690 + 720 = 4410.

Groundwater temperature for Springfield (from Fig. 8.56) is about 56°F. Assume that solar energy will be used to raise it to 115°F.

$$Q = 8.33 (4410 \text{ gpd}) (115 - 56) = 2,167,380 \text{ Btu}$$

**STEP 5.** *Desired percent solar.* For this best hour in July, assume a 100% solar contribution.

**STEP 6.** *Collector size.* Average daily insolation can be obtained from Appendix C as before:

Clear day total, 30° tilt, July: 2409 (Table C.12)

Clear day total, horizontal, July: 2534 (Table C.12)

Average day total, horizontal, July: 2058 (Table C.15)

Average July daily insolation, at 30° tilt
= 2409 × 2058/2534 = 1956

$$\text{collector area} = \frac{2,167,380 \text{ Btu} \times 100\%}{1956 \text{ Btu/ft}^2 \times 56\%}$$

$$= 1979 \text{ ft}^2$$

This size should be checked against average-month performance and adjusted as desired. Note that this is a ratio of about 1979/300 = 6.6 ft$^2$ of collector per student—lower than the general rule for typical residential DHW systems.

Note also that at the design guideline rate of 1 to 1.5 gal of storage per square foot of collector, a tank size of 2000 to 3000 gal is indicated. In Example 21.3, the minimum recovery rate required a storage tank of 8150 gal, and the fastest recovery rate required 3010 gal. Either seems sufficient for this size of collector array. ∎

*Swimming pool heating* is an especially attractive application for solar energy. A common design guideline for collector sizing is

collector area = 0.5 pool surface area

For summer ambient temperature operations, unglazed collectors are both the best-performing and the least expensive, as explained for Fig. 21.36. Another important consideration is a pool cover, which will not only conserve the pool's heat but also reduce water lost by evaporation. On very hot-dry days, water losses can reach 100 gpd from a 20 × 40-ft pool (380 L/day from a 6 × 12-m pool).

### (j) Heat Pump Water Heaters

The heat pump's use of the compressive refrigeration cycle was explained in Section 9.5 (also see Fig. 9.38). Air-water heat pumps are also used for DHW, as shown in Fig. 21.37. Because these devices remove heat from the air, they are usually installed either in normally overheated spaces (such as restaurant kitchens) or in unheated spaces such as garages or basements. (They can also remove heat from exhaust air, as explained in Chapter 5.) The spaces that contain these units will be cooled and dehumidified. Heat pumps require only a little more space than the simple hot-water storage tanks they serve. Some noise is created by the compressor and the fan that moves air across the evaporator.

Heat rejected from any refrigeration unit can be used for heating water via the heat pump. Applications include ice-making machines, refrigerated display units in grocery stores, walk-in freezers, and many others. Whenever constant refrigeration loads are present, there is an opportunity to utilize waste heat.

Graywater provides another opportunity for using a heat pump. Where filtered graywater is collected for recycling on site (as for irrigation systems), a water–water heat pump can heat a DHW storage tank as it lowers the graywater tank temperature. See Chapter 22 for graywater system design.

## 21.7 FIXTURES AND WATER CONSERVATION

Building design and fixture choice can affect water and energy consumption over the life of a building. It is the users who turn the faucets on and off but the designer who can encourage resource conservation through initial decisions, here centered largely in kitchens and bathrooms.

*Visible consumption* is one strategy to encourage the users of fixtures to conserve water. We see, hear,

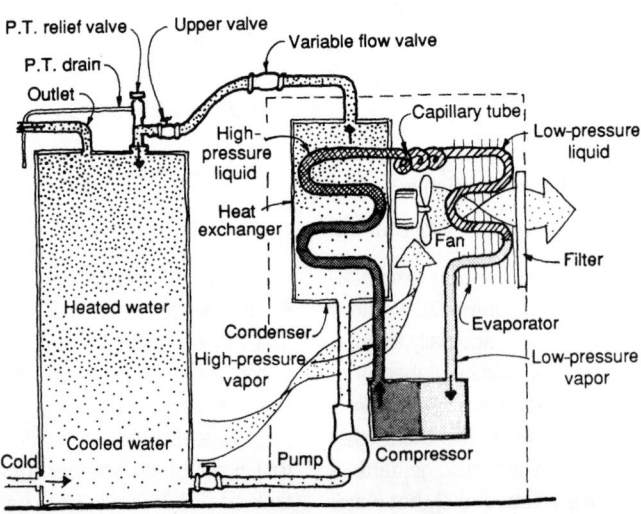

**Fig. 21.37** *Heat pump (air-water) used for heating DHW. (Adapted by permission from* Specifying Engineer, *October 1983. © 1983 by Cahners Publishing Co.)*

feel, and sometimes taste water as it issues from a fixture. If we could see *how much* water is involved in each use, we could more readily alter our pattern of use to conserve water where appropriate. Rainwater storage tanks outside bathroom windows, used to flush toilets, might have a visible indicator of the water level; the lower this level, the more that conservation is an imperative. The water-level indicator might be as simple as visible condensation on the tank's outer surface. Small, transparent tanks of clean supply water could be placed above lavatories, filled only when empty by means of a simple float valve. The user might thus be encouraged to use only a portion of the water already in the tank. Sufficient pressure at the fixture is achieved by such a tank's elevation above the fixture. The designer might also consider making the quantity of used (waste) water visible, but such water is often unsightly and the resulting quantity is "after the fact"—too late to conserve.

*Audible consumption* is another strategy, calling attention to the flow of supply (or waste) water. This is achieved most simply by slightly undersizing water supply pipes so that the water velocity becomes audible (at about 10 fps [3 m/s]). This strategy should be carefully considered; it is one thing to inform the water user, another to annoy. Audible flow is perhaps best justified on lines to exterior hose bibbs, where irrigation or car-washing hoses may inadvertently be left on.

Elimination of bodily wastes accounts for a significant portion of residential water usage (see Chapter 22), and bathing accounts for a significant portion of residential hot water usage (see Table 21.7). The choice of fixtures can have an impact on water conservation, and on energy conservation where hot water is involved. The bathroom is thus a site for potential resource conservation.

## (a) Physiology, Psychology, and Fixtures

One of the more challenging aspects of bathroom design is the frequent conflict between physiological design criteria and such psychological influences as cultural attitudes about bathroom activities. Physiological criteria change as slowly as evolution. Cultural attitudes can change very rapidly, even within a generation. The bathroom supports two primary human activities: cleansing and elimination. Today's common attitudes about these closely linked activi-

ties are that one is "clean," the other "dirty." One might be discussed (although rarely) among friends; the other is simply not fit for conversation. This conflict in attitudes influences the design of our plumbing fixtures. The toilet is the fixture used for elimination; it is the logical place to provide for the cleansing of the perineal zone that should immediately follow elimination. (This is especially true of public toilets, where stalls close off toilets from lavatories.) Yet very few toilets presently incorporate a cleansing feature: the mixture of cleansing and elimination in the same fixture too often seems abhorrent.

The brief summary presented in this section is based largely on a particularly revealing (and entertaining) study of the struggle between physiological and cultural criteria in fixture and bathroom design—Alexander Kira's *The Bathroom: Criteria for Design*, an expanded edition of which was published in 1976 by Viking Press.

## (b) Lavatories

A key issue here is the contrast between running water and a standing water body. Lavatories are used primarily for the cleansing of hands, face, and teeth—activities done quickly with running water that is wasted directly rather than being collected. Most lavatories are designed as collection bowls—perhaps a reflection of the days when washbasins of standing water, drawn and heated elsewhere, were brought to the bathing place. Most lavatories also have fittings that project out over the sink. Although these fittings have slowly evolved into very sleekly designed objects, most of them still dump running water directly into the drain, are hard to use as drinking fountains, and can wound those who try to wash their hair in the lavatory. In considering the prevalence of running as opposed to standing water, the role that running water could play in keeping lavatory surfaces clean, and the need for a drinking fountain when teeth are brushed, Kira proposed a quite different lavatory design (Fig. 21.38).

At full flow, lavatory faucets typically deliver 4 to 5 gpm (0.25 to 0.3 L/s). Newer low-flow faucets utilize a variety of devices to function as well (or better) with less water. Such devices include aerators (which add air bubbles to the stream, making it splash less and appear larger), flow restrictors, and mixing valves to control temperature. The lower

**Fig. 21.38** *Proposed lavatory that exploits running water. The water issues from a fountain-like stream for ease in drinking and in hair and face washing. The stream strikes the lavatory bowl in such a way as to minimize splashing outside the bowl, yet sets up a self-cleaning swirling action. A small repository at the back of the bowl, over the drain, can serve for standing water when desired. (From The Bathroom: New and Expanded Edition, © 1966, 1976 by Alexander Kira. By permission of Bantam Books, a division of Random House, Inc. All rights reserved.)*

flows achieved range from ½ to 2½ gpm (0.03 to 0.16 L/s). Another promising development is the foot-operated faucet, which frees the hands from having to control water flow, thus saving a few seconds of flow during each lavatory usage. These devices could be particularly helpful at kitchen sinks, where extensive washing of objects takes place.

### (c) Whole-Body Cleansing

The running/standing water contrast is even more clearly illustrated here. In cleansing, the ordinary sequence is *wet, soap, scrub, and rinse*. Standing water is good for wetting and very good for soaping/scrubbing; running water is superior for rinsing. The psychological aspect enters here in common attitudes toward tubs and showers. Tubs are often seen as places to relax in, to spend more time in, and perhaps to read in. Showers are viewed as quicker,

"no-nonsense," stand-up places. Yet each could benefit from some features of the other.

*Tubs* should be designed so that the reclining body is supported at the back; this requires a contoured surface (Fig. 21.39) rather than the ordinary straight-line design. It also requires braces for one's feet because otherwise the body will tend to float up and away from such a backrest. Tubs can be designed to accommodate persons of various leg lengths, as shown in Fig. 21.39*b*. There must also be a seat to give most of the body a chance to be out of the water and to facilitate safe entry into and exit from the tub. Especially needed is a handheld shower for the final rinse; soapy standing water leaves a scummy film on both people and fixtures.

*Showers* may seem very efficient, but cleaning would be more thorough and safe if the bather could turn off the water and sit for at least part of the soap/scrub activity, especially for the lower legs and feet. Showers with integral seats are now common.

Another consideration is the location of water controls (fittings). The user should be able to reach them easily from outside the tub or shower without wetting his or her arm, but also should be able to manipulate them from within the tub/shower even if temporarily blinded by soap.

Shower heads have been notorious for encouraging prodigal water usage; typical flow rates of 6 gpm (0.4 L/s) and maximum rates of 12 gpm (0.7 L/s) once were common. Even in a short (5-minute) shower, this rate of use could consume as much as 60 gal (227 L) of water, much of it heated. Many codes now require a limitation on showerhead flow; a flow of 2.5 gpm (0.2 L/s) is common. These flows can be designed into the shower head, or they can be achieved by cheap, simple flow restrictors in retrofit applications. Some utilities distribute flow restrictors free of charge. Most bathers notice no difference, either in enjoyment or in cleansing, when using restricted-flow shower heads.

### (d) Elimination

Several conflicts arise with toilets: the already-mentioned difficulty of combining cleansing with elimination, the issue of pure (high-grade) water for an impure (low-grade) purpose, and the conflict over the height of the toilet. A *lower toilet* is definitely of benefit to the average person, who will achieve far better bowel evacuation in a full squatting position.

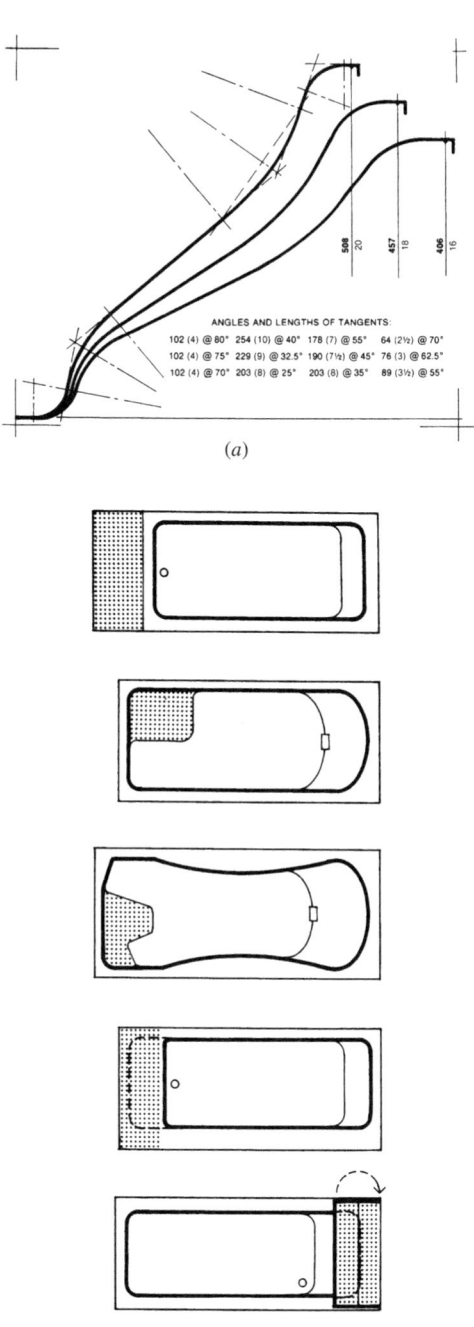

ANGLES AND LENGTHS OF TANGENTS:
102 (4) @ 80°  254 (10) @ 40°  178 (7) @ 55°  64 (2½) @ 70°
102 (4) @ 75°  229 (9) @ 32.5°  190 (7½) @ 45°  76 (3) @ 62.5°
102 (4) @ 70°  203 (8) @ 25°  203 (8) @ 35°  89 (3½) @ 55°

(a)

(b)

**Fig. 21.39** *Some considerations for tub design to facilitate relaxation during whole-body cleansing. (a) A contoured backrest allows comfortable reclining, with support of both the shoulder and lower back. Curve 2, with a median angle of 32.5°, promises comfort for most users. (b) Various tub plans, allowing long- and short-legged persons to brace their bodies against the contoured backrest. Raised seats are also provided to facilitate safe entry/exit. (From* The Bathroom: New and Expanded Edition, *© 1966, 1976 by Alexander Kira. By permission of Bantam Books, a division of Random House, Inc. All rights reserved.)*

If combined with a toilet seat contoured to best support the body during defecation, the proposed toilet in Fig. 21.40*a* would be physiologically superior to most of today's fixtures. There are problems with low height. The typical male stands while urinating, and a lower toilet presents a more difficult target. This can have serious maintenance consequences unless males can be induced to sit or a separate urinal is provided—an unlikely option for residential bathrooms. (In our culture, urinals look institutional; they are also expensive.) Another problem with a lower toilet is that the elderly and some handicapped people will have difficulty getting on and off the seat (see Fig. 21.51). Yet another problem arises from the fact that toilets are considered to be seats (for reading, toenail clipping, etc.). The low, squat-inducing toilet is decidedly not at a comfortable chair height.

As a result, a *higher toilet* with otherwise similar features is proposed (Fig. 21.40*b*). Note that both of these toilets feature openings whose shape differs from that of the conventional oval. This hourglass-shaped opening provides proper support for the body during defecation, and it allows a more generous opening, front and back, for proper perineal cleansing by hand. Another change is that there is no separate toilet seat; instead, these fixtures incorporate electric resistance heaters that warm the small portion of the toilet that contacts the body. Thus, the toilet becomes a consumer of energy—which is also used to heat the water that is provided for perineal cleansing. For those who insist on a separate seat, or for ordinary posture on conventional toilets, a physiologically sound contoured toilet seat is available (Fig. 21.41).

It is especially important that perineal cleansing be encouraged by the designers of bathrooms. The typical dry toilet tissue provided usually is not adequate for this task. In small, private bathrooms this problem is easily solved by placement of the toilet adjacent to the lavatory (or tub), where toilet tissue can be wetted with clean water. In public toilets separated by stalls, perineal cleansing requires either a clean water source built into the toilet or a separate *bidet* within the stall—a highly unlikely provision, given the extra floor space and the cost of the bidet. Again, cultural issues arise: bidets are quite common in the private bathrooms of Europe but a rarity in North America. Two common alternatives are (1) to build a cleansing source into the

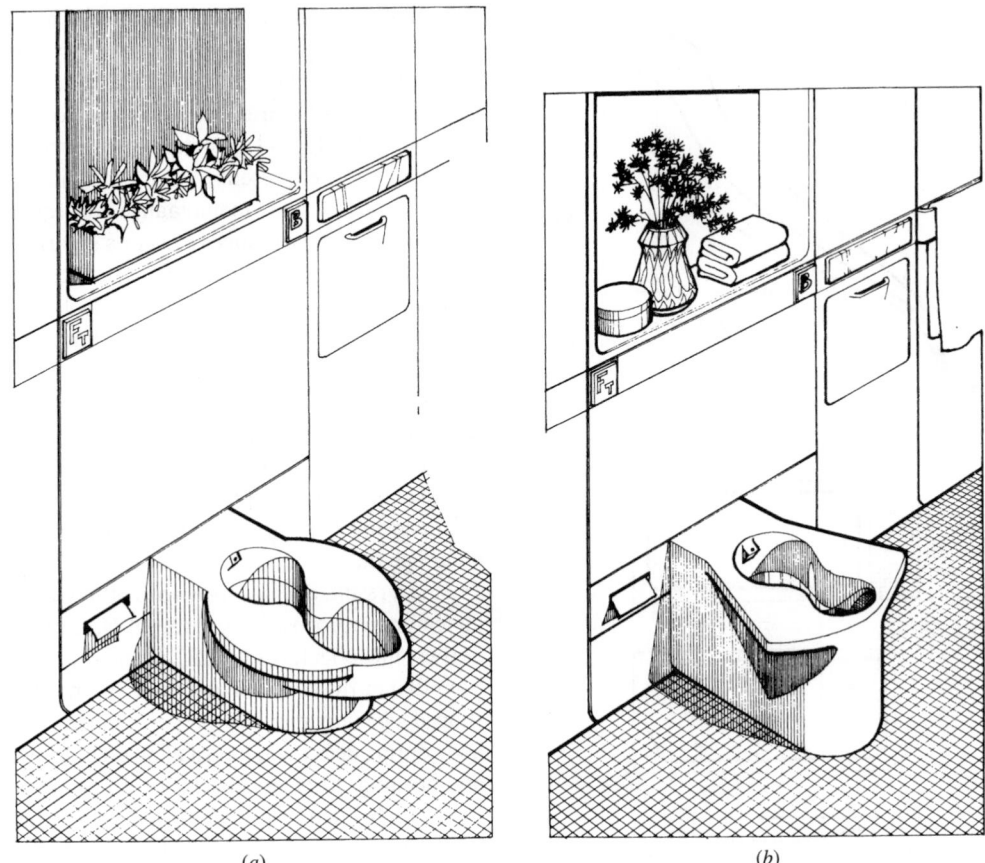

*(a)*                                                    *(b)*

**Fig. 21.40** *Two approaches to water closets (toilets) that encourage better posture during defecation and more thorough perineal cleansing. (a) A low toilet allows the best (full-squat) position. (b) The higher toilet is easier for the elderly and handicapped and better intercepts the urine from a standing male. (See Fig. 21.43 for a separate urinal design.) Both of these toilets eliminate the separate toilet seat and its maintenance problems; both use small amounts of energy to warm the toilet at its point of contact with the body and to heat the water for perineal cleansing. Such cleansing is also the reason for the larger front and back openings. The pushbuttons are labeled $F_T$ for the flush and B for the bidet (cleansing) function. (From* The Bathroom: New and Expanded Edition, *© 1966, 1976 by Alexander Kira. By permission of Bantam Books, a division of Random House, Inc. All rights reserved.)*

toilet itself (Fig. 21.42) or (2) to construct a toilet seat that provides for such cleansing. Several American manufacturers offer such seats, which can be easily adapted to existing toilets. Using recycled water in a toilet that also offers perineal cleansing complicates the high-grade water issue. Codes often require recycled water to be colored (blue or green), an aesthetic quality unappealing for skin contact even in one's nether regions.

The *urinal* is an answer to the problem posed by lower toilets to males, who stand while urinating. Although urinals are culturally acceptable in public toilet rooms, they have never achieved a place in private bathrooms. One solution proposed by Alexan-

der Kira is to build the home urinal into the wall (Fig. 21.43a); it could be pulled out for use, at the optimum mounting height, and at an optimum receiving shape to eliminate backsplash onto the user. When pushed back into the wall, it would flush automatically. A less satisfactory solution would be to modify the toilet by raising its back surface (Fig. 21.43b).

For the toilet (water closet), there are four major categories of fixtures, depending upon the amount of water used per flush. The *conventional* toilet uses 3.5 gal (13.2 L) or more per flush; the *watersaver* toilet uses 1.7 to 3.5 gal (6.4 to 13.2 L) per flush; the *low-consumption* toilet (also called

*Fig. 21.41* Toilet seat for a conventional water closet, which encourages better posture for defecation and somewhat better access for perineal cleansing. (Courtesy of American Standard, Inc.)

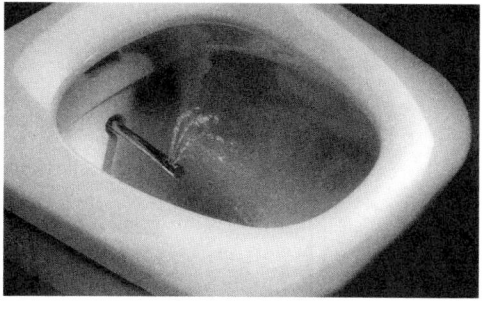

(*a*)

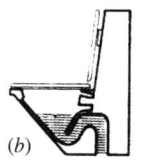    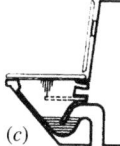    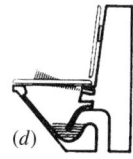

(*b*)            (*c*)            (*d*)

*Fig. 21.42* Provisions for perineal cleansing built into toilets. (a) An adjustable-intensity spray at warm (body) temperature continues for as long as a button is depressed in this Geberit ShowerToilet. (Courtesy of Geberit Manufacturing Inc., Michigan City, IN.) (b) Section through the toilet before/after the cleansing function. (c) At the initiation of perineal cleansing, a pipe extends and emits a controlled spray of warm water. (d) After the pipe is retracted, a warm jet of air issues from just below the seat. (From The Bathroom: New and Expanded Edition, © 1966, 1976 by Alexander Kira. By permission of Bantam Books, a division of Random House, Inc. All rights reserved.)

*ultra-low flush*) uses 1.6 gal (6 L) or less per flush; and the *waterless* toilet uses no water at all.

## (e) Conventional Toilets

These toilets may no longer be legally installed in the United States due to their high water consumption. They are still widely encountered in existing installations. In the common toilet, a sudden deluge of water removes human waste and simultaneously helps cleanse the toilet. Fast-moving water requires pressure and makes noise. The older flush-tank toilet (Fig. 21.44*a*) stores a smaller quantity of water (about 2.5 gal [9.5 L]) at sufficient height above the toilet bowl to achieve fast flow. Although it uses minimal amounts of water, it is noisy and its elevated tank has maximum visual impact. The more common (in North America) version is the two-piece toilet with the tank bolted to the bowl (Fig. 21.44*b*). With much less pressure available, this toilet requires 5 to 7 gal (19 to 26 L) per flush, but it is quieter. Newer *shallow-trap* models reduce the quantity needed to 3.5 gal (13.2 L). The maintenance problem posed by the seam between tank and toilet is eliminated in the one-piece toilet (Fig. 21.44*c*).

The cost of its low profile is an even greater need for water: 6 to 8 gal (23 to 30 L) per flush.

The fourth common alternative is the flush-valve toilet, which depends upon the building's water pressure rather than its own stored water. A noisy system requiring very high flow rates, it is seldom found in residences. However, it requires as little as 3 gal (11.4 L) per flush.

In addition to differing tank and bowl combinations, there are various types of flushing actions. Four common flush toilets are compared in Fig. 21.45.

*Washdown* toilets are no longer made in the United States, although many are still in use. This is the noisiest toilet and the most likely to become plugged because it has the smallest-diameter trap. It is usually found in the two-piece flush tank toilet; with an elevated tank, the flush required was only about 2.5 gal (9.5 L).

*Siphon jet* toilets are in widespread use in North America, particularly in residences. A small priming jet hurries the bowl's contents into the trap and

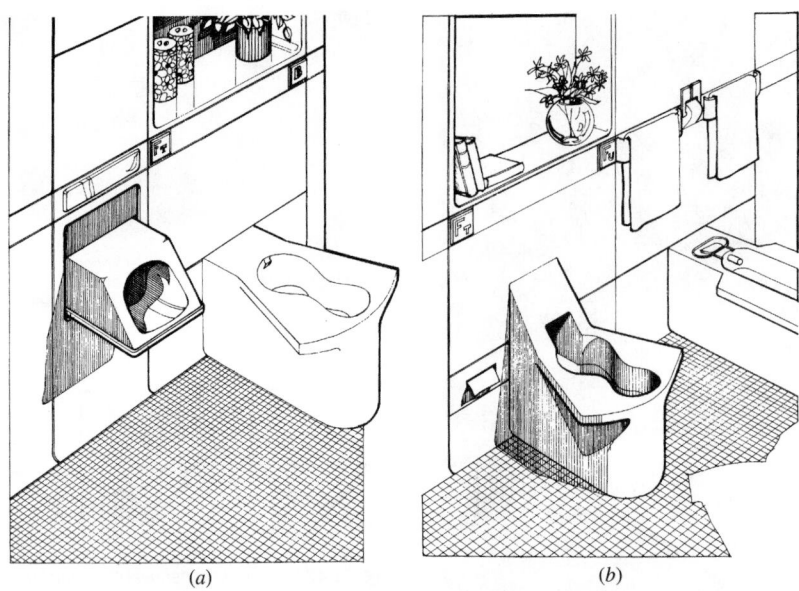

**Fig. 21.43** *Designs to accommodate male urination. (a) A built-in, tilt-out urinal for the home, visible only when in use. (b) A higher back for a toilet. A bidet for cleansing is shown to the right of the toilet. (From* The Bathroom: New and Expanded Edition, © *1966, 1976 by Alexander Kira. By permission of Bantam Books, a division of Random House, Inc. All rights reserved.)*

hastens the siphon action. With an elevated-tank (two-piece) toilet, this process requires a flush of about 3.75 gal (14.2 L). With the more common close-coupled two-piece toilet, it requires a flush of 5 to 7 gal (19 to 26.5 L). Siphon jet toilets are sometimes equipped with flush valves, which use less water (see the discussion of blowout toilets below).

*Siphon vortex* toilets are especially suitable for low-velocity water (often as a result of low pressure). They are therefore also the quietest, making them a favorite wherever bathrooms are adjacent to sleeping areas or other acoustically sensitive spaces. The water enters the bowl off-center in such a way as to form a vortex; this swirling action cleans the

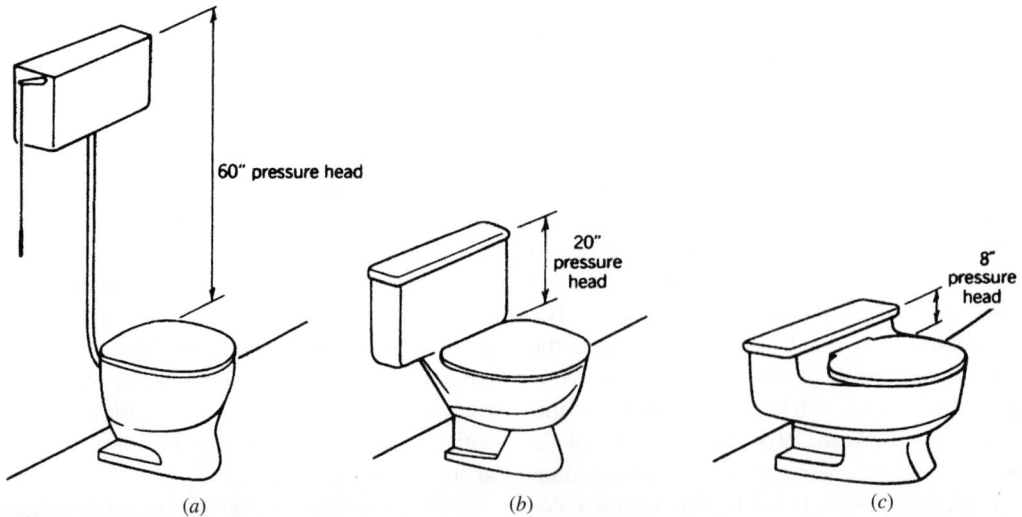

**Fig. 21.44** *Conventional flush tank toilets. (a) Two-piece flush tank toilet with an elevated wall-mounted tank and pull chain. (b) Two-piece flush tank toilet with the tank bolted to the bowl ("close-coupled"). (c) One-piece flush tank toilet with an integral tank. (From Milne, 1976.)*

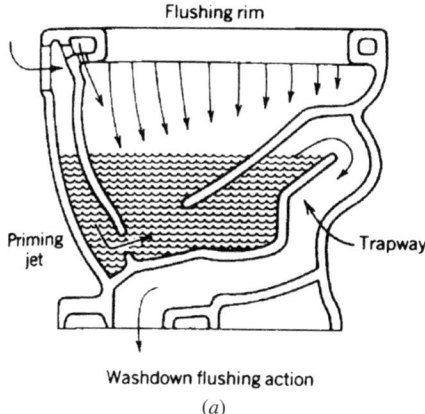

Washdown flushing action
(a)

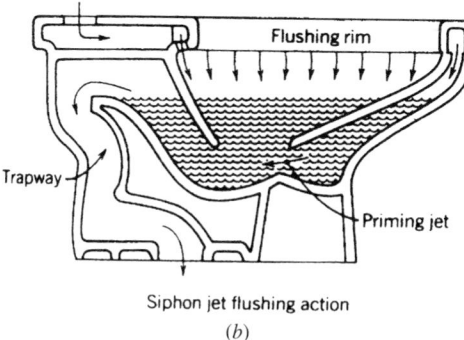

Siphon jet flushing action
(b)

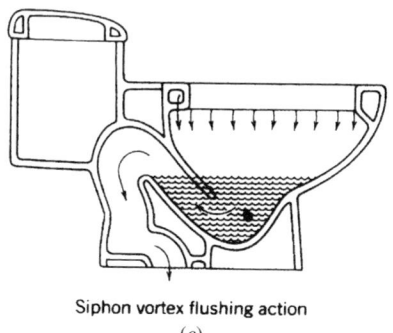

Siphon vortex flushing action
(c)

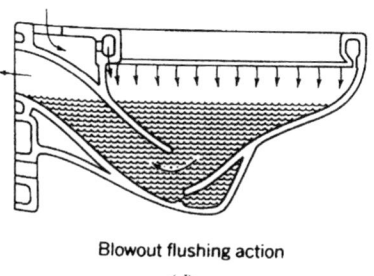

Blowout flushing action
(d)

**Fig. 21.45** *Four common flushing actions that are built into toilets. Flush tank toilets typically use either type (b) or type (c); flush valve toilets use either type (b) or type (d). (From Milne, 1976.)*

sides of the bowl and the trap, helping the siphon action in emptying both bowl and trap. The one-piece flush tank toilets usually have the siphon vortex flushing action, which typically requires 6 to 8 gal (23 to 30 L).

*Blowout* toilets combine very-high-velocity water and a simple trap to offer a noisy but very-low-maintenance toilet dependent upon flush valves rather than tanks. They are very common in commercial and institutional toilet rooms, where large water supply lines and high pressures are available. The high velocity of the water lowers the quantity required from 3 to 4 gal (11.4 to 15.1 L) per flush.

Architects can easily specify toilets that require less water per flush and yet have very little impact on the user. With effort on the part of both users and architects, water can be recycled, or rainwater collected, to supply toilets with lower-quality water.

### (f) Watersaver Toilets

These use 1.7 to 3.5 gal (6.4 to 13.2 L) per flush. Although these toilets use much less water than conventional ones, they may not meet the stricter water limits now established in many water-conscious states and municipalities in the United States (toilets must not exceed 1.6 gal (6 L) per flush). Many of these toilets represent a compromise between the conventional "water waster" and the noisy operation and/or smaller water surface area that typifies so many low-consumption models. Watersaver toilets tend to use a conventional flushing action, and hence need enough height between the tank and the bowl to provide sufficient water pressure within the bowl during the flush.

Toilets can be designed to use less water by varying several characteristics: the pressure of the water entering the bowl (generated either by the height of the tank above the bowl or by utilizing pressure from the water supply system or another source), the shape of the bowl, the way in which the bowl fills and empties, and the trap configuration. Watersaver toilets differ somewhat in most of these characteristics from a conventional toilet; more radical changes are needed to achieve a low-consumption toilet.

### (g) Low-Consumption Toilets

These use 1.6 gal (6 L) or less per flush. A heated controversy over these ultra-low-flush water closets developed in the late 1990s due in part to steep price increases and consumer discontent over performance. It appears that regulation got ahead of technology for a time, resulting in a need for repeated low-consumption flushes to accomplish what one higher-consumption flush could do. To remove fecal matter, however, two 1.6-gal flushes still approximately equal the old 3.5-gal flush; to flush urine, one 1.6-gal flush is adequate. Manufacturers are finding ways to change the entry point of water into the toilet bowl to increase swirl and thus waste removal power.

Several of these toilets achieve water conservation by using a *flushometer tank* (Fig. 21.46). Satisfactory flushing can be achieved with much less water by flushing with water entering at greater pressure. Instead of toilet tanks at ordinary air pressure, these tanks utilize water supply system pressure to compress air trapped within the tank. Water enters the bowl with much greater force with this combination of water under system pressure and compressed air. The flush is thorough, quick, and noisy. If the water supply pressure is greater than about 65 psi (448 kPa), some problems with excessive tank pressures may be expected. A pressure-reducing valve would be helpful in such cases.

**Fig. 21.46** *Example of a 1.5-gal-flush, pressure assisted, direct-fed siphon jet flush action flushometer tank, with a large 10 × 12-in. (250 × 300-mm) water surface area in the bowl: the Cadet Aquameter. (Courtesy of American Standard, Inc., Piscataway, NJ.)*

Another water-saving technique is to redesign the bowl to hold less water. The siphon jet, gravity-fed toilet whose tank is shown in Fig. 21.47 exposes a standing water surface of about 4½ × 6 in. (114 × 152 mm), much smaller than that in Fig. 21.46. Along with a very low 1.4-gal (5.3-L) flush, its trap is wide enough to pass a 2⅛-in. (54-mm) ball.

Far lower water consumption is achieved when a central compressed air system is combined with water supply system pressure. The Microphor flush toilet (Fig. 21.48) uses compressed air and 2 qt (1.9 L) of water per flush in a two-chambered toilet. When the flush lever is pressed, the water and waste in the bowl (upper chamber) is deposited into the secondary (lower) chamber, and 2 qt of water (direct from the supply pipe) washes down the sides of the bowl to await the next user. The now-closed secondary chamber then is pressurized with compressed air, and its contents are deposited into a conventional sewer line.

Air compressors are needed, along with compressed air lines. A small compressor (¼ to ½ hp [187 to 373 W]) with an accompanying air tank will operate up to three toilets. Although the compressor is noisy, the toilets themselves are no noisier than conventional toilets; they use the same plumbing lines. With such low flows, the designer may choose to increase the slope of waste lines; however, the 2-qt quantity has proven sufficient to carry the waste in existing installations.

In contrast, the Envirovac flush toilet (Fig. 21.49) uses a vacuum and 1.5 qt (1.4 L) per flush. Because a central sewage tank (kept under vacuum by a pump) must be used, the drain line from the toilet to this tank may run horizontally without a slope or may even run vertically upward from the toilet. This can have significant architectural advantages when vertical clearance is tight, toilets are being added within existing structures, or toilets must be located below the level of the public sewer, as in marinas or basements. In tall buildings, significant savings in power for water supply pumping are achieved. Easier approval for building permits, reduced water hookup fees, and lower monthly sewer charges can be substantial benefits. Furthermore, the central sewage holding tank can be flushed into the sewer at off-peak times—a benefit to the treatment plant. These water-saving, architectural, and water treatment advantages must be weighed against the cost of the toilets and the tank/pump combinations, the space required for

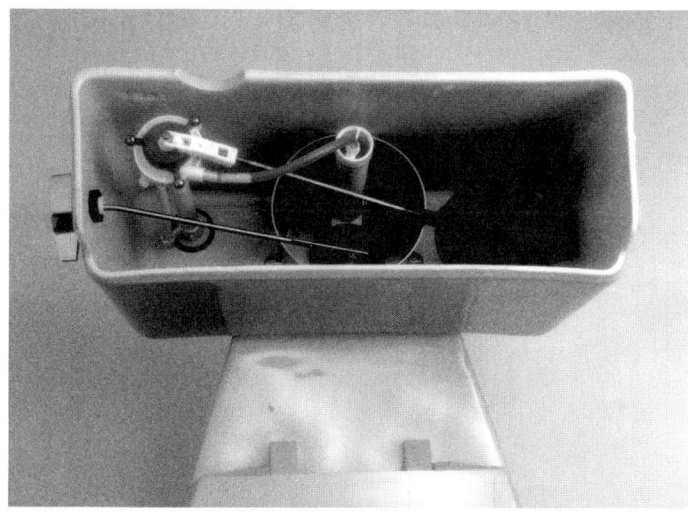

**Fig. 21.47** *A very low flow flush is achieved with a reservoir that meters about 1.4 gal (5.3 L) of water from the tank to the bowl during flushing. The standing water surface is about 4½ by 6 in. (115 by 150 mm). The Ultra-One/G. (Courtesy of Eljer Industries, Plano, TX.)*

the tank/pump, the power used by the pump, and the possibility that a power failure will halt system operation.

Vacuum systems can be installed for groups of buildings, such as subdivisions. Small pipe sizes, free-

dom from the need for continuous-sloping sewer lines, and conservation advantages for both water supply and treatment are the benefits. In addition, graywater (from kitchen, laundry, and bathing fixtures) can be kept separate from toilet water

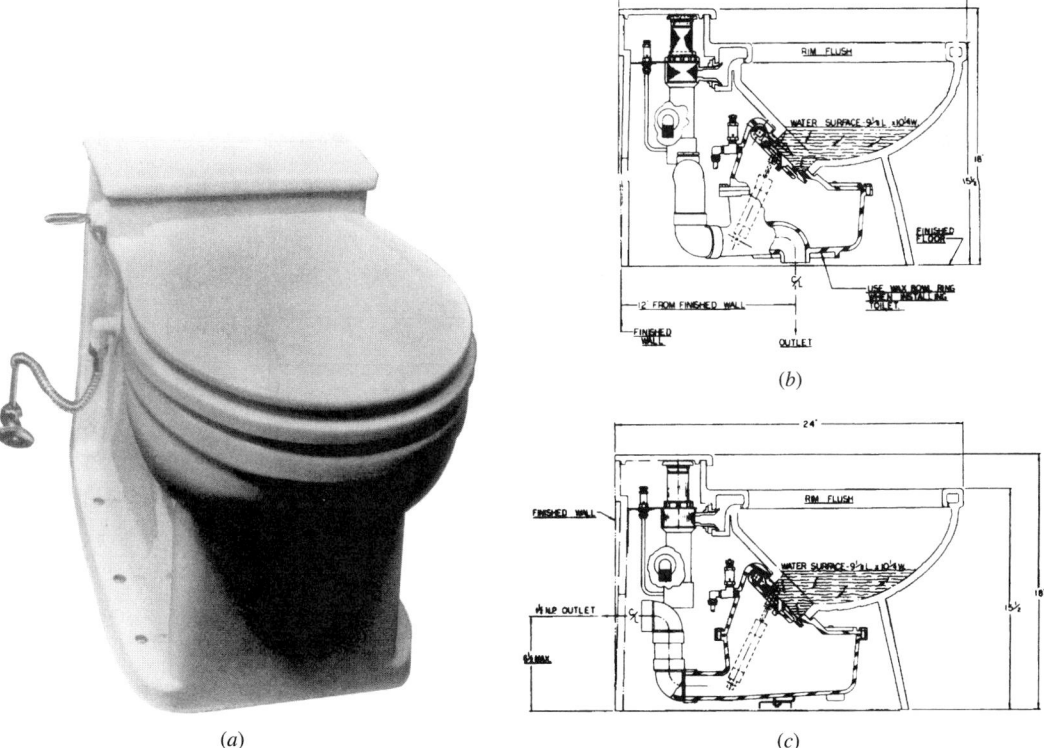

(a)             (b)

(c)

**Fig. 21.48** *(a) Compressed air is combined with a 2-qt (1.9-L) flush in the two-chamber Microphor toilet. (b) Section through the floor-discharge model. (c) Section through the wall-discharge model. (Courtesy of the Microphor Company, Willits, CA.)*

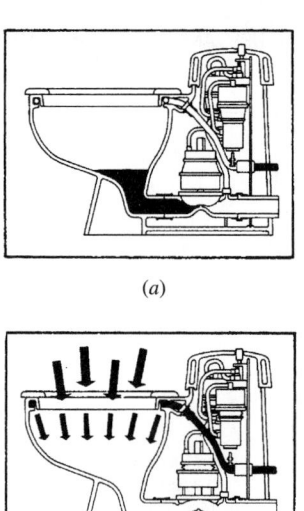

(a)

(b)

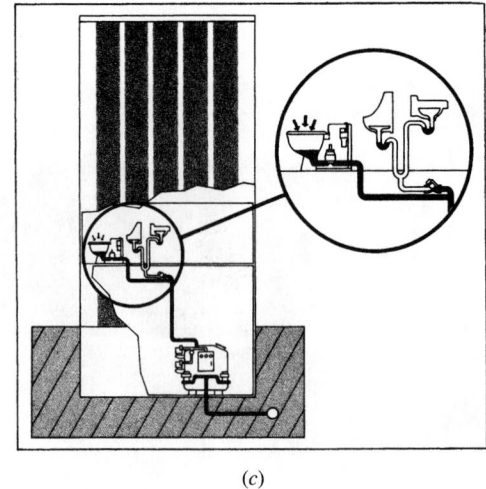

(c)

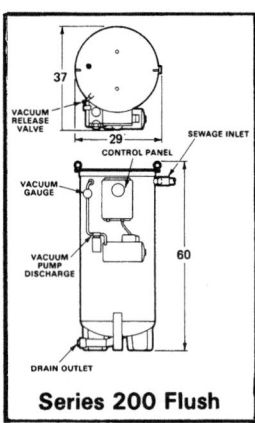

**Series 200 Flush**

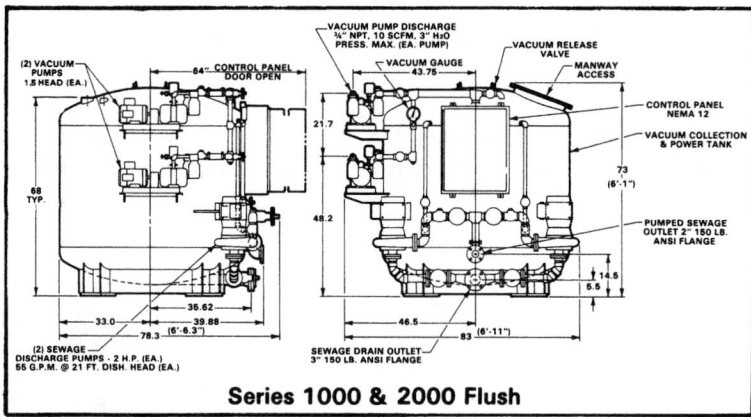

**Series 1000 & 2000 Flush**

(d)

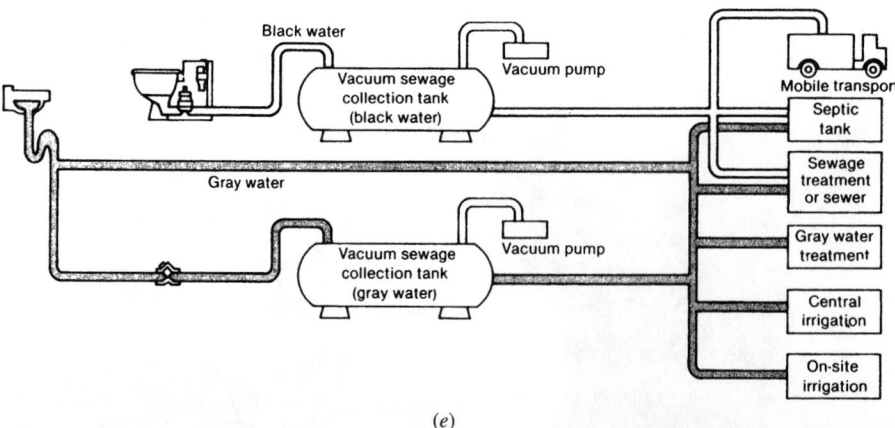

(e)

**Fig. 21.49** *The Envirovac system utilizes a vacuum and a 3-pt (1.4-L) flush. (a) Less than 3 pints of water remain in the bowl until the toilet is used. (b) When the pushbutton flush is pressed, the discharge valve opens for about 3 seconds; a rush of air into the evacuated piping carries along sewage, odor, and airborne bacteria. Simultaneously, the washdown water flow begins. The flow continues for 4 seconds after the discharge valve closes, so that most of the 3-pt flush remains in the bowl. (c) Toilets are directly connected to a central holding tank, kept under a vacuum by a pump. (d) Central tank/pump combinations can serve one house or a large building. (e) The separation of graywater and blackwater is another water-conserving opportunity. (Courtesy of Envirovac, Inc., Rockford, IL.)*

(blackwater) and thus made easier to recycle after moderate treatment.

### (h) Flushing Controls

Water conservation is encouraged by the *dual cycle* toilet, whose flushing mechanism allows a choice of fewer gallons for liquid wastes, more gallons for solid wastes. This simple mechanism is more common outside the United States; its handle is pushed up for liquid flushing and down for solids.

Another development is the *automatic flush*, common in public buildings and triggered by radiant heat from the pressure of a body at the fixture or by light reflected off the user and back to the control. This "touchless" approach seems to promise more for hygiene than for water conservation, although it does prevent the flush valve from being held open too long by a careless user.

In general, plumbing fixtures will consume less water as the supply water pressure is reduced. For this reason, *pressure-reducing valves* are becoming popular as water conservation devices (when they would not otherwise be required to protect fixtures from overpressure). Installed on the supply line to a building, they can save water throughout the system.

Several newer low-consumption toilets use vertical flush valve sleeves, where the flush handle is in the center of the top of the toilet tank. The handle is lifted to initiate the flush. An advantage is that the less time the handle is raised, the less water enters the bowl. Again, less water may be admitted to flush away liquid waste, more water for solid waste.

### (i) Waterless Toilets

Because these fixtures eliminate water entirely, they are described in Chapter 22, Section 22.1.

### (j) Appliances

Dishwashers, washing machines, and other appliances are big users of water—and of energy to heat it. Dishwashers use 12 to 18 gal (45 to 68 L) per cycle, much of it heated well beyond the 120°F (48°C) typical of a household hot water supply. Some models now allow shorter cycles, which can cut use to perhaps 7 gal (26.5 L).

Clothes washing machines use 40 to 55 gal (151 to 208 L) for full-sized loads. In the past, "suds

saver" features allowed soapy, hot wash water to be reused. Many newer washers allow for a wider selection of water quantities and temperatures—a feature that can save considerable water and energy. Horizontal-axis (front-loading) washers have greatly reduced the need for hot water per wash cycle.

## 21.8 FIXTURE ACCESSIBILITY AND PRIVACY

After plumbing fixtures designed for the human physique and for water conservation have been

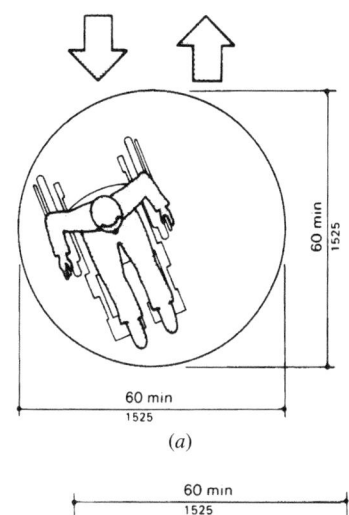

(a)

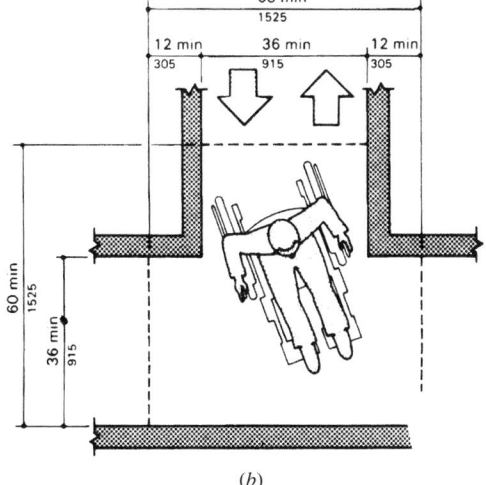

(b)

**Fig. 21.50** *Wheelchair turning space, minimum requirements. (a) A 60-in. (1525-mm)-diameter space. (b) T-shaped for 180° turns. Dashed lines indicate the minimum length of clear space required on each arm of the T-shaped space. (Reprinted by permission from* American National Standard A117.1-1986, © 1986 *by the American National Standards Institute.)*

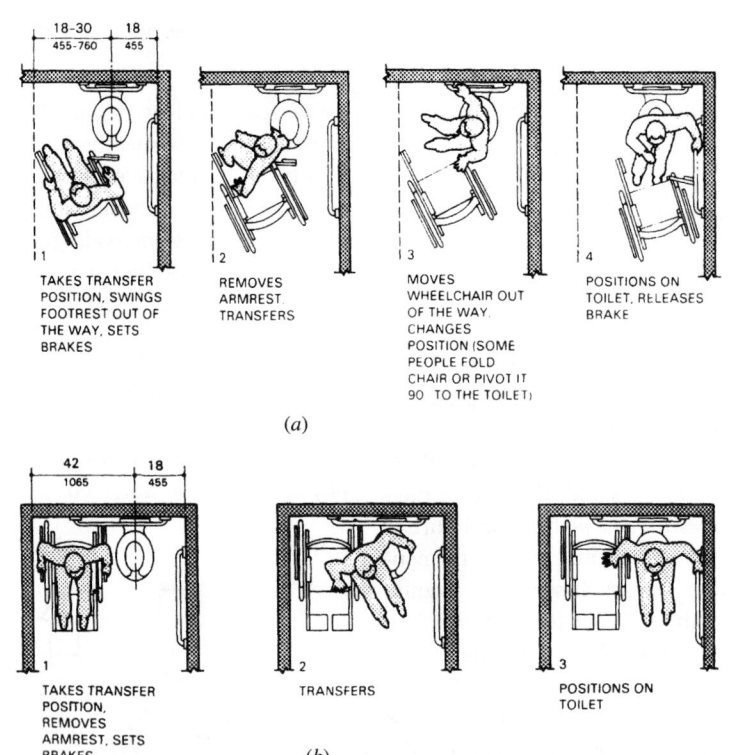

18-30 | 18
455-760 | 455

**1**
TAKES TRANSFER
POSITION, SWINGS
FOOTREST OUT OF
THE WAY, SETS
BRAKES

**2**
REMOVES
ARMREST
TRANSFERS

**3**
MOVES
WHEELCHAIR OUT
OF THE WAY,
CHANGES
POSITION (SOME
PEOPLE FOLD
CHAIR OR PIVOT IT
90 TO THE TOILET)

**4**
POSITIONS ON
TOILET, RELEASES
BRAKE

*(a)*

42 | 18
1065 | 455

**1**
TAKES TRANSFER
POSITION,
REMOVES
ARMREST, SETS
BRAKES

**2**
TRANSFERS

**3**
POSITIONS ON
TOILET

*(b)*

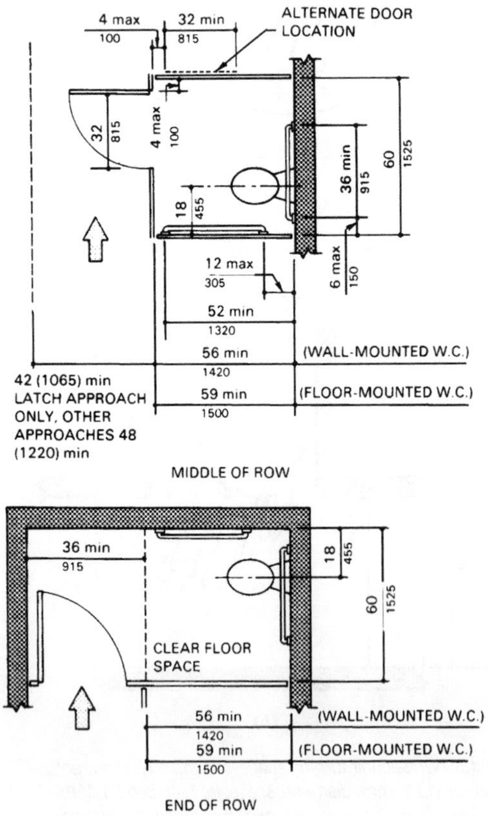

4 max / 100    32 min / 815    ALTERNATE DOOR LOCATION

32 / 815

4 max / 100

18 / 455

36 min / 915

60 / 1525

12 max / 305

6 max / 150

52 min / 1320

56 min / 1420   (WALL-MOUNTED W.C.)

42 (1065) min
LATCH APPROACH
ONLY, OTHER
APPROACHES 48
(1220) min

59 min / 1500   (FLOOR-MOUNTED W.C.)

MIDDLE OF ROW

36 min / 915

18 / 455

60 / 1525

CLEAR FLOOR
SPACE

56 min / 1420   (WALL-MOUNTED W.C.)
59 min / 1500   (FLOOR-MOUNTED W.C.)

END OF ROW

*(c)*

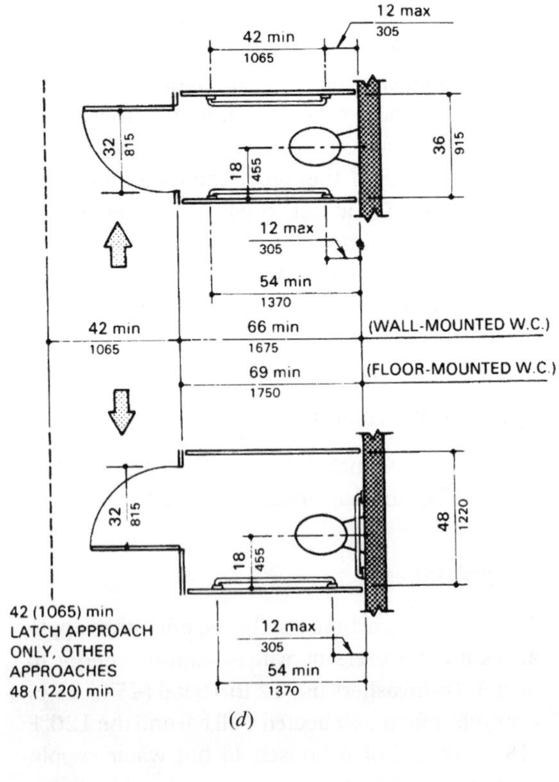

12 max / 305

42 min / 1065

32 / 815

18 / 455

36 / 915

12 max / 305

54 min / 1370

42 min / 1065    66 min / 1675   (WALL-MOUNTED W.C.)

69 min / 1750   (FLOOR-MOUNTED W.C.)

32 / 815

18 / 455

48 / 1220

42 (1065) min
LATCH APPROACH
ONLY, OTHER
APPROACHES
48 (1220) min

12 max / 305

54 min / 1370

*(d)*

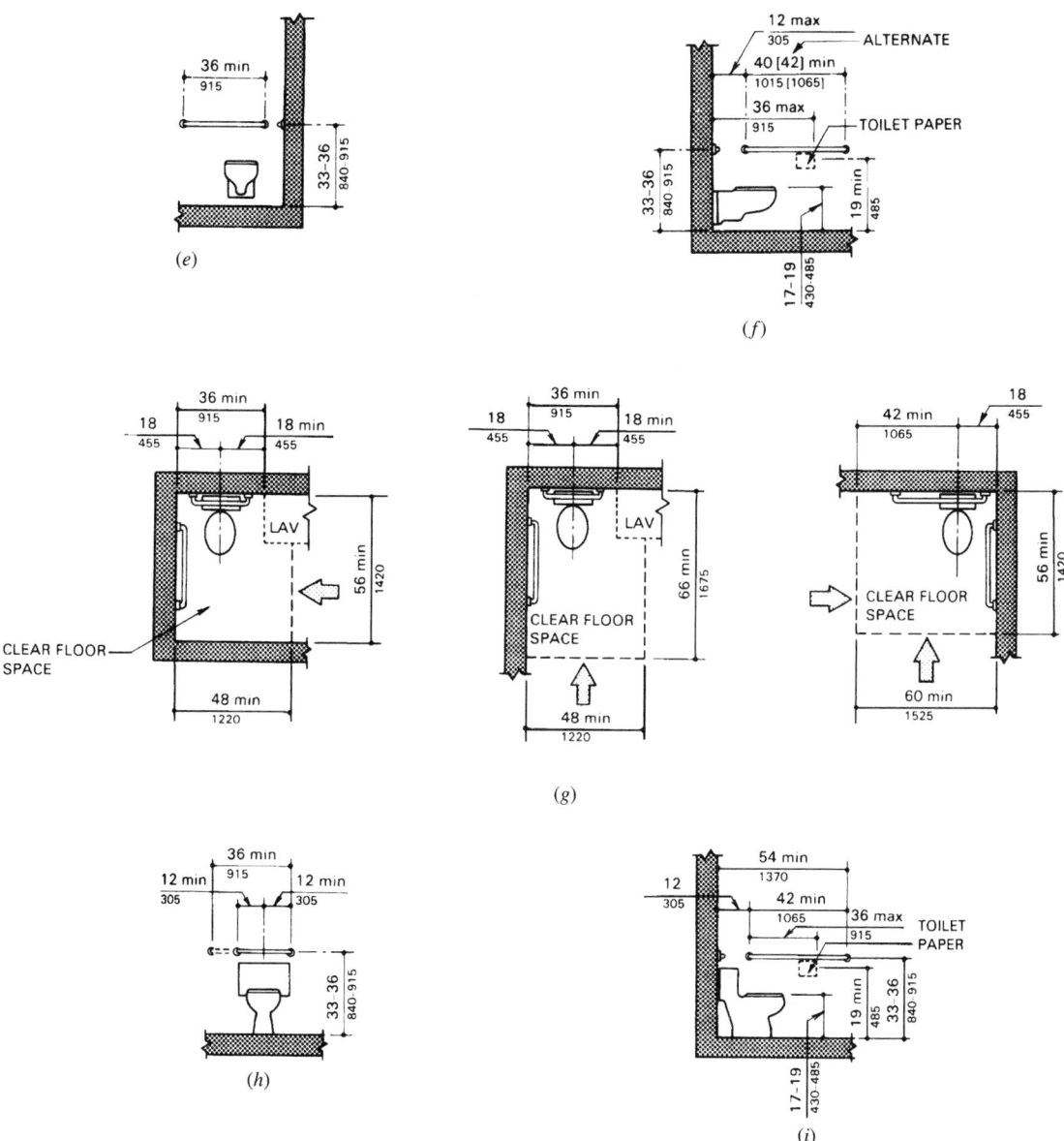

**Fig. 21.51** Clearances and grab bars for toilets. Two typical approaches (a,b) necessitate the stall clearances shown (c–f) as well as arrangements for toilets and urinals not within stalls (g–i). (Reprinted by permission from American National Standard A117.1-1986, © 1986 by the American National Standards Institute.)

WATER AND WASTE

chosen, they are placed in a space that has several unusual environmental needs. Much of the information in this section is reproduced with permission from *American National Standard for Buildings and Facilities A117.1-1986, Providing Accessibility and Usability for Physically Handicapped People*, copyright 1986 by the American National Standards Institute (ANSI). Copies of this standard

(updated as *Accessible and Useable Buildings and Facilities* in 2003) may be purchased from ANSI, 25 West 43rd Street, New York, NY 10036. Another useful guide is the *Americans with Disabilities Act (ADA) Accessibility Guidelines for Buildings and Facilities*, available from the U.S. Architectural and Transportation Barriers Compliance Board, Washington, DC.

## (a) Accessibility

Designers must allow room for wheelchair maneuvering into and within toilet rooms. The minimum clear floor area should be a circle 5 ft (1525 mm) in diameter. Minimum clearances for wheelchair maneuvering are shown in Fig. 21.50. Provisions for people in wheelchairs also influence the heights at which items such as light switches, electrical receptacles, paper towels, and water controls should be installed. In general, all objects to be reached by hand in toilet rooms should be placed more than 15 in. (380 mm) and less than 48 in. (1220 mm) above the floor.

Floor space requirements and mounting-height limitations also apply to the various plumbing fixtures in bathrooms, at least one of which (in each case) should be accessible to people in wheelchairs in most buildings. Requirements for drinking fountains (which often must be located outside toilet rooms), lavatories, bathtubs, and shower stalls include clearances beside the fixture (and clearance below lavatories) and grab bar locations; see the ANSI Standard for details.

Accessible toilets (water closets) have particularly large floor space requirements, as shown in Fig. 21.51. There is a conflict between the lower mounting height needed for proper defecation, the conventional height of 15 in. (380 mm), and the recommended height for handicapped users of 17 to 19 in. (430 to 485 mm). The doors on stalls designated for wheelchair access should swing out rather than in. Provision for a wheelchair sitting beside the out-swinging door must also be made. No doors should swing into the clearance space required for any fixture.

Where six or more water closets (or water closets plus urinals) are located within the same toilet room, *both* the "standard stall" (Fig. 21.51c) and one of the "alternate stalls" (Fig. 21.51d) are required. This impacts the design of public toilet rooms.

Urinals for handicapped users should either be of the floor-mounted stall type or, if wall-hung, have an elongated rim at a maximum of 17 in. (430 mm) above the floor. A clear floor space of 30×48 in. (760×1220 mm) is required in front of the urinal, and its flush controls should be no more than 44 in. (1120 mm) above the floor.

## (b) Privacy

A desire for visual privacy is easily understood and at least minimally provided by partial-height partitions. However, *acoustic* considerations are another important aspect of bathroom design. Because of their nonabsorbent surfaces, bathrooms are frequently the most reverberant spaces in a building. They encourage the singer or the whistler and may even serve as practice rooms for the family musician. Yet they can also intimidate a person who prefers to be quiet, such as a guest using the toilet while the rest of the dinner party sits quietly in the adjacent room.

Two common design responses are isolation and masking sound. Bathrooms can be separated from acoustically sensitive spaces by closets or hallways. With careful attention to sufficiently massive construction and/or other construction details, the doors, walls, ceilings, and floors of bathrooms can be constructed so as to reduce the passage of sound to acceptably low levels. (These considerations are discussed in Chapters 18 and 19.) Sound isolation also depends upon attention to detail: no cracks around the bathroom door, no back-to-back electrical outlets between bathrooms and adjacent spaces, no air grilles into ducts that have other grilles nearby, and no other open windows near a bathroom window. If such goals are unattainable or if additional acoustic security is desired, masking sound can be added as needed within the bathroom. All too often, the masking noise is provided by repeated flushings of the toilet (itself an embarrassing sound for some) or the running of water in the lavatory. However, a noisy ventilating fan would serve the purpose, as would a music source such as a radio. An elegant, expensive touch would be a recirculating fountain gracing one corner; it might provide splashing sounds only when the light or fan is on.

The final consideration in bathroom acoustics is the sound of water moving within pipes. Where such pipes are exposed or barely concealed within, or firmly attached to, thin walls, perceptible sounds are highly likely. Where water noise is undesirable, pipes should be wrapped, resiliently mounted, and/or located in a less acoustically critical wall. But remember that running water can also be a conservation signal as well as masking sound. Although many of us may have experienced this sudden acoustic signal that a bathroom is in use, few have experienced it quite like the occupants of the Aluminaire house on Long Island, cited by Reyner Banham (1969). Designed at the height of the modernist expression of mechanical services, this house featured a dining table cantilevered from the exposed waste stack serving the bathroom above.

## 21.9 WATER DISTRIBUTION

This section looks at ways to supply water throughout buildings at pressures sufficient to operate plumbing fixtures. Smaller buildings may be served simply by the pressure available in water mains (or pressure tanks fed by pumped wells). This is called *upfeed distribution*, because the water rises directly from mains to the plumbing fixtures. For taller buildings, several other options are available: *pumped upfeed* (in which pumps supply the additional pressure needed); *hydropneumatic* (in which pumps force water into sealed tanks, compressing the air within; this maintains the needed water pressure); and *downfeed* (in which pumps raise the water to storage tanks at the top of a building, and water then drops down to the plumbing fixtures).

In cities with municipal water supply systems, water is distributed through street mains at pressures varying at the main from about 50 psi (about 350 kPa) to about 70 psi (about 480 kPa). For low-rise buildings of two or three stories, these pressures are adequate to act against the static pressure of water standing in the vertical piping, overcome the frictional resistance of water flow in the pipes, and still deliver water at the pressure required to operate plumbing fixtures. The flow pressure required at fixtures varies from 5 to 20 psi (35 to 138 kPa), depending on the type of fixture—for example, a basin faucet, showerhead and faucet, or water closet. (Table 21.14 in Section 21.11 lists minimum flow rates and pressures for typical plumbing fixtures.)

### (a) Static Pressure

The pressure exerted at the bottom of a stationary "head" of water is related directly to its height. One cubic foot of water weighs 62.4 lb. Consider a "cube" of water 1 ft square and 1 ft high. Its weight (62.4 lb) rests on a bottom area of 1 ft$^2$ (144 in.$^2$). The static pressure at the bottom is 62.4/144 = 0.433 psi

(3 kPa). Reciprocally, 1 psi of pressure will *sustain* a static (stationary) column of water 1/0.433 = 2.3 ft in height. (In SI units, 1 m of head = 10 kPa.) When fixture pressure and pressure lost in friction-of-flow in pipes are considered, the problem becomes more complex. Example 21.6 in Section 21.11 illustrates this problem in the calculation of pipe size. For upfeed and downfeed distribution, the relationship of heights and static pressure is one controlling design factor.

### (b) Upfeed Distribution

In small, low buildings of moderate water demand, it is seldom difficult to achieve the proper flow pressure at fixtures by the use of an upfeed system. Pressure at the fixtures is usually more than required. When this causes an inconvenient splash, as at lavatory basins, a flow restrictor can be used in the faucet outlet.

Consider the typical upfeed system shown in Fig. 21.52, beginning at the point where supply water enters the building. In cold climates, water in the service entry pipe must not freeze. The pipe must therefore be below the *frost line* of frozen ground. This could vary from 0 to 7 ft (0 to 2 m), depending upon the geographical location. The onset of winter in cold climates requires the closing and draining of pipes supplying the hose bibbs (and other external piping) by means of a *stop-and-waste* valve. Houses left *unheated* in cold-winter weather must be entirely free of water that could freeze and burst the pipes. Note the drain valve at every low point in the system. House shutoff controls are usually located at the main, at the curb, and within the house.

Meters have recently taken on a new role: along with measuring the water quantity for which the occupant is to be charged, they now sometimes serve a restrictive function. During water shortages, they can be used to indicate water use in excess of established limits, beyond which fines are imposed and in some cases the water supply reduced by valves controlled by the water company.

In-house treatment is often performed to reduce water hardness that could clog piping and equipment or to neutralize acidity—a source of corrosion. During the short periods when the treatment tanks are valved off for backwashing or other servicing, the *bypass shutoff valve* is opened.

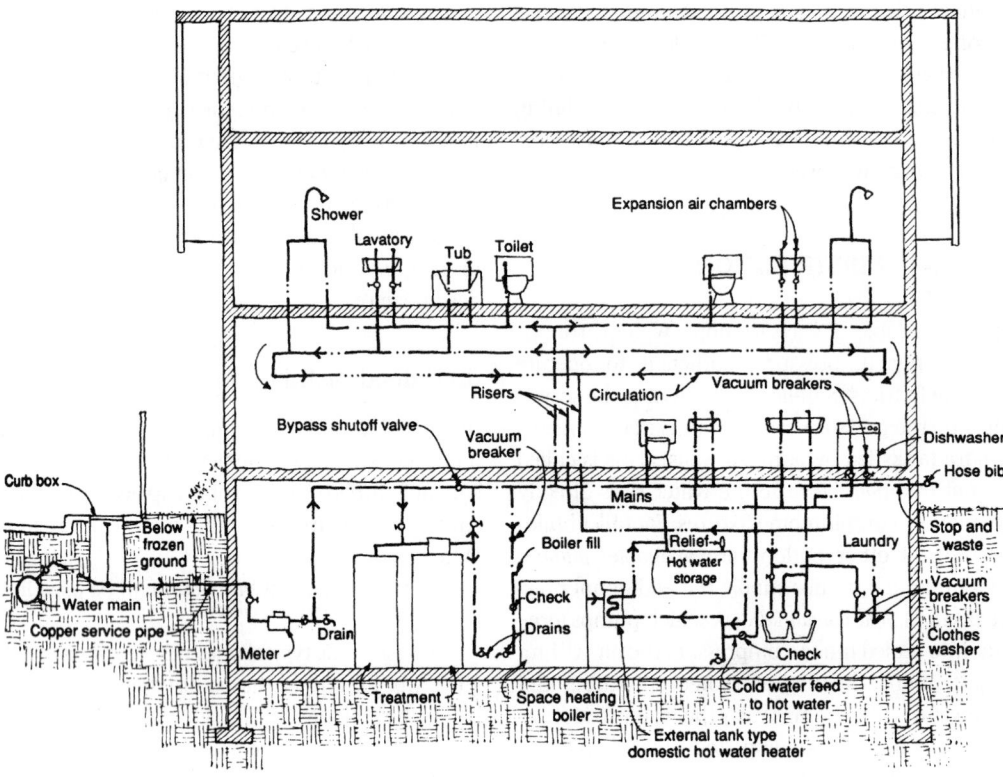

**Fig. 21.52** *Upfeed water distribution using pressure in street mains. A schematic section of the water services in a typical residence.*

From this point on, the water continues under pressure to

1. Supply makeup water to the space-heating boiler, as required
2. Supply water to and pressurize the cold water mains and branches, including the garden hose bibbs
3. Supply water to and pressurize the domestic hot water system through the hot water heater, the hot water storage tank, and the mains, branches, and circulating lines

The air-filled expansion chambers on cold water *runouts* absorb and reduce the shock of *water hammer*—the force exerted by decelerated flowing water that shakes and rattles pipes when faucets are shut off abruptly. On hot water runouts they perform the same function, as well as allowing for the expansion of the hot water as it increases in volume with increasing temperature. Vacuum breakers prevent backflow of polluted water into pipes carrying potable (hot or cold) water. Water from all fixtures

and appliances, such as dishwashers, clothes washers, and boilers, is thus isolated.

Figure 21.52 shows all parts lying in a two-dimensional plane. Obviously, in a real building, such a system is three-dimensional. For instance, economy of piping would suggest that, if possible, kitchen and lavatory, as well as the two upstairs bathrooms, be placed back-to-back.

## (c) Principles of Downfeed Distribution

Water pumped directly from the street main (or from a basement "suction tank" filled by gravity from the main) is lifted to a roof-storage tank. In cold climates, the water in an outdoor tank is kept at temperatures above freezing by heating coils in the tank. The fact that water pressure increases with distance below the tank water level is clearly shown by the construction of the tank (shown in Fig. 21.53). The iron rings, tensioned by adjustable threaded clamps, become more closely spaced toward the bottom of the tank, where the greater water pressure makes it

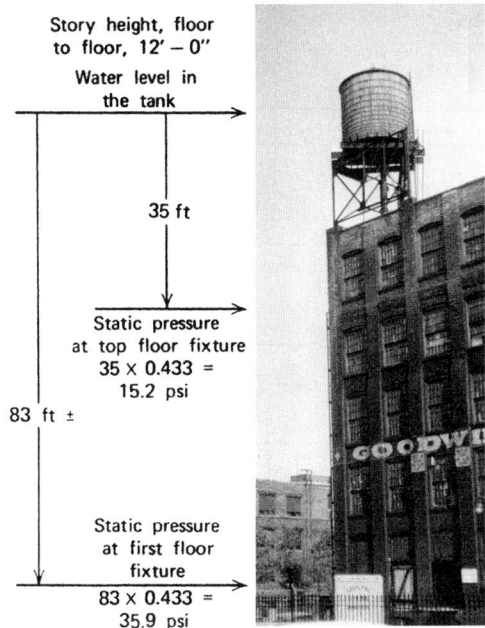

*Story height, floor to floor, 12′ − 0″*

*Water level in the tank*

35 ft

Static pressure at top floor fixture
35 × 0.433 =
15.2 psi

83 ft ±

Static pressure at first floor fixture
83 × 0.433 =
35.9 psi

**Fig. 21.53** *High-rise building of the 1870s with downfeed distribution. Although upfeed distribution from a street main is possible in buildings higher than two or three stories, when street main pressures drop below the usual 50 psi (345 kPa), combined with heavy use, very low pressure results in fixtures in the upper stories. Therefore, water is pumped from the main to elevated wooden roof tanks high enough to assure reasonable pressure at the top story and ample pressure at the bottom of the downfeed run.*

increasingly difficult to restrain the vertical wooden staves of this cylindrical barrel.

Tanks like these add interest, though perhaps not beauty, to a building's silhouette. A view-blocking screen enclosing this tank would have to be about 24 ft (7.3 m) high. Architecturally, such a rectangular lump on the roof would not be much of an improvement. Yet in the century following the appearance of such structures as the Goodwill Building, our technology has become more complex. Presently, for most high-rise buildings, including many 60 stories and more in height, an entire rooftop crowded with equipment and technical facilities is needed to serve the stories below (or the uppermost *zone* of a building). The items could include

Water storage tanks

Two-story penthouses over elevator banks

Chimneys

Numerous plumbing vents

Exhaust fans

Air-conditioning cooling towers

Cantilevered rolling rig to support a scaffold for exterior window washing

Perimeter track for a window washing rig

Photovoltaic cells and/or solar collectors for DHW

Thus, since the 1960s, tall buildings commonly have a *band* or *screen* two stories (or more) high above the structural roof. The best view locations are thus consigned to mechanical equipment. It might be said that it all began with the need for elevated water-storage tanks.

## (d) Tall Building Downfeed Distribution

Fig. 21.54*a* shows a medium-rise building in which one elevated tank can serve all of the lower floors. For taller buildings, it is advisable to separate groups of floors into *zones* with a maximum height (for plumbing pressure limits) of about 150 ft (about 45 m). This zone-height limitation is based upon the height-to-static pressure relationship. At the top of the zone (about 35 ft [10 m] below the storage tank), the minimum desirable pressure is probably at least 15 psi (103 kPa). At the bottom of the zone, the maximum desirable pressure is perhaps 80 psi (552 kPa); above this pressure, damage to fixtures might occur.

$$80 \text{ psi} - 15 \text{ psi} = 65 \text{ psi difference}$$
$$65 \text{ psi} \times 2.3 \text{ ft/psi} = \text{about } 150 \text{ ft}$$

In SI units:

$$552 \text{ kPa} - 103 \text{ kPa} = 449 \text{ kPa difference}$$
$$449 \text{ kPa} \times 0.1 \text{ m/kPa} = \text{about } 45 \text{ m}$$

With pressure-reducing valves at lower floors, however, these zones can be much higher, as shown in Fig. 21.54*b*.

Consider the system described in Fig. 21.54*a*, beginning with the elevated tank. The lower part of the tank often serves as a reserve space to hold a supply of water for a fire extinguishing system. In this case, only the water in the upper part is available for use as domestic (service) water. The amount stored must be enough to supplement what the pump will deliver during the several daily hours of high demand that occur in most buildings. The pump then continues, often for several hours, to replenish the house supply that had become partially depleted

WATER AND WASTE

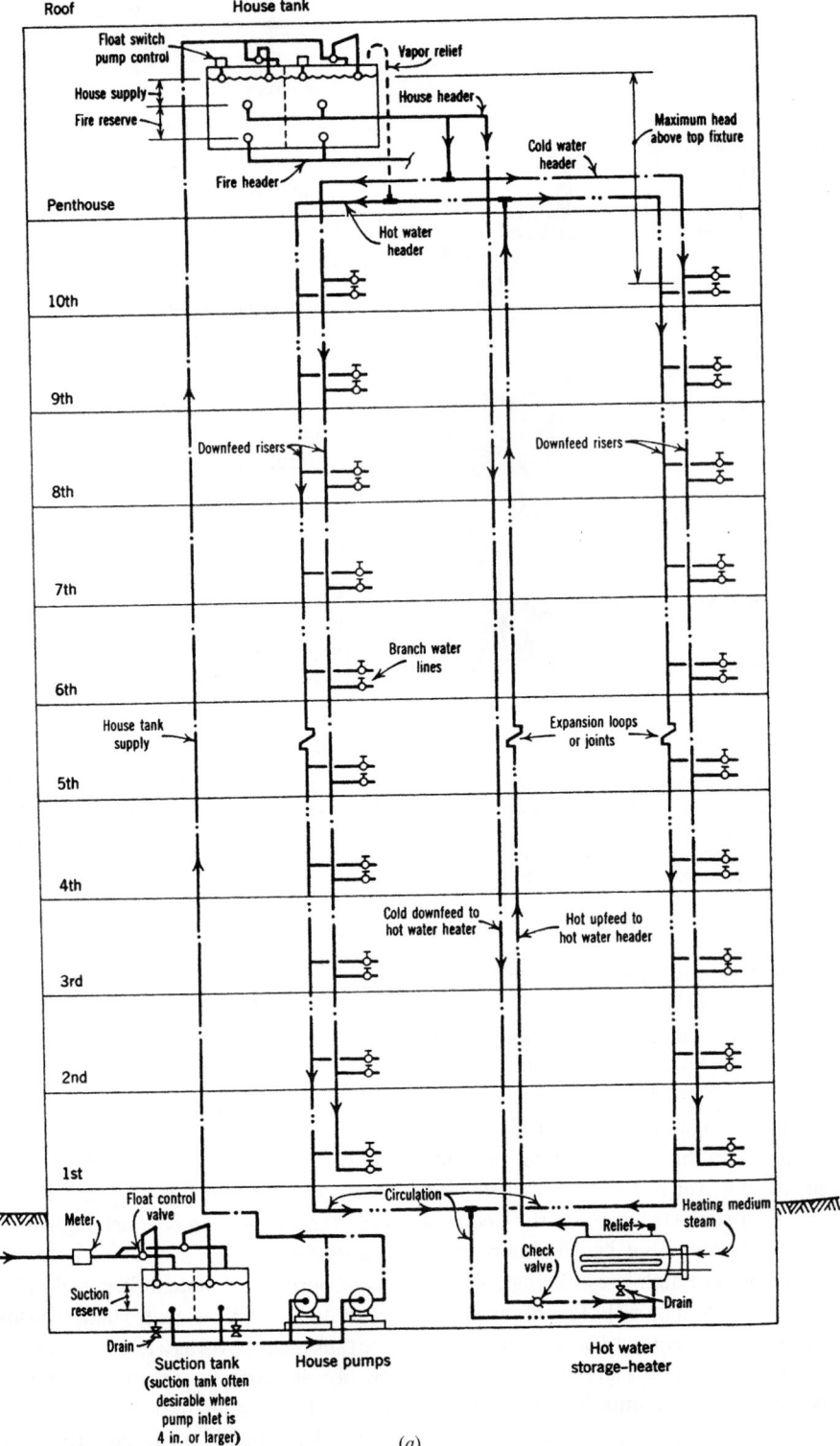

Roof
House tank
Float switch
pump control
Vapor relief
House supply
House header
Fire reserve
Maximum head
above top fixture
Cold water
header
Fire header
Penthouse
Hot water
header
10th
9th
Downfeed risers
Downfeed risers
8th
7th
6th
Branch water
lines
House tank
supply
Expansion loops
or joints
5th
4th
Cold downfeed to
hot water heater
Hot upfeed to
hot water header
3rd
2nd
1st
Circulation
Meter
Float control
valve
Heating medium
steam
Check
valve
Relief
Suction
reserve
Drain
Drain
Suction tank
(suction tank often
desirable when
pump inlet is
4 in. or larger)
House pumps
Hot water
storage–heater

(a)

**Fig. 21.54** (a) Downfeed water distribution, schematic section, part of the water services for a 10-story building. Hot water circulation moves from the hot upfeed in two directions at the hot water header, then down to the tank through the two downfeed hot water risers. For details of one type of steel house tank, a typical centrifugal house pump, expansion joints, and expansion in the hot water riser, see Figs. 21.55 and 21.56. (b) Downfeed water distribution, schematic section, part of the water services for a much taller zoned building. Zone tanks include a fire reserve, but standpipes are omitted from this drawing. For detail of the steam-type domestic water heater, see Fig. 21.24.

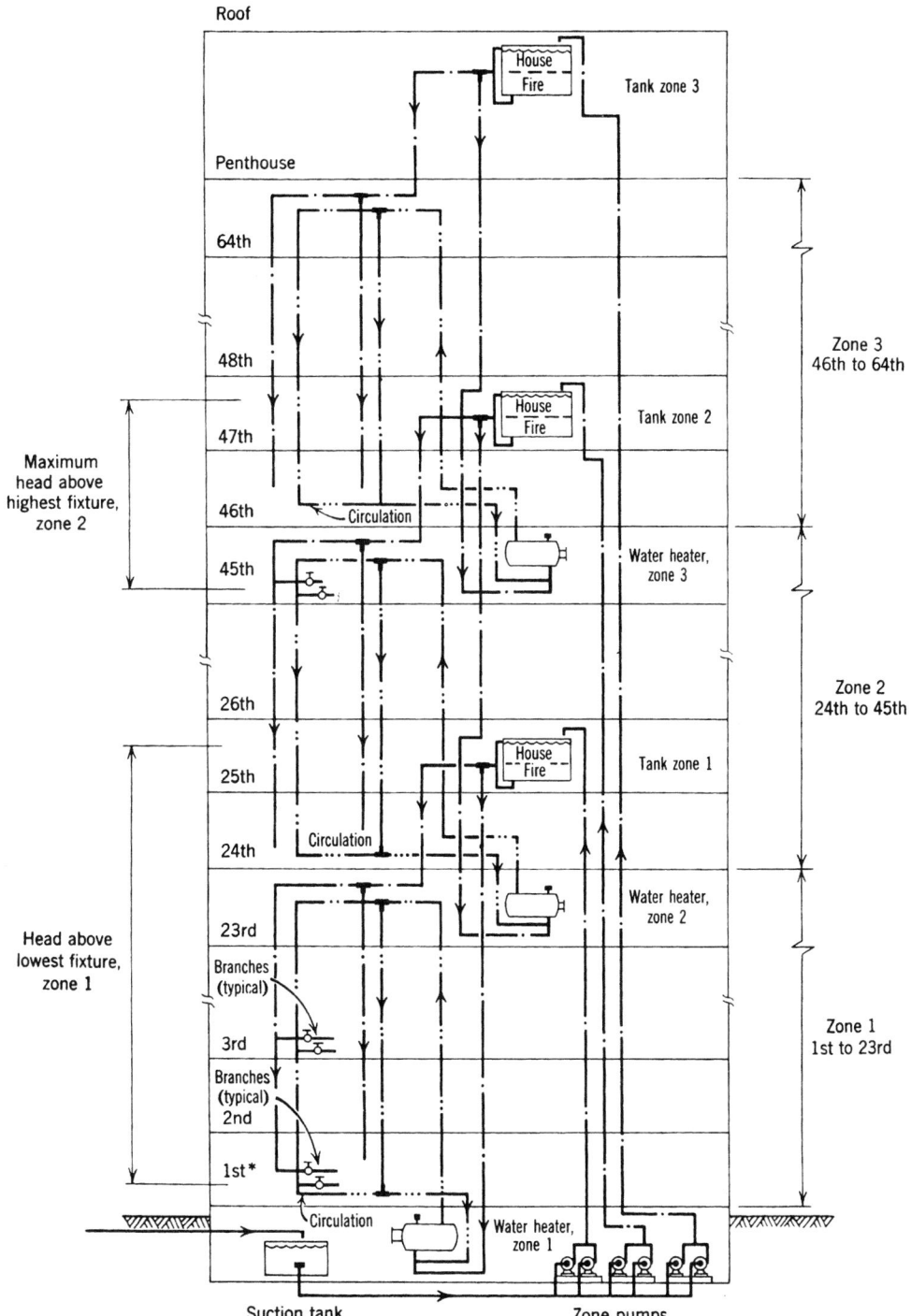

Roof

House
Fire
Tank zone 3

Penthouse

64th

48th

Zone 3
46th to 64th

47th

House
Fire
Tank zone 2

Maximum
head above
highest fixture,
zone 2

46th
Circulation

45th

Water heater,
zone 3

26th

Zone 2
24th to 45th

25th

House
Fire
Tank zone 1

24th
Circulation

Water heater,
zone 2

23rd

Head above
lowest fixture,
zone 1

Branches
(typical)

3rd

Zone 1
1st to 23rd

Branches
(typical)
2nd

1st*

Circulation

Water heater,
zone 1

Suction tank

Zone pumps

* In many cases a few lower floors are supplied from street mains

(b)

**Fig. 21.54** (Continued)

WATER AND WASTE

during the busy period. The suction tank is a buffer between the system and the street mains. It usually holds enough reserve to allow the pumps to make up the periodic depletion in the house tank. It refills automatically by flow from the street main that, consequently, will not suffer as much of a drop in pressure as it would if it were connected directly to the suction side of the house pumps. Neighboring water users are protected from the adverse effects of sudden demands within adjacent large buildings (Fig. 21.55).

House tanks and suction tanks are sometimes

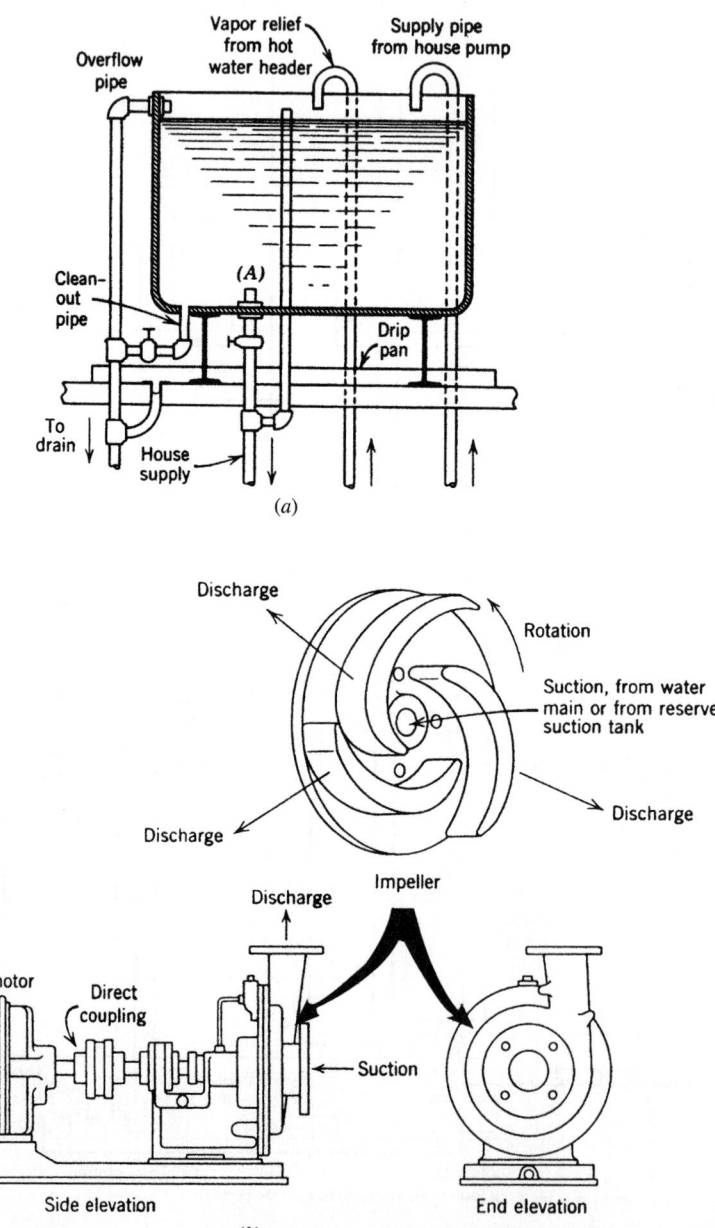

**Fig. 21.55** (a) House tank in an elevated position for downfeed by gravity. Sediment in the tank is drawn off through the clean-out pipe and is prevented from entering the house water supply by the pipe projection at A. Water for fire reserve could be provided by additional piping or by a separate tank. (b) Centrifugal house pump, commonly used to fill an elevated house tank at the top story or at an intermediate mechanical floor. The electric motor responds to a float switch pump control at the tank.

made of steel plate and divided vertically in half, each half having identical piping and controls. Hence, one-half of the tank can be cleaned out at a time of low demand without shutting down the entire system. One full-capacity pump is supplemented by an equivalent standby pump for alternate use. Because there is no suction lift below the pump, or any fixture pressure at the top of the house tank supply, the head against which the pump works is the sum of the distance from the suction tank water level to the top of the house tank and the feet of head equivalent to the friction loss in the tank supply pipe. For this kind of service, the vertical piping is on the order of 3 or 4 in. (76 or 102 mm) in diameter for large buildings. Sizes are established by formal calculations.

The house supply is fed by a short pipe from the house header to the cold water header that circles the top story and connects to many downfeed cold water risers. For simplicity, Fig. 21.54a shows only two risers and also omits many valves and controls.

Figure 21.54b is even more simplified. The circulating hot water originates as cold water at the house tank header, then takes quite a long route. Descending to the bottom of the hot water heater, it rises to seek its own level at the hot water header, becoming available there for hot water downfeed on demand. All of this occurs as flow below the general pressurizing effect of the house tank. In effect, when there is a cold- and hot-water demand on a story near the top of the building, the cold water makes a short trip down to the faucet, while the hot water goes through three vertical pipes instead of one.

This arrangement, with tank above and heater below, is used in multiple forms for very tall buildings. The zones are quite independent; the only common service is that provided by the general suction tank. With this zoning method, problems of pipe expansion, large pipe sizes, and high pressures in lower stories are minimized. Commonly, two and a half stories, or about 35 ft (about 10 m), comprise the minimum pressure head above the top fixture served by any zone tank. The static pressure created at the fixture is thus $35 \times 0.433 = 15$ psi ($10 \times 10 = 100$ kPa). If, during flow, not too much pressure is lost in friction, flushometers (flush valves) can be placed at this level, although flush tanks, because of their lower pressure demand, must often be accepted. The opposite problem occurs at the bottom of the zones, where excessive pressures must be reduced at the fixtures. In zone I of Fig. 21.54b, first-floor fixtures are below a head of 24½ stories, or about 149 psi (1027 kPa) of static pressure. Pressure-reducing valves must be used and fixture control valves must be throttled.

### (e) Pipe and Tube Expansion

The range of temperature experienced by hot water supply piping, from the normal indoor air temperature of about 70°F (21°C) to that of service hot water (which often exceeds 160°F [71°C]), can be 90F° (50C°), with resulting pipe expansion (Table 21.12). This longitudinal elongation of pipe, though negligibly small in houses, can be appreciable in a tall building. Two methods of allowing freedom for this longitudinal motion in long runs of expanding hot water piping are shown in Fig. 21.56a. These devices preclude the buildup of excessive stresses in the pipe material and the tendency of the pipes to buckle laterally.

**TABLE 21.12 Thermal Expansion of Pipe**

| Temperature Increase | | Approximate Expansion of Piping Material Inches per 100 ft (0.83 mm per m) of Length | | | | | |
|---|---|---|---|---|---|---|---|
| F° | C° | Carbon and Carbon Moly Steel | Cast Iron | Copper | Brass and Bronze | Wrought Iron | Plastic |
| 20 | 11.1 | 0.15 | 0.15 | 0.22 | 0.22 | 0.15 | 0.5 |
| 40 | 22.2 | 0.3 | 0.3 | 0.44 | 0.44 | 0.3 | 1.1 |
| 60 | 33.3 | 0.45 | 0.45 | 0.67 | 0.67 | 0.45 | 1.7 |
| 80 | 44.4 | 0.6 | 0.6 | 0.89 | 0.89 | 0.6 | 2.4 |
| 100 | 55.5 | 0.76 | 0.7 | 1.12 | 1.12 | 0.76 | 3.3 |
| 120 | 66.7 | 0.91 | 0.84 | 1.35 | 1.35 | 0.91 | 4.3 |
| 140 | 77.8 | 1.07 | 1.0 | 1.58 | 1.58 | 1.07 | 5.3 |
| 160 | 88.9 | 1.22 | 1.13 | 1.81 | 1.81 | 1.22 | 6.2 |
| 180 | 100 | 1.38 | 1.27 | 2.04 | 2.04 | 1.38 | — |
| 200 | 111.1 | 1.55 | 1.4 | 2.27 | 2.27 | 1.55 | — |

WATER AND WASTE

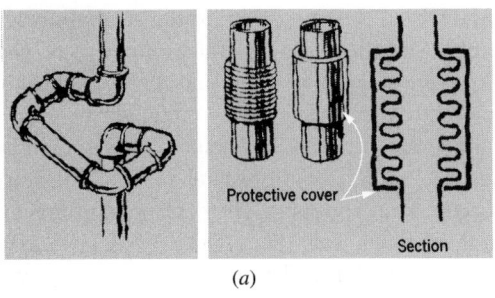

(a)

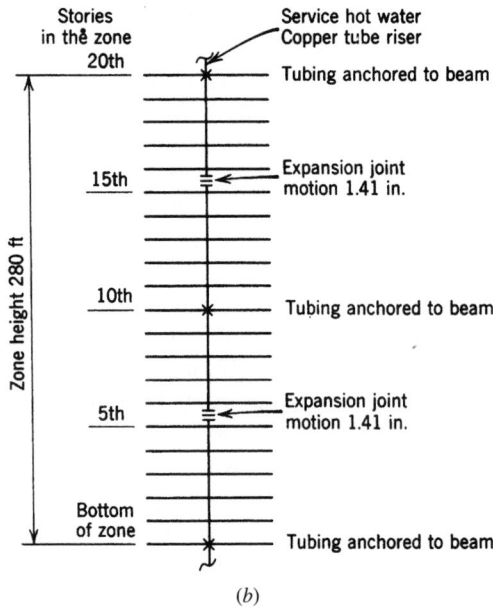

(b)

**Fig. 21.56** (a) Accommodation for the expansion of hot water piping or tubing. Left: Expansion joint of pipe and fittings. Right: A manufactured product in elevation and section. (b) Suggested scheme for location of the points of anchorage and expansion for service hot water tubing in a 20-story zone (Example 21.5).

**EXAMPLE 21.5** A 20-story zone in a tall building has a height of 280 ft. What will be the increase in length of a copper tube carrying "service hot water" (DHW) when its temperature increases from 70° to 160°F?

**SOLUTION**
The difference in temperature is 90F°. Approximate elongation per 100 ft for a 90F° increase is 1.01 in. (interpolating in Table 21.12).

expansion = 280 ft × (1.01 in./100 ft) = 2.82 in. ■

There are a number of ways of providing for this expansion. The one shown in Fig. 21.56b accepts this motion at two locations, which would make the expansion in each case 1.41 in. Equidistant anchorage to fix the tubing is provided at the bottom, the 10th floor, and the 20th floor. The support of the vertical riser at floors other than those at which it is anchored could consist of clamps of the type illustrated later in Fig. 21.61, supported on springs.

### (f) Pumped Upfeed Distribution

This distribution system (Fig. 21.57) is for medium-sized buildings—those too tall to rely on street main

pressure but not so tall as to necessitate heavy storage tanks on the roof. This equipment associated with this system can deliver water at rates varying from those needed for two or three faucets to full building demand while maintaining at each outlet a pressure within 2 psi (14 kPa) of the design pressure for that outlet.

The installation shown in Figs. 21.57a and 21.57b uses a triplex pump group. According to demand, one, two, or three pumps will operate. Because each pump is of the variable-speed type, virtually an infinite number of delivery rates can be achieved within the zero to maximum design range. The pumps operate in sequence. When a very small rate of demand occurs, the smallest or "jockey" pump starts in response to a low voltage impressed on its motor. All operations are triggered and adjusted by the pressure sensor at the base of the riser. The jockey pump continues to run until it has reached its maximum delivery rate, at which time the first of the larger pumps cuts in, joined by the other larger pump when required. Sequential operation of the three pumps, each increasing in delivery as called for by the sensor, meets the requirement for an increasing supply at nearly constant pressure.

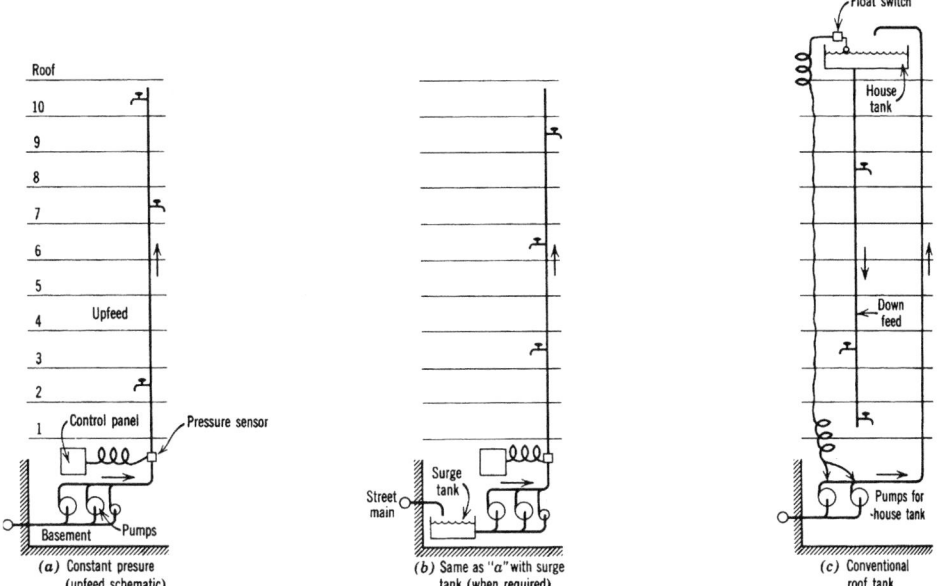

**Fig. 21.57** *Constant-pressure upfeed pumping* (a,b) *compared to gravity downfeed from a house tank on the roof* (c). *(By permission of Progressive Architecture.)*

Wear on the two large pumps is equalized by the assigning of "lead" and "lag" positions. For a period of 24 hours, one of the large pumps holds the lead position and starts after the jockey pump, giving the other large pump a smaller burden. The next day the rested pump takes over the more active assignment. All of this occurs automatically.

At full operation, this triplex unit can put a suction demand on the street main that can seriously reduce the available water pressure in the neighborhood (Fig. 21.57a). Therefore, utilities sometimes require that such a system draw from a surge tank, filled by casual flow from the street main, independent of the building requirements (Fig. 21.57b). This requirement is often imposed when analysis indicates a maximum building demand in excess of 400 gpm (25 L/s).

Obvious advantages of upfeed pumping are elimination of the house tank and the heavy structure that transmits its weight down to the footings, and elimination of the necessary periodic cleaning of the tank. A shortcoming is the lack of reserve storage, which could cause a serious problem during an electrical power failure. However, minimum flow during this kind of emergency can be arranged if a diesel or emergency standby motor is available to drive one of the pumps.

## 21.10 PIPING, TUBING, FITTINGS, AND CONTROLS

### (a) Piping, Tubing, and Fittings

The system of water supply piping or tubing should efficiently fulfill its purpose, be easily maintained, and interfere as little as possible with architectural form and function. Except in basements, in utility rooms, and at points of access to controls, the piping system is usually concealed. Stud-and-joist construction can provide space for concealment, but in concrete or masonry buildings, vertical and horizontal furred spaces must often be provided.

Water supply piping is subject to corrosion over time. When pipe materials corrode, they first lose some carrying capacity (due to increased wall roughness and perhaps buildup of materials) and ultimately fail. Sediment from corrosion can adversely impact plumbing fixtures as well. Steel piping is particularly subject to corrosion. In the nonferrous group, red brass and copper tubing are effective in providing corrosion resistance. Copper tubing is less expensive than brass, assembles more easily, and is not subject to dezincification (attack by acids on the zinc in brass). For use in handling aggressive waters, plastic is often a good choice. Like copper, it is light in weight and assembles with great ease.

For ferrous pipes and "iron pipe size" brass, threaded connections are used. The external, tapered thread on the pipe is covered with pipe compound and screws tight against the internal tapered thread of a coupling or other fitting. The solder–joint connection in copper depends on capillary attraction that draws the solder into a cylinder of clearance between the mating surfaces of tube and fitting. This occurs after the surfaces are polished and fluxed and the parts placed in final position. They are then heated, and molten solder is applied to the circular opening where the fitting edge surrounds the tube, with a small clearance. Solder is then drawn into the cylindrical connection. Solders are often tin-antimony alloys. This kind of joint permits the advantageous setting up of an entire tubing assembly without turning the parts (as in threaded installations) and before the soldering commences. For the same strength, copper tubing may have thinner walls than threaded pipes because no threads need to be cut into it. Its smooth interior surface offers low friction to flowing water. Although threaded- and solder–joint connections are the most common in small work, there are many other types (Figs. 21.58 and 21.59 and Table 21.13). Ferrous pipes in the larger sizes are often welded or connected by bolted flanges.

## (b) Plastic Pipe

Most of the plastic pipes and fittings now produced are synthetic resins derived from such materials as coal and petroleum. These corrosion-resistant materials are widely used in water supply piping, fittings, and drainage systems (see Chapter 22). The National Sanitation Foundation (NSF) tests and certifies plastic pipe; the NSF seal must appear on pipes that are to carry potable water.

Most of the materials used for piping are thermoplastics and will repeatedly soften under the application of heat. PVDC (polyvinylidene chloride) material can carry water at 180°F (82°C), but plastic pipe should not be subjected to temperatures higher than this. Expansion (see Table 21.12) is much greater than that of other piping materials for the same $\Delta t$ and affects the piping design. Quite shockproof, plastic is used in the plumbing systems of most mobile homes.

## (c) Valves and Controls

It is usually desirable to valve every riser, the branches that serve bathrooms or kitchens, and the runouts to individual fixtures. This facilitates repairs at any location with a minimum of shutdown within a system. Water treatment equipment will have a valved bypass (see Fig. 21.52). Pumps and other devices that may need repair should be able to be disconnected by unions (Fig. 21.59) after valves are closed.

A gate valve (Fig. 21.60a), with a retractable

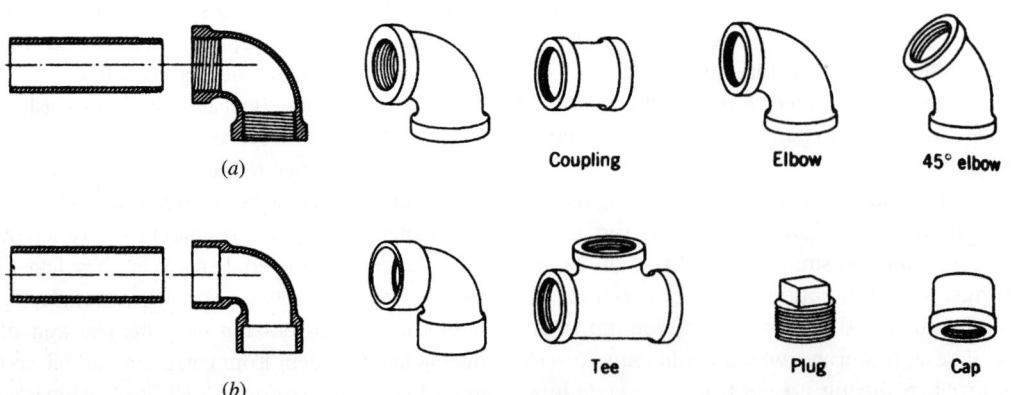

**Fig. 21.58** Methods of connecting pipes and fittings, and tubes and fittings. (a) Threaded: for ferrous pipe and fittings and for iron pipe size (IPS) brass. (b) Soldered: for copper tubing and fittings. A sliding fit similar to that of (b) is used for the solvent weld of plastic connections.

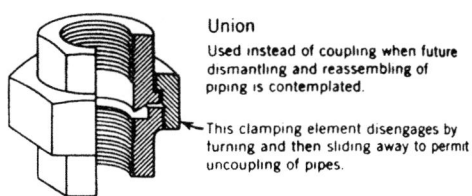

Union

Used instead of coupling when future dismantling and reassembling of piping is contemplated.

This clamping element disengages by turning and then sliding away to permit uncoupling of pipes.

**Fig. 21.59** *Example of threaded pipe fitting (a union) for ferrous or brass pipe. These and all common fittings are also available for solder–joint (copper) or solvent–weld (plastic) connections and usually for transition from one system or material to another.*

leaf machined to seal tightly against two sloping metal surfaces when closed, offers the least resistance to water flow when open. It is usually chosen for locations where it is left completely open most of the time. The compression-type globe valve (Fig. 21.60*b*) is usually used for the closing or throttling of flow near a point of occasional use. Faucets are usually of the compression type, as are drain

valves and hose connections. They are similar to an angle valve (Fig. 21.60*d*). When it is necessary to prevent flow in a direction opposite to that which is planned, a check valve (Fig. 21.60*c*) is introduced. The hinged leaf swings to permit flow in the direction of the arrow but closes against flow in the other direction.

### (d) Pipe Support

Water piping systems are heavy because of their water content and need regular support (Fig. 21.61). Vertical runs of piping should be supported at every story. Horizontal pipes should be supported at intervals of

6 ft (1829 mm) for ½-in. (12-mm) pipe;

8 ft (2438 mm) for ¾-in. or 1-in. (19- or 25-mm) pipe; and

10 ft (3048 mm) for 1¼-in. (32-mm) or larger pipe.

**WATER AND WASTE**

### TABLE 21.13  Water Supply Piping Materials

| PART A. WATER SERVICE ONLY[A] | |
|---|---|
| **Material** | **Connections[b]** |
| ABS (acrylonitrile butadiene styrene) plastic pipe | Mechanical with elastomeric seal (normally underground only); solvent cement; threaded joints |
| Asbestos–cement | Sleeve couplings of same material as pipe, sealed with elastomeric ring |
| Ductile iron water pipe | Depth of lead depends upon pipe size |
| PE (polyethylene) plastic pipe and tubing | Flared joints (see manufacturer), heat fusion, mechanical joints (see manufacturer) |
| PE-AL-PE (polyethylene/aluminum/polyethylene) pipe, PVC (polyvinyl chloride) plastic pipe | Mechanical with elastomeric seal (normally underground only); solvent cement; threaded joints (may reduce pressure rating). |
| PART B. WATER SERVICE AND DISTRIBUTION[c] | |
| **Material** | **Connections** |
| Brass pipe | Brazed, mechanical, threaded, or welded joints |
| CPVC (chlorinated polyvinyl chloride) plastic pipe (and tubing, indoors) | Mechanical (see manufacturer), solvent cement, or threaded joints (may reduce pressure rating) |
| Copper or copper-alloy pipe and tubing | Brazed, mechanical, soldered, threaded, or welded joints |
| Galvanized steel pipe[d] | Threaded, or mechanical joints with an elastomeric seal |
| PB (polybutylene) plastic pipe and tubing PEX (cross-linked polyethylene) plastic tubing PEX-AL-PEX (cross-linked polyethylene/cross-linked aluminum/polyethylene) pipe | Flared joints, heat-fusion, or mechanical joints (see manufacturer) |

[a]Materials as listed in the 1997 *International Plumbing Code*. Water service piping is for an outdoor, underground connection from the building to the main or supply tank.

[b]Joints between different piping materials require approved adapter fittings; dielectric or brass converter fittings required between copper (or copper alloy) and galvanized steel piping.

[c]Materials as listed in the 1997 *International Plumbing Code*. Water distribution piping is within the building.

[d]Sediment from corrosion over the life of the pipe.

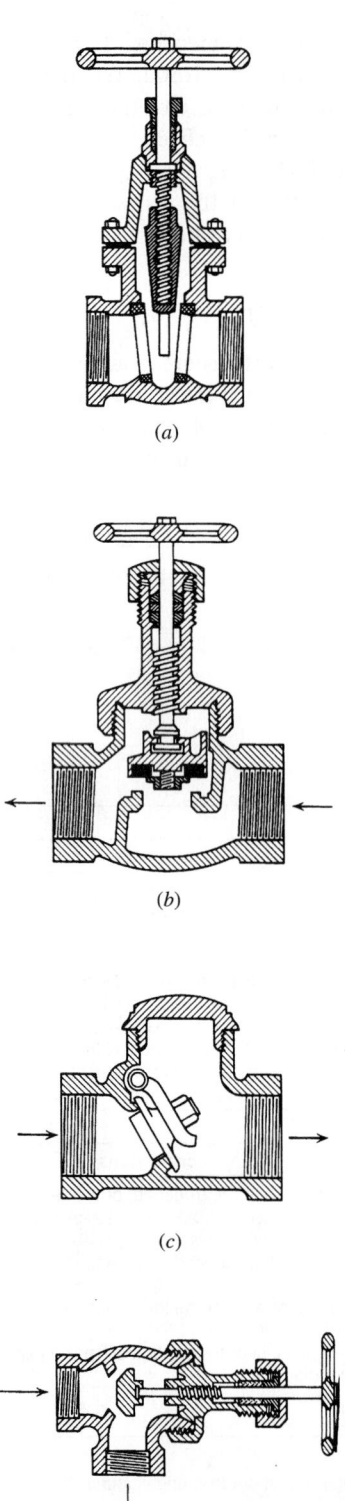

*(a)*

*(b)*

*(c)*

*(d)*

**Fig. 21.60** *Typical valves for water systems. (a) Gate valve. (b) Globe valve. (c) Check valve. (d) Angle valve.*

Although somewhat counterintuitive, the above recommendations reflect the ability of larger pipe to span greater distances between supports. Horizontal copper tubing should be supported at closer spacing than steel. Adequate positioning of horizontal runs is important to assure correct pitch and drainage. Hangers are adjustable for this purpose.

### (e) Shock and Hot Water Expansion

Expansion chambers were shown diagrammatically in Figure 21.52, and water hammer was explained in Section 21.9b. Expansion chambers are often made of capped lengths of vertical pipe about 2 ft (0.6 m) long at the fixture branches (Fig. 21.62a). They trap air, which absorbs the impact of the water surge. A somewhat better (more controllable) device is a *rechargeable air chamber* (Fig. 21.62b). By closing the valve and draining the water through the hose bibb while the petcock above is open to admit air, the chamber may be refilled with air. Closing the petcock and hose bibb and opening the valve completes the service operation and reconnects the device with the water system. Rechargeable chambers (instead of pipe extensions) are used on branch lines adjacent to groups of fixtures. Access for service must be provided. Perhaps the best device is the special shock absorber (Fig. 21.62c).

Air cushions also protect the water heater tank's relief valve against frequent operation, with the resultant leakage of hot water, as hot water periodically expands and contracts in closed systems.

### (f) Condensation or "Sweating"

The moisture that is always present in air can condense on the exterior surface of a cold pipe. Dropping off a pipe, it can create an unpleasantly wet condition, disfigure finished surfaces, and provide conditions amenable to mold growth. Groundwater in some parts of North America is 50°F (10°C) and colder (see Fig. 8.56). A pipe carrying such water might have a surface temperature of about 60°F (15°C). The psychrometric chart (Appendix F) indicates that at a summer air temperature of 85°F (29°C), condensation will occur on such a pipe when the relative humidity exceeds 40%. To avoid condensation damage, all cold water piping and fittings should be insulated. Glass fiber ½ to 1 in. (12 to 25 mm) thick is commonly chosen for this purpose.

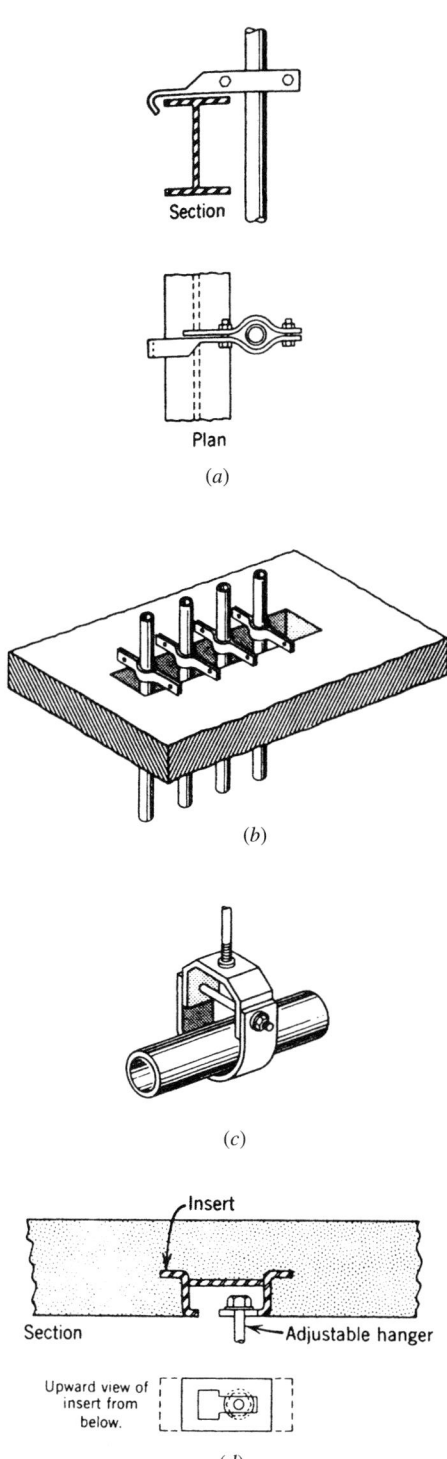

**Fig. 21.61** *Pipe supports. (a) Vertical riser supported at a steel beam. (b) Vertical riser group supported at a slot in a concrete slab. (c) Horizontal pipe hung from a slab above by an adjustable-length clevis hanger. (d) Typical metal insert in a concrete soffit to receive the hanger rod.*

A tight vapor retarder on the exterior surface of the insulation prevents water vapor from penetrating the insulation and reaching the cold surface. The insulation provides another important advantage: it retards heat flow from the warmer air to the water, thus preventing the water from becoming disagreeably warm.

### (g) Heat Conservation

Pipes carrying domestic hot water should be insulated to conserve energy used to heat the water and to assure a correct water temperature at the points of use. Parallel hot and cold water piping, even though insulated, should be separated to prevent heat exchange. The pipe insulation should have a $k = 0.22 - 0.28$ Btu in/h ft$^2$ °F, and its thickness should be a minimum of ½ in. (on pipes less than 1½-in. diameter) or 1 in. (on pipes 1½-in. diameter and above). In SI units, $k = 0.032 - 0.040$ W/m K, and the thickness should be a minimum of 1.3 mm (on pipes less than 25-mm diameter), or 2.5 mm (on pipes 25-mm diameter and above).

Storage tanks and water heaters are usually manufactured with integral insulation. Older devices, however, may have less insulation than today's energy concerns warrant. As a result, many older water heaters are retrofitted with added insulation. Some electric utilities find the savings from conservation so attractive (relative to the cost of building new generating facilities) that they provide water-heater wrapping as a free service to customers.

### 21.11  SIZING OF WATER PIPES

There must be sufficient pressure at fixtures to assure the user of a prompt and adequate flow of water. Municipal ordinances often state that the flow must be adequate to keep the fixtures clean and sanitary. These convenience and sanitation objectives result in prescribed pressures that must be maintained at the various fixtures to assure the proper flow rates listed in Table 21.14.

Minimum fixture pressures vary from 4 to 20 psi (28 to 138 kPa) for fixtures other than hose bibbs. Because the pressure in street mains is usually about 50 psi (345 kPa), it is possible to assure the minimum fixture pressure, provided that the water does not have to be lifted to too great a

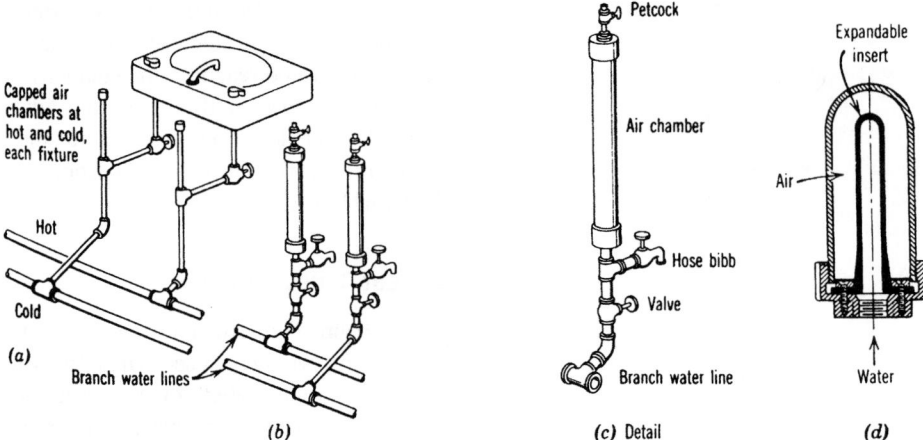

**Fig. 21.62** *Shock relief and expansion chambers. Air chambers cushion the shock of the water hammer when the fixture faucets are shut off abruptly. They also permit hot water to expand instead of periodically forcing open the hot water emergency pressure relief valve at the heater or tank. (a) Capped air chambers at each supply pipe of each fixture. (b,c) Rechargeable air chambers on hot and cold branch water lines (individual fixture chambers are omitted when these are used). (d) Special shock absorber.*

## TABLE 21.14  Flow and Pressure to Typical Plumbing Fixtures

| Fixture Served | Minimum Flow Rate (gpm)[a] | Minimum Pressure (psi)[b] | Maximum Flow Rate or Quantity |
|---|---|---|---|
| Bathtub | 4 | 8 | |
| Bidet | 2 | 4 | |
| Combination fixture | 4 | 8 | |
| Dishwasher, residential | 2.75 | 8 | |
| Drinking fountain | 0.75 | 8 | |
| Laundry tray | 4 | 8 | |
| Lavatory, private | 2 | 8 | 2.5 gpm at 80 psi |
| Lavatory, public | 2 | 8 | 0.5 gpm at 80 psi |
| Lavatory, public, metering or self-closing | 2 | 8 | 0.25 gallon per metering cycle |
| Shower head | 3 | 8 | 2.5 gpm at 80 psi |
| Shower head, temperature controlled | 3 | 20 | 2.5 gpm at 80 psi |
| Sink, residential | 2.5 | 8 | 2.5 gal at 60 psi |
| Sink, service | 3 | 8 | 2.5 gal at 60 psi |
| Urinal, valve | 15 | 15 | 1.5 gallons per flushing cycle[a] or 1.0 gallon per flushing cycle |
| Water closet, blow out, flushometer valve | 35 | 25 | 4 gallons per flushing cycle |
| Water closet, siphonic, flushometer valve | 25 | 15 | 4 gallons per flushing cycle[a] or 1.6 gallons per flushing cycle |
| Water closet, tank, close coupled | 3 | 8 | 1.6 gallons per flushing cycle |
| Water closet, tank, one piece | 6 | 20 | 1.6 gallons per flushing cycle |

*Source: International Plumbing Code.* © 1997, International Code Council, Inc., Falls Church VA. Reprinted with permission. All rights reserved.

For SI: 1 psi = 6.895 kPa, 1 gpm = 3.785 L/m.

[a]The higher maximum listed is for public use in places of assembly, and for patients, inmates, and residents in hospitals, nursing homes, sanitariums, prisons, asylums, and reformatories.

height and not too much pressure is lost by friction in distribution piping. Excessive friction results from piping that is too long in *developed length* (actual distance of water flow) or that interposes too many fittings (such as elbows and tees), or is too small in diameter.

The pressure losses in an upfeed system served by street main pressure are as follows. For A below, use the highest, most remote fixture from the main.

| | |
|---|---|
| Minimum fixture flow pressure | A |
| Pressure lost because of height | + B |
| Pressure lost by friction in piping | + C |
| Pressure lost by flow through meter | + D |
| *Total* required street main pressure | = E |

During design, items A, B, and E are known and are reasonably constant. A value for A can be found in Table 21.14. Street main pressure, E, is a characteristic of the local water supply and is obtained from the water utility. Item B, the pressure lost due to height, can be found by multiplying the height in feet by 0.433 (height in meters by 10) (see the discussion of static head in Section 21.9a). Item D, the pressure lost in flow through the water meter, depends upon flow and pipe size (Fig. 21.63), neither of which is yet known. Therefore, the value of item D is *estimated*. (For residences and small commercial buildings, meter size rarely exceeds 2 in. [50 mm].) Later, it must be checked and a recalculation made if

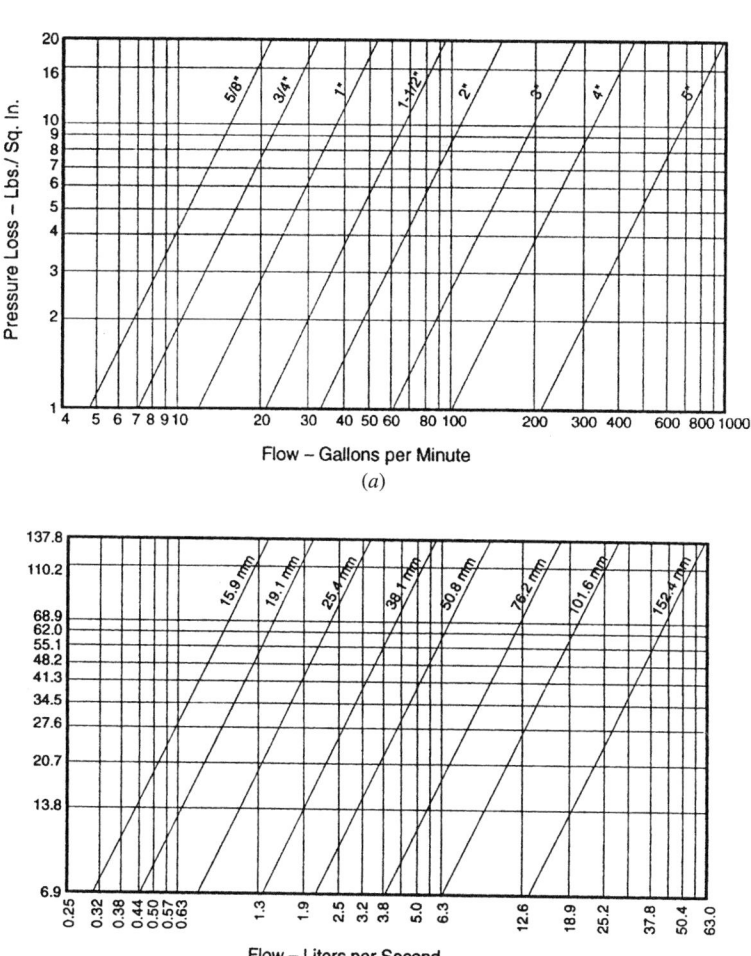

**Fig. 21.63** Pressure losses in disk-type water meters. (a) I-P units. (b) SI units. (Reprinted by permission from the 1997 Uniform Plumbing Code. © by the International Association of Plumbing and Mechanical Officials, Walnut, CA.)

necessary. This leaves one unknown, the value of C, where

$$C = E - A - B - D$$

Pipe size is based upon Fig. 21.64. Pipe diameter is determined by the point of intersection of a horizontal line representing flow and a vertical line expressing friction loss. To select a pipe size, one needs to know the probable flow and the *unit*-friction loss in the pipe and fittings. The noise created by water flow also must be considered. Flow above 10 fps (3.1 m/s) is usually too noisy; flow above 6 fps (1.8 m/s) may be too noisy in acoustical-critical locations.

Flow can be found for any pipe by first listing all fixtures to be served by that pipe, assigning them water supply fixture units (wsfu, listed in Table 21.15), and finding the total wsfu. This total is then converted to the likely demand flow in gpm from Fig. 21.65a (demand flow in L/s from Fig. 21.65b). These curves show that actual flow does *not* increase in direct proportion to an increase in fixture units. The larger an installation, the less likely that many fixtures will be operating concurrently. This reflects the statistical concept known as *diversity*.

To establish the desired friction loss, divide value C (pressure lost by friction in piping) by the *total equivalent length* of the piping. This length is the sum of the *developed length* (total linear distance of water travel) and the length equivalent of the fittings. For example, Table 21.16 shows that, in a

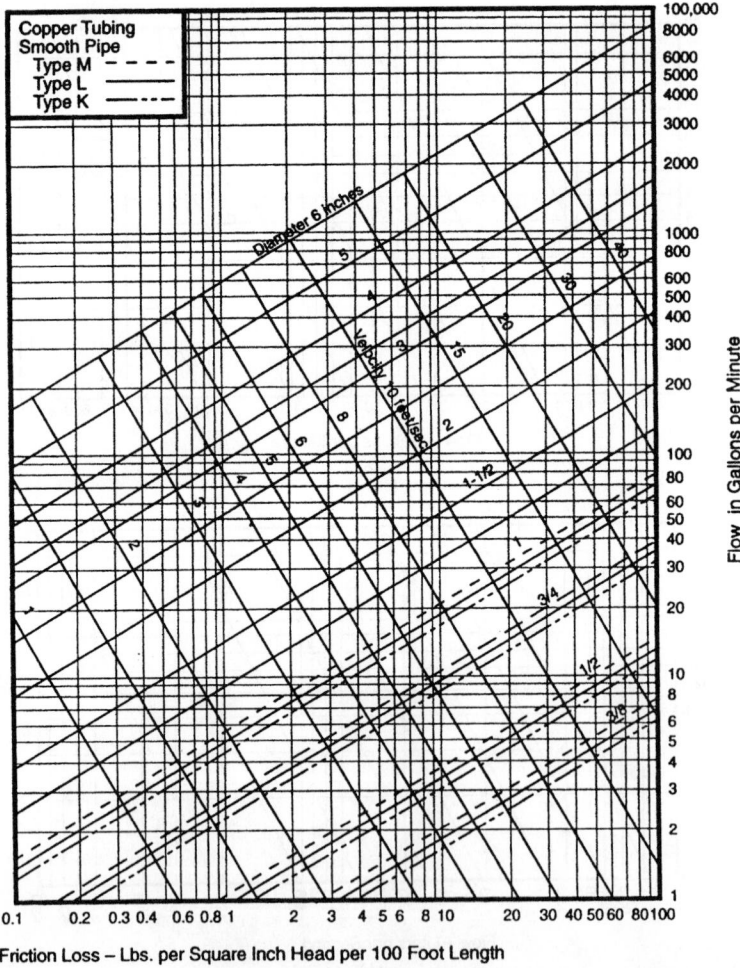

Friction Loss – Lbs. per Square Inch Head per 100 Foot Length

(a)

**Fig. 21.64** Friction loss chart for smooth pipe. Velocity is shown as an aid in noise control: above 10 fps (3m/s), moving water can be clearly heard within pipes. (a) I-P units. (b) SI units. (Reprinted by permission from the 1997 Uniform Plumbing Code. © by the International Association of Plumbing and Mechanical Officials, Walnut, CA.)

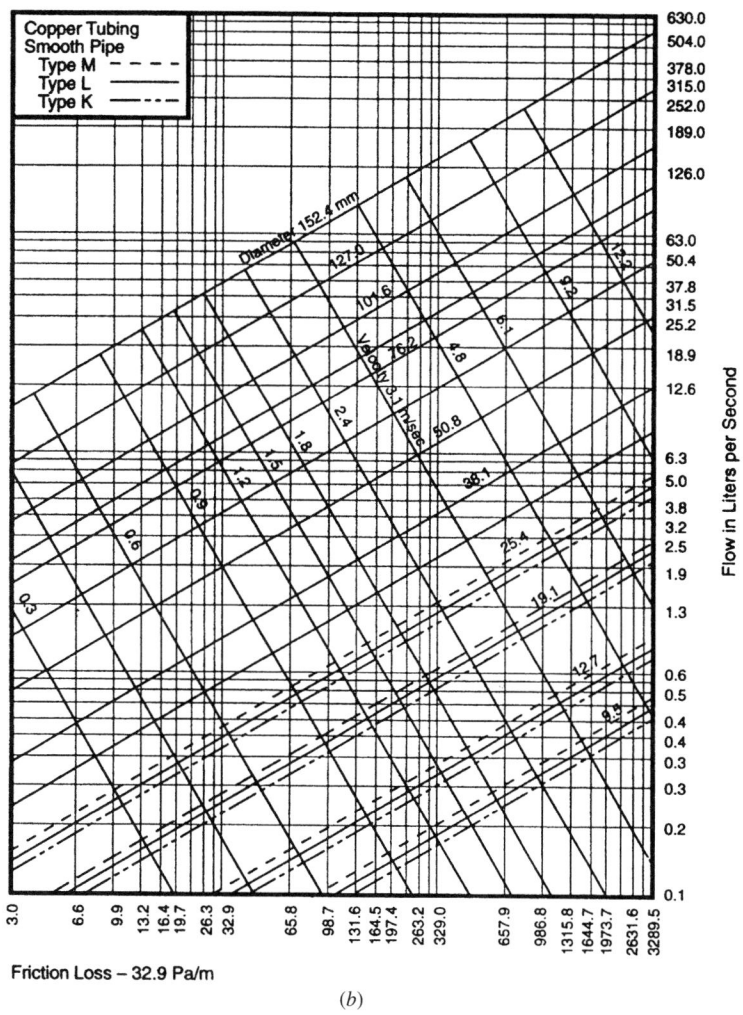

Friction Loss – 32.9 Pa/m

(*b*)

**Fig. 21.64** (Continued)

1-in.-diameter pipe run, a 90° ell causes a friction loss equivalent to that of 3 ft of straight pipe. The number and style of fittings are estimated, and the size of fittings is assumed. This may be puzzling, but it is a common trial and error engineering procedure that sometimes requires several recalculations.

**EXAMPLE 21.6** Using the following data—some of which have been arrived at by the assumptions referred to above—find the proper size for a metered water supply main.

| | |
|---|---|
| Street main pressure (minimum) | 50 psi |
| Height, topmost fixture above main | 30 ft |
| Topmost fixture type | Water closet with flush valve (1.6 gal per flush) |
| Fixture units in the system | 85 wsfu |
| Developed length of the piping (to the highest and most remote fixture) | 100 ft |
| Pipe length equivalent to fittings (commonly estimated at 50% of the developed length) | 50 ft |
| System uses predominantly | Flush valves |

**SOLUTION**

From the minimum street main pressure, subtract the sum of the fixture pressure, the static head, and the pressure lost in the meter. This sum is

| | *psi* |
|---|---|
| A: fixture pressure (Table 21.14) | 15 |
| B: static head 30 ft × 0.433 | 13 |
| D: pressure loss in meter (estimated, Fig. 21.63) | 8 |
| Subtotal | 36 |
| E: pressure in street main | 50 |
| (A + B + D) | −36 |
| E − (A + B + D) | 14 |

The pressure lost in 100 ft (DL) of piping plus the 50 ft of piping equivalent to the pressure lost by friction in the fittings therefore can total 14 psi. Total equivalent length is 150 ft. This procedure assures 15 psi at the critical fixture. The unit-friction loss, psi/100 ft of pipe, will be 14 psi × 100/150 (total equivalent length) = 9.33 psi/100 ft

From Fig. 21.65*a*, curve 1, a flush-valve system with 85 wsfu will have a probable flow (diversified demand) of about 64 gpm. Given this information, enter Fig. 21.64*a* horizontally at 64 gpm and vertically at 9.3 psi/100 ft. At the intersection of these lines, the pipe diameter and velocity are determined. Between 1½-in.- and 2-in.-diameter pipe:

velocity = 8 fps (less than 10 fps, so OK)

Therefore, a 2-in.-diameter supply pipe will be chosen with a 2-in. meter.

Now find the actual pressure loss in the 2-in. meter for a flow of 64 gpm. Figure 21.63*a* shows that this is about 4 psi. Because this is *less* than the 8 psi estimated, the pressure at the fixture will be slightly higher than the minimum needed. When a final system layout is made, the actual fittings are tabulated and the equivalent length of fittings is found. If this differs greatly from the 50 ft estimated, a recalculation is made. ■

WATER AND WASTE

**TABLE 21.15 Water Supply Fixture Units (WSFU)**

| Fixture | Occupancy | Type of Supply Control | *Load Values in WSFU* Cold | Hot | Total |
|---|---|---|---|---|---|
| Bathroom group | Private | Flush tank | 2.7 | 1.5 | 3.6 |
| Bathroom group | Private | Flush valve | 6 | 3 | 8 |
| Bathtub | Private | Faucet | 1 | 1 | 1.4 |
| Bathtub | Public | Faucet | 3 | 3 | 4 |
| Bidet | Private | Faucet | 1.5 | 1.5 | 2 |
| Combination fixture | Private | Faucet | 2.25 | 2.25 | 3 |
| Dishwashing machine | Private | Automatic | | 1.4 | 1.4 |
| Drinking fountain | Offices, etc. | ⅜″ (9.5 mm) valve | 0.25 | | 0.25 |
| Kitchen sink | Private | Faucet | 1 | 1 | 1.4 |
| Kitchen sink | Hotel, restaurant | Faucet | 3 | 3 | 4 |
| Laundry trays (1 to 3) | Private | Faucet | 1 | 1 | 1.4 |
| Lavatory | Private | Faucet | 0.5 | 0.5 | 0.7 |
| Lavatory | Public | Faucet | 1.5 | 1.5 | 2 |
| Service sink | Offices, etc. | Faucet | 2.25 | 2.25 | 3 |
| Shower head | Public | Mixing valve | 3 | 3 | 4 |
| Shower head | Private | Mixing valve | 1 | 1 | 1.4 |
| Urinal | Public | 1″ (25 mm) flush valve | 10 | | 10 |
| Urinal | Public | ¾″ (19 mm) flush valve | 5 | | 5 |
| Urinal | Public | Flush tank | 3 | | 3 |
| Washing machine (8 lbs) (3.6 kg) | Private | Automatic | 1 | 1 | 1.4 |
| Washing machine (8 lbs) (3.6 kg) | Public | Automatic | 2.25 | 2.25 | 3 |
| Washing machine (15 lbs) (6.8 kg) | Public | Automatic | 3 | 3 | 4 |
| Water closet | Private | Flush valve | 6 | | 6 |
| Water closet | Private | Flush tank | 2.2 | | 2.2 |
| Water closet | Public | Flush valve | 10 | | 10 |
| Water closet | Public | Flush tank | 5 | | 5 |
| Water closet | Public or private | Flushometer tank | 2 | | 2 |

*Source: International Plumbing Code.* © 1997, International Code Council, Falls Church, VA. Reprinted with permission. All rights reserved.

Fixture Units

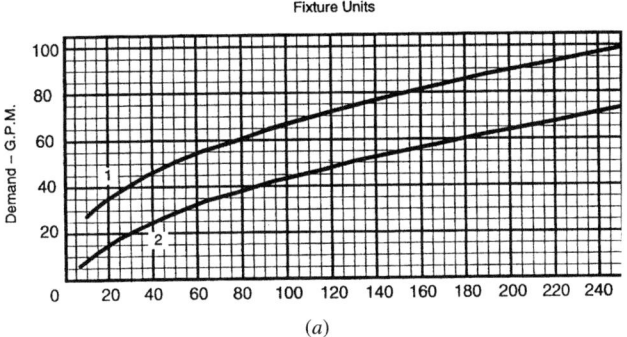

(a)

Fixture Units

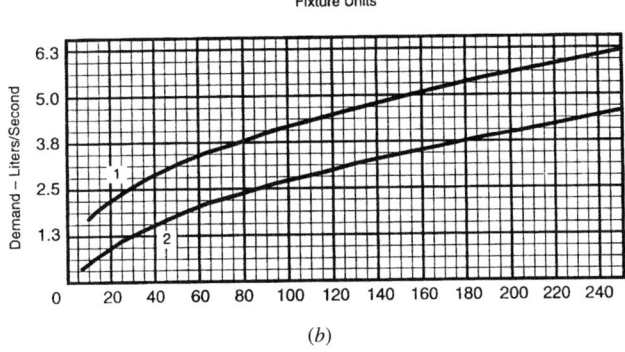

(b)

**Fig. 21.65** *Estimate curves for flow based upon total water supply fixture units. Curve 1 is for a system of predominantly flush valves. Curve 2 is for a system of predominantly flush tanks. (For fixture unit totals above 250, see the 1997* Uniform Plumbing Code.*) (a) I-P units. (b) SI units. (Reprinted by permission from the 1997* Uniform Plumbing Code. *© by the International Association of Plumbing and Mechanical Officials, Walnut, CA.)*

**EXAMPLE 21.7** Solve the pipe sizing problem posed in the previous example using SI units.

| | |
|---|---|
| Street main pressure (minimum) | 345 kPa |
| Height, topmost fixture above main | 10 m |
| Topmost fixture type | Water closet with flush valve |
| Fixture units in the system | 85 wsfu |
| Developed length of the piping (to the highest and most remote fixture) | 30 m |
| Pipe length equivalent to fittings (commonly estimated at 50% of developed length) | 15 m |
| System uses predominantly | Flush valves |

**SOLUTION**

From the minimum street main pressure, subtract the sum of the fixture pressure, the static head, and the pressure lost in the meter. This sum is

| | |
|---|---|
| A: fixture pressure (Table 21.14) | 103 kPa |
| B: static head 10 m × 10 kPa/m | 100 kPa |
| D: pressure loss in meter (estimated, Fig. 21.63) | 55 kPa |
| Subtotal | 258 kPa |
| E: pressure in street main | 345 kPa |
| (A + B + D) | −258 |
| E − (A + B + D) | 87 kPa |

The pressure lost in 30 m (developed length) of piping plus the 15 m of piping equivalent to the pressure lost by friction in the fittings therefore can total 87 kPa. Total equivalent length is 45 m. This procedure assures 103 kPa at the critical fixture. The unit-friction loss, kPa/100 m of pipe, will be

87 kPa × 100/45 (total equivalent length)
= 193 kPa/100 m

From Fig. 21.65b, curve 1, a flush-valve system with 85 wsfu will have a probable flow (diversified demand) of about 4 L/s. Given this information, enter Fig. 21.64b horizontally at 4 L/s and vertically at 193 kPa/100 m. At the intersection of these lines, the pipe diameter and velocity are determined. Between 38-mm- and 51-mm-diameter pipe:

velocity = about 2.4 m/s (less than 3 m/s, so OK)

Therefore, a 51-mm-diameter supply pipe will be chosen with a 51-mm meter. ∎

As in Example 21.6, the actual system pressures can be found using the exact pressure loss in a 51-mm meter.

## 21.12 IRRIGATION

The highest design priority for landscape watering is to ensure optimum rainfall retention (see Figs. 20.12 to 20.19). In the past, provision for watering

**TABLE 21.16 Allowance in Equivalent Length of Pipe for Friction Loss in Valves and Threaded Fittings[a]**

| | | | PART A. I-P UNITS | | | | |
|---|---|---|---|---|---|---|---|
| | | | *Equivalent Length of Pipe for Various Fittings* | | | | |
| Diameter of Fitting (in.) | 90° Standard Ell (ft) | 45° Standard Ell (ft) | Standard Tee 90° (ft) | Coupling or Straight Run of Tee (ft) | Gate Valve (ft) | Globe Valve (ft) | Angle Valve (ft) |
| ⅜ | 1 | 0.6 | 1.5 | 0.3 | 0.2 | 8 | 4 |
| ½ | 2 | 1.2 | 3 | 0.6 | 0.4 | 15 | 8 |
| ¾ | 2.5 | 1.5 | 4 | 0.8 | 0.5 | 20 | 12 |
| 1 | 3 | 1.8 | 5 | 0.9 | 0.6 | 25 | 15 |
| 1¼ | 4 | 2.4 | 6 | 1.2 | 0.8 | 35 | 18 |
| 1½ | 5 | 3 | 7 | 1.5 | 1.0 | 45 | 22 |
| 2 | 7 | 4 | 10 | 2 | 1.3 | 55 | 28 |
| 2½ | 8 | 5 | 12 | 2.5 | 1.6 | 65 | 34 |
| 3 | 10 | 6 | 15 | 3 | 2 | 80 | 40 |
| 3½ | 12 | 7 | 18 | 3.6 | 2.4 | 100 | 50 |
| 4 | 14 | 8 | 21 | 4.0 | 2.7 | 125 | 55 |
| 5 | 17 | 10 | 25 | 5 | 3.3 | 140 | 70 |
| 6 | 20 | 12 | 30 | 6 | 4 | 165 | 80 |
| | | | PART B. SI UNITS | | | | |
| | | | *Equivalent Length of Pipe for Various Fittings* | | | | |
| Diameter of Fitting (mm) | 90° Standard Elbow mm | 45° Standard Elbow (mm) | Standard Tee 90° (mm) | Coupling or Straight Run of Tee (mm) | Gate Valve (mm) | Globe Valve (mm) | Angle Valve (mm) |
| 9.5 | 305 | 183 | 457 | 91 | 61 | 2438 | 1219 |
| 12.7 | 610 | 366 | 914 | 183 | 122 | 4572 | 2438 |
| 19.1 | 762 | 457 | 1219 | 244 | 152 | 6096 | 3658 |
| 25.4 | 914 | 549 | 1524 | 274 | 183 | 7620 | 4572 |
| 32 | 1219 | 732 | 1829 | 366 | 244 | 10668 | 5486 |
| 38 | 1524 | 914 | 2134 | 457 | 305 | 13716 | 6706 |
| 51 | 2134 | 1219 | 3048 | 610 | 396 | 16764 | 8534 |
| 64 | 2438 | 1524 | 3658 | 762 | 488 | 19812 | 10363 |
| 76 | 3048 | 1829 | 4572 | 914 | 610 | 24384 | 12192 |
| 102 | 4267 | 2438 | 6401 | 1219 | 823 | 38100 | 16764 |
| 127 | 5182 | 3048 | 7620 | 1524 | 1006 | 42672 | 21336 |
| 152 | 6096 | 3658 | 9144 | 1829 | 1219 | 50292 | 24384 |

*Source:* Reprinted by permission from the 1997 *Uniform Plumbing Code.* © by the International Association of Plumbing and Mechanical Officials, Walnut, CA.

[a]Based on nonrecessed threaded fittings. Use one-half of these allowances for recessed threaded fittings or streamline solder fittings.

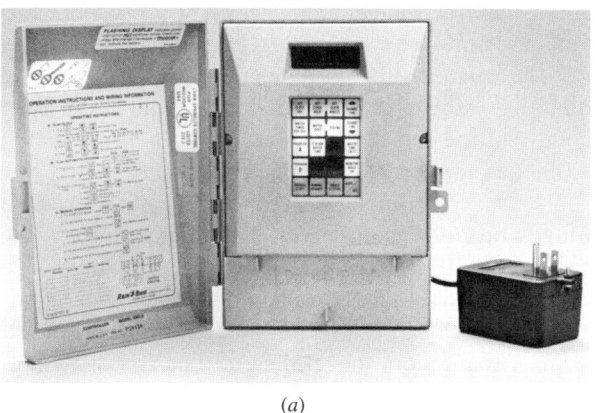

(a)

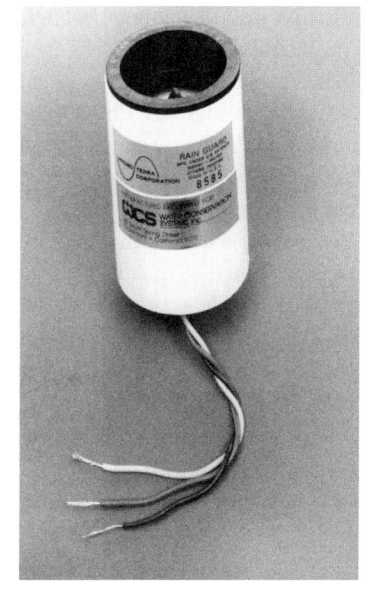

(b)

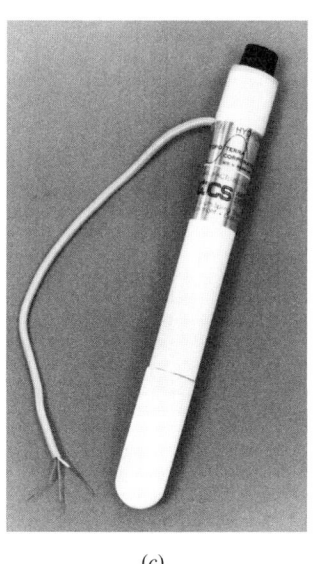

(c)

(d)

**Fig. 21.66** Components for automatic landscape sprinkling systems. (a) Programmable controllers feature multiple stations, up to 14-day sequences, and memory provisions in case of power failure. (Courtesy of Rain Bird Sales, Inc.) (b) A solid-state rain sensor automatically prevents needless watering during rainfall. (c) A tensiometer controls irrigation by monitoring the moisture content of the soil. (Courtesy of Water Conservation Systems, Inc.) (d) Bubblers are low-flow substitutes for sprinklers—a step toward drip irrigation. These can be pressure-compensating to permit constant flow. (Courtesy of Rain Bird Sales, Inc.)

the landscape around buildings frequently was limited to the installation of hose bibbs at building exteriors. In many areas of North America, half of the residential water usage (see Chapter 22) is for outdoor purposes. With increasing demands on a finite water supply, water-conserving irrigation equipment has become available.

*Lawn sprinklers* are relatively inefficient irrigating devices, as much of their water is lost to evaporation and runoff. It is commonly estimated that lawn sprinkling will provide ½ in. of water per hour per square foot of lawn (or 0.3 gph per ft$^2$) (0.005 L/s per m$^2$).

Landscape sprinkling is least efficient during the daytime, when hot sun and dry air combine to increase the rate of evaporation. However, nighttime sprinkling is rarely convenient for building custodians or homeowners. One solution to this problem is the use of *sprinkler timing devices* (Fig. 21.66a), which control electronic valves and a network of underground supply pipes and sprinkler heads that are permanently installed within landscaped areas. These timing devices typically include controls gov-

erning the length of the watering cycle, the time at which the cycle begins (usually before sunrise, when relative humidity is highest), and the number of days between cycles. A "rainy day switch" is often included so that irrigation can be discontinued during rainy weather. Rain sensors can be used to shut off irrigation automatically (Fig. 21.66b). Tighter control over the water–plant relationship can be obtained with *tensiometers* (Fig. 21.66c), which monitor the moisture content of the soil at the depth of the plants' root zone. These can be installed so as to override the automatic timing device, thus watering more—or less—frequently, depending upon the plants' needs. Instead of sprinklers, *bubblers* (Fig. 21.66d) can be installed, with very low flow rates and less evaporative water losses.

*Drip irrigation* takes an approach very different from that of the flooding method typical of sprinklers. From a network of plastic tubes, either just underground or on the surface, *emitters* (Fig. 21.67) slowly and steadily drip water onto the ground surface at each needy plant. Most of this water soaks into the soil at a rate that is better for most

(a)

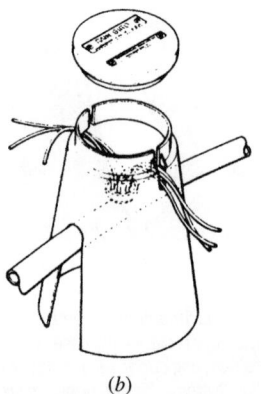

(b)

**Fig. 21.67** Equipment for drip irrigation. (a) On low-pressure lines, emitters are installed—one for each group of plants to be watered. The lines can be laid on the ground surface or just under the surface. (b) Simple emitter boxes allow easy access. (Courtesy of Rain Bird Sales, Inc.)

plants than the sudden, short flooding of intermittent sprinkling. Two requirements are especially important for drip systems: the water must not contain materials that can clog the small holes of the emitters, and the pressure must be low. Pressure-reducing valves at the source of the drip system are advisable; if necessary, filters should also be installed there.

Drip irrigation is not a universal solution to landscape watering; it is best for individual plants such as shrubs and small trees but is difficult to apply to large lawns. Where appropriate, it can achieve

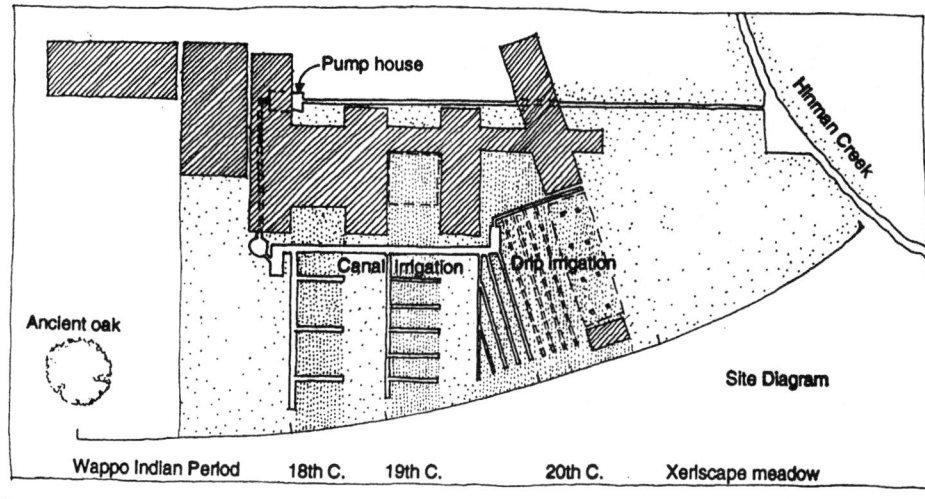

*(a)*

*(b)*

**Fig. 21.68** *The Napa Valley Museum shows the history of irrigation in northern California. (a) Site plan. Water, piped from behind a small dam in the adjacent creek, flows through open canals, then drip irrigation, eventually discharging to a field of native plants. (b) The canals help guide the visitor through the exhibit gardens. (Courtesy of Fernau and Hartman, architects, Berkeley, CA.)*

significant water conservation compared to sprinklers. A detailed discussion of soil texture, moisture, and plant growth is provided in Lowry (1988, Chap. 4).

Recycled or reclaimed water for irrigation, often provided by graywater systems and sometimes by stored rainwater, is gradually gaining acceptance in North American building codes. This topic is explored further in Section 22.9. Considered nonpotable, this water must sometimes be deliberately colored and/or be accompanied by warning signs at each outlet.

Irrigation, past and future, shaped the design of the Napa Valley Museum in California (Fig. 21.68). An adjacent creek provides water behind a small dam. A pump house carries water to a central observation tower, where visitors see the history of irrigation displayed in the context of the Napa Valley. The path of water in this model landscape invites visitors to explore canal irrigation of the eighteenth and nineteenth centuries, as well as advanced drip irrigation and xeriscape plantings of the twentieth century and beyond. (Xeriscape planting, as a part of the Hydrozone landscape concept, was shown in Fig. 20.21.)

## References

ANSI. 1986. *American National Standard for Buildings and Facilities A117.1-1986. Providing Accessibility and Usability for Physically Handicapped People*, American National Standards Institute, New York.

ASHRAE. 1995. *ASHRAE Handbook—HVAC Applications.* American Society of Heating, Refrigerating and Air-Conditioning Engineers, Atlanta, GA.

Banham, R. 1969. *The Architecture of the Well-tempered Environment.* Architectural Press, London.

Brown, G.Z., J. Reynolds, S. Ubbelohde. 1982. *Inside Out: Design Procedures for Passive Environmental Technologies.* John Wiley & Sons, New York.

HUD. 1977. *Intermediate Minimum Property Standards Supplement.* U.S. Department of Housing and Urban Development, Washington, DC.

IAPMO. 1997. *Uniform Plumbing Code.* International Association of Plumbing and Mechanical Officials, Walnut, CA.

ICC. 1997. *International Plumbing Code.* International Code Council, Falls Church, VA.

Kira, A. 1976. *The Bathroom, new and expanded ed.* Viking Press, New York.

Lowry, William P. 1988. *Atmospheric Ecology for Designers and Planners.* Van Nostrand Reinhold, New York.

Milne, M. 1976. *Residential Water Conservation.* U.S. Office of Water Research and Technology, Department of Commerce, NTIS.

National Drinking Water Clearinghouse, *Tech Brief* Series, U.S. Environmental Protection Agency, Washington, DC:
> June 1996. *Disinfection*
> September 1996. *Filtration*
> February 1997. *Corrosion Control*
> May 1997. *Ion Exchange and Demineralization*

Plante, R.H. 1983. *Solar Domestic Hot Water: A Practical Guide.* John Wiley & Sons, New York.

Reif, D.K. 1983. *Passive Solar Water Heaters.* Brick House Publishing, Andover, MA.

U.S. Architectural and Transportation Barriers Compliance Board. *Americans with Disabilities Act (ADA) Accessibility Guidelines for Buildings and Facilities.* Washington, DC.

U.S. Environmental Protection Agency. 1975. *Manual of Individual Water Supply Systems.* Washington, DC.

# CHAPTER 22

# Liquid Waste

IN CHAPTERS 20 AND 21, WE CONSIDERED the waste of resources inherent in the use of potable water to flush toilets. In this chapter we examine some consequences of this conventional approach to bodily waste removal, along with alternatives that use no water at all, and indeed that offer to convert waste to a useful resource. For typical U.S. residences, the potential impact of such alternatives on water usage and treatment is shown in Fig. 22.1.

Almost every plumbing fixture within buildings is provided with both water supply and waste pipes. Because the toilet is usually the largest user of water (Fig. 22.1), as well as one of the worst water polluters, this chapter begins with recycling/waterless alternatives to the flush toilet and urinals that were discussed in Section 21.7. Then more conventional systems using water to move waste, and conventional treatment systems from individual to large scale, are discussed. Discussions of graywater and stormwater treatment complete this chapter.

## 22.1 WATERLESS TOILETS AND URINALS

The array of waterless alternatives includes toilets in which chemicals or oil are substituted for water. These devices are commonly found in airplanes, vehicles, and boats, as well as in remote and environmentally sensitive areas. The chemicals must be frequently recharged and the waste products removed. Other waterless toilets temporarily treat the waste awaiting discharge to a sewer by freezing it, burning it, or otherwise packaging it so that it remains inoffensive. Obviously, such devices can become energy-intensive solutions to waterless waste disposal. Also, they are still dependent upon public sewer systems.

### (a) Composting Toilets

Once considered too unconventional to ever gain mainstream acceptance, these ecologically attrac-

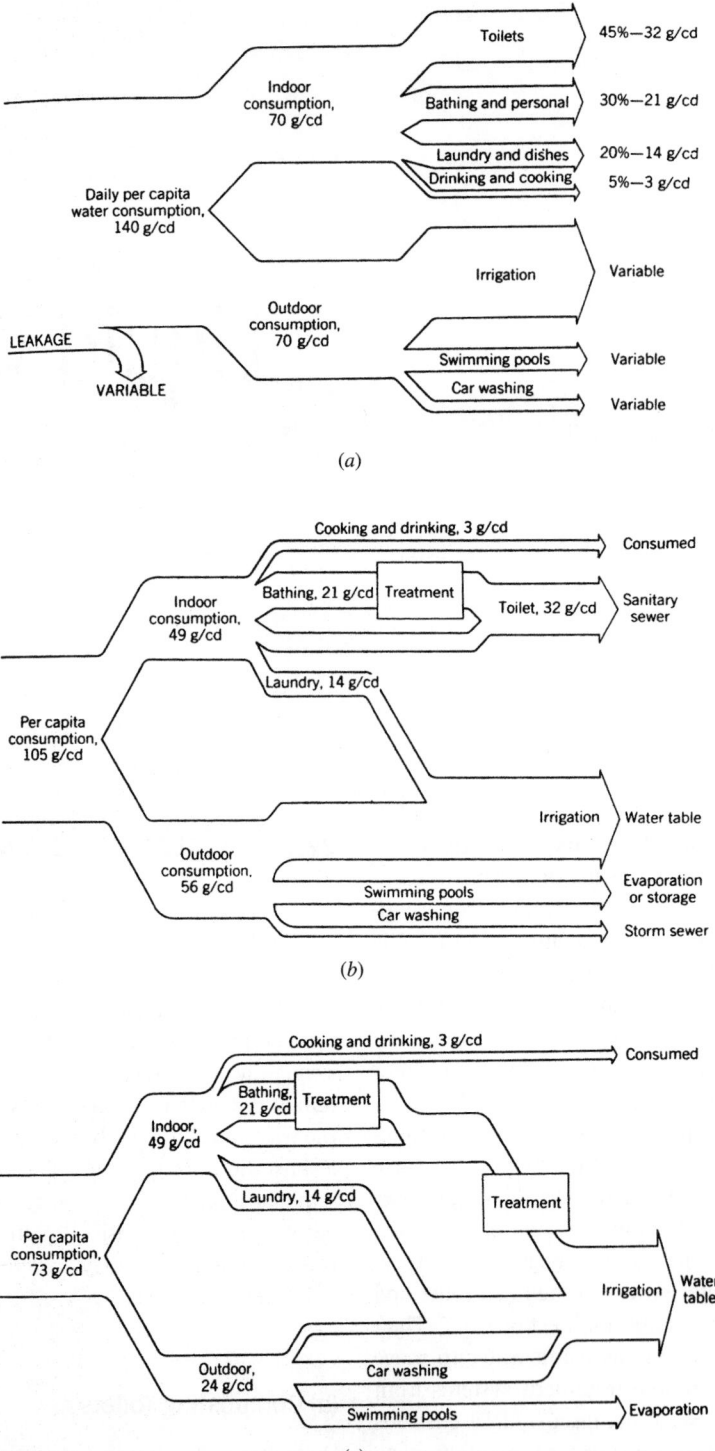

**Fig. 22.1** Opportunities for water conservation for the typical U.S. family. (a) At present, the average U.S. residential usage is about 140 g/cd (530 L/cd)—all of it potable. (b) With attention to recycling and the matching of water quality to usage, potable water usage could be cut to 105 g/cd (397 L/cd)—a 25% reduction. (c) With further on-site treatment and recycling, potable water usage drops to 73 g/cd (276 L/cd)—a 40% reduction—and the need for public sewage treatment (for residences) disappears. (From Milne, 1976.)

tive devices are now appearing not only in homes but within institutional buildings (Fig. 22.2). Perhaps several generations needed to pass between regular use of the outhouse and use of these composting toilets. They offer a self-contained and much less energy-intensive strategy for waterless toilets. Wherever water supplies are unreliable, or waste discharge is difficult or prohibited, composting toilets offer an alternative. Silent and immune from potential water damage (freezing pipes or overflowing), these systems rely upon *aerobic* digestion of waste (i.e., that which occurs in the presence of oxygen). Aerobic systems usually are essentially odor-free, and the exhaust air is rich in $CO_2$ and water vapor. In contrast, *anaerobic* decomposition (i.e., that which occurs without oxygen) is malodor-ous and produces methane gas as an important by-product. See Sections 22.7 and 22.8 for more about methane and anaerobic decomposition.

Waste material deposited in these toilets builds up into a mound, which retards the flow of air for decomposition. Wood shavings are commonly added with each use to encourage toppling of the mound and to add some carbon to the resulting compost. Some composting toilets include a means of regular mechanical stirring of the compost chamber, aiding aerobic decomposition.

Ventilation is important, both to reduce odors and to facilitate evaporation of excess moisture. A ventilation stack is an essential feature; it is often assisted by a very small (3- to 5-W) fan. Problems with insect hatchings within the composting chamber can be lessened by keeping vegetable scraps out of the chamber, as they are frequent sources of insect eggs.

The C.K. Choi Building at the University of British Columbia (Fig. 22.2) eliminated a connection to the city sewer system by a combination of composting toilets and a graywater treatment system. The composting toilets in this three-story, 30,000-ft$^2$ (2787-m$^2$) facility for the Institute of Asian Research are estimated to save 1500 gal (5680 L) of water per day. At an estimated 30,000 uses per year, these composting toilets are expected to produce an annual compost yield of about 4 ft$^3$ (0.1 m$^3$). In addition, their liquid end product will be added to the graywater treatment system at the rate of about 3 to 5 gal (11 to 19 L) per day. The composting chambers are located in a small basement below the toilet rooms at the center of the building.

The graywater includes waste from lavatories and sinks, as well as liquid from the composting toilets. This water drains into the subsurface "graywater trench," containing a variety of phragmite plant material. The roots support microbial life known for digesting and neutralizing bacteria. This graywater is then used for irrigation. Graywater systems are discussed further in Section 22.9.

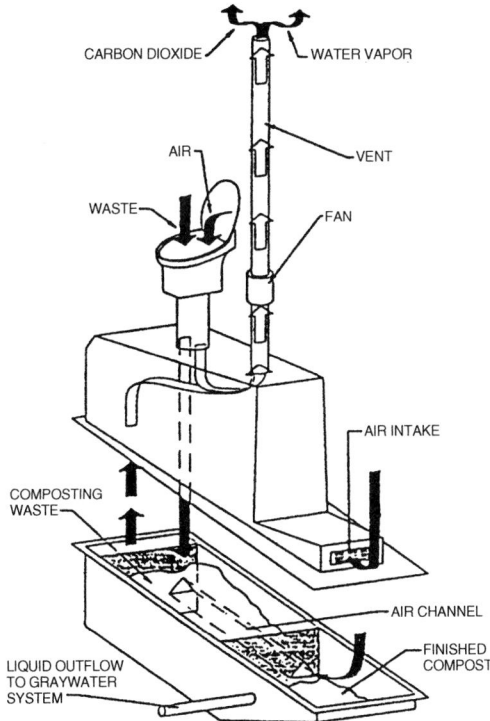

**Fig. 22.2** *Composting toilets save water and eliminate a sewer connection. They are used at the C.K. Choi Building, University of British Columbia, Canada, where the toilet rooms are concentrated near the center stairs, with a small basement below containing composting chambers and a rainwater cistern for irrigation. The composting toilets produce both dry humus and a liquid outflow to a graywater recycling system. (Courtesy of Matsuzaki Wright, architects, Vancouver, BC.)*

### (b) Vault-Type Composting Toilets

The *Clivus Multrum system* (Fig. 22.3) has a large decomposition chamber that must be below the toilet and the kitchen, from which it readily accepts

WATER AND WASTE

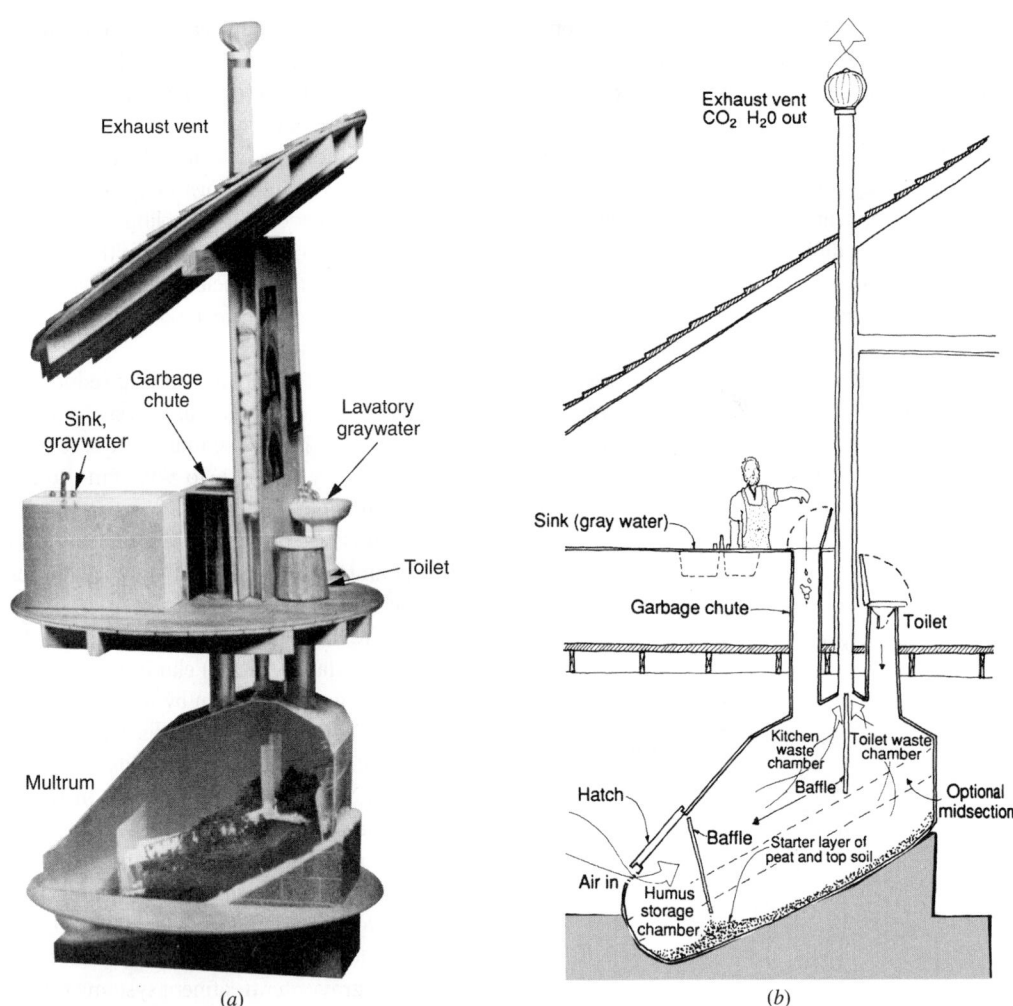

*(a)*              *(b)*

**Fig. 22.3** *Clivus Multrum System, with composting of both kitchen and toilet organic wastes. (a) The lower composting chamber requires 4 × 8 × 7 ft (1.2 × 2.4 × 2.1 m) of floor space, a generous supply of air (warm air speeds decomposition), and access for periodic removal of the garden-ready humus. (b) It must be arranged so that its upper end receives toilet wastes; as the bottom slopes downward, kitchen wastes are deposited on top of the decomposing toilet wastes. The vent stack assures continuous airflow, preventing odors from entering bathroom or kitchen. The toilet seat and cover must be kept closed when not in use to keep air flowing through the composting chamber and out of the stack. The toilet bowl (containing no water) swings open when the seat is occupied, exposing the composting chamber below. (Courtesy of Clivus Multrum USA, Inc.)*

organic waste. It must also be accessible to remove the humus—3 to 10 gal (11 to 38 L) of soil per person per year.

    The relatively large chamber (about the size of a Volkswagen "bug") and its low position/access requirement can have a significant impact on design. Also of significance is the device's appetite for fresh air. The more air it has, the speedier the process of aerobic decomposition and, therefore, the

less odor. In winter, this air could cause increased heat losses by infiltration in a building. The natural ventilation due to the stack effect (Section 8.6) is at least 15 cfm (7 L/s) in the Clivus Multrum. A ventilating fan in the stack will increase this rate significantly. However, if outdoor air is brought directly to the chamber, it could be too cold for proper decomposition; 98°F (36.7°C) is optimum. Solar heating of such input air offers one alternative. However,

(a)

(b)

*Fig. 22.4* On the shore of Crater Lake, Oregon, (a) a building with three Phoenix composting toilets serves visitors. An attached sunspace (b) contains and warms the composting chambers and contains an evaporator for liquid waste. The facility is closed all winter.

adding too much heat to compost piles encourages oxidation rather than aerobic decomposition; inferior compost results.

Another composting toilet application is shown on the shores of Oregon's Crater Lake National Park in Fig. 22.4. Consider the challenge: this facility is at the bottom of a 1-mile trail down from the crater rim; public sewers are miles away and hundreds of feet higher in elevation; severely cold winters and deep snow close the trail for more than half of the year; and Crater Lake is one of the clearest in the world, so water pollution is absolutely to be avoided. Three Phoenix composting toilet tanks sit within a solar-heated attached sunspace.

The *Phoenix* composting tanks (Fig. 22.5) feature rotatable tines that speed aerobic decomposition. They also can be fitted with an evaporation system, typically using a PV-driven small pump that delivers liquid to the top of a drip tank filled with plastic balls. This presents a very large surface area, and a small fan also aids evaporation.

## (c) Heater-Type Composting Toilets

Another approach puts the composting chamber within the toilet itself. The more compact the composting chamber, the more likely that heat must be added to speed decomposition and to evaporate liquid. Although electricity is an easy and compact way to heat, this energy used in treating waste reduces the compact toilet's ecological advantage.

In the most compact of *Sun-Mar*'s products (Fig. 22.6), a variable-diameter Bio-drum is turned by a fold-out hand crank at the front of this rather conventional-looking toilet. At this size, it is rated for continuous residential use by one adult or a family of two (for weekend/vacation use, by three adults or a family of four). Sun-Mar uses a three-chamber approach, separating the quick composting of waste and toilet paper, the evaporation of liquid, and the safe storage of the finished compost.

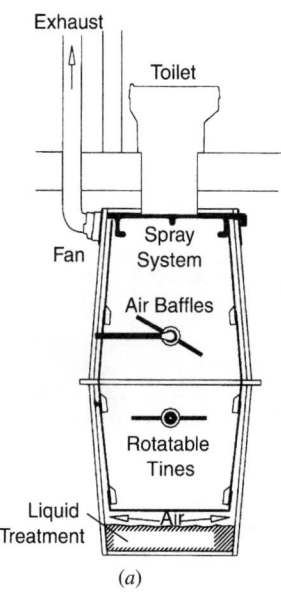

(a)　　　　　　　　　　　　　　　　　　　(b)

**Fig. 22.5** *The Phoenix composting toilet's tank serves either two toilets or one toilet and one food waste inlet. (a) Schematic section; rotatable tines mix waste and control the movement of compost to the access area for eventual removal. The fan exhausts air from the Phoenix tank to control odors. Passing behind the air baffles, air makes frequent contact with the compost pile, assuring aerobic decomposition. After passing through the compost pile and receiving secondary treatment in the bottom of the Phoenix, liquid is periodically sprayed on top of the compost pile to keep the pile moist and inoculate fresh waste with decomposing organisms. Each tank is 39 in. (1000 mm) wide and 61 in. (1550 mm) long, and requires another 5 ft (1525 mm) in front for maintenance access, and is available in three heights: 55 in. (1400 mm), 73 in. (1850 mm), and 92 in. (2350 mm). Another 12 in. (300 mm) above the top of the tank is convenient for maintenance. (b) Four chambers with their rotatable tines serve the Vogelsang High Sierra Camp at Yosemite National Park, California. (Courtesy of Advanced Composting Systems, Whitefish, MT.)*

**Fig. 22.6** *The Sun-Mar "Compact" composting toilet has a "Bio-drum"™ that is turned by a hand crank. The crank folds back into the body of the toilet after use. It also uses a 200-W electric heater to speed decomposition, a 25-W fan, and a 2-in. (50-mm) vent that attaches at the rear. The self-contained unit is designed for light residential use. (Courtesy of Sun-Mar Corporation, Tonawanda, NY.)*

### (d) Waterless Urinals

Urine is often kept out of composting toilets for several reasons. First, excess liquid in the chamber creates a discharge issue, complicating an otherwise simple end result of occasional harvesting of garden-ready humus. Second, urine is nearly sterile and an excellent fertilizer, and its separate capture could be useful. If not captured, urine is directed either to the graywater or to the wastewater system.

*Waterless No-Flush*™ urinals utilize a floating layer of BlueSeal® liquid that forms a barrier to sewer vapors but allows urine to readily pass through (Fig. 22.7). This liquid does not dissolve, mix, or react chemically with urine. It is more than 95% biodegradable and does not evaporate at

(a)

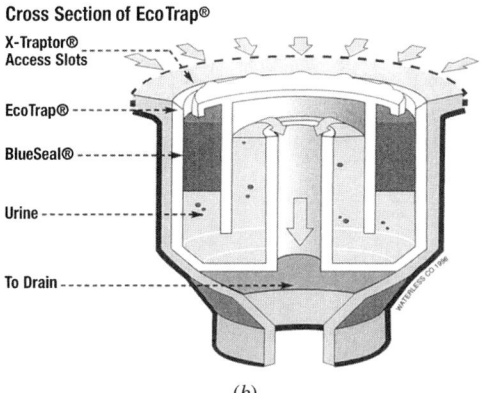

**Cross Section of EcoTrap®**

X-Traptor®
Access Slots

EcoTrap®

BlueSeal®

Urine

To Drain

(b)

**Fig. 22.7** *Using no water, (a) the Waterless No-Flush™ urinal employs a special liquid to allow urine to pass through (b) while keeping sewer gas trapped in the waste pipe. (Courtesy of the Waterless Company LLC, Del Mar, CA.)*

100°F (38°C). Because 1 quart (0.9 L) of this liquid lasts for 15,000 to 20,000 urinal uses, it replaces an estimated 15,000 to 60,000 gal (56,780 to 227,120 L) of water that would have been used by conventional flush urinals. The trap retains sediments and should be emptied occasionally, and the liquid is periodically replenished (3 oz [89 mL] per 1500 uses).

## 22.2 PRINCIPLES OF DRAINAGE

Early in the history of indoor plumbing, waste drainage was a simple matter; a pipe containing the wastewater led to a sewer (Fig. 22.8*a*). Before long, the noxious gases that were created by the anaerobic conditions in the sewer became a threat to the health of those indoors. Thus, the *trap* was invented (Fig. 22.8*b*) to block the pipe so that gases could not pass. However, as moving water filled the pipe downstream from the traps, the mass of water acted as a plunger, creating higher pressures in front of it and negative pressures behind. The positive pressures might force sewer gas through the water in other traps; worse, the negative pressures could suck (or siphon) the water from the trap, leaving it open to gas passage. A way to deal with these pressures was through the installation of *vents* so that the suction would draw air down through the vents rather than water from the traps (Fig. 22.8*c*). A typical arrangement of fixtures, traps, and vents is shown in Fig. 22.9.

### (a) Traps

The only separation between the unpleasant and dangerously unhealthy gases in sanitary drainage pipes and the air breathed by room occupants is the water caught in the fixture trap after each discharge from a fixture. Sufficient water must flow, especially in water closets, to keep this residual water clean. Traps are made of steel, cast iron, copper, plastic, or brass—except those in water closets and urinals, which are often made of vitreous china cast integrally with the fixture. The deeper the seal, the more resistance to siphonage but the greater the fouling area; therefore, a minimum depth of 2 in. (50 mm) and a maximum depth of 4 in. (100 mm) are common standards. All traps should be self-cleaning, that is, capable of being completely flushed each time the trap operates so that no sediment will remain inside to decompose.

There are a few exceptions to the rule that each fixture should have its own trap. Common exceptions include two laundry trays and a kitchen sink connected to a single trap; not more than three laundry trays using one trap; and three lavatories on a single trap. In the case of the laun-

WATER AND WASTE

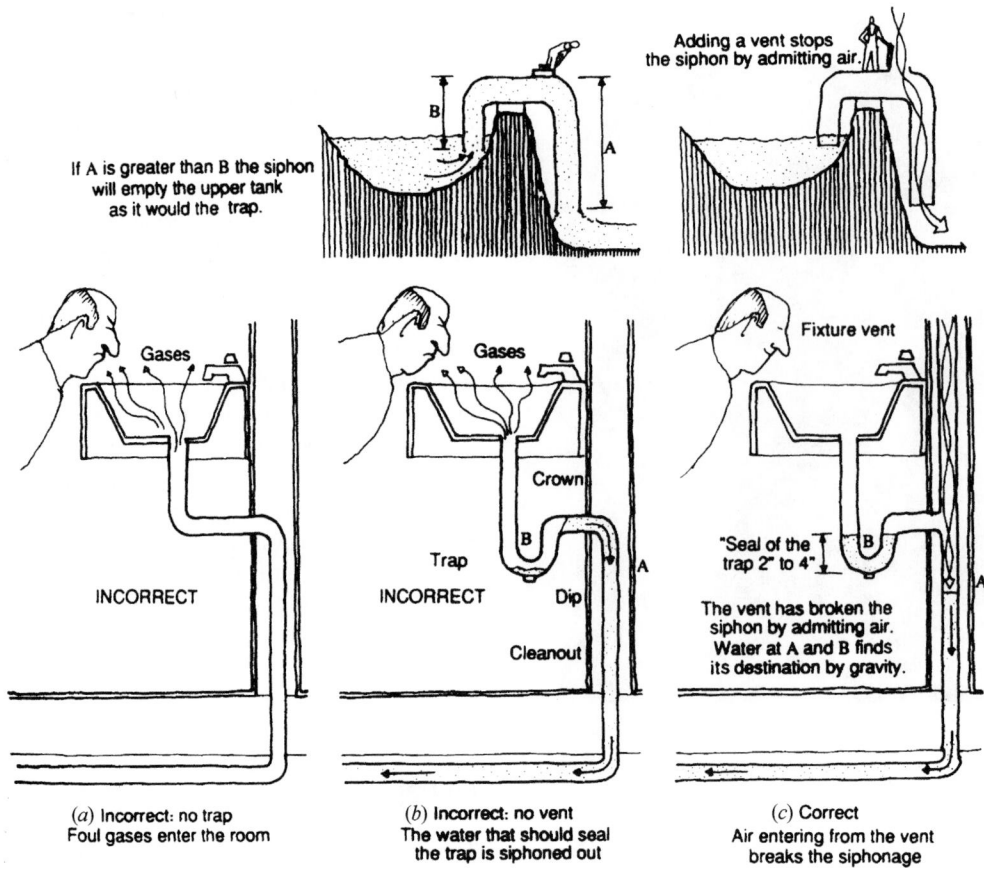

**Fig. 22.8** *Function of a trap and one of the several functions of a vent (preventing siphonage).*

dry trays and sink, the sink is equipped with the trap and set closest to the stack (see stack *b* in Fig. 22.9).

Traps are usually placed within 2 ft (610 mm) of the fixture and should be accessible for cleaning through a bottom opening that is otherwise closed by a plug. Overflow pipes from fixtures are connected to the inlet side of the trap. In long runs of horizontal pipe, so-called running traps are used only near the drains of floors, areas, or yards and should be provided with hand-hole cleanouts. "Island" sinks pose a special problem when the vent line cannot lead up from such an exposed location. The sink's waste line can be taken to a distant sump, which is then itself trapped and vented.

When fixtures are used very infrequently, the water in traps can evaporate into the air, break-

ing the seal of the trap. In contemplating the possible frequency of use, this fact should be kept in mind by the designer. Unoccupied residences (such as weekend or vacation homes) are likely candidates for sewer gas penetration through traps emptied by evaporation. Otherwise, evaporation to a dangerous degree rarely occurs, except in the case of floor drains. Used to carry away the water used in washing floors or drained from heating/cooling equipment, floor drains may often lose the water seal between infrequent operations. Some authorities are reluctant to approve floor drains connected to the building's sewer, requiring instead that they be separately connected to a dry well. In either case, the use of a special hose bibb, affording a source of water directly above the drain, is a wise precaution. It

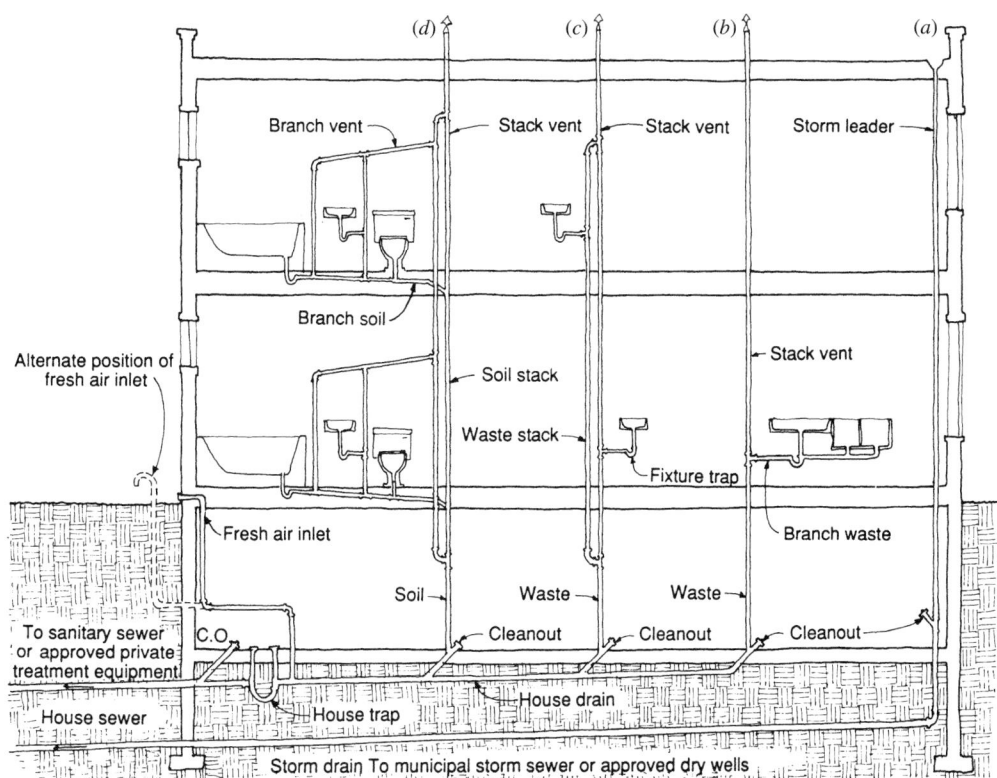

**Fig. 22.9** *Typical sanitary drainage system, which separates stormwater from sewage. This installation combines blackwater (from water closets) and graywater, although their separation would encourage graywater recycling. The house trap is optional under some plumbing codes and illegal under others (because sanitary sewers could be vented at street level through the fresh air inlet).*

can easily be used to manually refill the trap of the drain. Another strategy is to lead overflow lines from lavatories to the floor drain trap.

## (b) Vents

To admit air and discharge gases, soil and waste stacks are extended through roofs, and a system of air vents, largely paralleling the drainage system, is provided. As in the case of drainage stacks, the ventilating stacks extend through the roof or vent through the drainage stack. The functions of venting are often misunderstood. It is true that one important purpose is to ventilate the system by allowing air from the fresh-air inlet (or from the sewer, if there is no house trap or fresh-air inlet) to rise through the system and carry away offensive gases. This provides some purification for the piping. However, several other purposes are served by the vent piping. The

introduction of air near the fixture (and, in the case of circuit vents, at the branch soil line) breaks the possible siphonage of water out of the trap. Under other circumstances—namely, when drainage fluids descend to a fixture group through the soil stack—the foul gases, under pressure, could bubble through the trap seals of that group. The vent system provides a local escape for these gases. *Circuit venting*—which permits air and gases to pass in and out of the soil or waste branch instead of at each fixture, as in the case of continuous venting (individual fixture venting)—can prevent the siphonage of trap-seals or their penetration by gases (Fig. 22.10). Some codes may prohibit circuit venting.

## (c) Air Gaps and Vacuum Breakers

Nearly every plumbing fixture is supplied with pure water at one point, and most discharge contami-

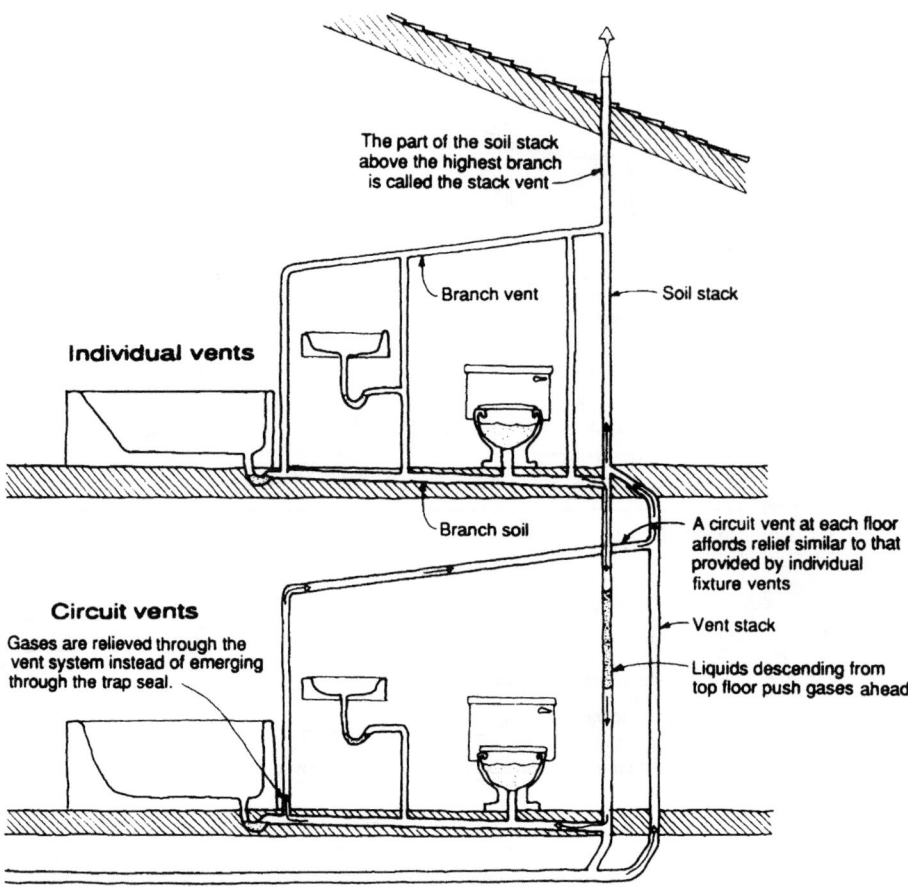

The part of the soil stack above the highest branch is called the stack vent

Branch vent

Soil stack

**Individual vents**

Branch soil

A circuit vent at each floor affords relief similar to that provided by individual fixture vents

Vent stack

**Circuit vents**

Gases are relieved through the vent system instead of emerging through the trap seal.

Liquids descending from top floor push gases ahead.

*Fig. 22.10* Sewer gas relief through vents. Gases pressurized by hydraulic action or by expansion due to putrefaction have an escape path through the vent system and will not enter the rooms. The upper floor shows individual vents (a vent at each fixture); the lower floor shows circuit vents, permitted by some codes, where one vent serves the entire branch.

nated fluids at another. The proximity of sewage to potable water at typical fixtures is inescapable; sewage could accidentally be siphoned into a pipe carrying potable water. Consider an improperly placed faucet whose outlet is below the rim of a fixture. If the fixture overflow is plugged and the fixture bowl full, the faucet can easily project into the foul drainage water. If, in this circumstance, the water piping is drained while the faucet is open, contaminated water could be drawn by suction into the water piping.

In water closets served by flushometers (flush valves), the water supply unavoidably enters the bowl below the rim. A vacuum breaker placed in the flushometer closes with water pressure but opens to admit air if there is suction in the water pipe. This

prevents siphonage in much the same way that a vent prevents trap siphonage (Fig. 22.11). The use of vacuum breakers at dishwashers and clothes washers was diagrammed in Figure 21.52. These are especially important locations because, in both of these appliances, pumps force the wastewater into the drain line.

## 22.3 PIPING, FITTINGS, AND ACCESSORIES

### (a) Piping and Fittings

The materials used within buildings for soil and waste piping and for venting include cast iron, cop-

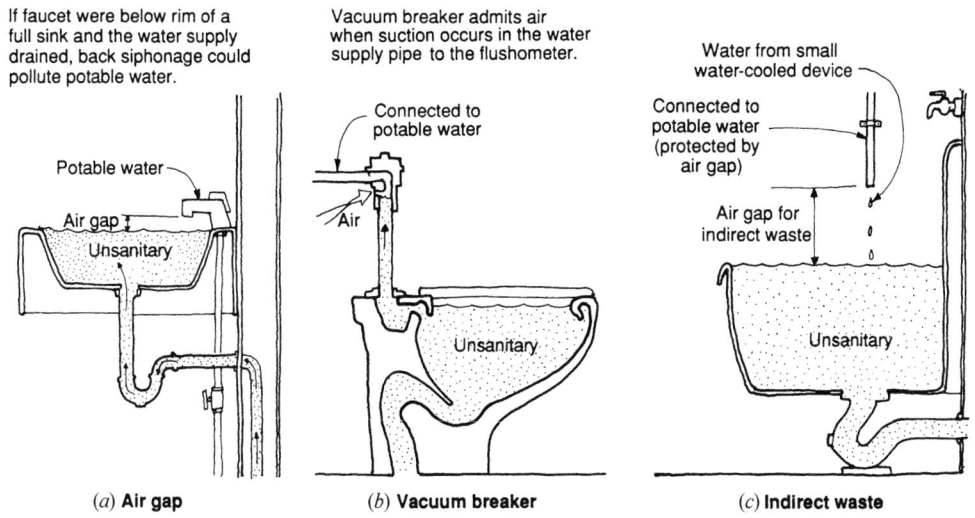

*If faucet were below rim of a full sink and the water supply drained, back siphonage could pollute potable water.*

Potable water

Air gap

Unsanitary

(*a*) **Air gap**

*Vacuum breaker admits air when suction occurs in the water supply pipe to the flushometer.*

Connected to potable water

Air

Unsanitary

(*b*) **Vacuum breaker**

Water from small water-cooled device

Connected to potable water (protected by air gap)

Air gap for indirect waste

Unsanitary

(*c*) **Indirect waste**

**Fig. 22.11** *Backflow preventers. Unsanitary fluid wastes cannot be siphoned into the potable water piping in these three examples.*

per, ABS plastic, and PVC type DWV pipe (Table 22.1, Fig. 22.12). Galvanized steel is sometimes chosen for vents and for tall stacks in high-rise structures. Sometimes, different materials are used in the same system. Where dissimilar metals are connected, dielectric unions are used to prevent corrosion due to electrolysis. Materials used underground for sewage disposal (depending upon local codes) include vitrified clay tile, cast iron, copper, asbestos-cement, ABS plastic, PVC type DWV, and concrete pipe.

***Cast Iron.*** Used first in Germany around 1562 and appearing in the United States about 1813, cast iron supplanted the tubing and culverts of earlier eras that employed clay, lead, bronze, and wood. Cast iron was thus the earliest of the modern materials used for piping. Its durability and resistance to corrosion make it appropriate for a wide range of uses, from small residential work to the stacks and branches of tall buildings.

Typical fittings for sanitary drainage appear in Figs. 22.12 and 22.13. In sanitary flow systems

### TABLE 22.1 Materials for Waste Piping

| Material | Aboveground DWV | Underground Building Drainage and Vent | Building Sewer |
|---|---|---|---|
| ABS (acrylonitrile butadiene styrene) | ✔ | ✔ | ✔ |
| Asbestos-cement | | ✔ | ✔ |
| Brass | ✔ | | |
| Cast-iron | ✔ | ✔ | ✔ |
| Concrete | | | ✔ |
| Copper (type) | ✔ (K, L, M, or DWV) | ✔ (K or L) | ✔ (K or L) |
| Galvanized steel | ✔ | | |
| Glass | ✔ | | |
| Polyolefin | ✔ | ✔ | |
| PVC (polyvinyl chloride) type DWV | ✔ | ✔ | ✔ |
| Vitrified clay | | | ✔ |

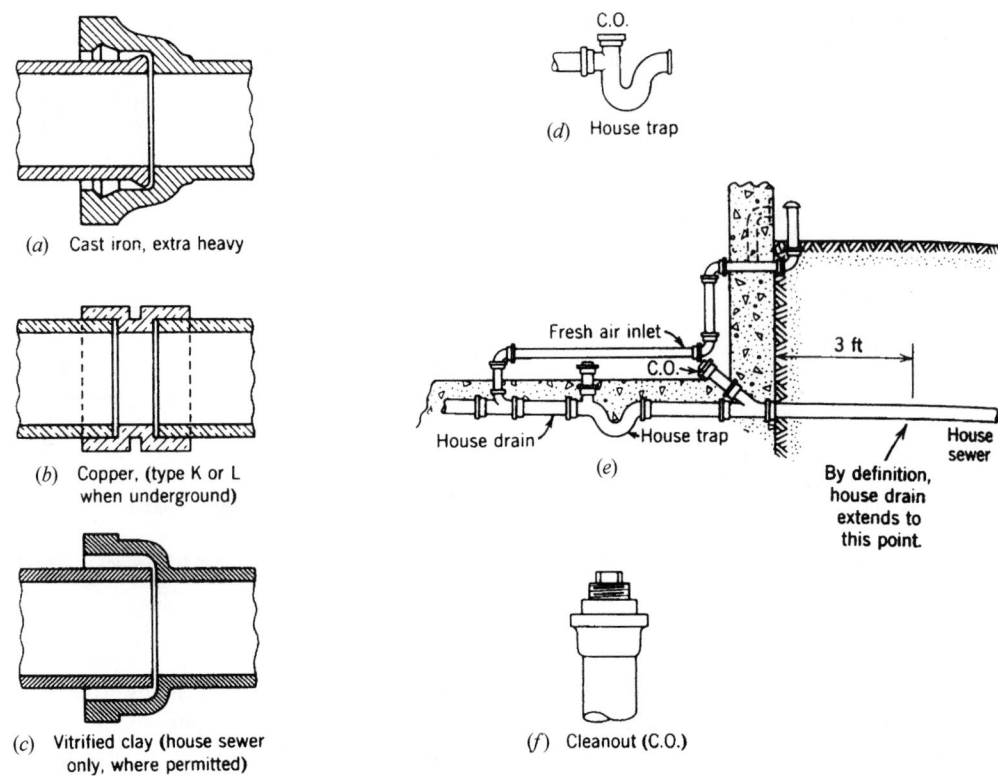

(a) Cast iron, extra heavy

(b) Copper, (type K or L when underground)

(c) Vitrified clay (house sewer only, where permitted)

(d) House trap

(e)

(f) Cleanout (C.O.)

**Fig. 22.12** Piping and fittings. (a) Connection of cast iron piping. (b) Coupling to connect copper tubing. (c) Connection of vitrified clay piping. (d) Detail of house trap fitting. (e) House drain, house trap with cleanouts and vent (fresh air inlet), and house sewer. (f) Cleanout showing removable threaded plug. For large buildings, the terms building drain, building sewer, and so on supplant house. Local codes differ on inclusion or omission of the house trap. Note: plastic pipe connections are similar to that in (b).

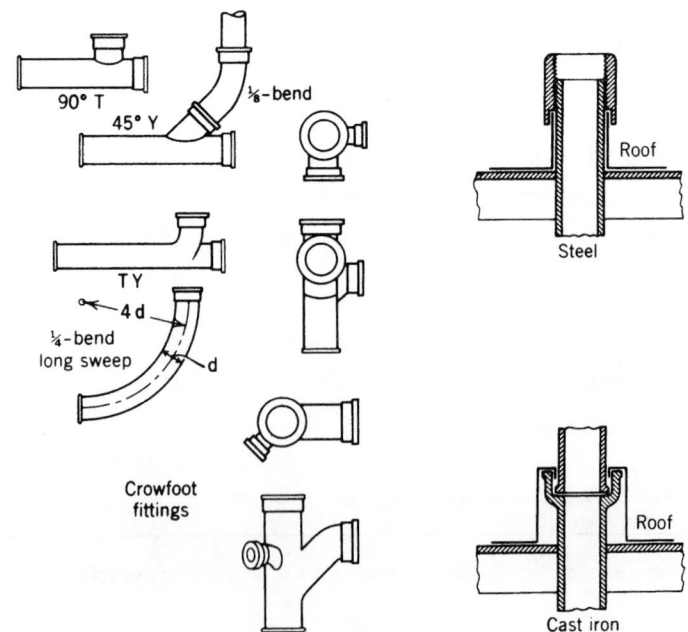

**Fig. 22.13** Principal types of cast-iron fittings and method of flashing at roofs for steel and cast iron.

composed of *any* material, changes in direction must be made with easy bends. To prevent clogging or fouling by the solid materials flowing in the piping, right-angle connections are not used. Thus, the choices in Fig. 22.13 would be for one-eighth bend plus a 45° Y, or a one-quarter bend long sweep. The top connection of the 90° T, in the position shown, would connect *only* to a vent.

The three cast-iron soil pipe joints shown in Fig. 22.14 are semirigid, watertight, and gastight connections of two or more pieces of pipe or fittings in a sanitary system. Types *b* and *c* provide a quieter plumbing system and slightly more flexible joints

than type *a*. See Fig. 22.15 for cast iron soil pipe and fittings in bathroom groups.

***Copper Tube and Fittings for DWV.*** There are several tube classifications for the copper products used in plumbing systems: types K, L, and M are for water supply systems, and type DWV is for drainage, waste, and vent installations. Connections between copper tubing and its couplings or fittings are made by a sliding fit (see Fig. 22.12*b*). Between the mating surfaces is a cylindrical capillary space filled with solder. The process of making the joint is a simple one, utilizing a flux, heat, and solder. Properly made,

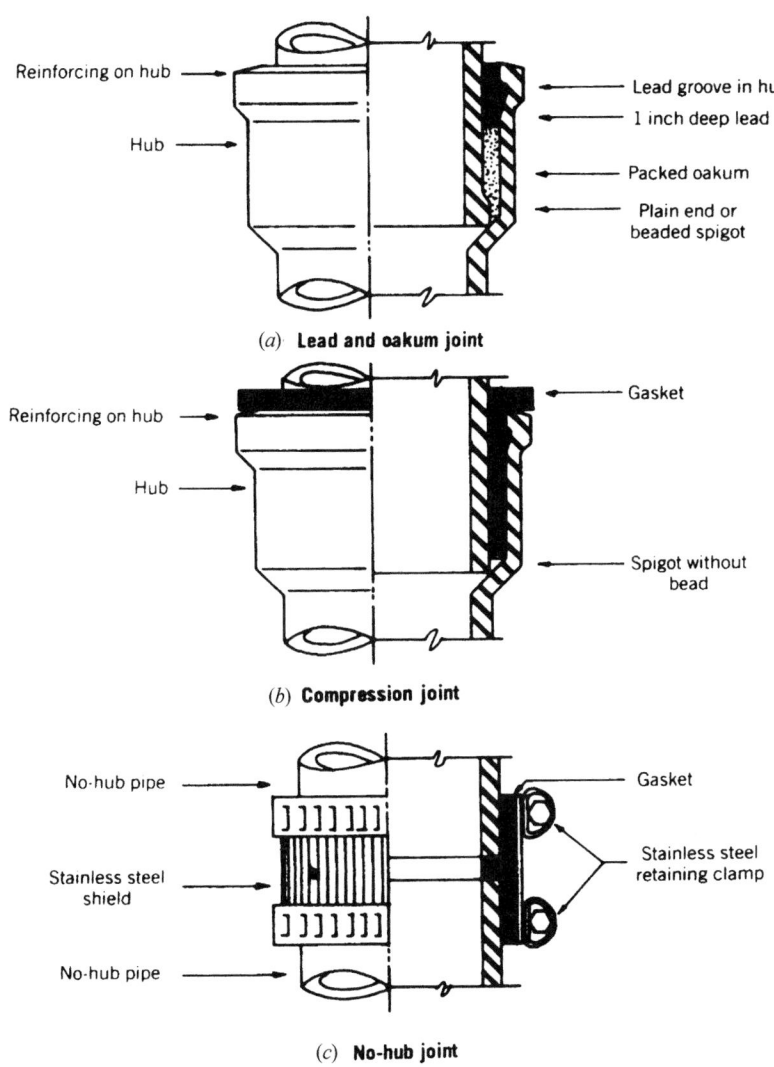

(*a*) **Lead and oakum joint**

(*b*) **Compression joint**

(*c*) **No-hub joint**

**Fig. 22.14** *Various joints used to connect cast-iron soil pipe and fittings. (Courtesy of the Cast Iron Soil Pipe Institute.)*

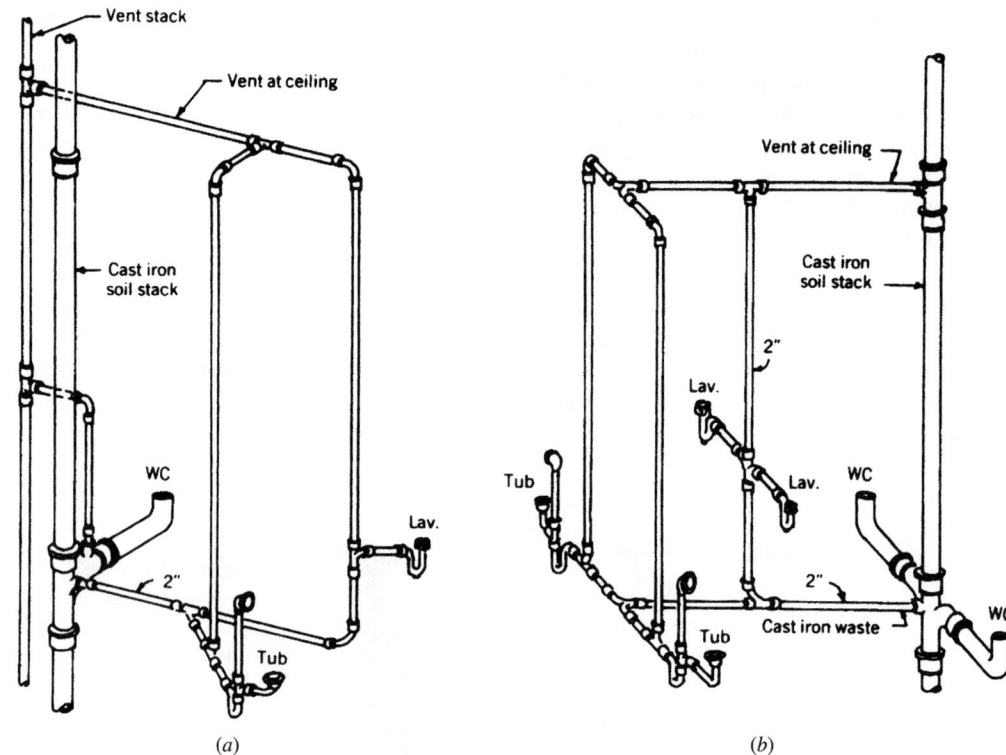

**Fig. 22.15** *Two typical piping arrangements for a water closet, lavatory, and tub. (a) For a multistory installation with each fixture vented; (b) for a single story, back-to-back installation. (Courtesy of the Cast Iron Soil Pipe Institute.)*

the joint will be airtight and capable of withstanding high pressure (although such pressure is not normal in waste lines). To undo the joint for repair or renovation, one simply reheats it until the solder melts. Like cast iron, copper has a history of use in ancient installations. Updated and highly developed in recent decades, it is now in widespread use.

*Plastic Materials for DWV.* Along with copper and cast iron, plastics are suitable for sanitary drainage systems. Table 22.1 lists the plastics suitable for drainage, waste, and vent, as well as building drains and sewers. Acrylonitrile-butadiene-styrene (ABS) is identified and further described by the labeling shown in Fig. 22.16. One of several steps used in making a "solvent-weld" connection is seen in Fig. 22.16, as is a method of support in wood frame construction. Figure 22.17 shows assembled bathroom piping in place. One advantage of plastic piping is its relatively light weight, encouraging some preassembly under better-controlled conditions, then carrying the piping to its place of installation.

## (b) Accessories

Among the many special devices that can form part of a plumbing system are floor drains, backwater valves, ejectors, and interceptors.

*Floor Drains.* When floors in buildings must be washed down after such operations as food preparation and cooking, floor drains are usually necessary. Figure 22.18 shows a typical floor drain. Because these drains are often connected to sanitary drainage systems (rather than dry wells) and, in long periods of disuse, might lose their trap-seals by evaporation, special precautions are necessary to preserve the trap-seal and avoid odors and unsanitary conditions in the room.

*Backwater Valves.* These devices (Fig. 22.19) are sometimes used when plumbing fixtures are installed at low elevations, such as in basements, or in other locations that are near the level of the sewer. They cannot be used to protect the entire

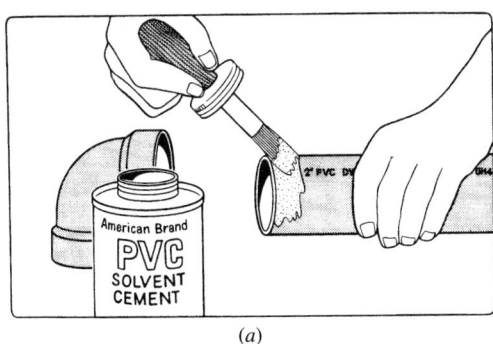

(a)

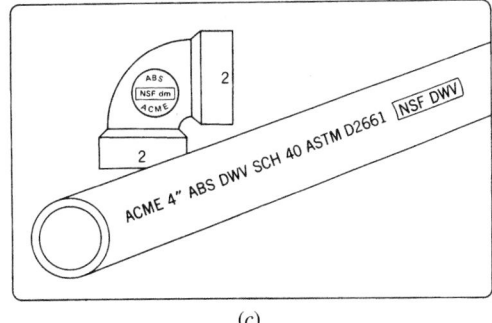

(c)

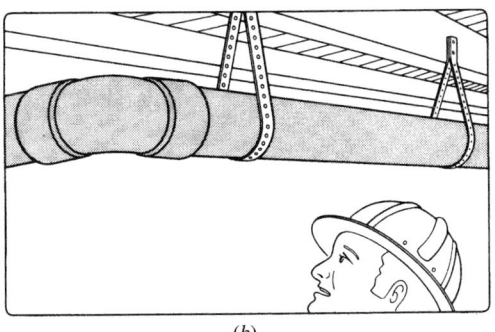

(b)

**Fig. 22.16** Details of the use of plastic pipe. (a) One of the steps in making a solvent weld of a plastic pipe to a plastic fitting. (b) In wood frame construction, plastic pipe assemblies can be supported by metal straps nailed to wood joists. The supports should be more closely spaced than those for metal piping because of plastic's increased flexibility. (c) Typical identification symbols on plastic pipe: ACME—the name of the manufacturer; 4 in.—diameter of the pipe; ABS—Acrylonitrile-butadiene-styrene, the material; DWV—suitable for drainage waste and vent; SCH 40—Schedule 40, this identifies the wall thickness of the pipe; ASTM D2661—the applicable American Society for Testing and Materials standard; NSF DWV—tested by the National Sanitation Foundation Testing Laboratory, the pipe meets or exceeds the current standards for sanitary service. (Courtesy of the Plastic Pipe Institute.)

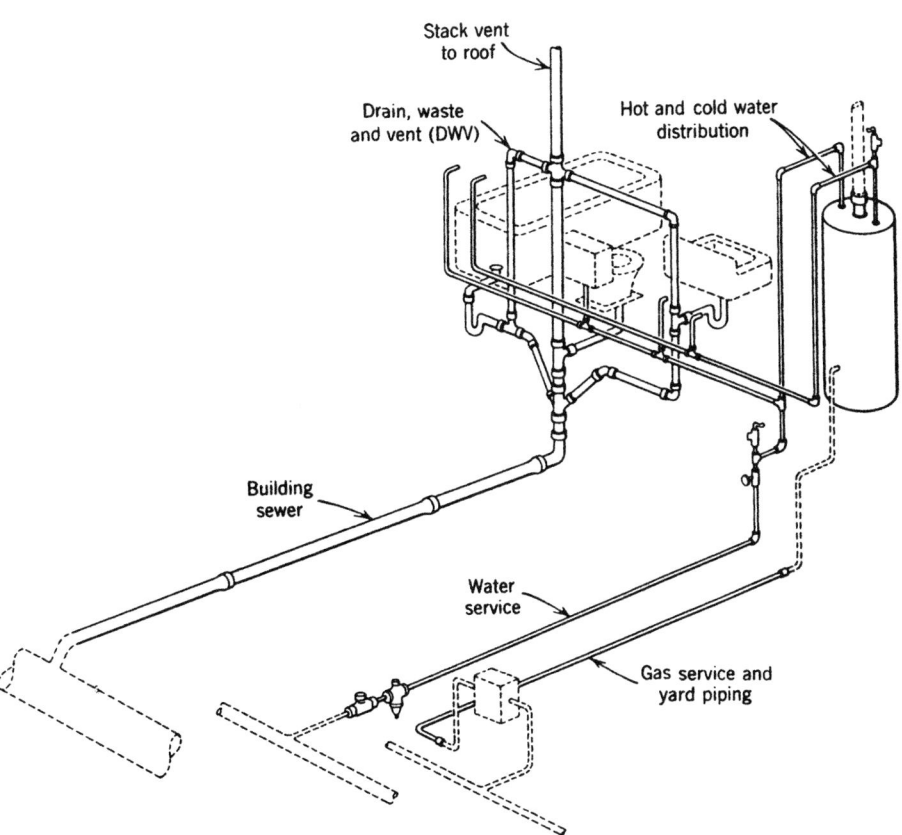

**Fig. 22.17** Plastic piping (solid lines) for water service, gas service, hot and cold water supply, and for drainage, waste, and vent. Gas service below grade can be PE, PB, or PVC. (Courtesy of the Plastic Pipe Institute.)

**995**

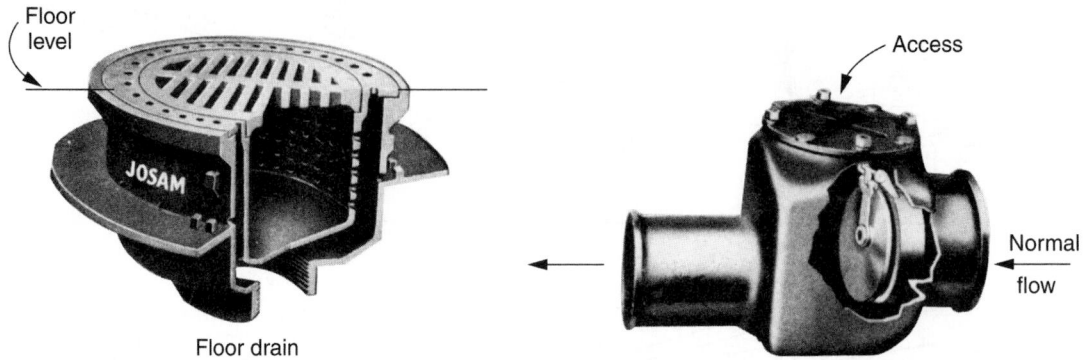

**Fig. 22.18** *Floor drain. (Courtesy of the Josam Manufacturing Company.)*

**Fig. 22.19** *Backwater valve, with access to facilitate unclogging and cleaning.*

plumbing system and should be used only when necessary. They must be accessible for maintenance. An alternative to the use of backwater valves is the sewage sump.

***Sewage Sumps and Ejectors.*** Whenever subsoil drainage, fixtures, or other equipment are situated below the level of public sewers, a sump pit or receptacle must be installed. Into this pit the drainage from the low fixtures may flow by gravity, and from it the contents are then lifted up into the building sewer. Sewer ejectors may be motor-driven centrifugal pumps (Fig. 22.20) or they may be operated by compressed air. The latter have no revolving

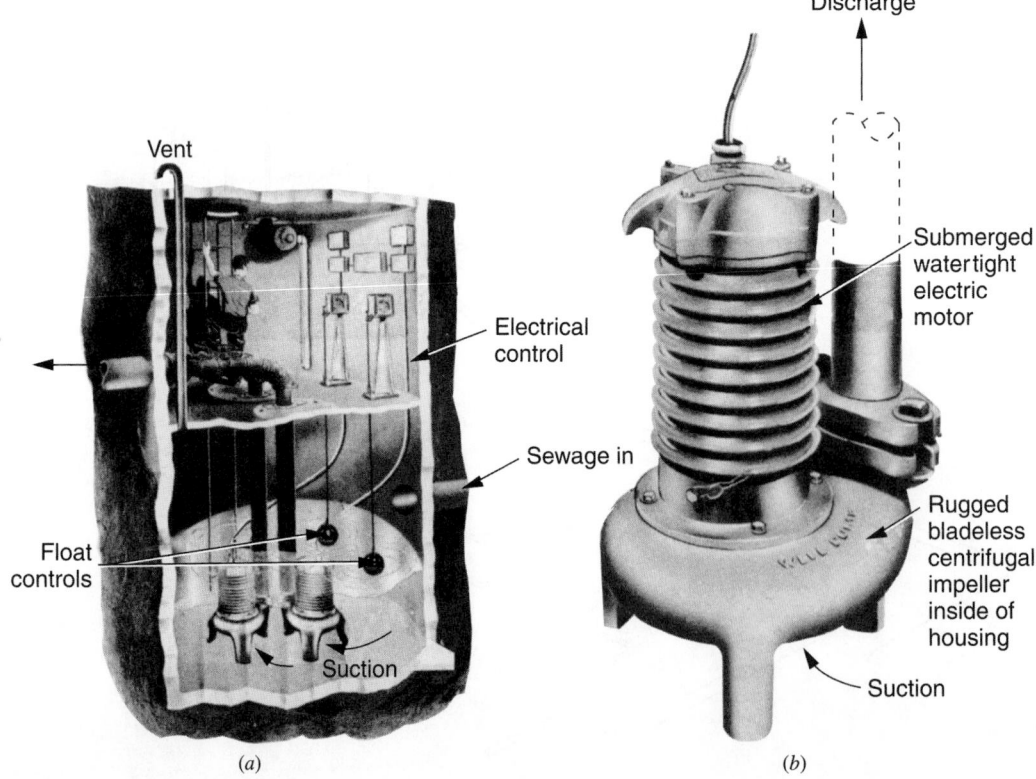

(a)                                   (b)

**Fig. 22.20** *Sump (a) and ejector pump (b). A submersible-type centrifugal pump for raising sewage to a higher level. Shown here for an outdoor subgrade sump installation, it may be used in basement applications within buildings. Venting must be carried to the roof. (Courtesy of the Weil Pump Company.)*

parts within the receptacle. An air compressor is started when the float within the sump reaches a certain level, and air at a pressure greater than 0.433 psi for each foot (10 kPa for each meter) of lift is delivered into the space above the liquid. The air pressure closes the inlet and opens the outlet check valves, expelling the contents of the sump and elevating it to the sewer.

***Interceptors.*** Sanitary drainage installations ultimately discharge their waste matter into private or public sewage treatment plants that attempt to digest or cope with anything that may come through the pipes. From any plumbing fixture to the end of the disposal process, all parts of systems should be openable through cleanouts and other points of access to relieve the clogging that will often occur in the piping (as well as in the septic tank or public disposal plant). Because it is impossible to control what will be discarded into the plumb-

ing drains, trouble can occasionally be expected. The problem can be reduced somewhat by devices known as *interceptors*, which catch foreign matter before it travels too far into the system.

Obviously, interceptors require periodic servicing. Interceptors for as many as 25 different kinds of extraneous material are listed by some manufacturers. They include devices to catch hair, grease, plaster, lubricating oil, glass grindings, or troublesome unwanted material from many industrial processes. One of the few interceptors that is sometimes used in homes, and more often in institutional kitchens, is the grease interceptor, or *grease trap*.

As the waste is passed from a kitchen sink through the circuitous path within the grease interceptor, the grease floats to the top, where it is trapped between baffles, while the more fluid wastes pass through at a lower level (Fig. 22.21). If not so intercepted, grease congeals within piping and thus physically retards the sewage digestion process.

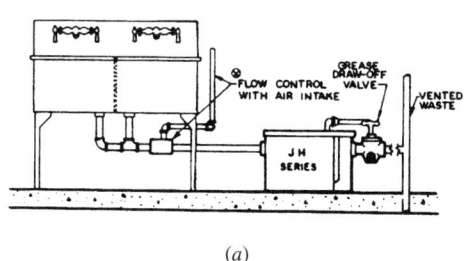

(a)

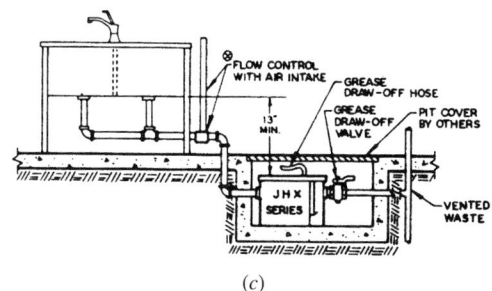

(c)

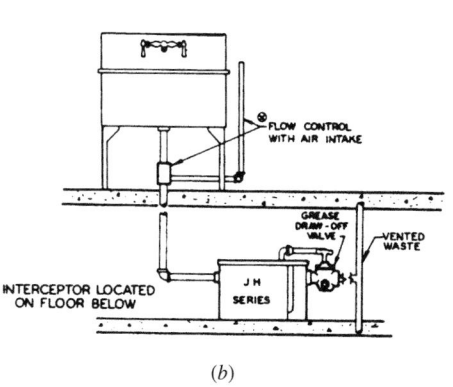

(b)

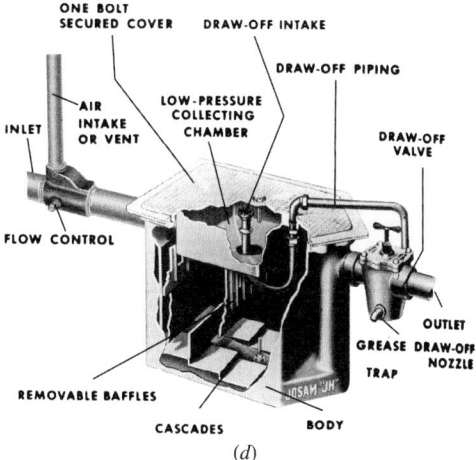

(d)

**Fig. 22.21** One type of institutional interceptor, a grease trap. Choice of three locations—adjacent to the sink (a), on the floor below (b), or in a pit (c). Periodic cleaning is both necessary and unpleasant; easy access is helpful. (d) Cutaway view with identification of component parts. (Courtesy of the Josam Manufacturing Company.)

### TABLE 22.2  Drainage Fixture Units (dfu)

| PART A. BY TYPE OF FIXTURE | | | |
|---|---|---|---|
| Fixture(s) | Drainage Fixture Units (dfu) | Minimum Trap Size in. | mm[a] |
| Automatic clothes washers: Commercial[b] | 3 | 2 | 51 |
| Residential | 2 | 2 | 51 |
| Bathroom group: Water closet, lavatory, bidet and bathtub or shower | 6 | — | — |
| Bathtub[c] (with or without overhead shower or whirlpool) | 2 | 1½ | 38 |
| Bidet | 2 | 1¼ | 32 |
| Combination sink and tray | 2 | 1½ | 38 |
| Dental lavatory | 1 | 1¼ | 32 |
| Dental unit or cuspidor | 1 | 1¼ | 32 |
| Dishwashing machine,[d] domestic | 2 | 1½ | 38 |
| Drinking fountain | ½ | 1¼ | 32 |
| Emergency floor drain | 0 | 2 | 51 |
| Floor drains | 2 | 2 | 51 |
| Kitchen sink, domestic | 2 | 1½ | 38 |
| Kitchen sink, domestic, with food waste grinder and/or dishwasher | 2 | 1½ | 38 |
| Laundry tray (1 or 2 compartments) | 2 | 1½ | 38 |
| Lavatory | 1 | 1¼ | 32 |
| Shower compartment, domestic | 2 | 2 | 51 |
| Sink | 2 | 1½ | 38 |
| Urinal | 4 | e | |
| Urinal, 1 gal (3.8 L) per flush or less | 2[f] | e | |
| Wash sink (circular or multiple) each set of faucets | 2 | 1½ | 38 |
| Water closet, flushometer tank, public or private | 4[f] | e | |
| Water closet, private installation | 4 | e | |
| Water closet, public installation | 6 | e | |

| PART B. BY SIZE OF TRAP | | |
|---|---|---|
| Fixture Drain or Trap Size in. | mm[a] | Drainage Fixture Unit (dfu) Value |
| 1¼ | 32 | 1 |
| 1½ | 38 | 2 |
| 2 | 51 | 3 |
| 2½ | 64 | 4 |
| 3 | 76 | 5 |
| 4 | 102 | 6 |

*Source:* Based on the *International Plumbing Code.* © 1997, International Code Council, Inc., Falls Church, VA. Reprinted with permission. All rights reserved.

[a]SI conversions provided by the author.

[b]For traps larger than 3 in. (76 mm), see Part B of this table.

[c]A showerhead over a bathtub or whirlpool bathtub attachments do not increase the dfu value.

[d]For ratings of fixtures not listed or for intermittent flows, see the 1997 *International Plumbing Code.*

[e]Trap size shall be consistent with fixture outlet size.

[f]For computing the loads on building drains and sewers, water closets or urinals shall not be rated at a lower dfu unless the lower values are confirmed by testing.

## 22.4 DESIGN OF RESIDENTIAL WASTE PIPING

In residential work, the piping assemblies may often be viewed as a "flag," where the mast is the soil stack, the horizontal top of the flag is the branch vent, the bottom is the soil or waste branch, and the outer edge is the vertical pipe of the last fixture. In frame construction, the flag usually fits into a 6-in. (150-mm) partition. Fixture branches project from the surface of the flag. There is considerable advantage in back-to-back planning of baths and kitchens; this allows the piping assembly to pick up the drainage of fixtures on both sides of it. When all the fixtures are on nearly the same level, it is unnecessary to have a separate vent stack standing beside the soil stack, as is often the case in multistory construction. Generally in one-story construction, the upper part of the soil stack forms a vent called a *stack vent*, to which the branch vents connect. A separate major vertical vent is called a *vent stack*.

The task of fitting the two plumbing "distribution trees" (supply and waste) into available horizontal and vertical chases can be difficult. This is especially true of the waste system, which has larger pipes, because it is not under pressure and must carry solids as well as liquid. To compound the problem, because of gravity flow requirements, these larger pipes must slope continuously down, from fixtures to the sewer. In residences, the "vertical chase" is sometimes nothing more than an extra-wide wall, increasing the designer's difficulty.

In residential applications and in other relatively small buildings, certain fairly standard minimum sizes (such as a 4-in. [102-mm] soil stack and building drain) are usually adequate. Horizontal fixture branches from *individual* fixtures should be the same size as the fixture trap. Horizontal fixture branches from *groups* of fixtures are sized by the drainage fixture units of the group. For individual fixture vents, the vent size is usually the same as the size of the fixture's horizontal branch. Under many codes, vertical vents that penetrate the roof must increase to a 4-in. (102-mm) size to prevent blocking by icing in freezing weather.

Tables 22.2 through 22.5 list minimum pipe sizes to carry waste and serve for venting.

### TABLE 22.3 Horizontal Fixture Branches and Stacks[a]

| Diameter of Pipe | | Horizontal Branch | Maximum Total Number of dfu Allowable | | |
|---|---|---|---|---|---|
| | | | Stacks[b] | | |
| in. | mm[c] | | One Branch Interval | Three Branch Intervals or Less | Greater than Three Branch Intervals |
| 1½ | 38 | 3 | 2 | 4 | 8 |
| 2 | 51 | 6 | 6 | 10 | 24 |
| 2½ | 64 | 12 | 9 | 20 | 42 |
| 3 | 76 | 20 | 20 | 48 | 72 |
| 4 | 102 | 160 | 90 | 240 | 500 |
| 5 | 127 | 360 | 200 | 540 | 1100 |
| 6 | 152 | 620 | 350 | 960 | 1900 |
| 8 | 203 | 1400 | 600 | 2200 | 3600 |
| 10 | 254 | 2500 | 1000 | 3800 | 5600 |
| 12 | 305 | 3900 | 1500 | 6000 | 8400 |
| 15 | 381 | 7000 | d | d | d |

*Source: Based on the International Plumbing Code.* © 1997, International Code Council, Inc., Falls Church, VA. Reprinted with permission. All rights reserved.

[a]Does not include branches of the building drain; see Table 22.5.

[b]Stacks shall be sized based on the total accumulated connected load at each story or branch interval. As the total accumulated connected load decreases, stacks may be reduced in size. Stack diameters shall not be reduced to less than one-half the diameter of the largest stack size required.

[c]SI conversions provided by the author.

[d]Sizing load based on design criteria.

## TABLE 22.4  Size and Developed Length of Stack Vents and Vent Stacks

| Diameter of Soil or Waste Stack in. (mm)[b] | Total Fixture Units Being Vented (dfu) | Maximum Developed Length[a] of Vent, feet (m)[b] Diameter of Vent, in. (mm)[b] | | | | | | | | | |
|---|---|---|---|---|---|---|---|---|---|---|---|
| | | 1¼ (32) | 1½ (38) | 2 (51) | 2½ (64) | 3 (76) | 4 (102) | 5 (127) | 6 (152) | 8 (203) | 10 (254) |
| 1¼ (32) | 2 | 30 / *9.1* | | | | | | | | | |
| 1½ (38) | 8 | 50 / *15.2* | 150 / *45.7* | | | | | | | | |
| 1½ (38) | 10 | 30 / *9.1* | 100 / *30.5* | | | | | | | | |
| 2 (51) | 12 | 30 / *9.1* | 75 / *22.9* | 200 / *61.0* | | | | | | | |
| 2 (51) | 20 | 26 / *7.9* | 50 / *15.2* | 150 / *45.7* | | | | | | | |
| 2½ (64) | 42 | | 30 / *9.1* | 100 / *30.5* | | | | | | | |
| 3 (76) | 10 | | 42 / *12.8* | 150 / *45.7* | 360 / *109.7* | 1040 / *317* | | | | | |
| 3 (76) | 21 | | 32 / *9.8* | 110 / *33.5* | 270 / *82.3* | 810 / *246.9* | | | | | |
| 3 (76) | 53 | | 27 / *8.2* | 94 / *28.7* | 230 / *70.1* | 680 / *207.3* | | | | | |
| 3 (76) | 102 | | 25 / *7.6* | 86 / *26.6* | 210 / *64.0* | 620 / *189.0* | | | | | |
| 4 (102) | 43 | | | 35 / *10.7* | 85 / *25.9* | 250 / *76.2* | 980 / *298.7* | | | | |
| 4 (102) | 140 | | | 27 / *8.2* | 65 / *19.8* | 200 / *61.0* | 750 / *228.6* | | | | |
| 4 (102) | 320 | | | 23 / *7.0* | 55 / *16.8* | 170 / *51.8* | 640 / *195.0* | | | | |
| 4 (102) | 540 | | | 21 / *6.4* | 50 / *15.2* | 150 / *45.7* | 580 / *176.8* | | | | |
| 5 (127) | 190 | | | | 28 / *8.5* | 82 / *25.0* | 320 / *97.5* | 990 / *301.8* | | | |
| 5 (127) | 490 | | | | 21 / *6.4* | 63 / *19.2* | 250 / *76.2* | 760 / *231.6* | | | |
| 5 (127) | 940 | | | | 18 / *5.5* | 53 / *16.2* | 210 / *64.0* | 670 / *204.2* | | | |
| 5 (127) | 1400 | | | | 16 / *4.9* | 49 / *14.9* | 190 / *57.9* | 590 / *179.8* | | | |
| 6 (152) | 500 | | | | 33 / *10.1* | 130 / *39.6* | 400 / *121.9* | 1000 / *304.8* | | | |
| 6 (152) | 1100 | | | | 26 / *7.9* | 100 / *30.5* | 310 / *94.5* | 780 / *237.7* | | | |
| 6 (152) | 2000 | | | | 22 / *6.7* | 84 / *25.6* | 260 / *79.2* | 660 / *201.2* | | | |
| 6 (152) | 2900 | | | | 20 / *6.1* | 77 / *23.5* | 240 / *73.2* | 600 / *182.9* | | | |
| 8 (203) | 1800 | | | | | 31 / *9.4* | 95 / *29.0* | 240 / *73.2* | 940 / *286.5* | | |
| 8 (203) | 3400 | | | | | 24 / *7.3* | 73 / *22.3* | 190 / *57.9* | 720 / *219.5* | | |
| 8 (203) | 5600 | | | | | 20 / *6.1* | 62 / *18.9* | 160 / *48.8* | 610 / *185.9* | | |
| 8 (203) | 7600 | | | | | 18 / *5.5* | 56 / *17.1* | 140 / *42.7* | 560 / *170.7* | | |
| 10 (254) | 4000 | | | | | | 31 / *9.4* | 78 / *23.8* | 310 / *94.5* | 960 / *292.6* | |
| 10 (254) | 7200 | | | | | | 24 / *7.3* | 60 / *18.3* | 240 / *73.2* | 740 / *225.6* | |
| 10 (254) | 11,000 | | | | | | 20 / *6.1* | 51 / *15.5* | 200 / *61.0* | 630 / *192.0* | |
| 10 (254) | 15,000 | | | | | | 18 / *5.5* | 46 / *14.0* | 180 / *54.9* | 570 / *173.7* | |

[a]The developed length is measured from the vent connection to the open air.

[b]SI conversions provided by the author.

## TABLE 22.5 Building Drains and Sewers

| Diameter of Pipe | | Maximum Number of dfu Connected to Any Portion of the Building Drain or Building Sewer, Including Branches of the Building Drain[a] Fall, in. per ft (% slope) | | | |
|---|---|---|---|---|---|
| (in.) | (mm)[b] | 1/16 (0.5%) | 1/8 (1.04%) | 1/4 (2.1%) | 1/2 (4.2%) |
| 2 | 51 | | | 21 | 26 |
| 2½ | 64 | | | 24 | 31 |
| 3 | 76 | | 36 | 42 | 50 |
| 4 | 102 | | 180 | 216 | 250 |
| 5 | 127 | | 390 | 480 | 575 |
| 6 | 152 | | 700 | 840 | 1,000 |
| 8 | 203 | 1,400 | 1,600 | 1,920 | 2,300 |
| 10 | 254 | 2,500 | 2,900 | 3,500 | 4,200 |
| 12 | 305 | 2,900 | 4,600 | 5,600 | 6,700 |
| 15 | 381 | 7,000 | 8,300 | 10,000 | 12,000 |

*Source:* Based on the *International Plumbing Code.* Copyright © 1997, International Code Council, Inc., Falls Church, VA. Reprinted with permission. All rights reserved.
[a]The minimum size of any building drain serving a water closet shall be 3 in. (76 mm).
[b]SI conversions provided by the author.

**EXAMPLE 22.1** Design, lay out, and size the piping for the sanitary drainage system for the house shown in Fig. 22.22.

### SOLUTION

The first step is to identify the locations where hot and cold water is needed at fixtures and where soil or waste drains must be provided. Figure 22.23*a* illustrates how this is done. A plan layout for the drains in both levels follows (Fig. 22.23*b*).

Next comes the plumbing section (Fig. 22.24). The local administrative authority usually requires this to

be submitted for approval. Sizes of all piping are determined from Tables 22.2 through 22.5. Drainage fixture units (dfu) for this system are summarized in Table 22.6 from data given in Table 22.2. ∎

Although Table 22.3 permits a 3-in. (76-mm) branch for up to 20 dfu, it is common practice to use a 4-in. (102-mm) branch for every water closet (or every group of water closets totaling 20 dfu or less). Because Table 22.3 allows 160 dfu for a 4-in. (102-mm) branch, this size is acceptable for any branch to which a water closet is connected. This size is also more than adequate for the stack in this example.

A water closet must have a 2-in. (51-mm) vent. Table 22.4 shows that a 2-in. (51-mm) vent, for developed vent lengths not exceeding 150 ft (45.7 m), will serve 20 dfu. This size is acceptable for all branch vents in this example.

The house drain should not be less than 4 in. (102 mm). From Table 22.5, at a ¼-in. fall per foot (2.1% slope), a 4-in. (102-mm) house drain will carry 216 dfu, which is more than adequate for our 26-dfu system.

There are prefabricated bathrooms in which one manufacturer assembles the piping for prese-lected fixtures (Fig. 22.25). Going further, there are a few examples of entirely one-piece bathrooms, which incorporate the maintenance advantage of having no seams between fixtures, walls, and floors.

## TABLE 22.6 Drainage Fixture Units (dfu) for Example 22.1

| Residential Fixture | Drainage Fixture Units (dfu) |
|---|---|
| Bar sink | 2 |
| Kitchen sink with dishwasher | 2 |
| Lavatory | 1 |
| Water closet | 4 |
| Automatic clothes washer | 2 |
| Master bathroom group | 6 |
| Extra lavatory | 1 |
| Shower | 2 |
| Lower floor bathroom group | 6 |
| **Total** | **26** |

Dfu values are from Table 22.2. The hose bibb drains to the ground; roof drainage is taken to dry wells.

WATER AND WASTE

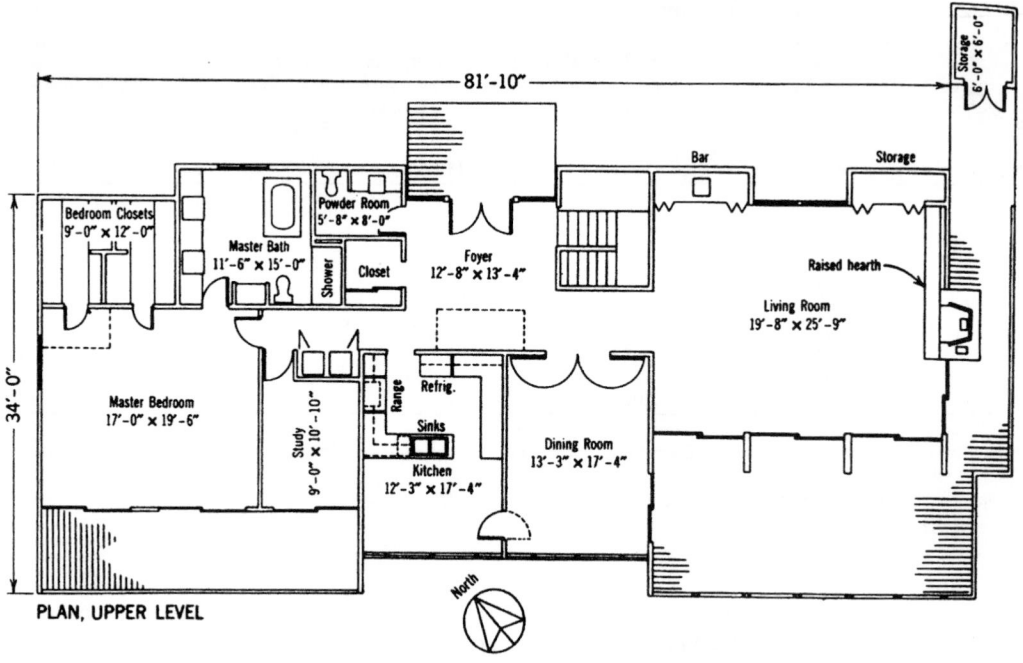

PLAN, UPPER LEVEL

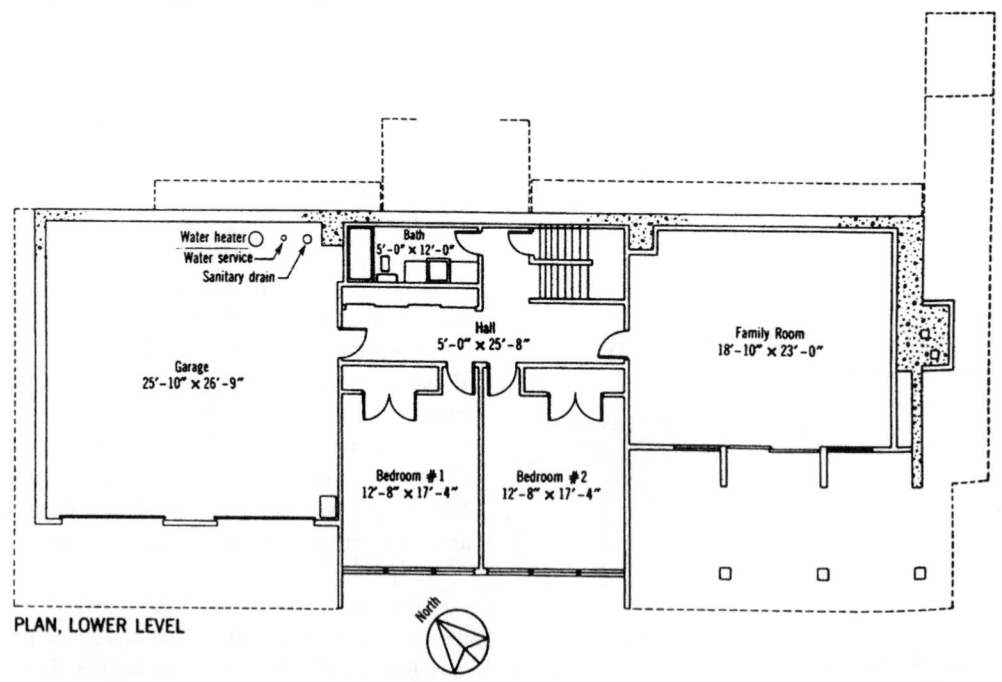

PLAN, LOWER LEVEL

**Fig. 22.22** Example 22.1. Floor plans of a house on Long Island, New York. (Budd Mogesen, architect and planner.)

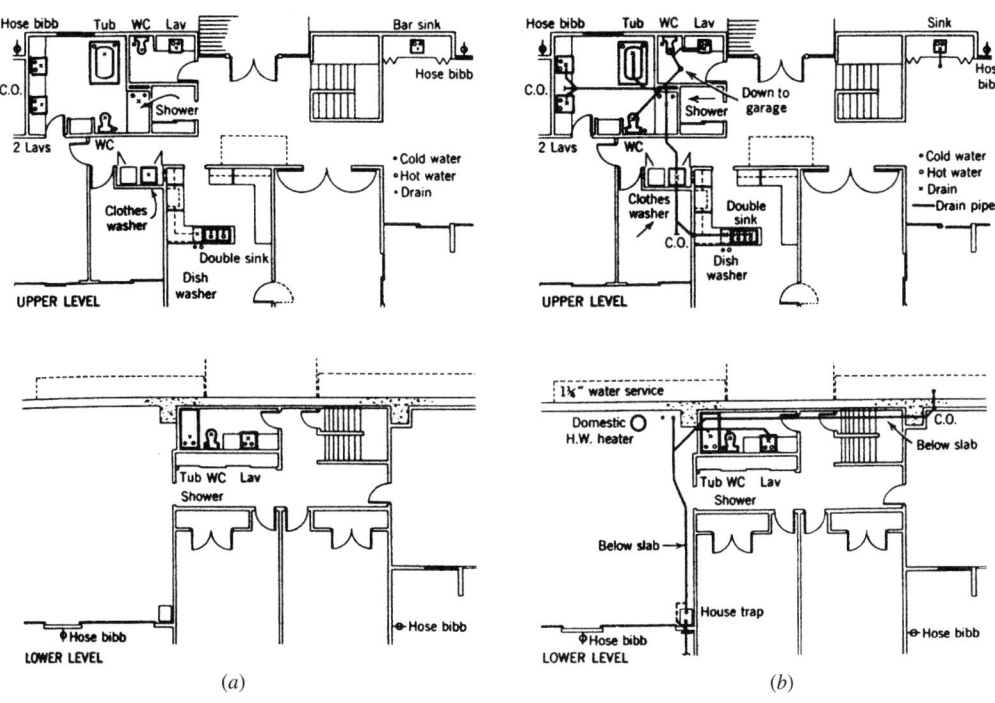

**Fig. 22.23** Example 22.1. (a) Plumbing requirements, water supply, and partial sanitary drainage. (b) Sanitary drainage plan.

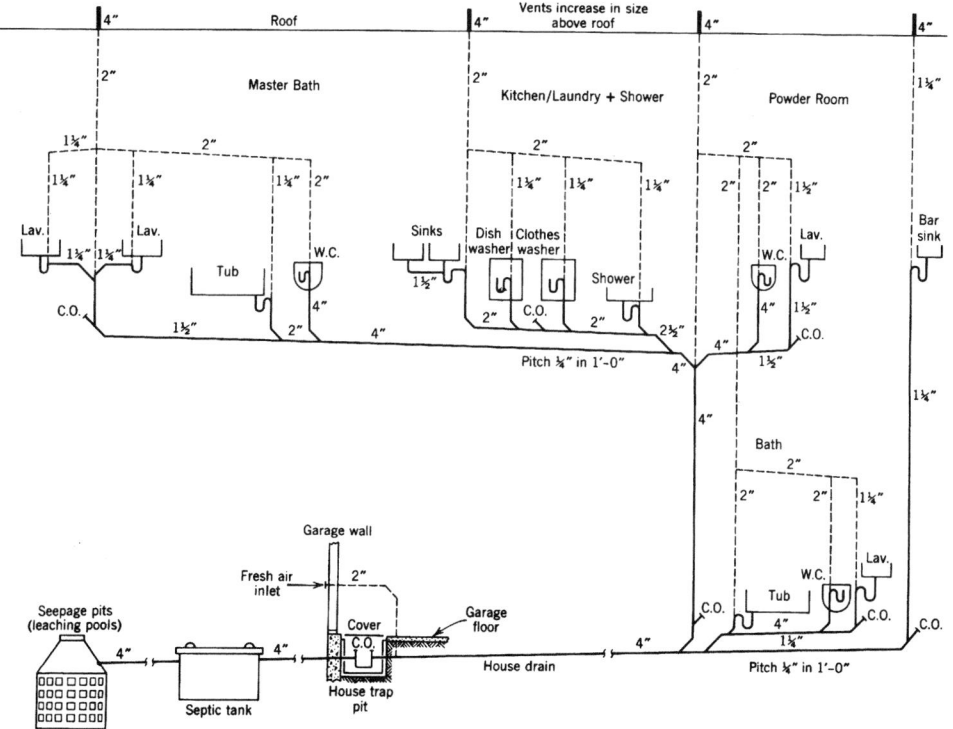

**Fig. 22.24** Example 22.1. Plumbing section. When every fixture is vented individually, as in this example, the method is known as continuous venting. In larger systems, batteries of fixtures may be vented by a loop or circuit vent. This reduces the amount of piping in the vent system (see Figs. 22.10 and Fig. 22.30). Note that septic tank discharge to a leaching pit is not permitted in many locations; see Section 22.6.

**Fig. 22.25** *Preassembled plumbing tree for manufactured housing, along with the prefabricated kitchen/bath in which it is installed. (Courtesy of Wausau Homes, Inc., and the Copper Development Association.)*

## 22.5 DESIGN OF LARGER-BUILDING WASTE PIPING

### (a) Basic Planning

Multistory construction, especially in office buildings, is often designed to be flexible and free of random partitions that would interfere with the periodic renovation of interior spaces and the relocating of dividing partitions. Building "cores" contain elevators, stairs, and shafts for plumbing, mechanical, and electrical equipment and are often placed in the central section of the building, freeing the surrounding areas for access to daylight. A hole in the floor for each pipe is often chosen in preference to a slot or shaft. This method usually interferes less with the floor construction (Fig. 22.26).

This, of course, makes fixture replacement expensive and difficult, and tends to put the burden of access to plumbing on adjacent rooms.

For the more ordinary bathroom, the two maintenance questions of greatest concern are (1) how easy is cleaning around and within the fixture? and (2) how accessible are those parts of the fixture most likely to need repair or replacement? Ease of cleaning is often determined more by the space around the fixture than by the design of the fixture itself. This is particularly true of toilets, where a generous amount of open floor on either side makes maintenance easy and therefore likely to be more frequent. Access to fixture parts may be more difficult. An access panel in the wall of the room behind the fixtures is often provided, encouraging speedy repair and replacement for fixtures at tubs, showers, and lavatories. Ideally, accessibility to all plumbing lines—whether via access panels in walls, trenches in concrete floors, exposed basement ceilings, or adequately deep crawl spaces—should be provided. The bathroom is likely to undergo thorough remodeling, including fixture replacement, as styles change and water/energy conservation becomes more important.

**Fig. 22.26** *Plumbing risers in a fireproof multistory building. Pipes, tubes, conduits, and ducts adjoin toilet rooms and utility spaces. Ventilation ducts and a master 5-in. (127-mm) copper hot-water riser are just to the left of center. Soil and vent stacks with hot and cold water supplies, all of copper, are to the right of this group. At the left, in a lighter tone, are the galvanized steel feeder conduit and distribution circuit conduits for a local electrical control panel box. Note that some pipes and tubes are supported at this floor by bolted clamps. After testing and before pipes are enclosed, covering will be completed. (Courtesy of Copper Development Association.)*

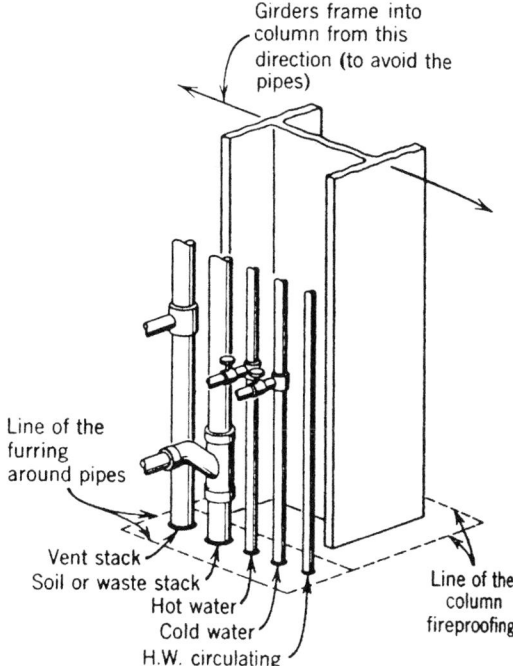

Girders frame into
column from this
direction (to avoid the
pipes)

Line of the
furring
around pipes

Vent stack
Soil or waste stack
Hot water
Cold water
H.W. circulating

Line of the
column
fireproofing

*Fig. 22.27* *Piping at a "wet" column. In large office buildings, there are usually several of these, remote from the core, in the general office area.*

Offices often need a single lavatory or a complete toilet room for executives at locations away from the central core of the building. The greater the horizontal distance from the core, the more vertical clearance will be needed to allow the drain to slope. When such vertical clearance becomes difficult, "wet" columns with a full complement of plumbing pipes offer a solution. If the pipes are to accompany a column in a steel building, structural coordination must be sought early in the planning if the pipes are to clear the structural framing of the floor (Fig. 22.27).

In some installations, the branch soil and waste piping perforates a floor and crosses below the slab to join the stack. Tubing has been developed, however, that sits above the structural slab, obviating the need for hung ceilings below (Fig. 22.28). A lightweight concrete fill is cast to cover the tubing, raising the floor by 5 or 6 in. (125 or 150 mm). This can create a raised floor in the toilet room—with attendant access problems—so the higher floor level is usually carried throughout the floor of the entire story, forming a convenient space into which the electrical conduit can be placed at a time later than would have been required if it were

to be placed in the structural slab. Such raised floors are becoming more common, along with the HVAC strategy of underfloor air distribution (see Section 10.4c).

### (b) Roughing-In

This is the process of getting all pipes installed, capped, and pressure-tested before the fixtures are installed. An example of roughing-in for an office building was shown in Fig. 22.28. The roughing-in of supply and waste piping for school lavatories is shown in Fig. 22.29.

Institutions such as schools have extensive requirements for durability and ease of maintenance. The fixtures are made of such resistant materials as stainless steel, chrome-plated cast brass, precast stone or terrazzo, or high-impact fiberglass. The fixture controls are designed to withstand heavy use—or misuse—and the fixtures are securely tied into the structure with concealed mounting hardware designed to resist extraordinary forces. (Some schools even move the lavatories into the hallway for better visual control of at least a part of the restroom facilities.)

(a)

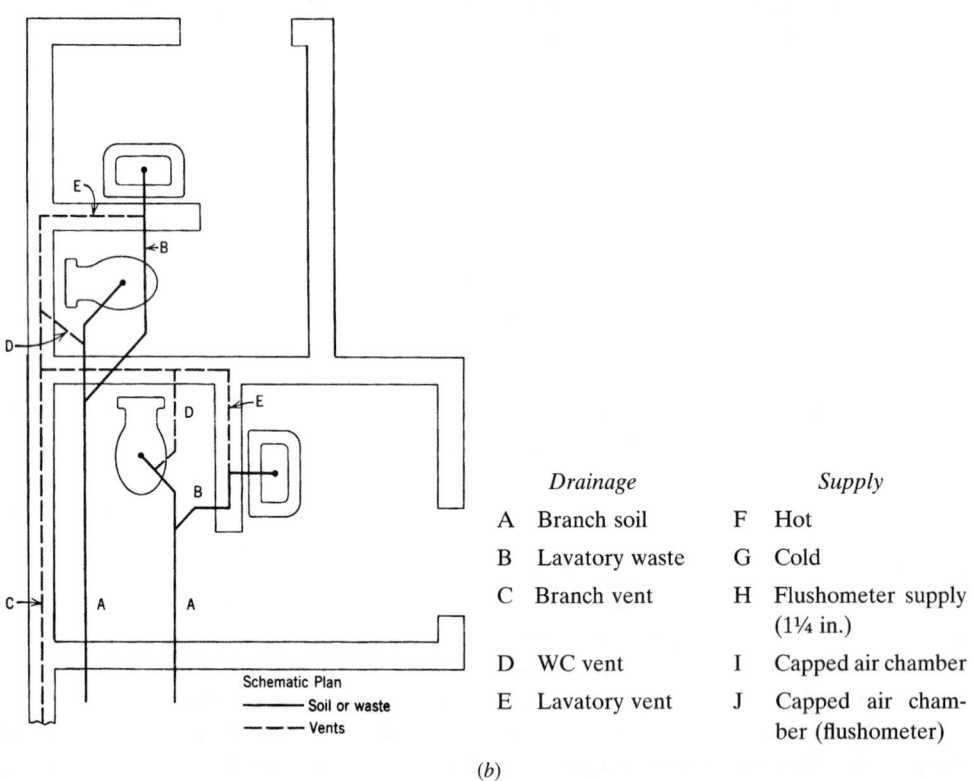

| Drainage | | Supply | |
|---|---|---|---|
| A | Branch soil | F | Hot |
| B | Lavatory waste | G | Cold |
| C | Branch vent | H | Flushometer supply (1¼ in.) |
| D | WC vent | I | Capped air chamber |
| E | Lavatory vent | J | Capped air chamber (flushometer) |

Schematic Plan
——— Soil or waste
- - - Vents

(b)

**Fig. 22.28** *(a) Horizontal waste piping above the structural slab, in an example of plumbing roughing for two lavatory rooms in a concrete office building. A lavatory and water closet (WC) in each room (see plan in b) are served by soil and waste branches below and vent branches above. Hot and cold water tubing with air chambers can be seen. Although the extensions of the water tubing above the two flushometer connections appear to connect into the horizontal vent branches, they actually do not; they are capped and merely touch the bottoms of the vent branches. Note that soil branches are above the structural slab. A fill of 5 or 6 in. (125 to 150 mm) is necessary to cover the tubing. All vertical tubing will be within the masonry block used to enclose the cubicles. (Courtesy of the Copper Development Association.)*

**Fig. 22.29** Roughing-in for four lavatories in the toilet room of a school. The waste branches have been capped and the system tested for leakage. The waste branches are cast iron, the vents galvanized steel, and the water supply lines are copper with soldered fittings. At the center of the supply array are vertical capped expansion and shock tubes, one for hot and one for cold supply water.

In prisons, extreme measures are taken to prevent plumbing fixtures from becoming weapons. Heavy-gauge stainless steel fixtures with nonremovable fittings are provided, at very high cost, for both the fixture and its tamperproof installation.

**EXAMPLE 22.2** Select sizes for drainage and vent piping for the plumbing in an office building for which the fixtures are shown in Fig. 22.30.

**SOLUTION**

Selected dfu values from Table 22.2 are applied to each section of the piping and totaled for each branch and stack, as well as for the building drain and the building sewer. Individual fixture branches should not be less than the size indicated in Table 22.3 for the minimum size of trap for each fixture. An example of a fixture-unit summary and sample sizes of individual branches that connect into a typical branch of the men's toilet group on any floor are shown in the following table.

| Features | Units per Fixture | Total Fixture Units | Diameter, Fixture Branch | |
|---|---|---|---|---|
| | | | in. | mm |
| 1 service sink (2-in. trap) | 3 | 3 | 3 | 76 |
| 3 lavatories | 1 | 3 | 3 | 76 |
| 3 urinals, washout | 4 | 12 | 2 | 51 |
| 3 water closets, valve operated | 6 | 18 | 4 | 102 |
| **Total fixture units,** men's toilet branch | | **36** | | |

Table 22.3 indicates that a 3-in. (76-mm) horizontal fixture branch is inadequate for the above group because it will handle only 20 dfu. Therefore, a 4-in. (102-mm) pipe is selected (also a wise minimum choice whenever WCs are served). Its capacity of 160 dfu will be more than enough for the 36 fixture units needed here. The same table shows that the soil stack diameter can be 4 in. (102 mm), and it is run thus for its entire height. Its capacity of 90 dfu per story (or interval) is sufficient for the 64 dfu that

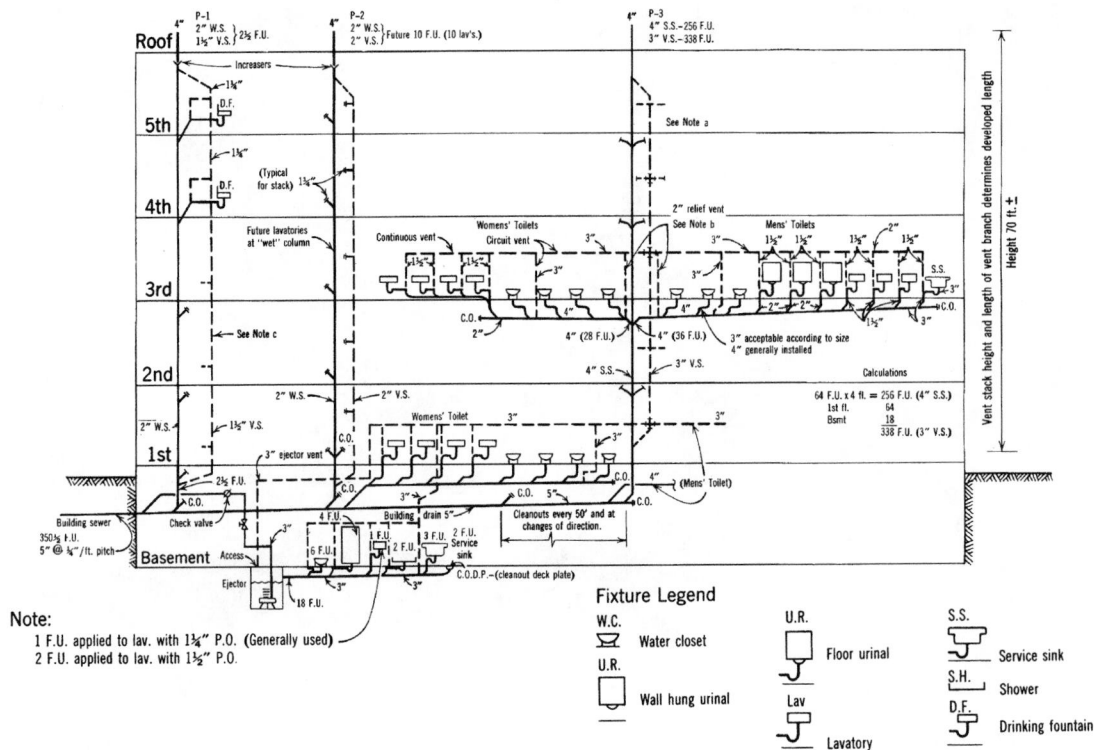

**Fig. 22.30** Example 22.2. Plumbing section of an office building. Circuit vents serve branch soil lines, but some codes require continuous venting (individual fixture vents). The house trap and fresh-air inlet are omitted from the building drain. Note that (1) a relief vent is not required on the top floor. (2) Men's and women's toilets on the third floor are typical and would be repeated on the first, second, fourth, and fifth floors. (3) Drinking fountains on the fifth and fourth floors would be repeated on the first, second, and third floors.

connect in at each T-Y connection. For three branch intervals, this size stack will carry 240 dfu (we need $3 \times 64 = 192$ dfu); for its entire height, it will carry 500 dfu (we need $5 \times 64 = 320$ dfu).

We need a vent stack serving up to 338 dfu, with a 4-in. (102-mm) soil stack, and at a maximum 70-ft (21.3-m) developed length. From Table 22.4, a 3-in. (76-mm) vent serving 540 dfu has a maximum developed length of 150 ft (45.7 m), well over our minimum requirements. All vent stack diameters will increase to 4 in. (102 mm) as they pass through the roof.

According to Table 22.5, the building drain and the building sewer at their pitch of ¼ in./ft should carry at least our total of 350 dfu. A 5-in. (127-mm) building drain or building sewer has a capacity of 480 dfu.

Although opinion may vary about the relative merits of continuous or circuit venting, either system, properly designed, will effectively prevent the siphoning of traps or relieve air pressures that could cause foul gases to bubble through the traps into the occupied space. Another system (below), especially suitable for high-rise buildings, eliminates the vent stack completely with equal effectiveness. ∎

## (c) The Sovent System

This essentially ventless system changes the nature of the effluent (discharge of wastes and sewage from the fixtures) instead of coping with the pressures and suctions that normal effluent would cause (see Figs. 22.31 and 22.32).

The "plunger" effect of a descending "slug" of water/waste within pipes was described in Section 22.2. If the effectiveness of the plunger can be reduced, the negative and positive pressures created by it will also be reduced. If their values can be brought down below the holding power of the several inches of water in the trap, no vents will be

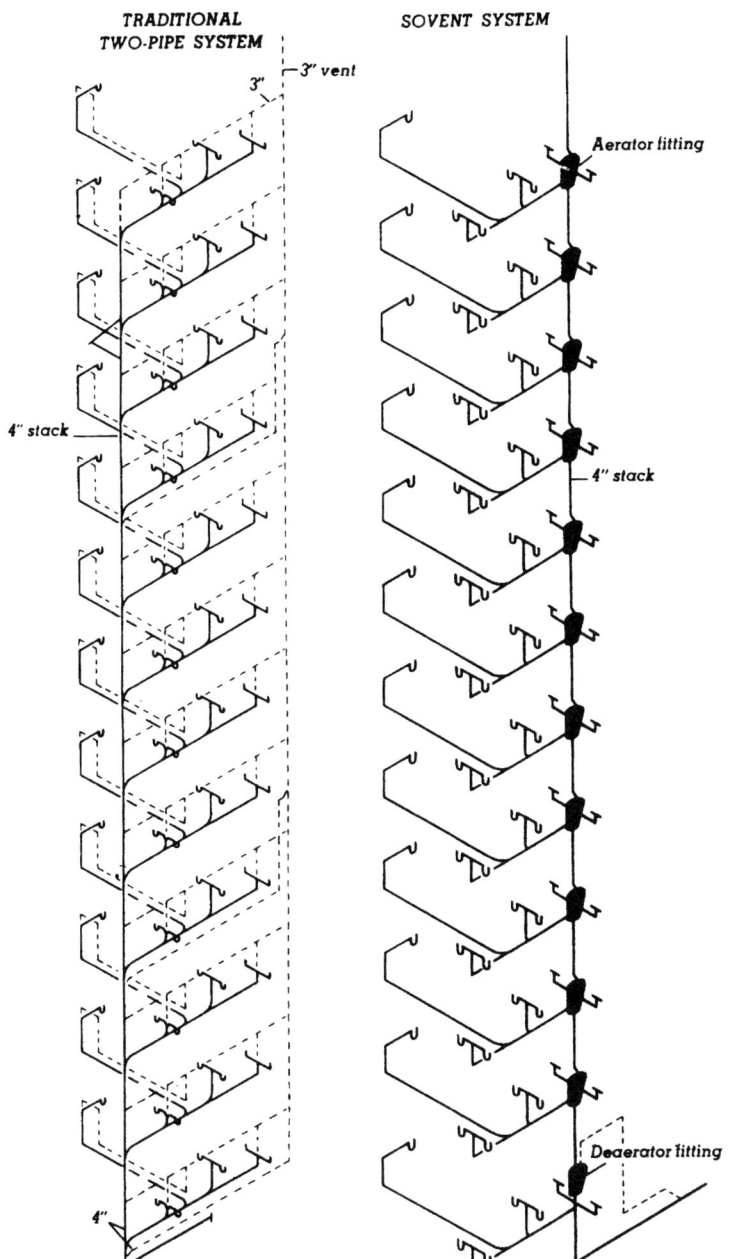

**Fig. 22.31** Comparing a two-pipe system for a 12-story stack serving an apartment grouping to the Sovent system. (Courtesy of the Copper Development Association.)

WATER AND WASTE

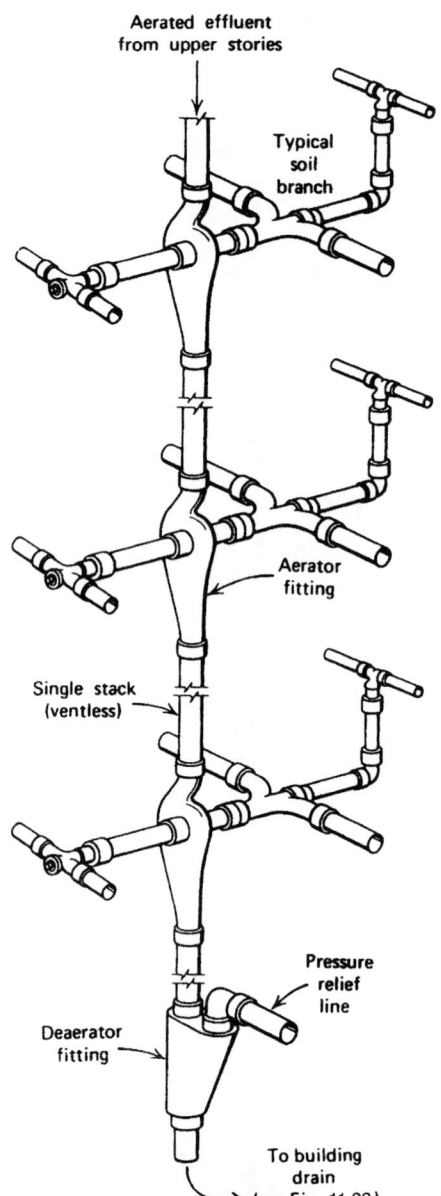

Aerated effluent
from upper stories

Typical
soil
branch

Aerator
fitting

Single stack
(ventless)

Pressure
relief
line

Deaerator
fitting

To building
drain
(see Fig. 11.33)

**Fig. 22.32** *The Sovent drainage stack consists of aerator fittings that join the horizontal branches to the stack at each floor level and a deaerator fitting at the bottom of the stack. The stack is open to the atmosphere above the roof at the top. (Courtesy of the Copper Development Association.)*

necessary. In the single-stack *Sovent* system illustrated in Fig. 22.32, this is done by dealing with the normal liquid effluent at each floor. Aeration there produces a *foam* that lacks the stack-filling tendency of the liquid effluent. Thus, through the creation of

a *soft* plunger, pressure variations in the single stack are minimized.

Tests have shown that the positive and negative pressures produced by normal liquid effluent during its descent (and relieved by the vent piping) are often about 5 to 12 in. (127 to 305 mm) water gauge. Obviously, if the vents were not provided, the 2 to 4 in. (51 to 102 mm) of water seal in the traps would be vulnerable to penetration by gases from pipes under positive pressure or siphonage of water seals into pipes that may be under negative pressure.

Figures 22.31, 22.32, and 22.33 illustrate the components and the action of the Sovent system. Effluent, already aerated and descending from upper stories, is diverted in the stack at each lower story. The aerator fitting there affords a passage for this diverted flow and also an air space into which the effluent from the local branch soil or waste can drop. Here it spatters, mixing with the air to form a rarefied mixture of air and liquid. Tests show that this mixture does not produce pressures, positive or negative, of more than 1 in. (25 mm) water gauge. Thus, a trap-seal of 2 in. (51 mm) or more is safe against siphonage or penetration.

At the foot of the single stack the aerated efflu-

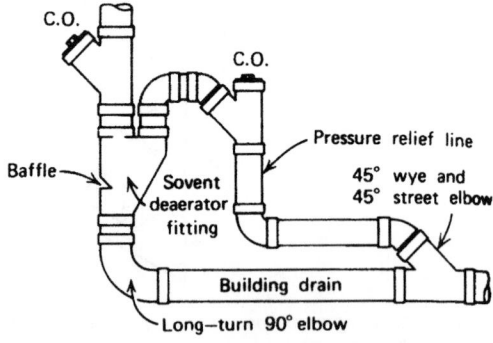

C.O.

C.O.

Pressure relief line

Baffle

Sovent
deaerator
fitting

45° wye and
45° street elbow

Building drain

Long–turn 90° elbow

**Fig. 22.33** *The deaerator consists of an air separation chamber with an internal nosepiece, a stack inlet, a pressure relief outlet at the top, and a stack outlet at the bottom. The deaerator fitting at the bottom of the stack functions in combination with the aerator fittings above to make the single stack self-venting. The deaerator is designed to overcome the tendency for the falling waste to build up excessive back pressure at the bottom of the stack when the flow decelerates at the bend into the horizontal drain. (Courtesy of the Copper Development Association.)*

ent is compacted—a process aided by a baffle in the path of the flow in the deaerator fitting (see Fig. 22.33). If not relieved, air piling up at this point could cause pressures in the stack at the first floor. An air-discharge pipe provides this relief of air from the deaerator fitting to the upper part of the building drain, above the liquid flow.

The Sovent system was invented by Fritz Sommer of Switzerland, who tested it in a 10-story drainage test tower. Since its introduction in 1962, it has been installed and used in hundreds of buildings in Europe and Africa. Canada used the Sovent method in the Habitat apartments at the 1967 Montreal exposition. Sovent was first granted U.S. code acceptance in 1968 in Richmond, California. Following this success, its code acceptance grew rapidly during the early 1970s. However, not all jurisdictions allow Sovent installations.

## 22.6 ON-SITE INDIVIDUAL BUILDING SEWAGE TREATMENT

The great majority of individual-building sewage treatment systems in North America use the septic tank (Fig. 22.34) as a *primary* treatment, where the settling of solids and anaerobic digestion take place. Subsequently, the effluent receives *secondary* treatment, which usually consists of a filtering process. Four common filtration systems are seepage pits, drain fields, mounds, and sand filters. Occasionally, a *tertiary* treatment (usually disinfection with chlorine) must be used, as when outflows from secondary treatment would flow directly into surface waterways.

### (a) Primary Treatment: Septic Tanks

These (Fig. 22.35) are commonly constructed of precast concrete; steel, fiberglass, and polyethylene tanks are also available. The sewage enters the first chamber, where solids sink to the bottom as sludge, and scum forms on the surface. Anaerobic decomposition proceeds, with production of methane gas. The liquid moves through the submerged opening in the middle of the tank to the second chamber, where finer solids continue to sink, and less scum forms on the surface. Finally, the effluent, about

70% purified, leaves the septic tank for secondary treatment.

The longer the sewage stays in the septic tank, the less polluted the effluent. This is why water conservation measures are so welcome with septic tank systems; the less the flow, the longer the water remains in the tank. The anaerobic decomposition process, while malodorous, is so thorough that the sludge needs only occasional removal—once in several years is about average for residences. The methane and its odor are kept within the tank, and access hatches are sometimes covered with soil (as in Fig. 22.35), even though this makes their occasional need to be located far more difficult.

Most systems with septic tanks will eventually experience failure, usually due to a breakdown in the secondary treatment rather than in the septic tank. With periodic removal of sludge, the septic tank itself is a reliable and simple primary treatment device. However, some types of domestic waste can disrupt the anaerobic process within the tank. The worst offenders are paints, varnishes, thinners, waste oil, photographic solutions, and pesticides. Also to be avoided are coffee grounds, dental floss, disposable diapers, kitty litter, sanitary napkins, tampons, cigarette butts, condoms, gauze bandages, fat and grease, and paper towels.

Septic tank sizes are commonly based on code requirements that consider the number of bedrooms in residences (Table 22.7) or the number of waste fixture units served (see Table 22.2). (As shown in Table 22.8, maximum size may also be related to the soil condition.) Sewage flow rates are also considered. Oversized septic tanks are more expensive to install, but they release cleaner effluent and prolong the life of the secondary treatment system.

The size of the secondary treatment system is usually based on the size of the septic tank it serves or on the expected total flow over a 24-hour period. (Flow can be estimated from Table 20.4.) Where effluent must be raised for secondary treatment, or when the total length of the secondary treatment disposal line exceeds 500 ft (154 m), a dosing tank (also called a *siphon* or *pumping chamber*; see Fig. 22.34c,d) is used to automatically discharge the septic tank's outflow chamber to the secondary treatment lines or sand filters.

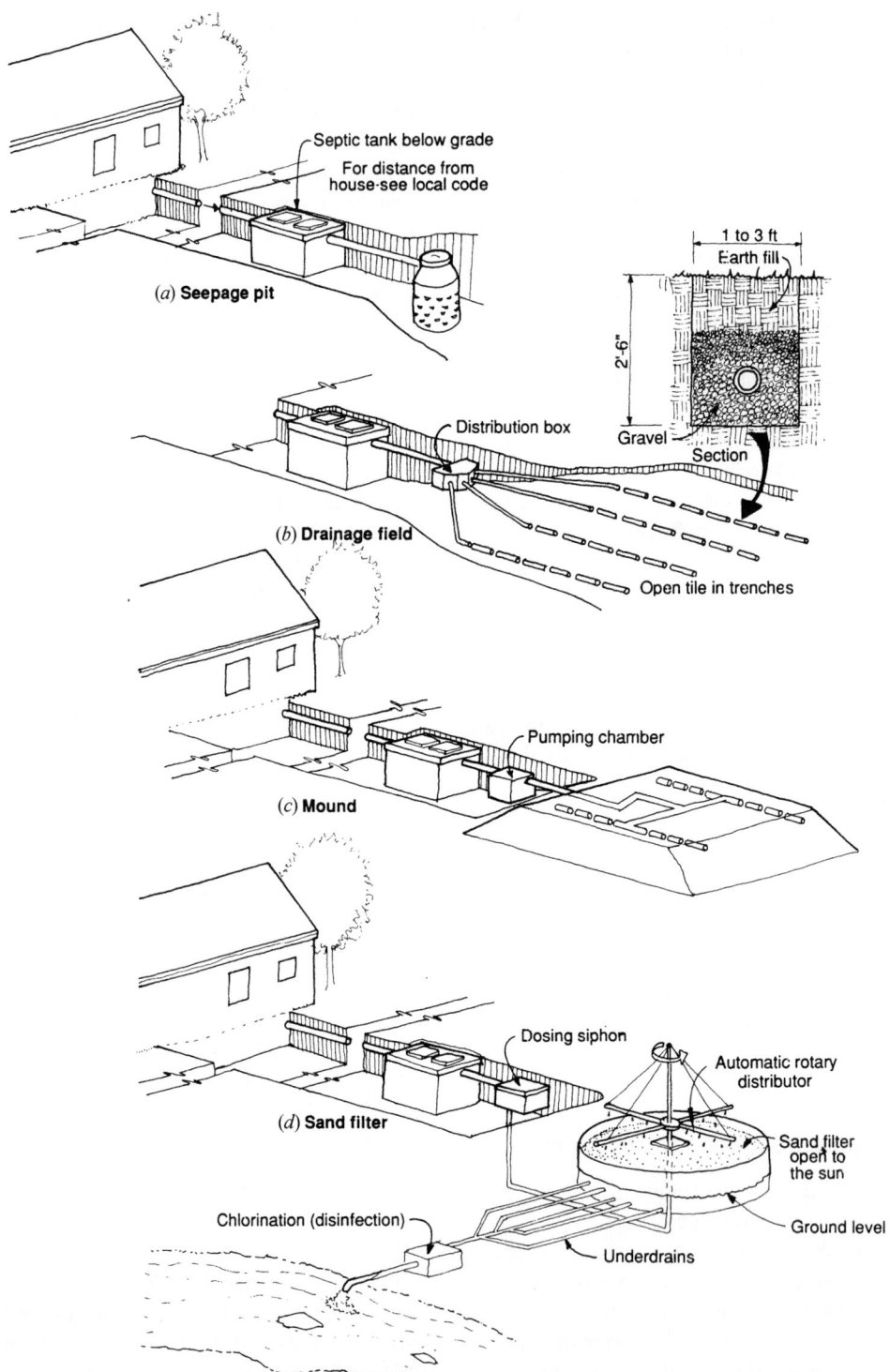

**Fig. 22.34** *For individual sewage treatment systems, septic tanks are commonly used for primary treatment. Four options for secondary treatment are shown here. (Tertiary treatment usually is required only for effluent discharge into waterways.) (a) Seepage pits are rarely allowed. (b) Drain fields are the most commonly used options. (c,d) Mounds and sand filters are more expensive to construct and are used where high water tables preclude the use of option (b).*

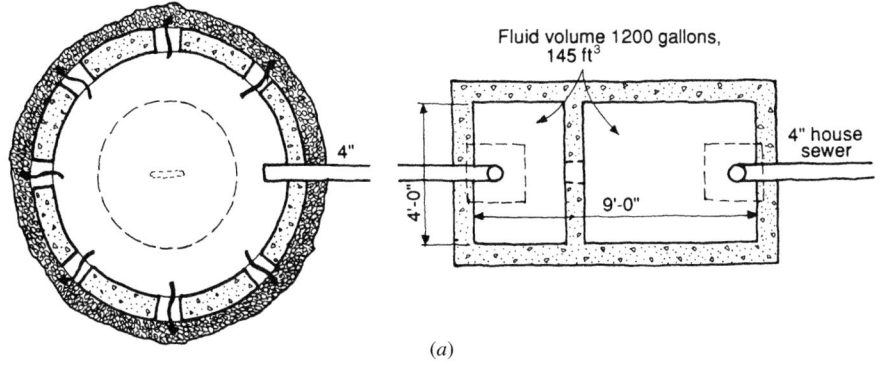

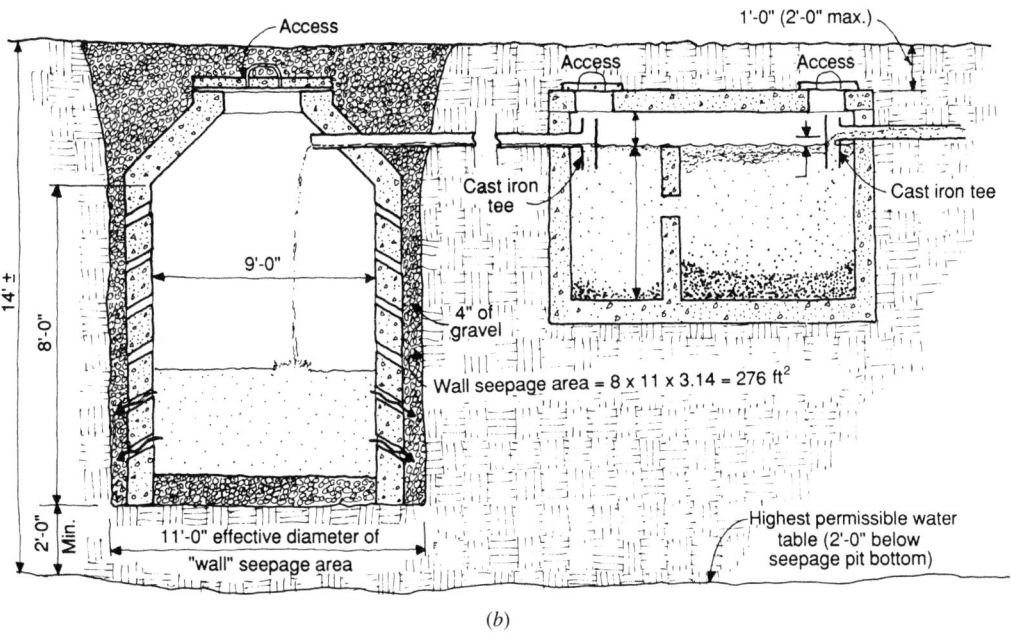

**Fig. 22.35** Plan (a) and section (b) of a septic tank and seepage pit for a four-bedroom house in sandy loam soil. A pit is suitable only when the earth is absorbent and the water table low (below the pit bottom). The drawing is not to scale.

**TABLE 22.7 Septic Tank Capacity[a]**

| Single Family Dwellings—Number of Bedrooms | Multiple-Dwelling Units or Apartments—One Bedroom Each | Other Uses: Maximum Fixture Units Served[b,c] | Minimum Septic Tank Capacity, Gal (L)[c] |
|---|---|---|---|
| 1 or 2 | | 15 | 750 (2,838) |
| 3 | | 20 | 1,000 (3,785) |
| 4 | 2 units | 25 | 1,200 (4,542) |
| 5 or 6 | 3 | 33 | 1,500 (5,678) |
| | 4 | 45 | 2,000 (7,570) |
| | 5 | 55 | 2,250 (8,516) |
| | 6 | 60 | 2,500 (9,463) |
| | 7 | 70 | 2,750 (10,409) |
| | 8 | 80 | 3,000 (11,355) |
| | 9 | 90 | 3,250 (12,301) |
| | 10 | 100 | 3,500 (13,248) |
| Extra bedroom: 150 gal (568 L) each. | | | |
| Extra dwelling units over 10: 250 gal (946 L) each. | | | |
| Extra fixture units over 100: 25 gal (95 L) per fixture unit. | | | |

*Source:* Reprinted by permission from the *Uniform Plumbing Code,* © 1997 by the International Association of Plumbing and Mechanical Officials.

[a]Septic tank sizes in this table include sludge storage capacity and the connection disposal of domestic food waste units without further volume increase.

[b]See Table 22.2.

[c]For larger or nonresidential installations in which sewage flow rate is known, size the septic tank as follows:

1. Flow up to 1500 gpd (5678 L/d): flow × 1.5 = septic tank capacity

2. Flow over 1500 gpd (5678 L/d): (flow × 0.75) + 1125 = septic tank capacity in gallons [(flow × 0.75) + 4258 = liters]

3. Secondary system shall be sized for total flow per 24 hours.

**TABLE 22.8 Septic Tank and Leaching Area Design Criteria for Five Typical Soils**

| Type of Soil | Required ft² of Leaching Area/100 Gal (m²/L) | Maximum Absorption Capacity, Gal/ft² of Leaching Area for a 24-h Period (L/m²) | Maximum Septic Tank Size Allowable | |
|---|---|---|---|---|
| | | | Gallons | Liters |
| 1. Coarse sand or gravel | 20 (0.005) | 5 (203.7) | 7,500 | 28,387 |
| 2. Fine sand | 25 (0.006) | 4 (162.9) | 7,500 | 28,387 |
| 3. Sandy loam or sandy clay | 40 (0.010) | 2.5 (101.9) | 5,000 | 18,925 |
| 4. Clay with considerable sand or gravel | 90 (0.022) | 1.10 (44.8) | 3,500 | 13,247 |
| 5. Clay with small amount of sand or gravel | 120 (0.029) | 0.83 (33.8) | 3,000 | 11,355 |

*Source:* Reprinted by permission from the *Uniform Plumbing Code,* © 1997 by the International Association of Plumbing and Mechanical Officials.

## (b) Primary Treatment: Aerobic Treatment Units

Active rather than passive aerobic treatment units (ATUs, Fig. 22.36) are an increasingly popular alternative to septic tanks for primary sewage treatment. They are frequently used to replace a septic tank in a troubled system; this "upgrade" can reju-venate an existing drain field and extend its life. ATUs depend on air bubbled through the sewage to achieve aerobic digestion, which is faster than anaerobic digestion; hence, they can be smaller than septic tanks. They are energy-intensive and require more maintenance than anaerobic tanks. A secondary treatment process is required as well. The

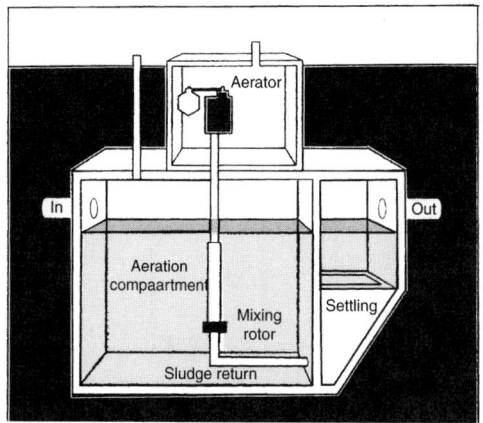

**Fig. 22.36** *Aerobic treatment unit (ATU), requiring more energy and maintenance, but with cleaner effluent than that of a passive anaerobic septic tank. (National Small Flows Clearinghouse.)*

effluent typically is less polluted than that of septic tanks; the biochemical oxygen demand (BOD) is reduced by 90% in ATUs compared to the 50% typical of septic tanks.

The sewage first enters an aeration chamber, where it is kept in turmoil so that air can continue to percolate through it. Either air is forced through the sewage by distribution lines fed by an air compres-

sor or the sewage is stirred by a variety of devices, depending on the manufacturer. After about a day's retention in the aeration chamber, the aerated wastewater enters a settling chamber, allowing remaining solids to settle and to be filtered out before the effluent leaves, for further treatment in a secondary process.

### (c) Secondary Treatment: Seepage Pits

Also called *cesspools* (Fig. 22.35), these follow treatment by either septic tanks or ATUs. They are increasingly rare, and are appropriate only in very porous soil, where the water table is at least 2 ft (0.6 m) below the bottom of the pit. Because these pits are commonly 10 to 15 ft (3 to 4.6 m) below the earth's surface, a very low water table may be required. Another common usage of precast seepage pits is for *dry wells* that receive runoff from paved areas during rainstorms.

Seepage pits are sized by the square footage of the wall area exposed to the earth—the *leaching area*, as listed in Table 22.8. Placement of them relative to buildings, water sources, waterways, and property lines is strictly controlled—see Table 22.9 and Fig. 22.37.

**WATER AND WASTE**

### TABLE 22.9  Location of On-Site Sewage Disposal Systems

| Minimum Horizontal Distance Clear Required from: | Building Sewer ft | mm | Septic Tank ft | mm | Disposal Field ft | mm | Seepage Pit (Cesspool) ft | mm |
|---|---|---|---|---|---|---|---|---|
| Buildings or structures[a] | 2 | 610 | 5 | 1,542 | 8 | 2,438 | 8 | 2,438 |
| Property line adjoining private property | Clear | | 5 | 1,542 | 5 | 1,542 | 8 | 2,438 |
| Water supply wells | 50[b] | 15,240 | 50 | 15,240 | 100 | 30.5 m | 150 | 45.7 m |
| Streams | 50 | 15,240 | 50 | 15,240 | 50 | 15,240 | 100 | 30.5 m |
| Trees | — | — | 10 | 3,048 | — | | 10 | 3,048 |
| Seepage pits or cesspools | — | — | 5 | 1,542 | 5 | 1,542 | 12 | 3,658 |
| Disposal field | — | — | 5 | 1,542 | 4[d] | 1,219 | 5 | 1,542 |
| On-site domestic water service line | 1 | 305 | 5 | 1,542 | 5 | 1,542 | 5 | 1,542 |
| Distribution box | — | — | — | | 5 | 1,542 | 5 | 1,542 |
| Pressure public water main | 10[c] | 3,048 | 10 | 3,048 | 10 | 3,048 | 10 | 3,048 |

*Source:* Reprinted by permission from the *1997 Uniform Plumbing Code.* © by the International Association of Plumbing and Mechanical Officials, Walnut, CA.

*Note:* When disposal fields and/or seepage pits are installed in sloping ground, the minimum horizontal distance between any part of the leaching system and the ground surface shall be 15 ft (4.6 m).

[a]Including porches and steps, whether covered or uncovered, breezeways, roofed porte-cocheres, roofed patios, carports, covered walks, covered driveways, and similar structures or appurtenances.

[b]All drainage piping shall clear domestic water supply wells by at least 50 ft (15.2 m). This distance may be reduced to not less than 25 ft (7.6 m) when the drainage piping is constructed of materials approved for use within a building.

[c]For parallel construction. For crossings, approval by the Health Department shall be required.

[d]Plus 2 ft (0.6 m) for each additional 1 ft (0.3 m) of depth in excess of 1 ft (0.3 m) below the bottom of the drain line.

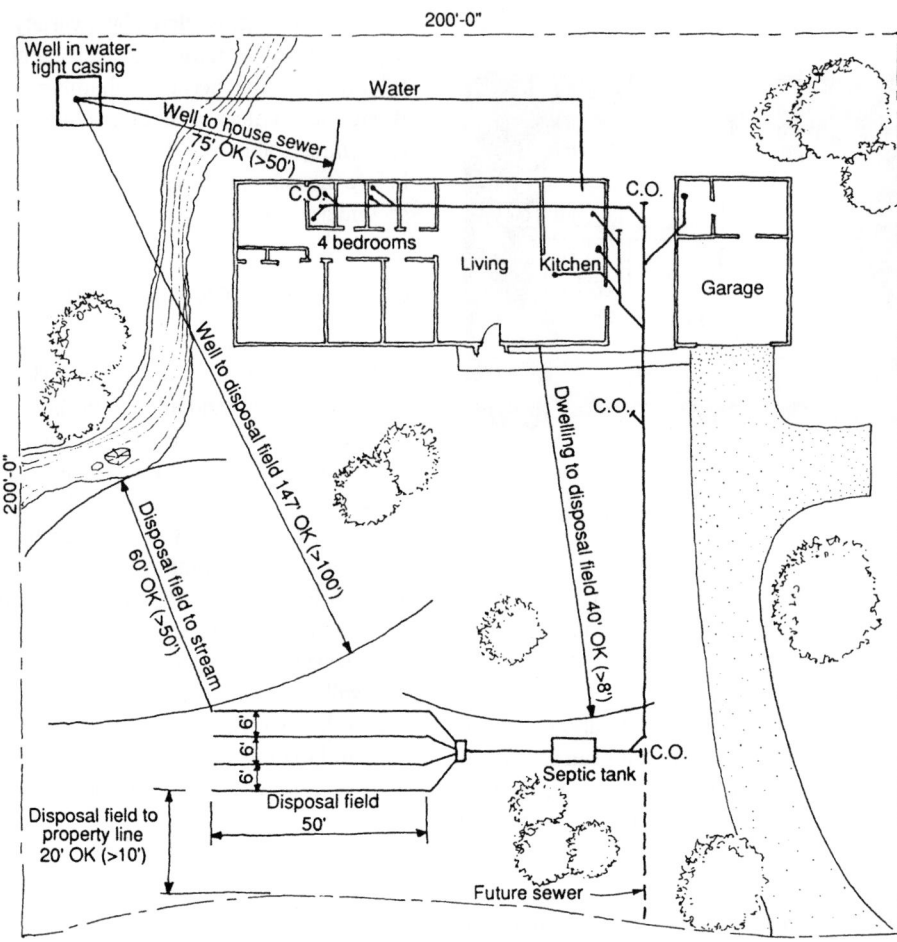

**Fig. 22.37** *Checking for required clearances on a suburban lot. Although the 40,000-ft² (3716-m²) lot is more than the minimum required, space should be left for a backup disposal field (of equal size) in case of failure of the original field.*

### (d) Disposal Fields

These are very commonly used as a secondary treatment method (Fig. 22.38) because they are relatively inexpensive to build and do not require a water table so deep—or soil so permeable—as seepage pits require. Drain lines consisting of either perforated pipes (of many approved materials) or square-edge agricultural tile, 4 in. (102 mm) or more in diameter (the ends of the tiles are separated by ¼-in. (6-mm) openings), are used. These lines are placed in shallow trenches, on a bed of gravel, and covered with gravel. The effluent runs out of these lines and stands in the interstices of the gravel until it seeps into the earth. In effect, the gravel provides spaces that act as a dry well to receive the fluids and accommodate them until they slowly sink into the ground.

Disposal fields are located according to regulations such as those given in Table 22.9 and sized in relation to total sewage flow and septic tank size (see Tables 22.8 and 22.10). Although Table 22.8 shows the maximum absorption capacities for soils based on gal/ft² (L/m²) over a 24-hour period, a common alternative is to use *percolation* tests. For such tests, a test pit is dug and water poured into it. The absorption in gal/ft² (L/m²) in a 24-hour period is determined from this test. Some codes require design sizes based on percolation test data. Codes

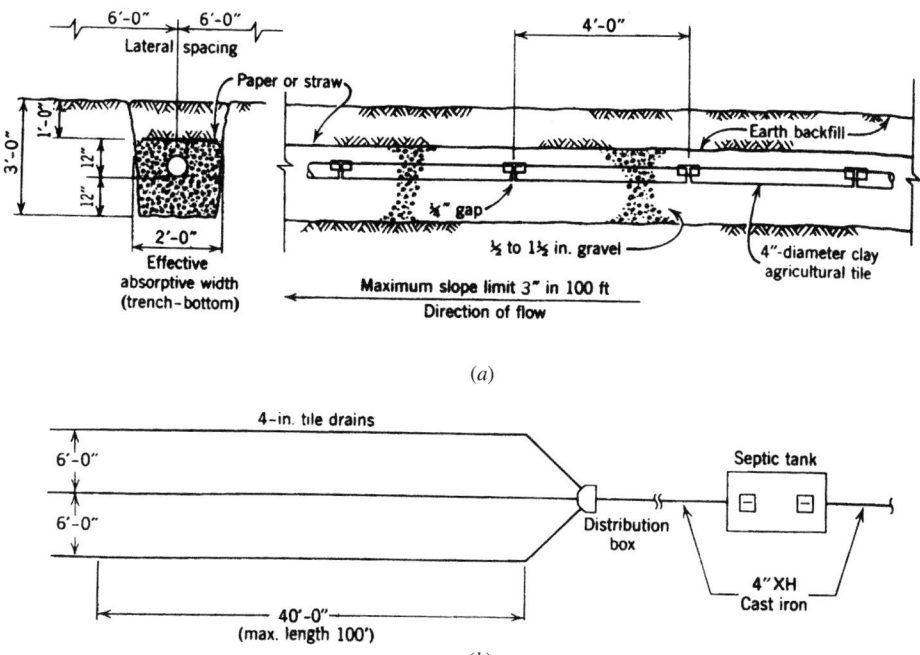

**Fig. 22.38** *Example 22.3. Tile drain field for a four-bedroom house. Although the drawings are not to scale, the dimensions show a required area of about 20 × 50 ft (6 × 15 m) on the lot. Because paving is not permitted over drain fields, sewage treatment on a small lot demands considerable space. (a) Transverse and longitudinal sections. (b) Schematic plan.*

often require that an area equal to the drainage field's required size be set aside for use in the event of failure in the original field.

Design guidelines for sizing disposal fields were presented in Section 20.3, but a more accurate method is shown here. The trench width, depth, and spacing shown in Fig. 22.38 are typical, but other combinations can be used, provided that they meet the requirements listed in Table 22.10.

### TABLE 22.10 Disposal Field Trenches

| PART A. DIMENSIONS | | |
|---|---|---|
| | **Minimum** | **Maximum** |
| Length of drain line(s) | — | 100 ft (30.5 m) |
| Bottom width of trench | 18 in. (457 mm) | 36 in. (0.9 m) |
| Spacing of lines, o.c.[a] | 6 ft (1.8 m) | — |
| Depth of earth cover over lines | 12 in. (305 mm) | — |
| | *Note:* 18 in. (457 mm) preferred | |
| Grade of lines | Level | 3 in./100 ft (25 mm/m) |
| Filter material | | |
|   Over drain lines | 2 in. (51 mm) | — |
|   Under drain lines[a] | 12 in. (305 mm) | —[d] |
| PART B. LEACHING AREAS | | |
| Trench bottom[b]: minimum 150 ft² (14 m²) per system<br>Trench side wall: minimum[c] 2 ft²/ft of length<br>  maximum[d] 6 ft²/ft of length | | |

*Source:* Adapted by permission from the *Uniform Plumbing Code,* © 1997 by the International Association of Plumbing and Mechanical Officials.

[a]Minimum spacing of drain lines: 4 ft (1.2 m) plus 2 ft (0.6 m) for *each* additional foot (0.3 m) of depth *beyond* 1 ft (0.3 m) below the bottom of the drain line.

[b]Exclusive of rock, clay, or other impervious formations.

[c]Based on the minimum 12-in. (305-mm) trench depth below drain tile.

[d]A maximum 36-in. (0.9-m) trench depth below the drain line can be counted when calculating the required absorption area.

WATER AND WASTE

**EXAMPLE 22.3** See Fig. 22.38. Design a septic tank and drain field system for a suburban residence under the following conditions:

| | |
|---|---|
| Bedrooms | 4 |
| Soil | Sandy loam |
| Depth to water table | 7 ft |

**SOLUTION**

Septic tank capacity (Table 22.7) = 1200 gal. Fluid volume of the tank (minimum required) is

$$1200 \text{ gal} \div 7.48 \text{ gal/ft}^3 = 160 \text{ ft}^3$$

Establish dimensions for the septic tank. Dimensions for the chosen tank (Fig. 22.35) are

$$4.5 \times 4.0 \times 9.0 = 162 \text{ ft}^3 \text{ volume}$$
$$162 > 160 \text{ (the minimum required)}$$

Drain field size is established from Table 22.8: sandy loam (soil type 3) requires 40 ft² of leaching area per 100 gal. With a septic tank of 1200 gal,

40 ft² × 1200 gal/100 gal
$$= 480 \text{ ft}^2 \text{ minimum absorption area}$$

The effective absorption area of the typical trench depth and spacing shown in Fig. 22.38 is

| | |
|---|---|
| Trench width | 2.0 ft²/ft of length |
| Trench sides | 2.0 ft²/ft of length (12 in. on each side) |
| Total absorption | 4.0 ft²/ft |

The required absorption area was determined to be 480 ft². Because the trench has 4.0 ft²/ft of length, the total trench length is

$$480 \text{ ft}^2/(4 \text{ ft}^2/\text{ft}) \text{ of length} = 120 \text{ ft}$$

A three-line disposal field is selected, so 120 ft/3 lines = 40 ft per line. Given the required clearances to the edges of the disposal field (Table 22.9) of 5 ft to property lines, the disposal field's area can be calculated as:

width:

5 ft + 12 ft (two spaces of 6 ft each) + 5 ft = 22 ft

length:

5 ft + 40-ft line + 5 ft = 50 ft

The site surface area, then, is 22 × 50 = 1100 ft². Note that double this area (for a second disposal field in the event of the first field's failure) is often required. ∎

### (e) Mounds with Leaching Beds

These (Fig. 22.39) represent a newer solution in the United States and thus may require special approval. The guidelines for leaching-bed sizing are essentially similar to those for drainage tile disposal fields. The absorption area for leaching beds, however, must be 50% greater than that required for trenches. The bottom of the leaching bed generally must be at least 5 ft (1.5 m) above the water table, although in water-scarce areas, officials may reduce this requirement.

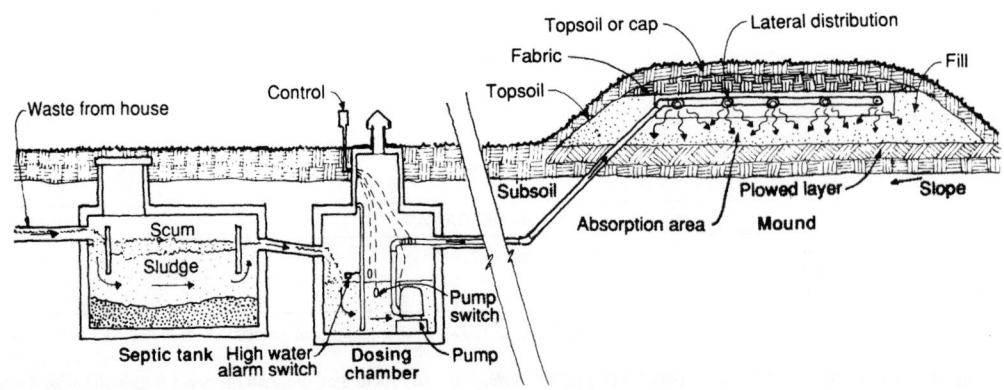

**Fig. 22.39** Mounds with leaching beds offer a disposal option when the water table is high. The system serves a two- or three-bedroom home. (Adapted from Converse, 1978.)

## (f) Buried Sand Filters

Slow sand filters for treating supply water were discussed in Section 21.2*d* and shown in Fig. 21.3. Buried sand filters (Fig. 22.40) work in a similar way, using primarily biological but also physical and chemical processes to clean wastewater. The medium is most commonly sand, although other locally available materials such as crushed glass, mineral tailings, bottom ash, and so on have been used. The grain size ranges from 0.3 to 3 mm in diameter.

Buried sand filters are one remedy for failed disposal fields. They are also used where high groundwater, shallow bedrock, or poor soil preclude use of the more simple disposal field. Although buried sand filters may require considerable site area, the ground surface can be used for lawns or other non-paved surface activities. More detailed information on the sand filter types discussed here is available from the National Small Flows Clearinghouse.

The layout is similar to that of the disposal fields described in Section 22.6*e*, but these filters, themselves 24 to 36 in. (610 to 914 mm) deep, usually require an excavation of 4 to 5 ft (1.2 to 1.5 m). The filter bed must be level and sited to avoid contact with groundwater and excess surface water runoff. Some authorities require the sand filter to be contained in an impermeable membrane liner. Underdrain pipes and a graded layer of washed gravel or crushed rock are placed on the bottom of the filter bed, with the finer gravel on top to keep the medium from washing into the underdrains. Then a layer of fine gravel first, with coarser gravel above, is placed around and over the distribution pipes. A geotextile fabric covers the top of the entire filter bed, which in turn is covered by backfill material. A properly constructed buried sand filter that receives properly digested waste from a well-maintained septic tank should last for up to 20 years without maintenance.

In general, sand filters are designed for normal dosing twice a day; they need time to allow the filter medium to drain between doses. The water that filters through the medium is collected in underdrains that are sloped toward an outlet and vented to the surface at the upstream end. This vent is a potential source of odors. The water is then taken either to a disinfection (usually chlorination) unit or to a disposal field for subsurface discharge.

Water from sand filters has been cleansed of many pathogens, such as harmful bacteria and viruses, and is therefore more likely to be approved for a down-line disposal field. Typical values for three types of sand filters are shown in Table 22.11. (Open and recirculating sand filters are discussed in the next section.)

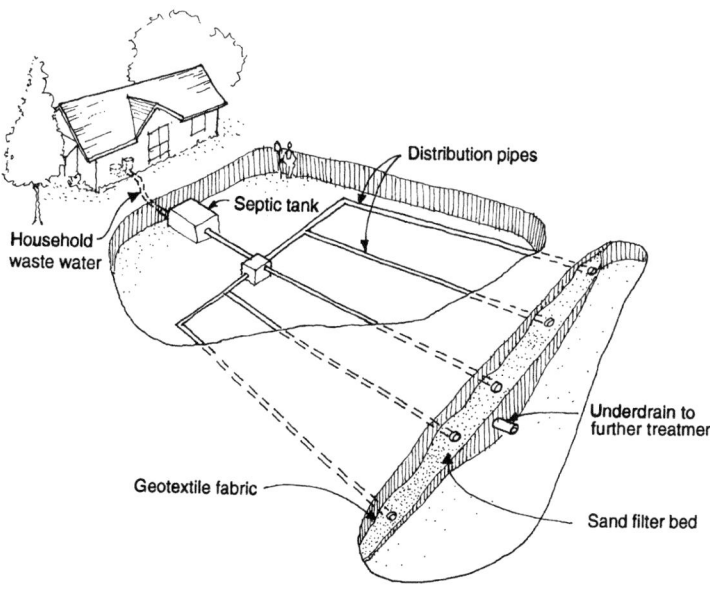

**Household waste water**

**Septic tank**

**Distribution pipes**

**Underdrain to further treatment**

**Geotextile fabric**

**Sand filter bed**

*Fig. 22.40* Buried sand filters appear to be like the common disposal fields, but are deeper and may require an impermeable membrane liner. (National Small Flows Clearinghouse.)

**TABLE 22.11 Sand Filter Design Values**[a]

| Design Factor | Buried Filters | Open Filters | Recirculating Filters |
|---|---|---|---|
| Pretreatment | All filters must be preceded by settling and removal of solids | | |
| Media | | | |
|   Materials | Washed, durable granular material | | |
|   Uniformity Coefficient | Less than 4.0 between smallest and largest size particles | | |
|   Depth | 24 to 36 in. (610 to 914 mm) | | |
|   Effective size | 0.3 to 1 mm | 0.3 to 1 mm | 0.8 to 3 mm |
|   Hydraulic loading | <1.5 gpd/ft$^2$ (<61 L/day m$^2$) | 2 to 5 gpd/ft$^2$ (82 to 204 L/day m$^2$) | 3 to 5 gpd/ft$^2$ [b] (122 to 204 L/day m$^2$) |
| Media temperature | >41°F (>5°C) | >41°F (>5°C) | >41°F (>5°C) |
| Dosing frequency | <2 per day | <2 per day | 5–10 min./30 min.[d] |
| Recirculation ratio[c] | (Not applicable) | (Not applicable) | 3:1 to 5:1 |

*Source:* National Small Flows Clearinghouse, *Pipeline,* Summer 1997, Vol. 8, No. 3.

[a]These values show typical design criteria and do not represent all possibilities.

[b]Forward flow only, not including recirculated effluent.

[c]Water recirculated: water discharged.

[d]Dosage lasts for 5 to 10 minutes every half hour.

## 22.7 ON-SITE MULTIPLE-BUILDING SEWAGE TREATMENT

The first approach involves septic tanks at each building for primary treatment. Then the combined outflows from the individual septic tanks can enter a secondary treatment process. Either of two more types of sand filters can be employed for this purpose. There are several advantages to this approach. The effluent lines from the septic tanks carry no solids; they can be much smaller in diameter and placed in shallower trenches than conventional sewer lines. The runs are far shorter, because distant centralized treatment plants are not involved. With sand filter secondary treatment, this septic tank wastewater is more likely to find a recycling use, such as irrigation.

### (a) Open Sand Filters

Also called *intermittent sand filters,* these are very similar in construction to slow sand filters (Fig. 21.3). They are nearly identical in performance to buried sand filters (Fig. 22.40); the difference is that these are at least partially aboveground. A detail of the filter medium is shown in Fig. 22.41. Table 22.11 shows that they are able to accept a somewhat higher flow, so they require less area than the buried filter, but the surface is no longer usable for any other purpose. (A removable cover for odor control is typical, despite their name.) They

are used on systems with flows of up to 120,000 gpd (454,240 L/day) of wastewater. Frequently, two such open filters are built so that one can rest while the other is in use. Alternatively, two filters may be used in series, with an even cleaner resulting final effluent.

### (b) Recirculating Sand Filters

These eliminate odors by ensuring an adequate supply of oxygen to the wastewater. The recirculating sand filter (Fig. 22.42) first receives water (from the septic tank or other primary treatment) into a recir-

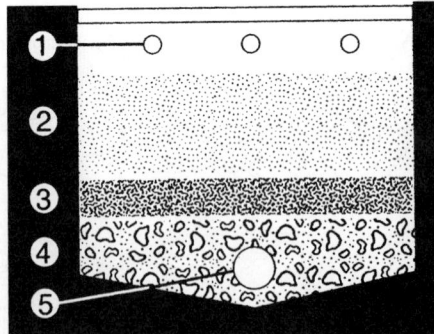

**Fig. 22.41** Open sand filter receives wastewater from above through (1) distribution pipes (or a thin layer of flooded wastewater). A filter medium (2) such as sand is on top, then (3) fine gravel, then (4) coarse gravel, surrounding the (5) underdrain. Slow sand filters are very similar. (National Small Flows Clearinghouse.)

culation tank. This tank has a pump, a timing mechanism, and float valves. Either in timed doses or when triggered by the float valve, doses of water are pumped to the sand filter. The treated wastewater collects in the underdrain, where the majority (75% to 80%) is directed back to the recirculation tank. Thus, mixing with the septic tank effluent and being pumped to the sand filter, the result is weaker (cleaner) effluent, containing more oxygen, as it enters the sand filter. This eliminates odor and allows a slightly larger filter grain size, making the system less prone to clogging. Table 22.11 compares sand filter types, showing the highest allowable flow rates per surface area for this recirculating sand filter.

At Stonehurst, a 47-lot subdivision in Contra Costa County, California, 1500-gal (5680-L) septic tanks serve each unit; each tank has a screened effluent vault, keeping solids out of the effluent lines. The effluent is taken through small-diameter, variable-grade sewers (located under the roadway) to a recirculating granular medium filter. (Homes lower than the roadway use effluent pumps at the septic tank outflows.) Outflow from the recirculating filter is then taken to a UV recirculation tank, where it is treated by UV radiation (see Fig. 21.6). From this tertiary treatment area, it is conducted to a subsurface drip irrigation system serving a community park. In winter, a 2.5-acre (10,117-m²) community soil absorption field is utilized.

### (c) Lagoons

These basins need sun, wind, and more land area than other methods, but are very simple to main-

tain and very low in energy use. They usually must be lined to prevent wastewater from polluting the groundwater below. Sometimes the outflow must undergo further treatment before release. If lagoons are not square or round, their length should not exceed three times their width. Outflow and inflow should be at opposite ends. Detailed information is available from the National Small Flows Clearinghouse (1997).

*Anaerobic lagoons* are usually used as the first of at least two lagoons in series. Working much like a septic tank, they hold wastewater for 20 to 50 days and are relatively deep, 8 to 15 ft (2.4 to 4.6 m). With a surface of floating scum that blocks oxygen from the wastewater, they are unsightly and malodorous.

*Aerobic lagoons* are shallower than other lagoons, so sunlight and oxygen from wind and air can better penetrate. Aerobic bacteria and algae do the work of cleaning the wastewater. These are best in warm climates where freezing is not a threat. They hold water for 3 to 50 days. The bottoms are either paved or lined to prevent weeds from growing within the lagoon.

*Aerated lagoons* take the aerobic lagoon a step further, actively stirring and adding oxygen to the lagoon. Because they speed aerobic action, they require shorter retention time and less land area than the passive aerobic lagoon.

*Facultative lagoons* (also called *stabilization, oxidation, photosynthetic,* or *aerobic–anaerobic ponds*) are the most common type used by small communities and individual households. They work in most climates and require no machinery. They require about 1 acre (0.4 hectare) for every 50

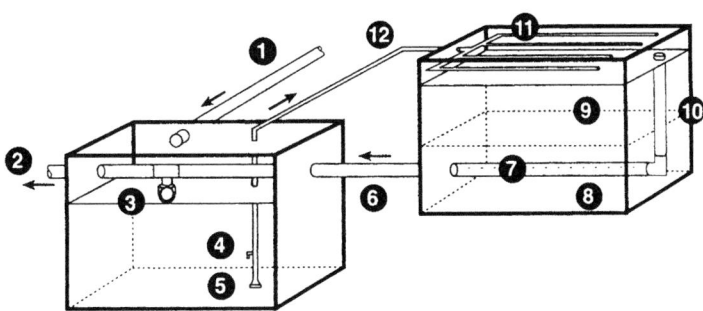

**Fig. 22.42** *Recirculating sand filters periodically distribute a cleaner, more oxygen-rich effluent to the sand filter. This eliminates odor and allows a larger medium grain size. (1) Effluent from pretreatment (such as a septic tank). (2) To disinfection/discharge. (3) Float valve. (4) Bleed line. (5) Submersible pump. (6) Recirculated effluent line. (7) Perforated underdrain piping. (8) Layered support gravel. (9) Filter media. (10) Riser pipe. (11) Distribution piping. (12) Recirculation pump discharge to recirculating filter. (National Small Flows Clearinghouse.)*

homes (or for every 200 people) they serve, are 3 to 8 ft (0.9 to 2.4 m) deep, and are designed to hold wastewater for 20 to 150 days (the longer period in cold weather).

Three layers form in these lagoons; the top aerobic zone, the middle facultative zone, and, at the bottom, the anaerobic zone. The depth of the surface aerobic zone depends on how much sunlight, wind, and rain can contribute oxygen; the deeper this zone is, the better it controls odors rising from the anaerobic zone. Therefore, these lagoons should be sited with full access to both sun and wind.

Down in the anaerobic zone, anaerobic bacteria convert sludge to gases, including hydrogen sulfide, ammonia, and methane. As these gases rise, they provide food for both the aerobic bacteria and the algae in the aerobic zone. Sludge accumulates more quickly in cold climates; sludge removal should be expected every 5 to 10 years.

### (d) Advanced Integrated Wastewater Pond System (AIWPS™)

At California Polytechnic State University, San Luis Obispo, there is an innovative proposal to create an integrated infrastructure facility combining wastewater treatment and solid waste processing on campus. In addition to the waste treatment and resources recovery for this campus of 18,000, the schematic design for the Energy Efficient Resource Recovery (E2R2) Facility includes an educational facility, a wildlife habitat, and park facilities. The wastewater and related wetlands proposals are shown in Fig. 22.43; the solid waste facility appears in Fig. 23.1.

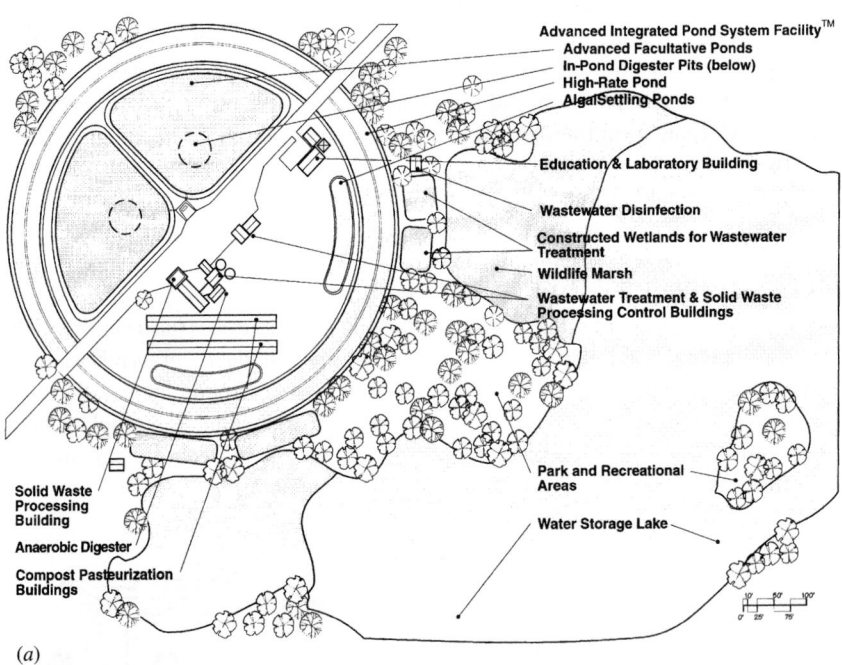

(a)

**Fig. 22.43** Schematic design proposal for wastewater treatment at the Energy Efficient Resource Recovery (E2R2) Facility at Cal Poly, San Luis Obispo. (a) Site plan. (b) Effluent enters the advanced facultative pond (AFP) through the in-pond digester, where solids settle, undergo methane fermentation, and are reduced to ash. Rising biogas is channeled by a proprietary submerged gas collector and passes through a highly aerobic surface layer where impurities are removed, yielding methane-rich biogas at the surface. (c) AFP effluent then goes to the high rate pond, where it is slowly moved along the shallow channels by an energy-efficient paddle wheel providing optimum solar exposure and enabling the profusely growing algae to supply abundant oxygen for oxidation of wastewater organics and disinfection of pathogens. Subsequently, effluent moves into the quiescent Algal Settling Pond where algae settles and is periodically removed for beneficial use. (d) Effluent then moves into constructed wetlands, where the plants in this system create an ideal environment for invertebrates and vertebrates (in free surface wetlands) and aerobic and anaerobic microbial populations to thrive and provide further treatment and effluent polishing. Finally, the effluent from the wetlands is disinfected by ozone or UV disinfection systems and stored in a lake that is part of a campus park and recreational facility. (Courtesy of Dr. William Oswald, Professor Emeritus of Civil and Environmental Engineering and Public Health at the University of California, Berkeley, and Principal of Oswald Engineering Associates, Inc.; Dr. Bailey Green, Research Engineer at the University of California, Berkeley, and Principal of Oswald Engineering Associates, Inc.; and Prof. Daniel Panetta, Architecture Department, Cal Poly, San Luis Obispo.)

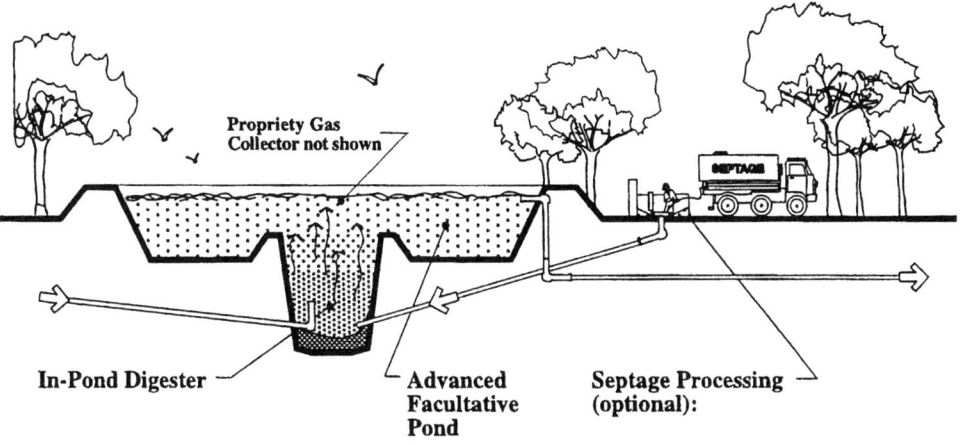

*Propriety Gas Collector not shown*

**In-Pond Digester**    **Advanced Facultative Pond**    **Septage Processing (optional):**

(*b*)

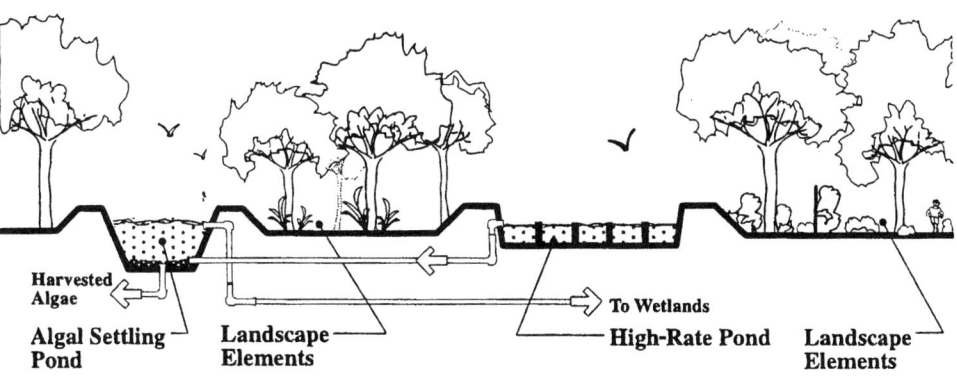

**Harvested Algae**

**Algal Settling Pond**    **Landscape Elements**    **To Wetlands**    **High-Rate Pond**    **Landscape Elements**

(*c*)

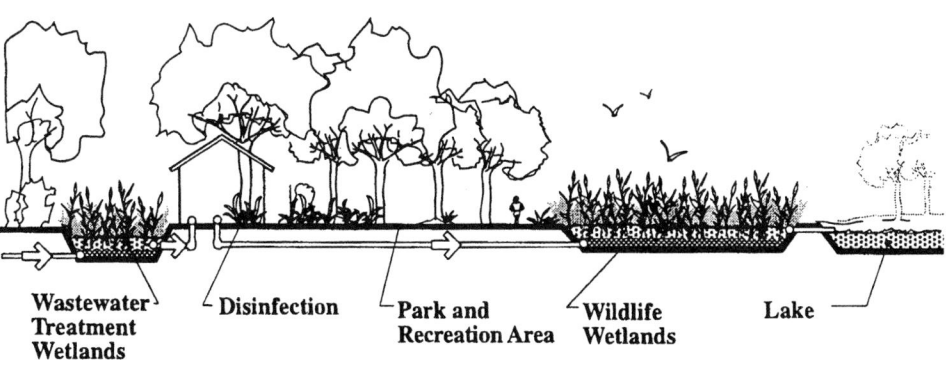

**Wastewater Treatment Wetlands**    **Disinfection**    **Park and Recreation Area**    **Wildlife Wetlands**    **Lake**

(*d*)

***Fig. 22.43*** (Continued)

Three ponds are used in the AIWPS facility (Oswald, 1990; Oswald et al., 1963). The first is the advanced facultative pond (AFP) (Fig. 22.43*a*), which includes an in-pond digester (IPD) or anaerobic fermentation pit at the bottom and an oxygen-rich aerobic surface water layer. Two AFPs used in parallel allow fail-safe operation. This design enables isolation and treatment of toxic substances in the event that such material is accidentally released into the community sewer system. Conventional sewage treatment plants typically cannot isolate or treat accidental toxic releases in wastewater due to their short processing time.

The raw sewage enters the system near the bottom of the IPD, where anaerobic bacteria slowly reduce biosolids to ash (Oswald et al., 1994; Green et al., 1995b). BOD is reduced to 60% to 80%, and minimal accumulation of refractory solids (sludge) has been demonstrated since the late 1960s.

Earthwork construction of the ponds allows them to be quite large and deep, without the substantially higher costs of concrete and steel construction typical of conventional wastewater treatment facilities. A berm surrounding the IPD prevents mixing of aerobic water with the influent below, whether by wind or thermal inversion. The aerobic upper layer oxidizes any malodorous biogases rising from the IPD. The biogas generated in the IPD is scrubbed as it passes through the water column and the aerobic surface layer. At the current campus population, between 3000 and 5000 ft$^3$ (85 and 142 m$^3$) per day of methane (CH$_4$) are estimated to be available for collection from the IPD and use as fuel (Green et al., 1995a).

The second pond is the high rate pond (HRP) designed to grow algae (Fig. 22.43*b*). The lined channels in this pond are approximately 1 m or less in water depth and include a paddle wheel (see Fig. 22.48) to slowly mix the primary effluent. The HRP design allows maximum oxygen production by algal photosynthesis due to optimum exposure to sunlight and abundant nutrients. The pond's aerobic and facultative bacteria oxidize most of the soluble and readily biodegradable BOD.

The next pond in the series is an algal settling pond (ASP) (Fig. 22.43*c*). The quiet waters of this pond allow 50% to 80% of the algae to settle over several days' residence time. The settled algae are periodically harvested, and the algal biomass does not produce objectionable odors. When properly disinfected, the harvested algae, rich in nitrogen and phosphorus, can be used as fish or livestock feed or as a nitrogen source in compost.

Near St. Helena, California, an AIWPS facility has functioned since the late 1960s. During that time period, no sludge removal has been necessary, attesting to the efficiency of the methane fermentation process. The facility is near several large wineries and is considered a good neighbor without odor problems (Oswald et al., 1970).

While a maturation pond usually follows the ASP in an AIWPS facility, various tertiary treatment components can subsequently be used to achieve advanced tertiary treatment: the highest wastewater reclamation and reuse standards required by the state of California (Green et al., 1996). In the proposed schematic design, the ASP effluent flows into a constructed wetland.

### (e) Constructed Wetlands

Wetlands as waste treatment are shown in Fig. 22.43*c*. In the schematic design, two parallel constructed wetlands are proposed, allowing a comparison between their outflows.

The free surface wetlands (FSW) consist of shallow open basins or channels that are lined to prevent seepage. Soil supports a succession of wetlands vegetation, nourished by a continuing flow of effluent from the initial wastewater treatment system(s) (such as a secondary AIWPS facility, septic tank, etc.). Although the water surface is exposed to beneficial air and sunlight, proper design is required to mitigate human contact with the treated effluent in FSWs. It is also necessary to address issues such as mosquitoes or other water-borne insects.

The subsurface flow wetlands utilize a lined treatment basin filled with a coarse medium such as large gravel or crushed rock. The large voids (if properly designed) allow the effluent to flow freely through rocks and plant root systems. The relationship between gravel size, slope of bed, and rate of flow is complex and is described in detail in EPA (1993). A layer of soil is used to cover the gravel bed, so subsequent human contacts with the effluent and insect problems are precluded.

In either type of constructed wetland system, saturated and nutrient-rich conditions encourage the growth of both aerobic and anaerobic microbes and certain invertebrates. The most common plants include varieties of *scripus* (bulrushes), *phragmites* (reeds), *typha* (cattails), *canna,* and *iris.* These plants not only provide large amounts of surface area for microbial activities, but also create an aerobic zone underwater by transporting air through their stems to their roots. A thin aerobic environment results among the root hairs.

Constructed wastelands become a vivid, colorful demonstration of how, in nature, there is no such thing as waste. After passage through a slow sand filter (see again Fig. 21.3) and final disinfection, the reclaimed water from the E2R2 facility can provide campus landscape irrigation water, be released into the campus creek to aid habitat restoration, or used for aquifer recharge. Arcata, California, completed its wetlands treatment system in 1986. This system draws several hundred species of birds, as its abundant vegetation removes nutrients from and allows habitat reuse of the city's treated wastewater.

## (f) Greenhouse Ecosystems

When constructed wetlands are moved indoors, greater control and less area become possible. As pioneered by marine biologist John Todd at Ocean Arks International, *Living Machines* are a series of tanks, each with its particular ecosystem, that form a "stream." The stream is followed by an indoor marsh; together they achieve a high degree of tertiary wastewater treatment.

These enclosed systems typically cost less to construct and about the same cost to maintain as conventional sewage treatment plants, but use less energy because photosynthesis (solar energy) and gravity flow largely replace energy-intensive pumps and large-scale aerators. These systems do not harm the environment with a final dose of chlorine, and they produce about one-quarter of the sludge of conventional systems. Because they are pleasant to look at and smell much like commercial greenhouses, they can be located within the neighborhoods they serve. This can save the huge costs associated with large-diameter, deep-trench sewer lines and pumping stations necessary with distant, centralized treatment plants.

A diagram of one greenhouse ecosystem is shown in Fig. 22.44. This Living Machine begins with an anaerobic bioreactor, in effect a septic tank where solids are reduced to sludge and

methane gas is produced. From here the wastewater passes to aerated tanks and then to *ecological fluidized beds,* each with a particular combination of bacteria, algae, snails, and fish. The further along the treatment process, the more advanced the ecosystem, with goldfish (carp) growing in the final clarifier. This process depends on the following food chain: (1) aerobic bacteria consume suspended organic matter and convert ammonia into nitrites and then nitrates; (2) algae and duckweed feed on the products of the bacterial action; (3) snails and zooplankton eat the algae, while floating duckweed provide shade that discourages algal growth in the later stages of the process; (4) fish eat the zooplankton and the snails. Plants range from water hyacinth in the early stages to an increasing variety of specialized "accumulators" that remove troublesome phosphorus, heavy metals, and so on. Papyrus, canna lilies, bald cypress, willows, and eucalyptus (among other plants) each have a role. When trees outgrow the greenhouse, they are transplanted outdoors—but they are long-term storage for the heavy metals they absorb, not final solutions, because heavy metals will be returned to the earth when the trees die and decompose. Plants can be sold to nurseries or ground up as compost; the fish also enrich the compost. The smaller fish (shiners) can be sold as bait fish. These are small enough to dart back and forth in the piping between tanks, choosing their water quality. If a toxic slug of waste enters the system, they swim to later, clearer tanks until the danger is past.

In Fig. 22.45*a*, this greenhouse solar aquatic system cleans the wastewater from a mobile home park on Vancouver Island, British Columbia. In Fig. 22.45*b*, a striking Living Machine design occupies the center of an urban block in Kolding, Denmark.

In milder climates and with specialized wastewater, these processes can be achieved without the greenhouse enclosure. The Sonoma Mountain Brewery's Living Machine (Fig. 22.46) near Sonoma, California, occupies about 1 acre (0.4 hectare) and treats a maximum flow of 7,800 gpd (29,525 L/day) of waste from the brewery process. Two streams move in parallel, allowing a comparison of their outflows as experiments with detailed ecosystems are conducted. A part-time operator manages this treatment plant. After an underground anaerobic holding tank, flow moves to:

**WATER AND WASTE**

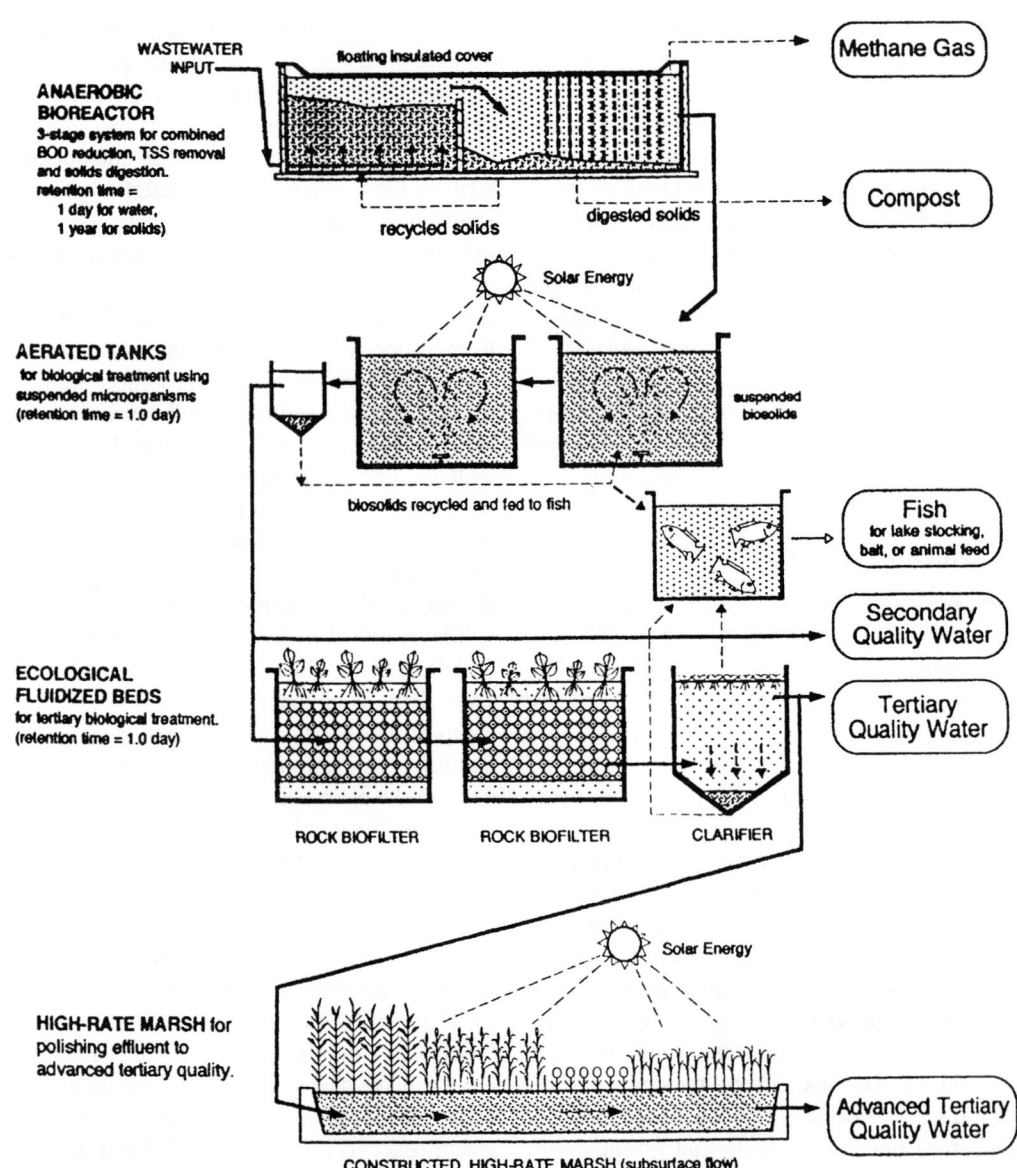

© Sunwater Systems, Inc.

**Fig. 22.44** *A Living Machine enclosed within a greenhouse uses a food chain in a series of ecosystems to treat wastewater. (Living Technologies, Inc.)*

(a)

(b)

**Fig. 22.45** (a) A greenhouse in a mobile home park in British Columbia encloses the 12 solar tanks and marsh of a Solar Aquatics System, designed by Ecological Engineering Associates and ECO-TEK, Inc. (Photo by Charles Rusch.) (b) In the middle of an urban block in Denmark, a Living Machine is a good neighbor. (Photo © 1996 by Robert Peña.)

Fig. 22.46 In the mild climate of California's Sonoma Valley, the waste from the Sonoma Mountain Brewery (a) is treated by an outdoor Living Machine (b). Flow is from left tanks to right tanks, then to a marsh. (c) Calla lilies and other plants lend beauty as well as function to this open aerobic tank. (Designed by Living Technologies, Burlington, VT.)

(a)

(b)

(c)

1. Closed aerobic reactors, each holding 2500 gal (9463 L). Air is blown through the tanks from the bottom (an adjacent small building encloses the blower); wastewater spends about 15 hours in these tanks. Next are—
2. Open aerobic tanks, four per stream. Plants, snails, and organisms can migrate among the agitated tanks, choosing their preferred level of nutrients ("pollution"). Water leaving the last of these tanks is much cleaner; water and sludge collected from the tanks are taken next to the—
3. Clarifier. Shaded by water hyacinth, algae cannot grow in the water at this advanced stage. In this quiet-water environment, solids settle out and are then pumped to an adjacent—
4. a. Open-air reed bed ("composter"), where they are dried and then composted. The compost will serve as nutrient for the hops used in the brewing process. Excess water percolates through an underlying sand bed and is returned to the beginning of the Living Machine streams. Meanwhile the clarified water passes next to—
   b. Ecological fluidized beds (two per stream), whose purpose is to further polish the wastewater. Water circulates many times through this section, pumped from the plant-shaded perimeter into a rock bed at the center. In this bed, microorganisms continue to clean the water. From the second ecological fluidized bed, water moves to an underground—
5. Irrigation pond, from where it is filtered through a—
6. Constructed wetland for further purification.

Finally, it is welcomed for irrigation of the hops and grapevines grown on site.

### (g) Pasveer Oxidation Stream

This method of sewage treatment (Figs. 22.47 to 22.49) has been adopted at the New York Institute of Technology. Serving the 450-acre campus at Old Westbury, Long Island, New York, this system provides an on-campus sewage treatment, which returns the puri-

fied effluent to the ground through 48 leaching wells located under the athletic field. The groundwater, thus restored, provides a contributing source of water for 400-ft (122-m)-deep wells, distantly located, that furnish part of the water supply for the campus buildings (see Fig. 22.47).

A few of the original small buildings, later converted to administration offices and classrooms, are still served by septic tanks and leaching fields. There was no public sewer near the campus, and in the 1970s the health authorities ruled out the use of septic tanks for the numerous additional buildings that were contemplated.

The oxidation stream process applied here was developed by the Netherlands Research Institute for Public Health Engineering and is now in operation in many U.S. locations as well as in Europe. It is considered to be a modified form of the *activated sludge* process (see later discussion of the Oceanside Plant in Section 22.8b), with aerobic digestion and periodic sludge removal.

The mechanical aerator (Fig. 22.48), which keeps the stream of sewage moving and provides the oxidation necessary for aerobic digestion, is an important feature. In this design, sludge-drying beds are placed on an island, surrounded by the continuously moving oxidation stream. Another feature of the system is its low profile (Fig. 22.49), readily screened by trees. The aerobic digestion process is not malodorous. The sounds are produced by the splashing water-wheel action of the mechanical aerators; no air compressor is required.

The plant has a full-time accredited operator; there are also two assistants and one relief operator. This design provides for an eventual population of 4330, with a 340,000-gpd (1,287,010-L/day) flow. The sludge was removed from the drying beds twice in the first five years of operation.

## 22.8 LARGER-SCALE SEWAGE TREATMENT SYSTEMS

Two examples of communitywide sewage treatment are presented here—one for a small, water-

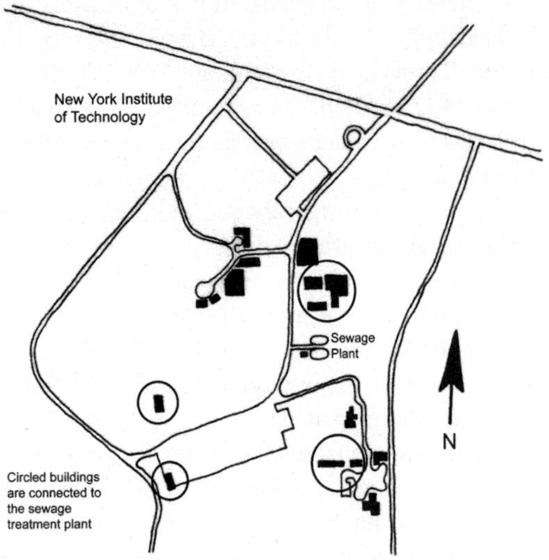

New York Institute
of Technology

Sewage
Plant

N

Circled buildings
are connected to
the sewage
treatment plant

*(a)*

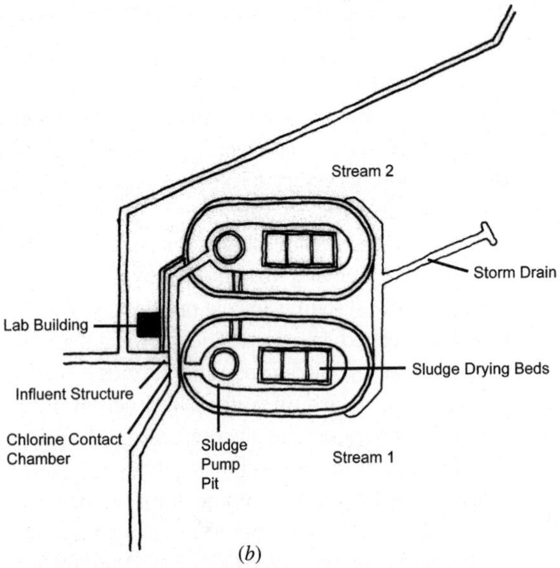

Stream 2

Storm Drain

Lab Building

Sludge Drying Beds

Influent Structure

Chlorine Contact
Chamber

Sludge
Pump
Pit

Stream 1

*(b)*

**Fig. 22.47** Sewage treatment plant, New York Institute of
Technology (NYIT). (a) The treatment plant's location is central
to all of the buildings it serves, an efficient choice. (b) Layout of
the treatment plant. Because of the plant's odorless operation,
its proximity to campus buildings poses no problem. (Courtesy
of Bogen Jenal, Engineers, P.C. Redrawn by Amanda Clegg.)

*(a)*

*(b)*

**Fig. 22.48** Details of NYIT treatment plant. (a) Rotor with a plastic "greenhouse" cover. (b) The rotor induces oxidation in this oval circulating stream. (A similar mechanism serves the high-rate pond in Fig. 22.43.)

(a)

(b)

**Fig. 22.49** *NYIT sewage treatment plant at completion. (a) The only projections above ground level are the office/laboratory and the plastic covers over the rotors. (b) The oxidation stream does not appear turbulent here, but it becomes so under the action of the mechanical aerator (rotor).*

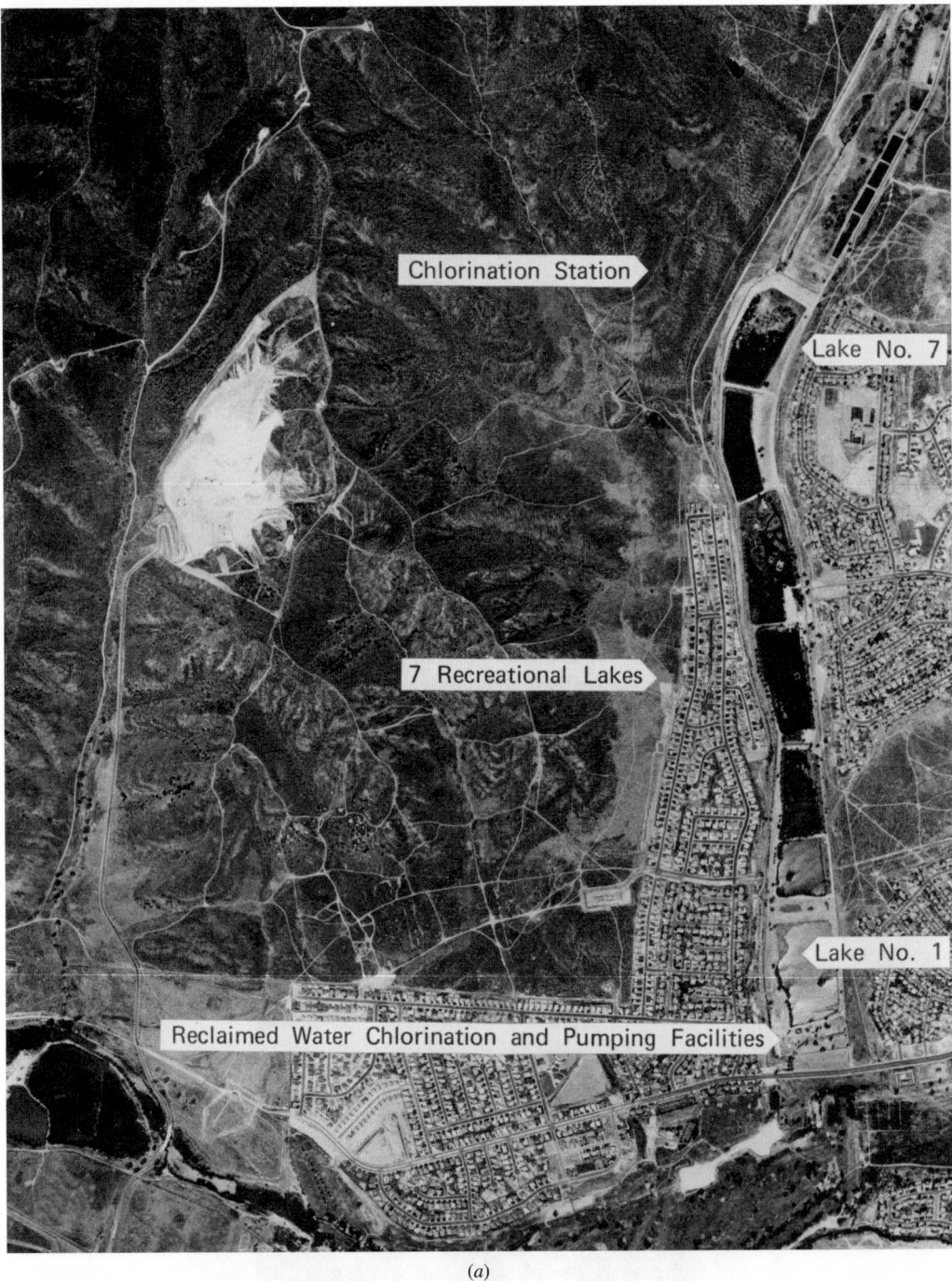

(a)

**Fig. 22.50** *Santee Water Reclamation Plant and Santee Park and Recreational Facilities, Padre Dam Municipal Water District, California. (a) Aerial photograph of the Santee Plant, including the seven recreational lakes. Note the location of the San Diego River at the bottom of the photograph. The water reclamation plant and the stabilization ponds do not appear in this aerial view. (b) Raw sewage from the community of Santee enters the treatment plant, which is located at the top of this diagram. The process then proceeds south to the point where reclaimed water is pumped to irrigation or recharges groundwater. Sludge is pumped to the San Diego Metro system. The plant is not burdened by stormwater, which is recharged to the ground locally in the community.*

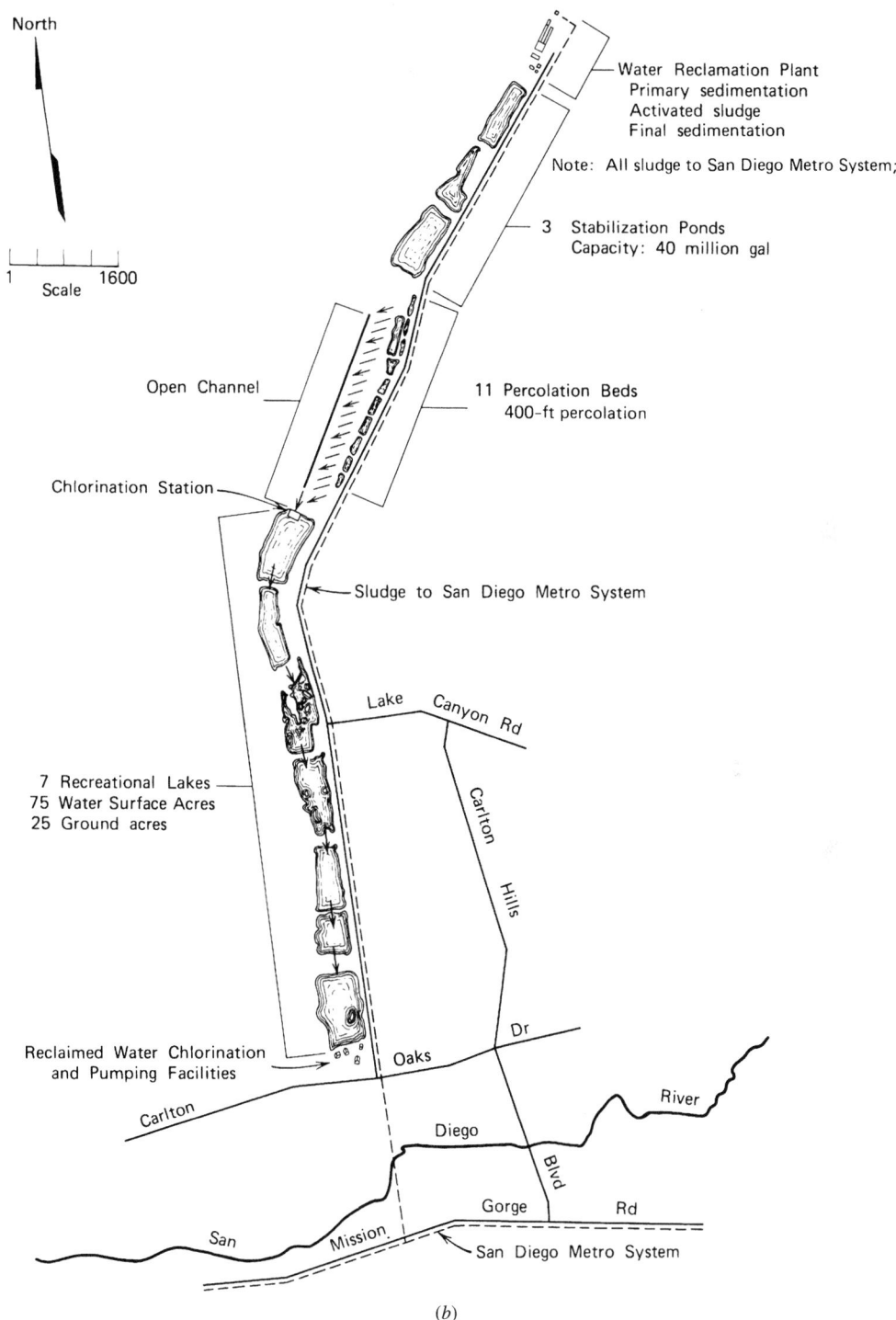

North

1  Scale  1600

Water Reclamation Plant
Primary sedimentation
Activated sludge
Final sedimentation

Note: All sludge to San Diego Metro System;

3  Stabilization Ponds
Capacity: 40 million gal

Open Channel

11  Percolation Beds
400-ft percolation

Chlorination Station

Sludge to San Diego Metro System

Lake  Canyon  Rd

Carlton  Hills

7  Recreational Lakes
75  Water Surface Acres
25  Ground acres

Dr

River

Reclaimed Water Chlorination
and Pumping Facilities

Oaks

Carlton

Diego

Blvd

Gorge  Rd

San  Mission

San Diego Metro System

(b)

**Fig. 22.50** (Continued)

scarce community near San Diego, California, and one for an oceanside urban neighborhood serving much of San Francisco.

### (a) Padre Dam Municipal Water District

The Santee Water Reclamation Plant (Fig. 22.50) represents innovations in water conservation and recycling during the 1960s and 1970s. The Padre Dam Municipal Water District is a region where rainfall is less than 15 in. (380 mm) per year. It has no local water supplies available, so water is imported 300 miles (480 km) from the Colorado River. If conventionally treated, the wastewater resulting from this flow would be discharged into the Pacific Ocean. Alternative sewage treatment (shown here) processes this fluid waste for secondary uses such as irrigation and recharging of groundwater. Many municipal potable water supplies are taken from rivers often heavily polluted with sewage and, after treatment, are pumped into the domestic water mains. Therefore, district officials here decided to install a system that would make use of purified wastewater for many secondary uses.

However, would such a concept be acceptable to the public? An effective public relations program concurrent with the technological advances has proved successful.

The project involved building a sewage treatment plant and utilizing seven pits left over from prior surface mining of sand and gravel. After partial purification of the sewage at the plant by the first two stages of a conventional activated sludge process, the effluent is discharged to form the lakes adjacent to the plant. This provides tertiary (oxidation) treatment, after which the cleaner effluent is pumped to a filter area at the north end of the complex, where it is further purified by flow through sand and gravel. Chlorination is performed, and the water then flows through a series of seven lakes, Nos. 7–6–5–4–3–2–1, in that order.

The appearance of new lakes in this semiarid region created initial interest that was rapidly augmented by a very clever program. At first, the lakes were fenced in. Then the fences were removed. Next, the seven lakes were made available for boating. They were stocked with fish, and careful studies were made that indicated that the fish were healthy and flourishing. Fishing was permitted, but for a while, all fish had to be returned to the water. Anglers were later permitted to keep and consume the fish. Finally, swimming was permitted. The overflow from the last lake, No. 1, discharges into Sycamore Canyon Creek and is used for irrigation of a golf course and the recharging of groundwater.

### (b) Oceanside Water Pollution Control Plant

Imagine the challenge of locating a major new sewage treatment plant on the shore of the Pacific, within the city of San Francisco, immediately adjacent to both the city zoo and a well-established neighborhood. This plant treats all the sewage in the western San Francisco watershed, an average dry weather flow of 21 million gal/day (79,491,720 L/day). The treated effluent is discharged through an ocean-bottom pipeline, 4.5 miles (7.2 km) offshore, with effluent and ocean water quality monitoring by an on-site laboratory. Sludge is shipped north to serve as the required daily cover for a landfill near Novato. The plant itself is concealed by large earth berms, through which entry and exit tunnels carry plant traffic. And the odor? It is completely contained within a huge underground complex of wastewater basins and channels, over which the zoo will build its new Mammal Conservation Center on 5 ft (1.5 m) of topsoil cover. Exhaust air from the underground plant is filtered through activated carbon and potassium permanganate before release.

The process is simplified in Fig. 22.51a. Incoming sewage passes through a bar screen and

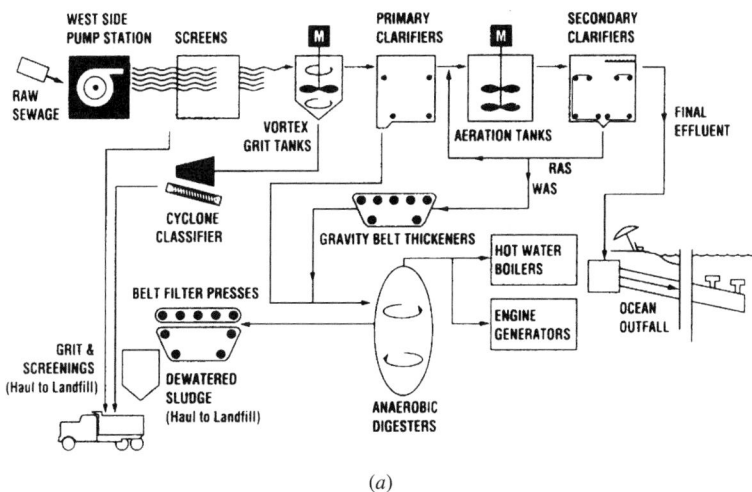

(a)

(b)

***Fig. 22.51*** *The Oceanside Water Pollution Control Plant, San Francisco, is largely underground, as well as hidden behind high earth berms. (a) Flow diagram. (Courtesy CH2M Hill and the City/County of San Francisco.) (b) View of some of the egg-shaped anaerobic digesters from within the berm walls. Biogas produced within the four 750,000-gal (28,375-m³) digesters supplies about 20% of the electricity used in the treatment plant.*

grit chambers, which remove the largest objects as well as grit, sand, and gravel (these are trucked to a landfill). Then the water flows to the primary clarifier, in which floating scum is skimmed off, and sludge settling on the bottom is pumped to the egg-shaped anaerobic digesters (Fig. 22.51b). The wastewater flows next to aeration tanks where bacteria grow fat, feeding off of and breaking down the remaining dissolved and suspended solids. Next is the secondary clarifier, where the fattened bacteria (activated sludge) sink to the bottom. Part of this activated sludge is returned to the aeration tanks to bolster the bacterial culture there. The remainder is thickened and pumped to the digesters.

Thus, the anaerobic digesters receive both primary sludge and thickened, activated sludge. As this sludge is stabilized and reduced in volume, biogas is produced: enough to heat water for both process and facility heating and to generate about one-fifth of the electricity used within the plant. Sludge leaving the digesters is dewatered in belt presses, then hauled to the landfill. During summer, biosolids are spread on agricultural land for soil enrichment.

Questions remain about such use of biosolids. Although a 1998 EPA national survey of sewage sludge found toxic chemicals at such low levels that regulation was deemed unnecessary, concentration of certain industries may produce local chemical or radioactive sludge pollutants at higher levels. Intercepting such wastes at their source, rather than allowing them into the sewage stream, seems an obvious imperative.

## 22.9 RECYCLING AND GRAYWATER

These chapters on water and waste have illustrated four common "grades" of water in buildings:

*Potable water* (usually treated, suitable for drinking)

*Rainwater*

*Graywater* (wastewater not from toilets or urinals)

*Blackwater* (water containing toilet or urinal waste)

Two more categories are:

*Dark graywater* (from washing machines with dirty diaper loads, kitchen sinks, and dishwashers; usually prohibited for reuse)

*Clearwater* (backwash water from reverse osmosis water treatment; condensation from a refrigerator or other cooling coils)

After considering the many blackwater sewage treatment complications described in the previous section, we might expect that graywater's reuse would, by contrast, be a simple matter. However, there are lingering prejudices against wastewater of any kind that complicate water recycling, if not prohibit it outright. This section presents some future directions in which we might well move if water conservation is to be taken seriously, followed by some code-mandated approaches.

*Rainwater*, given an adequate supply, could meet the need for many uses of water in and around buildings: bathing and laundry, toilet flushing, irrigation, and evaporative cooling, for example. Little or no treatment of rainwater is needed for such uses.

*Blackwater* requires much more extensive treatment, given the higher concentration of human waste in it. However, this may be the easiest waste to eliminate, given the waterless alternatives (Section 22.1).

*Graywater* (also spelled *greywater*) reuse opportunities are more limited than those of rainwater, because graywater carries increased threats from pathogens. For use in toilet flushing, it must undergo treatment before reuse. It is likely to contain soap, hair, and, occasionally, human waste (from soiled clothes in the laundry). If it is to be used in drip irrigation systems, filtering is essential. If kitchen wastes are to be included, grease and food solids add to the problems. Graywater may well be dirtier than rainwater, but its supply is likely far more predictable and constant. However, codes tend to limit the use of graywater to underground landscape irrigation for single-family houses.

The collection of these separate types of water poses a problem. In the early years of indoor plumbing, rainwater, graywater, and blackwater were mixed together, which severely overloaded sewage treatment facilities in rainy weather. Today, we usually mix graywater and blackwater. There are precedents for keeping them separate, however (Fig. 22.52).

### (a) Graywater and Future Recycling

Notwithstanding present code restrictions, the potential for residential graywater reuse (given

*(a)*

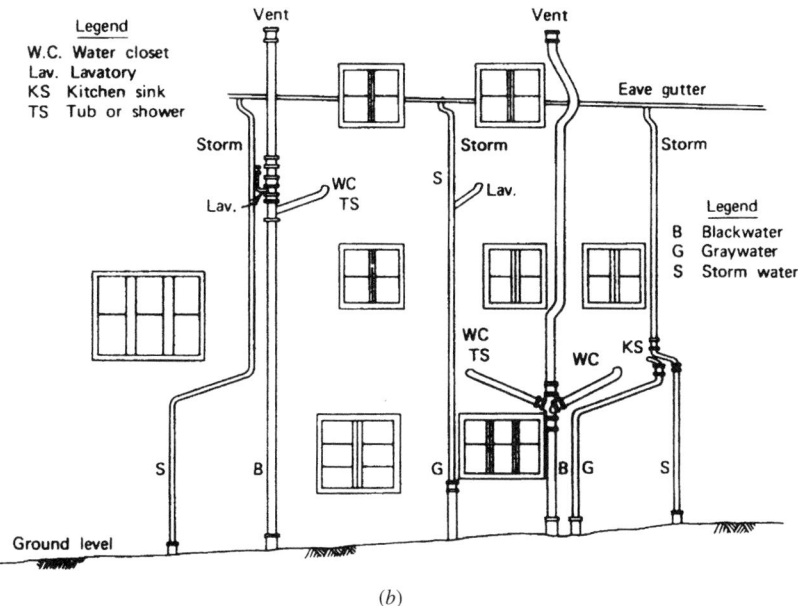

Legend
W.C. Water closet
Lav. Lavatory
KS Kitchen sink
TS Tub or shower

Vent

Vent

Eave gutter

Storm

Storm

Storm

Lav.

WC
TS

S

Lav.

Legend
B Blackwater
G Graywater
S Storm water

WC
TS

WC

KS

S

B

G

B G

S

Ground level

*(b)*

**Fig. 22.52** *Residence of the 27th Vicar of the Parish Church at Bibury (established 1086 A.D.) in the English Cotswolds. (a) Clearly, the plumbing was added after the construction of the residence—perhaps several hundred years later. In England the drainage system for an older building often appears on the outside of the building. (b) It seemed inappropriate to ask the vicar's permission to inspect his indoor facilities, so we have taken the liberty of assigning probable uses. Note that (1) offsets of the vertical piping are made with "easy bends" and are pitched down in the direction of the flow; (2) vents are located at high outdoor points; and (3) stacks carrying blackwater (from toilets) are of larger pipe size than those carrying graywater. In this sparsely settled region of the Cotswolds, the dispersal of stormwater to the ground and the private septic tank treatment of blackwater plus graywater can be satisfactory. In planning new systems, the possible separate treatment of blackwater, graywater, and stormwater is becoming an important consideration. (Photo by William McGuinness.)*

**1037**

separate waste collection systems) was shown earlier (Fig. 22.1b). Bathing's 21 g/cd (gallons per capita per day) can meet much of the conventional toilet's 32 g/cd and all of low-flush toilets' water needs. Laundry's 14 g/cd can help with irrigation; limiting contact between irrigation water and people seems prudent, hence the emphasis on underground distribution. Table 22.12 relates both the quantity and quality of residential graywater to its use in irrigation. Another potential benefit of separate graywater collection is that the water's heat content might be partially reclaimed (Table 22.13). In England, the Aquasaver Company diverts and cleans water from lavatories, baths, and showers for reuse in toilet flushing, clothes washing, car washing, and irrigation. This low-pressure system, installed behind panels in a house's bathroom, uses a pump to push graywater through a series of filters to remove soap, detergents, and other impurities and then to send it to a storage tank (in the attic or above points of use). Here it undergoes weir and trickle filtration and then treatment with nonhazardous cleaning agents. Finally, before reuse, it passes through a carbon filter network.

Storage raises a health issue. To avoid bacteria buildup in graywater, storage should be avoided and the filtered water taken directly to irrigation. However, storage allows for controlled dosage, allows solids to settle, allows for some potential heat recovery, provides for overflow to the sewer, and is often mandated by code.

Various stages of treatment are appropriate for different kinds of graywater (Fig. 22.53). Kitchen sink waste contains grease, which should be trapped (and periodically removed) before it can clog the filters and heat exchangers that serve graywater recycling systems. Similarly, lavatory showers and laundry waste contain lint and hair that must be intercepted quickly. Devices that do so, which are often called *interceptors*, were described in Fig. 22.21. However, code approval tends to sharply limit graywater recycling, assumes no filtering, and assigns its use.

**TABLE 22.12  Residential Graywater Sources for Irrigation**

| PART A. SOURCES WITH PUMPS | | |
|---|---|---|
| **Source** | **Average Outflow Water Quantity** | **Outflow Water Relative Quality** |
| Clothes washing machine | 30–50 gal (114–190 L) per load, top loader, 10 gal (38 L) per load, front loader. Average 1.5 loads per week per adult, 2.5 loads per week per child | Good: medium concentration of soaps, lint. With biocompatible cleaners, quality can improve to excellent; diapers degrade quality to poor. |
| Automatic dishwasher | 5–10 gal (19–38 L) per load | Poor: low to high quantity of solids depending on degree of pre-rinsing; high salt and pH from conventional cleaning compounds |
| PART B. GRAVITY FLOW SOURCES | | |
| **Source** | **Outflow Water Quantity** | **Outflow Water Quality** |
| Shower | 20 gal (76 L) per day per person, low-flow; 40 gal (151 L) per day per person, high-flow | Excellent: minimal concentration of soap and shampoo, but hair can clog distribution pumps and lines |
| Tub | 40 gal (151 L) per bath | Concentration in tub water likely less than shower water |
| Bathroom lavatory | 1–5 gal (4–19 L) per day per person | Good: concentration of soap, shaving cream, and toothpaste can be high |
| Kitchen sink | 5–15 gal (19–57 L) per day per person | Mixed: high in nutrients but also high in solids, grease, and soap |
| Reverse-osmosis water purifier | 3–5 gal per gallon of drinking water (3–5 L per L) | Excellent: usually no suspended solids; contains 25% more concentration of the same dissolved solids as tap water |
| Water softener backwash Softened water | 5% of indoor water use (All graywater from softened water source) | Very bad: extremely high in salt Poor: contains salt (unless potassium chloride is used in softening process) |

*Source:* Adapted from *Create an Oasis with Greywater,* by Art Ludwig, © 1997, Art Ludwig. Poor Richard's Press, Santa Maria, CA.

**TABLE 22.13  Heat Recovery from Graywater**

| Source | Volume/Use, Gal (L) | Flow Rate, Gal (L) | °F | °C | Quality | Coincident[a] |
|---|---|---|---|---|---|---|
| Kitchen sink | Up to 5 (up to 20) | 2.5 (10)/min | 85 | 30 | Poor | No |
| Lavatory | Up to 1.5 (up to 5) | 1.25 (5)/min | 85 | 30 | Fair | No |
| Bath | 32 (120) | 5 (20)/min | 100 | 37 | Good | No |
| Shower | 15 (50) | 2 (7)/min | 100 | 37 | Good | Yes |
| Washer | 40 (150) | 7 (25)/min | 60 | 15 | Moot | No |

*Source:* Glenn Nelson, "Graywater Heat Recovery," © *Solar Age,* August 1981. Reprinted by permission.

[a]Coincident flow of warm wastewater and input water to heater. If "no," a holding tank is needed for heat exchange. If "yes," a counter-current heat exchanger can be built into the drain line (see Fig. 22.53).

## (b) Subsurface Irrigation

In the 1997 *Uniform Building Code* (in its Appendix G), water *only from* bathroom lavatories, showers and tubs, and clothes washing machines and laundry tubs, and *only in* single-family residences, can *only be used for* subsurface "irrigation" on the same site as the residence. Such mandated trench constructions, in their zeal to avoid any surface graywater, may deliver very little water to surface plants.

Site conditions must meet the requirements of Table 22.14 and Fig. 22.54. The bottom of the trenches must be at least 5 feet (1.5 m) *above* the

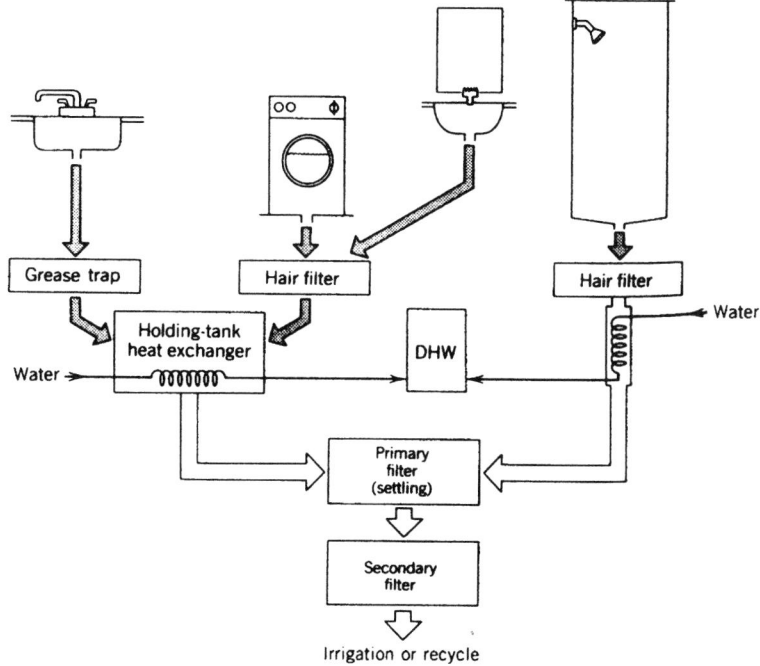

*Fig. 22.53* Sequence of water treatment and heat reclamation for domestic graywater, where such systems might be approved.

### TABLE 22.14 Location of Residential Graywater System

| From | To Holding Tank ft | To Holding Tank mm | To Irrigation/Disposal Field[a] ft | To Irrigation/Disposal Field[a] mm |
|---|---|---|---|---|
| Building structures[b] | 5[c] | 1,524 | 2[d] | 610 |
| Property line adjoining private property | 5 | 1,524 | 5 | 1,524 |
| Water supply wells[e] | 50 | 15,240 | 100 | 30,480 |
| Streams and lakes[e] | 50 | 15,240 | 50[f] | 15,240 |
| Sewage pits or cesspools | 5 | 1,524 | 5 | 1,524 |
| Disposal field and 100% expansion area | 5 | 1,524 | 4[g] | 1,219 |
| Septic tank | 0 | 0 | 5 | 1,524 |
| On-site domestic water service line | 5 | 1,524 | 5 | 1,524 |
| Pressurized public water main | 10 | 3,048 | 10[h] | 3,048 |

*Source:* Reprinted by permission from the *1997 Uniform Plumbing Code.* © by the International Association of Plumbing and Mechanical Officials, Walnut, CA.

[a]When irrigation/disposal fields are installed in sloping ground, the minimum horizontal distance between any part of the distribution system and the ground surface shall be fifteen (15) feet (4.6 m).

[b]Including porches and steps, whether covered or uncovered, breezeways, roofed porte-cocheres, roofed patios, carports, covered walks, covered driveways, and similar structures or appurtenances.

[c]The distance may be reduced to zero feet [meters] for above ground tanks when first approved by the Administrative Authority.

[d]Assumes a 45° (0.79 rad) angle from the foundation.

[e]Where special hazards are involved, the distance required shall be increased as may be directed by the Administrative Authority.

[f]These minimum clear horizontal distances shall also apply between the irrigation/disposal field and the ocean mean higher tide line.

[g]Plus two (2) feet (610 mm) for each additional foot of depth in excess of one (1) foot (305 mm) below the bottom of the drain line.

[h]For parallel construction. For crossings, approval by the Administrative Authority shall be required.

highest known seasonal groundwater. Flow estimates are as follows:

Number of occupants:
two for the first bedroom
one for each additional bedroom

Combined showers, bathtubs, and washbasins, flow per occupant: 25 gal/day (95 L/day)

Laundry, flow per occupant: 15 gal/day (57 L/day)

A holding tank is required with a minimum 50-gal (189-L) capacity. An unvalved overflow must connect to the building sewer (ahead of any septic tank).

The irrigation/disposal field must be divided into a minimum of three valved zones (allowing the occupants to better direct flow rates). The total area of this field is the aggregate length of the perforated pipe times the width of the proposed field. The required area is based on the estimated graywater flow (or size of the holding tank, whichever is larger), based upon the soil types and acceptance rates, as well as the trench design guidelines, shown in Table 22.15. If percolation tests are performed, the 24-hour percolation rate results must be within these limits:

Minimum, 0.83 gal/ft$^2$ (33.8 L/m$^2$)

Maximum, 5.12 gal/ft$^2$ (208.5 L/m$^2$)

One of several graywater system types from the 1997 *Uniform Building Code*, Appendix G, is shown in Fig. 22.55.

**EXAMPLE 22.4** A four-bedroom single-family residence is located on sandy clay soil more than 5 ft above the highest seasonal groundwater level. If the bathroom and laundry facilities contribute to a graywater system, how large an irrigation/disposal field is required?

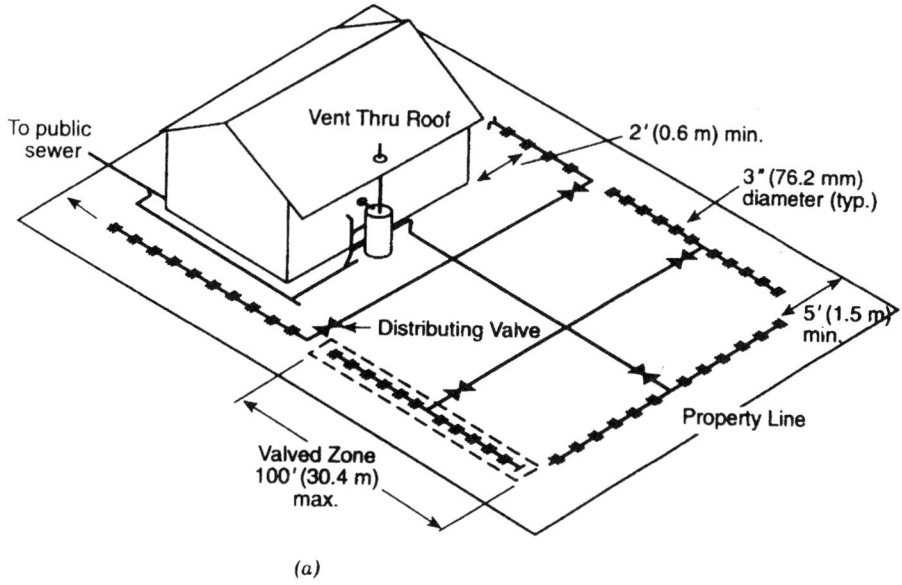

(a)

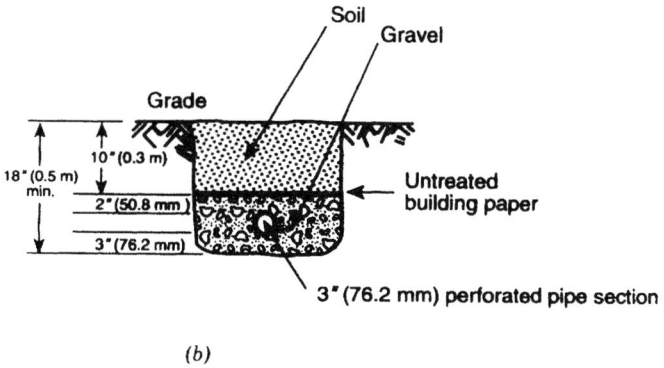

(b)

**Fig. 22.54** *Example of a residence whose daily graywater flow requires an irrigation/disposal field (a) using a valved zone as shown in the detail (b). In this example, five such valved zones are shown (a minimum of three such zones are required). (Reprinted by permission from the 1997* Uniform Plumbing Code. © *by the International Association of Plumbing and Mechanical Officials, Walnut, CA.)*

**TABLE 22.15  Graywater Design Criteria, Six Typical Soils**

| PART A. AREA OF IRRIGATION/LEACHING FIELD | | | | |
|---|---|---|---|---|
| | Minimum Area of Irrigation/Leaching per Estimated Graywater Discharge per Day | | Maximum Absorption Capacity of Irrigation/Leaching per 24-Hour Period | |
| Type of Soil | ft²/100 gal | m²/L | gal/ft² | L/m² |
| Coarse sand or gravel | 20 | 0.005 | 5 | 203.7 |
| Fine sand | 25 | 0.006 | 4 | 162.9 |
| Sandy loam | 40 | 0.010 | 2.5 | 101.8 |
| Sandy clay | 60 | 0.015 | 1.7 | 69.2 |
| Clay with considerable sand or gravel | 90 | 0.022 | 1.1 | 44.8 |
| Clay with small amounts of sand or gravel | 120 | 0.030 | 0.8 | 32.6 |

| PART B. IRRIGATION/LEACHING FIELD CONSTRUCTION | | | | |
|---|---|---|---|---|
| | Minimum | | Maximum | |
| | (I-P) | (SI) | (I-P) | (SI) |
| Number of valved zones | 3 | 3 | | |
| Number of lines per valved zone | 1 | 1 | | |
| Length of each perforated line | | | 100 ft | 30,840 mm |
| Bottom width of trench | 12 in. | 305 mm | 18 in. | 457 mm |
| Spacing of lines, center-to-center | 4 ft | 1219 mm | | |
| Depth of earth cover of lines | 10 in. | 254 mm | | |
| Depth of filter material cover of lines | 2 in. | 51 mm | | |
| Depth of filter material beneath lines | 3 in. | 76 mm | | |
| Grade of perforated lines | 3 in./100 ft | 2 mm/m | | |

*Source:* Adapted by permission from the *1997 Uniform Plumbing Code.* © by the International Association of Plumbing and Mechanical Officials, Walnut, CA.

**SOLUTION**

The total number of occupants = 2 + 1 + 1 + 1 = 5. Estimated graywater flow is

(25 + 15) gal/day per occupant × 5 occupants
= 200 gal/day

Required field size (Table 22.15) of each valved zone is

60 ft²/100 gal × 200 gal/day = 120 ft²

If each valved zone has two lines at the minimum spacing of 4 ft, the minimum width of the irrigation/disposal field is 4 ft; the minimum length of the two perforated lines is therefore

120 ft²/4 ft = 30 ft

Thus, three irrigation/disposal fields, each 4 ft by 30 ft, in trenches constructed as shown in Fig. 22.54, will meet code minimums. ∎

## 22.10 STORMWATER TREATMENT

Until relatively recently, stormwater runoff from buildings and paved surfaces was considered harmless and was conducted without treatment, and as rapidly as possible, to receiving streams, rivers, and lakes. However, almost three-quarters of the U.S. population lives on only about 7% of the land. And as our urban watersheds experience an increase in their percentage of paved or roofed land area, two problems develop. First, a heavy rain greatly increases the peaks in stream flows (because less rain is retained for later, slower release by unpaved surfaces); this accentuated peak–valley pattern of flows makes life more difficult for the aquatic flora and fauna. Second, pollutants (especially those accumulated on roadways) are washed into the stream; oil, gasoline, antifreeze, fragments of brake linings and tires, and so on are especially unwelcome deposits. Some design strategies to minimize these impacts, such as roof retention, porous pavement, and on-site groundwater recharging, were discussed in Section 20.6.

With minimal changes in street design, stormwater receptacles can be constructed not only to provide access for maintenance, but also to screen out sediment, capture oil and other floating contaminants, and provide a more even outflow (Fig. 22.56). But much more can be done.

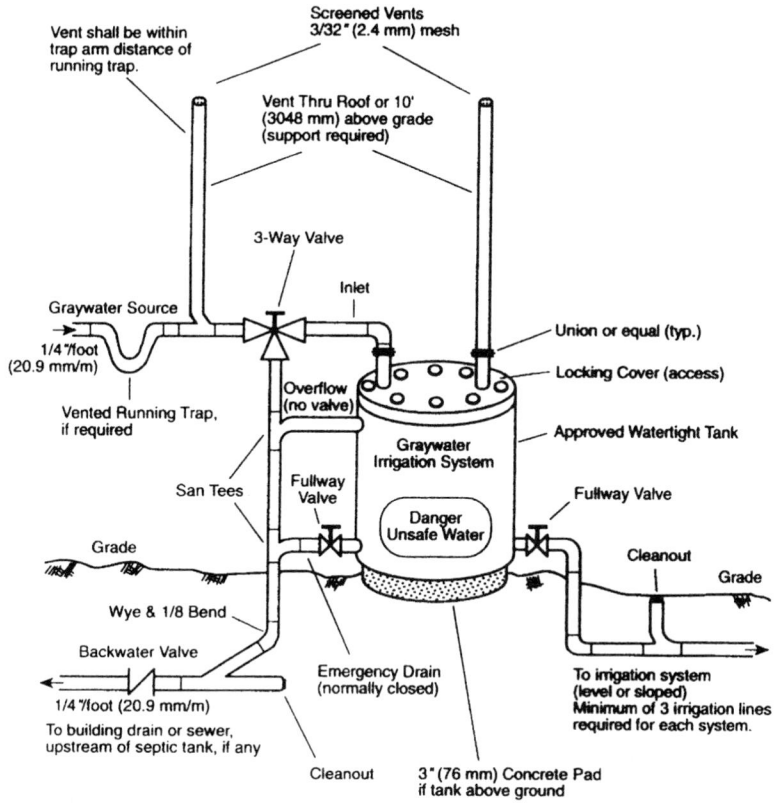

**Fig. 22.55** *Schematic diagram of a graywater system with an above-ground, gravity-emptied tank. (Reprinted by permission from the 1997 Uniform Plumbing Code. © by the International Association of Plumbing and Mechanical Officials, Walnut, CA.)*

*Phytoremediation* is the science of cleaning polluted soil and water with plants. (The constructed wetlands and algal ponds described in Sections 22.7 and 22.8 apply phytoremediation to sewage treatment.) Phytoremediation extracts, degrades, or contains contaminants. Plants that *extract* take up and accumulate contaminants in their shoots and leaves; when the vegetation is removed from the site, so is the contaminant. Plants that *degrade* break down contaminants (such as hydrocarbons and other organic compounds) until they are no longer toxic. Some plants degrade toxins in their root zone, whereas others use elements of organic toxins as food. Plants that *contain* essentially immobilize toxins in long-term storage. However, when the plants die and decompose, long-lived toxins may be released.

For treating stormwater, plants are typically incorporated in drainage strips between paved parking areas. The mix of plants is similar to those

cited earlier in wetland construction, with sedges and rushes especially effective at cleansing (see Fig. 22.2 for a sample trench). Shade trees could add welcome summer thermal benefits, as well as intercepting and storing some pollutants. The capacity of the planting strip should be sufficient to intercept the "first flush" of rain, which carries the heaviest load of pollutants. Underground drains intercept excess runoff and lead to a capture chamber where floatable contaminants can be periodically removed. The screened, regulated flow from the capture basin will be much gentler on the receiving stream.

A new community in Bellevue, Washington, incorporated a stormwater treatment facility in a 20-acre (8.1-hectare) public park. Beneath tennis courts, ballparks, picnic areas, and other playing fields, a 300,000-ft$^3$ (8495-m$^3$) sediment vault (surface area about 8 acres [3.2 hectares]) slows the flow of stormwater, encouraging sediment to

Grit/oil chambers

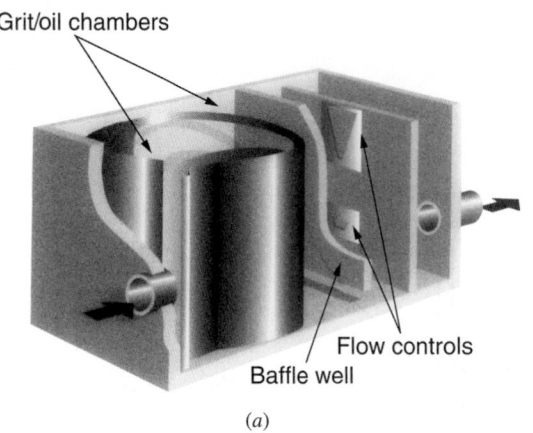

Flow controls

Baffle well

(a)

(b)

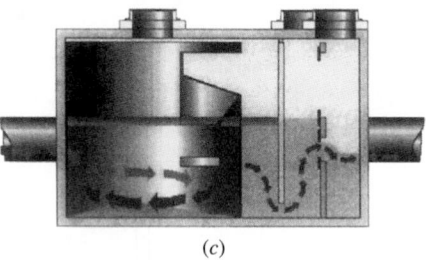

(c)

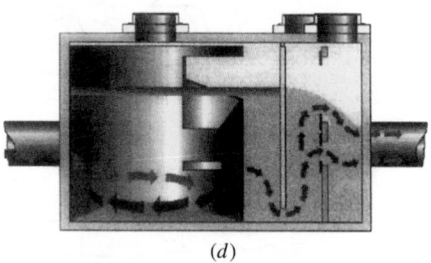

(d)

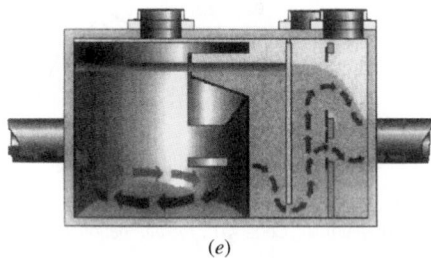

(e)

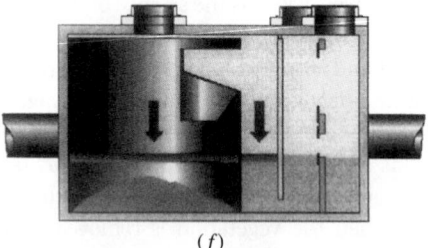

(f)

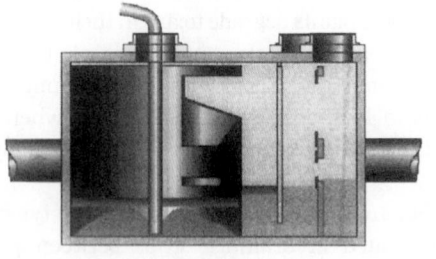

(g)

**Fig. 22.56** *This Vortechs™ Stormwater Treatment System (a) is designed to separate both floating oily contaminants and sediment from stormwater. (b) Rather like a septic tank, two chambers are separated by a baffle. Oil scum and sediment remain in the first chamber. (c) In a 2-month storm, water rises above the top of the inlet pipe, greatly reducing velocity and turbulence within the first chamber, encouraging sediment to precipitate to the bottom. (d) As flow increases, rising above the low-flow control (second chamber), oily contaminants accumulated from previous storms float upwards and more sediment falls out. (e) At the height of stormwater flow, usually designed for a 10-year storm, the high-level outlet begins to discharge. Sometimes a peak flow bypass is used. Oily scum and sediment continue to be trapped in the manhole. (f) As the flow subsides, water outflow is controlled, lessening the strain on the storm drainage system. (g) Cleanout, at the lowest water levels, is best done by a vacuum truck. As the grit chamber empties, oily liquids and floating debris drain back toward the inlet and can be removed along with the sediment. Note that the bottom of the baffle between the chambers always remains submerged. (Courtesy of Vortechnics Incorporated, Portland, ME.)*

drop out. The water then passes to a filter basin of vegetation, sand, and peat for the primary purpose of phosphorus removal. From there it is slowly released to the receiving stream, which then drains into a lake.

Building designers rarely encounter such communitywide pollution control strategies, but parking lots and on-site roadways are encountered regularly. Design for local pollution control, even for seemingly benign stormwater, is part of the architect's responsibilities.

## References

Converse, J., et al. 1978. *Design and Construction Manual for Mounds.* Small Scale Waste Management Project, University of Wisconsin Extension, Madison, WI.

EPA. 1993. *Subsurface Flow Constructed Wetlands for Wastewater Treatment.* EPA 832-R-93-008. U.S. Environmental Protection Agency, Office of Water. Washington, DC.

Green, F.B., L. Bernstone, T.J. Lundquist, J. Muir, R.B. Tresan, and W.J. Oswald. 1995a. "Methane Fermentation, Submerged Gas Collection, and the Fate of Carbon in Advanced Integrated Wastewater Pond Systems," in *Water Science and Technology,* Vol. 31, No. 12.

Green, F.B., T.J. Lundquist, and W.J. Oswald. 1995b. "Energetics of Advanced Integrated Wastewater Pond Systems," in *Water Science and Technology,* Vol. 31, No. 12.

Green, F.B., L. Bernstone, T.J. Lundquist, and W.J. Oswald. 1996. "Advanced Integrated Wastewater Pond Systems for Nitrogen Removal," in *Water Science and Technology,* Vol. 33, No. 7.

IAPMO. 1997. *Uniform Plumbing Code.* International Association of Plumbing and Mechanical Officials, Walnut, CA.

ICC. 1997. *International Plumbing Code.* International Code Council, Falls Church, VA.

Ludwig, A. 1997. *Create an Oasis with Greywater.* Poor Richard's Press, Santa Maria, CA.

Milne, M. 1976. *Residential Water Conservation.* U.S. Office of Water Research and Technology, Department of Commerce, NTIS, Springfield, VA.

National Small Flows Clearinghouse. 1997. "Lagoon Systems Can Provide Low-Cost Wastewater Treatment," in *Pipeline,* Vol. 8, No. 2. (National Small Flows Clearinghouse, Morgantown, WV.)

Nelson, G. 1981. "Graywater Heat Recovery," in *Solar Age,* August.

Oswald, W.J., C.G. Golueke, R.C. Cooper, H.K. Gee, and J.C. Bronson. 1963. "Water Reclamation, Algal Production and Methane Fermentation in Waste Ponds," in *International Journal of Air and Water Pollution,* Vol. 7.

Oswald, W.J., A. Meron, and M.D. Zabat. 1970. "Designing Waste Ponds to Meet Water Quality Criteria," in *Proceedings of the Second International Symposium on Waste Treatment Lagoons,* Kansas City, MO.

Oswald, W.J. 1990. "Advanced Integrated Waste-water Pond Systems," in *Supplying Water and Saving the Environment for Six Billion People: Proceedings of the 1990 American Society of Civil Engineers Convention,* Environmental Engineering Division, November 5–8, San Francisco.

Oswald, W.J., F.B. Green, and T.J. Lundquist. 1994. "Performance of Methane Fermentation Pits in Advanced Integrated Wastewater Pond Systems," in *Water Science and Technology,* Vol. 30, No. 12.

CHAPTER 23

# Solid Waste

FOR SOME DESIGNERS, the last building distribution system considered is the one involving the bulkiest items: the flow of supplies in and solid waste out. Because this system usually is not seen as consuming building energy or requiring specialized equipment, it ordinarily becomes a lower-priority system in the design process. Yet provisions for delivery of supplies, and especially for the collection and storage of solid wastes, can be more space-consuming than water/waste systems, can present a fire danger, and can create severe local environmental problems. The separation of solid waste for resource recovery involves significant energy and environmental consequences. Finally, mechanical equipment associated with solid waste is now more commonly installed.

## 23.1 WASTE AND RESOURCES

Since the late 1940s, there has been a marked increase in the amount of packaging material used for consumer products. Where shoppers once refilled reusable containers for bulk supplies at the market, for example, they now buy food in bags or cans that are discarded after use. This trend has increased spatial needs in the store, where shelves of cans are required instead of a bin of bulk products, and in the home, where such packaging soon turns into waste and must be stored until garbage collection day. Energy is required to make the boxes, bags, cans, and other containers; transport them; and

collect them as trash. Devices such as trash compactors add both space and energy requirements to the process. Landfills for garbage disposal fill more rapidly as solid-waste flows increase; methane and leachate from these landfills are potential environmental problems.

There may not be much that building designers can do about these increased packaging trends. But the solid wastes from buildings do contain important resources (Tables 23.1 and 23.2), and a designer can help society to recover those resources rather than bury them in landfills or dump them in the ocean. Furthermore, the buildings themselves can be designed for materials recovery upon remodeling or demolition, as was discussed in Sections 2.3 and 2.5.

The resources within solid waste can be divided into the high-grade resources represented by recyclable materials and the low-grade resources of heat and by-products obtainable from the burning or decomposing of combustible solid wastes.

### (a) High-Grade Resources

These include metals such as aluminum and steel, paper and paperboard, wood, and some plastics. For buildings, they pose the problem of storage while awaiting collection. For many of these materials, reuse in their present form is desirable, which eliminates compaction as a storage strategy. Table 23.1 compares uncompacted and compacted volumes of various recyclable materials.

## TABLE 23.1 Typical Recyclable Materials in Building Operation

| Category | Product | Recycling Label | Material Description | Conversion, Volume to Weight | |
|---|---|---|---|---|---|
| | | | | I-P Units | SI Units |
| Paper | Ledger paper, white letterhead | SWL | Sorted white ledger, high-grade white paper | Uncompacted: 1 yd³ = 500 lb Compacted: 1 yd³ = 750 lb | Uncompacted: 1 m³ = 297 kg Compacted: 1 m³ = 445 kg |
| | Computer paper | CPO | Computer printout | Uncompacted: 1 yd³ = 500–600 lb Compacted: 1 yd³ = 1000–1200 lb | Uncompacted: 1 m³ = 297–356 kg Compacted: 1 m³ = 593–712 kg |
| | Colored paper | SCL | Sorted color ledger | Uncompacted: 1 yd³ = 500 lb Compacted: 1 yd³ = 750 lb | Uncompacted: 1 m³ = 297 kg Compacted: 1 m³ = 445 kg |
| | Newspaper | Mix | Newsprint | Uncompacted: 1 yd³ = 350–500 lb Compacted: 1 yd³ = 750–1000 lb | Uncompacted: 1 m³ = 208–297 kg Compacted: 1 m³ = 445–593 kg |
| | Magazines | Mix | Clay-coated paper | Not available | Not available |
| | Telephone books | Mix | Mixed papers-adhesives | 1 book = 1–3 lb | 1 book = 0.5–1.4 kg |
| | Cereal boxes | Mix | Coated paperboard | Not available | Not available |
| | Shipping boxes | OCC | Old corrugated cardboard | Uncompacted: 1 yd³ = 285 lb Compacted: 1 yd³ = 500 lb | Uncompacted: 1 m³ = 169 kg Compacted: 1 m³ = 297 kg |
| Glass | Food jars, beverage bottles | 1 2 3 | Amber glass Green glass Clear glass | Loose, whole: 1 yd³ = 600 lb Manually crushed: 1 yd³ = 1000 lb Mechanically crushed: 1 yd³ = 1800 lb | Loose, whole: 1 m³ = 356 kg Manually crushed: 1 m³ = 593 kg Mechanically crushed: 1 m³ = 1068 kg |
| Plastic | Beverage containers | PET | Polyethylene terephthalate | Whole: 1 yd³ = 30 lb | Whole: 1 m³ = 18 kg |
| | Milk containers | HDPE | High-density polyethylene | Whole: 1 yd³ = 25 lb Crushed: 1 yd³ = 50 lb Compacted: 1 yd³ = 600 lb | Whole: 1 m³ = 15 kg Crushed: 1 m³ = 30 kg Compacted: 1 m³ = 356 kg |
| | "Clamshell" containers | — | Polystyrene plastic foam | Not available | Not available |
| | Film plastic | LDPE | Low-density polyethylene | Not available | Not available |
| Metals | Beverage cans | | Aluminum/ bimetal | Whole: 1 yd³ = 50–70 lb Crushed: 1 yd³ = 300–450 lb | Whole: 1 m³ = 30–42 kg Crushed: 1 m³ = 178–267 kg |
| | Food and beverage cans | | Steel with tin finish | Whole: 1 yd³ = 125–150 lb Crushed: 1 yd³ = 500–850 lb | Whole: 1 m³ = 74–89 kg Crushed: 1 m³ = 297–504 kg |
| Miscellaneous | Pallets | | Wood | Not available | Not available |
| | Food waste | | Organic solids and liquids | 55-gal drum—415 lb | (208 L drum = 188 kg) |
| | Yard waste | | Organic solids | Leaves, uncompacted: 1 yd³ = 250 lb Leaves, compacted: 1 yd³ = 450 lb Wood chips: 1 yd³ = 500 lb Grass clippings: 1 yd³ = 400 lb | Leaves, uncompacted: 1 m³ = 148 kg Leaves, compacted: 1 m³ = 267 kg Wood chips: 1 m³ = 297 kg Grass clippings: 1 m³ = 237 kg |
| | Used motor oil | | Petroleum product | 1 gallon = 71 lb | 1 L = 8.5 kg |
| | Tires | | Rubber | 1 passenger car = 20 lb 1 truck = 90 lb | 1 passenger car = 9 kg 1 truck = 41 kg |

*Source:* Reprinted by permission from AIA: Ramsey/Sleeper, *Architectural Graphic Standards,* 9th ed., © 1994 by John Wiley & Sons.

**TABLE 23.2  Solid Waste Sources in Building Operation**

| Classification | Occupancy | Types of Waste Generated (percent, where available) | Quantities of Waste Generated | |
|---|---|---|---|---|
| | | | I-P Units | SI Units |
| Residential | Studio or 1-bedroom apartment 2- or 3-bedroom apt. or single-family house | Newspaper (38/43)[a] Plastic (18/7)[a] Miscellaneous (13/18)[a] Metals (14/9)[a] Compost (10/15)[a] Glass (2/8)[a] | 1–1.5 yd$^3$/unit/month (200–250 lb) or 2.5 lb/ person/day 1.5–2 yd$^3$/unit/month (250–400 lb) or 2.5 lb/ person/day | 0.8–1.1 m$^3$/unit/month (91–113 kg) or 1.1 kg/person/day 1.1–1.5 m$^3$/unit/month (113–181 kg) or 1.1 kg/person/day |
| Commercial | Office | Plastics, compost, used oil, metals, and glass (30)[b] High-grade paper (29)[b] Mixed papers (23)[b] Newspaper (10)[b] Corrugated cardboard (8)[b] | 1.5 lb/employee/day or 1 yd$^3$/10,000 ft$^2$/day (includes 0.5 lb of high-grade paper/ employee/day) | 0.7 kg/employee/day, or 0.8 m$^3$/1,000 m$^2$/day (includes 0.2 kg of high-grade paper/ employee/day) |
| | Department store Wholesale/retail Shopping center | Corrugated cardboard, compost, wood pallets, high-grade paper, and plastic film | 1 yd$^3$/2,500 ft$^2$/day 70 lb/$1000 sales/day 2.5 lb/100 ft$^2$/day | 0.8 m$^3$/250 m$^2$/day 32 kg/$1000 sales/day 1.1 kg/100 ft$^2$/day |
| | Supermarkets | Corrugated cardboard, com-post, and wood pallets | 1 yd$^3$/2,500 ft$^2$/day | 0.8 m$^3$/250 m$^2$/day |
| | Restaurants/ entertainment | Compost (38)[b] Corrugated cardboard (11)[b] Newsprint (5)[b] High-grade paper (4)[b] | Cafeteria, 1 lb/meal Fast food, 200 lb/$1000 sales Restaurant, 1.5 lb/meal | Cafeteria, 0.45 kg/meal Fast food, 91 kg/ $1000 sales Restaurant, 0.7 kg/meal |
| | Drug stores | Corrugated cardboard and high-grade paper | 1 yd$^3$/2,000 ft$^2$/day | 0.8 m$^3$/200 m$^2$/day |
| | Banks, insurance companies | High-grade paper, mixed paper, and corrugated cardboard | 0.75 lb high-grade paper/ person/day (survey required) | 0.34 kg high-grade paper/person/day (survey required) |
| Hotels and Motels (not including restaurants) | High occupancy Average occupancy | Glass, aluminum, plastic, high-grade paper, news-paper, and corrugated cardboard | 0.5 yd$^3$/room/week 3.2 lb/room/day 0.17 yd$^3$/room/week 1.7 lb/room/day | 0.38 m$^3$/room/week 1.45 kg/room/day 0.13 m$^3$/room/week 0.8 kg/room/day |
| Institutional | Hospitals | Compost, high-grade paper, biomedical waste, corru-gated cardboard, glass, and plastics | 1 yd$^3$/5 occupied beds/day | 0.8 m$^3$/5 occupied beds/day |
| | Nursing homes Retirement homes | | 1 yd$^3$/15 persons/day 1 yd$^3$/20 persons/day | 0.8 m$^3$/15 persons/day 0.8 m$^3$/20 persons/day |
| Educational | Grade school High school Universities | High-grade paper, mixed paper, newspaper, corru-gated cardboard, compost, plastic, glass, and metals | 1 yd$^3$/8 rooms/day 1 yd$^3$/10 rooms/day (Survey required) | 0.8 m$^3$/8 rooms/day 0.8 m$^3$/10 rooms/day (Survey required) |

*Source:* Reprinted by permission from AIA: Ramsey/Sleeper, *Architectural Graphic Standards,* 9th ed., © 1994 by John Wiley & Sons.

[a]Percent by volume/percent by weight.

[b]Percent by volume.

Glass is especially suitable for recycling when it can be reused after simple washing, as in the case of beer and soft-drink bottles. Returnable bottles and cans have measurably decreased roadside litter in states that now require such containers. Newspaper is so readily stored and recycled that in many communities, charitable groups or service organizations recover newspapers for profit. Recycled paperboard can easily save 50% of the energy that would be required for pulp from virgin material.

The recovery of aluminum saves 96% of the energy necessary to produce it originally. Because the aluminum production process is dependent on electricity, the recycling of aluminum is even more attractive. Energy conservation in aluminum production developed to the point that only 82% as much energy was needed to make a pound of aluminum in 1982 as in 1972. More than a third of that reduction was attributable to aluminum recycling programs; the remainder came from improvements in the production process. As for steel, recycling can produce a 52% energy savings compared to the use of virgin material.

Wood has become more valuable as fiber than as fuel. Wood chips and scraps that once were burned or buried are now recycled to become oriented strand board (OSB) used in structural insulated panels (SIPs), window frames, and many other aspects of building and furniture construction.

Plastics are more difficult to recycle. Regulations and consumer preferences discourage the use of recycled plastics in food-related items. However, the plastics from items such as soft-drink containers, margarine tubs and lids, and milk jugs can be reprocessed into plastic pellets, which are cheaper than virgin plastic pellets. These pellets are made into nonfood items—toys, building products, sports products, and other things.

### (b) Low-Grade Resources

These resources include materials for which recycling is impractical but that are combustible. Low-grade resources include gaseous wastes, liquid and semiliquid wastes, and solid wastes. Industrial and commercial processes can generate wastes of all types, including some with very high heat content and some that are very toxic. Although recovering heat by burning such materials seems better than wasting them altogether, air pollution regulations have severely restricted simple trash burning as a means of solid waste disposal. Incinerators must meet increasingly strict regulations and, as a result, are very rarely installed in buildings. A better option is to compost them in landfills. The methane generated in landfills can then be used as a high-grade fuel.

Where large quantities of mixed trash, garbage, and other refuse are collected, special resource-recovery plants can be built to recover materials, produce useful steam for electricity generation, and reduce the flow of waste to landfill. In this energy-intensive process, the mixed garbage is shredded and blown through large "air classifiers" that separate the organic (burnable) wastes from metals and glass. The burnable wastes are then used, under controlled combustion, to generate electricity. The metals are further separated magnetically into ferrous and nonferrous classes; these and the glass are then recycled.

The proposed E2R2 integrated infrastructure facility at California Polytechnic State University, San Luis Obispo, combining wastewater treatment and solid waste processing, was introduced in Fig. 22.43. The solid waste facility is shown in Fig. 23.1.

The proposed schematic design for solid waste

processing (SWP) in the E2R2 facility involves transporting unsorted solid waste to a central processing facility located on campus. Materials will then be sorted using mechanical equipment and hand sorting. The composition of the solid waste stream is expected to be 70% organic materials, suitable for composting, and 815 tons (739 Mg) per year of recyclable materials (i.e., metals, glass, paper, and cardboard). Roughly 10% of the solid wastes are anticipated to be unsuitable for either recycling or composting. These "inert" materials will be disposed of at the local landfill.

Once separated, the design calls for the nonrecyclable, organic materials to be uniformly ground and combined with sludge (from the city of San Luis Obispo's wastewater treatment plant) or manure (from the university's farming operations) to achieve a proper carbon-to-nitrogen balance. Given the dynamic nature of the recyclable materials market, it is anticipated that, at times, some potentially recyclable paper will be included in the compost operations.

After mixing, the organic materials will be loaded into a high-solids anaerobic digester and undergo methane fermentation. The biogas generated by the fermentation process will be harvested and stored on site. Prior to use, the biogas will be scrubbed (processed) to remove impurities, making it suitable for use as fuel. A portion of the methane will be used to heat the digester, and the excess will power a micro fuel cell to generate electricity for campus use. Other potential uses include using methane in the campus's boiler, as an alternative to gasoline in the university's automobile and truck fleet, and even to power an ultra-high-temperature hazardous waste disposal system.

The anaerobic digestion process is estimated to generate 180,000 ft$^3$ (5098 m$^3$) per day of biogas. The energy content of the methane in this biogas is estimated at $35 \times 10^9$ Btu ($36.9 \times 10^9$ kJ) per year.

After being unloaded from the high-solids digester, the remaining biosolids will be pressed to recover bacteria-rich liquids, which are useful in inoculating the next batch of materials for the digester. The dewatered biosolids will then be mixed with algae from the adjacent AIWPS Facility to create a nutrient-rich compost. This compost will be available for use on the campus's landscape and its farm lands or for sale to local gardeners. Yearly compost production is anticipated to exceed 5000 tons (4535 Mg) air-dried weight.

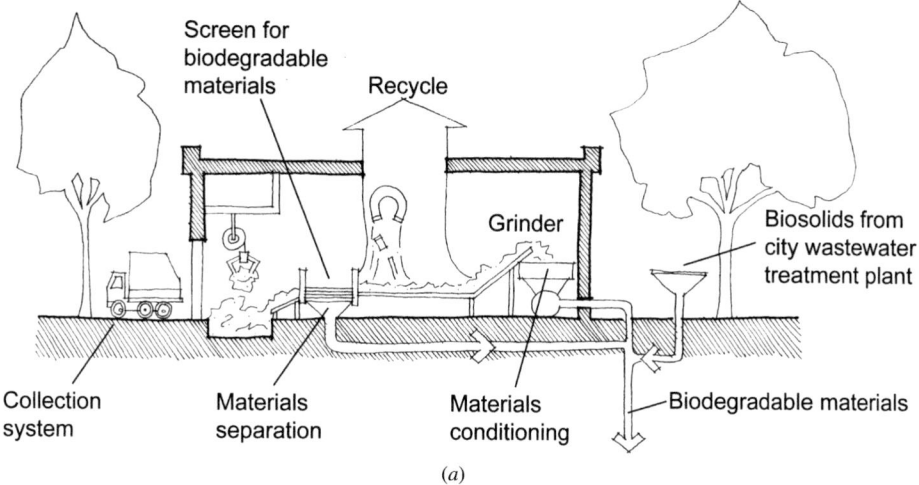

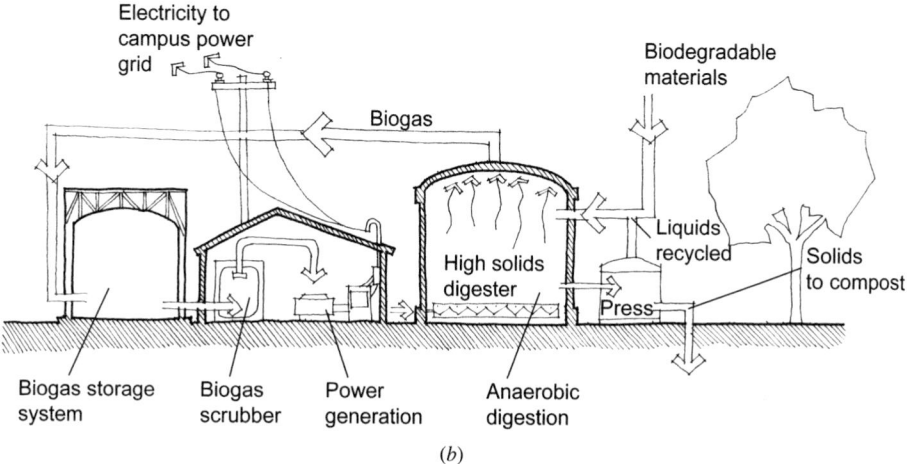

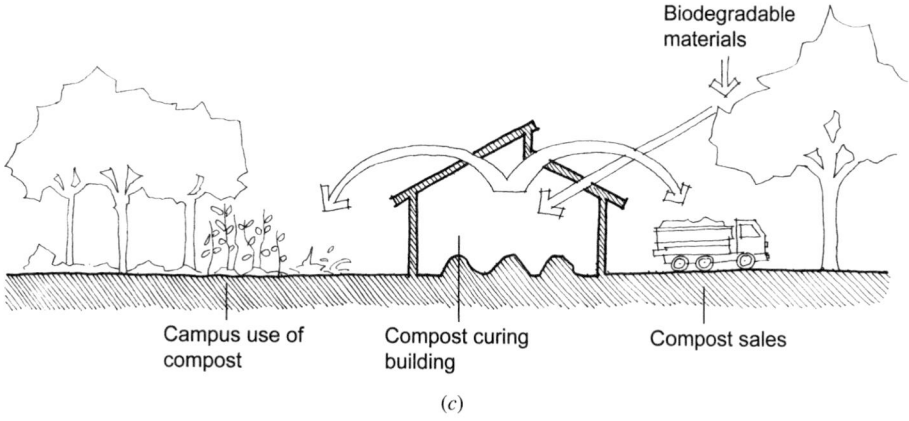

**Fig. 23.1** *Schematic design proposal for solid waste processing at the E2R2 Facility at Cal Poly State University, San Luis Obispo, California. (This works in conjunction with the proposed wastewater treatment facility; see Fig. 22.43.) Unsorted solid wastes are separated into recyclable, organic, and inert materials (a). The organic materials are ground, combined with sludge from the city's sewage treatment plant, and loaded into the high solids digester (b) to undergo anaerobic fermentation. The biogas generated by the fermentation process is scrubbed to remove impurities from the methane, making it suitable for use as a fuel to heat the digester and to power a micro fuel cell to generate electricity for the campus. Biosolids from the digester are pressed, and the liquid, rich in anaerobic bacteria, is then used to inoculate the next anaerobic fermentation cycle. Pressed biosolids are mixed with nutrient-rich algae from the adjacent AIWPS™ facility and pasteurized in compost curing buildings (c). Once cured, the compost can be used on landscaped areas or farm lands or sold to local gardeners. For the site plan, see Fig. 22.43d. (Courtesy of Prof. Douglas Williams, Agricultural Engineering Department, and Prof. Daniel Panetta, Architecture Department, Cal Poly, San Luis Obispo. Redrawn by Dain Carlson.)*

As an added touch, straw bale construction is proposed for the E2R2 facility buildings. Currently, plans call for an education and laboratory building, a shared control office for wastewater treatment and solid waste processing operations, a solid waste processing building, and on-site student housing. Straw bale construction is advantageous because of its affordability, as well as its exceptional insulation and noise mitigation characteristics. In addition, using rice straw for construction will help provide an alternative to burning waste straw from California's rice crop.

Conventional landfills accept unsorted garbage from a huge variety of sources. Over many years, anaerobic combustion in such enclosed landfills generates methane gas, a usable resource, as described previously. Fig. 23.2 shows an architectural opportunity for such landfill sites: a greenhouse heated by burning landfill (methane) gas. Electricity generated by landfill gas supplies the lights for the greenhouse. As a design guideline, a site with 1 million tons (907,000 Mg) of refuse in place should produce enough low-heat-value "landfill gas" to support

a cogeneration system that includes a greenhouse. Landfill gas, rather than natural (pipeline) gas, can also drive a fuel cell.

## 23.2 RESOURCE RECOVERY: CENTRAL OR LOCAL?

In general, the more thoroughly mixed the different types of solid waste, the harder it is to recover their high- and low-grade resources. From an energy conservation viewpoint, solid wastes should be kept locally as separate as possible; glass bottles should be washed and reused rather than broken and recycled, and unrecyclable but burnable solid wastes should be kept clean and dry until they can be burned or composted. The earlier different types of metals are separated, for example, the less energy spent later on to separate aluminum from steel, and so forth. Organic food wastes could be composted for use on site rather than ground up and added to the load on the sewage treatment system.

**Fig. 23.2** A 45,000-ft² (4180-m²) greenhouse built over part of a 70-acre (28-hectare) Michigan landfill utilizes "landfill gas" (including about a 65% methane component) both to power its space heaters and to generate electricity for lighting. Waste heat from the generator also contributes to space heating. (Photo by George P. Graff. Courtesy of Willow Run Farms, Inc., Ypsilanti, MI.)

However, there are two disadvantages to local solid waste separation (at the point of their discard)—one cultural, one physical. Keeping solid wastes separate requires somewhat more effort and time on the part of the consumer. Rather than dumping everything in a garbage can, the resource-conscious consumer will wash metal cans, remove both ends and flatten them, then deposit them in a container reserved for ferrous metal. Newspaper and box cardboard are stacked separately; returnable bottles and cans are kept apart from glass; compostable kitchen wastes are kept separate from garbage. Composting in urban apartments presents a unique space-odor challenge.

This disadvantage of slightly more work and time for waste separation is compounded by the physical disadvantage of the floor or cabinet space taken up by so many separate containers. In communities fortunate enough to have recycling-oriented garbage service, these containers can be carried out and lined up to await garbage collection. The garbage trucks used for this purpose are com-plex assemblages of various bins rather than simple massive caverns with a compactor. In many communities, however, each pile of recyclable materials at home must be dealt with separately, either collected by a specialized handler (nonprofit agencies are examples) or taken to a different collection point—a disadvantage in terms of both the time and energy used in transportation. Clearly, the designer's incentive to include recycling in building function will be stronger where recycling collection is well established.

The characteristics of local waste separation, then, are increased consumer time and building space requirements; see Tables 23.2 and 23.3. Central waste separation is characterized by energy-intensive (and noisy) industrial processes. Sections 23.3 and 23.4 offer a detailed look at the consequences of local waste separation for some common building types.

The massive quantities of solid waste generated in urban areas give rise to the terms *urban mines* and *urban forests*. To the extent that these solid

**WATER AND WASTE**

## TABLE 23.3  Space Planning for Solid Waste

| PART A. I-P UNITS | | | |
|---|---|---|---|
| | | **Exterior Area Required** | |
| **Occupancy** | **Building Area ft²** | **For Trash ft²** | **For Recyclable Materials, ft²** |
| Nonresidential | 0–5000 | 12 | 12 |
| | 5001–10,000 | 24 | 24 |
| | 10,001–25,000 | 48 | 48 |
| | 25,000+ | Each additional 25,000 ft² require 48 ft² each, trash and recyclables | |
| Multifamily Residential | 2–6 Units | 12 | 12 |
| | 7–15 Units | 24 | 24 |
| | 16–25 Units | 48 | 48 |
| | 25+ Units | Each additional 25 dwelling units require 48 ft² each, trash and recyclables | |
| PART B. SI UNITS | | | |
| | | **Exterior Area Required** | |
| **Occupancy** | **Building Area m²** | **For Trash m²** | **For Recyclable Materials, m²** |
| Nonresidential Buildings | 0–465 | 1.1 | 1.1 |
| | 466–929 | 2.2 | 2.2 |
| | 930–2323 | 4.5 | 4.5 |
| | 2323+ | Each additional 2323 m² require 4.5 m² each, trash and recyclables | |
| Multifamily Residential | 2–6 Units | 1.1 | 1.1 |
| | 7–15 Units | 2.2 | 2.2 |
| | 16–25 Units | 4.5 | 4.5 |
| | 25+ Units | Each additional 25 dwelling units require 4.5 m² each, trash and recyclables | |

*Source:* Reprinted by permission from AIA: Ramsey/Sleeper, *Architectural Graphic Standards,* 9th ed., © 1994 by John Wiley & Sons.

wastes can be turned into resources, savings can be realized: of energy in transportation, of land area for landfills, and of virgin materials replaced by these recycled products.

## 23.3 SOLID WASTE IN SMALL BUILDINGS

The choice between separation of waste or the mixing of garbage in one can is most often enjoyed by the occupants of small buildings. Where food preparation is involved, as in residences or restaurants, the collection of special containers can reach surprising complexity (Fig. 23.3).

As the point of origin for so many types of solid waste, the kitchen is often the location of waste separation and storage. There is an inherent conflict, however, between the kitchen's frequently hot, humid environment and the need for a cool, dry, well-aired place for solid wastes. This suggests that a space—pantry, air-lock entry, cabinet, or closet—

should be provided that opens to the kitchen on one side and to the outside on the other (Fig. 23.4). Both the daily deposits of solid waste and the weekly waste removal are made easy, and the near-outdoor conditions are better for the waste in most U.S. climates. It is also important that cleaning of the waste storage area be easy.

### (a) Garbage Disposer

This common appliance (Fig. 23.5) is usually installed below the kitchen sink, often in common with a dishwasher. It grinds up organic food scraps and sends them on through the sewer. This device is a boon to central garbage collection because it lightens the weight of the garbage can and adds less moisture to the garbage (which thus can be burned more efficiently). Finely chopped organic matter has a better chance to biodegrade at the wastewater treatment plant than in a tightly packed landfill. However, garbage disposer units require both water

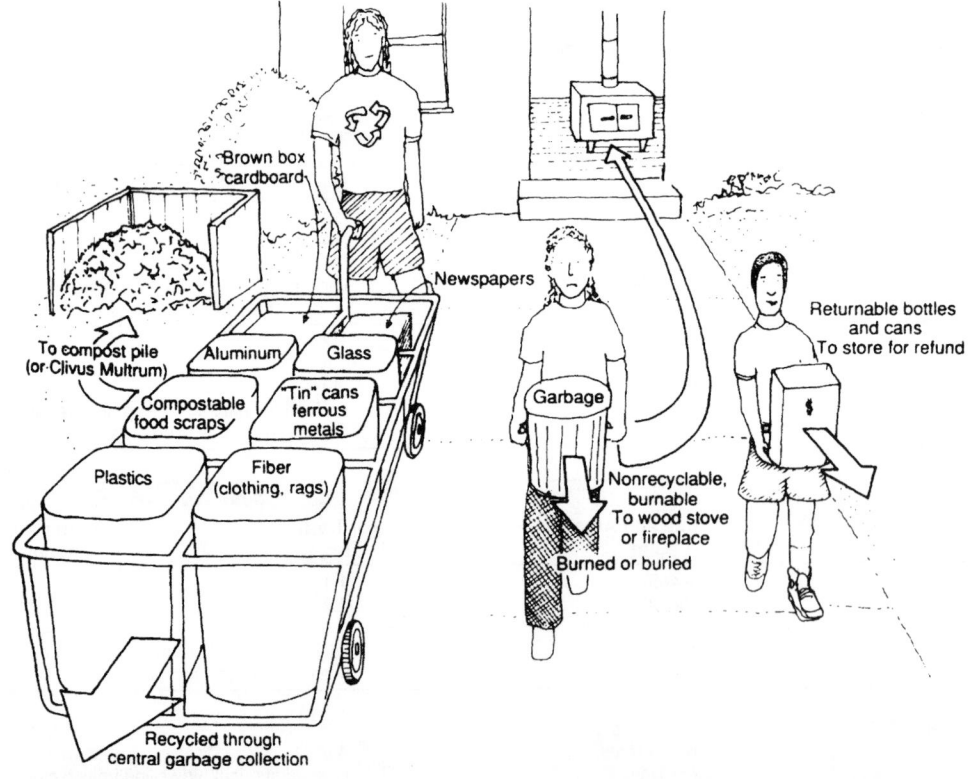

**Fig. 23.3** *Local separation of solid wastes can require many containers. In communities that offer recycling garbage collection service, the consumer can recycle without transporting individual containers to many different collection points. Some recycling services ask for glass to be separated into green, brown, and clear types; some offer to recycle white paper.*

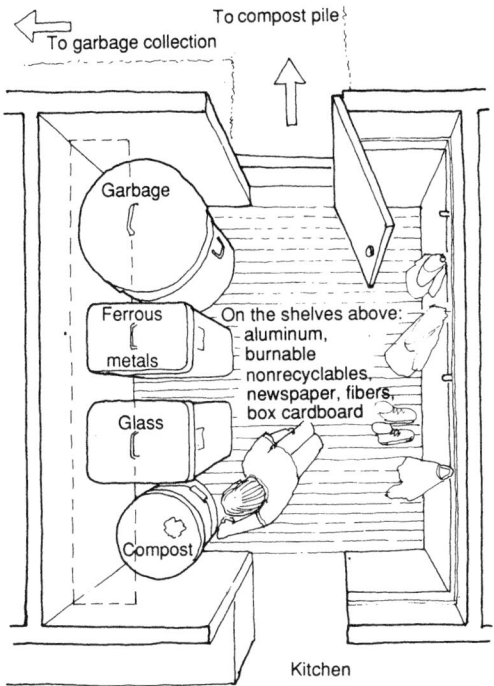

*Fig. 23.4* An entry vestibule (or airlock) to the kitchen can serve as a depository for solid wastes, as well as for coats, boots, and other items. Heavy, dirtier items belong on the floor, and many recyclables can be readily and conveniently stored on shelves. An outdoor hose bibb is useful for washing out soiled compost containers and garbage cans.

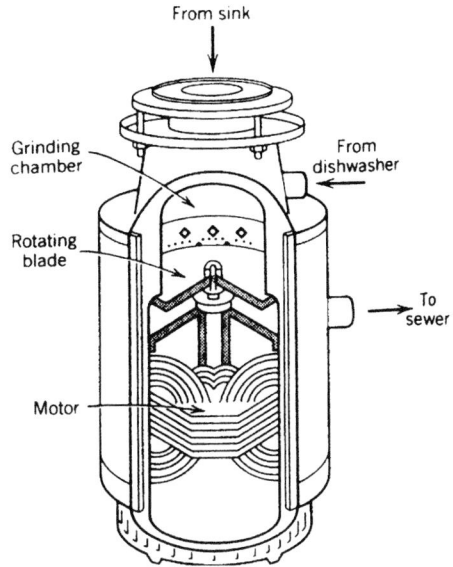

*Fig. 23.5* The garbage disposer (or grinder) diverts solid waste (food scraps) from landfills to sewage treatment plants. Occupying very little space, it saves garbage storage space and some of the user's time, but it requires more energy and much more water than treatment of the food scraps as garbage. Composting is another alternative. (From Milne, 1976.)

and energy. Water must be kept running during the grinding process to coagulate grease for chopping, wash the blades and keep them free of debris, and cool the grinder's motor. In all, 2 to 4 gal (7.5 to 15 L) are required for 1 minute of operation. Because more water and solid waste are deposited in the sewer system, moreover, the central sewage treatment plant requires more energy to operate.

### (b) Garbage Compactor

In some small buildings, this device allows a much less bulky storage arrangement. Used selectively, it can compact several of the stacks of items shown in Fig. 23.3, such as aluminum for recycling, ferrous metals, and box cardboard. Used indiscriminately, it can make the central process of garbage separation more difficult by crushing dissimilar items together. For a recycling-conscious single-family household, it is questionable whether the compactor will save much more storage volume than it takes for itself.

However, small stores and institutions with large quantities of a bulky waste (such as cardboard) could save considerable storage space.

### (c) Compost Pile

The alternative to the resource-hungry garbage disposal is the compost pile, familiar to most home gardeners as a source of excellent soil conditioners. Urban opportunities for composting seem very limited; yet, raised growing beds on a balcony or window flower boxes are logical recipients of a family's compost. The problem may not be so much what to do with the final product (humus) as where to locate the compost pile. Several self-contained composting containers are available.

The outdoor compost pile has several characteristics that challenge the designer. At its best, it is a frequently turned, quite warm, damp, well-aired source of rich humus (and red worms) for gardens; odors are noticeable only while the pile is turned. At its worst, it is a source of unpleasant odors, a breeding place for vermin, and a fast-food attraction for rats, dogs, and raccoons. Where odors are not objectionable, the heat generated in a frequently fed and

WATER AND WASTE

tended compost pile might be welcome against the exterior walls of residences. Clearly, these walls must have nonorganic exterior materials.

### (d) Storage Areas

As groups of residences are combined into large apartment complexes, the solid waste systems can make more significant demands on the designer. Approximate exterior storage areas are listed in Table 23.3. Where central storage compounds are provided, garbage cans should be fenced to ward off dogs and other marauders. Different bins for recyclable materials might be provided. Where space is limited, a single bin could accept specific materials on specified days—newspapers on Mondays, metals on Wednesdays—if collection schedules permit. A central compost pile could provide humus for the landscaping of the complex. (An incinerator, used to recover heat from nonrecyclable burnable wastes, is almost never approved for installation today.) This combination of space and equipment has special environmental needs, including garbage truck access, noise control, and location of both incinerator stacks and the compost pile with respect to prevailing winds.

### 23.4 SOLID WASTE IN LARGE BUILDINGS

In large buildings, it becomes more likely that solid wastes will be handled several times in the storage and collection process. This may inhibit the separated-waste approach because not only the employees who generate the waste, but also the custodians who collect and store it, must understand what items go into which bins. However, larger buildings tend to generate large and concentrated kinds of wastes, which makes recycling more attractive while also increasing the demand for storage space.

Consider the multistory office building. The office floor operations are likely to discard large quantities of white paper and smaller quantities of newspaper, box cardboard, and unrecyclable burnable trash (including floor sweepings). Much smaller quantities of food scraps (coffee grounds), metals, and glass are also generated. Given the high cost of rental space, the pressures are very high for a simple mixed-garbage can (rather than multiple separate bins). However, the cost of hauling trash to increasingly scarce landfills is now so high that many recycling programs pay for themselves both by avoiding landfill demands and by the value of recycled materials.

### (a) The Collection Process

In larger buildings, the collection of solid waste is typically a three-stage process (Fig. 23.6). The first stage is the generation of the waste itself by employees who might be provided with one wastebasket each. If each employee is expected to separate wastes, this suggests a redesigned receptacle. At the typical desk in an office building, white paper, recyclables, compostables, and garbage would be deposited in separate compartments. Waste separation at the point of origin thus may not require much more floor area, just more attention by the worker.

The second stage begins as custodians disconnect these individual baskets, dump them into separate bins on a collection cart, and reconnect the individual empty baskets for the next day's deposits. At various special-purpose stations, special wastebaskets can be supplied: white paper at the computer room and the copying machine, compostables and garbage at the employee lounge. Floor sweepings are added to garbage. When the cart is full, it is wheeled to the service closet, which is probably located within the core of the building. Here is a container for each category of waste, along with a service sink to wash out the garbage bin (and perhaps a paper shredder to be used selectively by employees).

The third stage begins at the ground floor of the service elevator, where white paper and other recyclables are perhaps shredded and stored until collection; compostables are stored or sent to a roof garden compost pile; and garbage is compacted and bagged. In the storage space, cool, dry, and fresh air is desirable. A sprinkler fire protection system is often required. The compactors and shredders are noisy and must be vibration-isolated from the floor. At the end of the third stage, a truck or van from the recycling center collects recyclables, and a garbage truck collects garbage bags.

### (b) Audubon House

This New York City office building (Fig. 23.7) utilizes several of the previously mentioned strategies and demonstrates some unusual design impacts of

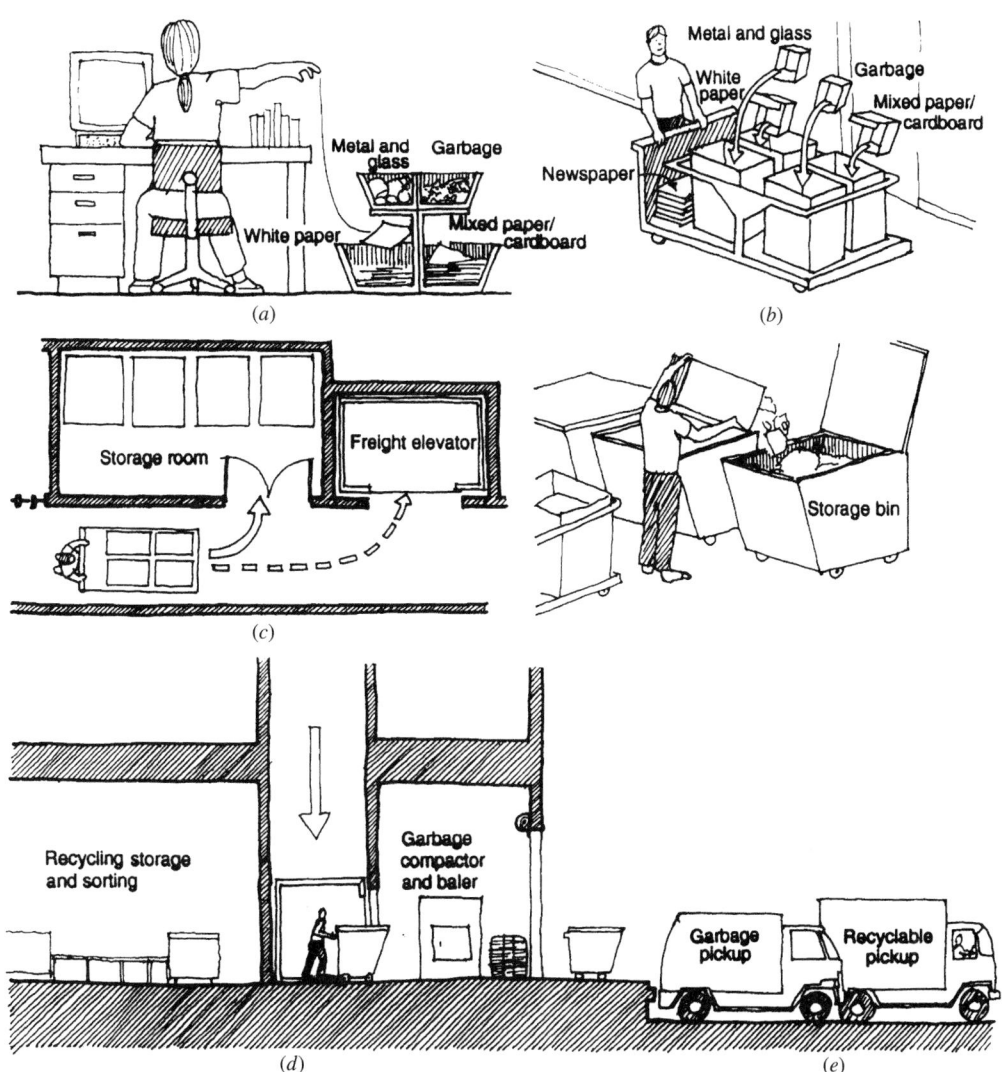

**Fig. 23.6** *Hypothetical three-stage solid waste collection process with four-category separation for office buildings. (a) At each workstation, a four-compartment waste receptacle is provided. White paper is the predominant waste product. (b) Custodians begin the second stage by collecting waste in separate bins. Floor sweepings are added to the garbage. (c) At the end of the second stage, the four categories of waste are deposited in separate bins. Service sinks, paper shredders, and other maintenance items can be incorporated in a service closet. (d) The third stage begins with compactors at the base of the service core. White paper is compacted, baled, and stored; recyclables are sorted and stored; garbage is compacted, baled, and stored. (e) At the end of the third stage, white paper, metal, plastic, and glass are collected for recycling and garbage is collected for separation at a central plant.*

planning for recycling in offices. Built at the end of the nineteenth century, this eight-story building was reused rather than demolished in its renovation near the end of the twentieth century as the offices of the National Audubon Society. Many new materials made from recycled materials were installed; four new vertical chutes conduct recyclable materials to a subbasement resource recovery center.

Details about this project are available in Audubon/ Croxton (1994).

Solid waste collection's first stage begins with two desktop recycled paper trays: one for paper for the employee's reuse, the other for recycling. Two wastebaskets are provided per employee: one for mixed paper, the other for unrecycled garbage. On each floor, central recycling points are gathered

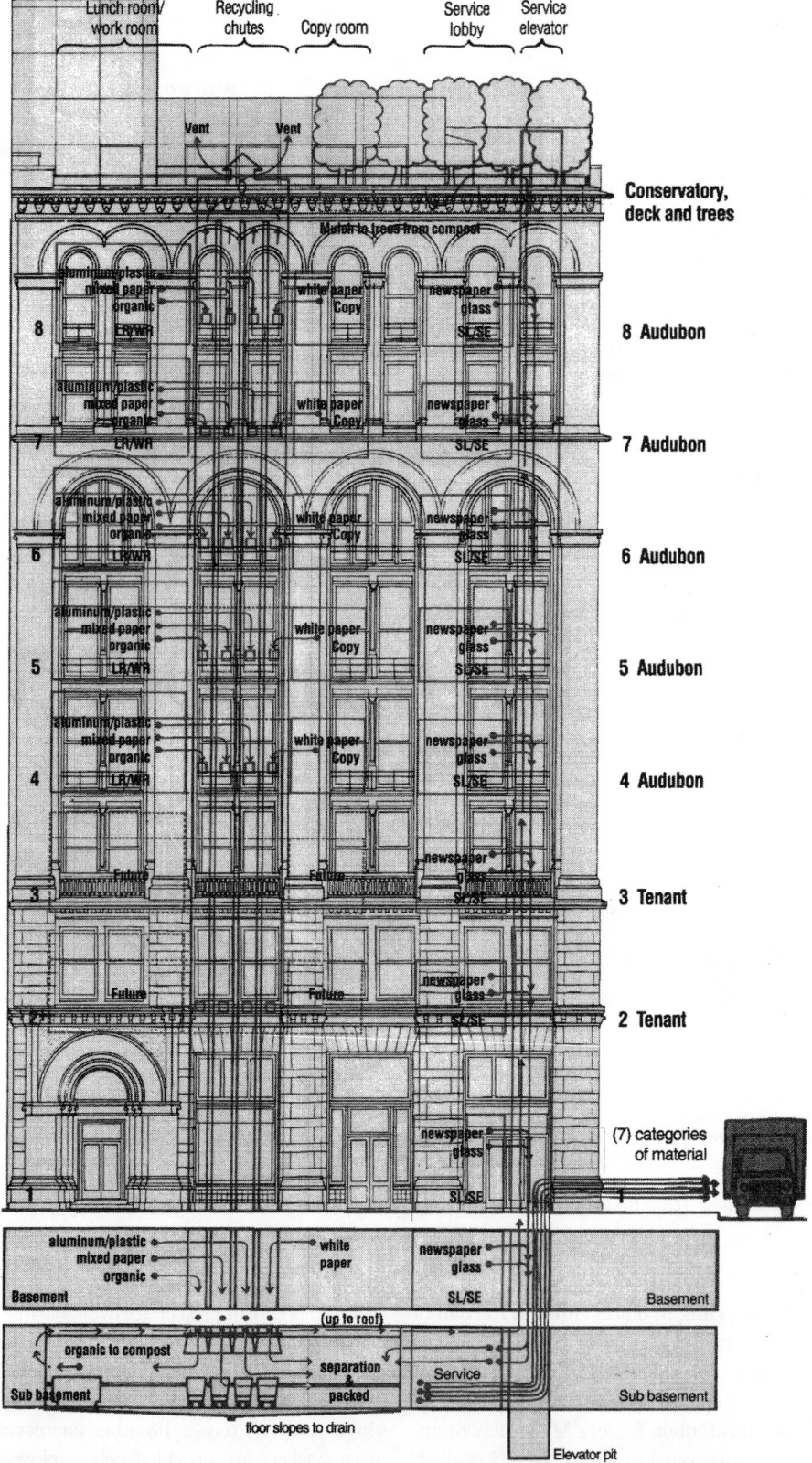

**Fig. 23.7** *Audubon House, New York City, renovated after its first 100 years, now features four 20-in. (508-mm)-diameter steel chutes that conduct recyclables to a subbasement resource recovery center. Organic materials become compost and are contributed to the roof garden. (Courtesy of the Croxton Collaborative Architects PC.)*

around four 20-in. (508-mm)-diameter steel chutes that run from top floor to subbasement. One of these chutes is for white (and computer) paper; the opening to this chute is in a vestibule for copy machines. The other three chute openings are in a "pantry" beside a small kitchen/break room and are for (1) organics (food wastes and soft soiled paper), (2) returnable plastic bottles and aluminum cans, and (3) mixed paper (colored paper, file folders, paperboard, Post-it notes). In addition, shelves in each pantry accept returnable glass bottles, coated papers (juice and milk cartons), magazines, and newspapers.

In Audubon House's second stage of collection, custodians pick up the wastebaskets from the work areas and the collection from the pantry shelves. These are taken to the subbasement recycling depot and sorted. In the third stage, large movable bins await deposits at the bottom of the four chutes. Glass bottles, newspapers, and so on from the pantry shelves are collected and boxed or baled. Recyclable materials and garbage are periodically taken up to the delivery dock to recycling and garbage collectors. The exception is the organic waste. This compostable material is first refrigerated until enough material accumulates. It is then screened (to remove the largest difficult-to-compost solids) and put into one of four composting "reactors," each with a 40-lb (18-kg) capacity. These are enclosed for odor control, yet are aerobic; air and heat, starter compost, and wood chips are provided to fast-start the process at about 140°F (60°C). Each reactor is vented through filters to control odors. After about 3 days, the second compost stage begins in which a steady flow of air sustains aerobic decomposition during a 3-month period. After this, the compost (greatly reduced in size and weight) has turned to humus and is taken to the roof garden.

## 23.5 EQUIPMENT FOR THE HANDLING OF SOLID WASTE

At Audubon House, four steel chutes of 20-in. (508-mm) diameter were inserted in the building's vertical core. Another approach is to use only one vertical chute, with a rotating receiving facility at the base (Fig. 23.8). With up to six receiving bins on a carousel and a control panel at each floor's open-

ing, considerable floor space is saved compared to six individual chutes. (Also, custodial workers have fewer recyclables to separate and move.) The user first checks the control panel setting (it will be at whatever bin the previous user had selected). After making the selection, the carousel below turns to position the appropriate bin. The access panel then opens (interlocked so that only one floor can be opened at one time), and the recyclable material (or garbage) can be deposited. The receiving room at the chute's base is about 12 ft (3.7 m) square, with a minimum height of 8 ft (2.4 m).

Where large amounts of solid waste are generated and little space is available in which to store it, various types of equipment can be used to change the volume or composition of the waste to facilitate its transportation and storage. The first step is to determine the probable daily flow of solid waste, as is done in Tables 23.1 and 23.2. When the extent of the problem is known, equipment can be selected.

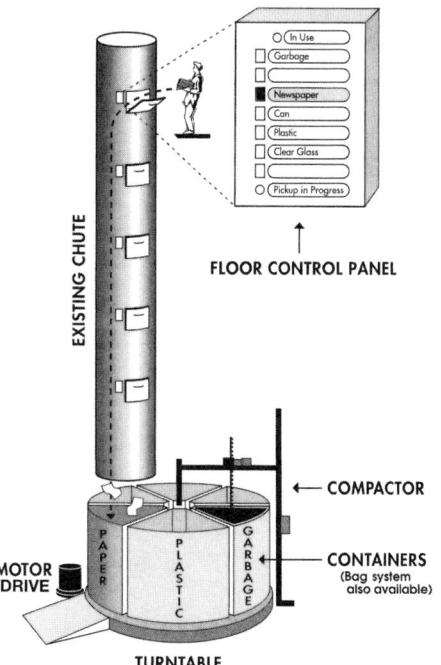

**Fig. 23.8** *One chute serves many different recyclables, as well as garbage, in this Hi-Rise Recycling System™. Interlocking access doors on each floor allow only one user at a time; the user selects the type of recyclable being deposited using the control panel at the opening. (Courtesy of Hi-Rise Recycling Systems, Inc., Miami, FL.)*

WATER AND WASTE

## (a) Compactors

Of the wide variety of compacting devices (Fig. 23.9), most are able to reduce the volume of solid waste to as little as 10% of the original volume. Among the many choices to be made are vertical versus horizontal compaction, automatic chute-fed versus manual freestanding, whether wastes are to be bagged or baled, and the final size of each unit of compacted waste. Compactors can be noisy; they can also be prone to fires as heat is generated in the compaction process. Many compactors have built-in sprays for both fire control and disinfecting. Access to wash water and a floor drain are highly desirable.

## (b) Vacuum Systems

The primary advantage of a vacuum system is that it reduces the building volume consumed by waste chutes or storage cans on each floor. A grinder-plus-evacuated-tube system is similar to the vacuum sewage systems described in Section 21.7 (Fig. 21.49). Vacuum conveying systems are now used in several industrial applications, such as ash and wood-chip transporting. Vacuum systems that use air, not water, as the transport medium are frequently used for linens in hotels and for trash. Separate vacuum systems for trash and linens are desirable, both to lessen the soiling of the linens and

to allow them to be delivered to a destination different from that for the trash.

The advantage of such pressurized systems is that lines can be small and the contents can be moved horizontally or even up. This allows far greater flexibility in design than is possible with gravity systems.

## (c) Summary

There are many options to be considered within the present waste-handling process of grind, crush, burn, and bury. As Earth's nonrenewable resources are stripped away, the recovery of materials from solid waste becomes increasingly attractive. The situation is similar to energy-and-design issues: after decades of energy-intensive building trends, designers are now giving space to daylighting atriums and allowing surfaces to act as thermal mass. A modest increase in floor space and equipment to encourage resource recovery from solid waste seems an increasingly worthwhile design investment.

## 23.6 THE SERVICE CORE

Medium- and high-rise buildings usually concentrate many building services within a core, from

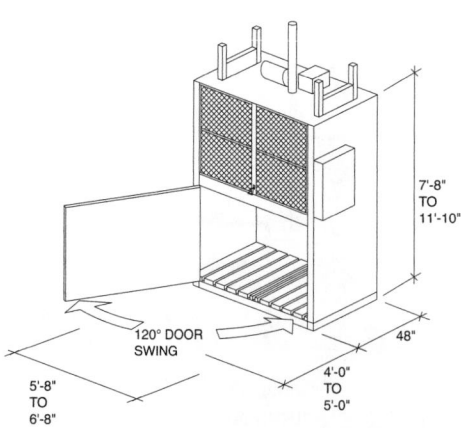

**Fig. 23.9** A refuse baler/compactor may save storage space where particular waste stream volumes are high. Cardboard, metal cans, and plastic bottles may be more attractive to recyclable haulers when their volume is thus reduced. (Reprinted by permission from AIA: Ramsey/Sleeper, Architectural Graphic Standards, 10th ed., © 2000 by John Wiley & Sons.)

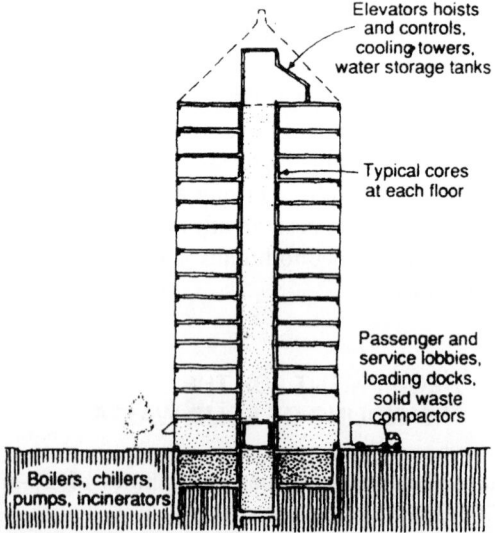

**Fig. 23.10** Schematic section through a core of a multistory building. At the top and bottom, floor space and ceiling height requirements increase. The consequences on the ground (or delivery/collection) floor are particularly great.

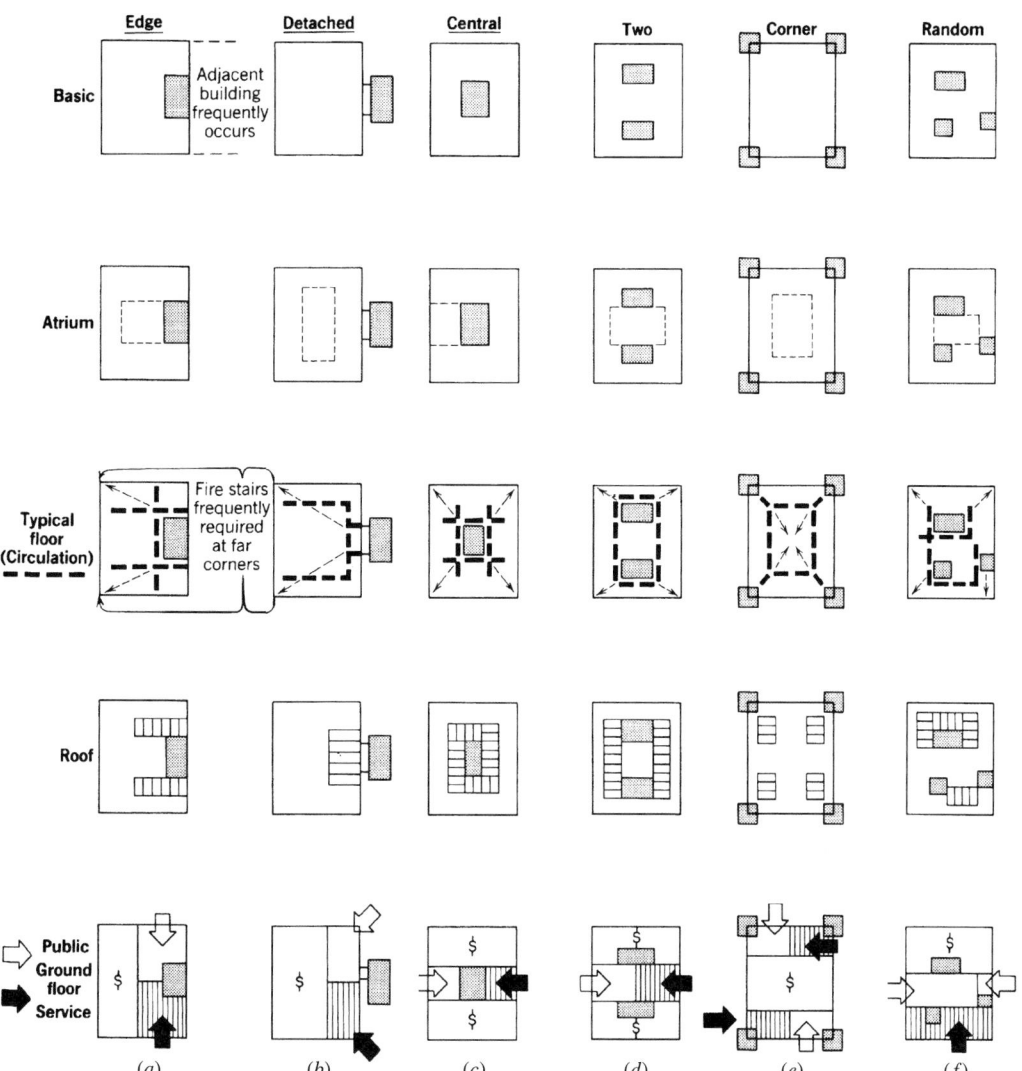

**Fig. 23.11** Comparison of core arrangements and building plans. (a,b) The edge and the detached cores give great flexibility to the rental floor area, with light and view for core spaces. (c) The central core expands readily at the roof and the ground floor, and has clear circulation and fairly flexible rental space. (d) Two-core is a popular, workable arrangement. (e) The corner cores give great flexibility to rental floors but are difficult at the roof and ground floor. (f) Random cores generally occur in low-rise buildings, in which the benefits of repetitious plans are minimal.

which services can be distributed as needed throughout each floor. The core typically contains stairs; passenger and service elevators; toilet rooms; service closets; mechanical, plumbing, and electrical chases; electrical/telephone closets with local switching capabilities; fire protection equipment; and supply closets. The size of such cores varies widely; for example, the taller the building, the more elevators.

Often, these service cores are identical in plan from one floor to the next. Alternatively, the vertical services can be identical (stairs, elevators, chases) and the arrangement of other elements (toilet rooms, supply closets) allowed to vary. Whether or not minor floor-plan variations occur, the cores usually depart radically from the typical plan—and ceiling height—at both roof and ground floor (Fig. 23.10). (Middle interruptions

**TABLE 23.4 Building Characteristics and Core Placement**

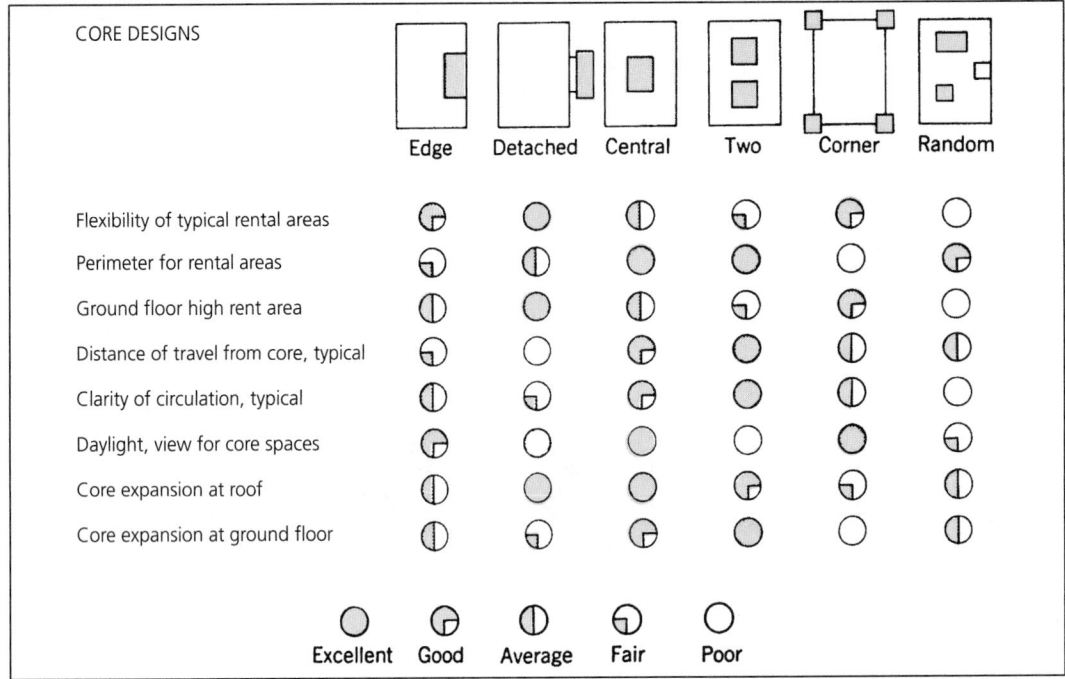

also occur where intermediate mechanical floors are provided.) Depending on the type of elevator machinery selected, the machine room ceiling height above the top floor served might be three times the ordinary floor height (see Chapter 31). On the delivery/loading floor, large trucks must be accommodated. For chillers, boilers, incinerators, and fans, higher ceilings may be required to accommodate the flues, ducts, and pipes to which they must be connected; see Section 10.3 for the dimensions required. Considering the entrance lobby as its extension, a building's service core can expand to nearly fill the ground floor. This core expansion typically happens on the roof and in the basement as well.

A service core can be related to the remaining service floor area in any of several ways (Fig. 23.11). A single central core is probably the most familiar arrangement (Occidental Chemical Offices, Fig. 5.11; Fox Plaza Los Angeles, Fig. 10.10; International Building, San Francisco, Fig. 10.11; CBS Tower, Fig. 10.59). Frequent variations include a core at the edge (Comstock Center, Fig. 10.63; Seeley

Mudd Library, Fig. 10.65) or one that is detached, two cores symmetrically placed, corner cores, or core services dispersed somewhat randomly (Iowa Public Service Building, Fig. 10.41).

Each core location has advantages and drawbacks. From the viewpoint of rentable space, important factors are the flexibility of the served (rental) floor area; the exposure of rental area to the perimeter, and thus to light, air, and view; and the extent to which ground floor high-rent space gets commercial exposure. For fire safety and occupant convenience, the distance of travel from the farthest rental floor area to the service core and the clarity of circulation within the rental area, are important. For the environment within the core—toilet rooms, stairs, elevator waiting areas—access to light, air, and view is desirable. For the mechanical services, ease of core area expansion on the roof, at the loading dock, and in the basement is important, as is the length of travel for ducts and other such elements within each floor. These factors are compared for various types of core designs in Table 23.4.

## References

Audubon/Croxton. 1994. *Audubon House: Building the Environmentally Responsible, Energy Efficient Office.* John Wiley & Sons, New York.

AIA: Ramsey/Sleeper. 1994. *Architectural Graphic Standards,* 9th ed. © John Wiley & Sons, New York.

AIA: Ramsey/Sleeper. 2000. *Architectural Graphic Standards,* 10th ed. John Wiley & Sons, New York.

Milne, M. 1976. *Residential Water Conservation.* U.S. Office of Water Research and Technology, Department of Commerce, NTIS, Springfield, VA.

**WATER AND WASTE**

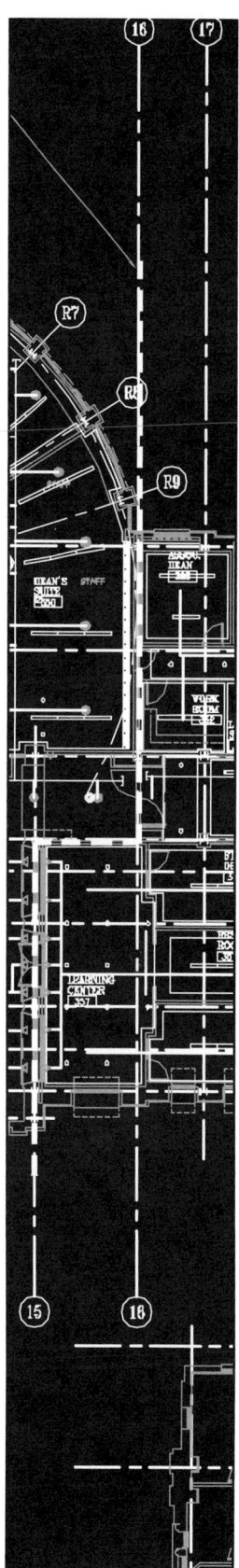

# PART VI

# FIRE PROTECTION

One of the most challenging aspects of building design is providing protection against the effects of fire. Designers may enjoy considering how their buildings can be made bearably cool on a stifling hot day or lighted and warm on a bitterly cold night. It is less pleasant to imagine a building burning and considering how first its occupants, then its contents—including the building itself—can be saved.

FIRE PROTECTION

To complicate this critically important concern, there are inherent conflicts between some optimum features of fire resistance and some common and useful features of design strategies such as daylighting, passive cooling, or forced-air HVAC systems. There are even potential design conflicts between systems for the safe evacuation of people and systems for the suppression of fire—because both people and fire thrive on oxygen. There is also some common ground among daylighting, passive and active climate control, acoustics, and fire protection. The challenge is to find the optimum balance.

Design for fire protection is an intriguing challenge. Many aspects of fire protection are rigidly governed by building codes that have evolved over the past 100 years in response to disasters and milestone events. On the other hand, most large buildings include some feature (such as an atrium) that is generally unique and not clearly addressed by the fire codes. Another challenge is to find an acceptable balance between respect for the codes and ingenuity to move beyond the codes when required.

# CHAPTER 24

# Fire Protection

JUST AS BUILDINGS CAN BE DESIGNED TO minimize heating, cooling, and lighting equipment and thus to reduce energy consumption, they can be designed to reduce the size of fire-fighting systems and retard the spread of smoke and fire. As the average size of fires within buildings continues to decline, the emphasis is shifting to minimizing water and smoke damage. Whole-building conflagrations, while spectacular, are relatively rare in large buildings in North America now that both automatic fire detection and fire-extinguishing systems are widespread.

This discussion of these topics in fire protection begins with basic design considerations for fire resistance. Smoke management (for safe evacuation and for limited smoke damage) is considered next, followed by fire suppression systems such as sprinklers and non-water-based approaches. Lightning protection is then discussed, along with the many fire detection and alarm systems that are keyed to the four stages of the typical building fire.

Throughout this chapter, the influence of the National Fire Protection Association (NFPA) is apparent. The periodically updated *Fire Protection Handbook* is an especially useful source of information for diagnosis, from which several illustrations have been reprinted in this text.

## FIRE RESISTANCE, EGRESS, AND EXTINGUISHMENT

### 24.1 DESIGN FOR FIRE RESISTANCE

Fire is a special kind of oxidation known as *combustion.* Oxidation, which has been discussed in previous chapters in terms of rust within water supply equipment and aerobic digestion in waste disposal systems, is a process in which molecules of a fuel are combined with molecules of oxygen, producing a mixture of gases and energy. When this occurs rapidly, as in a fire, energy is released as heat and light, and some gases become visible as smoke.

Fire has a triangle of needs: fuel, high temperature, and oxygen. If deprived of any of these needs, building fires will be extinguished. In general, this triangle's influence on building design is as follows. The *fuel* is the building's structure and contents; the designer controls the choice of structural and finish materials but rarely the final contents. The *temperatures* achieved in fires are well beyond the ability of building cooling systems to control, so special water systems (in the form of sprinklers) are often installed to deprive fire of the high temperatures it needs.

*Oxygen* may be denied to a fire partly by limitations on ventilation, but these can have serious safety consequences. Another design response is to install fire suppression systems that either cover the fuel (foam, dry chemicals) or displace oxygen with another gas—for example, carbon dioxide or "clean agents" that inhibit the chemical action of the flame itself.

## (a) Sources of Ignition

Buildings commonly contain three basic sources of ignition: chemical, electrical, and mechanical. (Much more rarely, nuclear sources are present.) In *chemical* combustion, most commonly known as *spontaneous combustion,* some chemicals reach ignition at ordinary temperatures within buildings. Chemical combustion depends on the rate of heat generation (related to the degree of saturation of combustible products by the chemicals involved), the air supply (enough to supply oxygen but not enough to lower the temperature), and the insulation provided by the immediate surroundings (again, the more insulation, the easier the attainment of combustion temperatures).

*Electrical* heat energy is most commonly supplied by resistance heating, a familiar process in many appliances and in space-heating equipment. Less common are electric ignition by induction, dialectic process, arcing, and static electricity. Lightning is an infrequent but enormously destructive electrical energy source.

*Mechanical* heat energy is produced by friction (including sparks), by overheating of machinery, and occasionally by the heat of compression.

## (b) Products of Combustion

Everyone has experienced to some degree the dangers of the *thermal* products of combustion (Fig. 24.1)—flame and heat. These visible and tactile elements of fire can cause burns, shock, dehydration, heat exhaustion, and fluid blockage of the respiratory tract, but they are responsible for only about one-quarter of the deaths resulting from building fires. Most fire deaths are caused by the *nonthermal* products—smoke (including gases). Smoke can usually be seen and smelled. Made up of droplets of flammable tars and small particles of carbon suspended in gases, it irritates the eyes and nasal passages, sometimes blinding and/or choking a person. Gases are especially dangerous because, without visible smoke, they are so often difficult to detect. Some gases are directly toxic, but all are dangerous because they displace oxygen. Common gases released in building fires include carbon monoxide and carbon dioxide.

*Carbon monoxide* is a deadly product of combustion and is often the most abundant. It is produced when insufficient oxygen is available to completely oxidize the burning material. It is more readily attached to hemoglobin molecules in red blood cells than is oxygen, thus depriving the brain and muscles of needed oxygen. This leads to irrational behavior and loss of consciousness, then to death.

*Carbon dioxide* is produced in large quantities, and rapidly overstimulates breathing, causing the lungs to swell. Other dangerous and common building-fire gases include hydrogen sulfide, sulfur dioxide, ammonia, oxides of nitrogen, cyanide, phosgene, and hydrogen chloride (from burning PVC).

In indoor fires, oxygen commonly becomes insufficient because the fire consumes it so rapidly. The normal concentration of oxygen in air is about 21%. At less than 17%, muscular coordination and judgment are diminished; at 14% down to 10%, people remain conscious but become irrational, and fatigue is rapid; at 10% down to 6%, collapse occurs, but revival is possible when increased oxygen is supplied. The technique of starving the fire of oxygen can therefore pose a threat to humans, both by increasing the chances of carbon monoxide production and by depriving people of oxygen.

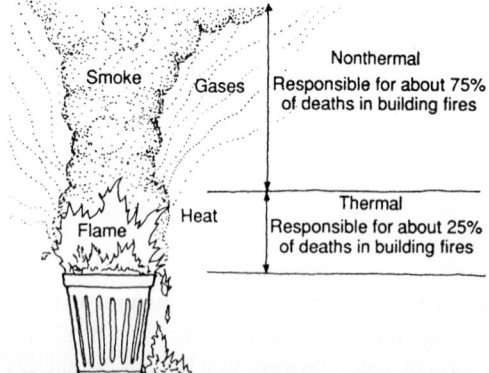

Smoke · Gases · Nonthermal Responsible for about 75% of deaths in building fires

Flame · Heat · Thermal Responsible for about 25% of deaths in building fires

**Fig. 24.1** *Although fire's thermal products—flame and heat—make strong visible and tactile impressions, fire's nonthermal products—smoke and gases—pose the greater threat to life.*

## (c) Objectives in Fire Safety

Many older buildings were designed at a time when building fires usually resulted in the loss of several adjacent structures; hence, the primary objective of fire fighting was to limit the conflagration to as few city blocks as possible. With the increased use of fire-resistant construction and code control of site and building planning, fires typically were confined to one building at a time. As fire-suppression systems came into common usage within buildings, it came to be expected that fires could be confined to one or two floors within a building. Now that automatic detection/suppression systems are technically advanced, fires can usually be confined to one room or to even smaller areas. In the United States, most fires are now extinguished with one to five sprinklers operating.

Three common objectives of building fire safety, in order of usual importance, are:

- Protection of life
- Protection of property
- Continuity of operation

Many of the elements of building fire safety are covered by building codes, but it is important to remember that codes specify the *minimum* acceptable protection. Designers can go much further than the codes require in order to enhance fire safety. It is also important to realize that codes *cannot* cover all aspects of fire safety, as there are too many variables in building design. Table 24.1 lists the stages of a fire and the factors that influence its growth. Many of these factors involve decisions made by building designers.

Codes typically *prescribe* design strategies that are *passive* means of limiting the spread of fire and protecting life: wall, floor, and ceiling constructions; maximum open floor areas; maximum distances to exits; and so on. Codes commonly allow

**TABLE 24.1  Major Factors Influencing Fire Growth**

| Realm | Approximate Ranges of Fire Sizes | Major Factors That Influence Growth |
|---|---|---|
| 1 Preburning | Overheat to ignition | Amount and duration of heat flux<br>Surface area receiving heat from material ignitability |
| 2 Initial burning | Ignition to radiation point (10-in. [254-mm] high flame) | Fuel continuity<br>Material ignitability<br>Thickness<br>Surface roughness<br>Thermal inertia of the fuel |
| 3 Vigorous burning | Radiation point to enclosure point (10-in. to 5-ft high flame [254 mm to 1.5 m]) | Interior finish<br>Fuel continuity<br>Feedback<br>Material ignitability<br>Thermal inertia of the fuel<br>Proximity of flames to walls |
| 4 Interactive burning | Enclosure point to ceiling point (5-ft [1.5-m] high flame to flame touching ceiling) | Interior finish<br>Fuel arrangement<br>Feedback<br>Height of fuels<br>Proximity of flames to walls<br>Ceiling height<br>Room insulation<br>Size and location of openings<br>HVAC operation |
| 5 Remote burning | Ceiling point to full room involvement | Fuel arrangement<br>Ceiling height<br>Length/width ratio<br>Room insulation<br>Size and location of openings<br>HVAC operations |

*Source:* Reprinted with permission from the *Fire Protection Handbook,* 18th ed., © 1997, National Fire Protection Association, Quincy, MA, 02269.

FIRE PROTECTION

some relaxing of such prescriptions when *active fire-suppression systems* (such as sprinklers) are designed into a building. Alternatively, a detailed *computer analysis* of fire spread and occupant evacuation within a given design may allow for even greater distances to exits, larger open floor areas, and alternate construction. This is a *performance-based*, rather than a prescriptive, approach to design. It requires close cooperation between designers and fire code enforcement agents.

### (d) Fire Safety and Other Environmental Control Systems

In some instances, the optimum design for fire safety will resemble the optimum design for lighting, thermal, acoustic, and water systems. Such matching characteristics include the following:

*Thermal mass*, which is useful for passive heating and cooling systems, for acoustic isolation of airborne sound, and for fire barriers (most thermally massive materials will not burn easily).

*High ceilings*, useful for daylight distribution and displacement ventilation, but also for collecting a large quantity of smoke before it reaches the occupant level and for allowing smoke and/or flames from a fire to be seen from a greater indoor distance.

*Windows*, for daylight, ventilation, and view, also allow access for fire fighting and rescue, provide escape routes, relieve smoke accumulation with fresh air, and thus relieve some of the stress of trapped occupants.

*Solid (noncombustible) overhangs over windows* not only provide sunshading, but also discourage the vertical spread of fire over the building face and can serve as emergency exterior places of refuge.

*Elevated water storage tanks* provide both adequate water pressure for plumbing fixtures and water for fire fighting in the first few minutes of a fire before fire fighters arrive.

In many other instances, there are potential design conflicts between building performance under ordinary conditions and building performance in the extraordinary event of a fire. Escalators invite shoppers to explore the upper levels of a department store or connect hotel lobbies to ballrooms; they are also efficient apertures for the vertical spread of fire and

smoke. (Special fire protection strategies at escalators are found in Figs. 33.9 to 33.13.) Daylighting and natural ventilation are best served by high ceilings and low partitions, which encourage light and air from the perimeter to pervade building interiors. However, if unchecked by sprinklers, fire and smoke can easily spread through such open plans. (However, smoke and fire can build up much more rapidly in small, enclosed rooms that retain heat.) Forced-air systems that heat, ventilate, and cool are also potential pathways for smoke and fire; this hazard is especially serious when the systems penetrate floors because vertically spreading fires are harder to fight than horizontally spreading ones. (However, carefully designed forced-air systems can aid in smoke management.) Windowless buildings that rely on electric light in place of daylight are especially dangerous in fires because fire fighters cannot easily evacuate occupants or gain access to the building. Sunscreens that completely cover windows are disadvantageous for the same reason; nonoperable windows, although considered advantageous for tightly controlled air conditioning, must be broken for fire evacuation/access. The higher the window, the greater the danger from falling shards of glass. Many excellent insulating materials will burn readily, and some will give off toxic gases. Some of the loveliest interior finishes are both flammable and deadly sources of gases in a fire. Table 24.2 summarizes the requirements for interior finishes.

This conflict, between interior comfort throughout a building's life and safety at the moment of its impending death by fire, is the principal reason why codes alone cannot assure fire safety. The dilemma of ordinary versus extraordinary performance is one that must be faced by the designer, owner, and occupants of buildings.

### (e) Protection of Life

The NFPA *Fire Protection Handbook* discusses human behavior in fires in great detail. Although panic behavior drives many code requirements, such behavior has been found to be rare. Designers should consider how building occupants make decisions in a fire. In the first phase, cues are detected—the smell of smoke, sounds associated with a fire (breaking glass, sirens, alarm bells), and, more rarely, the sight of flames. Open plans (with longer visible indoor distances) are more amenable to

**TABLE 24.2  Summary of Life Safety Code Requirements for Interior Finish**

| Occupancy | Exits | Access to Exits | Other Spaces |
|---|---|---|---|
| Assembly—New: >300 people | A | A or B | A or B |
| ≤300 people | A | A or B | A, B, or C |
| Assembly—Existing: >300 people | A | A or B | A or B |
| ≤300 people | A | A or B | A, B, or C |
| Educational—New | A | A or B | A or B; C on low partitions |
| Educational—Existing | A | A or B | A, B, or C |
| Day Care Centers—New | A | A | A or B |
| | I or II | I or II | NR |
| Day Care Centers—Existing | A or B | A or B | A or B |
| Group Day-Care Homes—New | A or B | A or B | A, B, or C |
| Group Day-Care Homes—Existing | A or B | A, B, or C | A, B, or C |
| Family Day-Care Homes | A or B | A, B, or C | A, B, or C |
| Health Care—New A. S. Mandatory | A or B | A or B | A or B |
| | | C on lower portion of corridor wall | C in small individual rooms |
| Health Care—Existing | A or B | A or B[a] | A or B[a] |
| Detention and Correctional—New | A | A | A, B, or C |
| | I | I | |
| Detention and Correctional—Existing | A or B | A or B | A, B, or C |
| | I or II | I or II | |
| Residential, Board and Care[b] | | | |
| Residential, Hotels and Dormitories—New | A | A or B | A, B, or C |
| | I or II | I or II | |
| Residential, Hotels and Dormitories—Existing | A or B | A or B | A, B, or C |
| | I or II | I or II | |
| Residential, Apartment Buildings—New | A | A or B | A, B, or C |
| | I or II | I or II | |
| Residential, Apartment Buildings—Existing | A or B | A or B | A, B, or C |
| | I or II | I or II | |
| Residential, 1- and 2-Family, Lodging or Rooming Houses | A, B, or C | A, B, or C | A, B, or C |
| Mercantile—New | A or B | A or B | A or B |
| Mercantile—Existing Class A or B | A or B | A or B | Ceiling A or B existing on walls A, B, or C |
| Mercantile—Existing Class C | A, B, or C | A, B, or C | A, B, or C |
| Office—New | A or B | A or B | A, B, or C |
| | I or II | I or II | |
| Office—Existing | A or B | A or B | A, B, or C |
| Industrial | A or B | A, B, or C | A, B, or C |
| Storage | A or B | A, B, or C | A, B, or C |

*Source:* Reprinted with permission from the *Fire Protection Handbook,* 18th ed., © 1997, National Fire Protection Association, Quincy, MA, 02269.

[a]See occupancy chapters of NFPA *101* for details.

[b]See Source, Chapters 22 and 23.

*Notes:* Class A Interior Wall and Ceiling Finish—flame spread 0–25, (new) smoke developed 0–450.

Class B Interior Wall and Ceiling Finish—flame spread 26–75, (new) smoke developed 0–450.

Class C Interior Wall and Ceiling Finish—flame spread 76–200, (new) smoke developed 0–450.

Class I Interior Floor Finish—critical radiant flux, minimum 0.45 Watts/cm$^2$.

Class II Interior Floor Finish—critical radiant flux, minimum 0.22 Watts/cm$^2$.

Automatic sprinklers—where a complete standard system of automatic sprinklers is installed, interior finish with flame spread rating not over Class C may be used in any location where Class B is normally specified, and with rating of Class B in any location where Class A is normally specified; similarly, Class II interior finish may be used in any location where Class I is normally specified, and no critical flux rating is required where Class II is normally specified. (This does not apply to new health care facilities.)

Exposed portions of structural members complying with the requirements of heavy timber construction may be permitted.

exposing such clues to a wider population. In the second phase, the occupants define the situation: Just how serious is this fire? The more numerous the cues, the more rapid the definition. How other people are reacting is influential, and in the absence of strong cues can lead to a group refusal to evacuate in the early stages of a fire. In the third phase, coping behavior begins: fight or flight. Clear exit pathways and access to fire fighting equipment are critical in this decision.

For most low-rise buildings, a reasonable goal is the evacuation of all occupants in the time interval between the detection of a fire and the arrival of the fire fighters. Designers can provide clearly defined pathways to exits (*exit access*) that can be kept relatively clear of smoke (Fig. 24.2). To accommodate a wheelchair, a minimum clear width of 32 in. (813 mm) is required. *Exits* can take a variety of forms. Vertical exits (Figs. 24.3 and 24.4) include smokeproof towers, exterior and interior stairs and ramps, and escalators that meet specific requirements. *Vertical exits do not include elevators;* they are too easily stalled or, worse, opened at the floor of a fire by malfunctioning signal equipment. Exits in the horizontal plane include doors leading directly

to the outside, 2-hour fire-rated enclosed hallways, and moving walks. Special *horizontal exits* are provided by internal firewalls penetrated by two fire doors—one swinging open in either direction. *Exit discharge* is the area outside the exits that leads to a public way and may still need protection in a fire.

Maximum allowable distances to exits are specified in Table 24.3. These are influenced by past experience with people exiting through smoke, especially when it may involve briefly traveling toward the perceived fire itself (as in the case of a dead-end corridor). When automatic fire suppression systems such as sprinklers are used, the allowable distances to exits are increased. Designers should remember, however, that at least 30% of building fire deaths result from fire cutting off the paths to exits.

Minimum egress widths per floor are found with the help of Table 24.4. First, calculate the floor area (net or gross, as specified in the table). Divide the floor area by the occupant load to determine the number of occupants for whom exits must be provided for that floor. Then calculate the exit capacity based on its clear width. Stairs are sized to meet the requirements of Table 24.5.

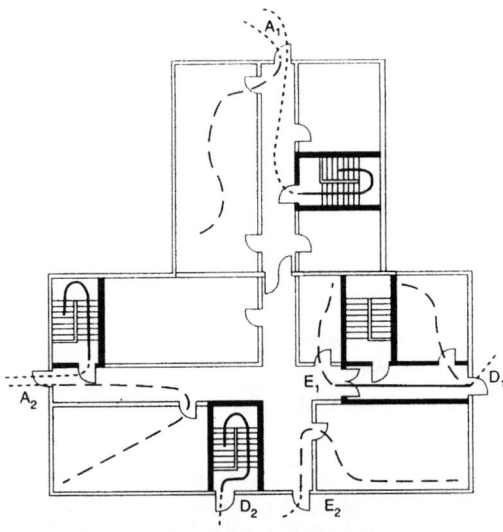

**Fig. 24.2** *Exit access, exit, and exit discharge on the first floor of a multistory building. Doors $A_1$, $A_2$, $E_1$, and $E_2$ are exits, and the path (dashed line) is the exit access. To the person emerging from the exit enclosure, doors $A_1$ and $A_2$ and the paths (dotted lines) are the exit discharge. Doors $D_1$ and $D_2$ are exit discharge doors. Solid-line paths are within the exit. (Reprinted with permission from the* Fire Protection Handbook, *18th ed., © 1997, National Fire Protection Association, Quincy, MA, 02269.)*

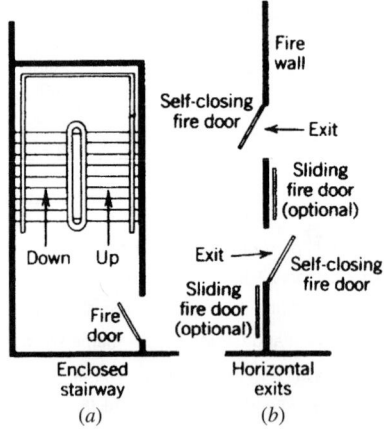

**Fig. 24.3** *Plan views of exits. (a) An enclosed stairway allows occupants on any floor above a fire to escape. A smokeproof tower is better, as it opens to the air at each floor, largely preventing accumulation of smoke in the stairway. (b) Horizontal exit through an interior firewall provides a quick refuge and lessens the need for a hasty flight down the stairs. Two wall openings are needed to facilitate exit in either direction. Fire-rated doors must be arranged to be self-closing or automatic-closing by smoke detection. (Reprinted with permission from the* Fire Protection Handbook, *18th ed., © 1997, National Fire Protection Association, Quincy, MA, 02269.)*

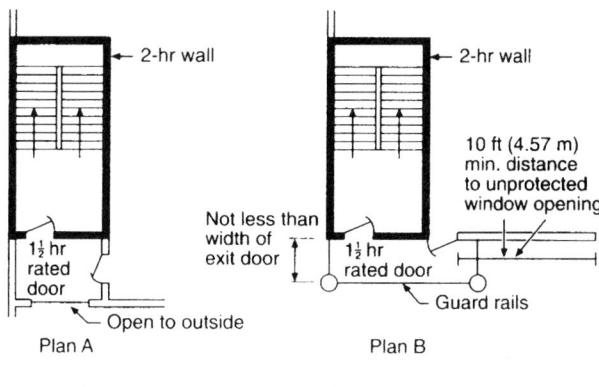

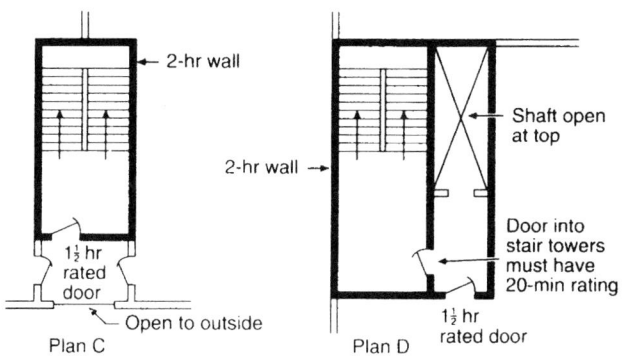

*Fig. 24.4* Four variations of smokeproof towers. Plan A has a vestibule opening from a corridor. Plan B shows an entrance by way of an outside balcony. Plan C could provide a stair tower entrance common to two areas. In plan D, smoke and gases entering the vestibule would be exhausted by a natural or induced draft in the open air shaft. In each case, a double entrance to the stair tower with at least one side open or vented is characteristic of this type of construction. Pressurization of the stair tower in the event of fire provides an attractive alternative for tall buildings and is a means of eliminating the entrance vestibule. (Reprinted with permission from the Fire Protection Handbook, 16th ed., © 1986. National Fire Protection Association, Quincy, MA, 02269.)

**EXAMPLE 24.1.** A multistory office building is 80 ft (24 m) wide by 300 ft (80 m) long. What exit capacity is required per floor?

**SOLUTION.**

The gross floor area = 80 × 300 = 2400 ft². From Table 24.4, "business" categories are based on one person per gross 100 ft².

The population per floor is, therefore, (2400 ft²)/(100 ft²/person) = 240 people.

Exit doors (to stairs): 240 people × 0.2 in./person = 48 in. total.

(One 34-in. clear door into each of two stairs = 2 doors × 34 = 68 in., more than the minimum.)

Stairs: 240 people × 0.3 in./person = 72 in. total.

(Two stairs at 44 in. each = 88 in., more than the minimum.) ∎

The building population as estimated for fire safety is usually much greater than the population for which HVAC, water, or elevator service is designed. The stairs must be designed to allow those already within the stairwell to continue down without interference from access doors on any floor. Stairs with direct access to outdoor air at each floor—so-called smokeproof towers—are the safest kind. A fire stair must allow fire fighters to move up while occupants are moving down. Another phenomenon is *reentry*, in which occupants who have exited decide to reenter despite the danger. (Rescue of family members, pets, and valuables is the likely reason.) This greatly complicates downward flow.

High-rise buildings present much more difficult problems because fire fighting equipment can ordinarily reach no higher than seven floors (about 90 ft [27 m]) and because, typically, only two exit stairways are provided. Downward flow rates in stairs were formerly assumed at about 45 persons/minute/22 in. (559 mm of width), but more recently, peak flows of only 24 persons/minute/22 in. (559 mm) have been observed. With a typical exit stair, a 15-story building housing 60 persons per floor per stair can be evacuated in about nine minutes. However, with the same stair in a 50-story building housing 240 persons per floor per stair, evacuation will take at least 2 hours, 11 minutes! When doors are held open at each floor to admit occupants fleeing the fire, smoke can readily enter the stairwell. Moreover, it is becoming increasingly

**TABLE 24.3 Common Path, Dead End, and Travel Distance Limits**

| Type of Occupancy | Common Path Limit Unsprinklered ft (m) | Common Path Limit Sprinklered ft (m) | Dead-End Limit Unsprinklered ft (m) | Dead-End Limit Sprinklered ft (m) | Travel Distance Limit Unsprinklered ft (m) | Travel Distance Limit Sprinklered ft (m) |
|---|---|---|---|---|---|---|
| **ASSEMBLY** | | | | | | |
| New | 20/75[a,b] (6.1/23) | 20/75[a,b] (6.1/23) | 20[b] (6.1) | 20[b] (6.1) | 150[c] (45) | 200[c] (60) |
| Existing | 20/75[a,b] (6.1/23) | 20/75[a,b] (6.1/23) | 20[b] (6.1) | 20[b] (6.1) | 150[c] (45) | 200[c] (60) |
| **EDUCATIONAL** | | | | | | |
| New | 75 (23) | 75 (23) | 20 (6.1) | 20 (6.1) | 150 (45) | 200 (60) |
| Existing | 75 (23) | 75 (23) | 20 (6.1) | 20 (6.1) | 150 (45) | 200 (60) |
| New Day-Care Center | NR[d,e] | NR[d,e] | 20 (6.1) | 20 (6.1) | 150 (45) | 200[f] (60) |
| Existing Day-Care Center | NR[d,e] | NR[d,e] | 20 (6.1) | 20 (6.1) | 150 (45) | 200[f] (60) |
| **HEALTH CARE** | | | | | | |
| New | NR[d] | NR[d] | 30 (9.1) | 30 (9.1) | NA[g] | 200[f] (60) |
| Existing | NR[d] | NR[d] | NR[d] | NR[d] | 150[f] (45) | 200[f] (60) |
| New Ambulatory Care | NR[d] | NR[d] | 30 (9.1) | 30 (9.1) | 150[f] (45) | 200[f] (60) |
| Existing Ambulatory Care | NR[d] | NR[d] | 50 (15) | 50 (15) | 150[f] (45) | 200[f] (60) |
| **DETENTION AND CORRECTIONAL** | | | | | | |
| New-Use Conditions II, III, IV | 50 (15) | 100 (30) | 50 (15) | 50 (15) | 150[f] (45) | 200[f] (60) |
| V | 50 (15) | 100 (30) | 20 (6.1) | 20 (6.1) | 150[f] (45) | 200[f] (60) |
| Existing-Use Conditions II, III, IV, V | 50[h] (15) | 100[h] (30) | NR[d] | NR[d] | 150[f] (45) | 200[f] (60) |
| **RESIDENTIAL** | | | | | | |
| Hotels and Dormitories | | | | | | |
| New | 35[i] (10.7) | 50[i] (15) | 35 (10.7) | 50 (15) | 175[f,i] (53) | 325[f,i] (99) |
| Existing | 35[i] (10.7) | 50[i] (15) | 50 (15) | 50 (15) | 175[f,i] (53) | 325[f,i] (99) |
| Apartments | | | | | | |
| New | 35[i] (10.7) | 50[i] (15) | 35 (10.7) | 50 (15) | 175[f,i] (53) | 325[f,i] (99) |
| Existing | 35[i] (10.7) | 50[i] (15) | 50 (15) | 50 (15) | 175[f,i] (53) | 325[f,i] (99) |
| Board and Care | | | | | | |
| Small, New and Existing | NR[d] | NR[d] | NR[d] | NR[d] | NR[d] | NR[d] |
| Large, New | NA | 125[k] (38.1) | NA | 15 (15) | NA | 325[f,i] (99) |
| Large, Existing | 110 (33.5) | 160 (48.8) | 50 (15) | 50 (15) | 175[f,i] (53) | 325[f,i] (99) |
| Lodging and Rooming Houses | NR[d] | NR[d] | NR[d] | NR[d] | NR[d] | NR[d] |
| One- and Two-Family Dwellings | NR[d] | NR[d] | NR[d] | NR[d] | NR[d] | NR[d] |
| **MERCANTILE** | | | | | | |
| Class A, B, C | | | | | | |
| New | 75 (23) | 100 (30) | 20 (6.1) | 50 (15) | 100 (30) | 200 (60) |
| Existing | 75 (23) | 100 (30) | 50 (15) | 50 (15) | 150 (45) | 200 (60) |
| Open Air Covered Mall | NR[d] | NR[d] | 0 (0) | 0 (0) | NR[d] | NR[d] |
| New | 75 (23) | 100 (30) | 20 (6.1) | 50 (15) | 100 (30) | 400[i] (120) |
| Existing | 75 (23) | 100 (30) | 50 (15) | 50 (15) | 150 (45) | 400[i] (120) |

| Occupancy | | | | | | |
|---|---|---|---|---|---|---|
| **BUSINESS** | | | | | | |
| New | 75[m] (23) | 100[m] (30) | 20 (6.1) | 50 (15) | 200 (60) | 300 (91) |
| Existing | 75[m] (23) | 100[m] (30) | 50 (15) | 50 (15) | 200 (60) | 300 (91) |
| **INDUSTRIAL** | | | | | | |
| General | 50 (15) | 100 (30) | 50 (15) | 50 (15) | 200[i] (60) | 250[n] (75) |
| Special Purpose | 50 (15) | 100 (30) | 50 (15) | 50 (15) | 300[p] (91) | 400[p] (122) |
| High Hazard | 0 (0) | 0 (0) | 0 (0) | 0 (0) | 75 (23) | 75 (23) |
| Aircraft Servicing Hangars, Ground Floor | 50[p] (15) | 50[p] (15) | 50[p] (15) | 50[p] (15) | 75 (23) | 75 (23) |
| Aircraft Servicing Hangars, Mezzanine Floor | 50[p] (15) | 50[p] (15) | 50[p] (15) | 50[p] (15) | 75 (23) | [o] |
| **STORAGE** | | | | | | |
| Low Hazard | NR[d] | NR[d] | NR[d] | NR[d] | NR[d] | NR[d] |
| Ordinary Hazard | 50 (15) | 100 (30) | 50 (15) | 100 (30) | 200 (60) | 400 (122) |
| High Hazard | 0 (0) | 0 (0) | 0 (0) | 0 (0) | 75 (23) | 75 (23) |
| Parking Garages, Open | 50 (15) | 50 (15) | 50 (15) | 50 (15) | 200 (60) | 300 (91) |
| Parking Garages, Enclosed | 50 (15) | 50 (15) | 50 (15) | 50 (15) | 150 (45) | 200 (60) |
| Aircraft Storage Hangars, Ground Floor | 50[p] (15) | 100[p] (30) | 50[p] (15) | 50[p] (15) | 75 (23) | |
| Aircraft Servicing Hangars, Mezzanine Floor | 50[p] (15) | 75[p] (23) | 50[p] (15) | 50[p] (15) | 75 (23) | 75 (23) |
| Underground Spaces in Grain Elevators | 50[p] (15) | 50[p] (15) | NR[d,p] | NR[d,p] | 200 (60) | 400 (122) |

*Source:* Adapted with permission from the *Fire Protection Handbook*, 18th ed., © 1997, National Fire Protection Association, Quincy, MA, 02269.

[a] 20 ft (6.1 m) for common path serving >50 persons; 75 ft (23 m) for common path serving <50 persons.

[b] See Source, Chapters 8 and 9, for special considerations for aisle accessways, aisles, and mezzanines.

[c] See Source, Chapters 8 and 9, for special considerations for smoke-protected assembly seating in arenas and stadia.

[d] No requirement.

[e] See Source, Sections 10–7 and 11–7 for requirement for second exit access based on room capacity or area.

[f] This dimension is for the total travel distance, assuming incremental portions have fully utilized their allowable maximums. For travel distance within the room, and from the room exit access door to the exit, see Source for the appropriate occupancy chapter.

[g] Not applicable.

[h] See Source, Chapter 15, for special considerations for existing common paths.

[i] This dimension is from the room/corridor or suite/corridor exit access door to the exit; thus it applies to corridor common path.

[j] See Source, appropriate occupancy chapter for special travel distance considerations for exterior ways of exit access.

[k] See Source, Section 22–3, for requirement for second exit access based on room area.

[l] See Source, Sections 24–4 and 25–4, for special travel distance considerations in covered malls considered pedestrian ways.

[m] See Source, Sections 26 and 27, for special common path considerations for single tenant spaces.

[n] See Source, Chapter 28, for industrial occupancy special travel distance considerations.

[o] See Source, Chapters 28 and 29, for special requirements on spacing of doors in aircraft hangars.

[p] See Source, Chapters 28 and 29, for special requirements if high hazard.

**TABLE 24.4 Summary of NFPA 101®, Life Safety Code®, Provisions for Occupant Load, and Exit Capacity**

| Occupancy | Occupant Load ft² (m²) per person | Level Components Doors, Corridors, Horizontal Exits, Ramps | | Stairs | |
|---|---|---|---|---|---|
| Assembly | | | | | |
| Less concentrated use without fixed seating | 15 Net (1.4) | 0.2 | | 0.3 | |
| Concentrated use without fixed seating | 7 Net (0.65) | 0.2 | | 0.3 | |
| Fixed seating | Actual number of seats | 0.2 | | 0.3 | |
| Educational | | | | | |
| Classrooms | 20 Net (1.9) | 0.2 | | 0.3 | |
| Shops and vocational | 50 Net (4.6) | 0.2 | | 0.3 | |
| Care centers | 35 Net (3.3) | 0.2 | | 0.3 | |
| Health care | | NAS | AS | NAS | AS |
| Sleeping departments | 120 Gross (11.1) | 0.5 | 0.2 | 0.6 | 0.3 |
| Treatment departments | 240 Gross (22.3) | 0.5 | 0.2 | 0.6 | 0.3 |
| Residential | 200 Gross (18.6) | 0.2 | | | |
| Board and care | 200 Gross (18.6) | 0.2 | | 0.3 | |
| Mercantile | | | | | |
| Street floor and sales basement | 30 Gross (3.7) | 0.2 | | 0.3 | |
| Multiple street floors (each) | 40 Gross (3.7) | 0.2 | | 0.3 | |
| Other floors | 60 Gross (5.6) | 0.2 | | 0.3 | |
| Storage–shipping | 300 Gross (27.9) | 0.2 | | 0.3 | |
| Malls | See Code | 0.2 | | 0.3 | |
| Business | 100 Gross (9.3) | 0.2 | | 0.3 | |
| Industrial | 100 Gross (9.3) | 0.2 | | 0.3 | |
| Detention and correctional | 120 Gross (11.1) | 0.2 | | 0.3 | |

*Source:* Adapted with permission from the *Fire Protection Handbook,* 18th ed., © 1997, National Fire Protection Association, Quincy, MA, 02269.

NAS = nonsprinklered; AS = sprinklered.

common for people to refuse to evacuate a high-rise building, placing their faith in fire-extinguishing systems instead of facing the daunting descent through many floors of stairs.

Because of the impracticality of rapid evacuation, larger buildings are required to provide *refuge areas* where smoke penetration is less likely. For example, stairs can be designed to hold *all* the occupants, in which case the design of discharge is based more on *capacity* than on *flow.* At maximum crowding, about 3 ft² (0.28 m²) per person is recommended. Again, smokeproof towers will be safer in such a design, although stairs can be pressurized with outdoor air. These details of smoke management are presented in the following section.

### (f) Property Protection

One of the earliest design concerns in this category is that the site should permit access for fire fighting

equipment. Ideally, fire trucks should be able to pull alongside each exterior wall. If accessibility is limited by adjacent buildings, and roadways alongside buildings and other measures are impractical, more reliance must be placed on internal fire-suppression systems. Another factor is the amount of time it will ordinarily take for fire fighters to reach a site. In congested urban areas or remote rural ones, the time that elapses between the alarm and the arrival of fire fighters can permit a fire to grow to unstoppable proportions. In these cases, emphasis again shifts to internal fire-suppression systems.

Another design concern is adequate water to fight the fire. Reliance on city water mains is not always a good solution. Elevated tanks on buildings can help in the early moments of fire fighting, but their capacity is soon exhausted. Some buildings therefore rely on lakes or enclosed reservoirs for a fire fighting water supply (see the discussion in Section 24.3).

FIRE PROTECTION

**TABLE 24.5 Requirements for Exit Stairs**

| | New Stairs | Existing Stairs | |
| | | Class A[a] | Class B[a] |
|---|---|---|---|
| Minimum width clear of all obstructions[b]: | | | |
|    Total occupant load[c] = 50 or more | 44 in. (1.12 m) | 44 in. (1.12 m) | 44 in. (1.12 m) |
|              = less than 50 | 36 in. (0.91 m) | 36 in. (0.91 m) | 36 in. (0.91 m) |
| Maximum height of risers | 7 in. (178 mm) | 7½ in. (191 mm) | 8 in. (203 mm) |
| Minimum height of risers | 4 in. (102 mm) | | |
| Minimum tread depth | 11 in. (279 mm) | 10 in. (244 mm) | 9 in. (229 mm) |
| Minimum headroom | 6 ft 8 in. (2.03 m) | 6 ft 8 in. (2.03 m) | 6 ft 8 in. (2.03 m) |
| Maximum height between landings | 12 ft (3.7 m) | 12 ft (3.7 m) | 12 ft (3.7 m) |
| Minimum dimension of landings in direction of travel | d | e | e |
| Doors opening immediately on stairs, without landing at least the width of door | No | No | No |

*Source:* Adapted with permission from the *Fire Protection Handbook,* 18th ed., © 1997, National Fire Protection Association, Quincy, MA, 02269.

[a]Class A and B stairs are defined as those having the tread and riser dimensions as listed in this table.

[b]Except projections not exceeding 3½ in. (89 mm) at and below handrail height on each side.

[c]Total occupant load includes all floors served by stairway.

[d]Every landing shall have a dimension, measured in the direction of travel, equal to the width of the stair. Such dimension need not exceed 4 ft (1.22 m) when the stair has a straight run.

[e]Stairways and intermediate landings shall continue with no decrease in width along the direction of exit travel.

*Exposure protection* is becoming common in areas where highly flammable surroundings pose a serious threat of fires originating *outside* a building. Candidates for exposure protection include buildings surrounded by older wooden buildings, bordered by lumberyards or other commercial activities that utilize highly flammable materials, or even bordered by open fields of dry grass or brush. Exposure protection guards against heat transfer by radiation and convective currents and against direct fire transfer via flying embers. Exposure protection begins with the use of nonflammable materials for the building's exterior. Erecting firewalls between the building and a fire-threatening neighbor is a more drastic, but sometimes necessary, step. Exposure protection sometimes includes external water sprinkler systems, in which a spray head is placed over (or under) each opening such as windows or doors. Sprinkler systems that soak the roof can play a cooling role on summer days, as well as an exposure protection role. Exterior doors can be chosen for their fire-delaying characteristics. Windows can also be protected by fire-rated shutters that are designed to close automatically at high temperatures (Fig. 24.5). Sometimes exterior protection is necessary to protect the *exit discharge* from flames

*Fig. 24.5* Rolling metal shutters, housed in valences above each window, punctuate the exterior of this downtown Pittsburgh building opening onto a narrow street. The shutters can protect the building's contents from an exterior fire, but can also protect the narrow street from a fire inside the building. Note the exterior fire escapes across this narrow street. These are no longer permitted, partly because smoke plumes rising from windows can make such stairs unusable long before heat from a fire becomes a threat.

originating *inside* the building, as with narrow alley-ways like that in Fig. 24.5 in downtown Pittsburgh.

*Compartmentation* has become increasingly important as buildings have become lightweight structures incorporating decreased fire resistance and open floor areas that encourage the spread of fires. Building codes establish the maximum floor areas permissible for various constructions and occupancies. If a building's floor area exceeds such limits, it must be subdivided by firewalls into areas that fall within the code limitations. Once again, sprinklers allow for increased floor area limits.

Openings in the firewalls must be protected by fire-rated doors and by fire dampers in forced-air systems (Fig. 24.6). As important as compartmentation is in buildings with large floor areas, how-ever, the vertical spread of fire (its natural path) poses a more serious problem. For this reason, compartmentation requirements around vertical openings are often especially strict.

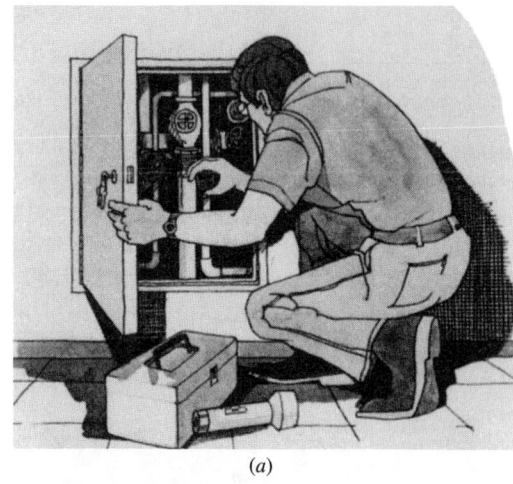

(a)

**Fig. 24.6** *Openings in firewalls are protected by various devices. (a) Access doors to controls and other elements within firewalls can be of fire-rated construction. (b) Locations of fire dampers in a typical forced-air system—at firewalls, partitions, ceilings, floors, and shaft enclosures. (Part b reprinted with permission from the* Fire Protection Handbook, *18th ed., copyright © 1997, National Fire Protection Association, Quincy, MA, 02269. Redrawn by Dain Carlson.)*

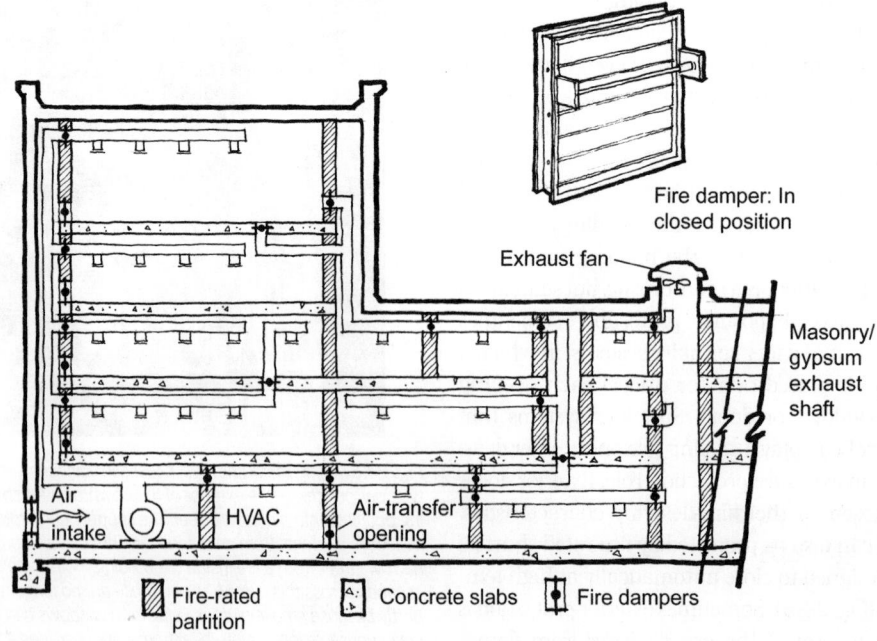

(b)

*Concealed spaces* are found in many contemporary buildings, especially over suspended ceilings but also behind walls, within pipe chases, in attics, under raised floors, and in other places. All such spaces can offer paths for the spread of fire. The designer can utilize noncombustible materials, as far as possible, in such spaces. Another important step is to include automatic fire detection and suppression systems in these uninhabited spaces—and often the use of oxygen-deprivation approaches to fire suppression. Another design response is compartmentation: using firestops (or firewalls) to break up otherwise continuous concealed spaces. Many codes, for example, require firestopping around each $1000\,\text{ft}^2$ ($93\,\text{m}^2$) of suspended ceiling area and every $2000\,\text{ft}^2$ ($186\,\text{m}^2$) of attic floor area.

*Structural protection,* another important requirement, allows a building to continue to stand during a fire and enables it to be salvaged rather than demolished after the fire. Codes require various protective layers for structural materials. In order of importance, it is usually most critical to protect columns, girders, beams, and, finally, the floor slab.

## (g) Continuity of Operations

For most building functions, it is desirable to minimize the disruption of operation that fire will cause. Design strategies to encourage continuity of operations include special fire alarm/suppression systems for especially critical operations areas (control rooms, for example), the design of HVAC systems to allow for 100% outside air (to aid in purging a building of smoke), and provision for the speedy removal of the water dumped on a fire from a sprinkler system. The floors in sprinkler-served buildings should be (yet rarely are) waterproof; waterproofing should also be carried up walls, columns, pipes, and other elements to a height of 4 to 6 in. (102 to 152 mm) above the floor. These provisions will help to minimize the water damage associated with fires.

## 24.2 SMOKE MANAGEMENT

As it has become increasingly evident that smoke kills more people in building fires than either heat or structural collapse, it has also become more common to design for smoke management as well as for resistance to fire and fire suppression. The objectives of smoke management are the same as those for fire resistance: to reduce deaths and property damage due to smoke and provide for continuity of operations with minimal smoke interference. Smoke management systems have been used for several generations; scientific laboratory buildings equipped with fume hoods offer a familiar example. Several options for smoke management are now available for ordinary building design.

### (a) Factors in Smoke Management

The heat of a fire produces air pressure and buoyancy that aid the spread of smoke well beyond the scene of the fire itself. In low buildings, the smoke is spread primarily by heat-induced convective air motion and by the differential air pressures caused by the expansion of gases as the temperature increases. In tall buildings, the stack effect complicates this pattern of smoke spread, encouraging the rapid rise of heated air within vertical shafts. Wind forces from outside are also more likely to be a factor in tall buildings. (Wind velocities are usually higher with increasing distance from the Earth's surface.) Forced-air systems, so common in high-rise buildings, can also contribute to the spread of smoke. The interactions among fire, the stack effect, wind, building geometry, and HVAC systems are complex. Several detailed calculation techniques are available, as explained in two publications, NFPA 92A (2000) and NFPA 92B (2005). These methods include building and testing a scale model, using a series of closed-form algebraic equations, and modeling a compartment fire with the help of a computer. In the last category, a more modest *zone* approach divides a fire into two zones (upper and lower), and the calculations can be done on a personal computer. The more ambitious *field* approach uses computational fluid dynamics (CFD), requiring a large-capacity computer but allowing the testing of wide variations in approaches to fire and smoke control; fire growth, smoke spread, structural behavior, and occupant evacuation can be examined together. Results can be surprising, leading at times to counterintuitive solutions.

### (b) Confinement

The most passive design response to smoke is to try to confine it to the fire area itself. Another confinement technique is to exclude smoke from specially protected areas, called *refuges*. Compartmentation is important in these approaches: where firewalls cannot be used, special smoke barriers (often called *curtain boards*) are suspended from the ceiling (Fig. 24.7) in an effort to trap the initial layer of hot air and smoke, and therefore to set off the fire detection/suppression system more quickly. As useful as these partial barriers can be in the fire's early stages, they quickly lose effectiveness as the smoke layer thickens or as air pressures force the smoke below the barrier. Even in firewalls with fire door protection, small cracks around such openings provide smoke paths, aided by the air pressures and velocities that fires can quickly produce. Therefore, the typical barrier is not very resistant to spreading smoke, and smoke control based solely on confinement must be closely linked to an effective early detection/suppression system such as a sprinkler system.

Compartmentation may not always be the best strategy. In the Potomac Mills Mall in Prince William County, Virginia, a 156,000-ft² (14,493-m²) addition to the shopping mall keeps the walls low and the ceiling high. The walls separating tenant spaces and public corridors are 12 ft (3.7 m) high, while the ceiling is at 17 ft (5.2 m), resulting in a large open volume of air. Smoke rising from a fire within a store can certainly spread well beyond that store, but will collect overhead in such a large space that evacuation of the mall can be completed well before the smoke reaches the occupied level. A computer model indicated that 20 minutes into a fire (and assuming *no* sprinkler operation), the smoke layer would have dropped to 12.3 ft (3.75 m) above the

floor. If the walls were extended to the ceiling, this same building would experience, in the same 20 minutes, within the origin store of the fire, smoke all the way down to the floor. In such a scheme, CFD analysis may well be required before approval is granted.

## (c) Dilution

For a limited time in a fire's early stages, the dilution of smoke with 100% outdoor air (provided by the HVAC system) may make conditions bearable during occupant evacuation. However, such large quantities of fresh air are needed in so short a time that dilution alone is rarely sufficient for smoke control, particularly when the smoke contains toxic fumes. When the dilution system is combined with both confinement and early detection/suppression systems, it becomes more attractive. Dilution systems that dump large quantities of outdoor air into refuge areas (including stairs) can create such high pressures within the refuge that smoke cannot enter through cracks around doors. However, when such doors are opened (as when more people enter the refuge), dilution systems that rely on the conventional HVAC system can rarely provide sufficient pressure to exclude smoke through the open doorway.

Fan-driven dilution systems raise several issues. Is the fan located where fire will not affect its performance and where its air intake can be kept free of smoke? Given the multiple possibilities in building fires, this is a difficult promise to make. "Dedicated" systems, independent of the building's HVAC system, are often favored to enable such emergency equipment to be isolated and because such huge quantities of air for relatively small spaces are well beyond the capability of the typical HVAC system. Where should the fresh air supply outlets be located? Most certainly, *below* the level of smoke accumulation so as not to drive smoke downward. (This may suggest using an under-floor air supply distribution system for such a purpose.) In a stair, how often should fresh air be supplied? Some fire engineers say every 3 floors, others every 10, but supplying all fresh air from either the top or the bottom is *not* recommended; it is too likely that open doors near that source would deplete fresh air for the rest of the stair. At what velocity should fresh air be supplied? The recommended maximum is 200 fpm (1 m/s) to avoid stirring the smoke layer and inviting its descent. The designer must consider that high rates

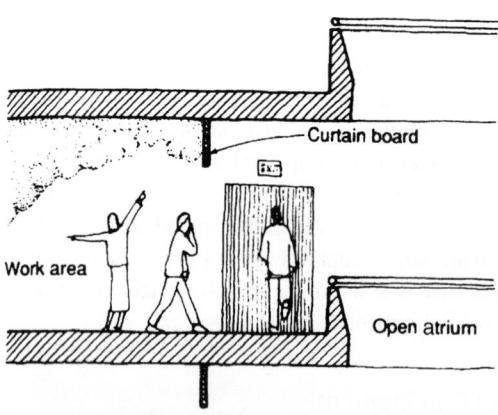

**Fig. 24.7** *Smoke barriers (curtain boards) are useful in a fire's early stages. By confining the initial layer of heated air and smoke produced by the fire, they help to slow the spread of smoke while making more likely the early detection and suppression of the fire.*

of flow at low velocities result in very large grilles located quite low within the space.

## (d) Exhaust

Special exhaust systems that function only in a fire are becoming more common. These systems employ both air velocity and air pressure to help control smoke. As with dilution systems, high velocities are avoided so that fresh air is not drawn up through the smoke layer. Such systems are particularly useful in large-volume atrium spaces, as in Fig. 24.8. Smoke accumulating at the ceiling is removed, while fresh air is supplied lower in these high spaces. If smoke can be removed at the same rate at which it is generated, the smoke layer can be kept from descending. Although they involve a greater initial expense, because they require special fans and special smoke-exhaust shafts, exhaust systems have several advantages over simple confinement or dilution approaches:

1. They can remove toxic gases from refuge areas (particularly when there is a dilution or outdoor air supply to such areas).
2. They help fire fighters by improving the air quality in the vicinity of the fire itself.
3. They can help control the direction that a fire takes by creating air currents that a fire will follow.
4. They remove the unburned but combustible gases from a fire before the latter can cause a *backdraft* or *flashover* (smoke explosion).
5. By creating higher air pressures in refuge areas and lower pressures in the fire zone, they keep smoke out of refuge areas even when doors are temporarily open.
6. With them, the tall-building stack effect, complicated by buoyancy and wind, is less likely to overcome the smoke management system.
7. They can help to remove smoke after the fire.

The two Petronas Towers in Kuala Lumpur, Malaysia, are each occupied by as many as 18,000 workers during the workday. As these are the tallest buildings in the world at the turn of the twenty-first century, most of these workers are well beyond the range of ordinary fire rescue operations. Smoke exhaust, stair pressurization, and vestibule pressurization systems are installed, and each floor has isolation dampers, controlled by the fire alarm system, wherever air ducts serve multiple floors. However, because each floor has its own air-handling system

(floor-by-floor fan rooms), vertical ductwork is minimized. There are both a main and a secondary fire command center for each of the towers and associated lower buildings.

## (e) HVAC Systems, Sprinklers, and Smoke

Two systems within the building must be closely coordinated with smoke exhaust systems: the HVAC system and the fire detection/suppression (usually sprinkler) system. As the fire detection system activates the smoke exhaust fans, it must also override the conventional HVAC system operation, usually by switching to 100% outside supply air and simultaneously blocking the return duct system. This process pressurizes each HVAC zone to form a barrier against the smoke (Fig. 24.9) and keeps smoke out of the return duct system. If the HVAC system is variable air volume (VAV), all supply control valves (dampers) must be moved to their full open position. If additional protection is needed for refuge areas, additional air supply to such areas can be provided, perhaps by a separate (smoke dilution) system described earlier.

Sprinkler systems can hamper the functioning of smoke exhaust systems, both by creating a curtain of water that inhibits the movement of smoke and by cooling the smoke, thus reducing its buoyancy. As less buoyant smoke descends, visibility is reduced and the danger of smoke inhalation increases, particularly in spaces with ordinary ceiling heights. The buoyancy of smoke is the factor relied on by smoke exhaust shafts, whose intakes are located at ceiling level within each zone. Although these two systems thus appear to work at cross purposes, suppression of the fire is obviously as important a goal as smoke management. Sprinkler systems potentially suppress fires so quickly that the size of the accompanying smoke exhaust systems can generally be reduced.

When the fire suppression system relies on oxygen displacement (such as carbon dioxide or "clean agents"), smoke exhaust systems clearly pose a threat; special attention is required if both systems are to be used. Again, oxygen displacement systems are often used in nonhabitable spaces, where smoke evacuation is less of an issue.

## (f) Automatic Ventilating Hatches

These heat-and-smoke venting devices (without fans) are often installed (Fig. 24.10) in smaller

FIRE PROTECTION

(*a*)

(*b*)

**Fig. 24.8** *The Rock and Roll Hall of Fame and Museum on Cleveland's lakefront (a) features a tall, pyramidal interior space. (b) At the apex, a large horizontal grille marks the smoke exhaust system intake. Smoke naturally accumulates here, and can be removed while special sources of fresh air operate at ground level around the perimeter. (When hot days alternate with cool nights, such a system could be used for night ventilation of thermal mass.)*

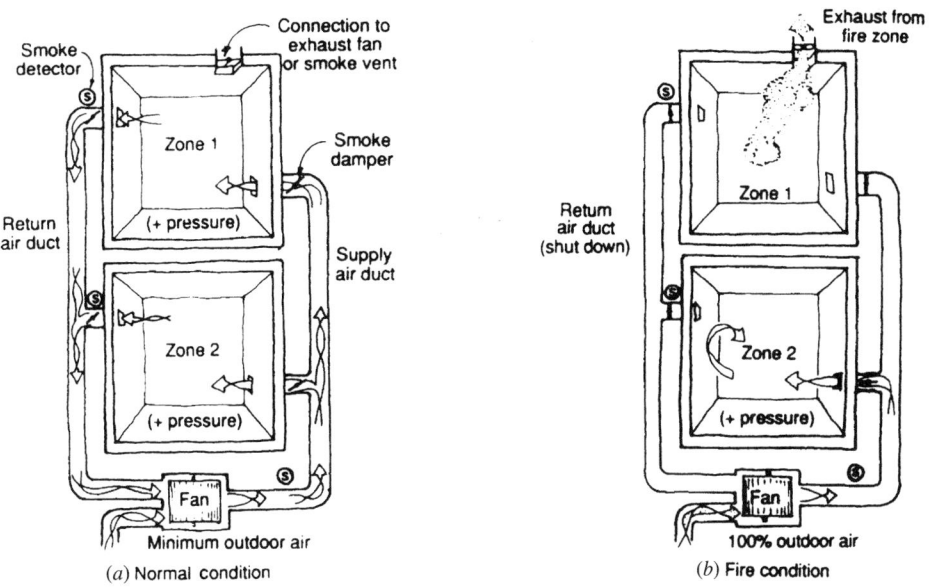

**Fig. 24.9** *Smoke management by a smoke exhaust system. (a) The conventional HVAC system is shown operating under normal minimum outdoor air supply conditions. (b) When smoke detectors trigger the smoke exhaust system, special fans begin to exhaust smoke through a special duct from the affected area only. Simultaneously, the conventional system undergoes two changes: All supply air becomes 100% outside air (for smoke dilution), and all return openings are closed off (to pressurize all nonfire areas, to keep smoke out of them and out of the return duct system.)*

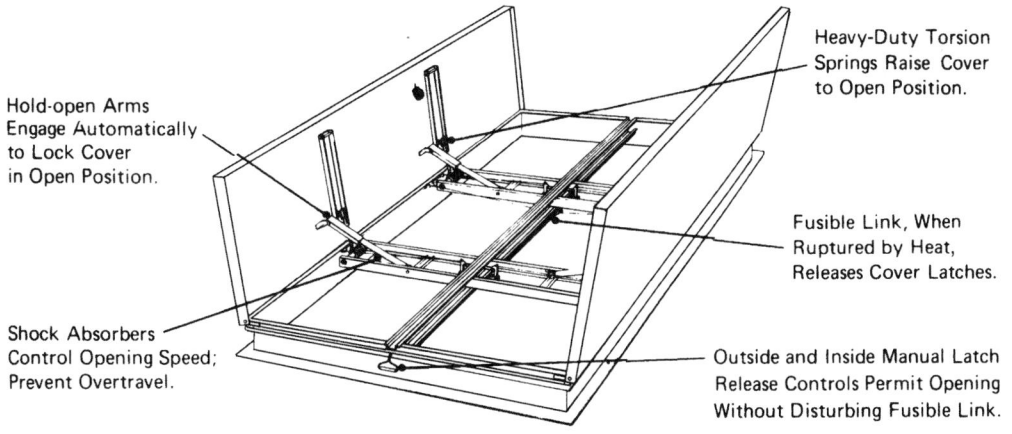

**Fig. 24.10** *Automatic heat-and-smoke roof hatch ventilators, primarily for one-story commercial and industrial buildings. (Courtesy of Inryco, Inc.)*

buildings or larger one-story buildings such as shopping malls. These hatches open individually as heat or smoke from the fire triggers their control devices. As a result, indoor conditions near the fire are improved for fire fighters, and fire fighters on the roof are quickly alerted to the location of the fire indoors.

In summary, designers can provide several components that aid in the exhaust of smoke and the provision of access and ventilation for fire fighters in almost any building: dedicated smoke exhaust and dilution systems, emergency controls on HVAC systems, operable windows and skylights (to facilitate ventilation and occupant escape), and smoke-and-heat hatches in roofs.

**FIRE PROTECTION**

## 24.3 WATER FOR FIRE SUPPRESSION

The most popular medium for building fire suppression is water, which is readily available and relatively low in cost. Water cools, smothers, emulsifies, and dilutes. As it turns to vapor, it removes 970 Btu/lb (2256 kJ/kg) at atmospheric pressure and its volume increases 1700 times—a process that helps push away the oxygen needed by the fire. Water has several disadvantages that sometimes preclude its use for fire suppression: it damages most contents of buildings, including interior surfaces; it conducts

electricity readily as a stream (less readily as a spray); many flammable oils float on the water's surface while continuing to burn; and as water vaporizes rapidly as steam, it can harm fire fighters. Where these factors are major considerations, other suppression media can be considered (see Section 24.4).

Fire-suppressing systems are commonly combined with smoke management systems; Fig. 24.11 shows a typical combination of provisions in one-story buildings. Taller buildings, especially those with multistory interior spaces (atriums), present a special challenge (Fig. 24.12). A typical approach is

FIRE PROTECTION

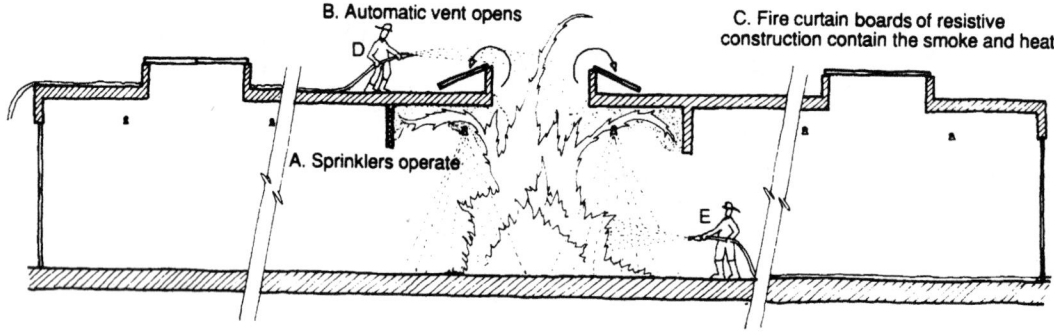

**Fig. 24.11** *Fire protection for large-area, one-story buildings showing the fire suppression (sprinkler) system (A) and the smoke management system (B,C). Locations D and E are relatively safe positions for fire fighters.*

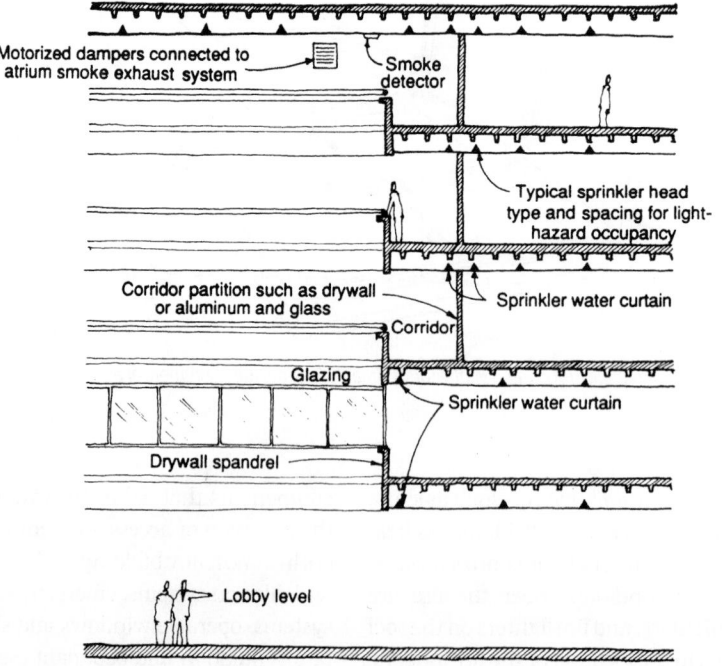

**Fig. 24.12** *Fire protection in an atrium-type office building: section showing the detection/suppression system and provisions for smoke management.*

to construct an approximately 6-ft (1.8-m)-deep curtain board (or fire spandrel) at the opening to the atrium on each floor. Smoke detectors, motorized dampers, and sprinklers can be combined in various ways. At the lobby level, the atrium floor level, and at any office floors open to the atrium, sprinklers at about 6 ft (1.8-m) o.c. (on center) will provide a water curtain at the edge of the atrium opening. Where glazing is used at the atrium's perimeter, the glazing frames sometimes can be designed for considerable thermal expansion, and sprinklers can provide a water curtain on the office side of the glazing. Fire-rated glazings and their frames are listed for both heat resistance and water pressure from a hose stream. Where balcony corridors adjoin the atrium, two sprinkler water curtains can be provided, one on either side of the partitions between corridor and office space. The smoke exhaust system often can provide six air changes per hour (ACH) for all spaces that open onto the atrium.

## (a) Standpipes and Hoses

Standpipes and hoses (Fig. 24.13) with a separate water reserve, upfeed pumping, or fire department connections are listed in three classes and five types. The major differences are whether the system is for first-aid or full-scale fire fighting and whether the system has an automatic water supply or a manual one.

*Class I* systems are for full-scale fire fighting, and are typically required in both sprinklered and unsprinklered buildings more than three stories high, as well as in malls. (This is based on the excessive time required to lay out hoses from outdoor fire connections.) They are for use by trained fire fighters using 2½-in. (64-mm) hose connections at designated locations. Class I is becoming the system of choice as use of the next two system types declines.

*Class II* systems are for first-aid fire fighting before the fire trucks arrive. They use 1½-in. (38-mm) hose connections and typically store a hose, nozzle, and hose rack (Fig. 24.14) in each specified location. The difficulty for untrained people to manage a 100-ft (30.5-m) hose containing a huge flow (100 gpm [378 L/min] or more) has led to a decline in the use of this system, although it can be found in large unsprinklered buildings and in special hazard areas such as

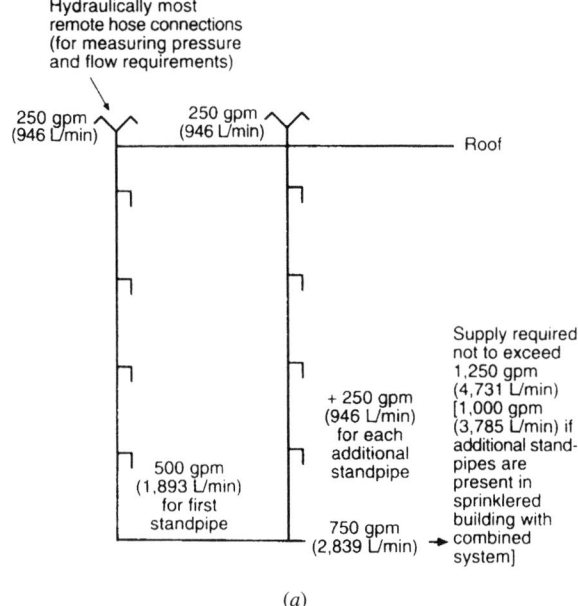

*(a)*

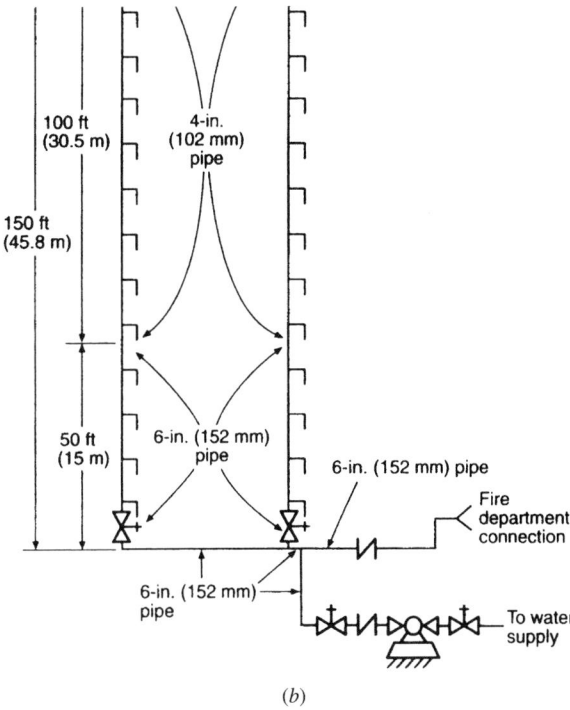

*(b)*

**Fig. 24.13** *Standpipe system variations. (a) Basic flow rate requirements for Class I and Class III systems. (b) Example of the pipe schedule sizing method for a Class I system. (c) Multiple-zone system. (Reprinted with permission from the* Fire Protection Handbook, 18th ed., © 1997, National *Fire Protection Association, Quincy, MA, 02269.)*

FIRE PROTECTION

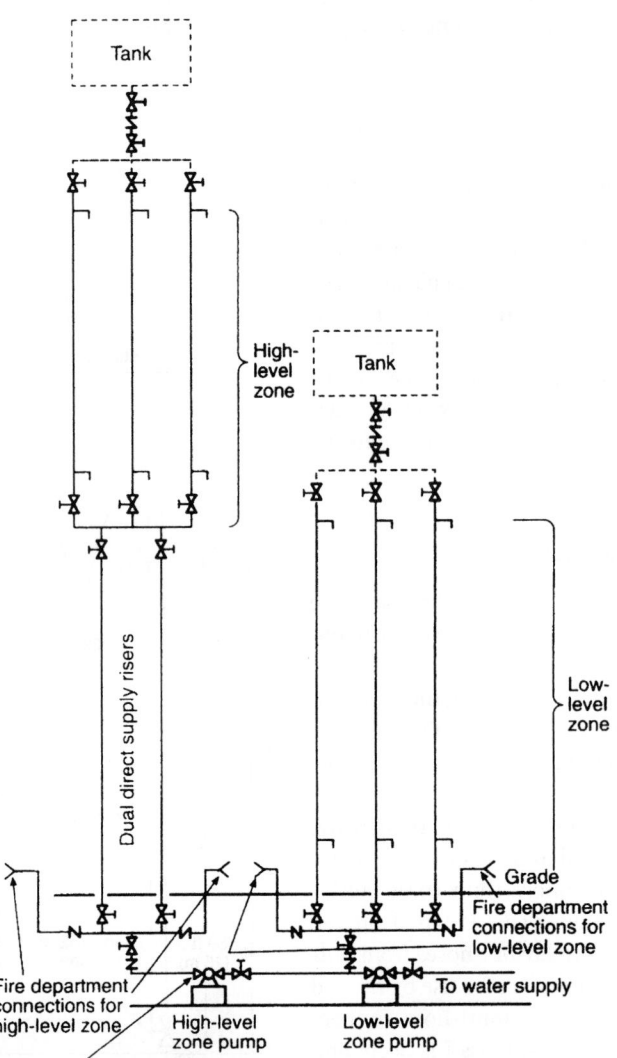

*(c)*

**Fig. 24.13** *(Continued)*

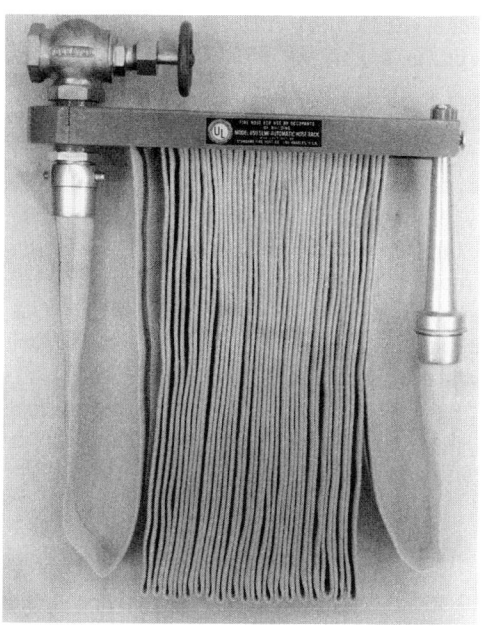

(a)

(b)

**Fig. 24.14** *Components of a standpipe and hose system. (a) Standard hose rack. (b) Hose rack and fire extinguisher in a cabinet with a glass door. (c) Siamese connection through which fire department pumping equipment supplies water to the standpipe system. The similar sprinkler siamese is marked "sprinkler." Color codes sometimes differentiate the two.*

(c)

stages. Many fire departments would rather the building occupants evacuate than try to manage such large hoses and water quantities. Significant water damage is a potential threat when hoses cannot be controlled.

*Class III* systems combine the characteristics of Classes I and II, thus serving both for first-aid and for full-scale fire fighting. Hose connections of both sizes are provided (with adapters) at each specified location. The disadvantages of Class II systems also apply to Class III, and their use is declining for the same reasons.

"*Combined*" systems are either Class I or III standpipe systems that also supply water to a sprinkler system.

The standpipe system types are as follows:

*Automatic-wet systems.* The pipes are filled with water and are connected to a water supply capable of automatically meeting the fire fighting demands. Water flows immediately on the opening of a hose valve.

*Automatic-dry systems.* The pipes are filled with pressurized air, and are connected to a water supply capable of automatically meeting the fire fighting demands. Through a device such as a dry-pipe valve, water replaces the air when a hose valve is opened.

*Semiautomatic-dry systems.* The pipes are filled with air and are connected to a water supply capable of automatically meeting the fire fighting demands. Through a device such as a deluge valve, water replaces the air when both a remote sensing device at the hose station and a hose valve are opened.

*Manual-dry systems.* The pipes are filled with air, and there is no connection to a water supply system other than that provided by the fire department (as in Fig. 24.14c).

*Manual-wet systems.* The pipes are filled with water, with a connection to a domestic water source that is used merely to fill and test the system. Water for fire fighting is provided by the fire department.

Minimum flow rates, minimum pressure, and maximum pressure all are involved in the design of standpipe systems. Two methods of sizing, the pipe schedule and the hydraulic method, are used. Neither is detailed here; see NFPA 14 (2003) for the procedures. The required supply flow rates are outlined in Fig. 24.13a. As a preliminary design guideline (Fig. 24.13b), Class I and III standpipes not exceeding 100 ft (30.5 m) in height must be a minimum of 4 in. (102 mm) nominal pipe size. Class I and III standpipes more than 100 ft (30.5 m) in height must be a minimum of 6 in. (152 mm) nominal pipe size (although the topmost 100 ft [30.5 m] may be a minimum of 4 in. [102 mm] nominal pipe size). For combined standpipe and sprinkler systems, regardless of height, a minimum of 6 in. (152 mm) nominal pipe size is required.

For Class I and III systems, a minimum hose pressure of 100 psi (690 kPa) is now required because of the widespread use of fog nozzles (rather than stream nozzles). The maximum hose pressure (also the maximum for sprinklers) is 175 psi (1207 kPa).

Water supply systems for fire fighting may use roof tanks or fire pumps. It is not practical to store enough water on the roof for a protracted fire fighting period, and it is usually assumed that a half-hour supply will be more than enough to provide for the short period it takes the fire engines to arrive. When the system is used by the fire department, its pumps are attached to the street siamese (Fig. 24.14c) to deliver water from street hydrants or the building's "secondary source." The check valve closest to the siamese in use opens and the check valves at the tank close to prevent the water from rising uselessly in the tank. After the engines are disconnected from the siamese, the water between the siamese and the adjacent check valve drains out through the ball drip so that it does not freeze.

An overhead tank is considered a most dependable source, but the height required can be architecturally undesirable and a considerable seismic disadvantage. In such a case, upfeed fire pumps (Fig. 24.13c) operating automatically to deliver water to higher stories may be used. Another option in this case is a pneumatic tank that delivers water by the

power of the air that is compressed in the upper portion of the tank.

A standpipe zone in a high-rise building is determined by hydraulic calculation so that maximum hose pressures are not exceeded. Fire standpipes and their hoses (for full-scale fire fighting) are now recommended to be located on the landings of fire stairs, from which personnel or fire fighters can approach a fire with charged hoses. This enables the hose connection to be made in a safe (relatively smoke-free) place, half a floor below the fire. Awkwardly, it means that the door to the stair will be held open by the hose, but the risk to occupants on the stairs is considered to be outweighed by the advantage to the fire fighters.

### (b) Sprinkler System Design Impacts

Unlike a fire hose, a sprinkler is likely to be already positioned above the point of a fire and is capable of being deployed in seconds, not minutes. Sprinkler systems are widely relied on as proven automatic fire suppressers. (They have come a very long way from the early water-filled bucket suspended by a black powder fuse; when the fuse was lit by a fire, it blew up the keg, dispersing the water and theoretically dousing the fire.)

However, designers must remember that the use of sprinklers does *not* give one a license to ignore fire code limitations, even though many codes are more lenient for sprinklered buildings. In addition, provision must be made for an adequate water supply, standby power for water pumping, and water drainage during and after the fire. Sprinklers require very large supply pipes and valves, and sometimes fire pumps—all of which are frequently considered unsightly. However, sprinklers afford opportunities for integration with energy-conserving HVAC systems (Section 24.3k). Also, the cost of sprinkler system installation can usually be recovered rather quickly through reductions in fire insurance premiums.

Automatic sprinkler systems usually consist of a horizontal pattern of pipes placed just below or within the ceilings of industrial buildings, warehouses, stores, theaters, offices, homes, and other structures in which a fire hazard requires their use. These pipes are provided with outlets and sprinkler heads constructed such that abnormally high temperatures will cause them to open automatically and emit a series of fine water sprays. NFPA 13 (2002) lists the detailed requirements for sprinkler systems (briefly referred to here) as well as numerous other regulations. In addition to pipes and sprinklers (discussed in the remainder of this section), several other system components are usually included.

*Alarm Gong.* An alarm gong mounted on the outside of the building warns of water flow through the alarm valve on the actuating of a sprinkler head. This warning gives the building personnel an opportunity to make additional fire fighting arrangements that can minimize loss and speed the termination of the fire; in this way, the sprinklers can be turned off to prevent excess water damage to building contents after the fire is out. Sprinkler alarms commonly are also connected to private regional supervisory offices that communicate promptly with municipal fire departments on the receipt of a signal. All public buildings, and other buildings as required, should be provided with fire detection and alarm systems that indicate, in the custodian's office, the location of the fire.

*Siamese Connections.* Siamese connections permit fire engines to pump into the sprinkler system in a manner similar to that used for standpipe systems (Fig. 24.14c).

*Provisions for Drainage.* Sprinkler heads can release a great deal of water, most of which will remain unvaporized and quickly collect at floor level. In addition to waterproofing the floors and at lower walls, columns, and other elements, provision should be made, where possible, for gravity drainage of water. Scuppers in exterior walls are preferable to floor drains, which are more easily clogged by debris. Scuppers are provided with hoods that protect against infiltration, birds, or insects.

*Water Supply.* In tall buildings, sprinklers (and standpipes) can be supplied with water from elevated storage tanks for domestic water. These tanks supply a constant pressure on the distribution lines, store sufficient water to balance supply and demand, prevent excessive starting and stopping of the pump, and provide a dependable fire reserve. The last factor has been critical in the calculation of fire insurance rates. When gravity tanks are used with sprinkler systems, they should provide enough

water to operate 25% of the sprinkler heads for 20 minutes. As in the case of standpipe and hose systems, this provision gives the fire department a chance to arrive and take over.

The principal objections to the use of tanks have been their unsightliness, the problem of freezing, and—in the case of large buildings—the tanks' tremendous weight, with resulting added structural costs.

Skidmore, Owings & Merrill, in their design for General Mills' central office building in suburban Minneapolis, turned to automatically controlled pumping systems instead of elevated storage (Fig. 24.15). For the General Mills building, the fire underwriters could have required at least 50,000 gal (189,250 L) of residual water in the tank for emergency purposes, which would have meant about 100,000 gal (378,500 L) of elevated storage. The alternative chosen was a reinforced-concrete structure placed underground to one side of the building and covered with 3 to 5 ft (0.9 to 1.5 m) of earth. Small vents rising from this reservoir blend in with the lawn and landscaped shrubbery above.

The saving in structural steel tended to make the overall cost comparable to that of an elevated tank of the same capacity. The savings were made possible chiefly by refined automatic-pump control. The underground reservoir eliminates the problems of an unsightly appearance and great weight, but the other usual advantages of an elevated tank—reliability in case of fire, minimum starting and stopping of motors, and the maintenance of pressure while balancing supply and demand—must be equaled in the automatic circuitry. This is not as simple as it might at first seem. Factors such as fluctuations in demand, friction within the pipes, elevations, starting surges from the pumps, and pressure-flow characteristics of the pumps themselves must be considered. Various combinations of these problems undoubtedly account to a great extent for the continued use of elevated tanks. The trend is toward the use of more sophisticated pump control.

A continuous flow from the deep-well pumps through both domestic and fire reservoirs prevents the water from becoming stale and rancid. The fire reservoir is given the necessary priority over the domestic reservoir by means of a simple weir. Even if the domestic reservoir were completely empty, the fire reservoir would remain full. Pressure for the

FIRE PROTECTION

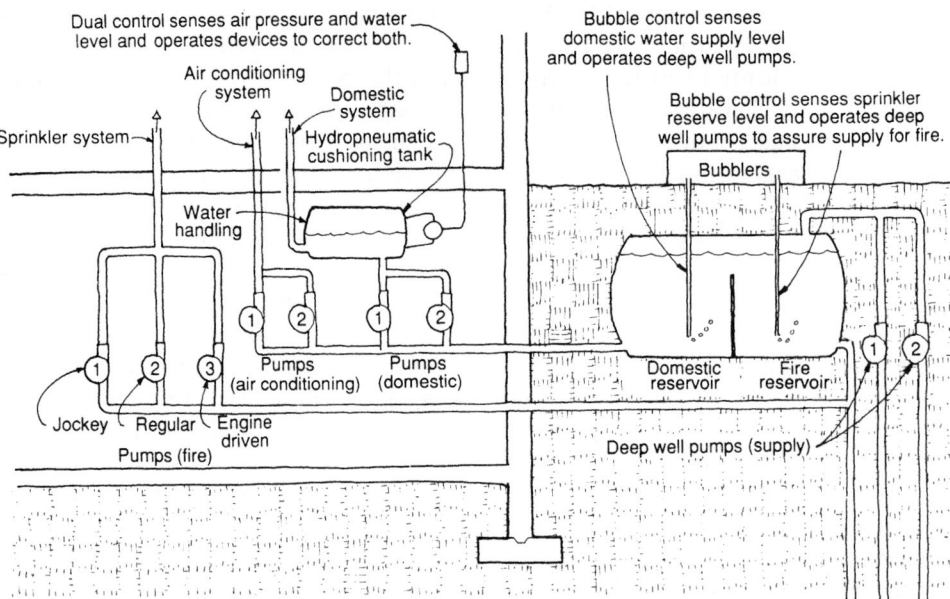

**Fig. 24.15** *Diagram of an upfeed pumping system to supply sprinklers and building demands for domestic water. In the General Mills Building, a large subsurface concrete water tank provides the secondary water source that reassures fire insurance carriers. Thus, the wells and their pumps constitute a primary water source. If the well yield fails, the concrete tank supply can be tapped. An additional resource provides engine-driven pumping that operates in the event of a power outage at an electric utility company. Automatic controls and piping for the storage tank reserve system are not shown in this illustration.*

sprinkler is supplied by a small 20-gpm (1.26-L/s) jockey pump. Signals from the sprinkler system bring in a 750-gpm (47.3-L/s) main pump. If this fails, a diesel engine–driven pump of equal capacity automatically takes over. Normally, the system automatically satisfies the heavy demands of air conditioning, fire control, and domestic water supply in this modern, rural, isolated office building.

*Valves.* Valves are required to allow the sprinkler system to be shut off for maintenance, system modification, or replacement of all sprinkler heads that have operated after a fire. Indicating valves of various types are required so that it is always immediately obvious whether this crucial valve is open (as it nearly always should be) or closed. Table 24.6 lists common failure modes in sprinkler systems; an improperly closed valve is the leading reason.

### (c) Sprinkler Construction, Orientation, and Rating

The common sprinkler head contains water by a plug or cap that is held tightly against the orifice by levers or other restraining devices. These, in turn, are held in place by the arms of the sprinkler body. In the past, the restraining device was usually a fusible metal link that melted at a predetermined temperature. Today, the restraining device is typically a glass bulb that contains both a colored liquid and an air bubble. As the liquid expands with the heat of a fire, it compresses the air bubble until it is absorbed. Then continued expansion ruptures the bulb (again, at a predetermined temperature) and releases the water in a solid stream through the orifice.

A deflector converts this solid stream of water into a spray, whose pattern is determined by the design of the deflector. Most sprinklers direct the spray down and horizontally rather than upward toward the ceiling. This yields better water distribution close to the sprinkler head and more effective coverage of burning material below. The spray pattern from the typical upright or pendant sprinkler is shown in Fig. 24.16.

Common types of sprinkler heads are upright (SSU), pendant (SSP), or sidewall types. Upright heads sit on top of the exposed supply piping. Pendant heads hang below piping, which thus can be concealed above suspended ceilings (Figs. 24.17a,b). The pendant heads themselves have a number of

**TABLE 24.6  Common Failure Modes for Automatic Sprinklers**

| Failure Mode | Potential Causes |
|---|---|
| Water supply valves are closed when sprinkler fuses | Inadequate valve supervision<br>Owner attitude<br>Maintenance policies |
| Water does not reach sprinkler | Dry pipe accelerator or exhauster malfunctions<br>Pre-action system malfunctions<br>Maintenance and inspection inadequate |
| Nozzle fails to open when expected | Fire rate of growth too fast<br>Response time and/or temperature of link inappropriate for the area protected<br>Sprinkler link protected from heat<br>Sprinkler link painted, taped, bagged, or corroded<br>Sprinkler skipping |
| Water cannot contact fuel (*Note:* The intent of this failure mode is to ensure that discharge is not interrupted in a manner that will prevent fire control by a sprinkler) | Fuel is protected<br>High piled storage is present<br>New construction (walls, ductwork, ceilings) obstructs water spray |
| Water discharge density is not sufficient | Discharge needs are insufficient for the type of fire and the rate of heat release<br>Change in combustible contents occurred<br>Number of sprinklers open is too great for the water supply<br>Water pressure too low<br>Water droplet size is inappropriate for the fire size |
| Enough water does not continue to flow | Water supply is inadequate because of original deficiencies, changes in water supply, or changes in the combustible contents<br>Pumps are inadequate or unreliable<br>Power supply malfunctions<br>System is disrupted |

*Source:* Reprinted with permission from the *Fire Protection Handbook,* 18th ed., © 1997, by the National Fire Protection Association, Quincy, MA, 02269.

**FIRE PROTECTION**

variations: recessed, flush, concealed, and ornamental pendant heads are available (Fig. 24.17*c*). Recessed heads have part of the sprinkler body and the deflector below the ceiling. Flush heads have only the thermosensitive element projecting below the ceiling. Concealed heads are entirely above a ceiling cover plate that falls away in a fire, exposing the thermosensitive element. Ornamental pendants are manufacturer-coated (never field-coated!) to match a desired decor.

Sidewall sprinklers (Fig. 24.17*d*) are usually located adjacent to one wall of smaller rooms— hotels, apartments, and so on—and throw a one-quarter-sphere spray of water entirely across such rooms. Therefore, only one sidewall sprinkler head per small room often is used.

The temperature at which a sprinkler is triggered is usually specified, as in Table 24.7; it should be at least 25F° (14C°) higher than the maximum ceiling temperature ordinarily expected. Ordinary sprinkler heads operate at between 135° and 170°F (57° and 77°C).

Buildings to be protected by sprinkler systems fall into several hazard groups, as described in Table 24.8. Newer *quick-response* sprinkler heads are now required throughout light hazard occupancies, including office buildings. These more thermally sensitive heads open sooner than ordinary heads, and thus tend to fight a fire with even fewer heads operating, even though they may sometimes open with extraordinary heat that is not fire-associated. They are considered superior for life protection because of their earlier operation.

Other special sprinkler models include an *extra large orifice* (for delivering large water quantities where water pressures are relatively low) and *multi-level* sprinklers (for use where other sprinklers are at a higher plane within the same space). These lower

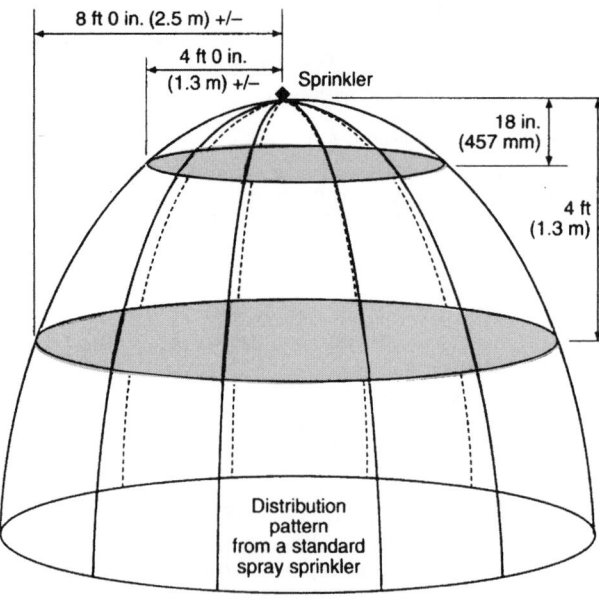

**Fig. 24.16** *Typical distribution pattern from a standard spray sprinkler. Obstructions (whether from structure or furnishings) are to be avoided within this volume. (Reprinted with permission from the Fire Protection Handbook, 18th ed., © 1997, National Fire Protection Association, Quincy, MA, 02269.)*

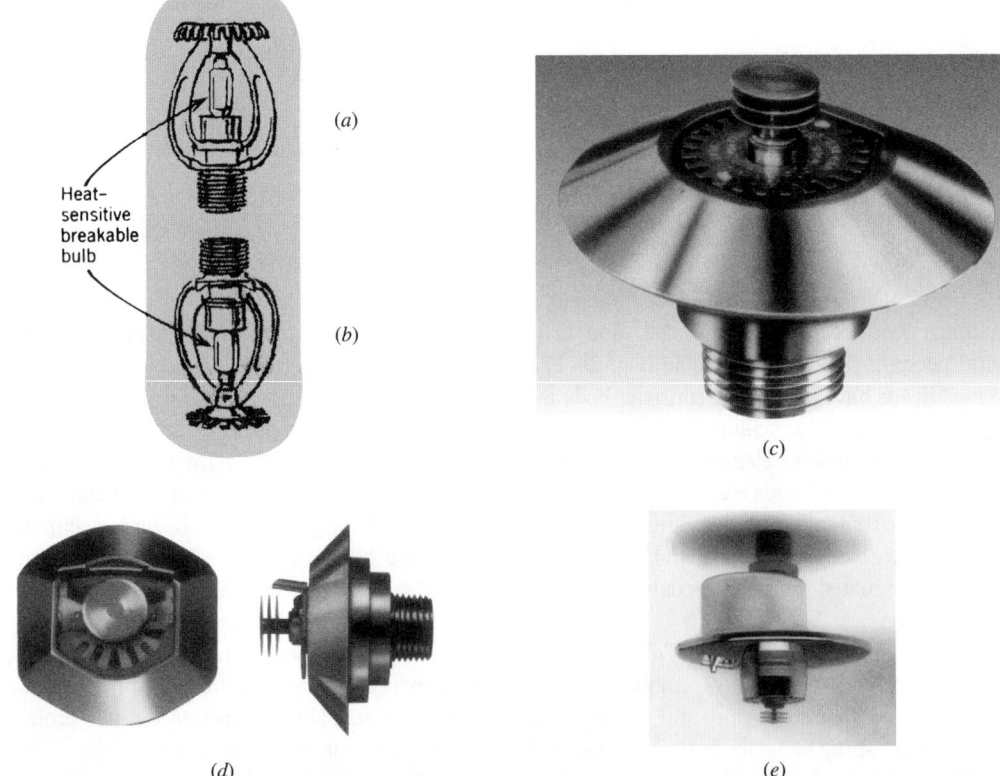

**Fig. 24.17** *Sprinkler heads. (a) Upright; it sits above exposed piping, just below the structural deck—where the hottest gases will first accumulate. (b) Pendant, projecting through a suspended ceiling in most installations. Both (a) and (b) use a quartzoid bulb, a transparent bulb with a colored liquid that ruptures at a preset temperature to release the water stream; heads can be manufactured as plated, polished, or colored—but never painted on site. (c) Pendant, styled for a more contemporary appearance. (d) Sidewall, front and side views. (e) Flow control pendant, which must be exposed within the space it serves. (Parts c–e courtesy of the Central Sprinkler Corporation.)*

**TABLE 24.7 Sprinkler Heads**

| DISCHARGE CHARACTERISTICS | | | | |
|---|---|---|---|---|
| **Nominal Orifice Size** | | | **Nominal Orifice Size Marked on Frame** | **Percent of Nominal ½-in. Discharge** |
| in. | mm | K Factor | | |
| ¼ | 6.4 | 1.3–1.5 | Yes | 25 |
| ⁵⁄₁₆ | 8.0 | 1.8–2.0 | Yes | 33.3 |
| ⅜ | 9.5 | 2.6–2.9 | Yes | 50 |
| ⁷⁄₁₆ | 11.0 | 4.0–4.4 | Yes | 75 |
| ½ | 12.7 | 5.3–5.8 | No | 100 |
| ¹⁷⁄₃₂ | 13.5 | 7.4–8.2 | No | 140 |
| ⅝ | 15.9 | 11.0–11.5 | Yes | 200 |
| ¾ | 19.0 | 13.5–14.5 | Yes | 250 |

| TEMPERATURE CLASSIFICATIONS | | | | | | |
|---|---|---|---|---|---|---|
| **Max. Ceiling Temp.** | | **Temperature Rating** | | **Temperature Classification** | **Color Code** | **Glass Bulb Colors** |
| °F | °C | °F | °C | | | |
| 100 | 38 | 135 to 170 | 57 to 77 | Ordinary | Uncolored or black | Orange or red |
| 150 | 66 | 175 to 225 | 79 to 107 | Intermediate | White | Yellow or green |
| 225 | 107 | 250 to 300 | 121 to 149 | High | Blue | Blue |
| 300 | 149 | 325 to 375 | 163 to 191 | Extra high | Red | Purple |
| 375 | 191 | 400 to 475 | 204 to 246 | Very extra high | Green | Black |
| 475 | 246 | 500 to 575 | 260 to 302 | Ultra high | Orange | Black |
| 625 | 329 | 650 | 343 | Ultra high | Orange | Black |

*Source:* Reprinted with permission from the *Fire Protection Handbook,* 18th ed., © 1997, National Fire Protection Association, Quincy, MA, 02269.

sprinklers may otherwise have their action retarded by water from the higher-elevation heads.

*Flow control sprinklers* close automatically once temperatures at the ceiling are sufficiently reduced. Ordinary sprinklers, once activated, continue to operate until a main valve is manually closed; this can waste large quantities of water and result in extensive water damage. The flow control sprinkler shown in Fig. 24.17*e* is designed to operate individually and in two stages. When the ambient temperature reaches 140°F (60°C), a fusible alloy melts and the sprinkler is readied for action. Water discharge, however, does not occur until the second stage, when the ambient temperature reaches 165°F (74°C). Later, when the ambient temperature falls below 95°F (35°C), the sprinkler returns to its first-stage condition: readied for action, but with no water discharge. Thus, if a fire reflash occurs, the sprinkler can again emit water, then again shut down as the temperature drops. Flow control sprinklers are not designed for use in dry-pipe systems (see Section 24.3*l*) and require special detailing when used in preaction systems (see Section 24.3*m*). They are not to be used in areas where nonfire ambi-

ent temperatures may exceed 100°F (37°C). These sprinklers are more visible because both the bottom and side elements within a space must be exposed.

### (d) Sprinkler Spacing and Hazard

The spacing of sprinkler heads and the sizing of their pipes are complex matters, and most sprinkler systems are designed by professionals working for sprinkler manufacturers. However, some guidelines for preliminary sprinkler location are included here. The first consideration is the degree of hazard faced by the occupants, as listed in Table 24.8. Once this is known, sprinklers and pipes can be approximately located in plan with the aid of Table 24.9. Maximum floor areas (per floor) to be protected by any single sprinkler system are shown, again by hazard, in Table 24.10. Within each space, sprinklers should be located so as to detect a fire readily and to discharge water over the greatest area (considering obstacles such as joists and beams, partial height partitions, etc.; see Fig. 24.16).

Piping for the sprinkler supply can be hydraulically designed, as shown in Section 21.9.

**TABLE 24.8 Relative Fire Hazard for Various Occupancies, as Related to Sprinkler Installations**

## CLASSIFICATION OF OCCUPANCIES

Occupancy classifications for this standard relate to sprinkler installations and their water supplies only. They are not intended to be a general classification of occupancy hazards. Examples in the listings below are intended to represent the norm for those occupancy types. Unusual or abnormal fuel loadings or combustible characteristics and susceptibility for changes in these characteristics, for a particular occupancy, are considerations that should be weighed in the selection and classification.

## LIGHT HAZARD OCCUPANCIES

Occupancies or portions of other occupancies where the quantity and/or combustibility of contents is low, and fires with relatively low rates of heat release are expected.

The Light Hazard classification is intended to encompass residential occupancies; however, this is not intended to preclude the use of listed sprinklers in residential occupancies or residential portions of other occupancies.

*Light Hazard Occupancies Include Occupancies Having Conditions Similar to:*

| | |
|---|---|
| Churches | Museums |
| Clubs | Nursing or convalescent homes |
| Eaves and overhangs[a] | Office, including data processing |
| Educational | Residential |
| Hospitals | Restaurant seating areas |
| Institutional | Theaters and Auditoriums, excluding stages and prosceniums |
| Libraries (except large stack rooms) | Unused attics |

[a]If combustible construction with no combustibles beneath.

## ORDINARY HAZARD OCCUPANCIES

Group 1: Occupancies or portions of other occupancies where combustibility is low, quantity of combustibles is moderate, stockpiles of combustibles do not exceed 8 ft (2.4 m), and fires with moderate rates of heat release are expected.

*Ordinary Hazard Occupancies (Group 1) Include Occupancies Having Conditions Similar to:*

| | |
|---|---|
| Automobile parking and showrooms | Electronic plants |
| Bakeries | Glass and glass product manufacturing |
| Beverage manufacturing | Laundries |
| Canneries | Restaurant service areas |
| Dairy products manufacturing and processing | |

Group 2: Occupancies or portions of other occupancies where quantity and combustibility of contents is moderate to high, stockpiles do not exceed 12 ft (3.7 m), and fires with moderate to high rates of heat release are expected.

*Ordinary Hazard Occupancies (Group 2) Include Occupancies Having Conditions Similar to:*

| | |
|---|---|
| Cereal mills | Paper and pulp mills |
| Chemical plants—ordinary | Paper process plants |
| Confectionery products | Piers and wharves |
| Distilleries | Post offices |
| Dry cleaners | Printing and publishing |
| Feed mills | Repair garages |
| Horse stables | Stages |
| Leather goods manufacturing | Textile manufacturing |
| Libraries—large stack areas | Tire manufacturing |
| Machine shops | Tobacco products manufacturing |
| Metal working | Wood machining |
| Mercantile | Wood product assembly |

## EXTRA HAZARD OCCUPANCIES

Occupancies or portions of other occupancies where quantity and combustibility of contents is very high and flammable and combustible liquids, dust, lint, or other materials are present, introducing the possibility of rapidly developing fires with high rates of heat release. Extra hazard occupancies involve a wide range of variables that may produce severe fires.

Extra Hazard Occupancies (Group 1) includes occupancies with little or no flammable or combustible liquids.

*Extra Hazard Occupancies (Group 1) Include Occupancies Having Conditions Similar to:*

| | |
|---|---|
| Aircraft hangars[b] | Printing using inks having flash points below 100°F (37.9°C) |
| Combustible hydraulic fluid use areas | Rubber reclaiming, compounding, drying, milling, vulcanizing |
| Die casting | Saw mills |
| Metal extruding | Textile picking, opening, blending, garnetting, carding, combining of cotton, synthetics, wool shoddy, or burlap |
| Plywood and particle board manufacturing | Upholstering with plastic foams |

Extra Hazard Occupancies (Group 2) includes occupancies with moderate to substantial amounts of flammable or combustible liquids or where shielding of combustibles is extensive.

*Extra Hazard Occupancies (Group 2) Include Occupancies Having Conditions Similar to:*

| | |
|---|---|
| Asphalt saturating | Open oil quenching |
| Flammable liquids spraying | Plastics processing |
| Flow coating | Solvent cleaning |
| Manufactured home or modular building assemblies[c] | Varnish and paint dipping |

[b]Except as governed by NFPA 409.

[c]Where finished enclosure is present and has combustible interiors.

## SPECIAL OCCUPANCY HAZARDS

See NFPA 13 for a listing of standards by occupancy.

*Source:* Adapted with permission from NFPA 13, *Installation of Sprinkler Systems*, © 1996, National Fire Protection Association, Quincy, MA. This reprinted material is not the complete and official position of the National Fire Protection Association on the referenced subject, which is represented only by the standard in its entirety.

**FIRE PROTECTION**

**TABLE 24.9 Protected Area, Maximum Spacing, and Distances Below Ceiling for Sprinklers**

PART A: UPRIGHT (SSU) AND PENDANT (SSP) SPRINKLERS

| Construction Type | Light Hazard Protection Area, ft² | Light Hazard Spacing (max.), ft | Ordinary Hazard Protection Area, ft² | Ordinary Hazard Spacing (max.), ft | Extra Hazard Protection Area, ft² | Extra Hazard Spacing (max.), ft | High-Piled Storage Protection Area, ft² | High-Piled Storage Spacing (max.), ft |
|---|---|---|---|---|---|---|---|---|
| Noncombustible obstructed and unobstructed, and combustible unobstructed | 225 | 15 | 130 | 15 | 100 | 12 | 100 | 12 |
| Combustible obstructed | 168 | 15 | 130 | 15 | 100 | 12 | 100 | 12 |

PART B: STANDARD SIDEWALL SPRAY SPRINKLERS

| | Light Hazard Combustible Finish | Light Hazard Noncombustible or Limited-Combustible Finish | Ordinary Hazard Combustible Finish | Ordinary Hazard Noncombustible or Limited-Combustible Finish |
|---|---|---|---|---|
| Maximum distance along the wall (S) | 14 ft | 14 ft | 10 ft | 10 ft |
| Maximum room width (L) | 12 ft | 14 ft | 10 ft | 10 ft |
| Maximum protection area | 120 ft² | 196 ft² | 80 ft² | 100 ft² |

PART C: EXTENDED COVERAGE (EC) UPRIGHT AND PENDANT SPRAY SPRINKLERS

| Construction Type | Light Hazard Protection Area, ft² | Light Hazard Spacing (max.), ft | Ordinary Hazard Protection Area, ft² | Ordinary Hazard Spacing (max.), ft | Extra Hazard Protection Area, ft² | Extra Hazard Spacing (max.), ft | High-Piled Storage Protection Area, ft² | High-Piled Storage Spacing (max.), ft |
|---|---|---|---|---|---|---|---|---|
| Unobstructed | 400, 324, 256 | 20, 18, 16 | 400, 324, 256, 196, 144 | 20, 18, 16, 14, 12 | 196, 144 | 14, 12 | 196, 144 | 14, 12 |
| Obstructed noncombustible (when specifically listed for such use) | 400, 324, 256 | 20, 18, 16 | 400, 324, 256, 196, 144 | 20, 18, 16, 14, 12 | 196, 144 | 14, 12 | 196 | 14 |
| Obstructed combustible | NA | NA | NA | NA | NA | NA | NA | NA |

PART D. EXTENDED COVERAGE (EC) SIDEWALL SPRINKLERS

| Construction Type | Light Hazard Protection Area, ft² | Light Hazard Spacing ft | Ordinary Hazard Protection Area, ft² | Ordinary Hazard Spacing, (L) ft |
|---|---|---|---|---|
| Unobstructed, smooth, flat | 400 | 28 | 400 | 24 |

PART E. LARGE-DROP SPRINKLERS

| Construction Type | Protection Area, ft² |
|---|---|
| Noncombustible unobstructed | 130 |
| Noncombustible obstructed | 130 |
| Combustible unobstructed | 130 |
| Combustible obstructed | 100 |

PART F. EARLY SUPPRESSION FAST-RESPONSE (ESFR) SPRINKLERS

| Construction Type | ESFR Sprinkler Up to 30 ft in Height Protection Area, ft² | Spacing (ft) | ESFR Sprinkler Up to 40 ft in Height Protection Area, ft² | Spacing (ft) |
|---|---|---|---|---|
| Noncombustible unobstructed | 100 | 12 | 100 | 10 |
| Noncombustible obstructed | 100 | 12 | 100 | 10 |
| Combustible unobstructed | 100 | 12 | 100 | 10 |
| Combustible obstructed | NA | NA | NA | NA |

PART G. MAXIMUM ALLOWABLE DEFLECTOR DISTANCES BELOW THE CEILING FOR VARIOUS SPRINKLERS[a]

| Sprinkler Type | Maximum Distance (in.) |
|---|---|
| SSU/SSP | 12 |
| Standard sidewall | 6 |
| EC upright and pendant | 12 |
| EC sidewall | 6 |
| Large drop | 8 |
| ESFR | 14 |

Source: Adapted with permission from the Fire Protection Handbook, 18th ed., © 1997, National Fire Protection Association, Quincy, MA, 02269.

[a] Greater distances are permitted based on construction type, special listings, or both.

Notes: For SI units: 1 ft² = 0.0929 m²; 1 ft = 0.305 m.
Obstructed: depth, spacing and openness of structural members impede heat flow to the sprinkler head and/or disrupt the spray pattern.
Unobstructed: structural members do not impede the operation of the sprinkler head.

FIRE PROTECTION

### TABLE 24.10  Maximum Floor Areas for Sprinkler Systems

The maximum floor area on any one floor to be protected by sprinklers supplied by any one sprinkler system riser (or combined system riser) is:

| Hazard | Area, ft² (m²) |
|---|---|
| Light Hazard | 52,000 (4,831) |
| Ordinary Hazard | 52,000 (4,831) |
| Extra Hazard (hydraulically calculated) | 40,000 (3,716) |
| High-Piled Storage | 40,000 (3,716) |

*Source:* Based on the *Fire Protection Handbook,* 18th ed., © 1997, National Fire Protection Association, Quincy, MA, 02269.
*Notes:* Floor area occupied by mezzanines shall not be included in the above areas. Where single systems protect extra hazard or high-piled storage and ordinary or light-hazard areas, the extra hazard or storage area coverage shall not exceed the floor area specified for that hazard and the total area coverage shall not exceed 52,000 ft² (4,831 m²).

A complicating factor is the expectation that only a small percentage of the sprinklers will actually open; more than 50% of the fires studied over a 49-year period were extinguished by two or fewer sprinklers. A detailed sizing procedure would consider both the available pressure at the highest sprinkler and the expected flow rate, which can vary from 500 to 5000 gpm (31.5 to 315 L/s). The sprinklers' actual performance is then determined by

$$Q = K \sqrt{p}$$

where

$Q$ = flow rate, gpm

$K$ = K factor, published for each sprinkler head by U.S. manufacturers (see Table 24.7)

$p$ = pressure, psi

In SI units:

$Q$ = flow rate, L/min

$K$ = K factor (14.3 × I-P K factor)

$p$ = pressure in bars

Sprinkler systems are usually designed for a maximum working pressure of 175 psi (1206 kPa). As a preliminary design guideline:

Light hazard systems need a minimum residual pressure of 15 psi (104 kPa) and 500–750 gpm (32–47 L/s) at the base of the system riser for 30 to 60 minutes.

Ordinary hazard systems need a minimum residual pressure of 20 psi (138 kPa) and 850–1500 gpm (54–95 L/s) at the base of the system riser for 60 to 90 minutes.

In the past, the *pipe schedule* design method made system layout and sizing fairly simple. Now

there are requirements for more complicated hydraulically designed systems, beyond the scope of this book, for pipe sizing.

## (e) Residential Sprinklers

Now that many codes require sprinklers in all residential occupancies, some new issues arise. The residential sprinkler is a fast-response device with a tested ability to enhance survivability in the room of fire origin and is thus listed for protection of dwelling units. It is sensitive to both smoldering and rapidly developing fires, opening quickly to fight a fire with only one or two heads operating. This is important because residences typically do not have a water supply with sufficient capacity for standard sprinkler systems. Toxic gases and smoke quickly fill the small rooms typical of residences; a fast response is important for life safety.

Most codes that otherwise require residential sprinklers in all areas make an exception for bathrooms no larger than 55 ft² (5.1 m²); for closets with the least dimension not exceeding 3 ft (0.9 m); for open porches, garages, and carports; for uninhabited attics and crawl spaces (if not used for storage); and for entrance foyers that are not the sole means of egress.

Residential sprinklers also have a special water distribution pattern capable of delivering water to the walls and high enough on the walls to prevent the fire from getting above the sprinklers. This water near the ceiling also tends to cool gases at the ceiling, reducing the likelihood of excessive sprinkler openings.

The added cost of residential sprinkler systems may be recovered in several ways. As with other buildings, reductions in fire insurance rates will be helpful, although in residences the payback time is rather long. Zoning could permit smaller parcels of land for sprinkler-protected residences, since separation between buildings is less important when sprinklers protect both buildings.

## (f) Quick-Response Sprinklers

All light hazard occupancies are now required to have quick-response (also called *fast-response*) sprinklers. These include hotels, motels, offices, and other buildings where faster sprinkler operation could enhance life safety. One measure of thermal sensitivity is the response time index (RTI), which indicates how fast the sprinkler can absorb heat from its surroundings sufficient to cause activation. It is expressed as the square root of meters-seconds. Quick- (fast-) response sprinklers have an RTI of 50 or less. Standard-response sprinklers have an RTI of 80 or more.

Because of the thermal lag in the glass bulb (or fusible link), the sprinkler body, and the water in sprinkler pipes, air temperature around the sprinkler may reach 1000°F (538°C) before a standard sprinkler, rated at 175°F (79°C), actually opens. The fast-response sprinkler's operating element has a smaller mass, enabling it to track the air temperature rise more quickly.

## (g) Early Suppression Fast-Response (ESFR) Sprinklers

See Table 24.9, part F. These are tested for their ability to suppress specific high-challenge fire hazards encountered in high-piled storage. They operate at a higher pressure and flow, so their water droplets depend on momentum rather than gravity to penetrate to the bottom of high-velocity fire plumes. These sprinkler heads require a minimum of 50 psi (345 kPa) and a minimum flow of 100 gpm (6.3 L/s). They have largely replaced the large-drop sprinklers (Table 24.9, Part E) that depend on the weight of the larger water droplet to penetrate the fire plume.

## (h) Extended Coverage Sprinklers

See Table 24.9, parts C and D. These are limited to a type of unobstructed construction consisting of flat, smooth ceilings of a slope not exceeding 2 in. per foot (158 mm/m). Note that a smooth ceiling means that luminaires and air grilles are flush or recessed, not suspended from the ceiling. However, such sprinklers can also be specifically listed for "noncombustible obstructed" construction, or as upright and pendant sprinklers within trusses or bar joists having web members not more than 1 in. (25 mm) thick, or where specifically listed for flat, smooth ceilings of a slope not exceeding 4 in. per foot (316 mm/m).

## (i) Future Developments

Now under development, *quick-response, early suppression* (QRES) sprinklers would use the same principles as the ESFR sprinkler, but with a smaller orifice suitable to lighter-hazard occupancies. This could be useful in a wide variety of business, mercantile, public assembly, and educational applications.

In future sprinkler systems, different types of sprays may be ejected from a single sprinkler head: one spray of larger droplets to penetrate the fire plume and thereby cool the burning surfaces as well as adjacent surfaces, and another, finer spray to cool the ceiling itself.

## (j) Wet-Pipe Systems

These are the most common and most simple systems, shown in Fig. 24.18. They are filled with water under pressure and are limited to spaces in which the air temperature does not fall below 40°F (4.4°C). (Wet-pipe systems that contain antifreeze, and admit ordinary water when a sprinkler head opens, are included in this category; the type of antifreeze is limited when potable water supplies the sprinkler system.) In the wet-pipe system, sprinklers in the affected area are opened by sensitive elements within the sprinkler heads themselves and immediately emit water.

A typical automatic wet-pipe sprinkler system is shown in Fig. 24.19. The building, a printing and publishing plant, is in the category of "Ordinary Hazard, Group 2" (see Table 24.8). The sprinkler design results in a nozzle spacing such that one nozzle (sprinkler head) takes care of 130 ft² (12 m²) of floor area.

An urban high-rise building with a wet-pipe system is the Transamerica Building (Fig. 24.20) in San Francisco. Two fire pumps can deliver 750 gpm at 275 psi (47.3 L/s at 1896 kPa) discharge

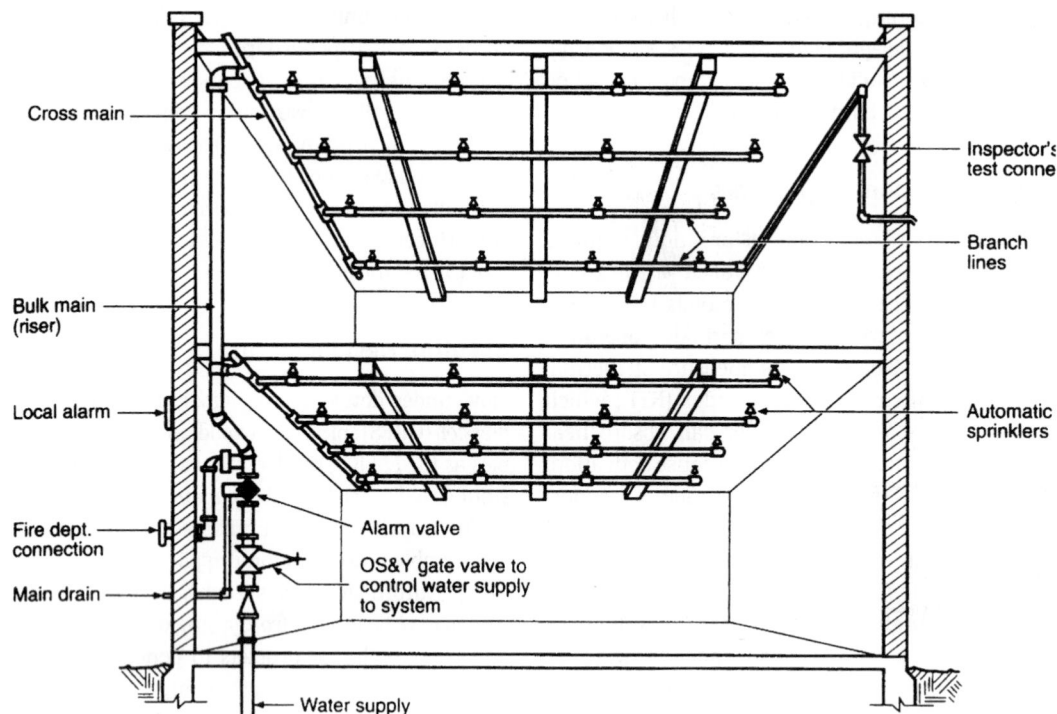

Cross main

Inspector's
test conne

Branch
lines

Bulk main
(riser)

Local alarm

Automatic
sprinklers

Fire dept.
connection

Alarm valve

OS&Y gate valve to
control water supply
to system

Main drain

Water supply

**Fig. 24.18** *Wet-pipe sprinkler system. This most typical system is under water pressure at all times so that water will be discharged immediately when an automatic sprinkler opens. The automatic alarm value shown causes a warning signal to sound when water flows through the sprinkler piping. OS&Y means "outside stem and yoke," describing a valve that clearly indicates the degree to which it is open or closed. (Reprinted with permission from the* Fire Protection Handbook, *18th ed., © 1997, National Fire Protection Association, Quincy, MA, 02269.)*

pressure; they draw from city mains at 50 psi (345 kPa) or from a 5000-gal (18,925-L) closed tank in the basement. These pumps feed into two 6-in. (15-mm)-diameter "express" risers, one in each stair tower, that rise the full height of this 48-story office building. These risers serve both the sprinkler system and fire department hose lines. In each of three 16-story zones, "local" 6-in. (150-mm)-diameter risers branch off to feed a 2-in. (50-mm) looped main at each floor.

The sprinkler piping is carried above suspended ceilings; pendant sprinkler locations are coordinated with the modular grid that also locates partitions, the air diffuser, and utility jacks. There are provisions for moving sprinkler heads as tenants change. At regular intervals, tees have been provided with one outlet stubbed and capped for future use. Typical office spaces are served by fully recessed, ½-in. orifice, 165°F (12.7-mm, 74°C) pendant sprinklers. Exposed pendant sprinklers are used in toilet rooms and service areas.

Although the typical sprinkler system is served by a single riser with a main line and branch lines, there are two variations that increase reliability by providing some redundancy in the supply lines. Figure 24.21a shows the gridded system, where each branch is served from either end, allowing each sprinkler head to receive water from either direction in the branch line. Figure 24.21b shows the loop system, where each branch line can receive water from either direction.

### (k) Circulating Closed-Loop Systems

These wet-pipe systems use the rather large sprinkler piping to circulate water for water source heat pumps. Water is not removed from this system, merely circulated. Water temperature in these systems must not exceed 120°F (49°C) or fall below 40°F (4°C). Such a system used for heating and cooling in a motel is shown in Fig. 24.22. An office building with a circulating closed-loop system was shown in Fig. 10.63.

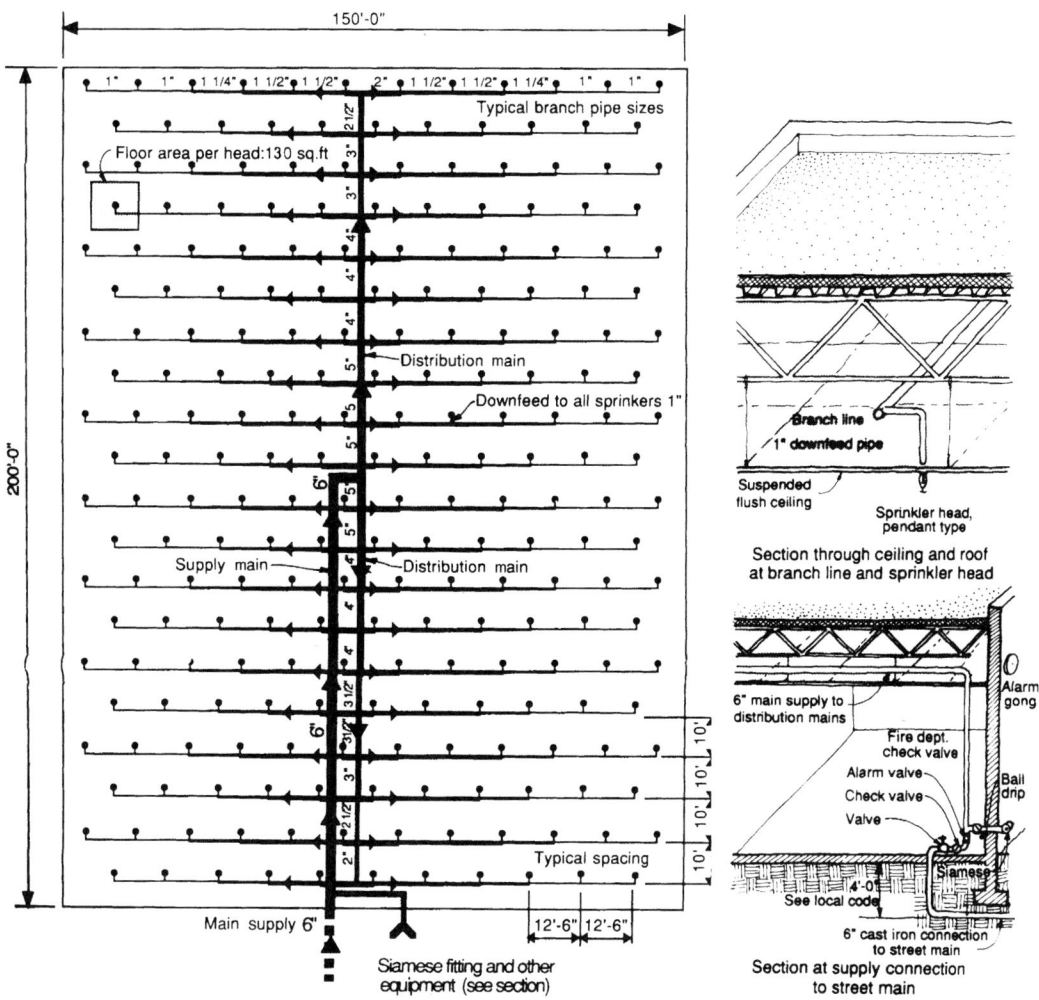

**Fig. 24.19** *Plan of a sprinklered industrial building, ordinary hazard occupancy, whose 30,000-ft² (2790-m²) floor area is protected by 230 sprinkler heads (average 130 ft² [12 m²] per sprinkler head). Sprinklers are closer together than the maximum allowed. Sprinklers (and standpipes) may use water from street mains when pressure is adequate. Either system may use pneumatic or gravity tanks. When the latter are used to supply both systems, an independent sprinkler reserve occupies the bottom and the fire standpipe supply occupies the top. Auxiliary fire engine feed by siamese should be provided in all cases. Auxiliary sources and standby pressurization may be required if street main adequacy is questionable.*

## (l) Dry-Pipe Systems

As shown in Fig. 24.23, these systems are filled with compressed air (or nitrogen) rather than with water. They are used in unheated areas, including loading docks and cold-storage areas. As soon as a sprinkler head opens, the compressed air rushes out, allowing water to enter the formerly dry-pipe system through a *dry-pipe valve*. It then functions like a wet-pipe system. The dry-pipe valve must be within a heated enclosure, since water under pressure is on one side of this

valve. Also, due to the delay in delivering water throughout a previously-dry piping system, a maximum system capacity of 750 gal (2839 L) is recommended.

Dry-pipe systems require a device to maintain design air pressure within the pipes; such air might be furnished from a compressor, from an air receiver tank, or from an existing pressurized air system (as in manufacturing operations). Dry-pipe systems also require a heated main control valve housing and the pitching of all piping to allow thorough drainage after usage.

### (m) Preaction Systems

In this system (Fig. 24.24), the pipes are filled with air that may or may not be under pressure. In addition to the sprinkler heads, either a heat- or a smoke-detection system is installed; it is more sensitive than the sprinkler head. Water is held back by the *preaction valve*. When the heat or smoke detectors are activated, they open the preaction valve, an alarm is sounded, and the water fills the pipes. It then functions like a wet-pipe system, with further water flow only on the opening of a sprinkler head.

A variation on this system, known as a *combined dry pipe-preaction* (or *double-interlock preaction*) system, has the pipes filled with compressed air. Heat or smoke detectors release the preaction valve and sound the alarm, but air pressure keeps water out of the piping until a sprinkler head opens.

Preaction systems are popular where the building's contents are especially subject to water damage—computer rooms, retail stores, museums, and so on—because the early alarm provided by water filling the piping often permits the fire to be found and extinguished manually, before any sprinklers open.

However, designers must remember that while the onset of the water spray is delayed, a fire can quickly grow in size. Typically, the 60-second delay in a preaction system could allow a fire to grow in area by at least 30%. Thus, about 30% more water would be required to extinguish it.

### (n) Deluge Systems

These systems (Fig. 24.25) have *open* sprinklers on dry pipes. As with preaction systems, a separate heat- or smoke-detection system is installed. These detectors control a *deluge valve*, which, once opened, floods the system with water, and *all* heads emit water. Obviously, huge quantities of water are thus released. Deluge systems are used where extremely rapid fire spread is expected—in aircraft hangars, for example, or other places where flammable liquid fires may break out.

### (o) Mist Systems

These are considered last because they introduce the next section, fire suppression alternatives to water. With a history of success in shipboard fires,

(*a*)

**Fig. 24.20** *Transamerica Building, San Francisco, California; William Pereira & Associates, Los Angeles, California, architects. (a) The pyramid form provides rentable floor areas ranging from 22,000 to 3000 ft² (2044 to 279 m²). (b) Sixth-floor sprinkler plan showing two risers, a looped feed main, and branch lines. (Courtesy of the Copper Development Association.)*

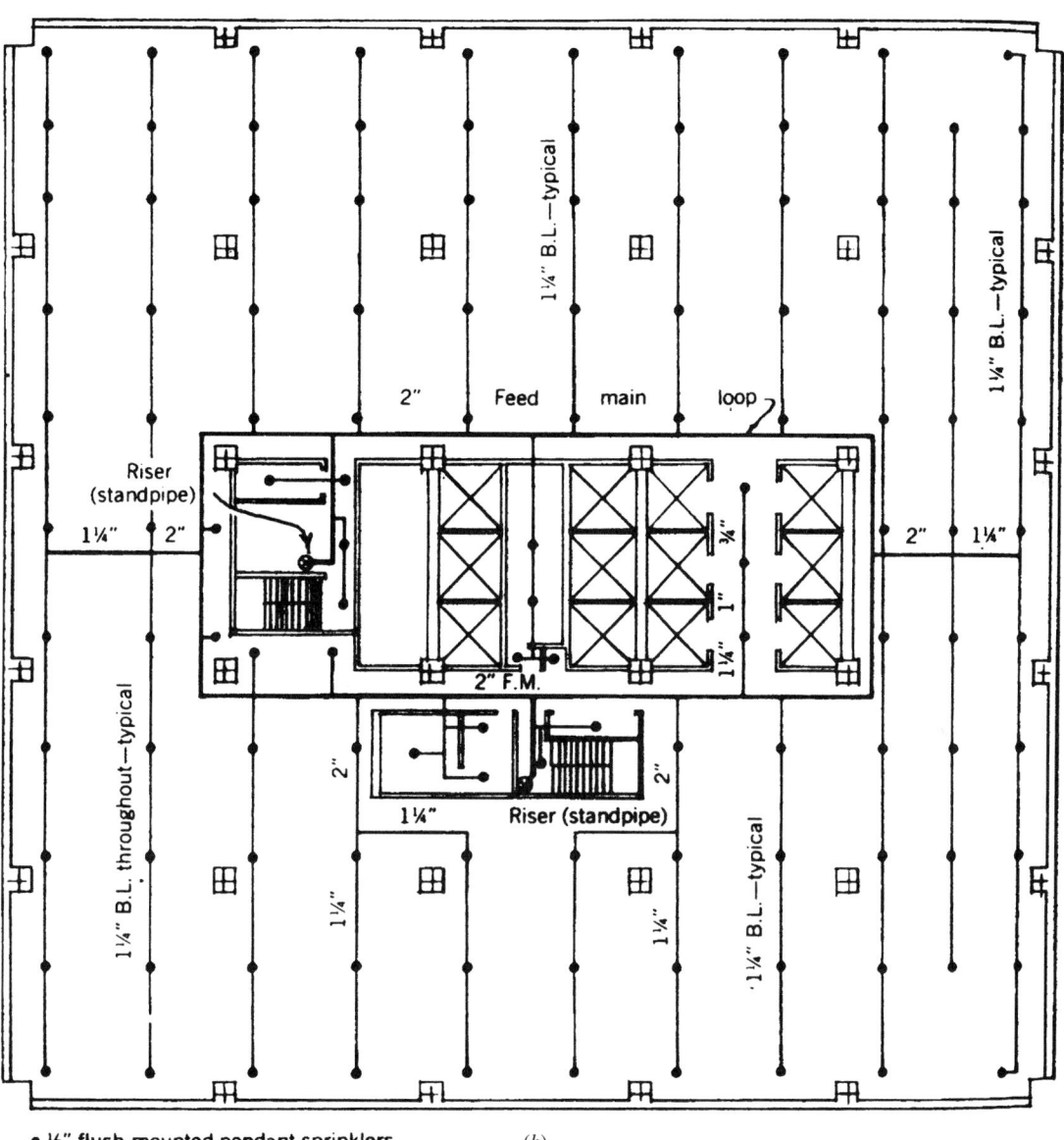

• ½" flush-mounted pendant sprinklers        (b)

**Fig. 24.20** (Continued)

they are being studied for much wider applications. NFPA 750 (2003) is the relevant standard.

As an equivalent to a sprinkler system, a mist system offers faster initiation of an alarm as well as more rapid response to a fire. Smaller volumes of water mean less water damage, and the mist can move more easily around obstructions.

As an alternative to halon or other clean agent gases (Section 24.4), mist systems do not pose a life-safety threat to fire fighters, are more tolerant of small amounts of ventilation, reduce the radiant heat transfer from the fire, and eliminate residues associated with many clean-agent gases. They also eliminate the expense of refilling a system with expensive clean-agent gases and allow a faster return to service after discharge.

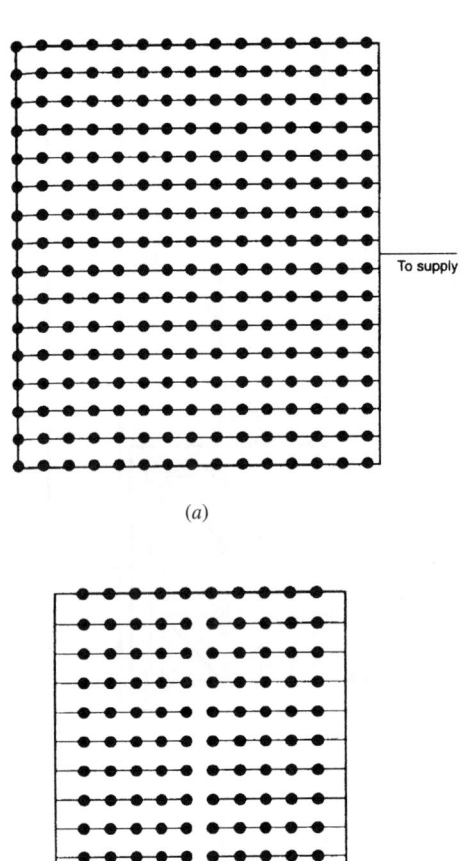

*(a)*

*(b)*

To supply

**Fig. 24.21** *Alternative piping arrangements that provide more than one path for water. (a) Gridded systems supply water from either end of a branch line, offering the potential for water to come from either side of each sprinkler. (b) Looped system, with the potential for each branch line to get water from either direction. (Reprinted with permission from NFPA 13, Installation of Sprinkler Systems, © 1996, National Fire Protection Association, Quincy, MA, 02269. This reprinted material is not the complete and official position of the National Fire Protection Association on the referenced subject, which is represented only by the standard in its entirety.)*

These systems produce a much smaller water droplet, thanks to inlet pressures ranging from 45 to 4100 psi (310 to 28,270 kPa), depending on the design of the sprinkler head. The heads are typically spaced closer together and have more sensitive thermal elements.

The NFPA *Fire Protection Handbook* (1997) suggests three classes of mist systems:

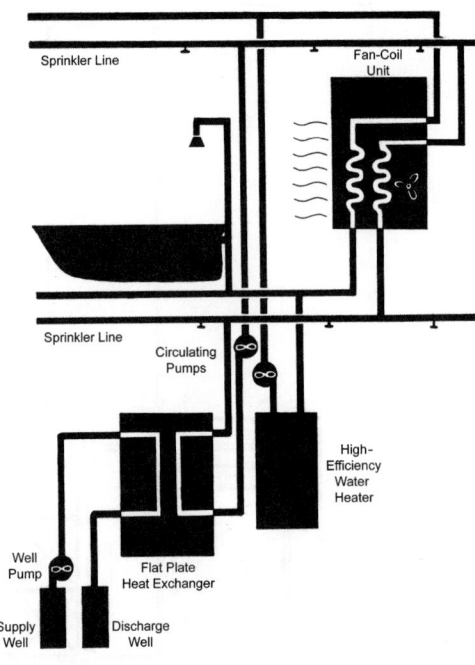

**Fig. 24.22** *Circulating closed-loop sprinkler system, whose water is used by heating/cooling systems. This application is for a motel; a ground source heat exchanger is supplemented by a water heater when required. (Courtesy of Montana Power. Redrawn by Amanda Clegg.) For another example of a circulating closed-loop sprinkler system, see Fig. 10.63.*

*Class I* mists have a droplet size ≤200 microns, the finest mists. This is achieved at the expense of flow rate and spray velocity, and thus requires significant input of energy to produce useful quantities. These mists are most suitable where enclosure reduces the need for spray momentum and fuel wetting is not critical (examples are liquid fuel fires and spray fires in enclosed spaces).

*Class II* mists have a droplet size from 200 to 400 microns. With larger drops, it is easier to achieve higher mass flow rates, and considerable surface wetting is possible. These mists are likely to be effective on fires involving ordinary combustibles.

*Class III* mists have a droplet size from 400 to 1000 microns. They can be generated by small-orifice sprinklers and fire hose fog nozzles, and deliver the highest mass flow rates.

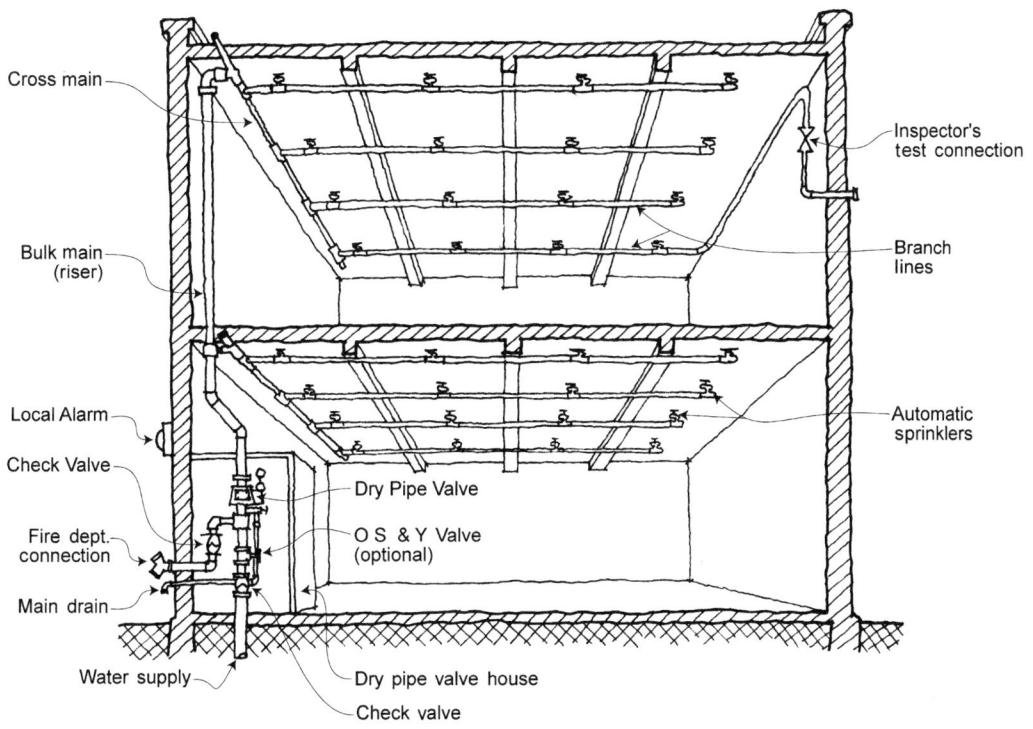

*Fig. 24.23* Dry-pipe sprinkler system for unheated occupancies. The system riser, cross mains, and branch lines are maintained at some air pressure. The dry-pipe valve and water supply must be within a heated enclosure. (Reprinted with permission from the Fire Protection Handbook, 18th ed., © 1997, National Fire Protection Association, Quincy, MA, 02269. Redrawn by Jonathan Meendering.)

Labels in figure: Cross main, Bulk main (riser), Local Alarm, Check Valve, Fire dept. connection, Main drain, Water supply, Inspector's test connection, Branch lines, Automatic sprinklers, Dry Pipe Valve, O S & Y Valve (optional), Dry pipe valve house, Check valve

The fire-extinguishing mechanisms of mists are shown in Fig. 24.26. Two primary mechanisms are heat extraction (through rapid evaporation of the finely divided water droplets) and oxygen displacement. Which mechanism dominates depends on whether the fire is poorly or well ventilated and the properties of the fuel. (Mist systems perform best in smaller enclosures with restricted ventilation, aided by heat entrapment and relatively easy oxygen displacement.) A third primary mechanism is the blocking of radiant heat. Secondary mechanisms in pool and spray fires are vapor/air dilution (mixing of water vapor and entrained air in the flammable vapor zone above the pool surface) and kinetic effects (the velocity of the flame front may be inhibited by small water droplets dispersed in a volume of flame).

Tests in 1998 by the Institute for Research in Construction, National Research Council of Can-

ada, demonstrated that cycling, rather than continuous, discharge from a mist system extinguished most test fires within a shorter time and ended all test fires with less water. A Class II spray with a twin-fluid (water and compressed air distributed separately, mixed at delivery) nozzle was used. The cycling discharges were "long water off" (50 seconds on, 30 seconds off) and "short water off" (30 seconds on, 20 seconds off). Compared to continuous discharge, cycling discharge produced higher room air temperatures near the ceiling and more rapid oxygen depletion. Both are threats to the contents and the occupants, yet both also make the mist system more efficient at extinguishing the fire. This resulted in less water used and, in most cases, a shorter fire extinction period.

After full-scale testing with resulting design guidelines, mist systems can be expected to replace

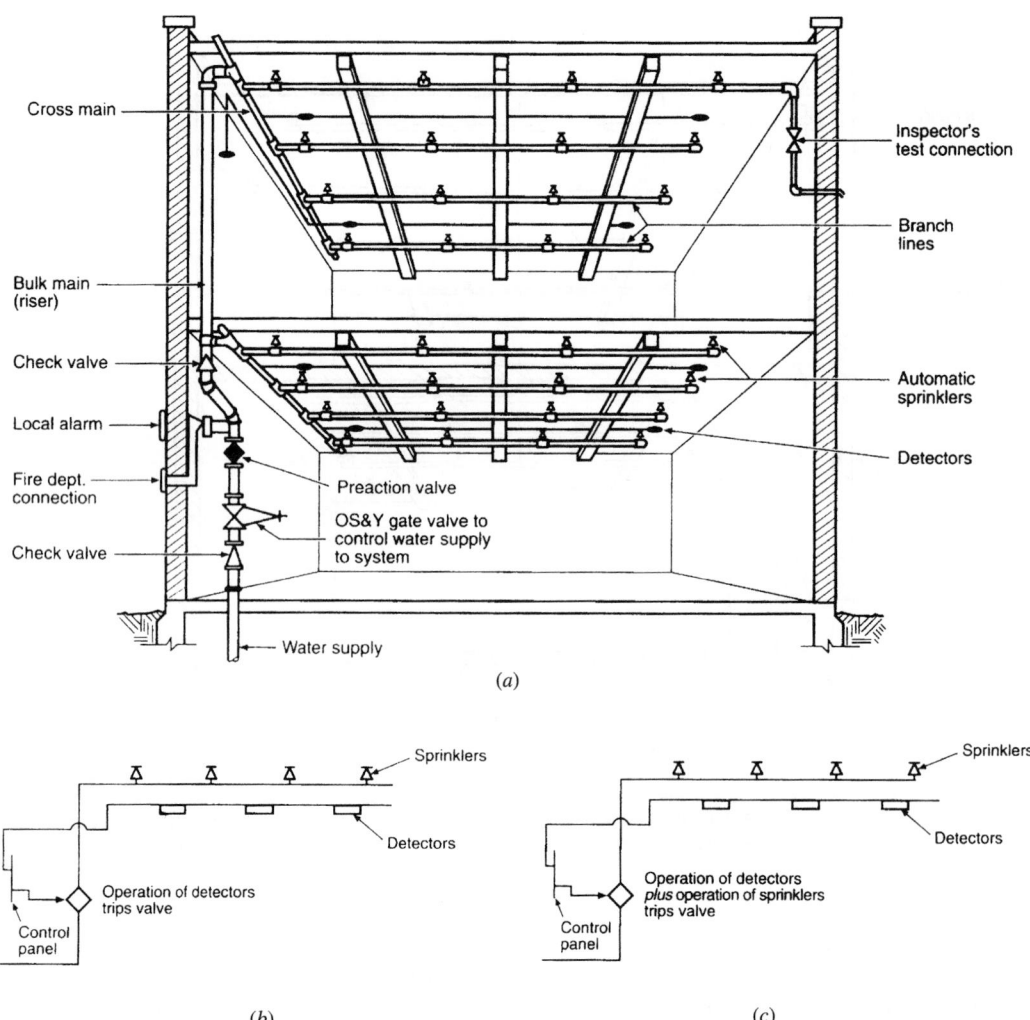

**Fig. 24.24** *(a) Preaction sprinkler system. Sprinkler piping contains air until the preaction valve allows water to enter the system piping. (b) In a preaction system, detectors open the valve. (c) In a double-interlock preaction system, both the detectors and then a sprinkler must open before the valve allows water to enter the system piping. (Reprinted with permission from the* Fire Protection Handbook, *18th ed., © 1997, National Fire Protection Association, Quincy, MA, 02269.)*

both standard sprinkler and clean agent systems for many applications.

## 24.4 OTHER FIRE SUPPRESSION METHODS

When water poses almost as much of a threat to a structure or its contents as does fire, a variety of other fire suppression methods are available. The most passive such measures are *intumescent* materials, which expand rapidly as they are touched by fire. This process creates air pockets that insulate a surface from the fire or swell a material until it blocks openings through which fire (or smoke) could have passed. Intumescent paints, caulks, and putties are available. Some intumescent materials come in ¼-in. (6.35-mm)-thick sheets, with various facing materials.

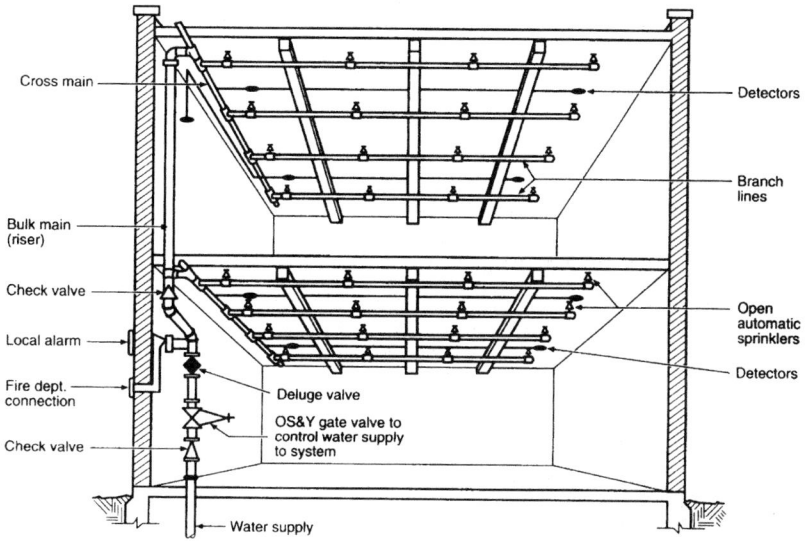

**Fig. 24.25** *Deluge system. All sprinklers are open; therefore, all operate when water is allowed to flow into the system piping. (Reprinted with permission from the* Fire Protection Handbook, *18th ed., © 1997, National Fire Protection Association, Quincy, MA, 02269.)*

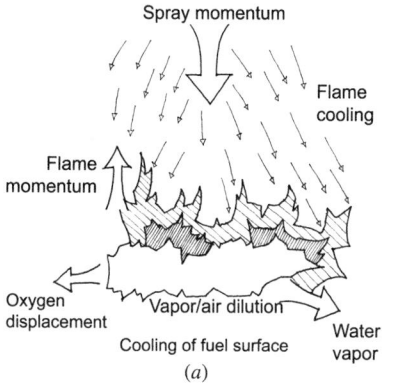

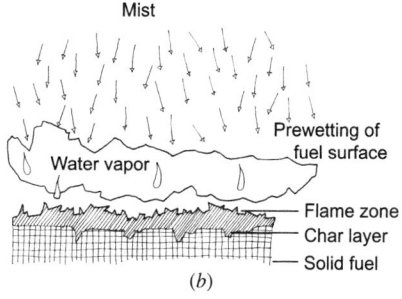

**Fig. 24.26** *Water mist fire-extinguishing mechanisms. (a) Interaction of mist with pool flame fire. (b) Mist and flame on a solid fuel with charring. (Reprinted with permission from the* Fire Protection Handbook, *18th ed., © 1997, National Fire Protection Association, Quincy, MA, 02269. Redrawn by Amanda Clegg.)*

## (a) The Rise and Fall of Halon 1301

Halogenated hydrocarbons, known commonly as *halons*, are gases (stored as liquids) in which one or more hydrogen atoms have been replaced by halogen atoms. Although the original hydrocarbons are often highly flammable gases, the substitution of halogen not only makes the gas itself inflammable, but also gives it a flame-extinguishing capability by interrupting the chemical chain reaction in a fire.

Until the mid-1990s, the most widely used of these agents was *Halon 1301*; the name signifies that its molecule contains one carbon atom, three fluorine atoms, no chlorine atoms, and one bromine atom. (When a fifth number is listed for halons, it refers to the number of iodine atoms.) Hydrogen atoms are not accounted for.

The advantages of Halon 1301 (and other halons) are that it can be released with relative safety in a flooding system in areas such as computer rooms, quickly extinguishing a fire with little harm to the contents, no oxygen displacement to threaten life support, little demonstrated harm to people, and no residue left on electronic components. It is also lightweight and space-saving relative to other fire suppressants, as seen in the following subsections. Halon 1301 was used on all commercial aircraft and in many special building applications, such as computer rooms, museums and libraries, telephone exchanges, and kitchens.

However, Halon 1301 was identified as a long-lived, significant threat to the Earth's protective stratospheric ozone layer. Its production was thus phased out as of 1994. The search for replacements for Halon 1301 (and 1211) has led to mists (discussed previously), foams, inerting gases, and clean agents (discussed later), as well as special application approaches beyond the scope of this book.

Inerting gas and clean agent replacements for Halon 1301, unlike water, actually protect the building's *contents* rather than its *structure*. They leave no sticky residue, such as most dry chemicals produce, and are therefore a popular choice where a clean fire-suppressing agent is required, people are present, and there are objects or processes of high value. Occasionally, they are used where water availability is low or where space cannot be found for systems that use other, bulkier fire suppressants.

### (b) Foams

Because foams (masses of gas-filled bubbles) are lighter than water and flammable liquids, they float on the surfaces of burning liquids, smothering and cooling the fire and sealing in vapors. Foams can be designed to inundate a surface or to fill cavities; they can be thin and rapidly spread or thick, tough, and heat-resistant. Some foams can be spread by foam-water sprinkler systems. Many foaming agents suited for various purposes are available. The NFPA *Fire Protection Handbook* presents a concise summary of foam-extinguishing agents and systems.

Foams are defined by their expansion ratio (final foam volume to original foam solution volume before air is added). *Low-expansion foam* has expansion up to 20:1 and is used principally to extinguish burning flammable or combustible liquid spill or tank fires. *Medium-expansion foam* has expansion from 20:1 to 200:1; *high-expansion foam* has expansion from 200:1 to 1000:1. These last two types are used for indoor fires in confined spaces to fill enclosures such as basement room areas or holds of ships. They are also (at about 500:1) used to control liquefied natural gas (LNG) spill fires and to help disperse the resulting vapor cloud.

Because foam breaks down and vaporizes its water under attack by heat and flame, it must be applied at a sufficient volume and rate to compen-

sate for water loss, leaving enough to produce a residual foam layer. Foam may easily be broken down by a water hose stream; turbulent air or rising combustion gases may divert foam from the burning area, especially if foam distribution occurs directly above the fire. Foam solutions are conductive, and therefore are not recommended for use on electrical fires. Fire fighter entry to a foam-filled passage requires self-contained breathing apparatus and a life line (vision and hearing are reduced).

An example of high-expansion foam is shown in Fig. 24.27, the North Central Airlines Hangar Building at the Metropolitan Airport in Detroit (Albert Kahn Associated Architects & Engineers, Inc.). The foam-generating equipment in this installation can fill the 38,400-ft$^2$ (3567-m$^2$) hangar with 1,400,000 ft$^3$ (39,644 m$^3$) of foam to a height of 36 ft (10.8 m) in less than 12 minutes. Automatic devices, on sensing an abnormal heat increase, operate the foam generators, open roof vents, start the smoke control exhaust fans in the vents, and transmit a fire alarm signal to the Airport Fire Department. Foam discharge is delayed 30 seconds while evacuation sirens sound to permit occupants to leave the fire area. Manual "override" controls allow personnel to start the system in the event of failure of the automatic controls or to stop it if the fire is small and controllable by other methods.

The high-expansion foam is created by wetting a nylon net with a mixture of water and a special detergent soap concentrate. A large blower directs an air current through the net, producing an avalanche of foam. Suds blanket the fire, attacking it in several ways. The water in the suds converts to steam, absorbing the heat of the fire. The expansion of the foam into steam reduces the oxygen content to about 7%, which is insufficient to support active combustion. A cooling effect is achieved by the wetting action of the breaking bubbles. The movement of air currents toward the fire to replace the rising hot gases draws the foam, supplied from the sides (rather than directly over the fire), to the center of the fire. There it blocks the airflow and cuts off the supply of oxygen. The fire, thus contained and diminished, can be approached by fire fighters for further control.

Aircraft are not harmed by the foam. Delicate machinery that might be injured by high-velocity streams of water is undamaged and left clean when the foam is rinsed away. The structure—in this case,

**Fig. 24.27** *Detergent foam (high-density) discharged from four units (two of which are seen in the illustration) after 3 minutes of operation. This installation at the North Central Airlines hangar building at Detroit Metropolitan Airport smothers fire, prevents its spread, and will not harm airplanes or machinery. The system can handle combustible liquid fires. (Note also the supplementary grid of upright sprinklers directly below the roof surface.)*

an open steel frame with a metal roof deck—is protected from excessive temperatures that might weaken it and cause it to collapse.

## (c) Carbon Dioxide ($CO_2$)

Among other inerting gases, $CO_2$ has long been used to prevent ignition of potentially flammable mixtures and extinguish fires involving flammable liquids or gases. Although inerting gases certainly help to extinguish fire by displacing oxygen, they are even more effective by acting as a "thermal ballast," absorbing combustion energy and reducing the temperature of the flame/vapor mixture below that necessary to sustain combustion.

Because of its oxygen displacement, $CO_2$ is often used in tightly confined spaces that are free of people or animals—for example, display cases, mechanical or electrical chases, and unventilated areas above suspended ceilings or below raised floors. However, it also finds applications in data centers, telecommunications equipment, and electrical equipment

rooms—in short, wherever water damage is a major concern.

$CO_2$ is stored in cylinders under great pressure as a liquid, requiring roughly four times as many cylinders as would Halon 1301 for the same hazard. When it is released as a gas, it absorbs about 120 Btu/lb to provide cooling as well as smothering action. $CO_2$ is noncombustible and will not react with most substances. It does not conduct electricity and will not normally damage sensitive electronic equipment. It has no residual clean-up associated with its use as a fire-suppressing agent. It spreads from discharge as a gas but will stratify over time. When properly vented, the gas escapes to the atmosphere after a fire has been extinguished.

The main problem with $CO_2$ is that it must be used at concentrations of 21% to 62% of that of the air, depending on the fire's fuel. However, at a $CO_2$ concentration of 9%, loss of consciousness will occur after a few minutes. This may allow enough time for an awake occupant to escape, but it does not help the fire fighters' (or trapped persons')

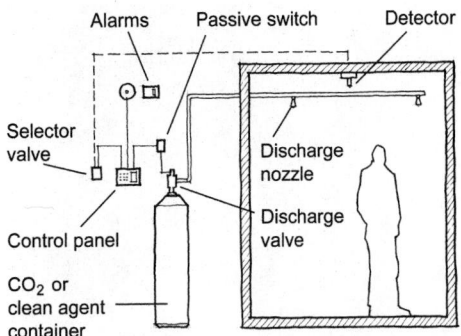

**Fig. 24.28** *Schematic of a $CO_2$ automatic fire detection/suppression system that provides total flooding of a space. Clean agent systems use similar components. (Drawing by Dain Carlson; © 2004 Walter Grondzik, all rights reserved.)*

environment. Another potential problem is that after the initial smothering and subsequent dissipation of the $CO_2$, smoldering embers might reignite.

A typical $CO_2$ automatic detection/suppression system (Fig. 24.28) has cylinders, a detection system, and a discharge valve that releases $CO_2$ into a piping system with discharge nozzles (rather similar to a sprinkler system).

### (d) Clean Agent Gases

An array of future replacement gases for Halon 1301 can be expected. Initially, HCFCs and HFCs are likely to predominate. Confined and vital spaces such as control rooms, computer and communication facilities, and emergency response centers are candidates for clean agent systems; consult the NFPA *Fire Protection Handbook* (2003) for system descriptions. Systems will resemble those for $CO_2$ (Fig. 24.28).

One of the earliest replacements to gain approval is FM-200® (heptafluoropropane, $CF_3$-CHF-$CF_3$, ASHRAE designation HFC-227ea). It works in a manner similar to that of halons but with zero ozone depletion potential. With a much shorter atmospheric lifetime than halons, it is also less threatening as a greenhouse gas contributing to global warming. It will typically require about 50% more cylinders than Halon 1301 for the same hazard.

FM-200 leaves no particulates or oily residue to damage electronic instruments, is electrically non-conductive and noncorrosive, is colorless, typically displaces only about 7% of the air in a space, and has acceptable toxicity at design concentration. This makes it relatively safe for fire fighters.

FM-200 decomposes to form halogen acids when exposed to open flames. Thermal decomposition products include hydrogen fluoride. With early warning and proper clean agent system installation, these by-products can be minimized.

### (e) Portable Fire Extinguishers

Most fires in buildings can be extinguished at an early stage with these common devices. Fire extinguishers are rated for the "class" of fire they are designed to fight; how many extinguishers are required and where they are located depend upon the hazard of the occupancy (as listed in Table 24.8). This material is a very short summary of NFPA 10, *Standard for Portable Fire Extinguishers* (2002).

Portable fire extinguishers are labeled as follows:

*Class 1A to 40A:* The numerals refer to the relative extinguishing potential (40A will extinguish 40 times as much as 1A). "A" refers to the contents: water, aqueous film-forming foam (AFFF), film-forming fluoroprotein foam (FFFP), and multipurpose dry chemical (ammonium-phosphate-base). (Halogenated agents are being phased out.) They are used on "Class A" fires: ordinary combustibles such as wood, cloth, paper, rubber, and many plastics, requiring the heat-absorbing, cooling effects of water, the coating effects of dry chemicals, or the interruption of the combustion chain reaction by dry chemicals.

*Class 5B to 40B:* The numerals refer to the approximate square footage of deep-layer liquid fire that an inexperienced operator can extinguish. "B" refers to the contents: smothering or flame-interrupting chemicals such as $CO_2$, dry chemicals, AFFF, or FFFP. (Halogenated agents are being phased out.) They are used on "Class B" fires with flammable or combustible liquids, flammable gases, greases, and similar materials that are fought by excluding oxygen, by inhibiting the release of combustible vapors, or by interrupting the combustion chain reaction.

*Class C:* Contents are nonelectrically conducting, such as $CO_2$ or dry chemicals. (Halogenated agents are being phased out.) They are used on "Class C" fires in live electrical equipment.

*Class A:B:C:* Multipurpose dry chemical extinguishers filled primarily with ammonium phosphate. Although indicated for all three classes of use, ammonium phosphate has the disadvantage of leaving an especially hard residue unless promptly and thoroughly cleaned as soon as the fire is extinguished.

*Class D:* Contents are dry powders, such as graphite or sodium chloride. They are used on a variety of combustible metals; the specific combustible metal for which the extinguisher is designed is printed on the extinguisher's nameplate.

A typical portable extinguisher is shown in Fig. 24.14*b*. The most common Class A extinguisher contains 2½ gal (9.5 L) of water, weighs about 30 lb (13.6 kg), and emits a stream from 30 to 40 ft (9 to 12 m). Fire extinguishers are rarely considered aesthetically pleasing, yet they should be located in conspicuous places along ordinary paths of egress (Fig. 24.29). The requirements for maximum floor area served and maximum actual path to extinguishers are shown in Table 24.11.

## 24.5 LIGHTNING PROTECTION

Lightning is nature's most destructive force: the average lighting discharge is estimated at 200 million V, 30,000 A, and courses through a grounded object in less than a thousandth of a second. "Cold" lightning bolts have ample current, voltage, and duration to shatter and kill but not to ignite combustibles. "Hot" bolts will ignite combustibles as well.

The decision on whether to protect a structure depends on an evaluation of these factors:

1. Frequency and severity of thunderstorms.
2. Value and nature of building and contents.
3. Hazard to building occupants.
4. Building exposure. Buildings in open and exposed areas are more susceptible to lightning than urban buildings, although very tall towers remain vulnerable.
5. Indirect effects. For example, loss of a water tower will seriously affect fire prevention and other services.

If a decision is reached to protect a building, it should be done completely and properly, with Underwriters Laboratories (UL) label equipment (Label A and B) and UL-approved installation (Label C). A partially protected building is in reality an improperly protected building, which might be worse than one with no protection at all.

The relevant standards are NFPA 780 *Standard for the Installation of Lightning Protection Systems* (2004), and UL Standard 96A, *Standard for Installa-*

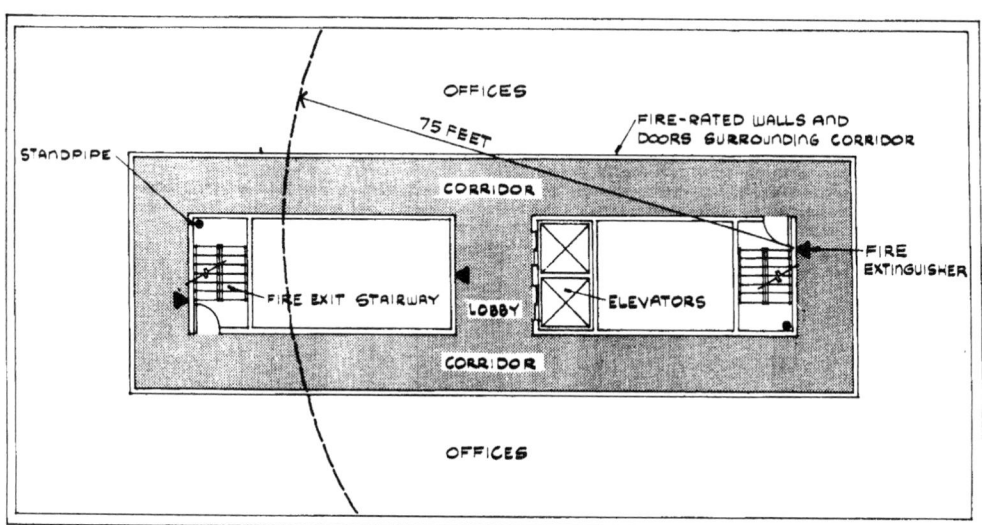

▲ FIRE EXTINGUISHER ( LOCATED SO THAT NO OCCUPANT IS MORE THAN 75 FT AWAY FROM AN EXTINGUISHER )

● STANDPIPE ( SERVES AS AN EXTENSION OF FIRE-FIGHTERS' HOSES TO GROUND-LEVEL HYDRANTS)

**Fig. 24.29** *Location of portable fire extinguishers for a typical multistory office building. Extinguishers should be easily reached and placed in conspicuous locations along normal paths of protected egress, away from potential fire hazards. Note that three locations are shown. (Reprinted by permission from Egan,* Concepts in Building Firesafety, *© 1978 by John Wiley & Sons.)*

**TABLE 24.11  Class A Fire Extinguishers**

| PART A. SIZE AND PLACEMENT FOR CLASS A HAZARDS | | | |
|---|---|---|---|
| | Occupancy | | |
| | **Light Hazard** | **Ordinary Hazard** | **Extra Hazard** |
| Minimum rated single extinguisher | 2-A[a] | 2-A | 4-A[b] |
| Maximum floor area per unit of A | 3,000 ft$^2$ | 1,500 ft$^2$ | 1,000 ft$^2$ |
| Maximum floor area for extinguisher | 11,250 ft$^{2c}$ | 11,250 ft$^{2c}$ | 11,250 ft$^{2c}$ |
| Maximum travel distance to extinguisher | 75 ft | 75 ft | 75 ft |

| PART B. MAXIMUM AREA TO BE PROTECTED PER EXTINGUISHER (FT2) | | | |
|---|---|---|---|
| | Occupancy | | |
| **Class A Rating on Extinguisher** | **Light Hazard** | **Ordinary Hazard** | **Extra Hazard** |
| 1A | — | — | — |
| 2A | 6,000 | 3,000 | — |
| 3A | 9,000 | 4,500 | — |
| 4A | 11,250 | 6,000 | 4,000 |
| 6A | 11,250 | 9,000 | 6,000 |
| 10A | 11,250 | 11,250 | 10,000 |
| 20A and more | 11,250 | 11,250 | 11,250 |

*Source:* Adapted with permission from the *Fire Protection Handbook,* 18th ed., © 1997, National Fire Protection Association, Quincy, MA, 02269.

[a]Up to two water-type extinguishers each with 1-A rating can be used to fulfill the requirements of one 2-A rated extinguisher for light hazard occupancies.

[b]Two 2½-gal. (9.5-L) water-type extinguishers can be used to fulfill the requirements of one 4-A rated extinguisher.

[c]This is the maximum area of a square that can be inscribed in a circle of 75-ft (22.8-m) radius where 75 ft is the maximum distance to a fire extinguisher.

*Note:* For SI units, 1 ft = 0.305 m; 1 ft$^2$ = 0.0929 m$^2$

*tion Requirements for Lightning Protection Systems.* (2001). The subject of lightning protection is complex, and the design of an adequate system is best left to specialists. Design considerations and available materials are discussed briefly later.

The basic principle in lightning protection is to provide a continuous metallic path to solid (low-resistance) ground for the lightning stroke because *there is no known method of protection that will prevent the occurrence of a lightning stroke.* This will prevent the stroke from passing through the nonconductive portions of a building, accompanied by intense heat and very large mechanical forces due to the high resistance of this path.

Any lightning protection system includes one (usually more) *air terminals* (pointed copper or aluminum rods projecting above a structure); *ground terminals* terminating in highly conductive soil (or in a network of buried wires if the soil is nonconductive, such as bedrock or very dry soil); and *down conductors* that connect air and ground terminals. The conductors may be exposed on the building's exterior or concealed within a structure.

Three techniques for protecting structures from lightning are shown in Fig. 24.30. All three offer the lightning stroke a good conductor above the level of the protected nonconductive structure. The lightning stroke is drawn to the conductor, then conducted harmlessly to the earth. (Trees can be similarly protected.)

### (a) Franklin Cone

A *Franklin cone* (Fig. 24.30a), named for Ben Franklin, is simply a mast with a conductor running straight to ground. A "cone of protection" is formed that protects the objects within it from strikes by absorbing the lightning stroke at the mast and grounding it harmlessly. The closer to the ground and the mast, the better the protection: buildings within an interior solid angle of 60° get excellent protection, those within an angle of 90° get good protection, and those within an angle of up to 126° get fair protection. (A more complex "geometric" version is described in the NFPA *Fire Protection Handbook.*)

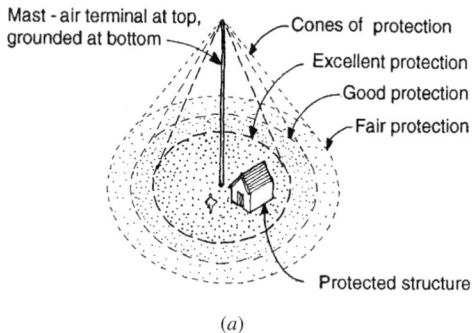

(a)

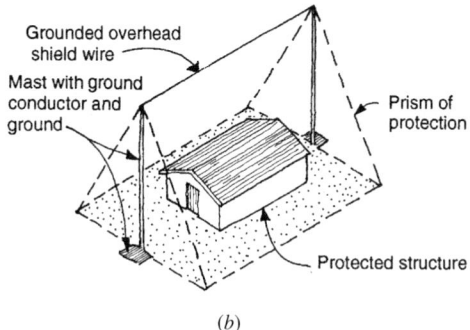

(b)

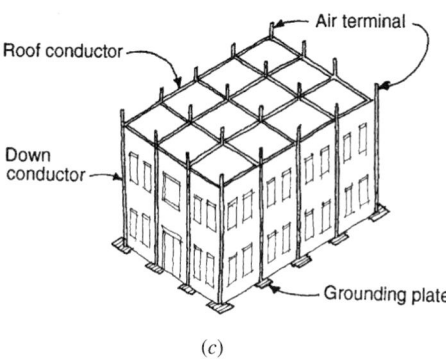

(c)

**Fig. 24.30** *Three approaches to lightning protection. (a) The Franklin cone is provided by a central mast that acts as an air terminal, conductor, and ground. (b) The overhead ground shield wire provides a triangular prism protection zone. (c) The Faraday cage wraps a protective mesh around a structure.*

### (b) Overhead Ground Shield Wire

A continuous air terminal, this shield (Fig. 24.30b) is linear and horizontal, so the protected volume is a triangular prism rather than a cone. This is most commonly used to protect overhead transmission lines.

For both the single mast and the overhead ground shield, it is recommended that they project

above any structure within the protected zone by at least 6 ft (1.8 m), and more height is better.

### (c) Faraday Cage

Named for Michael Faraday, this approach (Fig. 24.30c) depends on an open interconnected mesh covering a large, nonconducting mass (a building) to draw a lightning stroke and conduct it to Earth. This leaves large areas of the roof exposed, so the higher the air terminals and the denser the mesh, the better the protection. An office tower, with high air terminals at close intervals, is shown in Fig. 24.31. Because this is the most common method used to protect buildings, some details follow.

All metallic objects on the roof should be bonded to a looped conductor that joins the air terminals. These air terminals commonly are placed at both roof edges and ridges at intervals of about 20 ft (6 m). The conductors usually are made of copper and are best housed in a plastic pipe conduit. Ground

**Fig. 24.31** *Lighting rods (air terminals) enliven the silhouette of the Bank of America Corporate Center in Charlotte, North Carolina. (Photo by Dale Brentrup.)*

conductors, often called *counterpoise* conductors, are also installed to connect the earth terminals.

*Metal roof and siding* must not be substituted for air terminals and down conductors, unless constructed of ⅟₁₆-in. (4.8-mm) minimum sheet metal that has been made electrically continuous by bonding or an approved interlocking contact.

In *reinforced-concrete buildings,* concrete column steel reinforcing can be used for lightning conduction *if* the reinforcing steel is welded rather than merely tied. Where tied, and in buildings constructed with precast concrete panels, reinforcing steel (and other metal) within a few feet of lightning conductors should be bonded to them to avoid arcing. Such arcing, as noted previously, is accompanied by heat and mechanical forces that can produce major structural damage and fire. Precast concrete buildings present a challenge because the reinforcing steel within precast panels is not typically interconnected.

In *steel structure buildings,* the columns relatively easily become lightning conductors. Care must be taken to adequately bond both the tops and foundations of such columns to the air and ground terminals, respectively (Fig. 24.32).

### (d) Lightning Arresters

A special type of *surge arrester,* a lightning arrester is generally connected at one end to an overhead electrical line and at the other end to the ground. It thus suppresses any lightning voltage surges that appear in the electrical line so that equipment downstream on that line is not damaged. A lightning arrester operates only when it senses a voltage surge. At that point, it conducts the surge to the ground through its own low resistance and partially dissipates the energy of the surge in the body of the arrester. Once the surge is passed, the arrester returns to its high-impedance quiescent state, presenting an open-circuit connection to the normal voltage line to which it is connected.

### (e) High-Rise Buildings

Lightning protection systems are particularly important for very tall buildings. Figure 24.33 shows the schematic diagram for the John Hancock Center in Chicago, a structural steel building of unusual height. Intermediate loops at the mechanical equipment floors help to overcome the problem posed by the poorer conductivity of metal piping systems, which can become additional relatively high-impedance paths for lightning, with resultant destructive effects. The intermediate loops are connected to each outside column, as well as to the main electrical, plumbing, air conditioning, and fire protection risers. These loops are also connected to both the roof conductor and the ground counterpoise conductor.

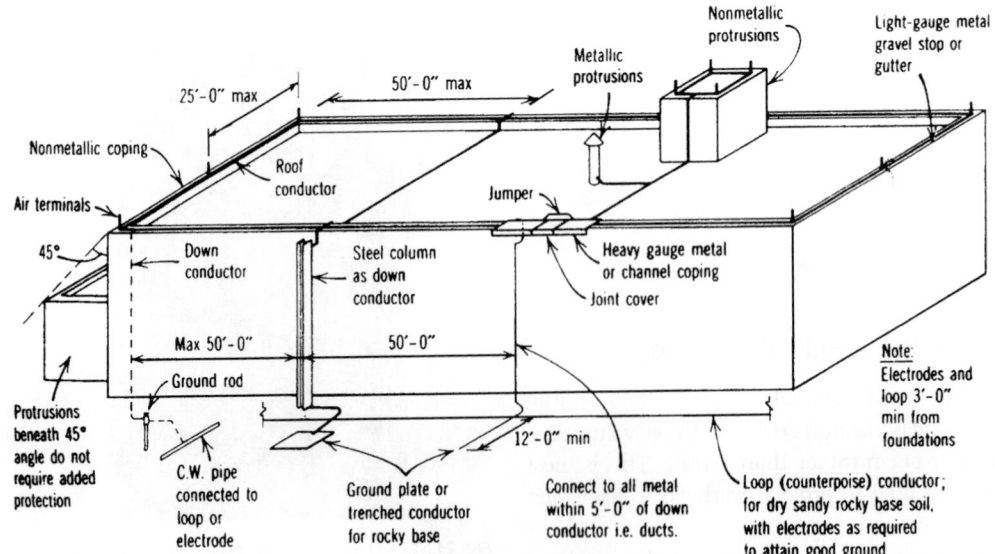

**Fig. 24.32** *Diagram of a typical lightning protection system.*

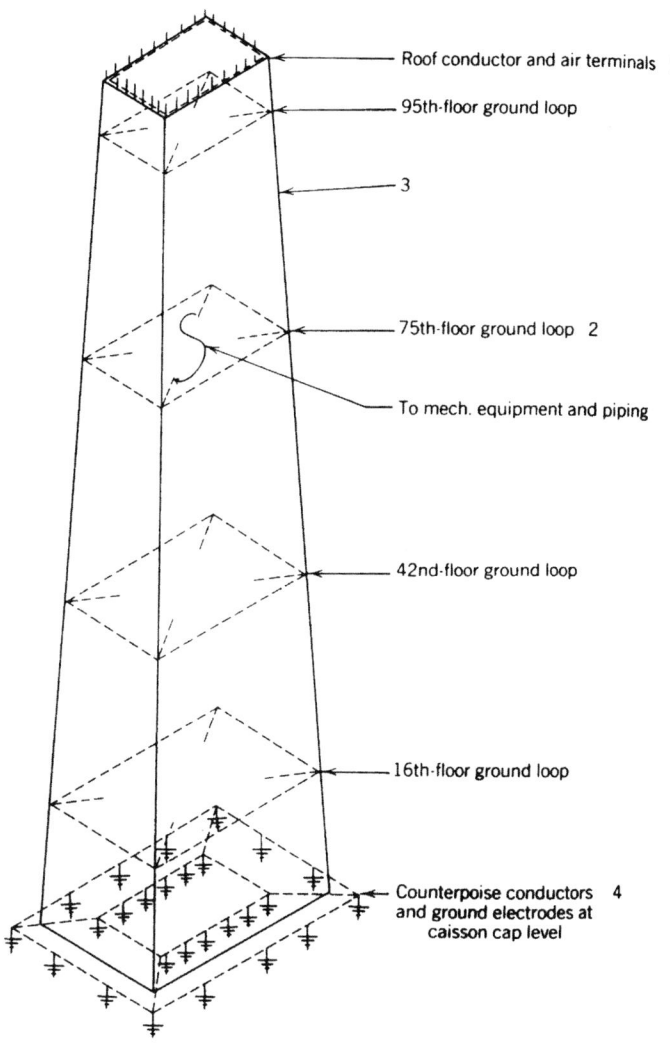

Roof conductor and air terminals 1

95th-floor ground loop

3

75th-floor ground loop 2

To mech. equipment and piping

42nd-floor ground loop

16th-floor ground loop

Counterpoise conductors 4
and ground electrodes at
caisson cap level

**Fig. 24.33** *Lightning protection for the John Hancock Center, Chicago. Air terminals with roof conductors (1) are bonded to building steel and looped to ensure that a single wiring break will not fault the system. Intermediate-floor ground loops (2) with connections to building mechanical equipment and piping protect against side-effect flashover to people or equipment. Building structural steel (3), properly bonded together, is used as a down conductor. Counterpoise conductors (4) cross-connect ground electrodes with ground conductors located below the minimum groundwater level to ensure adequate and uniform dissipation of the charge. (Reprinted by permission from Jensen (ed.),* Fire Protection for the Design Professional, *© 1975 by Van Nostrand Reinhold Company, Inc.)*

# FIRE ALARM SYSTEMS

## 24.6 GENERAL CONSIDERATIONS

A fire alarm system serves primarily to protect life and secondarily to prevent property loss. Because buildings vary in occupancy, flammability, type of construction, and value, the fire alarm system must be tailored to the needs of the specific facility. In schools, for instance, where the paramount consideration is rapid, orderly evacuation, the type of system used will generally be the same, although the means of automatic detection will vary with construction type, building height, specific area use, furnishings, and staffing. The fire alarm is part of the overall fire *protection* system design of the building. In particular, it overlaps with the design of safe egress and smoke/fire control in matters such as fan control and smoke venting, smoke door closers, rolling shutters, elevator capture, and the like. These automatic functions are initiated by operation of the alarm system but are designed in accordance with the overall fire protection plan.

Like other alarm systems, a fire alarm system has three basic parts: signal initiation, signal processing, and alarm indication (see Fig. 24.34). The signal initiation can be manual (pull stations or telephones) or automatic (fire and smoke detectors and/or waterflow switches). The alarm signal is processed by some sort of control equipment, which in turn activates audible and visible alarms and, in some cases, alerts a central fire station or municipal authorities.

## 24.7  FIRE CODES, AUTHORITIES, AND STANDARDS

There are probably more codes and standards in the area of fire protection than in any other area, with the possible exception of structural standards.

Although these codes are devoted primarily to fire protection and life safety, they also govern the type of fire alarm system to be installed, its components, and its installation.

NFPA codes (actually standards) are published by the National Fire Protection Association (NFPA), 1 Batterymarch Park, P.O. Box 9101, Quincy, MA 02269-9101. They bear directly on fire alarm system arrangements. Following is a list of NFPA codes, other codes, and the names of organizations concerned with standards and requirements.

*NFPA 101: Life Safety Code*—specifies the means of egress, the type of fire alarm system, and the type and location of fire and smoke detection equipment for all types of occupancies.

*NFPA 70: National Electrical Code;* particularly Art. 760—Fire Alarm Systems, which was completely rewritten for the 1996 edition.

*NFPA 72: National Fire Alarm Code.* This integrated code replaces the various individual sections that preceded it.

*NFPA 90A:* Standard for the installation of air conditioning and ventilating systems.

*NFPA 92B:* Smoke control systems in malls, atria,

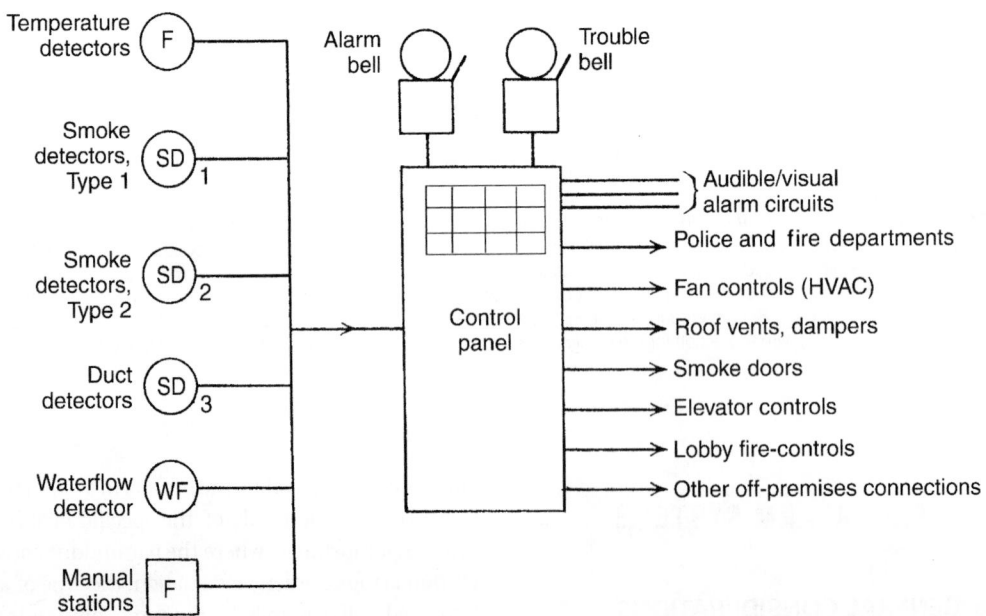

**Fig. 24.34** *Typical simplified arrangement of a building fire alarm system. Multiple circuits, zoning, and other accessories such as annunciators are not shown. In a protected premises (local alarm) system, the off-premises connections to police, fire department, and other central stations do not exist.*

and large areas. This code and 90A provide information on smoke detection in HVAC systems.

*NFPA 99:* Health Care Facilities.

*NFPA 110:* Emergency and Standby Power Systems.

*NFPA 111:* Stored Electrical Energy Emergency and Standby Power Systems.

*ANSI:* (American National Standards Institute)—1430 Broadway, New York, NY 10018. This organization does not develop standards; it accepts for review standards that are prepared by professional and technical societies and associations. Approval and listing by ANSI validates the technical acceptability of the work.

*CABO/ANSI 117.1:* Accessible and Usable Buildings and Facilities. This standard specifies requirements to make buildings completely accessible to and usable by people with physical disabilities of all sorts.

*SBCCI:* Also known as *SBC* (Standard Building Code), published by the Southern Building Code Congress—900 Montclair Road, Birmingham, AL 35213-1206. This code, generally used in the southern and southeastern areas of the United States, deals with auxiliary system controls such as door release, elevator capture, and so forth. In addition, SBC has a special section on high-rise buildings, covering fire communication, elevator control, and smoke alarms.

*BOCA:* Published by Building Officials and Code Administrators International, 4051 W. Flossmoor Rd., Country Club Hills, IL 60478-5795. Basic Fire Prevention Codes cover smoke detection and alarm audibility, plus requirements for high-rise buildings.

*UBC:* (Uniform Building Code), published by the International Conference of Building Officials (ICBO), 5360 Workman Mill Rd., Whittier, CA 90601-2298. This code, used generally in the western and southwestern areas of the United States, covers essentially the same material as the other codes. Like them, it relies heavily on the *Life Safety Code,* NFPA 101.

*HUD:* (Department of Housing and Urban Development)—covers requirements for residential buildings and care-type facilities. FHA minimum property standards are included here.

*UL:* (Underwriters Laboratories) performance standards, published by Underwriters Laboratories, Inc., 333 Pfingsten Road, Northbrook, IL 60062-2096. UL laboratories test equipment for inherent safety and for performance according to the manufacturer's specification. A product is "listed" if the equipment passes UL safety and performance tests.

### UL Detection Standards

*38:* Manually Activated Signaling Boxes

*217:* Smoke Detectors, Single and Multiple Stations

*268:* Smoke Detectors for Fire Protective Signaling Systems

*268A:* Smoke Detection for Duct Applications

*346:* Water Flow Indicators for Fire Protective Signaling Systems

*521:* Heat Detectors—Fire Protective Signaling Systems

*539:* Single and Multiple Station Heat Detectors

*985:* Household Fire Warning System Units

*1730:* Smoke Detectors, Monitors and Accessories for Individual Living Units of Multiple Residences and Hotels/Motels

### UL Alarm Standards

*228:* Door Closers with/without Integral Smoke Detectors

*464:* Audible Signal Devices

*827:* Central Station Alarm Services

*1480:* Speakers for Fire Protection

*1638:* Visual Signaling Applications—private mode, emergency, and general utility signals

*1971:* Signal Devices for Hearing Impaired

In addition, several insurance groups (e.g., Factory Mutual) issue standards that may apply to a specific facility, in addition to local codes, standards, and fire marshal regulations. As with other aspects of construction, the architect/designer must ascertain which regulations have jurisdiction before proceeding with the design. The specific recommendations given here are representative of current good practice. Actual design must be based on and must meet the requirements of *current* NFPA standards plus all codes having jurisdiction in the particular locale of the project.

**FIRE PROTECTION**

## 24.8 FIRE ALARM DEFINITIONS AND TERMS

The following short list can serve as a reference to terminology that a designer may encounter in fire alarm work.

**Addressable analog (smoke) detector** (also called an *intelligent detector*). An addressable (smoke) detector that continuously measures the concentration of smoke or other products of combustion in its test chamber. It continuously transmits an analog (voltage or current) signal to its (intelligent) control panel that is proportional to this measurement. It does not transmit an alarm signal (see Fig. 24.46).

**Addressable detector.** A smoke or fire detector that communicates an alarm condition along with an individual identification (address), thereby immediately establishing the alarm's location. It responds to polling (testing) when it senses its unique identification (address) signal, "replying" with a signal indicating OK, trouble, or alarm.

**Air sampling detector.** A system consisting of aspiration equipment and piping that draws air from the protected area to the detector, where analysis of the sampled air occurs.

**Alarm signal.** An audible and/or visual signal indicating a fire emergency, generally requiring immediate building evacuation.

**Alarm verification.** A technique for reducing false alarms. Common procedures include a requirement for alarming for a specific minimum time and a repeat alarm after initial reset.

**Automatic system.** A system in which an alarm-initiating device operates automatically to transmit or sound an alarm signal.

**Auxiliary fire alarm system.** See Section 24.9.

**Breakglass.** A false-alarm deterrent installed in manual fire alarm stations; a glass rod is placed across the pull-lever that breaks easily when the lever is pulled.

**Coded alarm signal.** An alarm signal that indicates the location of the fire alarm station operated by the pattern (code) of the audible signal.

**Coded system.** One in which not less than three rounds of coded alarm signals are sounded, after which the system may either be silenced or set to sound a continuous alarm.

**Control unit (fire alarm panel).** The controls, microprocessors, relays, switches, and associated circuits necessary to (1) furnish power to a fire alarm system, (2) receive and process signals from alarm-initiating devices and transmit them to indicating devices and accessory equipment, and (3) electrically supervise the system's circuitry.

**Drift compensation.** Adjustment function that compensates for detector sensitivity changes due to aging and environmental conditions. The adjustment may be automatic at the detector or manual/automatic at the control panel. Also referred to as *automatic gain control* in beam-type smoke detectors.

**Dual-coded system.** See Section 24.12.

**Intelligent control panel.** A control center that receives analog signals from detectors and, based on predetermined conditions (usually in software or a microprocessor), determines whether an alarm situation exists. Sensitivity control, polling, and remote diagnostic testing of connected detectors are other functions usually found in these panels.

**Listed, listing.** Inclusion of a device in a list published by a recognized testing organization, such as UL, indicating that the device meets the requirements of the referenced standard.

**Local fire alarm systems.** See Section 24.9.

**Manual system.** One in which the alarm-initiating device is operated manually to transmit or sound an alarm signal.

**Master-coded system.** See Section 24.12.

**Multiplex system.** An arrangement whereby a single pair of wires can carry more than one message (signal) at a time in either or both directions.

**Noncoded system.** See Section 24.12.

**Presignal system.** See Section 24.12.

**Private mode.** An alarm system installed in a location that has personnel specifically trained to act on receipt of an alarm signal.

**Proprietary fire alarm system.** See Section 24.9.

**Public mode.** An alarm system that alerts all of

the protected spaces and occupants audibly and/or visually.

**Rate-of-rise detector.** A temperature detector that alarms when the rate of temperature rise exceeds a specific design level, usually 15°F/minute.

**Remote-station fire alarm system.** See Section 24.9.

**Selective-coded system.** See Section 24.12.

**Station, fire alarm.** A manually operated alarm-initiating device. It may be equipped to generate a continuous signal (noncoded station) or a series of coded pulses (coded station).

**Supervised system.** A system in which a break or ground in the wiring that prevents the transmission of an alarm signal actuates a trouble signal.

**Trouble signal.** A signal indicating trouble of any nature, such as a circuit break or ground, occurring in the device or circuitry associated with a fire alarm system.

**Zone-coded system.** See Section 24.12.

## 24.9 TYPES OF FIRE ALARM SYSTEMS

Fire alarm systems can be classified according to several different criteria including location, application, connections, coding, and the degree of automation of the detection system. The last criterion is the one most often found in manufacturers' literature. To avoid confusion, we use the classification system found in the *National Fire Alarm Code* (NFPA 72). This code classifies fire alarm systems essentially by location and function:

- Household fire warning systems.
- Protected premises systems (local alarm).
- Off-premises systems (connections between local alarms and off-premises equipment and systems). This category includes auxiliary, remote station, proprietary, central station, and municipal fire alarm systems.

### (a) Household Fire Warning Systems

The term is self-explanatory. These systems are discussed at some length in Section 24.25.

### (b) Protected Premises Fire Alarm System

This arrangement (Fig. 24.34), as the name indicates, is intended to sound an alarm only in the protected premises. Action in response to an alarm must be taken locally, either manually or automatically. Thus, notification to the fire department must be manual, although fire suppression systems can be set into operation automatically. This arrangement is applicable to privately owned facilities. When the building is unoccupied, notification to the fire department can come only incidentally, perhaps from a passerby. Local systems and their components are discussed in detail in Section 24.11.

### (c) Auxiliary Fire Alarm System

This is simply a local system equipped with a direct connection to a municipal fire alarm box (Fig. 24.35). The received alarm signal is identical to that resulting from a manual alarm at that city box. Since the fire department is aware of all city box connections, arriving fire fighters would always check the protected premises. This type of system is usually applied to public buildings such as schools, government offices, museums, and the like.

### (d) Remote-Station Protective Signaling System

This system (see Fig. 24.36) is similar to the auxiliary system, except that the alarm or trouble signal is transmitted via a leased telephone line to a remote location (a police facility or a telephone answering

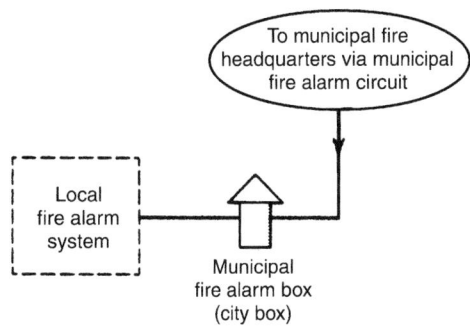

**Fig. 24.35** *An auxiliary fire alarm system is one that is hardwired to the municipal fire department, generally via a city box. This type of connection is usually restricted to public buildings.*

FIRE PROTECTION

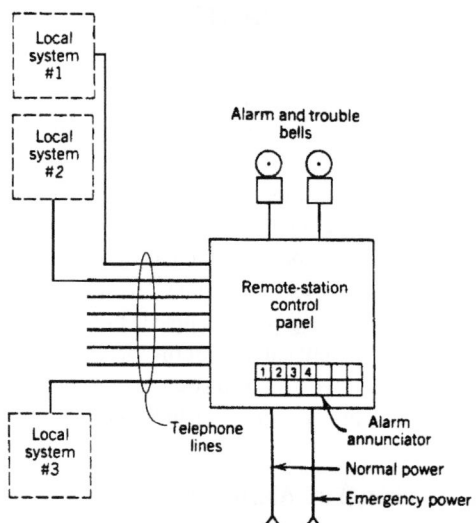

**Fig. 24.36** *A remote station system is one that transmits alarm, supervisory, and trouble signals to a remote location that is continuously attended. The required actions will be taken by those attending the remote stations.*

service) that is manned 24 hours a day. The notice of the alarm is then telephoned to the fire department. This arrangement is used in private buildings, such as stores and offices, that are unoccupied for considerable periods and for which reliance on passersby to turn in a fire signal is unacceptable. An audible

alarm circuit extended from a local system to a nearby building—as, for instance, from a store to a nearby residence—does *not* constitute a remote-station system unless all the requirements of NFPA 72 are met.

### (e) Proprietary Fire Alarm System

This system (see Fig. 24.37), which is applicable to large multibuilding facilities such as universities, manufacturing facilities, and the like, utilizes a dedicated central supervisory station to receive signals from all buildings. In a proprietary system, this station is on the site and is manned by persons associated with the facility. A common arrangement is for the station to be located in a guardhouse or similar supervisory location, from which point alarms are sent manually to a fire department and/or on-site fire brigades. Other actions that must be taken on receipt of alarm or trouble signals are performed by facility personnel at the central supervisory facility.

Information on the exact location or zone within each building at which an alarm occurs is transmitted to the central supervisory location. The central location has an audible alarm, some sort of visual display that indicates the alarm location, and a printer that makes a permanent record of each alarm. As mentioned, these central supervisory locations are frequently multipurpose, covering all

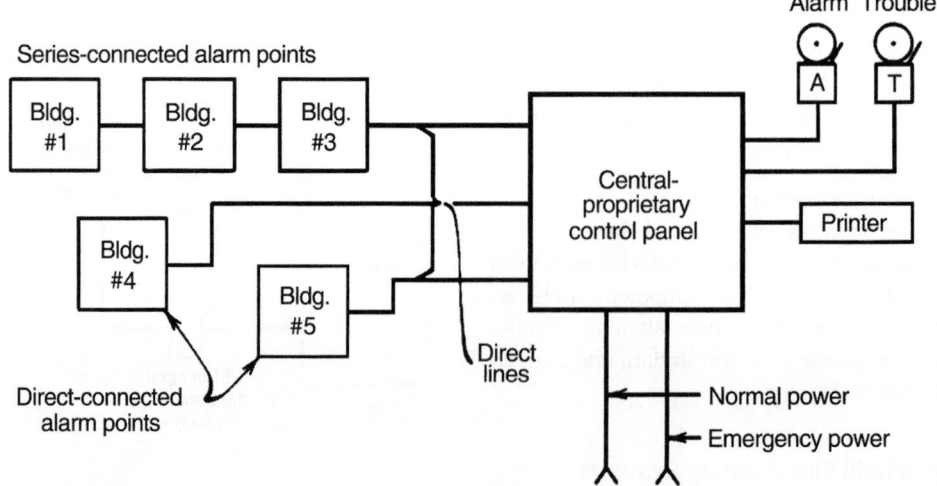

**Fig. 24.37** *Block diagram of a proprietary fire alarm system. The entire system, including all local devices, plus the central supervisory equipment, is the property of a single owner. The central supervisory installation is attended by facility personnel who are trained to respond appropriately to alarm, trouble, and supervisory signals.*

aspects of facility security plus, frequently, control functions such as energy management.

## (f) Central Station Fire Alarm System

This arrangement is similar to that described in Section 24.9*e* except that the system supervision and normally all of its equipment are owned and operated by a service company. In lieu of the multiple buildings of one facility shown in (*e*), the central station system supervises many individual unrelated local systems for a fee. Here too, operators of the central supervisory installation receive all signals from individual users and provide the required services, including alarm verification, fire reporting, and repairs when trouble signals are received. Here also, as with proprietary systems, central station consoles often supervise access control, intrusion alarms, and related systems.

## 24.10  CIRCUIT SUPERVISION

This very common term refers to the circuit arrangements in fire alarm systems that indicate a malfunction in the wiring of alarm (and other) devices by sounding a trouble bell. The trouble signal is separate and distinct from an alarm signal. The extent of circuit supervision required by fire codes varies with the type of facility and the specific

code having jurisdiction. Minimally, an open circuit (break) or a ground in any of the detector and manual station wiring will cause a trouble signal. In large, sophisticated systems, all of the wiring and circuitry is supervised, including the control panel(s), alarm device circuits, and even annunciator wiring.

In conventional systems (see Section 24.11), the supervision does not indicate where the fault lies, but only that a fault exists. It is up to the system operator to troubleshoot the system in order to locate the fault. Because this may be a protracted procedure, all codes require that the system wiring be such that a single break or a single nonsimultaneous ground will not prevent an alarm signal from being sounded beyond the break (Class B circuits) or on the entire circuit (Class A circuit).

## 24.11  CONVENTIONAL SYSTEMS

Manufacturers classify fire alarm systems by the type of information transmitted to the control panel by the detectors and the alarm devices. The simplest (and oldest) system is the conventional system (Fig. 24.38). A conventional system is one that uses detectors and manual stations that transmit an alarm signal only. When they are in standby or quiescent state, they do not transmit. Such detectors and manual stations are called *conventional units;*

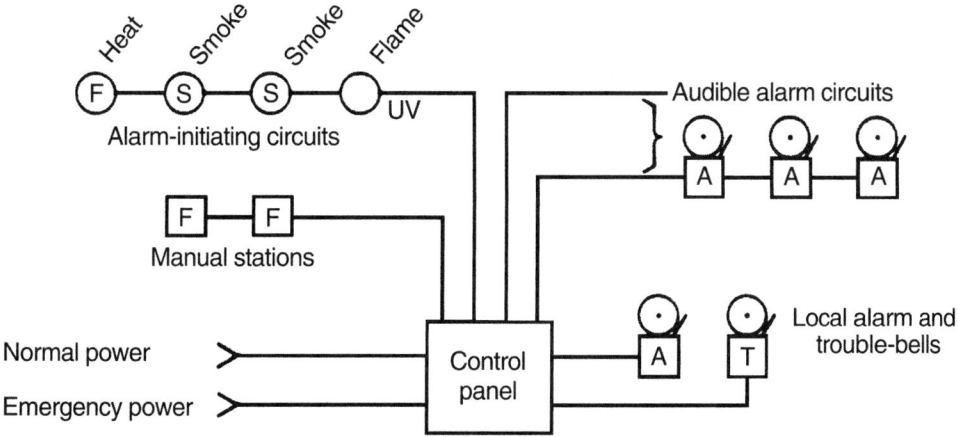

**Fig. 24.38** *Simplified schematic diagram of a conventional fire alarm system. Alarm-initiating devices are connected together on detector circuits, as are manual stations and audible/visual alarm devices. The control panel receives identical signals from all detectors, manual stations, and alarm-initiating circuits.*

hence the same name is used for the entire system. The signals transmitted by all detectors, and by manual stations as well, are identical and therefore indistinguishable from each other.

In theory, the number of alarm-initiating devices, such as fire and smoke detectors, that can be wired to a single detector circuit is unlimited. (In practice, there are electrical limitations.) As a result, when an alarm signal is received from such a multiple-device circuit, there is no way of knowing which of the devices alarmed and whether the alarm represents an actual fire condition or whether it is a false alarm caused by dirt, moisture, a puff of smoke, a short-circuit in the wiring, or a malfunction of a detector.

Smoke detectors of all types (ionization, photo) are subject to false alarms activated by particulate matter in the air. Detectors are, by nature, threshold devices; raising sensitivity increases false alarming, and decreasing sensitivity shortens the crucial early-warning period. False alarms are not merely nuisances; in facilities such as hospitals, and in places of public assembly such as theaters, public office buildings, and mass eating facilities, a false fire alarm can result in serious disruption, property loss, personal injury, and even death. As a result, conventional detectors in such buildings require continual maintenance and field sensitivity checks to minimize false alarms, which is an expensive and time-consuming procedure.

Because such facilities are so difficult and even dangerous to evacuate, most have fire and evacuation plans that call for some type of alarm verification before a general evacuation alarm is sounded. Because field experience has demonstrated that alarm verification can reduce false alarms appreciably without seriously degrading the early warning performance of combustion product detectors, recent issues of some fire codes now permit, or even require, alarm verification. This verification can be accomplished in conventional (hard-wired, zoned, nonaddressable) systems in a number of ways, including:

- Requiring activation of at least two cross-zoned detectors in a single area
- Requiring that a detector repeat its alarm after being reset (applicable only to systems with remote reset capabilities)
- Requiring that a smoke detector continue to

alarm for a minimum period of time, thus eliminating smoke puffs as the cause of an alarm
- Physically inspecting the site protected by the detector to visually determine whether the alarm is actual or false

This last technique for alarm verification is probably the most foolproof, but it is also the most difficult because it requires knowledge of the exact location of the alarming detector. This is possible only if detectors are grouped into zones and each zone is annunciated individually (Fig. 24.39). To assist in localizing areas in a building by annunciation, zones are wired to contain as few devices as possible, the ultimate arrangement being a separate zone for each detector. As can be seen in Fig. 24.39, as the number of zones increases, so does the necessary wiring and, concomitantly, the installation costs. Thus, a balance must be struck between two requirements: the need for more zones to localize and thereby simplify alarm identification and verification, and limitation of the increased cost of zoning. This balance depends on the type of facility being protected. All modern conventional systems are solid-state, and all but the smallest use a wiring system called *multiplexing*.

This system uses a time-sharing electronic technique to transmit and receive multiple signals on a single two-wire communications circuit. This has the great advantage of reducing the primary cost item in large hard-wired systems—individually wired circuits. The alarm and audible signal devices in standard multiplexed systems are the same as those in conventional hard-wired systems. Only the system wiring and the panel architecture change. Zones, and even coding, are substantially unaltered, although the reduction in wiring costs permits the use of smaller zones. Some multiplexed systems are hybrid, using multiplexed detector circuits and hard-wired audible/visual alarm circuits. An additional advantage of multiplex panels is that they can be reprogrammed readily, which was emphatically not the case with hard-wired relay panels.

## 24.12 SYSTEM CODING

### (a) Noncoded Systems

Noncoded systems are *continuous* ringing evacuation types using manual and automatic alarm initi-

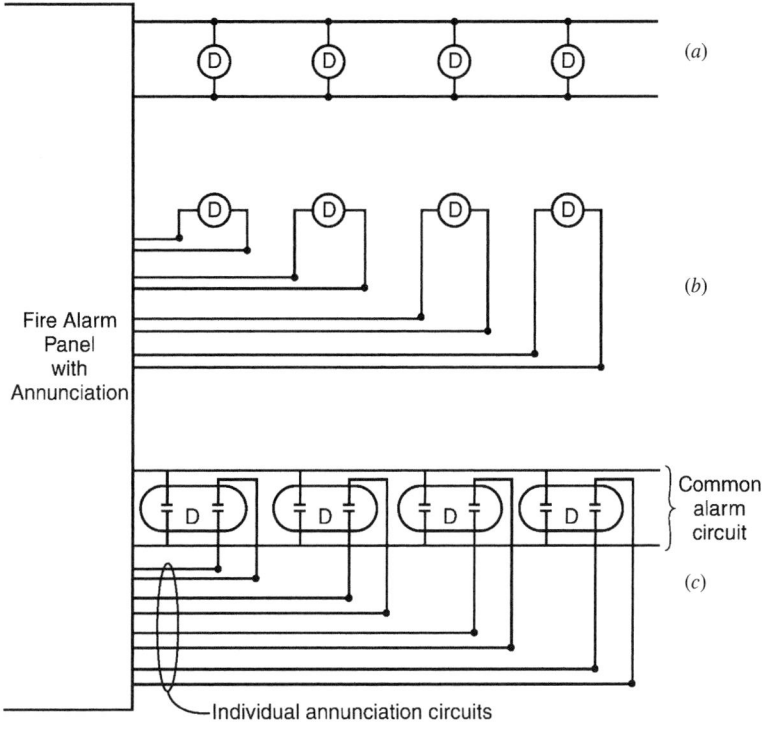

**Fig. 24.39** *Identification of detectors D can be accomplished in three ways. (a) Each circuit, with multiple detectors, is annunciated as a zone. This arrangement is applicable where detectors are close together. (b) Each detector is wired individually, comprising a separate zone. This is applicable for large or scattered detector spacing. (c) Multiple detectors are wired on a single alarm circuit but individually annunciated. This is applicable where the number of detector circuits in the panel is limited but individual detector identification is required.*

ation. If desired, the devices can be zoned and, if the system is sufficiently large, annunciation can be provided. Audible devices are continuous ringing, vibrating bells and horns. Visual devices are lights and strobes.

### (b) Master-Coded Systems

This system, also called *common-coded* and *fixed-coded*, generates four rounds of code that are sounded and flashed on all of the building alarm devices when any signal-initiating device operates. It utilizes a single code transmitter at the panel. Normally, the system stops after four rounds of code, although it can readily be arranged to sound and flash continuously thereafter. When the code is set to ring the bells at an even 108 strokes per minute, it is known as *march time* because of the rhythmic cadence. This beat aids the rapid, panic-free evacua-

tion of a building and therefore is frequently used in schools.

### (c) Zone-Coded Systems

Identification of the alarmed zone in a system can be accomplished with zone lights, with an annunciator, or by coding. In the first two cases, it is necessary to *go to the panel or annunciator* to determine the location of the operated device, which entails a possibly critical delay. All coded systems obviate this necessity by sounding the code on all the gongs in the building, thus immediately identifying the station and permitting the building staff to quickly investigate the cause of the alarm and take appropriate measures.

Therefore, if a *coded* system is desired, but by zone rather than by device (and this is less expensive by far), *noncoded* manual stations are used, along with automatic detectors, grouped by circuit into zones or individually zoned (Fig. 24.39). Each zone circuit trips a zone's transmitters *in the panel,* which in turn ring the zone's code on the single-stroke gongs or chimes. As with all coded systems, four rounds of coded signal are sounded, after which the system is silenced. In all coded systems, it is advisable to include a device that records all alarms in plain language, including the time of receipt and the code sounded.

### (d) Dual-Coded Systems

This arrangement is a combination of noncoded and zone-coded systems. When an alarm device operates, it initiates two separate functions—an identifying coded alarm and a continuous ringing evacuation alarm. The alarms are sounded simultaneously—the coded alarm in the building's maintenance office and the evacuation alarm on separate audible and visual devices throughout the building. A requisite to the application of this system is a continuously manned office in which the coded identifying signal can be received and acted on.

### (e) Selective-Coded Systems

These are fully coded systems in which all manual devices are individually coded and all automatic devices are arranged to trip code transmitters at the panel. Each manual station can be immediately identified by its distinctive code. Automatic devices may be grouped in any fashion desired and annunciated if desired. The combinations and circuitry are entirely in the hands of the designer. In large conventional systems, which fully selective-coded systems usually are, sprinkler transmitters and smoke detectors operate as integral subsystems of the main fire alarm panel.

### (f) Presignaling

When it is desired to alert only key personnel, a system called *presignaling* is used. Small bells or chimes are activated only at their work locations. Because these systems are always selectively coded, the personnel alerted can immediately investigate and, if necessary, manually turn in a general alarm by key operation of a station. Because of the delay involved, this type of system is used only in buildings where

evacuation is difficult and sufficient staff is available to immediately investigate the cause of an alarm.

## 24.13  SIGNAL PROCESSING

Once a hazard has been detected, a signal is transmitted to a fire control center and appropriate action taken. In conventional systems, the alarm signal is transmitted over dedicated conductors ("hard wiring") to a control panel, consisting of either electromechanical relays in older systems or solid-state switching circuitry in modern systems. The control panel, in turn, actuates audible and visible device circuits, illuminates annunciator panels, controls fans and door releases, and so on, all via dedicated wiring. This arrangement has the advantages of reliability and simplicity. As the system grows, however, the wiring becomes heavy, complex, and expensive; panels become large and bulky; and changes become difficult. Also, troubleshooting of system faults becomes time-consuming. Furthermore, minimizing false alarms is problematic because alarm verification in systems with large zones containing many alarm devices is very difficult within the extremely short time span permitted for this operation.

As a result of these problems, a system was developed in which every device can be individually identified and remotely checked, without separate wiring to each. It is called an *addressable system* because all the devices are individually addressable.

## 24.14  ADDRESSABLE FIRE ALARM SYSTEMS

An addressable fire alarm system is, by definition, one that uses addressable fire detection devices, both automatic and manual. These detectors are essentially identical to conventional detectors except that they are equipped with electronic circuitry, usually mounted in a special base, which makes each detector, in effect, a separate zone. Thus, in the event of an alarm, the control can require alarm confirmation from an adjacent detector or can require a repeat alarm from the same detector after (remote) reset.

Addressable devices are continuously polled from the control panel, and each replies to the poll by reporting its condition as OK–standby, alarm, or trouble. Because each device is identified by its address, an alarm or trouble signal from a detector can be quickly and easily confirmed. Detector identification is accomplished either visually on an annunciator panel or by alphanumeric readout on an LCD readout panel. Wiring is simplified; up to 100 detectors can be wired on a single multiplex line. Hardware costs of addressable systems are higher than those of a conventional system because of the additional electronics in the detector bases and the control panel. Maintenance costs, however, are somewhat lower because the panel's polling operation checks some (but not all) of a detector's electronics. Wiring costs of large systems are considerably lower than those of a conventional system. Measuring and calibrating the sensitivity of detectors is essentially the same as in a conventional system; the detector must be physically demounted and checked. This can be a major expense in a large system because periodic sensitivity checks are mandated by NFPA codes.

## 24.15 ADDRESSABLE ANALOG (INTELLIGENT) SYSTEMS

Addressable analog systems are known in the fire alarm industry as *intelligent* despite the fact that the word *intelligent* references to an inanimate device or system. To further confuse the issue, there is no accord among manufacturers on the exact meaning or use of the word when so applied. As a result, as of this writing, "very intelligent" and "extremely intelligent" devices have already appeared and other superlatives are sure to follow. We use the accurate technical term *addressable analog* in this book.

As mentioned previously, conventional and addressable detectors are identical except for the electronic identification coding of the latter. In function, both types test the adjacent space for temperature characteristics and/or products of combustion, compare the result to a preset threshold (adjustable), and latch into alarm position when the threshold is exceeded. The decision on whether to alarm is made at the detector, and the control panel acts only to receive and translate the alarm signal into an audiovisual buildingwide alarm. Expensive

periodic testing and adjustment of detectors in the field are essentially the same in both systems.

An analog detector gathers information from its sensor in analog form (voltage, current, or another variable characteristic proportional to the condition it senses) and transmits it to its control panel. That is, an analog detector acts only as a sensor; it does not make the threshold comparison leading to an alarm/no-alarm decision. That decision is made by a microprocessor at the control center. In modern systems, the data received from each detector are combined with other data including reports from adjacent detectors, "historical" data on the condition of the detectors' time patterns at each detector that might falsely indicate an alarm, plus other relevant information. Analysis of all of this information leads to the decision on whether to sound a general alarm. This arrangement has these advantages:

- False alarms are sharply reduced because of the alarm verification procedures at the panel.
- Threshold sensitivity is set at the panel, permitting adjustment without accessing the detector.
- A history of each detector's output is recorded, and sensitivity degradation can be compensated for automatically. This decreases maintenance costs by increasing the time period between cleanings.
- Any malfunction of a detector is noted immediately at the control panel, since all detectors are polled every few seconds.

Some manufacturers place an integral microprocessor in *each* detector that is capable of performing almost all of the functions described previously as panel functions. These detectors act as normal addressable analog detectors, transmitting analog signals to a control panel. However, in the event of a connection failure, the on-board microprocessor converts the unit to a stand-alone, self-contained detector/controller.

## 24.16 AUTOMATIC FIRE DETECTION: INCIPIENT STAGE

Fire authorities agree that *most* fires pass through four stages, the last of which is the visible flaming fire. These stages are shown in Fig. 24.40 as a

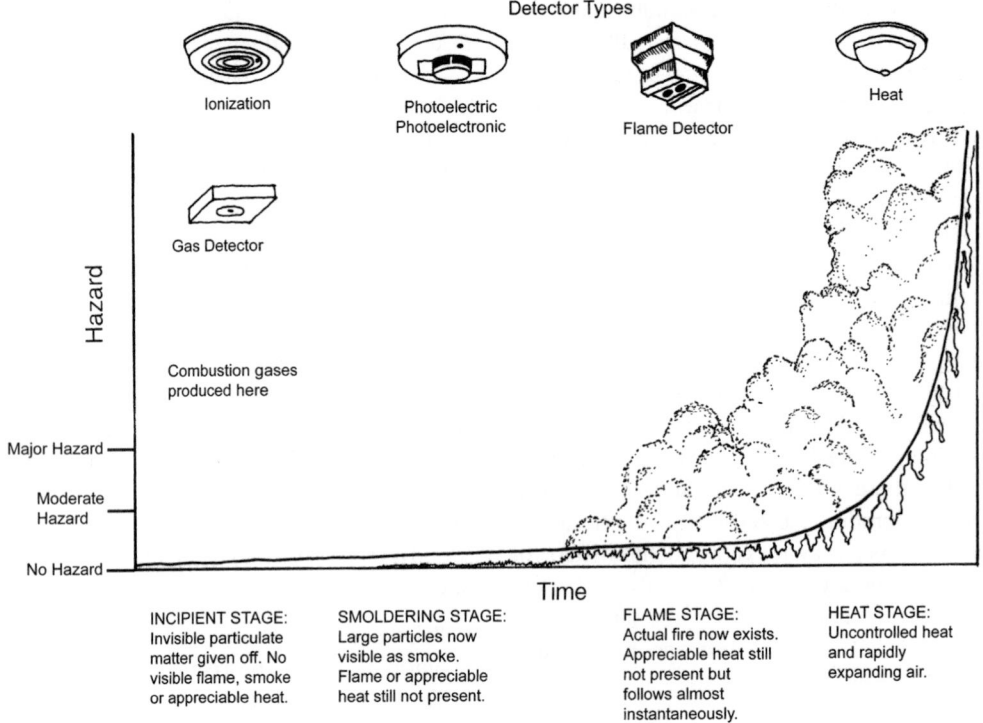

Detector Types

Ionization

Photoelectric
Photoelectronic

Flame Detector

Heat

Gas Detector

Hazard

Combustion gases
produced here

Major Hazard

Moderate
Hazard

No Hazard

Time

INCIPIENT STAGE:
Invisible particulate
matter given off. No
visible flame, smoke
or appreciable heat.

SMOLDERING STAGE:
Large particles now
visible as smoke.
Flame or appreciable
heat still not present.

FLAME STAGE:
Actual fire now exists.
Appreciable heat still
not present but
follows almost
instantaneously.

HEAT STAGE:
Uncontrolled heat
and rapidly
expanding air.

*Fig. 24.40 Four stages of a fire. Early detection of each stage requires a different type of detector that responds to the fire's particular characteristics (fire signature) at that stage. (Drawing by Amanda Clegg.)*

function of duration and degree of hazard. Also shown is the type of automatic detector recommended for detection at each stage.

In the incipient stage, the combustion products comprise a significant quantity of microscopic particles (0.01 to 1.0 micron), which are best detected by *ionization-type* detectors (Fig. 24.41). These detectors contain a small amount of radioactive material that serves to ionize the air between two charged surfaces, causing a current to flow. Combustion particles entering the detector chamber reduce air ion mobility, thus reducing current flow and increasing voltage. These changes are sensed, and the alarm is set off.

*Fig. 24.41 Addressable analog (intelligent) ionization-type smoke detector with an integral microprocessor that operates the unit's detection, diagnostic, and error-checking procedures. The detector is self-adjusting to compensate for sensitivity changes due to environmental conditions. Operating sensitivity is remotely adjustable from the control panel. (Photo courtesy of Cerberus Pyrotronics.)*

The response time of this type of detector depends on how rapidly the combustion particles can reach the detector—a factor that varies with room air currents and with the type of material "burning." Once the particles reach the detector, the response is essentially instantaneous. Therefore, ionization detectors are best applied indoors, in spaces with stagnant air or low air velocity (below

50 fpm [0.25 m/s]) and in which little visible smoke (large particles) is expected. (Some manufacturers make a special unit whose sensitivity *increases* with air velocity for application in spaces with air velocity of up to 500 fpm [2.5 m/s].)

Ionization detectors should not be installed on warm or hot ceilings, or in any other location where hot air concentrates, because the hot air prevents the combustion particles from reaching the detectors. As a corollary, ionization detector sensitivity is higher at low ambient temperature.

Because ionization detectors react to minute particles of combustion, they should not be applied in, or close to, areas where normal activities produce such particles, as the obvious result will be nuisance alarms. Thus, kitchens, bakeries, welding and brazing shops, workshops using open flames and burners, and areas with concentrated engine exhaust fumes are all spaces inappropriate for their application. Ionization detectors respond best to fast-burning flaming fires. Because dust settling in the ionization chambers reduces sensitivity, these units need periodic cleaning and recalibration.

As can be seen in Fig. 24.40, the incipient stage of a fire also produces changes in the gas content of the environment. These combustion gases, which are not normally present in the air, are detectable by devices known as *gas-sensing fire detectors*. The principle of operation of these detectors depends on the type. A semiconductor type reacts to a change in conductivity due to the presence of combustion gases, whereas a catalytic element type detects a change in the electrical resistance of the element caused by the presence of gases. Gas detectors are frequently used in conjunction with another type of detector so that both gases and particulate matter are detectable.

Coverage of incipient stage detectors varies from 150 to 900 ft² (14 to 83 m²) per unit, depending on the unit used, the type of combustible material in the space, and ambient conditions. Once the fire has passed the incipient stage and becomes smoky, or where even in the incipient stage the combustion products are the large particles typical of visible smoke, the ionization detector loses sensitivity and photoelectric detectors are recommended. Ionization, gas, and photoelectric detectors are classified as early-warning types.

## 24.17  AUTOMATIC FIRE DETECTION: SMOLDERING STAGE

### (a) Photoelectric Smoke Detection

Refer again to Fig. 24.40. The smoldering stage of a fire is characterized by particles up to 10 microns in size. Such particles, although small, are visible to the naked eye as smoke and are best detected by photometric means. The simplest type of *photoelectric smoke detector* operates on the principle of beam obscuration, as shown schematically in Fig. 24.42. A beam of light is directed onto a photosensor and a steady-state, no-smoke circuit condition is established. The presence of smoke in sufficient concentration partially obscures the beam, changing current flow in the photocell circuit and setting off an alarm response. Sensitivity is typically set at 0.5% to 4% obscuration per foot for gray smoke and 0.5% to 10% for black smoke to comply with UL Standard 1989.

In this design, accumulation of dust and dirt and the presence of heavy fumes from industrial processes cause gradual obscuration of both the photo cell and the lamp. This, combined with lamp aging and consequent light reduction, results in *increased* sensitivity, causing false alarms that are not only a nuisance, but also can constitute a hazard. To correct this condition, continuous maintenance and periodic recalibration are necessary. As a result, this design is no longer recommended for spot-type smoke detection but is used in beam detectors, as discussed next.

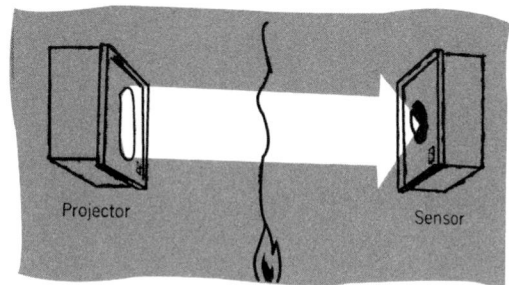

**Fig. 24.42** *Principle of smoke detection by beam obscuration. This design is used in two ways: as illustrated, with a projector in one location and a photocell at a remote location, or with the entire assembly in a single small housing, with the light source and receiver on opposite sides of a 2- to 3-in.- (50- to 75-mm-) wide smoke chamber.*

## (b) Projected Beam Photoelectric Smoke Detector

Beam-type photoelectric detectors consist of two separate units: a beam transmitter and a beam receiver, normally wall-mounted on opposite sides of a space, somewhat below the ceiling. They operate on the simple obscuration principle (Fig. 24.43). They are best applied in areas that do not lend themselves to the application of spot-type detectors. Among these are the following:

- High-ceiling areas such as atriums, churches, malls, auditoriums, and the like. In these spaces, ceiling-mounted spot-type detectors present serious maintenance problems. This is particularly true in spaces such as theaters and auditoriums.
- Spaces with medium- to high-velocity airflow at the ceiling level. This condition severely dilutes smoke entering the test chamber of a spot detector. Although beam detectors are also affected negatively, their long throw and wide "vision" angle make the dispersion and dilution problems much less serious.
- Closed areas with little airflow, resulting in a hot air layer at the ceiling that prevents smoke from reaching ceiling-mounted spot detectors (Fig. 24.44). Exposed ceiling beams also tend to trap hot air and reduce the effectiveness of ceiling spot detectors.

- Environments that militate against spot detector use, such as those that are extremely dirty, corrosive, humid, very hot, or very cold. Beam-type detectors can be physically shielded against these conditions in a manner that interferes only minimally with their light-beam transmission and reception characteristics.

Beam-type detectors have an effective throw (transmitter to receiver) ranging from 30 to 330 ft (9 to 100m), depending on ambient conditions, and can be spaced 30 to 60 ft (9 to 18 m) apart. This gives a coverage range of 900 to almost 20,000 ft$^2$ (84 to 1860 m$^2$), compared to spot detectors, whose *maximum* coverage is about 900 ft$^2$ (84 m$^2$). Furthermore, beam detectors are wall-mounted, which reduces maintenance problems.

Finally, beam detectors are usually drift-compensated (thus further reducing maintenance expense) and are arranged not to alarm with full obscuration, a condition that frequently occurs during routine building maintenance and repair work.

Disadvantages of beam-type detectors compared to spot units include:

- Response to the second (smoke) stage of a fire, and therefore not in the same early-warning category as ionization or gas detectors
- High cost
- Inapplicable to spaces that lack an unobstructed

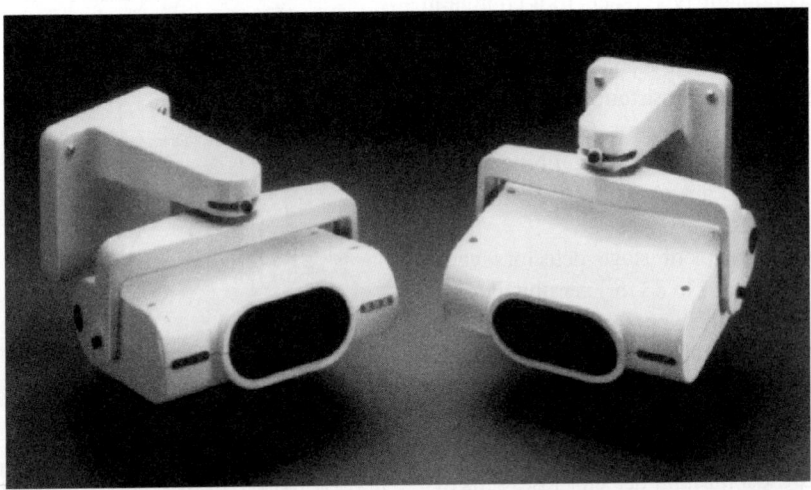

**Fig. 24.43** *This projected beam smoke detector utilizes an infrared (IR) beam so that light will not affect its operation. Two sensitivity levels are available that are selected in consonance with the distance between transmitter and receiver. Because heat attenuates or deflects the IR beam, the receiver translates this beam strength reduction as obscuration, identical to smoke obscuration. This effect makes the IR projected beam detector sensitive to high-heat flaming fires as well as smoky smoldering ones. (Photo courtesy of Notifier, division of Pittway.)*

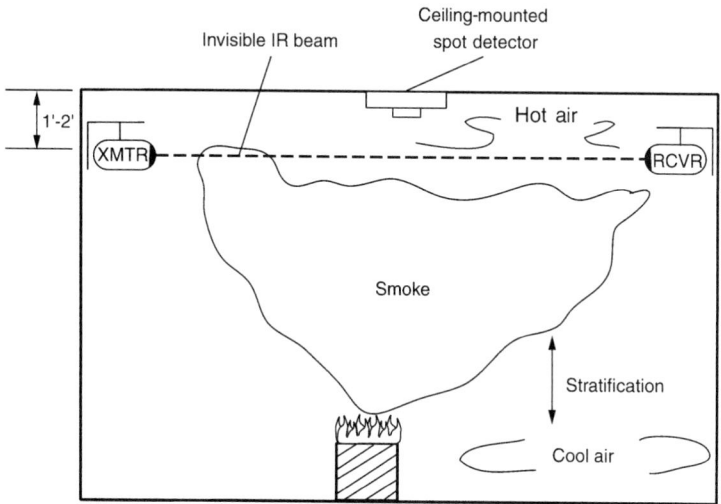

**Fig. 24.44** *Stratification of air in a closed space leads to a concentration of hot air at the ceiling that can prevent smoke from reaching a ceiling-mounted spot smoke detector. A projected beam detector mounted 1–2 ft (0.3–0.6 m) below the ceiling will alarm due to obscuration, as shown.*

view, such as spaces with pendant lighting fixtures, exposed ductwork, and the like

As with all types of detectors, the use of beam-type detectors in a particular application must be consonant with the requirements of *all* fire and administrative codes having jurisdiction in that area.

## (c) Scattered Light Photoelectric Smoke Detector

A second design, usually referred to as a *photoelectronic, scattered-light,* or *Tyndall-effect* detector, is illustrated in Figs. 24.45, 24.46, and 24.47. In this design, a beam of pulsed LED light is directed at a supervisory photocell, which serves to provide a baseline reference signal. The alarm photocell is shielded and normally receives no light. When smoke enters the unit, light is reflected from (scattered by) the smoke particles and strikes the alarm cell. This changes the cell's resistance and the resultant signal is amplified electronically, causing an alarm. An alternative design depends on refraction of light to set off the alarm. These designs are not sensitive to normal dust and dirt accumulation or to light source depreciation, and high sensitivity can be maintained without continual maintenance. Thus, they are usually found in commercial use and high-quality residential applications. Sensitivity is usually set at 1% obscuration per foot.

## (d) Laser Beam Photoelectric Detector

A very-high-sensitivity laser diode source, scattered-light type photoelectric smoke detector has been developed that is classified by its manufacturer as an early warning device (Fig. 24.48). The high sensitivity results from the ability of the very sharply focused laser beam and its associated control panel software to differentiate between a smoke particle and a dust particle. It is best applied in clean environments.

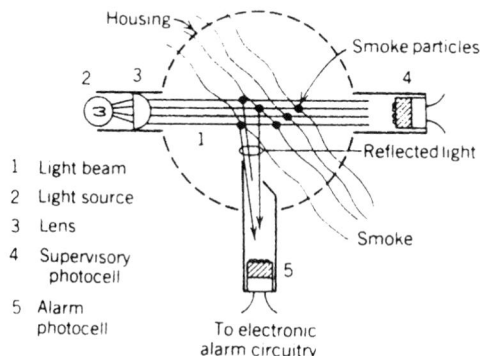

1 Light beam
2 Light source
3 Lens
4 Supervisory photocell
5 Alarm photocell

**Fig. 24.45** *Principle of operation of a scattered light photoelectronic spot smoke detector. A pulsed beam of light (1) from a source (2) is focused by a lens (3) on a supervisory cell (4). Smoke entering the unit reflects the light onto the alarm cell (5), changing its electrical characteristics and setting off an alarm. Photoelectric units with sensitivities of up to 2% per foot are usually classified as early-warning units.*

FIRE PROTECTION

**Fig. 24.46** *Addressable analog (intelligent) scattered-light-type photoelectric smoke detector with an on-board microprocessor that negates false-alarm signals caused by radio frequency and electromagnetic interference (RFI/EMI). It also validates trouble signals. If required, this unit can be equipped with a resettable heat detection element. The unit is approximately 5½ in. (140 mm) deep and 3 in. (75 mm) high. (Photo courtesy of Cerberus Pyrotronics.)*

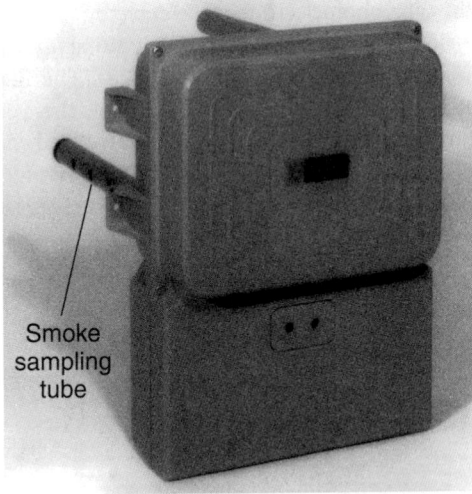

Smoke
sampling
tube

**Fig. 24.47** *This smoke detector, intended for duct mounting, can utilize either a photoelectronic or ionization-type head. The unit, which is usually mounted in return-air ducts rather than supply ducts to minimize smoke dilution, uses an air sampling tube that extends across the full width of the duct. Duct smoke detectors are intended primarily to prevent smoke dispersal throughout the building by causing shutdown of the air-handling system and by control of smoke damper operation. (Courtesy of Simplex.)*

**Fig. 24.48** *This unit is an addressable analog laser-beam type of scattered light photoelectric smoke detector. The sharply focused laser beam permits discrimination between smoke and dust particles, resulting in a high-sensitivity unit without false alarming. The unit is shown mounted on a sounder base containing a horn assembly, making it applicable to small, enclosed rooms. (Photo courtesy of Notifier, division of Pittway.)*

### (e) Air Sampling Detection System

Another system that takes advantage of the high sensitivity of laser beam–based photoelectric smoke detection is the air sampling system (Fig. 24.49). In principle the system is quite simple. Instead of waiting for air that may be carrying incipient smoke particles to reach the detector(s) by thermal air currents, this system samples air throughout the protected space by aspiration and brings it to the detector for testing. The aspiration and air conduction system is simply a piping system with holes at sampling points powered by a fan. The advantage of the system is that air throughout the space is sampled, thus, in effect, converting each sampling opening into a highly sensitive laser-beam detector.

Because the piping can be zoned into manageable areas that can readily be checked for alarm verification, the disadvantage of nonaddressability of the aspirating openings is avoided. Addition of a microprocessor to a multizone control unit, which adjusts the alarm sensitivity of each zone based on a history of environmental smoke levels, all but eliminates false/nuisance alarms.

### (f) Application of Photoelectric Detectors

Photoelectric smoke detectors of all designs will detect particles from about 0.2 to about 1000 microns. Thus, they are useful not only for smoldering fires but also for the smoky fires that character-

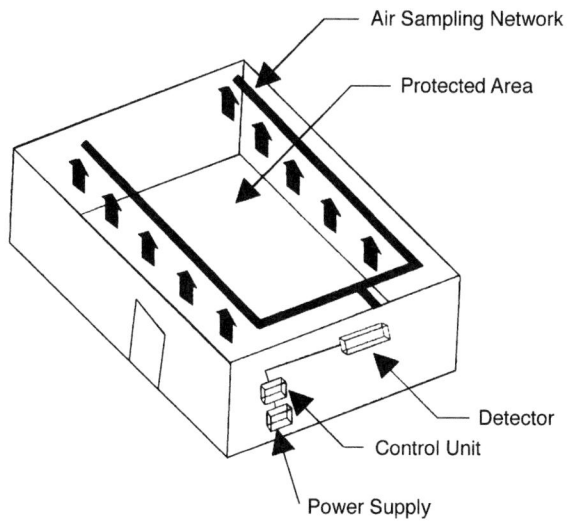

**Fig. 24.49** *In an air sampling system, air is aspirated at selected points in the protected area and piped to a highly sensitive laser beam photoelectric smoke detector for testing. Large areas are zone-piped so that trouble and alarm signals can be readily checked in the protected areas. (Drawing courtesy of Cerberus Pyrotronics.)*

ize the burning of certain plastics and chemicals. Also, particle agglomeration, which increases with the distance the smoke travels, does not reduce their sensitivity, as it does with ionization types.

*Maximum* recommended spacing for photoelectric detectors, as for other types, is given by UL and Factory Mutual Lab standards. Closer spacing is often mandated by the particular application or by the structure's characteristics. Manufacturers' recommendations should be obtained for all installations. In order to provide early-warning detection of a wider range of combustion products than is possible with either the photoelectronic or ionization-type detectors individually, several manufacturers produce a unit that combines a multichamber ionization detector with a photoelectronic detector.

## 24.18 AUTOMATIC FIRE DETECTION: FLAME STAGE

As noted in Fig. 24.40, the appearance of flame is followed almost instantaneously by heat buildup and the rapid spread of flame, with a concomitant large increase in hazard. Detection of flame is no longer "early warning," and the prime requirement for a detector at this stage is speed. It is self-evident that the actions taken as a result of a flame detection alarm, such as fire suppression and/or evacuation, must also be very rapid. This is not the case with early-warning equipment alarms, with which several minutes or more are normally available to investigate the authenticity of the alarm, *if desired,* before taking appropriate action.

Flame detectors are of two types: those that detect ultraviolet radiation and those that detect infrared radiation. Both types of radiation are present at the beginning of the flame stage.

*Ultraviolet (UV) radiation detectors* operate by optically detecting the UV radiation produced by flames, which is typically in the 170- to 290-nm range. (The entire UV range is 40 to 390 nm.) Hydrocarbon (organic material) fires in particular produce strong radiation in this range. A quartz filter in front of the UV sensor tube limits the detector's reception range to between 190 and 260 nm, which is below the visual range. This desensitizes it to sunlight and to incandescent and fluorescent sources. The detector need not "see" the flame directly; it can also detect UV radiation reflected from walls, ceilings, and the like. Obviously, the more direct the path, the stronger the radiation and the more rapidly the detector will respond. Because many devices—in particular, electric welders—produce UV radiation in sufficient quantity to activate the sensor, flame detectors are usually programmed to respond only to flickering sources at the usual flame flicker rate of 5 to 30 times per second.

In sum, UV detectors are long-range, very sensitive, react in milliseconds, and respond to most types of fires. They are best applied in highly flammable or explosive storage and work areas, either indoor or outdoor. Modern units are equipped with microprocessors that perform self-test diagnostics.

UV detectors are best applied to detect hydrocarbon-fueled fires such as burning alcohol, methane, propane, acetone, and all types of petroleum products. Their principal disadvantage is that they are "blinded" by thick black smoke.

*Infrared (IR) radiation detectors* are sensitive to radiation in the IR region, between 650 and 6500 nm. Most commercial units are filtered to be sensitive to radiation in the 3800- to 4300-nm range and to have maximum sensitivity at 4300 nm (4.3 microns), which corresponds to the radiation emitted by hot $CO_2$. They differ from ambient heat detectors in that they respond to radiant energy and are essentially optical detectors. IR units have about half the distance range of UV detectors, are sensitive (although not as sensitive as UV detectors), react in seconds, and must be programmed for flicker response to avoid false alarms, which can occur if they are exposed to other sources of IR radiation, such as sunlight. Because of interference problems that cause false alarms, IR units are normally applied to enclosed spaces such as sealed storage vaults and the like. IR flame detectors are best applied to fires that result in rapid flaming combustion and the production of $CO_2$. Typical materials in this risk category are petroleum products, wood and paper products, coal, and plastics.

Combination UV/IR detectors, such as the one illustrated in Fig. 24.50, cover a wide range of risk applications, including aircraft hangars, fueling stations of all sorts, tank farms, and flammable storage areas. The illustrated unit contains a microprocessor that performs self-diagnostics, not only of the optical system but also of the electronic circuitry and calibration. The presence in one unit of both UV and IR detectors permits the integral microprocessor to analyze the UV/IR ratio in any radiation detected and its time signature. This analysis is compared to known fire data in the microprocessor's memory in order to discriminate between actual fires and spurious radiation, thereby reducing false alarming.

## 24.19 AUTOMATIC FIRE DETECTION: HEAT STAGE

Again referring to Fig. 24.40, the heat stage is the last and most hazardous stage because, by this time, the fire is burning openly and producing great heat, incandescent air, and smoke. Spread of the fire depends on fuel, air currents, and the construction of the space in which the fire is burning. Detectors intended for use at this stage respond to heat and are referred to as *heat-actuated, thermal, thermostatic,* or simply *temperature detectors.* They act much like the fusible link in a sprinkler head. Effective application is restricted to locations where the subsequent alarm permits adequate countermeasures to be taken in time to prevent injury and minimize loss. Keep in mind that the heat stage *follows* the smoke stage and that smoke, not heat, is responsible for most casualties in fires.

Heat detectors have two designs: spot units, which are mounted in the center of the area that they protect, and linear units, which sense heat along their entire length. Both types respond to the high ambient temperatures caused by hot air convection from a fire. The linear type also senses the overheating of an object or surface with which it is in contact without the presence of fire.

### (a) Spot-Type Heat Detectors

Spot heat detectors are of two types: fixed-temperature units and rate-of-rise units. In the former, a set of contacts operates when a preset (nonadjustable) temperature is reached—usually 135°F or 185°F (57°C or 85°C). The rate-of-rise type

**Fig. 24.50** *The combined ultraviolet-infrared (UV/IR) flame detector has peak sensitivity at 200 and 4300 nm, respectively, which corresponds to the maximum emission points of a typical hydrocarbon (e.g., petroleum products) fire. This unit is equipped with a microprocessor that performs diagnostic procedures and ensures proper calibration. The detector, which measures approximately 5 × 5⅝ × 6 in. (127 × 144 × 152 mm) deep is suitable for use in specific class hazardous areas. (Photo courtesy of Megitt Avionics.)*

operates when the rate of ambient temperature change exceeds a predetermined amount (usually 15°F/minute), which is indicative of the heat stage of a fire. The rate-of-rise unit is normally combined with a fixed-temperature unit in a single housing.

The fixed-temperature unit is available in either a one-time nonrenewable design that utilizes a low-melting-point alloy plug or an automatic resetting unit that operates in the same fashion as a thermostat. For most applications, the resettable unit is preferred. Spot units are best applied in spaces that are separated from occupied areas and are subject to rapid-temperature-rise fires. Three different designs of heat detectors are shown in Figs. 24.51, 24.52, and 24.53.

**Fig. 24.51** *Detectors in the illustrated design are made with fixed-temperature, nonresetting elements that melt out at 130°F (54°C) or 200°F (93°C) and rate-of-rise elements that will alarm when the temperature rise is faster than 15°F/minute. The illustrated unit is a 200°F fixed-temperature-only model. It is applicable to spaces with unusually violent temperature fluctuations and ceiling temperatures between 100° and 150°F (37 and 65°C). (Photo courtesy of Cerberus Pyrotronics.)*

## (b) Linear Heat Detectors

In these devices, the entire length of the linear, cable-like element is heat sensitive. Its application is to long, narrow elements, the protection of which with spot detectors is impractical. Typical applications include cable trays, cable bundles of all sorts, conveyors, and large, long items of equipment. The heat-sensitive element is installed close to, or in direct continuous contact with, the protected item. As a result, any hot spot along the protected equipment will be detected.

The detectors themselves are available in a number of designs. One type uses a pair of steel wires under tension, held apart by thermoplastic insulation (Fig. 24.54). When exposed to its rated alarm temperature, the insulation melts and the wires come into contact, changing the current flow in the circuit to which they are connected and setting off an alarm. This type is available in a range of alarm temperatures (155°F [68°C], 190°F [88°C], 280°F [138°C]) and can be used in open- or closed-circuit configurations in lengths of up to 2500 ft (760 m). Meters are available that indicate the exact point of a fault, so that in long runs where overheating rather than fire is detected, countermeasures can be taken rapidly, without the delay caused by having to seek out the fault point. The major disadvantage of this type of detector is that it is nonrenewable; once it has faulted, the melted section must be cut out and replaced.

**FIRE PROTECTION**

(a)

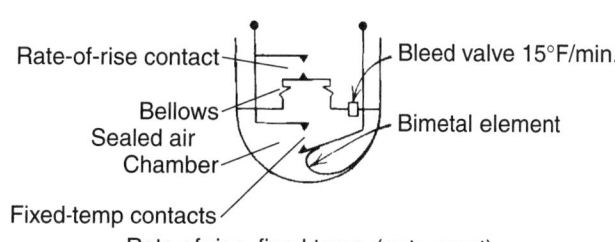

Rate-of-rise contact

Bellows
Sealed air
Chamber

Fixed-temp contacts

Bleed valve 15°F/min.

Bimetal element

Rate of rise, fixed temp (auto-reset)

(b)

**Fig. 24.52** *(a) A combination rate-of-rise and fixed-temperature detector. The detector and mounting plate fit on a standard 4-in. (100 mm-) square outlet box. (b) The principles of device operation are shown schematically. The rate-of-rise unit consists of an air chamber with a restricted bleed valve. Rapid temperature rise causes the bellows to expand before air is lost by bleeding, thereby setting off the alarm. The thermostatic element is a fixed-temperature, self-restoring bimetallic unit.*

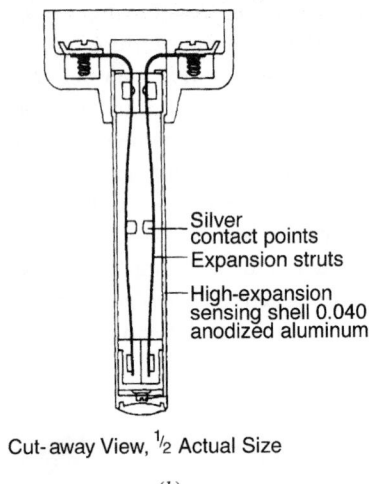

Cut-away View, ½ Actual Size

(a)                                                                     (b)

**Fig. 24.53** (a) A rate-of-temperature rise and fixed-temperature-type (135° or 200°F [57 or 93°C]) detector. Unlike the rate-of-rise detector in Fig. 24.52, this unit uses the principle of differential thermal expansion for both actions. (Photo courtesy of Cerebrus Pyrotronics.) (b) Cutaway of a detector with a different base than that shown in (a). The detector consists of an aluminum tubular shell containing two curved expansion struts under compression fitted with a pair of normally open opposed contact points that are insulated from the shell. The tubular shell and the struts have different coefficients of expansion. When subjected to a rapid heat rise, the tubular shell expands and lengthens slightly. The interior struts lengthen but at a slower rate than the shell. The rapid lengthening of the shell pulls the struts together, closing the contact points and initiating the alarm. When subjected to a slow heat rise, the tubular shell and the interior struts lengthen at approximately the same rate. At the detectors' set temperature point of 135°F or 200°F (57 or 93°C), the interior struts are fully extended, closing the contact points and initiating an alarm. (Drawing courtesy of Notifier, division of Pittway.)

Another type of linear detector is basically a linear thermistor—that is, a device whose electrical resistance varies with temperature (Fig. 24.55). The resistance is monitored at the control panel, and small changes are noted. Depending on the thermistor material selected, the detection range can be set anywhere from 70°F (21°C) to 1200°F (650°C). This type of linear detector is more sensitive, rapid, and expensive than the preceding type. When linear detectors are installed in direct contact

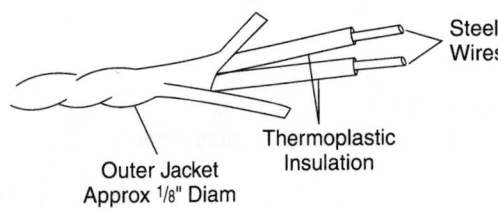

**Fig. 24.54** In this design of a linear heat detector, the thermoplastic insulation on the stiff, twisted steel conductors melts, allowing the two conductors to touch and thereby register an alarm.

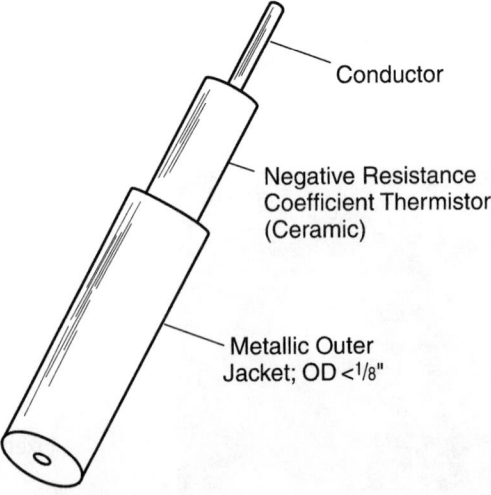

**Fig. 24.55** This linear heat detector is connected to an electrical control that monitors circuit current and thereby resistance. Any appreciable temperature rise along the entire cable length causes a sharp drop in the ceramic material's electrical resistance, which is immediately noted and pinpointed at the control point.

with the protected equipment, they detect surface temperature rather than air temperature and therefore act as early-warning devices. They are used extensively in this fashion to continuously monitor the surface temperature of such devices as transformers, switchgear, generators, and all sorts of hazardous equipment.

A third design for linear heat detectors uses optical fiber with a thermoplastic cladding (coating). Optical fiber transmits light by successive internal reflection at the fiber/cladding interface throughout the entire length of the fiber. Excessive temperature anywhere along the length of the fiber causes the cladding to melt and thereby appreciably changes the intensity of light received by the light sensor at the cable's end. This change is recorded and interpreted automatically as either a trouble signal or an overheat alarm signal. This design is available in lengths of up to 1 mile or more. Other linear sensor designs include variable electrical resistance polymers, phase change eutectic salt, and pneumatic tubing. Each type has its own advantages, drawbacks, and specific applicability.

Factors involved in the choice of a linear detector include:

- Physical characteristics and value of the protected elements
- Ease in locating the alarm area
- Requirements for resetting the system after an alarm
- Maintenance requirements
- Cost factors

## 24.20 SPECIAL TYPES OF FIRE DETECTORS

Several other sophisticated detectors are available for early-warning detection in the incipient and smoldering fire stages. Among the most sensitive is a unit based on the operating principle of a *Wilson cloud chamber* (see Figs. 24.56 and 24.57). When microscopic particles, such as those produced by the early stages of fire, are introduced into a saturated atmosphere (cloud chamber), they act as nuclei around which water vapor condenses to form visible droplets. The detector operates by continuously sampling air from the protected space and setting off an alarm when the cloud chamber indicates the presence of particulate matter.

(*a*)

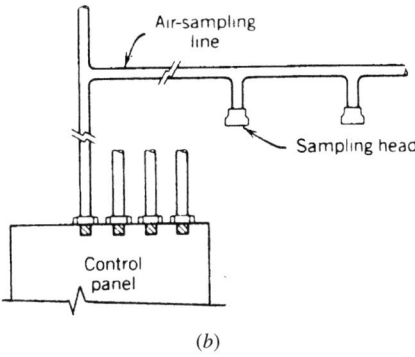
(*b*)

**Fig. 24.56** *This very early warning system detects microscopic combustion products in air aspirated from the protected area using a cloud chamber effect. (a) Tamperproof head housing, applicable to high-vandalism areas and correctional institutions. (b) Sampling heads are placed in all spaces to be monitored, including inside cabinets. (Courtesy of Protec Fire Detection, plc.)*

The continuous sampling procedure and the unit calibration make the system insensitive to dust and other noncombustion particulate matter and generally free of nuisance false alarms. The system's disadvantages are the need for piping and its high price in small installations. For large installations (30 or more detection points), the price is competitive with those of comparable systems. This detector is particularly applicable to high-value installations, such as museums, data-processing spaces, libraries, clean rooms, and facility control rooms, where its extremely high sensitivity may provide a critically important advance warning of an incipient fire.

## 24.21 FALSE ALARM MITIGATION

Smoke detectors of all types (ionization, photoelectric) are subject to false alarms due to moisture and

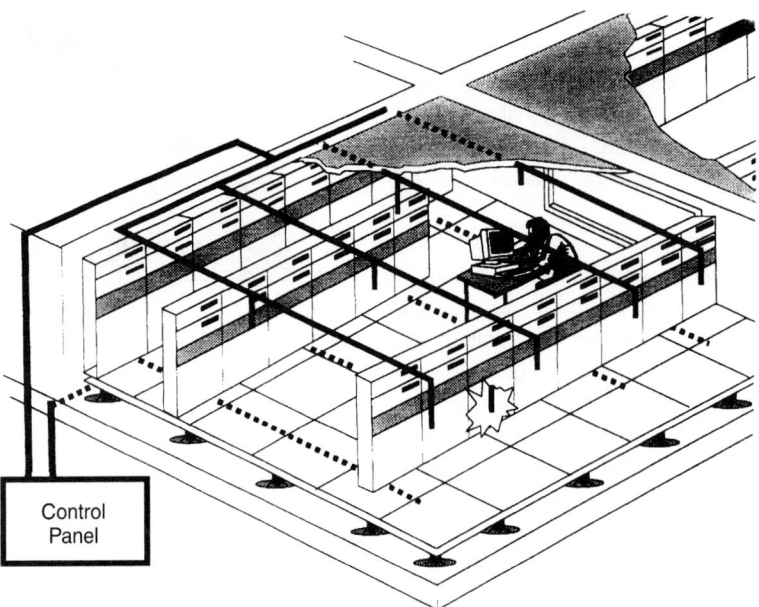

**Fig. 24.57** *Schematic diagram of a fire detection installation in a computer facility showing aspirating heads and piping. Note that air is sampled inside cabinets and under the raised floor in addition to sampling in the room itself. (Courtesy of Protec Fire Detection, plc.)*

Control
Panel

particulate matter in the air. Detectors are by nature threshold devices; raising their sensitivity increases the false alarming; decreasing their sensitivity shortens the crucial early-warning period. Much of the technological advancement since the late 1980s has been used to solve this intractable problem. Technology alone, however, will not entirely eliminate false alarms; the fire alarm system designer and the system user can contribute appreciably to minimizing this annoying and in certain circumstances dangerous occurrence by following some basic guidelines, including the following:

- Select appropriate smoke detectors. Choosing an ionization type where a photoelectric type is indicated, or vice versa, almost guarantees false alarms.
- Use detectors with drift compensation to permit accommodation of dirt/high temperature/humidity and air current ambient conditions, all of which increase false alarming.
- Where necessary, use radio-frequency interference (RFI) filters on detector circuits because line radio noise (spikes, surges) can cause alarms. This factor should be investigated in any installation with a heavy solid-state equipment load.

- Use control panels and detectors that have self-learning software and multispace-type algorithms that match sensitivity patterns to the spaces in which the detectors are installed.
- Use detector covers in areas where any type of construction work is taking place. This prevents fouling of detectors.
- When installing conventional systems, establish a regular maintenance and testing program that includes field testing and recalibration of all detectors.
- Avoid detector placement in areas where ambient conditions can cause problems. Where such placement is unavoidable, use compensating detectors, special maintenance, and appropriate verification procedures in the event of an alarm signal. These areas include:
  1. Kitchens, laundries, boiler rooms, shower rooms, and other spaces with high humidity and steam
  2. Repair shops, laboratories, and other areas where open flames are used in normal work
  3. Garages and engine test facilities where exhaust gases are present
  4. Smoking rooms and areas near spaces designated as smoking areas

5. Areas in which heavy accumulation of dust and dirt can be expected
6. Areas of high air movement, such as near loading docks and exit doors and near the discharge from diffusers or registers

## 24.22 MANUAL STATIONS

In contrast to automatic detectors, manual stations are operated by hand (Fig. 24.58). Manual stations serve to spread the alarm that has already been detected by other means, either human or automatic. In conventional systems, manual stations are either *coded* or *noncoded.* (See Section 24.12 for an explanation of coding.) If identification of the exact manual noncoded station operated is desirable, an annunciation panel can be added to the system; this is equivalent to using each station as a noncoded indicating zone. Because of wiring costs, annunciated systems become expensive; beyond 10 stations, coding should be considered (see Fig. 24.39).

When the system design is such that immediate *buildingwide* aural identification of the operated station is necessary, a coded station is used. The station code is received at the control panel, processed, and then transmitted audibly on the system gongs. Not less than three rounds of code, and normally four rounds, are transmitted. The code usually comprises three or four digits (e.g., 2-3-2) with a pause between the ringing groups and a longer pause between the rounds. The first number may identify the building floor, the second digit the wing, and the third digit the individual station. Establishment of codes is left to the user.

Manual stations are placed in the normal path of egress from a building so that an alarm may be turned in by a person as he or she exits. It is *imperative*, therefore, that stations be well marked and easily found. Architects who place fire alarm stations in nooks and corners and in camouflaged cabinets because they spoil the decor of the lobby are defeating the purpose of the system. Similarly, placement of bells *inside* hung ceilings because they are unattractive is not only foolish but dangerous and should *never* be done, regardless of the circumstances. Loss of property and even of life may result from such ill-conceived aesthetic considerations.

Addressable manual stations obviate the necessity for a code wheel for each manual station but

(a)

(b)

(c)

FIRE PROTECTION

**Fig. 24.58** (a) Manual fire alarm station with a break-glass rod. The unit is marked "local alarm," indicating that its operation will result in buildingwide evacuation signal but not a city alarm. Similar units are equipped with multiple sets of contacts for annunciation and other control functions or with addressable system electronics. Other similar units use a lock-open design that requires a key to reset; this avoids the necessity of replacing the broken glass rod. (b) A manual station, which is normally surface-mounted, can be enclosed in a well-marked recessed cabinet if desired for aesthetic reasons. (c) This addressable manual station latches in the down (operated) position when pulled and can only be reset by key. The manual switch can be polled from the control panel to verify that it has indeed been operated and that the alarm signal is not a circuit malfunction. The station is semi-recessed and measures approximately 4 in. (100 mm) wide by 5½ in. (140 mm) high. (Photo courtesy of Notifier, division of Pittway.)

do not eliminate coding of alarm bells if building-wide identification of the operated station is required. (Obviously, addressable manual stations are applicable only to addressable systems, not to conventional systems.) The purpose of gong coding is to permit rapid alarm verification by persons scattered about the building. An addressable manual station is identified only at the control panel, from which location a person must be sent to verify the alarm. In a small building, that is not a problem; in a large building, the time required to make such a verification is critical, because a manual alarm is already the result of a verified emergency unless it is a deliberate false alarm. The decision, then, as to whether to code the gongs must be made on the basis of data for that particular building. A typical addressable manual station is shown in Fig. 24.58*c*.

## 24.23 SPRINKLER ALARMS

Water flow switches are placed in sprinkler pipelines and operate when a sprinkler head goes off (Fig. 24.59). In electrical terms, a water flow switch is a set of contacts similar to a temperature detector. It can be used to trip a coded transmitter, setting off a sprinkler code; to show up on a sprinkler annunciator board, called a *sprinkler alarm panel*; or to act as a zone in a noncoded system. Wiring of water flow switches,

**Fig. 24.59** *Typical water flow indicator. The unit bolts onto a sprinkler pipe with the paddle inside the pipe. Any water motion deflects the paddle, causing a signal to be transmitted from the microswitch mounted in the box on top of the pipe. (Courtesy of Simplex/Potter.)*

or of switches activated by other extinguishing systems, is the same as that for a manual station.

## 24.24 AUDIBLE AND VISUAL ALARM DEVICES

The type and placement of audible and visual alarms in a (public) building must meet the requirements of NFPA 72 (*National Fire Alarm Code*) and the ADA (Americans with Disabilities Act). There are some disagreements between their requirements as of this writing. In such cases, the fire alarm system designer in concert with the building architect should settle these differences to the satisfaction of both authorities.

### (a) Audible Signals

In buildings where an evacuation signal will be sounded throughout the building (public mode) to alert all of its occupants, the requirements of NFPA 72 are normally used because they are more stringent than those of the ADA. There are several requirements, the most important of which to a building designer requires a minimum alarm sound level of 15 dBA above the average ambient sound level or 5 dB above the maximum sound level (whichever is greater) for at least 1 minute. This means that the designer must actually measure or predict, with a high degree of accuracy, the (anticipated) sound levels in the various portions of the building. Average sound level estimates are available from many sources, including the NFPA code itself; maximum sound levels are not. As a result, most designers use a considerable factor of safety to ensure compliance. The upper limit permissible in any area is 120 dBA. See Chapter 17 for technical information on architectural acoustics and sound levels.

The minimum private mode requirement is the greater of 10 dBA above the average ambient level or 5 dBA above the maximum ambient sound for at least 1 minute. Because this requirement generally applies to an enclosed space that houses the alarm panel or an adjacent space that is continuously occupied, the acoustic design is much simpler than for the public mode alarm.

The codes also specify the minimum required sound levels for sleeping areas and mechanical equipment rooms, in addition to directives regarding the location of audible devices. No purpose would be served by duplicating that information here, particularly because these codes, like others, are subject to constant review and revision. Suffice it to say that selection and placement of audible alarm devices is a highly technical task requiring acoustic analysis of spaces, occupancy, and the characteristics of available audible devices.

### (b) Visible Signals

The requirements for these signals, the primary purpose of which is to alert hearing-impaired people, are covered in NFPA 72, UL 1971, ANSI 117.1, ADA, and the various building code requirements. This is one of the areas in which, at this writing, discrepancies exist between ADA requirements and those of NFPA 72, UL 1971, and ANSI 117.1, which are in general agreement. All of the codes require that the visible alarm be a strobe light (usually a Xenon flashtube), flashing at less than 2 Hz to minimize problems for persons with photosensitive epilepsy. This does not, however, solve the problem, because a person viewing the light from more than one strobe (as at a corridor junction) can still be exposed to a higher flash rate unless all strobes are synchronized. Careful strobe placement can also help to minimize this problem.

### (c) Strobe Intensity

For nonsleeping areas, the ADA requires a strobe intensity of 75 cd and 50 ft (15 m) maximum unit spacing, whereas the other standards (UL, ANSI, NFPA) require an intensity dependent on the size of the room. For sleeping areas, the ADA requirement remains the same, whereas the other standards require a 110-cd source if wall-mounted and a 177-cd source if ceiling-mounted. For corridors, the ADA requirement remains 75 cd at 50 ft (15 m) strobe spacing, whereas the other standards have a requirement that varies with the corridor length.

Location of the strobes according to NFPA 72, Appendix B, must be such that the strobe light can be seen at any point in any space, regardless of the viewer's orientation, with the additional requirement that the maximum distance between strobes will not exceed 100 ft (30 m). A typical combination audiovisual fire alarm device is shown in Fig. 24.60.

**Fig. 24.60** *A combination audiovisual fire alarm signal is capable of sounding eight different warning tones that can be field selected. The 15-cd strobe flashes once every 1.5 seconds (2/3 Hz). The unit measures 4 in. (100 mm) square and can be mounted surface, semi-flush, or flush. (Photo courtesy of Notifier, division of Pittway Corp.)*

## 24.25 GENERAL RECOMMENDATIONS

The specific building system descriptions that follow are based on good present practice. Actual design must meet the requirements of the latest edition of NFPA standards plus other codes having jurisdiction. Special attention is directed to Article 760 of the *National Electrical Code* (NFPA 70). This article was completely rewritten and restructured in the 1996 edition of the code. It now covers fire alarm system wiring in its broadest sense (i.e., it now refers to "all circuits controlled and powered by the fire alarm system," including "fire detection and alarm notification, guard's tour, sprinkler . . . systems . . . safety functions, damper control and fan shutdown"). In office buildings the fire alarm system includes elevator capture and intercom, which, if powered and controlled by the fire alarm system, fall under Article 760.

## 24.26 RESIDENTIAL FIRE ALARMS

Refer to NFPA 72, *Household Fire Warning Equipment,* and NFPA 101, *Life Safety Code,* for detailed requirements. The system should provide sufficient time for the evacuation of the residents and for

**FIRE PROTECTION**

appropriate countermeasures to be initiated. The elements of the system are the various alarm-initiating devices, the wiring and control panel, and the audible alarm devices.

A good basic system requires the following:

- A listed smoke detector: outside and adjacent to each sleeping area, in each sleeping room, at the head of every stair, with at least one on every level, including the basement. Combined smoke/heat detectors should be installed in the boiler room, kitchen, garage, and attic. Listed heat detectors in the attic, kitchen, and boiler room are frequently set at 185°F (85°C) because of high ambient temperatures. Other units are set at 135°F (57°C).
- Ensure that an alarm in any detector produces an alarm in all audible and visual units.
- Control unit (central panel) annunciated to show the location of an alarmed device and arranged to shut off oil and gas lines and the attic fan (to prevent spread of smoke). Also, it should turn on lights both inside and out, operate an automatic dialer to ring a neighbor's phone or a commercial central station, and give a distinctive alarm sound when the phone is answered. An outside bell and some other device that transmits an alarm outside the residence are important.
- Backup power for the system—a supervised storage battery with a trickle charger.
- Wiring of all devices connected to a system control unit to be on supervised circuits that will sound a trouble alarm in the event of a fault. The trouble alarm should be distinct from the fire alarm.

Additional requirements for detectors, alarms, power supply, supervision signals, and other relevant items, including special requirements for fire alarm systems that are combined with other signal systems, are found in the current edition of the above-referenced standard.

## 24.27 MULTIPLE-DWELLING ALARM SYSTEMS

The exact requirements for the location and type of detectors and their action are specified in NFPA 101 and NFPA 72, and no purpose is served by repeating them here, particularly because they are constantly updated. Multiple dwellings include apartment houses, dormitories, hotels, motels, and rooming houses. The prime directive for these structures in the event of a fire emergency is early warning and orderly egress, with consideration of the fact that a fire emergency may occur when most of the building's occupants are asleep. Helpful design guidelines are as follows:

- Audible/visual alarms must be positioned so that all sleeping persons, including those with hearing and/or sight impairment, will be wakened. Code requirements for audible levels are minimum and are based on average spaces and furnishings. Unusually large or oddly shaped spaces should be designed individually.
- Consider the possibility of living rooms being used as sleeping areas on a regular basis.
- Provide smoke detection in corridors, service spaces, and utility and storage rooms.
- Battery-powered detectors may not be used because, unlike homeowners, apartment dwellers rely on the building's management for all building services, and periodic battery checks and replacement might not be carried out.
- Provide standby power to all fire alarm circuits.
- All alarms should be identifiable, either by addressing or by annunciation. Annunciators should be located at the system control panel in the building's management office or at the lobby desk in the case of hotels and dormitories. In all buildings, a lobby annunciator for the use of fire fighters is advantageous.
- An alarm light over the door of each apartment or suite to indicate the location of the alarm is desirable. This is particularly important if the central panel only provides zone annunciation.
- In high-rise buildings an emergency voice/alarm communication system should be provided.

## 24.28 COMMERCIAL AND INSTITUTIONAL BUILDING ALARM SYSTEMS

The requirements of these buildings are so varied that no specific recommendations can be made. A few suggestions, however, are in order:

- Presignaling, where permitted, is recommended for buildings that do not readily tolerate an evacuation alarm.

- In schools, particularly of the elementary grades, rapid, orderly evacuation of the building is the primary requirement. Consideration must be given to the uniqueness of the sound of the fire alarm gongs to allow no possibility of confusion with program gongs where the latter are used. Also, because regular fire drills are mandatory in most schools, the system must be arranged to provide this facility.
- Public buildings should have an auxiliary alarm connection to the fire department.
- For medium-sized buildings, an accurate cost estimate frequently indicates an advantage of addressable analog systems over addressable systems because of the high detector maintenance costs of the latter. Similarly, addressable systems frequently have an economic advantage over hard-wired systems because of the high labor costs for wiring.

Figures 24.61, 24.62, and 24.63 show a few modern system control panels.

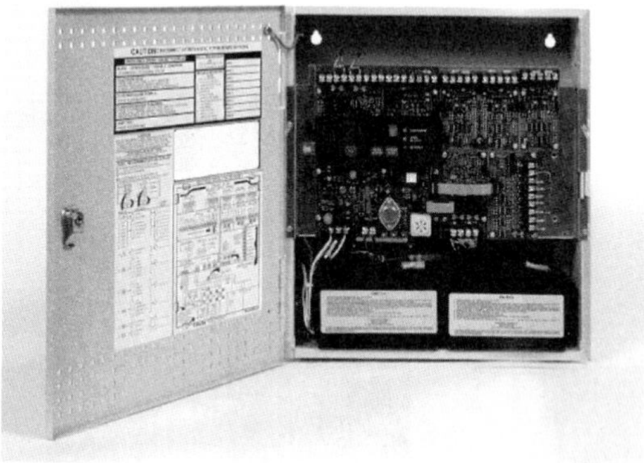

**Fig. 24.61** *Conventional hard-wired eight-zone (maximum) solid-state fire alarm control panel for supervised Class A or Class B circuits. A single annunciator light indicates the alarmed zone. Alarm verification for each zone is provided. Both audible and visible alarms can be used on two alarm-signal circuits. The audible evacuation signal is a slow (20-BPM) or fast (120-BPM) march time. Note the two standby batteries at the bottom of the cabinet. The unit measures approximately 16 in. (400 mm) wide and 17 in. (432 mm) high and is designed for surface mounting on 16-in. (400-mm) o.c. studs. (Photo courtesy of Simplex.)*

## 24.29 HIGH-RISE OFFICE BUILDING FIRE ALARM SYSTEMS

Sad experience has demonstrated that high-rise buildings, once thought to be fireproof, are emphatically not so. Indeed, due to their size, they have particularly severe fire protection problems, one of which is reliable communications during fire emergencies. As a result, fire codes now require that high-rise buildings be equipped with an emergency voice/alarm communications system. The system provides full control of transmission and building-wide distribution of all tones, alarm signals, and voice announcements on a selective or all-call basis. Specifically, it makes possible:

- Two-way active communications from the fire fighters' control command post (usually in the lobby) to at least one fire station per floor, all mechanical equipment rooms, elevator machine rooms, and air-handling (fan) rooms.
- Distribution to selected areas of alert tones, signals, and prerecorded messages on independent channels. (Obviously, a complete system of loudspeakers covering all areas of the building is an integral part of the system.)

**Fig. 24.62** *Control panel for a hard-wired conventional system that can accommodate up to 36 detection zones. The panel has a 2-line, 80-character alphanumeric display that indicates the zone alarm, trouble signals, and diagnostics information. The panel, which is 24 in. (610 mm) square, is microprocessor controlled and field programmable. A separate annunciator can be readily connected. (Photo courtesy of Simplex.)*

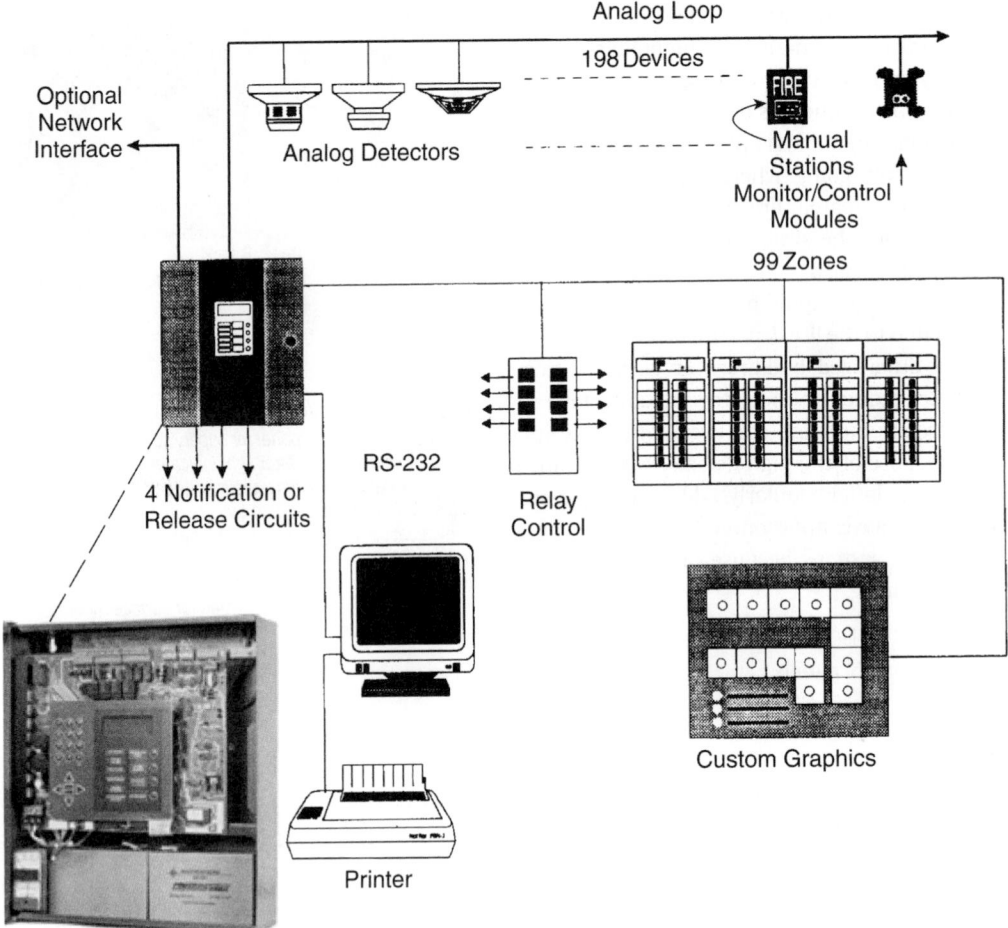

**Fig. 24.63** *System block diagram for a compact addressable analog fire system for a medium-sized building. The panel, which measures approximately 15 in. (380 mm) wide by 16 in. (400 mm) high by 6 in. (150 mm) deep, has a capacity of 99 analog detect and 99 monitor/control modules. Integral programming provides drift compensation and sensitivity adjustment for analog detectors, detector maintenance (dirt accumulation) alert, and automatic cyclic sensitivity adjustment, plus standard diagnostics. Custom graphics and the video display can be arranged to display floor plans, device locations, and the like. Relay control can be used for control of the building's air-handling unit, elevator interlock, and security/access system interlock. Notification circuits connect to the central station, remote station, proprietary alarms, or municipal alarms. (Diagram courtesy of Notifier, division of Pittway Corp.)*

- Communication with the fire department or central station.

In addition to communication functions, the (lobby) fire command post provides:

- Visual display (annunciation) of all fire alarm devices, including sprinkler valves and waterflow indicators
- Fire pump status indication
- Controls for any automatic stair door unlocking system (security access system)
- Emergency generator status

- Elevator location indicators plus operation and capture controls
- Control of smoke control devices (doors, dampers, etc.)
- Means for testing all circuits and devices

The exact equipment supplied depends on the building and the local fire code. The outstanding characteristics of this system are the communications system, the visual display of alarm locations, and the remote control of air-handling equipment (Figs. 24.64 and 24.65).

**Fig. 24.64** *Large addressable analog (intelligent) fire alarm control panel, suitable for network connection, computer interface, and operation in conjunction with an emergency voice/alarm communication system of the type used in multistory office buildings (see Fig. 24.65). The unit can accommodate 990 analog detectors and an equal number of monitor/control modules. It is illustrated with a 96-point integral annunciator, and additional annunciation can be connected. The integral control panel provides for polling, diagnostics, sensitivity control of all detectors, automatic detector drift compensation, alarm verification, front-panel programming, and options for system connections as shown in the block diagram of Fig. 24.63. (Photo courtesy of Notifier, division of Pittway Corp.)*

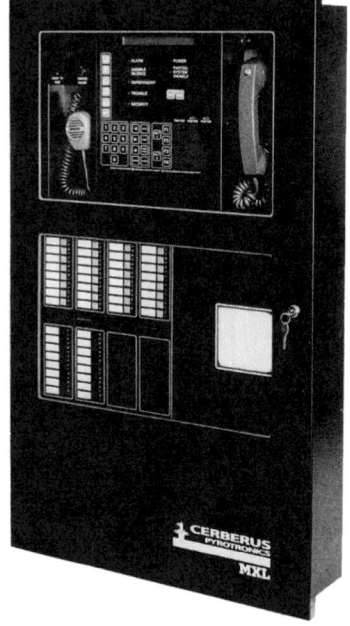

**Fig. 24.65** *A multiplex emergency voice alarm/communication system panel used in large high-rise buildings in conjunction with the fire alarm system panel (Fig. 24.64). The panel controls and feeds a multiplexed voice alarm system, as well as strobe light circuits and a complete firefighter's telephone intercom. It provides evacuation, alert, page, and auxiliary signals to all building speakers selectively. An integral annunciator indicates which audio devices are connected to which of the three available audio channels. All audio risers are fully supervised and power limited. (Photo courtesy of Cerberus Pyrotronics.)*

**FIRE PROTECTION**

## 24.30 INDUSTRIAL FACILITIES

In addition to manual stations at points of egress, these buildings use

- Temperature and smoke detectors in storage areas and laboratories
- Smoke and flame detectors in record rooms and continuous process laboratories
- Water flow switches on all sprinklers

The annunciators, control panel, and alarm register are best placed in the guardroom. If none is available, an auxiliary or remote-station circuit should be added to allow remote monitoring.

Because of the high ambient noise level in many plants, horns are substituted in such areas for bells and gongs, which might be inaudible.

In summary, the specific occupancy recom-

mendations given previously are representative of good practice at the present time. However, codes are constantly being changed in this particularly sensitive field, and they vary with the locale. In actual design situations, all codes and regulations having jurisdiction must be complied with.

## References

Various standards from NFPA, UL, ANSI, HUD, and the model code organizations dealing with fire alarm systems are presented in Section 24.7

U.S. Architectural and Transportation Barriers Compliance Board. *Americans with Disabilities Act (ADA) Accessibility Guidelines for Buildings and Facilities.* Washington, DC.

Egan, M.D. 1978. *Concepts in Building Firesafety.* John Wiley & Sons, New York.

Jensen, R., ed. 1975. *Fire Protection for the Design Professional.* Van Nostrand Reinhold, New York.

NFPA. 1997. *Fire Protection Handbook,* 18th ed. National Fire Protection Association, Quincy, MA.

NFPA. 2000. NFPA 92A: *Recommended Practice for Smoke-Control Systems.* National Fire Protection Association, Quincy, MA.

NFPA. 2002. NFPA 10: *Standard for Portable Fire Extinguishers.* National Fire Protection Association, Quincy, MA.

NFPA. 2002. NFPA 13: *Installation of Sprinkler Systems.* National Fire Protection Association, Quincy, MA.

NFPA. 2003. *Fire Protection Handbook,* 19th ed. National Fire Protection Association, Quincy, MA.

NFPA. 2003. NFPA 14: *Standard for the Installation of Standpipe and Hose Systems.* National Fire Protection Association, Quincy, MA.

NFPA. 2003. NFPA 750: *Standard on Water Mist Fire Protection Systems.* National Fire Protection Association, Quincy, MA.

NFPA. 2004. NFPA 780: *Standard for the Installation of Lightning Protection Systems.* National Fire Protection Association, Quincy, MA.

NFPA. 2005. NFPA 92B: *Standard for Smoke Management Systems in Malls, Atria, and Large Areas.* National Fire Protection Association, Quincy, MA.

UL. 2001. UL 96A: *Standard for Installation Requirements for Lightning Protection Systems.* Underwriters Laboratories, Inc., Northbrook, IL.

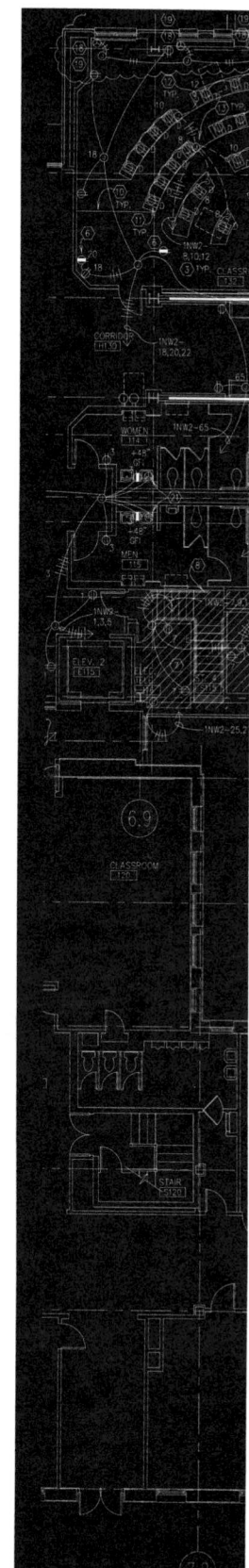

# ELECTRICITY

Electricity is the most prevalent form of energy in a modern building. It not only supplies electric outlets and electric lighting, but also provides the motive power for HVAC equipment, traction power for elevators and material transport, and power for all signal and communications equipment. An electric power failure can paralyze a facility. If properly designed, a facility can return to partial operation by virtue of emergency equipment that can furnish part of the facility's electricity needs for a limited time.

Given the complete dependence on electric power for normal operation that is characteristic of most modern buildings, designers must be familiar with the basic concepts and equipment of normal electrical systems. Chapter 25 reviews basic electrical relationships, with emphasis on electric circuits, power, energy, energy costs, and methods of energy management and electric load control. Chapter 26 describes electrical service, utilization, and emergency/standby power equipment. Also addressed are energy conservation considerations and economic factors. Chapter 27 introduces the concept of electrical equipment ratings and capacity, and continues with a description of modern wiring systems and their components. Chapter 28 draws on information given in the three preceding chapters to demonstrate straightforward design methods for building electrical systems. Chapter 29 presents information on photovoltaic (PV) systems, which are growing in interest among building designers.

# CHAPTER 25

# Principles of Electricity

HISTORICALLY, USABLE ENERGY WAS most often produced by burning a fossil fuel such as coal or oil. The resultant heat energy was used directly as heat and light or converted by machines into motion. Only since the end of the nineteenth century, however, has this heat in turn been used to create another very usable form of energy—electricity. The partial substitution of nuclear for fossil fuels has affected only the heat production portion of this process. Beyond that point, the heat is utilized in the same manner to drive generators that produce electricity. It is well to remember that in terms of natural resources, electricity is an expensive form of energy because the efficiency of the overall heat-to-electricity conversion, on a commercial scale, rarely exceeds 40%.

## 25.1 ELECTRIC ENERGY

Electricity is a form of energy that occurs naturally only in unusable forms such as lightning and other static discharges or in natural galvanic cells (which cause corrosion). The primary problem in the utilization of electric energy is that, unlike fuels or even heat, it cannot be readily stored and therefore must be generated and utilized in the same instant. This requires an entirely different concept of utilization than, for example, a heating system with its fuel source, burner, piping, and associated equipment.

The bulk of electric energy utilized today is in the form of *alternating current* (ac), produced by ac generators commonly called *alternators*. *Direct current* (dc) generators are utilized for special applications requiring large quantities of dc. In the building field, such a requirement was once almost universal for elevators because of the ease with which dc motors can be speed controlled. Today, however, because fine speed control of ac motors is practical, ac motors are the driver of choice for modern elevators, with attendant energy savings and a reduction in machinery space requirements. Smaller quantities of dc, furnished either by batteries, photovoltaic (PV) equipment, or rectifiers, are utilized for telephone and signal equipment, controls, and other specialized uses.

## 25.2 UNIT OF ELECTRIC CURRENT— THE AMPERE

Electricity flowing in a conductor is called *current*, which is measured in *amperes*, abbreviated *amp, amps,* or simply *A*. When current is used in an equation, it is usually represented by the letter $I$ or $i$.

It is convenient to establish an analogy between electrical systems and mechanical systems as an aid to comprehension. Current is a measure of flow and, as such, corresponds to water flow in a hydraulic system (Fig. 25.1). The analogy is not complete, however, because in the hydraulic system the velocity of water flow varies, whereas in the electric system the velocity of (electric) propagation is constant and may be considered instantaneous.

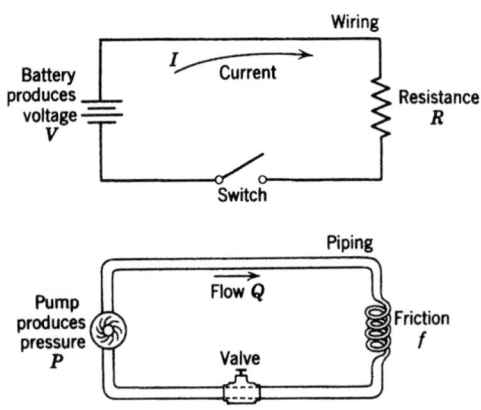

**Fig. 25.1** *Electric–hydraulic analogy. The circuits show that voltage is analogous to pressure, current to flow, friction to resistance, wiring to piping, and switches to valves.*

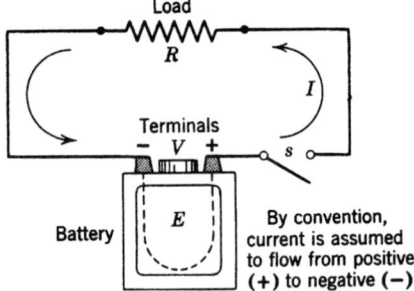

**Fig. 25.2** *Current flows in the electric circuit as a result of the voltage (potential difference) V that exists between the terminals of the battery. By convention, current direction is from positive (+) to negative (–) in the circuit (and from – to + inside the battery).*

## 25.3 UNIT OF ELECTRIC POTENTIAL—THE VOLT

The electron movement (and its concomitant energy, which constitutes electricity) is caused by creating a higher positive electric charge at one point on a conductor than exists at another point on that same conductor. This difference in charge can be created in a number of ways. The oldest and simplest method is by electrochemical action, as in a battery.

In the ordinary dry cell, or in a storage battery, chemical action causes positive charges (+) to collect on the positive terminal and electrons [i.e., negative charges (–)] to collect on the negative terminal. Assume for the moment that nothing is connected to the battery terminals. There is a tendency for flow between the electrified particles concentrated at the positive and negative terminals. *Potential difference* or *voltage* is the name given to this tendency, or force. It is analogous to pressure in a hydraulic or pneumatic system. Just as the pressure produced by a pump or blower causes water or air to flow in a connecting pipe, the potential (voltage) produced by a battery (or generator) causes current to flow in a conductor connecting the terminals between which a voltage exists (Fig. 25.2). The higher the voltage (pressure), the higher the current (flow) for a given resistance (friction). Other means of producing voltage, both direct (dc) and alternating (ac), are discussed in Section 25.8. The unit of voltage is the volt, abbreviated V or v.

## 25.4 UNIT OF ELECTRIC RESISTANCE—THE OHM

The flow of fluid in a hydraulic system is resisted by friction; the flow of current in an electric circuit is resisted by "resistance," which is the electrical term for friction. In a dc circuit this force is called *resistance* and is abbreviated R; in an ac circuit it is called *impedance* and is abbreviated Z. The unit of measurement is the *ohm*. The scientific convention of naming units after persons whose work is closely related to the field is followed here. Thus, the units ampere, volt, and ohm are derived from André Ampere, Alessandro Volta, and George Ohm.

Materials display different resistances to the flow of electric current. Metals generally have the least resistance and are therefore called *conductors*. The best conductors are the precious metals—silver, gold, and platinum—with copper and aluminum being only slightly inferior. Conversely, materials that resist the flow of current are called *insulators*. Glass, mica, rubber, oil, distilled water, porcelain, and certain synthetics exhibit this insulating property and are used to insulate electric conductors. Common examples are rubber and plastic wire coverings, porcelain lamp sockets, and oil-immersed switches.

## 25.5 OHM'S LAW

The current $I$ that will flow in a dc circuit is directly proportional to the voltage $V$ and inversely proportional to the resistance $R$ of the circuit.

Expressed as an equation, this is the basic form of Ohm's law:

$$I = \frac{V}{R} \qquad (25.1)$$

In ac circuits, the same relation holds true except that instead of dc resistance we use ac impedance. Ohm's law is frequently written in another form:

$$V = IR \qquad (25.2)$$

which expresses the mathematical relationship that voltage is the product of current and resistance. This form has no logical basis; therefore, we recommend remembering the form of Eq. 25.1. It clearly states the physical situation, that is, as a result of voltage $V$, a current $I$ is produced that is proportional to the electric voltage (pressure) $V$ and inversely proportional to the electric resistance (friction) $R$.

Example 25.1 illustrates the application of Ohm's law. The example chosen is applicable to both ac and dc because the load device is purely resistive. When a load is purely resistive, resistance and impedance are equal. This is more fully explained in the subsequent discussion of alternating current.

---

**EXAMPLE 25.1** An incandescent lamp having a hot resistance of 66 ohms is put into a socket that is connected to a 115-V supply. What current flows through the lamp (after it reaches operating temperature)?

**SOLUTION**

$$I = \frac{V}{R} = \frac{115 \text{ volts}}{66 \text{ ohms}} = 1.74 \text{ A}$$

(These figures correspond to a normal 200-W general service incandescent lamp.)  ∎

Hot resistance is mentioned in Example 25.1 because the electrical resistance of some materials changes with temperature. A typical example of this is a tungsten-filament lamp that when first turned on cold takes, for a fraction of a second, 10 to 15 times the steady-state hot filament current.

## 25.6  CIRCUIT ARRANGEMENTS

The two basic electric circuit arrangements are *series* and *parallel*. These concepts are the same for both dc

and ac. As previously, the focus is on purely resistive circuits so that circuit calculations are applicable to both dc and ac. In other than purely resistive circuits, ac circuit calculations are different and much more complicated than those for their dc counterparts.

## (a) Series Circuits

In a series arrangement the elements are connected one after another, that is, in series. Thus, resistances and voltages add. This is indicated graphically in Fig. 25.3. An electric circuit may be defined as a complete conducting path that carries current from a source of electricity to and through some electrical device (or load) and back to the source. A current can never flow unless there is a complete (closed) circuit. Due to the arrangement of components in a series circuit, *the current is the same in all parts of the circuit.* A somewhat more complicated circuit is shown in Fig. 25.4 and is analyzed in Example 25.2.

It is customary to refer to connection points on such wiring diagrams by letters, as $a$, $b$, $c$, $d$, and so on. The battery voltage may then be called $V_{ab} = 12$ V; the voltage across the load resistance, $V_{cd} = 11.5$ V; and the resistance of the two wires $r_{bc} + r_{da} = 0.04$ ohm. The positive and negative terminals of the battery are always shown.

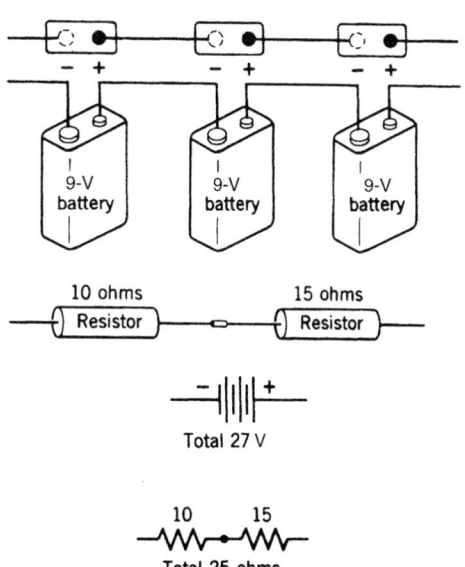

**Fig. 25.3** *Physical and graphic representation of a series connection of batteries and resistors.*

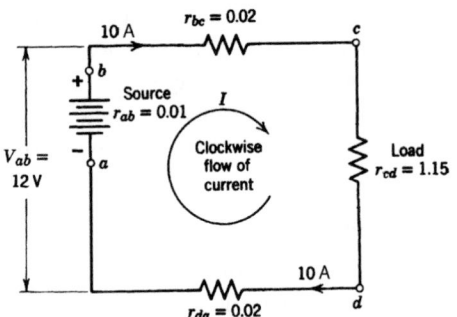

**Fig. 25.4** *A series circuit always contains a voltage source (here, the battery), and a load (R). This circuit also shows the resistance of the connecting wires for accuracy. Such resistances are normally neglected because they usually are very small compared to the load resistance.*

**EXAMPLE 25.2** The battery in Fig. 25.4 is rated at 12 V; the total line resistance (both wire segments) is 0.04 ohm; the battery internal resistance is 0.01 ohm; and the load resistance is 1.15 ohms. Determine (a) the current flowing in the circuit and (b) the voltage across the load ($V_{cd}$).

**SOLUTION**

(a) The current flowing is

$$I = \frac{V}{R} = \frac{V_{ab}}{r_{ab} + r_{bc} + r_{cd} + r_{da}}$$

$$= \frac{12}{0.01 + 0.02 + 1.15 + 0.02} = \frac{12}{1.2} = 10 \text{ A}$$

(b) The voltage drop across the load is

$$V_{cd} = I \times R_{cd} = 10 \times 1.15 = 11.5 \text{ V} \quad \blacksquare$$

Series circuits find very limited application in building wiring.

## (b) Parallel Circuits

When two or more branches or loads in a circuit are connected between the same two points, they are said to be connected in *parallel* or *multiple*. Such an arrangement and its hydraulic equivalent are shown in Fig. 25.5. From the circuit of Fig. 25.6 it should be apparent that the voltage across each load is the same, but the current in each load (branch) depends upon the resistance of that load. Parallel loads, in effect, constitute separate circuits. From this we conclude that in this arrangement, the total current in the circuit is the sum of the individual currents flowing in the branches—that is,

$$I_T = I_1 + I_2 + I_3$$

Notice in Fig. 25.6 that the total current flowing in the circuit is the sum of the currents in all the branches, but that the current in each branch is determined by a separate Ohm's law calculation. Thus, in the 10-ohm load, a 12-A current flows, and so forth.

The parallel connection is the standard arrangement in all building wiring. A typical lighting and receptacle arrangement for a large room is shown in Fig. 25.7. Here the lighting fixtures constitute one parallel (multiple) grouping, and the convenience wall outlets constitute a second parallel grouping. The fundamental principle to remember is that loads in parallel are additive for current and that each load has the same voltage across it.

One additional point is important to appreciate.

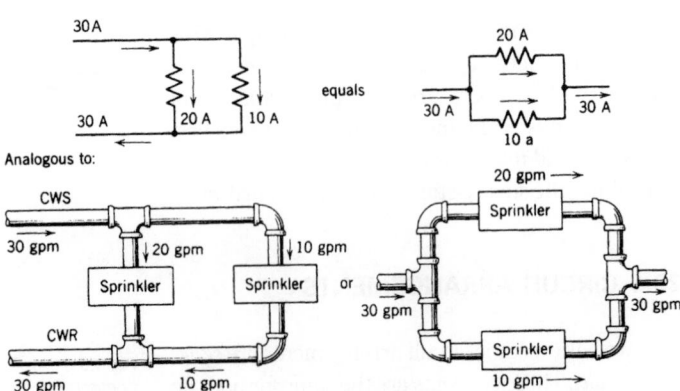

**Fig. 25.5** *In a parallel connection the flow divides between the branches, but the pressure is the same in each branch.*

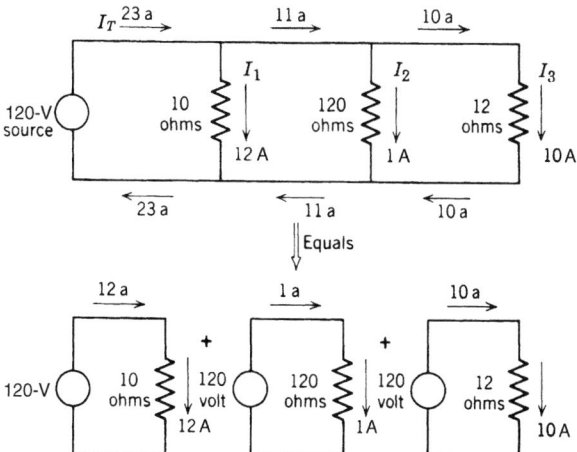

**Fig. 25.6** Note that loads connected in parallel are equivalent to separate circuits combined into a single circuit. Each load acts as an independent circuit unrelated to, and unaffected by, the other circuits.

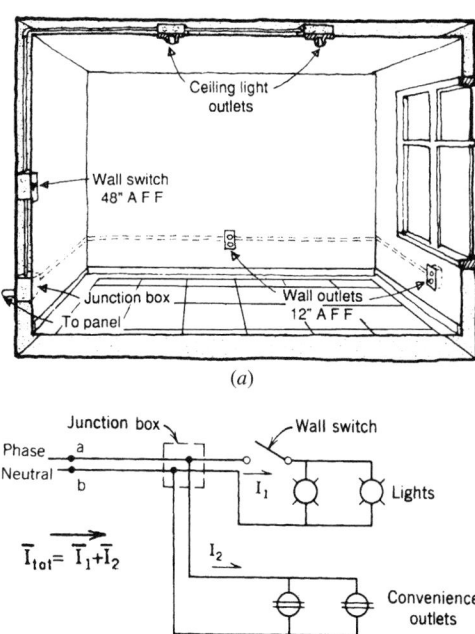

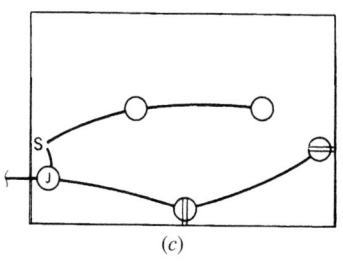

**Fig. 25.7** Parallel groupings of ceiling lamp outlets and wall outlets are in turn connected in parallel to each other. The circuit is shown (a) pictorially, (b) schematically, and (c) as on an electrical floor plan. (The horizontal bar over the currents $I_1$ and $I_2$ in (b) signifies the use of vectorial addition; see Example 25.5.)

Remember from Ohm's law that current is inversely proportional to resistance. Thus, as resistance drops, current rises. Now look at the circuit of Fig. 25.7. Normally, that circuit carries the (vector) sum of the currents in the two branches. (Vectorial addition is discussed in Section 25.11, Example 25.5c.) If, by some mischance, such as deterioration of the wiring insulation, a connection appears between points *a* and *b*, the circuit is *shortened* so that there is no resistance in parallel branch *ab*. The current rises instantly to a very high level, and we have a *short circuit*. If the circuit is properly protected, a fuse or circuit breaker opens and the circuit is deenergized. If not, the heat generated by the excessive current will probably start a fire.

## 25.7 DIRECT CURRENT AND ALTERNATING CURRENT

A flow of electric current that takes place at a constant time rate, practically unvarying and in the same direction around a circuit, is called a *direct current* (dc). Figure 25.8 shows dc voltages of 1.5 V positive polarity and 1.0 V negative polarity.

Whenever the flow of current is periodically varying in time and in direction, as indicated by the symmetrical positive and negative loops, or *sine waves*, in Fig. 25.9, it is called an *alternating current* (ac). The distance along the time axis spanned by a positive and a negative ac loop is called *one cycle*. The number of such cycles occurring in 1 second is known as the *frequency* of the ac current. Modern ac

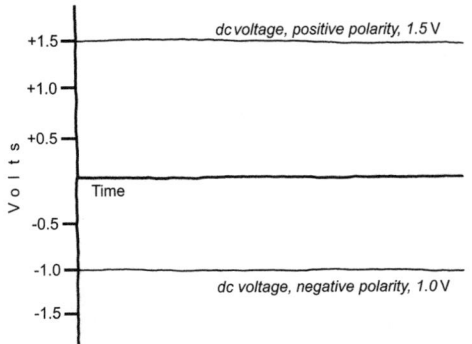

**Fig. 25.8** *Graphic representation of dc voltages with positive and negative polarity.*

systems in the United States and Canada operate at a frequency of 60 cycles per second, or 60 *hertz* (after Heinrich Hertz). This means that current at 60 hertz (abbreviated Hz) is delivered to the consumer. In Europe and much of Asia, 50 Hz is standard.

An ac circuit differs from a dc circuit in a number of important respects and, because the normal current supply is 60 Hz ac, it is important to understand ac terminology and usage. Instead of resistance, the corresponding parameter in an ac circuit is *impedance*, which is (also) measured in ohms.

Depending upon the circuit load, impedance can be markedly different from the dc resistance. For an ac circuit, Ohm's law is

$$I = \frac{V}{Z} \qquad (25.3)$$

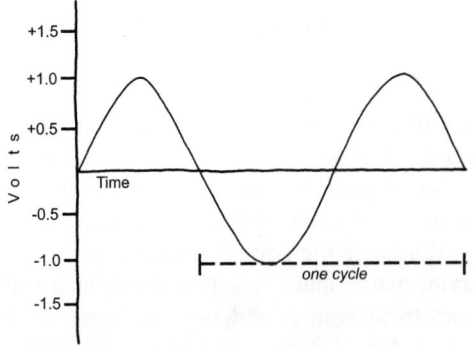

**Fig. 25.9** *Graphic representation of a pure single-frequency ac. The figure shown is a sine wave. Note that a complete cycle includes both the positive and negative loops. The number of full cycles in 1 second is defined as the frequency of the current (or voltage).*

where Z is the symbol for impedance. This book does not go into ac circuit calculations primarily because such calculations are not especially useful to the typical building designer. What *is* useful and important is an understanding of power and energy in both dc and ac circuits. These concepts are discussed in the following sections.

## 25.8 ELECTRIC POWER GENERATION—DC

With respect to the generation of large amounts of power, piezoelectric and thermoelectric effects can be ignored. Photovoltaic (PV) power generated from solar cells—as discussed in Chapter 29—is becoming increasingly important as a practical source of electricity for specialized uses and as part of green building efforts. This leaves the battery and the dc generator as everyday sources of dc electricity. Because the dc generator is in reality an ac generator with a device (commutator) attached that rectifies the ac to dc, the battery is still the major direct source of dc. (There are some special types of generators that produce dc *directly*, but their use to date has been extremely limited.) A discussion of the application of batteries for emergency and standby power supplies can be found in Chapter 28.

Another source of dc power is rectification of ac. This can be accomplished on any desired scale to provide as much dc power as there is available ac power. The principal application of dc in older buildings is for elevator motors and in all buildings, to a lesser extent, for standby power. Small amounts of dc are also used for controls and telephones.

## 25.9 ELECTRIC POWER GENERATION—AC

Alternating current is produced commercially by an ac generator, generally called an *alternator*. Its prime mover may be any type of engine or turbine. The process by which electricity is produced is illustrated in Fig. 25.10. It is based upon the fundamental discovery in 1831 by Michael Faraday of the principle of electromagnetic induction. Briefly, this principle states that when an electrical conductor is moved in a magnetic field, a voltage is induced in the conductor. The direction of movement determines the polarity of the induced voltage, as shown in Fig. 25.10.

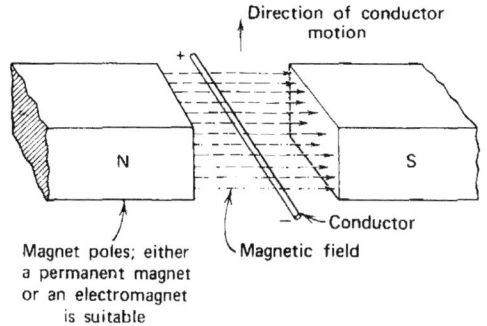

**Fig. 25.10** *The action fundamental to all generators. When a conductor of electricity moves through a magnetic field, a voltage is induced in the conductor, with polarity as shown.*

If a conductor is formed into a coil and rotated in a magnetic field, a voltage of alternating polarity is produced, that is, alternating current. It does not matter whether the conductor or the magnet moves; the motion of the conductor and the field with respect to each other produces the voltage (see Fig. 25.11). It is only one step (that of development) from this rudimentary ac generator to the large, powerful alternator that produces ac in a modern power plant. The frequency of the voltage generated is a function of the machine design and the speed at which it is driven. Normal generator frequency

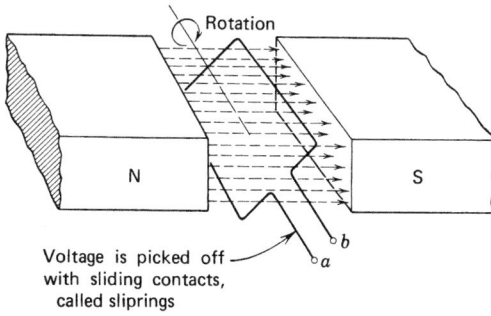

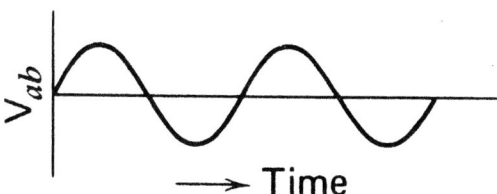

**Fig. 25.11** *Rotating a coil in a magnetic field produces an alternating sinusoidal voltage at terminals a and b because of the alternating polarity (see Fig. 25.10).*

in the United States, as noted previously, is 60 Hz, whereas in Europe and the Mideast it is 50 Hz.

## 25.10 POWER AND ENERGY

It is important, indeed imperative, that the distinction between power and energy be clearly understood, because all too frequently the terms are incorrectly used interchangeably. *Energy* is the technical term for the more common expression *work*. *Power* is the rate at which energy is used. Stated otherwise, it takes a fixed amount of energy to lift a weight a given distance either quickly or slowly, but the faster it is lifted, the more power that is required. Energy is the product of power and time, that is,

$$\text{energy (work)} = \text{power} \times \text{time} \qquad (25.4)$$

In practical terms, energy is synonymous with fuel and, therefore, with cost. Thus, energy can be expressed as barrels (liters) of oil, cubic feet (cubic meters) of gas, tons of coal, kilowatt-hours of electricity usage, and dollars of fuel cost. The outcome of energy efficiency efforts for buildings can be stated in terms of annual usage of oil, gas, and electricity or, alternatively, in terms of dollars of fuel or utility costs. In technical terms, energy is expressed in units of Btu (calories), foot-pounds (joules), and kilowatt-hours. Conversion factors between these (and other) units of energy are found in Appendix A. In terms of fuel, 1 kilowatt-hour (kWh) of energy is roughly equivalent to 0.5 lb (0.23 kg) of coal, 0.07 gal (0.26 L) of oil, 7.7 ft$^3$ (0.22 m$^3$) of natural gas, and 8200 gal (31,000 L) of dam water.

   *Power* is the rate at which energy is used or, alternatively, the rate at which work is done, because energy and work are synonymous. The term *power* implies continuity—that is, the use of energy at a particular rate, over a given, generally considerable, span of time. The concept of power necessarily involves the factor of time because it is, as stated previously, the *rate* at which work is done. Thus, multiplying power by time yields energy. Typical units of power in the I-P system are horsepower, Btu per hour, watt, and kilowatt. In the SI system, the corresponding units are joule per second, calorie per second, watt, and kilowatt. In physical terms, power is also the rate at which fuel (energy) is used. Thus, power can also be expressed as gallons (liters) of oil per hour, cubic feet (cubic meters) of gas per minute, and tons (kilograms) of coal per day. (When using fuel figures, it is important to note the assumed fuel-to-useful-energy conversion efficiency. When not stated, "energy content" [i.e., energy availability at 100% conversion efficiency] is assumed.)

## 25.11 POWER IN ELECTRIC CIRCUITS

The unit of electric power is the watt (W); 1000 watts is the commonly encountered kilowatt (kW). The power input in watts to any electrical device having a resistance $R$ and a current $I$ is given by the equation

$$W = I^2R = (I)(IR) \tag{25.5}$$

where $W$ is wattage. This is true for both dc and ac circuits. However, because the resistance of a device or load is generally not known, whereas the circuit voltage and current are known, it would be preferable to be able to calculate power using these latter two quantities. This can be done, but differently, for dc and ac.

In dc circuits, by Ohm's law, $V = IR$ and, because $W = I(IR)$ from Eq. 25.5, we obtain

$$W = VI \tag{25.6}$$

where

$W$ is watts

$R$ is ohms

$I$ is amperes

and $V$ is volts

In ac circuits, impedance consists of a combination of dc resistance and ac resistance (called *reactance*). Reactance causes a phase difference between voltage and current. This phase difference is represented by an angle, the cosine of which is called the *power factor* (*pf*). Power factor is extremely important in that it enables us to calculate power in an ac circuit. The ac power equation is similar to that for dc (see Eq. 25.6) with the addition of the ac power factor term; that is,

$$W = VI \times pf \tag{25.7}$$

If *pf* is not included in the equation, the product of voltage and current gives a quantity known as *volt-amperes*. In a purely resistive circuit, such as one with only electric heating elements, impedance equals resistance, power factor equals 1.0, and wattage

equals volt-amperage. A few examples should help make the application of these equations clear.

**EXAMPLE 25.3** Referring back to Example 25.1, calculate the power drawn using Eqs. 25.5, 25.6, and 25.7.

**SOLUTION**
Because an incandescent lamp is purely resistive and therefore has unity (1.0) power factor, it does not matter whether the circuit is ac or dc.
From Example 25.1,

$$R = 66 \text{ ohms}, I = 1.74 \text{ A}, V = 115 \text{ V}$$

1. In a dc circuit, we would use Eq. 25.6:
$$W = VI = 115 \times 1.74 = 200 \text{ W}$$

2. In an ac circuit, we would use Eq. 25.7:
$$W = VI \times pf = 115 \times 1.74 \times 1.0 = 200 \text{ W}$$

3. In either a dc or an ac circuit, we can use Eq. 25.5:
$$W = I^2R = (1.74)^2 \times 66 = 200 \text{ W} \quad \blacksquare$$

**EXAMPLE 25.4** Using the data given in Example 25.2 and Fig. 25.4, determine (a) the power loss in the wiring and (b) the power input to the load.

**SOLUTION**
(a) The total power loss in the wiring is
$$W = I^2R = I^2(r_{bc} + r_{da})$$
$$= (10)^2 \times 0.04 = 4 \text{ W}$$

(b) The power taken by the load is
$$W = I^2R = I^2R_{cd} = (10)^2 \times 1.15$$
$$= 115 \text{ W (or 0.115 kW)}.$$

Alternatively, we can find this power by multiplying voltage and current. The voltage on the load is
$$IR = 10 \times 1.15 = 11.5 \text{ V}$$
and
$$W = VI = 11.5 \times 10 = 115 \text{ W (or 0.115 kW)}. \quad \blacksquare$$

**EXAMPLE 25.5** Refer to Fig. 25.7. Assume a 150-W incandescent lamp at each ceiling outlet. Also assume the load connected to one convenience outlet to be a 10-A hair dryer with a power factor of 0.80.

Calculate the current and power in the two branches of the circuit, and the total circuit current, assuming a 120-V ac source.

**SOLUTION**

(a) In the circuit branch feeding the lamps, the power consumption is for two 150-W lamps; that is,

$$P = 2 \ (150) = 300 \ \text{W}$$

To calculate the current, we would use Eq. 25.7, which expresses power in an ac circuit:

$$W = VI \times pf$$

However, because incandescent lamps are (effectively) resistive loads, their power factor is 1.0. Therefore:

$$\text{Power} = VI$$
$$300 \ \text{W} = 120 \ \text{V} \times I$$
$$I = \frac{300 \ \text{W}}{120 \ \text{V}} = 2.5 \ \text{A}$$

(b) In the second branch, we have a 10-A, 0.8-*pf* load.

$$\text{Power in watts} = \text{volts} \times \text{amperes}$$
$$\times \text{power factor}$$
$$W = 120 \times 10 \times 0.8 = 960 \ \text{W}$$

However, the branch volt-amperes are

$$V \times A = 120 \times 10 = 1200 \ \text{VA}$$

This latter figure is important for sizing electrical equipment.

(c) To calculate the total current flowing from the panel to both branches of the circuit, we must combine a purely resistive current (lamp circuit) with a reactive one (dryer circuit). The exact value of the current is the *vectorial* sum of the two branch currents, which calculates to be 12.1 amperes. (Vectorial addition is a technique that considers the phase angle between two currents. Phase angle, or phase difference, was explained previously.) In normal circuit design practice, the *arithmetic* sum of 12.5 amps would be used. The error introduced is only 3.2%, and because it results in slightly oversized circuit components, it is on the safe side. Only where power factors are very low would a careful designer use vectorial addition. ∎

One further example at this point demonstrates the importance of power factor in normal situations.

**EXAMPLE 25.6** The nameplate of a single-phase motor shows the following data: 3 hp, 240 V, ac, 17 A. Assume an efficiency of 90%. Calculate the motor (and therefore circuit) power factor.

**SOLUTION**

From Appendix A the conversion 1 hp = 746 W is found. Therefore,

$$3 \ \text{hp} = 3 \times 746 = 2238 \ \text{W}$$

This, of course, represents the motor output. Because we know that for any device

$$\text{efficiency} = \frac{\text{output}}{\text{input}}$$

we have

$$\text{power input} = \frac{2238 \ \text{W}}{0.9} = 2487 \ \text{W}$$

However, for ac,

$$\text{power} = \text{volts} \times \text{amperes} \times \text{power factor}$$

so

$$2487 \ \text{W} = 240 \ \text{V} \times 17 \ \text{A} \times \text{power factor}$$

and

$$\text{power factor} = \frac{2487}{240 \times 17} = 0.61$$

Note the large difference between volt-amperes and watts:

$$V \times I = 240 \times 17 = 4080 \ \text{VA}$$
$$P = \text{as above} = 2487 \ \text{W}$$

where *P* designates power. This difference is important in circuit design, as we will see when we study the methods for sizing circuit components. ∎

## 25.12 ENERGY IN ELECTRIC CIRCUITS

Because energy = power × time, the amount of energy used is directly proportional to both the power of a system and the length of time it is in operation. Because power is expressed in watts or kilowatts and time in hours (seconds and minutes are too small for practical use), the units of energy used are watt-hours (Wh) or kilowatt-hours (kWh).

ELECTRICITY

**EXAMPLE 25.7**

(a) Find the daily energy consumption of the appliances listed if they are used daily for the length of time shown.

| | |
|---|---|
| Toaster (1340 W) | 15 min |
| Percolator (500 W) | 2 h |
| Fryer (1560 W) | ½ h |
| Iron (1400 W) | ½ h |

(b) Assuming that the average cost of energy is $0.12 per kilowatt-hour, find the daily operating cost.

**SOLUTION**

| | |
|---|---|
| (a) Toaster: | 1340 W = 1.34 kW × ¼ h |
| | = 0.335 kWh |
| Percolator: | 500 W = 0.5 kW × 2 h |
| | = 1.00 kWh |
| Fryer: | 1560 W = 1.56 kW × ½ h |
| | = 0.78 kWh |
| Iron: | 1400 W = 1.4 kW × ½ h |
| | = 0.70 kWh |
| | Total   2.815 kWh |

(b) The daily operating cost is 2.815 kWh × $0.12/ kWh = $0.3378 (say, 34 cents).  ■

The power used by a residential household varies with the time of day. A graph showing the power used by a typical American household during a normal weekday might look something like the one in Fig. 25.12. The average power demand of the household is obviously much lower than the maximum. The ratio between the two is called the *overall load factor* and runs between 20% and 30% for a typical household. The energy used by this household for the 24-hour period shown is represented by the *area* under the curve of Fig. 25.12. This can be determined only by integration because it varies continuously. This integration is exactly what a kilowatt-hour meter does, as explained in Section 25.15, which deals with electrical measurements.

**EXAMPLE 25.8** It has been estimated that the average power demand of an American household with an electric stove is 1.8 kW. Calculate the monthly electric bill of such a household, assuming a flat rate of $0.12 per kilowatt-hour.

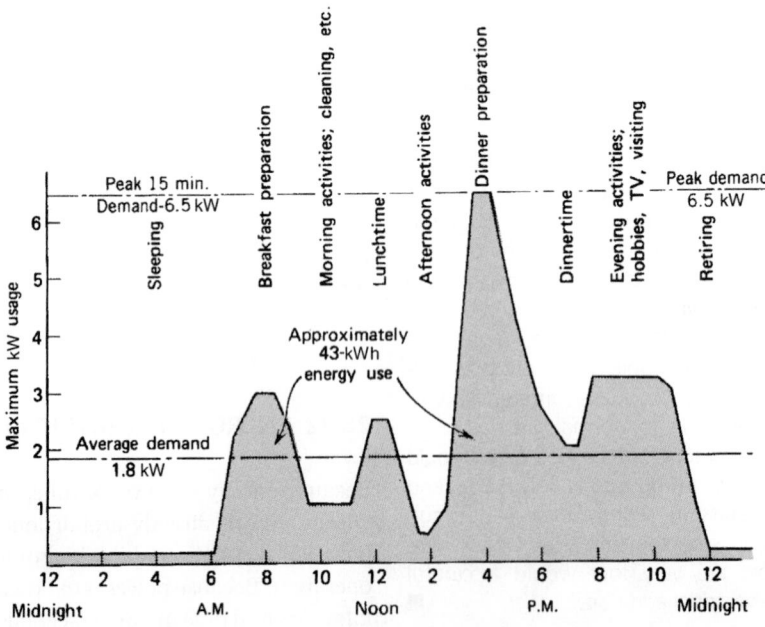

**Fig. 25.12** *Hypothetical graph of power usage for a typical U.S. household. Electric cooking is assumed. The area under the curve represents energy usage. Maximum kW demand (vertical axis) is based on a 15-minute integrated demand, thus eliminating spikes in demand, such as those caused by starting a refrigeration (air-conditioning) compressor. This curve has a 24-hour use of approximately 43 kWh, giving an average 24-hour demand of 1.8 kW. The ratio between this average demand and the peak demand of 6.5 kW is called the load factor, which here is 27.5%.*

## SOLUTION

Monthly energy use:

$$1.8 \text{ kW} \times \frac{24 \text{ h}}{\text{day}} \times \frac{30 \text{ days}}{\text{month}} = 1296 \text{ kWh per month}$$

Monthly electric power bill:

$$1296 \text{ kWh} \times \$0.12/\text{kWh} = \$155.52 \qquad \blacksquare$$

In Example 25.8 the bill was based on a *flat* rate of 12 cents per kWh because flat rates are very common for residential users. Such rates vary from a low of 5 to 6 cents per kWh in areas using hydroelectric power to as high as 17 cents per kWh in some major cities.

By the terms of their operating franchise, electric utilities must provide for a customer's *maximum* power demand, whereas their energy billing only compensates them for *average* demand, which is always lower. One technique used by electric utilities to adjust this condition is to levy a *demand charge* for power (kW) in addition to the normal energy (kWh) billing. This technique has long been the standard for industrial and commercial user rates, but it is still unusual for residential users. The demand charge is especially useful as a means of encouraging users to reduce their peak loads. In so doing, energy use is also somewhat reduced.

## 25.13 ELECTRIC DEMAND CHARGES

Electric utility companies normally levy a kW demand charge on all but individual residential and a few special category customers. Varying with the individual utility company involved, this monthly charge runs between $2 and $15 per kW of maximum average demand in any demand interval for that month. Demand intervals vary, usually being either 15 or 30 minutes (see Fig. 25.13).

Many utilities use a *sliding window* interval timing technique that starts a new interval every minute and updates the maximum interval demand accordingly. This enables them to find and bill for the maximum electric power demand in any 15- or 30-minute period in a month. Some companies also include a *ratchet* clause that levies a demand charge for a number of months based upon the maximum demand in any single month. This penalizes users with seasonal highs—that is, users with a low *yearly* load factor. The load factor is a measure of

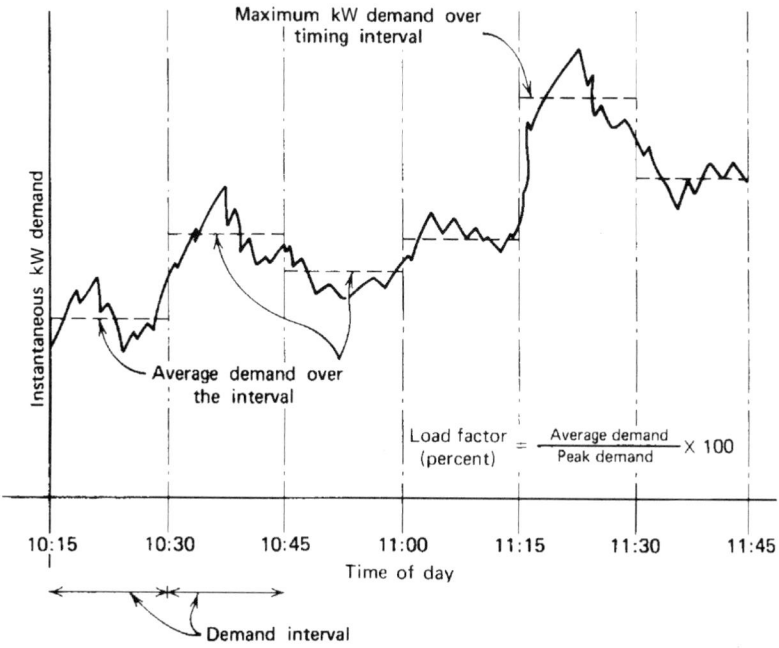

**Fig. 25.13** *Typical instantaneous load curve for a commercial facility. The utility demand meter records the average demand in each period (here, 15 minutes). The maximum interval demand—in this case, between 11:15 and 11:30—is used as the basis for monthly billing. A high load factor (utilization factor) indicates that little can be gained by demand control; a low load factor indicates the reverse.*

uniformity of power demand; a low load factor indicates short-time demand peaks for which the user is heavily charged. The justification for the imposition of a demand charge and the significance of load factors can best be demonstrated by an example.

Assume that a pottery manufacturer, whose average 8-hour daily load is 20 kW for lighting and pottery wheels, operates two 50-kW electric kilns twice a month for a 4-hour period each time. Further assume an energy charge of $0.10 per kWh. The total monthly *energy* bill for the operation of the two kilns only would be

$$\text{cost} = 2 \times 50 \text{ kW} \times 8 \text{ h} \times \frac{\$0.10}{\text{kWh}} = \$80$$

Thus, were it not for a demand charge, the utility company, which is required by law to supply the maximum customer demand, would have to provide and maintain 100 kW of generation, transmission, and distribution facilities in return for a payment that is the equivalent of 1.11 kW of average continuous load, that is,

$$\text{equivalent continuous load} = \frac{\$80.00}{720 \text{ hr/month} \times 0.10}$$
$$= 1.11 \text{ kW}$$

The user's load factor can be calculated readily. By definition,

$$\text{load factor} = \frac{\text{average power demand}}{\text{maximum power demand}} \quad (25.8)$$

For a given time interval (e.g., day, month, year) the average power demand is equal to the energy used in this period divided by the period length.

Average period power demand

$$= \frac{\text{kWh energy use}}{\text{hours of use}} \text{ for that time period.} \quad (25.9)$$

Therefore, the average *monthly* power demand equals the average monthly energy use divided by 720 hours. Substituting the monthly version of Eq. 25.9 as the general expression for the load factor (LF) in Eq. 25.8, we obtain

LF monthly

$$= \frac{\text{monthly kWh energy use} \div 720 \text{ hours}}{\text{maximum demand}} \quad (25.10)$$

For the case under consideration, the *monthly* load factor is

$$\text{LF} = \frac{\begin{array}{c}[23 \text{ days} \times 8 \text{ h} \times 20 \text{ kW} \\ + 2 \text{ days} \times 4 \text{ h} \times 100 \text{ kW}] \div 720 \text{ h}\end{array}}{120 \text{ kW}}$$

$$= \frac{(3680 + 800 \text{ kWh/month}) \div 720 \text{ h}}{120 \text{ kW}}$$

$$= 0.052 \text{ or } 5.2\%$$

This is a very poor load factor, which results in the levying of a high demand charge. Assuming an $8.00 per kW demand tariff, this pottery manufacturer would be billed, monthly, an additional

$$\text{demand charge} = 120 \text{ kW} \times \$8.00 = \$960.00$$

With an energy bill of only

$$\text{energy cost} = 4480 \text{ kWh} \times \$0.10 = \$448.00$$

this manufacturer is paying heavily for highly peaked power use.

Although the illustration selected is somewhat extreme in its type of power use, it is not uncommon to find demand charges of the same order of magnitude as energy charges. Obviously, it is impossible to eliminate demand charges entirely, but it is certainly possible and frequently very simple to reduce them. Such a step is in the interest of the user, the utility, and the public at large: the user—for economic reasons; the utility—to permit more efficient use of its facilities; and the general public—by avoiding unnecessary power plant construction and concomitant inefficient use of fuel at partial generator loading, and by overall reduction in fuel use. The last item is a secondary benefit of demand control.

The next section discusses user electric demand control, the primary function of which is to reduce electric *power* demand. This reduces demand charges and, incidentally (secondarily), somewhat reduces energy consumption. Electric demand control is different from *energy* management, which is primarily concerned with reduction of all types of energy use, including electricity. Electric demand control is frequently included as one part of an overall energy management system.

## 25.14 ELECTRIC DEMAND CONTROL

Electric demand control methods vary greatly in complexity and in degree of automation, but all basically perform the same task—enabling efficient utilization of available energy to produce a high load factor, resulting in a lowering of demand charges. An ancillary but important benefit is the improved utilization of building electrical power equipment, which normally runs underloaded. When demand control is incorporated during the design stage (instead of as a retrofit), it results in smaller equipment, a lower first cost, and less space allocated for equipment.

For a number of years beginning in the 1980s, many utilities offered their customers rebates that covered up to 40% of the cost of equipment and renovations that would reduce maximum demand *and* overall energy use. These programs, now essentially ended, acted as a clear financial incentive for the development of cost-effective demand control and energy conservation equipment and techniques that have since become widely adopted. The result has been a considerable reduction in nationwide per capita electric power and energy use. These rebates reduced the investment payback period to such an extent that many (perhaps most) large power users invested large sums in electric demand control and energy conservation and management equipment.

The advantage to the energy user is obvious: lower electric bills. The reasoning of the utility is equally simple. It costs $3500 to $5000 per kilowatt of generating capacity for new power plant construction (depending upon the location and required auxiliary construction), whereas rebates run from $150 to $1000 per kW conserved, with the larger amounts paid for by peak demand reduction. Because a utility must supply all the power demanded, it is very much in the interest of any utility that is generating power output near the maximum capacity of its equipment to reduce loads in general and peak loads in particular.

The oldest and simplest utility-sponsored demand control scheme, still in use today, is the time-of-day-dependent, variable utility rate schedule. This scheme encourages off-peak use of electricity by offering a lower energy rate for consumption during off-peak hours. Utilities offering off-peak rates generally install, at no charge to the user, a separate time-controlled circuit switch for use with time-deferrable loads such as water heaters, well pumps, battery chargers, and the like. The circuit is energized only during off-peak hours, which typically

are mid-afternoon and after midnight. One such arrangement is shown in Fig. 25.14. Note that the switch itself has no programming buttons. All programming is done by the utility with an auxiliary programmer (shown) that can be used to program many user switches.

The basic technique of *user* demand control is simple; electric loads are disconnected and reconnected in such a fashion that demand peaks are leveled off and the load factor is thereby improved. The extent to which a facility's electric loads can tolerate this type of switching is an indication of the potential effectiveness of a demand controller. An installation with a large *uninterruptible* load, such as from computers or office lighting, benefits minimally from demand control. Most industrial and commercial installations, however, contain a large percentage of interruptible loads (interruptions may be very short), and demand control systems frequently accomplish a 15% to 20% reduction in electric bills with a resultant short payback period on equipment investment.

The proliferation of demand control equipment has also produced a proliferation of nomenclature, including *load shedding control, automated load control, peak demand control,* and *programmable load control.* Descriptions that include the term *energy management* refer to devices whose primary

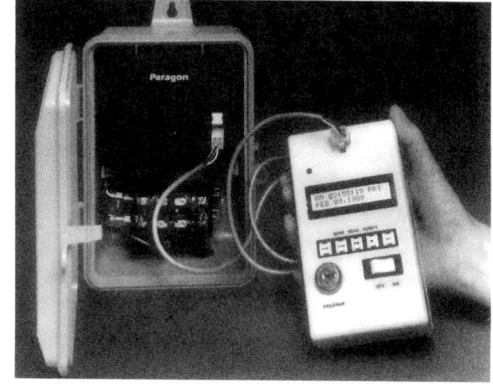

**Fig. 25.14** *Programmable electronic time control switch designed for time-of-use (off-peak) utility rate schedules. The programming device is separate from the switch and can be used by a utility to program many customer switches. The illustrated unit is arranged for 365-day scheduling, which permits full coordination with utility schedules that vary with the seasons of the year. Typical controlled loads include water heaters, thermal storage units, water and air accumulators, and any other load that either inherently or by design can be delayed for several hours. (Courtesy of Paragon Electric. Co., Inc.)*

function is the control of *energy* and that secondarily are equipped to provide electric demand control.

Demand control devices are intended to control *power*, which is *timed* energy use. Demand control produces energy savings as a secondary benefit. The expression *electric demand control* is used for simplicity in the following discussion (which also avoids knowingly using a term that refers to a particular manufacturer's equipment).

The various demand control schemes are discussed in some detail to enable the reader to differentiate them by recognizing their specific characteristics. A manufacturer's descriptive literature usually comprises a lengthy list of its equipment's abilities, none of its shortcomings, and little of its specific or comparative applicability. Furthermore, much equipment on the market today provides additional functions not directly related to demand control. That being so, a good understanding of the essentials of demand control schemes will greatly assist the prospective user.

### (a) Level 1—Load Scheduling and Duty-Cycle Control

This level is the simplest and most obvious, and is applicable to all types of facilities. The installation's electric loads are analyzed and then scheduled to restrict demand. Accordingly, large loads can be shifted to off-peak hours and controlled to avoid coincident operation. The user can also take advantage of special night and weekend utility rates for loads that do not require a specific time of operation, such as battery charging and transfer pumping. The demand control device used is essentially a programmable time switch (see Chapter 26) for a number of circuits or loads. It is not, strictly speaking, a demand controller in that no cognizance is taken of the actual continuous electric load. Instead, it operates on a preset timed duty cycle relying entirely on a prior analysis of the building loads. Typical applications of this device are control of HVAC loads, lighting loads, and process loads in small commercial, institutional, and industrial buildings.

### (b) Level 2—Demand Metering Alarm

If, in conjunction with a duty-cycle controller, some type of continuous demand metering is installed that goes into alarm mode when a predetermined demand level is exceeded, a basic load control system will have been established. The load analysis discussed previously would have to be extended to determine load priorities so that when the preset maximum demand load is exceeded and the alarm sounds, loads can be shed (disconnected) manually in a preestablished order of priority and subsequently reconnected, also in order of priority. This type of control is practical only for a limited size installation, inasmuch as the load switching activity is manual.

### (c) Level 3—Automatic Instantaneous Demand Control

This type of control (also called *rate control*) is, in effect, an automated version of the level 2 system described above. The unit accepts instantaneous kilowatt load information from the utility system, compares this information to the preset kilowatt limit, and acts automatically to disconnect and reconnect loads as required. These units *do not* recognize the utility's metering interval but act continuously on the basis of load comparison data. For this reason, these units are also referred to as *load comparator controllers*. Figure 25.15 will help explain the unit's operation.

Typical sheddable loads might include nonessential lighting, some cooling load, domestic hot water heating, snow melting, and the like. Nonsheddable loads (that do not tolerate even short interruption) might include essential lighting, elevators, communications equipment, computers, process control, emergency equipment, and the like. The nonsheddable loads are fed directly from the power line. The sheddable loads are fed through a panel of control relays that respond to on/off instructions from the demand controller. Although the resulting energy use with or without the controller is theoretically the same, in practice energy savings of 15% and more are common.

The principal drawback of this system is that the load-shedding is preprogrammed. This results in an inability to readily adapt to varying load patterns resulting from variable production schedules, time schedules, changes in weather, and so on. As a result of this limitation, this system is most useful in applications where operating modes do not change frequently and the facility is not very large. Thus, stores, supermarkets, warehouses, small industrial

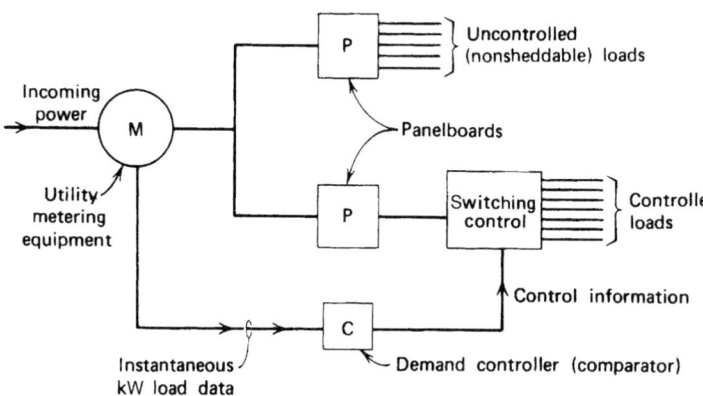

*Fig. 25.15* Block diagram of a system for automatic electric power demand control. The demand controller receives instantaneous load data from the metering equipment, compares them to preset limits, and disconnects and reconnects controllable loads automatically to keep kW demand (load) within these limits.

facilities, and commercial installations are well served with this type system if they have at least 20% sheddable loads and their connected electric load is at least 150 kilovolt-amperes (kVA).

This level of demand control, as well as levels 4 and 5 described later, is most often supplied as one of many functions of a larger building energy control system. Because this is so, it is necessary to ascertain that the type of demand control to be supplied by the overall building control hardware and software is what the electrical system designer intended.

### (d) Level 4—Ideal Curve Control

This control function operates by comparing the actual rate of *energy* usage to an ideal rate and controlling kilowatt demand by adjusting the total *energy* used within a metering interval. The utility company determines the demand over the demand interval by integrating the kilowatt-hour energy over the interval and dividing by the interval time. Thus, the user is actually given a block of energy (kWh) that can be utilized at any desired rate, not necessarily at a constant rate. The desirable rate of energy use is shown as the "ideal curve" on the typical usage curve shown in Fig. 25.16.

Shed points can be programmed independently for each load according to a predetermined priority, and priorities can be readily adjusted and rescheduled. Loads are normally shed only toward the end of an interval—when the permissible energy total for the interval is approached—and all loads are

*Fig. 25.16* Graph of the action of an ideal rate controller. Note that toward the latter part of the period the pattern has been established, and the actual rate of energy use corresponds closely to the ideal. This type of control is often simply a part of the software in a large, centralized building control system.

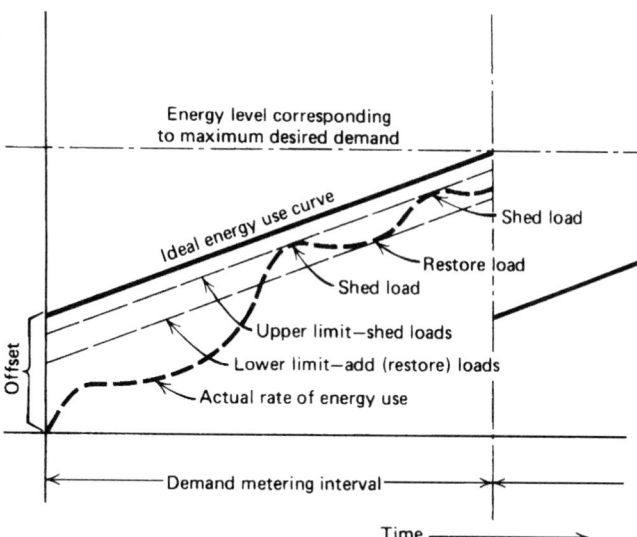

ELECTRICITY

restored at the beginning of an interval. Thus, during each interval, sheddable loads are off for only a few minutes at most. Controllers operating on the ideal curve principle are considerably more flexible than the kilowatt rate controller described in the level 3 system and are applicable to facilities of widely divergent load size, but with at least a 300-kW connected load. As with other controllers, the principal savings will be in demand charges, but almost always with considerable economies in energy billings.

### (e) Level 5—Forecasting Systems

These systems are by far the most sophisticated, the most expensive, and the most effective. They are best applied to large facilities where the number of loads, load patterns, and complexity of operation preclude the effective use of the preceding systems. Because of the large amount of load data that needs to be processed, these systems are usually installed as part of a computerized central control system. Details of operation are too complex to be described here, but the basic operation can be outlined. These units operate by continuously forecasting the amount of energy remaining in the demand interval based upon kilowatt-hour data received. They then examine the status and priority of each of the connected loads and decide on a course of action. Loads that in other systems are classified as nonsheddable are, in this system, controlled

because of the accuracy and rapidity of the control function. A pneumatic compressor, for instance, that supplies process air might, in lower-level control systems, be classified as noncontrollable, despite the fact that it has long off periods. With computer control, such a load would be classified as "delayable" or "inhibited" because a 30-second or 1-minute delay in activation after the pressure switch closes its contact is normally acceptable.

The advantage of these systems is that, if programmed properly, they can make small, accurate load changes throughout an interval, resulting in minimum load cycling and maximum efficiency.

## 25.15 ELECTRICAL MEASUREMENTS

Preceding sections have explained the fundamental electric quantities of voltage and current and defined the units involved as volts and amperes, respectively. As is true of all other physical quantities that are to be used in practical applications, a need exists for a simple means of measuring these quantities. This need was initially met by the development of the meter movement illustrated in Fig. 25.17.

Everyone at one time or another has felt the repulsion between like poles of two magnets held close together and, conversely, the attraction between opposite poles. This principle was used in the first basic meter movement: causing a deflection of the pointer as a result of the repulsion between the

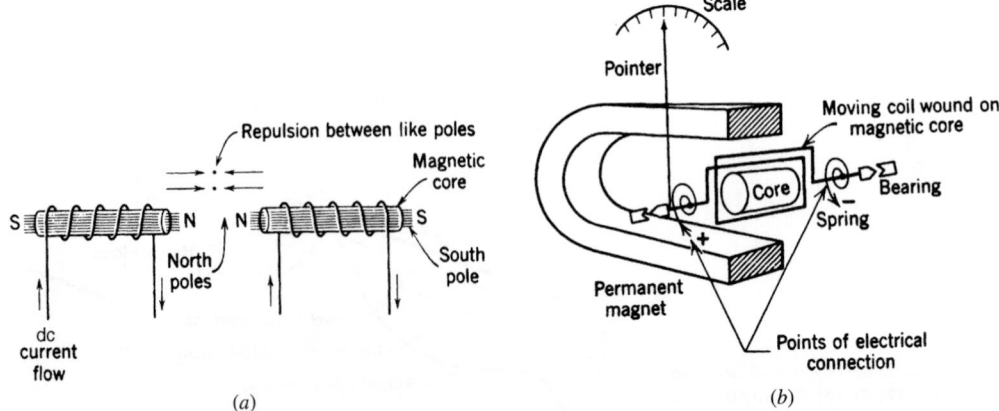

**Fig. 25.17** (a) Diagram showing the basic electromagnetic principle and interaction between electromagnets. Any iron core becomes an electromagnet when current flows in a coil wound around it, as shown. (b) The principle of the electromagnet is used in this basic meter movement. Current flowing in the movable coil forms an electromagnet whose field interacts (see a) with the permanent magnet's field, causing a deflection proportional to the current flow. (Courtesy of Wechsler, a division of Hughes Corp.)

field of a permanent magnet and an electromagnet. The electromagnet is formed when current flows in the coil, and its strength is proportional to the amount of measured current. Thus, a strong current causes a larger deflection of the needle and therefore a higher reading on the dial. A spring (see Fig. 25.17*b*) provides restraining torque on the pointer. To make this very sensitive basic unit usable for large currents (it is intrinsically a micro-ammeter, sensitive to millionths of an ampere), the device simply diverts, or *shunts* away, most of the current, allowing only a few microamperes to actually flow in the meter coil.

To use the same unit as a voltmeter, a large resistance called a *multiplier* is placed in series with the meter, thereby again limiting the current flowing to a few microamperes. The meter scale is then calibrated in volts. All dc meters are made in this fashion. Most ac meters operate on basically the same principle, except that instead of a permanent magnet, an electromagnet is used. Thus, when the polarity reverses, the deflecting force remains in the same direction. A dc meter connected to an ac circuit simply will not read because inertia prevents the needle from bouncing up and down 60 times a second.

The meters described above are conventional analog devices that read electrical units in proportion to mechanical forces exerted within the device—that is, by analogy. Modern electronics has produced a line of solid-state electrical meters (Fig. 25.18) that display the measured electrical values in analog mode (needle on a dial), digital mode, or both. They operate in a number of ways, all of which are different from that described previously and all of which are beyond the scope of this book.

The fact that a meter reads digitally does not necessarily mean that it is highly accurate. Accuracy depends upon the quality of the internal circuitry. Digital meters are obviously easier to read than analog meters because no visual interpretation or interpolation is involved. This advantage, plus the constantly declining cost of sophisticated electronics, will undoubtedly lead to solid-state digital meters replacing analog types, except for special applications.

In buildings, the measurement of current and voltage is generally not as important as the measurement of power and energy. To measure power, the fact that power is proportional to the product

(a)                    (b)

**Fig. 25.18** *(a) Solid-state clamp-on type meter with digital readout and automatic ranging. The latter feature eliminates the necessity to preselect a meter range and is particularly useful where the magnitude of current or voltage is unknown. The clamp-on feature permits use without wiring into, or otherwise disturbing, the circuit being measured. The meter measures approximately 8 × 3 × 1.5 in. (200 × 75 × 40 mm), weighs less than 1 lb (0.5 kg), and is battery-powered. Scales are 0.1–1000 amp ac, 0.1–1000 V ac, and 0.1–1000 ohms resistance, with ±2% accuracy. (b) Solid-state auto-ranging clamp-on ac meter with analog-type readout. It is similar in design to the meter in (a), but with somewhat larger range scales and additional features such as peak current measurement. (Photos courtesy of TIF Instruments, Inc.)*

of the voltage and current in a circuit is employed. Although actual construction is complex, the theory of operation of a conventional coil-type wattmeter is simple. The meter has two coils: a current coil that is similar in connection to an ammeter and a voltage coil that is similar in connection to a voltmeter (Fig. 25.19). By means of the physical

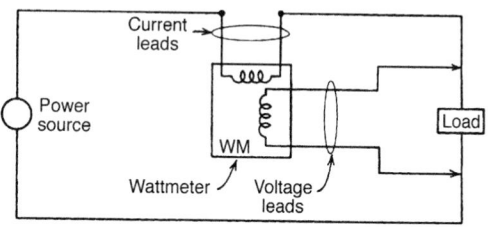

**Fig. 25.19** *Schematic arrangement of wattmeter connections. Note that the current coil is in series with the circuit load, whereas the voltage leads are in parallel.*

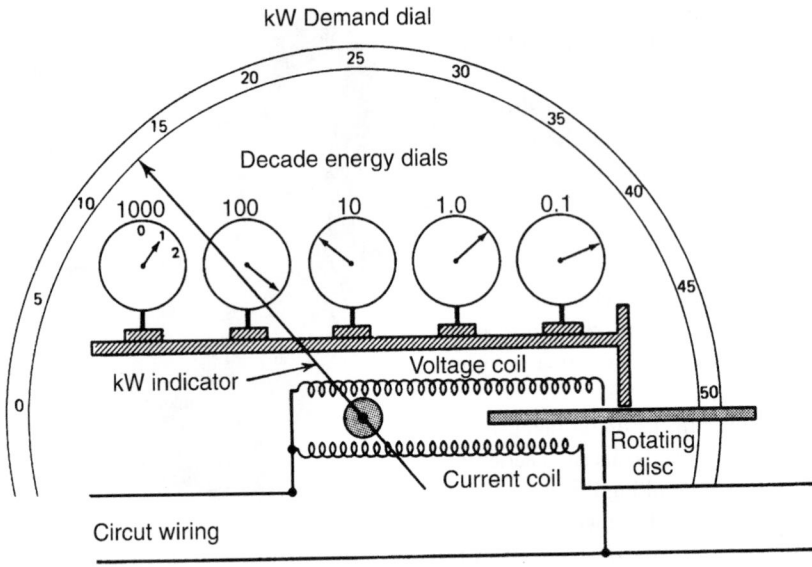

**Fig. 25.20** *Typical induction-type kilowatt-hour meter with kilowatt demand dial. Dials register total disc revolutions that are proportional to energy. Disc rotational speed is proportional to power. Note that the current coil is in series with the load and that the voltage coil is in parallel.*

(a)                                                                                  (b)

**Fig. 25.21** *(a) A modern multifunction solid-state kWh meter. This single-phase meter includes a mechanical register similar to that shown in Fig. 25.20 plus a large liquid crystal display. It can be arranged to provide either time-of-use billing or demand/time-of-use billing in addition to its energy measurement function. It can also be configured to provide load control, demand threshold alert, end-of-interval alert, and load profile recording. (b) Field programming of the meter shown in (a) is accomplished with a hand-held programming device. (Photos courtesy of Landis and Gyr.)*

coil arrangement, the meter deflection is proportional to the product of the two and therefore to the circuit power. The meter can be calibrated as desired, depending upon the size of the shunts and multipliers.

To measure energy, the factor of time must be introduced because

$$\text{energy} = \text{power} \times \text{time}.$$

Direct current energy meters are available but are not of general interest because of the rarity of dc power. Alternating current watt-hour meters are basically small motors, whose speed is proportional to the power being used. The number of rotations is counted on the dials, which are calibrated directly in kilowatt-hours. A diagram of the basic construction of an ac kilowatt-hour meter is shown in Fig. 25.20. As can be seen, kilowatt-hour energy consumption and maximum interval kilowatt demand can be read directly from the dials. If the numbers involved are too large (because of calibration), a multiplying factor is required to arrive at the proper kilowatt-hour consumption. This number is written directly on the meter nameplate, and the meter reading is multiplied by it to get the actual kilowatt-hours. Several types of solid-state multifunction

**Fig. 25.23** *This portable, programmable power analyzer unit measures approximately 13 × 7 × 12 in. (330 × 180 × 305 mm) and weighs 6 lb (2.7 kg). It is capable of measuring (and computing) 19 electrical parameters, among which are current, voltage, energy, power, power factor, and frequency, and of providing both an instantaneous digital readout and printed records (hard copy). Connections permit all functions to be addressed by remote control. (Courtesy of AEMC Corporation.)*

and energy meters are shown in Figs. 25.21, 25.22, and 25.23.

Continuous monitoring of the electrical characteristics of an entire distribution system, or of a portion fed from a specific switchboard, can be

**Fig. 25.22** *Digital power factor meter, which measures the power factor on single- and 3-phase circuits. In addition, the meter can measure 0–600 V ac and 0–2000 A ac. Connections to the circuits being measured are made with the clamp-on probes shown. A similar unit is available that measures power in both sinusoidal and nonsinusoidal waveforms in the range 0.1 W through 2000 kW. (Courtesy of AEMC Corporation.)*

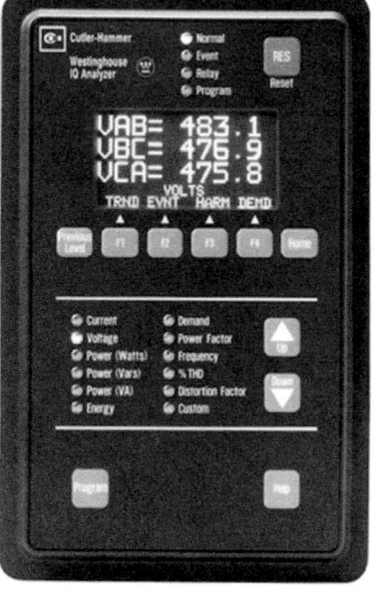

**Fig. 25.24** *Power system analysis unit designed to be mounted permanently into a system switchboard (see Chapter 26). This programmable unit can meter circuit parameters, measure energy and power use, analyze power quality (including harmonic analysis), and log events. (Photo courtesy of Cutler-Hammer.)*

readily accomplished by use of the solid-state device shown in Fig. 25.24. This programmable analyzer can be arranged to measure power usage, perform power quality and harmonic analyses (and the like), and display the results locally and remotely.

Because manual reading of kilowatt-hour meters of individual consumers is so labor-intensive a task—even without the routine job difficulties such as inaccessible meters, unfriendly dogs, inclement weather, and so on—utilities have long sought a more cost-effective means of determining consumer energy usage. To this end, with the benefit of modern microelectronics, a number of interesting, labor-saving kilowatt-hour meters have been developed. One type is equipped with a programmable electro-optical automatic meter reading system that can be activated locally or from a remote location. The meter data are transmitted electrically to a data-processing center, where they may be used by the utility to prepare subscribers' bills, prepare customer load profiles, and study, in combination with other such data, area load patterns, equipment loadings, and so on. The customer can use instantaneous data to control loads, as explained in the preceding section.

Another meter type is equipped with a miniature radio transmitter that can be remotely activated to transmit the current kilowatt-hour reading. The meter reader moves along a street and remotely activates a meter by entering a customer code into a digital pad. A special receiver not only receives the transmitted kilowatt-hour data but also encodes and records the data automatically. These and other new kilowatt-hour meters are obviously much more costly than the traditional units, but they are gradually being introduced because of large reductions in labor costs and increases in the quantity and quality of data made available.

ELECTRICITY

# CHAPTER 26

# Electrical Systems and Materials: Service and Utilization

THE FIRST STEP IN UNDERSTANDING building electrical systems is to examine the means by which electric service is brought into a structure.

## 26.1 ELECTRIC SERVICE

The codes and standards that apply to electric service include:

1. *National Electrical Code* © National Fire Protection Association; in particular, Section 230.
2. *National Electrical Safety Code,* published by the Institute of Electrical and Electronic Engineers (IEEE), 345 E. 47th Street, New York, NY 10017. This code is recognized as an American National Standard (ANSI). It deals with clearances for overhead lines, grounding methods, underground construction, and related matters.
3. Standards of the utility supplying electric service.

Public utility franchises require only that service be made available at the private property line. Thus, service is normally tapped onto the utility lines at a mutually agreeable point at or beyond the property line. The service tap may be a connection on a pole with an *overhead service drop* or an *underground service lateral* to the building, or a connection to an underground utility line with a service lateral to the building. Electrical construction work on private property is usually at the owner's expense.

Under certain conditions, the owner can influence the type of construction utilized by the electric utility company in conveying the electric service to the site. This is often the case in large tract developments and in places where owners are willing to share some of the cost of better grade construction. Also, in many areas, the utilities themselves have instituted "beautification" programs in an effort to decrease the objectionable appearance of much of their equipment.

Service from a utility line to a building may be run overhead or underground, depending on the following factors:

- Length of the service run
- Type of terrain
- Customer participation in the cost of service installation
- Service voltage
- Size and nature of the electric load
- Importance of appearance
- Local practices and ordinances
- Maintenance and service reliability
- Weather conditions
- Type of interbuilding distribution, if applicable

## 26.2 OVERHEAD SERVICE

The principal advantage (to the utility) of overhead electric lines is low cost. Depending on terrain and other factors, the cost saving of overhead compared to underground installation has ranged from 10% to 50% (the latter when compared to direct burial cable installation). This accounts for the majority of installations being overhead. In recent years, special techniques in underground installation have lowered that cost, making it a reasonable economic alternative when its advantages are considered.

Where the length of the service run is several hundred feet or more, voltages higher than the facility's utilization level may be involved. This weighs heavily in favor of overhead lines, particularly with voltages exceeding 5000 V. Similarly, when terrain is rocky and the electrical load is heavy, the cost of underground installation rises sharply. Because overhead lines are easily maintained and repaired and faults easily located, service continuity with overhead lines is generally acceptable. In areas with severe weather conditions, called *heavy loading areas*, where combinations of snow, wind, and ice increase the possibility of outages on overhead lines, underground service is preferable. This is particularly true when even short service outages cause hardship or financial loss. Reliability of overhead service can be improved markedly by taking service from two separate, and preferably separated, overhead lines. A final decision on the type and location of electrical service will be made after meetings among the building's architect, the electrical engineer, and the electric utility's technical personnel.

Overhead cables are of several types: bare, weatherproof, or preassembled aerial cable. Bare copper cables supported on porcelain or glass insulators on crossarms are normally used for high-voltage lines (2.4 kV and higher). Low-voltage circuits (600 V and below) are generally run on porcelain spool secondary racks using single-conductor (1/c) weatherproof cable. Pre-assembled aerial cable consists of three or four insulated cables wrapped together with a metallic tape and suspended by hooks from poles. This type of construction may be used for voltages of up to 15 kV (Fig. 26.1). It often proves to be more economical than crossarm or rack installation and more resistant to damage from severe weather conditions.

A typical detail of an overhead electric service entrance to a multiresidence building is shown in Fig. 26.2.

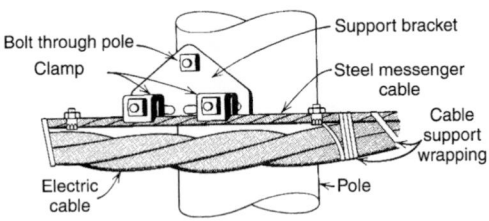

**Fig. 26.1** *Preassembled aerial messenger cables are carried by steel cables clamped to poles.*

## 26.3 UNDERGROUND SERVICE

The advantages of underground electric service are attractiveness (lack of overhead physical and visual clutter), service reliability, and long life. The principal disadvantage is high cost. To overcome this, utilities frequently use direct burial techniques that, by eliminating the raceway, reduce costs considerably. Because direct buried cable cannot be pulled out if it faults, as is the case with raceway-installed cable, restoration of service after a cable fault is time-consuming. It is recommended that the decision on which technique is to be used be based on the consideration of these factors:

- The cost premium for underground raceway installation, including handholes if required (see Section 26.4)
- The history of outages for direct burial installation by this installer, in the immediate area
- Cost and availability of repair service (utilities frequently will repair customer-owned underground service laterals *for a fee*).
- Impact of electric service outage in terms of time delays, inconvenience, necessity to dig up lawns and paved areas, and cost impact in the case of a commercial facility

## 26.4 UNDERGROUND WIRING

The methods available for underground wiring are:

- Direct burial (Fig. 26.3)
- Installation in Type I, concrete-encased duct (Fig. 26.4*a*)
- Installation in Type II, direct burial duct (Fig. 26.4*b*)

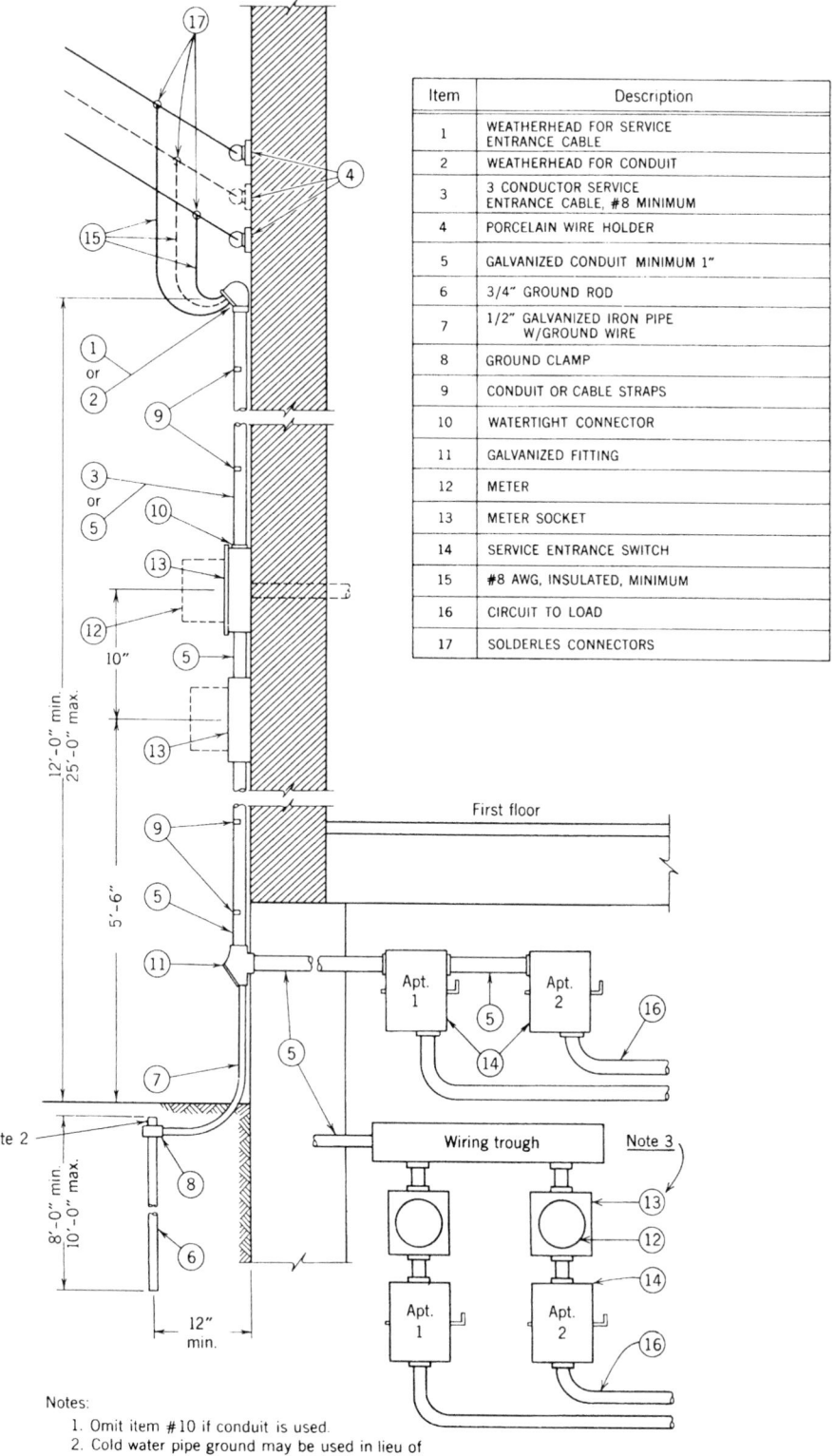

| Item | Description |
|------|-------------|
| 1 | WEATHERHEAD FOR SERVICE ENTRANCE CABLE |
| 2 | WEATHERHEAD FOR CONDUIT |
| 3 | 3 CONDUCTOR SERVICE ENTRANCE CABLE, #8 MINIMUM |
| 4 | PORCELAIN WIRE HOLDER |
| 5 | GALVANIZED CONDUIT MINIMUM 1" |
| 6 | 3/4" GROUND ROD |
| 7 | 1/2" GALVANIZED IRON PIPE W/GROUND WIRE |
| 8 | GROUND CLAMP |
| 9 | CONDUIT OR CABLE STRAPS |
| 10 | WATERTIGHT CONNECTOR |
| 11 | GALVANIZED FITTING |
| 12 | METER |
| 13 | METER SOCKET |
| 14 | SERVICE ENTRANCE SWITCH |
| 15 | #8 AWG, INSULATED, MINIMUM |
| 16 | CIRCUIT TO LOAD |
| 17 | SOLDERLES CONNECTORS |

Notes:
1. Omit item #10 if conduit is used.
2. Cold water pipe ground may be used in lieu of ground rod.
3. Meters may alternatively be placed inside the building.

**Fig. 26.2** *Detail of typical overhead electric service to a four-family residence. Note that meters in small residential buildings are frequently placed on the exterior of the building. Alternatively, they can be installed inside, provided that ready access is available or that some type of remote meter reading system is installed.*

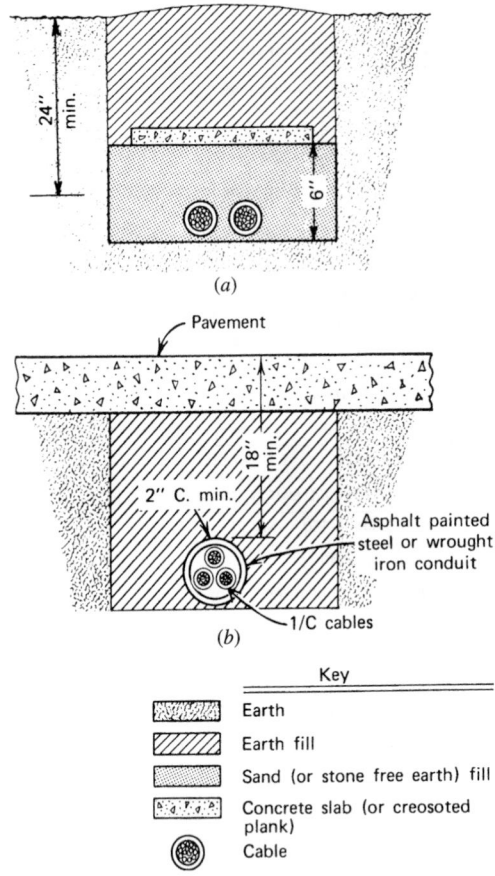

(a)

(b)

**Key**

Earth

Earth fill

Sand (or stone free earth) fill

Concrete slab (or creosoted plank)

Cable

**Fig. 26.3** (a) Technique for installation of direct-burial cable. (b) Under highways, streets, and other high-load areas, cable should be installed in metal conduit.

The first alternative offers low cost and ease of installation, with the disadvantage stated previously. The second offers high strength and permanence, but at the highest price of the three. The last offers median cost but little strength. It is applicable only for installations on undisturbed earth and/or under light paving.

Nonmetallic duct (conduit) intended for underground electrical use is commercially available in two wall thicknesses. NEMA (National Electrical Manufacturers Association) Type II with a heavy wall provides the physical protection required for direct burial installation with no concrete encasement. Type I is manufactured with a thinner wall and is intended for encasement in a minimum of 2 in. of concrete. Common trade names for asbestos-cement and fiber

Installing Type I Nonmetallic Underground Duct

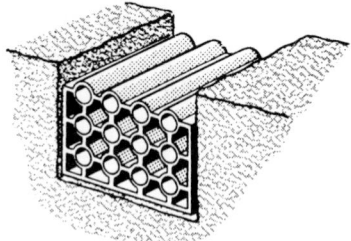

Duct tiers with separators

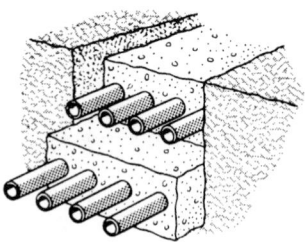

Tier-by-tier method in concrete

(a)

Installing Type II Nonmetallic Underground Duct

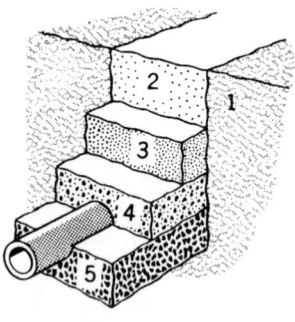

1.  Trench wall
2.  Ordinary backfill
3.  Selected backfill
4.  Selected backfill
5.  Bedding

(b)

**Fig. 26.4** Underground duct installation. (a) The concrete duct bank should have at least 6 in. (150 mm) of earth cover. (b) Heavy-wall duct should be buried at least 24 in. (600 mm) in ordinary traffic areas and 36 in. (900 mm) in areas subject to heavy traffic. Each layer is about 8 in. (200 mm) thick.

**ELECTRICITY**

ducts are Transite and Orangeburg. Plastic conduit is referred to as *PVC* or simply as *plastic.*

Nonmetallic conduit is most frequently used without concrete encasement for low-voltage and signal wiring and with encasement for high-voltage wiring. It offers several advantages over steel conduit for underground work, such as lower cost and freedom from corrosion.

When underground electric wiring is duct installed and the run extends over several hundred feet (the exact distance depending on the pulling tension), a pulling handhole or manhole is necessary. Handholes are used for low-voltage power and signal cables and for runs with a small number of cables. Manholes are used for high-voltage cables and where large duct banks must be accommodated. Precast handholes and manholes are readily available in most standard sizes and are usually cheaper than field-formed and poured units.

Cable used in underground wiring must be specially manufactured and approved for that purpose. Type SE is the basic service entrance cable, constructed with moisture- and flame-resistant covering. When it is provided with moistureproofing for underground use, the designation is SE type U, or simply USE. Underground cable for other than service runs is classified as type UF (underground feeder).

## 26.5 SERVICE EQUIPMENT

Referring to Fig. 27.1, note that interposed between the high-voltage incoming utility lines and the secondary service conductors is a block labeled "transformer." This equipment item is required whenever the building utilization voltage is different from the service voltage. It may be pole- or pad-mounted outside the building or installed in a room or vault inside the building, as discussed later.

## 26.6 TRANSFORMERS

A transformer is a device that changes or *transforms* alternating current (ac) of one voltage to alternating current of another voltage. Transformers used in building work consist essentially of an iron core on which are wound at least two coils: a primary (coil) winding and a secondary (coil) winding. A

voltage impressed on the primary winding induces (through the iron core) a voltage in the secondary winding in proportion to the ratio of turns in the two coils. Thus, a *step-down* transformer has a larger number of turns in its primary winding than in its secondary winding, and a *step-up* transformer has the reverse. In theory transformers are reversible, although in practice they are rarely used as such. Transformers cannot be used on dc.

A transformer typically is used to step down an incoming 4160-V service to 480-V distribution within a building. Another transformer is then used to step down the 480 V to 120 V for use on receptacle circuits. It is well to remember that ordinarily 120, 208, 240, 277, and 480 V are called *low* or *secondary voltages*, and 2400, 4160, 7200, 12,470 and 13,200 V are *high* or *primary voltages.*

Transformers are available in single-phase or three-phase construction. Transformer power capacity is rated in kilovolt-amperes (kVA). For a single-phase unit, this figure is the product of the full load current and the voltage. Because the voltages on the primary and secondary are different, so are the primary and secondary currents—because the kVA remains constant.

For example, a single-phase 100-kVA, 2400/120-V transformer will carry at full load

$$\text{primary current} = \frac{100,000 \text{ VA}}{2400 \text{ V}} = 41.6 \text{ A}$$

$$\text{secondary current} = \frac{100,000 \text{ VA}}{120 \text{ V}} = 832 \text{ A}$$

That is, the ratio of primary to secondary current is exactly the reverse of the ratio of primary to secondary voltage, because their product, which is the transformer's kVA rating, remains constant. Expressed in an equation,

primary current × primary voltage
= secondary current × secondary voltage

or

$$V_p I_p = V_s I_s \qquad (26.1)$$

where

$V_p$ is primary voltage
$I_p$ is primary current
$V_s$ is secondary voltage
$I_s$ is secondary current

ELECTRICITY

Heat is generated by the passage of current through the transformer coils due to the winding cable resistance. This heat is transferred to the unit's *cooling medium,* where it is radiated or otherwise disposed of. The unit's cooling medium is a characteristic of major importance. Transformers are either dry (air-cooled) or liquid filled. The choice depends upon the required electrical characteristics, the proposed physical location, and cost factors. Although detailed considerations are beyond the scope of this book, some selection criteria are presented in the following sections. In general, units rated above 5 kV are liquid-filled and units in the 600-V class are dry. Units installed indoors, except in vaults, are normally of the dry type, and are intended for general-purpose light and power circuits (Table 26.1). *Load center* transformers are installed in unit substations, both indoor and outdoor. *Distribution* transformers are mounted on a pole or on a concrete pad outdoors. *Substation* transformers are large and are always concrete-pad mounted.

The *insulation class* of a transformer affects its permissible temperature rise and operating temperature, and, as a direct result, its physical size, electrical power losses, overload capacity, and life.

The physical size of a transformer of a given kVA rating and voltage depends on the type of insulation used. In order of decreasing physical size and increasing operating temperature we have, for dry transformers, 105, 150, 185, and 220°C (221, 302, 365, and 428°F) systems that represent organic, inorganic (two types), and silicone insulating materials, respectively (Table 26.2).

The dimensional data in Table 26.3 vary considerably from one manufacturer to another. Therefore, the table is useful only for a concept of bulk volume and weight.

Although 220°C (428°F) system insulation transformers can withstand a 150°C (270°F) rise (see Table 26.2), some users specify 220°C (428°F) system insulation with a 115°C (207°F) rise or even a 80°C (144°F) rise; that is, a better grade of insulation is used with an *underrated* transformer. The advantages of this design are four:

• Longer life
• Higher overload capacity
• Lower operating cost
• Increased capability to withstand the heating effects of harmonics commonly found in commercial and industrial electrical systems (see Section 28.18)

A 220°C (428°F) system transformer operated at full load *continuously* (an unusual situation) has a short life—estimated on the basis of accelerated aging tests to be between 3 and 10 years. The same class insulation transformer (220°C [428°F] system) rated at 80°C (144°F) has a life expectancy of more than 100 years. With respect to operating cost, the same situations obtain here as with the high-temperature insulation discussed in Section 27.15 and studied in Table 27.5. The truism that one gets nothing for nothing can be expressed here as follows: in return for the smaller, lighter, cheaper, and hotter transformer (220°C [428°F] system), we

## TABLE 26.1 Typical Transformer Data

| Transformer Type (Application) | Maximum Capacity[a] (kVA) | Insulating (Cooling) Medium | Voltage Range | |
|---|---|---|---|---|
| | | | Primary | Secondary |
| General-purpose, dry type | 1000 | Air | 120–600 V | 120–600 V |
| Load center pad-mounted unit substation | 3000 | Air | 2.5–15 kV | 120–600 V |
| Load center pad-mounted unit substation | 5000 | Oil Silicone Fluid[b] Epoxy/resin[c] | 2.5–34.5 kV | 120–4800 V |
| Distribution system | 10,000 | Oil Silicone | 2.5–67 kV | 120–4800 V |

[a]Three-phase bank.
[b]High-fire-point paraffinic hydrocarbon fluid, manufactured expressly as a transformer dielectric-coolant.
[c]Solid dielectrics used for special installation, such as in high-hazard areas.

**TABLE 26.2 Air-Cooled Transformer Electrical Insulation Temperature Ratings (Based on 40°C Ambient)**

| Insulation Class (System)[a] | Insulation Type | Average Conductor Temperature Rise | Ambient Temperature | Hot-Spot Temperature Gradient | Total Maximum Temperature |
|---|---|---|---|---|---|
| 105°C | Organic (A) | 55°C | 40°C | 10°C | 105°C |
| 150°C | Mica, glass, resins (B) | 80°C | 40°C | 30°C | 150°C |
| 185°C | Asbestos (F) | 115°C | 40°C | 30°C | 185°C |
| 220°C | Silicones (H) | 150°C | 40°C | 30°C | 220°C |

[a]Modern terminology for insulation class uses *system* in lieu of *class*.

have higher losses, and therefore a higher operating cost and a shorter life.

It is good practice in buildings with a total transformer capacity in excess of 300 kVA to complete a calculation comparing the operating costs of various types of transformers in order to balance energy costs with necessary operating flexibility, reliability, and safety.

Such a calculation was performed assuming a daily/weekly operating cycle and loading that is representative of commercial use. The two transformers studied were the dry type, rated 750 kVA, one designed for a 80°C (144°F) rise and the second

**TABLE 26.3 Typical Dry-Type Transformer: Dimensions and Weights[a]**

| kVA Output Continuous | Temp. Rise (°C) | Approximate Dimensions (in.) H × W × D | Approx. Weight (lb) |
|---|---|---|---|
| PRIMARY: 480 V SECONDARY: 120/240 V | | | |
| Single-phase | | | |
| 3 | 115 | 13 × 8 × 8 | 60 |
| 5 | 115 | 14 × 12 × 12 | 120 |
| 10 | 115 | 15 × 12 × 12 | 150 |
| 15 | 150 | 24 × 16 × 16 | 200 |
| 25 | 150 | 24 × 16 × 16 | 230 |
| 50 | 150 | 30 × 20 × 20 | 430 |
| 100 | 150 | 40 × 24 × 24 | 650 |
| PRIMARY: 480 V SECONDARY: 208Y/120 V | | | |
| Three-phase | | | |
| 45 | 150 | 26 × 26 × 15 | 380 |
| 75 | 150 | 30 × 30 × 20 | 600 |
| 150 | 150 | 40 × 36 × 26 | 1100 |
| 225 | 150 | 44 × 36 × 26 | 1400 |

[a]Dimensions and weights vary among manufacturers. All figures increase with higher primary voltage and with lower temperature rise.

for a 150°C (270°F) temperature rise. Based on an energy cost of 7 cents per kWh, the lower energy waste of the more expensive 80°C (144°F) unit repays the first-cost differential in about 4 years. A higher electric energy cost reduces the payback time, and vice versa. Beyond the payback period the advantage is entirely on the side of the 80°C (144°F) unit. Using a 30-year life, 8% fixed capital cost, and 3% annual cost escalation, the life-cycle cost of the 80°C (144°F) unit is $10,000 less than that of the 150°C (270°F) unit. (At the assumed loading, even the 150°C (270°F) unit will probably last 30 years, the 80°C (144°F) unit much longer.)

Furthermore, with a total permissible hot-spot temperature of 220°C (428°F), the location of the 150°C (270°F) rise unit must be very carefully chosen because it can create a serious heat generation and radiation problem.

To summarize, a transformer is specified by type, phase, voltages, kVA rating, sound level, and insulation class. Thus, a 112.5-kVA, three-phase, 480/120–208-V, air-cooled, indoor, dry-type transformer with 220°C (428°F) insulation system and 115°C (207°F) rise, 45 db (decibel) maximum sound level, is an adequate transformer description. Sound ratings of transformers, as well as installation techniques and acoustical treatment, are discussed in Part IV.

## 26.7 TRANSFORMERS OUTDOORS

A service transformer bank is necessary, as explained above, when the facility utilization voltage is different from the utility voltage (Fig. 26.5a). The designer occasionally opts for a step-up, step-down arrangement when the service run is so long that the conductor cost when run at a low voltage would be excessive because of large cables. In such instances, the cost of the double transformer installation must be more than offset by the savings in feeder cost to be economically justifiable (Fig. 26.5*b*).

The advantages of an outdoor transformer installation are

- No building space required
- Reduced noise problem within the building
- Lower first cost
- Ease of maintenance and replacement
- No interior heat problem
- Opportunity to use low-cost, long-life, oil-filled units

However, it is frequently easier to find space indoors (preferably in a basement) than to find a suitable exterior location, and noise may be more disturbing from the available exterior spaces, such as courtyards, than from a basement. Costs can run high if long secondary voltage runs are required; heat can often be handled by louvers or areaways adjoining a basement or even used profitably if the transformer load is fairly constant. Also, because exposure to direct sunlight decreases a transformer's rating by increasing its temperature, a shaded spot may be dif-

ficult to find. Furthermore, an exterior transformer, other than pole-mounted, is a questionable choice in an area with a high incidence of vandalism, regardless of the sturdiness of construction.

Finally, the appearance of an exterior unit may be objectionable. This point has received much

(*a*)

(*b*)

**Fig. 26.6** (a) Pad-mounted exterior transformers are neat, compact, and, if sited properly, unobtrusive. Large units can be partially screened by shrubbery. (b) When the size is such that visual screening becomes a problem, consideration should be given to a structural screen such as a decorative brick wall, which also provides some screening. (Courtesy of General Electric, U.S.A.)

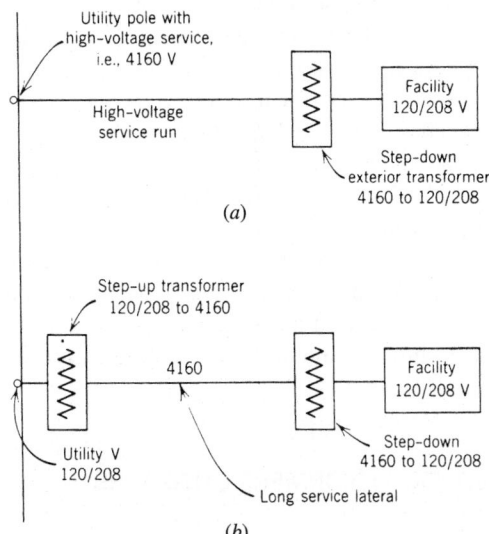

**Fig. 26.5** Service transformer arrangements. (a) High-voltage service with a step-down service transformer at the facility and (b) low-voltage service, with transformation at both ends of a long service run.

attention from manufacturers, and numerous designs have been developed that minimize the appearance problem (Fig. 26.6). The most popular type of exterior transformer installation is the pad mount. It has all of the advantages listed previously in addition to extreme simplicity of installation—it is simply set on a concrete pad. Consult manufacturers' catalogs for dimensional data.

## 26.8 TRANSFORMERS INDOORS: HEAT LOSS

When an indoor transformer installation is indicated, special consideration must be given to the transformer's heat-generating properties. Between 1% and 1½% of a transformer's rating, depending on the type, is converted to heat at full load. For a 750-kVA, dry-type, 150°C (270°F)-rise unit, 12 kW, or 41,000 Btu/h, of heat loss is generated at full load! Losses are lower for 80°C (144°F)-rise units. Liquid-filled units have approximately the same losses as 80°C (144°F)-rise dry units. Unless the heat can be used, sufficient ventilation—either natural or forced—must usually be provided to keep the room's ambient temperature from exceeding 40°C (104°F).

Ventilation by natural convection is most desirable, necessitating the location of the transformer room on an exterior wall (with an areaway if the room is located below grade). The size of the *free* area required for a ventilation opening is 3 in.$^2$/kVA (1935 mm$^2$/kVA) of capacity, plus an additional 1 in.$^2$/kVA (645 mm$^2$/kVA) for switchgear heat losses, if any. If a louver is used in the openings, the size of the opening must usually be doubled (total: 8 in.$^2$/kVA [5160 mm$^2$/kVA]) because most louvers have a 50% free area. For good convection, it is desirable to divide the louvered areas in half, placing one-half near the ceiling and the remaining one-half near the floor. To provide for equipment removal, if the areaway is large enough, it may be useful to add a louver area between the upper and lower openings and make the full louver removable. A bird screen is also desirable.

Because outside air temperature varies, it is advisable to use a temperature-controlled, adjustable louver. In extremely cold climates, heat loss from the electrical equipment may not be sufficient in winter to warm the room. In such instances, a unit

heater should be installed in the room with a thermostat set at 55°F (13°C).

## 26.9 TRANSFORMERS INDOORS: SELECTION

When transformers are installed indoors, they are subject to stringent *National Electrical Code* (*NEC*) regulations that are designed to make the installations intrinsically safe. These regulations are detailed in *NEC* Article 450, and no purpose is served in duplicating them here. Instead, we present the reader with the essential considerations involved.

### (a) Oil-Insulated Transformers

These present a fire hazard indoors because flammable oil can spread from a tank leak or rupture. To prevent this, most oil-filled transformers must be installed in a fire-resistant vault, the construction of which involves a heavy cost. (See *NEC* Article 450 for exceptions.) Advantages offsetting this cost are the oil-filled transformer's small size, low first cost, low losses, long life, excellent electrical characteristics, low noise level, and high overload capacity. Despite these, the vault requirement has had the effect of restricting oil-filled units to industrial facilities and other structures where electrical considerations favor its use.

### (b) "Less-Flammable" Liquid-Insulated Transformers

Transformers rated 35 kV or less that are insulated with a liquid whose fire point is not less than 300°C (572°F) ("less-flammable" liquid) may be installed indoors without a vault. A liquid confinement area is required, plus either a fire-extinguishing system or restrictions on the combustibility of the building's contents. The insulating liquids used are special types of silicones and hydrocarbons, and the transformer installation must meet all requirements for the specific liquid involved. Use of these units achieves the electrical and physical advantages of oil-insulated units without the cost penalty of vault construction and installation. The first cost of the units depends on the fluid type and the specific electrical characteristics required. In general, they are somewhat less expensive than nonflammable fluid-insulated units with similar electrical ratings.

### (c) Nonflammable Fluid-Filled Transformers

Certain nonflammable liquid coolants, called *askarels*, generally containing polychlorinated biphenyl (PCB), were once very widely used. However, since 1979, use of PCB coolants in new equipment has been banned by federal regulation due to their negative ecological impact, and existing units are gradually being phased out. Newer nonflammable fluid coolants have increased the price of these transformers considerably (Table 26.4). Such units have most of the advantages of oil-filled units and do *not* require a vault unless the voltage is very high. They do, however, require a sump or catch basin of sufficient capacity for all of the contained liquid. This and the high first cost are the negative aspects of this design.

### (d) Dry-Type Transformers

These are the units of choice in the majority of indoor installations, despite their shorter life, higher losses, high noise level, greater weight, and larger size than the liquid-filled units. The principal advantages are ease of installation and almost unrestricted choice of location. As explained previously, by using an underrated transformer (220°C [428°F] system, 80°C [144°F] rise), losses (heat) can be reduced and life extended. Also, for a price premium, the noise level can be reduced. Table 26.4 gives a comparison of the installed costs for these four classes of transformers.

## 26.10 TRANSFORMER VAULTS

A transformer vault is basically a fire-rated enclosure provided because of the possibility of fire due to rupture of an oil-filled transformer case. However, this must not be construed as implying that transformers are hazardous or delicate devices prone to faults. On the contrary, transformers are extremely tough, sturdy, long-lived, capable of sustaining large and prolonged overloads, and, indeed, among the most reliable elements of an electrical system. However, faults do occur, and an oil-filled transformer is a potential fire hazard.

Transformer vaults should be located, if practical, where they can be ventilated to the outside air without flues or ducts. The combined net area of all ventilating openings (gross area less screens, louvers, etc.) should be as explained in the preceding section: not less than 3 in.$^2$/kVA (1935 mm$^2$/kVA) of transformer but in no case less than 1 ft$^2$ (0.09 m$^2$). Further ventilation recommendations plus details of enclosure construction materials, fire rating, door and sill details, and other relevant and important information, are provided by *NEC* Article 450.

## 26.11 SERVICE EQUIPMENT ARRANGEMENTS AND METERING

Metering must be provided ahead (electrically) of the building's service entrance switch(es). The metering

**TABLE 26.4 Relative Installation Costs for a 300- to 1000-kVA Transformer**

| Transformer Type | Temperature Rise[a] (°C) | Relative Transformer Cost | Construction Cost | Total Relative First Cost |
|---|---|---|---|---|
| Oil filled | 65 | 1.00 | 50–100[b] | 1.50–2.00 |
| Less flammable and nonflammable fluids | 65 | 1.15–1.30 | 20–40[c] | 1.35–1.70 |
| Dry, ventilated | 80 | 1.65 | — | 1.65 |
| | 80 | 1.50 | — | 1.50 |
| | 150 | 1.35 | — | 1.35 |
| | 150 | 1.20 | — | 1.20 |

[a]Transformers of equal temperature rise have approximately equal life and equal losses.
[b]Cost of the vault depends on local labor costs and size of the transformer. The relative cost decreases with increasing transformer size.
[c]Cost of catch basin. As in the case of a vault, relative cost depends on labor rates and transformer size.

is either at the utility or the facility voltage, and either at the service point or inside the building. The choice is generally left to the owner with the understanding that interior meter equipment must be readily accessible to utility personnel. Although in increasingly large numbers of facilities actual meter reading is accomplished remotely, the meter equipment must still be available for inspection and service.

If high-voltage service is purchased, then the transformers and all equipment beyond the service connection must be furnished by the owner. Conversely, if low-voltage service is purchased, all equipment necessary to provide low voltage is furnished by the utility. Obviously, the electric service rates for low-voltage service are *higher* than those for high-voltage service in order to compensate the utility for the cost of providing and maintaining the step-down transformer and associated equipment. Therefore, it is often advisable for the owner of a large facility to investigate the economics of purchasing power at high voltage. Because many owners are not equipped to maintain high-voltage equipment, arrangements can sometimes be made to pay the utility to provide and maintain transformers while taking the cost advantage of high-voltage service.

For a single-occupant building or a building in which electric energy is included in the rental charge, only a single meter is necessary. Provision for such metering may be made in the main switchboard or the meter may be independently mounted. In both cases, the meter is furnished and installed by the utility company. Where submetering is required, such as in apartment houses, banks of meter sockets are installed to accommodate the multiple meters. Federal regulations forbid master metering in new multiple-dwelling constructions because it encourages energy waste (Fig. 26.7).

A low-voltage underground service detail as it would appear on a set of contract drawings, including relevant details, is given in Fig. 26.8. Note that here the service switch and meters are separately mounted.

## 26.12 SERVICE SWITCH(ES)

The purpose of the electric service switch(es) is to disconnect the normal service to the building. (See Section 28.20 for a discussion of emergency electric service.) Thus, in the event of fire, no electrical hazard will face firefighters. It is therefore obvious that this disconnecting apparatus must be located at a readily accessible spot near the point at which the service conductors enter the building. If such a location is not feasible, service conductors may be run in concrete encasement under the building and are considered "outside the building" up to the point at which they emerge from the floor in the building (see Fig. 26.8). At that point, the service disconnect must be installed. The service switch or, more accurately, the service-disconnecting means, may comprise one to six properly rated switches. These are frequently assembled into a switchboard. However, before discussing switchboards, a description of switches, circuit breakers, and fuses, the components of which switchboards are constructed, is in order.

**Fig. 26.7** *Three-section commercial submetering switchboard comprising a center service-entrance section rated 4000-A, 3-phase, 480-V maximum, equipped (usually) with a main switch or circuit breaker. Each of the end cubicles will accept six tenant meters, each on a 320-A 3-phase circuit, maximum, protected by an integral circuit breaker or fused switch. Tenant wiring enters the top or bottom of the board. Switchboards of this type can be installed on every floor of a large multitenant commercial office building. (Photo courtesy of Cutler-Hammer.)*

## 26.13 SWITCHES

In this and the following two sections we discuss traditional electrical switching devices, which close

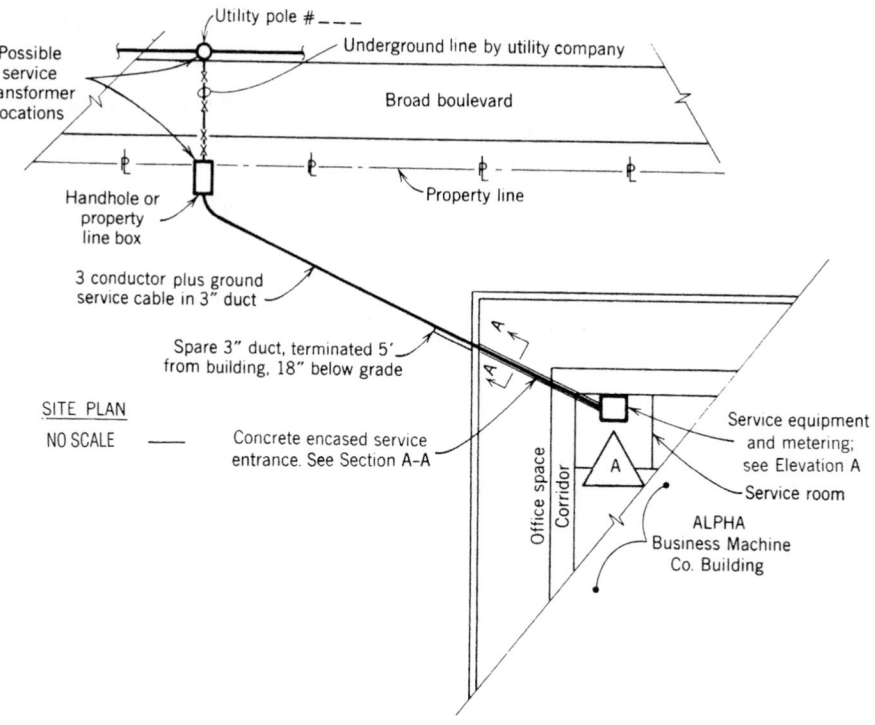

Utility pole # _ _ _

Underground line by utility company

Possible service transformer locations

Broad boulevard

Property line

Handhole or property line box

3 conductor plus ground service cable in 3" duct

Spare 3" duct, terminated 5' from building, 18" below grade

SITE PLAN
NO SCALE

Concrete encased service entrance. See Section A-A

Office space

Corridor

Service equipment and metering; see Elevation A

Service room

ALPHA Business Machine Co. Building

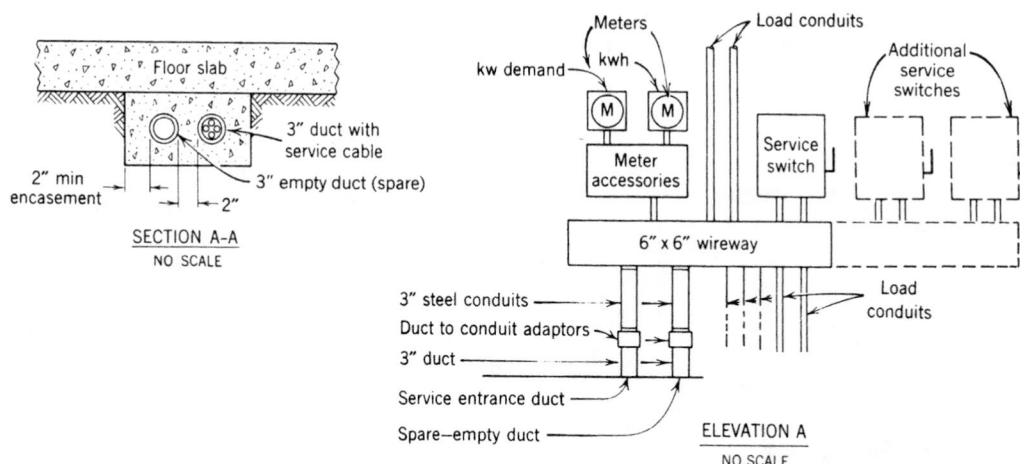

Floor slab

3" duct with service cable

2" min encasement

3" empty duct (spare)

2"

SECTION A-A
NO SCALE

Meters

kw demand

kwh

Load conduits

Additional service switches

Meter accessories

Service switch

6" x 6" wireway

Load conduits

3" steel conduits

Duct to conduit adaptors

3" duct

Service entrance duct

Spare—empty duct

ELEVATION A
NO SCALE

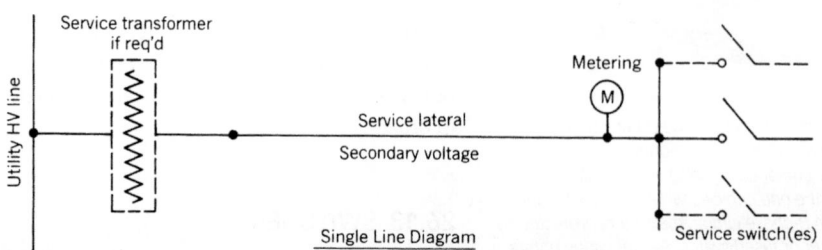

Service transformer if req'd

Utility HV line

Metering

Service lateral

Secondary voltage

Service switch(es)

Single Line Diagram

**Fig. 26.8** *Typical electrical plan and details of an underground low-voltage service (600-V class) to an industrial facility. Note that the portion of underground service beneath the building must be encased in concrete. A service transformer (if required) could be mounted on the utility pole or on a pad at the property line.*

**1178**

and open an electric circuit by physically moving two electrical conductors into contact with each other to close the circuit and physically separate them to open the circuit. The motive force can be supplied by hand, an electrical coil, a spring, or a motor, as discussed later, and the device name may change accordingly, but the action is the same. Solid-state switches perform the same electrical switching function but by a completely different process, without moving parts. They are discussed in Section 26.16.

An electrical switch is rated by current and voltage, duty, poles and throw, fusibility, and enclosure. The current rating is the amount of current that the switch can carry continuously and interrupt safely. Switches intended for motor control are also rated in horsepower. The voltage rating of a switch is, as for other electrical equipment, by voltage

(i)   Single-pole single-throw switch.

(ii)  Two-pole single-throw switch.

(iii) Three-pole and solid-neutral (3P and SN) switch.

(iv)  Single-pole double-throw switch (also called, in small sizes, a *3-way switch*).

(v)   Single-pole double-throw switch with center "off" position (in control work called a *hand-off-automatic switch*).

(vi)  Use of two single-pole, double-throw (3-way) switches for switching of a lighting circuit from two locations.

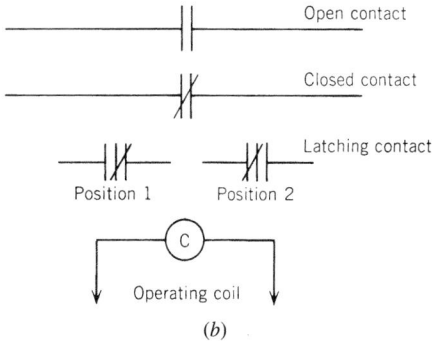

**Fig. 26.9** *(a) Typical switch configurations. Note that switches are always shown in the open position. (b) Graphic representation of switch (contact) positions in a contactor or relay. Note that a latching contactor shows the switch in one of two positions to indicate the latching nature of the contactor.*

class. Thus, a switch may be rated 250 V, 600 V, or 5 kV as required. Switches intended for normal use in lighting and power circuits are called *general duty safety switches.* Switches intended for frequent interrupting, high-fault currents, and ease of maintenance are rated HD for *heavy duty.*

An examination of Fig. 26.9*a* should clarify what is meant by the number of poles and throws of a switch. Unless otherwise noted, a switch is assumed to be single throw. Because the *NEC* states generally that the grounded neutral conductor of a circuit should not be broken, most switches carry the neutral through unbroken by means of a solid link within the switch. This gives rise to the term *solid neutral* (SN) *switch.* Switches are available in 1-, 2-, 3-, 4-, and 5-pole construction. Poles are indicated by a "P"; thus, 3-pole is written "3P," and so on.

A switch may be constructed with or without provision for fusing. If provided, the switch is fusible; if not, the switch is nonfusible. All individual enclosed switches are mounted in an appropriate cabinet. NEMA standardized the nomenclature and application of enclosures for all electrical control equipment, of which switches are only one item. These are detailed in Section 26.15. Summarizing, then, we could adequately describe a switch thus: switch, HD, 3P & SN, 200A/150AF, 600 V, in NEMA 12 enclosure. This "translates" as a heavy-duty switch, 3 poles and solid neutral, 200-amp rating with 150-amp fuses, 600-V rating, in an industrial use enclosure (Table 26.5). Such a switch is illustrated in Fig. 26.10.

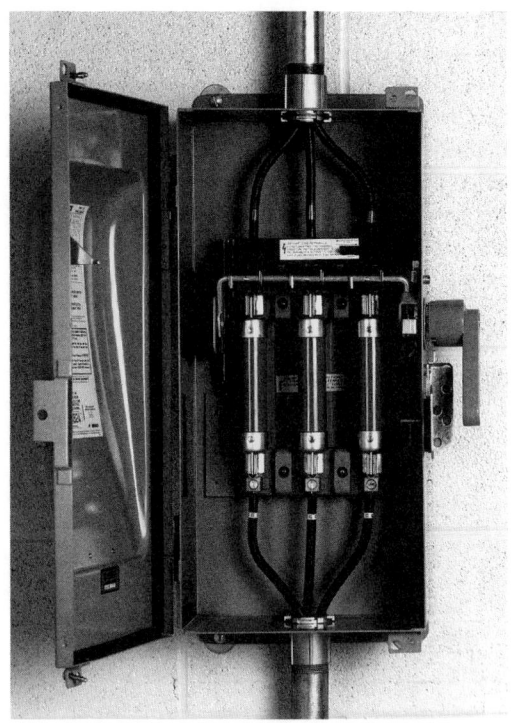

**Fig. 26.10** *The internal construction of a heavy-duty, fused, 3-pole, 3-wire switch in an industrial NEMA 12 enclosure. Note the gasketing on the inside of the cover, which seals the sheet steel box and makes it usable in environments containing lint, dust, dirt, sawdust, and the like. The external handle is a double-lockable indicating type for safety. (Courtesy of Siemens Energy & Automation, Inc.)*

## TABLE 26.5  Control Equipment Enclosures

| NEMA Designation: Type | Description | Application |
|---|---|---|
| 1 | General-purpose | Dry, indoor use |
| 2 | Dripproof | Indoor, subject to dripping |
| 3 | Dust-tight, rain-tight, and sleet-resistant | Indoor/outdoor, where subject to windblown dust and water |
| 3R | Rainproof and sleet-resistant | Outdoor, subject to falling rain, snow, and sleet |
| 3S | Dust-tight, rain-tight, and sleet-proof | Outdoor, subject to windblown water, dust, and sleet; most severe exterior duty |
| 4 | Watertight and dust-tight | Indoor/outdoor, subject to water from all directions; not sleetproof |
| 4X | Watertight, dust-tight, corrosion-resistant | Same as type 4 with added corrosion protection |
| 7–9 | Hazardous | Differing in application by class and group of hazardous use; see *NEC* |
| 12 | Industrial use, dust-tight, and drip-tight | Indoor only, general use, industrial and other "dirty" environments |

## 26.14 CONTACTORS

A contactor is a switch. Instead of a handle-operated, movable blade, a contactor uses *contact* blocks of silver-coated copper, which are forced together to *make* (close) or are separated to *break* (open) the circuit. The common wall light switch is a small, mechanically operated contactor. A relay is an electrically operated contactor. Most contactors are operated by means of an electromagnet that causes the contacts to close. They open either by spring action or by gravity.

Contactor terminology is somewhat different from that of a switch. Its condition when deenergized is its *normal* state. Thus, a contactor whose contacts are open when the coil is not energized is referred to as *normally open* (NO), whereas a contactor whose contacts are closed when deenergized is referred to as *normally closed* (NC). Units intended for motor connection are called *motor starters* and are discussed in Section 26.26. Current, voltage, and number of poles have the same nomenclature for contactors and relays as for switches.

The great advantage of contactors over switches is their facility for remote control. Switches must be manually thrown—or, in very special cases, thrown by a motor. The magnetic contactor, by contrast, is inherently a remotely controlled device, making it ideal for myriad control functions. It can be controlled by a manual or remote pushbutton or by automatic devices such as timers, float switches, thermostats, pressure switches, and so on. Because control can be both remote and automatic, the application of contactors is universal in control of lighting, heating, air conditioning, motors, and the like. A graphic representation of contactors is shown in Fig. 26.9*b*.

## 26.15 SPECIAL SWITCHES

Many special types of switches are available. Most of these types are beyond the scope of this book, except for the following, which are used extensively in building work.

### (a) Remote-Control (RC) Switches

A contactor that latches mechanically after being operated is known as a *remote-control switch*. It differs from a relay in that the latching operation performs the dual function of latching the contacts and disconnecting the electric circuit. To unlatch the contacts, the electric circuit must be reenergized. It is therefore also known as an *electrically operated mechanically held contactor*. This characteristic is advantageous where the position of the contacts will be maintained for a long period (as in a lighting control circuit) and it is undesirable to hold the contacts by continuous energizing of a coil, as would be required with an ordinary relay. A typical unit is shown in Fig. 26.11, and a typical application is shown in Fig. 28.13. A common application of these switches is for switching of a "block" load, such as exterior lighting, whole-floor or whole-building lighting, and the like.

### (b) Automatic Transfer Switch

This device, which is an essential part of all emergency and standby power arrangements, is basically a double throw switch—generally 3-pole—so

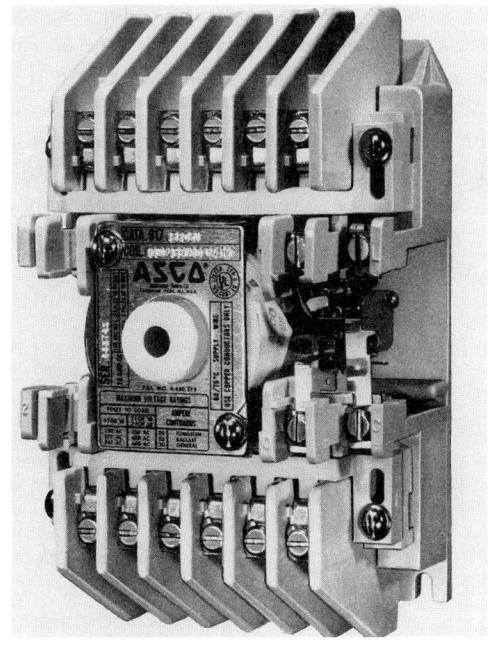

*Fig. 26.11* This remote-control switch (mechanically held, electrically operated contactor) is a 12-pole, 20-A, 600-V unit measuring approximately 7 in. H × 6 in. W × 4 in. D (178 × 150 × 100 mm), making it suitable for stud-wall mounting. Compact solid-state control modules are available that mount directly onto the unit and convert it to a flush, wall-mounted multicircuit programmable controller (see Section 26.16). (Courtesy of Automatic Switch Co.)

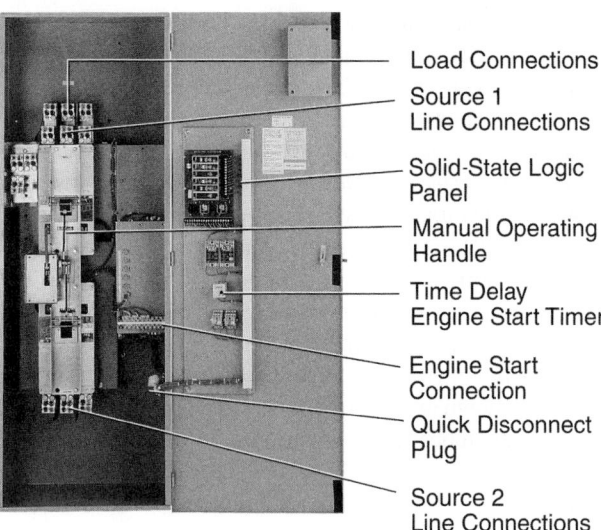

Load Connections

Source 1
Line Connections

Solid-State Logic
Panel

Manual Operating
Handle

Time Delay
Engine Start Timer

Engine Start
Connection

Quick Disconnect
Plug

Source 2
Line Connections

**Fig. 26.12** *Automatic transfer switch with solid-state logic. The unit can be operated manually if required. It is equipped with engine-starting contacts, auxiliary contacts for monitoring, electrical safety interlocks, and automatic differential voltage sensing on the normal source. An 800-A, 3-phase unit of this design, in a NEMA indoor sheet metal enclosure, measures 73 in. H × 26 in. W × 21 in. D (1.9 × 0.7 × 0.5 m) and weighs approximately 600 lb (272 kg). (Photo courtesy of Cutler-Hammer.)*

arranged that on failure of normal service it *automatically transfers* to the emergency service. When normal service is restored, it automatically retransfers to it. (Retransfer can also be arranged to have a minimum delay, or it can be entirely manual.) The switch control devices are voltage sensors that sense the condition of the service and operate the switch accordingly. Auxiliary devices can be built on to the basic switch, the most common of which are emergency generator starting equipment. Figure 26.12 illustrates a typical unit. Transfer switches used in uninterrupted power supplies are normally solid-state because they involve no switching time lapse.

### (c) Time-Controlled Switches

This category includes all switches whose operation is time-based. The timer can be either the familiar electromechanical device consisting of a low-speed miniature drive motor, to which some type of contact-making device is physically connected, or it can be a solid-state electronic timer, which in turn controls either a relay or a solid-state switch. The latter arrangement is the basis of all modern programmable time controls (Section 26.16), which find wide application in lighting control (Sections 15.14 and 30.27), energy management and automated building control (Sections 25.13 and 30.28), and clock and program systems (Section 30.14).

Motor-operated timers (other than hand-wound

types) depend for their accuracy on power-line frequency, have moving parts that wear, must be reset after power outages (some units have spring-wound reserve power motors), and become cumbersome and expensive with an increase in the number of

**Fig. 26.13** *Time switch arranged for three different types of control functions that are applicable to different types of load: (1) full photocell control—security lighting; (2) dusk "On," preset "Off"—signs, parking lots; (3) preset on, off—interiors. This unit's dial is a 7-day calendar type that permits setting a different schedule for every day of the week. The dial is not astronomical (solar) because all the turn-on functions are either photocell or time-of-day controlled. Solar dials are useful when photocell control is not used because sunrise and sunset times vary with the season. (Courtesy of Tork, Inc.)*

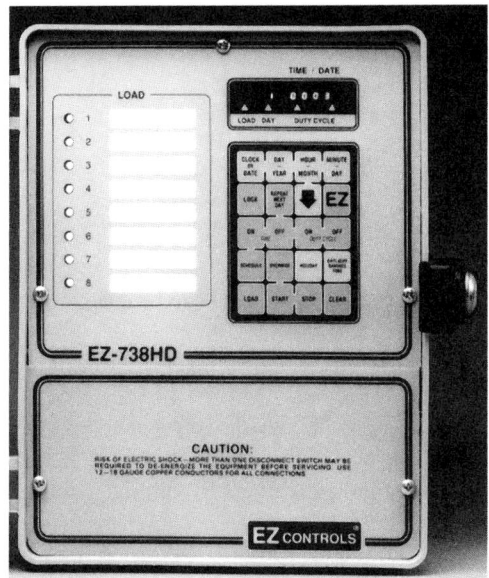

**Fig. 26.14** *Programmable time switch. This unit, which measures 9.5 in. H × 8 in. W × 4.5 in. D (240 × 200 × 115 mm) in its cabinet, permits independent programming of eight 20-A, 240-V circuits. Each circuit (channel) can be arranged for multiple on–off daily operations and can be programmed on a 7-day-a-week basis, with special 365-day functions such as automatic daylight savings time adjustment. (Courtesy of E.Z. Controls.)*

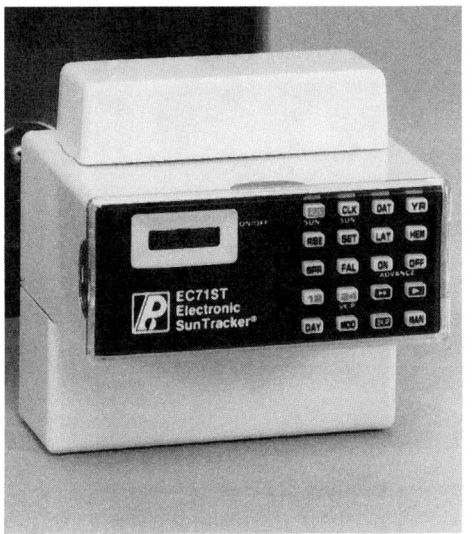

**Fig. 26.15** *This unit controls two 15-A, 240-V circuits. The built-in astronomical feature automatically adjusts the on–off setpoint according to sunset and sunrise throughout the year. (Courtesy of Paragon Electric Co., Inc.)*

controlled events. Electronic units are small, cool, quiet, independent of line frequency for accuracy (in many designs), carry through outages when standby-battery equipped, and have virtually unlimited event control capacity. They have largely replaced the older design except for the simplest applications (Figs. 26.13, 26.14, and 26.15).

## 26.16 SOLID-STATE SWITCHES, PROGRAMMABLE SWITCHES, MICROPROCESSORS, AND PROGRAMMABLE CONTROLLERS

A solid-state switch is an electronic device with a conducting state and a nonconducting state, corresponding to a conventional switch in its closed position and its open position. The change between the two states is accomplished by the application of a control signal (voltage). The change of state is instantaneous and noiseless, and occurs without arcing. The current ratings of solid-state switches are approximately the same as those of their mechanical counterparts, ranging from fractions of an ampere to hundreds of amperes. Voltage ratings are usually limited to secondary systems ratings (0 to 600 V) due to the nature of the semiconductors comprising the active elements of these switches, although higher-voltage equipment is available for special application.

If an electronic timing device is added to a solid-state switch, a time-controlled electronic switch is created that functions exactly like the electromechanical time-based switches described in Section 26.15. These switches have the advantages that their time control is independent of the utility line frequency and there are no moving parts to wear out. If a small, programmable memory circuit in the form of an erasable programmable read-only memory (EPROM) chip is added, we then have a programmable time switch. The simplest of these are single-circuit 15-A devices intended to replace common wall switches (see Fig. 26.48). A more complex unit corresponding roughly to the electromechanical unit shown in Fig. 26.13, but much more flexible, is shown in Fig. 26.14. Addition of a read-only memory (ROM) chip preprogrammed with astronomical information (sunrise–sunset) for the particular latitude involved converts the switch to a "sun tracker." Such a switch is particularly

**ELECTRICITY**

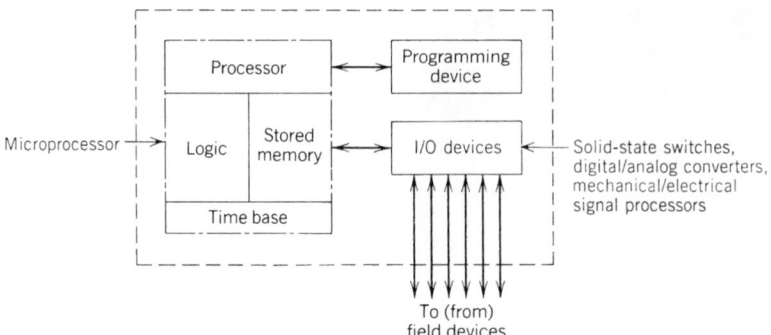

**Fig. 26.16** *Block diagram of a programmable controller. These devices are designed to supervise field conditions continuously and are therefore used in industrial applications as well as for lighting and environmental system control.*

applicable to control circuits with year-round dusk-to-dawn control (see Fig. 26.15).

If, in lieu of a programming device, which, until reprogrammed, gives fixed, invariable instructions to a (multichannel) switch, as is the case with a programmable time switch, we install a *microprocessor* [i.e., a logic/memory device programmed to *respond* according to a particular algorithm (control plan) to given input signals], we have constructed a programmable controller (Figs. 26.16, 26.17, and 26.18). By definition, a *programmable controller* is an electronic device that uses a programmable memory chip for internal storage of instructions that in turn implement spe-

cific functions such as logic, timing, counting, and so on.

The internal control/memory is frequently referred to as a *microprocessor,* whereas with respect to the overall programmable controller, it acts as a central processing unit (CPU). The programmable controller is therefore a special type of computer. It differs from a conventional computer in that its program is relatively short and exists in hardware (microprocessor), whereas a computer reads its program, which can be very lengthy, into its memory from software. Furthermore, the programmable controller is designed for a specific type of function, whereas the conventional computer is a general-use

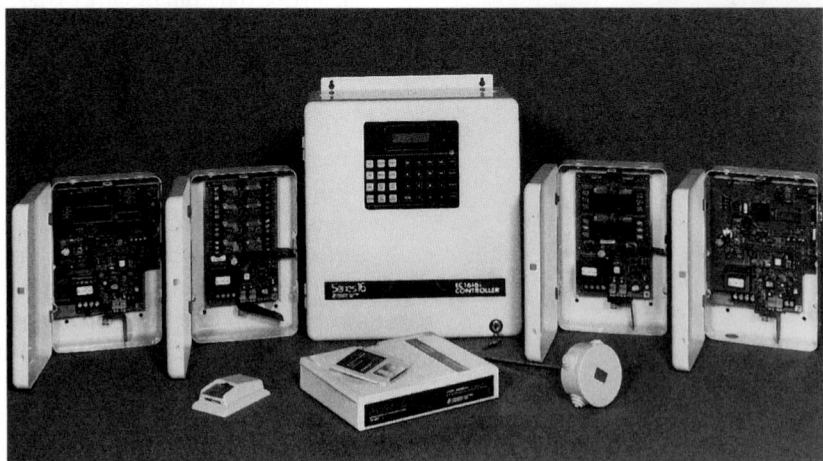

**Fig. 26.17** *Programmable controller (center) designed for use as an energy-monitoring and building system control device. The basic unit accepts up to 16 analog inputs, such as illuminance, temperature, kW demand load, time signals, central control signals, and the like. It processes data according to its current program (note the programming keypad) and can signal and control up to 16 outputs. Additional functions available include time-of-day switching, including astronomical considerations, and duty-cycle load switching for electric demand control. The controller is approximately 14 in. W × 16 in. H × 6 in. D (355 × 410 × 150 mm) and is arranged for surface mounting. On either side are accessory communication and relay modules. In the foreground are application software and typical sensors. (Photo courtesy of Paragon Electric Co., Inc.)*

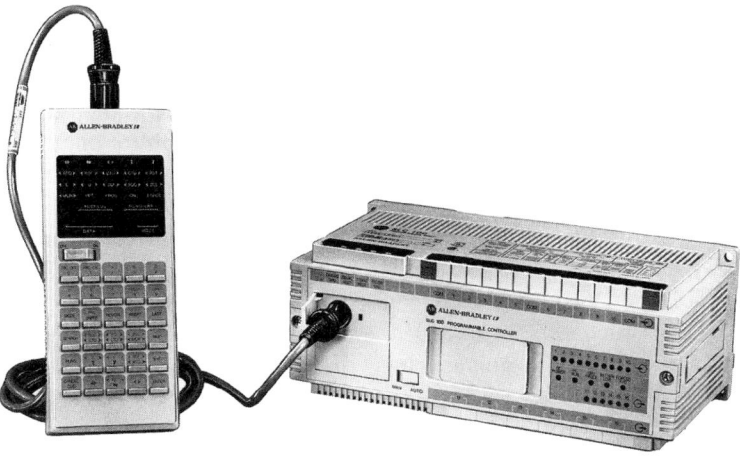

**Fig. 26.18** *General-purpose programmable controller. The processor unit contains the central processor (CPU), random access memory (RAM), and power supply. Peripherals include the illustrated hand-held programming unit and EEPROM (electrically erasable programmable read-only memory) module, and various computer/software interfaces (not shown). The unit illustrated, which measures approximately 10 × 9 × 3 in. (255 × 230 × 75 mm), has 16 I/O circuits and an 885-word memory. It is readily adaptable to a variety of automated control processes in manufacturing, building control, and material handling. (Courtesy of Allen-Bradley Co.)*

calculating (computing) device that processes almost any type of software.

Programmable controllers have all but replaced the ubiquitous *hard-wired* relay panel in applications such as industrial control, elevator control, process control, and the like, because of the ease with which such a controller can be reprogrammed (in lieu of rewiring a relay panel), the speed and accuracy of control, and the vastly increased complexity possible. See Chapter 30 for definitions and for an explanation of the application of these devices to building control.

Raintight NEMA 3R circuit breaker enclosure.

**Fig. 26.19** *Type 3R outdoor enclosure. This is the type usually intended when "weatherproof" is specified. (Courtesy of General Electric, U.S.A.)*

## 26.17 EQUIPMENT ENCLOSURES

Proper NEMA nonmenclature, descriptions, and applications for the more common enclosures are found in Table 26.5. It is important to note that there is no enclosure described as WP or weatherproof. Equipment intended for outdoor use should be specified:

1. In a type 3R enclosure to protect against rain (Fig. 26.19)
2. In a type 3S enclosure to protect against wind-driven rain and sleet
3. In a type 4 enclosure to protect against the above plus splashing and condensation

Note also that the type 12 industrial enclosure is similar to type 1 except that it is gasketed for dust and drip resistance and therefore is well applied in *all* "dirty" indoor environments, including commercial and institutional spaces (see Fig. 26.10).

## 26.18 CIRCUIT-PROTECTIVE DEVICES

To protect insulation, wiring, switches, and other apparatus from the destructive effects of overload and short-circuit currents, an automatic means for opening the circuit is required. The two most common devices employed to fulfill this function are the *fuse* and the *circuit breaker*, the latter frequently abbreviated *c/b*.

## (a) Fuses

The fuse is a simple device consisting of a *fusible* link or wire of low melting temperature that, when enclosed in an insulating fiber tube, is called a *cartridge fuse* and, when enclosed in a porcelain cup, is known as a *plug fuse*. Figure 26.20 shows common types of fuses. Plug fuses, such as those in residential use, are rated 5 to 30 A, 150 V to ground, maximum. Cartridge fuses of various designs are made up to 6000 A and 600 V.

## (b) Circuit Breakers

A circuit breaker is an electromechanical device that performs the same protective function as a fuse and, in addition, acts as a switch. Thus, it can be used in lieu of a switch-and-fuse combination to both protect and disconnect a circuit. Most circuit breakers are equipped with both thermal and magnetic trips. The thermal trip acts on overload, whereas the magnetic trip acts on short circuit. Both thermal and magnetic action have inverse-

time characteristics: that is, the heavier the overload, the faster the trip action. Modern circuit breakers in commercial and industrial applications are frequently equipped with solid-state electronic tripping control units, which provide fully adjustable overload, short-circuit, and ground-fault protection.

Air circuit breakers are available in two types: the molded-case breaker and the large air-circuit breaker. Molded-case breakers consist of a complete mechanism encased in a molded phenolic case. A light-duty molded case 50-A frame, plug-in circuit breaker is illustrated in Fig. 26.21*a*. The molded-case breaker shown in Fig. 26.21*b* has a 400-A frame and can be equipped with an adjustable electronic tripping unit, ground fault trip, plus other features formerly available only on large air-circuit breakers. The large air-circuit breaker is a more complicated and highly adjustable device that can be used in applications that preclude the use of molded-case breakers. A modern solid-state adjustable trip controller for a large air-circuit breaker is shown in Fig. 26.22, along with a circuit breaker with a similar solid-state trip unit.

All breakers can be equipped with remote trip

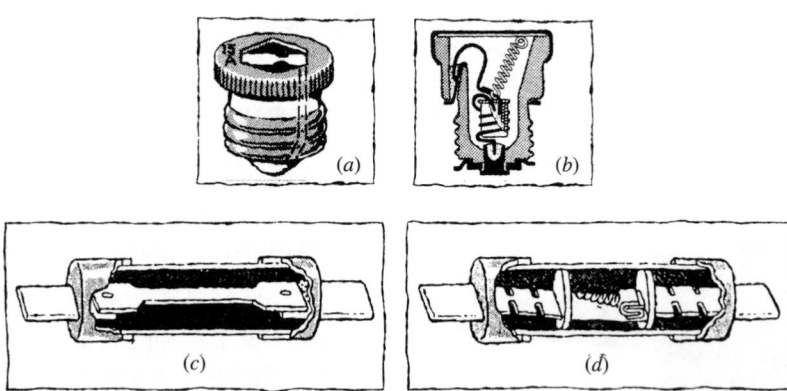

**Fig. 26.20** Plug-type fuses are made in two physical types. (a) The nonrenewable type with a standard (Edison) base screws directly into a standard socket. (b) Nonrenewable NEC type S fuses have a smaller base than type (a) and therefore require an adapter to fit into a standard socket. The adapter is current rated and nonremovable. This prevents deliberate or accidental use of a type S fuse of incorrect rating. Type S is required in new construction where plug fuses are used. Both fuses shown are nonrenewable and must be replaced after blowing. The type S fuse shown in (b) is a dual-element time-delay fuse. Cartridge fuses are available in a variety of designs. Illustrated are (c) the nonrenewable, single-element type and (d) the nonrenewable, dual-element, time-delay type. Because fuses are inherently very-fast-acting devices, time delay must be built into a fuse to prevent blowing on short-time overloads such as those caused by motor starting. A dual-element fuse as shown in (b) allows the heat generated by temporary overloads to be dissipated in the larger center metal element, preventing fuse blowing. If the overload reaches dangerous proportions, the metal will melt, releasing the spring and opening the circuit. The notched metal portions of the fuse element, at both ends of the dual center element, provide short-circuit protection. The time required to clear (blow) a fuse is generally inversely proportional to the amount of current. In renewable-cartridge fuses, the spent (blown) element is replaceable, reusing the original outer cartridge and blade connectors.

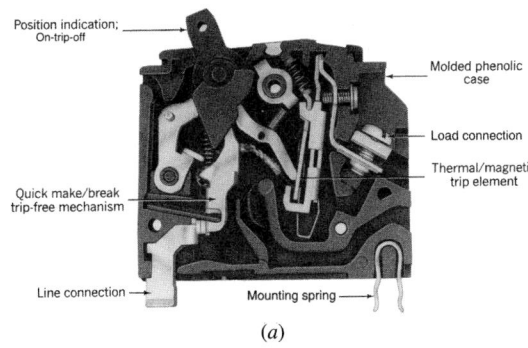

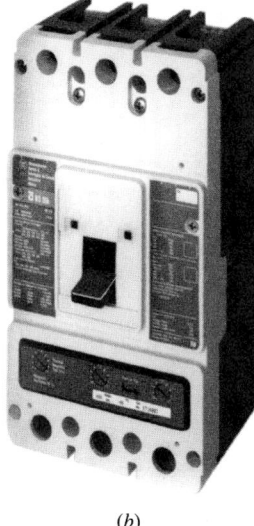

*(b)*

**Fig. 26.21** *(a) Essential elements of a plug-in-type molded case circuit breaker are shown in cutaway. Details vary with breakers designed for specific applications. The unit shown is typical of breakers in the 50- to 100-A size. (Photo courtesy of The Square D Company.) (b) Heavy-duty molded case circuit breaker, 400-A frame, 2- to 4-pole, 240–600 V. This circuit breaker is available with either a standard thermal-magnetic trip or an adjustable electronic trip with ground fault protection. Among the accessories available are auxiliary contacts, shunt trip, undervoltage release, and an electrical solenoid operator that permits remote control of the breaker. The unit is suitable for individual or panelboard mounting. A typical 3-pole unit measures 5½ in. W × 10 in. H × 4 in. D (140 × 255 × 100 mm) and weighs about 12 lb (5.5 kg). (Photo courtesy of Cutler-Hammer.)*

and auxiliary contacts, and all good breakers have trip-indicating handles and are *trip-free* (i.e., will trip out harmlessly if closed in on a short-circuited line). Low-voltage (600-V class) circuit breakers are available in frame sizes ranging from 50 to 4000 A and 1 to 3 poles. Their characteristics vary widely and are beyond the scope of this book.

## (c) Characteristics of Fuses and Circuit Breakers

Although both fuses and circuit breakers are circuit-protective devices, their characteristics differ markedly, as can be seen in the following tables.

### Fuses—Switch-and-Fuse Combination

| ADVANTAGES |
| --- |
| Simple and foolproof |
| Constant characteristics (no aging) |
| Initial economy |
| Very high interrupting capacity (IC) |
| No maintenance |
| Instantaneous; energy-limiting |

| DISADVANTAGES |
| --- |
| Fuses are single pole only |
| Necessity for storage of replacement fuses |
| Nonrenewable (one-time operation) |
| Nonadjustable |
| Nonindicating[a] |
| No electric or remote control[a] |
| Not trip-free |

### Circuit Breakers

| ADVANTAGES |
| --- |
| Usable as switches |
| Multipole |
| No replacement storage |
| Resettable |
| Indicates trip |
| Trip-free |
| Remote control |
| Adjustable |

| DISADVANTAGES |
| --- |
| Low to medium IC, except for special units |
| Periodic maintenance required |
| High initial cost |
| Complex construction changes with age |

[a]Can be accomplished with accessories.

These characteristics demonstrate that there is no relevance to the oft-posed question "Which are preferable—fuses or breakers?" The answer depends on the specific application involved and is often based on highly technical factors beyond this book's scope. Circuit breakers are generally used in lighting and appliance panelboards.

## 26.19  SWITCHBOARDS AND SWITCHGEAR

A switchboard is a large, free-standing assembly of switches, fuses, and/or circuit breakers, which normally provide switching and overcurrent protection

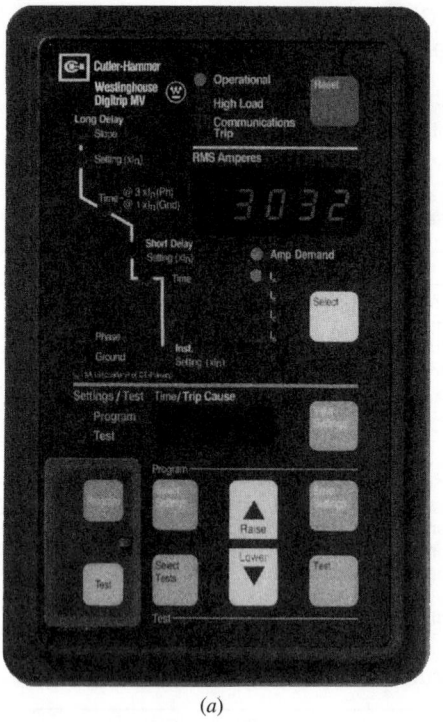

(a)

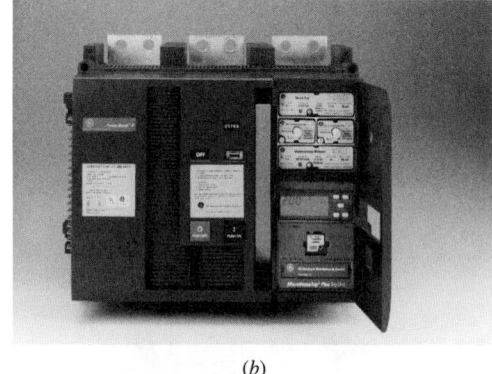

(b)

**Fig. 26.22** *(a) A multifunction microprocessor-based trip control unit for a large circuit breaker. It provides true root-mean-square (rms) sensing of phase and ground currents, selectable tripping characteristics, and trip alarm contacts. In addition, the unit can be used to display and transmit information on phase and ground currents. (Photo courtesy of Cutler-Hammer.) (b) Electrically operated large air circuit breaker with electronic microprocessor-controlled trip unit and additional remote-control accessories. The illustrated unit, rated 2000 A, is shown before stationary (nondrawout) mounting in a switchboard. (Courtesy of General Electric, U.S.A.)*

to a number of circuits connected to a single source. Metering and instruments are also often included. A switchboard may be represented in a single-line diagram, as in Fig. 26.23. In an electrical system, it serves to distribute, with adequate protection, bulk power into smaller "packages." Thus, by a hydraulic

analogy, the main buswork of the switchboard is equivalent to a main header, the switches to on/off valves, the fuses to flow-limiting devices, and the feeders to subheaders connected to the main header.

Modern switchboards (Figs. 26.24–26.27) are all deadfront; that is, they have all circuit breakers,

**Fig. 26.23** *One-line diagram of a typical switchboard. Switches are normally shown in the open position. Switches must be on the line (supply) side of fuses. Each line in a single-line diagram represents a 3-phase circuit. If circuit breakers were used, the entire board would be composed of units as depicted in circuit 6.*

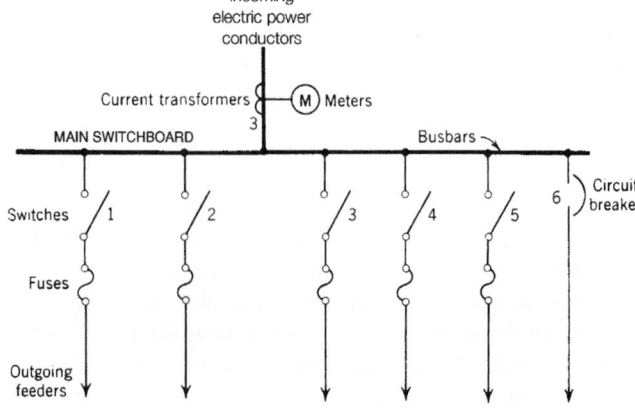

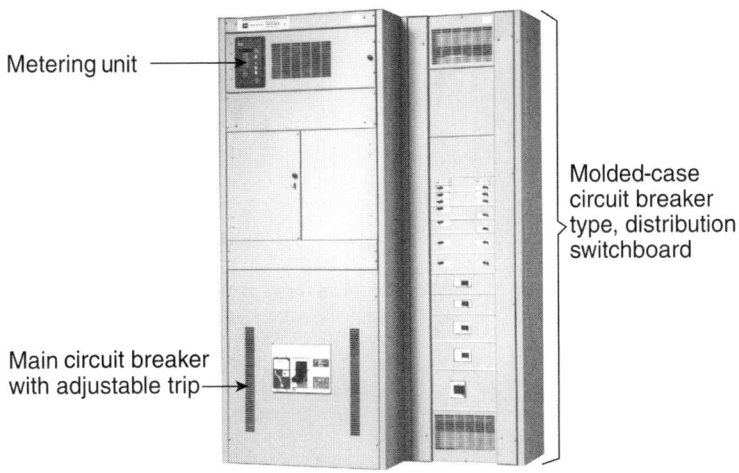

**Fig. 26.24** Free-standing, indoor-service, low-voltage (600-V) switchboard with group-mounted, fixed molded-case circuit breakers. The deeper section contains the service entrance main circuit breaker (maximum 3000 A) plus instrumentation. Section width varies from 30 to 45 in. (0.75 to 1.1 m) and depth from 18 to 36 in. (0.5 to 0.9 m), depending on the main circuit breaker rating. The smaller section contains branch circuit breakers. All units are 90 in. (2.3 m) high. (Photo courtesy of Cutler-Hammer.)

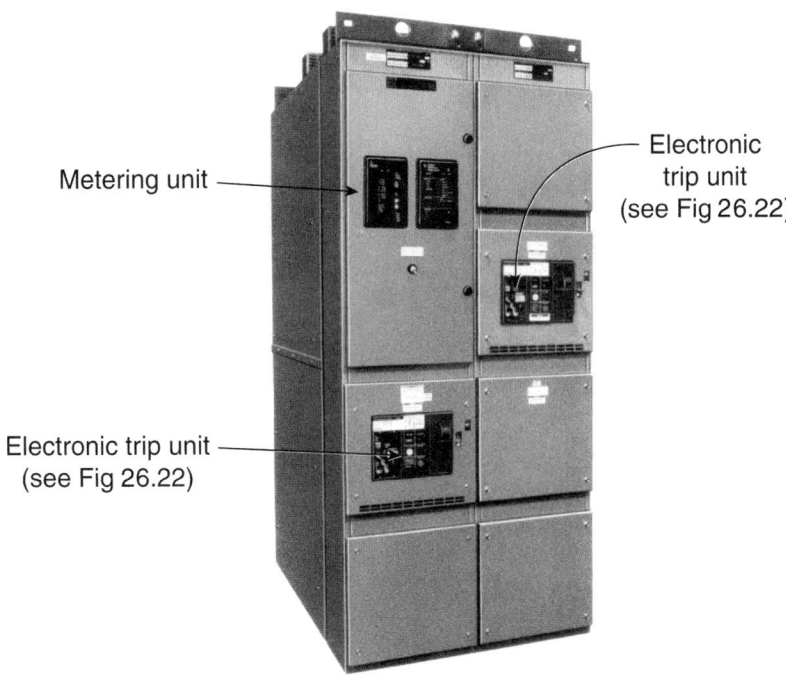

**Fig. 26.25** Metal-enclosed, indoor, low-voltage power switchboard with fixed or drawout large circuit breakers. Maximum bus capacity is 5000 A. Section dimensions vary with the size and type of contained equipment. All units are 90 in. (2.3 m) high. Each circuit breaker is equipped with a microprocessor-controlled trip unit similar to the unit shown in Fig. 26.22b. It can be seen clearly on the front of each unit. In addition, another programmable solid-state metering/analysis unit, similar to that shown in Fig. 29.15, can be seen at the left of the top compartment of the entry (left) cubicle of both this switchgear and the switchboard shown in Fig. 26.24. (Photo courtesy of Cutler-Hammer.)

Typical indoor assembly with a
breaker withdrawn on rails

*Fig. **26.26*** *Three sections of medium-voltage metal-clad drawout switchgear, suitable for indoor use as shown and for outdoor use with cabinet modification. Voltage ratings are 5–27 kV; current ratings are 1200–3000 A. Each indoor section measures 36 in. W × 95 in. H × 96 in. D (0.9 × 2.4 × 2.4 m) and contains two compartments for circuit breakers and/or instrumentation. Due to the drawout characteristic (shown), a 72-in. (1.8 m-) front aisle is required. Similar outdoor aisle-less switchgear measures 36 in. W × 115 in. H × 101 in. D (0.9 × 2.9 × 2.6 m). (Photo courtesy of Cutler-Hammer.)*

switches, fuses, and live parts completely enclosed in a metal structure. The operator controls all devices by means of push buttons and insulated handles on the front panel. When a switchboard has circuit breakers equipped with bayonet-type contacts and each is mounted in a movable drawer (like the drawers of a standard letter file), they are described as the *drawout* type. This drawout arrangement facilitates emergency replacements, inspection, and repairs and is illustrated in Fig. 26.26.

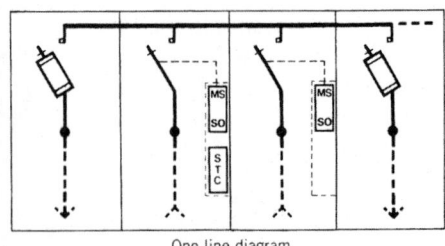

One line diagram

*Fig. **26.27*** *Four-bay (section), power-operated, high-voltage service entrance switchgear rated 13.8 kV, consisting of (see the single-line diagram inset and reading from left to right) a fused feeder section, two bays with parallel incoming lines and automatic transfer in the event of power failure, and a second switch and fuse feeder section. Fuses are electronic, with circuitry designed to provide a trip signal when excessive current is sensed rather than simply relying on the thermal characteristic of a fusible element. This gear, which measures 93 in. H × 172 in. W × 44 in. D (2.4 × 4.4 × 1.1 m) and weighs 8000 lb (3630 kg), is suitable for interior or exterior installation. (Courtesy of S&C Electric Co.)*

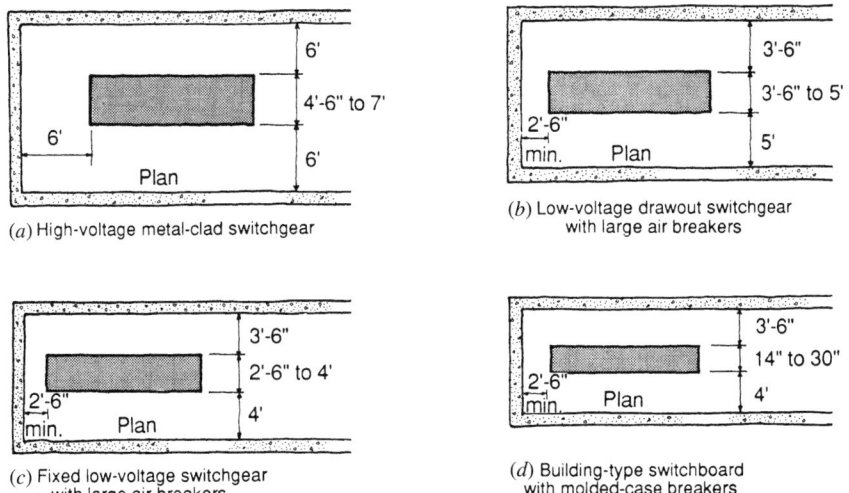

(a) High-voltage metal-clad switchgear

(b) Low-voltage drawout switchgear with large air breakers

(c) Fixed low-voltage switchgear with large air breakers

(d) Building-type switchboard with molded-case breakers

**Fig. 26.28** *Typical nominal switchgear space requirements. Each room should have two doors when switchgear is connected to high-capacity systems or operates at high voltage. Switchgear is shown in plan view. For detailed space requirements, see NEC Article 110.*

No clear distinction is made between the terms *switchboard* and *switchgear*, although generally, low-voltage switchboards with large circuit breakers and all high-voltage equipment (above 600 V) are referred to as *switchgear*. When molded case circuit breakers are utilized in a switchboard, it is often referred to as a *building-type switchboard*.

Recommended minimum space requirements for various types of switchgear are shown in Fig. 26.28. Working space around all types of electrical equipment must meet the requirements of *NEC* Article 110. Main metal-clad switchgear for commercial, industrial, and public buildings is almost invariably located in a basement and housed in separate well-ventilated electrical switchgear rooms. Smaller subdistribution switchboards require no special room. A wire screen enclosure to bar tampering or vandalism plus a large DANGER—HIGH VOLTAGE sign are usually adequate. The architect must provide adequate exits, hallways, and hatches for the installation and removal of all equipment. The specifications for switchgear should state the maximum overall dimensions of sections that can be accommodated in one piece.

When switchgear is to be installed outdoors, one of three methods is employed: build a small structure to enclosed normal indoor gear, utilize weatherproof outdoor gear, or utilize switchgear that is built into its own exterior enclosing structure, as seen in Fig. 26.27. These housings are equipped with heat and light and often prove to be the most economical choice.

## 26.20 UNIT SUBSTATIONS (TRANSFORMER LOAD CENTERS)

An assembly, comprising a primary voltage switch-and-fuse or circuit breaker, a step-down transformer, meters, controls, buswork, and secondary (low-voltage) switchgear, is known as a *unit substation* or a *load-center substation*. It is available for indoor or outdoor installation. Its function is to accept an incoming high-voltage power line, transform the high voltage to a level that can be utilized in the facility, and distribute the low-voltage power through low-voltage (secondary) switchgear. A dimensional physical sketch of a typical unit substation, along with its electrical single-line diagram, is shown in Fig. 26.29.

The location in the building of a unit substation is governed by the type of transformer utilized, as explained in the discussion on indoor transformer installations. For this reason, almost all indoor unit substations utilize dry-type (air-filled) transformers. Unit substations are utilized to effect the economies inherent in prefabricated construction with coordinated components. A basement location is most often selected, with ventilation

ELECTRICITY

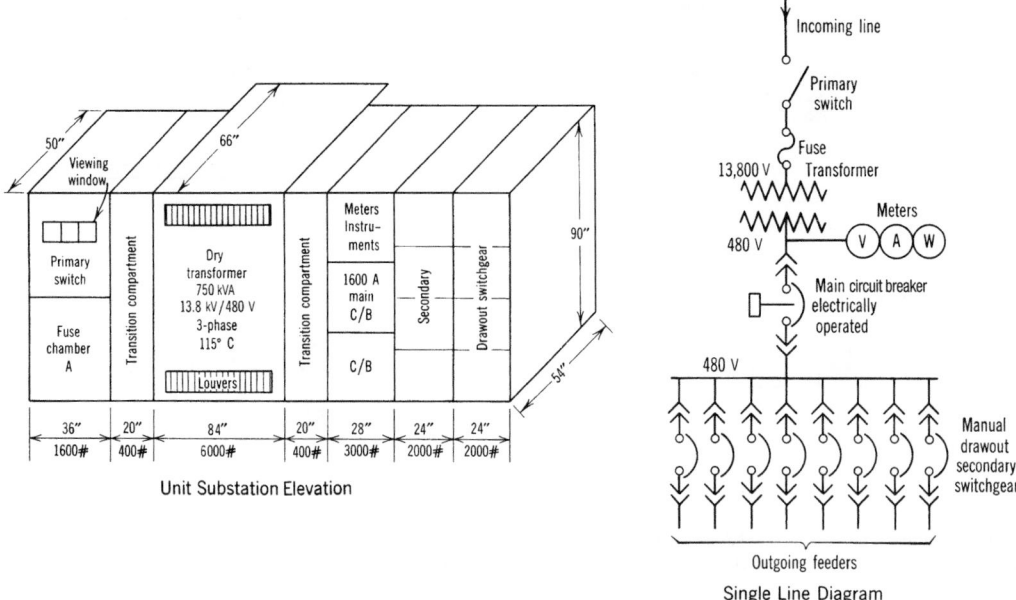

*Unit Substation Elevation*

*Single Line Diagram*

**Fig. 26.29** *Approximate sizes and weights of a typical large single-ended unit substation. Such a unit would supply a building with a maximum demand of 750 kVA. The incoming 13,800-V cables enter cubicle A and connect to the primary switch and fuses. The load side of the fuses connects to the transformer, which in turn connects to the secondary switchgear. The main secondary switchgear then feeds various switchboards and panelboards distributed throughout the building.*

requirements as detailed previously. Access should be restricted to authorized persons.

## 26.21 PANELBOARDS

An electrical panel, or *panelboard*, serves essentially the same function as a switchboard except on a smaller scale; that is, it accepts a relatively large block of power and distributes it in smaller blocks. Like the switchboard, it comprises main buses to which are connected circuit-protective devices (breakers or fuses) that feed smaller circuits. The panelboard level of the system is usually the final distribution point, feeding out to the branch circuits that contain the electrical utilization apparatus and devices, such as lighting, motors, and so on. Small panels, particularly in residential work, are frequently referred to as *load centers*.

The panelboard components—that is, the buses, breakers, and so on—are mounted inside an open metal cabinet called a *backbox*. The backbox is prefabricated with knockouts at the top, bottom, and sides to permit connection of conduits carrying circuit conductors (Fig. 26.30). The main feeders

that supply power to the panel's busbars enter through the large knockouts in the metal backbox, as in Fig. 26.30. Details of panelboard construction are presented in Fig. 26.31. A panel may be equipped with a main circuit breaker whose function is to disconnect the entire panel in the event of a major fault. Figures 26.30 and 26.31 show panelboards with only branch circuit devices and no main breaker or switch. These panels are described as "lugs in mains only," which means that the panel has only connectors (lugs) on the main busbars for connection of the main feeder cables and no main protective device. Three basic panelboard types are shown in Fig. 26.32.

The line terminal of each circuit protective device (breaker or fused switch) is connected to the busbars of the panelboard. The load terminal of the device then feeds the outgoing branch circuit. This is shown schematically in Fig. 26.33, which is a line-drawing representation of an electrical panelboard. Notice that the circuit breakers are arranged in two vertical rows, corresponding to actual construction. In the illustrated line drawing, a 3-phase panel is shown (i.e., three-phase busbars and a neutral bar). The busbars of the

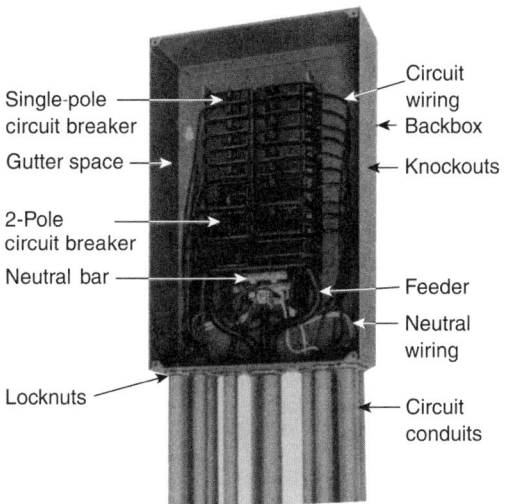

**Fig. 26.30** *A wired panelboard is shown with the cover removed. This panel is single-phase, 3-wire, meaning that it has two phase bars and a neutral (note the three heavy feeder cables). Lighting and appliance panels such as this are 4½–6 in. (115–150 mm) deep, 16–20 in. (400–500 mm) wide, and of sufficient height to accommodate the devices. The top device may be no higher than 78 in. (2 m) AFF and the bottom one no lower than 18 in. (450 mm) AFF to meet NEC requirements.*

**Fig. 26.31** *Cutaway of a typical lighting and appliance panelboard. Branch circuit protective devices are normally arranged in two vertical rows even in 3-phase panels. The circuit directory on the inside of the panelboard door gives a brief description of the loads connected to each branch circuit breaker.*

panelboard are energized by a feeder from a switchboard or large power distribution panel.

Panelboards are described and specified by type, bus arrangement, branch breakers, main breaker, voltage, and mounting—although not necessarily in that order. A typical description might be: lighting and appliance panel, 3-phase, 4-wire; 200A mains; main c/b, 225A frame, 150A trip. Branch breakers—all 100A frame; 8 ea. SP-20A, 4 ea. 2P-20A, 4 ea. 3P-20A; flush with hinged locked door.

## 26.22 PRINCIPLES OF ELECTRIC LOAD CONTROL

The key to energy conservation and electric demand limitation is load control. In the recent past, the extent to which both of these desirable operating functions were put into practice was essentially an economic decision: would the electric billing savings justify the expense of installing the required control? Today, and increasingly in the future, legislation resulting from environmental pressures mandates energy-use limitation. Already many states have enacted legislation requiring lighting control

in certain nonresidential buildings, and the trend toward additional and more stringent requirements is quite clear.

Lighting control strategies are discussed in Chapter 15. The essential control principle that applies to all loads, not only to lighting, is straightforward: the load is switched or modulated in response to a control signal (Fig. 26.34*a*). In terms of hardware, the process is somewhat more complex. Figure 26.34*b* shows in principle the equipment and wiring required in a common application: time control of exterior lighting, with photocell override. Figure 26.35 shows a more complex arrangement for area lighting control with local override, as might be installed in an area of a commercial office building. In each case, there are signal-initiation devices, one or more relays that performs the actual electrical switching, and the control and power wiring, in addition, of course, to the main power wiring from the panelboard.

Many users found that the payback from reduced electric billing was insufficient to cover the control equipment cost, particularly where local and/or remote override was required, and as a result relied on inefficient and generally ineffective manual control. That option, however, is limited today and will become even more so in the future.

ELECTRICITY

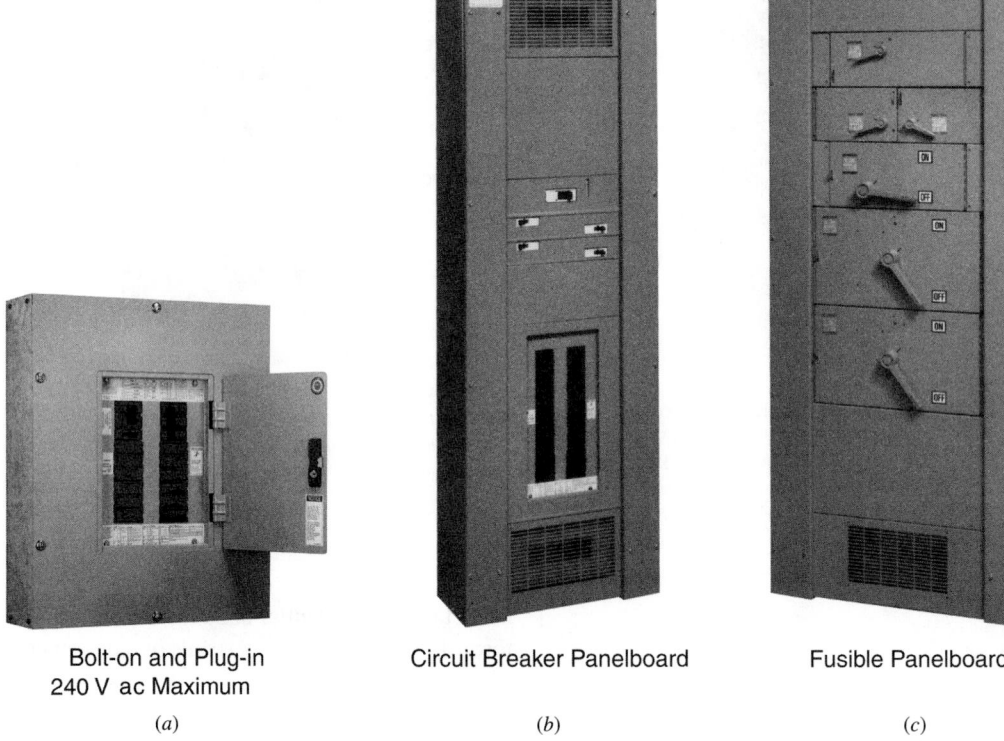

|  |  |  |
|---|---|---|
| Bolt-on and Plug-in<br>240 V ac Maximum | Circuit Breaker Panelboard | Fusible Panelboards |
| *(a)* | *(b)* | *(c)* |

**Fig. 26.32** *Three different designs of panelboards. (a) Circuit breaker–type panelboard using bolt-on and plug-in molded case circuit breakers 1- to 3-pole, 15–100 A. A main device of up to 400 A can be installed. The panel is rated for 240 V ac. The illustrated panel, which is arranged for surface mounting, measures 30 in. H × 20 in. W × 5¾ in. D (760 × 510 × 145 mm). (b) Circuit breaker distribution-type panelboard with maximum bus ratings of 1200 A and 600 V ac. Branch circuit breakers can be 15–100 A, single-pole and 15–1200 A, 2- or 3-pole. These panels are free-standing but can be wall-mounted, provided that the highest device does not exceed 6 ft, 6 in. (2 m) AFF. Similar units are available with a front door. Depending on the contents, cabinets can be 24–36 in. (610–915 mm) W × 60–90 in. H × 6–8 in. (150–200 mm) D. (c) Switch- and fuse-type distribution panelboard, free-standing, with ratings similar to those of the circuit breaker type in (b). Cabinets are 36–44 in. (915–1115 mm) W, 74–90 in. (1880–2290 mm) H, and 10½ in. (265 mm) D. (Photos courtesy of Cutler-Hammer.)*

As a result, a more economical way to accomplish the required control functions was developed, thanks in large measure to modern miniaturized programmable control elements. These can be installed directly in the lighting panelboard. Such a panel is known in the industry by the generic term *intelligent panelboard* and, of course, by various trade names that identify specific manufacturers.

## 26.23 INTELLIGENT PANELBOARDS

The idea behind the intelligent panelboard is straightforward. If many of the necessary electric load control and switching functions are incorporated into the panelboard by use of a compact centralized, programmable microprocessor, the corresponding external devices and the associated wiring are eliminated. The switching function is made possible by the use of motor- or solenoid-operated panel circuit breakers, thus taking advantage of their switching capability and eliminating the need for relays, contactors, and switches. The internal central controller permits local programmable control of each panel circuit breaker *individually*.

This central panel-mounted controller can also accept signal data from individual remote or network sources and can provide status reports, alarm

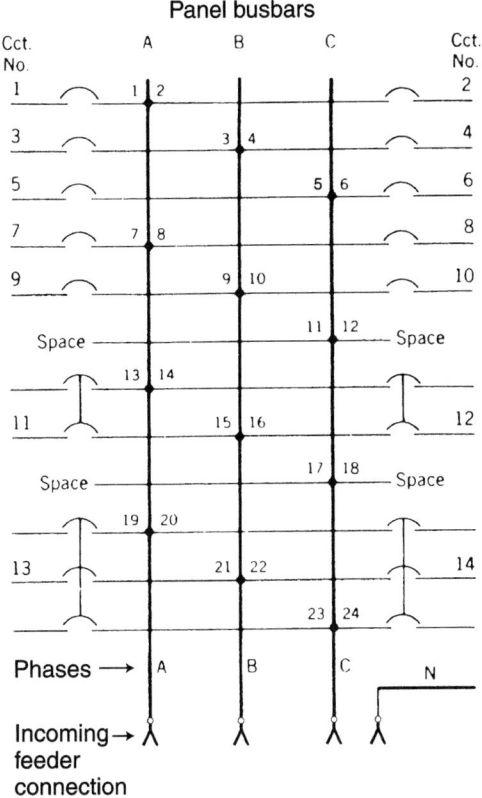

**Fig. 26.33** *Typical schematic diagram for a panel. Note that single-pole, 2-pole, and 3-pole circuit breakers are connected to 1-, 2-, and 3-phase buses, respectively. They supply 120-V 2-wire, 120/208-V 3-wire, and 120/208-V 3-phase 4-wire, respectively. This panelboard would be described as a 3-phase 4-wire lighting and appliance panel, with 10 SP, 2-2P, and 2-3P circuit breakers, lugs in mains only. The voltage of the panel and the ampere ratings of buses and all circuit breakers would then be specified.*

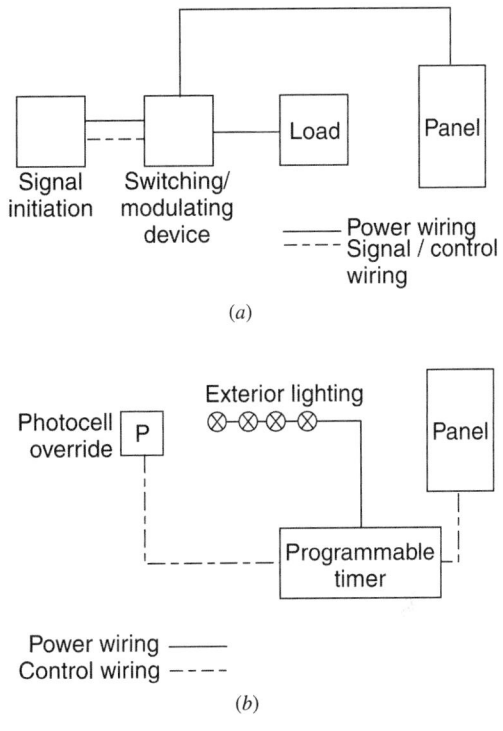

**Fig. 26.34** *(a) A low-voltage lighting control system always includes the elements shown. Signal initiation can come from a sensor such as a photocell or from a control device such as a timer, an override signal, a central computer signal, and the like. The signal activates the power-switching device, which in turn switches or modulates the load. The power source is always the associated lighting panelboard. (b) Schematic arrangement of a control system for exterior lighting. If the system is compact, a switching timer would be used. For a spread system, timer-controlled switching relays at each light would be used with attendant additional control wiring. Grouping of the lights into different control sequences (road, parking lot, building entrance, etc.) further complicates the wiring.*

**ELECTRICITY**

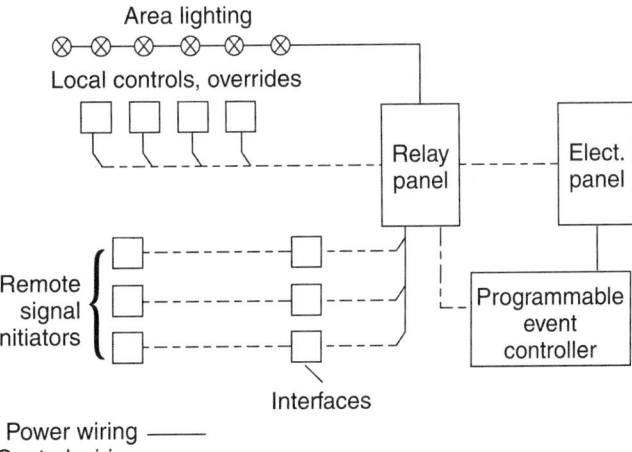

**Fig. 26.35** *Schematic diagram showing equipment and connections that might be used for control of a block of lighting in a large commercial building. The wiring can be even more complex and extensive if control of smaller lighting groups is desired or if control is desired from more than one location.*

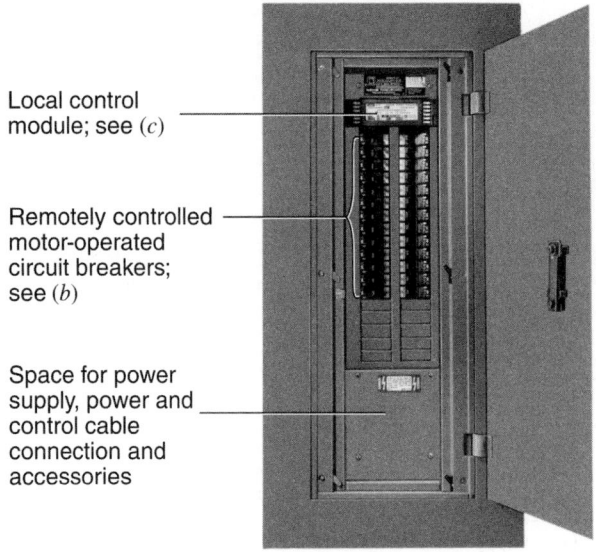

Local control module; see (c)

Remotely controlled motor-operated circuit breakers; see (b)

Space for power supply, power and control cable connection and accessories

(a) Intelligent panelboard

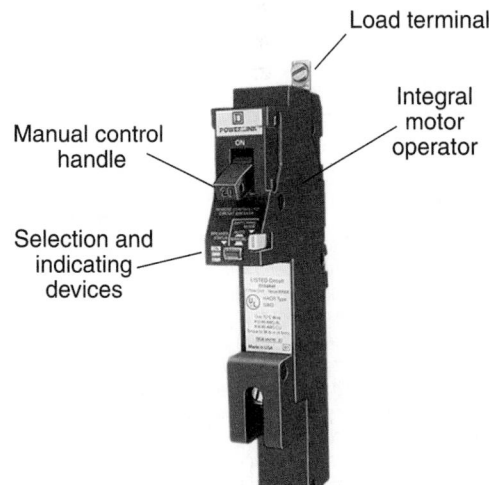

Load terminal

Manual control handle

Integral motor operator

Selection and indicating devices

(b) Single-pole motor-operated circuit breaker

Programming buttons

Status screen

(c) Control module

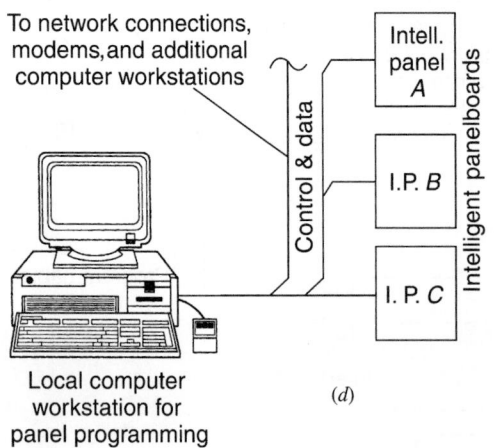

To network connections, modems, and additional computer workstations

Intell. panel *A*

I.P. *B*

I.P. *C*

Control & data

Intelligent panelboards

Local computer workstation for panel programming

(d)

**Fig. 26.36** (a) This intelligent panelboard contains (from top to bottom) a microprocessor control module, 26 single-pole, remote-controllable (motor-operated) circuit breakers, space for 10 additional poles, and a compartment containing a dc power supply (for circuit breaker operation) and accessories. (b) This motorized single-pole circuit breaker can be manually or remotely controlled. Electric operation latches the circuit breaker so that in the event of power failure, all breakers remain in their pre-outage condition. The illustrated unit is a 100-A frame, 20-A trip unit. A status window on each unit displays its condition, while electronic sensors in each unit sense voltages and indicate on the control panel the status of each unit. (c) The control module, mounted at the top of the panelboard, is equipped with programming buttons that allow control of each circuit breaker individually. The illustrated basic control module programs and controls essential functions including configuration, circuit breaker status, resetting, zoning, settings for hardware characteristics, override modes, screen selection, and self-diagnostics. Additional control modules (not shown) supply time-based programming, logging functions, and timed functions. The illustrated breaker status screen is one of five displays. The others are setup, program, override, and diagnostic screens. (d) Intelligent panels can be incorporated into a large control network by use of interface connections of various types (not shown). The illustration is a block diagram showing one of the networking possibilities. (All photos courtesy of Square D.)

signals, operational logs, and local bypass and override functions. Control cards in an adjacent cabinet, connected to the controller, can provide local switch override, daylight dimming and on-off control, telephone override energy control, electric demand control, and networking. With careful planning, the high initial cost of the intelligent panelboard can be offset by elimination of remote relays, relay panels, programmable time switches, remote-control switches, and all of their associated wiring. Simplified and improved facility operation, plus reduced maintenance costs and electric bills, are additional benefits that can be achieved with proper application and careful design. Figure 26.36 shows the essential elements of an intelligent panelboard and a typical network arrangement using the remote-control capabilities of such panels.

## 26.24 ELECTRIC MOTORS

Motors are frequently supplied with the equipment they drive (such as fans, blowers, and so on) as a complete package. The actual choice of motor is left to the driven-equipment supplier, because the supplier is presumed best qualified to select a motor that will optimally match the driven-equipment requirements for whose proper operation the supplier is responsible. In practice, however, the supplier is frequently guided primarily by the price. The designer, therefore, should be sufficiently knowledgeable so that the particular motor desired can be specified. The following sections are written with that purpose in mind, and are therefore primarily concerned with application.

### (a) Direct-Current Motors

These motors are not normally used in building work. The fine speed control for which they were primarily used previously is now available, more economically, with ac motors, as explained in Section 26.26.

### (b) Alternating-Current Motors

These motors fall into three general classifications: poly-phase induction motors, poly-phase synchronous motors, and single-phase motors. Within these categories there are further subdivisions. Of these

many types, most motors used in building equipment are squirrel-cage induction machines; therefore, this type is studied in some detail.

### (c) Squirrel-Cage Induction Motors

This motor owes its interesting name to an early design in which the rotor consisted of a group of bars welded together into a cylindrical cage-type shape. The design, invented by Nikola Tesla, is basically unchanged today except for refinements. Squirrel-cage motors are manufactured in four different NEMA designs to meet different application requirements. Of these the most common are:

*Type B:* standard design, high efficiency and power factor, normal torque; applicable to fans, blowers, and pumps

*Type C:* high starting torque, fair efficiency and power factor; applicable to compressors, conveyors, and other devices that start under load

A motor nameplate gives important information on a motor that is not self-evident:

1. *Type:* This is the manufacturer's designation and indicates primarily the enclosure. Common enclosures are open drip-proof, totally enclosed, fan-cooled, and weather-protected.
2. *Duty* (time rating): Continuous or intermittent.
3. *Service factor:* Permissible overload; generally 15%.
4. *kVA code:* Indicates by a letter the maximum starting current per horsepower. This is useful in selecting motor protective devices. More recently, actual locked rotor amperes have been given.
5. *Frame:* A NEMA standard number that indicates the motor's physical dimensions.
6. *Motor voltage:* The standard motor voltages are 208, 230/460, and 575 V. Induction motors generally operate satisfactorily ± 10% voltage. Only 208-V motors should be used on 208-V systems because the actual line voltage may be as low as 200 V. Using a 230-V on a 200-V motor will result in sharply reduced torque, increased temperature rise, and poor overload capacity.
7. *Motor nominal full-load efficiency:* A requirement of ANSI/ASHRAE/IESNA Standard 90.1 and EPACT.
8. *Design letter:* Indicates inherent motor design characteristics.

### (d) Electric Motor Energy Considerations

The Federal Energy Policy Act of 1992 (EPACT) requires that the most common designs of induction motors, accounting for 70% to 80% of all poly-phase induction motors sold in the United States, meet high-efficiency and power-factor standards. Specifically, these efficiency standards apply to single-speed, poly-phase, squirrel-cage induction motors, designs A and B, continuous rating, and operating at 230/460 V, 60 Hz. These motors are known as *premium efficiency* or *PE* motors.

The decision on whether to use a PE motor in applications not mandated by law is essentially an economic one. Many motor manufacturers make available straightforward software that enables a designer to rapidly make a detailed life-cycle cost comparison among motor types, with operating time, loading, first cost, energy and maintenance costs, and so on as the variables. As a guideline, motor applications involving low starting torque and frequent motor operation at partial load are likely candidates for short payback periods on the price premium of PE motors.

### 26.25 MOTOR CONTROL STANDARDS

Until recently, American designers and equipment manufacturers utilized motor control equipment manufactured to NEMA standards exclusively. It was also common practice in the construction industry to split the motor and motor-control package by having equipment manufacturers supply the motorized equipment while the electrical construction contractor supplied the motor-control equipment. Few coordination problems between controller and motor drive were encountered because control equipment sized and built to NEMA standards is large, heavy duty, and applicable to a wide range of motor sizes and characteristics.

The International Electrotechnical Commission (IEC) is a technical organization with 42 member nations, including the United States. It develops recommended standards for electrical equipment, including motor controls. The IEC design concept is one of close coordination between a motor and its controller. This application sensitivity, coupled with flexible ratings for a single piece of equipment, depending on its application, makes equipment built to

IEC standards smaller, lighter, and cheaper than similarly rated equipment built to NEMA standards. As a result, many U.S. manufacturers utilize IEC motor controls in conjunction with their motive equipment, particularly on international jobs. Because American architects, engineers, and contractors are involved in a large number of construction projects outside of the territorial United States, those responsible for designing, specifying, and purchasing electrical equipment in general and motor-control equipment in particular should be aware of the comparative characteristics of NEMA-based and IEC-based equipment.

### 26.26 MOTOR CONTROL

### (a) Fundamentals

A conventional ac motor controller is basically a contactor (see Section 26.14) designed to handle the heavy inrush currents encountered in starting an ac motor. Its function is twofold: to start and stop the motor and to protect it from overload. These two separate and distinct functions are accomplished by combining a set of contacts for on/off control with a set of thermal overload elements for overload protection in a single unit. When the contacts are operated by hand, the controller is called a *manual starter;* when the contacts are operated by a magnetic coil controlled by push buttons, thermostats, or other devices, the unit is known as a *magnetic controller* or, simply and more commonly, a *magnetic starter.*

Motors of 1 hp or less are generally controlled by a manual switch that contains an overload protection device. It is advisable to utilize such a device for all fractional horsepower motors.

Most starters are of the full-voltage across-the-line (ATL) type; that is, the contacts place the motor directly onto the line, and the motor starts up immediately. When such a procedure is undesirable because of voltage dip and flicker caused by the large inrush current or because of utility company limitations, a reduced voltage starter, sometimes called a *compensator,* is used. These units initially apply reduced voltage to the motor, thus reducing the starting inrush current and line voltage drop. This in turn reduces dimming of lights, flicker, and other undesirable effects.

Older reduced-voltage starters use a stepping

arrangement whereby voltage is increased incrementally, thus limiting inrush current. Unfortunately, starting torque is also limited, thus restricting the application of these starters. Solid-state starters have been available since the late 1980s that provide continuous (stepless), controlled motor starting. These units not only limit inrush current, but also, by adjusting the acceleration time, the required starting torque of the motor can be maintained. Such starters provide *soft* starts (i.e., starts that minimize the mechanical stresses caused by rapid application of accelerating torque). Additional advantages of these units are reduced size and weight, long life, more sophisticated motor protection, and additional operating functions such as jogging and reversal. Disadvantages are higher cost and possible radio frequency noise problems (Fig. 26.37).

## (b) Motor Speed Control

Many applications in HVAC, fluid piping, and industrial systems require speed control of electric motors.

Until recently, this constituted a problem whose solution was expensive, often space-consuming, and rarely energy efficient. The reason for this is that the common squirrel-cage induction motor is essentially a constant-speed device, where speed is determined by power line frequency and motor design. Speed drops slightly (slips) as the load increases. Until the advent of cheap electronic power equipment, speed control was usually accomplished by using a wound-rotor motor in lieu of a squirrel-cage unit. This drastically increases both motor and controller costs while providing only limited speed control.

Advances in power electronics have made practical a variable-voltage, variable-frequency (VVVF) controller that gives smooth, continuous speed control over a range exceeding 30 to 1 while maintaining motor torque. These highly reliable controllers (also known as *variable frequency drives*, VFD) provide considerable energy economies that usually result in rapid payback of the relatively high first cost of equipment. The essentials of the VVVF control scheme are shown in Fig. 26.38. Voltage control, in addition to frequency control, is necessary in order to maintain a constant voltage/frequency ratio, which is necessary for efficient, safe operation. An increase in this ratio results in overheating of the motor, whereas a decrease results in insufficient torque. Additional advances in power electronics have produced motor speed controllers with controlled speed ratios approaching 1000:1.

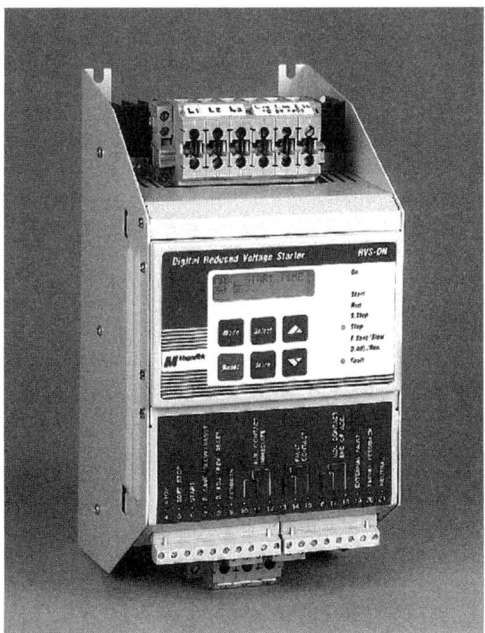

**Fig. 26.37** *Soft starter type for 3-phase induction motors of up to 1400 hp (1045 kW), shown without an enclosure. This microprocessor-based unit provides two selectable starting characteristics with stepless acceleration, plus additional control and protection functions including reversing control, over- and undervoltage and current protection, and two normal motor operating speeds. (Photo courtesy of MagneTek.)*

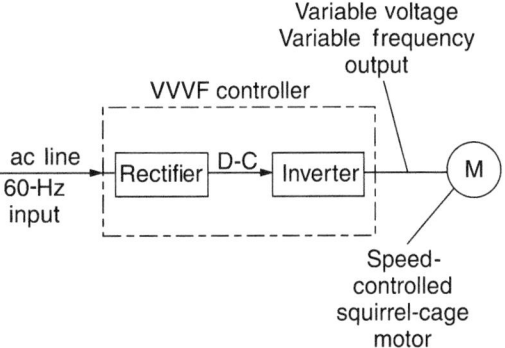

**Fig. 26.38** *Schematic representation of the action of a variable voltage and frequency (VVVF) motor speed controller. Line voltage at 60 Hz is rectified to dc and the inverted voltage to ac. Frequency and voltage control are built into the inverters, designed to maintain a constant voltage-to-frequency ratio. This constancy is necessary to prevent malfunction of the motor as its speed is varied.*

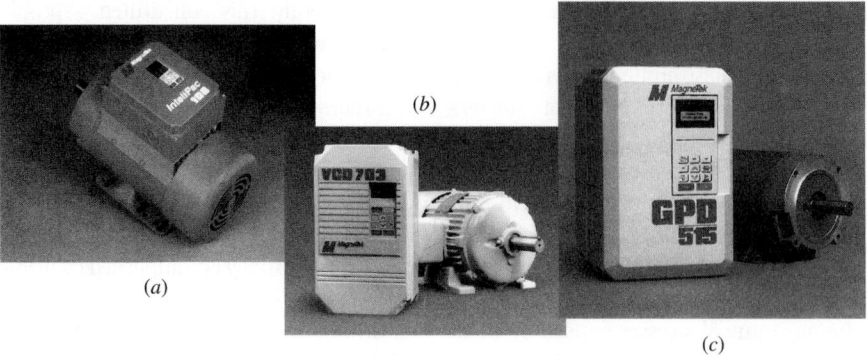

*(a)*

*(b)*

*(c)*

**Fig. 26.39** *Three modern programmable, microprocessor-controlled motor controllers. (a) Integrated motor and controller for motors of up to 1 hp (0.75 kW). The controller provides a soft start, a speed control range of 40:1, high efficiency, multiple speed settings, and full motor protection. (b) Controller for motors of up to 3-phase, 400 hp (300 kW); this type of unit provides programmable functions including speed control from standstill to double-rated speed, reversing, adjustable acceleration, and full motor protection. (c) Controller that can be programmed for three different control modes. Speed control range, regulation, and torque curves vary in each mode to suit the specific controlled load. The controller has low harmonic distortion plus fault diagnostic software, in addition to the normal overload and protective characteristics. It also has terminals for remote monitoring and control plus network control capabilities. A 20-hp (15 kW-), 230-V controller in a NEMA 1 (indoor) enclosure measures approximately 16 in. × 10 in. × 9 in. (400 × 250 × 230 mm) and weighs 24 lb (11 kg). (Photos courtesy of MagneTek.)*

One disadvantage of these speed controllers has been the production of line harmonics and radio noise of sufficient strength to constitute a serious engineering problem. At this writing, these problems have been mitigated, and will undoubtedly be pursued to the point that, except for highly sensitive areas, the interference will be negligible. At this point, however, the designer must investigate these effects for the particular application. Typical adjustable-frequency controllers are shown in Fig. 26.39.

## 26.27  MOTOR CONTROL EQUIPMENT

All the starters discussed previously are available in a wide range of sizes, voltages, and enclosure types. Every motor controller is required by the *NEC* to have a disconnecting means for safety reasons. Where convenient, this disconnect switch may be combined with the starter into a single unit known as a *combination starter.* A circuit breaker or fused switch is often used in such an arrangement, which then constitutes the branch circuit protection and disconnecting means (Fig. 26.40).

A typical, brief description of a conventional motor controller would be similar to the following: combination circuit-breaker type, across-the-line motor controller, NEMA size 2, three O.L. elements,

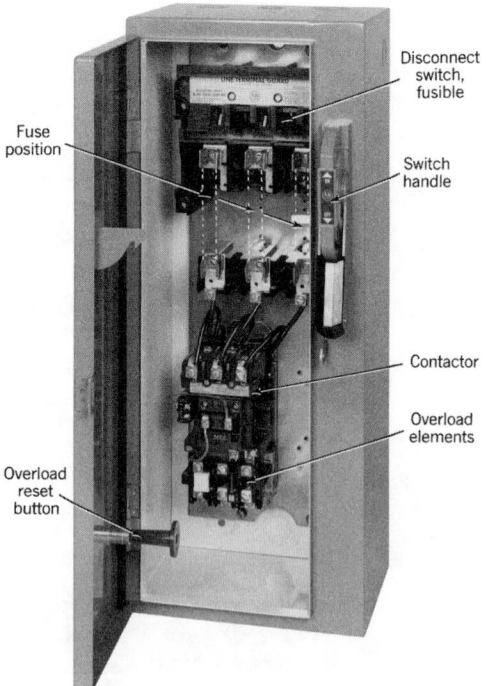

Disconnect switch, fusible

Fuse position

Switch handle

Contactor

Overload elements

Overload reset button

**Fig. 26.40** *Interior of a conventional (non-solid-state) combination fused switch-type, across-the-line motor controller. Note that the unit is essentially a switch and a starter wired together and installed in a single cabinet. Approximate dimensions of the illustrated unit are 20 in. W × 48 in. H × 9 in. D (510 × 1140 × 230 mm), with a 75-hp (56 kW) capacity. (Courtesy of Allen-Bradley Co.)*

ELECTRICITY

**TABLE 26.6 Rating and Approximate Dimensions[a] of ac Full-Voltage Conventional Single-Speed Motor Controllers, 3-Phase Combination Circuit Breaker Type[b]**

| NEMA Size[c] Designation | Maximum Horsepower[c] | Width | Height (in.) | Depth |
|---|---|---|---|---|
| 0 | 3 | 10 | 24 | 7 |
| 1 | 17½ | 10 | 24 | 7 |
| 2 | 15 | 10 | 24 | 7 |
| 3 | 30 | 20 | 24 | 9 |
| 4 | 50 | 20 | 48 | 9 |
| 5 | 100 | 20 | 56 | 11 |

[a]Dimensions vary among manufacturers.
[b]All starters are housed in a NEMA 1 indoor ventilated enclosure.
[c]Maximum hp that can be controlled at 208–230 V. Generally, when operating at 460 V, a starter one size smaller can be used.

208 V, in a NEMA 1 enclosure. Starter shall contain integral on/off push buttons.

Table 26.6 gives the approximate dimensions of one particular type of combination starter. As with other electrical equipment, a current manufacturer's catalog must be consulted for reliable data for *that* manufacturer. Dimensions vary somewhat among manufacturers producing a product that meets NEMA specifications. As noted previously, physical characteristics can vary greatly among items made to meet IEC specifications and those made to NEMA specifications for control equipment of similar rating.

In spaces where a group of motors are installed in close proximity to one another, it is often convenient from the control aspect, as well as economical, to combine the motor starters, disconnect switches, motor controls, and indicating devices into a single large assembly. Such an assembly is called a *motor control center* (MCC). Typical MCC types are shown in Fig. 26.41.

## 26.28 WIRING DEVICES: GENERAL DESCRIPTION

The general term *wiring devices* includes all devices that are normally installed in wall outlet boxes, including receptacles, switches, dimmers, fan controls, and so on. Attachment plugs, also called *caps*, and wall plates are also included in wiring devices.

The three grades of device quality, using NEMA and UL grading nomenclature, are hospital grade, federal specification grade, and UL general-purpose grade, in descending order of price and quality. Inasmuch as these grades correspond to published specification requirements, they are the only reliable gage of construction quality. Hospital-grade devices are identified by a green dot on the device face. These devices are built to withstand severe abuse while maintaining reliable operation and must meet UL requirements for this grade.

Manufacturers usually grade their devices as hospital grade, premium or industrial specification grade, commercial specification grade, and residential grade, in descending order of quality and price. Hospital grade is approximately the same among manufacturers because industry standards must be met. Industrial and commercial specification grades correspond roughly to federal specification grade and residential grade to UL general-purpose grade. Manufacturers' grades are qualitative, although reputable makers list the NEMA and UL specifications that each of their grades meets. It is only on this basis that a technical quality comparison can be made.

In application, industrial specification grade equipment is usually used in industrial and high-grade commercial construction; commercial specification grade in most educational and good residential buildings; and standard or residential grade in low-cost construction of all types. The grade of wiring devices should, as with all electrical equipment, be consistent with the quality of construction in the entire facility.

Although historically the term *wiring device* referred to devices that operate at line voltage (120, 208, 240, and 277 V) in a branch circuit, the term today also includes low-voltage lighting control devices, fan controls, and the like. The accepted criteria seem to be that only devices rated 30 amps or less, that can be mounted in a small wall box, are considered wiring devices. Communication attachments are not usually considered wiring devices even when they are supplied by the electrical contractor. They are part of *premise wiring*, which is the term commonly used to describe data and communication system wiring, raceways, and devices (see Fig. 27.22).

## 26.29 WIRING DEVICES: RECEPTACLES

A receptacle is, by *NEC* definition, "a contact device installed at the outlet for the connection of a single attachment plug." This usually takes the form of the

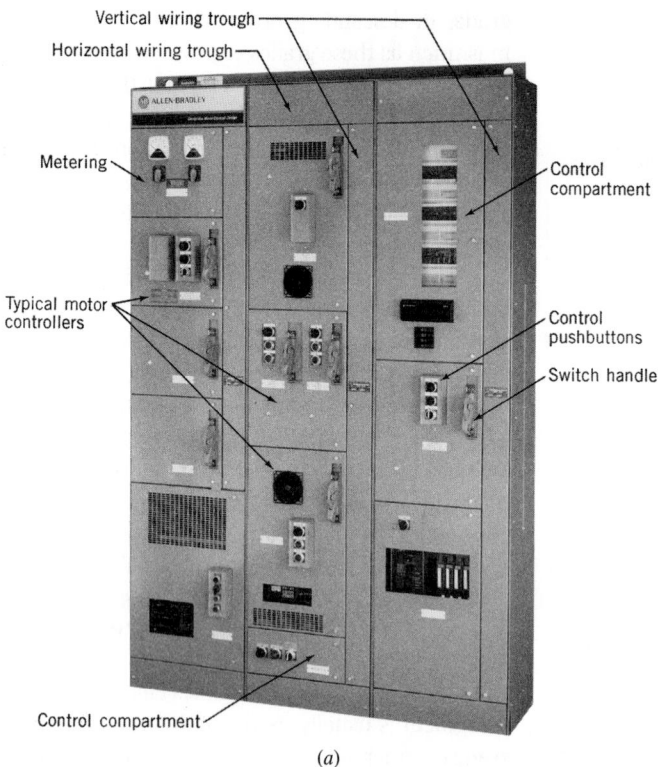

Vertical wiring trough

Horizontal wiring trough

Metering

Typical motor controllers

Control compartment

Control compartment

Control pushbuttons

Switch handle

*(a)*

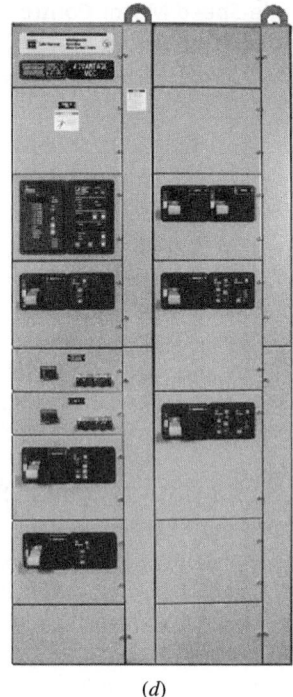

*(d)*

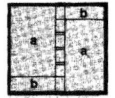

20" Back-to-back
Top view

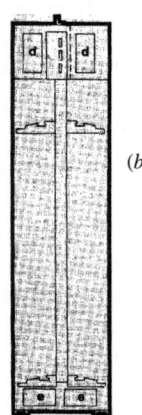

*(b)*

20" Back-to-back
Side view

a. Control Unit
b. Vertical Wireway
d. Top Horizontal Wireway
e. Bottom Horizontal Wireway

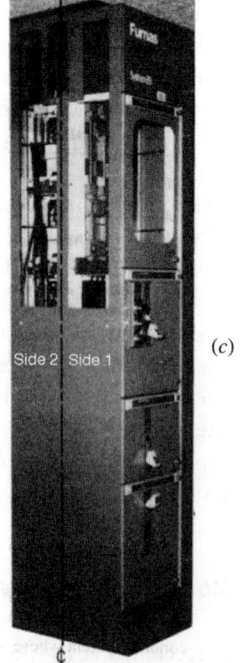

Side 2  Side 1

*(c)*

**Fig. 26.41** *Motor control centers (MCC) of various construc-tions. (a) Conventional MCC construction with magnetic motor starters, manual local control, analog metering, and conven-tional overload and protective equipment. (b) Back-to-back construction of conventional MCC adds only 5 in. (125 mm) to the basic 15-in. (380 mm) depth. (c) Units are 90 in. (2.3 m) high and 20 in. (500 mm) wide per section. Solid-state con-trollers can be installed in this unit, as well as in the MCC shown in (a), along with conventional magnetic contactors. (d) MCC that uses solid-state programmable controllers with heaterless overload protection. The unit also provides protection against phase loss, single phasing, and ground faults. Single side units in this construction are 90 in. (2.3 m) H × 20 in. (500 mm) W × 16–20 in. (400–500 mm) D (depending upon starter size). Back-to-back construction is 21 in. D. (Photos courtesy of Allen-Bradley [a], Siemens-Furnas [b,c], and Cutler-Hammer [d].)*

common wall outlet or, as illustrated later, larger and more complex devices. A comment here about terminology is required. The common wall outlet is properly called a *convenience receptacle outlet, a receptacle outlet,* or a *convenience outlet.* The term *wall plug,* which is heard so often, is incorrect. A *plug* is another name for the attachment device (cap) on the wire that carries electricity to the appliance. The device at the end of the line cord is plugged into the wall, hence *plug.*

Because by *NEC* definition a receptacle is a contact device for the connection of a single attachment plug, and because the normal wall conve-

nience receptacle takes two attachment plugs, it is properly called a *duplex convenience receptacle* or *duplex convenience outlet.* Most people shorten this term to *duplex receptacle* or *duplex outlet.*

Receptacles are identified by the number of poles, the number of wires, and whether the device is designed for connection of a separate equipment ground. (This ground may connect to a separate equipment grounding wire or to the metallic conduit system, according to system design.) The number of receptacle *poles* equals the number of current-carrying contacts, including the neutral. The equipment ground connection (pole) is not

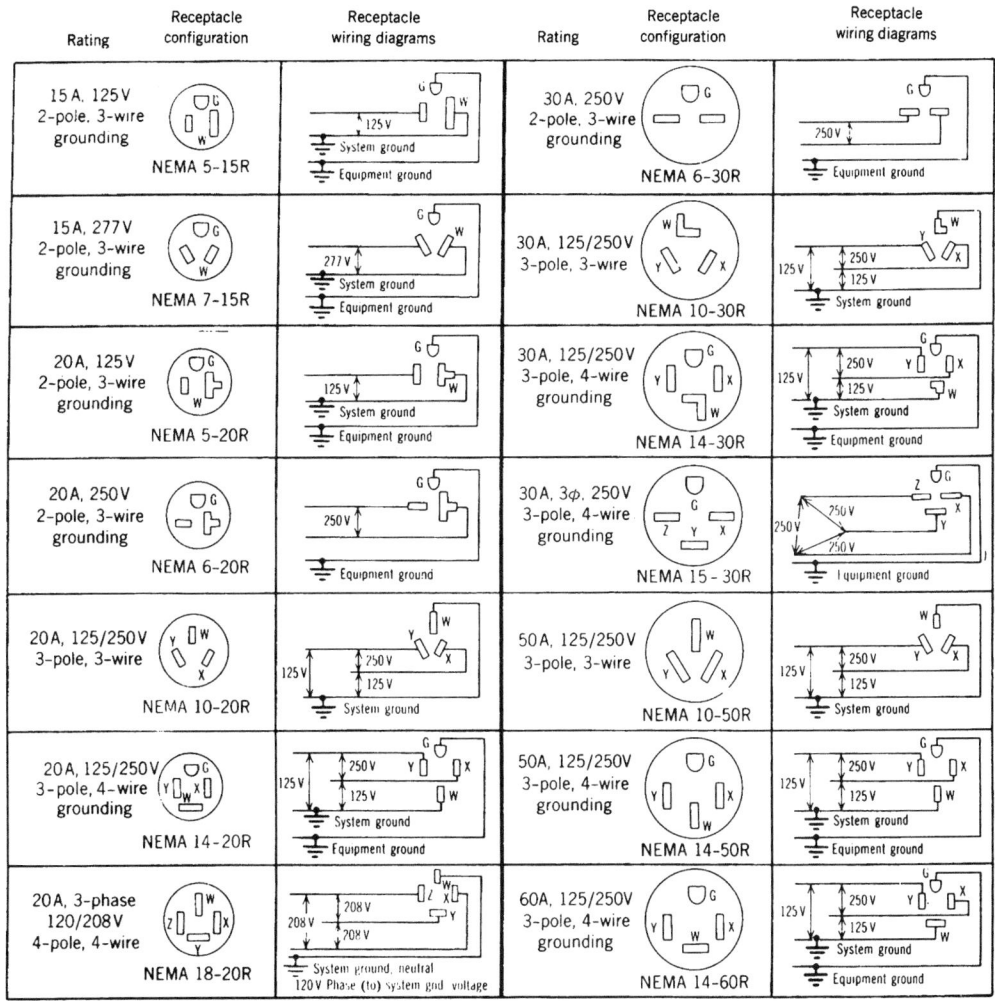

Note: "W" denotes white system ground or neutral
      "G" denotes green equipment ground.

**Fig. 26.42** Receptacle configuration chart with selected common general-purpose, nonlocking devices with associated NEMA designations.

counted because it does not (normally) carry current. The number of *wires* includes the equipment ground connection because it is wired. The equipment ground connection (grounding pole) should not be confused with the system ground (usually the neutral pole), nor may the wiring for the two be interchanged (see Section 28.3).

In a typical application, a receptacle for an electric dryer with a 4800-W, 208-V heating element and a ⅓-hp, 115-V motor would be NEMA 14-50R (Fig. 26.42). The motor would connect across W and X, the heater across X and Y, and the appliance case to G. Receptacles installed on standard 15- or 20-A branch circuits must be of the grounding type.

Receptacles connected to different voltages, frequencies, or current type (ac or dc) on the same premises must be polarized so that attachment plugs are not interchangeable. Figure 26.42 shows some of the standard receptacle configurations and their ratings. Figure 26.43 shows typical receptacle construction plus enclosures for use in damp and wet locations.

Receptacles are readily available from 10 to 400 A, 2 to 4 poles, and 125 to 600 V. In addition, special types such as locking, explosionproof, tamperproof, and decorative design units are made. Also, specific usage units such as range receptacles are available. All receptacles other than the normal 15/20-A, 3-wire, parallel-slot type should be

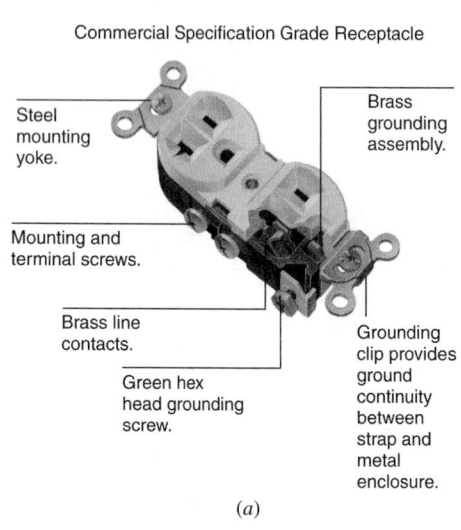

*(a)*

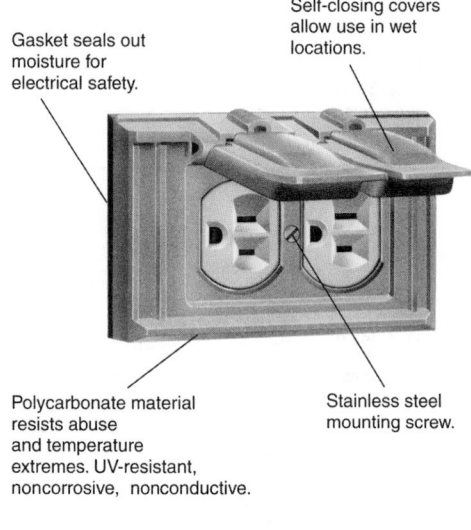

*(b)*

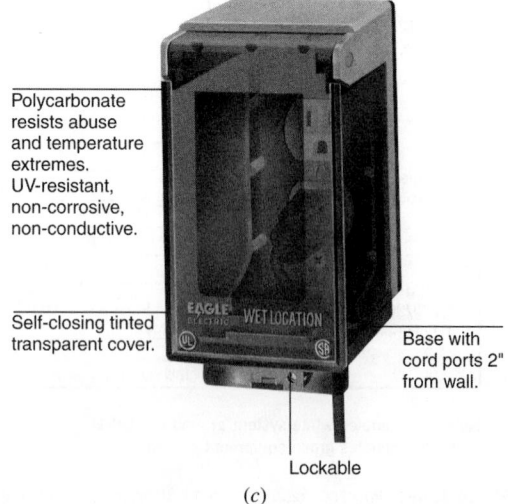

*(c)*

**Fig. 26.43** *(a) Cutaway of a commercial, specification-grade duplex convenience receptacle showing its construction. (b) Receptacles installed in damp locations, outdoors but not exposed to the rain, or in wet locations but used to energize portable equipment and continuously attended, should have a cover that is weatherproof when the attachment plug cap is removed, of which one type is illustrated. (NEC Article 410-57b(2).) (c) Receptacles in wet locations (that may be in use unattended) must have a weatherproof enclosure the integrity of which is not affected by the attachment of a plug. One such enclosure is illustrated. (NEC Article 410-57b(1).) (All photos courtesy of Eagle Electric Manufacturing Co., Inc.)*

specified to be furnished with at least two matching caps (plugs). A typical receptacle specification would be: receptacle, duplex, 2-pole, 3-wire, grounding type, 20 A, 250 V, federal specification grade, for indoor use. Receptacles are normally mounted vertically between 12 and 18 in. above the finished floor (AFF), except that in shops, labs, and other areas where tables are placed against the walls, 42 in. is the usual mounting height.

In addition to these types, several manufacturers produce a receptacle with built-in ground-fault circuit protection (Fig. 26.44). For a discussion of these ground-fault circuit interrupter (GFCI) receptacles, see Section 28.4.

The great sensitivity of modern electronic equipment to voltage surges and electrical *noise* (random, spurious electrical voltages) has led to the development of two special receptacles. Receptacles with built-in surge suppression protect the connected equipment from overvoltage spikes. Receptacles with an insulated equipment grounding terminal separate the device ground terminal from the system (raceway) ground because it has been determined that much of the unwanted electrical noise can be eliminated by such disconnection. The receptacle's insulated ground terminal is connected to an isolated green ground conductor that carries through the entire system but is connected to the system ground at the service entrance only; hence the name *isolated ground.* Isolated ground receptacles are identified by an orange triangle on their face (see Fig. 26.45).

*Fig. 26.45 Hospital-grade, insulated (isolated) ground receptacle used where heavy duty is expected, and both reliability and freedom from electronic noise are required. Hospital grade is denoted by the green dot on the upper left of the receptacle face and the insulated ground by the orange triangle on the upper right. (Some manufacturers make the entire receptacle orange.) (Photo courtesy of Leviton Manufacturing Co.)*

A receptacle with built-in surge suppression (see Section 26.37) is shown in Fig. 26.45.

## 26.30 WIRING DEVICES: SWITCHES

Switches of up to 30 A that can be outlet-box mounted fall into this category. Generally, ac-only switches are preferable to the ac/dc type because of their better construction. The usual ac switch rating is 15, 20, or 30 A at 120 or 120/277 V. Usual constructions are single-pole, 2-pole, 3-way, 4-way, momentary-contact, 2-circuit, maintained-contact SPDT and DPDT. Operating handles are toggle, key, push, touch, rocker, rotary, and tap-plate types. Mercury and ac quiet types are relatively noiseless; toggle, tumbler, and ac/dc types are generally not. A typical switch specification might be: switch, single-pole, ac, quiet type, federal specification grade, 15 A, 120 V, with press handles lighted when off, for side wiring only.

A switch incorporating a solidstate rectifier is readily available that gives high/off/low control for *incandescent lamps* and costs very little more than an ordinary switch. Typical applications are in areas where a lower illumination level is often acceptable and always desirable as an energy-conserving

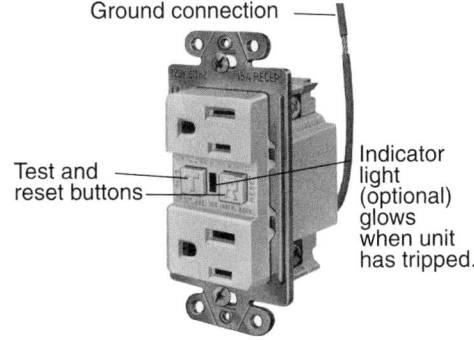

Ground connection

Test and reset buttons

Indicator light (optional) glows when unit has tripped.

*Fig. 26.44 Where it is desired to provide ground-fault protection at a single outlet only, and thereby localize power interruption (rather than using a GFCI circuit breaker on the entire circuit), a receptacle containing a ground-fault interrupter, as illustrated, may be used. A test button is provided to permit periodic testing, and a reset button reenergizes the receptacle after it has tripped on ground fault and the fault has been cleared. (Courtesy of Eagle Electric Manufacturing Co., Inc.)*

ELECTRICITY

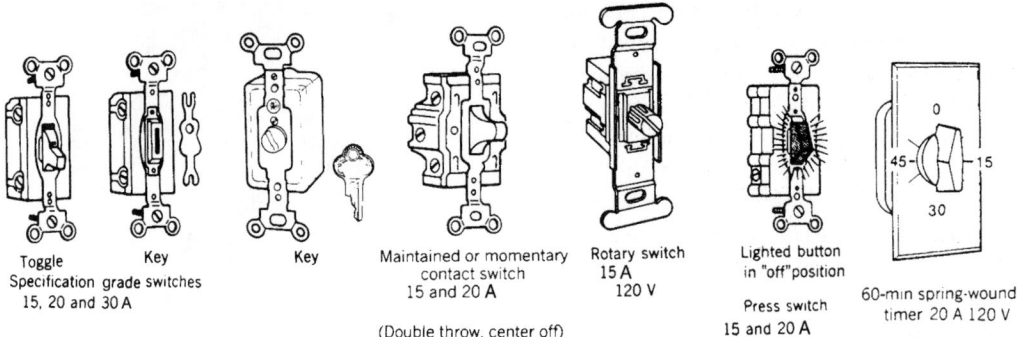

Toggle
Specification grade switches
15, 20 and 30 A

Key

Key

Maintained or momentary
contact switch
15 and 20 **A**

(Double throw, center off)

Rotary switch
15 A
120 V

Lighted button
in "off" position

Press switch
15 and 20 **A**

60-min spring-wound
timer 20 A 120 V

**Fig. 26.46** *A few of the most common wiring device switches, generally installed in a small wall box and used for control of lighting circuits.*

**ELECTRICITY**

## Industrial Specification Grade Switch

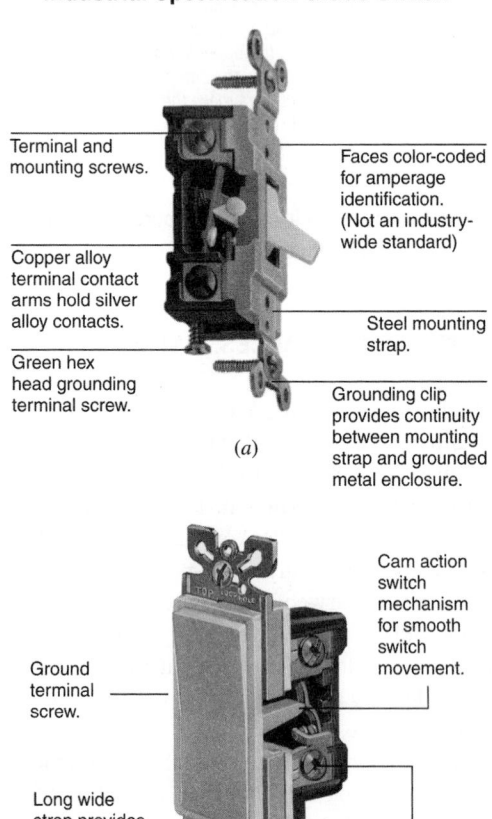

Terminal and
mounting screws.

Copper alloy
terminal contact
arms hold silver
alloy contacts.

Green hex
head grounding
terminal screw.

Faces color-coded
for amperage
identification.
(Not an industry-
wide standard)

Steel mounting
strap.

Grounding clip
provides continuity
between mounting
strap and grounded
metal enclosure.

(*a*)

Ground
terminal
screw.

Long wide
strap provides
good stable,
contact.

Cam action
switch
mechanism
for smooth
switch
movement.

Mounting and
terminal screws.

(*b*)

**Fig. 26.47** *Cutaway drawing of two common wall-switch designs showing operating mechanisms and construction details. (a) Conventional toggle design. (b) Rocker plate design. (Drawings courtesy of Eagle Electric Manufacturing Co.)*

measure. In high-security areas where the easily defeated normal key switch is inadequate, a tumbler lock–controlled unit can be used. Loads that can be timed out, such as bathroom heaters and ventilating fans, can be controlled by a spring-wound timer switch, as illustrated in Fig. 26.46 (which also illustrates a few common switch types). Figure 26.47 shows two common switch designs in cutaway, revealing their construction and operating mechanisms. A solid-state switch with adjustable time delay is shown in Fig. 26.48.

**Fig. 26.48** *Solid-state adjustable electronic switch timer that mounts in a common wall-switch box. The unit switches off, with delay adjustable from zero to 15 minutes in five discrete steps. Units such as these can control incandescent and fluorescent lamps and small motors. They are applicable where either direct manual control or occupancy sensors are inapplicable. (Photo courtesy of Leviton Manufacturing Co.)*

Thanks to miniature electronics, a programmable switch is available that fits into a wall-outlet box in lieu of an ordinary switch. The unit acts as a solid-state 15-A switch and can be readily programmed to switch the controlled circuit or device at preset times (see Fig. 26.48).

## 26.31 WIRING DEVICES: SPECIALTIES

The most common device in this group, which also includes pilot lights, fan controls, and other small motor controls, is the lighting dimmer. The original lighting dimmers were variable auto transformers. They were large, heavy, and expensive, and there-fore found application in theater lighting, displays, and the like. The advent of small electronic units (SCRs) made the modern wall-box dimmer possible. The simplest units are rotary, with limited capacity and applicable only to incandescent loads. One of their disadvantages, still found in inexpensive units, is the production of annoying radio frequency interference (RFI) and line harmonics. Newer models are available that provide preset control, combine dimming and switching, are usable for smooth fan speed control, can be remotely controlled (in addition to local control), provide automatic fade-out, and dim fluorescent lamps (provided with dimming ballasts). A few types are shown in Figs. 26.49 and 26.50.

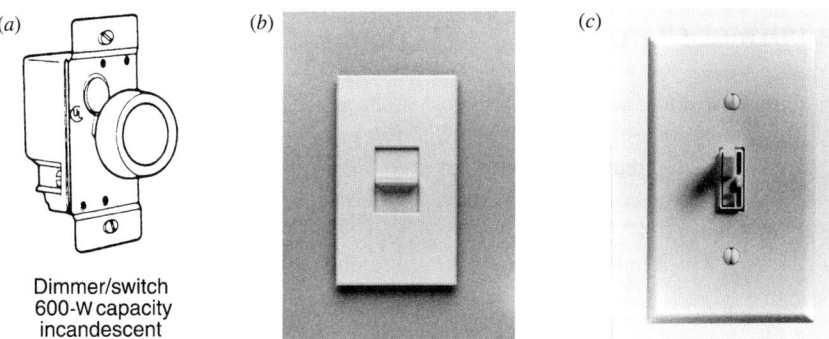

Dimmer/switch
600-W capacity
incandescent

**Fig. 26.49** *(a) Simple solid-state dimmer, suitable for a maximum 600-W incandescent load only. The switch permits level preset in a pushbutton design. Inexpensive units may cause considerable radio frequency interference (RFI). (b) Slide dimmer for a 600-W incandescent load, including on–off and level preset. This dimmer is available with a wireless receiver that responds to a hand-held infrared remote control. (c) Combined on–off switch and slide dimmer provides preset level control. Units similar to (b) and (c) are available for loads of up to 1000 W, fluorescent lamp control (with appropriate ballast), and fan speed control. (Photos b and c courtesy of LUTRON Electronics Co., Inc.)*

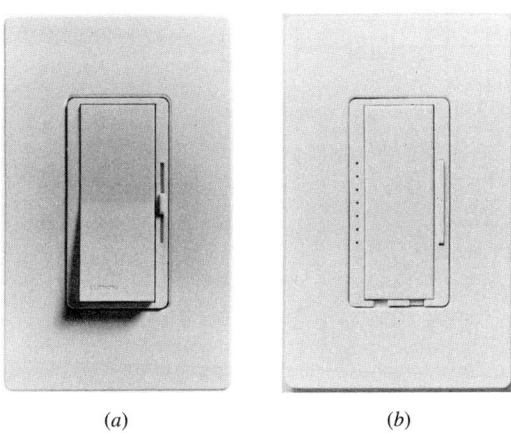

(a)        (b)

**Fig. 26.50** *(a) Combined tap switch and slide dimmer provides preset level control. A built-in night light indicates this design's residential application. (b) Multifunction tap switch in the center permits switching to full brightness, gradual brightening to preset, 10-second fade-out, and off. Rocker switch at the right controls levels. LED lights behind pinholes at the left provide an indication of the preset level. Both illustrated unit designs are available for incandescent and fluorescent (with special ballast) lamp control and fan speed control. Both are designed with RFI suppression. (Photos courtesy of LUTRON Electronics Co., Inc.)*

ELECTRICITY

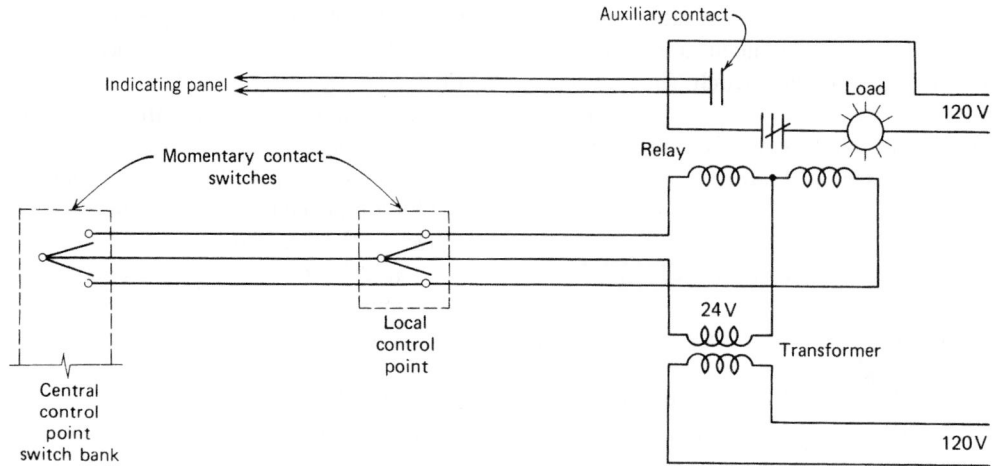

***Fig. 26.51*** *Low-voltage switching control. Multipoint control and central control are illustrated. The diagram shows the relays located at the load. Central relay cabinets are used in dense load areas.*

## 26.32 LOW-VOLTAGE SWITCHING

The switches illustrated and discussed in Section 26.30 are all full-voltage types; that is, they are wired directly into the load circuit and operate at line voltage and full current. A control scheme that uses light-duty, low-voltage (24-V) switches to control line voltage relays, which in turn do the actual

circuit switching, is illustrated in Figs. 26.51 and 26.52. The system is variously referred to as *low-voltage switching*, *remote control switching*, and *low-voltage control*.

Because the loads in this system are relay controlled, any number of switches and control devices can be wired in parallel to operate the relay and thereby effect the switching. Thus, the control can

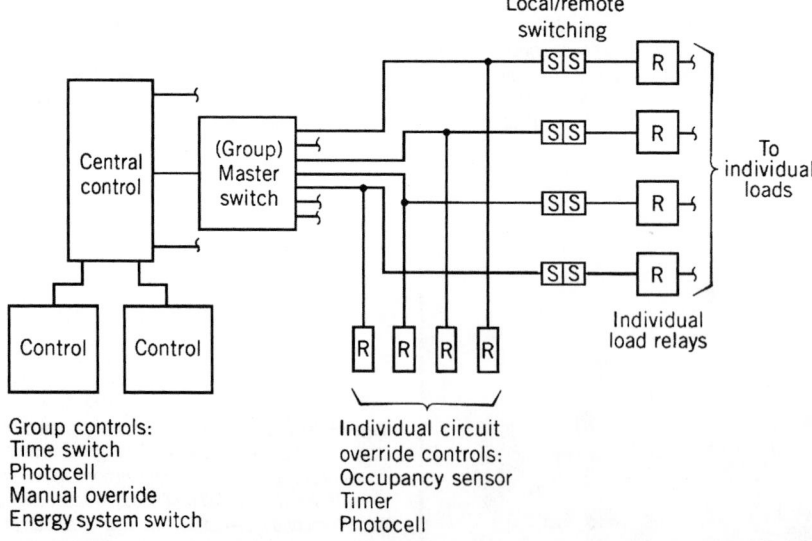

***Fig. 26.52*** *Schematic arrangement of a large low-voltage switching system. Individual loads can be switch-controlled from multiple locations, including a group (master) control point. In addition, local control devices such as photocells and occupancy sensors can override the local switches to maximize energy conservation. Central (group) control devices can perform the same override function at the group level.*

ELECTRICITY

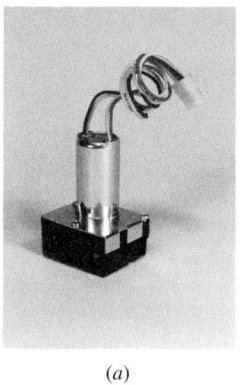
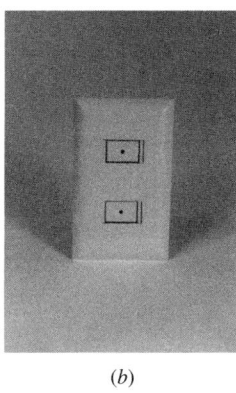

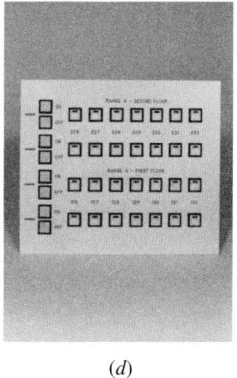

(*a*)　　　　　　　(*b*)　　　　　　　(*c*)　　　　　　　(*d*)

**Fig. 26.53** *Components of low-voltage switching and control systems for lighting and other electric loads. (a) The basic switching device is a single-pole, double-throw latching relay with a 24-V coil and 20-A, 120–277-V contacts. It is shown with a plug-in connector on its leads, making it suitable for rapid mounting in a relay panel (see Fig. 26.54). The same relay without the connector is suitable for hard-wiring at a controlled device. (b) Low-voltage (24-V) wall switches can be mounted single or ganged in groups of two to eight switches. Switches are available in button, rocker, key-operated, and lighted designs (as in b–d). (c) Switching station with pushbutton low-voltage switches, each with an LED to indicate switch position (on–off). (d) Switch bank with illuminated pushbutton switches identified with room number and floor. The on–off switches at the left of the bank are override switches for the corresponding horizontal line. (Photos courtesy of Touchplate Technologies, Inc.)*

be local, remote, and master, with individual override by local control devices such as occupancy sensors, and central control or override at a central controller by timers, daylight controllers, and so on. The advantages of this system of control over full-voltage switching are therefore:

- Flexibility of control location
- Individual load control override by local control devices (occupancy sensors, photocells; see Section 15.15)
- Group load override by central control devices such as timers, daylight controllers, and energy management systems
- Low cost of low-voltage, low-current wiring compared to full-voltage wire and conduit
- Inherent system flexibility and simplification of alterations
- Monitoring of the status of individual loads at a centralized control panel by the use of relays with auxiliary contacts (see Fig. 26.51)

The second and third points are particularly important from the energy conservation point of view, as it has been amply demonstrated that reliance on manual control for energy usage limitation yields disappointing results.

　The basic components of the entire system are the control relay and the switches, shown in Fig. 26.53. Small systems use hard wiring (physi-

cally wired connections), with the hard-wired relays installed either dispersed at each controlled light (or lighting groups) or grouped in a relay panel. When the number of relays exceeds about 20, these hard-wired panels become so densely wired that control changes become difficult and tedious. As a result, prewired computer- or microprocessor-controlled panels are used in large systems, where connections can be made and changed easily. The interior of one such panel is shown in Fig. 26.54. The architecture of a medium-sized computer-controlled system is

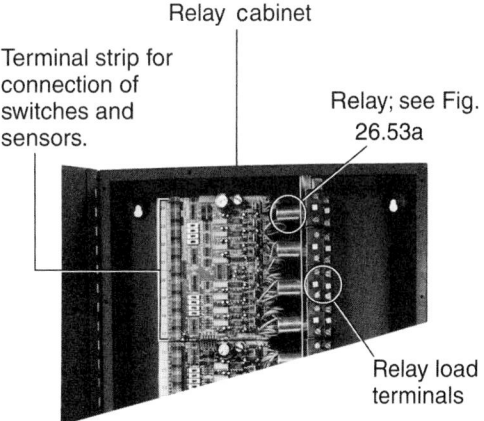

**Fig. 26.54** *Prewired relay cabinet (partial). Field wiring consists of control wiring to switches and sensors and power wiring to the controlled loads. (Photo courtesy of Touchplate Technologies, Inc.)*

ELECTRICITY

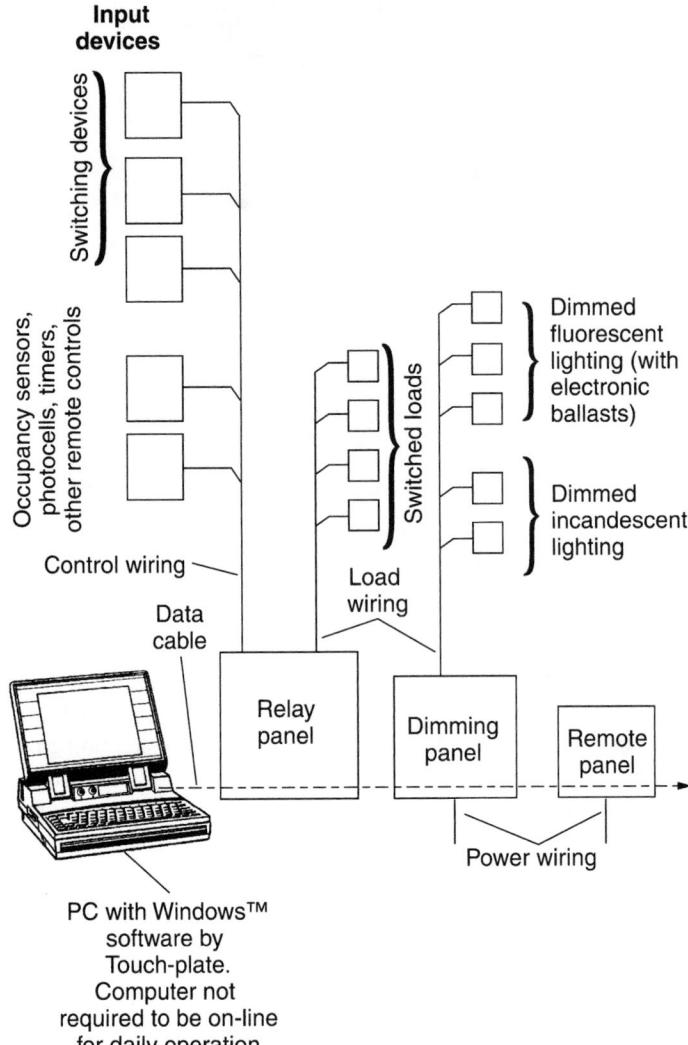

**Fig. 26.55** *Block diagram of a large, computer-controlled, low-voltage control system. The relay panel (Fig. 26.54) contains the incoming and outgoing power and control wiring, the switching relays, and the computer interface card. The system arrangement is controlled by computer software (see Fig. 26.56). (Courtesy of Touchplate Technologies, Inc.)*

shown in Fig. 26.55. Manufacturers have made software available that permits rapid, accurate design of complex, large systems. A typical computer screen in such a system is shown in Fig. 26.56.

## 26.33 WIRELESS SWITCHING AND CONTROL

Full-voltage local switching is described in Section 26.30, and low-voltage switching, both local and remote, is described in Section 26.32. The next logical step is remote wireless control and switching. These systems have undergone extensive development. They impinge on the design

considerations of architects and building engineers to the extent that facilities for their accommodation must be provided if such systems are to be used in a facility.

These system aspects, as well as system functioning and application to lighting control, are discussed in detail in Chapter 15.

## 26.34 POWER LINE CARRIER SYSTEMS

Although the advantages inherent in flexible, programmable load switching systems such as those described are undeniable, particularly in connection with energy management systems, their high

Dimmer control

Relay control Icons

Special software icons

Windows™ based lighting control for relay and dimmer systems.

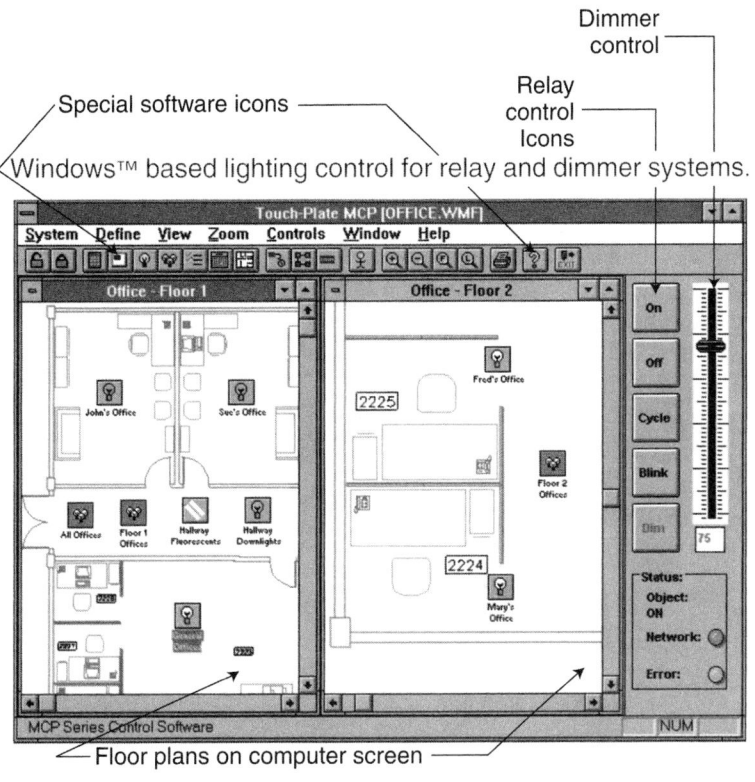

Floor plans on computer screen

**Fig. 26.56** *Typical computer screen of low-voltage control software showing facility floor plans (scanned in) and controlled loads. The icons at the top of the screen are customized and are used to control relays, make connections, check status, print, and log events by time and date. Other screens enable group and event definition, labeling, establishment of priorities, and other global sequential and individual control functions. (Courtesy of Touchplate Technologies, Inc.)*

first cost frequently makes the entire project uneconomical. This is particularly true in retrofitting large, complex facilities with energy management controls, where the cost of wiring, even when using low-voltage control wires, can amount to more than half of the total cost of the project.

As a result of the desire to minimize wiring costs, a system of control signal transmission that uses the electric power wiring (either new or existing) as its conductors was developed, thus obviating the need and expense of dedicated control wiring. The system is called a *power line carrier* (PLC) system because the power line carries the control signal. An ancillary benefit of such an arrangement is that not only can it be added to an existing system, it can also be removed, thus making possible and economical the leasing of energy management control systems.

The system operates by injecting into the power wiring a series of low-voltage, high-frequency,

binary-coded control signals, which then disperse over the entire power network. Only a receiver that is "tuned" to receive a particular code reacts, generally by operating a local device, which in turn connects, disconnects, dims, or brightens a particular load. In practice, the control signal generator can be a small, manually programmed controller, as in a residence, or a computer-operated energy management or lighting controller in a commercial facility. The receivers are normally of four types, most of which are designed to fit into an ordinary wiring device box:

1. A wall-switch module, which acts as a normal manual wall light switch but also contains a coded receiver and a fully rated relay (20 A, 125 V, or 277 V)
2. A wall receptacle module, which contains a 20-A receptacle along with a receiver and relay that serve to switch power on and off
3. A switching module, which can be connected

to activate contactors, small motors, and the like

4. A dimming module, which provides both local and remote control

Some of these components are shown in Fig. 26.57. Larger, multiple-circuit switching modules are also available for more extensive systems. Each miniature receiver can be manually tuned in the field (i.e., its code can be set and reset readily), giving the entire system extreme flexibility.

The system operates in the same manner as described for a low-voltage switching system or for an energy management or lighting control system, with the essential difference that no dedicated wiring is required. In retrofit jobs, existing wall switches and receptacle outlets are replaced with receiver-controlled units; in new installations, these devices are installed initially.

Due to the fact that the control signals travel over the power wiring and are therefore attenuated by high-resistance connections, poor grounds, and faulty wire insulation, PLC installations require a

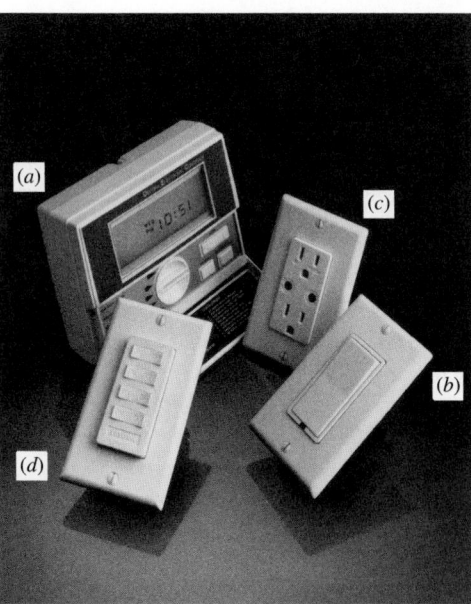

**Fig. 26.57** *Components of a residential-type PLC system. (a) Control unit, at which both manual and automatic (programmed) control of a maximum of 256 address codes are possible. (b) Wall-switch module has local dimming and switching functions in addition to being remotely controlled from the control center (a). (c) Duplex wall receptacle, either split-wired with local programming of one-half of the outlet or complete remote control. (d) Outlet box wall-mounted control switch that provides local group control (switching and/or dimming) for one to four devices (address codes). (Photo courtesy of Levington Manufacturing Co.)*

high-quality power wiring installation to operate properly. In retrofit work, preliminary tests occasionally indicate the impracticability of a PLC installation because of the condition of the power wiring system. Another problem encountered in PLC systems is interference with the PLC control signals from radio noise generated by faulty power equipment and by improperly shielded or grounded electronic equipment. These problems can frequently be overcome by careful testing, installation of filters and jumpers, and repair of improperly operating equipment. In addition to energy management and normal lighting control, PLC control can be applied to group switching of other loads, such as office machines, in addition to simplifying control of temporary light and power for special events.

## 26.35 POWER CONDITIONING

### (a) General Information

*Power conditioning* is a relatively new electrical term that describes the process of converting utility-supplied electrical power, which is frequently characterized by transient surges, spikes, radio frequency noise, and voltage fluctuation, to a pure (noiseless), accurately voltage-regulated sinusoidal waveform, normally referred to as *computer-grade power.*

Some form of power conditioning is required by all data processing and telecommunication equipment and by many other types of electronic equipment. This is due to their extreme sensitivity to fluctuations in the power supply voltage and particularly to random high-frequency voltages superimposed on the power supply voltage, known as *radio frequency interference* (RFI) or simply *radio noise.* In the case of data-processing equipment (computers and peripherals), radio noise can cause data errors; slow voltage fluctuations can cause overheating, data loss, and premature equipment failures; and large, rapid voltage fluctuations (spikes, transients) can cause equipment burnout and system collapse. The economic ramifications of such effects on large commercial data-handling systems are so severe that power conditioning has become almost universal. (Power conditioners must not be confused with uninterruptible power supplies [UPS], which function to maintain power during utility *failure,* although in many commercial installations, both functions are served by a single equipment package.)

## (b) Source of Disturbance

Utility power systems are designed to maintain voltage and frequency within certain limits on an average, long-term basis. For good technical reasons, utility equipment cannot react to the short-time disturbances that are so inimical to electronic equipment. Furthermore, much of the noise in utility lines is introduced by users and is therefore beyond the utility's point of control. The disturbances on the electrical power line can roughly be classified into three types: voltage variations, electrical noise, and transients.

1. *Voltage variations.* These can be caused by short-time current drain, such as during motor starts, deliberate voltage reduction (brownouts), load switching of EMS systems, excessive line voltage drop due to heavy loading or current leakage, and similar problems. These variations are *relatively* slow and long-lasting.

2. *Electrical noise.* This is simply low-amplitude, higher-frequency voltage that, when superimposed on the power line, results in a distorted waveform. Sources of such extraneous voltages are electronic equipment power supplies, lighting dimmers, solid-state motor controls, PLC systems, arc welding, switching transients caused by branch circuit switches and relays, and voltages induced by local magnetic fields.

3. *Transients.* These are large, short-duration voltage variations, also known as *surges* or *spikes.* They are the principal culprits in major equipment failures. They are caused by lightning strikes to overhead lines (even many miles from the affected installation), by electrical system faults, and by utility switching of high-voltage lines, transformers, capacitor banks, generators, and circuit reclosers.

## 26.36 POWER CONDITIONING EQUIPMENT

Each of the three types of problems described in the preceding section requires its own type of correction. Voltage variations are corrected with voltage regulators; noise problems are corrected by electrical isolation, filtering, and noise suppression; and transients are treated with surge suppressors. Each corrective device is a separate entity that can be applied individually to suit the particular problem encountered or in combination with another condi-

tioning device. The choice in specific installations is a highly technical one and should be made by the facility's electrical engineer after careful analysis of the quality of the utility service. A few general recommendations can be made.

1. All computer installations, including the smallest, should be protected from line transients with an appropriate surge suppressor. Multitap plug-in strips with integral surge suppressors are unsatisfactory unless they have surge current, clamping voltage, and surge-energy specifications, and these meet the installation requirements.

2. Major data-processing installations normally require all three types of line treatment. The most economical way to provide them is with a single integrated power conditioning unit.

3. Considerable improvement in the quality of electrical power can frequently be achieved by the simple expedient of running separate electrical feeders for sensitive loads.

4. Physical isolation of sensitive equipment areas is frequently helpful, particularly when disturbances are being induced by proximity to switching, arcing, and rectifying equipment. Discharge-type lighting, including fluorescent, mercury, sodium, and metal halide, can cause interference, especially when used with electronic ballasts.

5. It was found some years ago that much of the noise in electrical power systems was introduced through the (equipment) grounding pole of an electrical receptacle. Because this connection is required for electrical safety, it cannot be eliminated. It can, however, be separated from the wiring system ground and still maintain its function. Receptacles so constructed are color coded either entirely orange or have an orange triangle on the face (see Section 26.29). This "fix" is particularly effective where electronic dimmers, ballasts, and switching devices are present. Because the insulated ground receptacle (sometimes referred to as an *isolated* ground receptacle) is a special item, it must be so indicated on the construction contract documents. Furthermore, because these receptacles require special isolated wiring, the entire matter must be studied carefully during the electrical design stage and before issuance of construction contracts.

6. A problem related to equipment that produces radio noise is that this same equipment also produces harmonics that appear on the system neutral and can cause overloading. This subject is covered in Section 28.18.

**TABLE 26.7 Power Conditioning Equipment**

| Equipment Functions | | | | Dimensions (in.) | | |
| Voltage Regulation | Noise Suppression | Surge Suppression | KVA[a] | H | W | D |
|---|---|---|---|---|---|---|
| X | | | 5 | 15 | 20 | 20 |
| | | | 10 | 35 | 40 | 20 |
| | | | 20 | 40 | 40 | 30 |
| X | X | | 0.25 | 6 | 6 | 10 |
| | | | 0.5 | 7 | 7 | 12 |
| | | | 1.0 | 8 | 10 | 15 |
| | | | 3.0 | 10 | 12 | 20 |
| X | X | | 5 | 10 | 10 | 15 |
| | | | 10 | 20 | 30 | 15 |
| | | | 20 | 30 | 30 | 20 |
| X | X | X | 5 | 40 | 30 | 30 |
| | | | 10 | 50 | 35 | 30 |
| | | | 20 | 70 | 40 | 35 |
| | | X | NA[a] | | | |
| | | | 15 A | 3 | 3 | 16 |
| | | | 100 A | 10 | 12 | 12 |
| | | | 500 A | 15 | 15 | 12 |

[a]Surge suppressors are rated by transient energy-handling capability, and therefore a kVA rating is not applicable. However, because transient energy increases with the kVA size of protected equipment, a service ampere rating is given.

The issue of power quality is so important in modern commercial installations that the following steps should be taken in each such design:

- A detailed report should be obtained from the electric utility that will supply electric power showing power outages, voltage constancy, frequency control, and, if possible, oscillograms of waveforms at random intervals over the past year. Particularly important are periods of peak loading and periods of weather extremes.
- A power quality study should be made in the area in which the facility will be (or is) located. Sophisticated instrumentation for this purpose as well as specialists in this field are readily available (see Section 25.23).
- An in-depth study of all of the electrical equipment to be used in the building should be made for the purpose of:
  - Reducing the sources of electrical disturbances
  - Determining the tolerance or sensitivity of the data-processing equipment to be used. In the absence of reliable data for the second purpose, as is the case in speculative construction, a high degree of sensitivity must be assumed. In any case, for equipment that is known to produce radio noise, such as electronic lighting ballasts and dimmers, the specifications should include limitations plus a requirement for submission of laboratory tests on random samples.
- Many large power companies now offer "computer-grade power" that meets stated limitations

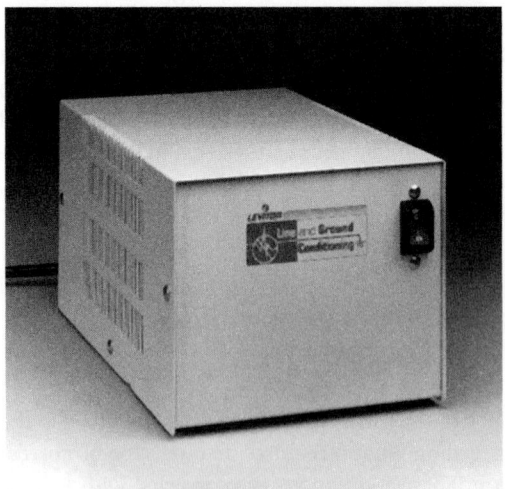

*Fig. 26.58* This single-phase power conditioner supplies noise filtering and surge suppression by use of an isolation transformer, ground noise filters, and category A and B surge suppression (see Section 26.37c). It is rated 5 A, 120 V, measures 6 in. H × 7 in. W × 11 in. D (150 × 180 × 280 mm), and weighs 18 lb (8 kg). A similar unit rated 20 A, 120 V measures 8 in. H × 8½ in. W × 14 in. D (200 × 215 × 355 mm). (Photo courtesy of Leviton Manufacturing Co.)

ELECTRICITY

to voltage fluctuation and the like. This type of service is desirable for all new office space or full building renovations. For individual office spaces, separate power conditioning equipment is required.

Table 26.7 and Fig. 26.58 give some idea of the physical characteristics of power conditioning equipment. Dimensions are typical and are indicative only of equipment bulk. Designs vary widely among manufacturers, as do, therefore, individual dimensions. Weights are not given for the same reasons. Voltage-regulating equipment containing regulating transformers is very heavy. Power conditioning equipment is frequently combined with UPS equipment (see Section 26.38).

## 26.37 SURGE SUPPRESSION

The full term for this aspect of power conditioning is *transient voltage surge suppression* (TVSS). Because transient voltages, also called *surges* or *spikes,* can cause major physical damage to computer systems and other types of electronic equipment, it is this aspect of system protection that has received the most attention (including a fair amount of highly imaginative advertising). Surge suppressors are available in a huge variety of designs and ratings, ranging from the ubiquitous cord-connected, multiple-outlet strip to large, 3-phase, service-entrance units. They all have a single purpose, however: to suppress (limit) a voltage surge to a level that the protected equipment can withstand without damage. This is done in two ways:

1. By placing one or more devices that present a high impedance in series with the incoming voltage transient, thereby limiting the let-through current.

2. By placing one or more devices across the incoming power line, in parallel with the protected load, that present a low impedance to the high transient voltage and thereby bypass (shunt away) the incoming wave's current. Many TVSS units use both of these techniques in a single "hybrid" unit.

### (a) Terminology

To be able to select and compare TVSS units and their ratings, a basic technical vocabulary is necessary.

**Avalanche diode.** A solid-state device placed in parallel with the protected load. It responds to overvoltage in nanoseconds by conducting (i.e., placing) its very low impedance across the incoming voltage. This results in a large ("avalanche") bypass current flow that serves to reduce the incoming voltage. It has a low clamping voltage (definition follows) and low energy absorption capacity. This latter characteristic causes "wear" and eventual failure.

**Clamping voltage.** The voltage at which shunting devices (such as the avalanche diode) begin to conduct and thereby clamp the incoming voltage surge. Therefore, the lower a device's clamping voltage, the better is its suppression action.

**EMI/RFI rejection.** A measure of the attenuation by a TVSS device of electromagnetic interference (EMI) and RFI. A TVSS device, due to its internal filters, always provides some radio noise attenuation. This characteristic is not of primary significance to a device used primarily for surge suppression.

**Gas tube surge protector.** A high-energy shunt device, typically found in high-capacity, good-quality TVSS units that are subject to large surges, such as those caused by lightning. It consists of a gas-filled tube with two electrodes. The tube conducts when a high-voltage surge causes its electrodes to arc over. Because of low arc impedance, the current drain and energy absorption are very high, thus clamping the incoming surge. Its principal drawback is its relatively slow response, which necessitates its use in conjunction with other, faster devices.

**Isolation transformer.** A transformer with two separate windings and no conductive electrical connection between them. By virtue of the inductive coupling of the windings, considerable attenuation of sharp voltage spikes is achieved. It can be used in high-capacity TVSS units in lieu of multiple solid-state components in parallel. It is also useful in attenuating high-frequency noise at any voltage level.

**Joule rating.** A *joule,* equal to 1 watt-second, is the SI unit of energy. A device's joule rating is a measure of its capacity to absorb and dissipate heat generated by its action. Although the joule rating is important where large amounts of energy are to be absorbed, as in (hard-wired) units at a service entrance or at a power distribution point, published ratings can be misleading because there is, at this writing, no industrywide standard calculation procedure for joule rating. As a result, the industry standard specification form (NEMA-LS 1) for TVSS devices does not include this rating.

**Maximum surge current.** A measure of a device's ability to divert and dissipate surge current without failing. It is a useful metric for TVSS devices used at service entrances and, for reasons explained previously, is more reliable than published joule ratings.

**Metal-oxide variable resistor (MOV).** A shunt device somewhat slower than an avalanche diode but with much higher energy absorption capability and higher clamping voltage than an avalanche diode. MOV is very widely used in TVSS designs. Like avalanche diodes, MOVs suffer degradation with use, thus limiting the useful life of the entire TVSS unit.

**Response time.** A measure of the rapidity with which a TVSS device begins to clamp a voltage surge, usually measured in microseconds or nanoseconds. It is not of great significance because both avalanche diodes and MOVs (found in all good-grade TVSS designs) are faster than the voltage buildup time of an incoming surge. Response time is not included in the NEMA standard specification, apparently because it is not considered to be a very important characteristic.

**Sinewave tracking.** A term of uncertain meaning found frequently in manufacturers' literature.

### (b) TVSS Operation

Figure 26.59 is intended to be self-explanatory. Each element acts to reduce the magnitude of the incoming voltage surge by either bypassing (shunting away) current or presenting a high impedance to the voltage surge. Both actions reduce the voltage. A commercial TVSS device may contain any or all of the illustrated components, single or in multiple, depending on the current and voltage it is expected to handle.

### (c) Standards

Standards applicable to surge suppression equipment are:

ANSI/IEEE C62.41.1: *Guide on the Surge Environment in Low-Voltage AC Power Circuits*

ANSI/IEEE C62.45: *Guide on Surge Testing for Equipment Connected to Low-Voltage AC Power Circuits*

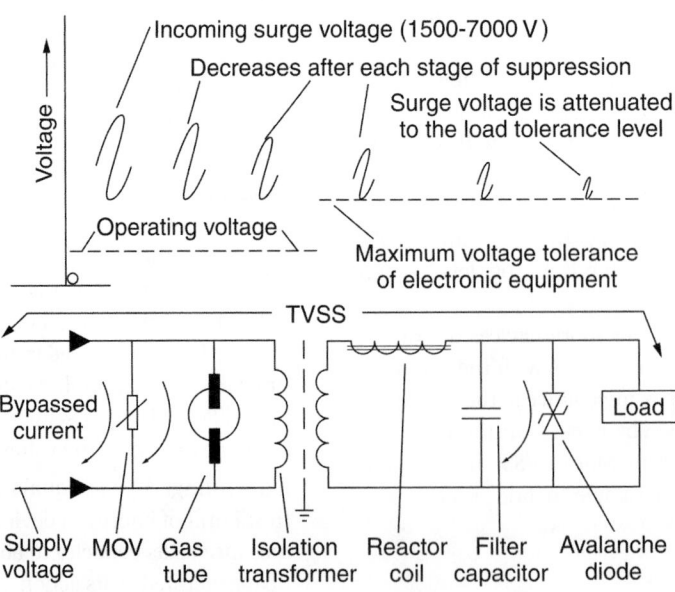

**Fig. 26.59** *All of the elements usually found in a surge suppression device are shown. Each acts to reduce the voltage surge, parallel elements by shunting away (bypassing) surge current and series elements by increasing the electrical impedance to the voltage surge.*

UL 1449: *Standard for Transient Voltage Surge Suppressors*

NEMA LS 1: *Low Voltage Surge Protection Devices*

Surge suppression equipment can be subjected to widely varied voltage stresses, depending on where in an electrical system a unit is installed. Standard C62.41 categorizes installation points according to the severity of service expected, that is, the (electrical) environment in which the surge suppression equipment operates.

*C62.41, category C* represents the most severe environment, corresponding to locations on the line side of the facility service switches. This environment is subject to the full force of lightning and switching surges and therefore requires robust equipment. As a voltage surge travels through a building's electrical system, it is attenuated by cables, transformers, and other electrical equipment. Therefore, the further along one travels in the building's electrical distribution system, the less severe the voltage surges become. All TVSS devices used in category C environments are hard-wired (Fig. 26.60).

*C62.41, category B* covers environments on the load side of the service equipment, including heavy feeder distribution panels and short branch circuits. Most TVSS devices in category B areas are hard-wired (Fig. 26.61).

*C62.41, category A*, which represents the least severe service, includes utilization outlets and long

**Fig. 26.61** *TVSS unit designed to be used at service and branch panelboards in residences and small commercial installations. The illustrated unit is rated single-phase, 120/240 V, 3-wire service, category B3. The unit measures 7 in. W × 7 in. H × 4 in. D (180 × 180 × 100 mm) and is equipped with indicating lights that indicate power and suppression status. (Photo courtesy of Leviton Manufacturing Co.)*

branch circuits (because the long cables attenuate the voltage surge). Most category A TVSS devices are arranged for loads to be plugged in directly. Devices used in category A environments include the common cord-connected multioutlet strip and surge suppression–equipped receptacles (Fig. 26.62).

Each of these three major categories is in turn subdivided into three minor categories: C3, C2, C1;

**Fig. 26.60** *Surge suppressor designed for use on the incoming service of a facility. It is rated for category C3 application: shunt type, 120/208 V, 3-phase, unlimited current. It is provided with RFI/EMI filtering and replaceable surge suppressor modules, and arranged to indicate the power and protection status of each phase. (Photo courtesy of Surge Control Ltd.)*

Lights indicate working condition and replacement requirement

**Fig. 26.62** *Typical category A utilization TVSS outlet equipped with indicating lights that show the condition of the surge suppressor circuitry. (Photo courtesy of Eagle Electric Co.)*

B3, B2, B1; and A3, A2, and A1, where 3 is the most severe and 1 the least. The differentiation depends on the electrical capacity of the facility's service and distribution system. Large systems have higher surges and less attenuation in the system, thus requiring larger and heavier TVSS equipment than smaller installations. The three major categories are shown graphically and in text in Fig. 26.63.

Standard C62.45 established the test procedures and test voltage levels to be used by manufacturers in testing and classifying their TVSS products in order to simulate the voltage stresses found in A, B, and C environments.

UL Standard 1449 establishes standards for surge suppressor safety and performance. An important aspect of this standard is that it establishes clamping voltage ratings. The ones of interest to building designers are the lowest four, which are 330, 400, 500, and 600 V. The UL standard requires testing for actual clamping levels, which are then rounded *upward* to one of the standard ratings. Thus, a TVSS that clamps at 240 V and one that clamps at 320 V would both be rated at 330 V, despite the fact that the 240-V unit gives better protection. This is particularly important for protecting computers and other electronic equipment equipped

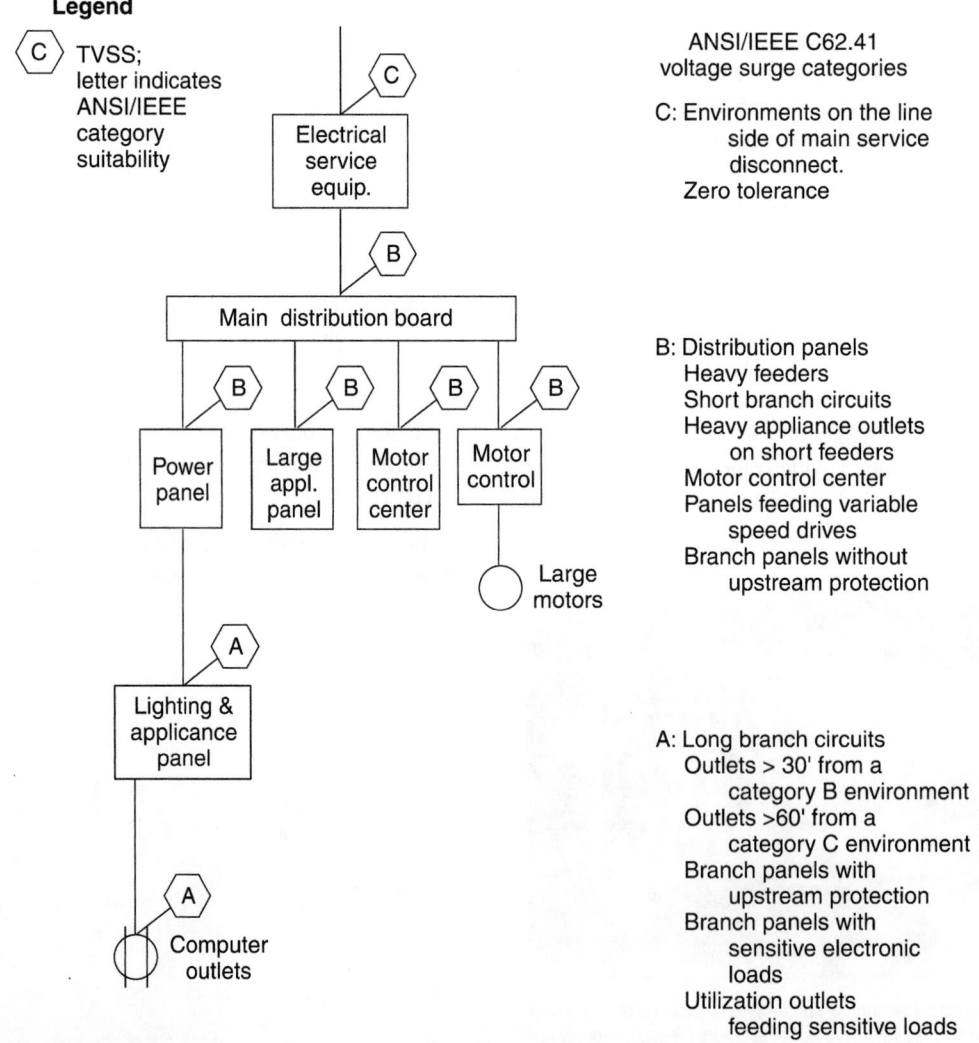

**Fig. 26.63** *Illustration based on ANSI/IEEE Standard C62.41 showing application of transient voltage suppression devices in a typical electrical facility, based on the three categories of electrical environments (C, B, A) defined by the standard.*

with switching power supplies, and requires comparison of *actual* clamping voltage of units rather than reliance on the voltage classification.

NEMA LS 1 standardizes the specification format for suppression units, and in so doing establishes a basis for comparison among various manufacturers' products. Important characteristics that must be specified include line voltage, maximum operating current and voltage, maximum surge current, and clamping voltage rating. (Actual clamping voltage is not supplied unless specifically requested, as it involves specific testing.)

All of the mentioned standards are voluntary industry and technical society publications. In 1997, the U.S. government developed a Performance Verification Test specification that established a quality grading system for cord-connected surge suppressors. This standard does not correspond readily to the C62.41 classifications, except in the B category. It is worth noting that the highest classification in this system is Class 1, Grade A, Mode 1, which corresponds to ANSI category B3, UL 1449 clamping voltage rating of 330 V, and line to neutral protection.

## (d) Application of TVSS Devices

Selection and application of surge suppression devices for even the simplest application requires considerable technical background, for which the foregoing material is intended. Important guidelines that assist in this area are:

- Every power level should have its own TVSS. Small systems can function with C62.41 categories B and A only. Large systems may require protection for two or three B levels (B3, B2, B1) and more than one A level.
- In the absence of accurate data on the surge tolerance of electronic equipment, use protection for the most sensitive level. At the utilization level, this is category A3 and U.S. government classification Grade A, Class 1, Mode 1.
- Mode 1 (also called Mode A) indicates protection line-to-line or line-to-neutral. This mode should be specified for stand-alone items. It is preferable to Mode 2 (or B), which is connected line-to-ground, because the latter can introduce random signals (contaminates) into the equipment grounding system.
- The primary criteria for selection and comparison are the *actual* clamping voltage and maximum surge current capacity. Other important characteristics are those listed in NEMA LS 1.
- TVSS units should be installed as close as possible to the loads being protected.
- Telephone and data lines must also be protected, particularly if the service is overhead. Most

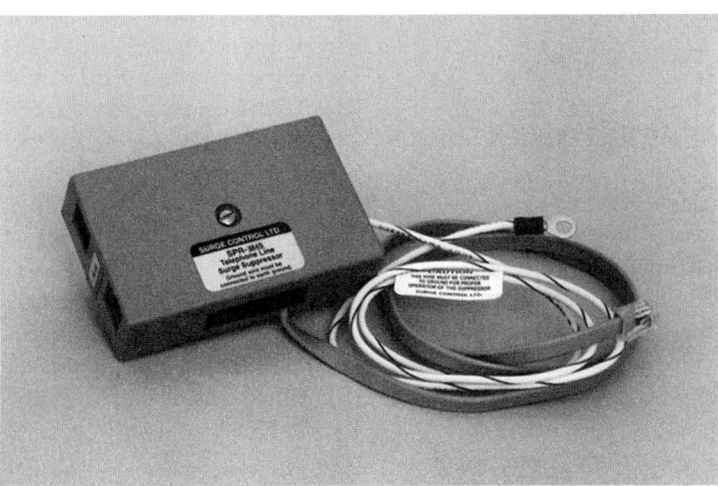

**Fig. 26.64** *Surge suppressor designed for use with standard snap-in low-voltage telephone and industry jacks. The illustrated unit is designed to protect telephone/telecommunication lines, both analog and digital. The units are rated 185 V, with a maximum surge current of 5000 A and a test clamping voltage of 240 V. The units can be used with telephones, modems, fax machines, and the like. They measure approximately 4 in. W × 2½ in. H × 1 in. D (100 × 64 × 25 mm). (Photo courtesy of Surge Control Ltd.)*

ELECTRICITY

ELECTRICITY

**Fig. 26.65** *TVSS device designed to protect video lines from voltage transients. The unit plugs into a standard outlet to establish a ground connection for current bypass. This unit is equipped with BNC connections; others have telephone, PL, and RS connectors for use on telephone, communication, and data lines. The illustrated unit has a clamping voltage of 15 V and a maximum surge current capacity of 5000 A. It measures approximately 2 in. W × 4 in. H × 1 in. D (50 × 100 × 25 mm). (Photo courtesy of Leviton Manufacturing Co.)*

telephone companies install an appropriate TVSS at the main distribution frame near the service entrance. This may not be sufficient to protect electronic equipment from disturbance that can cause distortion of data. A specialist should be consulted to determine the required protection levels and select appropriate equipment (Figs. 26.64 and 26.65).

## 26.38 UNINTERRUPTIBLE POWER SUPPLY

As explained previously, power conditioning equipment can supply clean utility power; it cannot, however, supply any power during a utility outage. That eventuality is covered by an alternate supply of power. Facilities that contain computers and/or data-processing equipment cannot tolerate power outages in excess of about 8 to 50 milliseconds (ms) without serious risk of data loss, with all the negative ramifications of such a loss.

An uninterruptible power supply (UPS) is an arrangement of normal and backup power supplies that transfer a facility's critical load from the normal to the backup supply in so short a time that no com-

puter malfunction results. This transfer time varies somewhat among different schemes and manufacturers but is always less than 8.3 ms, which is the minimum period of power outage that computers must tolerate without disturbance to meet the computer industry's manufacturing guidelines. This time period is double the maximum transfer time required by IEEE Standard 446, *Emergency and Standby Power,* of ¼ cycle at 60 Hz, or 4.16 ms. Thus, all computer systems fed from a UPS that meets this standard have a safety factor of at least 2 in transfer time.

The period of time after transfer that the equipment will run on the standby source depends on system design. In most cases it is 5 to 10 minutes, which is usually enough time to permit an orderly manual or automatic shutdown. Where shutdown is not a viable alternative, as with computer-controlled manufacturing processes, the standby power system can readily be designed to supply power indefinitely.

The selection of a UPS system for a facility is a complex process beyond the scope of this book. The material that follows is intended to provide sufficient familiarity with these systems to permit preliminary selection and planning.

### (a) Alternate Power Source

For most systems, battery backup is sufficient because, as noted above, only an orderly shutdown is required. Where this is not the case, a two-stage transfer can be used—the first transfer to a battery backup that can carry the load for up to 1½ hours (or more), depending on the load magnitude, and a second transfer to a long-term standby generator set. Thereafter, the availability of generated standby power is limited only by the fuel supply. Alternatively, the generator can be on-line, and the load picked up directly on the first transfer.

In large industrial installations, the standby source may also be a second utility line. In such cases, only a single transfer is required. Questions of service reliability and equipment redundancy are both technical and economic, and must be studied carefully for each individual installation.

### (b) Equipment Arrangement: Classic Standby and On-Line Topologies

Most UPS systems today are described by their manufacturers as on-line or simply as UPS rather than standby, primarily because there is no generally accepted industrywide definition of *on-line* and also because many systems are hybrid and do not fall easily into either category. Figure 26.66 shows the usual static (nonrotary) UPS equipment arrangement that applies, in principle, to both standby and on-line modes.

In the standby mode (Path *A*), utility power is normally passed through some power-conditioning equipment to shield against surges and remove random noise, and then, via position *A* in the static transfer switch, to the load. The ac line also provides a small current to the small battery charger that keeps the battery fully charged. The dc to ac inverter is open-circuited at the transfer switch and delivers no power. In the event of a utility power failure, the transfer switch moves to position *B,* and power flows from the battery through the inverter and transfer switch to the load (Path *B*). With proper equipment design, the transfer is usually accomplished smoothly, although an instantaneous change from no-load to full load on the inverter can sometimes cause a transfer voltage loss. For this reason, on-line schemes are preferred.

In the straightforward on-line arrangement shown in Fig. 26.67, Power Path A (normal) is through the rectifier and inverter to the load via position *B* in the transfer switch. The battery floats on the dc line. Failure of utility power causes the

power path to change to Path B (i.e., battery to inverter to load). The transfer switch remains in the *B* position. As a result, voltage to the load is undisturbed. If one of the Path *A* components fails, the transfer switch moves to position *A* and power is supplied directly from the utility lines via Path *C,* which is referred to as a *utility bypass.* The advantage of the on-line arrangement is that it eliminates reliance on the quality of utility power. Power conditioning and surge suppression are provided by the solid-state equipment.

The disadvantages of this classic on-line topology are high cost due to the full load capacity size of the charger/rectifier and an overall efficiency in Path *A* of about 75% to 80%. The resultant heat production can be problematic and is certainly expensive in terms of power costs.

### (c) Additional UPS Topologies

1. A money-saving arrangement called *on-line without bypass* is simply the bottom portion of Fig. 26.67. The first cost is reduced by elimination of the transfer switch and considerable cabling. Because failure of any of the components disables the system completely, as there is no backup power path, this system is not recommended for critical loads.

2. A system variously known as *line-interactive, standby line-interactive,* and *hybrid interactive* is shown

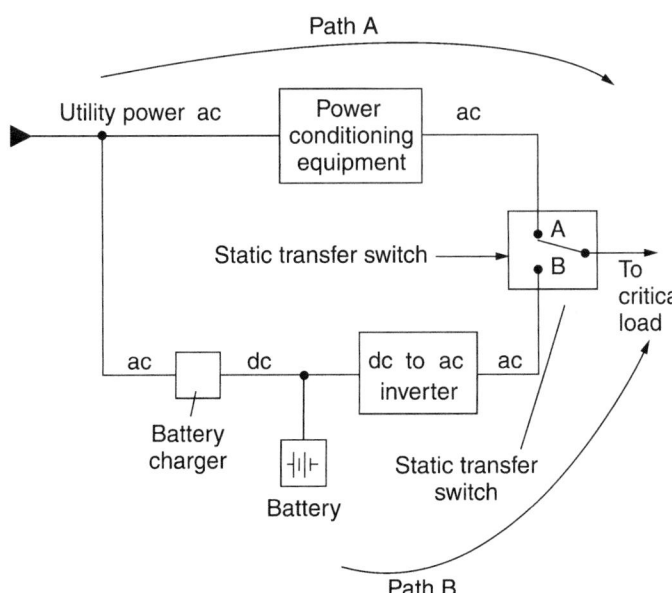

**Fig. 26.66** *In this UPS standby arrangement, power normally flows from the utility to the critical (computer) loads through position A of the transfer switch (Path A). The battery floats until a failure of utility power puts it on line through the inverter and position B in the solid-state, effectively instantaneous transfer switch (Path B). The inverter is normally inactive. The battery charger is small because it carries only charging current.*

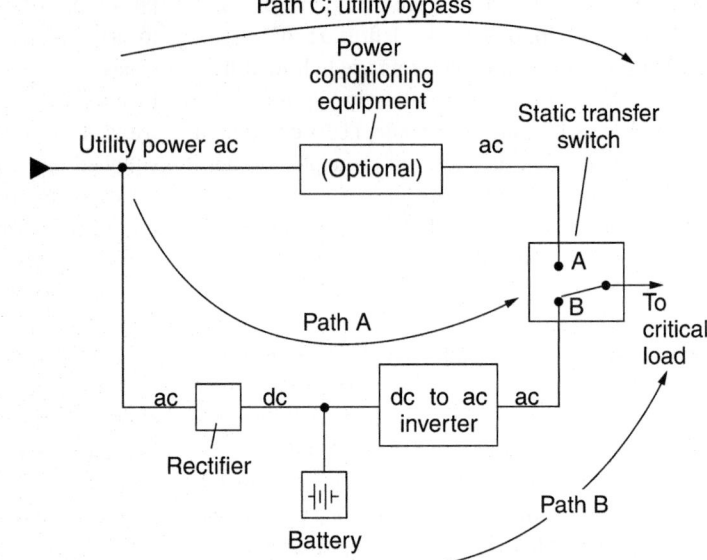

**Fig. 26.67** *In this UPS on-line mode, the normal power Path A is from the utility through a full-size rectifier to the load via an inverter. The battery floats, taking a small charging current from the rectifier. Utility failure brings the battery line through the inverter (Path B). In both instances, the transfer switch remains in position B and the load senses no change. Only in the event of UPS equipment failure is power taken directly from the utility (Path C).*

in Fig. 26.68. In its normal mode (Path *A*), power flows from the utility lines through power conditioning equipment and position *A* of the transfer switch to the load. (This is identical to the normal mode of the standby system shown in Fig. 26.66.) In addition, however, utility ac is tapped at the power conditioning equipment to feed the inverter that, operating in reverse as a rectifier (ac to dc), serves as a battery charger.

In the event of a utility power failure, power is drawn from the battery, and passing through the inverter, it is converted to ac. It then connects to the load via the *B* connection in the transfer switch.

The power path is indicated on Fig. 26.68 as Path *B*. The advantage of this topology is the elimination of a large rectifier by dual use of the inverter and elimination of transients caused by switching on the inverter at full load. The system is efficient, producing little heat, and is highly reliable.

## (d) System Selection and Comparison

The above system descriptions are intended to give the reader a topical familiarity with UPS equipment arrangements. Actual selection depends on the types and magnitudes of the critical loads,

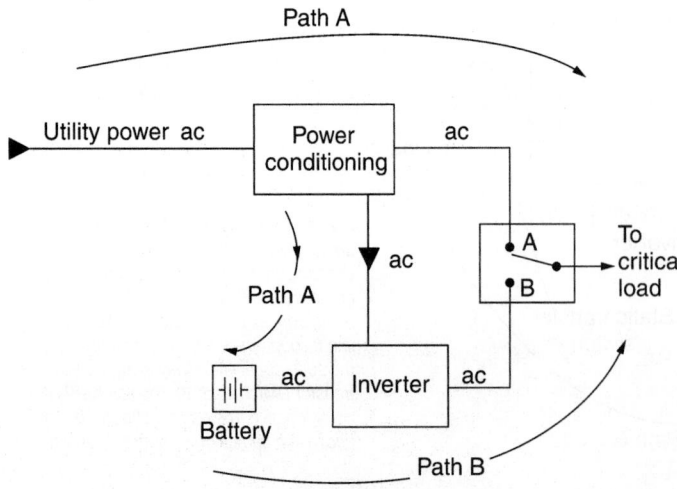

**Fig. 26.68** *In this UPS hybrid (line-interactive) mode, the normal power Path A takes power directly from the utility but keeps the inverter on line, working in reverse to furnish dc charging current to the battery. Utility failure causes a change to power Path B (i.e., from battery through the inverter now operating normally to invert dc to ac). The static switch changes to position B.*

**TABLE 26.8 Typical Dimensional Data for UPS Equipment**

| Topology | Capacity (kVA) | Dimensions H × W × D (in.) | Weight[a] (lb) | Full Load Run Time (min.) |
|---|---|---|---|---|
| Standby | 0.3 | 6 × 6 × 12 | 15 | 10 |
| Single phase | 0.6 | 7 × 6 × 14 | 25 | 5 |
| On-line | 3.0 | 30 × 12 × 24 | 250 | 20 |
| Single phase | 7.5 | 36 × 12 × 36 | 500 | 10 |
| Three-phase | 10.0 | 60 × 24 × 36 | 1200 | 15 |
| Interactive | 0.4 | 6 × 6 × 12 | 20 | 5 |
| Single phase | 1.0 | 8 × 8 × 15 | 40 | 5 |
| | 2.0 | 16 × 8 × 20 | 105 | 20 |
| | 3.5 | 18 × 10 × 24 | 140 | 10 |

[a]With batteries.

detailed outage histories, space considerations, cost factors, equipment and service redundancy, and overall system efficiency. The last characteristic can be a major decision factor where large loads are being supplied because overall system efficiencies vary from a high of about 95+% for classic standby systems (Fig. 26.66) to as low as 65% for some arrangements. Because all of the lost power turns to heat, a low-efficiency 100-kVA/80-kW system would produce about 27 kW or about 92,000 Btu/h! Unless this heat can be used effectively, power costs and ventilation arrangements should be factored into the selection process.

### (e) Ancillary Characteristics of UPS Systems

Programmable microprocessors and effective sensors and transducers have been added to basic UPS systems to produce what is known as *intelligent* UPS equipment. These additional devices now make available operating and diagnostic data including battery status, equipment operating temperatures, load data, waveform displays, on-line power quality analyses, operational and event logs, and the like. These data help an informed operator optimize system functioning, anticipate problems, and keep an accurate printed operational record. Table 26.8 gives some typical dimensional and run-time data for UPS units of different topologies.

## 26.39 EMERGENCY/STANDBY POWER EQUIPMENT

The *NEC* makes a clear distinction between emergency systems and standby systems, covering the former in Article 700 and the latter in Articles 701 and 702. The equipment, circuitry, and arrangement of both are similar; the purpose is somewhat different. The reader is also referred to NFPA Standard 110, *Emergency and Standby Power Systems*, which covers equipment requirements for these systems.

*Emergency* systems are intended to supply electric power to equipment essential for human *safety* on interruption of the normal power supply. Included in this classification are illumination in areas of assembly, to permit safe exiting and panic prevention, and such other vital functions as fire detection and alarm systems, elevators, fire pumps, public address and communication systems, and orderly shutdown or maintenance of hazardous processes.

*Standby* systems are divided into two categories: those legally required (Article 701) and optional systems (Article 702). The former are intended to power processes and systems (other than those classified as emergency systems) whose stoppage might create hazards or hamper firefighting operations. This classification is broader than the emergency system and could include HVAC systems, water supply equipment, and industrial processes whose interruption could cause a safety or health hazard. It is intended primarily as a safety measure.

*Optional* standby systems can cover any or all loads in a facility at the discretion of the owner, and are normally intended to protect property and prevent financial loss in the event of a normal service interruption. A typical application would be a critical industrial process or an ongoing research project.

*Health care facilities* are covered by a separate set of regulations: *NEC* Article 517 and NFPA

Standard 99, both of which are referenced as legally binding in the vast majority of jurisdictional codes. It is important to note that for both emergency and legally required standby systems, the *provision* of the system must be mandated by the authority having jurisdiction over construction, whether it is a local, state, or federal agency or a combination of these. The *NEC* dictates how the system is to be designed and constructed; its existence depends upon another authority.

A case in point is the fundamental item of exit lights. These are mandated by the NFPA *Life Safety Code* and Subpart E of OSHA regulations. They are not required by the *NEC*, and reference to *NEC*

will not assure their provision. The code(s) having authority and jurisdiction must specifically require an emergency and/or standby system by that nomenclature in order for *NEC* provisions to apply. The exact items of equipment to be powered are selected by the designer, keeping in mind the specific and general requirements of the codes having jurisdiction. The majority of codes make the provisions of the *NEC* and the relevant NFPA standards legally binding. Most codes require emergency systems; far fewer require standby systems, and then only for essential water and water-treatment systems and a few other essential uses. The designer must thoroughly investigate the matter of jurisdictional

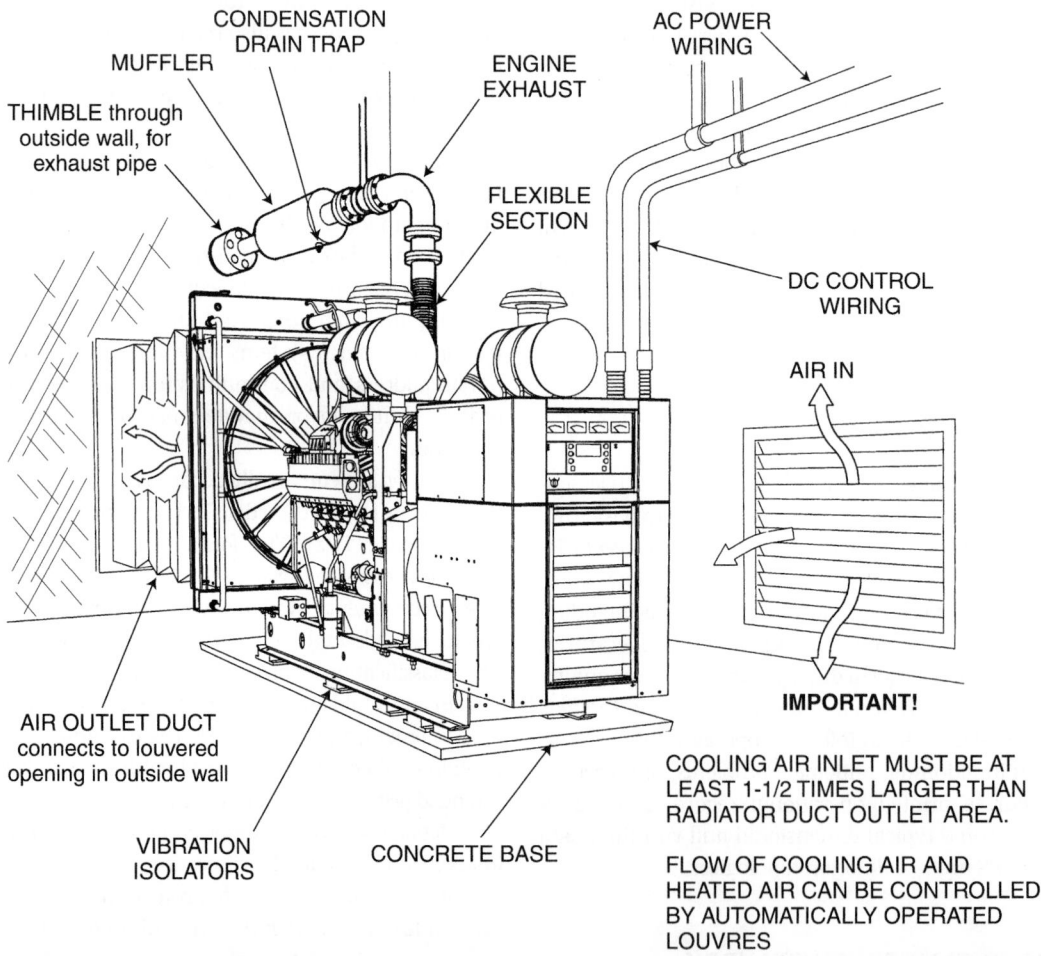

**Fig. 26.69** *Typical installation diagram of a radiator-cooled diesel engine–generator set. The set should be oriented so that the hot radiator cooling air vented through an outside wall flows in the same direction as prevailing winds. A solid masonry wind and noise barrier not less than 6 ft (1.8 m) from the radiator air outlet is recommended, as well as an elevated exhaust pipe outlet. (Illustration courtesy of ONAN Corp.)*

codes before considering emergency electric power systems and equipment.

System design arrangements are discussed in Sections 28.3 and 28.6. Emergency lighting equipment and system design is covered in Chapter 15. System equipment, which falls into two principal categories—that is, generator and battery installations—is discussed later. Optional standby systems normally use a fueled prime mover rather than batteries.

## (a) Engine–Generator Sets

An engine–generator set installation comprises basically three components: the fuel system, including storage, if necessary; the set itself, plus exhaust facilities; and the space housing the unit (Fig. 26.69). The principal advantages of the engine–generator set are unlimited kVA capacity, duration of power limited only by the size of the fuel tank, use for peak-load shaving, and, if properly maintained, indefinite life. The disadvantages are noise, vibration, the nuisance of exhaust piping, the need for constant maintenance and regular testing, and difficulties with fuel storage. Gasoline can be stored for only a year at most, and subsequent disposal is difficult. Diesel fuel keeps somewhat longer, but disposal is also difficult. Use of gas for a diesel or gas engine obviates the fuel storage problem but poses the alternate problem of availability of gas service during emergencies. In some large cities, steam is commercially available as an energy source. Here too, service reliability, particularly in the event of a widespread electric service failure, must be carefully investigated.

## (b) Battery Equipment

Storage batteries are often used to supply limited amounts of emergency power for lighting (see Section 16.31) and for UPS systems, as detailed previously. Such units are mounted in individual cabinets or in racks for large installations and are always provided with automatic charging equipment.

Battery types are undergoing intensive development. At this writing, the types principally in use are lead-acid, nickel-cadmium, lead-antimony, lead-calcium cells, and several alkaline types. The choice depends on the application. Installation requirements such as ventilation, gas detection, isolation, and the like depend entirely on the type of battery, size of the installation, and battery voltage. That being so, no general guidelines can be given except to follow *NEC* Article 480 requirements and consult a battery specialist for other data.

Batteries have the distinct advantage that they can be installed either in a central system with distribution of the battery power throughout the facility or in small package units around the building. Central systems are 24 to 125 V, dc or ac, and normally feed emergency lighting only. Individual packs are used often to supply ac power via built-in inverters. The great disadvantage of battery systems is their limited duration of power and the environmental effects upon disposal. The *NEC* requires that batteries maintain loads for a 1½-hour *minimum*, but larger capacity is frequently installed.

## 26.40 SYSTEM INSPECTION

Each electric wiring system is inspected at least twice by the local inspection authorities: once after raceways (roughing) have been installed and before the wiring and closing-in of walls and once after the entire job is complete. The purpose of these inspections is to determine whether the design, material, and installation techniques meet the national and local code requirements. Excellence of installation is the responsibility of the contractor. The designer, however, must be familiar with installation work and the equipment's physical characteristics in order to properly design an electrical system that will not present the contractor with unwarranted difficulties. The designer must understand and be aware of equipment substitutions by a contractor, who, having submitted a bid on the basis of plans and specifications, and should be required to supply the specified equipment.

# CHAPTER 27

# Electrical Systems and Materials: Wiring and Raceways

THE MAJOR COMPONENTS OF A BUILDING ELECTRICAL system can be arranged in three major categories: wiring and raceways, power-handling equipment, and utilization equipment. In the first category, we include conductors and raceways of all types; in the second, transformers, switchboards, panelboards, large switches, and circuit breakers; and in the last, actual utilization equipment such as lighting, motors, controls, and wiring devices. After some introduction applicable to all electrical materials, this chapter discusses in detail the items in the first of these three categories—that is, the wiring and raceway system. Chapter 26 covered most of the items in the next two categories, with the exception of lighting equipment, discussed in Part III. Signal systems, building control, and automation are discussed in Chapter 30.

## 27.1 SYSTEM COMPONENTS

Referring to Fig. 27.1, note the important fact that the power distribution equipment proceeds from the service point to the utilization points in a series of steps of decreasing circuit capacity. This is analogous to the arrangement of water supply systems and heating systems; the distribution equipment is largest at the supply point, narrowing in steps to reach the last utilization points. The reverse is true in collection systems such as those for drainage and venting; there, piping is smallest at the initial collec-

tion points, growing larger (in steps) as the quantities of fluid increase.

A normal single-line diagram does not differentiate by line weight between heavy and light (large and small) conductors (the heavier the conductor, the greater the amount of power being carried), but the single-line diagram of Fig. 27.1 does differentiate in order to show relative power levels in the system. This "size" differentiation is more clearly shown in Fig. 27.2, which is a pictorial representation of a system similar to that of Fig. 27.1—but in somewhat greater detail and omitting items beyond the panelboard. Only standard items are shown; special items, however common, such as emergency and uninterruptible power sources, energy controls, service transformers, and the like, are omitted for the sake of clarity.

## 27.2 NATIONAL ELECTRICAL CODE

The *National Electrical Code* (NEC) of the National Fire Protection Association (NFPA) defines the fundamental safety measures that must be followed in the selection, construction, and installation of all electrical equipment. This code is used by all inspectors, electrical designers, engineers, contractors, and operating personnel. Having been incorporated into the Occupational Safety and Health Act (OSHA) it has, in effect, the force of law. Frequent references are made to the *NEC* throughout this chapter. In

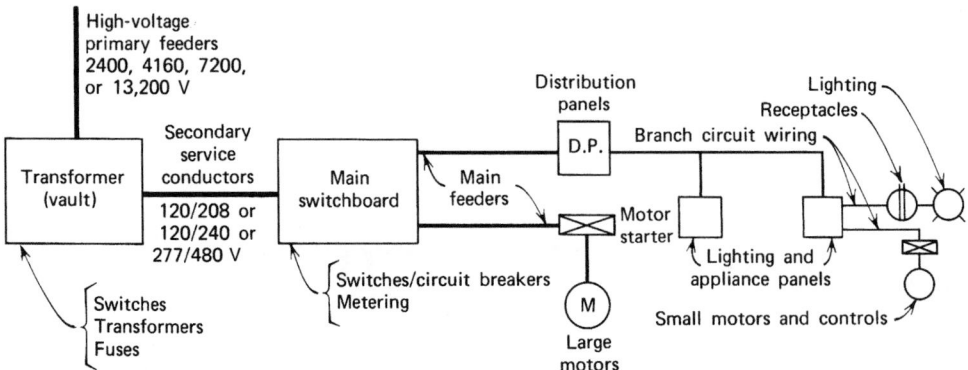

**Fig. 27.1** *Single-line diagram of a typical building electrical distribution system, from the incoming service to the utilization items at the end of the system. This is also referred to as a* block diagram *because the major components are shown as rectangles, or blocks. When this same type of information is presented showing the vertical spatial relationship between components, it is called a* riser diagram; *when electrical symbols are used in lieu of the blocks, it is called a* one-line or single-line diagram. *The connecting conductors between the major system components are drawn here to reflect size, and thereby power-handling capacity. In practice, all connecting lines are the same weight. Also, branch circuits are not usually shown; the typical diagram ends at the lighting and appliance panels.*

addition to the *NEC*, many large cities such as New York, Boston, and Washington, D.C., have their own electrical codes that, although similar to the *NEC*, contain numerous special requirements.

In order to ensure a minimum standard of intrinsic electrical safety for electrical equipment, a single agency was needed to establish standards and to test and inspect electrical equipment. Such an organization is Underwriters Laboratories (UL), which publishes lists of inspected and approved electrical equipment. These listings are universally accepted, and many local codes state that only electrical materials bearing the UL label (of approval) are acceptable.

## 27.3 ECONOMICS OF MATERIAL SELECTION

The selection of electrical materials involves not only choosing a material or assembly that is functionally adequate and, where necessary, visually satisfactory, but also the consideration of economic factors. This is necessary because usually many brands and types of equipment that will fulfill the construction need are available. In such cases, economic factors often decide the issue.

The decision is relatively simple when the various materials differ only slightly from each other and a straightforward first-cost comparison is all that is required. Often, however, the choice is not so simple, because the materials may vary considerably in characteristics other than functional suitability, and a more detailed cost study is required. Such economic analyses are frequently performed in comparisons among various HVAC systems. This happens less often when dealing with electrical systems because, from an energy point of view, electrical systems are more passive than HVAC systems, and it is energy costs that are often the decisive factor in economic analyses. We are speaking, of course, of life-cycle equipment costs (over the life of the *structure*) expressed in *present-value* dollars, or annual owning and operating costs, including equipment amortization costs. The type of analysis used depends on the situation. However, such comparisons are useful only when both the initial cost and the operating costs will be borne by the same individual—that is, an owner-operator. In the case of a speculative building venture, the first cost is usually the deciding factor.

It is often difficult to perform such cost analyses because accurate data on life and maintenance costs for electrical equipment may not be readily available. Still, even if a formal analysis cannot be done, the principle involved should be considered in the selection of all electrical material, bearing in mind its relative importance as energy-utilizing equipment. Detailed economic discussions on particular items are found with the related technical discussion. See Appendix I for an explanation of cost analysis techniques.

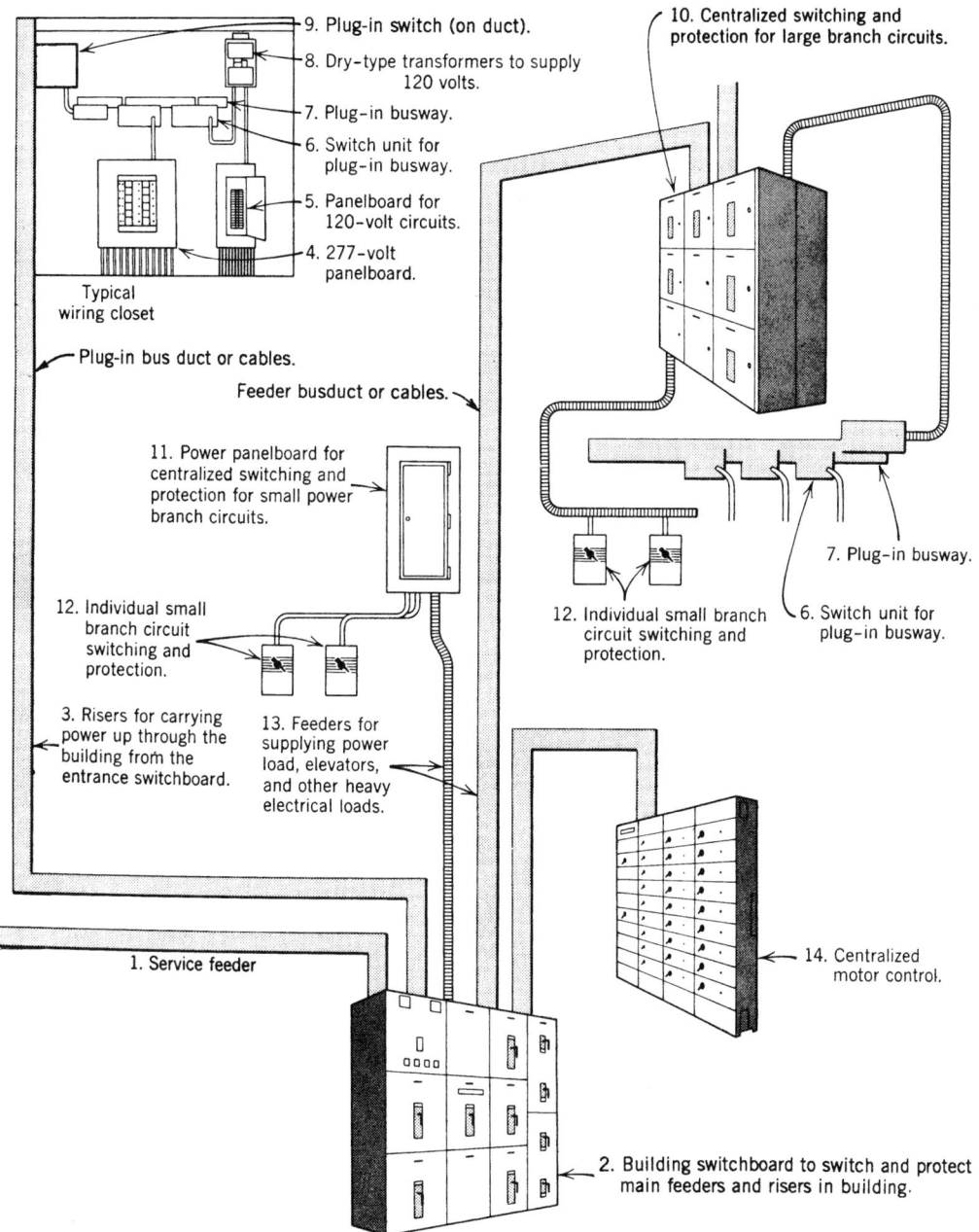

9. Plug-in switch (on duct).

8. Dry-type transformers to supply 120 volts.

7. Plug-in busway.

6. Switch unit for plug-in busway.

5. Panelboard for 120-volt circuits.

4. 277-volt panelboard.

Typical wiring closet

Plug-in bus duct or cables.

Feeder busduct or cables.

10. Centralized switching and protection for large branch circuits.

11. Power panelboard for centralized switching and protection for small power branch circuits.

7. Plug-in busway.

12. Individual small branch circuit switching and protection.

6. Switch unit for plug-in busway.

12. Individual small branch circuit switching and protection.

3. Risers for carrying power up through the building from the entrance switchboard.

13. Feeders for supplying power load, elevators, and other heavy electrical loads.

1. Service feeder

14. Centralized motor control.

2. Building switchboard to switch and protect main feeders and risers in building.

ELECTRICITY

**Fig. 27.2** *Diagrammatic pictorial representation of a typical building electrical power system with the relative power capacity indicated by the size of the conductors. This diagram does not extend beyond the local panelboard and includes only commonly used items. Note also that the entire wiring system is, in effect, jacketed in steel. (Courtesy of General Electric.)*

## 27.4 ENERGY CONSIDERATIONS

Energy costs are a major factor in economic analysis. However, energy considerations are at least as important in themselves because of environmental and natural resource considerations. This factor is examined in detail in the discussions of building energy budgets and their components, including lighting, elevators, and electric motors in all other systems.

## 27.5 ELECTRICAL EQUIPMENT RATINGS

All electrical equipment is rated for the normal service it is intended to perform. These ratings may be in voltage, current, duty, horsepower, kW, kVA, temperatures, enclosure, and so on. Ratings related to specific equipment are discussed in the sections that follow. The ratings that are specifically and characteristically electrical are those of voltage and current.

### (a) Voltage

The voltage rating of an item of electrical equipment is the maximum voltage that can safely be applied to the unit continuously. It frequently, but not always, corresponds to the voltage applied in normal use. Thus, an ordinary wall electrical receptacle is normally rated at 250 V maximum, although in normal use, only 120 V are applied to it. The rating is determined by the type and quantity of insulation used and the physical spacing between electrically energized parts.

### (b) Current

The current rating of an item is determined by the maximum operating temperature at which its components can operate at full load. That in turn depends on the type of insulation used. As a case in point, consider an electric motor. The current flowing in the motor windings causes a power loss ($I^2R$), which generates heat. If the windings are insulated with varnished cotton braid, with a maximum safe operating temperature of 65°C (150°F), the maximum permissible current (to which the horsepower rating of the motor is directly related) is the current that produces this operating temperature. If these same windings are insulated with a silicone or glass compound with a maximum operating tempera-

ture of 150°C (300°F), obviously more current can be carried safely, and the horsepower rating is consequently larger. Thus, we see that although a motor is rated in horsepower (or kW where SI units are used), a transformer is rated in kVA, and a cable (discussed later) is rated in amperes, the actual criterion on which all these ratings are based is the maximum permissible operating temperature of its insulation (and other components).

## 27.6 INTERIOR WIRING SYSTEMS

At this point, it is helpful to survey the different types of interior wiring systems before commencing a discussion of components. When the primary purpose of the system is to distribute electrical energy, it is referred to as an *electrical power system;* when the purpose is to transmit information, it is referred to as an *electrical signal* or a *communication system.* In this chapter, we deal with electrical power systems, except that the discussion of raceways covers equipment used by communication systems as well.

Due to the nature of electricity, its distribution within a structure for power use poses basically a single problem: how to construct a distribution system that *safely* provides the energy required at the location required. The safety consideration is all-important because even the smallest interior system is connected to the utility's powerful network, and the potential for physical damage, injury, and fire is always present. The solution to this problem is to isolate the electrical conductors from the structure, except at those specific points, such as wall receptacles, where contact is desired. This isolation is generally accomplished by insulating the conductors and placing them in protective raceways.

The principal types of interior wiring systems in use today are exposed insulated cables, insulated cables in open raceways, and insulated conductors in closed raceways.

### (a) Exposed Insulated Cables

This category includes (using the *NEC* nomenclature) cable types NM ("Romex") and AC ("BX"). (See Sections 27.11 and 27.12.) Also included are other types where the cable construction itself provides the necessary electrical insulation and mechanical protection.

## (b) Insulated Cables in Open Raceways (Trays)

This system is specifically intended for industrial application, and it relies on both the cable and the tray for safety.

## (c) Insulated Conductors in Closed Raceways

This system is the most general type and is applicable to all types of facilities. In general, the raceway is installed first and the wiring is pulled in or laid in later. The raceways themselves may be:

1. Buried in the structure—for example, conduit in the floor slab or underfloor duct. (See Sections 27.18, 27.19, 27.25, and 27.26.)
2. Attached to the structure—for example, all types of surface raceways, including conduit and wireways suspended above hung ceilings. (See Sections 27.23 and 27.30.)
3. Part of the structure—for example, cellular concrete and cellular metal floors. (See Sections 27.27 and 27.28.)

## (d) Combined Conductor and Enclosure

This category is intended to cover all types of factory-prepared and factory-constructed integral assemblies of conductor and enclosure. Included here are all types of busway, busduct, and cablebus (Section 27.15); flat cable assemblies and lighting track (Section 27.16); flat cable intended for undercarpet installation (Section 27.30); and manufactured wiring systems (Section 27.31).

## 27.7 CONDUCTORS

Electrical conductors (wiring) are the means by which the current is conducted through the electrical system, corresponding to the piping in the hydraulic analogy. The standard of the American wire and cable industry for round cross-section conductors is the American Wire Gauge (AWG). By convention, a single insulated conductor No. 6 AWG or larger, or several conductors of any size assembled into a single unit, are referred to as *cable*. Single conductors No. 8 AWG and smaller are called *wire*.

All wire sizes up to No. 0000 (also written as No. 4/0) are expressed in AWG. The AWG numbers run in *reverse* order to the size of the wire—that is, the smaller the AWG number, the larger the size. Thus, No. 10 is a heavier wire than No. 12 wire and is lighter (thinner) than No. 8 wire. The No. 4/0 size is the largest AWG designation, beyond which a different designation called *kcmil* (thousand circular mil) is used. The former designation for this unit was *MCM*, a term that is still used in many sources. In this designation, wire diameter *increases* with number; thus, 500 kcmil is a heavier wire (double the area) than 250 kcmil.

Outside of the United States, where SI units are in general use, conductor sizes are given simply as their diameter in millimeters. Table 27.1 gives dimensional and stranding data for common wire sizes and includes the millimeter equivalent of each size. This will prove useful in interfacing American gauges and SI sizes.

## 27.8 CONDUCTOR AMPACITY

Conductor current–carrying capacity, or *ampacity*, is determined, as explained previously, by the maximum safe operating temperature of the insulation. Heat generated as a result of current flow is dissipated into the environment. Thus, for a given environment (open-air, buried in earth, or enclosed), ampacity increases with increasing conductor size *and* with maximum permissible insulation temperature. These facts are clearly shown in Table 27.2. If more than three conductors are placed in a conduit, the resultant increase in temperature requires that the conductors be derated by the amount shown in Table 27.3.

Because heat dissipation from a conductor in free air is much greater than that from the same conductor enclosed in conduit or direct buried, its corresponding allowable ampacity is also greater. Conversely, if the ambient temperature around the conductor is higher than 30°C (86°F), on which all the ampacity tables are based, the permissible ampacity must be reduced.

Ampacity tables for conductors in free air, for cable types not shown in Table 27.2, and derating factors for high ambient temperatures are all found in the *NEC*. Typical ambient temperatures are given in Table 27.4.

**TABLE 27.1 Physical Properties of Bare Copper Conductors**

| Size AWG kcmil (MCM) | Area Circular Mils | Diameter (in.) Solid | Diameter (in.) Stranded | Diameter (mm) Solid | Diameter (mm) Stranded | dc Resistance at 25°C, (77°F) (Uncoated) ohms/1000 ft |
|---|---|---|---|---|---|---|
| 18 AWG | 1,620 | 0.040 | — | 1.02 | — | 7.77 |
| 16 | 2,580 | 0.051 | — | 1.29 | — | 4.89 |
| 14 | 4,110 | 0.064 | — | 1.63 | — | 3.07 |
| 12 | 6,530 | 0.081 | 0.092 | 2.05 | 2.34 | 1.93 |
| 10 | 10,380 | 0.102 | 0.116 | 2.59 | 4.06 | 1.21 |
| 8 | 16,510 | 0.128 | 0.146 | 3.26 | 3.71 | 0.764 |
| 6 | 26,240 | — | 0.184 | — | 4.11 | 0.491 |
| 4 | 41,740 | — | 0.232 | — | 5.18 | 0.308 |
| 2 | 66,360 | — | 0.292 | — | 6.55 | 0.194 |
| 1 | 83,690 | — | 0.332 | — | 7.34 | 0.154 |
| 0 (1/0) | 105,600 | — | 0.373 | — | 8.26 | 0.122 |
| 00 (2/0) | 133,100 | — | 0.418 | — | 9.27 | 0.097 |
| 000 (3/0) | 167,800 | — | 0.470 | — | 10.41 | 0.077 |
| 0000 (4/0) | 211,600 | — | 0.528 | — | 11.68 | 0.061 |
| 250 kcmil | 250,000 | — | 0.575 | — | 12.70 | 0.052 |
| 300 kcmil | 300,000 | — | 0.630 | — | 13.92 | 0.043 |
| 400 kcmil | 400,000 | — | 0.728 | — | 16.05 | 0.032 |
| 500 kcmil | 500,000 | — | 0.813 | — | 19.56 | 0.026 |

*Source:* Except for millimeter dimensions, this table was extracted from NFPA 70-1999, the *National Electrical Code.* © 1999, National Fire Protection Association, Quincy, MA 02269.

*Note:* This extracted material is not the complete and official position of the National Fire Protection Association on the referenced subject, which is represented only by the standard in its entirety.

## 27.9 CONDUCTOR INSULATION AND JACKETS

Most conductors are covered with some type of insulation that provides both electrical isolation and a degree of physical protection. Additional physical shielding, where necessary, is provided by a jacket over the insulation. Insulation is rated by voltage. Ordinary building wiring is rated for 300 or 600 V. Common types of building wire insulation are listed in Table 27.5 with their associated trade names, code letters, maximum permitted operating temperatures, and special provisions.

## 27.10 COPPER AND ALUMINUM CONDUCTORS

Aluminum has an inherent weight advantage over copper, with concomitant lower installation costs. Economy usually lies with copper in small- and medium-sized cable, because weight is not a problem and the smaller conduit required for the smaller copper conductors generally makes the combined installation cheaper. In the larger cable sizes, the aluminum weight advantage offsets the economy of smaller copper size and smaller conduit, and generally proves less expensive, particularly in areas of high labor cost such as urban areas.

Aluminum and copper both exhibit the low electrical resistivity necessary for a good electrical conductor. There are, however, difficulties inherent in splicing and terminating aluminum. These difficulties—which can be overcome with the use of proper equipment, techniques, and workmanship—stem from aluminum's cold-flow characteristic when under pressure (causing joints to loosen) and aluminum's oxide. This oxide, which forms within minutes on any exposed aluminum surface, is an adhesive, poorly conductive film that must be removed and prevented from reforming if a successful, long-life joint or termination is to be effected. If this is not done, the oxide causes a high-resistance joint with consequent excessive heat generation and possible incendiary effects. This oxide problem can be largely overcome by the use of copper-clad aluminum wire, but the cold-flow problem remains. Furthermore, when used in branch circuits, even if

**TABLE 27.2  Allowable Ampacities of Insulated Conductors Rated 0 through 2000 V, 60° to 90°C (140° to 194°F), Not More Than Three Current-Carrying Conductors in Raceway or Cable or Earth (Directly Buried), Based on Ambient Temperature of 30°C (86°F)**

| PART A | | | |
|---|---|---|---|
| | *Temperature Rating of Conductor*[a] | | |
| | **60°C (140°F)** | **75°C (167°F)** | **90°C (194°F)** |
| **Size AWG, kcmil** | **Type UF** | *Types RHW, THW, THWN, XHHW*[b] | **Types THWN, XHHW**[b] |
| 14 AWG[c] | 20 | 20 | 25 |
| 12[c] | 25 | 25 | 30 |
| 10[c] | 30 | 35 | 40 |
| 8 | 40 | 50 | 55 |
| 6 | 55 | 65 | 75 |
| 4 | 70 | 85 | 95 |
| 2 | 95 | 115 | 130 |
| 1 | 110 | 130 | 150 |
| 0 | 125 | 150 | 170 |
| 00 | 145 | 175 | 195 |
| 000 | 165 | 200 | 225 |
| 0000 | 195 | 230 | 260 |
| 250 kcmil | 215 | 255 | 290 |
| 300 | 240 | 285 | 320 |
| 350 | 260 | 310 | 350 |
| 400 | 280 | 335 | 380 |
| 500 | 320 | 380 | 430 |
| PART B | | | |
| *Ambient Temperature* | | *Correction Factors for Ambient Temperatures Other Than 30°C (86°F), Multiply the Allowable Ampacities Shown Above by the Appropriate Factor Shown Below.* | |
| **°C** | **°F** | | |
| 21–25 | 70–77 | 1.08 | 1.05 | 1.04 |
| 26–30 | 78–86 | 1.00 | 1.00 | 1.00 |
| 31–35 | 87–95 | .91 | .94 | .96 |
| 36–40 | 96–104 | .82 | .88 | .91 |
| 41–45 | 105–113 | .71 | .82 | .87 |

*Source:* This table was extracted from NFPA 70-1999, the *National Electrical Code.* © 1999, National Fire Protection Association, Quincy, MA 02269.

*Note:* This extracted material is not the complete and official position of the National Fire Protection Association on the referenced subject, which is represented only by the standard in its entirety.

[a]See Table 27.5.

[b]For dry and damp locations use the 90°C rating; for wet locations use the 75°C rating.

[c]Unless otherwise permitted by the *NEC,* the overcurrent protection for these conductors shall not exceed 15 A for 14 AWG, 20 A for 12 AWG, and 30 A for 10 AWG, after application of correction factors for ambient temperature and number of conductors in raceway.

properly installed initially, aluminum can create problems when wiring devices are replaced by unskilled homeowners.

As a result of a number of unfortunate incidents, some localities in the United States have banned the use of aluminum wire in branch circuitry. Heavy feeders are normally installed by experienced and skilled contractors, and the risk of a poor joint is minimized. We recommend that the use of aluminum conductors be restricted to sizes no smaller than No. 4 AWG, and that installation be permitted *only* by contractors who certify expertise in the specialized techniques involved. Also, local codes and electrical inspectors should be consulted. All references in this text, including all tables and illustrations, are to copper conductors. The following sections provide a brief description of the principal building wire types.

**ELECTRICITY**

### TABLE 27.3 Current-Carrying Capacity Derating Factors[a]

| Number of Current-Carrying Conductors in Raceway[b] | Derating Factor[c] |
|---|---|
| 4–6 | 0.80 |
| 7–9 | 0.70 |
| 10–20 | 0.50 |
| 21–30 | 0.45 |
| 31–40 | 0.40 |
| 41 and above | 0.35 |

*Source:* This table was extracted from NFPA 70-1999, the *National Electrical Code.* © 1999, National Fire Protection Association, Quincy, MA 02269.

*Note:* This extracted material is not the complete and official position of the National Fire Protection Association on the referenced subject, which is represented only by the standard in its entirety.

[a]These factors are to be applied after application of derating factors for ambient temperatures other than 30°C (86°F). See Table 27.2.

[b]Grounding and bonding conductors are not counted. Neutral conductors are counted in special cases. See 1999 *NEC,* Article 310-15(4).

[c]When a load diversity factor is applicable, derating factors as given in 1999 *NEC* Table B-310-11 shall be used.

### TABLE 27.4 Typical Ambient Temperatures

| Location | Temperature |
|---|---|
| Well-ventilated, normally heated buildings | 30°C (86°F) |
| Buildings with major heat sources such as power stations or industrial processes | 40°C (104°F) |
| Poorly ventilated spaces such as attics | 45°C (113°F) |
| Furnaces and boiler rooms       (min.) | 40°C (104°F) |
|                                            (max.) | 60°C (140°F) |
| Outdoors in shade in air | 40°C (104°F) |
| In thermal insulation | 45°C (113°F) |
| Direct solar exposure | 55°C (131°F) |

application and installation details and restrictions, see *NEC* Article 333, "Armored Cable."

A similar construction with much broader application (covered in *NEC* Article 334) is metal-clad (MC) cable. This cable may be used exposed or concealed and in cable trays, and, when covered with a moisture-impervious jacket, in wet and outdoor locations as well (Fig. 27.4).

## 27.11  FLEXIBLE ARMORED CABLE

Among the most common types of exposed wiring is *NEC* type AC armored cable, commonly known in the smaller sizes by the trade name *BX.* It is an assembly of insulated wires, bound together and enclosed in a protective armor made of a spiral-wound interlocking strip of steel tape (Fig. 27.3). The cable is installed with simple U-clamps or staples holding it against beams, walls, and so on. This type of installation is frequently used in residences and in the rewiring of existing buildings. Use of type AC cable is generally restricted to dry locations. For

## 27.12  NONMETALLIC SHEATHED CABLE (ROMEX)

In application, *NEC* types NM and NMC, also known by the trade name *Romex,* are restricted to small buildings—that is, residential and other structures not exceeding three floors above grade (Fig. 27.5). The plastic outer jacket, unlike the armor on type AC, makes type NM easier to handle but more vulnerable to physical damage. For application details and restrictions, see *NEC* Article 336, "Non-metallic Sheathed Cable." The typical installation technique is shown in Fig. 27.6.

### TABLE 27.5 Conductor Insulation and Application

| Trade Name | Type/Letter | Temperature | Application Provisions |
|---|---|---|---|
| Moisture and heat-resistant rubber | RHW | 75°C (167°F) | Dry and wet locations |
| Single conductor, underground feeder and branch-circuit | UF | 60°C (140°F) 75°C (167°F)[a] | Refer to NEC Article 339 |
| Moisture-resistant thermoplastic | TW | 60°C (140°F) | Dry and wet locations |
| Heat-resistant thermoplastic | THHN | 90°C (194°F) | Dry and damp locations |
| Moisture and heat-resistant thermoplastic | THW | 75°C (167°F) 90°C (194°F) | Dry and wet locations Special applications |
| Moisture and heat-resistant thermoplastic | THWN | 75°C (167°F) | Dry and wet locations |
| Moisture and heat-resistant cross-linked synthetic polyethylene | XHHW | 90°C (194°F) 75°C (167°F) | Dry and damp locations Wet locations |

*Source:* This table was extracted from NFPA 70-1999, the *National Electrical Code.* © 1999, National Fire Protection Association, Quincy, MA 02269. This extracted material is not the complete and official position of the National Fire Protection Association on the referenced subject, which is represented only by the standard in its entirety.

[a]For ampacity limitation, see *NEC* Article 339-5.

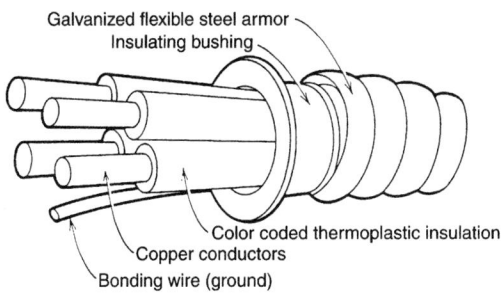

**Fig. 27.3** *Flexible armored cable (NEC type AC, trade name BX). Note the insulating bushing that is always installed on the end of the armor to protect the wires from damage from the sharp edges of the cut steel armor. (Courtesy of AFC Cable Systems.)*

## 27.13 CONDUCTORS FOR GENERAL WIRING

Under this heading (Article 310), the *NEC* lists the wire types that are generally installed in raceways and are referred to by the term *building wire.* The most common types are listed in Table 27.5. These wires consist of a copper conductor covered with insulation and, in some instances, with a jacket (Fig. 27.7).

## 27.14 SPECIAL CABLE TYPES

Although most building wiring is accomplished with plastic-insulated 300- and 600-V conductors of the types described in the preceding sections, some applications require the use of special cables. These include high-voltage cables, armored cables, corrosion-resistant jacketed cables, underground cables, and so on. The reader is referred to manufacturers' catalogs and the *NEC* for construction and application details. Service entrance cables and installation are discussed in Sections 26.1 and 26.4, which cover electric service.

## 27.15 BUSWAY/BUSDUCT/CABLEBUS

A busway (busduct) is an assembly of copper or aluminum bars in a rigid metallic housing (Fig. 27.8). Its use is almost always preferable, from an economic viewpoint, in two instances: when it is necessary to carry large amounts of current (power) and when it is necessary to tap onto an electrical power conductor at frequent intervals along its length.

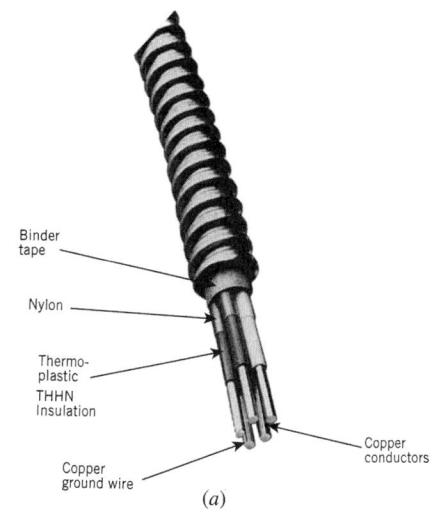

*(a)*

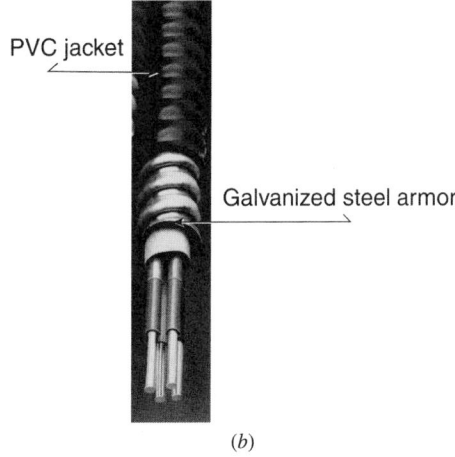

*(b)*

**Fig. 27.4** *(a) Metal-clad cable (NEC type MC, Article 334) with aluminum armor in lieu of the more common galvanized steel armor. Use is similar to that of the steel-armored cable, with the weight advantage of aluminum. Conductors are factory-installed, color-coded, and covered with type THHN insulation and nylon jacket. Cables of similar construction, using steel armor, are available for almost all power and control applications. (Courtesy of AFC Cable Systems.) (b) Jacketed-type MC cable has a wide variety of applications, including in wet locations, because of its water-impervious outer PVC jacket. Where specifically indicated by the manufacturer, this cable may be used for direct burial in earth or installed in concrete. (Courtesy of AFC Cable Systems.)*

In the case of a heavy current requirement, the alternatives to using busways are to use paralleled sets of round conductors or a single large conductor. Paralleled sets of conductors are almost always more expensive than a busway of similar current capacity because of the high installation cost of multiple conduits. Alternatively, using a single,

ELECTRICITY

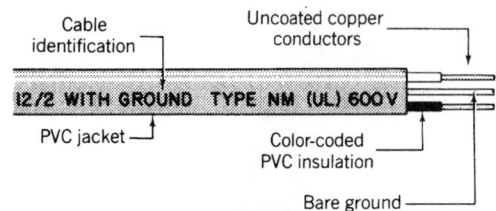

**Fig. 27.5** *Construction of typical NEC type NM cable. The illustrated cable is a two-conductor, No. 12 AWG with ground, insulated for 600 V. Normally shown are the manufacturer, cable trade name, and the letters (UL), which indicate listing of this product by Underwriters Laboratories, Inc. The ground wire is bare or covered, and the entire cable may be obtained flat (illustrated), oval, or round.*

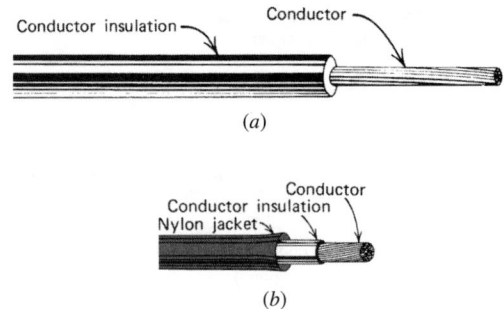

**Fig. 27.7** *(a) Typical construction of unjacketed building wire such as type THW (see Table 27.5). (b) The illustrated construction is typical for any nylon-jacketed cable such as THWN or THHN. (The first three letters indicate the type of insulation, and the final N indicates the nylon jacket.)*

large-diameter conductor becomes increasingly inefficient as cable size increases because large round conductors require more cross section per ampere of current-carrying capacity than do the flat conductors (busbars) used in busduct (Fig. 27.9).

Where many power tap-offs are required along an electric feeder run consisting of cable in conduit, costs become very high because of the large amount of expensive hand labor involved, since a connection must be made to each conductor in the run. The preferable alternative is to use "plug-in" busway to which connection can be made simply and rapidly with a plug-in device, similar to the insertion of a common plug into a wall receptacle. This has the additional advantage of convenience; connection and disconnection is simply a matter of inserting or withdrawing the plug-in device, whereas cable taps are permanent connections (Fig. 27.10).

A typical application of heavy-duty busduct might be a vertical feeder in a high-rise building connecting the basement switchboard to the penthouse machine room. The same building might also use heavy-duty plug-in busduct as vertical riser(s) with taps feeding individual floors (Fig. 27.2). Typical

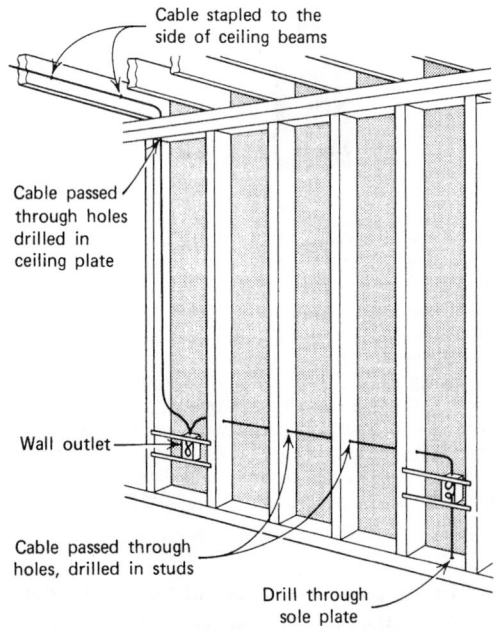

**Fig. 27.6** *Typical wiring technique using types NM (Romex) or AC (BX) in wood stud construction. With metal stud construction, BX cables are passed through precut openings in lieu of field-drilled holes. Where cables are exposed to damage from nails, screws, and other sources, protective metal cover plates are required. For details, see the installation limitations in the NEC.*

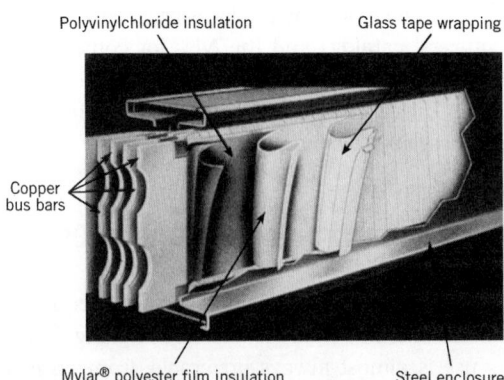

**Fig. 27.8** *Cutaway view showing construction of a typical feeder busduct. This design is highly compact and rigid, which gives desirable electrical characteristics as well as the advantage of small size. (Photo courtesy of Siemens Energy and Automation, Inc.)*

**Fig. 27.9** *Sectional view of a busduct (above) shows the tight assembly of insulated conductors within a metal housing. This design, unlike the ventilated type, can be mounted in any position because heat dissipation is by conduction from the busbars to the housing. The eight sets of cable shown (below) have the same current-carrying capacity as the busduct. (Reproduced by permission of Square D Company.)*

applications for light-duty plug-in busduct (70 to 100 A) could be any machine shop or workshop. The electrical supply to individual machine tools is made very simply and flexibly with a tap-on device (see Fig. 27.10).

Busduct is specified by type, material, number of buses, current capacity, and voltage (e.g., aluminum feeder busduct, 4-wire, 1000 A, 600 V, or copper plug-in busway, 100 A, 3-wire, 600 V). Feeder busduct (no plug-in capability) is available in ratings from 400 to 4000 A. Plug-in busway is available from 30 A for lighting or light-machinery circuits (see Section 27.16) to 3000 A. A wide variety of fittings and joints are available for all busways to permit easy installation (Fig. 27.11). Designs are available for indoor and outdoor application.

Cablebus is similar to ventilated busduct, except

that it uses insulated cables instead of busbars. The cables are rigidly mounted in an open space-frame. The advantage of this construction is that it carries the ampacity rating of its cables *in free air,* which is much higher than the conduit rating, thus giving a high amperes-per-dollar first-cost figure. Its principal disadvantages are bulkiness and difficulty in making tap-offs.

As an example of the type of economic analysis that should be made when considering an item as relatively simple as a heavy-current electrical feeder,

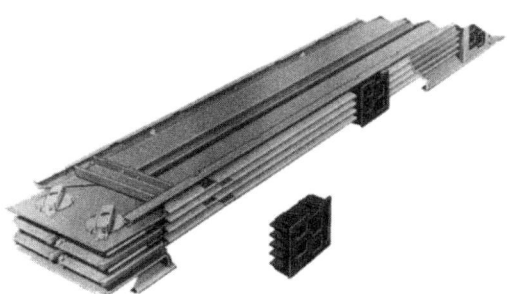

**Fig. 27.10** *Construction of one type of plug-in busduct. Plug-ins are evenly spaced on alternate sides to facilitate connection of plug-in breakers, switches, transformers, or cable taps. Housing is of sheet steel with openings for ventilation. The cover plate is not shown. (Courtesy of Square D Company.)*

**Fig. 27.11** *Typical installation of compact-design busduct. Note that the individual busducts are supported by channels hung from the ceiling and that the same hangers support more than one level of bus. Right-angle turns are easily made in the same plane (horizontal or vertical) and between vertical and horizontal planes. (Reproduced by permission of Square D Company.)*

**ELECTRICITY**

**TABLE 27.6 Life-Cycle Relative Cost Comparison of 2000-A, 208-V Feeder Installation**

| Feeder System Description[a] | Material Cost | Labor Cost[b] | Total First Cost | Power Loss per 100 ft[c] (kW) | Annual Energy Loss[d] (kWh) | Annual Energy Cost[e] | Life-Cycle Energy Cost[f] | Total Life-Cycle Cost[g] |
|---|---|---|---|---|---|---|---|---|
| Cable tray | 0.63 | 0.37 | 1.00 | 2.90 | 12,536 | 0.086 | 1.081 | 2.08 100% |
| Wire and conduit | 0.68 | 0.61 | 1.29 | 2.90 | 12,536 | 0.086 | 1.081 | 2.37 114% |
| Busduct | 1.14 | 0.24 | 1.38 | 5.60 | 24,207 | 0.166 | 2.087 | 3.47 167% |
| Cablebus | 0.69 | 0.45 | 1.14 | 5.91 | 25,547 | 0.174 | 2.203 | 3.34 161% |

[a]Equipment rating is 600 V, cable is copper, conduit is rigid steel, cable tray is aluminum.

[b]Labor cost includes overhead.

[c]Based on published resistivity data for cable and bus—assuming 80% demand (1600 A) and all conductors in the system equally loaded.

[d]Based on 80% demand, 12 hours per day, 360 days per year.

[e]Using $0.10 per kWh as the combined net rate, including demand charges.

[f]Using a 20-year life cycle, 8% fixed capital cost, and 3% annual escalation in energy cost.

[g]Sum of the fourth and eighth columns.

refer to Table 27.6, which shows the results of such a study in relative cost terms. Note that when considering the first cost alone, the advantage lies with cable tray (with interlocked armor cables) and cablebus. Adding the energy-loss consideration shifts the advantage to cable tray and wire in conduit. No general conclusion should be drawn from Table 27.6 regarding costs. A change in feeder length, number of taps, hours of operation, energy rates, or any of the other factors can shift the advantage to a different system. The point of our study is to demonstrate clearly that life-cycle costs and first costs often yield entirely different results and that, therefore, this type of study is required before an engineering decision can be made. (Life-cycle cost is taken to mean the present value of all costs over the installation's life cycle—in this case, 20 years for the system.)

Two additional items are worthy of note:

1. The very factors that yield a lower first cost operate to yield a higher operating cost. The smaller copper sizes in busduct and cablebus, permitted by high-temperature insulation and good ventilation, cause increased power loss because of their higher resistivity.
2. If the heat loss from the busduct or cablebus can be used to advantage, the related energy cost can be credited instead of being considered a total loss, and life-cycle costs can be changed considerably. Conversely, it can also affect the building cooling load.

## 27.16 LIGHT-DUTY BUSWAY, FLAT-CABLE ASSEMBLIES, AND LIGHTING TRACK

Special assemblies that act as light-duty (branch circuit) plug-in electrical feeders are widely used because of their simplicity of installation and, more importantly, because of their plug-in mode of connection.

### (a) Light-Duty Plug-In Busway

This construction, which may be used either for feeder or branch circuit application, is covered by the *NEC* general article on busways, with restrictions when applied as branch circuit wiring. Light-duty busways are rated from 20 to 60 A at 300 V, in 2- and 3-wire construction. A somewhat heavier design rated 60 A to 100 A at 600 V is available in 3- and 4-wire construction. Their application is principally for direct connection (with overcurrent protection) of light machinery and industrial lighting (Figs. 27.12 and 27.13).

### (b) Flat-Cable Assemblies

A specially designed cable (NEC Article 363; Type FC) consisting of two, three, or four No. 10 AWG conductors is field installed in a rigidly mounted standard 1⅜-in. square structural channel. Power-tap devices, installed where required, puncture the

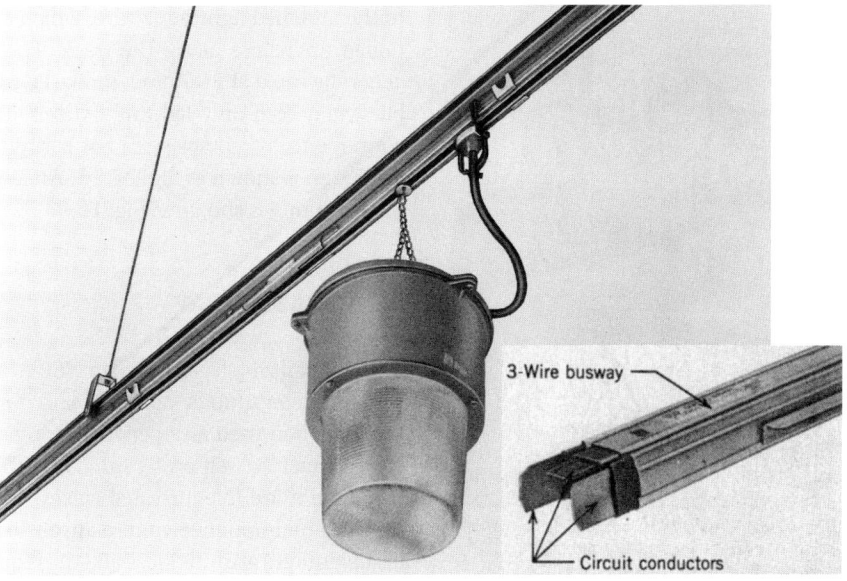

**Fig. 27.12** *Light-duty busway is rated 20–60 A, 300 V, and either 2- or 3-wire. Power takeoff devices twist into the bus to make contact with the circuit conductors. Within* NEC *restrictions of overcurrent protection, this busway may be used for standard and heavy-duty lighting fixtures and for other electrical devices such as electrically powered tools. (Courtesy of Siemens Energy and Automation, Inc.)*

insulation of one of the phase conductors and the neutral. Electrical connection is then made to the pigtail wires that extend from the tap devices. This connection can extend directly to the device or to an outlet box with a receptacle, which then acts as a disconnecting means for the electric device being served. In this fashion, lights, small motors, unit heaters, and other single-phase, light-duty devices can be served without the necessity of "hard" (conduit and cable) wiring. Figure 27.14 illustrates

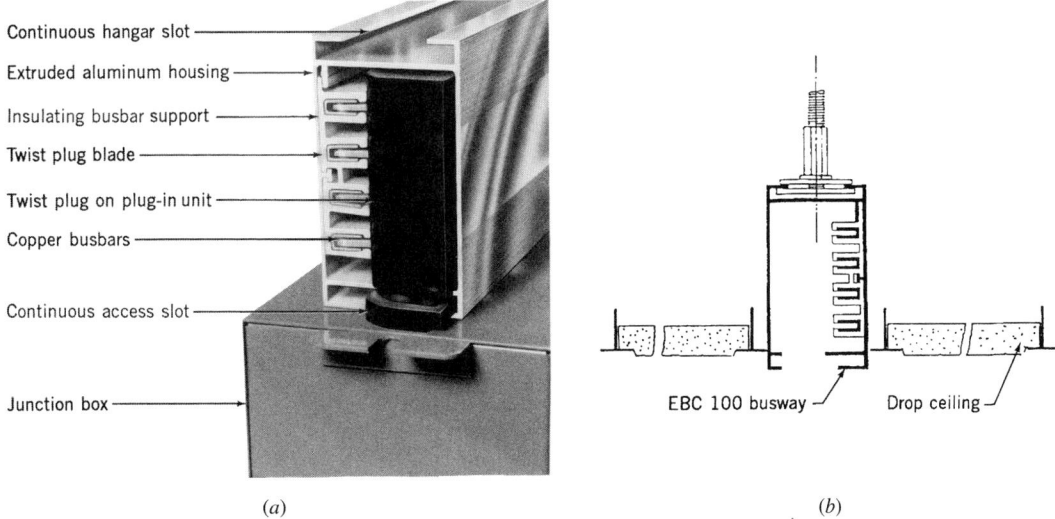

Continuous hangar slot

Extruded aluminum housing

Insulating busbar support

Twist plug blade

Twist plug on plug-in unit

Copper busbars

Continuous access slot

Junction box

EBC 100 busway       Drop ceiling

*(a)*                          *(b)*

**Fig. 27.13** *(a) Plug-in busway rated 100 A, 3-phase, 4-wire, 600 V measures approximately 2½ in. W × 4½ in. H (64 × 115 mm). The twist-in plug, which is integrally attached to a connection means (junction box in the illustration), is rated 30 to 100 A, single- or 3-phase, as required. The attached junction box (or receptacle, circuit breaker, or fuse box) then feeds the utilization device (e.g., heavy-duty lighting or machinery). (b) This bus can also be installed in a hung ceiling. (Courtesy of Universal Trolley.)*

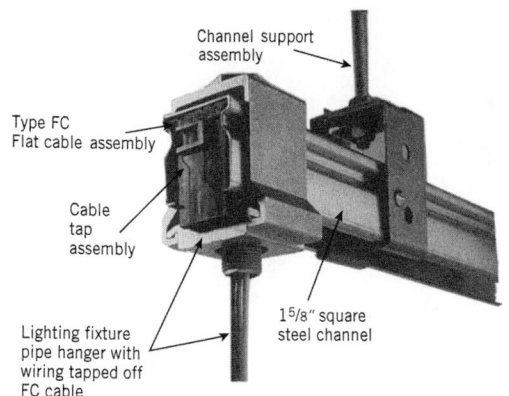

**Fig. 27.14** *A 3-phase, 4-wire flat cable assembly installed in a 1⅝-in. (41 mm) square steel channel is shown. Taps into the conductors are made by tightening the tap device. Taps can be made phase to ground to give 120 V and phase-to-phase to give 208 V. (If the cable is connected to a 277/480-V, 3-phase system, then the phase-to-ground and phase-to-phase voltages will be 277 and 480 V, respectively.) After a tap device is removed, the puncture made by the tap "heals" itself. (Courtesy of Chan-L-Wire/Wiremold Company.)*

similar equipment that is specifically intended to feed industrial lighting fixtures.

### (c) Lighting Track

This is a factory-assembled channel with conductors for one to four circuits *permanently* installed in the track (NEC Article 410-R). Power is taken from the track by special tap-off devices that contact the track's electrified conductors and carry the power to the attached lighting fixture, which can be positioned anywhere along the track. The tracks are generally rated at 20 A and, unlike FC cable assemblies, may feed only lighting fixtures. Taps to feed convenience receptacles are not permitted. A typical design is shown in Fig. 27.15. An application of track lighting is shown in Fig. 16.9.

### 27.17 CABLE TRAY

This system, which is covered in *NEC* Article 318, is simply a continuous open support for approved cables. When used as a general wiring system, the cables must be self-protected. The advantages of this system are free-air rated cables, easy installation and maintenance, and relatively low cost. The disadvantages are bulkiness and the required accessibility. Cable trays are used primarily in industrial applications.

### 27.18 DESIGN CONSIDERATIONS FOR RACEWAY SYSTEMS

The following sections deal with closed wiring raceways, which completes our discussion of raceways. The details of construction and application are not discussed here because of space limitation and because those data are readily available from manufacturers and the applicable *NEC* articles. However, enough material is provided for the reader to

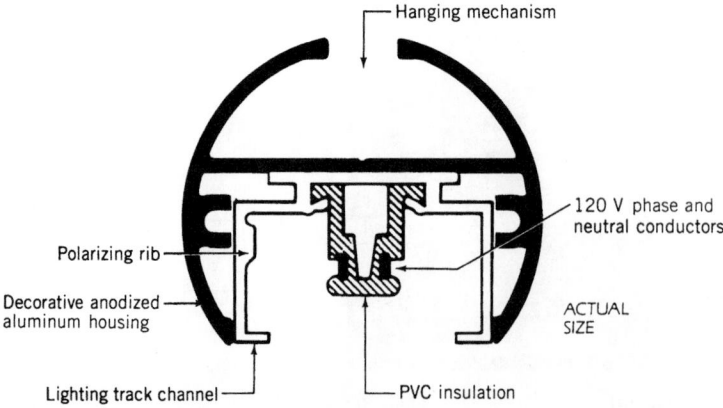

**Fig. 27.15** *Decorative lighting track. The circular housing can be obtained in a variety of finishes and hanger arrangements. The actual lighting track, shown full size, is readily available without the housing for direct surface mounting. Track is also available in a 4-conductor configuration (three 120-V conductors plus neutral). (Courtesy of Swivelier.)*

become familiar with the types of raceways, their common applications and limitations, and, where applicable, comparative characteristics.

Although our discussion in this chapter covers equipment used primarily in electrical power systems, empty raceways intended for signal, data, and communications wiring (by others) are normally provided under the electrical section of the construction contract. The raceway's function for such systems is largely the same as it is for power wiring: protection and isolation of the wiring.

Prior to the widespread use of computers in commercial establishments, the raceway space requirements for communication and signal wiring were easily established because such wiring consisted of small telephone cables plus miscellaneous signal and alarm wires. These requirements were also easily satisfied with empty conduits or cells in floor raceway systems. Today, when virtually every commercial establishment uses some type of data-processing equipment and communication networking becomes commonplace even in small facilities, raceways for communication cabling have become a major design item. They often far exceed in cross-sectional area the space required for power cabling. Their space requirements are sometimes so large that they, like ductwork, have architectural impact and must therefore be considered early in the design process.

Sizing of raceways for power wiring is an exact process (see Section 28.16), based as it is on maximum permissible temperatures for specific materials in a given environment. This is not so for communication cabling, even when a system's present requirements are known, because of the extremely rapid growth of networking and of data interchange. The problem is all the more difficult when designing commercial space for rental to an unknown client. The advisable approach in such cases is either to provide a reasonable amount of floor-level raceway space for main cabling (see Sections 27.25 and 27.26) and to rely on add-on systems such as under-carpet wiring (see Section 27.30) and surface or ceiling raceways (Section 27.31) for additional raceway area, or, alternatively, to use a structural system that provides virtually unlimited wiring space (Figs. 27.26 through 27.32). Because the latter is a major structural/architectural decision, it must be made at the preliminary stages of design.

Design considerations for power system raceways are discussed in the following sections and in Chapter 28. Design of raceways for data and communication system wiring includes the following considerations:

I. Number, type, and location of data-processing terminals.
II. Networking requirements
   A. The type of local area network largely determines the communication media (i.e., coaxial cable, shielded and unshielded wire, and fiber-optic cables), which affects the raceway space requirement. The cable type also determines the type of connectors needed (and their space requirements) and the type of floor outlets used for machine connection.
   B. Cable topology (i.e., interconnection arrangements). This item is frequently not within the domain of the architectural designer, although the raceway space availability seriously affects the cabling arrangement and vice versa.
   C. Requirement for interconnection of networks and connection to remote networks.
III. Number, location, and characteristics of major peripheral devices, such as mass storage, printing, and plotting.
IV. Location and type of major subsystems, such as computer-aided design/manufacturing spaces.
V. Location of presentation spaces that require interconnection to computer networks.

In view of these highly technical and rapidly changing requirements, engaging the services of a design consultant in this area is suggested.

## 27.19 STEEL CONDUIT

The purpose of conduit is to:

1. Protect the enclosed wiring from mechanical injury and damage from the surrounding atmosphere
2. Provide a grounded metal enclosure for the wiring in order to avoid a shock hazard
3. Provide a system ground path
4. Protect surroundings against a fire hazard as a

**ELECTRICITY**

**TABLE 27.7 Comparative Dimensions and Weights of Metallic Conduit[a]**

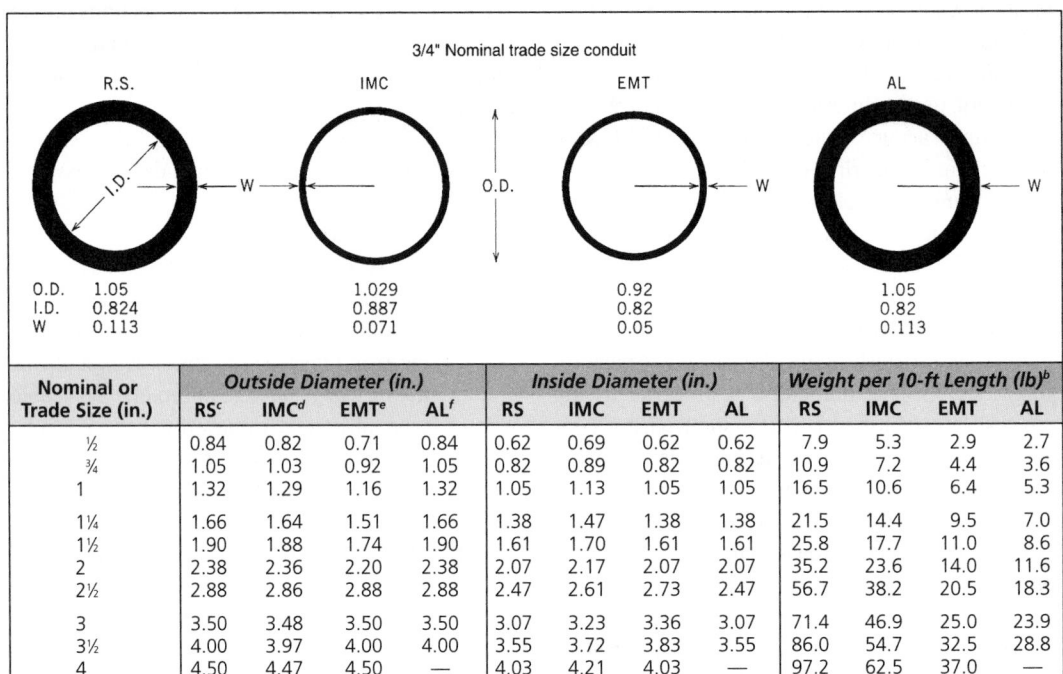

3/4" Nominal trade size conduit

| R.S. | IMC | EMT | AL |
|---|---|---|---|
| O.D. 1.05 | 1.029 | 0.92 | 1.05 |
| I.D. 0.824 | 0.887 | 0.82 | 0.82 |
| W 0.113 | 0.071 | 0.05 | 0.113 |

| Nominal or Trade Size (in.) | Outside Diameter (in.) | | | | Inside Diameter (in.) | | | | Weight per 10-ft Length (lb)[b] | | | |
|---|---|---|---|---|---|---|---|---|---|---|---|---|
| | RS[c] | IMC[d] | EMT[e] | AL[f] | RS | IMC | EMT | AL | RS | IMC | EMT | AL |
| ½ | 0.84 | 0.82 | 0.71 | 0.84 | 0.62 | 0.69 | 0.62 | 0.62 | 7.9 | 5.3 | 2.9 | 2.7 |
| ¾ | 1.05 | 1.03 | 0.92 | 1.05 | 0.82 | 0.89 | 0.82 | 0.82 | 10.9 | 7.2 | 4.4 | 3.6 |
| 1 | 1.32 | 1.29 | 1.16 | 1.32 | 1.05 | 1.13 | 1.05 | 1.05 | 16.5 | 10.6 | 6.4 | 5.3 |
| 1¼ | 1.66 | 1.64 | 1.51 | 1.66 | 1.38 | 1.47 | 1.38 | 1.38 | 21.5 | 14.4 | 9.5 | 7.0 |
| 1½ | 1.90 | 1.88 | 1.74 | 1.90 | 1.61 | 1.70 | 1.61 | 1.61 | 25.8 | 17.7 | 11.0 | 8.6 |
| 2 | 2.38 | 2.36 | 2.20 | 2.38 | 2.07 | 2.17 | 2.07 | 2.07 | 35.2 | 23.6 | 14.0 | 11.6 |
| 2½ | 2.88 | 2.86 | 2.88 | 2.88 | 2.47 | 2.61 | 2.73 | 2.47 | 56.7 | 38.2 | 20.5 | 18.3 |
| 3 | 3.50 | 3.48 | 3.50 | 3.50 | 3.07 | 3.23 | 3.36 | 3.07 | 71.4 | 46.9 | 25.0 | 23.9 |
| 3½ | 4.00 | 3.97 | 4.00 | 4.00 | 3.55 | 3.72 | 3.83 | 3.55 | 86.0 | 54.7 | 32.5 | 28.8 |
| 4 | 4.50 | 4.47 | 4.50 | — | 4.03 | 4.21 | 4.03 | — | 97.2 | 62.5 | 37.0 | — |

[a]Data vary slightly among manufacturers. O.D. is outside diameter; I.D. is inside diameter; W is wall thickness.
[b]Standard length including one coupling.
[c]Standard heavy-wall rigid steel conduit.
[d]Intermediate-weight steel conduit.
[e]Electric metallic tubing.
[f]Aluminum conduit.

ELECTRICITY

result of overheating or arcing of the enclosed conductors

5. Support the conductors

For these reasons, the *NEC* generally requires that all power wiring be enclosed in a rigid metallic corrosion-resistant conduit. To this latter end, steel conduit is manufactured in several types, among which are hot-dip galvanized, sherardized (coated with zinc dust), enameled, and plastic-covered.

There are three types of steel conduit that differ basically only in wall thickness. They are, in order of decreasing weight:

1. Heavy-wall steel conduit, also referred to simply as *rigid steel conduit;* covered by *NEC* Article 346
2. Intermediate metal conduit, usually referred to as *IMC;* covered by *NEC* Article 345
3. Electric metallic tubing, normally known as *EMT* or *thin-wall conduit;* covered by *NEC* Article 348

The differences are clearly shown in Table 27.7. Several types of heavy-wall conduit plus EMT are shown in Fig. 27.16. The equivalent millimeter sizes of conduits are given in Table 27.8.

Rigid conduit and IMC use the same threaded fittings. As a result of its thin wall, EMT is not threaded; instead, it uses set-screw and pressure fittings. The thinner walls of EMT and IMC yield a larger inside diameter (ID) and, therefore, easier wire pulling. The combination of lower weight and easier wire pulling gives EMT and IMC a distinct labor cost advantage over rigid conduit, which is further enhanced in jobs with a great deal of field bending and handling of conduit. Both, however, have application restrictions, which are detailed in the *NEC.*

Generally, no conduit smaller than ½-in. (13 mm) nominal trade diameter is used. Ordinary steel pipe may not be used for electric purposes, and all electric steel conduit is distinctively marked as such.

**TABLE 27.8 SI Equivalents of Rigid Steel
Conduit Sizes**

| Conduit Size (in.) | Dimensions (mm) | | |
|---|---|---|---|
| | Outside Diameter | Inside Diameter | Wall Thickness |
| ½ | 21.3 | 15.7 | 2.8 |
| ¾ | 26.7 | 20.8 | 3.0 |
| 1 | 33.5 | 26.7 | 3.4 |
| 1¼ | 42.2 | 35.1 | 3.5 |
| 1½ | 48.3 | 40.9 | 3.6 |
| 2 | 60.5 | 52.6 | 4.0 |
| 2½ | 73.2 | 62.7 | 5.2 |
| 3 | 88.9 | 78.0 | 5.5 |
| 3½ | 101.6 | 90.2 | 5.7 |
| 4 | 114.3 | 102.4 | 6.0 |

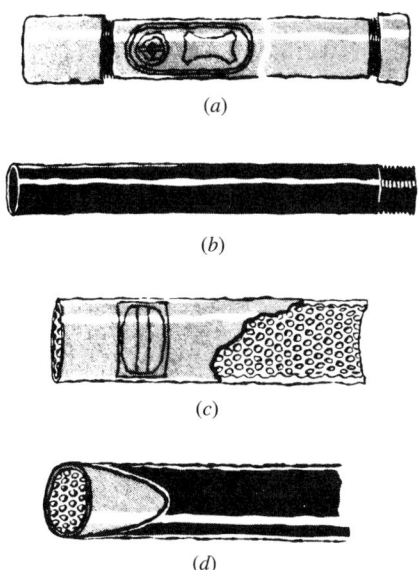

When steel conduit is installed in direct contact with the earth, it is advisable to use the hot-dip galvanized type and to coat the joints with asphaltum. If the earth is very wet, the entire conduit system should be coated with an appropriate waterproofing compound. Alternatively, a plastic-jacketed conduit can be used.

**Fig. 27.16** Steel conduits: (a) galvanized, heavy wall, rigid; (b) black enameled; (c) EMT thin wall; (d) plastic-coated conduit for use in highly corrosive atmospheres.

**Fig. 27.17** Typical overhead conduit bank installation. Note that due to field conditions the insert (a) for hanger rods was inadequate and an additional insert (b) was added. This conduit bank uses EMT, which has a pipe-wall thickness approximately one-third that of heavy wall rigid conduit. The resulting weight difference in a large bank such as this is very pronounced. EMT joints are made with set screw fittings (c). Note how individual conduits are fixed by clamps to the trapeze channel (d). (Courtesy of Republic Steel Corp.)

ELECTRICITY

Conduit is fastened to the structure in much the same way as pipe: with pipe straps and clamps. The vertical load at floor openings is taken with special support clamps. Trapeze mounting is common for conduit banks hung from the ceiling, as in Fig. 27.17.

Conduit size depends not only on the maximum permissible temperature of the contained conductors, but also on the number and diameter of the wires that may be drawn into the conduit without injuring the wire. The number and radius of bends in the conduit, as well as its total length, affect the degree of abrasion to the wiring insulation during installation. No wires should be installed until the conduit system has been inspected and approved.

For structural reasons, conduits in concrete slabs are run close to the bottom surface (in the portion of the slab in tension) or near the center. If a large number of conduits must be embedded, it may be necessary to increase the slab thickness. In many instances, the structural slab is covered with a concrete topping, in which conduit may be installed without affecting slab integrity. In all cases, local building codes should be consulted for limitations on embedded conduits. In any event, the top of any conduit shall be at least ¾ in. (19 mm) below the finished floor surface to prevent cracking. When heavy trucking is expected, this allowance should be increased to 1½ in. (38 mm) minimum.

In general, the following rules should be observed and included in all specifications for conduit work in concrete slabs:

1. Conduits shall have an outside diameter (OD) no greater than one-third the slab thickness, as measured at its thinnest point.
2. Conduits running parallel to each other shall be spaced not less than three times the OD of the largest conduit center-to-center.
3. Conduit crossings shall be as close to a right angle as possible.
4. Minimum cover over conduits shall be ¾ in. (19 mm).

## 27.20 ALUMINUM CONDUIT

The use of aluminum conduit has increased greatly in recent years because of the weight advantage that aluminum has over steel, being even lighter

than EMT. The savings in labor cost more than offsets the additional cost of the material itself. In addition, aluminum has better corrosion resistance in most atmospheres; it is nonmagnetic, giving a lower voltage drop; it is nonsparking; and, generally, it does not require painting.

Its major drawback is its deleterious effect on many types of concrete, causing spalling and cracking when embedded. Although manufacturers can demonstrate cases of embedding in concrete without harmful effect, this procedure should be avoided unless the concrete additives are rigidly controlled and the conduit is coated to prevent contact with the concrete. It is also inadvisable to bury aluminum in earth, with or without asphalt or another coating, because of the rapid corrosion often encountered. Other difficulties frequently encountered are freezing of threaded joints (because of thread deformation) and difficulty in obtaining electrical contact with grounding straps. With the exceptions noted, aluminum conduit may be used in all locations where steel conduit is used.

## 27.21 FLEXIBLE METAL CONDUIT

This type of conduit consists of an empty, spirally wound, interlocked armor steel or aluminum raceway. It is known to the trade as *Greenfield* and is covered in *NEC* Article 350. It is used principally for motor connections and other locations where vibration is present, where movement is encountered, or where physical obstructions make its use necessary. The acoustic and vibration isolation provided by flexible conduit is one of its most important applications. It should always be used in connections to motors, transformers, ballasts, and the like. A typical application is for wiring inside metal partitions. When covered with a liquid-tight plastic jacket, it is suitable for use in wet locations (Fig. 27.18). In this configuration, it is most often known by the trade name *Sealtite*.

## 27.22 NONMETALLIC CONDUIT

A separate classification of rigid conduit (*NEC* Article 347) covers raceways that are formed from such materials as fiber, asbestos-cement, soap-

**Fig. 27.18** *This is a particularly good application of liquid tight, flexible conduit because it provides weatherproofing and acoustical isolation of the vibration-producing equipment. (Courtesy of Electri-Flex Company.)*

stone, rigid polyvinyl chloride (PVC), and high-density polyethylene.

For use above ground, this conduit must be flame retardant, tough, and resistant to heat distortion, sunlight, and low-temperature effects. For use underground, the last two requirements are waived. Generally, nonmetallic conduit may be used without restriction in nonhazardous areas within the physical limitations of the material involved. Thus, plastic conduit has a temperature limitation, asbestos–cement has considerable physical strength limitations, and so on. As a result of these limitations, PVC conduit is the material of choice for indoor exposed use and asbestos–cement, fiber, and PVC plastic for outdoor and underground use. A separate ground wire *must* be provided because the ground provided by a metallic conduit is absent.

## 27.23 SURFACE METAL RACEWAYS (METALLIC AND NONMETALLIC)

These raceways are covered in *NEC* Article 352. Surface metal raceways and multioutlet assemblies

may be utilized only in dry, nonhazardous, noncorrosive locations and may generally contain only wiring operating below 300 V. Such raceways are normally installed exposed, in places not subject to physical injury.

The principal applications of surface metal raceways are:

1. Where economy in construction weighs very heavily in favor of surface raceways and where expansion is anticipated (Fig. 27.19)
2. Where outlets are required at frequent intervals and where rewiring is required or anticipated (Figs. 27.20 and 27.21)
3. Where access to equipment in the raceways is required and/or where necessary due to the nature of the wiring (Figs. 27.19, 27.22, 27.23, and 27.24)
4. Where we wish to avoid the extensive and expensive cutting and patching required to "bury" a raceway during rewiring (Fig. 27.25)

## 27.24 OUTLET AND DEVICE BOXES

These boxes are generally of galvanized stamped sheet metal. The most common sizes are the 4-in. (100 mm) square and 4-in. (100 mm) octagonal boxes used for fixtures, junctions, and devices and the 4- × 2⅛-in. (100 × 54 mm) box used for single devices where no splicing is required. Box depths

**Fig. 27.19** *Three-section nonmetallic baseboard-type raceway measures 4 in. H and 1.5 in. D (100 × 38 mm). The three sections are intended for telephone, power, and data cabling. The junction box on the right protrudes so as not to lessen the raceway wiring space. It is equipped with vertical dividers to keep the three types of wiring completely separated. The elevated wall box on the left contains telephone and data cable outlets and connects only to the low-voltage portions of the raceway. (Photo courtesy of Hubbell, Inc.)*

**ELECTRICITY**

**Fig. 27.20** *Multioutlet surface raceways find ready application in schools. Note the use of a surge and noise suppressor (foreground) to protect groups of computers from power line disturbances. (Courtesy of Walker Systems, A Wiremold Company.)*

**Fig. 27.21** *Large-capacity surface metal raceways are particularly useful for wiring in full-access floors (see Section 27.29) because of the heavy wiring (see wall box) and frequent rewiring. The perforated floor tile in the foreground is used to supply laminar airflow in this clean room at an integrated circuit manufacturer's facility. (Photos courtesy of Walker Systems, A Wiremold Company.)*

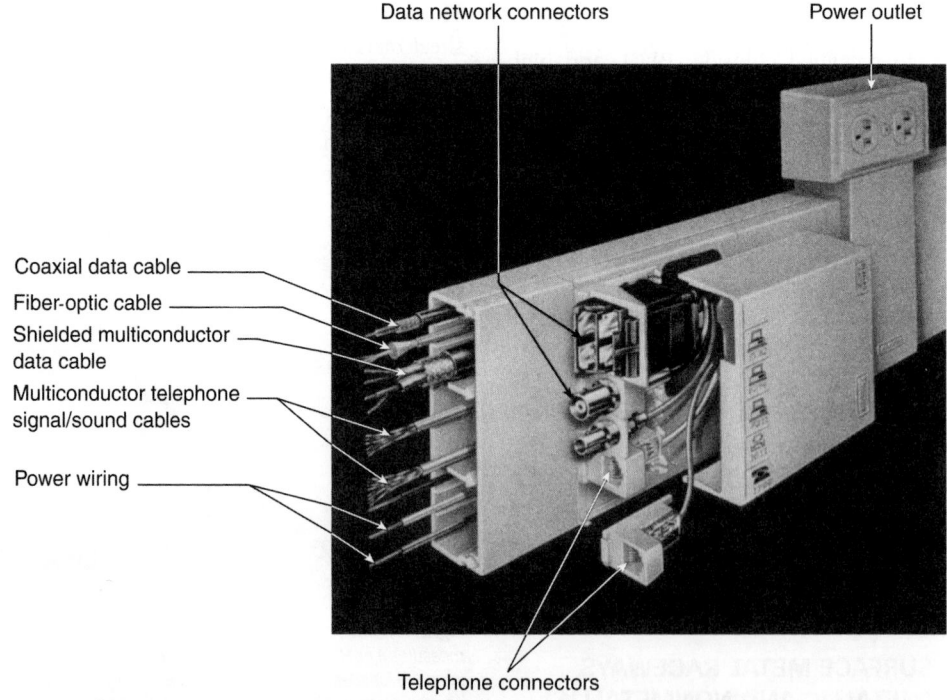

**Fig. 27.22** *Multichannel nonmetallic surface raceway with snap-in connector modules for data network, signal, and power wiring systems. (Similar metallic raceways are also available.) The raceway itself measures 4½ in. H × 1 in. D (115 × 25 mm). The internal dividers are movable or entirely removable, which permits varying the number and size of the wiring channels. The principal application of this type of wireway is in commercial occupancies using extensive desktop data-processing and communication equipment. (Courtesy of Panduit.)*

**Fig. 27.23** *The frequent wiring changes required for theater and exhibition lighting are easily made when the wiring is run in a suspended surface raceway. (Courtesy of Walker Systems, A Wiremold Company.)*

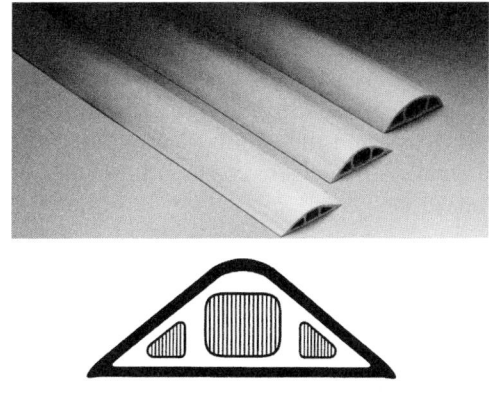

| FloorTrak | 2 | 3 | 4 |
|-----------|-----|-----|-----|
| Width | 2.68" | 2.985" | 2.975" |
| Height | .465" | .610" | .885" |
| Ctr. Hole | .50x.312" | .75x.500" | 1.0x.750" |

**Fig. 27.25** *Where low-voltage wire and cable cannot be concealed (buried), shallow, properly shaped floor track can be used. Such installations should avoid foot-traffic areas where possible to minimize trip hazards. (Courtesy of Hubbell Premise Wiring.)*

vary from 1½ to 3 in. (38 to 76 mm). Nonmetallic boxes may be used with NM and NMC cable and with nonmetallic conduit installations. In wet locations and for outdoor work, cast-iron or cast-aluminum boxes are recommended.

As a result of an *NEC* (Article 300-21) requirement that electrical penetrations in fire-rated floors be designed to maintain fire ratings, electrical manufacturers have produced a line of poke-through fittings to meet this need. (This requirement applies also to walls, ceilings, and partitions.) One such design is shown in Fig. 27.26. These electrical penetrations have become increasingly prevalent in

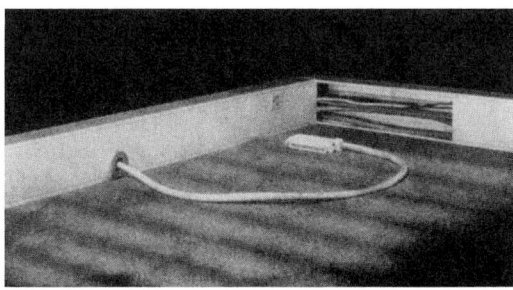

**Fig. 27.24** *The basic raceway illustrated is 1⅞ in. D × 3³⁄₁₆ in. W (48 × 81 mm). It is shown with a divider installed, permitting use of the top section for power wiring and the bottom section for low-voltage wiring. Because data and communication cables are frequently supplied with factory-installed terminations (as in the photo), a raceway where the cable can be laid in rather than pulled in is required. Also, terminal strips and other equipment can be installed in the low-voltage section of these large raceways, making the use of separate terminal cabinets unnecessary. (Courtesy of Walker Systems, A Wiremold Company.)*

Power & Low Tension outlets

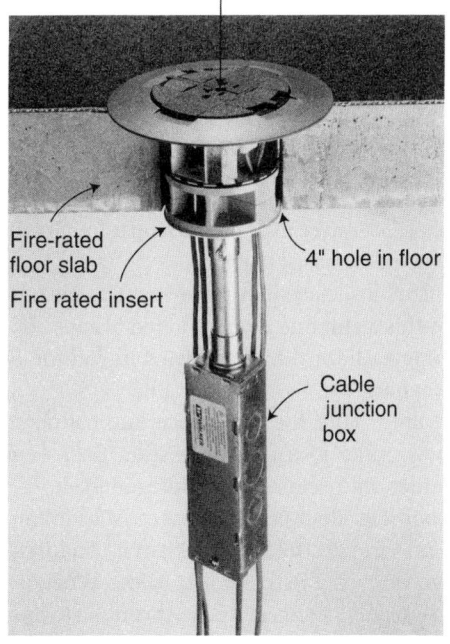

Fire-rated floor slab

Fire rated insert

4" hole in floor

Cable junction box

**Fig. 27.26** *Typical poke-through electrical fitting mounts in a 4-in. (100 mm) hole. It is wired from underneath with the required power, telephone, signal, and data cables. Power and low-voltage cables are separated as required by code. Units are available prewired or suitable for field wiring, and adaptable for varying floor thicknesses. The floor fitting is provided with power, telephone and data cable outlets as required for the specific installation. (Courtesy of Walker, Division of Wiremold.)*

**ELECTRICITY**

existing commercial spaces where the expanded need for desktop power and data wiring can be met most economically and rapidly by through-the-floor feeds from expanded wiring in the hung ceiling below. In addition, these fittings facilitate the electrical wiring relocations so common in rental office occupancies.

## 27.25  FLOOR RACEWAYS

In commercial spaces with large open floor areas, it is common practice to place desks and other workstations throughout the space, at considerable distances from permanent walls containing electrical services. Because in a modern office it must be assumed that each workstation requires electric power for a computer, desk lamp, and other common equipment plus a telephone line, a computer network connection, and possibly a data outlet, the problem of bringing these services to the workstations with a minimum of exposed wiring arises immediately. The required outlets can be installed on the floor adjacent to or under the workstation or, if half-height partitions are used, in these partitions.

To bring the various electrical and communication services to the user, in the absence of any sort of overall floor raceway system, the installing contractor has one of four choices:

1. Channel the floor and install a conduit in the chase, connecting it to the nearest wall outlet. Patch the chased portion.
2. Install a surface floor raceway. The usefulness of this technique is very limited because of the tripping hazard and problems in routine floor cleaning.
3. Drill through the floor twice and connect the new outlet to a nearby existing floor or wall outlet via a conduit on the underside of the floor slab. Floor penetrations must be fireproof.
4. Drill through the floor and run a conduit in or on the ceiling to the outlets below. When using this technique, special poke-through fittings are available that restore the fire rating of the slab (see Fig. 27.26). These fittings are designed to carry all the electric services normally required at a workstation. They can then be connected to a single-location multiservice floor outlet group, as in Fig. 27.26, or used to wire the partitions in a workstation, as in Fig. 27.27.

All four of these methods have serious disadvantages; method 1 is labor intensive, method 2 is unsightly and presents a safety hazard, methods 3 and 4 disturb the occupants below, and all four methods are inflexible and therefore unsuitable for spaces where changes in wiring and workstation location are anticipated. For these reasons, overall-access in-floor and underfloor raceway systems were developed and are widely used in high-grade commercial and institutional spaces.

The *NEC* recognizes three types of in-floor raceways:

Underfloor raceways—Article 354

Cellular metal floor raceways—Article 356

Cellular concrete floor raceways—Article 358

All three types are applicable to all types of structures, and none may be used in corrosive or hazardous areas. The fundamental difference between them is that underfloor raceways are added on to the structure, whereas cellular floor raceways are part of the structure itself—and therefore have a pronounced effect on the building's architecture. (Underfloor duct systems antedate poke-through fittings, which are a relatively recent development.)

## 27.26  UNDERFLOOR DUCT

These raceways may be installed beneath or flush with the floor. They find their widest application in office spaces because their use permits placement of power, data, and signal outlets close to desks and other furniture, regardless of furniture layout. Underfloor duct systems were widely employed until the introduction of what may be called *over-the-ceiling* ducts (in apposition to under-the-floor ducts) and flat-cable under-carpet wiring. These systems are discussed in Sections 27.30 and 27.31. The reason that alternate systems were developed is simply economic: underfloor duct systems are expensive and, because they are inflexible, being literally cast in concrete, they are therefore frequently underutilized in one area while being inadequate in another. However, before discussing the relative merits of systems, an understanding of what underfloor duct systems are and how they are assembled and utilized is necessary.

An underfloor duct system is simply an arrangement of parallel rectangular metal or heavy plastic raceways laid on the structural slab and cov-

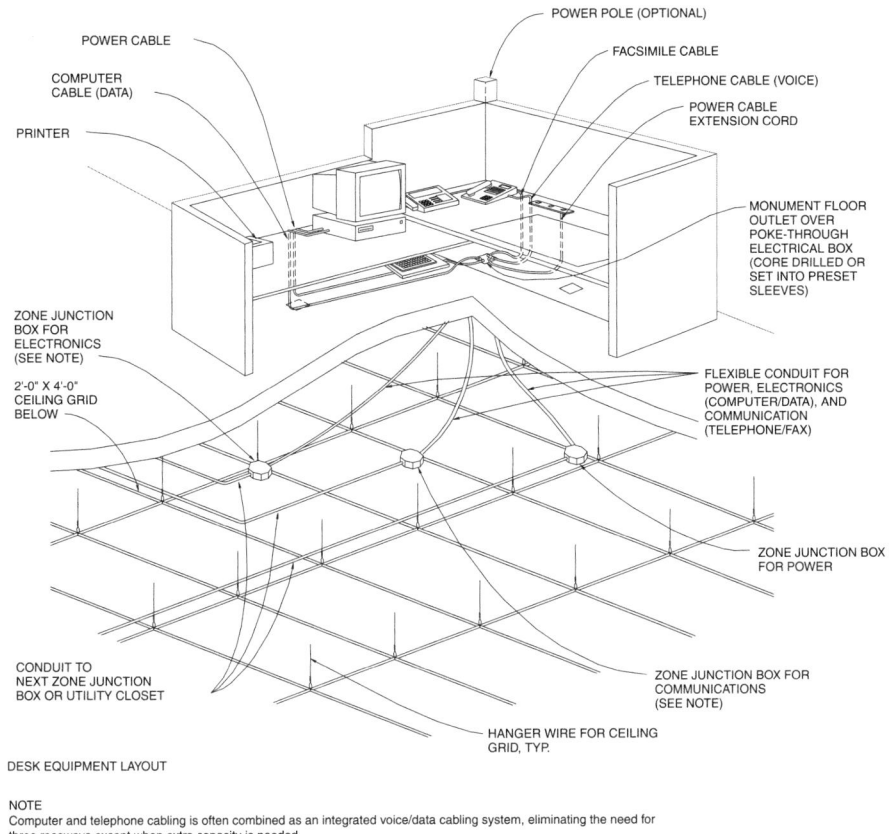

DESK EQUIPMENT LAYOUT

NOTE
Computer and telephone cabling is often combined as an integrated voice/data cabling system, eliminating the need for
three raceways except when extra capacity is needed.

**Fig. 27.27** *Typical application of a poke-through fitting to provide power, telephone, and data service to a modern workstation. This drawing shows the electrical services being tapped at junction boxes in a hung ceiling conduit system on the floor below. The ceiling wiring system can also be a raceway network (as in Fig. 27.50) in lieu of the hard-wiring shown here. (From AIA: Ramsey/Sleeper,* Architectural Graphic Standards, *10th ed., 2000, reprinted by permission of John Wiley & Sons.)*

ered with concrete fill. Access to the wiring in these *distribution ducts* is via inserts that connect to openings in the ducts and terminate in floor fittings for both power and signal/data wiring. Cable feeds to the distribution ducts are supplied by a second set of rectangular raceways called *feeder ducts,* usually laid at right angles to the distribution ducts.

In a *single-level* underfloor duct system, the distribution and feeder ducts are on the same level, and the interwiring between them is accomplished in junction boxes. The advantage of a single-level system is shallow concrete fill, normally 2½ to 3 in. (64 to 76 mm). The limiting factor of a single-level system is the junction box, which becomes more complex and multisectioned with an increasing number of ducts and wires. Newer systems utilize a one-piece triple-cell duct for both distribution and feeder ducts, with factory set inserts every 24 in. (610 mm) that

straddle all three cells at once (Fig. 27.28). By placing distribution ducts on 5-ft (1.5 m) centers with adequate crosswise feeder ducts and utilizing large flat junction boxes, a cost-effective installation adequate for all but the heaviest wiring demands can be assembled. For such areas, distribution ducts can be arranged to feed under-carpet cables (see Section 27.30).

Because the initial cost of a full underfloor system is high, an alternative arrangement utilizes *only* feeder ducts on approximately 25-ft (7.6 m) centers, with flat (under-carpet) cable box connectors spaced approximately every 20 ft (6.1 m) along the feeder ducts. The low-tension portion of this system relies completely on flat telephone and data transmission cables, including fiber-optic cables. Because these cables are generally precut and factory terminated, the system requirements must be

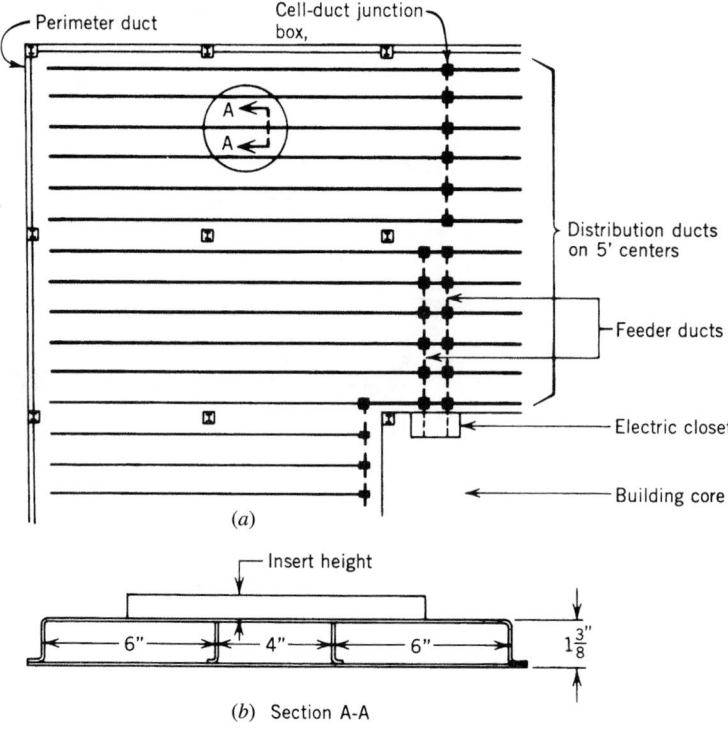

**Fig. 27.28** *Details of a single-level underfloor duct system utilizing three-section cell duct for both distribution and feeder ducts. (a) Portion of a typical large open floor space in a commercial facility. Distribution ducts may be placed as close as 5-ft (1.5 m) centers to satisfy dense desk spacing. (b) Three-cell distribution duct utilizes a 4-in.- (100 mm) wide center cell (4.9 in.² [3160 mm²]) for power and either 3-in.- (75 mm) wide (3.7 in.² [2386 mm²]) or 6-in.- (150 mm) wide (7.4 in.² [4773 mm²]) outer cells for signal and data cabling. Minimum concrete fill depth is 2½ in. (64 mm), resulting in a minimum 1-in. (25 mm) cover over the distribution ducts. Service fittings are flush with the floor. (Courtesy of Square D Company.)*

carefully analyzed (see Section 27.30) before committing to a complete under-carpet wiring system.

A two-level underfloor duct system is essentially the same as a single-level system except that the distribution ducts and feeder ducts are on different levels (Fig. 27.29). This arrangement has the advantages of simplifying junction boxes and of

giving the system unlimited feeder capacity but the distinct disadvantage of requiring a minimum of 3⅝ in. (92 mm) of concrete fill. This additional slab thickness can frequently be avoided by depressing part of the slab to accommodate feeder ducts run *under* the distribution ducts, as shown in Fig. 27.30.

A typical two-level system is illustrated in Fig. 27.31. Here, the feeder ducts run above the distribution ducts and intersect at a specially constructed

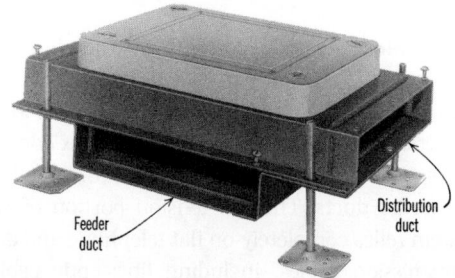

**Fig. 27.29** *Typical two-level junction box demonstrates the simplicity of the two-level system. (Courtesy of Square D Company.)*

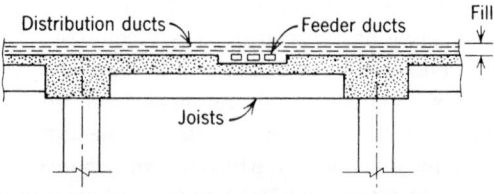

**Fig. 27.30** *Setting a two-level underfloor duct system. To avoid thickening fill, a depression in the slab can accept feeder ducts. Ducts would be run near the bay center to avoid the negative steel of joists near columns.*

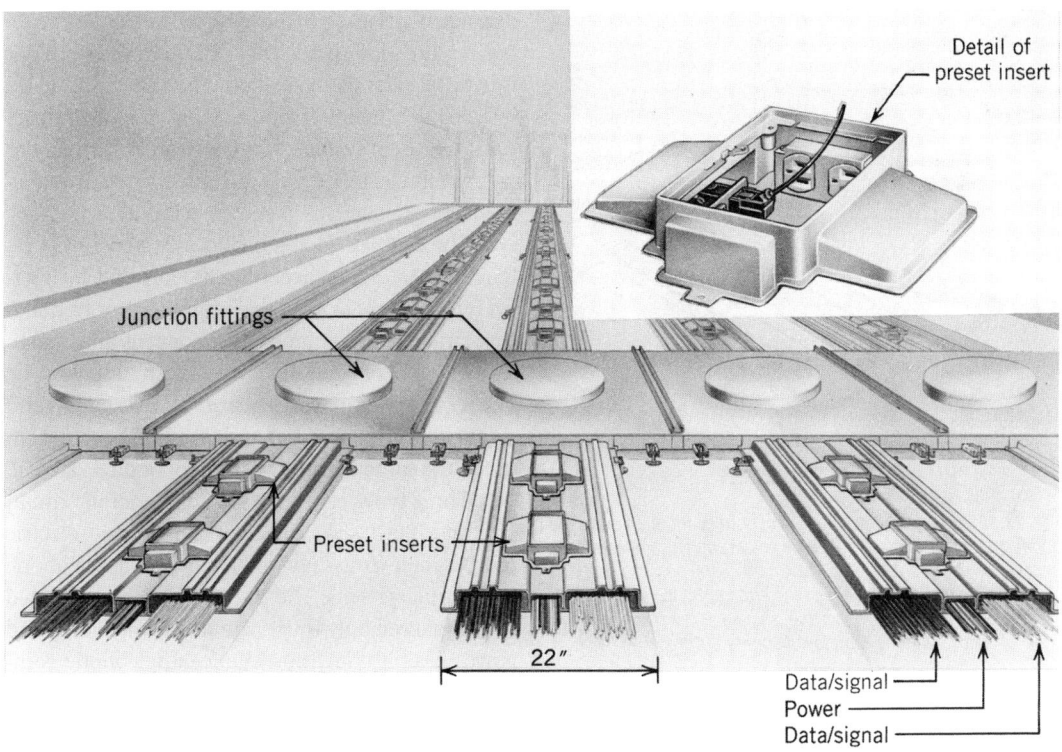

Detail of preset insert

Junction fittings

Preset inserts

22″

Data/signal
Power
Data/signal

**Fig. 27.31** *Modified two-level underfloor duct system requires 3 in. (75 mm) of concrete fill for 2-in.- (50 mm) high ducts and 4 in. (100 mm) of fill for 3-in.- (75 mm) high ducts. Preset inserts straddle all three cells and provide two duplex power receptacles plus connection to both low-tension cells. Distribution ducts recess into feeder-header ducts at junction boxes to reduce the overall height. See insert. (Courtesy of Walker Systems, A Wiremold Company.)*

junction fitting into which the distribution ducts partially recess in order to reduce overall system height. The required concrete fill is either 3 or 4 in. (75 or 100 mm), depending on the depth of the distribution cells (2 or 3 in. [50 or 75 mm]).

In all underfloor duct systems, the principal cable capacity bottleneck is usually the supply point to the *feeder* ducts. One solution to this problem uses a special feed arrangement at panels (Fig. 27.32). Another possibility is to subdivide large floor areas, supplying each via a system of multiple feed points arranged in closets or at wall panels. In such systems, care must be taken to ensure sufficient interconnection capacity between feed points because data networks are not only floorwide but frequently buildingwide.

Underfloor ducts may be cast into the structural slab in lieu of being installed in fill or topping, but the slab must be designed to accommodate them. The use of a fill or topping on the structural slab for underfloor duct has these advantages:

1. Ducts can be run in any direction, without conflict with structural elements.
2. Finishing is simplified.
3. Coordination is simplified.
4. Formwork and construction sequence are simplified.

The disadvantages are:

1. Additional concrete increases costs directly by increasing the weight. This is particularly expensive in seismic designs.
2. The building height may be increased.

In retrofit jobs where underfloor duct is decided on rather than one of the other floor or ceiling raceway systems, the ducts will obviously be placed in a new floor fill.

In conclusion, some general comments on the application of underfloor duct systems are in order. Underfloor duct systems are *expensive*. They can add 50% to the building's electric system cost, without

**ELECTRICITY**

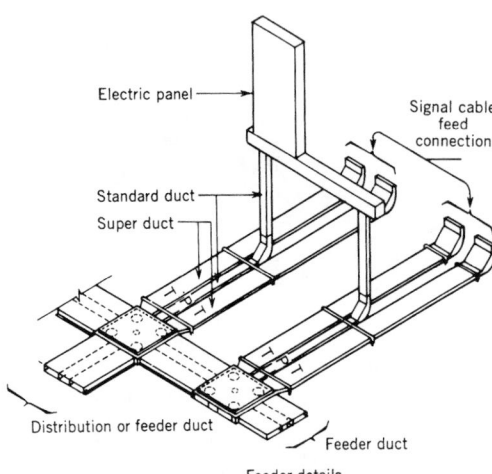

Electric panel

Signal cable feed connection

Standard duct

Super duct

Distribution or feeder duct

Feeder duct

Feeder details

**Fig. 27.32** *Due to the large capacity of both distribution ducts and feeder ducts, central cable feed points such as at electric closets can cause bottlenecks. Illustrated is one possible solution, consisting of a double-duct feed arrangement. Signal cable would feed in from cable boxes (not shown). P is a power duct; T a signal duct. (Diagrams courtesy of Square D Company.)*

consideration of the construction costs involved. To justify their use, therefore, the building should meet these criteria:

1. Open floor areas, with a requirement for outlets at locations removed from walls and partitions
2. Under-carpet wiring system is inapplicable
3. Outlets from ceiling systems are unacceptable
4. Frequent rearrangement of furniture and other items requiring electrical and signal service is anticipated

The facilities that may meet these criteria are prestigious office buildings, museums, galleries, and other display-case spaces, high-cost merchandising areas, and selected areas in industrial facilities. Bear in mind that even in high-cost office construction, underfloor duct systems are difficult to justify economically unless the furniture layout will change. In doubtful cases, alternate arrangements can be planned and an intelligent choice made after costs and the impact on the building structure are studied.

## 27.27 CELLULAR METAL FLOOR RACEWAY

The underfloor duct system described previously is best applied to rectilinear arrangements. Random arrangements, such as those found in office landscaping, require a fully accessible floor—if indeed the floor is to be used for electrification. This may be provided by a cellular (metal) floor that is an integrated structural/electrical system. The floor can be partially or completely electrified. One of the many structural element designs available is shown in Fig. 27.33.

The cellular floor is part of the structural system and is designed accordingly. Electrical wiring is fed into the cells from header ducts and/or trenches that run perpendicular to the floor cells and constitute a system of underfloor ducts in themselves. The header ducts in turn are fed from electric panels and signal data-transmission and telephone cabinets in much the same manner as underfloor ducts are fed.

Three types of wiring systems generally run in separate floor cells and header ducts—electric power, data-transmission wiring, and telephone and signal systems. The last two may be combined in a single cell only if the signal system voltage and power level are low and the local telephone company approves. A complete range of outlets and fittings is available.

## 27.28 PRECAST CELLULAR CONCRETE FLOOR RACEWAYS

This structural concrete system is similar to a cellular metal floor in application and has the same advantages: large capacity, versatility in that each cell is a potential raceway, and flexibility in outlet placement and movement. Here too, as with the metal cell constructions, the first cost is higher than that of standard underfloor duct installation, although the life-cycle cost is frequently lower, depending on space use.

A cell is defined in *NEC* Article 358 as a "single, enclosed, tubular space in a floor made of precast cellular concrete slabs, the direction of the cell being parallel to the direction of the floor member." Feed for these cells is provided, as with metal cellular floor construction, by header ducts. Although header ducts are normally installed in concrete fill above the hollow-core structural slab, a header arrangement with feed from the ceiling below is also entirely practical. As with the metallic cellular floor, the cells can be used for air distribution and even for piping, although these items are generally installed in a hung ceiling.

ELECTRICITY

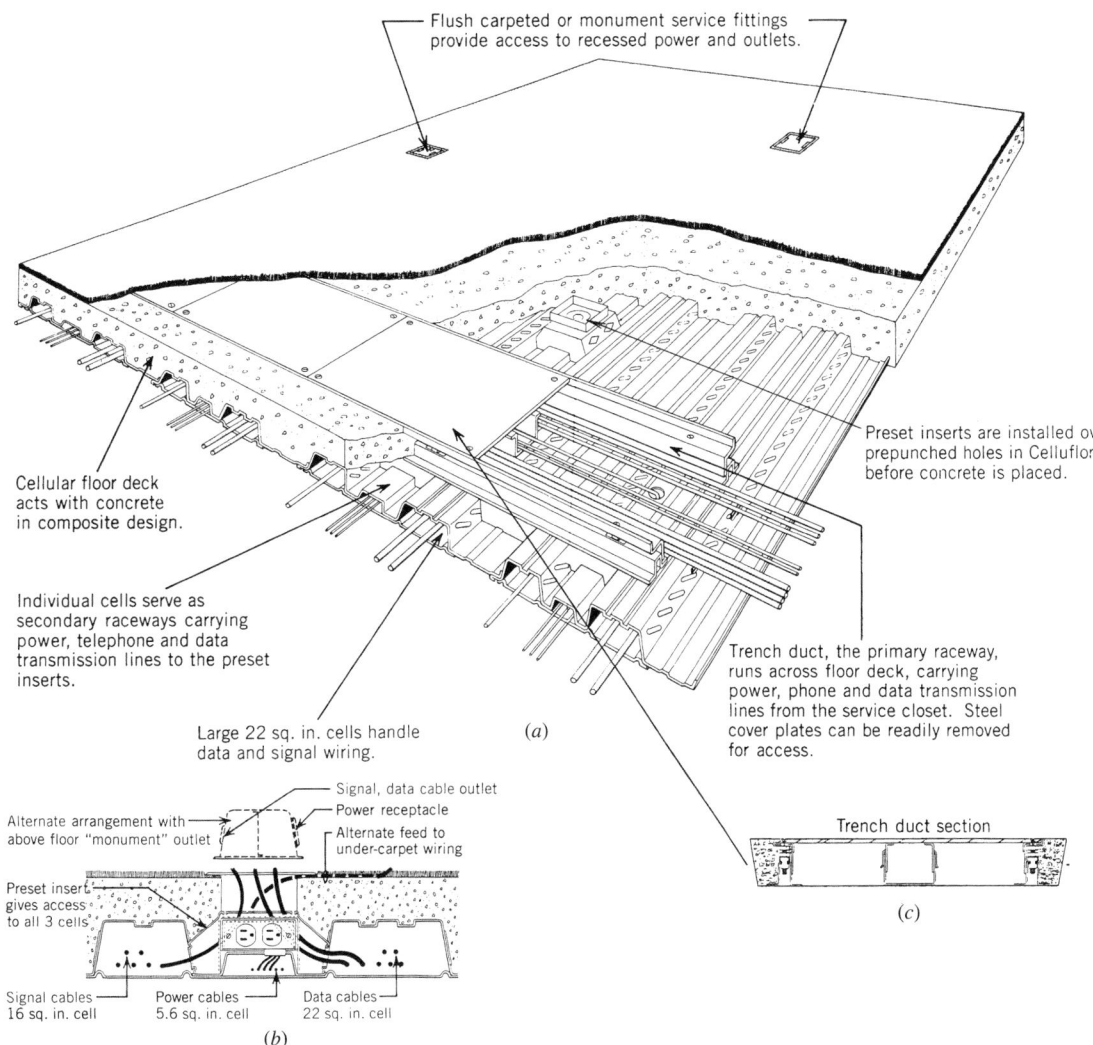

Flush carpeted or monument service fittings
provide access to recessed power and outlets.

Preset inserts are installed over
prepunched holes in Celluflor
before concrete is placed.

Cellular floor deck
acts with concrete
in composite design.

Individual cells serve as
secondary raceways carrying
power, telephone and data
transmission lines to the preset
inserts.

Large 22 sq. in. cells handle
data and signal wiring.

*(a)*

Trench duct, the primary raceway,
runs across floor deck, carrying
power, phone and data transmission
lines from the service closet. Steel
cover plates can be readily removed
for access.

Signal, data cable outlet
Power receptacle
Alternate arrangement with
above floor "monument" outlet
Alternate feed to
under-carpet wiring
Preset insert
gives access
to all 3 cells

Signal cables
16 sq. in. cell
Power cables
5.6 sq. in. cell
Data cables
22 sq. in. cell

*(b)*

Trench duct section

*(c)*

**Fig. 27.33** *(a) One of many designs for electrified cellular floors. The floor cells are available in many designs, depending primarily on the structural requirements. The trench (illustrated in c) that straddles the cells provides the electrical feeds through precut holes in the cells. The trench itself is completely accessible from the top and, when opened, exposes all the wiring and the cells below. (b) Activated preset insert. Note that the insert straddles the center (power) cell and provides access to the two adjoining low-tension wiring cells. Power and signal wiring are completely separated at all times by metal barriers. If desired, a standard surface "monument" fitting can be mounted on the floor or a connection can be made to under-carpet cables in lieu of the flush plate shown. When an insert is to be deactivated, the flush cover plate is simply replaced with a blank plate. (c) Section through a trench duct, which acts as a feeder for distribution ducts. The trench is available with or without bottom, in any required height, in widths from 9 to 36 in. (230 to 915 mm), and one, two, or three compartments, depending on floor cell design and cabling requirements. (Courtesy of Walker Systems, A Wiremold Company.)*

## 27.29 FULL-ACCESS FLOOR

This construction is applicable to spaces with very heavy cabling requirements, particularly if frequent recabling and reconnection are required. It provides rapid and complete access to an underfloor plenum. The system was originally developed for data-processing areas that require large, fully accessible

cable spaces and large quantities of conditioned air. The solution to both of these requirements is an infinite-access floor, usually constructed of lightweight die-cast aluminum panels supported on a network of adjustable steel or aluminum pedestals. Panels are available from $18 \times 18$ in. to $36 \times 36$ in. ($457 \times 457$ mm to $915 \times 915$ mm), and floor depth is normally 12 to 24 in. (305 to 610 mm), although taller pedestals are available. The subfloor space thus created can be used for cabling and also to carry conditioned air either in ducts or by using the entire space as an air plenum. (In the latter case, the wire and cable must be suitable for air plenum use; see Fig. 27.21.) The construction is usually fireproof. Obviously, sufficient ceiling height is necessary to accommodate the raised floor.

Where air requirements are limited or nonexistent and the floor is intended primarily for cabling, pedestals as short as 6 in. can be used, thus alleviating any ceiling height problems (Fig. 27.34). In such access floor spaces, use of multiservice distribution modules and modular wiring avoids cable tangles and reduces labor costs significantly (Figs. 27.34–27.37).

## 27.30 UNDER-CARPET WIRING SYSTEM

This system, which is covered in *NEC* Article 328, was originally developed as both an inexpensive alternative to an underfloor or cellular floor system and as a means for providing a flexible floor-level branch circuit wiring system. Essentially, the system consists of a factory-assembled flat cable (*NEC* type FCC), approved for floor installation *only* under carpet squares, plus the accessories necessary for connection to 120-V power outlets. The cable itself consists of three or more flat copper conductors, placed edge to edge and enclosed in an insulating material (Fig. 27.38). The entire assembly is covered with a grounded metal shield, which, like a

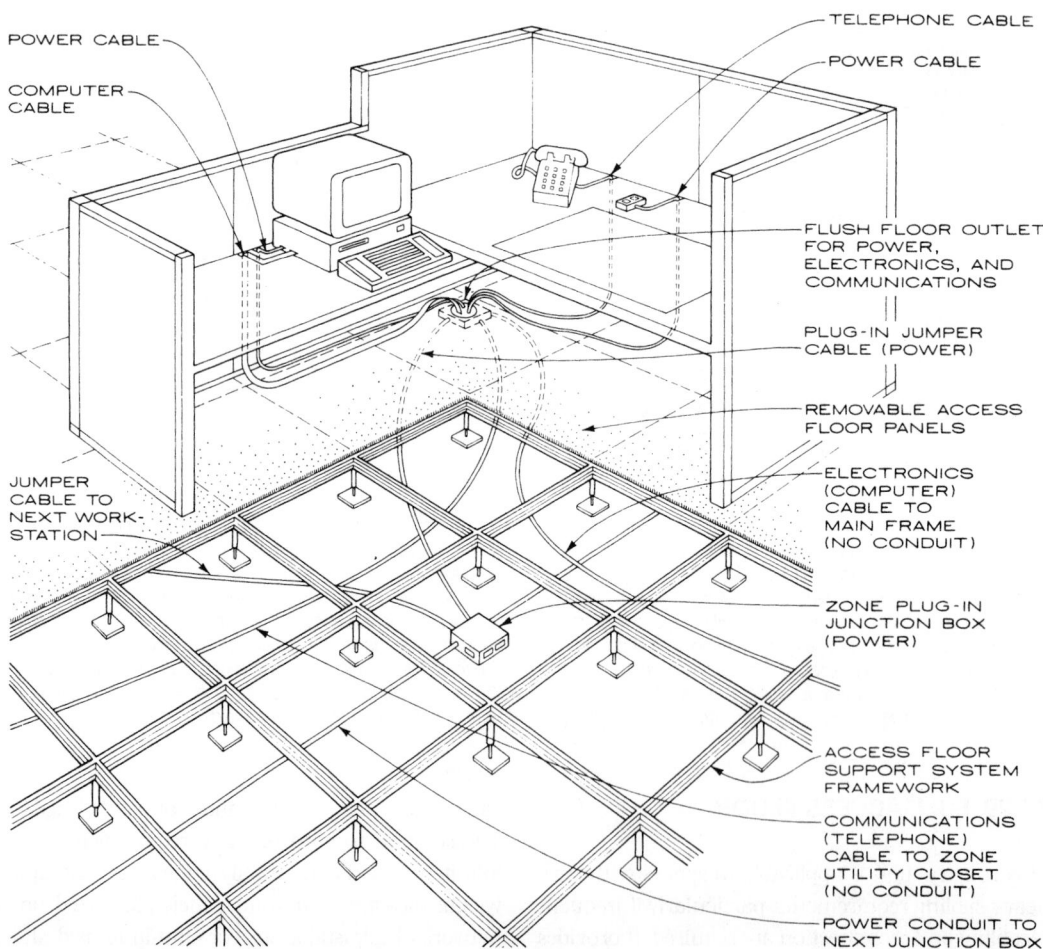

**Fig. 27.34** *Pictorial representation of a full-access floor designed to provide complete electrical, data and signal services to a modern workstation. The infinite access and unlimited space are ideal for heavily wired, rapidly changing workstations. (From AIA: Ramsey/Sleeper,* Architectural Graphic Standards, *10th ed., 2000, reprinted by permission of John Wiley & Sons.)*

**Fig. 27.35** Full-access floors provide infinite access to the myriad cables used in a modern commercial office. The modular wiring, using snap-on connectors and preassembled junction boxes, drastically reduces the labor cost of installation and the frequent cabling changes (see Fig. 27.39a). (Photo courtesy of Tate Access Floors, Inc.)

**Fig. 27.36** Reinforced nonferrous (aluminum) floors are used frequently in hi-tech applications requiring the large wiring capacity and convenience of full-access floors. Floor construction can be stringerless, as illustrated, or with stringers, as in Fig. 27.37. (Photo courtesy of Tate Access Floors, Inc.)

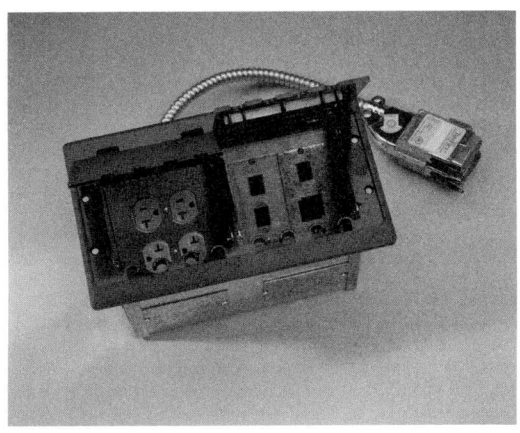

(a)

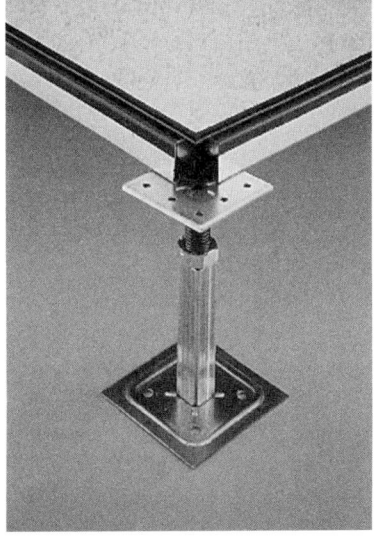

(b)

**Fig. 27.37** (a) Typical power/low-voltage floor box for use with a full-access floor. (b) Steel-bolted stringer support for access floor panels. Many stringer and stringerless designs are available (see Fig. 27.36). (Photos courtesy of Tate Access Floors, Inc.)

ELECTRICITY

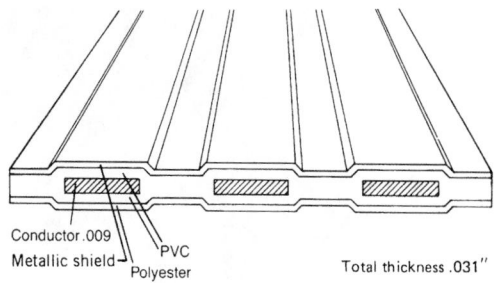

Conductor .009
Metallic shield
PVC
Polyester
Total thickness .031"

**Fig. 27.38** *Schematic section through one design of NEC type FCC under-carpet cable. The copper conductors illustrated are the equivalent of #12 AWG. The PVC acts as insulation, and the polyester as both insulation and physical protection. All designs require a metallic top shield and a metallic or nonmetallic bottom shield for physical protection.*

**Fig. 27.39** *Preterminated 25-pair under-carpet telephone cable. These cables are commercially available in lengths from 5 to 50 ft (1.5 to 15 m) in 5-ft. (1.5 m) increments. (Reprinted with permission of AMP.)*

metal conduit, provides both physical protection and a continuous electrical ground path. In addition, a bottom shield is required, which is usually heavy PVC or metal.

The cable, when properly installed on a hard, flat surface, is approximately 0.03 in. (0.8 mm) high, and thus is essentially undetectable when covered with carpet. Because the carpet squares are designed to be readily removable, the entire system can be repositioned to meet changing furniture layout requirements with a minimum of disruption and no structural work. The cable is designed to carry normal physical loads such as office traffic and furniture placement without affecting its electrical performance.

The attractiveness and simplicity of the system led to the development of similarly designed flat, low-tension cables for signal and communication wiring (Fig. 27.39) and, more recently, both electrical and fiber-optic cables and accessories for data transmission. Manufacturers also offer a complete line of junction fittings, connectors, adapters, and receptacles (Fig. 27.40).

The problems inherent in this type of on-the-floor wiring system, such as cable crossings, splicing, interfacing with round cable systems, interconnections at floor boxes and fittings, feed connections from cabinets, underfloor ducts, floor cells, and through-the-floor fittings, have all been solved by a full line of manufactured devices designed for specific usages.

Because under-carpet wiring systems are separate and distinct from wire and conduit systems, they, like underfloor duct systems, are usually shown on a separate electrical plan. A small typical plan of this

type is shown in Fig. 27.41. Figures 27.42 and 27.43 are photographs of essential portions of such an installation. Note that a complex system such as that shown in Fig. 27.43 requires recessing a floor box into the slab, which to an extent, obviates the essential simplicity and flexibility of an under-carpet system. Although these systems, at least in their simplest form, are particularly applicable to retrofit work, their low cost, combined with the inherent

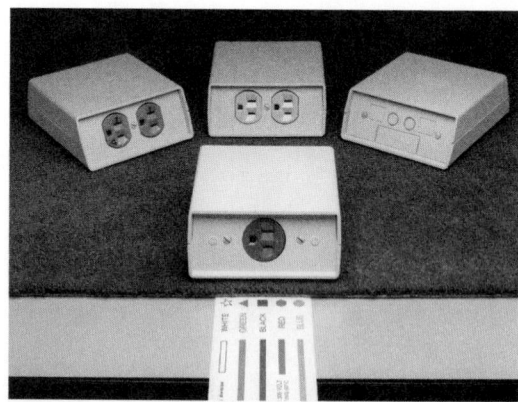

**Fig. 27.40** *Typical components of an under-carpet wiring system. The under-carpet FCC power cable is shown without the metallic top shield required in actual installation. It is a color-coded, five-conductor cable (neutral, equipment ground, and three circuit conductors or two circuit conducts and an isolated ground conductor). The floor outlets shown are front, single power outlet; rear (left to right), duplex power outlet (one of which has isolated ground); standard duplex power outlet; and combination data cable, communications, and telephone outlet. (Courtesy of Hubbell, Inc.)*

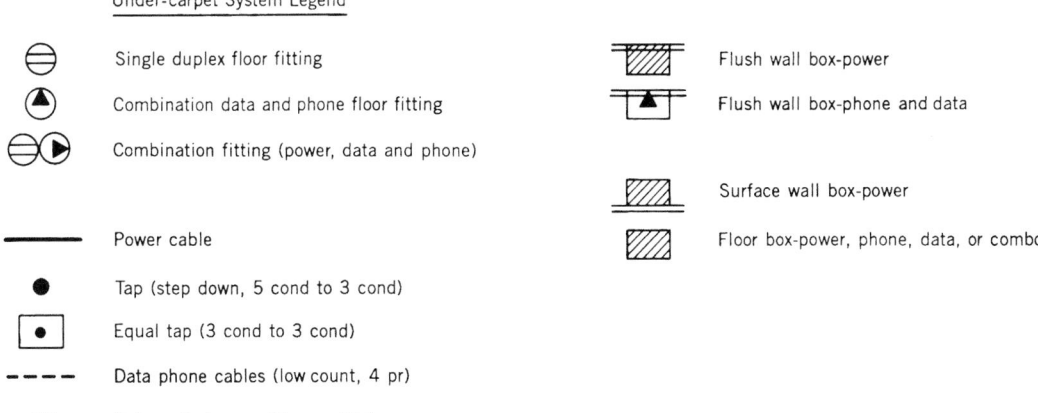

Under-carpet System Legend

⊖  Single duplex floor fitting

⦿  Combination data and phone floor fitting

⊖⦿  Combination fitting (power, data and phone)

———  Power cable

●  Tap (step down, 5 cond to 3 cond)

▣  Equal tap (3 cond to 3 cond)

- - - -  Data phone cables (low count, 4 pr)

--25--  Data and phone cables, multiple

▨  Flush wall box-power

▾  Flush wall box-phone and data

▨  Surface wall box-power

▨  Floor box-power, phone, data, or combo

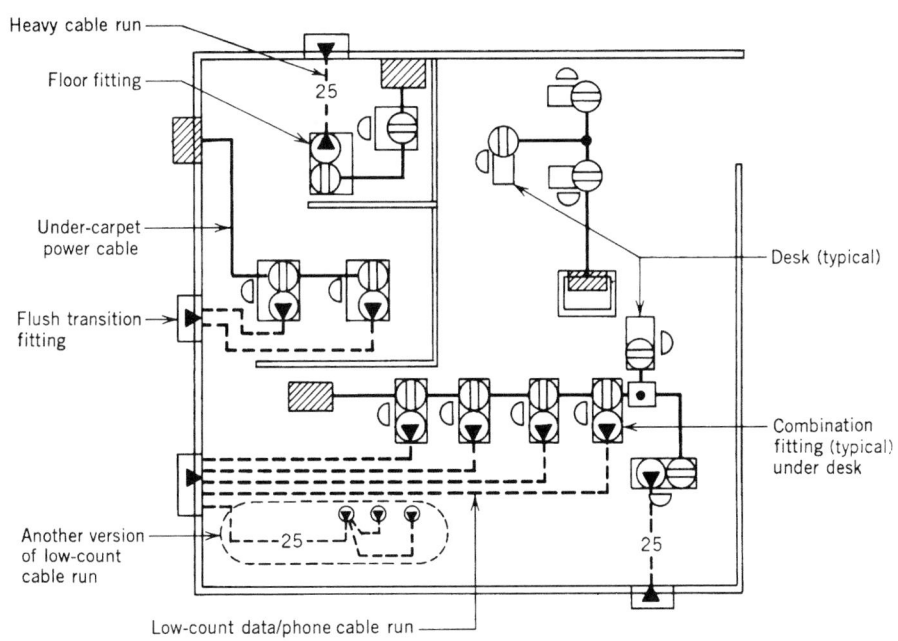

Heavy cable run

Floor fitting

Under-carpet power cable

Flush transition fitting

Another version of low-count cable run

Low-count data/phone cable run

Desk (typical)

Combination fitting (typical) under desk

**Fig. 27.41** *Typical layout of under-carpet power and low-tension/data cabling for a small office. A power, phone, and data cable connection on the floor is provided under or immediately adjacent to each desk (see Figs. 27.39, 27.40, 27.42, and 27.43 for typical details). (Reproduced with permission of AMP.)*

advantages of a flexible *floor-level* wiring system, particularly in open office areas, has made them a widely used first choice in new construction as well.

## 27.31 CEILING RACEWAYS AND MANUFACTURED WIRING SYSTEMS

The need for flexibility in a facility's electrical system coupled with the high cost of underfloor electrical

raceway systems encouraged the development of equivalent over-the-ceiling systems. These systems are actually more flexible than their underfloor counterparts because they energize lighting, provide power and telephone facilities, and even supply outlets for the floor above (see Figs. 27.26, 27.27, and 27.34), in addition to permitting very rapid layout changes at low cost. This last characteristic is particularly desirable in stores where frequent display changes necessitate corresponding electrical facility

**ELECTRICITY**

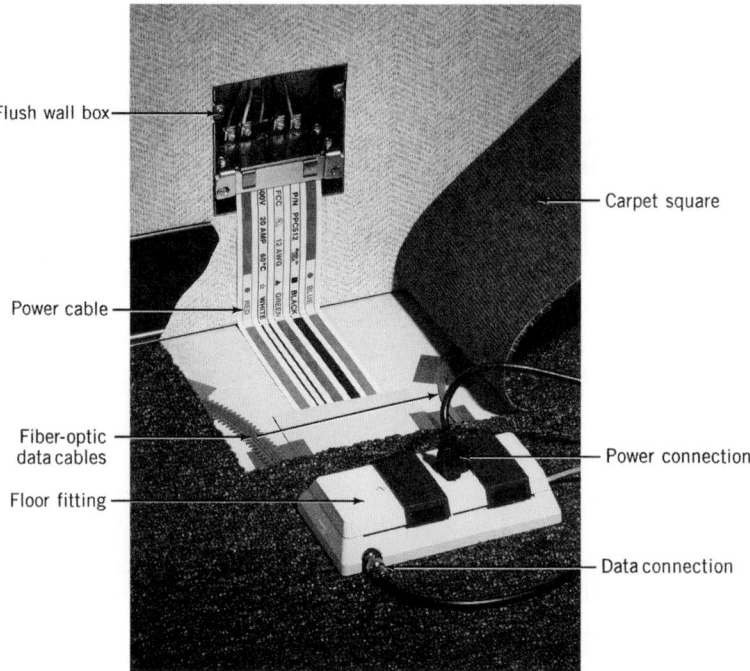

**Fig. 27.42** *Under-carpet flat power cables connect to round supply cables in a flush wall box and then extend to a combination power/low-tension/data floor fitting. Data cables are connected to the combination floor fitting from their system boxes, either individually or via other floor fittings. (Photo courtesy of Walker Systems, A Wiremold Company.)*

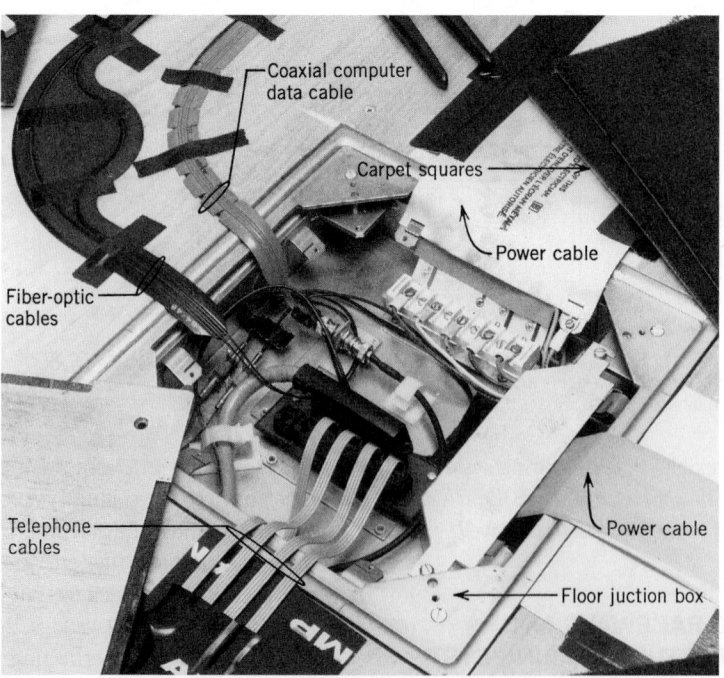

**Fig. 27.43** *Four service floor-junction boxes handle under-carpet power, telephone, low-tension electric, and FO cables. The floor box measures approximately 14 in. (356 mm) square and 2 in. (50 mm) deep. It can accommodate up to four 3-phase power circuits, eight 25-pair telephone/data cables, 24 coaxial cables, or, as illustrated, a mixture of cable types. (Reprinted with permission of AMP.)*

changes. Beyond the extreme flexibility made possible by the ceiling raceway system, it has the additional advantage that the system itself, not being cast in concrete like its underfloor counterpart, can be altered at will. Thus, not only layout changes (as mentioned previously) but also changes in the function of spaces can readily be accommodated. This is a particularly important characteristic in merchandising and educational facilities, where spaces can have their function repeatedly changed during the course of the building's life.

System details vary among manufacturers but the systems are essentially the same and, in principle, resemble underfloor systems. A typical system is constructed of metallic or nonmetallic surface-type raceways arranged in a tree formation (i.e., large trunk [header] raceways feed multiple smaller branch [distribution] raceways, and so on). The raceways are hung in the ceiling plenum from the concrete slab above. The hung ceiling must consist of lift-out panels because this type of wiring system is not permitted in spaces rendered inaccessible by the building structure. Header ducts (large area raceways) are fed from electrical panels and from signal, data, and telephone cabinets in the electrical and low-tension wiring service closets, respectively. Data headers are normally larger than the power header and can carry other low-voltage, low-power signal wiring as well. Distribution ducts (laterals) tap onto the headers. These laterals act as subdistribution raceways, feeding lighting fixtures and data, signal, telephone, and power outlets on the same floor and, via poke-through fittings, outlets on the floor above.

The standard method for extending wiring from the ceiling plenum raceways to floor-level or desk-level signal and power outlets uses vertical multisection raceways fed from the top (see Fig. 27.44). These service poles are available in a large variety of designs, finishes, and cross-sectional raceway areas and are easily installed in almost any location. They may be prewired and usually contain several power outlets, a telephone connection, and possibly data cable outlets. The result in a hung-ceiling office area or an exposed ceiling slab area (Figs. 27.45, 27.46, and 27.47) is certainly less elegant than that of a floor-level wiring system, but for most users it is satisfactory, and its low cost compared to any type of floor-duct system is a prime redeeming feature.

When electrical connections to poles, lighting fixtures, receptacles, and communication/data out-

**Fig. 27.44** *Typical floor-to-ceiling electrical/communication raceway poles. Units are available in a wide variety of shapes, sizes, and cross-sectional configurations in aluminum and steel. (Courtesy of Hubbell Premise Wiring.)*

lets are made with hard wiring, considerable field labor is required with a corresponding high cost. Furthermore, the relative permanence of such wiring lessens the inherent flexibility of the raceway system. To solve both problems, a number of manufacturers have developed a line of modular branch-circuit wiring elements. These, covered in *NEC* Article 604 under the very logical name *manufactured wiring systems,* consist of metal-clad or armored cable sets terminating in polarized plugs. The polarization prevents accidental interconnection of low-voltage, 120- and 277-V systems. Ceiling raceways can be equipped with matching receptacles, and connection to fixtures, poles, and other devices becomes a simple matter of plug insertion (Figs. 27.44, 27.48, and 27.49).

The result is a wiring system of extreme flexibility in which even extensive changes can be made very rapidly with minimal disruption and virtually no mess. Manufactured wiring systems are permitted in accessible areas only for obvious reasons. They are therefore also applicable to access floor spaces, as seen in Figs. 27.34 to 27.37. The additional cost of manufactured wiring elements is

**ELECTRICITY**

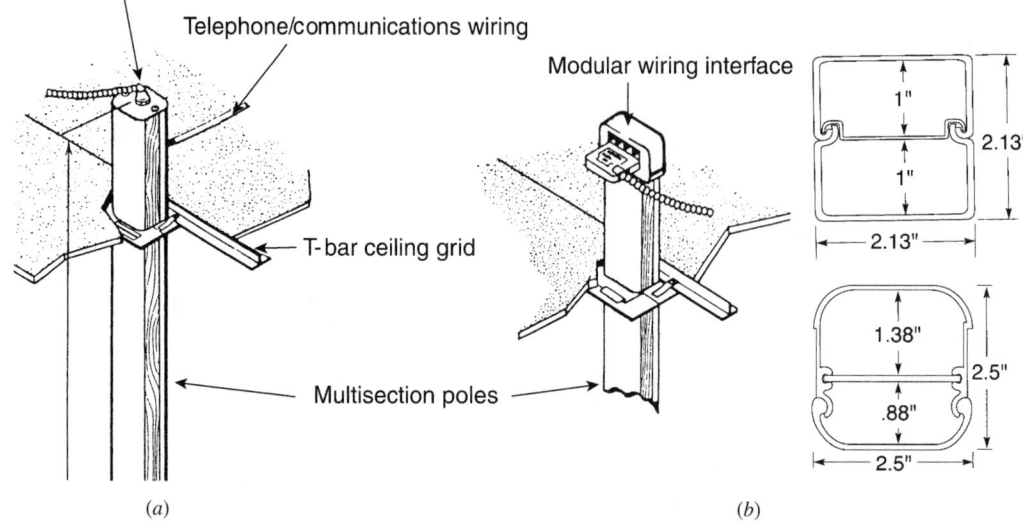

**Fig. 27.45** (a) Poles are fed at the top in the hung ceiling or from exposed ceiling raceways. These feeds can be made either conventional hard-wired or modular (plug-in), as shown. Modular connectors are used for power and low-tension (telephone, communication, data) wiring. (b) Two of the many cross-sectional configurations available are illustrated. Other designs divide the pole into three sections to suit the specific application. (Courtesy of Hubbell Premise Wiring.)

**Fig. 27.46** Power poles extend down from the ceiling to any desired height. In this library, power is required above the base cabinets, and the power pole is easily arranged to supply it. (Courtesy of Wiremold Company.)

**Fig. 27.47** The archetypical application of a power/signal pole is to provide electrical service to an island desk, as illustrated. This pole is equipped with telephone, data, and power outlets, all of which are in use. (Courtesy of Hubbell, Inc.)

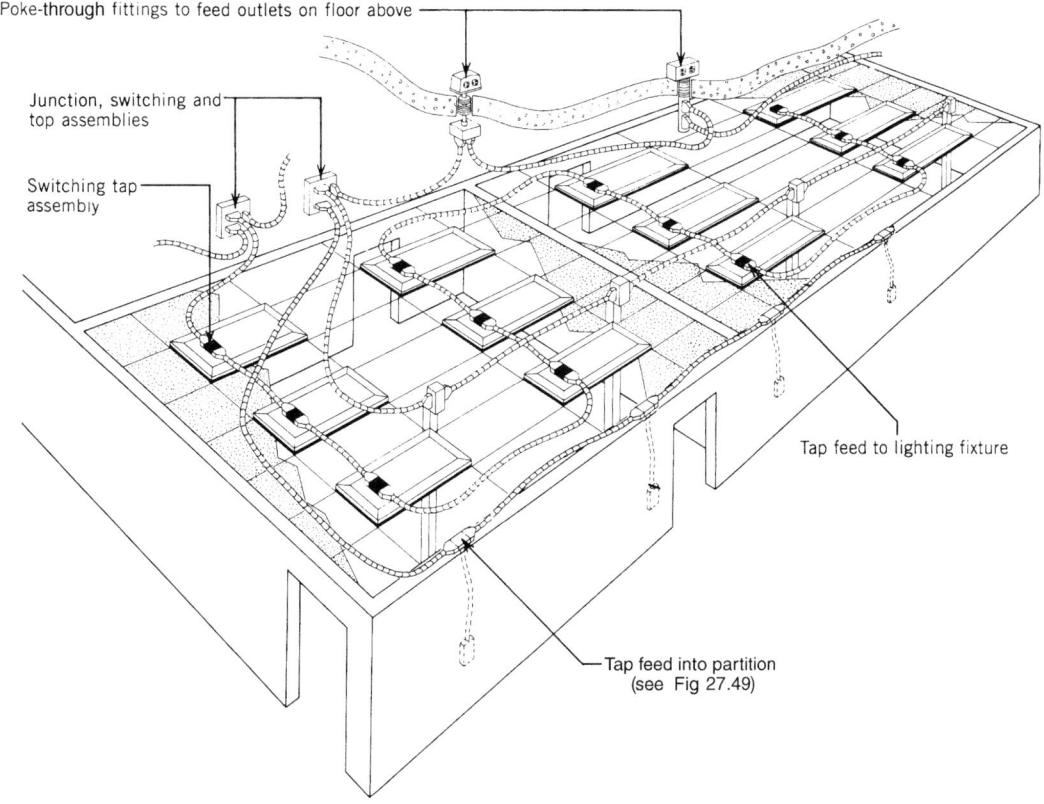

*Fig. 27.48* Manufactured modular wiring assemblies are used for tap connections to feed ceiling lighting fixtures or any other ceiling connection, as well as a complete range of junction, switching, tap, and poke-through units. (Courtesy of Walker Systems, A Wiremold Company.)

**ELECTRICITY**

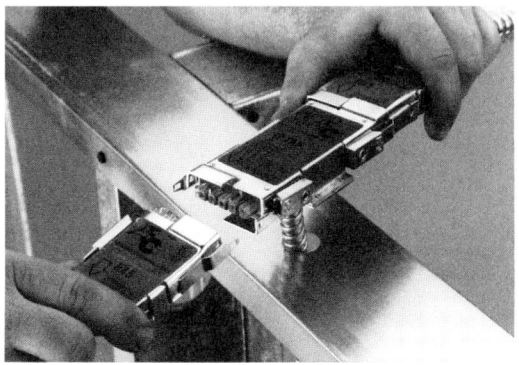

*Fig. 27.49* Manufactured modular wiring connectors simplify installation and reduce labor costs initially and in any subsequent rearrangements. (Courtesy of AFC Cable System.)

frequently offset by the labor savings even on initial installation and certainly after one or two field changes. Cable sets are available for power (120 V and 277 V), telephone, and all types of low-voltage signal equipment. The cables must be approved for use in conditioned air plenums and hung ceilings.

To take full advantage of the potential labor-cost savings inherent in the system, field labor must be minimized. This is accomplished by factory equipping all utilization equipment, including lighting fixtures, with appropriate plug-in connectors.

# Electric Wiring Design

IN WIRING DESIGN, THERE ARE NUMEROUS POSSIBLE solutions to each problem—some good, some fair, and some poor. Experience guides designers to a solution that best suits the job because it is their responsibility to establish the most economical design within the framework of the design intent and criteria. This chapter opens with a discussion of typical design criteria and continues with wiring design. The actual electrical design intended for construction must meet all the requirements of local and national building codes. To our knowledge, all building codes include NFPA 70, the *National Electrical Code* (*NEC*), as the minimum standard for all electrical design, and its requirements are thereby made legally binding. The electrical design data and procedures adduced in this chapter and in other sections of this book are based on *NEC* requirements and are, to the best of our knowledge, current. However, they are intended only as didactic materials. Code requirements change, as the *NEC* is revised every 3 years. The designer is therefore advised to obtain and study the latest edition of all codes having jurisdiction, including the *NEC*, to ensure that the electrical design meets the requirements of these codes.

## 28.1 GENERAL CONSIDERATIONS

### (a) Flexibility

Every wiring system should incorporate sufficient flexibility in branch circuitry, feeders, and panels to accommodate all probable patterns, arrangements, and locations of electric loads. The degree of flexibility to be incorporated depends largely on the type of facility. Thus, laboratories, research facilities, and small educational buildings require far more flexibility than do residential, office, and fixed-purpose industrial installations. As part of the design for flexibility, provision for expansion must be made, as experience has demonstrated that most facilities will grow, both physically and in electrical demand. Overdesign, however, is as bad as underdesign, being wasteful of money and resources, both initially and in operation.

### (b) Reliability

The reliability of electrical power within a facility is determined by two factors: the utility's service and the building's electrical system. The service record of the utility should be studied along with the economic impact of a power outage to determine whether and to what extent standby power equipment is justified (see Sections 26.39 and 28.21). Emergency equipment required for the safety of a building's occupants is determined by local, state, and national building codes.

Beyond the electrical service point, the reliability of power is entirely dependent on the wiring system. Here, too, economic studies must be made to determine the quality of equipment and the amount of redundancy (duplicate equipment) to be installed. The subject of reliability is complex, and we can state only a few general principles here.

1. The reliability of an electric system is only as good as that of its weakest element. Therefore, it may be necessary to provide redundancy at anticipated weak points in the system.
2. The electrical service and the building's distribution system act together. An extremely reliable (and expensive) service is of little use if the power cannot reach the desired points.
3. Critical loads within the facility should be pinpointed to determine how best to serve them reliably. This is done by establishing reliable power paths to them or by furnishing individual standby power packages for them. The latter course is often chosen for health care and other critical loads.

### (c) Safety

Although rigid adherence to the requirements of the *NEC* and other applicable NFPA codes ensures an initially safe electrical installation, the designer must constantly be alert to such factors as electric hazards caused by misuse or abuse of equipment or by equipment failure. Also, attention to the physical size of equipment eliminates the oft-encountered hazard caused by obstruction of access spaces, passages, closets, and walls with electric equipment. Finally, lightning protection can be subsumed under the heading of safety; this topic is discussed in Chapter 24.

### (d) Economic Factors

This can readily be divided into two frequently interrelated items: first cost and operating cost. All other factors being equal, the first cost depends largely on whether the owner is interested in the minimum first cost or the minimum owning cost. We have demonstrated that these two costs frequently stand in inverse relationship to one another (exceptions are mentioned later). Low first-cost equipment generally results in higher energy costs, higher maintenance costs, and shorter life.

The decision, however, is not purely an economic one, inasmuch as the electrical energy cost factor in the operating-cost equation is directly related to energy consumption, with one exception. That exception is the utility's demand charge, which is discussed in Section 25.12. Means for minimizing this cost are covered in detail in Section 25.13. Here, there occurs a coincidence of reduction to both first cost and operating costs. Load-leveling equipment permits the electrical distribution system to be sized without consideration of coincident load peaks, thus resulting in smaller equipment operating more efficiently—near its full-load capacity. All other reductions in electric energy cost flow directly from the corresponding reduction in energy consumption.

### (e) Energy Considerations

This factor is complex, involving considerations of energy codes and budgets, energy conservation techniques (see Section 28.5), and energy control.

Buildings constructed with governmental participation may be subject to energy budget limitations expressed in Btu/ft$^2$/year. Although the lion's share of this budget is taken by heating/cooling and lighting systems, the electrical distribution system is also subject to conformity to stated codes. Most important among these is ANSI/ASHRAE/IESNA Standard 90.1, a current copy of which is *de rigeur* in all design offices. Detailed energy conservation considerations are discussed in detail in Section 28.5. A discussion of the IESNA *Recommended Procedure for Lighting Power Limit Determination* is found in Chapter 15.

### (f) Space Allocations

The general impression that electrical equipment is small and easily concealed is accurate only for wire and conduit. Panels, motor control centers, busduct, distribution centers, switchboards, transformers, and so on can be large, bulky, noisy, and highly sensitive to tampering and vandalism. Thus, space allocations must be concerned with maintenance ease, ventilation, expandability, centrality (to limit the length of runs), limitation of access to authorized personnel, and noise limitation in addition to the fundamental consideration of space adequacy.

### (g) Special Considerations

These depend on the nature of the facility and may include items such as security, central and/or remote controls, interconnection with other facilities, and the like.

## 28.2 LOAD ESTIMATING

When initiating the electrical design of a facility, it is necessary to estimate the total building load in order to plan such spaces as transformer rooms, conduit chases, and electrical closets. This information is also required by the local power company well in advance of the start of construction. An exact load total can be made only after the design is completed, but because this is often several months later, a good preliminary estimate is required. Such an estimate can be made from the figures given in Table 28.1. These figures are averages. When it appears that a building will have heavier or lighter loads because of lighting levels, energy codes, load management equipment, or other factors, the figures should be modified accordingly. The figures in the individual categories should be added without application of demand or diversity factors in order to obtain the maximum load for which the building service equipment must be sized in the absence of electric load-control (load-leveling) equipment. (Non-

**TABLE 28.1 Electric Load Estimating[a]**

| I | II | III | IV | V | VI |
|---|---|---|---|---|---|
| | | Volt-Amperes per Square Foot[b] | | | Ten-Year Percent Load Growth |
| | | Misc. | Air Conditioning | | |
| Type of Occupancy | Lighting[c,d] | Power[e] | Electric | Nonelectric | |
| Auditorium | | | | | |
|   General | 1.0–2.0 | 0 | 12–20 | 5–8 | 20–40 |
|   Stage | 20–40 | 0.5 | | | |
| Art gallery | 2.0–4.0 | 0.5 | 5–7 | 2.0–3.2 | 20–40 |
| Bank | 1.5–2.5 | | 5–7 | 2.0–3.2 | 30–50 |
| Cafeteria | 1.0–1.6 | 0.5 | 6–10 | 2.5–4.5 | 20–40 |
| Church and synagogue | 1.0–3.0 | 0.5 | 5–7 | 2.0–3.2 | 10–30 |
| Computer area | 1.2–2.1 | 2.5[g] | 12–20 | 5–8 | 50–200 |
| Department store | | | | | |
|   Basement | 3–5 | 1.5 | | | |
|   Main floor | 2.0–3.5 | 1.5 | 5–7 | 2.0–3.2 | 50–100 |
|   Upper floor | | 2.0–3.5 | 1.0 | | |
| Dwelling (not hotel) 0–3000 ft² | 3.0 | 5.0 | — | — | 50–100 |
|   3000–120,000 | 0.4 | 0.15 | — | — | |
|   above 120,000 | 1.5–2.5 | 2.0 | — | — | |
| Garage (commercial) | | | — | — | 10–30 |
| Hospital | 1.0–3.5 | 1.5 | 5–7 | 2.0–3.2 | 40–80 |
| Hotel | 1.0–2.0 | 0.5 | | | |
|   Lobby | 1.0–1.5 | 1.0 | 5–8 | 2.0–3.5 | |
|   Rooms (no cooking) | 2.0–3.0 | 5–20 | 3–5 | 1.5–2.5 | 30–60 |
| Industrial loft building | 1.2–2.2 | 0.5 | — | — | 50–100 |
| Laboratories | 1.5–3.0 | 1.5 | 6–10 | 2.5–4.5 | 100–300 |
| Library | 1.0–2.0 | 0.5 | 5–7 | 2.2–3.2 | 30–40 |
| Medical center | 1.5–3.5 | 2.5 | 4–7 | 1.5–3.2 | 50–80 |
| Motel | 1.2–2.5 | 0.5 | — | — | 30–60 |
| Office building | 1.5–2.8 | 2.5 | 4–7 | 1.5–3.2 | 50–80 |
| Restaurant | | | 6–10 | 2.5–4.5 | 20–40 |
| School | 2.0–2.5 | 2.0 | 3.5–5.0 | 1.5–2.2 | 50–80 |
| Shops | 2.0–3.5 | 0.5 | | | |
|   Barber and beauty | 2.0–3.0 | 0.5 | | | |
|   Dress | 2.0–3.0 | 0.5 | 5–9 | 2–4 | |
|   Drug | 2.0–3.5 | 0.25 | | | 40–80 |
|   Five and ten | 0.25–1.0 | — | 4–7 | 1.5 to 3.2 | |
|   Hat, shoe, specialty | 0.3 | — | | | |
| Warehouse (storage) | 0.25 | — | — | — | 10–30 |
| In the above except single dwellings: | | | | | |
| Halls, closets, corridors, | 0.5 | — | — | — | |
| storage spaces | 0.25 | — | — | — | |

[a]Figures assume energy-conservation techniques applied.

[b]These figures do not include allowance for future loads.

[c]The figures given in Article 220 of the *NEC* are minimum figures for calculation of electric feeder sizes.

[d]See Section 15.5 for a discussion of power limits.

[e]These figures are based on experience and must be applied judiciously.

[f]Includes the loads of air-handling equipment and pumps.

[g]This figure does not include the power used by the computer.

coincident loads, such as heating and cooling in the same space, are not combined.)

At this point, an analysis can be made to determine the feasibility of incorporating electric load-control equipment into the facility (see Section 25.13). Input to this study includes the utility's complete rate schedule, including all penalty clauses, a detailed analysis of the building's equipment load patterns, and any external constraints such as maximum loads imposed by power and energy budgets. Equipment load patterns must be carefully analyzed because they determine a load's sheddability. Thus, for kitchen equipment, load interruption may be undesirable, but shifting of cooking time by half an hour is usually feasible. For HVAC equipment, building thermal inertia and "stretching" maximum and minimum temperatures and humidities permit considerable latitude without adverse effects. Also, as explained in Section 25.13, the degree and duration of load shedding are a function of the type of control equipment utilized. It is well to repeat here what is stated there—load control affects maximum demand with only minor effect on total energy consumption. The external constraints referred to are the energy budgets recommended or required by codes, legislation, and funding bodies.

After the load control analysis is complete, or simultaneously with it, a building energy consumption analysis must be performed. This may be done manually, although numerous computer programs are available that not only increase accuracy but also permit consideration of more factors. The results of this analysis indicate whether the target electrical energy budget is being met. If not, loads must be modified by reconsideration of projected systems and system criteria, by incorporating energy conservation devices and techniques into the electrical system, and by drawing up energy use guidelines that will be applied when the building is occupied. Because this last item depends for its success on the day-to-day voluntary actions of the building's occupants, experience has shown that it should not be considered as a major conservation source. Conservation measures are covered in Section 28.5.

The electrical loads in any facility can be categorized as follows:

1. Lighting
2. Miscellaneous power, which includes data-

processing equipment, convenience outlets, and small motors
3. Heating, ventilating, and air conditioning (HVAC)
4. Plumbing and other piping systems
5. Vertical transportation equipment and fixed material-handling equipment
6. Kitchen equipment
7. Special equipment

Category 1 is self-explanatory and is covered by column II of Table 28.1. Note, however, that lighting loads are carefully prescribed in ANSI/ASHRAE/IESNA Standard 90.1 and that these prescribed loads should be used where this standard has jurisdiction. Contradictions between recommendations in standards having jurisdiction must be adjudicated by local authorities before the design is established. Consult the current edition of ANSI/ASHRAE/IESNA Standard 90.1 for details of prescribed loads.

Category 2, column III, includes, in addition to receptacles and small motors, such items as desktop computers and data-processing terminals including all peripheral equipment, plus plug-in heaters, water fountains, and so on.

Category 3, column IV, includes all loads imposed by the HVAC equipment. Included therefore are fuel pumps, boiler motors, exhaust fans, and so forth. Also included in column IV, for air conditioning loads, are refrigeration compressors. This item is omitted in column V because the air conditioning utilizes absorption machines that do not use electricity for primary power. When air conditioning is not anticipated, the HVAC load is still appreciable because of heating and ventilating requirements. A rough estimate for this load would be two-thirds of the loads in column V.

Category 4 includes all loads associated with the water and sanitary systems, including house water pumps, air compressors and vacuum pumps, sump pumps and ejectors, well pumps and fire pumps, water heaters and pneumatic tubes, plus such special items as display fountain pumps. Also included in this category are electric loads connected with fixed systems such as cooking gas, medical gas piping, distilled water systems, and so on.

Because these loads vary widely with local conditions and with facility design as much as with the type of facility, an estimate cannot be made on a volt-amperes per square foot basis by type of building. For this reason, no figure is included in Table 28.1. If actual data cannot be used, it is helpful to remember that plumbing and piping electrical loads are relatively small, rarely exceeding 20% of the HVAC system electric load, although, for the most part, they are unrelated to it.

Category 5, vertical transportation, is related to square footage in some types of buildings but not in others. A close estimate of power and energy requirements can be made by the project's elevator consultant. Lacking this, a fair estimate can be made from the data given in Chapters 31–33. In addition to elevators, escalators, moving walkways, and ramps, these loads should include dumbwaiters, horizontal and vertical conveyors, trash and linen transport systems, automated container delivery, and fixed conveyors. These loads are readily available from the material transport consultant, who is frequently also the elevator consultant.

Category 6, kitchen equipment, is also not included in Table 28.1, although obviously present in all restaurants, most hospitals, and some office and religious-use buildings. The reason for this omission is that the primary power for the major load, the cooking equipment, may be either gas or electric. Other large energy-use equipment such as dishwashers can be electric, gas, or steam-fed. Furthermore, no correlation can be made between the facility type, area, and load, even if electrically powered, because population and schedule are also major factors. When kitchens are planned, a kitchen consultant or another experienced planner can supply an estimate of the electric power requirement.

Category 7, special equipment, is so variegated that no figures can be listed. Under this title are subsumed such items as laboratory equipment, shop loads, display area loads, floodlighting, canopy heaters, display window lighting, industrial loads, and so on. This load data must be gathered for individual items of equipment and added to the foregoing totals.

Table 28.2 gives a tabulation of service entrance size in amperes, based on single- and 3-phase service for typical occupancies. These figures are intended for quick estimate purposes and should be adjusted after the design is completed.

## 28.3 SYSTEM VOLTAGE

Several voltage systems are commonly available in the United States and Canada. (See *NEC* Article 230, "Services," for code requirements.)

### (a) 120-V, Single-Phase, 2-Wire

This (Fig. 28.1) is used for the smallest facilities, such as outbuildings, and isolated small loads of up to 6 kVA. The load is calculated by multiplying current and voltage. For 60-A service, which is the normal limit for this type of service, no more than 50 A are usually drawn. Thus

$$VA = 120 \times 50 = 6000 \text{ VA} = 6 \text{ kVA}$$

The nominal system voltage is 120 V, although it is also referred to as 110 V and 115 V.

**TABLE 28.2 Nominal Service Size in Amperes**

*Nominal service sizes are 100 A, 150 A, 200 A, 400 A, 600 A*

| Facility | Area in Square Feet | | | | Remarks |
|---|---|---|---|---|---|
| | 1,000 | 2,000 | 5,000 | 10,000 | |
| Single-phase, 120/240 V, 3-wire | | | | | |
| Residence | 100 A | 100 A | 150 A | — | Minimum 100 A |
| Store[a] | 100 A | 150 A | — | — | |
| School | 100 A | 100 A | 150 A | — | |
| Church[a] | 100 A | 150 A | — | — | |
| 3-phase, 120/208 V, 4-wire | | | | | |
| Apartment House | — | — | 100 A | 150 A | |
| Hospital[a] | — | — | 200 A | 400 A | |
| Office[a] | — | 100 A | 400 A | 600 A | |
| Store[a] | — | 100 A | 400 A | 600 A | |
| School | — | 100 A | 100 A | 200 A | |

[a]Fully air-conditioned using electric-driven compressors. Based on figures in Table 28.1.

ELECTRICITY

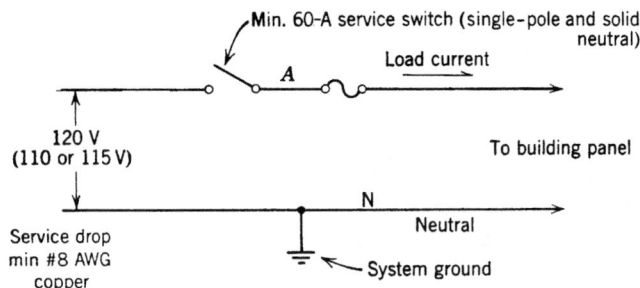

**Fig. 28.1** *Shown is a 120-V, single-phase, 2-wire service. This is also the arrangement of a typical branch circuit.*

### (b) 120/240-V, Single-Phase, 3-Wire

This system (Fig. 28.2) is for somewhat heavier loads. The code requires that all one-family residences with six or more 2-wire circuits or a net computed load (by code calculation) of 10 kVA or more have a minimum of 100-A, 3-wire service. Service disconnect for 100-A service would be a 100-A, 2-pole, solid neutral switch, fused at no more than 80% of the rating, or 80 A. This is usually written 100A, 2P & SN, 80AF. This service is used principally for residences, small stores, and other occupancies where the load does not exceed 80 A or 19.2 kVA. The load is calculated thus:

$$kVA = \frac{V \times I}{1000} = \frac{240 \times 80}{1000} = 19.2$$

Although it may appear otherwise, the neutral carries no more than full-load current. Note that each "hot leg" of the 3-wire system carries line current. Thus, total load can also be calculated:

load kVA = twice load on each line

Assuming a balanced 80-A load:

$$\text{total kVA} = 2 \times 80 \text{ A} \times 120 \text{ V}$$
$$= 2 \times 9600 \text{ VA} = 19{,}200 \text{ VA}$$
$$= 19.2 \text{ kVA}$$

If the loads were unbalanced with, say, 30 A in one line and 50 A in the other, the total load would be

$$120 \times 30 + 120 \times 50 = 3600 + 6000 = 9600 \text{ VA}$$
$$= 9.6 \text{ kVA}$$

Actual system voltages can be 120/240, 115/230, or 110/220, although 120/240 is the accepted industry standard. Loads at 120 V are connected between line and neutral; heavier loads, such as for a clothes dryer or an electric stove, are connected between phase lines at 240 V.

For example, to find the line currents caused by the three loads shown in Fig. 28.3:

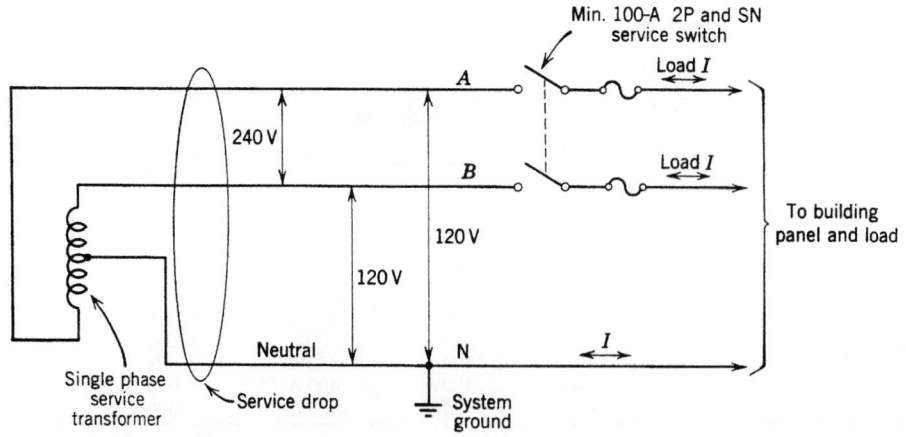

**Fig. 28.2** *Shown is a 120/240-V, single-phase, 3-wire service. The single-phase transformer, which is normally located on the power company's pole along a street or road, is center-tapped to establish a neutral. The neutral connection is always grounded.*

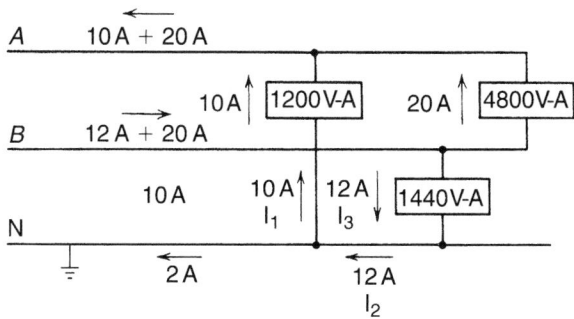

**Fig. 28.3** *An example 120/240 V, 3-wire distribution. Note that the neutral carries the difference in current between the A and B legs and therefore a maximum that is equal to the current in one of the legs (when the other is zero).*

1. 120-V, 1200-W iron, line A to neutral
2. 120-V, 1440-W hair dryer, line B to neutral
3. 240-V, 4800-W dryer, line A to line B

we calculate

$$I_1 = \frac{1200}{120} = 10 \text{ A}$$

$$I_2 = \frac{1440}{120} = 12 \text{ A}$$

$$I_3 = \frac{4800}{240} = 20 \text{ A}$$

Note that the neutral only carries the unbalance of 2 A (see Fig. 28.3).

Total current in line A = 20 + 10 = 30 A

Total current in line B = 20 + 12 = 32 A

Total current in neutral N = 2 A

Total load = 120(30) + 120(32) = 3600 + 3480 = 7440 VA

or

Loads are 1200 + 1440 + 4800 = 7440 VA

Because the loads are almost entirely resistive heating loads (small hair dryer motor and approximately a ⅙-hp dryer motor), the entire load has a power factor of almost 1.0. This, however, affects only energy calculations because equipment is sized, in general, by kVA capacity. The 120/240-V, single-phase system is derived from a center-tapped, single-phase transformer.

### (c) 120/208-V, Single-Phase, 3-Wire

Although this system (Fig. 28.4) appears similar to the one described in Section 28.3*b*, it is really part of a 3-phase system. It is most often found *within* a building that takes 3-phase service rather than constituting a service voltages arrangement. It is used to serve a load that does not require 3-phase, 4-wire but does require a voltage higher than 120 V to feed its load. Calculation of loads and line currents is considerably

ELECTRICITY

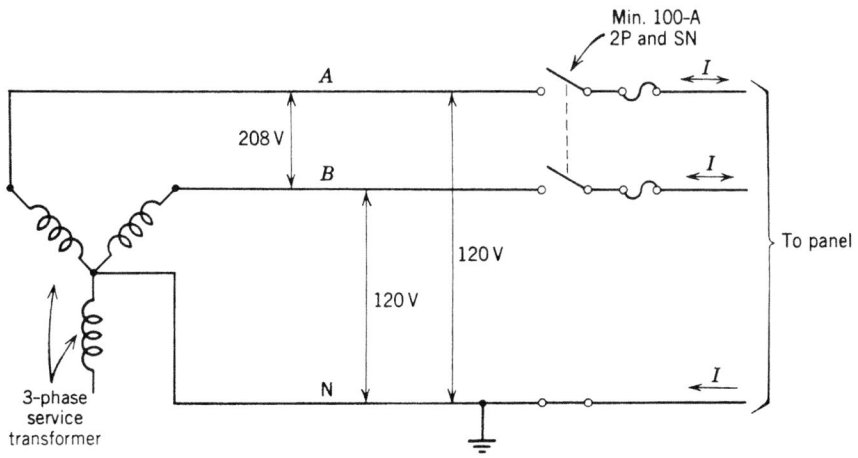

**Fig. 28.4** *Shown is a 120/208-V, single-phase, 3-wire service. This arrangement comprises two-thirds of the full 120/208-V, 3-phase, 4-wire connection shown in Fig. 28.5.*

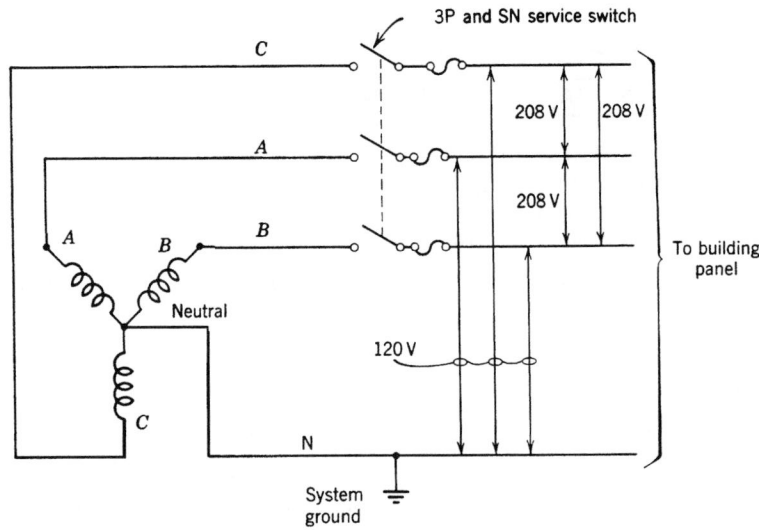

**Fig. 28.5** *Shown is a 120/208-V, 3-phase, 4-wire system. The neutral connection is connected to the system ground and is not broken by the service switch.*

more complex than in Section 28.3*b* because of the 120° phase displacement between phases A and B. Here, as before, the neutral carries no more than line current, regardless of whether the system is balanced.

### (d) 120/208-V, 3-Phase, 4-Wire

This system (Fig. 28.5) is a widely used 3-phase arrangement applicable to all facilities except very large ones. In the latter, lengths of feeders and sizes of loads become so great that a higher system voltage is required. In this system, 120-V loads such as lighting, computers and accessories, receptacles, and so on are fed at 120 V by connection between each phase leg (Fig. 28.6) and neutral. Motors larger than ½ hp and all 3-phase loads are fed at 208 V by connection to the 3-phase legs. Single-phase, 208-V loads such as heaters are accommodated as described in Section 28.3*c* by connection between two phase legs. Such loads are often referred to as *2-pole* loads, alluding to the 2-pole current breakers used to feed them.

### (e) 277/480-V, 3-Phase, 4-Wire

This system (Figs. 28.7 and 28.8) is applicable to large buildings (either horizontally or vertically) where lighting is principally fluorescent and/or HID

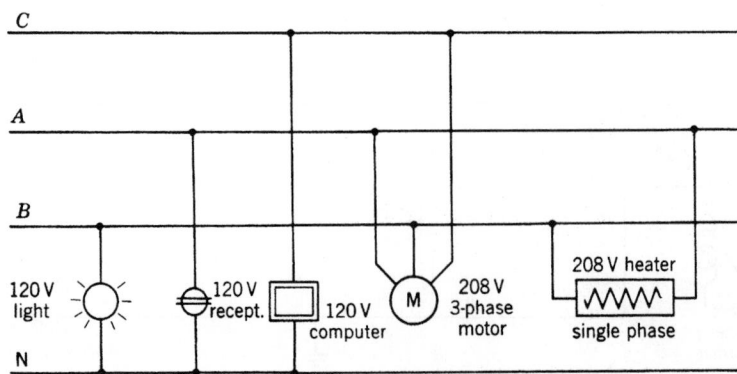

**Fig. 28.6** *The flexibility of the 120/208-V system is illustrated; this accounts for its wide usage. Loads are shown schematically. In practice, the loads are fed via protective devices in the panel. These are omitted here for clarity. A, B, C, and N represent the panel buses.*

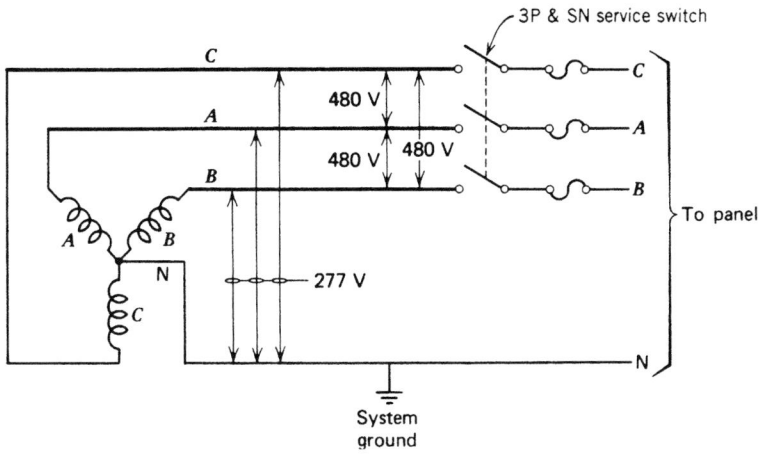

**Fig. 28.7** *Shown is a 277/480-V, 3-phase, 4-wire service system. The system is identical to the 120/208-V system shown in Fig. 28.5 except for the voltages.*

and the 120-V load does not exceed one-third of the total load. It provides 277 V for fluorescent and HID lighting and 480 V for machinery. Small (3- to 25-kVA) dry-type transformers are used to step down from 480 V for 120-V loads. This system is ideally suited to multistory office buildings and large single-level or multilevel industrial buildings. Cost savings are generated by the smaller feeder and conduit sizes and smaller switchgear, which more than offset the additional cost of step-down transformers for the 120-V load.

As an example of the economies possible with this system, let us consider the wiring required for a 15-kW heater. At 3-phase, 208 V:

$$I = \frac{15,000 \text{ W}}{\sqrt{3} \times 208 \text{ V}} = 42 \text{ A}$$

requiring No. 8 RHW wire (45-A capacity). At 480 V:

$$I = \frac{15,000 \text{ W}}{\sqrt{3} \times 480 \text{ V}} = 18 \text{ A}$$

requiring only No. 12 RHW wire (20-A capacity). This voltage system is also referred to as 265/460 V and 255/440 V.

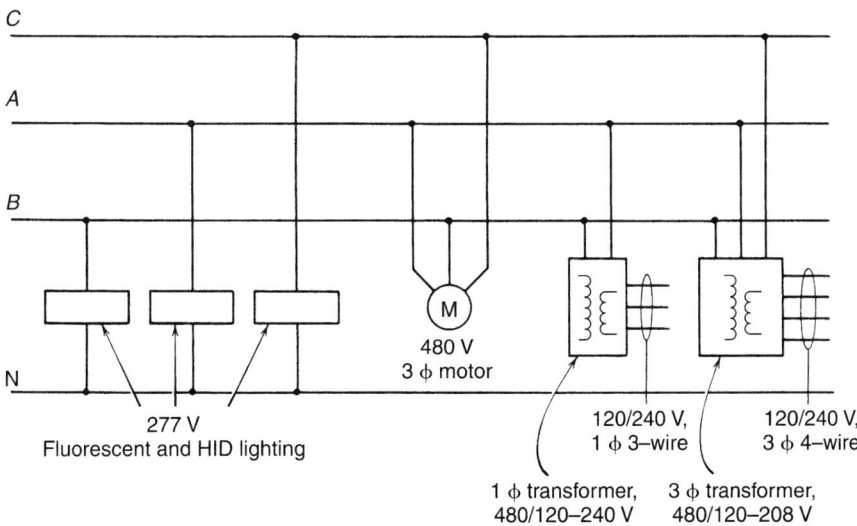

**Fig. 28.8** *Normal load arrangement for the 277/480 V system is illustrated. The lighting can be fluorescent or HID. Transformers, either single-phase or 3-phase, supply 120 V for receptacles and 208 V for loads requiring that voltage.*

**TABLE 28.3 Standard System and Utilization Voltages[a,b]**

| System Voltage (Transformers) | | Utilization Voltage[c] (Motors) | |
|---|---|---|---|
| Nominal | With 4% Drop[c] | Current Standard[d] | Obsolete Standard |
| 120 | 115.2 | 115 | 110 |
| 208[d] | 199.7 | 200 | 208 |
| 240[d] | 230.4 | 230 | 220 |
| 480 | 460.8 | 460 | 440 |
| 600 | 576.0 | 575 | 550 |

[a]To eliminate any confusion between system and utilization voltages, the current NEMA standards are tabulated above.

[b]When specifying transformers, use system voltages; for motors, use utilization voltage.

[c]Note that utilization voltage corresponds to a 4% drop from the system voltage, well within the normal motor tolerance.

[d]Motors for 208-V systems are rated 200 V. Motors for 240-V systems are rated 230 V. They cannot be used interchangeably without seriously affecting motor performance.

Voltages above 150 V to ground are generally avoided in residential branch circuits but may be used in commercial and industrial facilities within the guidelines established by the *NEC*.

### (f) 2400/4160-V, 3-Phase, 4-Wire

This system is used only in very large commercial buildings or in industrial buildings with machinery requiring these voltages. The cost of running these voltages within a building is high because of *NEC* requirements and the inherently higher cost of 5-kV equipment. A detailed cost and engineering analysis by a competent engineer is required for each case. Voltages above this level are widely used in large industrial plants but are beyond the scope of this book.

Reference was made earlier to the varied voltages assigned to the same voltage system. Thus, at the lowest level, we have 110, 115, and 120 V; at

the next level, 200, 208, 220, 230, and 240 V; at the next, 255, 265, and 277 V; and finally, 440, 465, and 480 V. These voltage differences arise because of the historical difference between transformer voltage standards, which establish the *system* voltage, and motor voltage standards, which govern *utilization* voltage (Table 28.3). Present motor voltage standards are established at a level that is consonant with the system voltage. Note the close correspondence between motor standard voltage and system voltage with a (normal) 4% feeder voltage drop. Thus, we see that on 240- and 480-V systems, 230- and 460-V motors are suitable.

Difficulties arise in application of 230- and 240-V motors to a 208-V system. Although motors will operate at plus or minus 10% voltage, 230- and 240-V motors should *not* be used on 208-V systems. Instead, motors specially wound for 200 V should be specified. A brief summary of the effects of undervoltage is given in Table 28.4.

**TABLE 28.4 Effects of Undervoltage on Utilization Equipment[a]**

| Load | 10% Undervoltage |
|---|---|
| Lighting | |
|    Incandescent | Output reduced 30% |
|    Fluorescent | Output reduced, poor start |
|    Mercury | Low output, poor start |
| Motors | 20% lower torque, hotter operation, reduced life, overloading |
| Heaters | 20% reduction in output |
| Small tools | Stalling, low power |

[a]Computers and peripherals are generally supplied with voltage-regulating power supplies that make them tolerant of ±10% variation in steady-state supply voltage. Conversely, they are highly intolerant of rapid supply voltage fluctuation.

## 28.4 GROUNDING AND GROUND-FAULT PROTECTION

The vast majority of secondary wiring systems are solidly grounded. The reasons for this arrangement are several and varied. Among them are:

1. To prevent sustained contact between the low-voltage secondary system and the high-voltage primary system in the event of an insulation failure. Such contact could cause a breakdown of the secondary system insulation and severely endanger the system's users.
2. To prevent single grounds from going unnoticed until a second ground occurs, which would extensively disable the secondary system.
3. To permit locating ground faults with ease.
4. To protect against voltage surges.

5. To establish a neutral at zero potential for safety and for reference.

Points 2 and 4 are highly technical, and a full explanation is beyond the scope of this book. Point 5 requires that the neutral in a single secondary system is:

1. Never interrupted by switches or other devices
2. Connected to ground only at one point—the service entrance
3. Color-coded white, natural gray, or by three continuous white stripes on any insulation color other than green, along the entire conductor length (in the United States), for easy recognition

A typical service-grounding diagram is given in Fig. 28.9. Universal acceptance and use of grounded secondary 120-V systems *introduces* another shock

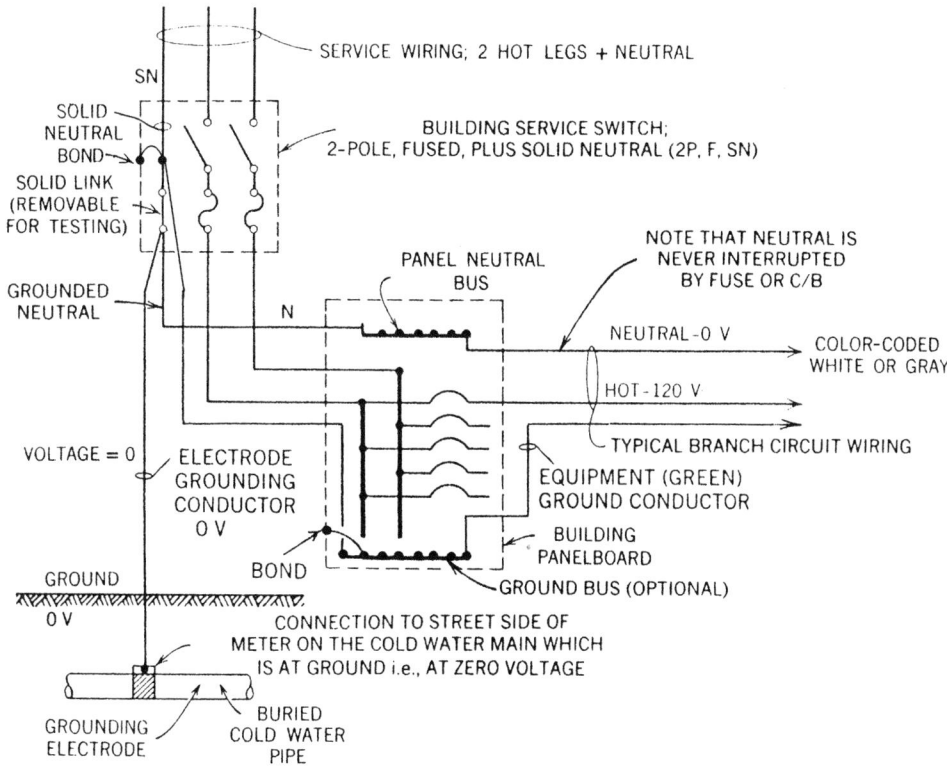

**Fig. 28.9** *Typical service-grounding arrangement. Note that the grounded neutral is unbroken throughout. The ground bus, if present, is separate and distinct from the neutral bus, and both are grounded at the service entrance point. For the sake of clarity, not all bonding connections, which are required to ground all non-current-carrying metal parts of the electrical system (boxes, enclosures, conduits, etc.) are shown.*

hazard while eliminating the dangers described previously. This is shown in Fig. 28.10a. An accidental fault within an appliance could connect the metal case of the appliance to the line. This can occur with such common devices as an electric saw, a clothes washer or dryer, or a food mixer. A person contacting the appliance housing and simultaneously a ground, such as a water pipe, would receive a nasty 120-V shock. If the contact were made with wet hands, the shock could be fatal. Unfortunately, however, until such an incident occurred, the internal fault would remain an unnoticed but incipient source of danger.

To eliminate this hazard, appliance manufacturers have always recommended that appliance housings be grounded to a cold-water pipe. In addition, the appliances are supplied with 3-wire plugs: two wires connected to the appliance and the third wire to the housing. To accommodate such plugs and to provide a ground path, the NEC requires all receptacles to be of the grounding type and all wiring systems to provide a ground path, separate and distinct

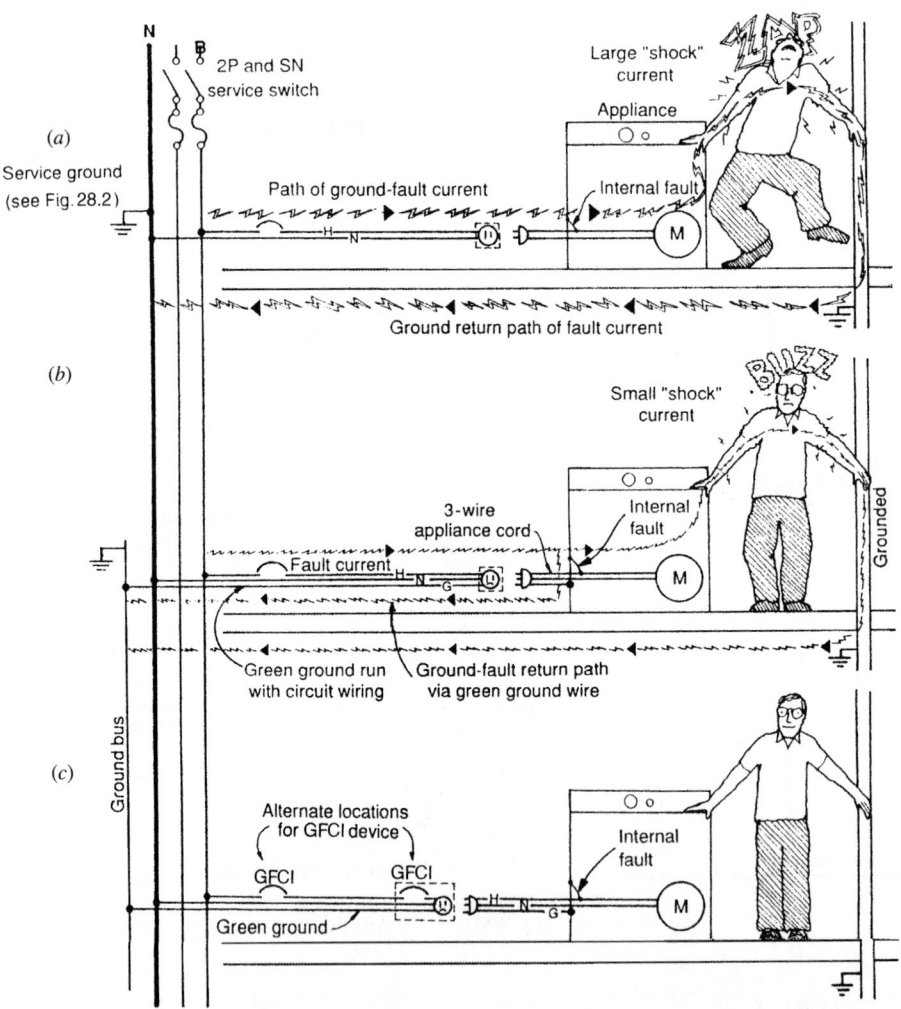

**Fig. 28.10** Three types of circuit arrangements. (a) A 2-wire, grounded-neutral circuit with no means of preventing shocks from ground faults. (b) A similar arrangement that includes a green ground wire, which considerably lessens the danger of ground-fault shocks. Note, however, that as long as the ground-fault current is below the rating of the branch circuit protective device (i.e., below 15 to 20 A), it will continue to flow within the appliance, causing overheating, arcing, and eventual destruction of the appliance. (c) The use of GFCIs, are mandatory for bathrooms, outdoor locations, swimming pool circuits, and other areas where ground faults are common and particularly hazardous.

from the neutral conductor (see Fig. 26.42). The result of such wiring is shown in Fig. 28.10b, where the ground current passes harmlessly through the internal fault, along the ground-wire path, and back to the panel. A person contacting the appliance housing establishes a parallel ground path. However, because this path is usually of much higher resistance than the ground-conductor path, only a very small current will flow. Wet hands materially reduce contact resistance, and shock current can increase to a dangerous level. If the ground current is sufficiently high, the branch circuit breaker or fuse opens, disconnecting the circuit.

When wiring systems are installed in metallic conduit, the conduit itself or the conduit plus a separate conductor within the conduit may be used as the grounding path. This latter method with the additional ground wire is far preferable, as explained in Fig. 28.10. When nonmetallic or flexible metallic wiring is used (Romex or BX), a separate grounding conductor run with the regular circuit conductors *must* be used.

All insulated ground conductors must have their covering colored green for identification as a grounding conductor. Many industrial installations install complete "green-ground" systems in an attempt to eliminate the shock hazard and reduce insulation failures. This has not been entirely successful for the reason alluded to previously; that is, in order to clear the ground fault, its current must be high enough to trip the branch circuit protective device. Otherwise, the ground fault continues to "leak," unnoticed by the system's protective devices. Unfortunately, ground faults are by nature low-current, leak-type faults, because they result from weak spots in insulation, dirt accumulation, and so on. Therefore, although the shock hazard is greatly reduced by the green-ground path, the fault continues to leak and arc until it becomes hot enough to cause a major breakdown, frequently accompanied by fire.

To eliminate this potentially dangerous situation, which occurs anytime there is a leak of current to ground in an electric circuit, the ground-fault circuit interrupter (GFCI) was developed (Fig. 28.10c). This device compares, with extreme precision, the current flowing in the hot and neutral legs of a circuit; if there is a difference, it indicates a ground fault and the device trips out. The rapidity of this action—approximately one-half

second—eliminates the possibility of a potentially dangerous shock hazard, which exists even in a properly grounded circuit, as in Fig. 28.10b, and all the more so for the circuit arrangement in Fig. 28.10a. The separate ground path required by the NEC (bonded metallic conduit system, green-ground, or both) serves to minimize the shock current taken by a person before the GFCI operates. The device can be applied at the panel to replace a normal circuit breaker (see Fig. 26.8b) or at an individual outlet to replace a normal receptacle device. It is advisable to use GFCI devices on all appliance circuits (see the NEC for locations where GFCI use is mandatory, such as outdoors and in kitchens and bathrooms). Application to lighting circuits is generally not required because fixtures are usually out of reach and are switch controlled. In mixed lighting and receptacle circuits, the GFCI is best applied at the outlet that is to be protected.

## 28.5 ENERGY CONSERVATION CONSIDERATIONS

Before proceeding with a detailed description of design procedure, many of the energy conservation ideas and techniques applicable to electrical distribution systems should be surveyed. This is done for ease of reference and cross-reference because the individual items appear throughout the lengthy design procedure. The following design procedures are intended to assist with meeting a design intent to conserve energy.

1. After establishing an energy budget based on projected loads and normal operation, set an energy reduction figure of 10% to 20% and meet this goal by using the techniques that follow. Annual energy consumption estimates are best made with the aid of one of the many computer programs available for this purpose.

2. Learn to recognize the energy-use characteristics of all equipment and systems specified (see Table 27.6). In general, select high-efficiency equipment (motors, transformers, etc.). If these are not identified as such, use materials and equipment with the lowest temperature rise because these have the lowest losses. This generally indicates the material of choice. When comparisons are close, a

detailed analysis may be necessary. Economic justification can be established with a life-cycle cost analysis. To avoid making a detailed cost analysis on every item, utilize one of the many available shortcut calculations for pay-back time (see Appendix I).

3. Provide electric load control equipment (demand control) either as part of an overall building control system or separately (see Sections 25.13 and 16.2).

4. In any multitenant residential building, provide individual user metering. (Metering of heating and cooling energy should be part of the HVAC contract.) All tenants should be made financially responsible for the energy they use. Exceptions to this rule would be made in the case of hotels, dormitories, and other transient facilities.

5. Where a choice of service voltages is available, consider the highest available. Similarly, consider the highest voltage in each class for interior distribution systems. This means 480 V in the 600-V class, 4 kV in the 5-kV class, and 13 kV in the 15-kV class. The result will be low line losses, small panelboards at the branch circuit level, and generally a lower electrical contract cost (see Section 28.3).

6. In any building, the maximum total voltage drop *shall not* exceed 3% in branch circuits or feeders, for a total of 5% to the farthest outlet, based on steady-state design load conditions.

7. Avoid the use of electric heating elements if alternatives are available. Electric heat is an inefficient use of natural resources because of the low overall efficiency of fuel-to-electricity conversion.

8. Provide metering points (for fixed or plug-in meters) throughout the system to permit accurate analysis of power and energy use. Meters, both instantaneous reading and recording types, provide essential data on equipment loading, load patterns, load coincidence, power factor, load voltage, power demand, and energy consumption. Analysis of these data indicates how to program and shift loads for maximum operational efficiency. A flexible design that permits this load shifting is assumed (see item 10).

9. Include provisions for power-factor correction in the system, both at devices and at the feeder level. Then, if metering (see item 8) indicates the necessity, add capacitors as required. High power factor reduces line losses, permits maximum utilization of equipment capacity, and avoids utility penalty charges. Utilization equipment rated greater than 1000 W and lighting equipment greater than 15 W, with an inductive reactance load component, should have a power factor of not less than 85% under rated load conditions. Utilization equipment with a power factor of less than 85% should be corrected to at least 90% under rated load conditions.

10. Size equipment as close as possible to the load. This normally results in maximum efficiency and a high power factor. Where the load varies considerably—for example, between day and night or on weekends—consider splitting the loads so that part of the equipment can be switched off when the load is low and the remaining load can be fed from a "night and weekend" feeder. Such a design permits one to shut down whole systems rather than operate at a very low load with concomitant high losses and low power factor. The design must be sufficiently flexible to permit shifting of loads between feeders if measurements on the operating facility indicate that this is desirable (see item 8). The aim is to operate equipment as close to the rated load as possible and to deenergize lightly loaded sections by shifting load to other, partly loaded equipment.

11. Use the most efficient type of control. This means solid-state control for motors, variable voltage variable frequency (VVVF) control for variable-speed equipment, remote switch control for blocks of lighting, electronic control systems for elevators, and so on.

12. Arrange automatic time controls for 24-hour loads, such as vent fans, water coolers, vending machines, and calculators.

13. Seal all electric riser shafts to avoid heat loss by stack action.

14. Generally select the coolest possible locations for electric equipment. Low ambient temperature (below 40°C) permits use of smaller equipment for the same load with concomitant lower cost and losses. If below-grade space is available, it is well suited for this purpose.

15. Provision for future expansion should be made by means of a design that accommodates additional equipment in lieu of oversiz-

ing equipment initially. Here again, the higher cost of two pieces of equipment can normally be justified by a detailed owning-and-operating cost analysis using realistic cost escalation and capital cost figures.

16. Energy-conservation techniques in lighting and lighting control are found in Sections 15.7 and 16.2 to 16.4.

17. Energy conservation techniques in vertical transportation are found in Sections 31.39, 31.40, 31.41, and 33.12.

## 28.6 DESIGN PROCEDURE

The steps involved in the electrical wiring design of any facility are outlined later in this section. These may in some instances be performed in a different order, or two or more steps may be combined, but the procedure normally used is the following:

1. Make an electrical load estimate based on areas involved, building data, and any other pertinent data (see Section 28.2).

2. In cooperation with the local electric utility, decide on the point of service entrance, type of service run, service voltage, metering location, and building utilization voltage (see Sections 26.1 to 26.11 and Section 28.5, item 5).

3. Determine with the client the proposed usage of all areas and information about all client-furnished equipment (including specific electric ratings and service connection requirements).

4. Determine from other consultants, such as for HVAC, plumbing, elevators, kitchens, and the like, the exact electrical rating of all the equipment in their realm of design. This determination is often made after conferences during which the electrical consultant makes valuable recommendations to these other specialists about the comparative characteristics and costs of equipment (see Section 28.5, items 2, 7, 9, and 10).

5. Determine the location and estimate the size of all required electric equipment spaces including switchboard rooms, emergency equipment spaces, electric closets, and so forth. Panelboards are normally located in closets but may be located in corridor walls or elsewhere. These decisions are necessary at this point to enable the architect to reserve the spaces for electrical equipment. Once the

design is accomplished in detail, the estimated space requirements can be checked and necessary adjustments made.

6. Design the lighting for the facility. This step is complex and involves continued interaction between the architect and the lighting designer. Coordination between daylighting and electric lighting elements is especially critical.

7. Depending on the type of facility, it may be necessary to separate the lighting layout from the layout plan(s) for other devices such as signaling, low-tension systems, and receptacles. When an underfloor wiring system is used, it is customary to show it on a separate plan to avoid clutter and confusion. This decision is made at the preliminary stage by the project engineer. Once the decision has been made as to how this is to be handled, the lighting fixture layout can be made.

8. On the plan(s), locate all electrical apparatus including receptacles, switches, motors, and other power-consuming apparatus. Underfloor, under-carpet, and over-ceiling wiring and raceway systems would be shown at this stage, in general on a separate plan.

9. On the plan(s), locate data-processing and signal apparatus such as telecommunication outlets, network connections, phone outlets, speakers, microphones, TV outlets, fire and smoke detectors, and so on. At this stage, provision is also made for load control wiring, building automatic control wiring, computer control, and the like. At this stage, the decision is also made as to whether intelligent panels will be used (see Section 26.23). Because some of these systems may be covered by separate contracts, the division of responsibility must be clearly defined. Often, only an empty conduit system and power outlets or sources are required in the electrical work contract.

    The material that follows deals with wiring design only; further discussion of signal equipment is reserved for Chapter 30.

10. Circuit all lighting, devices, and power equipment to the appropriate panels and prepare the panel schedules. Include in this step the circuitry for emergency equipment.

11. Compute panel loads.

12. Prepare the riser diagram. This includes the design of distribution panels, switchboards, and service equipment.

**ELECTRICITY**

13. Compute feeder sizes and all protective equipment ratings.
14. Check the preceding work.
15. Coordinate the electrical work with the other trades and with the architectural plans. This is not really a separate step, but rather a continuing process starting at step 9 and covering all subsequent stages of the work.

The material for items 1, 2, 3, 4, 6, 7, and 9 is covered elsewhere in this book. The remaining steps, that is, 5, 8, 10, 12, 13, and 15, are discussed in order below.

## 28.7 ELECTRICAL EQUIPMENT SPACES

The spaces required for electrical equipment in a facility vary greatly, depending on the design and the nature of the building. The working spaces required around major pieces of electrical switchgear and transformers were discussed previously (see Fig. 26.28). The *NEC* (in Article 110) further specifies the minimum working spaces required in front of electrical equipment.

### (a) Residences

In private residences, the service equipment and the building panelboard are generally incorporated into a single unit. The main disconnect(s) is usually installed as the main switch/breaker of the panel. A number of typical residential service-panel arrangements are shown in Fig. 28.11. The panel is normally placed in the garage, utility room, or basement. To minimize voltage drop, the panel should be placed as close to the major electrical loads as practical, without sacrificing valuable space or making the panel inaccessible. Frequently, a smaller panel can be subfed from the main panel to feed the kitchen and laundry loads. In apartments, panels are normally placed in the kitchen or the corridor immediately adjoining the kitchen. This location is chosen so that the panel circuit breaker can act as the required disconnecting means for most fixed appliances (see *NEC* Article 422C).

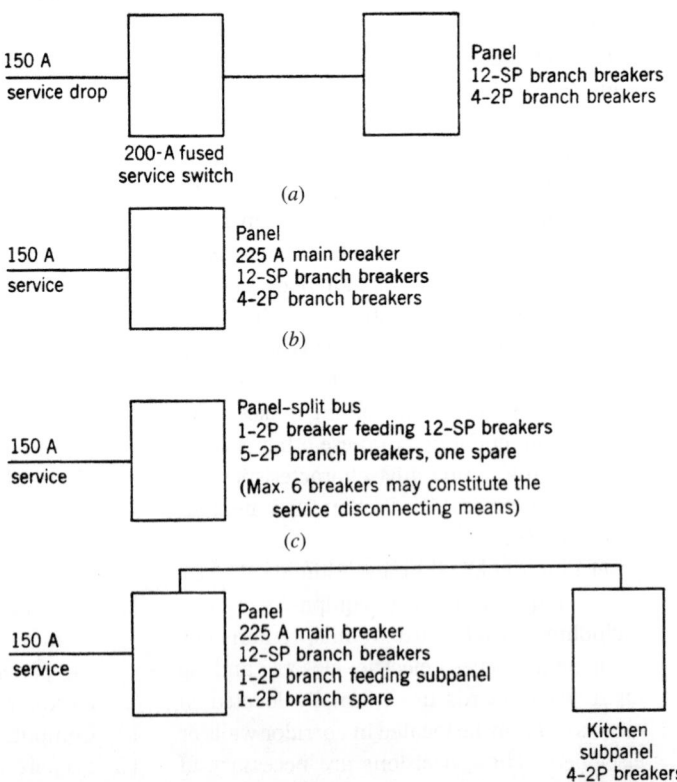

**Fig. 28.11** *Typical 150-A service arrangements, applicable to residences and small commercial buildings. The service switch can be a separate unit (a) or combined with the branch circuit panelboard (b–d). The panel may be a single unit (a,b), two units in a single enclosure, fed by separate service switches (c), or a central panel and a subfed single-use panel (d)—in this case, a kitchen subpanel.*

150 A
service drop

200-A fused
service switch

Panel
12-SP branch breakers
4-2P branch breakers

*(a)*

150 A
service

Panel
225 A main breaker
12-SP branch breakers
4-2P branch breakers

*(b)*

150 A
service

Panel–split bus
1-2P breaker feeding 12-SP breakers
5-2P branch breakers, one spare

(Max. 6 breakers may constitute the
service disconnecting means)

*(c)*

150 A
service

Panel
225 A main breaker
12-SP branch breakers
1-2P branch feeding subpanel
1-2P branch spare

Kitchen
subpanel
4-2P breakers
2-SP breakers

*(d)*

## (b) Commercial Spaces

The location of the required panelboards depends on their type and number and on availability of space. In the research building of which Fig. 28.12 is a partial plan, lighting panels are recessed into the corridor wall because the building is only two stories high and the panels can be vertically stacked and fed by a single conduit. If this building were six or more stories high, an electric closet (see Section 28.8) would be advisable to accommodate the panel and riser conduits. Of course, when panels are installed in finished areas such as corridors, flush mounting is required.

To limit the voltage drop on a branch circuit in accordance with code requirements, panelboards should be located so that no circuit exceeds 100 ft in length. If 15-A or 20-A branch circuits longer than this are unavoidable, No. 10 AWG wire should be used for runs of 100 to 150 ft and No. 8 AWG for longer circuits. These circuits are normally wired with No. 12 AWG wire.

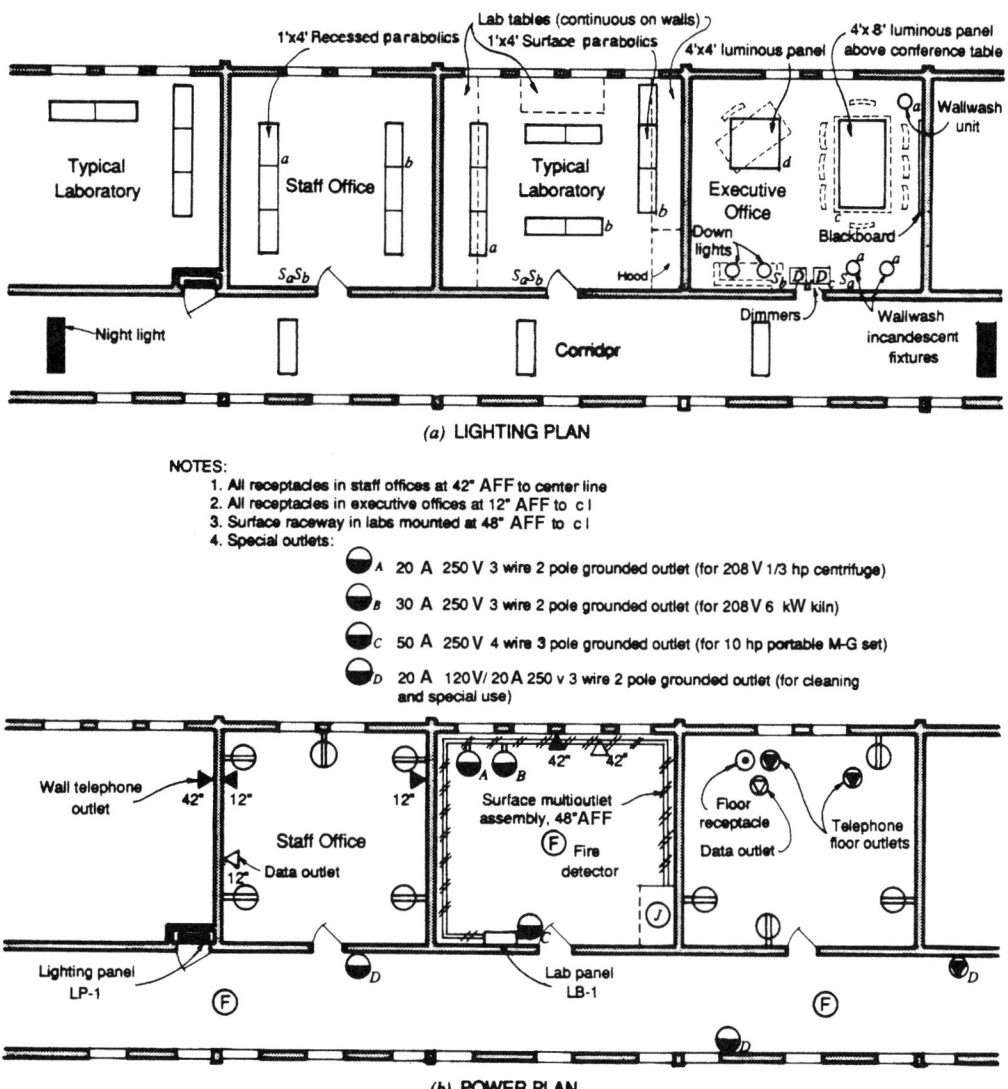

*(a)* LIGHTING PLAN

NOTES:
1. All receptacles in staff offices at 42" AFF to center line
2. All receptacles in executive offices at 12" AFF to c l
3. Surface raceway in labs mounted at 48" AFF to c l
4. Special outlets:

A  20 A  250 V 3 wire 2 pole grounded outlet (for 208 V 1/3 hp centrifuge)

B  30 A  250 V 3 wire 2 pole grounded outlet (for 208 V 6 kW kiln)

C  50 A  250 V 4 wire 3 pole grounded outlet (for 10 hp portable M-G set)

D  20 A  120V/ 20 A 250 v 3 wire 2 pole grounded outlet (for cleaning and special use)

*(b)* POWER PLAN

**Fig. 28.12** *Typical floor plans for lighting and power for a section of an office–laboratory building. Separate lighting (a) and power (b) plans are drawn for the sake of clarity. For circuited plans, see Figs. 28.21 and 28.22. Data outlets handle data-processing and telecommunications cabling.*

The laboratory between the two offices of Fig. 28.12 is intended to function as a self-contained unit and is therefore equipped with its own panel. Multioutlet assemblies, all wiring within the room, and the panel itself are surface-mounted to allow ready access to all components for the frequent rewiring encountered in laboratories. A main circuit breaker should be provided in such a panel to act as a main disconnect, whether required by code or not. Where panels are convenient to the load controlled, the panel circuit breakers may be used for switching.

Panels supplying large blocks of load simultaneously switched, such as auditorium house lights, lobby lights, large single-use office areas, store lighting, and the like, can be constructed with built-in electrically operated, mechanically held contactors to switch the entire load, with control at any desired remote location. These remote-control (RC) switches are discussed in Section 26.15. If only part of the panel's circuits is so arranged (i.e., for remotely controlled block switching), a split bus panel can be used, with the RC switch controlling only that part of the panel's load (Fig. 28.13).

Small offices, stores, and other small buildings frequently have lighting panels mounted in a convenient finished area and utilize the panels' circuit breakers for load switching. In large buildings, strategically located electric closets are provided to house all electrical supply equipment. Power panels and distribution panels are located as required by the loads fed through them.

In general, branch circuit panels, distribution panels, and switchboards are best located near the electrical load center. This minimizes feeder length and reduces voltage drop, making it the most economical arrangement.

Every completely enclosed switchgear room, emergency generator room, or transformer vault should be equipped with an emergency light source (see Section 16.36). In generator rooms, these should be battery-operated to give illumination for generator repairs in the event of generator failure during a power outage.

## 28.8 ELECTRICAL CLOSETS

In the design of the building electric system, particularly in multistory construction, it is often advantageous and convenient to group the electrical

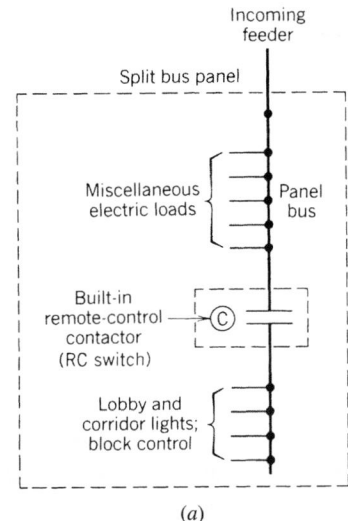

(a)

(b)

**Fig. 28.13** (a) Single-line diagram of a split bus panel. This design is used when a block of load is to be controlled as a unit. The control point is either remote, at the panel, or both. The contactor (RC switch) is electrically operated and mechanically latched. (b) RC switches (mechanically held contactors) are shown to the left of a bank of panels. The contactors were installed to provide photocell and timer control of department store lighting. (Photo courtesy of Automatic Switch Co.)

equipment in a small room called an *electrical closet* (Fig. 28.14). The shape of this space can be varied to fit the architectural requirements, and it should provide the following:

1. One or more locking doors.
2. Vertical stacking, above and below other electric closets, and located so as not to block conduits entering or leaving horizontally. Thus, locations on outside walls and adjoining shafts, columns, and stairs are poor.
3. Space free of other utilities such as piping or

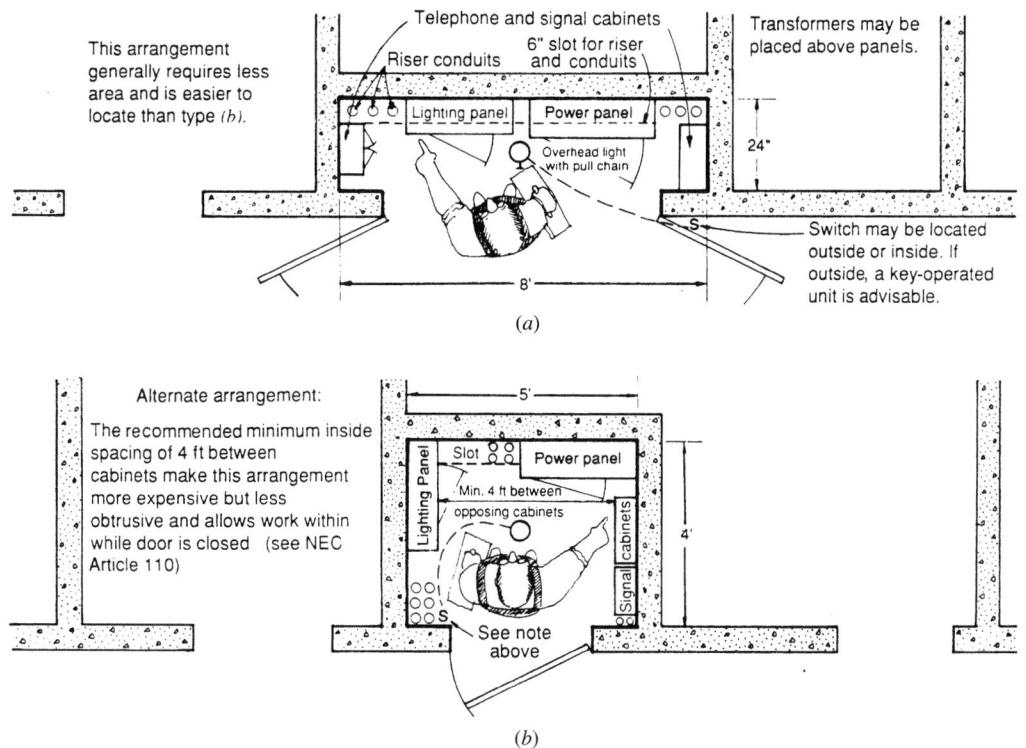

Telephone and signal cabinets
Riser conduits
6" slot for riser and conduits
Transformers may be placed above panels.

This arrangement generally requires less area and is easier to locate than type (b).

Lighting panel | Power panel

Overhead light with pull chain

24"

Switch may be located outside or inside. If outside, a key-operated unit is advisable.

8'

(a)

Alternate arrangement:

The recommended minimum inside spacing of 4 ft between cabinets make this arrangement more expensive but less obtrusive and allows work within while door is closed (see NEC Article 110)

5'

Slot | Power panel

Lighting Panel

Min. 4 ft between opposing cabinets

Signal cabinets

4'

See note above

(b)

**Fig. 28.14** Typical electric closets with some of the usual equipment. If warranted due to the amount of equipment, separate closets may be used for signal and telephone conduits and cabinets. Smaller spaces may be used, depending on the types of equipment cabinets and their arrangement (see NEC Article 110). In all cases, sufficient space must be provided to allow full opening of all cabinet doors. Empty wall space for future cabinets and equipment should also be considered.

ducts passing through the closet, either horizontally or vertically.

4. Sufficient wall space to mount all requisite and future panels, switches, transformers, telephone cabinets, and communication equipment. Wall cabinet space must be coordinated with raceway connections to underfloor ducts and over-the-ceiling raceways systems (see Sections 27.25 to 27.31).

5. Floor slots or sleeves of sufficient size for all present and future conduit or bus risers.

6. Sufficient floor space so that an electrician can work comfortably and safely on initial installation and repair.

7. Adequate illumination and ventilation.

## 28.9 EQUIPMENT LAYOUT

Wiring devices, principally comprising receptacles and switches, are located as required by the equipment to be served and by the anticipated area use.

Switches for control of lighting or receptacles are normally placed on the strike side of the door. Other devices such as plug-in strips on walls and special-purpose receptacles are shown and identified. Signal outlet locations are often noted but generally remain uncircuited on floor plans, a riser or floor raceway plan being utilized to show interconnections. These include fire-alarm equipment, telephone and intercom equipment, data and communications, radio and TV outlets, and so on. These devices may be identified by a special symbol or note where a standard symbol is not available.

As mentioned previously, lighting fixture outlets are normally placed on the same drawing as wiring devices unless the large number of the latter precludes showing the lighting without undue cluttering of the drawings. In such an event, the lighting is shown on one drawing and receptacles on another, with signals shown on the one less occupied. A ceiling or underfloor wiring system would necessitate such separation (see Figs. 27.41, 27.42, and 27.43). Motors, heaters, and other fixed and permanently wired equipment are shown and identified on the receptacle drawings (also called *power drawings,* in contrast to lighting drawings). Equipment furnished with a cord and plug is not normally shown. However, the receptacle intended for supplying such a device is shown and identified. A typical device layout is shown in Fig. 28.12. An abbreviated symbol list is given in Fig. 28.15. For a complete electrical symbol list, see Stein (1997).

## 28.10 APPLICATION OF OVERCURRENT EQUIPMENT

Before beginning an explanation of circuiting, it is necessary to explain the principles underlying overcurrent protection. As outlined in Chapter 26, the function of an overcurrent device is to open (interrupt) a circuit when the current rating of the equipment being protected is exceeded. All equipment must be protected in accordance with its current carrying capacity. Where ratings do not correspond exactly, the next larger standard size for protective devices up to 800 A may be used; above this use the next lower rating (see *NEC* Article 240-3 for further restrictions). The following general rules govern the application of overcurrent protection:

1. Overcurrent devices must be placed on the line or supply side of the equipment being protected (Fig. 28.16).
2. Overcurrent devices must be placed in all *ungrounded* conductors of the protected circuit (see Fig. 26.33).
3. In general, conductor sizes shall not be reduced in a circuit or tap unless the smallest-size wire is protected by the circuit overcurrent devices (Fig. 28.17).
4. Overcurrent devices shall be located so as to be readily accessible.

## 28.11 BRANCH CIRCUIT DESIGN

A branch circuit, by *NEC* definition, refers only to the circuit conductors, although for our purposes and in trade parlance it includes the protective device and the outlets served. Such circuits may be the multioutlet general-purpose type (Fig. 28.18*a*), the multioutlet appliance type (Fig. 28.18*b*), or the

single-outlet type intended for a specific piece of equipment (Fig. 28.18*c*). The multioutlet types are limited to 50 A in capacity, whereas the single-outlet type is governed in size only by the requirements of the item being served and may be 200 A or 300 A in size.

In its simplest form, a branch circuit comprises only two circuit wires. However, multiwire branch circuits carrying 2- to 3-phase wires plus a neutral are also widely used. Generally, each branch circuit should be sized for the load connected to it plus the load expansion that is expected. These general rules of good practice should be followed:

1. In all but the smallest installations, connect lighting, convenience receptacles, and appliances on separate (groups of) circuits, although this is not an *NEC* requirement.
2. General-purpose branch circuits should be 20 A and wired with No. 12 AWG wire. Switch legs may be No. 14 AWG if the lighting load permits and the wiring meets the tap requirements of *NEC* Article 240-21.
3. Limit the circuit load on 15-A and 20-A circuits to the values shown in Table 28.5. This provides the required building load expansion capability in the branch circuitry; that is, by bringing the loads on the branch circuits up to maximum, the additional building loads can be absorbed. However, because it is not always economical or feasible to expand existing circuits, panels are always equipped with spare breakers. These can also be utilized to pick up building load expansion, or the user may use a combination of the two techniques, expanding some existing circuits and adding new ones. (See Section 28.15 for a discussion of this subject.)

Because lighting and specific devices are circuited according to their nameplate rating, the only circuitry item left to the judgment of the designer is the number of convenience receptacles per circuit. The *NEC* specifies that plug outlets (convenience receptacles) be counted, in totaling loads, at 1.5 A each unless included in the load for general lighting. This point requires some clarification. The receptacle load for general illumination in dwellings is included in the 3 V-A/ft² specified by *NEC* Table 220-3(a) and found in Table 28.1. This is because receptacle outlets on general illumination circuits

ELECTRICAL SYMBOL LIST

## RACEWAYS

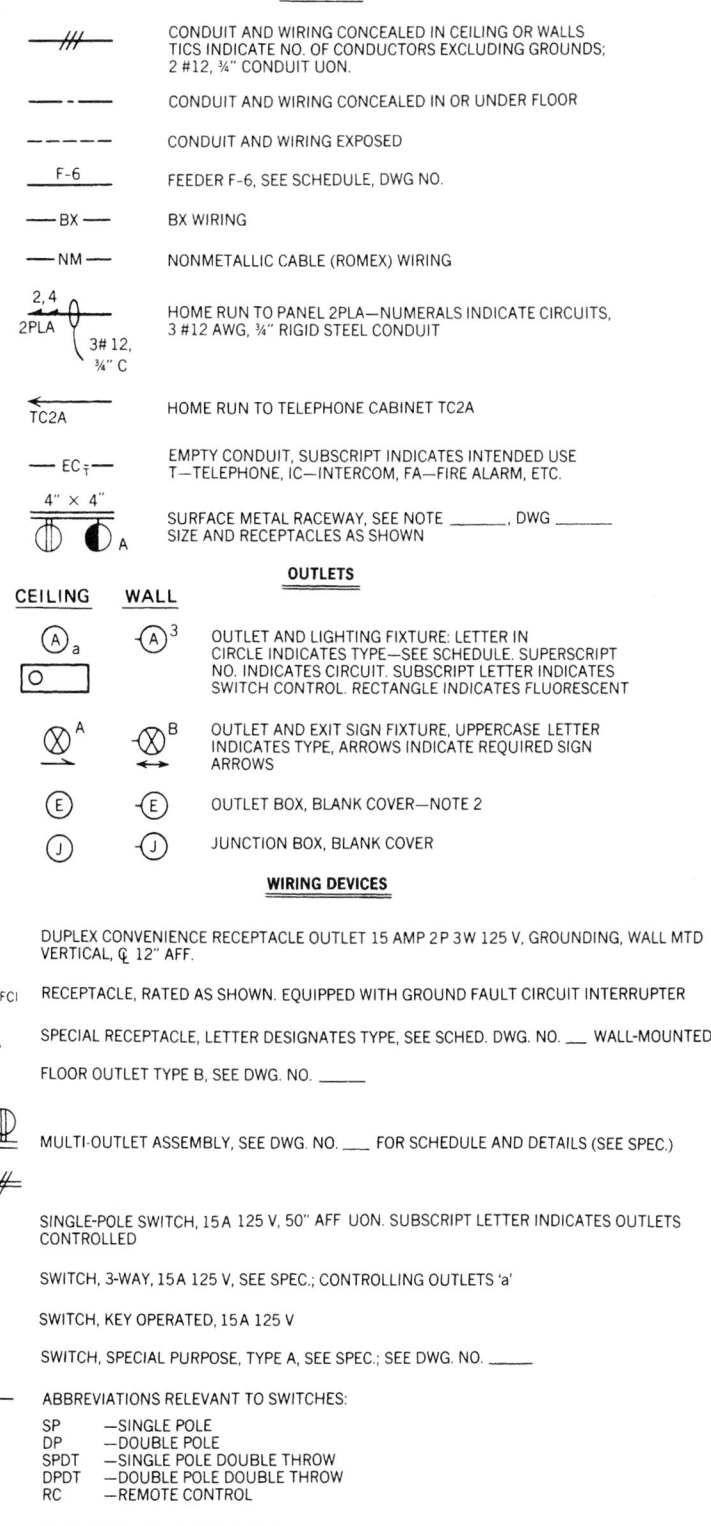

CONDUIT AND WIRING CONCEALED IN CEILING OR WALLS
TICS INDICATE NO. OF CONDUCTORS EXCLUDING GROUNDS;
2 #12, ¾" CONDUIT UON.

CONDUIT AND WIRING CONCEALED IN OR UNDER FLOOR

CONDUIT AND WIRING EXPOSED

FEEDER F-6, SEE SCHEDULE, DWG NO.

BX WIRING

NONMETALLIC CABLE (ROMEX) WIRING

HOME RUN TO PANEL 2PLA—NUMERALS INDICATE CIRCUITS,
3 #12 AWG, ¾" RIGID STEEL CONDUIT

HOME RUN TO TELEPHONE CABINET TC2A

EMPTY CONDUIT, SUBSCRIPT INDICATES INTENDED USE
T—TELEPHONE, IC—INTERCOM, FA—FIRE ALARM, ETC.

SURFACE METAL RACEWAY, SEE NOTE _____, DWG _____
SIZE AND RECEPTACLES AS SHOWN

## OUTLETS

CEILING    WALL

OUTLET AND LIGHTING FIXTURE: LETTER IN
CIRCLE INDICATES TYPE—SEE SCHEDULE. SUPERSCRIPT
NO. INDICATES CIRCUIT. SUBSCRIPT LETTER INDICATES
SWITCH CONTROL. RECTANGLE INDICATES FLUORESCENT

OUTLET AND EXIT SIGN FIXTURE, UPPERCASE LETTER
INDICATES TYPE, ARROWS INDICATE REQUIRED SIGN
ARROWS

OUTLET BOX, BLANK COVER—NOTE 2

JUNCTION BOX, BLANK COVER

## WIRING DEVICES

DUPLEX CONVENIENCE RECEPTACLE OUTLET 15 AMP 2 P 3 W 125 V, GROUNDING, WALL MTD
VERTICAL, ℄ 12" AFF.

RECEPTACLE, RATED AS SHOWN. EQUIPPED WITH GROUND FAULT CIRCUIT INTERRUPTER

SPECIAL RECEPTACLE, LETTER DESIGNATES TYPE, SEE SCHED. DWG. NO. __ WALL-MOUNTED

FLOOR OUTLET TYPE B, SEE DWG. NO. _____

MULTI-OUTLET ASSEMBLY, SEE DWG. NO. ___ FOR SCHEDULE AND DETAILS (SEE SPEC.)

SINGLE-POLE SWITCH, 15 A 125 V, 50" AFF UON. SUBSCRIPT LETTER INDICATES OUTLETS
CONTROLLED

SWITCH, 3-WAY, 15 A 125 V, SEE SPEC.; CONTROLLING OUTLETS 'a'

SWITCH, KEY OPERATED, 15 A 125 V

SWITCH, SPECIAL PURPOSE, TYPE A, SEE SPEC.; SEE DWG. NO. _____

ABBREVIATIONS RELEVANT TO SWITCHES:

SP   —SINGLE POLE
DP   —DOUBLE POLE
SPDT  —SINGLE POLE DOUBLE THROW
DPDT  —DOUBLE POLE DOUBLE THROW
RC   —REMOTE CONTROL

OUTLET-BOX-MOUNTED DIMMER

**Fig. 28.15** *Abbreviated electrical symbol list. For a complete list, see Stein (1997).*

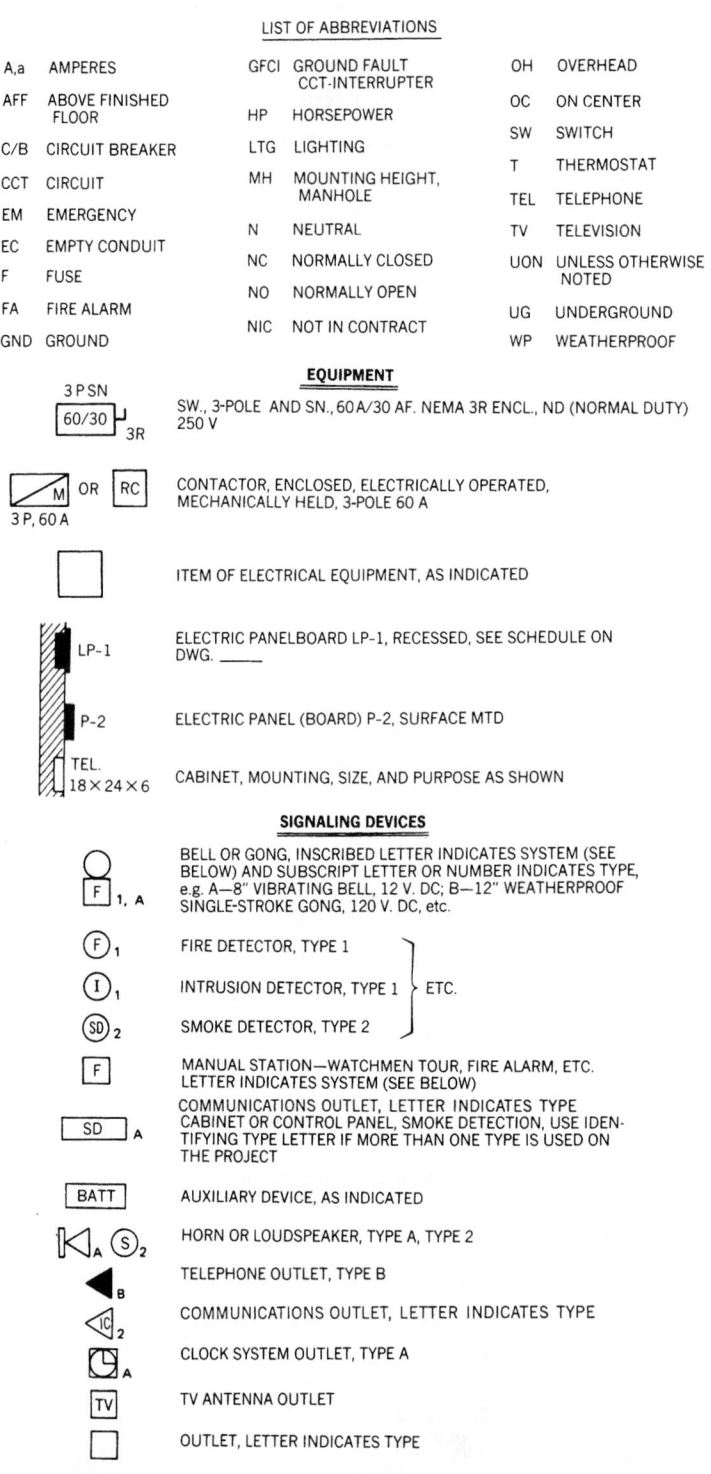

## LIST OF ABBREVIATIONS

| | | | | | |
|---|---|---|---|---|---|
| A,a | AMPERES | GFCI | GROUND FAULT CCT-INTERRUPTER | OH | OVERHEAD |
| AFF | ABOVE FINISHED FLOOR | HP | HORSEPOWER | OC | ON CENTER |
| C/B | CIRCUIT BREAKER | LTG | LIGHTING | SW | SWITCH |
| CCT | CIRCUIT | MH | MOUNTING HEIGHT, MANHOLE | T | THERMOSTAT |
| EM | EMERGENCY | | | TEL | TELEPHONE |
| EC | EMPTY CONDUIT | N | NEUTRAL | TV | TELEVISION |
| F | FUSE | NC | NORMALLY CLOSED | UON | UNLESS OTHERWISE NOTED |
| FA | FIRE ALARM | NO | NORMALLY OPEN | UG | UNDERGROUND |
| GND | GROUND | NIC | NOT IN CONTRACT | WP | WEATHERPROOF |

### EQUIPMENT

**3 P SN** / **60/30** / **3R** — SW., 3-POLE AND SN., 60 A/30 AF. NEMA 3R ENCL., ND (NORMAL DUTY) 250 V

**3 P, 60 A** [M] OR [RC] — CONTACTOR, ENCLOSED, ELECTRICALLY OPERATED, MECHANICALLY HELD, 3-POLE 60 A

□ — ITEM OF ELECTRICAL EQUIPMENT, AS INDICATED

LP-1 — ELECTRIC PANELBOARD LP-1, RECESSED, SEE SCHEDULE ON DWG. _____

P-2 — ELECTRIC PANEL (BOARD) P-2, SURFACE MTD

TEL. 18 × 24 × 6 — CABINET, MOUNTING, SIZE, AND PURPOSE AS SHOWN

### SIGNALING DEVICES

(F) 1, A — BELL OR GONG, INSCRIBED LETTER INDICATES SYSTEM (SEE BELOW) AND SUBSCRIPT LETTER OR NUMBER INDICATES TYPE, e.g. A—8" VIBRATING BELL, 12 V. DC; B—12" WEATHERPROOF SINGLE-STROKE GONG, 120 V. DC, etc.

(F) 1 — FIRE DETECTOR, TYPE 1

(I) 1 — INTRUSION DETECTOR, TYPE 1 ⎫ ETC.

(SD) 2 — SMOKE DETECTOR, TYPE 2 ⎭

[F] — MANUAL STATION—WATCHMEN TOUR, FIRE ALARM, ETC. LETTER INDICATES SYSTEM (SEE BELOW)

[SD] A — COMMUNICATIONS OUTLET, LETTER INDICATES TYPE CABINET OR CONTROL PANEL, SMOKE DETECTION, USE IDENTIFYING TYPE LETTER IF MORE THAN ONE TYPE IS USED ON THE PROJECT

[BATT] — AUXILIARY DEVICE, AS INDICATED

◁ A (S) 2 — HORN OR LOUDSPEAKER, TYPE A, TYPE 2

◀ B — TELEPHONE OUTLET, TYPE B

◁ IC 2 — COMMUNICATIONS OUTLET, LETTER INDICATES TYPE

◻ A — CLOCK SYSTEM OUTLET, TYPE A

[TV] — TV ANTENNA OUTLET

□ — OUTLET, LETTER INDICATES TYPE

**Fig. 28.15** (Continued)

ELECTRICITY

1284

## SYSTEM TYPES

| | |
|---|---|
| F, FA | FIRE ALARM |
| S, SD | SMOKE DETECTION |
| I, IA | INTRUSION ALARM |
| T, TEL | TELEPHONE |
| TV | TELEVISION |
| IC | INTERCOM |
| D | DATA, TELECOMMUNICATION |

## AUXILIARY DEVICES

| | |
|---|---|
| BATT | BATTERY |
| S, SP | SPEAKER, LOUDSPEAKER |
| TC | TELEPHONE CABINET |
| WF | WATERFLOW |

## MOTORS AND MOTOR CONTROL

COMBINATION TYPE MOTOR CONTROLLER; ATL STARTER PLUS FUSED DISCONNECT SWITCH, NEMA SIZE II, SEE SCHEDULE DWG.___ ___ ___

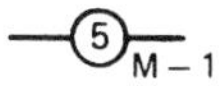

MOTOR #1, 5 HP, 3φ SQUIRREL CAGE UON

DEVICE 'T' (SEE LIST OF ABBREVIATIONS)

S_T    MANUAL MOTOR CONTROLLER WITH THERMAL ELEMENT.

ATL    ACROSS THE LINE STARTER—MAGNETIC

FS    FUSED SWITCH

CB    CIRCUIT BREAKER

FV    FULL VOLTAGE

MCC    MOTOR CONTROL CENTER

S    START BUTTON—MOMENTARY CONTACT

ST    STOP BUTTON—MOMENTARY CONTACT

## CONTROL DIAGRAMS AND WIRING DIAGRAMS

MOMENTARY CONTACT PUSH BUTTON—NO—(START)

MOMENTARY CONTACT PUSH BUTTON—NC—(STOP)

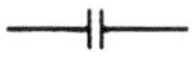

NORMALLY OPEN CONTACT—NO

NORMALLY CLOSED CONTACT—NC

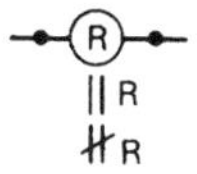

OPERATING COIL FOR RELAY OR OTHER MAGNETIC CONTROL DEVICE. WITH ONE NO AND ONE NC CONTACT. LETTERS NORMALLY USED ARE C, R FOR CONTROL COIL AND RELAY

POWER WIRING

CONTROL WIRING

## ONE LINE DIAGRAMS

CONDUCTOR, SIZE, AND TYPE INDICATED

CONDUCTORS CROSSING, CONNECTED

GROUND CONNECTION

GROUNDED WYE

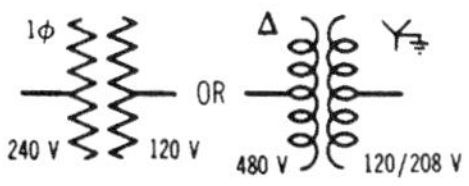

TRANSFORMER, 2 WINDING, 1φ INDICATES SINGLE PHASE; Δ-Y INDICATES 3-PHASE AND TYPE OF CONNECTION; NUMBERS INDICATE VOLTAGES.

FUSE, 30 A TYPE - - -

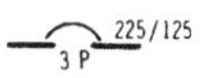

CIRCUIT BREAKER, MOLDED CASE, 3 POLE, 225 A FRAME/125 A TRIP

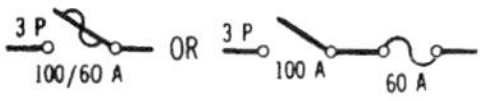

FUSED DISCONNECT SWITCH, 3 POLE, 100 A, 60 A FUSE

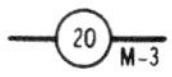

MOTOR, 3-PHASE SQUIRREL CAGE UON, 20 HP, MOTOR #3

**Fig. 28.15** (Continued)

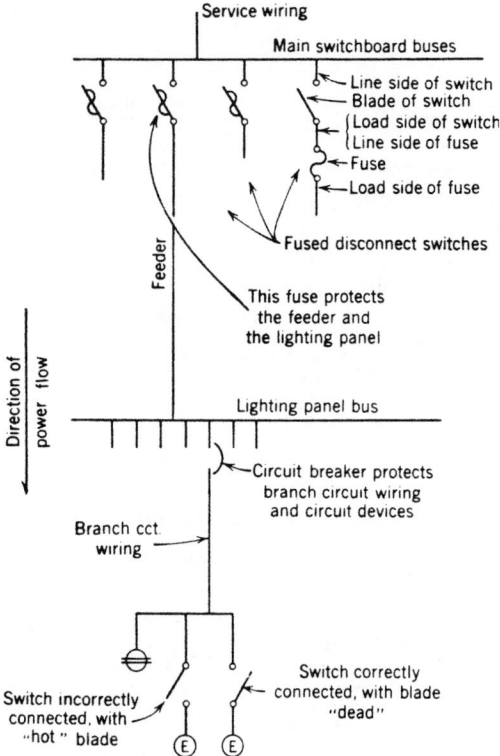

**Fig. 28.16** *Location of overcurrent protective equipment. Protective equipment should always be located at the point where the conductor receives its source of supply so that when it operates, the current supply is cut off.*

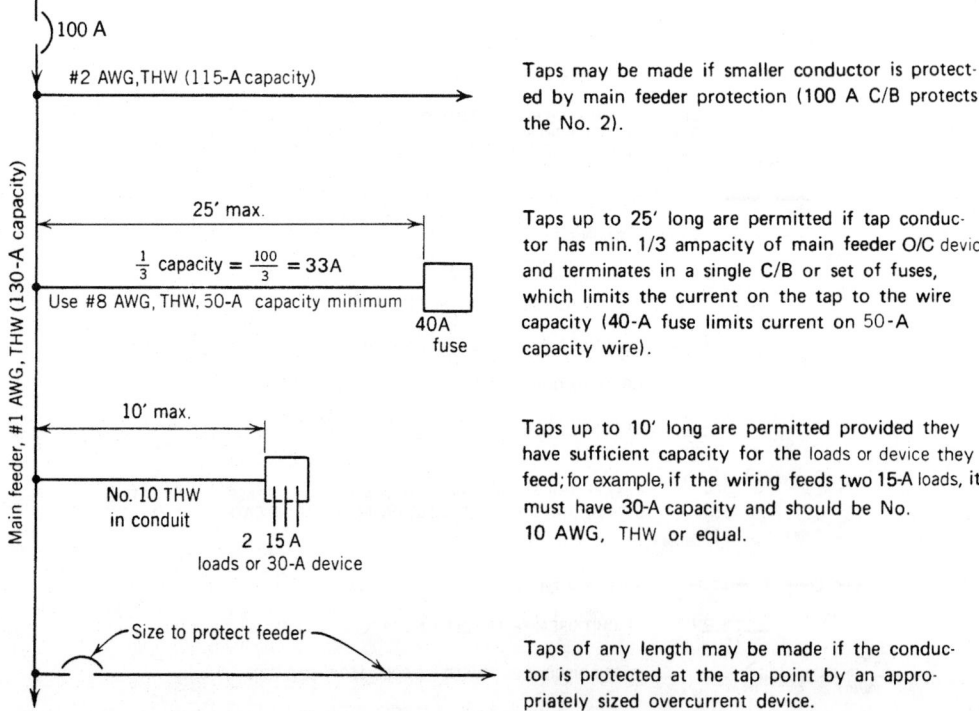

100 A

#2 AWG, THW (115-A capacity)

Taps may be made if smaller conductor is protected by main feeder protection (100 A C/B protects the No. 2).

25' max.

$\frac{1}{3}$ capacity = $\frac{100}{3}$ = 33A

Use #8 AWG, THW, 50-A capacity minimum

40A fuse

Taps up to 25' long are permitted if tap conductor has min. 1/3 ampacity of main feeder O/C device and terminates in a single C/B or set of fuses, which limits the current on the tap to the wire capacity (40-A fuse limits current on 50-A capacity wire).

10' max.

No. 10 THW in conduit

2  15 A loads or 30-A device

Taps up to 10' long are permitted provided they have sufficient capacity for the loads or device they feed; for example, if the wiring feeds two 15-A loads, it must have 30-A capacity and should be No. 10 AWG, THW or equal.

Main feeder, #1 AWG, THW (130-A capacity)

Size to protect feeder

Taps of any length may be made if the conductor is protected at the tap point by an appropriately sized overcurrent device.

**Fig. 28.17** *Some permissible tap arrangements. See NEC Article 240-21 for restrictions on taps in addition to those given in the figure.*

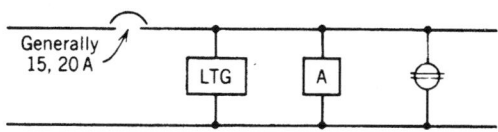

*(a)* General-purpose branch circuit. Supplies outlets for lighting and appliances, including convenience receptacles.

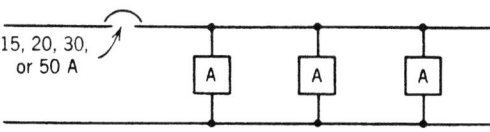

*(b)* Appliance branch circuit. Supplies outlets intended for feeding appliances. Fixed lighting not supplied.

*(c)* Individual branch circuit, designed to supply a single specific item.

**Fig. 28.18** Branch circuit types. See NEC Article 210 for detailed descriptions.

are assumed to supply illumination (i.e., lamps), and are therefore necessarily included in the lighting load. Application of the 1.5-A (180-V-A) per outlet criterion to such outlets would, in the opinion of the *NEC*, unnecessarily limit the number of outlets permitted on such circuits.

The same is true for appliance circuits. On both of these types, according to the *NEC*, the number of receptacle outlets is not limited. On 15- and 20-A receptacle circuits, which are neither general illumination nor small appliance branch circuits, the number of outlets is limited to 10 (1800 V-A) on a 15-A circuit and 13 (2340 V-A) on a 20-A circuit by applying the 1.5-A per outlet criterion.

It is our opinion that the distinction between circuit types is not completely clear, and that good practice dictates the limitation of convenience receptacles per circuit to a considerably smaller number. Thus, following the guidelines of Table 28.5 of 9- and 12-A loading on 15- and 20-A circuits, respectively, we would have by this method

$$\text{15-A circuit:} \quad \frac{9 \text{ A}}{1.5 \text{ A/outlet}} = 6 \text{ outlets per circuit}$$

$$\text{20-A circuit:} \quad \frac{12 \text{ A}}{1.5 \text{ A/outlet}} = 8 \text{ outlets per circuit}$$

These figures must be used judiciously. If the devices to be energized are small, these quantities may be used. Such would be the case in a business office, where only an electric pencil sharpener or a desk lamp (100 V-A) would be plugged in. However, for computer-equipped offices, service and repair facilities, and the like, plan for no more than four receptacles on a 20-A circuit—and preferably two or three receptacles should be used on a 15-A circuit. Of course, diversity of use is an all-important factor, and the more accurately it can be estimated, the better the design result.

A further note of caution is in order: Receptacles should be arranged, if at all possible, so that the loss of a single circuit does not deprive an entire area of power. That is, for the sake of reliability, circuitry should be alternated to give each space parts of different circuits.

*NEC* Table 210-24 specifies certain requirements for conductors, devices, and loads permissible on general-purpose branch circuits. These are excerpted and summarized in Table 28.6.

## 28.12 BRANCH CIRCUIT DESIGN GUIDELINES: RESIDENTIAL

1. The *NEC* (1999 edition) requires for residences sufficient circuitry to supply a load of 3 V-A/ft$^2$ in the building, excluding unfinished spaces such as porches, garages, and basements. Using Table 28.5, which already includes a 20% derating of circuits, to provide for continuous loading, this works out to 480 ft$^2$ on a 15-A circuit (1440 V-A) and 640 ft$^2$ on a 20-A circuit (1920 V-A). Allowing for some expansion, such as a finished basement or enclosing a porch to expand the living area, results in a good practice recommendation of 400–480 ft$^2$ per 15-A circuit and 530–640 ft$^2$ per 20 A circuit.

These good-practice figures normally provide enough circuits for all but the heaviest-loaded residences. However, if an actual design indicates the need for additional circuits, they obviously must be provided. The illustrative examples in *NEC* Chapter 20 are based on absolute minimum requirements.

ELECTRICITY

**TABLE 28.5  Recommended Branch Circuit Loads[a]**

| Size Circuit | Circuit[d] Amperes | Volt-Amperes[d] at 120 V | Volt-Amperes at 277 V |
|---|---|---|---|
| **0% Expansion[b,c]** | | | |
| 15 A | 12 | 1440 | — |
| 20 A | 16 | 1920 | 4440 |
| **25% Expansion[e]** | | | |
| 15 A | 9.6 | 1150 | — |
| 20 A | 12.8 | 1520 | 3600 |
| **50% Expansion[f]** | | | |
| 15 A | 8 | 960 | — |
| 20 A | 11 | 1300 | 3000 |

[a]The loading shown will provide the specified expansion in the branch circuits. Where branch circuit expansion is not practical, the required expansion can be obtained from panel spares.

[b]See Table 28.1 for anticipated load growth.

[c]For 0% expansion, initial load = 80% of the circuit rating, which is the maximum permissible for continuous loads. See Table 28.6, note 6.

[d]For branch circuits feeding utilization equipment such as air conditioners in addition to convenience outlets, see *NEC* Article 210-23.

[e]To accomplish 25% expansion, the circuit is derated to 80%, i.e., 80% of 12 A = 9.6 A, etc.

[f]To accomplish 50% expansion, the circuit is derated by ⅓, i.e., two-thirds of 12 A = 8 A, etc. (Then 50% expansion on 8 A yields 12 A.)

2. The *NEC* (Article 210-11) requires a minimum of two 20-A appliance branch circuits (Fig. 28.18*b*) to feed all the receptacle outlets in the kitchen, pantry, breakfast, and/or dining room and similar areas, and *only* these outlets. This is because any receptacle in these areas is a potential appliance outlet and must be fed and circuited as such. Permanently installed appliances such as a food disposal, dishwasher, fan hood, and the like may *not* be connected to these appliance circuits. An exception is made to permit clock outlets to be

**TABLE 28.6  Branch Circuit Requirements**

| | Branch Circuit Size | | | | |
|---|---|---|---|---|---|
| | **15 A** | **20 A** | **30 A** | **40 A** | **50 A** |
| Minimum size copper conductors | No. 14 | 12 | 10 | 8 | 6 |
| Minimum size taps | No. 14 | 14 | 14 | 12 | 12 |
| Overcurrent device rating | 15 A | 2s0 | 30 | 40 | 50 |
| Lampholders permitted | Any type | Any type | Heavy duty | Heavy duty | Heavy duty |
| Receptacle rating permitted (see note 7) | 15 A | 15 or 20 | 30 | 40 or 50 | 50 |
| Maximum load (see notes 6 & 8) | 15 | 20 | 30 | 40 | 50 |

*Notes:*

1. Wiring shall be types RHW, RHH, T, THW, TW, THWN, THHN, XHHW in raceway or cable.

2. On a 15-A circuit, the maximum single appliance shall draw 12 A. On a 20-A circuit, the maximum single appliance shall draw 16 A. If combined with lighting or portable appliances, any fixed appliance shall not draw more than 7.5 A on a 15-A circuit and 10 A on a 20-A circuit.

3. On a 30-A circuit, maximum single appliance draw shall be 24 A.

4. Heavy-duty lampholders are units rated not less than 750 W.

5. 30-, 40-, and 50-A circuits shall not be used for fixed lighting in residences.

6. When loads are connected for long periods (3 hours or more), the actual load shall not exceed 80% of the branch circuit rating. Conversely, continuous-type loads shall be figured at 125% of the actual load in all load calculations.

7. A single receptacle on an individual branch circuit shall have a rating not less than the circuit—for example, 15 A on a 15-A circuit. Also, 15-A receptacles on a 20-A circuit shall not supply a load greater than 12 A for appliances; 20-A receptacles on a 20-A circuit shall be limited to a 16-A load.

8. Rating of a single piece of cord-connected utilization equipment shall not exceed 24 A when connected to a 30-A circuit.

9. For additional restrictions and data on branch circuit use and application, see *NEC* Table 210-24.

fed from these circuits. Furthermore, the *NEC* requires that all kitchen outlets intended to serve countertop areas must be fed from at least two of these appliance circuits (these circuits may also feed appliance outlets in the other spaces specified previously). Thus, not all countertop workspace will be deenergized by the failure of a single circuit.

Although the code does not limit the number of appliance outlets to be wired into each circuit, good practice dictates that these receptacles be circuited with no more than four such outlets per 20-A circuit. This in turn usually requires more than the code minimum of two appliance circuits. Among other requirements, the *NEC* states that for kitchen and dining areas:

- No point on a wall behind a countertop shall be more than 24 in. from an outlet.
- All countertop convenience receptacles shall be of the GFCI type.

3. Locations utilized primarily or frequently for workshop-type activities, such as garages, utility rooms, and basements, should be provided with receptacles wired in appliance-type circuits (i.e., 20-A receptacles on 20-A circuits), with no more than four such receptacles to a circuit. Receptacles in garages, sheds, crawlspaces, below-grade finished or unfinished basements, and outdoors must be of the GFCI type. (For exceptions to this GFCI rule, see *NEC* Article 210-8.)

4. Additional circuits similar to appliance circuits (no fixed lighting outlets) should be furnished to supply one outlet in each bedroom of a house that is not centrally air-conditioned. Such outlets are intended for window air conditioners. (Good architectural and HVAC design provides a window arrangement, attic ventilation, insulation, sunscreening, and the like to obviate the necessity for these noisy energy users.)

5. The *NEC* requires that at least one 20-A appliance circuit supply the laundry outlets only. This requirement is good practice. If an electric clothes dryer is anticipated (and it should be unless it is definitely known that a gas dryer will be used), an individual branch circuit (distinct from the laundry outlets circuits and rated for the load) should be supplied to

serve this load via a heavy-duty receptacle. (Obviously, facilities for hanging clothes must be provided for those who prefer not to waste energy.)

6. Lay out convenience receptacles so that no point on a wall is more than 6 ft from an outlet. Use 20-A, grounding-type receptacles only. Do not combine receptacles and switches into a single outlet except where convenience of use dictates high mounting of receptacles.

7. Circuit the lighting and receptacles so that each room has parts of at least two circuits. This includes basements and garages.

8. Avoid placing all the lighting in a building on a single circuit.

9. Supply at least one 20-A wall-mounted receptacle adjacent to each bathroom basin location. Such receptacles must have GFCI protection. They should be fed from a 20-A circuit that energizes only such bathroom receptacles. Bathroom lighting, exhaust fan, heaters, or other outlets should not be connected to the bathroom receptacle circuit.

10. Provide at least two GFCI-protected and weatherproof receptacles on the outside of the house, one in front and one at the rear. Switch control of the outside receptacles from *inside* the house is an additional convenience.

11. In rooms without overhead lighting, provide switch control for one-half of a strategically located receptacle that is intended to supply a lamp (see Fig. 28.19 for the wiring arrangement in such a case).

12. Provide switch control for closet lights. Pull chains are a nuisance (but are considerably cheaper).

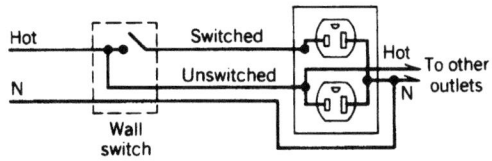

*Fig. 28.19* Split wiring of a duplex receptacle. The upper half is switch-controlled; the lower half is "hot" all the time. This allows wall switch control of a lamp or another device while maintaining part of the receptacle live for independent use. Notice that the receptacle is mounted with the grounding pole at the top. This is the safest way to install receptacles because a metallic item such as a paper clip falling on an inserted cap will contact the ground pole only.

ELECTRICITY

13. In bedrooms, supply two duplex outlets at either side of the bed location to accommodate electric blankets, clocks, radios, lamps, and other such appliances.
14. A disconnecting means, readily accessible and within sight of the controlled item, must be provided for electric ranges, cooktops, and ovens. It is good practice to utilize a small kitchen panel recessed into a kitchen wall to control the large kitchen appliances and provide the required disconnecting means.
15. Perimeter lighting, sensor control, and manual override can do much to lessen vandalism and discourage prowlers.
16. To accommodate the very rapidly increasing number of persons who work at home, at least partially, every study and work room or, in their absence, every large master bedroom should be equipped electrically to double as a home office. This room or area should be supplied *minimally* with these electrical services:
    • Six duplex 15-A or 20-A receptacles connected to at least two different circuits, one of which should serve no other outlets. All of these receptacles should be equipped with appropriate surge suppression (see Section 26.37).
    • An additional separate insulated and isolated ground wire, connected only at the service point, should be run to the box(es) containing two of the receptacles required in the preceding point, and there terminated, clearly marked, and labeled. These grounds are intended for possible use with IG receptacles, if it is found that the normal receptacles are too electrically noisy.
    • Two telephone jacks in recessed boxes should be provided in the area. In addition, an empty ¾-in. conduit from the telephone entry service point to an empty 4-in. square box in the area should be provided. An appropriate surge suppressor must be provided for the incoming telephone service lines.
    • For lighting design of this area, refer to the discussion on VDT lighting in Chapter 16.

A tabulation of residential electrical equipment, including recommended circuit and receptacle descriptions, is shown in Table 28.7. A complete residential wiring plan for a small house is shown in Fig. 28.20. Although residential plans are frequently left uncircuited, a completely circuited design is shown here for didactic purposes. The plan does not show the electrical equipment recommended for a home office that is described in item 16 for two reasons:

1. The plan is so compact that there is no single preferable location for the numerous outlets required.
2. Adding the required outlets to any selected area would clutter the plan to the point of illegibility.

## 28.13 BRANCH CIRCUIT DESIGN GUIDELINES: NONRESIDENTIAL

### (a) Schools

Because schools comprise an assembly of varied-use spaces, including instruction, lab, shop, assembly, office, and gymnasium, plus special areas such as swimming pools, photographic labs, and so on, it is not possible to generalize on branch circuit design considerations except for the following:

1. To accommodate the audio-visual equipment frequently used in the classroom, 20-A outlets wired two to a circuit are placed at the front and back of each such room. A similar receptacle, wired six or eight to a circuit, is placed on each remaining wall.
2. To accommodate computers, a detailed layout of the space is required showing the proposed computer area. A two-section surface-mounted or recessed raceway on the wall behind the row of computers should be provided. The power section will have two duplex 20-A receptacles at each computer station, wired on alternate circuits. The second section of the raceway is intended for low-tension wiring, such as network cabling, and wiring to computer peripherals.
3. Switching for lighting should provide:
    • High–low levels for energy conservation and to permit low-level lighting for screen viewing. With fluorescent lighting, this can be accomplished by alternate ballast wiring and switching, thus avoiding the high cost of dimming equipment.

**TABLE 28.7 Load, Circuit, and Receptacle Chart for Residential Electrical Equipment**

| Appliance | Typical Connected Volt-Amperes[a] | Volts | Wires[b] | NEMA Circuit Breaker or Fuse, amp | Outlets on Circuits | Device[d] and Configuration (see Fig. 26.42) |
|---|---|---|---|---|---|---|
| KITCHEN | | | | | | |
| Range[e,c,i] | 12,000 | 115/230 | 3 #6 | 60 | 1 | 14–60R |
| Oven (built-in)[c,i] | 4,500 | 115/230 | 3 #10 | 30 | 1 | 14–30R |
| Range tops[c,i] | 6,000 | 115/230 | 3 #10 | 30 | 1 | 14–30R |
| Dishwasher[c] | 1,200 | 115 | 2 #12 | 20 | 1 | 5–20R |
| Waste disposer[c] | 300 | 115 | 2 #12 | 20 | 1 | 5–20R |
| Microwave oven | 1,000 | 115 | 2 #12 | 20 | 1 or more | 5–20R |
| Refrigerator[f] | 300 | 115 | 2 #12 | 20 | 1 or more | 5–20R |
| Freezer[f] | 350 | 115 | 2 #12 | 20 | 1 or more | 5–20R |
| LAUNDRY | | | | | | |
| Washing machine | 1,200 | 115 | 2 #12 | 20 | 1 | 5–20R |
| Dryer[c,i] | 5,000 | 115/230 | 3 #10 | 30 | 1 | 14–30R |
| Hand iron[e] | 1,650 | 115 | 2 #12 | 20 | 1 | 5–20R |
| LIVING AREAS | | | | | | |
| Workshops[e,i] | 1,500 | 115 | 2 #12 | 20 | 1 or more | 5–20R |
| Portable heater[e] | 1,600 | 115 | 2 #12 | 20 | 1 | 5–20R |
| Television | 300 | 115 | 2 #12 | 20 | 1 or more | 5–20R |
| Audio center[g] | 350 | 115 | 2 #12 | 20 | 1 or more | 5–20R |
| VCR[g] | 150 | 115 | 2 #12 | 20 | 1 or more | 5–20R |
| Personal computer and peripherals[g,h] | 1,000 | 115 | 2 #12 | 20 | 1 or more | 5–20R |
| FIXED UTILITIES | | | | | | |
| Fixed lighting | 1,200 | 115 | 2 #12 | 20 | 1 or more | — |
| Air conditioner, ¾ hp[i,j] | 1,200 | 115 | 2 #12 | 20 or 30 | 1 | 5–20R 14–30R |
| Central air conditioner[c,i,j] | 5,000 | 115/230 | 3 #10 | 40 | 1 | — |
| Sump pump[j] | 300 | 115 | 2 #12 | 20 | 1 or more | — |
| Heating plant (i.e., forced-air furnace)[j,k] | 600 | 115 | 2 #12 | 20 | 1 | — |
| Attic fan[j] | 300 | 115 | 2 #12 | 20 | 1 or more | 5–20R |

[a]Wherever possible, use the actual equipment rating.

[b]Number of wires does not include equipment grounding wires. Ground wire is No. 12 AWG for a 20-amp circuit and No. 10 AWG for 30- and 50-amp circuits.

[c]For a discussion of disconnect requirements, see *NEC* Article 422.

[d]Equipment ground is provided in each receptacle.

[e]Heavy-duty appliances regularly used at one location should have a separate circuit. Only one such unit should be attached to a single circuit at the same time.

[f]A separate circuit serving only one other outlet is recommended.

[g]Surge protection recommended.

[h]Isolated ground may be required.

[i]A separate circuit is recommended.

[j]It is recommended that all motor-driven devices be protected by a local motor-protection element unless motor protection is built into the device.

[k]Connect through the disconnect switch equipped with a motor-protection element.

- Separate switching of the light fixtures on the window side of the room, which is often lighted sufficiently by daylight. Control should be initiated automatically by a photocell (see Section 16.3).

4. Provide appropriate outlets for all special equipment in labs, shops, cooking rooms, and the like.
5. Use heavy-duty devices and key-operated switches for public area lighting (corridors,

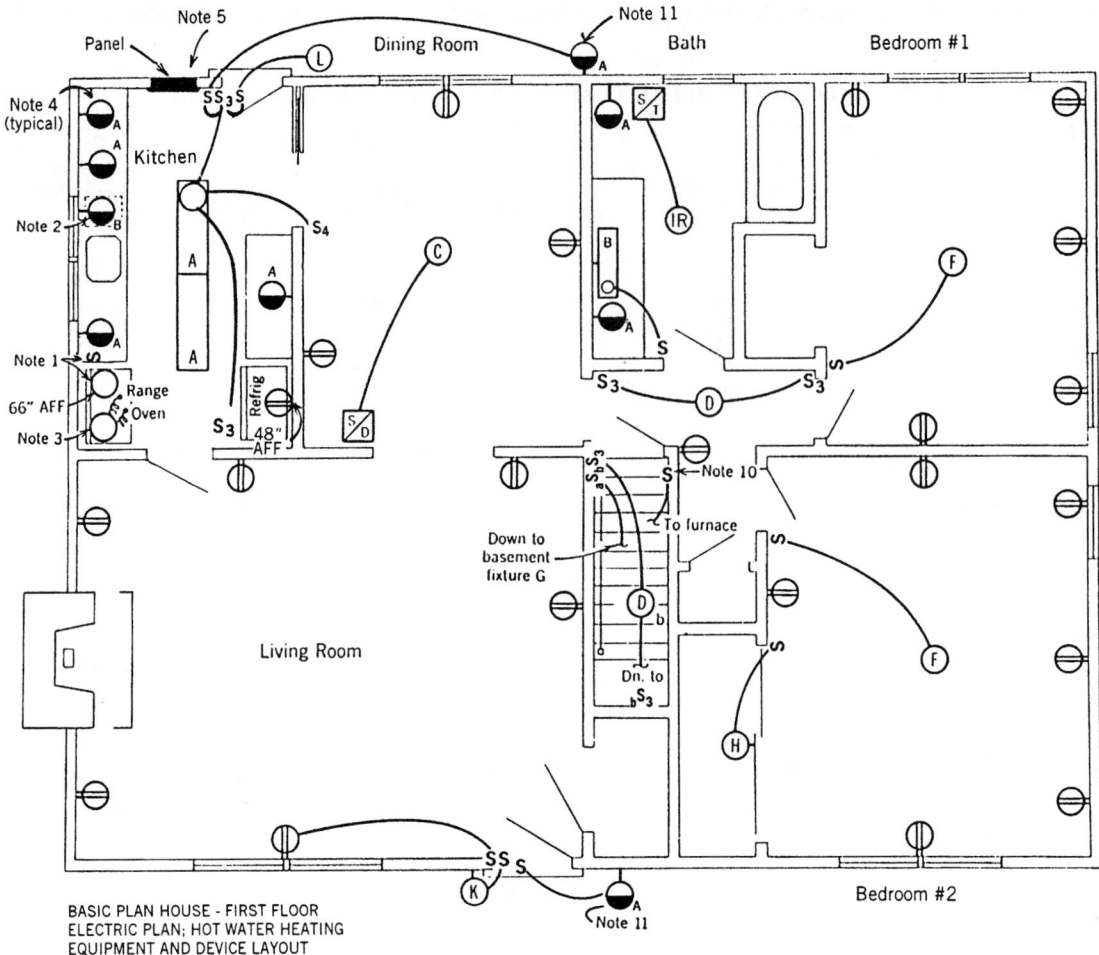

BASIC PLAN HOUSE - FIRST FLOOR
ELECTRIC PLAN; HOT WATER HEATING
EQUIPMENT AND DEVICE LAYOUT

Fixture Schedule: see Fig. 28.20 c
Symbol List: see Fig. 28.20 d
Circuited Plan: see Fig. 28.20 e, f
Plan, Schedule: see Fig. 28.20 g

Notes:

1. Switch and outlet for exhaust fan. Switch wall mtd. above counter-blacksplash. Outlet with blank cover mounted adjacent to fan wall opening. Separate switch may be omitted if fan is supplied with integral switch.

2. Dishwasher receptacle wall mtd. behind unit, 6" AFF.

3. Range and oven outlet boxes wall mtd., 36" AFF. Flexible connection to units.

4. Receptacles at countertop locations to be wall mounted 2" above backsplash.

5. Max. ht. of top c/b to be 78" AFF.

6. Wiring shown as run exposed indicates absence of finished ceiling in basement level. All BX to be run through framing members. Attachment below ceiling joists not permitted.

7. Connect to two type G fixtures ceiling mounted at $1/3$ points of crawl space.

8. Connect to one type G fixture at center of crawl space.

9. Connect to shutdown switch at top of stairs. Boiler control wiring by others. See Note 10.

10. Boiler wiring safety disconnect. Provide RED wall plate, clearly marked "BOILER ON—OFF."

11. Equipped with self-closing gasket WP cover.

(a)

**Fig. 28.20** *Electrical plan of a small house. (a) Layout of lighting and electrical devices for the main floor. (b) Layout of lighting and electrical devices for the basement. (c) Lighting fixture schedule. (d) Symbols and abbreviations. (e) Circuited plan, main floor of house. (f) Circuited plan, basement. (g) Panel schedule of the house.*

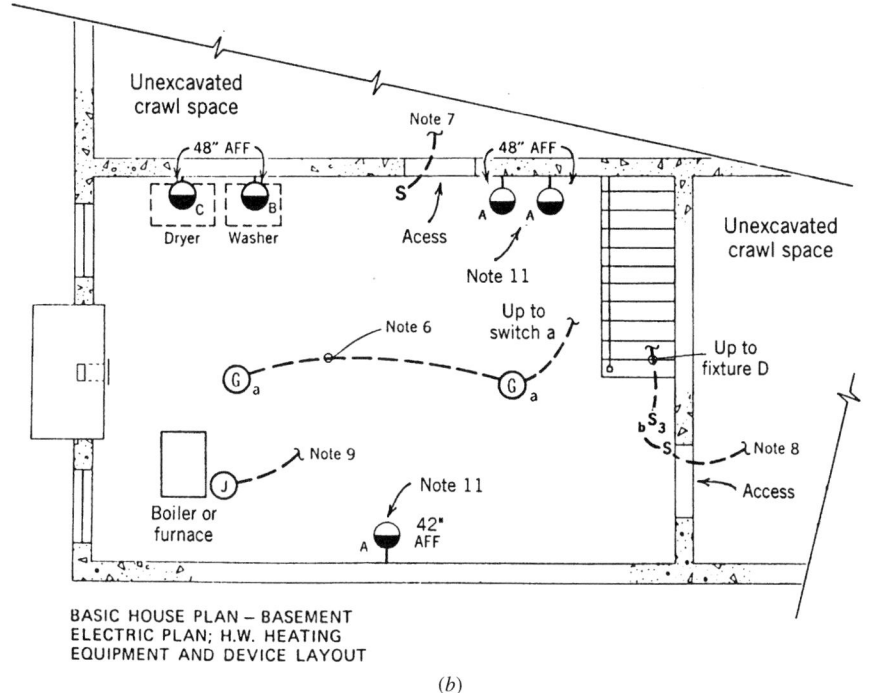

BASIC HOUSE PLAN – BASEMENT
ELECTRIC PLAN; H.W. HEATING
EQUIPMENT AND DEVICE LAYOUT

(b)

| LIGHTING FIXTURE SCHEDULE | | | |
|---|---|---|---|
| TYPE | DESCRIPTION | MANUFACTURER | REMARKS |
| A | 48" L X 12" W X 4" DEEP NOMINAL, 2 LAMP/FLUORES-CENT, WRAPAROUND ACRYLIC LENS, 40 W (NOM.) T8 LAMPS. SURFACE MTD. | BRITE–LITE CO. CAT. #2740/KFF OR EQUAL | 4" DEPTH MAXIMUM |
| B | 24" L, 1 LAMP 20 W FLUOR. FIXTURE, WRAPAROUND WHITE DIFFUSER, MOUNT ABOVE MEDICINE CABINET. | BRITE–LITE CO. CAT. #1/20/BFF OR EQUAL | MAX. MTG. HT. 78" TO ₵. |
| C | ADJUSTABLE HEIGHT PENDANT INCANDESCENT, 3–75 W MAX., BUILT-IN 3–POSITION SWITCH. | HOMELAMP CO. CAT. #3/75/DRP OR EQUAL | ——— |
| D | 10" D. DRUM-TYPE FIXTURE, WHITE GLASS DIFFUSER, CENTER LOCK-UP, 2–60 W INCAND. MAX., SURF. MTD. | BRITELITE CO. CAT. #2/60/HF OR EQUAL | 6" MAX. DEPTH. |
| F | 12" D. DRUM FIXTURE, CONCEALED HINGE ON OPAL GLASS DIFFUSER FOR RELAMPING WITHOUT GLASS REMOVAL, 2–75 W INCAND. MAX. SURFACE MTD. | DENMARK LIGHTING SPECIAL UNIT #374821 | NO SUBSTITUTION WILL BE ACCEPTED. |
| G | PORCELAIN LAMPHOLDER, PULL CHAIN WITH WIRE GUARD, 100 W. INCAND. SURF. MTD. | ——— | ——— |
| H | SAME AS TYPE G, EXCEPT W/O GUARD. | ——— | ——— |
| K | DECORATIVE OUTDOOR LANTERN, MAX. 150 W INCAND., WALL MTD. 84" AFF TO ₵. | TO BE CHOSEN BY OWNER | ——— |
| L | UTILITY OUTDOOR LIGHT, ANODIZED ALUMINUM BODY AND CYLINDRICAL OPAL GLASS DIFFUSER. 1–100 W INCAND. MAX. 84" AFF TO ₵. | UTIL–LITE CO. CAT. #1/100/BP OR EQUAL | IF VANDALISM IS OF CONCERN, SUBST. PLASTIC DIFFUSER. |
| IR | RECESSED FIXTURE WITH 150 W INFRARED HEAT LAMP | HEAT-LIGHT CO. | ——— |

(c)

**Fig. 28.20** (Continued)

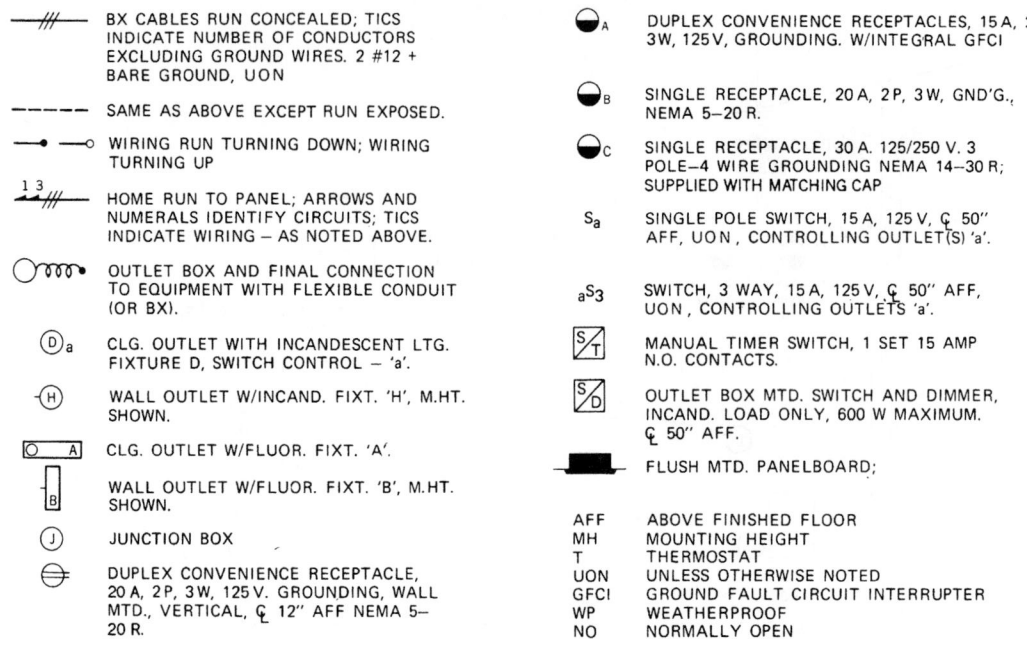

—///— BX CABLES RUN CONCEALED; TICS INDICATE NUMBER OF CONDUCTORS EXCLUDING GROUND WIRES. 2 #12 + BARE GROUND, UON

------ SAME AS ABOVE EXCEPT RUN EXPOSED.

—•—o WIRING RUN TURNING DOWN; WIRING TURNING UP

HOME RUN TO PANEL; ARROWS AND NUMERALS IDENTIFY CIRCUITS; TICS INDICATE WIRING — AS NOTED ABOVE.

OUTLET BOX AND FINAL CONNECTION TO EQUIPMENT WITH FLEXIBLE CONDUIT (OR BX).

(D)a CLG. OUTLET WITH INCANDESCENT LTG. FIXTURE D, SWITCH CONTROL — 'a'.

(H) WALL OUTLET W/INCAND. FIXT. 'H', M.HT. SHOWN.

[O   A] CLG. OUTLET W/FLUOR. FIXT. 'A'.

WALL OUTLET W/FLUOR. FIXT. 'B', M.HT. SHOWN.

(J) JUNCTION BOX

DUPLEX CONVENIENCE RECEPTACLE, 20 A, 2P, 3W, 125 V. GROUNDING, WALL MTD., VERTICAL, ₵ 12" AFF NEMA 5–20 R.

DUPLEX CONVENIENCE RECEPTACLES, 15 A, 2P, 3W, 125 V, GROUNDING. W/INTEGRAL GFCI

SINGLE RECEPTACLE, 20 A, 2P, 3W, GND'G., NEMA 5–20 R.

SINGLE RECEPTACLE, 30 A. 125/250 V. 3 POLE–4 WIRE GROUNDING NEMA 14–30 R; SUPPLIED WITH MATCHING CAP

Sa SINGLE POLE SWITCH, 15 A, 125 V, ₵ 50" AFF, UON, CONTROLLING OUTLET(S) 'a'.

aS3 SWITCH, 3 WAY, 15 A, 125 V, ₵ 50" AFF, UON, CONTROLLING OUTLETS 'a'.

[S/T] MANUAL TIMER SWITCH, 1 SET 15 AMP N.O. CONTACTS.

[S/D] OUTLET BOX MTD. SWITCH AND DIMMER, INCAND. LOAD ONLY, 600 W MAXIMUM. ₵ 50" AFF.

FLUSH MTD. PANELBOARD;

AFF   ABOVE FINISHED FLOOR
MH    MOUNTING HEIGHT
T     THERMOSTAT
UON   UNLESS OTHERWISE NOTED
GFCI  GROUND FAULT CIRCUIT INTERRUPTER
WP    WEATHERPROOF
NO    NORMALLY OPEN

(d)

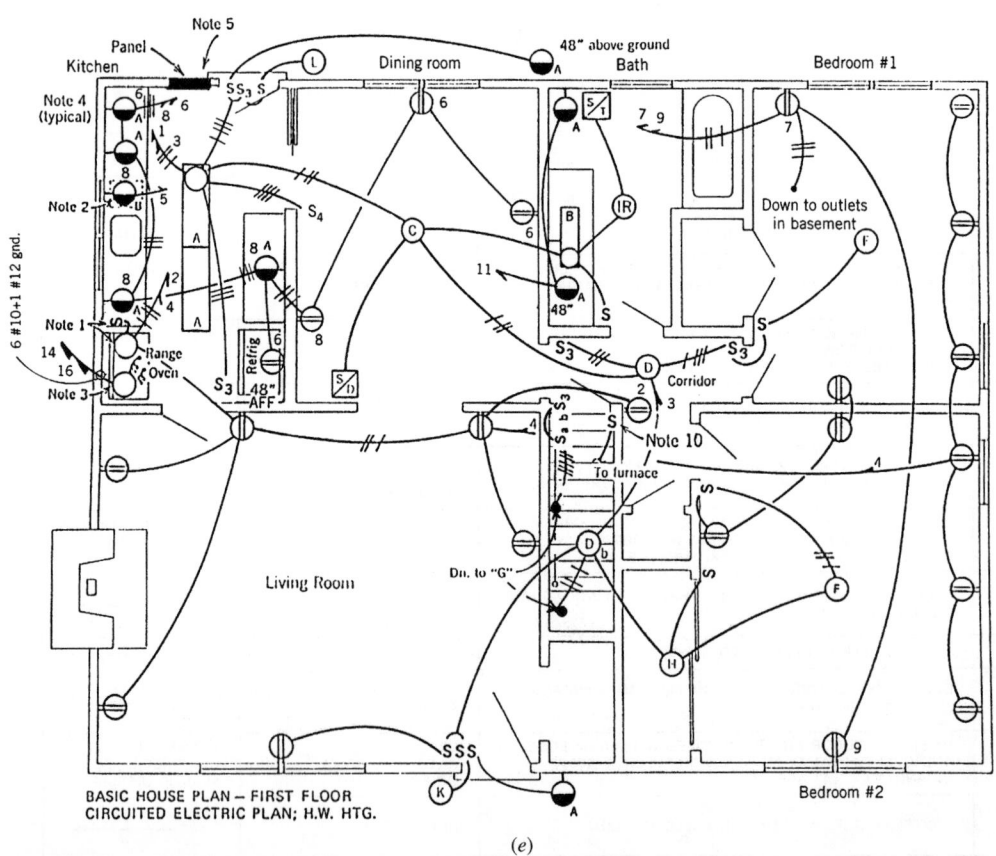

BASIC HOUSE PLAN – FIRST FLOOR
CIRCUITED ELECTRIC PLAN; H.W. HTG.

(e)

**Fig. 28.20** (Continued)

ELECTRICITY

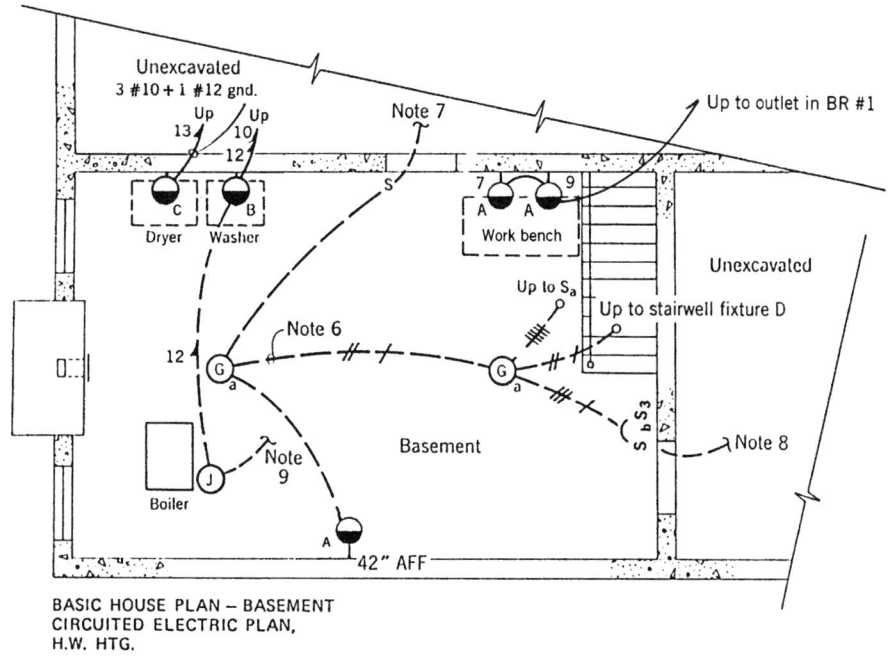

BASIC HOUSE PLAN – BASEMENT
CIRCUITED ELECTRIC PLAN,
H.W. HTG.

*(f)*

| Circ. no. | Description | LOAD V-A. | Circuit beakers | | Load V-A | Description | Circ. no. |
|---|---|---|---|---|---|---|---|
| | | | Panel schedule | | | | |
| 1 | LTG – { Kit. Dr., Br. #1, Hall, Outside, bath + Φ | 1045 1 R | 20   1 2 | 20 | 30 6 R | Outlets – LR. & corridor + Exh. fan | 2 |
| 3 | LTG – { Outside  Lr., Stair, Br. #2, Bsmt. + 6 Φ | 935 6 R | 20   A   3 4 | 20 | 6 R | Outlets – BR. 1 & 2 | 4 |
| 5 | Dishwasher | 1500 | 20   5 6 | 20 | 1500 | Outlets – Kit., Dr. | 6 |
| 7 | Outlets – BR 1, Bsmt | 2 R | 20   7 8 | 20 | 1500 | Outlets – Kit., Dr. | 8 |
| 9 | Outlets – Bsmt., BR 2 | 2 R | 20   9 10 | 20 | 1500 | Laundry outlet – Bsmt. | 10 |
| 11 | Bathroom outlets | 2 R | 20   11 12 | 20 | 1300 | Boiler or furnace | 12 |
| 13 | Spare | — | 20   13 14 | 30 A 2 P | 6Kw { | Range | 14 |
| 15 | Spare | — | 20   15 16 | | | | |
| 17 | Clothes dryer | 5Kw | 30 A 2 P   17 18 | 30 A 2 P | 4800w | Oven | 16 |
| | | | 19 20 | | | | |
| | Space for 2 – 1P or 1 – 2 pole | | 21 22 | | | Space for 1 – 2P or 2 – 1 pole | |
| | | | 23 24 | | | | |
| | Load total – phase A | | | | | Load total, phase B | |

Panel Data
Mains, GND. BUS:   150 A MNS., 60 A GND. BUS
Main C/B or ~~SW/F~~   150/100
Branch C/B INT. CAP. AMP. 10,000
Mounting – ~~Surf~~/recess
Remarks:   Front suitable for painting

Voltage   120/208
1 PH. 3 wire

*(g)*

**Fig. 28.20** *(Continued)*

etc.), plastic instead of glass in fixtures, and vandalproof equipment wherever possible. All panels *must* be locked and should be in locked closets.

6. The *NEC* requires sufficient branch circuitry to provide a minimum of 3 V-A/ft$^2$ for general lighting in schools (refer to *NEC* Article 220). Unlike the figure for residential occupancy, this figure does not include receptacles. Receptacles are calculated separately at 180 V-A each for ordinary convenience outlets.

### (b) Office Space

1. In small office spaces (less than 400 ft$^2$), provide either one convenience outlet for every 40 ft$^2$ or one outlet for every 10 linear ft of wall space, whichever is greater. In larger office spaces, provide one outlet every 100 to 125 ft$^2$ beyond the initial 400 ft$^2$ (10 outlets). These outlets are intended for miscellaneous electrical devices, in addition to anticipating the constantly increasing number of data system accessories. In addition, provide at *every* desk (on an adjacent wall, in a "power pole," or in the floor) at least one duplex 20-A receptacle for a computer. These receptacles should be circuited at no more than six to a 20-A branch circuit, and less if the equipment to be fed so dictates. Figure 28.21 shows one possible circuiting arrangement for the room layouts shown in Fig. 28.12. Although other arrange-

ments are possible, the net result is the same (see also Fig. 28.22).

2. Corridors should have a 20-A, 120-V outlet every 50 ft to supply cleaning and waxing machines.
3. As with all nonresidential buildings, convenience receptacles are figured at 180 V-A each.
4. Only specification grade equipment should be used.

### (c) Industrial Spaces

These areas are so specialized that no meaningful guidelines can be given.

### (d) Stores

In stores, good practice requires at least one convenience outlet receptacle for every 300 ft$^2$ in addition to outlets required for loads such as lamps, show windows, and demonstration appliances.

## 28.14 LOAD TABULATION

While circuiting the loads, a panel schedule is drawn up that lists the circuit numbers, load description, wattage (actually, volt-amperes), and the current rating and number of poles of the circuit-protective device feeding each circuit. Spare circuits are included to the extent that the designer considers them necessary and consonant with

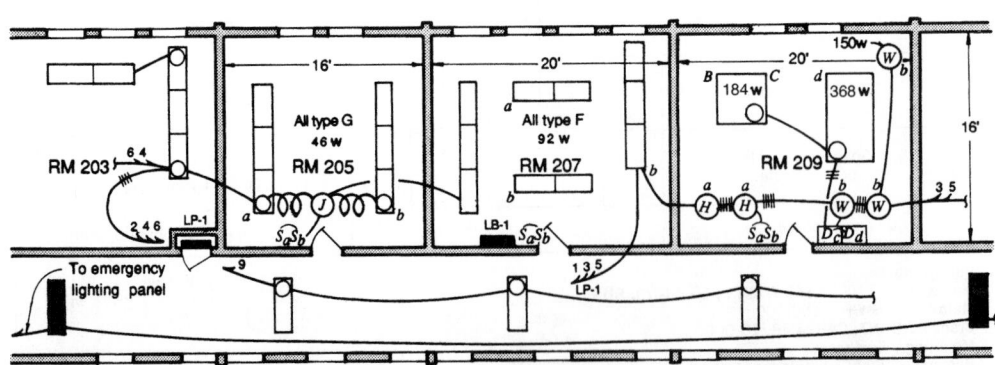

**Fig. 28.21** *Alternative methods of circuiting are shown in different rooms. Room 205 shows the actual junction box location, with flexible connections to the box at each fixture. Room 207 shows circuit numbers and switch designations only; the placement of junction boxes is understood, and conduit runs are omitted for clarity and because they usually are not representative of the actual installation. Room 209 shows an outlet box at each fixture, with schematic conduit connections. All of these methods are in common use.*

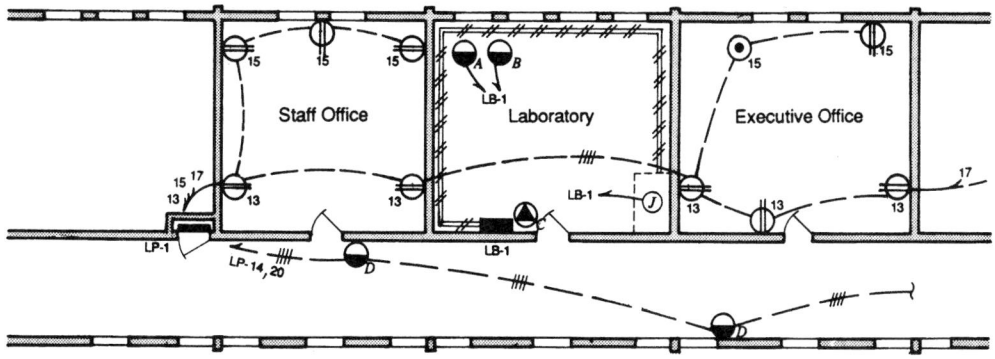

**Fig. 28.22** *Typical receptacle circuiting of several rooms in an office–laboratory building. Lighting (see Fig. 28.21) and power (receptacles) are shown on separate plans to avoid crowding. See Fig. 28.15 for symbols and Fig. 28.12 for notes. Lighting in offices is recessed; lighting in labs is surface-mounted for flexibility. Note the double circuiting of the type-D receptacles.*

economy, but normally no less than 20% of the number of active circuits. Finally, spaces are left for future circuit breakers in approximately the same quantity as the number of spare circuits, but always to round off the total number of circuits.

Panelboards are normally manufactured with an even number of poles. Thus, if a panel had, with spares, 21 poles, the designer would ask for (at least) three spaces, to give a 24-circuit box. A typical panel schedule is shown in Fig. 28.23, which serves the laboratory of Figs. 28.21 and 28.22.

In calculating panel loads, the following rules apply:

1. Each specific appliance, device, lighting fixture, or other load is taken at its nameplate rating, except certain kitchen and laundry appliances for which the *NEC* allows a demand factor (see *NEC* Article 220).
2. Each convenience outlet, in other than residential spaces, is counted as 1.5 A (180 V-A).
3. Loads for special areas and devices such as show window lighting, heavy-duty lampholders, and multioutlet assemblies are taken at the figures given in *NEC* Article 220.
4. Spare circuits are figured at approximately the same load as the average active circuits (1200 to 1500 V-A).
5. Spaces are not counted in the load.
6. Continuous loads such as lighting are calculated at 125% of their actual value. See *NEC* Articles 210-19(a) and 210-20(a), which refer to load calculations for conductor sizing and feeder overcurrent protection, respectively, for

combinations of continuous and noncontinuous loads. These calculations apply to panel load calculations as well and are best accomplished by adding 25% to the panelboard load, as shown in Fig. 28.23.

In calculating total panel loads, as shown in Fig. 28.23, no demand factor may be applied except as specifically stated in the *NEC* (see *NEC* Articles 220-11 and 220-13). This is true despite the knowledge that most often the usage will be such that the average load will be lower than the maximum demand (see Section 28.17 for multipanel feeders). If it is known that certain loads will not or cannot be used simultaneously, the load total should reflect only the larger of the two. Thus, heating and cooling loads are *generally* not concurrent. Nor is building night floodlighting concurrent with the business office load, but it may be with general interior lighting. Note in Fig. 28.23 that 2-pole loads (208 single-phase) appear in two columns. Similarly, 3-phase loads would appear in three columns. Also, note that the phase loads are *not equal*. It is the responsibility of the designer (or contractor) to circuit the loads so that the phases are as closely balanced in load as possible. If this is not done, one phase will carry considerably more current than the others. Because the panel feeder must be sized for the maximum phase current, this may lead to an oversized feeder and therefore a waste of money.

Once the loads have been tabulated, balanced and totaled by phase, the maximum current is calculated. A portion of the spare capacity available in

| | ELEC. PANEL-LP-I | | | |
|---|---|---|---|---|
| | 120/208 V 3φ 4W | LOAD IN WATTS | | |
| No. | LOAD | φA | φB | φC |
| 1 | Lighting | 1200 | | |
| 2 | Lighting | 1200 | | |
| 3 | Lighting | | 1450 | |
| 4 | Lighting | | 1050 | |
| 5 | Lighting | | | 1100 |
| 6 | Lighting | | | 1200 |
| 7 | Lighting | 800 | | |
| 8 | Lighting | 1100 | | |
| 9 | Lighting | | 700 | |
| 10 | Lighting | | 1050 | |
| 11 | Lighting | | | 1000 |
| 12 | Lighting | | | 1200 |
| 13 | Receptacle 2 | 900 | | |
| 14 | Receptacle 3 | 900 | | |
| 15 | Receptacle | | 900 | |
| 16 | Receptacle | | 900 | |
| 17 | Spare | | | 1200 |
| 18 | Spare | | | 1200 |
| 19 | Spare | 1200 | | |
| 20 | Receptacle 4 | | 1000 | |
| | | | | 1000 |
| 21 | Spare | 1200 | | |
| | | | 1200 | |
| 22-26 | Spaces only | | | |
| | | | | |
| | 25% addition for | | | |
| | continuos loads 5 | 1025 | 1063 | 1125 |
| | Phase totals | 9525 | 9313 | 9025 |
| | Panel total | | 27863 | V-A |
| | Max. φ current | | 77 amp | |
| | 25% spare capacity | ~ | 20 amp | (Future) |
| | | | | |
| | Max. Phase I | | 97 amps | |
| | | | | |

Main breaker  225A 3pole (see text
      example 17.1, section 17.18 )
Trip  100 A
Feeder size  4 #2 THW  in 1¼"C.

NOTES:
1. All C/B 1P; 50A frame, 20A trip except
   ccts 20 and 21, which are 2P, 50 AF/20 AT.
2. 5 receptacles @ 1.5A each.
3. Corridor receptacle; 120 V section.
4. Corridor receptacle; 208 V section.
5. See Section 17.15

**Fig. 28.23** Schedule for lighting panel LP-1 (see Fig. 28.22).

the branch circuits is added to the total as the basis for the calculation of the feeder load. This spare capacity, shown in Table 28.5, is between 25% and 50%. The exact amount to be added initially in feeder sizing is developed in the ensuing discussion.

## 28.15 SPARE CAPACITY

Load calculations for dwelling occupancies and other spaces are detailed in *NEC* Article 220, and examples are given in Chapter 20 of the *NEC*. Because these calculations are specialized (but rou-

tine) and are covered there in detail, they are not repeated here. Code requirements, it must be remembered, are minimum; quality design often exceeds these basic requirements.

Once the panel load totals have been determined as detailed previously, the next step is to size the conductors feeding the panel. To do this, an examination of the spare capacity of the panel and of the feeders is necessary in order that the system design be consistent, giving equal capacity for future growth in all of its components. Considering the panel circuitry first, let us examine the effect of load expansion, including spares and spaces (Table 28.8).

As noted in Section 28.11, spare capacity is built into the branch circuitry *and* into the panels. Most often, expansion is accomplished by additional loading on some circuits and by adding new circuits via spare circuit breakers in the panel. Table 28.8 gives the *ultimate* capacity of the panel—that is, fully loaded circuits and fully utilized spares and spaces. Because this ultimate capacity is rarely achieved, panel feeders need only be sized for initial loads as detailed in Section 28.14, and provision made for rewiring to meet anticipated expansion by one of the techniques listed in Section 28.16.

These results can be summarized as follows: For panels in buildings expecting limited expansion, for which branch circuits are loaded to 80% of capacity (25% *branch circuit* expansion; see Table 28.5, note *e*), the ultimate panel load without new conduit work (i.e., merely by filling out circuits), is 1.5 L, or 50% beyond the initial load. By adding breakers in the spaces, this load can be expanded to 75% beyond the initial load. The corresponding figures for panels that are lightly loaded (66% capacity, i.e., 50% branch circuit expansion, as in Table 28.5, note *f*), in anticipation of considerable load growth, are 80% to 110%, respectively. These results are summarized in Table 28.8.

## 28.16 FEEDER CAPACITY

To achieve economy, the panel feeder must accommodate the initial load plus some portion of the future load. Spare capacity in feeders (to accommodate a considerable portion of the panel spare capacity, as shown in Table 28.8) is provided by one or more of the following procedures:

**TABLE 28.8 Panel Initial and Expanded Loads**[a]

| | Panels in Facilities Expecting Limited Expansion; Circuits Initially Loaded to Give 25% Expandability (See Table 28.5) | Panels in Facilities Expecting Extensive Expansion; Circuits Initially Loaded to Give 50% Expandability (See Table 28.5) |
|---|---|---|
| Initial load | 100% | 100% |
| Initial plus spares | 120% | 120% |
| Load after all circuits including spares are loaded to maximum allowable | 150% | 180% |
| Load after utilizing 20% spaces also | 175% | 210% |

[a]For development of these figures, see McGuinness et al. (1980), pp. 690–691.

1. Provide feeder (and conduit) capacity initially to handle the entire eventual load. This method is most expensive, requiring an initial outlay for no return, and is rarely used.
2. Provide feeder for the initial load plus spare capacity, with properly sized conduit. Conduit is sized for type THW or RHW without covering. This method, as we shall see, yields limited spare capacity.
3. Provide feeder for the initial load plus spare capacity, with conduit oversized by one size. Size conduit for type THW wire, which is very widely used because of its attractive price and excellent electrical properties. Some additional costs are entailed here.

   If the initial wiring is done with type TW wire, the effect is approximately the same as that of oversizing the conduit by one size and

**TABLE 28.9 Maximum Number of Conductors in Trade Sizes of Rigid Metal Conduit**

| Type Letters | Conductor Size AWG, kcmil | Conduit Trade Size (in.) | | | | | | | | | |
|---|---|---|---|---|---|---|---|---|---|---|---|
| | | ½ | ¾ | 1 | 1¼ | 1½ | 2 | 2½ | 3 | 3½ | 4 |
| TW, THW, RHW, RHH (without outer covering), and THHW | 6 | 1 | 3 | 5 | 8 | 11 | 18 | | | | |
| | 4 | 1 | 1 | 3 | 6 | 8 | 14 | 20 | | | |
| | 2 | 1 | 1 | 2 | 4 | 6 | 10 | 14 | | | |
| | 1 | 1 | 1 | 1 | 3 | 4 | 7 | 10 | | | |
| | 0 | | 1 | 1 | 2 | 3 | 6 | 8 | 13 | | |
| | 00 | | 1 | 1 | 2 | 3 | 5 | 7 | 11 | | |
| | 000 | | 1 | 1 | 1 | 2 | 4 | 6 | 9 | 13 | |
| | 0000 | | | 1 | 1 | 1 | 3 | 5 | 8 | 10 | |
| | 250 | | | 1 | 1 | 1 | 3 | 4 | 6 | 8 | 11 |
| | 300 | | | 1 | 1 | 1 | 2 | 3 | 5 | 7 | 9 |
| | 350 | | | | 1 | 1 | 1 | 3 | 5 | 6 | 8 |
| | 400 | | | | 1 | 1 | 1 | 3 | 4 | 6 | 7 |
| | 500 | | | | 1 | 1 | 1 | 2 | 3 | 5 | 6 |
| | 600 | | | | 1 | 1 | 1 | 1 | 3 | 4 | 5 |
| THHN | 6 | 2 | 4 | 7 | 12 | 16 | 27 | | | | |
| | 4 | 1 | 2 | 4 | 7 | 10 | 16 | 24 | | | |
| | 2 | 1 | 1 | 3 | 5 | 7 | 12 | 17 | 26 | | |
| | 1 | 1 | 1 | 1 | 4 | 5 | 9 | 12 | 19 | 26 | |
| XHHW (4–500 kcmil) | 0 | 1 | 1 | 1 | 3 | 4 | 7 | 10 | 16 | | |
| | 00 | | 1 | 1 | 2 | 3 | 6 | 9 | 13 | | |
| | 000 | | 1 | 1 | 1 | 3 | 5 | 7 | 11 | 15 | |
| | 0000 | | 1 | 1 | 1 | 2 | 4 | 6 | 9 | 12 | 16 |
| | 250 | | | 1 | 1 | 1 | 3 | 5 | 7 | 10 | 13 |
| | 300 | | | 1 | 1 | 1 | 3 | 4 | 6 | 9 | 11 |
| | 350 | | | 1 | 1 | 1 | 2 | 3 | 6 | 7 | 10 |
| | 400 | | | 1 | 1 | 1 | 2 | 3 | 5 | 7 | 9 |
| | 500 | | | | 1 | 1 | 1 | 2 | 4 | 5 | 7 |
| | 600 | | | | 1 | 1 | 1 | 1 | 3 | 4 | |

*Source:* Extracted from the *National Electrical Code.*

wiring initially with THW wire. This is caused by the lower ampacity rating of TW compared to THW, resulting in a larger conduit size for the same initial ampacity. The reader can work out exact figures for each with the help of Tables 27.2a, and 27.2b, and conduit capacity in Table 28.9, or by using the *NEC* tables in Article 310 and Chapter 20.

4. Provide feeder for the initial load plus spare capacity and oversize conduit by *two* sizes. This yields most of the capacity necessary in facilities anticipating large expansion.

5. Provide for the initial load plus spare capacity, with an empty conduit for future use. This method is expensive because of the high conduit cost and is advisable only infrequently.

In procedures 2, 3, and 4, the future capacity beyond that initially supplied is handled by the use of larger-gauge wire in the existing conduit. To examine exactly what these alternatives provide in spare capacity, we have tabulated in Table 28.10 the maximum ampacity of various size conduits and the future capacity obtainable. Table 28.9 is taken directly from the *NEC*. Note from Table 28.10 that simply by rewiring, we can obtain up to 30% additional capacity, whereas if the conduit had been oversized, the additional capacity would be 35% to 146%.

Returning now to the question of how large to make the feeder for a given panel load, we must balance the future panel load, the initial cost of feeder, and the future capacity of existing conduit. It is best to avoid the installation of empty conduits because this is expensive. Rewiring, however, is relatively

inexpensive, and oversizing conduit is the method of choice if rapid expansion is anticipated.

Referring to Table 28.10, note that normal design uses THW cable. Design with T or TW is, in effect, a first step in oversizing conduit and is generally not economical. The second step is a deliberate oversizing of conduit that results in much increased conduit ampacity, as reflected by the figures. Using these figures in actual practice requires that the designer juggle cable and conduit costs against anticipated load growth to arrive at the most economical long-term solution. Applying these numbers to concepts previously developed, we have:

1. Buildings designed for 25% branch circuit expansion (see Table 28.5) have a panel capacity of 1.75 times the load (Table 28.8). If it is desired to design the panel feeder to carry the full expansion possible, then calculate the feeder on the basis of panel load plus 20% spares and oversize the feeder by 20%. This gives a feeder capacity of

$$1.20 \times 1.20 = 1.44 \times \text{initial load}$$

Table 28.10 indicates that rewiring adds another 20% on average. This gives

$$1.20 \times 1.44 = 1.73 \times \text{initial load}$$

which corresponds closely to the 1.75 desired.

2. For a building with 50% branch circuit expansion, utilizing the full panel space capacity gives us an ultimate panel capacity of 2.1 times the initial load (see Table 28.8). This is accomplished as follows: oversize the feeder by 15% and over-

**TABLE 28.10 Maximum Wire and Ampacity of a Conduit, and Ampacity Gain on Rewiring**

| | Initial Installation THW Cable | | Rewiring with XHHW | | | Capacity Increase Having Oversized Conduit | |
| | | | Using Original Conduit | | | | |
| Initial Conduit Size (in.) | Max.[a] Wire AWG/ kcmil | Max. Amperes | Maximum Wire | Maximum Amperes | Capacity Increase | One Size | Two Sizes |
|---|---|---|---|---|---|---|---|
| 1½ | 1 | 130 | 1/0 | 170 | 30% | 100% | 146% |
| 2 | 3/0 | 200 | 4/0 | 260 | 30% | 60% | 115% |
| 2½ | 250 | 255 | 300 | 320 | 25% | 69% | 100% |
| 3 | 400 | 335 | 500 | 430 | 28% | 35% | 81% |
| 3½ | 600 | 420 | 600 | 475 | 13% | 64% | — |

[a]Assuming four single conductors in conduit: 3-phase conductors plus neutral.

size the conduit by one size. The latter step gives an average expansion of 65%. Therefore,

feeder capacity = 1.15 × 1.20 × 1.65 = 2.3

which is approximately the desired figure.

If, as in the case of laboratories, more than 100% expansion is anticipated (see Table 28.1), conduit should be oversized by two sizes and initial wiring oversized by approximately 25%. Feeders thus arranged will handle the new panels required to meet the anticipated expansion.

Two factors should be carefully noted here. First, the smaller conduits offer the largest expandability, although, in dollars per amperes, they are more expensive. Second, in order to take advantage of spaces in a panel, conduit stubs should be taken from the panel and extended into hung ceilings or another procedure used to make the panel circuitry easily accessible in the future.

## 28.17 PANEL FEEDER LOAD CALCULATION

**EXAMPLE 28.1** Refer to Fig. 28.23. The panel is for a laboratory/office area. Because large expansion is anticipated, circuitry follows the bottom section of Table 28.5. The ultimate panel load would be 26 circuits at 1920 V-A = 50 kVA = 138 A. Thus, the initial feeder is sized for 115 A (4 #2 THW, 1¼ in. C), but rewiring with XHHW will allow as much as 150 A in the same conduit. A 225-A frame circuit breaker is chosen initially because eventually the trip will be raised to 150 A.  ■

The *NEC* in Article 220 specifies minimum volt-amperes per square foot (V-A/ft$^2$) figures for various occupancies, for lighting, and for miscellaneous power loads. Proper design procedure therefore requires that after detailed design of an area, the actual loading should be compared to these minima, and the larger of the figures, concerning the number of circuits and feeder load, be used. An example will help to make this clear.

**EXAMPLE 28.2** Assume a single floor of an office building of 100 × 200 ft. Assume also that 15% of the area is corridor and storage, equally divided

between the two. Calculate the load and feeder size. Assume a good-grade speculative construction venture.

### SOLUTION

| | |
|---|---|
| Office space | = 85% of 20,000 ft$^2$ |
| | = 17,000 ft$^2$ |
| Corridor and storage | = 15% of 20,000 ft$^2$ |
| | = 3,000 ft$^2$ |

The *NEC* specifies a minimum of 3½ V-A/ft$^2$ for lighting and 1 V-A/ft$^2$ for miscellaneous receptacles in office space. It further specifies ½ V-A/ft$^2$ minimum for corridors and ¼ V-A/ft$^2$ for storage. Therefore:

| Lighting | | |
|---|---|---|
| Office load: | 17,000 × 3½ | = 59.5 kVA |
| Corridor: | 1,500 × ½ | = 0.74 kVA |
| Storage: | 1,500 × ¼ | = 0.38 kVA |
| Minimum lighting load | | = 60.6 kVA |

| Receptacles |
|---|
| 17,000 × 1 V-A/ft$^2$ = 17 kVA |

These figures would then be compared to the actual design loads. Receptacles are counted at 180 V-A each, as noted in Section 28.14, item 2. If the code minima exceed the design load (as they well may if lighting is properly designed), panels must be equipped with sufficient additional circuits to make up the difference. The number of such circuits is up to the designer, because circuit loading is not specified.

For instance, suppose that the lighting design was accomplished at 2 V-A/ft$^2$. Panel circuits would have to be provided for the additional 1½ V-A/ft$^2$ as follows:

17,000 ft$^2$ × 1.5 V-A/ft$^2$ = 25.5 kVA

Assuming a 30% to 50% future load expansion, we would circuit the lighting loads at approximately 1300 V-A per circuit (see Table 28.5). Therefore, we would provide

$$\frac{25,500 \text{ V-A}}{1,300 \text{ V-A}} = 20 \text{ additional circuits}$$

With respect to the minimum feeder load, the *NEC* specifies that it be increased by 25% if loads are continuous (3 or more hours). This requirement allows for breakers to heat up in panels while carrying a continuous load and is waived for circuit breakers that are ambient compensated—that is, are rated to carry 100% load. Because we have established 80% of the breaker rating as the maximum load (see Table

ELECTRICITY

28.5), we have already accounted for this factor in circuitry but must utilize it in the feeder calculation. Assuming that the code minima are the design loads, then, for feeder calculation:

Lighting load = 60.6 kVA × 125% = 75.75 kVA
Receptacle load as above = 17.0 kVA
Minimum feeder load = 92.75 kVA
25% future load = 23.2 kVA
Design feeder load = 116 kVA

Because this load would be divided among several panels, the building electrical design might be such

that the panels are not all fed by one feeder (see Fig. 28.24). However, assuming they are, the feeder would be calculated in terms of 3-phase current thus:

$$I = \frac{kVA}{0.360} = \frac{116}{0.360} = 322 \text{ A}$$

Using THW cable, a minimum of 400 MCM would be required. Conduit would be a minimum of 3 in. but might be increased to 3½ or 4 in., according to the considerations of spare capacity discussed in the previous section. ■

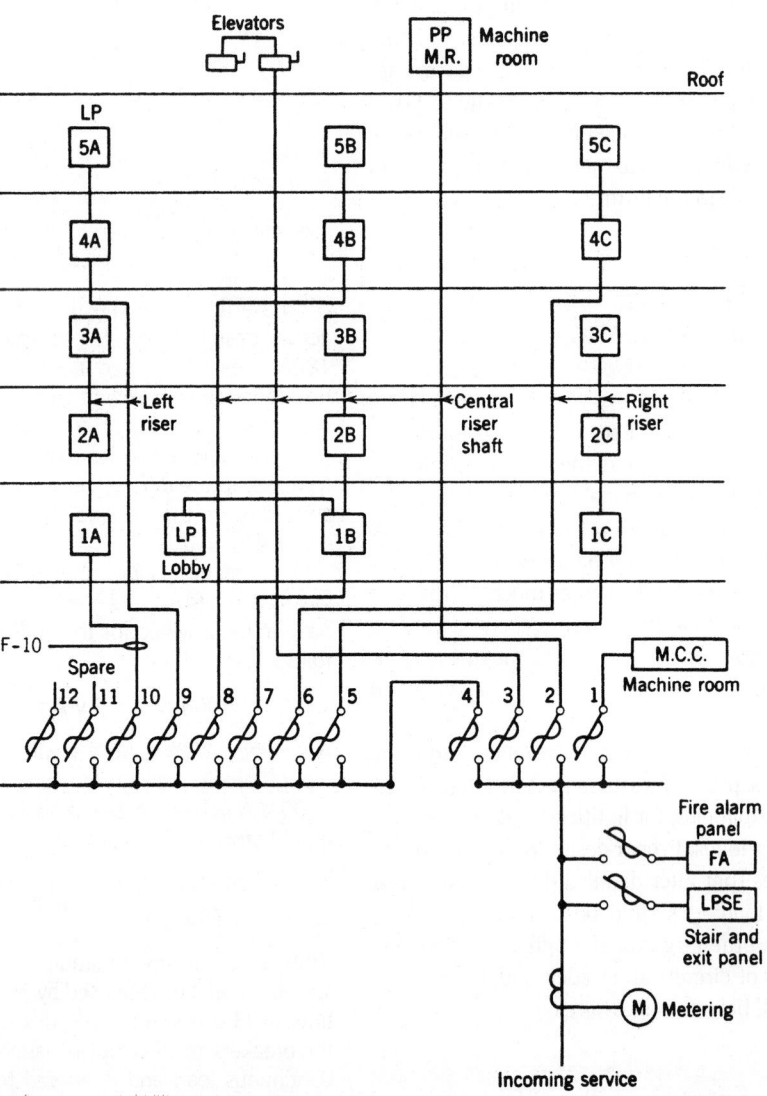

**Fig. 28.24** *Typical power riser diagram. Ordinarily, the main switchboard would be shown as a large rectangle with the feeders emanating from it, and a switchboard schedule would detail the contents. Here, because of the unusual bus arrangement, the main switchboard appears as it would on a single-line diagram.*

**TABLE 28.11 Current and Volt-Ampere Relationships**

| 120-V Single-Phase | 120/240-V 3-Wire | 120/208-V Single-Phase, 3-Wire | 120/208-V 3-Phase | 277/480-V 3-Phase | 277-V Single-Phase |
|---|---|---|---|---|---|
| $I = (V–A)/120$ | $I = (V–A)/240$ | $I = (V–A)/208$ | $I = (V–A)/360$ | $I = (V–A)/830$ | $I = (V–A)/277$ |

As an aid in computing currents, Table 28.11 lists the relevant relations. A note of caution is required regarding the use of computer programs to perform the previous calculations. The reader who has followed the discussion carefully understands that considerations of economy and anticipated expansion are combined in designing a feeder system. None of the programs that we have seen will make the necessary comparisons and calculations unless specifically directed, and this frequently requires a multistage calculation. Most programs are arranged to select feeders on the basis of current loads, plus a specified spare capacity, without regard to expansion or economic considerations. It is therefore important to understand the methods used by any particular computer program in order to apply it intelligently.

## 28.18 HARMONIC CURRENTS

A recent phenomenon, large harmonic currents, has been the cause of considerable difficulty in modern electrical installations. Without going into detail on this highly technical subject, a brief description of the problem and its causes can be given.

Conventional electrical loads such as lighting, resistive devices (heaters), motors, and the like are linear (i.e., the load impedance remains essentially constant, regardless of instantaneous voltage). This is not the case with most electronic equipment. Computers, modems, printers, electronic lighting ballasts, variable-speed motor drives, and solid-state equipment of all types are essentially nonlinear loads. As such, they produce harmonic currents, of which the odd-order ones are additive in the power system neutral conductor. The most troublesome of these are the third harmonic and its odd multiples (9th, 15th, 21st, ...). These currents can become so large in a modern computerized office installation with electronic lighting ballasts that instead of the neutral conductors carrying the unbalanced current in a 3-phase system (zero in a balanced sys-

tem), they actually carry *more* current than the phase wires.

Other serious negative effects of harmonic currents are:

- Deterioration of electronic equipment performance; continuous or sporadic computer malfunctions
- Overheating of neutral—possibly causing neutral burnout and resulting in equipment being subjected to severe voltage variations
- Overheating and premature failure of transformers—even when the transformer nameplate rating seems adequate
- Overheating of motors because of operation with a distorted voltage waveform
- Nuisance tripping of circuit breakers and adjustable-speed drives
- Telephone interference
- Capacitor fuse blowing

As the amount of electronic equipment in use increases (as it does continuously), the problem of destructive harmonic currents becomes progressively more severe.

In view of the fact that today, at least half of the electric load in a modern office-type facility is composed of nonlinear, harmonic-producing equipment, it follows that all such facilities, existing and in-planning, must take necessary corrective measures. In the past, these measures consisted of oversizing equipment to avoid overload burnout; adding passive harmonic filters (which act to reduce harmonic content) in the electric distribution system; using isolation transformers at sensitive loads; selecting power sources with low output impedance to minimize voltage distortion; using controls that are relatively insensitive to harmonic distortion; adding meters throughout the system that measure true rms voltage and current rather than the average figures shown by conventional meters; and other expensive and essentially passive power line conditioning (see Sections 26.35 and 26.36). In view of the increasing severity of the problem, computer-controlled variable power-conditioning equipment (called *active* line conditioning) has become available. This power-conditioning equipment operates in a fashion similar to that described for active noise cancellation in Section 19.28. The conditioner instantaneously and continuously analyzes the harmonic content of the line voltage and injects an equal but exactly out-of-phase voltage to cancel the harmonics and produce a pure sinusoidal voltage supply. The harmonic currents that are required by nonlinear loads are supplied by a digital signal generator. Other techniques are also used; their description is beyond the scope of this book.

In any rehab work, the electrical designer must obtain a detailed electrical system analysis of the existing system, performed by competent engineers experienced in the field of power quality. Many existing systems carry as much as 70% to 80% harmonic current and constitute a major system failure waiting to happen. A proper power quality study, performed with such instruments as true rms meters, harmonic analyzers, frequency selective voltmeters, and spectrum analyzers, will yield a true picture of the existing system and permits the electrical rehab work to be engineered with harmonic limitations as one of the important design parameters.

## 28.19 RISER DIAGRAMS

When all devices are circuited and panels are located and scheduled, we are ready to prepare a riser diagram. A typical diagram, shown in Fig. 27.2, represents a block version of a single-line diagram except that, as the name implies, vertical relationships are shown. All panels, feeders, switches, switchboards, and major components are shown up to, but not including, branch circuiting. This diagram is an electrical version of a vertical section taken through the building.

**EXAMPLE 28.3** Feeder F10 of Fig. 28.24 serves lighting panels 1A, 2A, and 3A. Calculate the required feeder size, considering loads, future expansion, and voltage drop.

### SOLUTION

The connected loads on these panels have been computed in accordance with the previous considerations and are:

LP-1A— 110 A

LP-2A— 125 A

LP-3A— $\dfrac{100\ \text{A}}{335\ \text{A}}$

These figures include connected load, spares, and a 25% future factor (the *NEC* requires that feeders be sized minimally for loads calculated by using the given *NEC* square foot loads plus any demand factors listed; see Article 220). If actual panel circuit loads are larger than the *NEC* minima, then they are the loads used in feeder calculations and any reasonable demand may be used, provided that at no time do the totals drop below the minima specified in the *NEC*. Therefore, we may apply diversity factors between panel loads in a judicious manner.

In office building work, typical diversity factors are as follows:

| Lighting Panels Fed from a Single Feeder | Diversity Factors |
| --- | --- |
| 1 or 2 | 1.00 |
| 3 or 4 | 1.09 |
| 5, 6, or 7 | 1.18 |
| 8, 9, or 10 | 1.33 |

Thus, the load on feeder F10, using 100% demand per panel and 1.09 diversity between panels, would be $335 \times 1.0/1.09 = 307$ A. ∎

Methods for handling future expansion were discussed previously. In this case, the feeder, before voltage drop considerations, would be (from Tables 27.2, 27.3, and 28.9) 4-350 MCM THW in 3-in. C. In this case, the figures in Table 28.10 can be misleading if not read carefully. A 3-in. conduit will take a maximum of 400 kcmil THW. However, because we are initially using only 350 kcmil that also requires a 3-in. conduit, the expansion possible by rewiring with XHHW is

$$\frac{\text{ampacity of 500 kcmil, XHHW in 3-in. C}}{\text{ampacity of 350 kcmil, THW in 3-in. C}}$$

$$= \frac{430\ \text{A}}{310\ \text{A}}$$

$$= 1.39\text{—i.e., 39\% expansion}$$

This should be sufficient.

The final consideration in sizing a feeder is voltage drop. The *NEC* suggests (does not require) that sizing branch circuit conductors for a maximum of 3% voltage drop to the farthest outlet, and sizing

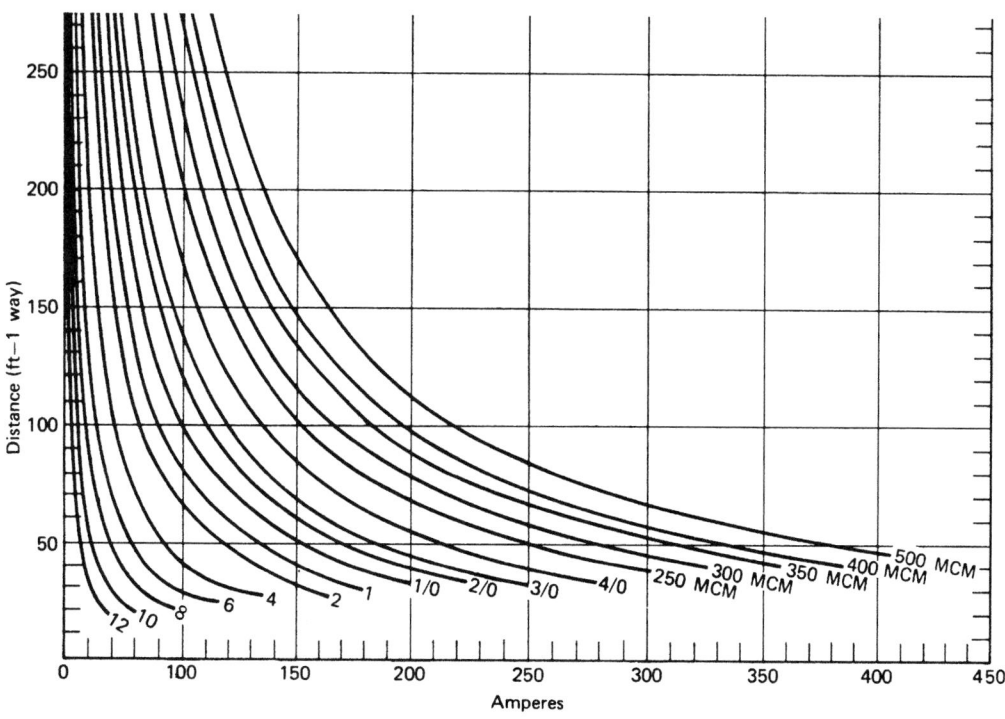

**Fig. 28.25** *Curves for determining voltage drop in copper cables. Curves show the maximum one-way circuit length for 1% voltage drop.*

feeders so that the maximum *combined* feeder and branch circuit voltage drop to the farthest outlet does not exceed 5% will give reasonably efficient operation. These figures apply to all types of loads.

We have always found that it makes economic sense to restrict branch circuit voltage drop to 2% to the farthest outlet while maintaining the overall 5% limitation. This allows a minimum 3% drop in feeders, which are frequently long and, because of size and ampacity inefficiency, relatively more expensive than small branch circuit wiring.

Many tables and curves are published by manufacturers from which voltage drop can be obtained. Such a set of curves is shown in Fig. 28.25, which shows the maximum length of run for a 1% drop. Applying these curves to Example 28.3 and assuming an 80-ft run gives the following.

From the curves, 307 A on 350 MCM cable gives a 1% drop in 50 ft. Therefore, the drop in 80 ft will be 1.6%, which is well within our criteria, assuming that branch circuits were limited to a 2% voltage drop.

In summary, then, feeders are sized in accordance with load (actual or square feet, whichever is larger) and voltage drop. Conduit may be oversized for large future load expansion.

## 28.20 SERVICE EQUIPMENT AND SWITCHBOARD DESIGN

The main switchboard shown in Fig. 28.24 constitutes a combination of service equipment and feeder switchboard. The service equipment portion of the board comprises the metering and the four main switches feeding the risers, motor control center (MCC), roof machine room, and elevators. The feeder board comprises switches 5 through 12. Such an arrangement is permissible inasmuch as the *NEC* allows up to six fused switches or circuit breakers to serve as the service disconnect means. This arrangement was chosen in order to separate to the largest extent possible the motor loads (elevators, air-conditioning equipment, basement power, etc.) from the lighting. Such a procedure minimizes lighting fluctuations resulting from motor starting and yields simpler maintenance. Also, the size of the main switch is reduced. This switchboard would

ELECTRICITY

be of the metal-clad dead-front type with switches or circuit breakers, as desired.

Other recommendations affecting service equipment are as follows:

- All equipment used for service, including cable, switches, meters, and so on, should be approved for that purpose.
- It is recommended that a minimum of 100-A, 3-wire, 120/240-V service be provided for all individual residences.
- No service switch smaller than 60 A or circuit breaker frame smaller than 100 A should be used.
- In multiple-occupancy buildings, tenants must have access to their own disconnect means.
- All building equipment should be connected on the load side of the service equipment, except that service fuses, metering, fire alarm and signal equipment, and equipment serving emergency systems may be connected ahead of the main disconnect (see Fig. 28.24).
- For additional information on electric service, see *NEC* Article 230.

## 28.21 EMERGENCY SYSTEMS

### (a) General Information

Some of the considerations relevant to power reliability were discussed in Section 28.1*b*, and a brief review of the equipment available to supply emergency power was presented in Section 26.39. Emergency lighting equipment is covered in Section 16.36. In this section, we discuss possible arrangements of an emergency power supply. The choice of arrangement and the size and type of equipment depend largely on the requirements of local codes, which determine the loads that must be supplied from the emergency system. Although we are using the term *emergency*, the concepts involved are equally applicable to standby systems. Where differences occur, they are pointed out.

The three classes of electrical power supply systems that are subsumed under the general category of emergency electric supply systems are:

- Emergency Systems: covered by *NEC* Article 700
- Legally Required Standby Systems: covered by *NEC* Article 701

- Optional Standby Systems: covered by *NEC* Article 702

The essential differences between the three classifications are concerned with the purpose, application, and duration of the specific nonnormal power supply. All three types are activated when normal power fails, but only the first two types are legally required when so specified by any governmental agency having jurisdiction. Optional standby systems are, as their name states, optional; they are entirely dependent on the requirements of the facility's owner or operator.

*Emergency systems* are those essential to human safety. As such, and depending on the type of facility involved, they may supply power for fire detection, lighting, alarm and communication systems, elevators, fire pumps, and such loads as ventilation, refrigeration, and industrial processes where power interruption would imperil human safety. Generally, emergency systems must pick up the loads automatically within 10 seconds of power interruption. Where the emergency power supply consists of batteries, they must have a full-load capacity of 90 minutes. Where the power supply is an engine–generator set, an on-site fuel supply that will suffice for a minimum of 2 hours of full load is required. The entire wiring system for the emergency power system must be separate and distinct from the normal power system, except, of course, where both systems are connected to the same item of equipment.

*Legally required standby systems* are those required to provide power for essentially the same types of loads as emergency systems, except that their absence does not involve immediate danger to human life. Such purposes as fire fighting, control of health hazards, long-term rescue operations, and industrial hazard prevention would be among the electrical loads served. For these systems, loads must be picked up automatically with 60 seconds of failure of normal power and maintained for the same minimum periods of time specified for emergency power systems. Wiring may be run in the same enclosures as those of the normal power system.

*Optional standby power systems* are intended essentially to minimize economic loss, and therefore can supply any load selected by the building's user. Connection of the selected standby source can be manual or automatic, and the duration of service is entirely at the user's discretion. No strictures are placed on the wiring system other than those applied to the normal electrical wiring system.

## (b) NFPA Codes

The *NEC* does not determine whether an emergency system is required; that is the function of the jurisdictional authorities and is generally covered by NFPA 101 (see the following list). Once a determination has been made, however, the equipment and installation must comply with the requirements of the *NEC*. Relevant NFPA codes include (latest issue):

NFPA 70 *National Electrical Code*
NFPA 99 *Standard for Health Care Facilities*
NFPA 101 *Life Safety Code*
NFPA 110 *Standard for Emergency and Standby Power Systems*
NFPA 111 *Standard on Stored Electrical Energy Emergency and Standby Power Systems*

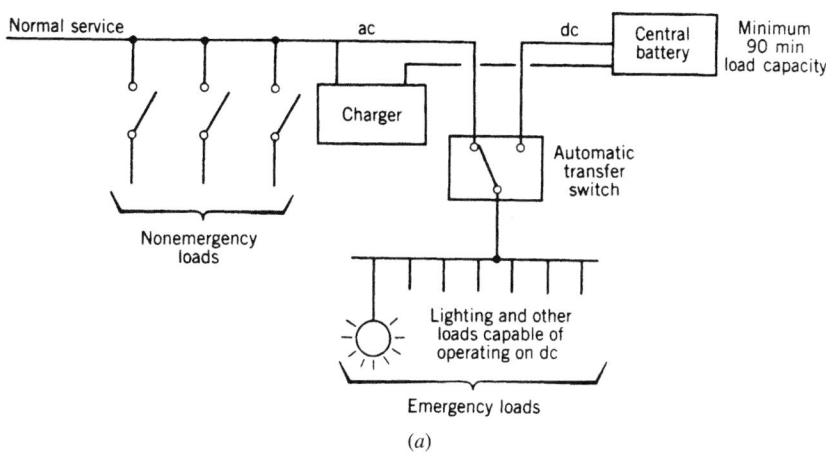

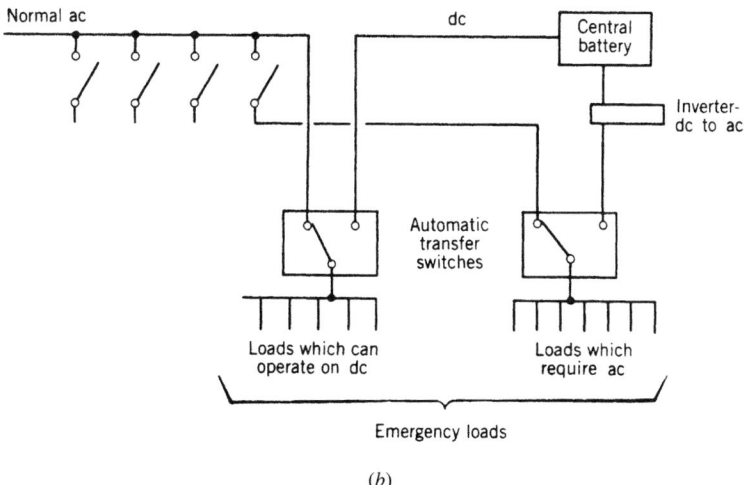

**Fig. 28.26** *Use of a central battery for emergency power supply. When all loads can be energized with dc, the arrangement of (a) is satisfactory. When ac as well as dc must be distributed, a central inverter is added, as in (b), and the ac and dc emergency loads are fed separately.*

ELECTRICITY

### (c) Technical Considerations

In general, when emergency power is discussed, it is assumed to be replacing normal power; that is, the assumption underlying most codes and ordinances is that power must be supplied to selected loads within the building because of utility power outages. In good-practice design, cognizance must also be taken of situations in which normal power has not failed and the outage is localized because of an equipment failure. That aspect of design—reliability—is left to the designer. Some of the arrangements discussed here differentiate between the nature of outages—that is, a utility or general outage versus an equipment or local outage. An exception to this generalization occurs with health-care facilities, where *NEC* in Article 517 specifies an internal electrical design that

largely covers both types of outages down to the distribution level. The interested reader should refer to the referenced *NEC* article for further information.

An emergency system includes all devices, wiring, raceways, and other electrical equipment, including the emergency source that is intended to supply electric power to the selected load. An important aspect of code requirements permits the use of a single power source for (1) emergency, (2) legally required standby, and (3) optional standby systems, provided that it is equipped with automatic selective load pickup and load-shedding equipment that will assure adequate power to the three types of systems in the order of priority stated (1, 2, 3).

1. Where emergency loads are light, a storage battery arranged to be connected automatically on

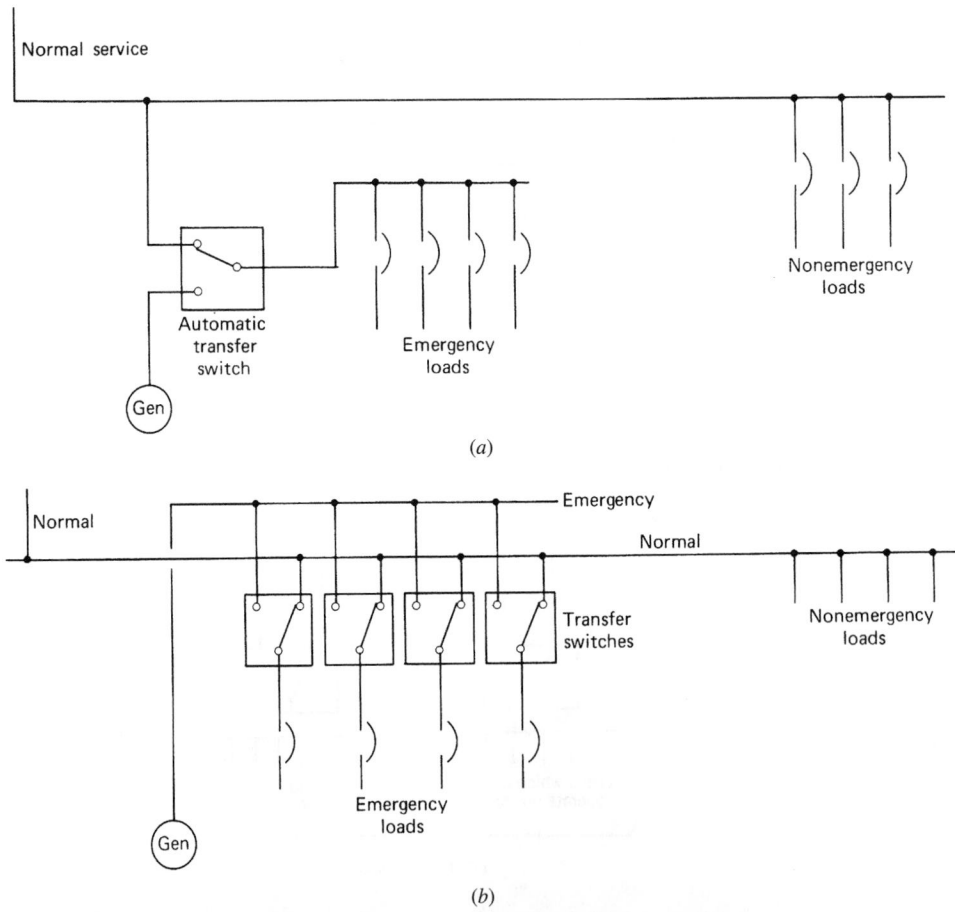

**Fig. 28.27** *Alternative arrangements of emergency/normal power feed. In (a) a single transfer switch serves the normal power and transfers to the generator on power failure. In (b) there are multiple smaller transfer switches, thus reducing the chance of a single equipment failure faulting out the entire emergency power system.*

power outage may be used. Where all emergency loads can be operated on dc, the arrangement of Fig. 28.26a is used. (Note that ac lighting can accept dc emergency power if equipped with a local inverter, as discussed in Section 16.36.) If ac is required, the arrangement of Fig. 28.26b can be utilized.

2. Where emergency loads are larger than can be practically supplied by a battery installation, and where a start-up and power transfer time of up to 10 seconds is tolerable, a generator set is employed. A combination of sources may be used in a single building. For instance, a generator can supply bulk power loads and a battery installation selected lighting loads.

The system can be arranged with a single transfer switch that senses normal power loss, as in Fig. 28.27a, or it can use multiple switches, each one of which will sense power loss *at its downstream location,* as in Fig. 28.27b. The latter system provides greater power reliability, provided that the design is such that the emergency power uses an independent power path to the transfer switches. Otherwise, a fault in a junction box or an item of equipment that interrupts normal power downstream will also prevent emergency power from reaching that point.

3. Many codes permit the use of two separate electric services in lieu of a normal service plus an emergency source, provided that the two sources are independent—that is, come from different utility transformers or feeders; enter the building at different points and, preferably, from different directions; and use separate service drops or laterals. The point is, of course, that the type of reliability desired can be obtained only by minimizing the possibility of a single event interrupting both services. The usual arrangement is for one such utility service to be normal and the other standby.

4. The least reliable arrangement is one in which the emergency loads are connected ahead of the main disconnects and are so arranged that a downstream fault within the building will not affect these items. This situation is illustrated in the riser diagram of Fig. 28.24, where the stair and exit panel, which supplies egress lighting, and the fire alarm panel are connected ahead of the building main disconnect and protected with their own fuses. This arrangement is permitted only for Legally Required Standby Systems (*NEC* Article 701) and not for Emergency Systems as defined in *NEC* Article 700. The items that may be connected ahead of the main service equipment are generally limited to emergency lighting, fire alarm, fire pumps, standby power equipment, and other alarm and protective equipment, each of which must be provided with a separate disconnect means and overcurrent protection (see *NEC* Article 230-82).

This power arrangement obviously can do nothing in the event of a power outage. It was once very common, but is now falling into disuse as a result of more stringent codes and its obvious shortcomings.

5. The *NEC* recognizes a category of equipment for emergency illumination called *unit equipment.* These devices, discussed and illustrated in Section 16.36, consist of individual self-contained packages with a battery, charger, and light source *permanently* mounted and wired at required locations. The panel device feeding these units should be clearly identified.

### References

Various standards from NFPA dealing with emergency power systems are presented in Section 28.21b.

McGuinness, W.J., B. Stein, and J.S. Reynolds. 1980. *Mechanical Equipment for Buildings,* 6th ed. John Wiley & Sons, New York.

Stein, B. 1997. *Building Technology, Mechanical and Electrical Systems,* 2nd ed. John Wiley & Sons, New York.

ELECTRICITY

CHAPTER 29

# Photovoltaic Systems

## 29.1 A CONTEXT FOR PHOTOVOLTAICS

A PHOTOVOLTAIC (PV) SYSTEM PRODUCES electricity through direct conversion of incident solar radiation (primarily radiation in the visible spectrum—light). This is an incredibly useful transformation process. Virtually every building of any size or function requires and consumes electricity. Solar radiation is a renewable resource, which will generally impinge upon a building in useful quantities in most climate zones (whether or not the building uses it for beneficial purposes). PV then seems an ideal design solution—providing a needed energy form from an otherwise often unused resource. In an ideal world, every building would have a PV system. Unfortunately, the economics of PV power production mitigate this idyllic picture. PV power production generally costs an owner more than utility-provided power, even considering reasonable life-cycle costing scenarios. In fact, the economics are currently such that building PV systems are the exception rather than the rule. There has long been hope that the economic picture would change through the evolving availability of much cheaper PV components. This has not really happened. Thus, any future shift in economics will likely occur mainly as the result of higher cost for competing electricity sources (utilities) and/or rebates or incentives from governments or utilities for distributed, greener power sources.

PV electric power systems, which until the middle to late 1980s were little more than a curios-

ity in the United States because of their very limited applicability and discouraging economics, have recently begun to make a noticeable impact on the commercial electrical power scene. This has come about for several reasons, the most important of which are federal government–financed research and development, and state and federal government–sponsored legislation in the electric utility field. The PV manufacturing technology project (PVMaT) involving the U.S. Department of Energy (DOE) and 20 private companies (launched in 1990), plus the 1992 Building Opportunities in the United States for Photovoltaics (PV:BONUS) program (also sponsored by the DOE) resulted in the reduction of PV module costs from upwards of $5.00 per watt to about $1.50–2.50. In addition, these initiatives spurred the development of new PV module materials, construction techniques, and product forms, increases in PV module efficiency, and a sharp overall decrease in the cost of power produced from about $0.50–$1.00 per watt to about half that amount, depending upon the system configuration. The legislative impetus has taken the form of utility deregulation and specific legislation requiring utilities to purchase power produced by small, private installations.

Several recent government efforts to foster the development of PV systems are underway. The (U.S.) Photovoltaic Industry Roadmap is an industry-led effort to help guide domestic PV research, technology, manufacturing, applications, markets, and policy. The Roadmap presents the intended direc-

ELECTRICITY

**1311**

tion of the PV industry, its critical partners, and U.S. government programs for the years 2000 to 2020. The U.S. PV program is associated with the International Energy Agency's (IEA) Task 16 effort focused upon PV systems for buildings. The potential market for building-integrated PV systems in the United States is enormous, and many companies are beginning to work at the development and commercialization of building-integrated PV components and systems. Increasing and substantial U.S. sales of PV roof shingles are indicative of this potential. Green building design efforts are also increasing interest in PV systems, with LEED new construction credit provided for renewable energy resources and for green power.

The 1978 federal Public Utilities Regulatory Policy Act (PURPA) required electric utilities to buy electric power at a price equal to the costs they avoid by not having to produce that power. The utilities generally interpret that "avoided cost" as fuel cost only, without considering other avoided costs such as additional plant construction. As a result, energy is often bought under this law at about $0.03/kWh by the same utilities that sell energy at $0.08–0.15/kWh. This spread obviously does not encourage the construction of grid-connected PV installations (see below). Subsequently, many states adopted net-metering laws that reduced this price unbalance. In this arrangement, the utility buys power during peak PV generation periods (see Fig. 29.1) at the same rate at which it sells power. At the end of a billing period (usually 1 month), any imbalance between utility-purchased and customer-purchased power is paid for at the above rates (i.e., if the customer used more power than he/she generated, that power is paid for at the conventional utility rate). Conversely (and less commonly), if the customer generated more power than was used, the utility would buy it at the avoided-cost rate.

The net metering arrangement is attractive economically to both utility and user. A utility buys power that offsets its peak afternoon load, which is particularly severe in warm weather, and sells power in its off-peak morning and evening periods. A residential customer sells power at the PV installation's peak output, which is also usually a period of low user consumption (Fig. 29.1), and buys it back at the same price in the evening, when PV output is low but power demand is high. The arrangement removes the necessity for an expensive battery installation that in a stand-alone system provides power during periods of low PV energy production. This advantage, combined with other advantages

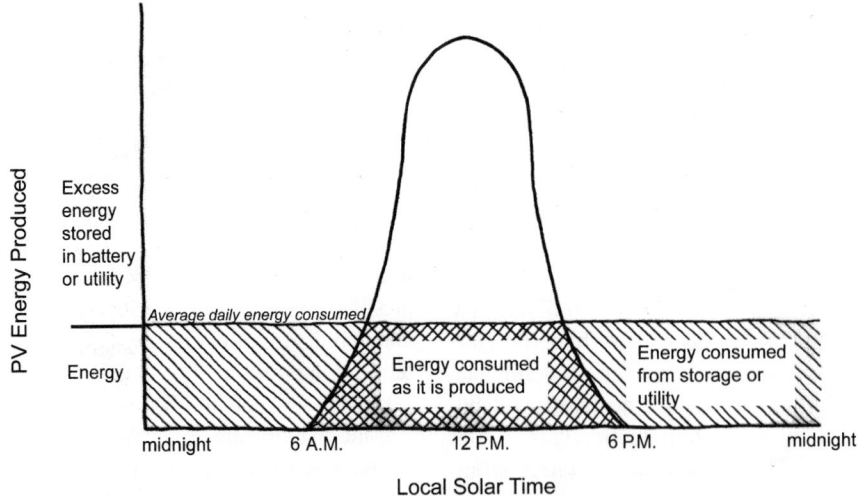

**Fig. 29.1** *Graph representing electricity produced and used by a residence equipped with a PV array. Daily energy use is shown as a uniform average over a 24-hour period to indicate overall quantity but not an actual usage pattern. The bell-shaped PV energy production curve is centered on solar noon and extends from sunrise (solar 6:00 A.M.) to sunset (solar 6:00 P.M.). The area of the bell curve above the average consumption line indicates electricity that would be purchased by a utility in a grid-connected system or stored in batteries in a stand-alone system. (From Harris, Miller, and Thomas, Solar Energy Systems Design, 1985, p. 715; reprinted with permission of John Wiley & Sons, Publishers. Redrawn by Amanda Clegg.)*

such as having a nonpolluting power source, has led to a large increase in grid-connected installations. Indeed, the attractiveness of having a green, environmentally beneficial electric power supply is so great that a number of electric utilities have instituted a (primarily residential) PV installation program in which the utility installs and maintains a PV system on the customer's premises (usually on the roof), for which the customer pays a nominal surcharge on the monthly electric power bill. The advantages of net billing, however, are available only to customers who initiate and maintain a PV system themselves.

## 29.2 TERMINOLOGY AND DEFINITIONS

A generally accepted terminology has developed in the PV field. The list that follows contains common terms and their accepted definitions. Particular attention should be paid to two terms that are often incorrectly used interchangeably: insolation and irradiance. *Irradiance* is the amount of solar *power* impinging on a specific area, usually measured in units of watts per square meter—and represents an instantaneous value. *Insolation* is the amount of solar *energy* received by a given area, measured in $Wh/m^2$, $kWh/m^2$, $Btu/ft^2$, or $MJ/m^2$ (with $1\ kWh/m^2 = 317.2\ Btu/ft^2$)—and represents a time-averaged value.

**Avoided cost.** The minimum amount an electric utility is required to pay an independent power producer under the Public Utilities Regulatory Policy Act of 1978.

**Balance of system (BOS).** All components other than the PV modules themselves, which include the system electronics, support structure, and storage.

**BIPV.** Building-integrated photovoltaic (system); a PV system with cells/modules incorporated as an integral part of the building envelope (rather than simply tacked on to the building shell).

**Blocking diode.** A diode used to block reverse flow of current into a PV source circuit.

**Interactive system.** A PV system that operates in parallel with the electric utility lines and may be designed to deliver power to the utility.

**Inverter.** Commonly known as a *power conver-*

*sion system* (PCS), an inverter is a device that changes direct-current (dc) to alternating-current (ac). An inverter is not the same as a power-conditioning unit (PCU).

**Panel.** A collection of modules mechanically fastened together, wired, and designed to provide a field-installable unit.

**Peak sun hours.** The number of hours per day at an irradiance of $1\ kW/m^2$ that is equivalent to the total daily insolation energy.

**PV array.** A mechanically integrated assembly of modules or panels with a support structure and other components, as required, including tracking apparatus where used, forming a power-producing unit.

**PV cell.** The basic PV device that generates electricity when exposed to solar radiation.

**PV module.** The smallest complete, environmentally protected assembly of PV cells and other components normally sold by a manufacturer; comprised of several (or many) PV cells.

**PV system.** All the components and subsystems that, in combination, convert solar energy into electrical energy suitable for connection to a load.

**Stand-alone system.** A PV system that supplies electrical energy without interconnection to any other power source.

**Thick-crystal photovoltaics.** The most common commercial type of PV material.

**Thin-film photovoltaics.** PV devices made of a semiconductor material, such as copper indium diselenide or amorphous silicon, a few micrometers thick deposited on substrates of glass, ceramic, or another compatible material.

## 29.3 PV CELLS

The production of an electric charge when solar radiation (primarily light) strikes some metallic surfaces has been a recognized physical phenomenon since the mid-nineteenth century. Around 1905, Einstein established a mathematical basis for what has come to be known as the *photoelectric effect*. The explanation of this effect, in simplified and qualitative terms, is approximately as follows. Light exhibits both wave characteristics and characteristics of a stream of energetic particles called *photons*. When a photon strikes a photoelectric metal surface, it dislodges a single electron from its normal orbit, causing a charge to appear. When pure silicon is exposed to an intense stream of photons, as from sunlight, a large number of electrons are dislodged from their orbits and proceed to wander about the crystal lattice structure of the silicon crystals. By a process called *doping*, impurities are deliberately added to pure silicon to create a P–N (positive–negative) junction, across which the electrons flow to create an electric current and give the doped silicon the properties of a semiconductor. If the semiconducting doped silicon is exposed to light and the negative and positive sides of the junction are connected through a load, an electron flow (a current) commences from the negative side of the junction to the positive side, doing work on the load. Work, of course, is energy. The energy comes from the fast-moving photons (light) and is imparted to the dislodged electrons by impact. The overall effect is to create a current flow proportional to the intensity of photon bombardment—that is, proportional to light intensity. Figure 29.2 is a schematic representation of this process.

The photons in solar radiation are not uniformly energetic. It has been found experimentally that a photon must have an energy of at least 1.08 electron-volts (corresponding to a wavelength of 1150 nm) to dislodge an electron. Photons at longer wavelengths do not have sufficient energy to create the photoelectric effect. Because shorter-wavelength photons are more energetic, the entire visible spectrum (light) is useful in PV action, the magnitude of current produced depending upon the intensity of light. According to Einstein's PV theory, a single photon can only dislodge a single electron; thus, any photon energy in excess of the minimum (dislodgement) level is dissipated as heat. For this reason, PV cells generate heat as well as electricity, and some means of heat collection or dissipation is required in practical PV array construction. Commercial PV cells operate at an overall insolation-to-electric energy conversion efficiency of around 8% to 12%. By layering different strata of semiconductor material, each of which is sensitive to a different limited frequency bandwidth (called *bandgap*), efficiencies as high as 35% to 40% can be achieved. Economic considerations, however, have limited today's commercial PV modules to a maximum of two layers with a maximum conversion efficiency of about 12%.

As mentioned previously, intensive research and development work on solar cell construction has yielded a new generation of PV materials that are variously referred to as *thin-film technology* and *polycrystalline technology*, using various copper and cadmium compounds and amorphous rather than large crystal silicon modules. These materials offer higher efficiencies, can be manufactured into large elements, and, because they do not require a glass cover, can be integrated into standard building ma-

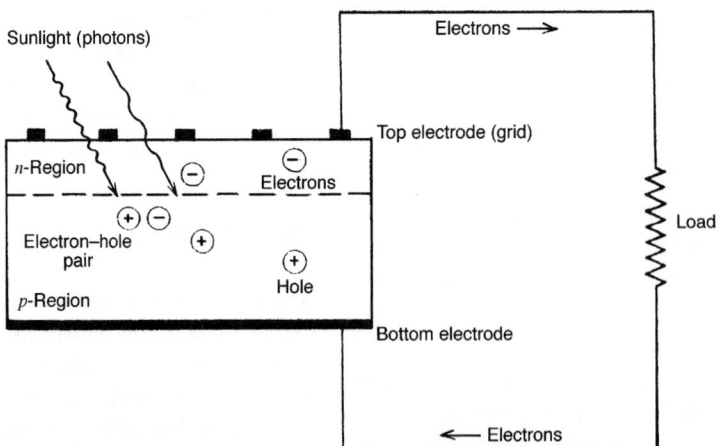

**Fig. 29.2** *Schematic drawing showing the conversion of solar energy to electrical energy by a PV cell. Photons from solar radiation dislodge electrons in the silicon semiconductor surface, creating a positively charged "hole." The N-P junction creates an electric field that "sweeps" electrons across the junction. When the top and bottom electrodes are electrically connected through a load, an electric current flows. This current is capable of doing work (supplying energy). (From Meyers, ed.,* Handbook of Energy Technologies, *1983, Wiley-Interscience; reprinted with permission.)*

terial formats (as building integrated PV elements). Typical data for two commercial products using this newer technology are:

***Amorphous Silicon Module***

Dimensions: 2 ft × 4 ft × 2 in.
   (610 mm × 1220 mm × 50 mm)
Weight: 33 lb (15 kg)
Electrical: 50 W, 72 V, 0.7 A

***Thin-Film Polycrystalline Module***

Dimensions: 2 ft × 4 ft × 2 in.
   (610 mm × 1220 mm × 50 mm)
Weight: 22 lb (10 kg)
Electrical: 75 W, 17 V, 4.5 A

These second-generation PV materials have high efficiency and a flexible physical format, distinct advantages over the older, large crystal silicon wafer elements. However, the latter are cheaper, *extremely* hardy, reliable, and long-lived, as has been demonstrated in many satellite and space vehicle applications. The newer units are initially somewhat unstable and have a shorter life than the large crystal silicon units.

The PV cell is the smallest electricity-producing unit in a PV system. Cells are not sold individually as commercial products but are arranged by the manufacturer into modules. A module is the smallest commercially available electricity-producing increment. The size of a module and the number of modules required to produce a desired electrical output depend upon the type of PV cell used, the manufacturer's product line, and the intent of the PV system

designer. If more than one module is required, as is often the case, multiple modules are assembled on site into arrays. A module is a self-contained product; an array is a field-assembled group of modules.

## 29.4 PV ARRAYS

A PV array is a complete and connected set of modules mounted and ready to deliver electricity. Two basic array arrangements exist: stationary and tracking. Building-mounted arrays are typically stationary and often have the advantage of not requiring a substantial support structure, although even a simple roof mounting on a pitched roof can be expensive due to the labor cost involved. The materials usually used for ground-mounted array structures are concrete, galvanized steel, and aluminum. Wood and painted iron are not recommended because of their high maintenance costs and relative short life. The design of an array mounting will depend upon the array size, weight, mounting angle, and wind loads and is therefore unique to each installation. Mounting on a wooden pole or steel pipe with a single support point is limited to relatively small arrays in low-wind areas because of the stress on the support arrangement. Several typical fixed mounting arrangements are shown in Figs. 29.3 through 29.5.

The angle, measured from the horizontal, at which a flat-plate collector is mounted is known as the *tilt angle*. For a stationary (nontracking) array, total *annual* insolation is maximum for a tilt angle equal to the latitude of the site. To maximize winter

ELECTRICITY

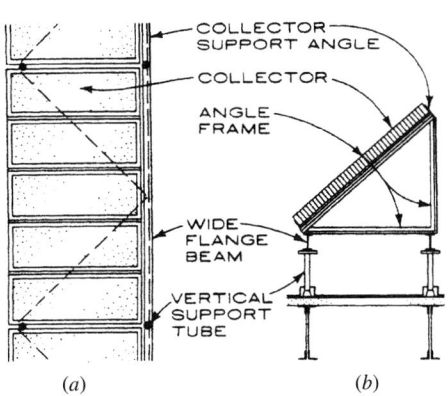
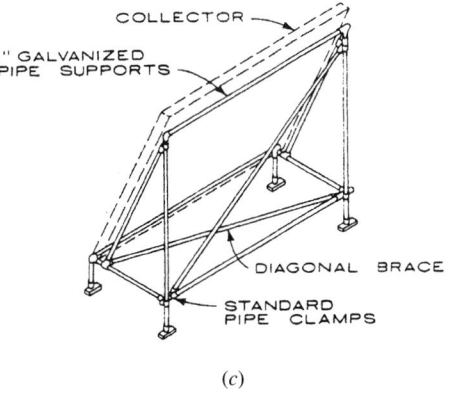

**Fig. 29.3** *Typical mounting techniques for PV arrays. (a) Plan of the arrangement for mounting large arrays on a horizontal surface. (b) Section through (a) showing mounting and arrays. (c) Simple pipe rack mounting for a small array. For all installations, ensure that wind and seismic loads, snow collection, and array cooling have been considered. (Diagrams reprinted with permission from* Architectural Graphic Standards, *10th ed., 2000, John Wiley and Sons, Publishers.)*

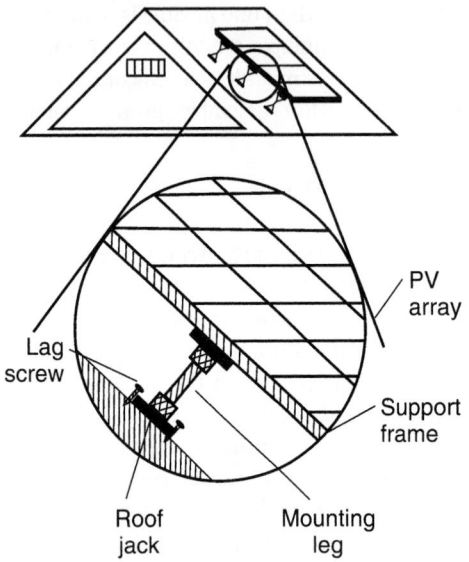

**Fig. 29.4** *Typical roof mount for a PV array. Array tilt and azimuth can be adjusted through the mount design. Exercise caution in snow areas so that accumulating snow does not block the array and does not slide off onto pedestrians. (From* Stand-Alone Photovoltaic Systems, *Sandia Laboratories, 1995.)*

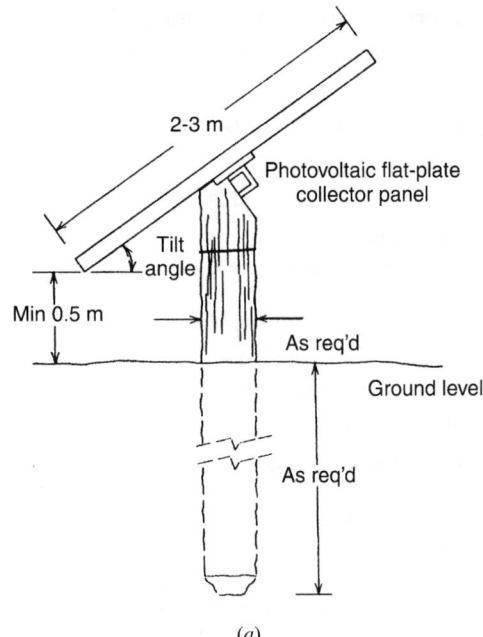

*(a)*

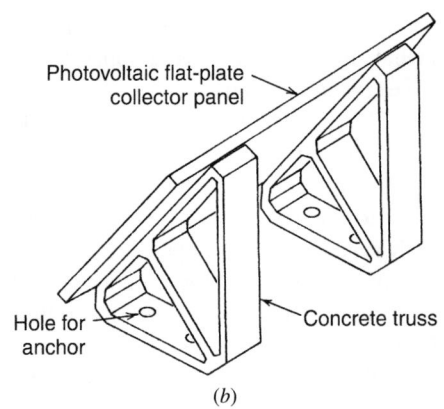

*(b)*

**Fig. 29.5** *Flat-plate PV collector support structures. (a) Wooden-pole torsion tube support. (b) Concrete truss support. (From Meyers, ed.,* Handbook of Energy Technologies, *1983, Wiley-Interscience; reprinted with permission.)*

insolation, when the sun is lowest, the tilt angle can be increased to latitude plus 15°. Conversely, to maximize summer insolation, when the sun is highest, the tilt angle can be reduced to latitude minus 15°. The tilt angle chosen for a stationary array depends upon PV output usage schedules, the amount of insolation for the months of the year, and economic considerations.

Tracking arrays follow the motion of the sun and thereby substantially increase the insolation per unit area of solar cells. Single-axis tracking follows the sun's motion from east to west while maintaining a constant array tilt angle. This type of tracking can increase insolation by 35% to 50% throughout the year and is most effective at low latitudes. Completely passive, sealed Freon, single-axis tracking drives are available for small to medium-sized pole-mounted arrays in low-wind areas. Where wind velocity is high, a motorized drive using PV-generated power may be necessary. Dual-axis tracking (following solar altitude and azimuth) for flat-plate collectors and altitude tracking for various designs of concentrating collectors are beyond the scope of this book.

## 29.5 PV SYSTEM TYPES AND APPLICATIONS

There are two basic types of PV systems: *stand-alone* and *grid-connected*. These names well describe their essential characteristic, which is their relationship to an external source of electricity such as a utility connection.

## (a) Stand-Alone Systems

A stand-alone system is just that; it is isolated from any outside electrical connections and is designed to do a carefully defined specific job. Stand-alone systems are the oldest PV installations because they were essentially the only solution to the problem of electrifying a remote and/or unattended load. Stand-alone installations typically supply electricity for sign lighting, railroad crossing lights, unattended pumps, lighthouses, unattended navigational aids, microwave repeaters, motor homes, sailboats and yachts, isolated small residences, and the like. The common characteristic of all these installations is the impossibility or impracticality of feeding them from a utility grid. When loads become larger than can be supplied by a practical solar array (from a physical or economic viewpoint), and particularly when the peak load is periodic rather than continuous, a fuel-powered generator is usually added to the system. Such a system is known as a *hybrid stand-alone system* and is used frequently for remote residences where the connection charge levied by the local utility is so large that a PV system is an economical and environmentally friendly solution to the problem of providing electricity. Schematic diagrams of stand-alone system types are shown in Figs. 29.6, 29.7, and 29.8.

The difference between the systems shown in Figs. 29.6 and 29.7 is the use of an energy storage medium—in most cases, a battery. The *direct-connected* system in Fig. 29.6 is applicable only with loads that tolerate variable power levels ranging from zero at night to maximum at solar noon. The most common of these is water pumping, using a dc motor to drive a positive displacement (piston) pump. Such installations are frequently used to fill elevated water tanks because a slow, interruptible fill rate does not adversely affect the water system's usability.

The vast majority of stand-alone systems are assembled as shown in Fig. 29.7, using a storage battery to store excess energy produced during peak insolation hours for use during periods of reduced solar resource (cloudy days) and full darkness (nighttime). With proper component selection, as is discussed in detail in subsequent sections, stand-alone PV systems can be used to adequately supply remote year-round residences, small mercantile establishments, and other off-grid loads.

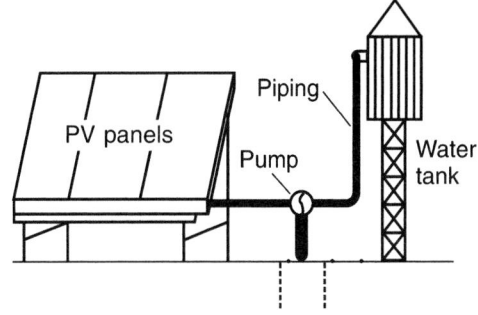

**Fig. 29.6** *Typical direct-connected PV installation feeds electric power to a positive-displacement pump that pumps water into an elevated tank.*

A stand-alone system may contain a dc-ac inverter if any of the loads require ac that cannot be supplied by a built-in inverter. The choice of whether to use a single central inverter or distributed smaller inverters is largely an economic decision. Fluorescent lamps, a reasonably efficient source of light, are available with integral inverter ballasts. Some kitchen appliances and power tools, however, are not readily available in a dc or battery-powered format, although that situation is rapidly changing and warrants careful investigation during system design.

The hybrid stand-alone system shown in Fig. 29.8 is similar to the stand-alone system in Fig. 29.7, with the addition of a small ac generator. This type of system, as noted previously, is used where operation

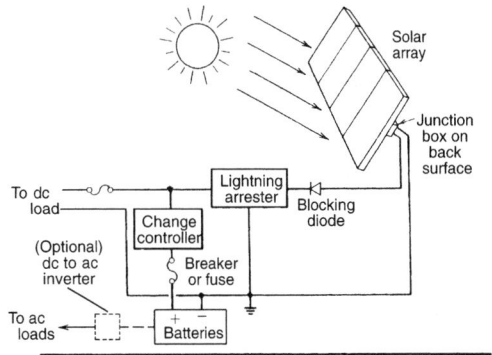

**Fig. 29.7** *Schematic diagram of the most common type of stand-alone PV installation. The blocking diode prevents reverse current flow when the battery voltage exceeds that of the PV array. The charge controller prevents overcharging of the battery and excessive current drain. If ac loads are present, a dc to ac inverter is added, sized as required. (From Harris, Miller, and Thomas,* Solar Energy Systems Design, *1985; reprinted with permission of John Wiley & Sons, Publishers.)*

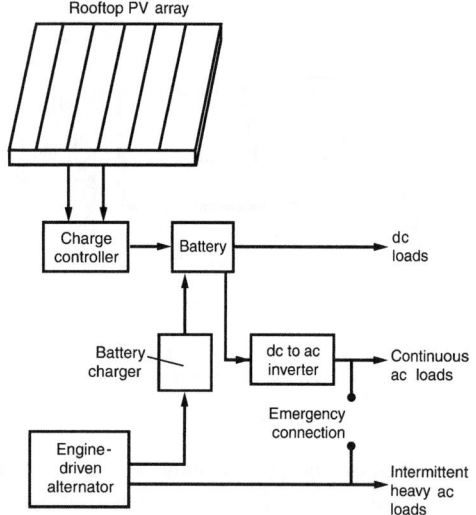

**Fig. 29.8** *One arrangement for a hybrid stand-alone PV system (control and safety devices are omitted for clarity). A diesel-, gasoline-, or propane-driven ac generator (alternator) supplies large, intermittent ac-only loads, whereas the PV array and battery supply dc loads and continuous ac loads through an inverter. During long, cloudy periods the alternator can charge the battery and, via an emergency connection, can also supply the regular ac loads. Other arrangements are possible, such as using a larger inverter to supply an all-ac system, with the alternator supplying intermittent heavy ac loads and all loads during low-battery periods.*

of a heavy ac load, such as a clothes washer or another relatively large ac motor–driven appliance, is necessary. Where these loads are fairly continuous, as with refrigeration and air-conditioning equipment, a larger PV system should be considered. A

continuously running engine–generator set is a source of noise and air pollution, and the ongoing consumption of a fossil fuel resource in such an installation makes it contradictory in spirit to a completely silent, pollution-free, renewable resource-driven, solar-powered PV installation.

## (b) Grid-Connected Systems

Prior to the advent of net-metering, the number of stand-alone PV systems in areas served by electric utility grids was small because of the overwhelming economic advantage of purchased power. Even today, with net-metering, the payback period upon an initial PV system investment seldom justifies an installation on purely economic grounds, although the continuous development work in PV materials may change that situation.

In a stand-alone arrangement, battery costs over the life of the system can exceed those of the PV array. Lead–acid batteries are relatively cheap but have a short life; nickel–cadmium batteries have a longer life but are much more expensive. By replacing the batteries that furnish energy storage and carryover in a stand-alone system with a utility connection that effectively serves the same purposes, and by taking advantage of the economies available from building-integrated PV elements (see Section 29.5c) in new construction, an economical PV system with reasonable payback can be constructed. A schematic diagram of one such system arrangement is shown in Fig. 29.9.

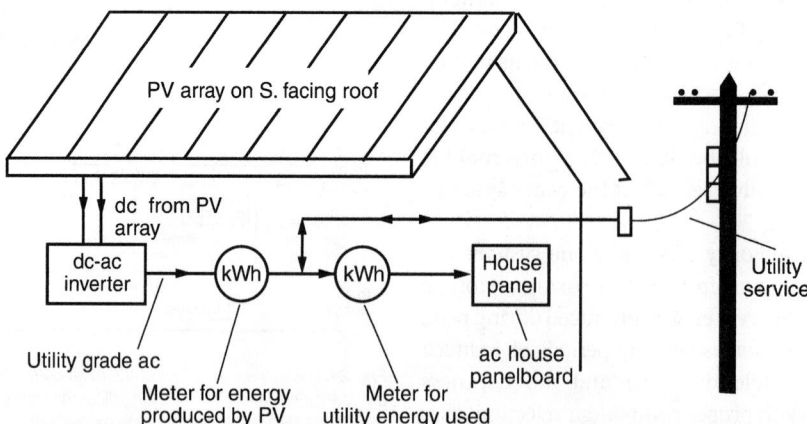

**Fig. 29.9** *Schematic drawing of a grid-connected PV system arranged for net-metering. Power from the array is converted from dc to utility-grade ac power (constant voltage and frequency; low harmonic content) by the inverter and fed to the utility connection through a kWh meter that measures the electricity sold to the utility. The second (utility company) meter measures the total power used by the structure. The difference in meter readings represents the positive or negative (credit) utility billing.*

The inverter shown in the diagram creates utility-grade ac from the PV array dc output and feeds it to the utility through an electric (kWh) meter. The inverter shown is equipped with a safety cutoff that instantly disconnects the PV system from the grid if a utility outage is sensed. This is a requirement of the electric utilities to ensure that their maintenance personnel are not endangered by PV electric power being fed into an apparently deenergized line.

The power company supplies house power through a second meter, as shown. In net-metering, the difference between the readings of the two meters is either billed to the consumer at the normal utility rate or credited to the consumer at the avoided-cost rate, depending on which meter reads higher. A consumer will notice no difference whatever in electric service, except in the lower, or possibly even negative, electric utility bills.

More recently, to encourage wider use of non-polluting PV power, some utilities have permitted the installation of small, individual PV modules in existing buildings. These units plug into convenience outlets in the building, supplying power to the building itself. Any excess electrical energy not required by the structure is fed back into the utility via their reversible energy meter. This arrangement is shown schematically in Fig. 29.10. Each PV module (panel) is equipped with a built-in, high-quality inverter that produces utility-grade power and is equipped with the safety devices described previously. A common commercially available unit of this type is rated at 100 W peak, 85 W nominal, 120 V, 60 Hz, measures approximately 2 ft × 4 ft × 4 in. (610 mm × 1220 mm × 50 mm) and weighs 25 lb (11 kg). The advantage gained with these units is that a user can begin with one or two modules and then expand as desired, without a relatively high first cost investment and without a centralized installation requiring professional construction and electrical personnel. Some power companies require separate disconnect switches for each PV unit rather than relying only on physical disconnection of the unit plug from the wall outlet.

## (c) Economic Considerations

Economic issues can be complex, and they are often the deciding factor in the choice between full PV, a PV-grid combination, and a conventional (grid) electric service. Among the cost factors are:

- Cost of a grid connection.
- Cost of power from the grid.
- Life-cycle cost of each component of the system over the life of the entire installation. This is necessary because the life span of individual components varies sharply; PV modules last about 20

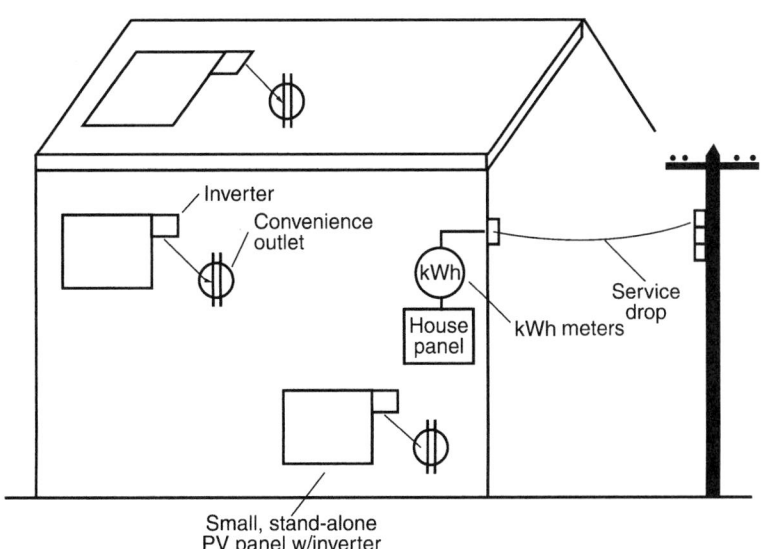

**Fig. 29.10** *Schematic drawing of a grid-connected PV system using small, distributed PV arrays connected to the house distribution system via convenience outlets. In this arrangement, the utility meter is unratcheted (reversible) so that excess PV-produced electricity reverses the power company's meter. Any net monthly credit is calculated at the avoided-cost rate.*

years, batteries about 5 years, electrical components 10 to 20 years (or more), PV module mountings 15 to 25 years, tracking mechanisms (if used) 5 to 10 years, and so on.

- Life-cycle maintenance costs (parts and labor) for each item.
- Financing costs.
- Cost credits for PV systems, including such items as use of a system battery as a UPS source for computers and peripherals, savings from avoided grid-power service interruptions, and construction cost credits when PV modules are used in lieu of building elements.

The last factor has taken on considerable importance in recent years with the development of building-integrated PV (BIPV) elements. A traditional PV installation consists of PV modules mounted near or on a building or facility. Often, the modules are roof-mounted if a roof surface of appropriate orientation (generally south-facing) and sufficient size is available. The development of thin-film amorphous silicon modules permits incorporation of PV cells into assemblies that can replace traditional building construction elements (typically in the form of roof panels, roofing tiles, wall panels, or skylights). These assemblies fulfill the dual function of producing electricity and weatherproofing the building envelope. In the jargon of design intent, they are *transformer* elements, replacing historically used barrier or filter elements. In such cases, only the difference in cost between standard construction and BIPV construction should be taken as an additional cost of the PV system.

## 29.6 PV SYSTEM BATTERIES

Batteries are required in a stand-alone system to store excess PV-produced electricity (from high production periods) for use during peak electricity consumption periods (the pattern illustrated in Fig. 29.1). For residential installations, there is usually almost no coincidence between these two time periods because PV output peaks at solar noon, whereas residential usage peaks in the evening and is low at noon. In comparison, commercial stand-alone PV installations such as off-grid stores, shops, cottage industries, and the like have a much higher coinci-

dence. This means that during normal, everyday use, the drain on the battery is shallow, and a cheaper battery will frequently suffice. These considerations are expanded in the system design discussions in Sections 29.8 and 29.9.

A PV system battery is usually expected to be capable of supplying most or all of an installation's electrical requirements for a given period—usually 3 days of cloudy weather if no other specific-use data are available. This means that the battery selected:

- Must be capable of repeated discharge to 20% capacity or less without injury
- Must be designed to accept slow, lengthy recharge

In addition, the battery:

- Should have long life
- Should be suitable for the physical conditions in which it will be used (i.e., space, ambient temperature, ventilation, and maintenance availability)

Pure lead automobile batteries (also known as *lead–acid* batteries) are not suitable for PV systems because they are designed for rapid, shallow discharge (engine starting) and very rapid recharge. If such a battery is repeatedly discharged more than about 20% (shallow discharge), or if it is not completely recharged after discharge, its life will be extremely short. Batteries made for use in electric vehicles such as golf carts and fork lifts are constructed with heavy plates and are designed for the deep-cycle discharge needed in PV systems. Most of these are flooded batteries, meaning that the plates must be completely covered with electrolyte to prevent damage. This requires frequent maintenance, making flooded batteries unsuitable for unattended sites.

Lead–calcium batteries are almost maintenance-free, rarely requiring addition of water or electrolyte. As a result, they are frequently used in unattended systems. They do not cycle as deeply as heavy-plate, deep-cycle units, and this must be considered during design.

Sealed batteries are completely maintenance-free. They are frequently constructed with a gel-type electrolyte that does not "boil" off on recharge and therefore needs no replacement. This makes them ideal for unattended sites, provided that they are specifically designed for deep-cycle service (80%

discharge) and are so labeled. Most maintenance-free batteries are shallow-cycle units and therefore are unsuitable. An important characteristic of sealed units is that they require very accurate charge and discharge control to prevent overcharge or overdischarge, or over-rapid charge or discharge. Any of these conditions severely shortens battery life. As a result, gel-cell batteries require the use of a high-quality charge controller.

Lead–antimony batteries are very sturdy and tolerate deep discharge and slow or rapid charge while maintaining their rated life. They require occasional water replenishment and adequate ventilation to disperse gases created during charging cycles.

Nickel–cadmium batteries are highly desirable in PV systems because they are maintenance-free, can be discharged to essentially 100% of capacity without injury or shortening of life, are much less temperature-sensitive than lead–acid batteries, and have a longer life. Their principal drawback is their high price. Nickel–cadmium batteries should not be shallow discharged because they tend to develop a "memory" of the discharge cycle and after many shallow cycles may not supply sufficient current for a deep discharge. Therefore, in essentially clear-sky climates, where a 3-day period of cloudiness is rare, nickel–cadmium batteries may not be applicable because storage sized for a 3-day system supply would normally be operating on shallow discharge.

Additional considerations in battery selection include:

*Temperature.* Lead–acid battery capacity declines with temperature. Nominal battery capacity ratings are referenced to a temperature of 77°F (25°C). Batteries operating at lower temperatures should be derated (in the absence of specific manufacturer's recommendations) as follows:

| Ambient Temperature | Derating Factor |
|---|---|
| 70°F (21°C) | 0.96 |
| 60°F (16°C) | 0.90 |
| 50°F (10°C) | 0.84 |
| 40°F (4°C) | 0.77 |
| 30°F (−1°C) | 0.71 |
| 20°F (−7°C) | 0.63 |

It is important to note that a discharged lead–acid battery contains essentially water and therefore freezes at about 30°F (−1°C). A fully charged battery will not freeze until the temperature dips below 0°F (−18°C). Nickel–cadmium batteries are much less affected by temperature.

*Voltage.* Battery terminal voltage varies with its state of charge, ranging for a 12-V battery at 77°F (25°C) from approximately 15 V at full charge to just over 12 V immediately after a deep discharge. If the battery is not charged immediately, voltage drops to a low of slightly over 11 V.

*Life.* A properly maintained deep-cycle lead–acid battery should last for a minimum of 5 years. A nickel–cadmium battery should last 1.5 to 2 times as long.

*Rating.* The manufacturer's stated ampere-hour (A-H) rating of a battery refers to a single continuous low-rate discharge under laboratory conditions. The actual field use A-H capacity of a battery depends heavily upon the discharge rate, and it is this figure, as calculated for the particular installation, that should be used to compare batteries of the same type. All reputable manufacturers have readily available discharge curves that show voltage and duration at different discharge rates.

The basic storage battery is rated at 12 V, although 6-V units are available. Figure 29.11 shows typical battery connection arrangements for the most common system voltages, which are 12, 24, 48, and 72 V. Table 29.1 shows some typical physical data for batteries.

## 29.7 BALANCE OF SYSTEM

The balance of system (BOS) consists of a charge controller, inverter if used, and accessories such as blocking diodes, a lightning arrester, an electrical grounding system, and other electrical specialty items. Refer to Figs. 29.7 through 29.9 for a sense of how these components fit into a PV system. Although they are important to overall system performance and safety, a full description of these items is beyond the scope of this chapter. Refer to the references at the end of this chapter for a few of the many excellent resources that address these components.

ELECTRICITY

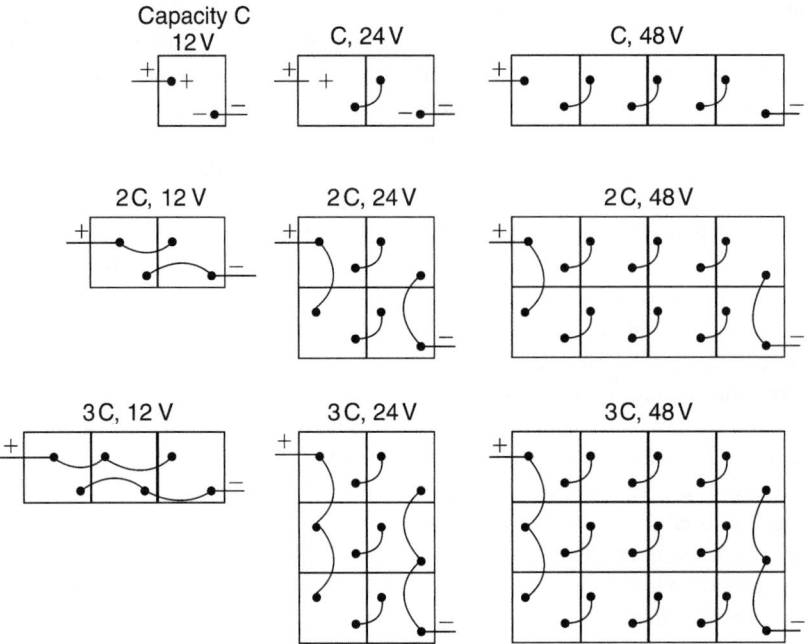

**Fig. 29.11** *Wiring arrangements for 12-V battery packs to obtain increased capacity, increased voltage, or both.*

## 29.8 DESIGN OF A STAND-ALONE PV SYSTEM

There is a fundamental difference in the procedures for array sizing and system arrangement between a stand-alone system and a grid-connected PV system. A stand-alone system must supply the entire electric load; there is no utility backup. Design decisions are primarily technical. A grid-connected system may supply part, or all, of the electric load. Design decisions are influenced by economics and owner intent, within the

context of technical feasibility. These two procedures are addressed separately in this and the following section.

Figure 29.7 shows the essentials of a stand-alone PV system. Because building electricity demands rarely exactly match the output of a PV system, the system operates successfully by using a PV array to charge a battery, which then supplies the building load. Any occasional coincidence between load and PV production is incidental to the design process. The three factors that establish the system requirements are:

**TABLE 29.1 Physical Characteristics of Typical 12-V[a] Batteries**

| Battery Type | Capacity in Amp-Hours | Dimensions of a 12-V Unit L × W × H in. (mm) | Weight lb (kg) |
|---|---|---|---|
| Deep-cycle Lead–acid | (6/20/100-hr discharge rates) 500/625/775 675/840/1025 850/1050/1275 | 33 × 6.5 × 24 (838 × 165 × 609) 40 × 6.5 × 24 (1016 × 165 × 609) 50 × 6.5 × 24 (1270 × 165 × 609) | 500 (227) 700 (318) 800 (363) |
| Sealed Gel-cells | (20-h discharge rate) 850 1050 1275 1500 | 40 × 8 × 24 (1016 × 203 × 609) 40 × 9 × 24 (1016 × 229 × 609) 40 × 10 × 24 (1016 × 254 × 609) 40 × 12 × 24 (1016 × 305 × 609) | 650 (295) 770 (349) 930 (422) 1100 (499) |

[a]Use two units for a 24-V battery and four units for a 48-V battery.

- The daily electricity usage
- The period of time for which the battery must supply the electric load without recharge
- The available insolation

The first factor is a function of building design. The second factor is a function of local weather conditions and would most often be chosen as the number of successive solid overcast or rainy days usually encountered. Designing for highly unusual weather results in a system that is oversized for normal (say, 95th percentile) weather patterns and, therefore, is essentially uneconomical. That scenario weighs heavily in favor of a hybrid system (see Fig. 29.8), which defeats, at least in part, the environmental benefits of the PV system.

The PV system design procedure is essentially straightforward and is as follows:

**STEP 1.** Determine the *average* daily energy usage in watt-hours.

**STEP 2.** Determine the maximum period for which the system battery must supply the entire electrical load. (In practice, this requirement can be reduced by switching off nonessential loads during unusually long sunless periods.)

**STEP 3.** Determine the required system battery capacity, which is equal to the daily load (Step 1) multiplied by the number of days that the battery must supply the load (Step 2). System losses are taken into account in this calculation.

**STEP 4.** Determine the preliminary size of the PV array. The array must supply all of the required electricity, that is, the facility's daily requirements (Step 1) plus all system losses. The preliminary array size is this capacity in terms of watt-hours. To determine the physical size of the array, the designer must determine the number of watt-hours produced per unit area of PV module. This involves knowledge of the site's solar radiation, the type and efficiency of the PV module used, mounting approach (stationary or tracking), and tilt angle (module angle measured from the horizontal). With these data in hand, the area of PV array required is simply the total load in watt-hours divided by the PV unit's energy production in watt-hours per unit area.

**STEP 5.** Check that the array selected is sufficiently large to charge the battery capacity estab-

lished in Step 3. Assuming, for instance, a 3-day period during which the battery must supply the entire load, about 3½ consecutive sunny days *at the same insolation* will be needed for full recharge (considering losses of approximately 20%). If the facility is to be used year-round and the PV array is sized on the basis of summer insolation, it is necessary to do a calculation for a winter month as well. Insolation data are usually available for both seasons, and average monthly insolation accounts for the appreciable energy available from cloudy skies as well as sunny skies. The result generally indicates a larger PV array requirement for winter months. In such cases, on purely economic grounds, the decision is often to use a small engine generator for battery-charging for winter use only. This is *not* the same as a hybrid system (Fig. 29.8), in which the generator is used year-round to pick up heavy ac loads. Some designers base the PV array size on *minimum* yearly (winter) insolation, reasoning that in other months any wasted overproduction from an array is the price paid for year-round operation without fossil fuels. Others use a median insolation figure, reasoning that power outages will be few and of short duration. Using summer insolation values will minimize the PV array size but probably will require inclusion of a small generator.

An illustrative example using the outlined design procedure should clarify the issues and calculations involved.

**EXAMPLE 29.1** Design a stand-alone PV system for a 650-ft² (60-m²) occupied-year-round, off-line cottage located near Prescott, Arizona. In addition to the usual appliances, an evaporative cooler is used during hot summer days. A PV array can be mounted on the south-facing sloping roof at any required angle. Determine the required battery and PV array sizes. State all assumptions.

**SOLUTION**

**STEP 1.** Determine the average electric energy consumption for the cottage. Although refrigerators that run on bottled gas are readily available, assume use of an electric refrigerator. Water heating is solar thermal, with a small gas booster heater. Cooking is with bottled gas.

| 24-V dc Appliances | Power Draw (W) | Daily Use (h) | Daily Energy Need (Wh) |
|---|---|---|---|
| Mid-sized refrigerator/freezer | 35 | 24 | 840 |
| Small color TV | 60 | 4 | 240 |
| Evaporative cooler[a] | 50 | 8 | 400 |
| Stereo system | 40 | 2 | 80 |
| Ceiling fans[a] (3) | 3 at 25 | 4 | 300 |
| Radios | 5 | 10 | 50 |
| Incandescent lighting | 50 | 8 | 400 |
| | | Daily use dc | 2310 Wh |

[a]Summer only.

| 120-V ac Appliances | Power Draw (W) | Daily Use (h) | Daily Energy Need (Wh) |
|---|---|---|---|
| Fluorescent lighting | 80 | 10 | 800 |
| Clothes washer | 500 | 0.5 | 250 |
| Vacuum cleaner | 600 | 0.25 | 150 |
| Kitchen appliances | 400 | 1 | 400 |
| Iron | 1000 | 0.25 | 250 |
| Miscellaneous | 100 | 3 | 300 |
| | | Daily use[a] ac | 2150 Wh |

[a]Maximum coincident ac load.

| Total daily energy usage: | |
|---|---|
| dc | 2310 |
| ac 2150 Wh × 1.15 (inverter loss) | 2473 |
| | 4783 Wh |
| Ampere-hours = (4783 Wh/24 V) = 199, say 200 A-H | |

**STEP 2.** Assume that investigation of local climate conditions indicates occasional periods throughout the year, but primarily in the winter, of 2–3 consecutive days of heavy overcast, clouds, and rain.

**STEP 3.** Sizing the system battery. Assume use of a deep-cycle flooded-cell lead battery with an appropriate charge controller. These batteries can supply about 80% of their rated ampere-hour capacity. Since the battery will be located indoors, but in an unheated room, it should be derated 15% for ambient temperature for a conservative design. Required battery size in ampere-hours equals:

$$\frac{3 \text{ days} \times 200 \text{ A-H daily load} \times 1.20 \text{ discharge factor}}{0.85 \text{ (temperature derating)}} = 850 \text{ A-H}$$

One type of 850-A-H, 24-V deep-cycle lead–acid battery (using two at 12-V units in series) measures approximately 40 in. long × 15 in. wide × 24 in. high (1020 mm × 380 mm × 610 mm) and weighs about 1500 lb (680 kg).

**STEP 4.** Sizing the PV array. Because the cottage will be occupied all year, and because the critical performance period is winter, when the sun is low, select a tilt of (latitude +15°) to maximize winter PV production. As the array is roof-mounted and will be relatively large, use a stationary rather than a tracking array. The array must supply the daily energy requirement of 4783 Wh or 4.8 kWh.

Referring to Fig. 29.12, which gives solar radiation data for the Prescott location, obtain the following data: average solar radiation for a PV array facing south at a tilt angle of latitude +15°:

January: 5.5 $kWh/m^2/day$; average monthly temperature 2.3°C (0.51 $kWh/ft^2/day$; 36°F)

June: 6.0 $kWh/m^2/day$; average monthly temperature 19.6°C (0.56 $kWh/ft^2/day$; 67°F)

Figure 29.13 was drawn using the computer program PVFORM, developed at Sandia National Laboratories. It gives daily average PV array production in $Wh/ft^2$ as a function of monthly average temperature and monthly average solar radiation in $kWh/m^2/day$. Using the January and June radiation figures stated previously, obtain from Fig. 29.13:

January: 52 $Wh/ft^2$ of a PV array (560 $Wh/m^2$)

June: 53 $Wh/ft^2$ of a PV array (570 $Wh/m^2$)

The PV array output is essentially constant throughout the year (with radiation availability and ambient temperature effects offsetting), and no difficulty will be encountered in an "off" season. If the PV modules to be used have an efficiency greater than 10%, then the array output would be increased proportionately.

The required area of PV array, assuming 10% conversion efficiency, is:

PV area =
(4783 Wh/day required) / (52 Wh/day/$ft^2$ output)
= 91 $ft^2$ (8.5 $m^2$)

**STEP 5.** Because the solar radiation in Prescott, Arizona, is essentially identical for summer and winter, PV production in 3 to 3½ clear days will fully recharge the battery, as explained previously.

**Summary.** The suggested PV system consists of approximately 90 $ft^2$ (8.4 $m^2$) of PV collector, an 850-A-H, 24-V battery, and a 1500-W inverter to handle the initial load plus the anticipated addition of small ac appliances. The summer-only DC cooling loads will likely be offset in winter by additional use of lighting and television and other in-cottage activities.

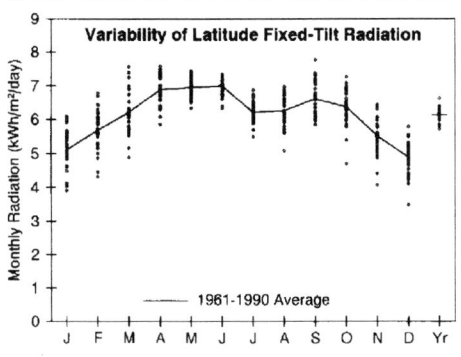

**Prescott, AZ**

**WBAN NO. 23184**

LATITUDE: 34.65° N
LONGITUDE: 112.43° W
ELEVATION: 1531 meters
MEAN PRESSURE: 847 millibars

STATION TYPE: Secondary

### Solar Radiation for Flat-Plate Collectors Facing South at a Fixed Tilt (kWh/m²/day), Uncertainty ±9%

| Tilt (°) | | Jan | Feb | Mar | Apr | May | June | July | Aug | Sept | Oct | Nov | Dec | Year |
|---|---|---|---|---|---|---|---|---|---|---|---|---|---|---|
| 0 | Average | 3.1 | 3.9 | 5.1 | 6.6 | 7.5 | 8.0 | 6.9 | 6.3 | 5.7 | 4.6 | 3.4 | 2.8 | 5.3 |
| | Min/Max | 2.6/3.5 | 3.2/4.5 | 4.2/6.1 | 5.7/7.1 | 6.8/8.0 | 7.2/8.5 | 6.1/7.7 | 5.1/7.0 | 5.1/6.7 | 3.6/5.2 | 2.8/3.9 | 2.2/3.2 | 5.1/5.7 |
| Latitude -15 | Average | 4.4 | 5.1 | 5.9 | 7.0 | 7.5 | 7.7 | 6.7 | 6.5 | 6.5 | 5.8 | 4.8 | 4.1 | 6.0 |
| | Min/Max | 3.5/5.1 | 4.0/6.0 | 4.7/7.2 | 6.0/7.7 | 6.8/8.0 | 6.9/8.1 | 5.9/7.5 | 5.3/7.2 | 5.7/7.6 | 4.4/6.6 | 3.7/5.5 | 3.0/4.8 | 5.6/6.5 |
| Latitude | Average | 5.1 | 5.7 | 6.2 | 6.9 | 7.0 | 7.0 | 6.2 | 6.3 | 6.6 | 6.4 | 5.5 | 4.9 | 6.1 |
| | Min/Max | 3.9/6.1 | 4.3/6.8 | 4.9/7.6 | 5.8/7.6 | 6.3/7.4 | 6.3/7.3 | 5.5/6.9 | 5.1/7.0 | 5.8/7.8 | 4.7/7.3 | 4.1/6.4 | 3.5/5.8 | 5.7/6.6 |
| Latitude +15 | Average | 5.5 | 5.9 | 6.1 | 6.4 | 6.1 | 6.0 | 5.4 | 5.7 | 6.4 | 6.5 | 5.9 | 5.4 | 5.9 |
| | Min/Max | 4.2/6.6 | 4.4/7.1 | 4.8/7.5 | 5.4/7.1 | 5.6/6.5 | 5.4/6.3 | 4.8/5.9 | 4.6/6.3 | 5.6/7.5 | 4.7/7.5 | 4.3/6.9 | 3.7/6.4 | 5.5/6.4 |
| 90 | Average | 5.0 | 4.9 | 4.3 | 3.5 | 2.6 | 2.3 | 2.3 | 2.9 | 4.0 | 5.0 | 5.2 | 5.0 | 3.9 |
| | Min/Max | 3.8/6.1 | 3.6/5.9 | 3.4/5.2 | 3.0/3.8 | 2.5/2.8 | 2.2/2.3 | 2.1/2.4 | 2.4/3.1 | 3.6/4.7 | 3.6/5.8 | 3.6/6.2 | 3.4/6.0 | 3.5/4.2 |

### Solar Radiation for 1-Axis Tracking Flat-Plate Collectors with a North-South Axis (kWh/m²/day), Uncertainty ±9%

| Axis Tilt (°) | | Jan | Feb | Mar | Apr | May | June | July | Aug | Sept | Oct | Nov | Dec | Year |
|---|---|---|---|---|---|---|---|---|---|---|---|---|---|---|
| 0 | Average | 4.8 | 5.9 | 7.4 | 9.4 | 10.6 | 11.3 | 9.3 | 8.8 | 8.4 | 7.1 | 5.4 | 4.4 | 7.7 |
| | Min/Max | 3.7/5.7 | 4.4/7.2 | 5.7/9.5 | 7.9/10.6 | 9.4/11.5 | 9.9/12.0 | 7.9/10.8 | 7.0/10.1 | 7.1/10.2 | 5.0/8.2 | 4.0/6.3 | 3.2/5.2 | 7.1/8.4 |
| Latitude -15 | Average | 5.7 | 6.8 | 8.1 | 9.8 | 10.6 | 11.2 | 9.3 | 9.0 | 9.0 | 8.0 | 6.3 | 5.4 | 8.3 |
| | Min/Max | 4.3/6.9 | 5.0/8.3 | 6.2/10.3 | 8.2/11.0 | 9.4/11.6 | 9.8/11.9 | 7.8/10.7 | 7.1/10.4 | 7.6/10.9 | 5.6/9.3 | 4.6/7.5 | 3.8/6.4 | 7.5/9.0 |
| Latitude | Average | 6.3 | 7.2 | 8.3 | 9.8 | 10.3 | 10.7 | 8.9 | 8.8 | 9.1 | 8.4 | 6.9 | 6.0 | 8.4 |
| | Min/Max | 4.7/7.6 | 5.3/8.8 | 6.3/10.6 | 8.1/11.0 | 9.2/11.3 | 9.4/11.4 | 7.6/10.4 | 7.0/10.2 | 7.7/11.1 | 5.8/9.8 | 4.9/8.2 | 4.1/7.2 | 7.6/9.2 |
| Latitude +15 | Average | 6.6 | 7.4 | 8.2 | 9.4 | 9.7 | 10.1 | 8.4 | 8.4 | 8.9 | 8.5 | 7.2 | 6.3 | 8.3 |
| | Min/Max | 4.9/8.1 | 5.3/9.1 | 6.2/10.6 | 7.7/10.6 | 8.6/10.6 | 8.8/10.7 | 7.1/9.7 | 6.7/9.8 | 7.5/10.9 | 5.8/9.9 | 5.1/8.6 | 4.3/7.6 | 7.5/9.0 |

### Average Climatic Conditions

| Element | Jan | Feb | Mar | Apr | May | June | July | Aug | Sept | Oct | Nov | Dec | Year |
|---|---|---|---|---|---|---|---|---|---|---|---|---|---|
| Temperature (°C) | 2.3 | 3.9 | 5.9 | 9.6 | 14.2 | 19.6 | 22.8 | 21.3 | 18.1 | 12.6 | 6.7 | 2.6 | 11.7 |
| Daily Minimum Temp | -5.6 | -4.4 | -2.2 | 0.8 | 5.1 | 9.9 | 14.4 | 13.2 | 9.3 | 3.4 | -1.9 | -5.4 | 3.1 |
| Daily Maximum Temp | 10.2 | 12.2 | 14.1 | 18.3 | 23.3 | 29.2 | 31.2 | 29.4 | 26.7 | 21.8 | 15.3 | 10.6 | 20.2 |
| Record Minimum Temp | -18.3 | -18.3 | -13.9 | -6.1 | -3.3 | 2.2 | 8.3 | 4.4 | 1.1 | -7.2 | -13.3 | -22.8 | -22.8 |
| Record Maximum Temp | 23.3 | 25.0 | 27.2 | 30.6 | 35.0 | 39.4 | 38.9 | 37.2 | 35.0 | 32.2 | 26.1 | 21.1 | 39.4 |
| HDD, Base 18.3°C | 496 | 403 | 384 | 262 | 138 | 32 | 0 | 0 | 41 | 182 | 350 | 489 | 2775 |
| CDD, Base 18.3°C | 0 | 0 | 0 | 0 | 11 | 68 | 139 | 96 | 32 | 4 | 0 | 0 | 351 |
| Relative Humidity (%) | 59 | 53 | 49 | 38 | 33 | 27 | 44 | 51 | 47 | 46 | 51 | 59 | 47 |
| Wind Speed (m/s) | 3.1 | 3.5 | 4.1 | 4.2 | 4.3 | 4.1 | 3.5 | 3.1 | 3.4 | 3.3 | 3.3 | 3.0 | 3.6 |

**Fig. 29.12** *Solar radiation data for Prescott, Arizona. (From Solar Radiation Data Manual, NREL, 1994.)*

**ELECTRICITY**

***Additional Considerations.*** Using accurate radiation and weather data for a site is necessary for the design of an efficient and workable system. Unfortunately, accurate radiation data, particularly for tilted surfaces, are not always available, and tilted-surface data are not easily derived from horizontal-surface or normal-to-the-sun data. PV installations are frequently planned for sites outside of large cities, and it is usually for these areas that a paucity of data exists. In such instances, data for nearby areas can be used but must be applied judiciously, as insolation conditions can vary sharply within a relatively small geographic area. ∎

## 29.9 DESIGN OF A GRID-CONNECTED PV SYSTEM

The design procedure for an interactive system (with feedback into the utility grid; see Fig. 29.9) differs from the typical stand-alone system design procedure described above. The principal differences are:

- The loads in an interactive system are all ac rather than a mix of ac and dc because the building is fed from the utility line or from the PV array dc-to-ac inverter; dc is not available to the loads.

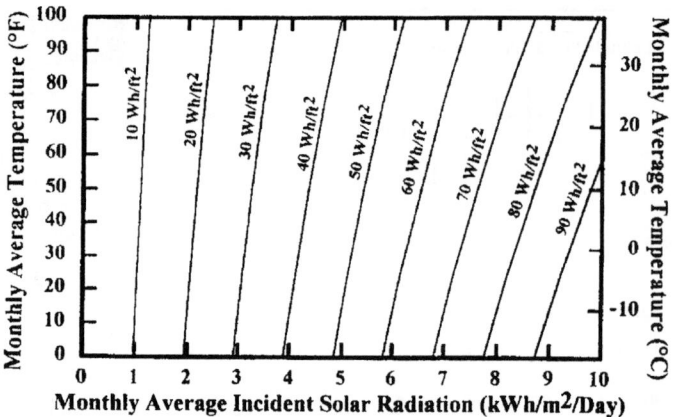

**Fig. 29.13** *Daily average PV array production, Wh/ft², as a function of monthly average temperature and incident solar radiation. Production was calculated using PVFORM for a 10% efficient solar cell array at 77°F (25°C) ambient temperature. (Reprinted with permission from "Design for PV," Association of Collegiate Schools of Architecture.)*

- The existence of a PV system has no bearing on the choice of fuel for appliances. Therefore, for instance, whether to use gas or electric cooking is a matter of fuel availability, costs, and owner preferences.
- Loads are generally heavier in interactive systems because one cannot run out of electric power, as with a stand-alone system. As a result, an all-electric house or a heavily automated house can be a candidate for a PV system as much as any other building.
- Load data can be obtained for a specific design or from general utility-supplied data similar to those shown in Figs. 29.14 and 29.15.
- Inasmuch as no battery or control equipment is used, space allocation in the structure for PV-related equipment is small and system maintenance is minimal. Also, system life is equivalent to PV array life. *Design note:* Some areas in North America have recently been hit with uncharacteristic ice storms that have caused unexpected and relatively lengthy power outages. Some owners with grid-connected PV systems in these areas have added a small battery to their systems to meet their essential needs (lights, heating, computer use) through such outages.

Assuming a net-metering arrangement, there is no advantage to coincidence of building electrical loads and peak PV production. This is emphatically not the case where PV production buyback is based upon avoided costs. In such situations, careful scrutiny of the utility's rate schedule may yield cost-saving opportunities in matching loads to lower-rate time periods. For nonresidential buildings, careful consideration of future electric loads should be reflected in architectural planning for future (larger) PV arrays. In this connection, the possibility of using distributed arrays, as in Fig. 29.10, should be studied.

The general design procedure for a grid-connected PV system is as follows:

**STEP 1.** Establish with the power company the metering arrangement to be used, both physical and financial. The utility can also generally supply reliable data on solar radiation, load patterns, PV array sizing and tracking feasibility, and other relevant design data.

**STEP 2.** Using a detailed cost analysis, determine the optimum size of the PV array in today's market. Although this analysis will involve technical data, the decision on an optimum size is essentially economic—trading utility cost savings against additional first costs for the PV system within a feasibility window established by the owner. This economic feasibility window will likely be influenced by qualitative factors such as a desire to reduce dependence on the "grid," a desire to be more environmentally friendly, or a desire to use a PV system to obtain green building certification points. Sensitivity analysis can be incorporated into the economic analysis to suggest ranges of reasonable system sizes given estimated changes in eco-

**Electrical Usage Profile for All-Electric House in Southern California during Winter**

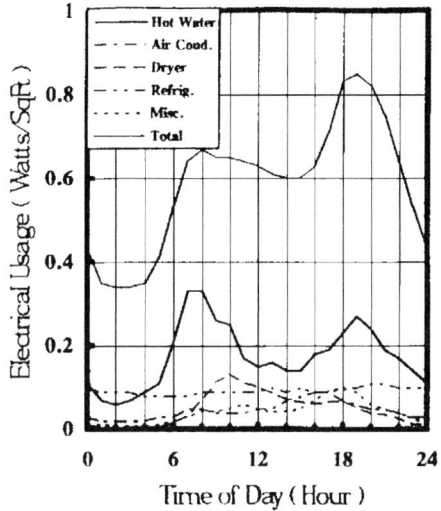

**Electrical Usage Profile for House in Southern California***

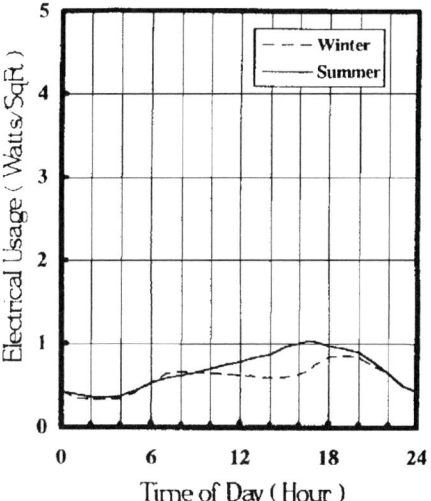

**Electrical Usage Profile for House in Midwest**

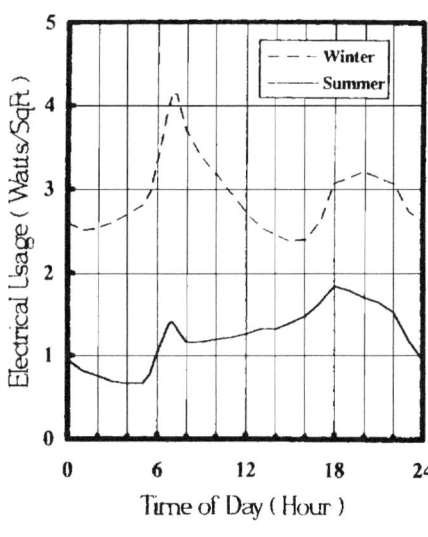

**Electrical Usage Profile for House in the Pacific Northwest**

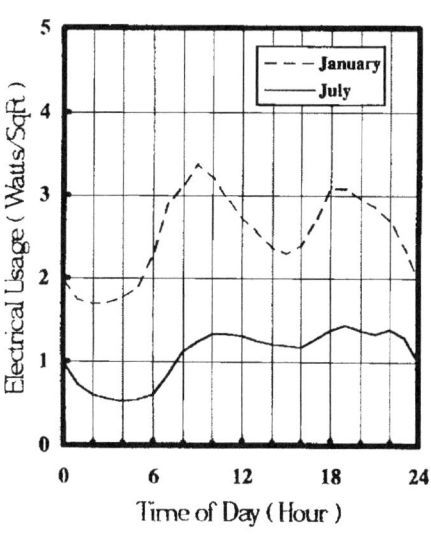

**Fig. 29.14** *Typical residential electricity usage profiles based on a 1500-ft² (140-m²) house. (Reprinted with permission from "Design for PV," Association of Collegiate Schools of Architecture.)*

nomic variables such as utility rates, tax credits, or future PV array costs.

**STEP 3.** Using specific data or generalized information similar to that given in Figs. 29.14 and 29.15, determine the required performance characteristics and physical size of the PV array established in Step 2.

In sizing PV systems, the need to provide charge to discharged batteries while simultaneously providing power to building loads should be considered. A system sized just large enough to meet all loads under design conditions will generally not have sufficient output to recharge depleted batteries.

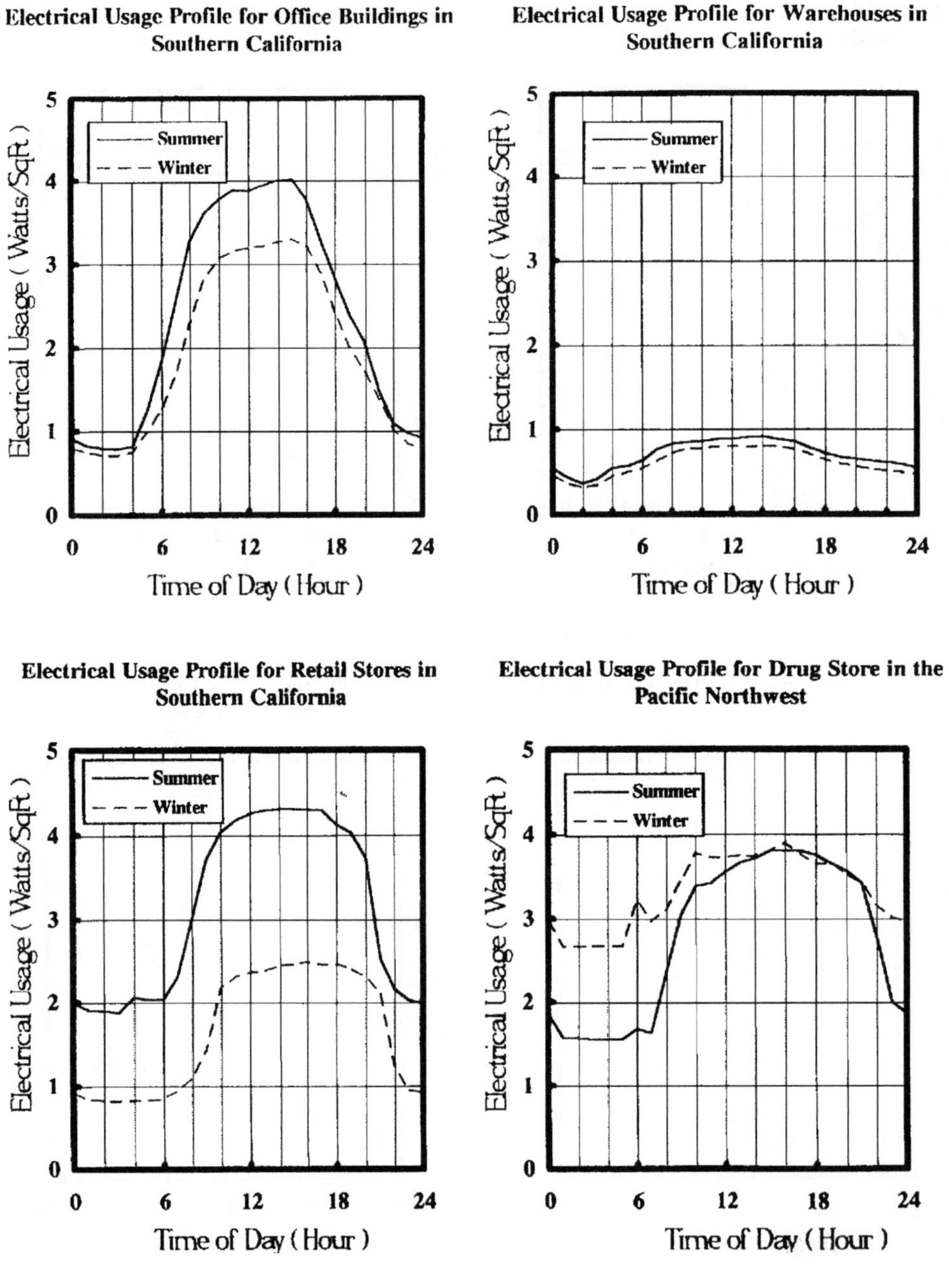
Electrical Usage Profile for Office Buildings in Southern California
Electrical Usage Profile for Warehouses in Southern California
Electrical Usage Profile for Retail Stores in Southern California
Electrical Usage Profile for Drug Store in the Pacific Northwest
Summer
Winter
Electrical Usage ( Watts/SqFt )
Time of Day ( Hour )
0
1
2
3
4
5
6
12
18
24

*Fig. 29.15* Typical electric load profiles for nonresidential U.S. buildings. (Reprinted with permission from "Design for PV," Association of Collegiate Schools of Architecture.)

## 29.10 CODES AND STANDARDS

PV installations are electrical systems and therefore fall under the purview of NFPA 70, the *National Electrical Code* (NEC). Article 690 of the *NEC*—"Solar Photovoltaic Systems"—is devoted specifically to such systems, and its requirements, plus those of applicable articles in the remainder of the *NEC*, must be observed in all areas where the *NEC* is mandated by local jurisdiction. In addition, for grid-connected systems, the specific requirements of the local utility must be followed, all equipment approved by them, and the final installation subject to their inspection.

Other codes and standards that contain important information and may be applicable if required by local authorities include (IEC = International Electrotechnical Commission; IEEE = Institute of Electrical and Electronics Engineers; UL = Underwriters Laboratories):

IEC 61194 Ed. 1.0 b:1992: *Characteristic Parameters of Stand-Alone Photovoltaic (PV) Systems*

IEC 61727 Ed. 1.0 b:1995: *Photovoltaic (PV) Systems—Characteristics of the Utility Interface*

IEC 61215-1993: *Crystalline Silicon Terrestrial Photovoltaic (PV) Modules—Design Qualification and Type Approval*

IEC 61173 Ed. 1.0 b:1992: *Overvoltage Protection for Photovoltaic (PV) Power Generating Systems*

IEC 61277 Ed. 1.0 B:1995: *Terrestrial Photovoltaic (PV) Power Generating Systems—General and Guide*

IEC 61702 Ed. 1.0 b:1995: *Rating of Direct Coupled Photovoltaic (PV) Pumping Systems*

IEEE 929-2000: *Recommended Practice for Utility Interface of Photovoltaic (PV) Systems*

IEEE 1013-2000: *Recommended Practice for Sizing Lead–Acid Batteries for Photovoltaic (PV) Systems*

IEEE 937-2000: *Recommended Practice for Installation and Maintenance of Lead–Acid Batteries for Photovoltaic (PV) Systems*

IEEE 1144-1996: *Recommended Practice for Sizing Nickel–Cadmium Batteries for Photovoltaic (PV) Systems*

IEEE 1145-1999: *Recommended Practice for Installation and Maintenance of Nickel–Cadmium Batteries for Photovoltaic (PV) Systems*

UL 1703-2002: *Standard for Flat-Plate Photovoltaic Modules and Panels*

UL 1741-1999: *Standard for Inverters, Converters, and Controllers for Use in Independent Power Systems*

## 29.11 PV INSTALLATIONS

PV power appears to be one of the dominant trends of the future. PV and buildings are too logical a connection to not develop into the norm. Even today, with unfavorable economics, numerous notable examples of PV installations exist. Several interesting examples are presented here.

Example 1: PV integrated with solar hot water collectors (Fig. 29.16). This is an interesting application that matches resource with need and obviates the need for storage of the PV system's energy production. The PV output drives a pump that circulates solar-heated water. The pump operates only when there is enough solar radiation to heat the water, which means there is also enough radiation to drive the pump via the PV module. No batteries or grid backup are required.

Example 2: An early example of BIPV (building integrated photovoltaics). A Georgia Power demonstration house (Fig. 29.17) in Atlanta that uses PV modules as the roofing (south-facing) for the house. The replacement of redundant elements (such as separate PV and roofing layers) and integration of components is a hallmark of BIPV.

Example 3: Photovoltaics at the Woods Hole Research Center (Fig. 29.18), Falmouth, Massachusetts. This green building features a substantial PV installation. The system is a net-metering arrangement (as described above). For further information, see the case study on this project presented in Chapter 1.

Example 4: PV on a solar racer (Fig. 29.19), Australia, 1983. An early example of TIPV (transportation integrated photovoltaics)? The lessons presented in this example actually apply well to buildings—reduce the load and optimize the collectors.

**Fig. 29.16** *Synergistic PV application—pumping hot water produced by solar collectors for a residence in the southwestern United States. (Photo ©2004 by Walter Grondzik; all rights reserved.)*

ELECTRICITY

*(a)*      *(b)*

**Fig. 29.17** *An early example of building-integrated PVs, Atlanta, Georgia. (a) South façade of a residence with a PV array. (b) View of the roof from inside the attic, thinking of the roof plane as a transformer rather than as a barrier. (Photos © 2004 by Walter Grondzik; all rights reserved.)*

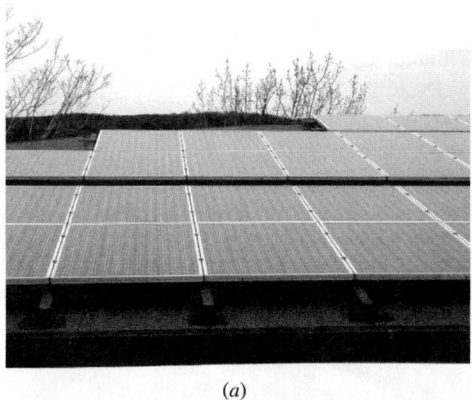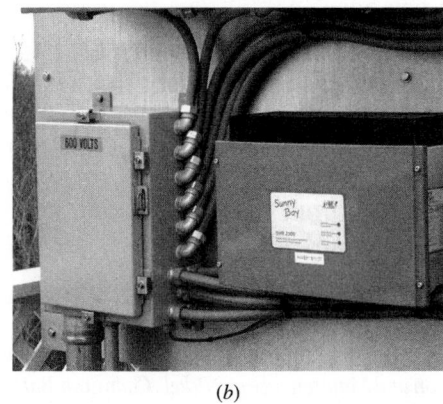

*(a)*      *(b)*

**Fig. 29.18** *PV at the Woods Hole Research Center, Falmouth, Massachusetts. (a) Low-slope PV array on the roof of the building. (b) An inverter (one of several) and a disconnect switch are also located on the roof. (Photos © 2004 by Walter Grondzik; all rights reserved.)*

**Fig. 29.19** *The PV car, a solar racer, after crossing Australia (not a minor feat). (Photo © 2004 by Walter Grondzik; all rights reserved.)*

**Lillis Business Complex, the University of Oregon**

**PROJECT BASICS**

- Location: Eugene, Oregon
- Latitude: 44.1 N; longitude: 123.2 W; elevation: 425 ft (130 m) above sea level
- Heating degree days: 4546 base 65°F (2528 base 18.3°C); cooling degree days: 300 base 65°F (167 base 18.3°C); annual precipitation: 43 in. (1092 mm); cloud cover: 75 clear days a year, 82 partly cloudy, and 209 cloudy
- Building type: institutional—classrooms, offices, common areas
- 196,500 ft$^2$ (18,255 m$^2$); four occupied stories
- Completed October 2003
- Client: University of Oregon
- Design team: SRG Partnership with Solar Design Associates (PV) and consultants

**Background.** The Lillis Business Complex is a new academic facility situated adjacent to a major circulation axis on the University of Oregon campus in Eugene. The building has the size, character, and location to serve as a teaching tool for thousands of students who walk by on any given day. North American universities as a rule (with notable exceptions) have not been in the forefront of the green design movement. This project is one of the exceptions.

**Context.** The Lillis Business Complex is an example of what can happen when various positive forces come into alignment on a project. In this case, an aware client seeking innovation partnered with a local utility promoting the use of alternative energy sources on a campus with an active Energy Studies in Buildings Laboratory (ESBL). The resulting synergy enabled this project to evolve into its final form without too much compromising of objectives. Utility support for the PV system was critical to its inclusion in the project; the utility will buy PV power output at $0.25 per kWh under a 10-year agreement.

**Design Intent.** The Lillis Business Complex was intended to be an example of innovation, particularly with respect to energy and the environment. An aggressive energy goal of 40% beyond the minimum efficiency required by code set a challenging benchmark for many design decisions.

Other green design features flowed from the intent to achieve LEED certification for the project.

**Design Criteria and Validation.** A desire for LEED certification provided the general criteria for the green features of the building (see Appendix G.3). The specific "40% better" energy benchmark provided the criterion for energy performance. Extensive modeling and analysis of concepts and systems was conducted by various consultants (including ESBL) during the course of the design process as a means of verifying design approaches and implementation methods. The building was commissioned.

**Key Design Features**

- The third-largest PV array in Oregon, with a peak capacity of about 45 kW of electricity
- A green roof, the first application of this design approach at the university
- Extensive use of daylighting, involving light shelves and solar shading devices
- Efficient electric lighting using dimmable fluorescent lamps with electronic ballasts and daylight sensor controls
- "Smart plugs" that are wired to occupancy sensors to turn off devices in unoccupied offices
- An integrated natural ventilation system to accommodate outdoor air needs (for acceptable IAQ), to cool the space and people, and to cool the thermal mass; ceiling fans augment the ventilation system
- Thermal mass to provide storage capacity for ventilation cooling and help the building heat up more slowly in summer and cool more slowly in winter

**Post-Occupancy Validation Methods and Performance Data.** The building is too recently occupied to provide much in the way of verifiable performance data (although a POE study is underway as of this writing). Instead, more information on the PV components is given here. The total PV system output is approximately 45 kW peak (dc). There are four distinct PV elements in the building. These include the PV cells integrated into the south-facing glass curtain wall at the entry, which produce about 6 kW (dc) and 2.7 kW (dc) of PV cells incorporated into skylights above the atrium. These two PV applications are good examples of BIPV. The PV cells are integrated into the glazing units such that these units are also the PV modules. In addition, another 6 kW (dc) of PV units are installed on sloping penthouse roof panels, and another 30 kW (dc) installed horizontally. Each element of the PV system is connected to separate inverter systems to maximize system output.

### FOR FURTHER INFORMATION

Brown, G.Z., et al. 2004. "A Lesson in Green," in *Solar Today,* Vol. 18, No. 2.

**Fig. 29.20** *South façade of the Lillis Business Complex, Eugene, Oregon. The atrium glazing incorporates PV cells in a recent example of building-integrated PVs. (Photo © 2004 by Alison Kwok; all rights reserved.)*

**Fig. 29.21** *Glazing panel on the Lillis Business Complex showing embedded PV cells. The glazing essentially acts as a power plant while still providing for views and daylighting. (Photo © 2004 by Alison Kwok; all rights reserved.)*

**Fig. 29.22** *Close-up view of an embedded polycrystalline PV cell at the Lillis Business Complex with interconnecting conductor strips. (Photo © 2004 by Alison Kwok; all rights reserved.)*

**Fig. 29.23** *View from the atrium of the Lillis Complex; the integrated PV cells provide partial shading and a lively lighting pattern. (Photo © 2004 by Alison Kwok; all rights reserved.)*

**Fig. 29.24** *Fisheye view of the Lillis Business Complex atrium. The south-facing panels of the rectangular skylights also contain embedded PV cells. (Photo © 2004 by Alison Kwok; all rights reserved.)*

## References

Various standards from IEC, IEEE, and UL dealing with PV systems are presented in Section 29.10.

ACSA. 1993. *Design for PV.* Association of Collegiate Schools of Architecture, Washington, DC.

CIBSE. 2000. *Understanding Building Integrated Photovoltaics.* The Chartered Institute of Building Service Engineers, London.

Florida Solar Energy Center. 2004. Photovoltaics and Distributed Generation. http://www.fsec.ucf.edu/pvt/pvbasics/index.htm/

Markvart, T. (ed.). 2000. *Solar Electricity,* 2nd ed. John Wiley & Sons, New York.

Messenger, R. and J. Ventre. 2003. *Photovoltaic Systems Engineering,* 2nd ed. CRC Press, Boca Raton, FL.

Sandia National Laboratories. 2004. Photovoltaic Systems Research & Development. http://www.sandia.gov/pv/

Strong, S. 1994. *The Solar Electric House: Energy for the Environmentally-Responsive, Energy-Independent Home.* Sustainability Press, Still River, MA.

U.S. Department of Energy. 2004. Energy Efficiency and Renewable Energy: Solar: Photovoltaics. http://www.eere.energy.gov/RE/solar_photovoltaics.html/

ELECTRICITY

# PART VIII

# SIGNAL SYSTEMS

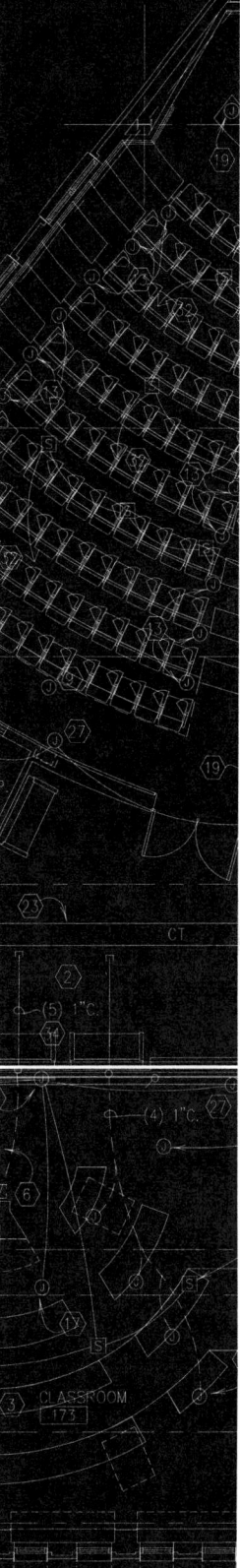

Modern buildings, from the simplest residence to the most complex industrial facility, depend extensively upon electrical, signal, alarm, control, and communication systems in their day-to-day functions. The systems discussed in this part of the book are those for security, music/sound, intercom, clock and program, paging, and building automation. (Fire and smoke detection and alarms are covered in Chapter 24.) Individual, single-purpose systems are discussed from the point of view of a user—that is, emphasizing system application rather than design.

Operating principles are presented, and application issues are related to available equipment. Although equipment is constantly being improved, operating and application principles remain substantially constant, thereby permitting the designer to readily adapt newly available hardware to existing system arrangements.

In terms of building automation, which is essentially a complex control system with sophisticated hardware, the discussion is limited to system architecture and application issues. A discussion of hardware, protocols, and software would be too detailed and would necessarily lag behind the times in a field that doubles its computing power, and therefore its control ability, roughly every 18 months.

# CHAPTER 30

# Signal Systems

## 30.1 INTRODUCTION

No AREA OF EQUIPMENT DESIGN AND APPLICATION to buildings has seen such rapid changes as that of signal equipment. This encompasses all signal, communication, and control equipment, the function of which is to assist in effecting proper building operation. Included are surveillance equipment such as that for fire and access control; audio and visual communication equipment such as the telephone, intercom, and television (both public and closed-circuit); and timing equipment such as clock and program equipment and all types of time-based controls. Specifically excepted is the entire area of data processing, which is not within the province of the building designer, except for equipment space allocation.

Clock and program equipment, which once was the exclusive interest of schools and some industrial facilities, is now incorporated into building mechanical equipment control systems. Closed-circuit TV, which once was limited to classroom and college use, is now commonplace in mercantile area surveillance systems. The hundreds of signals generated throughout a large facility are logged, channeled, and controlled by means of specially programmed computers and microprocessors. Signal systems that once were separate and distinct are now frequently combined and serve multiple purposes.

Obviously, a detailed study of the design and application of such equipment is beyond the scope of this book or, for that matter, of any single book. We attempt, however, to discuss the basic operation of the various systems, some of the equipment available, its application to different types of facilities, and the impact of these systems on the spaces within a structure. The types of facilities considered are single and multiple residences, schools, stores, office buildings, and industrial facilities. Hospitals and laboratories are combinations of these types, but they are too highly specialized to be discussed here.

## 30.2 PRINCIPLES OF INTRUSION DETECTION

To understand the design of intrusion detection systems, it is necessary first to understand the characteristics of the commonly available intrusion detectors (sensors) on which these systems are based. Once an intrusion alarm has been given by a sensor (as with fire detection), the signal must be processed and appropriate measures taken. This may include sounding loud alarms, turning on lights, sending signals to central proprietary or private surveillance services or police, and so on.

### (a) Sensors with Normally Open (NO) Contacts

These devices are no longer used in any reasonable system because the circuits are unsupervised. As a

**1337**

result, a defective circuit does not indicate trouble, and the system is thereby rendered ineffective.

### (b) Simple Normally Closed (NC) Contact Sensors

These come in a variety of designs, the most common of which are magnetic contacts for doors and windows, spring-loaded plunger contacts for doors and windows, window foil, and pressure/tension devices. They operate to transmit an alarm signal, as noted previously. Used in closed, supervised circuits, they provide a trouble signal when the sensors or circuits are damaged.

### (c) Mechanical Motion Detectors

Where window foil or fixed contacts are impractical, a mechanical motion detector can be used. This device is basically a spring-mounted contact suspended inside a second contact surface. Any appreciable motion of the surface on which the device is placed causes the contacts to "make" *momentarily,* turning in an alarm. These devices are very sensitive and can be activated by sonic booms, wind, and even a heavy truck passing by. For this reason, most such units are provided with sensitivity adjustment.

### (d) Photoelectric Devices

These devices operate on the simple principle of beam interruption. When the beam is received, a contact in the receiver is closed. Interruption of the beam causes the contact to open, setting off the alarm. Older devices of this design use a visible light beam and rely for concealment on the fact that (light) is invisible except when reflected from an intervening object. This is quite effective indoors, but outside dust, insects, birds, and so on show the location of the beam, permitting it to be circumvented. Birds and small animals set it off, too. Dispersion of light also limits the throw of the devices when used outside. Modern units use lasers or infrared (IR) beams, which are less easily detected and can be arranged to distinguish between intruders and other disturbances. These latter devices have a longer effective range for exterior use. When a laser beam is used, the signal can be picked up, amplified, and retransmitted in a different direction,

thus establishing a perimeter security "fence" from a single source.

### (e) Passive Infrared (PIR) "Presence" Detector

This device (Fig. 30.1*a*) acts on the principle that all objects emit IR radiation, or heat. The amount radiated depends primarily on the object's temperature and secondarily on its material, color, and texture. The PIR sensor uses a lens or mirror that focuses on a small area and concentrates the IR radiation collected from the conic volume of space between the device and that area in an IR sensor, forming, in effect, a conical beam of coverage called a *zone.* IR radiation in an area that is undisturbed (not necessarily unoccupied) changes very slowly (because object temperatures change very slowly). As a result, any rapid change in the IR reading of that zone indicates an object entering (or leaving) the space, and this triggers an alarm. (We have already seen how these detectors can be used as occupancy sensors to turn off lights when a space's occupant leaves.) See Section 15.17.

The ability to focus on a particular area is utilized to cover areas both horizontally and vertically by the simple expedient of using multiple lenses or mirrors to create multiple zones that cover any desired volume of space (see Fig. 15.28). Also, inasmuch as the IR detector is not intrinsically sensitive to motion, but only to heat, it is usable where motion in the monitored area is unavoidable. The principal disadvantage of PIR detectors is that rapid temperature changes caused by direct insolation, a cold breeze, a heater turning on, and the like can cause false alarms. PIR detectors can be, and are, applied as sensitive motion detectors by using a multibeam (zone) unit. Motion is detected as changes in the IR radiation of adjacent zones or in the radiation of a zone (target) with respect to the background, both of which characterize motion.

### (f) Motion Detectors

These devices (Fig. 30.1*b*), which operate at either microwave or ultrasonic frequencies, detect motion in the protected area by the Doppler effect. (This is the same effect that changes the perceived sound of a car horn or train whistle as the vehicle passes.) Any moving body changes the received frequency of

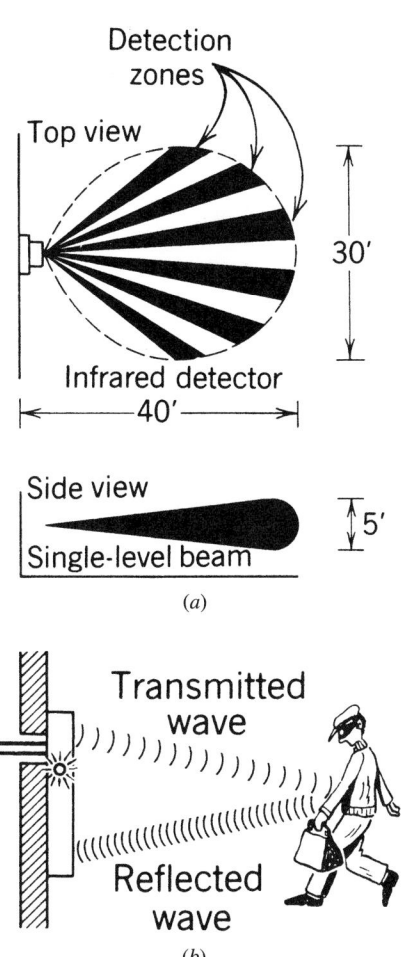

Detection zones

Top view

30′

Infrared detector

|← 40′ →|

Side view

5′

Single-level beam

(a)

Transmitted wave

Reflected wave

(b)

**Fig. 30.1** *(a) Passive infrared detectors give generally a 30 × 40-ft oval protective zone, starting as a narrow beam and widening with distance. Focusability permits exact coverage of any area in a space. Units are also readily available with multiple vertical sensitivity stacked beams that give the same type of high-level interbeam vertical sensitivity as is available from the multiple horizontal beams (zones). (b) Motion detectors depend upon the Doppler effect. They detect changes in the frequency of a signal reflected from a moving object. Sensitivity is highest when relative motion is greatest, that is, when the intruder is moving directly toward (or away from) the detector.*

the signal it reflects, and an alarm sounds. However, because the Doppler effect depends on relative motion between the source and the moving body, an intruder moving laterally may go undetected if sensitivity has been reduced to avoid false alarms. Therefore, units should be located so that the path of an intruder is as nearly as possible directly toward or away from the detector. Ultrasonic units are cheaper than microwave units but can be dis-

turbed by strong air turbulence and very loud noises. Microwave units are undisturbed by air or noise, but because they penetrate solids (like ordinary TV signals), they can be affected by motion outside the protected area.

### (g) Acoustic Detectors

These units alarm when the noise level exceeds a preset maximum. Alternatively, they can be arranged to respond to a particular range of frequencies corresponding to the noise of breaking glass, forced entry, or whatever is desired. Although applied principally in security systems, they can also be used as occupancy sensors for switching of lighting.

A selection of the devices described above is shown in Fig. 30.2.

As with fire detectors, a balance must be struck between the sensitivity of detectors and the nuisance of false alarms, as, unfortunately, increasing the former increases the latter as well. One very effective method of reducing nuisance alarms is to use multiple detectors with different technologies that, in effect, verify each other. Thus, as with occupancy detectors, units are available that combine PIR and ultrasonic detectors in a single housing. Such a unit does not transmit an alarm until both detectors indicate intrusion. This dual technique is applicable to area and perimeter protection as well as portal surveillance.

The previous discussion and that immediately following refer only to intrusion detection (i.e., indication of a human presence where it should not be). *Access control*, which is another area of security design entirely, deals with controlled entry to an area. It is generally more complex than intrusion control, and uses different technologies and a different philosophy of design (see Section 30.24).

## PRIVATE RESIDENTIAL SYSTEMS

### 30.3 GENERAL INFORMATION

Modern private residences utilize a variety of signal devices that greatly enhance their functional value. Indeed, automation in residences, which includes

**Fig. 30.2** *Typical intrusion detection equipment: (1) Passive infrared detector. Maximum sensitivity is a 2°C differential between target and background and a target motion of 2½ in./s. The unit is approximately 6 in. deep. These detectors are available with different coverage patterns and a maximum coverage of 2000 ft². (2) High-sensitivity, multizone, wide-angle passive infrared detector gives coverage of 17 zones horizontally and vertically. (3) Ultrasonic detectors arranged to give broad coverage of approximately 200 × 50 ft. Each unit is 11 × 5½ × 3½ in. (4) High-sensitivity, balanced signal type of ultrasonic detector differentiates between an intrusion signal and random environmental signals, thus reducing false alarms. (5) Unobtrusive, button-type ultrasonic sensor intended for ceiling mounting. (6–8) System control panel, annunciator, and mechanical alarm interface units. (Photos courtesy of Sentrol, an SLC Technologies Company.)*

control of and by a host of signal systems, is today a prime factor in valuation (see Sections 26.32–26.34 and 30.32). Figure 30.3 shows a conventional nonautomated residential plan providing what would be considered *minimally* adequate sound and signal equipment.

In general, all signal systems require a signal source; equipment to process the signal, including transmitting it; and finally, a means of indicating the signal, whether audibly, visually, or on a permanent hard copy. A complex system still falls into this threefold category except that the individual items of equipment and their functions become more sophisticated. In connection with the complexity of equipment, the reader is undoubtedly familiar with the sophisticated automated systems now on the residential market that handle security, fire alarm, time functions, lighting, and so on. They function

on dedicated wiring, control bus, or power line carrier (PLC) signals (high-frequency signals impressed on electric power wiring). As it is not our purpose to discuss automation or central control at this point, but rather the basic systems, the following descriptions consider the systems separately, despite the clear trend toward consolidated control.

Residential fire alarm systems are covered in Chapter 24. The components and design of such systems, including detectors, circuit arrangement, and the like, are also discussed there.

Table 30.1 lists the systems and equipment found in the residence shown in Fig. 30.3 by the threefold functions of source, processing, and indication. Note that the fire alarm, smoke detection, and intrusion alarm systems have been combined into a single system. This simplifies operation and avoids unnecessary equipment duplication. As we

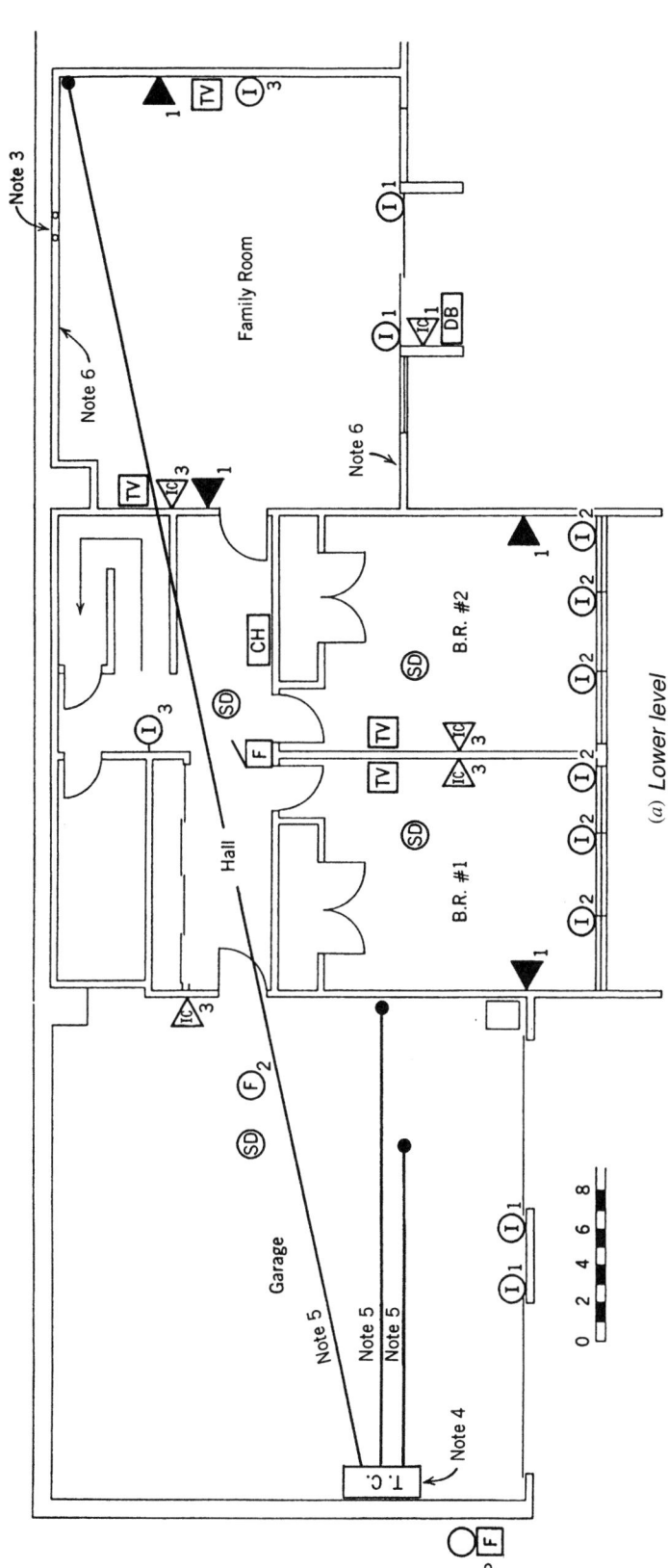

**Fig. 30.3** Signal systems in a residence. (a) Electrical signal equipment plan—lower level. (b) Electrical signal equipment plan—upper level. (c) Symbol list for signal equipment. (d) Signal equipment riser diagram.

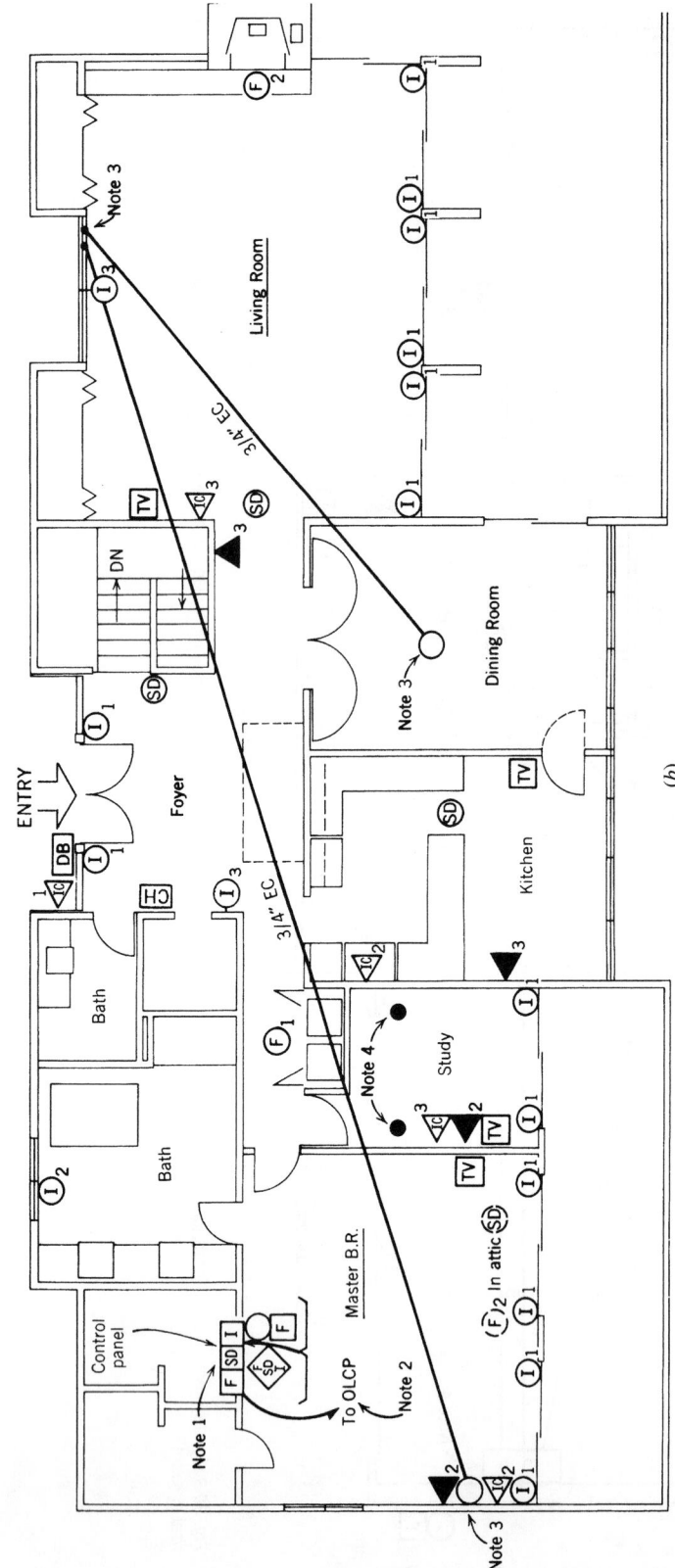

(b)

*Fig. 30.3* (Continued)

Notes:

1. The fire detection, smoke detection, and intrusion alarm devices all operate from a single control panel; see (d). The alarm bell is common. The annunciator indicates the device operated and its location.

2. The connection between the signal control panel and OLCP (outside lighting control panel) activates all outside lights when a signal device trips. Selected lights inside the house can also be connected to go on. See riser diagram (d).

3. Two ³/₄-in. empty plastic conduits, extending from 4-in. boxes in living room wall down to family room and terminating in 4-in. flush boxes. Boxes to be 18 in. AFF and fitted with blank covers. Also, extend a ³/₄-in. plastic EC from one 4-in. box in living room to 12-in. speaker backbox recessed in dining room ceiling. Locate in the field. From the second 4-in. box in living room extend a ³/₄-in. empty plastic conduit to an empty 4-in. box in the master bedroom, 18-in. AFF. Finish with blank cover.

4. Coordinate the location and size of the telephone entrance service cabinet (or box) with the 'phone company. Extend a ³/₄-in. empty plastic conduit from the telephone entrance cabinet to each of the signal raceways in the study. Extend one 1 ¼-in. or two ³/₄-in. empty plastic conduits from the cabinet to the empty signal raceway around the perimeter of the family room.

5. Provide a ³/₄-in. EC through the wall and capped at both ends, for entry of cables from an exterior satellite dish. Coordinate location with TV/CATV/satellite dish contractor.

6. Boiler room to contain smoke detector, fixed 190 F heat detector, and remote station intercom outlet.

Symbols for signal equipment

◇ Annunciator, custom design

○ | 6 in. ac vibrating bell, concealed in recessed box, with grill cloth cover, 84 in. AFF.

[CP] Central panel for fire alarm, smoke detector, and intrusion

[F] WP[F] 8-in. weatherproof bell or siren

[DB] Door bell

[CH] Chimes signal

⧄[F] Buzzer, ac, similar installation to above

▼1 Prewired phone outlet; jack 12 in. AFF.

F₁ Temp. detector; rate-of-rise and fixed temp., resettable

2 Prewired phone outlet; fixed, 12 in. AFF.

F₂ Temp. detector; fixed temp., 190°C

3 Prewired phone outlet; fixed wall outlet 60 in. AFF.

SD Smoke detector with resettable fixed temp. detector.

▽1 Intercom outlet, outdoor, W.P. 60 in. AFF.

①₁ Intrusion detector, magnetic door switch.

2 Intercom outlet, master station 60 in. AFF.

2 Intrusion detector; magnetic window switch.

3 Intercom outlet, remote station 60 in. AFF.

3 Intrusion detector; electronic, motion detector.

[TV] Prewired TV antenna outlet, 12 in. AFF.

[TC] Telephone cabinet

(c)

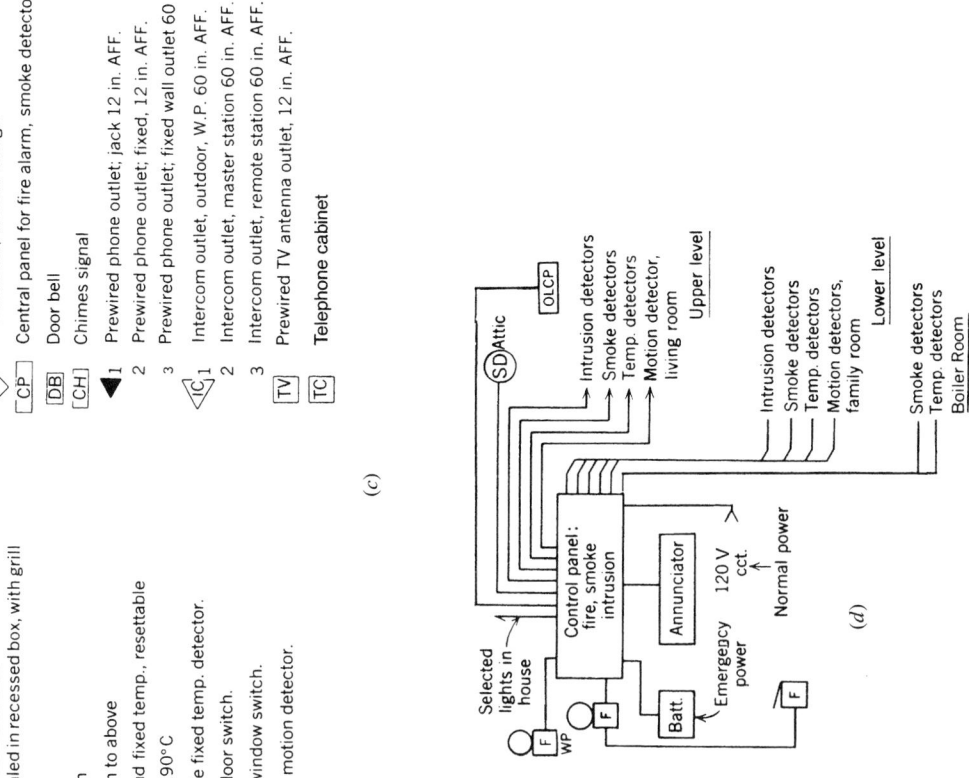

(d)

**Fig. 30.3** (Continued)

**TABLE 30.1 Elements of Residential Signal Systems**

| System Type | Signal Generator | Signal Processor[a] | Signal Transducer |
|---|---|---|---|
| Fire alarm | Temperature and smoke detectors | Control cabinet(s) | Bells, annunciator, buzzer, siren |
| Intrusion alarm | Door and window switches, motion detectors | Control cabinet | Bells, buzzer, annunciator, siren |
| Emergency call system | Pull, pushbutton | Control cabinet | Bells, annunciator, corridor lights |
| Intercom | Microphone, speaker–microphone | Amplifier | Speakers in various stations |

[a]Proper wiring and switching are included under this heading in all cases.

discuss the more complex systems, it will be seen that the basic functions remain unchanged.

As shown in Fig. 30.3, a single control panel can serve multiple residential systems. An annunciator, either integral with the panel or in a separate adjacent enclosure, displays the nature and location of the alarm device that has "tripped." A riser diagram for this residence is shown in Fig. 30.3d. The alarm devices themselves are not shown on the riser because they appear on the plans and duplication serves no useful purpose.

## 30.4 RESIDENTIAL INTRUSION ALARM SYSTEMS

Although any or all of the devices described in Section 30.2 may be used, residences normally utilize door and window magnetic switches, as well as PIR and/or motion detectors, as shown in Fig. 30.3. A manual switch at the end of a long cord is also often provided so that a resident may set off the alarm at will if an intruder is heard. If the system employs the same audible signal devices as the fire system, the sound should be distinctive so that the nature of the alarm can be discerned aurally. Intrusion alarm systems can be continuously supervised by connection with central stations of companies whose business is such supervision and that either respond directly to an alarm call or notify local police authorities of any illegal entry.

## 30.5 RESIDENTIAL INTERCOM SYSTEMS

The public demand for step-saving conveniences has resulted in wide acceptance of the home intercom. Although available with various features, the basic system comprises one or more masters and

several remote stations, one of which monitors the front door, allowing it to be answered from various points within the home. Where desired, a closed-circuit TV system can be added so that visual identification at entrances, in addition to voice communication, is effected. In general, master stations allow selective calling, whereas remote stations operating through the masters are nonselective. The systems are particularly useful when left in the open (monitor) position for remote "baby-sitting." The applicability of such systems to residences with outbuildings should be immediately apparent. As the wiring is low voltage and low power, multiconductor color-coded intercom cable is generally used, run concealed within walls, attics, and basements.

Systems are also available that impose voice signals on the house power wiring. This has the advantage of eliminating separate wiring and making remote stations portable—they are connected simply by plugging into a power outlet.

## 30.6 RESIDENTIAL TELECOMMUNICATION AND DATA SYSTEM

Prior to court decisions permitting users to install their own telephone equipment, the actual wiring within a structure was done only by the utility in the user's raceway system. For some years now, work beyond the service entrance may be done by the owner in a fashion similar to other signal work. In residential work, the telephone company normally follows the route of the electric service, entering the building overhead or underground as desired. In both cases, a separate service entrance means must be provided: if aerial, a sleeve through the wall; if underground, a separate entrance conduit. Unless a residence has many entering lines, no source of power is required for the telephone equipment.

Wiring of telephone instruments when in-stalled *after* completion of the residence consists of a single surface-mounted cable that, even if skillfully installed, is unsightly at best and completely objec-tionable at worst. Prewiring consists of running the cables on the wall framing and into empty device boxes to which instruments are later connected. The huge increase in the number of private residences with multiple phone lines, dedicated fax lines, and special high-speed data transfer lines connected to home office outlets has made telephone planning as much of a necessity as planning for any other sys-tem. This is not only true where residences are designed with a dedicated working-office space. The ease with which modern telecommunications has made it possible to work at home is resulting in an exponential increase in work-at-home situations, especially part-time positions. Failure to provide for multiple lines with adequate raceways is improper planning. In this regard, the locally franchised tele-phone company's technical representative can offer valuable planning advice. The system of raceways, boxes, and outlets dedicated to communications sys-tems of all sorts, generally excluding audio signals, has come to be known as *premise wiring.*

## 30.7  PREMISE WIRING

As noted above, premise wiring refers generally to communication system wiring, including raceways and outlets, and, paradoxically, frequently *not* includ-ing wiring. The "wiring" is installed by the various communication contractors and includes fiber-optic (FO) cabling under the category of wiring, despite the fact that wiring is traditionally understood to be metallic. The use of FO cables for most data cabling will undoubtedly increase in the near future because of their proven advantages in bandwidth, freedom from interference, and high-level security.

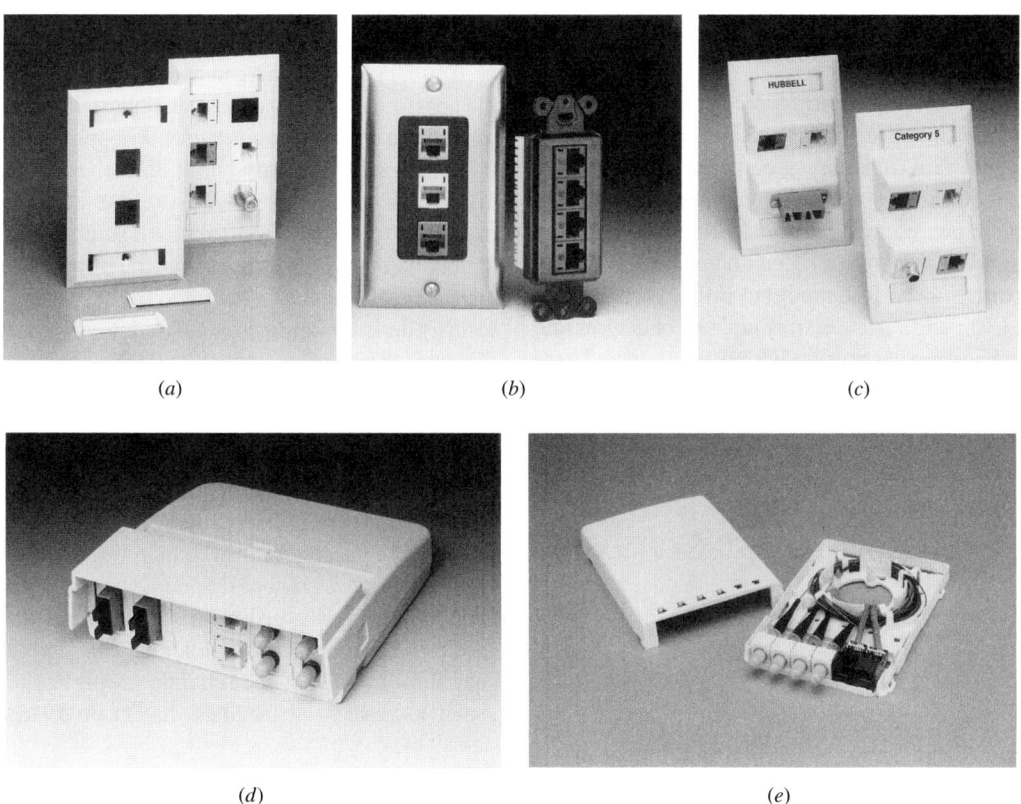

*(a)*     *(b)*     *(c)*

*(d)*     *(e)*

**Fig. 30.4** *Typical boxes, devices, and plates used in data and telecommunication wiring systems. (a) Single- and double-gang flush wall plates with various jacks and connectors. (b) Communication outlets styled to match standard rectangular electrical wiring device plates. (c) Angled faceplates with a variety of jacks and connectors. (d) Multimedia outlet that accepts both copper and fiber optic (FO) cable connections. (e) Multimedia FO outlet with a built-in storage spool for storage of spare FO cable. (Photos a–d courtesy of Hubbell Premise Wiring; photo e courtesy of Panduit Corp.)*

SIGNAL EQUIPMENT

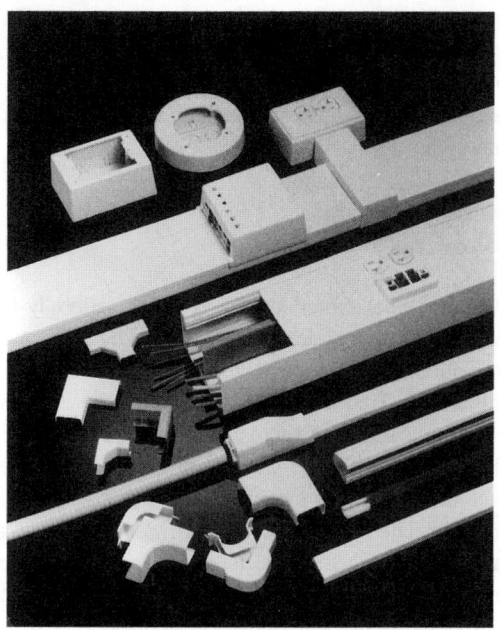

**Fig. 30.5** *A range of raceway sizes and shapes is available to meet all signal cabling requirements. (Photo courtesy of Panduit Corp.)*

Most premise wiring raceways are surface-mounted because access is frequently required and because data cables are often preterminated, making it difficult to pull them into recessed raceways. Finally, premise wiring raceways are usually large, and as such are much easier and more economical to install as surface-mounted units. Figures 30.4 and 30.5 show a few of the hundreds of devices and raceways used in premise wiring, also referred to as *wire management*. See also Fig. 27.22 for a cutaway view of a typical premise wiring raceway.

## MULTIPLE-DWELLING SYSTEMS

See Chapter 24 for a description of multiple-occupancy residential fire detection and alarm systems.

### 30.8 MULTIPLE-DWELLING ENTRY AND SECURITY SYSTEMS

Apartment houses and other large residences combine the functions of the access control doorbell system in the familiar lobby-to-apartment commu-

nication system. The most basic system is a series of pushbuttons in the lobby and an intercom speaker or telephone with which to communicate with residents. At the other end, the tenant has a speaker microphone plus a lobby door-opener button. This system can also be arranged to utilize the tenants' regular telephones. When the number of occupants is large, an alphabetical roster is added to the apartment-button panel to avoid the nuisance of scanning all the apartment names when the sought party's apartment number is not known. When the number is larger yet, a simple pushbutton per apartment arrangement becomes cumbersome, and is usually replaced by an alphabetical tenant register plus a dial or button phone. Closed-circuit television is frequently added to the lobby system, enabling the occupant not only to converse with but also see the caller. Such a system increases the electrical contract cost for an average apartment house by 5% to 7%.

In addition to the security provisions provided by the apartment-to-lobby audio and video connections, additional security and alarm devices have been used, such as emergency call buttons within the apartment. These perform any alarm functions required to deal with an intruder who manages to bypass the lobby security check. In geriatric housing designs, these buttons serve to *unlock* the apartment door to allow helpers to enter if summoned by lights and alarms. In luxury apartment buildings, apartment doors can be monitored from a central security desk and any unscheduled door movement subjected to immediate investigation. These systems are custom-designed to meet the needs and requirements of the owner.

A security problem applicable to all facilities, including residential ones, involves limiting entry in unsupervised areas to authorized persons (i.e., unsupervised access control). The problem of keys and locks is well known, despite advances in that field. More sophisticated means include magnetic cards and electronic combination locks, which, because of the ease of code change, are particularly useful in residential facilities that cater to transients.

Another aspect of security that is particularly appropriate to housing for the elderly and handicapped, although it is applicable to all housing installations, is the emergency call system. The purpose of this system is to alert *outsiders* to an emergency situation *inside* a closed apartment. This alarm system is essentially a way to call for help in time of illness or other distress. Many construction and housing codes include descriptions of the required equipment. Most often they prescribe a call initiation button in each bedroom *and* bathroom that registers an audible (alarm) and visible (annunciated) signal at a location that is monitored, locally or remotely, 24 hours a day. Additional signals are required in the floor corridor and at the apartment, the purpose of which is to alert immediate neighbors to the distress call.

## 30.9 MULTIPLE-DWELLING TELEVISION SYSTEMS

All modern multiple-dwelling residences supply each room with one or more TV/FM jack outlets. The TV jack may supply "house" signals from a rooftop satellite dish plus a house VCR or cable TV with pay TV as an option. As these systems are always subcontracted, the design requirements in new construction include a system of empty conduits connecting to cable pulling points in cabinets. Raceways should be sized liberally because of the constant expansion in the electronic entertainment field. As pointed out previously, the ideal raceway for signal cabling is 100% accessible (i.e., a surface-mounted raceway with a removable cover). However, the unsightliness of this solution plus the fact that, for most cabling, concealed raceways are adequate lead to their use in new work. In economy-type construction, only floor and wall sleeves are supplied and coaxial cables are run exposed in residential spaces. The cable contractor arranges for the electric power needed for any local amplification. Normally, a 15-A dedicated circuit is sufficient.

## 30.10 MULTIPLE-DWELLING TELEPHONE SYSTEMS

As in the small residence, the telephone service normally follows the same entrance path and method of entrance as the electric power service. For the sake of economy in underground construction, the two services often share the same trench, albeit in different raceways, and utilize twin manholes where these are required. Typical entrance arrange-

ments for any large building, residential or other, are shown in Fig. 30.6.

The service entrance space requirements vary with the size of the building and the telephone capacity. For a small apartment house of the garden or three-story type, a clear wall space of 4 to 6 ft is sufficient. A terminal (equipment) room is required only in large buildings (see Section 30.21). The telephone system is part of the overall telecommunications system utilized in modern residences, preparations for which fall into the premise wiring area discussed previously. As such, generalizations are not useful, and planning must consider the type of occupancy and future requirements.

Rental apartment buildings and dormitories differ from commercial structures in that the floor

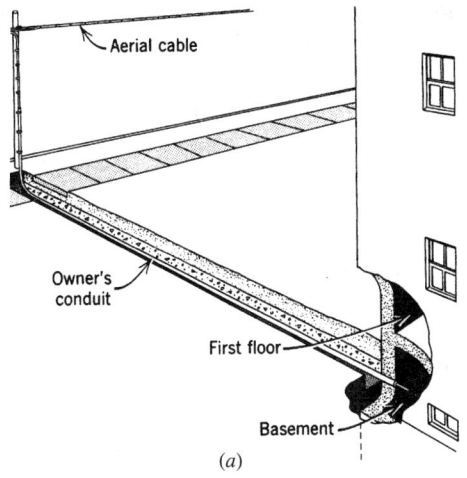

(a)

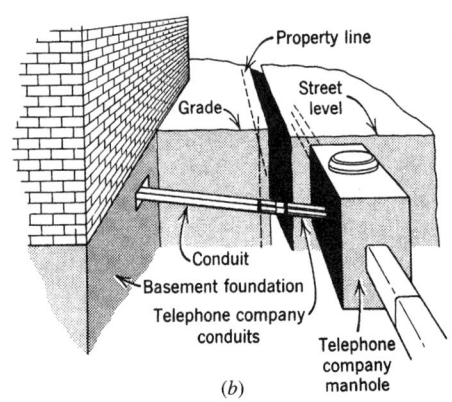

(b)

**Fig. 30.6** *Telephone cable may enter a building underground after originating on overhead lines (a) or via manholes (b).*

plans of all floors are similar, so that the arrangement of risers is relatively simple. It is common practice to utilize cable only in risers that extend through vertically aligned closets in apartments. To accommodate these cables, sleeves through the floor between closets are necessary. If a riser is located in a shaft other than a closet, conduit is normally utilized to allow for easy installation, protection, and repair. If the location is accessible, as in an alcove, only a sleeve is provided. When the riser is located outside the apartment, each dwelling unit is connected to the riser by a conduit with a junction box at either end. Beyond the apartment service point, the individual rooms can normally be prewired entirely without conduit or with only a few short sleeves. In condominium and coop structures, large raceways are run between the service entrance point to cabinets on each floor of the building. From terminal devices in these cabinets, the required service can be extended into each apartment. Prewiring of the apartment avoids the unsightliness of exposed cabling or of cabling within surface raceways.

**Fig. 30.7** *Punched-card coded key in common use in hotels. The punched hole coding is changed after each room use. The drawing of a key on the card is the size of an ordinary door key. (Photo by Stein.)*

## 30.11 HOTELS AND MOTELS

### (a) Security

Because of the transient population of these facilities and the need to provide maximum service and security to guests during their stay, the principal problem in addition to room access security is hotel equipment security (i.e., prevention of what is euphemistically known as *shrinkage*).

1. *Room access security.* The ineffectiveness of key locks in preventing undesired entry even in private homes is well known; in hotels, the key is little more than a psychological barrier. As a result, most modern hotels have installed electronic room locks whose opening device code is changed with every new guest. These locks may be of the coded pushbutton type, whose coding is changeable only from a central lock–security console; a magnetically or punched-hole coded-card type (Fig. 30.7); or a programmable electronic lock and coded-key type (Fig. 30.8). All of these systems have relative advantages and disadvantages, depending on the number of rooms, whether the installation is new construction or retrofit, and the average length of the guest stay.

2. *Equipment security.* The need to provide expected services requires that guest rooms be equipped with a television and possibly a VCR; meeting rooms with TV, VCR, projectors, and computer terminals; and that these and other devices be made available for guest use either as part of the room fee or on a rental basis. Theft control is a consultant specialty on its own and is beyond the scope of this book. One such system (equally applicable to equipment in hotels, schools, office buildings, and industrial facilities) senses the disconnection of the equipment from its power connection (wall plug) and transmits an alarm over the power lines to an annunciator at a selected control location. This arrangement has the advantage of alarming immediately on equipment removal, thus normally permitting appropriate retrieval action to be taken in time.

### (b) Telecommunication, Data

The telephone and data communication system in large modern hotels is important and complex.

(a)

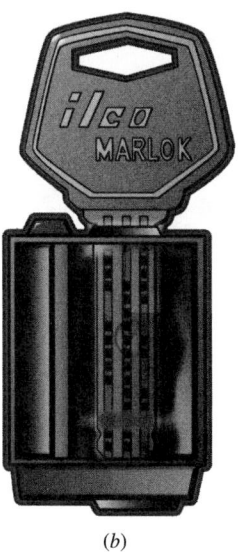

(b)

*Fig. 30.8* This electronic door-locking system uses a programmable electronic lock (a) with a nonduplicatable fixed-code metal key (b). The lock, which can be reprogrammed by the use of a portable programmer, reads the key code with an internal infrared reader. The keys, which can be programmed with up to 1.5 million codes, are discarded after use. Each lock accepts up to 3000 different codes and can be programmed in two modes: passage mode to admit a user during a specific time slot and time mode to accept specific keys during specific time slots. In addition, the locks can be audited to reveal access records, invalid attempts, activation/deactivation, and mechanical key override. (Photo courtesy of ILCO UNICAN Inc.)

Hotels catering to businesspersons must provide access to the Internet that can be used at the very least in rooms dedicated to that purpose and, increasingly, in guest rooms as well. Therefore, the building designer must provide adequate raceway (and cabling) facilities, as discussed in Section 27.18. The current practice of holding business meetings and technical conferences in hotels also requires that conference and meeting rooms be arranged for very heavy concentrations of electronic equipment that can be installed and rearranged in short order. This normally calls for some sort of access floor and modular cabling, as described in Sections 27.25 to 27.30.

# SCHOOL SYSTEMS

## 30.12 GENERAL INFORMATION

The proper operation of a modern school requires that flexible and efficient signal and communications equipment be available to the administrative and teaching staff. Such equipment, engineered to

meet the needs of the individual institution, promotes optimum utilization of staff and student time. School fire detection and alarm systems are discussed in Chapter 24.

## 30.13 SCHOOL SECURITY SYSTEMS

Although intrusion alarms and security systems were not historically normal school requirements, this situation has unfortunately changed. Sensing devices on doors and windows can be arranged both to trip local alarm devices and, via auxiliary circuits, to notify police headquarters. Often, vandals can be frightened off by having the alarm system actuate a protective lighting system that illuminates the building exterior and any interior areas desired, such as record rooms. A perimeter alarm detection system of the types described in Section 30.2 can be installed in particularly vandalism-prone areas to assist in preventing entry to school premises after hours. Although expensive, they are very frequently cost-effective.

An exit-control alarm is a type of exterior door security device applicable to schools (and other facilities) with doors that are locked from the out-

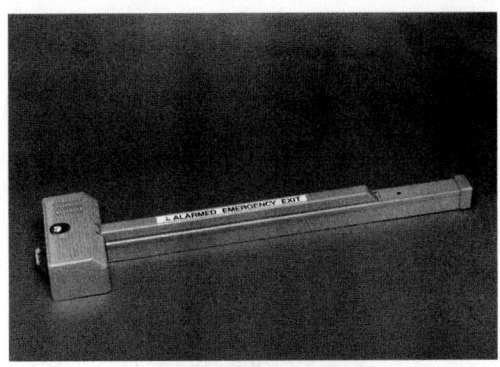

**Fig. 30.9** Delayed-egress exit alarm. When the bar is pushed an alarm sounds immediately, but egress is delayed 15 seconds to permit security personnel to investigate before the door release operates. In the event of fire or another emergency, the door releases instantly. The alarm can be bypassed and reset by key only. (Courtesy of NAPCO.)

**SIGNAL EQUIPMENT**

side but that must be openable from the inside in the event of an emergency. This device (Fig. 30.9) alarms as desired (i.e., audibly or visually, locally or remotely) when the controlled door is opened. It is normally equipped with a timed bypass that allows authorized personnel to key-operate the bypass but prevents the door from being held open without alarming. One model of door-opener alarm requires a continuous pressure on the door-opening bar of 10 to 20 seconds before the door opens, with the alarm being activated immediately on application of pressure. This arrangement (usable only where fire codes permit) allows sufficient time for the facility's staff to investigate the attempted exit before the door opens and is applicable only where exit control is the overriding consideration. As mentioned in Chapter 24, stair and exit door exit locks can be arranged to be centrally released from the fire control center.

## 30.14 SCHOOL CLOCK AND PROGRAM SYSTEMS

The clock system and the program system were once separate and distinct, sharing only the time-keeping facilities provided by the master clock. Now that electromechanical programming devices are effectively obsolete, the two systems are actually one, but the traditional two-system name remains. Referring to Fig. 30.10, we see that the heart of the system is the time base (electronic clock). This is the

same device that provides timing for all programmable switches and controllers. In a clock and program device it controls clock signals, audible devices, and, if desired, other switching functions.

The clock system may use analog clocks (with hands), digital units, or both. The former are usually locally powered, and a correction signal is periodically transmitted via dedicated wiring from the master clock at the controller. This same controller continuously transmits a binary-coded signal to digital "satellite" clocks, which can be either of the self-illuminated light-emitting diode (LED) type or of the liquid crystal display (LCD) type. LCD units are easily visible only in high ambient illumination and when viewed directly (not at an acute angle). LED units are best applied in areas of low ambient illuminance. Large-face analog clocks are easily visible in all ambient light situations.

The programming function of the controller serves to delineate audibly the various time periods into which the school day and week are divided. A single-circuit unit is utilized in an institution that operates entirely on one schedule, such as an elementary school on a morning period–lunch–afternoon period regimen. For a school employing different schedules for its various parts, the controller can provide multiple program schedules on different circuits, depending on its design. Controllers are user programmable and are normally provided with a crystal-controlled master clock to ensure accurate time keeping regardless of line frequency variation, a backup power source to maintain user programming and master clock local display, and various conveniences such as daylight saving time correction, security arrangements, event timers, and so on. If desired, the clock and program controllers can also be used for mechanical system control by the simple expedient of adding relays and switching devices, as shown in Fig. 30.10. Typical clock and program controllers are shown in Figs. 30.11 and 30.12.

The audible devices in a program system may be bells, gongs, buzzers, horns, or a tone reproduced on a classroom loudspeaker. The last system has the following advantages:

1. Clear audibility in each classroom, with an adjustable volume to accommodate both quiet and noisy areas.
2. No possibility of confusion between the pro-

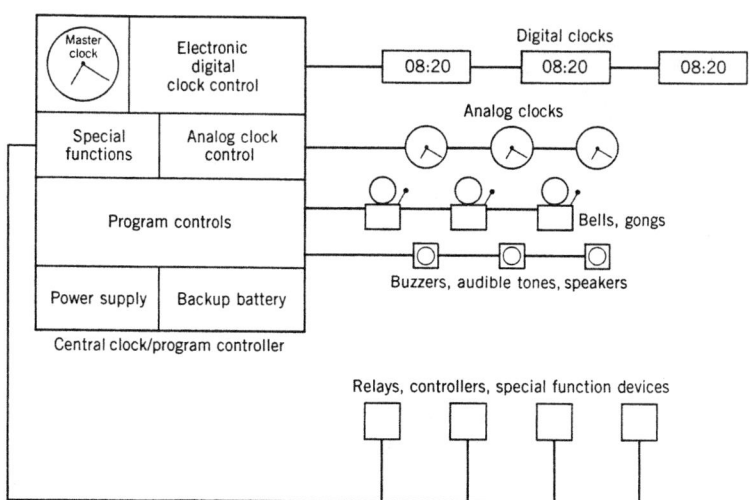

**Fig. 30.10** *A central clock and program device uses its electronic time base to control clocks of all types, time-based audible signals, and event controllers.*

gram tone and other signals such as fire alarm gongs.

3. Multiple use of the speaker unit for classroom sounds as well as program tones.

4. Complete flexibility of programming that is not possible with hall gongs. This is particularly desirable in schools with special programs for particular groups of students.

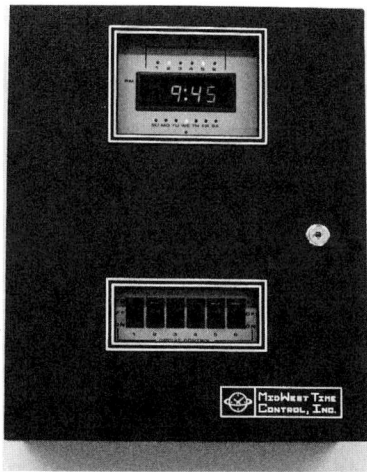

**Fig. 30.11** *Typical microprocessor-based master time clock and program device. This unit, which measures 14 × 12 × 5 in. deep, provides time correction signals to synchronous wired clocks of the digital or analog type and can be adapted to control minute impulse clocks. Programming of output circuits can be daily or weekly, and can accommodate daylight saving, holiday, and leap-year updates with a total program capacity in excess of 600 events. Programming is done locally or via a remote computer. This type of device is applicable to any facility requiring a clock and/or program system. (Courtesy of Midwest Time Control Inc.)*

## 30.15 SCHOOL INTERCOM SYSTEMS

Various types of intercom systems are available, depending on the needs of the school building involved. In small schools, a simple wired intercom system connecting the various offices is usually sufficient. This is supplemented by outside telephones in the administrative offices and a functional paging system that is normally part of the school's sound system arrangement. With larger buildings and correspondingly larger numbers of extensions and multiple-function demands, more sophisticated equipment is required. The unit illustrated in Fig. 30.13 is typical of modern school intercom equipment and is actually a private telephone system with considerable flexibility. Such a system is generally interfaced with the school sound system and may provide these functions:

1. Intercom between staff members and offices
2. Direct communication with classrooms, including selective and all-call capability
3. Paging zone, group call, and conference call functions
4. Interconnection with the outside phone system
5. Class-change signals via interface with the master clock system
6. School intercomputer communication via modems

A combination program/intercom controller is shown in Fig. 30.14. These systems use direct push-button dialing and programming, thus eliminating

**SIGNAL EQUIPMENT**

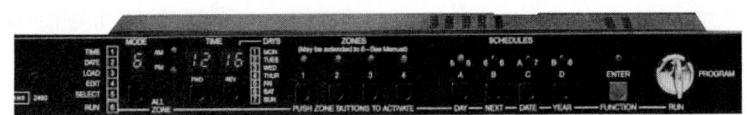

**Fig. 30.12** *User-programmable master clock and program controller. This unit has nonvolatile memory, eight program zones, automatic daylight saving and holiday adjustment, a maximum of four schedules, and is usable with most types of secondary clocks. (Courtesy of Rauland-Borg.)*

the need for switchboards and operators. All stations are coded with three-digit codes, and all switching is solid-state, minimizing maintenance problems. Such systems are adequate for all but the largest institutions.

## 30.16 SCHOOL SOUND SYSTEMS

An integrated sound-paging-radio system designed for school use offers several modes of operation and considerable flexibility. Its function is to provide a means for distributing recordings (CDs, tapes), broadcasts (AM/FM), or live sound to preselected areas of the school. Thus, a simple system might provide a CD player and single microphone input and a single channel to all the speakers in the school, whereas a complex system can be arranged to operate with three simultaneous input signals distributed to six different areas of the school (see Fig. 30.14).

A conventional system consists of a control console containing most of the input units, amplifiers, switching devices, and connections to the remote loudspeakers. The input units may comprise one or more AM/FM tuners, a VCR, CD/DVD player, tape deck, and microphones. One microphone is normally located at the console, with others in the principal's office, auditorium, school office, or other

selected locations. If desired, microphone outlets can be spotted around the school and a spare microphone and stand supplied that can be plugged in at any of these points (Fig. 30.15).

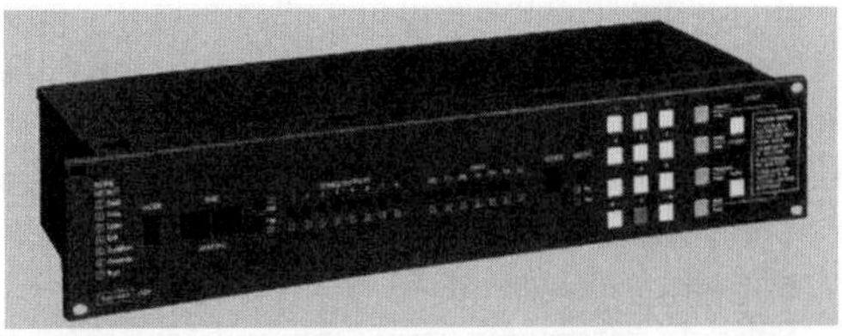

Loudspeakers, located in classrooms, gymnasium, auditorium, cafeteria, and outdoors, receive the amplified signal through the switching mechanisms located in the console. The function of these switches is to deliver the program material to the various loudspeaker circuits, called *program lines.* Thus, using a system with multiple amplifiers, music can be piped to the cafeteria, an educational broadcast program to selected classrooms, and instructions to an outdoor gym class or team during practice. An all-call feature also allows announcements to reach all speakers in the system simultaneously. The intercom system discussed previously can be interconnected with the sound system to allow conversation between classrooms and the console or other points. A small system can be contained in a compact desktop console (see Fig. 30.14). A large system frequently requires a console. A console is usually built in a desk arrangement, and it is advisable to provide sufficient space for it and for the person who operates it. Often an alcove of 30 to 50 ft$^2$ is reserved for it and a library of recordings.

Loudspeakers may be placed flush or in surface baffles at the discretion of the designer. Gymnasium, cafeteria, and auditorium units are normally flush-mounted in the ceiling. For large areas such as these, it is well to provide a volume control, enclosed in a recessed wall box with a locking cover. A common variation of this system uses separate subsystems for the cafeteria, auditorium, ball field, and other areas that utilize sound equipment frequently. These smaller systems have their own input, amplification, and control devices but utilize speakers in common with the central console. Normally, the console has an override feature that allows it to override local systems.

For a discussion of high-quality sound systems

**Fig. 30.13** *Intercom controller. The illustrated microcomputer design interfaces with the existing telephone system. In addition, it provides paging, tone distribution, and program distribution, with control from a telephone. The unit, which measures 19 in. × 9.5 in. × 3.4 in. high, weighs 9 lb, also has a built-in master clock. (Courtesy of Rauland-Borg.)*

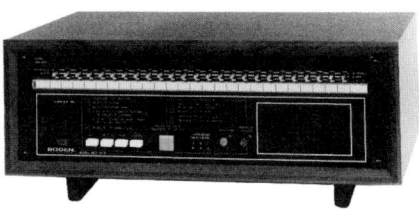

*Fig. 30.14* Compact program/intercom control center. This unit, which measures 20.5 × 11 × 8.5 in. high, has a separate power amplifier for intercom and program material. It distributes program material or provides intercom with 25 loudspeaker locations and has additional capabilities, such as emergency/all-call paging, tone calling, time signal, night bell, and telephone paging. Program material is accepted from separate external sources. (Courtesy of Bogen Communications Inc.)

required for recital halls in music schools and the like, refer to Sections 18.16 to 18.18.

## 30.17 SCHOOL ELECTRONIC TEACHING EQUIPMENT

This area, like so many others that utilize computers, is growing and changing so rapidly that almost anything written on it is immediately out of date.

We therefore simply outline the present and projected near-future uses of electronic teaching equipment. Figure 30.16 shows in block diagram form the arrangements possible with current technology.

### (a) Passive-Mode Usage

This category encompasses all recorded material, in whatever form, that is available to the student via some form of information retrieval technique. This includes printed, audio, and video material in conventional and electronic library forms.

### (b) Interactive Mode

Here students use a computer teaching terminal interactively to study at their own pace on a one-to-one basis, with the computer acting as a tutor. Modern teaching programs sense the students' weak points (as do good teachers) and emphasize them in teaching. The building designer must be aware of the rapid developments in this field at all educational levels, including elementary school, in order to make adequate electric power, cable raceway, lighting, and HVAC provisions.

**SIGNAL EQUIPMENT**

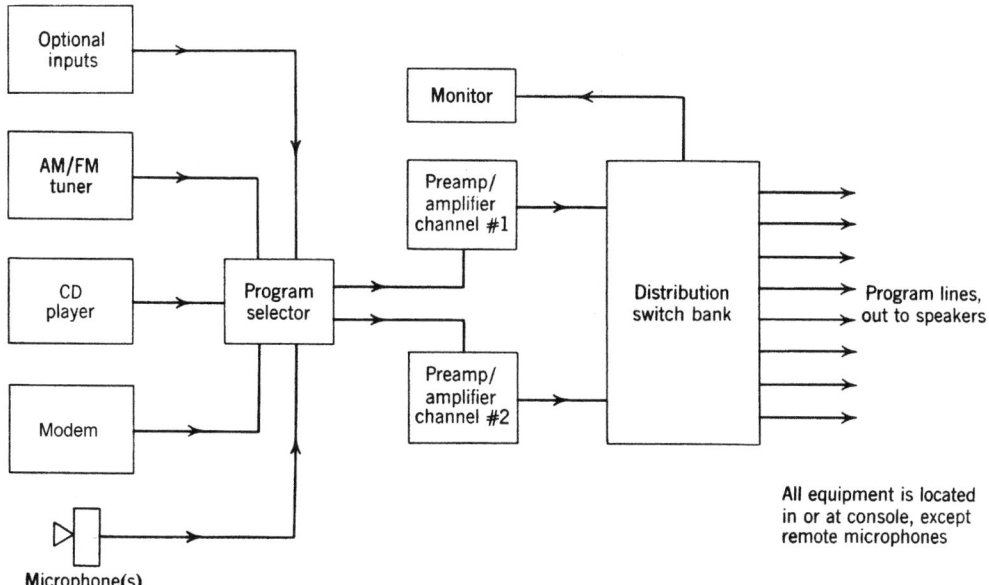

*Fig. 30.15* Block diagram of a two-channel sound system. Options that can be added to the console are a tape deck, private telephone communications, modem connection, and equipment for rebroadcast of signals between areas. An intercom and a master clock and program can also be incorporated or interconnected.

INPUT                                                    OUTPUT

Local Sources                                          Local Distribution

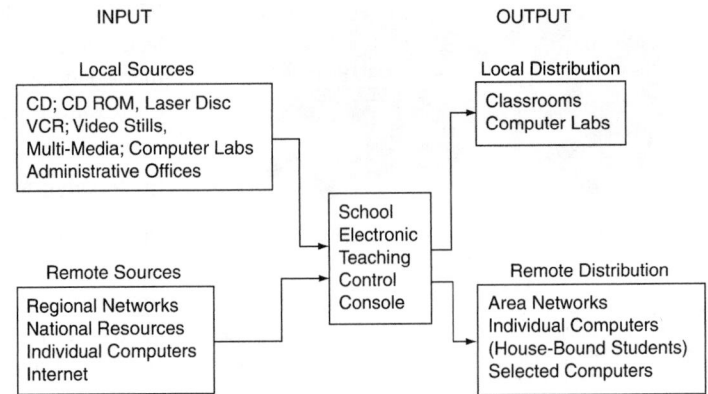

**Fig. 30.16** *Block diagram of connections to an electronic teaching console. Inputs change as technology changes.*

## OFFICE BUILDING SYSTEMS

### 30.18 GENERAL INFORMATION

This category includes systems found in all office-, professional-, and sales-type buildings. Such buildings house tenants with varying schedules and requirements. This must be considered in the design of the signal and communications systems for such buildings.

Although in many medium-sized and large buildings control, alarm, and security functions are combined in multiuse apparatus and consoles (discussed later in the chapter), the basic systems are essentially separate despite having shared-use equipment. We discuss them individually to demonstrate function and equipment, and combined to show economies and modern practice.

### 30.19 OFFICE BUILDING SECURITY SYSTEMS

Although automatic surveillance systems *are* applicable to office and mercantile occupancies, they are more frequently found in industrial facilities and are discussed under that heading. Office buildings normally utilize some type of manual watchman's tour system so that surveillance of unoccupied areas is conducted on a regular basis.

The simplest type is nonelectric and comprises a number of small cabinets, each containing a key, placed at intervals around the interior and exterior

of the building. The watchman uses these keys to operate a special clock that he carries about, thus recording the exact time at which he "clocked in" at any specific location. Alternatively, the clock is wall-mounted and the guard carries only a key (Fig. 30.17). A computerized version of this system is available that simplifies station check-in, automatically records guard visit data, and provides a hard-copy printout.

Guard tour systems are also available that permit constant supervision and are particularly effective where more than one person is on duty. Such systems show on a panel the location and progress of the watchman by means of lights that glow when the device at each location is operated.

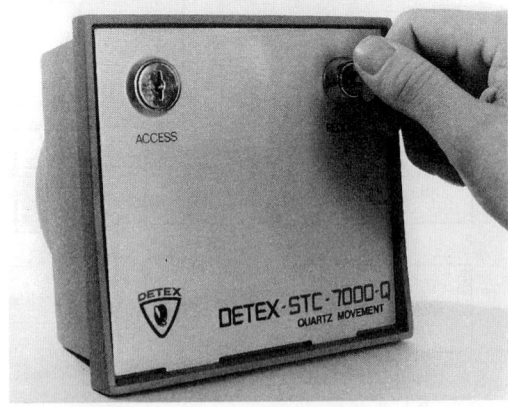

**Fig. 30.17** *Watchman's tour station. Here the clock and recording tape are inside. The station is operated by simply inserting a key, as shown, and giving it a quarter-turn. An alternative version of this system, commonly used, consists of a portable clock and tape operated by a key in a wall box at each station. (Courtesy of Detex Corp.)*

Because part of the effectiveness of these systems lies in the timing of the tour, a system can be arranged to sound an alarm if a particular station is not operated within a specific time period. Telephone jacks spaced at points along the guard's route allow the guard to communicate with the supervising office or other point without interrupting the scheduled tour. For protection of areas housing extremely valuable items or documents, an intrusion alarm system may be employed.

## 30.20 OFFICE BUILDING COMMUNICATIONS SYSTEMS

This planning item is composed of four parts, which are frequently melded into a single network:

1. Intraoffice voice communication, or intercom (Fig. 30.18)

*Fig. 30.18* This desktop multifunction intercom unit is usable in either handset or hands-free mode. In addition to standard intercom functions, each handset, which is connected to a central exchange, can provide additional features such as automatic redial, call conferencing, searching, all-call, group call, call transfer, call registration (stores incoming call numbers), privacy mode, special-meaning tones, and networking. The unit can also be used for voice and beeper paging. Exchanges, which can handle up to 4000 stations, are microcomputer-controlled, solid-state assemblies that vary in physical size according to the number of stations served. A typical large exchange might measure 60 in. W × 24 in. H × 20 in. D. (Courtesy of Dukane Corp., Communications Systems.)

2. Interoffice and intraoffice data communication using telephone and communication cabling
3. Outside-the-building communication via telephone company lines
4. Paging function

Today, in most locations, the user has the choice of purchasing or leasing as much of the system as is desired. In other words, the communication system, including cables, instruments, switching equipment, and so on, can be all privately owned, all supplied by the local telephone company, or almost any other arrangement desired. However, except for a small interoffice intercom, duplication is eliminated—that is, the same instruments and switching equipment are used for both intercom and outside connection.

What to buy, rent, or lease constitutes an economic decision because the required functions are satisfied by either private or telephone company equipment.

## 30.21 OFFICE BUILDING COMMUNICATIONS PLANNING

Planning for the telephone and other communications equipment in an office building is of prime importance because of the large amounts and critical locations of required space. Therefore, it must be done simultaneously with other space planning. Exact requirements for office space are generally unknown at design time, and even if they were known, planning would have to account for changes in space usage as well as increased communications and data transmission services. For this reason, all planning is based on usable office area. Planning is essentially for spaces only, from incoming service to final instrument, because cabling and equipment are furnished and installed either by a private telephone equipment supplier or by the telephone company.

The planning information that follows is applicable to office buildings with average telephone and data transmission loads. Buildings whose tenants include brokerage houses, insurance companies, headquarters of multibranch operations, and other heavy telephone and data transmission users may need more space. However advances in technology, including the use of FO cables, act to reduce equipment size and space

**SIGNAL EQUIPMENT**

requirements. As a result, the best course of action is to design for current requirements with a reasonable estimate for the future based on advice from experts in the field.

Space is required for:

1. A service entrance room that houses incoming cable (network cable), terminated empty conduits for expansion and data cables, a network cable splice box, connection (network interface) cabinets, and equipment required to interface the building's equipment, including telephones, modems, fax machines, and the like, to the main connection cabinets. This space is usually referred to as the *equipment room.*
2. Riser spaces (shafts) and riser closets. These are stacked vertically and carry main cables.
3. Satellite closets where required.
4. Horizontal distribution between closets and devices, including conduit, boxes and cabinets, underfloor raceways, and over-the-ceiling systems.

A block diagram of a typical system in shown in Fig. 30.19.

## (a) Service Entrance

At least two 4- or 5-in. conduits extend from the exterior service connection to the basement or ground floor service equipment room. One conduit will contain the service (network) cable; the other(s) is a spare. Spare conduits should be terminated 12 in. above the finished floor (AFF) of the equipment room, in a threaded fitting, and sealed against entrance of dirt, moisture, or rodents. The service entrance cable terminates in a splice box or cabinet, from which, in large buildings, additional cables extend to one or more wall cabinets, each measuring about 3 ft wide, 6 ft high, and 2 ft deep. Each cabinet can serve a maximum of 30,000 to 100,000 ft$^2$ of tenant area, depending on the density of service required. To accommodate these cabinets from which risers extend, a clear wall space of 6 ft for every cabinet should be provided. This wall area should be covered with ¾-in. marine plywood for cabinet mounting. A 1-in. PVC conduit should extend between cabinets from a suitable electrical system ground point, with a ground wire sized according to telephone company requirements. The room should be dry, ventilated, well lighted for close

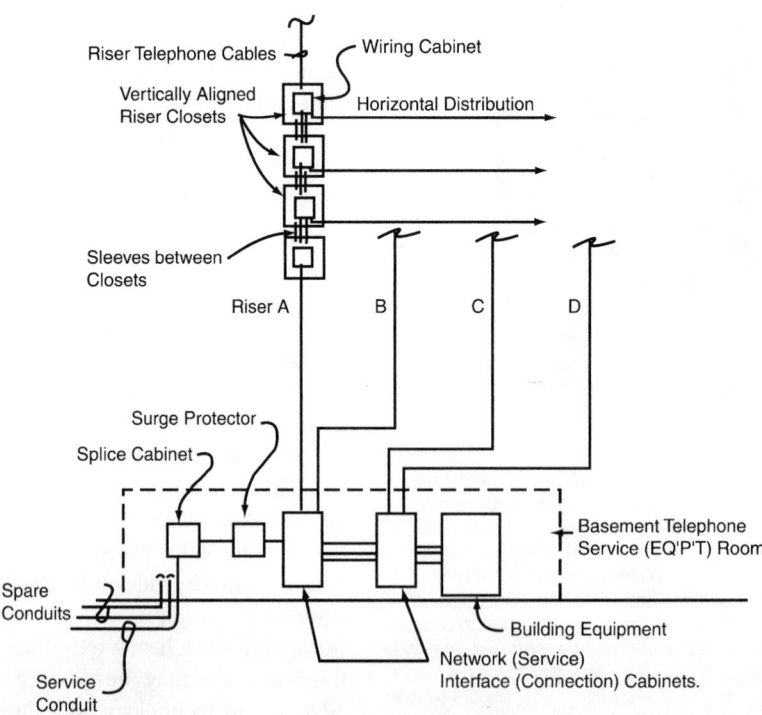

**Fig. 30.19** Riser diagram of the telecommunications system raceway and space arrangements for a typical commercial office building.

work in wiring and color recognition, and supplied with at least two 20-A duplex convenience outlets on a separate circuit.

In the event that additional connection, switching, or communication equipment is required, the project electrical engineer should obtain the equipment size and electrical requirements, supply the necessary space requirements to the architect, and design the electrical facility. In large equipment rooms, circulation and egress must be considered, as well as emergency light and power.

### (b) Riser Shafts

These accept the cables extending beyond the equipment room and carry them vertically through the building. Connection from the wall cabinets in the equipment room up to the riser shaft should preferably be in conduit.

The riser shafts provide means for the cables to extend vertically and terminate at each floor. Ideally, the riser spaces comprise a series of vertically aligned closets connected by ½ or 4-in. sleeves set in the floors and extending 1 in. above the floor. Sleeves are preferable to slots because they can easily be sealed and fireproofed; slots cannot. A minimum of four 4-in. conduit sleeves should be installed between closets vertically, plus an additional 4-in. sleeve for every additional 50,000 ft$^2$ of office space to be served. It is preferable to separate communications closets from electrical power closets. Where multiple risers are used, shafts should be interconnected by several 2-in. conduits to allow for interconnection of systems. Riser cables are also known as *backbone* cables.

### (c) Riser Closets

Here, cables from the riser system are interconnected to switching and power equipment, as well as to the cables that radiate from the closet to station locations throughout the floor area. These closets may also be called *zone closets* or *apparatus closets,* particularly if they function with an underfloor raceway system. The walls of the closet should be lined with plywood at least ¾ in. thick to support the weight of switching and connection panels, power equipment, terminals, connecting blocks, and other hardware. Each riser (apparatus) closet must be provided with a switched ceiling light and a separate 20-A, 120-V circuit with two duplex receptacles. A source of emergency power is desirable to avoid curtailment of telephone service during outages.

Riser closets should preferably have a minimum net area of 20 ft$^2$ (2 m$^2$) and a minimum clear wall of 5 ft for cabinet mounting. Each closet can supply telecommunication service for 8000 to 10,000 ft$^2$ (740 to 930 m$^2$) of floor area. Utilization telephone equipment should not be more than 250 ft from a closet.

### (d) Satellite Closets

Unlike riser (apparatus) closets, satellite closets do not contain switching and power equipment. Their primary use is to provide cable-connecting and -terminating facilities in large, complex facilities, where riser closet space is insufficient.

### (e) Auxiliary Equipment Rooms

Where extensive cross-connection is required or tenants utilize private switchboard (PBX) equipment, the required equipment is placed in a relatively small auxiliary equipment room. These spaces, which are actually small closets or alcoves, should contain a 20- to 30-A, 2P, 120/208-V circuit, a 20-A, 120-V outlet, a grounding point, good lighting, and, obviously, sufficient equipment space. Space requirements vary with each installation and are obtainable from the equipment supplier. Ventilation is essential, as is absorptive acoustic material on the ceiling and at least one wall. Connections between these spaces and other communication equipment closets should be via floor or ceiling ducts or multiple ½-in. conduits.

### (f) Horizontal Distribution

Cabling from riser, satellite, apparatus, and equipment closets to individual outlets and instruments can be underfloor (Sections 27.25–27.29), in ceilings and plenum spaces (Section 20.31), under-carpet (Section 20.30), or in surface raceways (Sections 27.23 and 30.7). Because of the large raceway volumes required, conduit is infrequently used. When underfloor raceways are used, a header capacity of 2 in.$^2$ per workstation is reasonable,

**SIGNAL EQUIPMENT**

based on one multiline telephone and one video display unit per station.

### (g) FO Cables

In installations with very heavy data transmission loads, in those using video systems, and/or in applications for which the high-security, low-noise, and broad-bandwidth characteristics of FO cables are desirable, they are used in lieu of copper cabling. Space requirements and connection accessories are obtainable from manufacturers. Because cables are often supplied with factory-applied terminations, raceways must be of the large-capacity, full-access type.

### 30.22  OFFICE BUILDING CONTROL AND AUTOMATION SYSTEMS

As the modern office building's mechanical and electrical systems increased in complexity, the need arose for a central point of supervision, control, and data collection from which to survey and control an entire building's functioning, plus increased automation at the utilization levels. From the supervisory location, water, air conditioning, heating, ventilating, electrical, and other systems could be controlled with great accuracy and convenience. Data on temperatures, pressures, flow, current, voltage, and all of the many parameters of mechanical systems could be made available instantly so that operational decisions could be made and automated procedures programmed. Also, all systems could be monitored here and all alarms instantly acted on automatically.

Such *supervisory control centers* are now installed as a matter of course in office buildings. They are equipped with computers that process the data received to arrive at operational decisions intended to optimize system performance. As such, they result in a considerable savings in operating and maintenance costs. These systems in their most generalized form are referred to as *building automation systems* (BAS) and are discussed at length in Sections 30.26 to 30.31.

Architecturally, BAS locations require good lighting, good ventilation, and extensive raceway space but little area. Because systems are tailored to a specific building, no guidelines can be stated for space requirements.

### INDUSTRIAL BUILDING SYSTEMS

### 30.23  GENERAL INFORMATION

All industrial facilities, ranging from the taxpayer loft housing a small hand-assembly plant to the immense steel manufacturing plant, require a variety of signal and communication equipment. Fire alarm systems for industrial buildings are discussed in Chapter 24. Audible alarms in industrial facilities for any building security system must be selected with the usually high ambient noise level in mind. See Figs. 30.20 and 30.21 for recommendations.

### 30.24  INDUSTRIAL BUILDING PERSONNEL ACCESS CONTROL

The design and engineering of personnel access control systems is a specialty that has burgeoned since the late 1980s and is now an independent profession. The large variety of identification technologies, plus the control and supervisory functions of computers, has greatly enhanced the access control capabilities of even a relatively small system. As a result, we do not attempt to discuss system design, but focus instead on the operation and application of the subsystems that comprise the whole.

The traditional access control question was always "Who is permitted to pass through a specific portal?" The means by which the identification is made that answers this question varies with the importance of access control at that portal. Thus, entrance to a large, multipurpose space used by many people must be rapid and must avoid the delay engendered by a physical barrier. The means commonly used is one or more guards who visually and in relatively cursory fashion inspect a badge with an ID photo. The next level of supervised, barrier-free access control may be a barrier-free "turnstile" that requires presentation of an access card to an electronic proximity reader. One such unit is illustrated

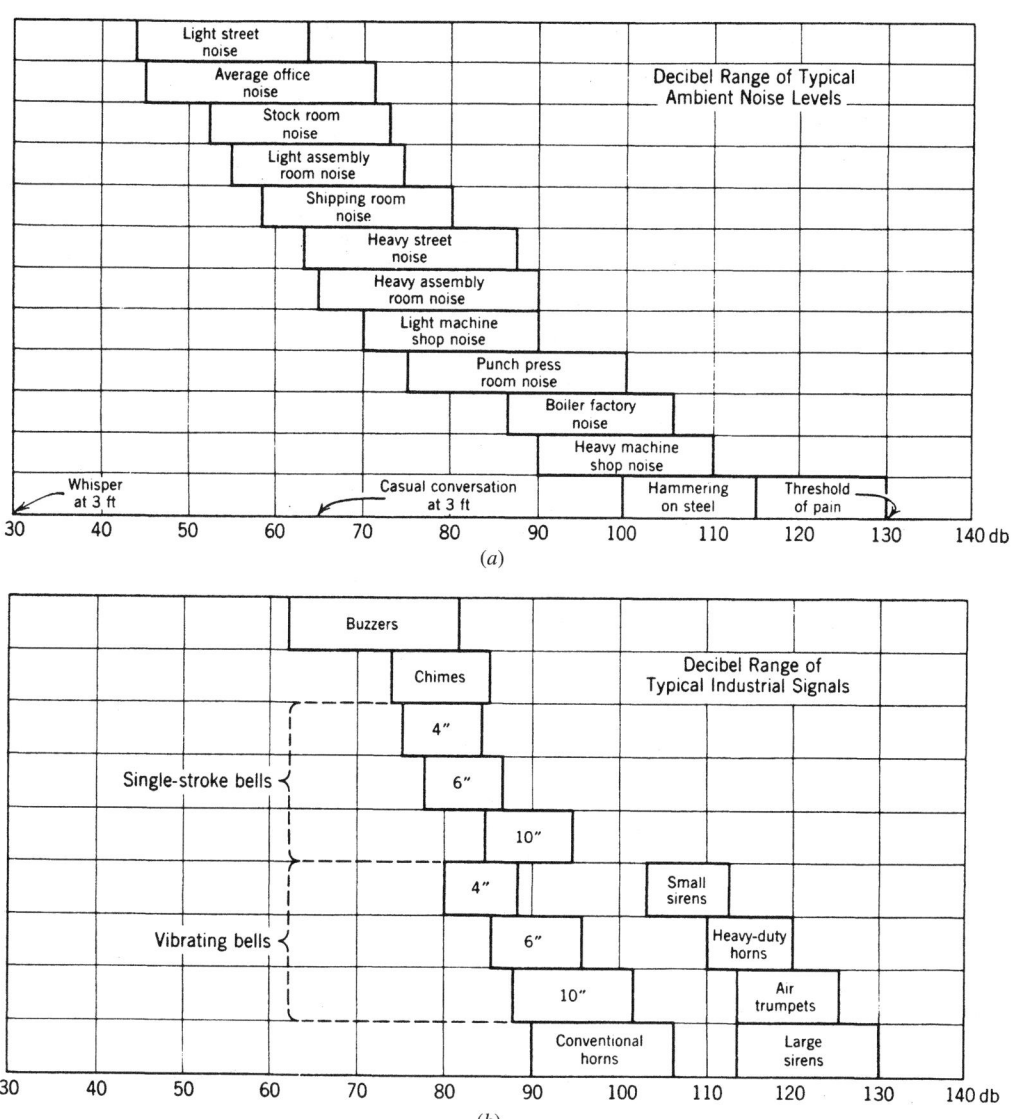

**Fig. 30.20** *(a) Decibel range of typical industrial noise situations. (b) The decibel rating of typical industrial signals may be compared with the anticipated ambient noise level (a) to facilitate selection.*

**SIGNAL EQUIPMENT**

in Fig. 30.22. This type of barrier is limited to a maximum pass-through rate of about one person per second. Because these two control methods are barrier-free, the action taken on alarm is to set up physical barriers and sound the appropriate alarms.

When a physical access barrier is involved, passage is slow, depending only partly on the identification process. Even with the most rapid electronic identification, a door or other portal closure must be physically released and operated, which can be very time-consuming, In the particular instance shown in Fig. 30.23, the identification process is very time-consuming as well, because it involves human intervention.

Unattended physical barriers can be released by a multitude of differing identification technologies, depending on their level of importance. At the lower end of the security scale are barriers that

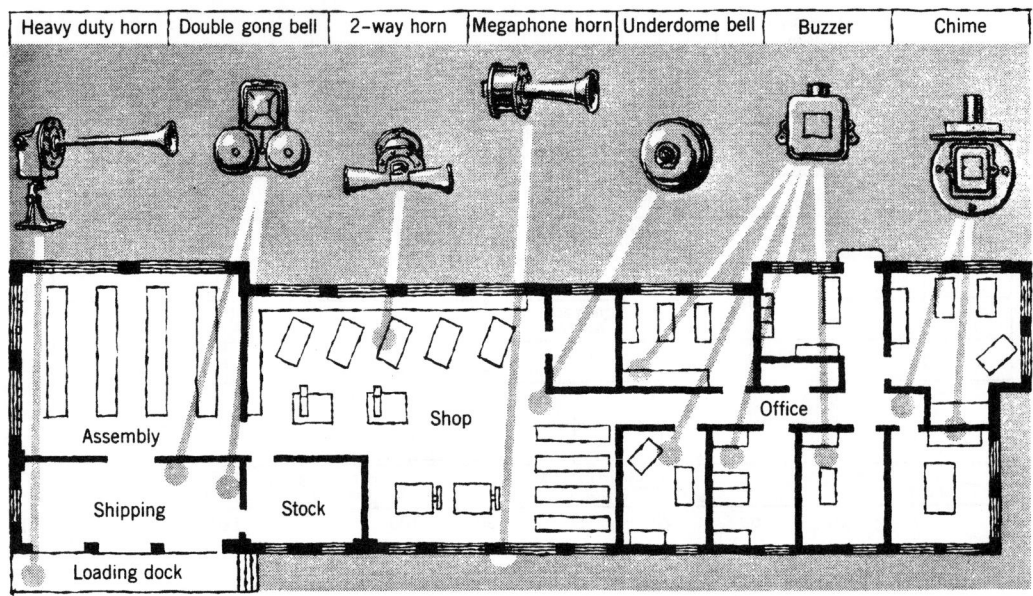

| Heavy duty horn | Double gong bell | 2-way horn | Megaphone horn | Underdome bell | Buzzer | Chime |

**Fig. 30.21** *Plan of a small industrial facility showing suggested locations of typical signals.*

require some knowledge and/or inspect an object presented by the user, such as a magnetic, bar-coded, or proximity-reader card (Fig. 30.24). Because locks can be controlled electronically and programmed remotely, they can be programmed to give access at certain times only, to specific groups of cards or individual cards only, or to prevent access to specific cards (e.g., cards reported lost) or for any other reason. The element of time-controlled access is relatively new; it integrates extremely well with intrusion security systems because portals can be easily coordinated with changing work schedules, both general and specific. Thus, a person can be barred entrance to an area with which that person should not be concerned at that time of day (Fig. 30.25).

The most sophisticated identification methods in use today are biometric: examining a specific physical characteristic that is unique to each person. This is considered a higher level of entry control because it identifies a person, not an object. Several of these methods are illustrated in Figs. 30.26 through 30.28.

In view of the plethora of available access-control technologies, access is being redefined beyond its original physical admittance meaning. Access control today is used to limit access to copy machines, fax machines, phone lines, and other office facilities frequently used by employees for other than purely business purposes.

## 30.25 INDUSTRIAL BUILDING SOUND AND PAGING SYSTEMS

The ubiquitous belt beepers and cellular phones are frequently ineffective in industrial facilities because of high ambient noise levels. In such areas, voice paging at high audio levels (see Figs. 30.20 and 30.21), radio-frequency paging calls on earphones or speakers built into sound-reducing ear protection (see Fig. 19.53), and paging calls using paging lights are used. Similar methods are employed in noisy areas for announcements and special voice messages. In less noisy areas such as offices adjoining manufacturing spaces, sound-reinforced conventional office-type systems are usually employed.

## AUTOMATION

## 30.26 GENERAL INFORMATION

It is important that the distinction between remote control and automation be clearly understood.

**Fig. 30.22** *Optical turnstile in an office lobby. These barrier-free access controls require presentation of a magnetic card to the proximity reader in the stanchion. Unauthorized access (or egress) causes an alarm that triggers audible and visual alarms and other measures as designed. This type of barrier-free access control is normally monitored by a guard positioned adjacent to the turnstile. See the desk at the left. (Photos courtesy of SMARTER Security Systems.)*

*Remote control* is simply a technique by which an action that can be performed manually locally is performed from a remote location by some intermediate means. The means may be low-voltage wiring (Section 26.32), a high-frequency wired signal (Section 26.34), a wireless signal such as the ubiquitous household TV/CD/VCR remote control, a similar IR lighting control unit (Fig. 15.24), or a radio-frequency wireless control (Fig. 16.6). In each case, however, we have a single-stage action that is manually initiated. When the signal initiation is nonmanual—that is, *automatic,* from a timing device (Section 26.16), a sensor such as a daylight or occupancy sensor (Section 15.17), or a programmable device such as a microprocessor or computer (Fig. 15.25)—we have an *automated* system. The

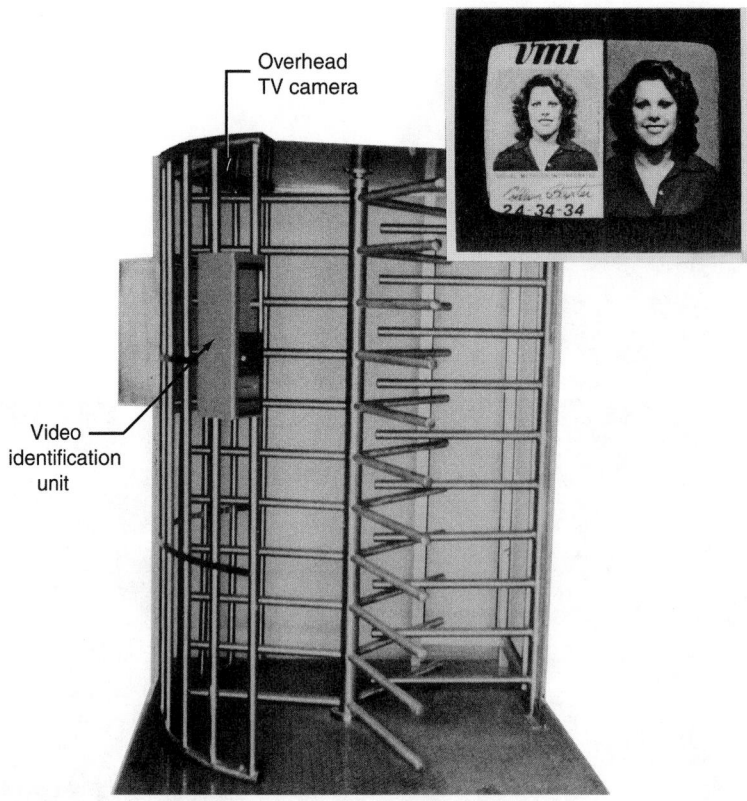

Overhead
TV camera

Video
identification
unit

**Fig. 30.23** *A person desiring entry at this high-security portal inserts her identification card and is viewed by a camera in the entrance device. At a remote location, images of the subject and her ID card are displayed on a screen (inset), while ID data from a memory bank are displayed simultaneously for comparison. Actual physical entry through the turnstile is controlled electrically and monitored by an overhead camera. (Photos courtesy of Visual Methods, Inc.)*

**Fig. 30.24** *This digital access lock has a programmable keypad and is arranged for remote or key override. Individuals or groups can be barred access by special local program codes. Multilevel user codes are available that act in the same fashion as multi-level master keys in cylinder locks. (Courtesy of NAPCO Security Systems.)*

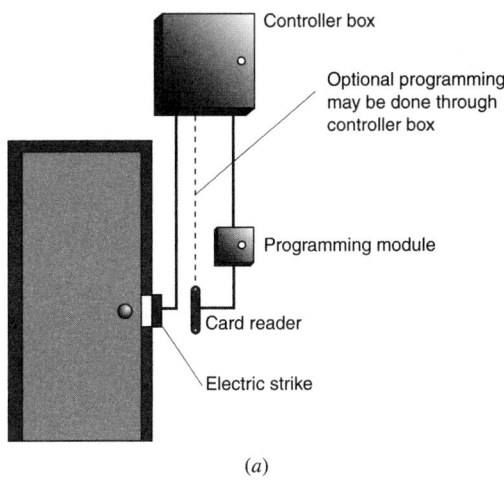

Controller box

Optional programming
may be done through
controller box

Programming module

Card reader

Electric strike

*(a)*

**Fig. 30.26** *This foolproof identification procedure relies on the fact that no two persons have identical retinal vascular patterns. The illustrated device, usually placed at an unmanned controlled-access portal, scans the entrant's retina, compares it to a previously recorded pattern, and then either permits or denies entry. (Photo courtesy of Eyedentify Inc.)*

*(b)*

**Fig. 30.25** *Stand-alone electronic locking system. (a) The arrangement and interconnection of equipment at each portal are shown. The card reader (swipe or insert type) is built into the door mechanism, which also contains the electric strike mechanism. (b) Each lock (swipe style on the left and insert type on the right) is individually programmed using the laptop computer seen in the background. Each lock can be programmed for up to 1500 users, each on a time-schedule basis, convering not only working hours, holidays, weekends, and the like, but also individual time schedules. All programming is performed locally, and the passage record of each lock can be downloaded easily. (Courtesy of Ilco-Unican Corp., Electronic Access Control Div.)*

ing scene-presets that activate dimmers and switches. Such a system is referred to as a *stand-alone* (automated) system. When several of these stand-alone systems are interconnected and supervised by a higher-level controller, each is referred to as a *subsystem* and the overall system is referred to as an *integrated control system*. Such an integrated system, when applied to the individual systems in a building, is referred to as a *building automation system* (BAS).

## 30.27 STAND-ALONE LIGHTING CONTROL SYSTEMS

A straightforward stand-alone lighting control system is shown in Fig. 30.29. The initial programming is entered from the laptop computer into the control panel using system (manufacturer's) software. The control panel contains the required microprocessors, timers, and output control devices to establish lighting levels and settings in the entire structure via dimming and switching panels. In addition, through interfaces (connection devices), the control panel reacts to signals from daylight and occupancy sensors in a preprogrammed fashion to switch and dim certain lights. Window shade control is two-way: they may close automatically when sunlight is detected in a specific area, causing lights to brighten, or they may open when direct sunlight is absent. In addition, a conference room arranged for audiovisual display may control lights and shades via a local lectern-mounted, scene-preset

system can be elementary, as just described, in which an automated signal controls a single action. It can also be very complex, with many levels, yet treat only a single function, such as an automated lighting system with sensors activating or overrid-

**SIGNAL EQUIPMENT**

(a)

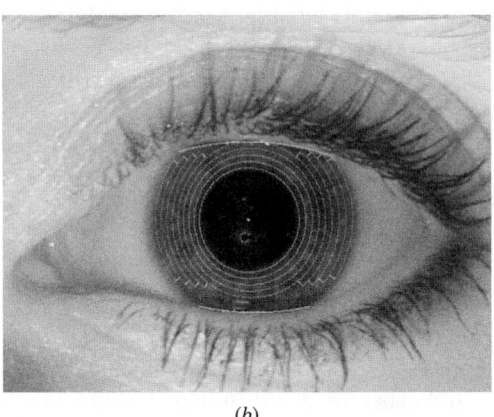

(b)

*Fig. 30.27* Iris scanning system. (a) When the system is used as an access control device at a portal, the entrant stands about 10 to 12 in. away from the scanner as the iris is scanned. The technology relies on the fact that each person's iris has unique and randomly formed features. (b) After scanning, the image is digitized (as shown) and compared to a computer record. The entire procedure is very rapid; entry access or denial is automatic. (Courtesy of IriScan Inc.)

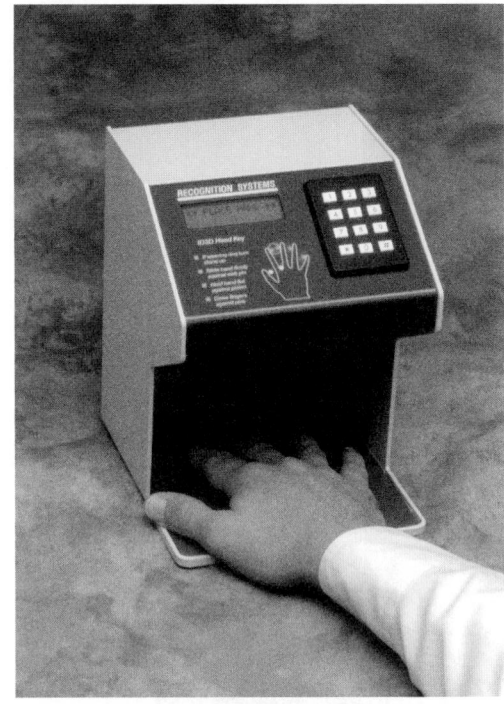

*Fig. 30.28* This biometric identification device takes more than 90 hand measurements that are then converted into a unique identification "template" that is stored in the system's computer. Once identification is made, it can be used for access control. Access points can be programmed to permit entry to authorized personnel on any time/date schedule desired. (Courtesy of Recognition Systems Inc.)

controller, and so on. When another stand-alone system such as a fire alarm system (which itself may be part of an overall security/safety system) sends an emergency signal, the control panel reacts as programmed. It may turn on all lights to full capacity, or it may brighten exit paths and dim other lights, and so forth. This interconnection between two stand-alone systems is characteristic of a BAS.

## 30.28 BUILDING AUTOMATION SYSTEMS

The clear trend in buildings, except for small or simple structures, is to use integrated system design plus centralized monitoring and control of building systems. The subsystems almost always included in a BAS are HVAC, energy management, and lighting control. Inclusion of security, life safety (fire alarm, fire control and suppression, plus emergency aspects of vertical transportation), material handling, and some aspects of communications depends on the specific needs of the building. This trend toward building automation, which previously was economically justifiable only in large owner-user facilities, is today not only economically feasible but also very nearly an economic necessity because of high labor costs and the relatively low cost of computer and microprocessor controls.

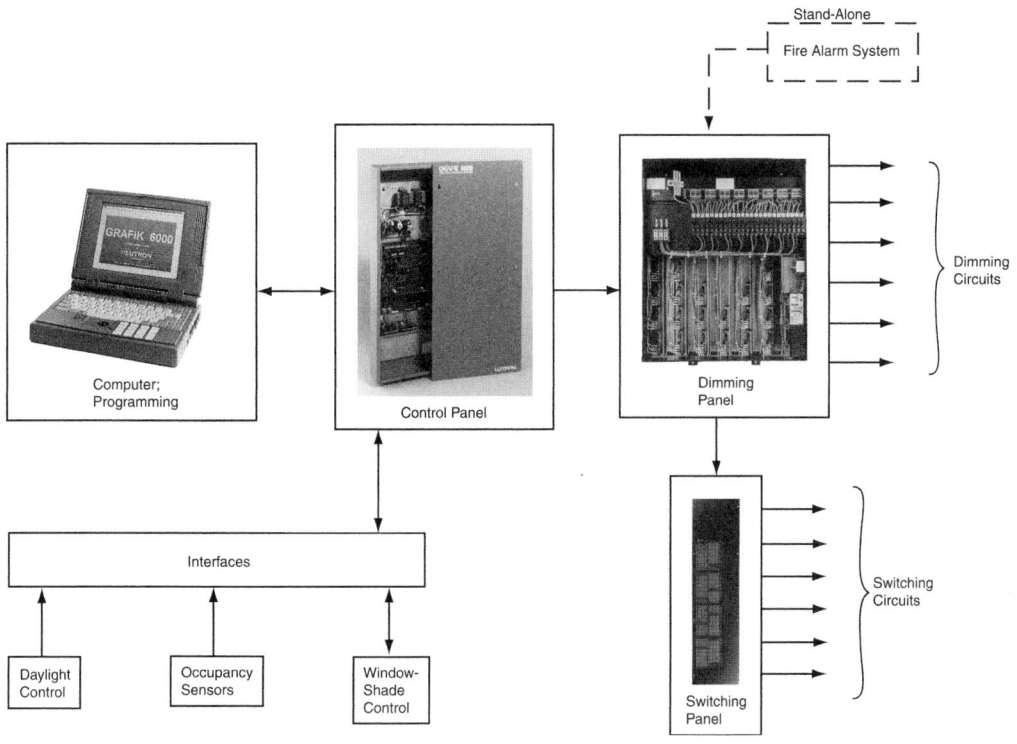

**Fig. 30.29** *Block diagram of a typical stand-alone automated lighting control system. A connection to another stand-alone system (fire alarm), shown dotted, would exist at the overall building control level. (Photos courtesy of Lutron Electronics.)*

What the rapid advances in microelectronics and computer technology have done is to make possible and practical detailed multipoint monitoring and control in real time, the result of which (assuming proper programming and BAS operation) is a highly cost-efficient and environmentally appropriate facility. Indeed, the advantages of such an arrangement are so great that retrofitting of existing buildings, which is obviously more costly than new construction, has become a major industry.

Because the technique of building automation has grown so rapidly in the past few years, the terminology surrounding it and its equipment suffers from a lack of standardization and uniformity. Thus, terms such as *building automation, intelligent buildings, computerized building control, integrated computer control, supervisory data systems, integrated building control, facilities management systems,* and so on are used almost interchangeably. We use the expression *building automation system* (BAS) in the discussion that follows. Inasmuch as every BAS is by definition programmed, any discussion necessarily involves basic control and computer terminology

in addition to that which specifically applies to a BAS. A short glossary should prove helpful in this regard.

## 30.29 GLOSSARY OF COMPUTER AND CONTROL TERMINOLOGY

**Address.** A number used to identify an I/O channel or a module.

**Algorithm.** A detailed description, often in flow-diagram form, of the method for solving a problem. The outline used for computer or microprocessor programming.

**Alphanumeric.** Characters, which are alphabet letters, symbols, or numbers.

**Analog data.** Numerical information about physical variables given in representative, continuously variable form (e.g., dial meters, temperature gauges).

**Artificial intelligence.** Branch of computer science concerned with computer programs that

**SIGNAL EQUIPMENT**

"learn" on the basis of experience and modify their own behavior accordingly.

**ASCII.** A 7-bit standard alphanumeric code established by ANSI. The acronym stands for American Standard Code for Information Interchange.

**Baud.** Rate at which information is transmitted in terms of bits per second.

**Binary.** A numeric system using only 1 and 0, normally used to represent the on/off circuit conditions.

**Bit.** The number 1 or 0. Contraction of *binary digit.*

**Building automation system (BAS).** Building-wide, computer-based system that monitors and controls selected aspects of building operations. The BAS is understood to include all of the equipment involved and its interconnections.

**Bus.** An electrical data channel.

**Byte.** A string of bits; 8 bits unless otherwise specified.

**Central processing unit (CPU).** That portion of a computer, microprocessor, or programmable controller that processes data and executes programs. The "brains" of the computer.

**Converter.** General term for a device that converts data or energy from one form to another. Data converters are analog to digital (ADC) or digital to analog (DAC). Energy-form converters (e.g., temperature or illuminance to voltage) are properly called *transducers.*

**Digital data.** Data in digital form, usually binary.

**Direct digital control (DDC).** Control arrangement that processes input data according to a programmed algorithm and yields a control function (e.g., the action of a local microprocessor controller or a computer control function).

**Duplex, duplex transmission.** Simultaneous bidirectional data transmission.

**EIA.** Electronic Industries Association. Establishes data communication standards (e.g., RS-232).

**Energy management system (EMS).** Computer-based system for monitoring and controlling facility energy use. Frequently part of a BAS.

**Hardware.** All physical equipment associated with a computer system, as opposed to software.

**I/O.** Input/output.

**LCD.** Acronym for liquid crystal display.

**LED.** Acronym for light-emitting diode.

**Load control, load management.** Arrangement whereby energy-using devices are monitored and controlled. Part of an EMS.

**Local area network (LAN).** Signal network providing intercomputer communications.

**Logic.** Factors involved in the design of a CPU.

**Microprocessor.** A CPU, usually in the form of a single integrated circuit (IC), preprogrammed to perform specific functions in response to specific input.

**Modem.** Device that connects a data-originating device to a communication line; from *modulator–demodulator.*

**Multiplex(er) (MX, MUX).** Use of a single communications line for simultaneous transmission of multiple signals.

**Nonvolatile memory.** Memory that remains intact when electric power is shut off.

**Operating system.** Software that controls input to, output from, and operation of the CPU.

**Parity.** Method of verifying the accuracy of recorded data.

**Point.** General term used to describe either a sensor/monitor location or a specific control signal in a BAS.

**Port.** An I/O connection point.

**Programmable controller.** Electronic device containing a microprocessor, a programming means or device, and I/O interfaces (see Section 26.16).

**Protocol.** Conventions governing the format and timing of signaling between communication devices (e.g., between computers in area networks).

**RAM, ROM, PROM, EPROM.** Random access memory, read-only memory, programmable ROM, erasable PROM.

**Real time.** Arrangement in which a computer receives and processes data without introduction of any deliberate time delay (i.e., effectively an immediate response).

**RS-232.** Standard for interconnecting digital communications devices, such as computers, printers, and monitors.

**Software.** Computer program; as opposed to hardware.

**Stand-alone.** Descriptive term for a control device or system that can perform independently and usually can also interconnect as a subsystem (controller) in a larger system.

**UPS.** Uninterruptible power supply.

**Volatile memory.** Memory that loses its information on loss of power.

**Word.** A group of bits that is treated as a single unit.

## 30.30 BAS ARRANGEMENT

The BAS described in this section is fairly recent. A block diagram is given in Fig. 30.30. Notice that information is transferred *vertically* between levels within a stand-alone system, whereas *horizontal* intersystem information transfer occurs only at the highest level (level 1), because that is the level at which systems are interconnected. (In practice, some interconnection also occurs at level 2.) This means that a smoke alarm from level 5 of the fire alarm system travels to level 1, where the central controller tells the other stand-alone systems to take the required action. These systems in turn then signal downwards, essentially to level 4, where appropriate commands are given to controls for supply fans, elevators, security, and so on. The stand-alone systems are essentially independent, and in practice cannot easily be interconnected because of the absence of standardization in wiring and computer protocols. The ostensibly simple problem of connector standardization has yet to be solved, as anyone who has done hands-on work in the communication field can attest, although the situation is much improved over the near chaos that once reigned. Fully or even partially integrated system buildings are at this writing in the planning stages only. These "intelligent buildings" are discussed in some detail in the next section.

Returning to the block diagram in Fig. 30.30 of a typical BAS: any specific BAS may contain more or fewer monitors, sensors, dependent and stand-alone local controllers, and interconnected systems, but the overall arrangement remains generally the same. Actually, depending on the

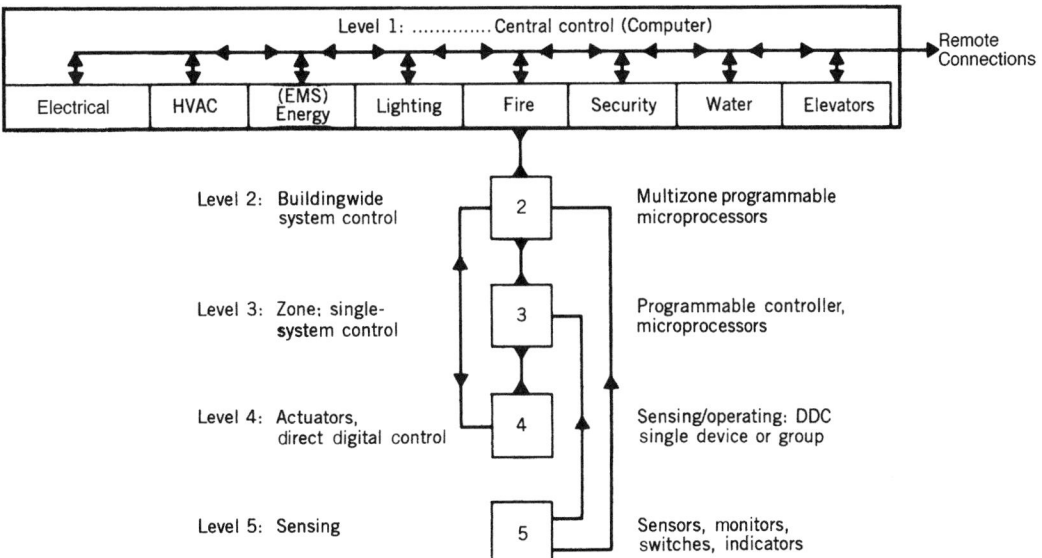

**Fig. 30.30** *Overall block diagram of a building automation system (BAS). In an actual system, the various levels and devices can be combined and telescoped by use of equipment that combines several level functions in a single housing.*

SIGNAL EQUIPMENT

equipment manufacturer and the system's complexity, some of the functions and levels can be telescoped or grouped. Our discussion deals with the general case. To concretize the discussion, we use fire control and evacuation system terminology as a relatively simple and straightforward example of overall system design, function, and interconnections.

Level 5 in the diagram is the lowest system level. It contains automatic operation passive sensors such as flame, smoke, ionization, temperature, and water-flow detectors; physical condition indicators such as smoke and fire-door position sensors; and manual fire alarm pull stations. All yield a digital signal (on–off), which is transmitted to level 3. As level 5 is purely a sensor/monitor level, it does not receive any control or operating signals.

Level 4 contains sensor/operators or actuators (i.e., devices that both transmit a condition signal and receive an operation signal). In our example, this level could contain smoke and fire-door actuators, smoke vent controls, sprinkler system valves, and fan control (override) switches. The last item can either reside at this level or can be controlled through the HVAC system via level 1, 2, or 3. Control at this level would involve hard-wiring of a single device or block.

Level 3 could represent a floor, zone, or area controller, with either relay circuitry or microprocessor control (programmable controller), depending on system complexity. This level receives information from levels 4 and 5, processes it, returns control information to level 4, operates specific local audible and visible alarms directly, and sends alarm and condition information to the level 2 building controller and to the master controller (computer) at level 1. It also receives and processes signals from levels 1 and 2, containing such information as alarm reset, alarm device operation for building emergency evacuation, and any reprogram or reset function.

Level 2 contains the building's central fire and evacuation system (life safety system) controls. This level can contain some of the level 1 programming logic relating to fire fighters' control of elevator operation and HVAC fan functioning, or it can simply act as an I/O device to level 1 for these functions, again depending on system design and complexity. In some facilities, these life safety system functions may be controlled from a separate console that is interconnected with the BAS. In a multibuilding facility, this level contains the individual building fire system central control and interconnections to other building fire system controllers, as well as external connections for city alarm and external supervisory services.

Level 1 is the central computer console. This level receives status reports on the individual devices from level 4 or 5; on areas, zones, and subsystems from level 3; and on the entire fire/evacuation system from level 2. Hard-copy printouts of all alarms and periodic status reports are generally made at this level, although they can also be made at level 2. Visual display terminals (VDTs) at level 1 can be used to view the status of any area and all the devices in the fire/evacuation system (Fig. 30.31 or, for any other system, Figs. 30.32 and 30.33) graphically or as a table (Fig. 30.34). Pictorial graphics for nontechnical end users are also possible (Fig. 30.35). At level 1, information is transferred to HVAC, elevator, security, and lighting systems in preprogrammed alarm modes, with manual override possible. A fire emergency scenario (Fig. 30.31) might interconnect with the HVAC, elevator, security, water, electrical, and lighting systems to activate exhaust and pressurizing fans and deactivate supply fans, place elevators in fire-evacuation mode, override access barriers to permit unhindered facility evacuation, connect fire pumps to emergency power feeders, disconnect high-voltage feeders, and activate emergency lighting, all via preprogrammed intersystem functions. The system operator at level 1 could then view any part of any system and reactivate or override preprogrammed functions as desired. With this system architecture, this is the level at which human supervisory intervention is planned. At other levels it occurs only if trouble arises, periodic maintenance is scheduled, or an alteration is required.

Returning to Fig. 30.30, we can consider a much more complex nonemergency system: energy. Here, level 5 senses and continuously monitors inside and outside temperature and humidity, duct air temperature, space pressurization, airflow, and the operating and load status of all fans, compressors, and refrigeration devices. Level 4 contains monitors and direct digital controls for valves, boilers, chillers, fans, humidifiers, dampers, cooling towers, and compressors (Fig. 30.36). Levels 2 and 3 contain electric load managers for duty cycling and demand limiting, HVAC controls for economizers,

**Fig. 30.31** *A workstation screen shows a selected building area, its zoning, all of the area's fire/evacuation system devices and their status, and the control system's present functioning. The illustrated screen shows an active alarm condition with messages: Elevators being recalled to lobby; Floor is now pressurized; Audio evacuation messages active; Call Fire Department. (Courtesy of Johnson Controls.)*

run-time optimizers, supply air and supply water reset, chilled and condenser water reset, enthalpy switching, load sequencing, and the like. These functions are performed either by a stand-alone EMS controller (see Fig. 26.15) or by individual demand, HVAC, and chiller load managers controlled by level 1 logic. Finally, level 1 contains the programming for all of the foregoing functions, any of which can be displayed at a level 1 workstation terminal at any time. The system operator can then reprogram any function after analyzing system performance (Fig. 30.31) at the level 1 computer and obtain immediate visual and hard-copy readout of the changes.

## 30.31 INTELLIGENT BUILDINGS

Although the term *intelligent building* seems to attribute a human characteristic to an inanimate object, the term's meaning (as defined by the Intelligent Buildings Institute in 1987) is, "a building which provides a productive and cost-effective environment through optimization of its four basic

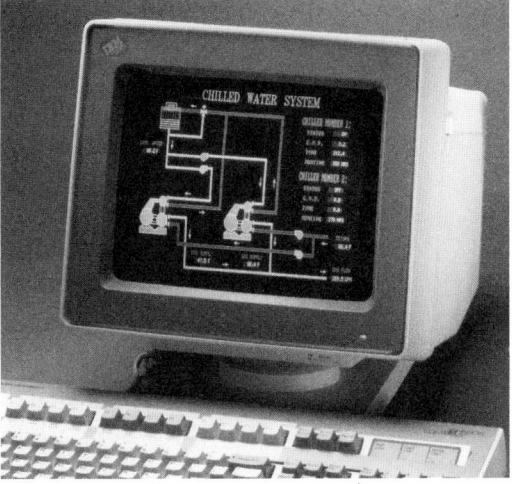

**Fig. 30.32** *Workstation screen shows a portion of a building system and the status of all components. Here we see the chiller status (on–off, COP, tons, run time), water temperature and flow in supply and return lines, and condenser water temperature. (Courtesy of Johnson Controls.)*

elements—structure, systems, services, and management—and the interrelationships between them. . . . Optimal building intelligence is the matching of solutions to occupant need." In the framework of such a definition, the intelligence of a building depends on the ability of the design team to understand the occupants' present needs and, based on study and interaction with the client, make an educated forecast of the occupants' future needs. Because a building's useful life depends

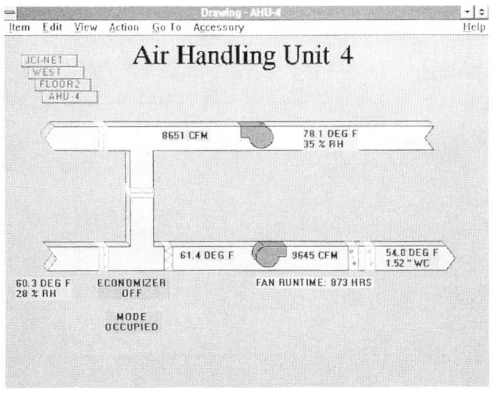

**Fig. 30.33** *Any workstation screen can display any portion of any system in the BAS. Here an air-handling unit (AHU) is displayed with data on airflow, air temperatures, and relative humidity in each duct section. In addition, cumulative fan run time, economizer status, and area occupancy status are displayed. (Courtesy of Johnson Controls.)*

SIGNAL EQUIPMENT

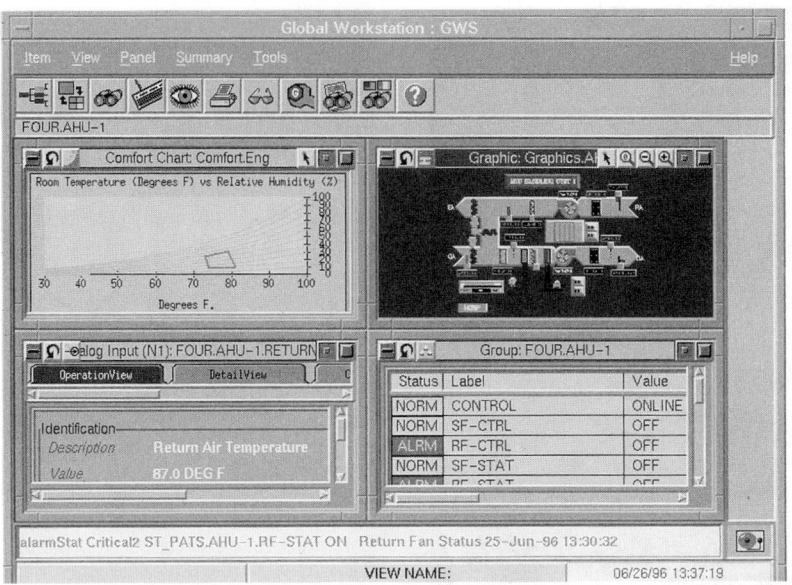

**Fig. 30.34** *Global workstation screen is arranged to display four separate data windows. Clockwise from upper left: (1) Comfort chart showing a plot of room psychrometric conditions on a graph of room temperature versus relative humidity. (2) Graphic screen showing the layout of an AHU with relevant data. (3) Status chart of items in AHU-1. (4) Temperature readout of return air for AHU-1. (Courtesy of Johnson Controls.)*

largely upon the accuracy of this forecast, building systems are being designed with open architecture that lends itself easily and economically to changes, both minor and major. The term *open architecture* is generally understood to mean, in terms of BAS, a system design that, at the very least, can do the following:

- Utilize equipment from any manufacturer whose equipment meets industry-wide specifications
- Readily reprogram equipment to serve a variety of functions
- Supply required system information at any I/O port
- Permit access to databases at any workstation

**SIGNAL EQUIPMENT**

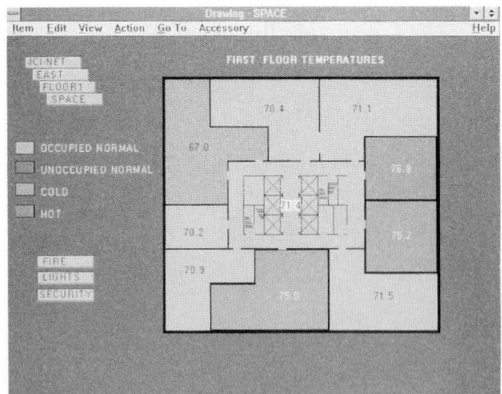

**Fig. 30.35** *Workstation screen showing a floor plan, indicating room status and temperature by color and in general descriptive terms—hot, cold, normal occupied, and normal unoccupied. A similar floor plan screen can be displayed for the status of fire alarm and security systems and the lighting system. (Courtesy of Johnson Controls.)*

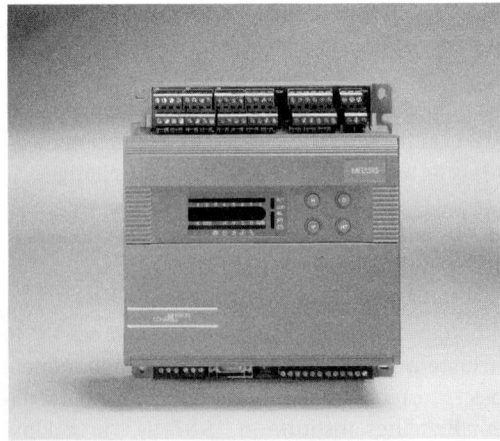

**Fig. 30.36** *Digital controller programmable for a variety of applications, such as chiller control, boiler control, or area environmental control. (Courtesy of Johnson Controls.)*

**Fig. 30.37** *Network control unit acts as a LAN platform, performing supervisory functions for a portion of a building. It connects to operator workstation(s) at the upper management level, to other network controllers at its level via the LAN, and to specific stand-alone controllers at the lower, control level. It can also be programmed to act as a high-quality industrial process controller. (Courtesy of Johnson Controls.)*

- Utilizes networks for information transfer at one or more levels

It is apparent that these requirements demand open protocols, standardized transmission media and connectors, and virtually unlimited cabling space. (The last requirement is easily fulfilled by using any of the underfloor or cellular floor raceway systems described in Chapter 27.) At this writing, proprietary interests and the prevalence of stand-alone system design are retarding the move to open architecture in the United States, whereas it is being adopted rapidly in Asian mercantile centers.

Referring to Fig. 30.30, which shows a prevalent BAS architecture, an open system would eliminate at least two of the levels, reducing the system to a three-level structure that *might* be structured as follows:

1. *Level 1*, the uppermost level, contains workstations connected to a main data bus. This level, which might be called the *management* or *information level*, has direct access to databases and information networks. It serves to establish operational programming and procedures. All points are accessible from any workstation. A typical workstation might appear as shown in Fig. 30.34.

2. *Level 2*, the intermediate level, might be called the *supervisory* or *network level*. Network controllers serve as a network platform both supervising stand-alone controllers from various systems at the third level and transferring information to other network controllers supervising other areas of the building via the LAN. In addition, the network controllers exchange information with one or more workstations. A network controller can also act as a process controller. A typical unit might appear as in Fig. 30.37.

**Fig. 30.38** *Fire control unit. This device controls intelligent addressable analog devices in an intelligent fire control system. (Courtesy of Johnson Controls.)*

**Fig. 30.39** *A unitary controller is designed to control single-zone HVAC subsystems such as heat pumps, fan coils, and the like. It is preset from the network level but can also be locally monitored and adjusted to satisfy local requirements. These adjustments are made with a portable plug-in control unit. (Courtesy of Johnson Controls.)*

3. *Level 3*, the control level, contains dedicated controllers in stand-alone systems. The units are highly flexible and can be locally reprogrammed to control the same function in a wide variety of different systems. Two such controllers are shown in Figs. 30.38 and 30.39.

## 30.32 INTELLIGENT RESIDENCES

The operational principles of remote control and automation discussed previously in this chapter, and also in Chapters 26 and 15, have found ready application in residential automation and control equipment. Residences so equipped and variously referred to as *intelligent, smart,* or *automated homes* appear increasingly in the real estate market news. The degree of automation in these houses varies considerably, the simplest being no more than a low-voltage control system of the type described in Section 26.32 with time-based programming using a relatively simple microprocessor. More complex systems use a touch-screen computer for control or a portable plug-in programmable microprocessor to set various portions of the system. The best (and most expensive) systems achieve a level of user–system integration not yet available in a complex BAS, in that the user's telephone can control any portion of the system directly. This is possible because only one level or, at most, two levels of control are involved, that is, the commands are on/off/set, where the last refers to a particular electronic contact. This contact in turn establishes a new preprogrammed condition such as night-setback, lighting preset for extended absence, reset of upper-level temperatures to daytime levels, and so forth. These systems also allow relatively straightforward reprogramming to meet the occupant's specific needs.

Because all new residential construction includes a degree of automation initially that will surely increase in the future, all residences, both single and multiple, should be provided with the cabling and/or raceway system that is the backbone of any automation system. At this writing, the cost of plastic optical cable (POC) is approximately competitive with copper cabling in large quantities. The decision as to what cabling and distribution system to install is technical and economic. At the very least, every room in a residence should be equipped with audio/video/control point outlets. In addition, a study, a recreation room, and at least one bedroom should be wired as an incipient fully equipped home office. This applies, of course, not only to anticipated automation wiring, but also to power outlets and lighting.

# TRANSPORTATION

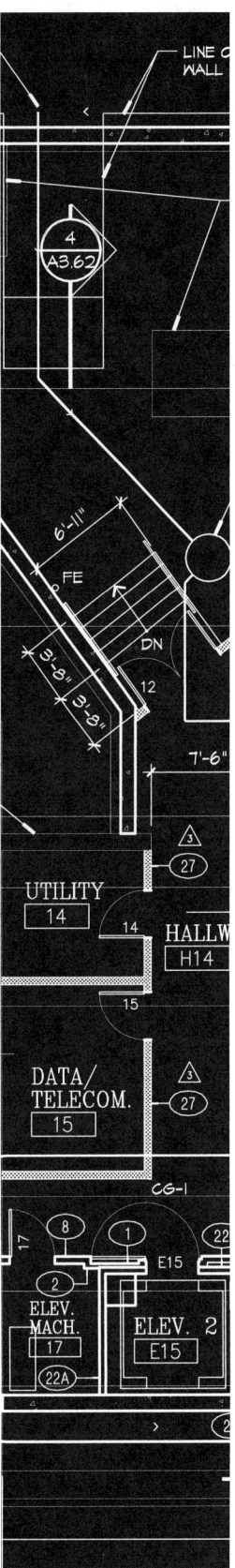

No systems in a multistory structure are taken for granted more consistently than elevators and escalators, which are relied upon to move people quickly and safely under all conditions, including emergencies. This movement should be rapid, trouble-free, and economical. Furthermore, because vertical transportation accounts for 10% to 15% of the construction budget in tall buildings, somewhat less of the building area, and somewhat more of the operating cost—and is a determining factor in building shape, core layout, and lobby design—elevator and escalator selection and design integration are major tasks for the architectural designer.

Chapter 31 introduces the subject with a description of components including traction equipment, cars, safety devices, and systems of control and supervision. Fundamental topics, such as codes and standards and requirements for the disabled, are also covered. Criteria are established for elevator performance, and extensive car performance data are presented that can be used in conjunction with design criteria to select the size, number, and speed of elevators to meet the design intent. Principles of elevator zoning are presented. The chapter continues with an analysis of spatial requirements for elevators, including shafts, lobbies, and core arrangements, and concludes with special considerations such as energy use, safety, security, and noise.

Chapter 32 first addresses important passenger elevator considerations above and beyond elevator selection. These include power and energy requirements, with an emphasis on energy conservation, fire–emergency operation, emergency power requirements, and passenger security. The next part of the chapter is devoted to special cases: freight elevators, hydraulic elevators, nonvertical designs, residential elevators, and chair lifts, with a discussion of equipment, capacities, selection, economy, and comparison (where applicable) to conventional design. The chapter also covers interesting design options including double-deck cars and observation cars.

Chapter 32 also describes material-handling devices: dumbwaiters, horizontal and vertical conveyors, pneumatic systems, container delivery systems, and automated self-propelled vehicles. This information responds to the need to plan for material handling in commercial and institutional buildings, rather than treating the subject as an afterthought, with resultant inefficiency, overloading of freight facilities, and misappropriation of passenger elevators.

Chapter 33 explains equipment components, capacities, sizes, speeds, and methods of selection, followed by special topics including lighting, fire protection, power and energy requirements, and budget estimating. Information on inclined walks (ramps) is also presented.

# CHAPTER 31

# Vertical Transportation: Passenger Elevators

## GENERAL INFORMATION

### 31.1 INTRODUCTION

OF THE MANY DECISIONS THAT MUST BE made by the designer of a multistory building, probably none is more important than the selection of the vertical transportation equipment—that is, the passenger, service, and freight elevators and the escalators. Not only do these items represent a major building expense, being in the case of a 25-story office building as much as 10% of the construction cost, but the quality of elevator service is also an important factor in a tenant's choice of space in competing buildings.

Although the final decision as to the type of equipment rests with the architect, the factors affecting it are so numerous that the building designer should consult with an elevator expert. This service is available from consultants in the field and from the major elevator and escalator manufacturers. The function of this chapter is to familiarize the architect and engineer with the nature and application of vertical transportation equipment in order to enable them to make preliminary design decisions and interact effectively with consultants.

### 31.2 PASSENGER ELEVATORS

This chapter is concerned primarily with the general-purpose traction elevator. Hydraulic elevators are covered in Chapter 32.

Ideal performance of an elevator installation provides minimum waiting time for a car at any floor level; comfortable acceleration; rapid transportation; smooth, rapid braking; accurate automatic leveling at landings; and rapid loading and unloading at all stops. The system must also provide quick, quiet power operation of doors; good floor and travel direction indication, both in the cars and at landings; easily operated car and landing call buttons (or other devices); smooth, quiet, and safe operation of all mechanical equipment for all conditions of loading; comfortable lighting; reliable emergency and security equipment; and a generally pleasant car atmosphere.

In addition to passenger-oriented service considerations, elevators have architectural impacts. Cars and shaftway doors must be treated in a manner consonant with the architectural unity of the building. More important, however, is the fact that shaftways are major space elements whose integration into the building is a prime factor in composition, as is the design of the elevator lobby.

TRANSPORTATION

## 31.3 CODES AND STANDARDS

Perhaps more than any other item of construction, elevators are governed by strict installation codes. The "bible" of the industry in the United States is the American Society of Mechanical Engineers ANSI/ASME Standard A17.1, *Safety Code for Elevators and Escalators*, the latest version of which should be in every architect's and engineer's working library. The code has legal force in most parts of the United States. In addition, ANSI/ASME Standard A17.3 covers existing elevators and escalators, and Standard A17.4 covers emergency evacuation of passengers from elevators. As with other building systems, some states and municipalities have their own elevator codes (Massachusetts, Wisconsin, Pennsylvania, New York City, Seattle, and Boston, among others) that are generally based upon, and more stringent than, the ANSI/ASME code.

In addition to the elevator code, other construction and installation codes have an influence on elevator work. Thus, NFPA 101, the *Life Safety Code*, states certain fire safety requirements; NFPA 70 (the *National Electrical Code*) governs some of the electrical aspects of elevator construction; and state and local laws add a multitude of requirements and restrictions bearing on fire safety, emergency power, security regulations, and special accommodations for handicapped persons. Provisions for the disabled are covered by ANSI A117.1 (*Accessible and Usable Buildings and Facilities*), a special industry code, by the requirements of the Americans with Disabilities Act (ADA), and, in most locations, by local law. Like most large industries, the elevator industry is self-regulating and standardized. The National Elevator Industry, Inc. (NEII) publishes standard elevator layouts for traction and hydraulic installations. Elevator consultants and elevator company representatives are normally knowledgeable about all of the codes and standards in force, but this does not relieve the architect-engineer of legal responsibility for the design. Therefore, we strongly recommend that in the preliminary planning stage all pertinent regulations concerning vertical transportation be acquired and studied.

## ELEVATOR EQUIPMENT

## 31.4 PRINCIPAL COMPONENTS

The car, cables, elevator machine, control equipment, counterweights, hoistway, rails, penthouse, and pit are the principal parts of a traction elevator installation. An idea of the functioning and orientation of these items of equipment can be obtained from Fig. 31.1.

The car is the only item with which the average passenger is familiar. Indeed, some of a building's prestige depends upon proper design of the car. Essentially, the car is a cage of some fire-resistant material supported on a structural frame, to the top member of which the lifting cables are fastened. By means of guide shoes on the side members, the car is guided in its vertical travel in the shaft. The car is provided with safety doors, operating-control equipment, floor-level indicators, illumination, emergency exits, and ventilation. It is designed for long life, quiet operation, and low maintenance.

The cables (ropes) that are connected to the cross-head (top beam of the elevator) and carry the weight of the car and its live load are made of groups of steel wires especially designed for this application. Four to eight cables, depending on car speed and capacity, are placed in parallel. Although multiple ropes are used primarily to increase the traction area on the drive sheaves, they also increase the elevator safety factor inasmuch as each rope is normally capable of supporting the entire load. The minimum factor of safety varies from 7.6 to 12.0 for passenger elevators and from 6.6 to 11.0 for freight elevators. The cables from the top of the car pass over a motor-driven cylindrical sheave at the traction machine (grooved for the cables) and then downward to the counterweight.

The counterweight is made up of cut steel plates stacked in a frame attached to the opposite ends of the cables to which the car is fastened. It is guided in its travel up and down the shaft by two guide rails typically installed on the back wall off the shaft. Its weight equals that of the empty car plus 40% of the rated live load. It serves several purposes: to provide adequate traction at the sheave for car lifting, reduce the size of the traction machine, and reduce power demand and energy cost. (These advantages come at the price of strengthening the overhead machine room floor, which must carry the additional structural load of the counterweight.)

Approximately 75% of the energy expended in lifting a car is returned to the system by regeneration when the car is lowered. Regeneration is the process in which the traction motor becomes a *generator* when the car is lowered and feeds power

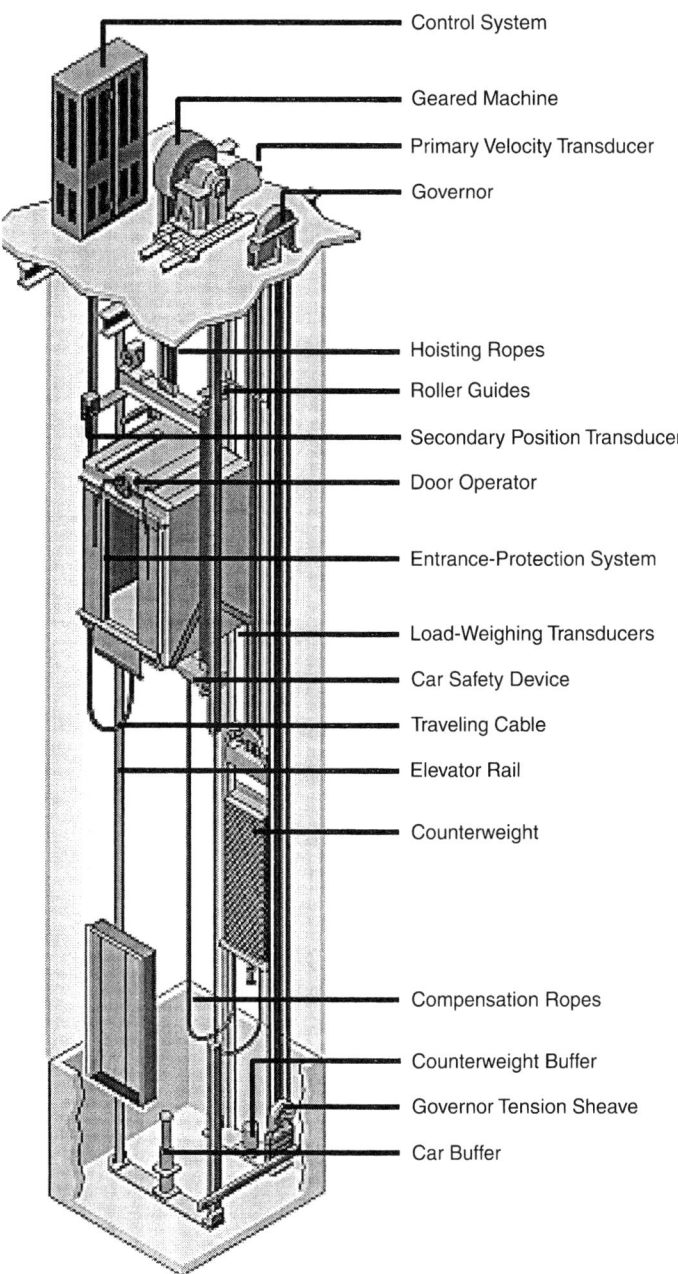

- Control System
- Geared Machine
- Primary Velocity Transducer
- Governor
- Hoisting Ropes
- Roller Guides
- Secondary Position Transducer
- Door Operator
- Entrance-Protection System
- Load-Weighing Transducers
- Car Safety Device
- Traveling Cable
- Elevator Rail
- Counterweight
- Compensation Ropes
- Counterweight Buffer
- Governor Tension Sheave
- Car Buffer

**Fig. 31.1** *Components of a geared elevator installation with solid-state control and motor drive. (Courtesy of Otis Elevator Co.)*

TRANSPORTATION

back into the electrical system. The lost energy appears as heat, primarily in the machine room. See Section 31.40 for a discussion of system energy requirements.

To compensate for the hoist rope weight, which becomes an important factor in high-rise elevators, cables are attached to the bottom of the car and the counterweight, thus equalizing loads regardless of the cab position. These cables can be seen in Fig. 31.1.

The elevator machine turns the sheave and lifts or lowers the car. It consists of a heavy structural frame on which are mounted the sheave and driving motor, the gears (if any), the brakes, the magnetic safety brake, and certain other auxiliaries. In many existing installations the elevator driving (traction) motor receives its energy from a separate motor–generator (m-g) set, which is in operation during the period that the particular elevator is available for handling traffic. This m-g set is properly considered a part of the elevator machine, although it may be located some distance from it. In modern installations, solid-state power and control equipment replaces the m-g set, as discussed in Section 31.17. A governor, which limits the car to safe speeds, is mounted on or near the elevator machine.

The control equipment is usually divided into three groups:

1. *Drive (motion) control* is concerned with the velocity, acceleration, position determination, and leveling of the car.
2. *Operating control* covers car door operation and functioning of car signals, including floor call buttons and all indicating devices.
3. *Supervisory control* is concerned with group operation of multiple-car installations.

The actual physical devices in these control systems were electromechanical in the past but are solid-state in modern installations. The indicating and control devices that are seen and used by the elevator user, including car and hallway buttons, lanterns, and audible devices, are all coordinated into the overall operational control scheme, which produces rapid, safe, and comfortable vertical transportation.

The *shaft*, or *hoistway*, is the vertical passageway for the car and counterweights. On its sidewalls are the car guide rails and certain mechanical and electrical auxiliaries of the control apparatus. At

the bottom of the shaft are the car and counterweight buffers. At the top is the structural platform on which the elevator machine rests. The elevator machine room (which may occupy one or two levels) is usually directly above the hoistway. It contains the traction machine and the m-g set or solid-state control that supplies energy to the elevator machine and control equipment. Machinery and control equipment are designed for quiet, vibration-free operation.

## 31.5 GEARLESS TRACTION MACHINES

A gearless traction machine consists of a dc or ac motor, the shaft of which is directly connected to a brake wheel and driving sheave. The elevator hoist ropes are placed around this sheave. The absence of gears means that the motor must run at the same relatively low speed as the driving sheave. As it is not economically practical to build motors for operation at very low speeds, a gearless machine is utilized for medium- and high-speed elevators—that is, 500 fpm and above. The motors range from 20 to 400 hp.

Gearless machines are generally utilized for passenger service, with car capacities of 2000 to 4000 lb, although specials of up to 10,000 lb have been built. Below 500 fpm, geared machines are used. At this writing, maximum car speeds are 2000 fpm. Faster drives have already been developed and will undoubtedly be an important factor in the development of ever taller, practical, workable buildings.

A gearless traction machine is considered superior to a geared machine because it is more efficient, quieter in operation, requires less maintenance, and has longer life. The decision as to whether these advantages are worth the additional cost involved is made only after a careful analysis. Generally, a gearless machine is chosen where the rise is more than 250 ft and very smooth, high-speed operation is desired. In the intermediate range of rise and speeds (i.e., 150–250 ft height and 400–500 fpm), excellent equipment, both geared and gearless, is available.

Because virtually every major elevator company operates internationally, and because outside the United States elevator size and speed are specified in kilograms and meters per second, it is useful to

**TABLE 31.1  Elevator Capacity and Speed; Approximate SI Equivalents**

| Weight | | Speed | |
|---|---|---|---|
| Standard SI (kg) | Approximate U.S. I-P (lb) | Standard SI (m/s) | Approximate U.S. I-P (fpm) |
| 1000 | 2000 | 1.0 | 200 |
| 1250 | 2500–3000 | 1.6 | 300 |
| 1600 | 3500 | 2.0 | 400 |
| 2000 | 4000 | 2.5 | 500 |
| | | 3.15 | 600 |
| | | 4.0 | 800 |
| | | 5.0 | 1000 |

have approximate conversion factors in mind. As the accurate equivalent for 1 kg is 2.2 lb and 1 m/s is 196.86 fpm, the factors of 2 and 200, respectively, give approximate I-P equivalents (Table 31.1).

## 31.6  GEARED TRACTION MACHINES

A geared traction machine has a worm and gear interposed between the driving motor and the hoisting sheave (Fig. 31.2). The driving motor can therefore be a smaller, cheaper, high-speed unit rather than the large, low-speed unit that a gearless installation would require.

The motor used in a geared installation, as in a gearless one, depends upon the type of drive system and may be either dc or ac. Characteristics of the various drive systems and their application are

**Fig. 31.2** Cutaway of a typical dc geared traction elevator machine. Note the grooves for multiple ropes in the traction sheave. (Photo courtesy of Schindler Elevator Corp.)

discussed in detail in Sections 31.12 to 31.15. Geared machines are used for car speeds of up to 450 fpm and a maximum rise of about 300 ft. With an appropriate drive and control system, a geared traction machine can give almost the same high-quality, accurate, smooth ride as is available from a gearless installation.

## 31.7  ARRANGEMENT OF ELEVATOR MACHINES, SHEAVES, AND ROPES

The simplest method of arranging vertical travel of a car is to pass a rope over a sheave and counterbalance the weight of the car by a counterweight. Then, by rotating the sheave, the car moves up or down and requires very little energy to do it. This is essentially the scheme that is used on a majority of high-speed passenger elevators, as illustrated in Fig. 31.3a.

When the four or more supporting ropes merely pass over the sheave T and connect directly to the counterweights, the lifting power is exerted by the sheave through the traction of the ropes in the parallel grooves on the sheave. This system is referred to as the *single-wrap traction elevator machine.* The function of sheave S is merely that of a guide pulley; it is called the *deflector sheave.*

The arrangement shown in Fig. 31.3b is called *double-wrap 1:1 roping.* It provides greater traction than the single-wrap arrangement and is used in many automatic high-speed installations.

A 1:1 roping arrangement (Fig. 31.3a,b,d) gives no mechanical advantage. The 2:1 roping (Fig. 31.3c) has a mechanical advantage of 2, which permits use of a high-speed, low-power, and, therefore, lower-cost traction machine. This arrangement is used for a wide variety of installations

**TRANSPORTATION**

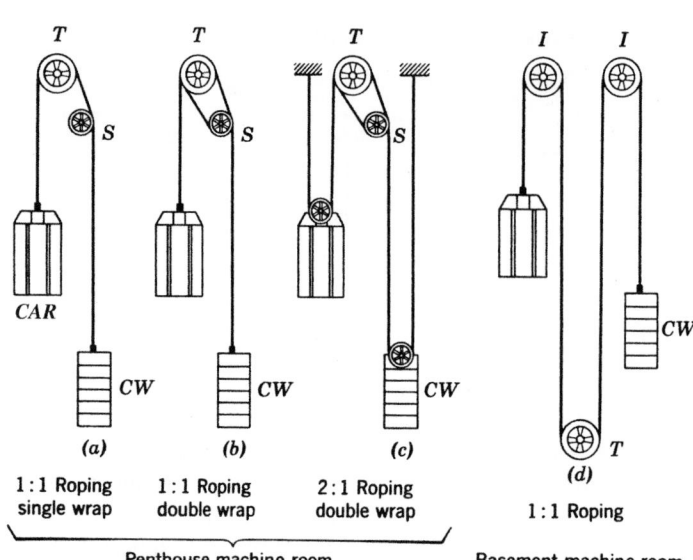

**Fig. 31.3** Elevator roping and sheave arrangements. (a) Basic single wrap rope arrangement. In (b) and (c), the rope passes over traction sheave T and sheave S, doubles back over T, and then extends past S to the counterweight CW. This double-wrap arrangement provides additional traction at the drive sheave. (d) Roping arrangement for a basement machine room.

varying from medium-speed (500–700 fpm) gearless passenger elevators to low-speed, heavy-duty freight units.

In types *a, b,* and *c* in Fig. 31.3, the elevator machines are located in a machine room penthouse at the top of the hoistway. If, for architectural or other reasons, it is desired to eliminate this penthouse, a basement machine room can be used with the roping shown in Fig. 31.3*d.* This arrangement uses geared traction equipment with speeds of up to 400 fpm. All the illustrated ropings are applicable to the full range of car capacities.

## 31.8 SAFETY DEVICES

The main brake of an elevator is mounted directly on the shaft of the elevator machine (see Fig. 31.2). The elevator is first slowed by dynamic braking of the motor, and the brake then operates to clamp the brake drum, thus holding the car still at the floor.

A dual safety system, designed to stop an elevator car automatically before its speed becomes excessive, is normally used. The device that acts first is a centrifugal governor or an electronic speed sensor that cuts off the power to the traction motor and sets the brake in the event of a limited over-

speed. This usually stops the car. Should the speed still increase, the governor actuates two safety rail clamps, which are mounted at the bottom of the car, one on either side. They clamp the guide rails by wedging action, bringing the car to a smooth stop (Fig. 31.4).

Oil or spring buffers are usually placed in the elevator pit. Their purpose is not to stop a falling car but to bring it to a partially cushioned stop if it overtravels the lower terminal. If a car overtravels (down or up), travel sensors deenergize the traction motor and set the main brake. Safety arrangements under emergency condition of fire or power failure are discussed in Section 31.43.

## 31.9 ELEVATOR DOORS

The choice of the car and hoistway door affects the speed and quality of elevator service considerably. Doors for passenger elevators are power-operated and are synchronized with the leveling controls so that the doors are fully opened by the time a car comes to a complete stop at the landing. The closing time, however, varies with the type of door and the size of the opening. For safety reasons, the kinetic energy of an automatic door is limited to 7 ft-lb and its closing pressure to 30 lb. To provide the fastest

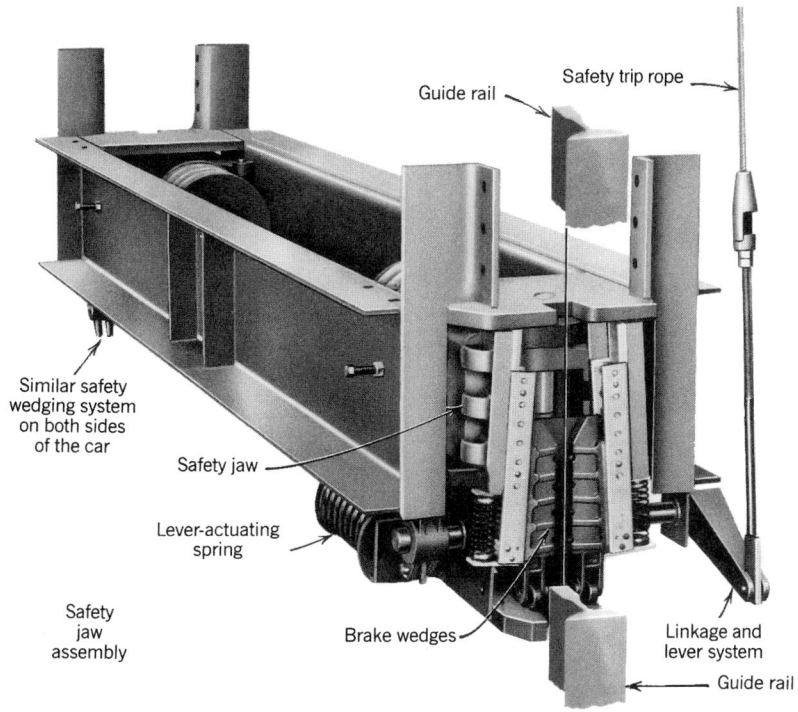

**Fig. 31.4** *Elevator safety devices. The governor or velocity transducer senses overspeed, clamping the safety trip rope and releasing the safety jaws, which exert a constant retarding force on the car rails, thus bringing the car to a gradual and safe stop. (Photo courtesy of Schindler Elevator Corp.)*

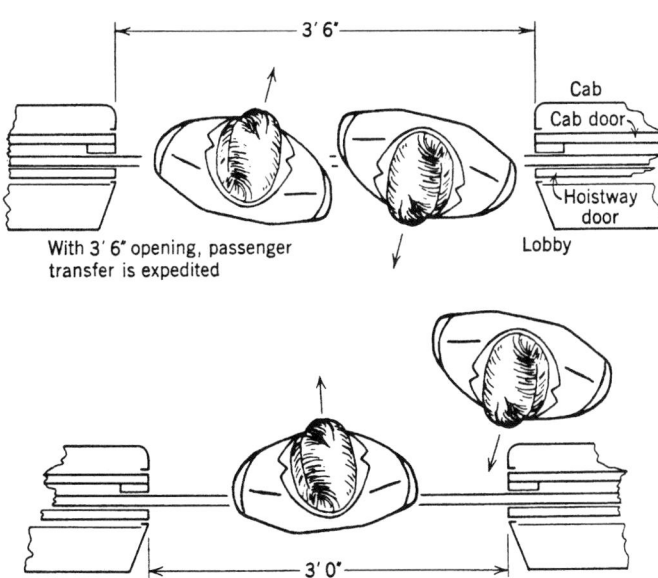

**Fig. 31.5** *Transfer of passengers with door openings of different widths. With openings smaller than 3 ft, 6 in. (42 in.), simultaneous loading and unloading is difficult and transfer time is lengthened. With a 42-in. opening, large people or people with bulky outerwear may brush against each other in passing. For complete isolation of passengers, a 48-in. opening may be necessary.*

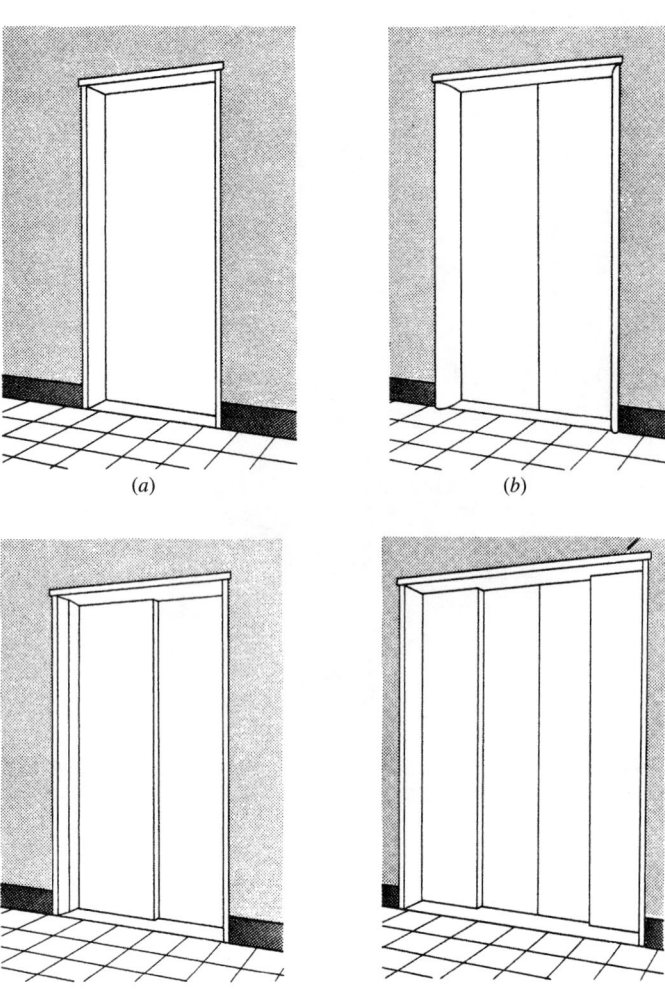

(a)

(b)

(c)

(d)

*Fig. 31.6* Typical hoistway doors and applications. (a) Single-slide door, 24 to 36 in. wide, for small commercial building or residential use. (b) Standard commercial 42-in. center-opening door for office building use or 48- to 60-in. center-opening for hospital or service car. (c) Two-speed, 42-in., general commercial use door. (d) Two-speed, center-opening, 60-in. department store door for freight, passenger, and nonautomatic service.

closing within this energy limitation, a center-opening door is used. Also, to reduce passenger transfer time and avoid discomfort, a clear opening of 3 ft, 6 in. (106.7 cm) is used in most commercial installations, which permits simultaneous loading and unloading without undue passenger contact. (Some consultants feel that simultaneous passenger transfer is practical only with a 48-in. clear opening; Fig. 31.5). When an opening narrower than 36 in. is used, loading is delayed until unloading is complete, and the speed and quality of service are thereby markedly reduced. Such small doors are applicable only in residential or small, light-traffic buildings. The available door types are shown in Fig. 31.6.

A two-speed door design is used where space conditions dictate or where a wide opening is required. The term *two-speed* reflects the fact that the two halves of the door must travel at different speeds to complete their travel simultaneously (see Fig. 31.6c).

Installations can be equipped with an electronic sensing device that detects passengers in a wide area on the landing in front of the car door rather than only directly in the door's path. Such detection, often accompanied by an audible signal, causes the car door to remain open for a predetermined length of time or a closing door to reverse. These devices are particularly useful in installations where passengers cannot approach the entrance or cannot enter the car quickly—for example, riders with baggage or holding children, people in wheelchairs, and employees moving bulky objects such as beds or carts in hospitals or wheeled objects in office and industrial facilities.

All automatic elevators, regardless of whether or not equipped with detection beams, are required by ANSI to have a safety edge device on the car doors that causes the car and hoistway doors, which operate in synchrony, to reopen when the safety edge meets any obstruction. Some cars doors are arranged to "nudge" when almost closed and/or after a specific time period. ADA requirements for the disabled are discussed in Section 31.11.

## 31.10  CARS AND SIGNALS

Possibly the only area in which the architect has a free hand in selecting equipment is the decor of the cars and the styling of hallway and car signals. A normal elevator specification is a functional one that describes the intended operation of the equip-

ment and normally includes an amount to cover optional decor of the cars. The type and functioning of signal equipment are also specified, but finish and styling are optional. Car interiors may be finished in wood paneling, plastic (Micarta or Formica), stainless steel, or almost any material desired. Floors may be tile, wood, or carpeting, as selected. Illumination may be from ceiling fixtures, coves, or a completely illuminated luminous ceiling of standard or special design. For each bank of elevators, it is wise to furnish at least one set of wall mats to protect wall finishes when cars are being used to move tenant furniture. This is especially important where no separate service car has been provided.

Car and hallway signals and lanterns should be designed to fulfill their basic functions, consider the needs of the handicapped, and be consonant with the decor of the cars and corridors. (For a discussion of the car control panel, see Section 31.24.) The hall buttons should indicate the desired direction of travel and by visual means confirm that a call has been placed. The hall lantern located at

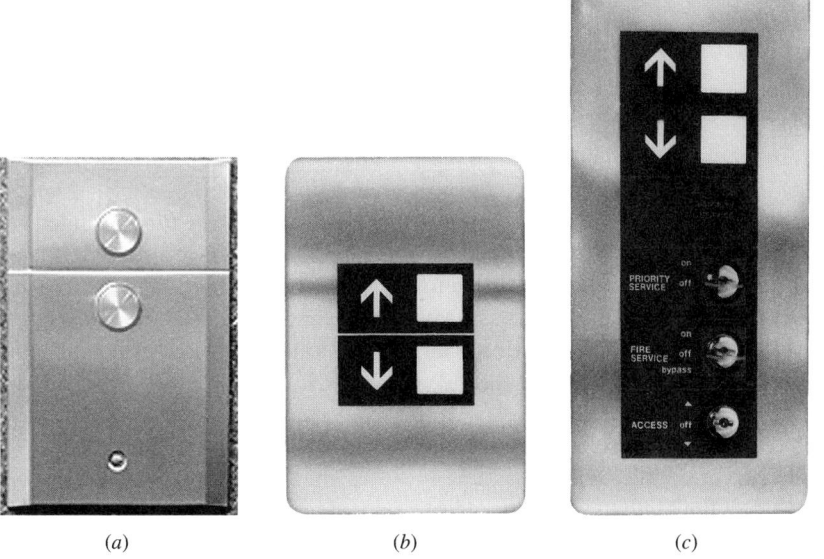

**Fig. 31.7** Typical hall stations. (a) Simple pushbutton station with an illuminated ring on each button to indicate call registration. (b) Large buttons show travel direction in addition to lighting up to indicate call registration. (c) Expanded hall stations may incorporate priority, emergency, or secure access controls as shown. (Photos courtesy of Otis Elevator Co.)

TRANSPORTATION

*(a)*

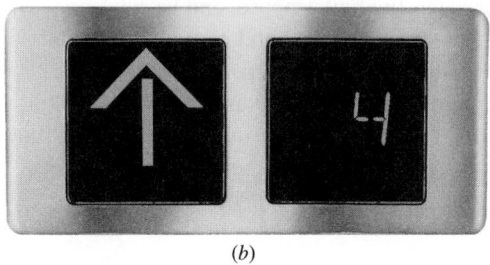

*(b)*

**Fig. 31.8** *Typical hall lanterns. (a) The top section is green and lights when a car approaches, indicating up travel. The bottom section is red, and when lighted indicates an approaching down car. (b) The lantern shows travel direction and present car location. (Photos courtesy of Otis Elevator Co.)*

each car entrance must visually indicate the direction of travel of an arriving elevator and preferably its present location (Figs. 31.7 and 31.8). An audible signal should announce its imminent arrival. This feature allows waiting passengers to move to the next arriving car in a bank and thereby speeds service. Hall stations can be equipped with special switches for fire, priority, and limited-access service, as required.

Within the car, the travel direction and present location can be indicated either with separate fixtures or by indicators built into the car panel. Most manufacturers can provide voice synthesizers built into the car panel that announce the floor, direction of travel, and any other desired message such as safety or emergency messages.

## 31.11 REQUIREMENTS FOR THE DISABLED

Elevator manufacturers follow ADA requirements as a minimum, and may add additional conveniences for the disabled as the specific facility design intent or local codes require. As with all aspects of design for buildings, no planning should begin until the project architect has verified and assembled all applicable codes and standards.

The basic physical limitations addressed are those of ambulation, sight, and hearing. Thus, to ease access for passengers in wheelchairs (or with walking aids), the ADA requires excellent car leveling, 36-in. minimum clear door opening (42-in. recommended), delayed door closing, detection beams that reopen the door *without contact* on sensing a passenger, inside car dimensions that permit turning a wheelchair, buttons and emergency controls within easy reach, and appropriate car furnishings.

For those with sight impairment, the ADA requires audible signals in addition to easily seen and recognized visual ones, both in the car and at landings, to indicate call registration, car approach, car landing, direction of travel, floor, car position, and so on. In this connection, a voice synthesizer is of invaluable assistance. In addition, Braille plates adjacent to car floor buttons and large, easily recognizable symbols adjacent to passenger-controlled emergency controls are to be used. Hearing-impaired passengers are assisted by the large visual signals required for sight-impaired passengers. In addition, call buttons must visually indicate that a call has been placed and the indication extinguished when the call has been answered. Many of these ideas also make life easier for passengers with full faculties. A note of caution is in order, however. Delayed door closing may increase travel time appreciably. In buildings with traffic peaks, it may be necessary to designate one or more specific elevators for use by the disabled during these periods.

The verbatim requirements of the ADA as stated in the *Federal Register,* Vol. 56, No. 144, July 26, 1991, are presented in Fig. 31.9, with accompanying illustrations as they appear in the *Federal Register.* References to illustrations within this figure are to the *Federal Register* figure numbers.

## ADA Accessibility Guidelines for Buildings and Facilities

### 4.10 Elevators

**4.10.1 General.** *Accessible* elevators shall be on an accessible route and shall comply with 4.10 and with the *ASME A17.1—1990. Safety Code for Elevators and Escalators. Freight elevators shall not be considered as meeting the requirements of this section unless the only elevators provided are used as combination passenger and freight elevators for the public and employees.*

**4.10.2 Automatic Operation.** Operation shall be automatic. Each car shall be equipped with a self-leveling feature that will automatically bring the car to floor landings within a tolerance of ½ in. (13 mm) under rated loading to zero loading conditions. This self-leveling feature shall be automatic and independent of the operating device and shall correct the overtravel or undertravel.

**4.10.3 Hall Call Buttons.** Call button in elevator lobbies and halls shall be centered at 42 in. (1065 mm) above the floor. Such call buttons shall have visual signals to indicate when each call is registered and when each call is answered. Call buttons shall be a minimum of ¾ in. (19 mm) in the smallest dimension. The button designating the up direction shall be on top. (See Fig. 20.) *Buttons shall be raised or flush. Objects mounted beneath hall call buttons shall not project into the elevator lobby more than 4 in. (100 mm).*

**4.10.4 Hall Lanterns.** A visible and audible signal shall be provided at each hoistway entrance to indicate which car is answering a call. Audible signals shall sound once for the up direction and twice for the down direction or shall have verbal annunciators that say "up" or "down". Visible signals shall have the following features:

(1) Hall lantern fixtures shall be mounted so that their centerline is at least 72 in. (1830 mm) above the lobby floor. (See Fig. 20.)

(2) Visual elements shall be at least 2½ in. (64 mm) in the smallest dimension.

(3) Signals shall be visible from the vicinity of the hall call button (see Fig. 20). In-car lanterns located in cars, visible from the vicinity of hall call

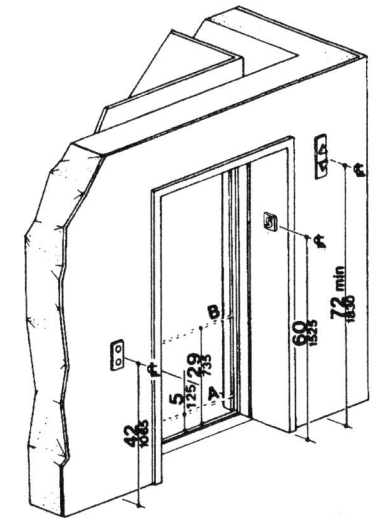

NOTE: The automatic door reopening device is activated if an object passes through either line A or line B. Line A and line B represent the vertical locations of the door reopening device not requiring contact.

**Fig. 20**
**Hoistway and Elevator Entrances**

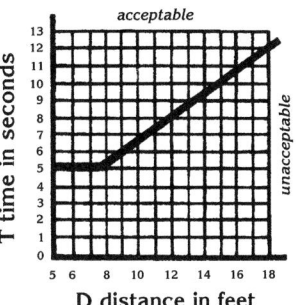

**Fig. 21**
**Graph of Timing Equation**

buttons, and conforming to the above requirements, shall be acceptable.

**4.10.5 Raised and Braille Characters on Hoistway Entrances.** All elevator hoistway entrances hall have *raised* and *Braille* floor designations provided on both jambs. The centerline of the characters shall be 60 in. (1525 mm)

**Fig. 31.9** *ADA requirements for elevators extracted from the* Federal Register, *July 26, 1991.*

**ADA Accessibility Guidelines for Buildings and Facilities** *(Continued)*

above *finish* floor. Such characters shall be 2 in. (50 mm) high and shall comply with 4.30.4. Permanently applied plates are acceptable if they are permanently fixed to the jambs (see Fig. 20).

**4.10.6 Door Protective and Reopening Device.** Elevator doors shall open and close automatically. They shall be provided with a reopening device that will stop and reopen a car door and hoistway door automatically if the door becomes obstructed by an object or person. The device shall be capable of completing these operations without requiring contact for an obstruction passing through the opening at heights of 5 in. and 29 in. (125 mm and 735 mm) above finish floor (see Fig. 20). Door reopening devices shall remain effective for at least 20 seconds. After such an interval, doors may close in accordance with the requirements of *ASME* A17.1—1990.

**4.10.7 Door and Signal Timing for Hall Calls.** The minimum acceptable time from notification that a car is answering a call until the doors of that car start to close shall be calculated from the following equation:

$$T = D/(1.5 \text{ ft/s}) \text{ or } T = D/(445 \text{ mm/s})$$

where *T* total time in seconds and *D* distance (in feet or millimeters) from a point in the lobby or corridor 60 in. (1525 mm) directly in front of the farthest call button controlling that car to the centerline of its hoistway door (see Fig. 21). For cars with in-car lanterns, *T* begins when the lantern is visible from the vicinity of hall call buttons and an audible signal is sounded. *The minimum acceptable notification time shall be 5 seconds.*

**4.10.8 Door Delay for Car Calls.** The minimum time for elevator doors to remain fully open in response to a car call shall be 3 seconds.

**4.10.9 Floor Plan of Elevator Cars.** The floor area of elevator cars shall provide space for wheelchair users to enter the car, maneuver within reach of controls, and exit from the car. Acceptable door opening and inside dimensions shall be as shown in Fig. 22. The clearance between the car platform sill and the edge of any hoistway landing shall be no greater than 1¼ in. (32 mm).

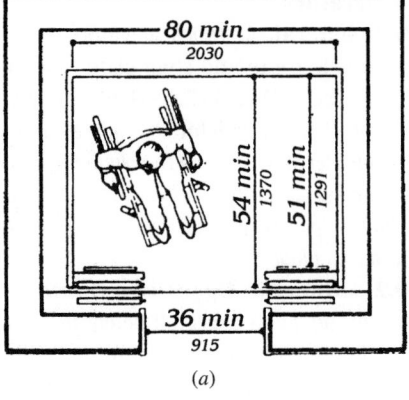

*(a)*

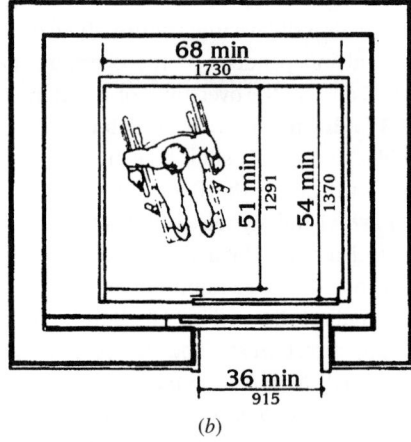

*(b)*

**Fig. 22**
**Minimum Dimensions of Elevator Cars**

**4.10.10 Floor Surfaces.** Floor surfaces shall comply with 4.5.

**4.10.11 Illumination Levels.** The level of illumination at the car controls, platform, and car threshold and landing sill shall be at least 5 footcandles (53.8 lux).

**4.10.12 Car Controls.** Elevator control panels shall have the following features:

(1) Buttons. All control buttons shall be at least ¾ in. (19 mm) in their smallest dimension. They *shall* be *raised* or flush.

(2) Tactile. *Braille*, and Visual Control Indicators. All control buttons shall be designated by

*Fig. 31.9 (Continued)*

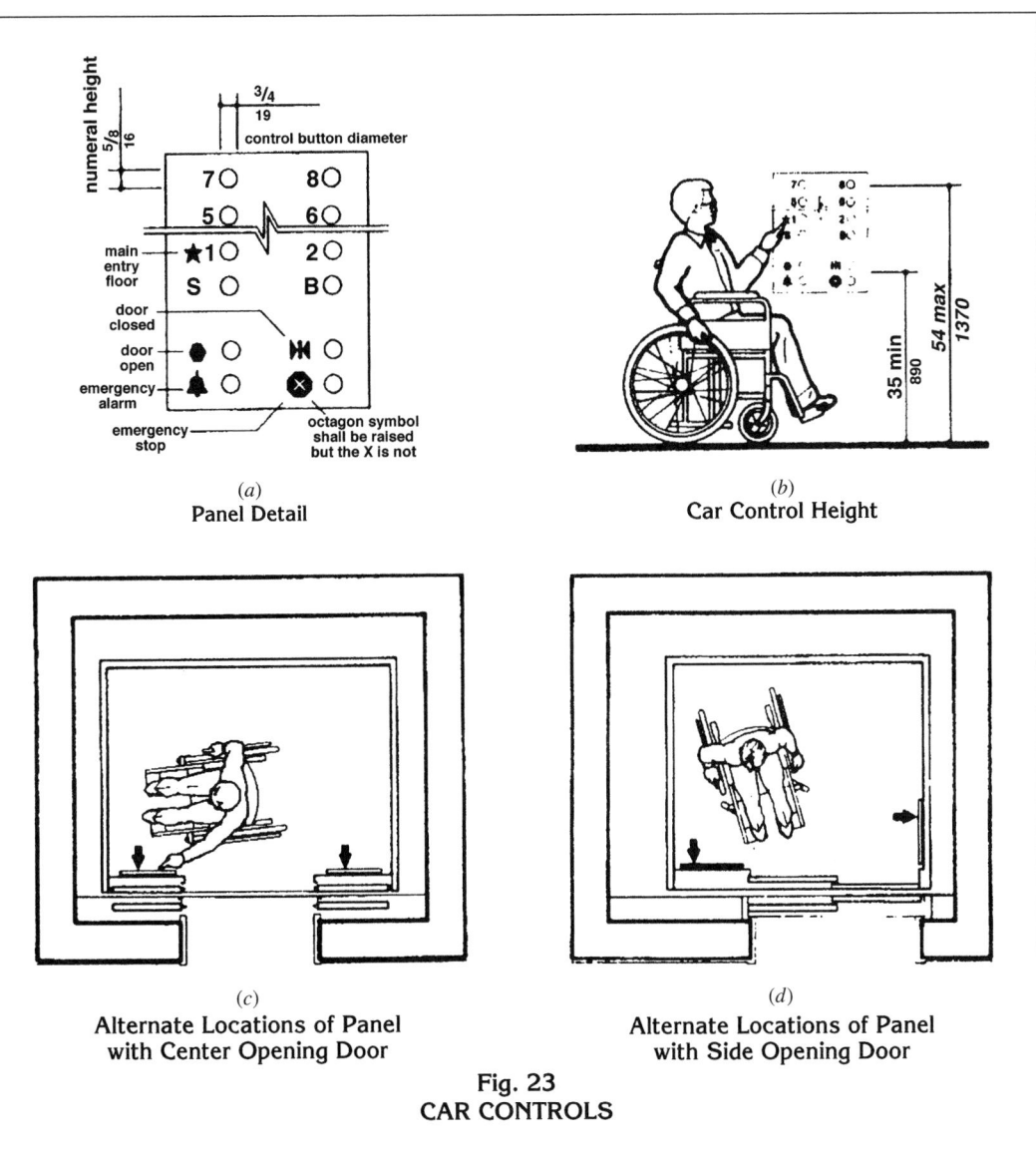

(a)
**Panel Detail**

(b)
**Car Control Height**

(c)
**Alternate Locations of Panel
with Center Opening Door**

(d)
**Alternate Locations of Panel
with Side Opening Door**

**Fig. 23
CAR CONTROLS**

*Braille* and by *raised* standard alphabet characters for letters, Arabic characters for numerals, or standard symbols as shown in Fig. 23a, and as required in *ASME* A17.1—1990. *Raised* and *Braille* characters and symbols shall comply with 4.30. The call button for the main entry floor shall be designated by a *raised* star at the left of the floor designation (see Fig. 23a), and as required in *ASME* A17.1—1990. *Raised* and *Braille* characters and symbols shall comply with 4.30. The call button for the main entry

floor shall be designated by a raised star at the left of the floor designation (see Fig. 23a). All raised designations for control buttons shall be placed immediately to the left of the button to which they apply. Applied plates, permanently attached, are an acceptable means to provide raised control designations. Floor buttons shall be provided with visual indicators to show when each call is registered. The visual indicators shall be extinguished when each call is answered.

**TRANSPORTATION**

*Fig. 31.9 (Continued)*

---

**ADA Accessibility Guidelines for Buildings and Facilities** *(Continued)*

(3) Height. All floor buttons shall be no higher than 54 in. (1370 mm) above the *finish floor for side approach and 48 in. (1220 mm) for front approach.* Emergency controls, including the emergency alarm and emergency stop, shall be grouped at the bottom of the panel and shall have their centerlines no less than 35 in. (890 mm) above the finish floor (see Fig. 23a and b).

**4.10.13 Car Position Indicators.** In elevator cars, a visual car position indicator shall be provided above the car control panel or over the door to show the position of the elevator in the hoistway. As the car passes or stops at a floor served by the elevators, the corresponding numerals shall illuminate, and an audible signal shall sound. Numerals shall be a minimum of ½ in. (13 mm) high. The audible signal shall be no less than 20 decibels with a frequency no higher than 1500 Hz. An automatic verbal announcement of the

floor number at which a car stops or which a car passes may be substituted for the audible signal.

**4.10.14 Emergency Communications.** If provided, emergency two-way communication systems between the elevator and a point outside the hoistway shall comply with *ASME A17.1—1990.* The highest operable part of a two-way communication system shall be a maximum of 48 in. (1220 mm) from the floor of the car. It shall be identified by a raised symbol and lettering complying with 4.30 and located adjacent to the device. If the system uses a handset then the length of the cord from the panel to the handset shall be at least 29 in. (735 mm). *If the system is located in a closed compartment the compartment door hardware shall conform to 4.27, Controls and Operating Mechanisms. The emergency inter-communication system shall not require voice communication.*

*Fig. 31.9* (Continued)

---

# ELEVATOR CAR CONTROL

## 31.12 DRIVE CONTROL

The movement of an elevator car and all of its parts is controlled by three different systems that combine and interact to provide a unified control system. The *supervisory system* controls a bank of elevators as a group and dictates which car answers which call. The *operational control* system determines when and where physical motion of a car and its doors should occur. It deals with the operation of the car doors and the integration of car buttons, lanterns, and passenger-operated devices into the overall control and indicating system. Operating control passes its information about car and door control to the *motion control* system (also known as *drive control*). Motion control determines the car's acceleration, velocity, braking, leveling, and regenerative braking, plus all aspects of door motion.

Elevator *car* control is separate and distinct from the control system that governs the functioning of a group of cars acting as a system. That arrangement is generally designated the *supervisory system* and is discussed separately.

Elevator car acceleration and deceleration are accomplished by controlling the speed of the motor that drives the elevator traction machine. This speed control can be accomplished in a number of ways, all of which are in use in elevator installations. They are:

- Thyristor control of an asynchronous (squirrel cage) ac traction motor
- Thyristor control of a dc traction motor
- Motor-generator (m-g) set control of a dc traction motor (Ward–Leonard system)
- Variable-voltage, variable-frequency (VVVF) control of an asynchronous ac traction motor

(Rheostatic control of single and multispeed ac motors is essentially obsolete and therefore is not discussed.)

We describe each system briefly and note its applicability, advantages, and disadvantages.

## 31.13 THYRISTOR CONTROL, AC AND DC

### (a) ac

High-power transistors make accurate speed control of inexpensive ac squirrel-cage motors practical by the simple expedient of supplying carefully regulated variable voltage to the motor. The resultant speed control is smooth and stepless, making it applicable to passenger elevator installations. However, the high motor slip caused by the constant frequency of the variable-voltage ac supply causes large thermal losses with resultant low operating efficiency. In addition, the "chopper," which is used to provide the necessary voltage control, introduces undesirable harmonics into the power system in considerable quantity. These harmonics cause undesirable radio noise and can cause system component overheating. Finally, the system's low power factor increases line losses and necessitates increased feeder sizes. As a result, this option, which is applicable to low and midrise passenger service (maximum rise 250 ft and maximum car speed of 350 fpm), using geared machines has been largely replaced by VVVF control (see Section 31.15).

### (b) dc

Excellent ride quality can be had by utilizing the dc version of thyristor control (see Fig. 31.10a) to supply variable voltage to a dc traction motor. This

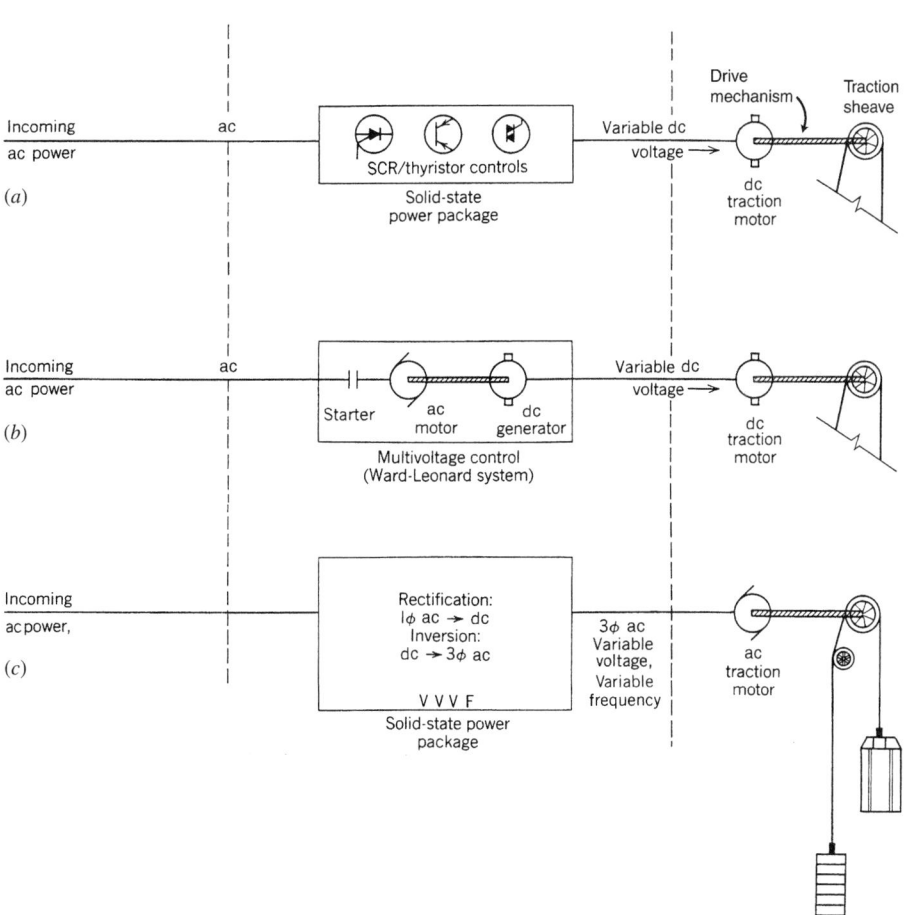

**Fig. 31.10** *Arrangement of electric speed-control equipment for elevator drives. (a) Solid-state, silicon-controlled rectifier (SCR) thyristor control produces variable-voltage dc and finds application in good-quality new and retrofit medium- and high-speed installations. (b) The traditional Ward–Leonard (m-g) system, which produces finely controlled, variable-voltage dc. This is the system in use today in the vast majority of existing high-quality installations. (c) Variable-frequency (VVVF) ac is energy efficient, highly accurate, and applicable to all types of installations.*

TRANSPORTATION

arrangement provides the high-quality ride and leveling characteristic of dc drives and removes the rise and speed limitations of its ac counterpart, but not the system's inherent low power factor, heavy line harmonics, and high machine room thermal losses. Line harmonics can be sharply reduced by the use of power line filters. This control system is widely used today, with both geared and gearless traction machines as applicable, for installations of all rises and car speeds. Its extremely smooth ride makes it one of the three systems of choice in high-quality construction. The other two systems are the traditional Ward–Leonard variable-voltage dc system and the VVVF ac system.

## 31.14 VARIABLE-VOLTAGE DC MOTOR CONTROL

Before the development of electronic motor control, the only practical way of obtaining the precise motor speed control necessary for smooth, stepless elevator operation was to impress a variable dc voltage on a dc traction motor. This variable dc voltage was obtained from an auxiliary m-g set comprising an ac motor and a dc generator. This arrangement (see Fig. 31.10b) is known as a *Ward–Leonard system* or as a *unit multivoltage* (UMV) drive. Although this arrangement has disadvantages, it is the classic high-quality elevator drive arrangement and is found in the vast majority of high-quality geared and gearless installations built before 1990. The disadvantages of this system—low overall efficiency, expensive machines, high thermal losses in machine rooms, high maintenance costs for three rotating machines, and high noise levels—were, until the development of a practical VVVF system, accepted as the price of a system that operates faultlessly, with supreme accuracy at all speeds, for long periods of time given proper maintenance. As noted previously, it is today still one of the three principal drive control systems being used in new installations. Its principal advantage over VVVF control is that it is "forgiving" of a less than ideal building power supply.

In modernization work, buildings with existing Ward–Leonard systems are frequently converted to solid-state dc thyristor control. This change maintains the excellent ride characteristics of variable-voltage dc traction machine installations while improving system performance by replacing the

auxiliary m-g set with a low-loss, low-maintenance, quiet solid-state thyristor drive control.

## 31.15 VARIABLE-VOLTAGE, VARIABLE-FREQUENCY AC MOTOR CONTROL

This system (see Fig. 31.10c), which is entirely solid-state, is in the opinion of many elevator professionals the drive system of choice for all high-quality new installations of any speed or rise. The system consists of a rectifier, which changes the incoming ac to dc, and an inverter, which creates variable-voltage, infinitely variable frequency, three-phase ac from the (rectified) dc. This ac is then applied to a standard squirrel-cage ac motor, which operates essentially at the speed corresponding to the frequency of the input. By maintaining a constant voltage-to-frequency ratio input to the traction motor, it is possible to provide continuously variable, highly accurate speed control at extremely high efficiencies throughout the full speed range of the motor. This is precisely what is required for high-quality geared and gearless installations, without the energy losses associated with the auxiliary m-g set of UMV control. The VVVF speed control system eliminates most of the disadvantages of the solid-state thyristor control system and has the following characteristics:

- Overall system efficiency is high at all motor speeds.
- Traction motors are economical single-speed squirrel-cage ac.
- Motors are 10% to 15% smaller than those of other drive systems.
- System power factor is close to unity.
- Line harmonics are lower than with thyristor controls.
- Speed control and leveling are equal to dc traction motor control.
- Equipment is 98% solid-state, thus requiring minimum maintenance.
- Electric energy use and peak loads are reduced, thus reducing electric billings by up to 35%.
- The system is applicable to all rises and speeds.
- Machine rooms are smaller, cooler, and quieter than those using Ward–Leonard UMV controls.

At this point, we can summarize the preceding discussion with tabular comparisons of the various drive systems in use today (Tables 31.2 and 31.3).

## TABLE 31.2 Comparative Energy Use of Various Motion Control Arrangements

| Energy Use | Relative Amount |
|---|---|
| Hydraulic elevators (discussed in Chapter 32) | 10 |
| UMV motion control; geared dc traction motor | 7–7.5 |
| UMV motion control; gearless dc traction motor | 6–6.5 |
| Thyristor control; geared dc traction motor | 5–5.5 |
| Thyristor control; gearless dc traction motor | 4–4.5 |
| Variable-voltage variable-frequency control, ac traction motor | 2.5–3.0 |

The exact degree of improvement possible when changing to solid-state equipment in a retrofit job depends on the quality of both the old and new systems. It is greatest when a microcomputer group control system replaces a relay-logic terminal dispatch system.

## 31.16 ELEVATOR OPERATING CONTROL

Assuming the system to be energized and at rest, registration of a call from a station in the lobby or an upper-floor corridor activates the system. The particular elevator that answers the call is selected by the supervisory system. In a UMV system the car's m-g set is started, whereas with solid-state motion control power is available immediately. Switching devices in the car control panel release the brake, energize the elevator motor, and accelerate the car to its rated speed. Reverse operations are initiated when decelerating and stopping (landing)

the car. When the car stops, the brake holds the sheave and elevator stationary.

The motion of a single car is determined by the action of three principal items of equipment: the car controller, the motion controls, and the system supervisory equipment. The action of the last equipment is discussed in Sections 31.17 to 31.21. The function of the car controller is to provide information on the car's exact location, car panel calls, and hall calls. This information is fed into the supervisory system and the motion control equipment, which in turn act to initiate all the procedures necessary to answer all calls via the individual car controller panels. The car controller panel also supplies the necessary signals to car and hall lanterns that indicate the car position and direction of travel.

## 31.17 SYSTEM CONTROL REQUIREMENTS

An operating system is one that provides the automatic response of a group of cars to calls for service. An effective system must take cognizance of all hall calls and car calls, car travel directions, and car position, in relation to each other and in relation to the call requirements, plus the trends of traffic. The last is required in order that the system anticipate demand rather than react to it. Because traffic and calls are never static, the control system that can satisfy all these demands in a large elevator system is necessarily an extremely sophisticated one. On small systems, the operating control is much simpler, as described in later sections. Throughout the discussion that follows, solid-state systems are assumed.

## TABLE 31.3 Comparative Characteristics of Elevator Drive Systems

| Type | Rise (ft) | Speed (fpm) | Control | Initial Cost | Operating Cost | Performance |
|---|---|---|---|---|---|---|
| Geared ac | 250 | 150–350 | Thyristor | Medium | Medium | Fair |
| | 300 | 150–450 | VVVF | High | Low | Excellent |
| Geared dc | 175 | 50–450 | UMV | High | High | Excellent |
| | 250 | 50–450 | Thyristor | Medium | Low | Very good |
| Gearless dc | Unlimited | 400–1200 | UMV | High | High | Excellent |
| | | | Thyristor | Medium | Low | Very good |
| Gearless ac | Unlimited | 400–2000 | VVVF | Medium | Low | Excellent |

*Note:* Life of equipment is generally indefinite except for the worm and gear of a geared unit, which have a 30- to 40-year life.

## 31.18 SINGLE AUTOMATIC PUSHBUTTON CONTROL

This system is the simplest of the passenger-operated automatic control schemes. It handles only one call at a time, providing an uninterrupted trip for each call. A single corridor button at each level can register a call only when the car is not in motion. This system is used only in private residences and for light-use freight elevators.

## 31.19 COLLECTIVE CONTROL

Cars stop at each floor that registers a call irrespective of direction, hence the term *collective*. This leads to slow and annoying service. As a result, this system is no longer used in new installations in the United States, although it is common in other countries.

## 31.20 SELECTIVE COLLECTIVE OPERATION

This type of collective operation is "selective" in that it is arranged to collect all waiting "up" calls on the trip up and all hall "down" calls on the trip down. The control system stores all calls until they are answered, and automatically reverses the direction of travel at the highest and lowest calls. when all calls have been cleared, the car will remain at the floor of its last stop awaiting the next call. Any hall button call will set the car into operation.

Selective collective control is standard in locations where service requirements are moderate, such as in apartment houses, small offices, and professional buildings. Because these locations often require more than one car, a group control scheme for up to three cars automatically assigns each hall call to the car best situated to answer it, prevents more than one car from answering a call, allows one car to be detached for freight duty, and automatically parks cars at the ground floor when they are not required.

Although selective collective control is in common use for residential and other light to moderate service requirement buildings, its inherent strong tendency toward bunching of cars can result in long waiting periods. This characteristic is particu-

larly annoying with groups of three cars. Frequently, a passenger arrives at a landing to find that all three cars have just passed, going in the same direction. The result is that service is only slightly better than that which would be rendered by a single car, except that the load (handling) capacity is greater. For this reason, operation of more than two cars with this system is not recommended, and operation of more than three cars is not feasible.

## 31.21 COMPUTERIZED SYSTEM CONTROL

Prior to the advent of computer supervisory control, large banks of elevators were controlled by lobby-based human "starters" who attempted, with limited success, to recognize and anticipate traffic patterns and thus speed service. However, due to the huge amount of information that had to be processed, service, particularly during heavy traffic, was frequently less than satisfactory. This situation changed with the advent of computerized, microprocessor-controlled operating systems.

A satisfactory control system must continuously monitor demand and control each car's motion in response to demand only; that is, it must analyze all the possibilities and answer each call in optimum fashion. The definition of *optimum* depends, of course, on the system design strategy, and this varies among manufacturers. Such a system is possible only with the aid of a central computer combined with programmable microprocessor-controlled peripherals because the amount of data that must be collected and instantaneously processed is enormous.

All manufacturers attempt to optimize the parameters by which system quality is measured—specifically, to minimize interval, hall waiting time, and average trip time. How this is accomplished depends on the relative weight assigned to each item of input data in an extremely complex computer algorithm. One manufacturer uses an algorithm that calculates a figure of merit for *each* car to answer *each* waiting hall call. The number arrived at represents the weighted sum of the prospective passenger hall waiting time (interval and hall wait time) and traveling passenger delay (average trip time). The car with the best figure of merit (minimum time) is assigned to the landing call. Another manufacturer uses a somewhat simpler dynamic call allocation algorithm that relies on the high-speed calculation capabilities of a central computer to make a last-moment decision as to which car answers which call. Another algorithm, in addition to analyzing each car's capability to answer a call, calculates the effect of any decision on the overall elevator service quality for the building and uses this factor as well in the final decision.

Still another algorithm uses a car dispatch program based on preprogrammed and learned traffic patterns modified by the history of the previous few minutes of operation. In the basic program, each car computes its own response time to a waiting hall call, considering its own position, velocity, and car calls, and compares it to that of all other cars in the group. The car that will provide optimum hall call service time (waiting time plus trip time) answers the hall call. To facilitate this type of car "bidding" for calls, some systems are arranged so that each car controller can act as a master group controller. At any one time, one car in the group acts as the master, but if it is taken out of service, another car controller becomes the master.

In addition, some programs use an artificial intelligence module to learn traffic data and history and continuously change the car dispatching mode. These changes consist in part of sectoring or zoning the building so that cars are grouped to provide optimal service. This type of system is particularly useful in buildings with single- or multifloor-use occupants with repetitive traffic patterns. Most algorithms permit overriding the program mode in response to crowd sensors or to particularly heavy, concentrated service demand such as might occur if the occupants of an entire floor left the building at an unusual hour.

The actual program logic for even a small group supervisory traffic control system is beyond the scope of this book, but its guidelines are not. An adequate system should:

- Be programmed initially for the anticipated service needs. These needs can be analyzed in existing buildings with computerized traffic analyzers (Fig. 31.11).
- Be reprogrammable to meet changes in building needs at nominal cost and with minimum shutdown time. It must be possible to detach cars from the system during testing, reprogramming,

and routine maintenance and to provide minimal service even during off-hours.
- Provide for priority calls (based on landing waiting times), statistical analysis of traffic in order to anticipate patterns, adaptive (zoned) car parking to meet specific needs, adjustable door timing based on the type of call (lobby, landing, car), backup dispatch means (in case of dispatch system failure), and automatic call cutout for constant (stuck) signals.
- Provide means for lobby and management office viewing (monitor) (Fig. 31.12) and obtaining hard copy (printer output) of elevator traffic information both stored and in real time (see Fig. 31.12). This data-handling equipment must provide a fault mode that stores, displays, and diagnoses system operating faults. The owner/manager unit may have on-site control and reprogramming capabilities if so required.
- Provide additional functions such as emergency power elevator selection and control (see Section 31.42), priority service, selective hall/car call cutout, swing (separate) car operation, and hoistway access controls for maintenance.
- Provide, where specifically required by the building's management, a riot control feature (access limitation at entrance levels), crossover floor operation in a zoned system, convention service (intense short-time usage at selected levels), and controlled access at given floors and for specific occupants.
- Be fully coordinated with the fire protection system in accordance with the local fire regulations (see Section 31.43).
- Act in consonance with the elevator security equipment so that operation of security/alarm devices initiates automatic elevator motion control procedures. This too must be coordinated with the local security authorities, and the automatic procedures must be subject to manual override (see Section 31.44).

Proper operation of the system should also result in:

1. All floors getting equal service, including the basement, if required
2. Proper handling of multifloor tenants in office buildings to permit efficient interfloor traffic
3. Appropriate action in emergencies, such as general or local circuit power failure or any type of abnormal car to signal operation

TRANSPORTATION

**Fig. 31.11** *Building elevator traffic studies can be performed quickly and accurately using a traffic analyzer with multiple inputs to feed data via appropriate software into a laptop computer (a). The unit illustrated has 64 inputs, which are scanned once per second, and is capable of unattended monitoring for a week. The accumulated data on call registration, response time, waiting time, maximum interval, cars in service, and so on, can be viewed visually on the computer screen or printed out as in (b). (Courtesy of UNITEC Parts Co.)*

*(a)*

COMPOSITE DN - AVERAGE WAITING TIME (SECONDS)

| TIME SLOT | | 0 | 5 | 10 | 15 | 20 |
|---|---|---|---|---|---|---|
| 14:49 | 14:54 | | | | | |
| 14:54 | 14:59 | | | | | |
| 14:59 | 15:04 | | | | | |
| 15:04 | 15:09 | | | | | |
| 15:09 | 15:14 | | | | | |
| 15:14 | 15:19 | | | | | |
| 15:19 | 15:24 | | | | | |
| 15:24 | 15:29 | | | | | |
| 15:29 | 15:34 | | | | | |
| 15:34 | 15:39 | | | | | |
| 15:39 | 15:44 | | | | | |
| 15:44 | 15:49 | | | | | |
| 15:49 | 15:54 | | | | | |

*(b)*

## 31.22 REHABILITATION WORK: PERFORMANCE PREDICTION

The primary reason for rehabilitating an existing elevator system is to improve its operating performance. Traction machinery has extremely long life, particularly in the gearless configuration. Thus, the usual rehabilitation project consists of replacing the m-g set, car control panel, and group controller with solid-state, microprocessor-controlled programmable equipment while retaining the original traction equipment. The car door operator, which controls door action, is usually also replaced because door opening and closing characteristics are an important factor in overall trip time and therefore in system performance.

Major manufacturers have developed computer-based elevator system simulators. Such programs enable architects and owners to input design (or as-is) building data, and to receive graphic and text output on the performance of proposed systems. Because such programs are interactive, the

Elevator bank displays

**Fig. 31.12** *Dedicated elevator display computer terminals are part of the extensive control and communication equipment installed at a lobby control desk in a New York City skyscraper. (Photo by B. Stein.)*

user can change the input data and the characteristics of the proposed equipment until the desired performance level is reached. These programs are particularly useful in modernization work because the owner sees in advance the operation of a proposed system *in his/her building* and can make appropriate decisions based upon good information.

## 31.23 LOBBY ELEVATOR PANEL

The traditional lobby elevator control and information panel for each elevator bank that was usually wall-mounted adjacent to the related elevators has become one or more computer monitor screens positioned at a lobby desk (Fig. 31.12) and/or in the building maintenance office. In addition, an information-only screen is frequently wall-mounted adjacent to the related elevators for the edification of waiting passengers.

The information displayed on the screen includes car locations, movement direction, waiting corridor calls, and any special status data. The control functions available at the computer terminal permit intervention to establish special types of operation including:

- Car movement without operating the usual audible and visual signals (*inconspicuous riser*)
- One or more cars removed from supervisory control and operated manually (*attendant or independent service*)

- Cars elected for night or weekend service while the other cars are shut down
- Car(s) assigned to a particular floor on a fixed- or priority-basis call (*convention feature or priority*)

Among other control functions are those concerned with emergency service, including the "fireman's return" feature required by ANSI and many local fire codes (Section 31.43) and the controls related to switching of power between cars in the event that operation on emergency generator power is necessary (Section 31.42).

In addition, means of two-way communication with each car and other selected locations are provided at the control center.

## 31.24 CAR OPERATING PANEL

A typical car operating panel is illustrated in Fig. 31.13. Every car panel (station) is equipped with full-access buttons for call registry, door-open, alarm, emergency stop, and fire fighter control. Also always provided is an intercom device that permits communication with the building control office. A door-close button is sometimes provided if extensive hand operation of the car is anticipated. It is activated only when the car is under manual control. Controls that do not concern the normal passenger are grouped in a locked compartment in the car panel. These include a hand-operation switch, light, fan, and power switches, and any

TRANSPORTATION

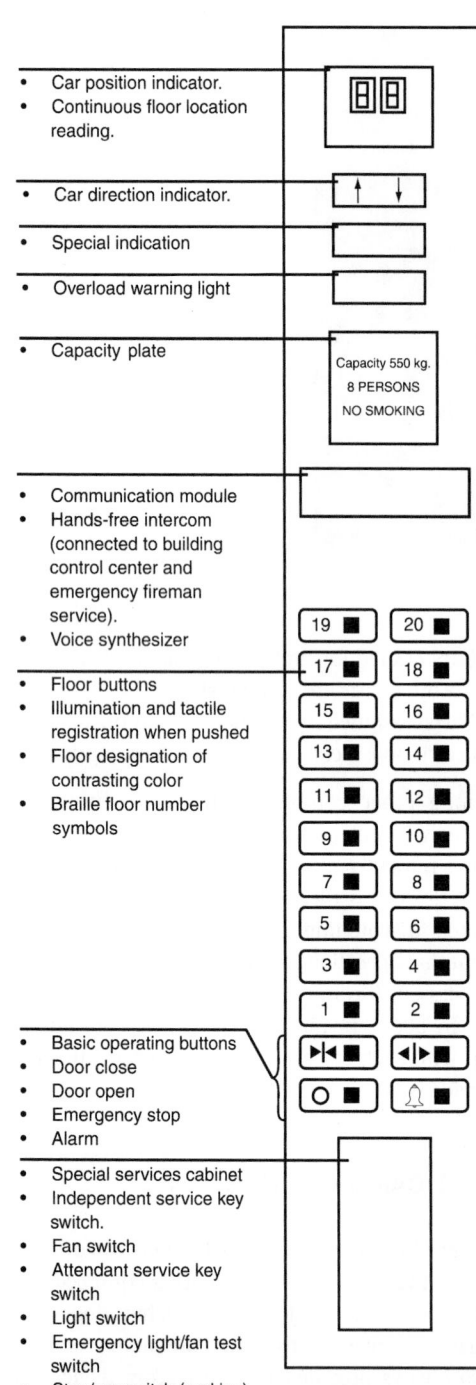

- Car position indicator.
- Continuous floor location reading.

- Car direction indicator.

- Special indication

- Overload warning light

- Capacity plate

  Capacity 550 kg.
  8 PERSONS
  NO SMOKING

- Communication module
- Hands-free intercom (connected to building control center and emergency fireman service).
- Voice synthesizer

- Floor buttons
- Illumination and tactile registration when pushed
- Floor designation of contrasting color
- Braille floor number symbols

- Basic operating buttons
- Door close
- Door open
- Emergency stop
- Alarm

- Special services cabinet
- Independent service key switch.
- Fan switch
- Attendant service key switch
- Light switch
- Emergency light/fan test switch
- Stop/run switch (parking)

**Fig. 31.13** *Typical car operating panel. Designs of these panels vary widely, but the essential components are as shown.*

special controls such as security and emergency devices. Finally, a compartment accessible only to technicians contains the devices controlling door motion, car signals, door and car position transducers, load-weighing control, door and platform detection beam equipment, speech synthesizer (if used), and visual displays.

## ELEVATOR SELECTION

### 31.25 GENERAL CONSIDERATIONS

The selection of elevators for any but the simplest buildings requires the simultaneous consideration of several factors: adequate elevator service for the intended building usage, the economics of elevator selection, and the architectural integration of spaces assigned to elevators, including lobbies, shafts, and machine rooms. In large buildings, many combinations are possible because these factors are interdependent. The selection of an optimum system for such buildings is most practical and accurate with the aid of a computer or simulator, and their use has become standard practice in the industry. Hand computation, following certain guidelines, can yield good results for small, straightforward buildings and reliable preliminary data for almost all buildings.

The criteria usually used in determining elevator service quality are

- Interval and average waiting time
- Handling capacity
- Travel time

The guidelines that are followed in the computation of elevator requirements, by hand or computer, result in a determination of these three criteria and their comparison with predetermined target values.

### 31.26 DEFINITIONS

A clear definition of important terms, including variant usages, follows.

**Average lobby time or average lobby waiting time.** Average time spent by a passenger between arriving in the lobby and leaving the lobby in a car.

**Handling capacity (HC).** Figure of merit for an elevator system, indicating the maximum number of passengers that can be handled in a given period—usually 5 minutes, thus the term *5-minute handling capacity.* When expressed as a percentage of the building's population, it is called *percent handling capacity* (PHC).

**Interval (I) or lobby dispatch time.** Average time between departure of cars from the lobby.

**Registration time.** Waiting time at an upper floor after registering a call.

**Round-trip time (RT).** Average time required for a car to make a round trip, starting from the lower terminal and returning to it. The time includes a statistically determined number of upper-floor stops in one direction and, *when calculating elevator requirements based on up-peak traffic,* an express return trip.

**Travel time or average trip time (AVTRP).** Average time spent by passengers from the moment they arrive in the lobby to the moment they leave the car at an upper floor.

**Zone.** Group of floors in a building that is considered as a unit with respect to elevator service. It may consist of a physical entity—a group of upper floors above and below which are blind shafts—or it may be a product of the elevator group control system, changing with system needs.

## 31.27 INTERVAL OR LOBBY DISPATCH TIME AND AVERAGE LOBBY WAITING TIME

In an ideal installation, at least from the riding public's point of view, a car would be waiting at the lower terminal on the rider's arrival or would be available after a short wait. Because cars leave the lobby separated in time by the *interval* (I) and passengers arrive at the lobby in random fashion, the average waiting time in the lobby should be half (50%) the interval. Field measurements show, however, that it is actually longer than this. The figure most often used in the industry is 60%—that is,

$$\text{average lobby waiting time} = 0.6 \times I$$

**TABLE 31.4 Recommended Elevator Intervals and Related[a] Lobby Waiting Time**

| Facility Type | Interval (sec) | Waiting Time[a] (sec) |
|---|---|---|
| OFFICE BUILDINGS | | |
| Excellent service | 15–24 | 9–14 |
| Good service | 25–29 | 15–17 |
| Fair service | 30–39 | 18–23 |
| Poor service | 40–49 | 24–29 |
| Unacceptable service | 50+ | 30+ |
| RESIDENTIAL | | |
| Prestige apartments | 50–70 | 30–42 |
| Middle-income apartments | 60–80 | 36–48 |
| Low-income apartments | 80–120 | 48–72 |
| Dormitories | 60–80 | 36–48 |
| Hotels—first quality | 30–50 | 18–30 |
| Hotels—second quality | 50–70 | 30–42 |

[a]Based on the relationship: waiting time = 0.6 × interval.

Table 31.4 lists intervals and their relative quality for office buildings and the related average waiting time based on the foregoing relationship. Because some control systems zone the building in such a way that some cars do not return to the lobby, the interval as a figure of merit may be somewhat misleading in such buildings. The table also lists recommended intervals for other types of buildings.

With intervals in the recommended range, riders are not conscious of any irksome delay in elevator service. Consciousness of delay is considered a major drawback in rental desirability and should be avoided for all conditions of traffic except during morning and evening peaks, when a certain delay is expected and therefore tolerated, however grudgingly. Even in peak periods, any modern group supervisory system will recognize a *timed-out* call—that is, a call with registration time exceeding 50 seconds—as a priority call. Priority calls are answered by the first available car, usually within 15 seconds. If a considerable amount of interfloor traffic is expected during peak periods, as may be the case when a large company occupies several upper floors, elevator capacity should be increased by 20% to 40% over that calculated to maintain proper intervals.

## 31.28 HANDLING CAPACITY

The frequency, or interval, with which a car appears at the lobby is one of the two factors that determine the passenger capacity of an elevator system. The other is obviously the size of the elevator car. The system's *handling capacity* is completely determined by these two factors—car size and interval—and is independent of the number of cars. This can be best understood by visualizing the system as a single set of doors that open periodically (interval) to remove a given number of passengers (car capacity) from the waiting group. Whether that set of doors represents a single car or many cars that take turns is immaterial. The only factors that determine the handling capacity are passenger load (car capacity) and frequency of loading (interval) (Table 31.5).

Note that cognizance is taken of the fact that during peak traffic periods, cars are not loaded to maximum capacity but only to about 80%—a figure determined by actual count in many existing installations.

As a convenient measure of capacity, the handling capacity of a system for 5 minutes is taken as a standard. This is because a 5-minute rush period is used as a measure of a system's ability to handle traffic. This may be expressed thus:

$$\text{handling capacity } (HC) = \text{passengers/car}$$
$$\times \text{cars/sec} \times 5 \text{ min}$$
$$\times 60 \text{ sec/min}$$

Because the number of cars per second is the reciprocal of the interval (e.g., 30 seconds between cars is the same as $\frac{1}{30}$th of a car per second), this equation reduces to

$$HC = \frac{\text{passengers/car}}{\text{interval}} \times 300$$

or

$$HC = \frac{300p}{I}$$

where $p$ is car loading (number of passengers/car). When the interval is 30 seconds, the system's

### TABLE 31.5 Car Passenger Capacity ($p$)

| Elevator Capacity (lb) | Maximum Passenger Capacity | Normal Passenger[a] Load per Trip |
|---|---|---|
| 2000 | 12 | 10 |
| 2500 | 17 | 13 |
| 3000 | 20 | 16 |
| 3500 | 23 | 19 |
| 4000 | 28 | 22 |

[a]The number of passengers carried on a trip during peak conditions is approximately 80% of the car capacity.

### TABLE 31.6 Minimum Percent Handling Capacities (PHC)

| Facility | Percent of Population to Be Carried in 5 Minutes |
|---|---|
| OFFICE BUILDINGS | |
| Center city | 12–14 |
| Investment | 11.5–13 |
| Single-purpose | 14–16 |
| RESIDENTIAL | |
| Prestige | 5–7 |
| Other | 6–8[a] |
| Dormitories | 10–11 |
| Hotels—first quality | 12–15 |
| Hotels—second quality | 10–12 |

[a]Due to more urgent traffic demands, particularly at the school and work exodus.

handling capacity is 10$p$, a convenient figure to remember.

To establish a figure of merit for building service, system $HC$ must be related to building size. This is normally done by establishing the minimum percentage of the building population that

### TABLE 31.7 Population of Typical Buildings for Estimating Elevator and Escalator Requirements

| Building Type | Net Area |
|---|---|
| OFFICE BUILDINGS | Square feet per person |
| Diversified (multiple tenancy) | |
| Normal | 110–130[a] |
| Prestige | 150–250 |
| Single tenancy | |
| Normal | 90–110 |
| Prestige | 130–200 |
| HOTELS | Persons per sleeping room |
| Normal use | 1.3 |
| Conventions | 1.9 |
| HOSPITALS | Visitors and staff per bed[b] |
| General private | 3 |
| General public (large wards) | 3–4 |
| APARTMENT HOUSES | Persons per bedroom |
| High-rental housing | 1.5 |
| Moderate-rental housing | 2.0 |
| Low-cost housing | 2.5–3.0 |

[a]Density may vary for different floors. The clerical and stenographic area may have a population density as high as 70 ft$^2$ per person.

[b]If visiting hours are restricted, the visitor population will determine elevator requirements. If visiting is not restricted to a certain few hours, staff requirements may determine elevator design. Where traffic is heavy, a combination of passenger cars and larger "hospital" cars should be used to provide optimum service.

## TABLE 31.8  Office Building Efficiency

| Building Height | Net Usable Area as Percentage of Gross Area |
|---|---|
| 0–10 floors | Approximately 80% |
| 0–20 floors | Floors 1–10 approximately 75% |
| | 11–20 approximately 80% |
| 0–30 floors | Floors 1–10 approximately 70% |
| | 11–20 approximately 75% |
| | 21–30 approximately 80% |
| 0–40 floors | Floors 1–10 approximately 70% |
| | 11–20 approximately 75% |
| | 21–30 approximately 80% |
| | 31–40 approximately 85% |

*Source:* Reprinted from G.R. Strakosch, *Vertical Transportation, Elevators and Escalators,* 2nd ed. Wiley, New York, 1983.

*Note:* Applicable to buildings with 15,000 to 20,000 gross square feet per floor.

the system must handle in 5-minutes, called PHC. A good system for a diversified office building will handle no less than 12% of the building population. Similar figures are shown in Table 31.6 for various types of facilities.

In planning a building's elevator requirements, its population must be estimated. This is particularly difficult in speculative-type, diversified-use buildings. However, based on rental cost, area, and building type, a fair estimate can be made. Population estimates for office buildings are based on net area—that is, actual available area for tenancy, Table 31.7 gives suggested density figures, and Table 31.8 gives average office building efficiency figures for use in calculating net area.

## 31.29  TRAVEL TIME OR AVERAGE TRIP TIME

The average trip time or time to destination is the sum of the lobby waiting time plus travel time to the median floor stop. Car round-trip time is also used as a criterion, but it is not as meaningful as trip time. In a commercial atmosphere, a trip of less than 1 minute is highly desirable, a 75-second trip is acceptable, a 90-second trip is annoying, and a 120-second trip is the limit of toleration. Obviously, in the more relaxed atmosphere of a residence, where interval alone can account for a minute or more of trip time, these maxima are revised upward.

From Fig. 31.14 we see that the 2000- and 2500-lb cars used in residential buildings can have a 17-story rise, even with a 60-second inter-

val, without excessive trip time. The 3500-lb car, however, which is almost universal in office buildings (Fig. 31.15), is limited to a maximum 16-floor local run before exceeding the 90-second limit and about 6 to 8 floors to stay within the 75-second criterion.

An important reservation on the foregoing statements must be noted. The curves of Figs. 31.14 and 31.15 are based on statistical calculations, empirical data, and field observations, as discussed in the next section. This being so, the average values that these curves give should be considered to be ±15%, and borderline cases can be shifted either way. Designs that show high travel time on paper frequently work out well in the field, because lobby loading is often less than 80%, upper-floor stops are less than the statistical figure due to groups of people going to the same floor, and staggered working hours relieve traffic peaks. Also, a feature called *high-call reversal* takes account of the fact that cars do not travel to the top on each trip, but reverse at the top call. This can reduce the average trip time by 5% to 10%. Finally, sophisticated solid-state traffic controls allow more rapid acceleration and deceleration without discomfort, variable door-closing time, and more efficient selection of landing call answers, all of which can further reduce the trip time another 5%.

## 31.30  ROUND-TRIP TIME

The figure for round-trip time during up-peak traffic conditions, used for calculating elevator requirements, is composed of the sum of four factors: (1) time to accelerate and decelerate; (2) time to open and close doors at all stops; (3) time to load and unload; and (4) running time (Figs. 31.16–31.18). Physically, round-trip time is the time from door opening at the lower terminal to door opening at the same terminal at the end of a round trip. Because the actual number of stops made by a car is unknown, a statistical probability figure is used, based upon the passenger capacity of the car and the number of local floors above the lower terminal. In calculating this round-trip time (*RT*), it is assumed that a car will depart the lower terminal when loaded. No intentional delay is included at either the lower or upper terminal. The *RT* thus calculated is a median figure, with any single actual

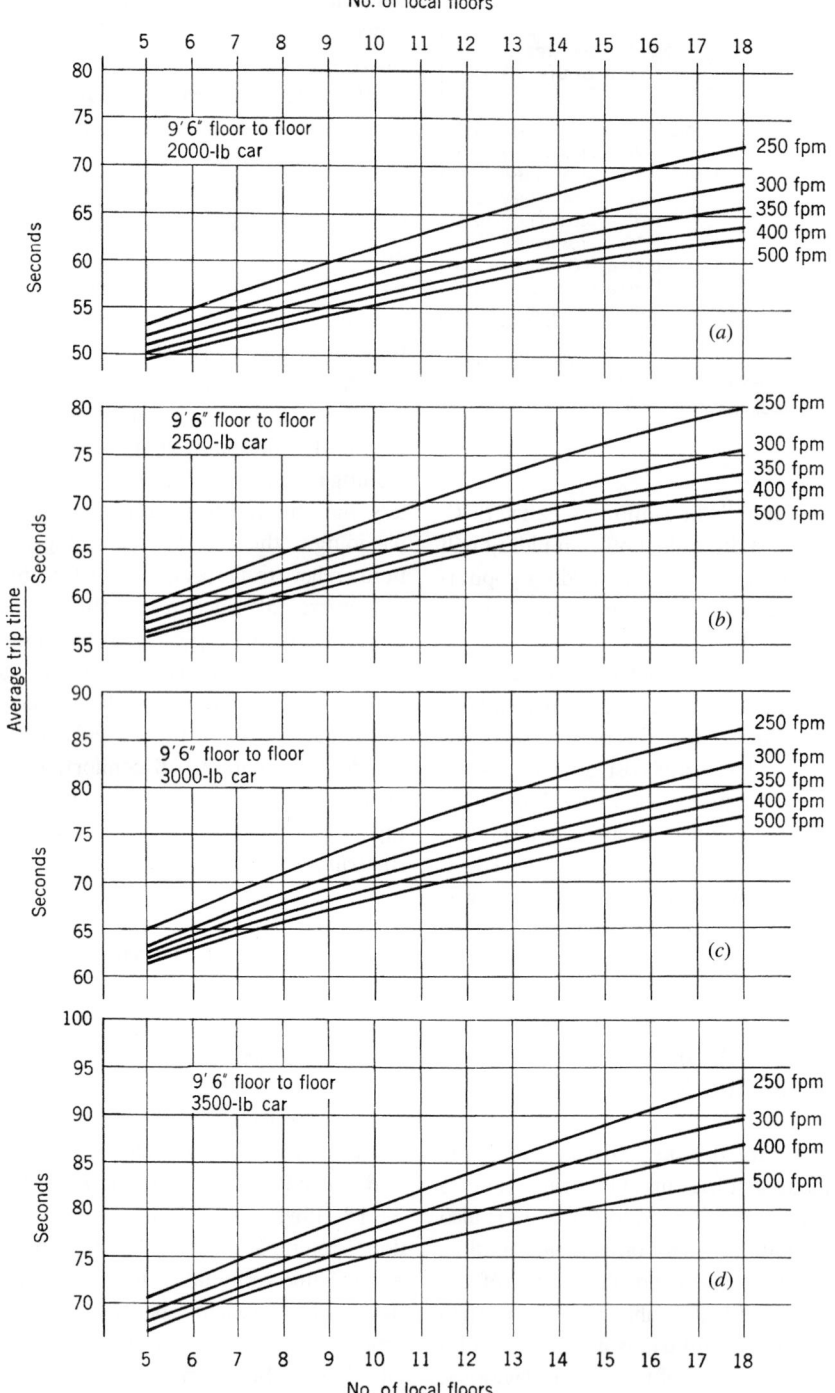

**Fig. 31.14** Plots of average trip time for various car speeds and capacities with a 9-ft, 6-in. floor height and a 30-second interval.

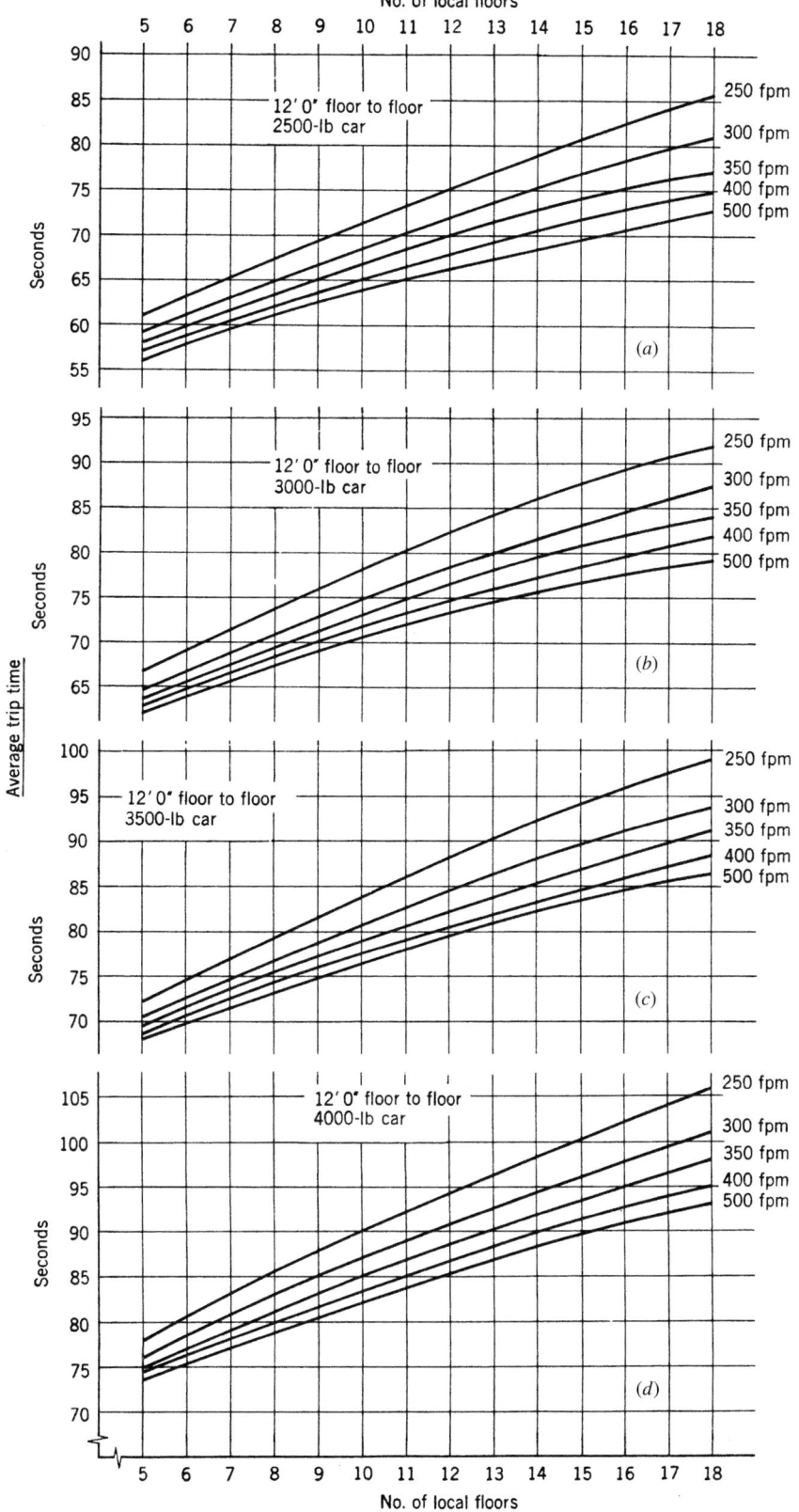

**Fig. 31.15** Plots of average trip time for various car speeds and capacities for a 12-ft floor height and a 30-second interval.

round trip taking more or less time. In detail, *RT* consists of the time expended in

1. Loading at the lobby
2. Door closing at the lobby
3. Accelerating from the terminal and from each stop
4. Decelerating at each stop
5. Passenger transfer at each stop
6. Door operations at each stop
7. Running time at rated speed between stops
8. Return express run from the last stop

These figures are obtained as follows:

1. *Field observations:* Items 1 and 5 are based on a 3-ft, 6-in. door opening. A smaller door opening increases passenger transfer time.
2. *Calculations:* Items 2, 3, 4, 6, 7, and 8. *Door-closing time* is based on a 3-ft, 6-in. center-opening door with adjustable speed.

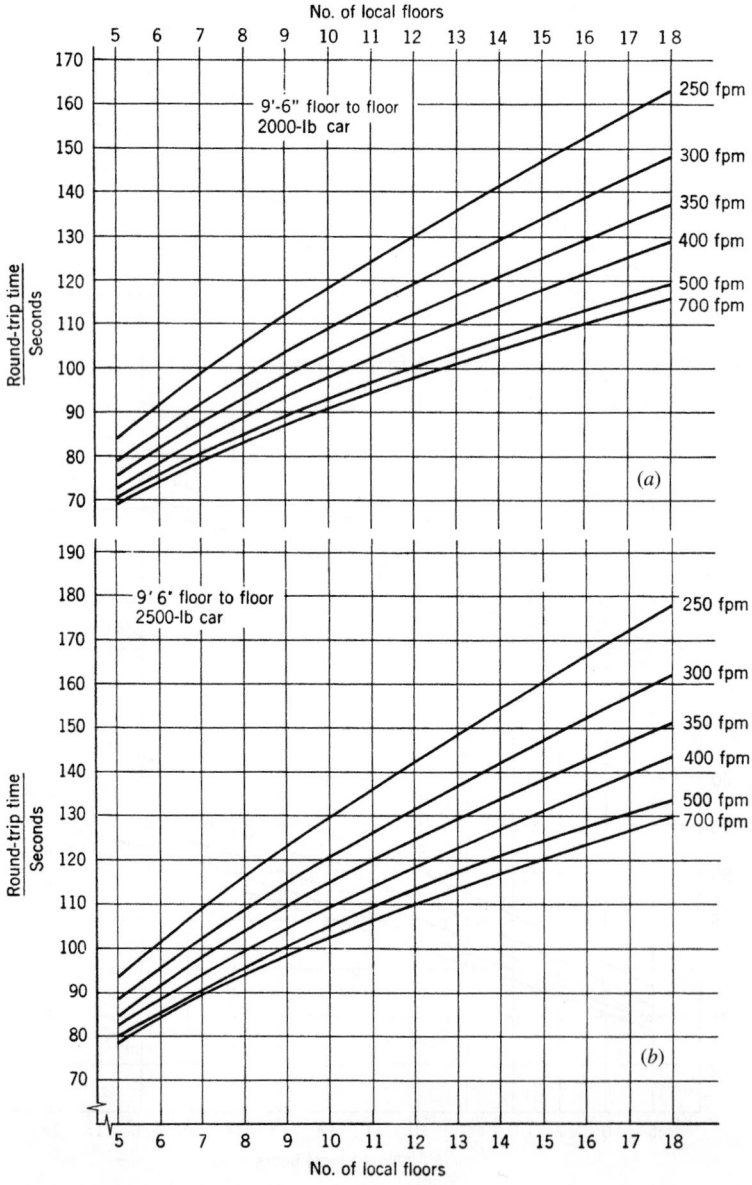

**Fig. 31.16** *Plots of round-trip time for various car speeds and capacities with a 9-ft, 6-in. floor height and a 30-second interval.*

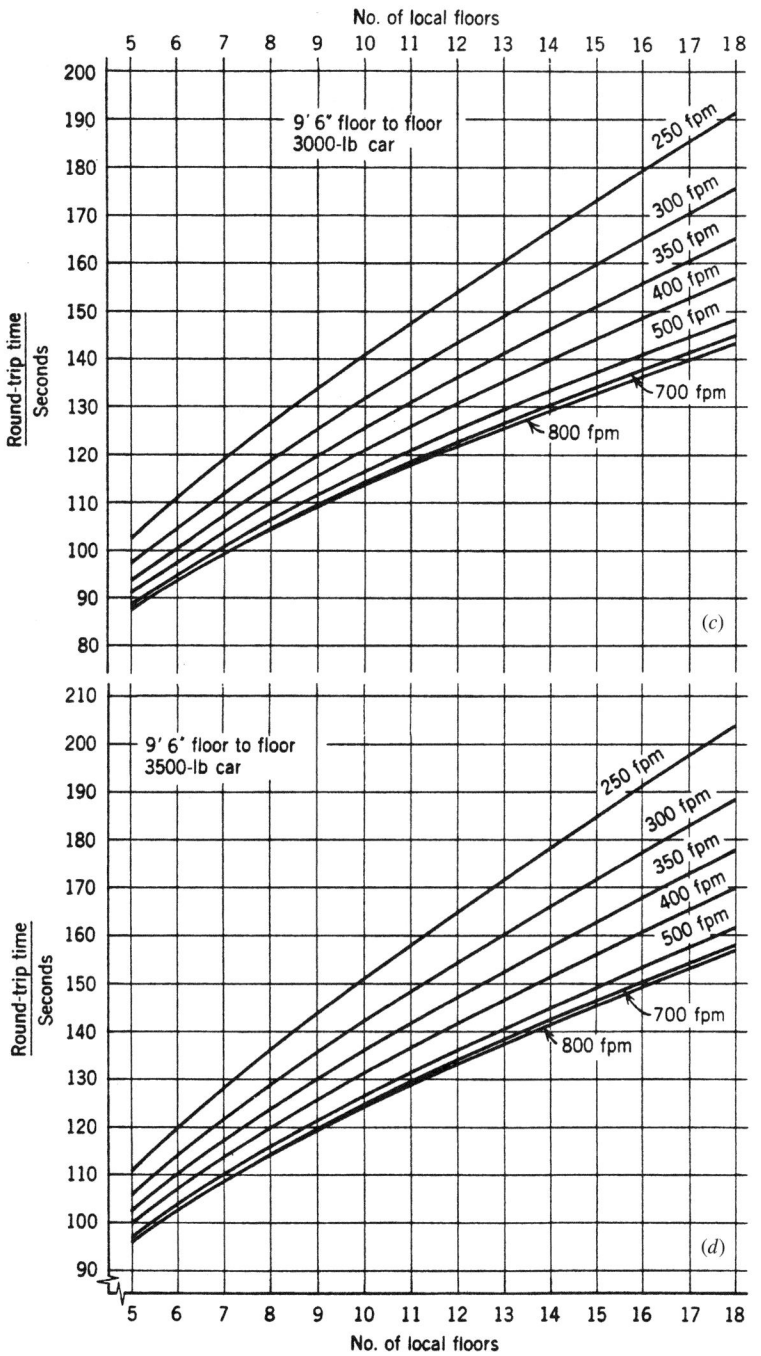

**Fig. 31.16** (Continued)

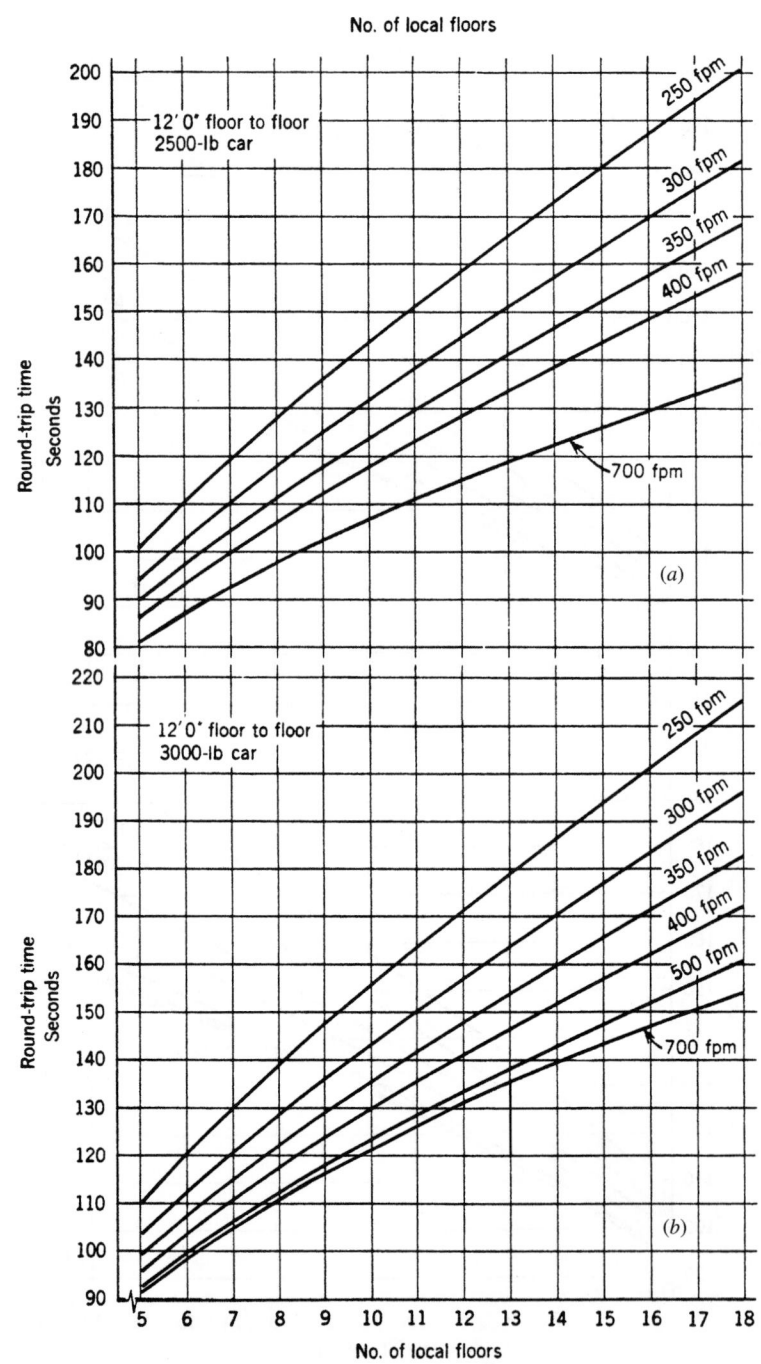

**Fig. 31.17** *Plots of round-trip time for various car speeds and capacities with a 12-ft floor height and a 30-second interval.*

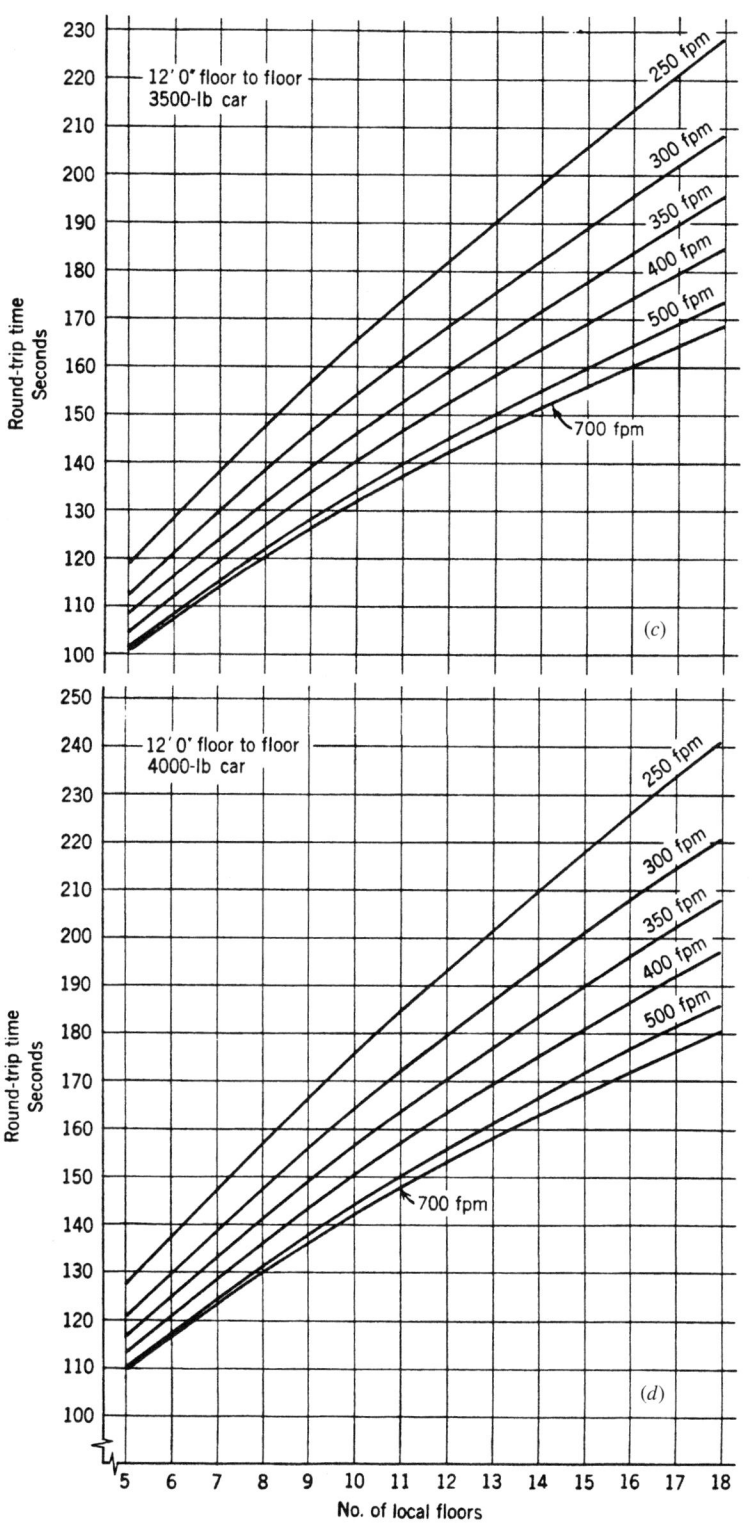

*Fig. 31.17* (Continued)

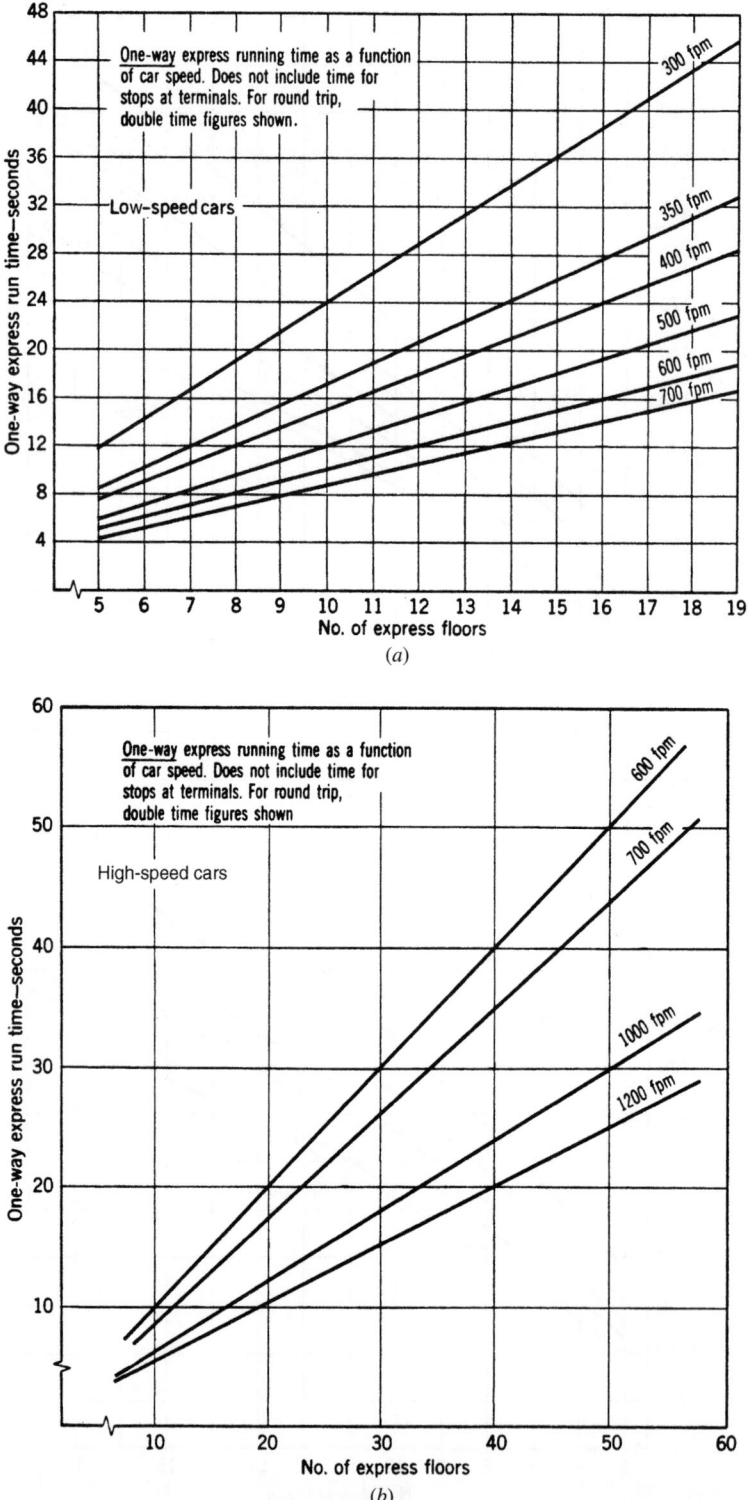

**Fig. 31.18** One-way express running time, not including terminal time; (a) low speed cars, (b) high speed cars.

*Acceleration and deceleration times* are calculated with a maximum of 4 ft/s/s because anything beyond that results in physical discomfort to the passengers.

*Running time at rated speed* takes place after the car has accelerated and before it begins to decelerate. If we consider that it takes between 20 and 30 ft to accelerate to 700 fpm, depending upon the rate of acceleration, we see that in local runs the car never gets to the rated speed. It simply accelerates and decelerates. The higher-speed equipment with its larger motor accelerates more quickly and gives some time advantage on the return express run, but it has no great time advantage overall. This accounts for the small reduction above 500 fpm in Figs. 31.16 and 31.17.

In calculating *RT* for cars in upper zones, it is necessary to know the time required to traverse the express floors. This may be obtained from Fig. 31.18. The times given there are for *one-way* express runs. Thus, to calculate *RT* for an upper-zone car, take the *RT* corresponding to the upper local floors and add *twice* the figure obtained for express run time from Fig. 31.18.

## 31.31 SYSTEM RELATIONSHIPS

The symbols that will be used in all elevator calculations are:

| | |
|---|---|
| $p$ | individual car capacity, equal to 80% of the maximum during peak hours |
| $h$ | 5-minute capacity of a single car |
| $N$ | number of cars in a system |
| $HC$ | system 5-minute handling capacity, expressed in number of persons |
| $RT$ | round-trip time, in seconds |
| $AVTRP$ | average trip time, in seconds |
| $I$ | interval, in seconds |
| $D$ | population density, in square feet per person |
| $PHC$ | percent of the population to be moved in 5 minutes, and expressed as a percentage |

Having considered the definition of interval, handling capacity, average trip time, and round trip time, we can demonstrate at this point the interrelationships between these quantities and the other equations governing the remaining factors that define elevator systems.

We previously established that handling capacity *HC* is determined by car capacity *p* and interval *I*:

$$HC = \frac{300p}{I} \qquad (31.1)$$

In a system comprising a single car, the interval (*I*) is obviously equal to the round-trip time (*RT*). In a system with more than one car, the interval is reduced in proportion to the number of cars. Thus,

$$I = \frac{RT}{N} \qquad (31.2)$$

The 5-minute handling capacity (*h*) of a single car is then

$$h = \frac{300p}{RT} \qquad (31.3)$$

if we remember that for a single car, its interval is its round-trip time. It follows then that if the handling capacity of a single car is *h*, then the handling capacity of *N* cars is *N* times as much. Thus,

$$HC = N \times h$$

or

$$N = \frac{HC}{h} \qquad (31.4)$$

## 31.32 CAR SPEED

The selection of the car speed to be used is a matter of trial and error, the final selection being that required to give an *RT* that in turn gives an acceptable interval. However, in order to establish a starting point, it has been found that a minimum car speed corresponding to a given building height—or, in elevator parlance, *rise*—can be established. Similarly, although car size can be selected at any value, it has been shown that for certain facility types, specific-size cars are indicated. These recommendations are given in Table 31.9.

It should also be borne in mind that elevator equipment falls into distinct speed categories. Thus, most manufacturers use geared equipment through 400 fpm (2 m/s) and gearless equipment thereafter. The next category is 500 fpm (2.5 m/s) gearless followed by 600 fpm (3.0 m/s) to 700 fpm (3.5 m/s), and so on. Therefore, it is wise to avoid moving into the next higher and more expensive equipment

**TABLE 31.9 Elevator Equipment Recommendations**

| Building Type | Car Capacity[a] | | Rise | | Minimum[a] Car Speed | |
|---|---|---|---|---|---|---|
| | lb | kg | ft | m | fpm | m/s |
| Office building | $\begin{cases} 2500 \\ 3000 \\ 3500 \end{cases}$ | $\begin{cases} 1250 \\ 1250 \\ 1600 \end{cases}$ | 0–125<br>126–225<br>226–275<br>276–375<br>Above 375 | 0–40<br>41–70<br>71–85<br>86–115<br>>115 | 350–400<br>500–600<br>700<br>800<br>1000 | 2.0<br>2.5<br>3.15<br>4.0<br>5.0 |
| Hotel | $\begin{cases} 2500 \\ 3000 \end{cases}$ | $\begin{cases} 1250 \\ 1250 \end{cases}$ | As above | | As above | |
| Hospital | $\begin{cases} 3500 \\ 4000 \end{cases}$ | $\begin{cases} 1600 \\ 2000 \end{cases}$ | 0–60<br>61–100<br>101–125<br>126–175<br>176–250<br>>250 | 0–20<br>21–30<br>31–40<br>41–55<br>56–75<br>>75 | 150<br>200–250<br>250–300<br>350–400<br>500–600<br>700 | 0.63<br>1.0<br>1.6<br>2.0<br>3.15<br>4.0 |
| Apartments | $\begin{cases} 2000 \\ 2500 \end{cases}$ | $\begin{cases} 1000 \\ 1250 \end{cases}$ | 0–75<br>76–125<br>126–200<br>>200 | 0–25<br>26–40<br>41–60<br>>60 | 100<br>200<br>250–300<br>350–400 | 0.63<br>1.0<br>1.6<br>2.0 |
| Stores | $\begin{cases} 3500 \\ 4000 \\ 5000 \end{cases}$ | $\begin{cases} 1600 \\ 2000 \\ 2500 \end{cases}$ | 0–100<br>101–150<br>151–200<br>>200 | 0–30<br>31–45<br>46–60<br>>60 | 200<br>250–300<br>350–400<br>500 | 1.0<br>1.6<br>2.0<br>2.5 |

[a]Car capacity is determined by building size, and car speed by rise.

category if possible. This may mean exceeding the recommended interval or dropping slightly low in handling capacity. It will be found, however, that this can be done without injury to the elevator system, provided that a good group supervisory control system is employed.

## 31.33 SINGLE-ZONE SYSTEMS

Having established the relationships that govern the design and performance of an elevator system comprising a single zone, it would be helpful to follow through an illustrative example.

**EXAMPLE 31.1** Office building, downtown, diversified use, 14 rentable floors above the lobby, each 12,000 ft² net. Floor-to-floor height—12 ft. Determine a workable elevator system.

### SOLUTION

From Table 31.6, recommended average *HC* is 13%. From Table 31.4, the maximum recommended interval is 25 seconds. From Table 31.7, average population density is 120 ft² per person.

*Trial 1*

Building population:

$$\frac{14 \text{ floors at } 12,000 \text{ ft}^2}{120 \text{ ft}^2 \text{ per person}} = 1400 \text{ persons}$$

Suggested minimum handling capacity:

$$PHC = 13\%$$
$$HC = 0.13 \times 1400 = 182 \text{ persons}$$
$$\text{rise} = 14 \text{ floors at } 12 \text{ ft} = 168 \text{ ft}$$

From Table 31.9, we select a car size of 3500 lb at 500 fpm.

3500 lb
500 fpm

Then, from Figs. 31.17c and 31.15c:

$$RT = 155 \text{ seconds} \qquad AVTRP = 82 \text{ seconds}$$

Single-car capacity: $h = 300p/RT$ (see Table 31.5 for *p*):

$$h = \frac{300 \,(19)}{155} = 36.8 \text{ persons}$$

$$N = \frac{HC}{h} = \frac{182}{36.8} = 4.9, \text{ say 5 cars}$$

$$I = \frac{RT}{5} = \frac{155}{5} = 31 \text{ seconds}$$

$$\text{actual } PHC = \frac{5\,(13\%)}{4.9} = 13\%$$

These figures are acceptable, but we should try faster cars to reduce the interval. We select 700 fpm.

*Trial 2*

3500 lb

700 fpm

$$RT = 151 \text{ seconds}$$
$$AVTRP = 81 \text{ seconds (by extrapolation)}$$

$$h = \frac{300\,(19)}{151} = 37.7 \text{ persons}$$

$$N = \frac{182}{37.7} = 4.8, \text{ say 5 cars}$$

$$I = \frac{RT}{N} = \frac{151}{5} \; 30 \text{ seconds}$$

$$\text{actual } PHC = \frac{5\,(13\%)}{4.8} = 13.5\%$$

This solution is only marginally better than the previous 500 fpm solution, and the increased cost would not be justified. A trial using smaller cars with shorter *RT* is called for.

*Trial 3*

3000-lb cars

500 fpm

$$RT = 143 \text{ seconds} \quad AVTRP = 76 \text{ seconds}$$

$$h = \frac{300\,(16)}{143} = 33.6 \text{ persons}$$

$$N = \frac{HC}{h} = \frac{182}{33.6} = 5.4 \text{ cars}$$

Using five cars:

$$I = \frac{RT}{N} = \frac{143}{5} = 28.4 \text{ seconds}$$

$$\text{actual } PHC = \frac{5}{5.4}\,(13\%) = 12\%$$

*Trial 4*

Using six 3000-lb, 500-fpm cars:

$$I = \frac{RT}{N} = \frac{143}{6} = 23.8 \text{ seconds}$$

and

$$PHC = \frac{6}{5.4}\,(13) = 14.4\%$$

Both solutions are acceptable.

Tabulating the calculation results, we have:

| Solution | Cars (lb) | Speed (fpm) | RT (s) | AVTRP (s) | I (s) | PHC (%) |
|---|---|---|---|---|---|---|
| 1 | 5 (3500) | 500 | 155 | 82 | 31 | 13 |
| 2 | 5 (3500) | 700 | 151 | 81 | 30 | 13.5 |
| 3 | 5 (3000) | 500 | 143 | 76 | 28.4 | 12 |
| 4 | 6 (3000) | 500 | 143 | 76 | 23.8 | 14.4 |

Solutions 1, 3, and 4 are acceptable. Solution 2 was discounted due to the high cost. Interestingly, solution 3, using smaller cars than the corresponding solution 1, and therefore being more economical, gives better results except for *HC*. Although the best solution is number 4, which gives excellent interval and *HC*, the additional cost of a sixth car plus the revenue loss from the rentable area occupied by the sixth shaft and the high additional cost of maintenance weigh heavily against this option. A trial with five 3000-lb cars at 700 fpm is in order, with the knowledge that a considerable cost increase would result because 700-fpm cars require gearless equipment, whereas 500-fpm cars are available in either geared or gearless format, both giving excellent service.

*Trial 5*

3000-lb cars

700 fpm

$$RT = 140 \text{ seconds} \quad AVTRP = 73 \text{ seconds}$$

$$h = \frac{300\,(16)}{140} = 34.3 \text{ persons}$$

$$N = \frac{182}{34.3} = 5.3; \text{ use 5 cars}$$

$$I = \frac{RT}{N} = \frac{140}{5} = 28 \text{ seconds}$$

$$PHC = \frac{5}{5.3}\,(13\%) = 12.3\%$$

As expected, the improvement over the performance at 500 fpm is very slight: an interval of 28 seconds rather than 28.4 seconds and a handling capacity of 12.3% versus 12%. The large increase in first cost for gearless equipment would not be justified. ∎

At this point, the final selection would be made on the basis of cost. When considering cost, note that first cost is the governing factor only in a speculative venture. With an owner-operator building, the cost comparison should be on a life-cycle basis.

**TABLE 31.10 Relative First Cost[a] Figures for Passenger Elevators of Various Speeds and Drive Systems**

| Car Size (lb) | Hydraulic (fpm) | Geared Traction (fpm) | | | Gearless Traction (fpm) | | | |
|---|---|---|---|---|---|---|---|---|
| | 100 | 200 | 350 | 500 | 500 | 700 | 1000 | 1200 |
| 2000 | 40 | 80 | 100 | 130 | 165 | 170 | 220 | 235 |
| 2500 | 43 | 85 | 115 | 145 | 175 | 180 | 235 | 250 |
| 3000 | 50 | 90 | 120 | 150 | 180 | 185 | 250 | 265 |
| 3500 | 58 | 95 | 125 | 155 | 190 | 195 | 265 | 275 |
| 4000 | 60 | 100 | 135 | 165 | 200 | 205 | 280 | 300 |
| 4500[b] | 70 | 120 | 150 | 185 | 225 | 230 | 300 | 325 |
| 5000[b] | 75 | 130 | 160 | 200 | 240 | 250 | 330 | 350 |

[a]Costs are ±10%; based on standard fixtures, cabs, and entrances, and average rise for the speed indicated.

[b]Service elevator or hospital elevator.

*Note:* See Table 31.9 for speed/rise recommendation.

Cost figures must reflect the impact of elevator space requirements on net rentable area in the building. Comparative cost figures are given in Table 31.10.

As mentioned earlier and shown in Fig. 31.11, planners of new buildings today generally take advantage of elevator selection software provided by consultants, manufacturers, or in-house capabilities. The results from one such program as applied to the problem just solved manually are shown in Fig. 31.19. This particular program was prepared by Otis Elevator Co. Note that the results in Fig. 31.19a correspond exactly to the final results obtained manually.

The correspondence is not surprising because the round-trip curves in Fig. 31.17 are based on a 3.3-second door time and a 4.0 ft/s² car acceleration, as shown in the figure. Note that high call reversal occurs at the top floor (13.5, i.e., 14) and that the number of up stops is 10 (9.7 from statistical calculations). The up-peak calculation assumes no counterflow traffic (i.e., an express down run and no interfloor traffic), as shown. These can be added, however, yielding very different results, as shown in Fig. 31.19b. We see there that counterflow traffic (down stops) of only 3% and interfloor traffic of 1%, both of which are reasonable, change the interval to only fair quality and handling capacity to unacceptable.

## 31.34 MULTIZONE SYSTEMS

In general, buildings with fewer than 15 stories are elevatored with a single zone (i.e., all cars serve all

floors), and buildings with more than 20 stories are split into two or more zones. The area in between—16 to 19 stories—can go either way, depending upon the population density and the required interval. Actually, a modern group supervisory system can automatically zone a building when traffic requires it. However, such an arrangement, although efficient, is expensive in terms of both equipment and construction because it does not take advantage of the considerable savings engendered by blind lower shaftways for upper-zone elevators. Analysis of multizone systems is complex and today is rarely done by hand. The interested reader is referred to Stein et al. (1986) for a detailed explanation of the technique involved. Most designers and consultants use one of the many available computerized simulation and selection programs. These have the advantage that, in addition to using basic criteria and building parameters, they can also consider the effect of variations in traffic control. Furthermore, the best of these programs can evaluate the engineering and economic impacts of such factors as varying rental rates for different floors, rental space, machine room and hoistway space, core layout, and the structural ramifications of the elevator system.

## 31.35 OTHER ELEVATOR SELECTION RECOMMENDATIONS

### (a) Office Buildings

All necessary design criteria can be selected from Tables 31.4 to 31.7. Supervisory group control

OtisPlan Single Group Performance

Single Deck Up Peak

name: Sample Problem
floors above lobby: 14
lobby height:  12.0ft
average floor height: 12.0ft
express zone height:     0.0ft
door time:  3.3sec
12ft flight time:  4.3sec
car acceleration: 4.00ft/sec/sec
high call reversal:  0.0
up probable stops:  0.0

(Use 0 to calculate high call
  reversal and probable stops)

population per floor:  100people
% counterflow: 0.0
% interfloor:  0.0
number of cars: 5
car speed:   500.0fpm
car capacity: 3000lbs
up car loading: 16people/car

added trip time:   .0sec

*** RESULTS ***

round trip time:        144.0sec
interval:               28.8sec
up handling capacity:   11.9%/5 min

high call reversal:     13.5
up probable stops:       9.7

(a)

OtisPlan Single Group Performance

Single Deck Up Peak

name: Sample Problem
floors above lobby: 14
lobby height:  12.0ft
average floor height: 12.0ft
express zone height:     0.0ft
door time:  3.3sec
12ft flight time:  4.3sec
car acceleration: 4.00ft/sec/sec
high call reversal:  0.0
up probable stops:  0.0

(Use 0 to calculate high call
  reversal and probable stops)

population per floor:  100people
% counterflow: 3.0
% interfloor:  1.0
number of cars: 5
car speed:   500.0fpm
car capacity: 3000lbs
up car loading: 16people/car

added trip time:  0.0sec

*** RESULTS ***

round trip time:        183.1sec
interval:               36.6sec
up handling capacity:   9.4%/5 min

high call reversal:     13.5
up probable stops:       9.7

(b)

**Fig. 31.19** Printout from a computerized elevator selection program. (a) Up-peak conditions, assuming no down stops (0% counter-flow) and no interfloor traffic. The results correspond very closely to those obtained manually in Example 31.1. (b) Addition of even light counterflow and interfloor traffic can seriously affect the system's performance, as can be seen from the round trip, interval, and handling capacity figures. (Courtesy of Otis Elevator Co.)

is normally microprocessor-based. Approximately 1 service car per 10 passenger cars should be provided or, alternatively, one car for every 300,000 ft² of net area. Service cars should be 5000 lb or larger without a dropped ceiling and, if also used for passenger service, equipped with wall pads. An oversized door (e.g., 4 ft, 0 in. or 4 ft, 6 in.) is particularly useful in handling furniture. Service elevators should have a shaftway door at every level plus easy access to the truck dock or other freight entrance as well as the lobby. These cars operate as service cars normally but can serve as passenger cars in peak periods to reduce congestion and delay. This fact is particularly useful in marginal service designs. See also Table 31.11 for approximate building costs.

TRANSPORTATION

**TABLE 31.11 Office Buildings: Cost of Elevator and Electric Work**

| Item | Number of Stories | | |
|---|---|---|---|
| | 20 | 35 | 60 |
| Elevator work | 10.9% | 11.9% | 12.2% |
| Electric work | 13.3% | 12.6% | 12.2% |

### (b) Apartment Buildings

Studies indicate that apartment building traffic depends not only on the population but also on the location and type of tenant. Buildings with many children experience a school-hour peak; buildings in midtown with predominantly adult tenancy exhibit evening peaks due to the home-coming working group and outgoing dinner traffic. Where two cars are required, the second car should function both as a service car and as a passenger car. The cars may be banked or separated, as desired. If a single car is used, it should be service elevator size.

Self-service collective control is the general choice, with provision for attendant control in prestigious buildings. With small cars and a short rise, a swing-type manual corridor door is acceptable; in larger installations, both the car and the corridor door should be the power-operated sliding type.

Service elevators must be large enough to handle bulky furniture and should therefore be at least 4000 lb, with a 48-in. door and a high ceiling. Hoistways must be isolated from sleeping rooms by lobbies or other space. Similarly, machine rooms must be isolated because the starting and stopping of motors and other machine room noises are a detriment to sound sleep. Security arrangements are discussed in Section 31.44.

### (c) Hospitals

As mentioned in Table 31.6, the governing factor in the determination of elevator requirements may be either normal hospital traffic or visitor traffic, depending on the visiting-hour schedule. Due to the large volume of vehicular traffic such as stretcher carts, wheelchairs, beds, linen carts, and laundry trucks, the elevator cars are much deeper than the normal passenger type. This type of car, when used for passenger service, holds more than 20 persons and therefore gives slow service. For this reason, it is occasionally advisable to utilize some normal pas-

senger cars in addition to hospital-size cars, particularly in large hospitals.

The use of tray and bulk carts in food service imposes a considerable load on the elevators before, during, and after meals, and passenger service is seriously disrupted. To reduce this congestion and delay, many architects and hospital administrators prefer the use of dumbwaiter cars or another of the many types of materials-handling systems that can handle a $15\frac{1}{2} \times 20$-in. food tray. These systems can also be used for transporting pharmaceuticals and other items, and are discussed in Sections 32.18 and 32.19.

Elevators should be grouped centrally, although separated by type of use. Car control is normally self-service collective.

The population of the hospital may be estimated from Table 31.7. Experience has shown that a carrying capacity of 45 passengers in a 5-minute period is adequate (estimating each vehicle as equivalent to 9 passengers).

Intervals should not exceed 1 minute. All recommendations for service to the handicapped should be adopted (see Section 31.11).

### (d) Retail Stores

Retail stores present a unique problem in vertical transportation inasmuch as the objective is partially to transport persons to specific levels and partially to expose the passengers to displayed merchandise. For this reason, modern stores rely heavily on escalators, with one or two elevators intended for use by staff and handicapped persons. When, for some reason, it is desired to equip a store with elevators, use the recommendations shown in Table 31.9, calculated for a load of 10% to 20% of the store's population. Control should be automatic, selective collective. Cars are arranged in a straight line to facilitate loading and waiting.

## PHYSICAL PROPERTIES AND SPATIAL REQUIREMENTS OF ELEVATORS

### 31.36 SHAFTS AND LOBBIES

The elevator lobbies and shafts form one of the major space factors with which the architect is concerned. The elevator lobby on each floor is the focal point from which the corridors radiate for access to all rooms, stairways, service rooms, and so forth. Such lobbies obviously must be located above each other. The ground-floor elevator lobby (also called the *lower terminal*) must be conveniently located with respect to the main entrances. The equipment within or adjacent to this area should include public telephones, a building directory, elevator indicators, and possibly a control desk (see Fig. 31.12).

All lobbies should be adequate in area for the peak-load gathering of passengers to ensure rapid and comfortable service to all. The number of people contributing to the period of peak load (15- to 20-minute peak) determines the required lobby area on the floor.

Not less than 5 ft$^2$ (0.5 m$^2$) of floor space per person should be provided at peak periods for waiting passengers at a given elevator or bank of elevators. The hallways leading to such lobbies should also provide at least 5 ft$^2$ (0.5 m$^2$) per person, approaching the lobby. Under self-adjusting relaxed conditions, density is about 7 ft$^2$ (0.65 m$^2$)

per person. However, in peak periods crowding occurs, reducing this amount 3 to 4 ft$^2$ per person (3–4 persons/m$^2$). An acceptable compromise is 5 ft$^2$ (0.5 m$^2$) per person.

The main lower terminal of elevator banks is generally on the street-floor level, although it may be on the mezzanine level when the elevations of the street entrances vary so that one side of the building is at mezzanine level, whereas another entrance is lower. Such a situation is ideal for the use of escalators, which economically and rapidly carry large numbers of persons between levels, thus making practical and efficient a single main lower elevator terminal. The upper terminal is usually the top floor of the building. Typical dimensional data and lobby arrangements are shown in Figs. 31.20 to 31.22.

## 31.37 DIMENSIONS AND WEIGHTS

Manufacturers and elevator consultants will supply standard layouts for elevators including dimensions, weights, and structural loads on request. Furthermore, to assist in preliminary design, major manufacturers have agreed on and publish a set of Standard Elevator Layouts via their trade organization, the National Elevator Industry, Inc. (NEII). One such standard is reproduced in Fig. 31.23 for 500- to 700-fpm gearless units in the full range of car capacities. These standards are available from the NEII.

As may be seen from Fig. 31.23, in providing for an elevator installation, it is necessary to consider such factors as the depth of the pit, the dimensions of the hoistway, the clearance from the top of the hoistway to the floor of the penthouse, the size of the penthouse, and the loads that must be carried by the supporting beams.

The penthouse floor (and the secondary level floor where required) are located above the shaft of each elevator and need approximately 1½ stories of additional height above the top of the support beam of a given elevator when it is standing at its top-floor location. The actual floor area required by the elevator traction machine and its controls is roughly two times the area of the elevator shaft itself. The machine room contains the bulk of the elevator machinery. Because some of this equipment must be moved for maintenance, it is advisable to furnish an overhead trolley beam

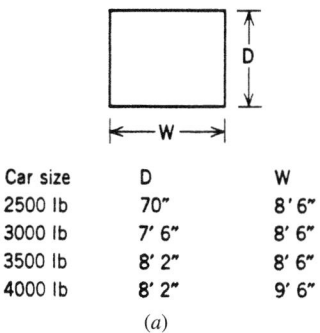

| Car size | D | W |
|---|---|---|
| 2500 lb | 70″ | 8′ 6″ |
| 3000 lb | 7′ 6″ | 8′ 6″ |
| 3500 lb | 8′ 2″ | 8′ 6″ |
| 4000 lb | 8′ 2″ | 9′ 6″ |

(*a*)

| Car size kg | speed m/s | D | W |
|---|---|---|---|
| | | mm | |
| 1000 | 2.5 | 2200 | 2400 |
| 1250 | 2.5 | 2200 | 2600 |
| 1600 | 2.5 | 2500 | 2600 |
| 2000 | 2.5 | 2500 | 2600 |
| 1000 | 4.0 | 2200 | 2400 |
| 1250 | 4.0 | 2200 | 2600 |
| 1600 | 4.0 | 2500 | 2600 |
| 1250 | 6.0 | 2250 | 2500 |

(*b*)

***Fig. 31.20*** *Rough hoistway dimensional data for use in schematic design. (a) I-P elevator sizes and dimensions. (b) SI elevator sizes and dimensions.*

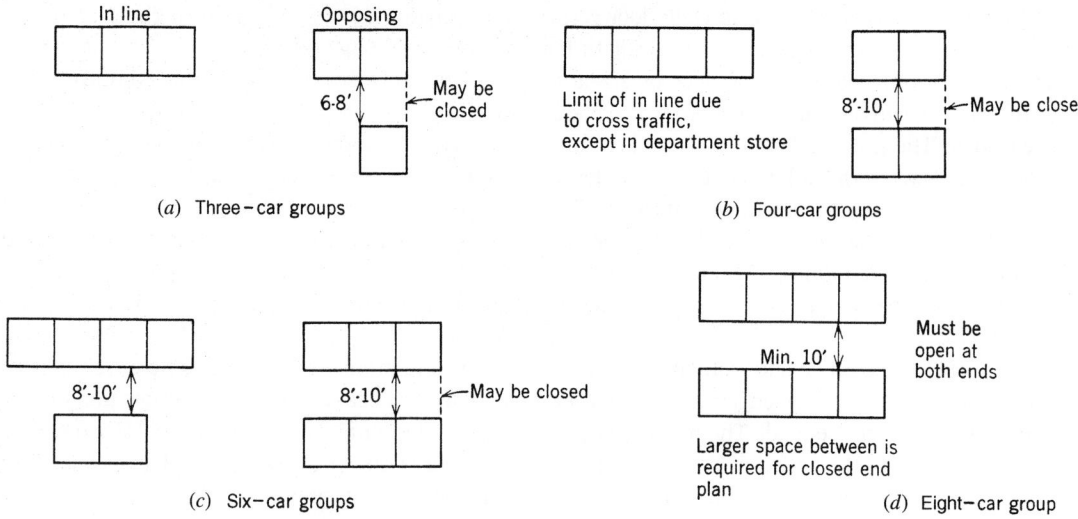

**Fig. 31.21** Lobby groupings for single zone systems: (a) three-, (b) four-, (c) six-, and (d) eight-car groups.

that can be used during installation as well. The maximum beam load is supplied by the elevator manufacturer.

Some typical machine room dimensional data are listed in Table 31.12, taken from actual installations. Because of the multiple drive options and the flexibility in equipment arrangement, no general conclusions can be drawn from these figures; they are listed simply to give a conceptual picture. A manufacturer's layout giving dimensional data

for the hoistway and machine room is shown in Fig. 31.24.

When penthouse space is not available and a hydraulic unit is not desired, a basement traction unit, also referred to as an *underslung arrangement*, can be used. These units are always low speed (100–350 fpm) and are therefore applicable where rise is limited and traffic is light to medium. Figure 31.25 shows a typical shaft section for this design with car and dimensional data.

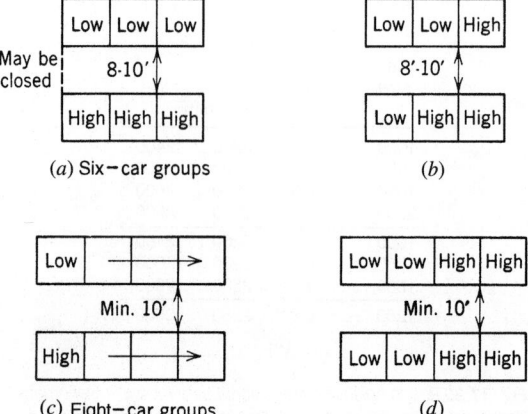

**Fig. 31.22** Lobby groupings for multiple zone systems. Arrangement (a) is preferable to (b), and (c) to (d). Groups with more than four cars in a row are not used because end-to-end walking time would excessively lengthen landing stops and hence total travel time.

# ELECTRIC PASSENGER ELEVATORS

RATED SPEEDS 500-700 fpm

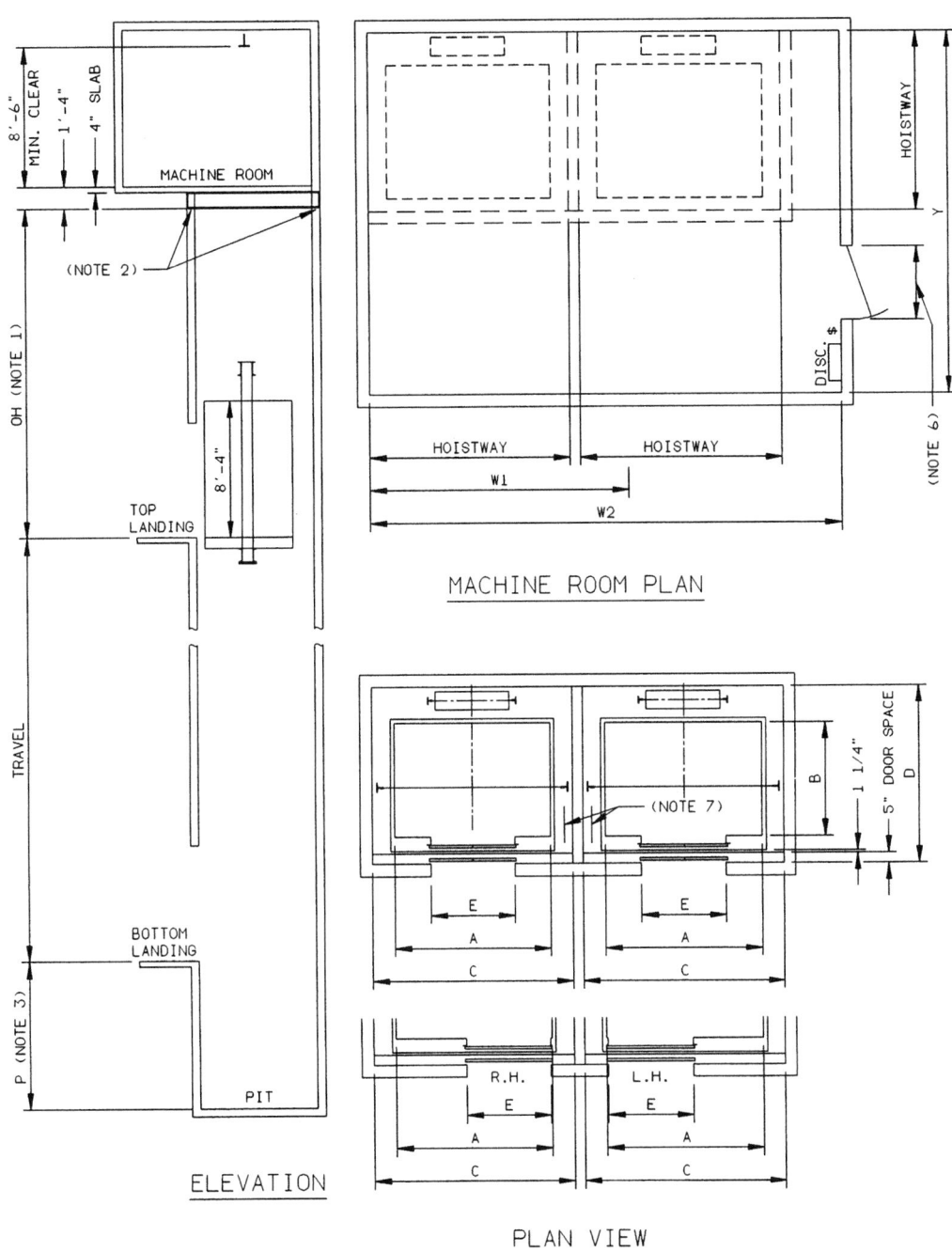

MACHINE ROOM PLAN

PLAN VIEW

ELEVATION

**Fig. 31.23** *Typical elevator installation dimensional data. (Reproduced with permission of National Elevator Industry, Inc., 185 Bridge Plaza North, Fort Lee, NJ 07024. Copyright 1992,* Vertical Transportation Standards, *7th edition.)*

| "OH" BOTTOM OF BEAM (Note 1) ft-in. | | | | | |
|---|---|---|---|---|---|
| SPEED (fpm) | RATED LOAD (lb) | | | | |
| | 2000 | 2500 | 3000 | 3500 | 4000 |
| 500 | 17-6 | 17-6 | 18-3 | 18-3 | 17-6 |
| 600 | 18-6 | 18-6 | 18-6 | 18-6 | 18-6 |
| 700 | 20-6 | 20-6 | 20-6 | 20-6 | 20-6 |

| CAR & HOISTWAY | | | | | | | | |
|---|---|---|---|---|---|---|---|---|
| RATED LOAD (lb) | AREA ■ | ft-in. (Note 11) | | | | | ENTRANCE (Note 10) | |
| | | A | B | C | D | E | TYPE | |
| 2000 | 24.2 | 5-8 | 4-3 | 7-4 | 6-11 | 3-0 | SSSO ▲ | SSCO |
| 2500 | 29.1 | 6-8 | 4-3 | 8-4 | 6-11 | 3-6 | SSSO ✳ | SSCO ▲ |
| 3000 | 33.7 | 6-8 | 4-7 | 8-4 | 7-5 | 3-6 | SSSO ✳ | SSCO ▲ |
| 3500 | 38.0 | 6-8 | 5-3 | 8-4 | 8-1 | 3-6 | SSSO ✳ | SSCO ▲ |
| 4000 | 42.2 | 7-8 | 5-3 | 9-6 | 8-1 | 4-0 | SSCO ✳ | |

| "P" PIT DEPTH (Note 3) ft-in. | | | | | |
|---|---|---|---|---|---|
| SPEED (fpm) | RATED LOAD (lb) | | | | |
| | 2000 | 2500 | 3000 | 3500 | 4000 |
| 500 | 10-1 | 10-1 | 10-1 | 10-1 | 10-1 |
| 600 | 11-5 | 11-5 | 11-5 | 11-5 | 11-5 |
| 700 | 11-5 | 11-5 | 11-5 | 11-5 | 11-5 |

| MACHINE ROOM ft-in. | | | | | |
|---|---|---|---|---|---|
| RATED LOAD (lb) | W1 WIDTH | W2 WIDTH | "Y" DEPTH FOR RATED SPEED (fpm) | | |
| | | | 500 | 600 | 700 |
| 2000 | 7-4 | 15-0 | 18-6 | 18-6 | 18-6 |
| 2500 | 8-4 | 17-0 | 18-6 | 18-6 | 18-6 |
| 3000 | 8-4 | 17-1 | 18-6 | 18-6 | 18-6 |
| 3500 | 8-4 | 17-1 | 18-6 | 18-6 | 18-6 |
| 4000 | 9-6 | 19-5 | 18-6 | 18-6 | 18-6 |

Notes
1. "OH" Dimensions are based on a 8'-4" overall car height.

2. Supports for elevator machine beams not by elevator supplier.

3. For travel greater than 400'-0" increased pit depth may be required. Consult elevator supplier.

4. 3'-6" x 7'-0" Recommended.

5. Pit ladder not by elevator supplier.

6. Dividing beams, not by elevator supplier, to be designed to sustain rail forces. Consult elevator supplier.

7. When compliance with seismic risk zone 2 or greater requirements is anticipated, provide additional hoistway space.

Notes
■ Maximum allowable inside car area in ft$^2$ per ASME A17.1, Rule 207.1.

✳ These car dimensions and entrance types provide wheelchair accessibility and accommodate an ambulance type stretcher (76 in. x 24 in.) in the horizontal position.

▲ These car dimensions and entrance types provide wheelchair accessibility.

**Fig. 31.23** (Continued)

## 31.38 STRUCTURAL STRESSES

For the purpose of structural design, it is necessary to know the overhead load that must be supported by the foundations, by structural columns extending up to the penthouse, and by the main beams that support the penthouse floor and subfloor. These loads (reactions) are supplied by manufacturers and usually include the actual dead weights of equipment when the elevator is not in motion, plus the added weight caused by the momentum of all moving parts and passengers when the elevator is at top speed and is suddenly stopped rapidly by the safety devices.

**TABLE 31.12 Typical Elevator Machine Room Dimensions**

| Bank Description | | Machine Room Dimensions (Rounded to Nearest Foot) |
|---|---|---|
| 2-2500 lb | 125 fpm | 16 × 18 |
| 3-2500 lb | 350 fpm | 19 × 26 |
| 3-2500 lb | 700 fpm | 21 × 26 |
| 6-3500 lb | 500 fpm | 26 × 29 |
| 6-3500 lb | 700 fpm | 27 × 27 |
| 6-3500 lb | 1000 fpm | 27 × 27 |
| 6-5000 lb | 450 fpm | 31 × 33 |

TRANSPORTATION

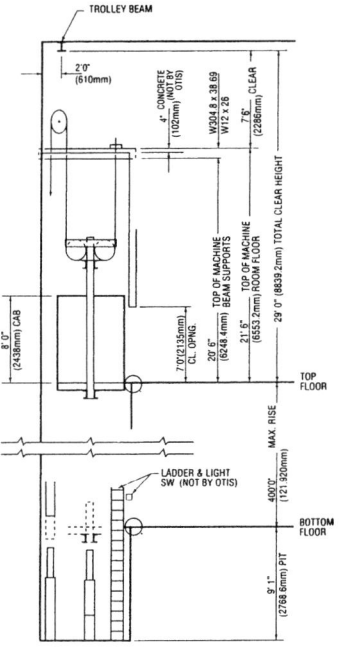

SECTIONAL ELEVATION

**HOISTWAY PLAN ( Below )**

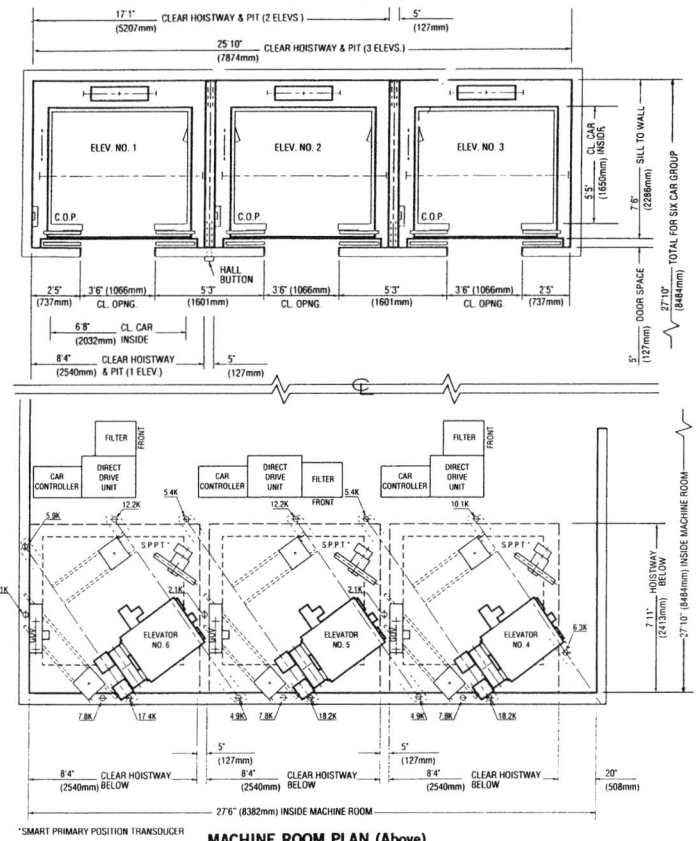

*SMART PRIMARY POSITION TRANSDUCER

**MACHINE ROOM PLAN (Above)**

**Fig. 31.24** *Manufacturer's layout data for a bank of six 3500-lb (1588-kg), 700-fpm (3.45-m/s) gearless passenger elevators. Equipment shown in the machine room is the thyristor control for dc traction machines. Because each controller provides group supervisory control in this design (Otis Elevonic 411), no separate group supervisory equipment is shown. No additional space would be required if a UMV drive with m-g sets were selected rather than thyristor control. (Courtesy of Otis Elevator Co.)*

**1417**

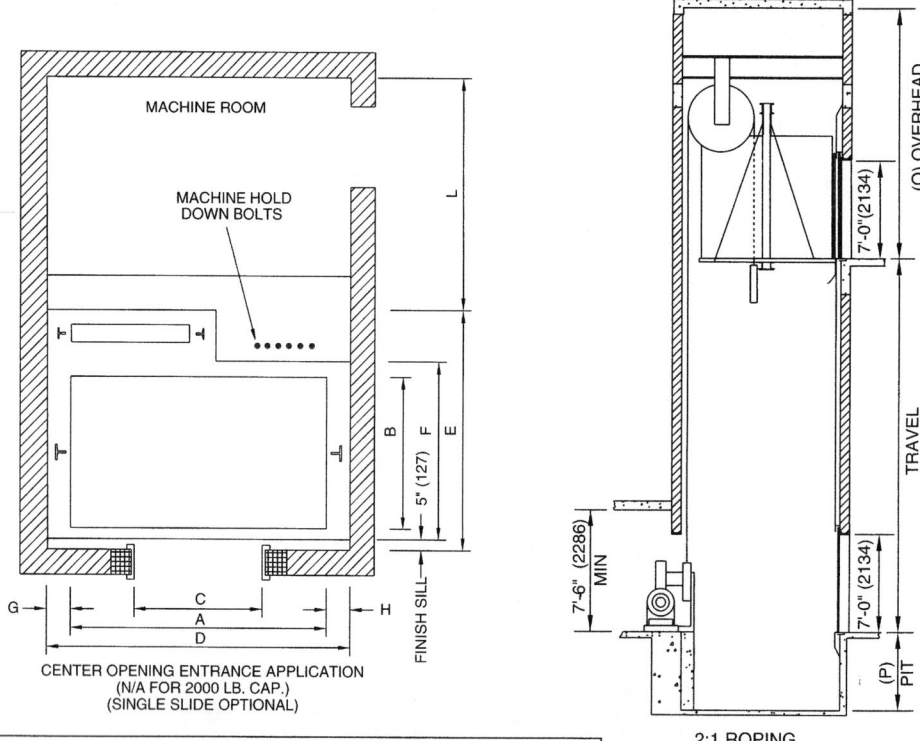

2:1 ROPING

## 2:1 ROPING ARRANGEMENT

Overhead dimension is reduced to a minimum. Greater pit depth is required due to the sheave arrangement as compared to a 1:1 installation.

### SIZES & CAPACITIES

| CAPACITY | 2000 lbs. (907 kg) | | 2500 lbs. (1134 kg) | | 3000 lbs. (1361 kg) | | 3500 lbs. (1588 kg) | |
|---|---|---|---|---|---|---|---|---|
| A | 6'-0" | (1829) | 7'-0" | (2134) | 7'-0" | (2134) | 7'-0" | (2134) |
| B | 5'-0" | (1524) | 5'-0" | (1524) | 5'-6" | (1676) | 6'-2" | (1880) |
| C | 3'-0" | (914) | 3'-6" | (1067) | 3'-6" | (1067) | 3'-6" | (1067) |
| D | 8'-0" | (2438) | 9'-0" | (2743) | 9'-0" | (2743) | 9'-0" | (2743) |
| E | 6'-10" | (2083) | 6'-10" | (2083) | 7'-4" | (2235) | 8'-0" | (2438) |
| F | 5'-4" | (1626) | 5'-4" | (1626) | 5'-10" | (1778) | 6'-6" | (1981) |
| G | 1'-0" | (305) | 1'-0" | (305) | 1'-0" | (305) | 1'-0" | (305) |
| H | 1'-0" | (305) | 1'-0" | (305) | 1'-0" | (305) | 1'-0" | (305) |

### MINIMUM MACHINE ROOM OVERHEAD & PIT DIMENSIONS

| SPEED | 100 FPM (0.51 m/s) | | 200 FPM (1.02 m/s) | | 250 FPM (1.27 m/s) | |
|---|---|---|---|---|---|---|
| L | 10'-6" | (3200) | 10'-6" | (3200) | 10'-6" | (3200) |
| O | 13'-0" | (3962) | 13'-2" | (4013) | 13'-7" | (4140) |
| P | 5'-6" | (1676) | 6'-6" | (1981) | 6'-11" | (2108) |

NOTES:

1. If a counterweight safety is applied, consult Montgomery KONE
2. Layouts and dimensions shown are for machine located at rear of hoistway.
3. Add 4" (102) to dimension D in seismic zones.
4. Other capacities, speeds, and arrangements are available.

**Fig. 31.25** *Typical data for a basement traction machine (underslung) arrangement, used where a penthouse is unavailable or undesirable. (Courtesy of Montgomery-KONE.)*

## POWER AND ENERGY

### 31.39 POWER REQUIREMENTS

The power required by an elevator drive is that needed to provide the necessary traction and to overcome friction. Because power is equal to the *rate* at which work is done, the elevator motor size is directly proportional to the speed of the system. In other words, it requires proportionately more power to lift a 3000-lb car at 700 fpm than at 200 fpm. This relationship is shown in Fig. 31.26, which shows the minimum size of a dc elevator traction motor as a function of speed for cars of different capacity. (For power data on hydraulic elevators, see Section 33.11.) As friction is higher in a geared machine than in a gearless unit, its traction motor must be larger for the same car speed. The size of the traction machine shown in Fig. 31.26 is independent of the power supply design (m-g set,

VVVF, thyristor control) because it is determined purely by traction system requirements. (In actual practice, however, traction motors with VVVF control are frequently smaller because they operate more efficiently.)

An elevator moves only about 50% of the time, the remainder being spent standing at various landings. As the number of cars in a bank increases, the probability of *all* the cars being in operation simultaneously decreases, resulting in a system demand factor of less than 1.0. The factor for different group sizes is shown in Fig. 31.26.

As an example of the use of the curves, consider a bank of five 3500-lb, 600-fpm units. From Fig. 31.26, each car requires 48 hp:

group demand factor = 0.67

total instantaneous power required
$$= 5 \times 48 \times 0.67 = 160 \text{ hp}$$

Note that this is the *traction motor* power requirements. If an m-g set with an overall efficiency of

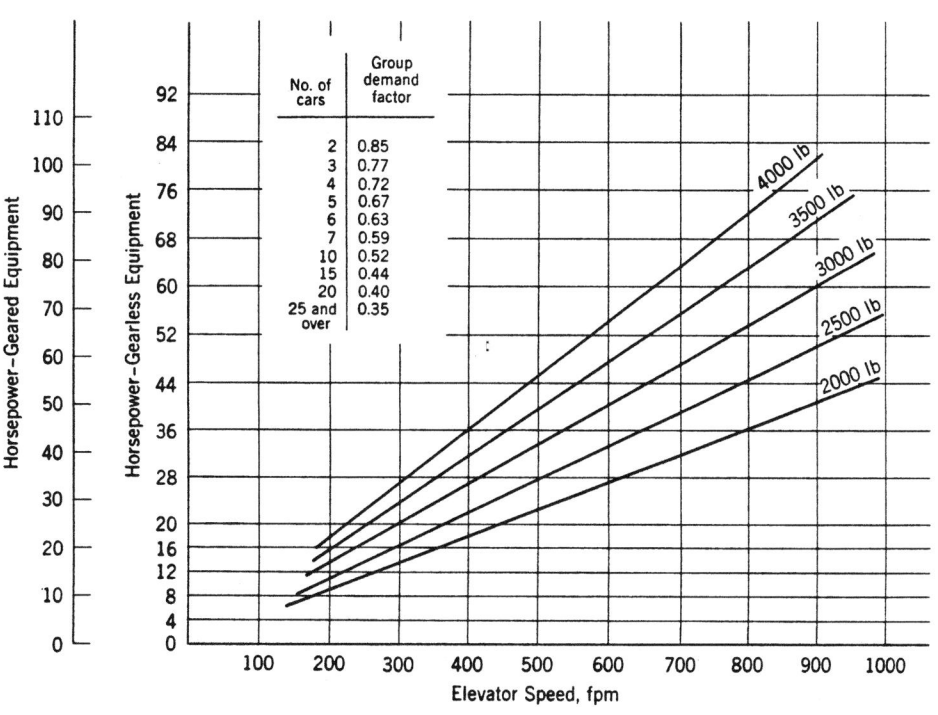

**Fig. 31.26** *Elevator traction motor power requirements per car. An m-g set drive (if used) is approximately 20% larger than a traction machine.*

80% is used to drive the traction motor, the elevator system power requirement is

$$\text{system power} = \frac{160 \text{ hp}}{80\% \text{ eff}} = 200 \text{ hp}$$

which must be provided by the building power system. If a solid-state power supply system with a (typical) efficiency of 92% is used, the system power requirement will be only

$$\text{system power} = \frac{160 \text{ hp}}{92\% \text{ eff}} = 174 \text{ hp}$$

which is a 13% reduction from the previously calculated 200-hp requirement.

## 31.40 ENERGY REQUIREMENTS

The energy used by an elevator is essentially the system friction, including the heat generated by the brakes plus the electrical losses in the traction motor and power supply equipment—rotary or solid state. The energy expended in raising the car and its passengers is simply stored as potential energy. It is *returned to the power system* when the car and passengers descend via the system of regenerative braking in use in almost all elevator systems. Refer to Fig. 31.27, which shows the approximate efficiencies of the components of a typical system. With these data, we are able to calculate a system's energy consumption.

**EXAMPLE 31.2** Given a system of five 3500-lb, 600-fpm (gearless) cars, calculate:

(a) Heat generated in the machine room during peak periods. Assume solid-state control.

(b) Approximate monthly energy cost using a combined demand/energy rate of $0.08/kWh.

### SOLUTION

(a) During peak periods, the traction motor operates approximately 50% of the time and is at standstill the other half. Assume that, while operating, it draws 90% of the full load (with a VVVF power supply this figure is reduced considerably). Therefore, for one car, from Fig. 31.26,

$$\text{traction motor} = 48 \text{ hp}$$

*Total loss per machine:*

In controls:

$$\frac{48 \text{ hp}}{0.9 \text{ eff}} \times 90\% \text{ load} \times 50\% \text{ operation} \times 10\% \text{ loss}$$
$$= 2.4 \text{ hp}$$

In traction motor:

$$48 \text{ hp} \times 90\% \text{ load} \times 50\% \text{ operation} \times 20\% \text{ loss}$$
$$= 4.32 \text{ hp}$$

$$\text{total} = 6.72 \text{ hp} = 17,100 \text{ Btu/h}$$

Because five elevators are operating, the total heat generated is

$$5 \times 17,100 \text{ Btu/h} = 85,500 \text{ Btu/h}$$

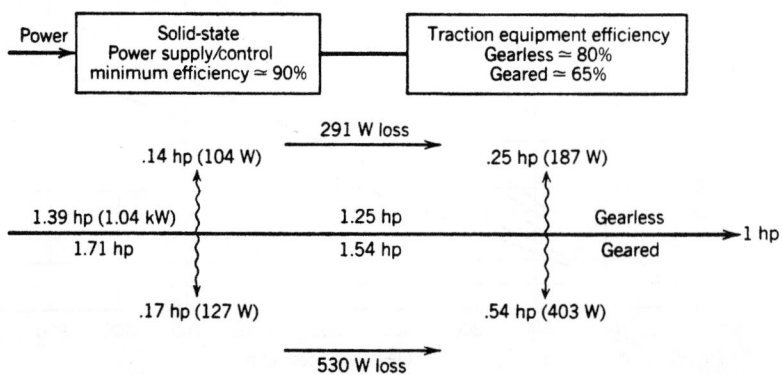

**Fig. 31.27** *Block diagram showing losses in the system per horsepower delivered to the elevator car and the equivalent wattages. Note that the losses in a geared system are almost double those of a gearless one. Figures shown are for solid-state thyristor controls.*

This is the rating of a home furnace. As a result, machine room temperatures in warm climates frequently reach 120°F! (No diversity is taken because all the machines are operating and the heating is additive; diversity is applicable only in calculating instantaneous load.)

As solid-state elevator equipment is much less tolerant of high ambient temperatures than were the electromechanical switches and relays previously used, the machine room should be held to a maximum dry bulb temperature of 90°F. (Temperatures above 90°F can result in unreliable elevator system performance.) This can sometimes be accomplished by thermostatically controlled forced ventilation, particularly if spill air from an air-conditioning system is present. However, because machine rooms are frequently on the building roof and exposed on all surfaces to direct solar radiation, air conditioning may be necessary. It is also important to prevent machine room temperature from dropping below 55°F. This can usually be done with one or more unit heaters, which will normally operate only during the winter, and then only on nights and weekends.

In actual design situations, accurate heat loss figures, which are available from manufacturers, would be used, along with accurate heat gain and heat loss calculations for the specific machine room being designed. Frequently, use of thermal insulation, thermostatically controlled louvers, and sunshading can ease the thermal load and result in appreciable financial and energy savings. Because of the high initial and operating costs of air conditioning, some manufacturers use control components that are tolerant of high temperatures. This point should be carefully examined with proposed elevator equipment manufacturers and the conclusions reflected in the project elevator specifications.

(b) To calculate the monthly energy cost, an estimate must be made of the total usage of the system. Assuming the system to be in an office building, a reasonable breakdown of operation during a 24-hour day would be

2 hours peak use
2 hours 70% of peak
6 hours 50% of peak
14 hours 10% of peak

This gives an average of 30% of peak load for the bank. Therefore, per car

$$energy = 30\% \times total\ losses \times 24\ hours$$
$$= 0.3 \times 6.72\ hp \times 24\ hours$$
$$= 48\ hp\text{-}h = 36\ kWh/day/car$$

Monthly cost would be

$$36\ \frac{kWh}{day} \times 25\ days \times \$0.08$$
$$= \$72/month/car$$
$$= \$360/month\ for\ the\ bank$$

This figure would be lower with a VVVF power supply and higher for a Ward–Leonard (m-g set) arrangement. ∎

## 31.41 ENERGY CONSERVATION

A reduction in energy consumption can be accomplished by implementing the following recommendations:

FOR EXISTING ELEVATORS
1. Increase the interval during nonpeak hours.
2. Replace m-g sets with a solid-state dc power supply or ac traction motors with a VVVF power supply. This conserves energy not only due to the higher efficiency of the power supply, but also because energy consumption of idling machines is eliminated.
3. Recycle machine room waste heat.
4. Shut down some units completely during off hours.

FOR A BUILDING IN THE PLANNING STAGE
1. Base the design on the maximum recommended trip time.
2. Use the lowest speeds possible within a type—that is, geared or gearless.
3. Use gearless equipment whenever possible.
4. After construction, implement the recommendations for existing elevators previously detailed.

Because elevator shafts have a powerful stack effect, measures should be taken to counteract this during the heating season.

## 31.42 EMERGENCY POWER

Major power failures and local brownouts have demonstrated forcefully the need for a standby or emergency power source of adequate size to operate a building's elevators. Few experiences are so harrowing as being trapped in the crowded confines of a small box suspended in a long vertical shaft, with little or no light, and complete strangers for companions.

A common misconception about elevators is that on failure of power, the cars will automatically descend to the nearest landing, where an exit is then possible. In actuality, the brake is set immediately on power outage and the car remains stationary. Hydraulic cars can be lowered by operation of a manual valve; *small* traction cars can be cranked to a landing by hand, but large cars are fixed in position. This is particularly bad for cars in blind shafts—that is, express shafts with no shaftway doors. In such cases, escape from the cars via a hatchway is not practical; when emergency power is not available, breaking through the shaftway walls is the only recourse.

In addition to simple inconvenience, loss of elevator service in facilities such as hospitals and mental and penal institutions constitutes a danger to life. For this reason, most codes require that emergency power be available in specific building types to operate at least one elevator at a time, and for elevator lighting and communications. Many installations separate the emergency power functions, providing a generator for elevator traction power and separate individual elevator battery packs for communications, lighting, and, preferably, the car fan. The last two items can be furnished by elevator manufacturers with the cars as an option.

The generator is normally sized to supply one elevator motor at a time, with manual or automatic switching arranged between unit controllers. Thus, each car in turn can be brought to a landing and thereafter a single car retained in service. Obviously, if it is desired to operate more than one car, a larger generator can be installed. This might well be the case in a multiwing building with critical service requirements, such as a hospital.

The amount of power required, the size of the emergency generator, and the equipment size necessary to absorb regenerative power are all data that can be furnished by the consulting engineer and the elevator manufacturer.

## SPECIAL CONSIDERATIONS

### 31.43 FIRE SAFETY

Most fire codes specify the procedures that the elevator control equipment must implement once a fire emergency has been initiated. Details vary somewhat, but in general the actions are these:

1. All cars close their doors and return nonstop to the lobby or another designated floor, where they park with the doors open. Thereafter, they are operable in manual mode only, by use of the fire fighter's key in the car panel.
2. All car and hall calls are canceled, and call-registered lights and directional arrows extinguished.
3. The fire emergency light or message panel in each car is activated to inform passengers of the nature of the alert and that cars are returning to a designated terminal.
4. Door sensors and in-car emergency stop switches are deactivated.
5. Traveling cars stop at the next landing without opening their doors and then proceed to the designated terminal.

The cars can then be used by trained personnel to transport fire personnel and equipment and for evacuation. In the event of a false alarm, the emergency procedure can be overridden at the (lobby) control point and the system returned to normal while the source of the alarm is located. (This is a particularly important feature in large buildings with automatic fire alarm systems containing hundreds of fire, smoke, and water-flow detectors.)

### 31.44 ELEVATOR SECURITY

Elevator security has two aspects: physical security of riders and consideration of the elevator as a portal in a building-access security system.

#### (a) Rider Security

This problem is particularly difficult inasmuch as the traveling elevator is an enclosed space that can be rendered inaccessible simply by pressing the emergency stop button. Thereafter, an attacker can escape at the floor of his choice. To reduce this danger to an extent, elevators are equipped with alarm buttons that alert residents and security personnel, if any. Every elevator, by code, must be equipped with communication equipment. A two-way communication system with "no-hands" operation in the car is particularly effective for security. When a closed-circuit TV monitor is added, utilizing a wide-angle camera in each car (Fig. 31.28), the security problem will have been addressed to a considerable extent. Obviously, using a communication and TV system presupposes continuous monitoring of the building security desk.

At least one major manufacturer now markets a device that alarms automatically on detecting sudden, violent motions or a sharp, pointed instrument. Handling of the alarm is problematic because an automatically locked door can be forced open manually. Furthermore, the advisability of locking a violent person in an elevator with potential victims is questionable.

### (b) Access Control

This is often a matter of restricting access to (and from) a floor or car. This can be accomplished by pushbutton combination locks or coded cards, the proper use of which permits access (see Chapter 30). However, if a second person happens to accompany

**Fig. 31.28** *Wide-angle TV camera intended for elevator car surveillance. A prominent printed warning in the car is an integral part of the system's effectiveness. (Photo courtesy of Visual Methods, Inc.)*

the authorized person, the effectiveness of this type of access barrier is seriously compromised. In sum, the most effective security system is a combination of automatic monitoring and access devices coupled with continuous supervision by persons who know the appropriate action to take in an emergency.

## 31.45 ELEVATOR NOISE

As already noted, elevator operation, with its rotating, sliding, and vibrating masses, can be a cause of serious noise disturbance to quiet areas such as sleeping rooms, libraries, and certain types of office space. Noise can be reduced by the appropriate application of vibration isolators (e.g., between guide rails and the structure) and by proper control, but primarily by placing noise-sensitive areas away from shafts and machine rooms. Furthermore, the clatter and whirring sound of the older machine room, caused by relays, step switches, m-g sets, and sliding contacts, can be entirely eliminated by the use of solid-state equipment.

## 31.46 ELEVATOR SPECIFICATIONS

Two basic types of specification for elevator equipment, as for other types of equipment, are utilized. The performance specifications describe job conditions and invite contractors to submit detailed proposals including those for full engineering. The burden of comparing proposals then falls on the owner, who—if competent to properly perform such an evaluation—would probably do better to utilize an equipment-type specification in the first place.

In recent years, the use of performance specifications has increased because of the advent of preengineered, premanufactured systems. These are supplied by the major manufacturers and have the following advantages:

1. Approximately 10% lower cost than a custom-designed system
2. A complete engineered and tested system whose performance and cost are known exactly
3. Rapid delivery
4. Minimum supervision required by the owner and architect

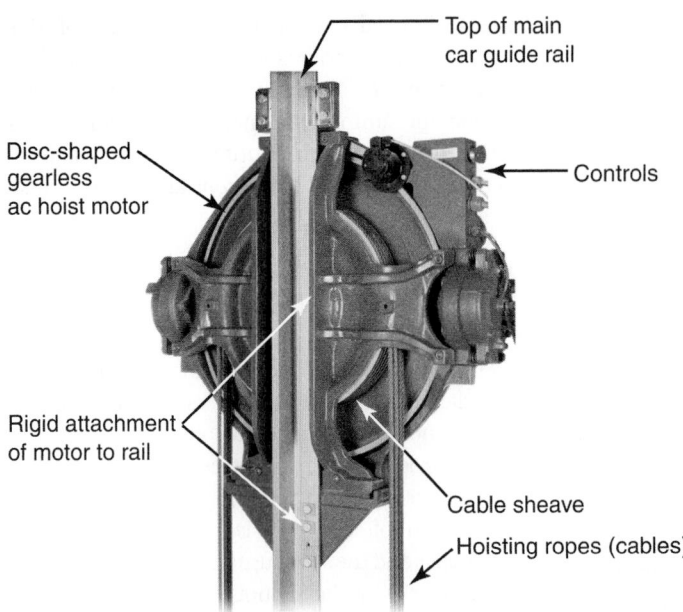

Top of main
car guide rail

Disc-shaped
gearless
ac hoist motor

Controls

Rigid attachment
of motor to rail

Cable sheave

Hoisting ropes (cables)

**Fig. 31.29** *Disc-shaped hoisting motor rigidly mounted on the elevator guide rail. The ac synchronous motor is connected directly to the hoisting cable drive sheave with no intervening gears. Brakes and controls are built into the assembly. (Courtesy of Montgomery-KONE.)*

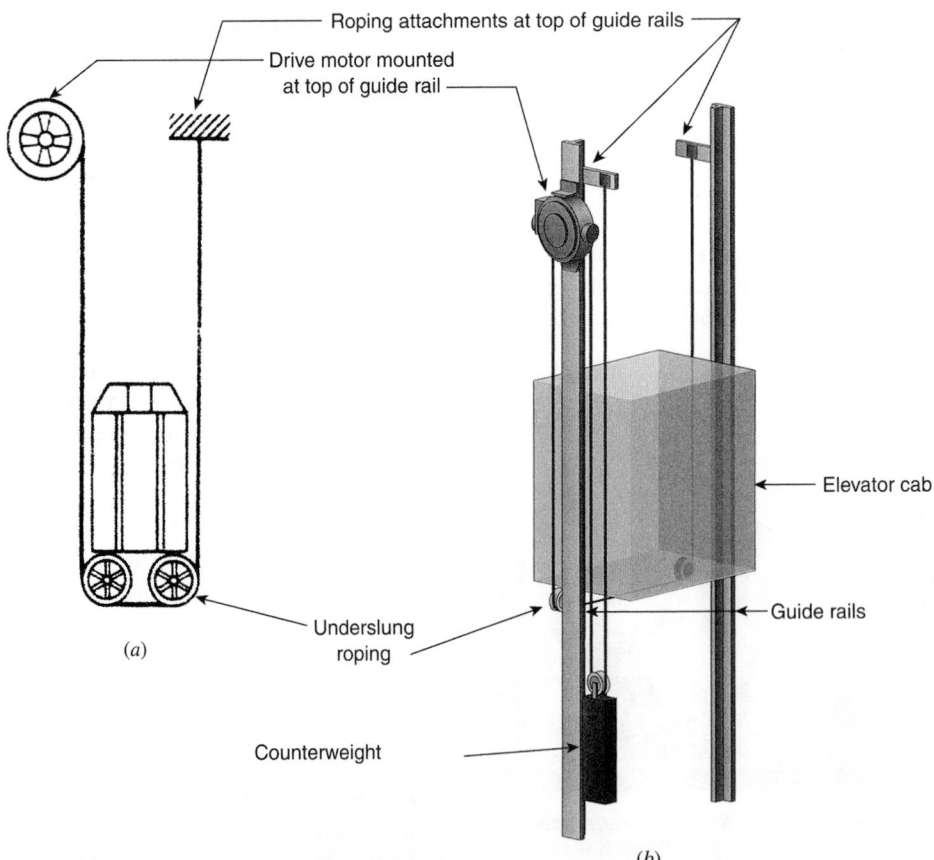

Roping attachments at top of guide rails

Drive motor mounted
at top of guide rail

Elevator cab

Guide rails

Underslung
roping

*(a)*

Counterweight

*(b)*

**Fig. 31.30** *Schematic (a) and pictorial (b) representations of the disc-motor-driven elevator arrangement. (Part b Courtesy of Montgomery-KONE.)*

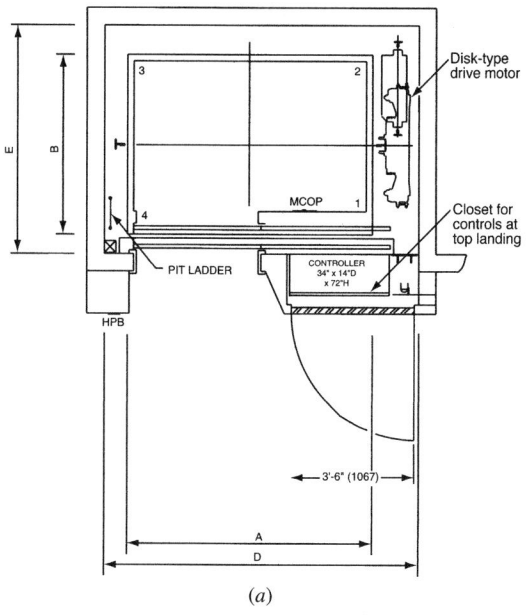

(a)

| MonoSpace Elevator System | |
|---|---|
| Speed | 200 FPM (1.0 m/s) |
| Capacity | 2500 lb. (1134 kg) |
| A | 7'-0" (2134) |
| B | 5'-0" (1524) |
| D | 9'-0" (2744) |
| E | 6'-4" (1931) |
| Entrance Types | Single Slide |
| Overhead (O) | 12'-9" (3887) min |
| Minimum Travel | 8'-4" (2540) |
| Maximum Travel | 80'-0" (24384) |
| Landings Served | 2 - 10 |
| Pit (P) | 5'-6" (1677) min |
| Machine Room (W x D x H) | 3'-9" x 16" x 8'-0" (1143 x 407 x 2439) |

(b)

**Fig. 31.31** (a) Section through the top of a hoistway showing dimensional data for a single 2500-lb, 200-fpm installation with a rise of up to 80 ft. Note that the drive motor occupies less than 2 ft in the width of the hoistway and that the elevator motion and operating controls (Section 31.4) are installed in a closet 42 in. wide and approximately 20 in. deep at the top landing. (b) Elevator system basic data for the simplex (single) unit shown in (a). (Courtesy of Montgomery-KONE.)

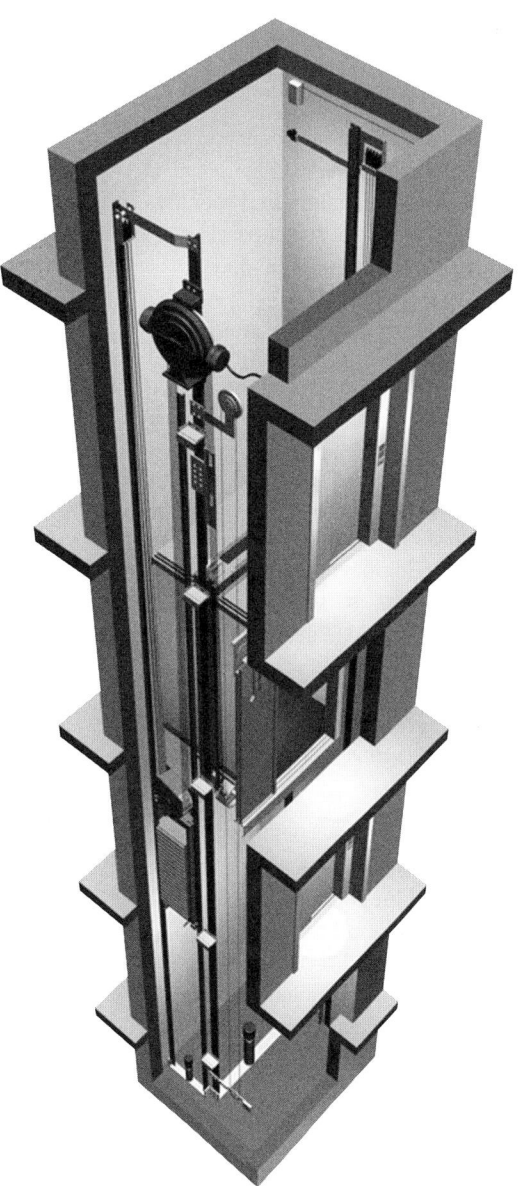

**Fig. 31.32** Pictorial representation of a disc-type traction motor hoistway showing the system's essentials. (Courtesy of Montgomery-KONE.)

TRANSPORTATION

If architects decide to use a custom-designed system, they must prepare detailed drawings and specifications. The specifications must include:

- Elevator type, rated load, and speed
- Maximum travel
- Number of landings and openings
- Type of control and supervisory system
- Details of car and shaft doors
- Signal equipment
- Characteristics of the power supply
- Finishes

The last item can be left as a dollar allowance for architectural treatment of the car interior. Because the selection of, and technical specifications for, elevators are specialized and complex, the services of an elevator consultant are usually required.

In addition to the technical portions of the specifications, it is imperative that the following items be covered in detail.

### (a) Owner's Responsibility

The *construction* contractor normally provides the following:

1. The hoistway, including a properly designed, lighted, drained, waterproofed, and ventilated machine room and pit
2. Access doors, ladders, and required guards
3. Guide rail bracket supports, and support for machine and sheave beams
4. Electric feeder terminating in a switch in the machine room
5. Hoistway outlets for light, power, and telephone
6. Temporary light and power during construction
7. Concrete machine foundations
8. Vents, holes, and other work to satisfy fire codes
9. All cutting, patching, and chasing of walls, beams, masonry, and so on
10. Coordination of all work
11. Any special work, as negotiated

### (b) Elevator Contractor's Responsibility

Provide a complete, working, tested, and approved system in accordance with specifications, plus any special work such as painting, special tests, work scheduling, and temporary elevator service.

### (c) Special Job Conditions

These include work restrictions, scheduling, penalties or bonuses, test reports, and the like.

In alteration and modernization work, the problems of coordination are more complex, and an elevator contractor experienced in this type of work must be selected. To this end, in all elevator contract work, bids should be solicited from parties named on qualified bidder lists. Part of an elevator contract comprises maintenance of the installation for a specific period after completion.

## 31.47 NOVEL DESIGNS

The elevator industry is constantly developing new equipment to improve the operation and safety of standard system designs. In addition, novel designs that are essentially different from standard traction arrangements are always being developed in an attempt to increase the efficiency of space use and to decrease the high cost of standard traction machinery. Among the interesting designs being developed in the first category is one that permits a car to travel horizontally in addition to its normal vertical motion, the purpose of which is to increase the number of cars using a single shaft. The second category includes a design using a linear motor (as opposed to a rotating unit) to supply traction power. These and other special designs are discussed in Chapter 32.

A recently developed interesting variation of the conventional traction design that effects a considerable space reduction is shown in Figs. 31.29–31.32. At this writing, its principal applications are in low-speed, low-rise installations now generally serviced by hydraulic elevators, but with higher speeds and rises in development. The novelty of the design lies in the use of a flat (disc-shaped), synchronous a-c gearless hoisting motor, which, due to its flat disc shape, can be mounted directly on the main car guide rail at one side of the shaft (see Fig. 31.29). This essentially obviates the need for a penthouse and a large machine room above the hoistway. Due to the traction motor's position at the side of the hoistway, the car is roped in an underslung arrangement, as shown in Fig. 31.30. Additional space economy is achieved by the use of a small drive controller built into an alcove at the top landing (see Fig. 31.31). The pictorial hoistway representation in Fig. 31.32 shows the equipment arrangement, demonstrating the absence of a penthouse and the limited machine room space requirement. An additional advantage of this arrangement is that the elevator loads and reactions are borne by the (stiffened) guide rail and transferred directly to the concrete pit below the bottom landing. This reduces the reactions borne by the machine room level in conventional traction design and results in reduced structural loads. Compared to hydraulic elevators, this design exhibits considerable energy economy due to its use of a gearless traction machine.

### References

Stein, B., J. Reynolds, and W. McGuinness. 1986. *Mechanical and Electrical Equipment for Buildings,* 7th ed. John Wiley & Sons, New York.

Strakosch, G.R. 1983. *Vertical Transportation, Elevators and Escalators,* 2nd ed. John Wiley & Sons, New York.

TRANSPORTATION

CHAPTER 32

# Vertical Transportation: Special Topics

CHAPTER 31 DEALT PRIMARILY with the construction, control, and selection of traction passenger elevators (i.e., units that are raised and lowered by cables attached to a crosshead beam above the cab). These elevators move in dedicated shafts: one per car, with the traction machine usually placed at the top of the shaft. This chapter discusses elevators with special shaft arrangements, special cars, lifting arrangements other than cable traction, and elevators designed primarily to carry freight rather than passengers. The chapter closes with a fairly extensive section devoted to material handling and movement in buildings using elevator-like traction cars and other means.

## SPECIAL SHAFT ARRANGEMENTS

The fact that each traction elevator in a building has a full-height basement-to-penthouse shaft that occupies valuable building space, and that the elevator only uses part of the space only part of the time, has always disturbed finance-conscious building operators. Development of high-speed drives and very sophisticated control systems has been spurred, in part, by the desire to increase the efficiency of elevator use of building space. An alternate approach to this problem lies in innovative use of shaft space so that more than one car can use a single shaft. A detailed analysis of these solutions is beyond our scope here; however, a rapid review is of interest.

## 32.1 SKY LOBBY ELEVATOR SYSTEM

For skyscraper buildings and high-rise multiple-use buildings (such as the John Hancock Center in Chicago)—which are, in effect, stacked multiple buildings—the elevator solution may involve transporting large groups of people from the street lobby to an upper lobby called a *sky lobby* or *plaza*. At this point, the passengers transfer to another elevator to continue their upward journey.

The traditional approach in tall, single-purpose buildings is to zone the structure and have banks of elevators serving groups of floors. The difficulty with this solution is that the upper-zone cars travel through a long, expensive blind shaft to reach the floors being served. The sky plaza approach stacks (figuratively) two (or more) shafts vertically with a lobby in between. The effective result is to have *two* cars operating in the equivalent of a single full-height shaft. To mitigate the annoyance felt by passengers resulting from a lengthened trip, the sky lobby is attractively appointed and may serve as an upper observation floor. Also, an elevator trip broken up by a lobby stop is thought to be less annoying than a somewhat shorter but still long, uninterrupted one. Finally, this plan is used principally where a clear differentiation in building use occurs, usually consisting of office areas below and residential units above, as in the John Hancock building. Thus, most lower-section occupants never use the sky lobby. Passengers headed for upper zones can use a sky lobby "shuttle" that travels from the entry

TRANSPORTATION

**1429**

level to the sky lobby and then continues the trip in an upper-section car.

Some designs also provide for a single-shaft upper-zone elevator group that serves upper-zone passengers who prefer not to use the sky lobby. A schematic drawing of the elevator arrangement in the John Hancock building should make the advantages of the system clear (Fig. 32.1).

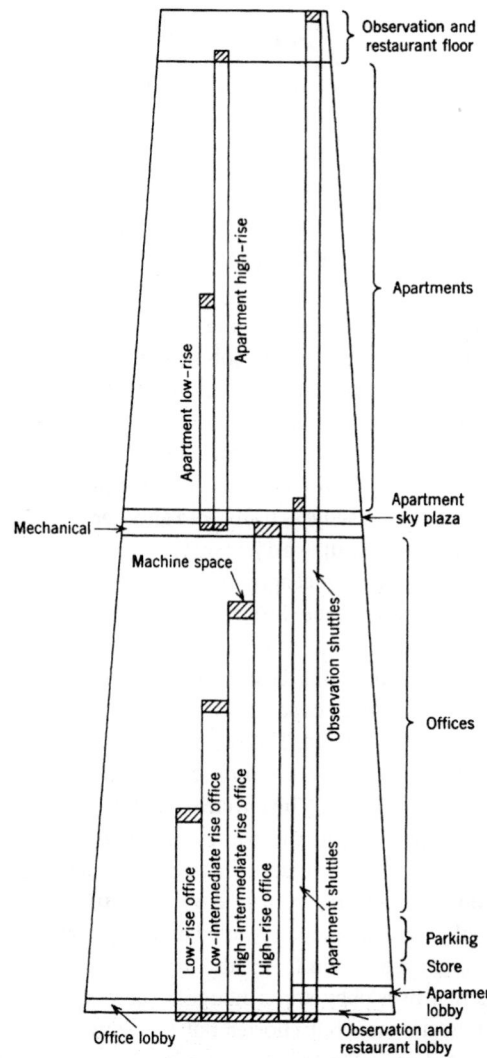

**Fig. 32.1** *The use of a sky plaza in this dual-purpose building (the John Hancock Center, Chicago) eliminates the blind shafts below the upper-section (residential) elevators. The lengthened trip time is well tolerated by riders destined for the upper residential section, whereas the lower office section remains unaffected and retains its short interval and trip time. (G.R. Strakosch,* Vertical Transportation, *2nd ed., John Wiley & Sons; reprinted with permission.)*

## 32.2 DOUBLE-DECK ELEVATORS

This is an old technique, recently revived and revised to answer the needs of tall buildings such as the Sears and Citicorp towers. Its principal purpose is to limit the otherwise prohibitively large amount of space occupied by elevator shafts (Fig. 32.2). The double-deck car increases shaft capacity, decreases the number of local stops, and increases the rental area available. This technique can also be combined with sky lobbies for further space economy, as was done in the Sears Tower.

## HYDRAULIC ELEVATORS

### 32.3 CONVENTIONAL PLUNGER-TYPE HYDRAULIC ELEVATORS

The elevators discussed thus far are traction types; that is, they are raised and lowered as a result of the tractive force of cables attached to or passing under the car. These cables in turn are raised and lowered by a motor-driven traction sheave. In contrast to these, the conventional hydraulic or plunger elevator is raised and lowered quite simply, by means of a movable rod (plunger) rigidly fixed to the bottom of the elevator car. The absence of cables, drums, traction motors, elaborate controllers, safety devices, and penthouse equipment makes this system inherently inexpensive and often the indicated choice for low-speed (up to 200 fpm), low-rise (up to 65 ft) applications, where construction of the plunger pit does not present difficulties and/or the absence of a penthouse is desirable.

As a matter of historical interest, the first hydraulic elevators used water as the system fluid, supplied at sufficient pressure from roof water tanks. The tanks in turn were kept full by building water pumps. All hydraulic elevators today use oil and obtain their motive power from a sealed oil-piping circuit powered by an oil pump.

The components of a typical hydraulic unit are shown in Fig. 32.3. This system operates the same way as a hydraulic automobile jack. Oil from a reservoir is pumped under the plunger, thereby raising it and the car. The pump is stopped during downward motion, the car being lowered by gravity and controlled by the action of bypass valves, which

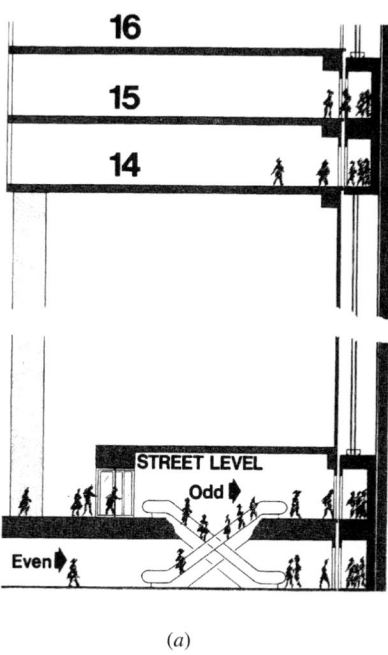

(a)

**Fig. 32.2** (a) A double-deck car serves to increase car capacity and decrease shaft space. Coincidence of calls in the upper and lower cars reduces the number of local stops made by the double-deck unit. (b) Graphical representation of the space saved in a 40-story building by the use of double-deck elevators. (Courtesy of Otis Elevator Co.)

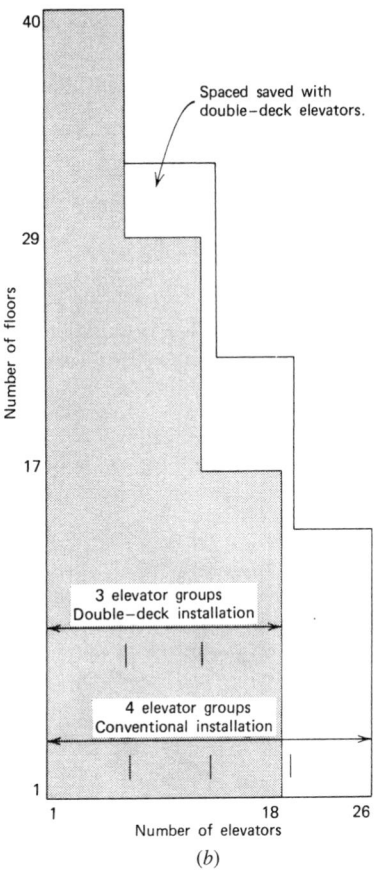

(b)

also control the positioning of the car during upward motion. Control systems normally used are similar to those for traction types—for example, collective, and selective collective. Similarly, door arrangements are the same as in traction types—that is, single-slide, center-opening, and two-speed. Automatic leveling is readily available and is standard on all automatic units.

From the point of view of architecture and construction, the major inherent advantage of hydraulic units is the absence of an overhead machine room, a penthouse, and traction equipment. In Fig. 32.3, we see that only the guide rails project above the car and, if these are camouflaged, the impression of a freestanding elevator car is given. This effect can be used to good advantage *inside* large, open spaces such as exist in shopping malls and stores; when it is combined with glass-enclosed, observation-type cars, the effect is striking (see Figs. 32.12 and 32.13).

Additional advantages of hydraulic elevators over traction units include:

- The elevator load is carried by the ground and not the structure. By contrast, traction units place a large structural load on the penthouse and machine room floors and on overhead steel as well.
- The hoistway is smaller due to the absence of a counterweight and its guide rails.
- Cars can be lowered manually by the operation of oil valves. This is particularly useful and important in the event of control equipment failure or power failure.
- There is essentially no limit to the load that can be lifted.

The major inherent *disadvantage* of the standard hydraulic elevator is its operating expense. Because it is not counterweighted, it requires a relatively large motor to drive the oil pump, and *all* the energy is lost in heat (see Table 31.2). As an example of the operating cost, consider a 3500-lb, 125-fpm hydraulic unit in a department store. Such a unit requires a 40-hp motor. Assuming the unit to

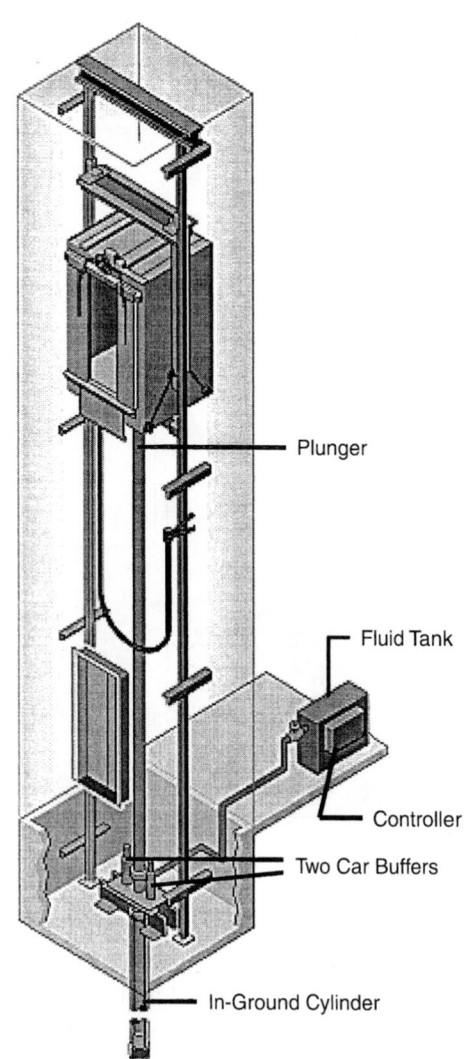

**Plunger**

**Fluid Tank**

**Controller**

**Two Car Buffers**

**In-Ground Cylinder**

***Fig. 32.3*** *Phantom view of a conventional "holed" hydraulic elevator, so called because the elevator car rests on a hydraulically activated steel plunger that descends into a hole in the ground. The hole is actually a buried hollow steel cylinder into which oil is pumped, under pressure, to raise the car. The oil pump is inside the fluid tank. (Illustration courtesy of Otis Elevator Co.)*

be in operation 10 hours a day, 6 days a week, and assuming a normal 60% time-in-operation figure, we have (remembering that the motor operates only in the up direction)

$$\text{energy used/day} = \frac{40 \text{ hp}}{0.82 \text{ eff}} \times \frac{0.746 \text{ kW}}{\text{hp}}$$
$$\times 60\% \times 10 \text{ h} \times \tfrac{1}{2}$$
$$= 110 \text{ kWh}$$

At $0.08/kWh, we have

$$\text{monthly energy cost} = 110 \times \frac{6 \text{ days}}{\text{week}}$$
$$\times \frac{4.33 \text{ weeks}}{\text{month}} \times \$0.08$$
$$= \$229/\text{month}$$

Compare this to the previously calculated (Section 31.40) monthly energy cost of $72 when using solid-state equipment and $126 with an m-g set power supply for a 3500-lb, 600-fpm traction car, for an appreciation of the value of a counterweight.

Other *disadvantages* of hydraulic units as compared to traction units include:

• They are limited to low-rise, low-speed applications.
• Ride quality is inferior to that of a good traction unit, although it is entirely acceptable for residential, mercantile, and industrial applications.
• Because oil viscosity changes with temperature, the ambient temperature of the space containing the pump and the oil storage tank must be controlled to maintain ride quality and performance.
• The high inrush current taken by the pump each time it starts, which is every time the elevator travels upward, requires a "stiff" power supply to avoid problems of light flicker and other undesirable line-voltage fluctuations.
• Noise from the pump and motor plus piping noise can be disturbing. This problem can be ameliorated by moving the pump mechanism (up to 50 ft from the elevator shaft).

Plunger-type hydraulic units have fallen into some disfavor recently because of problems caused by oil seepage into the ground. The plunger travels in a buried steel casing that, despite excellent butyl or other liquid-tight coatings, corrodes after an extended period and leaks oil. In many locations, this violates stringent EPA regulations dealing with groundwater pollution. Because repair of such an oil leak is expensive and entails extended elevator outage, other arrangements detailed in the next two sections have come into increasing use.

Hydraulic elevators are best applied to low-speed, low-rise applications such as office buildings and residential buildings up to four stories in height, low-rise department stores, malls, basement and garage shuttles, theater elevators, and stage lifts and freight applications of all sorts—in particular, those intended for very heavy loads. Another very

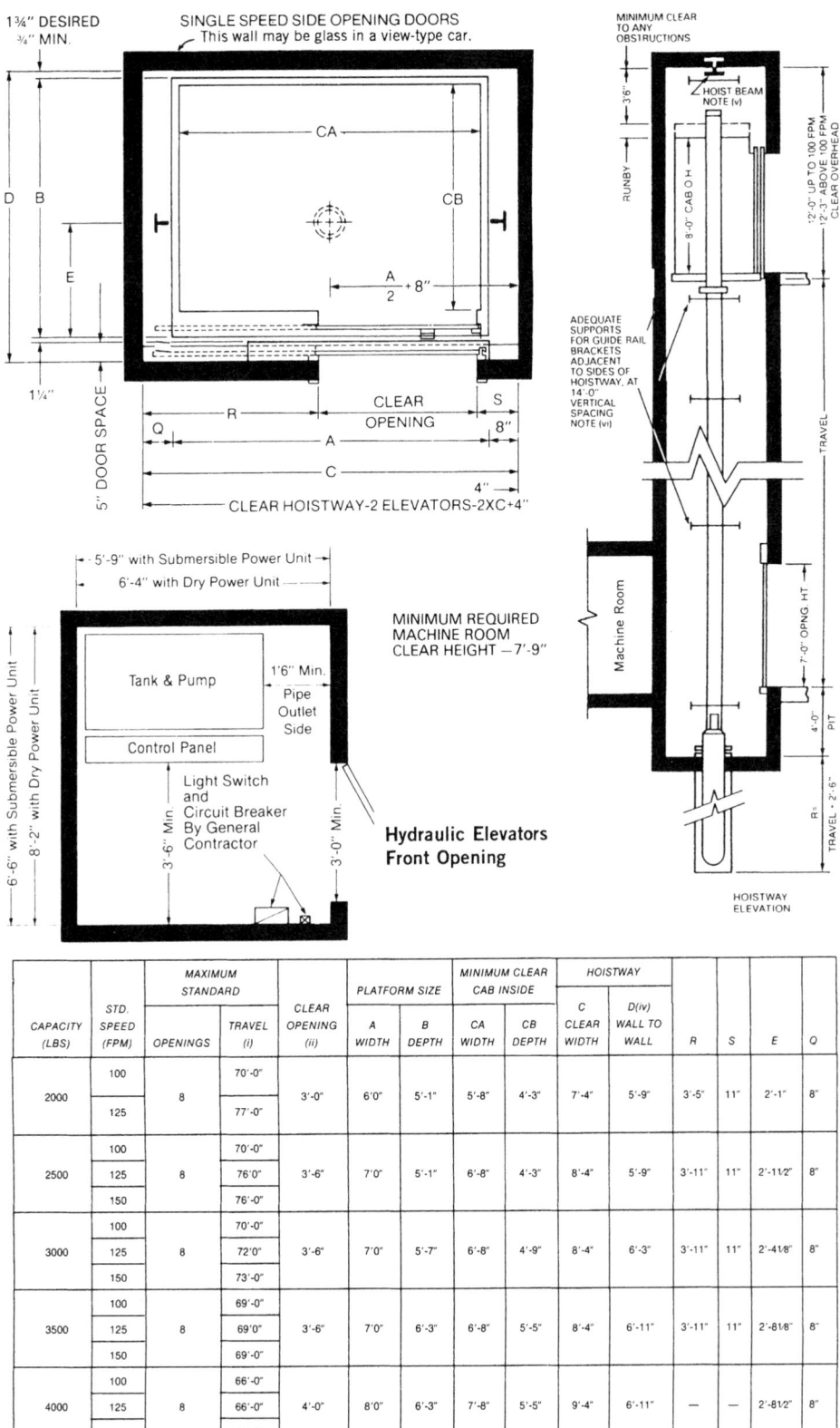

| | | MAXIMUM STANDARD | | | | PLATFORM SIZE | | MINIMUM CLEAR CAB INSIDE | | HOISTWAY | | | | | |
|---|---|---|---|---|---|---|---|---|---|---|---|---|---|---|---|
| CAPACITY (LBS) | STD. SPEED (FPM) | OPENINGS | TRAVEL (i) | CLEAR OPENING (ii) | | A WIDTH | B DEPTH | CA WIDTH | CB DEPTH | C CLEAR WIDTH | D(iv) WALL TO WALL | R | S | E | Q |
| 2000 | 100 | 8 | 70'-0" | 3'-0" | | 6'0" | 5'-1" | 5'-8" | 4'-3" | 7'-4" | 5'-9" | 3'-5" | 11" | 2'-1" | 8" |
| | 125 | | 77'-0" | | | | | | | | | | | | |
| 2500 | 100 | 8 | 70'-0" | 3'-6" | | 7'0" | 5'-1" | 6'-8" | 4'-3" | 8'-4" | 5'-9" | 3'-11" | 11" | 2'-11/2" | 8" |
| | 125 | | 76'0" | | | | | | | | | | | | |
| | 150 | | 76'-0" | | | | | | | | | | | | |
| 3000 | 100 | 8 | 70'-0" | 3'-6" | | 7'0" | 5'-7" | 6'-8" | 4'-9" | 8'-4" | 6'-3" | 3'-11" | 11" | 2'-41/8" | 8" |
| | 125 | | 72'0" | | | | | | | | | | | | |
| | 150 | | 73'-0" | | | | | | | | | | | | |
| 3500 | 100 | 8 | 69'-0" | 3'-6" | | 7'0" | 6'-3" | 6'-8" | 5'-5" | 8'-4" | 6'-11" | 3'-11" | 11" | 2'-81/8" | 8" |
| | 125 | | 69'0" | | | | | | | | | | | | |
| | 150 | | 69'-0" | | | | | | | | | | | | |
| 4000 | 100 | 8 | 66'-0" | 4'-0" | | 8'0" | 6'-3" | 7'-8" | 5'-5" | 9'-4" | 6'-11" | — | — | 2'-81/2" | 8" |
| | 125 | | 66'-0" | | | | | | | | | | | | |
| | 150 | | 66'-0" | | | | | | | | | | | | |

**Fig. 32.4** *Typical dimensional, capacity, and layout data for a conventional plunger-type hydraulic elevator. Door systems are single slide (SS) or center opening (CO). The car can be used as a viewing-type unit by utilizing a glass wall at the back of the car. (Extracted with permission from published data of Schindler Elevator Corp.)*

useful application is for the use of handicapped persons who cannot use escalators or negotiate stairs. A typical layout and dimensional data for standard plunger units are given in Fig. 32.4, along with capacities and application recommendations.

## 32.4 HOLE-LESS HYDRAULIC ELEVATORS

Where drilling a plunger hole presents difficulties, a hydraulic installation using a telescoping plunger or a roping arrangement can be installed. The telescoping jack design is shown schematically in Fig. 32.5 in a single-jack arrangement. As this causes a lateral stress in the building due to the cantilevered car, a dual-jack arrangement is used more frequently. When supported on both sides, as shown in Fig. 32.6, all of the vertical elevator load is transferred directly to the ground. The telescoping jack, in which all sections move simultaneously, is more complex than a simple plunger unit and requires more maintenance, although maintenance is simplified by the fact that the entire length of the jack is readily accessible. The ride in the telescoping jack arrangement is not as smooth as that on a straight plunger elevator due to the simultaneous movement of the jack's telescoping sections, causing a degree of jerk.

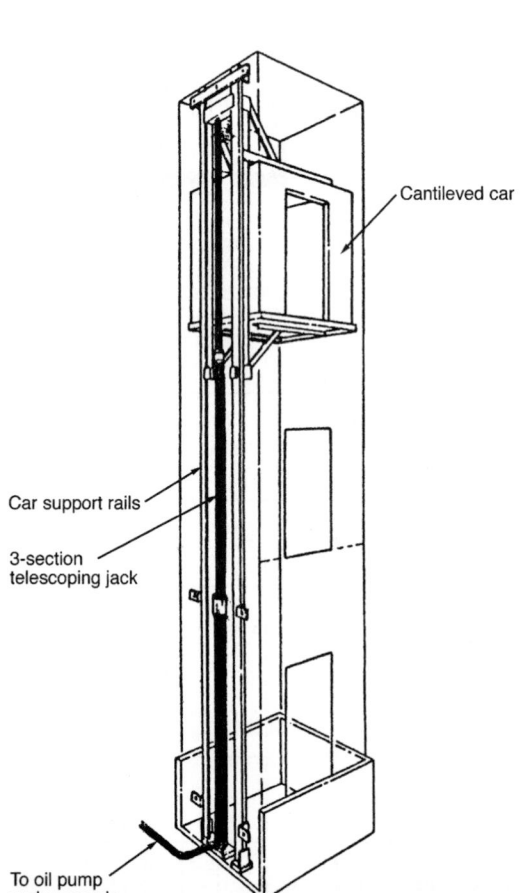

**Fig. 32.5** *Hole-less hydraulic elevator driven by a single telescoping jack. The cantilevered car exerts a lateral structural load on the building. This arrangement is used less than the two-jack mechanism shown in Fig. 32.6. (Courtesy of Otis Elevator Co.)*

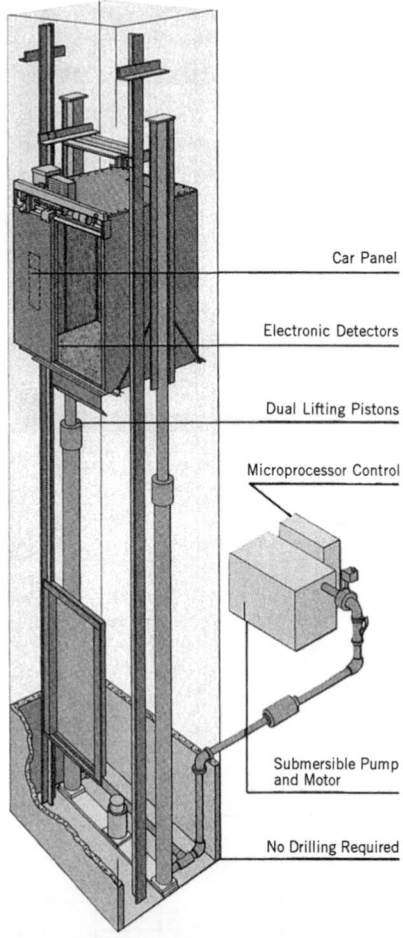

**Fig. 32.6** *Dual-jack hole-less hydraulic elevator. The balanced vertical load is borne by the ground only, with no building component. The unit illustrated is suitable only for a low-rise (two-stop) installation, with a car weighing no more than 2500 lb. (Courtesy of Otis Elevator Co.)*

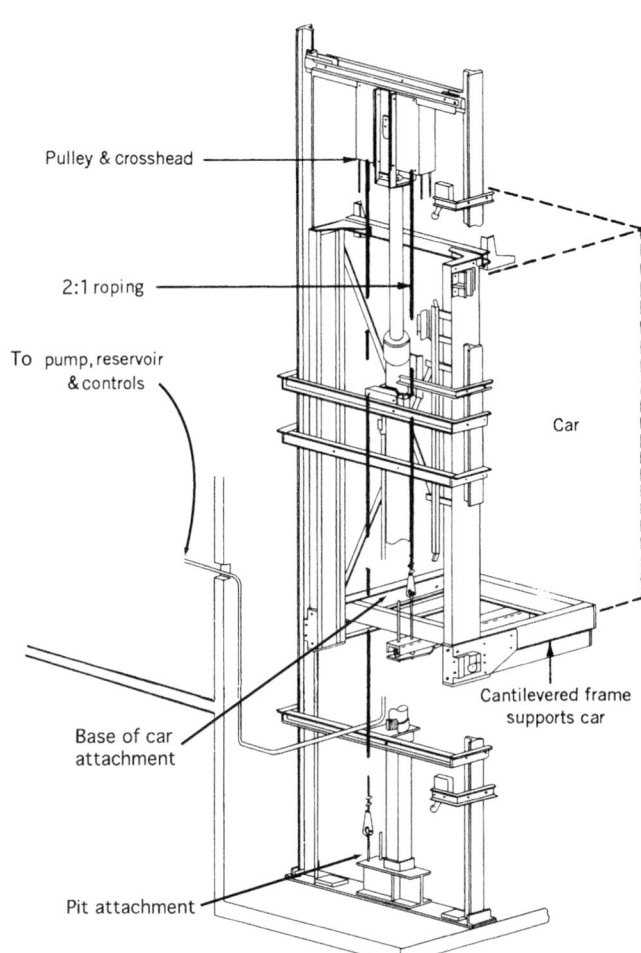

Pulley & crosshead

2:1 roping

To pump, reservoir & controls

Car

Cantilevered frame supports car

Base of car attachment

Pit attachment

**Fig. 32.7** *Low-rise residential-type elevator of the roped hydraulic type. The cantilevered car is lifted by cables from the cable crosshead, which is in turn lifted (and lowered) by the single-section telescoping piston (jack). The 2:1 roping arrangement lifts the car twice as far as the piston travels. The power unit, which includes the oil tank, pumps, and control, is usually mounted at the lower level. Control is automatic, including automatic leveling. Depending upon the specific design, the car is a 700- to 750-lb, 30- to 36-fpm unit, normally with a single-story rise. The hatchway is 16 to 26 ft$^2$, with the larger size used for a car intended to accommodate a wheelchair. (Courtesy of CEMCO, Corbett Elevator Co.)*

## 32.5 ROPED HYDRAULIC ELEVATORS

The roped hydraulic arrangement is simpler than the telescoping plunger unit because it uses only a single moving jack section compared to two or even three in the telescoping unit for the same rise. It accomplishes this by using 2:1 roping, which means that the car travels twice as far as the piston. This is accomplished by passing the rope over a pulley in the piston crosshead. One end of the rope is attached to a fixed point in the pit below the car, and the other end is attached to the base of the car (Fig. 32.7). The piston (plunger, jack) lifts the crosshead, which in turn lifts the car twice as far.

The arrangement shown in Fig. 32.7 uses a single jack and a cantilevered car. Other arrangements use two jacks to eliminate the lateral building load (Fig. 32.8). A 2:1 roping is standard in the United States. The simplicity and reliability of the single- or double-jack roped arrangement has made it in recent years by far the most common choice for low-rise, light- to medium-duty hydraulic elevators. Because it is a roped unit, it is equipped with a slack-rope safety in addition to the other safeties used on direct-connected hydraulics.

## FREIGHT ELEVATORS

### 32.6 GENERAL INFORMATION

The preceding material, which dealt with passenger traffic, had as its prime consideration the most effective solution to the problem of transporting a given number of persons vertically. The problem with

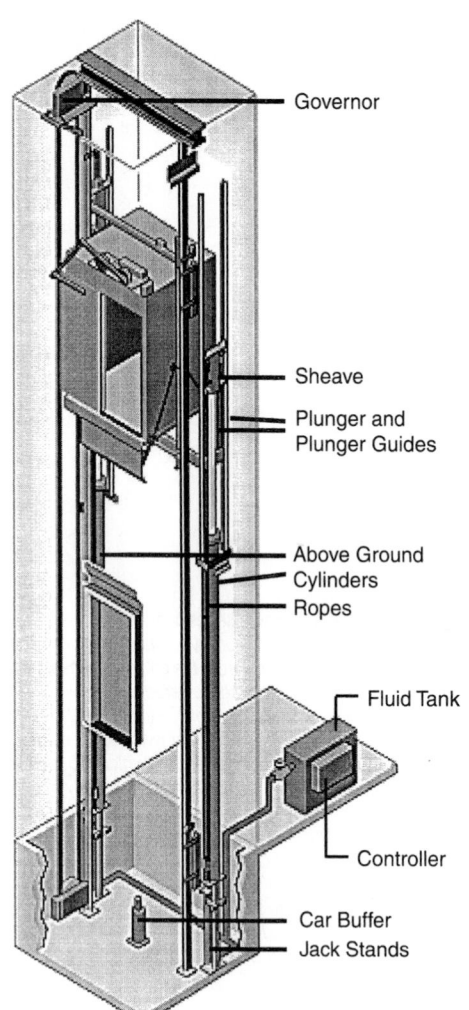

Governor

Sheave

Plunger and
Plunger Guides

Above Ground
Cylinders
Ropes

Fluid Tank

Controller

Car Buffer
Jack Stands

*Fig. 32.8 For heavier loads than can be readily handled by the cantilevered car in a single-jack design (Fig. 32.7), a balanced dual jack (cylinder) is used, as seen in this phantom view. This design, which can accomplish somewhat higher rise than a single-jack design, in addition to carrying much heavier loads, also requires a larger shaftway to accommodate the second cylinder and rails. (The second pulley and crosshead are omitted for clarity in the drawing.) (Courtesy of Otis Elevator Co.)*

respect to freight elevators is similar: to transport a given tonnage of freight efficiently, economically, and quickly. The service car in a facility can be considered to be a freight car but, if utilized for passenger duty at all, it must meet passenger service requirements. If passenger duty is not required or if much freight is to be handled, a car designed specifically for freight is used.

Factors to be considered in freight elevator selection, in addition to tonnage movement per hour, are

size of load, method of loading, travel, type of load, type of doors, and speed and capacity of cars. Due to the interrelation of these factors, the actual process of selection involves making assumptions on the basis of recommendations and then arriving at a solution, as is done for passenger elevators.

A detailed discussion of the selection of material-handling elevators is beyond the scope of this book because of the large number of considerations involved. Therefore, we restrict the following sections to descriptive material and recommendations. Also, inasmuch as freight elevators form such an important link in industrial processes, a careful and detailed material-flow study should be made before freight elevators are selected. Elevator manufacturers' representatives and materials-handling consultants can be very helpful in this regard.

## 32.7 FREIGHT CAR CAPACITY

Figure 32.9 is a section through a typical traction-type freight car shaft. Capacities corresponding to a specific platform size are due to the varying square-foot loads that are permissible. Cognizance of this is taken by the ASME Standard A17.1, which has established five load classifications for freight elevators:

*Class A.* General Freight Loading by hand truck. Single items may not exceed 25% of the car-rated load. The rated load is based on 50 pounds per square foot (psf) of net inside platform area.

*Class B.* Motor Vehicle Loading. The elevator car will carry automobiles or automobile trucks. The rating is based on a load of 30 psf of net inside platform area.

*Class C1.* Industrial truck loading; truck carried.

*Class C2.* Industrial truck loading; truck not carried.

*Class C3.* Concentrated loading; no truck used; increments greater than 25% rated capacity.

For classes C1, C2, and C3 the rated load is based on 50 psf. Cars have automatic leveling.

## 32.8 FREIGHT ELEVATOR DESCRIPTION

As speeds are generally between 50 and 200 fpm, a geared-type traction machine or a hydraulic unit is

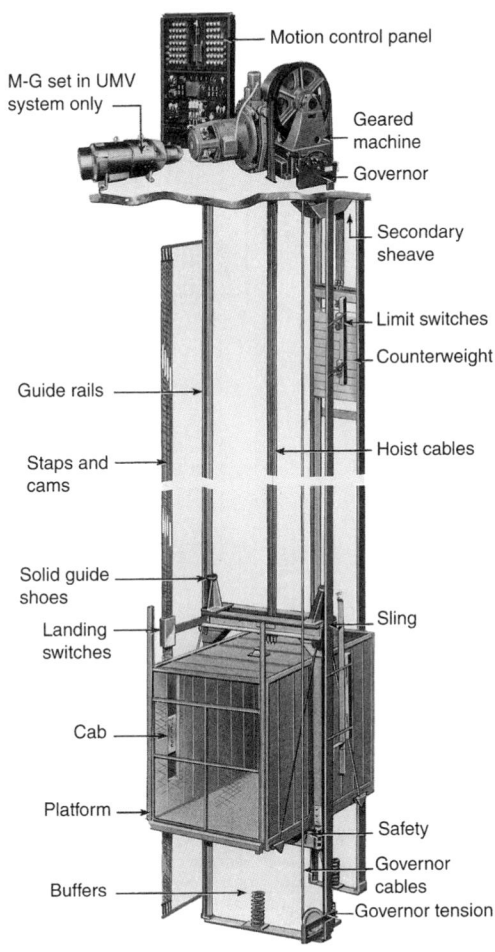

*Fig. 32.9 Components of a typical freight elevator installation utilizing a variable-voltage-controlled, geared traction machine. The sling that lifts the car is frequently arranged with double sheaves over which the hoisting ropes pass. This roping arrangement increases the mechanical advantage of the lifting ropes.*

similar to those for passenger elevators previously described.

General-purpose freight elevators, whether traction (Fig. 32.10) or hydraulic (Fig. 32.11), in load ranges of up to 20,000 lb, are standard design items applicable to all types of commercial and industrial buildings. Heavier units are individually engineered. Units of 20,000 lb and more require special safeties. As with passenger elevators, structural reactions for traction units are supplied by the manufacturer to the architect, who is responsible for providing adequate structural supports. This item is of great importance in larger car installations, because traction unit rails must be supported every few feet and additional steel provided to accomplish this.

## 32.9 FREIGHT ELEVATOR CARS, GATES, AND DOORS

Cars for freight service are normally built of heavy-gauge steel with a multilayer wooden floor, the entire unit being designed for hard service. Guarded ceiling light fixtures are required. Car gates slide vertically and are a minimum of 6 ft high. Hoistway doors are normally vertical lift, center-opening, manual or power-operated. Both car gate and hoistway doors are counterweighted and open fully to give complete floor and head clearance.

## 32.10 FREIGHT ELEVATOR COST DATA

The cost of a freight elevator installation, as with passenger elevator installation, is dependent upon many factors, principally capacity, type of control, use, and type of door operation.

As exact pricing, like actual selection, is outside the scope of our discussion, we recommend that a reputable manufacturer or elevator consultant be sought for such information. We can, however, make some general remarks on pricing, as follows:

1. Variable-voltage-controlled equipment, depending on the type, costs 20% to 50% more than rheostatically controlled equipment.
2. Above a basic two-floor rise, the cost increases linearly with rise.

used. The preferred system of control is collective, with a variable-voltage, dc supply, either (UMV) or VVVF. If the car is used infrequently (fewer than five trips a day), economy is very important, accurate leveling is not essential, and a rougher ride is tolerable, then a two-speed ac rheostatic control may be used.

For low-rise installations, a hydraulic unit is most often employed. These, like the variable-voltage dc traction units, provide accurate control, smooth operation, and accurate automatic leveling. Hydraulic units rarely exceed 60 ft in height and operate at speeds of up to 125 fpm. Accessories such as governors, safeties, and brakes are

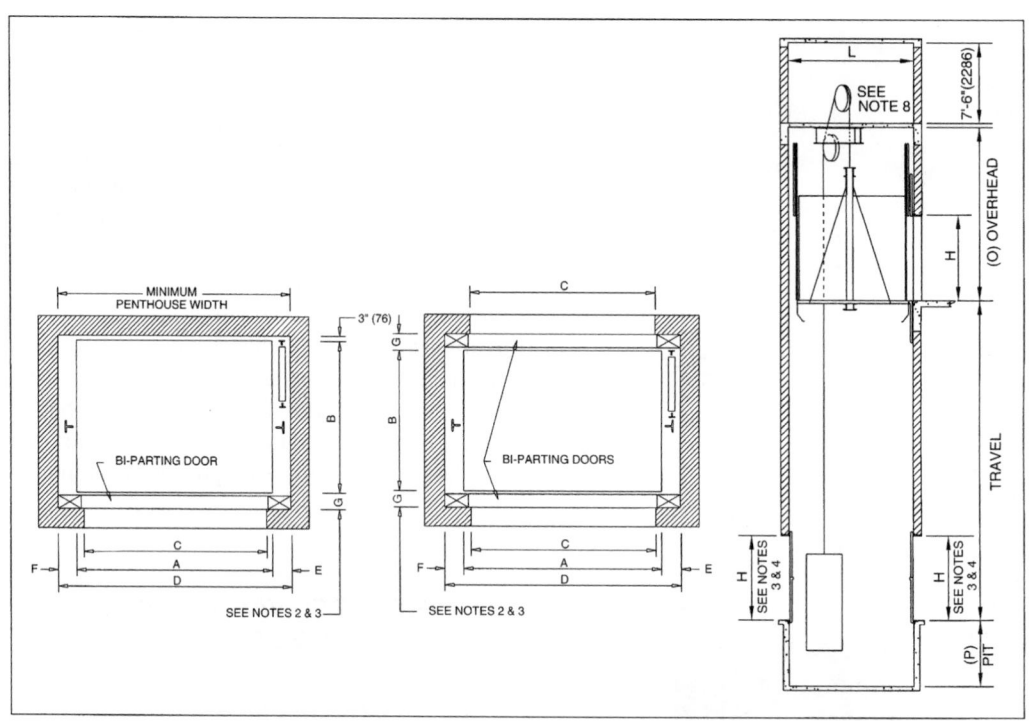

| | LIGHT AND MEDIUM DUTY FREIGHT ELEVATORS | | | | |
|---|---|---|---|---|---|
| CAPACITY | 2500 lbs. (1134 kg) | 3000 lbs. (1361 kg) | 4000 lbs. (1814 kg) | 6000 lbs. (2722 kg) | 8000 lbs. (3629 kg) |
| A | 5'-4" (1626) | 6'-4" (1930) | 6'-4" (1930) | 8'-4" (2540) | 8'-4" (2540) |
| B | 7'-0" (2134) | 8'-0" (2438) | 8'-0" (2438) | 10'-0" (3048) | 10'-0" (3048) |
| C | 5'-0" (1524) | 6'-0" (1829) | 6'-0" (1829) | 8'-0" (2438) | 8'-0" (2438) |
| D | 7'-10"(2388) | 8'-10"(2692) | 8'-10"(2692) | 10'-10"(3302) | 10'-10"(3302) |
| E | 1'-7" (483) | 1'-7" (483) | 1'-7" (483) | 1'-7" (483) | 1'-7" (483) |
| F | 0'-11" (279) | 0'-11" (279) | 0'-11" (279) | 0'-11" (279) | 0'-11" (279) |
| L | 13'-0" (3962) | 14'-0" (4267) | 14'-0" (4267) | 14'-0" (4267) | 14'-0" (4267) |

| | HEAVY DUTY POWER TRUCK LOADING FREIGHT ELEVATORS | | | | |
|---|---|---|---|---|---|
| CAPACITY | 10,000 lbs. (4536 kg) | 12,000 lbs. (5443 kg) | 16,000 lbs. (7258 kg) | 18,000 lbs. (8165 kg) | 20,000 lbs. (9072 kg) |
| A | 8'-4" (2540) | 10'-4" (3150) | 10'-4" (3150) | 10'-4" (3150) | 12'-4" (3759) |
| B | 12'-0" (3658) | 14'-0" (4267) | 14'-0" (4267) | 16'-0" (4877) | 20'-4" (6198) |
| C | 8'-0" (2438) | 10'-0" (3048) | 10'-0" (3048) | 10'-0" (3048) | 12'-0" (3658) |
| *D | 11'-4" (3454) | 13'-6" (4115) | 14'-0" (4267) | 14'-2" (4318) | 16'-6" (5029) |
| E | 1'-7" (483) | 1'-7" (483) | 1'-7" (483) | 1'-7" (483) | 1'-7" (483) |
| F | 0'-11" (279) | 0'-11" (279) | 0'-11" (279) | 0'-11" (279) | 0'-11" (279) |
| L | 14'-0" (4267) | 15'-0" (4572) | 15'-0" (4572) | 17'-0" (5182) | 21'-0" (6400) |

| MINIMUM OVERHEAD & PIT DIMENSIONS FOR LIGHT & MEDIUM DUTY FREIGHT ELEVATORS | | | | |
|---|---|---|---|---|
| SPEED | 50 FPM (0.25 m/s) | 75 FPM (0.38 m/s) | 100 FPM (0.51 m/s) | 200 FPM (1.02 m/s) |
| O | 16'-0"(4877) | 16'-0" (4877) | 16'-0"(4877) | 16'-0"(4877) |
| P | 5'-6"(1676) | 5'-6" (1676) | 5'-6"(1676) | 6'-0"(1829) |

NOTES:
2. Dimension G = 5" (127) for regular type counterbalanced hoistway doors and 6 3/4" (172) for pass type counterbalanced hoistway doors.
3. Pass type hoistway doors are required when floor heights are less than 11'-0" (3353) for 7'-0" (2134) openings and less than 12'-6" (3810) for 8'-0" (2438) openings.
4. Dimension H = 7'-0" (2134) on light and medium duty and 8'-0" (2438) (or as required) for heavy-duty doors. Doors higher than 8'-0" (2438) require additional overhead.
8. 2:1 Roping recommended for heavy duty power truck loading freight elevators
* D dimension includes space required when vertical columns are added (inside hoistway) for rail support. If no columns required, D=A+E+F.

**Fig. 32.10** *Typical dimensional data for traction-type freight elevators. (Courtesy of Montgomery-KONE.)*

3. Electric door operation can increase the cost of a car installation 10% to 25%.

As an example of *comparative* pricing, using a nominal 100 for an 8000-lb, 75-fpm, four-floor, manual-door car with ac rheostatic control and automatic leveling, the same car with variable-voltage control, 150 fpm, and electrically operated doors would cost approximately 180.

## SPECIAL ELEVATOR DESIGNS

### 32.11 OBSERVATION CARS

By placing a traction lifting mechanism *behind* the car, attaching the car at the back, and using a glass-enclosed, observation-style car, a spectacular unit

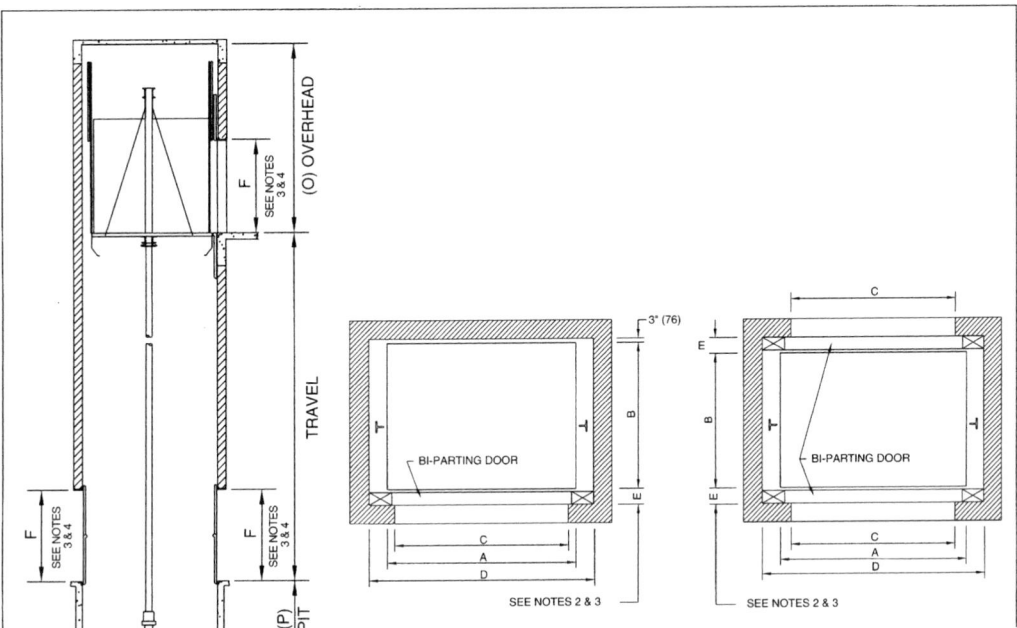

| LIGHT AND MEDIUM DUTY FREIGHT ELEVATORS | | | | | | |
|---|---|---|---|---|---|---|
| CAPACITY | 2000 lbs. (907 kg) | 3000 lbs. (1361 kg) | 4000 lbs. (1814 kg) | 5000 lbs. (2268 kg) | 6000 lbs. (2722kg) | 8000 lbs. (3629 kg) |
| A | 5'-0" (1524) | 5'-6" (1676) | 6'-6" (1981) | 8'-6" (2591) | 8'-6" (2591) | 8'-6" (2591) |
| B | 6'-0" (1829) | 7'-0" (2134) | 8'-0" (2438) | 10'-0" (3048) | 12'-0" (3658) | 12'-0" (3658) |
| C | 4'-8" (1422) | 5'-2" (1575) | 6'-2" (1880) | 8'-2" (2490) | 8'-2" (2490) | 8'-2" (2490) |
| D: manual doors | 6'-4" (1930) | 6'-10" (2083) | 7'-10" (2388) | 9'-10" (2997) | 10'-0" (3048) | 10'-6" (3200) |
| D: power doors | 6'-10" (2083) | 7'-4" (2235) | 8'-4" (2540) | 10'-4" (3150) | 10'-6" (3200) | 10'-6" (3200) |
| O: 7'-0" door ht. | 13'-2" (4013) | 13'-2" (4013) | 13'-2" (4013) | 13'-2" (4013) | 13'-2" (4013) | 13'-2" (4013) |
| P: 7'-0" door ht. | 4'-0" (2134) | 4'-0" (1219) | 4'-0" (1219) | 4'-6" (1372) | 4'-6" (1372) | 5'-0" (1524) |

| HEAVY DUTY POWER TRUCK LOADING FREIGHT ELEVATORS | | | | | |
|---|---|---|---|---|---|
| CAPACITY | 10,000 lbs. (4536 kg) | 12,000 lbs. (5443 kg) | 16,000 lbs. (7258 kg) | 18,000 lbs. (8165 kg) | 20,000 lbs. (9072 kg) |
| A | 10'-6" (3200) | 10'-6" (3200) | 10'-6" (3200) | 10'-6" (3200) | 12'-6" (3810) |
| B | 14'-0" (4267) | 14'-0" (4267) | 16'-0" (4877) | 16'-0" (4877) | 20'-0" (6096) |
| C | 10'-2" (3098) | 10'-2" (3098) | 10'-2" (3098) | 10'-2" (3098) | 12'-2" (3708) |
| D: manual doors | 12'-6" (3810) | 12'-6" (3810) | 12'-6" (3810) | 12'-6" (3810) | 14'-6" (4420) |
| D: power doors | 12'-6" (3810) | 12'-6" (3810) | 12'-6" (3810) | 12'-6" (3810) | 14'-6" (4420) |
| O: 7'-0" door ht. | 13'-2" (4013) | 13'-2" (4013) | 13'-2" (4013) | 13'-2" (4013) | 13'-2" (4013) |
| O: 8'-0" door ht. | 14'-2" (4318) | 14'-2" (4318) | 14'-2" (4318) | 14'-2" (4318) | 14'-2" (4318) |
| P: 7'-0" door ht. | 5'-0" (2134) | 6'-0" (1829) | 6'-0" (1829) | 6'-0" (1829) | 6'-0" (1829) |
| P: 8'-0" door ht. | 5'-0" (2438) | 6'-0" (1829) | 6'-0" (1829) | 6'-0" (1829) | 6'-0" (1829) |

NOTES:

2. Dimension E = 5" (127) for regular type counterbalanced hoistway doors and 6¾" (172) for pass type counterbalanced hoistway doors.

3. Pass type hoistway doors are required when floor heights are less than 11'-0" (3353) for 7'-0" (2134) openings and less than 12'-6" (3810) for 8'-0" (2438) openings.

4. Layout and dimensions shown for freight elevators based on bi-parting counterbalanced type hoistway doors – dimension F.

5. Dimensions O and P are minimums based on car speeds up to 150 FPM (0.76 m/s).

**Fig. 32.11** *Typical dimensional data for a hydraulic freight elevators. (Courtesy of Montgomery-KONE.)*

*(a)*

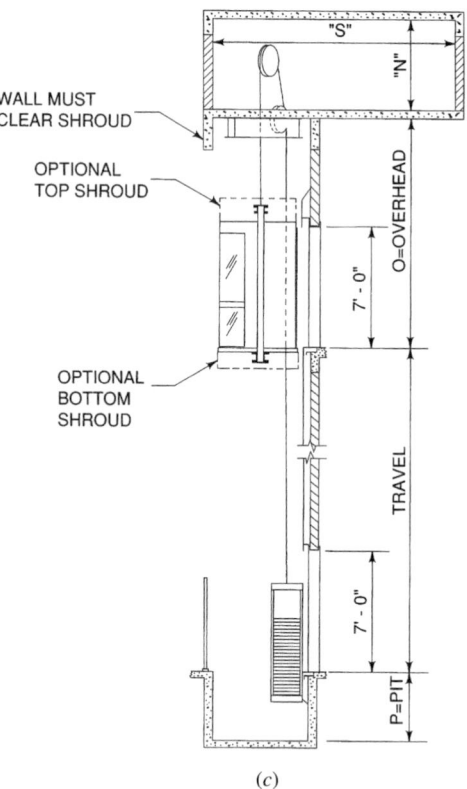

WALL MUST
CLEAR SHROUD

OPTIONAL
TOP SHROUD

OPTIONAL
BOTTOM
SHROUD

"S"

"N"

O=OVERHEAD

7' - 0"

TRAVEL

7' - 0"

P=PIT

*(c)*

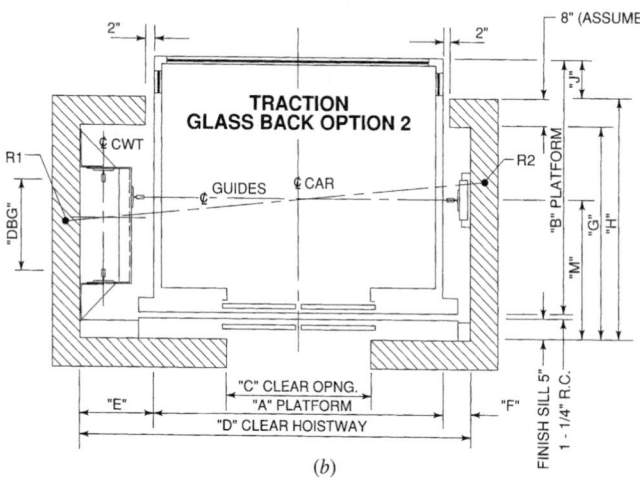

2"          2"

8" (ASSUMED)

**TRACTION
GLASS BACK OPTION 2**

₵ CWT

R1

₵ GUIDES    ₵ CAR

R2

"DBG"

"B" PLATFORM

"J"

"G"

"H"

"M"

"E"    "C" CLEAR OPNG.    "F"
       "A" PLATFORM
"D" CLEAR HOISTWAY

FINISH SILL 5"
1 - 1/4" R.C.

*(b)*

**Notes (Hydraulic and Traction)**
State and local code requirements may vary. Data is for
general application.

All layout details are based upon the use of center opening
type entrances.

Elevator machine room temperature should be maintained
between 65°F and 100°F.

Please consult Montgomery KONE for exact information
for working drawings and for designs which may vary from
those depicted.

*(d)*

**Fig. 32.12** *Details of a geared, traction-type observation elevator. (a) The glass-enclosed observation portion of this elevator faces the rear. The shaftway doors are seen in the wall behind the elevator. The traction ropes are also clearly seen above each elevator. (b) Plan drawing of the unit. Note that the counterweight is mounted at the side of the car so as to be essentially invisible and not detract from the relatively uncluttered view of the car in motion. (c) Vertical section of the installation, including the overhead machine room. (d) Notes applicable to drawings (b) and (c). (e) Dimensions applicable to drawings (b) and (c). (f) Vertical reactions required for structural design. (Courtesy of Montgomery-KONE.)*

TRANSPORTATION

| BASIC DIMENSIONS: TRACTION OBSERVATION ELEVATORS | | | | | |
|---|---|---|---|---|---|
| DIMENSION | CAPACITY | | | | |
| | 3000 lbs. | | 3500 lbs. | | 4500 lbs. |
| | Option 2 | | Option 2 | | Option 2 |
| "DBG" | 25" | | 31" | | 37" |
| A | 7' - 0" | | 7' - 0" | | 7' - 0" |
| B | 5' - 8" | | 6' - 6" | | 7' - 6" |
| C | 3' - 6" | | 3' - 6" | | 3' - 6" |
| D | 9' - 5" | | 9' - 5" | | 9' - 5" |
| E | 1' - 8" | | 1' - 8" | | 1' - 8" |
| F | 9" | | 9" | | 9" |
| G | 4' - 7¹/₂" | | 5' - 1¹/₂" | | 5' - 7¹/₂" |
| H | 5' - 3¹/₂" | | 5' - 9¹/₂ | | 6' - 3¹/₂" |
| J | 10³/₄" | | 1' - 2³/₄" | | 1' - 8³/₄" |
| K | NA | | NA | | NA |
| L | NA | | NA | | NA |
| M | 3' - 4¹/₂" | | 3' - 10¹/₂" | | 4' - 4¹/₂" |
| N | 7' - 6" | | 7' - 6" | | 7' - 6" |
| S | 13' - 8" | | 13' - 8" | | 13' - 8" |
| SPEED | 200 fpm | 350 fpm | 200 fpm | 350 fpm | 200 fpm | 350 fpm |
| *O(1) | 14' - 8" | 15' - 4" | 14' - 8" | 15' - 4" | 15' - 2" | 15' - 6" |
| **O(1) | 15' - 2" | 15' - 10" | 15' - 2" | 15' - 10" | 15' - 2" | 15' - 6" |
| *P | 5' - 0" | 5' - 6" | 5' - 0" | 5' - 6" | 5' - 4" | 5' - 10" |
| **P | 5' - 4" | 5' - 10" | 5' - 4" | 5' - 10" | 5' - 4" | 5' - 10" |

NOTE: * Dimension when no shroud is used.
     ** Dimension when 2' - 0" shroud is used.
    (1) Based on standard height cab (8' - 0" to underside of canopy).
    → The relationships between dimensions G, H, J and K are based on an assumed thickness of 8" for the rear wall of the hoistway.
    → The hoistway dimensions shown are based on a counterweight not requiring safeties.

(*e*)

| VERTICAL REACTIONS | TRACTION ELEVATOR CAPACITY | | | |
|---|---|---|---|---|
| | 2500 lbs. | 3000 lbs. | 3500 lbs. | 4500 lbs. |
| R1 | 17500 lbs. | 18900 lbs | 20900 lbs. | 24400 lbs. |
| R2 | 13900 lbs. | 15000 lbs. | 16300 lbs. | 19100 lbs. |

NOTE: Reactions include allowance for impact but DO NOT include weight of concrete slab. Reactions are for preliminary use only. Exact reactions will be provided when exact conditions are known.

(*f*)

**Fig. 32.12** (Continued)

can be constructed that becomes an attraction in itself. Basic construction and examples are seen in Figs. 32.12 and 32.13. If the back screen is treated properly, the car gives the impression of movement with no apparent motive force or machinery.

The same effect can be accomplished by using a single-jack hydraulic lift mechanism and a cantilevered car, as in Figs. 32.9 and 32.11. Observation cars, which are placed on the *outside* of a wall, do not require shaft space. Thus, in addition to an interesting and attractive appearance, they increase the amount of usable interior space, thereby effecting a considerable cost economy.

## 32.12 INCLINED ELEVATORS

Although elevators are normally conceived as traveling vertically, this is not necessarily so. Slant or inclined elevators have been constructed in numer-

(a)

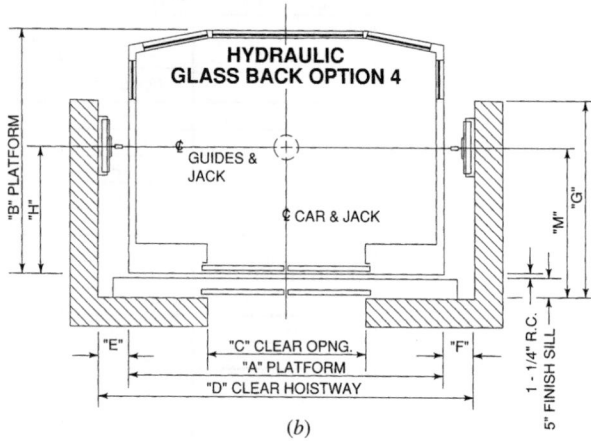

HYDRAULIC
GLASS BACK OPTION 4

GUIDES &
JACK

℄ CAR & JACK

"B" PLATFORM
"H"

"M"

"G"

1 - 1/4" R.C.

5" FINISH SILL

"E"     "C" CLEAR OPNG.     "F"
        "A" PLATFORM
        "D" CLEAR HOISTWAY

(b)

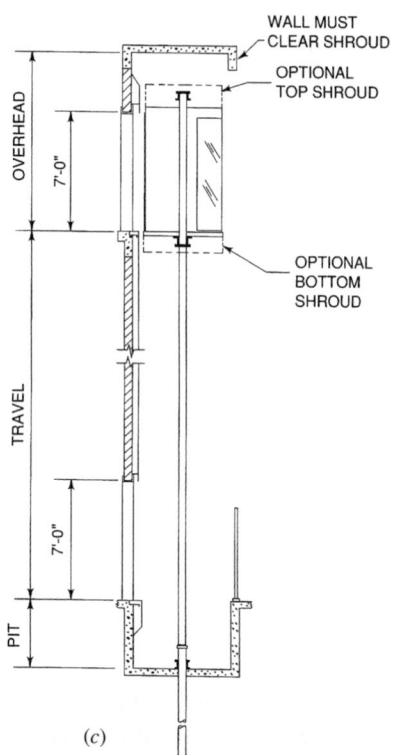

WALL MUST CLEAR SHROUD

OPTIONAL TOP SHROUD

OPTIONAL BOTTOM SHROUD

OVERHEAD

7'-0"

TRAVEL

7'-0"

PIT

(c)

| BASIC DIMENSIONS: HYDRAULIC OBSERVATION ELEVATORS | | | | |
|---|---|---|---|---|
| | CAPACITY | | | |
| | 2500 lbs. | 3000 lbs. | 3500 lbs. | |
| Dimension | OPTION 1,2,3,4 | OPTION 1,2,3,4 | OPTION 1,2,3,4 | OPTION 5 |
| A | 7' - 0" | 7' - 0" | 7' - 0" | 7' - 0" |
| B | 5' - 3" | 5' - 11" | 6' - 6" | 7' - 2" |
| C | 3' - 6" | 3' - 6" | 3' - 6" | 3' - 6" |
| D | 8' - 4" | 8' - 4" | 8' - 4" | 8' - 4" |
| E | 8" | 8" | 8" | 8" |
| F | 8" | 8" | 8" | 8" |
| G | 4' - 3" | 4' - 7" | 4' - 10$\frac{1}{2}$" | 4' - 10$\frac{1}{2}$" |
| H | 2' - 8$\frac{3}{4}$" | 3' - 0$\frac{3}{4}$" | 3' - 4$\frac{1}{4}$" | 3' - 4$\frac{1}{4}$" |
| M | 3' - 3" | 3' - 7" | 3' - 10$\frac{1}{2}$" | 3' - 10$\frac{1}{2}$" |

Basic dimensions outlined above are based upon conventional inground jack application.

(d)

| HYDRAULIC MACHINE ROOM SIZES (Typically recommended) | | |
|---|---|---|
| | CAPACITY | |
| SIZE | 2500 lbs. & 3000 lbs. | 3500lbs. |
| WIDTH | 9' - 0" | 9' - 6" |
| LENGTH | 6' - 8" | 6' - 9" |
| DOOR | 3' - 6" × 7' - 0" | |

NOTE: Hydraulic machine room location should be at the lowest landing adjacent to the hoistway. Consult your Montgomery KONE Professional for alternative locations and optimum sizes. Material above is typical for 150 FPM speed.

(e)

**Fig. 32.13** *Hydraulic observation elevator. (a) The hydraulic plunger is clearly visible below each car against the background of the shaftway doors. (b) Dimensioned sectional drawing of the car and its enclosure. (c) Vertical section of the elevator travel. (d) Dimensions applicable to drawings (b) and (c). (e) Dimensional data for the required machine room. Note that in drawing (c), the machine room is not shown because it can be remote from the elevators, connected only by piping (see Figs. 32.4 through 32.7). (Courtesy of Montgomery-KONE.)*

Acrylic dome enclosure

(a)

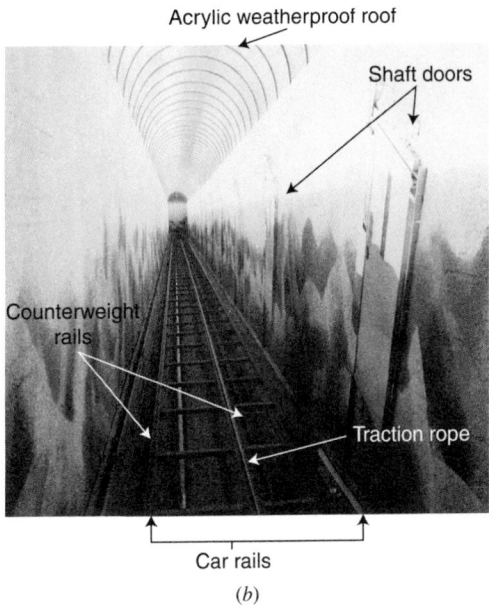

Acrylic weatherproof roof

Shaft doors

Counterweight rails

Traction rope

Car rails

(b)

**Fig. 32.14** *(a) Exterior inclined elevator enclosure, seen from inside the 14-story hotel it serves in Zichron Yaakov, Israel. The shaft is inclined at 26° and contains a single 10-passenger car. (b) Inclined elevator shaft containing rails (tracks) for the elevator, its counterweight, and the traction ropes. Shaft doors are swinging type, interlocked with the car. (Photos by B. Stein.)*

**Fig. 32.15** *The St. Louis Gateway Arch has a 10-passenger inclined elevator in each leg. Placement of doors, arrangement of the counterweight, and size of the shaft all depend upon the angle of incline. The car moves 82 ft horizontally during its 386 ft of total travel at an incline of approximately 12°. (Photo courtesy of Bethlehem Steel Corporation, which supplied the elevator rope for this installation.)*

ous locations where a building is built onto an inclined surface, usually a hillside or mountainside. One such elevator is shown in Fig. 32.14. Another well-known example is shown in Fig. 32.15. The motive mechanism used varies with the angle of inclination. In most instances, the car rides on inclined rails and is pulled up by a traction cable. It is counterweighted either by a weight riding on another set of rails (Fig. 32.14), in the case of a single car, or by the weight of another car in a two-car installation.

## 32.13 RACK AND PINION ELEVATORS

These operate on the straightforward principle of a rotating cogged-wheel pinion attached to a vertical rack. Rotation of the pinion forces the attached car up and down along the rack (see Fig. 32.16). The advantages of this system are its inherent simplicity and safety, unlimited rise, and low maintenance and operating costs, plus minimum space requirements. The last characteristic was primarily responsible for the selection of a rack and pinion design for

**Fig. 32.16** *The basis of the rack and pinion drive is simply a driven pinion on a stationary rack; rotation of the pinion forces the structure attached to the pinion to move along the rack. (Courtesy of Alimak Elevator Co.)*

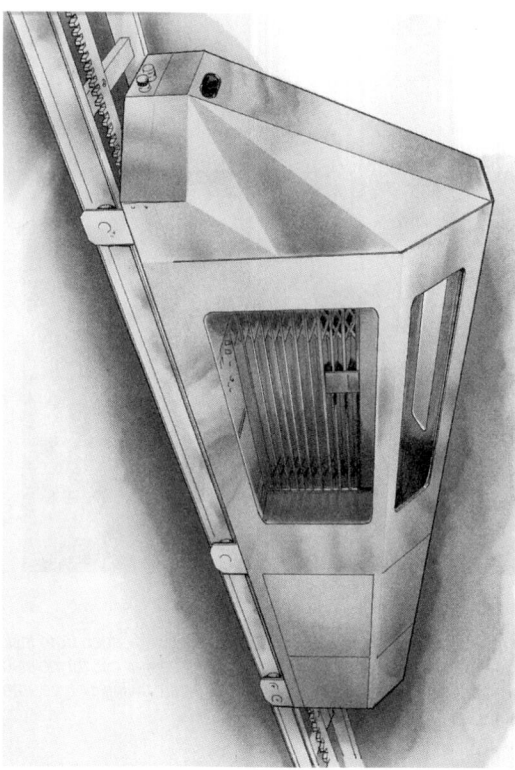

**Fig. 32.17** *The rack and pinion principle is used in the very small Statue of Liberty emergency elevator. The rack extends the entire length of the shaft and the car moves along the rack, driven by two motorized pinions below the passenger compartment. (Courtesy of Alimak Elevator Co.)*

the 210-ft-rise emergency elevator installed as part of the renovation and rehabilitation of the Statue of Liberty in 1986.

That elevator is used primarily to evacuate victims of heart attacks. Carrying them down the 171-step spiral staircase that connects the main upper landing to the observation platform in the crown is obviously not practical. The tight headroom available—less than 9 ft—ruled out a traction unit, as did the extremely tight shaftway, which measures only 2 ft, ½ in. × 4 ft, 10 in. The car itself (see Fig. 32.17) measures only 1 ft, 11 in. × 3 ft, 10 in. (outside dimensions). Space was so tight that the drive was divided between two motors, both of which are mounted under the passenger compartment. Rack and pinion lifts are used both indoors and outdoors primarily in industrial environments for both passengers and material vertical transport.

## 32.14 RESIDENTIAL ELEVATORS AND CHAIR LIFTS

Although the special needs of the handicapped have been widely recognized officially in legislated

requirements only since the late 1970s, the elevator industry has been providing for the handicapped for years on a voluntary basis. Small private-residence elevators can double as wheelchair lifts and are available in a wide range of designs, including winding-drum units (Figs. 32.18 and 32.19), roped hydraulics (Figs. 32.7 and 32.8), and worm and screw units. Standard traction designs are used infrequently due to the overhead space requirement. Similarly, standard hydraulics requiring a plunger bore hole are rarely used.

In recognition of the fact that most residential elevators (and elevators intended primarily for use of the disabled in public and private buildings) are low-speed, limited-load units, the ASME developed a new section for the elevator code (A17.1b part XXV) to cover such units. The section applies to Limited Use/Limited Application (LU/LA) elevators, defined as "a power passenger elevator where the use and application is limited by size, capacity, speed

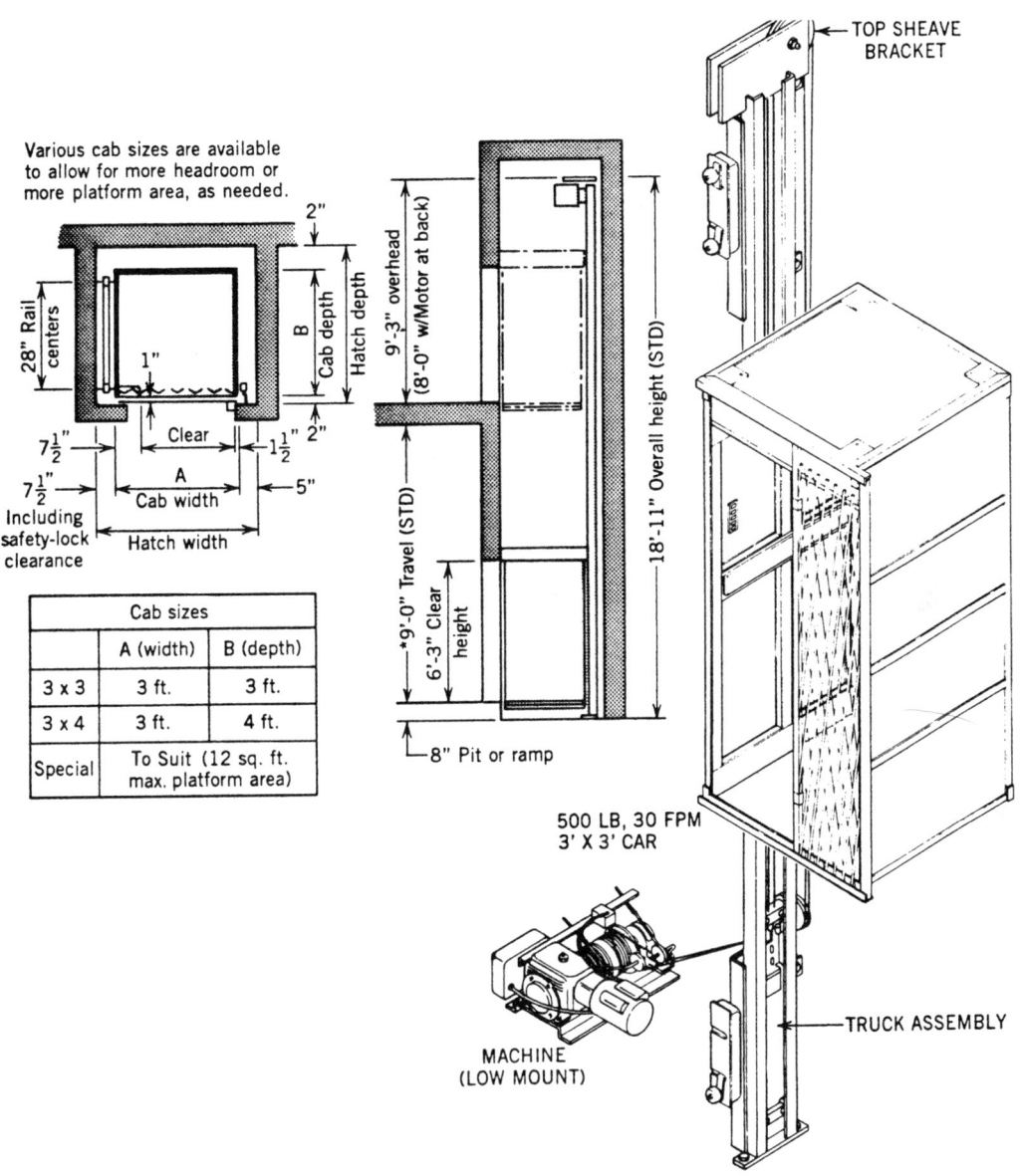

Various cab sizes are available
to allow for more headroom or
more platform area, as needed.

| Cab sizes | | |
|---|---|---|
| | A (width) | B (depth) |
| 3 x 3 | 3 ft. | 3 ft. |
| 3 x 4 | 3 ft. | 4 ft. |
| Special | To Suit (12 sq. ft. max. platform area) | |

500 LB, 30 FPM
3' X 3' CAR

MACHINE
(LOW MOUNT)

TOP SHEAVE
BRACKET

TRUCK ASSEMBLY

*Fig. 32.18* Isometric drawing of a typical residential elevator of the winding-drum design. Units of this type are usually limited to 500-lb car capacity, 40-ft rise, and 30-fpm speed. The car is rigidly attached in cantilever fashion to a rolling truck, which is raised and lowered by cables attached to the winding drum. The driving motor and winding drum shown here at the base of the assembly can be installed at any floor stop along the hoistway or overhead. Control is normally pushbutton automatic with limit-switch leveling. A 6-in. pit is required at the bottom of the hoistway. Door interlock, cable failure, and overrun safeties are standard items in such installations, which must meet elevator code requirements. (Courtesy of Waupaca Elevator Co., Inc.)

and rise, intended primarily to provide vertical transportation for people with physical disabilities." The new code section goes on to limit LU/LA elevators to a maximum size of 18 ft², a load of 1400 lb, a rise of 25 ft, and a speed of 30 fpm. Due to these limitations, safety requirements are different from those of standard traction elevators, drives may be different (e.g., winding-drum cable lifts), space requirements are smaller, and overall costs are considerably reduced. Elevators that exceed the limitations must conform to the requirements of the basic ANSI/ASME 17.1 elevator code.

TRANSPORTATION

**Fig. 32.19** *Winding-drum residential elevator specifically designed for use by handicapped persons. The car floor area is 12 ft², which accommodates a wheelchair plus another passenger. Note that the car is arranged with front and back access for maximum flexibility. (Courtesy of Waupaca Elevator Co., Inc.)*

Chair lifts and wheelchair platform lifts also come in a variety of designs. The chair lift shown in Fig. 32.20 uses a rack and pinion drive that permits it to negotiate a turn, as illustrated. For straight stair lifts, a winding-drum design similar to that used for residential elevators is often used. Other drives in use include worm gear and chain and screw worm and cog. The drive selection depends upon the load, rise, duty, and, of course, price. The wheelchair platform lift illustrated in Fig. 32.21 uses a ball screw drive. Other designs use some of the drives listed previously, including a roped-hydraulic drive for units requiring a large rise. The hydraulic scissor-jack type of wheelchair platform lift is most frequently found in public and institutional buildings because it is particularly applicable to short-rise, heavy-load, frequent-duty applications and requires minimal maintenance.

All of these units are covered by various sections of the elevator code and must be installed in accordance with code requirements, including safeties and controls. Residential units are normally arranged to operate on 120 V ac, although heavy-duty units may require 240 V service.

## 32.15 LINEAR ELEVATOR MOTOR DRIVE

A revolutionary elevator drive design developed by Otis utilizes a linear induction motor built into the counterweight frame. Motive power is therefore supplied at the counterweight, thus entirely eliminating the overhead traction machine and machine room. At this writing, the only installations using this design are in Japan. The inherent architectural, construction, energy usage, maintenance, and cost advantages of this elegant design will make it an extremely attractive option for American building designers once it has achieved the required code and jurisdictional approvals. However, because the code is predicated on the use of a traction machine, which in this design is entirely absent, an essentially new section of code must be prepared to cover this design, and in particular its safety aspects. So extensive a code revision probably will not be a short-term project. A schematic diagram of the design is shown in Fig. 32.22.

## MATERIAL HANDLING

## 32.16 GENERAL INFORMATION

The material-handling equipment discussed briefly here is that which finds application in commercial and institutional buildings. Industrial materials handling is an entirely separate subject not germane to our purpose.

The need to transport material within a building has always existed, and until about the late 1970s was done largely manually, with mechanical assistance. Thus, offices used messengers, and hospitals used dumbwaiters, service elevators, conveyors, and chutes. The single exception to this situation was (and still is) the extensive use of pneumatic tube systems in large stores, although in new facilities their use is declining. Today's sys-

(a)

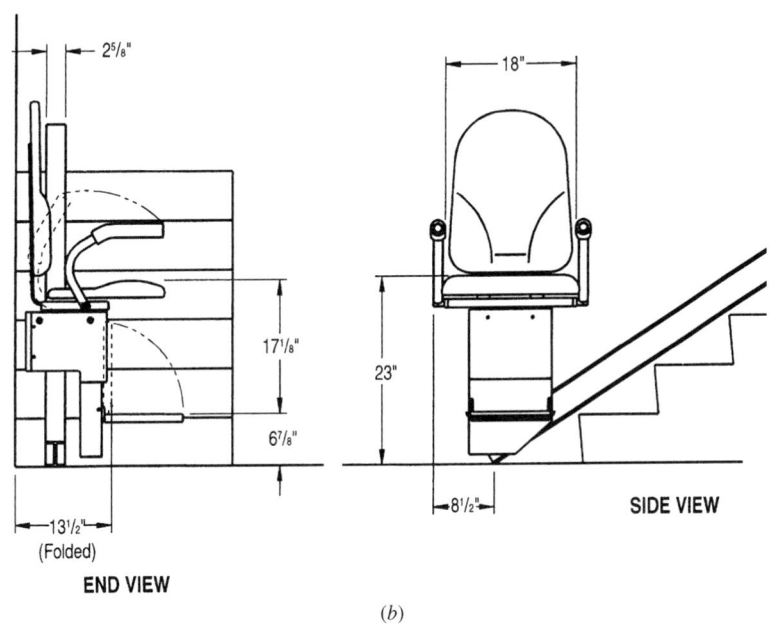

2⁵/₈"

18"

17¹/₈"

23"

6⁷/₈"

8¹/₂"

SIDE VIEW

13¹/₂"
(Folded)

END VIEW

(b)

*Fig. 32.20* (a) Chair lift with a rack and pinion drive permits the lift to negotiate a turn in the stairs. This unit is limited to a live load of 300 lb (135 kg), runs at 25 fpm (0.13 m/s), climbs at any incline up to 51°, and is driven by a ½-hp motor. Maximum rise is 60 ft. (b) Dimensional data for the chair lift shown. (Courtesy of Access Industries, Inc.)

tems accomplish the same end—that is, the transfer of materials—but automatically and, in general, much more rapidly. The first cost of these systems is frequently high, but the reduction in labor and the increase in speed generally yield a short payback period combined with a marked rise in efficiency.

Modern commercial material-handling systems can be grouped roughly into four categories:

1. *Elevator-type systems.* These are vertical-lift, car-type systems including the common dumbwaiter and ejection lifts, which are basically automated dumbwaiters.

(a)

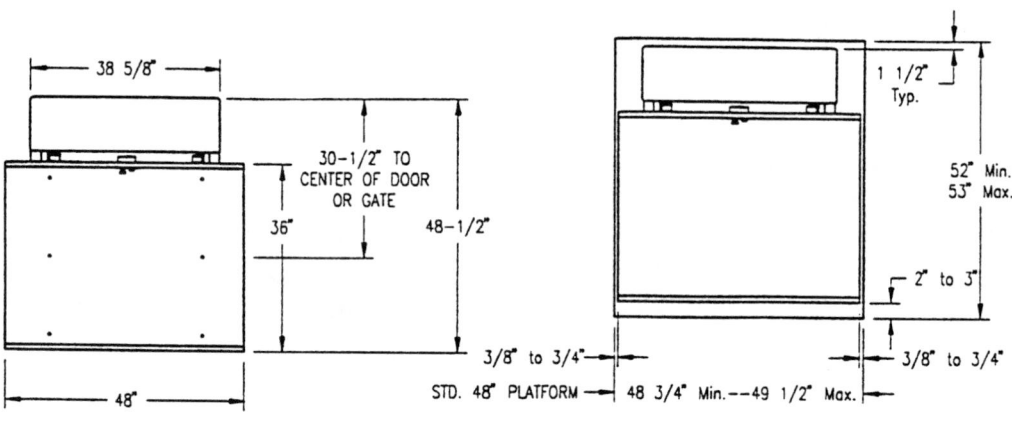

PLAN VIEW

HOISTWAY LAYOUT
(REF. ASME A17.1 PART XX)

(b)

**Fig. 32.21** *The wheelchair platform lift shown in (a) is designed for a maximum load of 750 lb (340 kg) and a maximum lifting height of 171 in. (4.34 m). The drive is a motor-driven recirculating ball screw unit for low-rise buildings and a 1:2 roped-hydraulic lift for high-rise buildings. Controls are mounted at the top of the right post of the gate frame. The unit is a straight-through design with gates at both ends. Dimensions are shown in (b). (Courtesy of Access Industries, Inc.)*

2. *Conveyor-type systems.* These include horizontal and vertical conveyors.
3. *Pneumatic systems.* These include sophisticated pneumatic tube systems and pneumatic trash and linen systems.

4. *Other systems.* Systems that do not fit easily into any of the previously mentioned categories include automated messenger carts and automatic track-type container delivery systems.

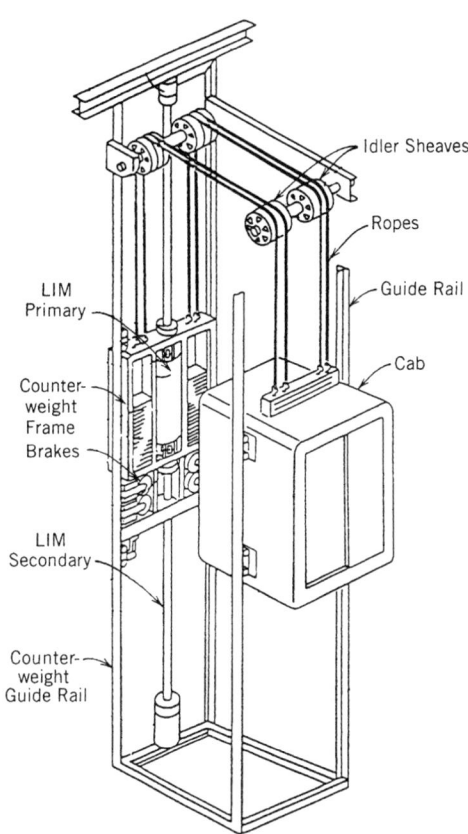

***Fig. 32.22*** *Schematic drawing of Otis Japan's linear induction motor (LIM) elevator design. Because the motor is built into the counterweight and supplies motive power linearly (vertically), the conventional overhead rotary traction machine is entirely eliminated, as is the machine room. The system operates more efficiently, quietly, and smoothly than the conventional design. (Courtesy of Otis Elevator.)*

## 32.17  MANUAL LOAD/UNLOAD DUMBWAITERS

Dumbwaiters often provide the most convenient and economical means of transporting relatively small articles between levels. In department stores, such units transport merchandise from stock areas to selling or pickup counters; in hospitals, dumbwaiters often transport food, drugs, linens, and other necessary small items. In multilevel restaurants and the like, dumbwaiters are almost always used for delivery of food from the kitchen and for return of soiled dishes.

Dumbwaiter cars are limited to a platform area of 9 ft² and a maximum height of 4 ft. The car may be, and frequently is, compartmented by shelves.

Normal speed ratings are 50 to 150 fpm, with a capacity of up to 500 lb. Cars may be of the traction (counterweighted) or drum (direct pickup) type. Control is normally "call and send" between two floors, although multibutton selector switch or central dispatching arrangements are available for applications with more than two floors. Loading may be floor, counter, or any other specified height (see Fig. 32.23 for typical layouts).

## 32.18  AUTOMATED DUMBWAITERS

These units are also known as *ejection lifts* because of the method of delivery (Fig. 32.24). They find their best application in institutions and other facilities that require rapid vertical movement of relatively large items. Thus, this device is ideally suited for delivery of food carts, linens, dishes, bulk-liquid containers, and so on. The load can be a cart (Fig. 32.25) or a basket (see Fig. 32.28) containing the items being transported. At the delivery terminal, the item must be picked up and transferred horizontally to its final destination if remote from the delivery point. Loading can be manual or automatic. Sophisticated ejection lifts use programmable controllers for automated loading, dispatch, and ejection; electronic sensors to determine whether space is available for a load; and automated return of the unloaded cart.

Payload capacity for cart systems is available up to 1000 lb and car speeds up to 350 fpm. Maximum cart size is approximately 32 in. W × 68 in. L × 70 in. H. The round-trip time for a 200-fpm unit with five loading stations is approximately 2 minutes. Major considerations for these units are their relatively high cost and the large shaft area required.

## 32.19  HORIZONTAL CONVEYORS

Although horizontal conveyors find their best application in industrial facilities, they are also usable in commercial buildings such as mail-order houses, which require a continuous flow of material. Restrictions in application stem from inflexible right-of-way requirements, noise generation, and a degree of danger if conveyors are left unprotected or exposed to unauthorized persons. The cost is relatively low, and the capacity is virtually unlimited.

**TRANSPORTATION**

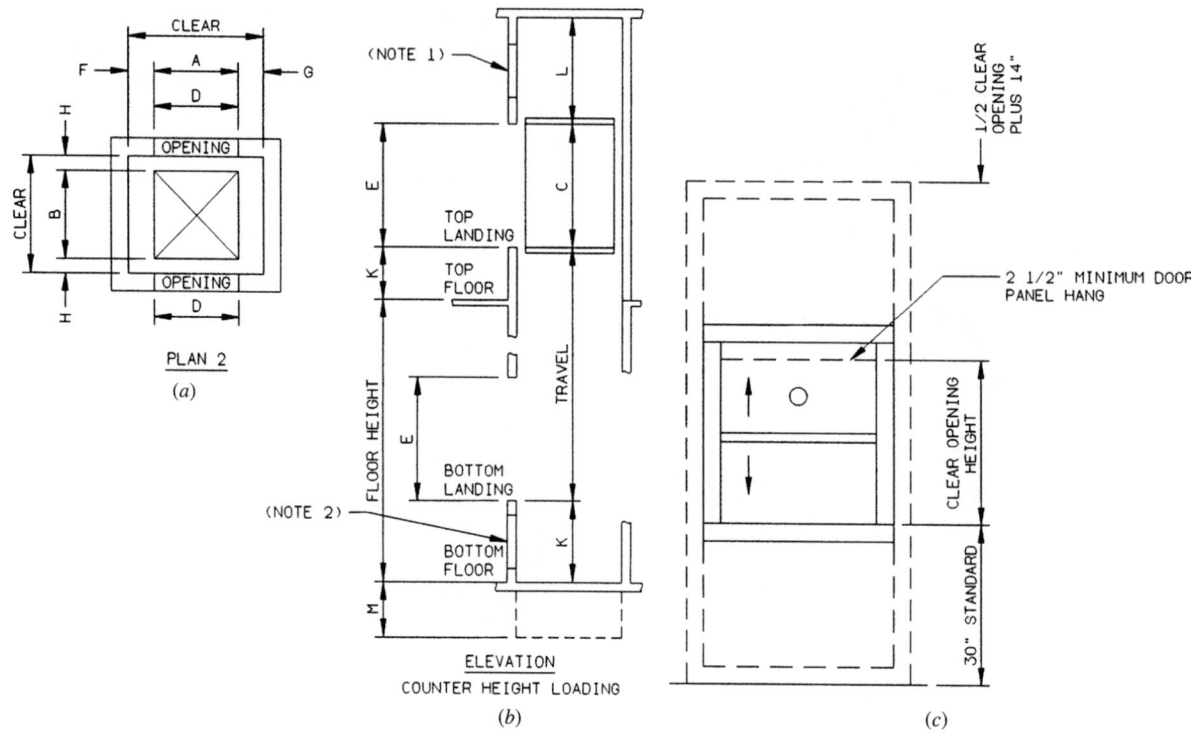

| RATED LOAD (lb) | VOL. ■ ft³ | COUNTER HEIGHT LOADING DIMENSIONS in. | | | | | | | | | | |
|---|---|---|---|---|---|---|---|---|---|---|---|---|
| | | CAR | | | DOORS ✶ | | HOISTWAY | | | | | |
| | | A WIDTH | B DEPTH | C HEIGHT | D WIDTH | E HEIGHT | F | G | H | J | K ▲ | L | M ✶✶ |
| PLAN 1 OR 2 (MACHINE OVERHEAD OR BELOW) RATED SPEED 50-150 fpm | | | | | | | | | | | | |
| APPLICATION – BULK SUPPLY, FOOD SERVICE (TRAY) & MATERIAL HANDLING | | | | | | | | | | | | |
| 100 | 8.33 | 20 | 24 | 30 | 20 | 30 | 6 1/2 | 7 | 5 | 3 | 30 ▲ | 48 | 6 |
| 200 | 12 | 24 | 24 | 36 | 36 | 24 | 6 1/2 | 7 | 5 | 3 | 30 ▲ | 48 | 12 |
| 300 | 17.5 | 30 | 24 | 42 | 30 | 42 | 6 1/2 | 7 | 5 | 3 | 30 ▲ | 48 | 18 |
| 400 | 30 | 36 | 30 | 48 | 48 | 48 | 6 1/2 | 7 | 5 | 3 | 30 ▲ | 48 | 24 |
| 500 | 36 | 36 | 36 | 48 | 48 | 48 | 6 1/2 | 7 | 5 | 3 | 30 ▲ | 48 | 24 |

✶ ONE OPENING MUST BE LARGER THAN THE CAR SIZE SO THE ASSEMBLED CAR CAN BE PLACED INTO, OR REMOVED FROM THE DUMBWAITER HOISTWAY.

✶✶ DEPTH OF PIT REQUIRED IF SLIDE DOWN TYPE DOORS ARE USED.

● 48 in. IF BI-PARTING HOISTWAY DOORS ARE USED.

▲ IF BASEMENT MACHINE (DRUM OR TRACTION) 42 in.

■ CAR SIZES ARE OPTIONAL UP TO 9 ft$^C$ OF FLOOR AREA AND UP TO 48 in. HIGH.

NOTES
1. 2'-0" X 2'-0" ACCESS DOOR IF MACHINE ABOVE.

2. 2'-0" X 2'-0" ACCESS DOOR IF MACHINE BELOW.

(d)

**Fig. 32.23** *Typical layout of a counter-height dumbwaiter with manual doors. The shaft openings (a, b) may be single (one side) or dual access. Doors can be biparting (c), slide up, or slide down. In addition to the sizes shown in the table (d), light-load cars rated 25 and 50 lb at 50 fpm are available in standard designs. (Reproduced with permission of National Elevator Industry, Inc., 185 Bridge Plaza N., Fort Lee, NJ 07024; copyrighted 1992, Vertical Transportation Standards.)*

**TRANSPORTATION**

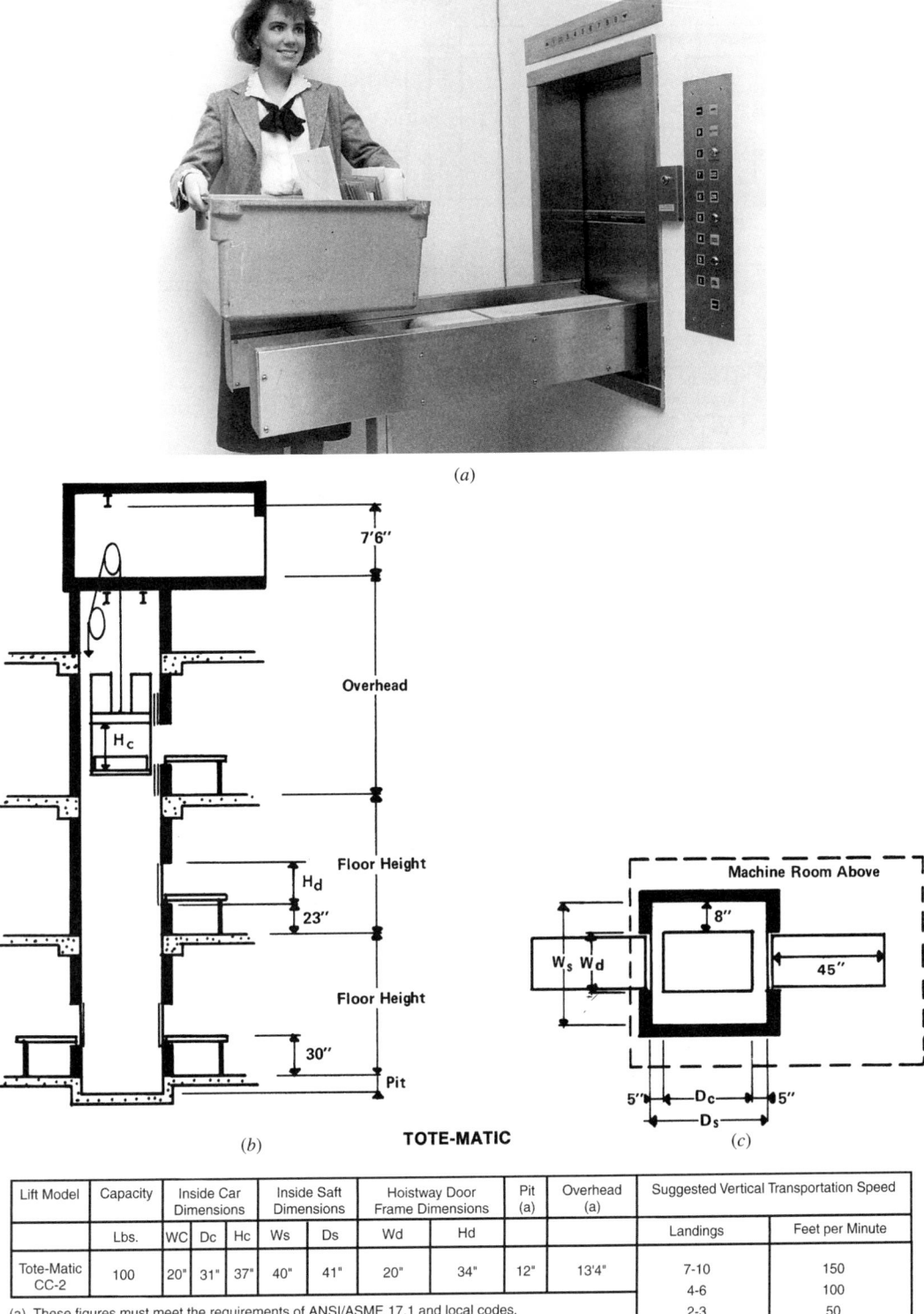

*(a)*

*(b)* **TOTE-MATIC** *(c)*

8"

45"

$W_s$ $W_d$

Machine Room Above

5" $D_c$ 5"

$D_s$

7'6"

Overhead

$H_c$

Floor Height

$H_d$

23"

Floor Height

30"

Pit

| Lift Model | Capacity | Inside Car Dimensions | | | Inside Saft Dimensions | | Hoistway Door Frame Dimensions | | Pit (a) | Overhead (a) | Suggested Vertical Transportation Speed | |
|---|---|---|---|---|---|---|---|---|---|---|---|---|
| | Lbs. | WC | Dc | Hc | Ws | Ds | Wd | Hd | | | Landings | Feet per Minute |
| Tote-Matic CC-2 | 100 | 20" | 31" | 37" | 40" | 41" | 20" | 34" | 12" | 13'4" | 7-10 | 150 |
| | | | | | | | | | | | 4-6 | 100 |
| | | | | | | | | | | | 2-3 | 50 |

(a) These figures must meet the requirements of ANSI/ASME 17.1 and local codes.

*(d)*

**Fig. 32.24** *This automated dumbwaiter or vertical ejection lift is designed to eject (unload) a container automatically at a preselected station. (Note the dispatching station to the right of the doors in a.) The tote box is carried on an ejection conveyor, which electrically senses arrival at its destination and effects ejection. Station doors are automatically electrically operated. The dumbwaiter car drive mechanism is a counterweighted traction drive. Car capacity is 100 lb, although containers are normally limited to 50 lb for ease of handling. Car speeds are normally 50 to 150 fpm (see table). A typical hoistway section (b) and plan (c) are shown. Vertical travel is not limited. Dimensions are provided in (d). (Courtesy of Courion Industries, Inc.)*

**1451**

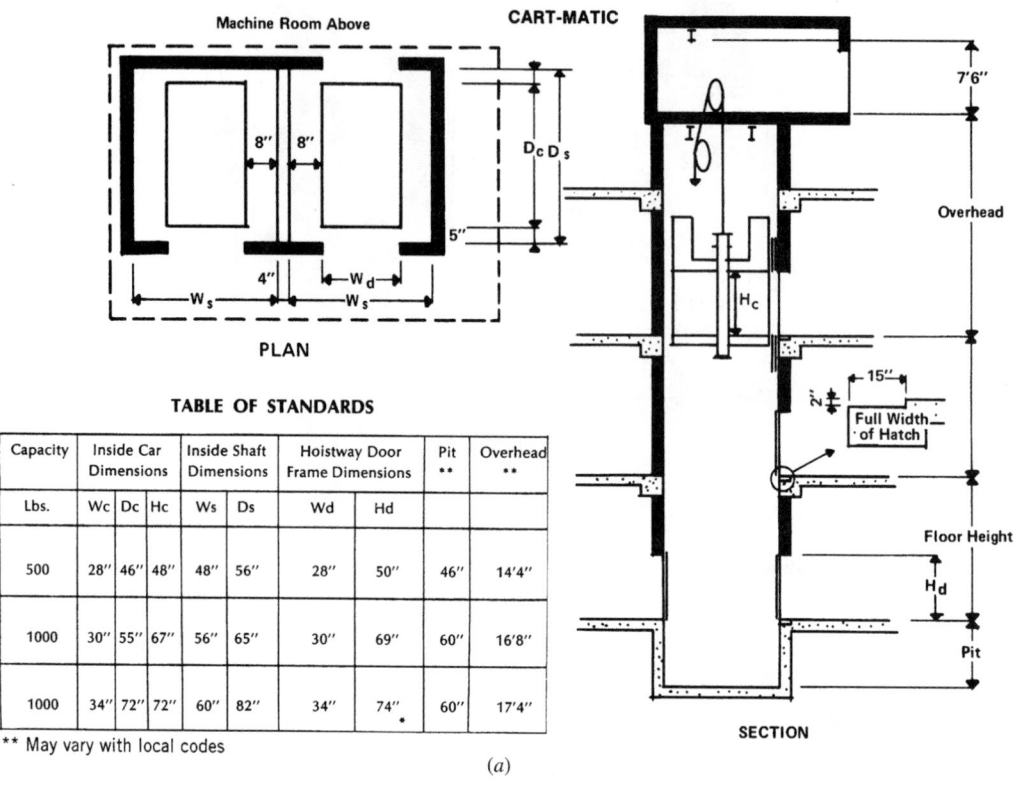

**CART-MATIC**

**PLAN AND SECTION**

Machine Room Above

PLAN

SECTION

**TABLE OF STANDARDS**

| Capacity | Inside Car Dimensions | | | Inside Shaft Dimensions | | Hoistway Door Frame Dimensions | | Pit ** | Overhead ** |
|---|---|---|---|---|---|---|---|---|---|
| Lbs. | Wc | Dc | Hc | Ws | Ds | Wd | Hd | | |
| 500 | 28" | 46" | 48" | 48" | 56" | 28" | 50" | 46" | 14'4" |
| 1000 | 30" | 55" | 67" | 56" | 65" | 30" | 69" | 60" | 16'8" |
| 1000 | 34" | 72" | 72" | 60" | 82" | 34" | 74" * | 60" | 17'4" |

** May vary with local codes

*(a)*

*(b)*

*(c)*

**Fig. 32.25** *(a) Plan, section, and dimensions for a vertical ejection lift system (automated dumbwaiter). (b) Open ejection-lift unit show-ing the cart ejection mechanism. Shaftway doors are vertical bi-parting. (c) The same unit being loaded with food carts and dispatched to the various floors of a hospital. At the upper floors the carts are rolled away by attendants. Later, the lifts are used to return soiled dishes and trays. (Courtesy of Courion Industries, Inc.)*

## 32.20  SELECTIVE VERTICAL CONVEYORS

The action of this system is similar to that of the automated dumbwaiter in that the system transfers vertically and automatically loads and unloads, but the similarity ends there. Vertical conveyors are constructed with a moving continuous-loop chain to which are attached carriages that pick up and deliver tote boxes (called *trays*). At sending and receiving stations, the operator places the items to be moved (up to 60 lb) in the tote box, "addresses" the box in one of several ways depending on the system, and places it at a pickup point (Fig. 32.26). The first empty carriage on the chain picks up the box and delivers it to its address. Functions in modern vertical conveyors are monitored by a microprocessor, which tracks system operation and furnishes maintenance data and operational diagnostics. Drawbacks of this system are the large shaft required, noise, and cumbersome arrangements when interfacing with horizontal conveyors. The cost is moderate.

## 32.21  PNEUMATIC TUBES

This well-tried system will undoubtedly continue in use where physical transfer of an item is required. Where information must be moved, electronic data reproduction has largely replaced the transfer of pieces of paper between two points. Pneumatic tube systems are available with a 2¼- to 6-in. range of tube diameters (special shapes are also used) and with single or multiple loops.

Whereas older systems were generally pressurized using a single large, noisy compressor, newer systems are computer-controlled, utilize a small blower in each of the zones, operate basically on vacuum but also on pressure, are relatively quiet, and are capable of being constructed in unlimited system length. Carriers travel at 25 fps. The computerized control center provides information and control of all system components, including status, traffic data, station assignments, scheduling, and the like. Overall, pneumatic tube systems perform their task reliably, rapidly, and efficiently at relatively low cost if installed during initial building construction. Typical system components are shown in Fig. 32.27.

## 32.22  PNEUMATIC TRASH AND LINEN SYSTEMS

Rapid movement of bagged or packaged trash and linen from numerous outlying stations to a central collecting point is usually the purpose of this system. (Health codes require separate tubes for trash and linen.) Linen systems are found generally in hospitals; trash systems in various facilities, frequently in conjunction with compactors. The system is basically a network of large pipes, negatively pressurized, with numerous loading stations throughout the building. Pipe sizes are 16, 18, or 20 in., operating at high static pressure. A system normally can handle only one unit load at a time, but moves it so quickly (20 to 30 fps) that system capacity is large and delays are not encountered. Material placed into a loading station is picked up as soon as the previous load clears. Compressors are large and very noisy, requiring considerable space allocation and acoustical isolation. In addition to the main vacuum system, a high-pressure air line is required to operate the doors, and sprinkler heads must be installed every few floors. Overall costs are low to moderate. For the specific task performed, a cheaper and more efficient transfer technique is difficult to find.

## 32.23  AUTOMATED CONTAINER DELIVERY SYSTEMS

This useful arrangement employs captive and secure containers locked onto a motorized carriage that, in turn, is locked onto the track system. Power for the motor in the carriage is picked off a third rail at 24 V dc. The entire assembly moves horizontally or vertically with equal ease. Containers move at a constant 120 fpm horizontally but more slowly on rises, depending upon the container load and the steepness of the rise. Containers are available in a number of shapes and volumes to suit the particular installation. Two standard sizes are approximately 18 in. × 6 in. × 13 in. narrow profile and 18 in. × 12 in. × 8 in. low profile. The normal payload is 20 lb, although for horizontal runs only, considerably heavier loads can be carried.

In simple single- or dual-track point-to-point or loop systems, right-of-way conflicts cannot occur and routing is simply a matter of address-

**TRANSPORTATION**

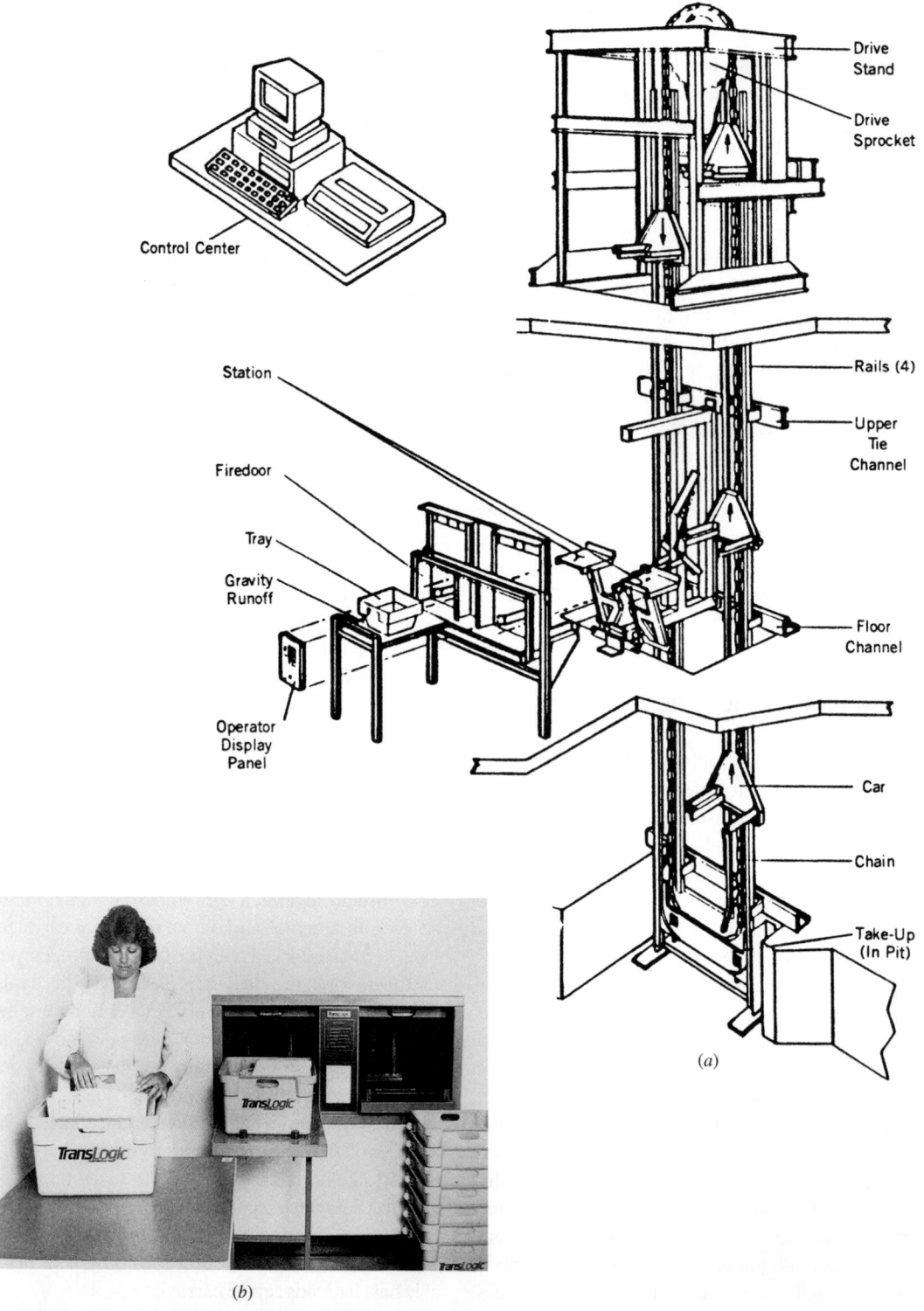

Control Center

Station

Firedoor

Tray

Gravity
Runoff

Operator
Display
Panel

Drive
Stand

Drive
Sprocket

Rails (4)

Upper
Tie
Channel

Floor
Channel

Car

Chain

Take-Up
(In Pit)

*(a)*

*(b)*

**Fig. 32.26** *(a) System diagram showing the operation of a selective vertical conveyor. Tote box addressing and dispatching can be accomplished remotely from the control center. (b) Typical sending and receiving terminal. Stations are arranged vertically in a common shaftway. Each station contains a keypad for addressing and dispatching boxes. (Courtesy of TransLogic Corp.)*

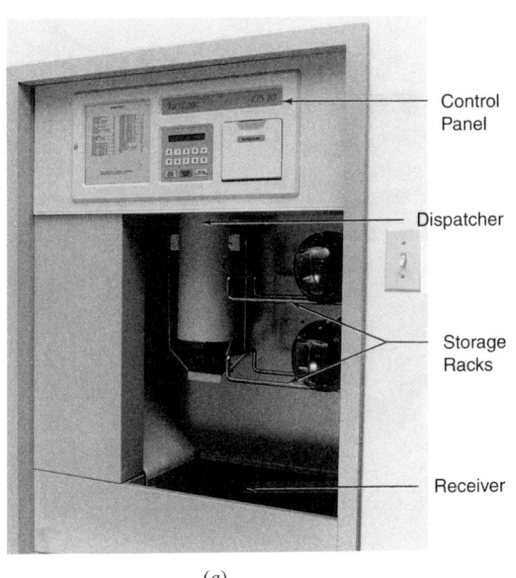

*(a)*

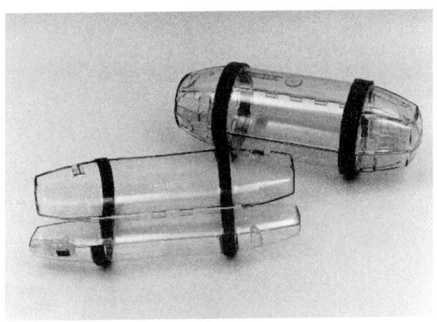

*(b)*

**Fig. 32.27** *(a) The recessed tube station illustrated is a full-facility combined sending and receiving point. Dispatching and receiving share the same pipe but operate independently and automatically. Arriving carriers (b) are decelerated by an air cushion and drop into the front bin on arrival. (b) Carriers (3¾ in. D × 15½ in. L and 5¾ in. D × 16 in L) are made of impact-resistant transparent polycarbonate plastic. Liners are used with fragile items. The bands around the carriers that maintain the air seal in the transport tubes are replaceable. (Courtesy of TransLogic Corp.)*

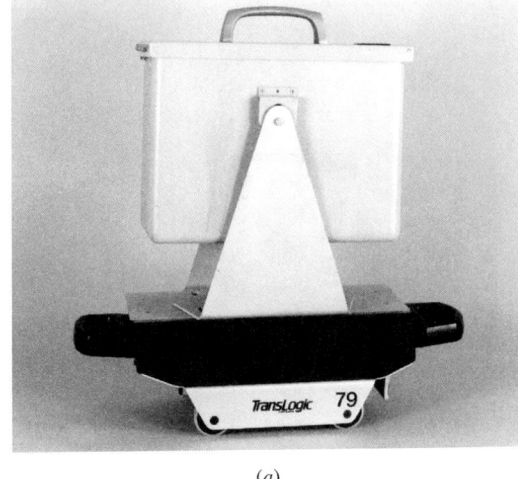

*(a)*

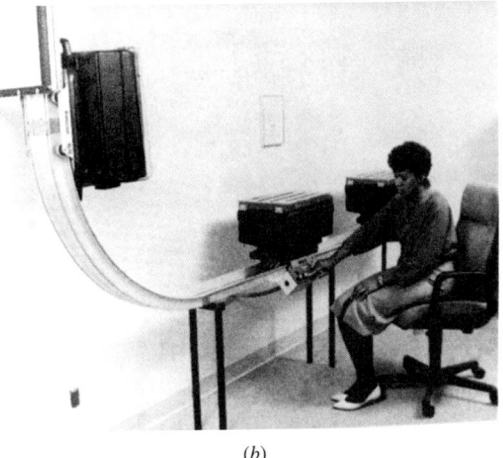

*(b)*

**Fig. 32.28** *(a) One of many designs of a carriage and container for an electric track vehicle system. This design uses a pivoted container that is maintained in a vertical position regardless of the carriage position, which may be horizontal, vertical sloped, or even inverted. (b) Dispatching and receiving station for a track vehicle system. (Courtesy of TransLogic Corp.)*

ing the car. For complex systems involving loops and branches, routing is decided by a central computerized controller that finds the shortest route for each car (up to 250 cars per system), side-tracks, parks cars, and, in effect, operates a miniature railroad-type system. The system is easily added onto a structure as a retrofit operation because of track flexibility and small size. Its major drawback is its high cost. Components are shown in Fig. 32.28.

## 32.24 AUTOMATED SELF-PROPELLED VEHICLES

These robot battery-powered vehicles follow a route determined by a passive guidance floor tape. They can be arranged to interface automatically with vertical transport means (elevators) so that the route can cover various levels in a facility. The floor tape, which can be installed below carpets, is entirely passive. It determines only the path that the

(a)

(b)

(c)

1 Pneumatic Safety Bumper

2 Sensor Panel
Contains metal detection sensors which confirm presence of the guide tape to the vehicle guidance system.

3 Steering System
Tricycle wheel design.

4 Batteries

5 Coupling Unit
Transcar tows trolleys. This device is activated to couple Transcar with trolley for transport. When trolley is delivered to destination, coupling unit retracts and vehicle drives from underneath trolley.

6 Drive Motors
Each drive wheel is powered by a 24 VDC motor.

7 Control Unit
Includes all electronic equipment

8 Charging System

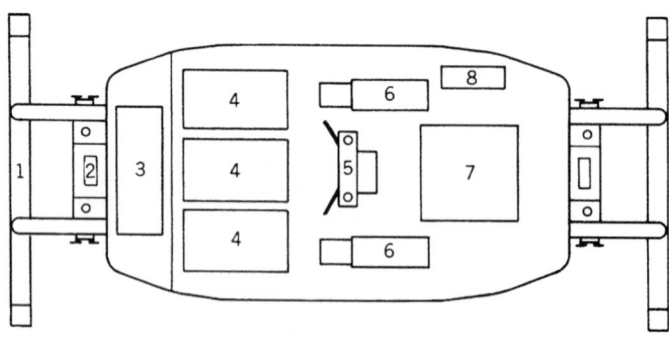

(d)

**TRANSPORTATION**

**Fig. 32.29** *An automated vehicle fits underneath its load, as seen in (a) and (c), which show two typical applications: automated food cart delivery in a hospital and automated mail service in a large office. The external programming device is seen in (b). Components of the automated vehicle are shown in the block diagram (d). (Courtesy of TransLogic Corp.)*

vehicle will follow. All sensing, instruction, motive power, and vehicle control are located on the vehicle itself, which can carry approximately a 300-lb payload and can operate a full 8-hour day without recharging its on-board batteries. Vehicle speeds are variable, ranging from 20 to 120 fpm. Programming of routes, stops, and vertical interfaces, changing of the cycle, and other functions is accomplished with an external programming device and an on-board programmable controller. Applications for an automated vehicle of this type are limitless: parts delivery and pickup in industrial facilities, food and supplies distribution in hospitals, and mail and document pickup and delivery in offices are among its basic functions. Typical units are shown in Fig. 32.29.

## 32.25 SUMMARY

The foregoing very brief overview simply describes the types of equipment available. For each facility

being planned the architect must study the material transfer problems, remembering that buildings not only handle and process but also generate material; an office building generates about 1 lb of waste per 100 ft$^2$ per day—a prodigious amount in today's large office structures. This type of dry waste can be compacted, bailed, and sold, unlike garbage and wet waste. In addition to considerations of the type of material being handled, there are factors of speed, scheduling, location of stations, labor and material costs, space requirements, noise generation, and energy requirements. To consider and evaluate all these factors in a large, complex facility is generally beyond the ability of the architect alone. Thus, expert advice from consultants who specialize in materials handling and from manufacturers' representatives should be sought.

**TRANSPORTATION**

# CHAPTER 33

# Moving Stairways and Walks

## MOVING ELECTRIC STAIRWAYS

### 33.1 GENERAL INFORMATION

THE MOVING STAIRWAY, ALSO REFERRED TO AS an *escalator* or an *electric stairway*, was first operated at the Paris Exposition in 1900. Its modern successors deliver passengers comfortably, rapidly, safely, and continuously at constant speed and usually with no delay at the boarding level. The annoyance of waiting for elevators is eliminated. Also, no time is lost by acceleration, retardation, leveling, and door operation, or by passenger interference in getting in or out of the cars, and so on. Instead of formal lobbies and hallways leading to a bank of elevators on each floor and a ride in a small, enclosed box, the electric stairway is always in motion, inviting passengers to ride on an open, airy, observation-type conveyance that can never trap them due to equipment or power failure. In contrast to the purely utilitarian function of an elevator, an escalator also has a decorative/design function, and its open, observation characteristic is frequently used to expose the rider to specific visual panoramas.

### 33.2 PARALLEL AND CRISSCROSS ARRANGEMENTS

Moving stairs can be constructed in three ways: two *parallel* arrangements and one *crisscross* arrangement. The crisscross arrangement, however, can be operated in two modes—spiral and walk-around—whereas the parallel arrangement defines usage by the physical arrangement of the stairs. The aptness of these descriptive terms becomes evident in the diagrams and photos that follow immediately.

The essential difference between the two plans is that in the crisscross arrangement, the upper and lower terminal entrances and exits to the up and down escalators are separated by the horizontal length of an escalator, whereas in either of the parallel arrangements the two escalators face in the same direction.

### (a) The Crisscross Arrangement

This is simpler to visualize and also more common. We therefore examine it first. Refer to Figs. 33.1*a* and 33.1*b*. Notice that the stair construction in both is *identical*; the difference occurs in the direction of operation of the second level of stairs. In the spiral crisscross arrangement of Fig. 33.1*a*, the rider begins an upward trip on stair L1A and by means of a 180° turn continues the trip uninterrupted on stair L2A, traveling, effectively in an upward spiral. The downward-traveling rider performs the same spiral trip on stairs L2B and L1B. This arrangement is rapid, pleasant, and very economical of space because the stairs nest into each other. It can be used for as many as five floors without excessive annoyance to the rider.

Now let us examine the arrangement of Fig. 33.1*b*. By reversing the direction of the second-level stairs L2A and L2B, the upward-traveling passenger must leave L1A at the first upper landing and walk

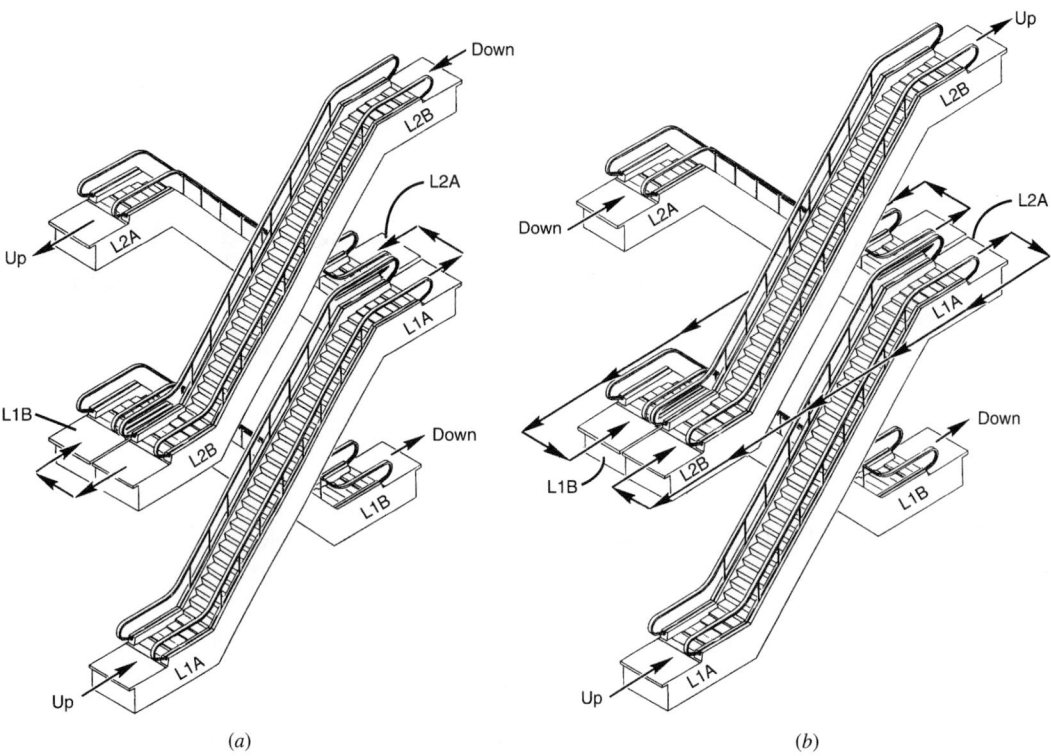

*(a)*                                                                  *(b)*

**Fig. 33.1** *Escalators constructed in a crisscross arrangement. (a) By operating stairs L1A and L2A up, a passenger from the lower ter-minal travels upward in a spiral, with only a turn-around at each level. The same is true for a down passenger. This system is called spi-ral operation, and the entire installation is referred to as crisscross spiral operation. (b) By reversing the stair direction of upper-level stairs L2A and L2B, passengers at each intermediate landing (only one shown) must traverse the entire horizontal stair length to reach the next stair traveling in the desired direction. (See arrows on the drawing.) As a result, this arrangement is known as crisscross walk-around operation. (Drawings courtesy of Otis Elevator Co.)*

around the entire length of the stair to continue the trip on L2B. Similarly, the downward-traveling passenger begins the trip on L2A but, at the first landing, must traverse the escalator length to reach the next down elevator L1B; hence the descriptive name *walk-around crisscross.* This arrangement obviously requires floor construction around the escalators, which is used in stores to display special sale merchandise. Indeed, this display purpose is the reason that stores force passengers to endure the extremely annoying forced walk-around.

A similar, but generally much shorter and therefore less objectionable walk-around characterizes the spiral crisscross plan, where the escalators are separated by distance $D$ (Fig. 33.2a). This distance is usually limited to about 10 ft because any greater distance places the trip-continuation escalator out of sight and causes not only annoyance at the enforced walk but also confusion and consequently resent-

ment. Separation of the escalators is frequently an architectural consideration and does have the advantage of easier mixing of riders entering at the various levels with riders making a continuous trip. Designers must, however, be continuously aware of the possibility of a negative reaction to separation of escalators, which can be further reinforced when:

1. Insufficient floor space is provided for the transit between escalators, causing crowding, pushing, and delay.
2. Insufficient elevator service is provided for passengers wishing to travel at least three floors. This forces people to make a multistory escalator trip, which can be wearying, particularly when carrying parcels. If such a trip is further lengthened by an enforced walk-around at each floor, it becomes a source of severe irritation, often sufficient to keep customers away from the store.

TRANSPORTATION

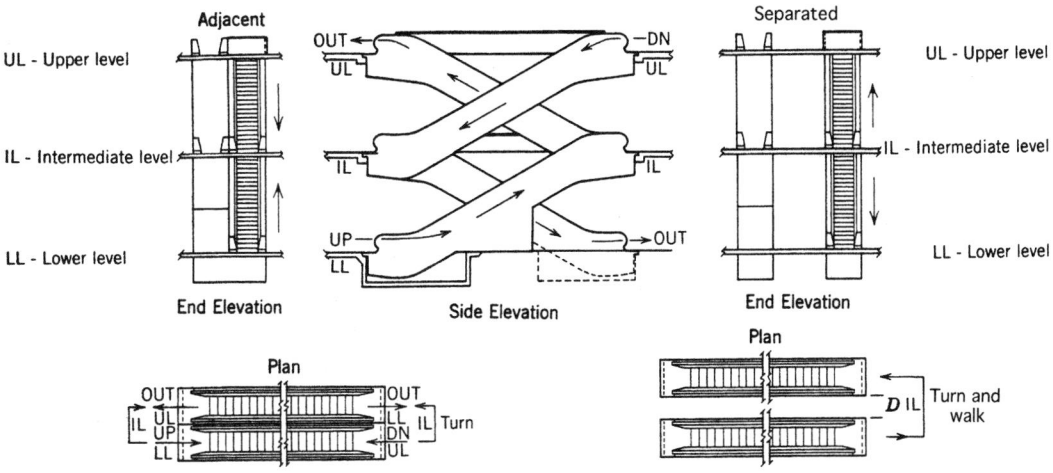

***Fig. 33.2*** *Side and end elevations and plan views of the crisscross escalator arrangements. In the spiral operation mode (see text), separating the escalators forces the rider to walk distance* D *to continue a trip. In the walk-around operation mode, escalators are frequently separated to provide a walk-around path.*

Figure 33.2 shows the plan and elevation of the crisscross arrangement for both spiral and walk-around operation, since, as shown, construction is identical for both. Figures 33.3 and 33.4 illustrate the crisscross plan.

## (b) Parallel Escalators

This configuration requires more floor space than the crisscross arrangement and is therefore used less often. As noted, it is constructed in either of two designs: the parallel spiral arrangement shown in

***Fig. 33.3*** *Bank of glass balustrade escalators with mirrored sides in a spiral operation, crisscross plan. Note the dividers at the turn-around at the intermediate level whose function is to guide traffic smoothly, either away from the escalator or to the turn point. This effectively eliminates undesirable bunching and crowding. (Photo courtesy of O&K Rolltreppen, Germany.)*

**Fig. 33.4** *Crisscross design applicable to either spiral or walk-around operation. The horizontal area in the right foreground can be used to display impulse-buying merchandise to walk-around riders during holiday seasons, after which operation can be returned to the spiral mode. (Photo courtesy of O&K Rolltreppen, Germany.)*

Fig. 33.5*a* and the stacked parallel arrangement, with forced walk-around, shown in Fig. 33.5*b*. The principal advantage of the parallel arrangement is its impressive appearance, as can be seen in Figs. 33.6 through 33.9. The stacked arrangement must be used with caution due to the inconvenience to the rider of an enforced long walk-around to continue the trip. This arrangement is found most often in mass purchase–type facilities and in malls. In department stores or malls this arrangement is much less objectionable, as many people are there to browse and window-shop rather than to purchase and leave, as is the case in a single-purpose store. In a multistory store, the inconvenience of this walk, which is frequently compounded by the crowds of people that normally congregate at special sale merchandise displays and counters, rapidly engenders annoyance. Thus, the stacked arrangement is seldom used above two floors (one such walk-around), and when it is used above two floors, riders can be expected to gravitate to elevators.

Escalators between two contiguous levels do not present the continued trip problem and therefore are frequently used in the parallel arrangement. This configuration is particularly common in public buildings, transportation terminals, and other heavy traffic areas, where the advantage of a single location for the entrance to the bank of escalators eliminates confusion and the safety hazard engendered by hesitant, confused riders in heavy traffic.

The consideration of division of rider traffic between elevators and escalators in a store is important. The general philosophy of store owners is to make escalators the primary means of vertical transportation for the obvious reason of merchandise exposure. As a result, elevators are frequently placed at one or both ends of a store, whereas the escalator banks are central. However, care must be taken to provide sufficient elevatoring in stores exceeding three floors, particularly in stores using the stacked or walk-around crisscross plans, because of the inconvenience of multifloor escalator trips discussed previously.

## 33.3 LOCATION

Because escalators are constantly moving and are generally part of a horizontal and vertical trip, they must be placed directly in the main line of traffic. This is in contrast to the elevator bank, which, being a vertical transportation unit, can be set off as an element on its own for people to approach and utilize. Escalators must therefore be placed in the area served, a totally and even dominating location. This allows potential riders to immediately:

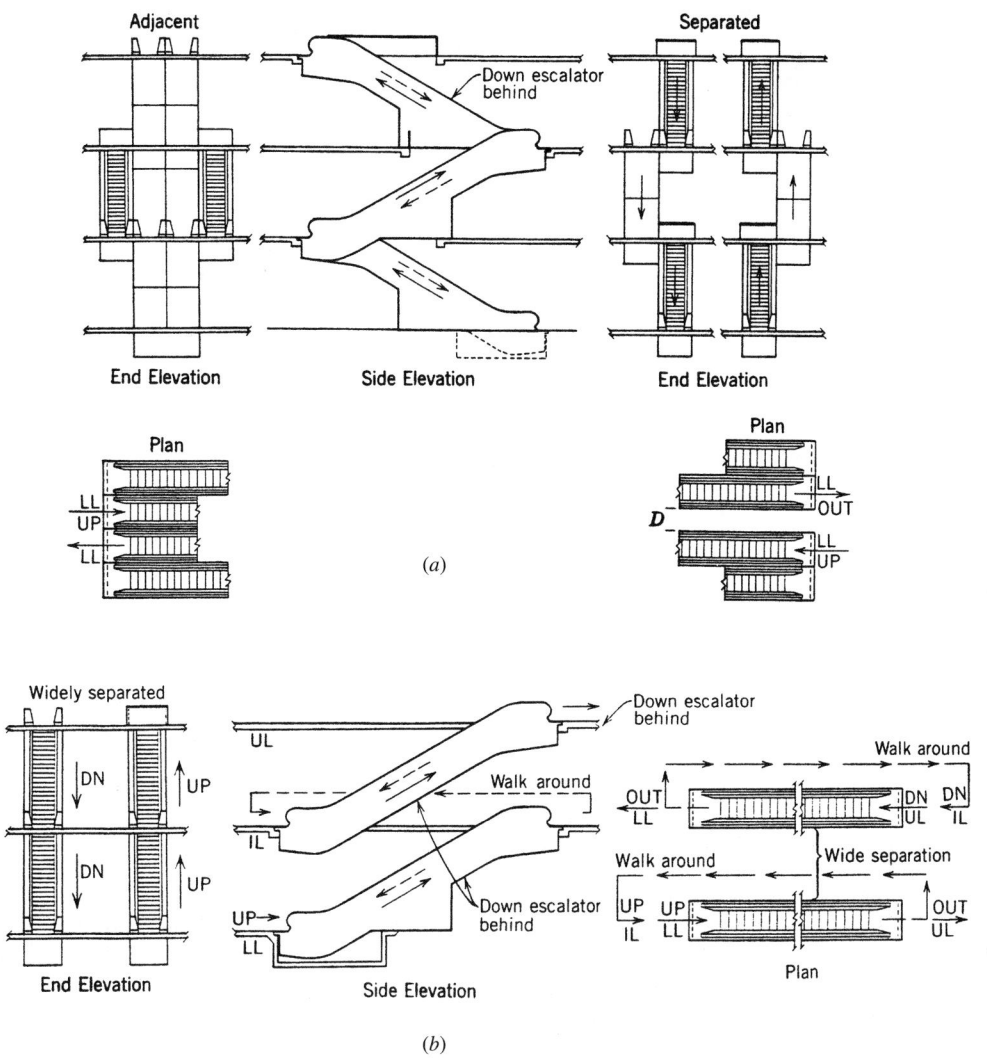

**Fig. 33.5** *Side and end elevations and plan views of escalators in (a) spiral parallel and (b) stacked parallel arrangements. In the spiral parallel arrangement (a) the separation has no effect on the rider, who simply makes a turn to continue in the same up or down direction. In the stacked arrangement (b), the rider must traverse the entire length of the escalator at each level in order to continue since the escalators are stacked vertically on each floor.*

1. Locate the escalators
2. Recognize the individual escalator's destination
3. Move easily and comfortably toward the escalator

One of the most effective ways to disorient traffic movement is to mark escalator destinations inadequately or poorly. The resultant milling about, false starts, and constant inquiries can be observed in numerous otherwise well-designed buildings, and often in stores where large displays block an originally open line-of-sight.

Sufficient lobby space must be provided at the base for queuing where anticipated, particularly at discharge points. A restricted or poorly marked discharge area causes passenger hesitation and traffic backup. Because the escalator discharges continuously, backup of traffic is dangerous and therefore intolerable. This is particularly crucial in theaters and stadiums, where even momentary hesitation during peak traffic periods can be disastrous. To avoid this, four design steps, in descending order of importance, are taken:

TRANSPORTATION

**Fig. 33.6** *An unusual design of stacked parallel escalators with a forced walk-around in a continuous trip. The exposed truss and drive elements, combined with a transparent balustrade, add considerable visual interest to the installation. (Photo courtesy of O&K Rolltreppen, Germany.)*

1. Provide well-marked escalators with sufficient traffic-carrying capacity.
2. Provide collecting space at intermediate landings so that pressure can be relieved.
3. Provide a physical divider at intermediate landing turnaround points that guides riders away from the discharge point, and provide adequate space (and time) for riders either to leave at that level or to follow the guide around and continue the trip (see Fig. 33.2).
4. Provide a slight setback for the next escalator so that the necessary 180° turn can readily be negotiated.

At the exit terminus, an escalator should discharge into an open area with no turns or choice of direction necessary. Where such is absolutely unavoidable, *large*, clear signs should make hesitation unnecessary. (See Fig. 33.8). The landing space beyond the escalator newels should be a minimum of 8 ft for 32-in. units and 10 ft for 48-in. units for a standard 100-fpm speed. For escalators that will be reversed to accommodate change in traffic direction, this landing space must be provided at the top and bottom.

**Fig. 33.7** *An unusual installation of parallel escalators adapted to a long four-story building. In the more common instance of a building that is taller than it is long, the escalators are stacked as in Fig. 33.6. Here the design is adapted to the large central court/atrium, with offset stepped landings and entrances to the buildings. (Photo courtesy of O&K Rolltreppen, Germany.)*

**Fig. 33.8** *Escalators in transportation terminals are subject to periods of very heavy use. To avoid crowding at the entrance to the up (left) moving stair, the balustrade is effectively extended with a glass-sided divider. Passengers on the down (right) escalator are guided by large overhead signs and a standing sign at the base and balustrade extension to avoid hesitation and bunching. Note also the large open area in front of both units. (Photo courtesy of O&K Rolltreppen, Germany.)*

**Fig. 33.9** *Among the longest moving stairways in Europe are those installed in the Stockholm subway system. This parallel bank of three escalators is 230 ft (70 m) long with a 108-ft (33-m) rise. Unlike American standards of 30° incline, 100-fpm speed, and 32- or 48-in. size, these units are at 27.3° incline, 147.6-fpm (45-m/min) speed, and 40-in. (1000-mm) width. (Photo courtesy of O&K Rolltreppen, Germany.)*

The crisscross arrangement has the advantages of lower cost, minimum floor space occupied, and the lowest structural requirements. The parallel arrangement, being less efficient and more expensive, has as a compensating virtue a very impressive appearance that strongly draws people to it. For this reason it is frequently employed, particularly in banks of three or four units, in transportation terminals (see Figs. 33.8 and 33.9). In such large installations, flexibility is maintained by operating all but one escalator in the direction of heaviest traffic. Reversibility of escalators provides this most desirable feature.

## 33.4 SIZE, SPEED, CAPACITY, AND RISE

Moving stairs are built according to manufacturers' and industry standards and are therefore available in standard designs. However, all major manufacturers also produce special designs for particular applications. The data that follow refer to standard designs.

All escalators in the United States are installed at an angle of 30° from the horizontal, with a minimum vertical clearance of 7 ft for escalator passengers. The 30° inclination means that the rise is equal to 57% of the unit's projected floor area for its inclined portion. The length of the horizontal portions of the stairway depends on the specific design. To meet ADA requirements, elongated newels with at least two horizontal treads before the landing plate are needed. Today, although the maximum linear speed permitted by the safety code (ANSI/ASME 17.1) is 125 fpm, the industry has standardized on a single speed of 100 fpm.

ANSI/ASME 17.1 now defines the *width* of an escalator as the width of the stair tread in inches. The previous width designation referred to the distance between balustrades, a figure difficult to define because of design variations. That measurement is now called *size*. The standard sizes and widths are:

| Escalator Size (in.) | Tread Width (in.) |
|---|---|
| 32 | 24 |
| 48 | 40 |

Table 33.1 lists theoretical, nominal (design), and observed average escalator passenger capacities. Maximum loads assume approximately 1¼ persons per tread for a 32-in. unit and almost 2 persons per tread for a 48-in. unit. In actuality, maximum capacity is approached only during peak-loads periods in transportation terminals and stadiums. At other

**TABLE 33.1 Escalator Passenger Capacity**

| Size (in.) | Tread Width (in.) | Speed (fpm) | Passengers per Hour | | |
| | | | Maximum[a] | Nominal[b] | Observed[c] |
|---|---|---|---|---|---|
| 32 | 24 | 100 | 5200 | 4000 | 2300 |
| 48 | 40 | 100 | 9000 | 6750 | 4500 |

[a]Theoretical maximum (see text).

[b]Heavy loading (see text).

[c]Average long-period loading.

times, a full (heavy) load is represented by the nominal capacity figure and an average (long period) load by the observed capacity figure. Although a 40-in. tread can indeed carry two persons, psychological factors, plus physical ones such as bulky clothing, packages, purses, and briefcases, mitigate against such loading. As a result, on a 40-in.-wide tread, one person uses each step in a diagonal pattern, and on 24-in. stairs one person occupies every other tread.

Some major manufacturers make two standard models (not styles) of moving stairs: a standard-duty unit intended for general indoor use, which provides low to medium rise, and a heavy-duty, sturdier unit intended for all-weather heavy-traffic use such as at transport terminals. The latter, because of heavier construction, can provide a greater maximum rise than standard units. Maximum rise for off-the-shelf design units varies among manufacturers. Approximate maximum figures are given in Table 33.2.

Specially designed units are available with rises of up to about 60 ft. In escalator design all the motive power is delivered at one point; that is, the drive motor drives the main chain, which drives the top sprocket, which drives the step chain, which pulls up the steps, causing the entire assembly to move. This arrangement is suitable for moderate rises of up to approximately 25 ft; beyond that, the design becomes increasingly inefficient. As the rise increases, the loads on all the drive components, including chains and sprockets, increase sharply. Furthermore, to accommodate the heavier equipment, truss width increases, as does wellway size and balustrade decks. For rises above 25 to 35 ft (depending on unit width), the drive motor is too large to fit inside the truss and requires a separate machine room below the truss, with attendant cost. These factors combine to limit standard escalators to a maximum rise of 60 ft (varies slightly among manufacturers).

## 33.5 COMPONENTS

The major components of a standard escalator installation are shown in Fig. 33.10. Safety devices are discussed in Section 33.6.

The truss is a welded steel frame that supports the entire apparatus (see Fig. 33.4). The tracks are steel angles attached to the truss on which the step rollers are guided, thus controlling the motion of the steps. The sprocket assemblies, chains, and machine provide the motive power for the unit, somewhat similar to the chain drive of a bicycle.

The handrail is driven by sheaves powered from the top sprocket assembly. It is synchronized with the tread motion to provide stability to riding passengers and support for entering and leaving passengers. Handrails disappear at inaccessible points at newels. The balustrade assembly is designed for maximum safety of persons stepping on or off the escalators.

Transparent balustrades are made of tempered glass and are frequently referred to as *crystal balustrades*. In these units, the handrail is pinch-driven within the truss. In addition to metal and glass as balustrade materials, fiberglass, wood, and various plastic materials are used.

**TABLE 33.2 Approximate Maximum Escalator Rise**

| Unit Size (in.) | Type | Supports | Maximum Rise (ft) |
|---|---|---|---|
| 32 | Standard | Ends | 24 |
| 48 | | | 16 |
| 32 | Standard | Ends and center | 30 |
| 48 | | | 20 |
| 32 | Heavy | Ends | 24 |
| 48 | | | 18 |
| 32 | Heavy | Ends and center | 40 |
| 48 | | | 20 |

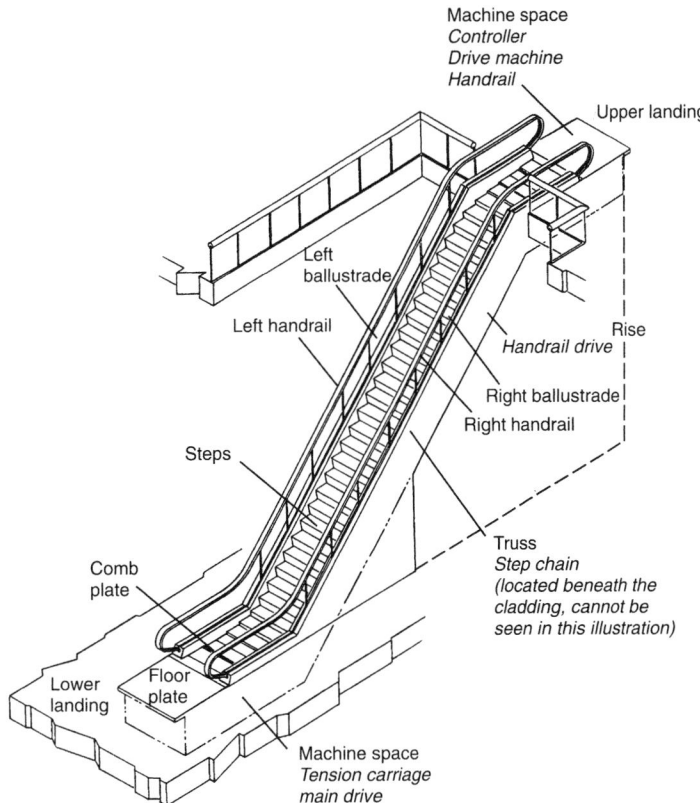

**Fig. 33.10** *View of a standard escalator showing the principal parts. (Drawing courtesy of Otis Elevator Co.)*

The control cabinet, which is normally located near the drive machine, contains malfunction indicators in addition to the drive controls. The cabinet may also contain a microprocessor malfunction analyzer and communication means for transmitting escalator operating conditions to a central control point. Operation of an emergency stop button that is wired to the controller and placed near or on the escalator housing at both ends stops the drive machine and applies the brake. Key-operated control switches at the top and bottom newels start, stop, and reverse the stairway.

## 33.6 SAFETY FEATURES

Protection of passengers during normal operation is assured by a number of safety features associated with moving stairways:

- Handrails and steps travel at exactly the same speed (100 fpm) to assure steadiness and balance and to aid stepping on or off the combplates.

- The steps are large and steady, and are designed to prevent slipping.
- Step design and step leveling with the combplates at each landing prevent tripping upon entering or leaving the escalator. This is accomplished with two or three (depending upon the manufacturer) horizontal steps at either end of the escalator.
- The balustrade is designed to prevent catching of passengers' clothing or packages. Close clearances provide safety near the combplates and step treads.
- Adequate illumination is provided by the building at all landings, at the combplates, and completely down all stairways. Some escalator designs provide built-in lighting, as discussed in Section 33.8.
- An automatic service brake will bring the stairway to a smooth stop if:
  - The drive chain or the step chain is broken or abnormally stretched
  - A foreign object is jammed into the handrail inlet, between the skirt guard and step, or between steps, causing them to separate

**TRANSPORTATION**

- A power failure occurs
- The emergency stop button is operated (one is located at either end of the escalator)
- Any of the fire safety system devices operates (see Section 33.7)
- A tread sags, rises, or breaks
- A drive motor malfunction occurs

In case of overspeed or underspeed, an automatic governor shuts down the escalator, prevents reversal of direction (up or down), and operates the service brake.

If the escalator is stopped by operation of a safety device, passengers can then walk the steps as they would any stationary stairway.

## 33.7 FIRE PROTECTION

Four methods of providing protection in case of fire near escalators are available: the rolling shutter, the smoke guard, the spray-nozzle curtain, and the sprinkler vent. One of these methods is required by code when more than two floors are pierced. Figure 33.11 illustrates clearly how the wellway at a given

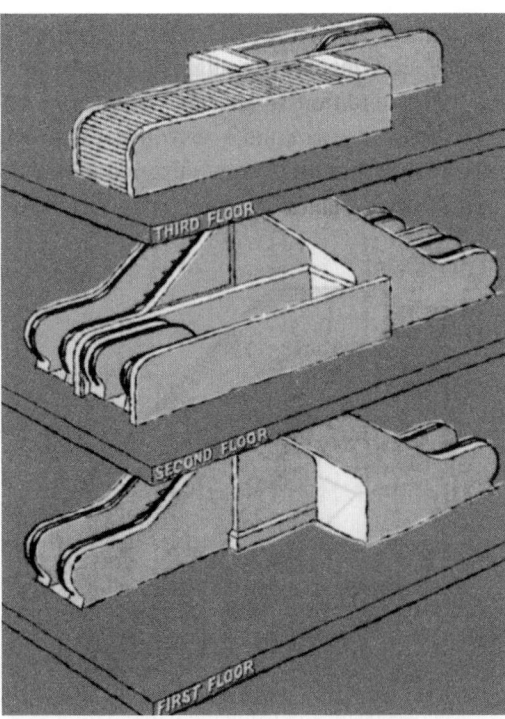

**Fig. 33.11** *Rolling-shutter method of wellway fire protection.*

floor level may be entirely closed off by the fire shutter, thus preventing draft and the spread of fire upward through escalator wells. Temperature and smoke detectors automatically actuate the motor-driven shutters. The shutter in Fig. 33.11 is shown at the third-floor level, but other shutters may be installed at the tops of horizontal wellway openings at any floor.

Figure 33.12 illustrates the smoke-guard method of protection. It consists of fireproof baffles surrounding the wellway and extending downward about 20 in. below the ceiling level. Smoke and flames rising upward to the escalator floor opening meet a curtain of water automatically released from sprinkler heads of the usual type, shown at the ceiling level. The baffle is a smoke and flame deflector. The vertical shields between adjacent sprinklers ensure that the spray from one will not cool the nearby thermal fuses, preventing the opening of adjacent sprinklers.

The spray-nozzle curtain of water (not shown) is similar to the smoke-guard protection. Here, closely spaced, high-velocity water nozzles form a compact water curtain to prevent smoke and flames from rising through the wellways. Automatic thermal or smoke relays open all nozzles simultaneously.

The sprinkler-vent fire control system is shown in Fig. 33.13. The fresh air intake housed on the roof contains a blower to drive air downward through escalator floor openings, while the exhaust fan on the roof creates a strong draft upward through an exhaust duct; this duct in turn draws air from the separate ducts just under the ceiling of each moving stairway floor opening. Three such separate wellway ducts are shown. Each duct has a number of smoke-pickup relays that automatically start the fresh air fans. The usual spray nozzles on the ceiling near the stairways aid in quenching the fire.

## 33.8 LIGHTING

Adequate illumination of a moving stairway, particularly at the landings, is important from decorative as well as safety standpoints. In a stairwell-type installation, where general-area lighting does not provide sufficient illumination for an escalator, lighting consonant with the adjacent illumination is installed on the ceiling above the escalator, with special emphasis on lighting the combplate. In Fig. 33.9, banks of fluorescent lamps are placed across

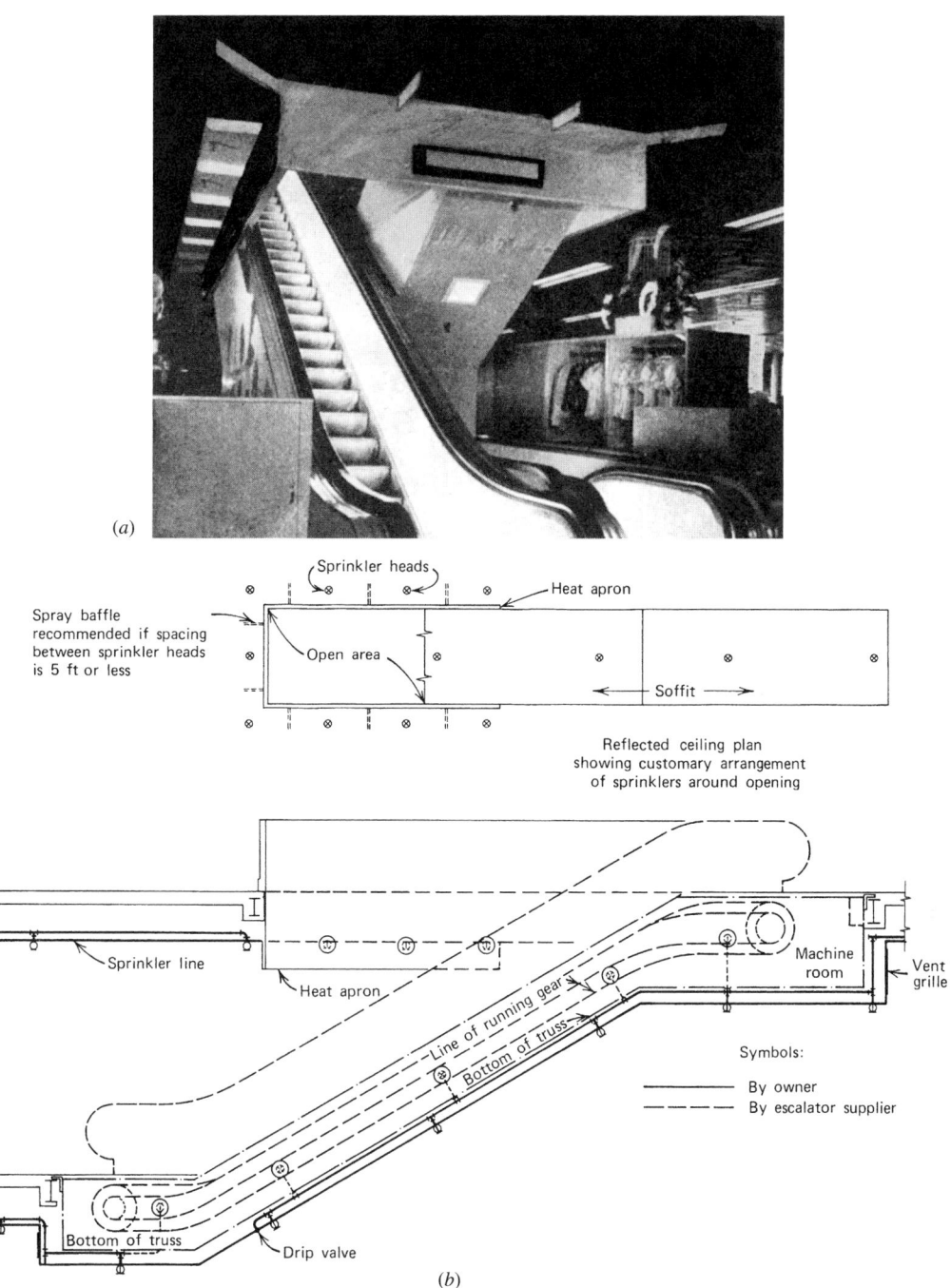

**Fig. 33.12** (a) Smoke-guard method of fire protection for a 32-in. moving stairway, crisscross type. The escalator floor opening (per floor) is approximately 4 ft, 4 in. × 14 ft, 8 in. (b) Reflected ceiling plan and section showing baffle and sprinkler layout. (Courtesy of Otis Elevator Co.)

the escalator bank at frequent intervals along the rise. Note the additional concentrated light at the combplate. Two different lighting treatments of similar installations are shown in Figs. 33.14 and 33.15 (see also the balustrade lighting in Fig. 33.16.)

## 33.9 APPLICATION

1. Main floor locations should be chosen in the direct flow of traffic to assure maximum use.
2. Vertical arrangements should be made to

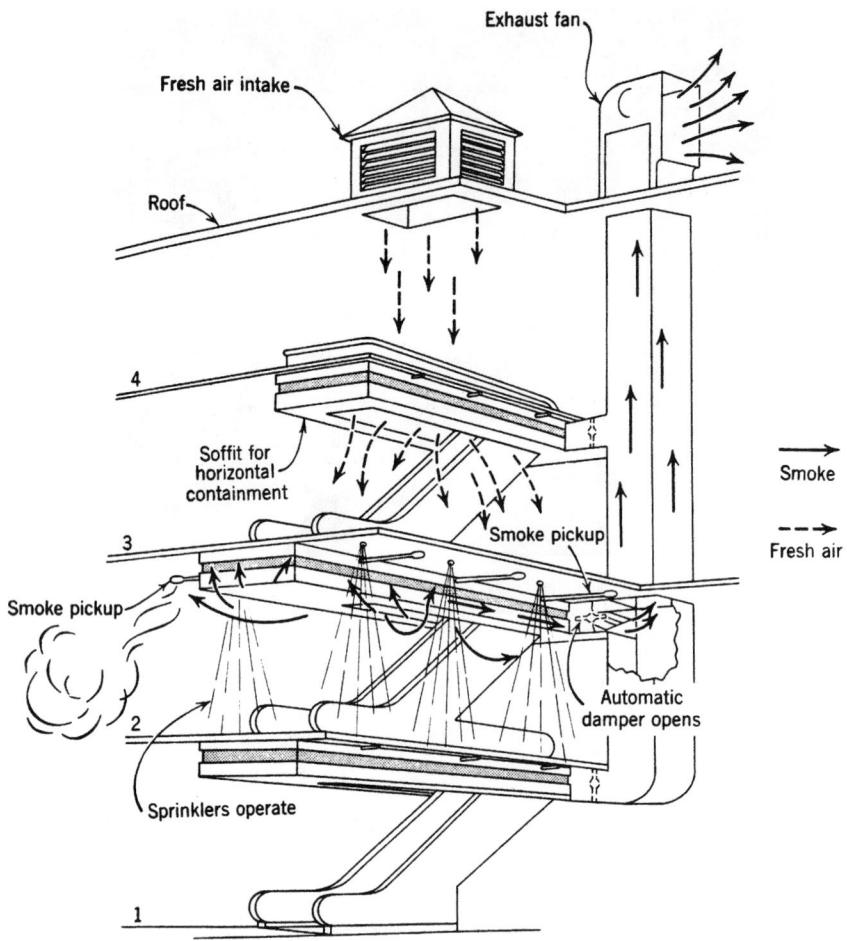

**Fig. 33.13** *Sprinkler vent fire protection for escalator openings. An exception (with control) to the rule against perforations in floors.*

**Fig. 33.14** *Stair lighting is supplied by a continuous fluorescent strip at the base of the balustrade in this California shopping mall. (Photo courtesy of Montgomery-KONE.)*

**Fig. 33.15** *A continuous fluorescent source is placed under the handrail of the crystal balustrade in this covered mall in West Germany. Note the additional light at the base for illuminating the combplate. (Photo courtesy of O&K Rolltreppen, Germany.)*

accomplish specific purposes, such as exposure of merchandise, maximum passenger capacity, and maximum accessibility to various areas.

3. Reversibility of an electric stairway should be considered in applications where major traffic flow is unidirectional. Light traffic in the reverse direction can be handled by a normal fixed stair adjacent to the escalator (see Fig. 33.16). Similarly, a bank of two escalators can operate either both up, both down, or one up and one down to

handle variable traffic conditions in such areas as office buildings and transportation terminals.

4. Exterior escalators can provide an attractive, interesting, and economical way of transporting people to elevated entry points in a building (see Fig. 33.16.)

## 33.10 ELEVATORS AND ESCALATORS

Elevators and escalators should be considered together—as a single solution in vertical transportation—for the particular facility being designed. In this connection, and particularly in the case of modernization, Fig. 33.17 provides an interesting comparison.

In certain facilities, there are times when demand for vertical transportation is so great that elevators are not a feasible solution. A prime example is a school building. During class change, virtually the entire building's population moves, with as many as 80% moving between floors. Since class change time is at most 10 minutes, the only reasonable solution is the combined use of fixed and moving stairs. In other buildings, such as multifloor stores, the escalator provides for short trips of one or two floors, and the elevator generally transports passengers traveling three or more stories.

A comparison of travel time between escalator and elevator is of interest. Using a normal speed of 100 fpm, a 12-ft floor requires 14.4 seconds for

**Fig. 33.16** *An all-weather exterior electric stairway is often the best solution to an elevated building entrance or exit. Note the elongated left balustrade, which prevents bunching at the escalator entrance, and the fluorescent fixture built into the balustrade, which provides nighttime step lighting. (Photo courtesy of O&K Rolltreppen, Germany.)*

TRANSPORTATION

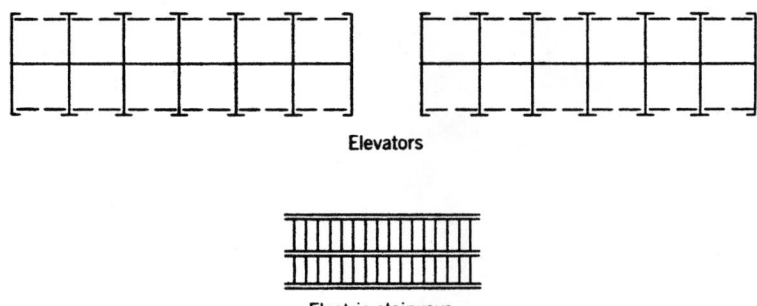

Elevators

Electric stairways

**Fig. 33.17** *Comparative space requirements for equivalent passenger-handling capacity. Note the substantial space savings offered by escalators.*

travel plus about 5 to 6 seconds to transfer to the next section, for a total of approximately 20 seconds. Thus, a four-story trip would take approximately 75 seconds. A similar elevator trip would take at most 60 seconds. The additional escalator time is not noticed in the activity of boarding, turning, and riding. However, a trip of more than four stories becomes tiresome, and all the more so when there is an enforced walk-around at each floor.

## 33.11 ELECTRIC POWER REQUIREMENTS

Standard North American electric stairways are driven by three-phase, 60-Hz, ac induction motors at standard voltages (208, 230, and 460 V). Approximate horsepower data for driving motors are shown in Table 33.3.

It is recommended that no more than four escalators be served by a single electric feeder, and further, that not all the escalators of an installation, whatever the number, be served from the same feeder.

Since obviously one cannot be trapped on an escalator, emergency power is rarely required. Ventilation for the machinery should be supplied for approximately 40% of the power, to be dissipated as heat. Thus, a 10-hp motor would require the dissipation of $0.40 \times 10 \times 2500$ (Btu/hp), or approximately 10,000 Btu/h.

## 33.12 SPECIAL-DESIGN ESCALATORS

Like elevators, escalators of nonstandard design are available on special order. These include units with slopes other than 30°, speeds other than 100 fpm, and rises beyond normal. The most unusual of these special designs is the curved escalator, which made its debut overseas and is now available in the United States. This design, which solves the very common problem of directing passengers to make a 90° turn as they leave the escalator, is an engineering tour-de-force. The entire drive mechanism had to be designed anew because the tread speed along the inside curve must be lower than that on the outside curve. Handrail drive was less of a problem since the handrail is narrow and therefore speed variation within it is minimal.

## 33.13 PRELIMINARY DESIGN DATA AND INSTALLATION DRAWINGS

At the preliminary design stage, the architect requires rough dimensional and structural data for escalators. Since at this stage of design a specific manufacturer for the stairs has not yet been selected, and because these data vary from one supplier to another, the information in Fig. 33.18 is

**TABLE 33.3 Typical Standard-Duty Escalator Motor Sizes**

| Escalator Size (in.) | Maximum Rise (ft) | Size of Motor (hp) |
|---|---|---|
| 32 | 14 | 5 |
| | 22 | 7½ |
| | 30 | 10 |
| 48 | 10 | 5 |
| | 15 | 10 |
| | 20 | 15 |

Range

| | | From | To |
|---|---|---|---|
| | A | 13'-2" | 16'-10" |
| | B | 7'-5" | 9'-4" |
| | C | 5'-9" | 7'-6" |
| | D | 3'-3" | 4'-2" |
| | E | 3'-3" | 3'-9" |
| | F | 12'-5" | 14'-10" |
| | G | 2'-11" | 3'-1" |
| | H | 2'-7" | 2'-9" |
| I | 32" | 3'-11" | 4'-6" |
| | 48" | 5'-3" | 5'-8" |
| | J | 0 | 1" |
| | K | 2'-11" | 3'-2" |
| L | 32" | 4' | 4'-4" |
| | 48" | 5'-3" | 5'-10" |
| M | 32" | 2' | 2' |
| | 48" | 3'-4" | 3'-4" |
| | N | 1'-9" | 2'-2" |
| | O | 3'-6" | 4'-0" |

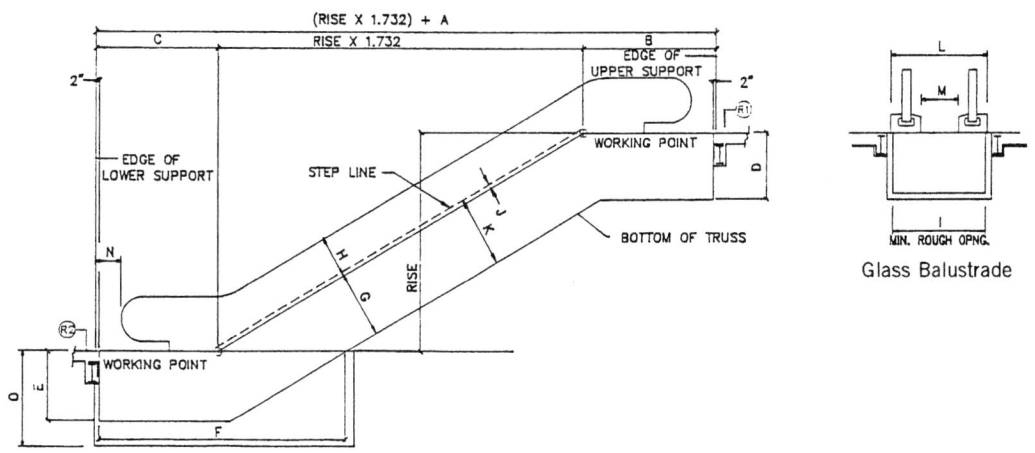

**Fig. 33.18** Escalator dimensional data in a range representing most of the major manufacturers. These data should be used for preliminary planning only.

given as a range that covers most of the major manufacturers. Similar data are available from the National Elevator Industry, Inc. Once construction contracts have been awarded and an escalator supplier selected, exact data on space requirements and structural reactions are made available by the vendor. On these, two "working points" are identified. From these two points all other measurements are made—that is, locating the centerline of the truss sections, placing the lower and upper landing truss support beams, and so on.

## 33.14 BUDGET ESTIMATING FOR ESCALATORS

The cost of an escalator includes the cost of the associated mechanical and electrical equipment, plus the shipping and installation charges. The manufacturer provides expert engineering and a field erector who supervises the installation.

A tabulation of *relative* escalator prices, on a base of 100, is given in Table 33.4. Prices for units with a rise above 35 ft increase very rapidly and

**TABLE 33.4 Relative<sup>a</sup> Escalator Prices**

| Rise (ft) | Size 32 in. | Size 48 in. |
|---|---|---|
| 14 | 100<sup>a,b</sup> | 118 |
| 16 | 103 | 121 |
| 18 | 105 | 124 |
| 20 | 108 | 127 |
| 22 | 111 | 130 |
| 24 | 113 | 133 |
| 26 | 115 | 135 |
| 28 | 118 | 138 |
| 30 | 120 | 141 |

<sup>a</sup>Base price figure is 100 for the shortest, narrowest, slowest unit (i.e., 14-ft rise, 32 in. wide). All units are 100 fpm. Add 5% for a glass balustrade for any unit.

depend upon the type of unit used. The designer is referred to suppliers for quotes on all units. To these figures must be added the cost of the contractor's work, wellway protection, lighting, outside balustrades, and plaster.

# MOVING WALKS AND RAMPS

## 33.15 GENERAL INFORMATION

Moving walks and ramps are different from moving stairways in application, function, construction, and capacity. Escalators have as their primary function the movement of large numbers of people vertically, when such vertical distance does not exceed approximately five stories, as noted above. The moving stair performs this specific transportation function extremely well with minimum cost, space, and maintenance.

When vertical transportation of wheeled vehicles and large parcels is required, the use of an electric stairway is at best awkward, if not impossible. For such functions and others discussed below, the moving ramp serves very well.

Unlike the elevator and escalator, the moving walk or ramp serves a dual function, that is, horizontal transportation only or a combination of horizontal and vertical transportation. For the purpose of our discussion, we will define a moving *walk* as one with an incline not exceeding 5° where the principal function is horizontal motion and inclined

motion is incidental to the horizontal. A moving *ramp* is a device with an incline limited by code to 15°, where vertical motion is generally more important than the horizontal component. It should be understood that the walk and ramp are physically the same device differently applied.

## 33.16 APPLICATION OF MOVING WALKS

The principal uses of moving walks, also known as *autowalks*, are to

1. Eliminate and/or accelerate burdensome walking
2. Eliminate congestion
3. Force movement
4. Easily transport large, bulky objects

Anyone who has walked the seemingly endless distances in a major airport, carrying or dragging a heavy suitcase, can appreciate the almost absolute necessity for a moving walkway. For this reason, transportation terminals have become major users of this item (Fig. 33.19). Other transportation terminals, such as rail and ship terminals, also can often find excellent applications for the moving walk, since much heavy and bulky luggage is moved in these areas.

The apparent distance compression that moving walks provide permits placement of parking areas remote from the pedestrians' destination. Thus, a store can extend its parking area with no annoyance to patrons who must make the long trip to their cars with bulky packages or shopping carts. These advantages are all the more appreciated by persons with a walking impediment.

A second application of walks, as noted previously, is the routing of traffic to avoid congestion, milling about, and lost time and motion. This is particularly applicable in transportation terminals, where persons are always traveling in opposite directions through the same and often restricted area, such as in "fingers" leading from airplanes to the main air terminal.

Moving walks are also useful to move people past a display window or some other point where congestion caused by stopping is undesirable. This "movement of objects" application demonstrates clearly that the moving walk is simply a large conveyor belt, regardless of its construction.

*Fig. 33.19* Twin autowalks in a transportation terminal. Of particular interest are the sloped entry and exit and the continuous fluorescent lighting beneath the handrails of the balustrades. (Photo courtesy of O&K Rolltreppen, Germany.)

## 33.17 APPLICATION OF MOVING RAMPS

The moving ramp that combines horizontal and vertical movement is principally applicable

1. To move persons and wheeled vehicles vertically
2. To move persons who lack the agility required to use an escalator
3. To vertically move large, bulky objects

Ramps have found an important field of application in multilevel stores where escalators are not feasible for shopping-cart users. Such stores may also utilize rooftop parking that is made accessible to cart users via a moving ramp. Since luggage carriers are not easily used on escalators, transportation terminals, which are almost always multilevel, also find extensive application for the moving ramps (Figs. 33.20 and 33.21).

## 33.18 SIZE, CAPACITY, AND SPEED

The speed, physical dimensions, and therefore passenger capacity of walks and ramps are not as extensively standardized as is the case with escalators. Manufacturers utilize several different tread widths, combined with various speeds and ramp angles of incline. The combinations are designed to

*Fig. 33.20* Unusual angular design of multiple ramps in a shopping mall in Brisbane, Australia. The use of ramps rather than escalators permits shopping carts and bulky packages to be moved with ease between the building levels. (Photo courtesy of O&K Rolltreppen, Germany.)

**TRANSPORTATION**

*Fig. 33.21* Multiple inclined ramp installation in a European department store (three up, two down). Of interest in this installation are the level entries, lighted balustrades, the ease with which shopping carts are handled on the wide pallet, and the use of the space between the transparent balustrades for merchandising. (Photo courtesy of O&K Rolltreppen, Germany.)

suit the application. Furthermore, since the maximum permissible walk ramp speed varies with the angle of slope, and with the design of the entering point, passenger capacity ratings vary with each design. By code, higher speeds are allowed for level entrance than for sloping entrance, for the obvious reason that the level entrance is easier to board. Tables 33.5 and 33.6 and Fig. 33.22 give data for typical units.

Since capacity varies with width, speed, and type of entrance, exact capacity figures must be obtained for each specific design. The maximum practical length at present is approximately 1000 ft, with longer units in design. Wide autowalk units (55 in.) are now extensively used in transportation terminals, since they permit passengers with luggage trolleys who walk on the moving walkway to easily pass other passengers with luggage carts, who prefer to stand still and move at the relatively slow rate of 100 to 130 fpm (about half of the normal walking speed). Recent installations of such wide autowalks are found at Heathrow, Gatwick, and Munich airports, among others.

### 33.19 COMPONENTS

Moving walks are manufactured in only one design: a derivative of the escalator, which uses a flattened

**TABLE 33.5  Maximum Permissible Operating Speeds of Moving Ramps**

| Angle of Incline | Maximum Speed fpm (m/s) | |
| --- | --- | --- |
| | Level Entrance | Sloping Entrance |
| 0–3° | 180 (0.9) | 180 (0.9) |
| 3–5° | 180 (0.9) | 160 (0.8) |
| 5–8° | 180 (0.9) | 140 (0.7) |
| 8–12° | 140 (0.7) | 130 (0.65) |
| 12–15° | 140 (0.7) | 125 (0.625) |

**TABLE 33.6  Moving Walks/Ramps: Commercial Ratings**

| Angle | Pallet Width in. (mm) | Ramp Speed fpm (m/s) |
| --- | --- | --- |
| 0–3° Walk | 32 (800) | 100 (0.5),  130 (0.65), 150 (0.75) |
| | 40 (1000) | 100 (0.5),  130 (0.65), 150 (0.75) |
| 10°, 11°, 12° Ramp | 62 (1400) | 90 (0.45), 100 (0.5),  130 (0.65) |
| | 32 (800) | 100 (0.5),  130 (0.65) |
| | 40 (1000) | 90 (0.45), 100 (0.5),  130 (0.65) |

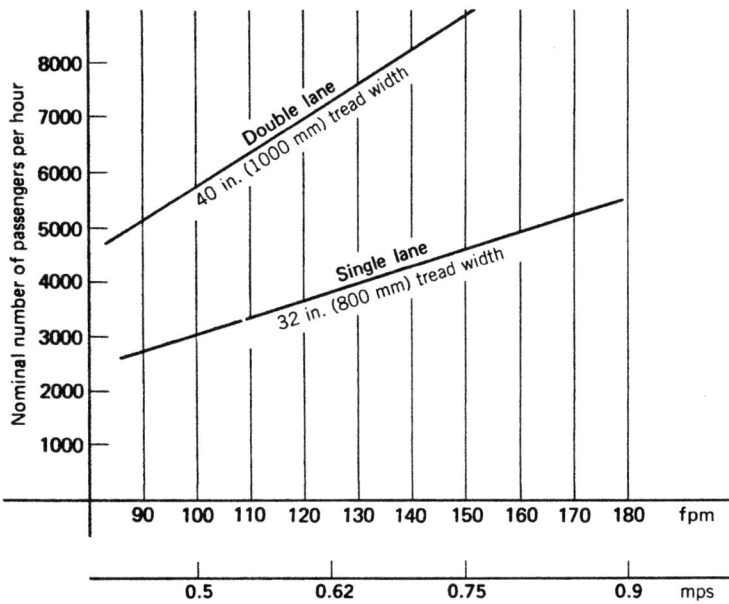

**Fig. 33.22** *The capacity of moving walks varies with speed, angle of incline, and tread width. The capacity shown is for the maximum incline permitted at that walk speed. Because of the requirement for handrail support, tread widths greater than double lane are not often utilized.*

pallet in place of a step. In all other respects—the drive mechanism, safeties, brake, handrails, and so on—the unit is similar to an escalator. Preliminary layout data and detailed plans are, as with moving stairs, available from vendors, who frequently are the same vendors that supply escalators.

# P A R T   X

# APPENDICES

The purpose of this part of the book is to present important reference and ancillary data (often extensive) in a separate, easily accessed area that does not detract from the flow of the text in the chapters. The appendices are:

A. Metrication, SI Units, and Conversions
B. Climatic Conditions for the United States, Canada, and Mexico
C. Solar and Daylighting Design Data
D. Solar Geometry
E. Thermal Properties of Materials and Assemblies

TING / TEMPERING / P

| HEATING CFM | SIZE (INCHES) HxL(No. Coils) | S |
|---|---|---|
| 9300 | 43x79(1) | |
| 9000 | 43x72(1) | |
| 4620 | 27x51(1) | |
| 4390 | 32x43(1) | |
| 2910 | 23x47(1) | |
| 940 | 12x20(1) | |
| 5830 | 35x40(1) | |
| 3640 | 45x22(1) | |
| 3740 | 45x22(1) | |
| 970 | 6x501) | |
| 1640 | 12x411) | |
| 950 | 9x34(1) | |
| 1010 | 12x30(1) | |
| 1040 | 9x37(1) | |
| 1030 | 9x37(1) | |
| 620 | 6x38(1) | |
| 890 | 6x44(1) | |
| 790 | 6x40(1) | |
| 780 | 6x40(1) | |
| 860 | 6x44(1) | |
| 4570 | 32x46(1) | |
| 2690 | 32x30(1) | |
| 2460 | 41x45(1) | |

LING COIL SCHEDUL

| COOLING CFM | SIZE (INCHES) HxL(No. Coils) | S |
|---|---|---|
| 9000 | 27x48(2) | |
| 1040 | 6x50(1) | |
| 2180 | 12x50(1) | |
| 1060 | 9x34(1) | |
| 1140 | 12x30(1) | |
| 16100 | 33x79(2) | |
| 13000 | 25x75(2) | |
| 1150 | 9x37(1) | |
| 1140 | 9x37(1) | |
| 780 | 6x38(1) | |
| 6200 | 35x49(1) | |
| 990 | 6x44(1) | |
| 1260 | 9x45(1) | |
| 1310 | 9x45(1) | |
| 1200 | 9x38(1) | |
| 6200 | 33x54(1) | |
| 4760 | 30x46(1) | |
| 940 | 12x31(1) | |
| 3640 | 45x26(1) | |
| 3740 | 45x26(1) | |

APPENDICES

F. Heating and Cooling Design Guidelines and Information

G. Standards/Guidelines for Energy- and Resource-Efficient Building Design

H. Annual Solar Performance

I. Economic Analysis

J. Lamp Data

K. Sound Transmission Data for Walls

L. Sound Transmission and Impact Insulation Data for Floor/Ceiling Constructions

M. Design Analysis Software

# Metrication, SI Units, and Conversions

## A.1 GENERAL COMMENTS ON SI UNITS

The building industry in the United States has been slow (glacially so) in adopting the metric (more accurately, Le Système International d'Unités or SI) system of units for many reasons. The rest of the world, however, is generally fully SI (or metric). Many professional societies use a mixture of I-P (inch-pound, previously "English" units) and SI units as a means of addressing their international audiences. Many older units are also so entrenched that changing them is an almost impossible task. This book attempts to provide dual units where feasible, but that is not always possible. For this reason, tables of conversions and approximations are presented below to enable the user to work with both systems. Also given below are some facts that should make use of the SI system a bit easier.

## A.2 SI NOMENCLATURE AND SYMBOLS

For a full discussion of the SI system see *AIA Metric Building and Construction Guide*, edited by S. Braybrooke (John Wiley & Sons, New York, 1980) and the ASHRAE publication "SI for HVAC&R" (May 1999) available from the ASHRAE web site (http://www.ashrae.org/). The units in common use include the basic SI staples—the meter, kilogram, and second (MKS)—plus a host of derived, supplementary, and non-SI units such as the pascal (pressure), radian (solid angle), and kilowatt-hour (energy). Also, multiple and submultiple units such as the liter, metric ton, and millibar are so common that they stand as separate units instead of being expressed as $10^{-3}$ m$^3$, $10^3$ kg, and $10^{-2}$ Pa.

Table A.1 lists SI unit prefixes with their accepted symbols. Symbols do not change in the plural. That is, 6 millimeters is written as 6 mm and 20 kilograms is written as 20 kg. All units and prefixes except Fahrenheit and Celsius are uncapitalized when written out (as in megaton or meter).

**TABLE A.1 Common SI Unit Prefixes**

| Multiples and Submultiples | Prefixes | Symbols |
|---|---|---|
| 1 000 000 000 = $10^9$ | giga | G |
| 1 000 000 = $10^6$ | mega | M[a] |
| 1 000 = $10^3$ | kilo | k[a] |
| 100 = $10^2$ | hecto[b] | h |
| 10 = 10 | deka | da |
| | | |
| 0.1 = $10^{-1}$ | deci | d |
| 0.01 = $10^{-2}$ | centi | c[a] |
| 0.001 = $10^{-3}$ | milli | m[a] |
| 0.000 001 = $10^{-6}$ | micro | μ[a] |
| 0.000 000 001 = $10^{-9}$ | nano | n |
| 0.000 000 000 001 = $10^{-12}$ | pico | p |

[a]Most commonly used.
[b]A hectare is a square hectometer (i.e., $10^4$ m$^2$).

## A.3 COMMON USAGE UNITS

1. *Length:* meter (m), kilometer (km), millimeter (mm), micrometer (μm), nanometer (nm)
2. *Area:* square meter (m²), hectare (ha)
3. *Volume:* cubic meter (m³), liter (L)
4. *Flow:* cubic meters per second (m³/s)
5. *Velocity—Airflow:* meters per second (m/s)
6. *Weight:* kilogram (kg), gram (g)

The SI system clearly differentiates between mass (kg) and force (kg · m/s²), the latter being given a separate name and symbol, newton (N). Weight is not used, because it is a force that depends upon acceleration and is therefore variable. However, the construction industry largely continues to use the terms *mass, weight,* and *force* interchangeably.

7. *Force:* newton (N), kilonewton (kN). A newton is the force required to accelerate 1 kg at 1.0 m/s².
8. *Pressure:* pascal (Pa), kilopascal (kPa). A pascal is a newton per square meter (N/m²).
9. *Energy, Work, Quantity of Heat:* joule (J), kilojoule (kJ), megajoule (MJ). A joule is a watt-second (W · s).
10. *Temperature:* degree Celsius (°C), degree kelvin (K).

SI temperature in degrees kelvin is not as commonly used as °C (*Celsius* is the accepted term; *Centigrade* is obsolete). The Celsius and kelvin scales are subdivided equally but start at different points—that is, 0 K is −273.15°C. Therefore, to determine kelvin from Celsius, simply add 273.15. Increments are equal because of equal subdivisions; that is, a change of 10 K is the same as a 10°C change. Because of its common usage in design work, a Fahrenheit/Celsius conversion chart is given in Fig. A.1.

11. *Illumination:* see Table 11.1.
12. *Acoustics:* see Table 19.16.
13. *CGS/MKS Conversions:* see Table 19.16.
14. *Abbreviations:* See Table A.2.
15. *Approximations:* See Table A.3.

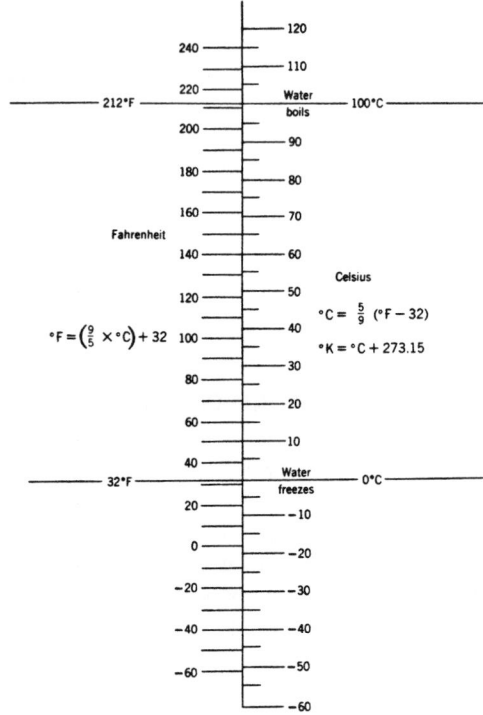

**Fig. A.1** *Conversion nomograph: Fahrenheit temperature–Celsius temperature.*

## A.4 CONVERSION FACTORS

Table A.4 is an alphabetized list of useful conversion factors. Decimal notation has been used in this table (versus scientific notation), that is, 0.00378 versus $3.78 \times 10^{-3}$ or 3.78 E-3.

**TABLE A.2 Typical Abbreviations: All Systems of Units**

| | | | |
|---|---|---|---|
| atmospheres | atm | gallons | gal |
| British thermal units | Btu | gallons per hour | gph, gal/h |
| British thermal units per hour | Btu/h | gallons per minute | gpm, gal/min |
| calorie | cal | grams | g |
| cubic feet | cf, ft³ | hectares | ha |
| cubic feet per minute | cfm, ft³/min | horsepower | hp |
| cubic feet per second | cfs, ft³/s | inches | in. |
| cubic meters | m³ | inches of mercury | in. Hg |
| feet | ft | joules | J |
| feet per second | fps, ft/s | kilocalories | kcal |

**TABLE A.2  Typical Abbreviations: All Systems of Units (*Continued*)**

| | | | |
|---|---|---|---|
| kilograms | kg | miles per hour | mph |
| kilograms per second | kg/s | millimeters | mm |
| kilojoules | kJ | millimeters of mercury | mm Hg |
| kilometers | km | newtons | N |
| kilometers per hour | kph, km/h | ounces | oz |
| | | | |
| kilonewtons | kN | pounds | lb |
| kilopascals | kPa | pounds of force | lbf |
| kilowatts | kW | pounds per cubic foot | lb/ft$^3$ |
| kilowatt-hours | kWh | second | sec, s |
| liters | L | square feet | ft$^2$ |
| | | | |
| liters per second | L/s | square inches | in.$^2$ |
| megajoules | MJ | square meters | m$^2$ |
| meganewtons | MN | watts | W |
| megapascals | MPa | watts per square meter | W/m$^2$ |
| meters | m | yards | yd |
| meters per second | m/s | | |

**TABLE A.3  Common Approximations**

| Approximate Common Equivalents | | | | | |
|---|---|---|---|---|---|
| 1 inch | = | 25 millimeters | 1 millimeter | = | 0.04 inch |
| 1 foot | = | 0.3 meter | 1 liter | = | 61 cubic inches |
| 1 yard | = | 0.9 meter | 1 meter | = | 3.3 feet |
| 1 mile | = | 1.6 kilometers | 1 meter | = | 1.1 yards |
| 1 square inch | = | 6.5 square centimeters | 1 kilometer | = | 0.6 mile |
| | | | | | |
| 1 square foot | = | 0.09 square meter | 1 square centimeter | = | 0.16 square inch |
| 1 square yard | = | 0.8 square meter | 1 square meter | = | 11 square feet |
| 1 acre | = | 0.4 hectare | 1 square meter | = | 1.2 square yards |
| 1 cubic inch | = | 16 cubic centimeters | 1 hectare | = | 2.5 acres |
| 1 cubic foot | = | 0.03 cubic meter | 1 cubic centimeter | = | 0.06 cubic inch |
| | | | | | |
| 1 cubic yard | = | 0.8 cubic meter | 1 cubic meter | = | 35 cubic feet |
| 1 quart | = | 1 liter | 1 cubic meter | = | 1.3 cubic yards |
| 1 gallon | = | 0.004 cubic meter | 1 liter | = | 1 quart |
| 1 ounce | = | 28 grams | 1 cubic meter | = | 250 gallons |
| 1 pound | = | 0.45 kilogram | 1 kilogram | = | 2.2 pounds |
| 1 horsepower | = | 0.75 kilowatt | 1 kilowatt | = | 1.3 horsepower |

**APPENDICES**

## TABLE A.4 Useful Conversion Factors: Alphabetized

| Multiply | By | To Get |
|---|---|---|
| acres | 4047 | square meters |
| atmospheres | 33.93 | feet of water |
| atmospheres | 29.92 | inches of mercury |
| atmospheres | 760.0 | millimeters of mercury |
| Btu (energy) | 0.252 | kilocalories |
| Btu (energy) | 1.055 | kilojoules |
| Btu/h (power) | 0.2931 | watts |
| Btu/h ft$^2$ (energy transfer) | 3.155 | watts per square meter |
| Btu/°F (heat capacity) | 1.897 | kilojoules per kelvin[a] |
| Btu/lb °F (specific heat) | 4.184 | kilojoules per kilogram per kelvin[a] |
| Btu/h°F ft (thermal conductivity[b]) | 1.729 | watts per kelvin[a] per meter |
| Btu/h °F ft$^2$ (conductance[c]) | 5.673 | watts per kelvin[a] per square meter |
| Btu/°F day (building load coefficient, BLC) | 0.022 | watts per kelvin[a] |
| Btu/°F day ft$^2$ (load-collector ratio, LCR) | 0.236 | watts per kelvin[a] per square meter |
| cubic feet | 0.02832 | cubic meters |
| cubic feet | 7.481 | gallons |
| cubic feet | 28.32 | liters |
| cubic feet per minute | 0.4719 | liters per second |
| cubic feet per second | 28.32 | liters per second |
| cubic inches | 16.39 | cubic centimeters |
| cubic meters | 35.32 | cubic feet |
| cubic meters | 1.308 | cubic yards |
| cubic meters | 264.2 | gallons |
| cubic yards | 0.765 | cubic meters |
| feet | 0.3048 | meters |
| feet | 304.8 | millimeters |
| feet per second | 0.3048 | meters per second |
| footcandle | 10.764 | lux |
| foot-pounds of force per second | 1.356 | watts |
| gallons | 3.785 | liters |
| gallons per hour | 1.052 | liters per second |
| gallons per minute | 0.0022 | cubic feet per second |
| gallons per minute | 0.06308 | liters per second |
| grams | 0.035 | ounces (avoirdupois) |
| hectares | 2.471 | acres |
| horsepower | 0.746 | kilowatts |
| horsepower | 746 | watts |
| inches | 25.4 | millimeters |
| inches of mercury | 0.033 | atmospheres |
| inches of mercury | 1.133 | feet of water |
| inches of mercury (60°F) (15.6 °C) | 3377 | newtons per square meter |
| inches of mercury | 0.491 | pounds per square inch |
| inches of water | 0.002458 | atmospheres |
| inches of water | 0.036 | pounds per square inch |

**TABLE A.4  Useful Conversion Factors: Alphabetized (*Continued*)**

| Multiply | By | To Get |
|---|---|---|
| inches of water (60°F) (15.6 °C) | 248.8 | newtons per square meter |
| kilocalories | 3.968 | British thermal units |
| kilocalories | 4190 | joules |
| kilograms | 2.205 | pounds |
| kilograms per cubic meter | 1.686 | pounds per cubic yard |
| kilograms per square meter | 0.0033 | feet of water |
| kilograms per square meter | 0.0029 | inches of mercury |
| kilograms per square meter | 0.205 | pounds per square foot |
| kilograms per square meter | 0.001422 | pounds per square inch |
| kilojoules | 0.948 | British thermal units |
| kilojoules per kilogram | 0.430 | British thermal units per pound |
| kilometers | 0.621 | miles |
| kilometers per hour | 0.621 | miles per hour |
| kilonewtons | 0.1004 | tons of force |
| kilonewtons | 224.8 | pounds of force |
| kilopascals | 20.89 | pounds of force per square foot |
| kilowatts | 1.341 | horsepower |
| kilowatt-hours | 3.6 | megajoules |
| liters | 0.03532 | cubic feet |
| liters | 61.02 | cubic inches |
| liters | 0.2642 | gallons |
| liters | 1.057 | quarts |
| liters per second | 2.119 | cubic feet per minute |
| liters per second | 951.0 | gallons per hour |
| liters per second | 15.85 | gallons per minute |
| megajoules | 0.278 | kilowatt-hours |
| meganewtons | 100.36 | tons of force |
| megapascals | 145.04 | pounds of force per square inch |
| megapascals | 9.324 | tons of force per square foot |
| meters | 3.281 | feet |
| meters | 1094 | yards |
| meters per second | 196.86 | feet per minute |
| meters per second | 2.237 | miles per hour |
| miles | 1.609 | kilometers |
| miles per hour | 1.609 | kilometers per hour |
| miles per hour | 0.447 | meters per second |
| milliliters | 0.061 | cubic inches |
| milliliters | 0.035 | fluid ounces |
| millimeters | 0.039 | inches |
| millimeters of mercury | 133.3 | newtons per square meter |
| million gallons per day | 18.94 | cubic meters per hour |
| newtons | 0.225 | pounds of force |
| ounces (avoirdupois) | 28.35 | grams |
| ounces (fluid) | 29.6 | milliliters |

**TABLE A.4 Useful Conversion Factors: Alphabetized (*Continued*)**

| Multiply | By | To Get |
|---|---|---|
| pounds | 0.4536 | kilograms |
| pounds of force | 4.448 | newtons |
| pounds of force per square foot | 47.88 | pascals |
| pounds of force per square inch | 6.895 | kilopascals |
| pounds per cubic foot | 16.02 | kilograms per cubic meter |
| pounds per cubic yard | 0.593 | kilograms per cubic meter |
| pounds per square foot | 4.882 | kilograms per square meter |
| square feet | 0.0929 | square meters |
| square inches | 645.2 | square millimeters |
| square kilometers | 0.386 | square miles |
| square meters | 10.76 | square feet |
| square meters | 1.196 | square yards |
| square miles | 2.590 | square kilometers |
| square yards | 0.836 | square meters |
| tons of force | 9.964 | kilonewtons |
| tons of force per square foot | 107.25 | kilopascals |
| tons of force per square inch | 15.44 | megapascals |
| torr (millimeters of mercury at 0°C [2 °F]) | 133.3 | newtons per square meter |
| watts | 3.412 | British thermal units per hour |
| watts | 0.738 | foot-pounds of force per second |
| watts per square meter | 0.317 | British thermal units per square foot |
| yards | 0.914 | meters |
| *Add your own conversion factors here.* | | |
| *Multiply* | *By* | *To Get* |
| *Multiply* | *By* | *To Get* |
| *Multiply* | *By* | *To Get* |

[a]K or °C.
[b]Thermal conductivity (k).
[c]Thermal conductance (C) or transmittance (U).

# Climatic Conditions for the United States, Canada, and Mexico

These climate data are taken from two ASHRAE publications, the 1997 *Handbook—Fundamentals* and ANSI/ASHRAE Standard 90.2-1993, *Energy Efficient Design of New Low-Rise Residential Buildings*, and are reprinted with permission of the American Society of Heating, Refrigerating and Air-Conditioning Engineers, Inc., Atlanta, Georgia.

STATION DESIGNATIONS

AP = airport

CO = urban areas, influenced by surroundings

No designation = semirural, similar to airport locations

## WINTER DESIGN CONDITIONS

- *HDD65* are annual heating degree-days at base 65°F (18.3°C), obtained from ANSI/ASHRAE Standard 90.2-1993: (HDD65 values are also given in Appendix C, and may vary slightly from these values because they are based upon data from a different set of years.)

- *Design dry-bulb* (97.5%) are outdoor temperatures (°F) that will, in a normal year, be the lowest experienced during 97.5% of the hours in a winter season defined as—

In the United States: December through February

In Canada: January

The implication of this set of values is that 2.5% of the winter hours will (statistically) see a temperature lower than the 97.5% design temperature; normally, this is quite acceptable (but this is a project-specific issue).

## SUMMER DESIGN CONDITIONS

- *Design dry-bulb and mean coincident wet-bulb* (2.5%) are outdoor temperatures (°F) that will, in a normal year, be the highest experienced during all but 2.5% of the hours in a summer season defined as—

In the United States: June through September

In Canada: July

- *Mean daily range* is the average difference (in F°) between daily maximum and minimum dry-bulb temperatures during the above-defined summer season.

- *Design wet-bulb* (2.5%) is the outdoor temperature (°F) that will, in a normal year, be the highest experienced during all but 2.5% of the hours in the summer season as defined above.

- *CDH74* are normal cooling degree-hours base 74°F (23.3°C), obtained from ANSI/ASHRAE Standard 90.2-1993.

Other stations: many more stations in each state or province, and for other countries, are found both in the current ASHRAE *Handbook—Fundamentals* and at various sites on the World Wide Web.

## INTERPRETATIONS BETWEEN STATIONS

As a general rule, weather data from listed stations can be adjusted for nearby locations in these ways:

• *Adjustment for elevation.* Elevations for many locations are given in Appendix C, Table C.15. For lower elevations, which tend to be warmer, *increase* the design temperatures:

Dry-bulb temperature: 1F° (0.56 C°) per 200 feet (60 m) of elevation

Wet-bulb temperature: 1F° (0.56 C°) per 500 feet (150 m) of elevation

• *Adjustment for air mass* (at coastlines). Along the West Coast, both dry-bulb and wet-bulb design temperatures increase with distance from the ocean. Along the Gulf Coast, dry-bulb temperatures increase for the first 200 to 300 miles (320 to 480 km) inland, while wet-bulb temperatures decrease slightly. (Beyond this 200- to 300-mile [320- to 480-km] belt, both dry-bulb and wet-bulb values decrease.)

### TABLE B.1 Outside Design Conditions: United States and Canada

| | | Winter | | Summer | | | |
|---|---|---|---|---|---|---|---|
| State and Station[a] | HDD65°F | Design Dry-Bulb (°F) (97.5%) | Design Dry-Bulb and Mean Coincident Wet-Bulb (°F) (2.5%) | Mean Daily Range (°F) | Design Wet-Bulb (°F) (2.5%) | CDH74°F |
| ALABAMA | | | | | | | |
| Auburn | 2614 | 22 | 93/76 | 21 | 78 | 19189 |
| Birmingham AP | 2876 | 21 | 94/75 | 21 | 77 | 20324 |
| Huntsville AP | 3289 | 16 | 93/74 | 23 | 77 | 18603 |
| Mobile AP | 1694 | 29 | 93/77 | 18 | 79 | 28912 |
| Montgomery AP | 2283 | 25 | 95/76 | 21 | 79 | 24564 |
| ALASKA | | | | | | | |
| Anchorage AP | 10825 | −18 | 68/58 | 15 | 59 | 35 |
| Fairbanks AP | 14280 | −47 | 78/60 | 24 | 62 | 752 |
| Juneau AP | 9113 | 1 | 70/58 | 15 | 59 | 34 |
| Nome AP | 14380 | −27 | 62/55 | 10 | 56 | 272 |
| ARIZONA | | | | | | | |
| Flagstaff AP | 7256 | 4 | 82/55 | 31 | 60 | 373 |
| Phoenix AP | 1444 | 34 | 107/71 | 27 | 75 | 54404 |
| Prescott AP | 4951 | 9 | 94/60 | 30 | 65 | 3828 |
| Tucson AP | 1737 | 32 | 102/66 | 26 | 71 | 35946 |
| Winslow AP | 4844 | 10 | 95/60 | 32 | 65 | 10636 |
| Yuma AP | 983 | 39 | 109/72 | 27 | 78 | 62507 |
| ARKANSAS | | | | | | | |
| Fayetteville AP | 4179 | 12 | 94/73 | 23 | 76 | 14956 |
| Fort Smith AP | 3482 | 17 | 98/76 | 24 | 79 | 23485 |
| Little Rock AP | 3161 | 20 | 96/77 | 22 | 79 | 23839 |
| Texarkana AP | 2506 | 23 | 96/77 | 21 | 79 | 27118 |
| CALIFORNIA | | | | | | | |
| Bakersfield AP | 2127 | 32 | 101/69 | 32 | 71 | 29954 |
| Barstow AP | 2585 | 29 | 104/68 | 37 | 71 | 27584 |

**TABLE B.1 Outside Design Conditions: United States and Canada (*Continued*)**

| State and Station[a] | HDD65°F | Winter Design Dry-Bulb (°F) (97.5%) | Summer Design Dry-Bulb and Mean Coincident Wet-Bulb (°F) (2.5%) | Mean Daily Range (°F) | Design Wet-Bulb (°F) (2.5%) | CDH74°F |
|---|---|---|---|---|---|---|
| Burbank AP | 1681 | 39 | 91/68 | 25 | 70 | 11400 |
| Eureka/Arcata AP | 4729 | 33 | 65/59 | 11 | 60 | 19 |
| Fresno AP | 2649 | 30 | 100/69 | 34 | 71 | 19366 |
| Long Beach AP | 1485 | 43 | 80/68 | 22 | 69 | 7755 |
| Los Angeles AP | 1595 | 43 | 80/68 | 15 | 69 | 4306 |
| Los Angeles CO | 1210 | 40 | 89/70 | 20 | 71 | 10575 |
| Needles AP | 1394 | 33 | 110/71 | 27 | 75 | 65218 |
| Oakland AP | 2880 | 36 | 80/63 | 19 | 64 | 435 |
| Pomona CO | 1973 | 30 | 99/69 | 36 | 72 | 10240 |
| Redding AP | 2548 | 31 | 102/67 | 32 | 69 | 27881 |
| Sacramento AP | 2775 | 32 | 98/70 | 36 | 71 | 10464 |
| San Diego AP | 1284 | 44 | 80/69 | 12 | 70 | 4643 |
| San Francisco AP | 3164 | 38 | 77/63 | 20 | 64 | 265 |
| San Francisco CO | 3078 | 40 | 71/62 | 14 | 62 | 216 |
| San Luis Obispo | 2503 | 35 | 88/70 | 26 | 71 | 1085 |
| Santa Barbara AP | 2493 | 36 | 77/66 | 24 | 67 | 894 |
| Santa Maria AP | 3064 | 33 | 76/63 | 23 | 64 | 131 |
| Santa Monica CO | 1873 | 43 | 80/68 | 16 | 69 | 1908 |
| COLORADO | | | | | | |
| Alamosa AP | 8718 | −16 | 82/57 | 35 | 61 | 46 |
| Boulder | 5466 | 8 | 91/59 | 27 | 63 | 7684 |
| Colorado Springs AP | 6353 | 2 | 88/57 | 30 | 62 | 3658 |
| Denver AP | 6023 | 1 | 91/59 | 28 | 63 | 5908 |
| Fort Collins | 6488 | −4 | 91/59 | 28 | 63 | 2946 |
| Grand Junction AP | 5684 | 7 | 94/59 | 29 | 63 | 12097 |
| Leadville | — | −4 | 81/51 | 30 | 55 | — |
| Pueblo AP | 5471 | 0 | 95/61 | 31 | 66 | 10983 |
| CONNECTICUT | | | | | | |
| Bridgeport AP | 5501 | 9 | 84/71 | 18 | 74 | 4977 |
| Hartford | 6197 | 7 | 88/73 | 22 | 75 | 4840 |
| New Haven AP | — | 7 | 84/73 | 17 | 75 | — |
| DELAWARE | | | | | | |
| Wilmington AP | 4989 | 14 | 89/74 | 20 | 76 | 8167 |
| DISTRICT OF COLUMBIA | | | | | | |
| Washington National AP | 4125 | 17 | 91/74 | 18 | 77 | 12389 |
| FLORIDA | | | | | | |
| Daytona Beach AP | 904 | 35 | 90/77 | 15 | 79 | 27107 |
| Fort Myers AP | 440 | 44 | 92/78 | 18 | 79 | 37444 |
| Gainesville AP | 1069 | 31 | 93/77 | 18 | 79 | 27679 |

**TABLE B.1 Outside Design Conditions: United States and Canada (*Continued*)**

| State and Station[a] | Winter | | Summer | | |
|---|---|---|---|---|---|
| | HDD65°F | Design Dry-Bulb (°F) (97.5%) | Design Dry-Bulb and Mean Coincident Wet-Bulb (°F) (2.5%) | Mean Daily Range (°F) | Design Wet-Bulb (°F) (2.5%) | CDH74°F |
| Jacksonville AP | 1402 | 32 | 94/77 | 19 | 79 | 24148 |
| Key West AP | 114 | 57 | 90/78 | 9 | 79 | 50236 |
| Miami AP | 198 | 47 | 90/77 | 15 | 79 | 39401 |
| Miami Beach CO | 132 | 48 | 89/77 | 10 | 79 | 39007 |
| Orlando AP | 660 | 38 | 93/76 | 17 | 78 | 33985 |
| Panama City, Tyndall AFB | — | 33 | 90/77 | 14 | 80 | — |
| Pensacola CO | 1570 | 29 | 93/77 | 14 | 79 | 29023 |
| Tallahassee AP | 1651 | 30 | 92/76 | 19 | 78 | 25185 |
| Tampa AP | 741 | 40 | 91/77 | 17 | 79 | 33677 |
| West Palm Beach AP | 262 | 45 | 91/78 | 16 | 79 | 35175 |
| GEORGIA | | | | | | |
| Athens | 2980 | 22 | 92/74 | 21 | 77 | 16147 |
| Atlanta AP | 3025 | 22 | 92/74 | 19 | 76 | 16803 |
| Augusta AP | 2571 | 23 | 95/76 | 19 | 79 | 19510 |
| Columbus, Lawson AFB | 2361 | 24 | 93/76 | 21 | 78 | 22115 |
| Macon AP | 2282 | 25 | 93/76 | 22 | 78 | 24443 |
| Rome AP | 3136 | 22 | 93/76 | 23 | 78 | 17745 |
| Savannah-Travis AP | 1921 | 27 | 93/77 | 20 | 79 | 22773 |
| Valdosta-Moody AFB | — | 31 | 94/77 | 20 | 79 | — |
| HAWAII | | | | | | |
| Hilo AP | 0 | 62 | 83/72 | 15 | 74 | 14770 |
| Honolulu AP | 0 | 63 | 86/73 | 12 | 75 | 36213 |
| Wahiawa | — | 59 | 85/72 | 14 | 74 | — |
| IDAHO | | | | | | |
| Boise AP | 5808 | 10 | 94/64 | 31 | 66 | 7979 |
| Lewiston AP | 5435 | 6 | 93/64 | 32 | 66 | 7923 |
| Moscow | 6691 | 0 | 87/62 | 32 | 64 | 1986 |
| Pocatello AP | 7131 | −1 | 91/60 | 35 | 63 | 3286 |
| ILLINOIS | | | | | | |
| Carbondale | 4568 | 7 | 93/77 | 21 | 79 | 14114 |
| Champaign/Urbana | 5766 | 2 | 92/74 | 21 | 77 | 9949 |
| Chicago, O'Hare AP | 6459 | −4 | 89/74 | 20 | 76 | 6606 |
| Chicago CO | 6013 | 2 | 91/74 | 15 | 77 | 8196 |
| Moline AP | 6504 | −4 | 91/75 | 23 | 77 | 8616 |
| Peoria AP | 6233 | −4 | 89/74 | 22 | 76 | 9503 |
| Rockford | 6955 | −4 | 89/73 | 24 | 76 | 6523 |
| Springfield AP | 5662 | 2 | 92/74 | 21 | 77 | 12438 |
| INDIANA | | | | | | |
| Evansville AP | 4734 | 9 | 93/75 | 22 | 78 | 14797 |
| Fort Wayne AP | 6324 | 1 | 89/72 | 24 | 75 | 6816 |

**TABLE B.1 Outside Design Conditions: United States and Canada (*Continued*)**

| State and Station[a] | HDD65°F | Winter Design Dry-Bulb (°F) (97.5%) | Summer Design Dry-Bulb and Mean Coincident Wet-Bulb (°F) (2.5%) | Mean Daily Range (°F) | Design Wet-Bulb (°F) (2.5%) | CDH74°F |
|---|---|---|---|---|---|---|
| Indianapolis AP | 5653 | 2 | 90/74 | 22 | 76 | 9082 |
| Lafayette | 6229 | 3 | 91/73 | 22 | 76 | 7728 |
| Muncie | — | 2 | 90/73 | 22 | 76 | — |
| South Bend AP | 6381 | 1 | 89/73 | 22 | 75 | 6606 |
| Terre Haute AP | 5527 | 4 | 92/74 | 22 | 77 | 9499 |
| IOWA | | | | | | |
| Ames | 6882 | –6 | 90/74 | 23 | 76 | 7453 |
| Burlington AP | 6094 | –3 | 91/75 | 22 | 77 | 8277 |
| Des Moines AP | 6563 | –5 | 91/74 | 23 | 77 | 10424 |
| Dubuque | 7379 | –7 | 88/73 | 22 | 75 | 4661 |
| Iowa City | 6378 | –6 | 89/76 | 22 | 78 | 10098 |
| Mason City AP | 7886 | –11 | 88/74 | 24 | 75 | 5584 |
| Sioux City AP | 6953 | –7 | 92/74 | 24 | 77 | 10117 |
| Waterloo | 7542 | –10 | 89/75 | 23 | 77 | 6641 |
| KANSAS | | | | | | |
| Dodge City AP | 5069 | 5 | 97/69 | 25 | 73 | 18470 |
| Goodland AP | 6703 | 0 | 96/65 | 31 | 70 | 10079 |
| Manhattan, Fort Riley | 5122 | 3 | 95/75 | 24 | 77 | 18826 |
| Topeka AP | 5323 | 4 | 96/75 | 24 | 78 | 16640 |
| Wichita AP | 4791 | 7 | 98/73 | 23 | 76 | 21208 |
| KENTUCKY | | | | | | |
| Bowling Green AP | 4314 | 10 | 92/75 | 21 | 77 | 14745 |
| Lexington AP | 4819 | 8 | 91/73 | 22 | 76 | 11208 |
| Louisville AP | 4529 | 10 | 93/74 | 23 | 77 | 13268 |
| LOUISIANA | | | | | | |
| Alexandria AP | 1960 | 27 | 94/77 | 20 | 79 | 27315 |
| Baton Rouge AP | 1673 | 29 | 93/77 | 19 | 80 | 26938 |
| Lafayette AP | 1560 | 30 | 94/78 | 18 | 80 | 28543 |
| Lake Charles AP | 1580 | 31 | 93/77 | 17 | 79 | 28573 |
| Monroe AP | 2406 | 25 | 96/76 | 20 | 79 | 26604 |
| New Orleans AP | 1311 | 33 | 92/78 | 16 | 80 | 32758 |
| Shreveport AP | 2271 | 25 | 96/76 | 20 | 79 | 28295 |
| MAINE | | | | | | |
| Bangor, Dow AFB | 7951 | –6 | 83/68 | 22 | 71 | 1232 |
| Caribou AP | 9621 | –13 | 81/67 | 21 | 69 | 920 |
| Portland | 7502 | –7 | 84/71 | 22 | 72 | 1104 |
| MARYLAND | | | | | | |
| Baltimore AP | 4714 | 13 | 91/75 | 21 | 77 | 9504 |
| Baltimore CO | 4086 | 17 | 89/76 | 17 | 78 | 15162 |
| Cumberland | 5108 | 10 | 89/74 | 22 | 76 | 7088 |

**TABLE B.1 Outside Design Conditions: United States and Canada (*Continued*)**

| State and Station[a] | HDD65°F | Winter<br>Design<br>Dry-Bulb<br>(°F)<br>(97.5%) | Summer<br>Design Dry-Bulb<br>and<br>Mean Coincident<br>Wet-Bulb<br>(°F)<br>(2.5%) | Mean<br>Daily<br>Range<br>(°F) | Design<br>Wet-Bulb<br>(°F)<br>(2.5%) | CDH74°F |
|---|---|---|---|---|---|---|
| Hagerstown | 5092 | 12 | 91/74 | 22 | 76 | 7346 |
| Salisbury | 4415 | 16 | 91/75 | 18 | 77 | 9166 |
| MASSACHUSETTS | | | | | | |
| Boston AP | 5596 | 9 | 88/71 | 16 | 74 | 5358 |
| Springfield | 5955 | 0 | 87/71 | 19 | 73 | 5205 |
| Worcester AP | 6951 | 4 | 84/70 | 18 | 72 | 1504 |
| MICHIGAN | | | | | | |
| Alpena AP | 8413 | −6 | 85/70 | 27 | 72 | 837 |
| Detroit | 6564 | 6 | 88/72 | 20 | 74 | 4889 |
| Escanaba | 8549 | −7 | 83/69 | 17 | 71 | 805 |
| Flint AP | 7067 | 1 | 87/72 | 25 | 74 | 2929 |
| Grand Rapids AP | 6927 | 5 | 88/72 | 24 | 74 | 4555 |
| Muskegon AP | 6926 | 6 | 84/70 | 21 | 73 | 2869 |
| Sault Ste. Marie AP | 9309 | −8 | 81/69 | 23 | 70 | 844 |
| Traverse City AP | 7800 | 1 | 86/71 | 22 | 73 | 2968 |
| MINNESOTA | | | | | | |
| Bemidji AP | 10207 | −26 | 85/69 | 24 | 71 | 2191 |
| Duluth AP | 9906 | −16 | 82/68 | 22 | 70 | 849 |
| International Falls AP | 10607 | −25 | 83/68 | 26 | 70 | 1614 |
| Minneapolis/St. Paul AP | 8010 | −12 | 89/73 | 22 | 75 | 6809 |
| Rochester AP | 8279 | −12 | 87/72 | 24 | 75 | 3901 |
| MISSISSIPPI | | | | | | |
| Columbus AFB | 2868 | 20 | 93/77 | 22 | 79 | 21829 |
| Jackson AP | 2390 | 25 | 95/76 | 21 | 78 | 25152 |
| Meridian AP | 2478 | 23 | 95/76 | 22 | 79 | 23827 |
| Vicksburg CO | 2200 | 26 | 95/78 | 21 | 80 | 24595 |
| MISSOURI | | | | | | |
| Columbia AP | 5214 | 4 | 94/74 | 22 | 77 | 14475 |
| Kansas City AP | 4814 | 6 | 96/74 | 20 | 77 | 20256 |
| St Louis AP | 4948 | 6 | 94/75 | 21 | 77 | 17843 |
| St Louis CO | — | 8 | 94/75 | 18 | 77 | — |
| Springfield AP | 4665 | 9 | 93/74 | 23 | 77 | 16262 |
| MONTANA | | | | | | |
| Billings AP | 7220 | −10 | 91/64 | 31 | 66 | 6167 |
| Bozeman | 9876 | −14 | 87/60 | 32 | 62 | 407 |
| Cut Bank AP | 9161 | −20 | 85/61 | 35 | 62 | 1378 |
| Glasgow AP | 8952 | −18 | 89/63 | 29 | 66 | 4762 |
| Great Falls AP | 7784 | −15 | 88/60 | 28 | 62 | 3574 |
| Helena AP | 8186 | −16 | 88/60 | 32 | 62 | 2547 |

**TABLE B.1 Outside Design Conditions: United States and Canada (*Continued*)**

| State and Station[a] | HDD65°F | Winter Design Dry-Bulb (°F) (97.5%) | Summer Design Dry-Bulb and Mean Coincident Wet-Bulb (°F) (2.5%) | Mean Daily Range (°F) | Design Wet-Bulb (°F) (2.5%) | CDH74°F |
|---|---|---|---|---|---|---|
| Kalispell AP | 8363 | −7 | 87/61 | 34 | 63 | 1210 |
| Lewistown AP | 8623 | −16 | 87/61 | 30 | 63 | 2127 |
| Miles City AP | 7907 | −15 | 95/66 | 30 | 68 | 10094 |
| Missoula AP | 7844 | −6 | 88/61 | 36 | 63 | 1068 |
| NEBRASKA | | | | | | |
| Grand Island AP | 6491 | −3 | 94/71 | 28 | 74 | 11957 |
| Lincoln CO | 5974 | −2 | 95/74 | 24 | 77 | 14607 |
| North Platte AP | 6914 | −4 | 94/69 | 28 | 72 | 8492 |
| Omaha AP | 6201 | −3 | 91/75 | 22 | 77 | 13180 |
| Scottsbluff AP | 6711 | −3 | 92/65 | 31 | 68 | 7339 |
| NEVADA | | | | | | |
| Elko AP | 7252 | −2 | 92/59 | 42 | 62 | 3840 |
| Ely AP | 7705 | −4 | 87/56 | 39 | 59 | 659 |
| Las Vegas AP | 2535 | 28 | 106/65 | 30 | 70 | 43153 |
| Lovelock AP | 5843 | 12 | 96/63 | 42 | 65 | 7845 |
| Reno AP | 6032 | 10 | 92/60 | 45 | 62 | 2192 |
| Tonopah AP | 5754 | 10 | 92/59 | 40 | 62 | 5888 |
| Winnemucca AP | 6417 | 3 | 94/60 | 42 | 62 | 4284 |
| NEW HAMPSHIRE | | | | | | |
| Berlin | — | −9 | 84/69 | 22 | 71 | — |
| Concord AP | 7483 | −3 | 87/70 | 26 | 73 | 1993 |
| Portsmouth, Pease AFB | 6919 | 2 | 85/71 | 22 | 74 | 2239 |
| NEW JERSEY | | | | | | |
| Atlantic City CO | 4766 | 13 | 89/74 | 18 | 77 | 5880 |
| Newark AP | 4977 | 14 | 91/73 | 20 | 76 | 9107 |
| Trenton CO | 4953 | 14 | 88/74 | 19 | 76 | 7393 |
| NEW MEXICO | | | | | | |
| Albuquerque AP | 4415 | 16 | 94/61 | 27 | 65 | 11012 |
| Farmington AP | 5734 | 6 | 93/62 | 30 | 65 | 5047 |
| Las Cruces | 3120 | 20 | 96/64 | 30 | 68 | 14507 |
| Los Alamos | 6389 | 9 | 87/60 | 32 | 61 | 1235 |
| Raton AP | — | 1 | 89/60 | 34 | 64 | — |
| Roswell, Walker AFB | 3131 | 18 | 98/66 | 33 | 70 | 20216 |
| Silver City AP | — | 10 | 94/60 | 30 | 64 | — |
| Tucumcari AP | 3938 | 13 | 97/66 | 28 | 69 | 14754 |
| NEW YORK | | | | | | |
| Albany AP | 6929 | −1 | 88/72 | 23 | 74 | 2998 |
| Binghamton AP | 7346 | 1 | 83/69 | 20 | 72 | 1646 |
| Buffalo AP | 6799 | 6 | 85/70 | 21 | 73 | 3044 |

APPENDICES

**TABLE B.1 Outside Design Conditions: United States and Canada (*Continued*)**

| State and Station[a] | HDD65°F | Winter Design Dry-Bulb (°F) (97.5%) | Summer Design Dry-Bulb and Mean Coincident Wet-Bulb (°F) (2.5%) | Mean Daily Range (°F) | Design Wet-Bulb (°F) (2.5%) | CDH74°F |
|---|---|---|---|---|---|---|
| Ithaca | 7182 | 0 | 85/71 | 24 | 73 | 1583 |
| Massena AP | 8184 | −8 | 83/69 | 20 | 72 | 2012 |
| NYC—Central Park | 4869 | 15 | 89/73 | 17 | 75 | 9537 |
| NYC—Kennedy AP | 5171 | 15 | 87/72 | 16 | 75 | 7634 |
| Rochester AP | 6718 | 5 | 88/71 | 22 | 73 | 3764 |
| Syracuse AP | 6789 | 2 | 87/71 | 20 | 73 | 3516 |
| NORTH CAROLINA | | | | | | |
| Asheville AP | 4300 | 14 | 87/72 | 21 | 74 | 6292 |
| Charlotte AP | 3348 | 22 | 93/74 | 20 | 76 | 15159 |
| Greensboro AP | 3877 | 18 | 91/73 | 21 | 76 | 11020 |
| Raleigh/Durham AP | 3538 | 20 | 92/75 | 20 | 77 | 11845 |
| Wilmington AP | 2473 | 26 | 91/78 | 18 | 80 | 17648 |
| NORTH DAKOTA | | | | | | |
| Bismarck AP | 9080 | −19 | 91/68 | 27 | 71 | 4554 |
| Fargo AP | 9349 | −18 | 89/71 | 25 | 74 | 4284 |
| Minot AP | 9425 | −20 | 89/67 | 25 | 70 | 4157 |
| Williston | 9252 | −21 | 88/67 | 25 | 70 | 4034 |
| OHIO | | | | | | |
| Akron-Canton AP | 6248 | 6 | 86/71 | 21 | 73 | 4808 |
| Cincinnati CO | 4958 | 6 | 90/72 | 21 | 75 | 10698 |
| Cleveland AP | 6179 | 5 | 88/72 | 22 | 74 | 4772 |
| Columbus AP | 5692 | 5 | 90/73 | 24 | 75 | 7490 |
| Dayton AP | 5260 | 4 | 89/72 | 20 | 75 | 11400 |
| Mansfield AP | 6254 | 5 | 87/72 | 22 | 74 | 4948 |
| Toledo AP | 6576 | 1 | 88/73 | 25 | 75 | 5081 |
| Youngstown AP | 6562 | 4 | 86/71 | 23 | 73 | 2975 |
| OKLAHOMA | | | | | | |
| Norman | — | 13 | 96/74 | 24 | 76 | — |
| Oklahoma City AP | 3742 | 13 | 97/74 | 23 | 77 | 22978 |
| Stillwater | 3802 | 13 | 96/74 | 24 | 76 | 23621 |
| Tulsa AP | 3741 | 13 | 98/75 | 22 | 78 | 26468 |
| OREGON | | | | | | |
| Astoria AP | 5250 | 29 | 71/62 | 16 | 63 | 30 |
| Bend | 7082 | 4 | 87/60 | 33 | 62 | 647 |
| Eugene AP | 4803 | 22 | 89/66 | 31 | 67 | 1296 |
| Medford AP | 4803 | 23 | 94/67 | 35 | 68 | 6151 |
| Pendleton AP | 5275 | 5 | 93/64 | 29 | 65 | 8121 |
| Portland AP | 4693 | 23 | 85/67 | 23 | 67 | 1851 |
| Portland CO | 4417 | 24 | 86/67 | 21 | 67 | 2105 |
| Salem AP | 4978 | 23 | 88/66 | 31 | 68 | 1081 |

**TABLE B.1 Outside Design Conditions: United States and Canada (*Continued*)**

| State and Station[a] | HDD65°F | Winter Design Dry-Bulb (°F) (97.5%) | Design Dry-Bulb and Mean Coincident Wet-Bulb (°F) (2.5%) | Summer Mean Daily Range (°F) | Design Wet-Bulb (°F) (2.5%) | CDH74°F |
|---|---|---|---|---|---|---|
| PENNSYLVANIA | | | | | | |
| Allentown AP | 5815 | 9 | 88/72 | 22 | 75 | 5804 |
| Erie AP | 6768 | 9 | 85/72 | 18 | 74 | 2222 |
| Harrisburg AP | 5339 | 11 | 91/74 | 21 | 76 | 9071 |
| Philadelphia AP | 4954 | 14 | 90/74 | 21 | 76 | 8896 |
| Pittsburgh AP | 5957 | 5 | 86/71 | 22 | 73 | 5009 |
| Pittsburgh CO | — | 7 | 88/71 | 19 | 73 | — |
| Scranton/Wilkes-Barre | 6332 | 5 | 87/71 | 19 | 73 | 3774 |
| State College | 6252 | 7 | 87/71 | 23 | 73 | 3484 |
| RHODE ISLAND | | | | | | |
| Newport | 6122 | 9 | 85/72 | 16 | 75 | 1682 |
| Providence AP | 5909 | 9 | 86/72 | 19 | 74 | 3631 |
| SOUTH CAROLINA | | | | | | |
| Anderson | 2960 | 23 | 92/74 | 21 | 76 | 16014 |
| Charleston AFB | 2146 | 27 | 91/78 | 18 | 80 | 20315 |
| Charleston CO | 1868 | 28 | 92/78 | 13 | 80 | 23283 |
| Columbia AP | 2632 | 24 | 95/75 | 22 | 78 | 21805 |
| Spartanburg AP | 3246 | 22 | 91/74 | 20 | 76 | 14069 |
| SOUTH DAKOTA | | | | | | |
| Huron AP | 8109 | −14 | 93/72 | 28 | 75 | 8677 |
| Pierre AP | 7580 | −10 | 95/71 | 29 | 74 | 10444 |
| Rapid City AP | 7308 | −7 | 92/65 | 28 | 69 | 8286 |
| Sioux Falls AP | 7890 | −11 | 91/72 | 24 | 75 | 8638 |
| TENNESSEE | | | | | | |
| Bristol-Tri City AP | 4367 | 14 | 89/72 | 22 | 75 | 8954 |
| Chattanooga AP | 3591 | 18 | 93/74 | 22 | 77 | 17017 |
| Knoxville AP | 3666 | 19 | 92/73 | 21 | 76 | 14214 |
| Memphis AP | 3214 | 18 | 95/76 | 21 | 79 | 24504 |
| Nashville AP | 3767 | 14 | 94/74 | 21 | 77 | 18543 |
| TEXAS | | | | | | |
| Abilene AP | 2628 | 20 | 99/71 | 22 | 74 | 31904 |
| Amarillo AP | 4240 | 11 | 95/67 | 26 | 70 | 15718 |
| Austin AP | 1759 | 28 | 98/74 | 22 | 77 | 35180 |
| Brownsville AP | 607 | 39 | 93/77 | 18 | 79 | 42529 |
| Bryan AP | 1784 | 29 | 96/76 | 20 | 78 | 33178 |
| Corpus Christi AP | 970 | 35 | 94/78 | 19 | 80 | 42515 |
| Dallas AP | 2303 | 22 | 100/75 | 20 | 78 | 36697 |
| Del Rio, Laughlin AFB | 1511 | 31 | 98/73 | 24 | 77 | 42032 |
| El Paso AP | 2672 | 24 | 98/64 | 27 | 68 | 22966 |
| Fort Worth AP | 2420 | 22 | 99/74 | 22 | 77 | 36294 |

**TABLE B.1  Outside Design Conditions: United States and Canada (*Continued*)**

| State and Station[a] | HDD65°F | Winter Design Dry-Bulb (°F) (97.5%) | Summer Design Dry-Bulb and Mean Coincident Wet-Bulb (°F) (2.5%) | Mean Daily Range (°F) | Design Wet-Bulb (°F) (2.5%) | CDH74°F |
|---|---|---|---|---|---|---|
| Galveston AP | 1252 | 36 | 89/79 | 10 | 80 | 31911 |
| Houston AP | 1548 | 32 | 94/77 | 18 | 79 | 30474 |
| Houston CO | — | 33 | 95/77 | 18 | 79 | — |
| Laredo AFB | 927 | 36 | 101/73 | 23 | 78 | 52560 |
| Lubbock AP | 3522 | 15 | 96/69 | 26 | 72 | 18218 |
| Lufkin AP | 1929 | 29 | 97/76 | 20 | 79 | 30370 |
| Midland AP | 2664 | 21 | 98/69 | 26 | 72 | 24201 |
| Port Arthur AP | 1476 | 31 | 93/78 | 19 | 80 | 31667 |
| San Angelo, Goodfellow AFB | 2325 | 22 | 99/71 | 24 | 74 | 32735 |
| San Antonio AP | 1605 | 30 | 97/73 | 19 | 76 | 36179 |
| Sherman Perrin AFB | 2943 | 20 | 98/75 | 22 | 77 | 29686 |
| Waco AP | 2128 | 26 | 99/75 | 22 | 78 | 36671 |
| Wichita Falls AP | 3017 | 18 | 101/73 | 24 | 76 | 34510 |
| UTAH | | | | | | |
| Cedar City AP | 6002 | 5 | 91/60 | 32 | 63 | 5036 |
| Salt Lake City AP | 5805 | 8 | 95/62 | 32 | 65 | 9898 |
| Vernal AP | 7671 | 0 | 89/60 | 32 | 63 | 2678 |
| VERMONT | | | | | | |
| Barre | 8529 | −11 | 81/69 | 23 | 71 | 1377 |
| Burlington AP | 7956 | −7 | 85/70 | 23 | 72 | 2562 |
| VIRGINIA | | | | | | |
| Charlottesville | 4195 | 18 | 91/74 | 23 | 76 | 10287 |
| Norfolk AP | 3458 | 22 | 91/76 | 18 | 78 | 13677 |
| Richmond AP | 3969 | 17 | 92/76 | 21 | 78 | 12260 |
| Roanoke AP | 4321 | 16 | 91/72 | 23 | 74 | 9275 |
| WASHINGTON | | | | | | |
| Bellingham AP | 5726 | 15 | 77/65 | 19 | 65 | 185 |
| Olympia AP | 5710 | 22 | 83/65 | 32 | 66 | 341 |
| Seattle CO | 4684 | 27 | 82/66 | 19 | 67 | 897 |
| Seattle–Tacoma AP | 5122 | 26 | 80/64 | 22 | 64 | 1050 |
| Spokane AP | 6885 | 2 | 90/63 | 28 | 64 | 3453 |
| Walla Walla AP | 5057 | 7 | 94/66 | 27 | 67 | 9511 |
| Yakima AP | 6035 | 5 | 93/65 | 36 | 66 | 4118 |
| WEST VIRGINIA | | | | | | |
| Beckley | 5585 | 4 | 81/69 | 22 | 71 | 2408 |
| Charleston AP | 4705 | 11 | 90/73 | 20 | 75 | 8790 |
| Huntington CO | 4687 | 10 | 91/74 | 22 | 77 | 11185 |
| Wheeling | 5456 | 5 | 86/71 | 21 | 73 | 6793 |

**TABLE B.1 Outside Design Conditions: United States and Canada (*Continued*)**

| | | Winter | | Summer | | |
| State and Station[a] | HDD65°F | Design Dry-Bulb (°F) (97.5%) | Design Dry-Bulb and Mean Coincident Wet-Bulb (°F) (2.5%) | Mean Daily Range (°F) | Design Wet-Bulb (°F) (2.5%) | CDH74°F |
|---|---|---|---|---|---|---|
| WISCONSIN | | | | | | |
| Ashland | 9067 | −16 | 82/68 | 23 | 70 | 1288 |
| Eau Claire AP | 8466 | −11 | 89/73 | 23 | 75 | 3886 |
| Green Bay AP | 8146 | −9 | 85/72 | 23 | 74 | 2453 |
| La Cross AP | 7544 | −9 | 88/73 | 22 | 75 | 6795 |
| Madison AP | 7643 | −7 | 88/73 | 22 | 75 | 3343 |
| Milwaukee AP | 7327 | −4 | 87/73 | 21 | 74 | 3313 |
| WYOMING | | | | | | |
| Casper AP | 7649 | −5 | 90/57 | 31 | 61 | 4370 |
| Cheyenne AP | 7315 | −1 | 86/58 | 30 | 62 | 2087 |
| Lander AP | 7915 | −11 | 88/61 | 32 | 63 | 3940 |
| Laramie AP | 8933 | −6 | 81/56 | 28 | 60 | 428 |
| Rock Springs AP | 8358 | −3 | 84/55 | 32 | 58 | 1043 |
| Sheridan AP | 7852 | −8 | 91/62 | 32 | 65 | 4115 |
| ALBERTA | | | | | | |
| Calgary AP | 9657 | −23 | 81/61 | 25 | 63 | 1176 |
| Edmonton AP | 10782 | −25 | 82/65 | 23 | 66 | 1137 |
| Medicine Hat AP | 8762 | −24 | 90/65 | 28 | 68 | 3291 |
| BRITISH COLUMBIA | | | | | | |
| Dawson Creek | 11212 | −33 | 79/63 | 26 | 64 | 427 |
| Nanaimo | 5834 | 20 | 80/65 | 21 | 66 | 560 |
| Prince George AP | 9677 | −28 | 80/62 | 26 | 64 | 431 |
| Prince Rupert CO | 7398 | 2 | 63/57 | 12 | 58 | 3 |
| Vancouver AP | 5454 | 19 | 77/66 | 17 | 67 | 221 |
| Victoria CO | 5607 | 23 | 73/62 | 16 | 62 | 42 |
| MANITOBA | | | | | | |
| Flin Flon | 12551 | −37 | 81/66 | 19 | 68 | 1424 |
| Winnipeg AP | 10661 | −27 | 86/71 | 22 | 73 | 2615 |
| NEW BRUNSWICK | | | | | | |
| Fredericton AP | 8530 | −11 | 85/69 | 23 | 71 | 1000 |
| Saint John AP | 8582 | −8 | 77/65 | 19 | 68 | 49 |
| NEWFOUNDLAND | | | | | | |
| Corner Brook | 8550 | 0 | 73/63 | 17 | 66 | 505 |
| St. John's AP | 8683 | 7 | 75/65 | 18 | 67 | 1163 |
| NORTHWEST TERR. | | | | | | |
| Fort Smith AP | 14013 | −45 | 81/64 | 24 | 66 | 724 |
| Inuvik | 18182 | −53 | 77/60 | 21 | 62 | 1055 |
| Yellowknife AP | 15352 | −46 | 77/61 | 16 | 63 | 655 |

**TABLE B.1 Outside Design Conditions: United States and Canada (*Continued*)**

| State and Station[a] | | Winter | | Summer | | | |
|---|---|---|---|---|---|---|---|
| | HDD65°F | Design Dry-Bulb (°F) (97.5%) | Design Dry-Bulb and Mean Coincident Wet-Bulb (°F) (2.5%) | Mean Daily Range (°F) | Design Wet-Bulb (°F) (2.5%) | CDH74°F |
| NOVA SCOTIA | | | | | | |
| Halifax AP | 7963 | 5 | 76/65 | 16 | 67 | 332 |
| Sydney AP | 8172 | 3 | 80/68 | 19 | 70 | 668 |
| Yarmouth AP | 7315 | 9 | 72/64 | 15 | 66 | 14 |
| ONTARIO | | | | | | |
| Kitchener | 7463 | −2 | 85/72 | 23 | 74 | 2447 |
| Ottawa AP | 8442 | −13 | 87/71 | 21 | 73 | 2051 |
| Sudbury AP | 9812 | −19 | 83/67 | 22 | 70 | 1222 |
| Toronto AP | 7457 | −1 | 87/72 | 20 | 74 | 2676 |
| PRINCE EDWARD ISLAND | | | | | | |
| Charlottetown AP | 8438 | −4 | 78/68 | 16 | 70 | 409 |
| QUEBEC | | | | | | |
| Chicoutimi | 9781 | −22 | 83/68 | 20 | 70 | 1135 |
| Montréal AP | 8167 | −10 | 85/72 | 17 | 74 | 2035 |
| Québec AP | 9297 | −14 | 84/70 | 20 | 72 | 882 |
| Sept Iles AP | 11077 | −21 | 73/61 | 17 | 65 | 225 |
| Val d'Or AP | 11158 | −27 | 83/68 | 22 | 70 | 1269 |
| SASKATCHEWAN | | | | | | |
| Prince Albert AP | 11806 | −35 | 84/66 | 25 | 68 | 502 |
| Regina AP | 10577 | −29 | 88/68 | 26 | 70 | 1973 |
| Saskatoon AP | 10912 | −31 | 86/66 | 26 | 68 | 1439 |
| YUKON TERRITORY | | | | | | |
| Whitehorse AP | 12578 | −43 | 77/58 | 22 | 59 | 454 |

[a]U.S. solar data are available from the National Climatic Data Center, Federal Building, Asheville, NC 28801. Canadian solar data are available from The Canadian Climate Center, Atmospheric Environment Service, 4905 Dufferin St., Downsview, Ontario M3H 5T4.

**TABLE B.2 Outdoor Design Conditions: Mexico and Puerto Rico**

| Country and Station | Winter[a] | | Summer[b] | | | |
|---|---|---|---|---|---|---|
| | HDD65°F | Design Dry-Bulb (°F) 97.5% | Design Dry-Bulb (°F) 2.5% | Mean Daily Range (°F) | Design Wet-Bulb (°F) 2.5% | CDH74°F |
| MEXICO | | | | | | |
| Guadalajara | — | 42 | 91 | 29 | 67 | — |
| Mérida | — | 61 | 95 | 21 | 79 | — |
| Mexico City | — | 39 | 81 | 25 | 60 | — |
| Monterrey | — | 41 | 95 | 20 | 78 | — |
| Vera Cruz | — | 62 | 89 | 20 | 83 | — |
| PUERTO RICO | | | | | | |
| San Juan | 0 | 68 | 88 | 11 | 80 | 55224 |

[a]For the 3-month winter period.
[b]For the 4-month summer period.

# Solar and Daylighting Design Data

This appendix contains the following information.

### Solar Intensity and Solar Heat Gain Factors (Tables C.1–C.10)

These values are listed for five latitudes (16, 32, 40, 48, and 64 °N) in both I-P and SI units (the tables are so labeled). Solar heat gain factors represent the solar radiation gain through one layer of double-strength sheet (DSA) glass. The data are based on an "average cloudless" day at the respective latitudes. Tables C.1 through C.10 are copyrighted by the American Society of Heating, Refrigerating and Air-Conditioning Engineers, Inc., Atlanta, GA, and are reprinted by permission from the 1989 *ASHRAE Handbook—Fundamentals.*

### Solar Position and Clear Day Global Irradiance (Insolation) (Tables C.11–C.14)

These values are listed for four latitudes (32, 40, 48, and 64 °N). The data are presented in I-P units for the following tilt angles (degrees above the horizontal): latitude −10°, latitude, latitude +10°, lati-

tude +20°, and vertical. To convert to SI units: Btu/h ft$^2$ × 3.152 = W/m$^2$.

### Elevation, Latitude, Average Horizontal Insolation, Average Vertical Insolation, Average Temperature, and Heating Degree Days (Table C.15)

These values are listed in I-P units for January and July, along with yearly totals, for a number of locations (Fig. C.1). To convert horizontal surface (HS) and vertical surface (VS) insolation values to SI units: Btu/h ft$^2$ × 3.152 = W/m$^2$. To convert average temperature (TA) to Celsius: (°F − 32)/1.8 = °C. To convert DD(F) to DD(C): DD(F) × ⅝ = DD(C) or DD(K).

### Daylighting Coefficient of Utilization Values for Various Window and Sky Configurations (Tables C.16–C.21)

### Reflectances of Materials used to Construct Daylighting Models (Table C.22)

## TABLE C.1  Solar Intensity and Solar Heat Gain Factors[a] for 16°N Latitude (I-P Units)[b]

| Date | Solar Time A.M. | Direct Normal Btu/h ft² | N | NNE | NE | ENE | E | ESE | SE | SSE | S | SSW | SW | WSW | W | WNW | NW | NNW | HOR | Solar Time P.M. |
|---|---|---|---|---|---|---|---|---|---|---|---|---|---|---|---|---|---|---|---|---|
| Jan 21 | 7 | 141 | 5 | 6 | 44 | 92 | 124 | 134 | 126 | 96 | 49 | 6 | 5 | 5 | 5 | 5 | 5 | 5 | 14 | 5 |
|  | 8 | 262 | 14 | 15 | 55 | 147 | 210 | 240 | 233 | 189 | 114 | 25 | 14 | 14 | 14 | 14 | 14 | 14 | 79 | 4 |
|  | 9 | 300 | 21 | 21 | 32 | 122 | 200 | 244 | 251 | 219 | 152 | 58 | 22 | 21 | 21 | 21 | 21 | 21 | 150 | 3 |
|  | 10 | 317 | 26 | 26 | 27 | 66 | 150 | 209 | 233 | 223 | 178 | 102 | 31 | 26 | 26 | 26 | 26 | 26 | 203 | 2 |
|  | 11 | 325 | 29 | 29 | 29 | 31 | 77 | 148 | 195 | 210 | 194 | 146 | 75 | 31 | 29 | 29 | 29 | 29 | 236 | 1 |
|  | 12 | 327 | 30 | 30 | 30 | 30 | 32 | 73 | 139 | 184 | 199 | 184 | 138 | 72 | 32 | 30 | 30 | 30 | 248 | 12 |
|  | HALF-DAY TOTALS |  | 110 | 112 | 196 | 461 | 760 | 1000 | 1096 | 1020 | 781 | 426 | 211 | 127 | 111 | 110 | 110 | 110 | 805 |  |
| Feb 21 | 7 | 182 | 8 | 17 | 84 | 138 | 169 | 172 | 150 | 103 | 36 | 8 | 8 | 8 | 8 | 8 | 8 | 8 | 25 | 5 |
|  | 8 | 273 | 17 | 19 | 96 | 180 | 231 | 247 | 224 | 166 | 77 | 18 | 17 | 17 | 17 | 17 | 17 | 17 | 101 | 4 |
|  | 9 | 305 | 23 | 24 | 64 | 153 | 214 | 242 | 233 | 188 | 110 | 30 | 23 | 23 | 23 | 23 | 23 | 23 | 174 | 3 |
|  | 10 | 319 | 28 | 29 | 33 | 92 | 161 | 202 | 211 | 188 | 134 | 61 | 30 | 28 | 28 | 28 | 28 | 28 | 229 | 2 |
|  | 11 | 326 | 32 | 32 | 32 | 37 | 83 | 136 | 167 | 172 | 149 | 102 | 49 | 33 | 32 | 32 | 32 | 32 | 263 | 1 |
|  | 12 | 328 | 33 | 33 | 33 | 33 | 34 | 60 | 107 | 142 | 154 | 142 | 106 | 60 | 34 | 33 | 33 | 33 | 275 | 12 |
|  | HALF-DAY TOTALS |  | 124 | 137 | 321 | 609 | 865 | 1023 | 1034 | 885 | 582 | 287 | 174 | 132 | 124 | 124 | 124 | 124 | 930 |  |
| Mar 21 | 7 | 201 | 11 | 53 | 124 | 172 | 192 | 183 | 145 | 82 | 15 | 10 | 10 | 10 | 10 | 10 | 10 | 10 | 40 | 5 |
|  | 8 | 272 | 20 | 50 | 140 | 205 | 239 | 235 | 195 | 123 | 35 | 19 | 19 | 19 | 19 | 19 | 19 | 19 | 120 | 4 |
|  | 9 | 299 | 26 | 35 | 109 | 179 | 218 | 225 | 197 | 138 | 57 | 27 | 26 | 26 | 26 | 26 | 26 | 26 | 192 | 3 |
|  | 10 | 312 | 31 | 33 | 61 | 120 | 165 | 182 | 172 | 134 | 76 | 34 | 32 | 31 | 31 | 31 | 31 | 31 | 247 | 2 |
|  | 11 | 318 | 34 | 35 | 36 | 53 | 87 | 114 | 125 | 116 | 89 | 55 | 36 | 35 | 34 | 34 | 34 | 34 | 280 | 1 |
|  | 12 | 320 | 35 | 35 | 36 | 36 | 37 | 47 | 69 | 87 | 93 | 86 | 68 | 47 | 37 | 38 | 36 | 35 | 291 | 12 |
|  | HALF-DAY TOTALS |  | 141 | 226 | 494 | 755 | 928 | 975 | 879 | 643 | 319 | 187 | 153 | 142 | 139 | 139 | 139 | 139 | 1025 |  |
| Apr 21 | 6 | 14 | 2 | 8 | 12 | 14 | 14 | 12 | 8 | 2 | 1 | 1 | 1 | 1 | 1 | 1 | 1 | 1 | 1 | 6 |
|  | 7 | 197 | 24 | 94 | 153 | 187 | 191 | 167 | 117 | 45 | 14 | 13 | 13 | 13 | 13 | 13 | 13 | 13 | 53 | 5 |
|  | 8 | 256 | 27 | 99 | 172 | 216 | 227 | 204 | 150 | 69 | 24 | 22 | 22 | 22 | 22 | 22 | 22 | 22 | 131 | 4 |
|  | 9 | 280 | 31 | 79 | 149 | 193 | 208 | 193 | 147 | 77 | 31 | 29 | 29 | 29 | 29 | 29 | 29 | 29 | 197 | 3 |
|  | 10 | 293 | 35 | 54 | 102 | 141 | 158 | 151 | 120 | 73 | 37 | 34 | 33 | 33 | 33 | 33 | 33 | 33 | 249 | 2 |
|  | 11 | 299 | 38 | 40 | 54 | 72 | 86 | 88 | 78 | 60 | 43 | 38 | 38 | 36 | 36 | 36 | 36 | 37 | 279 | 1 |
|  | 12 | 301 | 39 | 39 | 39 | 40 | 40 | 41 | 43 | 45 | 45 | 45 | 43 | 41 | 40 | 39 | 39 | 39 | 289 | 12 |
|  | HALF-DAY TOTALS |  | 179 | 403 | 674 | 859 | 922 | 851 | 653 | 352 | 174 | 159 | 157 | 156 | 155 | 155 | 155 | 156 | 1057 |  |
| May 21 | 6 | 44 | 14 | 30 | 41 | 45 | 43 | 34 | 19 | 4 | 3 | 3 | 3 | 3 | 3 | 3 | 3 | 3 | 5 | 6 |
|  | 7 | 193 | 50 | 120 | 168 | 191 | 185 | 150 | 92 | 24 | 16 | 16 | 16 | 16 | 16 | 16 | 16 | 17 | 62 | 5 |
|  | 8 | 244 | 52 | 132 | 189 | 218 | 215 | 179 | 115 | 38 | 25 | 24 | 24 | 24 | 24 | 24 | 24 | 25 | 135 | 4 |
|  | 9 | 268 | 49 | 116 | 171 | 198 | 197 | 167 | 109 | 45 | 32 | 30 | 30 | 30 | 30 | 30 | 30 | 32 | 197 | 3 |
|  | 10 | 280 | 47 | 89 | 130 | 151 | 150 | 126 | 84 | 44 | 37 | 35 | 35 | 35 | 35 | 35 | 35 | 37 | 245 | 2 |
|  | 11 | 286 | 47 | 63 | 79 | 87 | 83 | 70 | 52 | 41 | 40 | 39 | 38 | 38 | 38 | 39 | 39 | 41 | 273 | 1 |
|  | 12 | 288 | 46 | 46 | 44 | 43 | 42 | 41 | 41 | 41 | 41 | 41 | 41 | 42 | 43 | 44 | 46 | 46 | 282 | 12 |
|  | HALF-DAY TOTALS |  | 283 | 575 | 804 | 916 | 897 | 748 | 493 | 217 | 172 | 167 | 167 | 167 | 167 | 168 | 169 | 176 | 1058 |  |
| Jun 21 | 6 | 53 | 20 | 39 | 52 | 55 | 51 | 39 | 20 | 4 | 4 | 4 | 4 | 4 | 4 | 4 | 4 | 4 | 7 | 6 |
|  | 7 | 188 | 62 | 128 | 172 | 190 | 179 | 141 | 80 | 20 | 16 | 16 | 16 | 16 | 16 | 16 | 16 | 18 | 64 | 5 |
|  | 8 | 238 | 66 | 142 | 194 | 217 | 207 | 167 | 99 | 31 | 25 | 25 | 25 | 25 | 25 | 25 | 25 | 27 | 135 | 4 |
|  | 9 | 261 | 63 | 130 | 178 | 198 | 190 | 154 | 93 | 37 | 31 | 31 | 31 | 31 | 31 | 31 | 31 | 33 | 194 | 3 |
|  | 10 | 273 | 59 | 104 | 140 | 154 | 145 | 115 | 70 | 39 | 37 | 36 | 36 | 36 | 36 | 36 | 36 | 38 | 241 | 2 |
|  | 11 | 279 | 57 | 76 | 90 | 92 | 82 | 63 | 46 | 41 | 40 | 39 | 39 | 39 | 39 | 40 | 41 | 43 | 268 | 1 |
|  | 12 | 281 | 57 | 55 | 50 | 45 | 43 | 42 | 41 | 41 | 41 | 41 | 41 | 42 | 42 | 45 | 50 | 55 | 277 | 12 |
|  | HALF-DAY TOTALS |  | 356 | 648 | 850 | 929 | 876 | 700 | 430 | 194 | 174 | 171 | 171 | 171 | 172 | 173 | 176 | 190 | 1049 |  |
| Jul 21 | 6 | 41 | 14 | 29 | 39 | 42 | 40 | 31 | 18 | 4 | 3 | 3 | 3 | 3 | 3 | 3 | 3 | 3 | 6 | 6 |
|  | 7 | 184 | 51 | 118 | 164 | 185 | 179 | 145 | 88 | 23 | 16 | 16 | 16 | 16 | 16 | 16 | 16 | 17 | 62 | 5 |
|  | 8 | 236 | 55 | 132 | 187 | 214 | 210 | 174 | 111 | 37 | 25 | 25 | 25 | 25 | 25 | 25 | 25 | 26 | 133 | 4 |
|  | 9 | 259 | 52 | 117 | 170 | 196 | 193 | 163 | 106 | 44 | 32 | 31 | 31 | 31 | 31 | 31 | 31 | 33 | 194 | 3 |
|  | 10 | 272 | 50 | 92 | 131 | 151 | 148 | 123 | 81 | 44 | 38 | 36 | 36 | 36 | 36 | 36 | 36 | 38 | 241 | 2 |
|  | 11 | 278 | 49 | 66 | 81 | 88 | 83 | 69 | 52 | 42 | 41 | 40 | 39 | 39 | 39 | 40 | 40 | 42 | 289 | 1 |
|  | 12 | 279 | 49 | 48 | 46 | 44 | 43 | 42 | 42 | 42 | 42 | 42 | 42 | 42 | 43 | 44 | 46 | 48 | 277 | 12 |
|  | HALF-DAY TOTALS |  | 296 | 580 | 799 | 903 | 878 | 729 | 478 | 215 | 176 | 172 | 171 | 171 | 171 | 172 | 173 | 182 | 1043 |  |
| Aug 21 | 6 | 11 | 2 | 7 | 10 | 12 | 12 | 10 | 6 | 2 | 1 | 1 | 1 | 1 | 1 | 1 | 1 | 1 | 1 | 6 |
|  | 7 | 180 | 26 | 92 | 145 | 176 | 180 | 156 | 109 | 42 | 15 | 14 | 14 | 14 | 14 | 14 | 14 | 14 | 53 | 5 |
|  | 8 | 240 | 30 | 100 | 168 | 209 | 219 | 196 | 143 | 65 | 25 | 23 | 23 | 23 | 23 | 23 | 23 | 23 | 128 | 4 |
|  | 9 | 266 | 33 | 82 | 148 | 190 | 203 | 187 | 142 | 74 | 33 | 30 | 30 | 30 | 30 | 30 | 30 | 30 | 193 | 3 |
|  | 10 | 279 | 37 | 58 | 104 | 140 | 155 | 147 | 117 | 71 | 39 | 36 | 35 | 35 | 35 | 35 | 35 | 35 | 243 | 2 |
|  | 11 | 285 | 40 | 43 | 57 | 75 | 86 | 87 | 76 | 59 | 44 | 40 | 39 | 38 | 38 | 38 | 38 | 39 | 273 | 1 |
|  | 12 | 287 | 41 | 41 | 41 | 42 | 42 | 42 | 44 | 44 | 45 | 46 | 45 | 44 | 43 | 42 | 41 | 41 | 282 | 12 |
|  | HALF-DAY TOTALS |  | 191 | 410 | 666 | 837 | 891 | 817 | 624 | 339 | 180 | 167 | 165 | 164 | 163 | 163 | 163 | 164 | 1033 |  |
| Sep 21 | 7 | 179 | 12 | 50 | 114 | 158 | 176 | 168 | 133 | 76 | 15 | 11 | 11 | 11 | 11 | 11 | 11 | 11 | 39 | 5 |
|  | 8 | 253 | 21 | 49 | 134 | 196 | 227 | 224 | 186 | 119 | 36 | 20 | 20 | 20 | 20 | 20 | 20 | 20 | 116 | 4 |
|  | 9 | 281 | 28 | 36 | 106 | 173 | 211 | 217 | 191 | 134 | 57 | 28 | 27 | 27 | 27 | 27 | 27 | 27 | 185 | 3 |
|  | 10 | 295 | 32 | 34 | 61 | 118 | 161 | 178 | 168 | 132 | 76 | 35 | 33 | 32 | 32 | 32 | 32 | 32 | 236 | 2 |
|  | 11 | 302 | 35 | 36 | 37 | 54 | 86 | 113 | 123 | 114 | 88 | 56 | 38 | 36 | 35 | 35 | 35 | 35 | 271 | 1 |
|  | 12 | 304 | 36 | 36 | 37 | 38 | 39 | 49 | 69 | 86 | 93 | 86 | 69 | 48 | 39 | 38 | 37 | 36 | 282 | 12 |
|  | HALF-DAY TOTALS |  | 146 | 226 | 475 | 722 | 885 | 931 | 842 | 622 | 319 | 192 | 159 | 148 | 145 | 144 | 144 | 144 | 991 |  |
| Oct 21 | 7 | 166 | 8 | 18 | 79 | 128 | 156 | 159 | 139 | 95 | 33 | 9 | 8 | 8 | 8 | 8 | 8 | 8 | 25 | 5 |
|  | 8 | 259 | 17 | 20 | 95 | 174 | 223 | 237 | 215 | 159 | 74 | 19 | 17 | 17 | 17 | 17 | 17 | 17 | 99 | 4 |
|  | 9 | 292 | 24 | 25 | 65 | 150 | 209 | 235 | 225 | 187 | 106 | 31 | 24 | 24 | 24 | 24 | 24 | 24 | 170 | 3 |
|  | 10 | 307 | 29 | 30 | 34 | 92 | 158 | 197 | 205 | 183 | 130 | 60 | 31 | 29 | 29 | 29 | 29 | 29 | 224 | 2 |
|  | 11 | 314 | 32 | 32 | 33 | 39 | 83 | 133 | 163 | 167 | 145 | 100 | 49 | 34 | 32 | 32 | 32 | 32 | 258 | 1 |
|  | 12 | 316 | 33 | 33 | 33 | 34 | 35 | 60 | 105 | 139 | 150 | 138 | 104 | 60 | 35 | 34 | 33 | 33 | 270 | 12 |
|  | HALF-DAY TOTALS |  | 127 | 141 | 318 | 592 | 836 | 986 | 996 | 852 | 563 | 283 | 175 | 136 | 128 | 127 | 127 | 127 | 911 |  |
| Nov 21 | 7 | 134 | 5 | 6 | 43 | 89 | 119 | 129 | 120 | 92 | 47 | 6 | 5 | 5 | 5 | 5 | 5 | 5 | 14 | 5 |
|  | 8 | 255 | 15 | 15 | 55 | 145 | 206 | 235 | 228 | 185 | 111 | 25 | 15 | 15 | 15 | 15 | 15 | 15 | 78 | 4 |
|  | 9 | 295 | 21 | 21 | 33 | 121 | 197 | 241 | 247 | 215 | 150 | 57 | 22 | 21 | 21 | 21 | 21 | 21 | 149 | 3 |
|  | 10 | 312 | 26 | 26 | 28 | 67 | 147 | 206 | 230 | 220 | 176 | 100 | 31 | 26 | 26 | 26 | 26 | 26 | 201 | 2 |
|  | 11 | 320 | 29 | 29 | 29 | 31 | 77 | 146 | 192 | 207 | 191 | 144 | 74 | 31 | 29 | 29 | 29 | 29 | 234 | 1 |
|  | 12 | 322 | 30 | 30 | 30 | 30 | 32 | 72 | 137 | 181 | 196 | 181 | 137 | 72 | 32 | 30 | 30 | 30 | 246 | 12 |
|  | HALF-DAY TOTALS |  | 112 | 113 | 197 | 456 | 749 | 983 | 1077 | 1001 | 767 | 420 | 210 | 128 | 112 | 112 | 112 | 112 | 799 |  |
| Dec 21 | 7 | 118 | 4 | 5 | 30 | 72 | 101 | 112 | 107 | 85 | 48 | 7 | 4 | 4 | 4 | 4 | 4 | 4 | 10 | 5 |
|  | 8 | 255 | 13 | 14 | 41 | 132 | 198 | 233 | 231 | 193 | 124 | 33 | 13 | 13 | 13 | 13 | 13 | 13 | 69 | 4 |
|  | 9 | 297 | 20 | 20 | 25 | 108 | 191 | 241 | 254 | 227 | 165 | 72 | 21 | 20 | 20 | 20 | 20 | 20 | 138 | 3 |
|  | 10 | 315 | 25 | 25 | 26 | 56 | 144 | 208 | 239 | 233 | 192 | 117 | 35 | 25 | 25 | 25 | 25 | 25 | 191 | 2 |
|  | 11 | 323 | 28 | 28 | 28 | 29 | 73 | 150 | 202 | 221 | 207 | 161 | 86 | 30 | 28 | 28 | 28 | 28 | 223 | 1 |
|  | 12 | 325 | 29 | 29 | 29 | 29 | 30 | 77 | 149 | 197 | 212 | 196 | 149 | 76 | 30 | 29 | 29 | 29 | 234 | 12 |
|  | HALF-DAY TOTALS |  | 104 | 105 | 159 | 402 | 710 | 975 | 1099 | 1050 | 836 | 484 | 228 | 125 | 105 | 104 | 104 | 104 | 748 |  |
|  |  |  | N | NNW | NW | WNW | W | WSW | SW | SSW | S | SSE | SE | ESE | E | ENE | NE | NNE | HOR | PM |

[a]Total solar heat gains for DSA glass (based on a ground reflectance of 0.20).
[b]Half-day totals computed by Simpson's rule, time interval = 10 minutes.

## TABLE C.2  Solar Intensity and Solar Heat Gain Factors[a] for 32°N Latitude (I-P Units)[b]

| Date | Solar Time A.M. | Direct Normal Btu/h ft² | N | NNE | NE | ENE | E | ESE | SE | SSE | S | SSW | SW | WSW | W | WNW | NW | NNW | HOR | Solar Time P.M. |
|---|---|---|---|---|---|---|---|---|---|---|---|---|---|---|---|---|---|---|---|---|
| Jan 21 | 7 | 1 | 0 | 0 | 0 | 1 | 1 | 1 | 1 | 1 | 0 | 0 | 0 | 0 | 0 | 0 | 0 | 0 | 0 | 5 |
| | 8 | 203 | 9 | 9 | 29 | 105 | 160 | 189 | 189 | 159 | 103 | 28 | 9 | 9 | 9 | 9 | 9 | 9 | 32 | 4 |
| | 9 | 269 | 15 | 15 | 15 | 91 | 175 | 229 | 246 | 225 | 169 | 82 | 17 | 15 | 15 | 15 | 15 | 15 | 88 | 3 |
| | 10 | 295 | 20 | 20 | 20 | 41 | 135 | 209 | 249 | 250 | 212 | 141 | 46 | 20 | 20 | 20 | 20 | 20 | 136 | 2 |
| | 11 | 306 | 23 | 23 | 23 | 24 | 68 | 159 | 221 | 249 | 238 | 191 | 110 | 29 | 23 | 23 | 23 | 23 | 166 | 1 |
| | 12 | 310 | 24 | 24 | 24 | 24 | 25 | 88 | 174 | 228 | 246 | 228 | 174 | 88 | 25 | 24 | 24 | 24 | 176 | 12 |
| | HALF-DAY TOTALS | | 79 | 79 | 107 | 284 | 570 | 856 | 1015 | 1014 | 853 | 553 | 264 | 112 | 80 | 79 | 79 | 79 | 512 | |
| Feb 21 | 7 | 112 | 4 | 7 | 47 | 82 | 102 | 106 | 95 | 67 | 26 | 4 | 4 | 4 | 4 | 4 | 4 | 4 | 9 | 5 |
| | 8 | 245 | 13 | 14 | 65 | 149 | 205 | 228 | 216 | 170 | 95 | 17 | 13 | 13 | 13 | 13 | 13 | 13 | 64 | 4 |
| | 9 | 287 | 19 | 19 | 32 | 122 | 199 | 242 | 248 | 216 | 149 | 55 | 20 | 19 | 19 | 19 | 19 | 19 | 127 | 3 |
| | 10 | 305 | 24 | 24 | 25 | 62 | 151 | 213 | 241 | 232 | 189 | 112 | 31 | 24 | 24 | 24 | 24 | 24 | 176 | 2 |
| | 11 | 314 | 26 | 26 | 26 | 28 | 76 | 156 | 208 | 227 | 212 | 165 | 87 | 28 | 26 | 26 | 26 | 26 | 207 | 1 |
| | 12 | 316 | 27 | 27 | 27 | 27 | 29 | 79 | 155 | 204 | 221 | 204 | 155 | 79 | 29 | 27 | 27 | 27 | 217 | 12 |
| | HALF-DAY TOTALS | | 100 | 103 | 201 | 445 | 735 | 978 | 1080 | 1010 | 780 | 452 | 228 | 122 | 100 | 100 | 100 | 100 | 691 | |
| Mar 21 | 7 | 185 | 10 | 37 | 105 | 153 | 176 | 173 | 142 | 88 | 20 | 9 | 9 | 9 | 9 | 9 | 9 | 9 | 32 | 5 |
| | 8 | 260 | 17 | 25 | 107 | 183 | 227 | 237 | 209 | 150 | 62 | 18 | 17 | 17 | 17 | 17 | 17 | 17 | 100 | 4 |
| | 9 | 290 | 23 | 25 | 64 | 151 | 210 | 237 | 227 | 183 | 107 | 30 | 23 | 23 | 23 | 23 | 23 | 23 | 164 | 3 |
| | 10 | 304 | 28 | 28 | 30 | 87 | 158 | 202 | 215 | 195 | 144 | 70 | 29 | 28 | 28 | 28 | 28 | 28 | 211 | 2 |
| | 11 | 311 | 31 | 31 | 31 | 34 | 82 | 142 | 179 | 188 | 168 | 120 | 59 | 32 | 31 | 31 | 31 | 31 | 242 | 1 |
| | 12 | 313 | 32 | 32 | 32 | 32 | 33 | 66 | 122 | 162 | 176 | 162 | 122 | 66 | 33 | 32 | 32 | 32 | 252 | 12 |
| | HALF-DAY TOTALS | | 124 | 162 | 359 | 629 | 875 | 1033 | 1041 | 888 | 589 | 326 | 193 | 136 | 125 | 124 | 124 | 124 | 874 | |
| Apr 21 | 6 | 66 | 9 | 35 | 54 | 65 | 66 | 56 | 38 | 12 | 4 | 3 | 3 | 3 | 3 | 3 | 3 | 3 | 7 | 6 |
| | 7 | 206 | 17 | 80 | 146 | 188 | 200 | 182 | 136 | 65 | 16 | 14 | 14 | 14 | 14 | 14 | 14 | 14 | 61 | 5 |
| | 8 | 255 | 23 | 61 | 144 | 200 | 227 | 219 | 177 | 107 | 30 | 22 | 22 | 22 | 22 | 22 | 22 | 22 | 129 | 4 |
| | 9 | 278 | 28 | 36 | 103 | 168 | 206 | 212 | 187 | 133 | 58 | 29 | 28 | 28 | 28 | 28 | 28 | 28 | 188 | 3 |
| | 10 | 290 | 32 | 34 | 52 | 108 | 155 | 177 | 172 | 141 | 87 | 39 | 33 | 32 | 32 | 32 | 32 | 32 | 233 | 2 |
| | 11 | 295 | 35 | 35 | 36 | 47 | 83 | 118 | 135 | 132 | 108 | 70 | 40 | 36 | 35 | 35 | 35 | 35 | 262 | 1 |
| | 12 | 297 | 36 | 36 | 36 | 37 | 38 | 53 | 82 | 106 | 115 | 106 | 82 | 53 | 38 | 37 | 36 | 36 | 271 | 12 |
| | HALF-DAY TOTALS | | 161 | 296 | 550 | 792 | 952 | 992 | 889 | 645 | 360 | 228 | 177 | 157 | 153 | 152 | 152 | 152 | 1015 | |
| May 21 | 6 | 119 | 33 | 77 | 108 | 121 | 116 | 94 | 56 | 13 | 8 | 8 | 8 | 8 | 8 | 8 | 8 | 9 | 21 | 6 |
| | 7 | 211 | 36 | 111 | 170 | 202 | 204 | 174 | 118 | 42 | 19 | 18 | 18 | 18 | 18 | 18 | 18 | 19 | 81 | 5 |
| | 8 | 250 | 29 | 94 | 165 | 208 | 220 | 199 | 149 | 73 | 27 | 25 | 25 | 25 | 25 | 25 | 25 | 25 | 146 | 4 |
| | 9 | 269 | 33 | 61 | 128 | 177 | 198 | 190 | 155 | 93 | 37 | 32 | 31 | 31 | 31 | 31 | 31 | 31 | 201 | 3 |
| | 10 | 280 | 36 | 40 | 76 | 121 | 150 | 156 | 138 | 99 | 54 | 37 | 35 | 35 | 35 | 35 | 35 | 35 | 243 | 2 |
| | 11 | 285 | 38 | 39 | 42 | 59 | 83 | 99 | 102 | 90 | 68 | 47 | 40 | 39 | 37 | 37 | 37 | 37 | 269 | 1 |
| | 12 | 286 | 38 | 39 | 40 | 40 | 41 | 47 | 59 | 70 | 74 | 70 | 59 | 47 | 41 | 40 | 40 | 39 | 277 | 12 |
| | HALF-DAY TOTALS | | 222 | 438 | 702 | 900 | 985 | 933 | 747 | 447 | 250 | 199 | 183 | 177 | 175 | 174 | 174 | 175 | 1098 | |
| Jun 21 | 6 | 131 | 44 | 92 | 123 | 135 | 127 | 99 | 55 | 12 | 10 | 10 | 10 | 10 | 10 | 10 | 10 | 11 | 28 | 6 |
| | 7 | 210 | 47 | 122 | 176 | 204 | 201 | 168 | 108 | 35 | 20 | 20 | 20 | 20 | 20 | 20 | 20 | 21 | 88 | 5 |
| | 8 | 245 | 36 | 106 | 171 | 208 | 214 | 189 | 135 | 60 | 28 | 27 | 27 | 27 | 27 | 27 | 27 | 27 | 151 | 4 |
| | 9 | 264 | 35 | 74 | 137 | 178 | 193 | 180 | 139 | 77 | 35 | 32 | 32 | 32 | 32 | 32 | 32 | 32 | 204 | 3 |
| | 10 | 274 | 38 | 47 | 86 | 125 | 146 | 145 | 123 | 83 | 45 | 38 | 36 | 36 | 36 | 36 | 36 | 36 | 244 | 2 |
| | 11 | 279 | 40 | 41 | 47 | 64 | 82 | 91 | 89 | 75 | 56 | 43 | 41 | 40 | 39 | 39 | 39 | 39 | 269 | 1 |
| | 12 | 280 | 41 | 41 | 41 | 42 | 42 | 46 | 52 | 58 | 60 | 58 | 52 | 46 | 42 | 42 | 41 | 41 | 276 | 12 |
| | HALF-DAY TOTALS | | 261 | 504 | 762 | 935 | 985 | 897 | 678 | 372 | 225 | 197 | 189 | 185 | 184 | 184 | 183 | 186 | 1122 | |
| Jul 21 | 6 | 113 | 34 | 76 | 105 | 117 | 113 | 90 | 53 | 12 | 9 | 9 | 9 | 9 | 9 | 9 | 9 | 9 | 22 | 6 |
| | 7 | 203 | 38 | 111 | 167 | 198 | 198 | 169 | 114 | 41 | 20 | 19 | 19 | 19 | 19 | 19 | 19 | 19 | 81 | 5 |
| | 8 | 241 | 31 | 95 | 163 | 204 | 215 | 194 | 145 | 70 | 28 | 26 | 26 | 26 | 26 | 26 | 26 | 26 | 145 | 4 |
| | 9 | 261 | 34 | 64 | 129 | 175 | 195 | 186 | 150 | 90 | 37 | 32 | 32 | 32 | 32 | 32 | 32 | 32 | 198 | 3 |
| | 10 | 271 | 37 | 42 | 78 | 121 | 148 | 153 | 134 | 96 | 53 | 38 | 36 | 36 | 36 | 36 | 36 | 36 | 240 | 2 |
| | 11 | 277 | 39 | 40 | 43 | 60 | 83 | 98 | 99 | 88 | 66 | 47 | 41 | 40 | 38 | 38 | 38 | 38 | 265 | 1 |
| | 12 | 279 | 40 | 40 | 41 | 41 | 42 | 48 | 58 | 68 | 72 | 68 | 58 | 48 | 42 | 41 | 41 | 40 | 273 | 12 |
| | HALF-DAY TOTALS | | 231 | 444 | 701 | 890 | 967 | 912 | 726 | 433 | 248 | 202 | 187 | 182 | 180 | 179 | 179 | 180 | 1088 | |
| Aug 21 | 6 | 59 | 10 | 33 | 50 | 60 | 60 | 51 | 34 | 11 | 4 | 4 | 4 | 4 | 4 | 4 | 4 | 4 | 8 | 6 |
| | 7 | 190 | 19 | 79 | 141 | 179 | 190 | 172 | 128 | 61 | 17 | 15 | 15 | 15 | 15 | 15 | 15 | 15 | 61 | 5 |
| | 8 | 240 | 25 | 63 | 141 | 195 | 219 | 210 | 170 | 102 | 31 | 23 | 23 | 23 | 23 | 23 | 23 | 23 | 128 | 4 |
| | 9 | 263 | 30 | 39 | 104 | 166 | 200 | 206 | 181 | 127 | 57 | 31 | 29 | 29 | 29 | 29 | 29 | 29 | 185 | 3 |
| | 10 | 276 | 34 | 36 | 55 | 109 | 153 | 173 | 167 | 136 | 84 | 40 | 35 | 34 | 34 | 34 | 34 | 34 | 229 | 2 |
| | 11 | 282 | 36 | 37 | 39 | 50 | 84 | 116 | 131 | 127 | 104 | 69 | 41 | 38 | 36 | 36 | 36 | 36 | 256 | 1 |
| | 12 | 284 | 37 | 37 | 37 | 39 | 40 | 54 | 81 | 103 | 111 | 103 | 81 | 54 | 40 | 39 | 37 | 37 | 265 | 12 |
| | HALF-DAY TOTALS | | 171 | 303 | 546 | 774 | 922 | 955 | 854 | 618 | 352 | 231 | 184 | 166 | 162 | 161 | 160 | 160 | 999 | |
| Sep 21 | 7 | 163 | 10 | 35 | 96 | 139 | 159 | 156 | 128 | 80 | 20 | 10 | 10 | 10 | 10 | 10 | 10 | 10 | 31 | 5 |
| | 8 | 240 | 18 | 26 | 103 | 173 | 215 | 224 | 198 | 143 | 60 | 19 | 18 | 18 | 18 | 18 | 18 | 18 | 96 | 4 |
| | 9 | 272 | 24 | 26 | 64 | 146 | 202 | 227 | 218 | 177 | 105 | 31 | 24 | 24 | 24 | 24 | 24 | 24 | 158 | 3 |
| | 10 | 287 | 29 | 29 | 32 | 86 | 154 | 196 | 208 | 189 | 141 | 70 | 31 | 29 | 29 | 29 | 29 | 29 | 204 | 2 |
| | 11 | 294 | 32 | 32 | 32 | 36 | 81 | 139 | 174 | 182 | 163 | 118 | 59 | 34 | 32 | 32 | 32 | 32 | 234 | 1 |
| | 12 | 296 | 33 | 33 | 33 | 33 | 35 | 66 | 120 | 158 | 171 | 158 | 120 | 66 | 35 | 33 | 33 | 33 | 244 | 12 |
| | HALF-DAY TOTALS | | 130 | 164 | 345 | 598 | 831 | 982 | 993 | 852 | 574 | 325 | 197 | 142 | 130 | 129 | 129 | 129 | 845 | |
| Oct 21 | 7 | 99 | 4 | 7 | 43 | 74 | 92 | 96 | 85 | 60 | 24 | 5 | 4 | 4 | 4 | 4 | 4 | 4 | 10 | 5 |
| | 8 | 229 | 13 | 15 | 63 | 143 | 195 | 217 | 206 | 162 | 90 | 17 | 13 | 13 | 13 | 13 | 13 | 13 | 63 | 4 |
| | 9 | 273 | 20 | 20 | 33 | 120 | 193 | 234 | 239 | 208 | 144 | 54 | 21 | 20 | 20 | 20 | 20 | 20 | 125 | 3 |
| | 10 | 293 | 24 | 24 | 26 | 62 | 147 | 207 | 234 | 225 | 183 | 109 | 32 | 24 | 24 | 24 | 24 | 24 | 173 | 2 |
| | 11 | 302 | 27 | 27 | 27 | 29 | 76 | 152 | 203 | 221 | 207 | 160 | 85 | 29 | 27 | 27 | 27 | 27 | 203 | 1 |
| | 12 | 304 | 28 | 28 | 28 | 28 | 30 | 78 | 151 | 199 | 215 | 199 | 151 | 78 | 30 | 28 | 28 | 28 | 213 | 12 |
| | HALF-DAY TOTALS | | 103 | 106 | 200 | 433 | 708 | 941 | 1038 | 972 | 753 | 441 | 226 | 125 | 104 | 103 | 103 | 103 | 679 | |
| Nov 21 | 7 | 2 | 0 | 0 | 0 | 1 | 1 | 1 | 1 | 1 | 1 | 0 | 0 | 0 | 0 | 0 | 0 | 0 | 0 | 5 |
| | 8 | 196 | 9 | 9 | 29 | 103 | 156 | 184 | 184 | 155 | 100 | 27 | 9 | 9 | 9 | 9 | 9 | 9 | 32 | 4 |
| | 9 | 263 | 16 | 16 | 17 | 90 | 173 | 225 | 241 | 221 | 166 | 80 | 17 | 16 | 16 | 16 | 16 | 16 | 88 | 3 |
| | 10 | 289 | 20 | 20 | 21 | 41 | 134 | 206 | 245 | 246 | 209 | 138 | 45 | 21 | 20 | 20 | 20 | 20 | 136 | 2 |
| | 11 | 301 | 23 | 23 | 23 | 24 | 67 | 157 | 218 | 245 | 234 | 188 | 109 | 29 | 23 | 23 | 23 | 23 | 165 | 1 |
| | 12 | 304 | 24 | 24 | 24 | 24 | 25 | 87 | 171 | 224 | 243 | 224 | 171 | 87 | 25 | 24 | 24 | 24 | 175 | 12 |
| | HALF-DAY TOTALS | | 80 | 81 | 108 | 282 | 561 | 841 | 996 | 995 | 838 | 544 | 261 | 113 | 81 | 80 | 80 | 80 | 509 | |
| Dec 21 | 8 | 176 | 7 | 7 | 19 | 84 | 135 | 163 | 166 | 143 | 97 | 31 | 7 | 7 | 7 | 7 | 7 | 7 | 22 | 4 |
| | 9 | 257 | 14 | 14 | 15 | 77 | 162 | 218 | 238 | 222 | 171 | 89 | 15 | 14 | 14 | 14 | 14 | 14 | 72 | 3 |
| | 10 | 288 | 18 | 18 | 18 | 34 | 127 | 204 | 246 | 251 | 216 | 148 | 52 | 19 | 18 | 18 | 18 | 18 | 119 | 2 |
| | 11 | 301 | 21 | 21 | 21 | 22 | 63 | 157 | 222 | 252 | 243 | 197 | 116 | 29 | 21 | 21 | 21 | 21 | 148 | 1 |
| | 12 | 304 | 22 | 22 | 22 | 22 | 23 | 89 | 177 | 232 | 252 | 232 | 177 | 89 | 23 | 22 | 22 | 22 | 158 | 12 |
| | HALF-DAY TOTALS | | 71 | 71 | 84 | 227 | 500 | 792 | 965 | 986 | 852 | 578 | 275 | 107 | 71 | 71 | 71 | 71 | 440 | |
| | | | N | NNW | NW | WNW | W | WSW | SW | SSW | S | SSE | SE | ESE | E | ENE | NE | NNE | HOR | PM |

[a]Total solar heat gains for DSA glass (based on a ground reflectance of 0.20).
[b]Half-day totals computed by Simpson's rule, time interval = 10 minutes.

## TABLE C.3 Solar Intensity and Solar Heat Gain Factors[a] for 40°N Latitude (I-P Units)[b]

| Date | Solar Time A.M. | Direct Normal Btu/h ft² | N | NNE | NE | ENE | E | ESE | SE | SSE | S | SSW | SW | WSW | W | WNW | NW | NNW | HOR | Solar Time P.M. |
|---|---|---|---|---|---|---|---|---|---|---|---|---|---|---|---|---|---|---|---|---|
| Jan 21 | 8 | 142 | 5 | 5 | 17 | 71 | 111 | 132 | 133 | 114 | 75 | 22 | 6 | 5 | 5 | 5 | 5 | 5 | 14 | 4 |
|  | 9 | 239 | 12 | 12 | 13 | 74 | 154 | 205 | 224 | 209 | 160 | 82 | 13 | 12 | 12 | 12 | 12 | 12 | 55 | 3 |
|  | 10 | 274 | 16 | 16 | 16 | 31 | 124 | 199 | 241 | 246 | 213 | 146 | 51 | 17 | 16 | 16 | 16 | 16 | 96 | 2 |
|  | 11 | 289 | 19 | 19 | 19 | 20 | 61 | 156 | 222 | 252 | 244 | 198 | 118 | 28 | 19 | 19 | 19 | 19 | 124 | 1 |
|  | 12 | 294 | 20 | 20 | 20 | 20 | 21 | 90 | 179 | 234 | 254 | 234 | 179 | 90 | 21 | 20 | 20 | 20 | 133 | 12 |
|  | HALF-DAY TOTALS |  | 61 | 61 | 73 | 199 | 452 | 734 | 904 | 932 | 813 | 561 | 273 | 101 | 62 | 61 | 61 | 61 | 354 |  |
| Feb 21 | 7 | 55 | 2 | 3 | 23 | 40 | 51 | 53 | 47 | 34 | 14 | 2 | 2 | 2 | 2 | 2 | 2 | 2 | 4 | 5 |
|  | 8 | 219 | 10 | 11 | 50 | 129 | 183 | 206 | 199 | 160 | 94 | 18 | 10 | 10 | 10 | 10 | 10 | 10 | 43 | 4 |
|  | 9 | 271 | 16 | 16 | 22 | 107 | 186 | 234 | 245 | 218 | 157 | 66 | 17 | 16 | 16 | 16 | 16 | 16 | 98 | 3 |
|  | 10 | 294 | 21 | 21 | 21 | 49 | 143 | 211 | 246 | 243 | 203 | 129 | 38 | 21 | 21 | 21 | 21 | 21 | 143 | 2 |
|  | 11 | 304 | 23 | 23 | 23 | 24 | 71 | 160 | 219 | 244 | 231 | 184 | 103 | 27 | 23 | 23 | 23 | 23 | 171 | 1 |
|  | 12 | 307 | 24 | 24 | 24 | 24 | 25 | 86 | 170 | 222 | 241 | 222 | 170 | 86 | 25 | 24 | 24 | 24 | 180 | 12 |
|  | HALF-DAY TOTALS |  | 84 | 86 | 152 | 361 | 648 | 916 | 1049 | 1015 | 821 | 508 | 250 | 114 | 85 | 84 | 84 | 84 | 548 |  |
| Mar 21 | 7 | 171 | 9 | 29 | 93 | 140 | 163 | 161 | 135 | 86 | 22 | 8 | 8 | 8 | 8 | 8 | 8 | 8 | 26 | 5 |
|  | 8 | 250 | 16 | 18 | 91 | 169 | 218 | 232 | 211 | 157 | 74 | 17 | 16 | 16 | 16 | 16 | 16 | 16 | 85 | 4 |
|  | 9 | 282 | 21 | 22 | 47 | 136 | 203 | 238 | 236 | 198 | 128 | 40 | 22 | 21 | 21 | 21 | 21 | 21 | 143 | 3 |
|  | 10 | 297 | 25 | 25 | 27 | 72 | 153 | 207 | 229 | 216 | 171 | 95 | 29 | 25 | 25 | 25 | 25 | 25 | 186 | 2 |
|  | 11 | 305 | 28 | 28 | 28 | 30 | 78 | 151 | 198 | 213 | 197 | 150 | 77 | 30 | 28 | 28 | 28 | 28 | 213 | 1 |
|  | 12 | 307 | 29 | 29 | 29 | 29 | 31 | 75 | 145 | 191 | 206 | 191 | 145 | 75 | 31 | 29 | 29 | 29 | 223 | 12 |
|  | HALF-DAY TOTALS |  | 114 | 139 | 302 | 563 | 832 | 1035 | 1087 | 968 | 694 | 403 | 220 | 132 | 114 | 113 | 113 | 113 | 764 |  |
| Apr 21 | 6 | 89 | 11 | 46 | 72 | 87 | 88 | 76 | 52 | 18 | 5 | 5 | 5 | 5 | 5 | 5 | 5 | 5 | 11 | 6 |
|  | 7 | 206 | 16 | 71 | 140 | 185 | 201 | 186 | 143 | 75 | 16 | 14 | 14 | 14 | 14 | 14 | 14 | 14 | 61 | 5 |
|  | 8 | 252 | 22 | 44 | 128 | 190 | 224 | 223 | 188 | 124 | 41 | 22 | 21 | 21 | 21 | 21 | 21 | 21 | 123 | 4 |
|  | 9 | 274 | 27 | 29 | 80 | 155 | 202 | 219 | 203 | 156 | 83 | 29 | 27 | 27 | 27 | 27 | 27 | 27 | 177 | 3 |
|  | 10 | 286 | 31 | 31 | 37 | 92 | 152 | 187 | 193 | 170 | 121 | 56 | 32 | 31 | 31 | 31 | 31 | 41 | 217 | 2 |
|  | 11 | 292 | 33 | 33 | 34 | 39 | 81 | 130 | 160 | 166 | 146 | 102 | 52 | 35 | 33 | 33 | 33 | 33 | 243 | 1 |
|  | 12 | 293 | 34 | 34 | 34 | 34 | 36 | 62 | 108 | 142 | 154 | 142 | 108 | 62 | 36 | 34 | 34 | 34 | 252 | 12 |
|  | HALF-DAY TOTALS |  | 154 | 265 | 501 | 758 | 957 | 1051 | 994 | 782 | 488 | 296 | 199 | 157 | 148 | 147 | 147 | 147 | 957 |  |
| May 21 | 5 | 1 | 0 | 1 | 1 | 1 | 1 | 1 | 0 | 0 | 0 | 0 | 0 | 0 | 0 | 0 | 0 | 0 | 0 | 7 |
|  | 6 | 144 | 36 | 90 | 128 | 145 | 141 | 115 | 71 | 18 | 10 | 10 | 10 | 10 | 10 | 10 | 10 | 11 | 31 | 6 |
|  | 7 | 216 | 28 | 102 | 165 | 202 | 209 | 184 | 131 | 54 | 20 | 19 | 19 | 19 | 19 | 19 | 19 | 19 | 87 | 5 |
|  | 8 | 250 | 27 | 73 | 149 | 199 | 220 | 208 | 164 | 93 | 29 | 25 | 25 | 25 | 25 | 25 | 25 | 25 | 146 | 4 |
|  | 9 | 267 | 31 | 42 | 105 | 164 | 197 | 200 | 175 | 121 | 53 | 32 | 30 | 30 | 30 | 30 | 30 | 30 | 195 | 3 |
|  | 10 | 277 | 34 | 36 | 54 | 105 | 148 | 168 | 163 | 133 | 83 | 40 | 35 | 34 | 34 | 34 | 34 | 34 | 234 | 2 |
|  | 11 | 283 | 36 | 36 | 38 | 48 | 81 | 113 | 130 | 127 | 105 | 70 | 42 | 38 | 36 | 36 | 36 | 36 | 257 | 1 |
|  | 12 | 284 | 37 | 37 | 37 | 38 | 40 | 54 | 82 | 114 | 113 | 104 | 82 | 54 | 40 | 38 | 37 | 37 | 265 | 12 |
|  | HALF-DAY TOTALS |  | 215 | 404 | 666 | 893 | 1024 | 1025 | 881 | 601 | 358 | 247 | 200 | 180 | 176 | 175 | 174 | 175 | 1083 |  |
| Jun 21 | 5 | 22 | 10 | 17 | 21 | 22 | 20 | 14 | 6 | 2 | 1 | 1 | 1 | 1 | 1 | 1 | 1 | 2 | 3 | 7 |
|  | 6 | 155 | 48 | 104 | 143 | 159 | 151 | 121 | 70 | 17 | 13 | 13 | 13 | 13 | 13 | 13 | 13 | 14 | 40 | 6 |
|  | 7 | 216 | 37 | 113 | 172 | 205 | 207 | 178 | 122 | 46 | 22 | 21 | 21 | 21 | 21 | 21 | 21 | 21 | 97 | 5 |
|  | 8 | 246 | 30 | 85 | 156 | 201 | 216 | 199 | 152 | 80 | 29 | 27 | 27 | 27 | 27 | 27 | 27 | 27 | 153 | 4 |
|  | 9 | 263 | 33 | 51 | 114 | 166 | 192 | 190 | 161 | 105 | 45 | 33 | 32 | 32 | 32 | 32 | 32 | 32 | 201 | 3 |
|  | 10 | 272 | 35 | 38 | 63 | 109 | 145 | 158 | 148 | 116 | 69 | 39 | 36 | 35 | 35 | 35 | 35 | 35 | 238 | 2 |
|  | 11 | 277 | 38 | 39 | 40 | 52 | 81 | 105 | 116 | 110 | 88 | 60 | 41 | 39 | 38 | 38 | 38 | 38 | 260 | 1 |
|  | 12 | 279 | 38 | 38 | 40 | 41 | 52 | 72 | 89 | 95 | 89 | 72 | 52 | 41 | 40 | 38 | 38 | 38 | 267 | 12 |
|  | HALF-DAY TOTALS |  | 253 | 470 | 734 | 941 | 1038 | 999 | 818 | 523 | 315 | 236 | 204 | 191 | 188 | 187 | 186 | 188 | 1126 |  |
| Jul 21 | 5 | 2 | 1 | 2 | 2 | 2 | 2 | 1 | 1 | 0 | 0 | 0 | 0 | 0 | 0 | 0 | 0 | 0 | 0 | 7 |
|  | 6 | 138 | 37 | 89 | 125 | 142 | 137 | 112 | 68 | 18 | 11 | 11 | 11 | 11 | 11 | 11 | 11 | 12 | 32 | 6 |
|  | 7 | 208 | 30 | 102 | 163 | 198 | 204 | 179 | 127 | 53 | 21 | 20 | 20 | 20 | 20 | 20 | 20 | 20 | 88 | 5 |
|  | 8 | 241 | 28 | 75 | 148 | 196 | 216 | 203 | 160 | 90 | 30 | 26 | 26 | 26 | 26 | 26 | 26 | 26 | 145 | 4 |
|  | 9 | 259 | 32 | 44 | 106 | 163 | 193 | 196 | 170 | 118 | 52 | 33 | 31 | 31 | 31 | 31 | 31 | 31 | 194 | 3 |
|  | 10 | 269 | 35 | 37 | 56 | 106 | 146 | 165 | 159 | 129 | 81 | 41 | 36 | 35 | 35 | 35 | 35 | 35 | 231 | 2 |
|  | 11 | 275 | 37 | 38 | 40 | 50 | 81 | 111 | 127 | 123 | 102 | 69 | 43 | 39 | 37 | 37 | 37 | 37 | 254 | 1 |
|  | 12 | 276 | 38 | 38 | 38 | 40 | 41 | 55 | 80 | 101 | 109 | 101 | 80 | 55 | 41 | 40 | 38 | 38 | 262 | 12 |
|  | HALF-DAY TOTALS |  | 223 | 411 | 666 | 885 | 1008 | 1003 | 956 | 584 | 352 | 248 | 204 | 186 | 181 | 180 | 180 | 181 | 1076 |  |
| Aug 21 | 6 | 81 | 12 | 44 | 68 | 81 | 82 | 71 | 48 | 17 | 6 | 5 | 5 | 5 | 5 | 5 | 5 | 5 | 12 | 6 |
|  | 7 | 191 | 17 | 71 | 135 | 177 | 191 | 177 | 135 | 70 | 17 | 16 | 16 | 16 | 16 | 16 | 16 | 16 | 62 | 5 |
|  | 8 | 237 | 24 | 47 | 126 | 185 | 216 | 214 | 180 | 118 | 41 | 23 | 23 | 23 | 23 | 23 | 23 | 23 | 122 | 4 |
|  | 9 | 260 | 28 | 31 | 82 | 153 | 197 | 212 | 196 | 151 | 80 | 31 | 28 | 28 | 28 | 28 | 28 | 28 | 174 | 3 |
|  | 10 | 272 | 32 | 33 | 40 | 93 | 150 | 182 | 187 | 165 | 116 | 56 | 34 | 32 | 32 | 32 | 32 | 32 | 214 | 2 |
|  | 11 | 278 | 35 | 35 | 36 | 41 | 81 | 128 | 156 | 160 | 141 | 99 | 52 | 37 | 35 | 35 | 35 | 35 | 239 | 1 |
|  | 12 | 280 | 35 | 35 | 35 | 36 | 38 | 63 | 106 | 138 | 149 | 138 | 106 | 63 | 38 | 36 | 35 | 35 | 247 | 12 |
|  | HALF-DAY TOTALS |  | 164 | 273 | 498 | 741 | 928 | 1013 | 956 | 751 | 474 | 296 | 205 | 166 | 157 | 156 | 156 | 156 | 946 |  |
| Sep 21 | 7 | 149 | 9 | 27 | 84 | 125 | 146 | 144 | 121 | 77 | 21 | 9 | 9 | 9 | 9 | 9 | 9 | 9 | 25 | 5 |
|  | 8 | 230 | 17 | 19 | 87 | 160 | 205 | 218 | 199 | 148 | 71 | 18 | 17 | 17 | 17 | 17 | 17 | 17 | 82 | 4 |
|  | 9 | 263 | 22 | 23 | 47 | 131 | 194 | 227 | 226 | 190 | 124 | 41 | 23 | 22 | 22 | 22 | 22 | 22 | 138 | 3 |
|  | 10 | 280 | 27 | 27 | 28 | 71 | 148 | 200 | 221 | 209 | 165 | 93 | 30 | 27 | 27 | 27 | 27 | 27 | 180 | 2 |
|  | 11 | 287 | 29 | 29 | 29 | 31 | 78 | 147 | 192 | 207 | 191 | 146 | 77 | 31 | 29 | 29 | 29 | 29 | 206 | 1 |
|  | 12 | 290 | 30 | 30 | 30 | 30 | 32 | 75 | 142 | 185 | 200 | 185 | 142 | 75 | 32 | 30 | 30 | 30 | 215 | 12 |
|  | HALF-DAY TOTALS |  | 119 | 142 | 291 | 534 | 787 | 980 | 1033 | 925 | 672 | 396 | 222 | 137 | 119 | 118 | 118 | 118 | 738 |  |
| Oct 21 | 7 | 48 | 2 | 3 | 20 | 36 | 45 | 47 | 42 | 30 | 12 | 2 | 2 | 2 | 2 | 2 | 2 | 2 | 4 | 5 |
|  | 8 | 204 | 11 | 12 | 49 | 123 | 173 | 195 | 188 | 151 | 89 | 18 | 11 | 11 | 11 | 11 | 11 | 11 | 43 | 4 |
|  | 9 | 257 | 17 | 17 | 23 | 104 | 180 | 225 | 235 | 209 | 151 | 64 | 18 | 17 | 17 | 17 | 17 | 17 | 97 | 3 |
|  | 10 | 280 | 21 | 21 | 22 | 50 | 139 | 205 | 238 | 235 | 196 | 125 | 38 | 22 | 21 | 21 | 21 | 21 | 140 | 2 |
|  | 11 | 291 | 24 | 24 | 24 | 25 | 71 | 156 | 212 | 236 | 224 | 178 | 101 | 28 | 24 | 24 | 24 | 24 | 168 | 1 |
|  | 12 | 294 | 25 | 25 | 25 | 25 | 25 | 85 | 165 | 216 | 234 | 216 | 165 | 85 | 27 | 25 | 25 | 25 | 177 | 12 |
|  | HALF-DAY TOTALS |  | 88 | 89 | 152 | 351 | 623 | 878 | 1006 | 974 | 791 | 493 | 247 | 117 | 89 | 88 | 88 | 88 | 540 |  |
| Nov 21 | 8 | 136 | 5 | 5 | 18 | 69 | 108 | 128 | 129 | 110 | 72 | 21 | 6 | 5 | 5 | 5 | 5 | 5 | 14 | 4 |
|  | 9 | 232 | 12 | 12 | 13 | 73 | 151 | 201 | 219 | 204 | 156 | 80 | 13 | 12 | 12 | 12 | 12 | 12 | 55 | 3 |
|  | 10 | 268 | 16 | 16 | 16 | 31 | 122 | 196 | 237 | 242 | 209 | 143 | 50 | 17 | 16 | 16 | 16 | 18 | 96 | 2 |
|  | 11 | 283 | 19 | 19 | 19 | 20 | 61 | 154 | 218 | 248 | 240 | 194 | 116 | 28 | 19 | 19 | 19 | 19 | 123 | 1 |
|  | 12 | 288 | 20 | 20 | 20 | 20 | 21 | 89 | 176 | 231 | 250 | 231 | 176 | 89 | 21 | 20 | 20 | 20 | 132 | 12 |
|  | HALF-DAY TOTALS |  | 63 | 63 | 75 | 198 | 445 | 721 | 887 | 914 | 798 | 551 | 269 | 101 | 63 | 63 | 63 | 63 | 354 |  |
| Dec 21 | 8 | 89 | 3 | 3 | 8 | 41 | 67 | 82 | 84 | 73 | 50 | 17 | 3 | 3 | 3 | 3 | 3 | 3 | 6 | 4 |
|  | 9 | 217 | 10 | 10 | 11 | 60 | 135 | 185 | 205 | 194 | 151 | 83 | 13 | 10 | 10 | 10 | 10 | 10 | 39 | 3 |
|  | 10 | 261 | 14 | 14 | 14 | 25 | 113 | 188 | 232 | 239 | 210 | 146 | 55 | 15 | 14 | 14 | 14 | 14 | 77 | 2 |
|  | 11 | 280 | 17 | 17 | 17 | 17 | 56 | 151 | 217 | 249 | 242 | 198 | 120 | 28 | 17 | 17 | 17 | 17 | 104 | 1 |
|  | 12 | 285 | 18 | 18 | 18 | 18 | 19 | 89 | 178 | 233 | 253 | 233 | 178 | 89 | 19 | 18 | 18 | 18 | 113 | 12 |
|  | HALF-DAY TOTALS |  | 52 | 52 | 56 | 146 | 374 | 649 | 822 | 867 | 775 | 557 | 276 | 94 | 53 | 52 | 52 | 52 | 282 |  |
|  |  |  | N | NNW | NW | WNW | W | WSW | SW | SSW | S | SSE | SE | ESE | E | ENE | NE | NNE | HOR | PM |

[a] Total solar heat gains for DSA glass (based on a ground reflectance of 0.20).
[b] Half-day totals computed by Simpson's rule, time interval = 10 minutes.

## TABLE C.4 Solar Intensity and Solar Heat Gain Factors[a] for 48°N Latitude (I-P Units)[b]

| Date | Solar Time A.M. | Direct Normal Btu/h ft² | N | NNE | NE | ENE | E | ESE | SE | SSE | S | SSW | SW | WSW | W | WNW | NW | NNW | HOR | Solar Time P.M. |
|---|---|---|---|---|---|---|---|---|---|---|---|---|---|---|---|---|---|---|---|---|
| Jan 21 | 8 | 37 | 1 | 1 | 4 | 18 | 29 | 34 | 35 | 30 | 20 | 6 | 1 | 1 | 1 | 1 | 1 | 2 | 4 | 4 |
| | 9 | 185 | 8 | 8 | 8 | 53 | 118 | 160 | 176 | 166 | 129 | 69 | 10 | 8 | 8 | 8 | 8 | 8 | 25 | 3 |
| | 10 | 239 | 12 | 12 | 12 | 22 | 106 | 175 | 216 | 223 | 195 | 136 | 50 | 12 | 12 | 12 | 12 | 12 | 55 | 2 |
| | 11 | 261 | 14 | 14 | 14 | 15 | 53 | 144 | 208 | 239 | 233 | 190 | 116 | 26 | 14 | 14 | 14 | 14 | 77 | 1 |
| | 12 | 267 | 15 | 15 | 15 | 15 | 16 | 86 | 171 | 226 | 245 | 226 | 171 | 86 | 16 | 15 | 15 | 15 | 85 | 12 |
| | HALF-DAY TOTALS | | 43 | 43 | 46 | 117 | 316 | 567 | 729 | 776 | 701 | 512 | 259 | 85 | 43 | 43 | 43 | 43 | 203 | |
| Feb 21 | 7 | 4 | 0 | 0 | 1 | 3 | 3 | 3 | 3 | 2 | 1 | 0 | 0 | 0 | 0 | 0 | 0 | 0 | 0 | 5 |
| | 8 | 180 | 8 | 8 | 36 | 103 | 149 | 170 | 166 | 136 | 82 | 17 | 8 | 8 | 8 | 8 | 8 | 8 | 25 | 4 |
| | 9 | 247 | 13 | 13 | 16 | 90 | 168 | 216 | 230 | 209 | 155 | 71 | 14 | 13 | 13 | 13 | 13 | 13 | 66 | 3 |
| | 10 | 275 | 17 | 17 | 17 | 38 | 131 | 203 | 242 | 244 | 207 | 138 | 44 | 18 | 17 | 17 | 17 | 17 | 105 | 2 |
| | 11 | 288 | 19 | 19 | 19 | 20 | 65 | 158 | 221 | 249 | 239 | 192 | 113 | 27 | 19 | 19 | 19 | 19 | 130 | 1 |
| | 12 | 292 | 20 | 20 | 20 | 20 | 22 | 89 | 176 | 231 | 250 | 231 | 176 | 89 | 22 | 20 | 20 | 20 | 138 | 12 |
| | HALF-DAY TOTALS | | 68 | 68 | 107 | 274 | 541 | 816 | 968 | 967 | 813 | 531 | 261 | 104 | 68 | 68 | 68 | 68 | 395 | |
| Mar 21 | 7 | 153 | 7 | 22 | 80 | 123 | 145 | 145 | 123 | 80 | 23 | 7 | 7 | 7 | 7 | 7 | 7 | 7 | 20 | 5 |
| | 8 | 236 | 14 | 15 | 76 | 154 | 204 | 222 | 206 | 158 | 82 | 15 | 14 | 14 | 14 | 14 | 14 | 14 | 68 | 4 |
| | 9 | 270 | 19 | 19 | 3 | 121 | 193 | 234 | 239 | 207 | 142 | 52 | 20 | 19 | 19 | 19 | 19 | 19 | 118 | 3 |
| | 10 | 287 | 23 | 23 | 24 | 58 | 146 | 208 | 237 | 231 | 189 | 115 | 33 | 23 | 23 | 23 | 23 | 23 | 156 | 2 |
| | 11 | 295 | 25 | 25 | 25 | 26 | 74 | 156 | 210 | 232 | 218 | 172 | 94 | 28 | 25 | 25 | 25 | 25 | 180 | 1 |
| | 12 | 298 | 26 | 26 | 26 | 26 | 27 | 83 | 161 | 211 | 228 | 211 | 161 | 83 | 27 | 26 | 26 | 26 | 188 | 12 |
| | HALF-DAY TOTALS | | 100 | 118 | 250 | 494 | 775 | 1012 | 1100 | 1014 | 767 | 465 | 244 | 126 | 101 | 100 | 100 | 100 | 636 | |
| Apr 21 | 6 | 108 | 12 | 53 | 86 | 105 | 107 | 93 | 64 | 23 | 6 | 6 | 6 | 6 | 6 | 6 | 6 | 6 | 15 | 6 |
| | 7 | 205 | 15 | 61 | 132 | 180 | 199 | 189 | 148 | 84 | 18 | 14 | 14 | 14 | 14 | 14 | 14 | 14 | 60 | 5 |
| | 8 | 247 | 20 | 32 | 111 | 179 | 219 | 225 | 196 | 138 | 55 | 21 | 20 | 20 | 20 | 20 | 20 | 20 | 114 | 4 |
| | 9 | 268 | 25 | 26 | 60 | 141 | 197 | 223 | 215 | 176 | 106 | 33 | 25 | 25 | 25 | 25 | 25 | 25 | 161 | 3 |
| | 10 | 280 | 28 | 28 | 31 | 77 | 148 | 193 | 209 | 194 | 150 | 80 | 31 | 28 | 28 | 28 | 28 | 28 | 196 | 2 |
| | 11 | 286 | 31 | 31 | 31 | 33 | 78 | 140 | 181 | 193 | 177 | 133 | 69 | 33 | 31 | 31 | 31 | 31 | 218 | 1 |
| | 12 | 288 | 31 | 31 | 31 | 31 | 34 | 71 | 131 | 172 | 186 | 172 | 131 | 71 | 34 | 31 | 31 | 31 | 226 | 12 |
| | HALF-DAY TOTALS | | 147 | 242 | 461 | 724 | 957 | 1098 | 1081 | 895 | 605 | 370 | 226 | 156 | 141 | 140 | 140 | 140 | 875 | |
| May 21 | 5 | 41 | 17 | 31 | 40 | 42 | 39 | 29 | 14 | 3 | 3 | 3 | 3 | 3 | 3 | 3 | 3 | 3 | 5 | 7 |
| | 6 | 162 | 35 | 97 | 141 | 162 | 160 | 133 | 85 | 24 | 12 | 12 | 12 | 12 | 12 | 12 | 12 | 13 | 40 | 6 |
| | 7 | 219 | 23 | 90 | 158 | 200 | 212 | 191 | 142 | 68 | 21 | 19 | 19 | 19 | 19 | 19 | 19 | 19 | 91 | 5 |
| | 8 | 248 | 26 | 54 | 132 | 190 | 218 | 214 | 178 | 113 | 38 | 25 | 25 | 25 | 25 | 25 | 25 | 25 | 142 | 4 |
| | 9 | 264 | 29 | 32 | 82 | 151 | 194 | 208 | 192 | 147 | 77 | 32 | 29 | 29 | 29 | 29 | 29 | 29 | 185 | 3 |
| | 10 | 274 | 33 | 34 | 39 | 90 | 145 | 178 | 184 | 163 | 116 | 57 | 35 | 33 | 33 | 33 | 33 | 33 | 219 | 2 |
| | 11 | 279 | 35 | 35 | 36 | 40 | 79 | 126 | 155 | 160 | 142 | 101 | 54 | 37 | 35 | 35 | 35 | 35 | 240 | 1 |
| | 12 | 280 | 35 | 35 | 35 | 36 | 38 | 63 | 107 | 139 | 150 | 139 | 107 | 63 | 38 | 36 | 35 | 35 | 247 | 12 |
| | HALF-DAY TOTALS | | 215 | 388 | 645 | 893 | 1065 | 1114 | 1007 | 749 | 483 | 316 | 225 | 184 | 174 | 173 | 173 | 174 | 1045 | |
| Jun 21 | 5 | 77 | 35 | 61 | 76 | 80 | 72 | 53 | 24 | 6 | 5 | 5 | 5 | 5 | 5 | 5 | 5 | 8 | 12 | 7 |
| | 6 | 172 | 46 | 110 | 155 | 175 | 169 | 138 | 84 | 22 | 14 | 14 | 14 | 14 | 14 | 14 | 14 | 16 | 51 | 6 |
| | 7 | 220 | 29 | 101 | 165 | 204 | 211 | 187 | 135 | 60 | 23 | 21 | 21 | 21 | 21 | 21 | 21 | 21 | 103 | 5 |
| | 8 | 246 | 29 | 64 | 139 | 191 | 215 | 206 | 168 | 101 | 34 | 27 | 27 | 27 | 27 | 27 | 27 | 27 | 152 | 4 |
| | 9 | 261 | 31 | 36 | 91 | 153 | 190 | 199 | 180 | 133 | 66 | 33 | 31 | 31 | 31 | 31 | 31 | 31 | 193 | 3 |
| | 10 | 269 | 34 | 36 | 45 | 94 | 143 | 169 | 171 | 148 | 101 | 50 | 36 | 34 | 34 | 34 | 34 | 34 | 225 | 2 |
| | 11 | 274 | 36 | 36 | 38 | 44 | 79 | 118 | 142 | 145 | 126 | 88 | 49 | 38 | 36 | 36 | 36 | 36 | 246 | 1 |
| | 12 | 275 | 37 | 37 | 37 | 38 | 40 | 60 | 96 | 124 | 134 | 124 | 96 | 60 | 40 | 38 | 37 | 37 | 252 | 12 |
| | HALF-DAY TOTALS | | 257 | 459 | 722 | 955 | 1095 | 1102 | 955 | 678 | 436 | 299 | 228 | 197 | 189 | 188 | 188 | 191 | 1108 | |
| Jul 21 | 5 | 43 | 18 | 33 | 42 | 45 | 41 | 30 | 15 | 3 | 3 | 3 | 3 | 3 | 3 | 3 | 3 | 4 | 6 | 7 |
| | 6 | 156 | 37 | 96 | 138 | 159 | 156 | 129 | 82 | 24 | 13 | 13 | 13 | 13 | 13 | 13 | 13 | 14 | 41 | 6 |
| | 7 | 211 | 25 | 90 | 156 | 196 | 207 | 186 | 138 | 66 | 22 | 20 | 20 | 20 | 20 | 20 | 20 | 20 | 92 | 5 |
| | 8 | 240 | 27 | 56 | 132 | 187 | 214 | 209 | 174 | 110 | 38 | 26 | 26 | 26 | 26 | 26 | 26 | 26 | 142 | 4 |
| | 9 | 256 | 30 | 34 | 83 | 149 | 191 | 204 | 187 | 143 | 75 | 33 | 30 | 30 | 30 | 30 | 30 | 30 | 184 | 3 |
| | 10 | 266 | 34 | 35 | 41 | 90 | 143 | 174 | 180 | 158 | 113 | 56 | 36 | 34 | 34 | 34 | 34 | 34 | 217 | 2 |
| | 11 | 271 | 36 | 36 | 37 | 42 | 79 | 124 | 151 | 156 | 138 | 99 | 54 | 38 | 36 | 36 | 36 | 36 | 237 | 1 |
| | 12 | 272 | 36 | 36 | 36 | 37 | 39 | 63 | 104 | 136 | 146 | 136 | 104 | 63 | 39 | 37 | 36 | 36 | 244 | 12 |
| | HALF-DAY TOTALS | | 223 | 395 | 646 | 886 | 1050 | 1092 | 983 | 730 | 474 | 315 | 229 | 190 | 181 | 179 | 179 | 180 | 1042 | |
| Aug 21 | 6 | 99 | 13 | 51 | 81 | 98 | 100 | 87 | 60 | 22 | 7 | 7 | 7 | 7 | 7 | 7 | 7 | 7 | 16 | 6 |
| | 7 | 190 | 17 | 61 | 128 | 172 | 190 | 179 | 141 | 79 | 19 | 15 | 15 | 15 | 15 | 15 | 15 | 15 | 61 | 5 |
| | 8 | 232 | 22 | 34 | 110 | 174 | 211 | 216 | 188 | 132 | 53 | 23 | 22 | 22 | 22 | 22 | 22 | 22 | 114 | 4 |
| | 9 | 154 | 27 | 28 | 63 | 139 | 192 | 216 | 108 | 169 | 102 | 34 | 27 | 27 | 27 | 27 | 27 | 27 | 159 | 3 |
| | 10 | 266 | 30 | 30 | 33 | 78 | 145 | 188 | 203 | 188 | 144 | 78 | 33 | 30 | 30 | 30 | 30 | 30 | 193 | 2 |
| | 11 | 272 | 32 | 32 | 32 | 36 | 78 | 137 | 175 | 187 | 171 | 129 | 68 | 35 | 32 | 32 | 32 | 32 | 215 | 1 |
| | 12 | 274 | 33 | 33 | 33 | 33 | 36 | 71 | 128 | 167 | 189 | 167 | 128 | 71 | 36 | 33 | 33 | 33 | 223 | 12 |
| | HALF-DAY TOTALS | | 157 | 251 | 459 | 709 | 929 | 1060 | 1040 | 862 | 587 | 366 | 231 | 165 | 151 | 149 | 149 | 149 | 869 | |
| Sep 21 | 7 | 131 | 8 | 21 | 71 | 108 | 128 | 128 | 108 | 71 | 21 | 8 | 7 | 7 | 7 | 7 | 7 | 7 | 20 | 5 |
| | 8 | 215 | 15 | 16 | 72 | 144 | 191 | 207 | 193 | 148 | 77 | 16 | 15 | 15 | 15 | 15 | 15 | 15 | 65 | 4 |
| | 9 | 251 | 20 | 20 | 34 | 116 | 184 | 223 | 227 | 197 | 136 | 52 | 21 | 20 | 20 | 20 | 20 | 20 | 114 | 3 |
| | 10 | 269 | 24 | 24 | 25 | 58 | 141 | 200 | 228 | 221 | 182 | 112 | 34 | 24 | 24 | 24 | 24 | 24 | 151 | 2 |
| | 11 | 278 | 26 | 26 | 26 | 28 | 73 | 151 | 203 | 223 | 210 | 166 | 92 | 29 | 26 | 26 | 26 | 26 | 174 | 1 |
| | 12 | 280 | 27 | 27 | 27 | 27 | 29 | 82 | 156 | 204 | 220 | 204 | 156 | 82 | 29 | 27 | 27 | 27 | 182 | 12 |
| | HALF-DAY TOTALS | | 105 | 121 | 240 | 465 | 729 | 953 | 1040 | 963 | 737 | 453 | 243 | 131 | 106 | 105 | 105 | 105 | 614 | |
| Oct 21 | 7 | 4 | 0 | 0 | 2 | 3 | 4 | 4 | 3 | 2 | 1 | 0 | 0 | 0 | 0 | 0 | 0 | 0 | 0 | 5 |
| | 8 | 165 | 8 | 9 | 35 | 96 | 139 | 159 | 155 | 126 | 77 | 16 | 8 | 8 | 8 | 8 | 8 | 8 | 25 | 4 |
| | 9 | 233 | 14 | 14 | 16 | 88 | 161 | 207 | 220 | 199 | 148 | 68 | 15 | 14 | 14 | 14 | 14 | 14 | 66 | 3 |
| | 10 | 262 | 18 | 18 | 18 | 39 | 128 | 196 | 233 | 234 | 199 | 133 | 43 | 18 | 18 | 18 | 18 | 18 | 104 | 2 |
| | 11 | 274 | 20 | 20 | 20 | 21 | 64 | 153 | 213 | 241 | 231 | 186 | 109 | 27 | 20 | 20 | 20 | 20 | 128 | 1 |
| | 12 | 278 | 21 | 21 | 21 | 21 | 23 | 87 | 171 | 223 | 242 | 223 | 171 | 87 | 23 | 21 | 21 | 21 | 136 | 12 |
| | HALF-DAY TOTALS | | 71 | 71 | 108 | 266 | 519 | 780 | 925 | 925 | 779 | 513 | 256 | 106 | 72 | 71 | 71 | 71 | 391 | |
| Nov 21 | 8 | 36 | 1 | 1 | 4 | 18 | 29 | 34 | 35 | 30 | 20 | 6 | 1 | 1 | 1 | 1 | 1 | 2 | 4 | 4 |
| | 9 | 179 | 8 | 8 | 8 | 52 | 115 | 156 | 171 | 161 | 125 | 67 | 10 | 8 | 8 | 8 | 8 | 8 | 26 | 3 |
| | 10 | 233 | 12 | 12 | 12 | 22 | 104 | 172 | 212 | 218 | 191 | 133 | 49 | 13 | 12 | 12 | 12 | 12 | 55 | 2 |
| | 11 | 255 | 15 | 15 | 15 | 15 | 52 | 142 | 204 | 234 | 228 | 186 | 114 | 26 | 15 | 15 | 15 | 15 | 77 | 1 |
| | 12 | 261 | 15 | 15 | 15 | 15 | 17 | 85 | 168 | 222 | 240 | 222 | 168 | 85 | 17 | 15 | 15 | 15 | 85 | 12 |
| | HALF-DAY TOTALS | | 44 | 44 | 47 | 117 | 310 | 555 | 713 | 760 | 686 | 502 | 255 | 85 | 44 | 44 | 44 | 44 | 204 | |
| Dec 21 | 9 | 140 | 5 | 5 | 6 | 36 | 86 | 120 | 133 | 127 | 100 | 56 | 8 | 5 | 5 | 5 | 5 | 5 | 13 | 3 |
| | 10 | 214 | 10 | 10 | 10 | 16 | 91 | 156 | 194 | 201 | 179 | 126 | 49 | 10 | 10 | 10 | 10 | 10 | 38 | 2 |
| | 11 | 242 | 12 | 12 | 12 | 13 | 46 | 134 | 195 | 225 | 220 | 180 | 111 | 25 | 12 | 12 | 12 | 12 | 57 | 1 |
| | 12 | 250 | 13 | 13 | 13 | 13 | 14 | 81 | 163 | 215 | 233 | 215 | 168 | 81 | 14 | 13 | 13 | 13 | 65 | 12 |
| | HALF-DAY TOTALS | | 33 | 33 | 34 | 73 | 233 | 458 | 610 | 665 | 616 | 468 | 247 | 76 | 34 | 33 | 33 | 33 | 141 | |
| | | | N | NNW | NW | WNW | W | WSW | SW | SSW | S | SSE | SE | ESE | E | ENE | NE | NNE | HOR | PM |

[a] Total solar heat gains for DSA glass (based on a ground reflectance of 0.20).
[b] Half-day totals computed by Simpson's rule, time interval = 10 minutes.

### TABLE C.5  Solar Intensity and Solar Heat Gain Factors[a] for 64°N Latitude (I-P Units)[b]

| Date | Solar Time A.M. | Direct Normal Btu/h ft² | N | NNE | NE | ENE | E | ESE | SE | SSE | S | SSW | SW | WSW | W | WNW | NW | NNW | HOR | Solar Time P.M. |
|---|---|---|---|---|---|---|---|---|---|---|---|---|---|---|---|---|---|---|---|---|
| Jan 21 | 10 | 22 | 1 | 1 | 1 | 1 | 9 | 16 | 20 | 21 | 19 | 13 | 5 | 1 | 1 | 1 | 1 | 1 | 1 | 2 |
| | 11 | 81 | 3 | 3 | 3 | 3 | 15 | 45 | 67 | 77 | 75 | 62 | 38 | 8 | 3 | 3 | 3 | 3 | 6 | 1 |
| | 12 | 100 | 3 | 3 | 3 | 3 | 4 | 33 | 67 | 89 | 96 | 89 | 67 | 33 | 4 | 3 | 3 | 3 | 8 | 12 |
| | HALF-DAY TOTALS | | 5 | 5 | 5 | 6 | 25 | 79 | 121 | 142 | 141 | 119 | 75 | 23 | 5 | 5 | 5 | 5 | 11 | |
| Feb 21 | 8 | 18 | 1 | 1 | 3 | 10 | 15 | 17 | 17 | 14 | 9 | 2 | 1 | 1 | 1 | 1 | 1 | 1 | 1 | 4 |
| | 9 | 134 | 5 | 5 | 6 | 43 | 89 | 118 | 128 | 119 | 90 | 45 | 6 | 5 | 5 | 5 | 5 | 5 | 13 | 3 |
| | 10 | 190 | 8 | 8 | 8 | 18 | 87 | 144 | 176 | 180 | 157 | 108 | 38 | 9 | 8 | 8 | 8 | 8 | 28 | 2 |
| | 11 | 215 | 10 | 10 | 10 | 10 | 44 | 122 | 177 | 202 | 197 | 160 | 97 | 20 | 10 | 10 | 10 | 10 | 41 | 1 |
| | 12 | 222 | 11 | 11 | 11 | 11 | 12 | 73 | 147 | 194 | 210 | 194 | 147 | 73 | 12 | 11 | 11 | 11 | 45 | 12 |
| | HALF-DAY TOTALS | | 29 | 30 | 33 | 89 | 244 | 446 | 578 | 617 | 560 | 411 | 212 | 66 | 30 | 29 | 29 | 29 | 106 | |
| Mar 21 | 7 | 95 | 4 | 11 | 47 | 74 | 90 | 91 | 79 | 53 | 17 | 4 | 4 | 4 | 4 | 4 | 4 | 4 | 9 | 5 |
| | 8 | 185 | 9 | 10 | 46 | 113 | 158 | 177 | 170 | 135 | 78 | 14 | 9 | 9 | 9 | 9 | 9 | 9 | 32 | 4 |
| | 9 | 227 | 13 | 13 | 16 | 88 | 159 | 203 | 215 | 194 | 143 | 64 | 14 | 13 | 13 | 13 | 13 | 13 | 59 | 3 |
| | 10 | 249 | 16 | 16 | 16 | 35 | 122 | 190 | 226 | 228 | 194 | 130 | 42 | 16 | 16 | 16 | 16 | 16 | 84 | 2 |
| | 11 | 260 | 17 | 17 | 17 | 18 | 60 | 148 | 209 | 236 | 228 | 184 | 109 | 25 | 17 | 17 | 17 | 17 | 99 | 1 |
| | 12 | 263 | 18 | 18 | 18 | 18 | 19 | 85 | 168 | 221 | 239 | 221 | 168 | 85 | 19 | 18 | 18 | 18 | 105 | 12 |
| | HALF-DAY TOTALS | | 68 | 74 | 150 | 334 | 596 | 854 | 984 | 958 | 779 | 504 | 257 | 104 | 68 | 68 | 68 | 68 | 335 | |
| Apr 21 | 5 | 27 | 8 | 18 | 24 | 27 | 26 | 20 | 12 | 2 | 1 | 1 | 1 | 1 | 1 | 1 | 1 | 1 | 2 | 7 |
| | 6 | 133 | 12 | 59 | 102 | 127 | 132 | 118 | 84 | 35 | 8 | 8 | 8 | 8 | 8 | 8 | 8 | 8 | 21 | 6 |
| | 7 | 194 | 14 | 41 | 113 | 163 | 189 | 185 | 153 | 96 | 25 | 13 | 13 | 13 | 13 | 13 | 13 | 13 | 51 | 5 |
| | 8 | 228 | 17 | 19 | 79 | 153 | 201 | 217 | 201 | 153 | 79 | 19 | 17 | 17 | 17 | 17 | 17 | 17 | 85 | 4 |
| | 9 | 248 | 21 | 21 | 32 | 111 | 180 | 219 | 225 | 197 | 138 | 55 | 22 | 21 | 21 | 21 | 21 | 21 | 116 | 3 |
| | 10 | 260 | 23 | 23 | 24 | 51 | 134 | 194 | 225 | 221 | 185 | 118 | 38 | 24 | 23 | 23 | 23 | 23 | 140 | 2 |
| | 11 | 266 | 24 | 24 | 24 | 26 | 68 | 148 | 202 | 225 | 214 | 171 | 99 | 29 | 24 | 24 | 24 | 24 | 155 | 1 |
| | 12 | 268 | 25 | 25 | 25 | 25 | 27 | 83 | 159 | 208 | 224 | 208 | 159 | 83 | 27 | 25 | 25 | 25 | 160 | 12 |
| | HALF-DAY TOTALS | | 131 | 218 | 410 | 671 | 943 | 1150 | 1186 | 1036 | 763 | 487 | 273 | 149 | 121 | 120 | 120 | 120 | 651 | |
| May 21 | 4 | 51 | 30 | 44 | 51 | 51 | 43 | 28 | 8 | 3 | 3 | 3 | 3 | 3 | 3 | 3 | 3 | 10 | 6 | 8 |
| | 5 | 132 | 48 | 95 | 125 | 135 | 125 | 96 | 50 | 11 | 9 | 9 | 9 | 9 | 9 | 9 | 9 | 11 | 26 | 7 |
| | 6 | 185 | 28 | 97 | 150 | 181 | 183 | 158 | 109 | 40 | 15 | 15 | 15 | 15 | 15 | 15 | 15 | 15 | 55 | 6 |
| | 7 | 218 | 21 | 63 | 138 | 189 | 211 | 201 | 161 | 94 | 24 | 19 | 19 | 19 | 19 | 19 | 19 | 19 | 90 | 5 |
| | 8 | 239 | 23 | 28 | 97 | 167 | 209 | 220 | 198 | 146 | 68 | 25 | 23 | 23 | 23 | 23 | 23 | 23 | 124 | 4 |
| | 9 | 252 | 26 | 27 | 45 | 122 | 183 | 215 | 215 | 184 | 123 | 46 | 27 | 26 | 26 | 26 | 26 | 26 | 152 | 3 |
| | 10 | 261 | 28 | 28 | 30 | 61 | 135 | 188 | 212 | 205 | 167 | 102 | 36 | 28 | 28 | 28 | 28 | 28 | 174 | 2 |
| | 11 | 265 | 30 | 30 | 30 | 32 | 72 | 141 | 188 | 207 | 195 | 154 | 87 | 33 | 30 | 30 | 30 | 30 | 188 | 1 |
| | 12 | 267 | 30 | 30 | 30 | 30 | 33 | 78 | 146 | 189 | 204 | 189 | 146 | 78 | 33 | 30 | 30 | 30 | 192 | 12 |
| | HALF-DAY TOTALS | | 247 | 425 | 680 | 950 | 1177 | 1291 | 1218 | 985 | 708 | 465 | 288 | 191 | 169 | 168 | 168 | 176 | 911 | |
| Jun 21 | 4 | 93 | 53 | 83 | 96 | 94 | 78 | 50 | 14 | 7 | 7 | 7 | 7 | 7 | 7 | 7 | 7 | 21 | 16 | 8 |
| | 5 | 154 | 62 | 114 | 148 | 158 | 145 | 110 | 55 | 14 | 12 | 12 | 12 | 12 | 12 | 12 | 12 | 14 | 39 | 7 |
| | 6 | 194 | 36 | 107 | 162 | 191 | 192 | 163 | 110 | 39 | 18 | 17 | 17 | 17 | 17 | 17 | 17 | 18 | 71 | 6 |
| | 7 | 221 | 24 | 71 | 145 | 193 | 213 | 200 | 158 | 89 | 25 | 22 | 22 | 22 | 22 | 22 | 22 | 22 | 105 | 5 |
| | 8 | 239 | 25 | 33 | 104 | 170 | 208 | 216 | 192 | 139 | 62 | 27 | 25 | 25 | 25 | 25 | 25 | 25 | 137 | 4 |
| | 9 | 251 | 28 | 29 | 51 | 124 | 181 | 210 | 208 | 175 | 115 | 43 | 29 | 28 | 28 | 28 | 28 | 28 | 165 | 3 |
| | 10 | 258 | 30 | 30 | 32 | 65 | 134 | 183 | 204 | 195 | 157 | 94 | 36 | 30 | 30 | 30 | 30 | 30 | 186 | 2 |
| | 11 | 262 | 32 | 32 | 32 | 34 | 72 | 137 | 180 | 196 | 184 | 144 | 82 | 35 | 32 | 32 | 32 | 32 | 199 | 1 |
| | 12 | 263 | 32 | 32 | 32 | 32 | 35 | 76 | 138 | 179 | 193 | 179 | 138 | 76 | 35 | 32 | 32 | 32 | 203 | 12 |
| | HALF-DAY TOTALS | | 322 | 533 | 801 | 1061 | 1253 | 1317 | 1195 | 946 | 679 | 455 | 296 | 211 | 192 | 190 | 191 | 213 | 1021 | |
| Jul 21 | 4 | 53 | 32 | 47 | 55 | 54 | 46 | 29 | 9 | 4 | 4 | 4 | 4 | 4 | 4 | 4 | 4 | 11 | 8 | 8 |
| | 5 | 128 | 49 | 94 | 123 | 133 | 124 | 95 | 50 | 11 | 10 | 10 | 10 | 10 | 10 | 10 | 10 | 11 | 28 | 7 |
| | 6 | 179 | 30 | 96 | 148 | 177 | 180 | 155 | 106 | 39 | 16 | 15 | 15 | 15 | 15 | 15 | 15 | 15 | 57 | 6 |
| | 7 | 211 | 22 | 64 | 137 | 186 | 207 | 197 | 157 | 92 | 25 | 20 | 20 | 20 | 20 | 20 | 20 | 20 | 92 | 5 |
| | 8 | 231 | 24 | 30 | 97 | 165 | 205 | 215 | 193 | 142 | 67 | 26 | 24 | 24 | 24 | 24 | 24 | 24 | 124 | 4 |
| | 9 | 245 | 27 | 28 | 47 | 121 | 180 | 211 | 211 | 179 | 120 | 46 | 28 | 27 | 27 | 27 | 27 | 27 | 152 | 3 |
| | 10 | 253 | 29 | 29 | 31 | 62 | 134 | 185 | 208 | 200 | 164 | 100 | 37 | 29 | 29 | 29 | 29 | 29 | 174 | 2 |
| | 11 | 257 | 31 | 31 | 31 | 33 | 72 | 139 | 185 | 202 | 191 | 151 | 86 | 34 | 31 | 31 | 31 | 31 | 187 | 1 |
| | 12 | 259 | 31 | 31 | 31 | 31 | 34 | 78 | 143 | 185 | 200 | 185 | 143 | 78 | 34 | 31 | 31 | 31 | 192 | 12 |
| | HALF-DAY TOTALS | | 258 | 434 | 684 | 946 | 1163 | 1269 | 1193 | 965 | 697 | 462 | 292 | 198 | 177 | 175 | 175 | 185 | 918 | |
| Aug 21 | 5 | 29 | 9 | 20 | 27 | 30 | 28 | 22 | 13 | 2 | 2 | 2 | 2 | 2 | 2 | 2 | 2 | 2 | 3 | 7 |
| | 6 | 123 | 13 | 58 | 97 | 121 | 125 | 111 | 80 | 34 | 9 | 9 | 9 | 9 | 9 | 9 | 9 | 9 | 23 | 6 |
| | 7 | 181 | 15 | 42 | 109 | 157 | 180 | 176 | 145 | 92 | 26 | 14 | 14 | 14 | 14 | 14 | 14 | 14 | 53 | 5 |
| | 8 | 214 | 19 | 21 | 78 | 148 | 193 | 208 | 192 | 147 | 76 | 21 | 19 | 19 | 19 | 19 | 19 | 19 | 87 | 4 |
| | 9 | 234 | 22 | 22 | 34 | 109 | 174 | 211 | 217 | 189 | 133 | 55 | 23 | 22 | 22 | 22 | 22 | 22 | 117 | 3 |
| | 10 | 246 | 25 | 25 | 26 | 52 | 131 | 188 | 217 | 214 | 178 | 114 | 39 | 25 | 25 | 25 | 25 | 25 | 140 | 2 |
| | 11 | 252 | 26 | 26 | 26 | 28 | 69 | 144 | 196 | 217 | 207 | 166 | 97 | 31 | 26 | 26 | 26 | 26 | 154 | 1 |
| | 12 | 254 | 27 | 27 | 27 | 27 | 29 | 82 | 155 | 201 | 217 | 201 | 155 | 82 | 29 | 27 | 27 | 27 | 159 | 12 |
| | HALF-DAY TOTALS | | 142 | 226 | 410 | 657 | 914 | 1109 | 1141 | 997 | 740 | 478 | 275 | 158 | 131 | 130 | 130 | 130 | 656 | |
| Sep 21 | 7 | 77 | 4 | 10 | 39 | 62 | 74 | 75 | 65 | 44 | 15 | 4 | 4 | 4 | 4 | 4 | 4 | 4 | 8 | 5 |
| | 8 | 163 | 10 | 10 | 43 | 103 | 143 | 160 | 154 | 123 | 71 | 14 | 10 | 10 | 10 | 10 | 10 | 10 | 31 | 4 |
| | 9 | 206 | 14 | 14 | 17 | 83 | 148 | 189 | 200 | 181 | 133 | 61 | 15 | 14 | 14 | 14 | 14 | 14 | 57 | 3 |
| | 10 | 229 | 16 | 16 | 17 | 35 | 116 | 179 | 213 | 214 | 183 | 123 | 41 | 17 | 16 | 16 | 16 | 16 | 81 | 2 |
| | 11 | 240 | 18 | 18 | 18 | 19 | 59 | 141 | 198 | 224 | 216 | 174 | 104 | 26 | 18 | 18 | 18 | 18 | 96 | 1 |
| | 12 | 244 | 19 | 19 | 19 | 19 | 21 | 82 | 160 | 209 | 227 | 209 | 160 | 82 | 21 | 19 | 19 | 19 | 101 | 12 |
| | HALF-DAY TOTALS | | 71 | 77 | 142 | 307 | 547 | 787 | 910 | 891 | 731 | 480 | 249 | 106 | 72 | 71 | 71 | 71 | 324 | |
| Oct 21 | 8 | 17 | 1 | 1 | 3 | 10 | 14 | 16 | 16 | 13 | 8 | 2 | 1 | 1 | 1 | 1 | 1 | 1 | 1 | 4 |
| | 9 | 122 | 5 | 5 | 6 | 40 | 82 | 109 | 118 | 110 | 83 | 42 | 6 | 5 | 5 | 5 | 5 | 5 | 13 | 3 |
| | 10 | 176 | 9 | 9 | 9 | 18 | 83 | 135 | 165 | 169 | 147 | 102 | 36 | 9 | 9 | 9 | 9 | 9 | 29 | 2 |
| | 11 | 201 | 11 | 11 | 11 | 11 | 43 | 116 | 167 | 191 | 186 | 152 | 92 | 20 | 11 | 11 | 11 | 11 | 41 | 1 |
| | 12 | 208 | 11 | 11 | 11 | 11 | 13 | 70 | 140 | 184 | 199 | 184 | 140 | 70 | 13 | 11 | 11 | 11 | 46 | 12 |
| | HALF-DAY TOTALS | | 31 | 31 | 34 | 86 | 231 | 460 | 542 | 580 | 527 | 388 | 202 | 66 | 32 | 31 | 31 | 31 | 108 | |
| Nov 21 | 10 | 23 | 1 | 1 | 1 | 1 | 10 | 17 | 21 | 22 | 20 | 14 | 5 | 1 | 1 | 1 | 1 | 1 | 1 | 2 |
| | 11 | 79 | 3 | 3 | 3 | 3 | 15 | 44 | 65 | 75 | 74 | 61 | 37 | 8 | 3 | 3 | 3 | 3 | 6 | 1 |
| | 12 | 97 | 4 | 4 | 4 | 4 | 4 | 32 | 66 | 87 | 93 | 87 | 66 | 32 | 4 | 4 | 4 | 4 | 8 | 12 |
| | HALF-DAY TOTALS | | 5 | 5 | 5 | 6 | 26 | 79 | 120 | 141 | 140 | 117 | 74 | 23 | 6 | 5 | 5 | 5 | 11 | |
| Dec 21 | 11 | 4 | 0 | 0 | 0 | 0 | 1 | 2 | 3 | 4 | 4 | 3 | 2 | 0 | 0 | 0 | 0 | 0 | 0 | 1 |
| | 12 | 16 | 0 | 0 | 0 | 0 | 1 | 5 | 11 | 14 | 15 | 14 | 11 | 5 | 1 | 0 | 0 | 0 | 1 | 12 |
| | HALF-DAY TOTALS | | 0 | 0 | 0 | 0 | 1 | 5 | 9 | 11 | 11 | 10 | 7 | 3 | 0 | 0 | 0 | 0 | 1 | |
| | | | N | NNW | NW | WNW | W | WSW | SW | SSW | S | SSE | SE | ESE | E | ENE | NE | NNE | HOR | PM |

[a] Total solar heat gains for DSA glass (based on a ground reflectance of 0.20).
[b] Half-day totals computed by Simpson's rule, time interval = 10 minutes.

**TABLE C.6 Solar Intensity and Solar Heat Gain Factors[a] for 16°N Latitude (SI Units)[b]**

| Date | Solar Time A.M. | Direct Normal W/m² | N | NNE | NE | ENE | E | ESE | SE | SSE | S | SSW | SW | WSW | W | WNW | NW | NNW | HOR | Solar Time P.M. |
|---|---|---|---|---|---|---|---|---|---|---|---|---|---|---|---|---|---|---|---|---|
| Jan 21 | 7 | 445 | 17 | 19 | 138 | 291 | 390 | 424 | 397 | 303 | 155 | 19 | 17 | 17 | 17 | 17 | 17 | 17 | 43 | 5 |
| | 8 | 827 | 45 | 48 | 174 | 463 | 662 | 757 | 734 | 596 | 359 | 79 | 45 | 45 | 45 | 45 | 45 | 45 | 249 | 4 |
| | 9 | 948 | 67 | 67 | 102 | 384 | 630 | 770 | 791 | 690 | 481 | 183 | 69 | 67 | 67 | 67 | 67 | 67 | 472 | 3 |
| | 10 | 1001 | 82 | 82 | 86 | 209 | 474 | 658 | 737 | 704 | 563 | 321 | 97 | 82 | 82 | 82 | 82 | 82 | 640 | 2 |
| | 11 | 1025 | 92 | 92 | 92 | 96 | 242 | 467 | 614 | 663 | 612 | 462 | 236 | 96 | 92 | 92 | 92 | 92 | 745 | 1 |
| | 12 | 1032 | 95 | 95 | 95 | 95 | 100 | 228 | 438 | 580 | 628 | 580 | 438 | 228 | 100 | 95 | 95 | 95 | 782 | 12 |
| HALF-DAY TOTALS | | | 348 | 352 | 618 | 1453 | 2398 | 3153 | 3458 | 3217 | 2465 | 1344 | 666 | 401 | 350 | 348 | 348 | 348 | 2539 | |
| Feb 21 | 7 | 575 | 24 | 55 | 265 | 435 | 532 | 544 | 474 | 326 | 113 | 26 | 24 | 24 | 24 | 24 | 24 | 24 | 80 | 5 |
| | 8 | 862 | 53 | 60 | 304 | 567 | 729 | 778 | 706 | 525 | 244 | 56 | 53 | 53 | 53 | 53 | 53 | 53 | 319 | 4 |
| | 9 | 961 | 74 | 77 | 202 | 482 | 676 | 763 | 733 | 592 | 347 | 96 | 74 | 74 | 74 | 74 | 74 | 74 | 549 | 3 |
| | 10 | 1006 | 90 | 91 | 104 | 292 | 508 | 636 | 665 | 593 | 423 | 193 | 94 | 90 | 90 | 90 | 90 | 90 | 722 | 2 |
| | 11 | 1027 | 99 | 99 | 102 | 118 | 262 | 428 | 527 | 542 | 471 | 323 | 154 | 103 | 99 | 99 | 99 | 99 | 831 | 1 |
| | 12 | 1034 | 103 | 103 | 103 | 105 | 108 | 189 | 336 | 448 | 487 | 448 | 336 | 189 | 108 | 105 | 103 | 103 | 868 | 12 |
| HALF-DAY TOTALS | | | 390 | 431 | 1013 | 1922 | 2730 | 3228 | 3263 | 2792 | 1836 | 906 | 547 | 417 | 393 | 390 | 390 | 390 | 2933 | |
| Mar 21 | 7 | 634 | 36 | 167 | 393 | 544 | 606 | 578 | 458 | 260 | 47 | 33 | 33 | 33 | 33 | 33 | 33 | 33 | 126 | 5 |
| | 8 | 857 | 63 | 157 | 442 | 648 | 752 | 741 | 615 | 390 | 111 | 61 | 61 | 61 | 61 | 61 | 61 | 61 | 380 | 4 |
| | 9 | 943 | 84 | 110 | 343 | 565 | 689 | 709 | 622 | 435 | 180 | 86 | 82 | 82 | 82 | 82 | 82 | 82 | 606 | 3 |
| | 10 | 983 | 98 | 103 | 191 | 379 | 519 | 575 | 543 | 424 | 240 | 107 | 100 | 98 | 98 | 98 | 98 | 98 | 778 | 2 |
| | 11 | 1003 | 108 | 110 | 113 | 166 | 273 | 361 | 395 | 366 | 281 | 173 | 114 | 110 | 108 | 108 | 108 | 108 | 885 | 1 |
| | 12 | 1008 | 111 | 111 | 112 | 114 | 117 | 149 | 216 | 273 | 295 | 273 | 216 | 149 | 117 | 114 | 112 | 111 | 919 | 12 |
| HALF-DAY TOTALS | | | 443 | 712 | 1558 | 2383 | 2926 | 3076 | 2773 | 2028 | 1006 | 588 | 483 | 448 | 440 | 483 | 437 | 437 | 3233 | |
| Apr 21 | 6 | 44 | 7 | 24 | 37 | 43 | 43 | 37 | 24 | 7 | 2 | 2 | 2 | 2 | 2 | 2 | 12 | 2 | 4 | 6 |
| | 7 | 622 | 75 | 298 | 482 | 589 | 604 | 528 | 369 | 141 | 45 | 42 | 42 | 42 | 42 | 42 | 42 | 42 | 169 | 5 |
| | 8 | 807 | 85 | 312 | 543 | 682 | 718 | 644 | 473 | 217 | 74 | 70 | 70 | 70 | 70 | 70 | 70 | 70 | 413 | 4 |
| | 9 | 885 | 97 | 248 | 469 | 610 | 657 | 608 | 465 | 244 | 97 | 90 | 90 | 90 | 90 | 90 | 90 | 90 | 623 | 3 |
| | 10 | 924 | 112 | 171 | 321 | 444 | 499 | 476 | 380 | 231 | 118 | 109 | 166 | 106 | 106 | 106 | 106 | 106 | 784 | 2 |
| | 11 | 942 | 120 | 127 | 169 | 228 | 270 | 276 | 245 | 189 | 136 | 121 | 118 | 115 | 115 | 115 | 115 | 118 | 882 | 1 |
| | 12 | 948 | 123 | 123 | 124 | 125 | 126 | 129 | 135 | 141 | 143 | 141 | 135 | 129 | 126 | 125 | 124 | 123 | 911 | 12 |
| HALF-DAY TOTALS | | | 565 | 1272 | 2127 | 2711 | 2909 | 2684 | 2059 | 1111 | 547 | 503 | 494 | 491 | 490 | 489 | 490 | 492 | 3333 | |
| May 21 | 6 | 138 | 43 | 94 | 128 | 141 | 134 | 106 | 59 | 11 | 9 | 9 | 9 | 9 | 9 | 9 | 9 | 10 | 195 | 6 |
| | 7 | 608 | 157 | 378 | 531 | 603 | 583 | 474 | 290 | 76 | 49 | 49 | 49 | 49 | 49 | 49 | 49 | 55 | 17 | 5 |
| | 8 | 771 | 165 | 415 | 598 | 689 | 677 | 564 | 362 | 121 | 78 | 76 | 76 | 76 | 76 | 76 | 76 | 80 | 425 | 4 |
| | 9 | 845 | 156 | 366 | 539 | 626 | 622 | 526 | 344 | 141 | 100 | 96 | 96 | 96 | 96 | 96 | 96 | 100 | 621 | 3 |
| | 10 | 883 | 149 | 281 | 410 | 478 | 474 | 398 | 264 | 139 | 116 | 111 | 111 | 111 | 111 | 111 | 111 | 116 | 772 | 2 |
| | 11 | 902 | 147 | 198 | 248 | 273 | 262 | 220 | 166 | 130 | 126 | 123 | 120 | 120 | 120 | 122 | 124 | 128 | 862 | 1 |
| | 12 | 907 | 146 | 144 | 140 | 134 | 132 | 131 | 130 | 129 | 129 | 129 | 130 | 131 | 132 | 134 | 140 | 144 | 890 | 12 |
| HALF-DAY TOTALS | | | 893 | 1813 | 2537 | 2891 | 2829 | 2360 | 1555 | 685 | 541 | 528 | 527 | 526 | 527 | 529 | 533 | 557 | 3338 | |
| Jun 21 | 6 | 168 | 64 | 124 | 163 | 175 | 162 | 123 | 64 | 13 | 12 | 12 | 12 | 12 | 12 | 12 | 12 | 13 | 24 | 6 |
| | 7 | 593 | 195 | 404 | 543 | 598 | 565 | 445 | 252 | 63 | 52 | 52 | 52 | 52 | 52 | 52 | 52 | 57 | 203 | 5 |
| | 8 | 750 | 209 | 449 | 612 | 684 | 653 | 526 | 313 | 98 | 79 | 79 | 79 | 79 | 79 | 79 | 79 | 84 | 425 | 4 |
| | 9 | 823 | 199 | 409 | 560 | 626 | 601 | 487 | 294 | 118 | 98 | 98 | 98 | 98 | 98 | 98 | 98 | 105 | 613 | 3 |
| | 10 | 861 | 187 | 329 | 441 | 486 | 459 | 363 | 222 | 125 | 116 | 113 | 113 | 113 | 113 | 113 | 113 | 121 | 759 | 2 |
| | 11 | 879 | 181 | 241 | 283 | 290 | 258 | 200 | 146 | 130 | 126 | 122 | 122 | 122 | 122 | 125 | 128 | 136 | 847 | 1 |
| | 12 | 885 | 179 | 173 | 158 | 142 | 134 | 132 | 130 | 129 | 129 | 129 | 130 | 132 | 134 | 142 | 158 | 173 | 873 | 12 |
| HALF-DAY TOTALS | | | 1122 | 2043 | 2683 | 2930 | 2763 | 2207 | 1357 | 612 | 548 | 540 | 540 | 540 | 542 | 545 | 555 | 599 | 3308 | |
| Jul 21 | 6 | 128 | 43 | 90 | 122 | 134 | 127 | 99 | 55 | 11 | 9 | 9 | 9 | 9 | 9 | 9 | 9 | 10 | 17 | 6 |
| | 7 | 579 | 161 | 373 | 518 | 585 | 564 | 456 | 277 | 74 | 51 | 51 | 51 | 51 | 51 | 51 | 51 | 55 | 194 | 5 |
| | 8 | 743 | 173 | 415 | 590 | 676 | 661 | 549 | 349 | 110 | 78 | 78 | 78 | 78 | 78 | 78 | 78 | 83 | 419 | 4 |
| | 9 | 818 | 165 | 371 | 537 | 618 | 610 | 513 | 334 | 138 | 102 | 99 | 99 | 99 | 99 | 99 | 99 | 104 | 611 | 3 |
| | 10 | 857 | 157 | 289 | 413 | 475 | 467 | 389 | 257 | 138 | 118 | 113 | 113 | 113 | 113 | 113 | 113 | 120 | 759 | 2 |
| | 11 | 876 | 154 | 207 | 255 | 277 | 262 | 218 | 164 | 133 | 128 | 125 | 122 | 122 | 122 | 125 | 128 | 131 | 848 | 1 |
| | 12 | 882 | 154 | 151 | 145 | 138 | 135 | 134 | 133 | 132 | 132 | 132 | 133 | 134 | 135 | 138 | 145 | 151 | 875 | 12 |
| HALF-DAY TOTALS | | | 933 | 1829 | 2521 | 2848 | 2770 | 2298 | 1507 | 680 | 554 | 541 | 539 | 539 | 540 | 542 | 547 | 575 | 3289 | |
| Aug 21 | 6 | 136 | 7 | 21 | 31 | 37 | 36 | 31 | 20 | 6 | 2 | 2 | 2 | 2 | 2 | 2 | 2 | 2 | 4 | 6 |
| | 7 | 569 | 81 | 289 | 458 | 555 | 567 | 493 | 343 | 131 | 48 | 45 | 45 | 45 | 45 | 45 | 45 | 45 | 167 | 5 |
| | 8 | 757 | 94 | 315 | 531 | 660 | 691 | 617 | 451 | 206 | 79 | 74 | 74 | 74 | 74 | 74 | 74 | 74 | 404 | 4 |
| | 9 | 838 | 104 | 258 | 467 | 598 | 640 | 589 | 448 | 235 | 103 | 95 | 95 | 95 | 95 | 95 | 95 | 95 | 608 | 3 |
| | 10 | 879 | 118 | 183 | 328 | 443 | 490 | 464 | 368 | 224 | 122 | 114 | 110 | 110 | 110 | 110 | 110 | 110 | 766 | 2 |
| | 11 | 899 | 127 | 136 | 180 | 235 | 271 | 273 | 240 | 185 | 138 | 127 | 124 | 120 | 120 | 120 | 120 | 124 | 860 | 1 |
| | 12 | 905 | 129 | 130 | 130 | 131 | 132 | 134 | 139 | 143 | 145 | 143 | 139 | 134 | 132 | 131 | 130 | 130 | 889 | 12 |
| HALF-DAY TOTALS | | | 603 | 1294 | 2099 | 2640 | 2810 | 2578 | 1969 | 1069 | 568 | 528 | 519 | 516 | 515 | 514 | 515 | 518 | 3258 | |
| Sep 21 | 7 | 565 | 38 | 157 | 360 | 497 | 554 | 529 | 419 | 240 | 48 | 35 | 35 | 35 | 35 | 35 | 35 | 35 | 122 | 5 |
| | 8 | 797 | 67 | 156 | 424 | 618 | 716 | 705 | 587 | 374 | 113 | 64 | 64 | 64 | 64 | 64 | 64 | 64 | 367 | 4 |
| | 9 | 887 | 87 | 114 | 335 | 547 | 665 | 684 | 602 | 424 | 181 | 90 | 86 | 86 | 86 | 86 | 86 | 86 | 585 | 3 |
| | 10 | 931 | 101 | 107 | 193 | 372 | 507 | 560 | 529 | 415 | 240 | 111 | 103 | 101 | 101 | 101 | 101 | 101 | 752 | 2 |
| | 11 | 952 | 111 | 114 | 118 | 169 | 272 | 356 | 389 | 361 | 279 | 176 | 118 | 114 | 111 | 111 | 111 | 111 | 856 | 1 |
| | 12 | 958 | 114 | 114 | 116 | 118 | 121 | 153 | 217 | 272 | 293 | 272 | 217 | 153 | 121 | 118 | 116 | 114 | 889 | 12 |
| HALF-DAY TOTALS | | | 461 | 712 | 1500 | 2276 | 2791 | 2937 | 2658 | 1963 | 1007 | 605 | 501 | 466 | 457 | 455 | 454 | 454 | 3126 | |
| Oct 21 | 7 | 524 | 25 | 56 | 249 | 404 | 492 | 502 | 437 | 300 | 105 | 27 | 25 | 25 | 25 | 25 | 25 | 25 | 79 | 5 |
| | 8 | 816 | 55 | 64 | 299 | 548 | 702 | 747 | 677 | 502 | 234 | 59 | 55 | 55 | 55 | 55 | 55 | 55 | 313 | 4 |
| | 9 | 920 | 76 | 80 | 204 | 473 | 659 | 741 | 711 | 573 | 336 | 97 | 76 | 76 | 76 | 76 | 76 | 76 | 537 | 3 |
| | 10 | 968 | 92 | 94 | 108 | 291 | 499 | 621 | 647 | 577 | 412 | 189 | 97 | 92 | 92 | 92 | 92 | 92 | 707 | 2 |
| | 11 | 991 | 102 | 105 | 122 | 261 | 420 | 515 | 528 | 459 | 315 | 154 | 106 | 102 | 102 | 102 | 102 | 102 | 814 | 1 |
| | 12 | 997 | 105 | 105 | 105 | 107 | 111 | 189 | 330 | 437 | 474 | 437 | 330 | 189 | 111 | 107 | 105 | 105 | 850 | 12 |
| HALF-DAY TOTALS | | | 402 | 444 | 1002 | 1869 | 2637 | 3110 | 3141 | 2689 | 1776 | 891 | 553 | 428 | 404 | 402 | 402 | 402 | 2872 | |
| Nov 21 | 7 | 423 | 17 | 19 | 134 | 280 | 375 | 406 | 379 | 289 | 147 | 19 | 17 | 17 | 17 | 17 | 17 | 17 | 43 | 5 |
| | 8 | 806 | 46 | 49 | 174 | 456 | 651 | 742 | 719 | 583 | 350 | 78 | 46 | 46 | 46 | 46 | 46 | 46 | 247 | 4 |
| | 9 | 929 | 67 | 67 | 105 | 382 | 623 | 779 | 779 | 679 | 472 | 180 | 70 | 67 | 67 | 67 | 67 | 67 | 468 | 3 |
| | 10 | 984 | 83 | 83 | 87 | 210 | 470 | 651 | 727 | 694 | 554 | 316 | 98 | 83 | 83 | 83 | 83 | 83 | 635 | 2 |
| | 11 | 1009 | 92 | 92 | 92 | 98 | 242 | 462 | 607 | 654 | 603 | 455 | 234 | 98 | 92 | 92 | 92 | 92 | 740 | 1 |
| | 12 | 1016 | 96 | 96 | 96 | 96 | 101 | 227 | 432 | 572 | 619 | 572 | 432 | 227 | 101 | 96 | 96 | 96 | 775 | 12 |
| HALF-DAY TOTALS | | | 352 | 357 | 620 | 1438 | 2363 | 3101 | 3399 | 3159 | 2421 | 1324 | 664 | 404 | 355 | 352 | 352 | 352 | 2520 | |
| Dec 21 | 7 | 372 | 13 | 14 | 93 | 228 | 318 | 354 | 339 | 268 | 150 | 21 | 13 | 13 | 13 | 13 | 13 | 13 | 31 | 5 |
| | 8 | 803 | 42 | 43 | 129 | 416 | 625 | 734 | 728 | 609 | 390 | 105 | 42 | 42 | 42 | 42 | 42 | 42 | 219 | 4 |
| | 9 | 936 | 63 | 63 | 78 | 341 | 604 | 761 | 800 | 716 | 520 | 226 | 66 | 63 | 63 | 63 | 63 | 63 | 436 | 3 |
| | 10 | 993 | 78 | 78 | 81 | 178 | 445 | 657 | 753 | 735 | 605 | 369 | 112 | 79 | 78 | 78 | 78 | 78 | 602 | 2 |
| | 11 | 1019 | 87 | 87 | 87 | 92 | 231 | 474 | 638 | 698 | 654 | 506 | 270 | 94 | 87 | 87 | 87 | 87 | 704 | 1 |
| | 12 | 1026 | 91 | 91 | 91 | 91 | 95 | 241 | 469 | 620 | 670 | 620 | 469 | 241 | 95 | 91 | 91 | 91 | 739 | 12 |
| HALF-DAY TOTALS | | | 328 | 328 | 501 | 1269 | 2241 | 3076 | 3468 | 3312 | 2638 | 1527 | 721 | 393 | 331 | 328 | 328 | 328 | 2361 | |
| | | | N | NNW | NW | WNW | W | WSW | SW | SSW | S | SSE | SE | ESE | E | ENE | NE | NNE | HOR | PM |

[a] Total solar heat gains for DSA glass (based on a ground reflectance of 0.20).
[b] Half-day totals computed by Simpson's rule, time interval = 10 minutes.

**TABLE C.7  Solar Intensity and Solar Heat Gain Factors[a] for 32°N Latitude (SI Units)[b]**

| Date | Solar Time A.M. | Direct Normal W/m² | N | NNE | NE | ENE | E | ESE | SE | SSE | S | SSW | SW | WSW | W | WNW | NW | NNW | HOR | Solar Time P.M. |
|---|---|---|---|---|---|---|---|---|---|---|---|---|---|---|---|---|---|---|---|---|
| Jan 21 | 7 | 4 | 0 | 0 | 1 | 3 | 4 | 4 | 4 | 3 | 2 | 0 | 0 | 0 | 0 | 0 | 0 | 0 | 0 | 5 |
| | 8 | 640 | 28 | 29 | 93 | 330 | 505 | 597 | 596 | 502 | 326 | 88 | 29 | 28 | 28 | 28 | 28 | 28 | 102 | 4 |
| | 9 | 849 | 48 | 48 | 53 | 286 | 553 | 721 | 775 | 711 | 534 | 258 | 52 | 48 | 48 | 48 | 48 | 48 | 278 | 3 |
| | 10 | 931 | 63 | 63 | 64 | 129 | 427 | 659 | 784 | 788 | 670 | 444 | 144 | 64 | 63 | 63 | 63 | 63 | 430 | 2 |
| | 11 | 967 | 71 | 71 | 71 | 75 | 213 | 502 | 698 | 784 | 750 | 602 | 347 | 90 | 71 | 71 | 71 | 71 | 524 | 1 |
| | 12 | 977 | 74 | 74 | 74 | 74 | 79 | 277 | 548 | 718 | 777 | 718 | 548 | 277 | 79 | 74 | 74 | 74 | 556 | 12 |
| | HALF-DAY TOTALS | | 249 | 250 | 338 | 897 | 1797 | 2700 | 3201 | 3198 | 2692 | 1746 | 831 | 353 | 252 | 249 | 249 | 249 | 1614 | |
| Feb 21 | 7 | 352 | 13 | 23 | 148 | 258 | 323 | 336 | 299 | 212 | 83 | 14 | 13 | 13 | 13 | 13 | 13 | 13 | 30 | 5 |
| | 8 | 771 | 41 | 43 | 204 | 469 | 646 | 719 | 683 | 538 | 300 | 53 | 41 | 41 | 41 | 41 | 41 | 41 | 200 | 4 |
| | 9 | 905 | 60 | 60 | 101 | 386 | 627 | 764 | 783 | 680 | 471 | 174 | 63 | 60 | 60 | 60 | 60 | 60 | 402 | 3 |
| | 10 | 964 | 75 | 75 | 78 | 195 | 475 | 670 | 760 | 733 | 595 | 352 | 99 | 75 | 75 | 75 | 75 | 75 | 556 | 2 |
| | 11 | 990 | 83 | 83 | 83 | 88 | 240 | 491 | 656 | 717 | 670 | 519 | 275 | 89 | 83 | 83 | 83 | 83 | 652 | 1 |
| | 12 | 998 | 86 | 86 | 86 | 86 | 91 | 250 | 489 | 644 | 696 | 643 | 489 | 250 | 91 | 86 | 86 | 86 | 684 | 12 |
| | HALF-DAY TOTALS | | 314 | 324 | 635 | 1405 | 2318 | 3086 | 3408 | 3188 | 2461 | 1426 | 718 | 384 | 316 | 314 | 314 | 314 | 2181 | |
| Mar 21 | 7 | 583 | 30 | 116 | 332 | 483 | 556 | 545 | 447 | 276 | 63 | 29 | 29 | 29 | 29 | 29 | 29 | 29 | 100 | 5 |
| | 8 | 821 | 54 | 78 | 339 | 576 | 717 | 746 | 661 | 473 | 195 | 57 | 54 | 54 | 54 | 54 | 54 | 54 | 315 | 4 |
| | 9 | 914 | 74 | 77 | 203 | 475 | 662 | 746 | 716 | 578 | 339 | 95 | 74 | 74 | 74 | 74 | 74 | 74 | 516 | 3 |
| | 10 | 959 | 88 | 88 | 96 | 273 | 499 | 637 | 677 | 615 | 456 | 221 | 93 | 88 | 88 | 88 | 88 | 88 | 667 | 2 |
| | 11 | 980 | 96 | 96 | 98 | 108 | 258 | 447 | 564 | 592 | 529 | 380 | 185 | 101 | 96 | 96 | 96 | 96 | 762 | 1 |
| | 12 | 987 | 99 | 99 | 99 | 99 | 105 | 209 | 386 | 511 | 554 | 511 | 386 | 209 | 105 | 99 | 99 | 99 | 795 | 12 |
| | HALF-DAY TOTALS | | 393 | 512 | 1131 | 1985 | 2761 | 3258 | 3284 | 2801 | 1858 | 1030 | 609 | 429 | 394 | 391 | 391 | 391 | 2757 | |
| Apr 21 | 6 | 210 | 29 | 110 | 172 | 205 | 207 | 177 | 119 | 38 | 11 | 11 | 11 | 11 | 11 | 11 | 11 | 11 | 23 | 6 |
| | 7 | 649 | 53 | 253 | 462 | 593 | 631 | 575 | 428 | 204 | 49 | 45 | 45 | 45 | 45 | 45 | 45 | 45 | 191 | 5 |
| | 8 | 804 | 73 | 192 | 453 | 632 | 715 | 689 | 559 | 337 | 95 | 69 | 69 | 69 | 69 | 69 | 69 | 69 | 408 | 4 |
| | 9 | 876 | 89 | 113 | 324 | 532 | 649 | 669 | 591 | 418 | 183 | 92 | 87 | 87 | 87 | 87 | 87 | 87 | 593 | 3 |
| | 10 | 913 | 101 | 106 | 165 | 342 | 490 | 599 | 543 | 445 | 275 | 123 | 104 | 101 | 101 | 101 | 101 | 101 | 736 | 2 |
| | 11 | 932 | 109 | 109 | 115 | 149 | 263 | 371 | 426 | 415 | 340 | 222 | 126 | 113 | 109 | 109 | 109 | 109 | 825 | 1 |
| | 12 | 937 | 112 | 112 | 112 | 116 | 120 | 167 | 260 | 335 | 363 | 335 | 260 | 167 | 120 | 116 | 112 | 112 | 854 | 12 |
| | HALF-DAY TOTALS | | 508 | 932 | 1734 | 2500 | 3003 | 3129 | 2806 | 2033 | 1135 | 721 | 559 | 496 | 482 | 479 | 479 | 479 | 3203 | |
| May 21 | 6 | 374 | 104 | 244 | 340 | 381 | 367 | 290 | 175 | 39 | 26 | 26 | 26 | 26 | 26 | 26 | 26 | 28 | 67 | 6 |
| | 7 | 666 | 112 | 350 | 535 | 638 | 643 | 550 | 374 | 134 | 60 | 57 | 57 | 57 | 57 | 57 | 57 | 59 | 256 | 5 |
| | 8 | 787 | 93 | 295 | 519 | 655 | 694 | 629 | 470 | 229 | 86 | 80 | 80 | 80 | 80 | 80 | 80 | 80 | 461 | 4 |
| | 9 | 849 | 104 | 194 | 404 | 558 | 625 | 601 | 488 | 294 | 117 | 100 | 97 | 97 | 97 | 97 | 97 | 97 | 633 | 3 |
| | 10 | 882 | 114 | 127 | 240 | 383 | 473 | 491 | 435 | 313 | 170 | 117 | 110 | 110 | 110 | 110 | 110 | 110 | 766 | 2 |
| | 11 | 898 | 121 | 124 | 132 | 186 | 260 | 312 | 322 | 285 | 215 | 147 | 126 | 122 | 118 | 118 | 118 | 118 | 847 | 1 |
| | 12 | 904 | 121 | 123 | 125 | 127 | 130 | 148 | 186 | 220 | 223 | 220 | 186 | 148 | 130 | 127 | 125 | 123 | 873 | 12 |
| | HALF-DAY TOTALS | | 700 | 1382 | 2214 | 2840 | 3106 | 2943 | 2356 | 1409 | 788 | 627 | 577 | 558 | 551 | 550 | 548 | 551 | 3464 | |
| Jun 21 | 6 | 412 | 140 | 289 | 388 | 425 | 401 | 314 | 174 | 39 | 32 | 32 | 32 | 32 | 32 | 32 | 32 | 35 | 89 | 6 |
| | 7 | 662 | 148 | 384 | 556 | 644 | 634 | 529 | 342 | 109 | 62 | 62 | 62 | 62 | 62 | 62 | 62 | 65 | 279 | 5 |
| | 8 | 773 | 115 | 335 | 540 | 656 | 677 | 597 | 427 | 188 | 89 | 84 | 84 | 84 | 84 | 84 | 84 | 84 | 476 | 4 |
| | 9 | 831 | 110 | 234 | 431 | 563 | 609 | 567 | 440 | 244 | 111 | 101 | 101 | 101 | 101 | 101 | 101 | 101 | 642 | 3 |
| | 10 | 863 | 120 | 150 | 272 | 395 | 461 | 458 | 387 | 261 | 143 | 120 | 114 | 114 | 114 | 114 | 114 | 114 | 770 | 2 |
| | 11 | 880 | 126 | 130 | 148 | 202 | 258 | 288 | 280 | 237 | 177 | 135 | 128 | 125 | 122 | 122 | 122 | 122 | 847 | 1 |
| | 12 | 885 | 128 | 129 | 130 | 132 | 134 | 148 | 164 | 183 | 191 | 183 | 164 | 144 | 134 | 132 | 130 | 129 | 871 | 12 |
| | HALF-DAY TOTALS | | 824 | 1589 | 2403 | 2950 | 3108 | 2829 | 2137 | 1174 | 710 | 621 | 595 | 585 | 582 | 579 | 579 | 586 | 3538 | |
| Jul 21 | 6 | 358 | 106 | 240 | 332 | 370 | 355 | 285 | 168 | 39 | 27 | 27 | 27 | 27 | 27 | 27 | 27 | 30 | 70 | 6 |
| | 7 | 640 | 118 | 349 | 526 | 623 | 626 | 534 | 361 | 129 | 62 | 59 | 59 | 59 | 59 | 59 | 59 | 61 | 257 | 5 |
| | 8 | 761 | 98 | 300 | 515 | 645 | 680 | 613 | 456 | 221 | 89 | 82 | 82 | 82 | 82 | 82 | 82 | 82 | 457 | 4 |
| | 9 | 823 | 107 | 202 | 405 | 553 | 615 | 588 | 475 | 284 | 117 | 100 | 100 | 100 | 100 | 100 | 100 | 100 | 626 | 3 |
| | 10 | 856 | 118 | 133 | 247 | 383 | 467 | 481 | 424 | 303 | 167 | 120 | 113 | 113 | 113 | 113 | 113 | 113 | 757 | 2 |
| | 11 | 873 | 124 | 128 | 137 | 191 | 261 | 308 | 314 | 277 | 210 | 147 | 129 | 125 | 121 | 121 | 121 | 121 | 836 | 1 |
| | 12 | 879 | 126 | 127 | 128 | 130 | 133 | 150 | 184 | 214 | 226 | 214 | 184 | 150 | 133 | 130 | 128 | 127 | 861 | 12 |
| | HALF-DAY TOTALS | | 728 | 1402 | 2210 | 2809 | 3052 | 2877 | 2290 | 1365 | 783 | 637 | 591 | 574 | 568 | 566 | 565 | 568 | 3431 | |
| Aug 21 | 6 | 187 | 30 | 103 | 159 | 188 | 189 | 162 | 108 | 35 | 12 | 12 | 12 | 12 | 12 | 12 | 12 | 12 | 25 | 6 |
| | 7 | 599 | 59 | 250 | 444 | 565 | 598 | 542 | 402 | 192 | 53 | 49 | 49 | 49 | 49 | 49 | 49 | 49 | 192 | 5 |
| | 8 | 756 | 79 | 200 | 446 | 614 | 690 | 662 | 536 | 321 | 96 | 74 | 74 | 74 | 74 | 74 | 74 | 74 | 403 | 4 |
| | 9 | 830 | 95 | 124 | 328 | 523 | 632 | 648 | 570 | 402 | 179 | 98 | 93 | 93 | 93 | 93 | 93 | 93 | 583 | 3 |
| | 10 | 869 | 106 | 113 | 175 | 344 | 482 | 544 | 526 | 429 | 266 | 126 | 110 | 106 | 106 | 106 | 106 | 106 | 722 | 2 |
| | 11 | 889 | 115 | 117 | 122 | 157 | 264 | 365 | 414 | 401 | 328 | 217 | 131 | 119 | 115 | 115 | 115 | 115 | 808 | 1 |
| | 12 | 894 | 117 | 117 | 117 | 122 | 126 | 170 | 255 | 324 | 350 | 324 | 255 | 170 | 126 | 122 | 117 | 117 | 837 | 12 |
| | HALF-DAY TOTALS | | 539 | 957 | 1721 | 2443 | 2910 | 3014 | 2694 | 1950 | 1109 | 729 | 581 | 524 | 510 | 507 | 506 | 506 | 3152 | |
| Sep 21 | 7 | 514 | 32 | 109 | 302 | 437 | 502 | 492 | 405 | 252 | 62 | 30 | 30 | 30 | 30 | 30 | 30 | 30 | 97 | 5 |
| | 8 | 758 | 57 | 81 | 343 | 546 | 679 | 706 | 626 | 450 | 190 | 61 | 57 | 57 | 57 | 57 | 57 | 57 | 304 | 4 |
| | 9 | 857 | 77 | 82 | 201 | 459 | 637 | 717 | 688 | 557 | 330 | 99 | 77 | 77 | 77 | 77 | 77 | 77 | 498 | 3 |
| | 10 | 905 | 91 | 91 | 101 | 270 | 485 | 617 | 655 | 596 | 444 | 220 | 97 | 91 | 91 | 91 | 91 | 91 | 645 | 2 |
| | 11 | 928 | 100 | 100 | 102 | 113 | 257 | 437 | 549 | 576 | 516 | 373 | 187 | 106 | 100 | 100 | 100 | 100 | 737 | 1 |
| | 12 | 935 | 103 | 103 | 103 | 103 | 110 | 210 | 379 | 499 | 540 | 499 | 379 | 210 | 110 | 103 | 103 | 103 | 769 | 12 |
| | HALF-DAY TOTALS | | 409 | 518 | 1089 | 1888 | 2621 | 3097 | 3132 | 2689 | 1812 | 1025 | 620 | 446 | 411 | 408 | 408 | 408 | 2664 | |
| Oct 21 | 7 | 312 | 13 | 24 | 136 | 233 | 291 | 301 | 268 | 190 | 75 | 14 | 13 | 13 | 13 | 13 | 13 | 13 | 30 | 5 |
| | 8 | 724 | 42 | 46 | 200 | 450 | 616 | 684 | 649 | 511 | 285 | 54 | 42 | 42 | 42 | 42 | 42 | 42 | 198 | 4 |
| | 9 | 862 | 63 | 63 | 104 | 378 | 608 | 738 | 755 | 655 | 454 | 170 | 65 | 63 | 63 | 63 | 63 | 63 | 395 | 3 |
| | 10 | 923 | 77 | 77 | 81 | 197 | 465 | 652 | 737 | 711 | 577 | 342 | 101 | 77 | 77 | 77 | 77 | 77 | 546 | 2 |
| | 11 | 952 | 86 | 86 | 86 | 91 | 239 | 481 | 639 | 697 | 651 | 505 | 269 | 93 | 86 | 86 | 86 | 86 | 639 | 1 |
| | 12 | 960 | 89 | 89 | 89 | 89 | 95 | 247 | 478 | 626 | 677 | 626 | 478 | 247 | 95 | 89 | 89 | 89 | 671 | 12 |
| | HALF-DAY TOTALS | | 325 | 335 | 632 | 1365 | 2234 | 2968 | 3275 | 3067 | 2376 | 1390 | 714 | 394 | 328 | 325 | 325 | 325 | 2143 | |
| Nov 21 | 7 | 5 | 0 | 0 | 1 | 3 | 4 | 5 | 4 | 3 | 2 | 0 | 0 | 0 | 0 | 0 | 0 | 0 | 0 | 5 |
| | 8 | 619 | 29 | 29 | 93 | 323 | 493 | 581 | 579 | 488 | 316 | 86 | 29 | 29 | 29 | 29 | 29 | 29 | 102 | 4 |
| | 9 | 829 | 49 | 49 | 55 | 284 | 545 | 709 | 761 | 697 | 523 | 253 | 49 | 49 | 49 | 49 | 49 | 49 | 277 | 3 |
| | 10 | 912 | 64 | 64 | 65 | 131 | 423 | 650 | 772 | 775 | 659 | 437 | 142 | 65 | 64 | 64 | 64 | 64 | 428 | 2 |
| | 11 | 949 | 72 | 72 | 72 | 76 | 213 | 496 | 688 | 773 | 739 | 593 | 342 | 91 | 72 | 72 | 72 | 72 | 521 | 1 |
| | 12 | 959 | 75 | 75 | 75 | 75 | 80 | 274 | 541 | 708 | 766 | 708 | 541 | 274 | 80 | 75 | 75 | 75 | 553 | 12 |
| | HALF-DAY TOTALS | | 253 | 254 | 342 | 890 | 1769 | 2653 | 3143 | 3138 | 2643 | 1717 | 824 | 355 | 256 | 253 | 253 | 253 | 607 | |
| Dec 21 | 8 | 556 | 22 | 22 | 59 | 265 | 426 | 515 | 523 | 452 | 305 | 98 | 23 | 22 | 22 | 22 | 22 | 22 | 69 | 4 |
| | 9 | 812 | 43 | 43 | 46 | 244 | 512 | 686 | 751 | 701 | 539 | 282 | 48 | 43 | 43 | 43 | 43 | 43 | 228 | 3 |
| | 10 | 908 | 57 | 57 | 57 | 106 | 402 | 642 | 777 | 792 | 683 | 466 | 163 | 59 | 57 | 57 | 57 | 57 | 375 | 2 |
| | 11 | 949 | 66 | 66 | 66 | 69 | 200 | 497 | 700 | 795 | 766 | 620 | 367 | 92 | 66 | 66 | 66 | 66 | 468 | 1 |
| | 12 | 960 | 69 | 69 | 69 | 69 | 73 | 281 | 559 | 733 | 794 | 733 | 559 | 281 | 73 | 69 | 69 | 69 | 500 | 12 |
| | HALF-DAY TOTALS | | 223 | 223 | 265 | 717 | 1577 | 2499 | 3043 | 3111 | 2687 | 1823 | 868 | 339 | 225 | 223 | 223 | 223 | 1389 | |
| | | | N | NNW | NW | WNW | W | WSW | SW | SSW | S | SSE | SE | ESE | E | ENE | NE | NNE | HOR | PM |

[a]Total solar heat gains for DSA glass (based on a ground reflectance of 0.20).
[b]Half-day totals computed by Simpson's rule, time interval = 10 minutes.

## TABLE C.8  Solar Intensity and Solar Heat Gain Factors[a] for 40°N Latitude (SI Units)[b]

| Date | Solar Time A.M. | Direct Normal W/m² | N | NNE | NE | ENE | E | ESE | SE | SSE | S | SSW | SW | WSW | W | WNW | NW | NNW | HOR | Solar Time P.M. |
|------|------|------|-----|-----|-----|-----|-----|-----|-----|-----|-----|-----|-----|-----|-----|-----|-----|-----|-----|------|
| Jan 21 | 8 | 446 | 17 | 17 | 55 | 223 | 350 | 417 | 420 | 358 | 236 | 60 | 17 | 17 | 17 | 17 | 17 | 17 | 44 | 4 |
| | 9 | 753 | 37 | 37 | 41 | 233 | 485 | 648 | 706 | 658 | 504 | 260 | 42 | 37 | 37 | 37 | 37 | 37 | 173 | 3 |
| | 10 | 865 | 51 | 51 | 51 | 97 | 390 | 627 | 761 | 776 | 671 | 460 | 161 | 53 | 51 | 51 | 51 | 51 | 303 | 2 |
| | 11 | 913 | 59 | 59 | 59 | 62 | 193 | 493 | 699 | 796 | 769 | 623 | 372 | 89 | 59 | 59 | 59 | 59 | 390 | 1 |
| | 12 | 926 | 62 | 62 | 62 | 62 | 66 | 293 | 563 | 740 | 802 | 740 | 563 | 283 | 62 | 62 | 62 | 62 | 419 | 12 |
| | HALF-DAY TOTALS | | 194 | 194 | 231 | 628 | 1426 | 2316 | 2852 | 2941 | 2566 | 1770 | 860 | 318 | 196 | 194 | 194 | 194 | 1117 | |
| Feb 21 | 7 | 175 | 6 | 10 | 71 | 127 | 160 | 167 | 150 | 107 | 43 | 6 | 6 | 6 | 6 | 6 | 6 | 6 | 11 | 5 |
| | 8 | 692 | 33 | 35 | 158 | 407 | 576 | 651 | 628 | 505 | 296 | 56 | 33 | 33 | 33 | 33 | 33 | 33 | 136 | 4 |
| | 9 | 854 | 52 | 52 | 70 | 337 | 587 | 738 | 773 | 689 | 496 | 209 | 55 | 52 | 52 | 52 | 52 | 52 | 309 | 3 |
| | 10 | 926 | 65 | 65 | 67 | 155 | 450 | 666 | 777 | 768 | 641 | 408 | 120 | 66 | 65 | 65 | 65 | 65 | 451 | 2 |
| | 11 | 958 | 73 | 73 | 73 | 77 | 224 | 504 | 690 | 769 | 730 | 579 | 325 | 86 | 73 | 73 | 73 | 73 | 538 | 1 |
| | 12 | 967 | 76 | 76 | 76 | 76 | 80 | 271 | 536 | 702 | 760 | 702 | 536 | 271 | 80 | 76 | 76 | 76 | 568 | 12 |
| | HALF-DAY TOTALS | | 267 | 271 | 478 | 1140 | 2043 | 2888 | 3308 | 3202 | 2591 | 1602 | 790 | 361 | 269 | 267 | 267 | 267 | 1730 | |
| Mar 21 | 7 | 540 | 27 | 92 | 295 | 441 | 514 | 509 | 425 | 271 | 69 | 26 | 26 | 26 | 26 | 26 | 26 | 26 | 83 | 5 |
| | 8 | 789 | 50 | 57 | 288 | 534 | 686 | 732 | 665 | 494 | 232 | 54 | 50 | 50 | 50 | 50 | 50 | 50 | 268 | 4 |
| | 9 | 889 | 67 | 70 | 147 | 429 | 640 | 749 | 744 | 625 | 404 | 127 | 69 | 67 | 67 | 67 | 67 | 67 | 450 | 3 |
| | 10 | 938 | 80 | 80 | 85 | 226 | 482 | 653 | 722 | 682 | 539 | 299 | 91 | 80 | 80 | 80 | 80 | 80 | 587 | 2 |
| | 11 | 961 | 88 | 88 | 88 | 94 | 247 | 476 | 623 | 673 | 622 | 473 | 244 | 94 | 88 | 88 | 88 | 85 | 673 | 1 |
| | 12 | 968 | 91 | 91 | 91 | 91 | 97 | 238 | 458 | 602 | 650 | 602 | 458 | 238 | 97 | 91 | 91 | 91 | 702 | 12 |
| | HALF-DAY TOTALS | | 358 | 440 | 954 | 1777 | 2626 | 3265 | 3429 | 3055 | 2191 | 1270 | 694 | 417 | 360 | 357 | 357 | 357 | 2411 | |
| Apr 21 | 6 | 282 | 36 | 144 | 228 | 275 | 279 | 241 | 164 | 56 | 16 | 15 | 15 | 15 | 15 | 15 | 15 | 15 | 34 | 6 |
| | 7 | 651 | 50 | 223 | 442 | 584 | 633 | 588 | 451 | 255 | 51 | 45 | 45 | 45 | 45 | 45 | 45 | 45 | 193 | 5 |
| | 8 | 795 | 69 | 140 | 402 | 601 | 706 | 703 | 594 | 391 | 130 | 69 | 67 | 67 | 67 | 67 | 67 | 67 | 389 | 4 |
| | 9 | 865 | 84 | 91 | 254 | 488 | 638 | 691 | 640 | 494 | 260 | 91 | 84 | 84 | 84 | 84 | 84 | 84 | 557 | 3 |
| | 10 | 901 | 96 | 99 | 117 | 291 | 480 | 589 | 608 | 538 | 381 | 177 | 101 | 96 | 96 | 96 | 96 | 96 | 685 | 2 |
| | 11 | 920 | 104 | 107 | 107 | 122 | 255 | 411 | 506 | 522 | 459 | 323 | 164 | 109 | 104 | 104 | 104 | 104 | 766 | 1 |
| | 12 | 926 | 106 | 106 | 106 | 109 | 114 | 196 | 341 | 448 | 486 | 448 | 341 | 196 | 114 | 109 | 106 | 106 | 794 | 12 |
| | HALF-DAY TOTALS | | 487 | 835 | 1580 | 2390 | 3020 | 3314 | 3135 | 2466 | 1539 | 935 | 628 | 495 | 467 | 464 | 464 | 464 | 3020 | |
| May 21 | 5 | 3 | 1 | 2 | 3 | 3 | 3 | 2 | 1 | 0 | 0 | 0 | 0 | 0 | 0 | 0 | 0 | 0 | 0 | 7 |
| | 6 | 453 | 113 | 284 | 403 | 458 | 446 | 364 | 223 | 56 | 33 | 33 | 33 | 33 | 33 | 33 | 33 | 35 | 96 | 6 |
| | 7 | 681 | 90 | 320 | 520 | 638 | 659 | 580 | 412 | 172 | 63 | 59 | 59 | 59 | 59 | 59 | 59 | 59 | 276 | 5 |
| | 8 | 787 | 86 | 230 | 471 | 629 | 694 | 655 | 519 | 295 | 92 | 80 | 80 | 80 | 80 | 80 | 80 | 80 | 461 | 4 |
| | 9 | 844 | 99 | 131 | 330 | 518 | 620 | 632 | 551 | 382 | 168 | 101 | 96 | 96 | 96 | 96 | 96 | 96 | 616 | 3 |
| | 10 | 875 | 107 | 114 | 171 | 332 | 467 | 529 | 513 | 419 | 262 | 127 | 111 | 107 | 107 | 107 | 107 | 107 | 737 | 2 |
| | 11 | 891 | 115 | 115 | 121 | 152 | 256 | 357 | 409 | 400 | 331 | 222 | 133 | 120 | 115 | 115 | 115 | 115 | 812 | 1 |
| | 12 | 896 | 117 | 117 | 117 | 121 | 126 | 171 | 258 | 329 | 355 | 329 | 258 | 171 | 126 | 121 | 117 | 117 | 836 | 12 |
| | HALF-DAY TOTALS | | 679 | 1275 | 2102 | 2818 | 3231 | 3232 | 2778 | 1897 | 1129 | 780 | 630 | 568 | 554 | 550 | 550 | 552 | 3418 | |
| Jun 21 | 5 | 68 | 32 | 54 | 68 | 70 | 63 | 46 | 20 | 5 | 4 | 4 | 4 | 4 | 4 | 4 | 4 | 7 | 8 | 7 |
| | 6 | 488 | 150 | 329 | 450 | 501 | 478 | 380 | 222 | 53 | 39 | 39 | 39 | 39 | 39 | 39 | 39 | 43 | 126 | 6 |
| | 7 | 681 | 118 | 355 | 543 | 648 | 654 | 562 | 385 | 145 | 68 | 65 | 65 | 65 | 65 | 65 | 65 | 67 | 306 | 5 |
| | 8 | 776 | 94 | 268 | 492 | 633 | 680 | 626 | 480 | 252 | 93 | 85 | 85 | 85 | 85 | 85 | 85 | 85 | 484 | 4 |
| | 9 | 829 | 105 | 160 | 359 | 524 | 607 | 601 | 507 | 332 | 142 | 104 | 100 | 100 | 100 | 100 | 100 | 100 | 633 | 3 |
| | 10 | 859 | 112 | 121 | 197 | 345 | 457 | 500 | 468 | 366 | 218 | 122 | 115 | 112 | 112 | 112 | 112 | 112 | 750 | 2 |
| | 11 | 874 | 119 | 123 | 128 | 166 | 254 | 332 | 367 | 347 | 279 | 188 | 130 | 124 | 119 | 119 | 119 | 119 | 821 | 1 |
| | 12 | 879 | 121 | 121 | 121 | 127 | 131 | 163 | 227 | 281 | 301 | 281 | 227 | 163 | 131 | 127 | 121 | 121 | 844 | 12 |
| | HALF-DAY TOTALS | | 799 | 1483 | 2314 | 2968 | 3275 | 3151 | 2580 | 1649 | 995 | 743 | 642 | 602 | 592 | 588 | 587 | 594 | 3551 | |
| Jul 21 | 5 | 7 | 3 | 5 | 7 | 7 | 6 | 5 | 2 | 0 | 0 | 0 | 0 | 0 | 0 | 0 | 0 | 1 | 1 | 7 |
| | 6 | 435 | 116 | 281 | 395 | 447 | 433 | 352 | 216 | 55 | 34 | 34 | 34 | 34 | 34 | 34 | 34 | 37 | 100 | 6 |
| | 7 | 656 | 95 | 321 | 513 | 625 | 643 | 564 | 400 | 166 | 66 | 62 | 62 | 62 | 62 | 62 | 62 | 62 | 278 | 5 |
| | 8 | 762 | 90 | 236 | 468 | 620 | 680 | 639 | 505 | 285 | 94 | 83 | 83 | 83 | 83 | 83 | 83 | 83 | 459 | 4 |
| | 9 | 818 | 102 | 138 | 333 | 513 | 610 | 618 | 537 | 371 | 165 | 104 | 99 | 99 | 99 | 99 | 99 | 99 | 611 | 3 |
| | 10 | 850 | 110 | 118 | 177 | 333 | 462 | 519 | 501 | 407 | 255 | 129 | 114 | 110 | 110 | 110 | 110 | 110 | 729 | 2 |
| | 11 | 866 | 117 | 120 | 125 | 157 | 256 | 352 | 400 | 389 | 321 | 217 | 135 | 123 | 117 | 117 | 117 | 117 | 802 | 1 |
| | 12 | 871 | 120 | 120 | 120 | 125 | 130 | 172 | 253 | 320 | 345 | 320 | 253 | 172 | 130 | 125 | 120 | 120 | 826 | 12 |
| | HALF-DAY TOTALS | | 705 | 1296 | 2102 | 2792 | 3180 | 3164 | 2707 | 1842 | 1110 | 783 | 643 | 586 | 572 | 568 | 567 | 570 | 3395 | |
| Aug 21 | 6 | 255 | 38 | 137 | 214 | 256 | 259 | 223 | 151 | 52 | 18 | 17 | 17 | 17 | 17 | 17 | 17 | 17 | 38 | 6 |
| | 7 | 603 | 55 | 223 | 426 | 557 | 602 | 557 | 426 | 222 | 55 | 49 | 49 | 49 | 49 | 49 | 49 | 49 | 196 | 5 |
| | 8 | 747 | 75 | 149 | 397 | 584 | 681 | 676 | 569 | 374 | 128 | 74 | 72 | 72 | 72 | 72 | 72 | 72 | 386 | 4 |
| | 9 | 819 | 89 | 97 | 259 | 481 | 621 | 669 | 618 | 475 | 251 | 97 | 89 | 89 | 89 | 89 | 89 | 89 | 549 | 3 |
| | 10 | 857 | 102 | 105 | 126 | 294 | 472 | 574 | 590 | 519 | 367 | 173 | 107 | 102 | 102 | 102 | 102 | 102 | 674 | 2 |
| | 11 | 876 | 109 | 109 | 113 | 130 | 257 | 403 | 492 | 505 | 443 | 313 | 165 | 116 | 109 | 109 | 109 | 109 | 753 | 1 |
| | 12 | 882 | 112 | 112 | 112 | 115 | 120 | 197 | 333 | 434 | 470 | 434 | 333 | 197 | 120 | 115 | 112 | 112 | 780 | 12 |
| | HALF-DAY TOTALS | | 518 | 861 | 1571 | 2338 | 2929 | 3196 | 3015 | 2370 | 1496 | 932 | 647 | 524 | 496 | 492 | 492 | 492 | 2983 | |
| Sep 21 | 7 | 472 | 28 | 87 | 265 | 395 | 460 | 456 | 381 | 244 | 66 | 27 | 27 | 27 | 27 | 27 | 27 | 27 | 80 | 5 |
| | 8 | 725 | 52 | 61 | 275 | 504 | 646 | 689 | 626 | 467 | 224 | 57 | 52 | 52 | 52 | 52 | 52 | 52 | 258 | 4 |
| | 9 | 830 | 71 | 73 | 148 | 413 | 613 | 717 | 712 | 599 | 391 | 129 | 73 | 71 | 71 | 71 | 71 | 71 | 434 | 3 |
| | 10 | 882 | 84 | 84 | 89 | 224 | 468 | 631 | 697 | 659 | 522 | 293 | 96 | 84 | 84 | 84 | 84 | 84 | 567 | 2 |
| | 11 | 906 | 92 | 92 | 92 | 99 | 245 | 463 | 604 | 652 | 603 | 461 | 242 | 99 | 92 | 92 | 92 | 92 | 651 | 1 |
| | 12 | 914 | 95 | 95 | 95 | 95 | 101 | 237 | 446 | 584 | 631 | 584 | 446 | 237 | 101 | 95 | 95 | 95 | 679 | 12 |
| | HALF-DAY TOTALS | | 374 | 447 | 917 | 1683 | 2484 | 3092 | 3257 | 2918 | 2119 | 1250 | 699 | 432 | 376 | 373 | 373 | 373 | 2329 | |
| Oct 21 | 7 | 153 | 6 | 10 | 64 | 113 | 142 | 148 | 132 | 94 | 38 | 7 | 6 | 6 | 6 | 6 | 6 | 6 | 12 | 5 |
| | 8 | 644 | 35 | 37 | 155 | 387 | 545 | 615 | 592 | 476 | 280 | 56 | 35 | 35 | 35 | 35 | 35 | 35 | 136 | 4 |
| | 9 | 811 | 54 | 54 | 73 | 329 | 567 | 710 | 743 | 661 | 476 | 202 | 57 | 54 | 54 | 54 | 54 | 54 | 305 | 3 |
| | 10 | 884 | 67 | 67 | 70 | 157 | 439 | 646 | 752 | 742 | 619 | 395 | 120 | 69 | 67 | 67 | 67 | 67 | 443 | 2 |
| | 11 | 917 | 76 | 76 | 76 | 80 | 223 | 491 | 670 | 745 | 707 | 562 | 317 | 89 | 76 | 76 | 76 | 76 | 529 | 1 |
| | 12 | 927 | 78 | 78 | 78 | 78 | 84 | 267 | 521 | 681 | 737 | 681 | 521 | 267 | 84 | 78 | 78 | 78 | 558 | 12 |
| | HALF-DAY TOTALS | | 277 | 282 | 479 | 1106 | 1965 | 2771 | 3173 | 3074 | 2494 | 1555 | 780 | 369 | 280 | 277 | 277 | 277 | 1704 | |
| Nov 21 | 8 | 430 | 17 | 17 | 55 | 217 | 339 | 404 | 406 | 346 | 228 | 66 | 18 | 17 | 17 | 17 | 17 | 17 | 44 | 4 |
| | 9 | 733 | 38 | 38 | 42 | 231 | 476 | 634 | 691 | 643 | 492 | 254 | 43 | 38 | 38 | 38 | 38 | 38 | 173 | 3 |
| | 10 | 846 | 52 | 52 | 52 | 99 | 385 | 617 | 748 | 763 | 659 | 450 | 159 | 54 | 52 | 52 | 52 | 52 | 303 | 2 |
| | 11 | 894 | 60 | 60 | 60 | 63 | 192 | 486 | 688 | 783 | 757 | 613 | 367 | 89 | 60 | 60 | 60 | 60 | 388 | 1 |
| | 12 | 908 | 63 | 63 | 63 | 63 | 67 | 280 | 555 | 728 | 789 | 728 | 555 | 280 | 67 | 63 | 63 | 63 | 418 | 12 |
| | HALF-DAY TOTALS | | 197 | 198 | 235 | 623 | 1403 | 2273 | 2797 | 2884 | 2516 | 1738 | 850 | 319 | 199 | 197 | 197 | 197 | 1115 | |
| Dec 21 | 8 | 279 | 9 | 9 | 25 | 129 | 212 | 259 | 264 | 230 | 157 | 52 | 10 | 9 | 9 | 9 | 9 | 9 | 20 | 4 |
| | 9 | 685 | 31 | 31 | 33 | 188 | 427 | 584 | 646 | 611 | 477 | 260 | 40 | 31 | 31 | 31 | 31 | 31 | 124 | 3 |
| | 10 | 825 | 45 | 45 | 45 | 78 | 358 | 594 | 732 | 755 | 661 | 462 | 173 | 47 | 45 | 45 | 45 | 45 | 244 | 2 |
| | 11 | 882 | 53 | 53 | 53 | 55 | 177 | 477 | 685 | 786 | 765 | 624 | 379 | 90 | 53 | 53 | 53 | 53 | 327 | 1 |
| | 12 | 898 | 56 | 56 | 56 | 56 | 60 | 281 | 560 | 736 | 798 | 738 | 560 | 281 | 60 | 56 | 56 | 56 | 356 | 12 |
| | HALF-DAY TOTALS | | 165 | 165 | 178 | 461 | 1180 | 2046 | 2594 | 2736 | 2446 | 1757 | 870 | 298 | 167 | 165 | 165 | 165 | 891 | |
| | | | N | NNW | NW | WNW | W | WSW | SW | SSW | S | SSE | SE | ESE | E | ENE | NE | NNE | HOR | PM |

[a] Total solar heat gains for DSA glass (based on a ground reflectance of 0.20).
[b] Half-day totals computed by Simpson's rule, time interval = 10 minutes.

## TABLE C.9  Solar Intensity and Solar Heat Gain Factors[a] for 48°N Latitude (SI Units)[b]

| Date | Solar Time A.M. | Direct Normal W/m² | N | NNE | NE | ENE | E | ESE | SE | SSE | S | SSW | SW | WSW | W | WNW | NW | NNW | HOR | Solar Time P.M. |
|---|---|---|---|---|---|---|---|---|---|---|---|---|---|---|---|---|---|---|---|---|
| Jan 21 | 8 | 116 | 4 | 4 | 13 | 57 | 90 | 109 | 110 | 94 | 63 | 18 | 4 | 4 | 4 | 4 | 4 | 4 | 7 | 4 |
|  | 9 | 584 | 24 | 24 | 26 | 168 | 371 | 505 | 555 | 523 | 406 | 217 | 30 | 24 | 24 | 24 | 24 | 24 | 79 | 3 |
|  | 10 | 754 | 37 | 37 | 37 | 69 | 333 | 554 | 682 | 702 | 615 | 429 | 159 | 39 | 37 | 37 | 37 | 37 | 174 | 2 |
|  | 11 | 823 | 45 | 45 | 45 | 47 | 166 | 455 | 656 | 753 | 734 | 598 | 365 | 83 | 45 | 45 | 45 | 45 | 244 | 1 |
|  | 12 | 842 | 48 | 48 | 48 | 48 | 51 | 271 | 541 | 713 | 772 | 713 | 541 | 271 | 51 | 48 | 48 | 48 | 269 | 12 |
|  | HALF-DAY TOTALS |  | 134 | 134 | 144 | 368 | 996 | 1788 | 2298 | 2448 | 2212 | 1616 | 817 | 267 | 136 | 134 | 134 | 134 | 639 |  |
| Feb 21 | 7 | 11 | 0 | 1 | 5 | 8 | 10 | 11 | 10 | 7 | 0 | 0 | 0 | 0 | 0 | 0 | 0 | 0 | 1 | 5 |
|  | 8 | 568 | 24 | 25 | 114 | 324 | 470 | 537 | 524 | 428 | 259 | 52 | 24 | 24 | 24 | 24 | 24 | 24 | 78 | 4 |
|  | 9 | 780 | 42 | 42 | 49 | 284 | 530 | 683 | 727 | 660 | 488 | 225 | 45 | 42 | 42 | 42 | 42 | 42 | 210 | 3 |
|  | 10 | 869 | 54 | 54 | 55 | 120 | 415 | 641 | 764 | 768 | 653 | 434 | 137 | 55 | 54 | 54 | 54 | 54 | 331 | 2 |
|  | 11 | 908 | 61 | 61 | 61 | 64 | 205 | 498 | 696 | 787 | 755 | 607 | 355 | 84 | 61 | 61 | 61 | 61 | 409 | 1 |
|  | 12 | 920 | 63 | 63 | 63 | 63 | 68 | 280 | 555 | 728 | 790 | 728 | 555 | 280 | 68 | 63 | 63 | 63 | 435 | 12 |
|  | HALF-DAY TOTALS |  | 214 | 216 | 337 | 864 | 1708 | 2575 | 3054 | 3051 | 2565 | 1677 | 825 | 328 | 216 | 214 | 214 | 214 | 1246 |  |
| Mar 21 | 7 | 482 | 23 | 71 | 253 | 387 | 458 | 458 | 388 | 253 | 71 | 23 | 22 | 22 | 22 | 22 | 22 | 22 | 64 | 5 |
|  | 8 | 744 | 44 | 48 | 239 | 486 | 644 | 701 | 651 | 498 | 257 | 49 | 44 | 44 | 44 | 44 | 44 | 44 | 214 | 4 |
|  | 9 | 853 | 60 | 60 | 104 | 381 | 609 | 738 | 753 | 652 | 449 | 164 | 62 | 60 | 60 | 60 | 60 | 60 | 371 | 3 |
|  | 10 | 906 | 72 | 72 | 75 | 183 | 460 | 656 | 749 | 728 | 596 | 363 | 104 | 72 | 72 | 72 | 72 | 72 | 493 | 2 |
|  | 11 | 932 | 79 | 79 | 79 | 83 | 232 | 492 | 663 | 731 | 689 | 541 | 297 | 88 | 79 | 79 | 79 | 79 | 568 | 1 |
|  | 12 | 939 | 81 | 81 | 81 | 81 | 86 | 261 | 509 | 666 | 720 | 666 | 509 | 261 | 86 | 81 | 81 | 81 | 594 | 12 |
|  | HALF-DAY TOTALS |  | 317 | 372 | 790 | 1558 | 2446 | 3193 | 3471 | 3199 | 2420 | 1466 | 769 | 399 | 319 | 316 | 316 | 316 | 2006 |  |
| Apr 21 | 6 | 340 | 39 | 167 | 271 | 330 | 337 | 294 | 203 | 74 | 20 | 19 | 19 | 19 | 19 | 19 | 19 | 19 | 46 | 6 |
|  | 7 | 646 | 49 | 191 | 417 | 567 | 628 | 595 | 468 | 264 | 56 | 45 | 45 | 45 | 45 | 45 | 45 | 45 | 189 | 5 |
|  | 8 | 779 | 64 | 99 | 350 | 566 | 690 | 709 | 619 | 435 | 173 | 67 | 64 | 64 | 64 | 64 | 64 | 64 | 359 | 4 |
|  | 9 | 847 | 79 | 83 | 191 | 444 | 621 | 703 | 679 | 554 | 335 | 105 | 79 | 79 | 79 | 79 | 79 | 79 | 507 | 3 |
|  | 10 | 884 | 90 | 90 | 97 | 244 | 466 | 610 | 660 | 613 | 472 | 251 | 97 | 90 | 90 | 90 | 90 | 90 | 618 | 2 |
|  | 11 | 902 | 97 | 97 | 97 | 105 | 245 | 443 | 570 | 609 | 558 | 419 | 217 | 103 | 97 | 97 | 97 | 97 | 689 | 1 |
|  | 12 | 908 | 99 | 99 | 99 | 99 | 106 | 224 | 414 | 543 | 587 | 543 | 414 | 224 | 106 | 99 | 99 | 99 | 713 | 12 |
|  | HALF-DAY TOTALS |  | 463 | 765 | 1454 | 2285 | 3019 | 3465 | 3411 | 2825 | 1908 | 1169 | 713 | 492 | 445 | 442 | 442 | 442 | 2761 |  |
| May 21 | 5 | 129 | 52 | 97 | 125 | 133 | 122 | 91 | 44 | 9 | 8 | 8 | 8 | 8 | 8 | 8 | 8 | 10 | 16 | 7 |
|  | 6 | 511 | 112 | 305 | 443 | 512 | 504 | 418 | 267 | 75 | 38 | 38 | 38 | 38 | 38 | 38 | 38 | 40 | 125 | 6 |
|  | 7 | 690 | 73 | 283 | 498 | 631 | 668 | 604 | 448 | 214 | 66 | 61 | 61 | 61 | 61 | 61 | 61 | 61 | 287 | 5 |
|  | 8 | 782 | 83 | 170 | 418 | 599 | 689 | 675 | 562 | 358 | 120 | 79 | 79 | 79 | 79 | 79 | 79 | 79 | 449 | 4 |
|  | 9 | 834 | 93 | 101 | 259 | 475 | 611 | 656 | 605 | 463 | 243 | 100 | 93 | 93 | 93 | 93 | 93 | 93 | 585 | 3 |
|  | 10 | 864 | 103 | 106 | 124 | 283 | 458 | 561 | 579 | 513 | 366 | 179 | 109 | 103 | 103 | 103 | 103 | 103 | 690 | 2 |
|  | 11 | 879 | 109 | 109 | 113 | 127 | 249 | 396 | 488 | 505 | 447 | 320 | 170 | 116 | 109 | 109 | 109 | 109 | 756 | 1 |
|  | 12 | 884 | 111 | 111 | 111 | 114 | 120 | 198 | 336 | 439 | 474 | 439 | 336 | 198 | 120 | 114 | 111 | 111 | 778 | 12 |
|  | HALF-DAY TOTALS |  | 679 | 1224 | 2035 | 2817 | 3360 | 3515 | 3176 | 2363 | 1525 | 997 | 709 | 579 | 550 | 546 | 549 |  | 3297 |  |
| Jun 21 | 5 | 243 | 111 | 192 | 241 | 252 | 228 | 166 | 75 | 19 | 17 | 17 | 17 | 17 | 17 | 17 | 17 | 24 | 38 | 7 |
|  | 6 | 544 | 146 | 348 | 488 | 552 | 534 | 434 | 266 | 70 | 46 | 46 | 46 | 46 | 46 | 46 | 46 | 49 | 162 | 6 |
|  | 7 | 693 | 93 | 317 | 521 | 642 | 467 | 591 | 427 | 188 | 72 | 67 | 67 | 67 | 67 | 67 | 67 | 67 | 324 | 5 |
|  | 8 | 775 | 90 | 203 | 440 | 604 | 678 | 651 | 529 | 320 | 107 | 84 | 84 | 84 | 84 | 84 | 84 | 84 | 479 | 4 |
|  | 9 | 822 | 98 | 114 | 286 | 482 | 600 | 629 | 567 | 418 | 208 | 105 | 98 | 98 | 98 | 98 | 98 | 98 | 610 | 3 |
|  | 10 | 849 | 108 | 113 | 142 | 296 | 450 | 534 | 540 | 465 | 319 | 157 | 114 | 108 | 108 | 108 | 108 | 108 | 711 | 2 |
|  | 11 | 864 | 114 | 114 | 120 | 138 | 248 | 373 | 449 | 456 | 396 | 279 | 156 | 121 | 114 | 114 | 114 | 114 | 775 | 1 |
|  | 12 | 869 | 116 | 116 | 116 | 120 | 126 | 189 | 303 | 391 | 423 | 391 | 303 | 189 | 126 | 120 | 116 | 116 | 796 | 12 |
|  | HALF-DAY TOTALS |  | 811 | 1446 | 2279 | 3013 | 3454 | 3477 | 3013 | 2140 | 1376 | 942 | 718 | 620 | 597 | 593 | 592 | 602 | 3495 |  |
| Jul 21 | 5 | 135 | 57 | 104 | 133 | 141 | 129 | 96 | 46 | 11 | 9 | 9 | 9 | 9 | 9 | 9 | 9 | 12 | 18 | 7 |
|  | 6 | 492 | 116 | 302 | 436 | 501 | 492 | 407 | 259 | 75 | 40 | 40 | 40 | 40 | 40 | 40 | 40 | 43 | 130 | 6 |
|  | 7 | 666 | 78 | 285 | 492 | 619 | 653 | 588 | 436 | 207 | 69 | 63 | 63 | 63 | 63 | 63 | 63 | 63 | 290 | 5 |
|  | 8 | 757 | 86 | 176 | 416 | 590 | 675 | 660 | 547 | 348 | 119 | 81 | 81 | 81 | 81 | 81 | 81 | 81 | 448 | 4 |
|  | 9 | 809 | 96 | 106 | 263 | 471 | 601 | 643 | 591 | 450 | 237 | 104 | 96 | 96 | 96 | 96 | 96 | 96 | 582 | 3 |
|  | 10 | 839 | 106 | 109 | 130 | 285 | 453 | 550 | 566 | 500 | 356 | 177 | 112 | 106 | 106 | 106 | 106 | 106 | 684 | 2 |
|  | 11 | 855 | 112 | 112 | 117 | 132 | 249 | 390 | 477 | 492 | 435 | 312 | 169 | 119 | 112 | 112 | 112 | 112 | 749 | 1 |
|  | 12 | 859 | 114 | 114 | 114 | 117 | 124 | 189 | 329 | 428 | 462 | 428 | 329 | 198 | 124 | 117 | 114 | 114 | 771 | 12 |
|  | HALF-DAY TOTALS |  | 705 | 1247 | 2039 | 2795 | 3311 | 3446 | 3101 | 2303 | 1496 | 993 | 721 | 598 | 570 | 565 | 565 | 569 | 3287 |  |
| Aug 21 | 6 | 311 | 42 | 161 | 256 | 310 | 316 | 273 | 190 | 70 | 22 | 21 | 21 | 21 | 21 | 21 | 21 | 21 | 51 | 6 |
|  | 7 | 599 | 53 | 193 | 403 | 543 | 598 | 565 | 444 | 250 | 59 | 49 | 49 | 49 | 49 | 49 | 49 | 49 | 193 | 5 |
|  | 8 | 732 | 69 | 108 | 347 | 549 | 665 | 681 | 593 | 417 | 168 | 73 | 69 | 69 | 69 | 69 | 69 | 69 | 358 | 4 |
|  | 9 | 801 | 84 | 90 | 198 | 437 | 605 | 681 | 655 | 534 | 323 | 108 | 84 | 84 | 84 | 84 | 84 | 84 | 502 | 3 |
|  | 10 | 839 | 95 | 95 | 104 | 247 | 453 | 593 | 639 | 592 | 456 | 245 | 104 | 95 | 95 | 95 | 95 | 95 | 610 | 2 |
|  | 11 | 858 | 102 | 102 | 102 | 112 | 247 | 433 | 553 | 589 | 539 | 406 | 215 | 110 | 102 | 102 | 102 | 102 | 679 | 1 |
|  | 12 | 864 | 104 | 104 | 104 | 104 | 113 | 224 | 404 | 526 | 568 | 526 | 404 | 224 | 113 | 104 | 104 | 104 | 702 | 12 |
|  | HALF-DAY TOTALS |  | 496 | 790 | 1449 | 2237 | 2929 | 3343 | 3282 | 2720 | 1852 | 1156 | 728 | 521 | 475 | 471 | 471 | 471 | 2741 |  |
| Sep 21 | 7 | 414 | 24 | 66 | 224 | 342 | 403 | 403 | 342 | 224 | 66 | 24 | 23 | 23 | 23 | 23 | 23 | 23 | 62 | 5 |
|  | 8 | 678 | 46 | 51 | 228 | 455 | 602 | 654 | 608 | 467 | 244 | 52 | 46 | 46 | 46 | 46 | 46 | 46 | 206 | 4 |
|  | 9 | 792 | 63 | 63 | 107 | 366 | 581 | 702 | 716 | 621 | 430 | 163 | 66 | 63 | 63 | 63 | 63 | 63 | 358 | 3 |
|  | 10 | 848 | 75 | 75 | 79 | 182 | 444 | 630 | 719 | 699 | 573 | 353 | 107 | 75 | 75 | 75 | 75 | 75 | 476 | 2 |
|  | 11 | 876 | 82 | 82 | 82 | 88 | 230 | 476 | 639 | 704 | 664 | 524 | 292 | 93 | 82 | 82 | 82 | 82 | 549 | 1 |
|  | 12 | 884 | 84 | 84 | 84 | 84 | 91 | 257 | 494 | 643 | 695 | 643 | 494 | 257 | 91 | 84 | 84 | 84 | 574 | 12 |
|  | HALF-DAY TOTALS |  | 332 | 381 | 758 | 1467 | 2300 | 3007 | 3280 | 3039 | 2324 | 1429 | 766 | 334 | 331 | 331 | 331 | 1937 |  |  |
| Oct 21 | 7 | 12 | 0 | 1 | 5 | 9 | 11 | 12 | 11 | 8 | 3 | 0 | 0 | 0 | 0 | 0 | 0 | 0 | 1 | 5 |
|  | 8 | 522 | 25 | 27 | 111 | 304 | 439 | 501 | 489 | 398 | 242 | 51 | 25 | 25 | 25 | 25 | 25 | 25 | 79 | 4 |
|  | 9 | 734 | 44 | 44 | 52 | 276 | 508 | 652 | 693 | 629 | 466 | 216 | 47 | 44 | 44 | 44 | 44 | 44 | 208 | 3 |
|  | 10 | 825 | 56 | 56 | 58 | 122 | 403 | 618 | 736 | 739 | 628 | 418 | 136 | 58 | 56 | 56 | 56 | 56 | 327 | 2 |
|  | 11 | 866 | 64 | 64 | 64 | 67 | 203 | 483 | 673 | 760 | 729 | 587 | 345 | 86 | 64 | 64 | 64 | 64 | 403 | 1 |
|  | 12 | 878 | 66 | 66 | 66 | 66 | 71 | 274 | 538 | 704 | 763 | 704 | 538 | 274 | 71 | 66 | 66 | 66 | 429 | 12 |
|  | HALF-DAY TOTALS |  | 223 | 225 | 340 | 838 | 1637 | 2461 | 2918 | 2918 | 2459 | 1619 | 808 | 334 | 226 | 223 | 223 | 223 | 1233 |  |
| Nov 21 | 8 | 115 | 4 | 4 | 13 | 57 | 90 | 108 | 109 | 93 | 62 | 18 | 4 | 4 | 4 | 4 | 4 | 4 | 7 | 4 |
|  | 9 | 565 | 25 | 25 | 27 | 165 | 363 | 492 | 540 | 509 | 394 | 211 | 31 | 25 | 25 | 25 | 25 | 25 | 81 | 3 |
|  | 10 | 735 | 38 | 38 | 38 | 70 | 328 | 543 | 668 | 688 | 602 | 420 | 156 | 40 | 38 | 38 | 38 | 38 | 175 | 2 |
|  | 11 | 804 | 46 | 46 | 46 | 48 | 165 | 448 | 645 | 739 | 720 | 587 | 358 | 83 | 46 | 46 | 46 | 46 | 244 | 1 |
|  | 12 | 823 | 48 | 48 | 48 | 48 | 53 | 267 | 531 | 700 | 758 | 700 | 531 | 267 | 53 | 48 | 48 | 48 | 269 | 12 |
|  | HALF-DAY TOTALS |  | 138 | 138 | 147 | 368 | 979 | 1752 | 2250 | 2396 | 2165 | 1583 | 804 | 268 | 140 | 138 | 138 | 138 | 642 |  |
| Dec 21 | 9 | 442 | 16 | 16 | 18 | 113 | 272 | 378 | 420 | 401 | 316 | 176 | 25 | 16 | 16 | 16 | 16 | 16 | 42 | 3 |
|  | 10 | 676 | 30 | 30 | 30 | 51 | 286 | 491 | 613 | 636 | 563 | 398 | 156 | 32 | 30 | 30 | 30 | 30 | 119 | 2 |
|  | 11 | 765 | 38 | 38 | 38 | 40 | 146 | 421 | 615 | 708 | 694 | 569 | 351 | 80 | 38 | 38 | 38 | 38 | 181 | 1 |
|  | 12 | 789 | 41 | 41 | 41 | 41 | 44 | 256 | 514 | 679 | 734 | 679 | 514 | 256 | 44 | 41 | 41 | 41 | 204 | 12 |
|  | HALF-DAY TOTALS |  | 105 | 105 | 107 | 231 | 735 | 1444 | 1924 | 2098 | 1944 | 1477 | 778 | 239 | 107 | 105 | 105 | 105 | 444 |  |
|  |  |  | N | NNW | NW | WNW | W | WSW | SW | SSW | S | SSE | SE | ESE | E | ENE | NE | NNE | HOR | PM |

[a] Total solar heat gains for DSA glass (based on a ground reflectance of 0.20).

[b] Half-day totals computed by Simpson's rule, time interval = 10 minutes.

**TABLE C.10  Solar Intensity and Solar Heat Gain Factors[a] for 64°N Latitude (SI Units)[b]**

| Date | Solar Time A.M. | Direct Normal W/m² | Solar Heat Gain Factors, W/m² | | | | | | | | | | | | | | | | | Solar Time P.M. |
|---|---|---|---|---|---|---|---|---|---|---|---|---|---|---|---|---|---|---|---|---|
| | | | N | NNE | NE | ENE | E | ESE | SE | SSE | S | SSW | SW | WSW | W | WNW | NW | NNW | HOR | |
| Jan 21 | 10 | 69 | 2 | 2 | 2 | 4 | 29 | 51 | 63 | 66 | 58 | 41 | 16 | 2 | 2 | 2 | 2 | 2 | 4 | 2 |
| | 11 | 256 | 9 | 9 | 9 | 46 | 143 | 210 | 242 | 238 | 196 | 120 | 24 | 9 | 9 | 9 | 9 | 9 | 18 | 1 |
| | 12 | 316 | 11 | 11 | 11 | 11 | 12 | 104 | 211 | 280 | 302 | 280 | 211 | 104 | 12 | 11 | 11 | 11 | 24 | 12 |
| | HALF-DAY TOTALS | | 16 | 16 | 16 | 18 | 80 | 250 | 382 | 449 | 446 | 374 | 237 | 72 | 17 | 16 | 16 | 16 | 33 | |
| Feb 21 | 8 | 56 | 2 | 2 | 9 | 31 | 46 | 53 | 53 | 44 | 27 | 6 | 2 | 2 | 2 | 2 | 2 | 2 | 3 | 4 |
| | 9 | 422 | 16 | 16 | 18 | 136 | 280 | 373 | 403 | 375 | 285 | 143 | 18 | 16 | 16 | 16 | 16 | 16 | 41 | 3 |
| | 10 | 601 | 26 | 26 | 26 | 57 | 276 | 453 | 554 | 567 | 495 | 341 | 119 | 28 | 26 | 26 | 26 | 26 | 90 | 2 |
| | 11 | 679 | 32 | 32 | 32 | 34 | 139 | 386 | 558 | 638 | 622 | 506 | 307 | 64 | 32 | 32 | 32 | 32 | 128 | 1 |
| | 12 | 701 | 34 | 34 | 34 | 34 | 37 | 231 | 464 | 613 | 662 | 613 | 464 | 231 | 37 | 34 | 34 | 34 | 143 | 12 |
| | HALF-DAY TOTALS | | 93 | 93 | 103 | 279 | 770 | 1408 | 1822 | 1947 | 1767 | 1298 | 668 | 210 | 95 | 93 | 93 | 93 | 334 | |
| Mar 21 | 7 | 300 | 12 | 34 | 147 | 235 | 284 | 288 | 249 | 168 | 54 | 13 | 12 | 12 | 12 | 12 | 12 | 12 | 27 | 5 |
| | 8 | 582 | 29 | 31 | 147 | 357 | 499 | 559 | 536 | 427 | 246 | 44 | 29 | 29 | 29 | 29 | 29 | 29 | 100 | 4 |
| | 9 | 717 | 41 | 41 | 51 | 277 | 502 | 642 | 679 | 613 | 450 | 203 | 44 | 41 | 41 | 41 | 41 | 41 | 187 | 3 |
| | 10 | 786 | 49 | 49 | 50 | 111 | 386 | 598 | 714 | 719 | 613 | 410 | 132 | 51 | 49 | 49 | 49 | 49 | 264 | 2 |
| | 11 | 821 | 54 | 54 | 54 | 57 | 190 | 468 | 658 | 746 | 718 | 579 | 344 | 80 | 54 | 54 | 54 | 54 | 314 | 1 |
| | 12 | 831 | 56 | 56 | 56 | 56 | 61 | 269 | 530 | 696 | 754 | 696 | 530 | 269 | 61 | 56 | 56 | 56 | 331 | 12 |
| | HALF-DAY TOTALS | | 213 | 234 | 473 | 1052 | 1879 | 2695 | 3103 | 3021 | 2457 | 1591 | 811 | 330 | 215 | 213 | 213 | 213 | 1057 | |
| Apr 21 | 5 | 85 | 24 | 56 | 77 | 85 | 81 | 64 | 36 | 6 | 4 | 4 | 4 | 4 | 4 | 4 | 4 | 5 | 8 | 7 |
| | 6 | 419 | 38 | 187 | 320 | 400 | 416 | 371 | 266 | 111 | 26 | 24 | 24 | 24 | 24 | 24 | 24 | 24 | 67 | 6 |
| | 7 | 613 | 43 | 129 | 355 | 516 | 595 | 584 | 482 | 303 | 80 | 41 | 41 | 41 | 41 | 41 | 41 | 41 | 162 | 5 |
| | 8 | 720 | 54 | 60 | 249 | 483 | 633 | 685 | 634 | 484 | 250 | 60 | 54 | 54 | 54 | 54 | 54 | 54 | 269 | 4 |
| | 9 | 783 | 65 | 65 | 100 | 351 | 567 | 691 | 710 | 620 | 436 | 174 | 68 | 65 | 65 | 65 | 65 | 65 | 367 | 3 |
| | 10 | 820 | 72 | 72 | 76 | 161 | 421 | 613 | 709 | 698 | 582 | 372 | 119 | 74 | 72 | 72 | 72 | 72 | 441 | 2 |
| | 11 | 839 | 77 | 77 | 77 | 82 | 215 | 468 | 637 | 710 | 676 | 540 | 312 | 93 | 77 | 77 | 77 | 77 | 448 | 1 |
| | 12 | 846 | 79 | 79 | 79 | 79 | 85 | 262 | 503 | 655 | 708 | 655 | 503 | 262 | 85 | 79 | 79 | 79 | 504 | 12 |
| | HALF-DAY TOTALS | | 413 | 687 | 1294 | 2116 | 2974 | 3628 | 3741 | 3268 | 2408 | 1537 | 861 | 471 | 381 | 378 | 378 | 378 | 2053 | |
| May 21 | 4 | 160 | 94 | 139 | 162 | 161 | 135 | 87 | 25 | 10 | 10 | 10 | 10 | 10 | 10 | 10 | 11 | 31 | 20 | 8 |
| | 5 | 416 | 152 | 298 | 393 | 425 | 395 | 302 | 158 | 34 | 29 | 29 | 29 | 29 | 29 | 29 | 29 | 33 | 81 | 7 |
| | 6 | 584 | 90 | 304 | 474 | 570 | 579 | 499 | 343 | 126 | 49 | 46 | 46 | 46 | 46 | 46 | 46 | 46 | 175 | 6 |
| | 7 | 688 | 65 | 199 | 436 | 596 | 665 | 634 | 507 | 297 | 78 | 60 | 60 | 60 | 60 | 60 | 60 | 60 | 284 | 5 |
| | 8 | 754 | 72 | 89 | 307 | 527 | 660 | 694 | 623 | 459 | 215 | 78 | 72 | 72 | 72 | 72 | 72 | 72 | 390 | 4 |
| | 9 | 796 | 82 | 85 | 143 | 384 | 577 | 678 | 679 | 579 | 389 | 147 | 85 | 82 | 82 | 82 | 82 | 82 | 481 | 3 |
| | 10 | 822 | 89 | 89 | 94 | 193 | 427 | 593 | 668 | 645 | 528 | 322 | 114 | 89 | 89 | 89 | 89 | 89 | 549 | 2 |
| | 11 | 836 | 93 | 93 | 93 | 100 | 226 | 446 | 594 | 652 | 614 | 485 | 275 | 105 | 93 | 93 | 93 | 93 | 592 | 1 |
| | 12 | 841 | 95 | 95 | 95 | 95 | 103 | 248 | 460 | 597 | 644 | 597 | 460 | 248 | 103 | 95 | 95 | 95 | 606 | 12 |
| | HALF-DAY TOTALS | | 780 | 1341 | 2145 | 2998 | 3712 | 4073 | 3841 | 3108 | 2234 | 1467 | 909 | 603 | 533 | 529 | 530 | 554 | 2875 | |
| Jun 21 | 4 | 294 | 181 | 263 | 302 | 297 | 247 | 156 | 43 | 21 | 21 | 21 | 21 | 21 | 21 | 21 | 23 | 67 | 50 | 8 |
| | 5 | 485 | 195 | 360 | 466 | 498 | 458 | 346 | 175 | 44 | 39 | 39 | 39 | 39 | 39 | 39 | 39 | 45 | 124 | 7 |
| | 6 | 614 | 113 | 338 | 510 | 603 | 605 | 516 | 347 | 122 | 57 | 55 | 55 | 55 | 55 | 55 | 55 | 57 | 223 | 6 |
| | 7 | 698 | 74 | 225 | 457 | 610 | 672 | 632 | 498 | 282 | 80 | 68 | 68 | 68 | 68 | 68 | 68 | 68 | 331 | 5 |
| | 8 | 754 | 79 | 105 | 327 | 535 | 657 | 682 | 605 | 438 | 197 | 85 | 79 | 79 | 79 | 79 | 79 | 79 | 433 | 4 |
| | 9 | 791 | 89 | 93 | 161 | 393 | 572 | 662 | 655 | 552 | 362 | 137 | 92 | 89 | 89 | 89 | 89 | 89 | 520 | 3 |
| | 10 | 814 | 96 | 96 | 102 | 205 | 423 | 577 | 642 | 615 | 497 | 297 | 114 | 96 | 96 | 96 | 96 | 96 | 585 | 2 |
| | 11 | 827 | 100 | 100 | 100 | 108 | 228 | 431 | 568 | 619 | 581 | 456 | 258 | 110 | 100 | 100 | 100 | 100 | 627 | 1 |
| | 12 | 831 | 101 | 101 | 101 | 101 | 110 | 240 | 436 | 566 | 610 | 566 | 436 | 240 | 110 | 101 | 101 | 101 | 641 | 12 |
| | HALF-DAY TOTALS | | 1016 | 1080 | 2526 | 3347 | 3953 | 4154 | 3769 | 2985 | 2142 | 1436 | 935 | 667 | 605 | 601 | 603 | 673 | 3219 | |
| Jul 21 | 4 | 168 | 101 | 149 | 172 | 172 | 144 | 93 | 27 | 12 | 12 | 12 | 12 | 12 | 12 | 12 | 12 | 35 | 24 | 8 |
| | 5 | 405 | 154 | 297 | 389 | 420 | 390 | 298 | 156 | 36 | 32 | 32 | 32 | 32 | 32 | 32 | 32 | 36 | 88 | 7 |
| | 6 | 564 | 94 | 303 | 467 | 560 | 567 | 488 | 335 | 124 | 51 | 49 | 49 | 49 | 49 | 49 | 49 | 49 | 181 | 6 |
| | 7 | 665 | 69 | 201 | 431 | 585 | 652 | 620 | 495 | 290 | 80 | 63 | 63 | 63 | 63 | 63 | 63 | 63 | 289 | 5 |
| | 8 | 730 | 75 | 94 | 307 | 520 | 648 | 680 | 609 | 449 | 211 | 81 | 75 | 75 | 75 | 75 | 75 | 75 | 393 | 4 |
| | 9 | 771 | 85 | 88 | 148 | 382 | 567 | 665 | 664 | 566 | 380 | 146 | 88 | 85 | 85 | 85 | 85 | 85 | 481 | 3 |
| | 10 | 797 | 92 | 92 | 98 | 196 | 422 | 583 | 655 | 631 | 516 | 315 | 116 | 92 | 92 | 92 | 92 | 92 | 548 | 2 |
| | 11 | 812 | 96 | 96 | 96 | 104 | 226 | 439 | 582 | 638 | 601 | 475 | 271 | 108 | 96 | 96 | 96 | 96 | 590 | 1 |
| | 12 | 816 | 98 | 98 | 98 | 98 | 107 | 246 | 452 | 585 | 630 | 585 | 452 | 246 | 107 | 98 | 98 | 98 | 604 | 12 |
| | HALF-DAY TOTALS | | 814 | 1370 | 2157 | 2985 | 3669 | 4004 | 3763 | 3046 | 2198 | 1457 | 920 | 626 | 557 | 553 | 554 | 582 | 2985 | |
| Aug 21 | 5 | 92 | 28 | 62 | 85 | 94 | 89 | 71 | 40 | 8 | 6 | 6 | 6 | 6 | 6 | 6 | 6 | 6 | 11 | 7 |
| | 6 | 388 | 42 | 182 | 306 | 380 | 395 | 352 | 252 | 107 | 30 | 27 | 27 | 27 | 27 | 27 | 27 | 27 | 73 | 6 |
| | 7 | 570 | 48 | 132 | 344 | 494 | 567 | 555 | 458 | 289 | 81 | 45 | 45 | 45 | 45 | 45 | 45 | 45 | 168 | 5 |
| | 8 | 675 | 59 | 66 | 247 | 468 | 609 | 657 | 607 | 464 | 241 | 66 | 59 | 59 | 59 | 59 | 59 | 59 | 273 | 4 |
| | 9 | 737 | 70 | 70 | 107 | 345 | 549 | 666 | 683 | 597 | 420 | 172 | 74 | 70 | 70 | 70 | 70 | 70 | 368 | 3 |
| | 10 | 775 | 78 | 78 | 82 | 165 | 412 | 594 | 685 | 674 | 562 | 361 | 122 | 80 | 78 | 78 | 78 | 78 | 440 | 2 |
| | 11 | 794 | 82 | 82 | 82 | 89 | 216 | 456 | 617 | 686 | 653 | 523 | 305 | 99 | 82 | 82 | 82 | 82 | 485 | 1 |
| | 12 | 801 | 84 | 84 | 84 | 84 | 92 | 260 | 489 | 635 | 684 | 633 | 489 | 260 | 92 | 84 | 84 | 84 | 501 | 12 |
| | HALF-DAY TOTALS | | 447 | 714 | 1932 | 2073 | 2884 | 3498 | 3600 | 3147 | 2333 | 1509 | 869 | 500 | 413 | 409 | 409 | 410 | 2069 | |
| Sep 21 | 7 | 242 | 12 | 30 | 122 | 194 | 234 | 238 | 206 | 139 | 47 | 13 | 12 | 12 | 12 | 12 | 12 | 12 | 26 | 5 |
| | 8 | 513 | 30 | 33 | 136 | 324 | 451 | 505 | 484 | 387 | 225 | 45 | 30 | 30 | 30 | 30 | 30 | 30 | 97 | 4 |
| | 9 | 651 | 43 | 43 | 54 | 261 | 468 | 596 | 631 | 570 | 420 | 193 | 47 | 43 | 43 | 43 | 43 | 43 | 181 | 3 |
| | 10 | 722 | 52 | 52 | 53 | 111 | 366 | 563 | 672 | 677 | 577 | 388 | 131 | 54 | 52 | 52 | 52 | 52 | 255 | 2 |
| | 11 | 758 | 57 | 57 | 57 | 61 | 185 | 445 | 623 | 706 | 680 | 549 | 329 | 83 | 57 | 57 | 57 | 57 | 303 | 1 |
| | 12 | 769 | 59 | 59 | 59 | 59 | 65 | 260 | 504 | 660 | 715 | 660 | 504 | 260 | 65 | 59 | 59 | 59 | 320 | 12 |
| | HALF-DAY TOTALS | | 224 | 241 | 448 | 968 | 1727 | 2484 | 2821 | 2810 | 2305 | 1513 | 787 | 336 | 227 | 224 | 224 | 224 | 1021 | |
| Oct 21 | 8 | 54 | 2 | 2 | 10 | 30 | 45 | 52 | 51 | 42 | 26 | 6 | 2 | 2 | 2 | 2 | 2 | 2 | 4 | 4 |
| | 9 | 383 | 17 | 17 | 19 | 127 | 259 | 345 | 372 | 346 | 263 | 133 | 19 | 17 | 17 | 17 | 17 | 17 | 42 | 3 |
| | 10 | 556 | 28 | 28 | 28 | 58 | 261 | 426 | 520 | 532 | 465 | 321 | 115 | 29 | 28 | 28 | 28 | 28 | 91 | 2 |
| | 11 | 633 | 34 | 34 | 34 | 36 | 135 | 367 | 528 | 603 | 588 | 479 | 292 | 64 | 34 | 34 | 34 | 34 | 130 | 1 |
| | 12 | 655 | 36 | 36 | 36 | 36 | 40 | 222 | 441 | 581 | 628 | 581 | 441 | 222 | 40 | 36 | 36 | 36 | 144 | 12 |
| | HALF-DAY TOTALS | | 98 | 98 | 109 | 273 | 728 | 1324 | 1711 | 1829 | 1661 | 1225 | 638 | 209 | 100 | 98 | 98 | 98 | 339 | |
| Nov 21 | 10 | 72 | 2 | 2 | 2 | 5 | 31 | 53 | 67 | 69 | 62 | 43 | 17 | 3 | 2 | 2 | 2 | 2 | 4 | 2 |
| | 11 | 250 | 9 | 9 | 9 | 10 | 46 | 140 | 207 | 238 | 234 | 192 | 118 | 24 | 9 | 9 | 9 | 9 | 19 | 1 |
| | 12 | 307 | 11 | 11 | 11 | 11 | 13 | 102 | 207 | 274 | 295 | 274 | 207 | 102 | 13 | 11 | 11 | 11 | 25 | 12 |
| | HALF-DAY TOTALS | | 17 | 17 | 17 | 19 | 81 | 248 | 378 | 444 | 440 | 369 | 234 | 71 | 18 | 17 | 17 | 17 | 35 | |
| Dec 21 | 11 | 12 | 0 | 0 | 0 | 0 | 2 | 7 | 10 | 12 | 11 | 9 | 6 | 1 | 0 | 0 | 0 | 0 | 1 | 1 |
| | 12 | 51 | 1 | 1 | 1 | 1 | 2 | 16 | 34 | 45 | 48 | 45 | 34 | 16 | 2 | 1 | 1 | 1 | 3 | 12 |
| | HALF-DAY TOTALS | | 1 | 1 | 1 | 1 | 3 | 16 | 28 | 35 | 36 | 32 | 22 | 8 | 1 | 1 | 1 | 1 | 2 | |
| | | | N | NNW | NW | WNW | W | WSW | SW | SSW | S | SSE | SE | ESE | E | ENE | NE | NNE | HOR | PM |

[a] Total solar heat gains for DSA glass (based on a ground reflectance of 0.20).
[b] Half-day totals computed by Simpson's rule, time interval = 10 minutes.

**TABLE C.11 Solar Position and Clear Day Insolation, 32°N Latitude**

| Date | Solar Time A.M. | Solar Time P.M. | Solar Position Alt | Solar Position Azm | Direct Normal | Horiz | 22 | 32 | 42 | 52 | 90 |
|------|------|------|------|------|------|------|------|------|------|------|------|
| Jan 21 | 7 | 5 | 1.4 | 65.2 | 1 | 0 | 0 | 0 | 0 | 1 | 1 |
| | 8 | 4 | 12.5 | 56.5 | 203 | 56 | 93 | 106 | 116 | 123 | 115 |
| | 9 | 3 | 22.5 | 46.0 | 269 | 118 | 175 | 193 | 206 | 212 | 181 |
| | 10 | 2 | 30.6 | 33.1 | 295 | 167 | 235 | 256 | 269 | 274 | 221 |
| | 11 | 1 | 36.1 | 17.5 | 306 | 198 | 273 | 295 | 308 | 312 | 245 |
| | 12 | | 38.0 | 0.0 | 310 | 209 | 285 | 308 | 321 | 324 | 253 |
| | Surface Daily Totals | | | | 2458 | 1288 | 1839 | 2008 | 2118 | 2166 | 1779 |
| Feb 21 | 7 | 5 | 7.1 | 73.5 | 121 | 22 | 34 | 37 | 40 | 42 | 38 |
| | 8 | 4 | 19.0 | 64.4 | 247 | 95 | 127 | 136 | 140 | 141 | 108 |
| | 9 | 3 | 29.9 | 53.4 | 288 | 161 | 206 | 217 | 222 | 220 | 158 |
| | 10 | 2 | 39.1 | 39.4 | 306 | 212 | 266 | 278 | 283 | 279 | 193 |
| | 11 | 1 | 45.6 | 21.4 | 315 | 244 | 304 | 317 | 321 | 315 | 214 |
| | 12 | | 48.0 | 0.0 | 317 | 255 | 316 | 330 | 334 | 328 | 222 |
| | Surface Daily Totals | | | | 2872 | 1724 | 2188 | 2300 | 2345 | 2322 | 1644 |
| Mar 21 | 7 | 5 | 12.7 | 81.9 | 185 | 54 | 60 | 60 | 59 | 56 | 32 |
| | 8 | 4 | 25.1 | 73.0 | 260 | 129 | 146 | 147 | 144 | 137 | 78 |
| | 9 | 3 | 36.8 | 62.1 | 290 | 194 | 222 | 224 | 220 | 209 | 119 |
| | 10 | 2 | 47.3 | 47.5 | 304 | 245 | 280 | 283 | 278 | 265 | 150 |
| | 11 | 1 | 55.0 | 26.8 | 311 | 277 | 317 | 321 | 315 | 300 | 170 |
| | 12 | | 58.0 | 0.0 | 313 | 287 | 329 | 333 | 327 | 312 | 177 |
| | Surface Daily Totals | | | | 3012 | 2084 | 2378 | 2403 | 2358 | 2246 | 1276 |
| Apr 21 | 6 | 6 | 6.1 | 99.9 | 66 | 14 | 9 | 6 | 6 | 5 | 3 |
| | 7 | 5 | 18.8 | 92.2 | 206 | 86 | 78 | 71 | 62 | 51 | 10 |
| | 8 | 4 | 31.5 | 84.0 | 255 | 158 | 156 | 148 | 136 | 120 | 35 |
| | 9 | 3 | 43.9 | 74.2 | 278 | 220 | 225 | 217 | 203 | 183 | 68 |
| | 10 | 2 | 55.7 | 60.3 | 290 | 267 | 279 | 272 | 256 | 234 | 95 |
| | 11 | 1 | 65.4 | 37.5 | 295 | 297 | 313 | 306 | 290 | 265 | 112 |
| | 12 | | 69.6 | 0.0 | 297 | 307 | 325 | 318 | 301 | 276 | 118 |
| | Surface Daily Totals | | | | 3076 | 2390 | 2444 | 2356 | 2206 | 1994 | 764 |
| May 21 | 6 | 6 | 10.4 | 107.2 | 119 | 36 | 21 | 13 | 13 | 12 | 7 |
| | 7 | 5 | 22.8 | 100.1 | 211 | 107 | 88 | 75 | 60 | 44 | 13 |
| | 8 | 4 | 35.4 | 92.9 | 250 | 175 | 159 | 145 | 127 | 105 | 15 |
| | 9 | 3 | 48.1 | 84.7 | 269 | 233 | 223 | 209 | 188 | 163 | 33 |
| | 10 | 2 | 60.6 | 73.3 | 280 | 277 | 273 | 259 | 237 | 208 | 56 |
| | 11 | 1 | 72.0 | 51.9 | 285 | 305 | 305 | 290 | 268 | 237 | 72 |
| | 12 | | 78.0 | 0.0 | 286 | 315 | 315 | 301 | 278 | 247 | 77 |
| | Surface Daily Totals | | | | 3112 | 2582 | 2454 | 2284 | 2064 | 1788 | 469 |
| Jun 21 | 6 | 6 | 12.2 | 110.2 | 131 | 45 | 26 | 16 | 15 | 14 | 9 |
| | 7 | 5 | 24.3 | 103.4 | 210 | 115 | 91 | 76 | 59 | 41 | 14 |
| | 8 | 4 | 36.9 | 96.8 | 245 | 180 | 159 | 143 | 122 | 99 | 16 |
| | 9 | 3 | 49.6 | 89.4 | 264 | 236 | 221 | 204 | 181 | 153 | 19 |
| | 10 | 2 | 62.2 | 79.7 | 274 | 279 | 268 | 251 | 227 | 197 | 41 |
| | 11 | 1 | 74.2 | 60.9 | 279 | 306 | 299 | 282 | 257 | 224 | 56 |
| | 12 | | 81.5 | 0.0 | 280 | 315 | 309 | 292 | 267 | 234 | 60 |
| | Surface Daily Totals | | | | 3084 | 2634 | 2436 | 2234 | 1990 | 1690 | 370 |

**TABLE C.11 Solar Position and Clear Day Insolation, 32°N Latitude (*continued*)**

| Date | A.M. | P.M. | Alt | Azm | Direct Normal | Horiz | 22 | 32 | 42 | 52 | 90 |
|---|---|---|---|---|---|---|---|---|---|---|---|
| | Solar Time | | Solar Position | | Global Irradiance (Btu/h ft²) | | | | | | |
| | | | | | | | South-Facing Elevation Angle | | | | |
| Jul 21 | 6 | 6 | 10.7 | 107.7 | 113 | 37 | 22 | 14 | 13 | 12 | 8 |
| | 7 | 5 | 23.1 | 100.6 | 203 | 107 | 87 | 75 | 60 | 44 | 14 |
| | 8 | 4 | 35.7 | 93.6 | 241 | 174 | 158 | 143 | 125 | 104 | 16 |
| | 9 | 3 | 48.4 | 85.5 | 261 | 230 | 220 | 205 | 185 | 159 | 31 |
| | 10 | 2 | 60.9 | 74.3 | 271 | 274 | 269 | 254 | 232 | 204 | 54 |
| | 11 | 1 | 72.4 | 53.3 | 277 | 302 | 300 | 285 | 262 | 232 | 69 |
| | | 12 | 78.6 | 0.0 | 279 | 311 | 310 | 296 | 273 | 242 | 74 |
| | | | Surface Daily Totals | | 3012 | 2558 | 2422 | 2250 | 2030 | 1754 | 458 |
| Aug 21 | 6 | 6 | 6.5 | 100.5 | 59 | 14 | 9 | 7 | 6 | 6 | 4 |
| | 7 | 5 | 19.1 | 92.8 | 190 | 85 | 77 | 69 | 60 | 50 | 12 |
| | 8 | 4 | 31.8 | 84.7 | 240 | 156 | 152 | 144 | 132 | 116 | 33 |
| | 9 | 3 | 44.3 | 75.0 | 263 | 216 | 220 | 212 | 197 | 178 | 65 |
| | 10 | 2 | 56.1 | 61.3 | 276 | 262 | 272 | 264 | 249 | 226 | 91 |
| | 11 | 1 | 66.0 | 38.4 | 282 | 292 | 305 | 298 | 281 | 257 | 107 |
| | | 12 | 70.3 | 0.0 | 284 | 302 | 317 | 309 | 292 | 268 | 113 |
| | | | Surface Daily Totals | | 2902 | 2352 | 2388 | 2296 | 2144 | 1934 | 736 |
| Sep 21 | 7 | 5 | 12.7 | 81.9 | 163 | 51 | 56 | 56 | 55 | 52 | 30 |
| | 8 | 4 | 25.1 | 73.0 | 240 | 124 | 140 | 141 | 138 | 131 | 75 |
| | 9 | 3 | 36.8 | 62.1 | 272 | 188 | 213 | 215 | 211 | 201 | 114 |
| | 10 | 2 | 47.3 | 47.5 | 287 | 237 | 270 | 273 | 268 | 255 | 145 |
| | 11 | 1 | 55.0 | 26.8 | 294 | 268 | 306 | 309 | 303 | 289 | 164 |
| | | 12 | 58.0 | 0.0 | 296 | 278 | 318 | 321 | 315 | 300 | 171 |
| | | | Surface Daily Totals | | 2808 | 2014 | 2288 | 2308 | 2264 | 2154 | 1226 |
| Oct 21 | 7 | 5 | 6.8 | 73.1 | 99 | 19 | 29 | 32 | 34 | 36 | 32 |
| | 8 | 4 | 18.7 | 64.0 | 229 | 90 | 120 | 128 | 133 | 134 | 104 |
| | 9 | 3 | 29.5 | 53.0 | 273 | 155 | 198 | 208 | 213 | 212 | 153 |
| | 10 | 2 | 38.7 | 39.1 | 293 | 204 | 257 | 269 | 273 | 270 | 188 |
| | 11 | 1 | 45.1 | 21.1 | 302 | 236 | 294 | 307 | 311 | 306 | 209 |
| | | 12 | 47.5 | 0.0 | 304 | 247 | 306 | 320 | 324 | 318 | 217 |
| | | | Surface Daily Totals | | 2696 | 1654 | 2100 | 2208 | 2252 | 2232 | 1588 |
| Nov 21 | 7 | 5 | 1.5 | 65.4 | 2 | 0 | 0 | 0 | 1 | 1 | 1 |
| | 8 | 4 | 12.7 | 56.6 | 196 | 55 | 91 | 104 | 113 | 119 | 111 |
| | 9 | 3 | 22.6 | 46.1 | 263 | 118 | 173 | 190 | 202 | 208 | 176 |
| | 10 | 2 | 30.8 | 33.2 | 289 | 166 | 233 | 252 | 265 | 270 | 217 |
| | 11 | 1 | 36.2 | 17.6 | 301 | 197 | 270 | 291 | 303 | 307 | 241 |
| | | 12 | 38.2 | 0.0 | 304 | 207 | 282 | 304 | 316 | 320 | 249 |
| | | | Surface Daily Totals | | 2406 | 1280 | 1816 | 1980 | 2084 | 2130 | 1742 |
| Dec 21 | 8 | 4 | 10.3 | 53.8 | 176 | 41 | 77 | 90 | 101 | 108 | 107 |
| | 9 | 3 | 19.8 | 43.6 | 257 | 102 | 161 | 180 | 195 | 204 | 183 |
| | 10 | 2 | 27.6 | 31.2 | 288 | 150 | 221 | 244 | 259 | 267 | 226 |
| | 11 | 1 | 32.7 | 16.4 | 301 | 180 | 258 | 282 | 298 | 305 | 251 |
| | | 12 | 34.6 | 0.0 | 304 | 190 | 271 | 295 | 311 | 318 | 259 |
| | | | Surface Daily Totals | | 2348 | 1136 | 1704 | 1888 | 2016 | 2086 | 1794 |

*Source:* Reprinted by permission from Peter J. Lunde, *Solar Thermal Engineering,* © 1980, John Wiley & Sons, New York.
*Note:* 1 Btu/h ft² × 3.152 = W/m².

APPENDICES

**TABLE C.12 Solar Position and Clear Day Insolation, 40° N Latitude**

| Date | Solar Time A.M. | Solar Time P.M. | Alt | Azm | Direct Normal | Horiz | 22 | 32 | 42 | 52 | 90 |
|---|---|---|---|---|---|---|---|---|---|---|---|
| Jan 21 | 8 | 4 | 8.1 | 55.3 | 142 | 28 | 65 | 74 | 81 | 85 | 84 |
| | 9 | 3 | 16.8 | 44.0 | 239 | 83 | 155 | 171 | 182 | 187 | 171 |
| | 10 | 2 | 23.8 | 30.9 | 274 | 127 | 218 | 237 | 249 | 254 | 223 |
| | 11 | 1 | 28.4 | 16.0 | 289 | 154 | 257 | 277 | 290 | 293 | 253 |
| | 12 | | 30.0 | 0.0 | 294 | 164 | 270 | 291 | 303 | 306 | 263 |
| | Surface Daily Totals | | | | 2182 | 948 | 1660 | 1810 | 1906 | 1944 | 1726 |
| Feb 21 | 7 | 5 | 4.8 | 72.7 | 69 | 10 | 19 | 21 | 23 | 24 | 22 |
| | 8 | 4 | 15.4 | 62.2 | 224 | 73 | 114 | 122 | 126 | 127 | 107 |
| | 9 | 3 | 25.0 | 50.2 | 274 | 132 | 195 | 205 | 209 | 208 | 167 |
| | 10 | 2 | 32.8 | 35.9 | 295 | 178 | 256 | 267 | 271 | 267 | 210 |
| | 11 | 1 | 38.1 | 18.9 | 305 | 206 | 293 | 306 | 310 | 304 | 236 |
| | 12 | | 40.0 | 0.0 | 308 | 216 | 306 | 319 | 323 | 317 | 245 |
| | Surface Daily Totals | | | | 2640 | 1414 | 2060 | 2162 | 2202 | 2176 | 1730 |
| Mar 21 | 7 | 5 | 11.4 | 80.2 | 171 | 46 | 55 | 55 | 54 | 51 | 35 |
| | 8 | 4 | 22.5 | 69.6 | 250 | 114 | 140 | 141 | 138 | 131 | 89 |
| | 9 | 3 | 32.8 | 57.3 | 282 | 173 | 215 | 217 | 213 | 202 | 138 |
| | 10 | 2 | 41.6 | 41.9 | 297 | 218 | 273 | 276 | 271 | 258 | 176 |
| | 11 | 1 | 47.7 | 22.6 | 305 | 247 | 310 | 313 | 307 | 293 | 200 |
| | 12 | | 50.0 | 0.0 | 307 | 257 | 322 | 326 | 320 | 305 | 208 |
| | Surface Daily Totals | | | | 2916 | 1852 | 2308 | 2330 | 2284 | 2174 | 1484 |
| Apr 21 | 6 | 6 | 7.4 | 98.9 | 89 | 20 | 11 | 8 | 7 | 7 | 4 |
| | 7 | 5 | 18.9 | 89.5 | 206 | 87 | 77 | 70 | 61 | 50 | 12 |
| | 8 | 4 | 30.3 | 79.3 | 252 | 152 | 153 | 145 | 133 | 117 | 53 |
| | 9 | 3 | 41.3 | 67.2 | 274 | 207 | 221 | 213 | 199 | 179 | 93 |
| | 10 | 2 | 51.2 | 51.4 | 286 | 250 | 275 | 267 | 252 | 229 | 126 |
| | 11 | 1 | 58.7 | 29.2 | 292 | 277 | 308 | 301 | 285 | 260 | 147 |
| | 12 | | 61.6 | 0.0 | 293 | 287 | 320 | 313 | 296 | 271 | 154 |
| | Surface Daily Totals | | | | 3092 | 2274 | 2412 | 2320 | 2168 | 1956 | 1022 |
| May 21 | 5 | 7 | 1.9 | 114.7 | 1 | 0 | 0 | 0 | 0 | 0 | 0 |
| | 6 | 6 | 12.7 | 105.6 | 144 | 49 | 25 | 15 | 14 | 13 | 9 |
| | 7 | 5 | 24.0 | 96.6 | 216 | 114 | 89 | 76 | 60 | 44 | 13 |
| | 8 | 4 | 35.4 | 87.2 | 250 | 175 | 158 | 144 | 125 | 104 | 25 |
| | 9 | 3 | 46.8 | 76.0 | 267 | 227 | 221 | 206 | 186 | 160 | 60 |
| | 10 | 2 | 57.5 | 60.9 | 277 | 267 | 270 | 255 | 233 | 205 | 89 |
| | 11 | 1 | 66.2 | 37.1 | 283 | 293 | 301 | 287 | 264 | 234 | 108 |
| | 12 | | 70.0 | 0.0 | 284 | 301 | 312 | 297 | 274 | 243 | 114 |
| | Surface Daily Totals | | | | 3160 | 2552 | 2442 | 2264 | 2040 | 1760 | 724 |
| Jun 21 | 5 | 7 | 4.2 | 117.3 | 22 | 4 | 3 | 3 | 2 | 2 | 1 |
| | 6 | 6 | 14.8 | 108.4 | 155 | 60 | 30 | 18 | 17 | 16 | 10 |
| | 7 | 5 | 26.0 | 99.7 | 216 | 123 | 92 | 77 | 59 | 40 | 14 |
| | 8 | 4 | 37.4 | 90.7 | 246 | 182 | 159 | 142 | 121 | 97 | 16 |
| | 9 | 3 | 48.8 | 80.2 | 263 | 233 | 219 | 202 | 179 | 151 | 47 |
| | 10 | 2 | 59.8 | 65.8 | 272 | 272 | 266 | 248 | 224 | 193 | 74 |
| | 11 | 1 | 69.2 | 41.9 | 277 | 296 | 296 | 278 | 253 | 221 | 92 |
| | 12 | | 73.5 | 0.0 | 279 | 304 | 306 | 289 | 263 | 230 | 98 |
| | Surface Daily Totals | | | | 3180 | 2648 | 2434 | 2224 | 1974 | 1670 | 610 |

**TABLE C.12 Solar Position and Clear Day Insolation, 40° N Latitude (*continued*)**

| Date | A.M. | P.M. | Alt | Azm | Direct Normal | Horiz | 22 | 32 | 42 | 52 | 90 |
|---|---|---|---|---|---|---|---|---|---|---|---|
| Jul 21 | 5 | 7 | 2.3 | 115.2 | 2 | 0 | 0 | 0 | 0 | 0 | 0 |
| | 6 | 6 | 13.1 | 106.1 | 138 | 50 | 26 | 17 | 15 | 14 | 9 |
| | 7 | 5 | 24.3 | 97.2 | 208 | 114 | 89 | 75 | 60 | 44 | 14 |
| | 8 | 4 | 35.8 | 87.8 | 241 | 174 | 157 | 142 | 124 | 102 | 24 |
| | 9 | 3 | 47.2 | 76.7 | 259 | 225 | 218 | 203 | 182 | 157 | 58 |
| | 10 | 2 | 57.9 | 61.7 | 269 | 265 | 266 | 251 | 229 | 200 | 86 |
| | 11 | 1 | 66.7 | 37.9 | 275 | 290 | 296 | 281 | 258 | 228 | 104 |
| | 12 | | 70.6 | 0.0 | 276 | 298 | 307 | 292 | 269 | 238 | 111 |
| | Surface Daily Totals | | | | 3062 | 2534 | 2409 | 2230 | 2006 | 1728 | 702 |
| Aug 21 | 6 | 6 | 7.9 | 99.5 | 81 | 21 | 12 | 9 | 8 | 7 | 5 |
| | 7 | 5 | 19.3 | 90.0 | 191 | 87 | 76 | 69 | 60 | 49 | 12 |
| | 8 | 4 | 30.7 | 79.9 | 237 | 150 | 150 | 141 | 129 | 113 | 50 |
| | 9 | 3 | 41.8 | 67.9 | 260 | 205 | 216 | 207 | 193 | 173 | 89 |
| | 10 | 2 | 51.7 | 52.1 | 272 | 246 | 267 | 259 | 244 | 221 | 120 |
| | 11 | 1 | 59.3 | 29.7 | 278 | 273 | 300 | 292 | 276 | 252 | 140 |
| | 12 | | 62.3 | 0.0 | 280 | 282 | 311 | 303 | 287 | 262 | 147 |
| | Surface Daily Totals | | | | 2916 | 2244 | 2354 | 2258 | 2104 | 1894 | 978 |
| Sep 21 | 7 | 5 | 11.4 | 80.2 | 149 | 43 | 51 | 51 | 49 | 47 | 32 |
| | 8 | 4 | 22.5 | 69.6 | 230 | 109 | 133 | 134 | 131 | 124 | 84 |
| | 9 | 3 | 32.8 | 57.3 | 263 | 167 | 206 | 208 | 203 | 193 | 132 |
| | 10 | 2 | 41.6 | 41.9 | 280 | 211 | 262 | 265 | 260 | 247 | 168 |
| | 11 | 1 | 47.7 | 22.6 | 287 | 239 | 298 | 301 | 295 | 281 | 192 |
| | 12 | | 50.0 | 0.0 | 290 | 249 | 310 | 313 | 307 | 292 | 200 |
| | Surface Daily Totals | | | | 2708 | 1788 | 2210 | 2228 | 2182 | 2074 | 1416 |
| Oct 21 | 7 | 5 | 4.5 | 72.3 | 48 | 7 | 14 | 15 | 17 | 17 | 16 |
| | 8 | 4 | 15.0 | 61.9 | 204 | 68 | 106 | 113 | 117 | 118 | 100 |
| | 9 | 3 | 24.5 | 49.8 | 257 | 126 | 185 | 195 | 200 | 198 | 160 |
| | 10 | 2 | 32.4 | 35.6 | 280 | 170 | 245 | 257 | 261 | 257 | 203 |
| | 11 | 1 | 37.6 | 18.7 | 291 | 199 | 283 | 295 | 299 | 294 | 229 |
| | 12 | | 39.5 | 0.0 | 294 | 208 | 295 | 308 | 312 | 306 | 238 |
| | Surface Daily Totals | | | | 2454 | 1348 | 1962 | 2060 | 2098 | 2074 | 1654 |
| Nov 21 | 8 | 4 | 8.2 | 55.4 | 136 | 28 | 63 | 72 | 78 | 82 | 81 |
| | 9 | 3 | 17.0 | 44.1 | 232 | 82 | 152 | 167 | 178 | 183 | 167 |
| | 10 | 2 | 24.0 | 31.0 | 268 | 126 | 215 | 233 | 245 | 249 | 219 |
| | 11 | 1 | 28.6 | 16.1 | 283 | 153 | 254 | 273 | 285 | 288 | 248 |
| | 12 | | 30.2 | 0.0 | 288 | 163 | 267 | 287 | 298 | 301 | 258 |
| | Surface Daily Totals | | | | 2128 | 942 | 1636 | 1778 | 1870 | 1908 | 1686 |
| Dec 21 | 8 | 4 | 5.5 | 53.0 | 89 | 14 | 39 | 45 | 50 | 54 | 56 |
| | 9 | 3 | 14.0 | 41.9 | 217 | 65 | 135 | 152 | 164 | 171 | 163 |
| | 10 | 2 | 20.7 | 29.4 | 261 | 107 | 200 | 221 | 235 | 242 | 221 |
| | 11 | 1 | 25.0 | 15.2 | 280 | 134 | 239 | 262 | 276 | 283 | 252 |
| | 12 | | 26.6 | 0.0 | 285 | 143 | 253 | 275 | 290 | 296 | 263 |
| | Surface Daily Totals | | | | 1978 | 782 | 1480 | 1634 | 1740 | 1796 | 1646 |

*Source:* Reprinted by permission from Peter J. Lunde, *Solar Thermal Engineering,* © 1980, John Wiley & Sons, New York.
*Note:* 1 Btu/h ft$^2$ × 3.152 = W/m$^2$.

APPENDICES

**TABLE C.13 Solar Position and Clear Day Insolation, 48° N Latitude**

| Date | Solar Time A.M. | Solar Time P.M. | Solar Position Alt | Solar Position Azm | Direct Normal | Global Irradiance (Btu/h ft²) Horiz | South-Facing Elevation Angle 22 | 32 | 42 | 52 | 90 |
|------|------|------|------|------|------|------|------|------|------|------|------|
| Jan 21 | 8 | 4 | 3.5 | 54.6 | 37 | 4 | 17 | 19 | 21 | 22 | 22 |
| | 9 | 3 | 11.0 | 42.6 | 185 | 46 | 120 | 132 | 140 | 145 | 139 |
| | 10 | 2 | 16.9 | 29.4 | 239 | 83 | 190 | 206 | 216 | 220 | 206 |
| | 11 | 1 | 20.7 | 15.1 | 261 | 107 | 231 | 249 | 260 | 263 | 243 |
| | | 12 | 22.0 | 0.0 | 267 | 115 | 245 | 264 | 275 | 278 | 255 |
| | Surface Daily Totals | | | | 1710 | 596 | 1360 | 1478 | 1550 | 1578 | 1478 |
| Feb 21 | 7 | 5 | 2.4 | 72.2 | 12 | 1 | 3 | 4 | 4 | 4 | 4 |
| | 8 | 4 | 11.6 | 60.5 | 188 | 49 | 95 | 102 | 105 | 106 | 96 |
| | 9 | 3 | 19.7 | 47.7 | 251 | 100 | 178 | 187 | 191 | 190 | 167 |
| | 10 | 2 | 26.2 | 33.3 | 278 | 139 | 240 | 251 | 255 | 251 | 217 |
| | 11 | 1 | 30.5 | 17.2 | 290 | 165 | 278 | 290 | 294 | 288 | 247 |
| | | 12 | 32.0 | 0.0 | 293 | 173 | 291 | 304 | 307 | 301 | 258 |
| | Surface Daily Totals | | | | 2330 | 1080 | 1880 | 1972 | 2024 | 1978 | 1720 |
| Mar 21 | 7 | 5 | 10.0 | 78.7 | 153 | 37 | 49 | 49 | 47 | 45 | 35 |
| | 8 | 4 | 19.5 | 66.8 | 236 | 96 | 131 | 132 | 129 | 122 | 96 |
| | 9 | 3 | 28.2 | 53.4 | 270 | 147 | 205 | 207 | 203 | 193 | 152 |
| | 10 | 2 | 35.4 | 37.8 | 287 | 187 | 263 | 266 | 261 | 248 | 195 |
| | 11 | 1 | 40.3 | 19.8 | 295 | 212 | 300 | 303 | 297 | 283 | 223 |
| | | 12 | 42.0 | 0.0 | 298 | 220 | 312 | 315 | 309 | 294 | 232 |
| | Surface Daily Totals | | | | 2780 | 1578 | 2208 | 2228 | 2182 | 2074 | 1632 |
| Apr 21 | 6 | 6 | 8.6 | 97.8 | 108 | 27 | 13 | 9 | 8 | 7 | 5 |
| | 7 | 5 | 18.6 | 86.7 | 205 | 85 | 76 | 68 | 59 | 48 | 21 |
| | 8 | 4 | 28.5 | 74.9 | 247 | 142 | 149 | 141 | 129 | 113 | 69 |
| | 9 | 3 | 37.8 | 61.2 | 268 | 191 | 216 | 208 | 194 | 174 | 115 |
| | 10 | 2 | 45.8 | 44.6 | 280 | 228 | 268 | 260 | 245 | 223 | 152 |
| | 11 | 1 | 51.5 | 24.0 | 286 | 252 | 301 | 294 | 278 | 254 | 177 |
| | | 12 | 53.6 | 0.0 | 288 | 260 | 313 | 305 | 289 | 264 | 185 |
| | Surface Daily Totals | | | | 3076 | 2106 | 2358 | 2266 | 2114 | 1902 | 1262 |
| May 21 | 5 | 7 | 5.2 | 114.3 | 41 | 9 | 4 | 4 | 4 | 3 | 2 |
| | 6 | 6 | 14.7 | 103.7 | 162 | 61 | 27 | 16 | 15 | 13 | 10 |
| | 7 | 5 | 24.6 | 93.0 | 219 | 118 | 89 | 75 | 60 | 43 | 13 |
| | 8 | 4 | 34.7 | 81.6 | 248 | 171 | 156 | 142 | 123 | 101 | 45 |
| | 9 | 3 | 44.3 | 68.3 | 264 | 217 | 217 | 202 | 182 | 156 | 86 |
| | 10 | 2 | 53.0 | 51.3 | 274 | 252 | 265 | 251 | 229 | 200 | 120 |
| | 11 | 1 | 59.5 | 28.6 | 279 | 274 | 296 | 281 | 258 | 228 | 141 |
| | | 12 | 62.0 | 0.0 | 280 | 281 | 306 | 292 | 269 | 238 | 149 |
| | Surface Daily Totals | | | | 3254 | 2482 | 2418 | 2234 | 2010 | 1728 | 982 |
| Jun 21 | 5 | 7 | 7.9 | 116.5 | 77 | 21 | 9 | 9 | 8 | 7 | 5 |
| | 6 | 6 | 17.2 | 106.2 | 172 | 74 | 33 | 19 | 18 | 16 | 12 |
| | 7 | 5 | 27.0 | 95.8 | 220 | 129 | 93 | 77 | 59 | 39 | 15 |
| | 8 | 4 | 37.1 | 84.6 | 246 | 181 | 157 | 140 | 119 | 95 | 35 |
| | 9 | 3 | 46.9 | 71.6 | 261 | 225 | 216 | 198 | 175 | 147 | 74 |
| | 10 | 2 | 55.8 | 54.8 | 269 | 259 | 262 | 244 | 220 | 189 | 105 |
| | 11 | 1 | 62.7 | 31.2 | 274 | 280 | 291 | 273 | 248 | 216 | 126 |
| | | 12 | 65.5 | 0.0 | 275 | 287 | 301 | 283 | 258 | 225 | 133 |
| | Surface Daily Totals | | | | 3312 | 2626 | 2420 | 2204 | 1950 | 1644 | 874 |

**TABLE C.13  Solar Position and Clear Day Insolation, 48° N Latitude (*continued*)**

| Date | A.M. (Solar Time) | P.M. (Solar Time) | Alt (Solar Position) | Azm (Solar Position) | Direct Normal | Horiz | 22 | 32 | 42 | 52 | 90 |
|------|------|------|------|------|------|------|------|------|------|------|------|
| Jul 21 | 5 | 7 | 5.7 | 114.7 | 43 | 10 | 5 | 5 | 4 | 4 | 3 |
|  | 6 | 6 | 15.2 | 104.1 | 156 | 62 | 28 | 18 | 16 | 15 | 11 |
|  | 7 | 5 | 25.1 | 93.5 | 211 | 118 | 89 | 75 | 59 | 42 | 14 |
|  | 8 | 4 | 35.1 | 82.1 | 240 | 171 | 154 | 140 | 121 | 99 | 43 |
|  | 9 | 3 | 44.8 | 68.8 | 256 | 215 | 214 | 199 | 178 | 153 | 83 |
|  | 10 | 2 | 53.5 | 51.9 | 266 | 250 | 261 | 246 | 224 | 195 | 116 |
|  | 11 | 1 | 60.1 | 29.0 | 271 | 272 | 291 | 276 | 253 | 223 | 137 |
|  | 12 | | 62.6 | 0.0 | 272 | 279 | 301 | 286 | 263 | 232 | 144 |
|  | Surface Daily Totals | | | | 3158 | 2474 | 2386 | 2200 | 1974 | 1694 | 956 |
| Aug 21 | 6 | 6 | 9.1 | 98.3 | 99 | 28 | 14 | 10 | 9 | 8 | 6 |
|  | 7 | 5 | 19.1 | 87.2 | 190 | 85 | 75 | 67 | 58 | 47 | 20 |
|  | 8 | 4 | 29.0 | 75.4 | 232 | 141 | 145 | 137 | 125 | 109 | 65 |
|  | 9 | 3 | 38.4 | 61.8 | 254 | 189 | 210 | 201 | 187 | 168 | 110 |
|  | 10 | 2 | 46.4 | 45.1 | 266 | 225 | 260 | 252 | 237 | 214 | 146 |
|  | 11 | 1 | 52.2 | 24.3 | 272 | 248 | 293 | 285 | 268 | 244 | 169 |
|  | 12 | | 54.3 | 0.0 | 274 | 256 | 304 | 296 | 279 | 255 | 177 |
|  | Surface Daily Totals | | | | 2898 | 2086 | 2300 | 2200 | 2046 | 1836 | 1208 |
| Sep 21 | 7 | 5 | 10.0 | 78.7 | 131 | 35 | 44 | 44 | 43 | 40 | 31 |
|  | 8 | 4 | 19.5 | 66.8 | 215 | 92 | 124 | 124 | 121 | 115 | 90 |
|  | 9 | 3 | 28.2 | 53.4 | 251 | 142 | 196 | 197 | 193 | 183 | 143 |
|  | 10 | 2 | 35.4 | 37.8 | 269 | 181 | 251 | 254 | 248 | 236 | 185 |
|  | 11 | 1 | 40.3 | 19.8 | 278 | 205 | 287 | 289 | 284 | 269 | 212 |
|  | 12 | | 42.0 | 0.0 | 280 | 213 | 299 | 302 | 296 | 281 | 221 |
|  | Surface Daily Totals | | | | 2568 | 1522 | 2102 | 2118 | 2070 | 1966 | 1546 |
| Oct 21 | 7 | 5 | 2.0 | 71.9 | 4 | 0 | 1 | 1 | 1 | 1 | 1 |
|  | 8 | 4 | 11.2 | 60.2 | 165 | 44 | 86 | 91 | 95 | 95 | 87 |
|  | 9 | 3 | 19.3 | 47.4 | 233 | 94 | 167 | 176 | 180 | 178 | 157 |
|  | 10 | 2 | 25.7 | 33.1 | 262 | 133 | 228 | 239 | 242 | 239 | 207 |
|  | 11 | 1 | 30.0 | 17.1 | 274 | 157 | 266 | 277 | 281 | 276 | 237 |
|  | 12 | | 31.5 | 0.0 | 278 | 166 | 279 | 291 | 294 | 288 | 247 |
|  | Surface Daily Totals | | | | 2154 | 1022 | 1774 | 1860 | 1890 | 1866 | 1626 |
| Nov 21 | 8 | 4 | 3.6 | 54.7 | 36 | 5 | 17 | 19 | 21 | 22 | 22 |
|  | 9 | 3 | 11.2 | 42.7 | 179 | 46 | 117 | 129 | 137 | 141 | 135 |
|  | 10 | 2 | 17.1 | 29.5 | 233 | 83 | 186 | 202 | 212 | 215 | 201 |
|  | 11 | 1 | 20.9 | 15.1 | 255 | 107 | 227 | 245 | 255 | 258 | 238 |
|  | 12 | | 22.2 | 0.0 | 261 | 115 | 241 | 259 | 270 | 272 | 250 |
|  | Surface Daily Totals | | | | 1668 | 596 | 1336 | 1448 | 1518 | 1544 | 1442 |
| Dec 21 | 9 | 3 | 8.0 | 40.9 | 140 | 27 | 87 | 98 | 105 | 110 | 109 |
|  | 10 | 2 | 13.6 | 28.2 | 214 | 63 | 164 | 180 | 192 | 197 | 190 |
|  | 11 | 1 | 17.3 | 14.4 | 242 | 86 | 207 | 226 | 239 | 244 | 231 |
|  | 12 | | 18.6 | 0.0 | 250 | 94 | 222 | 241 | 254 | 260 | 244 |
|  | Surface Daily Totals | | | | 1444 | 446 | 1136 | 1250 | 1326 | 1364 | 1304 |

*Source:* Reprinted by permission from Peter J Lunde, *Solar Thermal Engineering,* © 1980, John Wiley & Sons, New York.
*Note:* 1 Btu/h ft$^2$ × 3.152 = W/m$^2$.

**TABLE C.14  Solar Position and Clear Day Insolation, 64° N Latitude**

| Date | A.M. | P.M. | Alt | Azm | Direct Normal | Horiz | 22 | 32 | 42 | 52 | 90 |
|------|------|------|-----|-----|--------------|-------|----|----|----|----|----|
| | | | *Solar Time* | | | | | *Global Irradiance (Btu/h ft²)* | | | | |
| Jan 21 | 10 | 2 | 2.8 | 28.1 | 22 | 2 | 17 | 19 | 20 | 20 | 20 |
| | 11 | 1 | 5.2 | 14.1 | 81 | 12 | 72 | 77 | 80 | 81 | 81 |
| | | 12 | 6.0 | 0.0 | 100 | 16 | 91 | 98 | 102 | 103 | 103 |
| | | | Surface Daily Totals | | 306 | 45 | 268 | 290 | 302 | 306 | 304 |
| Feb 21 | 8 | 4 | 3.4 | 58.7 | 35 | 4 | 17 | 19 | 19 | 19 | 19 |
| | 9 | 3 | 8.6 | 44.8 | 147 | 31 | 103 | 108 | 111 | 110 | 107 |
| | 10 | 2 | 12.6 | 30.3 | 199 | 55 | 170 | 178 | 181 | 178 | 173 |
| | 11 | 1 | 15.1 | 15.3 | 222 | 71 | 212 | 220 | 223 | 219 | 213 |
| | | 12 | 16.0 | 0.0 | 228 | 77 | 225 | 235 | 237 | 232 | 226 |
| | | | Surface Daily Totals | | 1432 | 400 | 1230 | 1286 | 1302 | 1282 | 1252 |
| Mar 21 | 7 | 5 | 6.5 | 76.5 | 95 | 18 | 30 | 29 | 29 | 27 | 25 |
| | 8 | 4 | 12.7 | 62.6 | 185 | 54 | 101 | 102 | 99 | 94 | 89 |
| | 9 | 3 | 18.1 | 48.1 | 227 | 87 | 171 | 172 | 169 | 160 | 153 |
| | 10 | 2 | 22.3 | 32.7 | 249 | 112 | 227 | 229 | 224 | 213 | 203 |
| | 11 | 1 | 25.1 | 16.6 | 260 | 129 | 262 | 265 | 259 | 246 | 235 |
| | | 12 | 26.0 | 0.0 | 263 | 134 | 274 | 277 | 271 | 258 | 246 |
| | | | Surface Daily Totals | | 2296 | 932 | 1856 | 1870 | 1830 | 1736 | 1656 |
| Apr 21 | 5 | 7 | 4.0 | 108.5 | 27 | 5 | 2 | 2 | 2 | 1 | 1 |
| | 6 | 6 | 10.4 | 95.1 | 133 | 37 | 15 | 9 | 8 | 7 | 6 |
| | 7 | 5 | 17.0 | 81.6 | 194 | 76 | 70 | 63 | 54 | 43 | 37 |
| | 8 | 4 | 23.3 | 67.5 | 228 | 112 | 136 | 128 | 116 | 102 | 91 |
| | 9 | 3 | 29.0 | 52.3 | 248 | 144 | 197 | 189 | 176 | 158 | 145 |
| | 10 | 2 | 33.5 | 36.0 | 260 | 169 | 246 | 239 | 224 | 203 | 188 |
| | 11 | 1 | 36.5 | 18.4 | 266 | 184 | 278 | 270 | 255 | 233 | 216 |
| | | 12 | 37.6 | 0.0 | 268 | 190 | 289 | 281 | 266 | 243 | 225 |
| | | | Surface Daily Totals | | 2982 | 1644 | 2176 | 2082 | 1936 | 1736 | 1594 |
| May 21 | 4 | 8 | 5.8 | 125.1 | 51 | 11 | 5 | 4 | 4 | 3 | 3 |
| | 5 | 7 | 11.6 | 112.1 | 132 | 42 | 13 | 11 | 10 | 9 | 8 |
| | 6 | 6 | 17.9 | 99.1 | 185 | 79 | 29 | 16 | 14 | 12 | 11 |
| | 7 | 5 | 24.5 | 85.7 | 218 | 117 | 86 | 72 | 56 | 39 | 28 |
| | 8 | 4 | 30.9 | 71.5 | 239 | 152 | 148 | 133 | 115 | 94 | 80 |
| | 9 | 3 | 36.8 | 56.1 | 252 | 182 | 204 | 190 | 170 | 145 | 128 |
| | 10 | 2 | 41.6 | 38.9 | 261 | 205 | 249 | 235 | 213 | 186 | 167 |
| | 11 | 1 | 44.9 | 20.1 | 265 | 219 | 278 | 264 | 242 | 213 | 193 |
| | | 12 | 46.0 | 0.0 | 267 | 224 | 288 | 274 | 251 | 222 | 201 |
| | | | Surface Daily Totals | | 3470 | 2236 | 2312 | 2124 | 1898 | 1624 | 1436 |
| Jun 21 | 3 | 9 | 4.2 | 139.4 | 21 | 4 | 2 | 2 | 2 | 2 | 1 |
| | 4 | 8 | 9.0 | 126.4 | 93 | 27 | 10 | 9 | 8 | 7 | 6 |
| | 5 | 7 | 14.7 | 113.6 | 154 | 60 | 16 | 15 | 13 | 11 | 10 |
| | 6 | 6 | 21.0 | 100.8 | 194 | 96 | 34 | 19 | 17 | 14 | 13 |
| | 7 | 5 | 27.5 | 87.5 | 221 | 132 | 91 | 74 | 55 | 36 | 23 |
| | 8 | 4 | 34.0 | 73.3 | 239 | 166 | 150 | 133 | 112 | 88 | 73 |
| | 9 | 3 | 39.9 | 57.8 | 251 | 195 | 204 | 187 | 164 | 137 | 119 |
| | 10 | 2 | 44.9 | 40.4 | 258 | 217 | 247 | 230 | 206 | 177 | 157 |
| | 11 | 1 | 48.3 | 20.9 | 262 | 231 | 275 | 258 | 233 | 202 | 181 |
| | | 12 | 49.5 | 0.0 | 263 | 235 | 284 | 267 | 242 | 211 | 189 |
| | | | Surface Daily Totals | | 3650 | 2488 | 2342 | 2118 | 1862 | 1558 | 1356 |

**TABLE C.14  Solar Position and Clear Day Insolation, 64° N Latitude (*continued*)**

| Date | A.M. | P.M. | Alt | Azm | Direct Normal | Horiz | 22 | 32 | 42 | 52 | 90 |
|------|------|------|-----|-----|---------------|-------|----|----|----|----|----|
| | | | Solar Time | Solar Position | | Global Irradiance (Btu/h ft²) South-Facing Elevation Angle | | | | | |
| Jul 21 | 4 | 8 | 6.4 | 125.3 | 53 | 13 | 6 | 5 | 5 | 4 | 4 |
| | 5 | 7 | 12.1 | 112.4 | 128 | 44 | 14 | 13 | 11 | 10 | 9 |
| | 6 | 6 | 18.4 | 99.4 | 179 | 81 | 30 | 17 | 16 | 13 | 12 |
| | 7 | 5 | 25.0 | 86.0 | 211 | 118 | 86 | 72 | 56 | 38 | 28 |
| | 8 | 4 | 31.4 | 71.8 | 231 | 152 | 146 | 131 | 113 | 91 | 77 |
| | 9 | 3 | 37.3 | 56.3 | 245 | 182 | 201 | 186 | 166 | 141 | 124 |
| | 10 | 2 | 42.2 | 39.2 | 253 | 204 | 245 | 230 | 208 | 181 | 162 |
| | 11 | 1 | 45.4 | 20.2 | 257 | 218 | 273 | 258 | 236 | 207 | 187 |
| | 12 | | 46.6 | 0.0 | 259 | 223 | 282 | 267 | 245 | 216 | 195 |
| | Surface Daily Totals | | | | 3372 | 2248 | 2280 | 2090 | 1864 | 1588 | 1400 |
| Aug 21 | 5 | 7 | 4.6 | 108.8 | 29 | 6 | 3 | 3 | 2 | 2 | 2 |
| | 6 | 6 | 11.0 | 95.5 | 123 | 39 | 16 | 11 | 10 | 8 | 7 |
| | 7 | 5 | 17.6 | 81.9 | 181 | 77 | 69 | 61 | 52 | 42 | 35 |
| | 8 | 4 | 23.9 | 67.8 | 214 | 113 | 131 | 123 | 112 | 97 | 87 |
| | 9 | 3 | 29.6 | 52.6 | 234 | 144 | 190 | 182 | 169 | 150 | 138 |
| | 10 | 2 | 34.2 | 36.2 | 246 | 168 | 237 | 229 | 215 | 194 | 179 |
| | 11 | 1 | 37.2 | 18.5 | 252 | 183 | 268 | 260 | 244 | 222 | 205 |
| | 12 | | 38.3 | 0.0 | 254 | 188 | 278 | 270 | 255 | 232 | 215 |
| | Surface Daily Totals | | | | 2808 | 1646 | 2108 | 2008 | 1860 | 1662 | 1522 |
| Sep 21 | 7 | 5 | 6.5 | 76.5 | 77 | 16 | 25 | 24 | 24 | 23 | 21 |
| | 8 | 4 | 12.7 | 62.6 | 163 | 51 | 92 | 92 | 90 | 85 | 81 |
| | 9 | 3 | 18.1 | 48.1 | 206 | 83 | 159 | 159 | 156 | 147 | 141 |
| | 10 | 2 | 22.3 | 32.7 | 229 | 108 | 212 | 213 | 209 | 198 | 189 |
| | 11 | 1 | 25.1 | 16.6 | 240 | 124 | 246 | 248 | 243 | 230 | 220 |
| | 12 | | 26.0 | 0.0 | 244 | 129 | 258 | 260 | 254 | 241 | 230 |
| | Surface Daily Totals | | | | 2074 | 892 | 1726 | 1736 | 1696 | 1608 | 1532 |
| Oct 21 | 8 | 4 | 3.0 | 58.5 | 17 | 2 | 9 | 9 | 10 | 10 | 10 |
| | 9 | 3 | 8.1 | 44.6 | 122 | 26 | 86 | 91 | 93 | 92 | 90 |
| | 10 | 2 | 12.1 | 30.2 | 176 | 50 | 152 | 159 | 161 | 159 | 155 |
| | 11 | 1 | 14.6 | 15.2 | 201 | 65 | 193 | 201 | 203 | 200 | 195 |
| | 12 | | 15.5 | 0.0 | 208 | 71 | 207 | 215 | 217 | 213 | 208 |
| | Surface Daily Totals | | | | 1238 | 358 | 1088 | 1136 | 1152 | 1134 | 1106 |
| Nov 21 | 10 | 2 | 3.0 | 28.1 | 23 | 3 | 18 | 20 | 21 | 21 | 21 |
| | 11 | 1 | 5.4 | 14.2 | 79 | 12 | 70 | 76 | 79 | 80 | 79 |
| | 12 | | 6.2 | 0.0 | 97 | 17 | 89 | 96 | 100 | 101 | 100 |
| | Surface Daily Totals | | | | 302 | 46 | 266 | 286 | 298 | 302 | 300 |
| Dec 21 | 11 | 1 | 1.8 | 13.7 | 4 | 0 | 3 | 4 | 4 | 4 | 4 |
| | 12 | | 2.6 | 0.0 | 16 | 2 | 14 | 15 | 16 | 17 | 17 |
| | Surface Daily Totals | | | | 24 | 2 | 20 | 22 | 24 | 24 | 24 |

*Source:* Reprinted by permission from Peter J. Lunde, *Solar Thermal Engineering,* © 1980, John Wiley & Sons, New York.
*Note:* 1 Btu/h ft² × 3.152 = W/m².

**APPENDICES**

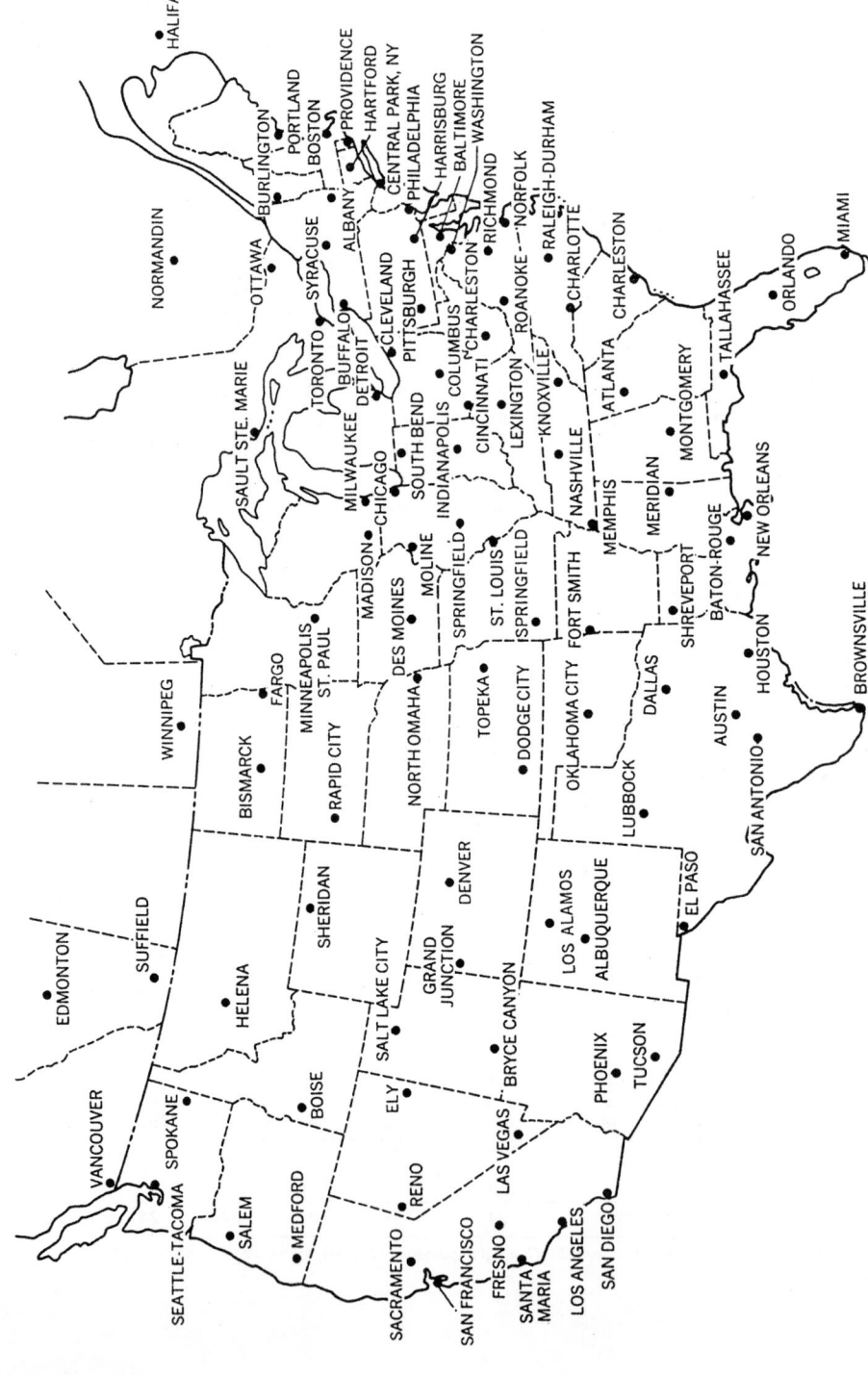

**Fig. C.1** U.S. and Canadian locations for which solar data are given in Table C.15. (Basic temperature data for many more locations are found in Appendix B.) (Adapted by permission from J.D. Balcomb et al., Passive Solar Design Handbook, Vol. 3, © 1983, American Solar Energy Society, Inc., Boulder, CO.)

## TABLE C.15  Average Insolation, Temperature, and DD Data

Elevation in feet, latitude in degrees north latitude, HS (horizontal surface) and VS (vertical surface) insolation in Btu/day ft², TA in degrees F, D50, D55, D60, and D65 in degree days F (for these various "base" temperatures).

### United States

**MONTGOMERY, ALABAMA**  Elev 203  Lat 32.3

|     | HS | VS | TA | D50 | D55 | D60 | D65 |
|-----|----|----|----|-----|-----|-----|-----|
| Jan | 752 | 896 | 48 | 148 | 256 | 394 | 556 |
| Jul | 1841 | 820 | 81 | 0 | 0 | 0 | 0 |
| Yr | 1390 | 946 | 65 | 445 | 866 | 1474 | 2269 |

**PHOENIX, ARIZONA**  Elev 1112  Lat 33.4

|     | HS | VS | TA | D50 | D55 | D60 | D65 |
|-----|----|----|----|-----|-----|-----|-----|
| Jan | 1021 | 1462 | 51 | 78 | 162 | 285 | 428 |
| Jul | 2486 | 964 | 91 | 0 | 0 | 0 | 0 |
| Yr | 1371 | 1326 | 70 | 187 | 459 | 919 | 1552 |

**TUCSON, ARIZONA**  Elev 2556  Lat 32.1

|     | HS | VS | TA | D50 | D55 | D60 | D65 |
|-----|----|----|----|-----|-----|-----|-----|
| Jan | 1099 | 1539 | 51 | 80 | 166 | 292 | 442 |
| Jul | 2341 | 922 | 86 | 0 | 0 | 0 | 0 |
| Yr | 1874 | 1307 | 68 | 214 | 525 | 1036 | 1752 |

**FORT SMITH, ARKANSAS**  Elev 463  Lat 35.3

|     | HS | VS | TA | D50 | D55 | D60 | D65 |
|-----|----|----|----|-----|-----|-----|-----|
| Jan | 744 | 996 | 39 | 346 | 497 | 651 | 806 |
| Jul | 2065 | 908 | 82 | 0 | 0 | 0 | 0 |
| Yr | 1406 | 1013 | 61 | 996 | 1622 | 2405 | 3336 |

**FRESNO, CALIFORNIA**  Elev 328  Lat 36.8

|     | HS | VS | TA | D50 | D55 | D60 | D65 |
|-----|----|----|----|-----|-----|-----|-----|
| Jan | 657 | 886 | 45 | 176 | 308 | 457 | 611 |
| Jul | 2685 | 1076 | 81 | 0 | 0 | 0 | 0 |
| Yr | 1714 | 1210 | 62 | 507 | 1021 | 1741 | 2650 |

**LOS ANGELES, CALIFORNIA**  Elev 105  Lat 33.9

|     | HS | VS | TA | D50 | D55 | D60 | D65 |
|-----|----|----|----|-----|-----|-----|-----|
| Jan | 926 | 1293 | 55 | 21 | 83 | 186 | 331 |
| Jul | 2307 | 942 | 69 | 0 | 1 | 5 | 19 |
| Yr | 1596 | 1157 | 62 | 64 | 299 | 849 | 1819 |

**SACRAMENTO, CALIFORNIA**  Elev 26  Lat 38.5

|     | HS | VS | TA | D50 | D55 | D60 | D65 |
|-----|----|----|----|-----|-----|-----|-----|
| Jan | 597 | 829 | 45 | 183 | 315 | 464 | 617 |
| Jul | 2688 | 1131 | 75 | 0 | 0 | 0 | 0 |
| Yr | 1646 | 1198 | 60 | 554 | 1097 | 1871 | 2843 |

**SAN DIEGO, CALIFORNIA**  Elev 30  Lat 32.7

|     | HS | VS | TA | D50 | D55 | D60 | D65 |
|-----|----|----|----|-----|-----|-----|-----|
| Jan | 976 | 1325 | 55 | 9 | 58 | 160 | 314 |
| Jul | 2186 | 902 | 70 | 0 | 0 | 1 | 6 |
| Yr | 1600 | 1151 | 63 | 23 | 170 | 623 | 1507 |

**SAN FRANCISCO, CALIFORNIA**  Elev 16  Lat 37.6

|     | HS | VS | TA | D50 | D55 | D60 | D65 |
|-----|----|----|----|-----|-----|-----|-----|
| Jan | 708 | 1023 | 48 | 82 | 210 | 363 | 518 |
| Jul | 2392 | 1034 | 63 | 0 | 2 | 21 | 93 |
| Yr | 1556 | 1156 | 57 | 202 | 705 | 1643 | 3042 |

**SANTA MARIA, CALIFORNIA**  Elev 236  Lat 34.9

|     | HS | VS | TA | D50 | D55 | D60 | D65 |
|-----|----|----|----|-----|-----|-----|-----|
| Jan | 854 | 1198 | 51 | 51 | 150 | 296 | 450 |
| Jul | 2341 | 965 | 62 | 0 | 3 | 30 | 112 |
| Yr | 1610 | 1172 | 57 | 155 | 624 | 1604 | 3053 |

**DENVER, COLORADO**  Elev 5331  Lat 39.7

|     | HS | VS | TA | D50 | D55 | D60 | D65 |
|-----|----|----|----|-----|-----|-----|-----|
| Jan | 840 | 1465 | 30 | 623 | 778 | 933 | 1088 |
| Jul | 2273 | 1053 | 73 | 0 | 0 | 0 | 0 |
| Yr | 1570 | 1334 | 50 | 2592 | 3588 | 4733 | 6016 |

**GRAND JUNCTION, COLO**  Elev 4839  Lat 39.1

|     | HS | VS | TA | D50 | D55 | D60 | D65 |
|-----|----|----|----|-----|-----|-----|-----|
| Jan | 791 | 1296 | 27 | 726 | 880 | 1035 | 1190 |
| Jul | 2465 | 1094 | 79 | 0 | 0 | 0 | 0 |
| Yr | 1661 | 1346 | 53 | 2514 | 3412 | 4434 | 5605 |

**HARTFORD, CONNECTICUT**  Elev 180  Lat 41.9

|     | HS | VS | TA | D50 | D55 | D60 | D65 |
|-----|----|----|----|-----|-----|-----|-----|
| Jan | 477 | 694 | 25 | 781 | 936 | 1091 | 1246 |
| Jul | 1649 | 861 | 73 | 0 | 0 | 0 | 0 |
| Yr | 1060 | 835 | 49 | 2971 | 3948 | 5075 | 6350 |

**WASHINGTON, DC**  Elev 289  Lat 38.9

|     | HS | VS | TA | D50 | D55 | D60 | D65 |
|-----|----|----|----|-----|-----|-----|-----|
| Jan | 572 | 793 | 32 | 555 | 710 | 865 | 1020 |
| Jul | 1817 | 883 | 75 | 0 | 0 | 0 | 0 |
| Yr | 1210 | 912 | 54 | 2004 | 2869 | 3864 | 5010 |

**MIAMI, FLORIDA**  Elev 7  Lat 25.8

|     | HS | VS | TA | D50 | D55 | D60 | D65 |
|-----|----|----|----|-----|-----|-----|-----|
| Jan | 1057 | 1121 | 67 | 1 | 4 | 18 | 53 |
| Jul | 1763 | 787 | 82 | 0 | 0 | 0 | 0 |
| Yr | 1474 | 941 | 76 | 3 | 14 | 55 | 206 |

**ORLANDO, FLORIDA**  Elev 118  Lat 28.5

|     | HS | VS | TA | D50 | D55 | D60 | D65 |
|-----|----|----|----|-----|-----|-----|-----|
| Jan | 999 | 1151 | 60 | 13 | 42 | 105 | 197 |
| Jul | 1801 | 795 | 81 | 0 | 0 | 0 | 0 |
| Yr | 1488 | 984 | 72 | 39 | 126 | 348 | 733 |

**TALLAHASSEE, FLORIDA**  Elev 69  Lat 30.4

|     | HS | VS | TA | D50 | D55 | D60 | D65 |
|-----|----|----|----|-----|-----|-----|-----|
| Jan | 877 | 1033 | 53 | 73 | 150 | 256 | 408 |
| Jul | 1748 | 786 | 81 | 0 | 0 | 0 | 0 |
| Yr | 1434 | 969 | 68 | 215 | 501 | 951 | 1563 |

**ATLANTA, GEORGIA**  Elev 1033  Lat 33.6

|     | HS | VS | TA | D50 | D55 | D60 | D65 |
|-----|----|----|----|-----|-----|-----|-----|
| Jan | 718 | 884 | 42 | 246 | 393 | 546 | 701 |
| Jul | 1812 | 821 | 78 | 0 | 0 | 0 | 0 |
| Yr | 1347 | 941 | 61 | 758 | 1362 | 2150 | 3095 |

**BOISE, IDAHO**  Elev 2867  Lat 43.6

|     | HS | VS | TA | D50 | D55 | D60 | D65 |
|-----|----|----|----|-----|-----|-----|-----|
| Jan | 485 | 770 | 29 | 651 | 806 | 961 | 1116 |
| Jul | 2613 | 1309 | 75 | 0 | 0 | 0 | 0 |
| Yr | 1499 | 1255 | 51 | 2420 | 3395 | 4536 | 5833 |

**CHICAGO, ILLINOIS**  Elev 623  Lat 41.8

|     | HS | VS | TA | D50 | D55 | D60 | D65 |
|-----|----|----|----|-----|-----|-----|-----|
| Jan | 507 | 756 | 24 | 797 | 952 | 1107 | 1262 |
| Jul | 1944 | 984 | 75 | 0 | 0 | 0 | 0 |
| Yr | 1217 | 960 | 51 | 2954 | 3881 | 4940 | 6127 |

**MOLINE, ILLINOIS**  Elev 594  Lat 41.4

|     | HS | VS | TA | D50 | D55 | D60 | D65 |
|-----|----|----|----|-----|-----|-----|-----|
| Jan | 535 | 803 | 22 | 884 | 1039 | 1194 | 1349 |
| Jul | 1939 | 974 | 75 | 0 | 0 | 0 | 0 |
| Yr | 1226 | 973 | 50 | 3191 | 4117 | 5178 | 6395 |

**SPRINGFIELD, ILLINOIS**  Elev 614  Lat 39.8

|     | HS | VS | TA | D50 | D55 | D60 | D65 |
|-----|----|----|----|-----|-----|-----|-----|
| Jan | 585 | 852 | 27 | 723 | 877 | 1032 | 1187 |
| Jul | 2058 | 984 | 76 | 0 | 0 | 0 | 0 |
| Yr | 1304 | 1003 | 53 | 2558 | 3434 | 4425 | 5558 |

### TABLE C.15  Average Insolation, Temperature, and DD Data (*continued*)

Elevation in feet, latitude in degrees north latitude, HS (horizontal surface) and VS (vertical surface) insolation in Btu/day ft², TA in degrees F, D50, D55, D60, and D65 in degree days F (for these various "base" temperatures).

**United States**

| INDIANAPOLIS, INDIANA | | | | Elev 807 | Lat 39.7 | |
|---|---|---|---|---|---|---|
| | HS | VS | TA | D50 | D55 | D60 | D65 |
| Jan | 496 | 668 | 28 | 685 | 840 | 995 | 1150 |
| Jul | 1806 | 891 | 75 | 0 | 0 | 0 | 0 |
| Yr | 1167 | 873 | 52 | 2511 | 3403 | 4421 | 5577 |

| SAULT STE. MARIE, MICHIGAN | | | | Elev 725 | Lat 46.5 | |
|---|---|---|---|---|---|---|
| | HS | VS | TA | D50 | D55 | D60 | D65 |
| Jan | 325 | 492 | 14 | 1110 | 1265 | 1420 | 1575 |
| Jul | 1835 | 1045 | 64 | 1 | 7 | 33 | 96 |
| Yr | 1044 | 861 | 40 | 4969 | 6198 | 7607 | 9193 |

| SOUTH BEND, INDIANA | | | | Elev 774 | Lat 41.7 | |
|---|---|---|---|---|---|---|
| | HS | VS | TA | D50 | D55 | D60 | D65 |
| Jan | 416 | 566 | 24 | 806 | 961 | 1116 | 1271 |
| Jul | 1852 | 944 | 72 | 0 | 1 | 4 | 6 |
| Yr | 1140 | 864 | 49 | 3112 | 4084 | 5199 | 6462 |

| MINNEAPOLIS, MINNESOTA | | | | Elev 837 | Lat 44.9 | |
|---|---|---|---|---|---|---|
| | HS | VS | TA | D50 | D55 | D60 | D65 |
| Jan | 464 | 768 | 12 | 1172 | 1327 | 1482 | 1637 |
| Jul | 1970 | 1071 | 72 | 0 | 1 | 5 | 11 |
| Yr | 1172 | 996 | 44 | 4584 | 5631 | 6824 | 8159 |

| DES MOINES, IOWA | | | | Elev 965 | Lat 41.5 | |
|---|---|---|---|---|---|---|
| | HS | VS | TA | D50 | D55 | D60 | D65 |
| Jan | 581 | 912 | 19 | 949 | 1104 | 1259 | 1414 |
| Jul | 2097 | 1037 | 75 | 0 | 0 | 0 | 0 |
| Yr | 1314 | 1065 | 49 | 3491 | 4435 | 5510 | 6710 |

| MERIDIAN, MISSISSIPPI | | | | Elev 308 | Lat 32.3 | |
|---|---|---|---|---|---|---|
| | HS | VS | TA | D50 | D55 | D60 | D65 |
| Jan | 744 | 883 | 47 | 163 | 274 | 413 | 575 |
| Jul | 1823 | 815 | 81 | 0 | 0 | 0 | 0 |
| Yr | 1371 | 933 | 65 | 510 | 955 | 1582 | 2388 |

| DODGE CITY, KANSAS | | | | Elev 2582 | Lat 37.8 | |
|---|---|---|---|---|---|---|
| | HS | VS | TA | D50 | D55 | D60 | D65 |
| Jan | 827 | 1303 | 31 | 596 | 750 | 905 | 1060 |
| Jul | 2295 | 1013 | 79 | 0 | 0 | 0 | 0 |
| Yr | 1562 | 1232 | 55 | 2131 | 2980 | 3945 | 5046 |

| SAINT LOUIS, MISSOURI | | | | Elev 564 | Lat 38.7 | |
|---|---|---|---|---|---|---|
| | HS | VS | TA | D50 | D55 | D60 | D65 |
| Jan | 627 | 898 | 31 | 581 | 735 | 890 | 1045 |
| Jul | 2049 | 959 | 79 | 0 | 0 | 0 | 0 |
| Yr | 1329 | 1006 | 56 | 1961 | 2762 | 3686 | 4750 |

| TOPEKA, KANSAS | | | | Elev 886 | Lat 39.1 | |
|---|---|---|---|---|---|---|
| | HS | VS | TA | D50 | D55 | D60 | D65 |
| Jan | 681 | 1033 | 28 | 683 | 837 | 992 | 1147 |
| Jul | 2128 | 993 | 78 | 0 | 0 | 0 | 0 |
| Yr | 1387 | 1036 | 54 | 2325 | 3175 | 4137 | 5243 |

| SPRINGFIELD, MISSOURI | | | | Elev 1270 | Lat 37.2 | |
|---|---|---|---|---|---|---|
| | HS | VS | TA | D50 | D55 | D60 | D65 |
| Jan | 684 | 956 | 33 | 534 | 686 | 840 | 995 |
| Jul | 2063 | 936 | 78 | 0 | 0 | 0 | 0 |
| Yr | 1364 | 1011 | 56 | 1848 | 2618 | 3522 | 4571 |

| LEXINGTON, KENTUCKY | | | | Elev 988 | Lat 38.0 | |
|---|---|---|---|---|---|---|
| | HS | VS | TA | D50 | D55 | D60 | D65 |
| Jan | 546 | 714 | 33 | 531 | 685 | 840 | 995 |
| Jul | 1850 | 881 | 76 | 0 | 0 | 0 | 0 |
| Yr | 1221 | 892 | 55 | 1865 | 2686 | 3632 | 4729 |

| HELENA, MONTANA | | | | Elev 3898 | Lat 46.6 | |
|---|---|---|---|---|---|---|
| | HS | VS | TA | D50 | D55 | D60 | D65 |
| Jan | 419 | 719 | 18 | 989 | 1144 | 1299 | 1454 |
| Jul | 2334 | 1312 | 68 | 1 | 3 | 12 | 33 |
| Yr | 1266 | 1134 | 43 | 4151 | 5342 | 6689 | 8190 |

| BATON ROUGE, LOUISIANA | | | | Elev 75 | Lat 30.5 | |
|---|---|---|---|---|---|---|
| | HS | VS | TA | D50 | D55 | D60 | D65 |
| Jan | 785 | 889 | 51 | 90 | 174 | 294 | 451 |
| Jul | 1746 | 786 | 82 | 0 | 0 | 0 | 0 |
| Yr | 1380 | 913 | 67 | 232 | 530 | 1006 | 1670 |

| NORTH OMAHA, NEBRASKA | | | | Elev 1325 | Lat 41.4 | |
|---|---|---|---|---|---|---|
| | HS | VS | TA | D50 | D55 | D60 | D65 |
| Jan | 634 | 1034 | 20 | 924 | 1079 | 1234 | 1389 |
| Jul | 2106 | 1038 | 75 | 0 | 1 | 3 | 7 |
| Yr | 1323 | 1078 | 49 | 3369 | 4309 | 5381 | 6601 |

| NEW ORLEANS, LOUISIANA | | | | Elev 10 | Lat 30.0 | |
|---|---|---|---|---|---|---|
| | HS | VS | TA | D50 | D55 | D60 | D65 |
| Jan | 835 | 950 | 53 | 73 | 150 | 252 | 403 |
| Jul | 1813 | 801 | 82 | 0 | 0 | 0 | 0 |
| Yr | 1438 | 943 | 68 | 197 | 465 | 887 | 1465 |

| ELY, NEVADA | | | | Elev 6253 | Lat 39.3 | |
|---|---|---|---|---|---|---|
| | HS | VS | TA | D50 | D55 | D60 | D65 |
| Jan | 819 | 1380 | 24 | 818 | 973 | 1128 | 1283 |
| Jul | 2447 | 1094 | 67 | 0 | 2 | 11 | 23 |
| Yr | 1675 | 1391 | 44 | 3716 | 4922 | 6291 | 7814 |

| SHREVEPORT, LOUISIANA | | | | Elev 259 | Lat 32.5 | |
|---|---|---|---|---|---|---|
| | HS | VS | TA | D50 | D55 | D60 | D65 |
| Jan | 762 | 920 | 47 | 154 | 264 | 403 | 552 |
| Jul | 2014 | 864 | 83 | 0 | 0 | 0 | 0 |
| Yr | 1428 | 973 | 66 | 428 | 832 | 1415 | 2167 |

| LAS VEGAS, NEVADA | | | | Elev 2178 | Lat 36.1 | |
|---|---|---|---|---|---|---|
| | HS | VS | TA | D50 | D55 | D60 | D65 |
| Jan | 978 | 1553 | 44 | 216 | 346 | 493 | 645 |
| Jul | 2588 | 1039 | 90 | 0 | 0 | 0 | 0 |
| Yr | 1866 | 1431 | 66 | 631 | 1129 | 1788 | 2601 |

| PORTLAND, MAINE | | | | Elev 62 | Lat 43.6 | |
|---|---|---|---|---|---|---|
| | HS | VS | TA | D50 | D55 | D60 | D65 |
| Jan | 450 | 689 | 22 | 884 | 1039 | 1194 | 1349 |
| Jul | 1659 | 894 | 68 | 0 | 1 | 5 | 27 |
| Yr | 1052 | 857 | 45 | 3648 | 4758 | 6039 | 7498 |

| RENO, NEVADA | | | | Elev 4400 | Lat 39.5 | |
|---|---|---|---|---|---|---|
| | HS | VS | TA | D50 | D55 | D60 | D65 |
| Jan | 800 | 1345 | 32 | 561 | 716 | 871 | 1026 |
| Jul | 2692 | 1167 | 69 | 0 | 1 | 5 | 17 |
| Yr | 1764 | 1439 | 49 | 2292 | 3345 | 4590 | 6022 |

| DETROIT, MICHIGAN | | | | Elev 627 | Lat 42.4 | |
|---|---|---|---|---|---|---|
| | HS | VS | TA | D50 | D55 | D60 | D65 |
| Jan | 417 | 585 | 26 | 760 | 915 | 1070 | 1225 |
| Jul | 1835 | 951 | 73 | 0 | 0 | 0 | 0 |
| Yr | 1122 | 869 | 50 | 2931 | 3890 | 4986 | 6228 |

| ALBUQUERQUE, NEW MEXICO | | | | Elev 5312 | Lat 35.0 | |
|---|---|---|---|---|---|---|
| | HS | VS | TA | D50 | D55 | D60 | D65 |
| Jan | 1016 | 1562 | 35 | 459 | 614 | 769 | 924 |
| Jul | 2489 | 995 | 79 | 0 | 0 | 0 | 0 |
| Yr | 1830 | 1379 | 57 | 1497 | 2292 | 3216 | 4292 |

**TABLE C.15  Average Insolation, Temperature, and DD Data (*continued*)**

Elevation in feet, latitude in degrees north latitude, HS (horizontal surface) and VS (vertical surface) insolation in Btu/day ft², TA in degrees F, D50, D55, D60, and D65 in degree days F (for these various "base" temperatures).

**United States**

| LOS ALAMOS, NEW MEXICO | | | | Elev 7380 | | Lat 35.9 |
|---|---|---|---|---|---|---|
| | HS | VS | TA | D50 | D55 | D60 | D65 |
| Jan | 893 | 1340 | 29 | 650 | 804 | 958 | 1113 |
| Jul | 2051 | 913 | 68 | 0 | 0 | 0 | 0 |
| Yr | 1579 | 1244 | 48 | 2822 | 3851 | 5043 | 6437 |

| COLUMBUS, OHIO | | | | Elev 833 | | Lat 40.0 |
|---|---|---|---|---|---|---|
| | HS | VS | TA | D50 | D55 | D60 | D65 |
| Jan | 459 | 606 | 28 | 670 | 825 | 980 | 1135 |
| Jul | 1755 | 876 | 74 | 0 | 0 | 0 | 0 |
| Yr | 1128 | 834 | 52 | 2524 | 3438 | 4491 | 5702 |

| BUFFALO, NEW YORK | | | | Elev 705 | | Lat 42.9 |
|---|---|---|---|---|---|---|
| | HS | VS | TA | D50 | D55 | D60 | D65 |
| Jan | 349 | 465 | 24 | 815 | 970 | 1125 | 1280 |
| Jul | 1776 | 935 | 70 | 0 | 0 | 3 | 12 |
| Yr | 1037 | 780 | 47 | 3322 | 4363 | 5551 | 6927 |

| OKLAHOMA CITY, OKLAHOMA | | | | Elev 1302 | | Lat 35.4 |
|---|---|---|---|---|---|---|
| | HS | VS | TA | D50 | D55 | D60 | D65 |
| Jan | 801 | 1114 | 37 | 413 | 565 | 719 | 874 |
| Jul | 2128 | 925 | 82 | 0 | 0 | 0 | 0 |
| Yr | 1463 | 1073 | 60 | 1232 | 1903 | 2734 | 3695 |

| NEW YORK, NEW YORK | | | | Elev 187 | | Lat 40.8 |
|---|---|---|---|---|---|---|
| | HS | VS | TA | D50 | D55 | D60 | D65 |
| Jan | 500 | 708 | 32 | 552 | 707 | 862 | 1017 |
| Jul | 1688 | 861 | 77 | 0 | 0 | 0 | 0 |
| Yr | 1101 | 849 | 55 | 1931 | 2759 | 3737 | 4848 |

| MEDFORD, OREGON | | | | Elev 1299 | | Lat 42.4 |
|---|---|---|---|---|---|---|
| | HS | VS | TA | D50 | D55 | D60 | D65 |
| Jan | 407 | 565 | 37 | 417 | 571 | 725 | 880 |
| Jul | 2475 | 1207 | 72 | 0 | 1 | 3 | 11 |
| Yr | 1356 | 1033 | 53 | 1576 | 2505 | 3621 | 4930 |

| SYRACUSE, NEW YORK | | | | Elev 407 | | Lat 43.1 |
|---|---|---|---|---|---|---|
| | HS | VS | TA | D50 | D55 | D60 | D65 |
| Jan | 385 | 538 | 24 | 818 | 973 | 1128 | 1283 |
| Jul | 1758 | 931 | 72 | 0 | 1 | 3 | 11 |
| Yr | 1037 | 791 | 48 | 3215 | 4218 | 5366 | 6678 |

| SALEM, OREGON | | | | Elev 200 | | Lat 44.9 |
|---|---|---|---|---|---|---|
| | HS | VS | TA | D50 | D55 | D60 | D65 |
| Jan | 332 | 471 | 39 | 348 | 502 | 657 | 812 |
| Jul | 2142 | 1154 | 67 | 0 | 1 | 7 | 43 |
| Yr | 1130 | 897 | 52 | 1265 | 2220 | 3411 | 4852 |

| ALBANY, NEW YORK | | | | Elev 292 | | Lat 42.7 |
|---|---|---|---|---|---|---|
| | HS | VS | TA | D50 | D55 | D60 | D65 |
| Jan | 456 | 674 | 22 | 884 | 1039 | 1194 | 1349 |
| Jul | 1725 | 908 | 72 | 0 | 1 | 3 | 9 |
| Yr | 1068 | 843 | 48 | 3424 | 4428 | 5586 | 6888 |

| HARRISBURG, PENNSYLVANIA | | | | Elev 348 | | Lat 40.2 |
|---|---|---|---|---|---|---|
| | HS | VS | TA | D50 | D55 | D60 | D65 |
| Jan | 536 | 763 | 30 | 617 | 772 | 927 | 1082 |
| Jul | 1764 | 883 | 76 | 0 | 0 | 0 | 0 |
| Yr | 1152 | 887 | 53 | 2221 | 3093 | 4086 | 5224 |

| CHARLOTTE, NORTH CAROLINA | | | | Elev 768 | | Lat 35.2 |
|---|---|---|---|---|---|---|
| | HS | VS | TA | D50 | D55 | D60 | D65 |
| Jan | 719 | 944 | 42 | 255 | 402 | 555 | 710 |
| Jul | 1831 | 841 | 79 | 0 | 0 | 0 | 0 |
| Yr | 1346 | 981 | 61 | 828 | 1451 | 2257 | 3218 |

| PHILADELPHIA, PENNSYLVANIA | | | | Elev 30 | | Lat 39.9 |
|---|---|---|---|---|---|---|
| | HS | VS | TA | D50 | D55 | D60 | D65 |
| Jan | 555 | 792 | 32 | 549 | 704 | 859 | 1014 |
| Jul | 1758 | 876 | 77 | 0 | 0 | 0 | 0 |
| Yr | 1170 | 905 | 55 | 1935 | 2775 | 3749 | 4865 |

| RALEIGH-DURHAM, NORTH CAROLINA | | | | Elev 440 | | Lat 35.9 |
|---|---|---|---|---|---|---|
| | HS | VS | TA | D50 | D55 | D60 | D65 |
| Jan | 694 | 924 | 41 | 300 | 451 | 605 | 760 |
| Jul | 1776 | 832 | 78 | 0 | 0 | 0 | 0 |
| Yr | 1297 | 955 | 59 | 990 | 1659 | 2509 | 3514 |

| PITTSBURGH, PENNSYLVANIA | | | | Elev 1224 | | Lat 40.5 |
|---|---|---|---|---|---|---|
| | HS | VS | TA | D50 | D55 | D60 | D65 |
| Jan | 424 | 553 | 28 | 679 | 834 | 989 | 1144 |
| Jul | 1689 | 857 | 72 | 0 | 1 | 3 | 7 |
| Yr | 1071 | 793 | 50 | 2635 | 3574 | 4669 | 5930 |

| BISMARCK, NORTH DAKOTA | | | | Elev 1647 | | Lat 46.8 |
|---|---|---|---|---|---|---|
| | HS | VS | TA | D50 | D55 | D60 | D65 |
| Jan | 467 | 847 | 8 | 1296 | 1451 | 1606 | 1761 |
| Jul | 2184 | 1241 | 71 | 0 | 2 | 8 | 18 |
| Yr | 1251 | 1145 | 41 | 5235 | 6364 | 7627 | 9044 |

| PROVIDENCE, RHODE ISLAND | | | | Elev 62 | | Lat 41.7 |
|---|---|---|---|---|---|---|
| | HS | VS | TA | D50 | D55 | D60 | D65 |
| Jan | 506 | 750 | 28 | 670 | 825 | 980 | 1135 |
| Jul | 1695 | 878 | 72 | 0 | 0 | 0 | 0 |
| Yr | 1114 | 884 | 50 | 2566 | 3543 | 4669 | 5972 |

| FARGO, NORTH DAKOTA | | | | Elev 899 | | Lat 46.9 |
|---|---|---|---|---|---|---|
| | HS | VS | TA | D50 | D55 | D60 | D65 |
| Jan | 415 | 720 | 6 | 1367 | 1522 | 1677 | 1832 |
| Jul | 2120 | 1210 | 71 | 0 | 1 | 5 | 13 |
| Yr | 1206 | 1075 | 41 | 5485 | 6607 | 7858 | 9271 |

| CHARLESTON, SOUTH CAROLINA | | | | Elev 39 | | Lat 32.9 |
|---|---|---|---|---|---|---|
| | HS | VS | TA | D50 | D55 | D60 | D65 |
| Jan | 744 | 904 | 49 | 120 | 222 | 360 | 521 |
| Jul | 1799 | 813 | 80 | 0 | 0 | 0 | 0 |
| Yr | 1346 | 940 | 65 | 360 | 756 | 1355 | 2146 |

| CINCINNATI, OHIO | | | | Elev 889 | | Lat 39.1 |
|---|---|---|---|---|---|---|
| | HS | VS | TA | D50 | D55 | D60 | D65 |
| Jan | 500 | 659 | 31 | 587 | 741 | 896 | 1051 |
| Jul | 1771 | 869 | 76 | 0 | 0 | 0 | 0 |
| Yr | 1160 | 858 | 54 | 2117 | 2973 | 3951 | 5070 |

| RAPID CITY, SOUTH DAKOTA | | | | Elev 3169 | | Lat 44.0 |
|---|---|---|---|---|---|---|
| | HS | VS | TA | D50 | D55 | D60 | D65 |
| Jan | 542 | 928 | 22 | 871 | 1026 | 1181 | 1336 |
| Jul | 2223 | 1161 | 73 | 1 | 2 | 8 | 13 |
| Yr | 1344 | 1177 | 47 | 3681 | 4749 | 5965 | 7324 |

| CLEVELAND, OHIO | | | | Elev 804 | | Lat 41.4 |
|---|---|---|---|---|---|---|
| | HS | VS | TA | D50 | D55 | D60 | D65 |
| Jan | 388 | 507 | 27 | 716 | 871 | 1026 | 1181 |
| Jul | 1828 | 929 | 71 | 0 | 1 | 4 | 9 |
| Yr | 1093 | 808 | 50 | 2804 | 3768 | 4879 | 6154 |

| KNOXVILLE, TENNESSEE | | | | Elev 981 | | Lat 35.8 |
|---|---|---|---|---|---|---|
| | HS | VS | TA | D50 | D55 | D60 | D65 |
| Jan | 621 | 785 | 41 | 302 | 449 | 602 | 756 |
| Jul | 1804 | 839 | 78 | 0 | 0 | 0 | 0 |
| Yr | 1275 | 909 | 60 | 1018 | 1671 | 2504 | 3478 |

## TABLE C.15  Average Insolation, Temperature, and DD Data (*continued*)

Elevation in feet, latitude in degrees north latitude, HS (horizontal surface) and VS (vertical surface) insolation in Btu/day ft², TA in degrees F, D50, D55, D60, and D65 in degree days F (for these various "base" temperatures).

### United States

**MEMPHIS, TENNESSEE** — Elev 285 — Lat 35.0

|     | HS | VS | TA | D50 | D55 | D60 | D65 |
|-----|----|----|----|-----|-----|-----|-----|
| Jan | 683 | 870 | 41 | 312 | 455 | 606 | 760 |
| Jul | 1972 | 879 | 82 | 0 | 0 | 0 | 0 |
| Yr | 1368 | 965 | 62 | 988 | 1588 | 2357 | 3227 |

**BURLINGTON, VERMONT** — Elev 341 — Lat 44.5

|     | HS | VS | TA | D50 | D55 | D60 | D65 |
|-----|----|----|----|-----|-----|-----|-----|
| Jan | 385 | 572 | 17 | 1029 | 1184 | 1339 | 1494 |
| Jul | 1721 | 941 | 70 | 0 | 1 | 4 | 20 |
| Yr | 1023 | 815 | 44 | 4142 | 5230 | 6464 | 7876 |

**NASHVILLE, TENNESSEE** — Elev 591 — Lat 36.1

|     | HS | VS | TA | D50 | D55 | D60 | D65 |
|-----|----|----|----|-----|-----|-----|-----|
| Jan | 580 | 721 | 38 | 369 | 519 | 673 | 828 |
| Jul | 1891 | 869 | 80 | 0 | 0 | 0 | 0 |
| Yr | 1272 | 891 | 59 | 1195 | 1874 | 2720 | 3696 |

**NORFOLK, VIRGINIA** — Elev 30 — Lat 36.9

|     | HS | VS | TA | D50 | D55 | D60 | D65 |
|-----|----|----|----|-----|-----|-----|-----|
| Jan | 678 | 932 | 41 | 300 | 450 | 605 | 760 |
| Jul | 1853 | 868 | 78 | 0 | 0 | 0 | 0 |
| Yr | 1327 | 990 | 59 | 974 | 1646 | 2489 | 3488 |

**AUSTIN, TEXAS** — Elev 620 — Lat 30.3

|     | HS | VS | TA | D50 | D55 | D60 | D65 |
|-----|----|----|----|-----|-----|-----|-----|
| Jan | 864 | 1008 | 50 | 116 | 207 | 333 | 483 |
| Jul | 2105 | 865 | 85 | 0 | 0 | 0 | 0 |
| Yr | 1478 | 974 | 68 | 289 | 602 | 1088 | 1737 |

**RICHMOND, VIRGINIA** — Elev 164 — Lat 37.5

|     | HS | VS | TA | D50 | D55 | D60 | D65 |
|-----|----|----|----|-----|-----|-----|-----|
| Jan | 632 | 863 | 38 | 390 | 543 | 698 | 853 |
| Jul | 1774 | 849 | 78 | 0 | 0 | 0 | 0 |
| Yr | 1250 | 936 | 58 | 1296 | 2021 | 2909 | 3939 |

**BROWNSVILLE, TEXAS** — Elev 20 — Lat 25.9

|     | HS | VS | TA | D50 | D55 | D60 | D65 |
|-----|----|----|----|-----|-----|-----|-----|
| Jan | 913 | 923 | 60 | 18 | 51 | 116 | 225 |
| Jul | 2212 | 867 | 84 | 0 | 0 | 0 | 0 |
| Yr | 1550 | 917 | 74 | 44 | 127 | 325 | 650 |

**ROANOKE, VIRGINIA** — Elev 1175 — Lat 37.3

|     | HS | VS | TA | D50 | D55 | D60 | D65 |
|-----|----|----|----|-----|-----|-----|-----|
| Jan | 660 | 911 | 36 | 423 | 577 | 832 | 887 |
| Jul | 1796 | 854 | 75 | 0 | 0 | 0 | 0 |
| Yr | 1271 | 958 | 56 | 1486 | 2277 | 3211 | 4307 |

**DALLAS, TEXAS** — Elev 489 — Lat 32.8

|     | HS | VS | TA | D50 | D55 | D60 | D65 |
|-----|----|----|----|-----|-----|-----|-----|
| Jan | 821 | 1035 | 45 | 189 | 312 | 457 | 608 |
| Jul | 2122 | 890 | 86 | 0 | 0 | 0 | 0 |
| Yr | 1470 | 1014 | 66 | 505 | 943 | 1543 | 2290 |

**SEATTLE-TACOMA, WASHINGTON** — Elev 400 — Lat 47.4

|     | HS | VS | TA | D50 | D55 | D60 | D65 |
|-----|----|----|----|-----|-----|-----|-----|
| Jan | 262 | 378 | 38 | 367 | 521 | 676 | 831 |
| Jul | 2248 | 1299 | 65 | 0 | 2 | 16 | 80 |
| Yr | 1056 | 857 | 51 | 1386 | 2393 | 3662 | 5185 |

**EL PASO, TEXAS** — Elev 3917 — Lat 31.8

|     | HS | VS | TA | D50 | D55 | D60 | D65 |
|-----|----|----|----|-----|-----|-----|-----|
| Jan | 1125 | 1572 | 44 | 210 | 355 | 509 | 663 |
| Jul | 2450 | 934 | 82 | 0 | 0 | 0 | 0 |
| Yr | 1901 | 1327 | 63 | 561 | 1102 | 1810 | 2678 |

**SPOKANE, WASHINGTON** — Elev 2365 — Lat 47.6

|     | HS | VS | TA | D50 | D55 | D60 | D65 |
|-----|----|----|----|-----|-----|-----|-----|
| Jan | 315 | 496 | 25 | 763 | 918 | 1073 | 1228 |
| Jul | 2357 | 1368 | 70 | 0 | 2 | 7 | 21 |
| Yr | 1227 | 1068 | 47 | 3061 | 4150 | 5411 | 6835 |

**HOUSTON, TEXAS** — Elev 108 — Lat 30.0

|     | HS | VS | TA | D50 | D55 | D60 | D65 |
|-----|----|----|----|-----|-----|-----|-----|
| Jan | 772 | 852 | 52 | 71 | 150 | 263 | 416 |
| Jul | 1828 | 805 | 83 | 0 | 0 | 0 | 0 |
| Yr | 1353 | 884 | 69 | 161 | 409 | 825 | 1434 |

**CHARLESTON, WEST VIRGINIA** — Elev 951 — Lat 38.4

|     | HS | VS | TA | D50 | D55 | D60 | D65 |
|-----|----|----|----|-----|-----|-----|-----|
| Jan | 498 | 638 | 35 | 483 | 636 | 791 | 946 |
| Jul | 1682 | 827 | 75 | 0 | 0 | 0 | 0 |
| Yr | 1125 | 822 | 55 | 1726 | 2540 | 3488 | 4590 |

**LUBBOCK, TEXAS** — Elev 3241 — Lat 33.6

|     | HS | VS | TA | D50 | D55 | D60 | D65 |
|-----|----|----|----|-----|-----|-----|-----|
| Jan | 1031 | 1497 | 39 | 343 | 494 | 648 | 803 |
| Jul | 2412 | 956 | 80 | 0 | 0 | 0 | 0 |
| Yr | 1768 | 1279 | 60 | 1069 | 1739 | 2582 | 3545 |

**MADISON, WISCONSIN** — Elev 860 — Lat 43.1

|     | HS | VS | TA | D50 | D55 | D60 | D65 |
|-----|----|----|----|-----|-----|-----|-----|
| Jan | 515 | 822 | 17 | 1029 | 1184 | 1339 | 1494 |
| Jul | 1934 | 1009 | 70 | 0 | 1 | 5 | 14 |
| Yr | 1193 | 978 | 45 | 4086 | 5143 | 6352 | 7730 |

**SAN ANTONIO, TEXAS** — Elev 794 — Lat 29.5

|     | HS | VS | TA | D50 | D55 | D60 | D65 |
|-----|----|----|----|-----|-----|-----|-----|
| Jan | 895 | 1026 | 51 | 93 | 179 | 302 | 451 |
| Jul | 2121 | 863 | 85 | 0 | 0 | 0 | 0 |
| Yr | 1501 | 973 | 69 | 213 | 490 | 941 | 1570 |

**MILWAUKEE, WISCONSIN** — Elev 692 — Lat 42.9

|     | HS | VS | TA | D50 | D55 | D60 | D65 |
|-----|----|----|----|-----|-----|-----|-----|
| Jan | 479 | 731 | 19 | 949 | 1104 | 1259 | 1414 |
| Jul | 1962 | 1017 | 70 | 0 | 1 | 4 | 15 |
| Yr | 1194 | 957 | 46 | 3774 | 4833 | 6045 | 7444 |

**BRYCE CANYON, UTAH** — Elev 7588 — Lat 37.7

|     | HS | VS | TA | D50 | D55 | D60 | D65 |
|-----|----|----|----|-----|-----|-----|-----|
| Jan | 914 | 1511 | 20 | 936 | 1091 | 1246 | 1401 |
| Jul | 2424 | 1044 | 62 | 1 | 8 | 47 | 128 |
| Yr | 1742 | 1404 | 40 | 4693 | 6106 | 7675 | 9133 |

**SHERIDAN, WYOMING** — Elev 3966 — Lat 44.8

|     | HS | VS | TA | D50 | D55 | D60 | D65 |
|-----|----|----|----|-----|-----|-----|-----|
| Jan | 517 | 900 | 21 | 899 | 1054 | 1209 | 1364 |
| Jul | 2329 | 1237 | 70 | 1 | 2 | 9 | 28 |
| Yr | 1333 | 1170 | 45 | 3860 | 4991 | 6279 | 7708 |

**SALT LAKE CITY, UTAH** — Elev 4226 — Lat 40.8

|     | HS | VS | TA | D50 | D55 | D60 | D65 |
|-----|----|----|----|-----|-----|-----|-----|
| Jan | 639 | 1017 | 28 | 683 | 837 | 992 | 1147 |
| Jul | 2590 | 1186 | 77 | 0 | 0 | 0 | 0 |
| Yr | 1606 | 1301 | 51 | 2648 | 3612 | 4725 | 5983 |

**TABLE C.15  Average Insolation, Temperature, and DD Data (*continued*)**

| Elevation in feet, latitude in degrees north latitude, HS (horizontal surface) and VS (vertical surface) insolation in Btu/day ft², TA in degrees F, D50, D55, D60, and D65 in degree days F (for these various "base" temperatures). |
| --- |

**Canada**

| EDMONTON, ALBERTA | | | | Elev 2220 | | Lat 53.6 | | HALIFAX, NOVA SCOTIA | | | | Elev 136 | | Lat 44.6 |
|---|---|---|---|---|---|---|---|---|---|---|---|---|---|---|
| | HS | VS | TA | D50 | D55 | D60 | D65 | | HS | VS | TA | D50 | D55 | D60 | D65 |
| Jan | 324 | 746 | 4 | 1421 | 1574 | 1728 | 1883 | Jan | 456 | 737 | 26 | 752 | 900 | 1051 | 1204 |
| Jul | 1977 | 1378 | 62 | 1 | 7 | 38 | 117 | Jul | 1694 | 929 | 65 | 0 | 1 | 8 | 57 |
| Yr | 1114 | 1205 | 36 | 6317 | 7563 | 9016 | 10650 | Yr | 1076 | 907 | 46 | 3457 | 4500 | 5746 | 7211 |

| SUFFIELD, ALBERTA | | | | Elev 2549 | | Lat 50.3 | | OTTAWA, ONTARIO | | | | Elev 377 | | Lat 45.5 |
|---|---|---|---|---|---|---|---|---|---|---|---|---|---|---|
| | HS | VS | TA | D50 | D55 | D60 | D65 | | HS | VS | TA | D50 | D55 | D60 | D65 |
| Jan | 433 | 937 | 7 | 1333 | 1486 | 1640 | 1794 | Jan | 510 | 914 | 13 | 1169 | 1320 | 1473 | 1627 |
| Jul | 2173 | 1377 | 67 | 0 | 2 | 11 | 49 | Jul | 1875 | 1040 | 69 | 0 | 1 | 4 | 23 |
| Yr | 1239 | 1269 | 40 | 5500 | 6637 | 7923 | 9393 | Yr | 1158 | 1015 | 43 | 4912 | 5965 | 7158 | 8529 |

| VANCOUVER, BRITISH COLUMBIA | | | | Elev 310 | | Lat 37.5 | | TORONTO, ONTARIO | | | | Elev 443 | | Lat 43.7 |
|---|---|---|---|---|---|---|---|---|---|---|---|---|---|---|
| | HS | VS | TA | D50 | D55 | D60 | D65 | | HS | VS | TA | D50 | D55 | D60 | D65 |
| Jan | 254 | 395 | 37 | 425 | 572 | 724 | 878 | Jan | 487 | 777 | 22 | 891 | 1041 | 1194 | 1348 |
| Jul | 2021 | 1239 | 63 | 0 | 1 | 13 | 82 | Jul | 1958 | 1035 | 70 | 0 | 0 | 3 | 18 |
| Yr | 1060 | 916 | 50 | 1791 | 2781 | 4041 | 5588 | Yr | 1171 | 948 | 46 | 3842 | 4853 | 6013 | 7343 |

| WINNIPEG, MANITOBA | | | | Elev 820 | | Lat 49.95 | | NORMANDIN, QUEBEC | | | | Elev 450 | | Lat 48.8 |
|---|---|---|---|---|---|---|---|---|---|---|---|---|---|---|
| | HS | VS | TA | D50 | D55 | D60 | D65 | | HS | VS | TA | D50 | D55 | D60 | D65 |
| Jan | 461 | 1011 | 0 | 1588 | 1740 | 1893 | 2047 | Jan | 454 | 921 | 0 | 1564 | 1719 | 1873 | 2028 |
| Jul | 2025 | 1264 | 67 | 0 | 2 | 9 | 45 | Jul | 1707 | 1031 | 62 | 1 | 7 | 40 | 118 |
| Yr | 1190 | 1199 | 36 | 6925 | 8062 | 9338 | 10790 | Yr | 1092 | 1053 | 34 | 7037 | 8308 | 9762 | 11376 |

*Source:* Reprinted by permission from J.D. Balcomb et al., *Passive Solar Design Handbook,* Vol. 3, © 1983, American Solar Energy Society, Inc., Boulder, CO.

**TABLE C.16  Coefficient of Utilization from Window Without Blinds; Sky Component $E_{xvk}/E_{xhk} = 0.75$**

| Room Depth/ | | Window Width/Window Height | | | | | | | |
|---|---|---|---|---|---|---|---|---|---|
| Window Height | Depth | .5 | 1 | 2 | 3 | 4 | 6 | 8 | Infinite |
| 1 | 10 | .824 | .864 | .870 | .873 | .875 | .879 | .880 | .883 |
| | 30 | .547 | .711 | .777 | .789 | .793 | .798 | .799 | .801 |
| | 50 | .355 | .526 | .635 | .659 | .666 | .669 | .670 | .672 |
| | 70 | .243 | .386 | .505 | .538 | .548 | .544 | .545 | .547 |
| | 90 | .185 | .304 | .418 | .451 | .464 | .444 | .446 | .447 |
| 2 | 10 | .667 | .781 | .809 | .812 | .813 | .815 | .816 | .824 |
| | 30 | .269 | .416 | .519 | .544 | .551 | .556 | .557 | .563 |
| | 50 | .122 | .204 | .287 | .319 | .331 | .339 | .341 | .345 |
| | 70 | .068 | .116 | .173 | .201 | .214 | .223 | .226 | .229 |
| | 90 | .050 | .084 | .127 | .151 | .164 | .167 | .171 | .172 |
| 3 | 10 | .522 | .681 | .739 | .746 | .747 | .749 | .747 | .766 |
| | 30 | .139 | .232 | .320 | .350 | .360 | .366 | .364 | .373 |
| | 50 | .053 | .092 | .139 | .163 | .174 | .183 | .182 | .187 |
| | 70 | .031 | .053 | .081 | .097 | .106 | .116 | .116 | .119 |
| | 90 | .025 | .041 | .061 | .074 | .082 | .089 | .090 | .092 |
| 4 | 10 | .405 | .576 | .658 | .670 | .673 | .675 | .674 | .707 |
| | 30 | .075 | .134 | .197 | .224 | .235 | .243 | .243 | .255 |
| | 50 | .028 | .050 | .078 | .094 | .104 | .112 | .114 | .119 |
| | 70 | .018 | .031 | .048 | .059 | .065 | .073 | .074 | .078 |
| | 90 | .016 | .026 | .040 | .048 | .053 | .059 | .061 | .064 |
| 6 | 10 | .242 | .392 | .494 | .516 | .521 | .524 | .523 | .588 |
| | 30 | .027 | .054 | .086 | .102 | .111 | .119 | .120 | .135 |
| | 50 | .011 | .023 | .036 | .044 | .049 | .055 | .056 | .063 |
| | 70 | .009 | .018 | .027 | .032 | .035 | .040 | .041 | .046 |
| | 90 | .008 | .016 | .023 | .028 | .031 | .034 | .035 | .040 |
| 8 | 10 | .147 | .257 | .352 | .380 | .387 | .391 | .392 | .482 |
| | 30 | .012 | .026 | .043 | .054 | .060 | .067 | .070 | .086 |
| | 50 | .006 | .013 | .021 | .026 | .029 | .033 | .035 | .043 |
| | 70 | .005 | .011 | .017 | .021 | .023 | .026 | .027 | .034 |
| | 90 | .004 | .010 | .015 | .019 | .021 | .023 | .025 | .030 |
| 10 | 10 | .092 | .168 | .248 | .275 | .284 | .290 | .291 | .395 |
| | 30 | .006 | .014 | .026 | .032 | .036 | .041 | .044 | .059 |
| | 50 | .003 | .008 | .014 | .017 | .019 | .022 | .024 | .032 |
| | 70 | .003 | .007 | .012 | .014 | .016 | .018 | .019 | .026 |
| | 90 | .003 | .006 | .011 | .013 | .015 | .016 | .017 | .024 |

*Source:* IES RP-23-1989; reprinted with permission of Illuminating Engineering Society of North America.

APPENDICES

**TABLE C.17  Coefficient of Utilization from Window Without Blinds; Sky Component $E_{xvk}/E_{xhk} = 1.00$**

| Room Depth/ Window Height | Depth | *Window Width/Window Height* | | | | | | | |
|---|---|---|---|---|---|---|---|---|---|
| | | .5 | 1 | 2 | 3 | 4 | 6 | 8 | Infinite |
| 1 | 10 | .671 | .704 | .711 | .715 | .717 | .726 | .726 | .728 |
| | 30 | .458 | .595 | .654 | .668 | .672 | .682 | .683 | .685 |
| | 50 | .313 | .462 | .563 | .589 | .598 | .607 | .608 | .610 |
| | 70 | .227 | .362 | .478 | .515 | .527 | .530 | .532 | .534 |
| | 90 | .186 | .306 | .424 | .465 | .481 | .468 | .471 | .472 |
| 2 | 10 | .545 | .636 | .658 | .660 | .661 | .665 | .666 | .672 |
| | 30 | .239 | .367 | .459 | .484 | .491 | .499 | .501 | .506 |
| | 50 | .121 | .203 | .286 | .320 | .335 | .348 | .351 | .355 |
| | 70 | .074 | .128 | .192 | .226 | .243 | .259 | .264 | .267 |
| | 90 | .058 | .101 | .156 | .188 | .207 | .215 | .221 | .223 |
| 3 | 10 | .431 | .561 | .607 | .613 | .614 | .616 | .615 | .631 |
| | 30 | .133 | .223 | .306 | .337 | .348 | .357 | .357 | .366 |
| | 50 | .058 | .103 | .155 | .183 | .197 | .211 | .213 | .218 |
| | 70 | .037 | .064 | .098 | .119 | .132 | .147 | .150 | .154 |
| | 90 | .030 | .051 | .079 | .098 | .110 | .122 | .126 | .129 |
| 4 | 10 | .339 | .482 | .549 | .560 | .563 | .566 | .565 | .593 |
| | 30 | .078 | .139 | .204 | .234 | .247 | .258 | .260 | .272 |
| | 50 | .033 | .060 | .094 | .114 | .126 | .139 | .143 | .150 |
| | 70 | .022 | .039 | .061 | .074 | .083 | .095 | .099 | .104 |
| | 90 | .019 | .032 | .050 | .061 | .070 | .080 | .084 | .089 |
| 6 | 10 | .211 | .343 | .433 | .453 | .458 | .461 | .461 | .518 |
| | 30 | .033 | .065 | .103 | .123 | .135 | .145 | .148 | .167 |
| | 50 | .015 | .029 | .047 | .057 | .064 | .073 | .077 | .086 |
| | 70 | .011 | .021 | .033 | .040 | .045 | .051 | .054 | .060 |
| | 90 | .010 | .019 | .028 | .034 | .038 | .044 | .046 | .052 |
| 8 | 10 | .135 | .238 | .326 | .353 | .362 | .366 | .367 | .452 |
| | 30 | .016 | .034 | .058 | .072 | .080 | .090 | .094 | .116 |
| | 50 | .008 | .017 | .027 | .034 | .039 | .045 | .048 | .059 |
| | 70 | .006 | .013 | .021 | .026 | .028 | .032 | .035 | .043 |
| | 90 | .005 | .012 | .019 | .023 | .025 | .029 | .031 | .038 |
| 10 | 10 | .090 | .165 | .244 | .272 | .283 | .290 | .291 | .395 |
| | 30 | .009 | .020 | .036 | .045 | .052 | .060 | .064 | .087 |
| | 50 | .005 | .010 | .019 | .023 | .026 | .030 | .033 | .044 |
| | 70 | .004 | .009 | .015 | .018 | .020 | .023 | .025 | .033 |
| | 90 | .003 | .008 | .014 | .016 | .018 | .020 | .022 | .030 |

*Source:* IES RP-23-1989; reprinted with permission of Illuminating Engineering Society of North America.

**TABLE C.18 Coefficient of Utilization from Window Without Blinds; Sky Component $E_{xvk}/E_{xhk} = 1.25$**

| Room Depth/ Window Height | Depth | Window Width/Window Height | | | | | | | |
|---|---|---|---|---|---|---|---|---|---|
| | | .5 | 1 | 2 | 3 | 4 | 6 | 8 | Infinite |
| 1 | 10 | .578 | .607 | .614 | .619 | .621 | .633 | .634 | .635 |
| | 30 | .405 | .525 | .580 | .594 | .599 | .612 | .614 | .615 |
| | 50 | .287 | .423 | .519 | .547 | .556 | .569 | .571 | .573 |
| | 70 | .218 | .347 | .461 | .501 | .515 | .522 | .525 | .526 |
| | 90 | .186 | .307 | .428 | .473 | .491 | .483 | .486 | .487 |
| 2 | 10 | .472 | .549 | .566 | .569 | .570 | .574 | .575 | .581 |
| | 30 | .221 | .337 | .422 | .447 | .456 | .465 | .467 | .472 |
| | 50 | .120 | .202 | .285 | .321 | .337 | .353 | .357 | .361 |
| | 70 | .078 | .136 | .204 | .242 | .261 | .281 | .287 | .290 |
| | 90 | .064 | .112 | .174 | .211 | .233 | .244 | .251 | .253 |
| 3 | 10 | .377 | .488 | .527 | .533 | .534 | .536 | .536 | .549 |
| | 30 | .130 | .217 | .298 | .329 | .341 | .352 | .353 | .362 |
| | 50 | .062 | .110 | .165 | .195 | .211 | .228 | .231 | .237 |
| | 70 | .040 | .070 | .109 | .132 | .147 | .166 | .171 | .175 |
| | 90 | .033 | .057 | .090 | .112 | .127 | .142 | .148 | .152 |
| 4 | 10 | .300 | .424 | .484 | .494 | .497 | .499 | .499 | .524 |
| | 30 | .080 | .143 | .209 | .240 | .255 | .267 | .269 | .283 |
| | 50 | .036 | .066 | .104 | .126 | .140 | .156 | .160 | .168 |
| | 70 | .024 | .043 | .068 | .083 | .094 | .109 | .115 | .120 |
| | 90 | .021 | .036 | .056 | .070 | .080 | .092 | .099 | .103 |
| 6 | 10 | .193 | .314 | .395 | .415 | .420 | .423 | .423 | .476 |
| | 30 | .036 | .071 | .113 | .136 | .149 | .161 | .165 | .186 |
| | 50 | .017 | .033 | .053 | .065 | .074 | .084 | .089 | .100 |
| | 70 | .012 | .024 | .037 | .045 | .050 | .058 | .061 | .069 |
| | 90 | .011 | .021 | .031 | .038 | .043 | .049 | .053 | .060 |
| 8 | 10 | .128 | .226 | .310 | .337 | .346 | .351 | .352 | .433 |
| | 30 | .019 | .039 | .066 | .082 | .092 | .104 | .109 | .134 |
| | 50 | .009 | .019 | .031 | .040 | .045 | .052 | .056 | .069 |
| | 70 | .007 | .015 | .023 | .029 | .032 | .037 | .040 | .049 |
| | 90 | .006 | .013 | .021 | .025 | .028 | .032 | .035 | .043 |
| 10 | 10 | .088 | .164 | .241 | .270 | .282 | .290 | .291 | .396 |
| | 30 | .011 | .024 | .043 | .054 | .062 | .071 | .076 | .103 |
| | 50 | .005 | .012 | .022 | .026 | .030 | .035 | .038 | .052 |
| | 70 | .004 | .010 | .017 | .020 | .023 | .026 | .028 | .038 |
| | 90 | .004 | .009 | .016 | .018 | .020 | .023 | .025 | .034 |

*Source:* IES RP-23-1989; reprinted with permission of Illuminating Engineering Society of North America.

**TABLE C.19  Coefficient of Utilization from Window Without Blinds; Sky Component $E_{xvk}/E_{xhk} = 1.50$**

| Room Depth/ Window Height | Depth | Window Width/Window Height | | | | | | | |
|---|---|---|---|---|---|---|---|---|---|
| | | .5 | 1 | 2 | 3 | 4 | 6 | 8 | Infinite |
| 1 | 10 | .503 | .528 | .536 | .541 | .544 | .557 | .558 | .559 |
| | 30 | .359 | .464 | .514 | .528 | .534 | .549 | .550 | .552 |
| | 50 | .261 | .384 | .471 | .499 | .508 | .524 | .526 | .527 |
| | 70 | .204 | .325 | .432 | .470 | .485 | .497 | .499 | .500 |
| | 90 | .179 | .295 | .412 | .456 | .475 | .474 | .477 | .478 |
| 2 | 10 | .412 | .477 | .490 | .492 | .493 | .498 | .499 | .505 |
| | 30 | .201 | .304 | .379 | .402 | .410 | .422 | .424 | .429 |
| | 50 | .115 | .192 | .269 | .304 | .320 | .339 | .343 | .347 |
| | 70 | .078 | .136 | .204 | .241 | .261 | .286 | .292 | .295 |
| | 90 | .066 | .117 | .183 | .221 | .246 | .262 | .271 | .273 |
| 3 | 10 | .331 | .426 | .458 | .461 | .462 | .465 | .465 | .477 |
| | 30 | .121 | .202 | .275 | .304 | .316 | .327 | .329 | .337 |
| | 50 | .062 | .109 | .164 | .193 | .209 | .228 | .232 | .238 |
| | 70 | .041 | .073 | .114 | .138 | .154 | .176 | .183 | .188 |
| | 90 | .035 | .062 | .099 | .123 | .141 | .159 | .169 | .173 |
| 4 | 10 | .265 | .372 | .422 | .430 | .433 | .435 | .435 | .456 |
| | 30 | .077 | .137 | .199 | .229 | .243 | .256 | .259 | .272 |
| | 50 | .037 | .069 | .107 | .130 | .144 | .161 | .167 | .175 |
| | 70 | .026 | .046 | .073 | .089 | .101 | .119 | .126 | .132 |
| | 90 | .022 | .039 | .063 | .078 | .090 | .106 | .114 | .120 |
| 6 | 10 | .173 | .281 | .351 | .368 | .373 | .375 | .375 | .422 |
| | 30 | .037 | .073 | .115 | .137 | .151 | .164 | .168 | .189 |
| | 50 | .018 | .036 | .058 | .071 | .080 | .092 | .098 | .110 |
| | 70 | .013 | .026 | .040 | .049 | .056 | .064 | .069 | .078 |
| | 90 | .012 | .023 | .035 | .043 | .048 | .057 | .062 | .070 |
| 8 | 10 | .117 | .207 | .282 | .305 | .314 | .319 | .320 | .393 |
| | 30 | .020 | .042 | .071 | .087 | .098 | .111 | .116 | .143 |
| | 50 | .010 | .021 | .035 | .044 | .050 | .058 | .063 | .078 |
| | 70 | .007 | .016 | .026 | .032 | .036 | .041 | .045 | .055 |
| | 90 | .076 | .014 | .023 | .028 | .031 | .036 | .040 | .049 |
| 10 | 10 | .082 | .153 | .224 | .250 | .262 | .269 | .271 | .368 |
| | 30 | .012 | .026 | .047 | .059 | .068 | .078 | .084 | .114 |
| | 50 | .006 | .014 | .024 | .030 | .034 | .040 | .044 | .060 |
| | 70 | .005 | .011 | .019 | .022 | .025 | .029 | .032 | .043 |
| | 90 | .004 | .010 | .017 | .020 | .023 | .026 | .028 | .038 |

*Source:* IES RP-23-1989; reprinted with permission of Illuminating Engineering Society of North America.

**TABLE C.20 Coefficient of Utilization from Window Without Blinds; Sky Component $E_{xvk}/E_{xhk}$ = 1.75**

| Room Depth/ Window Height | Depth | Window Width/Window Height | | | | | | | |
|---|---|---|---|---|---|---|---|---|---|
| | | .5 | 1 | 2 | 3 | 4 | 6 | 8 | Infinite |
| 1 | 10 | .435 | .457 | .465 | .471 | .474 | .486 | .488 | .489 |
| | 30 | .317 | .407 | .452 | .466 | .471 | .486 | .488 | .489 |
| | 50 | .234 | .343 | .422 | .447 | .456 | .472 | .475 | .476 |
| | 70 | .187 | .297 | .395 | .430 | .445 | .458 | .461 | .462 |
| | 90 | .168 | .276 | .384 | .426 | .444 | .447 | .450 | .451 |
| 2 | 10 | .357 | .412 | .422 | .424 | .424 | .430 | .431 | .436 |
| | 30 | .180 | .271 | .335 | .356 | .363 | .375 | .378 | .381 |
| | 50 | .106 | .177 | .246 | .278 | .293 | .313 | .318 | .321 |
| | 70 | .074 | .130 | .194 | .229 | .249 | .274 | .282 | .284 |
| | 90 | .065 | .116 | .181 | .219 | .244 | .264 | .273 | .276 |
| 3 | 10 | .288 | .369 | .394 | .397 | .397 | .400 | .401 | .411 |
| | 30 | .110 | .183 | .247 | .272 | .282 | .294 | .296 | .304 |
| | 50 | .058 | .104 | .154 | .181 | .196 | .215 | .221 | .226 |
| | 70 | .040 | .072 | .112 | .136 | .152 | .176 | .184 | .188 |
| | 90 | .035 | .063 | .101 | .126 | .144 | .166 | .177 | .182 |
| 4 | 10 | .232 | .324 | .365 | .371 | .373 | .375 | .375 | .394 |
| | 30 | .071 | .127 | .183 | .209 | .222 | .235 | .238 | .250 |
| | 50 | .036 | .067 | .104 | .125 | .139 | .157 | .163 | .171 |
| | 70 | .025 | .046 | .072 | .089 | .101 | .119 | .127 | .134 |
| | 90 | .022 | .041 | .065 | .082 | .095 | .114 | .124 | .130 |
| 6 | 10 | .153 | .247 | .307 | .320 | .324 | .326 | .327 | .367 |
| | 30 | .035 | .070 | .109 | .130 | .143 | .155 | .160 | .180 |
| | 50 | .018 | .036 | .058 | .071 | .080 | .091 | .098 | .110 |
| | 70 | .013 | .026 | .041 | .051 | .058 | .067 | .073 | .082 |
| | 90 | .012 | .023 | .037 | .046 | .052 | .062 | .069 | .078 |
| 8 | 10 | .104 | .184 | .249 | .269 | .276 | .281 | .282 | .346 |
| | 30 | .020 | .042 | .070 | .086 | .096 | .109 | .115 | .141 |
| | 50 | .010 | .022 | .036 | .046 | .052 | .060 | .066 | .081 |
| | 70 | .008 | .017 | .027 | .033 | .038 | .044 | .048 | .059 |
| | 90 | .007 | .015 | .024 | .030 | .034 | .040 | .044 | .054 |
| 10 | 10 | .074 | .138 | .201 | .223 | .233 | .240 | .242 | .328 |
| | 30 | .012 | .027 | .048 | .059 | .067 | .078 | .084 | .114 |
| | 50 | .006 | .014 | .026 | .032 | .036 | .043 | .047 | .064 |
| | 70 | .005 | .011 | .020 | .024 | .027 | .031 | .034 | .046 |
| | 90 | .004 | .010 | .018 | .022 | .024 | .028 | .031 | .042 |

*Source:* IES RP-23-1989; reprinted with permission of Illuminating Engineering Society of North America.

**TABLE C.21  Coefficient of Utilization from Window Without Blinds; Ground Component**

| Room Depth/ Window Height | Depth | Window Width/Window Height | | | | | | | |
|---|---|---|---|---|---|---|---|---|---|
| | | .5 | 1 | 2 | 3 | 4 | 6 | 8 | Infinite |
| 1 | 10 | .105 | .137 | .177 | .197 | .207 | .208 | .210 | .211 |
| | 30 | .116 | .157 | .203 | .225 | .235 | .241 | .243 | .244 |
| | 50 | .110 | .165 | .217 | .241 | .252 | .267 | .269 | .270 |
| | 70 | .101 | .162 | .217 | .243 | .253 | .283 | .285 | .286 |
| | 90 | .091 | .146 | .199 | .230 | .239 | .290 | .292 | .293 |
| 2 | 10 | .095 | .124 | .160 | .178 | .186 | .186 | .189 | .191 |
| | 30 | .082 | .132 | .179 | .201 | .212 | .219 | .222 | .225 |
| | 50 | .062 | .113 | .165 | .189 | .202 | .214 | .218 | .220 |
| | 70 | .051 | .093 | .141 | .165 | .179 | .194 | .198 | .200 |
| | 90 | .045 | .079 | .118 | .140 | .153 | .179 | .183 | .185 |
| 3 | 10 | .088 | .120 | .157 | .175 | .183 | .185 | .163 | .167 |
| | 30 | .059 | .107 | .154 | .176 | .187 | .198 | .193 | .198 |
| | 50 | .039 | .074 | .114 | .134 | .146 | .157 | .163 | .170 |
| | 70 | .031 | .055 | .085 | .101 | .111 | .122 | .127 | .130 |
| | 90 | .028 | .047 | .070 | .083 | .092 | .107 | .113 | .115 |
| 4 | 10 | .073 | .113 | .154 | .174 | .183 | .187 | .176 | .184 |
| | 30 | .040 | .082 | .127 | .148 | .159 | .170 | .177 | .185 |
| | 50 | .025 | .049 | .078 | .094 | .103 | .113 | .117 | .123 |
| | 70 | .020 | .036 | .054 | .065 | .071 | .079 | .083 | .087 |
| | 90 | .019 | .032 | .046 | .054 | .060 | .069 | .073 | .076 |
| 6 | 10 | .056 | .106 | .143 | .164 | .175 | .184 | .173 | .194 |
| | 30 | .021 | .050 | .081 | .098 | .107 | .117 | .123 | .138 |
| | 50 | .013 | .027 | .041 | .049 | .054 | .060 | .064 | .072 |
| | 70 | .011 | .021 | .029 | .033 | .035 | .039 | .041 | .046 |
| | 90 | .011 | .020 | .026 | .030 | .032 | .035 | .037 | .042 |
| 8 | 10 | .036 | .082 | .122 | .143 | .156 | .166 | .170 | .208 |
| | 30 | .011 | .029 | .050 | .062 | .070 | .078 | .082 | .101 |
| | 50 | .007 | .016 | .024 | .028 | .031 | .035 | .038 | .046 |
| | 70 | .006 | .013 | .018 | .020 | .021 | .023 | .025 | .030 |
| | 90 | .006 | .013 | .017 | .019 | .020 | .022 | .023 | .028 |
| 10 | 10 | .024 | .061 | .109 | .120 | .131 | .144 | .147 | .200 |
| | 30 | .006 | .017 | .034 | .040 | .046 | .053 | .056 | .076 |
| | 50 | .004 | .010 | .016 | .018 | .020 | .023 | .024 | .033 |
| | 70 | .004 | .009 | .013 | .014 | .015 | .016 | .016 | .022 |
| | 90 | .004 | .009 | .013 | .013 | .014 | .015 | .016 | .021 |

*Source:* IES RP-23-1989; reprinted with permission of Illuminating Engineering Society of North America.

**Table C.22 Reflectances of Daylighting Modeling Materials**

| Materials | Number | Reflectance (%) |
|---|---|---|
| **Crescent Board Color** | | |
| Raven Black | 989 | 6.7 |
| Newport Blue | 977 | 9.1 |
| Wine | 907A | 12.6 |
| Williamsburg Green | 988 | 12.7 |
| Volcano Blue | 1081 | 15.5 |
| Russet | 996 | 15.7 |
| Las Cruces Purple | 1076 | 16.1 |
| Marine Blue | 1082 | 16.5 |
| Madeira Red | 1075 | 19.4 |
| Avocado | 1084 | 21.6 |
| Persimmon | 1087 | 26.4 |
| Gibraltar Grey | 1074 | 28.5 |
| Burnt Orange | 1077 | 33.5 |
| Moss Point Green | 1001 | 35.0 |
| Bar Harbor Grey | 976 | 38.5 |
| Bimini Blue | 1080 | 43.3 |
| Mist Grey | 1002 | 43.5 |
| Suntan | 1062 | 45.8 |
| Stone Grey | 975 | 48.5 |
| French Blue | 972 | 54.2 |
| Cameo Rose | 973 | 56.0 |
| Biscay Blue | 1073 | 56.0 |
| Sauterne | 1089 | 59.4 |
| Lime | 910A | 59.7 |
| Sandstone | 1061 | 61.0 |
| Madagascar Pink | 1078 | 73.7 |
| Mist | 1088 | 76.3 |
| Daffodil | 971 | 88.8 |
| Yellow | 902A | 93.1 |
| **Mat Board** | | |
| Artic White | 3297 | 91.4 |
| **Linen Board** | | |
| Cream Linen | 2961A | 91.0 |
| French Grey | 2962A | 76.0 |
| **Museum Board** | | |
| 2-ply White | 1150 | 96.5 |
| 2-ply Antique | 1157 | 91.5 |
| 2-ply Cream | 1152 | 94.2 |

*Source:* Technical Reference Sheet; The Lighting Design Lab, Seattle, WA.

# Solar Geometry

This appendix provides information on various aspects of solar geometry, including three sets of sun-path diagrams (sun peg charts, horizontal projection charts, and vertical projection charts).

*Solar Altitude and Azimuth Data for 30, 34, 38, 42, 44, and 48°N Latitudes (Table D.1)*

*Sun Peg Charts for 28, 32, 36, 40, 44, 48, and 52°N Latitudes (Figure D.1)*

*Horizontal Projection (Equidistant) Sunpath Charts for 24, 28, 32, 36, 40, 44, 48, and 52°N Latitudes (Figure D.2)*

## D.1 SOLAR ALTITUDE AND AZIMUTH DATA FOR 30, 34, 38, 42, 44, AND 48°N LATITUDES

**TABLE D.1 Typical Solar Altitude and Azimuth Data as a Function of Date and Time of Day**

| Latitude (°N) | | Date | Solar Time[a] AM: 6 PM: 6 | 7 5 | 8 4 | 9 3 | 10 2 | 11 1 | Noon |
|---|---|---|---|---|---|---|---|---|---|
| 30 | Altitude | June 21 | 12 | 24 | 37 | 50 | 63 | 75 | 83 |
| | | Mar.–Sept. 21 | — | 13 | 26 | 38 | 49 | 57 | 60 |
| | | Dec. 21 | — | — | 12 | 21 | 29 | 35 | 37 |
| | Azimuth | June 21 | 111 | 104 | 99 | 92 | 84 | 67 | 0 |
| | | Mar.–Sept. 21 | 90 | 83 | 74 | 64 | 49 | 28 | 0 |
| | | Dec. 21 | — | 60 | 54 | 44 | 32 | 17 | 0 |
| 34 | Altitude | June 21 | 13 | 25 | 37 | 50 | 62 | 74 | 79 |
| | | Mar.–Sept. 21 | — | 12 | 25 | 36 | 46 | 53 | 56 |
| | | Dec. 21 | — | — | 9 | 18 | 26 | 31 | 33 |
| | Azimuth | June 21 | 110 | 103 | 95 | 90 | 78 | 58 | 0 |
| | | Mar.–Sept. 21 | 90 | 82 | 72 | 61 | 46 | 26 | 0 |
| | | Dec. 21 | — | — | 54 | 43 | 30 | 16 | 0 |
| 38 | Altitude | June 21 | 14 | 26 | 37 | 49 | 61 | 71 | 75 |
| | | Mar.–Sept. 21 | — | 12 | 23 | 34 | 43 | 50 | 52 |
| | | Dec. 21 | — | — | 7 | 16 | 23 | 27 | 28 |
| | Azimuth | June 21 | 109 | 101 | 90 | 83 | 70 | 46 | 0 |
| | | Mar.–Sept. 21 | 90 | 81 | 71 | 58 | 43 | 24 | 0 |
| | | Dec. 21 | — | — | 54 | 43 | 30 | 16 | 0 |
| 42 | Altitude | June 21 | 16 | 26 | 38 | 49 | 60 | 68 | 71 |
| | | Mar.–Sept. 21 | — | 11 | 22 | 32 | 40 | 46 | 48 |
| | | Dec. 21 | — | — | 4 | 13 | 19 | 23 | 25 |
| | Azimuth | June 21 | 108 | 99 | 89 | 78 | 63 | 39 | 0 |
| | | Mar.–Sept. 21 | 90 | 80 | 69 | 56 | 41 | 22 | 0 |
| | | Dec. 21 | — | — | 53 | 42 | 29 | 15 | 0 |
| 46 | Altitude | June 21 | 17 | 27 | 37 | 48 | 57 | 65 | 67 |
| | | Mar.–Sept. 21 | — | 10 | 20 | 30 | 37 | 42 | 44 |
| | | Dec. 21 | — | — | 2 | 10 | 15 | 20 | 21 |
| | Azimuth | June 21 | 107 | 97 | 88 | 74 | 58 | 34 | 0 |
| | | Mar.–Sept. 21 | 90 | 79 | 67 | 54 | 39 | 21 | 0 |
| | | Dec. 21 | — | — | 52 | 41 | 28 | 14 | 0 |
| 48 | Altitude | June 21 | 17 | 27 | 37 | 47 | 56 | 63 | 65 |
| | | Mar.–Sept. 21 | — | 10 | 20 | 29 | 36 | 40 | 42 |
| | | Dec. 21 | — | — | 1 | 8 | 14 | 17 | 19 |
| | Azimuth | June 21 | 106 | 95 | 85 | 72 | 55 | 31 | 0 |
| | | Mar.–Sept. 21 | 90 | 79 | 67 | 53 | 38 | 20 | 0 |
| | | Dec. 21 | — | — | 52 | 41 | 28 | 14 | 0 |

[a]Solar time and clock time do not usually coincide. They are related by the equation of time (see Section 6.2).

## D.2 SUN PEG CHARTS FOR 28, 32, 36, 40, 44, 48, AND 52°N LATITUDES

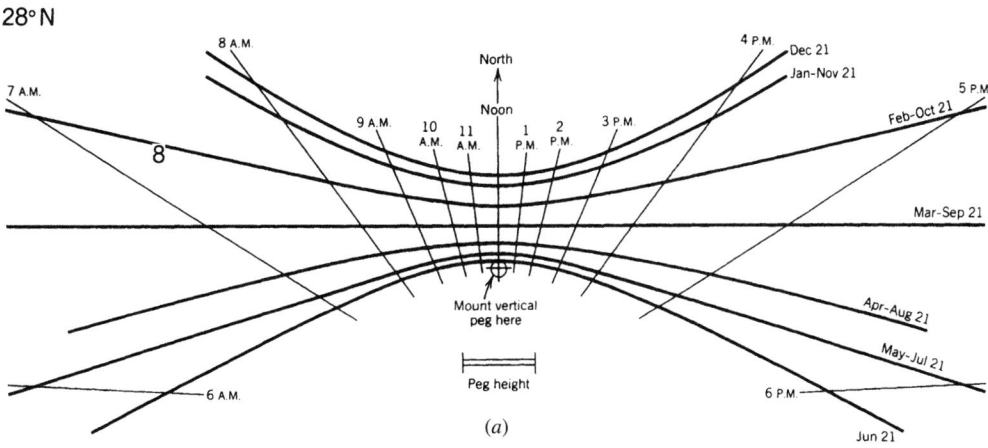

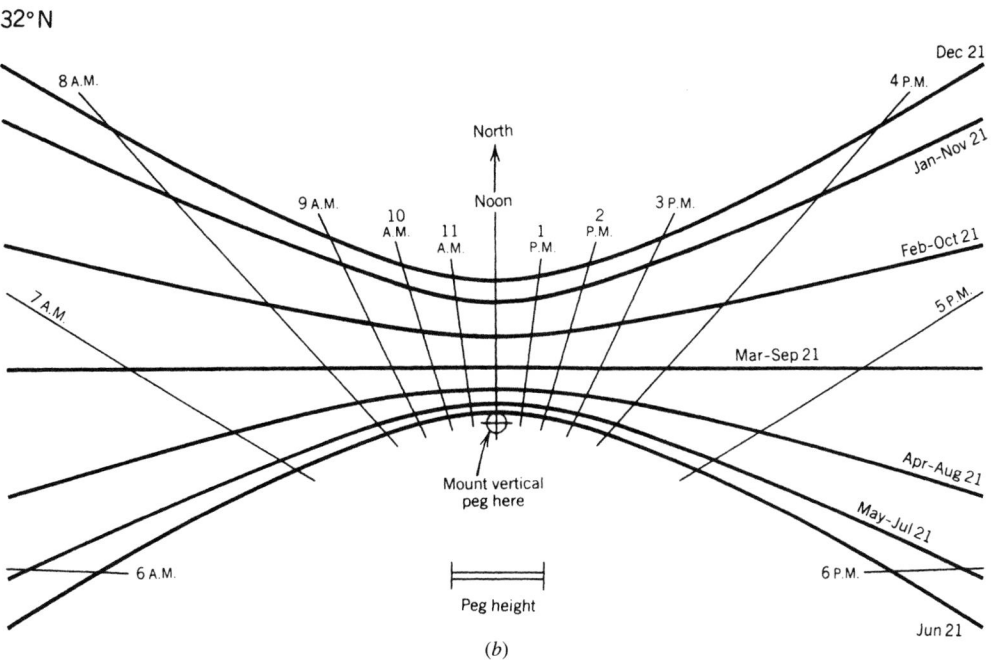

**Fig. D.1** *(a–g) Sun peg shadow charts. These charts show the exact positions of solar radiation and shadow on a model of any scale, on any date, at any time of day between shortly after sunrise and shortly before sunset. Latitudes as indicated in the upper-left-hand corner of each chart. See Chapter 6 for instructions on the use of these charts.*

36° N

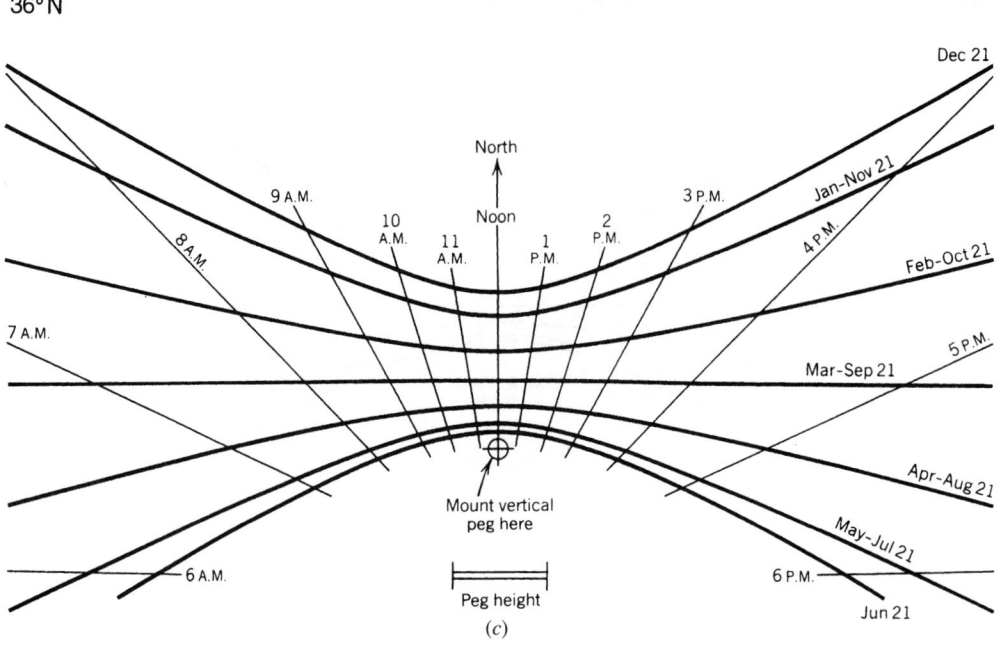

(c)

40° N

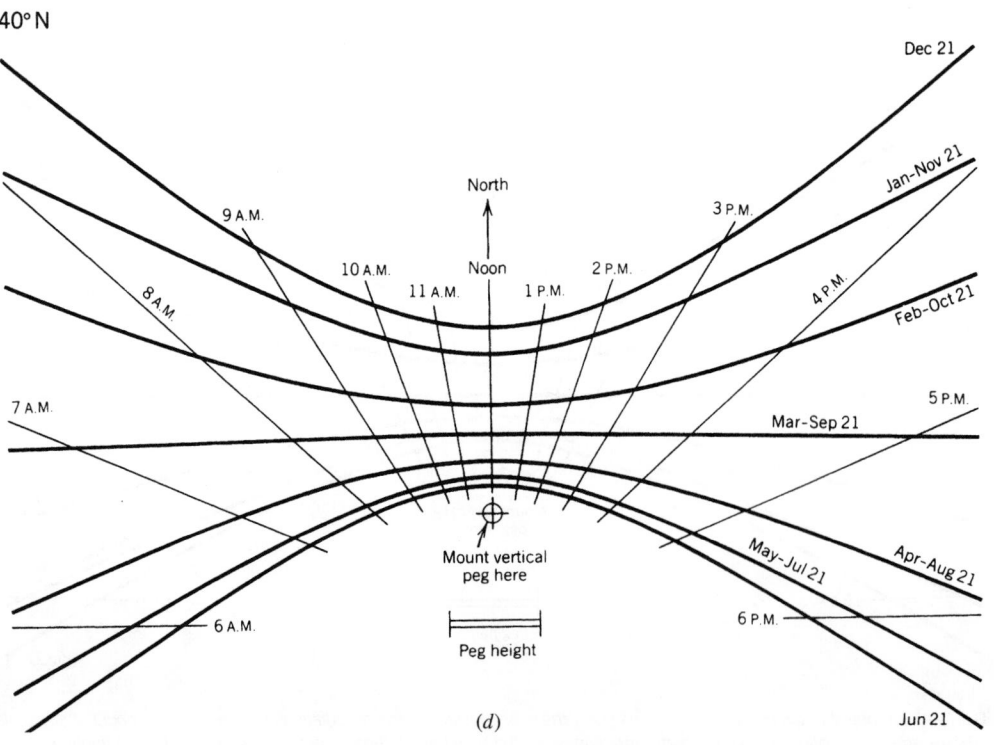

(d)

**Fig. D.1** (continued)

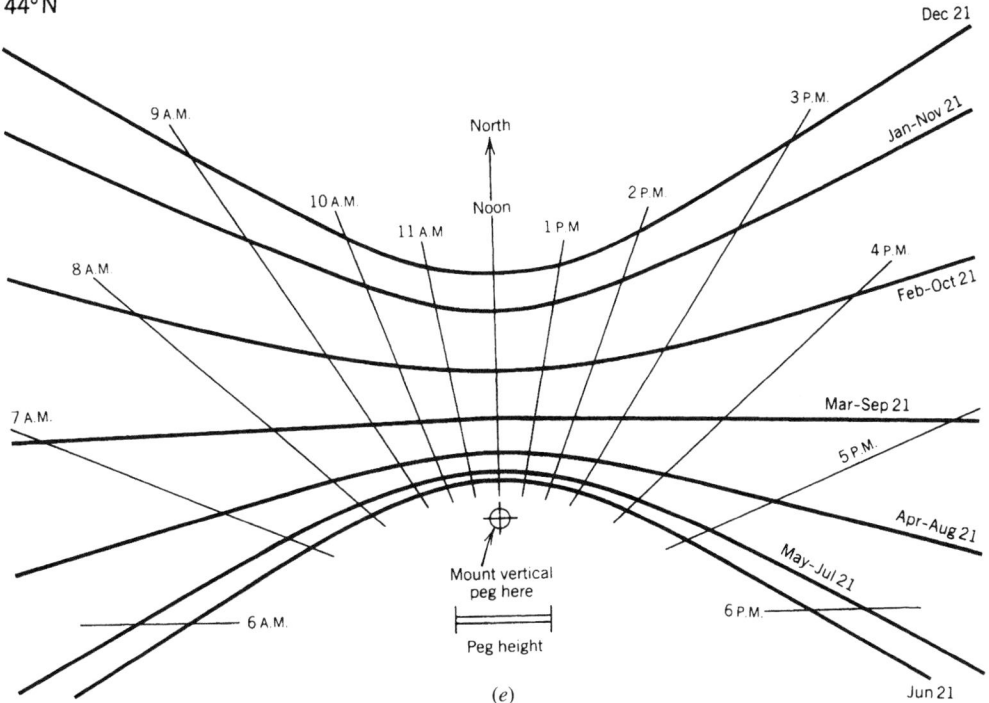

44° N

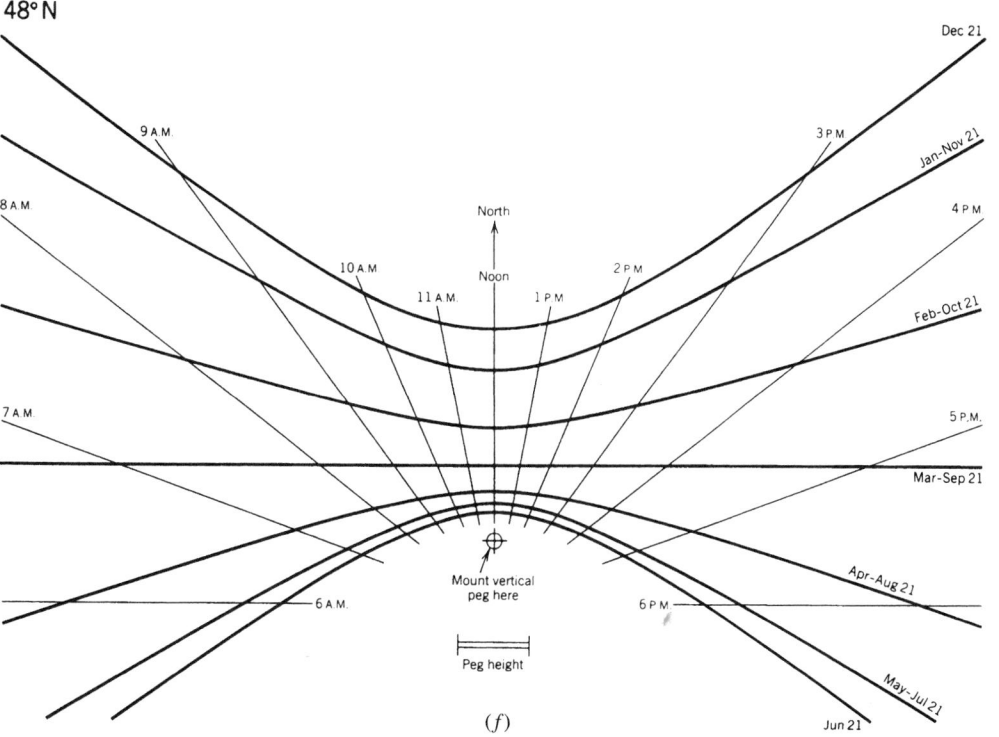

48° N

**Fig. D.1** (continued)

(g)

*Fig. D.1* (continued)

## D.3 HORIZONTAL PROJECTION (EQUIDISTANT) SUNPATH CHARTS FOR 24, 28, 32, 36, 40, 44, 48, AND 52°N LATITUDES

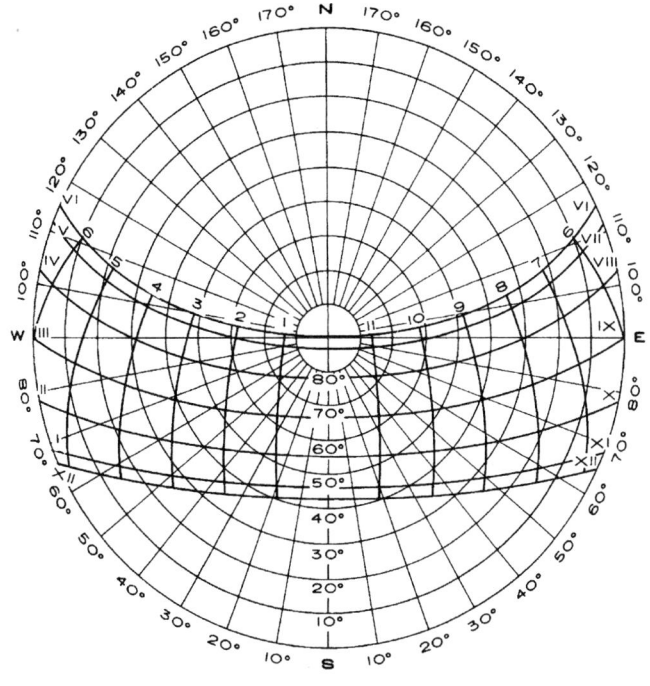

(*a*) **24°N LATITUDE**

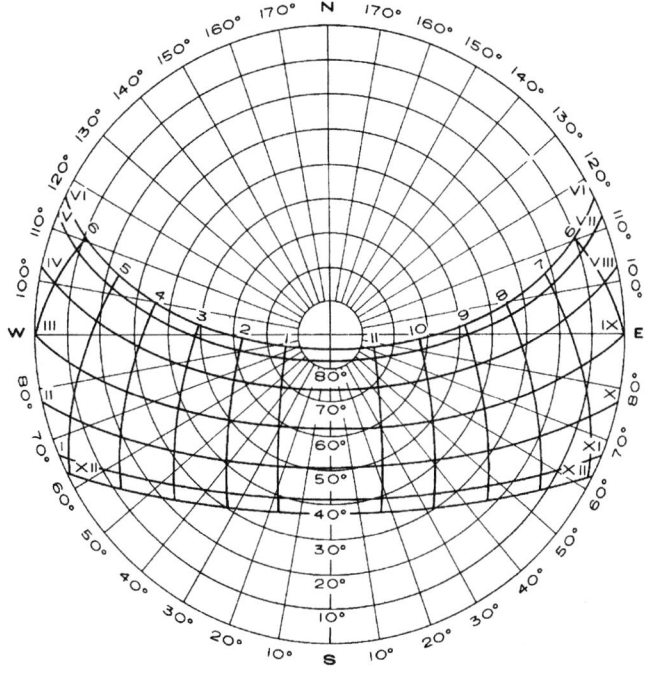

(*b*) **28°N LATITUDE**

***Fig. D.2*** *(a–h) Set of horizontal projection (equidistant) sunpath charts at 4° latitude intervals. Latitudes as indicated below each chart. (Reprinted with permission from* Architectural Graphic Standards, *8th ed., 1988, John Wiley & Sons, Publishers.)*

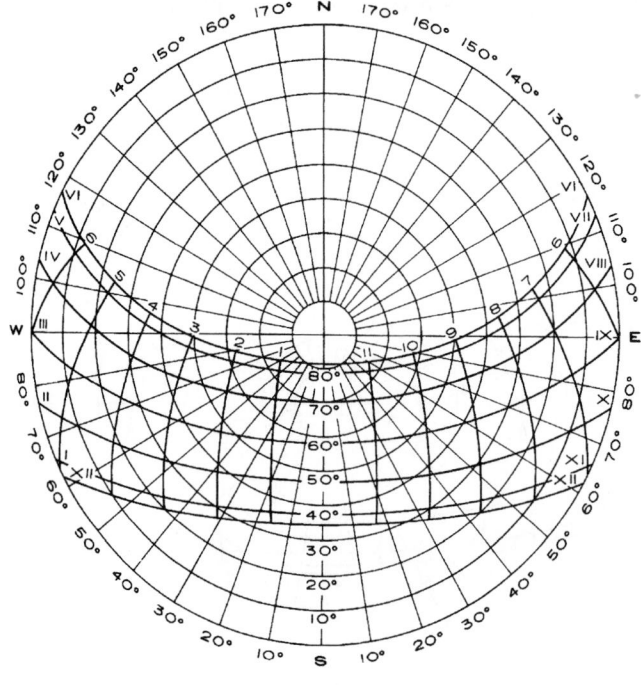

(c)   **32°N LATITUDE**

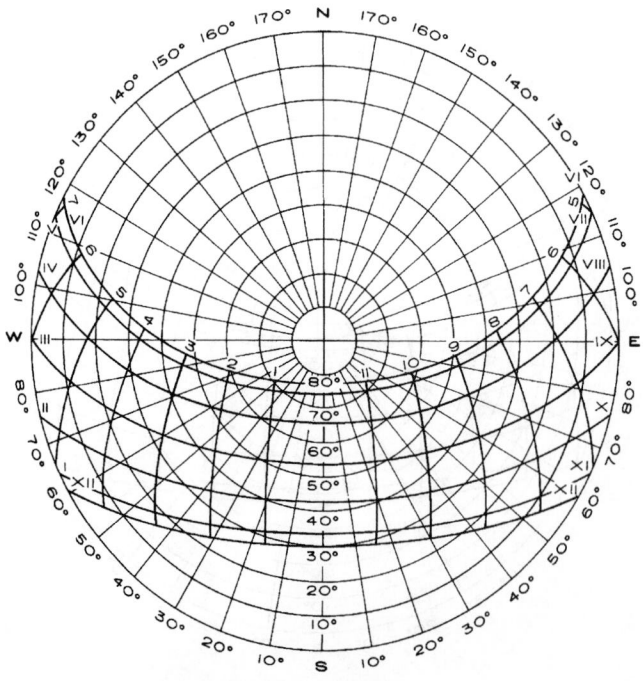

(d)   **36°N LATITUDE**

**Fig. D.2** (continued)

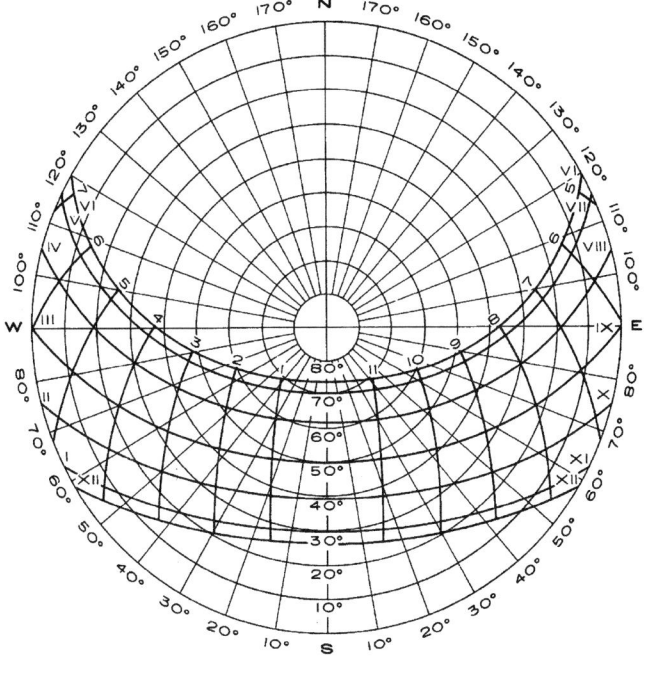

(e)  **40°N LATITUDE**

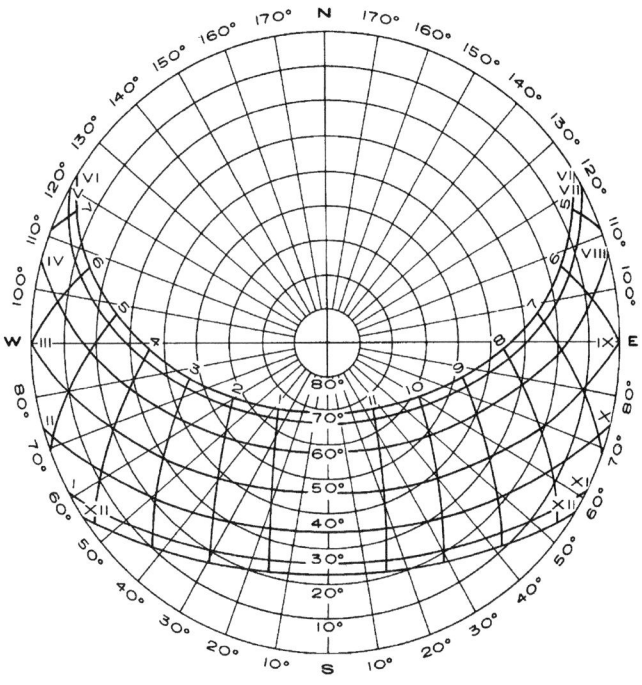

(f)  **44°N LATITUDE**

**Fig. D.2** (continued)

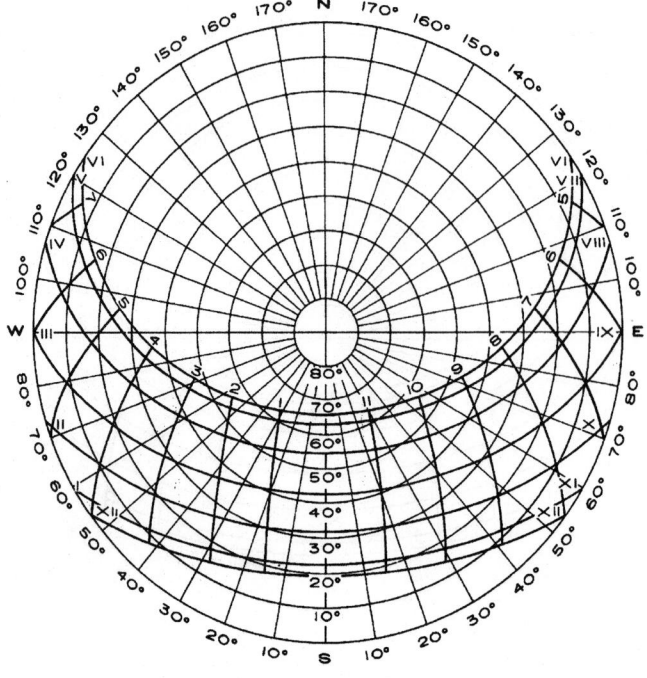

(*g*) **48°N LATITUDE**

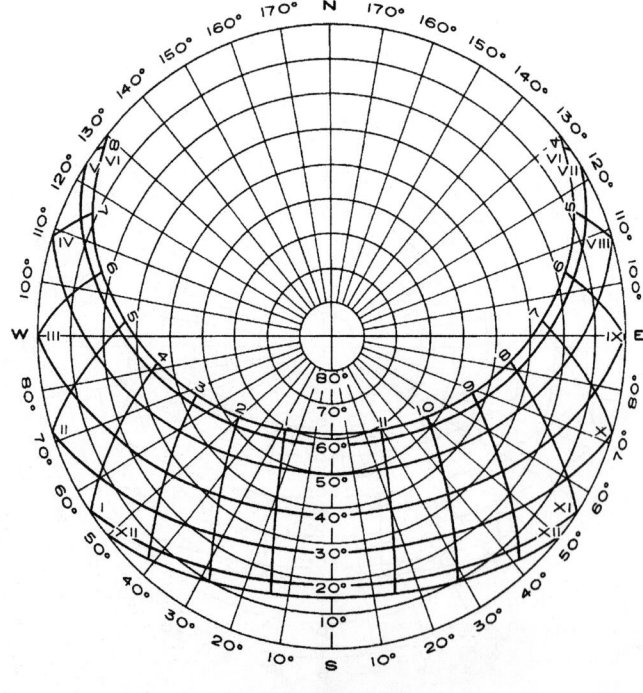

(*h*) **52°N LATITUDE**

***Fig. D.2*** (continued)

## D.4 VERTICAL PROJECTION SUNPATH CHARTS FOR 28, 32, 36, 40, 44, 48, 52, AND 56°N LATITUDES

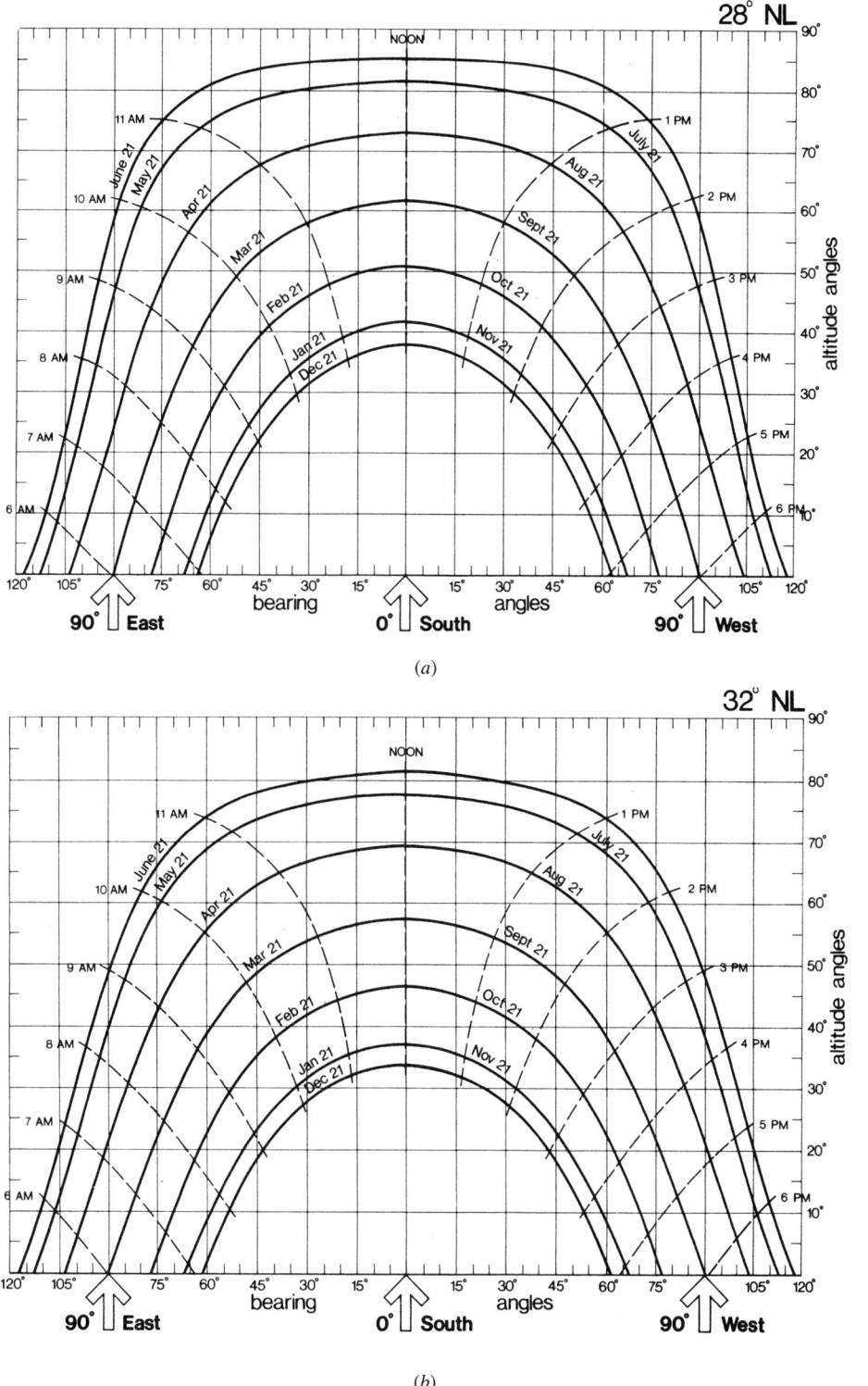

**Fig. D.3** *(a–h) Set of vertical projection sunpath charts at 4° latitude intervals. Latitudes as indicated in the upper right-hand corner of each chart. (© Edward Mazria,* The Passive Solar Energy Book; *used with permission.)*

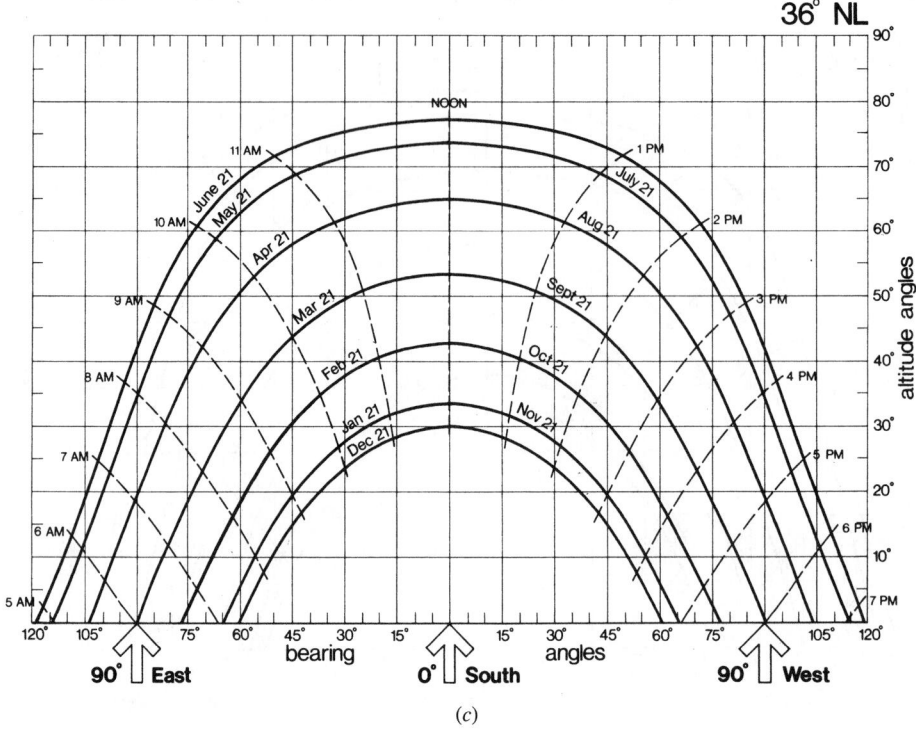

(c)

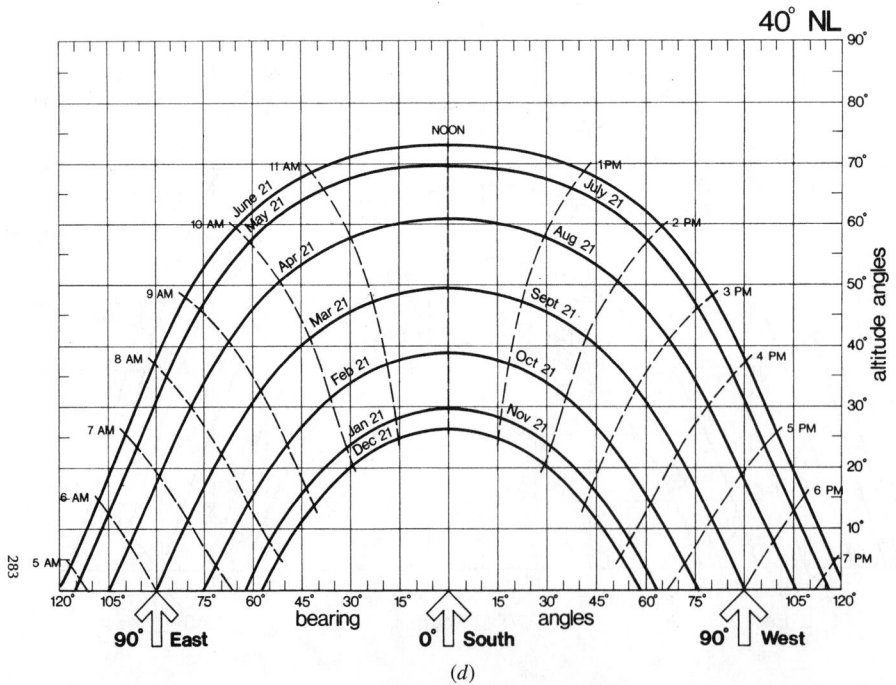

(d)

**Fig. D.3** (continued)

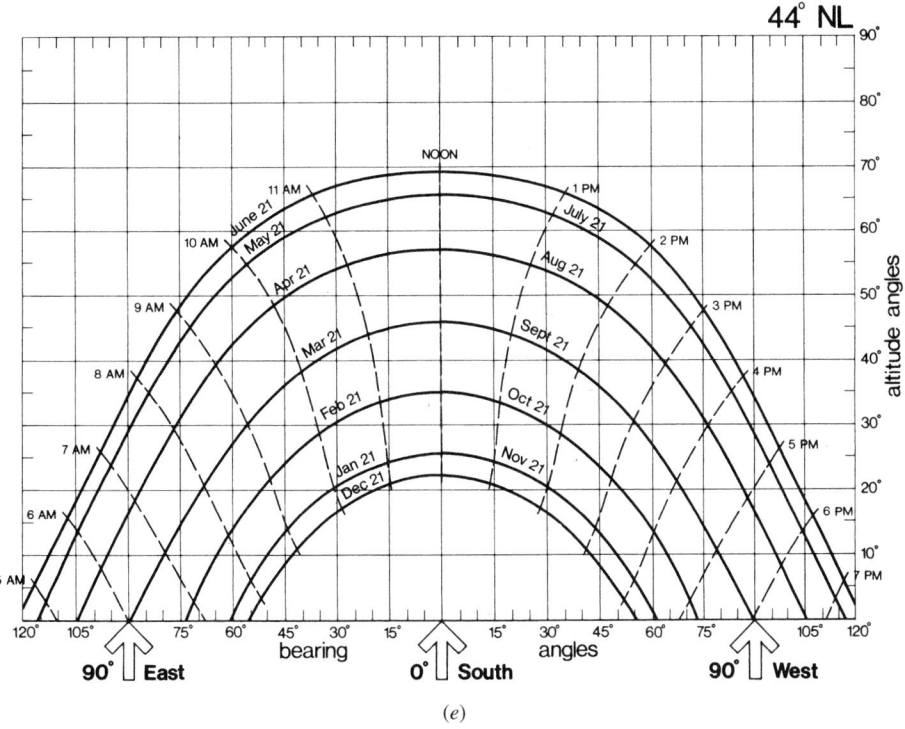

(e)

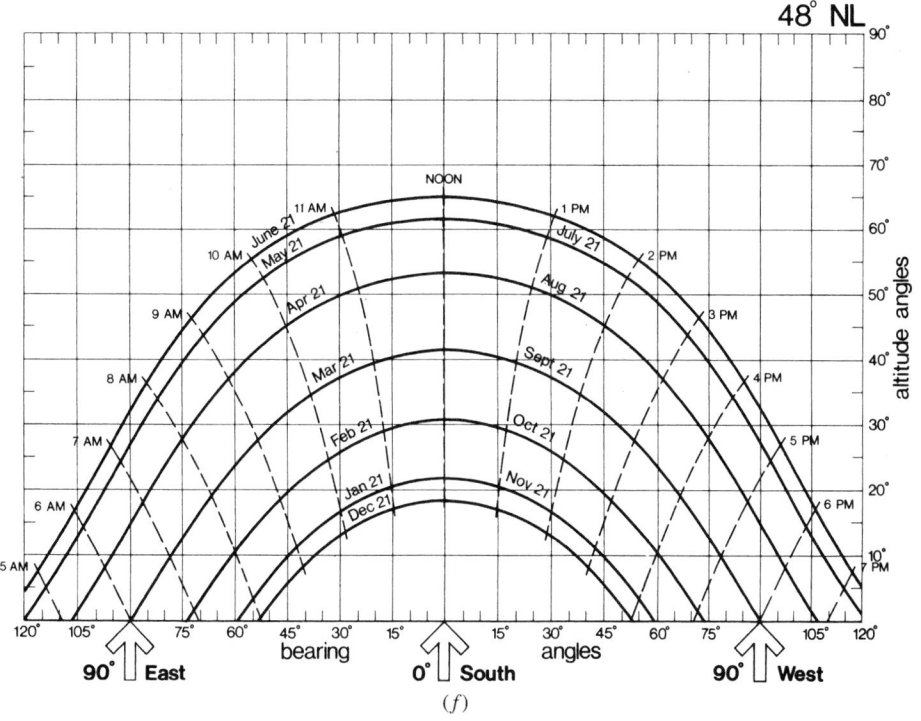

(f)

**Fig. D.3** (continued)

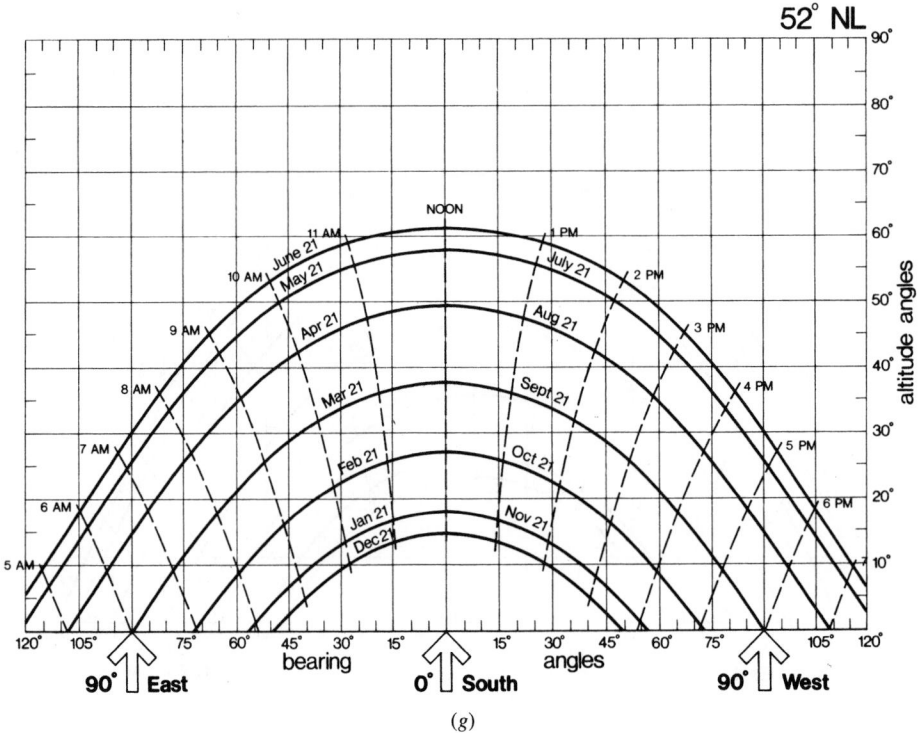

(g)

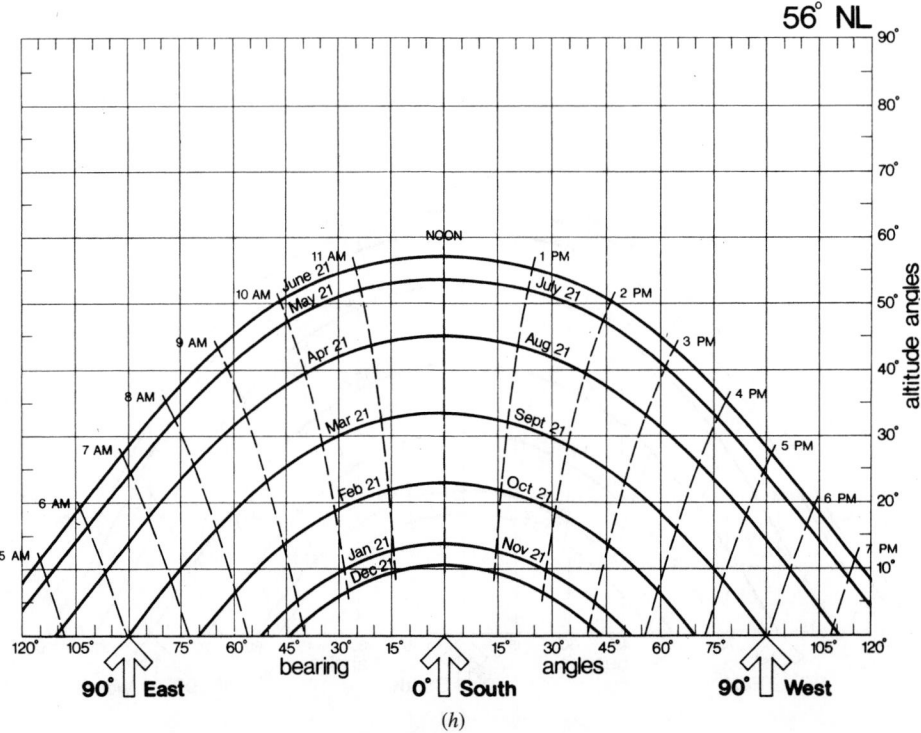

(h)

**Fig. D.3** (continued)

# Thermal Properties of Materials and Assemblies

The tables in this appendix provide information on thermal (and occasionally visual) properties for a range of generic building materials and assemblies. The information provided herein would typically be used early in the design process before specific products have been selected and/or to determine the properties of assemblies that are fabricated from generic materials (such as gypsum wallboard, bricks, and concrete masonry units). When specific products (such as insulations, windows, or skylights) have been selected for use, data for those specific materials and assemblies should be obtained from appropriate manufacturers' resources (catalogs or WWW sites) rather than from these tables.

**APPENDICES**

**TABLE E.1 Thermal Properties of Conventional Building and Insulating Materials[a]**

| Description | Density lb/ft³ (kg/m³) | Conductivity[b] (k) Btu-in / h ft² °F (W/m K) | Conductance (C) Btu / h ft² °F (W/m² K) | I-P Resistance[c] (R) Per Inch Thickness (1/k) °F ft² h / Btu-in | For Thickness Listed (1/C) °F ft² h / Btu | Specific Heat Btu / lb °F (kJ/kg K) | SI Resistance[c] (R) Per Meter Thickness (1/k) K m / W | For Thickness Listed (1/C) K m² / W |
|---|---|---|---|---|---|---|---|---|
| **Building Board** | | | | | | | | |
| Asbestos-cement board | 120 (1900) | 4.0 (0.58) | — | 0.25 | — | 0.24 (1.00) | 1.73 | — |
| Asbestos-cement board 0.125 in. 3.2 mm | 120 (1900) | — | 33.00 (187.4) | — | 0.03 | — | — | 0.005 |
| Asbestos-cement board 0.25 in. 6.4 mm | 120 (1900) | — | 16.50 (93.7) | — | 0.06 | — | — | 0.011 |
| Gypsum or plaster board 0.375 in. 9.5 mm | 50 (800) | — | 3.10 (17.6) | — | 0.32 | 0.26 (1.09) | — | 0.056 |
| Gypsum or plaster board 0.5 in. 12.7 mm | 50 (800) | — | 2.22 (12.6) | — | 0.45 | — | — | 0.079 |
| Gypsum or plaster board 0.625 in. 15.9 mm | 50 (800) | — | 1.78 (10.1) | — | 0.56 | — | — | 0.099 |
| Plywood (Douglas Fir)[d] | 34 (540) | 0.80 (0.12) | — | 1.25 | — | 0.29 (1.21) | 8.66 | — |
| Plywood (Douglas Fir) 0.25 in. 6.4 mm | 34 (540) | — | 3.20 (18.2) | — | 0.31 | — | — | 0.055 |
| Plywood (Douglas Fir) 0.375 in. 9.5 mm | 34 (540) | — | 2.13 (12.1) | — | 0.47 | — | — | 0.083 |
| Plywood (Douglas Fir) 0.5 in. 12.7 mm | 34 (540) | — | 1.60 (9.1) | — | 0.62 | — | — | 0.11 |
| Plywood (Douglas Fir) 0.625 in. 15.9 mm | 34 (540) | — | 1.29 (7.3) | — | 0.77 | — | — | 0.14 |

**TABLE E.1 Thermal Properties of Conventional Building and Insulating Materials[a] (Continued)**

| Description | Density lb/ft³ (kg/m³) | Conductivity[b] (k) Btu-in / h ft² °F (W/m K) | Conductance (C) Btu / h ft² °F (W/m² K) | I-P Resistance[c] (R) Per Inch Thickness (1/k) °F ft² h / Btu-in | I-P Resistance[c] (R) For Thickness Listed (1/C) °F ft² h / Btu | Specific Heat Btu / lb °F (kJ/kg K) | SI Resistance[c] (R) Per Meter Thickness (1/k) K m / W | SI Resistance[c] (R) For Thickness Listed (1/C) K m² / W |
|---|---|---|---|---|---|---|---|---|
| Plywood or wood panels | 0.75 in. 19.0 mm | 34 (540) | — | 1.07 (6.1) | — | 0.93 | 0.29 (1.21) | — | 0.16 |
| Vegetable fiber board | | | | | | | | | |
| Sheathing, regular density[e] | 0.5 in. 12.7 mm | 18 (290) | — | 0.76 (4.3) | — | 1.32 | 0.31 (1.3) | — | 0.23 |
| | 0.78125 in. 19.8 mm | 18 (290) | — | 0.49 (2.8) | — | 2.06 | | — | 0.36 |
| Sheathing intermediate density[e] | 0.5 in. 12.7 mm | 22 (350) | — | 0.92 (5.2) | — | 1.09 | 0.31 (1.30) | — | 0.19 |
| Nail-base sheathing[e] | 0.5 in. 12.7 mm | 25 (400) | — | 0.94 (5.3) | — | 1.06 | 0.31 (1.30) | — | 0.19 |
| Shingle backer | 0.375 in. 9.5 mm | 18 (290) | — | 1.06 (6.0) | — | 0.94 | 0.31 (1.30) | — | 0.17 |
| Shingle backer | 0.3125 in. 7.9 mm | 18 (290) | — | 1.28 (7.3) | — | 0.78 | | — | 0.14 |
| Sound deadening board | 0.5 in. 12.7 mm | 15 (240) | — | 0.74 (4.2) | — | 1.35 | 0.30 (1.26) | — | 0.24 |
| Tile and lay-in panels, plain or acoustic | | 18 (290) | 0.40 (0.058) | — | 2.50 (17.0) | — | 0.14 (0.59) | 17.0 | — |
| | 0.5 in. 12.7 mm | 18 (290) | — | 0.80 (4.5) | — | 1.25 | | — | 0.22 |
| | 0.75 in. 19.0 mm | 18 (290) | — | 0.53 (3.0) | — | 1.89 | | — | 0.33 |

Continuation of building materials thermal properties table (column headings appear on the preceding page).

| Description | Density lb/ft³ (kg/m³) | Conductivity k | Conductance C | Resistance R, per inch 1/k | Resistance R, for thickness 1/C | Specific Heat | |
|---|---|---|---|---|---|---|---|
| Laminated paperboard | 30 (480) | 0.50 (0.072) | — | 2.00 (13.9) | — | 0.33 (1.38) | — |
| Homogeneous board from repulped paper | 30 (480) | 0.50 (0.072) | — | 2.00 (13.9) | — | 0.28 (1.17) | — |
| Hardboard[e] | | | | | | | |
| Medium density | 50 (800) | 0.73 (0.105) | — | 1.37 (9.50) | — | 0.31 (1.30) | — |
| High density, service-tempered grade and service grade | 55 (880) | 0.82 (0.118) | — | 1.22 (8.46) | — | 0.32 (1.34) | — |
| High density, standard-tempered grade | 63 (1010) | 1.00 (0.144) | — | 1.00 (6.93) | — | 0.32 (1.34) | — |
| Particleboard[e] | | | | | | | |
| Low density | 37 (590) | 0.71 (0.102) | — | 1.41 (9.77) | — | 0.31 (1.30) | — |
| Medium density | 50 (800) | 0.94 (0.135) | — | 1.06 (7.35) | — | 0.31 (1.30) | — |
| High density<br>0.625 in.<br>15.9 mm | 62 (1000) | 0.5 (0.17) | 1.18 | — (5.90) | 0.85 | 1.30 | — |
| Underlayment<br>0.75 in<br>19.0 mm | 40 (640) | — | 1.22 (6.9) | — | 0.82 | 0.29 (1.21) | — |
| Waferboard | 37 (590) | 0.63 | — | 1.59 (11.0) | — | — | 0.14 |
| Wood subfloor | — | — | 1.06 (6.0) | — | 0.94 | 0.33 (1.38) | — |
| **Building Membrane** | | | | | | | |
| Vapor—permeable felt | — | — | 16.70 (94.9) | — | 0.06 (0.011) | — | — |
| Vapor—seal, 2 layers of mopped 15-lb (0.73 kg/m²) felt | — | — | 8.35 (47.4) | — | 0.12 (0.21) | — | — |
| Vapor—seal, plastic film | — | — | Negl. | — | Negl. (Negl.) | — | 0.17 |

**TABLE E.1 Thermal Properties of Conventional Building and Insulating Materials[a]** *(Continued)*

| Description | Density lb/ft³ (kg/m³) | Conductivity[b] (k) Btu-in / h ft² °F (W/m K) | Conductance (C) Btu / h ft² °F (W/m² K) | I-P Resistance[c] (R) Per Inch Thickness (1/k) °F ft² h / Btu-in | I-P Resistance[c] (R) For Thickness Listed (1/C) °F ft² h / Btu | Specific Heat Btu / lb °F (kJ/kg K) | SI Resistance[c] (R) Per Meter Thickness (1/k) K m / W | SI Resistance[c] (R) For Thickness Listed (1/C) K m² / W |
|---|---|---|---|---|---|---|---|---|
| **Finish Flooring Materials** | | | | | | | | |
| Carpet and fibrous pad | — | 0.48 (2.73) | — | 2.08 | 0.34 (1.42) | — | 0.37 | |
| Carpet and rubber pad | — | 0.81 (4.60) | — | 1.23 | 0.33 (1.38) | — | 0.22 | |
| Cork tile 0.125 in. 3.2 mm | — | — | 3.60 (20.4) | — | 0.28 | 0.48 (2.01) | — | 0.049 |
| Terrazzo 1 in. 25.0 mm | — | — | 12.50 (71.0) | — | 0.08 | 0.19 (0.80) | — | 0.014 |
| Tile—asphalt, linoleum, vinyl, rubber | — | — | 20.00 (113.6) | — | 0.05 | 0.30 (1.26) | — | 0.009 |
| Vinyl asbestos | | | | | | 0.24 (1.01) | | |
| Ceramic | | | | | | 0.19 (0.80) | | |
| Wood, hardwood finish 0.75 in. 19 mm | — | — (8.35) | 1.47 | — | 0.68 | — | — | 0.12 |
| **Insulating Materials** | | | | | | | | |
| ***Blanket and Batt[f,g]*** | | | | | | | | |
| Mineral fiber, fibrous form processed from rock, slag, or glass | | | | | | | | |
| Approx. 3–4 in. 75–100 mm | 0.4–2.0 (6.4–32.0) | — | 0.091 (0.52) | — | 11 | — | — | 1.94 |

| Description | Density, lb/ft³ (kg/m³) | Conductivity k | Conductance C | Resistance (1/k) | Resistance (1/C) | Specific heat | Resistance (SI) | Resistance (SI) |
|---|---|---|---|---|---|---|---|---|
| Approx. 3.5 in. 90 mm | 0.4–2.0 (6.4–32.0) | — | 0.077 (0.44) | — | 13 | — | — | 2.29 |
| Approx. 3.5 in. 90 mm | 1.2–1.6 (19.0–26.0) | — | 0.067 (0.38) | — | 15 | — | — | 2.63 |
| Approx. 5.5–6.5 in. 140–165 mm | 0.4–2.0 (6.4–32.0) | — | 0.053 (0.30) | — | 19 | — | — | 3.32 |
| Approx. 5.5 in. 140 mm | 0.6–1.0 (10.0–16.0) | — | 0.048 (0.27) | — | 21 | — | — | 3.67 |
| Approx. 6–7.5 in. 150–190 mm | 0.4–2.0 (6.4–32.0) | — | 0.045 (0.26) | — | 22 | — | — | 3.91 |
| Approx. 8.25–10 in. 210–250 mm | 0.4–2.0 (6.4–32.0) | — | 0.033 (0.19) | — | 30 | — | — | 5.34 |
| Approx. 10–13 in. 250–330 mm | 0.4–2.0 (6.4–32.0) | — | 0.026 (0.15) | — | 38 | — | — | 6.77 |
| *Board and Slabs* | | | | | | | | |
| Cellular glass | 8.0 (136) | 0.33 (0.050) | — | 3.03 | — | 0.18 (0.75) | 19.8 | — |
| Glass fiber, organic bonded | 4.0–9.0 (64–140) | 0.25 (0.036) | — | 4.00 | — | 0.23 (0.96) | 27.7 | — |
| Expanded perlite, organic bonded | 1.0 (16) | 0.36 (0.052) | — | 2.78 | — | 0.30 (1.26) | 19.3 | — |
| Expanded rubber (rigid) | 4.5 (72) | 0.22 (0.032) | — | 4.55 | — | 0.40 (1.68) | 31.6 | — |
| Expanded polystyrene, extruded (smooth surface)(CFC-12 exp.) | 1.8–3.5 (29–56) | 0.20 (0.029) | — | 5.00 | — | 0.29 (1.21) | 34.7 | — |
| Expanded polystyrene, extruded (smooth skin surface) (HCFC-142b exp.)[h] | 1.8–3.5 (29–56) | 0.20 (0.029) | — | 5.00 | — | 0.29 (1.21) | 34.7 | — |
| Expanded polystyrene, molded beads | 1.0 (16) | 0.26 (0.037) | — | 3.85 | — | — | 26.7 | — |

**TABLE E.1 Thermal Properties of Conventional Building and Insulating Materials[a] (Continued)**

| Description | Density lb/ft³ (kg/m³) | Conductivity[b] (k) Btu-in / h ft² °F (W/m K) | Conductance (C) Btu / h ft² °F (W/m² K) | I-P Resistance[c] (R) Per Inch Thickness (1/k) °F ft² h / Btu-in | I-P Resistance[c] (R) For Thickness Listed (1/C) °F ft² h / Btu | Specific Heat Btu / lb °F (kJ/kg K) | SI Resistance[c] (R) Per Meter Thickness (1/k) K m / W | SI Resistance[c] (R) For Thickness Listed (1/C) K m² / W |
|---|---|---|---|---|---|---|---|---|
| Expanded polystyrene, molded beads | 1.25 (20) | 0.25 (0.036) | — | 4.00 | — | — | 27.7 | — |
|  | 1.5 (24) | 0.24 (0.035) | — | 4.17 | — | — | 28.9 | — |
|  | 1.75 (28) | 0.24 (0.035) | — | 4.17 | — | — | 28.9 | — |
|  | 2.0 (32) | 0.23 (0.033) | — | 4.35 | — | — | 30.2 | — |
| Cellular polyurethane/polyisocyanurate[i] (CFC-11 exp.) (unfaced) | 1.5 (24) | 0.16–0.18 (0.023–0.026) | — | 6.25–5.56 | — | 0.38 (1.59) | 43.3–38.5 | — |
| Cellular polyisocyanurate[j] (CFC-11 exp.) (gas-permeable facers) | 1.5–2.5 (24–40) | 0.16–0.18 (0.023–0.026) | — | 6.25–5.56 | — | 0.22 (0.92) | 43.3–38.5 | — |
| Cellular polyisocyanurate[j] (CFC-11 exp.) (gas-impermeable facers) | 2.0 (32) | 0.14 (0.020) | — | 7.04 | — | 0.22 (0.92) | 48.8 | — |
| Cellular phenolic (closed cell) (CFC-11, CFC-113 exp.)[k] | 3.0 (32) | 0.12 (0.017) | — | 8.20 | — | — | 56.8 | — |
| Cellular phenolic (open cell) | 1.8–2.2 (29–35) | 0.23 (0.033) | — | 4.40 | — | — | 30.5 | — |
| Mineral fiber with resin binder | 15.0 (240) | 0.29 (0.042) | — | 3.45 | — | 0.17 | 23.9 | — |
| Mineral fiberboard, wet felted |  |  |  |  |  |  |  |  |
| Core or roof insulation | 16–17 (260–270) | 0.34 (0.049) | — | 2.94 | — | — | 20.4 | — |

| | | | | | | | | |
|---|---|---|---|---|---|---|---|---|
| Acoustical tile | | 18.0 (290) | 0.35 (0.050) | — | 2.86 | — | 0.19 (0.80) | 19.8 | — |
| Acoustical tile | | 21.0 (340) | 0.37 (0.053) | — | 2.70 | — | — | 18.7 | — |
| Mineral fiberboard, wet molded | | | | | | | | | |
| Acoustical tile[1] | | 23.0 (370) | 0.42 (0.060) | — | 2.38 | — | 0.14 (0.59) | 16.5 | — |
| Wood or cane fiberboard | | | | | | | | | |
| Acoustical tile[1] | 0.5 in. 12.7 mm | — | — | 0.80 4.5 | — | 1.25 | 0.31 (1.30) | — | 0.22 |
| Acoustical tile[1] | 0.75 in. 19 mm | — | — | 0.53 3.0 | — | 1.89 | — | — | 0.33 |
| Interior finish (plank, tile) | | 15.0 (240) | 0.35 (0.050) | — | 2.86 | — | 0.32 (1.34) | 19.8 | — |
| Cement fiber slabs (shredded wood with Portland cement binder) | | 25–27.0 (400–430) | 0.50–0.53 (0.072–0.076) | — | 2.0–1.89 (13.9–13.1) | — | — | 13.9–13.1 | — |
| Cement fiber slabs (shredded wood with magnesia oxysulfide binder) | | 22.0 (350) | 0.57 (0.082) | — | 1.75 | — | 0.31 (1.30) | 12.1 | — |
| *Loose Fill* | | | | | | | | | |
| Cellulosic insulation (milled paper or wood pulp) | | 2.3–3.2 (37–51) | 0.27–0.32 (0.039–0.046) | — | 3.70–3.13 | — | 0.33 (1.38) | 25.6–21.7 | — |
| Perlite, expanded | | 2.0–4.1 (32–66) | 0.27–0.31 (0.039–0.045) | — | 3.7–3.3 | — | 0.26 (1.09) | 25.6–22.9 | — |
| | | 4.1–7.4 (66–120) | 0.31–0.36 (0.045–0.052) | — | 3.3–2.8 | — | — | 22.9–19.4 | — |
| | | 7.4–11.0 (120–180) | 0.36–0.42 (0.052–0.060) | — | 2.8–2.4 | — | — | 19.4–16.6 | — |
| Mineral fiber (rock, slag, or glass)[9] | | | | | | | | | |
| Approx. 3.75–5 in. 95–130 mm | | 0.6–2.0 (9.6–32) | — | — | — | 11.0 | 0.17 (0.71) | — | 1.94 |

TABLE E.1 Thermal Properties of Conventional Building and Insulating Materials[a] *(Continued)*

| Description | Density lb/ft³ (kg/m³) | Conductivity[b] (k) Btu-in / h ft² °F (W/m K) | Conductance (C) Btu / h ft² °F (W/m² K) | I-P Resistance[c] (R) Per Inch Thickness (1/k) °F ft² h / Btu-in | I-P Resistance[c] (R) For Thickness Listed (1/C) °F ft² h / Btu | Specific Heat Btu / lb °F (kJ/kg K) | SI Resistance[c] (R) Per Meter Thickness (1/k) K m / W | SI Resistance[c] (R) For Thickness Listed (1/C) K m² / W |
|---|---|---|---|---|---|---|---|---|
| Mineral fiber (rock, slag, or glass)[g] | | | | | | | | |
| Approx. 6.5–8.75 in. 170–220 mm | 0.6–2.0 (9.6–32) | — | — | — | 19.0 | — | — | 3.35 |
| Approx. 7.5–10 in. 190–250 mm | 0.6–2.0 (9.6–32) | — | — | — | 22.0 | — | — | 3.87 |
| Approx. 10.3–13.7 in. 260–350 mm | 0.6–2.0 (9.6–32) | — | — | — | 30.0 | — | — | 5.28 |
| Mineral fiber (rock, slag, or glass)[g] Approx. 3.5 in. (90 mm) (closed sidewall application) | 2.0–3.5 (32–56) | — | — | — | 12.0–14.0 | — | — | 2.1–2.5 |
| Vermiculite, exfoliated | 7.0–8.2 (110–130) | 0.47 0.068 | — | 2.13 | — | 0.32 (1.34) | 14.8 | — |
|  | 4.0–6.0 (64–96) | 0.44 0.063 | — | 2.27 | — | — | 15.7 | — |
| *Spray Applied* | | | | | | | | |
| Polyurethane foam | 1.5–2.5 (24–40) | 0.16–0.18 0.023–0.026 | — | 6.25–5.56 | — | — | 43.3–38.5 | — |
| Ureaformaldehyde foam | 0.7–1.6 (11–26) | 0.22–0.28 0.032–0.040 | — | 4.55–3.57 | — | — | 31.5–24.7 | — |
| Cellulosic fiber | 3.5–6.0 (56–96) | 0.29–0.34 0.042–0.049 | — | 3.45–2.94 | — | — | 23.9–20.4 | — |
| Glass fiber | 3.5–4.5 (56–72) | 0.26–0.27 0.038–0.039 | — | 3.85–3.70 | — | — | 26.7–25.6 | — |

*Reflective Insulation*

| Material | | k | C | R (1/k) | | R (1/C) | | Specific Heat |
|---|---|---|---|---|---|---|---|---|
| Reflective material (ε < 0.5) in center of ¾ in. (20 mm) cavity forms two ⅜ in. (10 mm) vertical air spaces^m | — | — | 0.31 (1.76) | — | — | 3.2 | 0.57 | — |

**Metals**

(See 2001 *ASHRAE Handbook—Fundamentals*)

**Roofing**

| Material | Density | k | C | R (1/k) | | R (1/C) | | Specific Heat |
|---|---|---|---|---|---|---|---|---|
| Asbestos-cement shingles | 120 (1900) | — | 4.76 (27.0) | — | — | 0.21 | 0.037 | 0.24 (1.0) |
| Asphalt roll roofing | 70 (1100) | — | 6.50 (36.9) | — | — | 0.15 | 0.026 | 0.36 (1.51) |
| Asphalt shingles | 70 (1100) | — | 2.27 (12.9) | — | — | 0.44 | 0.077 | 0.30 (1.26) |
| Built-up roofing, 0.375 in. (10 mm) | 70 (1100) | — | 3.00 (17.0) | — | — | 0.33 | 0.058 | 0.35 (1.46) |
| Slate, 0.5 in. (13 mm) | — | — | 20.00 (114.0) | — | — | 0.05 | 0.009 | 0.30 (1.26) |
| Wood shingles, plain and plastic film faced | — | — | 1.06 (6.0) | — | — | 0.94 | 0.166 | 0.31 (1.30) |

**Plastering Materials**

| Material | Density | k | C | R (1/k) | | R (1/C) | | Specific Heat |
|---|---|---|---|---|---|---|---|---|
| Cement plaster, sand aggregate | 116 (1860) | 5.0 (0.72) | — | 0.20 | 1.39 | — | — | 0.20 (0.84) |
| Sand aggregate, 0.375 in. (10 mm) | — | — | 13.3 (75.5) | — | — | 0.08 | 0.013 | 0.20 (0.84) |
| Sand aggregate, 0.75 in. (20 mm) | — | — | 6.66 (37.8) | — | — | 0.15 | 0.026 | 0.20 (0.84) |
| Gypsum plaster | | | | | | | | |
| Lightweight aggregate, 0.5 in. (13 mm) | 45 (720) | — | 3.12 (17.7) | — | — | 0.32 | 0.056 | — |
| Lightweight aggregate, 0.625 in. (16 mm) | 45 (720) | — | 2.67 (15.2) | — | — | 0.39 | 0.066 | — |
| Lightweight aggregate on metal lath, 0.75 in. (19 mm) | — | — | 2.13 (12.1) | — | — | 0.47 | 0.083 | — |

**TABLE E.1 Thermal Properties of Conventional Building and Insulating Materials[a] (Continued)**

| Description | Density lb/ft³ (kg/m³) | Conductivity[b] (k) Btu-in / h ft² °F (W/m K) | Conductance (C) Btu / h ft² °F (W/m² K) | I-P Resistance[c] (R) Per Inch Thickness (1/k) °F ft² h / Btu-in | I-P Resistance[c] (R) For Thickness Listed (1/C) °F ft² h / Btu | Specific Heat Btu / lb °F (kJ/kg K) | SI Resistance[c] (R) Per Meter Thickness (1/k) K m / W | SI Resistance[c] (R) For Thickness Listed (1/C) K m² / W |
|---|---|---|---|---|---|---|---|---|
| Perlite aggregate | 45 (720) | 1.5 (0.22) | — | 0.67 | — | 0.32 (1.34) | 4.64 | — |
| Sand aggregate | 105 (1680) | 5.6 (0.81) | — | 0.18 | — | 0.20 (0.84) | 1.25 | — |
| Sand aggregate | 0.5 in. 13 mm | 105 (1680) | — | 11.10 (63.0) | — | 0.09 | — | — | 0.016 |
| Sand aggregate | 0.625 in. 16 mm | 105 (1680) | — | 9.10 (51.7) | — | 0.11 | — | — | 0.019 |
| Sand aggregate on metal lath | 0.75 in. 19 mm | — | — | 7.70 (43.7) | — | 0.13 | — | — | 0.023 |
| Vermiculite aggregate | 45 (720) | 1.7 (0.24) | — | 0.59 | — | — | 4.09 | — |
| **Masonry Materials** | | | | | | | | |
| *Masonry Units* | | | | | | | | |
| Brick, fired clay | 150 (2400) | 8.4–10.2 (1.21–1.47) | — | 0.12–0.10 | — | — | 0.83–0.68 | — |
| | 140 (2240) | 7.4–9.0 (1.07–1.30) | — | 0.14–0.11 | — | — | 0.94–0.77 | — |
| | 130 (2080) | 6.4–7.8 (0.92–1.12) | — | 0.16–0.12 | — | — | 1.08–0.89 | — |
| | 120 (1920) | 5.6–6.8 (0.81–0.98) | — | 0.18–0.15 | — | 0.19 (0.79) | 1.24–1.02 | — |

| | | lb/ft³ (kg/m³) | | | | | | | |
|---|---|---|---|---|---|---|---|---|---|
| **Clay tile, hollow** | | | | | | | | | |
| 1 cell deep | 3 in. / 75 mm | 110 (1760) | 4.9–5.9 (0.71–0.85) | 1.25 (7.10) | 0.20–0.17 | 0.80 | 0.21 (0.88) | 1.42–1.18 | 0.14 |
| 1 cell deep | 4 in. / 100 mm | 100 (1600) | 4.2–5.1 (0.71–0.85) | 0.90 (5.11) | 0.24–0.20 | 1.11 | — | 1.65–1.36 | 0.20 |
| 2 cells deep | 6 in. / 150 mm | 90 (1440) | 3.6–4.3 (0.52–0.62) | 0.66 (3.75) | 0.28–0.24 | 1.52 | — | 1.93–1.61 | 0.27 |
| 2 cells deep | 8 in. / 200 mm | 80 (1280) | 3.0–3.7 (0.43–0.53) | 0.54 (3.07) | 0.33–0.27 | 1.85 | — | 2.31–1.87 | 0.33 |
| 2 cells deep | 10 in. / 250 mm | 70 (1120) | 2.5–3.1 (0.36–0.45) | 0.45 (2.56) | 0.40–0.33 | 2.22 | — | 2.77–2.23 | 0.39 |
| 3 cells deep | 12 in. / 300 mm | — | — | 0.40 (2.27) | — | 2.50 | — | — | 0.44 |
| **Concrete blocks[n,o]** | | | | | | | | | |
| **Limestone aggregate** | | | | | | | | | |
| 8 in., 36 lb, 138 lb/ft³ concrete, 200 mm, 16.3 kg, 2210 kg/m³ 2 cores | | — | — | — | — | — | — | — | — |
| Same with perlite filled cores | | — | — | 0.48 (2.73) | — | 2.1 | — | — | 0.37 |

**TABLE E.1 Thermal Properties of Conventional Building and Insulating Materials^a (Continued)**

| Description | Density lb/ft³ (kg/m³) | Conductivity^b (k) Btu-in / h ft² °F (W/m K) | Conductance (C) Btu / h ft² °F (W/m² K) | I-P Resistance^c (R) | | Specific Heat Btu / lb °F (kJ/kg K) | SI Resistance^c (R) | |
|---|---|---|---|---|---|---|---|---|
| | | | | Per Inch Thickness (1/k) °F ft² h / Btu-in | For Thickness Listed (1/C) °F ft² h / Btu | | Per Meter Thickness (1/k) K m / W | For Thickness Listed (1/C) K m² / W |
| Limestone aggregate | | | | | | | | |
| 12 in., 55 lb, 138 lb/ft³ concrete, 300 mm, 25 kg, 2210 kg/m³ 2 cores | — | — | — | — | — | — | — | — |
| Same with perlite filled cores | — | — | 0.27 (1.53) | — | 3.7 | — | — | 0.65 |
| Normal weight aggregate (sand and gravel) | | | | | | | | |
| 8 in., 33–36 lb, 126–136 lb/ft³ 200 mm, 15–16 kg, 2020–2180 kg/m³ concrete, 2 or 3 cores | — | — | 0.90–1.03 (5.1–5.8) | — | 1.11–0.97 | 0.22 (0.92) | — | 0.20–0.17 |
| Same with perlite filled cores | — | — | 0.50 (2.84) | — | 2.0 | — | — | 0.35 |
| Same with vermiculite filled cores | — | — | 0.52–0.73 (3.0–4.1) | — | 1.92–1.37 | — | — | 0.34–0.24 |
| 12 in., 50 lb, 125 lb/ft³ concrete, 300 mm, 22.7 kg, 2000 kg/m³ 2 cores | — | — | 0.81 (4.60) | — | 1.23 | 0.22 (0.92) | — | 0.217 |
| Medium weight aggregate (combinations of normal-weight and lightweight aggregate) | | | | | | | | |
| 8 in., 26–29 lb, 97–112 lb/ft³ 200 mm, 15–16 kg, 2020–2180 kg/m³ concrete, 2 or 3 cores | — | — | 0.58–0.78 (3.3–4.4) | — | 1.71–1.28 | — | — | 0.30–0.22 |
| Same with perlite filled cores | — | — | 0.27–0.44 (1.5–2.5) | — | 3.7–2.3 | — | — | 0.65–0.41 |

| Description | | Conductivity / Resistance | | | | | |
|---|---|---|---|---|---|---|---|
| Same with vermiculite filled cores | — | 0.30 (1.70) | — | — | 3.3 | — | 0.58 |
| Same with molded EPS (beads) filled cores | — | 0.32 (1.82) | — | — | 3.2 | — | 0.56 |
| Same with molded EPS inserts in cores | — | 0.37 (2.10) | — | — | 2.7 | — | 0.47 |
| Lightweight aggregate (expanded shale, clay, slate or slag, pumice) | | | | | | | |
| 6 in., 16–17 lb, 85–87 lb/ft³ 150 mm, 7.3–7.7 kg, 1360–1390 kg/m³ concrete, 2 or 3 cores | — | 0.52–0.61 (3.0–3.5) | — | — | 1.93–1.65 | — | 0.34–0.29 |
| Same with perlite filled cores | — | 0.24 (1.36) | — | — | 4.2 | — | 0.74 |
| Same with vermiculite filled cores | — | 0.33 (1.87) | — | — | 3.0 | — | 0.53 |
| 8 in., 19–22 lb, 72–86 lb/ft³ 200 mm, 8.6–10 kg, 1150–1380 kg/m³ concrete | — | 0.32–0.54 (1.8–3.1) | — | — | 3.2–1.90 | 0.21 (0.88) | 0.56–0.33 |
| Same with perlite filled cores | — | 0.15–0.23 (0.9–1.3) | — | — | 6.8–4.4 | — | 1.20–0.77 |
| Same with vermiculite filled cores | — | 0.19–0.26 (1.1–1.5) | — | — | 5.3–3.9 | — | 0.93–0.69 |
| Same with molded EPS (beads) filled cores | — | 0.21 (1.19) | — | — | 4.8 | — | 0.85 |
| Same with UF foam-filled cores | — | 0.22 (1.25) | — | — | 4.5 | — | 0.79 |
| Same with molded EPS inserts in cores | — | 0.29 (1.65) | — | — | 3.5 | — | 0.62 |
| 12 in., 32–36 lb, 80–90 lb/ft³ concrete, 300 mm, 14.5–16.3 kg, 1280–1440 kg/m³ 2 or 3 cores | — | 0.38–0.44 (2.2–2.5) | — | — | 2.6–2.3 | — | 0.46–0.40 |
| Same with perlite filled cores | — | 0.11–0.16 (0.6–0.9) | — | — | 9.2–6.3 | — | 1.6–1.1 |

**TABLE E.1 Thermal Properties of Conventional Building and Insulating Materials^a (Continued)**

| Description | Density lb/ft³ (kg/m³) | Conductivity^b (k) Btu-in / h ft² °F (W/m K) | Conductance (C) Btu / h ft² °F (W/m² K) | I-P Resistance^c (R) Per Inch Thickness (1/k) °F ft² h / Btu-in | For Thickness Listed (1/C) °F ft² h / Btu | Specific Heat Btu / lb °F (kJ/kg K) | SI Resistance^c (R) Per Meter Thickness (1/k) K m / W | For Thickness Listed (1/C) K m² / W |
|---|---|---|---|---|---|---|---|---|
| Lightweight aggregate (expanded shale, clay, slate or slag, pumice) | | | | | | | | |
| Same with vermiculite filled cores | — | — | 0.17 (0.97) | — | 5.8 | — | — | 1.0 |
| Stone, lime, or sand | 180 (2880) | 72 (10.4) | — | 0.01 | — | — | 0.10 | — |
| Quartzitic and sandstone | 160 (2560) | 43 (6.2) | — | 0.02 | — | — | 0.16 | — |
| | 140 (2240) | 24 (3.5) | — | 0.04 | — | — | 0.29 | — |
| | 120 (1920) | 13 (1.9) | — | 0.08 | — | 0.19 | 0.53 | — |
| Calcitic, dolomitic, limestone, marble, and granite | 180 (2880) | 30 (4.3) | — | 0.03 | — | — | 0.23 | — |
| | 160 (2560) | 22 (3.2) | — | 0.05 | — | — | 0.32 | — |
| | 140 (2240) | 16 (2.3) | — | 0.06 | — | — | 0.43 | — |
| | 120 (1920) | 11 (1.6) | — | 0.09 | — | 0.19 | 0.63 | — |
| | 100 (1600) | 8 (1.1) | — | 0.13 | — | — | 0.90 | — |

| | Density lb/ft³ (kg/m³) | | | | | | | |
|---|---|---|---|---|---|---|---|---|
| Gypsum partition tile | | | | | | | | |
| 3 × 12 × 30 in., solid (75 × 300 × 760 mm) | — | — | 0.79 (4.50) | — | 1.26 | 0.19 (0.79) | — | 0.222 |
| 3 × 12 × 30 in., 4 cells (75 × 300 × 760 mm) | — | — | 0.74 (4.20) | — | 1.35 | — | — | 0.238 |
| 4 × 12 × 30 in., 3 cells (100 × 300 × 760 mm) | — | — | 0.60 (3.40) | — | 1.67 | — | — | 0.294 |
| *Concretes*[o] | | | | | | | | |
| Sand and gravel or stone aggregate concretes | 150 (2400) | 10.0–20.0 (1.4–2.9) | — | 0.10–0.05 | — | — | 0.69–0.35 | — |
| (concretes with more than 50% quartz or quartzite sand have conductivities in the higher end of the range) | 140 (2240) | 9.0–18.0 (1.3–2.6) | — | 0.11–0.06 | — | 0.19–0.24 (0.8–1.0) | 0.77–0.39 | — |
| | 130 (2080) | 7.0–13.0 (1.0–1.9) | — | 0.14–0.08 | — | — | 0.99–0.53 | — |
| Limestone concretes | 140 (2240) | 11.1 (1.60) | — | 0.09 | — | — | 0.62 | — |
| | 120 (1920) | 7.9 (1.14) | — | 0.13 | — | — | 0.88 | — |
| | 100 (1600) | 5.5 (0.79) | — | 0.18 | — | — | 1.26 | — |
| Gypsum-fiber concrete (87.5% gypsum, 12.5% wood chips) | 51 (816) | 1.66 (0.24) | — | 0.60 | — | 0.21 (0.88) | 4.18 | — |
| Cement/lime, mortar, and stucco | 120 (1920) | 9.7 (1.40) | — | 0.10 | — | — | 0.71 | — |
| | 100 (1600) | 6.7 (0.97) | — | 0.15 | — | — | 1.04 | — |
| | 80 (1280) | 4.5 (0.65) | — | 0.22 | — | — | 1.54 | — |

**TABLE E.1 Thermal Properties of Conventional Building and Insulating Materials[a] (Continued)**

| Description | Density lb/ft³ (kg/m³) | Conductivity[b] (k) Btu-in / h ft² °F (W/m K) | Conductance (C) Btu / h ft² °F (W/m² K) | I-P Resistance[c] (R) | | Specific Heat Btu / lb °F (kJ/kg K) | SI Resistance[c] (R) | |
|---|---|---|---|---|---|---|---|---|
| | | | | Per Inch Thickness (1/k) °F ft² h / Btu-in | For Thickness Listed (1/C) °F ft² h / Btu | | Per Meter Thickness (1/k) K m / W | For Thickness Listed (1/C) K m² / W |
| Lightweight aggregate concretes | | | | | | | | |
| Expanded shale, clay, or slate; expanded slags; cinders; pumice (with density up to 100 lb/ft³); and scoria (sanded concretes have conductivities in the higher end of the range) | 120 (1920) | 6.4–9.1 (0.9–1.3) | — | 0.16–0.11 | — | — | 1.08–0.76 | — |
| | 100 (1600) | 4.7–6.2 (0.68–0.89) | — | 0.21–0.16 | — | 0.20 (0.84) | 1.48–1.12 | — |
| | 80 (1280) | 3.3–4.1 (0.48–0.59) | — | 0.30–0.24 | — | 0.20 (0.84) | 2.10–1.69 | — |
| | 60 (960) | 2.1–2.5 (0.30–0.36) | — | 0.48–0.40 | — | — | 3.30–2.77 | — |
| | 40 (640) | 1.3 (0.18) | — | 0.78 | — | — | 5.40 | — |
| Perlite, vermiculite, and polystyrene beads | 50 (800) | 1.8–1.9 (0.26–0.27) | — | 0.55–0.53 | — | — | 3.81–3.68 | — |
| | 40 (640) | 1.4–1.5 (0.20–0.22) | — | 0.71–0.67 | — | 0.15–0.23 (0.63–0.96) | 4.92–4.65 | — |
| | 30 (480) | 1.1 (0.16) | — | 0.91 | — | — | 6.31 | — |
| | 20 (320) | 0.8 (0.12) | — | 1.25 | — | — | 8.67 | — |
| Foam concretes | 120 (1920) | 5.4 (0.75) | — | 0.19 | — | — | 1.32 | — |
| | 100 (1600) | 4.1 (0.60) | — | 0.24 | — | — | 1.66 | — |

| Material | | | | | | | | |
|---|---|---|---|---|---|---|---|---|
| Foam concretes and cellular concretes | 80 (1280) | 3.0 (0.44) | — | 0.33 | — | — | 2.29 | — |
| | 70 (1120) | 2.5 (0.36) | — | 0.40 | — | — | 2.77 | — |
| | 60 (960) | 2.1 (0.30) | — | 0.48 | — | — | 3.33 | — |
| | 40 (640) | 1.4 (0.20) | — | 0.71 | — | — | 4.92 | — |
| | 20 (320) | 0.8 (0.12) | — | 1.25 | — | — | 8.67 | — |

**Siding Materials (on flat surface)**

*Shingles*

| Material | | | | | | | | |
|---|---|---|---|---|---|---|---|---|
| Asbestos-cement | 120 (1900) | — | 4.75 (27.0) | — | 0.21 | — | — | 0.037 |
| Wood, 16 in., 7.5 exposure (400 mm, 190 mm exp.) | — | — | 1.15 (6.53) | — | 0.87 | 0.31 (1.30) | — | 0.15 |
| Wood, double, 16-in., 12-in. exposure (400 mm, 300 mm exp.) | — | — | 0.84 (4.77) | — | 1.19 | 0.28 (1.17) | — | 0.21 |
| Wood, plus ins. backer board, 0.312 in. (8 mm) | — | — | 0.71 (4.03) | — | 1.40 | 0.31 (1.30) | — | 0.25 |

*Siding*

| Material | | | | | | | | |
|---|---|---|---|---|---|---|---|---|
| Asbestos-cement, 0.25 in. (6.4 mm) lapped | — | — | 4.76 (27.0) | — | 0.21 | 0.24 (1.01) | — | 0.037 |
| Asphalt roll siding | — | — | 6.50 (36.9) | — | 0.15 | 0.35 (1.47) | — | 0.026 |
| Asphalt insulating siding (0.5 in. [12.7 mm] bed.) | — | — | 0.69 (3.92) | — | 1.46 | 0.35 (1.47) | — | 0.26 |
| Hardboard siding, 0.4375 in. (11 mm) | — | — | 1.49 (8.46) | — | 0.67 | 0.28 (1.17) | — | 0.12 |
| Wood, drop, 1 × 8 in. (25 × 200 mm) | — | — | 1.27 (7.21) | — | 0.79 | 0.28 (1.17) | — | 0.14 |

**TABLE E.1 Thermal Properties of Conventional Building and Insulating Materials[a] (Continued)**

| Description | Density lb/ft³ (kg/m³) | I-P Resistance[c] (R) | | | | | SI Resistance[c] (R) | |
|---|---|---|---|---|---|---|---|---|
| | | Conductivity[b] (k) Btu-in / h ft² °F (W/m K) | Conductance (C) Btu / h ft² °F (W/m² K) | Per Inch Thickness (1/k) °F ft² h / Btu-in | For Thickness Listed (1/C) °F ft² h / Btu | Specific Heat Btu / lb °F (kJ/kg K) | Per Meter Thickness (1/k) K m / W | For Thickness Listed (1/C) K m² / W |
| Wood, bevel, 0.5 × 8 in., lapped (13 × 200 mm) | — | — | 1.23 (6.98) | — | 0.81 | 0.28 (1.17) | — | 0.14 |
| Wood, bevel, 0.75 × 10 in., lapped (19 × 250 mm) | — | — | 0.95 (5.40) | — | 1.05 | 0.28 (1.17) | — | 0.18 |
| Wood, plywood, 0.375 in. (9.5 mm) lapped | — | — | 1.69 (9.60) | — | 0.59 | 0.29 (1.22) | — | 0.10 |
| Aluminum, steel, or vinyl[p,q] over sheathing | | | | | | | | |
| Hollow-backed | — | — | 1.64 (9.31) | — | 0.61 | 0.29[q/] (1.22) | — | 0.11 |
| Insulating-board backed nominal 0.375 in. (9.5 mm) | — | — | 0.55 (3.12) | — | 1.82 | 0.32 (1.34) | — | 0.32 |
| Insulating-board backed nominal 0.375 in. (9.5 mm), foil backed | — | — | 0.34 (1.93) | — | 2.96 | — | — | 0.52 |
| Architectural (soda-lime float) glass | 158 (2528) | 6.9 | 56.8 | — | — | 0.21 (0.84) | — | 0.018 |
| **Woods (12% moisture content)[e,r]** | | | | | | | | |
| *Hardwoods* | | | | | | | | |
| Oak | 41.2–46.8 (659–749) | 1.12–1.25 (0.16–0.18) | — | 0.89–0.80 | — | 0.39[s] (1.63) | 6.2–5.5 | — |
| Birch | 42.6–45.4 (682–726) | 1.16–1.22 (0.167–0.176) | — | 0.87–0.82 | — | | 6.0–5.7 | — |

| Species | | | | | | | | |
|---|---|---|---|---|---|---|---|---|
| Maple | 39.8–44.0 (637–704) | 1.09–1.19 (0.157–0.171) | — | — | 0.92–0.84 | — | 6.4–5.8 | — |
| Ash | 38.4–41.9 (614–670) | 1.06–1.14 (0.153–0.164) | — | — | 0.94–0.88 | — | 6.5–6.1 | — |
| *Softwoods* | | | | | $0.39^s$ (1.63) | | | |
| Southern Pine | 35.6–441.2 (570–659) | 1.00–1.12 (0.144–0.161) | — | — | 1.00–0.89 | — | 6.9–6.2 | — |
| Douglas Fir-Larch | 33.5–36.3 (536–581) | 0.95–1.01 (0.13–0.145) | — | — | 1.06–0.99 | — | 7.3–6.9 | — |
| Southern Cypress | 31.4–32.1 (502–514) | 0.90–0.92 (0.130–0.132) | — | — | 1.11–1.09 | — | 7.7–7.6 | — |
| Hem-Fir, Spruce-Pine-Fir | 24.5–31.4 (392–502) | 0.74–0.90 (0.107–0.130) | — | — | 1.35–1.11 | — | 9.3–7.7 | — |
| West Coast Woods, Cedars | 21.7–31.4 (347–502) | 0.68–0.90 (0.098–0.130) | — | — | 1.48–1.11 | — | 10.3–7.7 | — |
| California Redwood | 24.5–28.0 (392–448) | 0.74–0.82 (0.107–0.118) | — | — | 1.35–1.22 | — | 9.4–8.5 | — |

*Source:* Reprinted with permission of the American Society of Heating, Refrigerating and Air-Conditioning Engineers, Inc. from 2001 *ASHRAE Handbook—Fundamentals.*

*Note:* The SI units for various properties were appended to the ASHRAE I-P data by the authors.

[a]Values are for a mean temperature of 75°F (24°C). Representative values for dry materials are intended as design (not specification) values for materials in normal use. Thermal values of insulating materials may differ from design values depending on their *in situ* properties (e.g., density, moisture content, orientation) and variability experienced during manufacture. For properties of a particular product, use the value supplied by the manufacturer or by unbiased tests.

[b]To obtain thermal conductivities in Btu/h·ft·°F, divide the k-factor by 12 in./ft.

[c]Resistance values are the reciprocals of C before rounding off C to two decimal places.

[d]From Lewis (1967), *Thermal Conductivity of Wood-Base Fiber and Particle Panel Materials.* Forest Products Laboratory, Research Paper FPL 77, June.

[e]U.S. Dept. of Agriculture (1974), *Wood Handbook.* Handbook No. 72.

[f]Does not include paper backing and facing, if any. Where insulation forms a boundary (reflective or otherwise) of an airspace, see Tables E.3 and E.4/E.5 for the insulating value of an airspace with the appropriate effective emittance and temperature conditions of the space.

[g]Conductivity varies with fiber diameter. (See Chapter 23, Factors Affecting Thermal Performance, of *2001 ASHRAE Handbook—Fundamentals.*) Batt, blanket, and loose-fill mineral fiber insulations are manufactured to achieve specified R-values, the most common of which are listed in the table. Due to differences in manufacturing processes and materials, the product thicknesses, densities and thermal conductivities vary over considerable ranges for a specified R-value.

[h]This material is relatively new and data are based on limited testing.

[i]For additional information, see Society of Plastics Engineers (SPI) *Bulletin U108.* Values are for aged, unfaced board stock. For change in conductivity with age of expanded polyurethane/polyisocyanurate, see Chapter 23, Factors Affecting Thermal Performance, of *2001 ASHRAE Handbook—Fundamentals.*

[j]Values are for aged products with gas-impermeable facers on the two major surfaces. An aluminum foil facer of 0.001 in. (25 μm) thickness or greater is generally considered impermeable to gases. For change in conductivity with age of expanded polyisocyanurate, see Chapter 23, Factors Affecting Thermal Performance, of *2001 ASHRAE Handbook—Fundamentals.*

[k]Cellular phenolic insulation may no longer be manufactured. These thermal conductivity and resistance values do not represent aged insulation, which may have higher thermal conductivity and lower thermal resistance.

**TABLE E.1 Thermal Properties of Conventional Building and Insulating Materials** [a] **(Continued)**

[l] Insulating values of acoustical tile vary, depending on density of the board and on type, size, and depth of perforations.

[m] Cavity is framed with 0.75-in. (20-mm) wood furring strips. Caution should be used in applying this value for other framing materials. The reported value was derived from tests and applies to the reflective path only. The effect of studs or furring strips must be included in determining the overall performance of the wall.

[n] Values for fully grouted block may be approximated using values for concrete with a similar unit weight.

[o] Values for concrete block and concrete are at moisture contents representative of normal use.

[p] Values for metal or vinyl siding applied over flat surfaces vary widely, depending on amount of ventilation of airspace beneath the siding; whether airspace is reflective or nonreflective; and on thickness, type, and application of insulating backing used. Values are averages for use as design guides, and were obtained from several guarded hot box tests (ASTM C 236) or calibrated hot box (ASTM C 976) on hollow-backed types and types made using backing-boards of wood fiber, foamed plastic, and glass fiber. Departures of ±50% or more from these values may occur.

[q] Vinyl specific heat = 0.25 Btu/lb °F (1.0 kJ/kg K).

[r] See Adams (1971. Supporting Cryogenic Equipment with Wood. *Chemical Engineering* (May): pp. 156–158), MacLean (1941. Thermal Conductivity of Wood. *ASHVE Transactions 47:323*), and Wilkes (1979. Thermophysical Properties Data Base Activities at Owens-Corning Fiberglas. *Proceedings of the ASHRAE/DOE-ORNL Conference, Thermal Performance of the Exterior Envelopes of Buildings*, ASHRAE SP 28:662–77). The conductivity values listed are for heat transfer across the grain. The thermal conductivity of wood varies linearly with the density, and the density ranges listed are those normally found for the wood species given. If the density of the wood species is not known, use the mean conductivity value. For extrapolation to other moisture contents, refer to the equation presented in footnote "s" on page 25.9 of the 2001 *ASHRAE Handbook—Fundamentals*.

[s] For an empirical equation for the specific heat of moist wood at 75°F (24°C), refer to the equation presented in footnote "r" on page 25.9 of the *2001 ASHRAE Handbook—Fundamentals*.

**TABLE E.2 Thermal Properties of Alternative[a] Building and Insulating Materials**

| Description | Density lb/ft³ (kg/m³) | Conductivity (k) Btu-in/h ft² °F | Conductance (C) Btu/h ft² °F | I-P Resistance Per Inch Thickness (1/k) h ft² °F/Btu-in. | I-P Resistance for Thickness Listed h ft² °F/Btu | SI Resistance Per Meter Thickness (1/k) K m/W | SI Resistance for Thickness Listed K m²/W |
|---|---|---|---|---|---|---|---|
| **Building Materials** | | | | | | | |
| Adobe[b] typically 10–24 in. (250–610 mm) | 100± (1600) | 3.3 | | 0.3 | | 2.1 | |
| Cob[c] typically 18–36 in. (460–915 mm) | | 4 | | 0.25 | | 1.7 | |
| Rammed earth[d] 18 in. (460 mm) typical | 100± (1600) | 4 | | 0.25 | | 1.7 | |
| Straw bale[e] typically 16–24 in. (400–610 mm) | 5–10 (80–160) | 0.67–0.33 | | 1.5–3.0 | | 10.4–20.8 | |
| SIP (EPS)[f] Various thicknesses | | 0.25 | | 4 | | 27.7 | |
| SIP (XPS)[g] Various thicknesses | | 0.20 | | 5 | | 34.7 | |
| SIP (POLY/ISO)[h] Various thicknesses | | 0.17–0.14 | | 6–7 | | 41.6–48.5 | |
| **Insulating Materials** | | | | | | | |
| Air-Krete[i] | 2.2 (35) | 0.26 | | 3.9 | | 27.0 | |
| Insulating Forms[j] | | | 0.06–0.03 | | 17–40 | | 3.0–7.0 |

[a]It is generally difficult to obtain reliable and consistent information on the thermal properties of alternative building materials that are not sold as a proprietary product. The data in this table have been obtained from a variety of sources (as noted); note the wide range of values and lack of data for some properties (such as specific heat).
[b]Adobe: http://www.eere.energy.gov/EE/strawhouse/house-of-straw.html
[c]Rammed earth: http://www.toolbase.org/
[d]Cob: http://www.toolbase.org/
[e]Straw bale: http://www.buildinggreen.com/news/r-value.cfm; recent tests suggest lower R-values than previously reported, with a typical R per inch of 1.5 for a per-bale R around 26–28.
[f]Structural insulated panel with expanded polystyrene insulation: http://www.eere.energy.gov/consumerinfo/factsheets/bd1.html
[g]Structural insulated panel with extruded polystyrene insulation: http://www.eere.energy.gov/consumerinfo/factsheets/bd1.html
[h]Structural insulated panel with polyurethane/isocyanurate insulation: http://www.eere.energy.gov/consumerinfo/factsheets/bd1.html; R-values are for aged panels (as opposed to new panel R-values of 7–9 as typically reported).
[i]Air-krete: http://www.airkrete.com/
[j]Insulating concrete forms come in a variety of configurations; the listed R-17 value is for a 9 inch wall section (4 in. polystyrene and 5 in. poured concrete); http://www.forms.org/product_info/product_benefits.html. Another manufacturer produces forms that provide R-values of 22, 32, and 40.

**TABLE E.3 Thermal Properties of Surface Air Films and Air Spaces**

Part A. Surface Conductances[a] and Resistances[b] for Surface Air Films

| | | Surface Emittance, $\varepsilon$ | | | | | | | | | | |
|---|---|---|---|---|---|---|---|---|---|---|---|---|
| | | I-P Units[c] | | | | | | SI Units[d] | | | | |
| | | Nonreflective | | Reflective | | | | Nonreflective | | Reflective | | |
| | | $\varepsilon = 0.90$ | | $\varepsilon = 0.20$ | | $\varepsilon = 0.05$ | | $\varepsilon = 0.90$ | | $\varepsilon = 0.20$ | | $\varepsilon = 0.05$ |
| Position of Surface | Direction of Heat Flow | $h_i$ | R | $h_i$ | R | $h_i$ | R | $h_i$ | R | $h_i$ | R | $h_i$ R |
| *Still Air* | | | | | | | | | | | | |
| Horizontal | Upward | 1.63 | 0.61 | 0.91 | 1.10 | 0.76 | 1.32 | 9.26 | 0.11 | 5.17 | 0.19 | 4.32 0.23 |
| Sloping–45° | Upward | 1.60 | 0.62 | 0.88 | 1.14 | 0.73 | 1.37 | 9.09 | 0.11 | 5.00 | 0.20 | 4.15 0.24 |
| Vertical | Horizontal | 1.46 | 0.68 | 0.74 | 1.35 | 0.59 | 1.70 | 8.29 | 0.12 | 4.20 | 0.24 | 3.35 0.30 |
| Sloping–45° | Downward | 1.32 | 0.76 | 0.60 | 1.67 | 0.45 | 2.22 | 7.50 | 0.13 | 3.41 | 0.29 | 2.56 0.39 |
| Horizontal | Downward | 1.08 | 0.92 | 0.37 | 2.70 | 0.22 | 4.55 | 6.13 | 0.16 | 2.10 | 0.48 | 1.25 0.80 |
| *Moving Air (any position)* | | $h_o$ | R | | | | | $h_o$ | R | | | |
| Winter Wind | | | | | | | | | | | | |
| 15 mph (6.7 m/s) | Any | 6.00 | 0.17 | | | | | 34.0 | 0.030 | | | |
| Summer Wind | | | | | | | | | | | | |
| 7.5 mph (3.4 m/s) | Any | 4.00 | 0.25 | | | | | 22.7 | 0.044 | | | |

Part B. Emittance Values of Various Surfaces and Effective Emittances of Air Spaces

| | Effective Emittance $\varepsilon_{eff}$ of Air Space | | |
|---|---|---|---|
| | | One Surface | |
| | Average | Emittance $\varepsilon$; | Both Surfaces |
| | Emittance $\varepsilon$ | Other Surface 0.9 | Emittance $\varepsilon$ |
| Aluminum foil, bright | 0.05 | 0.05 | 0.03 |
| Aluminum foil with condensate just visible | 0.30 | 0.29 | — |
| Aluminum foil with condensate clearly visible | 0.70 | 0.65 | — |
| Aluminum sheet | 0.12 | 0.12 | 0.06 |
| Aluminum-coated paper, polished | 0.20 | 0.20 | 0.11 |
| Steel, galvanized, bright | 0.25 | 0.24 | 0.15 |
| Aluminum paint | 0.50 | 0.47 | 0.35 |
| Building materials: wood, paper, masonry, nonmetallic paints | 0.90 | 0.82 | 0.82 |
| Regular glass | 0.84 | 0.77 | 0.72 |

*Source:* Reprinted with permission of the American Society of Heating, Refrigerating and Air-Conditioning Engineers, Inc. from the 2001 *ASHRAE Handbook—Fundamentals.*
[a]Conductances are for surfaces of the stated emittance facing virtual blackbody surroundings at the same temperature as the ambient air. Values are based on a surface–air temperature difference of 10F° (5.5C°) and for surface temperatures of 70°F (21°C).
[b]No surface has both an air space resistance value and a surface resistance value.
[c]I-P units: surface conductance $h_i$ and $h_o$ measured in Btu/h ft$^2$ °F; resistance R measured in h ft$^2$ °F/Btu.
[d]SI units: surface conductance $h_i$ and $h_o$ measured in W/m$^2$ K; resistance R measured in m$^2$ K/W.

# TABLE E.4 Thermal Resistances of Plane[a] Air Spaces (I-P Units)

All resistance values expressed in ft² °F·h/Btu. Values apply only to air spaces of uniform thickness bounded by plane, smooth, parallel surfaces with no leakage of air to or from the space. These conditions are not normally present in standard building construction. When accurate values are required, use overall U-factors determined for a particular construction. Thermal resistance values for multiple air spaces must be based on careful estimates of mean temperature differences for each air space.

| Position of Air Space | Direction of Heat Flow | | Air Space | | 0.5-in. Air Space[d] Value of E[b,c] | | | | | 0.75-in. Air Space[d] Value of E[b,c] | | | | |
|---|---|---|---|---|---|---|---|---|---|---|---|---|---|---|
| | | | Mean Temp,[b] °F | Temp Diff,[b] °F | 0.03 | 0.05 | 0.2 | 0.5 | 0.82 | 0.03 | 0.05 | 0.2 | 0.5 | 0.82 |
| Horiz. | Up | ↑ | 90 | 10 | 2.13 | 2.03 | 1.51 | 0.99 | 0.73 | 2.34 | 2.22 | 1.61 | 1.04 | 0.75 |
| | | | 50 | 30 | 1.62 | 1.57 | 1.29 | 0.96 | 0.75 | 1.71 | 1.66 | 1.35 | 0.99 | 0.77 |
| | | | 50 | 10 | 2.13 | 2.05 | 1.60 | 1.11 | 0.84 | 2.30 | 2.21 | 1.70 | 1.16 | 0.87 |
| | | | 0 | 20 | 1.73 | 1.70 | 1.45 | 1.12 | 0.91 | 1.83 | 1.79 | 1.52 | 1.16 | 0.93 |
| | | | 0 | 10 | 2.10 | 2.04 | 1.70 | 1.27 | 1.00 | 2.23 | 2.16 | 1.78 | 1.31 | 1.02 |
| | | | -50 | 20 | 1.69 | 1.66 | 1.49 | 1.23 | 1.04 | 1.77 | 1.74 | 1.55 | 1.27 | 1.07 |
| | | | -50 | 10 | 2.04 | 2.00 | 1.75 | 1.40 | 1.16 | 2.16 | 2.11 | 1.84 | 1.46 | 1.20 |
| 45° Slope | Up | ↗ | 90 | 10 | 2.44 | 2.31 | 1.65 | 1.06 | 0.76 | 2.96 | 2.78 | 1.88 | 1.15 | 0.81 |
| | | | 50 | 30 | 2.06 | 1.98 | 1.56 | 1.10 | 0.83 | 1.99 | 1.92 | 1.52 | 1.08 | 0.82 |
| | | | 50 | 10 | 2.55 | 2.44 | 1.83 | 1.22 | 0.90 | 2.90 | 2.75 | 2.00 | 1.29 | 0.94 |
| | | | 0 | 20 | 2.20 | 2.14 | 1.76 | 1.30 | 1.02 | 2.13 | 2.07 | 1.72 | 1.28 | 1.00 |
| | | | 0 | 10 | 2.63 | 2.54 | 2.03 | 1.44 | 1.10 | 2.72 | 2.62 | 2.08 | 1.47 | 1.12 |
| | | | -50 | 20 | 2.08 | 2.04 | 1.78 | 1.42 | 1.17 | 2.05 | 2.01 | 1.76 | 1.41 | 1.16 |
| | | | -50 | 10 | 2.62 | 2.56 | 2.17 | 1.66 | 1.33 | 2.53 | 2.47 | 2.10 | 1.62 | 1.30 |
| Vertical | Horiz. | → | 90 | 10 | 2.47 | 2.34 | 1.67 | 1.06 | 0.77 | 3.50 | 3.24 | 2.08 | 1.22 | 0.84 |
| | | | 50 | 30 | 2.57 | 2.46 | 1.84 | 1.23 | 0.90 | 2.91 | 2.77 | 2.01 | 1.30 | 0.94 |
| | | | 50 | 10 | 2.66 | 2.54 | 1.88 | 1.24 | 0.91 | 3.70 | 3.46 | 2.35 | 1.43 | 1.01 |
| | | | 0 | 20 | 2.82 | 2.72 | 2.14 | 1.50 | 1.13 | 3.14 | 3.02 | 2.32 | 1.58 | 1.18 |
| | | | 0 | 10 | 2.93 | 2.82 | 2.20 | 1.53 | 1.15 | 3.77 | 3.59 | 2.64 | 1.73 | 1.26 |
| | | | -50 | 20 | 2.90 | 2.82 | 2.35 | 1.76 | 1.39 | 2.90 | 2.83 | 2.36 | 1.77 | 1.39 |
| | | | -50 | 10 | 3.20 | 3.10 | 2.54 | 1.87 | 1.46 | 3.72 | 3.60 | 2.87 | 2.04 | 1.56 |
| 45° Slope | Down | ↘ | 90 | 10 | 2.48 | 2.34 | 1.67 | 1.06 | 0.77 | 3.53 | 3.27 | 2.10 | 1.22 | 0.84 |
| | | | 50 | 30 | 2.64 | 2.52 | 1.87 | 1.24 | 0.91 | 3.43 | 3.23 | 2.24 | 1.39 | 0.99 |
| | | | 50 | 10 | 2.67 | 2.55 | 1.89 | 1.25 | 0.92 | 3.81 | 3.57 | 2.40 | 1.45 | 1.02 |
| | | | 0 | 20 | 2.91 | 2.80 | 2.19 | 1.52 | 1.15 | 3.75 | 3.57 | 2.63 | 1.72 | 1.26 |
| | | | 0 | 10 | 2.94 | 2.83 | 2.21 | 1.53 | 1.15 | 4.12 | 3.91 | 2.81 | 1.80 | 1.30 |
| | | | -50 | 20 | 3.16 | 3.07 | 2.52 | 1.86 | 1.45 | 3.78 | 3.65 | 2.90 | 2.05 | 1.57 |
| | | | -50 | 10 | 3.26 | 3.16 | 2.58 | 1.89 | 1.47 | 4.35 | 4.18 | 3.22 | 2.21 | 1.66 |

**APPENDICES**

**TABLE E.4 Thermal Resistances of Plane[a] Air Spaces (I-P Units) (Continued)**

| Position of Air Space | Direction of Heat Flow | Air Space Mean Temp,[b] °F | Temp Diff,[b] °F | 0.5-in. Air Space[d] Value of E[b,c] | | | | | 0.75-in. Air Space[d] Value of E[b,c] | | | | |
|---|---|---|---|---|---|---|---|---|---|---|---|---|---|
| | | | | 0.03 | 0.05 | 0.2 | 0.5 | 0.82 | 0.03 | 0.05 | 0.2 | 0.5 | 0.82 |
| Horiz. | Down ↓ | 90 | 10 | 2.48 | 2.34 | 1.67 | 1.06 | 0.77 | 3.55 | 3.29 | 2.10 | 1.22 | 0.85 |
| | | 50 | 30 | 2.66 | 2.54 | 1.88 | 1.24 | 0.91 | 3.77 | 3.52 | 2.38 | 1.44 | 1.02 |
| | | 50 | 10 | 2.67 | 2.55 | 1.89 | 1.25 | 0.92 | 3.84 | 3.59 | 2.41 | 1.45 | 1.02 |
| | | 0 | 20 | 2.94 | 2.83 | 2.20 | 1.53 | 1.15 | 4.18 | 3.96 | 2.83 | 1.81 | 1.30 |
| | | 0 | 10 | 2.96 | 2.85 | 2.22 | 1.53 | 1.16 | 4.25 | 4.02 | 2.87 | 1.82 | 1.31 |
| | | −50 | 20 | 3.25 | 3.15 | 2.58 | 1.89 | 1.47 | 4.60 | 4.41 | 3.36 | 2.28 | 1.69 |
| | | −50 | 10 | 3.28 | 3.18 | 2.60 | 1.90 | 1.47 | 4.71 | 4.51 | 3.42 | 2.30 | 1.71 |

| Position of Air Space | Direction of Heat Flow | Air Space Mean Temp,[b] °F | Temp Diff,[b] °F | 1.5-in. Air Space[d] Value of E[b,c] | | | | | 3.5-in. Air Space[d] Value of E[b,c] | | | | |
|---|---|---|---|---|---|---|---|---|---|---|---|---|---|
| | | | | 0.03 | 0.05 | 0.2 | 0.5 | 0.82 | 0.03 | 0.05 | 0.2 | 0.5 | 0.82 |
| Horiz. | Up ↑ | 90 | 10 | 2.55 | 2.41 | 1.71 | 1.08 | 0.77 | 2.84 | 2.66 | 1.83 | 1.13 | 0.80 |
| | | 50 | 30 | 1.87 | 1.81 | 1.45 | 1.04 | 0.80 | 2.09 | 2.01 | 1.58 | 1.10 | 0.84 |
| | | 50 | 10 | 2.50 | 2.40 | 1.81 | 1.21 | 0.89 | 2.80 | 2.66 | 1.95 | 1.28 | 0.93 |
| | | 0 | 20 | 2.01 | 1.95 | 1.63 | 1.23 | 0.97 | 2.25 | 2.18 | 1.79 | 1.32 | 1.03 |
| | | 0 | 10 | 2.43 | 2.35 | 1.90 | 1.38 | 1.06 | 2.71 | 2.62 | 2.07 | 1.47 | 1.12 |
| | | −50 | 20 | 1.94 | 1.91 | 1.68 | 1.36 | 1.13 | 2.19 | 2.14 | 1.86 | 1.47 | 1.20 |
| | | −50 | 10 | 2.37 | 2.31 | 1.99 | 1.55 | 1.26 | 2.65 | 2.58 | 2.18 | 1.67 | 1.33 |
| 45° Slope | Up ↗ | 90 | 10 | 2.92 | 2.73 | 1.86 | 1.14 | 0.80 | 3.18 | 2.96 | 1.97 | 1.18 | 0.82 |
| | | 50 | 30 | 2.14 | 2.06 | 1.61 | 1.12 | 0.84 | 2.26 | 2.17 | 1.67 | 1.15 | 0.86 |
| | | 50 | 10 | 2.88 | 2.74 | 1.99 | 1.29 | 0.94 | 3.12 | 2.95 | 2.10 | 1.34 | 0.96 |
| | | 0 | 20 | 2.30 | 2.23 | 1.82 | 1.34 | 1.04 | 2.42 | 2.35 | 1.90 | 1.38 | 1.06 |
| | | 0 | 10 | 2.79 | 2.69 | 2.12 | 1.49 | 1.13 | 2.98 | 2.87 | 2.23 | 1.54 | 1.16 |
| | | −50 | 20 | 2.22 | 2.17 | 1.88 | 1.49 | 1.21 | 2.34 | 2.29 | 1.97 | 1.54 | 1.25 |
| | | −50 | 10 | 2.71 | 2.64 | 2.23 | 1.69 | 1.35 | 2.87 | 2.79 | 2.33 | 1.75 | 1.39 |
| Vertical | Horiz. → | 90 | 10 | 3.99 | 3.66 | 2.25 | 1.27 | 0.87 | 3.69 | 3.40 | 2.15 | 1.24 | 0.85 |
| | | 50 | 30 | 2.58 | 2.46 | 1.84 | 1.23 | 0.90 | 2.67 | 2.55 | 1.89 | 1.25 | 0.91 |
| | | 50 | 10 | 3.79 | 3.55 | 2.39 | 1.45 | 1.02 | 3.63 | 3.40 | 2.32 | 1.42 | 1.01 |
| | | 0 | 20 | 2.76 | 2.66 | 2.10 | 1.48 | 1.12 | 2.88 | 2.78 | 2.17 | 1.51 | 1.14 |
| | | 0 | 10 | 3.51 | 3.35 | 2.51 | 1.67 | 1.23 | 3.49 | 3.33 | 2.50 | 1.67 | 1.23 |
| | | −50 | 20 | 2.64 | 2.58 | 2.18 | 1.66 | 1.33 | 2.82 | 2.75 | 2.30 | 1.73 | 1.37 |
| | | −50 | 10 | 3.31 | 3.21 | 2.62 | 1.91 | 1.48 | 3.40 | 3.30 | 2.67 | 1.94 | 1.50 |

| Position of Air Space | Direction of Heat Flow | $t_m$ | Δt | | | 0.5-in. Air Space | | | | | 3.5-in. Air Space | | |
|---|---|---|---|---|---|---|---|---|---|---|---|---|---|
| 45° Slope | Down ↗ | 90 | 10 | 5.07 | 4.55 | 2.56 | 1.36 | 0.91 | 4.81 | 4.33 | 2.49 | 1.34 | 0.90 |
| | | 50 | 30 | 3.58 | 3.36 | 2.31 | 1.42 | 1.00 | 3.51 | 3.30 | 2.28 | 1.40 | 1.00 |
| | | 50 | 10 | 5.10 | 4.66 | 2.85 | 1.60 | 1.09 | 4.74 | 4.36 | 2.73 | 1.57 | 1.08 |
| | | 0 | 20 | 3.85 | 3.66 | 2.68 | 1.74 | 1.27 | 3.81 | 3.63 | 2.66 | 1.74 | 1.27 |
| | | 0 | 10 | 4.92 | 4.62 | 3.16 | 1.94 | 1.37 | 4.59 | 4.32 | 3.02 | 1.88 | 1.34 |
| | | −50 | 20 | 3.62 | 3.50 | 2.80 | 2.01 | 1.54 | 3.77 | 3.64 | 2.90 | 2.05 | 1.57 |
| | | −50 | 10 | 4.67 | 4.47 | 3.40 | 2.29 | 1.70 | 4.50 | 4.32 | 3.31 | 2.25 | 1.68 |
| Horiz. | Down ↓ | 90 | 10 | 6.09 | 5.35 | 2.79 | 1.43 | 0.94 | 10.07 | 8.19 | 3.41 | 1.57 | 1.00 |
| | | 50 | 30 | 6.27 | 5.63 | 3.18 | 1.70 | 1.14 | 9.60 | 8.17 | 3.86 | 1.88 | 1.22 |
| | | 50 | 10 | 6.61 | 5.90 | 3.27 | 1.73 | 1.15 | 11.15 | 9.27 | 4.09 | 1.93 | 1.24 |
| | | 0 | 20 | 7.03 | 6.43 | 3.91 | 2.19 | 1.49 | 10.90 | 9.52 | 4.87 | 2.47 | 1.62 |
| | | 0 | 10 | 7.31 | 6.66 | 4.00 | 2.22 | 1.51 | 11.97 | 10.32 | 5.08 | 2.52 | 1.64 |
| | | −50 | 20 | 7.73 | 7.20 | 4.77 | 2.85 | 1.99 | 11.64 | 10.49 | 6.02 | 3.25 | 2.18 |
| | | −50 | 10 | 8.09 | 7.52 | 4.91 | 2.89 | 2.01 | 12.98 | 11.56 | 6.36 | 3.34 | 2.22 |

*Source:* Reprinted with permission of the American Society of Heating, Refrigerating and Air-Conditioning Engineers, Inc. from the 2001 *ASHRAE Handbook—Fundamentals*.

[a]Thermal resistance values were determined from the relation $R = 1/C$, where $C = h_c + \varepsilon_{eff}\, h_r$; $h_r$ is the radiation coefficient $= 0.0068\varepsilon_{eff}[(t_m + 460)/100]^3$; $h_c$ is the conduction–convection coefficient; $\varepsilon_{eff}$ is the effective emittance; and $t_m$ is the mean temperature of the air space. For extrapolation from this table to air spaces less than 0.5 in. (as in insulating window glass), assume that $h_c = 0.159(1 + 0.0016t_m)/l$, where $l$ is the air space thickness in inches and $h_c$ is heat transfer through the air space only.

[b]Interpolation is permissible for other values of mean temperature, temperature difference, and effective emittance $\varepsilon_{eff}$. Interpolation and moderate extrapolation for air spaces greater than 3.5 in. are also permissible.

[c]Effective emittance, $\varepsilon_{eff}$, of the air space is given by $1/\varepsilon_{eff} = 1/\varepsilon_1 + 1/\varepsilon_2 - 1$, where $\varepsilon_1$ and $\varepsilon_2$ are emittances of the surfaces of the air space (from Table E.3).

[d]A single resistance value cannot account for multiple air spaces. Each air space requires a separate resistance calculation that applies only for the established boundary conditions.

[e]Resistances of horizontal spaces with heat flow downward are substantially independent of temperature difference.

APPENDICES

## TABLE E.5 Thermal Resistances of Plane[a] Air Spaces (SI Units)

All resistance values expressed in K m²/W. Values apply only to air spaces of uniform thickness bounded by plane, smooth, parallel surfaces with no leakage of air to or from the space. These conditions are not normally present in standard building construction. When accurate values are required, use overall U-factors determined for a particular construction. Thermal resistance values for multiple air spaces must be based on careful estimates of mean temperature differences for each air space.

| Position of Air Space | Direction of Heat Flow | | Air Space | | 13-mm Air Space[d] | | | | | 20-mm Air Space[d] | | | | |
|---|---|---|---|---|---|---|---|---|---|---|---|---|---|---|
| | | | Mean Temp,[b] °C | Temp Diff,[b] °C | Value of E[b,c] | | | | | Value of E[b,c] | | | | |
| | | | | | 0.03 | 0.05 | 0.2 | 0.5 | 0.82 | 0.03 | 0.05 | 0.2 | 0.5 | 0.82 |
| Horiz. | Up | ↑ | 32.2 | 5.6 | 0.37 | 0.36 | 0.27 | 0.17 | 0.13 | 0.41 | 0.39 | 0.28 | 0.18 | 0.13 |
| | | | 10.0 | 16.7 | 0.29 | 0.28 | 0.23 | 0.17 | 0.13 | 0.30 | 0.29 | 0.24 | 0.17 | 0.14 |
| | | | 10.0 | 5.6 | 0.37 | 0.36 | 0.28 | 0.30 | 0.15 | 0.40 | 0.39 | 0.30 | 0.20 | 0.15 |
| | | | -17.8 | 11.1 | 0.30 | 0.30 | 0.26 | 0.20 | 0.16 | 0.32 | 0.32 | 0.27 | 0.23 | 0.16 |
| | | | -17.8 | 5.6 | 0.37 | 0.36 | 0.30 | 0.22 | 0.18 | 0.39 | 0.38 | 0.31 | 0.23 | 0.18 |
| | | | -45.6 | 11.1 | 0.30 | 0.29 | 0.26 | 0.22 | 0.18 | 0.31 | 0.31 | 0.27 | 0.22 | 0.19 |
| | | | -45.6 | 5.6 | 0.36 | 0.35 | 0.31 | 0.25 | 0.20 | 0.38 | 0.37 | 0.32 | 0.26 | 0.21 |
| 45° Slope | Up | ↗ | 32.2 | 5.6 | 0.43 | 0.41 | 0.29 | 0.19 | 0.13 | 0.52 | 0.49 | 0.33 | 0.20 | 0.14 |
| | | | 10.0 | 16.7 | 0.36 | 0.35 | 0.27 | 0.19 | 0.15 | 0.35 | 0.34 | 0.27 | 0.19 | 0.14 |
| | | | 10.0 | 5.6 | 0.45 | 0.43 | 0.32 | 0.21 | 0.16 | 0.51 | 0.48 | 0.35 | 0.23 | 0.17 |
| | | | -17.8 | 11.1 | 0.39 | 0.38 | 0.31 | 0.23 | 0.18 | 0.37 | 0.36 | 0.30 | 0.23 | 0.18 |
| | | | -17.8 | 5.6 | 0.46 | 0.45 | 0.36 | 0.25 | 0.19 | 0.48 | 0.46 | 0.37 | 0.26 | 0.20 |
| | | | -45.6 | 11.1 | 0.37 | 0.36 | 0.31 | 0.25 | 0.21 | 0.36 | 0.35 | 0.31 | 0.25 | 0.20 |
| | | | -45.6 | 5.6 | 0.46 | 0.45 | 0.38 | 0.29 | 0.23 | 0.45 | 0.43 | 0.37 | 0.29 | 0.23 |
| Vertical | Horiz. | → | 32.2 | 5.6 | 0.43 | 0.41 | 0.29 | 0.19 | 0.14 | 0.62 | 0.57 | 0.37 | 0.21 | 0.15 |
| | | | 10 | 16.7 | 0.45 | 0.43 | 0.32 | 0.22 | 0.16 | 0.51 | 0.49 | 0.35 | 0.23 | 0.17 |
| | | | 10 | 5.6 | 0.47 | 0.45 | 0.33 | 0.22 | 0.16 | 0.65 | 0.61 | 0.41 | 0.25 | 0.18 |
| | | | -17.8 | 11.1 | 0.50 | 0.48 | 0.38 | 0.26 | 0.20 | 0.55 | 0.53 | 0.41 | 0.28 | 0.21 |
| | | | -17.8 | 5.6 | 0.52 | 0.50 | 0.39 | 0.27 | 0.20 | 0.66 | 0.63 | 0.46 | 0.30 | 0.22 |
| | | | -45.6 | 11.1 | 0.51 | 0.50 | 0.41 | 0.31 | 0.24 | 0.51 | 0.50 | 0.42 | 0.31 | 0.24 |
| | | | -45.6 | 5.6 | 0.56 | 0.55 | 0.45 | 0.33 | 0.26 | 0.65 | 0.63 | 0.51 | 0.36 | 0.27 |
| 45° Slope | Down | ↘ | 32.2 | 5.6 | 0.44 | 0.41 | 0.29 | 0.19 | 0.14 | 0.62 | 0.58 | 0.37 | 0.21 | 0.15 |
| | | | 10.0 | 16.7 | 0.46 | 0.44 | 0.33 | 0.22 | 0.16 | 0.60 | 0.57 | 0.39 | 0.24 | 0.17 |
| | | | 10.0 | 5.6 | 0.47 | 0.45 | 0.33 | 0.22 | 0.16 | 0.67 | 0.63 | 0.42 | 0.26 | 0.18 |
| | | | -17.8 | 11.1 | 0.51 | 0.49 | 0.39 | 0.27 | 0.20 | 0.66 | 0.63 | 0.46 | 0.30 | 0.22 |
| | | | -17.8 | 5.6 | 0.52 | 0.50 | 0.39 | 0.27 | 0.20 | 0.73 | 0.69 | 0.49 | 0.32 | 0.23 |
| | | | -45.6 | 11.1 | 0.56 | 0.54 | 0.44 | 0.33 | 0.25 | 0.67 | 0.64 | 0.51 | 0.36 | 0.28 |
| | | | -45.6 | 5.6 | 0.57 | 0.56 | 0.45 | 0.33 | 0.26 | 0.77 | 0.74 | 0.57 | 0.39 | 0.29 |

| Position of Air Space | Direction of Heat Flow | | Air Space Mean Temp,[b] °C | Air Space Temp Diff,[b] °C | 40-mm Air Space[d] Value of E[b,c] | | | | | 90-mm Air Space[d] Value of E[b,c] | | | | |
|---|---|---|---|---|---|---|---|---|---|---|---|---|---|---|
| | | | | | 0.03 | 0.05 | 0.2 | 0.5 | 0.82 | 0.03 | 0.05 | 0.2 | 0.5 | 0.82 |
| Horiz. | Down | → | 32.2 | 5.6 | 0.44 | 0.41 | 0.29 | 0.19 | 0.14 | 0.62 | 0.58 | 0.37 | 0.21 | 0.15 |
| | | | 10.0 | 16.7 | 0.47 | 0.45 | 0.33 | 0.22 | 0.16 | 0.66 | 0.62 | 0.42 | 0.25 | 0.18 |
| | | | 10.0 | 5.6 | 0.47 | 0.45 | 0.33 | 0.22 | 0.16 | 0.68 | 0.63 | 0.42 | 0.26 | 0.18 |
| | | | −17.8 | 11.1 | 0.52 | 0.50 | 0.39 | 0.27 | 0.20 | 0.74 | 0.70 | 0.50 | 0.32 | 0.23 |
| | | | −17.8 | 5.6 | 0.52 | 0.50 | 0.39 | 0.27 | 0.20 | 0.75 | 0.71 | 0.51 | 0.32 | 0.23 |
| | | | −45.6 | 11.1 | 0.57 | 0.55 | 0.45 | 0.33 | 0.26 | 0.81 | 0.78 | 0.59 | 0.40 | 0.30 |
| | | | −45.6 | 5.6 | 0.58 | 0.56 | 0.46 | 0.33 | 0.26 | 0.83 | 0.79 | 0.60 | 0.40 | 0.30 |
| Horiz. | Up | ↑ | 32.2 | 5.6 | 0.45 | 0.42 | 0.30 | 0.19 | 0.14 | 0.50 | 0.47 | 0.32 | 0.20 | 0.14 |
| | | | 10.0 | 16.7 | 0.33 | 0.32 | 0.26 | 0.18 | 0.14 | 0.27 | 0.35 | 0.28 | 0.19 | 0.15 |
| | | | 10.0 | 5.6 | 0.44 | 0.42 | 0.32 | 0.21 | 0.16 | 0.49 | 0.47 | 0.34 | 0.23 | 0.16 |
| | | | −17.8 | 11.1 | 0.35 | 0.34 | 0.29 | 0.22 | 0.17 | 0.40 | 0.38 | 0.32 | 0.23 | 0.18 |
| | | | −17.8 | 5.6 | 0.43 | 0.41 | 0.33 | 0.24 | 0.19 | 0.48 | 0.46 | 0.36 | 0.26 | 0.20 |
| | | | −45.6 | 11.1 | 0.34 | 0.34 | 0.30 | 0.24 | 0.20 | 0.39 | 0.38 | 0.33 | 0.26 | 0.21 |
| | | | −45.6 | 5.6 | 0.42 | 0.41 | 0.35 | 0.27 | 0.22 | 0.47 | 0.45 | 0.38 | 0.29 | 0.23 |
| 45° Slope | Up | ↗ | 32.2 | 5.6 | 0.51 | 0.48 | 0.33 | 0.20 | 0.14 | 0.56 | 0.52 | 0.35 | 0.21 | 0.14 |
| | | | 10.0 | 16.7 | 0.38 | 0.36 | 0.28 | 0.20 | 0.15 | 0.40 | 0.38 | 0.29 | 0.20 | 0.15 |
| | | | 10.0 | 5.6 | 0.51 | 0.48 | 0.35 | 0.23 | 0.17 | 0.55 | 0.52 | 0.37 | 0.24 | 0.17 |
| | | | −17.8 | 11.1 | 0.40 | 0.39 | 0.32 | 0.24 | 0.18 | 0.43 | 0.41 | 0.33 | 0.24 | 0.19 |
| | | | −17.8 | 5.6 | 0.49 | 0.47 | 0.37 | 0.26 | 0.20 | 0.52 | 0.51 | 0.39 | 0.27 | 0.20 |
| | | | −45.6 | 11.1 | 0.39 | 0.38 | 0.33 | 0.26 | 0.21 | 0.41 | 0.40 | 0.35 | 0.27 | 0.22 |
| | | | −45.6 | 5.6 | 0.48 | 0.46 | 0.39 | 0.30 | 0.24 | 0.51 | 0.49 | 0.41 | 0.31 | 0.24 |
| Vertical | Horiz. | → | 32.2 | 5.6 | 0.70 | 0.64 | 0.40 | 0.22 | 0.15 | 0.65 | 0.60 | 0.38 | 0.22 | 0.15 |
| | | | 10.0 | 16.7 | 0.45 | 0.43 | 0.32 | 0.22 | 0.16 | 0.47 | 0.45 | 0.33 | 0.22 | 0.16 |
| | | | 10.0 | 5.6 | 0.67 | 0.62 | 0.42 | 0.26 | 0.18 | 0.64 | 0.60 | 0.41 | 0.25 | 0.18 |
| | | | −17.8 | 11.1 | 0.49 | 0.47 | 0.37 | 0.26 | 0.20 | 0.51 | 0.49 | 0.38 | 0.27 | 0.20 |
| | | | −17.8 | 5.6 | 0.62 | 0.59 | 0.44 | 0.29 | 0.22 | 0.61 | 0.59 | 0.44 | 0.29 | 0.22 |
| | | | −45.6 | 11.1 | 0.46 | 0.45 | 0.38 | 0.29 | 0.23 | 0.50 | 0.48 | 0.40 | 0.30 | 0.24 |
| | | | −45.6 | 5.6 | 0.58 | 0.56 | 0.46 | 0.34 | 0.26 | 0.60 | 0.58 | 0.47 | 0.34 | 0.26 |
| 45° Slope | Down | ↗ | 32.2 | 5.6 | 0.89 | 0.80 | 0.45 | 0.24 | 0.16 | 0.85 | 0.76 | 0.44 | 0.24 | 0.16 |
| | | | 10.0 | 16.7 | 0.63 | 0.59 | 0.41 | 0.25 | 0.18 | 0.62 | 0.58 | 0.40 | 0.25 | 0.18 |
| | | | 10.0 | 5.6 | 0.90 | 0.82 | 0.50 | 0.28 | 0.19 | 0.83 | 0.77 | 0.48 | 0.28 | 0.19 |
| | | | −17.8 | 11.1 | 0.68 | 0.64 | 0.47 | 0.31 | 0.22 | 0.67 | 0.64 | 0.47 | 0.31 | 0.22 |
| | | | −17.8 | 5.6 | 0.87 | 0.81 | 0.56 | 0.34 | 0.24 | 0.81 | 0.76 | 0.53 | 0.33 | 0.24 |
| | | | −45.6 | 1.1 | 0.64 | 0.62 | 0.49 | 0.35 | 0.27 | 0.66 | 0.64 | 0.51 | 0.36 | 0.28 |
| | | | −45.6 | 5.6 | 0.82 | 0.79 | 0.60 | 0.40 | 0.30 | 0.79 | 0.76 | 0.58 | 0.40 | 0.30 |

**TABLE E.5 Thermal Resistances of Plane[a] Air Spaces (SI Units) (Continued)**

| Position of Air Space | Direction of Heat Flow | Air Space Mean Temp.,[b] °C | Air Space Temp. Diff.,[b] °C | 40-mm Air Space[d] Value of E[b,c] | | | | | | 90-mm Air Space[d] Value of E[b,c] | | | | | |
|---|---|---|---|---|---|---|---|---|---|---|---|---|---|---|---|
| | | | | 0.03 | 0.05 | 0.2 | 0.5 | 0.82 | | 0.03 | 0.05 | 0.2 | 0.5 | 0.82 |
| Horiz. | Down → | 32.2 | 5.6 | 1.07 | 0.94 | 0.49 | 0.25 | 0.17 | | 1.77 | 1.44 | 0.60 | 0.28 | 0.18 |
| | | 10.0 | 16.7 | 1.10 | 0.99 | 0.56 | 0.30 | 0.20 | | 1.69 | 1.44 | 0.68 | 0.33 | 0.21 |
| | | 10.0 | 5.6 | 1.16 | 1.04 | 0.58 | 0.30 | 0.20 | | 1.96 | 1.63 | 0.72 | 0.34 | 0.22 |
| | | -17.8 | 11.1 | 1.24 | 1.13 | 0.69 | 0.39 | 0.26 | | 1.92 | 1.68 | 0.86 | 0.4 | 0.29 |
| | | -17.8 | 5.6 | 1.29 | 1.17 | 0.70 | 0.39 | 0.27 | | 2.11 | 1.82 | 0.89 | 0.44 | 0.29 |
| | | -45.6 | 11.1 | 1.36 | 1.27 | 0.84 | 0.50 | 0.35 | | 2.05 | 1.85 | 1.06 | 0.57 | 0.38 |
| | | -45.6 | 5.6 | 1.42 | 1.32 | 0.86 | 0.51 | 0.35 | | 2.28 | 2.03 | 1.12 | 0.59 | 0.39 |

*Source:* I-P table (E.4) reprinted with permission of the American Society of Heating, Refrigerating and Air-Conditioning Engineers, Inc. from the 2001 *ASHRAE Handbook—Fundamentals*; SI unit table (E.5) developed by the authors.

[a]Thermal resistance values were determined from the relation R = 1/C, where C = $h_c$ + $\varepsilon_{eff}$ $h_r$, where $h_r$ is the conduction–convection coefficient; $\varepsilon_{eff}$ $h_r$ is the radiation coefficient = 0.227$\varepsilon_{eff}$[($t_m$ + 273)/100]$^3$; and $t_m$ is the mean temperature of the air space. For extrapolation from this table to air spaces less than 12.5 mm (as in insulating window glass), assume that $h_c$ = 21.8 (1 + 0.00274$t_m$)/*l*, where *l* is the air space thickness in inches and $h_c$ is the heat transfer in W/m$^2$ K through the air space only.

[b]Interpolation is permissible for other values of mean temperature, temperature difference, and effective emittance $\varepsilon_{eff}$. Interpolation and moderate extrapolation for air spaces greater than 90 mm are also permissible.

[c]Effective emittance, $\varepsilon_{eff}$, of the air space is given by 1/$\varepsilon_{eff}$ = 1/$\varepsilon_1$ + 1/$\varepsilon_2$ − 1, where $\varepsilon_1$ and $\varepsilon_2$ are emittances of the surfaces of the air space (from Table E.3).

[d]A single resistance value cannot account for multiple air spaces. Each air space requires a separate resistance calculation that applies only for the established boundary conditions.

[e]Resistances of horizontal spaces with heat flow downward are substantially independent of temperature difference.

**TABLE E.6 U-Factors for Common Wall, Roof, and Floor Assemblies**

| Assembly | Basic Construction | Other Thermal Components | Insulation R-value °F ft² h/Btu | Insulation R-value K m²/W | Assembly U-factor Btu/°F ft² h | Assembly U-factor W/K m² |
|---|---|---|---|---|---|---|
| Wall | Wood studs, nominal 2 in. × 4 in., 16 in. o.c. (50 mm × 100 mm, 400 mm o.c.) | Exterior air film, stucco, exterior gypsum board, interior gypsum board, interior air film | 11 | 1.94 | 0.096 | 0.55 |
| | As above | Above plus R-4 (SI: R-0.7) continuous insulation | 15 | 2.64 | 0.068 | 0.39 |
| | Wood studs, nominal 2 in. × 6 in., 24 in. o.c. (50 mm × 150 mm, 400 mm o.c.) | | 18 | 3.17 | 0.065 | 0.37 |
| | Steel studs, nominal 2 in. × 4 in., 16 in. o.c. (50 mm × 100 mm, 400 mm o.c.) | | 11 | 1.94 | 0.132 | 0.75 |
| | As above | Above plus R-4 (SI: R-0.7) continuous insulation | 15 | 2.64 | 0.087 | 0.49 |
| | 6 in. concrete masonry unit at 115 lb/ft³ (150 mm, at 1840 kg/m³) | Partly grouted, cells insulated | N/A | N/A | 0.41 | 2.33 |
| | As above | Above with R-4 (SI: R-0.7) continuous insulation with 1-in. (25-mm) framing and interior gypsum board | 4 | 0.70 | 0.17 | 0.97 |
| Roof | Standard wood joists | Semiexterior air film, gypsum board, interior air film | 30 | 5.28 | 0.034 | 0.19 |
| | Steel joists | Exterior air film, metal deck, interior air film | 30 | 5.28 | 0.041 | 0.23 |
| | System with insulation entirely above roof deck | Exterior air film, metal deck, interior air film | 30 | 5.28 | 0.032 | 0.18 |

**TABLE E.6 U-Factors for Common Wall, Roof, and Floor Assemblies** *(Continued)*

| Assembly | Basic Construction | Other Thermal Components | Insulation R-value °F ft² h/Btu | Insulation R-value K m²/W | Assembly U-factor Btu/°F ft² h | Assembly U-factor W/K m² |
|---|---|---|---|---|---|---|
| Floor | Nominal 6 in. wood joists (150 mm) | Semiexterior air film, wood subfloor, carpet and pad, interior air film | 11 | 1.94 | 0.074 | 0.42 |
| | Steel floor joists | Semiexterior air film, metal deck, concrete slab, carpet and pad, interior air film | 11 | 1.94 | 0.078 | 0.44 |

*Source:* Extracted and used with permission of the American Society of Heating, Refrigerating and Air-Conditioning Engineers, Inc. from ANSI/ASHRAE/IESNA Standard 90.1-2001: *Energy Standard for Buildings Except Low-Rise Residential Buildings.*
*Note:* The SI units for resistance and U-factor were appended to the ASHRAE I-P data by the authors.

**TABLE E.7 U-Factors for Walls in Passive Solar Heating Systems**

| | Whole-Assembly U-Factor, Winter | | | | | |
|---|---|---|---|---|---|---|
| | No Night Insulation | | With Night Insulation[a] of Specified R | | | |
| | Btu | W | R-4 Btu | SI:R-0.7 W | R-9 Btu | SI:R-1.58 W |
| Component | °F ft² h | K m² | °F ft² h | K m² | °F ft² h | K m² |
| Direct gain[b] | 0.55 | 3.1 | 0.30 | 1.7 | 0.24 | 1.4 |
| Trombe wall, 18-in. (460-mm) thick[b] | 0.22 | 1.2 | 0.15 | 0.9 | 0.12 | 0.7 |
| Water wall[b] | 0.33 | 1.9 | 0.20 | 1.1 | 0.17 | 1.0 |

[a]These are daily average U-factors rather than the instantaneous U-factors generally presented in these tables. Insulation is assumed to be in place from 5:00 P.M. to 8:00 A.M., solar time, for Trombe and water walls and 5:00 P.M. to 7:00 A.M. for direct gain.
[b]From J.D. Balcomb et al., *Passive Solar Design Handbook, Volume Two: Passive Solar Design Analysis* (1980). U.S. Department of Energy, Washington, DC. SI values were appended to the I-P data by the authors.

**TABLE E.8 Effective R-Values for Wall, Roof, and Floor Systems with Steel Framing**

| Construction | Spacing of Framing | Cavity Insulation R-value I-P °F ft² h / Btu | SI K m² / W | Correction Factor[a,b] | Effective Framing/ Insulation R-value I-P °F ft² h / Btu | SI K m² / W |
|---|---|---|---|---|---|---|
| **Walls** | | | | | | |
| Nominal Cavity Depth (Stud Size) | | | | | | |
| 4 in. (100 mm) | 16 in. (400 mm) o.c. | 11 | 1.94 | 0.50 | 5.5 | 0.96 |
| | | 13 | 2.29 | 0.46 | 6.0 | 1.06 |
| | | 15 | 2.64 | 0.43 | 6.4 | 1.13 |
| 4 in. (100 mm) | 24 in. (600 mm) o.c. | 11 | 1.94 | 0.60 | 6.6 | 1.16 |
| | | 13 | 2.29 | 0.55 | 7.2 | 1.27 |
| | | 15 | 2.64 | 0.52 | 7.8 | 1.37 |
| 6 in. (150 mm) | 16 in. (400 mm) o.c. | 19 | 3.35 | 0.37 | 7.1 | 1.25 |
| | | 21 | 3.70 | 0.35 | 7.4 | 1.30 |
| 6 in. (150 mm) | 24 in. (600 mm) o.c. | 19 | 3.35 | 0.45 | 8.6 | 1.52 |
| | | 21 | 3.70 | 0.43 | 9.0 | 1.59 |
| 8 in. (200 mm) | 16 in. (400 mm) o.c. | 25 | 4.41 | 0.31 | 7.8 | 1.37 |
| 8 in. (200 mm) | 24 in. (600 mm) o.c. | 25 | 4.41 | 0.38 | 9.6 | 1.69 |
| **Roofs/Floors** | | | | | | |
| Insulation R | 4 ft (1.2 m) o.c. | | | | | |
| 11 | | 11 | 1.94 | 0.91 | 10.0 | 1.76 |
| 13 | | 13 | 2.29 | 0.90 | 11.7 | 2.06 |
| 15 | | 15 | 2.64 | 0.88 | 13.2 | 2.32 |
| 19 | | 19 | 3.35 | 0.86 | 16.3 | 2.87 |
| 21 | | 21 | 3.70 | 0.84 | 17.6 | 3.10 |
| 25 | | 25 | 4.41 | 0.81 | 20.2 | 3.56 |
| 30 | | 30 | 5.28 | 0.79 | 23.7 | 4.17 |
| 35 | | 35 | 6.16 | 0.76 | 26.6 | 4.68 |
| 40 | | 40 | 7.04 | 0.73 | 29.2 | 5.14 |
| 45 | | 45 | 7.92 | 0.71 | 32.0 | 5.63 |
| 50 | | 50 | 8.80 | 0.69 | 34.5 | 6.07 |
| 55 | | 55 | 9.68 | 0.67 | 36.9 | 6.49 |

*Source:* Reprinted with permission of the American Society of Heating, Refrigerating and Air-Conditioning Engineers, Inc. from Tables A-21 and A-20 of ANSI/ASHRAE/IESNA Standard 90.1-2001, *Energy Standard for Buildings Except Low-Rise Residential Buildings* and Table 5-1 of ANSI/ASHRAE Standard 90.2-2001, *Energy-Efficient Design of Low-Rise Residential Buildings.*
*Note:* The SI units for resistance were appended to the ASHRAE I-P data by the authors.
[a]Correction factors for walls are from Standard 90.2.
[b]Roof/floor correction factors are from Standard 90.1 and are based upon 4-ft (1.2-m) o.c. metal trusses that penetrate the insulation and 0.66-in. (16.8-mm) crossbars every 1 ft (0.3 m).

APPENDICES

**TABLE E.9 Comparison of U-Factors for Framed Walls and Structural Insulated Panels (SIP)**

| Component | Cavity R-Value[a] | | Center-of-Insulated Whole-Wall R-Value[b] | | Wall U-Factor[c] | |
|---|---|---|---|---|---|---|
| | I-P $\frac{°F\,ft^2\,h}{Btu}$ | SI $\frac{K\,m^2}{W}$ | I-P $\frac{°F\,ft^2\,h}{Btu}$ | SI $\frac{K\,m^2}{W}$ | I-P $\frac{Btu}{°F\,ft^2\,h}$ | SI $\frac{W}{k\,m^2}$ |
| 2-in. × 4-in. insulated stud wall at 16-in. o.c. (50 × 100 mm at 400 mm) | 13.6 | 2.40 | 9.6 | 1.69 | 0.104 | 0.59 |
| 2-in. × 6-in. insulated stud wall at 24-in. o.c. (50 × 150 mm at 600 mm) | 19.6 | 3.46 | 13.7 | 2.41 | 0.073 | 0.41 |
| 6½-in. SIP (165 mm) | 25.1 | 4.35 | 21.6 | 3.80 | 0.046 | 0.26 |

*Source:* Oak Ridge National Laboratory (in *Environmental Building News,* 7(5) May 1998).
[a]Center-of-insulated-cavity R-value is the maximum total R-value through a cross section of the wall system, including insulation, sheathing, drywall, etc.
[b]Whole-wall R-value is an average value that includes adjustments for the framing as well as some openings in a standardized wall system.
[c]U-factors calculated by the authors and appended to Oak Ridge R-value data; SI dimensions also appended by the authors.

## TABLE E.10 Transmission Coefficients (U-Factors) for Wood and Steel Doors

| Nominal Door Thickness | | | Wood Storm Door[a] | | Metal Storm Door[b] | | No Storm Door | |
|---|---|---|---|---|---|---|---|---|
| in. | mm | | Btu/h ft² °F | W/m² K | Btu/h ft² °F | W/m² K | Btu/h ft² °F | W/m² K |
| **Unglazed Wood Doors[c,d]** | | | | | | | | |
| 1⅜ | 35 | Panel door with ⁷⁄₁₆-in. (11-mm) panels[e] | 0.33 | 1.87 | 0.37 | 2.10 | 0.57 | 3.24 |
| 1⅜ | 35 | Hollow core flush door | 0.30 | 1.70 | 0.32 | 1.82 | 0.47 | 2.67 |
| 1⅜ | 35 | Solid core flush door | 0.26 | 1.48 | 0.28 | 1.59 | 0.39 | 2.21 |
| 1¾ | 45 | Panel door with ⁷⁄₁₆-in. (11-mm) panels | 0.32 | 1.82 | 0.36 | 2.04 | 0.54 | 3.07 |
| 1¾ | 45 | Hollow core flush door | 0.29 | 1.65 | 0.32 | 1.82 | 0.46 | 2.61 |
| 1¾ | 45 | Panel door with 1⅛-in. (29-mm) panels | 0.26 | 1.48 | 0.28 | 1.59 | 0.39 | 2.21 |
| 1¾ | 45 | Solid core flush door | — | — | 0.26 | 1.48 | 0.40 | 2.27 |
| 2¼ | 57 | Solid core flush door | 0.20 | 1.14 | 0.21 | 1.19 | 0.27 | 1.53 |
| **Unglazed Steel Doors[d]** | | | | | | | | |
| 1¾ | 45 | Fiberglass or mineral wool core with steel stiffeners, no thermal break[f] | — | — | — | — | 0.60 | 3.41 |
| 1¾ | 45 | Paper honeycomb core without thermal break[f] | — | — | — | — | 0.56 | 3.18 |
| 1¾ | 45 | Solid urethane foam core without thermal break[c] | — | — | — | — | 0.40 | 2.27 |
| 1¾ | 45 | Solid fire-rated mineral fiberboard core without thermal break[f] | — | — | — | — | 0.38 | 2.16 |
| 1¾ | 45 | Polystyrene core without thermal break[f] (18-gage [1.31-mm] commercial steel) | — | — | — | — | 0.35 | 1.99 |
| 1¾ | 45 | Polyurethane core without thermal break[f] (24-gage[0.70-mm] residential steel) | — | — | — | — | 0.29 | 1.65 |
| 1¾ | 45 | Polyurethane core with thermal break and wood perimeter[f] (24-gage [0.70-mm] residential steel) | — | — | — | — | 0.20 | 1.14 |
| 1¾ | 45 | Solid urethane foam core with thermal break[c] | — | — | 0.16 | 0.91 | 0.20 | 1.14 |

*Source:* Reprinted with permission of the American Society of Heating, Refrigerating and Air-Conditioning Engineers, Inc. from the 2001 *ASHRAE Handbook—Fundamentals* (Chapter 25, Table 6).
*Note:* U-factors listed are for exterior doors with no glazing, except for the storm doors, which are in addition to the main exterior door. Any glazing area in exterior doors should be included with the appropriate glass type and analyzed as a window (see Table E.14). Interpolation and moderate extrapolation are permitted for door thicknesses other than those specified.
[a]Values for wood storm door are approximately 50% glass.
[b]Values for metal storm door are for any percentage of glass area.
[c]Values are based on a nominal 32-in. × 80-in. (810-mm × 2030-mm) door size with no glazing.
[d]Outside air conditions: 15 mph (24 km/h) wind speed, 0°F (−18°C) air temperature; inside air conditions: natural convection, 70°F (21°C) air temperature.
[e]A 55% panel area.
[f]ASTM C 236 hot box data on a nominal 3-ft × 7-ft (910-mm × 2130-mm) door size with no glazing.

## TABLE E.11 Heat Loss Coefficients ($F_2$) for Slab-on-Grade Floors

| Construction[a] | Insulation | Btu/h °F ft Perimeter Degree Days (65°F base) | | | W/K m Perimeter Degree Days (18°C base) | | |
|---|---|---|---|---|---|---|---|
| | | 2950 | 5350 | 7433 | 1640 | 2970 | 4130 |
| (a) Block wall, 8 in. (200 mm), brick facing | Uninsulated | 0.62 | 0.68 | 0.72 | 1.07 | 1.17 | 1.24 |
| | Insulated from edge to footer: R-5.4 h ft² °F/Btu (SI: R-0.95 m² K/W) | 0.48 | 0.50 | 0.56 | 0.83 | 0.86 | 0.97 |
| (b) Block wall, 4 in. (100 mm), brick facing | Uninsulated | 0.80 | 0.84 | 0.93 | 1.38 | 1.45 | 1.61 |
| | Insulated from edge to footer: R-5.4 h ft² °F/Btu (SI: R-0.95 m² K/W) | 0.47 | 0.49 | 0.54 | 0.81 | 0.85 | 0.93 |
| (c) Metal stud wall, stucco | Uninsulated | 1.15 | 1.20 | 1.34 | 1.99 | 2.07 | 2.32 |
| | Insulated from edge to footer: R-5.4 h ft² °F/Btu (SI: R-0.95 m² K/W) | 0.51 | 0.53 | 0.58 | 0.88 | 0.92 | 1.00 |
| (d) Poured concrete wall with duct near perimeter[b] | Uninsulated | 1.84 | 2.12 | 2.73 | 3.18 | 3.67 | 4.72 |
| | Insulated from edge to footer, and 3 ft. [910 mm] under floor slab R-5.4 h ft² °F/Btu (SI: R-0.95 m² K/W) | 0.64 | 0.72 | 0.90 | 1.11 | 1.24 | 1.56 |

*Source:* Reprinted with permission of the American Society of Heating, Refrigerating and Air-Conditioning Engineers, Inc. from 2001 *ASHRAE Handbook—Fundamentals.* The SI units for $F_2$ shown in the last three columns were appended to the ASHRAE I-P data by the authors.

[a]See Fig. E.1 for illustrations of the listed constructions.

[b]Weighted average temperature of heating duct was assumed to be 110°F (43°C) during the heating season (outdoor air temperature less than 65°F [18°C]).

*Note:* To use this table:

$$q = F_2 P \Delta t$$

where
$q$ = heat loss through perimeter (Btu/h or W)
$F_2$ = heat loss coefficients from above
$P$ = perimeter of exposed slab edge (ft or m)
$\Delta t$ = temperature difference between indoor and outdoor air (°F or °C).
Do not assume additional losses from the slab to the earth below. Heat gains are assumed to be nonexistent.

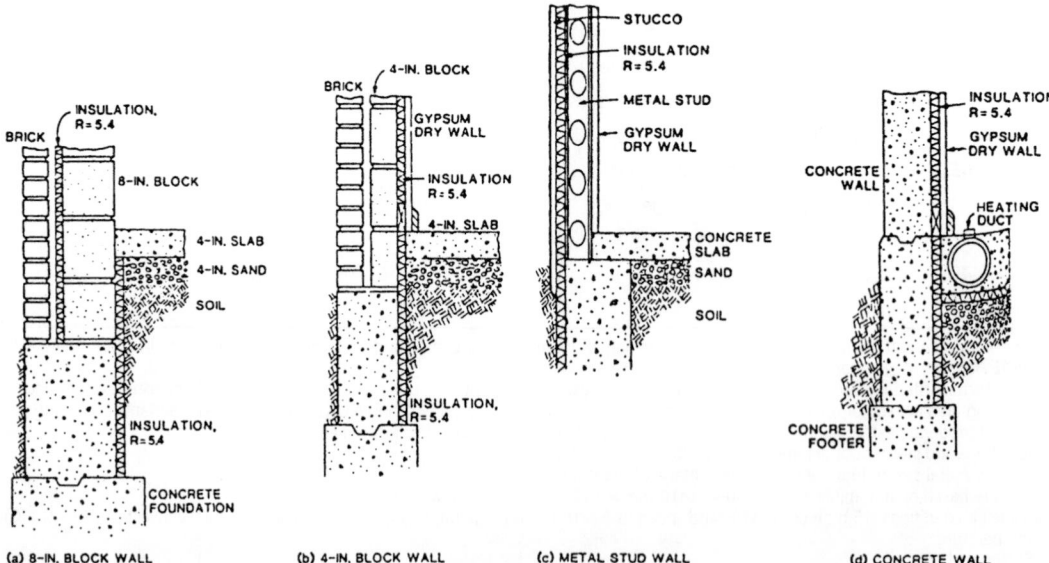

**Fig. E.1** *Insulation and construction configurations for Table E.11 data (heat loss coefficients for slab-on-grade floors). Reprinted with permission of the American Society of Heating, Refrigerating and Air-Conditioning Engineers, Inc. from 2001* ASHRAE Handbook—Fundamentals.

**TABLE E.12 Heat Flow Coefficients ($F_2$) for Slab-on-Grade Floors with Various Insulation Strategies**

| R-Value, Position, and Width (or Depth) of Insulation | $F_2$ (Btu/h ft °F) | (W/m K) |
|---|---|---|
| Uninsulated slab | 0.73 | 1.26 |
| R-5 (SI: R-0.88) Horizontal insulation, 2 ft (0.6 m), no thermal break | 0.70 | 1.21 |
| R-10 (SI: R-1.76) Horizontal insulation, 2 ft (0.6 m), no thermal break | 0.70 | 1.21 |
| R-15 (SI: R-2.64) Horizontal insulation, 2 ft (0.6 m), no thermal break | 0.69 | 1.19 |
| R-5 (SI: R-0.88) Horizontal insulation, 4 ft (1.2 m), no thermal break | 0.67 | 1.16 |
| R-10 (SI: R-1.76) Horizontal insulation, 4 ft (1.2 m), no thermal break | 0.64 | 1.11 |
| R-15 (SI: R-2.64) Horizontal insulation, 4 ft (1.2 m), no thermal break | 0.63 | 1.09 |
| R-5 (SI: R-0.88) Vertical insulation, 2 ft (0.6 m) | 0.58 | 1.00 |
| R-10 (SI: R-1.76) Vertical insulation, 2 ft (0.6 m) | 0.54 | 0.93 |
| R-15 (SI: R-2.64) Vertical insulation, 2 ft (0.6 m) | 0.52 | 0.90 |
| R-5 (SI: R-0.88) Vertical insulation, 4 ft (1.2 m) | 0.54 | 0.93 |
| R-10 (SI: R-1.76) Vertical insulation, 4 ft (1.2 m) | 0.48 | 0.83 |
| R-15 (SI: R-2.65) Vertical insulation, 4 ft (1.2 m) | 0.45 | 0.78 |
| R-10 (SI: R-1.76) Fully insulated slab (insulated under entire slab as well as around edge) | 0.36 | 0.62 |

Source: *Super Good Cents Heat Loss Reference, Vol. 1,* Ecotope Group, for Bonneville Power Administration, 1988.
Insulation is extruded polystyrene, R = 5.0 h ft$^2$ °F/Btu-in. (SI: R = 34.7 m K/W).
Soil conductivity is 0.75 Btu/h ft °F (1.30 W/K m).
No thermal break at edge of slab, where so indicated. If a thermal break is provided with horizontal insulation, use the corresponding value for vertical insulation.
Values assume an unheated slab (as per data in Table A-16 of ANSI/ASHRAE/IESNA Standard 90.1-2001, *Energy Standard for Buildings Except Low-Rise Residential Buildings*); $F_2$ values increase substantially when slab is heated.

**TABLE E.13 Heat Flow through Below-Grade Walls and Floors[a]**

| Part A. Basement Walls | | | | | | | | | | |
|---|---|---|---|---|---|---|---|---|---|---|
| SI Heat Loss[b] (W/m²K) | | | | Depth | | Path Length Through Soil | | I-P Heat Loss[b] (Btu/h ft² °F) | | |
| Uninsulated | R = 0.73 | R = 1.47 | R = 2.20 | m | (ft) | m | (ft) | Uninsulated | R = 4.17 | R = 8.34 | R = 12.5 |
| 2.33 | 0.86 | 0.53 | 0.38 | 0–0.3 | (0–1) | 0.2 | (0.68) | 0.410 | 0.152 | 0.093 | 0.067 |
| 1.26 | 0.66 | 0.45 | 0.36 | 0.3–0.6 | (1–2) | 0.69 | (2.27) | 0.222 | 0.116 | 0.079 | 0.059 |
| 0.88 | 0.53 | 0.38 | 0.30 | 0.6–0.9 | (2–3) | 1.18 | (3.88) | 0.155 | 0.094 | 0.068 | 0.053 |
| 0.67 | 0.45 | 0.34 | 0.27 | 0.9–1.2 | (3–4) | 1.68 | (5.52) | 0.119 | 0.079 | 0.060 | 0.048 |
| 0.54 | 0.39 | 0.30 | 0.25 | 1.2–1.5 | (4–5) | 2.15 | (7.05) | 0.096 | 0.069 | 0.053 | 0.044 |
| 0.45 | 0.34 | 0.27 | 0.23 | 1.5–1.8 | (5–6) | 2.64 | (8.65) | 0.079 | 0.060 | 0.048 | 0.040 |
| 0.39 | 0.30 | 0.25 | 0.21 | 1.8–2.1 | (6–7) | 3.13 | (10.28) | 0.069 | 0.054 | 0.044 | 0.037 |

| Part B. Basement Floors | | | | | | | | | |
|---|---|---|---|---|---|---|---|---|---|
| SI Heat Loss[b] (W/m²K), Width of House in Meters | | | | Depth of Foundation Wall Below Grade | | I-P Heat Loss[b] (Btu/h ft² °F), Width of House in Feet | | | |
| 6 | 7.3 | 8.5 | 9.7 | m | (ft) | 20 | 24 | 28 | 32 |
| 0.18 | 0.16 | 0.15 | 0.13 | 1.5 | (5) | 0.032 | 0.029 | 0.026 | 0.023 |
| 0.17 | 0.15 | 0.14 | 0.12 | 1.8 | (6) | 0.030 | 0.027 | 0.025 | 0.022 |
| 0.16 | 0.15 | 0.13 | 0.12 | 2.1 | (7) | 0.029 | 0.026 | 0.023 | 0.021 |

Source: Reprinted with permission of the American Society of Heating, Refrigerating and Air-Conditioning Engineers, Inc. from the 2001 *ASHRAE Handbook—Fundamentals*.
[a]Only heat losses are assumed because the interior temperature is normally higher than the ground temperature.
[b]Soil conductivity assumed to be 1.38 W/m K (9.6 Btu-in/h ft² °F).

**TABLE E.14 U-Factors of Representative Window Assemblies**

| Glazing System Description | Aluminum without Thermal Break Btu/h ft² °F (W/m² K) | Aluminum with Thermal Break Btu/h ft² °F (W/m² K) | Wood/ Vinyl Btu/h ft² °F (W/m² K) | Insulated Fiberglass/ Vinyl Btu/h ft² °F (W/m² K) |
|---|---|---|---|---|
| Single glazing with uncoated ⅛ in. [3.2 mm] clear pane | 1.27 (7.21) | 1.08 (6.13) | 0.89 (5.03) | 0.81 (4.60) |
| Single glazing with uncoated ¼ in. [6.4 mm] acrylic/polycarbonate pane | 1.14 (6.47) | 0.96 (5.45) | 0.78 (4.43) | 0.71 (4.03) |
| Double glazing with ⅛ in. [3.2 mm] panes: uncoated clear I clear with ¼ in. [6.4 mm] air space | 0.87 (4.94) | 0.65 (3.69) | 0.55 (3.12) | 0.49 (2.78) |
| Double glazing with ⅛ in. [3.2 mm] panes: uncoated clear I clear with ½ in. [13 mm] air space | 0.81 (4.60) | 0.60 (3.41) | 0.51 (2.90) | 0.44 (2.50) |
| Double glazing with ⅛ in. [3.2 mm] panes: uncoated clear I low-e (0.2) on surface[b] 3 with ½ in. [13 mm] air space | 0.71 (4.03) | 0.51 (2.90) | 0.42 (2.39) | 0.36 (2.04) |
| Triple glazing with ⅛ in. [3.2 mm] panes: uncoated clear I clear with ½ in. [13 mm] air spaces | 0.67 (3.80) | 0.46 (2.61) | 0.39 (2.21) | 0.34 (1.93) |
| Triple glazing with ⅛ in. [3.2 mm] panes: uncoated clear I low-e (0.2) on surfaces[b] 3 and 5 with ½ in. [13 mm] air spaces | 0.58 (3.29) | 0.38 (2.16) | 0.31 (1.76) | 0.27 (1.53) |
| Quadruple glazing with ⅛ in. [3.2 mm] panes: uncoated clear I low-e (0.1) on surfaces[b] 3 and 5 with ½ in. [13 mm] air spaces | 0.54 (3.07) | 0.34 (1.93) | 0.28 (1.59) | 0.24 (1.36) |

*Source:* Excerpted with permission of the American Society of Heating, Refrigerating and Air-Conditioning Engineers, Inc. from the 2001 *ASHRAE Handbook—Fundamentals* (Chapter 30, Table 4).
[a]Based upon an operable 3-ft × 5-ft (0.9-m × 1.5-m) aluminum-framed window.
[b]Glazing surfaces are numbered starting with the surface closest to the sun; thus, surface 2 would be the inner surface of an exterior pane of glass.

**TABLE E.15 Representative Window Characteristics[a]**

| Glazing Description and Reference Number | Layers of Glazing and Spaces (outside to inside) | Frame (and spacer) | *Total Window U-Factor* Btu/h ft² °F | *Total Window U-Factor* W/m²K | SHGC[b] | VT[c] | LSG[d] | FHR[e] | FCR[e] | | | | |
|---|---|---|---|---|---|---|---|---|---|---|---|---|---|
| 1. Single-glazed clear | ⅛ in. (3 mm) clear | Aluminum, no thermal break | 1.30 | *7.38* | 0.79 | 0.69 | 0.87 | 0 | 0 | [illegible] | [illegible] | [illegible] | [illegible] |
| 2. Single-glazed bronze | ⅛ in. (3 mm) bronze | Aluminum, no thermal break | 1.30 | *7.38* | 0.69 | 0.52 | 0.75 | −2 | 8 | [illegible] | [illegible] | [illegible] | [illegible] |
| 3. Double-glazed clear | ⅛ in. (3 mm) clear<br>½ in. (13 mm) air<br>⅛ in. (3 mm) clear | Aluminum, thermal break (aluminum) | 0.64 | *3.63* | 0.65 | 0.62 | 0.95 | 19 | 12 | 0.37 | *0.57* | 0.56 | *2.85* |
| 4. Double-glazed bronze | ⅛ in. (3 mm) bronze<br>½ in. (13 mm) air<br>⅛ in. (3 mm) clear | Aluminum, thermal break (aluminum) | 0.64 | *3.63* | 0.55 | 0.47 | 0.85 | 17 | 20 | 0.37 | *0.57* | 0.56 | *2.85* |
| 5. Double-glazed clear | ⅛ in. (3 mm) clear<br>½ in. (13 mm) air<br>⅛ in. (3 mm) clear | Wood or vinyl (aluminum) | 0.49 | *2.78* | 0.58 | 0.57 | 0.98 | 24 | 18 | 0.37 | *0.57* | 0.56 | *2.85* |
| 6. Double-glazed bronze | ⅛ in. (3 mm) bronze<br>½ in. (13 mm) air<br>⅛ in. (3 mm) clear | Wood or vinyl (aluminum) | 0.49 | *2.78* | 0.48 | 0.43 | 0.90 | 22 | 25 | 0.37 | *0.57* | 0.56 | *2.85* |
| 7. Double-glazed low-ε[f] | ⅛ in. (3 mm) clear<br>½ in. (13 mm) argon<br>⅛ in. (3 mm) low-ε 0.20 | Wood or vinyl (stainless) | 0.33 | *1.87* | 0.55 | 0.52 | 0.95 | 32 | 19 | 0.10 | *0.16* | 0.15 | *0.76* |
| 8. Double-glazed low-ε[f] | ⅛ in. (3 mm) low-ε 0.08<br>½ in. (13 mm) argon<br>⅛ in. (3 mm) clear | Wood or vinyl (stainless) | 0.30 | *1.70* | 0.44 | 0.56 | 1.27 | 32 | 27 | 0.10 | *0.16* | 0.15 | *0.76* |

Visual Transmitance
air leakage

**TABLE E.15 Representative Window Characteristics^a (Continued)**

| Glazing Description and Reference Number | Layers of Glazing and Spaces (outside to inside) | Frame (and spacer) | Total Window U-Factor | | Total Window | | | | | Air Leakage | | | |
|---|---|---|---|---|---|---|---|---|---|---|---|---|---|
| | | | Btu/h ft² °F | W/m²K | SHGC^b | VT^c | LSG^d | FHR^e | FCR^e | cfm/lin ft | L/s m | cfm/ft² | L/s m² |
| 9. Double-glazed spectrally selective^f | ⅛ in. (3 mm) low-ε 0.04 / ½ in. (13 mm) argon / ⅛ in. (3 mm) clear | Wood or vinyl (stainless) | 0.29 | 1.65 | 0.31 | 0.51 | 1.65 | 30 | 36 | 0.10 | 0.16 | 0.15 | 0.76 |
| 10. Double-glazed spectrally selective^f | ⅛ in. (3 mm) low-ε 0.10 / ½ in. (13 mm) argon / ⅛ in. (3 mm) clear | Wood or vinyl (stainless) | 0.31 | 1.76 | 0.26 | 0.31 | 1.19 | 27 | 40 | 0.10 | 0.16 | 0.15 | 0.76 |
| 11. Triple-glazed low-ε^f superwindow | ⅛ in. (3 mm) low-ε 0.08 / ½ in. (13 mm) krypton / ⅛ in. (3 mm) clear / ½ in. (13 mm) krypton / ⅛ in. (3 mm) low-ε 0.08 | Insulated vinyl (insulated) | 0.15 | 0.85 | 0.37 | 0.48 | 1.30 | 38 | 33 | 0.05 | 0.08 | 0.08 | 0.41 |
| 12. Triple-glazed clear | ⅛ in. (3 mm) clear / ½ in. (13 mm) air / ⅛ in. (3 mm) clear / ½ in. (13 mm) air / ⅛ in. (3 mm) clear | Wood or vinyl (stainless) | 0.34 | 1.93 | 0.52 | 0.53 | 1.02 | 32 | 22 | 0.10 | 0.16 | 0.15 | 0.76 |

*Source: Residential Windows: A Guide to New Technologies and Energy Performance*, by John Carmody, Stephen Selkowitz, and Lisa Heschong. © 1996 by John Carmody, Stephen Selkowitz and Lisa Heschong. Adapted by permission of W.W. Norton & Company, New York. (SI units appended by the authors.)

^aBased upon a casement window, 2 ft × 4 ft (610 mm × 1220 mm).
^bSolar heat gain coefficient (higher numbers mean more solar heat flow).
^cVisible transmittance (higher numbers mean more light transmitted).
^dLight-to-solar gain ratio, LSG = VT/SHGC. For typical center-of-glass values for VT and SHGC, see Table E.18.
^eFHR (fenestration heating rating) and FCR (fenestration cooling rating) are heating season and cooling season (respectively) estimates of the percentage of energy saved in a typical residential application compared to using window 1.
^fLow-ε ratings indicate what percentage of the long-wavelength radiant energy is admitted; 0.20 therefore admits 20% and reflects 80%.

**TABLE E.16 Representative Skylight U-Factors[a]**

| | Part A. I-P Units (Btu/h ft² °F) | | | | | | |
|---|---|---|---|---|---|---|---|
| | *Manufactured Skylight[b]* | | | | *Site-Assembled Glazing[c]* | | |
| | Aluminum, Thermal Break | | Reinforced Vinyl/ Aluminum- Clad Wood | Wood/ Vinyl | Aluminum, Thermal Break | | Structural Glazing |
| Glazing | No | Yes | | | No | Yes | |
| Single glazed | 1.98 | 1.89 | 1.75 | 1.47 | 1.36 | 1.25 | 1.25 |
| Double glazed, ½-in. gap | 1.30 | 1.10 | 1.04 | 0.84 | 0.81 | 0.69 | 0.65 |
| Double glazed, ½-in. gap, argon low-ε 0.20 on surface[d] 2 or 3 | 1.15 | 0.95 | 0.89 | 0.68 | 0.66 | 0.55 | 0.51 |
| Double glazed, ½-in. gap, argon low-ε 0.10 on surface[d] 2 or 3 | 1.13 | 0.93 | 0.87 | 0.67 | 0.65 | 0.53 | 0.49 |
| Double glazed, ½-in. gap, low-ε 0.05 on surface[d] 2 or 3 | 1.11 | 0.91 | 0.85 | 0.65 | 0.63 | 0.52 | 0.47 |
| Triple glazed, ½-in. gaps | 1.10 | 0.87 | 0.81 | 0.61 | 0.62 | 0.51 | 0.45 |
| Triple glazed, ½-in. gaps, argon low-ε 0.10 on surfaces[d] 2 or 3 and 4 or 5 | 0.95 | 0.72 | 0.67 | 0.47 | 0.48 | 0.37 | 0.31 |
| | Part B. SI Units (W/m² K) | | | | | | |
| | *Manufactured Skylight[b]* | | | | *Site-Assembled Glazing[c]* | | |
| | Aluminum, Thermal Break | | Reinforced Vinyl/ Aluminum- Clad Wood | Wood/ Vinyl | Aluminum, Thermal Break | | Structural Glazing |
| Glazing | No | Yes | | | No | Yes | |
| Single glazed | 11.24 | 10.73 | 9.94 | 8.35 | 7.72 | 7.10 | 7.10 |
| Double glazed, 12.7 mm gap | 7.38 | 6.25 | 5.91 | 4.77 | 4.60 | 3.92 | 3.69 |
| Double glazed, 12.7 mm gap, argon, low-ε 0.20 on surface[d] 2 or 3 | 6.53 | 5.39 | 5.05 | 3.86 | 3.75 | 3.12 | 2.90 |
| Double glazed, 12.7 mm gap, argon, low-ε 0.10 on surface[d] 2 or 3 | 6.42 | 5.28 | 4.94 | 3.80 | 3.69 | 3.01 | 2.78 |
| Double glazed, 12.7 mm gap, low-ε 0.05 on surface[d] 2 or 3 | 6.30 | 5.17 | 4.83 | 3.69 | 3.58 | 2.95 | 2.67 |
| Triple glazed, 12.7 mm gaps | 6.25 | 4.94 | 4.60 | 3.46 | 3.52 | 2.90 | 2.56 |
| Triple glazed, 12.7-mm gaps, argon, low-ε 0.10 on surfaces[d] 2 or 3 and 4 or 5 | 5.39 | 4.09 | 3.80 | 2.67 | 2.73 | 2.10 | 1.76 |

*Source:* Excerpted with permission of the American Society of Heating, Refrigerating and Air-Conditioning Engineers, Inc. from the 2001 *ASHRAE Handbook—Fundamentals.* The SI values were appended to ASHRAE's I-P data by the authors.
[a]All glazings are ⅛-in. (3-mm) thickness. Winter conditions, 15-mph (24-km/h) wind with 0°F (−18°C) outside, 70°F (21°C) inside, no solar radiation. Glazing is sloped at 20° from the horizontal.
[b]Skylight is 2 ft × 4 ft (600 mm × 1200 mm) with frame width as follows: aluminum (with or without thermal break), 0.7 in. (18 mm); wood/vinyl/aluminum-clad, 0.9 in. (23 mm); structural glazing, not applicable.
[c]Site-assembled glazing size is 4 ft × 4 ft (1200 mm × 1200 mm) with aluminum frame width of 2.25 in. (57 mm), structural glazing width of 2.5 in. (64 mm).
[d]Glazing layer surfaces are numbered from outside to inside.

**TABLE E.17 Solar Heat Gain Coefficients (SHGC) for Plastic Domed Horizontal Skylights**

| Dome | Light Diffuser (Translucent) | Curb (See Sketch Below) | | Solar Heat Gain Coefficient | U-Factor at Center of Skylight[b] |
|---|---|---|---|---|---|
| | | Height, in. (mm) | Width-to-Height Ratio | | |
| Clear ($\tau = 0.86^a$) | Yes ($\tau = 0.58$) | 0 (0) | $\infty$ | 0.53 | 0.46 (2.6) |
| | | 9 (230) | 5 | 0.50 | 0.43 (2.4) |
| | | 18 (460) | 2.5 | 0.44 | 0.40 (2.3) |
| Clear ($\tau = 0.86$) | None | 0 (0) | $\infty$ | 0.86 | 0.80 (4.5) |
| | | 9 (230) | 5 | 0.77 | 0.75 (4.3) |
| | | 18 (460) | 2.5 | 0.70 | 0.70 (4.0) |
| Translucent ($\tau = 0.52$) | None | 0 (0) | $\infty$ | 0.50 | 0.80 (4.5) |
| | | 18 (460) | 2.5 | 0.40 | 0.70 (4.0) |
| Translucent ($\tau = 0.27$) | None | 0 (0) | $\infty$ | 0.30 | 0.80 (4.5) |
| | | 9 (230) | 5 | 0.26 | 0.75 (4.3) |
| | | 18 (460) | 2.5 | 0.24 | 0.70 (4.0) |

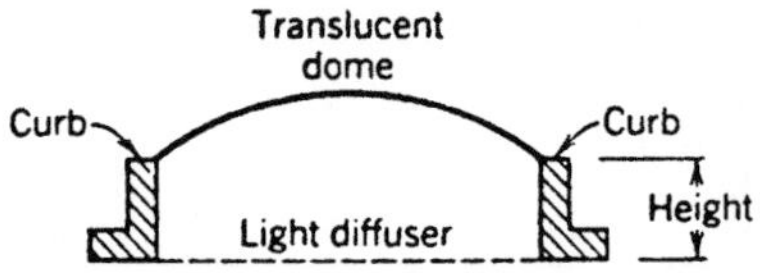

*Source:* Excerpted with permission of the American Society of Heating, Refrigerating and Air-Conditioning Engineers, Inc. from the 2001 *ASHRAE Handbook—Fundamentals* (solar heat gain coefficients) and the 1981 *ASHRAE Handbook—Fundamentals* (U-factors).
[a]$\tau$ refers to the transmittance of the dome or diffusing element, as noted.
[b]U-factor units: Btu/h ft$^2$ °F (W/m$^2$ K).

Visual transmittance

## resentative Glazings and Window Assemblies

| | Center of Glazing VT | Center of Glazing SHGC[a] | Window Assembly VT[a,b] | Window Assembly SHGC[a,b] |
|---|---|---|---|---|
| ept as noted | | | | |
| | 0.90 | 0.86 | 0.77 | 0.75 |
| | 0.88 | 0.81 | 0.75 | 0.71 |
| | 0.54 | 0.62 | 0.45 | 0.54 |
| | 0.76 | 0.60 | 0.65 | 0.53 |
| Uncoated gray | 0.46 | 0.59 | 0.39 | 0.53 |
| Reflective stainless steel on clear (8%) | 0.08 | 0.19 | 0.07 | 0.18 |
| Reflective stainless steel on clear (20%) | 0.20 | 0.31 | 0.17 | 0.28 |
| **Double Glazings—all with ¼ in. (6.4 mm) panes** | | | | |
| Uncoated clear I clear | 0.78 | 0.70 | 0.66 | 0.61 |
| Uncoated bronze I clear | 0.47 | 0.49 | 0.40 | 0.44 |
| Uncoated high-performance green I clear | 0.59 | 0.39 | 0.50 | 0.35 |
| Reflective stainless steel on clear (8%) I clear | 0.07 | 0.13 | 0.06 | 0.13 |
| Reflective titanium on clear (20%) I clear | 0.18 | 0.21 | 0.15 | 0.20 |
| Clear I low-e (0.2) on surface[c] 3 | 0.73 | 0.65 | 0.62 | 0.57 |
| Low-e (0.2) on surface 2 I clear | 0.73 | 0.60 | 0.62 | 0.53 |
| Low-e (0.1) on surface 2 I clear | 0.72 | 0.60 | 0.61 | 0.45 |
| Low-e (0.05) on surface 2 I clear | 0.70 | 0.37 | 0.60 | 0.34 |
| **Triple Glazings—all with ¼ in. (6.4 mm) panes** | | | | |
| Clear I clear I clear | 0.70 | 0.61 | 0.60 | 0.54 |
| High-performance green I clear I clear | 0.53 | 0.32 | 0.45 | 0.31 |
| Low-e (0.2) on surface 2 I clear I clear | 0.64 | 0.53 | 0.54 | 0.47 |
| Low-e (0.1) on surface 2 I clear I low-e (0.1) on surface 5 | 0.59 | 0.36 | 0.50 | 0.33 |
| Low-e (0.05) on surface 2 I low-e (0.05) on surface 4 I clear | 0.55 | 0.26 | 0.47 | 0.24 |

*Source:* Excerpted with permission of the American Society of Heating, Refrigerating and Air-Conditioning Engineers, Inc. from the 2001 *ASHRAE Handbook—Fundamentals* (Chapter 30, Table 13).
[a]Values are for a normal angle of solar incidence, see the 2001 *ASHRAE Handbook—Fundamentals* for SHGC values at other incidence angles.
[b]Based upon an operable 3-ft × 5-ft (0.9-m × 1.5-m) aluminum-framed window.
[c]Glazing surfaces are numbered starting with the surface closest to the sun; thus, surface 2 would be the inner surface of an exterior pane of glass.

**TABLE E.19 Solar Optical Properties of Transparent Plastics**

| Type of Plastic | Transmittance | | SC[a] |
| | Visible | Solar | |
| --- | --- | --- | --- |
| Acrylic | | | |
|   Clear | 0.92 | 0.85 | 0.98 |
|   Gray tint | 0.16 | 0.27 | 0.52 |
|   Gray tint | 0.33 | 0.41 | 0.63 |
|   Gray tint | 0.45 | 0.55 | 0.74 |
|   Gray tint | 0.59 | 0.62 | 0.80 |
|   Gray tint | 0.76 | 0.74 | 0.89 |
|   Bronze tint | 0.27 | 0.35 | 0.58 |
|   Bronze tint | 0.49 | 0.56 | 0.75 |
|   Bronze tint | 0.61 | 0.62 | 0.80 |
|   Bronze tint | 0.75 | 0.75 | 0.90 |
|   Reflective[b] | 0.14 | 0.12 | 0.21 |
| Polycarbonate | | | |
|   Clear, ⅛ in. (3 mm) | 0.88 | 0.82 | 0.98 |
|   Gray or bronze, ⅛ in. (3 mm) | 0.50 | 0.57 | 0.74 |

*Source:* Reprinted with permission of the American Society of Heating, Refrigerating and Air-Conditioning Engineers, Inc. from the 1997 *ASHRAE Handbook—Fundamentals.*
[a]Shading coefficient; multiply SC by 0.87 for an approximate estimation of SHGC (solar heat gain coefficient).
[b]Aluminum metalized polyester film on plastic.

**TABLE E.20 Approximate Shading Coefficients (SC) of External Shading Devices**

| | |
| --- | --- |
| Awnings | |
|   Of venetian blind type, ⅔ drawn | 0.43 |
|   Of venetian blind type, fully drawn | 0.15 |
|   Dark or medium canvas | 0.25 |
| Shading screens | 0.28–0.23 |
| Louvers, movable | 0.15–0.10 |
| Overhang: continuous, completely shading window | 0.25 |
| Dense tree casting heavy shade | 0.25 |

*Source:* Adapted from V. Olgyay *Design with Climate,* 1963. Princeton University Press, Princeton, NJ.

Plastic transmittance

**...ouvered Sun Screens**

| | Group 3[c] | | Group 4[d] | | Group 5[e] | | Group 6[f] | |
|---|---|---|---|---|---|---|---|---|
| | Trans-mittance | SC | Trans-mittance | SC | Trans-mittance | SC | Trans-mittance | SC |
| | 0.40 | 0.51 | 0.48 | 0.59 | 0.15 | 0.27 | 0.26 | 0.45 |
| | 0.32 | 0.42 | 0.39 | 0.50 | 0.04 | 0.11 | 0.20 | 0.35 |
| | 0.21 | 0.31 | 0.28 | 0.38 | 0.03 | 0.10 | 0.13 | 0.26 |
| | 0.07 | 0.18 | 0.20 | 0.30 | 0.03 | 0.10 | 0.04 | 0.13 |

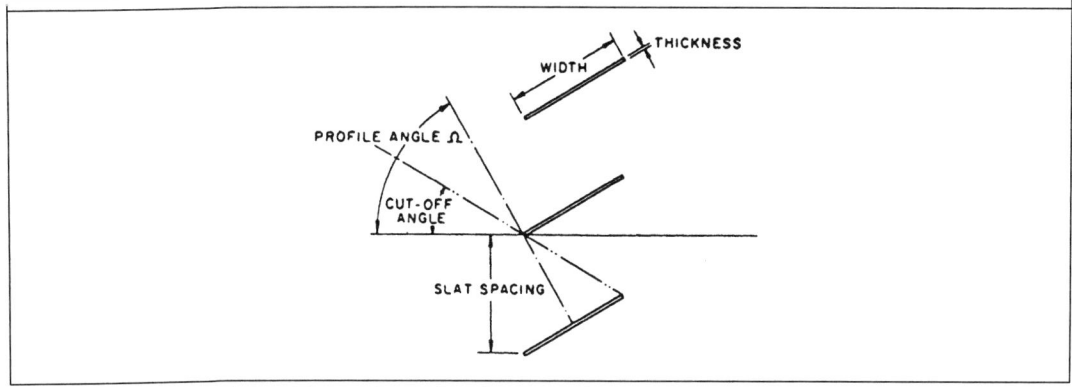

*Source:* Reprinted with permission of the American Society of Heating, Refrigerating and Air-Conditioning Engineers, Inc., from the 1997 *ASHRAE Handbook—Fundamentals.*

*Note:* The 2001 *ASHRAE Handbook—Fundamentals* presents a different and more complex methodology for determining solar gain through externally shaded glazing systems—involving the variables $F_u$ (unshaded fraction of glazing) and EAC (exterior solar attenuation coefficient). The methodology and associated data are not presented here, as they are less intuitive and less suited to hand calculations than the older SC data presented above. Refer to the 2001 *Handbook* for this new information.

[a]*Group 1:* Black, width over spacing ratio 1.15/1; 23 louvers/in. (1.1 mm between louvers), with ¼-in. (6.4-mm) single clear glass.

[b]*Group 2:* Light color; high reflectance, otherwise same as group 1.

[c]*Group 3:* Black or dark color; width over spacing ratio 0.85/1; 17 louvers/in. (1.5 mm between louvers), with ¼-in. (6.4-mm) single clear glass.

[d]*Group 4:* Light color or unpainted aluminum; high reflectance; otherwise same as group 3.

[e]*Group 5:* Same as group 1, except two lights of ¼-in. (6.4-mm) clear glass with ½-in. (12.7-mm) air space.

[f]*Group 6:* Same as group 3, except two lights of ¼ in. (6.4 mm) clear glass with ½-in. (12.7-mm) air space.

**TABLE E.22 Performance Characteristics of Adjustable Sunshading**

| Shading Type | Typical Orientations | Solar Heat Rejection | Daylight Distribution | Ventilation | View | Winter Night Insulation |
|---|---|---|---|---|---|---|
| Roll-down Shades Solid | East and west, where it is rolled down daily during hours of direct sun exposure. | Blocks direct, diffuse, and reflected radiation. Heat builds up between shade and glass unless vented. | Translucent shades diffuse light evenly but may become a source of glare due to brightness. If shades are colored, incoming light will be colored. | Blocks ventilation air flow. | Blocks view. | Can be effective if air flow around the shade is blocked (providing dead air space). |
| Slatted | East and west, where it is rolled down daily during hours of direct sun exposure. | Blocks most direct, diffuse, and reflected radiation. Heat can convect through slats into space. | A thinly striped pattern of direct sun is admitted. Opaque slats will heavily filter daylight. | Greatly reduces ventilation air flow. | Substantially obstructs and filters view. | No value unless slats can be fitted together to close openings and trap air between shade and glass. |
| Fold-out Retractable Awning | Any orientation. On south, awnings can be deep and be left in position for weeks at a time. | Blocks direct sun; admits most diffuse and reflected radiation. Heat collected on awning easily dissipated outside of building. | Translucent awnings evenly diffuse sunlight. (Colored awnings will color light). Awnings can redirect ground-reflected light into building for distribution by ceiling. | Blocks view of sky; permits view of ground and some surroundings (depending upon positioning). | Reduces ventilation air flow somewhat, tends to direct wind upward into space. | If awning folds back against window, it can provide some insulation value—but likely limited. |

**TABLE E.22 Performance Characteristics of Adjustable Sunshading** *(Continued)*

| Shading Type | Typical Orientations | Solar Heat Rejection | Daylight Distribution | Ventilation | View | Winter Night Insulation |
|---|---|---|---|---|---|---|
| Pivoting Louvers<br><br>Horizontal | South; also used on other elevations (although less effective). | Blocks direct sun; admits ground-reflected and some diffuse radiation. | Skylight can be reflected to ceiling plane for distribution into building. Ground-reflected light can be reflected into space. | Reduces ventilation air flow somewhat (based upon louver spacing); directs wind upward into space. | Blocks sky view; filters ground and surrounding views (depending upon angle of louvers). | Closed louvers can provide some insulation value (depending upon tightness of closure). |
| Vertical | East and west; also used on north. Not as effective on south as horizontal louvers. | Blocks direct sun; admits most ground-reflected and diffuse radiation. | Blocks direct sunlight; admits most ground-reflected and diffuse light. | Reduces ventilation air flow somewhat (to extent of solididity); directs wind sideways into space. | Filters view, especially to sides. | Closed louvers can provide some insulation value (depending upon tightness of closure). |

APPENDICES

**TABLE E.23 Shading Coefficients (SC) for Glazing with Integral or Interior Shading**

**Part A. SC for Insulating Glass[a] with Indoor Shading by Venetian Blinds or Roller Shades**

| Type of Glass | Nominal Thickness, Each Light | Solar Transmittance[b] | | Type of Shading | | | | | |
|---|---|---|---|---|---|---|---|---|---|
| | | | | Venetian Blinds[c] | | Roller Shade | | | |
| | | | | | | Opaque | | Translucent | |
| | | Outer Pane | Inner Pane | Medium | Light | Dark | White | Light | |
| Clear out | ⅛ in. (3.2 mm) | 0.87 | 0.87 | 0.62 | 0.58 (45° open) | 0.71 | 0.35 | 0.40 | |
| Clear in | | | | 0.63 | 0.58 (closed)[d] | | | | |
| Heat-absorbing[e] out | ¼ in. (6.4 mm) | 0.46 | 0.80 | 0.39 | 0.36 | 0.40 | 0.22 | 0.30 | |
| Clear in | | | | | | | | | |
| Reflective coated glass | | | | | | | | | |
| SC[f] = 0.20 | | | | 0.19 | 0.18 | | | | |
| 0.30 | | | | 0.27 | 0.26 | | | | |
| 0.40 | | | | 0.34 | 0.33 | | | | |

**Part B. Shading Coefficients for Double Glazing with Between-Glass Shading**

| Type of Glass | Nominal Thickness, Each Pane | Solar Transmittance[b] | | Type of Shading | | | Description of Air Space |
|---|---|---|---|---|---|---|---|
| | | | | Venetian Blinds | | Louvered Sun Screen | |
| | | Outer Pane | Inner Pane | Light | Medium | | |
| Clear out | ⅛ in. (3.2 mm) | 0.87 | 0.87 | 0.33 | 0.36 | 0.43 | Shade in contact with glass or shade separated from glass by airspace |
| Clear in | | | | | | | |

| | | | | | | | |
|---|---|---|---|---|---|---|---|
| Clear out<br>Clear in | ¼ in. (6.4 mm) | 0.80 | 0.80 | Shade in contact with glass; voids filled with plastic | — | — | 0.49 |
| Heat-absorbing[e] out | | | | Shade in contact with glass or shade separated from glass by airspace | 0.28 | 0.30 | 0.37 |
| Clear in | ¼ in. (6 mm) | 0.46 | 0.80 | Shade in contact with glass; voids filled with plastic | — | — | 0.41 |

*Source:* Reprinted with permission of the American Society of Heating, Refrigerating and Air-Conditioning Engineers, Inc. from the 1997 *ASHRAE Handbook—Fundamentals*.

*Note:* The 2001 *ASHRAE Handbook—Fundamentals* presents a different and more complex methodology for determining solar gain through internally and integrally shaded glazing systems—involving the variables IAC and BAC (interior and between-glass solar attenuation coefficients, respectively). The methodology and associated data are not presented here, as they are less intuitive and less suited to hand calculations than the older SC data presented above. Refer to the 2001 *Handbook* for this new information.

[a]Factory-manufactured units with ³⁄₁₆-in. (4.8-mm), ¼-in. (6.4-mm), ½-in. (12.7 mm) air space.
[b]Refer to manufacturers' literature for exact values.
[c]For vertical blinds with opaque, white, or beige louvers, tightly closed, SC is approximately the same as for opaque white roller shades.
[d]Use these values only when operation is automated for solar gain reduction (as opposed to daylight use).
[e]Refers to bronze or green-tinted, heat-absorbing glass.
[f]SC for glass with no shading device.

**TABLE E.24 Shading Coefficients (SC) for Single and Insulating Glass with Draperies**

| Glazing | Glass Trans. | Glass SC[a] | Range of Shading Coefficients — Drapery Fabrics[b] | |
|---|---|---|---|---|
| | | | High Transmittance, Low Reflectance[c] | Low Transmittance, High Reflectance[c] |
| Single Glass | | | | |
| ⅛ in. (3.2 mm) clear | 0.86 | 1.00 | 0.87 | 0.37 |
| ¼ in. (6.4 mm) clear | 0.80 | 0.95 | 0.80 | 0.35 |
| ½ in. (12.7 mm) clear | 0.71 | 0.88 | 0.74 | 0.35 |
| ¼ in. (6.4 mm) heat absorbing | 0.46 | 0.67 | 0.57 | 0.33 |
| ½ in. (12.7 mm) heat absorbing | 0.24 | 0.50 | 0.43 | 0.30 |
| Reflective coated (see manufacturers' | — | 0.60 | 0.57 | 0.33 |
| literature for exact values) | — | 0.50 | 0.46 | 0.31 |
| | — | 0.40 | 0.36 | 0.26 |
| | — | 0.30 | 0.25 | 0.20 |
| Insulating Glass, ½ in. (12.7 mm) air space | | | | |
| Clear out and clear in | 0.64 | 0.83 | 0.66 | 0.35 |
| Heat absorbing out and clear in | 0.37 | 0.55 | 0.49 | 0.32 |
| Reflective coated (see manufacturers' | — | 0.40 | 0.38 | 0.28 |
| literature for exact | — | 0.30 | 0.29 | 0.24 |
| values) | — | 0.20 | 0.19 | 0.15 |

*Source:* Based upon the 1997 *ASHRAE Handbook—Fundamentals,* published by the American Society of Heating, Refrigerating and Air-Conditioning Engineers, Inc., Atlanta, GA.
*Note:* The 2001 *ASHRAE Handbook—Fundamentals* presents a different and more complex methodology for determining solar gain through internally shaded glazing systems—involving the variable IAC (interior solar attenuation coefficient). The methodology and associated data are not presented here, as they are less intuitive and less suited to hand calculations than the older SC data presented above. Refer to the 2001 *Handbook* for this new information.
[a]For glass alone, with no drapery.
[b]Draperies of 100% fullness, loose hanging.
[c]See the 1997 *ASHRAE Handbook—Fundamentals,* Chapter 27, Table 29, for more detailed listings.

Ventilation rates *(handwritten)*

**TABLE E.25 Minimum Ventilation Rates in Breathing Zone (for Buildings Except Low-Rise Residential)**

| Occupancy Category | People Outdoor Air Rate $R_P$ cfm/Person | People Outdoor Air Rate $R_P$ L/s Person | Area Outdoor Air Rate $R_A$ cfm/ft² | Area Outdoor Air Rate $R_A$ L/s m² | Notes | Occupant Density #/1000 ft² | D… cfm/person | D… L/s person |
|---|---|---|---|---|---|---|---|---|
| **Correctional Facilities** | | | | | | | | |
| Cell | 5 | 2.5 | 0.12 | 0.6 | | 25 | | |
| Day room | 5 | 2.5 | 0.06 | 0.3 | | 30 | | |
| Guard stations | 5 | 2.5 | 0.06 | 0.3 | | 15 | | |
| Booking/waiting | 7.5 | 3.8 | 0.06 | 0.3 | | 50 | | |
| **Educational Facilities** | | | | | | | | |
| Daycare (through age 4) | 10 | 5 | 0.18 | 0.9 | | 25 | | 6.7 |
| Classrooms (ages 5–8) | 10 | 5 | 0.12 | 0.6 | | 25 | | 4.3 |
| Classrooms (age 9 plus) | 10 | 5 | 0.12 | 0.6 | | 35 | 13 | 4.0 |
| Lecture classroom | 7.5 | 3.8 | 0.06 | 0.3 | | 65 | 8 | 9.5 |
| Lecture hall (fixed seats) | 7.5 | 3.8 | 0.06 | 0.3 | | 150 | 8 | 8.6 |
| Art classroom | 10 | 5 | 0.18 | 0.9 | | 20 | 19 | 9.5 |
| Science laboratories | 10 | 5 | 0.18 | 0.9 | | 25 | 17 | 9.5 |
| Wood/metal shop | 10 | 5 | 0.18 | 0.9 | | 20 | 19 | 7.4 |
| Computer lab | 10 | 5 | 0.12 | 0.6 | | 25 | 15 | 7.4 |
| Media center | 10 | 5 | 0.12 | 0.6 | A | 25 | 15 | 5.9 |
| Music/theater/dance | 10 | 5 | 0.06 | 0.3 | | 35 | 12 | 4.1 |
| Multi-use assembly | 7.5 | 3.8 | 0.06 | 0.3 | | 100 | 8 | |
| **Food and Beverage Service** | | | | | | | | |
| Restaurant dining rooms | 7.5 | 3.8 | 0.18 | 0.9 | | 70 | 10 | 5.1 |
| Cafeteria/fast food dining | 7.5 | 3.8 | 0.18 | 0.9 | | 100 | 9 | 4.7 |
| Bars, cocktail lounges | 7.5 | 3.8 | 0.18 | 0.9 | | 100 | 9 | 4.7 |
| **General** | | | | | | | | |
| Conference/meeting | 5 | 2.5 | 0.06 | 0.3 | | 50 | 6 | 3.1 |
| Corridors | — | — | 0.06 | 0.3 | | — | | |
| Storage rooms | — | — | 0.12 | 0.6 | B | — | | |

TABLE E.25 Minimum Ventilation Rates in Breathing Zone (for Buildings Except Low-Rise Residential) (Continued)

| Occupancy Category | People Outdoor Air Rate | | | | Area Outdoor Air Rate | | | Default Values | | | |
|---|---|---|---|---|---|---|---|---|---|---|---|
| | $R_P$ | | | | $R_A$ | | Notes | Occupant Density | Combined Outdoor Air Rate | | |
| | cfm/Person | L/s Person | | | cfm/ft² | L/s m² | | #/1000 ft² | cfm/Person | | L/s Person |
| **Hotels, Motels, Resorts, Dormitories** | | | | | | | | | | | |
| Bedroom/living room | 5 | 2.5 | | | 0.06 | 0.3 | | 10 | 11 | | 5.5 |
| Barracks sleeping areas | 5 | 2.5 | | | 0.06 | 0.3 | | 20 | 8 | | 4 |
| Lobbies/prefunction | 7.5 | 3.8 | | | 0.06 | 0.3 | | 30 | 10 | | 4.8 |
| Multi-purpose assembly | 5 | 2.5 | | | 0.06 | 0.3 | | 120 | 6 | | 2.8 |
| **Office Buildings** | | | | | | | | | | | |
| Office space | 5 | 2.5 | | | 0.06 | 0.3 | | 5 | 17 | | 8.5 |
| Reception areas | 5 | 2.5 | | | 0.06 | 0.3 | | 30 | 7 | | 3.5 |
| Telephone/data entry | 5 | 2.5 | | | 0.06 | 0.3 | | 60 | 6 | | 3 |
| Main entry lobbies | 5 | 2.5 | | | 0.06 | 0.3 | | 10 | 11 | | 5.5 |
| **Miscellaneous Spaces** | | | | | | | | | | | |
| Bank vaults/safe deposit | 5 | 2.5 | | | 0.06 | 0.3 | | 5 | 17 | | 8.5 |
| Computer (not printing) | 5 | 2.5 | | | 0.06 | 0.3 | | 4 | 20 | | 10 |
| Pharmacy (prep. area) | 5 | 2.5 | | | 0.18 | 0.9 | | 10 | 23 | | 11.5 |
| Photo studios | 5 | 2.5 | | | 0.12 | 0.6 | | 10 | 17 | | 8.5 |
| Shipping/receiving | — | — | | | 0.12 | 0.6 | B | — | | | |
| Transportation waiting | 7.5 | 3.8 | | | 0.06 | 0.3 | | 100 | 8 | | 4.1 |
| Warehouses | — | — | | | 0.06 | 0.3 | B | — | | | |
| **Public Assembly Spaces** | | | | | | | | | | | |
| Auditorium seating area | 5.0 | 2.5 | | | 0.06 | 0.3 | | 150 | 5 | | 2.7 |
| Places of religious worship | 5.0 | 2.5 | | | 0.06 | 0.3 | | 120 | 6 | | 2.8 |
| Courtrooms | 5.0 | 2.5 | | | 0.06 | 0.3 | | 70 | 6 | | 2.9 |
| Legislative chambers | 5.0 | 2.5 | | | 0.06 | 0.3 | | 50 | 6 | | 3.1 |
| Libraries | 5.0 | 2.5 | | | 0.12 | 0.6 | | 10 | 17 | | 8.5 |
| Lobbies | 5.0 | 2.5 | | | 0.06 | 0.3 | | 150 | 5 | | 2.7 |
| Museums (children's) | 7.5 | 3.8 | | | 0.12 | 0.6 | | 40 | 11 | | 5.3 |
| Museums/galleries | 7.5 | 3.8 | | | 0.06 | 0.3 | | 40 | 9 | | 4.6 |

| | | | | | Notes | | | |
|---|---|---|---|---|---|---|---|---|
| **Retail** | | | | | | | | |
| Sales (except as below) | 7.5 | 3.8 | 0.12 | 0.6 | | 15 | 16 | 7.8 |
| Mall common areas | 7.5 | 3.8 | 0.06 | 0.3 | | 40 | 9 | 4.6 |
| Barber shop | 7.5 | 3.8 | 0.06 | 0.3 | | 25 | 10 | 5 |
| Beauty and nail salons | 20 | 10 | 0.12 | 0.6 | | 25 | 25 | 12.4 |
| Pet shops (animal areas) | 7.5 | 3.8 | 0.18 | 0.9 | | 10 | 26 | 12.8 |
| Supermarket | 7.5 | 3.8 | 0.06 | 0.3 | | 8 | 15 | 7.6 |
| Coin-operated laundries | 7.5 | 3.8 | 0.06 | 0.3 | | 20 | 11 | 5.3 |
| **Sports and Entertainment** | | | | | | | | |
| Sports arena (play area) | — | — | 0.3 | 1.5 | | — | | |
| Gym, stadium (play area) | — | — | 0.3 | 1.5 | | 30 | | |
| Spectator areas | 7.5 | 3.8 | 0.06 | 0.3 | | 150 | 8 | 4 |
| Swimming (pool and deck) | — | — | 0.48 | 2.4 | C | — | | |
| Disco/dance floors | 20 | 10 | 0.06 | 0.3 | | 100 | 21 | 10.3 |
| Health club/aerobics room | 20 | 10 | 0.06 | 0.3 | | 40 | 22 | 10.8 |
| Health club/weight rooms | 20 | 10 | 0.06 | 0.3 | | 10 | 26 | 13 |
| Bowling alley (seating) | 10 | 5 | 0.12 | 0.6 | | 40 | 13 | 6.5 |
| Gambling casinos | 7.5 | 3.8 | 0.18 | 0.9 | | 120 | 9 | 4.6 |
| Game arcades | 7.5 | 3.8 | 0.18 | 0.9 | | 20 | 17 | 8.3 |
| Stages, studios | 10 | 5 | 0.06 | 0.3 | D | 70 | 11 | 5.4 |

*Source:* Reprinted with permission of the American Society of Heating, Refrigerating and Air-Conditioning Engineers, Inc. from *Addendum n to ANSI / ASHRAE Standard 62-2001, Ventilation for Acceptable Indoor Air Quality.*

*General Notes*

1. *Related Requirements:* The rates in this table are based on all other requirements of Standard 62-2001 (with addenda) being met.
2. *Smoking:* This table applies to no-smoking areas. Rates for smoking-permitted spaces must be determined using other means.
3. *Air Density:* Volumetric airflow rates are based on an air density of 1.2 $kg_{da}/m^3$ (0.075 $lb_{da}/ft^2$), which corresponds to dry air at a barometric pressure of 101.3 kPa (1 atm) and an air temperature of 21°C (70°F). Rates may be adjusted for actual density, but such adjustment is not required for compliance with this standard.
4. *Default Occupant Density:* The default occupant density shall be used when actual occupant density is not known.
5. *Default Combined Outdoor Air Rate (per person):* This rate is based on the default occupant density.
6. *Unlisted Occupancies:* If the occupancy category for a proposed space or zone is not listed, the requirements for the listed occupancy category that is most similar in terms of occupant density, activities, and building construction shall be used.
7. *Residential Facilities, Health Care Facilities, and Vehicles:* Rates shall be determined in accordance with Appendix E (of Standard 62-2001).

*Item-Specific Notes:*

A. For high school and college libraries, use values shown for Public Spaces—Libraries.
B. Rate may not be sufficient when stored materials include those having potentially harmful emissions.
C. Rate does not allow for humidity control. Additional ventilation or dehumidification may be required to remove moisture.
D. Rate does not include special exhaust for stage effects, e.g., dry ice vapors, smoke.

**TABLE E.26 Recommended Ventilation and Exhaust Air Requirements—Low-Rise Residential**

| Part A: Ventilation Air (cfm [L/s]) | | | | | |
|---|---|---|---|---|---|
| Floor Area | 0–1 | 2–3 | 4–5 | 6–7 | >7 |
| ft² (m²) | Bedroom | Bedrooms | Bedrooms | Bedrooms | Bedrooms |
| <1500 (<139) | 30 (14) | 45 (21) | 60 (28) | 75 (35) | 90 (42) |
| 1501–3000 (139.1–279) | 45 (21) | 60 (28) | 75 (35) | 90 (42) | 105 (50) |
| 3001–4500 (279.1–418) | 60 (28) | 75 (35) | 90 (42) | 105 (50) | 120 (57) |
| 4501–6000 (418.1–557) | 75 (35) | 90 (42) | 105 (50) | 120 (57) | 135 (64) |
| 6001–7500 (557.1–697) | 90 (42) | 105 (50) | 120 (57) | 135 (64) | 150 (71) |
| >7500 (>697) | 105 (50) | 120 (57) | 135 (64) | 150 (71) | 165 (78) |
| Part B: Exhaust Air | | | | | |

If **continuous**—local ventilation exhaust air flow rates:

  Kitchen: 5 air changes per hour (based upon kitchen volume)

  Bathroom: 20 cfm [10 L/s]

If **intermittent**—local ventilation exhaust air flow rates:

  Kitchen: 100 cfm (50 L/s) (vented range hood required if exhaust fan flow rate is less than 5 kitchen air changes per hour)

  Bathroom: 50 cfm (25 L/s)

*Source:* Reprinted with permission of the American Society of Heating, Refrigerating and Air-Conditioning Engineers, Inc. from ASHRAE Standard 62.2-2003, *Ventilation and Acceptable Indoor Air Quality in Low-Rise Residential Buildings.* See Standard 62.2 for definitions, assumptions, implementation, and exceptions.

**TABLE E.27 Estimated Overall Infiltration Rates for Small Buildings**

| Part A. Construction Types | |
|---|---|
| Construction Type | Description |
| Tight | Good multifamily residential construction with close-fitting doors, windows, and framing is considered tight. New houses with full vapor retarder, no fireplace, well-fitted windows, weather-stripped doors, one-story, and less than 1500 ft² (140 m²) floor area fall into this category. |
| Medium | Medium structures include new two-story frame houses or one-story houses more than 10 years old with average maintenance, a floor area greater than 1500 ft² (140 m²), average-fit windows and doors, and a fireplace with damper and glass closure. Below-average multifamily construction falls in this category. |
| Loose | Loose structures are poorly constructed single and multifamily residences with poorly fitted windows and doors. Examples include houses more than 20 years old, of average maintenance, having a fireplace without damper or glass closure, or having more than an average number of vented appliances. Average manufactured homes are in this category. |

**Part B. Design Infiltration Rate (ACH) for Winter: Indoors 68°F (20°C); Wind Speed = 15 mph (6.7 m/s)**

| Construction Type | Winter Outdoor Design Temperature | | | | | | | | | |
|---|---|---|---|---|---|---|---|---|---|---|
| | °F: 50 | 40 | 30 | 20 | 10 | 0 | −10 | −20 | −30 | −40 |
| | °C: 10 | 4 | −1 | −7 | −12 | −18 | −23 | −29 | −34 | −40 |
| Tight | 0.41 | 0.43 | 0.45 | 0.47 | 0.49 | 0.51 | 0.53 | 0.55 | 0.57 | 0.59 |
| Medium | 0.69 | 0.73 | 0.77 | 0.81 | 0.85 | 0.89 | 0.93 | 0.97 | 1.00 | 1.05 |
| Loose | 1.11 | 1.15 | 1.20 | 1.23 | 1.27 | 1.30 | 1.35 | 1.40 | 1.43 | 1.47 |

**Part C. Design Infiltration Rate (ACH) for Summer: Indoors 75°F (24°C); Wind Speed = 7.5 mph (3.4 m/s)**

| Construction Type | Summer Outdoor Design Temperature | | | | | |
|---|---|---|---|---|---|---|
| | °F: 85 | 90 | 95 | 100 | 105 | 110 |
| | °C: 29 | 32 | 35 | 38 | 41 | 43 |
| Tight | 0.33 | 0.34 | 0.35 | 0.36 | 0.37 | 0.38 |
| Medium | 0.46 | 0.48 | 0.50 | 0.52 | 0.54 | 0.56 |
| Loose | 0.68 | 0.70 | 0.72 | 0.74 | 0.76 | 0.78 |

**Part D. Infiltration Rates per Unit Floor Area**

| Ceiling Height | Air Flow | Air Changes per Hour | | | | | | | | | | | | | | | | |
|---|---|---|---|---|---|---|---|---|---|---|---|---|---|---|---|---|---|---|
| | | 0.3 | 0.4 | 0.5 | 0.6 | 0.7 | 0.8 | 0.9 | 1.0 | 1.1 | 1.2 | 1.3 | 1.4 | 1.5 | 1.6 | 1.7 | 1.8 | 1.9 | 2.0 |
| 7.5 ft | cfm/ft² | 0.04 | 0.05 | 0.06 | 0.08 | 0.09 | 0.10 | 0.11 | 0.13 | 0.14 | 0.15 | 0.16 | 0.18 | 0.19 | 0.20 | 0.21 | 0.23 | 0.24 | 0.25 |
| 2.3 m | L/s m² | 0.20 | 0.25 | 0.31 | 0.41 | 0.46 | 0.51 | 0.56 | 0.66 | 0.71 | 0.76 | 0.81 | 0.91 | 0.97 | 1.02 | 1.07 | 1.17 | 1.22 | 1.27 |
| 8.0 ft | cfm/ft² | 0.04 | 0.05 | 0.07 | 0.08 | 0.09 | 0.11 | 0.12 | 0.13 | 0.15 | 0.16 | 0.17 | 0.19 | 0.20 | 0.21 | 0.23 | 0.24 | 0.26 | 0.27 |
| 2.4 m | L/s m² | 0.20 | 0.25 | 0.36 | 0.41 | 0.46 | 0.56 | 0.61 | 0.66 | 0.76 | 0.81 | 0.86 | 0.97 | 1.02 | 1.07 | 1.17 | 1.22 | 1.32 | 1.37 |
| 8.5 ft | cfm/ft² | 0.04 | 0.06 | 0.07 | 0.09 | 0.10 | 0.11 | 0.13 | 0.14 | 0.16 | 0.17 | 0.18 | 0.20 | 0.21 | 0.23 | 0.24 | 0.26 | 0.27 | 0.28 |
| 2.6 m | L/s m² | 0.20 | 0.31 | 0.36 | 0.46 | 0.51 | 0.56 | 0.66 | 0.71 | 0.81 | 0.86 | 0.91 | 1.02 | 1.07 | 1.17 | 1.22 | 1.32 | 1.37 | 1.42 |
| 9.0 ft | cfm/ft² | 0.05 | 0.06 | 0.08 | 0.09 | 0.11 | 0.12 | 0.14 | 0.15 | 0.17 | 0.18 | 0.20 | 0.21 | 0.23 | 0.24 | 0.26 | 0.27 | 0.29 | 0.30 |
| 2.7 m | L/s m² | 0.25 | 0.31 | 0.41 | 0.46 | 0.56 | 0.61 | 0.71 | 0.76 | 0.86 | 0.91 | 1.02 | 1.07 | 1.17 | 1.22 | 1.32 | 1.37 | 1.47 | 1.52 |

**TABLE E.27 Estimated Overall Infiltration Rates for Small Buildings** *(Continued)*

| Part E. Infiltration Heat Flow Rates per Unit Floor Area | | | | | | | | | | | | | | | | | | | |
|---|---|---|---|---|---|---|---|---|---|---|---|---|---|---|---|---|---|---|---|
| Ceiling Height | Units | | | | | | | | | | | | | | | | | | | |
| 7.5 ft | Btu/h ft² °F | 0.04 | 0.05 | 0.07 | 0.08 | 0.09 | 0.11 | 0.12 | 0.14 | 0.15 | 0.16 | 0.18 | 0.20 | 0.20 | 0.22 | 0.23 | 0.24 | 0.26 | 0.27 |
| 2.3 m | W/m² K | 0.23 | 0.28 | 0.40 | 0.45 | 0.51 | 0.63 | 0.68 | 0.80 | 0.85 | 0.91 | 1.02 | 1.14 | 1.14 | 1.25 | 1.31 | 1.36 | 1.48 | 1.53 |
| 8.0 ft | Btu/h ft² °F | 0.04 | 0.06 | 0.07 | 0.09 | 0.10 | 0.12 | 0.13 | 0.14 | 0.16 | 0.17 | 0.19 | 0.22 | 0.22 | 0.23 | 0.24 | 0.26 | 0.27 | 0.29 |
| 2.4 m | W/m² K | 0.23 | 0.34 | 0.40 | 0.51 | 0.57 | 0.68 | 0.74 | 0.80 | 0.91 | 0.97 | 1.08 | 1.25 | 1.25 | 1.31 | 1.36 | 1.48 | 1.53 | 1.65 |
| 8.5 ft | Btu/h ft² °F | 0.05 | 0.06 | 0.08 | 0.09 | 0.11 | 0.12 | 0.14 | 0.15 | 0.17 | 0.18 | 0.20 | 0.23 | 0.23 | 0.24 | 0.26 | 0.28 | 0.29 | 0.30 |
| 2.4 m | W/m² K | 0.28 | 0.34 | 0.45 | 0.51 | 0.63 | 0.68 | 0.80 | 0.85 | 0.97 | 1.02 | 1.14 | 1.31 | 1.31 | 1.36 | 1.48 | 1.59 | 1.65 | 1.70 |
| 9.0 ft | Btu/h ft² °F | 0.05 | 0.06 | 0.08 | 0.10 | 0.11 | 0.13 | 0.15 | 0.16 | 0.18 | 0.19 | 0.21 | 0.24 | 0.24 | 0.26 | 0.28 | 0.29 | 0.31 | 0.32 |
| 2.4 m | W/m² K | 0.28 | 0.34 | 0.45 | 0.57 | 0.63 | 0.74 | 0.85 | 0.91 | 1.02 | 1.08 | 1.19 | 1.36 | 1.36 | 1.48 | 1.59 | 1.65 | 1.76 | 1.82 |

*Sources:* Parts A through C reprinted with permission of the American Society of Heating, Refrigerating and Air-Conditioning Engineers, Inc. from the 2001 *ASHRAE Handbook—Fundamentals;* Parts D and E, from ASHRAE *Cooling and Heating Load Calculation Manual,* 1979; with SI units appended by the authors.

**TABLE E.28 Approximate Infiltration Through Doors and Windows of Small Buildings**

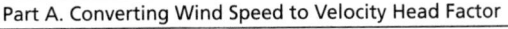

Part A. Converting Wind Speed to Velocity Head Factor

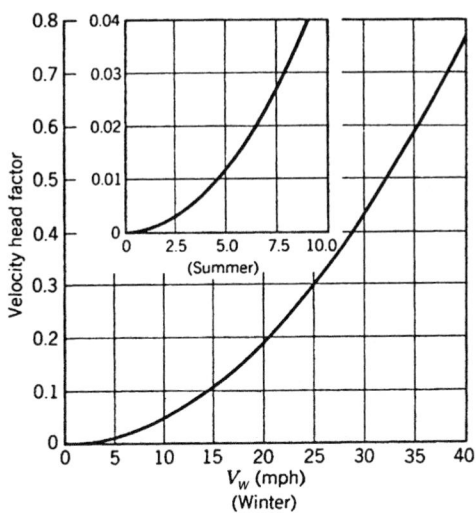

*Note:*  Design conditions are often assumed as:

Winter wind $V_w$ = 15 mph (6.7 m/s) corresponding to a velocity head factor of 0.105

Summer wind $V_w$ = 7.5 mph (3.4 m/s) corresponding to a velocity head factor of 0.028

Part B. Infiltration Rates for Velocity Head Factors

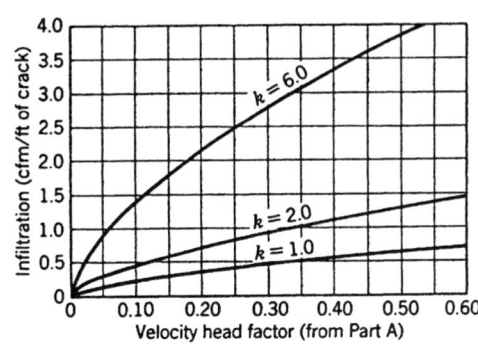

*Note:* Enter this graph with velocity head factor (from Part A) to find infiltration rate in cfm/ft of crack (using values of *k* found in Part C or D). L/s per m = (cfm/ft) (0.646).

**TABLE E.28 Approximate Infiltration Through Doors and Windows of Small Buildings** *(Continued)*

| Part C. Classifications of Windows for Infiltration | | |
|---|---|---|
| Window Fit | Wood<br>Double-Hung<br>(Locked) | Other Types |
| Tight, $k = 1.0$ | Weather stripped; average gap (1/64 in. [0.4 mm] crack) | Wood casement and awning windows; weather stripped.<br>Metal casement windows; weather stripped. |
| Average, $k = 2.0$ | Non-weather stripped; average gap (1/64 in. [0.4 mm] crack) or weather stripped; large gap (3/32 in. [2.4 mm] crack) | All types of vertical and horizontal sliding windows; weather stripped. If average gap (1/64 in. [0.4 mm] crack), this could be tight-fitting window.<br>Metal casement windows; non-weather stripped. If large gap (3/32 in. [2.4 mm] crack), this could be a loose-fitting window. |
| Loose, $k = 6.0$ | Non-weather stripped; large gap (3/32 in. [2.4 mm] crack) | Vertical and horizontal sliding windows; non-weather stripped. |
| Part D. Classification of Residential-type Doors for Infiltration | | |
| Door Fit | Comments | |
| Tight, $k = 1.0$ | Very small perimeter gap and perfect fit weather stripping—often characteristic of new doors. | |
| Average, $k = 2.0$ | Small perimeter gap having stop trim fitting properly around door; weather stripped. | |
| Loose, $k = 6.0$ | Large perimeter gap having poor fitting stop trim; weather stripped.<br>or<br>Small perimeter gap; no weather stripping. | |

*Source:* Reprinted with permission of the American Society of Heating, Refrigerating and Air-Conditioning Engineers, Inc., Atlanta, GA, from *Cooling and Heating Load Calculation Manual,* 2nd edition, 1992. SI units appended by the authors.

# Heating and Cooling Design Guidelines and Information

This appendix provides data to support the design and analysis processes presented in Chapter 8.

## F.1 GLAZING AREAS FOR PASSIVE SOLAR BUILDINGS

**TABLE F.1** Design Guidelines for Passive Solar Glazing Area

| Location | Area of Solar Glazing[a] as Ratio of Floor Area | | Approximate SSF Values | | | |
|---|---|---|---|---|---|---|
| | | | Standard Performance[b] | | Superior Performance[c] | |
| | Low | High | Low | High | Low | High |
| Birmingham, Alabama | 0.09 | 0.18 | 22 | 37 | 34 | 58 |
| Mobile, Alabama | 0.06 | 0.12 | 26 | 44 | 34 | 60 |
| Montgomery, Alabama | 0.07 | 0.15 | 24 | 41 | 34 | 59 |
| Phoenix, Arizona | 0.06 | 0.12 | 37 | 60 | 48 | 75 |
| Prescott, Arizona | 0.10 | 0.20 | 29 | 48 | 44 | 72 |
| Tucson, Arizona | 0.06 | 0.12 | 35 | 57 | 45 | 73 |
| Winslow, Arizona | 0.12 | 0.24 | 30 | 47 | 48 | 74 |
| Yuma, Arizona | 0.04 | 0.09 | 43 | 66 | 51 | 78 |
| Fort Smith, Arkansas | 0.10 | 0.20 | 24 | 39 | 38 | 64 |
| Little Rock, Arkansas | 0.10 | 0.19 | 23 | 38 | 37 | 62 |
| Bakersfield, California | 0.08 | 0.15 | 31 | 50 | 42 | 67 |
| Daggett, California | 0.07 | 0.15 | 35 | 56 | 46 | 73 |
| Fresno, California | 0.09 | 0.17 | 29 | 46 | 41 | 65 |
| Long Beach, California | 0.05 | 0.10 | 35 | 58 | 44 | 72 |
| Los Angeles, California | 0.05 | 0.09 | 36 | 58 | 44 | 72 |
| Mount Shasta, California | 0.11 | 0.21 | 24 | 38 | 42 | 67 |
| Needles, California | 0.06 | 0.12 | 39 | 61 | 49 | 76 |
| Oakland, California | 0.07 | 0.15 | 35 | 55 | 46 | 72 |
| Red Bluff, California | 0.09 | 0.18 | 29 | 46 | 41 | 65 |
| Sacramento, California | 0.09 | 0.18 | 29 | 47 | 41 | 66 |
| San Diego, California | 0.04 | 0.09 | 37 | 61 | 46 | 74 |
| San Francisco, California | 0.06 | 0.13 | 34 | 54 | 45 | 71 |
| Santa Maria, California | 0.05 | 0.11 | 31 | 53 | 42 | 69 |
| Colorado Springs, Colorado | 0.12 | 0.24 | 27 | 42 | 47 | 74 |
| Denver, Colorado | 0.12 | 0.23 | 27 | 43 | 47 | 74 |
| Eagle, Colorado | 0.14 | 0.29 | 25 | 35 | 53 | 77 |
| Grand Junction, Colorado | 0.13 | 0.27 | 29 | 43 | 50 | 76 |
| Pueblo, Colorado | 0.11 | 0.23 | 29 | 45 | 48 | 75 |
| Hartford, Connecticut | 0.17 | 0.35 | 14 | 19 | 40 | 64 |
| Wilmington, Delaware | 0.15 | 0.29 | 19 | 30 | 39 | 63 |
| Washington, DC | 0.12 | 0.23 | 18 | 28 | 37 | 61 |
| Apalachicola, Florida | 0.05 | 0.10 | 28 | 47 | 36 | 61 |
| Daytona Beach, Florida | 0.04 | 0.08 | 30 | 51 | 36 | 63 |
| Jacksonville, Florida | 0.05 | 0.09 | 27 | 47 | 35 | 62 |
| Miami, Florida | 0.01 | 0.02 | 27 | 48 | 31 | 54 |
| Orlando, Florida | 0.03 | 0.06 | 30 | 52 | 37 | 63 |
| Tallahassee, Florida | 0.05 | 0.11 | 26 | 45 | 35 | 60 |
| Tampa, Florida | 0.03 | 0.06 | 30 | 52 | 36 | 63 |
| West Palm Beach, Florida | 0.01 | 0.03 | 30 | 51 | 34 | 59 |
| Atlanta, Georgia | 0.08 | 0.17 | 22 | 36 | 34 | 58 |
| Augusta, Georgia | 0.08 | 0.16 | 24 | 40 | 35 | 60 |
| Macon, Georgia | 0.07 | 0.15 | 25 | 41 | 35 | 59 |
| Savannah, Georgia | 0.06 | 0.13 | 25 | 43 | 35 | 60 |
| Boise, Idaho | 0.14 | 0.28 | 27 | 38 | 48 | 71 |
| Lewiston, Idaho | 0.15 | 0.29 | 22 | 29 | 44 | 65 |
| Pocatello, Idaho | 0.13 | 0.26 | 25 | 35 | 51 | 74 |
| Chicago, Illinois | 0.17 | 0.35 | 17 | 23 | 43 | 67 |
| Moline, Illinois | 0.20 | 0.39 | 17 | 22 | 46 | 70 |
| Springfield, Illinois | 0.15 | 0.30 | 19 | 28 | 42 | 67 |
| Evansville, Indiana | 0.14 | 0.27 | 19 | 29 | 37 | 61 |
| Fort Wayne, Indiana | 0.16 | 0.33 | 13 | 17 | 37 | 60 |
| Indianapolis, Indiana | 0.14 | 0.28 | 15 | 21 | 37 | 60 |
| South Bend, Indiana | 0.18 | 0.35 | 12 | 15 | 39 | 61 |
| Burlington, Iowa | 0.18 | 0.36 | 20 | 27 | 47 | 71 |
| Des Moines, Iowa | 0.21 | 0.43 | 19 | 25 | 50 | 75 |
| Mason City, Iowa | 0.22 | 0.44 | 18 | 19 | 56 | 79 |
| Sioux City, Iowa | 0.23 | 0.46 | 20 | 24 | 53 | 76 |
| Dodge City, Kansas | 0.12 | 0.23 | 27 | 42 | 46 | 73 |
| Goodland, Kansas | 0.13 | 0.27 | 26 | 39 | 47 | 74 |
| Topeka, Kansas | 0.14 | 0.28 | 24 | 35 | 45 | 71 |
| Wichita, Kansas | 0.14 | 0.28 | 26 | 41 | 45 | 72 |
| Lexington, Kentucky | 0.13 | 0.27 | 17 | 26 | 35 | 58 |
| Louisville, Kentucky | 0.13 | 0.27 | 18 | 27 | 35 | 59 |
| Baton Rouge, Louisiana | 0.06 | 0.12 | 26 | 43 | 34 | 59 |

**TABLE F.1** Design Guidelines for Passive Solar Glazing Area *(Continued)*

| Location | Area of Solar Glazing[a] as Ratio of Floor Area | | Approximate SSF Values | | | |
|---|---|---|---|---|---|---|
| | | | Standard Performance[b] | | Superior Performance[c] | |
| | Low | High | Low | High | Low | High |
| Lake Charles, Louisiana | 0.06 | 0.11 | 24 | 41 | 32 | 57 |
| New Orleans, Louisiana | 0.05 | 0.11 | 27 | 46 | 35 | 61 |
| Shreveport, Louisiana | 0.08 | 0.15 | 26 | 43 | 36 | 61 |
| Caribou, Maine | 0.25 | 0.50 | — | NR[c] | 53 | 74 |
| Portland, Maine | 0.17 | 0.34 | 14 | 17 | 45 | 69 |
| Baltimore, Maryland | 0.14 | 0.27 | 19 | 30 | 38 | 62 |
| Boston, Massachusetts | 0.15 | 0.29 | 17 | 25 | 40 | 64 |
| Alpena, Michigan | 0.21 | 0.42 | — | NR | 47 | 69 |
| Detroit, Michigan | 0.17 | 0.34 | 13 | 17 | 39 | 61 |
| Flint, Michigan | 0.15 | 0.31 | 11 | 12 | 40 | 62 |
| Grand Rapids, Michigan | 0.19 | 0.38 | 12 | 13 | 39 | 61 |
| Sault Ste. Marie, Michigan | 0.25 | 0.50 | — | NR | 50 | 70 |
| Traverse City, Michigan | 0.18 | 0.36 | — | NR | 42 | 62 |
| Duluth, Minnesota | 0.25 | 0.50 | — | NR | 50 | 70 |
| International Falls, Minnesota | 0.25 | 0.50 | — | NR | 47 | 66 |
| Minneapolis–St. Paul, Minnesota | 0.25 | 0.50 | — | NR | 55 | 76 |
| Rochester, Minnesota | 0.24 | 0.49 | — | NR | 54 | 76 |
| Jackson, Mississippi | 0.08 | 0.15 | 24 | 40 | 34 | 59 |
| Meridian, Mississippi | 0.08 | 0.15 | 23 | 39 | 34 | 58 |
| Columbia, Missouri | 0.13 | 0.26 | 20 | 30 | 41 | 66 |
| Kansas City, Missouri | 0.14 | 0.29 | 22 | 32 | 44 | 70 |
| Saint Louis, Missouri | 0.15 | 0.29 | 21 | 33 | 41 | 65 |
| Springfield, Missouri | 0.13 | 0.26 | 22 | 34 | 40 | 65 |
| Billings, Montana | 0.16 | 0.32 | 24 | 31 | 53 | 76 |
| Cut Bank, Montana | 0.24 | 0.49 | 22 | 23 | 62 | 81 |
| Dillon, Montana | 0.16 | 0.32 | 24 | 32 | 54 | 77 |
| Glasgow, Montana | 0.25 | 0.50 | — | NR | 55 | 75 |
| Great Falls, Montana | 0.18 | 0.37 | 23 | 28 | 56 | 77 |
| Helena, Montana | 0.20 | 0.39 | 21 | 25 | 55 | 77 |
| Lewistown, Montana | 0.19 | 0.38 | 21 | 25 | 54 | 76 |
| Miles City, Montana | 0.23 | 0.47 | 21 | 23 | 60 | 80 |
| Missoula, Montana | 0.18 | 0.36 | 15 | 16 | 47 | 68 |
| Grand Island, Nebraska | 0.18 | 0.36 | 24 | 33 | 51 | 76 |
| North Omaha, Nebraska | 0.20 | 0.40 | 21 | 29 | 51 | 76 |
| North Platte, Nebraska | 0.17 | 0.34 | 25 | 36 | 50 | 76 |
| Scotts Bluff, Nebraska | 0.16 | 0.31 | 24 | 36 | 49 | 74 |
| Elko, Nevada | 0.12 | 0.25 | 27 | 39 | 52 | 76 |
| Ely, Nevada | 0.12 | 0.23 | 27 | 41 | 50 | 77 |
| Las Vegas, Nevada | 0.09 | 0.18 | 35 | 56 | 48 | 75 |
| Lovelock, Nevada | 0.13 | 0.25 | 32 | 48 | 53 | 78 |
| Reno, Nevada | 0.11 | 0.22 | 31 | 48 | 49 | 76 |
| Tonopah, Nevada | 0.11 | 0.23 | 31 | 48 | 51 | 77 |
| Winnemucca, Nevada | 0.13 | 0.26 | 28 | 42 | 49 | 75 |
| Concord, New Hampshire | 0.17 | 0.34 | 13 | 15 | 45 | 68 |
| Newark, New Jersey | 0.13 | 0.25 | 19 | 29 | 39 | 64 |
| Albuquerque, New Mexico | 0.11 | 0.22 | 29 | 47 | 46 | 73 |
| Clayton, New Mexico | 0.10 | 0.20 | 28 | 45 | 45 | 73 |
| Farmington, New Mexico | 0.12 | 0.24 | 29 | 45 | 49 | 76 |
| Los Alamos, New Mexico | 0.11 | 0.22 | 25 | 40 | 44 | 72 |
| Roswell, New Mexico | 0.10 | 0.19 | 30 | 49 | 45 | 73 |
| Truth or Consequences, New Mexico | 0.09 | 0.17 | 32 | 51 | 46 | 73 |
| Tucumcari, New Mexico | 0.10 | 0.20 | 30 | 48 | 45 | 73 |
| Zuñi, New Mexico | 0.11 | 0.21 | 27 | 43 | 45 | 73 |
| Albany, New York | 0.21 | 0.41 | 13 | 15 | 43 | 66 |
| Binghamton, New York | 0.15 | 0.30 | — | NR | 35 | 56 |
| Buffalo, New York | 0.19 | 0.37 | — | NR | 36 | 57 |
| Massena, New York | 0.25 | 0.50 | — | NR | 50 | 71 |
| New York (Central Park), New York | 0.15 | 0.30 | 16 | 25 | 36 | 59 |
| Rochester, New York | 0.18 | 0.37 | — | NR | 37 | 58 |
| Syracuse, New York | 0.19 | 0.38 | — | NR | 37 | 59 |
| Asheville, North Carolina | 0.10 | 0.20 | 21 | 35 | 36 | 61 |
| Cape Hatteras, North Carolina | 0.09 | 0.17 | 24 | 40 | 36 | 60 |
| Charlotte, North Carolina | 0.08 | 0.17 | 23 | 38 | 36 | 60 |
| Greensboro, North Carolina | 0.10 | 0.20 | 23 | 37 | 37 | 63 |

**TABLE F.1 Design Guidelines for Passive Solar Glazing Area** *(Continued)*

| Location | Area of Solar Glazing[a] as Ratio of Floor Area | | Approximate SSF Values | | | |
|---|---|---|---|---|---|---|
| | | | Standard Performance[b] | | Superior Performance[c] | |
| | Low | High | Low | High | Low | High |
| Raleigh–Durham, North Carolina | 0.09 | 0.19 | 22 | 37 | 36 | 61 |
| Bismarck, North Dakota | 0.25 | 0.50 | — NR — | | 56 | 77 |
| Fargo, North Dakota | 0.25 | 0.50 | — NR — | | 51 | 72 |
| Minot, North Dakota | 0.25 | 0.50 | — NR — | | 52 | 72 |
| Akron–Canton, Ohio | 0.15 | 0.31 | 12 | 16 | 35 | 57 |
| Cincinnati, Ohio | 0.12 | 0.24 | 15 | 23 | 35 | 57 |
| Cleveland, Ohio | 0.15 | 0.31 | 11 | 14 | 34 | 55 |
| Columbus, Ohio | 0.14 | 0.28 | 13 | 18 | 35 | 57 |
| Dayton, Ohio | 0.14 | 0.28 | 14 | 20 | 36 | 59 |
| Toledo, Ohio | 0.17 | 0.34 | 13 | 17 | 38 | 61 |
| Youngstown, Ohio | 0.16 | 0.32 | — NR — | | 34 | 54 |
| Oklahoma City, Oklahoma | 0.11 | 0.22 | 25 | 41 | 41 | 67 |
| Tulsa, Oklahoma | 0.11 | 0.22 | 24 | 38 | 40 | 65 |
| Astoria, Oregon | 0.09 | 0.19 | 21 | 34 | 37 | 60 |
| Burns, Oregon | 0.13 | 0.25 | 23 | 32 | 47 | 71 |
| Medford, Oregon | 0.12 | 0.24 | 21 | 32 | 38 | 60 |
| North Bend, Oregon | 0.09 | 0.17 | 25 | 42 | 38 | 64 |
| Pendleton, Oregon | 0.14 | 0.27 | 22 | 30 | 43 | 64 |
| Portland, Oregon | 0.13 | 0.26 | 21 | 31 | 38 | 60 |
| Redmond, Oregon | 0.13 | 0.27 | 26 | 38 | 47 | 71 |
| Salem, Oregon | 0.12 | 0.24 | 21 | 32 | 37 | 59 |
| Allentown, Pennsylvania | 0.15 | 0.29 | 16 | 24 | 39 | 63 |
| Erie, Pennsylvania | 0.17 | 0.34 | — NR — | | 35 | 55 |
| Harrisburg, Pennsylvania | 0.13 | 0.26 | 17 | 26 | 38 | 62 |
| Philadelphia, Pennsylvania | 0.15 | 0.29 | 19 | 29 | 38 | 62 |
| Pittsburgh, Pennsylvania | 0.14 | 0.28 | 12 | 16 | 33 | 55 |
| Wilkes Barre–Scranton, Pennsylvania | 0.16 | 0.32 | 13 | 18 | 37 | 60 |
| Providence, Rhode Island | 0.15 | 0.30 | 17 | 24 | 40 | 64 |
| Charleston, South Carolina | 0.07 | 0.14 | 25 | 41 | 34 | 59 |
| Columbia, South Carolina | 0.08 | 0.17 | 25 | 41 | 36 | 61 |
| Greenville–Spartanburg, South Carolina | 0.08 | 0.17 | 23 | 38 | 36 | 60 |
| Huron, South Dakota | 0.25 | 0.50 | — NR — | | 58 | 79 |
| Pierre, South Dakota | 0.22 | 0.43 | 21 | 23 | 58 | 80 |
| Rapid City, South Dakota | 0.15 | 0.30 | 23 | 32 | 51 | 76 |
| Sioux Falls, South Dakota | 0.22 | 0.45 | 18 | 19 | 57 | 79 |
| Chattanooga, Tennessee | 0.09 | 0.19 | 19 | 32 | 33 | 56 |
| Knoxville, Tennessee | 0.09 | 0.18 | 20 | 33 | 33 | 56 |
| Memphis, Tennessee | 0.09 | 0.19 | 22 | 36 | 36 | 60 |
| Nashville, Tennessee | 0.10 | 0.21 | 19 | 30 | 33 | 55 |
| Abilene, Texas | 0.09 | 0.18 | 29 | 47 | 41 | 68 |
| Amarillo, Texas | 0.11 | 0.22 | 29 | 46 | 45 | 72 |
| Austin, Texas | 0.06 | 0.13 | 27 | 46 | 37 | 63 |
| Brownsville, Texas | 0.03 | 0.06 | 27 | 46 | 32 | 57 |
| Corpus Christi, Texas | 0.05 | 0.09 | 29 | 49 | 36 | 63 |
| Dallas, Texas | 0.08 | 0.17 | 27 | 44 | 38 | 64 |
| Del Rio, Texas | 0.06 | 0.12 | 30 | 50 | 39 | 66 |
| El Paso, Texas | 0.09 | 0.17 | 32 | 53 | 45 | 72 |
| Fort Worth, Texas | 0.09 | 0.17 | 26 | 44 | 38 | 64 |
| Houston, Texas | 0.06 | 0.11 | 25 | 43 | 34 | 59 |
| Laredo, Texas | 0.05 | 0.09 | 31 | 52 | 39 | 64 |
| Lubbock, Texas | 0.09 | 0.19 | 30 | 49 | 44 | 72 |
| Lufkin, Texas | 0.07 | 0.14 | 26 | 43 | 35 | 61 |
| Midland–Odessa, Texas | 0.09 | 0.18 | 32 | 52 | 44 | 72 |
| Port Arthur, Texas | 0.06 | 0.11 | 26 | 44 | 34 | 60 |
| San Angelo, Texas | 0.08 | 0.15 | 29 | 48 | 40 | 67 |
| San Antonio, Texas | 0.06 | 0.12 | 28 | 48 | 38 | 64 |
| Sherman, Texas | 0.10 | 0.20 | 25 | 41 | 38 | 64 |
| Waco, Texas | 0.08 | 0.15 | 27 | 45 | 38 | 64 |
| Wichita Falls, Texas | 0.10 | 0.20 | 27 | 45 | 41 | 67 |
| Bryce Canyon, Utah | 0.13 | 0.25 | 26 | 39 | 52 | 78 |
| Cedar City, Utah | 0.12 | 0.24 | 28 | 43 | 48 | 75 |
| Salt Lake City, Utah | 0.13 | 0.26 | 27 | 39 | 48 | 72 |
| Burlington, Vermont | 0.22 | 0.43 | — NR — | | 46 | 68 |

# TABLE F.1 Design Guidelines for Passive Solar Glazing Area *(Continued)*

| Location | Area of Solar Glazing[a] as Ratio of Floor Area | | Approximate SSF Values | | | |
| | | | Standard Performance[b] | | Superior Performance[c] | |
| | Low | High | Low | High | Low | High |
|---|---|---|---|---|---|---|
| Norfolk, Virginia | 0.09 | 0.19 | 23 | 38 | 37 | 62 |
| Richmond, Virginia | 0.11 | 0.22 | 21 | 34 | 37 | 61 |
| Roanoke, Virginia | 0.11 | 0.23 | 21 | 34 | 37 | 61 |
| Olympia, Washington | 0.12 | 0.23 | 20 | 29 | 38 | 59 |
| Seattle-Tacoma, Washington | 0.11 | 0.22 | 21 | 30 | 39 | 59 |
| Spokane, Washington | 0.20 | 0.39 | 20 | 24 | 48 | 68 |
| Yakima, Washington | 0.18 | 0.36 | 24 | 31 | 49 | 70 |
| Charleston, West Virginia | 0.13 | 0.25 | 16 | 24 | 32 | 54 |
| Huntington, West Virginia | 0.13 | 0.25 | 17 | 27 | 34 | 57 |
| Eau Claire, Wisconsin | 0.25 | 0.50 | — | NR — | 53 | 75 |
| Green Bay, Wisconsin | 0.23 | 0.46 | — | NR — | 53 | 75 |
| La Crosse, Wisconsin | 0.21 | 0.43 | — | NR — | 52 | 75 |
| Madison, Wisconsin | 0.20 | 0.40 | 15 | 17 | 51 | 74 |
| Milwaukee, Wisconsin | 0.18 | 0.35 | 15 | 18 | 48 | 71 |
| Casper, Wyoming | 0.13 | 0.26 | 27 | 39 | 53 | 78 |
| Cheyenne, Wyoming | 0.11 | 0.21 | 25 | 39 | 47 | 74 |
| Rock Springs, Wyoming | 0.14 | 0.28 | 26 | 38 | 54 | 79 |
| Sheridan, Wyoming | 0.16 | 0.31 | 22 | 30 | 52 | 75 |
| *CANADA* | | | | | | |
| Edmonton, Alberta | 0.25 | 0.50 | — | NR — | 54 | 72 |
| Suffield, Alberta | 0.25 | 0.50 | 28 | 30 | 67 | 85 |
| Nanaimo, British Columbia | 0.13 | 0.26 | 26 | 35 | 45 | 66 |
| Vancouver, British Columbia | 0.13 | 0.26 | 20 | 28 | 40 | 60 |
| Winnipeg, Manitoba | 0.25 | 0.50 | — | NR — | 54 | 74 |
| Dartmouth, Nova Scotia | 0.14 | 0.28 | 17 | 24 | 45 | 70 |
| Moosonee, Ontario | 0.25 | 0.50 | — | NR — | 48 | 67 |
| Ottawa, Ontario | 0.25 | 0.50 | — | NR — | 59 | 80 |
| Toronto, Ontario | 0.18 | 0.36 | 17 | 23 | 44 | 68 |
| Normandin, Quebec | 0.25 | 0.50 | — | NR — | 54 | 74 |

*Source:* Adapted from J. D. Balcomb et al. (1980). *Passive Solar Design Handbook,* Vol. 2 (Passive Solar Design Analysis), U.S. Department of Energy, Washington, DC.

NR = not recommended.

[a]Due south–facing openings are assumed.

[b]Double-glazed, clear glass (approximately equal to window 3, Table E.15).

[c]Either movable window insulation of R-9, in place from 5:30 P.M. to 7:30 A.M., solar time, or superwindows with an overall U-factor near 0.30 (approximately equal to windows 7 or 12, Table E.15).

## F.2 THERMAL MASS FOR PASSIVE SOLAR BUILDINGS

### TABLE F.2 Design Guidelines for Passive Solar Thermal Mass

| Expected Solar Savings Fraction (SSF), % | Thermal Storage by Weight/Collector Area | | | | Recommended Effective[a] Thermal Storage Area Per Unit Area of Solar Collection Area | |
| | Water | | Masonry | | Water Surface Area[b] / Collector Surface Area | Masonry Surface Area[c] / Collector Surface Area |
| | lb/ft² | kg/m² | lb/ft² | kg/m² | | |
|---|---|---|---|---|---|---|
| 10 | 6 | 29 | 30 | 147 | 0.1 | 0.7 |
| 20 | 12 | 59 | 60 | 293 | 0.2 | 1.5 |
| 30 | 18 | 88 | 90 | 440 | 0.3 | 2.2 |
| 40 | 24 | 117 | 120 | 586 | 0.4 | 2.9 |
| 50 | 30 | 147 | 150 | 733 | 0.5 | 3.7 |
| 60 | 36 | 176 | 180 | 879 | 0.6 | 4.4 |
| 70 | 42 | 205 | 210 | 1026 | 0.7 | 5.1 |
| 80 | 48 | 234 | 240 | 1172 | 0.8 | 5.9 |
| 90 | 54 | 264 | 270 | 1319 | 0.9 | 6.6 |

*Source:* Adapted from J. D. Balcomb et al. (1980). *Passive Solar Design Handbook,* Vol. 2 (Passive Solar Design Analysis), U.S. Department of Energy, Washington, DC.

[a]Effective area is that area exposed at some point to direct sun during a clear winter day.

[b]For a water container 12 in. (300 mm) thick.

[c]For a 4-in. (100-mm)-thick brick, density 123 lb/ft³ (1970 kg/m³).

APPENDICES

## F.3 ESTIMATING SUMMER HEAT GAINS

### TABLE F.3 Approximate Summer Heat Gains from Occupants, Equipment, Lighting, and Envelope

| Part A. Internal Heat Sources—People and Equipment | | | | | | | | |
|---|---|---|---|---|---|---|---|---|
| | Area per Person[a] | | Sensible Heat Gain (Btu/h ft² of Floor Area) | | | Sensible Heat Gain (W/m² of Floor Area) | | |
| **Function** | ft² | m² | People[b] | Equipment[c] | Total | People[b] | Equipment[c] | Total |
| Office, U.S.[c] | 180–100 | 16.7–9.3 | 1.3–2.3 | 0.4–1.1 | 1.7–3.4 | 4.1–7.3 | 1.2–3.4 | 5.3–10.7 |
| Office, Europe[e] | | | 1–1.6 | 2.2–4.2 | 3.2–5.8 | 3–5 | 7–13.1 | 10–18.1 |
| School: elementary, U.S. | 100–20 | 9.3–1.9 | 2.3–11.5 | 0–0.6 | 2.3–12.1 | 7.3–36.3 | 0–2.0 | 7.3–38.3 |
| Schools, Europe[e] | | | 3.8–8.0 | 0–0.6 | 3.8–8.6 | 12–25.2 | 0–2.0 | 12.0–27.2 |
| School: secondary, college | 150–100 | 13.9–9.3 | 1.7–2.6 | 0–0.6 | 1.7–3.2 | 5.4–8.2 | 0–2.0 | 5.4–10.2 |
| Health care | | | | | | | | |
|   Sleeping (hospital) | 240 | 22.3 | 0.9 | 0.6[e] | 1.5 | 2.8 | 2.0[e] | 4.8 |
|   In-patient (clinic) | 120 | 11.1 | 1.9 | Varies | 1.9+ | 6.0 | Varies | 6.0+ |
| Assembly: fixed seats | 15 | 1.4 | 14.0 | — | 14.0 | 44.2 | — | 44.2 |
|   standing space, concentrated use | 15–7 | 1.4–0.7 | 21.0–45.0 | 0–0.5 | 21.0–45.5 | 66.3–142.0 | 0–1.6 | 66.3–143.6 |
| Restaurant:[f] | | | | | | | | |
|   Fast food: dining area | 15 | 1.4 | 17 | 3.4 | 20.4 | 53.6 | 10.7 | 64.3 |
|     Kitchen, refrigeration | | | | 17.1 | 17.1 | | 54.0 | 54.0 |
|   Sit-down: dining area | 25 | 2.3 | 10.2 | 5.1 | 15.3 | 32.2 | 16.1 | 48.3 |
|     Kitchen, refrigeration | | | | 7.2 | 7.2 | | 22.7 | 22.7 |
| Mercantile: street floor | 50–30 | 4.7–2.8 | 6.3–10.5 | 3.4 | 9.7–13.9 | 19.9–33.1 | 10.7 | 30.6–43.8 |
|   Other sales floors | 60–50 | 5.6–4.7 | 5.3–6.3 | 3.4 | 8.7–9.7 | 16.7–19.9 | 10.7 | 27.4–30.6 |
| Shopping center, Europe[e] | | | 3.2 | 0.3–1.3 | 3.5–4.5 | 10 | 1.0–4.0 | 11.0–14.0 |
| Warehouse | 1000–300 | 92.9–27.9 | 0.4–1.2 | — | 0.4–1.2 | 1.3–3.8 | — | 1.3–3.8 |
| Hotels, nursing homes | 300–200 | 27.9–18.6 | 0.8–1.2 | 3.4 | 4.2–4.6 | 2.5–3.8 | 10.7 | 13.2–14.5 |
| Apartments[g] | 300–200 | 27.9–18.6 | 0.8–1.2 | See note g | See note g | 2.5–3.8 | See note g | See note g |

| Part B. Internal Heat Sources—Electric Lighting | | | | | | |
|---|---|---|---|---|---|---|
| | Sensible Heat Gain[h] (Btu/h ft² of Floor Area) | | | Sensible Heat Gain[h] (W/m² of Floor Area) | | |
| **Function** | DF<1 | 1<DF<4[h] | DF>4[h] | DF<1 | 1<DF<4[h] | DF>4[h] |
| Office | 5.1 | 2.0 | 0.5 | 16.1 | 6.3 | 1.6 |
| School: elementary | 6.3–6.8 | 2.5–2.7 | 0.6–0.7 | 19.9–21.5 | 7.9–8.5 | 1.9–2.2 |
| School: secondary, college | 6.3–6.8 | 2.5–2.7 | 0.6–0.7 | 19.9–21.5 | 7.9–8.5 | 1.9–2.2 |
| Health care | | | | | | |
|   Sleeping (hospital) | 6.8 | 2.7 | 0.7 | 21.5 | 8.5 | 2.2 |
|   In-patient (clinic) | 6.8 | 2.7 | 0.7 | 21.5 | 8.5 | 2.2 |
| Assembly | 3.8 | 1.5 | 0.4 | 12.0 | 4.7 | 1.3 |
| Restaurants[i] | 6.3 | 2.5 | 0.6 | 19.9 | 7.9 | 1.9 |
| Mercantile | 5.1–6.8 | 2.0–2.7 | 0.5–0.7 | 16.1–21.5 | 6.3–8.5 | 1.6–2.2 |
| Warehouse | 2.4 | 1.0 | 0.2 | 7.6 | 3.2 | 0.6 |
| Hotels, nursing homes | 6.8 | 2.7 | 0.7 | 21.5 | 8.5 | 2.2 |
| Apartments[g] | Up to 6.8 | Up to 2.7 | Up to 0.7 | Up to 21.5 | Up to 8.5 | Up to 2.2 |

| Part C. Heat Gain through Envelope[j] | | | | |
|---|---|---|---|---|
| | (Btu/h ft² of Floor Area) Outdoor Design Temperature | | (W/m² of Floor Area) Outdoor Design Temperature | |
| | 90°F | 100°F | 32°C | 38°C |
| I. Gains through externally shaded windows[k]: Find ratio, $\frac{\text{total window area}}{\text{total floor area}}$, then multiply by | 16 | 21 | 50 | 66 |
| II. Gains through opaque walls: Find ratio, $\frac{\text{total opaque wall area}}{\text{total floor area}} \times (U_{wall})$, then multiply by | 15 | 25 | 8 | 14 |
| III. Gains through roofs: Find ratio, $\frac{\text{total opaque roof area}}{\text{total floor area}} \times (U_{roof})$, then multiply by | 35 | 45 | 19 | 25 |

Summer gains

ns **from Occupants, Equipment, Lighting, and Envelope**

Part D. Summary Gains

uildings

t ventilation

**:ooling load, in Btu/h ft² (W/m²), of floor area.**

buildings

t ventilation

**Total gains: Add Parts A, B, C, and E to obtain cooling load, in Btu/h ft² (W/m²), of floor area.**

| Part E. Gains from Infiltration/Ventilation of "Closed" Buildings | | Outdoor Design Temperature | | Outdoor Design Temperature | |
|---|---|---|---|---|---|
| | | 90°F | 100°F | 32°C | 38°C |
| **Infiltration:** | | | | | |
| Find ratio, $\dfrac{\text{total window + opaque wall area}}{\text{total floor area}}$ , then multiply by | | 1.0 | 1.9 | 3.2 | 6.0 |
| **Ventilation:** | | | | | |
| Find known $\dfrac{\text{total cfm of outdoor air}}{\text{total floor area, ft}^2}$ , then multiply by | | 16 | 27 | — | — |
| Find known $\dfrac{\text{total L/s of outdoor air}}{\text{total floor area, m}^2}$ , then multiply by | | — | — | 9.9 | 16.8 |

[a]Lower density from elevator population estimates, Chapter 31; higher density from fire population estimates, Chapter 24. Lower density for whole-building estimates; higher density for single-space estimates.

[b]Sensible gains per adult from Table F.8 for activities as stated.

[c]The typical "miscellaneous" load of 1 W/ft² produces 3.4 Btu/h ft².

[d]P. Komor, "Space Cooling Demand from Office Plug Loads," in *ASHRAE Journal,* December 1997. This is considerably lower than the typical 2 to 4 W/ft² for offices used by many designers.

[e]Adapted from S. R. Hastings (1994). *Passive Solar Commercial and Institutional Buildings,* International Energy Agency, John Wiley & Sons, Chichester, UK.

[f]Based on total area of restaurant + kitchen. From ACSA (1994). *Design with PV.* Association of Collegiate Schools of Architecture, Washington, DC., data from Electric Power Research Institute.

[g]Residential internal gains are often assumed at 230 Btu/h per occupant plus 1200 to 1600 Btu/h total from appliances (68 W per person, plus 350 to 470 W total from appliances).

[h]Adapted from Northwest Power Planning Council (1983). *Maximum Lighting Standards,* Portland, OR. Values shown for DF ≥ 1 assume automatic dimming in the presence of daylight.

[i]Lighting is often much lower in sit-down than in fast-food restaurants.

[j]Averaged from the more specific data found in Table F.5.

[k]If windows are not externally shaded, see Table F.6 for multipliers, which will vary by orientation.

## F.4 PASSIVE SOLAR BUILDING CHARACTERISTICS

**TABLE F.4 Design Data for Some Early Passive Solar Buildings**

| Name, Function Location | Floor Area (ft²) | Area Ratio, South glass / Floor Area | System Type (see Apx. H) | Approx. LCR | | | Approx. SSF | |
|---|---|---|---|---|---|---|---|---|
| | | | | ACH = 1.0 | ACH = 0.5 | ACH = 1.0 | ACH = 1.0 | ACH = 0.5 |
| Dove Publication, Pecos, N.M. Office Warehouse | 2660 5040 | 0.37 0.11 | DGB1 and WWB4 DGC1 | 37[a] | 25[a] | | 43%[a] | 51%[a] |
| Karen Terry House, Santa Fe, N.M. | 850 | 0.45 | DGC1 | 22 | 17 | | 52% | 60% |
| Kelbaugh House, Princeton, N.J. | 1640[b] | 0.49[c] | TWE2, DGB1, and SSB4 | 13 | 9 | | 62% | 72% |
| First Village, Unit 1, Santa Fe, N.M. | 1800[b] | 0.22[c] | SSE1 | 49 | 31 | | 38% | 52% |

[a]Treating office and warehouse as one large zone.
[b]Not including sunspace floor area.
[c]Includes sunspace south glazing but not sunspace floor area.

**TABLE F.5 Design Equivalent Temperature Differences (DETD)**

**Part A. Mass Walls, Roofs, and Floors**

| | 29.4°C | | 32.2 | | | 35.0 | | | 37.7 | | 40.5 | 43.3 | 85°F | | 90 | | | 95 | | | 100 | | 105 | 110 |
|---|---|---|---|---|---|---|---|---|---|---|---|---|---|---|---|---|---|---|---|---|---|---|---|---|
| **Daily Temperature Range[a]** (Outdoor Design Temperature) | L | M | L | M | H | L | M | H | M | H | H | H | L | M | L | M | H | L | M | H | M | H | H | H |
| *Walls* | | | | | | | | | | | | | | | | | | | | | | | | |
| 1. Masonry walls, 200-mm (8-in.) block or brick | 5.7 | 3.5 | 8.5 | 6.3 | 3.5 | 11.3 | 9.1 | 6.3 | 11.8 | 9.1 | 11.8 | 14.6 | 10.3 | 6.3 | 15.3 | 11.3 | 6.3 | 20.3 | 16.3 | 11.3 | 21.3 | 16.3 | 21.3 | 26.3 |
| 2. Partitions, frame masonry | 5.0 | 2.7 | 7.7 | 5.5 | 2.7 | 10.5 | 8.3 | 5.5 | 11.1 | 8.3 | 11.1 | 13.8 | 9.0 | 5.0 | 14.0 | 10.0 | 5.0 | 19.0 | 15.0 | 10.0 | 20.0 | 15.0 | 20.0 | 25.0 |
| | 1.4 | 0.0 | 4.2 | 1.9 | 0.0 | 6.9 | 4.7 | 1.9 | 7.5 | 4.7 | 7.5 | 10.3 | 2.5 | 0 | 7.5 | 3.5 | 0 | 12.5 | 8.5 | 3.5 | 13.5 | 8.5 | 13.5 | 18.5 |
| *Ceilings and Roofs* | | | | | | | | | | | | | | | | | | | | | | | | |
| 1. Ceilings under naturally vented attic or vented flat roof—dark | 21.1 | 18.8 | 23.8 | 21.6 | 18.8 | 26.6 | 24.4 | 21.6 | 27.2 | 24.4 | 27.2 | 30.0 | 38.0 | 34.0 | 43.0 | 39.0 | 34.0 | 48.0 | 44.0 | 39.0 | 49.0 | 44.0 | 49.0 | 54.0 |
| —light | 16.6 | 14.9 | 19.4 | 17.2 | 14.4 | 22.2 | 20.0 | 17.2 | 22.7 | 20.0 | 22.7 | 25.5 | 30.0 | 26.0 | 35.0 | 31.0 | 26.0 | 40.0 | 36.0 | 31.0 | 41.0 | 36.0 | 41.0 | 46.0 |
| 2. Built-up roof, no ceiling—dark | 21.1 | 18.8 | 23.3 | 21.6 | 18.8 | 26.6 | 24.4 | 21.6 | 27.2 | 24.4 | 27.2 | 30.0 | 38.0 | 34.0 | 43.0 | 39.0 | 34.0 | 48.0 | 44.0 | 39.0 | 49.0 | 44.0 | 49.0 | 54.0 |
| —light | 16.6 | 14.9 | 19.4 | 17.2 | 14.4 | 22.2 | 20.2 | 17.2 | 22.7 | 20.0 | 22.7 | 25.5 | 30.0 | 26.0 | 35.0 | 31.0 | 26.0 | 40.0 | 36.0 | 31.0 | 41.0 | 36.0 | 41.0 | 46.0 |
| 3. Ceilings under unconditioned rooms | 5.0 | 2.7 | 7.7 | 5.5 | 2.7 | 10.5 | 8.3 | 5.5 | 11.1 | 8.3 | 11.1 | 13.8 | 9.0 | 5.0 | 14.0 | 10.0 | 5.0 | 19.0 | 15.0 | 10.0 | 20.0 | 15.0 | 20.0 | 25.0 |
| *Floors* | | | | | | | | | | | | | | | | | | | | | | | | |
| 1. Over unconditioned rooms | 5.0 | 2.7 | 7.7 | 5.5 | 2.7 | 10.5 | 8.3 | 5.5 | 11.1 | 8.3 | 11.1 | 13.8 | 9.0 | 5.0 | 14.0 | 10.0 | 5.0 | 19.0 | 15.0 | 10.0 | 20.0 | 15.0 | 20.0 | 25.0 |
| 2. Over basement, enclosed crawl space or concrete slab on ground | 0.0 | 0.0 | 0.0 | 0.0 | 0.0 | 0.0 | 0.0 | 0.0 | 0.0 | 0.0 | 0.0 | 0.0 | 0 | 0 | 0 | 0 | 0 | 0 | 0 | 0 | 0 | 0 | 0 | 0 |
| 3. Over open crawl space | 5.0 | 2.7 | 7.7 | 5.5 | 2.7 | 10.5 | 8.3 | 5.5 | 11.1 | 8.3 | 11.1 | 13.8 | 9.0 | 5.0 | 14.0 | 10.0 | 5.0 | 19.0 | 15.0 | 10.0 | 20.0 | 15.0 | 20.0 | 25.0 |

APPENDICES

**TABLE F.5 Design Equivalent Temperature Differences (DETD) (Continued)**

### Part B. Frame Walls and Doors

**Outdoor Design Temperature / Daily Temperature Range[a]**

| Orientation | 29.4°C L | M | H | 32 L | M | H | 35 L | M | H | 38 L | M | H | 41 L | M | H | 43 L | M | H |
|---|---|---|---|---|---|---|---|---|---|---|---|---|---|---|---|---|---|---|
| North | 4 | 2 |  | 7 | 4 | 2 | 10 | 7 | 4 |  | 10 | 7 |  |  | 10 |  |  | 13 |
| NE and NW | 8 | 5 |  | 11 | 8 | 5 | 13 | 11 | 7 |  | 13 | 11 |  |  | 13 |  |  | 16 |
| East and West | 10 | 7 |  | 13 | 10 | 7 | 16 | 13 | 10 |  | 16 | 13 |  |  | 16 |  |  | 18 |
| SE and SW | 9 | 6 |  | 12 | 9 | 6 | 14 | 12 | 9 |  | 14 | 12 |  |  | 14 |  |  | 17 |
| South | 6 | 3 |  | 9 | 6 | 3 | 12 | 9 | 6 |  | 12 | 9 |  |  | 12 |  |  | 14 |

| Orientation | 85°F L | M | H | 90 L | M | H | 95 L | M | H | 100 L | M | H | 105 L | M | H | 110 L | M | H |
|---|---|---|---|---|---|---|---|---|---|---|---|---|---|---|---|---|---|---|
| North | 8 | 3 |  | 13 | 8 | 3 | 18 | 13 | 8 |  | 18 | 13 |  |  | 18 |  |  | 23 |
| NE and NW | 14 | 9 |  | 19 | 14 | 9 | 24 | 19 | 14 |  | 24 | 19 |  |  | 24 |  |  | 29 |
| East and West | 18 | 13 |  | 23 | 18 | 13 | 28 | 23 | 18 |  | 28 | 23 |  |  | 28 |  |  | 33 |
| SE and SW | 16 | 11 |  | 21 | 16 | 11 | 26 | 21 | 16 |  | 26 | 21 |  |  | 26 |  |  | 31 |
| South | 11 | 6 |  | 16 | 11 | 6 | 21 | 16 | 11 |  | 21 | 16 |  |  | 21 |  |  | 26 |

*Source:* Part A, 1981 *ASHRAE Handbook—Fundamentals;* Part B, 1997 *ASHRAE Handbook—Fundamentals;* both © by the American Society of Heating, Refrigerating and Air-Conditioning Engineers, Inc., Atlanta, GA.

[a] Daily temperature range: L (low), M (medium), H (high).

From Appendix B, mean daily range column: L is less than 16°F (9°C); M is 16 to 25°F (9 to 14°C); H is greater than 25°F (14°C).

# F.6 HEAT GAINS (COOLING LOADS) THROUGH GLASS

**TABLE F.6 Design Cooling Load Factors Through Glass**

Part A: SI: W/m²

| Outdoor | Regular Single Glass | | | | | | Regular Double Glass | | | | | | Heat-Absorbing Double Glass | | | | | | Clear Triple Glass | | |
|---|---|---|---|---|---|---|---|---|---|---|---|---|---|---|---|---|---|---|---|---|---|
| Design Temp.[a] | 29.4 | 32.2 | 35.0 | 37.7 | 40.5 | 43.3 | 29.4 | 32.2 | 35.0 | 37.7 | 40.5 | 43.3 | 29.4 | 32.2 | 35.0 | 37.7 | 40.5 | 43.3 | 29.4 | 32.2 | 35.0 |
| **No Awnings or Inside Shading** | | | | | | | | | | | | | | | | | | | | | |
| North | 72.6 | 85.2 | 97.8 | 110.4 | 123.0 | 138.8 | 59.9 | 66.2 | 75.8 | 82.0 | 88.3 | 94.7 | 37.9 | 44.1 | 53.7 | 60.0 | 66.2 | 72.6 | 53.7 | 60.0 | 63.1 |
| NE and NW | 176.6 | 189.3 | 202.0 | 214.6 | 227.1 | 243.0 | 145.1 | 151.4 | 161.0 | 167.2 | 173.6 | 179.9 | 85.1 | 91.4 | 101.0 | 107.2 | 113.6 | 119.9 | 132.6 | 135.7 | 138.9 |
| East and west | 255.5 | 268.1 | 280.8 | 293.4 | 306.0 | 321.9 | 214.6 | 220.9 | 230.3 | 236.7 | 243.0 | 249.2 | 132.6 | 138.9 | 148.2 | 154.6 | 160.9 | 167.2 | 195.7 | 198.8 | 202.6 |
| SE and SW | 220.9 | 233.4 | 246.0 | 258.8 | 271.3 | 287.1 | 186.1 | 192.4 | 202.0 | 208.2 | 214.6 | 220.9 | 110.4 | 116.8 | 126.2 | 132.6 | 138.9 | 145.1 | 167.2 | 173.6 | 176.7 |
| South | 126.2 | 138.9 | 151.4 | 164.0 | 176.7 | 192.4 | 104.1 | 110.4 | 119.9 | 126.2 | 132.6 | 138.9 | 66.0 | 72.6 | 82.0 | 88.3 | 94.7 | 101.0 | 94.7 | 97.9 | 104.1 |
| Horiz. skylight | 504.8 | 517.4 | 530.0 | 542.7 | 555.2 | 571.0 | 438.6 | 444.9 | 454.3 | 460.7 | 467.0 | 473.0 | 280.8 | 287.1 | 296.6 | 302.9 | 309.1 | 315.6 | 397.6 | 400.7 | 407.0 |
| **Draperies or Venetian Blinds** | | | | | | | | | | | | | | | | | | | | | |
| North | 47.3 | 60.0 | 72.6 | 85.1 | 97.9 | 113.6 | 37.9 | 44.1 | 53.7 | 60.0 | 66.2 | 72.6 | 28.3 | 34.7 | 44.1 | 50.4 | 56.8 | 63.1 | 34.7 | 37.9 | 44.2 |
| NE and NW | 101.0 | 113.6 | 126.2 | 138.9 | 151.4 | 167.2 | 85.1 | 91.4 | 101.0 | 107.2 | 113.6 | 119.9 | 63.1 | 69.4 | 78.9 | 85.1 | 91.4 | 97.9 | 75.8 | 82.0 | 85.1 |
| East and west | 151.4 | 164.0 | 176.7 | 189.3 | 202.0 | 217.7 | 132.6 | 138.9 | 148.2 | 154.6 | 161.0 | 167.2 | 94.7 | 101.0 | 110.4 | 116.8 | 123.0 | 129.3 | 119.9 | 123.0 | 129.3 |
| E and SW | 126.2 | 138.9 | 151.4 | 164.0 | 176.7 | 192.4 | 110.4 | 116.8 | 126.2 | 132.6 | 138.9 | 145.1 | 75.8 | 82.0 | 91.4 | 97.9 | 104.1 | 110.4 | 101.0 | 104.1 | 107.2 |
| South | 72.5 | 85.1 | 97.9 | 110.4 | 123.0 | 138.9 | 63.1 | 69.4 | 78.9 | 85.1 | 91.4 | 97.9 | 47.3 | 53.7 | 63.1 | 69.4 | 75.8 | 82.0 | 56.8 | 60.0 | 66.2 |
| **Roller Shades Half-Drawn** | | | | | | | | | | | | | | | | | | | | | |
| North | 56.8 | 69.4 | 82.0 | 92.7 | 102.2 | 123.0 | 47.3 | 53.7 | 63.1 | 69.4 | 75.8 | 82.0 | 31.6 | 37.9 | 47.3 | 53.7 | 60.0 | 66.2 | 41.0 | 44.2 | 47.3 |
| NE and NW | 126.2 | 138.9 | 151.4 | 164.0 | 176.7 | 192.4 | 119.9 | 126.2 | 135.7 | 142.0 | 148.2 | 154.6 | 75.8 | 82.0 | 91.4 | 97.9 | 104.1 | 110.4 | 107.2 | 110.4 | 110.4 |
| East and west | 192.5 | 205.0 | 217.7 | 230.3 | 243.0 | 258.8 | 170.3 | 176.7 | 186.1 | 192.4 | 198.8 | 205.0 | 110.4 | 116.8 | 126.2 | 132.5 | 138.9 | 145.1 | 154.1 | 154.6 | 157.8 |
| SE and SW | 164.0 | 176.7 | 189.3 | 202.0 | 214.6 | 230.3 | 145.1 | 151.4 | 161.0 | 167.2 | 173.6 | 179.9 | 94.7 | 101.0 | 110.4 | 116.8 | 123.0 | 129.3 | 129.3 | 132.6 | 135.7 |
| South | 91.4 | 104.1 | 116.8 | 129.3 | 142.0 | 157.8 | 85.1 | 91.4 | 101.0 | 107.2 | 113.6 | 119.9 | 56.8 | 63.1 | 72.6 | 78.9 | 85.1 | 91.4 | 78.9 | 82.0 | 82.0 |
| **Awnings[b]** | | | | | | | | | | | | | | | | | | | | | |
| North | 63.1 | 75.8 | 88.3 | 101.0 | 113.6 | 129.3 | 41.0 | 56.8 | 60.0 | 63.1 | 69.4 | 75.8 | 31.6 | 34.7 | 37.9 | 47.3 | 50.4 | 56.8 | 34.7 | 37.9 | 41.0 |
| NE and NW | 66.2 | 78.9 | 91.4 | 104.1 | 116.8 | 132.6 | 44.2 | 60.0 | 63.1 | 66.2 | 72.6 | 78.9 | 34.7 | 41.0 | 44.2 | 50.4 | 56.8 | 60.0 | 37.9 | 41.0 | 44.2 |
| East and west | 69.4 | 82.0 | 94.7 | 107.2 | 119.9 | 135.7 | 44.2 | 60.0 | 63.1 | 66.2 | 72.6 | 78.9 | 37.9 | 44.2 | 53.6 | 60.0 | 66.2 | 72.6 | 37.9 | 41.0 | 44.2 |
| SE and SW | 66.2 | 78.9 | 91.4 | 104.1 | 116.8 | 132.6 | 44.2 | 60.0 | 63.1 | 66.2 | 72.6 | 78.9 | 34.7 | 41.0 | 44.2 | 50.4 | 56.8 | 60.0 | 37.9 | 41.0 | 44.2 |
| South | 66.2 | 75.8 | 88.3 | 101.0 | 113.6 | 129.3 | 41.0 | 56.8 | 60.0 | 63.1 | 69.4 | 75.8 | 34.7 | 37.9 | 41.0 | 50.4 | 56.8 | 56.8 | 34.7 | 37.9 | 41.0 |

**TABLE F.6  Design Cooling Load Factors Through Glass (Continued)**

### Part B. I-P: Btu/h ft²

| Outdoor Design Temp.[a] | Regular Single Glass | | | | | | Regular Double Glass | | | | | | Heat-Absorbing Double Glass | | | | | | Clear Triple Glass | | |
|---|---|---|---|---|---|---|---|---|---|---|---|---|---|---|---|---|---|---|---|---|---|
| | 85 | 90 | 95 | 100 | 105 | 110 | 85 | 90 | 95 | 100 | 105 | 110 | 85 | 90 | 95 | 100 | 105 | 110 | 85 | 90 | 95 |
| **No Awnings or Inside Shading** | | | | | | | | | | | | | | | | | | | | | |
| North | 23 | 27 | 31 | 35 | 39 | 44 | 19 | 21 | 24 | 26 | 28 | 30 | 12 | 14 | 17 | 19 | 21 | 23 | 17 | 19 | 20 |
| NE and NW | 56 | 60 | 64 | 68 | 72 | 77 | 46 | 48 | 51 | 53 | 55 | 57 | 27 | 29 | 32 | 34 | 36 | 38 | 42 | 43 | 44 |
| East and west | 81 | 85 | 89 | 93 | 97 | 102 | 68 | 70 | 73 | 75 | 77 | 79 | 42 | 44 | 47 | 49 | 51 | 53 | 62 | 63 | 64 |
| SE and SW | 70 | 74 | 78 | 82 | 86 | 91 | 59 | 61 | 64 | 66 | 68 | 70 | 35 | 37 | 40 | 42 | 44 | 46 | 53 | 55 | 56 |
| South | 40 | 44 | 48 | 52 | 56 | 61 | 33 | 35 | 38 | 40 | 42 | 44 | 19 | 21 | 24 | 26 | 28 | 30 | 30 | 31 | 33 |
| Horiz. skylight | 160 | 164 | 168 | 172 | 176 | 181 | 139 | 141 | 144 | 146 | 148 | 150 | 89 | 91 | 94 | 96 | 98 | 100 | 126 | 127 | 129 |
| **Draperies or Venetian Blinds** | | | | | | | | | | | | | | | | | | | | | |
| North | 15 | 19 | 23 | 27 | 31 | 36 | 12 | 14 | 17 | 19 | 21 | 23 | 9 | 11 | 14 | 16 | 18 | 20 | 11 | 12 | 14 |
| NE and NW | 32 | 36 | 40 | 44 | 48 | 53 | 27 | 29 | 32 | 34 | 36 | 38 | 20 | 22 | 25 | 27 | 29 | 31 | 24 | 26 | 27 |
| East and west | 48 | 52 | 56 | 60 | 64 | 69 | 42 | 44 | 47 | 49 | 51 | 53 | 30 | 32 | 35 | 37 | 39 | 41 | 38 | 39 | 41 |
| SE and SW | 40 | 44 | 48 | 52 | 56 | 61 | 35 | 37 | 40 | 42 | 44 | 46 | 24 | 26 | 29 | 31 | 33 | 35 | 32 | 33 | 34 |
| South | 23 | 27 | 31 | 35 | 39 | 44 | 20 | 22 | 25 | 27 | 29 | 31 | 15 | 17 | 20 | 22 | 24 | 26 | 18 | 19 | 21 |
| **Roller Shades Half-Drawn** | | | | | | | | | | | | | | | | | | | | | |
| North | 18 | 22 | 26 | 30 | 34 | 39 | 15 | 17 | 20 | 22 | 24 | 26 | 10 | 12 | 15 | 17 | 19 | 21 | 13 | 14 | 15 |
| NE and NW | 40 | 44 | 48 | 52 | 56 | 61 | 38 | 40 | 43 | 45 | 47 | 49 | 24 | 26 | 29 | 31 | 33 | 35 | 34 | 35 | 35 |
| East and west | 61 | 65 | 69 | 73 | 77 | 82 | 54 | 56 | 59 | 61 | 63 | 65 | 35 | 37 | 40 | 42 | 44 | 46 | 49 | 49 | 50 |
| SE and SW | 52 | 56 | 60 | 64 | 68 | 73 | 46 | 48 | 51 | 53 | 55 | 57 | 30 | 32 | 35 | 37 | 39 | 41 | 41 | 42 | 43 |
| South | 29 | 33 | 37 | 41 | 45 | 50 | 27 | 29 | 32 | 34 | 36 | 38 | 18 | 20 | 23 | 25 | 27 | 29 | 25 | 26 | 26 |
| **Awnings[b]** | | | | | | | | | | | | | | | | | | | | | |
| North | 20 | 24 | 28 | 32 | 36 | 41 | 15 | 18 | 20 | 22 | 23 | 24 | 10 | 12 | 15 | 17 | 19 | 21 | 11 | 12 | 13 |
| NE and NW | 21 | 25 | 29 | 33 | 37 | 42 | 16 | 19 | 21 | 23 | 25 | 25 | 11 | 13 | 16 | 18 | 20 | 22 | 12 | 13 | 14 |
| East and west | 22 | 26 | 30 | 34 | 38 | 43 | 16 | 19 | 21 | 23 | 25 | 25 | 12 | 14 | 17 | 19 | 21 | 23 | 12 | 13 | 14 |
| SE and SW | 21 | 25 | 29 | 33 | 37 | 42 | 16 | 19 | 21 | 23 | 25 | 25 | 11 | 13 | 16 | 18 | 20 | 22 | 12 | 13 | 14 |
| South | 21 | 24 | 28 | 32 | 36 | 41 | 15 | 18 | 20 | 22 | 24 | 24 | 11 | 13 | 16 | 18 | 20 | 22 | 11 | 12 | 13 |

*Source:* © by the American Society of Heating, Refrigerating and Air-Conditioning Engineers, Inc., Atlanta, GA. Reprinted by permission from the 1981 *ASHRAE Handbook—Fundamentals.*

[a] Based on indoor design temperature of 75°F (23.8°C) and outdoor design temperatures as indicated. Interpolate to obtain factors for outdoor design temperatures other than those given.

[b] For other external shading devices that completely shade the glass at any orientation, use the values for "Awnings, north."

## F.7  HEAT GAINS (COOLING LOADS) DUE TO INFILTRATION/VENTILATION

**TABLE F.7  Sensible Cooling Load Factors Due to Infiltration and Ventilation**

| °C: 29.4 | 32.2 | 35.0 | 37.7 | 41.5 | 43.3 | Units | Design Temperature | Units | °F: 85 | 90 | 95 | 100 | 105 | 110 |
|---|---|---|---|---|---|---|---|---|---|---|---|---|---|---|
| 2.2 | 3.5 | 4.7 | 6.0 | 6.9 | 8.2 | W/m² | Infiltration, per gross exposed wall area | Btu/h ft² | 0.7 | 1.1 | 1.5 | 1.9 | 2.2 | 2.6 |
| 6.8 | 9.9 | 13.6 | 16.7 | 19.8 | 23.6 | W per L/s | Mechanical ventilation | Btu/h per cfm | 11.0 | 16.0 | 22.0 | 27.0 | 32.0 | 38.0 |

*Source:* © by the American Society of Heating, Refrigerating and Air-Conditioning Engineers, Inc., Atlanta, GA. Reprinted by permission from the 1981 *ASHRAE Handbook—Fundamentals.*

## F.8  HEAT GAINS FROM BUILDING OCCUPANTS

**TABLE F.8  Rates of Heat Gain from Occupants of Conditioned Spaces**

| Activity | Location | Total Heat Gain W Adult Male | Adjusted[b] | Sensible[a] Heat | Latent[a] Heat | Btu/h Adult Male | Adjusted[b] | Sensible[a] Heat | Latent[a] Heat |
|---|---|---|---|---|---|---|---|---|---|
| Seated at theater | Theater, matinee | 115 | 95 | 65 | 30 | 390 | 330 | 225 | 105 |
| Seated at theater, night | Theater, night | 115 | 105 | 70 | 35 | 390 | 350 | 245 | 105 |
| Seated, very light work | Offices, hotels, apartments | 130 | 115 | 70 | 45 | 450 | 400 | 245 | 155 |
| Moderately active office work | Offices, hotels, apartments | 140 | 130 | 75 | 55 | 475 | 450 | 250 | 200 |
| Standing, light work; walking | Department or retail store | 160 | 130 | 75 | 55 | 550 | 450 | 250 | 200 |
| Walking, standing | Drug store, bank | 160 | 145 | 75 | 70 | 550 | 500 | 250 | 250 |
| Sedentary work | Restaurant[c] | 170 | 160 | 80 | 80 | 590 | 550 | 275 | 275 |
| Light bench work | Factory | 235 | 220 | 80 | 140 | 800 | 750 | 275 | 475 |
| Moderate dancing | Dance hall | 265 | 250 | 90 | 160 | 900 | 850 | 305 | 545 |
| Walking 4.8 km/h (3 mph), light machine work | Factory | 295 | 295 | 110 | 185 | 1000 | 1000 | 375 | 625 |
| Bowling[d] | Bowling alley | 440 | 425 | 170 | 255 | 1500 | 1450 | 580 | 870 |
| Heavy work | Factory | 440 | 425 | 170 | 255 | 1500 | 1450 | 580 | 870 |
| Heavy machine work, lifting | Factory | 470 | 470 | 185 | 285 | 1600 | 1600 | 635 | 965 |
| Athletics | Gymnasium | 585 | 525 | 210 | 315 | 2000 | 1800 | 710 | 1090 |

*Source:* Reprinted with permission of the American Society of Heating, Refrigerating and Air-Conditioning Engineers, Inc., Atlanta, GA from the 1997 *ASHRAE Handbook—Fundamentals.*

[a]All values are rounded to the nearest 5 W (and 5 Btu/h). Based on 75°F (24°C) room dry bulb temperature. For 80°F (27°C) room dry bulb temperature, the total heat remains the same but the sensible heat values should be *decreased* by approximately 20% and the latent heat values *increased* accordingly.

[b]Adjusted heat gain based on the normal percentage of men, women, and children for the application listed, assuming that the gain from an adult female is 85% (and from children 75%) of that from an adult male.

[c]Adjusted heat gain includes 60 Btu/h (18 W) for food per individual: 50% sensible, 50% latent.

[d]Assume only one person per alley actually bowling and all others as sitting, standing, or walking slowly.

APPENDICES

## F.9 HEAT GAINS FROM OFFICE EQUIPMENT

### TABLE F.9  Rates of Heat Gain from Office Equipment

| Appliance | Maximum Input Rating (W) | Recommended Rate of Heat Gain | |
|---|---|---|---|
| | | W | Btu/h |
| Check processing workstation, 12 pockets | 4,800 | 2,460 | 8,410 |
| **Computer Devices** | | | |
| Communication/transmission | 1,800–4,600 | 1,640–2,810 | 5,600–9,600 |
| Disk drives/mass storage | 1,000–10,000 | 1,000–6,570 | 3,412–22,420 |
| Microcomputer | 100–600 | 90–530 | 300–1,800 |
| Minicomputer | 2,200–6,600 | 2,200–6,600 | 7,500–15,000 |
| Optical reader | 3,000–6,000 | 2,350–4,980 | 8,000–17,000 |
| Plotters | 75 | 63 | 214 |
| **Printers** | | | |
| Letter quality, 30–45 characters/min. | 350 | 292 | 1,000 |
| Line, high speed, 5000 or more lines/min | 1,000–5,300 | 730–3,810 | 2,500–13,000 |
| Line, low speed 300–600 lines/min | 450 | 376 | 1,280 |
| Tape drives | 1,200–6,500 | 1,000–4,700 | 3,500–15,000 |
| Terminal | 90–200 | 80–180 | 270–600 |
| **Copiers/Duplicators** | | | |
| Copiers, large; 30–67[a] copies/min | 1,700–6,600 | 1,700–6,600 | 5,800–22,500 |
| Copiers, small; 6–30[a] copies/min | 460–1,700 | 460–1,700 | 1,570–5,800 |
| Feeder | 30 | 30 | 100 |
| Microfilm printer | 450 | 450 | 1,540 |
| Sorter/collator | 60–600 | 60–600 | 200–2,050 |
| **Audio Equipment** | | | |
| Cassette recorders/players | 60 | 60 | 200 |
| Receiver/tuner | 100 | 100 | 340 |
| Signal analyzer | 60–650 | 60–650 | 90–2,220 |
| **Mail Processing** | | | |
| Folding machine | 125 | 80 | 270 |
| Inserting machine, 3600–6800 pieces/h | 600–3,300 | 390–2,150 | 1,330–7,340 |
| Labelling machine, 1,500–30,000 pieces/h | 600–6,600 | 390–4,300 | 1,330–14,700 |
| Postage meter | 230 | 150 | 510 |
| **Vending Machines** | | | |
| Cigarette | 72 | 72 | 250 |
| Cold food/beverage | 1,150–1,920 | 575–960 | 1,960–3,280 |
| Hot beverage | 1,725 | 862 | 2,940 |
| Snack | 240–275 | 240–275 | 820–940 |
| **Miscellaneous** | | | |
| Barcode printer | 440 | 370 | 1,260 |
| Cash registers | 60 | 48 | 160 |
| Coffee maker, 10 cups | 1,500 | 1,050 sensible | 3,580 sensible |
| | | 450 latent | 1,540 latent |
| Microfiche reader/printer | 1,150 | 1,150 | 3,920 |
| Microwave oven, 28 L (1 ft³) | 600 | 400 | 1,360 |
| Paper shredder | 250–3,000 | 200–2,420 | 680–8,250 |
| Water cooler, 30 L/h (32 qt/h) | 1,750 | 1,750 | 5,970 |

*Source:* Excerpted with permission of the American Society of Heating, Refrigerating and Air-Conditioning Engineers, Inc., Atlanta, GA, from the 1997 *ASHRAE Handbook—Fundamentals.*

[a]Input of power is not proportional to capacity.

## F.10  HEAT GAINS FROM APPLIANCES

**TABLE F.10  Rate of Heat Gain from Miscellaneous Appliances[a]**

| Miscellaneous Data | Manufacturer's Rating Watts | Recommended Rate of Heat Gain (W) | | | Appliance | Miscellaneous Data | Manufacturer's Rating Btu/h | Recommended Rate of Heat Gain (Btu/h) | | |
|---|---|---|---|---|---|---|---|---|---|---|
| | | Sensible | Latent | Total | | | | Sensible | Latent | Total |
| **Electrical Appliances** | | | | | | | | | | |
| Blower type | 1580 | 675 | 120 | 785 | Hair dryer | Blower type | 5,400 | 2,300 | 400 | 2,700 |
| Helmet type 60 heaters at 25 W | 700 | 550 | 100 | 650 | Hair dryer | Helmet type | 2,400 | 1,870 | 330 | 2,200 |
| 91.44-cm normal use | 1500 | 250 | 50 | 300 | Permanent wave machine | 60 heaters at 25 W / 36-in. normal use | 5,000 | 850 | 150 | 1,000 |
| 1.27-cm diameter | | 28 | | 28 | Neon sign, per linear meter of tube[b] | 0.5-in. diameter | | 30 | | 30 |
| 0.95-cm diameter | | 56 | | 56 | | 0.375-in. diameter | | 60 | | 60 |
| | 1100 | 190 | 350 | 540 | Sterilizer, instrument | | 3,750 | 650 | 1200 | 1,850 |
| | 202 | 102 | | 102 | Magnetic card type-writer | | 690 | 350 | | |
| Running | 1760 | 1760 | | 1760 | Small copier | Running | 6,000 | 6,000 | | |
| Standby | 880 | 880 | | 880 | | Standby | 3,000 | 3,000 | | |
| Running | 3515 | 3515 | | 3515 | Large copier | Running | 12,000 | 12,000 | | |
| Standby | 1760 | 1760 | | 1760 | | Standby | 6,000 | 6,000 | | |
| **Gas-Burning Appliances** | | | | | | | | | | |
| | | | | | *Lab Burners* | | | | | |
| 1.1-cm barrel | 880 | 495 | 125 | 620 | Bunsen | 0.4375-in. barrel | 3,000 | 1,680 | 430 | 2,100 |
| 3.8 cm wide | 1465 | 820 | 205 | 1025 | Fishtail | 1.5 in. wide | 5,000 | 2,800 | 700 | 3,500 |
| 2.54-cm diameter | 1760 | 985 | 245 | 1230 | Meeker | 1-in. diameter | 6,000 | 3,360 | 840 | 4,200 |
| Mantle type | 585 | 530 | 60 | 590 | Gas light, per burner | Mantle type | 2,000 | 1,800 | 200 | 2,000 |

*Source:* © by the American Society of Heating, Refrigerating and Air-Conditioning Engineers, Inc., Atlanta, GA. Reprinted by permission from the 1981 *ASHRAE Handbook—Fundamentals.*

[a]For residential appliances, see Table 8.6.

[b]I-P (Btu/h) values are per linear foot of tube.

## F.11 CLIMATE DATA FOR BUILDING COOLING

**TABLE F.11 Cooling Climate Data[a] for Some North American Cities**

| City, State, or Province | I-P Units, °F CDD50 | No. Hours 8 A.M.–4 P.M. (°F) 55 < DB < 69 (°C) 13 < DB < 21 | SI Units, °C CDD10 |
|---|---|---|---|
| Albuquerque, New Mexico | 3908 | 703 | 2171 |
| Albany, New York | 2525 | 605 | 1403 |
| Amarillo, Texas | 4128 | 680 | 2293 |
| Atlanta, Georgia | 5038 | 749 | 2799 |
| Baltimore, Maryland | 3683 | 593 | 2046 |
| Birmingham, Alabama | 5206 | 760 | 2892 |
| Brownsville, Texas | 8777 | 422 | 4876 |
| Boise, Idaho | 2807 | 647 | 1559 |
| Boston, Massachusetts | 2897 | 713 | 1609 |
| Buffalo, New York | 2468 | 697 | 1371 |
| Burlington, Vermont | 2228 | 637 | 1238 |
| Calgary, Alberta | 1167 | NA | 648 |
| Charleston, South Carolina | 5722 | 767 | 3179 |
| Cheyenne, Wyoming | 1886 | 608 | 1048 |
| Chicago, Illinois (O'Hare) | 2941 | 613 | 1634 |
| Columbia, Missouri | 3752 | 633 | 2084 |
| Columbus, Ohio | 3119 | 708 | 1733 |
| Denver, Colorado | 2732 | 739 | 1518 |
| Detroit, Michigan | 3199 | 632 | 1777 |
| Dodge City, Kansas | 5001 | 637 | 2272 |
| El Paso, Texas | 5488 | 735 | 3049 |
| Fort Wayne, Indiana | 3077 | 601 | 1709 |
| Fort Worth, Texas | 6174 | 772 | 3430 |
| Fresno, California | 5350 | 785 | 2972 |
| Great Falls, Montana | 1993 | 641 | 1107 |
| Honolulu, Hawaii | 9949 | 69 | 5527 |
| Houston, Texas | 7215 | 703 | 4008 |
| Jackson, Mississippi | 5900 | 640 | 3278 |
| Jacksonville, Florida | 6847 | 674 | 3804 |
| Lake Charles, Louisiana | 6813 | 668 | 3785 |
| Las Vegas, Nevada | 6745 | 719 | 3747 |
| Los Angeles, California | 4777 | 1849 | 2654 |
| Louisville, Kentucky | 4000 | 636 | 2222 |
| Lubbock, Texas | 4833 | 743 | 2685 |
| Madison, Wisconsin | 2389 | 658 | 1327 |
| Medford, Oregon | 2989 | 749 | 1661 |
| Memphis, Tennessee | 5467 | 851 | 3037 |
| Miami, Florida | 9474 | 259 | 5263 |
| Minneapolis, Minnesota | 2680 | 566 | 1489 |
| Montreal, Quebec | 2146 | NA | 1192 |
| Nashville, Tennessee | 4689 | 749 | 2605 |
| New Orleans, Louisiana | 6910 | 789 | 3893 |
| New York (Central Park), New York | 3643 | 790 | 2019 |
| Norfolk, Virginia | 4478 | 685 | 2488 |
| Oklahoma City, Oklahoma | 4972 | 733 | 2762 |
| Omaha, Nebraska | 3618 | 586 | 2010 |
| Ottawa, Ontario | 2045 | NA | 1136 |
| Philadelphia, Pennsylvania | 3623 | 646 | 2013 |
| Phoenix, Arizona | 8425 | 746 | 4681 |
| Pittsburgh, Pennsylvania | 2836 | 700 | 1576 |
| Portland, Maine | 1943 | 665 | 1079 |
| Portland, Oregon | 2517 | 1060 | 1398 |
| Raleigh, North Carolina | 4499 | 740 | 2499 |

**TABLE F.11 Cooling Climate Data[a] for Some North American Cities *(Continued)***

| City, State, or Province | I-P Units, °F CDD50 | No. Hours 8 A.M.–4 P.M. (°F) 55 < DB < 69 (°C) 13 < DB < 21 | SI Units, °C CDD10 |
|---|---|---|---|
| Regina, Saskatchewan | 1620 | NA | 900 |
| Richmond, Virginia | 4223 | 716 | 2346 |
| Sacramento, California | 4474 | 990 | 2486 |
| Salt Lake City, Utah | 3276 | 586 | 1820 |
| San Antonio, Texas | 7170 | 690 | 3983 |
| San Diego, California | 5223 | 1911 | 2902 |
| San Francisco, California | 2883 | 1796 | 1602 |
| Saint Louis, Missouri | 4193 | 614 | 2329 |
| Seattle–Tacoma, Washington | 2021 | 982 | 1123 |
| Sioux City, Iowa | 3149 | 602 | 1749 |
| Tampa, Florida | 8239 | 592 | 4577 |
| Topeka, Kansas | 3880 | 608 | 2156 |
| Toronto, Ontario | 2370 | NA | 1317 |
| Tucson, Arizona | 6921 | 716 | 3845 |
| Tulsa, Oklahoma | 5150 | 591 | 2861 |
| Vancouver, British Columbia | 1536 | NA | 853 |
| Washington, DC | 3734 | 657 | 2074 |
| Winnipeg, Manitoba | 1784 | NA | 991 |

*Source:* ANSI / ASHRAE Standard 90.1-1989, *Energy Efficient Design of New Buildings Except Low-Rise Residential Buildings,* © by the American Society of Heating, Refrigerating and Air-Conditioning Engineers, Inc., Atlanta, GA.

NA = not available

[a]CDH74 (cooling degree hours base 74°F); values are found in Appendix B.

## F.12 DESIGN DATA FOR EARTH TUBES

**TABLE F.12 Earth Tube Applications[a]**

### Part A. I-P Units

| Soil, No. of Tubes, Arrangement | Tube Diameter (in.) | Tube Length (ft) | Depth[b] (ft) | Temperatures (°F) Soil | Temperatures (°F) Intake Air[c] | Air Velocity (ft/min) | Total Flow Rate (ft³/h) | Flow Per Tube (ft³/h) | Peak Sensible Cooling (Btu/h ft) |
|---|---|---|---|---|---|---|---|---|---|
| Sandy soil; 8 tubes, radial | 10 | 100 | 8 | 73° | 67–105° | 220 | 57,600 | 7,200 | 20 |
| Silt loam, wet; 9 tubes, radial (3 trenches) | 5 | 150 | 8–11.5 | 73° | 62–95° | 500 | 60,000 | 4,000 | 11 |
| Silt loam; 22 tubes, parallel at 4 ft | 10 | 180 | 8–9 | 72° | 70–94° | 500 | 359,700 | 16,350 | 30 |
| Silt loam; 5 tubes, parallel | 12 | 260 | 10 | 63° | 60–92° | 530 | 123,600 | 24,720 | 40 |

### Part B. SI Units

| Soil, No. of Tubes, Arrangement | Tube Diameter (cm.) | Tube Length (m) | Depth[b] (m) | Temperatures (°C) Soil | Temperatures (°C) Intake Air[c] | Air Velocity (m/s) | Total Flow Rate (m³/h) | Flow Per Tube (m³/h) | Peak Sensible Cooling (W/m) |
|---|---|---|---|---|---|---|---|---|---|
| Sandy soil; 8 tubes, radial | 25 | 30.5 | 2.4 | 23° | 19.4–40.6° | 1.1 | 1,632 | 204 | 19.7 |
| Silt loam, wet; 9 tubes, radial (3 trenches) | 13 | 46 | 2.4–3.5 | 23° | 16.7–35° | 2.5 | 1,080 | 120 | 10.6 |
| Silt loam; 22 tubes, parallel at 1.2m | 25 | 55 | 2.4–2.7 | 22° | 21.1–34.4° | 2.5 | 10,200 | 463 | 29.5 |
| Silt loam; 5 tubes, parallel | 30 | 79 | 3 | 17° | 15.6–33.3° | 2.7 | 3,500 | 696 | 38.8 |

*Source:* Givoni, B. (1994). *Passive and Low Energy Cooling of Buildings.* Van Nostrand Reinhold, New York. I-P conversions added by the author.

[a]Central Illinois farm applications, summertime.

[b]Tubes often slope to drain the condensation that occurs as air reaches the dewpoint within the tube.

[c]Intake air temperature range during the period of study.

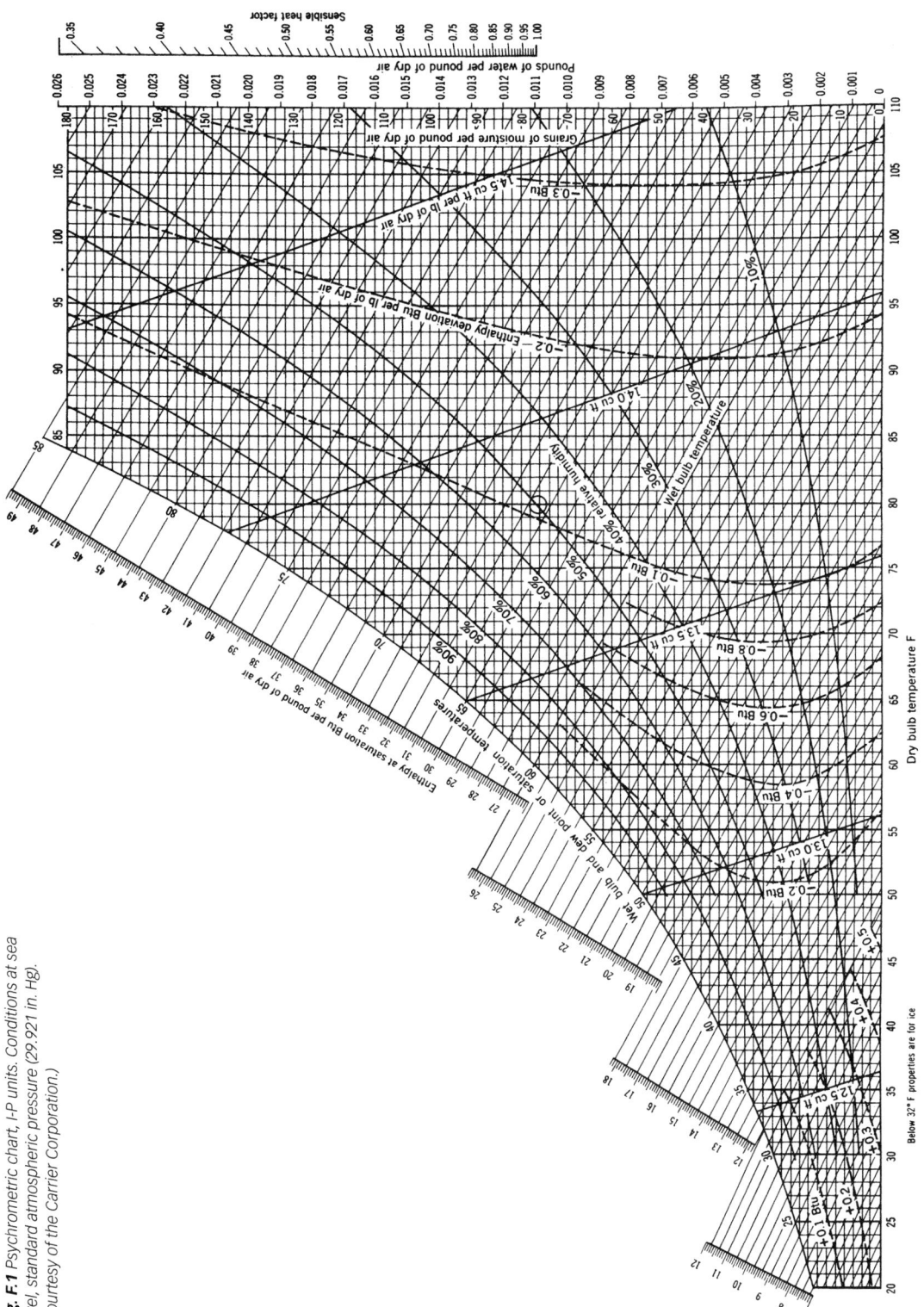

**Fig. F.1** Psychrometric chart, I-P units. Conditions at sea level, standard atmospheric pressure (29.921 in. Hg). (Courtesy of the Carrier Corporation.)

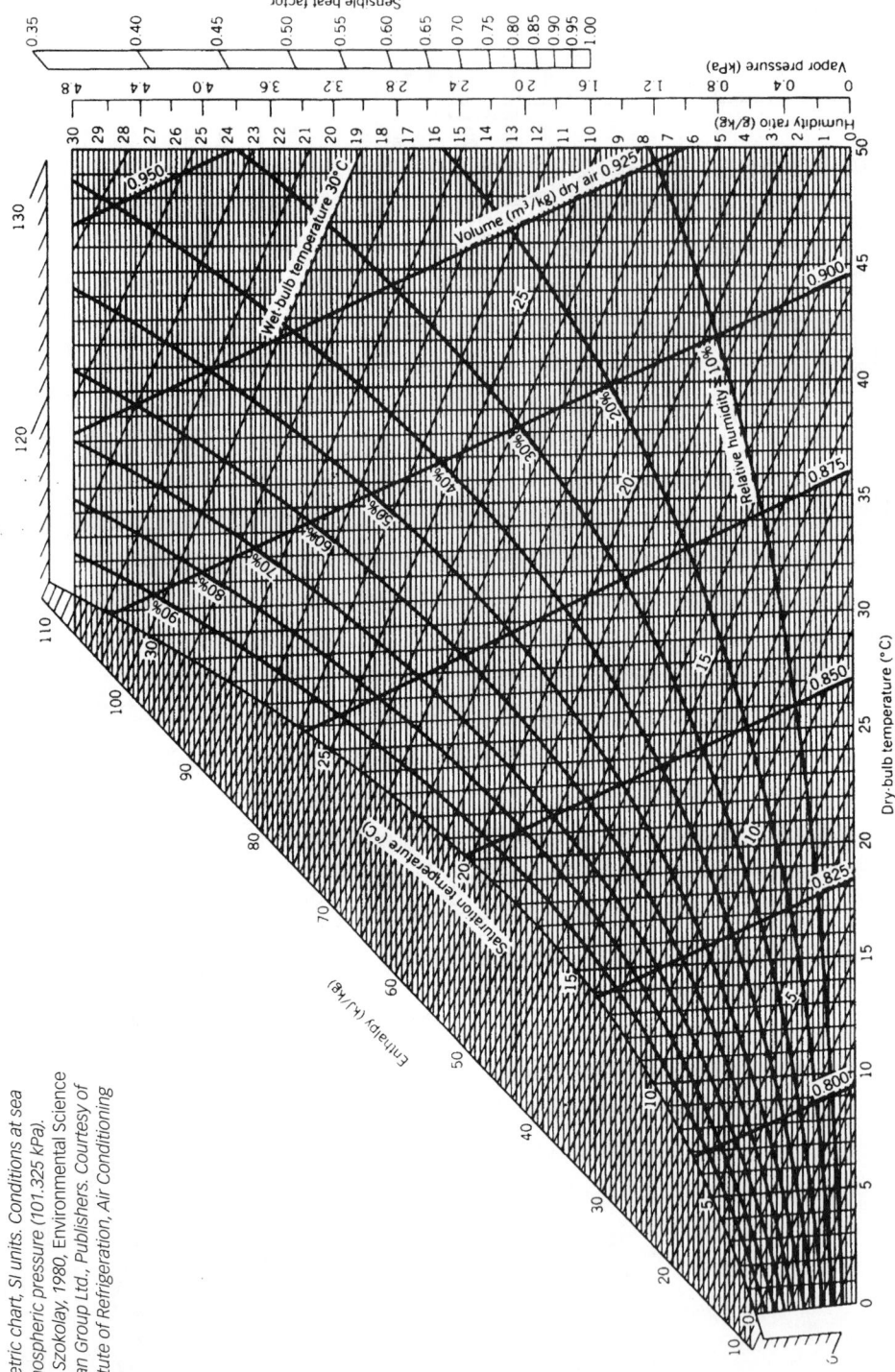

**Fig. F.2** Psychrometric chart, SI units. Conditions at sea level, standard atmospheric pressure (101.325 kPa). (Adapted from S.V. Szokolay, 1980, Environmental Science Handbook, Longman Group Ltd., Publishers. Courtesy of the Australian Institute of Refrigeration, Air Conditioning and Heating.)

# Standards/Guidelines for Energy- and Resource-Efficient Building Design

This appendix provides extracts from ASHRAE Standard 90.1, ASHRAE Standard 90.2, and the U.S. Green Building Council (USGBC) LEED rating system for new construction. The purpose of these extracts is to indicate the nature of these guidance documents and their format. For full details of each of these documents, obtain complete information from ASHRAE (www.ashrae.org) or USGBC (www.usgbc.org).

### G.1 SAMPLE OF PRESCRIPTIVE BUILDING ENVELOPE REQUIREMENTS EXTRACTED FROM ANSI/ASHRAE/ IESNA STANDARD 90.1-2001: *ENERGY STANDARD FOR BUILDINGS EXCEPT LOW-RISE RESIDENTIAL BUILDINGS*

**TABLE G.1 ASHRAE 90.1: Building Envelope Requirements (for HDD65: 3601-5400, CDD50: 3601+)**

| | Semiheated | | Nonresidential | | Residential | |
|---|---|---|---|---|---|---|
| Opaque Elements | Assembly Maximum | Insulation Min. R-Value | Assembly Maximum | Insulation Min. R-Value | Assembly Maximum | Insulation Min. R-Value |
| *Roofs* | | | | | | |
| Insulation entirely above deck | U-0.063 | R-15.0 ci | U-0.063 | R-15.0 ci | U-0.218 | R-3.8 ci |
| Metal building | U-0.065 | R-19.0 | U-0.065 | R-19.0 | U-0.097 | R-10.0 |
| Attic and other | U-0.034 | R-30.0 | U-0.027 | R-38.0 | U-0.081 | R-13.0 |
| *Walls, Above Grade* | | | | | | |
| Mass | U-0.051[a] | R-5.7 ci[a] | U-0.104 | R-9.5 ci[a] | U-0.580 | NR |
| Metal building | U-0.113 | R-13.0 | U-0.113 | R-13.0 | U-0.134 | R-10.0 |
| Steel framed | U-0.124 | R-13.0 | U-0.064 | R-13.0 + R-7.5 ci | U-0.124 | R-13.0 |
| Wood framed and other | U-0.089 | R-13.0 | U-0.089 | R-13.0 | U-0.089 | R-13.0 |
| *Wall, Below Grade* | | | | | | |
| Below grade wall | C-1.140 | NR | C-1.140 | NR | C-1.140 | NR |
| *Floors* | | | | | | |
| Mass | U-0.107 | R-6.3 ci | U-0.087 | R-8.3 ci | U-0.322 | NR |
| Steel joist | U-0.052 | R-19.0 | U-0.038 | R-30.0 | U-0.069 | R-13.0 |
| Wood framed and other | U-0.051 | R-19.0 | U-0.033 | R-30.0 | U-0.066 | R-13.0 |
| *Slab-on-Grade Floors* | | | | | | |
| Unheated | F-0.730 | NR | F-0.730 | NR | F-0.730 | NR |
| Heated | F-0.950 | R-7.5 for 24 in. | F-0.840 | R-10 for 36 in. | F-1.020 | R-7.5 for 12 in. |
| *Opaque Doors* | | | | | | |
| Swinging | U-0.700 | | U-0.700 | | U-0.700 | |
| Non-swinging | U-1.450 | | U-0.500 | | U-1.450 | |

**TABLE G.1 ASHRAE 90.1: Building Envelope Requirements (for HDD65: 3601-5400, CDD50: 3601+)** *(Continued)*

| Fenestration | *Semiheated* Assembly Max. U (Fixed/ Operable) | *Semiheated* Assembly Max. SHGC (All Orientations/North-Oriented) | *Nonresidential* Assembly Max. U (Fixed/ Operable) | *Nonresidential* Assembly Max. SHGC (All Orientations/North-Oriented) | *Residential* Assembly Max. U (Fixed/ Operable) | *Residential* Assembly Max. SHGC (All Orientations/North-Oriented) |
|---|---|---|---|---|---|---|
| *Vertical Glazing, % of Wall* | | | | | | |
| 0–10.0% | $U_{fixed}$-0.57 | $SHGC_{all}$-0.39 | $U_{fixed}$-0.57 | $SHGC_{all}$-0.39 | $U_{fixed}$-1.22 | $SHGC_{all}$-NR |
|  | $U_{oper}$-0.67 | $SHGC_{north}$-0.49 | $U_{oper}$-0.67 | $SHGC_{north}$-0.49 | $U_{oper}$-1.27 | $SHGC_{north}$-NR |
| 10.1–20.0% | $U_{fixed}$-0.57 | $SHGC_{all}$-0.39 | $U_{fixed}$-0.57 | $SHGC_{all}$-0.39 | $U_{fixed}$-1.22 | $SHGC_{all}$-NR |
|  | $U_{oper}$-0.67 | $SHGC_{north}$-0.49 | $U_{oper}$-0.67 | $SHGC_{north}$-0.49 | $U_{oper}$-1.27 | $SHGC_{north}$-NR |
| 20.1–30.0% | $U_{fixed}$-0.57 | $SHGC_{all}$-0.39 | $U_{fixed}$-0.57 | $SHGC_{all}$-0.39 | $U_{fixed}$-1.22 | $SHGC_{all}$-NR |
|  | $U_{oper}$-0.67 | $SHGC_{north}$-0.49 | $U_{oper}$-0.67 | $SHGC_{north}$-0.49 | $U_{oper}$-1.27 | $SHGC_{north}$-NR |
| 30.1–40.0% | $U_{fixed}$-0.57 | $SHGC_{all}$-0.39 | $U_{fixed}$-0.57 | $SHGC_{all}$-0.39 | $U_{fixed}$-1.22 | $SHGC_{all}$-NR |
|  | $U_{oper}$-0.67 | $SHGC_{north}$-0.49 | $U_{oper}$-0.67 | $SHGC_{north}$-0.49 | $U_{oper}$-1.27 | $SHGC_{north}$-NR |
| 40.1–50.0% | $U_{fixed}$-0.46 | $SHGC_{all}$-0.25 | $U_{fixed}$-0.46 | $SHGC_{all}$-0.25 | $U_{fixed}$-0.98 | $SHGC_{all}$-NR |
|  | $U_{oper}$-0.47 | $SHGC_{north}$-0.36 | $U_{oper}$-0.47 | $SHGC_{north}$-0.36 | $U_{oper}$-1.02 | $SHGC_{north}$-NR |
| *Skylight with Curb, Glass, % of Roof* | | | | | | |
| 0–2.0% | $U_{all}$-1.17 | $SHGC_{all}$-0.49 | $U_{all}$-0.98 | $SHGC_{all}$-0.36 | $U_{all}$-1.98 | $SHGC_{all}$-NR |
| 2.1–5.0% | $U_{all}$-1.17 | $SHGC_{all}$-0.39 | $U_{all}$-0.98 | $SHGC_{all}$-0.19 | $U_{all}$-1.98 | $SHGC_{all}$-NR |
| *Skylight with Curb, Plastic, % of Roof* | | | | | | |
| 0–2.0% | $U_{all}$-1.30 | $SHGC_{all}$-0.65 | $U_{all}$-1.30 | $SHGC_{all}$-0.62 | $U_{all}$-1.90 | $SHGC_{all}$-NR |
| 2.1–5.0% | $U_{all}$-1.30 | $SHGC_{all}$-0.34 | $U_{all}$-1.30 | $SHGC_{all}$-0.27 | $U_{all}$-1.90 | $SHGC_{all}$-NR |
| *Skylight without Curb, All, % of Roof* | | | | | | |
| 0–2.0% | $U_{all}$-0.69 | $SHGC_{all}$-0.49 | $U_{all}$-0.58 | $SHGC_{all}$-0.36 | $U_{all}$-1.36 | $SHGC_{all}$-NR |
| 2.1–5.0% | $U_{all}$-0.69 | $SHGC_{all}$-0.39 | $U_{all}$-0.58 | $SHGC_{all}$-0.19 | $U_{all}$-1.36 | $SHGC_{all}$-NR |

*Source:* Reprinted by permission of the American Society of Heating, Refrigerating and Air-Conditioning Engineers, Inc., from ANSI/ASHRAE/IESNA Standard 90.1-2001, *Energy Standard for Buildings Except Low-Rise Residential Buildings* (I-P Edition).
[a]An exception to Standard 90.1 5.3.1.2a applies.
NR = no requirement.
ci = continuous insulation.

# G.2 SAMPLES OF PRESCRIPTIVE BUILDING ENVELOPE REQUIREMENTS FROM ANSI/ASHRAE STANDARD 90.2-2001: *ENERGY-EFFICIENT DESIGN OF LOW-RISE RESIDENTIAL BUILDINGS*

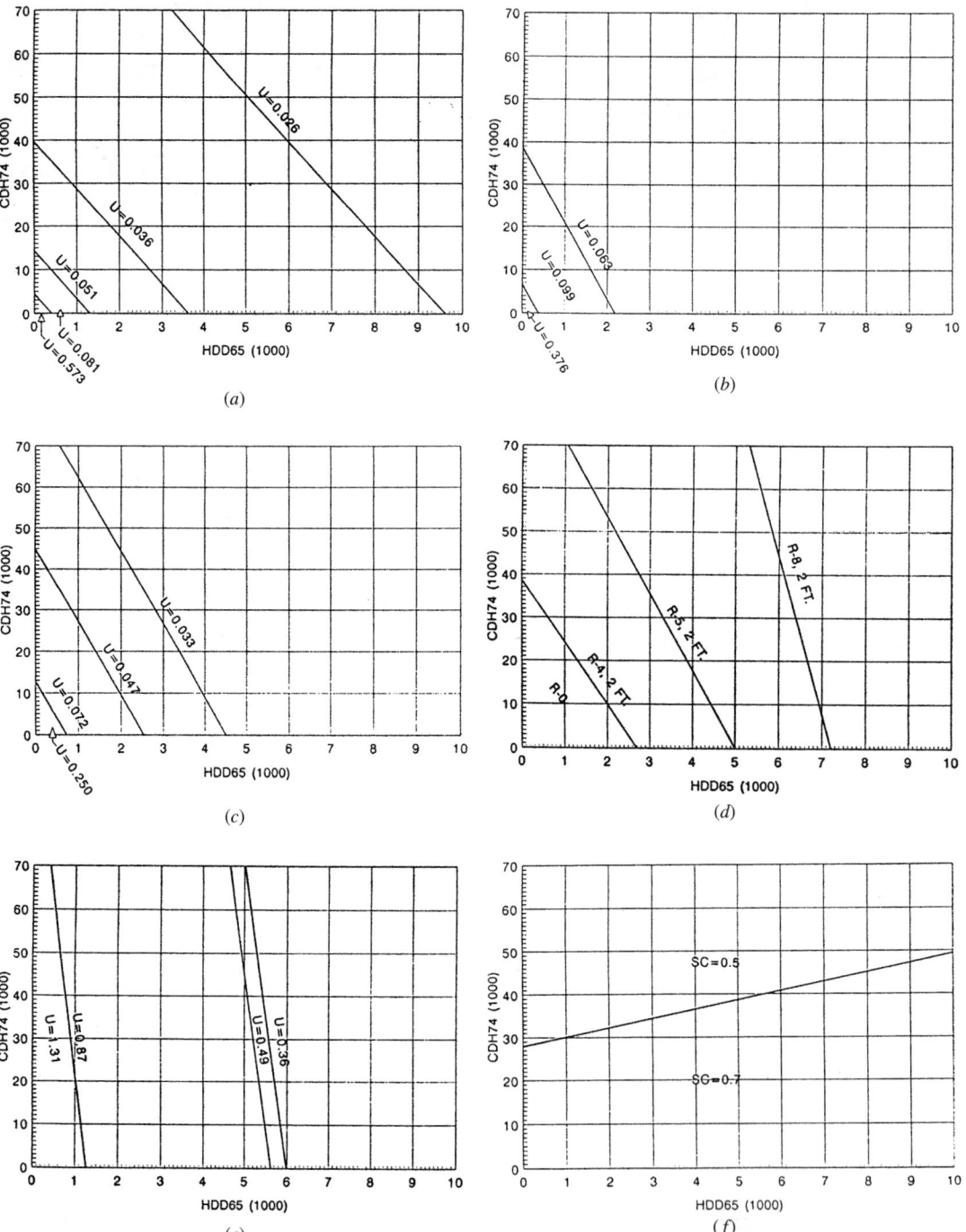

**Fig. G.1** *Prescriptive building envelope requirements extracted from ASHRAE Standard 90.2. (a) maximum U-factors for ceilings with attics (with ductwork distribution within the conditioned space); (b) maximum U-factors for above-grade frame wall and band joists (with ductwork distribution within the conditioned space); (c) maximum U-factors for wood frame floors over unconditioned space—vented crawl spaces, basements, enclosed garages or porches (with ductwork distribution within the conditioned space); (d) minimum R-values for slab-on-grade floors (with ductwork distribution within the conditioned space); (e) maximum fenestration U-factors—including framing (with ductwork distribution within the conditioned space); (f) maximum fenestration shading coefficients (with duct-work distribution within the conditioned space). (Reprinted with permission of the American Society of Heating, Refrigerating and Air-Conditioning Engineers, Inc., from ANSI/ASHRAE Standard 90.2-2001, Energy-Efficient Design of New Low-Rise Residential Buildings.)*

## G.3 PROJECT CHECKLIST FROM LEED-NC, VERSION 2.1 (REV. 3/14/03)

### TABLE G.2 LEED-NC Project Checklist

**LEED (Leadership in Energy & Environmental Design) Green Building Rating System for New Construction & Major Renovations (LEED-NC) Version 2.1**

| Project Checklist |
|---|
| **Sustainable Sites** |
| Prerequisite 1   Erosion & Sedimentation Control |
| Credit 1   Site Selection |
| Credit 2   Development Density |
| Credit 3   Brownfield Redevelopment |
| Credit 4   Alternative Transportation |
| Credit 5   Reduced Site Disturbance |
| Credit 6   Stormwater Management |
| Credit 7   Heat Island Effect |
| Credit 8   Light Pollution Reduction |
| **Water Efficiency** |
| Credit 1   Water Efficient Landscaping |
| Credit 2   Innovative Wastewater Technologies |
| Credit 3   Water Use Reduction |
| **Energy & Atmosphere** |
| Prerequisite 1   Fundamental Building Systems Commissioning |
| Prerequisite 2   Minimum Energy Performance |
| Prerequisite 3   CFC Reduction in HVAC&R Equipment |
| Credit 1   Optimize Energy Performance |
| Credit 2   Renewable Energy |
| Credit 3   Additional Commissioning |
| Credit 4   Ozone Depletion |
| Credit 5   Measurement & Verification |
| Credit 6   Green Power |
| **Materials & Resources** |
| Prerequisite 1   Storage & Collection of Recyclables |
| Credit 1   Building Reuse |
| Credit 2   Construction Waste Management |
| Credit 3   Resource Reuse |
| Credit 4   Recycled Content |
| Credit 5   Local/Regional Materials |
| Credit 6   Rapidly Renewable Materials |
| Credit 7   Certified Wood |
| **Indoor Environmental Quality** |
| Prerequisite 1   Minimum IAQ Performance |
| Prerequisite 2   Environmental Tobacco Smoke (ETS) Control |
| Credit 1   Carbon Dioxide ($CO_2$) Monitoring |
| Credit 2   Ventilation Effectiveness |
| Credit 3   Construction IAQ Management Plan |
| Credit 4   Low-Emitting Materials |
| Credit 5   Indoor Chemical & Pollutant Source Control |
| Credit 6   Controllability of Systems |
| Credit 7   Thermal Comfort |
| Credit 8   Daylight & Views |
| **Innovation & Design Process** |
| Credit 1   Innovation in Design |
| Credit 2   LEED Accredited Professional |

*Source:* Extracted from LEED-NC Version 2.1 rating system (March 2003); used with permission of the U.S. Green Building Council (www.usgbc.org).

# Annual Solar Performance

This appendix lists 30 passive solar heating systems (3 water wall, 9 Trombé wall, 9 direct gain, and 9 sunspace) for the U.S. and Canadian locations shown in Fig. C.1. *Passive Solar Heating Analysis* lists a total of 94 systems for about twice as many locations. The specifications for all 94 systems are listed in Table H.1, so the original source may be consulted if a proposed passive system is not represented by one of those in this appendix.

**TABLE H.1 Characteristics of Selected Passive Solar Heating Systems**

| | | | | | |
|---|---|---|---|---|---|
| Part A. **Water Wall Systems** | | | | | |
| Designation | Thermal Storage Capacity[a] (Btu/ft$^2$ °F) | Wall Thickness (in.) | No. of Glazings | Wall Surface | Night Insulation |
| WW-A1 | 15.6 | 3 | 2 | Normal | No |
| WW-A2 | 31.2 | 6 | 2 | Normal | No |
| WW-A3[b] | 46.8 | 9 | 2 | Normal | No |
| WW-A4 | 62.4 | 12 | 2 | Normal | No |
| WW-A5 | 93.6 | 18 | 2 | Normal | No |
| WW-A6 | 124.8 | 24 | 2 | Normal | No |
| WW-B1 | 46.8 | 9 | 1 | Normal | No |
| WW-B2 | 46.8 | 9 | 3 | Normal | No |
| WW-B3 | 46.8 | 9 | 1 | Normal | Yes |
| WW-B4[b] | 46.8 | 9 | 2 | Normal | Yes |
| WW-B5 | 46.8 | 9 | 3 | Normal | Yes |
| WW-C1 | 46.8 | 9 | 1 | Selective | No |
| WW-C2[b] | 46.8 | 9 | 2 | Selective | No |
| WW-C3 | 46.8 | 9 | 1 | Selective | Yes |
| WW-C4 | 46.8 | 9 | 2 | Selective | Yes |
| Part B. **Trombe Wall Systems: Vented** | | | | | |
| Designation | Thermal Storage Capacity[c] (Btu/ft$^2$ °F) | Wall Thickness[c] (in.) | $\rho ck$[d] (Btu$^2$/h ft$^4$ °F$^2$) | No. of Glazings | Wall Surface | Night Insulation |
| TW-A1[b] | 15 | 6 | 30 | 2 | Normal | No |
| TW-A2[b] | 22.5 | 9 | 30 | 2 | Normal | No |
| TW-A3[b] | 30 | 12 | 30 | 2 | Normal | No |
| TW-A4[b] | 45 | 18 | 30 | 2 | Normal | No |
| TW-B1 | 15 | 6 | 15 | 2 | Normal | No |
| TW-B2 | 22.5 | 9 | 15 | 2 | Normal | No |
| TW-B3[b] | 30 | 12 | 15 | 2 | Normal | No |
| TW-B4 | 45 | 18 | 15 | 2 | Normal | No |
| TW-C1 | 15 | 6 | 7.5 | 2 | Normal | No |
| TW-C2 | 22.5 | 9 | 7.5 | 2 | Normal | No |
| TW-C3 | 30 | 12 | 7.5 | 2 | Normal | No |
| TW-C4 | 45 | 18 | 7.5 | 2 | Normal | No |
| TW-D1 | 30 | 12 | 30 | 1 | Normal | No |
| TW-D2 | 30 | 12 | 30 | 3 | Normal | No |
| TW-D3 | 30 | 12 | 30 | 1 | Normal | Yes |
| TW-D4[b] | 30 | 12 | 30 | 2 | Normal | Yes |
| TW-D5 | 30 | 12 | 30 | 3 | Normal | Yes |
| TW-E1 | 30 | 12 | 30 | 1 | Selective | No |
| TW-E2[b] | 30 | 12 | 30 | 2 | Selective | No |
| TW-E3 | 30 | 12 | 30 | 1 | Selective | Yes |
| TW-E4 | 30 | 12 | 30 | 2 | Selective | Yes |

*Note:* Vented systems may be made to perform as unvented systems by sealing both top and bottom vent openings.

**TABLE H.1 Characteristics of Selected Passive Solar Heating Systems (*Continued*)**

| | | | | | | |
|---|---|---|---|---|---|---|
| Part C. **Trombe Wall Systems: Unvented** | | | | | | |
| Designation | Thermal Storage Capacity[c] (Btu/ft² °F) | Wall Thickness[c] (in.) | $\rho ck^d$ (Btu²/h ft⁴ °F²) | No. of Glazings | Wall Surface | Night Insulation |
| TW-F1 | 15 | 6 | 30 | 2 | Normal | No |
| TW-F2 | 22.5 | 9 | 30 | 2 | Normal | No |
| TW-F3[b] | 30 | 12 | 30 | 2 | Normal | No |
| TW-F4 | 45 | 18 | 30 | 2 | Normal | No |
| TW-G1 | 15 | 6 | 15 | 2 | Normal | No |
| TW-G2 | 22.5 | 9 | 15 | 2 | Normal | No |
| TW-G3 | 30 | 12 | 15 | 2 | Normal | No |
| TW-G4 | 45 | 18 | 15 | 2 | Normal | No |
| TW-H1 | 15 | 6 | 7.5 | 2 | Normal | No |
| TW-H2 | 22.5 | 9 | 7.5 | 2 | Normal | No |
| TW-H3 | 30 | 12 | 7.5 | 2 | Normal | No |
| TW-H4 | 45 | 18 | 7.5 | 2 | Normal | No |
| TW-I1 | 30 | 12 | 30 | 1 | Normal | No |
| TW-I2 | 30 | 12 | 30 | 3 | Normal | No |
| TW-I3 | 30 | 12 | 30 | 1 | Normal | Yes |
| TW-I4 | 30 | 12 | 30 | 2 | Normal | Yes |
| TW-I5 | 30 | 12 | 30 | 3 | Normal | Yes |
| TW-J1 | 30 | 12 | 30 | 1 | Selective | No |
| TW-J2[b] | 30 | 12 | 30 | 2 | Selective | No |
| TW-J3 | 30 | 12 | 30 | 1 | Selective | Yes |
| TW-J4 | 30 | 12 | 30 | 2 | Selective | Yes |
| Part D. **Direct-Gain Systems** | | | | | | |
| Designation | Thermal Storage Capacity[c] (Btu/ft² °F) | Mass Thickness[c] (in.) | Ratio of Mass to Glazing Area | No. of Glazings | | Night Insulation |
| DG-A1[b] | 30 | 2 | 6 | 2 | | No |
| DG-A2[b] | 30 | 2 | 6 | 3 | | No |
| DG-A3[b] | 30 | 2 | 6 | 2 | | Yes |
| DG-B1[b] | 45 | 6 | 3 | 2 | | No |
| DG-B2[b] | 45 | 6 | 3 | 3 | | No |
| DG-B3[b] | 45 | 6 | 3 | 2 | | Yes |
| DG-C1[b] | 60 | 4 | 6 | 2 | | No |
| DG-C2[b] | 60 | 4 | 6 | 3 | | No |
| DG-C3[b] | 60 | 4 | 6 | 2 | | Yes |

**TABLE H.1 Characteristics of Selected Passive Solar Heating Systems** *(Continued)*

| Part E. Sunspace Systems | | | | | |
|---|---|---|---|---|---|
| Designation | Type | Tilt (Degrees) | Common Wall[f] | End Walls[g] | Night Insulation |
| SS-A1[b] | Attached | 50 | Masonry | Opaque | No |
| SS-A2 | Attached | 50 | Masonry | Opaque | Yes |
| SS-A3 | Attached | 50 | Masonry | Glazed | No |
| SS-A4 | Attached | 50 | Masonry | Glazed | Yes |
| SS-A5 | Attached | 50 | Insulated | Opaque | No |
| SS-A6 | Attached | 50 | Insulated | Opaque | Yes |
| SS-A7 | Attached | 50 | Insulated | Glazed | No |
| SS-A8 | Attached | 50 | Insulated | Glazed | Yes |
| SS-B1[b] | Attached | 90/30 | Masonry | Opaque | No |
| SS-B2[b] | Attached | 90/30 | Masonry | Opaque | Yes |
| SS-B3[b] | Attached | 90/30 | Masonry | Glazed | No |
| SS-B4 | Attached | 90/30 | Masonry | Glazed | Yes |
| SS-B5 | Attached | 90/30 | Insulated | Opaque | No |
| SS-B6 | Attached | 90/30 | Insulated | Opaque | Yes |
| SS-B7 | Attached | 90/30 | Insulated | Glazed | No |
| SS-B8 | Attached | 90/30 | Insulated | Glazed | Yes |
| SS-C1[b] | Semi-enclosed | 90 | Masonry | Common | No |
| SS-C2[b] | Semi-enclosed | 90 | Masonry | Common | Yes |
| SS-C3 | Semi-enclosed | 90 | Insulated | Common | No |
| SS-C4 | Semi-enclosed | 90 | Insulated | Common | Yes |
| SS-D1 | Semi-enclosed | 50 | Masonry | Common | No |
| SS-D2 | Semi-enclosed | 50 | Masonry | Common | Yes |
| SS-D3 | Semi-enclosed | 50 | Insulated | Common | No |
| SS-D4 | Semi-enclosed | 50 | Insulated | Common | Yes |
| SS-E1[b] | Semi-enclosed | 90/30 | Masonry | Common | No |
| SS-E2[b] | Semi-enclosed | 90/30 | Masonry | Common | Yes |
| SS-E3[b] | Semi-enclosed | 90/30 | Insulated | Common | No |
| SS-E4 | Semi-enclosed | 90/30 | Insulated | Common | Yes |

*Source:* Reprinted by permission of the American Society of Heating, Refrigerating and Air-Conditioning Engineers, Inc., from *Passive Solar Heating Analysis,* 1984, by J.D. Balcomb et al.

[a]Per unit of projected area.

[b]Listed in this book.

[c]The thermal storage capacity is per projected area of south aperture (ft$^2$ of glazing) or, equivalently, the quantity $\rho ct$. The wall thickness ($t$) is listed only as an approximate guide by assuming that $\rho c = 30$ Btu/ft$^3$ °F ($\rho$ of 150 lb/ft$^3$ and c of 0.2 Btu/lb °F), typical of ordinary concrete. Example: for DGA1 thermal storage capacity, (30 Btu/ft$^3$ °F) × (2 in./12 in/ft) × (6 ft$^2$ of mass/1 ft$^2$ of glazing) = 30 Btu/ft$^2$ °F

[d]$\rho ck$ is the product of the density (lb/ft$^3$), specific heat (Btu/lb °F), and thermal conductivity (Btu/h ft °F); see Table H.2.

[e]See Fig. H.1 for additional description.

[f]Common walls should provide closable openings that total at minimum 6% (maximum 15%) of the wall area. Insulated common walls should have minimum R-10 insulation and thermal mass in water at a minimum of 3.8 gal/ft$^2$ (0.5 ft$^3$/ft$^2$) of projected area. Masonry common walls should have a minimum 3:1 (mass:glass) area ratio.

[g]Opaque end walls are strongly recommended; glazed end walls perform more poorly in most winter climates and are subject to severe summer overheating.

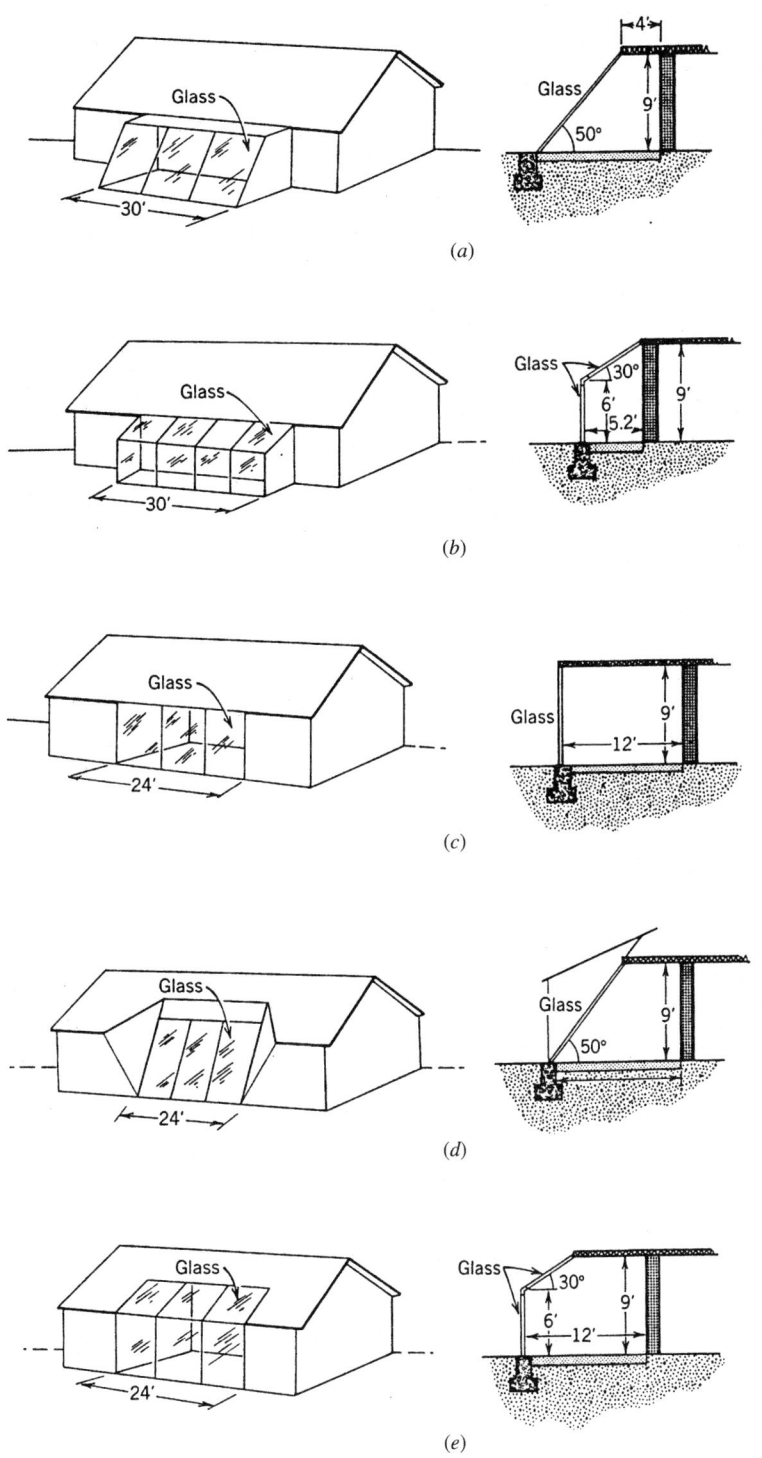

**Fig. H.1** *Types of sunspaces, described in Table H.1, Part E. Types (a) and (b) are considered attached to the building; types (c), (d), and (e) are considered semienclosed by the building. The architectural detail at the sides of type (d) is insignificant; no shading of the sunspace by the building was accounted for in the performance estimates in Table H.2. (Reprinted by permission from J.D. Balcomb et al.,* Passive Solar Heating Analysis, *© 1984, American Society of Heating, Refrigerating and Air-Conditioning Engineers, Inc., Atlanta, GA.)*

**TABLE H.2 Thermal Properties of Various Materials**

| Material | Source[a] | Density, ρ (lb/ft³) | Specific Heat, c (Btu/lb °F) | Heat Capacity, ρc (Btu/ft³ °F) | Conductivity, k (Btu/h °F ft) | Diffusivity, k/ρc (ft²/h) | ρck (Btu²/h ft⁴ °F²) |
|---|---|---|---|---|---|---|---|
| Water | 1 | 62.3 | 1.0 | 62.3 | | | |
| Concrete | 2, 3 | 144 | 0.21 | 30.3 | 0.89 | 0.029 | 27.0 |
| Concrete block | 2, 3 | | | | | | |
| Heavy weight | | 135 | 0.21 | 28.4 | 0.74 | 0.026 | 21.0 |
| Medium weight | | 105 | 0.22 | 23.1 | 0.41 | 0.018 | 9.5 |
| Light weight | | 85 | 0.23 | 19.6 | 0.27 | 0.014 | 5.3 |
| Brick | 1, 4 | | | | | | |
| Paving | | 135 | 0.19 | 25.7 | 0.75 | 0.029 | 19.3 |
| Face | | 130 | 0.19 | 24.7 | 0.75 | 0.030 | 18.5 |
| Building | | 120 | 0.19 | 22.8 | 0.42 | 0.018 | 9.6 |
| Mortar or grout | 1 | 116 | 0.20 | 23.2 | 0.42 | 0.018 | 9.7 |
| Adobe | 5 | 80 | 0.20 | 16.0 | 0.38 | 0.024 | 6.1 |
| | | 100 | 0.20 | 20.0 | 0.75 | 0.038 | 15.0 |
| Gypsum or plasterboard | 1 | 50 | 0.26 | 13.0 | 0.097 | 0.058 | 9.8 |
| Douglas fir plywood | 1 | 34 | 0.29 | 9.9 | 0.067 | 0.007 | 0.7 |
| Hardwood | 1 | 45 | 0.30 | 13.5 | 0.092 | 0.007 | 1.2 |

Source: Reprinted by permission from *Passive Solar Heating Analysis*, © 1984, American Society of Heating, Refrigerating and Air-Conditioning Engineers, Inc., Atlanta, GA.

Note: Concrete and masonry products can be manufactured in a variety of densities. The values listed are for the most commonly available products. The thermal properties of concrete and concrete block vary with density. Empirical correlations between these properties are[2] $k = 0.05e^{0.02\rho}$ and [3]$c = 1/(3.934 + 0.006\rho)$.

[a]1. ASHRAE 1981 Handbook—Fundamentals (American Society of Heating, Refrigerating and Air-Conditioning Engineers, Inc., Atlanta, GA).

2. "Calculating the Steady State U-Value of Concrete Masonry Walls," Expanded Shale, Clay, and Slate Institute, 4905 Del Ray Avenue, Bethesda, MD.

3. D. Whiting, A. Litvin, and S.E. Goodwin, "Specific Heat of Selected Concretes," *Research and Development Bulletin RD058.01B*, Portland Cement Association, 5420 Old Orchard Road, Skokie, IL.

4. "Brick Passive Solar Heating Systems. Material Properties—Part IV," *Technical Notes on Brick Construction 43D*, Brick Institute of America, 1750 Old Meadow Road, Mc Lean, VA, September/October 1980.

5. Benjamin T. Rogers, private communication; and W.L. Sibbitt, measurements at Los Alamos National Laboratory. Note that the thermal conductivity of adobe varies greatly with moisture content, which is part of the reason for the wide range quoted.

## TABLE H.3 Annual Passive Heating Performance: SSF

For each location, degree days listed are DD65. For a description of SSF (solar savings fraction) see Fig. 8.8; the LCR (load–collector ratio) is described in Section 8.10. LCR units are Btu/DD ft$^2$.

United States

### MONTGOMERY, ALABAMA SSF (%) — 2272 DD65

| Type | LCR = 200 | 100 | 70 | 50 | 40 | 30 | 25 | 20 |
|---|---|---|---|---|---|---|---|---|
| WWA3 | 17 | 29 | 38 | 48 | 55 | 64 | 70 | 77 |
| WWB4 | 18 | 36 | 48 | 60 | 68 | 78 | 83 | 89 |
| WWC2 | 18 | 36 | 47 | 59 | 67 | 76 | 82 | 88 |
| TWA1 | 17 | 24 | 29 | 35 | 40 | 46 | 51 | 57 |
| TWA2 | 16 | 26 | 33 | 41 | 46 | 54 | 60 | 67 |
| TWA3 | 15 | 26 | 34 | 43 | 49 | 58 | 63 | 70 |
| TWA4 | 14 | 26 | 34 | 43 | 49 | 58 | 64 | 72 |
| TWB3 | 14 | 24 | 31 | 39 | 45 | 54 | 60 | 67 |
| TWD4 | 17 | 33 | 43 | 54 | 62 | 72 | 78 | 84 |
| TWE2 | 18 | 34 | 44 | 55 | 63 | 72 | 78 | 84 |
| TWF3 | 12 | 23 | 30 | 39 | 45 | 54 | 60 | 67 |
| TWJ2 | 15 | 31 | 41 | 52 | 60 | 70 | 76 | 82 |
| DGA1 | 12 | 21 | 28 | 34 | 39 | 45 | 49 | 53 |
| DGA2 | 13 | 23 | 31 | 39 | 45 | 53 | 58 | 64 |
| DGA3 | 15 | 27 | 37 | 47 | 55 | 64 | 70 | 77 |
| DGB1 | 12 | 22 | 29 | 37 | 43 | 51 | 56 | 62 |
| DGB2 | 13 | 24 | 32 | 41 | 49 | 58 | 65 | 71 |
| DGB3 | 15 | 28 | 38 | 49 | 57 | 68 | 75 | 82 |
| DGC1 | 14 | 25 | 34 | 43 | 50 | 60 | 66 | 72 |
| DGC2 | 15 | 27 | 36 | 47 | 55 | 66 | 72 | 79 |
| DGC3 | 18 | 32 | 42 | 55 | 64 | 75 | 81 | 87 |
| SSA1 | 21 | 32 | 40 | 48 | 55 | 63 | 68 | 75 |
| SSB1 | 17 | 28 | 35 | 42 | 48 | 56 | 62 | 68 |
| SSB2 | 21 | 35 | 44 | 54 | 61 | 70 | 76 | 82 |
| SSB3 | 17 | 26 | 33 | 40 | 46 | 54 | 59 | 66 |
| SSC1 | 15 | 26 | 34 | 43 | 49 | 58 | 64 | 71 |
| SSC2 | 17 | 31 | 41 | 51 | 58 | 68 | 74 | 80 |
| SSE1 | 20 | 33 | 42 | 51 | 58 | 67 | 72 | 78 |
| SSE2 | 23 | 40 | 51 | 62 | 69 | 78 | 83 | 88 |
| SSE3 | 18 | 28 | 35 | 43 | 48 | 56 | 62 | 68 |

### TUCSON, ARIZONA SSF (%) — 1758 DD65

| Type | LCR = 200 | 100 | 70 | 50 | 40 | 30 | 25 | 20 |
|---|---|---|---|---|---|---|---|---|
| WWA3 | 31 | 52 | 64 | 75 | 82 | 89 | 93 | 96 |
| WWB4 | 37 | 62 | 75 | 85 | 91 | 96 | 98 | 99 |
| WWC2 | 37 | 61 | 74 | 84 | 90 | 95 | 97 | 99 |
| TWA1 | 25 | 38 | 47 | 56 | 63 | 72 | 78 | 84 |
| TWA2 | 27 | 44 | 55 | 65 | 73 | 81 | 86 | 91 |
| TWA3 | 28 | 46 | 58 | 69 | 76 | 84 | 89 | 93 |
| TWA4 | 27 | 46 | 58 | 69 | 77 | 85 | 89 | 94 |
| TWB3 | 25 | 42 | 53 | 64 | 72 | 81 | 86 | 91 |
| TWD4 | 33 | 56 | 69 | 80 | 86 | 93 | 96 | 98 |
| TWE2 | 35 | 57 | 70 | 81 | 87 | 93 | 96 | 98 |
| TWF3 | 24 | 42 | 53 | 65 | 73 | 81 | 86 | 91 |
| TWJ2 | 31 | 54 | 67 | 78 | 85 | 92 | 95 | 97 |
| DGA1 | 24 | 41 | 52 | 62 | 68 | 76 | 80 | 84 |
| DGA2 | 24 | 43 | 54 | 65 | 72 | 81 | 85 | 89 |
| DGA3 | 27 | 49 | 62 | 74 | 80 | 87 | 90 | 93 |
| DGB1 | 24 | 44 | 58 | 69 | 76 | 83 | 87 | 91 |
| DGB2 | 25 | 45 | 59 | 72 | 79 | 87 | 91 | 94 |
| DGB3 | 27 | 50 | 65 | 78 | 85 | 91 | 94 | 96 |
| DGC1 | 28 | 51 | 66 | 78 | 84 | 90 | 93 | 95 |
| DGC2 | 28 | 51 | 66 | 79 | 85 | 92 | 95 | 97 |
| DGC3 | 31 | 56 | 72 | 84 | 90 | 95 | 96 | 98 |
| SSA1 | 33 | 51 | 62 | 72 | 79 | 86 | 90 | 94 |
| SSB1 | 29 | 45 | 55 | 66 | 73 | 81 | 86 | 91 |
| SSB2 | 35 | 55 | 67 | 77 | 84 | 90 | 94 | 97 |
| SSB3 | 27 | 43 | 53 | 64 | 71 | 79 | 84 | 89 |
| SSC1 | 28 | 47 | 59 | 70 | 77 | 86 | 90 | 94 |
| SSC2 | 32 | 54 | 66 | 78 | 84 | 91 | 94 | 97 |
| SSE1 | 35 | 55 | 67 | 77 | 83 | 90 | 93 | 96 |
| SSE2 | 41 | 64 | 76 | 85 | 91 | 95 | 97 | 99 |
| SSE3 | 30 | 46 | 56 | 67 | 73 | 82 | 86 | 91 |

### PHOENIX, ARIZONA SSF (%) — 1556 DD65

| Type | LCR = 200 | 100 | 70 | 50 | 40 | 30 | 25 | 20 |
|---|---|---|---|---|---|---|---|---|
| WWA3 | 33 | 54 | 66 | 77 | 83 | 90 | 93 | 96 |
| WWB4 | 39 | 64 | 76 | 86 | 91 | 96 | 98 | 99 |
| WWC2 | 39 | 63 | 75 | 85 | 90 | 95 | 97 | 99 |
| TWA1 | 26 | 40 | 49 | 58 | 65 | 73 | 79 | 84 |
| TWA2 | 29 | 46 | 56 | 67 | 74 | 82 | 87 | 91 |
| TWA3 | 29 | 48 | 59 | 70 | 77 | 85 | 89 | 93 |
| TWA4 | 29 | 48 | 60 | 71 | 78 | 86 | 90 | 94 |
| TWB3 | 26 | 44 | 55 | 66 | 73 | 82 | 86 | 91 |
| TWD4 | 35 | 58 | 70 | 81 | 87 | 93 | 96 | 98 |
| TWE2 | 36 | 59 | 71 | 82 | 88 | 93 | 96 | 98 |
| TWF3 | 26 | 44 | 55 | 67 | 74 | 82 | 87 | 92 |
| TWJ2 | 33 | 56 | 68 | 79 | 86 | 92 | 95 | 97 |
| DGA1 | 26 | 43 | 54 | 64 | 70 | 77 | 81 | 85 |
| DGA2 | 26 | 45 | 56 | 67 | 74 | 82 | 86 | 90 |
| DGA3 | 29 | 51 | 64 | 75 | 81 | 87 | 91 | 93 |
| DGB1 | 27 | 47 | 60 | 71 | 77 | 84 | 87 | 91 |
| DGB2 | 27 | 48 | 61 | 73 | 80 | 88 | 91 | 94 |
| DGB3 | 29 | 53 | 67 | 79 | 85 | 91 | 94 | 96 |
| DGC1 | 31 | 54 | 58 | 79 | 85 | 90 | 93 | 95 |
| DGC2 | 30 | 54 | 68 | 80 | 86 | 92 | 95 | 97 |
| DGC3 | 33 | 59 | 74 | 85 | 90 | 95 | 96 | 98 |
| SSA1 | 34 | 52 | 62 | 73 | 79 | 86 | 90 | 94 |
| SSB1 | 29 | 46 | 56 | 67 | 73 | 81 | 86 | 90 |
| SSB2 | 36 | 56 | 67 | 78 | 84 | 90 | 93 | 96 |
| SSB3 | 28 | 44 | 54 | 64 | 71 | 79 | 84 | 89 |
| SSC1 | 30 | 49 | 61 | 72 | 78 | 86 | 90 | 94 |
| SSC2 | 34 | 56 | 68 | 79 | 85 | 91 | 94 | 97 |
| SSE1 | 36 | 56 | 67 | 77 | 83 | 90 | 93 | 96 |
| SSE2 | 42 | 64 | 76 | 85 | 90 | 95 | 97 | 99 |
| SSE3 | 30 | 47 | 57 | 67 | 74 | 82 | 86 | 91 |

### FORT SMITH, ARKANSAS SSF (%) — 3338 DD65

| Type | LCR = 200 | 100 | 70 | 50 | 40 | 30 | 25 | 20 |
|---|---|---|---|---|---|---|---|---|
| WWA3 | 11 | 21 | 29 | 37 | 43 | 51 | 57 | 64 |
| WWB4 | 12 | 27 | 37 | 49 | 57 | 67 | 73 | 80 |
| WWC2 | 12 | 27 | 37 | 48 | 55 | 65 | 72 | 79 |
| TWA1 | 14 | 19 | 23 | 27 | 31 | 36 | 40 | 45 |
| TWA2 | 12 | 20 | 25 | 31 | 36 | 43 | 48 | 54 |
| TWA3 | 11 | 19 | 26 | 33 | 38 | 46 | 51 | 58 |
| TWA4 | 10 | 19 | 25 | 33 | 38 | 46 | 52 | 59 |
| TWB3 | 10 | 18 | 23 | 30 | 35 | 43 | 48 | 54 |
| TWD4 | 12 | 25 | 34 | 44 | 51 | 61 | 67 | 75 |
| TWE2 | 13 | 26 | 35 | 44 | 52 | 61 | 67 | 75 |
| TWF3 | 8 | 16 | 22 | 29 | 34 | 42 | 47 | 54 |
| TWJ2 | 10 | 23 | 32 | 41 | 48 | 58 | 65 | 72 |
| DGA1 | 8 | 14 | 19 | 24 | 28 | 33 | 36 | 39 |
| DGA2 | 9 | 17 | 23 | 29 | 35 | 42 | 46 | 52 |
| DGA3 | 11 | 21 | 28 | 37 | 44 | 53 | 59 | 66 |
| DGB1 | 8 | 15 | 19 | 25 | 30 | 36 | 40 | 46 |
| DGB2 | 9 | 17 | 23 | 31 | 36 | 45 | 51 | 58 |
| DGB3 | 11 | 21 | 29 | 38 | 45 | 56 | 63 | 71 |
| DGC1 | 10 | 18 | 24 | 30 | 36 | 44 | 50 | 56 |
| DGC2 | 11 | 20 | 27 | 35 | 42 | 52 | 59 | 66 |
| DGC3 | 13 | 24 | 32 | 43 | 51 | 63 | 70 | 78 |
| SSA1 | 16 | 25 | 31 | 38 | 43 | 50 | 55 | 62 |
| SSB1 | 13 | 21 | 26 | 33 | 37 | 44 | 49 | 55 |
| SSB2 | 15 | 27 | 35 | 43 | 50 | 59 | 64 | 71 |
| SSB3 | 12 | 19 | 25 | 30 | 35 | 41 | 46 | 52 |
| SSC1 | 10 | 19 | 25 | 32 | 37 | 45 | 51 | 57 |
| SSC2 | 11 | 23 | 31 | 40 | 47 | 56 | 62 | 69 |
| SSE1 | 14 | 24 | 31 | 39 | 45 | 53 | 58 | 65 |
| SSE2 | 16 | 30 | 39 | 50 | 57 | 66 | 72 | 79 |
| SSE3 | 14 | 21 | 27 | 33 | 37 | 44 | 49 | 55 |

## TABLE H.3 Annual Passive Heating Performance: SSF *(Continued)*

### FRESNO, CALIFORNIA SSF (%) — 2657 DD65

| Type | LCR = 200 | 100 | 70 | 50 | 40 | 30 | 25 | 20 |
|---|---|---|---|---|---|---|---|---|
| WWA3 | 18 | 31 | 40 | 49 | 56 | 64 | 69 | 75 |
| WWB4 | 20 | 38 | 49 | 61 | 68 | 77 | 82 | 87 |
| WWC2 | 20 | 38 | 49 | 60 | 67 | 76 | 81 | 86 |
| TWA1 | 18 | 26 | 31 | 37 | 41 | 48 | 52 | 58 |
| TWA2 | 18 | 28 | 35 | 42 | 48 | 55 | 60 | 67 |
| TWA3 | 17 | 28 | 36 | 44 | 50 | 58 | 64 | 70 |
| TWA4 | 16 | 28 | 36 | 44 | 51 | 59 | 64 | 71 |
| TWB3 | 15 | 26 | 33 | 41 | 47 | 55 | 60 | 67 |
| TWD4 | 19 | 35 | 45 | 56 | 63 | 72 | 77 | 83 |
| TWE2 | 20 | 36 | 46 | 56 | 63 | 72 | 77 | 83 |
| TWF3 | 14 | 25 | 32 | 40 | 47 | 55 | 60 | 66 |
| TWJ2 | 17 | 33 | 43 | 53 | 60 | 69 | 75 | 81 |
| DGA1 | 14 | 23 | 30 | 36 | 41 | 47 | 50 | 54 |
| DGA2 | 14 | 25 | 33 | 41 | 47 | 54 | 59 | 65 |
| DGA3 | 17 | 30 | 39 | 49 | 56 | 65 | 70 | 76 |
| DGB1 | 14 | 24 | 31 | 39 | 45 | 52 | 56 | 61 |
| DGB2 | 15 | 26 | 35 | 44 | 50 | 59 | 64 | 70 |
| DGB3 | 17 | 30 | 40 | 51 | 58 | 68 | 74 | 80 |
| DGC1 | 16 | 28 | 37 | 46 | 52 | 60 | 64 | 70 |
| DGC2 | 17 | 30 | 39 | 49 | 56 | 65 | 71 | 77 |
| DGC3 | 19 | 34 | 45 | 56 | 64 | 74 | 79 | 85 |
| SSA1 | 22 | 34 | 41 | 49 | 54 | 61 | 66 | 72 |
| SSB1 | 19 | 29 | 36 | 43 | 49 | 56 | 61 | 66 |
| SSB2 | 22 | 36 | 45 | 54 | 60 | 69 | 74 | 79 |
| SSB3 | 18 | 28 | 34 | 41 | 46 | 53 | 58 | 64 |
| SSC1 | 16 | 28 | 36 | 44 | 50 | 59 | 64 | 70 |
| SSC2 | 18 | 33 | 42 | 52 | 59 | 68 | 73 | 79 |
| SSE1 | 22 | 34 | 42 | 51 | 57 | 64 | 69 | 75 |
| SSE2 | 25 | 41 | 51 | 61 | 67 | 75 | 80 | 85 |
| SSE3 | 20 | 30 | 36 | 44 | 49 | 56 | 60 | 66 |

### SACRAMENTO, CALIFORNIA SSF (%) — 2845 DD65

| Type | LCR = 200 | 100 | 70 | 50 | 40 | 30 | 25 | 20 |
|---|---|---|---|---|---|---|---|---|
| WWA3 | 18 | 31 | 39 | 48 | 55 | 63 | 68 | 74 |
| WWB4 | 19 | 37 | 49 | 60 | 67 | 76 | 81 | 87 |
| WWC2 | 20 | 37 | 48 | 59 | 66 | 75 | 80 | 85 |
| TWA1 | 18 | 25 | 30 | 36 | 41 | 47 | 51 | 57 |
| TWA2 | 17 | 27 | 34 | 42 | 47 | 55 | 60 | 66 |
| TWA3 | 16 | 28 | 35 | 44 | 50 | 58 | 63 | 69 |
| TWA4 | 15 | 27 | 35 | 44 | 50 | 58 | 64 | 70 |
| TWB3 | 15 | 25 | 32 | 40 | 46 | 54 | 60 | 66 |
| TWD4 | 18 | 34 | 44 | 55 | 62 | 71 | 76 | 82 |
| TWE2 | 20 | 35 | 45 | 56 | 63 | 71 | 77 | 82 |
| TWF3 | 13 | 24 | 32 | 40 | 46 | 54 | 59 | 66 |
| TWJ2 | 17 | 32 | 42 | 53 | 60 | 69 | 74 | 80 |
| DGA1 | 13 | 23 | 29 | 36 | 40 | 46 | 49 | 53 |
| DGA2 | 14 | 25 | 32 | 40 | 46 | 54 | 58 | 64 |
| DGA3 | 16 | 29 | 38 | 48 | 55 | 64 | 69 | 75 |
| DGB1 | 13 | 23 | 31 | 39 | 44 | 51 | 55 | 60 |
| DGB2 | 14 | 25 | 34 | 43 | 50 | 58 | 64 | 70 |
| DGB3 | 16 | 30 | 39 | 50 | 58 | 67 | 73 | 79 |
| DGC1 | 16 | 27 | 36 | 45 | 51 | 59 | 63 | 69 |
| DGC2 | 16 | 29 | 39 | 49 | 56 | 65 | 70 | 76 |
| DGC3 | 19 | 33 | 44 | 56 | 63 | 73 | 79 | 85 |
| SSA1 | 22 | 33 | 41 | 48 | 54 | 61 | 65 | 71 |
| SSB1 | 19 | 29 | 36 | 43 | 48 | 55 | 60 | 66 |
| SSB2 | 22 | 36 | 45 | 54 | 60 | 68 | 73 | 79 |
| SSB3 | 18 | 28 | 34 | 41 | 46 | 53 | 57 | 63 |
| SSC1 | 16 | 27 | 35 | 44 | 50 | 58 | 63 | 69 |
| SSC2 | 18 | 32 | 42 | 51 | 58 | 67 | 72 | 78 |
| SSE1 | 21 | 34 | 42 | 50 | 56 | 64 | 68 | 74 |
| SSE2 | 24 | 41 | 50 | 60 | 67 | 75 | 79 | 84 |
| SSE3 | 20 | 30 | 36 | 43 | 48 | 55 | 60 | 65 |

### LOS ANGELES, CALIFORNIA SSF (%) — 1793 DD65

| Type | LCR = 200 | 100 | 70 | 50 | 40 | 30 | 25 | 20 |
|---|---|---|---|---|---|---|---|---|
| WWA3 | 37 | 58 | 70 | 81 | 87 | 93 | 95 | 98 |
| WWB4 | 43 | 68 | 80 | 89 | 93 | 97 | 99 | 99 |
| WWC2 | 42 | 67 | 79 | 88 | 93 | 97 | 98 | 99 |
| TWA1 | 29 | 43 | 52 | 62 | 69 | 77 | 82 | 88 |
| TWA2 | 32 | 49 | 60 | 71 | 78 | 86 | 90 | 94 |
| TWA3 | 33 | 52 | 63 | 74 | 81 | 88 | 92 | 96 |
| TWA4 | 32 | 52 | 64 | 75 | 82 | 89 | 93 | 96 |
| TWB3 | 29 | 48 | 59 | 70 | 77 | 85 | 90 | 94 |
| TWD4 | 39 | 62 | 74 | 85 | 90 | 95 | 97 | 99 |
| TWE2 | 40 | 63 | 75 | 85 | 91 | 95 | 97 | 99 |
| TWF3 | 29 | 48 | 59 | 71 | 78 | 86 | 90 | 94 |
| TWJ2 | 37 | 60 | 72 | 83 | 89 | 94 | 97 | 98 |
| DGA1 | 29 | 47 | 58 | 68 | 74 | 81 | 85 | 89 |
| DGA2 | 29 | 49 | 60 | 71 | 78 | 85 | 89 | 93 |
| DGA3 | 32 | 55 | 68 | 79 | 84 | 90 | 93 | 95 |
| DGB1 | 30 | 52 | 65 | 76 | 82 | 88 | 91 | 94 |
| DGB2 | 30 | 52 | 66 | 78 | 84 | 91 | 94 | 96 |
| DGB3 | 33 | 57 | 72 | 83 | 89 | 94 | 96 | 97 |
| DGC1 | 34 | 59 | 73 | 84 | 89 | 93 | 95 | 96 |
| DGC2 | 34 | 58 | 73 | 84 | 90 | 95 | 97 | 98 |
| DGC3 | 37 | 63 | 78 | 89 | 93 | 96 | 97 | 98 |
| SSA1 | 41 | 60 | 71 | 80 | 86 | 92 | 95 | 97 |
| SSB1 | 35 | 53 | 64 | 74 | 81 | 88 | 91 | 95 |
| SSB2 | 42 | 64 | 75 | 84 | 89 | 94 | 97 | 98 |
| SSB3 | 34 | 52 | 63 | 73 | 79 | 87 | 90 | 94 |
| SSC1 | 33 | 53 | 65 | 76 | 83 | 90 | 93 | 96 |
| SSC2 | 38 | 60 | 72 | 83 | 88 | 94 | 96 | 98 |
| SSE1 | 43 | 64 | 75 | 84 | 90 | 94 | 97 | 98 |
| SSE2 | 50 | 73 | 83 | 91 | 95 | 98 | 99 | 100 |
| SSE3 | 36 | 54 | 65 | 75 | 81 | 88 | 92 | 95 |

### SAN DIEGO, CALIFORNIA SSF (%) — 1512 DD65

| Type | LCR = 200 | 100 | 70 | 50 | 40 | 30 | 25 | 20 |
|---|---|---|---|---|---|---|---|---|
| WWA3 | 39 | 61 | 73 | 83 | 89 | 94 | 97 | 98 |
| WWB4 | 46 | 71 | 83 | 91 | 95 | 98 | 99 | 100 |
| WWC2 | 45 | 70 | 82 | 91 | 95 | 98 | 99 | 100 |
| TWA1 | 30 | 45 | 55 | 65 | 72 | 80 | 85 | 90 |
| TWA2 | 34 | 52 | 64 | 74 | 81 | 88 | 92 | 95 |
| TWA3 | 35 | 55 | 67 | 78 | 84 | 91 | 94 | 97 |
| TWA4 | 34 | 55 | 67 | 78 | 85 | 91 | 94 | 97 |
| TWB3 | 31 | 51 | 62 | 73 | 80 | 88 | 92 | 95 |
| TWD4 | 41 | 65 | 77 | 87 | 92 | 96 | 98 | 99 |
| TWE2 | 43 | 66 | 78 | 88 | 92 | 97 | 98 | 99 |
| TWF3 | 31 | 51 | 63 | 74 | 81 | 89 | 92 | 96 |
| TWJ2 | 39 | 63 | 76 | 86 | 91 | 96 | 98 | 99 |
| DGA1 | 31 | 51 | 62 | 72 | 78 | 84 | 88 | 91 |
| DGA2 | 31 | 52 | 64 | 74 | 81 | 88 | 91 | 95 |
| DGA3 | 34 | 58 | 71 | 81 | 87 | 92 | 94 | 95 |
| DGB1 | 32 | 56 | 69 | 79 | 85 | 90 | 93 | 95 |
| DGB2 | 32 | 56 | 70 | 81 | 87 | 93 | 95 | 97 |
| DGB3 | 35 | 61 | 75 | 86 | 91 | 95 | 96 | 97 |
| DGC1 | 37 | 63 | 77 | 87 | 91 | 95 | 96 | 97 |
| DGC2 | 36 | 63 | 77 | 87 | 92 | 96 | 98 | 99 |
| DGC3 | 39 | 68 | 82 | 91 | 94 | 97 | 98 | 98 |
| SSA1 | 43 | 62 | 73 | 82 | 88 | 93 | 95 | 98 |
| SSB1 | 37 | 56 | 67 | 77 | 83 | 89 | 93 | 96 |
| SSB2 | 45 | 66 | 77 | 86 | 91 | 95 | 97 | 99 |
| SSB3 | 36 | 54 | 65 | 75 | 81 | 88 | 92 | 95 |
| SSC1 | 36 | 56 | 68 | 79 | 85 | 92 | 95 | 97 |
| SSC2 | 40 | 63 | 75 | 85 | 91 | 95 | 97 | 99 |
| SSE1 | 45 | 67 | 78 | 86 | 91 | 95 | 97 | 99 |
| SSE2 | 52 | 75 | 85 | 92 | 96 | 98 | 99 | 100 |
| SSE3 | 38 | 57 | 67 | 77 | 83 | 90 | 93 | 96 |

**TABLE H.3 Annual Passive Heating Performance: SSF** *(Continued)*

| SAN FRANCISCO, CALIFORNIA SSF (%) | | | | | | | 3050 DD65 | | DENVER, COLORADO SSF (%) | | | | | | | 6022 DD65 |
|---|---|---|---|---|---|---|---|---|---|---|---|---|---|---|---|---|
| Type | LCR = 200 | 100 | 70 | 50 | 40 | 30 | 25 | 20 | Type | LCR = 100 | 70 | 50 | 40 | 30 | 25 | 20 | 15 |
| WWA3 | 24 | 41 | 51 | 62 | 69 | 77 | 82 | 88 | WWA3 | 21 | 29 | 37 | 43 | 52 | 58 | 66 | 75 |
| WWB4 | 27 | 49 | 61 | 73 | 80 | 87 | 91 | 95 | WWB4 | 27 | 38 | 50 | 58 | 68 | 75 | 82 | 89 |
| WWC2 | 27 | 48 | 61 | 72 | 79 | 87 | 91 | 94 | WWC2 | 27 | 37 | 48 | 56 | 67 | 73 | 80 | 88 |
| TWA1 | 21 | 31 | 38 | 46 | 51 | 59 | 64 | 71 | TWA1 | 19 | 23 | 27 | 31 | 37 | 41 | 46 | 53 |
| TWA2 | 22 | 35 | 44 | 53 | 59 | 68 | 73 | 80 | TWA2 | 19 | 25 | 31 | 36 | 44 | 49 | 55 | 64 |
| TWA3 | 21 | 36 | 46 | 56 | 63 | 71 | 77 | 83 | TWA3 | 19 | 26 | 33 | 39 | 47 | 52 | 59 | 69 |
| TWA4 | 20 | 36 | 46 | 56 | 63 | 72 | 78 | 84 | TWA4 | 18 | 25 | 33 | 39 | 47 | 53 | 61 | 70 |
| TWB3 | 19 | 33 | 42 | 52 | 59 | 68 | 73 | 79 | TWB3 | 17 | 23 | 30 | 36 | 43 | 49 | 56 | 65 |
| TWD4 | 25 | 44 | 56 | 67 | 75 | 83 | 87 | 92 | TWD4 | 25 | 34 | 44 | 52 | 62 | 68 | 76 | 85 |
| TWE2 | 26 | 46 | 57 | 68 | 75 | 83 | 88 | 92 | TWE2 | 26 | 35 | 45 | 52 | 62 | 69 | 76 | 85 |
| TWF3 | 18 | 32 | 42 | 52 | 59 | 68 | 73 | 80 | TWF3 | 16 | 22 | 29 | 35 | 43 | 48 | 55 | 65 |
| TWJ2 | 23 | 42 | 54 | 65 | 73 | 81 | 86 | 91 | TWJ2 | 23 | 32 | 42 | 49 | 59 | 66 | 74 | 83 |
| DGA1 | 18 | 31 | 40 | 48 | 54 | 61 | 65 | 69 | DGA1 | 14 | 19 | 25 | 28 | 33 | 37 | 40 | 45 |
| DGA2 | 18 | 33 | 42 | 52 | 59 | 67 | 72 | 78 | DGA2 | 16 | 22 | 30 | 35 | 43 | 47 | 53 | 61 |
| DGA3 | 21 | 38 | 49 | 60 | 68 | 76 | 81 | 86 | DGA3 | 20 | 28 | 37 | 44 | 54 | 61 | 68 | 76 |
| DGB1 | 18 | 33 | 43 | 53 | 60 | 68 | 72 | 77 | DGB1 | 14 | 19 | 25 | 30 | 37 | 42 | 48 | 54 |
| DGB2 | 19 | 34 | 45 | 56 | 64 | 73 | 79 | 84 | DGB2 | 17 | 23 | 31 | 37 | 46 | 52 | 60 | 68 |
| DGB3 | 21 | 39 | 51 | 63 | 72 | 81 | 85 | 90 | DGB3 | 21 | 29 | 38 | 46 | 57 | 65 | 73 | 82 |
| DGC1 | 21 | 38 | 49 | 61 | 68 | 76 | 81 | 85 | DGC1 | 17 | 24 | 31 | 37 | 46 | 52 | 59 | 66 |
| DGC2 | 22 | 39 | 51 | 63 | 71 | 80 | 85 | 89 | DGC2 | 20 | 27 | 36 | 43 | 53 | 60 | 68 | 77 |
| DGC3 | 24 | 43 | 57 | 70 | 78 | 86 | 90 | 94 | DGC3 | 24 | 33 | 43 | 52 | 64 | 72 | 80 | 88 |
| SSA1 | 30 | 45 | 54 | 63 | 68 | 76 | 80 | 85 | SSA1 | 25 | 31 | 38 | 44 | 51 | 56 | 63 | 71 |
| SSB1 | 25 | 39 | 48 | 57 | 63 | 70 | 75 | 81 | SSB1 | 21 | 26 | 33 | 38 | 45 | 50 | 56 | 65 |
| SSB2 | 30 | 47 | 58 | 67 | 74 | 81 | 86 | 90 | SSB2 | 27 | 35 | 44 | 50 | 59 | 65 | 72 | 81 |
| SSB3 | 24 | 38 | 46 | 55 | 61 | 68 | 73 | 79 | SSB3 | 19 | 25 | 31 | 35 | 42 | 46 | 52 | 61 |
| SSC1 | 21 | 37 | 47 | 57 | 64 | 72 | 78 | 84 | SSC1 | 19 | 25 | 33 | 38 | 46 | 52 | 59 | 68 |
| SSC2 | 24 | 42 | 54 | 64 | 72 | 80 | 85 | 90 | SSC2 | 23 | 31 | 41 | 48 | 57 | 63 | 71 | 80 |
| SSE1 | 30 | 47 | 57 | 66 | 72 | 79 | 84 | 88 | SSE1 | 24 | 31 | 39 | 45 | 54 | 59 | 66 | 74 |
| SSE2 | 35 | 54 | 65 | 75 | 81 | 87 | 91 | 94 | SSE2 | 30 | 40 | 50 | 57 | 67 | 73 | 80 | 87 |
| SSE3 | 26 | 40 | 48 | 57 | 63 | 71 | 75 | 81 | SSE3 | 21 | 27 | 33 | 38 | 45 | 49 | 56 | 64 |

| SANTA MARIA, CALIFORNIA SSF (%) | | | | | | | 3061 DD65 | | GRAND JUNCTION, COLORADO SSF (%) | | | | | | | 5607 DD65 |
|---|---|---|---|---|---|---|---|---|---|---|---|---|---|---|---|---|
| Type | LCR = 200 | 100 | 70 | 50 | 40 | 30 | 25 | 20 | Type | LCR = 100 | 70 | 50 | 40 | 30 | 25 | 20 | 15 |
| WWA3 | 25 | 44 | 55 | 67 | 74 | 82 | 87 | 92 | WWA3 | 20 | 27 | 35 | 41 | 49 | 55 | 62 | 71 |
| WWB4 | 29 | 53 | 66 | 78 | 84 | 91 | 95 | 97 | WWB4 | 26 | 36 | 47 | 55 | 65 | 71 | 79 | 87 |
| WWC2 | 29 | 52 | 65 | 77 | 84 | 91 | 94 | 97 | WWC2 | 26 | 35 | 46 | 54 | 64 | 70 | 77 | 85 |
| TWA1 | 21 | 32 | 40 | 49 | 55 | 64 | 69 | 76 | TWA1 | 19 | 22 | 26 | 30 | 35 | 38 | 43 | 50 |
| TWA2 | 22 | 37 | 47 | 57 | 64 | 73 | 79 | 85 | TWA2 | 19 | 24 | 30 | 35 | 41 | 46 | 52 | 61 |
| TWA3 | 22 | 39 | 49 | 60 | 67 | 77 | 82 | 88 | TWA3 | 19 | 25 | 31 | 37 | 44 | 49 | 56 | 65 |
| TWA4 | 22 | 38 | 49 | 61 | 68 | 77 | 83 | 88 | TWA4 | 18 | 24 | 31 | 37 | 45 | 50 | 57 | 66 |
| TWB3 | 20 | 35 | 45 | 56 | 64 | 73 | 78 | 84 | TWB3 | 17 | 22 | 29 | 34 | 41 | 46 | 52 | 61 |
| TWD4 | 27 | 48 | 60 | 72 | 79 | 87 | 91 | 95 | TWD4 | 24 | 32 | 42 | 49 | 59 | 65 | 73 | 81 |
| TWE2 | 28 | 49 | 61 | 73 | 80 | 88 | 92 | 95 | TWE2 | 25 | 33 | 43 | 50 | 59 | 66 | 73 | 82 |
| TWF3 | 19 | 35 | 45 | 56 | 64 | 73 | 79 | 85 | TWF3 | 15 | 21 | 28 | 33 | 40 | 45 | 52 | 61 |
| TWJ2 | 25 | 46 | 58 | 70 | 77 | 86 | 90 | 94 | TWJ2 | 22 | 30 | 40 | 47 | 57 | 63 | 70 | 79 |
| DGA1 | 19 | 34 | 43 | 52 | 59 | 66 | 71 | 75 | DGA1 | 14 | 18 | 23 | 26 | 31 | 34 | 37 | 41 |
| DGA2 | 19 | 35 | 46 | 56 | 64 | 72 | 77 | 83 | DGA2 | 16 | 22 | 28 | 33 | 40 | 45 | 50 | 57 |
| DGA3 | 22 | 40 | 53 | 65 | 73 | 81 | 85 | 89 | DGA3 | 20 | 27 | 36 | 42 | 52 | 58 | 65 | 73 |
| DGB1 | 19 | 35 | 47 | 59 | 66 | 74 | 79 | 83 | DGB1 | 14 | 18 | 24 | 28 | 34 | 38 | 43 | 49 |
| DGB2 | 20 | 37 | 49 | 62 | 70 | 79 | 84 | 89 | DGB2 | 16 | 22 | 29 | 35 | 43 | 49 | 56 | 64 |
| DGB3 | 22 | 41 | 55 | 69 | 77 | 86 | 90 | 93 | DGB3 | 20 | 28 | 37 | 44 | 54 | 61 | 69 | 78 |
| DGC1 | 22 | 41 | 54 | 67 | 75 | 83 | 87 | 90 | DGC1 | 17 | 22 | 29 | 34 | 42 | 47 | 53 | 61 |
| DGC2 | 23 | 42 | 55 | 69 | 77 | 85 | 90 | 93 | DGC2 | 19 | 26 | 34 | 41 | 50 | 56 | 64 | 73 |
| DGC3 | 25 | 47 | 62 | 76 | 83 | 91 | 94 | 96 | DGC3 | 23 | 31 | 41 | 49 | 61 | 68 | 76 | 85 |
| SSA1 | 32 | 49 | 60 | 70 | 76 | 83 | 87 | 92 | SSA1 | 24 | 29 | 36 | 41 | 48 | 52 | 58 | 66 |
| SSB1 | 27 | 43 | 53 | 63 | 70 | 78 | 83 | 88 | SSB1 | 20 | 25 | 31 | 35 | 42 | 46 | 52 | 60 |
| SSB2 | 33 | 52 | 64 | 74 | 80 | 88 | 91 | 95 | SSB2 | 25 | 33 | 41 | 47 | 56 | 62 | 68 | 77 |
| SSB3 | 26 | 41 | 51 | 61 | 68 | 76 | 81 | 86 | SSB3 | 18 | 23 | 29 | 33 | 39 | 43 | 48 | 56 |
| SSC1 | 23 | 40 | 50 | 61 | 69 | 78 | 83 | 89 | SSC1 | 18 | 24 | 31 | 36 | 43 | 49 | 55 | 64 |
| SSC2 | 26 | 46 | 58 | 69 | 77 | 85 | 89 | 94 | SSC2 | 22 | 30 | 38 | 45 | 54 | 60 | 67 | 76 |
| SSE1 | 33 | 52 | 63 | 73 | 80 | 87 | 91 | 94 | SSE1 | 23 | 29 | 37 | 42 | 49 | 54 | 61 | 69 |
| SSE2 | 38 | 61 | 72 | 82 | 88 | 93 | 96 | 98 | SSE2 | 28 | 37 | 47 | 54 | 63 | 69 | 75 | 83 |
| SSE3 | 28 | 43 | 53 | 63 | 70 | 76 | 83 | 88 | SSE3 | 20 | 25 | 31 | 35 | 41 | 46 | 51 | 59 |

**TABLE H.3 Annual Passive Heating Performance: SSF *(Continued)***

| HARTFORD, CONNECTICUT SSF (%) | | | | | | | | 6354 DD65 | | MIAMI, FLORIDA SSF (%) | | | | | | | | 206 DD65 |
|---|---|---|---|---|---|---|---|---|---|---|---|---|---|---|---|---|---|---|
| Type | LCR = 100 | 70 | 50 | 40 | 30 | 25 | 20 | 15 | | Type | LCR = 200 | 100 | 70 | 50 | 40 | 30 | 25 | 20 |
| WWA3 | 7 | 10 | 13 | 16 | 19 | 22 | 25 | 30 | | WWA3 | 82 | 96 | 99 | 100 | 100 | 100 | 100 | 100 |
| WWB4 | 9 | 16 | 23 | 28 | 36 | 41 | 49 | 58 | | WWB4 | 89 | 98 | 100 | 100 | 100 | 100 | 100 | 100 |
| WWC2 | 10 | 15 | 22 | 27 | 35 | 40 | 46 | 55 | | WWC2 | 88 | 98 | 100 | 100 | 100 | 100 | 100 | 100 |
| TWA1 | 12 | 13 | 14 | 15 | 16 | 17 | 18 | 20 | | TWA1 | 64 | 84 | 92 | 97 | 98 | 100 | 100 | 100 |
| TWA2 | 9 | 11 | 13 | 15 | 18 | 20 | 22 | 25 | | TWA2 | 73 | 91 | 96 | 99 | 100 | 100 | 100 | 100 |
| TWA3 | 8 | 10 | 13 | 15 | 18 | 21 | 24 | 28 | | TWA3 | 76 | 93 | 97 | 99 | 100 | 100 | 100 | 100 |
| TWA4 | 6 | 9 | 12 | 15 | 18 | 21 | 24 | 29 | | TWA4 | 76 | 93 | 98 | 99 | 100 | 100 | 100 | 100 |
| TWB3 | 7 | 10 | 12 | 15 | 18 | 20 | 23 | 28 | | TWB3 | 71 | 90 | 96 | 99 | 100 | 100 | 100 | 100 |
| TWD4 | 10 | 15 | 21 | 25 | 32 | 37 | 43 | 52 | | TWD4 | 84 | 97 | 99 | 100 | 100 | 100 | 100 | 100 |
| TWE2 | 10 | 15 | 21 | 26 | 32 | 37 | 43 | 51 | | TWE2 | 85 | 97 | 99 | 100 | 100 | 100 | 100 | 100 |
| TWF3 | 5 | 7 | 10 | 12 | 15 | 17 | 20 | 24 | | TWF3 | 72 | 91 | 96 | 99 | 100 | 100 | 100 | 100 |
| TWJ2 | 8 | 13 | 19 | 23 | 30 | 34 | 40 | 49 | | TWJ2 | 83 | 96 | 99 | 100 | 100 | 100 | 100 | 100 |
| DGA1 | 4 | 5 | 6 | 6 | 7 | 7 | 6 | 5 | | DGA1 | 75 | 92 | 97 | 99 | 100 | 100 | 100 | 100 |
| DGA2 | 7 | 9 | 12 | 14 | 17 | 19 | 22 | 26 | | DGA2 | 74 | 92 | 97 | 99 | 100 | 100 | 100 | 100 |
| DGA3 | 10 | 14 | 18 | 22 | 28 | 32 | 37 | 45 | | DGA3 | 79 | 93 | 96 | 97 | 97 | 97 | 97 | 97 |
| DGB1 | 4 | 5 | 6 | 6 | 7 | 7 | 7 | 7 | | DGB1 | 81 | 95 | 98 | 99 | 99 | 99 | 99 | 99 |
| DGB2 | 7 | 9 | 12 | 15 | 18 | 20 | 23 | 28 | | DGB2 | 80 | 96 | 99 | 100 | 100 | 100 | 100 | 100 |
| DGB3 | 10 | 14 | 19 | 23 | 29 | 33 | 39 | 48 | | DGB3 | 83 | 96 | 98 | 98 | 98 | 98 | 98 | 98 |
| DGC1 | 5 | 7 | 9 | 10 | 12 | 13 | 14 | 15 | | DGC1 | 88 | 97 | 98 | 99 | 99 | 99 | 98 | 98 |
| DGC2 | 9 | 12 | 15 | 18 | 23 | 26 | 30 | 36 | | DGC2 | 86 | 98 | 99 | 100 | 100 | 100 | 100 | 100 |
| DGC3 | 12 | 17 | 22 | 27 | 33 | 38 | 45 | 54 | | DGC3 | 88 | 98 | 98 | 99 | 99 | 98 | 98 | 98 |
| SSA1 | 13 | 16 | 19 | 21 | 25 | 27 | 30 | 34 | | SSA1 | 81 | 95 | 99 | 100 | 100 | 100 | 100 | 100 |
| SSB1 | 11 | 13 | 15 | 17 | 20 | 22 | 25 | 28 | | SSB1 | 75 | 92 | 97 | 99 | 100 | 100 | 100 | 100 |
| SSB2 | 14 | 18 | 24 | 27 | 33 | 37 | 43 | 50 | | SSB2 | 84 | 97 | 99 | 100 | 100 | 100 | 100 | 100 |
| SSB3 | 10 | 11 | 14 | 15 | 17 | 19 | 21 | 23 | | SSB3 | 74 | 92 | 97 | 99 | 100 | 100 | 100 | 100 |
| SSC1 | 6 | 8 | 11 | 13 | 15 | 17 | 20 | 23 | | SSC1 | 78 | 94 | 98 | 99 | 100 | 100 | 100 | 100 |
| SSC2 | 8 | 12 | 17 | 21 | 26 | 30 | 35 | 42 | | SSC2 | 83 | 96 | 99 | 100 | 100 | 100 | 100 | 100 |
| SSE1 | 10 | 13 | 16 | 18 | 21 | 23 | 26 | 29 | | SSE1 | 85 | 97 | 99 | 100 | 100 | 100 | 100 | 100 |
| SSE2 | 13 | 18 | 24 | 29 | 35 | 40 | 45 | 53 | | SSE2 | 91 | 99 | 100 | 100 | 100 | 100 | 100 | 100 |
| SSE3 | 11 | 13 | 15 | 17 | 19 | 21 | 23 | 26 | | SSE3 | 76 | 93 | 97 | 99 | 100 | 100 | 100 | 100 |

| WASHINGTON, DC SSF (%) | | | | | | | | 5010 DD65 | | ORLANDO, FLORIDA SSF (%) | | | | | | | | 733 DD65 |
|---|---|---|---|---|---|---|---|---|---|---|---|---|---|---|---|---|---|---|
| Type | LCR = 100 | 70 | 50 | 40 | 30 | 25 | 20 | 15 | | Type | LCR = 200 | 100 | 70 | 50 | 40 | 30 | 25 | 20 |
| WWA3 | 12 | 16 | 22 | 25 | 31 | 35 | 40 | 48 | | WWA3 | 45 | 68 | 80 | 89 | 93 | 97 | 98 | 99 |
| WWB4 | 15 | 23 | 32 | 39 | 48 | 54 | 62 | 72 | | WWB4 | 52 | 78 | 88 | 95 | 97 | 99 | 100 | 100 |
| WWC2 | 15 | 23 | 31 | 38 | 46 | 52 | 60 | 69 | | WWC2 | 52 | 77 | 87 | 94 | 97 | 99 | 100 | 100 |
| TWA1 | 14 | 16 | 18 | 20 | 23 | 25 | 28 | 32 | | TWA1 | 34 | 51 | 61 | 72 | 78 | 86 | 90 | 94 |
| TWA2 | 13 | 16 | 20 | 22 | 27 | 30 | 34 | 40 | | TWA2 | 38 | 59 | 70 | 80 | 86 | 92 | 95 | 98 |
| TWA3 | 12 | 15 | 20 | 23 | 28 | 32 | 37 | 43 | | TWA3 | 40 | 62 | 73 | 83 | 89 | 94 | 97 | 98 |
| TWA4 | 10 | 14 | 19 | 23 | 29 | 33 | 38 | 45 | | TWA4 | 39 | 62 | 74 | 84 | 90 | 95 | 97 | 99 |
| TWB3 | 11 | 14 | 18 | 22 | 27 | 30 | 35 | 42 | | TWB3 | 36 | 57 | 69 | 80 | 86 | 92 | 95 | 97 |
| TWD4 | 15 | 21 | 29 | 35 | 43 | 49 | 56 | 65 | | TWD4 | 47 | 72 | 83 | 91 | 95 | 98 | 99 | 100 |
| TWE2 | 15 | 22 | 29 | 35 | 43 | 49 | 56 | 65 | | TWE2 | 49 | 73 | 84 | 92 | 96 | 98 | 99 | 100 |
| TWF3 | 9 | 12 | 16 | 20 | 25 | 28 | 33 | 40 | | TWF3 | 36 | 58 | 70 | 80 | 87 | 93 | 95 | 98 |
| TWJ2 | 13 | 19 | 27 | 32 | 40 | 46 | 53 | 62 | | TWJ2 | 45 | 70 | 82 | 90 | 94 | 98 | 99 | 100 |
| DGA1 | 7 | 9 | 12 | 14 | 16 | 17 | 18 | 19 | | DGA1 | 37 | 58 | 69 | 78 | 84 | 90 | 93 | 95 |
| DGA2 | 10 | 13 | 18 | 21 | 26 | 29 | 33 | 39 | | DGA2 | 36 | 59 | 71 | 81 | 86 | 92 | 95 | 97 |
| DGA3 | 13 | 18 | 25 | 29 | 37 | 42 | 48 | 57 | | DGA3 | 40 | 66 | 77 | 86 | 90 | 94 | 95 | 96 |
| DGB1 | 7 | 9 | 12 | 14 | 17 | 19 | 21 | 23 | | DGB1 | 38 | 64 | 76 | 85 | 90 | 94 | 96 | 97 |
| DGB2 | 10 | 14 | 18 | 22 | 27 | 31 | 36 | 42 | | DGB2 | 38 | 64 | 77 | 87 | 92 | 96 | 98 | 99 |
| DGB3 | 14 | 19 | 25 | 31 | 38 | 44 | 51 | 61 | | DGB3 | 41 | 69 | 82 | 90 | 94 | 97 | 97 | 98 |
| DGC1 | 9 | 12 | 16 | 19 | 23 | 25 | 29 | 33 | | DGC1 | 44 | 72 | 84 | 91 | 94 | 97 | 97 | 98 |
| DGC2 | 12 | 17 | 22 | 26 | 33 | 37 | 43 | 51 | | DGC2 | 42 | 71 | 83 | 92 | 95 | 98 | 99 | 100 |
| DGC3 | 16 | 22 | 29 | 35 | 43 | 50 | 58 | 68 | | DGC3 | 46 | 76 | 88 | 94 | 96 | 98 | 98 | 98 |
| SSA1 | 17 | 21 | 26 | 29 | 34 | 38 | 42 | 49 | | SSA1 | 46 | 67 | 78 | 87 | 92 | 96 | 98 | 99 |
| SSB1 | 14 | 17 | 21 | 25 | 29 | 32 | 36 | 42 | | SSB1 | 40 | 61 | 72 | 82 | 88 | 93 | 96 | 98 |
| SSB2 | 18 | 24 | 31 | 36 | 43 | 48 | 54 | 63 | | SSB2 | 48 | 71 | 82 | 90 | 94 | 98 | 99 | 100 |
| SSB3 | 13 | 16 | 20 | 22 | 26 | 29 | 33 | 38 | | SSB3 | 39 | 59 | 71 | 81 | 86 | 92 | 95 | 98 |
| SSC1 | 10 | 14 | 18 | 21 | 26 | 30 | 34 | 40 | | SSC1 | 41 | 63 | 75 | 85 | 90 | 95 | 97 | 99 |
| SSC2 | 13 | 19 | 25 | 30 | 37 | 42 | 48 | 57 | | SSC2 | 46 | 70 | 81 | 90 | 94 | 98 | 99 | 100 |
| SSE1 | 15 | 19 | 24 | 28 | 33 | 37 | 41 | 48 | | SSE1 | 50 | 72 | 83 | 91 | 95 | 98 | 90 | 100 |
| SSE2 | 19 | 26 | 34 | 39 | 47 | 53 | 59 | 68 | | SSE2 | 57 | 80 | 90 | 95 | 98 | 99 | 100 | 100 |
| SSE3 | 15 | 18 | 22 | 24 | 29 | 31 | 35 | 41 | | SSE3 | 41 | 62 | 73 | 82 | 88 | 93 | 96 | 98 |

## TABLE H.3 Annual Passive Heating Performance: SSF *(Continued)*

### TALLAHASSEE, FLORIDA SSF (%) — 1563 DD65

| Type | LCR = 200 | 100 | 70 | 50 | 40 | 30 | 25 | 20 |
|---|---|---|---|---|---|---|---|---|
| WWA3 | 24 | 41 | 52 | 63 | 70 | 79 | 84 | 90 |
| WWB4 | 27 | 49 | 62 | 74 | 82 | 89 | 93 | 96 |
| WWC2 | 27 | 49 | 61 | 73 | 81 | 88 | 92 | 96 |
| TWA1 | 21 | 31 | 38 | 46 | 52 | 60 | 65 | 72 |
| TWA2 | 22 | 35 | 44 | 53 | 60 | 70 | 75 | 82 |
| TWA3 | 22 | 36 | 46 | 56 | 64 | 73 | 79 | 85 |
| TWA4 | 21 | 36 | 46 | 57 | 64 | 74 | 79 | 86 |
| TWB3 | 19 | 33 | 42 | 52 | 59 | 69 | 75 | 81 |
| TWD4 | 25 | 45 | 57 | 69 | 76 | 85 | 89 | 93 |
| TWE2 | 26 | 46 | 58 | 69 | 77 | 85 | 90 | 94 |
| TWF3 | 18 | 32 | 42 | 52 | 60 | 69 | 75 | 82 |
| TWJ2 | 23 | 42 | 54 | 66 | 74 | 83 | 88 | 92 |
| DGA1 | 18 | 31 | 40 | 48 | 54 | 62 | 66 | 71 |
| DGA2 | 18 | 33 | 42 | 53 | 60 | 68 | 74 | 79 |
| DGA3 | 21 | 37 | 49 | 62 | 69 | 78 | 83 | 87 |
| DGB1 | 18 | 32 | 43 | 54 | 61 | 70 | 75 | 80 |
| DGB2 | 19 | 34 | 45 | 58 | 66 | 75 | 81 | 86 |
| DGB3 | 21 | 38 | 51 | 65 | 73 | 83 | 87 | 92 |
| DGC1 | 21 | 38 | 49 | 62 | 71 | 79 | 83 | 88 |
| DGC2 | 21 | 38 | 51 | 65 | 73 | 82 | 87 | 91 |
| DGC3 | 24 | 43 | 57 | 72 | 80 | 88 | 92 | 95 |
| SSA1 | 27 | 42 | 52 | 62 | 69 | 77 | 82 | 88 |
| SSB1 | 23 | 36 | 46 | 55 | 62 | 71 | 77 | 83 |
| SSB2 | 28 | 45 | 56 | 67 | 74 | 83 | 87 | 92 |
| SSB3 | 22 | 35 | 44 | 53 | 60 | 69 | 74 | 81 |
| SSC1 | 21 | 37 | 47 | 57 | 65 | 74 | 80 | 86 |
| SSC2 | 24 | 42 | 54 | 66 | 73 | 82 | 87 | 92 |
| SSE1 | 28 | 45 | 55 | 66 | 73 | 81 | 86 | 91 |
| SSE2 | 32 | 53 | 65 | 76 | 83 | 90 | 93 | 96 |
| SSE3 | 24 | 37 | 46 | 56 | 63 | 71 | 77 | 83 |

### BOISE, IDAHO SSF (%) — 5837 DD65

| Type | LCR = 100 | 70 | 50 | 40 | 30 | 25 | 20 | 15 |
|---|---|---|---|---|---|---|---|---|
| WWA3 | 18 | 23 | 29 | 34 | 40 | 45 | 50 | 57 |
| WWB4 | 22 | 31 | 40 | 47 | 56 | 62 | 69 | 77 |
| WWC2 | 22 | 30 | 39 | 46 | 54 | 60 | 67 | 75 |
| TWA1 | 17 | 20 | 24 | 26 | 30 | 33 | 36 | 41 |
| TWA2 | 17 | 21 | 26 | 30 | 35 | 38 | 43 | 49 |
| TWA3 | 16 | 21 | 27 | 31 | 37 | 41 | 46 | 53 |
| TWA4 | 15 | 21 | 26 | 31 | 37 | 41 | 47 | 54 |
| TWB3 | 15 | 20 | 25 | 29 | 35 | 39 | 44 | 50 |
| TWD4 | 20 | 28 | 36 | 43 | 51 | 56 | 63 | 71 |
| TWE2 | 21 | 29 | 37 | 43 | 51 | 56 | 63 | 71 |
| TWF3 | 13 | 18 | 23 | 27 | 33 | 37 | 42 | 49 |
| TWJ2 | 19 | 26 | 34 | 40 | 48 | 54 | 61 | 69 |
| DGA1 | 12 | 15 | 19 | 21 | 25 | 26 | 28 | 30 |
| DGA2 | 14 | 19 | 24 | 28 | 34 | 38 | 42 | 48 |
| DGA3 | 18 | 24 | 32 | 37 | 45 | 50 | 56 | 64 |
| DGB1 | 12 | 16 | 20 | 23 | 27 | 29 | 32 | 35 |
| DGB2 | 15 | 20 | 26 | 30 | 36 | 40 | 46 | 52 |
| DGB3 | 19 | 25 | 33 | 39 | 47 | 52 | 59 | 68 |
| DGC1 | 15 | 19 | 25 | 28 | 33 | 37 | 41 | 45 |
| DGC2 | 17 | 23 | 30 | 35 | 42 | 47 | 53 | 60 |
| DGC3 | 21 | 28 | 37 | 43 | 52 | 58 | 65 | 74 |
| SSA1 | 22 | 26 | 32 | 35 | 41 | 44 | 48 | 54 |
| SSB1 | 18 | 22 | 27 | 31 | 36 | 39 | 43 | 49 |
| SSB2 | 23 | 30 | 37 | 42 | 49 | 54 | 59 | 67 |
| SSB3 | 17 | 21 | 25 | 28 | 33 | 36 | 39 | 44 |
| SSC1 | 15 | 20 | 26 | 30 | 35 | 39 | 44 | 50 |
| SSC2 | 18 | 25 | 33 | 38 | 46 | 51 | 57 | 65 |
| SSE1 | 20 | 25 | 31 | 35 | 40 | 44 | 48 | 54 |
| SSE2 | 25 | 32 | 40 | 46 | 53 | 58 | 64 | 71 |
| SSE3 | 19 | 23 | 27 | 31 | 35 | 38 | 42 | 47 |

### ATLANTA, GEORGIA SSF (%) — 3105 DD65

| Type | LCR = 200 | 100 | 70 | 50 | 40 | 30 | 25 | 20 |
|---|---|---|---|---|---|---|---|---|
| WWA3 | 12 | 22 | 29 | 37 | 43 | 52 | 57 | 65 |
| WWB4 | 12 | 28 | 38 | 49 | 57 | 67 | 74 | 81 |
| WWC2 | 13 | 28 | 38 | 48 | 56 | 66 | 72 | 79 |
| TWA1 | 15 | 20 | 24 | 28 | 32 | 37 | 41 | 46 |
| TWA2 | 13 | 20 | 26 | 32 | 37 | 44 | 49 | 55 |
| TWA3 | 12 | 20 | 26 | 33 | 39 | 47 | 52 | 59 |
| TWA4 | 10 | 19 | 26 | 33 | 39 | 47 | 53 | 60 |
| TWB3 | 10 | 18 | 24 | 31 | 36 | 43 | 49 | 55 |
| TWD4 | 12 | 25 | 35 | 45 | 52 | 61 | 68 | 75 |
| TWE2 | 13 | 26 | 35 | 45 | 52 | 62 | 68 | 75 |
| TWF3 | 9 | 17 | 23 | 30 | 35 | 43 | 48 | 55 |
| TWJ2 | 11 | 23 | 32 | 42 | 49 | 59 | 65 | 73 |
| DGA1 | 9 | 15 | 20 | 25 | 29 | 34 | 37 | 40 |
| DGA2 | 10 | 17 | 23 | 30 | 35 | 43 | 47 | 53 |
| DGA3 | 12 | 21 | 29 | 38 | 44 | 54 | 60 | 67 |
| DGB1 | 9 | 15 | 20 | 26 | 31 | 37 | 42 | 47 |
| DGB2 | 10 | 18 | 24 | 32 | 37 | 46 | 52 | 59 |
| DGB3 | 12 | 22 | 30 | 39 | 46 | 57 | 64 | 72 |
| DGC1 | 10 | 19 | 25 | 32 | 37 | 45 | 50 | 57 |
| DGC2 | 11 | 21 | 28 | 36 | 43 | 53 | 59 | 67 |
| DGC3 | 13 | 25 | 33 | 44 | 52 | 63 | 71 | 79 |
| SSA1 | 17 | 26 | 32 | 40 | 45 | 52 | 57 | 64 |
| SSB1 | 14 | 22 | 28 | 34 | 39 | 46 | 51 | 57 |
| SSB2 | 16 | 28 | 36 | 45 | 52 | 60 | 66 | 73 |
| SSB3 | 13 | 21 | 26 | 32 | 37 | 43 | 48 | 54 |
| SSC1 | 11 | 20 | 26 | 33 | 38 | 46 | 51 | 58 |
| SSC2 | 12 | 24 | 32 | 41 | 47 | 57 | 63 | 70 |
| SSE1 | 15 | 26 | 33 | 41 | 47 | 55 | 60 | 67 |
| SSE2 | 17 | 32 | 41 | 51 | 58 | 68 | 74 | 80 |
| SSE3 | 15 | 23 | 28 | 34 | 39 | 46 | 51 | 57 |

### CHICAGO, ILLINOIS SSF (%) — 6125 DD65

| Type | LCR = 100 | 70 | 50 | 40 | 30 | 25 | 20 | 15 |
|---|---|---|---|---|---|---|---|---|
| WWA3 | 9 | 13 | 16 | 19 | 23 | 26 | 30 | 35 |
| WWB4 | 12 | 19 | 26 | 32 | 40 | 46 | 53 | 62 |
| WWC2 | 12 | 18 | 25 | 31 | 38 | 44 | 51 | 60 |
| TWA1 | 13 | 14 | 16 | 17 | 19 | 20 | 22 | 24 |
| TWA2 | 11 | 13 | 16 | 18 | 21 | 23 | 26 | 30 |
| TWA3 | 9 | 12 | 16 | 18 | 22 | 25 | 28 | 33 |
| TWA4 | 8 | 11 | 15 | 18 | 22 | 25 | 29 | 34 |
| TWB3 | 9 | 11 | 15 | 17 | 21 | 24 | 27 | 32 |
| TWD4 | 12 | 17 | 24 | 29 | 36 | 41 | 47 | 56 |
| TWE2 | 12 | 18 | 24 | 29 | 36 | 40 | 47 | 55 |
| TWF3 | 7 | 9 | 13 | 15 | 19 | 21 | 25 | 29 |
| TWJ2 | 10 | 15 | 22 | 26 | 33 | 38 | 44 | 53 |
| DGA1 | 5 | 7 | 8 | 9 | 10 | 10 | 11 | 10 |
| DGA2 | 8 | 11 | 14 | 17 | 20 | 23 | 26 | 30 |
| DGA3 | 12 | 16 | 21 | 25 | 31 | 35 | 41 | 48 |
| DGB1 | 5 | 7 | 8 | 9 | 11 | 11 | 12 | 12 |
| DGB2 | 8 | 11 | 15 | 17 | 21 | 24 | 28 | 33 |
| DGB3 | 12 | 16 | 22 | 26 | 32 | 37 | 43 | 52 |
| DGC1 | 7 | 9 | 12 | 13 | 16 | 17 | 19 | 21 |
| DGC2 | 10 | 14 | 18 | 21 | 26 | 30 | 34 | 41 |
| DGC3 | 14 | 19 | 25 | 30 | 37 | 42 | 49 | 58 |
| SSA1 | 15 | 18 | 21 | 24 | 27 | 30 | 33 | 38 |
| SSB1 | 12 | 15 | 18 | 20 | 23 | 25 | 28 | 32 |
| SSB2 | 16 | 20 | 26 | 30 | 36 | 40 | 46 | 53 |
| SSB3 | 11 | 13 | 16 | 17 | 20 | 22 | 24 | 27 |
| SSC1 | 8 | 11 | 14 | 16 | 19 | 21 | 24 | 28 |
| SSC2 | 10 | 15 | 20 | 24 | 30 | 34 | 39 | 46 |
| SSE1 | 12 | 15 | 18 | 21 | 25 | 27 | 30 | 34 |
| SSE2 | 15 | 21 | 27 | 32 | 39 | 43 | 49 | 57 |
| SSE3 | 12 | 15 | 17 | 19 | 22 | 24 | 27 | 30 |

**TABLE H.3 Annual Passive Heating Performance: SSF** *(Continued)*

| MOLINE, ILLINOIS SSF (%) | | | | | | | 6399 DD65 | |
|---|---|---|---|---|---|---|---|---|
| Type | LCR = 100 | 70 | 50 | 40 | 30 | 25 | 20 | 15 |
| WWA3 | 9 | 13 | 16 | 19 | 24 | 26 | 30 | 36 |
| WWB4 | 12 | 19 | 26 | 32 | 40 | 46 | 53 | 63 |
| WWC2 | 12 | 18 | 25 | 31 | 39 | 44 | 51 | 60 |
| TWA1 | 13 | 14 | 16 | 17 | 19 | 20 | 22 | 24 |
| TWA2 | 11 | 13 | 16 | 18 | 21 | 23 | 26 | 30 |
| TWA3 | 9 | 12 | 16 | 18 | 22 | 25 | 28 | 33 |
| TWA4 | 8 | 11 | 15 | 18 | 22 | 25 | 29 | 34 |
| TWB3 | 9 | 11 | 15 | 17 | 21 | 24 | 27 | 32 |
| TWD4 | 12 | 17 | 24 | 29 | 36 | 41 | 47 | 56 |
| TWE2 | 12 | 18 | 24 | 29 | 36 | 41 | 47 | 55 |
| TWF3 | 7 | 9 | 13 | 15 | 19 | 21 | 25 | 29 |
| TWJ2 | 10 | 15 | 22 | 26 | 33 | 38 | 44 | 53 |
| DGA1 | 5 | 7 | 8 | 9 | 10 | 11 | 11 | 10 |
| DGA2 | 8 | 11 | 14 | 17 | 20 | 23 | 26 | 30 |
| DGA3 | 12 | 16 | 21 | 25 | 31 | 35 | 41 | 49 |
| DGB1 | 5 | 7 | 8 | 9 | 11 | 11 | 12 | 12 |
| DGB2 | 8 | 11 | 15 | 17 | 21 | 24 | 28 | 33 |
| DGB3 | 12 | 16 | 22 | 26 | 32 | 37 | 43 | 52 |
| DGC1 | 7 | 9 | 12 | 13 | 16 | 17 | 19 | 21 |
| DGC2 | 10 | 14 | 18 | 21 | 26 | 29 | 34 | 41 |
| DGC3 | 14 | 19 | 25 | 30 | 37 | 42 | 49 | 59 |
| SSA1 | 15 | 18 | 21 | 24 | 27 | 30 | 33 | 38 |
| SSB1 | 12 | 14 | 17 | 20 | 23 | 25 | 28 | 32 |
| SSB2 | 15 | 20 | 26 | 30 | 36 | 40 | 46 | 53 |
| SSB3 | 11 | 13 | 15 | 17 | 20 | 22 | 24 | 27 |
| SSC1 | 8 | 11 | 14 | 16 | 19 | 21 | 24 | 29 |
| SSC2 | 10 | 15 | 20 | 24 | 30 | 34 | 39 | 47 |
| SSE1 | 12 | 15 | 18 | 21 | 24 | 27 | 30 | 34 |
| SSE2 | 15 | 21 | 27 | 32 | 38 | 43 | 49 | 57 |
| SSE3 | 12 | 15 | 17 | 19 | 22 | 24 | 27 | 30 |

| INDIANAPOLIS, INDIANA SSF (%) | | | | | | | 5585 DD65 | |
|---|---|---|---|---|---|---|---|---|
| Type | LCR = 100 | 70 | 50 | 40 | 30 | 25 | 20 | 15 |
| WWA3 | 9 | 12 | 16 | 19 | 22 | 25 | 29 | 34 |
| WWB4 | 11 | 18 | 25 | 31 | 39 | 45 | 52 | 61 |
| WWC2 | 11 | 18 | 25 | 30 | 37 | 43 | 49 | 58 |
| TWA1 | 12 | 14 | 15 | 17 | 18 | 19 | 21 | 23 |
| TWA2 | 10 | 13 | 15 | 17 | 20 | 22 | 25 | 29 |
| TWA3 | 9 | 12 | 15 | 18 | 21 | 24 | 27 | 31 |
| TWA4 | 8 | 11 | 14 | 17 | 21 | 24 | 28 | 33 |
| TWB3 | 8 | 11 | 14 | 17 | 20 | 23 | 26 | 31 |
| TWD4 | 11 | 17 | 23 | 28 | 35 | 40 | 46 | 55 |
| TWE2 | 12 | 17 | 23 | 28 | 35 | 39 | 46 | 54 |
| TWF3 | 6 | 9 | 12 | 14 | 18 | 20 | 23 | 28 |
| TWJ2 | 10 | 15 | 21 | 26 | 32 | 37 | 43 | 52 |
| DGA1 | 5 | 6 | 8 | 8 | 9 | 10 | 10 | 9 |
| DGA2 | 8 | 11 | 14 | 16 | 20 | 22 | 25 | 29 |
| DGA3 | 11 | 15 | 20 | 24 | 30 | 34 | 40 | 47 |
| DGB1 | 5 | 6 | 8 | 9 | 10 | 10 | 11 | 11 |
| DGB2 | 8 | 11 | 14 | 17 | 20 | 23 | 26 | 31 |
| DGB3 | 12 | 16 | 21 | 25 | 31 | 36 | 42 | 51 |
| DGC1 | 7 | 9 | 11 | 13 | 15 | 16 | 17 | 19 |
| DGC2 | 10 | 13 | 17 | 21 | 25 | 28 | 33 | 39 |
| DGC3 | 13 | 18 | 24 | 29 | 36 | 41 | 48 | 57 |
| SSA1 | 15 | 18 | 21 | 24 | 27 | 30 | 33 | 38 |
| SSB1 | 12 | 15 | 17 | 20 | 23 | 25 | 28 | 32 |
| SSB2 | 15 | 20 | 26 | 30 | 36 | 40 | 46 | 53 |
| SSB3 | 11 | 13 | 16 | 17 | 20 | 22 | 24 | 27 |
| SSC1 | 8 | 10 | 13 | 15 | 18 | 20 | 23 | 27 |
| SSC2 | 10 | 14 | 19 | 23 | 29 | 33 | 38 | 45 |
| SSE1 | 12 | 15 | 18 | 21 | 24 | 27 | 30 | 34 |
| SSE2 | 15 | 21 | 27 | 32 | 38 | 43 | 49 | 56 |
| SSE3 | 12 | 15 | 17 | 19 | 22 | 24 | 26 | 30 |

| SPRINGFIELD, ILLINOIS SSF (%) | | | | | | | 5563 DD65 | |
|---|---|---|---|---|---|---|---|---|
| Type | LCR = 100 | 70 | 50 | 40 | 30 | 25 | 20 | 15 |
| WWA3 | 12 | 16 | 21 | 24 | 30 | 33 | 38 | 45 |
| WWB4 | 15 | 23 | 31 | 38 | 46 | 53 | 60 | 70 |
| WWC2 | 15 | 22 | 30 | 36 | 45 | 50 | 58 | 67 |
| TWA1 | 14 | 16 | 18 | 20 | 22 | 24 | 27 | 30 |
| TWA2 | 13 | 16 | 19 | 22 | 26 | 29 | 32 | 38 |
| TWA3 | 12 | 15 | 19 | 22 | 27 | 30 | 35 | 41 |
| TWA4 | 10 | 14 | 19 | 22 | 27 | 31 | 36 | 43 |
| TWB3 | 11 | 14 | 18 | 21 | 26 | 29 | 33 | 39 |
| TWD4 | 14 | 21 | 28 | 34 | 42 | 47 | 54 | 63 |
| TWE2 | 15 | 21 | 28 | 34 | 42 | 47 | 54 | 63 |
| TWF3 | 9 | 12 | 16 | 19 | 24 | 27 | 31 | 37 |
| TWJ2 | 13 | 19 | 26 | 31 | 39 | 44 | 51 | 60 |
| DGA1 | 7 | 9 | 11 | 13 | 15 | 16 | 17 | 17 |
| DGA2 | 10 | 13 | 17 | 20 | 25 | 28 | 32 | 37 |
| DGA3 | 14 | 18 | 24 | 28 | 35 | 40 | 47 | 55 |
| DGB1 | 7 | 9 | 12 | 13 | 16 | 17 | 19 | 21 |
| DGB2 | 10 | 14 | 18 | 21 | 26 | 29 | 34 | 40 |
| DGB3 | 14 | 19 | 25 | 30 | 37 | 42 | 49 | 59 |
| DGC1 | 9 | 12 | 15 | 18 | 21 | 23 | 26 | 30 |
| DGC2 | 12 | 16 | 21 | 25 | 31 | 35 | 41 | 49 |
| DGC3 | 16 | 22 | 28 | 34 | 42 | 48 | 56 | 66 |
| SSA1 | 17 | 20 | 25 | 28 | 32 | 36 | 40 | 46 |
| SSB1 | 14 | 17 | 21 | 23 | 28 | 30 | 34 | 40 |
| SSB2 | 18 | 23 | 29 | 34 | 41 | 46 | 52 | 60 |
| SSB3 | 13 | 15 | 19 | 21 | 25 | 27 | 30 | 35 |
| SSC1 | 10 | 14 | 17 | 20 | 25 | 28 | 32 | 38 |
| SSC2 | 13 | 18 | 24 | 29 | 36 | 40 | 46 | 55 |
| SSE1 | 14 | 18 | 23 | 26 | 31 | 34 | 38 | 44 |
| SSE2 | 18 | 25 | 32 | 37 | 45 | 50 | 57 | 65 |
| SSE3 | 14 | 17 | 21 | 23 | 27 | 29 | 33 | 38 |

| SOUTH BEND, INDIANA SSF (%) | | | | | | | 6465 DD65 | |
|---|---|---|---|---|---|---|---|---|
| Type | LCR = 100 | 70 | 50 | 40 | 30 | 25 | 20 | 15 |
| WWA3 | 7 | 9 | 12 | 14 | 16 | 18 | 20 | 23 |
| WWB4 | 9 | 14 | 21 | 26 | 33 | 38 | 44 | 52 |
| WWC2 | 9 | 14 | 20 | 25 | 31 | 36 | 41 | 49 |
| TWA1 | 12 | 12 | 13 | 14 | 15 | 16 | 16 | 17 |
| TWA2 | 9 | 11 | 13 | 14 | 16 | 17 | 19 | 21 |
| TWA3 | 8 | 10 | 12 | 14 | 16 | 18 | 20 | 23 |
| TWA4 | 6 | 9 | 11 | 13 | 16 | 18 | 21 | 24 |
| TWB3 | 7 | 9 | 12 | 13 | 16 | 18 | 20 | 23 |
| TWD4 | 9 | 14 | 19 | 23 | 29 | 34 | 39 | 47 |
| TWE2 | 10 | 14 | 19 | 23 | 29 | 33 | 38 | 45 |
| TWF3 | 5 | 7 | 9 | 11 | 13 | 15 | 17 | 20 |
| TWJ2 | 8 | 12 | 17 | 21 | 27 | 31 | 36 | 43 |
| DGA1 | 4 | 4 | 5 | 5 | 5 | 5 | 4 | 1 |
| DGA2 | 7 | 9 | 11 | 13 | 16 | 17 | 20 | 22 |
| DGA3 | 10 | 14 | 18 | 21 | 26 | 29 | 34 | 41 |
| DGB1 | 4 | 4 | 5 | 5 | 5 | 5 | 5 | 3 |
| DGB2 | 7 | 9 | 12 | 14 | 16 | 18 | 21 | 24 |
| DGB3 | 10 | 14 | 19 | 22 | 27 | 31 | 36 | 44 |
| DGC1 | 5 | 7 | 8 | 9 | 10 | 10 | 10 | 10 |
| DGC2 | 9 | 11 | 15 | 17 | 21 | 23 | 26 | 31 |
| DGC3 | 12 | 16 | 21 | 25 | 31 | 35 | 41 | 49 |
| SSA1 | 13 | 15 | 18 | 20 | 23 | 25 | 27 | 29 |
| SSB1 | 11 | 13 | 15 | 17 | 19 | 20 | 22 | 25 |
| SSB2 | 14 | 18 | 23 | 26 | 31 | 35 | 39 | 46 |
| SSB3 | 10 | 11 | 13 | 14 | 16 | 17 | 18 | 19 |
| SSC1 | 6 | 8 | 10 | 11 | 13 | 14 | 15 | 17 |
| SSC2 | 8 | 11 | 15 | 18 | 23 | 26 | 30 | 36 |
| SSE1 | 10 | 12 | 15 | 16 | 18 | 20 | 21 | 23 |
| SSE2 | 13 | 17 | 23 | 26 | 32 | 36 | 40 | 47 |
| SSE3 | 11 | 13 | 15 | 16 | 18 | 19 | 20 | 22 |

**TABLE H.3 Annual Passive Heating Performance: SSF** *(Continued)*

| DES MOINES, IOWA SSF (%) | | | | | | | 6718 DD65 | |
|---|---|---|---|---|---|---|---|---|
| Type | LCR = 100 | 70 | 50 | 40 | 30 | 25 | 20 | 15 |
| WWA3 | 11 | 14 | 19 | 22 | 27 | 30 | 34 | 41 |
| WWB4 | 13 | 21 | 29 | 35 | 43 | 49 | 57 | 66 |
| WWC2 | 14 | 20 | 28 | 34 | 42 | 47 | 54 | 64 |
| TWA1 | 13 | 15 | 17 | 18 | 21 | 22 | 24 | 27 |
| TWA2 | 12 | 14 | 17 | 20 | 23 | 26 | 29 | 34 |
| TWA3 | 11 | 14 | 17 | 20 | 24 | 27 | 31 | 37 |
| TWA4 | 9 | 13 | 17 | 20 | 25 | 28 | 32 | 39 |
| TWB3 | 10 | 13 | 16 | 19 | 23 | 26 | 30 | 36 |
| TWD4 | 13 | 19 | 26 | 31 | 39 | 44 | 51 | 60 |
| TWE2 | 14 | 20 | 26 | 31 | 39 | 44 | 50 | 59 |
| TWF3 | 8 | 11 | 14 | 17 | 21 | 24 | 28 | 33 |
| TWJ2 | 11 | 17 | 24 | 29 | 36 | 41 | 48 | 57 |
| DGA1 | 6 | 8 | 10 | 11 | 12 | 13 | 14 | 14 |
| DGA2 | 9 | 12 | 16 | 18 | 23 | 25 | 29 | 34 |
| DGA3 | 13 | 17 | 22 | 27 | 33 | 38 | 44 | 52 |
| DGB1 | 6 | 8 | 10 | 11 | 13 | 14 | 15 | 17 |
| DGB2 | 9 | 13 | 16 | 19 | 23 | 27 | 31 | 36 |
| DGB3 | 13 | 18 | 23 | 28 | 34 | 39 | 46 | 56 |
| DGC1 | 8 | 11 | 14 | 15 | 18 | 20 | 22 | 26 |
| DGC2 | 11 | 15 | 20 | 23 | 29 | 32 | 38 | 45 |
| DGC3 | 15 | 20 | 26 | 31 | 39 | 45 | 52 | 62 |
| SSA1 | 16 | 19 | 23 | 25 | 29 | 32 | 36 | 41 |
| SSB1 | 13 | 16 | 19 | 21 | 25 | 27 | 31 | 35 |
| SSB2 | 16 | 22 | 27 | 32 | 38 | 43 | 48 | 56 |
| SSB3 | 12 | 14 | 17 | 19 | 22 | 24 | 27 | 30 |
| SSC1 | 9 | 12 | 16 | 18 | 22 | 25 | 28 | 33 |
| SSC2 | 12 | 16 | 22 | 26 | 33 | 37 | 43 | 51 |
| SSE1 | 13 | 16 | 20 | 23 | 27 | 30 | 34 | 39 |
| SSE2 | 16 | 22 | 29 | 34 | 41 | 46 | 52 | 61 |
| SSE3 | 13 | 16 | 19 | 21 | 24 | 26 | 29 | 33 |

| TOPEKA, KANSAS SSF (%) | | | | | | | 5243 DD65 | |
|---|---|---|---|---|---|---|---|---|
| Type | LCR = 700 | 70 | 50 | 40 | 30 | 25 | 20 | 15 |
| WWA3 | 16 | 21 | 27 | 32 | 38 | 43 | 49 | 57 |
| WWB4 | 20 | 28 | 38 | 45 | 55 | 61 | 69 | 78 |
| WWC2 | 20 | 28 | 37 | 44 | 53 | 60 | 67 | 76 |
| TWA1 | 16 | 19 | 22 | 24 | 28 | 31 | 34 | 39 |
| TWA2 | 15 | 19 | 24 | 27 | 33 | 36 | 41 | 48 |
| TWA3 | 15 | 19 | 24 | 29 | 35 | 39 | 44 | 52 |
| TWA4 | 14 | 18 | 24 | 29 | 35 | 40 | 45 | 54 |
| TWB3 | 13 | 18 | 23 | 27 | 32 | 36 | 42 | 50 |
| TWD4 | 18 | 26 | 34 | 41 | 49 | 55 | 63 | 72 |
| TWE2 | 19 | 27 | 35 | 41 | 50 | 55 | 63 | 72 |
| TWF3 | 12 | 16 | 21 | 25 | 31 | 35 | 40 | 48 |
| TWJ2 | 17 | 24 | 32 | 38 | 47 | 53 | 60 | 70 |
| DGA1 | 10 | 13 | 16 | 19 | 22 | 24 | 26 | 28 |
| DGA2 | 13 | 17 | 22 | 26 | 32 | 35 | 40 | 46 |
| DGA3 | 16 | 22 | 29 | 34 | 42 | 48 | 55 | 64 |
| DGB1 | 10 | 13 | 17 | 19 | 23 | 26 | 29 | 34 |
| DGB2 | 13 | 17 | 23 | 27 | 33 | 38 | 43 | 51 |
| DGB3 | 17 | 23 | 30 | 36 | 44 | 50 | 58 | 69 |
| DGC1 | 13 | 17 | 21 | 25 | 30 | 34 | 38 | 44 |
| DGC2 | 15 | 21 | 27 | 32 | 39 | 44 | 51 | 60 |
| DGC3 | 19 | 26 | 34 | 40 | 50 | 57 | 65 | 76 |
| SSA1 | 20 | 24 | 30 | 33 | 39 | 43 | 48 | 55 |
| SSB1 | 16 | 20 | 25 | 29 | 34 | 38 | 42 | 49 |
| SSB2 | 21 | 28 | 35 | 40 | 48 | 53 | 60 | 68 |
| SSB3 | 15 | 19 | 23 | 26 | 31 | 34 | 38 | 44 |
| SSC1 | 14 | 18 | 23 | 27 | 33 | 37 | 43 | 50 |
| SSC2 | 17 | 23 | 30 | 36 | 44 | 49 | 56 | 65 |
| SSE1 | 18 | 23 | 29 | 33 | 39 | 43 | 49 | 56 |
| SSE2 | 23 | 30 | 39 | 45 | 53 | 59 | 66 | 74 |
| SSE3 | 17 | 21 | 25 | 29 | 33 | 37 | 41 | 48 |

| DODGE CITY, KANSAS SSF (%) | | | | | | | 5053 DD65 | |
|---|---|---|---|---|---|---|---|---|
| Type | LCR = 100 | 70 | 50 | 40 | 30 | 25 | 20 | 15 |
| WWA3 | 21 | 27 | 35 | 41 | 50 | 55 | 63 | 72 |
| WWB4 | 26 | 36 | 47 | 55 | 66 | 72 | 80 | 88 |
| WWC2 | 26 | 36 | 46 | 54 | 64 | 71 | 78 | 86 |
| TWA1 | 19 | 22 | 27 | 30 | 35 | 39 | 44 | 51 |
| TWA2 | 19 | 24 | 30 | 35 | 42 | 47 | 53 | 61 |
| TWA3 | 19 | 25 | 32 | 37 | 44 | 50 | 57 | 66 |
| TWA4 | 18 | 24 | 31 | 37 | 45 | 51 | 58 | 67 |
| TWB3 | 17 | 23 | 29 | 34 | 41 | 46 | 53 | 62 |
| TWD4 | 24 | 33 | 43 | 50 | 60 | 66 | 74 | 82 |
| TWE2 | 25 | 33 | 43 | 50 | 60 | 66 | 74 | 83 |
| TWF3 | 16 | 21 | 28 | 33 | 40 | 46 | 52 | 62 |
| TWJ2 | 22 | 30 | 40 | 47 | 57 | 63 | 71 | 80 |
| DGA1 | 14 | 18 | 23 | 27 | 31 | 34 | 38 | 42 |
| DGA2 | 16 | 22 | 28 | 33 | 41 | 45 | 51 | 58 |
| DGA3 | 20 | 27 | 36 | 42 | 52 | 58 | 65 | 74 |
| DGB1 | 14 | 18 | 24 | 28 | 35 | 39 | 44 | 50 |
| DGB2 | 17 | 22 | 29 | 35 | 44 | 49 | 57 | 65 |
| DGB3 | 20 | 28 | 37 | 44 | 55 | 62 | 70 | 79 |
| DGC1 | 17 | 22 | 29 | 34 | 42 | 48 | 55 | 62 |
| DGC2 | 19 | 26 | 34 | 41 | 51 | 57 | 65 | 74 |
| DGC3 | 23 | 31 | 41 | 50 | 61 | 69 | 77 | 86 |
| SSA1 | 24 | 30 | 37 | 42 | 49 | 54 | 60 | 68 |
| SSB1 | 20 | 25 | 31 | 36 | 43 | 47 | 53 | 62 |
| SSB2 | 26 | 33 | 42 | 48 | 57 | 63 | 70 | 78 |
| SSB3 | 19 | 24 | 29 | 33 | 40 | 44 | 50 | 57 |
| SSC1 | 18 | 24 | 31 | 36 | 44 | 49 | 56 | 65 |
| SSC2 | 22 | 30 | 39 | 45 | 55 | 61 | 68 | 77 |
| SSE1 | 23 | 30 | 37 | 43 | 51 | 56 | 63 | 71 |
| SSE2 | 29 | 38 | 48 | 55 | 64 | 70 | 77 | 85 |
| SSE3 | 21 | 26 | 32 | 36 | 42 | 47 | 53 | 61 |

| LEXINGTON, KENTUCKY SSF (%) | | | | | | | 4732 DD65 | |
|---|---|---|---|---|---|---|---|---|
| Type | LCR = 100 | 70 | 50 | 40 | 30 | 25 | 20 | 15 |
| WWA3 | 12 | 16 | 21 | 25 | 30 | 34 | 39 | 45 |
| WWB4 | 15 | 23 | 31 | 38 | 47 | 53 | 60 | 70 |
| WWC2 | 15 | 22 | 30 | 37 | 45 | 51 | 58 | 67 |
| TWA1 | 14 | 16 | 18 | 20 | 23 | 25 | 27 | 31 |
| TWA2 | 13 | 16 | 19 | 22 | 26 | 29 | 33 | 38 |
| TWA3 | 12 | 15 | 19 | 23 | 27 | 31 | 35 | 41 |
| TWA4 | 11 | 14 | 19 | 23 | 28 | 31 | 36 | 43 |
| TWB3 | 11 | 14 | 18 | 21 | 26 | 29 | 34 | 40 |
| TWD4 | 15 | 21 | 28 | 34 | 42 | 47 | 54 | 64 |
| TWE2 | 15 | 22 | 29 | 34 | 42 | 47 | 54 | 63 |
| TWF3 | 9 | 12 | 16 | 19 | 24 | 27 | 32 | 38 |
| TWJ2 | 13 | 19 | 26 | 32 | 39 | 45 | 51 | 61 |
| DGA1 | 7 | 9 | 12 | 13 | 15 | 16 | 17 | 18 |
| DGA2 | 10 | 14 | 17 | 21 | 25 | 28 | 32 | 37 |
| DGA3 | 14 | 19 | 24 | 29 | 36 | 41 | 47 | 55 |
| DGB1 | 7 | 10 | 12 | 14 | 16 | 17 | 19 | 21 |
| DGB2 | 10 | 14 | 18 | 21 | 26 | 30 | 34 | 40 |
| DGB3 | 14 | 19 | 25 | 30 | 37 | 43 | 50 | 59 |
| DGC1 | 10 | 13 | 16 | 18 | 21 | 24 | 27 | 31 |
| DGC2 | 12 | 17 | 22 | 26 | 31 | 36 | 41 | 49 |
| DGC3 | 16 | 22 | 29 | 34 | 42 | 48 | 56 | 66 |
| SSA1 | 17 | 21 | 25 | 29 | 34 | 37 | 41 | 47 |
| SSB1 | 14 | 18 | 21 | 24 | 29 | 32 | 35 | 41 |
| SSB2 | 18 | 24 | 30 | 35 | 42 | 47 | 53 | 61 |
| SSB3 | 13 | 16 | 20 | 22 | 26 | 28 | 32 | 36 |
| SSC1 | 10 | 14 | 18 | 21 | 25 | 28 | 32 | 38 |
| SSC2 | 13 | 18 | 25 | 29 | 36 | 41 | 47 | 55 |
| SSE1 | 15 | 19 | 24 | 27 | 32 | 35 | 40 | 46 |
| SSE2 | 19 | 26 | 33 | 38 | 46 | 51 | 58 | 66 |
| SSE3 | 15 | 18 | 21 | 24 | 28 | 31 | 34 | 39 |

**TABLE H.3 Annual Passive Heating Performance: SSF *(Continued)***

### BATON ROUGE, LOUISIANA SSF (%) — 1670 DD65

| Type | LCR = 200 | 100 | 70 | 50 | 40 | 30 | 25 | 20 |
|---|---|---|---|---|---|---|---|---|
| WWA3 | 21 | 35 | 45 | 56 | 63 | 72 | 78 | 84 |
| WWB4 | 23 | 43 | 56 | 68 | 75 | 84 | 89 | 93 |
| WWC2 | 23 | 43 | 55 | 67 | 74 | 83 | 88 | 92 |
| TWA1 | 19 | 28 | 34 | 41 | 46 | 54 | 59 | 65 |
| TWA2 | 19 | 31 | 39 | 47 | 54 | 63 | 68 | 75 |
| TWA3 | 19 | 32 | 40 | 50 | 57 | 66 | 72 | 79 |
| TWA4 | 18 | 31 | 40 | 50 | 57 | 67 | 73 | 79 |
| TWB3 | 17 | 29 | 37 | 46 | 53 | 62 | 68 | 75 |
| TWD4 | 22 | 39 | 51 | 62 | 70 | 79 | 84 | 90 |
| TWE2 | 23 | 40 | 52 | 63 | 70 | 80 | 85 | 90 |
| TWF3 | 16 | 28 | 37 | 46 | 53 | 62 | 68 | 75 |
| TWJ2 | 20 | 37 | 48 | 60 | 68 | 77 | 82 | 88 |
| DGA1 | 15 | 27 | 34 | 42 | 47 | 54 | 58 | 63 |
| DGA2 | 16 | 28 | 37 | 46 | 53 | 61 | 67 | 73 |
| DGA3 | 18 | 33 | 43 | 55 | 63 | 72 | 77 | 83 |
| DGB1 | 15 | 27 | 36 | 46 | 53 | 62 | 67 | 72 |
| DGB2 | 16 | 29 | 39 | 50 | 58 | 68 | 74 | 80 |
| DGB3 | 19 | 33 | 45 | 57 | 66 | 76 | 82 | 87 |
| DGC1 | 18 | 32 | 42 | 53 | 61 | 71 | 76 | 81 |
| DGC2 | 19 | 33 | 44 | 56 | 65 | 75 | 80 | 86 |
| DGC3 | 21 | 38 | 50 | 64 | 73 | 83 | 88 | 92 |
| SSA1 | 24 | 38 | 47 | 56 | 63 | 71 | 76 | 82 |
| SSB1 | 21 | 33 | 41 | 50 | 56 | 65 | 70 | 77 |
| SSB2 | 25 | 41 | 51 | 62 | 69 | 77 | 83 | 88 |
| SSB3 | 20 | 31 | 39 | 48 | 54 | 63 | 68 | 75 |
| SSC1 | 19 | 32 | 41 | 51 | 58 | 67 | 73 | 79 |
| SSC2 | 21 | 37 | 48 | 59 | 66 | 76 | 81 | 87 |
| SSE1 | 24 | 40 | 49 | 60 | 67 | 75 | 80 | 86 |
| SSE2 | 28 | 47 | 59 | 70 | 77 | 85 | 89 | 93 |
| SSE3 | 22 | 33 | 41 | 50 | 56 | 65 | 70 | 77 |

### SHREVEPORT, LOUISIANA SSF (%) — 2175 DD65

| Type | LCR = 200 | 100 | 70 | 50 | 40 | 30 | 25 | 20 |
|---|---|---|---|---|---|---|---|---|
| WWA3 | 17 | 30 | 39 | 49 | 56 | 65 | 71 | 78 |
| WWB4 | 19 | 38 | 49 | 61 | 69 | 79 | 84 | 90 |
| WWC2 | 19 | 37 | 49 | 60 | 68 | 78 | 83 | 89 |
| TWA1 | 17 | 25 | 30 | 36 | 41 | 48 | 52 | 58 |
| TWA2 | 17 | 27 | 34 | 42 | 48 | 56 | 61 | 68 |
| TWA3 | 16 | 27 | 35 | 44 | 50 | 59 | 65 | 72 |
| TWA4 | 15 | 27 | 35 | 44 | 51 | 60 | 66 | 73 |
| TWB3 | 14 | 25 | 32 | 40 | 47 | 55 | 61 | 68 |
| TWD4 | 18 | 34 | 45 | 56 | 63 | 73 | 79 | 85 |
| TWE2 | 19 | 35 | 46 | 57 | 64 | 74 | 79 | 86 |
| TWF3 | 13 | 24 | 31 | 40 | 46 | 55 | 61 | 68 |
| TWJ2 | 16 | 32 | 42 | 53 | 61 | 71 | 77 | 84 |
| DGA1 | 13 | 22 | 29 | 36 | 41 | 47 | 51 | 55 |
| DGA2 | 13 | 24 | 32 | 40 | 47 | 55 | 60 | 66 |
| DGA3 | 16 | 28 | 38 | 48 | 56 | 66 | 72 | 78 |
| DGB1 | 13 | 22 | 30 | 38 | 45 | 53 | 58 | 64 |
| DGB2 | 14 | 25 | 33 | 43 | 50 | 60 | 66 | 73 |
| DGB3 | 16 | 29 | 39 | 50 | 59 | 70 | 76 | 83 |
| DGC1 | 15 | 26 | 35 | 45 | 52 | 62 | 68 | 74 |
| DGC2 | 16 | 28 | 38 | 49 | 57 | 68 | 74 | 80 |
| DGC3 | 18 | 32 | 44 | 56 | 66 | 77 | 83 | 88 |
| SSA1 | 21 | 33 | 41 | 49 | 56 | 64 | 70 | 76 |
| SSB1 | 18 | 28 | 35 | 43 | 49 | 58 | 63 | 70 |
| SSB2 | 21 | 36 | 45 | 55 | 62 | 71 | 77 | 83 |
| SSB3 | 17 | 27 | 34 | 41 | 47 | 55 | 60 | 67 |
| SSC1 | 15 | 27 | 35 | 44 | 51 | 60 | 66 | 73 |
| SSC2 | 17 | 32 | 42 | 52 | 60 | 69 | 75 | 82 |
| SSE1 | 21 | 34 | 43 | 52 | 59 | 68 | 73 | 80 |
| SSE2 | 24 | 41 | 52 | 63 | 70 | 79 | 84 | 89 |
| SSE3 | 19 | 29 | 36 | 44 | 50 | 58 | 63 | 70 |

### NEW ORLEANS, LOUISIANA SSF (%) — 1465 DD65

| Type | LCR = 200 | 100 | 70 | 50 | 40 | 30 | 25 | 20 |
|---|---|---|---|---|---|---|---|---|
| WWA3 | 24 | 41 | 52 | 63 | 70 | 79 | 84 | 89 |
| WWB4 | 28 | 50 | 63 | 74 | 81 | 89 | 93 | 96 |
| WWC2 | 28 | 49 | 62 | 73 | 80 | 88 | 92 | 95 |
| TWA1 | 21 | 31 | 38 | 46 | 52 | 60 | 65 | 72 |
| TWA2 | 22 | 35 | 44 | 54 | 60 | 70 | 75 | 81 |
| TWA3 | 22 | 36 | 46 | 57 | 64 | 73 | 79 | 85 |
| TWA4 | 21 | 36 | 46 | 57 | 64 | 74 | 79 | 85 |
| TWB3 | 20 | 33 | 42 | 52 | 60 | 69 | 75 | 81 |
| TWD4 | 25 | 45 | 57 | 69 | 76 | 85 | 89 | 93 |
| TWE2 | 27 | 46 | 58 | 69 | 77 | 85 | 89 | 94 |
| TWF3 | 19 | 33 | 42 | 53 | 60 | 69 | 75 | 82 |
| TWJ2 | 24 | 43 | 55 | 66 | 74 | 83 | 88 | 92 |
| DGA1 | 18 | 32 | 40 | 49 | 55 | 62 | 66 | 71 |
| DGA2 | 19 | 33 | 43 | 53 | 60 | 68 | 73 | 79 |
| DGA3 | 21 | 38 | 50 | 62 | 69 | 78 | 83 | 87 |
| DGB1 | 18 | 33 | 43 | 54 | 61 | 70 | 75 | 80 |
| DGB2 | 19 | 34 | 45 | 58 | 66 | 75 | 80 | 86 |
| DGB3 | 22 | 39 | 51 | 65 | 73 | 83 | 87 | 91 |
| DGC1 | 21 | 38 | 50 | 62 | 70 | 79 | 83 | 87 |
| DGC2 | 22 | 39 | 51 | 65 | 73 | 82 | 87 | 91 |
| DGC3 | 24 | 44 | 57 | 72 | 80 | 88 | 92 | 95 |
| SSA1 | 28 | 43 | 52 | 62 | 69 | 78 | 82 | 88 |
| SSB1 | 24 | 37 | 46 | 56 | 63 | 72 | 77 | 83 |
| SSB2 | 28 | 46 | 57 | 68 | 75 | 83 | 87 | 92 |
| SSB3 | 23 | 36 | 44 | 54 | 61 | 69 | 75 | 81 |
| SSC1 | 22 | 37 | 47 | 58 | 65 | 74 | 79 | 85 |
| SSC2 | 25 | 43 | 54 | 66 | 73 | 82 | 87 | 91 |
| SSE1 | 28 | 45 | 55 | 67 | 74 | 82 | 86 | 91 |
| SSE2 | 33 | 53 | 66 | 76 | 83 | 90 | 93 | 96 |
| SSE3 | 25 | 38 | 47 | 56 | 63 | 72 | 77 | 83 |

### PORTLAND, MAINE SSF (%) — 7499 DD65

| Type | LCR = 100 | 70 | 50 | 40 | 30 | 25 | 20 | 15 |
|---|---|---|---|---|---|---|---|---|
| WWA3 | 7 | 10 | 13 | 15 | 18 | 21 | 24 | 28 |
| WWB4 | 9 | 15 | 22 | 28 | 35 | 41 | 47 | 57 |
| WWC2 | 9 | 15 | 21 | 26 | 33 | 38 | 45 | 53 |
| TWA1 | 12 | 13 | 14 | 15 | 16 | 17 | 18 | 19 |
| TWA2 | 9 | 11 | 13 | 15 | 17 | 19 | 21 | 24 |
| TWA3 | 8 | 10 | 13 | 15 | 18 | 20 | 22 | 26 |
| TWA4 | 6 | 9 | 12 | 14 | 18 | 20 | 23 | 28 |
| TWB3 | 7 | 10 | 12 | 14 | 17 | 19 | 22 | 26 |
| TWD4 | 10 | 15 | 20 | 25 | 31 | 36 | 42 | 50 |
| TWE2 | 10 | 15 | 21 | 25 | 31 | 35 | 41 | 49 |
| TWF3 | 5 | 7 | 10 | 12 | 15 | 17 | 19 | 23 |
| TWJ2 | 8 | 13 | 18 | 22 | 29 | 33 | 39 | 47 |
| DGA1 | 4 | 5 | 5 | 6 | 6 | 6 | 6 | 4 |
| DGA2 | 7 | 9 | 12 | 14 | 17 | 19 | 21 | 25 |
| DGA3 | 10 | 14 | 18 | 22 | 27 | 31 | 36 | 43 |
| DGB1 | 4 | 5 | 5 | 6 | 6 | 7 | 6 | 5 |
| DGB2 | 7 | 9 | 12 | 14 | 17 | 20 | 22 | 26 |
| DGB3 | 11 | 14 | 19 | 23 | 28 | 33 | 38 | 46 |
| DGC1 | 5 | 7 | 8 | 10 | 11 | 12 | 13 | 13 |
| DGC2 | 9 | 12 | 15 | 18 | 22 | 25 | 29 | 34 |
| DGC3 | 12 | 16 | 22 | 26 | 32 | 37 | 43 | 53 |
| SSA1 | 13 | 16 | 19 | 21 | 24 | 26 | 29 | 32 |
| SSB1 | 11 | 13 | 15 | 17 | 20 | 22 | 24 | 27 |
| SSB2 | 14 | 18 | 23 | 27 | 33 | 37 | 42 | 49 |
| SSB3 | 10 | 11 | 13 | 15 | 17 | 18 | 20 | 22 |
| SSC1 | 6 | 8 | 10 | 12 | 14 | 16 | 18 | 21 |
| SSC2 | 8 | 12 | 16 | 20 | 25 | 28 | 33 | 40 |
| SSE1 | 10 | 13 | 15 | 17 | 20 | 22 | 24 | 27 |
| SSE2 | 13 | 18 | 24 | 28 | 34 | 38 | 44 | 51 |
| SSE3 | 11 | 13 | 15 | 17 | 19 | 20 | 22 | 25 |

**TABLE H.3 Annual Passive Heating Performance: SSF** *(Continued)*

| BALTIMORE, MARYLAND SSF (%) | | | | | | | 4731 DD65 | |
|---|---|---|---|---|---|---|---|---|
| Type | LCR = 100 | 70 | 50 | 40 | 30 | 25 | 20 | 15 |
| WWA3 | 13 | 18 | 24 | 28 | 35 | 39 | 45 | 53 |
| WWB4 | 17 | 26 | 35 | 42 | 52 | 58 | 66 | 75 |
| WWC2 | 17 | 25 | 34 | 41 | 50 | 56 | 64 | 73 |
| TWA1 | 15 | 17 | 20 | 22 | 25 | 28 | 31 | 36 |
| TWA2 | 14 | 17 | 21 | 25 | 30 | 33 | 38 | 44 |
| TWA3 | 13 | 17 | 22 | 26 | 31 | 35 | 41 | 48 |
| TWA4 | 12 | 16 | 22 | 26 | 32 | 36 | 42 | 50 |
| TWB3 | 12 | 16 | 20 | 24 | 29 | 33 | 38 | 46 |
| TWD4 | 16 | 23 | 31 | 38 | 46 | 52 | 59 | 69 |
| TWE2 | 17 | 24 | 32 | 38 | 46 | 52 | 59 | 69 |
| TWF3 | 10 | 14 | 19 | 22 | 28 | 32 | 37 | 44 |
| TWJ2 | 14 | 21 | 29 | 35 | 44 | 49 | 57 | 66 |
| DGA1 | 8 | 11 | 14 | 16 | 19 | 20 | 22 | 24 |
| DGA2 | 11 | 15 | 19 | 23 | 28 | 32 | 37 | 42 |
| DGA3 | 15 | 20 | 26 | 32 | 39 | 45 | 52 | 60 |
| DGB1 | 8 | 11 | 14 | 17 | 20 | 22 | 25 | 29 |
| DGB2 | 11 | 15 | 20 | 24 | 30 | 34 | 40 | 47 |
| DGB3 | 15 | 21 | 27 | 33 | 41 | 47 | 55 | 65 |
| DGC1 | 11 | 14 | 18 | 22 | 26 | 29 | 34 | 39 |
| DGC2 | 13 | 18 | 24 | 29 | 36 | 41 | 47 | 56 |
| DGC3 | 17 | 24 | 31 | 37 | 47 | 53 | 62 | 72 |
| SSA1 | 18 | 22 | 27 | 31 | 37 | 41 | 46 | 52 |
| SSB1 | 15 | 19 | 23 | 27 | 31 | 35 | 40 | 46 |
| SSB2 | 19 | 26 | 33 | 38 | 45 | 51 | 57 | 66 |
| SSB3 | 14 | 17 | 21 | 24 | 29 | 32 | 36 | 41 |
| SSC1 | 12 | 16 | 20 | 24 | 30 | 33 | 38 | 46 |
| SSC2 | 15 | 21 | 28 | 33 | 41 | 46 | 52 | 61 |
| SSE1 | 16 | 21 | 26 | 31 | 36 | 40 | 46 | 53 |
| SSE2 | 21 | 28 | 36 | 42 | 51 | 56 | 63 | 72 |
| SSE3 | 16 | 19 | 23 | 26 | 31 | 34 | 38 | 44 |

| BOSTON, MASSACHUSETTS SSF (%) | | | | | | | 5622 DD65 | |
|---|---|---|---|---|---|---|---|---|
| Type | LCR = 100 | 70 | 50 | 40 | 30 | 25 | 20 | 15 |
| WWA3 | 10 | 14 | 18 | 21 | 26 | 29 | 33 | 40 |
| WWB4 | 13 | 20 | 28 | 34 | 43 | 49 | 56 | 66 |
| WWC2 | 13 | 20 | 27 | 33 | 41 | 47 | 54 | 63 |
| TWA1 | 13 | 15 | 16 | 18 | 20 | 21 | 23 | 26 |
| TWA2 | 11 | 14 | 17 | 19 | 23 | 25 | 28 | 33 |
| TWA3 | 10 | 13 | 17 | 20 | 24 | 27 | 31 | 36 |
| TWA4 | 9 | 12 | 16 | 19 | 24 | 27 | 32 | 38 |
| TWB3 | 9 | 12 | 16 | 18 | 23 | 26 | 29 | 35 |
| TWD4 | 13 | 18 | 25 | 31 | 38 | 43 | 50 | 59 |
| TWE2 | 13 | 19 | 26 | 31 | 38 | 43 | 50 | 58 |
| TWF3 | 7 | 10 | 14 | 16 | 20 | 23 | 27 | 33 |
| TWJ2 | 11 | 17 | 23 | 28 | 35 | 41 | 47 | 56 |
| DGA1 | 6 | 7 | 9 | 10 | 12 | 12 | 13 | 13 |
| DGA2 | 9 | 12 | 15 | 18 | 22 | 25 | 28 | 33 |
| DGA3 | 12 | 17 | 22 | 26 | 32 | 37 | 43 | 51 |
| DGB1 | 6 | 7 | 9 | 11 | 12 | 13 | 14 | 16 |
| DGB2 | 9 | 12 | 16 | 18 | 23 | 26 | 30 | 36 |
| DGB3 | 13 | 17 | 23 | 27 | 34 | 39 | 46 | 55 |
| DGC1 | 8 | 10 | 13 | 15 | 17 | 19 | 22 | 25 |
| DGC2 | 11 | 15 | 19 | 23 | 28 | 32 | 37 | 44 |
| DGC3 | 14 | 20 | 26 | 31 | 38 | 44 | 51 | 62 |
| SSA1 | 15 | 19 | 22 | 25 | 29 | 32 | 36 | 41 |
| SSB1 | 13 | 15 | 19 | 21 | 25 | 27 | 31 | 36 |
| SSB2 | 16 | 21 | 27 | 32 | 38 | 43 | 48 | 56 |
| SSB3 | 11 | 14 | 17 | 19 | 22 | 24 | 27 | 31 |
| SSC1 | 9 | 12 | 15 | 17 | 21 | 24 | 27 | 32 |
| SSC2 | 11 | 16 | 21 | 26 | 32 | 36 | 42 | 50 |
| SSE1 | 13 | 16 | 20 | 23 | 27 | 30 | 34 | 39 |
| SSE2 | 16 | 22 | 29 | 34 | 41 | 46 | 52 | 61 |
| SSE3 | 13 | 16 | 19 | 21 | 24 | 26 | 29 | 33 |

| DETROIT, MICHIGAN SSF (%) | | | | | | | 6234 DD65 | |
|---|---|---|---|---|---|---|---|---|
| Type | LCR = 100 | 70 | 50 | 40 | 30 | 25 | 20 | 15 |
| WWA3 | 7 | 10 | 13 | 15 | 18 | 20 | 23 | 26 |
| WWB4 | 9 | 15 | 22 | 27 | 35 | 40 | 46 | 55 |
| WWC2 | 9 | 15 | 21 | 26 | 33 | 38 | 44 | 52 |
| TWA1 | 12 | 13 | 14 | 15 | 16 | 16 | 17 | 18 |
| TWA2 | 9 | 11 | 13 | 15 | 17 | 19 | 21 | 23 |
| TWA3 | 8 | 10 | 13 | 15 | 17 | 19 | 22 | 25 |
| TWA4 | 7 | 9 | 12 | 14 | 17 | 20 | 23 | 27 |
| TWB3 | 7 | 10 | 12 | 14 | 17 | 19 | 22 | 26 |
| TWD4 | 10 | 15 | 20 | 25 | 31 | 35 | 41 | 49 |
| TWE2 | 10 | 15 | 20 | 25 | 31 | 35 | 40 | 48 |
| TWF3 | 5 | 7 | 10 | 12 | 14 | 16 | 19 | 22 |
| TWJ2 | 8 | 13 | 18 | 22 | 28 | 33 | 38 | 46 |
| DGA1 | 4 | 5 | 6 | 6 | 6 | 6 | 5 | 4 |
| DGA2 | 7 | 9 | 12 | 14 | 17 | 19 | 21 | 24 |
| DGA3 | 10 | 14 | 18 | 22 | 27 | 31 | 36 | 43 |
| DGB1 | 4 | 5 | 6 | 6 | 7 | 7 | 6 | 5 |
| DGB2 | 7 | 9 | 12 | 14 | 17 | 20 | 22 | 26 |
| DGB3 | 11 | 14 | 19 | 23 | 28 | 32 | 38 | 46 |
| DGC1 | 5 | 7 | 9 | 10 | 11 | 12 | 12 | 13 |
| DGC2 | 9 | 12 | 15 | 18 | 22 | 25 | 28 | 33 |
| DGC3 | 12 | 17 | 22 | 26 | 32 | 37 | 43 | 52 |
| SSA1 | 13 | 16 | 19 | 21 | 24 | 26 | 28 | 32 |
| SSB1 | 11 | 13 | 15 | 17 | 20 | 21 | 24 | 26 |
| SSB2 | 14 | 18 | 23 | 27 | 32 | 36 | 41 | 48 |
| SSB3 | 10 | 11 | 13 | 15 | 17 | 18 | 20 | 21 |
| SSC1 | 6 | 8 | 10 | 12 | 14 | 16 | 18 | 20 |
| SSC2 | 8 | 12 | 16 | 20 | 25 | 28 | 32 | 39 |
| SSE1 | 10 | 13 | 15 | 17 | 20 | 21 | 23 | 25 |
| SSE2 | 13 | 18 | 24 | 28 | 33 | 37 | 43 | 49 |
| SSE3 | 11 | 13 | 15 | 17 | 19 | 20 | 22 | 24 |

| SAULT STE. MARIE, MICHIGAN SSF (%) | | | | | | | 9201 DD65 | |
|---|---|---|---|---|---|---|---|---|
| Type | LCR = 100 | 70 | 50 | 40 | 30 | 25 | 20 | 15 |
| WWA3 | 5 | 7 | 8 | 10 | 11 | 12 | 13 | 13 |
| WWB4 | 6 | 11 | 17 | 21 | 27 | 31 | 37 | 44 |
| WWC2 | 7 | 11 | 16 | 20 | 25 | 29 | 34 | 40 |
| TWA1 | 10 | 11 | 11 | 12 | 12 | 12 | 12 | 11 |
| TWA2 | 8 | 9 | 10 | 11 | 12 | 13 | 13 | 14 |
| TWA3 | 6 | 7 | 9 | 10 | 12 | 13 | 14 | 15 |
| TWA4 | 4 | 6 | 8 | 10 | 12 | 13 | 14 | 16 |
| TWB3 | 6 | 7 | 9 | 10 | 12 | 13 | 15 | 16 |
| TWD4 | 7 | 11 | 16 | 19 | 25 | 28 | 33 | 39 |
| TWE2 | 8 | 12 | 16 | 19 | 24 | 27 | 31 | 37 |
| TWF3 | 3 | 5 | 6 | 8 | 9 | 10 | 11 | 12 |
| TWJ2 | 6 | 9 | 14 | 17 | 22 | 25 | 30 | 36 |
| DGA1 | 2 | 3 | 3 | 2 | 2 | 1 | 0 | -4 |
| DGA2 | 5 | 7 | 9 | 11 | 13 | 14 | 15 | 17 |
| DGA3 | 9 | 12 | 15 | 18 | 23 | 26 | 30 | 35 |
| DGB1 | 2 | 2 | 3 | 2 | 2 | 1 | 0 | -2 |
| DGB2 | 5 | 7 | 9 | 11 | 13 | 14 | 16 | 18 |
| DGB3 | 9 | 12 | 16 | 19 | 24 | 27 | 32 | 38 |
| DGC1 | 4 | 5 | 5 | 6 | 6 | 6 | 5 | 3 |
| DGC2 | 7 | 9 | 12 | 14 | 17 | 19 | 22 | 25 |
| DGC3 | 10 | 14 | 19 | 22 | 28 | 31 | 36 | 43 |
| SSA1 | 12 | 13 | 15 | 17 | 19 | 20 | 21 | 22 |
| SSB1 | 9 | 11 | 12 | 14 | 15 | 16 | 17 | 18 |
| SSB2 | 12 | 16 | 20 | 23 | 27 | 30 | 34 | 39 |
| SSB3 | 8 | 9 | 11 | 11 | 12 | 13 | 13 | 12 |
| SSC1 | 4 | 5 | 7 | 7 | 8 | 8 | 8 | 7 |
| SSC2 | 6 | 9 | 12 | 14 | 18 | 20 | 23 | 27 |
| SSE1 | 8 | 9 | 11 | 12 | 13 | 13 | 13 | 11 |
| SSE2 | 10 | 14 | 18 | 22 | 26 | 29 | 32 | 36 |
| SSE3 | 10 | 11 | 12 | 13 | 14 | 15 | 15 | 14 |

## TABLE H.3 Annual Passive Heating Performance: SSF *(Continued)*

| MINNEAPOLIS, MINNESOTA SSF (%) | | | | | | | 8165 DD65 | |
|---|---|---|---|---|---|---|---|---|
| Type | LCR = 100 | 70 | 50 | 40 | 30 | 25 | 20 | 15 |
| WWA3 | 7 | 9 | 12 | 14 | 17 | 19 | 21 | 24 |
| WWB4 | 9 | 14 | 21 | 26 | 33 | 38 | 45 | 54 |
| WWC2 | 9 | 14 | 20 | 25 | 32 | 36 | 42 | 50 |
| TWA1 | 11 | 12 | 13 | 14 | 15 | 16 | 16 | 17 |
| TWA2 | 9 | 11 | 13 | 14 | 16 | 18 | 19 | 21 |
| TWA3 | 7 | 10 | 12 | 14 | 16 | 18 | 20 | 23 |
| TWA4 | 6 | 8 | 11 | 13 | 16 | 19 | 21 | 25 |
| TWB3 | 7 | 9 | 12 | 13 | 16 | 18 | 21 | 24 |
| TWD4 | 9 | 14 | 19 | 24 | 30 | 34 | 40 | 48 |
| TWE2 | 10 | 14 | 20 | 24 | 29 | 34 | 39 | 46 |
| TWF3 | 5 | 7 | 9 | 11 | 13 | 15 | 17 | 20 |
| TWJ2 | 7 | 12 | 17 | 21 | 27 | 31 | 37 | 44 |
| DGA1 | 4 | 4 | 5 | 5 | 5 | 5 | 4 | 2 |
| DGA2 | 7 | 9 | 11 | 13 | 16 | 18 | 20 | 23 |
| DGA3 | 10 | 13 | 18 | 21 | 26 | 30 | 35 | 41 |
| DGB1 | 3 | 4 | 5 | 5 | 6 | 5 | 5 | 3 |
| DGB2 | 7 | 9 | 12 | 14 | 17 | 19 | 21 | 25 |
| DGB3 | 10 | 14 | 19 | 22 | 28 | 31 | 37 | 44 |
| DGC1 | 5 | 6 | 8 | 9 | 10 | 11 | 11 | 11 |
| DGC2 | 8 | 11 | 15 | 17 | 21 | 24 | 27 | 32 |
| DGC3 | 12 | 16 | 21 | 25 | 31 | 36 | 42 | 50 |
| SSA1 | 12 | 15 | 17 | 19 | 21 | 23 | 25 | 28 |
| SSB1 | 10 | 12 | 14 | 16 | 18 | 19 | 21 | 23 |
| SSB2 | 13 | 17 | 22 | 25 | 30 | 34 | 38 | 45 |
| SSB3 | 9 | 10 | 12 | 13 | 15 | 16 | 17 | 18 |
| SSC1 | 6 | 8 | 10 | 11 | 13 | 14 | 16 | 18 |
| SSC2 | 8 | 11 | 15 | 19 | 23 | 27 | 31 | 37 |
| SSE1 | 9 | 11 | 13 | 15 | 17 | 18 | 20 | 21 |
| SSE2 | 12 | 16 | 21 | 25 | 31 | 34 | 39 | 46 |
| SSE3 | 10 | 12 | 14 | 15 | 17 | 18 | 19 | 20 |

| SAINT LOUIS, MISSOURI SSF (%) | | | | | | | 4754 DD65 | |
|---|---|---|---|---|---|---|---|---|
| Type | LCR = 100 | 70 | 50 | 40 | 30 | 25 | 20 | 15 |
| WWA3 | 15 | 20 | 26 | 30 | 36 | 41 | 47 | 55 |
| WWB4 | 19 | 27 | 37 | 44 | 53 | 59 | 67 | 76 |
| WWC2 | 19 | 27 | 36 | 42 | 51 | 57 | 65 | 74 |
| TWA1 | 16 | 18 | 21 | 23 | 27 | 29 | 32 | 37 |
| TWA2 | 15 | 19 | 23 | 26 | 31 | 35 | 39 | 46 |
| TWA3 | 14 | 18 | 23 | 27 | 33 | 37 | 42 | 50 |
| TWA4 | 13 | 18 | 23 | 27 | 33 | 38 | 43 | 51 |
| TWB3 | 13 | 17 | 22 | 25 | 31 | 35 | 40 | 47 |
| TWD4 | 18 | 25 | 33 | 39 | 48 | 53 | 61 | 70 |
| TWE2 | 18 | 26 | 33 | 39 | 48 | 53 | 61 | 70 |
| TWF3 | 11 | 15 | 20 | 24 | 29 | 33 | 38 | 46 |
| TWJ2 | 16 | 23 | 31 | 37 | 45 | 51 | 58 | 67 |
| DGA1 | 9 | 12 | 15 | 18 | 20 | 22 | 24 | 26 |
| DGA2 | 12 | 16 | 21 | 25 | 30 | 34 | 38 | 44 |
| DGA3 | 16 | 21 | 28 | 33 | 41 | 46 | 53 | 62 |
| DGB1 | 10 | 13 | 16 | 18 | 22 | 24 | 27 | 31 |
| DGB2 | 12 | 17 | 22 | 26 | 32 | 36 | 41 | 49 |
| DGB3 | 16 | 22 | 29 | 34 | 43 | 48 | 56 | 66 |
| DGC1 | 12 | 16 | 20 | 23 | 28 | 31 | 35 | 41 |
| DGC2 | 15 | 20 | 26 | 30 | 37 | 42 | 49 | 57 |
| DGC3 | 19 | 25 | 33 | 39 | 48 | 54 | 63 | 73 |
| SSA1 | 19 | 24 | 29 | 32 | 38 | 42 | 47 | 53 |
| SSB1 | 16 | 20 | 24 | 28 | 33 | 36 | 41 | 47 |
| SSB2 | 21 | 27 | 34 | 39 | 46 | 52 | 58 | 67 |
| SSB3 | 15 | 18 | 22 | 25 | 30 | 33 | 37 | 43 |
| SSC1 | 13 | 17 | 22 | 26 | 31 | 35 | 40 | 47 |
| SSC2 | 16 | 22 | 29 | 34 | 42 | 47 | 54 | 63 |
| SSE1 | 18 | 22 | 28 | 32 | 38 | 42 | 47 | 54 |
| SSE2 | 22 | 29 | 37 | 43 | 52 | 57 | 64 | 73 |
| SSE3 | 17 | 20 | 24 | 28 | 32 | 35 | 40 | 46 |

| MERIDIAN, MISSISSIPPI SSF (%) | | | | | | | 2393 DD65 | |
|---|---|---|---|---|---|---|---|---|
| Type | LCR = 200 | 100 | 70 | 50 | 40 | 30 | 25 | 20 |
| WWA3 | 15 | 28 | 36 | 45 | 52 | 61 | 67 | 74 |
| WWB4 | 16 | 34 | 46 | 58 | 66 | 75 | 81 | 87 |
| WWC2 | 17 | 34 | 45 | 56 | 64 | 74 | 80 | 86 |
| TWA1 | 16 | 23 | 28 | 33 | 38 | 44 | 49 | 54 |
| TWA2 | 15 | 24 | 31 | 38 | 44 | 52 | 57 | 64 |
| TWA3 | 14 | 25 | 32 | 41 | 47 | 55 | 61 | 68 |
| TWA4 | 13 | 24 | 32 | 41 | 47 | 56 | 62 | 69 |
| TWB3 | 13 | 22 | 29 | 37 | 43 | 52 | 57 | 64 |
| TWD4 | 16 | 31 | 41 | 52 | 60 | 70 | 76 | 82 |
| TWE2 | 17 | 32 | 42 | 53 | 60 | 70 | 76 | 83 |
| TWF3 | 11 | 21 | 29 | 37 | 43 | 51 | 57 | 64 |
| TWJ2 | 14 | 29 | 39 | 50 | 57 | 67 | 73 | 80 |
| DGA1 | 11 | 20 | 26 | 32 | 37 | 43 | 46 | 50 |
| DGA2 | 12 | 22 | 29 | 37 | 43 | 51 | 56 | 62 |
| DGA3 | 14 | 26 | 35 | 45 | 52 | 62 | 68 | 75 |
| DGB1 | 11 | 20 | 27 | 34 | 40 | 48 | 53 | 59 |
| DGB2 | 12 | 22 | 30 | 39 | 46 | 56 | 62 | 69 |
| DGB3 | 14 | 26 | 36 | 46 | 55 | 66 | 72 | 80 |
| DGC1 | 13 | 24 | 31 | 41 | 47 | 57 | 63 | 69 |
| DGC2 | 14 | 26 | 34 | 45 | 52 | 63 | 69 | 76 |
| DGC3 | 16 | 30 | 40 | 52 | 61 | 73 | 79 | 86 |
| SSA1 | 20 | 31 | 38 | 46 | 52 | 61 | 66 | 72 |
| SSB1 | 16 | 26 | 33 | 41 | 46 | 54 | 59 | 66 |
| SSB2 | 20 | 33 | 42 | 52 | 59 | 68 | 74 | 80 |
| SSB3 | 16 | 25 | 31 | 39 | 44 | 52 | 57 | 63 |
| SSC1 | 14 | 24 | 32 | 41 | 47 | 55 | 61 | 68 |
| SSC2 | 15 | 29 | 38 | 49 | 56 | 65 | 71 | 78 |
| SSE1 | 19 | 31 | 40 | 49 | 55 | 64 | 70 | 76 |
| SSE2 | 22 | 38 | 49 | 60 | 67 | 76 | 81 | 87 |
| SSE3 | 17 | 27 | 33 | 41 | 46 | 54 | 59 | 66 |

| SPRINGFIELD, MISSOURI SSF (%) | | | | | | | 4571 DD65 | |
|---|---|---|---|---|---|---|---|---|
| Type | LCR = 100 | 70 | 50 | 40 | 30 | 25 | 20 | 15 |
| WWA3 | 16 | 22 | 28 | 33 | 40 | 45 | 52 | 60 |
| WWB4 | 21 | 30 | 40 | 47 | 57 | 63 | 71 | 80 |
| WWC2 | 21 | 29 | 39 | 46 | 55 | 62 | 69 | 78 |
| TWA1 | 17 | 19 | 23 | 25 | 29 | 32 | 36 | 41 |
| TWA2 | 16 | 20 | 25 | 29 | 34 | 38 | 43 | 51 |
| TWA3 | 15 | 20 | 26 | 30 | 36 | 41 | 46 | 55 |
| TWA4 | 14 | 19 | 25 | 30 | 37 | 41 | 48 | 56 |
| TWB3 | 14 | 18 | 24 | 28 | 34 | 38 | 44 | 52 |
| TWD4 | 19 | 27 | 36 | 42 | 51 | 57 | 65 | 74 |
| TWE2 | 20 | 28 | 36 | 43 | 51 | 57 | 65 | 74 |
| TWF3 | 12 | 17 | 22 | 26 | 33 | 37 | 43 | 51 |
| TWJ2 | 18 | 25 | 33 | 40 | 49 | 55 | 62 | 72 |
| DGA1 | 11 | 14 | 18 | 20 | 24 | 26 | 28 | 30 |
| DGA2 | 13 | 18 | 23 | 27 | 33 | 37 | 42 | 48 |
| DGA3 | 17 | 23 | 30 | 36 | 44 | 50 | 57 | 66 |
| DGB1 | 11 | 14 | 18 | 21 | 25 | 28 | 32 | 37 |
| DGB2 | 14 | 18 | 24 | 28 | 35 | 39 | 46 | 54 |
| DGB3 | 17 | 24 | 31 | 37 | 46 | 52 | 60 | 71 |
| DGC1 | 13 | 18 | 22 | 26 | 32 | 36 | 41 | 47 |
| DGC2 | 10 | 21 | 28 | 33 | 41 | 46 | 54 | 63 |
| DGC3 | 20 | 27 | 35 | 42 | 52 | 59 | 67 | 78 |
| SSA1 | 21 | 26 | 31 | 36 | 42 | 46 | 51 | 59 |
| SSB1 | 17 | 22 | 27 | 30 | 36 | 40 | 45 | 52 |
| SSB2 | 22 | 29 | 37 | 42 | 50 | 55 | 62 | 71 |
| SSB3 | 16 | 20 | 25 | 28 | 33 | 37 | 41 | 48 |
| SSC1 | 14 | 19 | 25 | 29 | 35 | 39 | 45 | 53 |
| SSC2 | 18 | 24 | 32 | 38 | 46 | 51 | 58 | 68 |
| SSE1 | 19 | 25 | 31 | 36 | 42 | 47 | 52 | 60 |
| SSE2 | 24 | 32 | 41 | 47 | 56 | 62 | 69 | 77 |
| SSE3 | 18 | 22 | 27 | 30 | 36 | 39 | 44 | 51 |

**TABLE H.3 Annual Passive Heating Performance: SSF** *(Continued)*

| HELENA, MONTANA SSF (%) | | | | | | | 8197 DD65 | |
|---|---|---|---|---|---|---|---|---|
| Type | LCR = 100 | 70 | 50 | 40 | 30 | 25 | 20 | 15 |
| WWA3 | 11 | 14 | 19 | 22 | 27 | 30 | 34 | 39 |
| WWB4 | 14 | 21 | 29 | 35 | 43 | 49 | 55 | 64 |
| WWC2 | 14 | 20 | 28 | 34 | 41 | 47 | 53 | 62 |
| TWA1 | 13 | 15 | 17 | 19 | 21 | 22 | 24 | 27 |
| TWA2 | 12 | 14 | 18 | 20 | 24 | 26 | 29 | 33 |
| TWA3 | 11 | 14 | 18 | 20 | 25 | 28 | 31 | 36 |
| TWA4 | 9 | 13 | 17 | 20 | 25 | 28 | 32 | 38 |
| TWB3 | 10 | 13 | 16 | 19 | 23 | 26 | 30 | 35 |
| TWD4 | 13 | 19 | 26 | 31 | 39 | 44 | 50 | 58 |
| TWE2 | 14 | 20 | 26 | 31 | 39 | 43 | 49 | 57 |
| TWF3 | 8 | 11 | 14 | 17 | 21 | 24 | 28 | 33 |
| TWJ2 | 12 | 17 | 24 | 29 | 36 | 41 | 47 | 55 |
| DGA1 | 6 | 8 | 10 | 11 | 13 | 13 | 14 | 14 |
| DGA2 | 9 | 12 | 16 | 19 | 23 | 26 | 29 | 33 |
| DGA3 | 13 | 17 | 23 | 27 | 34 | 38 | 44 | 51 |
| DGB1 | 6 | 8 | 10 | 12 | 14 | 15 | 16 | 17 |
| DGB2 | 9 | 13 | 17 | 20 | 24 | 27 | 31 | 37 |
| DGB3 | 13 | 18 | 24 | 28 | 35 | 40 | 46 | 55 |
| DGC1 | 8 | 11 | 14 | 16 | 19 | 21 | 23 | 26 |
| DGC2 | 11 | 15 | 20 | 24 | 30 | 33 | 38 | 44 |
| DGC3 | 15 | 20 | 27 | 32 | 40 | 45 | 52 | 61 |
| SSA1 | 16 | 19 | 23 | 25 | 29 | 32 | 35 | 39 |
| SSB1 | 13 | 16 | 19 | 21 | 25 | 27 | 30 | 34 |
| SSB2 | 17 | 22 | 27 | 32 | 38 | 42 | 47 | 54 |
| SSB3 | 12 | 14 | 17 | 19 | 22 | 24 | 26 | 29 |
| SSC1 | 9 | 12 | 16 | 18 | 22 | 25 | 28 | 32 |
| SSC2 | 12 | 17 | 22 | 27 | 33 | 37 | 42 | 49 |
| SSE1 | 13 | 17 | 20 | 23 | 27 | 29 | 32 | 35 |
| SSE2 | 17 | 23 | 29 | 34 | 40 | 45 | 50 | 57 |
| SSE3 | 13 | 16 | 19 | 21 | 24 | 26 | 29 | 32 |

| ELY, NEVADA SSF (%) | | | | | | | 7818 DD65 | |
|---|---|---|---|---|---|---|---|---|
| Type | LCR = 100 | 70 | 50 | 40 | 30 | 25 | 20 | 15 |
| WWA3 | 18 | 24 | 32 | 37 | 46 | 51 | 58 | 67 |
| WWB4 | 23 | 33 | 44 | 52 | 62 | 69 | 76 | 85 |
| WWC2 | 23 | 32 | 43 | 50 | 60 | 67 | 75 | 83 |
| TWA1 | 17 | 20 | 24 | 29 | 32 | 35 | 40 | 47 |
| TWA2 | 17 | 21 | 27 | 31 | 38 | 43 | 49 | 57 |
| TWA3 | 16 | 22 | 28 | 33 | 41 | 46 | 52 | 61 |
| TWA4 | 15 | 21 | 28 | 33 | 41 | 46 | 53 | 63 |
| TWB3 | 15 | 20 | 26 | 31 | 38 | 43 | 49 | 58 |
| TWD4 | 21 | 29 | 39 | 46 | 56 | 62 | 70 | 79 |
| TWE2 | 22 | 30 | 40 | 47 | 56 | 63 | 70 | 79 |
| TWF3 | 13 | 18 | 25 | 29 | 37 | 42 | 48 | 57 |
| TWJ2 | 19 | 27 | 37 | 44 | 53 | 60 | 68 | 77 |
| DGA1 | 11 | 15 | 20 | 23 | 27 | 30 | 33 | 37 |
| DGA2 | 14 | 19 | 25 | 30 | 37 | 41 | 47 | 54 |
| DGA3 | 18 | 24 | 32 | 39 | 48 | 55 | 62 | 71 |
| DGB1 | 11 | 15 | 20 | 24 | 30 | 34 | 39 | 45 |
| DGB2 | 14 | 19 | 26 | 31 | 39 | 45 | 52 | 61 |
| DGB3 | 18 | 25 | 33 | 40 | 51 | 58 | 66 | 76 |
| DGC1 | 14 | 19 | 25 | 30 | 38 | 43 | 49 | 57 |
| DGC2 | 17 | 23 | 31 | 37 | 46 | 53 | 60 | 70 |
| DGC3 | 21 | 29 | 38 | 46 | 57 | 65 | 74 | 83 |
| SSA1 | 23 | 29 | 36 | 40 | 48 | 52 | 58 | 66 |
| SSB1 | 19 | 24 | 30 | 34 | 41 | 46 | 51 | 59 |
| SSB2 | 25 | 32 | 41 | 47 | 55 | 61 | 68 | 76 |
| SSB3 | 18 | 22 | 28 | 32 | 38 | 42 | 48 | 55 |
| SSC1 | 16 | 21 | 27 | 32 | 40 | 45 | 51 | 60 |
| SSC2 | 19 | 27 | 35 | 42 | 51 | 57 | 64 | 74 |
| SSE1 | 22 | 28 | 36 | 41 | 49 | 54 | 60 | 68 |
| SSE2 | 27 | 37 | 46 | 53 | 62 | 68 | 75 | 83 |
| SSE3 | 20 | 24 | 30 | 34 | 41 | 45 | 51 | 58 |

| OMAHA, NEBRASKA SSF (%) | | | | | | | 6606 DD65 | |
|---|---|---|---|---|---|---|---|---|
| Type | LCR = 100 | 70 | 50 | 40 | 30 | 25 | 20 | 15 |
| WWA3 | 12 | 16 | 21 | 24 | 30 | 33 | 38 | 45 |
| WWB4 | 15 | 22 | 31 | 37 | 47 | 53 | 60 | 70 |
| WWC2 | 15 | 22 | 30 | 36 | 45 | 51 | 58 | 68 |
| TWA1 | 14 | 16 | 18 | 20 | 22 | 24 | 27 | 30 |
| TWA2 | 13 | 15 | 19 | 22 | 26 | 29 | 32 | 38 |
| TWA3 | 11 | 15 | 19 | 22 | 27 | 30 | 35 | 41 |
| TWA4 | 10 | 14 | 19 | 22 | 27 | 31 | 36 | 43 |
| TWB3 | 11 | 14 | 18 | 21 | 25 | 29 | 33 | 40 |
| TWD4 | 14 | 21 | 28 | 34 | 42 | 47 | 54 | 64 |
| TWE2 | 15 | 21 | 28 | 34 | 42 | 47 | 54 | 63 |
| TWF3 | 9 | 12 | 16 | 19 | 24 | 27 | 31 | 37 |
| TWJ2 | 13 | 19 | 26 | 31 | 39 | 49 | 51 | 61 |
| DGA1 | 7 | 9 | 11 | 13 | 15 | 16 | 17 | 17 |
| DGA2 | 10 | 13 | 17 | 20 | 25 | 28 | 32 | 37 |
| DGA3 | 13 | 18 | 24 | 28 | 35 | 40 | 47 | 55 |
| DGB1 | 7 | 9 | 11 | 13 | 15 | 17 | 19 | 21 |
| DGB2 | 10 | 14 | 18 | 21 | 26 | 29 | 34 | 40 |
| DGB3 | 14 | 19 | 25 | 29 | 37 | 42 | 49 | 59 |
| DGC1 | 9 | 12 | 15 | 18 | 21 | 23 | 26 | 30 |
| DGC2 | 12 | 16 | 21 | 25 | 31 | 35 | 41 | 49 |
| DGC3 | 16 | 21 | 28 | 33 | 42 | 48 | 56 | 66 |
| SSA1 | 16 | 20 | 24 | 27 | 32 | 35 | 39 | 45 |
| SSB1 | 14 | 17 | 20 | 23 | 27 | 30 | 34 | 39 |
| SSB2 | 18 | 23 | 29 | 34 | 40 | 45 | 51 | 59 |
| SSB3 | 12 | 15 | 18 | 21 | 24 | 26 | 29 | 34 |
| SSC1 | 10 | 13 | 17 | 20 | 25 | 28 | 32 | 38 |
| SSC2 | 13 | 18 | 29 | 29 | 36 | 40 | 47 | 55 |
| SSE1 | 14 | 18 | 22 | 25 | 30 | 33 | 37 | 40 |
| SSE2 | 18 | 24 | 31 | 37 | 44 | 49 | 56 | 65 |
| SSE3 | 14 | 17 | 20 | 23 | 26 | 29 | 32 | 37 |

| RENO, NEVADA SSF (%) | | | | | | | 6027 DD65 | |
|---|---|---|---|---|---|---|---|---|
| Type | LCR = 100 | 70 | 50 | 40 | 30 | 25 | 20 | 15 |
| WWA3 | 25 | 33 | 41 | 48 | 57 | 63 | 70 | 78 |
| WWB4 | 31 | 42 | 54 | 62 | 72 | 78 | 84 | 91 |
| WWC2 | 31 | 41 | 53 | 60 | 70 | 76 | 83 | 90 |
| TWA1 | 21 | 25 | 30 | 35 | 41 | 45 | 50 | 58 |
| TWA2 | 22 | 28 | 35 | 40 | 48 | 53 | 60 | 69 |
| TWA3 | 22 | 29 | 37 | 43 | 51 | 57 | 64 | 73 |
| TWA4 | 21 | 29 | 37 | 43 | 52 | 58 | 65 | 74 |
| TWB3 | 20 | 26 | 34 | 40 | 48 | 53 | 60 | 69 |
| TWD4 | 28 | 38 | 48 | 56 | 66 | 72 | 79 | 87 |
| TWE2 | 29 | 39 | 49 | 57 | 66 | 72 | 79 | 87 |
| TWF3 | 19 | 25 | 33 | 39 | 47 | 53 | 60 | 69 |
| TWJ2 | 26 | 36 | 46 | 54 | 64 | 70 | 77 | 85 |
| DGA1 | 17 | 23 | 29 | 33 | 38 | 42 | 46 | 50 |
| DGA2 | 19 | 26 | 34 | 39 | 47 | 52 | 58 | 65 |
| DGA3 | 23 | 32 | 41 | 49 | 58 | 64 | 71 | 79 |
| DGB1 | 17 | 23 | 30 | 36 | 43 | 48 | 53 | 60 |
| DGB2 | 20 | 27 | 35 | 42 | 51 | 57 | 64 | 72 |
| DGB3 | 24 | 32 | 43 | 51 | 62 | 69 | 76 | 84 |
| DGC1 | 21 | 28 | 36 | 43 | 52 | 57 | 64 | 71 |
| DGC2 | 23 | 31 | 41 | 48 | 59 | 65 | 72 | 80 |
| DGC3 | 27 | 37 | 48 | 57 | 69 | 75 | 83 | 89 |
| SSA1 | 29 | 36 | 43 | 49 | 56 | 61 | 67 | 75 |
| SSB1 | 24 | 30 | 37 | 43 | 50 | 55 | 61 | 69 |
| SSB2 | 31 | 39 | 48 | 55 | 64 | 69 | 76 | 83 |
| SSB3 | 23 | 29 | 35 | 40 | 47 | 52 | 58 | 65 |
| SSC1 | 22 | 29 | 37 | 43 | 51 | 57 | 64 | 73 |
| SSC2 | 26 | 35 | 45 | 52 | 61 | 67 | 75 | 83 |
| SSE1 | 29 | 36 | 45 | 51 | 59 | 64 | 70 | 78 |
| SSE2 | 35 | 45 | 55 | 62 | 71 | 76 | 82 | 89 |
| SSE3 | 25 | 31 | 38 | 43 | 50 | 55 | 61 | 68 |

## TABLE H.3 Annual Passive Heating Performance: SSF *(Continued)*

**LAS VEGAS, NEVADA SSF (%)**  2602 DD65

| Type | LCR = 200 | 100 | 70 | 50 | 40 | 30 | 25 | 20 |
|------|-----------|-----|----|----|----|----|----|----|
| WWA3 | 24 | 41 | 52 | 63 | 71 | 80 | 85 | 90 |
| WWB4 | 28 | 50 | 63 | 75 | 82 | 89 | 93 | 96 |
| WWC2 | 28 | 49 | 62 | 74 | 81 | 88 | 92 | 96 |
| TWA1 | 21 | 31 | 38 | 46 | 52 | 61 | 66 | 73 |
| TWA2 | 22 | 35 | 44 | 54 | 61 | 70 | 76 | 82 |
| TWA3 | 22 | 37 | 47 | 57 | 64 | 74 | 79 | 85 |
| TWA4 | 21 | 36 | 47 | 58 | 64 | 74 | 80 | 86 |
| TWB3 | 19 | 33 | 43 | 53 | 60 | 69 | 75 | 82 |
| TWD4 | 25 | 45 | 57 | 69 | 77 | 85 | 89 | 93 |
| TWE2 | 27 | 46 | 58 | 70 | 77 | 85 | 90 | 94 |
| TWF3 | 18 | 33 | 43 | 53 | 60 | 70 | 76 | 82 |
| TWJ2 | 24 | 43 | 55 | 67 | 75 | 83 | 88 | 92 |
| DGA1 | 18 | 32 | 40 | 49 | 55 | 63 | 67 | 72 |
| DGA2 | 18 | 33 | 43 | 53 | 60 | 69 | 74 | 80 |
| DGA3 | 21 | 38 | 50 | 62 | 70 | 78 | 83 | 88 |
| DGB1 | 18 | 33 | 44 | 55 | 62 | 71 | 75 | 80 |
| DGB2 | 19 | 35 | 46 | 58 | 67 | 76 | 81 | 86 |
| DGB3 | 21 | 39 | 52 | 65 | 74 | 83 | 87 | 92 |
| DGC1 | 21 | 38 | 51 | 63 | 71 | 79 | 84 | 88 |
| DGC2 | 21 | 39 | 52 | 65 | 74 | 83 | 87 | 91 |
| DGC3 | 24 | 44 | 58 | 72 | 80 | 88 | 92 | 95 |
| SSA1 | 26 | 40 | 50 | 59 | 66 | 74 | 79 | 85 |
| SSB1 | 22 | 35 | 44 | 53 | 60 | 69 | 74 | 80 |
| SSB2 | 27 | 44 | 54 | 65 | 72 | 81 | 85 | 90 |
| SSB3 | 21 | 34 | 42 | 51 | 57 | 66 | 71 | 78 |
| SSC1 | 22 | 37 | 47 | 58 | 65 | 75 | 80 | 86 |
| SSC2 | 24 | 43 | 55 | 66 | 73 | 82 | 87 | 92 |
| SSE1 | 27 | 43 | 53 | 64 | 71 | 79 | 84 | 89 |
| SSE2 | 31 | 51 | 63 | 74 | 80 | 88 | 91 | 95 |
| SSE3 | 23 | 36 | 45 | 54 | 60 | 69 | 74 | 80 |

**LOS ALAMOS, NEW MEXICO SSF (%)**  6437 DD65

| Type | LCR = 100 | 70 | 50 | 40 | 30 | 25 | 20 | 15 |
|------|-----------|----|----|----|----|----|----|----|
| WWA3 | 20 | 27 | 35 | 41 | 49 | 55 | 62 | 72 |
| WWB4 | 25 | 36 | 47 | 55 | 66 | 72 | 80 | 88 |
| WWC2 | 25 | 35 | 46 | 54 | 64 | 71 | 78 | 86 |
| TWA1 | 18 | 22 | 26 | 29 | 34 | 38 | 43 | 50 |
| TWA2 | 18 | 23 | 29 | 34 | 41 | 46 | 52 | 61 |
| TWA3 | 18 | 24 | 31 | 36 | 44 | 49 | 56 | 66 |
| TWA4 | 17 | 23 | 31 | 36 | 45 | 50 | 57 | 67 |
| TWB3 | 16 | 22 | 28 | 33 | 41 | 46 | 53 | 62 |
| TWD4 | 23 | 32 | 42 | 49 | 59 | 66 | 74 | 83 |
| TWE2 | 24 | 33 | 43 | 50 | 60 | 66 | 74 | 83 |
| TWF3 | 15 | 20 | 27 | 32 | 40 | 45 | 52 | 62 |
| TWJ2 | 21 | 30 | 40 | 47 | 57 | 63 | 71 | 80 |
| DGA1 | 13 | 17 | 22 | 26 | 31 | 34 | 37 | 41 |
| DGA2 | 15 | 21 | 28 | 33 | 40 | 45 | 50 | 58 |
| DGA3 | 19 | 26 | 35 | 42 | 52 | 58 | 65 | 74 |
| DGB1 | 13 | 18 | 23 | 27 | 34 | 38 | 44 | 50 |
| DGB2 | 16 | 21 | 28 | 34 | 43 | 49 | 56 | 65 |
| DGB3 | 20 | 27 | 36 | 43 | 54 | 62 | 70 | 80 |
| DGC1 | 16 | 22 | 28 | 34 | 42 | 48 | 55 | 63 |
| DGC2 | 18 | 25 | 33 | 40 | 50 | 57 | 65 | 74 |
| DGC3 | 22 | 31 | 41 | 49 | 61 | 69 | 77 | 86 |
| SSA1 | 25 | 31 | 38 | 44 | 51 | 56 | 63 | 71 |
| SSB1 | 21 | 26 | 32 | 37 | 44 | 49 | 56 | 64 |
| SSB2 | 27 | 34 | 43 | 50 | 59 | 65 | 72 | 80 |
| SSB3 | 19 | 24 | 30 | 35 | 42 | 46 | 52 | 60 |
| SSC1 | 17 | 23 | 30 | 36 | 43 | 49 | 56 | 65 |
| SSC2 | 21 | 29 | 38 | 45 | 54 | 61 | 68 | 78 |
| SSE1 | 24 | 31 | 39 | 45 | 53 | 59 | 65 | 74 |
| SSE2 | 30 | 39 | 50 | 57 | 67 | 73 | 79 | 87 |
| SSE3 | 21 | 27 | 33 | 37 | 44 | 49 | 55 | 63 |

**ALBUQUERQUE, N. MEXICO SSF (%)**  4293 DD65

| Type | LCR = 100 | 70 | 50 | 40 | 30 | 25 | 20 | 15 |
|------|-----------|----|----|----|----|----|----|----|
| WWA3 | 28 | 37 | 46 | 54 | 63 | 69 | 76 | 85 |
| WWB4 | 35 | 47 | 59 | 67 | 77 | 83 | 89 | 94 |
| WWC2 | 34 | 46 | 58 | 66 | 76 | 82 | 88 | 94 |
| TWA1 | 22 | 27 | 33 | 38 | 45 | 50 | 56 | 64 |
| TWA2 | 24 | 31 | 39 | 45 | 53 | 59 | 66 | 75 |
| TWA3 | 25 | 32 | 41 | 48 | 57 | 63 | 70 | 79 |
| TWA4 | 24 | 32 | 41 | 48 | 58 | 64 | 71 | 80 |
| TWB3 | 22 | 29 | 38 | 44 | 53 | 59 | 66 | 75 |
| TWD4 | 31 | 42 | 53 | 61 | 71 | 77 | 84 | 91 |
| TWE2 | 32 | 43 | 54 | 62 | 72 | 78 | 84 | 91 |
| TWF3 | 21 | 29 | 37 | 44 | 53 | 59 | 66 | 76 |
| TWJ2 | 29 | 40 | 51 | 59 | 69 | 75 | 82 | 90 |
| DGA1 | 19 | 26 | 33 | 37 | 44 | 48 | 52 | 58 |
| DGA2 | 21 | 29 | 38 | 44 | 52 | 57 | 64 | 71 |
| DGA3 | 25 | 35 | 46 | 54 | 64 | 70 | 76 | 84 |
| DGB1 | 20 | 27 | 35 | 41 | 50 | 55 | 61 | 68 |
| DGB2 | 22 | 30 | 40 | 47 | 58 | 64 | 71 | 79 |
| DGB3 | 26 | 36 | 47 | 56 | 68 | 75 | 82 | 89 |
| DGC1 | 23 | 32 | 42 | 49 | 60 | 66 | 72 | 79 |
| DGC2 | 25 | 34 | 46 | 54 | 65 | 72 | 79 | 86 |
| DGC3 | 30 | 40 | 53 | 63 | 75 | 82 | 88 | 93 |
| SSA1 | 30 | 38 | 46 | 52 | 61 | 66 | 73 | 81 |
| SSB1 | 25 | 32 | 40 | 46 | 54 | 60 | 67 | 75 |
| SSB2 | 32 | 42 | 52 | 59 | 68 | 74 | 81 | 88 |
| SSB3 | 24 | 30 | 38 | 43 | 51 | 57 | 63 | 72 |
| SSC1 | 24 | 32 | 41 | 48 | 57 | 63 | 71 | 79 |
| SSC2 | 29 | 39 | 50 | 57 | 67 | 73 | 80 | 88 |
| SSE1 | 31 | 39 | 49 | 55 | 64 | 70 | 77 | 85 |
| SSE2 | 38 | 49 | 60 | 67 | 76 | 82 | 88 | 93 |
| SSE3 | 26 | 33 | 40 | 46 | 54 | 60 | 66 | 75 |

**ALBANY, NEW YORK SSF (%)**  6891 DD65

| Type | LCR = 100 | 70 | 50 | 40 | 30 | 25 | 20 | 15 |
|------|-----------|----|----|----|----|----|----|----|
| WWA3 | 7 | 9 | 12 | 14 | 17 | 18 | 21 | 24 |
| WWB4 | 8 | 14 | 21 | 26 | 33 | 38 | 45 | 54 |
| WWC2 | 9 | 14 | 20 | 25 | 31 | 36 | 42 | 51 |
| TWA1 | 11 | 12 | 13 | 14 | 15 | 15 | 16 | 17 |
| TWA2 | 9 | 10 | 12 | 14 | 16 | 17 | 19 | 21 |
| TWA3 | 7 | 9 | 12 | 13 | 16 | 18 | 20 | 23 |
| TWA4 | 6 | 8 | 11 | 13 | 16 | 18 | 21 | 25 |
| TWB3 | 7 | 9 | 11 | 13 | 16 | 18 | 20 | 24 |
| TWD4 | 9 | 13 | 19 | 23 | 30 | 34 | 40 | 48 |
| TWE2 | 9 | 14 | 19 | 23 | 29 | 33 | 39 | 46 |
| TWF3 | 5 | 7 | 9 | 11 | 13 | 15 | 17 | 20 |
| TWJ2 | 7 | 12 | 17 | 21 | 27 | 31 | 37 | 45 |
| DGA1 | 3 | 4 | 5 | 5 | 5 | 4 | 4 | 2 |
| DGA2 | 6 | 8 | 11 | 13 | 15 | 17 | 20 | 23 |
| DGA3 | 10 | 13 | 17 | 21 | 26 | 29 | 34 | 41 |
| DGB1 | 3 | 4 | 4 | 5 | 5 | 5 | 4 | 3 |
| DGB2 | 6 | 9 | 11 | 13 | 16 | 18 | 21 | 24 |
| DGB3 | 10 | 14 | 18 | 22 | 27 | 31 | 36 | 44 |
| DGC1 | 5 | 6 | 7 | 8 | 10 | 10 | 10 | 10 |
| DGC2 | 8 | 11 | 14 | 17 | 20 | 23 | 27 | 32 |
| DGC3 | 12 | 16 | 21 | 25 | 31 | 35 | 41 | 50 |
| SSA1 | 12 | 15 | 17 | 19 | 22 | 24 | 27 | 30 |
| SSB1 | 10 | 12 | 14 | 16 | 18 | 20 | 22 | 25 |
| SSB2 | 13 | 17 | 22 | 26 | 31 | 35 | 39 | 46 |
| SSB3 | 9 | 11 | 12 | 14 | 15 | 16 | 18 | 19 |
| SSC1 | 6 | 7 | 9 | 11 | 13 | 14 | 16 | 18 |
| SSC2 | 7 | 11 | 15 | 18 | 23 | 26 | 31 | 37 |
| SSE1 | 9 | 11 | 14 | 16 | 18 | 20 | 21 | 23 |
| SSE2 | 12 | 17 | 22 | 26 | 32 | 36 | 41 | 48 |
| SSE3 | 11 | 12 | 14 | 16 | 17 | 19 | 20 | 22 |

**TABLE H.3 Annual Passive Heating Performance: SSF** *(Continued)*

| BUFFALO, NEW YORK SSF (%) | | | | | | 6931 DD65 | | |
|---|---|---|---|---|---|---|---|---|
| Type | LCR = 100 | 70 | 50 | 40 | 30 | 25 | 20 | 15 |
| WWA3 | 5 | 6 | 8 | 9 | 11 | 11 | 12 | 13 |
| WWB4 | 6 | 11 | 16 | 21 | 27 | 31 | 37 | 44 |
| WWC2 | 6 | 11 | 16 | 19 | 25 | 29 | 34 | 41 |
| TWA1 | 10 | 11 | 11 | 12 | 12 | 12 | 11 | 11 |
| TWA2 | 8 | 9 | 10 | 11 | 12 | 13 | 13 | 13 |
| TWA3 | 6 | 7 | 9 | 10 | 11 | 12 | 13 | 14 |
| TWA4 | 4 | 6 | 8 | 9 | 11 | 12 | 14 | 16 |
| TWB3 | 6 | 7 | 9 | 10 | 12 | 13 | 14 | 16 |
| TWD4 | 7 | 11 | 15 | 19 | 24 | 28 | 32 | 39 |
| TWE2 | 8 | 11 | 15 | 19 | 23 | 27 | 31 | 37 |
| TWF3 | 3 | 5 | 6 | 7 | 9 | 10 | 11 | 12 |
| TWJ2 | 5 | 9 | 13 | 17 | 22 | 25 | 29 | 36 |
| DGA1 | 2 | 2 | 2 | 2 | 1 | 0 | −1 | −4 |
| DGA2 | 5 | 7 | 9 | 10 | 12 | 13 | 14 | 16 |
| DGA3 | 9 | 12 | 15 | 18 | 22 | 25 | 29 | 34 |
| DGB1 | 2 | 2 | 2 | 2 | 1 | 0 | 0 | −4 |
| DGB2 | 5 | 7 | 9 | 10 | 12 | 14 | 15 | 17 |
| DGB3 | 9 | 12 | 16 | 19 | 23 | 27 | 31 | 37 |
| DGC1 | 4 | 4 | 5 | 5 | 5 | 5 | 4 | 2 |
| DGC2 | 7 | 9 | 12 | 14 | 16 | 18 | 21 | 24 |
| DGC3 | 10 | 14 | 18 | 22 | 27 | 31 | 35 | 42 |
| SSA1 | 11 | 13 | 15 | 17 | 19 | 20 | 21 | 22 |
| SSB1 | 9 | 11 | 12 | 13 | 15 | 16 | 17 | 18 |
| SSB2 | 12 | 15 | 19 | 23 | 27 | 30 | 34 | 40 |
| SSB3 | 8 | 9 | 11 | 11 | 12 | 13 | 13 | 13 |
| SSC1 | 4 | 5 | 6 | 7 | 8 | 8 | 8 | 7 |
| SSC2 | 6 | 8 | 12 | 14 | 17 | 20 | 23 | 27 |
| SSE1 | 8 | 9 | 11 | 12 | 13 | 13 | 14 | 13 |
| SSE2 | 10 | 14 | 18 | 22 | 26 | 29 | 33 | 38 |
| SSE3 | 10 | 11 | 12 | 13 | 14 | 15 | 15 | 15 |

| SYRACUSE, NEW YORK SSF (%) | | | | | | 6681 DD65 | | |
|---|---|---|---|---|---|---|---|---|
| Type | LCR = 100 | 70 | 50 | 40 | 30 | 25 | 20 | 15 |
| WWA3 | 5 | 7 | 9 | 10 | 12 | 13 | 14 | 15 |
| WWB4 | 6 | 12 | 17 | 22 | 28 | 33 | 38 | 46 |
| WWC2 | 7 | 11 | 17 | 21 | 26 | 30 | 36 | 43 |
| TWA1 | 11 | 11 | 12 | 12 | 12 | 13 | 12 | 12 |
| TWA2 | 8 | 9 | 10 | 11 | 13 | 13 | 14 | 15 |
| TWA3 | 6 | 8 | 9 | 11 | 12 | 14 | 15 | 16 |
| TWA4 | 5 | 7 | 9 | 10 | 12 | 14 | 15 | 18 |
| TWB3 | 6 | 7 | 9 | 11 | 13 | 14 | 16 | 18 |
| TWD4 | 7 | 11 | 16 | 20 | 25 | 29 | 34 | 41 |
| TWE2 | 8 | 12 | 16 | 20 | 25 | 28 | 33 | 39 |
| TWF3 | 4 | 5 | 7 | 8 | 10 | 11 | 12 | 13 |
| TWJ2 | 6 | 10 | 14 | 18 | 23 | 26 | 31 | 38 |
| DGA1 | 2 | 3 | 3 | 3 | 2 | 1 | 0 | −3 |
| DGA2 | 6 | 7 | 9 | 11 | 13 | 14 | 16 | 17 |
| DGA3 | 9 | 12 | 16 | 18 | 23 | 26 | 30 | 36 |
| DGB1 | 2 | 3 | 3 | 3 | 2 | 1 | 0 | −2 |
| DGB2 | 6 | 7 | 9 | 11 | 13 | 15 | 16 | 19 |
| DGB3 | 9 | 12 | 16 | 19 | 24 | 28 | 32 | 38 |
| DGC1 | 4 | 5 | 5 | 6 | 6 | 6 | 5 | −4 |
| DGC2 | 7 | 9 | 12 | 14 | 17 | 19 | 22 | 25 |
| DGC3 | 11 | 14 | 19 | 22 | 28 | 31 | 36 | 44 |
| SSA1 | 12 | 14 | 16 | 17 | 19 | 20 | 22 | 24 |
| SSB1 | 9 | 11 | 13 | 14 | 16 | 17 | 18 | 19 |
| SSB2 | 12 | 16 | 20 | 23 | 28 | 31 | 35 | 41 |
| SSB3 | 8 | 10 | 11 | 12 | 13 | 13 | 14 | 14 |
| SSC1 | 4 | 6 | 7 | 8 | 9 | 9 | 9 | 9 |
| SSC2 | 6 | 9 | 12 | 15 | 19 | 21 | 24 | 29 |
| SSE1 | 8 | 10 | 11 | 13 | 14 | 14 | 15 | 15 |
| SSE2 | 10 | 14 | 19 | 22 | 27 | 30 | 34 | 40 |
| SSE3 | 10 | 11 | 12 | 13 | 15 | 15 | 16 | 16 |

| NEW YORK, NEW YORK SSF (%) | | | | | | 4851 DD65 | | |
|---|---|---|---|---|---|---|---|---|
| Type | LCR = 100 | 70 | 50 | 40 | 30 | 25 | 20 | 15 |
| WWA3 | 10 | 14 | 19 | 22 | 27 | 30 | 35 | 41 |
| WWB4 | 13 | 21 | 29 | 35 | 44 | 50 | 57 | 67 |
| WWC2 | 13 | 20 | 28 | 34 | 42 | 48 | 55 | 64 |
| TWA1 | 13 | 15 | 17 | 18 | 21 | 22 | 24 | 27 |
| TWA2 | 11 | 14 | 17 | 20 | 23 | 26 | 30 | 35 |
| TWA3 | 10 | 14 | 17 | 20 | 25 | 28 | 32 | 38 |
| TWA4 | 9 | 13 | 17 | 20 | 25 | 28 | 33 | 39 |
| TWB3 | 9 | 13 | 16 | 19 | 23 | 26 | 30 | 36 |
| TWD4 | 13 | 19 | 26 | 31 | 39 | 44 | 51 | 60 |
| TWE2 | 14 | 20 | 26 | 32 | 39 | 44 | 51 | 60 |
| TWF3 | 7 | 11 | 14 | 17 | 21 | 24 | 28 | 34 |
| TWJ2 | 11 | 17 | 24 | 29 | 36 | 42 | 48 | 57 |
| DGA1 | 6 | 8 | 10 | 11 | 12 | 13 | 14 | 14 |
| DGA2 | 9 | 12 | 16 | 18 | 23 | 25 | 29 | 34 |
| DGA3 | 12 | 17 | 22 | 27 | 33 | 38 | 44 | 52 |
| DGB1 | 6 | 8 | 10 | 11 | 13 | 14 | 16 | 17 |
| DGB2 | 9 | 12 | 16 | 19 | 23 | 27 | 31 | 37 |
| DGB3 | 13 | 17 | 23 | 28 | 35 | 40 | 47 | 56 |
| DGC1 | 8 | 11 | 13 | 15 | 18 | 20 | 23 | 26 |
| DGC2 | 11 | 15 | 20 | 23 | 29 | 33 | 38 | 45 |
| DGC3 | 15 | 20 | 26 | 31 | 39 | 45 | 53 | 63 |
| SSA1 | 16 | 19 | 23 | 26 | 30 | 33 | 37 | 43 |
| SSB1 | 13 | 16 | 19 | 22 | 26 | 28 | 32 | 37 |
| SSB2 | 17 | 22 | 28 | 33 | 39 | 44 | 50 | 58 |
| SSB3 | 12 | 14 | 17 | 20 | 23 | 25 | 28 | 32 |
| SSC1 | 9 | 12 | 15 | 18 | 22 | 25 | 29 | 34 |
| SSC2 | 11 | 16 | 22 | 27 | 33 | 37 | 43 | 52 |
| SSE1 | 13 | 17 | 21 | 24 | 28 | 31 | 35 | 41 |
| SSE2 | 17 | 23 | 30 | 35 | 43 | 48 | 54 | 62 |
| SSE3 | 13 | 16 | 19 | 22 | 25 | 27 | 31 | 35 |

| CHARLOTTE, NORTH CAROLINA SSF (%) | | | | | | | 3226 DD65 | |
|---|---|---|---|---|---|---|---|---|
| Type | LCR = 200 | 100 | 70 | 50 | 40 | 30 | 25 | 20 |
| WWA3 | 13 | 23 | 31 | 39 | 45 | 54 | 59 | 66 |
| WWB4 | 13 | 29 | 40 | 51 | 59 | 69 | 75 | 82 |
| WWC2 | 13 | 29 | 39 | 50 | 58 | 67 | 74 | 81 |
| TWA1 | 15 | 20 | 24 | 29 | 33 | 38 | 42 | 47 |
| TWA2 | 13 | 21 | 27 | 33 | 38 | 45 | 50 | 57 |
| TWA3 | 12 | 21 | 27 | 35 | 40 | 48 | 54 | 60 |
| TWA4 | 11 | 20 | 27 | 35 | 41 | 49 | 54 | 61 |
| TWB3 | 11 | 19 | 25 | 32 | 37 | 45 | 50 | 57 |
| TWD4 | 13 | 26 | 36 | 46 | 53 | 63 | 69 | 76 |
| TWE2 | 14 | 27 | 37 | 47 | 54 | 63 | 69 | 77 |
| TWF3 | 9 | 18 | 24 | 31 | 36 | 44 | 50 | 56 |
| TWJ2 | 11 | 24 | 34 | 44 | 51 | 61 | 67 | 74 |
| DGA1 | 9 | 16 | 21 | 26 | 30 | 35 | 39 | 42 |
| DGA2 | 10 | 18 | 24 | 31 | 37 | 44 | 49 | 55 |
| DGA3 | 12 | 22 | 30 | 39 | 46 | 55 | 61 | 68 |
| DGB1 | 9 | 16 | 21 | 28 | 32 | 39 | 44 | 49 |
| DGB2 | 10 | 19 | 25 | 33 | 39 | 48 | 54 | 61 |
| DGB3 | 12 | 23 | 31 | 40 | 48 | 58 | 65 | 73 |
| DGC1 | 11 | 20 | 26 | 33 | 39 | 47 | 53 | 59 |
| DGC2 | 12 | 22 | 29 | 38 | 45 | 55 | 61 | 69 |
| DGC3 | 14 | 26 | 35 | 45 | 53 | 65 | 72 | 80 |
| SSA1 | 17 | 26 | 33 | 40 | 45 | 53 | 58 | 64 |
| SSB1 | 14 | 22 | 28 | 35 | 40 | 47 | 52 | 58 |
| SSB2 | 17 | 29 | 37 | 46 | 52 | 61 | 67 | 74 |
| SSB3 | 14 | 21 | 26 | 33 | 37 | 44 | 49 | 55 |
| SSC1 | 11 | 20 | 27 | 34 | 40 | 48 | 53 | 60 |
| SSC2 | 12 | 25 | 33 | 42 | 49 | 58 | 64 | 72 |
| SSE1 | 16 | 26 | 34 | 42 | 47 | 55 | 61 | 68 |
| SSE2 | 18 | 32 | 42 | 52 | 59 | 69 | 74 | 81 |
| SSE3 | 15 | 23 | 29 | 35 | 40 | 47 | 51 | 57 |

**APPENDICES**

**TABLE H.3 Annual Passive Heating Performance: SSF** *(Continued)*

| RALEIGH-DURHAM, N. CAROLINA SSF (%) | | | | | | | 3520 DD65 | |
|---|---|---|---|---|---|---|---|---|
| Type | LCR = 200 | 100 | 70 | 50 | 40 | 30 | 25 | 20 |
| WWA3 | 11 | 21 | 28 | 36 | 41 | 50 | 55 | 62 |
| WWB4 | 11 | 26 | 37 | 48 | 55 | 66 | 72 | 79 |
| WWC2 | 12 | 26 | 36 | 46 | 54 | 64 | 70 | 77 |
| TWA1 | 14 | 19 | 22 | 27 | 30 | 35 | 39 | 44 |
| TWA2 | 12 | 19 | 24 | 30 | 35 | 42 | 47 | 53 |
| TWA3 | 11 | 19 | 25 | 32 | 37 | 45 | 50 | 56 |
| TWA4 | 9 | 18 | 24 | 32 | 37 | 45 | 51 | 57 |
| TWB3 | 10 | 17 | 23 | 29 | 34 | 41 | 46 | 53 |
| TWD4 | 11 | 24 | 33 | 43 | 50 | 60 | 66 | 73 |
| TWE2 | 12 | 25 | 34 | 43 | 50 | 60 | 66 | 73 |
| TWF3 | 8 | 16 | 21 | 28 | 33 | 41 | 46 | 52 |
| TWJ2 | 10 | 22 | 31 | 40 | 47 | 57 | 63 | 71 |
| DGA1 | 8 | 14 | 19 | 23 | 27 | 32 | 35 | 38 |
| DGA2 | 9 | 16 | 22 | 29 | 34 | 41 | 45 | 51 |
| DGA3 | 11 | 20 | 27 | 36 | 43 | 52 | 58 | 65 |
| DGB1 | 8 | 14 | 19 | 24 | 29 | 35 | 39 | 44 |
| DGB2 | 9 | 17 | 23 | 30 | 35 | 44 | 49 | 56 |
| DGB3 | 11 | 21 | 28 | 37 | 44 | 55 | 61 | 70 |
| DGC1 | 9 | 17 | 23 | 30 | 35 | 42 | 48 | 54 |
| DGC2 | 10 | 20 | 27 | 35 | 41 | 50 | 57 | 64 |
| DGC3 | 13 | 24 | 32 | 42 | 50 | 61 | 68 | 77 |
| SSA1 | 16 | 24 | 30 | 37 | 42 | 50 | 54 | 61 |
| SSB1 | 13 | 21 | 26 | 32 | 37 | 44 | 48 | 54 |
| SSB2 | 15 | 26 | 34 | 43 | 49 | 58 | 63 | 70 |
| SSB3 | 12 | 19 | 24 | 30 | 34 | 41 | 45 | 51 |
| SSC1 | 10 | 18 | 24 | 31 | 36 | 44 | 49 | 56 |
| SSC2 | 11 | 22 | 30 | 39 | 46 | 55 | 61 | 68 |
| SSE1 | 14 | 24 | 31 | 38 | 44 | 52 | 57 | 63 |
| SSE2 | 16 | 30 | 39 | 49 | 56 | 65 | 71 | 77 |
| SSE3 | 14 | 21 | 26 | 32 | 37 | 43 | 48 | 54 |

| FARGO, NORTH DAKOTA SSF (%) | | | | | | | 9278 DD65 | |
|---|---|---|---|---|---|---|---|---|
| Type | LCR = 100 | 70 | 50 | 40 | 30 | 25 | 20 | 15 |
| WWA3 | 6 | 8 | 11 | 12 | 15 | 16 | 18 | 20 |
| WWB4 | 8 | 13 | 20 | 24 | 31 | 36 | 42 | 50 |
| WWC2 | 8 | 13 | 19 | 23 | 29 | 33 | 39 | 46 |
| TWA1 | 11 | 12 | 13 | 13 | 14 | 14 | 15 | 15 |
| TWA2 | 9 | 10 | 12 | 13 | 15 | 16 | 17 | 18 |
| TWA3 | 7 | 9 | 11 | 13 | 15 | 16 | 18 | 20 |
| TWA4 | 6 | 8 | 10 | 12 | 15 | 16 | 19 | 21 |
| TWB3 | 7 | 8 | 11 | 12 | 15 | 16 | 18 | 21 |
| TWD4 | 8 | 13 | 18 | 22 | 28 | 32 | 37 | 44 |
| TWE2 | 9 | 13 | 18 | 22 | 27 | 31 | 36 | 43 |
| TWF3 | 4 | 6 | 8 | 10 | 12 | 13 | 15 | 17 |
| TWJ2 | 7 | 11 | 16 | 20 | 25 | 29 | 34 | 41 |
| DGA1 | 3 | 4 | 4 | 4 | 4 | 3 | 2 | 0 |
| DGA2 | 6 | 8 | 11 | 12 | 15 | 16 | 18 | 20 |
| DGA3 | 10 | 13 | 17 | 20 | 25 | 28 | 33 | 39 |
| DGB1 | 3 | 4 | 4 | 4 | 4 | 4 | 3 | 1 |
| DGB2 | 6 | 8 | 11 | 13 | 15 | 17 | 19 | 22 |
| DGB3 | 10 | 13 | 18 | 21 | 26 | 30 | 35 | 41 |
| DGC1 | 5 | 6 | 7 | 8 | 9 | 9 | 9 | 8 |
| DGC2 | 8 | 11 | 14 | 16 | 20 | 22 | 25 | 29 |
| DGC3 | 11 | 15 | 20 | 24 | 30 | 34 | 39 | 47 |
| SSA1 | 12 | 14 | 16 | 17 | 19 | 21 | 22 | 24 |
| SSB1 | 9 | 11 | 13 | 14 | 16 | 17 | 18 | 20 |
| SSB2 | 12 | 16 | 20 | 23 | 28 | 31 | 36 | 41 |
| SSB3 | 8 | 10 | 11 | 12 | 13 | 14 | 14 | 14 |
| SSC1 | 5 | 7 | 8 | 10 | 11 | 12 | 13 | 14 |
| SSC2 | 7 | 10 | 14 | 17 | 21 | 24 | 28 | 33 |
| SSE1 | 8 | 10 | 12 | 13 | 14 | 15 | 16 | 15 |
| SSE2 | 11 | 15 | 20 | 23 | 28 | 31 | 35 | 41 |
| SSE3 | 10 | 11 | 13 | 14 | 15 | 16 | 16 | 17 |

| BISMARCK, NORTH DAKOTA SSF (%) | | | | | | | 9057 DD65 | |
|---|---|---|---|---|---|---|---|---|
| Type | LCR = 100 | 70 | 50 | 40 | 30 | 25 | 20 | 15 |
| WWA3 | 8 | 11 | 15 | 17 | 21 | 23 | 26 | 30 |
| WWB4 | 11 | 17 | 24 | 30 | 37 | 43 | 49 | 58 |
| WWC2 | 11 | 17 | 23 | 28 | 36 | 40 | 47 | 55 |
| TWA1 | 12 | 13 | 15 | 16 | 17 | 18 | 19 | 21 |
| TWA2 | 10 | 12 | 15 | 16 | 19 | 21 | 23 | 26 |
| TWA3 | 9 | 11 | 14 | 16 | 20 | 22 | 25 | 29 |
| TWA4 | 7 | 10 | 14 | 16 | 20 | 22 | 26 | 30 |
| TWB3 | 8 | 10 | 13 | 16 | 19 | 21 | 24 | 29 |
| TWD4 | 11 | 16 | 22 | 27 | 33 | 38 | 44 | 52 |
| TWE2 | 11 | 16 | 22 | 27 | 33 | 37 | 43 | 51 |
| TWF3 | 6 | 8 | 11 | 13 | 17 | 19 | 21 | 25 |
| TWJ2 | 9 | 14 | 20 | 24 | 31 | 35 | 41 | 49 |
| DGA1 | 5 | 6 | 7 | 8 | 8 | 8 | 8 | 7 |
| DGA2 | 8 | 10 | 13 | 15 | 19 | 21 | 24 | 27 |
| DGA3 | 11 | 15 | 20 | 23 | 29 | 33 | 38 | 45 |
| DGB1 | 4 | 6 | 7 | 8 | 9 | 9 | 9 | 9 |
| DGB2 | 8 | 10 | 14 | 16 | 20 | 22 | 25 | 29 |
| DGB3 | 11 | 15 | 21 | 25 | 30 | 35 | 40 | 48 |
| DGC1 | 6 | 8 | 10 | 12 | 14 | 15 | 16 | 17 |
| DGC2 | 10 | 13 | 17 | 20 | 24 | 27 | 31 | 37 |
| DGC3 | 13 | 18 | 24 | 28 | 35 | 39 | 46 | 55 |
| SSA1 | 13 | 16 | 19 | 21 | 24 | 26 | 28 | 31 |
| SSB1 | 11 | 13 | 15 | 17 | 20 | 22 | 24 | 27 |
| SSB2 | 14 | 18 | 23 | 27 | 33 | 36 | 41 | 48 |
| SSB3 | 10 | 11 | 13 | 15 | 17 | 18 | 20 | 21 |
| SSC1 | 7 | 9 | 12 | 14 | 17 | 18 | 21 | 24 |
| SSC2 | 9 | 13 | 18 | 22 | 27 | 31 | 35 | 42 |
| SSE1 | 10 | 13 | 16 | 18 | 20 | 22 | 24 | 26 |
| SSE2 | 13 | 18 | 24 | 28 | 34 | 38 | 43 | 50 |
| SSE3 | 11 | 13 | 15 | 17 | 19 | 20 | 22 | 24 |

| CINCINNATI, OHIO SSF (%) | | | | | | | 5079 DD65 | |
|---|---|---|---|---|---|---|---|---|
| Type | LCR = 100 | 70 | 50 | 40 | 30 | 25 | 20 | 15 |
| WWA3 | 10 | 13 | 17 | 21 | 25 | 28 | 32 | 38 |
| WWB4 | 13 | 20 | 27 | 33 | 42 | 47 | 55 | 64 |
| WWC2 | 13 | 19 | 27 | 32 | 40 | 45 | 52 | 62 |
| TWA1 | 13 | 14 | 16 | 18 | 20 | 21 | 23 | 25 |
| TWA2 | 11 | 14 | 17 | 19 | 22 | 25 | 28 | 32 |
| TWA3 | 10 | 13 | 16 | 19 | 23 | 26 | 30 | 35 |
| TWA4 | 9 | 12 | 16 | 19 | 23 | 26 | 31 | 36 |
| TWB3 | 9 | 12 | 15 | 18 | 22 | 25 | 29 | 34 |
| TWD4 | 12 | 18 | 25 | 30 | 37 | 42 | 49 | 58 |
| TWE2 | 13 | 19 | 25 | 30 | 37 | 42 | 48 | 57 |
| TWF3 | 7 | 10 | 13 | 16 | 20 | 23 | 26 | 31 |
| TWJ2 | 11 | 16 | 23 | 28 | 35 | 40 | 46 | 55 |
| DGA1 | 6 | 7 | 9 | 10 | 11 | 12 | 12 | 12 |
| DGA2 | 9 | 12 | 15 | 18 | 21 | 24 | 27 | 32 |
| DGA3 | 12 | 16 | 22 | 26 | 32 | 36 | 42 | 50 |
| DGB1 | 6 | 7 | 9 | 10 | 12 | 13 | 14 | 14 |
| DGB2 | 9 | 12 | 15 | 18 | 22 | 25 | 29 | 34 |
| DGB3 | 12 | 17 | 23 | 27 | 33 | 38 | 44 | 53 |
| DGC1 | 8 | 10 | 13 | 15 | 17 | 18 | 20 | 23 |
| DGC2 | 11 | 15 | 19 | 22 | 27 | 31 | 36 | 42 |
| DGC3 | 14 | 20 | 26 | 31 | 38 | 43 | 50 | 60 |
| SSA1 | 15 | 19 | 23 | 25 | 29 | 32 | 36 | 41 |
| SSB1 | 13 | 16 | 19 | 21 | 25 | 27 | 31 | 35 |
| SSB2 | 16 | 22 | 27 | 32 | 38 | 42 | 48 | 56 |
| SSB3 | 12 | 14 | 17 | 19 | 22 | 24 | 27 | 30 |
| SSC1 | 8 | 11 | 15 | 17 | 21 | 23 | 26 | 31 |
| SSC2 | 11 | 16 | 21 | 25 | 31 | 35 | 41 | 49 |
| SSE1 | 13 | 16 | 20 | 23 | 27 | 30 | 33 | 38 |
| SSE2 | 16 | 22 | 29 | 34 | 41 | 46 | 52 | 60 |
| SSE3 | 13 | 16 | 19 | 21 | 24 | 26 | 29 | 33 |

**TABLE H.3 Annual Passive Heating Performance: SSF** *(Continued)*

| CLEVELAND, OHIO SSF (%) | | | | | | | 6160 DD65 | |
|---|---|---|---|---|---|---|---|---|
| Type | LCR = 100 | 70 | 50 | 40 | 30 | 25 | 20 | 15 |
| WWA3 | 6 | 9 | 11 | 13 | 15 | 16 | 18 | 20 |
| WWB4 | 8 | 13 | 20 | 24 | 31 | 36 | 42 | 50 |
| WWC2 | 8 | 13 | 19 | 23 | 29 | 34 | 39 | 47 |
| TWA1 | 11 | 12 | 13 | 14 | 14 | 15 | 15 | 15 |
| TWA2 | 9 | 10 | 12 | 13 | 15 | 16 | 17 | 19 |
| TWA3 | 7 | 9 | 11 | 13 | 15 | 16 | 18 | 20 |
| TWA4 | 6 | 8 | 10 | 12 | 15 | 17 | 19 | 22 |
| TWB3 | 7 | 9 | 11 | 13 | 15 | 16 | 19 | 21 |
| TWD4 | 9 | 13 | 18 | 22 | 28 | 32 | 37 | 45 |
| TWE2 | 9 | 14 | 18 | 22 | 28 | 31 | 36 | 43 |
| TWF3 | 5 | 6 | 8 | 10 | 12 | 13 | 15 | 17 |
| TWJ2 | 7 | 11 | 16 | 20 | 25 | 29 | 34 | 41 |
| DGA1 | 3 | 4 | 4 | 4 | 4 | 4 | 2 | 0 |
| DGA2 | 7 | 8 | 11 | 12 | 15 | 16 | 18 | 21 |
| DGA3 | 10 | 13 | 17 | 20 | 25 | 28 | 33 | 39 |
| DGB1 | 3 | 4 | 4 | 5 | 4 | 4 | 3 | 1 |
| DGB2 | 7 | 9 | 11 | 13 | 15 | 17 | 19 | 22 |
| DGB3 | 10 | 14 | 18 | 21 | 26 | 30 | 35 | 42 |
| DGC1 | 5 | 6 | 7 | 8 | 9 | 9 | 9 | 8 |
| DGC2 | 8 | 11 | 14 | 16 | 20 | 22 | 25 | 29 |
| DGC3 | 12 | 16 | 21 | 24 | 30 | 34 | 39 | 47 |
| SSA1 | 13 | 15 | 18 | 19 | 22 | 24 | 26 | 28 |
| SSB1 | 10 | 12 | 14 | 16 | 18 | 19 | 21 | 23 |
| SSB2 | 13 | 17 | 22 | 25 | 30 | 34 | 38 | 44 |
| SSB3 | 9 | 11 | 13 | 14 | 15 | 16 | 17 | 18 |
| SSC1 | 5 | 7 | 9 | 10 | 11 | 12 | 13 | 14 |
| SSC2 | 7 | 11 | 14 | 17 | 21 | 24 | 28 | 33 |
| SSE1 | 9 | 12 | 14 | 15 | 17 | 18 | 19 | 20 |
| SSE2 | 12 | 17 | 22 | 25 | 31 | 34 | 39 | 45 |
| SSE3 | 11 | 13 | 14 | 16 | 17 | 18 | 19 | 20 |

| OKLAHOMA CITY, OKLAHOMA SSF (%) | | | | | | | 3699 DD65 | |
|---|---|---|---|---|---|---|---|---|
| Type | LCR = 200 | 100 | 70 | 50 | 40 | 30 | 25 | 20 |
| WWA3 | 12 | 22 | 29 | 37 | 44 | 52 | 58 | 65 |
| WWB4 | 12 | 28 | 38 | 50 | 58 | 68 | 74 | 81 |
| WWC2 | 13 | 27 | 38 | 49 | 56 | 66 | 73 | 80 |
| TWA1 | 14 | 20 | 23 | 28 | 32 | 37 | 41 | 46 |
| TWA2 | 13 | 20 | 26 | 32 | 37 | 44 | 49 | 55 |
| TWA3 | 12 | 20 | 26 | 33 | 39 | 47 | 52 | 59 |
| TWA4 | 10 | 19 | 26 | 33 | 39 | 47 | 53 | 60 |
| TWB3 | 10 | 18 | 24 | 31 | 36 | 44 | 49 | 56 |
| TWD4 | 12 | 25 | 34 | 45 | 52 | 62 | 68 | 76 |
| TWE2 | 13 | 26 | 35 | 45 | 53 | 62 | 68 | 76 |
| TWF3 | 9 | 17 | 23 | 30 | 35 | 43 | 48 | 55 |
| TWJ2 | 11 | 23 | 32 | 42 | 49 | 59 | 66 | 73 |
| DGA1 | 8 | 15 | 20 | 25 | 29 | 34 | 37 | 41 |
| DGA2 | 9 | 17 | 23 | 30 | 35 | 43 | 47 | 53 |
| DGA3 | 11 | 21 | 29 | 38 | 45 | 54 | 60 | 67 |
| DGB1 | 8 | 15 | 20 | 26 | 31 | 37 | 42 | 47 |
| DGB2 | 10 | 18 | 24 | 31 | 37 | 46 | 52 | 59 |
| DGB3 | 12 | 22 | 29 | 39 | 46 | 57 | 64 | 72 |
| DGC1 | 10 | 18 | 24 | 32 | 37 | 46 | 51 | 58 |
| DGC2 | 11 | 21 | 28 | 36 | 43 | 53 | 60 | 68 |
| DGC3 | 13 | 25 | 33 | 44 | 52 | 64 | 71 | 79 |
| SSA1 | 16 | 25 | 32 | 39 | 44 | 51 | 57 | 63 |
| SSB1 | 14 | 21 | 27 | 33 | 38 | 45 | 50 | 56 |
| SSB2 | 16 | 27 | 35 | 44 | 51 | 60 | 65 | 72 |
| SSB3 | 13 | 20 | 25 | 31 | 36 | 42 | 47 | 53 |
| SSC1 | 11 | 19 | 26 | 33 | 38 | 46 | 52 | 59 |
| SSC2 | 12 | 23 | 32 | 41 | 48 | 57 | 63 | 71 |
| SSE1 | 15 | 25 | 32 | 40 | 46 | 54 | 59 | 66 |
| SSE2 | 17 | 31 | 40 | 51 | 58 | 67 | 73 | 80 |
| SSE3 | 15 | 22 | 27 | 34 | 38 | 45 | 50 | 56 |

| COLUMBUS, OHIO SSF (%) | | | | | | | 5705 DD65 | |
|---|---|---|---|---|---|---|---|---|
| Type | LCR = 100 | 70 | 50 | 40 | 30 | 25 | 20 | 15 |
| WWA3 | 8 | 11 | 14 | 16 | 20 | 22 | 25 | 29 |
| WWB4 | 10 | 16 | 23 | 29 | 36 | 41 | 48 | 57 |
| WWC2 | 10 | 16 | 22 | 27 | 34 | 39 | 46 | 54 |
| TWA1 | 12 | 13 | 14 | 15 | 17 | 17 | 18 | 20 |
| TWA2 | 10 | 12 | 14 | 16 | 18 | 20 | 22 | 25 |
| TWA3 | 8 | 11 | 13 | 16 | 19 | 21 | 24 | 27 |
| TWA4 | 7 | 10 | 13 | 15 | 19 | 21 | 24 | 29 |
| TWB3 | 8 | 10 | 13 | 15 | 18 | 20 | 23 | 27 |
| TWD4 | 10 | 15 | 21 | 26 | 32 | 37 | 43 | 51 |
| TWE2 | 11 | 16 | 21 | 26 | 32 | 36 | 42 | 50 |
| TWF3 | 6 | 8 | 11 | 13 | 16 | 18 | 20 | 24 |
| TWJ2 | 9 | 13 | 19 | 23 | 30 | 34 | 40 | 48 |
| DGA1 | 4 | 5 | 6 | 7 | 7 | 7 | 7 | 5 |
| DGA2 | 7 | 10 | 12 | 15 | 18 | 20 | 22 | 26 |
| DGA3 | 11 | 14 | 19 | 23 | 28 | 32 | 37 | 44 |
| DGB1 | 4 | 5 | 6 | 7 | 8 | 8 | 8 | 7 |
| DGB2 | 7 | 10 | 13 | 15 | 18 | 21 | 24 | 28 |
| DGB3 | 11 | 15 | 20 | 24 | 29 | 33 | 39 | 47 |
| DGC1 | 6 | 8 | 10 | 11 | 12 | 13 | 14 | 15 |
| DGC2 | 9 | 12 | 16 | 19 | 23 | 26 | 30 | 35 |
| DGC3 | 13 | 17 | 23 | 27 | 33 | 38 | 44 | 53 |
| SSA1 | 14 | 17 | 20 | 22 | 25 | 27 | 30 | 34 |
| SSB1 | 11 | 13 | 16 | 18 | 21 | 23 | 25 | 29 |
| SSB2 | 15 | 19 | 24 | 28 | 34 | 38 | 43 | 50 |
| SSB3 | 10 | 12 | 14 | 16 | 18 | 20 | 22 | 24 |
| SSC1 | 7 | 9 | 11 | 13 | 16 | 17 | 19 | 22 |
| SSC2 | 9 | 13 | 17 | 21 | 26 | 30 | 34 | 41 |
| SSE1 | 11 | 14 | 17 | 19 | 22 | 24 | 26 | 29 |
| SSE2 | 14 | 19 | 25 | 29 | 36 | 40 | 45 | 53 |
| SSE3 | 12 | 14 | 16 | 18 | 20 | 22 | 24 | 26 |

| MEDFORD, OREGON SSF (%) | | | | | | | 4935 DD65 | |
|---|---|---|---|---|---|---|---|---|
| Type | LCR = 100 | 70 | 50 | 40 | 30 | 25 | 20 | 15 |
| WWA3 | 16 | 21 | 26 | 31 | 36 | 40 | 45 | 51 |
| WWB4 | 20 | 28 | 37 | 43 | 52 | 58 | 64 | 72 |
| WWC2 | 20 | 28 | 36 | 42 | 50 | 56 | 62 | 70 |
| TWA1 | 17 | 19 | 22 | 24 | 28 | 30 | 33 | 37 |
| TWA2 | 16 | 20 | 24 | 27 | 32 | 35 | 39 | 44 |
| TWA3 | 15 | 19 | 24 | 28 | 33 | 37 | 41 | 47 |
| TWA4 | 14 | 19 | 24 | 28 | 34 | 37 | 42 | 49 |
| TWB3 | 14 | 18 | 23 | 26 | 31 | 35 | 40 | 46 |
| TWD4 | 19 | 26 | 34 | 39 | 47 | 52 | 59 | 67 |
| TWE2 | 20 | 26 | 34 | 40 | 47 | 52 | 59 | 66 |
| TWF3 | 12 | 16 | 21 | 25 | 30 | 33 | 38 | 44 |
| TWJ2 | 17 | 24 | 31 | 37 | 45 | 50 | 56 | 64 |
| DGA1 | 11 | 14 | 17 | 19 | 21 | 23 | 24 | 25 |
| DGA2 | 13 | 17 | 22 | 26 | 31 | 34 | 38 | 43 |
| DGA3 | 17 | 23 | 29 | 34 | 41 | 46 | 52 | 59 |
| DGB1 | 11 | 14 | 18 | 20 | 23 | 25 | 28 | 30 |
| DGB2 | 14 | 18 | 23 | 27 | 33 | 37 | 41 | 47 |
| DGB3 | 17 | 23 | 30 | 36 | 44 | 49 | 55 | 63 |
| DGC1 | 13 | 17 | 22 | 25 | 30 | 32 | 36 | 39 |
| DGC2 | 16 | 21 | 27 | 32 | 39 | 43 | 48 | 55 |
| DGC3 | 20 | 26 | 34 | 40 | 49 | 54 | 61 | 69 |
| SSA1 | 21 | 26 | 30 | 34 | 39 | 42 | 46 | 51 |
| SSB1 | 18 | 22 | 26 | 29 | 34 | 37 | 41 | 46 |
| SSB2 | 22 | 28 | 35 | 40 | 47 | 51 | 57 | 64 |
| SSB3 | 17 | 20 | 24 | 27 | 31 | 34 | 37 | 42 |
| SSC1 | 14 | 18 | 23 | 27 | 32 | 35 | 39 | 44 |
| SSC2 | 17 | 23 | 30 | 35 | 42 | 46 | 52 | 59 |
| SSE1 | 19 | 24 | 29 | 33 | 38 | 41 | 45 | 50 |
| SSE2 | 24 | 31 | 38 | 44 | 51 | 55 | 61 | 68 |
| SSE3 | 18 | 22 | 26 | 29 | 33 | 36 | 40 | 44 |

### TABLE H.3 Annual Passive Heating Performance: SSF *(Continued)*

| SALEM, OREGON SSF (%) | | | | | | 4854 DD65 | | |
|---|---|---|---|---|---|---|---|---|
| Type | LCR = 100 | 70 | 50 | 40 | 30 | 25 | 20 | 15 |
| WWA3 | 15 | 20 | 25 | 29 | 35 | 38 | 43 | 48 |
| WWB4 | 19 | 27 | 36 | 42 | 50 | 56 | 63 | 71 |
| WWC2 | 19 | 27 | 35 | 41 | 49 | 54 | 60 | 68 |
| TWA1 | 16 | 19 | 22 | 24 | 27 | 29 | 31 | 35 |
| TWA2 | 16 | 19 | 23 | 26 | 30 | 33 | 37 | 42 |
| TWA3 | 15 | 19 | 23 | 27 | 32 | 35 | 40 | 45 |
| TWA4 | 14 | 18 | 23 | 27 | 32 | 36 | 40 | 47 |
| TWB3 | 14 | 18 | 22 | 25 | 30 | 34 | 38 | 44 |
| TWD4 | 18 | 25 | 32 | 38 | 46 | 51 | 57 | 65 |
| TWE2 | 19 | 26 | 33 | 38 | 46 | 51 | 57 | 65 |
| TWF3 | 12 | 16 | 20 | 24 | 28 | 32 | 36 | 42 |
| TWJ2 | 16 | 23 | 30 | 36 | 43 | 48 | 55 | 63 |
| DGA1 | 10 | 13 | 16 | 18 | 20 | 21 | 23 | 23 |
| DGA2 | 13 | 17 | 22 | 25 | 30 | 33 | 37 | 41 |
| DGA3 | 17 | 22 | 28 | 33 | 40 | 45 | 51 | 58 |
| DGB1 | 11 | 13 | 17 | 19 | 22 | 24 | 26 | 28 |
| DGB2 | 13 | 18 | 22 | 26 | 32 | 35 | 40 | 45 |
| DGB3 | 17 | 23 | 29 | 35 | 42 | 47 | 54 | 62 |
| DGC1 | 13 | 16 | 21 | 24 | 28 | 30 | 33 | 37 |
| DGC2 | 16 | 21 | 26 | 31 | 37 | 41 | 46 | 53 |
| DGC3 | 19 | 26 | 33 | 39 | 47 | 53 | 59 | 68 |
| SSA1 | 21 | 25 | 29 | 33 | 37 | 40 | 44 | 49 |
| SSB1 | 17 | 21 | 25 | 28 | 33 | 36 | 39 | 44 |
| SSB2 | 22 | 28 | 34 | 39 | 46 | 50 | 55 | 62 |
| SSB3 | 16 | 20 | 23 | 26 | 30 | 33 | 36 | 40 |
| SSC1 | 14 | 18 | 22 | 25 | 30 | 33 | 37 | 42 |
| SSC2 | 17 | 22 | 29 | 33 | 40 | 45 | 50 | 57 |
| SSE1 | 19 | 23 | 28 | 32 | 36 | 39 | 43 | 47 |
| SSE2 | 23 | 30 | 37 | 42 | 49 | 54 | 59 | 66 |
| SSE3 | 18 | 22 | 25 | 28 | 32 | 35 | 38 | 42 |

| PHILADELPHIA, PENNSYLVANIA SSF (%) | | | | | | 4875 DD65 | | |
|---|---|---|---|---|---|---|---|---|
| Type | LCR = 100 | 70 | 50 | 40 | 30 | 25 | 20 | 15 |
| WWA3 | 12 | 17 | 22 | 26 | 32 | 36 | 41 | 49 |
| WWB4 | 16 | 24 | 33 | 40 | 49 | 55 | 63 | 72 |
| WWC2 | 16 | 23 | 32 | 38 | 47 | 53 | 61 | 70 |
| TWA1 | 14 | 16 | 19 | 21 | 24 | 26 | 29 | 33 |
| TWA2 | 13 | 16 | 20 | 23 | 27 | 31 | 35 | 41 |
| TWA3 | 12 | 16 | 20 | 24 | 29 | 33 | 38 | 44 |
| TWA4 | 11 | 15 | 20 | 24 | 29 | 33 | 39 | 46 |
| TWB3 | 11 | 15 | 19 | 22 | 27 | 31 | 36 | 42 |
| TWD4 | 15 | 22 | 30 | 35 | 44 | 49 | 57 | 66 |
| TWE2 | 16 | 22 | 30 | 36 | 44 | 49 | 56 | 66 |
| TWF3 | 9 | 13 | 17 | 20 | 25 | 29 | 34 | 40 |
| TWJ2 | 13 | 20 | 27 | 33 | 41 | 47 | 54 | 63 |
| DGA1 | 8 | 10 | 12 | 14 | 17 | 18 | 19 | 20 |
| DGA2 | 10 | 14 | 18 | 22 | 26 | 30 | 34 | 39 |
| DGA3 | 14 | 19 | 25 | 30 | 37 | 42 | 49 | 58 |
| DGB1 | 7 | 10 | 13 | 15 | 17 | 19 | 22 | 24 |
| DGB2 | 11 | 14 | 19 | 22 | 28 | 31 | 37 | 43 |
| DGB3 | 14 | 20 | 26 | 31 | 39 | 44 | 52 | 62 |
| DGC1 | 10 | 13 | 17 | 19 | 23 | 26 | 30 | 34 |
| DGC2 | 13 | 17 | 23 | 27 | 33 | 38 | 44 | 52 |
| DGC3 | 16 | 22 | 29 | 35 | 44 | 50 | 58 | 69 |
| SSA1 | 17 | 21 | 26 | 29 | 34 | 38 | 43 | 49 |
| SSB1 | 14 | 18 | 22 | 25 | 29 | 33 | 37 | 43 |
| SSB2 | 19 | 24 | 31 | 36 | 43 | 48 | 54 | 63 |
| SSB3 | 13 | 16 | 20 | 22 | 26 | 29 | 33 | 38 |
| SSC1 | 11 | 14 | 19 | 22 | 27 | 30 | 35 | 42 |
| SSC2 | 14 | 19 | 26 | 31 | 38 | 43 | 49 | 58 |
| SSE1 | 15 | 20 | 25 | 28 | 34 | 37 | 42 | 48 |
| SSE2 | 19 | 26 | 34 | 40 | 48 | 53 | 60 | 68 |
| SSE3 | 15 | 18 | 22 | 25 | 29 | 32 | 36 | 41 |

| HARRISBURG, PENNSYLVANIA SSF (%) | | | | | | 5226 DD65 | | |
|---|---|---|---|---|---|---|---|---|
| Type | LCR = 100 | 70 | 50 | 40 | 30 | 25 | 20 | 15 |
| WWA3 | 11 | 14 | 19 | 22 | 28 | 31 | 36 | 42 |
| WWB4 | 14 | 21 | 29 | 36 | 45 | 51 | 58 | 68 |
| WWC2 | 14 | 20 | 28 | 34 | 43 | 48 | 56 | 65 |
| TWA1 | 13 | 15 | 17 | 19 | 21 | 23 | 25 | 28 |
| TWA2 | 12 | 14 | 18 | 20 | 24 | 27 | 30 | 35 |
| TWA3 | 10 | 14 | 18 | 21 | 25 | 28 | 33 | 39 |
| TWA4 | 9 | 13 | 17 | 20 | 25 | 29 | 34 | 40 |
| TWB3 | 10 | 13 | 16 | 19 | 24 | 27 | 31 | 37 |
| TWD4 | 13 | 19 | 26 | 32 | 40 | 45 | 52 | 61 |
| TWE2 | 14 | 20 | 27 | 32 | 40 | 45 | 52 | 61 |
| TWF3 | 8 | 11 | 14 | 17 | 22 | 25 | 29 | 35 |
| TWJ2 | 11 | 17 | 24 | 29 | 37 | 42 | 49 | 58 |
| DGA1 | 6 | 8 | 10 | 11 | 13 | 14 | 15 | 15 |
| DGA2 | 9 | 12 | 16 | 19 | 23 | 26 | 30 | 35 |
| DGA3 | 12 | 17 | 23 | 27 | 34 | 38 | 45 | 53 |
| DGB1 | 6 | 8 | 10 | 12 | 14 | 15 | 16 | 18 |
| DGB2 | 9 | 12 | 16 | 19 | 24 | 27 | 32 | 38 |
| DGB3 | 13 | 18 | 23 | 28 | 35 | 40 | 47 | 57 |
| DGC1 | 8 | 11 | 14 | 16 | 19 | 21 | 24 | 27 |
| DGC2 | 11 | 15 | 20 | 24 | 29 | 33 | 39 | 46 |
| DGC3 | 15 | 20 | 27 | 32 | 40 | 46 | 53 | 64 |
| SSA1 | 16 | 19 | 23 | 26 | 31 | 34 | 38 | 44 |
| SSB1 | 13 | 16 | 19 | 22 | 26 | 29 | 32 | 38 |
| SSB2 | 17 | 22 | 28 | 33 | 40 | 44 | 50 | 58 |
| SSB3 | 12 | 14 | 17 | 20 | 23 | 25 | 28 | 33 |
| SSC1 | 9 | 12 | 16 | 19 | 23 | 26 | 30 | 35 |
| SSC2 | 12 | 17 | 22 | 27 | 34 | 38 | 44 | 52 |
| SSE1 | 13 | 17 | 21 | 24 | 29 | 32 | 36 | 42 |
| SSE2 | 17 | 23 | 30 | 36 | 43 | 48 | 55 | 63 |
| SSE3 | 13 | 16 | 19 | 22 | 25 | 28 | 31 | 36 |

| PITTSBURGH, PENNSYLVANIA SSF (%) | | | | | | 5937 DD65 | | |
|---|---|---|---|---|---|---|---|---|
| Type | LCR = 100 | 70 | 50 | 40 | 30 | 25 | 20 | 15 |
| WWA3 | 7 | 9 | 12 | 14 | 17 | 19 | 21 | 24 |
| WWB4 | 9 | 15 | 21 | 26 | 33 | 38 | 45 | 53 |
| WWC2 | 9 | 14 | 20 | 25 | 32 | 36 | 42 | 50 |
| TWA1 | 12 | 12 | 13 | 14 | 15 | 16 | 16 | 17 |
| TWA2 | 9 | 11 | 13 | 14 | 16 | 18 | 19 | 22 |
| TWA3 | 8 | 10 | 12 | 14 | 17 | 18 | 21 | 23 |
| TWA4 | 6 | 9 | 11 | 14 | 17 | 19 | 21 | 25 |
| TWB3 | 7 | 9 | 12 | 14 | 16 | 18 | 21 | 24 |
| TWD4 | 9 | 14 | 20 | 24 | 30 | 34 | 40 | 48 |
| TWE2 | 10 | 14 | 20 | 24 | 30 | 34 | 39 | 46 |
| TWF3 | 5 | 7 | 9 | 11 | 14 | 15 | 17 | 20 |
| TWJ2 | 8 | 12 | 17 | 21 | 27 | 31 | 37 | 44 |
| DGA1 | 4 | 5 | 5 | 5 | 5 | 5 | 4 | 2 |
| DGA2 | 7 | 9 | 11 | 13 | 16 | 18 | 20 | 23 |
| DGA3 | 10 | 14 | 18 | 21 | 26 | 30 | 35 | 41 |
| DGB1 | 4 | 4 | 5 | 6 | 6 | 6 | 5 | 3 |
| DGB2 | 7 | 9 | 12 | 14 | 17 | 19 | 21 | 25 |
| DGB3 | 10 | 14 | 19 | 22 | 28 | 32 | 37 | 44 |
| DGC1 | 5 | 7 | 8 | 9 | 10 | 11 | 11 | 11 |
| DGC2 | 9 | 11 | 15 | 17 | 21 | 24 | 27 | 32 |
| DGC3 | 12 | 16 | 21 | 26 | 32 | 36 | 42 | 50 |
| SSA1 | 13 | 16 | 19 | 21 | 23 | 25 | 28 | 31 |
| SSB1 | 11 | 13 | 15 | 17 | 19 | 21 | 23 | 26 |
| SSB2 | 14 | 18 | 23 | 27 | 32 | 36 | 40 | 47 |
| SSB3 | 10 | 11 | 13 | 15 | 17 | 18 | 19 | 21 |
| SSC1 | 6 | 8 | 10 | 11 | 13 | 15 | 16 | 18 |
| SSC2 | 8 | 12 | 16 | 19 | 23 | 27 | 31 | 37 |
| SSE1 | 10 | 12 | 15 | 17 | 19 | 21 | 23 | 24 |
| SSE2 | 13 | 18 | 23 | 27 | 33 | 37 | 42 | 48 |
| SSE3 | 11 | 13 | 15 | 17 | 19 | 20 | 21 | 23 |

**TABLE H.3 Annual Passive Heating Performance: SSF** *(Continued)*

### PROVIDENCE, RHODE ISLAND SSF (%) — 5974 DD65

| Type | LCR = 100 | 70 | 50 | 40 | 30 | 25 | 20 | 15 |
|---|---|---|---|---|---|---|---|---|
| WWA3 | 10 | 13 | 18 | 21 | 26 | 29 | 33 | 40 |
| WWB4 | 13 | 20 | 28 | 34 | 43 | 49 | 56 | 66 |
| WWC2 | 13 | 19 | 27 | 33 | 41 | 47 | 54 | 63 |
| TWA1 | 13 | 14 | 16 | 18 | 20 | 21 | 23 | 26 |
| TWA2 | 11 | 14 | 17 | 19 | 22 | 25 | 28 | 33 |
| TWA3 | 10 | 13 | 16 | 19 | 24 | 27 | 31 | 36 |
| TWA4 | 9 | 12 | 16 | 19 | 24 | 27 | 31 | 38 |
| TWB3 | 9 | 12 | 15 | 18 | 22 | 25 | 29 | 35 |
| TWD4 | 12 | 18 | 25 | 30 | 38 | 43 | 50 | 59 |
| TWE2 | 13 | 19 | 25 | 30 | 38 | 43 | 50 | 59 |
| TWF3 | 7 | 10 | 13 | 16 | 20 | 23 | 27 | 32 |
| TWJ2 | 11 | 16 | 23 | 28 | 35 | 40 | 47 | 56 |
| DGA1 | 6 | 7 | 9 | 10 | 11 | 12 | 13 | 13 |
| DGA2 | 8 | 11 | 15 | 18 | 22 | 24 | 28 | 33 |
| DGA3 | 12 | 16 | 21 | 26 | 32 | 37 | 43 | 51 |
| DGB1 | 5 | 7 | 9 | 10 | 12 | 13 | 14 | 15 |
| DGB2 | 9 | 12 | 15 | 18 | 22 | 26 | 30 | 36 |
| DGB3 | 12 | 17 | 22 | 27 | 34 | 39 | 45 | 55 |
| DGC1 | 7 | 10 | 12 | 14 | 17 | 19 | 21 | 25 |
| DGC2 | 10 | 14 | 19 | 22 | 28 | 31 | 37 | 44 |
| DGC3 | 14 | 19 | 26 | 31 | 38 | 44 | 52 | 62 |
| SSA1 | 15 | 19 | 22 | 25 | 30 | 33 | 36 | 42 |
| SSB1 | 12 | 15 | 19 | 21 | 25 | 28 | 31 | 36 |
| SSB2 | 16 | 21 | 27 | 32 | 38 | 43 | 49 | 57 |
| SSB3 | 11 | 14 | 17 | 19 | 22 | 24 | 27 | 31 |
| SSC1 | 8 | 11 | 15 | 17 | 21 | 24 | 27 | 32 |
| SSC2 | 11 | 16 | 21 | 25 | 32 | 36 | 42 | 50 |
| SSE1 | 13 | 16 | 20 | 23 | 27 | 30 | 34 | 39 |
| SSE2 | 16 | 22 | 29 | 34 | 42 | 47 | 53 | 61 |
| SSE3 | 13 | 16 | 19 | 21 | 24 | 27 | 30 | 34 |

### RAPID CITY, SOUTH DAKOTA SSF (%) — 7332 DD65

| Type | LCR = 100 | 70 | 50 | 40 | 30 | 25 | 20 | 15 |
|---|---|---|---|---|---|---|---|---|
| WWA3 | 13 | 18 | 23 | 27 | 33 | 37 | 42 | 50 |
| WWB4 | 17 | 25 | 34 | 40 | 50 | 56 | 63 | 73 |
| WWC2 | 17 | 24 | 33 | 39 | 48 | 54 | 61 | 70 |
| TWA1 | 15 | 17 | 19 | 21 | 24 | 27 | 30 | 34 |
| TWA2 | 13 | 17 | 21 | 24 | 28 | 32 | 36 | 42 |
| TWA3 | 13 | 17 | 21 | 25 | 30 | 34 | 38 | 45 |
| TWA4 | 11 | 16 | 21 | 25 | 30 | 34 | 39 | 47 |
| TWB3 | 11 | 15 | 20 | 23 | 28 | 32 | 37 | 43 |
| TWD4 | 16 | 23 | 30 | 36 | 45 | 50 | 57 | 66 |
| TWE2 | 17 | 23 | 31 | 37 | 45 | 50 | 57 | 66 |
| TWF3 | 10 | 13 | 18 | 21 | 26 | 30 | 35 | 41 |
| TWJ2 | 14 | 21 | 28 | 34 | 42 | 47 | 54 | 64 |
| DGA1 | 8 | 11 | 13 | 15 | 18 | 19 | 20 | 21 |
| DGA2 | 11 | 15 | 19 | 22 | 27 | 31 | 35 | 40 |
| DGA3 | 14 | 20 | 26 | 31 | 38 | 43 | 50 | 58 |
| DGB1 | 8 | 11 | 14 | 16 | 19 | 21 | 23 | 26 |
| DGB2 | 11 | 15 | 20 | 32 | 29 | 33 | 38 | 44 |
| DGB3 | 15 | 20 | 27 | 32 | 40 | 45 | 53 | 62 |
| DGC1 | 10 | 14 | 18 | 21 | 25 | 27 | 31 | 35 |
| DGC2 | 13 | 18 | 24 | 28 | 34 | 39 | 45 | 53 |
| DGC3 | 17 | 23 | 31 | 36 | 45 | 51 | 59 | 69 |
| SSA1 | 18 | 22 | 26 | 30 | 34 | 38 | 42 | 48 |
| SSB1 | 15 | 19 | 22 | 25 | 30 | 33 | 37 | 42 |
| SSB2 | 19 | 25 | 31 | 36 | 43 | 48 | 54 | 62 |
| SSB3 | 13 | 16 | 20 | 23 | 26 | 29 | 33 | 37 |
| SSC1 | 11 | 15 | 20 | 23 | 28 | 31 | 36 | 42 |
| SSC2 | 14 | 20 | 27 | 32 | 39 | 44 | 50 | 59 |
| SSE1 | 16 | 20 | 25 | 28 | 33 | 37 | 41 | 47 |
| SSE2 | 20 | 27 | 34 | 40 | 47 | 52 | 59 | 67 |
| SSE3 | 15 | 18 | 22 | 25 | 29 | 32 | 35 | 40 |

### CHARLESTON, SOUTH CAROLINA SSF (%) — 2150 DD65

| Type | LCR = 200 | 100 | 70 | 50 | 40 | 30 | 25 | 20 |
|---|---|---|---|---|---|---|---|---|
| WWA3 | 18 | 31 | 40 | 50 | 57 | 66 | 72 | 79 |
| WWB4 | 19 | 38 | 50 | 62 | 70 | 79 | 85 | 90 |
| WWC2 | 20 | 38 | 49 | 61 | 69 | 78 | 84 | 89 |
| TWA1 | 17 | 25 | 30 | 37 | 41 | 48 | 53 | 59 |
| TWA2 | 17 | 27 | 34 | 42 | 48 | 57 | 62 | 69 |
| TWA3 | 16 | 28 | 36 | 45 | 51 | 60 | 66 | 73 |
| TWA4 | 15 | 27 | 36 | 45 | 52 | 61 | 67 | 74 |
| TWB3 | 14 | 25 | 33 | 41 | 47 | 56 | 62 | 69 |
| TWD4 | 18 | 35 | 45 | 56 | 64 | 74 | 79 | 86 |
| TWE2 | 19 | 36 | 46 | 57 | 65 | 74 | 80 | 86 |
| TWF3 | 13 | 24 | 32 | 41 | 47 | 56 | 62 | 69 |
| TWJ2 | 17 | 32 | 43 | 54 | 62 | 72 | 77 | 84 |
| DGA1 | 13 | 23 | 29 | 36 | 41 | 48 | 51 | 56 |
| DGA2 | 13 | 24 | 32 | 41 | 47 | 55 | 60 | 67 |
| DGA3 | 16 | 29 | 38 | 49 | 57 | 66 | 72 | 78 |
| DGB1 | 13 | 23 | 31 | 39 | 45 | 54 | 59 | 65 |
| DGB2 | 14 | 25 | 34 | 43 | 51 | 61 | 67 | 74 |
| DGB3 | 16 | 29 | 39 | 51 | 60 | 71 | 77 | 83 |
| DGC1 | 15 | 27 | 36 | 46 | 53 | 63 | 69 | 75 |
| DGC2 | 16 | 29 | 38 | 49 | 58 | 68 | 75 | 81 |
| DGC3 | 18 | 33 | 44 | 57 | 66 | 77 | 83 | 89 |
| SSA1 | 22 | 33 | 41 | 50 | 56 | 65 | 70 | 76 |
| SSB1 | 18 | 29 | 36 | 44 | 50 | 58 | 64 | 70 |
| SSB2 | 22 | 36 | 46 | 56 | 63 | 72 | 77 | 83 |
| SSB3 | 17 | 27 | 34 | 42 | 48 | 56 | 61 | 68 |
| SSC1 | 16 | 28 | 36 | 45 | 51 | 60 | 66 | 73 |
| SSC2 | 18 | 32 | 42 | 53 | 60 | 70 | 76 | 82 |
| SSE1 | 21 | 34 | 43 | 53 | 60 | 69 | 74 | 80 |
| SSE2 | 24 | 42 | 53 | 64 | 71 | 80 | 85 | 90 |
| SSE3 | 19 | 29 | 36 | 44 | 50 | 58 | 64 | 70 |

### MEMPHIS, TENNESSEE SSF (%) — 3237 DD65

| Type | LCR = 200 | 100 | 70 | 50 | 40 | 30 | 25 | 20 |
|---|---|---|---|---|---|---|---|---|
| WWA3 | 11 | 21 | 27 | 35 | 41 | 49 | 54 | 61 |
| WWB4 | 11 | 26 | 36 | 47 | 54 | 65 | 71 | 78 |
| WWC2 | 12 | 26 | 35 | 46 | 53 | 63 | 69 | 77 |
| TWA1 | 14 | 19 | 22 | 27 | 30 | 35 | 38 | 43 |
| TWA2 | 12 | 19 | 24 | 30 | 35 | 41 | 46 | 52 |
| TWA3 | 11 | 19 | 25 | 31 | 36 | 44 | 49 | 55 |
| TWA4 | 9 | 18 | 24 | 31 | 37 | 44 | 50 | 56 |
| TWB3 | 10 | 17 | 23 | 29 | 34 | 41 | 46 | 52 |
| TWD4 | 11 | 24 | 32 | 42 | 49 | 59 | 65 | 72 |
| TWE2 | 12 | 25 | 33 | 43 | 50 | 59 | 65 | 72 |
| TWF3 | 8 | 16 | 21 | 28 | 33 | 40 | 45 | 51 |
| TWJ2 | 10 | 22 | 30 | 40 | 47 | 56 | 62 | 70 |
| DGA1 | 8 | 14 | 18 | 23 | 26 | 31 | 34 | 37 |
| DGA2 | 9 | 16 | 22 | 28 | 33 | 40 | 44 | 50 |
| DGA3 | 11 | 20 | 27 | 36 | 42 | 51 | 57 | 64 |
| DGB1 | 8 | 14 | 19 | 24 | 28 | 34 | 38 | 43 |
| DGB2 | 9 | 17 | 22 | 29 | 35 | 43 | 48 | 55 |
| DGB3 | 11 | 21 | 28 | 37 | 43 | 54 | 60 | 69 |
| DGC1 | 10 | 17 | 23 | 29 | 34 | 41 | 46 | 52 |
| DGC2 | 11 | 20 | 26 | 34 | 40 | 49 | 56 | 63 |
| DGC3 | 13 | 24 | 31 | 41 | 49 | 60 | 67 | 76 |
| SSA1 | 16 | 24 | 30 | 37 | 42 | 49 | 54 | 60 |
| SSB1 | 13 | 21 | 26 | 32 | 36 | 43 | 47 | 53 |
| SSB2 | 15 | 26 | 34 | 42 | 49 | 57 | 63 | 70 |
| SSB3 | 12 | 19 | 24 | 30 | 34 | 40 | 44 | 50 |
| SSC1 | 10 | 18 | 24 | 31 | 36 | 43 | 48 | 54 |
| SSC2 | 11 | 22 | 30 | 38 | 45 | 54 | 59 | 67 |
| SSE1 | 14 | 24 | 30 | 38 | 43 | 51 | 56 | 62 |
| SSE2 | 16 | 29 | 38 | 48 | 55 | 64 | 70 | 77 |
| SSE3 | 14 | 21 | 26 | 32 | 36 | 43 | 47 | 53 |

**TABLE H.3 Annual Passive Heating Performance: SSF** *(Continued)*

**KNOXVILLE, TENNESSEE SSF (%)**     3486 DD65

| Type | LCR = 200 | 100 | 70 | 50 | 40 | 30 | 25 | 20 |
|---|---|---|---|---|---|---|---|---|
| WWA3 | 10 | 18 | 24 | 31 | 36 | 44 | 49 | 55 |
| WWB4 | 9 | 23 | 32 | 43 | 50 | 60 | 66 | 74 |
| WWC2 | 10 | 23 | 32 | 42 | 49 | 58 | 65 | 72 |
| TWA1 | 13 | 17 | 21 | 24 | 27 | 32 | 35 | 39 |
| TWA2 | 11 | 17 | 22 | 27 | 31 | 37 | 41 | 47 |
| TWA3 | 10 | 17 | 22 | 28 | 33 | 39 | 44 | 50 |
| TWA4 | 8 | 16 | 21 | 28 | 33 | 40 | 45 | 51 |
| TWB3 | 9 | 15 | 20 | 26 | 30 | 37 | 41 | 47 |
| TWD4 | 10 | 21 | 29 | 38 | 45 | 54 | 60 | 68 |
| TWE2 | 11 | 22 | 30 | 39 | 46 | 55 | 60 | 68 |
| TWF3 | 7 | 14 | 19 | 25 | 29 | 36 | 40 | 46 |
| TWJ2 | 8 | 19 | 27 | 36 | 43 | 52 | 58 | 65 |
| DGA1 | 7 | 12 | 16 | 20 | 23 | 27 | 29 | 32 |
| DGA2 | 8 | 15 | 19 | 25 | 30 | 36 | 40 | 45 |
| DGA3 | 10 | 18 | 25 | 32 | 38 | 47 | 53 | 60 |
| DGB1 | 7 | 12 | 16 | 21 | 24 | 29 | 32 | 36 |
| DGB2 | 8 | 15 | 20 | 26 | 31 | 38 | 43 | 49 |
| DGB3 | 10 | 19 | 26 | 33 | 40 | 49 | 55 | 64 |
| DGC1 | 8 | 15 | 20 | 25 | 29 | 36 | 40 | 45 |
| DGC2 | 9 | 18 | 24 | 31 | 36 | 44 | 50 | 57 |
| DGC3 | 12 | 22 | 29 | 38 | 45 | 55 | 62 | 71 |
| SSA1 | 15 | 22 | 28 | 34 | 38 | 45 | 49 | 55 |
| SSB1 | 12 | 19 | 24 | 29 | 33 | 39 | 43 | 49 |
| SSB2 | 14 | 24 | 31 | 39 | 45 | 53 | 59 | 65 |
| SSB3 | 12 | 18 | 22 | 27 | 31 | 36 | 40 | 45 |
| SSC1 | 9 | 16 | 21 | 27 | 32 | 38 | 43 | 49 |
| SSC2 | 9 | 19 | 27 | 35 | 41 | 49 | 55 | 62 |
| SSE1 | 13 | 21 | 27 | 34 | 39 | 48 | 51 | 57 |
| SSE2 | 14 | 27 | 35 | 44 | 51 | 60 | 65 | 72 |
| SSE3 | 13 | 19 | 24 | 29 | 33 | 39 | 43 | 48 |

**AUSTIN, TEXAS SSF (%)**     1737 DD65

| Type | LCR = 200 | 100 | 70 | 50 | 40 | 30 | 25 | 20 |
|---|---|---|---|---|---|---|---|---|
| WWA3 | 21 | 37 | 47 | 58 | 65 | 75 | 80 | 86 |
| WWB4 | 24 | 45 | 58 | 70 | 77 | 86 | 90 | 94 |
| WWC2 | 24 | 44 | 57 | 69 | 76 | 85 | 89 | 94 |
| TWA1 | 20 | 29 | 35 | 42 | 48 | 56 | 61 | 67 |
| TWA2 | 20 | 32 | 40 | 49 | 56 | 65 | 70 | 77 |
| TWA3 | 19 | 33 | 42 | 52 | 59 | 68 | 74 | 81 |
| TWA4 | 18 | 32 | 42 | 52 | 60 | 69 | 75 | 82 |
| TWB3 | 17 | 30 | 38 | 48 | 55 | 64 | 70 | 77 |
| TWD4 | 22 | 41 | 52 | 64 | 72 | 81 | 86 | 91 |
| TWE2 | 24 | 42 | 53 | 65 | 73 | 81 | 86 | 91 |
| TWF3 | 16 | 29 | 38 | 48 | 55 | 64 | 70 | 77 |
| TWJ2 | 21 | 39 | 50 | 62 | 70 | 79 | 84 | 90 |
| DGA1 | 16 | 28 | 36 | 44 | 49 | 57 | 61 | 66 |
| DGA2 | 16 | 29 | 38 | 48 | 55 | 64 | 69 | 75 |
| DGA3 | 19 | 34 | 45 | 57 | 65 | 74 | 79 | 84 |
| DGB1 | 16 | 29 | 38 | 48 | 56 | 64 | 69 | 75 |
| DGB2 | 17 | 30 | 41 | 52 | 60 | 70 | 76 | 82 |
| DGB3 | 19 | 35 | 46 | 60 | 68 | 79 | 84 | 89 |
| DGC1 | 19 | 33 | 44 | 56 | 64 | 74 | 79 | 84 |
| DGC2 | 19 | 34 | 46 | 59 | 68 | 77 | 83 | 88 |
| DGC3 | 22 | 39 | 52 | 66 | 75 | 85 | 89 | 93 |
| SSA1 | 25 | 39 | 48 | 58 | 65 | 73 | 78 | 84 |
| SSB1 | 21 | 34 | 42 | 51 | 58 | 67 | 72 | 79 |
| SSB2 | 26 | 42 | 53 | 63 | 71 | 79 | 84 | 89 |
| SSB3 | 20 | 32 | 40 | 49 | 56 | 65 | 70 | 77 |
| SSC1 | 19 | 33 | 42 | 53 | 60 | 69 | 75 | 81 |
| SSC2 | 22 | 39 | 50 | 61 | 68 | 78 | 83 | 89 |
| SSE1 | 25 | 41 | 51 | 62 | 69 | 77 | 82 | 88 |
| SSE2 | 29 | 49 | 61 | 72 | 79 | 87 | 91 | 94 |
| SSE3 | 22 | 34 | 43 | 52 | 58 | 67 | 72 | 79 |

**NASHVILLE, TENNESSEE SSF (%)**     3704 DD65

| Type | LCR = 200 | 100 | 70 | 50 | 40 | 30 | 25 | 20 |
|---|---|---|---|---|---|---|---|---|
| WWA3 | 8 | 15 | 20 | 26 | 31 | 37 | 42 | 47 |
| WWB4 | 7 | 19 | 28 | 37 | 44 | 54 | 60 | 68 |
| WWC2 | 8 | 19 | 27 | 36 | 43 | 52 | 58 | 66 |
| TWA1 | 12 | 16 | 18 | 21 | 24 | 27 | 30 | 33 |
| TWA2 | 10 | 15 | 19 | 23 | 27 | 32 | 35 | 40 |
| TWA3 | 8 | 14 | 19 | 24 | 28 | 34 | 38 | 43 |
| TWA4 | 7 | 13 | 18 | 24 | 28 | 34 | 38 | 44 |
| TWB3 | 7 | 13 | 17 | 22 | 26 | 32 | 35 | 41 |
| TWD4 | 8 | 18 | 25 | 34 | 40 | 48 | 54 | 61 |
| TWE2 | 9 | 19 | 26 | 34 | 40 | 48 | 54 | 61 |
| TWF3 | 6 | 11 | 16 | 21 | 24 | 30 | 34 | 39 |
| TWJ2 | 6 | 16 | 23 | 31 | 37 | 46 | 51 | 59 |
| DGA1 | 6 | 10 | 13 | 16 | 18 | 21 | 23 | 25 |
| DGA2 | 7 | 12 | 17 | 21 | 25 | 31 | 34 | 39 |
| DGA3 | 9 | 16 | 22 | 28 | 34 | 41 | 47 | 54 |
| DGB1 | 6 | 10 | 13 | 16 | 19 | 22 | 25 | 28 |
| DGB2 | 7 | 13 | 17 | 22 | 26 | 32 | 36 | 42 |
| DGB3 | 9 | 17 | 22 | 29 | 35 | 43 | 49 | 57 |
| DGC1 | 7 | 12 | 16 | 20 | 24 | 29 | 32 | 36 |
| DGC2 | 8 | 15 | 20 | 26 | 31 | 38 | 43 | 49 |
| DGC3 | 10 | 19 | 25 | 33 | 39 | 49 | 55 | 64 |
| SSA1 | 13 | 20 | 25 | 30 | 34 | 39 | 43 | 49 |
| SSB1 | 11 | 17 | 21 | 25 | 29 | 34 | 38 | 42 |
| SSB2 | 12 | 21 | 28 | 35 | 40 | 48 | 53 | 60 |
| SSB3 | 10 | 15 | 19 | 23 | 27 | 31 | 35 | 39 |
| SSC1 | 7 | 13 | 18 | 23 | 26 | 32 | 36 | 41 |
| SSC2 | 8 | 16 | 23 | 30 | 35 | 43 | 48 | 55 |
| SSE1 | 11 | 18 | 23 | 29 | 33 | 39 | 44 | 49 |
| SSE2 | 12 | 23 | 31 | 39 | 45 | 53 | 59 | 66 |
| SSE3 | 12 | 17 | 21 | 25 | 29 | 34 | 37 | 41 |

**BROWNSVILLE, TEXAS SSF (%)**     650 DD65

| Type | LCR = 200 | 100 | 70 | 50 | 40 | 30 | 25 | 20 |
|---|---|---|---|---|---|---|---|---|
| WWA3 | 39 | 60 | 72 | 82 | 88 | 94 | 96 | 98 |
| WWB4 | 45 | 70 | 82 | 90 | 94 | 98 | 99 | 100 |
| WWC2 | 45 | 69 | 81 | 90 | 94 | 97 | 99 | 99 |
| TWA1 | 30 | 45 | 54 | 64 | 71 | 79 | 84 | 89 |
| TWA2 | 33 | 51 | 63 | 73 | 80 | 87 | 91 | 95 |
| TWA3 | 34 | 54 | 66 | 76 | 83 | 90 | 93 | 96 |
| TWA4 | 34 | 54 | 66 | 77 | 84 | 90 | 94 | 97 |
| TWB3 | 31 | 50 | 61 | 72 | 79 | 87 | 91 | 95 |
| TWD4 | 41 | 64 | 76 | 86 | 91 | 96 | 98 | 99 |
| TWE2 | 42 | 65 | 77 | 87 | 92 | 96 | 98 | 99 |
| TWF3 | 30 | 50 | 62 | 73 | 80 | 88 | 91 | 95 |
| TWJ2 | 39 | 62 | 75 | 85 | 90 | 95 | 97 | 99 |
| DGA1 | 31 | 50 | 61 | 70 | 76 | 83 | 87 | 90 |
| DGA2 | 31 | 51 | 63 | 73 | 80 | 87 | 90 | 94 |
| DGA3 | 34 | 57 | 70 | 80 | 86 | 91 | 93 | 95 |
| DGB1 | 32 | 54 | 67 | 78 | 84 | 89 | 92 | 94 |
| DGB2 | 32 | 55 | 69 | 80 | 86 | 92 | 95 | 97 |
| DGB3 | 34 | 60 | 74 | 85 | 90 | 94 | 96 | 97 |
| DGC1 | 36 | 62 | 76 | 85 | 90 | 94 | 95 | 97 |
| DGC2 | 35 | 61 | 75 | 86 | 91 | 95 | 97 | 99 |
| DGC3 | 39 | 66 | 80 | 90 | 94 | 97 | 98 | 98 |
| SSA1 | 42 | 62 | 73 | 82 | 87 | 93 | 96 | 98 |
| SSB1 | 36 | 55 | 66 | 76 | 83 | 89 | 93 | 96 |
| SSB2 | 44 | 65 | 77 | 86 | 91 | 95 | 97 | 99 |
| SSB3 | 35 | 54 | 65 | 75 | 81 | 88 | 92 | 95 |
| SSC1 | 35 | 55 | 67 | 78 | 84 | 91 | 94 | 97 |
| SSC2 | 39 | 62 | 74 | 84 | 90 | 95 | 97 | 99 |
| SSE1 | 45 | 66 | 77 | 86 | 91 | 95 | 97 | 99 |
| SSE2 | 51 | 74 | 85 | 92 | 96 | 98 | 99 | 100 |
| SSE3 | 37 | 56 | 67 | 77 | 83 | 90 | 93 | 96 |

**TABLE H.3 Annual Passive Heating Performance: SSF** *(Continued)*

| DALLAS, TEXAS SSF (%) Type | LCR = 200 | 100 | 70 | 50 | 40 | 30 | 25 | 20 |
|---|---|---|---|---|---|---|---|---|
| WWA3 | 17 | 30 | 39 | 49 | 56 | 66 | 72 | 79 |
| WWB4 | 19 | 37 | 49 | 62 | 70 | 79 | 85 | 90 |
| WWC2 | 19 | 37 | 49 | 61 | 69 | 78 | 84 | 89 |
| TWA1 | 17 | 24 | 30 | 36 | 41 | 48 | 52 | 59 |
| TWA2 | 16 | 26 | 33 | 42 | 48 | 56 | 62 | 69 |
| TWA3 | 16 | 27 | 35 | 44 | 50 | 60 | 65 | 73 |
| TWA4 | 14 | 26 | 35 | 44 | 51 | 60 | 66 | 74 |
| TWB3 | 14 | 24 | 32 | 40 | 47 | 56 | 62 | 69 |
| TWD4 | 18 | 34 | 45 | 56 | 64 | 74 | 79 | 86 |
| TWE2 | 19 | 35 | 46 | 57 | 65 | 74 | 80 | 86 |
| TWF3 | 13 | 23 | 31 | 40 | 46 | 56 | 62 | 69 |
| TWJ2 | 16 | 32 | 42 | 54 | 61 | 71 | 77 | 84 |
| DGA1 | 12 | 22 | 29 | 36 | 41 | 47 | 51 | 56 |
| DGA2 | 13 | 24 | 31 | 40 | 47 | 55 | 60 | 66 |
| DGA3 | 15 | 28 | 38 | 49 | 56 | 66 | 72 | 78 |
| DGB1 | 12 | 22 | 30 | 38 | 45 | 54 | 59 | 64 |
| DGB2 | 13 | 24 | 33 | 43 | 51 | 61 | 67 | 74 |
| DGB3 | 16 | 28 | 38 | 50 | 59 | 71 | 77 | 83 |
| DGC1 | 15 | 26 | 35 | 45 | 53 | 63 | 69 | 75 |
| DGC2 | 15 | 28 | 38 | 49 | 57 | 68 | 74 | 81 |
| DGC3 | 18 | 32 | 43 | 57 | 66 | 77 | 83 | 89 |
| SSA1 | 21 | 32 | 40 | 49 | 56 | 64 | 70 | 76 |
| SSB1 | 17 | 28 | 35 | 43 | 49 | 58 | 63 | 70 |
| SSB2 | 21 | 35 | 45 | 55 | 62 | 71 | 77 | 83 |
| SSB3 | 17 | 26 | 33 | 41 | 47 | 55 | 60 | 67 |
| SSC1 | 15 | 27 | 35 | 44 | 51 | 60 | 66 | 73 |
| SSC2 | 17 | 32 | 42 | 52 | 60 | 70 | 76 | 82 |
| SSE1 | 20 | 33 | 42 | 52 | 59 | 68 | 74 | 80 |
| SSE2 | 23 | 41 | 52 | 63 | 70 | 79 | 84 | 90 |
| SSE3 | 18 | 28 | 35 | 43 | 49 | 58 | 63 | 70 |

| HOUSTON, TEXAS SSF (%) Type | LCR = 200 | 100 | 70 | 50 | 40 | 30 | 25 | 20 |
|---|---|---|---|---|---|---|---|---|
| WWA3 | 21 | 37 | 47 | 58 | 65 | 74 | 80 | 86 |
| WWB4 | 24 | 45 | 58 | 70 | 77 | 86 | 90 | 94 |
| WWC2 | 24 | 44 | 57 | 69 | 76 | 85 | 89 | 93 |
| TWA1 | 20 | 29 | 35 | 42 | 48 | 55 | 61 | 67 |
| TWA2 | 20 | 32 | 40 | 49 | 56 | 65 | 70 | 77 |
| TWA3 | 20 | 33 | 42 | 52 | 59 | 68 | 74 | 80 |
| TWA4 | 18 | 32 | 42 | 52 | 59 | 69 | 75 | 81 |
| TWB3 | 18 | 30 | 38 | 48 | 55 | 64 | 70 | 77 |
| TWD4 | 22 | 41 | 52 | 64 | 72 | 81 | 86 | 91 |
| TWE2 | 24 | 42 | 53 | 65 | 72 | 81 | 86 | 91 |
| TWF3 | 16 | 29 | 38 | 48 | 55 | 64 | 70 | 77 |
| TWJ2 | 21 | 38 | 50 | 62 | 69 | 79 | 84 | 89 |
| DGA1 | 16 | 28 | 36 | 44 | 49 | 56 | 61 | 65 |
| DGA2 | 16 | 29 | 38 | 48 | 55 | 63 | 69 | 74 |
| DGA3 | 19 | 34 | 45 | 57 | 64 | 74 | 79 | 84 |
| DGB1 | 16 | 28 | 38 | 48 | 55 | 64 | 69 | 74 |
| DGB2 | 17 | 30 | 40 | 52 | 60 | 70 | 75 | 81 |
| DGB3 | 19 | 35 | 46 | 59 | 68 | 78 | 83 | 89 |
| DGC1 | 18 | 33 | 44 | 56 | 64 | 73 | 78 | 83 |
| DGC2 | 19 | 34 | 46 | 59 | 67 | 77 | 82 | 88 |
| DGC3 | 21 | 39 | 52 | 66 | 75 | 84 | 89 | 93 |
| SSA1 | 25 | 39 | 48 | 58 | 65 | 73 | 78 | 84 |
| SSB1 | 21 | 34 | 42 | 52 | 58 | 67 | 72 | 79 |
| SSB2 | 26 | 42 | 53 | 64 | 71 | 79 | 84 | 89 |
| SSB3 | 20 | 32 | 41 | 50 | 56 | 65 | 70 | 77 |
| SSC1 | 19 | 33 | 42 | 52 | 60 | 69 | 75 | 81 |
| SSC2 | 22 | 38 | 49 | 61 | 68 | 77 | 83 | 88 |
| SSE1 | 25 | 41 | 51 | 62 | 69 | 77 | 82 | 88 |
| SSE2 | 29 | 49 | 61 | 72 | 79 | 87 | 90 | 94 |
| SSE3 | 22 | 34 | 43 | 52 | 59 | 67 | 73 | 79 |

DALLAS, TEXAS 2297 DD65; HOUSTON, TEXAS 1434 DD65

| EL PASO, TEXAS SSF (%) Type | LCR = 200 | 100 | 70 | 50 | 40 | 30 | 25 | 20 |
|---|---|---|---|---|---|---|---|---|
| WWA3 | 22 | 38 | 49 | 60 | 68 | 77 | 83 | 88 |
| WWB4 | 25 | 47 | 60 | 72 | 80 | 88 | 92 | 95 |
| WWC2 | 25 | 46 | 59 | 71 | 79 | 87 | 91 | 95 |
| TWA1 | 20 | 29 | 36 | 44 | 49 | 58 | 63 | 70 |
| TWA2 | 20 | 33 | 41 | 51 | 58 | 67 | 73 | 80 |
| TWA3 | 20 | 34 | 44 | 54 | 61 | 71 | 77 | 83 |
| TWA4 | 19 | 34 | 44 | 54 | 62 | 72 | 77 | 84 |
| TWB3 | 18 | 31 | 40 | 50 | 57 | 67 | 73 | 79 |
| TWD4 | 23 | 42 | 54 | 66 | 74 | 83 | 88 | 93 |
| TWE2 | 24 | 43 | 55 | 67 | 75 | 84 | 88 | 93 |
| TWF3 | 16 | 30 | 40 | 50 | 57 | 67 | 73 | 80 |
| TWJ2 | 21 | 40 | 52 | 64 | 72 | 81 | 86 | 91 |
| DGA1 | 16 | 29 | 37 | 46 | 52 | 59 | 64 | 69 |
| DGA2 | 17 | 30 | 40 | 50 | 57 | 66 | 71 | 77 |
| DGA3 | 19 | 35 | 47 | 59 | 67 | 76 | 81 | 86 |
| DGB1 | 16 | 30 | 40 | 51 | 58 | 67 | 72 | 78 |
| DGB2 | 17 | 31 | 42 | 55 | 63 | 73 | 79 | 84 |
| DGB3 | 19 | 36 | 48 | 62 | 71 | 81 | 86 | 91 |
| DGC1 | 19 | 35 | 47 | 59 | 68 | 77 | 82 | 86 |
| DGC2 | 20 | 36 | 48 | 62 | 70 | 80 | 85 | 90 |
| DGC3 | 22 | 40 | 55 | 69 | 78 | 87 | 91 | 94 |
| SSA1 | 25 | 39 | 49 | 58 | 65 | 74 | 79 | 85 |
| SSB1 | 21 | 34 | 43 | 52 | 59 | 68 | 73 | 80 |
| SSB2 | 26 | 42 | 53 | 64 | 71 | 80 | 85 | 90 |
| SSB3 | 20 | 32 | 41 | 50 | 56 | 65 | 71 | 77 |
| SSC1 | 20 | 34 | 44 | 55 | 62 | 72 | 77 | 84 |
| SSC2 | 22 | 40 | 51 | 63 | 71 | 80 | 85 | 90 |
| SSE1 | 25 | 41 | 52 | 62 | 69 | 78 | 83 | 89 |
| SSE2 | 29 | 49 | 61 | 73 | 80 | 87 | 91 | 95 |
| SSE3 | 22 | 35 | 43 | 52 | 59 | 68 | 73 | 80 |

| LUBBOCK, TEXAS SSF (%) Type | LCR = 200 | 100 | 70 | 50 | 40 | 30 | 25 | 20 |
|---|---|---|---|---|---|---|---|---|
| WWA3 | 17 | 30 | 40 | 50 | 57 | 67 | 73 | 80 |
| WWB4 | 18 | 38 | 50 | 63 | 71 | 80 | 86 | 91 |
| WWC2 | 19 | 37 | 49 | 61 | 69 | 79 | 85 | 90 |
| TWA1 | 17 | 24 | 30 | 36 | 41 | 48 | 53 | 60 |
| TWA2 | 16 | 26 | 34 | 42 | 48 | 57 | 63 | 70 |
| TWA3 | 15 | 27 | 35 | 44 | 51 | 60 | 67 | 74 |
| TWA4 | 14 | 26 | 35 | 45 | 52 | 61 | 67 | 75 |
| TWB3 | 14 | 24 | 32 | 41 | 47 | 56 | 62 | 70 |
| TWD4 | 17 | 34 | 45 | 57 | 65 | 75 | 80 | 87 |
| TWE2 | 19 | 35 | 46 | 58 | 65 | 75 | 81 | 87 |
| TWF3 | 12 | 23 | 31 | 40 | 47 | 56 | 63 | 70 |
| TWJ2 | 16 | 32 | 43 | 54 | 62 | 72 | 78 | 85 |
| DGA1 | 12 | 22 | 29 | 36 | 41 | 48 | 52 | 57 |
| DGA2 | 13 | 23 | 32 | 41 | 47 | 56 | 61 | 67 |
| DGA3 | 15 | 28 | 38 | 49 | 57 | 67 | 73 | 79 |
| DGB1 | 12 | 22 | 30 | 39 | 46 | 55 | 60 | 66 |
| DGB2 | 13 | 24 | 33 | 43 | 51 | 62 | 68 | 75 |
| DGB3 | 15 | 28 | 39 | 51 | 60 | 72 | 78 | 84 |
| DGC1 | 14 | 26 | 35 | 46 | 54 | 65 | 70 | 76 |
| DGC2 | 15 | 28 | 38 | 50 | 59 | 70 | 76 | 82 |
| DGC3 | 17 | 32 | 44 | 57 | 67 | 79 | 84 | 90 |
| SSA1 | 21 | 33 | 41 | 50 | 56 | 65 | 70 | 77 |
| SSB1 | 17 | 28 | 35 | 43 | 50 | 58 | 64 | 71 |
| SSB2 | 21 | 35 | 45 | 55 | 63 | 72 | 77 | 84 |
| SSB3 | 16 | 26 | 33 | 41 | 47 | 55 | 61 | 68 |
| SSC1 | 15 | 27 | 35 | 45 | 52 | 61 | 67 | 74 |
| SSC2 | 17 | 32 | 42 | 53 | 61 | 71 | 77 | 83 |
| SSE1 | 20 | 34 | 43 | 53 | 60 | 69 | 74 | 81 |
| SSE2 | 23 | 41 | 52 | 63 | 71 | 80 | 85 | 90 |
| SSE3 | 18 | 28 | 36 | 44 | 50 | 58 | 64 | 70 |

EL PASO, TEXAS 2685 DD65; LUBBOCK, TEXAS 3551 DD65

## TABLE H.3 Annual Passive Heating Performance: SSF *(Continued)*

**SAN ANTONIO, TEXAS SSF (%)** — 1570 DD65

| Type | LCR = 200 | 100 | 70 | 50 | 40 | 30 | 25 | 20 |
|------|-----|-----|-----|-----|-----|-----|-----|-----|
| WWA3 | 23 | 40 | 50 | 61 | 69 | 78 | 83 | 88 |
| WWB4 | 26 | 48 | 61 | 73 | 80 | 88 | 92 | 95 |
| WWC2 | 26 | 47 | 60 | 72 | 79 | 87 | 91 | 95 |
| TWA1 | 21 | 30 | 37 | 45 | 50 | 59 | 64 | 71 |
| TWA2 | 21 | 34 | 42 | 52 | 59 | 68 | 74 | 80 |
| TWA3 | 21 | 35 | 45 | 55 | 62 | 71 | 77 | 83 |
| TWA4 | 20 | 35 | 45 | 55 | 63 | 72 | 78 | 84 |
| TWB3 | 19 | 32 | 41 | 51 | 58 | 67 | 73 | 80 |
| TWD4 | 24 | 43 | 55 | 67 | 75 | 83 | 88 | 93 |
| TWE2 | 26 | 44 | 56 | 68 | 75 | 84 | 88 | 93 |
| TWF3 | 18 | 31 | 41 | 51 | 58 | 68 | 74 | 80 |
| TWJ2 | 22 | 41 | 53 | 65 | 73 | 82 | 87 | 91 |
| DGA1 | 17 | 30 | 38 | 47 | 53 | 60 | 64 | 69 |
| DGA2 | 18 | 32 | 41 | 51 | 58 | 67 | 72 | 78 |
| DGA3 | 20 | 36 | 48 | 60 | 68 | 77 | 81 | 86 |
| DGB1 | 17 | 31 | 41 | 52 | 59 | 68 | 73 | 78 |
| DGB2 | 18 | 33 | 44 | 56 | 64 | 74 | 79 | 85 |
| DGB3 | 20 | 37 | 50 | 63 | 72 | 81 | 86 | 91 |
| DGC1 | 20 | 36 | 48 | 60 | 68 | 77 | 82 | 86 |
| DGC2 | 21 | 37 | 49 | 63 | 71 | 80 | 85 | 90 |
| DGC3 | 23 | 42 | 56 | 70 | 78 | 87 | 91 | 94 |
| SSA1 | 27 | 41 | 51 | 61 | 68 | 76 | 81 | 87 |
| SSB1 | 23 | 36 | 45 | 54 | 61 | 70 | 75 | 82 |
| SSB2 | 27 | 45 | 55 | 66 | 73 | 82 | 86 | 91 |
| SSB3 | 22 | 34 | 43 | 52 | 59 | 68 | 73 | 80 |
| SSC1 | 21 | 35 | 45 | 56 | 63 | 72 | 78 | 84 |
| SSC2 | 23 | 41 | 53 | 64 | 71 | 80 | 85 | 91 |
| SSE1 | 27 | 44 | 54 | 65 | 72 | 80 | 85 | 90 |
| SSE2 | 31 | 52 | 64 | 75 | 82 | 89 | 92 | 96 |
| SSE3 | 24 | 37 | 45 | 55 | 62 | 70 | 76 | 82 |

**SALT LAKE CITY, UTAH SSF (%)** — 5988 DD65

| Type | LCR = 100 | 70 | 50 | 40 | 30 | 25 | 20 | 15 |
|------|-----|-----|-----|-----|-----|-----|-----|-----|
| WWA3 | 19 | 25 | 32 | 37 | 44 | 49 | 55 | 64 |
| WWB4 | 23 | 33 | 43 | 50 | 60 | 66 | 73 | 82 |
| WWC2 | 23 | 32 | 42 | 49 | 59 | 65 | 72 | 80 |
| TWA1 | 18 | 21 | 25 | 28 | 32 | 35 | 39 | 45 |
| TWA2 | 18 | 22 | 28 | 32 | 38 | 42 | 47 | 54 |
| TWA3 | 17 | 22 | 29 | 33 | 40 | 44 | 50 | 58 |
| TWA4 | 16 | 22 | 28 | 33 | 40 | 45 | 51 | 60 |
| TWB3 | 16 | 21 | 26 | 31 | 37 | 42 | 48 | 55 |
| TWD4 | 22 | 30 | 39 | 45 | 55 | 60 | 68 | 76 |
| TWE2 | 23 | 31 | 39 | 46 | 55 | 61 | 68 | 76 |
| TWF3 | 14 | 19 | 25 | 30 | 36 | 41 | 46 | 54 |
| TWJ2 | 20 | 28 | 37 | 43 | 52 | 58 | 65 | 74 |
| DGA1 | 12 | 16 | 20 | 23 | 27 | 30 | 32 | 35 |
| DGA2 | 15 | 20 | 26 | 30 | 36 | 41 | 46 | 52 |
| DGA3 | 19 | 25 | 33 | 39 | 47 | 53 | 60 | 68 |
| DGB1 | 13 | 17 | 21 | 25 | 30 | 33 | 37 | 41 |
| DGB2 | 15 | 21 | 27 | 32 | 39 | 44 | 50 | 58 |
| DGB3 | 19 | 26 | 34 | 40 | 50 | 56 | 64 | 73 |
| DGC1 | 15 | 20 | 26 | 31 | 37 | 41 | 46 | 52 |
| DGC2 | 18 | 24 | 31 | 37 | 45 | 51 | 58 | 66 |
| DGC3 | 22 | 29 | 38 | 46 | 56 | 62 | 70 | 79 |
| SSA1 | 23 | 28 | 34 | 38 | 44 | 48 | 54 | 60 |
| SSB1 | 19 | 24 | 29 | 33 | 39 | 43 | 48 | 55 |
| SSB2 | 24 | 31 | 39 | 45 | 52 | 58 | 64 | 72 |
| SSB3 | 18 | 22 | 27 | 31 | 36 | 39 | 44 | 50 |
| SSC1 | 16 | 22 | 28 | 32 | 39 | 43 | 49 | 57 |
| SSC2 | 20 | 27 | 35 | 41 | 49 | 55 | 62 | 70 |
| SSE1 | 22 | 27 | 34 | 38 | 45 | 49 | 55 | 62 |
| SSE2 | 27 | 35 | 44 | 50 | 58 | 63 | 70 | 78 |
| SSE3 | 20 | 24 | 29 | 33 | 38 | 42 | 47 | 53 |

**BRYCE CANYON, UTAH SSF (%)** — 9136 DD65

| Type | LCR = 100 | 70 | 50 | 40 | 30 | 25 | 20 | 15 |
|------|-----|-----|-----|-----|-----|-----|-----|-----|
| WWA3 | 16 | 22 | 30 | 35 | 43 | 59 | 56 | 65 |
| WWB4 | 21 | 31 | 42 | 49 | 60 | 67 | 75 | 84 |
| WWC2 | 21 | 30 | 40 | 48 | 58 | 65 | 73 | 82 |
| TWA1 | 16 | 19 | 22 | 25 | 30 | 33 | 38 | 44 |
| TWA2 | 15 | 20 | 25 | 29 | 36 | 40 | 46 | 55 |
| TWA3 | 15 | 20 | 26 | 31 | 38 | 43 | 50 | 59 |
| TWA4 | 14 | 19 | 26 | 31 | 39 | 44 | 51 | 60 |
| TWB3 | 13 | 18 | 24 | 29 | 35 | 40 | 47 | 55 |
| TWD4 | 19 | 28 | 37 | 44 | 54 | 60 | 68 | 78 |
| TWE2 | 20 | 28 | 38 | 45 | 54 | 60 | 68 | 78 |
| TWF3 | 12 | 17 | 23 | 27 | 34 | 39 | 46 | 55 |
| TWJ2 | 17 | 25 | 35 | 42 | 51 | 58 | 65 | 75 |
| DGA1 | 10 | 14 | 18 | 21 | 25 | 28 | 30 | 34 |
| DGA2 | 12 | 17 | 23 | 28 | 35 | 39 | 44 | 51 |
| DGA3 | 16 | 23 | 31 | 37 | 46 | 52 | 60 | 69 |
| DGB1 | 10 | 14 | 18 | 22 | 27 | 31 | 36 | 41 |
| DGB2 | 13 | 18 | 24 | 29 | 37 | 42 | 49 | 58 |
| DGB3 | 17 | 23 | 32 | 38 | 48 | 55 | 64 | 74 |
| DGC1 | 13 | 17 | 23 | 28 | 35 | 39 | 46 | 54 |
| DGC2 | 15 | 21 | 29 | 35 | 43 | 50 | 58 | 67 |
| DGC3 | 19 | 27 | 36 | 43 | 55 | 62 | 72 | 81 |
| SSA1 | 23 | 29 | 35 | 40 | 47 | 52 | 58 | 66 |
| SSB1 | 18 | 23 | 29 | 34 | 40 | 45 | 51 | 59 |
| SSB2 | 24 | 32 | 40 | 46 | 55 | 61 | 68 | 76 |
| SSB3 | 17 | 22 | 27 | 32 | 38 | 42 | 47 | 55 |
| SSC1 | 14 | 19 | 25 | 30 | 37 | 42 | 49 | 58 |
| SSC2 | 18 | 25 | 33 | 40 | 49 | 55 | 62 | 72 |
| SSE1 | 21 | 28 | 35 | 41 | 48 | 53 | 60 | 68 |
| SSE2 | 27 | 36 | 46 | 53 | 62 | 68 | 75 | 83 |
| SSE3 | 19 | 24 | 30 | 34 | 40 | 45 | 50 | 58 |

**BURLINGTON, VERMONT SSF (%)** — 7878 DD65

| Type | LCR = 100 | 70 | 50 | 40 | 30 | 25 | 20 | 15 |
|------|-----|-----|-----|-----|-----|-----|-----|-----|
| WWA3 | 4 | 6 | 8 | 9 | 10 | 11 | 12 | 13 |
| WWB4 | 6 | 10 | 16 | 20 | 26 | 31 | 36 | 44 |
| WWC2 | 6 | 10 | 15 | 19 | 25 | 29 | 34 | 41 |
| TWA1 | 10 | 11 | 11 | 11 | 11 | 11 | 11 | 10 |
| TWA2 | 7 | 8 | 9 | 10 | 11 | 12 | 12 | 13 |
| TWA3 | 6 | 7 | 8 | 10 | 11 | 12 | 13 | 14 |
| TWA4 | 4 | 6 | 8 | 9 | 11 | 12 | 14 | 15 |
| TWB3 | 5 | 7 | 8 | 10 | 11 | 12 | 14 | 16 |
| TWD4 | 7 | 10 | 15 | 19 | 24 | 27 | 32 | 39 |
| TWE2 | 7 | 11 | 15 | 18 | 23 | 26 | 31 | 37 |
| TWF3 | 3 | 4 | 6 | 7 | 8 | 9 | 10 | 11 |
| TWJ2 | 5 | 9 | 13 | 16 | 21 | 25 | 29 | 35 |
| DGA1 | 2 | 2 | 2 | 1 | 1 | 0 | -1 | -5 |
| DGA2 | 5 | 7 | 8 | 10 | 11 | 13 | 14 | 16 |
| DGA3 | 8 | 11 | 15 | 17 | 22 | 25 | 29 | 34 |
| DGB1 | 2 | 2 | 2 | 1 | 1 | 0 | -1 | -4 |
| DGB2 | 5 | 7 | 9 | 10 | 12 | 13 | 15 | 17 |
| DGB3 | 9 | 12 | 15 | 18 | 23 | 26 | 31 | 37 |
| DGC1 | 3 | 4 | 4 | 4 | 5 | 4 | 3 | 2 |
| DGC2 | 7 | 9 | 11 | 13 | 16 | 18 | 20 | 23 |
| DGC3 | 10 | 13 | 18 | 21 | 26 | 30 | 35 | 42 |
| SSA1 | 11 | 13 | 14 | 16 | 17 | 18 | 20 | 21 |
| SSB1 | 9 | 10 | 11 | 13 | 14 | 15 | 16 | 17 |
| SSB2 | 11 | 15 | 19 | 22 | 26 | 29 | 33 | 38 |
| SSB3 | 8 | 9 | 10 | 10 | 11 | 11 | 12 | 11 |
| SSC1 | 4 | 5 | 6 | 7 | 7 | 7 | 7 | 7 |
| SSC2 | 5 | 8 | 11 | 14 | 17 | 19 | 22 | 27 |
| SSE1 | 7 | 8 | 10 | 11 | 12 | 12 | 12 | 11 |
| SSE2 | 9 | 13 | 17 | 21 | 25 | 28 | 32 | 37 |
| SSE3 | 9 | 10 | 11 | 12 | 13 | 13 | 14 | 14 |

**TABLE H.3 Annual Passive Heating Performance: SSF** *(Continued)*

| NORFOLK, VIRGINIA SSF (%) | | | | | | | 3493 DD65 | |
|---|---|---|---|---|---|---|---|---|
| Type | LCR = 200 | 100 | 70 | 50 | 40 | 30 | 25 | 20 |
| WWA3 | 12 | 22 | 29 | 37 | 42 | 51 | 56 | 63 |
| WWB4 | 12 | 27 | 37 | 49 | 56 | 67 | 73 | 80 |
| WWC2 | 12 | 27 | 37 | 47 | 55 | 65 | 71 | 78 |
| TWA1 | 14 | 19 | 23 | 27 | 31 | 36 | 40 | 45 |
| TWA2 | 12 | 20 | 25 | 31 | 36 | 43 | 48 | 54 |
| TWA3 | 11 | 20 | 26 | 33 | 38 | 46 | 51 | 57 |
| TWA4 | 10 | 19 | 25 | 33 | 38 | 46 | 52 | 59 |
| TWB3 | 10 | 18 | 23 | 30 | 35 | 42 | 48 | 54 |
| TWD4 | 12 | 25 | 34 | 44 | 51 | 61 | 67 | 74 |
| TWE2 | 13 | 26 | 35 | 44 | 51 | 61 | 67 | 74 |
| TWF3 | 8 | 16 | 22 | 29 | 34 | 42 | 47 | 53 |
| TWJ2 | 10 | 23 | 32 | 41 | 48 | 58 | 64 | 72 |
| DGA1 | 8 | 15 | 19 | 24 | 28 | 33 | 36 | 39 |
| DGA2 | 9 | 17 | 23 | 29 | 35 | 42 | 46 | 52 |
| DGA3 | 11 | 21 | 28 | 37 | 44 | 53 | 59 | 66 |
| DGB1 | 8 | 15 | 20 | 25 | 30 | 36 | 40 | 45 |
| DGB2 | 9 | 17 | 23 | 31 | 36 | 45 | 50 | 58 |
| DGB3 | 11 | 21 | 29 | 38 | 45 | 56 | 63 | 71 |
| DGC1 | 10 | 18 | 24 | 31 | 36 | 44 | 49 | 56 |
| DGC2 | 11 | 20 | 27 | 36 | 42 | 52 | 58 | 66 |
| DGC3 | 13 | 24 | 33 | 43 | 51 | 62 | 70 | 78 |
| SSA1 | 16 | 25 | 31 | 38 | 43 | 50 | 55 | 62 |
| SSB1 | 13 | 21 | 26 | 33 | 38 | 44 | 49 | 55 |
| SSB2 | 16 | 27 | 35 | 44 | 50 | 59 | 64 | 71 |
| SSB3 | 13 | 20 | 25 | 31 | 35 | 42 | 46 | 52 |
| SSC1 | 10 | 19 | 25 | 32 | 37 | 45 | 50 | 57 |
| SSC2 | 11 | 23 | 31 | 40 | 46 | 56 | 62 | 69 |
| SSE1 | 15 | 25 | 31 | 39 | 45 | 53 | 58 | 64 |
| SSE2 | 16 | 30 | 40 | 50 | 57 | 66 | 72 | 78 |
| SSE3 | 14 | 22 | 27 | 33 | 38 | 44 | 49 | 55 |

| ROANOKE, VIRGINIA SSF (%) | | | | | | | 4314 DD65 | |
|---|---|---|---|---|---|---|---|---|
| Type | LCR = 100 | 70 | 50 | 40 | 30 | 25 | 20 | 15 |
| WWA3 | 17 | 23 | 30 | 35 | 42 | 47 | 54 | 63 |
| WWB4 | 22 | 31 | 41 | 49 | 59 | 65 | 73 | 82 |
| WWC2 | 21 | 30 | 40 | 47 | 57 | 64 | 71 | 80 |
| TWA1 | 17 | 20 | 23 | 26 | 30 | 33 | 37 | 43 |
| TWA2 | 16 | 21 | 26 | 30 | 36 | 40 | 45 | 53 |
| TWA3 | 16 | 21 | 27 | 31 | 38 | 43 | 49 | 57 |
| TWA4 | 15 | 20 | 26 | 31 | 38 | 43 | 50 | 59 |
| TWB3 | 14 | 19 | 25 | 29 | 35 | 40 | 46 | 54 |
| TWD4 | 20 | 28 | 37 | 44 | 53 | 59 | 67 | 76 |
| TWE2 | 21 | 29 | 38 | 44 | 53 | 59 | 67 | 76 |
| TWF3 | 13 | 17 | 23 | 28 | 34 | 39 | 45 | 53 |
| TWJ2 | 18 | 26 | 35 | 41 | 50 | 57 | 64 | 74 |
| DGA1 | 11 | 14 | 18 | 21 | 25 | 27 | 30 | 33 |
| DGA2 | 13 | 18 | 24 | 28 | 35 | 39 | 44 | 50 |
| DGA3 | 17 | 23 | 31 | 37 | 46 | 52 | 59 | 67 |
| DGB1 | 11 | 15 | 19 | 22 | 27 | 30 | 35 | 40 |
| DGB2 | 14 | 19 | 25 | 30 | 37 | 42 | 48 | 56 |
| DGB3 | 18 | 24 | 32 | 38 | 48 | 54 | 63 | 73 |
| DGC1 | 14 | 18 | 24 | 28 | 34 | 38 | 44 | 51 |
| DGC2 | 16 | 22 | 29 | 35 | 43 | 49 | 56 | 65 |
| DGC3 | 20 | 28 | 36 | 43 | 54 | 61 | 70 | 79 |
| SSA1 | 21 | 27 | 32 | 37 | 43 | 48 | 54 | 61 |
| SSB1 | 18 | 22 | 28 | 32 | 38 | 42 | 47 | 55 |
| SSB2 | 23 | 30 | 38 | 44 | 52 | 57 | 64 | 73 |
| SSB3 | 17 | 21 | 26 | 29 | 35 | 39 | 44 | 50 |
| SSC1 | 15 | 20 | 26 | 30 | 37 | 41 | 47 | 56 |
| SSC2 | 18 | 25 | 33 | 39 | 48 | 53 | 61 | 70 |
| SSE1 | 20 | 26 | 33 | 37 | 44 | 49 | 55 | 63 |
| SSE2 | 25 | 34 | 43 | 49 | 58 | 64 | 71 | 79 |
| SSE3 | 18 | 23 | 28 | 32 | 37 | 41 | 46 | 53 |

| RICHMOND, VIRGINIA SSF (%) | | | | | | | 3941 DD65 | |
|---|---|---|---|---|---|---|---|---|
| Type | LCR = 200 | 100 | 70 | 50 | 40 | 30 | 25 | 20 |
| WWA3 | 9 | 17 | 23 | 30 | 35 | 43 | 48 | 54 |
| WWB4 | 8 | 22 | 31 | 42 | 49 | 59 | 65 | 73 |
| WWC2 | 9 | 22 | 31 | 41 | 48 | 57 | 64 | 71 |
| TWA1 | 13 | 17 | 20 | 24 | 26 | 31 | 34 | 38 |
| TWA2 | 11 | 17 | 21 | 26 | 30 | 36 | 40 | 46 |
| TWA3 | 9 | 16 | 21 | 27 | 32 | 38 | 43 | 49 |
| TWA4 | 8 | 15 | 21 | 27 | 32 | 39 | 44 | 50 |
| TWB3 | 8 | 15 | 19 | 25 | 29 | 36 | 40 | 46 |
| TWD4 | 9 | 20 | 28 | 37 | 44 | 53 | 59 | 67 |
| TWE2 | 10 | 21 | 29 | 38 | 45 | 54 | 60 | 67 |
| TWF3 | 6 | 13 | 18 | 24 | 28 | 35 | 39 | 45 |
| TWJ2 | 8 | 19 | 26 | 35 | 42 | 51 | 57 | 64 |
| DGA1 | 6 | 11 | 15 | 19 | 22 | 26 | 28 | 30 |
| DGA2 | 7 | 14 | 19 | 24 | 29 | 35 | 39 | 44 |
| DGA3 | 9 | 18 | 24 | 31 | 37 | 46 | 52 | 59 |
| DGB1 | 6 | 11 | 15 | 20 | 23 | 28 | 31 | 35 |
| DGB2 | 8 | 14 | 19 | 25 | 30 | 37 | 42 | 48 |
| DGB3 | 10 | 18 | 25 | 33 | 39 | 48 | 55 | 63 |
| DGC1 | 8 | 14 | 19 | 24 | 29 | 35 | 39 | 44 |
| DGC2 | 9 | 17 | 23 | 30 | 35 | 43 | 49 | 56 |
| DGC3 | 11 | 21 | 28 | 37 | 44 | 54 | 61 | 70 |
| SSA1 | 14 | 21 | 27 | 33 | 37 | 43 | 48 | 53 |
| SSB1 | 12 | 18 | 23 | 28 | 32 | 38 | 42 | 47 |
| SSB2 | 13 | 23 | 30 | 38 | 44 | 52 | 57 | 64 |
| SSB3 | 11 | 17 | 21 | 26 | 29 | 35 | 39 | 44 |
| SSC1 | 8 | 15 | 20 | 26 | 31 | 37 | 42 | 48 |
| SSC2 | 9 | 19 | 26 | 34 | 40 | 48 | 54 | 61 |
| SSE1 | 12 | 20 | 26 | 33 | 37 | 44 | 49 | 55 |
| SSE2 | 13 | 25 | 34 | 43 | 49 | 58 | 64 | 71 |
| SSE3 | 13 | 19 | 23 | 28 | 32 | 37 | 41 | 46 |

| SEATTLE-TACOMA, WASHINGTON SSF (%) | | | | | | | 5188 DD65 | |
|---|---|---|---|---|---|---|---|---|
| Type | LCR = 100 | 70 | 50 | 40 | 30 | 25 | 20 | 15 |
| WWA3 | 14 | 18 | 23 | 26 | 30 | 33 | 36 | 41 |
| WWB4 | 17 | 25 | 32 | 38 | 46 | 51 | 57 | 64 |
| WWC2 | 17 | 24 | 32 | 37 | 44 | 49 | 55 | 62 |
| TWA1 | 16 | 18 | 20 | 22 | 25 | 26 | 28 | 31 |
| TWA2 | 14 | 18 | 21 | 24 | 27 | 30 | 33 | 36 |
| TWA3 | 14 | 17 | 21 | 24 | 29 | 31 | 35 | 39 |
| TWA4 | 12 | 17 | 21 | 24 | 29 | 32 | 35 | 40 |
| TWB3 | 13 | 16 | 20 | 23 | 27 | 30 | 34 | 38 |
| TWD4 | 17 | 23 | 30 | 35 | 42 | 46 | 52 | 59 |
| TWE2 | 17 | 24 | 30 | 35 | 42 | 46 | 51 | 58 |
| TWF3 | 11 | 14 | 18 | 21 | 25 | 28 | 31 | 36 |
| TWJ2 | 15 | 21 | 28 | 33 | 39 | 44 | 49 | 57 |
| DGA1 | 9 | 12 | 14 | 16 | 17 | 18 | 18 | 18 |
| DGA2 | 12 | 16 | 20 | 23 | 27 | 30 | 33 | 37 |
| DGA3 | 16 | 21 | 27 | 31 | 37 | 41 | 46 | 53 |
| DGB1 | 10 | 12 | 15 | 17 | 19 | 20 | 21 | 22 |
| DGB2 | 13 | 16 | 21 | 24 | 29 | 32 | 35 | 40 |
| DGB3 | 16 | 22 | 28 | 33 | 39 | 44 | 49 | 57 |
| DGC1 | 12 | 15 | 19 | 21 | 24 | 26 | 28 | 30 |
| DGC2 | 15 | 19 | 25 | 28 | 34 | 37 | 42 | 47 |
| DGC3 | 19 | 24 | 31 | 36 | 44 | 49 | 54 | 62 |
| SSA1 | 19 | 23 | 27 | 30 | 34 | 36 | 39 | 43 |
| SSB1 | 16 | 20 | 23 | 26 | 30 | 32 | 35 | 38 |
| SSB2 | 21 | 26 | 32 | 36 | 42 | 46 | 51 | 57 |
| SSB3 | 15 | 18 | 22 | 24 | 27 | 29 | 31 | 34 |
| SSC1 | 12 | 16 | 20 | 22 | 26 | 28 | 31 | 35 |
| SSC2 | 15 | 20 | 26 | 30 | 36 | 40 | 45 | 51 |
| SSE1 | 17 | 21 | 25 | 28 | 32 | 34 | 38 | 39 |
| SSE2 | 21 | 27 | 34 | 38 | 44 | 48 | 53 | 58 |
| SSE3 | 17 | 20 | 23 | 26 | 29 | 31 | 33 | 36 |

## TABLE H.3 Annual Passive Heating Performance: SSF *(Continued)*

| SPOKANE, WASHINGTON SSF (%) | | | | | | | | 6839 DD65 |
|---|---|---|---|---|---|---|---|---|
| Type | LCR = 100 | 70 | 50 | 40 | 30 | 25 | 20 | 15 |
| WWA3 | 11 | 14 | 18 | 21 | 24 | 27 | 29 | 32 |
| WWB4 | 14 | 20 | 28 | 33 | 40 | 45 | 51 | 58 |
| WWC2 | 14 | 20 | 27 | 32 | 38 | 43 | 48 | 55 |
| TWA1 | 14 | 15 | 17 | 19 | 21 | 22 | 23 | 25 |
| TWA2 | 12 | 15 | 18 | 20 | 23 | 24 | 27 | 29 |
| TWA3 | 11 | 14 | 17 | 20 | 23 | 25 | 28 | 31 |
| TWA4 | 10 | 13 | 17 | 20 | 23 | 26 | 29 | 33 |
| TWB3 | 10 | 13 | 17 | 19 | 22 | 25 | 28 | 32 |
| TWD4 | 13 | 19 | 25 | 30 | 36 | 41 | 46 | 53 |
| TWE2 | 14 | 20 | 26 | 30 | 36 | 40 | 45 | 52 |
| TWF3 | 8 | 11 | 14 | 17 | 20 | 22 | 25 | 28 |
| TWJ2 | 12 | 17 | 23 | 28 | 34 | 38 | 43 | 50 |
| DGA1 | 7 | 9 | 10 | 11 | 12 | 12 | 12 | 11 |
| DGA2 | 10 | 13 | 16 | 19 | 22 | 25 | 27 | 30 |
| DGA3 | 13 | 18 | 23 | 27 | 33 | 36 | 41 | 47 |
| DGB1 | 7 | 9 | 11 | 12 | 14 | 14 | 14 | 14 |
| DGB2 | 10 | 13 | 17 | 20 | 24 | 26 | 29 | 33 |
| DGB3 | 14 | 18 | 24 | 28 | 34 | 39 | 44 | 51 |
| DGC1 | 9 | 12 | 14 | 16 | 19 | 20 | 21 | 21 |
| DGC2 | 12 | 16 | 20 | 24 | 29 | 32 | 35 | 40 |
| DGC3 | 16 | 21 | 27 | 32 | 39 | 43 | 49 | 56 |
| SSA1 | 16 | 19 | 22 | 25 | 28 | 30 | 32 | 35 |
| SSB1 | 13 | 16 | 19 | 21 | 24 | 26 | 28 | 30 |
| SSB2 | 17 | 22 | 27 | 31 | 36 | 40 | 44 | 50 |
| SSB3 | 12 | 15 | 17 | 19 | 21 | 22 | 24 | 25 |
| SSC1 | 10 | 12 | 15 | 17 | 20 | 22 | 24 | 26 |
| SSC2 | 12 | 17 | 22 | 25 | 30 | 34 | 38 | 43 |
| SSE1 | 13 | 17 | 20 | 22 | 24 | 26 | 27 | 29 |
| SSE2 | 17 | 22 | 28 | 32 | 37 | 41 | 45 | 50 |
| SSE3 | 14 | 16 | 19 | 21 | 23 | 25 | 26 | 28 |

| MADISON, WISCONSIN SSF (%) | | | | | | | | 7731 DD65 |
|---|---|---|---|---|---|---|---|---|
| Type | LCR = 100 | 70 | 50 | 40 | 30 | 25 | 20 | 15 |
| WWA3 | 8 | 11 | 14 | 17 | 20 | 23 | 26 | 30 |
| WWB4 | 10 | 16 | 24 | 29 | 37 | 42 | 49 | 58 |
| WWC2 | 10 | 16 | 23 | 28 | 35 | 40 | 47 | 55 |
| TWA1 | 12 | 13 | 14 | 15 | 17 | 18 | 19 | 20 |
| TWA2 | 10 | 12 | 14 | 16 | 19 | 20 | 23 | 26 |
| TWA3 | 8 | 11 | 14 | 16 | 19 | 21 | 24 | 28 |
| TWA4 | 7 | 10 | 13 | 16 | 19 | 22 | 25 | 30 |
| TWB3 | 8 | 10 | 13 | 15 | 18 | 21 | 24 | 28 |
| TWD4 | 10 | 16 | 22 | 26 | 33 | 38 | 44 | 52 |
| TWE2 | 11 | 16 | 22 | 26 | 33 | 37 | 43 | 51 |
| TWF3 | 6 | 8 | 11 | 13 | 16 | 18 | 21 | 25 |
| TWJ2 | 9 | 14 | 19 | 24 | 30 | 35 | 41 | 49 |
| DGA1 | 4 | 5 | 6 | 7 | 8 | 8 | 7 | 6 |
| DGA2 | 7 | 10 | 13 | 15 | 18 | 20 | 23 | 26 |
| DGA3 | 11 | 15 | 19 | 23 | 28 | 32 | 38 | 45 |
| DGB1 | 4 | 5 | 6 | 7 | 8 | 8 | 8 | 8 |
| DGB2 | 8 | 10 | 13 | 15 | 19 | 21 | 24 | 29 |
| DGB3 | 11 | 15 | 20 | 24 | 30 | 34 | 40 | 48 |
| DGC1 | 6 | 8 | 10 | 11 | 13 | 14 | 15 | 16 |
| DGC2 | 9 | 12 | 16 | 19 | 23 | 26 | 30 | 36 |
| DGC3 | 13 | 17 | 23 | 27 | 34 | 39 | 45 | 55 |
| SSA1 | 14 | 16 | 19 | 22 | 25 | 27 | 30 | 33 |
| SSB1 | 11 | 13 | 16 | 18 | 21 | 22 | 25 | 28 |
| SSB2 | 14 | 19 | 24 | 28 | 33 | 37 | 42 | 50 |
| SSB3 | 10 | 12 | 14 | 16 | 18 | 19 | 21 | 23 |
| SSC1 | 7 | 9 | 12 | 13 | 16 | 18 | 20 | 23 |
| SSC2 | 9 | 13 | 18 | 21 | 27 | 30 | 35 | 42 |
| SSE1 | 11 | 13 | 16 | 18 | 21 | 23 | 25 | 28 |
| SSE2 | 14 | 19 | 25 | 29 | 35 | 39 | 45 | 52 |
| SSE3 | 12 | 14 | 16 | 17 | 20 | 21 | 23 | 26 |

| CHARLESTON, WEST VIRGINIA SSF (%) | | | | | | | | 4594 DD65 |
|---|---|---|---|---|---|---|---|---|
| Type | LCR = 100 | 70 | 50 | 40 | 30 | 25 | 20 | 15 |
| WWA3 | 11 | 15 | 20 | 23 | 28 | 31 | 36 | 42 |
| WWB4 | 14 | 21 | 30 | 36 | 45 | 50 | 58 | 67 |
| WWC2 | 14 | 21 | 29 | 35 | 43 | 48 | 55 | 65 |
| TWA1 | 14 | 15 | 18 | 19 | 21 | 23 | 25 | 28 |
| TWA2 | 12 | 15 | 18 | 21 | 24 | 27 | 30 | 35 |
| TWA3 | 11 | 14 | 18 | 21 | 26 | 29 | 33 | 38 |
| TWA4 | 10 | 13 | 18 | 21 | 26 | 29 | 34 | 40 |
| TWB3 | 10 | 13 | 17 | 20 | 24 | 27 | 31 | 37 |
| TWD4 | 14 | 20 | 27 | 32 | 40 | 45 | 52 | 61 |
| TWE2 | 15 | 20 | 27 | 32 | 40 | 45 | 51 | 60 |
| TWF3 | 8 | 11 | 15 | 18 | 22 | 25 | 29 | 35 |
| TWJ2 | 12 | 18 | 25 | 30 | 37 | 42 | 49 | 58 |
| DGA1 | 7 | 9 | 11 | 12 | 14 | 14 | 15 | 15 |
| DGA2 | 10 | 13 | 16 | 19 | 24 | 26 | 30 | 35 |
| DGA3 | 13 | 18 | 23 | 28 | 34 | 39 | 45 | 53 |
| DGB1 | 7 | 9 | 11 | 12 | 14 | 15 | 17 | 18 |
| DGB2 | 10 | 13 | 17 | 20 | 25 | 28 | 32 | 38 |
| DGB3 | 13 | 18 | 24 | 29 | 36 | 41 | 47 | 56 |
| DGC1 | 9 | 12 | 15 | 17 | 20 | 22 | 24 | 27 |
| DGC2 | 12 | 16 | 21 | 24 | 30 | 34 | 39 | 46 |
| DGC3 | 15 | 21 | 28 | 33 | 40 | 46 | 53 | 63 |
| SSA1 | 17 | 20 | 24 | 27 | 32 | 35 | 39 | 44 |
| SSB1 | 14 | 17 | 20 | 23 | 27 | 30 | 33 | 38 |
| SSB2 | 18 | 23 | 29 | 34 | 40 | 45 | 51 | 59 |
| SSB3 | 13 | 15 | 19 | 21 | 24 | 27 | 30 | 34 |
| SSC1 | 10 | 13 | 16 | 19 | 23 | 26 | 30 | 35 |
| SSC2 | 12 | 17 | 23 | 27 | 34 | 38 | 44 | 52 |
| SSE1 | 14 | 18 | 22 | 26 | 30 | 33 | 37 | 42 |
| SSE2 | 18 | 24 | 31 | 37 | 44 | 49 | 55 | 63 |
| SSE3 | 14 | 17 | 20 | 23 | 26 | 29 | 32 | 36 |

| MILWAUKEE, WISCONSIN SSF (%) | | | | | | | | 7445 DD65 |
|---|---|---|---|---|---|---|---|---|
| Type | LCR = 100 | 70 | 50 | 40 | 30 | 25 | 20 | 15 |
| WWA3 | 8 | 11 | 14 | 16 | 20 | 22 | 25 | 30 |
| WWB4 | 10 | 16 | 23 | 29 | 37 | 42 | 49 | 58 |
| WWC2 | 10 | 16 | 23 | 28 | 35 | 40 | 46 | 55 |
| TWA1 | 12 | 13 | 15 | 15 | 17 | 18 | 19 | 20 |
| TWA2 | 10 | 12 | 14 | 16 | 18 | 20 | 22 | 26 |
| TWA3 | 9 | 11 | 14 | 16 | 19 | 21 | 24 | 28 |
| TWA4 | 7 | 10 | 13 | 15 | 19 | 22 | 25 | 29 |
| TWB3 | 8 | 10 | 13 | 15 | 18 | 21 | 24 | 28 |
| TWD4 | 10 | 16 | 21 | 26 | 33 | 37 | 43 | 52 |
| TWE2 | 11 | 16 | 22 | 26 | 32 | 37 | 43 | 51 |
| TWF3 | 6 | 8 | 11 | 13 | 16 | 18 | 21 | 25 |
| TWJ2 | 9 | 14 | 19 | 24 | 30 | 35 | 40 | 49 |
| DGA1 | 4 | 6 | 6 | 7 | 8 | 8 | 7 | 6 |
| DGA2 | 7 | 10 | 13 | 15 | 18 | 20 | 23 | 26 |
| DGA3 | 11 | 15 | 19 | 23 | 28 | 32 | 37 | 45 |
| DGB1 | 4 | 5 | 6 | 7 | 8 | 8 | 8 | 8 |
| DGB2 | 8 | 10 | 13 | 15 | 19 | 21 | 24 | 28 |
| DGB3 | 11 | 15 | 20 | 24 | 30 | 34 | 40 | 48 |
| DGC1 | 6 | 8 | 10 | 11 | 13 | 14 | 15 | 16 |
| DGC2 | 9 | 12 | 16 | 19 | 23 | 26 | 30 | 36 |
| DGC3 | 13 | 17 | 23 | 27 | 34 | 39 | 45 | 54 |
| SSA1 | 14 | 17 | 20 | 22 | 25 | 27 | 30 | 34 |
| SSB1 | 11 | 14 | 16 | 18 | 21 | 23 | 25 | 29 |
| SSB2 | 15 | 19 | 24 | 28 | 34 | 38 | 43 | 50 |
| SSB3 | 10 | 12 | 14 | 16 | 18 | 19 | 21 | 23 |
| SSC1 | 7 | 9 | 12 | 13 | 16 | 18 | 20 | 23 |
| SSC2 | 9 | 13 | 18 | 21 | 26 | 30 | 35 | 41 |
| SSE1 | 11 | 14 | 16 | 19 | 21 | 23 | 26 | 28 |
| SSE2 | 14 | 19 | 25 | 29 | 35 | 39 | 45 | 52 |
| SSE3 | 12 | 14 | 16 | 18 | 20 | 22 | 24 | 26 |

**TABLE H.3 Annual Passive Heating Performance: SSF** *(Continued)*

| SHERIDAN, WYOMING SSF (%) | | | | | | | 7717 DD65 | |
|---|---|---|---|---|---|---|---|---|
| Type | LCR = 100 | 70 | 50 | 40 | 30 | 25 | 20 | 15 |
| WWA3 | 12 | 17 | 22 | 26 | 31 | 35 | 40 | 47 |
| WWB4 | 16 | 24 | 32 | 39 | 48 | 54 | 61 | 71 |
| WWC2 | 16 | 23 | 31 | 38 | 46 | 52 | 59 | 68 |
| TWA1 | 14 | 16 | 19 | 21 | 23 | 25 | 28 | 32 |
| TWA2 | 13 | 16 | 20 | 23 | 27 | 30 | 34 | 40 |
| TWA3 | 12 | 16 | 20 | 23 | 28 | 32 | 36 | 43 |
| TWA4 | 11 | 15 | 20 | 23 | 29 | 33 | 37 | 44 |
| TWB3 | 11 | 14 | 19 | 22 | 27 | 30 | 35 | 41 |
| TWD4 | 15 | 22 | 29 | 35 | 43 | 48 | 55 | 65 |
| TWE2 | 16 | 22 | 30 | 35 | 43 | 48 | 55 | 64 |
| TWF3 | 9 | 13 | 17 | 20 | 25 | 28 | 33 | 39 |
| TWJ2 | 13 | 20 | 27 | 32 | 40 | 46 | 53 | 62 |
| DGA1 | 7 | 10 | 12 | 14 | 16 | 17 | 19 | 19 |
| DGA2 | 10 | 14 | 18 | 21 | 26 | 29 | 33 | 38 |
| DGA3 | 14 | 19 | 25 | 30 | 37 | 42 | 48 | 56 |
| DGB1 | 7 | 10 | 13 | 15 | 17 | 19 | 21 | 23 |
| DGB2 | 10 | 14 | 19 | 22 | 27 | 31 | 36 | 42 |
| DGB3 | 14 | 19 | 26 | 31 | 38 | 44 | 51 | 60 |
| DGC1 | 10 | 13 | 17 | 19 | 23 | 26 | 29 | 33 |
| DGC2 | 13 | 17 | 23 | 27 | 33 | 37 | 43 | 51 |
| DGC3 | 16 | 22 | 30 | 35 | 43 | 49 | 57 | 67 |
| SSA1 | 17 | 21 | 25 | 28 | 33 | 36 | 40 | 46 |
| SSB1 | 14 | 17 | 21 | 24 | 28 | 31 | 35 | 40 |
| SSB2 | 18 | 24 | 30 | 35 | 42 | 46 | 52 | 60 |
| SSB3 | 13 | 16 | 19 | 22 | 25 | 28 | 31 | 35 |
| SSC1 | 11 | 14 | 18 | 22 | 26 | 30 | 34 | 40 |
| SSC2 | 13 | 19 | 25 | 30 | 37 | 42 | 48 | 56 |
| SSE1 | 15 | 19 | 24 | 27 | 32 | 35 | 39 | 44 |
| SSE2 | 19 | 25 | 33 | 38 | 46 | 50 | 57 | 65 |
| SSE3 | 15 | 18 | 21 | 24 | 28 | 30 | 34 | 38 |

| SUFFIELD, ALBERTA SSF (%) | | | | | | | 9423 DD65 | |
|---|---|---|---|---|---|---|---|---|
| Type | LCR = 100 | 70 | 50 | 40 | 30 | 25 | 20 | 15 |
| WWA3 | 11 | 15 | 19 | 23 | 27 | 30 | 34 | 39 |
| WWB4 | 14 | 21 | 29 | 35 | 44 | 49 | 56 | 65 |
| WWC2 | 14 | 21 | 28 | 34 | 42 | 47 | 53 | 62 |
| TWA1 | 14 | 15 | 17 | 19 | 21 | 23 | 25 | 28 |
| TWA2 | 12 | 15 | 18 | 21 | 24 | 27 | 30 | 34 |
| TWA3 | 11 | 14 | 18 | 21 | 25 | 28 | 32 | 37 |
| TWA4 | 10 | 13 | 18 | 21 | 25 | 29 | 33 | 38 |
| TWB3 | 10 | 13 | 17 | 20 | 24 | 27 | 31 | 36 |
| TWD4 | 14 | 20 | 27 | 32 | 39 | 44 | 50 | 59 |
| TWE2 | 14 | 20 | 27 | 32 | 39 | 44 | 50 | 58 |
| TWF3 | 8 | 11 | 15 | 18 | 22 | 25 | 28 | 33 |
| TWJ2 | 12 | 18 | 24 | 29 | 37 | 41 | 48 | 56 |
| DGA1 | 7 | 9 | 11 | 12 | 13 | 14 | 15 | 14 |
| DGA2 | 10 | 13 | 17 | 19 | 24 | 26 | 30 | 34 |
| DGA3 | 13 | 18 | 23 | 28 | 34 | 38 | 44 | 51 |
| DGB1 | 7 | 9 | 11 | 13 | 15 | 16 | 17 | 18 |
| DGB2 | 10 | 13 | 17 | 20 | 25 | 28 | 32 | 37 |
| DGB3 | 13 | 18 | 24 | 29 | 36 | 41 | 47 | 55 |
| DGC1 | 9 | 12 | 15 | 17 | 20 | 22 | 24 | 26 |
| DGC2 | 12 | 16 | 21 | 25 | 30 | 34 | 39 | 45 |
| DGC3 | 15 | 21 | 28 | 33 | 40 | 46 | 52 | 61 |
| SSA1 | 15 | 18 | 22 | 24 | 28 | 30 | 33 | 37 |
| SSB1 | 12 | 15 | 18 | 21 | 24 | 26 | 29 | 32 |
| SSB2 | 16 | 21 | 27 | 31 | 37 | 41 | 46 | 53 |
| SSB3 | 11 | 14 | 16 | 18 | 21 | 22 | 24 | 27 |
| SSC1 | 10 | 13 | 16 | 19 | 23 | 25 | 28 | 33 |
| SSC2 | 12 | 17 | 23 | 27 | 33 | 37 | 43 | 50 |
| SSE1 | 13 | 16 | 19 | 22 | 25 | 27 | 30 | 33 |
| SSE2 | 16 | 22 | 28 | 32 | 39 | 43 | 48 | 55 |
| SSE3 | 13 | 16 | 18 | 20 | 23 | 25 | 27 | 30 |

## Canada

| EDMONTON, ALBERTA SSF (%) | | | | | | | 10671 DD65 | |
|---|---|---|---|---|---|---|---|---|
| Type | LCR = 100 | 70 | 50 | 40 | 30 | 25 | 20 | 15 |
| WWA3 | 9 | 12 | 16 | 18 | 21 | 23 | 26 | 28 |
| WWB4 | 11 | 18 | 25 | 30 | 37 | 42 | 48 | 56 |
| WWC2 | 12 | 17 | 24 | 29 | 36 | 40 | 46 | 53 |
| TWA1 | 12 | 14 | 15 | 17 | 18 | 19 | 20 | 22 |
| TWA2 | 10 | 13 | 15 | 17 | 20 | 22 | 24 | 26 |
| TWA3 | 9 | 12 | 15 | 17 | 20 | 23 | 25 | 28 |
| TWA4 | 8 | 11 | 14 | 17 | 21 | 23 | 26 | 29 |
| TWB3 | 8 | 11 | 14 | 17 | 20 | 22 | 25 | 28 |
| TWD4 | 11 | 17 | 23 | 27 | 34 | 38 | 43 | 50 |
| TWE2 | 12 | 17 | 23 | 27 | 33 | 37 | 42 | 49 |
| TWF3 | 6 | 9 | 12 | 14 | 17 | 19 | 22 | 25 |
| TWJ2 | 10 | 15 | 21 | 25 | 31 | 35 | 41 | 47 |
| DGA1 | 5 | 7 | 8 | 9 | 9 | 9 | 9 | 7 |
| DGA2 | 8 | 11 | 14 | 17 | 20 | 22 | 25 | 28 |
| DGA3 | 12 | 16 | 21 | 25 | 30 | 34 | 39 | 45 |
| DGB1 | 5 | 7 | 8 | 9 | 10 | 11 | 11 | 10 |
| DGB2 | 8 | 11 | 15 | 17 | 21 | 24 | 27 | 30 |
| DGB3 | 12 | 16 | 22 | 26 | 32 | 36 | 41 | 48 |
| DGC1 | 7 | 9 | 12 | 13 | 15 | 17 | 18 | 18 |
| DGC2 | 10 | 14 | 18 | 21 | 26 | 29 | 33 | 37 |
| DGC3 | 14 | 19 | 25 | 29 | 36 | 41 | 46 | 53 |
| SSA1 | 13 | 16 | 19 | 21 | 23 | 25 | 27 | 28 |
| SSB1 | 11 | 13 | 16 | 17 | 20 | 21 | 23 | 25 |
| SSB2 | 14 | 19 | 23 | 27 | 32 | 35 | 40 | 45 |
| SSB3 | 10 | 12 | 14 | 15 | 17 | 18 | 18 | 19 |
| SSC1 | 8 | 10 | 13 | 15 | 17 | 19 | 21 | 22 |
| SSC2 | 10 | 14 | 19 | 23 | 28 | 31 | 35 | 40 |
| SSE1 | 10 | 13 | 15 | 17 | 19 | 20 | 21 | 21 |
| SSE2 | 13 | 18 | 23 | 27 | 32 | 36 | 40 | 45 |
| SSE3 | 11 | 13 | 16 | 17 | 19 | 20 | 21 | 22 |

| VANCOUVER, BRITISH COL. SSF (%) | | | | | | | 5607 DD65 | |
|---|---|---|---|---|---|---|---|---|
| Type | LCR = 100 | 70 | 50 | 40 | 30 | 25 | 20 | 15 |
| WWA3 | 15 | 19 | 24 | 27 | 31 | 34 | 37 | 41 |
| WWB4 | 18 | 25 | 33 | 39 | 46 | 51 | 57 | 65 |
| WWC2 | 18 | 25 | 32 | 38 | 45 | 49 | 55 | 62 |
| TWA1 | 16 | 18 | 21 | 23 | 25 | 27 | 29 | 32 |
| TWA2 | 15 | 18 | 22 | 25 | 28 | 31 | 34 | 37 |
| TWA3 | 14 | 18 | 22 | 25 | 29 | 32 | 35 | 40 |
| TWA4 | 13 | 17 | 22 | 25 | 30 | 33 | 36 | 41 |
| TWB3 | 13 | 17 | 21 | 24 | 28 | 31 | 35 | 39 |
| TWD4 | 17 | 24 | 31 | 36 | 43 | 47 | 53 | 60 |
| TWE2 | 18 | 24 | 31 | 36 | 42 | 47 | 52 | 59 |
| TWF3 | 11 | 15 | 19 | 22 | 26 | 29 | 32 | 37 |
| TWJ2 | 16 | 22 | 29 | 33 | 40 | 45 | 50 | 57 |
| DGA1 | 10 | 12 | 15 | 17 | 18 | 19 | 20 | 19 |
| DGA2 | 12 | 16 | 21 | 24 | 28 | 31 | 34 | 37 |
| DGA3 | 16 | 22 | 28 | 32 | 38 | 42 | 47 | 53 |
| DGB1 | 10 | 13 | 16 | 18 | 20 | 21 | 22 | 23 |
| DGB2 | 13 | 17 | 22 | 25 | 30 | 33 | 36 | 41 |
| DGB3 | 17 | 22 | 29 | 33 | 40 | 45 | 50 | 57 |
| DGC1 | 13 | 16 | 20 | 22 | 26 | 27 | 29 | 31 |
| DGC2 | 15 | 20 | 25 | 29 | 35 | 38 | 42 | 48 |
| DGC3 | 19 | 25 | 32 | 37 | 45 | 49 | 55 | 62 |
| SSA1 | 20 | 23 | 27 | 30 | 34 | 36 | 39 | 42 |
| SSB1 | 16 | 20 | 24 | 26 | 30 | 32 | 35 | 38 |
| SSB2 | 21 | 26 | 32 | 36 | 42 | 46 | 51 | 57 |
| SSB3 | 15 | 19 | 22 | 24 | 27 | 29 | 31 | 34 |
| SSC1 | 13 | 17 | 21 | 23 | 27 | 29 | 32 | 35 |
| SSC2 | 16 | 21 | 27 | 31 | 37 | 41 | 45 | 51 |
| SSE1 | 18 | 22 | 26 | 28 | 32 | 34 | 36 | 38 |
| SSE2 | 21 | 28 | 34 | 38 | 44 | 48 | 52 | 58 |
| SSE3 | 17 | 20 | 24 | 26 | 29 | 31 | 33 | 36 |

**TABLE H.3 Annual Passive Heating Performance: SSF** *(Continued)*

| WINNIPEG, MANITOBA **SSF (%)** | | | | | | 10776 DD65 | | |
|---|---|---|---|---|---|---|---|---|
| Type | LCR = 100 | 70 | 50 | 40 | 30 | 25 | 20 | 15 |
| WWA3 | 8 | 10 | 13 | 16 | 19 | 21 | 24 | 28 |
| WWB4 | 10 | 16 | 23 | 28 | 36 | 41 | 48 | 57 |
| WWC2 | 10 | 16 | 22 | 27 | 34 | 39 | 45 | 54 |
| TWA1 | 12 | 13 | 14 | 15 | 16 | 17 | 18 | 19 |
| TWA2 | 10 | 11 | 14 | 15 | 18 | 20 | 22 | 25 |
| TWA3 | 8 | 10 | 13 | 15 | 18 | 20 | 23 | 27 |
| TWA4 | 7 | 9 | 13 | 15 | 18 | 21 | 24 | 28 |
| TWB3 | 8 | 10 | 13 | 15 | 18 | 20 | 23 | 27 |
| TWD4 | 10 | 15 | 21 | 25 | 32 | 37 | 43 | 51 |
| TWE2 | 11 | 16 | 21 | 25 | 32 | 36 | 42 | 49 |
| TWF3 | 5 | 8 | 10 | 12 | 15 | 17 | 20 | 24 |
| TWJ2 | 8 | 13 | 19 | 23 | 29 | 34 | 40 | 47 |
| DGA1 | 4 | 5 | 6 | 7 | 7 | 7 | 6 | 5 |
| DGA2 | 7 | 9 | 12 | 14 | 18 | 20 | 22 | 26 |
| DGA3 | 10 | 14 | 19 | 22 | 28 | 32 | 37 | 44 |
| DGB1 | 4 | 5 | 6 | 7 | 7 | 8 | 8 | 7 |
| DGB2 | 7 | 10 | 13 | 15 | 18 | 21 | 24 | 28 |
| DGB3 | 11 | 15 | 19 | 23 | 29 | 34 | 39 | 47 |
| DGC1 | 6 | 7 | 9 | 10 | 12 | 13 | 14 | 15 |
| DGC2 | 9 | 12 | 16 | 19 | 23 | 26 | 30 | 35 |
| DGC3 | 12 | 17 | 22 | 27 | 34 | 38 | 45 | 53 |
| SSA1 | 12 | 15 | 17 | 19 | 22 | 23 | 26 | 28 |
| SSB1 | 10 | 12 | 14 | 16 | 18 | 20 | 22 | 24 |
| SSB2 | 13 | 17 | 22 | 25 | 31 | 34 | 39 | 46 |
| SSB3 | 9 | 10 | 12 | 13 | 15 | 16 | 17 | 18 |
| SSC1 | 6 | 9 | 11 | 13 | 15 | 17 | 19 | 21 |
| SSC2 | 9 | 12 | 17 | 20 | 26 | 29 | 34 | 40 |
| SSE1 | 9 | 11 | 14 | 15 | 18 | 19 | 21 | 22 |
| SSE2 | 12 | 17 | 22 | 26 | 31 | 35 | 40 | 46 |
| SSE3 | 11 | 12 | 14 | 15 | 17 | 18 | 20 | 21 |

| OTTAWA, ONTARIO **SSF (%)** | | | | | | 8461 DD65 | | |
|---|---|---|---|---|---|---|---|---|
| Type | LCR = 100 | 70 | 50 | 40 | 30 | 25 | 20 | 15 |
| WWA3 | 8 | 11 | 14 | 17 | 20 | 23 | 26 | 31 |
| WWB4 | 10 | 16 | 24 | 29 | 37 | 43 | 50 | 59 |
| WWC2 | 10 | 16 | 23 | 28 | 36 | 41 | 47 | 56 |
| TWA1 | 12 | 13 | 14 | 15 | 17 | 18 | 19 | 21 |
| TWA2 | 10 | 12 | 14 | 16 | 19 | 20 | 23 | 26 |
| TWA3 | 8 | 11 | 14 | 16 | 19 | 22 | 25 | 29 |
| TWA4 | 7 | 10 | 13 | 16 | 19 | 22 | 26 | 31 |
| TWB3 | 8 | 10 | 13 | 15 | 19 | 21 | 24 | 29 |
| TWD4 | 10 | 15 | 22 | 26 | 33 | 38 | 44 | 53 |
| TWE2 | 11 | 16 | 22 | 26 | 33 | 38 | 44 | 52 |
| TWF3 | 6 | 8 | 11 | 13 | 16 | 18 | 21 | 26 |
| TWJ2 | 9 | 14 | 19 | 24 | 31 | 35 | 41 | 50 |
| DGA1 | 4 | 5 | 6 | 7 | 8 | 8 | 7 | 6 |
| DGA2 | 7 | 10 | 13 | 15 | 18 | 20 | 23 | 27 |
| DGA3 | 11 | 14 | 19 | 23 | 28 | 33 | 38 | 46 |
| DGB1 | 4 | 5 | 6 | 7 | 8 | 8 | 9 | 8 |
| DGB2 | 7 | 10 | 13 | 15 | 19 | 21 | 25 | 29 |
| DGB3 | 11 | 15 | 20 | 24 | 30 | 34 | 40 | 49 |
| DGC1 | 6 | 7 | 9 | 11 | 13 | 14 | 15 | 17 |
| DGC2 | 9 | 12 | 16 | 19 | 24 | 27 | 31 | 37 |
| DGC3 | 13 | 17 | 23 | 27 | 34 | 39 | 46 | 55 |
| SSA1 | 13 | 16 | 19 | 21 | 24 | 26 | 29 | 33 |
| SSB1 | 11 | 13 | 15 | 17 | 20 | 22 | 24 | 28 |
| SSB2 | 14 | 18 | 23 | 27 | 33 | 37 | 42 | 49 |
| SSB3 | 10 | 11 | 13 | 15 | 17 | 18 | 20 | 22 |
| SSC1 | 7 | 9 | 11 | 13 | 16 | 18 | 21 | 24 |
| SSC2 | 9 | 13 | 18 | 21 | 27 | 31 | 36 | 43 |
| SSE1 | 10 | 13 | 16 | 18 | 21 | 23 | 25 | 28 |
| SSE2 | 13 | 18 | 24 | 28 | 35 | 39 | 44 | 52 |
| SSE3 | 11 | 13 | 15 | 17 | 19 | 21 | 23 | 25 |

| HALIFAX, NOVA SCOTIA **SSF (%)** | | | | | | 7167 DD65 | | |
|---|---|---|---|---|---|---|---|---|
| Type | LCR = 100 | 70 | 50 | 40 | 30 | 25 | 20 | 15 |
| WWA3 | 10 | 13 | 17 | 20 | 25 | 28 | 32 | 38 |
| WWB4 | 12 | 19 | 27 | 33 | 42 | 47 | 55 | 64 |
| WWC2 | 12 | 19 | 26 | 32 | 40 | 45 | 52 | 61 |
| TWA1 | 13 | 14 | 16 | 17 | 19 | 21 | 23 | 25 |
| TWA2 | 11 | 13 | 16 | 19 | 22 | 24 | 27 | 32 |
| TWA3 | 10 | 13 | 16 | 19 | 23 | 26 | 29 | 35 |
| TWA4 | 8 | 12 | 16 | 19 | 23 | 26 | 30 | 36 |
| TWB3 | 9 | 12 | 15 | 18 | 22 | 24 | 28 | 34 |
| TWD4 | 12 | 18 | 24 | 30 | 37 | 42 | 49 | 58 |
| TWE2 | 13 | 18 | 25 | 30 | 37 | 42 | 48 | 57 |
| TWF3 | 7 | 10 | 13 | 16 | 19 | 22 | 26 | 31 |
| TWJ2 | 10 | 16 | 22 | 27 | 34 | 39 | 46 | 55 |
| DGA1 | 6 | 7 | 9 | 10 | 11 | 11 | 12 | 11 |
| DGA2 | 8 | 11 | 15 | 17 | 21 | 24 | 27 | 31 |
| DGA3 | 12 | 16 | 21 | 25 | 32 | 36 | 42 | 50 |
| DGB1 | 5 | 7 | 9 | 10 | 11 | 12 | 13 | 14 |
| DGB2 | 9 | 12 | 15 | 18 | 22 | 25 | 29 | 34 |
| DGB3 | 12 | 17 | 22 | 26 | 33 | 38 | 45 | 54 |
| DGC1 | 7 | 10 | 12 | 14 | 16 | 18 | 20 | 23 |
| DGC2 | 10 | 14 | 18 | 22 | 27 | 31 | 36 | 43 |
| DGC3 | 14 | 19 | 25 | 30 | 38 | 43 | 51 | 60 |
| SSA1 | 15 | 19 | 22 | 25 | 29 | 32 | 36 | 41 |
| SSB1 | 13 | 15 | 19 | 21 | 25 | 27 | 30 | 35 |
| SSB2 | 16 | 21 | 27 | 31 | 38 | 42 | 48 | 56 |
| SSB3 | 11 | 14 | 17 | 19 | 22 | 24 | 26 | 30 |
| SSC1 | 8 | 11 | 14 | 17 | 20 | 23 | 26 | 30 |
| SSC2 | 11 | 15 | 21 | 25 | 31 | 35 | 41 | 48 |
| SSE1 | 13 | 16 | 20 | 23 | 27 | 29 | 33 | 37 |
| SSE2 | 16 | 22 | 29 | 34 | 41 | 46 | 52 | 59 |
| SSE3 | 13 | 16 | 19 | 21 | 24 | 26 | 29 | 33 |

| TORONTO, ONTARIO **SSF (%)** | | | | | | 7280 DD65 | | |
|---|---|---|---|---|---|---|---|---|
| Type | LCR = 100 | 70 | 50 | 40 | 30 | 25 | 20 | 15 |
| WWA3 | 8 | 11 | 14 | 17 | 21 | 23 | 27 | 31 |
| WWB4 | 10 | 17 | 24 | 30 | 38 | 43 | 50 | 59 |
| WWC2 | 11 | 16 | 23 | 28 | 36 | 41 | 47 | 56 |
| TWA1 | 12 | 13 | 15 | 16 | 17 | 18 | 19 | 21 |
| TWA2 | 10 | 12 | 14 | 16 | 19 | 21 | 23 | 27 |
| TWA3 | 9 | 11 | 14 | 16 | 20 | 22 | 25 | 29 |
| TWA4 | 7 | 10 | 13 | 16 | 20 | 22 | 26 | 31 |
| TWB3 | 8 | 10 | 13 | 16 | 19 | 21 | 24 | 29 |
| TWD4 | 11 | 16 | 22 | 27 | 33 | 38 | 44 | 53 |
| TWE2 | 11 | 16 | 22 | 27 | 33 | 38 | 44 | 52 |
| TWF3 | 6 | 8 | 11 | 13 | 16 | 19 | 22 | 26 |
| TWJ2 | 9 | 14 | 20 | 24 | 31 | 35 | 41 | 50 |
| DGA1 | 5 | 6 | 7 | 7 | 8 | 8 | 8 | 7 |
| DGA2 | 8 | 10 | 13 | 15 | 18 | 21 | 23 | 27 |
| DGA3 | 11 | 15 | 19 | 23 | 29 | 33 | 38 | 46 |
| DGB1 | 4 | 5 | 7 | 7 | 8 | 9 | 9 | 9 |
| DGB2 | 8 | 10 | 13 | 16 | 19 | 22 | 25 | 29 |
| DGB3 | 11 | 15 | 20 | 24 | 30 | 35 | 41 | 49 |
| DGC1 | 6 | 8 | 10 | 11 | 13 | 14 | 16 | 17 |
| DGC2 | 9 | 12 | 16 | 19 | 24 | 27 | 31 | 37 |
| DGC3 | 13 | 17 | 23 | 28 | 34 | 39 | 46 | 55 |
| SSA1 | 14 | 17 | 20 | 22 | 25 | 28 | 31 | 35 |
| SSB1 | 11 | 14 | 16 | 18 | 21 | 23 | 26 | 29 |
| SSB2 | 15 | 19 | 24 | 28 | 34 | 38 | 43 | 51 |
| SSB3 | 10 | 12 | 14 | 16 | 18 | 20 | 22 | 24 |
| SSC1 | 7 | 9 | 12 | 14 | 17 | 18 | 21 | 24 |
| SSC2 | 9 | 13 | 18 | 22 | 27 | 31 | 36 | 43 |
| SSE1 | 11 | 14 | 17 | 19 | 22 | 24 | 27 | 30 |
| SSE2 | 14 | 19 | 25 | 30 | 36 | 40 | 46 | 59 |
| SSE3 | 12 | 14 | 16 | 18 | 20 | 22 | 24 | 27 |

**TABLE H.3 Annual Passive Heating Performance: SSF** *(Continued)*

| NORMANDIN, QUEBEC | SSF (%) | | | | | 11364 DD65 | | |
|---|---|---|---|---|---|---|---|---|
| Type | LCR = 100 | 70 | 50 | 40 | 30 | 25 | 20 | 15 |
| WWA3 | 6 | 9 | 11 | 14 | 17 | 19 | 21 | 25 |
| WWB4 | 8 | 14 | 21 | 26 | 33 | 39 | 45 | 54 |
| WWC2 | 8 | 14 | 20 | 25 | 32 | 36 | 43 | 51 |
| TWA1 | 11 | 12 | 13 | 13 | 14 | 15 | 16 | 17 |
| TWA2 | 8 | 10 | 12 | 13 | 16 | 17 | 19 | 22 |
| TWA3 | 7 | 9 | 11 | 13 | 16 | 18 | 20 | 24 |
| TWA4 | 5 | 8 | 11 | 13 | 16 | 18 | 21 | 25 |
| TWB3 | 6 | 8 | 11 | 13 | 16 | 18 | 20 | 24 |
| TWD4 | 8 | 13 | 19 | 23 | 30 | 34 | 40 | 48 |
| TWE2 | 9 | 14 | 19 | 23 | 29 | 34 | 39 | 47 |
| TWF3 | 4 | 6 | 8 | 10 | 13 | 15 | 17 | 20 |
| TWJ2 | 7 | 11 | 17 | 21 | 27 | 31 | 37 | 45 |
| DGA1 | 3 | 3 | 4 | 5 | 5 | 4 | 4 | 2 |
| DGA2 | 6 | 8 | 10 | 13 | 15 | 17 | 20 | 23 |
| DGA3 | 9 | 13 | 17 | 20 | 26 | 30 | 35 | 42 |
| DGB1 | 3 | 3 | 4 | 4 | 5 | 5 | 5 | 4 |
| DGB2 | 6 | 8 | 11 | 13 | 16 | 18 | 21 | 25 |
| DGB3 | 10 | 13 | 18 | 21 | 27 | 31 | 37 | 45 |
| DGC1 | 4 | 6 | 7 | 8 | 9 | 10 | 11 | 12 |
| DGC2 | 8 | 10 | 14 | 16 | 20 | 23 | 27 | 33 |
| DGC3 | 11 | 15 | 20 | 25 | 31 | 36 | 42 | 51 |
| SSA1 | 12 | 14 | 16 | 18 | 21 | 23 | 25 | 27 |
| SSB1 | 9 | 11 | 13 | 15 | 17 | 19 | 20 | 23 |
| SSB2 | 12 | 16 | 21 | 24 | 30 | 33 | 38 | 44 |
| SSB3 | 8 | 10 | 11 | 12 | 14 | 15 | 16 | 17 |
| SSC1 | 5 | 7 | 9 | 11 | 13 | 14 | 16 | 18 |
| SSC2 | 7 | 11 | 15 | 18 | 23 | 27 | 31 | 37 |
| SSE1 | 8 | 10 | 13 | 14 | 16 | 18 | 19 | 20 |
| SSE2 | 11 | 16 | 21 | 25 | 30 | 34 | 39 | 45 |
| SSE3 | 10 | 11 | 13 | 14 | 16 | 17 | 19 | 20 |

# Economic Analysis

## I.1 ECONOMIC DECISION MAKING

As has been emphasized in this book, economic factors are a major consideration in almost every decision in the building design process. The discussion of economic analysis below is necessarily brief but covers the essential principles. For in-depth information on economic analyses of the design and construction process, see the references at the end of this appendix. Publications of the National Institute of Science and Technology (NIST; formerly The National Bureau of Standards) by Marshall and Ruegg (1980a,b) and Ruegg et al. (1978) are key sources for the following information.

The most common economic analysis questions are:

1. How can the cost of different systems *that produce the same result* be compared? This may apply to the purchase of a single item such as a motor, to the lighting of a room, or to the choice of the type of HVAC installation for an entire building. The principle involved is the same.

2. Assuming that improving a system, existing or proposed, will reduce operating costs and that many different types of improvement are possible, what preliminary economic guidelines can be established to determine whether any proposed investment appears cost effective? Stated otherwise, how can the initial simple rate of return of a proposed investment be determined so that it can be compared to the *minimum expected rate of return* on the investment?

3. Having determined in answer to question 2 that a number of different proposals, each with its own investment and payback (savings) figures, are apparently cost effective, how can these be compared to determine which one is *most* cost effective? Put another way, which proposal has the maximum *net benefit or savings?* This is the type of decision that must be made when comparing energy conservation projects such as lighting control systems.

4. Having established which of the proposals is most cost effective, how can the *actual* rate of return of the investment be determined? This figure, often referred to as *internal rate of return* (IRR), can then be compared to an expected (or required) rate of return to determine whether an investment is economically desirable.

5. What is the *payback period* of an investment? These questions are considered individually in the discussion that follows.

## I.2 LIFE-CYCLE COST

Cost comparisons made on the basis of first cost are legitimate only when operating/replacement costs are nominal or hold no concern for an owner. Proposals for a concrete slab might represent the former situation and speculative construction for immediate sale the latter. In such cases, cost comparison is simple—and based upon first cost. When operating/maintenance/replacement costs are involved that vary from system to system, however, the only reasonable way to determine and compare overall (true) system costs is by *life-cycle costing.* The basic theory and calculation techniques are well documented (see Fuller and Peterson, 1995; AIA, 1977; Marshal and Ruegg, 1980a,b; Ruegg et al., 1978), so they will be reviewed only briefly here. For discipline-specific information, see ASHRAE (2003) and IESNA (2000).

Stated simply, life-cycle cost represents the total cost of an item or system over its entire life cycle—it is the sum of the first cost and all future costs, less salvage. When the life cycle of a system being studied does not correspond to that of the building in which it is installed (as is usually the case), the designer must decide which life cycle to use. For instance, a lighting system may have an estimated life of 15 to 20 years, whereas the building life is usually at least double that figure. It is often wiser to conduct the analysis over the shorter time period, since (almost certainly) changes in technology will present an entirely different picture when the time comes to replace the system. This is true of most mechanical/electrical systems being designed today.

Life-cycle cost, in equation form, is

$$LCC = IC + MC + AC + OC$$

where

$LCC$ = life-cycle cost

$IC$   = investment cost (i.e., first cost minus salvage)

$MC$ = maintenance and repair cost

$AC$  = amortization (replacement) cost (not always included in life-cycle calculations)

$OC$  = operating costs, including labor and energy

All costs are typically expressed in terms of present value, which is explained below. Costs can also be annualized. The present value approach tends to give a clearer answer to the economic decision questions posed above. When calculating costs, the designer must include related ancillary costs. For instance, in comparing lighting systems, the costs must reflect the impact of lighting on the HVAC system (capacity and energy use) and the wiring system. For lighting system cost analysis information, see IESNA.

When the life-cycle analysis period is short—say, 2 years maximum—the last three components of Eq. I.1 can be assumed to be constant without introducing an unacceptable error. For longer peri-

ods, however, this is obviously not a valid assumption because of escalating labor and energy costs. Since these two factors weigh heavily in all economic comparisons, and since they generally apply in different proportions in the systems being considered, escalation of such costs will change the economic balance over the equipment life span. It is therefore necessary to conduct an analysis that reflects this escalation.

The present value of future payments of 1.00, made at the *end* of each of a series of $n$ periods, is

$$PV = \frac{(1 + i)^n - 1}{i\,(1 + i)^n} \qquad (I.2)$$

where

$PV$ = present value

$i$   = interest (discount) rate expressed as a decimal (e.g., 8% = 0.08)

$n$  = number of periods

The factor $i$ is more accurately the discount rate rather than the simple interest rate, the difference being that the latter reflects inflation, either actual or anticipated. If, however, the future payments escalate, the expression becomes

$$PV = \frac{1 - \left(\dfrac{K}{1 + i}\right)^n}{\dfrac{1 + i}{K} - 1} \qquad (I.3)$$

where $K$ is the escalation rate per period, and the remaining items are as in Eq. I.2. Thus, for a 5% annual escalation, $K = 1.05$, and so on. The values of $PV$ are tabulated in Table I.1 for interest (discount) rates of 8%, 10%, and 12% and for escalation rates of 3%, 5%, 8%, and 10%. Other values can be calculated from the equations.

It is apparent that the result of a life-cycle cost analysis depends heavily on accurate forecasting when using escalations and that comparative results can readily shift not only with escalation estimates of the various components but also with the length of the life cycle being analyzed.

## I.3  INITIAL (SIMPLE) RATE OF RETURN

Addressing the second question in Section I.1, is a proposed cost-saving investment worth examining in detail? That is, is the initial (simple) rate of return sufficient? The difference between initial and *actual* rates of return is that the former is simply

$$\text{initial rate of return} = \frac{\text{annual savings}}{\text{investment}}$$

whereas the actual (internal) rate of return depends upon the life of a project and the *total* savings accrued during the life cycle. The calculation method for the latter is given in Section I.5.

Initial rate of return is a useful criterion in that it can be very easily calculated and then compared to a *minimum* desired rate of return on investment. The *minimum* rate is emphasized, since the actual rate of return is always lower than the simple initial rate of return, except where savings escalate during the project life cycle. If a preliminary comparison indicates that the simple rate of return exceeds the minimum by a margin of a few percentage points, it is likely that a project is economically feasible, and a detailed analysis can be undertaken.

## I.4 COST-EFFECTIVENESS COMPARISON

This analysis is undertaken in answer to question 3; that is, several projects are being considered, all of which meet the stated criteria. Which one is most desirable from a cost-effectiveness viewpoint? (If only one project is under consideration, the analysis proceeds directly to the IRR calculation described in the next section).

The difference between life-cycle project cost and life-cycle project savings, that is, net lifetime savings, is a measure of cost effectiveness. Therefore, to compare the cost effectiveness of two projects, it is necessary simply to calculate this differential. This can be expressed as an equation:

net savings = life-cycle savings − life-cycle cost

Since this is true for all projects, a comparative differential can be calculated by using the differential in each term of the life-cycle cost equation, that is,

$$\Delta_{net} = \Delta_{savings} - \Delta_{costs}$$

Taking a very simple example, assume the following conditions.

|  | Project A | Project B |
|---|---|---|
| Investment | $10,000 | $15,000 |
| Life cycle | 10 years | 8 years |
| Annual savings | $2,000 | $3,200 |
| Initial rate of return | 20% | 32.3% |
| Discount rate | 12% | 12% |

Since project B has an apparently higher rate of return, we would formulate the differential as B minus A. Using Table I.1*a*, the present worth of lifetime savings for project B (without escalation) is $3200 × (4.97) or $15,904, and that of project A is $2000 × (5.65) or $11,120. Therefore, the differential cost benefit (CB) of B minus A is

$$\Delta_{\text{life savings}} \text{ minus } \Delta_{\text{project costs}}$$

or

$$\Delta_{CB} = (\$15,904 - \$11,120) - (\$15,000 - \$10,000)$$

$$= \$4784 - 5000 = -\$216$$

indicating that project A is more cost effective despite the apparently higher rate of return of project B. If we include escalation in costs, the life savings of the longer life project (A) will shrink more than the savings of shorter life project B. Using Table I.1*d*, the projects are about equal, with 3% annual escalation. With higher rates of escalation, project B is more cost effective.

## I.5 INTERNAL RATE OF RETURN (IRR)

Internal rate of return represents the rate of return at which lifetime savings are exactly equal to lifetime costs. IRR cannot be calculated directly, but only by trial and error. Assume a rate of return, calculate the present value (PV) of savings, and compare PV to lifetime costs. Because of this difficulty, IRR is frequently neglected and initial rate of return is used instead, although, as discussed above, this can readily give misleading results. To illustrate the calculation of IRR, return to the very simple example above and calculate IRR for projects A and B.

For project A, a present worth factor of $10,000/$2000 = 5 is needed, in 10 years. Referring to Table I.1*a*, note that at a discount (interest) rate of 15%, the present worth factor is 5.019, indicating that IRR is slightly above this level. (It is actually 15.1%.) For project B an 8-year present worth factor of $15,000/$3200, or 4.69, is needed. Referring again to Table I.1*a*, note that this factor falls between 12% and 15%. A series of trials with Eq. I.2 will eventually arrive at 13.7%, which is 1.4% *less* than the same factor for project A, confirming the conclusion reached in Section I.4.

## I.6 PAYBACK PERIOD

The payback period referred to in most proposals is the reciprocal of the initial rate of return and is usually referred to as the *simple* payback period. However, just as the actual rate of return differs from the simple rate (and is usually *lower*), the actual or *discount payback period* is different from, and usually *longer* than, the simple one. This period is, logically, the period of time required for the accumulated net savings to equal the initial investment, with all figures expressed in present value dollars. To illustrate, return to the example given above.

1. *Proposal A.* As in the IRR calculation, a present worth factor of $10,000/$2000, or 5, is needed at a 12% discount rate. From Table I.1a, find this to be about 8.1 years. This compares to a simple payback period of 5 years. Note that this is *not* the reciprocal of IRR.
2. *Proposal B.* Here a present worth factor of $15,000/$3200, or 4.7, is needed. At a 12%

discount rate, from Table I.1a, this is about 7.3 years. This compares to a simple payback period of 4.7 years.

Note from these results the very important fact that a shorter payback period does not necessarily indicate a more cost-effective investment. Here, proposal A has a better return precisely because it yields good savings for a longer period. Summing up the results of the simple study, we have:

|  | Project A | Project B |
|---|---|---|
| Initial rate of return | 20% | 21.3% |
| Internal rate of return | 15.1% | 13.7% |
| Simple payback period | 5 years | 4.7 years |
| Discounted payback period | 8.1 years | 7.3 years |

This shows very clearly that although initial rate of return and simple payback are indications of a proposal's value, they are useless in comparative studies.

**TABLE I.1a  Present Value of n Future Payments Beginning at the End of the First Period**

| n | Interest (Discount) Rate per Period | | | | | | |
|---|---|---|---|---|---|---|---|
|  | 6% | 8% | 10% | 12% | 15% | 20% | 25% |
| 1 | 0.94 | 0.93 | 0.91 | 0.89 | 0.87 | 0.83 | 0.80 |
| 2 | 1.83 | 1.78 | 1.74 | 1.69 | 1.63 | 1.53 | 1.44 |
| 3 | 2.67 | 2.58 | 2.49 | 2.40 | 2.28 | 2.11 | 1.95 |
| 4 | 3.47 | 3.31 | 3.17 | 3.04 | 2.86 | 2.59 | 2.36 |
| 5 | 4.21 | 3.99 | 3.79 | 3.61 | 3.35 | 2.99 | 2.69 |
| 6 | 4.92 | 4.62 | 4.36 | 4.11 | 3.78 | 3.33 | 2.95 |
| 7 | 5.58 | 5.21 | 4.87 | 4.56 | 4.16 | 3.61 | 3.16 |
| 8 | 6.21 | 5.75 | 5.34 | 4.97 | 4.49 | 3.84 | 3.33 |
| 9 | 6.80 | 6.25 | 5.76 | 5.33 | 4.77 | 4.03 | 3.46 |
| 10 | 7.36 | 6.71 | 6.15 | 5.65 | 5.02 | 4.19 | 3.57 |
| 11 | 7.89 | 7.14 | 6.50 | 5.94 | 5.23 | 4.33 | 3.66 |
| 12 | 8.83 | 7.54 | 6.81 | 6.19 | 5.42 | 4.44 | 3.73 |
| 13 | 8.85 | 7.90 | 7.10 | 6.42 | 5.58 | 4.53 | 3.78 |
| 14 | 9.30 | 8.24 | 7.37 | 6.63 | 5.72 | 4.61 | 3.82 |
| 15 | 9.71 | 8.56 | 7.61 | 6.81 | 5.85 | 4.68 | 3.86 |
| 16 | 10.11 | 8.85 | 7.82 | 6.97 | 5.95 | 4.73 | 3.89 |
| 17 | 10.48 | 9.12 | 8.02 | 7.12 | 6.05 | 4.78 | 3.91 |
| 18 | 10.83 | 9.37 | 8.20 | 7.25 | 6.13 | 4.81 | 3.93 |
| 19 | 11.16 | 9.60 | 8.37 | 7.37 | 6.20 | 4.84 | 3.94 |
| 20 | 11.47 | 9.82 | 8.51 | 7.47 | 6.26 | 4.87 | 3.95 |
| 21 | 11.76 | 10.02 | 8.65 | 7.56 | 6.31 | 4.89 | 3.96 |
| 22 | 12.04 | 10.20 | 8.77 | 7.65 | 6.36 | 4.91 | 3.97 |
| 23 | 12.30 | 10.37 | 8.88 | 7.72 | 6.40 | 4.93 | 3.98 |
| 24 | 12.55 | 10.53 | 8.99 | 7.78 | 6.43 | 4.94 | 3.98 |
| 25 | 12.78 | 10.68 | 9.08 | 7.84 | 6.46 | 4.95 | 3.99 |

**TABLE I.1b** Present Value of n Future Payments Beginning at the End of the First Period and Escalating at K per Period—8% Discount

| | Interest (Discount) Rate per Period 8% | | | |
|---|---|---|---|---|
| | Annual (Periodic) Escalation Rate K | | | |
| n | 1.03 | 1.05 | 1.08 | 1.10 |
| 1 | 0.95 | 0.97 | 1.00 | 1.02 |
| 2 | 1.86 | 1.92 | 2.00 | 2.06 |
| 3 | 2.73 | 2.84 | 3.00 | 3.11 |
| 4 | 3.56 | 3.73 | 4.00 | 4.19 |
| 5 | 4.35 | 4.60 | 5.00 | 5.28 |
| 6 | 5.10 | 5.44 | 6.00 | 6.40 |
| 7 | 5.82 | 6.26 | 7.00 | 7.54 |
| 8 | 6.50 | 7.06 | 8.00 | 8.70 |
| 9 | 7.15 | 7.84 | 9.00 | 9.88 |
| 10 | 7.78 | 8.59 | 10.00 | 11.08 |
| 11 | 8.37 | 9.33 | 11.00 | 12.30 |
| 12 | 8.94 | 10.04 | 12.00 | 13.55 |
| 13 | 9.48 | 10.73 | 13.00 | 14.82 |
| 14 | 9.99 | 11.41 | 14.00 | 16.11 |
| 15 | 10.48 | 12.06 | 15.00 | 17.43 |
| 16 | 10.95 | 12.70 | 16.00 | 18.77 |
| 17 | 11.40 | 13.32 | 17.00 | 20.13 |
| 18 | 11.82 | 13.92 | 18.00 | 21.53 |
| 19 | 12.23 | 14.51 | 19.00 | 22.94 |
| 20 | 12.62 | 15.08 | 20.00 | 24.39 |
| 21 | 12.99 | 15.63 | 21.00 | 25.86 |
| 22 | 13.34 | 16.17 | 22.00 | 27.35 |
| 23 | 13.68 | 16.69 | 23.00 | 28.88 |
| 24 | 14.00 | 17.20 | 24.00 | 30.43 |
| 25 | 14.30 | 17.69 | 25.00 | 32.01 |

**TABLE I.1c** Present Value of n Future Payments Beginning at the End of the First Period and Escalating at K per Period—10% Discount

| | Interest (Discount) Rate per Period 10% | | | |
|---|---|---|---|---|
| | Annual (Periodic) Escalation Rate K | | | |
| n | 1.03 | 1.05 | 1.08 | 1.10 |
| 1 | 0.94 | 0.96 | 0.98 | 1.00 |
| 2 | 1.81 | 1.87 | 1.95 | 2.00 |
| 3 | 2.63 | 2.74 | 2.89 | 3.00 |
| 4 | 3.40 | 3.57 | 3.82 | 4.00 |
| 5 | 4.12 | 4.36 | 4.73 | 5.00 |
| 6 | 4.80 | 5.12 | 5.63 | 6.00 |
| 7 | 5.43 | 5.84 | 6.51 | 7.00 |
| 8 | 6.02 | 6.53 | 7.37 | 8.00 |
| 9 | 6.57 | 7.18 | 8.22 | 9.00 |
| 10 | 7.09 | 7.81 | 9.05 | 10.00 |
| 11 | 7.58 | 8.41 | 9.87 | 11.00 |
| 12 | 8.03 | 8.98 | 10.67 | 12.00 |
| 13 | 8.46 | 9.53 | 11.46 | 13.00 |
| 14 | 8.85 | 10.05 | 12.23 | 14.00 |
| 15 | 9.23 | 10.55 | 12.99 | 15.00 |
| 16 | 9.58 | 11.27 | 13.74 | 16.00 |
| 17 | 9.90 | 11.48 | 14.47 | 17.00 |
| 18 | 10.21 | 11.91 | 15.19 | 18.00 |
| 19 | 10.50 | 12.32 | 15.90 | 19.00 |
| 20 | 10.76 | 12.72 | 16.59 | 20.00 |
| 21 | 11.02 | 13.09 | 17.27 | 21.00 |
| 22 | 11.25 | 13.45 | 17.94 | 22.00 |
| 23 | 11.47 | 13.80 | 18.59 | 23.00 |
| 24 | 11.68 | 14.12 | 19.24 | 24.00 |
| 25 | 11.87 | 14.44 | 19.87 | 25.00 |

**TABLE I.1d** Present Value of n Future Payments Beginning at the End of First Period and Escalating at K per Period—12% Discount

| | Interest (Discount) Rate per Period 12% | | | |
|---|---|---|---|---|
| | Annual (Periodic) Escalation Rate K | | | |
| n | 1.03 | 1.05 | 1.08 | 1.10 |
| 1 | .92 | .94 | .96 | .98 |
| 2 | 1.77 | 1.82 | 1.89 | 1.95 |
| 3 | 2.54 | 2.64 | 2.79 | 2.89 |
| 4 | 3.26 | 3.41 | 3.66 | 3.82 |
| 5 | 3.92 | 4.14 | 4.49 | 4.74 |
| 6 | 4.52 | 4.82 | 5.29 | 5.64 |
| 7 | 5.08 | 5.45 | 6.07 | 6.52 |
| 8 | 5.59 | 6.05 | 6.82 | 7.38 |
| 9 | 6.06 | 6.61 | 7.54 | 8.23 |
| 10 | 6.49 | 7.13 | 8.23 | 9.07 |
| 11 | 6.89 | 7.63 | 8.90 | 9.89 |
| 12 | 7.26 | 8.09 | 9.55 | 10.69 |
| 13 | 7.59 | 8.52 | 10.17 | 11.49 |
| 14 | 7.90 | 8.92 | 10.77 | 12.26 |
| 15 | 8.19 | 9.30 | 11.35 | 13.03 |
| 16 | 8.45 | 9.66 | 11.91 | 13.78 |
| 17 | 8.69 | 9.99 | 12.45 | 14.51 |
| 18 | 8.91 | 10.31 | 12.97 | 15.23 |
| 19 | 9.11 | 10.60 | 13.47 | 15.95 |
| 20 | 9.30 | 10.87 | 13.95 | 16.64 |
| 21 | 9.47 | 11.13 | 14.42 | 17.34 |
| 22 | 9.63 | 11.37 | 14.87 | 18.00 |
| 23 | 9.78 | 11.60 | 15.30 | 18.66 |
| 24 | 9.91 | 11.81 | 15.72 | 19.31 |
| 25 | 10.04 | 12.01 | 16.12 | 19.95 |

### REFERENCES

AIA. 1977. *Life Cycle Cost Analysis—A Guide for Architects.* American Institute of Architects, Washington, DC.

ASHRAE. 2003. *ASHRAE Handbook—HVAC Applications,* Chapter 36. American Society of Heating, Refrigerating and Air-Conditioning Engineers, Inc., Atlanta, GA.

Fuller, S.K. and S.R. Peterson. 1995. *Life-Cycle Costing Manual for the Federal Energy Management Program.* National Institute of Standards and Technology. Gaithersburg, MD. http://fire.nist.gov/bfrlpubs/

IESNA. 2000. *Lighting Handbook,* 9th ed., Chapter 25. Illuminating Engineering Society of North America, New York.

Marshall, H.E. and R.T. Ruegg. 1980a. *Energy Conservation in Buildings: An Economics Guidebook for Investment Decision.* National Bureau of Standards Handbook 132, NBS, Washington, DC.

Marshall, H.E. and R.T. Ruegg. 1980b. *Simplified Energy Design Economics.* National Bureau of Standards Special Publication 544, NBS, Washington, DC.

Ruegg, R.T., et al. 1978. *Life Cycle Costing.* National Bureau of Standards Publication 113, NBS, Washington, DC.

*Incandescent characteristics*

# Lamp Data

This appendix provides technical information on incandescent lamps.

## J.1 INCANDESCENT LAMP CHARACTERISTICS

**TABLE J.1  Typical Incandescent Lamp Data (Listing a Few of the Many Sizes and Types of 115-, 120-, and 125-V Lamps)**

| Lamp Watts[a] | Average Rated Life (h) | Approximate Color Temperature[b] (K) | Initial Lumens | Lumens per Watt[c] | Shape of Bulb[d] | Base | Description |
|---|---|---|---|---|---|---|---|
| 15 | 2,500 | — | 126 | 8.4 | A-15 | Med | Long life |
| 25 | 1,000 | — | 240 | 9.6 | A-19 | Med | Rough service |
| 25 | 2,500 | 2,500 | 232 | 9.3 | A-19 | Med | — |
| 40 | 1,500 | — | 495 | 12.4 | A-19 | Med | — |
| 50 | 2,000 | — | 525 | 10.5 | ER30 | Med | — |
| 60 | 1,000 | 2,800 | 890 | 14.8 | A-19 | Med | — |
| 60 | 2,500 | — | 800 | 13.3 | A-19 | Med | Long life |
| 75 | 750 | — | 1,220 | 16.3 | A-19 | Med | — |
| 100 | 750 | 2,870 | 1,650 | 16.5 | A-19 | Med | — |
| 100 | 750 | — | 1,750 | 17.5 | A-21 | Med | — |
| 100 | 2,500 | — | 1,500 | 15.0 | A-19 | Med | Long life |
| 135 | 750 | — | 2,425 | 18.0 | A-21 | Med | Economy |
| 150 | 750 | 2,900 | 2,810 | 18.7 | A-21 | Med | — |
| 150 | 750 | — | 2,600 | 17.3 | PS-25 | Med | — |
| 200 | 750 | 2,930 | 4,000 | 20.0 | A-23 | Med | — |
| 300 | 750 | 2,940 | 6,300 | 21.0 | PS-25 | Med | — |
| 500 | 1,000 | 3,000 | 10,850 | 21.7 | PS-25 | Mogul | — |

[a]Figures in this column designate the input watts thus: 60 = 60 W.
[b]See Section 11.34.
[c]Luminous efficacy, in lumens per watt, increases with filament temperature and, therefore, with wattage.
[d]See Fig. 12.8.

## J.2 INCANDESCENT LAMP DIMENSIONAL DATA

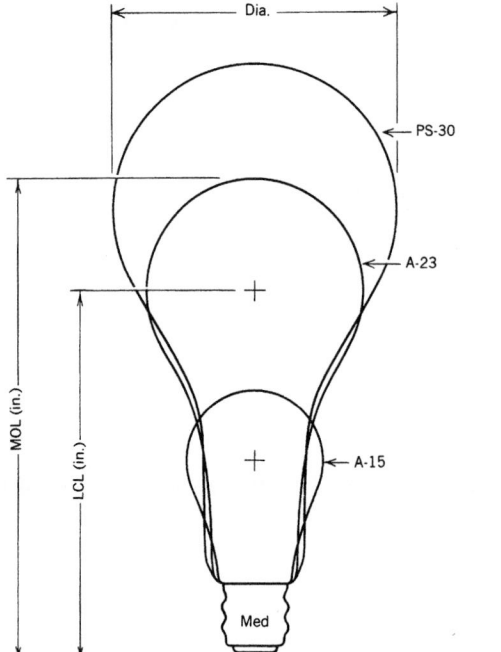

Bulb diameter is given in 1/8 in. Example: An A-19 bulb has a diameter of 19/8 in. or 2 3/8 in.

MOL: Maximum overall length: This figure refers to the maximum length of the bulb.

LCL: Light center length: This dimension, important when designing reflectors, is measured from the filament to a point that varies with base type. See Fig. 19.31.

| | A – STANDARD SHAPE | | | | | | | | | PS – PEAR SHAPE | | | | | | | |
|---|---|---|---|---|---|---|---|---|---|---|---|---|---|---|---|---|---|
| WATTS | 15 | 25 | 40 | 60 | 75 | 100 | 100 | 150 | 150 | 150 | 200 | 300 | 300 | 500 | 750 | 1000 | 1500 |
| BULB | A–15 | A–19$_1$ | A–19$_2$ | A–19$_3$ | A–19$_3$ | A–19$_3$ | A–21$_1$ | A–21$_2$ | A–23 | PS–25 | PS–30 | PS–30 | PS–35 | PS–40 | PS–52 | PS–52 | PS–52 |
| DIAMETER" | 1$^7$/8 | 2$^3$/8 | 2$^3$/8 | 2$^3$/8 | 2$^3$/8 | 2$^3$/8 | 2$^5$/8 | 2$^5$/8 | 2$^7$/8 | 3$^1$/8 | 3$^3$/4 | 3$^3$/4 | 4$^3$/8 | 5 | 6$^1$/2 | 6$^1$/2 | 6$^1$/2 |
| M.O.L." | 3$^1$/2 | 3$^7$/8 | 4$^1$/4 | 4$^7$/16 | 4$^7$/16 | 4$^7$/16 | 5$^1$/4 | 5$^1$/2 | 6$^3$/16 | 6$^{15}$/16 | 8$^1$/16 | 8$^1$/16 | 9$^3$/8 | 9$^3$/4 | 13 | 13 | 13 |
| L.C.L." | 2$^3$/8 | 2$^1$/2 | 2$^{15}$/16 | 3$^1$/8 | 3$^1$/8 | 3$^1$/8 | 3$^7$/8 | 4 | 4$^5$/8 | 5$^3$/16 | 6 | 6 | 7 | 7 | 9$^1$/2 | 9$^1$/2 | 9$^1$/2 |
| BASE | MED | MED | MED | MED | MED | MED | MED | MED | MED | MED | MED | MED | MOG | MOG | MOG | MOG | MOG |
| STANDARD FINISH | IF | IF | IF W | IF | IF | IF | IF | IF | CL IF | CL IF | IF | IF | IF | IF | CL IF | CL IF | CL IF |

CL – CLEAR    IF – INSIDE FROSTED

**Fig. J.1** *Typical dimensional data for common general-service incandescent lamps.*

## J.3  INCANDESCENT LAMP PERFORMANCE DATA

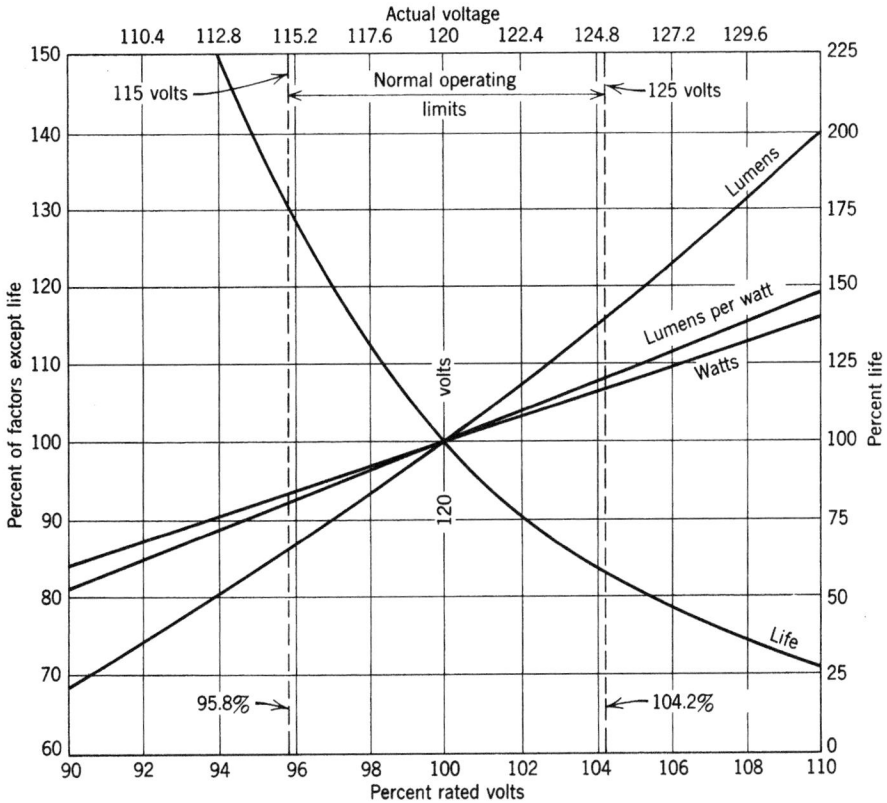

**Fig. J.2** *Characteristics of a standard 120-V general-service incandescent lamp as a function of voltage.*

# Sound Transmission Data for Walls

To use Appendix K:

1. Find the desired construction Type Code in Index K.1.
2. Find the desired STC corresponding to the selected construction Type Code.
3. Refer to Table K.1 for details of wall construction.
4. Refer to Index K.2 for wall thickness, weight, STC, fire rating, and the transmission losses at standard octave midpoints.

### EXAMPLE

An interior masonry partition with an STC between 50 and 55 is desired. Of particular interest is the TL at 1000 Hz.

1. From Index K.1, find Type Code "c."
2. From Index K.1, note that constructions W6, W12b, W15b, and W22b are suitable for the proposed use.
3. From Table K.1, decide that construction W15b is most suitable for the proposed use.
4. From Index K.2, find that construction W15b has a transmission loss of 55 dB at 1000 Hz.

## INDEX K.1 Sound Transmission Class: Walls

| Type Code: | | | | | |
|---|---|---|---|---|---|
| a. Wooden stud | | f. Plaster | | j. Fiber board | |
| b. Metal stud | | g. Gypsum wallboard | | k. Lead | |
| c. Masonry | | h. With resilient element | | l. Gypsum core board | |
| d. Concrete | | i. Absorbent blankets or fill | | m. Double wall | |
| e. Staggered stud | | | | | |

| STC | Type | Item No. | STC | Type | Item No. |
|---|---|---|---|---|---|
| 63 | d,f | W4 | 46 | a,f | W30a |
| 62[a] | c,f,j,m | W23 | 46 | a,f | W31a |
| 56 | c | W7 | 46 | a,f,k | W31c |
| 56[a] | c,f | W8 | 46 | a,e,g,i | W37 |
| 55[a] | b,g,i | W63 | 45 | c,d | W10b |
| 54[a] | c,f,m | W22b | 45 | a,f,i | W38 |
| 54 | b,f,h | W52 | 45 | b,f,h | W50a |
| 53[a] | d,f | W2 | 45 | g,l,m | W85a |
| 53 | c,f,h | W15b | 44 | c | W12a |
| 52[a] | c,f | W6 | 44 | a,e,g | W34a |
| 51 | b,f | W44a | 44 | a,e,g | W35a |
| 51 | b,g,h,i | W67 | 44 | a,f,h | W40a |
| 50 | c,k | W12b | 43 | c,d, | W10a |
| 50 | b,g,i | W60 | 43[a] | d,j,m | W21 |
| 50 | g,i,l,m | W85b | 43 | a,e,f | W35a |
| 49[a] | c,f,m | W22a | 43 | a,e,f | W36a |
| 48 | c,d | W9 | 43 | b,f,k | W43b |
| 48 | a,e,f,i | W36b | 42[a] | c,f | W5 |
| 48 | b,f,k | W43c | 42 | a,f | W14a |
| 47 | d | W1 | 41 | b,f | W43a |
| 47 | a,g,k | W28b | 41 | b,g | W55a |
| 47 | a,f,k | W31b | 40 | c,f | W13 |
| 47 | a,g,h | W39a | 40 | a,g | W32a |
| 46 | c,f | W11a | 39 | a,g | W28a |
| 46 | c,f,h | W15a | 36 | g,l | W80 |

[a]Field measurement.

## INDEX K.2 Sound Transmission Loss: Walls

| Designation | Thickness (in.) | Weight (lb/ft²) | Transmission Loss (dB) at frequency | | | | | | STC | Fire Rating (hr) |
|---|---|---|---|---|---|---|---|---|---|---|
| | | | 125 | 250 | 500 | 1K | 2K | 4K | | |
| W1 | 3 | 39 | 35 | 40 | 44 | 52 | 58 | 64 | 47 | ½ |
| W2 | 7 | 80 | 39 | 42 | 50 | 58 | 64 | — | 53 | 3 |
| W4 | Approx. 16 | 184 | 50 | 54 | 59 | 65 | 71 | 68 | 63 | 4+ |
| W5 | 5½ | 55 | 34 | 34 | 41 | 50 | 66 | — | 42 | 2.5 |
| W6 | 10 | 100 | 41 | 43 | 49 | 55 | 57 | — | 52 | 4+ |
| W7 | 12 | 121 | 45 | 45 | 53 | 58 | 60 | 61 | 56 | 4+ |
| W8 | 25 | 280 | 50 | 53 | 52 | 58 | 61 | — | 56 | 4+ |
| W9 | 12 | 79 | 46.5 | 44 | 46 | 52 | 54 | 56 | 48 | 4 |
| W10a | 6 | 34 | 32 | 33 | 40 | 47 | 51 | 48 | 43 | 1 |
| W10b | 6 | 34 | 37 | 36 | 42 | 49 | 55 | 58 | 45 | 1 |
| W11a | 5¼ | 35.8 | 36 | 37 | 44 | 51 | 55 | 62 | 46 | 2 |
| W12a | 3¾ | 26.1 | 40 | 40 | 40 | 48 | 55 | 56 | 44 | 1.5 |
| W12b | 5 | 31 | 41 | 46 | 46 | 56 | 63 | 67 | 50 | 1.5 |
| W13 | 4 | 21.5 | 39 | 34 | 38 | 43 | 48 | 46 | 40 | 3 |
| W14a | 5 | 23.4 | 37 | 42 | 39 | 44 | 49 | 49 | 42 | 4 |
| W15a | 5 | 27 | 38 | 37 | 44 | 51 | 56 | 59 | 46 | 3 |
| W15b | 6 | 31 | 45 | 44 | 50 | 55 | 56 | 59 | 53 | 4 |
| W20 | 7 | 22.9 | 32 | 46 | 49 | 53 | 58 | 66 | 52 | 3 |
| W21 | 10¼ | 37 | 41 | 42 | 46 | 51 | 52 | — | 43 | Not available |
| W22a | 12 | 100 | 37 | 41 | 48 | 60 | 60.5 | — | 49 | 4+ |
| W22b | 12 | 100 | 40 | 44 | 55 | 67.5 | 70 | — | 54 | 4+ |
| W23 | 18 | 120 | 48 | 54 | 58 | 64 | 69 | — | 62 | 4+ |
| W28a | 5 | 6 | 21 | 28 | 35 | 42 | 45 | 41 | 39 | 0.5 |
| W28b | Approx. 5⅛ | 12 | 27 | 37 | 43 | 52 | 56 | — | 47 | 0.5 |
| W30a | 5¾ | 13.4 to 15.7 | 32 | 37 | 42 | 47 | 47 | 63 | 46 | 0.75 |
| W31a | 5¾ | 13.4 to 15.7 | 32 | 37 | 42 | 48 | 48 | 63 | 46 | 0.75 |
| W31b | Approx. 5⅞ | | 33 | 41 | 45 | 52 | 55 | 65 | 47 | 0.75 |
| W31c | Approx. 5⅞ | 17–19 | 32.5 | 40 | 43 | 47 | 50 | 62 | 46 | 0.75 |
| W32a | 5½ | 8.2 | 27 | 31 | 39 | 45 | 52.5 | 48 | 40 | 1 |
| W34a | 5 | 6.2 | 36 | 36 | 40 | 47 | 52 | 45 | 44 | 0.5 |
| W35a | 6½ | 13.4 | 41 | 41 | 46 | 49 | 41 | 54 | 44 | 1.5 |
| W35b | 5¾ | 15.6 | 48 | 46 | 48 | 48 | 48 | 59 | 43 | 1 |
| W36a | 6¼ | 11.1 | 36 | 33 | 42 | 42 | 41 | 51 | 43 | 0.75 |
| W36b | 6¼ | 12.8 | 37 | 37 | 49 | 50 | 52 | 66 | 48 | 1 |
| W37 | 5¾ | 13.8 | 39 | 40 | 42 | 47.5 | 55 | 51.5 | 46 | 0.5 |
| W38 | 5⅜ | 14.2 | 39 | 45 | 48 | 50 | 44 | 54 | 45 | 1 |
| W39a | 6¼ | 6.7 | 30 | 40 | 46 | 50 | 49 | 49 | 47 | 1 |
| W40a | Approx. 6½ | 14.4 | 43 | 41 | 48 | 50 | 42 | 56 | 44 | 1 |
| W43a | 3⅜ | 12.3 | 27 | 37 | 43 | 46 | 39 | 47 | 41 | 0.75 |
| W43b | 3½ | 15.2 | 35 | 43 | 45 | 47 | 48 | 58 | 43 | 0.75 |
| W43c | Approx. 3½ | 18.2 | 36 | 45 | 47 | 50 | 53 | 61 | 48 | 0.75 |
| W44a | 5 | 15.7 | 34 | 38 | 47 | 50 | 52 | 58 | 51 | 1 |
| W50a | 5¼ | 13 | 35 | 46 | 48 | 51 | 48 | 43 | 45 | 0.75 |
| W52 | 5 | 19 | 50 | 52 | 55 | 56 | 52 | 60 | 54 | 1 |
| W55a | 4⅞ | 6 | 29 | 36 | 40 | 46 | 40 | 46 | 41 | 1 |
| W60 | 3½ | 5.4 | 34 | 40 | 47 | 50 | 53 | 54 | 50 | 1 |
| W63 | 6⅛ | 11.5 | 36 | 47 | 51 | 57 | 57 | 62 | 55 | 2 |
| W67 | 6½ | 11.3 | 41 | 46 | 49 | 51 | 50 | 60 | 51 | 2 |
| W80 | 2¼ | 10.2 | 34 | 34 | 37 | 38 | 39 | 45 | 36 | 1 |
| W85a | 5⅛ | 14.6 | 36 | 35 | 45 | 51 | 53 | 57 | 45 | 3 |
| W85b | 6 | 12.8 | 37 | 37 | 54 | 56 | 56 | 62 | 50 | 3 |

**APPENDICES**

**TABLE K.1 Acoustic Characteristics of Walls: For Other Data See Index K.2**

| Designation | Description | Section Sketch |
|---|---|---|
| **Solid Concrete** | | |
| W1 | 3-in.-thick solid concrete wall poured in situ in test opening. All surface cavities were sealed with thin mortar mix. | |
| W2 | 6-in.-thick concrete wall with ½-in.-thick layer of plaster on both sides. | W 1, W 2 |
| W4 | Wall of 4, 6, and 8 × 8 × 16 in. sand and gravel aggregate solid concrete blocks; on each side, ¼- to ½-in.-thick layer of cement gypsum plaster and sand. | W 4 |
| W5 | 4½-in.-thick brick wall with ½-in.-thick layer of plaster on each side. | W 5, W 6 |
| W6 | 9-in.-thick brick wall with ½-in.-thick layer of plaster on each side. | |
| W7 | 12-in.-thick brick wall. | W 7 |
| W8 | 24-in.-thick stone wall with ½-in.-thick layer of plaster on both sides. | W 8 |

**TABLE K.1 Acoustic Characteristics of Walls: For Other Data See Index K.2** *(Continued)*

| Designation | Description | Section Sketch |
|---|---|---|

**Hollow Concrete Block**

W9    12-in. wall made of hollow 8 × 8 × 12 in. and
       8 × 4 × 16 in. concrete blocks.

W10a   6-in. hollow concrete blocks constructed with
       vertical mortar joints staggered.

W10b   Similar to W10a except wall painted.

**Cinder Block**

W11a   4 × 8 × 16-in. hollow cinder blocks; on each side,
       ⅝ in. of sanded gypsum plaster.

**Cement Block**

W12a   3⅝ × 7¾ × 13½-in. lightweight-aggregate cement
       blocks with ½-in. mortar joints; three coats of
       masonry paint applied to each side of partition.

W12b   Same as W12a, except that 1 × 2-in. furring strips
       were nailed vertically to partition on one side; ¹⁄₁₆-in.
       layer of lead, 3.94 lb/ft², nailed to furring strips,
       ¼-in. plywood-covered lead with joints caulked.

**Hollow Gypsum Block**

W13    3-in. hollow gypsum blocks cemented together
       with ⅜-in. mortar joints; on each side, ½-in. sanded
       gypsum plaster.

W14a   4-in. hollow gypsum blocks cemented together
       with ⅜-in. mortar joints; on each side, ½-in. sanded
       gypsum plaster.

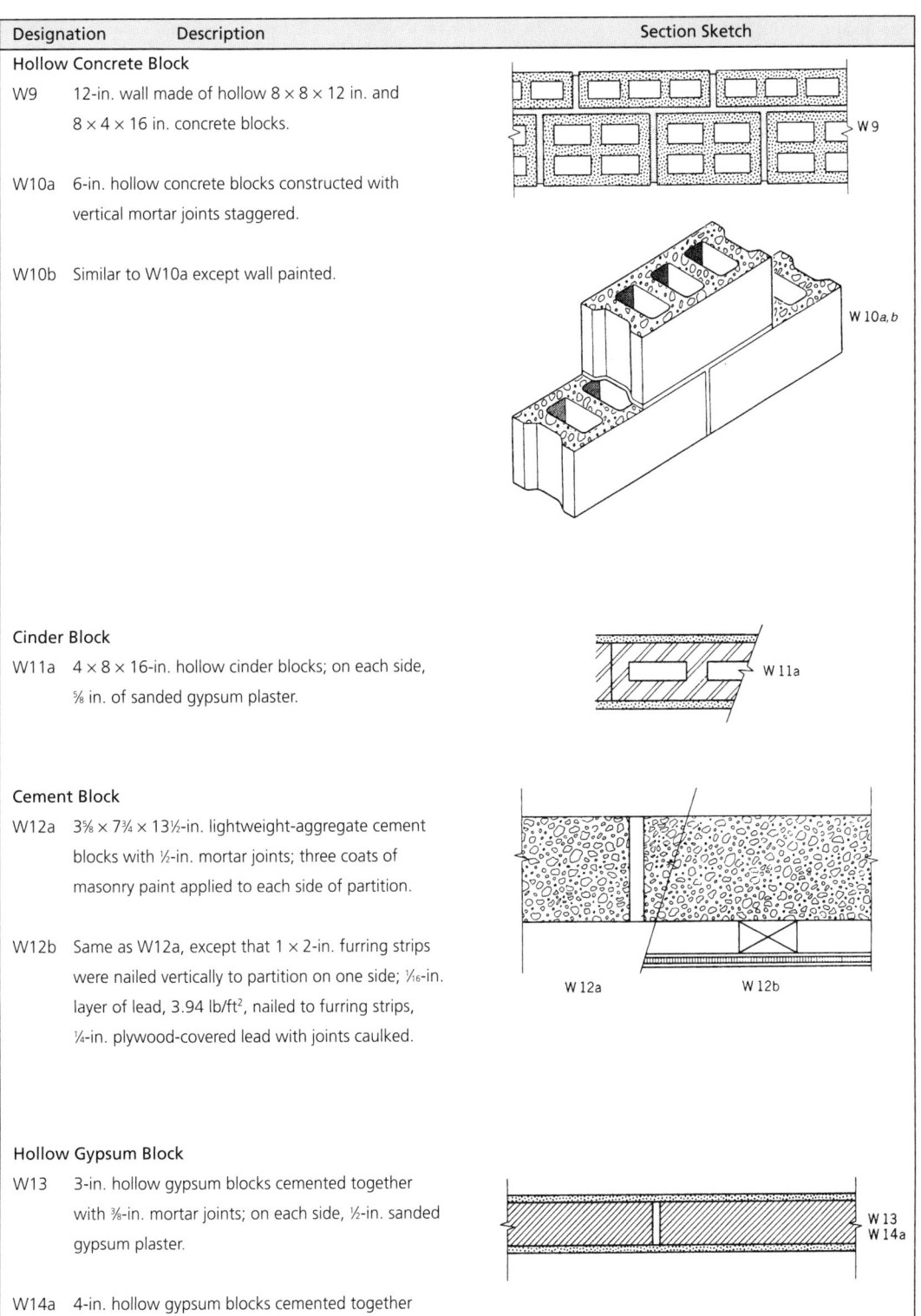

**TABLE K.1 Acoustic Characteristics of Walls: For Other Data See Index K.2** *(Continued)*

| Designation | Description | Section Sketch |
|---|---|---|

**Hollow Gypsum Block, Resilient on One Side, Plaster on Both Sides**

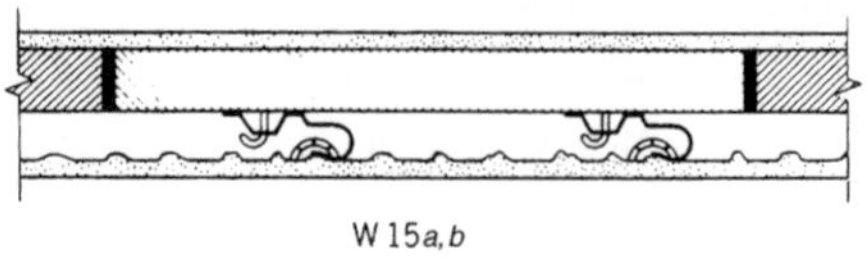

W15a    3 × 12 × 30-in. hollow gypsum blocks with ½-in. mortar joints. On one side, ⁷⁄₁₆-in. sanded gypsum plaster; on the other side, resilient clips, spaced 18 in. apart on centers vertically and 16 in. apart on centers horizontally, held to ¾- in. metal channels 16 in. on centers, to which expanded metal lath was wire-tied; ¹¹⁄₁₆-in. sanded gypsum plaster; ¹⁄₁₆-in. white-coat finish applied to both sides.

W 15*a, b*

W15b    Similar to W15a except that 4 × 12 × 30-in. gypsum blocks were used.

**Hollow Concrete**

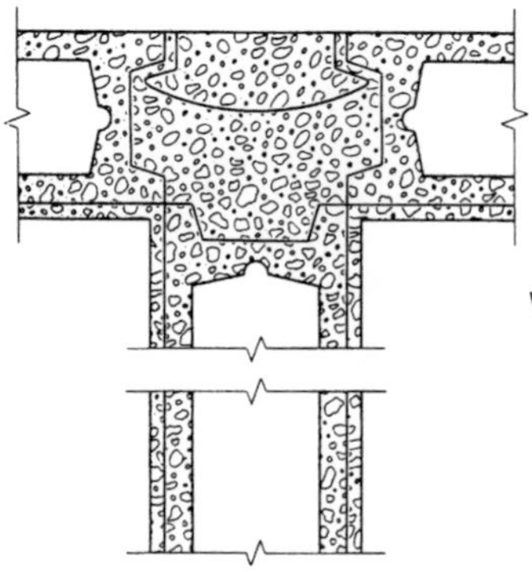

W21    Precast concrete hollow wall panels with in situ concrete posts and beams. Panels have 1½-in.-thick concrete shells with 6¼-in. air space between them. Layer of fiberboard ½ in. thick is adhered to the exposed surfaces of the panel.

W 21

**Double Walls**

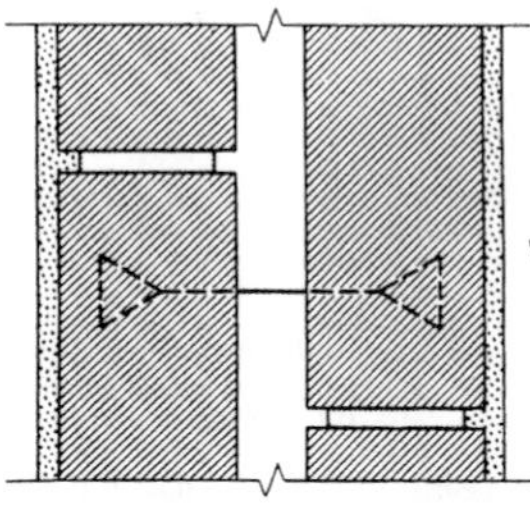

W22a    Double wall with 4½-in.-thick brick leaves separated by a 2-in. cavity (wire ties between leaves); ½-in. plaster on exposed sides.

W22b    Similar to W22a but no wire ties between the leaves.

W 22

W23    Double wall with 4½-in.-thick brick leaves, 6-in. cavity (no ties); on exposed sides, ½-in. plaster on 1-in.-thick wood-wool slabs mortared to the brick walls.

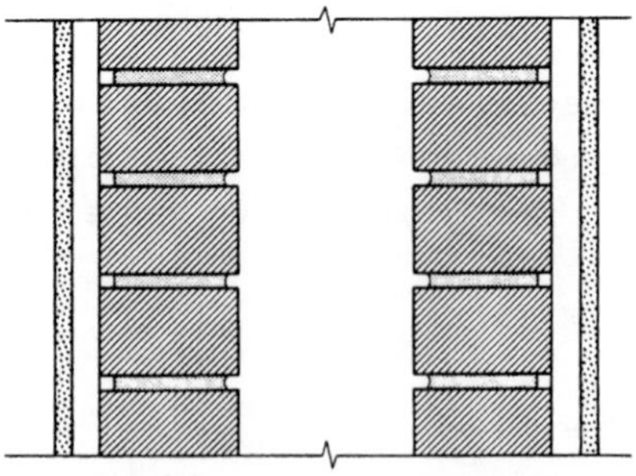

W 23

**Wood Stud Walls**

W28a    2 × 4-in. wooden studs, 16 in. on centers, ½-in. gypsum wallboard nailed to each side. All joints taped and finished.

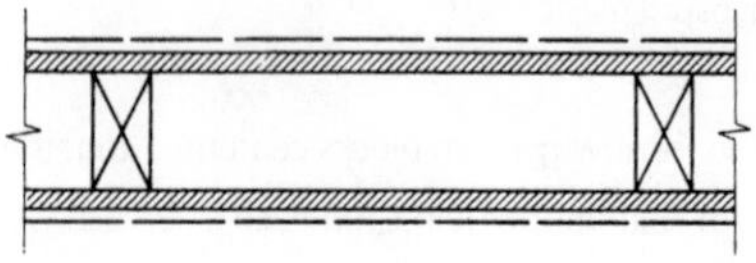

W 28*a, b*

W28b    Similar to W28a except that a layer of lead, 2.95 lb/ft² , was laminated to each side of the panel.

**TABLE K.1 Acoustic Characteristics of Walls: For Other Data See Index K.2** *(Continued)*

| Designation | Description | Section Sketch |
|---|---|---|
| W30a | 2 × 4-in. wooden studs, 16 in. on centers, attached to 2 × 4-in. wooden floor and ceiling plates, ⅜-in. gypsum lath nailed to studs on both sides, ½-in. sanded plaster with white-coat finish. | W 30a W 31a,b,c W 32a |
| W31a | 2 × 4-in. wooden studs, 16 in. on centers, ⅜-in. gypsum lath nailed to studs on both sides, ½-in. sanded plaster with white-coat finish. | |
| W31b | Similar to W31a except that a 0.065-in.-thick layer of lead weighing 3.85 lb/ft$^2$ was laminated to each side of the panel. | |
| W31c | Similar to W31a except that a 0.13-in.-thick layer of lead weighing 7.9 lb/ft$^2$ was laminated to one side of the panel. | |
| W32a | 2 × 4-in. wooden studs, 16 in. on centers; on each side, two layers of ⅜-in. gypsum wallboard were cemented together; joints in exposed surfaces taped and finished. | |

**Staggered Wood Stud Walls**

| | | |
|---|---|---|
| W34a | 2 × 3-in. wooden studs, 16 in. on centers, staggered 8 in. on centers, attached to 2 × 4-in. wooden plates at ceiling and floor; ½-in. gypsum wallboard nailed 7 in. on centers on both sides to studs. All joints taped and finished. | W 34a |
| W35a | 2 × 3-in. wooden studs, 16 in. on centers, staggered 8 in. on centers (attached to 2 × 4-in. wooden plates at floor and ceiling); two layers of ⅝-in. tapered-edge gypsum wallboard, first layer nailed 7 in. on centers, second layer nailed 16 in. on centers. All exposed joints taped and finished. | W 35a      W 35b |
| W35b | Similar to W35a except that the wall was constructed with ⅜-in. perforated gypsum lath and ½-in. sanded gypsum plaster with white-coat finish. | |

**TABLE K.1 Acoustic Characteristics of Walls: For Other Data See Index K.2 *(Continued)***

| Designation | Description | Section Sketch |
|---|---|---|
| W36a | 2 × 4-in. wooden studs, 16 in. on center, staggered 8 in. on center and offset ½ in. On each side, ⅜-in. gypsum lath nailed to studs, ½-in. gypsum vermiculite plaster, machine applied, and a hand-applied white-coat finish. | W 36a |
| W36b | Same as W36a except that the space between the studs contained vermiculite fill with a density of 6.3 lb/ft³. | W 36b |
| W37 | 2 × 4-in. wooden studs, 16 in. on center, staggered 8 in. on center, attached to a 2 × 4¾-in. wooden floor and ceiling plates; ½-in. gypsum wallboard nailed on both sides to studs, 0.9-in. wood-fiber wool blanket stapled on the inside of one side of the wall. All joints taped and finished. | W 37 |

**Slotted Wood Studs**

| | | |
|---|---|---|
| W38 | 2 × 4-in. slotted wooden studs, 16 in. on centers, attached to 2 × 4-in. wooden floor and ceiling plates; ⅜-in. gypsum lath nailed 7 in. on center to studs, ½-in. gypsum plaster with white-coat finish applied to both sides. 3-in. mineral fiber batts stapled between studs. | W 38 |

**TABLE K.1 Acoustic Characteristics of Walls: For Other Data See Index K.2** *(Continued)*

| Designation | Description | Section Sketch |
|---|---|---|
| **Wood Studs; Resilient Mounting** | | |
| W39a | 2 × 4-in. wooden studs, 16 in. on centers, attached to 2 × 4-in. wooden floor and ceiling plates; resilient channels nailed horizontally to both sides of studs 24 in. on center, ⅝-in. gypsum wallboard screwed 12 in. on center to channels. All joints taped and finished. | 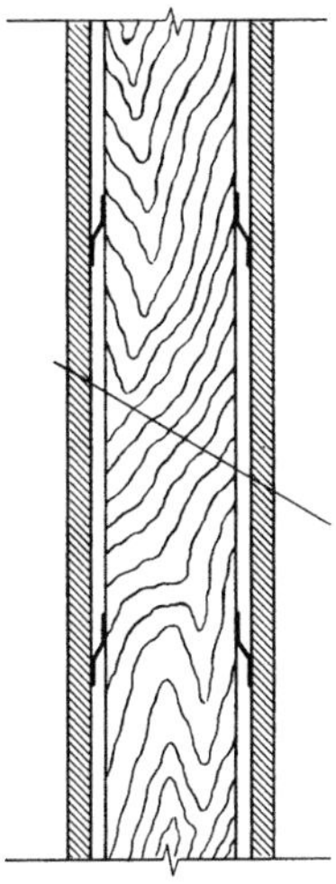 W 39a |
| W40a | 2 × 4-in. wooden studs, 16 in. on center; resilient clips, nailed to studs on both sides, held ⅜-in. gypsum lath, ½-in. sanded gypsum plaster with white-coat finish. |  W 40a |
| **Steel Truss Stud Wall** | | |
| W43a | 1⅝-in. steel truss studs; ⅜-in. gypsum lath, ½-in. plaster on both sides. | 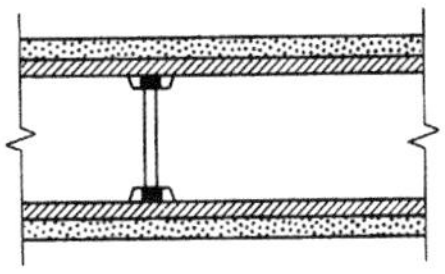 W 43a,b,c  W 44a |
| W43b | Similar to W43a except that a layer of lead, 2.95 lb/ft², was laminated to one side of the partition. | |
| W43c | Similar to W43a except that a layer of lead, 2.95 lb/ft², was laminated to each side of the partition. | |
| W44a | 3¼-in. steel truss studs, 24 in. on center, attached to metal floor and ceiling tracks; on both sides, ⅜-in. perforated gypsum lath attached with wire clips wire-tied to studs, ½-in. sanded gypsum plaster. | |
| **Steel Truss Studs; Resilient Mounting** | | |
| W50a | 2½-in. steel truss studs 16 in. on center, ⅜-in. gypsum lath attached with resilient clips to studs, ½-in. plaster applied to both sides. | 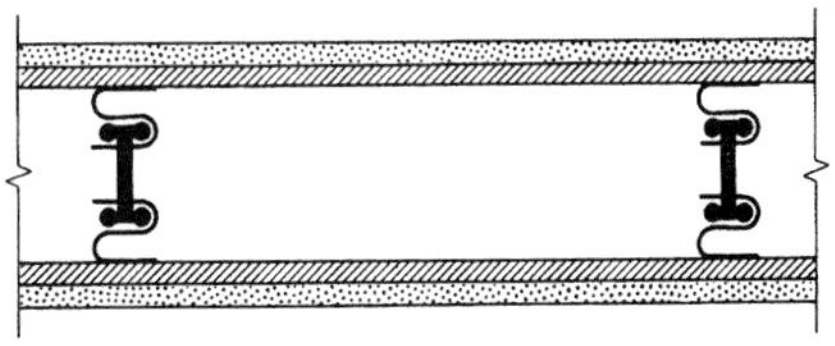 W 50a |

**APPENDICES**

**TABLE K.1 Acoustic Characteristics of Walls: For Other Data See Index K.2 *(Continued)***

| Designation | Description | Section Sketch |
|---|---|---|
| W52 | 3¼-in. steel truss studs, 16 in. on center; on each side, resilient clips fastened 16 in. on center to studs, ¼-in. metal rod wiretied to clips, diamond mesh metal lath wiretied to metal rods, ¾-in. sanded gypsum plaster. | 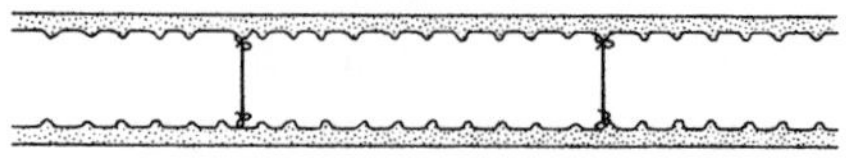 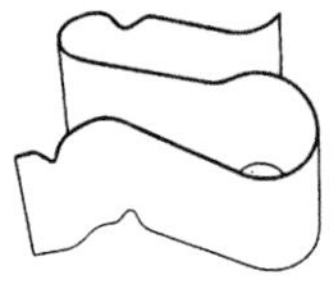 |
| **Metal Channel Stud Wall** | | |
| W55a | 3⅝-in. metal channel studs, 24 in. on center, set into 3⅝-in. metal floor and ceiling runners; ⅝-in. gypsum wallboard screwed to studs on both sides. All joints taped and finished. | 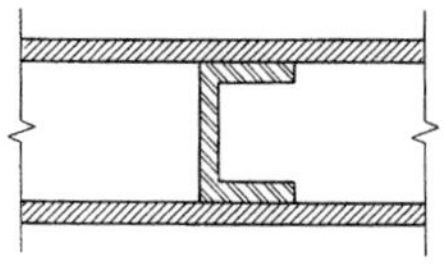 |
| W60 | 2½-in. metal channel studs, 24 in. on center, set in 2½- in. metal floor and ceiling runners; ½-in. vinyl-coated gypsum wallboard adhesively attached and screwed to studs on both sides. All joints sealed with caulking compound. Aluminum batten strips screwed 12 in. on center to gypsum board at joints; top and bottom finished with aluminum ceiling and base trim. 2-in. mineral fiber blankets hung between studs. | 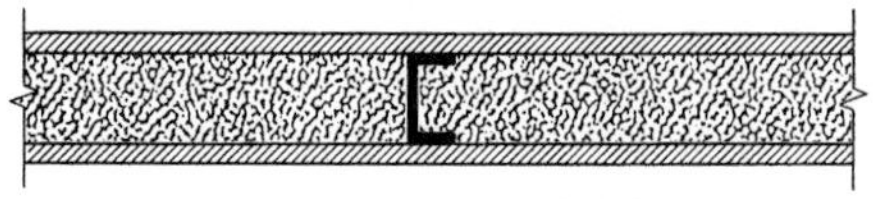 |
| W63 | 3⅝-in. metal channel studs, 24 in. on center, set into 3⅝-in. metal runners, which were attached through continuous beads of nonsetting resilient caulking compound to floor and ceiling, respectively. Two layers of ⅝-in. gypsum wallboard attached to both sides of studs. First layer screwed 8 in. on center at joints and 12 in. on center in field; second layer laminated and screwed 24 in. on center to first layer, with joints staggered 24 in. 1½-in.-thick mineral fiber felt, 3 lb/ft³, stapled between studs. All exposed joints taped and finished. The ¼-in. clearance around the perimeter closed with a nonsetting resilient caulking compound. | 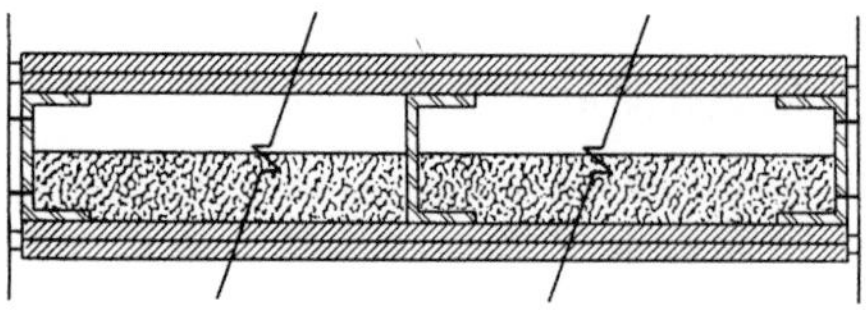 |
| W67 | 3⅝-in. metal channel studs, 24 in. on center, set in 3⅝-in. metal floor and ceiling runners; ⅝-in. gypsum wallboard screwed to studs on both sides. On one side, resilient channels screwed horizontally, 24 in. on center to inner layer; ⅝-in. gypsum wallboard screwed to channels. On the other side, ⅝-in. gypsum wallboard laminated directly to inner layer. 3-in. mineral fiber blankets hung between studs. All exposed joints taped and finished. | 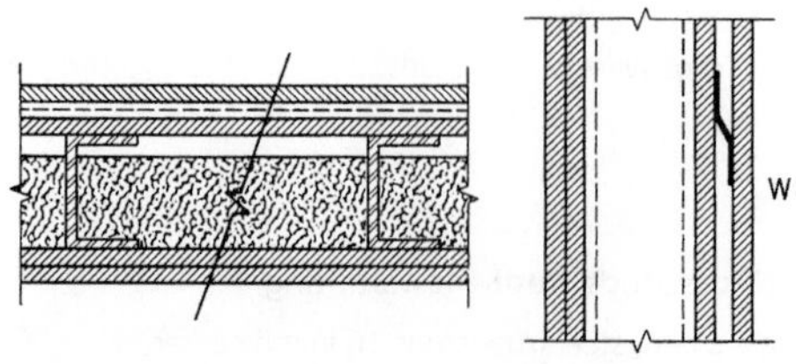 |

**TABLE K.1 Acoustic Characteristics of Walls: For Other Data See Index K.2** *(Continued)*

| Designation | Description | Section Sketch |
|---|---|---|
| **Gypsum Partitions** | | |
| W80 | 24-in.-wide panels constructed of 1 × 24-in. gypsum core board offset 1½ in. at edges to form tongue-and-groove edge; ⅝-in. vinyl-faced gypsum wallboard laminated to both sides of core board. Panels inserted into two-piece metal floor and ceiling tracks. Gypsum-to-gypsum screws at ¼ and ½ points along vertical edges of face boards. | 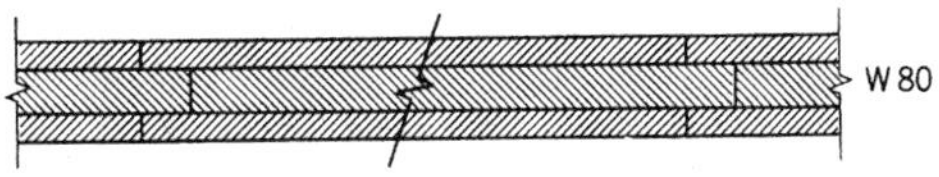 W 80 |
| W85a | Double wall with 1⅜-in. air space. Each leaf consisted of 24-in.-wide panels of ⅝-in. gypsum core board strips, 7½ and 4⅜ in. wide, offset 1½ in. at edges to form tongue and groove; ⅝-in., vinyl-faced, gypsum wallboard laminated to both sides of core board strips. Panels screwed 12 in. on center to 1¼ × 1-in. angle floor and ceiling runners. | 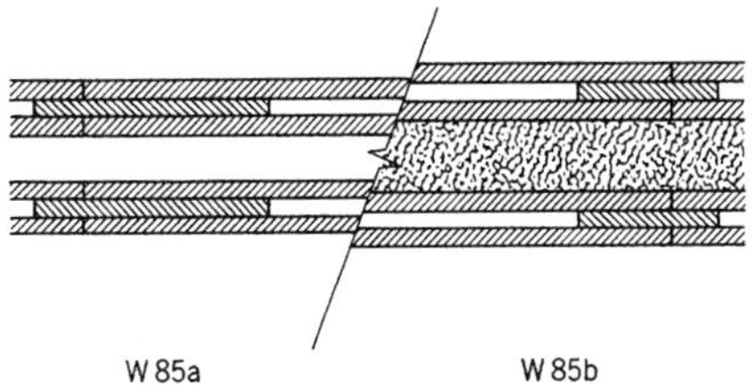 W 85a    W 85b |
| W85b | Similar to W85a except that space between leaves was 2⅛ in. and contained 2-in. mineral fiber blankets stapled to one leaf. ¼-in. perimeter clearance closed with a nonsetting resilient caulking compound. Vertical face layer joints sealed with joint compound. | |

*Source:* All data extracted from *A Guide to Airborne, Impact and Structure-Borne Noise Control in Multi-Family Dwellings,* HUD/FHA/NBS 1971. U.S. Department of Housing and Urban Development, Washington, DC.

# Sound Transmission and Impact Insulation Data for Floor/Ceiling Constructions

To use Appendix L:

1. Find the desired construction Type Code in Index L.1 or L.2. Index L.1 lists the constructions by STC. Index L.2 lists the constructions by IIC.
2. Find desired STC/IIC ratings corresponding to the selected Type Code.
3. Refer to Table L.1 for details of construction.
4. Refer to Index L.3 for thickness, weight, STC, IIC, fire rating, and transmission loss at standard octave midpoints.

## EXAMPLE

A standard (simple) wooden joist floor-ceiling construction is required with a minimum IIC of 55. Any STC above 35 is acceptable.

1. Since IIC is the determining factor, refer to Index L.2. The basic Type Code is "a."
2. From Index L.2, note that none of the listed "a" constructions gives an IIC of 55. However, from Section 19.24, note that the addition of carpeting will add 10 to 27 points to the IIC, making items F34, F38a, F39a, and F30 all suitable. This selection avoids special and resilient constructions because of the requirement for standard (simple) construction.
3. From Table L.1, select construction F30 as being most appropriate for application of carpeting.
4. From Index L.3 find (without carpet): thickness 9½ in., STC 34, IIC 32 (+ carpet), fire rating ¼ hour. With high-pile carpeting on a foam rubber pad, this construction will have an IIC > 55, STC > 35.

**INDEX L.1 Sound Transmission Class: Floor/Ceiling Constructions**

| Type Code | |
|---|---|
| a. Wooden joist | f. With resilient elements |
| b. Metal joist | h. With carpeting |
| c. Concrete or masonry | i. With absorbent blankets |
| d. Plaster ceiling | j. With separate ceiling joists |
| e. Gypsum board ceiling | |

| STC | IIC | Type | Item No. | STC | IIC | Type | Item No. |
|---|---|---|---|---|---|---|---|
| 55[a] | 57[a] | c,d,f | F14 | 46[a] | 42[a] | c,d | F23 |
| 54[a] | 64[a] | c,d,f | F17b | 45 | 44 | a,e,f | F44 |
| 52 | 80 | a,e,i,j | F48 | 44 | 42 | c,f | F3-2d |
| 51[a] | 53[a] | c,d,f | F10 | 44 | 41 | c | F3-1a |
| 51[a] | 48[a] | c,d | F7a | 44 | 80 | c,h | F2-1a |
| 50[a] | 53[a] | c,e,f | F25 | 44 | 29 | c | F1-c |
| 50[a] | 51[a] | c,e | F27 | 44 | 25 | c | F1a |
| 50[a] | 48[a] | c,d | F16 | 43[a] | 43[a] | a,d | F32b |
| 49[a] | 48[a] | c,d | F9 | 42[a] | 32[a] | c | F22 |
| 48[a] | 47[a] | c,d | F12 | 40 | 32 | a,e,i | F40a |
| 48 | 33 | b,c,d | F60a | 39[a] | 37[a] | a,e | F34 |
| 47 | 62 | b,c,e | F58 | 37 | 33 | a,e | F38a |
| 47[a] | 42[a] | c,d | F24 | 37 | 32 | a,e | F39a |
| 47 | 59 | b,c,d,h | F57b | 34[a] | 32[a] | a,e | F30 |
| 47 | 37 | b,c,d | F57a | 29[a] | 32[a] | a,e | F35a |
| 46 | 74 | b,c,d | F60c | 29 | 56 | a,e,h | F35b |

[a]Field measurement.

**INDEX L.2  Impact Insulation Class: Floor/Ceiling Constructions**

| Type Code | |
|---|---|
| a. Wooden joist | f. With resilient ceiling element |
| b. Metal joist | g. With resilient floor element |
| c. Concrete or masonry | h. With carpeting |
| d. Plaster ceiling | i. With absorbent blankets |
| e. Gypsum board ceiling | j. With separate ceiling joists |

| IIC | STC | Type | Item No. | IIC | STC | Type | Item No. |
|---|---|---|---|---|---|---|---|
| 80 | 52 | a,e,h,i,j | F48 | 44 | 45 | a,e,f | F44 |
| 80 | 44[b] | c,h | F2-1a | 43[a] | 43[a] | a,d,g | F32b |
| 74 | 46 | b,c,d,h | F60c | 42[a] | 47[a] | c,d | F24 |
| 64[a] | 54[a] | c,d,g | F17b | 42[a] | 46[a] | c,d | F23 |
| 62 | 47 | b,c,e,h | F58 | 42 | 44[b] | c,g | F3-2(d) |
| 59 | 47[b] | b,c,d,h | F57b | 41 | 44[b] | c | F3-1(a) |
| 57[a] | 55[a] | c,d,g | F14 | 37 | 47 | b,c,d | F57a |
| 56[a] | 29[b] | a,e,h | F35b | 37[a] | 39[a] | a,e | F34 |
| 53[a] | 51[a] | c,d,g | F10 | 33 | 48 | b,c,d | F60a |
| 53[a] | 50[a] | c,e,g | F25 | 33 | 37 | a,e | F38a |
| 51[a] | 50[a] | c,e | F27 | 32[a] | 42[a] | c | F22 |
| 48[a] | 51[a] | c,d | F7a | 32 | 40 | a,e,i | F40a |
| 48[a] | 50[a] | c,d | F16 | 32 | 37 | a,e | F39a |
| 48[a] | 49[a] | c,d,g | F9 | 32[a] | 34[a] | a,e | F30 |
| 47[a] | 48[a] | c,d | F12 | 32[a] | 29[a] | a,e | F35a |
| | | | | 29 | 44[b] | c | F1c |
| | | | | 25 | 44 | c | F1a |

[a]Field measurement.
[b]Estimated on the basis of similar structures.

**INDEX L.3 Floor/Ceiling Sound Transmission and Construction Data[a]**

| Designation | Thickness (in.) | Weight (lb/ft²) | Transmission Loss (dB) at frequency | | | | | | STC | IIC | Fire Rating (h) |
|---|---|---|---|---|---|---|---|---|---|---|---|
| | | | 125 | 250 | 500 | 1K | 2K | 4K | | | |
| F1a | 4 | 53 | 47 | 42 | 45 | 56 | 58 | 66 | 44 | 25 | 1 |
| F7a | 5¼ | 61 | 41 | 42 | 47 | 54 | 59 | 63 | 51 | 48 | 2 |
| F9 | 6⅝ | 65 | 40 | 42 | 46 | 48 | 57 | 62 | 49 | 48 | 2 |
| F10 | 8¼ | 90 | 38 | 43 | 47 | 53 | 60 | — | 51 | 53 | 2½ |
| F12 | 10 | 62 | 38 | 40 | 44 | 51 | 56 | 59 | 48 | 47 | 3 |
| F14 | 9½ | 83 | 38 | 44 | 52 | 55 | 60 | — | 55 | 57 | 3 |
| F16 | 8⅛ | 65 | 40 | 42 | 46 | 52 | 58 | — | 50 | 48 | 2 |
| F17b | 9¼ | 57 | 38 | 46 | 52 | 59 | 64 | — | 54 | 64 | 2 |
| F22 | 6¼ | 28 | 34 | 34 | 38 | 45 | 55 | 61 | 42 | 32 | ¾ |
| F23 | 9½ | 45 | 33 | 37 | 43 | 52 | 58 | 62 | 46 | 42 | ¾ |
| F24 | 10¼ | 65 | 34 | 37 | 43 | 52 | 57 | — | 47 | 42 | ¾ |
| F25 | 10 | 45 | 30 | 38 | 46 | 58 | 64 | — | 50 | 53 | ¾ |
| F27 | 7⅝ | 50 | 36 | 40 | 47 | 54 | 58 | — | 50 | 51 | ¾ |
| F30 | 9½ | 7 | 19 | 24 | 31 | 35 | 45 | — | 34 | 32 | ¼ |
| F32b | 11 | 12 | 30 | 31 | 41 | 47 | 52 | — | 43 | 43 | ¾ |
| F34 | 10¼ | 9.9 | 15 | 32 | 44 | 48 | 54 | 53 | 39 | 37 | 1 |
| F35a | 10 | 9.2 | 14 | 17 | 30 | 44 | 47 | 52 | 29 | 32 | — |
| F38a | 11¾ | 9 | 30 | 38 | 36 | 43 | 48 | 49 | 37 | 33 | ½ |
| F39a | 11⅞ | 9.5 | 22 | 32 | 36 | 45 | 49 | 56 | 37 | 32 | 1 |
| F40a | 11⅞ | 10 | 25 | 36 | 38 | 46 | 51 | 57 | 40 | 32 | 1 |
| F44 | 10½ | 10.1 | 43 | 41 | 41 | 52 | 50 | 60 | 45 | 44 | ¾ |
| F48 | 12⅜ | 10.7 | 35 | 42 | 52 | 56 | 69 | 74 | 52 | 80 | ¾ |
| F57a | 18⁹⁄₁₆ | 23.2 | 33 | 44 | 45 | 46 | 58 | 62 | 47 | 37 | 3 |
| F58 | 21½ | 20.4 | 27 | 37 | 45 | 54 | 60 | 65 | 47 | 62 | 1 |
| F60a | 11 | 38.2 | 42 | 44 | 44 | 51 | 51 | 61 | 48 | 33 | 1½ |
| F60c | 11⅝ | 39 | 39 | 43 | 44 | 52 | 52 | 65 | 46 | 74 | — |

[a]Material extracted from *A Guide to Airborne, Impact and Structure-Borne Noise Control in Multifamily Dwellings*, HUD/FHA/NBS, 1971. U.S. Department of Housing and Urban Development, Washington, DC.

**TABLE L.1  Acoustic Characteristics of Floors: For Acoustic Data see Index L.3**

| Code[a] | Description | Section Sketch |
|---|---|---|

**Reinforced Concrete Slab**

F1a  4-in.-thick reinforced concrete slab, isolated from support structure. Concrete was reinforced with 6 × 6-in. No. 6 AWG reinforcing mesh placed at the centerline horizontal plane of the slab. All surface cavities were sealed with a thin mortar mix.

F1c  Same as F1a except that ⅛-in.-thick vinyl tile was adhered to concrete.

F 1a          F 1c

**Reinforced Concrete with Floor Coverings**

**See also F1c above.**

F2-1(a)  4-in.-thick reinforced concrete slab with carpeting and pad. The carpeting was of ¼-in. wool loop pile with ⅛-in. woven jute backing, 0.49 lb/ft$^2$; the foam rubber pad was ¼ in. thick and weighed 0.53 lb/ft$^2$.

F 2-1

F3-1(a)  4-in. reinforced concrete slab with ½ × 9 × 9-in. oak blocks, 1.8 lb/ft$^2$, set in mastic.

F3-2(d)  4-in. concrete slab with ⅛-in. cork.

F 3

F7a  4⅜-in.-thick reinforced concrete slab. On the floor side, ¾-in.-thick, sand-cement screed with ⅛-in. linoleum floor covering. On the ceiling side, ⅜-in. layer of plaster.

F 7a

F9  4⅜-in.-thick reinforced concrete slab. On the floor side, ½-in.-thick layer of bitumen with ½-in.-thick soft wood fiberboard, which was covered with a thin layer of bitumen with sand and a ¾-in.-thick sand-cement screed. On the ceiling side, ⅜-in. layer of plaster.

F 9

**Reinforced Concrete Slab, Floating Floor**

F10  5-in.-thick reinforced concrete. On the floor side, 1½-in.-thick wire mesh reinforced sand-cement screed floating on ½-in.-thick bitumen-bonded, glass-wool quilt covered with building paper. On the screed, ½-in.-thick pitch-mastic with a linoleum floor covering. On the ceiling side, ½-in. layer of plaster.

F 10

F12  4⅜-in.-thick reinforced concrete slab. On the floor side, ¾-in.-thick sand-cement screed. On the ceiling side, brick wire mesh, suspended 4 in. with wire hangers, held ⅞-in. gypsum plaster.

F 12

**TABLE L.1 Acoustic Characteristics of Floors: For Acoustic Data see Index L.3** *(Continued)*

| Code[a] | Description | Section Sketch |
|---|---|---|
| F14 | 6-in.-thick reinforced concrete slab. On the floor side, ¾-in.-thick tongue-and-groove wood flooring nailed to 1½ × 2-in. wooden battens, 16 in. on centers, floating on 1-in.-thick glass-wool quilt. On the ceiling side, ½-in. layer of plaster. | 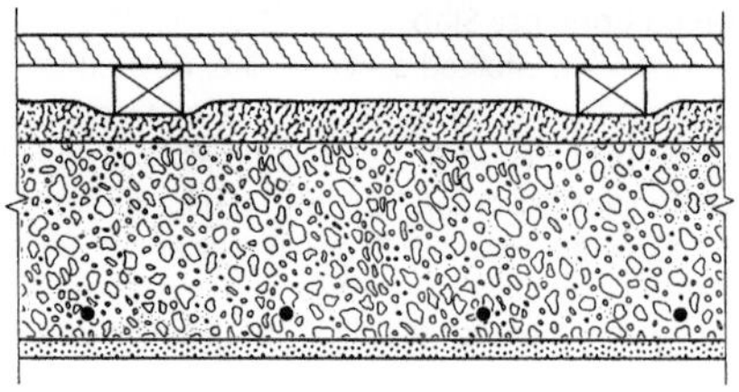 F 14 |

**Concrete with Hollow Blocks**

| | | |
|---|---|---|
| F16 | 5 × 10-in. hollow masonry blocks, 14 in. on center, with spaces between blocks filled with 5-in.-thick reinforced concrete. On the floor side, ⅞-in.-thick wood blocks adhered to 1½-in.-thick sand-cement screed. On the ceiling side, ¾-in. layer of plaster. | 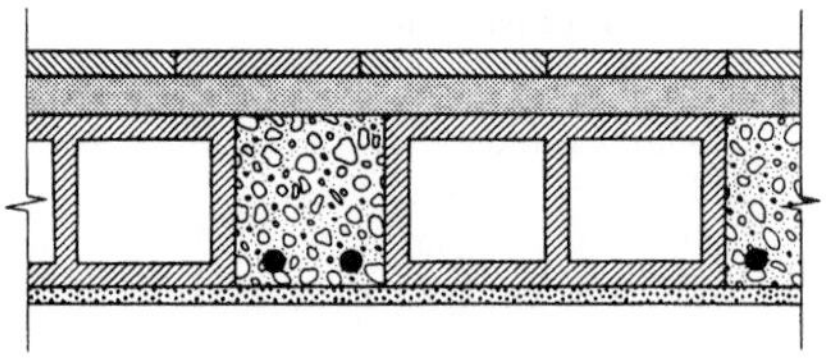 F 16 |
| F17b | 4 × 12½-in. hollow masonry blocks, 15½ in. on center, with spaces between blocks filled with 4-in.- thick reinforced concrete. On the floor side, 2-in.- thick sand-cement screed; linoleum on 1-in.-thick wood flooring nailed to 1 × 2-in. wooden battens, spaced 15½ in. on centers, floating on a glass-wool quilt approximately 1 in. thick. On the ceiling side, ¾-in. layer of plaster. | 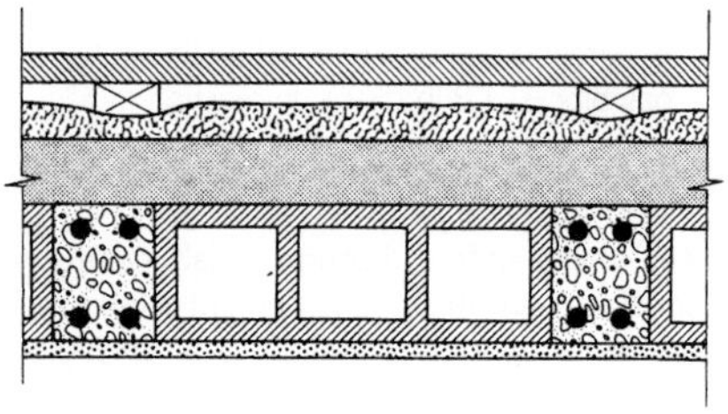 F 17b |

**Concrete Channel Slab**

| | | |
|---|---|---|
| F22 | Prefabricated concrete channel slabs mortared together 20 in. on center. Each slab had a 3-in.-deep trapezoidal channel with bases of 11 and 14¾ in. On the floor side, ¾-in.-thick sand-cement finish. | 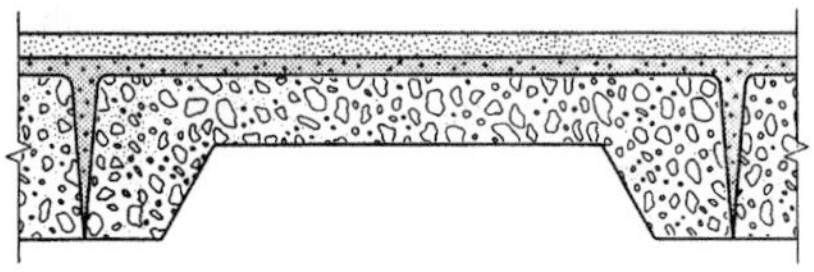 F 22 |

**Ribbed Concrete**

| | | |
|---|---|---|
| F23 | 7¼-in. ribbed concrete floor. Ribs were 5¼ × 3¾ in., spaced 21 in. on center, with 1 × 2-in. wooden nailing strips cast into ends. On the floor side, the slab was 2 in. thick with a ¾-in.-thick sand-cement screed. On the ceiling side, ⅝-in.-thick wooden laths nailed to nailing strips held ⅝-in.-thick plaster. | 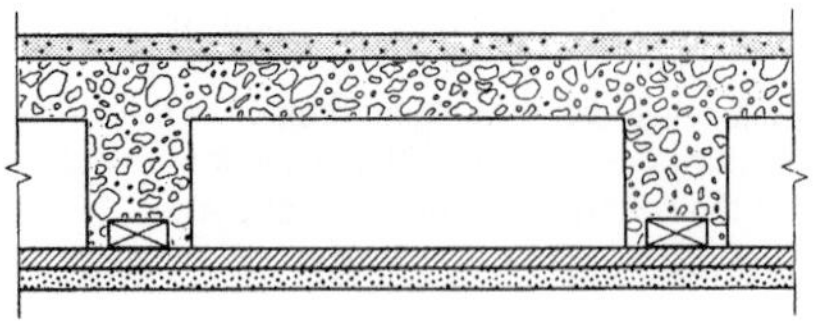 F 23 |

**Concrete Channel Beam**

| | | |
|---|---|---|
| F24 | 7-in. precast trapezoidal concrete channel beams, 14 in. on center, with spaces between beams filled with sand-cement mix. On the floor side, 1½-in.-thick sand-cement screed with 1-in.-thick wood-block floor covering. On the ceiling side, approximately ¾-in.-thick layer of plaster on expanded metal lath. | 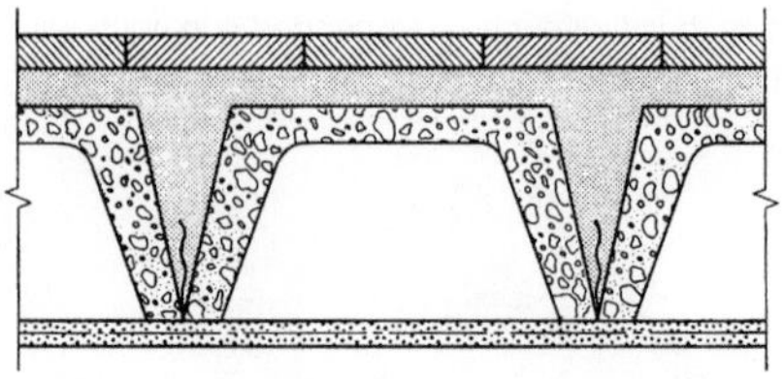 F 24 |

**TABLE L.1  Acoustic Characteristics of Floors: For Acoustic Data see Index L.3 *(Continued)***

| Code[a] | Description | Section Sketch |
|---|---|---|
| | **Precast Concrete Beam, Floating Floor** | |
| F25 | 5-in. precast concrete channel beams, 14½ in. on center, with spaces between beams filled with a sand-cement mix. On the floor side, ⅞-in.-thick tongue-and-groove wood flooring nailed to 1 × 2-in. wooden battens, 20 in. on center, on approximately 1-in.-thick glass-wool quilt on ¾-in.-thick sand- cement screed. On the ceiling side, ⅛-in. layer of plaster on ⅜-in. gypsum wallboard nailed to 1 × 2-in. wooden battens spaced 14½ in. on centers. | 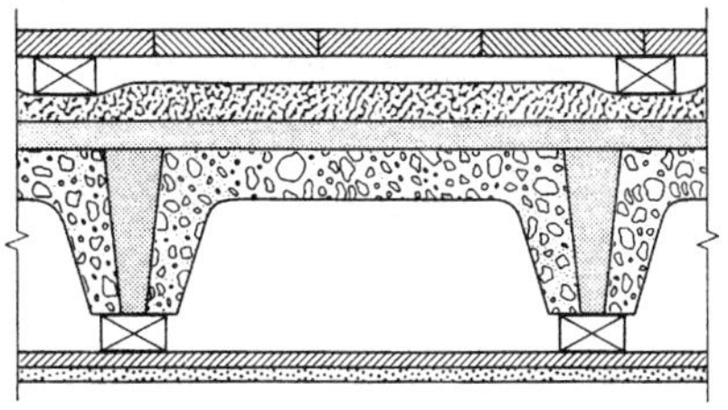 F 25 |
| | **Hollow Concrete Beam** | |
| F27 | 5-in. precast trapezoidal hollow concrete beams, 14½ in. on center, with bases of 14 and 12½ in. Spaces between beams filled with sand-cement mix. On the floor side, 1-in.-thick sand-cement screed with 3/16-in. cork tile floor covering. On the ceiling side, ⅜-in.-thick gypsum wallboard attached to 1 × 2- in. wooden battens held by metal clips. | 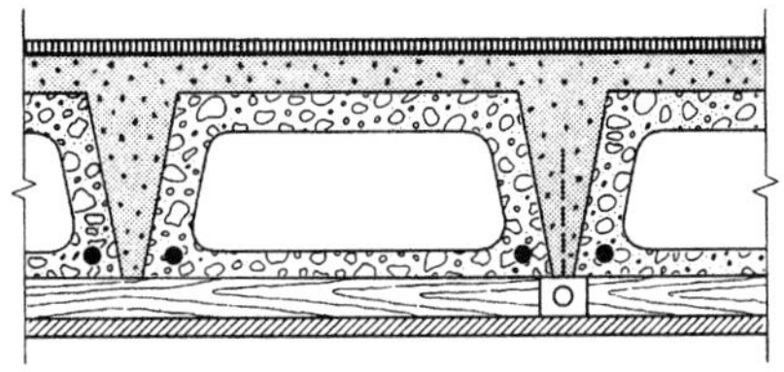 F 27 |
| | **Wooden Joist** | |
| F30 | 2 × 8-in. wooden joists, 16 in. on center. On the floor side, ⅞-in. tongue-and-groove flooring nailed to joists; on the ceiling side, ⅜-in. gypsum wallboard nailed to joists with joints sealed. | 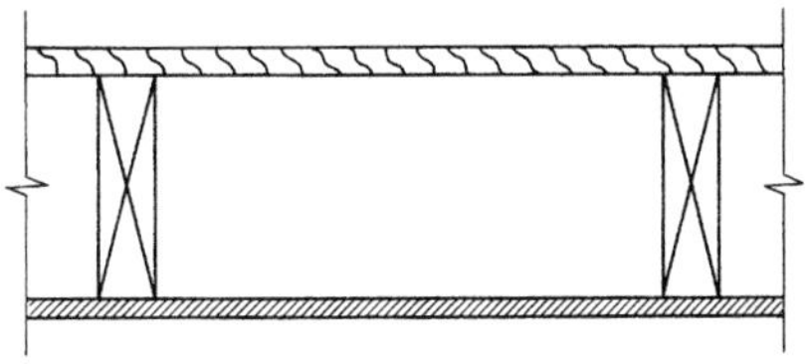 F 30 |
| F32 | 2 × 8-in. wooden joists 18 in. on center. On the floor side, ⅞-in. tongue-and-groove wood flooring nailed to joists. On the ceiling side, 1-in. battens nailed through glass-wool quilt approximately 1 in. thick; ½-in. layer plaster on ¼-in.-thick wood lath. | 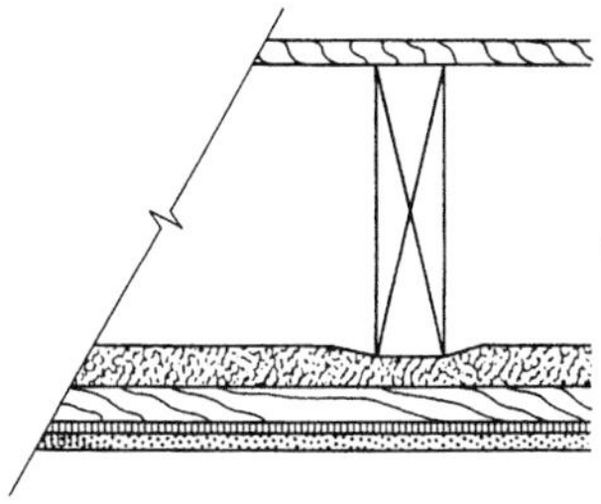 F 32 |
| F34 | 2 × 8-in. wooden joists, 16 in. on center. On the floor side, ½-in.-thick C-D plywood nailed 8 in. on center to joists, 25/32-in.-thick hardwood flooring on plywood. On the ceiling side, ½-in.-thick gypsum wallboard nailed 6 in. on center to joists. All joints taped and finished; ceiling tile adhered to gypsum board. | 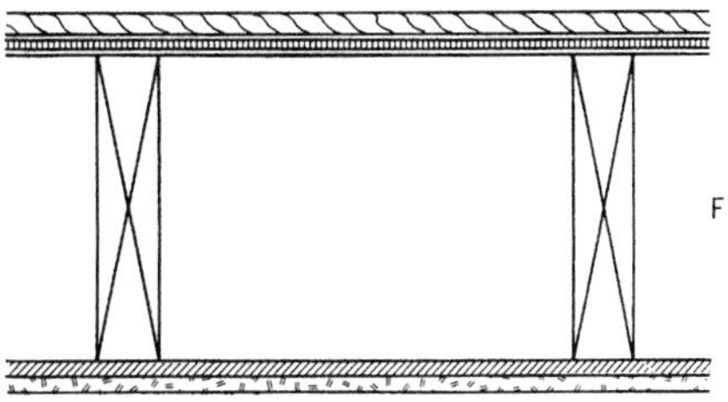 F 34 |

APPENDICES

**TABLE L.1 Acoustic Characteristics of Floors: For Acoustic Data see Index L.3 *(Continued)***

| Codeᵃ | Description | Section Sketch |
|---|---|---|
| F35a | 2 × 8-in. wooden joists, 16 in. on center. On the floor side, 1½-in.-thick tongue-and-groove wood fiber-board nailed to joists, vinyl tile floor covering. On the ceiling side, ½-in.-thick gypsum wallboard nailed 6 in. on center to joists. All joints taped and finished. | 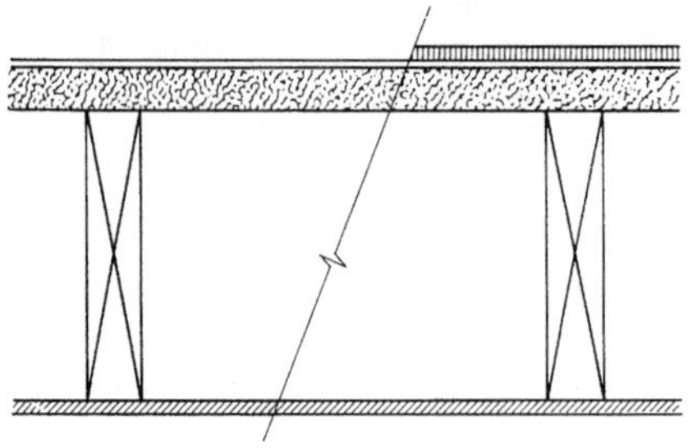 |
| F35b | Similar to F35a except that fiberboard was covered with carpet and pad. | |
| F38a | 2 × 10-in. wooden floor joists spaced 16 in. on center. ⅝-in. fir plywood subfloor nailed to joists 8 in. on center; ½-in. plywood underlayment nailed to subfloor with joints staggered to miss joints of the subfloor; ⅛ × 9 × 9-in. vinyl asbestos tile glued to underlayment. On the ceiling side, ½-in. gypsum wallboard nailed 12 in. on center, with all joints and nailheads taped and finished. | 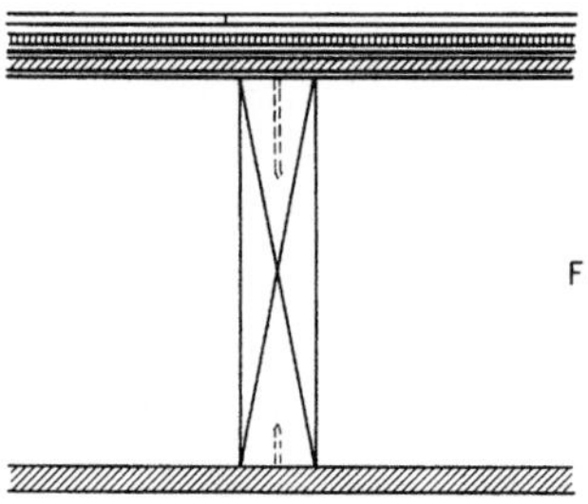 |
| F39a | 2 × 10-in. wooden joists, 16 in. on center. On the floor side, ½-in.-thick plywood subfloor nailed 6 in. on center along edges and 10 in. on center in field, building paper underlayment, ²⁵⁄₃₂ × 2¼-in. oak wood flooring nailed at each joist intersection and midway between joists. On the ceiling side, ⅝-in.-thick gypsum wallboard, nailed 6 in. on center to joists. All joints taped and finished. | 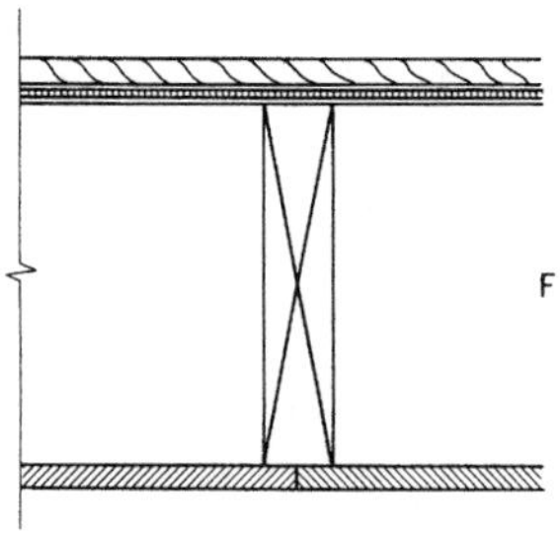 |

**Wooden Joist, Resilient Ceiling**

| | | |
|---|---|---|
| F44 | 2 × 8-in. wooden joists 16 in. on center. On the floor side, ¾-in.-thick wood subfloor, layer of building paper, and ¾-in.-thick tongue-and-groove fir finish flooring. On the ceiling side, resilient runners bridged across joists and nailed 12 in. on center to joists; ⅝-in.-thick gypsum wallboard screwed to resilient runners. All joints taped and finished. |  |

**Wooden Joist with Insulation**

| | | |
|---|---|---|
| F40a | 2 × 10-in. wooden joists 16 in. on center with 3-in.-thick mineral fiber batts stapled between joists. On the floor side, ½-in.-thick plywood subfloor nailed 6 in. on center along edges and 10 in. on center in field, building paper underlayment, ²⁵⁄₃₂ × 2¼-in. oak wood flooring nailed at each joist intersection and midway between joists. On the ceiling side, ⅝-in.- thick gypsum wallboard nailed 6 in. on center to joists. All joints taped and finished. | 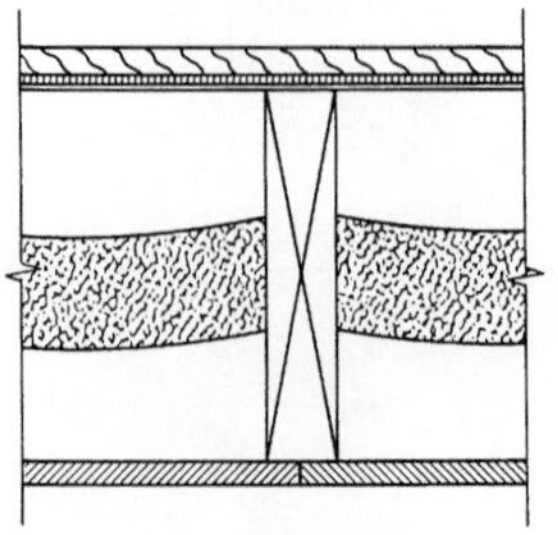 |

**TABLE L.1 Acoustic Characteristics of Floors: For Acoustic Data see Index L.3** ***(Continued)***

| Code[a] | Description | Section Sketch |
|---|---|---|
| F48 | 2 × 8-in. wooden joists, 16 in. on center. On the floor side, 1⅛-in.-thick regular C-D rough plywood nailed 6 in. on center along periphery and 16 in. on center at other bearings, plywood covered with an all-hair pad (40 oz/yd$^2$) and all-wool pile (44 oz/yd$^2$) carpet. The total weight of the carpet was 4.14 lb/yd$^2$ and the total thickness was ⅜ in. On the ceiling side, 2 × 4-in. wooden joists, 16 in. on center, staggered 8 in. on center relative to the floor joists, 3-in.-thick fibered glass blankets stapled between ceiling joists, ⅝-in.-thick gypsum wallboard nailed to ceiling joists. All joints taped and finished; entire periphery of panel caulked and sealed. The ceiling was supported independently of the floor structure. | 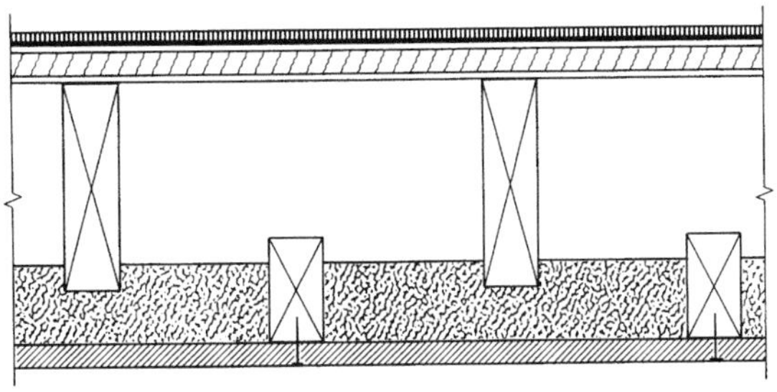 F 48 |
| **Steel Joist with Concrete Floor** | | |
| F57a | 2½-in.-thick perlite concrete, 72 lb/ft$^3$ on 28-gauge corrugated steel units supported by 14-in. steel bar joints; ⅛-in.-thick asphalt tile cemented to concrete. On the ceiling side, ¾-in. furring channels, 13½ in. on center, wiretied to joists, 3.4 lb/yd$^2$ diamond mesh metal lath wire-tied to furring channels, 9/16-in. coat of plaster with 1/16-in. white-coat finish. | 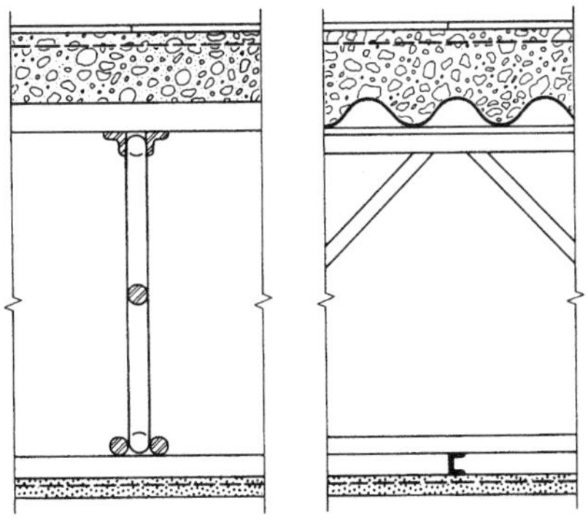 F 57*a, b* |
| F57b | Same as F57a except for carpet and pad in lieu of asphalt tile. | |
| F58 | 18-in. steel joists, 16 in. on center. On the floor side, ⅝-in.-thick C-D rough plywood nailed to joists, 1⅝- in.-thick foamed concrete, 100 lb/ft$^3$, slab constructed on the plywood; concrete covered with an all-hair pad (40 oz/yd$^2$) and an all-wool pile (44 oz/yd$^2$) carpet. Total weight of the carpet, 4.14 lb/yd$^2$; total thickness, ⅜ in. On the ceiling side, ⅝- in.-thick gypsum wallboard nailed to joists. All joints taped and finished; entire periphery of panel caulked and sealed. | 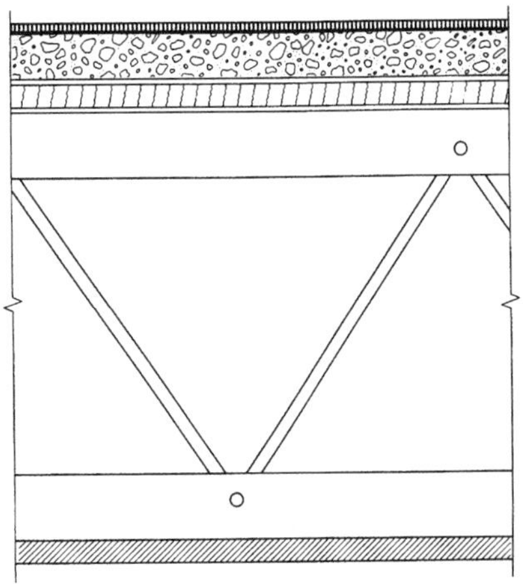 F 58 |

APPENDICES

**TABLE L.1  Acoustic Characteristics of Floors: For Acoustic Data see Index L.3 *(Continued)***

| Code[a] | Description | Section Sketch |
|---|---|---|
| F60a | 7-in. steel bar joists spaced 27 in. on center. On the floor side, ⅜-in. metal rib lath attached to top of joists and 2-in.-thick poured concrete floor. On the ceiling side, ¾-in. metal furring channels wiretied to joists 16 in. on centers; ⅜ × 16 × 48-in. plain gypsum lath held with wire clips and sheet metal end joint clips; ⁷⁄₁₆-in. sanded gypsum plaster and ¹⁄₁₆- in. white-coat finish. | F 60a,c |
| F60c | Structure F60a with nylon carpeting and foam rubber pad placed on the floor. The carpet pad had an uncompressed thickness of ¼ in. backed with a oven jute fiber cloth. The carpet had ⅛-in. woven backing and ¼-in. looped pile spaced 7 loops per inch with a total thickness of ⅜ in. | |

[a]Material extracted from *A Guide to Airborne, Impact and Structure-Borne Noise Control in Multifamily Dwellings,* HUD/FHA/NBS, 1971. U.S. Department of Housing and Urban Development, Washington, DC.

# Design Analysis Software

This book emphasizes manual (or hand) calcula- tion procedures, those simple enough to require only a hand-held calculator and an informed (and patient) user. This emphasis is based upon the belief that such calculations provide the most ex- plicit means of understanding the variables (and relationships among variables) that influence build- ing performance. In practice, many analyses are conducted using computer software programs. Today, these programs generally run on personal computers. This appendix presents a very short list (drawn from a larger and rapidly expanding array) of software programs for personal comput- ers that may be used in the design of mechanical and electrical systems (both passive and active) for buildings.

Updated and improved versions of existing pro- grams appear as often as do new programs; any list of such programs quickly becomes obsolete. Thus, the descriptions presented here avoid version numbers and other transient information and instead describe the basics of each program. Mailing addresses are generally included; current World Wide Web URLs should be readily found using Internet search engines.

Excellent computer programs have been devel- oped at many universities and research labs. Apolo- gies to those whose programs do not appear here; the following list attempts to name the programs most widely used (or readily available) in North America.

### (a) Building Products Life-Cycle Assessment

- *BEES.* Aids in the selection of building products by generating an overall score. It weighs the environmental and economic life-cycle perfor- mance scores, with relative importance scales specified by the user. Global warming, acidifica- tion, nutrification, natural resource depletion, IAQ, and solid waste impacts are included.

    National Institute of Standards and Technol- ogy (NIST), Green Buildings Program, 100 Bureau Drive, Gaithersburg, MD 20899-0001.

### (b) Design Strategies and Climate Analysis

- *Climate Consultant.* A detailed database that can be queried by users, this program plots climate for every hour of the year on a variety of charts and provides commentary.

    Research Professor Murray Milne, UCLA, Department of Architecture and Urban Design, Los Angeles, CA 90095.

- *Green Building Advisor.* Provide a project location, building type, and size, and this program re- sponds with "moderately" and "strongly" sug- gested design strategies, relevant case studies, an information library, and a products directory. This is a tool for brainstorming rather than for detailed design decisions.

    Environmental Building News, 28 Birge St., Suite 30, Brattleboro, VT 05301.

- *Solar 2.* Graphically displays the performance

of a window with any combination of fins and overhang, as well as remote objects such as walls of an exterior courtyard or a distant building; for Windows 95.

Research Professor Murray Milne, UCLA, Department of Architecture and Urban Design, Los Angeles, CA 90095.

### (c) Heating, Cooling, and Energy Performance

- *Building Energy Software Tools Directory.* An extensive online directory of software tools, compiled by the U.S. Department of Energy. Well worth browsing.

  http://www.eere.energy.gov/buildings/tools_directory/

- *Building Design Advisor.* Provides energy modeling in a CAD-like environment, simulating energy and daylighting performance from CAD drawings.

  Lawrence Berkeley Laboratory, 1 Cyclotron Road, Berkeley, CA 94720.

- *EnergyPlus.* This program aims to combine the best features of BLAST and DOE-2, providing a detailed HVAC system performance analysis that includes passive energy sources and strategies. Considerable operator training is required.

  For information: http://www.eere.energy.gov/buildings/energyplus/getting.html.

- *Energy Scheming.* Using schematic drawings generated on-screen (Mac), this program calculates thermal performance for 4 months (representing the seasons), with passive strategies included. Design advice accompanies the performance information. Also available in a Windows version.

  Prof. G.Z. Brown, Energy Studies in Buildings Laboratory, Department of Architecture, University of Oregon, Eugene, OR 97403-1206.

- *Energy 10.* Provides energy simulations for smaller ($\leq$10,000 ft$^2$ [929 m$^2$]) commercial and institutional buildings. Input can be extremely simple (floor area, location, function), and a "shoebox" generic building is generated with design alternatives and energy comparisons. A detailed "weathermaker" climate data feature allows for adjustments to address local climates and provides "Olgyay"-style climate analyses (a hidden gem of a feature).

Sustainable Buildings Industry Council, 1112 16th St. NW, Suite 240, Washington, DC, 20036.

- *ENVSTD.* Determines compliance with the envelope trade-off aspects of ANSI/ASHRAE/IESNA 90.1, *Energy Standard for Buildings Except Low-Rise Residential Buildings.* The software is updated as the standard is updated.

  ASHRAE, 1791 Tullie Circle NE, Atlanta, GA 30329-5478.

- *Opaque.* Graphically displays the performance of a wall or roof of any composition, any color, any orientation, and at any latitude.

  Research Professor Murray Milne, UCLA, Department of Architecture and Urban Design, Los Angeles, CA 90095.

- *SOLAR 5.* Using information about climate, orientation, geometry, and shading configurations, users get a three-dimensional plot of building energy performance based upon hourly TMY data for over 230 sites. Almost all passive heating and cooling strategies can be simulated. Most HVAC system parameters can also be modeled.

  Research Professor Murray Milne, UCLA, Department of Architecture and Urban Design, Los Angeles, CA 90095.

### (d) Environmental Assessment

- *BREEAM* (Building Research Establishment Environmental Assessment Method). This well-tested UK program, in use since 1990, rates designs and building management policies, awarding certificates of overall performance (pass, good, very good, or excellent). Priority areas can be targeted early in design, making this potentially useful as a design tool.

  Building Research Establishment, Marston, Watford, WD2 7JR, United Kingdom.

- *Green Building Assessment Tool (GBTool).* Developed for use in the 1998 Green Buildings Challenge, this analysis tool is being further developed at various levels of detail and includes a weighting system with defaults that can be modified to reflect regional priorities. The "narrative" that accompanies the tool makes for an interesting comparison with the U.S. Green Building Council LEED program.

International Initiative for a Sustainable Built Environment; http://www.iisbe.org/iisbe/gbc2k5/gbc2k5-start.htm/

### (e) Solar Water Heating

- *F-Chart.* These programs allow analysis of various collector and system types (including some passive heating strategies).

  F-Chart Software, 4406 Fox Bluff Road, Middleton, WI 53562.

### (f) Fire Safety

- *Fire-Safe Building Design for Architects and Designers.* A self-paced, interactive CD-ROM combining graphics, text, narration, animation, and video, allows the user to manipulate graphics, make materials selections, consult supporting fire safety tables, review videos, and explore a library of static and animated appendices. Feedback is provided, encouraging improvements to fire safety during design.

  National Fire Academy, USFA/FEMA, 16825 South Seton Ave., Emmitsburg, MD 21727.

### (g) Daylighting and Electric Lighting

- *Visual.* This software includes several lighting calculation tools and three-dimensional modeling capabilities intended to provide comprehensive analysis for lighting design projects.

  Lithonia Lighting: http://www.lithonia.com/software/Lightware7/Visual/
- *AGI32.* Performs point-by-point calculations of direct or reflected light on any real surface or imaginary plane. Can be used to predict/quantify the distribution of electric light or daylight. Ray tracing, daylight factor, and glare rating calculations are included.

  Lighting Analysts, Inc.: http://www.agi32.com/
- *Lumen Designer.* Models all types of architectural spaces using a "designer wizard" or a CAD interface. Provides automated tools, a product database, a library of objects and materials, space visualization and rendering, and analytic output.

Lighting Technologies, Inc.: http://www.lighting-technologies.com/
- *Autodesk® VIZ 2005.* Three-dimensional modeling, rendering, and presentation software with a materials library and sophisticated rendering routines. Provides the ability to import luminaire data or select from a built-in fixture library and to define sky conditions for daylighting.

  Autodesk: http://usa.autodesk.com/
- *Desktop Radiance.* A software package that integrates the Radiance Synthetic Imaging System with AutoCAD Release 14 for lighting modeling. Includes libraries of materials, glazings, luminaires, and furnishings.

  Lawrence Berkeley National Laboratory: http://radsite.lbl.gov/deskrad/
- *FormZ RadioZity.* A version of formZ that includes radiosity-based rendering, permitting accurate simulation of the distribution of light in a space.

  auto·des·sys, Inc.: http://www.formz.com/products/formz_radiozity.html/

### (h) Life-Cycle Costing

- *Building Life-Cycle Cost Program (BLCC).* Developed by the National Institute of Standards and Technology (NIST) for the Federal Energy Management Program (FEMP), BLCC provides a computer-based structure for life-cycle cost analyses in conformance with *Handbook 135* (*Life-Cycle Costing Manual for the Federal Energy Management Program*).

  http://www.eere.energy.gov/femp/information/download_blcc.cfm/

### (i) Multipurpose Suites

- *Ecotect.* A multifaceted, multipurpose program that includes analysis tools for lighting, energy, acoustics, airflow, and other design issues. This program is approaching the idea of "one-stop" shopping for building analysis assistance.

  Square One Research PTY LTD: http://www.squ1.com/ecotect/ecotect.html/

# Index